Annals of Mathematics Studies
Number 193

# The $p$-adic Simpson Correspondence

Ahmed Abbes
Michel Gros
Takeshi Tsuji

PRINCETON UNIVERSITY PRESS
PRINCETON AND OXFORD
2016

Published by Princeton University Press, 41 William Street,
Princeton, New Jersey 08540
In the United Kingdom: Princeton University Press, 6 Oxford Street,
Woodstock, Oxfordshire OX20 1TW

press.princeton.edu



Library of Congress Cataloging-in-Publication Data

Abbes, Ahmed, 1970-
The p-adic Simpson correspondence / Ahmed Abbes, Michel Gros, Takeshi Tsuji.
pages cm. – (Annals of mathematics studies : number 193)
Includes bibliographical references and index.
ISBN 978-0-691-17028-2 (hardcover : alk. paper) – ISBN 978-0-691-17029-9 (pbk. : alk. paper) 1. Hodge theory, 2. p-adic fields, 3. Arithmetical algebraic geometry. I. Gros, Michel, 1956- II. Tsuji, Takeshi, 1967- III. Title.
QA179.A23 2016
512'.2–dc23
2015031778

British Library Cataloging-in-Publication Data is available

This book has been composed in LaTeX.

Printed on acid-free paper. ∞

The publisher would like to acknowledge the authors of this volume for providing the camera-ready copy from which this book was printed.

Printed in the United States of America

1 3 5 7 9 10 8 6 4 2

*In Memoriam Hyodo Osamu*

# Contents

# Foreword

In 1965, generalizing and clarifying a result of Weil, Narasimhan and Seshadri **[56]** established a bijective correspondence between the set of equivalence classes of unitary irreducible representations of the fundamental group of a compact Riemann surface $X$ of genus $\geq 2$ and the set of isomorphism classes of stable vector bundles of degree zero on $X$. The correspondence was then extended to all projective and smooth complex varieties by Donaldson **[21]**. The analogue for arbitrary linear representations is due to Simpson; to obtain a correspondence of the same type as Narasimhan and Seshadri, we need an additional structure on the vector bundle. This led to the notion of a Higgs bundle that was first introduced by Hitchin for algebraic curves. If $X$ is a smooth scheme over a field $K$, a *Higgs bundle* on $X$ is a pair $(M, \theta)$ consisting of a locally free $\mathscr{O}_X$-module of finite type $M$ and an $\mathscr{O}_X$-linear morphism $\theta\colon M \to M \otimes_{\mathscr{O}_X} \Omega^1_{X/K}$ such that $\theta \wedge \theta = 0$. Simpson's main result **[67, 68, 69, 70]** establishes an equivalence between the category of (complex) finite-dimensional linear representations of the fundamental group of a smooth and projective complex variety and the category of semi-stable Higgs bundles with vanishing Chern classes (cf. **[54]**).

Simpson's results and the important developments they inspired have led, in recent years, to the search for a $p$-adic analogue. Looking back, the first examples of such a construction (which did not yet use the terminology of Higgs bundles) can be found in the work of Hyodo **[43]**, who had treated the conceptually important case of $p$-adic variations of Hodge structures, called *Hodge–Tate local systems*. At present, the most advanced approach to such a correspondence is due to Faltings **[27, 28]**. It aims to describe all $p$-adic representations of the geometric fundamental group of a smooth algebraic variety over a $p$-adic field in terms of Higgs bundles. The constructions use several tools that he developed to establish the existence of Hodge–Tate decompositions **[24]**, in particular his theory of almost étale extensions **[26]**. Once completed, this *p-adic Simpson correspondence* should thus naturally provide the best Hodge–Tate type statements in $p$-adic Hodge theory. But at present, Faltings' construction seems satisfying only for curves, and even in that case, many fundamental questions remain open.

In this volume, we undertake a systematic development of the *p-adic Simpson correspondence* started by Faltings following two new approaches, one by the first two authors (A.A. and M.G.), the other by the third author (T.T.). The need to resume and develop Faltings' construction was felt given the number of results sketched in a rather short and extremely dense article **[27]**. This correspondence applies to objects more general than $p$-adic representations of the geometric fundamental group, introduced by Faltings and called *generalized representations*. We focus mainly on those $p$-adically close to the trivial representation, qualified by him as *small*. Though some of our constructions extend beyond this setting, let us make clear right away that we do not, however, discuss the descent techniques that enabled Faltings, in the case of curves, to get rid of this smallness condition.

Independently of the work of Faltings, Deninger and Werner **[19, 20, 18]** have developed a partial analogue to the theory of Narasimhan and Seshadri for $p$-adic curves, which should correspond to Higgs bundles with vanishing Higgs field in the $p$-adic Simpson correspondence. On the other hand, also inspired by the complex case, Ogus and Vologodsky **[59]** have introduced a correspondence between modules with integrable connection and Higgs modules for varieties in characteristic $p$. That work, in turn, inspired the first approach developed in this volume for the $p$-adic Simpson correspondence. Let us, however, note that unlike the complex case where a link between the two variants of the Simpson correspondence is established (by definition) through the Riemann–Hilbert correspondence, the link between the $p$-adic and the modulo $p$ Simpson correspondences is not, at present, known. Several works exploring this direction are in progress **[52, 61, 66]**.

Let us now give some indications on the structure of this volume. The first approach is presented in Chapters I, II, and III. Chapter I provides an overview of this approach and can also serve as an introduction to the general theme of this volume. In Chapter II, we study the case of an affine scheme of a particular type, qualified also as *small* by Faltings. We introduce the notion of *Dolbeault generalized representation* and the companion notion of *solvable Higgs module*, and then construct a natural equivalence between these two categories. We prove that this approach generalizes simultaneously Faltings' construction for small generalized representations and Hyodo's theory of $p$-adic variations of Hodge–Tate structures. In Chapter III, we address the global aspects of the theory. We introduce the *Higgs–Tate algebra*, which is the main novelty of this approach compared to that of Faltings, the notion of *Dolbeault module* that globalizes that of Dolbeault generalized representation, and the companion notion of *solvable Higgs bundle*. The main result is the equivalence between the category of Dolbeault modules and that of solvable Higgs bundles. We also prove the compatibility of this equivalence with the natural cohomologies. The general construction is obtained from the affine case by a gluing technique relying on the *Faltings topos*, developed in a more general context in Chapter VI.

This first approach, like Faltings' original one, requires the datum of a deformation of the scheme over a $p$-adic infinitesimal thickening of order 1 introduced by Fontaine. The second approach, developed in Chapter IV of this volume, avoids this additional datum. For this purpose, we introduce a crystalline type topos, and replace the notion of Higgs bundles by that of *Higgs (iso)crystals*. The link between these two notions uses *Higgs envelopes* and calls to mind the link between classical crystals and modules with integrable connections. The main result is the construction of a fully faithful functor from the category of Higgs (iso)crystals satisfying an *overconvergence* condition to that of *small* generalized representations. We also prove the compatibility of this functor with the natural cohomologies. Finally, we compare the period rings used in the two approaches developed in this volume, showing the compatibility of the two constructions.

The last part of the volume, consisting of Chapters V and VI, contains results of wider interest in $p$-adic Hodge theory. Chapter V provides a concise introduction to Faltings' theory of *almost étale extensions*, a tool that has become essential in many questions in arithmetic geometry, even beyond $p$-adic Hodge theory. The point of view adopted here is closer to Faltings' original one than the more systematic development given later by Gabber and Ramero. Chapter VI is devoted to the *Faltings topos*. Though it is the general framework for Faltings' approach in $p$-adic Hodge theory, this topos remains relatively unexplored. We present a new approach to it based on a generalization of Deligne's *covanishing topos*. Along the way, we correct the original definition of Faltings.

The reader will find at the end of the volume the facsimile of Faltings' article, which is reprinted from *Advances in Mathematics* 198(2), Faltings, Gerd, "A $p$-adic Simpson Correspondence," pp. 847–862, Copyright 2005, with permission from Elsevier. We thank the author warmly for having allowed us to reproduce it.

Ahmed Abbes, Michel Gros, and Takeshi Tsuji

September 2014

# The $p$-adic Simpson Correspondence

CHAPTER I

# Representations of the fundamental group and the torsor of deformations. An overview

Ahmed Abbes and Michel Gros

## I.1. Introduction

**I.1.1.** We develop a new approach to the $p$-adic Simpson correspondence, closely related to Faltings' original approach [**27**], and inspired by the work of Ogus and Vologodsky [**59**] on an analogue in characteristic $p$ of the complex Simpson correspondence. Before giving the details of this approach in Chapters II and III, we give a summary in this introductory chapter.

**I.1.2.** Let $K$ be a complete discrete valuation ring of characteristic 0, with perfect residue field of characteristic $p>0$, $\mathscr{O}_K$ the valuation ring of $K$, $\overline{K}$ an algebraic closure of $K$, and $\mathscr{O}_{\overline{K}}$ the integral closure of $\mathscr{O}_K$ in $\overline{K}$. Let $X$ be a smooth $\mathscr{O}_K$-scheme of finite type with integral generic geometric fiber $X_{\overline{K}}$, $\overline{x}$ a geometric point of $X_{\overline{K}}$, and $\mathfrak{X}$ the formal scheme $p$-adic completion of $X\otimes_{\mathscr{O}_K}\mathscr{O}_{\overline{K}}$. In this work, we consider a more general smooth logarithmic situation (cf. II.6.2 and III.4.7). Nevertheless, to simplify the presentation, we restrict ourselves in this introductory chapter to the smooth case in the usual sense. We are looking for a functor from the category of $p$-adic representations of the geometric fundamental group $\pi_1(X_{\overline{K}},\overline{x})$ (that is, the finite-dimensional continuous linear $\mathbb{Q}_p$-representations of $\pi_1(X_{\overline{K}},\overline{x})$) to the category of Higgs $\mathscr{O}_{\mathfrak{X}}[\frac{1}{p}]$-bundles (that is, the pairs $(\mathscr{M},\theta)$ consisting of a locally projective $\mathscr{O}_{\mathfrak{X}}[\frac{1}{p}]$-module of finite type $\mathscr{M}$ and an $\mathscr{O}_{\mathfrak{X}}[\frac{1}{p}]$-linear morphism $\theta\colon \mathscr{M}\to\mathscr{M}\otimes_{\mathscr{O}_X}\Omega^1_{X/\mathscr{O}_K}$ such that $\theta\wedge\theta=0$). Following Faltings' strategy, which at present has been only partly achieved, this functor should extend to a strictly larger category than that of the $p$-adic representations of $\pi_1(X_{\overline{K}},\overline{x})$, called category of *generalized representations* of $\pi_1(X_{\overline{K}},\overline{x})$. It would then be an *equivalence of categories* between this new category and the category of Higgs $\mathscr{O}_{\mathfrak{X}}[\frac{1}{p}]$-bundles. The main motivation for the present work is the construction of such an equivalence of categories. When $X_K$ is a proper and smooth curve over $K$, Faltings shows that the Higgs bundles associated with the "true" $p$-adic representations of $\pi_1(X_{\overline{K}},\overline{x})$ are semi-stable of slope zero and expresses the hope that all semi-stable Higgs bundles of slope zero are obtained this way. This statement, which would correspond to the difficult part of Simpson's result in the complex case, seems out of reach at present.

**I.1.3.** The notion of generalized representations is due to Faltings. They are, in simplified terms, continuous $p$-adic *semi-linear* representations of $\pi_1(X_{\overline{K}},\overline{x})$ on modules over a certain $p$-adic ring endowed with a continuous action of $\pi_1(X_{\overline{K}},\overline{x})$. Faltings' approach in [**27**] to construct a functor $\mathscr{H}$ from the category of these generalized representations to the category of Higgs bundles consists of two steps. He first defines $\mathscr{H}$ for the generalized representations that are $p$-adically close to the trivial representation, which he calls *small*. He carries out this step in arbitrary dimension. In the second step, achieved only for curves, he extends the functor $\mathscr{H}$ to all generalized representations of $\pi_1(X_{\overline{K}}, \overline{x})$ by descent. Indeed, every generalized representation becomes small over a finite étale cover of $X_{\overline{K}}$.

**I.1.4.** Our new approach, which works in arbitrary dimension, allows us to define the functor $\mathscr{H}$ on the category of generalized representations of $\pi_1(X_{\overline{K}}, \overline{x})$ satisfying an admissibility condition à la Fontaine, called *Dolbeault* generalized representations. For this purpose, we introduce a family of period rings that we call *Higgs–Tate algebras*, and that are the main novelty of our approach compared to that of Faltings. We show that the admissibility condition for rational coefficients corresponds to the smallness condition of Faltings; but it is strictly more general for integral coefficients. Note that Faltings' construction for small rational coefficients is limited to curves and that it presents a number of difficulties that can be avoided with our approach.

**I.1.5.** We proceed in two steps. We first study in Chapter II the case of an affine scheme of a certain type, called also *small* by Faltings. We then tackle in Chapter III the global aspects of the theory. The general construction is obtained from the affine case using a gluing technique presenting unexpected difficulties. To do this we will use the *Faltings topos*, a fibered variant of Deligne's notion of *covanishing topos*, which we develop in Chapter VI.

**I.1.6.** This introductory chapter offers, in a geometric situation simplified for the clarity of the exposition, a detailed summary of the global steps leading to our main results. Let us take a quick look at its contents. We begin, in I.3, with a short aside on small generalized representations in the affine case, which will be used as intermediary for the study of Dolbeault representations. Section I.4 summarizes the local study conducted in Chapter II. We introduce the notion of generalized Dolbeault representation for a small affine scheme and the companion notion of solvable Higgs module, and then construct a natural equivalence between these two categories. We in fact develop two variants, an integral one and a more subtle rational one. We establish links between these notions and Faltings smallness conditions. We also link this to Hyodo's theory [**43**]. The global aspects of the theory developed in Chapter III are summarized in Sections I.5 and I.6. After a short introduction to Faltings' ringed topos in I.5, we introduce the *Higgs–Tate algebras* (I.5.13). The notion of *Dolbeault module* that globalizes that of generalized Dolbeault representation and the companion notion of *solvable Higgs bundle* are defined in I.6.13. Our main result (I.6.18) is the equivalence of these two categories. For the proof of this result, we need acyclicity statements for the Higgs–Tate algebras that we give in I.6.5 and I.6.8, which also allow us to show the compatibility of this equivalence with the relevant cohomologies on each side (I.6.19). We also study the functoriality of the various introduced properties by étale morphisms (I.6.21), as well as their local character for the étale topology (I.6.22, I.6.23, I.6.24). Finally, we return in this global situation to the logical links (I.6.26, I.6.27, I.6.28), for a Higgs bundle, between smallness (I.6.25) and solvability.

At the beginning of Chapters II and III, the reader will find a detailed description of their structure. Chapter VI, which is of separate interest, has its own introduction.

**Acknowledgments.** This work could obviously not have existed without the work of G. Faltings, and first and foremost, that on the $p$-adic Simpson correspondence [**27**]. We would like to convey our deep gratitude to him. The genesis of this work immediately followed a workshop held in Rennes in 2008–2009 on his article [**27**]. We benefited, on that occasion, from the text of O. Brinon's talk [**13**] and from the work of T. Tsuji [**75**] presenting his own approach to the $p$-adic Simpson correspondence. These two texts have been extremely useful to us and we are grateful to their authors for having made them available to us spontaneously. We also thank O. Brinon, G. Faltings, and T. Tsuji for all the exchanges we had with them on questions related to this work, and A. Ogus for the clarifying discussions we had with him on his work with V. Vologodsky **[59]**. We thank Reinie Erné warmly for translating, with great skill and under tight deadlines, Chapters I–III and VI of this volume, keeping in mind our stylistic preferences. The first author (A.A.) thanks the Centre Émile Borel, the Institut des Hautes Études Scientifiques, and the University of Tokyo for their hospitality. He also thanks those who followed the course he gave on this subject at the University of Tokyo during the fall of 2010 and the winter of 2011, whose questions and remarks have been precious for perfecting this work. The second author (M.G.) thanks the Institut des Hautes Études Scientifiques and the University of Tokyo for their hospitality. Finally, we thank the participants of the summer school *Higgs bundles on p-adic curves and representation theory* that took place in Mainz in September 2012, during which our main results were presented, for their remarks and their stimulating interest. This work was supported by the ANR program $p$-adic Hodge theory and beyond (ThéHopaD) ANR-11-BS01-005.

## I.2. Notation and conventions

*All rings in this chapter have an identity element; all ring homomorphisms map the identity element to the identity element. We mostly consider commutative rings, and rings are assumed to be commutative unless stated otherwise; in particular, when we take a ringed topos $(X, A)$, the ring $A$ is assumed to be commutative unless stated otherwise.*

**I.2.1.** In this introduction, $K$ denotes a complete discrete valuation ring of characteristic 0, with perfect residue field $k$ of characteristic $p > 0$, $\mathscr{O}_K$ the valuation ring of $K$, $\overline{K}$ an algebraic closure of $K$, $\mathscr{O}_{\overline{K}}$ the integral closure of $\mathscr{O}_K$ in $\overline{K}$, $\mathscr{O}_C$ the $p$-adic Hausdorff completion of $\mathscr{O}_{\overline{K}}$, and $C$ the field of fractions of $\mathscr{O}_C$. From I.5 on, we will assume that $k$ is algebraically closed. We set $S = \mathrm{Spec}(\mathscr{O}_K)$, $\overline{S} = \mathrm{Spec}(\mathscr{O}_{\overline{K}})$, and $\check{\overline{S}} = \mathrm{Spec}(\mathscr{O}_C)$. We denote by $s$ (resp. $\eta$, resp. $\overline{\eta}$) the closed point of $S$ (resp. the generic point of $S$, resp. the generic point of $\overline{S}$). For any integer $n \geq 1$ and any $S$-scheme $X$, we set $S_n = \mathrm{Spec}(\mathscr{O}_K/p^n\mathscr{O}_K)$,

$$X_n = X \times_S S_n, \quad \overline{X} = X \times_S \overline{S}, \quad \text{and} \quad \check{\overline{X}} = X \times_S \check{\overline{S}}. \tag{I.2.1.1}$$

For any abelian group $M$, we denote by $\widehat{M}$ its $p$-adic Hausdorff completion.

**I.2.2.** Let $G$ be a profinite group and $A$ a topological ring endowed with a continuous action of $G$ by ring homomorphisms. An *$A$-representation* of $G$ consists of an $A$-module $M$ and an $A$-semi-linear action of $G$ on $M$, that is, such that for all $g \in G$, $a \in A$, and $m \in M$, we have $g(am) = g(a)g(m)$. We say that the $A$-representation is *continuous* if $M$ is a topological $A$-module and if the action of $G$ on $M$ is continuous. Let $M$, $N$ be two $A$-representations (resp. two continuous $A$-representations) of $G$. A morphism from $M$ to $N$ is a $G$-equivariant and $A$-linear (resp. $G$-equivariant, continuous, and $A$-linear) morphism from $M$ to $N$.

**I.2.3.** Let $(X, A)$ be a ringed topos and $E$ an $A$-module. A *Higgs $A$-module with coefficients in $E$* is a pair $(M, \theta)$ consisting of an $A$-module $M$ and an $A$-linear morphism $\theta\colon M \to M \otimes_A E$ such that $\theta \wedge \theta = 0$ (cf. II.2.8). Following Simpson (**[68]** p. 24), we call *Dolbeault* complex of $(M, \theta)$ and denote by $\mathbb{K}^\bullet(M, \theta)$ the complex of cochains of $A$-modules

$$M \to M \otimes_A E \to M \otimes_A \wedge^2 E \ldots \tag{I.2.3.1}$$

deduced from $\theta$ (cf. II.2.8.2).

**I.2.4.** Let $(X, A)$ be a ringed topos, $B$ an $A$-algebra, $M$ a $B$-module, and $\lambda \in \Gamma(X, A)$. A $\lambda$-*connection* on $M$ with respect to the extension $B/A$ consists of an $A$-linear morphism

$$\nabla\colon M \to \Omega^1_{B/A} \otimes_B M \tag{I.2.4.1}$$

such that for all local sections $x$ of $B$ and $s$ of $M$, we have

$$\nabla(xs) = \lambda d(x) \otimes s + x\nabla(s). \tag{I.2.4.2}$$

It is *integrable* if $\nabla \circ \nabla = 0$ (cf. II.2.10). We will leave the extension $B/A$ out of the terminology when there is no risk of confusion.

Let $(M, \nabla)$, $(M', \nabla')$ be two modules with $\lambda$-connections. A morphism from $(M, \nabla)$ to $(M', \nabla')$ is a $B$-linear morphism $u\colon M \to M'$ such that $(\mathrm{id} \otimes u) \circ \nabla = \nabla' \circ u$.

Classically, 1-connections are called *connections*. Integrable 0-connections are the Higgs $B$-fields with coefficients in $\Omega^1_{B/A}$.

**Remark I.2.5.** Let $(X, A)$ be a ringed topos, $B$ an $A$-algebra, $\lambda \in \Gamma(X, A)$, and $(M, \nabla)$ a module with $\lambda$-connection with respect to the extension $B/A$. Suppose that there exist an $A$-module $E$ and a $B$-linear isomorphism $\gamma\colon E \otimes_A B \xrightarrow{\sim} \Omega^1_{B/A}$ such that for every local section $\omega$ of $E$, we have $d(\gamma(\omega \otimes 1)) = 0$. The $\lambda$-connection $\nabla$ is integrable if and only if the morphism $\theta\colon M \to E \otimes_A M$ induced by $\nabla$ and $\gamma$ is a Higgs $A$-field on $M$ with coefficients in $E$ (cf. II.2.12).

**I.2.6.** If $\mathscr{C}$ is an additive category, we denote by $\mathscr{C}_{\mathbb{Q}}$ and call *category of objects of $\mathscr{C}$ up to isogeny* the category with the same objects as $\mathscr{C}$, and such that the set of morphisms between two objects is given by

$$\mathrm{Hom}_{\mathscr{C}_{\mathbb{Q}}}(E, F) = \mathrm{Hom}_{\mathscr{C}}(E, F) \otimes_{\mathbb{Z}} \mathbb{Q}. \tag{I.2.6.1}$$

The category $\mathscr{C}_{\mathbb{Q}}$ is none other than the localized category of $\mathscr{C}$ with respect to the multiplicative system of the *isogenies* of $\mathscr{C}$ (cf. III.6.1). We denote by

$$\mathscr{C} \to \mathscr{C}_{\mathbb{Q}}, \quad M \mapsto M_{\mathbb{Q}} \tag{I.2.6.2}$$

the localization functor. If $\mathscr{C}$ is an abelian category, the category $\mathscr{C}_{\mathbb{Q}}$ is abelian and the localization functor (I.2.6.2) is exact. Indeed, $\mathscr{C}_{\mathbb{Q}}$ identifies canonically with the quotient of $\mathscr{C}$ by the thick subcategory of objects of finite exponent (III.6.1.4).

**I.2.7.** Let $(X, A)$ be a ringed topos. We denote by $\mathbf{Mod}(A)$ the category of $A$-modules of $X$ and by $\mathbf{Mod}_{\mathbb{Q}}(A)$, instead of $\mathbf{Mod}(A)_{\mathbb{Q}}$, the category of $A$-modules up to isogeny (I.2.6). The tensor product of $A$-modules induces a bifunctor

$$\mathbf{Mod}_{\mathbb{Q}}(A) \times \mathbf{Mod}_{\mathbb{Q}}(A) \to \mathbf{Mod}_{\mathbb{Q}}(A), \quad (M, N) \mapsto M \otimes_{A_{\mathbb{Q}}} N \tag{I.2.7.1}$$

making $\mathbf{Mod}_{\mathbb{Q}}(A)$ into a symmetric monoidal category with $A_{\mathbb{Q}}$ as unit object. The objects of $\mathbf{Mod}_{\mathbb{Q}}(A)$ will also be called $A_{\mathbb{Q}}$-*modules*. This terminology is justified by considering $A_{\mathbb{Q}}$ as a monoid of $\mathbf{Mod}_{\mathbb{Q}}(A)$.

**I.2.8.** Let $(X, A)$ be a ringed topos and $E$ an $A$-module. We call *Higgs $A$-isogeny with coefficients in $E$* a quadruple

$$(M, N, u\colon M \to N, \theta\colon M \to N \otimes_A E) \tag{I.2.8.1}$$

consisting of two $A$-modules $M$ and $N$ and two $A$-linear morphisms $u$ and $\theta$ satisfying the following property: there exist an integer $n \neq 0$ and an $A$-linear morphism $v\colon N \to M$ such that $v \circ u = n \cdot \mathrm{id}_M$, $u \circ v = n \cdot \mathrm{id}_N$, and that $(M, (v \otimes \mathrm{id}_E) \circ \theta)$ and $(N, \theta \circ v)$ are Higgs $A$-modules with coefficients in $E$ (I.2.3). Note that $u$ induces an isogeny of

Higgs modules from $(M, (v \otimes \mathrm{id}_E) \circ \theta)$ to $(N, \theta \circ v)$ (III.6.1), whence the terminology. Let $(M, N, u, \theta)$, $(M', N', u', \theta')$ be two Higgs $A$-isogenies with coefficients in $E$. A morphism from $(M, N, u, \theta)$ to $(M', N', u', \theta')$ consists of two $A$-linear morphisms $\alpha\colon M \to M'$ and $\beta\colon N \to N'$ such that $\beta \circ u = u' \circ \alpha$ and $(\beta \otimes \mathrm{id}_E) \circ \theta = \theta' \circ \alpha$. We denote by $\mathbf{HI}(A, E)$ the category of Higgs $A$-isogenies with coefficients in $E$. It is an additive category. We denote by $\mathbf{HI}_{\mathbb{Q}}(A, E)$ the category of objects of $\mathbf{HI}(A, E)$ up to isogeny.

**I.2.9.** Let $(X, A)$ be a ringed topos, $B$ an $A$-algebra, and $\lambda \in \Gamma(X, A)$. We call *$\lambda$-isoconnection with respect to the extension $B/A$* (or simply *$\lambda$-isoconnection* when there is no risk of confusion) a quadruple

$$(M, N, u\colon M \to N, \nabla\colon M \to \Omega^1_{B/A} \otimes_B N) \tag{I.2.9.1}$$

where $M$ and $N$ are $B$-modules, $u$ is an isogeny of $B$-modules (III.6.1), and $\nabla$ is an $A$-linear morphism such that for all local sections $x$ of $B$ and $t$ of $M$, we have

$$\nabla(xt) = \lambda d(x) \otimes u(t) + x\nabla(t). \tag{I.2.9.2}$$

For every $B$-linear morphism $v\colon N \to M$ for which there exists an integer $n$ such that $u \circ v = n \cdot \mathrm{id}_N$ and $v \circ u = n \cdot \mathrm{id}_M$, the pairs $(M, (\mathrm{id} \otimes v) \circ \nabla)$ and $(N, \nabla \circ v)$ are modules with $(n\lambda)$-connections (I.2.2), and $u$ is a morphism from $(M, (\mathrm{id} \otimes v) \circ \nabla)$ to $(N, \nabla \circ v)$. We call the $\lambda$-isoconnection $(M, N, u, \nabla)$ *integrable* if there exist a $B$-linear morphism $v\colon N \to M$ and an integer $n \neq 0$ such that $u \circ v = n \cdot \mathrm{id}_N$, $v \circ u = n \cdot \mathrm{id}_M$, and that the $(n\lambda)$-connections $(\mathrm{id} \otimes v) \circ \nabla$ on $M$ and $\nabla \circ v$ on $N$ are integrable.

Let $(M, N, u, \nabla)$ and $(M', N', u', \nabla')$ be two $\lambda$-isoconnections. A morphism from $(M, N, u, \nabla)$ to $(M', N', u', \nabla')$ consists of two $B$-linear morphisms $\alpha\colon M \to M'$ and $\beta\colon N \to N'$ such that $\beta \circ u = u' \circ \alpha$ and $(\mathrm{id} \otimes \beta) \circ \nabla = \nabla' \circ \alpha$.

## I.3. Small generalized representations

**I.3.1.** In this section, we fix a smooth affine $S$-scheme $X = \mathrm{Spec}(R)$ such that $X_{\overline{\eta}}$ is connected and $X_s$ is nonempty, an integer $d \geq 1$, and an étale $S$-morphism

$$X \to \mathbb{G}^d_{m,S} = \mathrm{Spec}(\mathscr{O}_K[T_1^{\pm 1}, \dots, T_d^{\pm 1}]). \tag{I.3.1.1}$$

This is the typical example of a Faltings' small affine scheme. The assumption that $X_{\overline{\eta}}$ is connected is not necessary but allows us to simplify the presentation. The reader will recognize the logarithmic nature of the datum (I.3.1.1). Following **[27]**, we consider in this work a more general smooth logarithmic situation, which turns out to be necessary even for defining the $p$-adic Simpson correspondence for a proper smooth curve over $S$. Indeed, in the second step of the descent, we will need to consider finite covers of its generic fiber, which brings us to the case of a semi-stable scheme over $S$. Nevertheless, to simplify the presentation, we will restrict ourselves in this introduction to the smooth case in the usual sense (cf. II.6.2 for the logarithmic smooth affine case). We denote by $t_i$ the image of $T_i$ in $R$ $(1 \leq i \leq d)$, and we set

$$R_1 = R \otimes_{\mathscr{O}_K} \mathscr{O}_{\overline{K}}. \tag{I.3.1.2}$$

**I.3.2.** Let $\overline{y}$ be a geometric point of $X_{\overline{\eta}}$ and $(V_i)_{i \in I}$ a universal cover of $X_{\overline{\eta}}$ at $\overline{y}$. We denote by $\Delta$ the geometric fundamental group $\pi_1(X_{\overline{\eta}}, \overline{y})$. For every $i \in I$, we denote by $\overline{X}_i = \mathrm{Spec}(R_i)$ the integral closure of $\overline{X}$ in $V_i$, and we set

$$\overline{R} = \varinjlim_{i \in I} R_i. \tag{I.3.2.1}$$

In this context, the *generalized representations* of $\Delta$ are the continuous $\widehat{\overline{R}}$-representations of $\Delta$ with values in projective $\widehat{\overline{R}}$-modules of finite type, endowed with their $p$-adic topologies (I.2.2). Such a representation $M$ is called *small* if $M$ is a free $\widehat{\overline{R}}$-module of finite type having a basis made up of elements that are $\Delta$-invariant modulo $p^{2\alpha}M$ for a rational number $\alpha > \frac{1}{p-1}$. The main property of the small generalized representations of $\Delta$ is their good behavior under descent for certain quotients of $\Delta$ isomorphic to $\mathbb{Z}_p(1)^d$. Let us fix such a quotient $\Delta_\infty$ by choosing, for every $1 \leq i \leq d$, a compatible system $(t_i^{(n)})_{n\in\mathbb{N}}$ of $p^n$th roots of $t_i$ in $\overline{R}$. We define the notion of small $\widehat{R_1}$-representation of $\Delta_\infty$ similarly. The functor

$$M \mapsto M \otimes_{\widehat{R_1}} \widehat{\overline{R}} \tag{I.3.2.2}$$

from the category of small $\widehat{R_1}$-representations of $\Delta_\infty$ to that of small $\widehat{\overline{R}}$-representations of $\Delta$ is then an equivalence of categories (cf. II.14.4). This is a consequence of Faltings' almost purity theorem (cf. II.6.16; **[26]** § 2b).

**I.3.3.** If $(M, \varphi)$ is a small $\widehat{R_1}$-representation of $\Delta_\infty$, we can consider the logarithm of $\varphi$, which is a homomorphism from $\Delta_\infty$ to $\operatorname{End}_{\widehat{R_1}}(M)$. By fixing a $\mathbb{Z}_p$-basis $\zeta$ of $\mathbb{Z}_p(1)$, the latter can be written uniquely as

$$\log(\varphi) = \sum_{i=1}^{d} \theta_i \otimes \chi_i \otimes \zeta^{-1}, \tag{I.3.3.1}$$

where $\zeta^{-1}$ is the dual basis of $\mathbb{Z}_p(-1)$, $\chi_i$ is the character of $\Delta_\infty$ with values in $\mathbb{Z}_p(1)$ that gives its action on the system $(t_i^{(n)})_{n\in\mathbb{N}}$, and $\theta_i$ is an $\widehat{R_1}$-linear endomorphism of $M$. We immediately see that

$$\theta = \sum_{i=1}^{d} \theta_i \otimes d\log(t_i) \otimes \zeta^{-1} \tag{I.3.3.2}$$

is a Higgs $\widehat{R_1}$-field on $M$ with coefficients in $\Omega^1_{R/\mathscr{O}_K} \otimes_R \widehat{R}_1(-1)$ (I.2.3) (to simplify we will say with coefficients in $\Omega^1_{R/\mathscr{O}_K}(-1)$). The resulting correspondence $(M, \varphi) \mapsto (M, \theta)$ is in fact an *equivalence of categories* between the category of small $\widehat{R_1}$-representations of $\Delta_\infty$ and that of *small* Higgs $\widehat{R_1}$-modules with coefficients in $\Omega^1_{R/\mathscr{O}_K}(-1)$ (that is, the category of Higgs $\widehat{R_1}$-modules with coefficients in $\Omega^1_{R/\mathscr{O}_K}(-1)$ whose underlying $\widehat{R_1}$-module is free of finite type and whose Higgs field is zero modulo $p^{2\alpha}$ for a rational number $\alpha > \frac{1}{p-1}$). Combining this with the previous descent statement (I.3.2.2), we obtain an equivalence between the category of small $\widehat{\overline{R}}$-representations of $\Delta$ and that of small Higgs $\widehat{R_1}$-modules with coefficients in $\Omega^1_{R/\mathscr{O}_K}(-1)$. The disadvantage of this construction is its dependence on the $(t_i^{(n)})_{n\in\mathbb{N}}$ $(1 \leq i \leq d)$, which excludes any globalization. To remedy this defect, Faltings proposes another equivalent definition that depends on another choice that can be globalized easily. Our approach, which is the object of the remainder of this introduction, was inspired by this construction.

## I.4. The torsor of deformations

**I.4.1.** In this section, we are given a smooth affine $S$-scheme $X = \operatorname{Spec}(R)$ such that $X_{\overline{\eta}}$ is connected, $X_s$ is nonempty, and that there exist an integer $d \geq 1$ and an étale $S$-morphism $X \to \mathbb{G}^d_{m,S}$ (but we do not fix such a morphism). We also fix a geometric point $\overline{y}$ of $X_{\overline{\eta}}$ and a universal cover $(V_i)_{i\in I}$ of $X_{\overline{\eta}}$ at $\overline{y}$, and we use the notation of I.3.2: $\Delta = \pi_1(X_{\overline{\eta}}, \overline{y})$, $R_1 = R \otimes_{\mathscr{O}_K} \mathscr{O}_{\overline{K}}$, and $\overline{R}$ (I.3.2.1).

**I.4.2.** Recall that Fontaine associates functorially with each $\mathbb{Z}_{(p)}$-algebra $A$ the ring

(I.4.2.1) $$\mathscr{R}_A = \varprojlim_{x\mapsto x^p} A/pA$$

and a homomorphism $\theta$ from the ring $\mathrm{W}(\mathscr{R}_A)$ of Witt vectors of $\mathscr{R}_A$ to the $p$-adic Hausdorff completion $\widehat{A}$ of $A$ (cf. II.9.3). We set

(I.4.2.2) $$\mathscr{A}_2(A) = \mathrm{W}(\mathscr{R}_A)/\ker(\theta)^2$$

and denote also by $\theta\colon \mathscr{A}_2(\mathscr{R}_A) \to \widehat{A}$ the homomorphism induced by $\theta$.

For the remainder of this chapter, we fix a sequence $(p_n)_{n\in\mathbb{N}}$ of elements of $\mathscr{O}_{\overline{K}}$ such that $p_0 = p$ and $p_{n+1}^p = p_n$ for every $n \geq 0$. We denote by $\underline{p}$ the element of $\mathscr{R}_{\mathscr{O}_{\overline{K}}}$ induced by the sequence $(p_n)_{n\in\mathbb{N}}$ and set

(I.4.2.3) $$\xi = [\underline{p}] - p \in \mathrm{W}(\mathscr{R}_{\mathscr{O}_{\overline{K}}}),$$

where $[\ ]$ is the multiplicative representative. The sequence

(I.4.2.4) $$0 \longrightarrow \mathrm{W}(\mathscr{R}_{\mathscr{O}_{\overline{K}}}) \xrightarrow{\cdot\xi} \mathrm{W}(\mathscr{R}_{\mathscr{O}_{\overline{K}}}) \xrightarrow{\theta} \mathscr{O}_C \longrightarrow 0$$

is exact (II.9.5). It induces an exact sequence

(I.4.2.5) $$0 \longrightarrow \mathscr{O}_C \xrightarrow{\cdot\xi} \mathscr{A}_2(\mathscr{O}_{\overline{K}}) \xrightarrow{\theta} \mathscr{O}_C \longrightarrow 0,$$

where $\cdot\xi$ again denotes the morphism deduced from the morphism of multiplication by $\xi$ in $\mathscr{A}_2(\mathscr{O}_{\overline{K}})$. The ideal $\ker(\theta)$ of $\mathscr{A}_2(\mathscr{O}_{\overline{K}})$ has square zero. It is a free $\mathscr{O}_C$-module with basis $\xi$. It will be denoted by $\xi\mathscr{O}_C$. Note that unlike $\xi$, this module does not depend on the choice of the sequence $(p_n)_{n\in\mathbb{N}}$. We denote by $\xi^{-1}\mathscr{O}_C$ the dual $\mathscr{O}_C$-module of $\xi\mathscr{O}_C$. For every $\mathscr{O}_C$-module $M$, we denote the $\mathscr{O}_C$-modules $M\otimes_{\mathscr{O}_C}(\xi\mathscr{O}_C)$ and $M\otimes_{\mathscr{O}_C}(\xi^{-1}\mathscr{O}_C)$ simply by $\xi M$ and $\xi^{-1}M$, respectively.

Likewise, we have an exact sequence (II.9.11.2)

(I.4.2.6) $$0 \longrightarrow \widehat{\overline{R}} \xrightarrow{\cdot\xi} \mathscr{A}_2(\overline{R}) \xrightarrow{\theta} \widehat{\overline{R}} \longrightarrow 0.$$

The ideal $\ker(\theta)$ of $\mathscr{A}_2(\overline{R})$ has square zero. It is a free $\widehat{\overline{R}}$-module with basis $\xi$, canonically isomorphic to $\xi\widehat{\overline{R}}$. The group $\Delta$ acts by functoriality on $\mathscr{A}_2(\overline{R})$.

We set $\mathscr{A}_2(\overline{S}) = \mathrm{Spec}(\mathscr{A}_2(\mathscr{O}_{\overline{K}}))$, $Y = \mathrm{Spec}(\overline{R})$, $\widehat{Y} = \mathrm{Spec}(\widehat{\overline{R}})$, and $\mathscr{A}_2(Y) = \mathrm{Spec}(\mathscr{A}_2(\overline{R}))$.

**I.4.3.** From now on, we fix a smooth $\mathscr{A}_2(\overline{S})$-deformation $\widetilde{X}$ of $\check{\overline{X}}$, that is, a smooth $\mathscr{A}_2(\overline{S})$-scheme $\widetilde{X}$ that fits into a Cartesian diagram

(I.4.3.1) $$\begin{array}{ccc} \check{\overline{X}} & \longrightarrow & \widetilde{X} \\ \downarrow & & \downarrow \\ \check{\overline{S}} & \longrightarrow & \mathscr{A}_2(\overline{S}) \end{array}$$

This additional datum replaces the datum of an étale $S$-morphism $X \to \mathbb{G}^d_{m,S}$; in fact, such a morphism provides a deformation.

We set

(I.4.3.2) $$\mathrm{T} = \mathrm{Hom}_{\widehat{\overline{R}}}(\Omega^1_{R/\mathscr{O}_K} \otimes_R \widehat{\overline{R}}, \xi\widehat{\overline{R}}).$$

We identify the dual $\widehat{\overline{R}}$-module with $\xi^{-1}\Omega^1_{R/\mathscr{O}_K} \otimes_R \widehat{\overline{R}}$ (I.4.2) and denote by $\mathbf{T}$ the associated $\widehat{Y}$-vector bundle, in other words,

$$\mathbf{T} = \mathrm{Spec}(\mathrm{Sym}_{\widehat{\overline{R}}}(\xi^{-1}\Omega^1_{R/\mathscr{O}_K} \otimes_R \widehat{\overline{R}})). \tag{I.4.3.3}$$

Let $U$ be an open subscheme of $\widehat{Y}$ and $\widetilde{U}$ the open subscheme of $\mathscr{A}_2(Y)$ defined by $U$. We denote by $\mathscr{L}(U)$ the set of morphisms represented by dotted arrows that complete the diagram

(I.4.3.4)

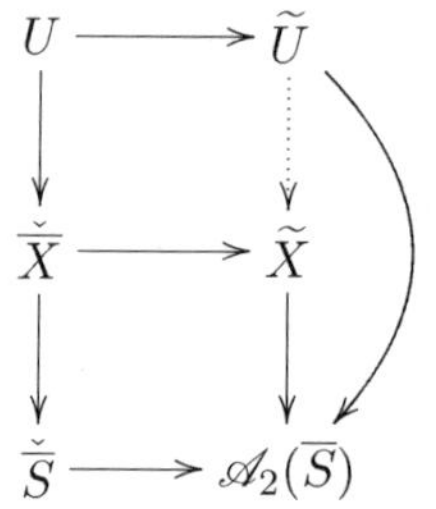

in such a way that it remains commutative. The functor $U \mapsto \mathscr{L}(U)$ is a $T$-torsor for the Zariski topology of $\widehat{Y}$. We denote by $\mathscr{F}$ the $\widehat{\overline{R}}$-module of affine functions on $\mathscr{L}$ (cf. II.4.9). The latter fits into a canonical exact sequence (II.4.9.1)

$$0 \to \widehat{\overline{R}} \to \mathscr{F} \to \xi^{-1}\Omega^1_{R/\mathscr{O}_K} \otimes_R \widehat{\overline{R}} \to 0. \tag{I.4.3.5}$$

This sequence induces for every integer $n \geq 1$ an exact sequence

$$0 \to \mathrm{Sym}^{n-1}_{\widehat{\overline{R}}}(\mathscr{F}) \to \mathrm{Sym}^{n}_{\widehat{\overline{R}}}(\mathscr{F}) \to \mathrm{Sym}^{n}_{\widehat{\overline{R}}}(\xi^{-1}\Omega^1_{R/\mathscr{O}_K} \otimes_R \widehat{\overline{R}}) \to 0. \tag{I.4.3.6}$$

The $\widehat{\overline{R}}$-modules $(\mathrm{Sym}^{n}_{\widehat{\overline{R}}}(\mathscr{F}))_{n\in\mathbb{N}}$ therefore form a filtered direct system whose direct limit

$$\mathscr{C} = \varinjlim_{n\geq 0} \mathrm{Sym}^{n}_{\widehat{\overline{R}}}(\mathscr{F}) \tag{I.4.3.7}$$

is naturally endowed with a structure of $\widehat{\overline{R}}$-algebra. By II.4.10, the $\widehat{Y}$-scheme

$$\mathbf{L} = \mathrm{Spec}(\mathscr{C}) \tag{I.4.3.8}$$

is naturally a principal homogeneous $\mathbf{T}$-bundle on $\widehat{Y}$ that canonically represents $\mathscr{L}$.

The natural action of $\Delta$ on the scheme $\mathscr{A}_2(Y)$ induces an $\widehat{\overline{R}}$-semi-linear action of $\Delta$ on $\mathscr{F}$, such that the morphisms in sequence (I.4.3.5) are $\Delta$-equivariant. From this we deduce an action of $\Delta$ on $\mathscr{C}$ by ring automorphisms, compatible with its action on $\widehat{\overline{R}}$, which we call *canonical action*. These actions are continuous for the $p$-adic topologies (II.12.4). The $\widehat{\overline{R}}$-algebra $\mathscr{C}$, endowed with the canonical action of $\Delta$, is called the *Higgs–Tate algebra* associated with $\widetilde{X}$. The $\widehat{\overline{R}}$-representation $\mathscr{F}$ of $\Delta$ is called the *Higgs–Tate extension* associated with $\widetilde{X}$.

**I.4.4.** Let $(M,\theta)$ be a *small* Higgs $\widehat{R_1}$-module with coefficients in $\xi^{-1}\Omega^1_{R/\mathscr{O}_K}$ (that is, a Higgs $\widehat{R_1}$-module with coefficients in $\xi^{-1}\Omega^1_{R/\mathscr{O}_K} \otimes_R \widehat{R_1}$ whose underlying $\widehat{R_1}$-module is free of finite type and whose Higgs field is zero modulo $p^\alpha$ for a rational number $\alpha > \frac{1}{p-1}$) and let $\psi \in \mathscr{L}(\widehat{Y})$. For every $\sigma \in \Delta$, we denote by ${}^\sigma\psi$ the section of $\mathscr{L}(\widehat{Y})$

defined by the commutative diagram

(I.4.4.1)
$$\begin{array}{ccc} \mathbf{L} & \xrightarrow{\sigma} & \mathbf{L} \\ {\scriptstyle\psi}\uparrow\downarrow & & \downarrow\uparrow{\scriptstyle {}^{\sigma}\psi} \\ \widehat{Y} & \xrightarrow{\sigma} & \widehat{Y} \end{array}$$

The difference $D_\sigma = \psi - {}^{\sigma}\psi$ is an element of $\mathrm{Hom}_{\widehat{\overline{R}}}(\xi^{-1}\Omega^1_{R/\mathscr{O}_K} \otimes_R \widehat{\overline{R}}, \widehat{\overline{R}})$. The endomorphism $\exp((D_\sigma \otimes \mathrm{id}_M) \circ \theta)$ of $M \otimes_{\widehat{R_1}} \widehat{\overline{R}}$ is well-defined, in view of the smallness of $\theta$. We then obtain a small $\widehat{\overline{R}}$-representation of $\Delta$ on $M \otimes_{\widehat{R_1}} \widehat{\overline{R}}$. The resulting correspondence is in fact an equivalence of categories from the category of small Higgs $\widehat{R_1}$-modules with coefficients in $\xi^{-1}\Omega^1_{R/\mathscr{O}_K}$ to that of small $\widehat{\overline{R}}$-representations of $\Delta$. It is essentially a quasi-inverse of the equivalence of categories defined in I.3.3.

To avoid the choice of a section $\psi$ of $\mathscr{L}(\widehat{Y})$, we can carry out the base change from $\widehat{\overline{R}}$ to $\mathscr{C}$ and use the diagonal embedding of $\mathbf{L}$. In this setting, the previous construction can be interpreted following the classic scheme of correspondences introduced by Fontaine (or even the more classic complex analytic Riemann–Hilbert correspondence) by taking for period ring making the link between generalized representations and Higgs modules a weak $p$-adic completion $\mathscr{C}^\dagger$ of $\mathscr{C}$ (the completion is made necessary by the exponential). With this ring is naturally associated a notion of admissibility; it is the notion of *generalized Dolbeault representation.* Before developing this approach, we will say a few words about the ring $\mathscr{C}$ that can itself play the role of period ring between the generalized representations and Higgs modules. Indeed, $\mathscr{C}$ is an *integral model of the Hyodo ring* (cf. (I.4.6.1) and II.15.6), which explains the link between our approach and that of Hyodo.

**I.4.5.** Recall that Faltings has defined a canonical extension of $\widehat{\overline{R}}$-representations of $\pi_1(X, \overline{y})$

(I.4.5.1)
$$0 \to \rho^{-1}\widehat{\overline{R}} \to \mathscr{E} \to \Omega^1_{R/\mathscr{O}_K} \otimes_R \widehat{\overline{R}}(-1) \to 0,$$

where $\rho$ is an element of $\mathscr{O}_{\overline{K}}$ of valuation $\geq \frac{1}{p-1}$ that plays an important role in his approach to $p$-adic Hodge theory (cf. II.7.22). We show in II.10.19 that there exists a $\Delta$-equivariant and $\widehat{\overline{R}}$-linear morphism

(I.4.5.2)
$$p^{-\frac{1}{p-1}}\mathscr{F} \to \mathscr{E}$$

that fits into a commutative diagram

(I.4.5.3)
$$\begin{array}{ccccccccc} 0 & \longrightarrow & p^{-\frac{1}{p-1}}\widehat{\overline{R}} & \longrightarrow & p^{-\frac{1}{p-1}}\mathscr{F} & \longrightarrow & p^{-\frac{1}{p-1}}\xi^{-1}\Omega^1_{R/\mathscr{O}_K} \otimes_R \widehat{\overline{R}} & \longrightarrow & 0 \\ & & \downarrow & & \downarrow & & \downarrow{\scriptstyle -c} & & \\ 0 & \longrightarrow & \rho^{-1}\widehat{\overline{R}} & \longrightarrow & \mathscr{E} & \longrightarrow & \Omega^1_{R/\mathscr{O}_K} \otimes_R \widehat{\overline{R}}(-1) & \longrightarrow & 0 \end{array}$$

where $c$ is the isomorphism induced by the canonical isomorphism $\widehat{\overline{R}}(1) \xrightarrow{\sim} p^{\frac{1}{p-1}}\xi\widehat{\overline{R}}$ (II.9.18). The morphism (I.4.5.2) is canonical if we take for $\widetilde{X}$ the deformation induced by an étale $S$-morphism $X \to \mathbb{G}^d_{m,S}$. It is important to note that in the logarithmic setting that will be considered in this work, the Faltings extension changes form slightly because the factor $\rho^{-1}\widehat{\overline{R}}$ is replaced by $(\pi\rho)^{-1}\widehat{\overline{R}}$, where $\pi$ is a uniformizer for $R$.

**I.4.6.** Taking Faltings extension $\mathscr{E}$ (I.4.5.1) as a starting point, Hyodo **[43]** defines an $\widehat{\overline{R}}$-algebra $\mathscr{C}_{\mathrm{HT}}$ using a direct limit analogous to (I.4.3.7). Note that $p$ being invertible in $\mathscr{C}_{\mathrm{HT}}$, this is equivalent to beginning with $\mathscr{E}\otimes_{\mathbb{Z}_p}\mathbb{Q}_p$, which corresponds to Hyodo's original definition. The morphism (I.4.5.2) therefore induces a $\Delta$-equivariant isomorphism of $\widehat{\overline{R}}$-algebras

$$\mathscr{C}[\frac{1}{p}] \xrightarrow{\sim} \mathscr{C}_{\mathrm{HT}}. \tag{I.4.6.1}$$

For every continuous $\mathbb{Q}_p$-representation $V$ of $\Gamma=\pi_1(X,\overline{y})$ and every integer $i$, Hyodo defines the $\widehat{R}[\frac{1}{p}]$-module $\mathrm{D}^i(V)$ by setting

$$\mathrm{D}^i(V)=(V\otimes_{\mathbb{Q}_p}\mathscr{C}_{\mathrm{HT}}(i))^{\Gamma}. \tag{I.4.6.2}$$

The representation $V$ is called *Hodge–Tate* if it satisfies the following conditions:

(i) $V$ is a $\mathbb{Q}_p$-vector space of finite dimension, endowed with the $p$-adic topology.

(ii) The canonical morphism

$$\oplus_{i\in\mathbb{Z}}\mathrm{D}^i(V)\otimes_{\widehat{R}[\frac{1}{p}]}\mathscr{C}_{\mathrm{HT}}(-i)\to V\otimes_{\mathbb{Q}_p}\mathscr{C}_{\mathrm{HT}} \tag{I.4.6.3}$$

is an isomorphism.

**I.4.7.** For any rational number $r\geq 0$, we denote by $\mathscr{F}^{(r)}$ the $\widehat{\overline{R}}$-representation of $\Delta$ deduced from $\mathscr{F}$ by inverse image under the morphism of multiplication by $p^r$ on $\xi^{-1}\Omega^1_{R/\mathscr{O}_K}\otimes_R\widehat{\overline{R}}$, so that we have an exact sequence

$$0\to\widehat{\overline{R}}\to\mathscr{F}^{(r)}\to\xi^{-1}\Omega^1_{R/\mathscr{O}_K}\otimes_R\widehat{\overline{R}}\to 0. \tag{I.4.7.1}$$

For every integer $n\geq 1$, this sequence induces an exact sequence

$$0\to\mathrm{Sym}^{n-1}_{\widehat{\overline{R}}}(\mathscr{F}^{(r)})\to\mathrm{Sym}^{n}_{\widehat{\overline{R}}}(\mathscr{F}^{(r)})\to\mathrm{Sym}^{n}_{\widehat{\overline{R}}}(\xi^{-1}\Omega^1_{R/\mathscr{O}_K}\otimes_R\widehat{\overline{R}})\to 0. \tag{I.4.7.2}$$

The $\widehat{\overline{R}}$-modules $(\mathrm{Sym}^n_{\widehat{\overline{R}}}(\mathscr{F}^{(r)}))_{n\in\mathbb{N}}$ therefore form a filtered direct system, whose direct limit

$$\mathscr{C}^{(r)}=\lim_{\substack{\longrightarrow\\ n\geq 0}}\mathrm{S}^n_{\widehat{\overline{R}}}(\mathscr{F}^{(r)}) \tag{I.4.7.3}$$

is naturally endowed with a structure of $\widehat{\overline{R}}$-algebra. The action of $\Delta$ on $\mathscr{F}^{(r)}$ induces an action on $\mathscr{C}^{(r)}$ by ring automorphisms, compatible with its action on $\widehat{\overline{R}}$, which we call *canonical action*. The $\widehat{\overline{R}}$-algebra $\mathscr{C}^{(r)}$ endowed with this action is called the *Higgs–Tate algebra of thickness* $r$ associated with $\widetilde{X}$. We denote by $\widehat{\mathscr{C}}^{(r)}$ the $p$-adic Hausdorff completion of $\mathscr{C}^{(r)}$ that we always assume endowed with the $p$-adic topology.

For all rational numbers $r\geq r'\geq 0$, we have an injective and $\Delta$-equivariant canonical $\widehat{\overline{R}}$-homomorphism $\alpha^{r,r'}\colon\mathscr{C}^{(r')}\to\mathscr{C}^{(r)}$. One easily verifies that the induced homomorphism $h^{r,r'}_{\alpha}\colon\widehat{\mathscr{C}}^{(r')}\to\widehat{\mathscr{C}}^{(r)}$ is injective. We set

$$\mathscr{C}^{\dagger}=\lim_{\substack{\longrightarrow\\ r\in\mathbb{Q}_{>0}}}\widehat{\mathscr{C}}^{(r)}, \tag{I.4.7.4}$$

which we identify with a sub-$\widehat{\overline{R}}$-algebra of $\widehat{\mathscr{C}}=\widehat{\mathscr{C}}^{(0)}$. The group $\Delta$ acts naturally on $\mathscr{C}^{\dagger}$ by ring automorphisms, in a manner compatible with its actions on $\widehat{\overline{R}}$ and on $\widehat{\mathscr{C}}$.

We denote by

$$d_{\mathscr{C}^{(r)}}\colon\mathscr{C}^{(r)}\to\xi^{-1}\Omega^1_{R/\mathscr{O}_K}\otimes_R\mathscr{C}^{(r)} \tag{I.4.7.5}$$

the universal $\widehat{\overline{R}}$-derivation of $\mathscr{C}^{(r)}$ and by

$$d_{\widehat{\mathscr{C}}^{(r)}} : \widehat{\mathscr{C}}^{(r)} \to \xi^{-1}\Omega^1_{R/\mathscr{O}_K} \otimes_R \widehat{\mathscr{C}}^{(r)} \tag{I.4.7.6}$$

its extension to the completions (note that the $R$-module $\Omega^1_{R/\mathscr{O}_K}$ is free of finite type). The derivations $d_{\mathscr{C}^{(r)}}$ and $d_{\widehat{\mathscr{C}}^{(r)}}$ are $\Delta$-equivariant. They are also Higgs $\widehat{\overline{R}}$-fields with coefficients in $\xi^{-1}\Omega^1_{R/\mathscr{O}_K}$ because $\xi^{-1}\Omega^1_{R/\mathscr{O}_K} \otimes_R \widehat{\overline{R}} = d_{\mathscr{C}^{(r)}}(\mathscr{F}^{(r)}) \subset d_{\mathscr{C}^{(r)}}(\mathscr{C}^{(r)})$ (cf. I.2.5).

For all rational numbers $r \geq r' \geq 0$, we have

$$p^{r'}(\mathrm{id} \times \alpha^{r,r'}) \circ d_{\mathscr{C}^{(r')}} = p^r d_{\mathscr{C}^{(r)}} \circ \alpha^{r,r'}. \tag{I.4.7.7}$$

The derivations $p^r d_{\widehat{\mathscr{C}}^{(r)}}$ therefore induce an $\widehat{\overline{R}}$-derivation

$$d_{\mathscr{C}^\dagger} : \mathscr{C}^\dagger \to \xi^{-1}\Omega^1_{R/\mathscr{O}_K} \otimes_R \mathscr{C}^\dagger, \tag{I.4.7.8}$$

that is none other than the restriction op $d_{\widehat{\mathscr{C}}}$ to $\mathscr{C}^\dagger$.

**I.4.8.** For any $\widehat{\overline{R}}$-representation $M$ of $\Delta$, we denote by $\mathbb{H}(M)$ the Higgs $\widehat{R_1}$-module with coefficients in $\xi^{-1}\Omega^1_{R/\mathscr{O}_K}$ defined by

$$\mathbb{H}(M) = (M \otimes_{\widehat{\overline{R}}} \mathscr{C}^\dagger)^\Delta \tag{I.4.8.1}$$

and by the Higgs field induced by $d_{\mathscr{C}^\dagger}$. For every Higgs $\widehat{R_1}$-module $(N, \theta)$ with coefficients in $\xi^{-1}\Omega^1_{R/\mathscr{O}_K}$, we denote by $\mathbb{V}(N)$ the $\widehat{\overline{R}}$-representation of $\Delta$ defined by

$$\mathbb{V}(N) = (N \otimes_{\widehat{R_1}} \mathscr{C}^\dagger)^{\theta_{\mathrm{tot}}=0}, \tag{I.4.8.2}$$

where $\theta_{\mathrm{tot}} = \theta \otimes \mathrm{id} + \mathrm{id} \otimes d_{\mathscr{C}^\dagger}$, and by the action of $\Delta$ induced by its canonical action on $\mathscr{C}^\dagger$. In order to make the most of these functors we establish acyclicity results for $\mathscr{C}^\dagger \otimes_{\mathbb{Z}_p} \mathbb{Q}_p$ for the Dolbeault cohomology (II.12.3) and for the continuous cohomology of $\Delta$ (II.12.5), slightly generalizing earlier results of Tsuji (cf. IV).

A continuous $\widehat{\overline{R}}$-representation $M$ of $\Delta$ is called *Dolbeault* if it satisfies the following conditions (cf. II.12.11):

(i) $M$ is a projective $\widehat{\overline{R}}$-module of finite type, endowed with the $p$-adic topology;
(ii) $\mathbb{H}(M)$ is a projective $\widehat{R_1}$-module of finite type;
(iii) the canonical $\mathscr{C}^\dagger$-linear morphism

$$\mathbb{H}(M) \otimes_{\widehat{R_1}} \mathscr{C}^\dagger \to M \otimes_{\widehat{\overline{R}}} \mathscr{C}^\dagger \tag{I.4.8.3}$$

is an isomorphism.

A Higgs $\widehat{R_1}$-module $(N, \theta)$ with coefficients in $\xi^{-1}\Omega^1_{R/\mathscr{O}_K}$ is called *solvable* if it satisfies the following conditions (cf. II.12.12):

(i) $N$ is a projective $\widehat{R_1}$-module of finite type;
(ii) $\mathbb{V}(N)$ is a projective $\widehat{\overline{R}}$-module of finite type;
(iii) the canonical $\mathscr{C}^\dagger$-linear morphism

$$\mathbb{V}(N) \otimes_{\widehat{\overline{R}}} \mathscr{C}^\dagger \to N \otimes_{\widehat{R_1}} \mathscr{C}^\dagger \tag{I.4.8.4}$$

is an isomorphism.

One immediately sees that the functors $\mathbb{V}$ and $\mathbb{H}$ induce equivalences of categories quasi-inverse to each other between the category of Dolbeault $\widehat{\overline{R}}$-representations of $\Delta$ and that of solvable Higgs $\widehat{R_1}$-modules with coefficients in $\xi^{-1}\Omega^1_{R/\mathscr{O}_K}$ (II.12.15).

We show that small $\widehat{\overline{R}}$-representations of $\Delta$ are Dolbeault (II.14.6), that small Higgs $\widehat{R_1}$-modules are solvable (II.13.20), and that $\mathbb{V}$ and $\mathbb{H}$ induce equivalences of categories quasi-inverse to each other between the categories of these objects (II.14.7). We in fact recover the correspondence defined in I.3.3, up to renormalization (cf. II.13.18).

**I.4.9.** We define the notions of *Dolbeault* $\widehat{\overline{R}}[\frac{1}{p}]$-representation of $\Delta$ and *solvable* Higgs $\widehat{R_1}[\frac{1}{p}]$-module with coefficients in $\xi^{-1}\Omega^1_{R/\mathscr{O}_K}$ by copying the definitions given in the integral case (cf. II.12.16 and II.12.18). We show that the functors $\mathbb{V}$ and $\mathbb{H}$ induce equivalences of categories quasi-inverse to each other between the category of Dolbeault $\widehat{\overline{R}}[\frac{1}{p}]$-representations of $\Delta$ and that of solvable Higgs $\widehat{R_1}[\frac{1}{p}]$-modules with coefficients in $\xi^{-1}\Omega^1_{R/\mathscr{O}_K}$ (II.12.24). This result is slightly more delicate than its integral analogue (I.4.8).

Unlike the integral case, the rational admissibility conditions can be interpreted in terms of divisibility conditions. More precisely, we say that a continuous $\widehat{\overline{R}}[\frac{1}{p}]$-representation $M$ of $\Delta$ is *small* if it satisfies the following conditions:

(i) $M$ is a projective $\widehat{\overline{R}}[\frac{1}{p}]$-module of finite type, endowed with a $p$-adic topology (II.2.2);

(ii) there exist a rational number $\alpha > \frac{2}{p-1}$ and a sub-$\widehat{\overline{R}}$-module $M^\circ$ of $M$ of finite type, stable under $\Delta$, generated by a finite number of elements $\Delta$-invariant modulo $p^\alpha M^\circ$, and that generates $M$ over $\widehat{\overline{R}}[\frac{1}{p}]$.

We say that a Higgs $\widehat{R_1}[\frac{1}{p}]$-module $(N, \theta)$ with coefficients in $\xi^{-1}\Omega^1_{R/\mathscr{O}_K}$ is *small* if it satisfies the following conditions:

(i) $N$ is a projective $\widehat{R_1}[\frac{1}{p}]$-module of finite type;

(ii) there exist a rational number $\beta > \frac{1}{p-1}$ and a sub-$\widehat{R_1}$-module $N^\circ$ of $N$ of finite type that generates $N$ over $\widehat{R_1}[\frac{1}{p}]$, such that we have

(I.4.9.1) $$\theta(N^\circ) \subset p^\beta \xi^{-1} N^\circ \otimes_R \Omega^1_{R/\mathscr{O}_K}.$$

**Proposition I.4.10** (cf. II.13.25). *A Higgs $\widehat{R_1}[\frac{1}{p}]$-module with coefficients in $\xi^{-1}\Omega^1_{R/\mathscr{O}_K}$ is solvable if and only if it is small.*

**Proposition I.4.11** (cf. II.13.26). *Every Dolbeault $\widehat{\overline{R}}[\frac{1}{p}]$-representation of $\Delta$ is small.*

We prove that the converse implication is equivalent to a descent property for small $\widehat{\overline{R}}[\frac{1}{p}]$-representations of $\Delta$ (II.14.8).

**Proposition I.4.12** (cf. II.12.26). *Let $M$ be a Dolbeault $\widehat{\overline{R}}[\frac{1}{p}]$-representation of $\Delta$ and $(\mathbb{H}(M), \theta)$ the associated Higgs $\widehat{R_1}[\frac{1}{p}]$-module with coefficients in $\xi^{-1}\Omega^1_{R/\mathscr{O}_K}$. We then have a functorial canonical isomorphism in $\mathbf{D}^+(\mathbf{Mod}(\widehat{R_1}[\frac{1}{p}]))$*

(I.4.12.1) $$\mathrm{C}^\bullet_{\mathrm{cont}}(\Delta, M) \xrightarrow{\sim} \mathbb{K}^\bullet(\mathbb{H}(M), \theta),$$

*where $\mathrm{C}^\bullet_{\mathrm{cont}}(\Delta, M)$ is the complex of continuous cochains of $\Delta$ with values in $M$ and $\mathbb{K}^\bullet(\mathbb{H}(M), \theta)$ is the Dolbeault complex* (I.2.3).

This statement was proved by Faltings for small representations ([**27**] § 3) and by Tsuji (IV.5.3.2).

**I.4.13.** It follows from (I.4.6.1) that if $V$ is a Hodge–Tate $\mathbb{Q}_p$-representation of $\Gamma$, then $V \otimes_{\mathbb{Z}_p} \widehat{\overline{R}}$ is a Dolbeault $\widehat{\overline{R}}[\frac{1}{p}]$-representation of $\Delta$; we have a functorial $\widehat{R_1}$-linear isomorphism

$$\mathbb{H}(V \otimes_{\mathbb{Z}_p} \widehat{\overline{R}}) \xrightarrow{\sim} \oplus_{i \in \mathbb{Z}} \mathrm{D}^i(V) \otimes_{\widehat{R}} \widehat{R_1}(-1), \tag{I.4.13.1}$$

and the Higgs field on $\mathbb{H}(V \otimes_{\mathbb{Z}_p} \widehat{\overline{R}})$ is induced by the $\widehat{R}$-linear morphisms

$$\mathrm{D}^i(V) \to \mathrm{D}^{i-1}(V) \otimes_R \Omega^1_{R/\mathscr{O}_K} \tag{I.4.13.2}$$

deduced from the universal derivation of $\mathscr{C}_{\mathrm{HT}}$ over $\widehat{\overline{R}}[\frac{1}{p}]$ (cf. II.15.7). Moreover, the isomorphism (I.4.13.1) is canonical if we take for $\widetilde{X}$ the deformation induced by an étale $S$-morphism $X \to \mathbb{G}^d_{m,S}$.

**I.4.14.** Hyodo ([**43**] 3.6) has proved that if $f\colon Y \to X$ is a proper and smooth morphism, for every integer $m \geq 0$, the sheaf $\mathrm{R}^m f_{\eta *}(\mathbb{Q}_p)$ is Hodge–Tate of weight between 0 and $m$; for every $0 \leq i \leq m$, we have a canonical isomorphism

$$\mathrm{D}^i(\mathrm{R}^m f_{\eta *}(\mathbb{Q}_p)) \xrightarrow{\sim} (\mathrm{R}^{m-i} f_{\eta *}(\Omega^i_{Y/X})) \otimes_R \widehat{R}, \tag{I.4.14.1}$$

and the morphism (I.4.13.2) is induced by the Kodaira–Spencer class of $f$. It follows that the Higgs bundle associated with $\mathrm{R}^m f_{\eta *}(\mathbb{Q}_p)$ is equal to the vector bundle

$$\oplus_{0 \leq i \leq m} \mathrm{R}^{m-i} f_{\eta *}(\Omega^i_{Y/X}), \tag{I.4.14.2}$$

endowed with the Higgs field $\theta$ defined by the Kodaira–Spencer class of $f$.

## I.5. Faltings ringed topos

**I.5.1.** We will tackle in Chapter III the global aspects of the theory in a logarithmic setting. However, in order to maintain a simplified presentation, we again restrict ourselves here to the smooth case in the usual sense (cf. III.4.7 for the smooth logarithmic case). In the remainder of this introduction, we suppose that $k$ is *algebraically closed* and we denote by $X$ a smooth $S$-scheme of finite type. From I.5.12 on, we will moreover suppose that there exists a smooth $\mathscr{A}_2(\overline{S})$-deformation $\widetilde{X}$ of $\check{\overline{X}}$ that we will fix.

**I.5.2.** The first difficulty we encounter in gluing the local construction described in I.4 is the sheafification of the notion of generalized representation. To do this, we use the *Faltings topos*, a fibered variant of Deligne's notion of *covanishing topos* that we develop in Chapter VI. We denote by $E$ the category of morphisms of schemes $V \to U$ over the canonical morphism $X_{\overline{\eta}} \to X$, that is, the commutative diagrams

$$\begin{array}{ccc} V & \longrightarrow & U \\ \big\downarrow & & \big\downarrow \\ X_{\overline{\eta}} & \longrightarrow & X \end{array} \tag{I.5.2.1}$$

such that the morphism $U \to X$ is étale and that the morphism $V \to U_{\overline{\eta}}$ is finite étale. It is fibered over the category $\mathbf{\acute{E}t}_{/X}$ of étale $X$-schemes, by the functor

$$\pi\colon E \to \mathbf{\acute{E}t}_{/X}, \quad (V \to U) \mapsto U. \tag{I.5.2.2}$$

The fiber of $\pi$ over an étale $X$-scheme $U$ is the category $\mathbf{\acute{E}t}_{\mathrm{f}/U_{\overline{\eta}}}$ of finite étale schemes over $U_{\overline{\eta}}$, which we endow with the étale topology. We denote by $U_{\overline{\eta},\mathrm{f\acute{e}t}}$ the topos of sheaves of sets on $\mathbf{\acute{E}t}_{\mathrm{f}/U_{\overline{\eta}}}$ (cf. VI.9.2). If $U_{\overline{\eta}}$ is connected and if $\overline{y}$ is a geometric point of $U_{\overline{\eta}}$, denoting by $\mathbf{B}_{\pi_1(U_{\overline{\eta}},\overline{y})}$ the classifying topos of the fundamental group $\pi_1(U_{\overline{\eta}},\overline{y})$, we have a canonical equivalence of categories (VI.9.8.4)

$$\nu_{\overline{y}}\colon U_{\overline{\eta},\mathrm{f\acute{e}t}} \xrightarrow{\sim} \mathbf{B}_{\pi_1(U_{\overline{\eta}},\overline{y})}. \tag{I.5.2.3}$$

We endow $E$ with the *covanishing topology* generated by the coverings $\{(V_i \to U_i) \to (V \to U)\}_{i\in I}$ of the following two types:

(v) $U_i = U$ for every $i \in I$, and $(V_i \to V)_{i\in I}$ is a covering;
(c) $(U_i \to U)_{i\in I}$ is a covering and $V_i = U_i \times_U V$ for every $i \in I$.

The resulting covanishing site $E$ is also called *Faltings site* of $X$. We denote by $\widetilde{E}$ and call *Faltings topos* of $X$ the topos of sheaves of sets on $E$. We refer to Chapter VI for a detailed study of this topos. Let us give a practical and simple description of $\widetilde{E}$.

**Proposition I.5.3** (cf. VI.5.10). *Giving a sheaf $F$ on $E$ is equivalent to giving, for every object $U$ of $\mathbf{\acute{E}t}_{/X}$, a sheaf $F_U$ of $U_{\overline{\eta},\mathrm{f\acute{e}t}}$, and for every morphism $f\colon U' \to U$ of $\mathbf{\acute{E}t}_{/X}$, a morphism $F_U \to f_{\mathrm{f\acute{e}t}*}(F_{U'})$, these morphisms being subject to compatibility relations such that for every covering family $(f_n\colon U_n \to U)_{n\in\Sigma}$ of $\mathbf{\acute{E}t}_{/X}$, if for any $(m,n)\in\Sigma^2$, we set $U_{mn} = U_m \times_U U_n$ and denote by $f_{mn}\colon U_{mn} \to U$ the canonical morphism, the sequence of morphisms of sheaves of $U_{\overline{\eta},\mathrm{f\acute{e}t}}$*

$$F_U \to \prod_{n\in\Sigma} (f_{n,\overline{\eta}})_{\mathrm{f\acute{e}t}*}(F_{U_n}) \rightrightarrows \prod_{(m,n)\in\Sigma^2} (f_{mn,\overline{\eta}})_{\mathrm{f\acute{e}t}*}(F_{U_{mn}}) \tag{I.5.3.1}$$

*is exact.*

From now on, we will identify every sheaf $F$ on $E$ with the associated functor $\{U \mapsto F_U\}$, the sheaf $F_U$ being the restriction of $F$ to the fiber $\mathbf{\acute{E}t}_{\mathrm{f}/U_{\overline{\eta}}}$ of $\pi$ over $U$.

**I.5.4.** The canonical injection functor $\mathbf{\acute{E}t}_{\mathrm{f}/X_{\overline{\eta}}} \to E$ is continuous and left exact (VI.5.32). It therefore defines a morphism of topos

$$\beta\colon \widetilde{E} \to Y_{\mathrm{f\acute{e}t}}. \tag{I.5.4.1}$$

Likewise, the functor

$$\sigma^+\colon \mathbf{\acute{E}t}_{/X} \to E, \quad U \mapsto (U_{\overline{\eta}} \to U) \tag{I.5.4.2}$$

is continuous and left exact (VI.5.32). It therefore defines a morphism of topos

$$\sigma\colon \widetilde{E} \to X_{\mathrm{\acute{e}t}}. \tag{I.5.4.3}$$

**I.5.5.** Let $\overline{x}$ be a geometric point of $X$ and $X'$ the strict localization of $X$ at $\overline{x}$. We denote by $E'$ the Faltings site associated with $X'$, by $\widetilde{E}'$ the topos of sheaves of sets on $E'$, and by

$$\beta'\colon \widetilde{E}' \to X'_{\overline{\eta},\mathrm{f\acute{e}t}} \tag{I.5.5.1}$$

the canonical morphism (I.5.4.1). We prove in VI.10.27 that the functor $\beta'_*$ is exact. This property is crucial for the study of the main sheaves of the Faltings topos considered in this work. The canonical morphism $X' \to X$ induces, by functoriality, a morphism of topos (VI.10.12)

$$\Phi\colon \widetilde{E}' \to \widetilde{E}. \tag{I.5.5.2}$$

We denote by

$$\varphi_{\overline{x}}\colon \widetilde{E} \to X'_{\overline{\eta},\mathrm{f\acute{e}t}} \tag{I.5.5.3}$$

the composed functor $\beta'_* \circ \Phi^*$.

We denote by $\mathfrak{V}_{\overline{x}}$ the category of $\overline{x}$-pointed étale $X$-schemes, or, equivalently, the category of neighborhoods of $\overline{x}$ in the site $\mathbf{\acute{E}t}_{/X}$. For every object $(U, \xi\colon \overline{x} \to U)$ of $\mathfrak{V}_{\overline{x}}$, we denote also by $\xi\colon X' \to U$ the $X$-morphism induced by $\xi$. We prove in VI.10.37 that for every sheaf $F = \{U \mapsto F_U\}$ of $\widetilde{E}$, we have a functorial canonical isomorphism

$$\varphi_{\overline{x}}(F) \xrightarrow{\sim} \varinjlim_{U \in \mathfrak{V}_{\overline{x}}^{\circ}} (\xi_{\overline{\eta}})^*_{\text{fét}}(F_U). \tag{I.5.5.4}$$

Assume that $\overline{x}$ is over $s$. We prove (III.3.7) that $\overline{X}'$ is normal and strictly local (and in particular integral). Let $\overline{y}$ be a geometric point of $X'_{\overline{\eta}}$ (which is integral), $\mathbf{B}_{\pi_1(X'_{\overline{\eta}}, \overline{y})}$ the classifying topos of the fundamental group $\pi_1(X'_{\overline{\eta}}, \overline{y})$, and

$$\nu_{\overline{y}}\colon X'_{\overline{\eta},\text{fét}} \xrightarrow{\sim} \mathbf{B}_{\pi_1(X'_{\overline{\eta}}, \overline{y})} \tag{I.5.5.5}$$

the fiber functor at $\overline{y}$ (VI.9.8.4). The composed functor

$$\widetilde{E} \xrightarrow{\varphi_{\overline{x}}} X'_{\overline{\eta},\text{fét}} \xrightarrow{\nu_{\overline{y}}} \mathbf{B}_{\pi_1(X'_{\overline{\eta}}, \overline{y})} \longrightarrow \mathbf{Set}, \tag{I.5.5.6}$$

where the last arrow is the forgetful functor of the action of $\pi_1(X'_{\overline{\eta}}, \overline{y})$, is a fiber functor (VI.10.31 and VI.9.9). It corresponds to a point of geometric origin of the topos $\widetilde{E}$, denoted by $\rho(\overline{y} \rightsquigarrow \overline{x})$ (cf. III.8.6).

**Theorem I.5.6** (cf. VI.10.30). *Under the assumptions of* I.5.5, *for every abelian sheaf* $F$ *of* $\widetilde{E}$ *and every integer* $i \geq 0$, *we have a functorial canonical isomorphism* (I.5.4.3)

$$\mathrm{R}^i\sigma_*(F)_{\overline{x}} \xrightarrow{\sim} \mathrm{H}^i(X'_{\overline{\eta},\text{fét}}, \varphi_{\overline{x}}(F)). \tag{I.5.6.1}$$

**Corollary I.5.7.** *We keep the assumptions of* I.5.5 *and moreover assume that* $x$ *is over* $s$. *Then, for every abelian sheaf* $F$ *of* $\widetilde{E}$ *and for every integer* $i \geq 0$, *we have a canonical functorial isomorphism*

$$\mathrm{R}^i\sigma_*(F)_{\overline{x}} \xrightarrow{\sim} \mathrm{H}^i(\pi_1(X'_{\overline{\eta}}, \overline{y}), \nu_{\overline{y}}(\varphi_{\overline{x}}(F))). \tag{I.5.7.1}$$

**Proposition I.5.8** (cf. VI.10.32). *When* $\overline{x}$ *goes through the set of geometric points of* $X$, *the family of functors* $\varphi_{\overline{x}}$ (I.5.5.3) *is conservative.*

**I.5.9.** For every object $(V \to U)$ of $E$, we denote by $\overline{U}^V$ the integral closure of $\overline{U} = U \times_S \overline{S}$ in $V$ and we set

$$\overline{\mathscr{B}}(V \to U) = \Gamma(\overline{U}^V, \mathscr{O}_{\overline{U}^V}). \tag{I.5.9.1}$$

We thus define a presheaf of rings on $E$, which turns out to be a *sheaf* (III.8.16). Note that $\overline{\mathscr{B}}$ is not in general a sheaf for the topology of $E$ originally defined by Faltings in ([**26**] page 214) (cf. III.8.18). For every $U \in \mathrm{Ob}(\mathbf{\acute{E}t}_{/X})$, we denote by $\overline{\mathscr{B}}_U$ the restriction of $\overline{\mathscr{B}}$ to the fiber $\mathbf{\acute{E}t}_{\mathrm{f}/U_{\overline{\eta}}}$ of $\pi$ over $U$, so that $\overline{\mathscr{B}} = \{U \to \overline{\mathscr{B}}_U\}$. In I.5.10 below, we give an explicit description of this sheaf. For any integer $n \geq 0$, we set

$$\overline{\mathscr{B}}_n = \overline{\mathscr{B}}/p^n\overline{\mathscr{B}}, \tag{I.5.9.2}$$

$$\overline{\mathscr{B}}_{U,n} = \overline{\mathscr{B}}_U/p^n\overline{\mathscr{B}}_U. \tag{I.5.9.3}$$

Note that the correspondence $\{U \mapsto \overline{\mathscr{B}}_{U,n}\}$ naturally forms a presheaf on $E$ whose associated sheaf is canonically isomorphic to $\overline{\mathscr{B}}_n$. It is in general difficult, if not impossible, to describe explicitly the restrictions of $\overline{\mathscr{B}}_n$ to the fibers of the functor $\pi$ (I.5.2.2). However, its images by the fiber functors (I.5.5.6) are accessible (III.10.8.5).

We denote by $\hbar\colon \overline{X} \to X$ the canonical projection (I.2.1.1) and by

$$\hbar_*(\mathscr{O}_{\overline{X}}) \to \sigma_*(\overline{\mathscr{B}}) \tag{I.5.9.4}$$

the homomorphism defined for every $U \in \mathrm{Ob}(\mathbf{\acute{E}t}_{/X})$ by the canonical homomorphism

$$\Gamma(\overline{U}, \mathscr{O}_{\overline{U}}) \to \Gamma(\overline{U}^{U_{\overline{\eta}}}, \mathscr{O}_{\overline{U}^{U_{\overline{\eta}}}}). \tag{I.5.9.5}$$

Unless explicitly stated otherwise, we consider $\sigma\colon \widetilde{E} \to X_{\text{ét}}$ (I.5.4.3) as a morphism of ringed topos (by $\overline{\mathscr{B}}$ and $\hbar_*(\mathscr{O}_{\overline{X}})$, respectively).

**I.5.10.** Let $U$ be an object of $\mathbf{\acute{E}t}_{/X}$, $\overline{y}$ a geometric point of $U_{\overline{\eta}}$, and $V$ the connected component of $U_{\overline{\eta}}$ containing $\overline{y}$. We denote by $\mathbf{B}_{\pi_1(V,\overline{y})}$ the classifying topos of the fundamental group $\pi_1(V,\overline{y})$, by $(V_i)_{i\in I}$ the normalized universal cover of $V$ at $\overline{y}$ (VI.9.8), and by

$$\nu_{\overline{y}}\colon V_{\text{fét}} \xrightarrow{\sim} \mathbf{B}_{\pi_1(V,\overline{y})}, \quad F \mapsto \varinjlim_{i\in I^\circ} F(V_i) \tag{I.5.10.1}$$

the fiber functor at $\overline{y}$. For every $i \in I$, $(V_i \to U)$ is naturally an object of $E$. We can therefore consider the inverse system of schemes $(\overline{U}^{V_i})_{i\in I}$. We set

$$\overline{R}^{\overline{y}}_U = \varinjlim_{i\in I^\circ} \Gamma(\overline{U}^{V_i}, \mathscr{O}_{\overline{U}^{V_i}}), \tag{I.5.10.2}$$

which is a ring of $\mathbf{B}_{\pi_1(V,\overline{y})}$. By III.8.15, we have a canonical isomorphism of $\mathbf{B}_{\pi_1(V,\overline{y})}$

$$\nu_{\overline{y}}(\overline{\mathscr{B}}_U|V) \xrightarrow{\sim} \overline{R}^{\overline{y}}_U. \tag{I.5.10.3}$$

**I.5.11.** Since $X_\eta$ is a subobject of the final object $X$ of $X_{\text{ét}}$, $\sigma^*(X_\eta)$ is a subobject of the final object of $\widetilde{E}$. We denote by

$$\gamma\colon \widetilde{E}_{/\sigma^*(X_\eta)} \to \widetilde{E} \tag{I.5.11.1}$$

the localization morphism of $\widetilde{E}$ at $\sigma^*(X_\eta)$. We denote by $\widetilde{E}_s$ the closed subtopos of $\widetilde{E}$ complement of $\sigma^*(X_\eta)$, that is, the full subcategory of $\widetilde{E}$ made up of the sheaves $F$ such that $\gamma^*(F)$ is a final object of $\widetilde{E}_{/\sigma^*(X_\eta)}$, and by

$$\delta\colon \widetilde{E}_s \to \widetilde{E} \tag{I.5.11.2}$$

the canonical embedding, that is, the morphism of topos such that $\delta_*\colon \widetilde{E}_s \to \widetilde{E}$ is the canonical injection functor. There exists a morphism

$$\sigma_s\colon \widetilde{E}_s \to X_{s,\text{ét}}, \tag{I.5.11.3}$$

unique up to isomorphism, such that the diagram

$$\begin{array}{ccc} \widetilde{E}_s & \xrightarrow{\sigma_s} & X_{s,\text{ét}} \\ {\scriptstyle\delta}\downarrow & & \downarrow{\scriptstyle\iota_{\text{ét}}} \\ \widetilde{E} & \xrightarrow{\sigma} & X_{\text{ét}} \end{array} \tag{I.5.11.4}$$

where $\iota\colon X_s \to X$ is the canonical injection, is commutative up to isomorphism (cf. III.9.8).

For every integer $n \geq 1$, if we identify the étale topos of $X_s$ and $\overline{X}_n$ (I.2.1.1) ($k$ being algebraically closed), the morphism $\sigma_s$ and the homomorphism (I.5.9.4) induce a morphism of ringed topos (III.9.9)

$$\sigma_n\colon (\widetilde{E}_s, \overline{\mathscr{B}}_n) \to (X_{s,\text{ét}}, \mathscr{O}_{\overline{X}_n}). \tag{I.5.11.5}$$

**I.5.12.** For the remainder of this introduction, we assume that there exists a smooth $\mathscr{A}_2(\overline{S})$-deformation $\widetilde{X}$ of $\check{\overline{X}}$ that we fix (cf. (I.2.1.1) and I.4.2 for the notation):

$$\begin{array}{ccc} \check{\overline{X}} & \longrightarrow & \widetilde{X} \\ \downarrow & & \downarrow \\ \check{\overline{S}} & \longrightarrow & \mathscr{A}_2(\overline{S}) \end{array} \tag{I.5.12.1}$$

Let $Y = \mathrm{Spec}(R)$ be a connected affine object of $\mathbf{\acute{E}t}_{/X}$ admitting an étale $S$-morphism to $\mathbb{G}^d_{m,S}$ for an integer $d \geq 1$ and such that $Y_s \neq \emptyset$ (in other words, $Y$ satisfies the conditions of I.4.1) and $\widetilde{Y} \to \widetilde{X}$ the unique étale morphism that lifts $\check{\overline{Y}} \to \check{\overline{X}}$. For any geometric point $\overline{y}$ of $Y_{\overline{\eta}}$, we denote by $Y'_{\overline{\eta}}$ the connected component of $Y_{\overline{\eta}}$ containing $\overline{y}$, by

$$\nu_{\overline{y}}\colon Y'_{\overline{\eta},\mathrm{f\acute{e}t}} \xrightarrow{\sim} \mathbf{B}_{\pi_1(Y'_{\overline{\eta}},\overline{y})} \tag{I.5.12.2}$$

the fiber functor at $\overline{y}$, and by $\mathscr{F}^{\overline{y}}_Y$ the Higgs–Tate $\widehat{\overline{R}}^{\overline{y}}_Y$-extension associated with $(Y, \widetilde{Y})$ (I.4.3).

For every integer $n \geq 0$, there exists a $\overline{\mathscr{B}}_{Y,n}$-module $\mathscr{F}_{Y,n}$, where $\overline{\mathscr{B}}_{Y,n} = \overline{\mathscr{B}}_Y/p^n\overline{\mathscr{B}}_Y$ (I.5.9.3), unique up to canonical isomorphism, such that for every geometric point $\overline{y}$ of $Y_{\overline{\eta}}$, we have a canonical isomorphism of $\overline{R}^{\overline{y}}_Y$-modules (I.5.10.3)

$$\nu_{\overline{y}}(\mathscr{F}_{Y,n}|Y'_{\overline{\eta}}) \xrightarrow{\sim} \mathscr{F}^{\overline{y}}_Y/p^n\mathscr{F}^{\overline{y}}_Y. \tag{I.5.12.3}$$

The exact sequence (I.4.3.5) induces a canonical exact sequence of $\overline{\mathscr{B}}_{Y,n}$-modules

$$0 \to \overline{\mathscr{B}}_{Y,n} \to \mathscr{F}_{Y,n} \to \xi^{-1}\Omega^1_{X/S}(Y) \otimes_{\mathscr{O}_X(Y)} \overline{\mathscr{B}}_{Y,n} \to 0. \tag{I.5.12.4}$$

For every rational number $r \geq 0$, we denote by $\mathscr{F}^{(r)}_{Y,n}$ the extension of $\overline{\mathscr{B}}_{Y,n}$-modules of $Y_{\overline{\eta},\mathrm{f\acute{e}t}}$ deduced from $\mathscr{F}_{Y,n}$ by inverse image under the morphism of multiplication by $p^r$ on $\xi^{-1}\Omega^1_{X/S}(Y) \otimes_{\mathscr{O}_X(Y)} \overline{\mathscr{B}}_{Y,n}$, so that we have a canonical exact sequence of $\overline{\mathscr{B}}_{Y,n}$-modules

$$0 \to \overline{\mathscr{B}}_{Y,n} \to \mathscr{F}^{(r)}_{Y,n} \to \xi^{-1}\Omega^1_{X/S}(Y) \otimes_{\mathscr{O}_X(Y)} \overline{\mathscr{B}}_{Y,n} \to 0. \tag{I.5.12.5}$$

This induces, for every integer $m \geq 1$, an exact sequence of $\overline{\mathscr{B}}_{Y,n}$-modules

$$0 \to \mathrm{S}^{m-1}_{\overline{\mathscr{B}}_{Y,n}}(\mathscr{F}^{(r)}_{Y,n}) \to \mathrm{S}^{m}_{\overline{\mathscr{B}}_{Y,n}}(\mathscr{F}^{(r)}_{Y,n}) \to \mathrm{S}^{m}_{\overline{\mathscr{B}}_{Y,n}}(\xi^{-1}\Omega^1_{X/S}(Y) \otimes_{\mathscr{O}_X(Y)} \overline{\mathscr{B}}_{Y,n}) \to 0.$$

The $\overline{\mathscr{B}}_{Y,n}$-modules $(\mathrm{S}^{m}_{\overline{\mathscr{B}}_{Y,n}}(\mathscr{F}^{(r)}_{Y,n}))_{m\in\mathbb{N}}$ therefore form a direct system whose direct limit

$$\mathscr{C}^{(r)}_{Y,n} = \varinjlim_{m\geq 0} \mathrm{S}^{m}_{\overline{\mathscr{B}}_{Y,n}}(\mathscr{F}^{(r)}_{Y,n}) \tag{I.5.12.6}$$

is naturally endowed with a structure of $\overline{\mathscr{B}}_{Y,n}$-algebra of $Y_{\overline{\eta},\mathrm{f\acute{e}t}}$.

For all rational numbers $r \geq r' \geq 0$, we have a canonical $\overline{\mathscr{B}}_{Y,n}$-linear morphism

$$\mathsf{a}^{r,r'}_{Y,n}\colon \mathscr{F}^{(r)}_{Y,n} \to \mathscr{F}^{(r')}_{Y,n} \tag{I.5.12.7}$$

that lifts the morphism of multiplication by $p^{r'-r}$ on $\xi^{-1}\Omega^1_{X/S}(Y) \otimes_{\mathscr{O}_X(Y)} \overline{\mathscr{B}}_{Y,n}$ and that extends the identity on $\overline{\mathscr{B}}_{Y,n}$ (I.5.12.5). It induces a homomorphism of $\overline{\mathscr{B}}_{Y,n}$-algebras

$$\alpha^{r,r'}_{Y,n}\colon \mathscr{C}^{(r)}_{Y,n} \to \mathscr{C}^{(r')}_{Y,n}. \tag{I.5.12.8}$$

Note that $\mathscr{C}^{(r)}_{Y,n}$ and $\mathscr{F}^{(r)}_{Y,n}$ depend on the choice of the deformation $\widetilde{X}$.

We extend the previous definitions to connected affine objects $Y$ of $\mathbf{\acute{E}t}_{/X}$ such that $Y_s = \emptyset$ by setting $\mathscr{C}^{(r)}_{Y,n} = \mathscr{F}^{(r)}_{Y,n} = 0$.

**I.5.13.** Let $n$ be an integer $\geq 0$ and $r$ a rational number $\geq 0$. The correspondences $\{Y \mapsto \mathscr{F}^{(r)}_{Y,n}\}$ and $\{Y \mapsto \mathscr{C}^{(r)}_{Y,n}\}$ naturally form presheaves on the full subcategory of $E$ made up of the objects $(V \to Y)$ such that $Y$ is affine, connected, and admits an étale morphism to $\mathbb{G}^d_{m,S}$ for an integer $d \geq 1$ (cf. III.10.19). Since this subcategory is clearly topologically generating in $E$, we can consider the associated sheaves in $\widetilde{E}$

(I.5.13.1) $$\mathscr{F}^{(r)}_n = \{Y \mapsto \mathscr{F}^{(r)}_{Y,n}\}^a,$$

(I.5.13.2) $$\mathscr{C}^{(r)}_n = \{Y \mapsto \mathscr{C}^{(r)}_{Y,n}\}^a.$$

These are in fact a $\overline{\mathscr{B}}_n$-module and a $\overline{\mathscr{B}}_n$-algebra of $\widetilde{E}_s$ (III.10.22). We call $\mathscr{F}^{(r)}_n$ the *Higgs–Tate* $\overline{\mathscr{B}}_n$*-extension of thickness* $r$ and call $\mathscr{C}^{(r)}_n$ the *Higgs–Tate* $\overline{\mathscr{B}}_n$*-algebra of thickness* $r$ associated with $\widetilde{X}$. We have a canonical exact sequence of $\overline{\mathscr{B}}_n$-modules (I.5.11.5)

(I.5.13.3) $$0 \to \overline{\mathscr{B}}_n \to \mathscr{F}^{(r)}_n \to \sigma^*_n(\xi^{-1}\Omega^1_{\overline{X}_n/\overline{S}_n}) \to 0.$$

In III.10.29 we describe explicitly the images of $\mathscr{F}^{(r)}_n$ and $\mathscr{C}^{(r)}_n$ by the fiber functors (I.5.5.3).

For all rational numbers $r \geq r' \geq 0$, the homomorphisms (I.5.12.8) induce a homomorphism of $\overline{\mathscr{B}}_n$-algebras

(I.5.13.4) $$\alpha^{r,r'}_n : \mathscr{C}^{(r)}_n \to \mathscr{C}^{(r')}_n.$$

For all rational numbers $r \geq r' \geq r'' \geq 0$, we have

(I.5.13.5) $$\alpha^{r,r''}_n = \alpha^{r',r''}_n \circ \alpha^{r,r'}_n.$$

We have a canonical $\mathscr{C}^{(r)}_n$-linear isomorphism

(I.5.13.6) $$\Omega^1_{\mathscr{C}^{(r)}_n/\overline{\mathscr{B}}_n} \xrightarrow{\sim} \sigma^*_n(\xi^{-1}\Omega^1_{\overline{X}_n/\overline{S}_n}) \otimes_{\overline{\mathscr{B}}_n} \mathscr{C}^{(r)}_n.$$

The universal $\overline{\mathscr{B}}_n$-derivation of $\mathscr{C}^{(r)}_n$ corresponds through this isomorphism to the unique $\overline{\mathscr{B}}_n$-derivation

(I.5.13.7) $$d^{(r)}_n : \mathscr{C}^{(r)}_n \to \sigma^*_n(\xi^{-1}\Omega^1_{\overline{X}_n/\overline{S}_n}) \otimes_{\overline{\mathscr{B}}_n} \mathscr{C}^{(r)}_n$$

that extends the canonical morphism $\mathscr{F}^{(r)}_n \to \sigma^*_n(\xi^{-1}\Omega^1_{\overline{X}_n/\overline{S}_n})$. For all rational numbers $r \geq r' \geq 0$, we have

(I.5.13.8) $$p^{r-r'}(\mathrm{id} \otimes \alpha^{r,r'}_n) \circ d^{(r)}_n = d^{(r')}_n \circ \alpha^{r,r'}_n.$$

**I.5.14.** The inverse systems of objects of $\widetilde{E}_s$ (I.5.4) indexed by the ordered set of natural numbers $\mathbb{N}$ form a topos that we denote by $\widetilde{E}^{\mathbb{N}^\circ}_s$. Section III.7 contains useful results for this type of topos. We denote by $\breve{\overline{\mathscr{B}}}$ the ring $(\overline{\mathscr{B}}_{n+1})_{n\in\mathbb{N}}$ of $\widetilde{E}^{\mathbb{N}^\circ}_s$, by $\mathscr{O}_{\breve{\overline{X}}}$ the ring $(\mathscr{O}_{\overline{X}_{n+1}})_{n\in\mathbb{N}}$ of $X^{\mathbb{N}^\circ}_{s,\text{ét}}$, and by $\xi^{-1}\Omega^1_{\breve{\overline{X}}/\breve{\overline{S}}}$ the $\mathscr{O}_{\breve{\overline{X}}}$-module $(\xi^{-1}\Omega^1_{\overline{X}_{n+1}/\overline{S}_{n+1}})_{n\in\mathbb{N}}$ of $X^{\mathbb{N}^\circ}_{s,\text{ét}}$. The morphisms $(\sigma_{n+1})_{n\in\mathbb{N}}$ (I.5.11.5) induce a morphism of ringed topos

(I.5.14.1) $$\breve{\sigma} : (\widetilde{E}^{\mathbb{N}^\circ}_s, \breve{\overline{\mathscr{B}}}) \to (X^{\mathbb{N}^\circ}_{s,\text{ét}}, \mathscr{O}_{\breve{\overline{X}}}).$$

We say that a $\breve{\overline{\mathscr{B}}}$-module $(M_{n+1})_{n\in\mathbb{N}}$ of $\widetilde{E}^{\mathbb{N}^\circ}_s$ is *adic* if for all integers $m$ and $n$ such that $m \geq n \geq 1$, the morphism $M_m \otimes_{\overline{\mathscr{B}}_m} \overline{\mathscr{B}}_n \to M_n$ deduced from the transition morphism $M_m \to M_n$ is an isomorphism.

Let $r$ be a rational number $\geq 0$. For all integers $m \geq n \geq 1$, we have a canonical $\overline{\mathscr{B}}_m$-linear morphism $\mathscr{F}_m^{(r)} \to \mathscr{F}_n^{(r)}$ and a canonical homomorphism of $\overline{\mathscr{B}}_m$-algebras $\mathscr{C}_m^{(r)} \to \mathscr{C}_n^{(r)}$ such that the induced morphisms

$$\mathscr{F}_m^{(r)} \otimes_{\overline{\mathscr{B}}_m} \overline{\mathscr{B}}_n \to \mathscr{F}_n^{(r)} \quad \text{and} \quad \mathscr{C}_m^{(r)} \otimes_{\overline{\mathscr{B}}_m} \overline{\mathscr{B}}_n \to \mathscr{C}_n^{(r)} \tag{I.5.14.2}$$

are isomorphisms. These morphisms form compatible systems when $m$ and $n$ vary, so that $(\mathscr{F}_{n+1}^{(r)})_{n\in\mathbb{N}}$ and $(\mathscr{C}_{n+1}^{(r)})_{n\in\mathbb{N}}$ are inverse systems. We call *Higgs–Tate* $\breve{\overline{\mathscr{B}}}$*-extension of thickness* $r$ associated with $\widetilde{X}$, and denote by $\breve{\mathscr{F}}^{(r)}$, the $\breve{\overline{\mathscr{B}}}$-module $(\mathscr{F}_{n+1}^{(r)})_{n\in\mathbb{N}}$ of $\widetilde{E}_s^{\mathbb{N}^\circ}$. We call *Higgs–Tate* $\breve{\overline{\mathscr{B}}}$*-algebra of thickness* $r$ associated with $\widetilde{X}$ and denote by $\breve{\mathscr{C}}^{(r)}$, the $\breve{\overline{\mathscr{B}}}$-algebra $(\mathscr{C}_{n+1}^{(r)})_{n\in\mathbb{N}}$ of $\widetilde{E}_s^{\mathbb{N}^\circ}$. These are adic $\breve{\overline{\mathscr{B}}}$-modules. We have an exact sequence of $\breve{\overline{\mathscr{B}}}$-modules

$$0 \to \breve{\overline{\mathscr{B}}} \to \breve{\mathscr{F}}^{(r)} \to \breve{\sigma}^*(\xi^{-1}\Omega^1_{\breve{\overline{X}}/\breve{\overline{S}}}) \to 0. \tag{I.5.14.3}$$

For all rational numbers $r \geq r' \geq 0$, the homomorphisms $(\alpha_n^{r,r'})_{n\in\mathbb{N}}$ induce a homomorphism of $\breve{\overline{\mathscr{B}}}$-algebras

$$\breve{\alpha}^{r,r'} : \breve{\mathscr{C}}^{(r)} \to \breve{\mathscr{C}}^{(r')}. \tag{I.5.14.4}$$

For all rational numbers $r \geq r' \geq r'' \geq 0$, we have

$$\breve{\alpha}^{r,r''} = \breve{\alpha}^{r',r''} \circ \breve{\alpha}^{r,r'}. \tag{I.5.14.5}$$

The derivations $(d_{n+1}^{(r)})_{n\in\mathbb{N}}$ define a morphism

$$\breve{d}^{(r)} : \breve{\mathscr{C}}^{(r)} \to \breve{\sigma}^*(\xi^{-1}\Omega^1_{\breve{\overline{X}}/\breve{\overline{S}}}) \otimes_{\breve{\overline{\mathscr{B}}}} \breve{\mathscr{C}}^{(r)} \tag{I.5.14.6}$$

that is none other than the universal $\breve{\overline{\mathscr{B}}}$-derivation of $\breve{\mathscr{C}}^{(r)}$. For all rational numbers $r \geq r' \geq 0$, we have

$$p^{r-r'}(\mathrm{id} \otimes \breve{\alpha}^{r,r'}) \circ \breve{d}^{(r)} = \breve{d}^{(r')} \circ \breve{\alpha}^{r,r'}. \tag{I.5.14.7}$$

## I.6. Dolbeault modules

**I.6.1.** We keep the hypotheses and notation of I.5 in this section. We set $\mathscr{S} = \mathrm{Spf}(\mathscr{O}_C)$ and denote by $\mathfrak{X}$ the formal scheme $p$-adic completion of $\overline{X}$ and by $\xi^{-1}\Omega^1_{\mathfrak{X}/\mathscr{S}}$ the $p$-adic completion of the $\mathscr{O}_{\overline{X}}$-module $\xi^{-1}\Omega^1_{\overline{X}/\overline{S}} = \xi^{-1}\Omega^1_{X/S} \otimes_{\mathscr{O}_X} \mathscr{O}_{\overline{X}}$. We denote by

$$\breve{u} \colon (X_{s,\text{ét}}^{\mathbb{N}^\circ}, \mathscr{O}_{\breve{\overline{X}}}) \to (X_{s,\text{zar}}^{\mathbb{N}^\circ}, \mathscr{O}_{\breve{\overline{X}}}) \tag{I.6.1.1}$$

the canonical morphism of ringed topos (I.5.14 and III.2.9), by

$$\lambda \colon (X_{s,\text{zar}}^{\mathbb{N}^\circ}, \mathscr{O}_{\breve{\overline{X}}}) \to (X_{s,\text{zar}}, \mathscr{O}_{\mathfrak{X}}) \tag{I.6.1.2}$$

the morphism of ringed topos whose corresponding direct image functor is the inverse limit (III.7.4), and by

$$\top \colon (\widetilde{E}_s^{\mathbb{N}^\circ}, \breve{\overline{\mathscr{B}}}) \to (\mathfrak{X}_{\text{zar}}, \mathscr{O}_{\mathfrak{X}}) \tag{I.6.1.3}$$

the composed morphism $\lambda \circ \breve{u} \circ \breve{\sigma}$ (I.5.14.1). For modules, we use the notation $\top^{-1}$ to denote the inverse image in the sense of abelian sheaves, and we keep the notation $\top^*$ for the inverse image in the sense of modules; we do likewise for $\breve{\sigma}$.

We denote by

$$\delta \colon \xi^{-1}\Omega^1_{\mathfrak{X}/\mathscr{S}} \to \mathrm{R}^1\top_*(\breve{\overline{\mathscr{B}}}) \tag{I.6.1.4}$$

the $\mathscr{O}_{\mathfrak{X}}$-linear morphism of $X_{s,\mathrm{zar}}$ composed of the adjunction morphism (III.11.2.5)

$$\xi^{-1}\Omega^1_{\mathfrak{X}/\mathscr{S}} \to \top_*(\breve{\sigma}^*(\xi^{-1}\Omega^1_{\breve{X}/\breve{S}})) \tag{I.6.1.5}$$

and the boundary map of the long exact sequence of cohomology deduced from the canonical exact sequence

$$0 \to \breve{\overline{\mathscr{B}}} \to \breve{\mathscr{F}} \to \breve{\sigma}^*(\xi^{-1}\Omega^1_{\breve{X}/\breve{S}}) \to 0. \tag{I.6.1.6}$$

Note that the morphism

$$\top^*(\xi^{-1}\Omega^1_{\mathfrak{X}/\mathscr{S}}) \to \breve{\sigma}^*(\xi^{-1}\Omega^1_{\breve{X}/\breve{S}}), \tag{I.6.1.7}$$

adjoint to (I.6.1.5), is an isomorphism (III.11.2.6).

**Theorem I.6.2** (cf. III.11.8). *There exists a unique isomorphism of graded $\mathscr{O}_{\mathfrak{X}}[\frac{1}{p}]$-algebras*

$$\wedge\,(\xi^{-1}\Omega^1_{\mathfrak{X}/\mathscr{S}}[\frac{1}{p}]) \xrightarrow{\sim} \oplus_{i\geq 0}\mathrm{R}^i\top_*(\breve{\overline{\mathscr{B}}})[\frac{1}{p}] \tag{I.6.2.1}$$

*whose component in degree one is the morphism $\delta\otimes_{\mathbb{Z}_p}\mathbb{Q}_p$ (I.6.1.4).*

This statement is the key step in Faltings' approach in $p$-adic Hodge theory. We encounter it here and there in different forms. Its local Galois form (II.8.21) is a consequence of Faltings' almost purity theorem (II.6.16). The global statement has an integral variant (III.11.3) that follows from the local case by localization (I.5.7).

**I.6.3.** The canonical exact sequence (I.6.1.6) induces, for every integer $m\geq 1$, an exact sequence

$$0 \to \mathrm{Sym}^{m-1}_{\breve{\overline{\mathscr{B}}}}(\breve{\mathscr{F}}) \to \mathrm{Sym}^{m}_{\breve{\overline{\mathscr{B}}}}(\breve{\mathscr{F}}) \to \breve{\sigma}^*(\mathrm{Sym}^m_{\mathscr{O}_{\breve{X}}}(\xi^{-1}\Omega^1_{\breve{X}/\breve{S}})) \to 0. \tag{I.6.3.1}$$

**Proposition I.6.4** (cf. III.11.12). *Let $m$ be an integer $\geq 1$. Then:*

(i) *The morphism*

$$\top_*(\mathrm{Sym}^{m-1}_{\breve{\overline{\mathscr{B}}}}(\breve{\mathscr{F}}))[\frac{1}{p}] \to \top_*(\mathrm{Sym}^{m}_{\breve{\overline{\mathscr{B}}}}(\breve{\mathscr{F}}))[\frac{1}{p}] \tag{I.6.4.1}$$

*induced by (I.6.3.1) is an isomorphism.*

(ii) *For every integer $q\geq 1$, the morphism*

$$\mathrm{R}^q\top_*(\mathrm{Sym}^{m-1}_{\breve{\overline{\mathscr{B}}}}(\breve{\mathscr{F}}))[\frac{1}{p}] \to \mathrm{R}^q\top_*(\mathrm{Sym}^{m}_{\breve{\overline{\mathscr{B}}}}(\breve{\mathscr{F}}))[\frac{1}{p}] \tag{I.6.4.2}$$

*induced by (I.6.3.1) is zero.*

The local Galois variant of this statement is due to Hyodo ([**43**] 1.2). It is the main ingredient in the definition of Hodge–Tate local systems.

**Proposition I.6.5** (cf. III.11.18). *The canonical homomorphism*

$$\mathscr{O}_{\mathfrak{X}}[\frac{1}{p}] \to \varinjlim_{r\in\mathbb{Q}_{>0}} \top_*(\breve{\mathscr{C}}^{(r)})[\frac{1}{p}] \tag{I.6.5.1}$$

*is an isomorphism, and for every integer $q\geq 1$,*

$$\varinjlim_{r\in\mathbb{Q}_{>0}} \mathrm{R}^q\top_*(\breve{\mathscr{C}}^{(r)})[\frac{1}{p}] = 0. \tag{I.6.5.2}$$

The local Galois variant of this statement (II.12.5) is mainly due to Tsuji (IV.5.3.4).

**I.6.6.** We denote by $\mathbf{Mod}(\breve{\overline{\mathscr{B}}})$ the category of $\breve{\overline{\mathscr{B}}}$-modules of $\widetilde{E}_s^{\mathbb{N}^\circ}$, by $\mathbf{Mod}^{\mathrm{ad}}(\breve{\overline{\mathscr{B}}})$ (resp. $\mathbf{Mod}^{\mathrm{aft}}(\breve{\overline{\mathscr{B}}})$) the full subcategory made up of the adic $\breve{\overline{\mathscr{B}}}$-modules (resp. the adic $\breve{\overline{\mathscr{B}}}$-modules of finite type) (I.5.14), and by $\mathbf{Mod}_{\mathbb{Q}}(\breve{\overline{\mathscr{B}}})$ (resp. $\mathbf{Mod}_{\mathbb{Q}}^{\mathrm{ad}}(\breve{\overline{\mathscr{B}}})$, resp. $\mathbf{Mod}_{\mathbb{Q}}^{\mathrm{aft}}(\breve{\overline{\mathscr{B}}})$) the category of objects of $\mathbf{Mod}(\breve{\overline{\mathscr{B}}})$ (resp. $\mathbf{Mod}^{\mathrm{ad}}(\breve{\overline{\mathscr{B}}})$, resp. $\mathbf{Mod}^{\mathrm{aft}}(\breve{\overline{\mathscr{B}}})$) up to isogeny (I.2.6). The category $\mathbf{Mod}_{\mathbb{Q}}(\breve{\overline{\mathscr{B}}})$ is abelian and the canonical functors

$$\mathbf{Mod}_{\mathbb{Q}}^{\mathrm{aft}}(\breve{\overline{\mathscr{B}}}) \to \mathbf{Mod}_{\mathbb{Q}}^{\mathrm{ad}}(\breve{\overline{\mathscr{B}}}) \to \mathbf{Mod}_{\mathbb{Q}}(\breve{\overline{\mathscr{B}}}) \tag{I.6.6.1}$$

are fully faithful. We denote by $\mathbf{Mod}^{\mathrm{coh}}(\mathscr{O}_{\mathfrak{X}})$ (resp. $\mathbf{Mod}^{\mathrm{coh}}(\mathscr{O}_{\mathfrak{X}}[\frac{1}{p}])$) the category of coherent $\mathscr{O}_{\mathfrak{X}}$-modules (resp. $\mathscr{O}_{\mathfrak{X}}[\frac{1}{p}]$-modules) of $X_{s,\mathrm{zar}}$ and by $\mathbf{Mod}_{\mathbb{Q}}^{\mathrm{coh}}(\mathscr{O}_{\mathfrak{X}})$ the category of coherent $\mathscr{O}_{\mathfrak{X}}$-modules up to isogeny. By III.6.16, the canonical functor

$$\mathbf{Mod}^{\mathrm{coh}}(\mathscr{O}_{\mathfrak{X}}) \to \mathbf{Mod}^{\mathrm{coh}}(\mathscr{O}_{\mathfrak{X}}[\frac{1}{p}]), \quad \mathscr{F} \mapsto \mathscr{F}_{\mathbb{Q}_p} = \mathscr{F} \otimes_{\mathbb{Z}_p} \mathbb{Q}_p, \tag{I.6.6.2}$$

induces an equivalence of abelian categories

$$\mathbf{Mod}_{\mathbb{Q}}^{\mathrm{coh}}(\mathscr{O}_{\mathfrak{X}}) \xrightarrow{\sim} \mathbf{Mod}^{\mathrm{coh}}(\mathscr{O}_{\mathfrak{X}}[\frac{1}{p}]). \tag{I.6.6.3}$$

**I.6.7.** For every rational number $r \geq 0$, we denote also by

$$\breve{d}^{(r)} \colon \breve{\mathscr{C}}^{(r)} \to \mathrm{T}^*(\xi^{-1}\Omega^1_{\mathfrak{X}/\mathscr{S}}) \otimes_{\breve{\overline{\mathscr{B}}}} \breve{\mathscr{C}}^{(r)} \tag{I.6.7.1}$$

the $\breve{\overline{\mathscr{B}}}$-derivation induced by $d^{(r)}$ (I.5.14.6) and the isomorphism (I.6.1.7), that we identify with the universal $\breve{\overline{\mathscr{B}}}$-derivation of $\breve{\mathscr{C}}^{(r)}$. This is a Higgs $\breve{\overline{\mathscr{B}}}$-field with coefficients in $\mathrm{T}^*(\xi^{-1}\Omega^1_{\mathfrak{X}/\mathscr{S}})$ (I.2.5). We denote by $\mathbb{K}^\bullet(\breve{\mathscr{C}}^{(r)}, p^r\breve{d}^{(r)})$ the Dolbeault complex of the Higgs $\breve{\overline{\mathscr{B}}}$-module $(\breve{\mathscr{C}}^{(r)}, p^r\breve{d}^{(r)})$ (I.2.3) and by $\mathbb{K}^\bullet_{\mathbb{Q}}(\breve{\mathscr{C}}^{(r)}, p^r\breve{d}^{(r)})$ its image in $\mathbf{Mod}_{\mathbb{Q}}(\breve{\overline{\mathscr{B}}})$. By (I.5.14.7), for all rational numbers $r \geq r' \geq 0$, the homomorphism $\breve{\alpha}^{r,r'}$ (I.5.14.4) induces a morphism of complexes

$$\breve{\nu}^{r,r'} \colon \mathbb{K}^\bullet(\breve{\mathscr{C}}^{(r)}, p^r\breve{d}^{(r)}) \to \mathbb{K}^\bullet(\breve{\mathscr{C}}^{(r')}, p^{r'}\breve{d}^{(r')}). \tag{I.6.7.2}$$

**Proposition I.6.8** (cf. III.11.24). *The canonical morphism*

$$\breve{\overline{\mathscr{B}}}_{\mathbb{Q}} \to \varinjlim_{r \in \mathbb{Q}_{>0}} \mathrm{H}^0(\mathbb{K}^\bullet_{\mathbb{Q}}(\breve{\mathscr{C}}^{(r)}, p^r\breve{d}^{(r)})) \tag{I.6.8.1}$$

*is an isomorphism, and for every integer $q \geq 1$,*

$$\varinjlim_{r \in \mathbb{Q}_{>0}} \mathrm{H}^q(\mathbb{K}^\bullet_{\mathbb{Q}}(\breve{\mathscr{C}}^{(r)}, p^r\breve{d}^{(r)})) = 0. \tag{I.6.8.2}$$

Observe that filtered direct limits are not a priori representable in the category $\mathbf{Mod}_{\mathbb{Q}}(\breve{\overline{\mathscr{B}}})$.

**I.6.9.** The functor $\top_*$ (I.6.1.3) induces an additive and left exact functor that we denote also by

$$\top_* \colon \mathbf{Mod}_{\mathbb{Q}}(\breve{\overline{\mathscr{B}}}) \to \mathbf{Mod}(\mathscr{O}_{\mathfrak{X}}[\frac{1}{p}]). \tag{I.6.9.1}$$

For every integer $q \geq 0$, we denote by

$$\mathrm{R}^q\top_* \colon \mathbf{Mod}_{\mathbb{Q}}(\breve{\overline{\mathscr{B}}}) \to \mathbf{Mod}(\mathscr{O}_{\mathfrak{X}}[\frac{1}{p}]) \tag{I.6.9.2}$$

the $q$th right derived functor of $\top_*$. By (I.6.6.3), the inverse image functor $\top^*$ induces an additive functor that we denote also by

(I.6.9.3) $$\top^*\colon \mathbf{Mod}^{\rm coh}(\mathscr{O}_{\mathfrak{X}}[\frac{1}{p}]) \to \mathbf{Mod}^{\rm aft}_{\mathbb{Q}}(\breve{\overline{\mathscr{B}}}).$$

For every coherent $\mathscr{O}_{\mathfrak{X}}[\frac{1}{p}]$-module $\mathscr{F}$ and every $\breve{\overline{\mathscr{B}}}_{\mathbb{Q}}$-module $\mathscr{G}$, we have a bifunctorial canonical homomorphism

(I.6.9.4) $$\mathrm{Hom}_{\breve{\overline{\mathscr{B}}}_{\mathbb{Q}}}(\top^*(\mathscr{F}),\mathscr{G}) \to \mathrm{Hom}_{\mathscr{O}_{\mathfrak{X}}[\frac{1}{p}]}(\mathscr{F},\top_*(\mathscr{G}))$$

that is injective (III.12.1.5).

**I.6.10.** We denote by $\mathbf{HI}(\mathscr{O}_{\mathfrak{X}},\xi^{-1}\Omega^1_{\mathfrak{X}/\mathscr{S}})$ the category of Higgs $\mathscr{O}_{\mathfrak{X}}$-isogenies with coefficients in $\xi^{-1}\Omega^1_{\mathfrak{X}/\mathscr{S}}$ (I.2.8) and by $\mathbf{HI}^{\rm coh}(\mathscr{O}_{\mathfrak{X}},\xi^{-1}\Omega^1_{\mathfrak{X}/\mathscr{S}})$ the full subcategory made up of the quadruples $(\mathscr{M},\mathscr{N},u,\theta)$ such that $\mathscr{M}$ and $\mathscr{N}$ are coherent $\mathscr{O}_{\mathfrak{X}}$-modules. These are additive categories. We denote by $\mathbf{HI}_{\mathbb{Q}}(\mathscr{O}_{\mathfrak{X}},\xi^{-1}\Omega^1_{\mathfrak{X}/\mathscr{S}})$ (resp. $\mathbf{HI}^{\rm coh}_{\mathbb{Q}}(\mathscr{O}_{\mathfrak{X}},\xi^{-1}\Omega^1_{\mathfrak{X}/\mathscr{S}})$) the category of objects of $\mathbf{HI}(\mathscr{O}_{\mathfrak{X}},\xi^{-1}\Omega^1_{\mathfrak{X}/\mathscr{S}})$ (resp. $\mathbf{HI}^{\rm coh}(\mathscr{O}_{\mathfrak{X}},\xi^{-1}\Omega^1_{\mathfrak{X}/\mathscr{S}})$) up to isogeny (I.2.6).

By Higgs $\mathscr{O}_{\mathfrak{X}}[\frac{1}{p}]$-module with coefficients in $\xi^{-1}\Omega^1_{\mathfrak{X}/\mathscr{S}}$, we will mean a Higgs $\mathscr{O}_{\mathfrak{X}}[\frac{1}{p}]$-module with coefficients in $\xi^{-1}\Omega^1_{\mathfrak{X}/\mathscr{S}}[\frac{1}{p}]$ (I.2.3). We denote by $\mathbf{HM}(\mathscr{O}_{\mathfrak{X}}[\frac{1}{p}],\xi^{-1}\Omega^1_{\mathfrak{X}/\mathscr{S}})$ the category of such modules and by $\mathbf{HM}^{\rm coh}(\mathscr{O}_{\mathfrak{X}}[\frac{1}{p}],\xi^{-1}\Omega^1_{\mathfrak{X}/\mathscr{S}})$ the full subcategory made up of the Higgs modules whose underlying $\mathscr{O}_{\mathfrak{X}}[\frac{1}{p}]$-module is coherent.

The functor

(I.6.10.1) $$\begin{array}{ccc}\mathbf{HI}(\mathscr{O}_{\mathfrak{X}},\xi^{-1}\Omega^1_{\mathfrak{X}/\mathscr{S}}) & \to & \mathbf{HM}(\mathscr{O}_{\mathfrak{X}}[\frac{1}{p}],\xi^{-1}\Omega^1_{\mathfrak{X}/\mathscr{S}}) \\ (\mathscr{M},\mathscr{N},u,\theta) & \mapsto & (\mathscr{M}_{\mathbb{Q}_p},(\mathrm{id}\otimes u^{-1}_{\mathbb{Q}_p})\circ\theta_{\mathbb{Q}_p})\end{array}$$

induces a functor

(I.6.10.2) $$\mathbf{HI}_{\mathbb{Q}}(\mathscr{O}_{\mathfrak{X}},\xi^{-1}\Omega^1_{\mathfrak{X}/\mathscr{S}}) \to \mathbf{HM}(\mathscr{O}_{\mathfrak{X}}[\frac{1}{p}],\xi^{-1}\Omega^1_{\mathfrak{X}/\mathscr{S}}).$$

By III.6.21, this induces an equivalence of categories

(I.6.10.3) $$\mathbf{HI}^{\rm coh}_{\mathbb{Q}}(\mathscr{O}_{\mathfrak{X}},\xi^{-1}\Omega^1_{\mathfrak{X}/\mathscr{S}}) \xrightarrow{\sim} \mathbf{HM}^{\rm coh}(\mathscr{O}_{\mathfrak{X}}[\frac{1}{p}],\xi^{-1}\Omega^1_{\mathfrak{X}/\mathscr{S}}).$$

**Definition I.6.11.** We call *Higgs $\mathscr{O}_{\mathfrak{X}}[\frac{1}{p}]$-bundle with coefficients in* $\xi^{-1}\Omega^1_{\mathfrak{X}/\mathscr{S}}$ any Higgs $\mathscr{O}_{\mathfrak{X}}[\frac{1}{p}]$-module with coefficients in $\xi^{-1}\Omega^1_{\mathfrak{X}/\mathscr{S}}$ whose underlying $\mathscr{O}_{\mathfrak{X}}[\frac{1}{p}]$-module is locally projective of finite type (III.2.8).

**I.6.12.** Let $r$ be a rational number $\geq 0$. We denote by $\Xi^r$ the category of integrable $p^r$-isoconnections with respect to the extension $\breve{\mathscr{C}}^{(r)}/\breve{\overline{\mathscr{B}}}$ (I.2.9) and by $\Xi^r_{\mathbb{Q}}$ the category of objects of $\Xi^r$ up to isogeny (I.2.6). We denote by $\mathfrak{S}^r$ the functor

(I.6.12.1) $$\mathfrak{S}^r\colon \mathbf{Mod}(\breve{\overline{\mathscr{B}}}) \to \Xi^r, \quad \mathscr{M} \mapsto (\breve{\mathscr{C}}^{(r)}\otimes_{\breve{\overline{\mathscr{B}}}}\mathscr{M},\breve{\mathscr{C}}^{(r)}\otimes_{\breve{\overline{\mathscr{B}}}}\mathscr{M},\mathrm{id},p^r\breve{d}^{(r)}\otimes\mathrm{id}).$$

This induces a functor that we denote also by

(I.6.12.2) $$\mathfrak{S}^r\colon \mathbf{Mod}_{\mathbb{Q}}(\breve{\overline{\mathscr{B}}}) \to \Xi^r_{\mathbb{Q}}.$$

We denote by $\mathscr{K}^r$ the functor

(I.6.12.3) $$\mathscr{K}^r\colon \Xi^r \to \mathbf{Mod}(\breve{\overline{\mathscr{B}}}), \quad (\mathscr{F},\mathscr{G},u,\nabla)\mapsto \ker(\nabla).$$

This induces a functor that we denote also by

(I.6.12.4) $$\mathscr{K}^r\colon \Xi^r_{\mathbb{Q}} \to \mathbf{Mod}_{\mathbb{Q}}(\breve{\overline{\mathscr{B}}}).$$

It is clear that (I.6.12.1) is a left adjoint of (I.6.12.3). Consequently, (I.6.12.2) is a left adjoint of (I.6.12.4).

If $(\mathscr{N},\mathscr{N}',v,\theta)$ is a Higgs $\mathscr{O}_{\mathfrak{X}}$-isogeny with coefficients in $\xi^{-1}\Omega^1_{\mathfrak{X}/\mathscr{S}}$,

$$(\text{I.6.12.5})\quad (\breve{\mathscr{C}}^{(r)}\otimes_{\overline{\breve{\mathscr{B}}}}\top^*(\mathscr{N}),\breve{\mathscr{C}}^{(r)}\otimes_{\overline{\breve{\mathscr{B}}}}\top^*(\mathscr{N}'),\mathrm{id}\otimes_{\overline{\breve{\mathscr{B}}}}\top^*(v),\mathrm{id}\otimes\top^*(\theta)+p^r\breve{d}^{(r)}\otimes\top^*(v))$$

is an object of $\Xi^r$ (III.6.12). We thus obtain a functor (I.6.10)

$$(\text{I.6.12.6})\qquad \top^{r+}\colon \mathbf{HI}(\mathscr{O}_{\mathfrak{X}},\xi^{-1}\Omega^1_{\mathfrak{X}/\mathscr{S}})\to\Xi^r.$$

By (I.6.10.3), this induces a functor that we denote also by

$$(\text{I.6.12.7})\qquad \top^{r+}\colon \mathbf{HM}^{\mathrm{coh}}(\mathscr{O}_{\mathfrak{X}}[\frac{1}{p}],\xi^{-1}\Omega^1_{\mathfrak{X}/\mathscr{S}})\to\Xi^r_{\mathbb{Q}}.$$

Let $(\mathscr{F},\mathscr{G},u,\nabla)$ be an object of $\Xi^r$. By the projection formula (III.12.4), $\nabla$ induces an $\mathscr{O}_{\mathfrak{X}}$-linear morphism

$$(\text{I.6.12.8})\qquad \top_*(\nabla)\colon \top_*(\mathscr{F})\to\xi^{-1}\Omega^1_{\mathfrak{X}/\mathscr{S}}\otimes_{\mathscr{O}_{\mathfrak{X}}}\top_*(\mathscr{G}).$$

We immediately see that $(\top_*(\mathscr{F}),\top_*(\mathscr{G}),\top_*(u),\top_*(\nabla))$ is a Higgs $\mathscr{O}_{\mathfrak{X}}$-isogeny with coefficients in $\xi^{-1}\Omega^1_{\mathfrak{X}/\mathscr{S}}$. We thus obtain a functor

$$(\text{I.6.12.9})\qquad \top^r_+\colon \Xi^r\to\mathbf{HI}(\mathscr{O}_{\mathfrak{X}},\xi^{-1}\Omega^1_{\mathfrak{X}/\mathscr{S}})$$

that is clearly a right adjoint of (I.6.12.6). The composition of the functors (I.6.12.9) and (I.6.10.1) induces a functor that we denote also by

$$(\text{I.6.12.10})\qquad \top^r_+\colon \Xi^r_{\mathbb{Q}}\to\mathbf{HM}(\mathscr{O}_{\mathfrak{X}}[\frac{1}{p}],\xi^{-1}\Omega^1_{\mathfrak{X}/\mathscr{S}}).$$

**Definition I.6.13** (cf. III.12.10). Let $\mathscr{M}$ be an object of $\mathbf{Mod}^{\mathrm{aft}}_{\mathbb{Q}}(\breve{\overline{\mathscr{B}}})$ and $\mathscr{N}$ a Higgs $\mathscr{O}_{\mathfrak{X}}[\frac{1}{p}]$-bundle with coefficients in $\xi^{-1}\Omega^1_{\mathfrak{X}/\mathscr{S}}$.

(i) Let $r>0$ be a rational number. We say that $\mathscr{M}$ and $\mathscr{N}$ are *$r$-associated* if there exists an isomorphism of $\Xi^r_{\mathbb{Q}}$

$$(\text{I.6.13.1})\qquad \alpha\colon \top^{r+}(\mathscr{N})\xrightarrow{\sim}\mathfrak{S}^r(\mathscr{M}).$$

We then also say that the triple $(\mathscr{M},\mathscr{N},\alpha)$ is *$r$-admissible.*

(ii) We say that $\mathscr{M}$ and $\mathscr{N}$ are *associated* if there exists a rational number $r>0$ such that $\mathscr{M}$ and $\mathscr{N}$ are $r$-associated.

Note that for all rational numbers $r\geq r'>0$, if $\mathscr{M}$ and $\mathscr{N}$ are $r$-associated, they are also $r'$-associated.

**Definition I.6.14** (cf. III.12.11). (i) We call *Dolbeault $\breve{\overline{\mathscr{B}}}_{\mathbb{Q}}$-module* any object of the category $\mathbf{Mod}^{\mathrm{aft}}_{\mathbb{Q}}(\breve{\overline{\mathscr{B}}})$ for which there exists an associated Higgs $\mathscr{O}_{\mathfrak{X}}[\frac{1}{p}]$-bundle with coefficients in $\xi^{-1}\Omega^1_{\mathfrak{X}/\mathscr{S}}$.

(ii) We say that a Higgs $\mathscr{O}_{\mathfrak{X}}[\frac{1}{p}]$-bundle with coefficients in $\xi^{-1}\Omega^1_{\mathfrak{X}/\mathscr{S}}$ is *solvable* if it admits an associated Dolbeault module.

We denote by $\mathbf{Mod}^{\mathrm{Dolb}}_{\mathbb{Q}}(\breve{\overline{\mathscr{B}}})$ the full subcategory of $\mathbf{Mod}^{\mathrm{aft}}_{\mathbb{Q}}(\breve{\overline{\mathscr{B}}})$ made up of the Dolbeault $\breve{\overline{\mathscr{B}}}_{\mathbb{Q}}$-modules and by $\mathbf{HM}^{\mathrm{sol}}(\mathscr{O}_{\mathfrak{X}}[\frac{1}{p}],\xi^{-1}\Omega^1_{\mathfrak{X}/\mathscr{S}})$ the full subcategory of $\mathbf{HM}(\mathscr{O}_{\mathfrak{X}}[\frac{1}{p}],\xi^{-1}\Omega^1_{\mathfrak{X}/\mathscr{S}})$ made up of the solvable Higgs $\mathscr{O}_{\mathfrak{X}}[\frac{1}{p}]$-bundles with coefficients in $\xi^{-1}\Omega^1_{\mathfrak{X}/\mathscr{S}}$.

**I.6.15.** For every $\breve{\overline{\mathscr{B}}}_{\mathbb{Q}}$-module $\mathscr{M}$ and all rational numbers $r \geq r' \geq 0$, we have a canonical morphism of $\mathbf{HM}(\mathscr{O}_{\mathfrak{X}}[\frac{1}{p}], \xi^{-1}\Omega^1_{\mathfrak{X}/\mathscr{S}})$

$$\mathsf{T}^r_+(\mathfrak{S}^r(\mathscr{M})) \to \mathsf{T}^{r'}_+(\mathfrak{S}^{r'}(\mathscr{M})). \tag{I.6.15.1}$$

We thus obtain a filtered direct system $(\mathsf{T}^r_+(\mathfrak{S}^r(\mathscr{M})))_{r\in\mathbb{Q}_{\geq 0}}$. We denote by $\mathscr{H}$ the functor

$$\mathscr{H}\colon \mathbf{Mod}_{\mathbb{Q}}(\breve{\overline{\mathscr{B}}}) \to \mathbf{HM}(\mathscr{O}_{\mathfrak{X}}[\frac{1}{p}], \xi^{-1}\Omega^1_{\mathfrak{X}/\mathscr{S}}), \quad \mathscr{M} \mapsto \varinjlim_{r\in\mathbb{Q}_{>0}} \mathsf{T}^r_+(\mathfrak{S}^r(\mathscr{M})). \tag{I.6.15.2}$$

For every object $\mathscr{N}$ of $\mathbf{HM}(\mathscr{O}_{\mathfrak{X}}[\frac{1}{p}], \xi^{-1}\Omega^1_{\mathfrak{X}/\mathscr{S}})$ and all rational numbers $r \geq r' \geq 0$, we have a canonical morphism of $\mathbf{Mod}_{\mathbb{Q}}(\breve{\overline{\mathscr{B}}})$

$$\mathscr{K}^r(\mathsf{T}^{r+}(\mathscr{N})) \to \mathscr{K}^{r'}(\mathsf{T}^{r'+}(\mathscr{N})). \tag{I.6.15.3}$$

We thus obtain a filtered direct system $(\mathscr{K}^r(\mathsf{T}^{r+}(\mathscr{N})))_{r\geq 0}$. Note, however, that filtered direct limits are not a priori representable in the category $\mathbf{Mod}_{\mathbb{Q}}(\breve{\overline{\mathscr{B}}})$.

**Proposition I.6.16** (cf. III.12.18). *For every Dolbeault $\breve{\overline{\mathscr{B}}}_{\mathbb{Q}}$-module $\mathscr{M}$, $\mathscr{H}(\mathscr{M})$* (I.6.15.2) *is a solvable Higgs $\mathscr{O}_{\mathfrak{X}}[\frac{1}{p}]$-bundle associated with $\mathscr{M}$. In particular, $\mathscr{H}$ induces a functor that we denote also by*

$$\mathscr{H}\colon \mathbf{Mod}^{\mathrm{Dolb}}_{\mathbb{Q}}(\breve{\overline{\mathscr{B}}}) \to \mathbf{HM}^{\mathrm{sol}}(\mathscr{O}_{\mathfrak{X}}[\frac{1}{p}], \xi^{-1}\Omega^1_{\mathfrak{X}/\mathscr{S}}), \quad \mathscr{M} \mapsto \mathscr{H}(\mathscr{M}). \tag{I.6.16.1}$$

**Proposition I.6.17** (cf. III.12.23). *We have a functor*

$$\mathscr{V}\colon \mathbf{HM}^{\mathrm{sol}}(\mathscr{O}_{\mathfrak{X}}[\frac{1}{p}], \xi^{-1}\Omega^1_{\mathfrak{X}/\mathscr{S}}) \to \mathbf{Mod}^{\mathrm{Dolb}}_{\mathbb{Q}}(\breve{\overline{\mathscr{B}}}), \quad \mathscr{N} \mapsto \varinjlim_{r\in\mathbb{Q}_{>0}} \mathscr{K}^r(\mathsf{T}^{r+}(\mathscr{N})). \tag{I.6.17.1}$$

*Moreover, for every object $\mathscr{N}$ of $\mathbf{HM}^{\mathrm{sol}}(\mathscr{O}_{\mathfrak{X}}[\frac{1}{p}], \xi^{-1}\Omega^1_{\mathfrak{X}/\mathscr{S}})$, $\mathscr{V}(\mathscr{N})$ is associated with $\mathscr{N}$.*

**Theorem I.6.18** (cf. III.12.26). *The functors* (I.6.16.1) *and* (I.6.17.1)

$$\mathbf{Mod}^{\mathrm{Dolb}}_{\mathbb{Q}}(\breve{\overline{\mathscr{B}}}) \underset{\mathscr{V}}{\overset{\mathscr{H}}{\rightleftarrows}} \mathbf{HM}^{\mathrm{sol}}(\mathscr{O}_{\mathfrak{X}}[\tfrac{1}{p}], \xi^{-1}\Omega^1_{\mathfrak{X}/\mathscr{S}}) \tag{I.6.18.1}$$

*are equivalences of categories quasi-inverse to each other.*

**Theorem I.6.19** (cf. III.12.34). *Let $\mathscr{M}$ be a Dolbeault $\breve{\overline{\mathscr{B}}}_{\mathbb{Q}}$-module and $q \geq 0$ an integer. We denote by $\mathbb{K}^\bullet(\mathscr{H}(\mathscr{M}))$ the Dolbeault complex of the Higgs $\mathscr{O}_{\mathfrak{X}}[\frac{1}{p}]$-bundle $\mathscr{H}(\mathscr{M})$* (I.2.3)*. We then have a functorial canonical isomorphism of $\mathscr{O}_{\mathfrak{X}}[\frac{1}{p}]$-modules* (I.6.9.2)

$$\mathrm{R}^q\mathsf{T}_*(\mathscr{M}) \xrightarrow{\sim} \mathrm{H}^q(\mathbb{K}^\bullet(\mathscr{H}(\mathscr{M}))). \tag{I.6.19.1}$$

**I.6.20.** Let $g\colon X' \to X$ be an étale morphism. There exists essentially a unique étale morphism $\widetilde{g}\colon \widetilde{X}' \to \widetilde{X}$ that fits into a Cartesian diagram (I.2.1.1)

$$\begin{array}{ccc} \check{\overline{X}}' & \longrightarrow & \widetilde{X}' \\ {\scriptstyle \check{\overline{g}}}\downarrow & & \downarrow{\scriptstyle \widetilde{g}} \\ \check{\overline{X}} & \longrightarrow & \widetilde{X} \end{array}, \tag{I.6.20.1}$$

so that $\widetilde{X}'$ is a smooth $\mathscr{A}_2(\overline{S})$-deformation of $\check{\overline{X}}'$. We associate with $(X', \widetilde{X}')$ objects analogous to those defined earlier for $(X, \widetilde{X})$, which we will denote by the same symbols equipped with an exponent $'$. The morphism $g$ defines by functoriality a morphism of ringed topos (III.8.20)

$$\Phi\colon (\widetilde{E}', \overline{\mathscr{B}}') \to (\widetilde{E}, \overline{\mathscr{B}}). \tag{I.6.20.2}$$

We prove in III.8.21 that $\Phi$ identifies with a localization morphism of $(\widetilde{E}, \overline{\mathscr{B}})$ at $\sigma^*(X')$. Furthermore, $\Phi$ induces a morphism of ringed topos

$$\breve{\Phi}\colon (\widetilde{E}_s'^{\mathbb{N}^\circ}, \breve{\overline{\mathscr{B}}}') \to (\widetilde{E}_s^{\mathbb{N}^\circ}, \breve{\overline{\mathscr{B}}}). \tag{I.6.20.3}$$

We denote by $\mathfrak{g}\colon \mathfrak{X}' \to \mathfrak{X}$ the extension of $\overline{g}\colon \overline{X}' \to \overline{X}$ to the $p$-adic completions.

**Proposition I.6.21** (cf. III.14.9). *Under the assumptions of* I.6.20, *let moreover* $\mathscr{M}$ *be a Dolbeault* $\breve{\overline{\mathscr{B}}}_{\mathbb{Q}}$*-module and* $\mathscr{N}$ *a solvable Higgs* $\mathscr{O}_{\mathfrak{X}}[\frac{1}{p}]$*-bundle with coefficients in* $\xi^{-1}\Omega^1_{\mathfrak{X}/\mathscr{S}}$. *Then* $\breve{\Phi}^*(\mathscr{M})$ *is a Dolbeault* $\breve{\overline{\mathscr{B}}}'_{\mathbb{Q}}$*-module and* $\mathfrak{g}^*(\mathscr{N})$ *is a solvable Higgs* $\mathscr{O}_{\mathfrak{X}'}[\frac{1}{p}]$*-bundle with coefficients in* $\xi^{-1}\Omega^1_{\mathfrak{X}'/\mathscr{S}}$. *If, moreover,* $\mathscr{M}$ *and* $\mathscr{N}$ *are associated, then* $\breve{\Phi}^*(\mathscr{M})$ *and* $\mathfrak{g}^*(\mathscr{N})$ *are associated.*

We in fact prove that the diagrams of functors

$$\begin{array}{ccccc}
\mathbf{Mod}^{\mathrm{Dolb}}_{\mathbb{Q}}(\breve{\overline{\mathscr{B}}}) & \xrightarrow{\mathscr{H}} & \mathbf{HM}^{\mathrm{sol}}(\mathscr{O}_{\mathfrak{X}}[\frac{1}{p}], \xi^{-1}\Omega^1_{\mathfrak{X}/\mathscr{S}}) & \xrightarrow{\mathscr{V}} & \mathbf{Mod}^{\mathrm{Dolb}}_{\mathbb{Q}}(\breve{\overline{\mathscr{B}}}) \\
\big\downarrow{\scriptstyle \breve{\Phi}^*} & & \big\downarrow{\scriptstyle \mathfrak{g}^*} & & \big\downarrow{\scriptstyle \breve{\Phi}^*} \\
\mathbf{Mod}^{\mathrm{Dolb}}_{\mathbb{Q}}(\breve{\overline{\mathscr{B}}}') & \xrightarrow{\mathscr{H}'} & \mathbf{HM}^{\mathrm{sol}}(\mathscr{O}_{\mathfrak{X}'}[\frac{1}{p}], \xi^{-1}\Omega^1_{\mathfrak{X}'/\mathscr{S}}) & \xrightarrow{\mathscr{V}'} & \mathbf{Mod}^{\mathrm{Dolb}}_{\mathbb{Q}}(\breve{\overline{\mathscr{B}}}')
\end{array} \tag{I.6.21.1}$$

are commutative up to canonical isomorphisms (III.14.11).

**I.6.22.** There exists a unique morphism of topos

$$\psi\colon \widetilde{E}_s^{\mathbb{N}^\circ} \to X_{\text{ét}} \tag{I.6.22.1}$$

such that for every $U \in \mathrm{Ob}(\mathbf{\acute{E}t}_{/X})$, $\psi^*(U)$ is the constant inverse system $(\sigma_s^*(U_s))_{\mathbb{N}}$ (I.5.11.3). We denote by $\mathbf{\acute{E}t}_{\mathrm{coh}/X}$ the full subcategory of $\mathbf{\acute{E}t}_{/X}$ made up of étale schemes of finite presentation over $X$. We have a canonical fibered category

$$\mathrm{MOD}_{\mathbb{Q}}(\breve{\overline{\mathscr{B}}}) \to \mathbf{\acute{E}t}_{\mathrm{coh}/X} \tag{I.6.22.2}$$

whose fiber over an object $U$ of $\mathbf{\acute{E}t}_{\mathrm{coh}/X}$ is the category $\mathbf{Mod}_{\mathbb{Q}}(\breve{\overline{\mathscr{B}}}|\psi^*(U))$ and the inverse image functor under a morphism $U' \to U$ of $\mathbf{\acute{E}t}_{\mathrm{coh}/X}$ is the restriction functor (I.6.20.2)

$$\mathbf{Mod}_{\mathbb{Q}}(\breve{\overline{\mathscr{B}}}|\psi^*(U)) \to \mathbf{Mod}_{\mathbb{Q}}(\breve{\overline{\mathscr{B}}}|\psi^*(U')), \quad \mathscr{M} \mapsto \mathscr{M}|\psi^*(U'). \tag{I.6.22.3}$$

By I.6.21, it induces a fibered category

$$\mathrm{MOD}^{\mathrm{Dolb}}_{\mathbb{Q}}(\breve{\overline{\mathscr{B}}}) \to \mathbf{\acute{E}t}_{\mathrm{coh}/X} \tag{I.6.22.4}$$

whose fiber over an object $U$ of $\mathbf{\acute{E}t}_{\mathrm{coh}/X}$ is the category $\mathbf{Mod}^{\mathrm{Dolb}}_{\mathbb{Q}}(\breve{\overline{\mathscr{B}}}|\psi^*(U))$.

**Proposition I.6.23** (cf. III.15.4). *Let* $\mathscr{M}$ *be an object of* $\mathbf{Mod}^{\mathrm{aft}}_{\mathbb{Q}}(\breve{\overline{\mathscr{B}}})$ *and* $(U_i)_{i\in I}$ *a covering of* $\mathbf{\acute{E}t}_{\mathrm{coh}/X}$. *Then* $\mathscr{M}$ *is Dolbeault if and only if for every* $i \in I$, *the* $(\breve{\overline{\mathscr{B}}}|\psi^*(U_i))_{\mathbb{Q}}$*-module* $\mathscr{M}|\psi^*(U_i)$ *is Dolbeault.*

**Proposition I.6.24** (cf. III.15.5). *The following conditions are equivalent:*

(i) *The fibered category* (I.6.22.4)

$$\mathrm{MOD}^{\mathrm{Dolb}}_{\mathbb{Q}}(\breve{\overline{\mathscr{B}}}) \to \mathbf{\acute{E}t}_{\mathrm{coh}/X} \tag{I.6.24.1}$$

*is a stack* ([**35**] II 1.2.1).

(ii) *For every covering* $(U_i \to U)_{i\in I}$ *of* $\mathbf{\acute{E}t}_{\mathrm{coh}/X}$*, denoting by* $\mathscr{U}$ *(resp.* $\mathscr{U}_i$*, for* $i \in I$*) the formal p-adic completion of* $\overline{U}$ *(resp.* $\overline{U}_i$*), a Higgs* $\mathscr{O}_{\mathscr{U}}[\frac{1}{p}]$*-bundle* $\mathscr{N}$ *with coefficients in* $\xi^{-1}\Omega^1_{\mathscr{U}/\mathscr{S}}$ *is solvable if and only if for every* $i \in I$*, the Higgs* $\mathscr{O}_{\mathscr{U}_i}[\frac{1}{p}]$*-bundle* $\mathscr{N} \otimes_{\mathscr{O}_{\mathscr{U}}} \mathscr{O}_{\mathscr{U}_i}$ *with coefficients in* $\xi^{-1}\Omega^1_{\mathscr{U}_i/\mathscr{S}}$ *is solvable.*

**Definition I.6.25** (cf. III.15.6)**.** Let $(\mathscr{N},\theta)$ be Higgs $\mathscr{O}_{\mathfrak{X}}[\frac{1}{p}]$-bundle with coefficients in $\xi^{-1}\Omega^1_{\mathfrak{X}/\mathscr{S}}$.

(i) We say that $(\mathscr{N},\theta)$ is *small* if there exist a coherent sub-$\mathscr{O}_{\mathfrak{X}}$-module $\mathfrak{N}$ of $\mathscr{N}$ that generates it over $\mathscr{O}_{\mathfrak{X}}[\frac{1}{p}]$ and a rational number $\varepsilon > \frac{1}{p-1}$ such that

$$\theta(\mathfrak{N}) \subset p^{\varepsilon}\xi^{-1}\Omega^1_{\mathfrak{X}/\mathscr{S}} \otimes_{\mathscr{O}_{\mathfrak{X}}} \mathfrak{N}. \tag{I.6.25.1}$$

(ii) We say that $(\mathscr{N},\theta)$ is *locally small* if there exists an open covering $(U_i)_{i\in I}$ of $X_s$ such that for every $i \in I$, $(\mathscr{N}|U_i, \theta|U_i)$ is small.

**Proposition I.6.26** (cf. III.15.8)**.** *Every solvable Higgs* $\mathscr{O}_{\mathfrak{X}}[\frac{1}{p}]$*-bundle* $(\mathscr{N},\theta)$ *with coefficients in* $\xi^{-1}\Omega^1_{\mathfrak{X}/\mathscr{S}}$ *is locally small.*

**Proposition I.6.27** (cf. III.15.9)**.** *Suppose that* $X$ *is affine and connected, and that it admits an étale* $S$*-morphism to* $\mathbb{G}^d_{m,S}$ *for an integer* $d \geq 1$*. Then, every small Higgs* $\mathscr{O}_{\mathfrak{X}}[\frac{1}{p}]$*-bundle with coefficients in* $\xi^{-1}\Omega^1_{\mathfrak{X}/\mathscr{S}}$ *is solvable.*

**Corollary I.6.28** (cf. III.15.10)**.** *Under the conditions of* I.6.24*, every locally small Higgs* $\mathscr{O}_{\mathfrak{X}}[\frac{1}{p}]$*-bundle with coefficients in* $\xi^{-1}\Omega^1_{\mathfrak{X}/\mathscr{S}}$ *is solvable.*

CHAPTER II

# Representations of the fundamental group and the torsor of deformations. Local study

AHMED ABBES AND MICHEL GROS

## II.1. Introduction

The current chapter is devoted to the construction and study of the *p-adic Simpson correspondence*, following the general approach summarized in Chapter I, for an affine logarithmic scheme of a certain type (II.6.2). Section II.2 contains the main notation and general conventions, in particular, those related to Higgs modules (II.2.8). Section II.3 contains several results on the continuous cohomology of profinite groups and, in particular, for lack of a good reference, a treatment of the Künneth formula adapted to the situation. Section II.4 recalls and details the existing relations both between torsors for the Zariski topology and principal homogeneous bundles, and between the associated equivariant notions (under an abstract group). Next, in Section II.5 we recall a few notions from logarithmic geometry that will play an important role in this work, in order to fix the notation and give reference points for readers unfamiliar with this theory. In Section II.6, we introduce the logarithmic setting (II.6.2), the rings (II.6.7), and the Galois groups (II.6.10) used throughout this chapter, and then establish some of their properties (II.6.6, II.6.8, and II.6.15). We then recall Faltings' almost purity theorem (II.6.16) and a number of corollaries (II.6.17)–(II.6.25). Section II.7 is devoted to the Faltings extension; it contains two variants (II.7.17.2) and (II.7.22.2). For the convenience of the reader, in Section II.8, we present, in some detail, the computation of the Galois cohomology due to Faltings, which is central to his approach of $p$-adic Hodge theory. In Section II.9, we introduce Fontaine infinitesimal $p$-adic thickenings, and, following Tsuji, endow them with logarithmic structures (II.9.11). The most novel part of the chapter begins at Section II.10, with the introduction of the *Higgs–Tate torsor* (II.10.3). Theorem II.10.18 establishes an important link between this torsor and the Faltings extension. Section II.11 is devoted to the study of the de Rham (II.11.4) and Galois (II.11.7) cohomology of various algebras associated with this torsor. In Section II.12, we define the main functors (II.12.8.2) and (II.12.9.2) that link the category of generalized representations to that of Higgs modules. In it we also introduce the notions of Dolbeault representation and of solvable Higgs module. We in fact develop two variants, an integral one (II.12.11 and II.12.12) and a more subtle rational one (II.12.16 and II.12.18). We then show that for each variant, these notions lead to two equivalent categories (II.12.15 and II.12.24). Section II.13 is devoted to studying small representations and small Higgs modules following Faltings' approach in **[27]**. We also establish links between these notions and the notions of Dolbeault representation and solvable Higgs module (II.13.20, II.13.25, and II.13.26). Section II.14 contains a descent statement for small representations due to Faltings (II.14.4). From this we deduce new links between the different notions of representations and Higgs modules introduced previously (II.14.6,

II.14.8, and II.14.16). The last section makes the link (II.15.7) with Hyodo's theory for Hodge–Tate representations.

## II.2. Notation and conventions

*All rings in this chapter have an identity element; all ring homomorphisms map the identity element to the identity element. We mostly consider commutative rings, and rings are assumed to be commutative unless stated otherwise; in particular, when we take a ringed topos $(X, A)$, the ring $A$ is assumed to be commutative unless stated otherwise.*

**II.2.1.** In this chapter, $p$ denotes a prime number, $K$ a complete discrete valuation ring of characteristic 0, with perfect residue field $k$ of characteristic $p$, and $\overline{K}$ an algebraic closure of $K$. We denote by $\mathscr{O}_K$ the valuation ring of $K$, by $\mathscr{O}_{\overline{K}}$ the integral closure of $\mathscr{O}_K$ in $\overline{K}$, by $\mathfrak{m}_{\overline{K}}$ the maximal ideal of $\mathscr{O}_{\overline{K}}$, by $\overline{k}$ the residue field of $\mathscr{O}_{\overline{K}}$, and by $v$ the valuation of $\overline{K}$ normalized so that $v(p) = 1$. We denote by $\mathscr{O}_C$ the $p$-adic Hausdorff completion of $\mathscr{O}_{\overline{K}}$, by $C$ its field of fractions, and by $\mathfrak{m}_C$ its maximal ideal.

We choose a compatible system $(\beta_n)_{n>0}$ of $n$th roots of $p$ in $\mathscr{O}_{\overline{K}}$. For any rational number $\varepsilon > 0$, we set $p^\varepsilon = (\beta_n)^{\varepsilon n}$, where $n$ is a positive integer such that $\varepsilon n$ is an integer.

We denote by $G_K = \mathrm{Gal}(\overline{K}/K)$ the Galois group of $\overline{K}$ over $K$ and by $\widehat{\mathbb{Z}}(1)$ and $\mathbb{Z}_p(1)$ the $\mathbb{Z}[G_K]$-modules

$$\widehat{\mathbb{Z}}(1) = \varprojlim_{n\geq 1} \mu_n(\mathscr{O}_{\overline{K}}), \tag{II.2.1.1}$$

$$\mathbb{Z}_p(1) = \varprojlim_{n\geq 0} \mu_{p^n}(\mathscr{O}_{\overline{K}}), \tag{II.2.1.2}$$

where $\mu_n(\mathscr{O}_{\overline{K}})$ denotes the subgroup of $n$th roots of unity in $\mathscr{O}_{\overline{K}}$. For any $\mathbb{Z}_p[G_K]$-module $M$ and any integer $n$, we set $M(n) = M \otimes_{\mathbb{Z}_p} \mathbb{Z}_p(1)^{\otimes n}$.

For any abelian group $A$, we denote by $\widehat{A}$ its $p$-adic Hausdorff completion.

**II.2.2.** We endow $\mathbb{Z}_p$ with the $p$-adic topology and do the same for every adic $\mathbb{Z}_p$-algebra (that is, every $\mathbb{Z}_p$-algebra that is separated and complete for the $p$-adic topology). Let $A$ be an adic $\mathbb{Z}_p$-algebra and $i\colon A \to A[\frac{1}{p}]$ the canonical homomorphism. We call *$p$-adic topology* on $A[\frac{1}{p}]$ the unique topology compatible with its structure of additive group for which the subgroups $i(p^n A)$, for $n \in \mathbb{N}$, form a fundamental system of neighborhoods of 0 (**[12]** III § 1.2 Prop. 1). It makes $A[\frac{1}{p}]$ into a topological ring. Let $M$ be an $A[\frac{1}{p}]$-module of finite type and $M^\circ$ a sub-$A$-module of $M$ of finite type that generates it over $A[\frac{1}{p}]$. We call *$p$-adic topology* on $M$ the unique topology compatible with its structure of additive group for which the subgroups $p^n M^\circ$, for $n \in \mathbb{N}$, form a fundamental system of neighborhoods of 0. This topology does not depend on the choice of $M^\circ$. Indeed, if $M'$ is another sub-$A$-module of $M$ of finite type that generates it over $A[\frac{1}{p}]$, then there exists an $m \geq 0$ such that $p^m M^\circ \subset M'$ and $p^m M' \subset M^\circ$. It is clear that $M$ is a topological $A[\frac{1}{p}]$-module.

**II.2.3.** Let $A$ be a ring and $n$ an integer $\geq 1$. We denote by $\mathrm{W}(A)$ (resp. $\mathrm{W}_n(A)$) the ring of Witt vectors (resp. of Witt vectors of length $n$) with respect to $p$ with coefficients in $A$. We have a ring homomorphism

$$\begin{array}{rccc} \Phi_n\colon & \mathrm{W}_n(A) & \to & A, \\ & (x_1, \dots, x_n) & \mapsto & x_1^{p^{n-1}} + p x_2^{p^{n-2}} + \cdots + p^{n-1} x_n. \end{array} \tag{II.2.3.1}$$

called the $n$th ghost component. We also have at our disposal the restriction, the shift, and the Frobenius morphisms

$$\mathrm{R}\colon \mathrm{W}_{n+1}(A) \to \mathrm{W}_n(A), \tag{II.2.3.2}$$

$$\mathrm{V}\colon \mathrm{W}_n(A) \to \mathrm{W}_{n+1}(A), \tag{II.2.3.3}$$

$$\mathrm{F}\colon \mathrm{W}_{n+1}(A) \to \mathrm{W}_n(A). \tag{II.2.3.4}$$

When $A$ is of characteristic $p$, F induces an endomorphism of $\mathrm{W}_n(A)$ that we denote also by F.

**II.2.4.** For any abelian category $\mathbf{A}$, we denote by $\mathbf{D}(\mathbf{A})$ its derived category and by $\mathbf{D}^-(\mathbf{A})$, $\mathbf{D}^+(\mathbf{A})$, and $\mathbf{D}^{\mathrm{b}}(\mathbf{A})$ the full subcategories of $\mathbf{D}(\mathbf{A})$ of complexes with cohomology bounded from above, from below, and from both sides, respectively. Unless mentioned otherwise, complexes in $\mathbf{A}$ have a differential of degree $+1$, the degree being written as an exponent.

**II.2.5.** Let $(X, A)$ be a ringed topos. We denote by $\mathbf{Mod}(A)$ or $\mathbf{Mod}(A, X)$ the category of $A$-modules of $X$. If $M$ is an $A$-module, we denote by $\mathrm{S}_A(M)$ (resp. $\wedge_A(M)$, resp. $\Gamma_A(M)$) the symmetric algebra (resp. the exterior algebra, resp. the divided power algebra) of $M$ (**[45]** I 4.2.2.6) and for any integer $n \geq 0$, by $\mathrm{S}^n_A(M)$ (resp. $\wedge^n_A(M)$, resp. $\Gamma^n_A(M)$) its homogeneous part of degree $n$. We will leave the ring $A$ out of the notation when there is no risk of confusion. Forming these algebras commutes with localizing over an object of $X$.

**II.2.6.** Let $A$ be a ring, $L$ an $A$-module, and $u\colon L \to A$ a linear form. For any $x \in \wedge(L)$, we denote by $d_u(x)$ the inner product of $x$ and $u$ (**[9]** III § 11.7 Example p. 161). By (loc. cit. p. 162), we have

$$d_u(x_1 \wedge \cdots \wedge x_n) = \sum_{i=1}^{n} (-1)^{i+1} u(x_i) x_1 \wedge \cdots \wedge x_{i-1} \wedge x_{i+1} \wedge \cdots \wedge x_n \tag{II.2.6.1}$$

for $x_1, \ldots, x_n \in L$. The map $d_u\colon \wedge(L) \to \wedge(L)$ is an anti-derivation of degree $-1$ and square 0 (**[9]** III § 11.8 Example p. 165). The algebra $\wedge(L)$ endowed with the anti-derivation $d_u$ is called the *Koszul* algebra (or complex) of $u$. We denote it by $\mathbb{K}^A_\bullet(u)$; we therefore have $\mathbb{K}^A_n(u) = \wedge^n L$ and the differentials of $\mathbb{K}^A_\bullet(u)$ are of degree $-1$ (cf. **[10]** § 9.1).

For any complex of $A$-modules $C$, we define a chain complex (**[10]** § 5.1)

$$\mathbb{K}^A_\bullet(u, C) = \mathbb{K}^A_\bullet(u) \otimes_A C \tag{II.2.6.2}$$

and a cochain complex

$$\mathbb{K}^\bullet_A(u, C) = \mathrm{Homgr}_A(\mathbb{K}^A_\bullet(u), C). \tag{II.2.6.3}$$

By (**[10]** § 9.1 Cor. 2 to Prop. 1), if $\mathrm{Ann}(C)$ is the annihilator of $C$, then $u(L) + \mathrm{Ann}(C)$ annihilates $\mathrm{H}^*(\mathbb{K}^\bullet_A(u, C))$ and $\mathrm{H}_*(\mathbb{K}^A_\bullet(u, C))$.

Suppose that $L$ is the direct sum of $L_1, \ldots, L_r$ and denote by $u_i\colon L_i \to A$ the restriction of $u$ to $L_i$. Then the canonical isomorphism (**[9]** III § 7.7 Prop. 10)

$$^g\!\otimes_{1 \leq i \leq r} \wedge(L_i) \xrightarrow{\sim} \wedge(L) \tag{II.2.6.4}$$

is an isomorphism of complexes $\otimes_{1 \leq i \leq r} \mathbb{K}^A_\bullet(u_i) \xrightarrow{\sim} \mathbb{K}^A_\bullet(u)$, where the symbol $^g\!\otimes$ denotes the left tensor product (cf. **[9]** III § 4.7 Remarks p. 49).

Since $d_u$ is an anti-derivation, taking the product in the algebra $\wedge(L)$ induces a morphism of complexes

$$\mathbb{K}^A_\bullet(u) \otimes_A \mathbb{K}^A_\bullet(u) \to \mathbb{K}^A_\bullet(u). \tag{II.2.6.5}$$

Supposing $L$ projective of rank $n$ and composing with the canonical morphism $\mathbb{K}^A_\bullet(u) \to \wedge^n L[-n]$, we deduce from this a morphism of complexes

(II.2.6.6) $$\mathbb{K}^A_\bullet(u) \to \mathrm{Homgr}_A(\mathbb{K}^A_\bullet(u), \wedge^n L[-n])$$

that is bijective ([**9**] III § 7.8 p. 87). For any complex of $A$-modules $C$, we deduce from this an isomorphism of complexes ([**10**] § 9.1 p. 149)

(II.2.6.7) $$\mathbb{K}^A_\bullet(u, C) \xrightarrow{\sim} \mathbb{K}^\bullet_A(u, C \otimes_A \wedge^n L[-n]).$$

Taking the homology, we therefore have, for every integer $i$, an isomorphism

(II.2.6.8) $$\mathrm{H}_i(\mathbb{K}^A_\bullet(u, C)) \xrightarrow{\sim} \mathrm{H}^{n-i}(\mathbb{K}^\bullet_A(u, C \otimes_A \wedge^n L)).$$

**II.2.7.** Let $A$ be a ring, $L$ an $A$-module, and $u\colon \mathrm{S}(L) \otimes_A L \to \mathrm{S}(L)$ the linear form defined by $u(s \otimes x) = sx$ for every $s \in \mathrm{S}(L)$ and $x \in L$. The canonical isomorphism ([**9**] III § 7.5 Prop. 8)

(II.2.7.1) $$\wedge\,(\mathrm{S}(L) \otimes_A L) \xrightarrow{\sim} \mathrm{S}(L) \otimes_A \wedge(L)$$

transforms the differential of the Koszul complex $\mathbb{K}^{\mathrm{S}(L)}_\bullet(u)$ (II.2.6) into the map

(II.2.7.2) $$d\colon \mathrm{S}(L) \otimes_A \wedge(L) \to \mathrm{S}(L) \otimes_A \wedge(L)$$

that is defined, for every $x_1, \dots, x_n, y_1, \dots, y_m \in L$, by

(II.2.7.3)
$$d((x_1 \dots x_n) \otimes (y_1 \wedge \cdots \wedge y_m)) = \sum_{i=1}^{m} (-1)^{i+1} y_i x_1 \dots x_n \otimes (y_1 \wedge \cdots \wedge y_{i-1} \wedge y_{i+1} \wedge \cdots \wedge y_m).$$

For any complex $C$ of $\mathrm{S}(L)$-modules, we set

(II.2.7.4) $$\mathbb{K}^{\mathrm{S}(L)}_\bullet(C) = \mathbb{K}^{\mathrm{S}(L)}_\bullet(u, C),$$

(II.2.7.5) $$\mathbb{K}^\bullet_{\mathrm{S}(L)}(C) = \mathbb{K}^\bullet_{\mathrm{S}(L)}(u, C)$$

(since the morphism $u$ is canonical, it can be left out of the notation).

Let $L'$ be an $A$-module and $u'\colon \mathrm{S}(L') \otimes_A L' \to \mathrm{S}(L')$ the linear form such that $u'(s' \otimes x') = s'x'$ for $s' \in \mathrm{S}(L')$ and $x' \in L'$. The isomorphism (II.2.6.4) induces an isomorphism

(II.2.7.6) $$(\mathrm{S}(L) \otimes_A \Lambda(L))\,{}^g\boxtimes_A(\mathrm{S}(L') \otimes_A \Lambda(L')) \xrightarrow{\sim} \mathrm{S}(L \oplus L') \otimes_A \Lambda(L \oplus L'),$$

where the left exterior tensor product is taken with respect to the canonical co-Cartesian diagram

$$\begin{array}{ccc} A & \longrightarrow & \mathrm{S}(L) \\ \downarrow & & \downarrow \\ \mathrm{S}(L') & \longrightarrow & \mathrm{S}(L \oplus L') \end{array}$$

It follows from (II.2.7.3) that (II.2.7.6) is an isomorphism of complexes

(II.2.7.7) $$\mathbb{K}^{\mathrm{S}(L)}_\bullet(u)\,{}^g\boxtimes_A \mathbb{K}^{\mathrm{S}(L')}_\bullet(u') \xrightarrow{\sim} \mathbb{K}^{\mathrm{S}(L\oplus L')}_\bullet(u \oplus u').$$

For any complex $C$ of $\mathrm{S}(L \oplus L')$-modules, this leads to isomorphisms of complexes

(II.2.7.8) $$\mathbb{K}^{\mathrm{S}(L\oplus L')}_\bullet(C) \xrightarrow{\sim} \mathbb{K}^{\mathrm{S}(L)}_\bullet(\mathbb{K}^{\mathrm{S}(L')}_\bullet(C)),$$

(II.2.7.9) $$\mathbb{K}^\bullet_{\mathrm{S}(L\oplus L')}(C) \xrightarrow{\sim} \mathbb{K}^\bullet_{\mathrm{S}(L)}(\mathbb{K}^\bullet_{\mathrm{S}(L')}(C)).$$

**Definition II.2.8.** Let $(X, A)$ be a ringed topos and $E$ an $A$-module.

(i) We call *Higgs $A$-module with coefficients in $E$* a pair $(M, \theta)$ consisting of an $A$-module $M$ and an $A$-linear morphism

$$\theta\colon M \to M \otimes_A E \tag{II.2.8.1}$$

such that $\theta \wedge \theta = 0$. We also say that $\theta$ is a *Higgs $A$-field* on $M$ with coefficients in $E$.

(ii) If $(M_1, \theta_1)$ and $(M_2, \theta_2)$ are two Higgs $A$-modules, a morphism from $(M_1, \theta_1)$ to $(M_2, \theta_2)$ is an $A$-linear morphism $u\colon M_1 \to M_2$ such that $(u \otimes \mathrm{id}_E) \circ \theta_1 = \theta_2 \circ u$.

The Higgs $A$-modules with coefficients in $E$ form a category that we denote by $\mathbf{HM}(A, E)$. Let us complete the terminology and make a few remarks.

II.2.8.2. Let $(M, \theta)$ be a Higgs $A$-module with coefficients in $E$. For each $i \geq 1$,

$$\theta_i\colon M \otimes_A \wedge^i E \to M \otimes_A \wedge^{i+1} E \tag{II.2.8.3}$$

is the $A$-linear morphism defined, for all local sections $m$ of $M$ and $\omega$ of $\wedge^i E$, by $\theta_i(m \otimes \omega) = \theta(m) \wedge \omega$. We have $\theta_{i+1} \circ \theta_i = 0$. Following Simpson ([**68**] p. 24), the *Dolbeault* complex of $(M, \theta)$, denoted by $\mathbb{K}^\bullet(M, \theta)$, is the cochain complex of $A$-modules

$$M \xrightarrow{\theta} M \otimes_A E \xrightarrow{\theta_1} M \otimes_A \wedge^2 E \ldots, \tag{II.2.8.4}$$

where $M$ is placed in degree zero and the differentials are of degree one.

II.2.8.5. Let $(M, \theta)$ be a Higgs $A$-module with coefficients in $E$ with $M$ a locally free $A$-module of finite type. Consider, for an integer $i \geq 1$, the composition

$$\wedge^i M \xrightarrow{\wedge^i \theta} \wedge^i (M \otimes_A E) \longrightarrow \wedge^i M \otimes_A \mathrm{S}^i E, \tag{II.2.8.6}$$

where the second arrow is the canonical morphism ([**45**] V 4.5). We call *ith characteristic invariant* of $\theta$, and denote by $\lambda_i(\theta)$, the trace of the morphism (II.2.8.6) viewed as a section of $\Gamma(X, \mathrm{S}^i E)$.

II.2.8.7. Let $(M_1, \theta_1)$ and $(M_2, \theta_2)$ be two Higgs $A$-modules with coefficients in $E$. We call *total* Higgs field on $M_1 \otimes_A M_2$ the $A$-linear morphism

$$\theta_{\mathrm{tot}}\colon M_1 \otimes_A M_2 \to M_1 \otimes_A M_2 \otimes_A E \tag{II.2.8.8}$$

defined by

$$\theta_{\mathrm{tot}} = \theta_1 \otimes \mathrm{id}_{M_2} + \mathrm{id}_{M_1} \otimes \theta_2. \tag{II.2.8.9}$$

We call $(M_1 \otimes_A M_2, \theta_{\mathrm{tot}})$ the *tensor product* of $(M_1, \theta_1)$ and $(M_2, \theta_2)$.

II.2.8.10. Suppose that $E$ is locally free of finite type over $A$; let $F = \mathscr{H}om_A(E, A)$. Since for every $A$-module $M$, the canonical morphism

$$\mathscr{E}nd_A(M) \otimes_A E \to \mathscr{H}om_A(M, M \otimes_A E) \tag{II.2.8.11}$$

is an isomorphism, giving a Higgs $A$-field $\theta$ on $M$ is equivalent to giving a structure of $\mathrm{S}(F)$-module on $M$ that is compatible with its structure of $A$-module. On the other hand, by ([**9**] § 11.5 Prop. 7), the $A$-module $\mathscr{H}om_A(\wedge(F), A)$ can be identified with the graded dual algebra of $\wedge(F)$ and we have a canonical isomorphism of graded algebras

$$\wedge(E) \to \mathscr{H}om_A(\wedge(F), A). \tag{II.2.8.12}$$

One verifies that this induces an isomorphism of complexes of $A$-modules (II.2.7.5)

$$\mathbb{K}^\bullet(M, \theta) \xrightarrow{\sim} \mathbb{K}^\bullet_{\mathrm{S}(F)}(M). \tag{II.2.8.13}$$

**II.2.9.** Let $f\colon (X',A')\to(X,A)$ be a morphism of ringed topos, $E$ an $A$-module, $E'$ an $A'$-module, and $\gamma\colon f^*(E)\to E'$ an $A'$-linear morphism. Let $(M,\theta)$ be a Higgs $A$-module with coefficients in $E$. The composition

$$\theta'\colon\ f^*(M)\xrightarrow{f^*(\theta)} f^*(M)\otimes_{A'} f^*(E)\xrightarrow{\mathrm{id}\otimes\gamma} f^*(M)\otimes_{A'} E' \tag{II.2.9.1}$$

is then a Higgs $A'$-field with coefficients in $E'$. We call the Higgs $A'$-module $(f^*(M),\theta')$ the *inverse image* of $(M,\theta)$ under $(f,\gamma)$.

**II.2.10.** Let $(X,A)$ be a ringed topos, $B$ an $A$-algebra, $\Omega^1_{B/A}$ the $B$-module of Kähler differential forms of $B$ over $A$ ([**45**] II 1.1.2), and

$$\Omega_{B/A}=\oplus_{n\in\mathbb{N}}\Omega^n_{B/A}=\wedge_B(\Omega^1_{B/A}) \tag{II.2.10.1}$$

the exterior algebra of $\Omega^1_{B/A}$. Then there exists a unique $A$-anti-derivation $d\colon\Omega_{B/A}\to\Omega_{B/A}$ of degree one and square zero that extends the universal $A$-derivation $d\colon B\to\Omega^1_{B/A}$. This follows, for example, from ([**10**] § 2.10 Prop. 13) by noting that $\Omega^1_{B/A}$ is the sheaf associated with the presheaf $U\mapsto\Omega^1_{B(U)/A(U)}$ $(U\in\mathrm{Ob}(X))$.

Let $M$ be a $B$-module and $\lambda\in\Gamma(X,A)$. A $\lambda$*-connection* on $M$ with respect to the extension $B/A$ is an $A$-linear morphism

$$\nabla\colon M\to\Omega^1_{B/A}\otimes_B M \tag{II.2.10.2}$$

such that for all local sections $x$ of $B$ and $s$ of $M$, we have

$$\nabla(xs)=\lambda d(x)\otimes s+x\nabla(s). \tag{II.2.10.3}$$

We also say that $(M,\nabla)$ is a $B$-module with $\lambda$-connection with respect to the extension $B/A$. We will leave the extension $B/A$ out of the terminology when there is no risk of confusion. The morphism $\nabla$ extends into a unique graded $A$-linear morphism

$$\nabla\colon\Omega_{B/A}\otimes_B M\to\Omega_{B/A}\otimes_B M \tag{II.2.10.4}$$

of degree one such that for all local sections $\omega$ of $\Omega^i_{B/A}$ and $s$ of $\Omega^j_{B/A}\otimes_B M$ $(i,j\in\mathbb{N})$, we have

$$\nabla(\omega\wedge s)=\lambda d(\omega)\wedge s+(-1)^i\omega\wedge\nabla(s). \tag{II.2.10.5}$$

Iterating this formula gives

$$\nabla\circ\nabla(\omega\wedge s)=\omega\wedge\nabla\circ\nabla(s). \tag{II.2.10.6}$$

We say that $\nabla$ is *integrable* if $\nabla\circ\nabla=0$.

Let $(M,\nabla)$ and $(M',\nabla')$ be two modules with $\lambda$-connections. A morphism from $(M,\nabla)$ to $(M',\nabla')$ is a $B$-linear morphism $u\colon M\to M'$ such that $(\mathrm{id}\otimes u)\circ\nabla=\nabla'\circ u$.

Classically, 1-connections are called *connections*. Integrable 0-connections are Higgs $B$-fields with coefficients in $\Omega^1_{B/A}$ (II.2.8).

**II.2.11.** Let $f\colon(X',A')\to(X,A)$ be a morphism of ringed topos, $B$ an $A$-algebra, $B'$ an $A'$-algebra, $\alpha\colon f^*(B)\to B'$ a homomorphism of $A'$-algebras, $\lambda\in\Gamma(X,A)$, and $(M,\nabla)$ a module with $\lambda$-connection with respect to the extension $B/A$. We denote by $\lambda'$ the canonical image of $\lambda$ in $\Gamma(X',A')$, by $d'\colon B'\to\Omega^1_{B'/A'}$ the universal $A'$-derivation of $B'$, and by

$$\gamma\colon f^*(\Omega^1_{B/A})\to\Omega^1_{B'/A'} \tag{II.2.11.1}$$

the canonical $\alpha$-linear morphism. We immediately see that $f^*(\nabla)$ is a $\lambda'$-connection on $f^*(M)$ with respect to the extension $f^*(B)/A'$. It is integrable if $\nabla$ is. Moreover, there exists a unique $A'$-linear morphism

$$\nabla'\colon B' \otimes_{f^*(B)} f^*(M) \to \Omega^1_{B'/A'} \otimes_{f^*(B)} f^*(M) \tag{II.2.11.2}$$

such that for all local sections $x'$ of $B'$ and $t$ of $f^*(M)$, we have

$$\nabla'(x' \otimes t) = \lambda' d'(x') \otimes t + x'(\gamma \otimes \mathrm{id})(f^*(\nabla)(t)). \tag{II.2.11.3}$$

This is a $\lambda'$-connection on $B' \otimes_{f^*(B)} f^*(M)$ with respect to the extension $B'/A'$. It is integrable if $\nabla$ is.

**II.2.12.** Let $(X, A)$ be a ringed topos, $B$ an $A$-algebra, $\lambda \in \Gamma(X, A)$, and $(M, \nabla)$ a module with $\lambda$-connection with respect to the extension $B/A$. Suppose that there exist an $A$-module $E$ and a $B$-linear isomorphism $\gamma\colon E \otimes_A B \xrightarrow{\sim} \Omega^1_{B/A}$ such that for every local section $\omega$ of $E$, we have $d(\gamma(\omega \otimes 1)) = 0$. Let $\vartheta\colon M \to E \otimes_A M$ be the morphism induced by $\nabla$ and $\gamma$. Then the $\lambda$-connection $\nabla$ is integrable if and only if $\vartheta$ is a Higgs $A$-field on $M$ with coefficients in $E$. Indeed, the diagram

$$\begin{array}{ccc} E \otimes_A M & \xrightarrow{-\mathrm{id}\wedge\vartheta} & (\wedge^2 E) \otimes_A M \\ \downarrow & & \downarrow \\ \Omega^1_{B/A} \otimes_B M & \xrightarrow{\nabla} & \Omega^2_{B/A} \otimes_B M \end{array} \tag{II.2.12.1}$$

where the vertical arrows are the isomorphisms induced by $\gamma$, is clearly commutative.

**II.2.13.** Let $(X, A)$ be a ringed topos, $B$ an $A$-algebra, $\lambda \in \Gamma(X, A)$, and $(M, \nabla)$ a module with integrable $\lambda$-connection with respect to the extension $B/A$. Suppose that there exist an $A$-module $E$ and a $B$-isomorphism $\gamma\colon E \otimes_A B \xrightarrow{\sim} \Omega^1_{B/A}$ such that for every local section $\omega$ of $E$, we have $d(\gamma(\omega \otimes 1)) = 0$ (cf. II.2.12). Let $(N, \theta)$ be a Higgs $A$-module with coefficients in $E$. There exists a unique $A$-linear morphism

$$\nabla'\colon M \otimes_A N \to \Omega^1_{B/A} \otimes_B M \otimes_A N \tag{II.2.13.1}$$

such that for all local sections $x$ of $M$ and $y$ of $N$, we have

$$\nabla'(x \otimes y) = \nabla(x) \otimes_A y + (\gamma \otimes_B \mathrm{id}_{M \otimes_A N})(x \otimes_A \theta(y)). \tag{II.2.13.2}$$

This is an integrable $\lambda$-connection on $M \otimes_A N$ with respect to the extension $B/A$.

**II.2.14.** Let $A$ be an adic ring, $I$ an ideal of definition of $A$, and $B$ an adic $A$-algebra; that is, an $A$-algebra $B$ that is separated and complete for the $(IB)$-adic topology. Recall that the canonical topology on the $B$-module $\Omega^1_{B/A}$ is deduced from that of $B$ ([**42**] 0.20.4.5). We denote by $\widehat{\Omega}^1_{B/A}$ the Hausdorff completion of $\Omega^1_{B/A}$ and by

$$d\colon B \to \widehat{\Omega}^1_{B/A} \tag{II.2.14.1}$$

the universal continuous $A$-derivation of $B$. Let $M$ be a $B$-module that is separated and complete for the $(IB)$-adic topology, and $\lambda \in A$. An *adic* (or *I-adic*) *$\lambda$-connection* on $M$ with respect to the extension $B/A$ is an $A$-linear morphism

$$\nabla\colon M \to \widehat{\Omega}^1_{B/A} \widehat{\otimes}_B M \tag{II.2.14.2}$$

such that for all $x \in B$ and $t \in M$, we have

$$\nabla(xt) = \lambda d(x) \widehat{\otimes} t + x\nabla(t). \tag{II.2.14.3}$$

We also say that $(M, \nabla)$ is a $B$-module with adic (or $I$-adic) $\lambda$-connection with respect to the extension $B/A$. For every integer $n \geq 1$, we set $A_n = A/I^n$, $B_n = B \otimes_A A_n$, and $M_n = M \otimes_A A_n$. We denote by $\overline{\lambda}_n$ the class of $\lambda$ in $A_n$. Then $\nabla$ induces a (usual) $\overline{\lambda}_n$-connection with respect to the extension $B_n/A_n$:

$$\nabla_n : M_n \to \Omega^1_{B_n/A_n} \otimes_{B_n} M_n. \tag{II.2.14.4}$$

Moreover, $\nabla$ can be identified with the inverse limit of the morphisms $\nabla_n$. We say that $\nabla$ is *integrable* if $\nabla_n$ is integrable for every $n \geq 1$.

Let $(M, \nabla)$ and $(M', \nabla')$ be two modules with adic $\lambda$-connections. A morphism from $(M, \nabla)$ to $(M', \nabla')$ is a $B$-linear morphism $u : M \to M'$ such that $(\mathrm{id} \widehat{\otimes} u) \circ \nabla = \nabla' \circ u$.

**II.2.15.** Let $A$ be an adic ring, $\lambda \in A$, $B$ an adic $A$-algebra, $B'$ an adic $B$-algebra, and $(M, \nabla)$ a $B$-module with adic $\lambda$-connection with respect to the extension $B/A$. We denote by $d' : B' \to \widehat{\Omega}^1_{B'/A}$ the universal continuous $A$-derivation of $B'$ and by

$$\gamma : \widehat{\Omega}^1_{B/A} \to \widehat{\Omega}^1_{B'/A} \tag{II.2.15.1}$$

the canonical morphism. There exists a unique $A$-linear morphism

$$\nabla' : B' \widehat{\otimes}_B M \to \widehat{\Omega}^1_{B'/A} \widehat{\otimes}_B M \tag{II.2.15.2}$$

such that for all $x' \in B'$ and $t \in M$, we have

$$\nabla'(x' \widehat{\otimes} t) = \lambda d'(x') \widehat{\otimes} t + x'(\gamma \widehat{\otimes} \mathrm{id}_M)(\nabla(t)). \tag{II.2.15.3}$$

This is an adic $\lambda$-connection on $B' \widehat{\otimes}_B M$ with respect to the extension $B'/A$. It is integrable if $\nabla$ is.

**II.2.16.** Let $A$ be an adic ring, $I$ an ideal of definition of $A$, $\lambda \in A$, $B$ an adic $A$-algebra, and $(M, \nabla)$ a $B$-module with adic $\lambda$-connection with respect to the extension $B/A$. For every integer $n \geq 1$, we set $A_n = A/I^n$ and $B_n = B \otimes_A A_n$. We assume that the following conditions are satisfied:

(i) The ideal $I$ is of finite type over $A$ and $\Omega^1_{B_1/A_1}$ is a $B_1$-module of finite type.

(ii) There exist a *free* $A$-module *of finite type* $E$ and a $B$-linear isomorphism $\gamma : E \otimes_A B \xrightarrow{\sim} \widehat{\Omega}^1_{B/A}$ such that $\gamma(E) \subset d(B)$.

Note that the topology on $\widehat{\Omega}^1_{B/A}$ is the $(IB)$-adic topology and that we have $\widehat{\Omega}^1_{B/A} \otimes_A A_n = \Omega^1_{B_n/A_n}$ for all $n \geq 1$ (**[11]** III § 2.12 Cor. 1 to Prop. 14). Moreover, we have $\widehat{\Omega}^1_{B/A} \widehat{\otimes}_B M = \widehat{\Omega}^1_{B/A} \otimes_B M = E \otimes_A M$. Let $\vartheta : M \to E \otimes_A M$ be the morphism induced by $\nabla$ and $\gamma$. Then the adic $\lambda$-connection $\nabla$ is integrable if and only if $\vartheta$ is a Higgs $A$-field on $M$ with coefficients in $E$. Indeed, $\vartheta$ is a Higgs $A$-field on $M$ with coefficients in $E$ if and only if for every $n \geq 1$, $\vartheta \otimes_A A_n$ is a Higgs $A_n$-field on $M \otimes_A A_n$ with coefficients in $E \otimes_A A_n$. The desired statement now follows from II.2.12.

**II.2.17.** Let $A$ be an adic ring, $\lambda \in A$, $B$ an adic $A$-algebra, and $(M, \nabla)$ a $B$-module with integrable adic $\lambda$-connection with respect to the extension $B/A$. Suppose that conditions (i) and (ii) of II.2.16 are satisfied. Let $(N, \theta)$ be a Higgs $A$-module with coefficients in $E$; we endow $N$ with the topology induced by that on $A$. Taking the limit of II.2.13, we see that there exists a unique $A$-linear morphism

$$\nabla' : M \widehat{\otimes}_A N \to \widehat{\Omega}^1_{B/A} \widehat{\otimes}_B M \widehat{\otimes}_A N \tag{II.2.17.1}$$

such that for all $x \in M$ and $y \in N$, we have

$$\nabla'(x \widehat{\otimes} y) = \nabla(x) \widehat{\otimes}_A y + (\gamma \otimes_B \mathrm{id}_{M \widehat{\otimes}_A N})(x \widehat{\otimes}_A \theta(y)). \tag{II.2.17.2}$$

This is an integrable adic $\lambda$-connection on $M \widehat{\otimes}_A N$ with respect to the extension $B/A$.

**II.2.18.** Let $X$ be a smooth projective complex variety. A harmonic bundle on $X$ is a triple $(M, D, \langle\ ,\ \rangle)$, where $M$ is a complex vector bundle of class $\mathscr{C}^\infty$ on $X$, $D$ is an integrable connection on $M$, and $\langle\ ,\ \rangle$ is a Hermitian metric on $M$ satisfying a condition described below ([**54**] § 1). There is a unique way to write $D$ as a sum $\nabla + \alpha$ with $\nabla$ a Hermitian connection and $\alpha$ a differential form of degree one with values in $\mathrm{End}_{\mathscr{O}_X}(M)$ that is auto-adjoint with respect to $\langle\ ,\ \rangle$. We decompose $\nabla$ and $\alpha$ using their types:

$$\nabla = \partial + \overline{\partial}, \quad \alpha = \theta + \theta^*, \tag{II.2.18.1}$$

where $\partial$ and $\overline{\partial}$ are of type $(1,0)$ and $(0,1)$, respectively, and $\theta$ is a differential form of type $(1,0)$ with values in $\mathrm{End}_{\mathscr{O}_X}(M)$. The required condition is that the operator $D'' = \overline{\partial} + \theta$ is integrable (that is, $D''^2 = 0$). This condition is equivalent to saying that $\overline{\partial}^2 = 0$, $\overline{\partial}\theta = 0$, and $\theta \wedge \theta = 0$. Thus the operator $\overline{\partial}$ defines on $M$ a structure of holomorphic vector bundle and $\theta$ is a Higgs field on $M$ with coefficients in $\Omega^1_{X/\mathbb{C}}$. The Dolbeault complex

$$0 \to \mathrm{A}^0(M) \xrightarrow{D''} \mathrm{A}^1(M) \xrightarrow{D''} \dots \tag{II.2.18.2}$$

of $(M, \theta)$ is obtained by extending $D''$ to the differential forms of class $\mathscr{C}^\infty$. By restriction to the holomorphic differential forms, it induces the complex $\mathbb{K}(M, \theta)$; this explains the terminology.

## II.3. Results on continuous cohomology of profinite groups

**II.3.1.** Let $G$ be a profinite group and $A$ a topological ring endowed with a continuous action of $G$ by ring homomorphisms. An *$A$-representation* of $G$ is an $A$-module $M$ endowed with an $A$-semi-linear action of $G$; that is, an action such that for all $g \in G$, $a \in A$, and $m \in M$, we have $g(am) = g(a)g(m)$. We say that the $A$-representation is *continuous* if $M$ is a topological $A$-module and the action of $G$ on $M$ is continuous. Let $M$, $N$ be two $A$-representations (resp. two continuous $A$-representations) of $G$. A morphism from $M$ to $N$ is an $A$-linear and $G$-equivariant (resp. $A$-linear, continuous, and $G$-equivariant) morphism from $M$ to $N$. We denote by $\mathbf{Rep}_A(G)$ (resp. $\mathbf{Rep}_A^{\mathrm{cont}}(G)$) the category of $A$-representations (resp. continuous $A$-representations) of $G$. If $M$ and $N$ are two $A$-representations of $G$, then the $A$-modules $M \otimes_A N$ and $\mathrm{Hom}_A(M, N)$ are naturally $A$-representations of $G$.

Suppose that the action of $G$ on $A$ is trivial. The objects of $\mathbf{Rep}_A^{\mathrm{cont}}(G)$ are then also called *topological $A$-$G$-modules.* A topological $A$-$G$-module whose topology is discrete is called a *discrete $A$-$G$-module.* We denote by $\mathbf{Rep}_A^{\mathrm{disc}}(G)$ the full subcategory of $\mathbf{Rep}_A^{\mathrm{cont}}(G)$ made up of discrete $A$-$G$-modules.

If $R$ is a ring without topology, it is understood that the topological $R$-$G$-modules are defined with respect to the discrete topology on $R$ (and the trivial action of $G$ on $R$).

**II.3.2.** Let $A$ be a ring, $G$ a profinite group, and $M$ an $A$-module endowed with the discrete topology. The *induced $A$-$G$-module* of $M$, denoted by $\mathrm{Ind}_{A,G}(M)$, is the $A$-module of continuous maps from $G$ to $M$, endowed with the action of $G$ defined, for all $f \in \mathrm{Ind}_{A,G}(M)$ and $g \in G$, by

$$(g \cdot f)(x) = f(x \cdot g). \tag{II.3.2.1}$$

It is a discrete $A$-$G$-module. We thus define an exact functor

$$\mathrm{Ind}_{A,G} \colon \mathbf{Mod}(A) \to \mathbf{Rep}_A^{\mathrm{disc}}(G), \quad M \mapsto \mathrm{Ind}_{A,G}(M) \tag{II.3.2.2}$$

that is a right adjoint to the forgetful functor for the action of $G$ (V.11.1). It therefore transforms injective $A$-modules into injective $A$-$G$-modules. The category $\mathbf{Rep}_A^{\mathrm{disc}}(G)$ has enough injective objects. An object of $\mathbf{Rep}_A^{\mathrm{disc}}(G)$ is injective if and only if it is a direct summand of an object of the form $\mathrm{Ind}_{A,G}(I)$, where $I$ is an injective $A$-module (V.11.2). A discrete $A$-$G$-module is called *induced* if it is isomorphic to the induced $A$-$G$-module of an $A$-module.

We denote by $\Gamma(G,-)$ the left exact functor

(II.3.2.3) $$\Gamma(G,-)\colon \mathbf{Rep}_A^{\mathrm{disc}}(G) \to \mathbf{Mod}(A), \quad M \mapsto M^G,$$

and by

(II.3.2.4) $$\mathrm{R}\Gamma(G,-)\colon \mathbf{D}^+(\mathbf{Rep}_A^{\mathrm{disc}}(G)) \to \mathbf{D}^+(\mathbf{Mod}(A)),$$

(II.3.2.5) $$\mathrm{H}^q(G,-)\colon \mathbf{Rep}_A^{\mathrm{disc}}(G) \to \mathbf{Mod}(A), \quad (q \geq 0),$$

its right derived functors (cf. II.2.4).

**II.3.3.** Let $A$ be a ring, $G$ a profinite group, and $H$ a normal closed subgroup of $G$. Then the groups $H$ and $G/H$ are profinite. We again denote by $\Gamma(H,-)$ the left exact functor

(II.3.3.1) $$\Gamma(H,-)\colon \mathbf{Rep}_A^{\mathrm{disc}}(G) \to \mathbf{Rep}_A^{\mathrm{disc}}(G/H), \quad M \mapsto M^H,$$

and by $\mathrm{R}\Gamma(H,-)$ and $\mathrm{H}^q(H,-)$ $(q \geq 0)$ its right derived functors. This abuse of notation is justified by the fact that the diagram

(II.3.3.2) $$\begin{array}{ccc} \mathbf{D}^+(\mathbf{Rep}_A^{\mathrm{disc}}(G)) & \xrightarrow{\mathrm{R}\Gamma(H,-)} & \mathbf{D}^+(\mathbf{Rep}_A^{\mathrm{disc}}(G/H)) \\ \downarrow & & \downarrow \\ \mathbf{D}^+(\mathbf{Rep}_A^{\mathrm{disc}}(H)) & \xrightarrow{\mathrm{R}\Gamma(H,-)} & \mathbf{D}^+(\mathbf{Mod}(A)) \end{array}$$

where the vertical arrows are induced by the forgetful functors, is commutative up to canonical isomorphism (V.11.5).

**Proposition II.3.4** (cf. V.11.7)**.** *Let $A$ be a ring, $G$ a profinite group, $H$ a normal closed subgroup of $G$, and $M$ a discrete $A$-$G$-module. Then we have a canonical functorial isomorphism of* $\mathbf{D}^+(\mathbf{Mod}(A))$

(II.3.4.1) $$\mathrm{R}\Gamma(G/H, \mathrm{R}\Gamma(H,M)) \xrightarrow{\sim} \mathrm{R}\Gamma(G,M).$$

**Remark II.3.5.** Let $A$ be a ring, $G$ a profinite group, $H$ a normal closed subgroup of $G$, and $M$ a discrete $A$-$G$-module. Then for every integer $n \geq 0$, the restriction

(II.3.5.1) $$\rho_n\colon \mathrm{H}^n(G,M) \to \mathrm{H}^n(H,M)$$

coincides with the composition

(II.3.5.2)
$$u_n\colon \mathrm{H}^n(G,M) \xrightarrow{\sim} \mathrm{H}^n(G/H, \tau_{\leq n}\mathrm{R}\Gamma(H,M)) \to \Gamma(G/H, \mathrm{H}^n(H,M)) \to \mathrm{H}^n(H,M),$$

where $\tau_{\leq n}\mathrm{R}\Gamma(H,M)$ is the canonical filtration of $\mathrm{R}\Gamma(H,M)$ (**[15]** 1.4.6), the first arrow is induced by the isomorphism (II.3.4.1), and the other arrows are the canonical morphisms. Indeed, $(\rho_n)$ and $(u_n)$ are two morphisms of universal $\partial$-functors that coincide in degree zero.

**II.3.6.** We denote by $\mathfrak{D}$ the category whose objects are the ordered sets $[n] = \{0,\dots,n\}$ (for $n \in \mathbb{N}$) and whose morphisms are the increasing maps. We keep the notation $\Delta$ for a Galois group that will come up in (II.6.10). For all $n \in \mathbb{N}$ and $i \in [n]$, we denote by $d_n^i\colon [n-1] \to [n]$ the increasing injection that forgets $i$, and by $s_n^i\colon [n+1] \to [n]$ the increasing surjection that repeats $i$. When there is no risk of confusion, we leave the index $n$ out of the notation $d_n^i$ and $s_n^i$.

Let $\mathscr{A}$ be an additive category, $q$ an integer $\geq 1$, and $X$ a $q$-cosimplicial object of $\mathscr{A}$ ([**45**] I 1.1). We call *diagonal subobject* of $X$, and denote by $\Delta X$, the cosimplicial object of $\mathscr{A}$ defined by $[n] \mapsto X^{n,\dots,n}$. We call *cochain complex* of $X$ the complex of $q$-uple cochains $\widetilde{X}$ defined for all $(n_1,\dots,n_q) \in \mathbb{N}^q$ by $\widetilde{X}^{n_1,\dots,n_q} = X^{n_1,\dots,n_q}$ and, for $1 \leq i \leq q$, by the differential

$$d^i = \sum_j (-1)^j X(\mathrm{id},\dots,d^j,\dots,\mathrm{id}), \tag{II.3.6.1}$$

where $d^j$ is placed in $i$th position ([**45**] I 1.2.2). For every complex of $q$-uple cochains $M$ of $\mathscr{A}$, we call *associated simple complex* the cochain complex $\int M$ defined, for all $n \in \mathbb{N}$, by

$$(\int M)^n = \oplus_{\sum_{i=1}^q n_i = n} M^{n_1,\dots,n_q}, \tag{II.3.6.2}$$

where the differential $d$, when restricted to $M^{n_1,\dots,n_q}$, is given by $\sum_j (-1)^{\sum_{i<j} n_i} d^j$ ([**45**] I 1.2.1).

By the Eilenberg–Zilber–Cartier theorem ([**45**] I 1.2.2), for every 2-cosimplicial object $X$ of $\mathscr{A}$, there exist two functorial homomorphisms, the "Alexander–Whitney" arrow

$$\int \widetilde{X} \to (\Delta X)^{\sim} \tag{II.3.6.3}$$

and the "shuffle map"

$$(\Delta X)^{\sim} \to \int \widetilde{X}, \tag{II.3.6.4}$$

which induce the identity in degree zero, and are inverse to each other up to functorial homotopy. Notice that the direction of the arrows is reversed with respect to loc. cit. which treats simplicial objects. More precisely, the arrow (II.3.6.3) sends $X^{m,n}$ to $X^{m+n,m+n}$ by taking $X(d^{m+n}\dots d^{m+1}, d^0 \dots d^0)$, where $d^0$ is composed $m$ times, and the arrow (II.3.6.4) sends $X^{m+n,m+n}$ to $X^{m,n}$ by

$$\sum_{(\mu,\nu)} \varepsilon(\mu,\nu) X(s^{\nu_1}\dots s^{\nu_n}, s^{\mu_1}\dots s^{\mu_m}), \tag{II.3.6.5}$$

where the sum is taken over all $(m,n)$-shuffles of the set $[m+n-1]$, and $\varepsilon(\mu,\nu)$ is the sign of the shuffle.

**II.3.7.** Let $A$ be a ring and $X$, $Y$ two cosimplicial $A$-modules. We denote by $X \otimes_A Y$ the 2-cosimplicial $A$-module defined by $(m,n) \mapsto X^m \otimes_A Y^n$ and by $X \otimes_A^{\Delta} Y$ the diagonal subobject $\Delta(X \otimes_A Y)$. Note that the simple complex associated with the cochain complex of $X \otimes_A Y$ is none other than the tensor product $\widetilde{X} \otimes_A \widetilde{Y}$ of the complexes of $A$-modules $\widetilde{X}$ and $\widetilde{Y}$ (cf. [**10**] § 4.1). We therefore have the "Alexander-Whitney" arrow

$$\widetilde{X} \otimes_A \widetilde{Y} \to (X \otimes_A^{\Delta} Y)^{\sim}, \tag{II.3.7.1}$$

and the "shuffle map"

$$(X \otimes_A^{\Delta} Y)^{\sim} \to \widetilde{X} \otimes_A \widetilde{Y}. \tag{II.3.7.2}$$

**II.3.8.** Let $A$ be a topological ring, $G$ a profinite group, and $M$ a topological $A$-$G$-module. We associate with $M$ a bifunctorial cosimplicial $A[G]$-module $\mathrm{K}^\bullet(G,M)$ (II.3.6) that is contravariant in $G$ and covariant in $M$, as follows: for every integer $n \in \mathbb{N}$, $\mathrm{K}^n(G,M)$ is the $A$-module of continuous maps from $G^{[n]}$ to $M$ (II.3.6), endowed with the action of $G$ defined, for all $g \in G$ and $f \in \mathrm{K}^n(G,M)$, by

(II.3.8.1) $$(g \cdot f)(g_0,\dots,g_n) = g \cdot f(g^{-1}g_0,\dots,g^{-1}g_n).$$

Every morphism $\delta\colon [n] \to [m]$ in $\mathfrak{D}$ (II.3.6) induces a morphism of $A[G]$-modules

(II.3.8.2) $$\mathrm{K}^n(G,M) \to \mathrm{K}^m(G,M)$$

defined by the composition with the map

(II.3.8.3) $$G^{[m]} \to G^{[n]}, \quad (g_0,\dots,g_m) \mapsto (g_{\delta(0)},\dots,g_{\delta(n)}).$$

The cochain complex $\widetilde{\mathrm{K}}^\bullet(G,M)$ associated with $\mathrm{K}^\bullet(G,M)$ (II.3.6) is called the *complex of continuous homogeneous cochains of $G$ with values in $M$*.

We call complex of *continuous nonhomogeneous cochains* of $G$ with values in $M$, and denote by $\mathrm{C}^\bullet_{\mathrm{cont}}(G,M)$, the complex of $A$-modules defined as follows: for all $n \in \mathbb{N}$, $\mathrm{C}^n_{\mathrm{cont}}(G,M)$ is the set of all continuous maps from $G^n$ to $M$, and the differential $d\colon \mathrm{C}^n_{\mathrm{cont}}(G,M) \to \mathrm{C}^{n+1}_{\mathrm{cont}}(G,M)$ is defined by the formula

(II.3.8.4) $$\begin{aligned} d(f)(g_1,\dots,g_{n+1}) &= g_1 \cdot f(g_2,\dots,g_{n+1}) \\ &\quad + \sum_{i=1}^{n} (-1)^i f(g_1,\dots,g_ig_{i+1},\dots,g_{n+1}) \\ &\quad + (-1)^{n+1} f(g_1,\dots,g_n). \end{aligned}$$

The cohomology groups of $\mathrm{C}^\bullet_{\mathrm{cont}}(G,M)$ are denoted by $\mathrm{H}^\bullet_{\mathrm{cont}}(G,M)$ and are called the *continuous cohomology* groups of $G$ with values in $M$ (cf. **[72]** § 2). The continuous maps (for every $n \in \mathbb{N}$)

(II.3.8.5) $$G^n \to G^{[n]}, \quad (g_1,\dots,g_n) \mapsto (1, g_1, g_1g_2, \dots, g_1\cdots g_n)$$

induce an isomorphism of complexes of $A$-modules

(II.3.8.6) $$\widetilde{\mathrm{K}}^\bullet(G,M)^G \xrightarrow{\sim} \mathrm{C}^\bullet_{\mathrm{cont}}(G,M).$$

Suppose that $M$ is discrete. Then for every $n \in \mathbb{N}$, $\mathrm{K}^n(G,M)$ is an induced discrete $A$-$G$-module (V.11.9). We denote by $\varepsilon\colon M \to \mathrm{K}^0(G,M)$ the morphism of discrete $A$-$G$-modules defined, for all $m \in M$, by $\varepsilon(m)(g_0) = m$. We have $d^0 \circ \varepsilon = 0$, where $d^0$ is the differential of degree zero of the complex $\widetilde{\mathrm{K}}^\bullet(G,M)$. The augmented complex of $A$-modules $M \xrightarrow{\varepsilon} \widetilde{\mathrm{K}}^\bullet(G,M)$ is homotopic to 0 (V.11.8). From this, we deduce a functorial canonical isomorphism of $\mathbf{D}^+(\mathbf{Mod}(A))$

(II.3.8.7) $$\mathrm{R}\Gamma(G,M) \xrightarrow{\sim} \mathrm{C}^\bullet_{\mathrm{cont}}(G,M).$$

We can therefore leave out the index "cont" from the notation $\mathrm{C}^\bullet_{\mathrm{cont}}$ and $\mathrm{H}^\bullet_{\mathrm{cont}}$ without causing any ambiguity.

**II.3.9.** For any abelian category $\mathscr{A}$, we denote by $\mathscr{A}^{\mathbb{N}}$ the category of inverse systems in $\mathscr{A}$ indexed by the ordered set $\mathbb{N}$. Then $\mathscr{A}^{\mathbb{N}}$ is an abelian category whose kernels and cokernels can be computed component-wise. If $\mathscr{A}$ has enough injectives, the same holds for $\mathscr{A}^{\mathbb{N}}$ (**[47]** 1.1); an object $(A_n, d_n)_{n\in\mathbb{N}}$ in $\mathscr{A}^{\mathbb{N}}$ is injective if and only if for all $n \in \mathbb{N}$, $A_n$ is injective and $d_n\colon A_{n+1} \to A_n$ has a left inverse.

Let $h\colon \mathscr{A} \to \mathscr{B}$ be a left exact functor between abelian categories. We denote by $h^{\mathbb{N}}\colon \mathscr{A}^{\mathbb{N}} \to \mathscr{B}^{\mathbb{N}}$ its natural extension. Assume that $\mathscr{A}$ has enough injectives. Then

$\mathrm{R}^i h^{\mathbb{N}} = (\mathrm{R}^i h)^{\mathbb{N}}$ for every $i \geq 0$ ([**47**] 1.2). Assume, moreover, that inverse limits indexed by $\mathbb{N}$ are representable in $\mathscr{B}$. Then we denote by

(II.3.9.1) $$\varprojlim_n h \colon \mathscr{A}^{\mathbb{N}} \to \mathscr{B}$$

the functor that sends $(A_n, d_n)_{n \in \mathbb{N}}$ to the inverse limit

(II.3.9.2) $$\varprojlim_n (h(A_n), h(d_n)),$$

and by $\mathrm{R}^+(\varprojlim_n h)$ its right derived functor. If $h$ maps injectives to injectives, then the same holds for $h^{\mathbb{N}}$. If, moreover, the right derived functor $\mathrm{R}^+(\varprojlim_n)$ of the functor

(II.3.9.3) $$\varprojlim_n \colon \mathscr{B}^{\mathbb{N}} \to \mathscr{B}$$

exists, then we have a canonical isomorphism

(II.3.9.4) $$\mathrm{R}^+(\varprojlim_n h) \xrightarrow{\sim} \mathrm{R}^+(\varprojlim_n) \circ (\mathrm{R}^+ h)^{\mathbb{N}}.$$

**II.3.10.** Let $A$ be a ring and $G$ a profinite group. Inverse limits in $\mathbf{Mod}(A)$ are representable: the functor

(II.3.10.1) $$\varprojlim_n \colon \mathbf{Mod}(A)^{\mathbb{N}} \to \mathbf{Mod}(A)$$

admits a right derived functor and we have $\mathrm{R}^i(\varprojlim_n) = 0$ for every $i \geq 2$ (cf. [**47**] 1.4 and [**63**] 2.1). If $(M_n)_{n \in \mathbb{N}}$ is an inverse system of $A$-modules satisfying the Mittag–Leffler condition, then we have ([**47**] 1.15 and [**63**] 3.1)

(II.3.10.2) $$\mathrm{R}^1 \varprojlim_n M_n = 0.$$

We denote by

(II.3.10.3) $$\Gamma(G, -) \colon (\mathbf{Rep}_A^{\mathrm{disc}}(G))^{\mathbb{N}} \to \mathbf{Mod}(A)$$

the functor $\varprojlim_n \Gamma(G, -)$ (II.3.9.1) and by $\mathrm{R}^+\Gamma(G, -)$ its right derived functor. For every inverse system $(M_n)_{n \in \mathbb{N}}$ of $\mathbf{Rep}_A^{\mathrm{disc}}(G)$ and every integer $i \geq 0$, we have, by (II.3.9.4), an exact sequence

(II.3.10.4) $$0 \to \mathrm{R}^1 \varprojlim_n \mathrm{H}^{i-1}(G, M_n) \to \mathrm{H}^i(G, (M_n)_{n \in \mathbb{N}}) \to \varprojlim_n \mathrm{H}^i(G, M_n) \to 0.$$

Suppose that $(M_n)_{n \in \mathbb{N}}$ satisfies the Mittag–Leffler condition and let $M$ be its inverse limit as a topological $A$-$G$-module. By ([**47**] 2.2), there is a canonical isomorphism of $\mathbf{D}^+(\mathbf{Mod}(A))$

(II.3.10.5) $$\mathrm{C}^\bullet_{\mathrm{cont}}(G, M) \xrightarrow{\sim} \mathrm{R}^+\Gamma(G, (M_n)_{n \in \mathbb{N}}).$$

**II.3.11.** Let $A$ be a ring and $G$, $H$ two groups. If $M$ is an $A[G]$-module and $N$ is an $A[H]$-module, then we denote by $M \boxtimes N$ the $A[G \times H]$-module with underlying $A$-module $M \otimes_A N$ such that for all $(g,h) \in G \times H$ and $(x,y) \in M \times N$, we have

$$(g,h) \cdot (x \otimes y) = (g \cdot x) \otimes (h \cdot y). \tag{II.3.11.1}$$

If $G$ and $H$ are profinite, then if $M$ is a discrete $A$-$G$-module and $N$ is a discrete $A$-$H$-module, $M \boxtimes N$ is a discrete $A$-$(G \times H)$-module.

If $C$ is a complex of $A[G]$-modules and $D$ is a complex of $A[H]$-modules, then we denote by $C \boxtimes D$ the simple complex associated with the bicomplex of $A[G \times H]$-modules defined by $(m,n) \mapsto C^m \boxtimes D^n$.

If $X$ is a cosimplicial $A[G]$-module and $Y$ is a cosimplicial $A[H]$-module, then we denote by $X \boxtimes Y$ the 2-cosimplicial $A[G \times H]$-module defined by $(m,n) \mapsto X^m \boxtimes Y^n$ and by $X \boxtimes^\Delta Y$ the diagonal subobject $\Delta(X \boxtimes Y)$ (II.3.6).

**II.3.12.** Let $A$ be a topological ring, $G$ and $H$ two profinite groups, $M$ a topological $A$-$G$-module, and $N$ a topological $A$-$H$-module. With the notation of II.3.8 and II.3.11, for every $n \in \mathbb{N}$, we have a morphism of $A[G \times H]$-modules

$$\mathrm{K}^n(G,M) \boxtimes \mathrm{K}^n(H,N) \to \mathrm{K}^n(G \times H, M \boxtimes N) \tag{II.3.12.1}$$

defined, for $\varphi \in \mathrm{K}^n(G,M)$ and $\psi \in \mathrm{K}^n(H,N)$, by

$$\varphi \otimes \psi \mapsto (((g_0,h_0),\dots,(g_n,h_n)) \mapsto \varphi(g_0,\dots,g_n) \otimes \psi(h_0,\dots,h_n)). \tag{II.3.12.2}$$

We thus obtain a morphism of cosimplicial $A[G \times H]$-modules (II.3.6)

$$\mathrm{K}^\bullet(G,M) \boxtimes^\Delta \mathrm{K}^\bullet(H,N) \to \mathrm{K}^\bullet(G \times H, M \boxtimes N). \tag{II.3.12.3}$$

From this we deduce a morphism of cosimplicial $A$-modules

$$\mathrm{K}^\bullet(G,M)^G \otimes_A^\Delta \mathrm{K}^\bullet(H,N)^H \to \mathrm{K}^\bullet(G \times H, M \boxtimes N)^{G \times H}. \tag{II.3.12.4}$$

Taking the associated cochain complexes and composing with the arrow (II.3.7.1), we obtain a morphism of complexes of $A$-modules (II.3.8.6)

$$\mathrm{C}^\bullet_{\mathrm{cont}}(G,M) \otimes_A \mathrm{C}^\bullet_{\mathrm{cont}}(H,N) \to \mathrm{C}^\bullet_{\mathrm{cont}}(G \times H, M \boxtimes N). \tag{II.3.12.5}$$

For all integers $m,n \geq 0$, this induces the *cross product*

$$\mathrm{H}^m_{\mathrm{cont}}(G,M) \otimes_A \mathrm{H}^n_{\mathrm{cont}}(H,N) \to \mathrm{H}^{m+n}_{\mathrm{cont}}(G \times H, M \boxtimes N), \quad x \otimes y \mapsto x \times y. \tag{II.3.12.6}$$

The morphisms (II.3.12.3), (II.3.12.5), and (II.3.12.6) are functorial, contravariant in $G$ and $H$, and covariant in $M$ and $N$.

**II.3.13.** We keep the assumptions of II.3.12 and moreover suppose that $G = H$. Composing the morphism (II.3.12.5) with the morphism

$$\mathrm{C}^\bullet_{\mathrm{cont}}(G \times G, M \boxtimes N) \to \mathrm{C}^\bullet_{\mathrm{cont}}(G, M \otimes N) \tag{II.3.13.1}$$

induced by the diagonal homomorphism $G \to G \times G$, we obtain a morphism of complexes of $A$-modules

$$\mathrm{C}^\bullet_{\mathrm{cont}}(G,M) \otimes_A \mathrm{C}^\bullet_{\mathrm{cont}}(G,N) \to \mathrm{C}^\bullet_{\mathrm{cont}}(G, M \otimes N). \tag{II.3.13.2}$$

For all integers $m,n \geq 0$, this induces the *cup product*

$$\mathrm{H}^m_{\mathrm{cont}}(G,M) \otimes_A \mathrm{H}^n_{\mathrm{cont}}(G,N) \to \mathrm{H}^{m+n}_{\mathrm{cont}}(G, M \otimes_A N), \quad x \otimes y \mapsto x \cup y. \tag{II.3.13.3}$$

The morphisms (II.3.13.2) and (II.3.13.3) are functorial, contravariant in $G$, and covariant in $M$ and $N$.

**Remark II.3.14.** Under the assumptions of II.3.12, for all $m, n \geq 0$, the diagram

$$\begin{array}{ccc} \mathrm{H}^m_{\mathrm{cont}}(G,M)\otimes_A \mathrm{H}^n_{\mathrm{cont}}(H,N) & & \\ \downarrow{\scriptstyle \pi_1^*\times\pi_2^*} & \searrow^{\times} & \\ \mathrm{H}^m_{\mathrm{cont}}(G\times H,M)\otimes_A \mathrm{H}^n_{\mathrm{cont}}(G\times H,N) & \xrightarrow{\cup} & \mathrm{H}^{m+n}_{\mathrm{cont}}(G\times H, M\boxtimes N) \end{array} \tag{II.3.14.1}$$

where $\pi_1\colon G\times H\to G$ and $\pi_2\colon G\times H\to H$ are the canonical projections, is commutative. This follows from the functoriality of the cross product and the fact that the composition

$$G\times H \xrightarrow{\delta} (G\times H)\times(G\times H) \xrightarrow{\pi_1\times\pi_2} G\times H\ , \tag{II.3.14.2}$$

where $\delta$ is the diagonal, is the identity.

**Lemma II.3.15.** *Let $A$ be a ring, $G$ a profinite group, $M$ a discrete $A$-$G$-module, and $N$ a flat $A$-module. Then we have a bifunctorial canonical isomorphism of* $\mathbf{D}^+(\mathbf{Mod}(A))$

$$\mathrm{R}\Gamma(G,M)\otimes^{\mathrm{L}}_A N \xrightarrow{\sim} \mathrm{R}\Gamma(G,M\otimes_A N). \tag{II.3.15.1}$$

Let us first show that for every integer $q\geq 0$, we have an isomorphism

$$\mathrm{H}^q(G,M)\otimes_A N \xrightarrow{\sim} \mathrm{H}^q(G,M\otimes_A N). \tag{II.3.15.2}$$

We have a canonical isomorphism

$$\mathrm{H}^q(G,M) \xrightarrow{\sim} \varinjlim_{H} \mathrm{H}^q(G/H,M^H), \tag{II.3.15.3}$$

where the limit is taken over the normal open subgroups $H$ of $G$, and likewise for $M\otimes_A N$. On the other hand, since $N$ is $A$-flat, for every subgroup $H$ of $G$, the canonical morphism

$$M^H\otimes_A N \to (M\otimes_A N)^H \tag{II.3.15.4}$$

is an isomorphism. We can therefore reduce to the case where $G$ is finite. Let

$$\cdots\to P_i\to P_{i-1}\to\cdots\to P_0\to\mathbb{Z}\to 0 \tag{II.3.15.5}$$

be a resolution of the $\mathbb{Z}[G]$-module $\mathbb{Z}$ by free $\mathbb{Z}[G]$-modules of finite type. For every $i\geq 0$, the canonical morphism

$$\mathrm{Hom}_{\mathbb{Z}[G]}(P_i,M)\otimes_A N\to \mathrm{Hom}_{\mathbb{Z}[G]}(P_i,M\otimes_A N) \tag{II.3.15.6}$$

is then an isomorphism; this leads to the desired conclusion because $\mathrm{H}^*(G,M)$ is the cohomology of the cochain complex $\mathrm{Hom}_{\mathbb{Z}[G]}(P_\bullet,M)$.

With the notation of II.3.8, the complex $\widetilde{\mathrm{K}}(G,M)\otimes_A N$ is a resolution of $M\otimes_A N$ by discrete and $G$-acyclic $A$-$G$-modules, by virtue of (II.3.15.2). On the other hand, the canonical morphism

$$\mathrm{C}^\bullet(G,M)\otimes_A N\to(\widetilde{\mathrm{K}}(G,M)\otimes_A N)^G \tag{II.3.15.7}$$

is an isomorphism by (II.3.15.4). The lemma follows.

**Lemma II.3.16.** *Let $A$ be a Dedekind ring, $G$ a profinite group, $M$ an $A$-$G$-module that is discrete, $G$-acyclic, and flat as an $A$-module, and $N$ an $A$-module. Then $M\otimes_A N$ is $G$-acyclic and the canonical morphism*

$$M^G\otimes_A N\to(M\otimes_A N)^G \tag{II.3.16.1}$$

*is an isomorphism.*

We can restrict to the case where $N$ is of finite type ([**2**] VI 5.3). Then, there exist two projective $A$-modules of finite type $N_1$ and $N_2$ and an exact sequence $0 \to N_2 \to N_1 \to N \to 0$. We can thus reduce to the case where $N$ is projective of finite type, and even to the case where $N$ is free of finite type, in which case the lemma is immediate.

**Lemma II.3.17.** *Let $A$ be a Dedekind ring and $C$, $D$ two complexes of $A$-modules such that $D$ is $A$-flat (that is, all its components are $A$-flat). If $C$ or $D$ is acyclic, the same holds for $C \otimes_A D$.*

This follows from the Künneth formula ([**10**] § 4.7 Thm. 3).

**Lemma II.3.18.** *Let $A$ be a Dedekind ring, $C$, $C'$, and $D$ complexes of $A$-modules, and $u\colon C \to C'$ a quasi-isomorphism. If $D$ is $A$-flat or $C$ and $C'$ are $A$-flat, then $u \otimes \mathrm{id}\colon C \otimes_A D \to C' \otimes_A D$ is a quasi-isomorphism.*

This follows from II.3.17 and ([**10**] § 4.3 Lem. 2).

**II.3.19.** Let $A$ be a Dedekind ring, $G$ and $H$ two profinite groups, $M$ a discrete $A$-$G$-module whose underlying $A$-module is flat, and $N$ a discrete $A$-$H$-module. Since the complex $\widetilde{\mathrm{K}}^\bullet(G,M)$ is $A$-flat (II.3.8), $\widetilde{\mathrm{K}}^\bullet(G,M) \boxtimes \widetilde{\mathrm{K}}^\bullet(H,N)$ (II.3.11) is a resolution of $M \boxtimes N$ by discrete and $(G \times H)$-acyclic $A$-$(G \times H)$-modules, by virtue of II.3.4, II.3.15, II.3.16, and II.3.18. It is also an acyclic resolution of $M \boxtimes N$ for the functor (II.3.3.1)

$$\Gamma(H,-)\colon \mathbf{Rep}_A^{\mathrm{disc}}(G \times H) \to \mathbf{Rep}_A^{\mathrm{disc}}(G), \quad L \mapsto L^H, \tag{II.3.19.1}$$

by II.3.15 and (II.3.3.2). We have a canonical isomorphism

$$\widetilde{\mathrm{K}}^\bullet(G,M) \otimes_A \mathrm{C}^\bullet(H,N) \xrightarrow{\sim} (\widetilde{\mathrm{K}}^\bullet(G,M) \otimes_A \widetilde{\mathrm{K}}^\bullet(H,N))^H. \tag{II.3.19.2}$$

Moreover, the canonical morphism

$$M \otimes_A \mathrm{C}^\bullet(H,N) \to \widetilde{\mathrm{K}}^\bullet(G,M) \otimes_A \mathrm{C}^\bullet(H,N) \tag{II.3.19.3}$$

is a quasi-isomorphism (II.3.18); $\widetilde{\mathrm{K}}^\bullet(G,M) \otimes_A \mathrm{C}^\bullet(H,N)$ is a complex of discrete and $G$-acyclic $A$-$G$-modules and the canonical morphism

$$\mathrm{C}^\bullet(G,M) \otimes_A \mathrm{C}^\bullet(H,N) \to (\widetilde{\mathrm{K}}^\bullet(G,M) \otimes_A \mathrm{C}^\bullet(H,N))^G \tag{II.3.19.4}$$

is an isomorphism by II.3.16. Consequently, the canonical morphism (II.3.12.5)

$$\mathrm{C}^\bullet(G,M) \otimes_A \mathrm{C}^\bullet(H,N) \to \mathrm{C}^\bullet(G \times H, M \boxtimes N) \tag{II.3.19.5}$$

is a quasi-isomorphism. The composition of (II.3.19.5) and the inverse of (II.3.19.4) then induces an isomorphism of $\mathbf{D}^+(\mathbf{Mod}(A))$

$$\mathrm{R}\Gamma(G, M \otimes_A^{\mathrm{L}} \mathrm{R}\Gamma(H,N)) \xrightarrow{\sim} \mathrm{R}\Gamma(G \times H, M \boxtimes N). \tag{II.3.19.6}$$

In view of (II.3.15.1), we can deduce from this an isomorphism of $\mathbf{D}^+(\mathbf{Mod}(A))$

$$\mathrm{R}\Gamma(G, \mathrm{R}\Gamma(H, M \boxtimes N)) \xrightarrow{\sim} \mathrm{R}\Gamma(G \times H, M \boxtimes N). \tag{II.3.19.7}$$

This is equal to the isomorphism (II.3.4.1).

**Lemma II.3.20.** *Let $G$ be a profinite group isomorphic to $\mathbb{Z}_p$, $\gamma$ a topological generator of $G$, $A$ a ring, and $M$ a discrete $A$-$G$-module whose elements are all $p$-primary torsion. Then we have an exact sequence of discrete $A$-$G$-modules*

$$0 \longrightarrow M \xrightarrow{\varepsilon} \mathrm{Ind}_{A,G}(M) \xrightarrow{d_\gamma} \mathrm{Ind}_{A,G}(M) \longrightarrow 0, \tag{II.3.20.1}$$

*where for all $x \in M$, $f \in \mathrm{Ind}_{A,G}(M)$, and $g \in G$, we have $\varepsilon(x)(g) = g \cdot x$ and $d_\gamma(f)(g) = \gamma f(\gamma^{-1}g) - f(g)$.*

Only the surjectivity of $d_\gamma$ needs to be proved. Let $f\colon G \to M$ be continuous map. Since $G$ is compact, there exists an integer $n \geq 0$ such that $f$ factors through $G/\gamma^{p^n\mathbb{Z}_p}$, $p^n f(G) = 0$, and for every $x \in f(G)$, we have $\gamma^{p^n} \cdot x = x$. We denote by $h\colon \mathbb{Z}_{>0} \to M$ the map defined, for every $a \in \mathbb{Z}_{>0}$, by

$$h(a) = -\sum_{i=1}^{a} \gamma^{a-i} \cdot f(\gamma^i). \tag{II.3.20.2}$$

This factors through $\mathbb{Z}/p^{2n}\mathbb{Z}$. Composing this with the surjective homomorphism $G \to \mathbb{Z}/p^{2n}\mathbb{Z}$ that sends $\gamma$ onto the class of 1 gives a continuous map $\widetilde{h}\colon G \to M$ such that $d_\gamma(\widetilde{h}) = f$.

**Lemma II.3.21.** *We keep the assumptions of* II.3.20 *and moreover denote by*

$$\partial_M \colon M \to \mathrm{H}^1(G, M) \tag{II.3.21.1}$$

*the boundary map of the exact sequence* (II.3.20.1). *Then:*

(i) *For every $x \in M$, there exists a unique continuous crossed homomorphism $\nu_x\colon G \to M$ such that $\nu_x(\gamma) = x$.*
(ii) *For every $x \in M$, the class of $\nu_x$ in $\mathrm{H}^1(G, M)$ is equal to $-\partial_M(x)$.*

(i) Indeed, assume that such a continuous crossed homomorphism $\nu_x$ exists. Then for every integer $a \geq 1$, we have

$$\nu_x(\gamma^a) = (\gamma^{a-1} + \cdots + 1) \cdot x. \tag{II.3.21.2}$$

Let $n$ be an integer $\geq 0$ such that $p^n x = 0$ and $\gamma^{p^n} \cdot x = x$. For every $g \in G$, we have

$$\nu_x(\gamma^{p^{2n}} g) = g \cdot \nu_x(\gamma^{p^{2n}}) + \nu_x(g) = \nu_x(g). \tag{II.3.21.3}$$

By continuity, $\nu_x$ therefore factors through the canonical homomorphism $G \to G/\gamma^{p^{2n}\mathbb{Z}_p}$. On the other hand, the map

$$\mathbb{Z}_{>0} \to M, \quad a \mapsto (\gamma^{a-1} + \cdots + 1) \cdot x \tag{II.3.21.4}$$

factors through $\mathbb{Z}/p^{2n}\mathbb{Z}$. Composing this with the surjective homomorphism $G \to \mathbb{Z}/p^{2n}\mathbb{Z}$ that sends $\gamma$ onto the class of 1 gives a continuous crossed homomorphism from $G$ to $M$. The statement follows.

(ii) For every $g \in G$, we have (II.3.20.1)

$$d_\gamma(\nu_x)(g) = \gamma \cdot \nu_x(\gamma^{-1} g) - \nu_x(g) = -\nu_x(\gamma) = -x. \tag{II.3.21.5}$$

**Lemma II.3.22.** *Let $G$ be a profinite group isomorphic to $\mathbb{Z}_p$, $\gamma$ a topological generator of $G$, $A$ a ring with $p$-primary torsion, $M$ a discrete $A$-$G$-module, and $x \in M^G$. Then we have*

$$\partial_M(x) = \partial_A(1) \cup x, \tag{II.3.22.1}$$

*where $\partial_M$ is the boundary map of the exact sequence* (II.3.20.1) *and $\cup$ is the cup product.*

This follows from II.3.21. Indeed, let $\nu_x\colon G \to M$ and $\mu_1\colon G \to A$ be the continuous crossed homomorphisms such that $\nu_x(\gamma) = x$ and $\mu_1(\gamma) = 1$. For every integer $a \geq 1$, we have $\nu_x(\gamma^a) = ax = \mu_1(\gamma^a)x$ by (II.3.21.2). It follows that $\nu_x(g) = \mu_1(g)x$ for every $g \in G$. The lemma follows.

**Proposition II.3.23.** *Let $n$ be an integer $\geq 1$ and for every integer $1 \leq i \leq n$, let $G_i$ be a profinite group isomorphic to $\mathbb{Z}_p$ and $\gamma_i$ a topological generator of $G_i$. Set $G = \prod_{i=1}^n G_i$. Let $A$ be a ring and $M$ a discrete $A$-$G$-module whose elements are all $p$-primary torsion. By induction, we define a complex of discrete $A$-$G$-modules $K_i^\bullet$, for integers $0 \leq i \leq n$,* *as follows. We set* $K_0^\bullet = M[0]$ *and for every* $1 \leq i \leq n$, $K_i^\bullet$ *is the fiber of the morphism* $\gamma_i - 1\colon K_{i-1}^\bullet \to K_{i-1}^\bullet$. *Then there exists a canonical isomorphism of* $\mathbf{D}^+(\mathbf{Mod}(A))$

(II.3.23.1) $$\mathrm{R}\Gamma(G, M) \xrightarrow{\sim} K_n^\bullet.$$

For every integer $0 \leq j \leq n$, we set $G_{\leq j} = \prod_{1 \leq i \leq j} G_i$ and $G_{>j} = \prod_{j<i\leq n}^n G_i$. Let us show by induction that for every integer $0 \leq j \leq n$, we have a canonical isomorphism of $\mathbf{D}^+(\mathbf{Rep}_A^{\mathrm{disc}}(G_{>j}))$

(II.3.23.2) $$\mathrm{R}\Gamma(G_{\leq j}, M) \xrightarrow{\sim} K_j^\bullet.$$

The assertion is immediate for $j = 0$. Suppose that the assertion has been proved for every integer $0 \leq j \leq n-1$. We denote by $\mathrm{Ind}_{A,G_{j+1}}(K_j^\bullet)$ the image of the complex $K_j^\bullet$ by the functor $\mathrm{Ind}_{A,G_{j+1}}$ (II.3.2.2). We endow it with the action of $G_{>j}$ defined, for all $q \in \mathbb{Z}$, $f \in \mathrm{Ind}_{A,G_{j+1}}(K_j^q)$, $g = (g_0, g_1) \in G_{>j} = G_{j+1} \times G_{>j+1}$, and $x \in G_{j+1}$, by

(II.3.23.3) $$(g \cdot f)(x) = g_1 \cdot f(x \cdot g_0) \in K_j^q.$$

It immediately follows from II.3.20 that we have an exact sequence of complexes of $\mathbf{Rep}_A^{\mathrm{disc}}(G_{>j})$

(II.3.23.4) $$0 \longrightarrow K_j^\bullet \xrightarrow{\varepsilon_j} \mathrm{Ind}_{A,G_{j+1}}(K_j^\bullet) \xrightarrow{d_{\gamma_{j+1}}} \mathrm{Ind}_{A,G_{j+1}}(K_j^\bullet) \longrightarrow 0\ ,$$

where for all $q \in \mathbb{Z}$, $c \in K_j^q$, $f \in \mathrm{Ind}_{A,G_{j+1}}(K_j^q)$ and $x \in G_{j+1}$, we have

(II.3.23.5) $$\varepsilon_j^q(c)(x) = x \cdot c,$$

(II.3.23.6) $$d_{\gamma_{j+1}}^q(f)(x) = \gamma_{j+1} f(\gamma_{j+1}^{-1} x) - f(x).$$

By the induction hypothesis, we therefore have a distinguished triangle

(II.3.23.7) $$\mathrm{R}\Gamma(G_{\leq j}, M) \longrightarrow \mathrm{Ind}_{A,G_{j+1}}(K_j^\bullet) \xrightarrow{d_{\gamma_{j+1}}} \mathrm{Ind}_{A,G_{j+1}}(K_j^\bullet) \xrightarrow{+1}$$

in $\mathbf{D}^+(\mathbf{Rep}_A^{\mathrm{disc}}(G_{>j}))$. By II.3.2 and (II.3.3.2), for every integer $q$, the morphism

(II.3.23.8) $$K_j^q \to \Gamma(G_{j+1}, \mathrm{Ind}_{A,G_{j+1}}(K_j^q)), \quad u \mapsto (x \mapsto u)$$

induces an isomorphism of $\mathbf{D}^+(\mathbf{Rep}_A^{\mathrm{disc}}(G_{>j+1}))$

(II.3.23.9) $$K_j^q[0] \xrightarrow{\sim} \mathrm{R}\Gamma(G_{j+1}, \mathrm{Ind}_{A,G_{j+1}}(K_j^q)).$$

By virtue of II.3.4, the triangle (II.3.23.7) therefore induces a distinguished triangle in $\mathbf{D}^+(\mathbf{Rep}_A^{\mathrm{disc}}(G_{>j+1}))$

(II.3.23.10) $$\mathrm{R}\Gamma(G_{\leq j+1}, M) \longrightarrow K_j^\bullet \xrightarrow{\gamma_{j+1}-1} K_j^\bullet \xrightarrow{+1}$$

The proposition follows.

**Corollary II.3.24.** *For every integer* $d \geq 1$, *the* $p$*-cohomological dimension of the profinite group* $\mathbb{Z}_p^d$ *is equal to* $d$.

For every $p$-primary torsion discrete $\mathbb{Z}$-$\mathbb{Z}_p^d$-module $M$, the associated complex $K_d^\bullet$ in II.3.23 is concentrated in the degrees $[0, d]$. The $p$-cohomological dimension of the profinite group $\mathbb{Z}_p^d$ is therefore lesser than or equal to $d$ by virtue of II.3.23 and (**[65]** I Prop. 11). For the trivial discrete $\mathbb{Z}$-$\mathbb{Z}_p^d$-module $\mathbb{F}_p$, the differentials of the associated complex $K_d^\bullet$ are zero and $\mathrm{H}^d(\mathbb{Z}_p^d, \mathbb{F}_p)$ is therefore isomorphic to $\mathbb{F}_p$ (II.3.23.1); the corollary follows.

**Corollary II.3.25.** *Let $n$ be an integer $\geq 1$, $G$ a profinite group isomorphic to $\mathbb{Z}_p^n$ with $\mathbb{Z}_p$-basis $e_1, \dots, e_n$, $A$ a topological ring, and $M$ a topological $A$-$G$-module. We denote by $\varphi\colon G \to \mathrm{Aut}_A(M)$ the representation of $G$ over $M$, by $\mathrm{S}_A(G)$ the symmetric algebra of the $A$-module $G \otimes_{\mathbb{Z}} A$, and by $M^{\triangleright}$ the $\mathrm{S}_A(G)$-module with underlying $A$-module $M$ such that for every $1 \leq i \leq n$, the action of $e_i$ on $M$ is given by $\varphi(e_i) - \mathrm{id}_M$. Suppose that one of the following conditions is satisfied:*

(i) *$M$ is a $p$-primary torsion discrete $A$-$G$-module.*

(ii) *$M$ is endowed with the $p$-adic topology, for which it is complete and separated.*

*Then we have a canonical isomorphism of* $\mathbf{D}^+(\mathbf{Mod}(A))$

(II.3.25.1) $$\mathrm{C}^{\bullet}_{\mathrm{cont}}(G, M) \xrightarrow{\sim} \mathbb{K}^{\bullet}_{\mathrm{S}_A(G)}(M^{\triangleright})$$

*that is functorial in $M$, where the complex on the left is defined in* (II.3.8) *and that on the right is defined in* (II.2.7.5).

Case (i) follows from II.3.23 and (II.2.7.9). Let us consider case (ii). For every $r \geq 0$, we set $M_r = M/p^rM$. By virtue of (II.3.10.5), we have a canonical isomorphism of $\mathbf{D}^+(\mathbf{Mod}(A))$

(II.3.25.2) $$\mathrm{C}^{\bullet}_{\mathrm{cont}}(G, M) \xrightarrow{\sim} \mathrm{R}^+\Gamma(G, (M_r)_{r\in\mathbb{N}}).$$

On the other hand, by case (i), we have a compatible system of isomorphisms

(II.3.25.3) $$\mathrm{R}\Gamma(G, M_r) \xrightarrow{\sim} \mathbb{K}^{\bullet}_{\mathrm{S}_A(G)}(M^{\triangleright}/p^rM^{\triangleright})$$

of $\mathbf{D}^+(\mathbf{Mod}(A))$. Since for every integer $n$, the inverse system $(\mathbb{K}^{n}_{\mathrm{S}_A(G)}(M^{\triangleright}/p^rM^{\triangleright}))_{r\geq 0}$ satisfies the Mittag–Leffler condition, in view of (II.3.9.4) and (II.3.10.2), we obtain an isomorphism

(II.3.25.4) $$\mathrm{C}^{\bullet}_{\mathrm{cont}}(G, M) \xrightarrow{\sim} \varprojlim_{r\geq 0} \mathbb{K}^{\bullet}_{\mathrm{S}_A(G)}(M^{\triangleright}/p^rM^{\triangleright}).$$

The corollary follows because for every integer $n$, $\mathbb{K}^{n}_{\mathrm{S}_A(G)}(M^{\triangleright})$ is complete and separated for the $p$-adic topology.

**Remark II.3.26.** We keep the assumptions of II.3.25. By II.2.8.10, the $\mathrm{S}_A(G)$-module $M^{\triangleright}$ corresponds to a Higgs $A$-field $\theta$ on $M$ with coefficients in $\mathrm{Hom}_{\mathbb{Z}}(G, A)$, and the complex $\mathbb{K}^{\bullet}_{\mathrm{S}_A(G)}(M^{\triangleright})$ can be identified with the Dolbeault complex of $(M, \theta)$.

**II.3.27.** Let $G$ be a profinite group isomorphic to $\mathbb{Z}_p$, $\gamma$ a topological generator of $G$, $H$ a profinite group, $A$ a $p$-primary torsion ring, and $M$ a discrete $A$-$H$-module that we also view as a discrete $A$-$(G \times H)$-module through the canonical projection $G \times H \to H$. We denote by $\mathrm{Ind}_{A,G}(\mathrm{C}^{\bullet}(H, M))$ the image of the complex of continuous cochains $\mathrm{C}^{\bullet}(H, M)$ (II.3.8) by the functor $\mathrm{Ind}_{A,G}$ (II.3.2.2). We endow $\mathrm{C}^{\bullet}(H, M)$ with the trivial action of $G$. By II.3.20, we have an exact sequence of complexes of $\mathbf{Rep}^{\mathrm{disc}}_A(G)$

(II.3.27.1) $$0 \longrightarrow \mathrm{C}^{\bullet}(H, M) \xrightarrow{\varepsilon^{\bullet}} \mathrm{Ind}_{A,G}(\mathrm{C}^{\bullet}(H, M)) \xrightarrow{d^{\bullet}_{\gamma}} \mathrm{Ind}_{A,G}(\mathrm{C}^{\bullet}(H, M)) \longrightarrow 0,$$

where for all $q \in \mathbb{Z}$, $c \in \mathrm{C}^q(H, M)$, $f \in \mathrm{Ind}_{A,G}(\mathrm{C}^q(H, M))$ and $x \in G$, we have

(II.3.27.2) $$\varepsilon^q(c)(x) = x,$$

(II.3.27.3) $$d^q_{\gamma}(f)(x) = \gamma f(\gamma^{-1}x) - f(x).$$

For every integer $q \geq 0$, $\varepsilon^q$ induces an isomorphism of $\mathbf{D}^+(\mathbf{Mod}(A))$

(II.3.27.4) $$\mathrm{C}^q(H, M)[0] \xrightarrow{\sim} \mathrm{R}\Gamma(G, \mathrm{Ind}_{A,G}(\mathrm{C}^q(H, M))).$$

By virtue of II.3.4, (II.3.27.1) therefore induces a distinguished triangle in $\mathbf{D}^+(\mathbf{Mod}(A))$

(II.3.27.5) $$\mathrm{R}\Gamma(G\times H,M)\longrightarrow \mathrm{C}^\bullet(H,M)\xrightarrow{0}\mathrm{C}^\bullet(H,M)\xrightarrow{+1}$$

where 0 is the zero morphism. From this we deduce, for every integer $n\geq 0$, an exact sequence

(II.3.27.6) $$0\longrightarrow \mathrm{H}^{n-1}(H,M)\xrightarrow{\alpha_n}\mathrm{H}^n(G\times H,M)\xrightarrow{\beta_n}\mathrm{H}^n(H,M)\longrightarrow 0.$$

Moreover, again by II.3.20, we have a canonical exact sequence (II.3.20.1)

(II.3.27.7) $$0\to A\to \mathrm{Ind}_{A,G}(A)\to \mathrm{Ind}_{A,G}(A)\to 0.$$

It induces an isomorphism

(II.3.27.8) $$\partial_A\colon A\xrightarrow{\sim}\mathrm{H}^1(G,A).$$

**Proposition II.3.28.** *Under the assumptions of* II.3.27, *with $n$ an integer $\geq 0$, we have:*

(i) *$\beta_n$ is the restriction morphism with respect to the canonical injection $H\to G\times H$.*

(ii) *For every $x\in \mathrm{H}^{n-1}(H,M)$, we have*

(II.3.28.1) $$\alpha_n(x)=\partial_A(1)\times x,$$

*where the cross product is defined in* (II.3.12.6).

We denote by $\tau_{\leq n}\mathrm{C}^\bullet(H,M)$ the canonical filtration of $\mathrm{C}^\bullet(H,M)$ ([**15**] 1.4.6). By II.3.20, we have a commutative diagram of complexes of $\mathbf{Rep}^{\mathrm{disc}}_A(G)$

(II.3.28.2)

$$\begin{array}{ccccccccc}0&\longrightarrow&\tau_{\leq n}\mathrm{C}^\bullet(H,M)&\xrightarrow{\psi_n}&\mathrm{Ind}_{A,G}(\tau_{\leq n}\mathrm{C}^\bullet(H,M))&\xrightarrow{\phi_n}&\mathrm{Ind}_{A,G}(\tau_{\leq n}\mathrm{C}^\bullet(H,M))&\longrightarrow&0\\ &&\downarrow{\scriptstyle u_n}&&\downarrow{\scriptstyle \mathrm{Ind}_{A,G}(u_n)}&&\downarrow{\scriptstyle \mathrm{Ind}_{A,G}(u_n)}&&\\ 0&\longrightarrow&\mathrm{H}^n(H,M)[-n]&\longrightarrow&\mathrm{Ind}_{A,G}(\mathrm{H}^n(H,M))[-n]&\longrightarrow&\mathrm{Ind}_{A,G}(\mathrm{H}^n(H,M))[-n]&\longrightarrow&0\end{array}$$

where $u_n$ is the canonical morphism and the horizontal lines are the exact sequences defined as in (II.3.27.1). In fact, since the functor $\mathrm{Ind}_{A,G}$ is exact, the top horizontal line can be deduced from the exact sequence (II.3.27.1) by applying the functor $\tau_{\leq n}$.

(i) We clearly have $\beta_n=\mathrm{H}^n(G,\mathrm{Ind}_{A,G}(u_n)\circ\psi_n)$. In view of (II.3.28.2), we deduce from this that $\beta_n=\mathrm{H}^n(G,u_n)$. The statement therefore follows from II.3.5.

(ii) It follows by induction from II.3.24 that for every integer $i\geq 2$, we have

(II.3.28.3) $$\mathrm{H}^n(G,\tau_{\leq n-i}\mathrm{C}^\bullet(H,M))=0.$$

Consequently, the canonical morphism

(II.3.28.4) $$\mathrm{H}^n(G,\tau_{\leq n-1}\mathrm{C}^\bullet(H,M))\to \mathrm{H}^1(G,\mathrm{H}^{n-1}(H,M))$$

is an isomorphism. In view of (II.3.4.1), we deduce from this a canonical morphism

(II.3.28.5) $$v_n\colon \mathrm{H}^1(G,\mathrm{H}^{n-1}(H,M))\to \mathrm{H}^n(G\times H,M).$$

Let us moreover consider the commutative diagram

(II.3.28.6) $$\begin{array}{ccc}\mathrm{Ind}_{A,G}(\tau_{\leq n-1}\mathrm{C}^\bullet(H,M))&\xrightarrow{\delta_{\leq n-1}}&\tau_{\leq n-1}\mathrm{C}^\bullet(H,M)[1]\\ \downarrow{\scriptstyle \mathrm{Ind}_{A,G}(u_{n-1})}&&\downarrow{\scriptstyle u_{n-1}[1]}\\ \mathrm{Ind}_{A,G}(\mathrm{H}^{n-1}(H,M))[-n+1]&\xrightarrow{\delta_{n-1}}&\mathrm{H}^{n-1}(H,M)[-n+2]\end{array}$$

induced by the diagram (II.3.28.2) (for $n-1$ instead of $n$). Let

(II.3.28.7) $$\partial_{n-1} = \mathrm{H}^{n-1}(G, \delta_{n-1}) \colon \mathrm{H}^{n-1}(H, M) \to \mathrm{H}^1(G, \mathrm{H}^{n-1}(H, M)),$$

which is in fact an isomorphism. Since (II.3.28.4) is an isomorphism, we have

(II.3.28.8) $$\alpha_n = v_n \circ \mathrm{H}^{n-1}(G, u_{n-1}[1] \circ \delta_{\leq n-1}) = v_n \circ \partial_{n-1}.$$

By II.3.22, for every $x \in \mathrm{H}^{n-1}(H, M)$, we have

(II.3.28.9) $$\partial_{n-1}(x) = \partial_A(1) \cup x.$$

It therefore suffices to show that the diagram

(II.3.28.10) $$\begin{array}{ccc} \mathrm{H}^1(G, A) \otimes_A \mathrm{H}^{n-1}(H, M) & & \\ \downarrow{\scriptstyle \cup} & \searrow{\scriptstyle \times} & \\ \mathrm{H}^1(G, \mathrm{H}^{n-1}(H, M)) & \xrightarrow{v_n} & \mathrm{H}^n(G \times H, M) \end{array}$$

where the cup product on the left is defined in (II.3.13.3) and the cross product on the right is defined in (II.3.12.6), is commutative. By II.3.25, we have a canonical isomorphism

(II.3.28.11) $$\partial_{\mathbb{Z}_p} \colon \mathbb{Z}_p \xrightarrow{\sim} \mathrm{H}^1(G, \mathbb{Z}_p)$$

that is compatible with the isomorphism $\partial_A$ (II.3.27.8) through the canonical homomorphism $\mathbb{Z}_p \to A$. It therefore also suffices to show that the diagram

(II.3.28.12) $$\begin{array}{ccc} \mathrm{H}^1(G, \mathbb{Z}_p) \otimes_{\mathbb{Z}_p} \mathrm{H}^{n-1}(H, M) & & \\ \downarrow{\scriptstyle \cup} & \searrow{\scriptstyle \times} & \\ \mathrm{H}^1(G, \mathrm{H}^{n-1}(H, M)) & \xrightarrow{v_n} & \mathrm{H}^n(G \times H, M) \end{array}$$

is commutative. By II.3.19, we have a canonical isomorphism of $\mathbf{D}^+(\mathbf{Mod}(A))$

(II.3.28.13) $$\mathrm{R}\Gamma(G, \tau_{\leq n-1}\mathrm{C}^\bullet(H, M)) \xrightarrow{\sim} \mathrm{C}^\bullet(G, \mathbb{Z}_p) \otimes_{\mathbb{Z}_p} \tau_{\leq n-1}\mathrm{C}^\bullet(H, M),$$

and in view of (II.3.28.4), the morphism $v_n$ can be obtained by applying the functor $\mathrm{H}^n$ to the morphism

(II.3.28.14) $$\mathrm{C}^\bullet(G, \mathbb{Z}_p) \otimes_{\mathbb{Z}_p} \tau_{\leq n-1}\mathrm{C}^\bullet(H, M) \to \mathrm{C}^\bullet(G \times H, M)$$

deduced from (II.3.12.5). The commutativity of (II.3.28.12) follows.

**Remark II.3.29.** Proposition II.3.28 gives a "Künneth formula" for the cohomology of products of profinite groups with values in certain discrete modules that we have not been able to find in the literature. We have only treated a simple case, where one of the groups is $\mathbb{Z}_p$. This is the only case we will need in this work. For the general case, **[48]** contains a weaker statement that does not study the compatibility with the cross product.

**Proposition II.3.30.** *Let $A$ be a $\mathbb{Z}_p$-algebra that is complete and separated for the $p$-adic topology and $d$ an integer $\geq 1$. Then there exists a unique isomorphism of graded $A$-algebras*

(II.3.30.1) $$\wedge\, (\mathrm{H}^1_{\mathrm{cont}}(\mathbb{Z}_p^d, A)) \xrightarrow{\sim} \oplus_{n \geq 0} \mathrm{H}^n_{\mathrm{cont}}(\mathbb{Z}_p^d, A),$$

*where the right-hand side is endowed with the cup product, which is the identity in degree one.*

Let us first consider the case where $A$ is $p$-primary torsion. We proceed by induction on $d$. The case $d = 1$ is immediate in view of II.3.24 and the canonical isomorphism $A \xrightarrow{\sim} \mathrm{H}^1_{\mathrm{cont}}(\mathbb{Z}_p, A)$. Suppose $d \geq 2$ and that the statement has been proved for $d-1$. Let $G = \mathbb{Z}_p$ and $H = \mathbb{Z}_p^{d-1}$. It follows from II.3.28 and II.3.14 that the cross product (II.3.12.6) induces an isomorphism of graded $A$-algebras

$$(A \oplus \mathrm{H}^1(G, A))^g\otimes_A(\oplus_{n\geq 0}\mathrm{H}^n(H, A)) \xrightarrow{\sim} \oplus_{n\geq 0}\mathrm{H}^n(G \times H, A), \tag{II.3.30.2}$$

where the symbol ${}^g\otimes_A$ denotes the left tensor product (cf. [9] III § 4.7 Remarks p. 49). The desired statement now follows using the induction hypothesis. Also note that we have a canonical isomorphism $A^d \xrightarrow{\sim} \mathrm{H}^1_{\mathrm{cont}}(\mathbb{Z}_p^d, A)$.

Let us now consider the general case. It follows from the above that for every integer $n \geq 0$, the inverse system $(\mathrm{H}^n(\mathbb{Z}_p^d, A/p^rA))_{r\in\mathbb{N}}$ satisfies the Mittag–Leffler condition. By virtue of (II.3.10.2), (II.3.10.4), and (II.3.10.5), the canonical morphism

$$\mathrm{H}^n_{\mathrm{cont}}(\mathbb{Z}_p^d, A) \to \varprojlim_{r\geq 0} \mathrm{H}^n(\mathbb{Z}_p^d, A/p^rA) \tag{II.3.30.3}$$

is therefore an isomorphism. The proposition then follows from the earlier case.

**II.3.31.** Let $G$ be a profinite group. A *$G$-set* is a discrete topological space endowed with a continuous action of $G$. The $G$-sets naturally form a category. A $G$-group $M$ is a group in this category. We associate with it its subgroup of $G$-invariants $\mathrm{H}^0(G, M) = M^G$ and its first cohomology set $\mathrm{H}^1(G, M)$; we refer to ([65] I § 5) for the definition and main properties of this pointed set.

**II.3.32.** Let $G$ be a profinite group, $A$ a ring endowed with the discrete topology and a continuous action of $G$, $M$ a free $A$-module of rank $r \geq 1$ endowed with the discrete topology, and $(e_1, \dots, e_r)$ a basis of $M$ over $A$. We denote by $\mathrm{Mat}_r(A)$ the $A$-algebra of square $r$ by $r$ matrices with coefficients in $A$ and by $\mathrm{GL}_r(A)$ the group of invertible elements of $\mathrm{Mat}_r(A)$. Note that $\mathrm{GL}_r(A)$ is naturally a $G$-group. Giving a continuous $A$-representation $\rho$ of $G$ over $M$ is equivalent to giving, for every $g \in G$, an element $U_g$ of $\mathrm{GL}_r(A)$ such that the map $g \mapsto U_g$ is continuous and that for every $g, h \in G$, we have

$$U_{gh} = U_g \cdot {}^gU_h. \tag{II.3.32.1}$$

The matrix $U_g$ gives the coordinates of the vectors $e_1, \dots, e_r$ in the basis $g(e_1), \dots, g(e_r)$. Changing the basis $(e_i)_{1\leq i\leq r}$ changes the cocycle $g \mapsto U_g$ into a cohomologous cocycle. The map that sends $\rho$ to the class $[\rho]$ of the cocycle $g \mapsto U_g$ in $\mathrm{H}^1(G, \mathrm{GL}_r(A))$ is a bijection from the set of isomorphism classes of continuous $A$-representations of $G$ over $M$ to the set $\mathrm{H}^1(G, \mathrm{GL}_r(A))$. The $A$-representation of $G$ over $M$ that fixes the $(e_i)_{1\leq i\leq r}$ corresponds to the neutral element of $\mathrm{H}^1(G, \mathrm{GL}_r(A))$.

Let $a \in A^G$ be such that $a$ is nilpotent in $A$ and $\rho$ an $A$-representation of $G$ over $M$ such that $\rho(g)(e_i) - e_i \in aM$ for all $g \in G$ and $1 \leq i \leq r$. Then the cocycle $g \mapsto U_g$ defined above takes on values in the subgroup $\mathrm{id}_r + a\mathrm{Mat}_r(A)$ of $\mathrm{GL}_r(A)$. If we change the basis $(e_i)_{1\leq i\leq r}$ to a basis $(e'_i)_{1\leq i\leq r}$ such that $e'_i - e_i \in aM$ for every $1 \leq i \leq r$, then the cocycle $g \mapsto U_g$ transforms into a cohomologous cocycle in $\mathrm{id}_r + a\mathrm{Mat}_r(A)$. The map that sends $\rho$ to the class $[\rho]$ of the cocycle $g \mapsto U_g$ in $\mathrm{H}^1(G, \mathrm{id}_r + a\mathrm{Mat}_r(A))$ is a bijection from the set of isomorphism classes of continuous $A$-representations of $G$ over $M$ that fix $e_i$ modulo $aM$ for every $1 \leq i \leq r$, modulo the isomorphisms that fix $e_i$ modulo $aM$ for every $1 \leq i \leq r$, to the set $\mathrm{H}^1(G, \mathrm{id}_r + a\mathrm{Mat}_r(A))$.

**II.3.33.** Let $G$ be a profinite group, $A$ a ring endowed with an action of $G$, $a \in A^G$, and $m, n, q, r$ integers $\geq 1$ such that $q \geq n \geq m$ and $m + n \geq q$. Suppose that the action of $G$ on $A$ is continuous for the $a$-adic topology and that the multiplication by $a^n$ in $A$ induces an isomorphism

(II.3.33.1) $$A/a^{q-n}A \xrightarrow{\sim} a^nA/a^qA.$$

The second condition is satisfied, for example, if $a$ is not a zero divisor in $A$. Consider the canonical exact sequence of $G$-groups

(II.3.33.2) $$1 \to \mathrm{id}_r + a^n\mathrm{Mat}_r(A/a^qA) \to \mathrm{id}_r + a^m\mathrm{Mat}_r(A/a^qA) \to \mathrm{id}_r + a^m\mathrm{Mat}_r(A/a^nA) \to 1.$$

Then $\mathrm{id}_r + a^n\mathrm{Mat}_r(A/a^qA)$ is contained in the center of $\mathrm{id}_r + a^m\mathrm{Mat}_r(A/a^qA)$. On the other hand, by (II.3.33.1), we have a canonical isomorphism of abelian $G$-groups

(II.3.33.3) $$\mathrm{Mat}_r(A/a^{q-n}A) \xrightarrow{\sim} \mathrm{id}_r + a^n\mathrm{Mat}_r(A/a^qA).$$

By ([**65**] I prop. 43), we have a canonical exact sequence of pointed sets

(II.3.33.4) $$\mathrm{H}^1(G, \mathrm{Mat}_r(A/a^{q-n}A)) \longrightarrow \mathrm{H}^1(G, \mathrm{id}_r + a^m\mathrm{Mat}_r(A/a^qA)) \longrightarrow$$
$$\mathrm{H}^1(G, \mathrm{id}_r + a^m\mathrm{Mat}_r(A/a^nA)) \xrightarrow{\partial} \mathrm{H}^2(G, \mathrm{Mat}_r(A/a^{q-n}A)).$$

**II.3.34.** We keep the assumptions of II.3.33 and let $N$ be a free $(A/a^qA)$-module of rank $r$ with basis $f_1, \ldots, f_r$ and $\rho, \rho'$ two continuous representations of $G$ over $N$ (endowed with the discrete topology) such that $\rho(g)(f_i) - f_i \in a^mN$ and $\rho'(g)(f_i) - f_i \in a^mN$ for all $g \in G$ and $1 \leq i \leq r$. We denote by $g \mapsto V_g$ and $g \mapsto V'_g$ the cocycles of $G$ with values in $\mathrm{id}_r + a^m\mathrm{Mat}_r(A/a^qA)$ associated with $N$ and $N'$, respectively, by the choice of the basis $(f_i)_{1 \leq i \leq d}$. By virtue of (II.3.33.4) and ([**65**] I Prop. 42), the following conditions are equivalent:

(i) There exists an $A$-linear automorphism

(II.3.34.1) $$u\colon N/a^nN \xrightarrow{\sim} N/a^nN$$

such that $u \circ \rho(g) = \rho'(g) \circ u$ for every $g \in G$ and $u(f_i) - f_i \in a^mN$ for every $1 \leq i \leq r$.

(ii) There exist a cocycle $g \mapsto W_g$ of $G$ with values in $\mathrm{Mat}_r(A/a^{q-n}A)$ and a matrix $U \in \mathrm{id}_r + a^m\mathrm{Mat}_r(A/a^qA)$ such that for every $g \in G$, we have (II.3.33.3)

(II.3.34.2) $$V'_g = U^{-1}(\mathrm{id}_r + a^nW_g)V_g{}^gU.$$

If these conditions hold, we will say that $\rho'$ is deduced from $\rho$ by twisting by the cocycle $g \mapsto W_g$. Two cohomologous cocycles define representations that are isomorphic by an isomorphism that is compatible with (II.3.34.1). We will then also say that $\rho'$ is deduced from $\rho$ by twisting by the class $\mathfrak{c} \in \mathrm{H}^1(G, \mathrm{Mat}_r(A/a^{q-n}A))$ of the cocycle $g \mapsto W_g$.

Let $n', m', q'$ be integers $\geq 1$ such that $n \geq n' \geq m'$, $m \geq m' \geq q - n$, and $q' = q - n + n'$. Suppose that multiplication by $a^{n'}$ in $A$ induces an isomorphism

(II.3.34.3) $$A/a^{q'-n'}A \xrightarrow{\sim} a^{n'}A/a^{q'}A.$$

Then we have a commutative diagram

(II.3.34.4) $$\begin{array}{ccccc}
\mathrm{Mat}_r(A/a^{q-n}A) & \longrightarrow & \mathrm{id}_r + a^m\mathrm{Mat}_r(A/a^qA) & \longrightarrow & \mathrm{id}_r + a^m\mathrm{Mat}_r(A/a^nA) \\
\Big\downarrow{\scriptstyle \cdot a^{n-n'}} & & \Big\downarrow{\scriptstyle \alpha} & & \Big\downarrow{\scriptstyle \beta} \\
\mathrm{Mat}_r(A/a^{q'-n'}A) & \longrightarrow & \mathrm{id}_r + a^{m'}\mathrm{Mat}_r(A/a^{q'}A) & \longrightarrow & \mathrm{id}_r + a^{m'}\mathrm{Mat}_r(A/a^{n'}A)
\end{array}$$

where the lines correspond to the exact sequences (II.3.33.2) and $\alpha$ and $\beta$ are induced by the reduction homomorphisms $A/a^qA \to A/a^{q'}A$ and $A/a^nA \to A/a^{n'}A$. Consequently, if $\rho'$ is deduced from $\rho$ by twisting by a class $\mathfrak{c} \in \mathrm{H}^1(G, \mathrm{Mat}_r(A/a^{q-n}A))$, the representation over $N/a^{q'}N$ induced by $\rho'$ is obtained from the representation over $N/a^{q'}N$ induced by $\rho$ by twisting by the class

(II.3.34.5) $$a^{n-n'} \cdot \mathfrak{c} \in \mathrm{H}^1(G, \mathrm{Mat}_r(A/a^{q'-n'}A)).$$

**II.3.35.** We keep the assumptions of II.3.33 and let $M$ be a free $(A/a^nA)$-module of rank $r$ with basis $e_1, \dots, e_r$ and $\rho$ a continuous representation of $G$ over $M$ (endowed with the discrete topology) such that $\rho(g)(e_i) - e_i \in a^mM$ for every $g \in G$ and every $1 \leq i \leq r$. We denote by $g \mapsto U_g$ the cocycle of $G$ with values in $\mathrm{id}_r + a^m\mathrm{Mat}_r(A/a^nA)$ associated with $M$ through the choice of the basis $(e_i)_{1 \leq i \leq d}$, and by $[M]$ its class in $\mathrm{H}^1(G, \mathrm{id}_r + a^m\mathrm{Mat}_r(A/a^nA))$. By virtue of (II.3.33.4) and ([**65**] I prop. 41), the class

(II.3.35.1) $$\partial([M]) \in \mathrm{H}^2(G, \mathrm{Mat}_r(A/a^{q-n}A))$$

is the obstruction to lifting $M$ to a continuous $(A/a^qA)$-representation of $G$ with underlying $(A/a^qA)$-module that is free of finite type.

Let $n'$ be an integer such that $n \geq n' \geq m$, $q' = q-n+n'$ and that the multiplication by $a^{n'}$ in $A$ induces an isomorphism

(II.3.35.2) $$A/a^{q'-n'}A \xrightarrow{\sim} a^{n'}A/a^{q'}A.$$

We denote by

(II.3.35.3) $$\partial' \colon \mathrm{H}^1(G, \mathrm{id}_r + a^m\mathrm{Mat}_r(A/a^{n'}A)) \to \mathrm{H}^2(G, \mathrm{End}_A(M/a^{q'-n'}M))$$

the boundary map defined as in (II.3.33.4). By functoriality, we have

(II.3.35.4) $$\partial'([M/a^{n'}M]) = a^{n-n'}\partial([M]).$$

## II.4. Objects with group actions

**II.4.1.** Let **Sch** be the category of schemes. We choose a normalized cleavage of the category of arrows of **Sch** ([**37**] VI § 11); in other words, for every morphism of schemes $f \colon X' \to X$, we choose a base change functor

(II.4.1.1) $$f^\bullet \colon \mathbf{Sch}_{/X} \to \mathbf{Sch}_{/X'}, \quad Y \mapsto Y \times_X X'$$

such that for every scheme $X$, $f = \mathrm{id}_X$ implies $f^\bullet = \mathrm{id}_{\mathbf{Sch}_{/X}}$. For any scheme $X$, we denote by $X_{\mathrm{zar}}$ the Zariski topos of $X$ and by

(II.4.1.2) $$\varphi_X \colon \mathbf{Sch}_{/X} \to X_{\mathrm{zar}}, \quad Y \mapsto \mathrm{Hom}_X(-, Y)$$

the canonical functor.

**Remark II.4.2.** If $f \colon X' \to X$ is a morphism of schemes, the diagram

(II.4.2.1) $$\begin{array}{ccc} \mathbf{Sch}_{/X} & \xrightarrow{\varphi_X} & X_{\mathrm{zar}} \\ {\scriptstyle f^\bullet}\downarrow & & \downarrow{\scriptstyle f^*} \\ \mathbf{Sch}_{/X'} & \xrightarrow{\varphi_{X'}} & X'_{\mathrm{zar}} \end{array}$$

where the horizontal arrows are the canonical functors (II.4.1.2), is not commutative in general. Nevertheless, we have a canonical morphism of functors

(II.4.2.2) $$f^* \circ \varphi_X \to \varphi_{X'} \circ f^\bullet.$$

If, moreover, $f$ is an open immersion, then the morphism (II.4.2.2) is an isomorphism. Indeed, every open subscheme of $X'$ is an open subscheme of $X$ and $f^*$ is the restriction functor. In this case, the diagram (II.4.2.1) is therefore commutative, up to canonical isomorphism.

**II.4.3.** Let $X$ be a scheme and $G$ an $X$-group scheme. In this chapter, a *principal homogeneous $G$-bundle over $X$* will mean a (right) $G$-pseudo-torsor of $\mathbf{Sch}_{/X}$ that is locally trivial for the Zariski topology on $X$ ([**35**] III 1.1.5). We denote by $\mathbf{PHB}(G/X)$ the category of principal homogeneous $G$-bundles over $X$ and by $\mathbf{Tors}(\varphi_X(G), X_{\mathrm{zar}})$ the category of (right) $\varphi_X(G)$-torsors of $X_{\mathrm{zar}}$ ([**35**] III 1.4.1). The functor $\varphi_X$ (II.4.1.2) induces a functor that we denote also by

(II.4.3.1) $$\varphi_X\colon \mathbf{PHB}(G/X) \to \mathbf{Tors}(\varphi_X(G), X_{\mathrm{zar}}), \quad Y \mapsto \mathrm{Hom}_X(-, Y).$$

**Proposition II.4.4.** *Let $X$ be a scheme and $G$ an $X$-group scheme. Then:*

(i) *The functor* (II.4.3.1) *is fully faithful.*

(ii) *If, moreover, $X$ is coherent (that is, quasi-compact and quasi-separated) and $G$ is affine over $X$, then the functor* (II.4.3.1) *is an equivalence of categories.*

(i) Let $Y, Z$ be two objects of $\mathbf{PHB}(G/X)$. We denote by $\mathrm{Hom}_G(Y, Z)$ the set of morphisms from $Y$ to $Z$ in $\mathbf{PHB}(G/X)$ and by $\mathrm{Hom}_{\varphi_X(G)}(\varphi_X(Y), \varphi_X(Z))$ the set of morphisms from $\varphi_X(Y)$ to $\varphi_X(Z)$ in $\mathbf{Tors}(\varphi_X(G), X_{\mathrm{zar}})$. Let us show that the map

(II.4.4.1) $$\mathrm{Hom}_G(Y, Z) \to \mathrm{Hom}_{\varphi_X(G)}(\varphi_X(Y), \varphi_X(Z))$$

induced by the functor $\varphi_X$ (II.4.3.1) is bijective. Let $(U_i)_{i\in I}$ be a Zariski open covering of $X$. For every $i \in I$, we set $G_i = G \times_X U_i$, $Y_i = Y \times_X U_i$, and $Z_i = Z \times_X U_i$. For every $(i, j) \in I^2$, we set $U_{ij} = U_i \cap U_j$, $G_{ij} = G \times_X U_{ij}$, $Y_{ij} = Y \times_X U_{ij}$, and $Z_{ij} = Z \times_X U_{ij}$. We then have a commutative diagram of maps of sets

(II.4.4.2)
$$\begin{array}{ccc}
\mathrm{Hom}_G(Y, Z) & \longrightarrow & \mathrm{Hom}_{\varphi_X(G)}(\varphi_X(Y), \varphi_X(Z)) \\
\downarrow & & \downarrow \\
\prod_{i\in I} \mathrm{Hom}_{G_i}(Y_i, Z_i) & \longrightarrow & \prod_{i\in I} \mathrm{Hom}_{\varphi_{X_i}(G_i)}(\varphi_{X_i}(Y_i), \varphi_{X_i}(Z_i)) \\
\downarrow\downarrow & & \downarrow\downarrow \\
\prod_{(i,j)\in I^2} \mathrm{Hom}_{G_{ij}}(Y_{ij}, Z_{ij}) & \longrightarrow & \prod_{(i,j)\in I^2} \mathrm{Hom}_{\varphi_{X_{ij}}(G_{ij})}(\varphi_{X_{ij}}(Y_{ij}), \varphi_{X_{ij}}(Z_{ij}))
\end{array}$$

where the horizontal arrows are induced by the functor (II.4.3.1) and the vertical arrows are defined by restriction (II.4.2). The vertical columns are clearly exact, so we may restrict to the case $Y = G$. We then identify the source and the target of the map (II.4.4.1) with the set $\mathrm{Hom}_X(X, Z)$, giving the desired statement.

(ii) It suffices to show that the functor in question is essentially surjective. Let $F$ be an object of $\mathbf{Tors}(\varphi_X(G), X_{\mathrm{zar}})$ and $(U_i)_{i\in I}$ a covering of $X$ by affine open sets such that for every $i \in I$, $F|U_i$ is trivial. Let $X' = \sqcup_{i\in I} U_i$, $X'' = X' \times_X X'$, $G' = G \times_X X'$, $G'' = G \times_X X''$, and denote by $f\colon X' \to X$ the canonical morphism and by $\mathrm{pr}_1, \mathrm{pr}_2\colon X'' \to X'$ the canonical projections. Since $X$ is quasi-compact, we may assume that $I$ is finite. Since $X$ is quasi-separated, $f$ is then quasi-compact. We have an isomorphism $\varphi_{X'}(G') \xrightarrow{\sim} f^*(F)$ of $\mathbf{Tors}(\varphi_{X'}(G'), X'_{\mathrm{zar}})$. The canonical descent datum on $f^*(F)$ induces a descent datum on $\varphi_{X'}(G')$ with respect to $f$ (as a $\varphi_{X'}(G')$-torsor of $X'_{\mathrm{zar}}$); that is, an isomorphism of $\mathbf{Tors}(\varphi_{X''}(G''), X''_{\mathrm{zar}})$

(II.4.4.3) $$\psi\colon \mathrm{pr}_1^*(\varphi_{X'}(G')) \xrightarrow{\sim} \mathrm{pr}_2^*(\varphi_{X'}(G'))$$

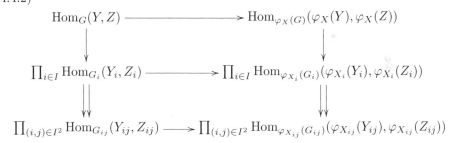

that satisfies a cocycle relation. Note that $\mathrm{pr}_1^*(\varphi_{X'}(G'))$ and $\mathrm{pr}_2^*(\varphi_{X'}(G'))$ are canonically isomorphic to $\varphi_{X''}(G'')$ (II.4.2) and that $\psi$ is in general different from the trivial descent datum (induced by $\varphi_X(G)$). In view of statement (i), $\psi$ induces a descent datum on $G'$ with respect to $f$ (as a principal homogeneous $G'$-bundle over $X'$); that is, an isomorphism of $\mathbf{PHB}(G''/X'')$

(II.4.4.4) $$\varphi\colon \mathrm{pr}_1^\bullet(G') \xrightarrow{\sim} \mathrm{pr}_2^\bullet(G')$$

satisfying a cocycle relation. By virtue of ([**37**] VIII 2.1), there exists a principal homogeneous $G$-bundle $Y$ over $X$ that corresponds to $\varphi$. Since the $G$-torsor $\varphi_X(Y)$ of $X_{\mathrm{zar}}$ corresponds to the descent datum $\psi$, there exists an isomorphism $\varphi_X(Y) \xrightarrow{\sim} F$ of $\mathbf{Tors}(\varphi_X(G), X_{\mathrm{zar}})$, giving the desired statement.

**II.4.5.** Let $f\colon X' \to X$ be a morphism of schemes, T an $\mathscr{O}_X$-module, and $\mathscr{L}$ a T-torsor of $X_{\mathrm{zar}}$. From now on, for $\mathscr{O}_X$-modules, we will use the notation $f^{-1}$ to denote the inverse image in the sense of abelian sheaves and will keep the notation $f^*$ for the inverse image in the sense of modules. The *affine inverse image* of $\mathscr{L}$ under $f$, denoted by $f^+(\mathscr{L})$, is the $f^*(\mathrm{T})$-torsor of $X'_{\mathrm{zar}}$ deduced from the $f^{-1}(\mathrm{T})$-torsor $f^*(\mathscr{L})$ by extending its structural group by the canonical homomorphism $f^{-1}(\mathrm{T}) \to f^*(\mathrm{T})$,

(II.4.5.1) $$f^+(\mathscr{L}) = f^*(\mathscr{L}) \wedge^{f^{-1}(\mathrm{T})} f^*(\mathrm{T});$$

in other words, the quotient of $f^*(\mathscr{L}) \times f^*(\mathrm{T})$ by the diagonal action of $f^{-1}(\mathrm{T})$ ([**35**] III 1.4.6). Let $\mathrm{T}'$ be an $\mathscr{O}_X$-module, $\mathscr{L}'$ a $\mathrm{T}'$-torsor of $X_{\mathrm{zar}}$, $u\colon \mathrm{T} \to \mathrm{T}'$ an $\mathscr{O}_X$-linear morphism, and $v\colon \mathscr{L} \to \mathscr{L}'$ a $u$-equivariant morphism of $X_{\mathrm{zar}}$. By ([**35**] III 1.3.6), there exists a unique $f^*(u)$-equivariant morphism

(II.4.5.2) $$f^+(v)\colon f^+(\mathscr{L}) \to f^+(\mathscr{L}')$$

that fits into the commutative diagram

(II.4.5.3) $$\begin{array}{ccc} f^*(\mathscr{L}) & \xrightarrow{f^*(v)} & f^*(\mathscr{L}') \\ \downarrow & & \downarrow \\ f^+(\mathscr{L}) & \xrightarrow{f^+(v)} & f^+(\mathscr{L}') \end{array}$$

where the vertical arrows are the canonical morphisms. The resulting correspondence $(\mathrm{T}, \mathscr{L}) \mapsto (f^*(\mathrm{T}), f^+(\mathscr{L}))$ is a functor from the category of torsors of $X_{\mathrm{zar}}$ under an $\mathscr{O}_X$-module to the category of torsors of $X'_{\mathrm{zar}}$ under an $\mathscr{O}_{X'}$-module.

Let $g\colon X'' \to X'$ be a morphism of schemes and $h = f \circ g\colon X'' \to X$. We have a canonical isomorphism of functors

(II.4.5.4) $$h^* \xrightarrow{\sim} g^* \circ f^*.$$

Since $g^*$ commutes with direct limits, we have a canonical isomorphism

(II.4.5.5) $$g^*(f^+(\mathscr{L})) \xrightarrow{\sim} g^*(f^*(\mathscr{L})) \wedge^{h^{-1}(\mathrm{T})} g^{-1}(f^*(\mathrm{T})).$$

In view of ([**35**] III 1.3.5), this induces a canonical $h^*(\mathrm{T})$-equivariant isomorphism

(II.4.5.6) $$g^+(f^+(\mathscr{L})) \xrightarrow{\sim} h^+(\mathscr{L}).$$

One immediately verifies that this isomorphism is functorial and that it satisfies a cocycle relation of the type ([**37**] VI 7.4 B)) for the composition of three morphisms of schemes.

**II.4.6.** Let $f\colon X' \to X$ be a morphism of schemes, T a locally free $\mathscr{O}_X$-module of finite type, $\Omega = \mathscr{H}om_{\mathscr{O}_X}(\mathrm{T}, \mathscr{O}_X)$ its dual, $\mathbf{T} = \mathrm{Spec}(\mathrm{S}_{\mathscr{O}_X}(\Omega))$ the corresponding $X$-bundle, and $\mathbf{L}$ a principal homogeneous $\mathbf{T}$-bundle over $X$. We have a commutative diagram of group morphisms of $X'_{\mathrm{zar}}$

(II.4.6.1)
$$\begin{array}{ccc} f^{-1}(\mathrm{T}) & \xrightarrow{\sim} & f^*(\varphi_X(\mathbf{T})) \\ \downarrow & & \downarrow \\ f^*(\mathrm{T}) & \xrightarrow{\sim} & \varphi_{X'}(\mathbf{T} \times_X X') \end{array}$$

where the vertical arrow on the right is the morphism (II.4.2.2) and the other arrows are the canonical morphisms. Moreover, we have an $f^{-1}(\mathrm{T})$-equivariant canonical morphism (II.4.2.2)

(II.4.6.2)
$$f^*(\varphi_X(\mathbf{L})) \to \varphi_{X'}(\mathbf{L} \times_X X').$$

By ([**35**] III 1.3.6), this induces an isomorphim of $f^*(\mathrm{T})$-torsors

(II.4.6.3)
$$f^+(\varphi_X(\mathbf{L})) \xrightarrow{\sim} \varphi_{X'}(\mathbf{L} \times_X X').$$

**II.4.7.** Let $X$ be a scheme, T a locally free $\mathscr{O}_X$-module of finite type, $\Omega = \mathscr{H}om_{\mathscr{O}_X}(\mathrm{T}, \mathscr{O}_X)$ its dual, and $\mathscr{L}$ a T-torsor of $X_{\mathrm{zar}}$. An *affine function on* $\mathscr{L}$ is a morphism $f\colon \mathscr{L} \to \mathscr{O}_X$ of $X_{\mathrm{zar}}$ satisfying the following equivalent conditions:

(i) For every open subscheme $U$ of $X$ and every $s \in \mathscr{L}(U)$, the map

(II.4.7.1)
$$\mathrm{T}(U) \to \mathscr{O}_X(U), \quad t \mapsto f(s+t) - f(s)$$

is $\mathscr{O}_X(U)$-linear.

(ii) There exists a section $\omega \in \Omega(X)$, called the *linear term* of $f$, such that for every open subscheme $U$ of $X$ and all $s \in \mathscr{L}(U)$ and $t \in \mathrm{T}(U)$, we have

(II.4.7.2)
$$f(s+t) = f(s) + \omega(t).$$

We then also say that the morphism $f$ is *affine*. Indeed, (ii) clearly implies (i). Conversely, assume that condition (i) is satisfied. Let $(U_i)_{i \in I}$ be an open covering of $X$ such that for every $i \in I$, there exists an $s_i \in \mathscr{L}(U_i)$. Then there exist $\omega_i \in \Omega(U_i)$ such that for every open subscheme $U$ of $U_i$ and every $t \in \mathrm{T}(U)$, we have

(II.4.7.3)
$$f(s_i + t) = f(s_i) + \omega_i(t) \in \mathscr{O}_X(U).$$

For every $(i,j) \in I^2$, let $t_{ij} \in \mathrm{T}(U_i \cap U_j)$ be such that $s_i = s_j + t_{ij} \in \mathscr{L}(U_i \cap U_j)$. For every open subscheme $U$ of $U_i \cap U_j$ and every $t \in \mathrm{T}(U)$, we have

(II.4.7.4)
$$\begin{aligned} f(s_i) + \omega_i(t) &= f(s_i + t) = f(s_j + t_{ij} + t) \\ &= f(s_j) + \omega_j(t_{ij}) + \omega_j(t) = f(s_i) + \omega_j(t) \in \mathscr{O}_X(U). \end{aligned}$$

We therefore have $\omega_i | U_i \cap U_j = \omega_j | U_i \cap U_j$. Consequently, the sections $(\omega_i)_{i \in I}$ glue to define a section $\omega \in \Omega(X)$. Let $U$ be an open subscheme $X$, $s \in \mathscr{L}(U)$, $t \in \mathrm{T}(U)$. Let us show that $f(s+t) = f(s) + \omega(t)$. We may assume that there exists $i \in I$ such that $U \subset U_i$. Let $t' \in \mathrm{T}(U)$ be such that $s = s_i + t'$. Then we have

(II.4.7.5)
$$f(s+t) = f(s_i + t' + t) = f(s_i) + \omega(t') + \omega(t) = f(s) + \omega(t).$$

**Remark II.4.8.** Condition II.4.7(ii) corresponds to saying that there exists $\omega \in \Omega(X)$ such that $f$ is T-equivariant when we endow $\mathscr{O}_X$ with the structure of T-object defined by $\omega$, or more precisely, by the morphism

(II.4.8.1)
$$\mathrm{T} \times \mathscr{O}_X \to \mathscr{O}_X, \quad (t, x) \mapsto \omega(t) + x.$$

**II.4.9.** Let $X$ be a scheme, $\mathrm{T}$ a locally free $\mathscr{O}_X$-module of finite type, $\Omega = \mathscr{H}om_{\mathscr{O}_X}(\mathrm{T}, \mathscr{O}_X)$ its dual, and $\mathscr{L}$ a T-torsor of $X_{\mathrm{zar}}$. Condition II.4.7(i) is clearly local for the Zariski topology on $X$, so we denote by $\mathscr{F}$ the subsheaf of $\mathscr{H}om_{X_{\mathrm{zar}}}(\mathscr{L}, \mathscr{O}_X)$ consisting of affine functions on $\mathscr{L}$; in other words, for every open subscheme $U$ of $X$, $\mathscr{F}(U)$ is the set of affine functions on $\mathscr{L}|U$. We call $\mathscr{F}$ the sheaf of *affine functions on* $\mathscr{L}$. It is naturally endowed with a structure of $\mathscr{O}_X$-module. We have a canonical $\mathscr{O}_X$-linear morphism $c\colon \mathscr{O}_X \to \mathscr{F}$ whose image consists of the constant functions. The "linear term" defines an $\mathscr{O}_X$-linear morphism $\nu\colon \mathscr{F} \to \Omega$. One verifies that the sequence

(II.4.9.1) $$0 \longrightarrow \mathscr{O}_X \xrightarrow{c} \mathscr{F} \xrightarrow{\nu} \Omega \longrightarrow 0$$

is exact; to do this, we may assume that $\mathscr{L}$ is trivial. By ([**45**] I 4.3.1.7), the sequence induces, for every integer $n \geq 1$, an exact sequence (II.2.5)

(II.4.9.2) $$0 \to \mathrm{S}^{n-1}_{\mathscr{O}_X}(\mathscr{F}) \to \mathrm{S}^n_{\mathscr{O}_X}(\mathscr{F}) \to \mathrm{S}^n_{\mathscr{O}_X}(\Omega) \to 0.$$

The $\mathscr{O}_X$-modules $(\mathrm{S}^n_{\mathscr{O}_X}(\mathscr{F}))_{n\in\mathbb{N}}$ therefore form a direct system, whose direct limit

(II.4.9.3) $$\mathscr{C} = \varinjlim_{n\geq 0} \mathrm{S}^n_{\mathscr{O}_X}(\mathscr{F})$$

is naturally endowed with a structure of $\mathscr{O}_X$-algebra. For any integer $n \geq 0$, the canonical morphism $\mathrm{S}^n_{\mathscr{O}_X}(\mathscr{F}) \to \mathscr{C}$ is injective. Note that for every $\mathscr{O}_X$-algebra $\mathscr{B}$, if $u\colon \mathscr{F} \to \mathscr{B}$ is an $\mathscr{O}_X$-linear morphism such that $u \circ c$ is the structural homomorphism, there exists a unique homomorphism of $\mathscr{O}_X$-algebras $\mathscr{C} \to \mathscr{B}$ that extends $u$.

There exists a unique homomorphism of $\mathscr{O}_X$-algebras

(II.4.9.4) $$\mu\colon \mathscr{C} \to \mathrm{S}_{\mathscr{O}_X}(\Omega) \otimes_{\mathscr{O}_X} \mathscr{C}$$

such that for every local section $x$ of $\mathscr{F}$, we have

(II.4.9.5) $$\mu(x) = \nu(x) \otimes 1 + 1 \otimes x.$$

We denote by $\mathbf{T} = \mathrm{Spec}(\mathrm{S}_{\mathscr{O}_X}(\Omega))$ the $X$-vector bundle associated with $\Omega$ and set

(II.4.9.6) $$\mathbf{L} = \mathrm{Spec}(\mathscr{C}).$$

We have a canonical isomorphism of groups of $X_{\mathrm{zar}}$

(II.4.9.7) $$\mathrm{T} \xrightarrow{\sim} \varphi_X(\mathbf{T}).$$

For any $s \in \mathscr{L}(X)$, the morphism $\rho_s\colon \mathscr{F} \to \mathscr{O}_X$ that sends a local section $f$ of $\mathscr{F}$ to $f(s)$ is a section of the exact sequence (II.4.9.1). It extends to a unique homomorphism of $\mathscr{O}_X$-algebras $\varrho_s\colon \mathscr{C} \to \mathscr{O}_X$ that induces a section $\sigma_s \in \mathbf{L}(X)$. The map $s \mapsto \sigma_s$ defines a morphism of $X_{\mathrm{zar}}$

(II.4.9.8) $$\iota\colon \mathscr{L} \to \varphi_X(\mathbf{L}).$$

**Proposition II.4.10.** *Under the assumptions of* II.4.9, *the morphism* $\mathbf{T} \times_X \mathbf{L} \to \mathbf{L}$ *induced by* $\mu$ (II.4.9.4) *makes* $\mathbf{L}$ *into a principal homogeneous* $\mathbf{T}$*-bundle over* $X$*, and the canonical morphism* $\iota\colon \mathscr{L} \to \varphi_X(\mathbf{L})$ (II.4.9.8) *is an isomorphism of* T*-torsors.*

The questions are local, so we may restrict to the case where $\mathscr{L}$ is trivial. For $s \in \mathscr{L}(X)$, let $\rho_s\colon \mathscr{F} \to \mathscr{O}_X$ be the associated splitting of the exact sequence (II.4.9.1). The morphism $\lambda_s\colon \Omega \to \mathscr{F}$ deduced from $\mathrm{id}_{\mathscr{F}} - c \circ \rho_s$ extends to an isomorphism of $\mathscr{O}_X$-algebras

(II.4.10.1) $$\psi\colon \mathrm{S}_{\mathscr{O}_X}(\Omega) \to \mathscr{C}$$

that is compatible with the filtrations $(\oplus_{0\leq i\leq n}\mathrm{S}^i_{\mathscr{O}_X}(\Omega))_{n\in\mathbb{N}}$ and $(\mathrm{S}^n_{\mathscr{O}_X}(\mathscr{F}))_{n\in\mathbb{N}}$. It follows from (II.4.9.2) that $\psi$ is an isomorphism. Denote by

(II.4.10.2) $$\delta\colon \mathrm{S}_{\mathscr{O}_X}(\Omega) \to \mathrm{S}_{\mathscr{O}_X}(\Omega) \otimes_{\mathscr{O}_X} \mathrm{S}_{\mathscr{O}_X}(\Omega)$$

the homomorphism of $\mathscr{O}_X$-algebras such that for any local section $\omega$ of $\Omega$, we have $\delta(\omega) = \omega \otimes 1 + 1 \otimes \omega$. We immediately see that the diagram

(II.4.10.3)
$$\begin{array}{ccc} \mathrm{S}_{\mathscr{O}_X}(\Omega) & \xrightarrow{\psi} & \mathscr{C} \\ \downarrow{\scriptstyle\delta} & & \downarrow{\scriptstyle\mu} \\ \mathrm{S}_{\mathscr{O}_X}(\Omega) \otimes_{\mathscr{O}_X} \mathrm{S}_{\mathscr{O}_X}(\Omega) & \xrightarrow{\mathrm{id}\otimes\psi} & \mathrm{S}_{\mathscr{O}_X}(\Omega) \otimes_{\mathscr{O}_X} \mathscr{C} \end{array}$$

is commutative. Consequently, the morphism $\mathbf{T} \times_X \mathbf{L} \to \mathbf{L}$ induced by $\mu$ makes $\mathbf{L}$ into a principal homogeneous $\mathbf{T}$-bundle over $X$, and the morphism

(II.4.10.4)
$$\Psi \colon \mathbf{L} \to \mathbf{T}$$

induced by $\psi$ is an isomorphism of $\mathbf{T}$-torsors. Since $\rho_s \circ \lambda_s = 0$, we have $\Psi(\iota(s)) = 0$ in $\mathbf{T}(X)$. For any $t \in \mathrm{T}(X)$, we have $\rho_{s+t} = \rho_s + t \circ \nu$ and therefore $\rho_{s+t} \circ \lambda_s = t \circ \nu \circ \lambda_s = t$. It follows that $\Psi(\iota(s+t)) = t$ in $\mathbf{T}(X)$, and consequently that

(II.4.10.5)
$$\iota(s+t) = \iota(s) + t.$$

Hence $\iota$ is a morphism of T-torsors and therefore an isomorphism.

**Definition II.4.11.** Under the assumptions of II.4.9, we say that $\mathbf{L}$ is the canonical principal homogeneous $\mathbf{T}$-bundle over $X$ that *represents* $\mathscr{L}$.

By II.4.4(i), there exists at most one principal homogeneous $\mathbf{T}$-bundle over $X$ that represents $\mathscr{L}$, up to canonical isomorphism. The construction in II.4.9 gives a canonical one.

**II.4.12.** Let $X$ be a scheme, T and T′ two locally free $\mathscr{O}_X$-modules of finite type, $\mathscr{L}$ a T-torsor of $X_{\mathrm{zar}}$, and $\mathscr{L}'$ a T′-torsor of $X_{\mathrm{zar}}$. We set $\Omega = \mathscr{H}om_{\mathscr{O}_X}(\mathrm{T}, \mathscr{O}_X)$, $\Omega' = \mathscr{H}om_{\mathscr{O}_X}(\mathrm{T}', \mathscr{O}_X)$, $\mathbf{T} = \mathrm{Spec}(\mathrm{S}_{\mathscr{O}_X}(\Omega))$, and $\mathbf{T}' = \mathrm{Spec}(\mathrm{S}_{\mathscr{O}_X}(\Omega'))$. We denote by $\mathscr{F}$ the sheaf of affine functions on $\mathscr{L}$ (II.4.9), by $\mathscr{F}'$ the sheaf of affine functions on $\mathscr{L}'$, by $\mathbf{L}$ the canonical principal homogeneous $\mathbf{T}$-bundle over $X$ that represents $\mathscr{L}$ (II.4.11), and by $\mathbf{L}'$ the canonical principal homogeneous $\mathbf{T}'$-bundle over $X$ that represents $\mathscr{L}'$. Let $u \colon \mathrm{T} \to \mathrm{T}'$ be an $\mathscr{O}_X$-linear morphism, $u^\vee \colon \Omega' \to \Omega$ the dual morphism of $u$, and $v \colon \mathscr{L} \to \mathscr{L}'$ a $u$-equivariant morphism of $X_{\mathrm{zar}}$. If $h \colon \mathscr{L}' \to \mathscr{O}_X$ is an affine function with linear term $\omega' \in \Omega'(X)$, then the composition $h' = h \circ v \colon \mathscr{L} \to \mathscr{O}_X$ is affine, with linear term $u^\vee(\omega')$. The resulting correspondence $h \mapsto h'$ induces an $\mathscr{O}_X$-linear morphism

(II.4.12.1)
$$w \colon \mathscr{F}' \to \mathscr{F}$$

that fits into a commutative diagram

(II.4.12.2)
$$\begin{array}{ccccccccc} 0 & \to & \mathscr{O}_X & \to & \mathscr{F}' & \to & \Omega' & \to & 0 \\ & & \| & & \downarrow{\scriptstyle w} & & \downarrow{\scriptstyle u^\vee} & & \\ 0 & \to & \mathscr{O}_X & \to & \mathscr{F} & \to & \Omega & \to & 0 \end{array}$$

where the lines are the canonical exact sequences (II.4.9.1).

The morphism $u^\vee$ induces a morphism of $X$-group schemes $\alpha \colon \mathbf{T} \to \mathbf{T}'$. The morphism $w$ induces an $\alpha$-equivariant $X$-morphism

(II.4.12.3)
$$\beta \colon \mathbf{L} \to \mathbf{L}'.$$

The diagram

(II.4.12.4)
$$\begin{array}{ccc} \mathscr{L} & \xrightarrow{v} & \mathscr{L}' \\ \downarrow{\scriptstyle \iota} & & \downarrow{\scriptstyle \iota'} \\ \varphi_X(\mathbf{L}) & \xrightarrow{\varphi_X(\beta)} & \varphi_X(\mathbf{L}') \end{array}$$

where $\iota$ and $\iota'$ are the canonical isomorphisms (II.4.9.8), is clearly commutative.

The correspondence that sends a T-torsor of $X_{\rm zar}$ to the canonical principal homogeneous $\mathbf{T}$-bundle over $X$ that represents it therefore defines a functor

(II.4.12.5)
$$\mathbf{Tors}(\mathrm{T}, X_{\rm zar}) \to \mathbf{PHB}(\mathbf{T}/X), \quad \mathscr{L} \mapsto \mathbf{L}.$$

This is a quasi-inverse of the functor (II.4.3.1)

(II.4.12.6)
$$\varphi_X \colon \mathbf{PHB}(\mathbf{T}/X) \to \mathbf{Tors}(\mathrm{T}, X_{\rm zar}), \quad \mathbf{L} \mapsto \varphi_X(\mathbf{L}),$$

by virtue of II.4.10, II.4.4(i), and (II.4.12.4).

**II.4.13.** Let $f\colon X' \to X$ be a morphism of schemes, T a locally free $\mathscr{O}_X$-module of finite type, and $\mathscr{L}$ a T-torsor of $X_{\rm zar}$. We set $\Omega = \mathscr{H}om_{\mathscr{O}_X}(\mathrm{T}, \mathscr{O}_X)$ and $\mathbf{T} = \mathrm{Spec}(\mathrm{S}_{\mathscr{O}_X}(\Omega))$. We denote by $\mathscr{F}$ the sheaf of affine functions on $\mathscr{L}$ (II.4.7), by $\mathscr{F}^+$ the sheaf of affine functions on $f^+(\mathscr{L})$ (II.4.5), by $\mathbf{L}$ the canonical principal homogeneous $\mathbf{T}$-bundle over $X$ that represents $\mathscr{L}$ (II.4.11), and by $\mathbf{L}^+$ the canonical principal homogeneous $(\mathbf{T} \times_X X')$-bundle over $X'$ that represents $f^+(\mathscr{L})$. Let $\ell\colon \mathscr{L} \to \mathscr{O}_X$ be an affine morphism, $\omega \in \Omega(X)$ its linear term, and $\omega' = f^*(\omega) \in f^*(\Omega)(X')$. Endowing $\mathscr{O}_{X'}$ with the structure of $f^*(\mathrm{T})$-object defined by $\omega'$ (II.4.8), there exists a unique $f^*(\mathrm{T})$-equivariant morphism $\ell'\colon f^+(\mathscr{L}) \to \mathscr{O}_{X'}$ that fits into the commutative diagram

(II.4.13.1)
$$\begin{array}{ccc} f^*(\mathscr{L}) & \xrightarrow{f^*(\ell)} & f^{-1}(\mathscr{O}_X) \\ \downarrow & & \downarrow \\ f^+(\mathscr{L}) & \xrightarrow{\ell'} & \mathscr{O}_{X'} \end{array}$$

where the vertical arrows are the canonical morphisms ([**35**] III 1.3.6). The morphism $h'$ is therefore affine, with linear term $\omega'$. The resulting correspondence $\ell \mapsto \ell'$ induces an $\mathscr{O}_X$-linear morphism

(II.4.13.2)
$$\lambda_\sharp \colon \mathscr{F} \to f_*(\mathscr{F}^+)$$

that fits into a commutative diagram

(II.4.13.3)
$$\begin{array}{ccccc} \mathscr{O}_X & \longrightarrow & \mathscr{F} & \longrightarrow & \Omega \\ \downarrow & & \downarrow{\scriptstyle \lambda_\sharp} & & \downarrow \\ f_*(\mathscr{O}_{X'}) & \longrightarrow & f_*(\mathscr{F}^+) & \longrightarrow & f_*(f^*(\Omega)) \end{array}$$

where the other arrows are the canonical morphisms. The adjoint morphism

(II.4.13.4)
$$\lambda\colon f^*(\mathscr{F}) \to \mathscr{F}^+$$

therefore fits into a commutative diagram

(II.4.13.5)
$$\begin{array}{ccccccccc} 0 & \longrightarrow & \mathscr{O}_{X'} & \longrightarrow & f^*(\mathscr{F}) & \longrightarrow & f^*(\Omega) & \longrightarrow & 0 \\ & & \| & & \downarrow{\scriptstyle \lambda} & & \| & & \\ 0 & \longrightarrow & \mathscr{O}_{X'} & \longrightarrow & \mathscr{F}^+ & \longrightarrow & f^*(\Omega) & \longrightarrow & 0 \end{array}$$

where the lines are the canonical exact sequences (II.4.9.1). Consequently, $\lambda$ is an isomorphism. We deduce from this an isomorphism of principal homogeneous $(\mathbf{T}\times_X X')$-bundles

(II.4.13.6) $$\mathbf{L}^+ \xrightarrow{\sim} \mathbf{L}\times_X X'.$$

The diagram

(II.4.13.7) $$\begin{array}{ccc} f^*(\mathscr{L}) & \xrightarrow{a} & f^+(\mathscr{L}) \\ {\scriptstyle f^*(\iota)}\downarrow & & \downarrow{\scriptstyle \iota^+} \\ f^*(\varphi_X(\mathbf{L})) & \xrightarrow{b} & \varphi_{X'}(\mathbf{L}^+) \end{array}$$

where $\iota$ and $\iota^+$ are the canonical isomorphisms (II.4.9.8), $a$ is the canonical morphism, and $b$ is the morphism induced by (II.4.2.2) and (II.4.13.6), is commutative. Indeed, it suffices to show that the diagram

(II.4.13.8) $$\begin{array}{ccc} \mathscr{L} & \xrightarrow{a_\sharp} & f_*(\mathscr{L}^+) \\ {\scriptstyle \iota}\downarrow & & \downarrow{\scriptstyle f_*(\iota^+)} \\ \varphi_X(\mathbf{L}) & \xrightarrow{b_\sharp} & f_*(\varphi_{X'}(\mathbf{L}^+)) \end{array}$$

where $a_\sharp$ and $b_\sharp$ are adjoints of $a$ and $b$, is commutative, or that the diagram deduced from this one by evaluating the sheaves on $X$ is commutative. For $s\in\mathscr{L}(X)$, let $\rho_s\colon \mathscr{F}\to\mathscr{O}_X$ be the morphism that sends a local section $\ell$ of $\mathscr{F}$ to $\ell(s)$. Let us set $s'=a_\sharp(s)\in\mathscr{L}^+(X')$ and let $\rho_{s'}\colon \mathscr{F}^+\to\mathscr{O}_{X'}$ be the morphism that sends a local section $\ell'$ of $\mathscr{F}^+$ to $\ell'(s')$. It immediately follows from the definition of the morphism $\lambda_\sharp$ (II.4.13.2) that the diagram

(II.4.13.9) $$\begin{array}{ccc} \mathscr{F} & \xrightarrow{\lambda_\sharp} & f_*(\mathscr{F}^+) \\ {\scriptstyle \rho_s}\downarrow & & \downarrow{\scriptstyle f_*(\rho_{s'})} \\ \mathscr{O}_X & \longrightarrow & f_*(\mathscr{O}_{X'}) \end{array}$$

is commutative. We deduce from this the relation $\rho_{s'}\circ\lambda=f^*(\rho_s)$, which immediately implies the commutativity of the diagram (II.4.13.8).

Let $g\colon X''\to X'$ be a morphism of schemes and $h=f\circ g\colon X''\to X$. We denote by $\mathscr{F}^\dagger$ the sheaf of affine functions on $h^+(\mathscr{L})$. We then have a canonical isomorphism

(II.4.13.10) $$\theta\colon h^*(\mathscr{F})\xrightarrow{\sim}\mathscr{F}^\dagger.$$

In view of (II.4.5.6), we also have a canonical isomorphism

(II.4.13.11) $$\lambda^+\colon g^*(\mathscr{F}^+)\xrightarrow{\sim}\mathscr{F}^\dagger.$$

One immediately verifies that

(II.4.13.12) $$\theta=\lambda^+\circ g^*(\lambda).$$

**II.4.14.** Let $f\colon X'\to X$ be a morphism of schemes, $\mathrm{T}$ a locally free $\mathscr{O}_X$-module of finite type, $\mathrm{T}'$ a locally free $\mathscr{O}_{X'}$-module of finite type, $\mathscr{L}$ a $\mathrm{T}$-torsor of $X_{\mathrm{zar}}$, and $\mathscr{L}'$ a $\mathrm{T}'$-torsor of $X'_{\mathrm{zar}}$. Set $\Omega=\mathscr{H}om_{\mathscr{O}_X}(\mathrm{T},\mathscr{O}_X)$, $\Omega'=\mathscr{H}om_{\mathscr{O}_{X'}}(\mathrm{T}',\mathscr{O}_{X'})$, $\mathbf{T}=\mathrm{Spec}(\mathrm{S}_{\mathscr{O}_X}(\Omega))$, and $\mathbf{T}'=\mathrm{Spec}(\mathrm{S}_{\mathscr{O}_{X'}}(\Omega'))$. We denote by $\mathscr{F}$ the sheaf of affine functions on $\mathscr{L}$ (II.4.9), by $\mathscr{F}'$ the sheaf of affine functions on $\mathscr{L}'$, by $\mathbf{L}$ the canonical principal homogeneous $\mathbf{T}$-bundle over $X$ that represents $\mathscr{L}$ (II.4.11), and by $\mathbf{L}'$ the canonical principal homogeneous $\mathbf{T}'$-bundle over $X'$ that represents $\mathscr{L}'$. The sheaf $f_*(\mathscr{L}')$ is naturally an $f_*(\mathrm{T}')$-object of $X_{\mathrm{zar}}$. Let $u\colon \mathrm{T}\to f_*(\mathrm{T}')$ be an $\mathscr{O}_X$-linear morphism and $v\colon\mathscr{L}\to f_*(\mathscr{L}')$ a $u$-equivariant morphism. We denote by $\gamma\colon f^{-1}(\mathrm{T})\to f^*(\mathrm{T})$ the canonical morphism, by

$u^\sharp\colon f^*(\mathrm{T}) \to \mathrm{T}'$ the adjoint morphism of $u$, by $u^\vee\colon \Omega' \to f^*(\Omega)$ the dual morphism of $u^\sharp$, and by $v^\sharp\colon f^*(\mathscr{L}) \to \mathscr{L}'$ the adjoint morphism of $v$. Since $v^\sharp$ is $(u^\sharp \circ \gamma)$-equivariant, it factors uniquely through a $u^\sharp$-equivariant morphism

(II.4.14.1) $$v^+\colon f^+(\mathscr{L}) \to \mathscr{L}'.$$

By (II.4.12.1) and (II.4.13.4), this induces an $\mathscr{O}_{X'}$-linear morphism

(II.4.14.2) $$w\colon \mathscr{F}' \to f^*(\mathscr{F})$$

that fits into a commutative diagram

(II.4.14.3) $$\begin{array}{ccccccccc} 0 & \longrightarrow & \mathscr{O}_{X'} & \longrightarrow & \mathscr{F}' & \longrightarrow & \Omega' & \longrightarrow & 0 \\ & & \| & & \downarrow{\scriptstyle w} & & \downarrow{\scriptstyle u^\vee} & & \\ 0 & \longrightarrow & \mathscr{O}_{X'} & \longrightarrow & f^*(\mathscr{F}) & \longrightarrow & f^*(\Omega) & \longrightarrow & 0 \end{array}$$

where the lines are the canonical exact sequences (II.4.9.1).

The morphism $u^\vee$ induces a morphism of $X'$-group schemes

(II.4.14.4) $$\alpha\colon \mathbf{T} \times_X X' \to \mathbf{T}'.$$

The morphism $w$ induces an $\alpha$-equivariant $X'$-morphism

(II.4.14.5) $$\beta\colon \mathbf{L} \times_X X' \to \mathbf{L}'.$$

It follows from (II.4.12.4) and (II.4.13.7) that the diagram

(II.4.14.6) $$\begin{array}{ccc} f^*(\mathscr{L}) & \xrightarrow{v^\sharp} & \mathscr{L}' \\ {\scriptstyle f^*(\iota)}\downarrow & & \downarrow{\scriptstyle \iota'} \\ f^*(\varphi_X(\mathbf{L})) & \xrightarrow{\delta} & \varphi_{X'}(\mathbf{L}') \end{array}$$

where $\iota$ and $\iota'$ are the canonical isomorphisms (II.4.9.8) and $\delta$ is the morphism induced by (II.4.2.2) and $\beta$, is commutative.

**II.4.15.** We keep the assumptions of II.4.14 and moreover let $g\colon X'' \to X'$ be a morphism of schemes, $\mathrm{T}''$ a locally free $\mathscr{O}_{X''}$-module of finite type, $\mathscr{L}''$ a $\mathrm{T}''$-torsor of $X''_{\mathrm{zar}}$, and $\mathscr{F}''$ the sheaf of affine functions on $\mathscr{L}''$. Let $u'\colon \mathrm{T}' \to g_*(\mathrm{T}'')$ be an $\mathscr{O}_{X'}$-linear morphism and $v'\colon \mathscr{L}' \to g_*(\mathscr{L}'')$ a $u'$-equivariant morphism. By II.4.14, the pair $(u', v')$ induces an $\mathscr{O}_{X''}$-linear morphism

(II.4.15.1) $$w'\colon \mathscr{F}'' \to g^*(\mathscr{F}').$$

Likewise, the pair $(f_*(u') \circ u, f_*(v') \circ v)$ induces an $\mathscr{O}_{X''}$-linear morphism

(II.4.15.2) $$t\colon \mathscr{F}'' \to g^*(f^*(\mathscr{F})).$$

We then have

(II.4.15.3) $$t = g^*(w) \circ w'.$$

This follows from (II.4.13.12) and from a chase in the commutative diagram

$$\text{(II.4.15.4)}\qquad \begin{array}{ccccccc}
g^*(f^*(\mathscr{L})) & & \xrightarrow{\quad g^*(v^\sharp)\quad} & & g^*(\mathscr{L}') & \xrightarrow{\quad v'^\sharp \quad} & \mathscr{L}'' \\
& \searrow & & \nearrow_{g^*(v^+)} & & \searrow & \nearrow_{v'^+} \\
& & g^*(f^+(\mathscr{L})) & & & & g^+(\mathscr{L}') \\
& & & \searrow & & \nearrow_{g^+(v^+)} & \\
& & & & g^+(f^+(\mathscr{L})), & &
\end{array}$$

where $v'^\sharp$ is the adjoint morphism of $v'$, $v'^+$ is the morphism induced by $v'^\sharp$, and the unlabeled arrows are the canonical morphisms.

**II.4.16.** Let $\mathscr{C}$, $\mathscr{F}$ be two categories,

$$\text{(II.4.16.1)}\qquad \pi\colon \mathscr{F} \to \mathscr{C}$$

a fibration ([**37**] VI 6.1), and $\Delta$ an (abstract) group. For any $X \in \mathrm{Ob}(\mathscr{C})$, we denote by $\mathscr{F}_X$ the fiber of $\pi$ over $X$ ([**37**] VI § 4). We denote also by $\Delta$ the groupoid associated with $\Delta$ (that is, the category having only one object, with morphisms class $\Delta$).

Let $X$ be an object of $\mathscr{C}$ and $\varphi$ a left action of $\Delta$ on $X$; in other words, $\varphi\colon \Delta \to \mathscr{C}$ is a functor that sends the unique object of $\Delta$ to $X$. For $\sigma \in \Delta$, we (abusively) denote by $\sigma$ the automorphism $\varphi(\sigma)$ of $X$. The base change of $\pi$ by $\varphi$ ([**37**] VI § 3)

$$\text{(II.4.16.2)}\qquad \pi_\Delta\colon \mathscr{F}_\Delta \to \Delta$$

is a fibration. The category $\mathscr{F}_\Delta$ has the same objects as the fiber $\mathscr{F}_X$, but has more morphisms. The Cartesian sections of $\pi_\Delta$ are called the *$\Delta$-equivariant $X$-objects* of $\mathscr{F}$ (or *$\Delta$-equivariant objects* of $\mathscr{F}_X$). Giving a $\Delta$-equivariant $X$-object of $\mathscr{F}$ is therefore equivalent to giving an object $Y$ of $\mathscr{F}_X$ and a left action of $\Delta$ on $Y$ (viewed as an object of $\mathscr{F}$) that is compatible with its action on $X$ by the functor $\pi$. The $\Delta$-equivariant $X$-objects of $\mathscr{F}$ naturally form a category, namely the category of Cartesian sections of $\pi_\Delta$ ([**2**] VI 6.10).

Let us choose a normalized cleavage of $\mathscr{F}$ over $\mathscr{C}$ ([**37**] VI § 7; cf. also [**1**] 1.1.2); in other words, let us choose, for every morphism $f\colon Z \to Y$ in $\mathscr{C}$, an inverse image functor

$$\text{(II.4.16.3)}\qquad f^*\colon \mathscr{F}_Y \to \mathscr{F}_Z,$$

such that for any $Y \in \mathrm{Ob}(\mathscr{C})$, $f = \mathrm{id}_Y$ implies that $f^* = \mathrm{id}_{\mathscr{F}_Y}$. For every pair of composable morphisms $(f, g)$ in $\mathscr{C}$, we have a canonical isomorphism of functors

$$\text{(II.4.16.4)}\qquad c_{g,f}\colon g^* f^* \xrightarrow{\sim} (fg)^*$$

satisfying compatibility relations ([**37**] VI 7.4). By ([**37**] VI § 12), giving a $\Delta$-equivariant object of $\mathscr{F}_X$ is equivalent to giving an object $Y$ of $\mathscr{F}_X$ and for every $\sigma \in \Delta$, an isomorphism

$$\text{(II.4.16.5)}\qquad \tau^Y_\sigma\colon Y \xrightarrow{\sim} \sigma^*(Y)$$

such that $\tau^Y_{\mathrm{id}} = \mathrm{id}_Y$ and for any $(\sigma, \sigma') \in \Delta^2$, we have

$$\text{(II.4.16.6)}\qquad \tau^Y_{\sigma\sigma'} = c_{\sigma',\sigma} \circ (\sigma'^* * \tau^Y_\sigma) \circ \tau^Y_{\sigma'}.$$

We will leave it to the reader to explicitly describe the morphisms of $\Delta$-equivariant $X$-objects of $\mathscr{F}$.

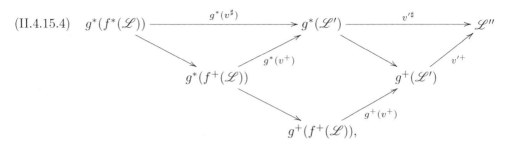

**II.4.17.** Let $\mathscr{C}$ be a category in which fibered products are representable and $X$ an object of $\mathscr{C}$ endowed with a left action of an (abstract) group $\Delta$. We denote also by $\Delta$ the groupoid associated with $\Delta$, and denote by $\varphi\colon \Delta \to \mathscr{C}$ the action of $\Delta$ on $X$. We denote by $\mathbf{Fl}(\mathscr{C})$ the category of arrows of $\mathscr{C}$ and by

(II.4.17.1) $$\mathbf{Fl}(\mathscr{C}) \to \mathscr{C}$$

the target functor, which is a fibration ([**37**] VI § 11 a)). We deduce from this, by base change by $\varphi$ ([**37**] VI § 3), a fibration

(II.4.17.2) $$\mathbf{Fl}(\mathscr{C})_\Delta \to \Delta$$

whose Cartesian sections are called the *$\Delta$-equivariant $X$-objects* of $\mathscr{C}$ (or *$\Delta$-equivariant objects* of $\mathscr{C}_{/X}$). Giving such an object is equivalent to giving an object $Y$ of $\mathscr{C}_{/X}$ and a left action of $\Delta$ on $Y$ viewed as an object of $\mathscr{C}$ that is compatible with its action on $X$.

We define the category $\mathbf{Gr}(\mathscr{C})$ as follows. The objects of $\mathbf{Gr}(\mathscr{C})$ are pairs consisting of a morphism $G \to Y$ of $\mathscr{C}$ and a structure of group of $\mathscr{C}_{/Y}$ on $G$, in the sense of ([**35**] III 1.1.1). We will leave the group structure out of the notation. Let $G \to Y$ and $G' \to Y'$ be two objects of $\mathbf{Gr}(\mathscr{C})$. A morphism from $G' \to Y'$ to $G \to Y$ is a commutative diagram of $\mathscr{C}$

(II.4.17.3) $$\begin{array}{ccc} G' & \longrightarrow & G \\ \downarrow & & \downarrow \\ Y' & \longrightarrow & Y \end{array}$$

such that the induced morphism $G' \to G \times_Y Y'$ is a morphism of groups of $\mathscr{C}_{/Y'}$. The target functor

(II.4.17.4) $$\mathbf{Gr}(\mathscr{C}) \to \mathscr{C}, \quad (G \to Y) \mapsto Y$$

is a fibration; its fiber over an object $Y$ of $\mathscr{C}$ is the category of groups of $\mathscr{C}_{/Y}$. We deduce from it by base change by $\varphi$ a fibration

(II.4.17.5) $$\mathbf{Gr}(\mathscr{C})_\Delta \to \Delta,$$

whose Cartesian sections are called the *$\Delta$-equivariant $X$-groups* of $\mathscr{C}$ (or *$\Delta$-equivariant groups* of $\mathscr{C}_{/X}$). Giving such an object is equivalent to giving a group $G$ of $\mathscr{C}_{/X}$ and a left action of $\Delta$ on $G$ viewed as an object in $\mathscr{C}$ that is compatible with its action on $X$ and with the group structure on $G$ in a sense that we will not go into.

We define the category $\mathbf{Op}(\mathscr{C})$ as follows (cf. [**35**] III 1.1.6). The objects of $\mathbf{Op}(\mathscr{C})$ are triples consisting of an object $G \to Y$ of $\mathbf{Gr}(\mathscr{C})$, a morphism $Z \to Y$ of $\mathscr{C}$, and a (right) action $m$ of $G$ on $Z$ over $Y$; that is, a $Y$-morphism $m\colon Z \times_Y G \to Z$ subject to the usual algebraic conditions (cf. [**35**] III 1.1.2). Let $(G \to Y, Z \to Y)$ and $(G' \to Y', Z' \to Y')$ be two objects of $\mathbf{Op}(\mathscr{C})$. A morphism from $(G \to Y, Z \to Y, m)$ to $(G' \to Y', Z' \to Y', m')$ consists of two commutative diagrams of $\mathscr{C}$

(II.4.17.6) $$\begin{array}{ccc} G' & \longrightarrow & G \\ \downarrow & & \downarrow \\ Y' & \longrightarrow & Y \end{array} \qquad \begin{array}{ccc} Z' & \longrightarrow & Z \\ \downarrow & & \downarrow \\ Y' & \longrightarrow & Y \end{array}$$

such that the diagram

(II.4.17.7) $$\begin{array}{ccc} Z' \times_{Y'} G' & \longrightarrow & Z \times_Y G \times_Y Y' \\ {\scriptstyle m'}\downarrow & & \downarrow{\scriptstyle m \times_Y Y'} \\ Z' & \longrightarrow & Z \times_Y Y' \end{array}$$

is commutative. The functor

$$\mathbf{Op}(\mathscr{C}) \to \mathscr{C}, \quad (G \to Y, Z \to Y, m) \mapsto Y \tag{II.4.17.8}$$

is a fibration. Its fiber over an object $Y$ of $\mathscr{C}$ is the category of objects with (right) group actions of $\mathscr{C}_{/Y}$. We deduce from this by base change by $\varphi$ a fibration

$$\mathbf{Op}(\mathscr{C})_\Delta \to \Delta \tag{II.4.17.9}$$

whose Cartesian sections are called the $\Delta$-*equivariant* $X$-*objects with group actions* of $\mathscr{C}$ (or $\Delta$-*equivariant objects with group actions* of $\mathscr{C}_{/X}$). Giving such an object is equivalent to giving a $\Delta$-equivariant group $G$ of $\mathscr{C}_{/X}$, a $\Delta$-equivariant object $Y$ of $\mathscr{C}_{/X}$, and an action $m$ of $G$ on $Y$ that is compatible with the $\Delta$-equivariant structures in a sense we will not go into. We also say that $(Y, m)$ is a $\Delta$-*equivariant* $G$-*object of* $\mathscr{C}_{/X}$.

**II.4.18.** Let $X$ be a scheme endowed with a left action by an (abstract) group $\Delta$. We define the $\Delta$-*equivariant objects of* $X_{\text{zar}}$ by taking for $\mathscr{F}$ in (II.4.16) the cleaved and normalized fibered category

$$\mathscr{X} \to \mathbf{Sch} \tag{II.4.18.1}$$

obtained by associating with any scheme $Y$ the topos $Y_{\text{zar}}$ and with any morphism of schemes $f\colon Z \to Y$ the inverse image functor $f^*\colon Y_{\text{zar}} \to Z_{\text{zar}}$. We define the $\Delta$-*equivariant groups of* $X_{\text{zar}}$ by taking for $\mathscr{F}$ in (II.4.16) the cleaved and normalized fibered category

$$\mathscr{G} \to \mathbf{Sch} \tag{II.4.18.2}$$

obtained by associating with any scheme $Y$ the category of groups of $Y_{\text{zar}}$ and with any morphism of schemes $f\colon Z \to Y$ the inverse image functor $f^*$. We define the $\Delta$-*equivariant* $\mathscr{O}_X$-*modules of* $X_{\text{zar}}$ by taking for $\mathscr{F}$ in (II.4.16) the cleaved and normalized fibered category

$$\mathscr{M} \to \mathbf{Sch} \tag{II.4.18.3}$$

obtained by associating with any scheme $Y$ the category of $\mathscr{O}_Y$-modules of $Y_{\text{zar}}$ and with any morphism of schemes $f\colon Z \to Y$ the inverse image functor in the sense of modules. Note that since the inverse images of an $\mathscr{O}_X$-module under an automorphism of $X$ in the sense of modules and in the sense of abelian groups are equal, every $\Delta$-equivariant $\mathscr{O}_X$-module of $X_{\text{zar}}$ is also a $\Delta$-equivariant group.

We define the category $\mathscr{P}$ as follows. The objects of $\mathscr{P}$ are the triples $(Y, G, P)$, where $Y$ is a scheme, $G$ is a group of $Y_{\text{zar}}$, and $P$ is a (right) $G$-object of $Y_{\text{zar}}$ ([**35**] III 1.1.2). Let $(Y, G, P)$ and $(Y', G', P')$ be two objects of $\mathscr{P}$. A morphism from $(Y', G', P')$ to $(Y, G, P)$ consists of a morphism of schemes $f\colon Y' \to Y$, a group morphism $\gamma\colon G' \to f^*(G)$ of $Y'_{\text{zar}}$, and a $\gamma$-equivariant morphism $\delta\colon P' \to f^*(P)$ of $Y'_{\text{zar}}$. The functor

$$\mathscr{P} \to \mathbf{Sch}, \quad (Y, G, P) \mapsto Y \tag{II.4.18.4}$$

is a fibration. Taking for $\mathscr{F}$ in (II.4.16) the fibered category above, we obtain the notion of $\Delta$-*equivariant objects with group actions of* $X_{\text{zar}}$. Giving such an object is equivalent to giving a $\Delta$-equivariant group $G$ of $X_{\text{zar}}$, a $\Delta$-equivariant object $P$ of $X_{\text{zar}}$, and an action $m$ of $G$ on $P$ that is compatible with the $\Delta$-equivariant structures in a sense that we will not go into. We also say that $P$ is a $\Delta$-*equivariant* $G$-*object of* $X_{\text{zar}}$. If, moreover, $P$ is a $G$-torsor of $X_{\text{zar}}$, then we also say that it is a $\Delta$-*equivariant* $G$-*torsor of* $X_{\text{zar}}$.

**Remarks II.4.19.** Let $X$ and $X'$ be two schemes endowed with a left action by an (abstract) group $\Delta$, and $f\colon X' \to X$ a $\Delta$-equivariant morphism.

(i) For every $\Delta$-equivariant object with group actions $(G, P)$ of $X_{\text{zar}}$, $(f^*(G), f^*(P))$ is naturally a $\Delta$-equivariant object with group actions of $X'_{\text{zar}}$.

(ii) For every $\Delta$-equivariant object with group action $(G', P')$ of $X'_{\rm zar}$, $(f_*(G'), f_*(P'))$ is naturally a $\Delta$-equivariant object with group action of $X_{\rm zar}$. Indeed, for every $\sigma \in \Delta$, the base change morphism for the Zariski topos $\sigma^* f_* \to f_*\sigma^*$, deduced from the relation $f\sigma = \sigma f$ ([**1**] 1.2.2), is an isomorphism. The statement follows using ([**1**] 1.2.4(i)).

**II.4.20.** Let $X$ be a scheme endowed with a left action by an (abstract) group $\Delta$ and $G$ a $\Delta$-equivariant $X$-group scheme. A *$\Delta$-equivariant principal homogeneous $G$-bundle over $X$* is a $\Delta$-equivariant $G$-object $(Y, m)$ of $\mathbf{Sch}_{/X}$ (II.4.17) such that $Y$ is a principal homogeneous $G$-bundle over $X$ (II.4.3).

In view of II.4.2, the functor (II.4.1.2)

$$\varphi_X\colon \mathbf{Sch}_{/X} \to X_{\rm zar}, \quad Y \mapsto \mathrm{Hom}_X(-, Y) \tag{II.4.20.1}$$

transforms $\Delta$-equivariant $X$-schemes (resp. $\Delta$-equivariant $X$-group schemes) (II.4.17) into $\Delta$-equivariant objects (resp. groups) of $X_{\rm zar}$ (II.4.18). Likewise, we set $\underline{G} = \varphi_X(G)$. The functor (II.4.3.1)

$$\varphi_X\colon \mathbf{PHB}(G/X) \to \mathbf{Tors}(\underline{G}, X_{\rm zar}), \quad Y \mapsto \varphi_X(Y) \tag{II.4.20.2}$$

transforms $\Delta$-equivariant principal homogeneous $G$-bundles over $X$ into $\Delta$-equivariant $\underline{G}$-torsors of $X_{\rm zar}$. Conversely, let $Y$ be a principal homogeneous $G$-bundle over $X$ and $\underline{Y} = \varphi_X(Y)$. Giving a $\Delta$-equivariant structure on the $\underline{G}$-torsor $\underline{Y}$ determines on $Y$ a unique structure of $\Delta$-equivariant principal homogeneous $G$-bundle over $X$. Indeed, for any $\sigma \in \Delta$, let

$$\tau^G_\sigma\colon G \xrightarrow{\sim} \sigma^\bullet(G), \tag{II.4.20.3}$$

$$\tau^{\underline{Y}}_\sigma\colon \underline{Y} \xrightarrow{\sim} \sigma^*(\underline{Y}) \tag{II.4.20.4}$$

be the isomorphisms induced by the $\Delta$-equivariant structures of $G$ and $\underline{Y}$ (II.4.16.5), respectively. By definition, $\tau^G_\sigma$ is an isomorphism of $X$-group schemes. By II.4.2, it induces a group isomorphism of $X_{\rm zar}$

$$\tau^{\underline{G}}_\sigma\colon \underline{G} \xrightarrow{\sim} \sigma^*(\underline{G}). \tag{II.4.20.5}$$

The isomorphisms $(\tau^{\underline{G}}_\sigma)_{\sigma\in\Delta}$ make $\underline{G}$ into a $\Delta$-equivariant group of $X_{\rm zar}$. For any $\sigma \in \Delta$, $\tau^{\underline{Y}}_\sigma$ is an isomorphism of $\underline{G}$-torsors, where $\sigma^*(\underline{Y})$ is viewed as a $\underline{G}$-torsor via $\tau^{\underline{G}}_\sigma$. By II.4.4(i), $\tau^{\underline{Y}}_\sigma$ is the image by the functor (II.4.3.1) of an isomorphism of $\mathbf{PHB}(G/X)$

$$\tau^Y_\sigma\colon Y \xrightarrow{\sim} \sigma^\bullet(Y), \tag{II.4.20.6}$$

where $\sigma^\bullet(Y)$ is seen as a principal homogeneous $G$-bundle via $\tau^G_\sigma$. The isomorphisms $(\tau^Y_\sigma)_{\sigma\in\Delta}$ satisfy the compatibility relations (II.4.16.6). They make $Y$ into a $\Delta$-equivariant principal homogeneous $G$-bundle over $X$ (cf. II.4.16).

**II.4.21.** Let $X$ be a scheme endowed with a left action by an (abstract) group $\Delta$, $\mathrm{T}$ a locally free $\Delta$-equivariant $\mathscr{O}_X$-module of finite type (II.4.18), and $\mathscr{L}$ a $\Delta$-equivariant $\mathrm{T}$-torsor of $X_{\rm zar}$. Set $\Omega = \mathscr{H}om_{\mathscr{O}_X}(\mathrm{T}, \mathscr{O}_X)$ and $\mathbf{T} = \mathrm{Spec}(\mathrm{S}_{\mathscr{O}_X}(\Omega))$. We denote by $\mathscr{F}$ the sheaf of affine functions on $\mathscr{L}$ (II.4.9) and by $\mathbf{L}$ the canonical principal homogeneous $\mathbf{T}$-bundle that represents $\mathscr{L}$ (II.4.11). For any $\sigma \in \Delta$, we denote by

$$\tau^{\mathrm{T}}_\sigma\colon \mathrm{T} \xrightarrow{\sim} \sigma^*(\mathrm{T}) \quad \text{and} \quad \tau^{\mathscr{L}}_\sigma\colon \mathscr{L} \xrightarrow{\sim} \sigma^*(\mathscr{L}) \tag{II.4.21.1}$$

the isomorphisms that define the $\Delta$-equivariant structures on $\mathrm{T}$ and $\mathscr{L}$, respectively (cf. II.4.16). By II.4.14, the inverses of $\tau^{\mathrm{T}}_\sigma$ and $\tau^{\mathscr{L}}_\sigma$ induce an $\mathscr{O}_X$-linear morphism

$$\tau^{\mathscr{F}}_\sigma\colon \mathscr{F} \to \sigma^*(\mathscr{F}) \tag{II.4.21.2}$$

that fits into a commutative diagram

(II.4.21.3)
$$\begin{CD} 0 @>>> \mathscr{O}_X @>>> \mathscr{F} @>>> \Omega @>>> 0 \\ @. @| @VV{\tau_\sigma^{\mathscr{F}}}V @VV{\tau_\sigma^{\Omega}}V @. \\ 0 @>>> \mathscr{O}_X @>>> \sigma^*(\mathscr{F}) @>>> \sigma^*(\Omega) @>>> 0 \end{CD}$$

where the lines are the canonical exact sequences (II.4.9.1) and $\tau_\sigma^{\Omega}$ is the inverse of the isomorphism induced by $\tau_\sigma^{\mathrm{T}}$. It follows that $\tau_\sigma^{\mathscr{F}}$ is an isomorphism. By (II.4.15.3), the $(\tau_\sigma^{\mathscr{F}})_{\sigma\in\Delta}$ satisfy compatibility relations (II.4.16.6). They therefore make $\mathscr{F}$ into a $\Delta$-equivariant $\mathscr{O}_X$-module. Likewise, the $(\tau_\sigma^{\Omega})_{\sigma\in\Delta}$ make $\Omega$ into a $\Delta$-equivariant $\mathscr{O}_X$-module. We deduce from this on $\mathbf{T}$ a $\Delta$-equivariant structure of $X$-group scheme and on $\mathbf{L}$ a structure of $\Delta$-equivariant principal homogeneous $\mathbf{T}$-bundle over $X$ (II.4.20). By II.4.10 and (II.4.14.6), we have an isomorphism of objects with $\Delta$-equivariant group actions of $X_{\mathrm{zar}}$

(II.4.21.4)
$$(\mathrm{T},\mathscr{L}) \xrightarrow{\sim} (\varphi_X(\mathbf{T}),\varphi_X(\mathbf{L})).$$

**II.4.22.** Let $X$ and $X'$ be two schemes endowed with a left action by an (abstract) group $\Delta$, $f\colon X'\to X$ a $\Delta$-equivariant morphism, T a locally free $\Delta$-equivariant $\mathscr{O}_X$-module of finite type, T$'$ a locally free $\Delta$-equivariant $\mathscr{O}_{X'}$-module of finite type, $\mathscr{L}$ a $\Delta$-equivariant T-torsor of $X_{\mathrm{zar}}$, and $\mathscr{L}'$ a $\Delta$-equivariant T$'$-torsor of $X'_{\mathrm{zar}}$. We denote by $\mathscr{F}$ the sheaf of affine functions on $\mathscr{L}$ (II.4.9) and by $\mathscr{F}'$ the sheaf of affine functions on $\mathscr{L}'$. The pair $(f_*(\mathrm{T}'),f_*(\mathscr{L}'))$ is naturally a $\Delta$-equivariant object with group actions of $X_{\mathrm{zar}}$ (II.4.19). Let $u\colon \mathrm{T}\to f_*(\mathrm{T}')$ be a $\Delta$-equivariant $\mathscr{O}_X$-linear morphism and $v\colon \mathscr{L}\to f_*(\mathscr{L}')$ a $\Delta$-equivariant and $u$-equivariant morphism. The pair $(u,v)$ induces an $\mathscr{O}_{X'}$-linear morphism (II.4.14.2)

(II.4.22.1)
$$w\colon \mathscr{F}'\to f^*(\mathscr{F}).$$

It immediately follows from (II.4.15.3) that $w$ is $\Delta$-equivariant when we endow $\mathscr{F}$ and $\mathscr{F}'$ with the canonical $\Delta$-equivariant structures (II.4.21).

Let $u^\sharp\colon f^*(\mathrm{T})\to \mathrm{T}'$ be the adjoint of $u$, $v^\sharp\colon f^*(\mathscr{L})\to\mathscr{L}'$ the adjoint of $v$, and $v^+\colon f^+(\mathscr{L})\to\mathscr{L}'$ the morphism induced by $v^\sharp$ (II.4.14.1). If $u^\sharp$ is an isomorphism, then $v^+$ is an isomorphism of $f^*(\mathrm{T})$-torsors and $w$ is an isomorphism by virtue of II.4.13.

## II.5. Logarithmic geometry lexicon

We recall a few notions of logarithmic geometry that will play an important role in this work, in order to fix the notation and to give readers unfamiliar with this theory points of reference. We refer to **[50, 51, 33, 58]** for the systematic development of the theory.

**II.5.1.** By *monoid* we will mean a commutative monoid with an identity element. Homomorphisms of monoids are assumed to map the identity element to the identity element. If $M$ is a monoid, then we denote by $M^{\mathrm{gp}}$ the associated group, by $M^\times$ the group of invertible elements of $M$, by $M^\sharp$ the set of orbits $M/M^\times$ (which is also the quotient of $M$ by $M^\times$ in the category of monoids), and by $i_M\colon M\to M^{\mathrm{gp}}$ the canonical homomorphism. We set $M^{\mathrm{int}}=i_M(M)$ and

(II.5.1.1)
$$M^{\mathrm{sat}}=\{x\in M^{\mathrm{gp}}|x^n\in i_M(M) \text{ for an integer } n\geq 1\}.$$

We say that a monoid $M$ is *integral* if the canonical homomorphism $i_M\colon M\to M^{\mathrm{gp}}$ is injective, that $M$ is *fine* if it is integral and of finite type, that $M$ is *saturated* if it is integral and isomorphic to $M^{\mathrm{sat}}$, and that $M$ is *toric* if it is fine and saturated and if $M^{\mathrm{gp}}$ is free over $\mathbb{Z}$.

If $M$ is integral, then $M^\sharp$ is integral, and $M$ is saturated if and only if $M^\sharp$ is saturated.

A morphism of monoids $u\colon M \to N$ is said to be *strict* if the induced morphism $u^\sharp\colon M^\sharp \to N^\sharp$ is an isomorphism.

**II.5.2.** Let $u\colon M \to N$ be a morphism of integral monoids. We say that $u$ is *exact* if the diagram

(II.5.2.1)
$$\begin{array}{ccc} M & \xrightarrow{u} & N \\ \downarrow & & \downarrow \\ M^{\rm gp} & \xrightarrow{u^{\rm gp}} & N^{\rm gp} \end{array}$$

is Cartesian. We say that $u$ is *integral* if for every integral monoid $M'$ and every homomorphism $v\colon M \to M'$, the amalgamated sum $M' \oplus_M N$ is integral. We say that $u$ is *saturated* if it is integral and if for every saturated monoid $M'$ and every homomorphism $v\colon M \to M'$, the amalgamated sum $M' \oplus_M N$ is saturated.

**II.5.3.** Let $T$ be a topos. We denote by $\mathbf{Mon}_T$ the category of (commutative and unitary) monoids of $T$ and by $\mathbf{Ab}_T$ the category of abelian groups of $T$. The canonical injection functor from $\mathbf{Ab}_T$ to $\mathbf{Mon}_T$ admits a right adjoint

(II.5.3.1)
$$\mathbf{Mon}_T \to \mathbf{Ab}_T, \quad \mathscr{M} \mapsto \mathscr{M}^\times.$$

It is immediate that for any $U \in \mathrm{Ob}(T)$, the adjunction morphism $\mathscr{M}^\times(U) \to \mathscr{M}(U)$ induces an isomorphism $\mathscr{M}^\times(U) \xrightarrow{\sim} \mathscr{M}(U)^\times$. We say that a monoid $\mathscr{M}$ of $T$ is *sharp* if $\mathscr{M}^\times = 1_T$. The canonical injection functor from the full subcategory of sharp monoids of $T$ to $\mathbf{Mon}_T$ admits a left adjoint

(II.5.3.2)
$$\mathscr{M} \mapsto \mathscr{M}^\sharp = \mathscr{M}/\mathscr{M}^\times.$$

The canonical injection functor from $\mathbf{Ab}_T$ to $\mathbf{Mon}_T$ admits a left adjoint

(II.5.3.3)
$$\mathbf{Mon}_T \to \mathbf{Ab}_T, \quad \mathscr{M} \mapsto \mathscr{M}^{\rm gp}.$$

We say that a monoid $\mathscr{M}$ of $T$ is *integral* if the adjunction morphism $\mathscr{M} \to \mathscr{M}^{\rm gp}$ is a monomorphism. We denote by $\mathbf{Mon}_{T,\rm int}$ the full subcategory of $\mathbf{Mon}_T$ made up of integral monoids of $T$. The canonical injection functor from $\mathbf{Mon}_{T,\rm int}$ to $\mathbf{Mon}_T$ admits a left adjoint

(II.5.3.4)
$$\mathbf{Mon}_T \to \mathbf{Mon}_{T,\rm int}, \quad \mathscr{M} \to \mathscr{M}^{\rm int}.$$

**II.5.4.** Let $\mathscr{C}$ be a site, $\widetilde{\mathscr{C}}$ the topos of sheaves of sets on $\mathscr{C}$ (relative to a fixed universe). For every presheaf of monoids $\mathscr{P}$ on $\mathscr{C}$, we denote by $\mathscr{P}^{\rm gp}$ (resp. $\mathscr{P}^{\rm int}$) the presheaf of monoids on $\mathscr{C}$ that sends $U \in \mathrm{Ob}(\mathscr{C})$ to the monoid $\mathscr{P}(U)^{\rm gp}$ (resp. $\mathscr{P}(U)^{\rm int}$) and by $\mathscr{P}^a$ the sheaf of monoids associated with $\mathscr{P}$. We then have a functorial canonical isomorphism

(II.5.4.1)
$$(\mathscr{P}^{\rm gp})^a \xrightarrow{\sim} (\mathscr{P}^a)^{\rm gp}.$$

Since the functor $\mathscr{P} \mapsto \mathscr{P}^a$ is exact, it transforms the presheaves of integral monoids on $\mathscr{C}$ into integral monoids of $\widetilde{\mathscr{C}}$. We have a functorial canonical isomorphism

(II.5.4.2)
$$(\mathscr{P}^{\rm int})^a \xrightarrow{\sim} (\mathscr{P}^a)^{\rm int}.$$

Consequently, a monoid $\mathscr{M}$ of $\widetilde{\mathscr{C}}$ is integral if and only if for every $U \in \mathrm{Ob}(\mathscr{C})$, the monoid $\mathscr{M}(U)$ is integral.

**II.5.5.** Let $T$ be a topos. We say that a morphism of integral monoids $u\colon \mathscr{M} \to \mathscr{N}$ of $T$ is *exact* if the canonical diagram

$$\begin{array}{ccc} \mathscr{M} & \xrightarrow{u} & \mathscr{N} \\ \downarrow & & \downarrow \\ \mathscr{M}^{\mathrm{gp}} & \xrightarrow{u^{\mathrm{gp}}} & \mathscr{N}^{\mathrm{gp}} \end{array} \tag{II.5.5.1}$$

is Cartesian. It amounts to requiring, for every $U \in \mathrm{Ob}(T)$, the homomorphism

$$u(U)\colon \mathscr{M}(U) \to \mathscr{N}(U)$$

to be exact. Indeed, if the diagram (II.5.5.1) is Cartesian, then the same holds for the diagrams obtained by taking its value at any $U \in \mathrm{Ob}(T)$. Consequently, the diagram

$$\begin{array}{ccc} \mathscr{M}(U) & \xrightarrow{u} & \mathscr{N}(U) \\ \downarrow & & \downarrow \\ \mathscr{M}(U)^{\mathrm{gp}} & \xrightarrow{u^{\mathrm{gp}}} & \mathscr{N}(U)^{\mathrm{gp}} \end{array} \tag{II.5.5.2}$$

is Cartesian because the canonical morphism $\mathscr{M}(U)^{\mathrm{gp}} \to \mathscr{M}^{\mathrm{gp}}(U)$ is injective (and the same holds for $\mathscr{N}$). The converse follows from (II.5.4.1) and from the exactness of the functor $\mathscr{P} \mapsto \mathscr{P}^a$.

Let $\mathscr{M}$ be an integral monoid of $T$ and $n$ an integer $\geq 1$. We say that $\mathscr{M}$ is *$n$-saturated* if the endomorphism defined by taking the $n$th power in $\mathscr{M}$ is exact. We say that $\mathscr{M}$ is *saturated* if it is $n$-saturated for every integer $n \geq 1$. It amounts to requiring, for every $U \in \mathrm{Ob}(T)$, the monoid $\mathscr{M}(U)$ to be saturated. We denote by $\mathbf{Mon}_{T,\mathrm{sat}}$ the full subcategory of $\mathbf{Mon}_T$ made up of saturated monoids of $T$. The canonical injection functor from $\mathbf{Mon}_{T,\mathrm{sat}}$ to $\mathbf{Mon}_T$ admits a left adjoint

$$\mathbf{Mon}_T \to \mathbf{Mon}_{T,\mathrm{sat}}, \quad \mathscr{M} \mapsto \mathscr{M}^{\mathrm{sat}}. \tag{II.5.5.3}$$

**II.5.6.** Let $\mathscr{C}$ be a site, $\widetilde{\mathscr{C}}$ the topos of sheaves of sets on $\mathscr{C}$ (relative to a fixed universe). For every presheaf of monoids $\mathscr{P}$ on $\mathscr{C}$, we denote by $\mathscr{P}^{\mathrm{sat}}$ the presheaf of saturated monoids defined, for $U \in \mathrm{Ob}(\mathscr{C})$, by $U \mapsto \mathscr{P}(U)^{\mathrm{sat}}$. Then the functor "associated sheaf of monoids," $\mathscr{P} \mapsto \mathscr{P}^a$, transforms the presheaves of saturated monoids on $\mathscr{C}$ into saturated monoids of $\widetilde{\mathscr{C}}$, and we have a functorial canonical isomorphism

$$(\mathscr{P}^{\mathrm{sat}})^a \xrightarrow{\sim} (\mathscr{P}^a)^{\mathrm{sat}}. \tag{II.5.6.1}$$

Consequently, a monoid $\mathscr{M}$ of $\widetilde{\mathscr{C}}$ is saturated if and only if for every $U \in \mathrm{Ob}(\mathscr{C})$, the monoid $\mathscr{M}(U)$ is saturated.

**II.5.7.** Let $f\colon T' \to T$ be a morphism of topos and $\mathscr{M}$ a monoid of $T$.

(i) If $\mathscr{M}$ is integral (resp. saturated), then the same holds for $f^*(\mathscr{M})$.

(ii) We have functorial canonical isomorphisms

$$f^*(\mathscr{M})^{\mathrm{gp}} \xrightarrow{\sim} f^*(\mathscr{M}^{\mathrm{gp}}), \tag{II.5.7.1}$$

$$f^*(\mathscr{M})^{\mathrm{int}} \xrightarrow{\sim} f^*(\mathscr{M}^{\mathrm{int}}). \tag{II.5.7.2}$$

If $\mathscr{M}$ is moreover integral, then we have a canonical isomorphism

$$f^*(\mathscr{M})^{\mathrm{sat}} \xrightarrow{\sim} f^*(\mathscr{M}^{\mathrm{sat}}). \tag{II.5.7.3}$$

(iii) If $u\colon \mathscr{M} \to \mathscr{N}$ is an exact morphism of integral monoids of $T$, then the morphism $f^*(u)\colon f^*(\mathscr{M}) \to f^*(\mathscr{N})$ is exact.

**II.5.8.** Let $T$ be a topos.

(i) If $M$ is an integral (resp. saturated) monoid, then the constant sheaf of monoids $M_T$ with value $M$ on $T$ is integral (resp. saturated).

(ii) Suppose that $T$ has enough points. A monoid $\mathscr{M}$ of $T$ is integral (resp. saturated) if and only if for every point $p$ of $T$, the monoid $\mathscr{M}_p$ is integral (resp. saturated). A morphism of integral monoids $u\colon \mathscr{M} \to \mathscr{N}$ of $T$ is exact if and only if for every point $p$ of $T$, the homomorphism $u_p\colon \mathscr{M}_p \to \mathscr{N}_p$ is exact.

(iii) Let $\mathscr{M}$ be an integral monoid of $T$. Then $\mathscr{M}^\sharp$ is integral and $\mathscr{M}$ is saturated if and only if $\mathscr{M}^\sharp$ is saturated.

**II.5.9.** A *prelogarithmic* structure on a scheme $X$ is a pair $(\mathscr{P}, \beta)$, where $\mathscr{P}$ is a sheaf of abelian monoids on the étale site of $X$ and $\beta$ is a homomorphism from $\mathscr{P}$ to the multiplicative monoid $\mathscr{O}_X$. A prelogarithmic structure $(\mathscr{P}, \beta)$ is called *logarithmic* if $\beta$ induces an isomorphism $\beta^{-1}(\mathscr{O}_X^\times) \xrightarrow{\sim} \mathscr{O}_X^\times$. The prelogarithmic structures on $X$ naturally form a category that contains the full subcategory of logarithmic structures on $X$. The canonical injection from the category of logarithmic structures on $X$ to the category of prelogarithmic structures on $X$ admits a left adjoint. It associates with a prelogarithmic structure $(\mathscr{P}, \beta)$ the logarithmic structure $(\mathscr{M}, \alpha)$, where $\mathscr{M}$ is defined by the co-Cartesian diagram

(II.5.9.1)
$$\begin{array}{ccc} \beta^{-1}(\mathscr{O}_X^\times) & \longrightarrow & \mathscr{P} \\ \downarrow & & \downarrow \\ \mathscr{O}_X^\times & \longrightarrow & \mathscr{M} \end{array}$$

We say that $(\mathscr{M}, \alpha)$ is the logarithmic structure *associated* with $(\mathscr{P}, \beta)$.

**II.5.10.** Let $f\colon X \to Y$ be a morphism of schemes. For sheaves of monoids, we use the notation $f^{-1}$ to denote the inverse image in the sense of sheaves of monoids and keep the notation $f^*$ for the inverse image in the sense of logarithmic structures, defined as follows. The *inverse image* under $f$ of a logarithmic structure $(\mathscr{M}, \alpha)$ on $Y$ is the logarithmic structure $(f^*(\mathscr{M}), \beta)$ on $X$ associated with the prelogarithmic structure defined by the composition $f^{-1}(\mathscr{M}) \to f^{-1}(\mathscr{O}_Y) \to \mathscr{O}_X$. It immediately follows from the definition that the canonical homomorphism

(II.5.10.1)
$$f^{-1}(\mathscr{M}^\sharp) \to (f^*(\mathscr{M}))^\sharp$$

is an isomorphism.

**II.5.11.** A *prelogarithmic* (resp. *logarithmic*) scheme is a triple $(X, \mathscr{M}_X, \alpha_X)$ consisting of a scheme $X$ and a prelogarithmic (resp. logarithmic) structure $(\mathscr{M}_X, \alpha_X)$ on $X$. When there is no risk of confusion, we will leave $\alpha_X$, and even $\mathscr{M}_X$, out of the notation. A morphism of prelogarithmic (resp. logarithmic) schemes $(X, \mathscr{M}_X, \alpha_X) \to (Y, \mathscr{M}_Y, \alpha_Y)$ is a pair $(f, f^\flat)$ consisting of a morphism of schemes $f\colon X \to Y$ and a homomorphism $f^\flat\colon f^{-1}(\mathscr{M}_Y) \to \mathscr{M}_X$ such that the diagram

(II.5.11.1)
$$\begin{array}{ccc} f^{-1}(\mathscr{M}_Y) & \xrightarrow{f^{-1}(\alpha_Y)} & f^{-1}(\mathscr{O}_Y) \\ {\scriptstyle f^\flat}\downarrow & & \downarrow \\ \mathscr{M}_X & \xrightarrow{\alpha_X} & \mathscr{O}_X \end{array}$$

is commutative.

We say that a logarithmic scheme $(X, \mathscr{M}_X, \alpha_X)$ is *integral* (resp. *saturated*) if $\mathscr{M}_X$ is integral (resp. saturated). We say that a morphism of logarithmic schemes

$$f\colon (X, \mathscr{M}_X, \alpha_X) \to (Y, \mathscr{M}_Y, \alpha_Y)$$

is *strict* if $(\mathscr{M}_X, \alpha_X)$ is the inverse image of $(\mathscr{M}_Y, \alpha_Y)$ under $f$, or, equivalently, if the canonical homomorphism $f^{-1}(\mathscr{M}_Y^\sharp) \to \mathscr{M}_X^\sharp$ is an isomorphism.

**II.5.12.** Let $M$ be a monoid. For every integer $n \geq 1$, we (abusively) denote by $\varpi_n\colon M \to M$ the Frobenius homomorphism of order $n$ of $M$ (that is, taking the $n$th power in $M$ in the multiplicative notation).

For any ring $R$, we denote by $R[M]$ the $R$-algebra of $M$ and by $e\colon M \to R[M]$ the canonical homomorphism, where $R[M]$ is seen as a multiplicative monoid. For any $x \in M$, we will write $e^x$ instead of $e(x)$.

We denote by $\mathbf{A}_M$ the scheme $\mathrm{Spec}(\mathbb{Z}[M])$ endowed with the logarithmic structure associated with the prelogarithmic structure defined by $e\colon M \to \mathbb{Z}[M]$.

For any homomorphism of monoids $\vartheta\colon M \to N$, we denote by $\mathbf{A}_\vartheta\colon \mathbf{A}_N \to \mathbf{A}_M$ the associated morphism of logarithmic schemes.

**II.5.13.** Let $(X, \mathscr{M}_X)$ be a logarithmic scheme, $M$ a monoid, and $M_X$ the constant (étale) sheaf of monoids with value $M$ on $X$. The following data are equivalent:

(i) a homomorphism $\gamma\colon M \to \Gamma(X, \mathscr{M}_X)$;
(ii) a homomorphism $\widetilde{\gamma}\colon M_X \to \mathscr{M}_X$;
(iii) a morphism of logarithmic schemes $\gamma^a\colon (X, \mathscr{M}_X) \to \mathbf{A}_M$.

Moreover, the following conditions are equivalent:

(a) $\mathscr{M}_X$ is associated with the prelogarithmic structure it induces on $M_X$;
(b) the morphism $\gamma^a\colon (X, \mathscr{M}_X) \to \mathbf{A}_M$ is strict.

We then say that $(M, \gamma)$ is a *chart* for $(X, \mathscr{M}_X)$. We say that the chart $(M, \gamma)$ is *coherent* (resp. *integral*, resp. *fine*, resp. *saturated*, resp. *toric*) if the monoid $M$ is of finite type (resp. integral, resp. fine, resp. saturated, resp. toric).

**II.5.14.** Let $f\colon (X, \mathscr{M}_X) \to (Y, \mathscr{M}_Y)$ be a morphism of logarithmic schemes. A *chart* for $f$ is a triple $((M, \gamma), (N, \delta), \vartheta\colon N \to M)$ consisting of a chart $(M, \gamma)$ for $(X, \mathscr{M}_X)$, a chart $(N, \delta)$ for $(Y, \mathscr{M}_Y)$, and a homomorphism $\vartheta\colon N \to M$ such that the diagram

(II.5.14.1)
$$\begin{array}{ccc} N & \xrightarrow{\delta} & \Gamma(Y, \mathscr{M}_Y) \\ {\scriptstyle\vartheta}\downarrow & & \downarrow{\scriptstyle f^\flat} \\ M & \xrightarrow{\gamma} & \Gamma(X, \mathscr{M}_X) \end{array}$$

is commutative, or equivalently, such that the associated diagram of morphisms of logarithmic schemes (II.5.12)

(II.5.14.2)
$$\begin{array}{ccc} (X, \mathscr{M}_X) & \xrightarrow{\gamma^a} & \mathbf{A}_M \\ {\scriptstyle f}\downarrow & & \downarrow{\scriptstyle \mathbf{A}_\vartheta} \\ (Y, \mathscr{M}_Y) & \xrightarrow{\delta^a} & \mathbf{A}_N \end{array}$$

is commutative. We say that the chart $((M, \gamma), (N, \delta), \vartheta\colon N \to M)$ is *coherent* (resp. *fine*) if $M$ and $N$ are of finite type (resp. fine).

**II.5.15.** Let $(X, \mathscr{M}_X)$ be a logarithmic scheme. We say that $(X, \mathscr{M}_X)$ is *coherent* if every geometric point $\overline{x}$ of $X$ admits an étale neighborhood $U$ in $X$ such that $(U, \mathscr{M}_X|U)$ admits a coherent chart. We say that $(X, \mathscr{M}_X)$ is *fine* if it is coherent and integral.

The logarithmic scheme $(X, \mathscr{M}_X)$ is fine (resp. fine and saturated) if and only if every geometric point $\overline{x}$ of $X$ admits an étale neighborhood $U$ in $X$ such that $(U, \mathscr{M}_X|U)$ admits a fine (resp. fine and saturated) chart.

We say that $(X, \mathscr{M}_X)$ is *toric* if every geometric point $\overline{x}$ of $X$ admits an étale neighborhood $U$ in $X$ such that $(U, \mathscr{M}_X|U)$ admits a toric chart.

**Lemma II.5.16** ([**73**] 1.3.2). *For every fine and saturated logarithmic scheme $(X, \mathscr{M}_X)$, the following conditions are equivalent:*

(i) *There exists a fine and saturated chart $\gamma\colon P \to \Gamma(X, \mathscr{M}_X)$ for $(X, \mathscr{M}_X)$* (II.5.13) *such that the composition*

(II.5.16.1)
$$P \to \Gamma(X, \mathscr{M}_X) \to \Gamma(X, \mathscr{M}_X)/\Gamma(X, \mathscr{O}_X^\times)$$

*is an isomorphism.*

(ii) *There exists a coherent chart $\gamma\colon P \to \Gamma(X, \mathscr{M}_X)$ for $(X, \mathscr{M}_X)$ such that the composition* (II.5.16.1) *is surjective.*

(iii) *The monoid $\Gamma(X, \mathscr{M}_X)/\Gamma(X, \mathscr{O}_X^\times)$ is of finite type and the identity of $\Gamma(X, \mathscr{M}_X)$ is a chart for $(X, \mathscr{M}_X)$.*

The implication (i)⇒(ii) is clear. Let us show (ii)⇒(iii). We set $Q = \Gamma(X, \mathscr{M}_X)$ and denote by $\mathscr{P}$ and $\mathscr{Q}$ the logarithmic structures on $X$ associated with the prelogarithmic structures defined by the homomorphism $\gamma$ and the identity of $\Gamma(X, \mathscr{M}_X)$, respectively, and by $\theta\colon \mathscr{P} \to \mathscr{Q}$ the homomorphism induced by $\gamma$. Since the composition $\mathscr{P} \xrightarrow{\theta} \mathscr{Q} \to \mathscr{M}$ is an isomorphism, it suffices to show that $\theta$ is surjective. Let $\overline{x}$ be a geometric point of $X$. We denote by $\beta\colon Q \to \mathscr{M}_{\overline{x}}$ the canonical homomorphism and by $\alpha\colon P \to \mathscr{M}_{\overline{x}}$ the composition $\beta \circ \gamma$. We then have a commutative diagram of homomorphisms of monoids

(II.5.16.2)
$$\begin{array}{ccc} P/\alpha^{-1}(\mathscr{O}_{X,\overline{x}}^\times) & \xrightarrow{\overline{\gamma}} & Q/\beta^{-1}(\mathscr{O}_{X,\overline{x}}^\times) \\ \downarrow & & \downarrow \\ \mathscr{P}_{\overline{x}}/\mathscr{O}_{X,\overline{x}}^\times & \xrightarrow{\theta_{\overline{x}}^\sharp} & \mathscr{Q}_{\overline{x}}/\mathscr{O}_{X,\overline{x}}^\times \end{array}$$

where $\overline{\gamma}$ (resp. $\theta_{\overline{x}}^\sharp$) is induced by $\gamma$ (resp. $\theta_{\overline{x}}$) and the vertical arrows are the canonical isomorphisms. Since $\Gamma(X, \mathscr{O}_X^\times) \subset \beta^{-1}(\mathscr{O}_{X,\overline{x}}^\times)$, $\overline{\gamma}$ is surjective. Consequently, $\theta_{\overline{x}}^\sharp$ is surjective. Hence $\theta_{\overline{x}}$ is surjective, giving the statement. Finally, let us show (iii)⇒(i). Let $P = \Gamma(X, \mathscr{M}_X)/\Gamma(X, \mathscr{O}_X^\times)$. Since $\Gamma(X, \mathscr{M}_X)$ is saturated, $P$ is fine and saturated. The torsion subgroup of $P^{\rm gp}$ is therefore contained in $P$. Since $P$ is sharp, $P^{\rm gp}$ is a free abelian group of finite type. Let $\delta\colon P^{\rm gp} \to \Gamma(X, \mathscr{M}_X)^{\rm gp}$ be a section of the canonical homomorphism $\Gamma(X, \mathscr{M}_X)^{\rm gp} \to P^{\rm gp}$. The restriction of $\delta$ to $P$ induces a homomorphism $\gamma\colon P \to \Gamma(X, \mathscr{M}_X)$ that satisfies the conditions by virtue of ([**73**] 1.3.1).

**Lemma II.5.17** ([**73**] 1.3.3). *Let $(X, \mathscr{M}_X)$ be a fine and saturated logarithmic scheme whose underlying scheme $X$ is noetherian, $x \in X$, and $\gamma\colon P \to \Gamma(X, \mathscr{M}_X)$ a fine and saturated chart for $(X, \mathscr{M}_X)$. Then there exists a (Zariski) open neighborhood $U$ of $x$ in $X$ such that for every (Zariski) open neighborhood $V$ of $x$ in $U$, the logarithmic scheme $(V, \mathscr{M}_X|V)$ satisfies the equivalent conditions of* (II.5.16).

We denote by $\mathrm{Spec}(P)$ the set of prime ideals of $P$ (cf. [**51**] 5.1), by $P_X$ the constant (étale) sheaf on $X$ with value $P$, and by $\widetilde{\gamma}\colon P_X \to \mathscr{M}_X$ the homomorphism induced by $\gamma$.

For every geometric point $\overline{y}$ of $X$, we set

(II.5.17.1) $$\mathfrak{p}_{\overline{y}} = P - \widetilde{\gamma}_{\overline{y}}^{-1}(\mathscr{O}_{X,\overline{y}}^{\times}) \subset (P_X)_{\overline{y}} = P,$$

which is a prime ideal of $P$. For every $t \in P$, we denote by $Y_t$ the cosupport of the image of $t$ in $\Gamma(X, \mathscr{M}_X/\mathscr{O}_X^{\times})$, or equivalently, since $\mathscr{M}_X$ is integral, the cosupport of the image of $t$ in $\Gamma(X, \mathscr{M}_X^{\mathrm{gp}}/\mathscr{O}_X^{\times})$ ([**2**] IV 8.5.2). It is a subobject of the final object of the étale topos of $X$ that we identify with an open subset of $X$ ([**2**] VIII 6.1). We set $X_t = X - Y_t$. One immediately verifies that the fiber of $X_t$ over a geometric point $\overline{y}$ of $X$ is nonempty if and only if $t \in \mathfrak{p}_{\overline{y}}$. For every $\mathfrak{p} \in \mathrm{Spec}(P)$, we set $X_{\mathfrak{p}} = \cap_{t \in \mathfrak{p}} X_t$. The fiber of $X_{\mathfrak{p}}$ over a geometric point $\overline{y}$ of $X$ is nonempty if and only if $\mathfrak{p} \subset \mathfrak{p}_{\overline{y}}$. Since $X$ is noetherian and $\mathrm{Spec}(P)$ is finite (cf. [**51**] 5.5), there exists a (Zariski) open neighborhood $U$ of $x$ in $X$ with the following property. For every $\mathfrak{p} \in \mathrm{Spec}(P)$ such that $U_{\mathfrak{p}} = U \cap X_{\mathfrak{p}} \neq \emptyset$, $x$ belongs to all irreducible components of $U_{\mathfrak{p}}$. Since every (Zariski) open neighborhood $V$ of $x$ in $U$ has the same property, it suffices to prove that $(U, \mathscr{M}_X|U)$ satisfies the equivalent conditions of II.5.16.

Let $\overline{x}$ be a geometric point of $X$ over $x$. Since the composition of the canonical homomorphisms

(II.5.17.2) $$P \to \Gamma(U, \mathscr{M}_X)/\Gamma(U, \mathscr{O}_X^{\times}) \xrightarrow{u} \Gamma(U, \mathscr{M}_X/\mathscr{O}_X^{\times}) \xrightarrow{v} \mathscr{M}_{X,\overline{x}}/\mathscr{O}_{X,\overline{x}}^{\times}$$

is surjective and the morphism $u$ is injective, it suffices to show that $v$ is injective. Let $a, b \in \Gamma(U, \mathscr{M}_X/\mathscr{O}_X^{\times})$ be such that $v(a) = v(b)$ and $\overline{y}$ a geometric point of $U$. Then there exist a geometric point $\overline{z}$ of $U$ and a specialization map $\varphi \colon \overline{z} \rightsquigarrow \overline{y}$ such that the canonical image $z$ of $\overline{z}$ in $U$ is a generic point of $U_{\mathfrak{p}_{\overline{y}}}$. We have $\widetilde{\gamma}_{\overline{z}} = \varphi^* \circ \widetilde{\gamma}_{\overline{y}}$, where $\varphi^* \colon \mathscr{M}_{X,\overline{y}} \to \mathscr{M}_{X,\overline{z}}$ is the specialization homomorphism associated with $\varphi$ ([**2**] VIII 7.7). It follows that $\mathfrak{p}_{\overline{z}} \subset \mathfrak{p}_{\overline{y}}$. Since $z \in U_{\mathfrak{p}_{\overline{y}}}$, we have $\mathfrak{p}_{\overline{y}} \subset \mathfrak{p}_{\overline{z}}$ and therefore $\mathfrak{p}_{\overline{y}} = \mathfrak{p}_{\overline{z}}$. The commutative diagram

(II.5.17.3) $$\begin{array}{ccc} P/\widetilde{\gamma}_{\overline{y}}^{-1}(\mathscr{O}_{X,\overline{y}}^{\times}) & \xrightarrow{\widetilde{\gamma}_{\overline{y}}} & \mathscr{M}_{X,\overline{y}}/\mathscr{O}_{X,\overline{y}}^{\times} \\ \| & & \downarrow{\scriptstyle \varphi^*} \\ P/\widetilde{\gamma}_{\overline{z}}^{-1}(\mathscr{O}_{X,\overline{z}}^{\times}) & \xrightarrow{\widetilde{\gamma}_{\overline{z}}} & \mathscr{M}_{X,\overline{z}}/\mathscr{O}_{X,\overline{z}}^{\times} \end{array}$$

where the horizontal arrows are isomorphisms then shows that

$$\varphi^* \colon \mathscr{M}_{X,\overline{y}}/\mathscr{O}_{X,\overline{y}}^{\times} \to \mathscr{M}_{X,\overline{z}}/\mathscr{O}_{X,\overline{z}}^{\times}$$

is an isomorphism. Since $\overline{x}$ is a specialization of $\overline{z}$ in $U$, the canonical images $a_{\overline{z}}$ and $b_{\overline{z}}$ of $a$ and $b$ in $\mathscr{M}_{X,\overline{z}}/\mathscr{O}_{X,\overline{z}}^{\times}$ are equal. Hence the canonical images $a_{\overline{y}}$ and $b_{\overline{y}}$ of $a$ and $b$ in $\mathscr{M}_{X,\overline{y}}/\mathscr{O}_{X,\overline{y}}^{\times}$ are also equal. It follows that $a = b$, giving the desired result.

**II.5.18.** Let $f \colon (X, \mathscr{M}_X) \to (Y, \mathscr{M}_Y)$ be a morphism of integral logarithmic schemes. We say that $f$ is *integral* (resp. *saturated*) if for every geometric point $\overline{x}$ of $X$, the homomorphism $\mathscr{M}_{Y,f(\overline{x})} \to \mathscr{M}_{X,\overline{x}}$ is integral (resp. saturated), or, equivalently, if the homomorphism $\mathscr{M}_{Y,f(\overline{x})}^{\sharp} \to \mathscr{M}_{X,\overline{x}}^{\sharp}$ is integral (resp. saturated).

**II.5.19.** A morphism $f \colon X \to Y$ of integral (resp. fine) logarithmic schemes is integral if and only if for every integral (resp. fine) logarithmic scheme $Z$ and every morphism $Z \to Y$, the fibered product $Z \times_Y X$ in the category of logarithmic schemes is integral (resp. fine) ([**50**] 4.3.1).

**II.5.20.** An integral morphism $f\colon X \to Y$ of fine and saturated logarithmic schemes is saturated if and only if for every fine and saturated logarithmic scheme $Z$ and every morphism $Z \to Y$, the fibered product $Z \times_Y X$ in the category of logarithmic schemes is fine and saturated ([**74**] II 2.13 page 24).

**II.5.21.** Let $f\colon (X, \mathscr{M}_X) \to (Y, \mathscr{M}_Y)$ be a morphism of prelogarithmic schemes. We set

$$\Omega^1_{(X,\mathscr{M}_X)/(Y,\mathscr{M}_Y)} = \frac{\Omega^1_{X/Y} \oplus (\mathscr{O}_X \otimes_{\mathbb{Z}} \mathscr{M}_X^{\mathrm{gp}})}{\mathscr{F}}, \tag{II.5.21.1}$$

where $\Omega^1_{X/Y}$ is the $\mathscr{O}_X$-module of relative 1-differentials of $X$ over $Y$ and $\mathscr{F}$ is the sub-$\mathscr{O}_X$-module locally generated by the sections of the form

(i) $(d(\alpha_X(a)), 0) - (0, \alpha_X(a) \otimes a)$ for every local section $a$ of $\mathscr{M}_X$;
(ii) $(0, 1 \otimes a)$ for every local section $a$ of the image of the morphism $f^\flat\colon f^{-1}(\mathscr{M}_Y) \to \mathscr{M}_X$.

We also denote by

$$d\colon \mathscr{O}_X \to \Omega^1_{(X,\mathscr{M}_X)/(Y,\mathscr{M}_Y)} \tag{II.5.21.2}$$

the morphism induced by the universal derivation $d\colon \mathscr{O}_X \to \Omega^1_{X/Y}$ and we denote by

$$d\log\colon \mathscr{M}_X \to \Omega^1_{(X,\mathscr{M}_X)/(Y,\mathscr{M}_Y)} \tag{II.5.21.3}$$

the homomorphism defined, for a local section $a$ of $\mathscr{M}_X$, by

$$d\log(a) = 1 \otimes a. \tag{II.5.21.4}$$

Then the triple $(\Omega^1_{(X,\mathscr{M}_X)/(Y,\mathscr{M}_Y)}, d, d\log)$ is universal for logarithmic derivations. We call $\Omega^1_{(X,\mathscr{M}_X)/(Y,\mathscr{M}_Y)}$ the $\mathscr{O}_X$-module of *logarithmic* 1-*differentials* of $(X, \mathscr{M}_X)$ over $(Y, \mathscr{M}_Y)$ (or of $f$). It satisfies the same functoriality properties as the module $\Omega^1_{X/Y}$.

If $\mathscr{M}_X^a$ (resp. $\mathscr{M}_Y^a$) denotes the logarithmic structure on $X$ (resp. $Y$) associated with $\mathscr{M}_X$ (resp. $\mathscr{M}_Y$), then $f$ induces a morphism $f^a\colon (X, \mathscr{M}_X^a) \to (Y, \mathscr{M}_Y^a)$, and we have a canonical isomorphism

$$\Omega^1_{(X,\mathscr{M}_X)/(Y,\mathscr{M}_Y)} \xrightarrow{\sim} \Omega^1_{(X,\mathscr{M}_X^a)/(Y,\mathscr{M}_Y^a)}. \tag{II.5.21.5}$$

If $f$ is a morphism of coherent logarithmic schemes, then $\Omega^1_{(X,\mathscr{M}_X)/(Y,\mathscr{M}_Y)}$ is a quasi-coherent $\mathscr{O}_X$-module. If, moreover, the morphism of schemes underlying $f$ is locally of finite presentation, then $\Omega^1_{(X,\mathscr{M}_X)/(Y,\mathscr{M}_Y)}$ is an $\mathscr{O}_X$-module of finite presentation.

If $f$ is a strict morphism of logarithmic schemes, then the canonical morphism

$$\Omega^1_{X/Y} \to \Omega^1_{(X,\mathscr{M}_X)/(Y,\mathscr{M}_Y)} \tag{II.5.21.6}$$

is an isomorphism.

**II.5.22.** A morphism of logarithmic schemes $f\colon (X, \mathscr{M}_X) \to (Y, \mathscr{M}_Y)$ is a *closed immersion* (resp. *exact closed immersion*) if the underlying morphism of schemes $X \to Y$ is a closed immersion and if the morphism $f^*(\mathscr{M}_Y) \to \mathscr{M}_X$ is an epimorphism (resp. an isomorphism).

**II.5.23.** Consider a commutative diagram of morphisms of fine logarithmic schemes

(II.5.23.1)
$$\begin{array}{ccc} (Z', \mathscr{M}_{Z'}) & \xrightarrow{u'} & (X, \mathscr{M}_X) \\ {\scriptstyle j}\downarrow & \overset{u}{\nearrow} & \downarrow{\scriptstyle f} \\ (Z, \mathscr{M}_Z) & \xrightarrow{g} & (Y, \mathscr{M}_Y) \end{array}$$

where $Z'$ is a closed subscheme of $Z$ defined by an ideal $\mathscr{I}$ of $\mathscr{O}_Z$ with square zero and $j$ is an exact closed immersion. We denote by $\mathrm{P}_f(j, u')$ the set of $(Y, \mathscr{M}_Y)$-morphisms $u\colon (Z, \mathscr{M}_Z) \to (X, \mathscr{M}_X)$ such that $u' = u \circ j$. Then $\mathrm{P}_f(j, u')$ is a pseudo-torsor under

(II.5.23.2)
$$\mathrm{Hom}_{\mathscr{O}_{Z'}}(u'^*\Omega^1_{(X,\mathscr{M}_X)/(Y,\mathscr{M}_Y)}, \mathscr{I}).$$

More precisely ([**50**] 3.9), if $\mathrm{P}_f(j, u')$ is nonempty, then every element $u \in \mathrm{P}_f(j, u')$ uniquely determines an isomorphism

(II.5.23.3)
$$\varphi_u\colon \mathrm{P}_f(j, u') \xrightarrow{\sim} \mathrm{Hom}_{\mathscr{O}_{Z'}}(u'^*\Omega^1_{(X,\mathscr{M}_X)/(Y,\mathscr{M}_Y)}, \mathscr{I})$$

such that for every $v \in \mathrm{P}_f(j, u')$, every local section $a$ of $\mathscr{O}_X$, and every local section $b$ of $\mathscr{M}_X$, we have

(II.5.23.4)
$$\varphi_u(v)(u'^*(da)) = v^\sharp(v^{-1}(a)) - u^\sharp(u^{-1}(a)),$$

(II.5.23.5)
$$\varphi_u(v)(u'^*(d\log b)) = \beta - 1,$$

where $\beta$ is the unique local section of $1 + \mathscr{I} \subset \mathscr{O}_Z^\times \subset \mathscr{M}_Z$ such that $v^\flat(v^{-1}(b)) = \beta \cdot u^\flat(u^{-1}(b))$.

**II.5.24.** Let $f\colon X \to Y$ be a morphism of fine logarithmic schemes. We say that $f$ is *formally smooth* (resp. *formally étale*) if for every logarithmic scheme $Y'$ whose underlying scheme is affine, every exact closed immersion with nilpotent ideal $j\colon Y'_0 \to Y'$, and every morphism $Y' \to Y$, the map

(II.5.24.1)
$$\mathrm{Hom}_Y(Y', X) \to \mathrm{Hom}(Y'_0, X)$$

deduced from $j$ is surjective (resp. bijective). We say that $f$ is *smooth* (resp. *étale*) if it is formally smooth (resp. formally étale), if the logarithmic schemes $X$ and $Y$ are coherent, and if the morphism of schemes underlying $f$ is locally of finite presentation.

**II.5.25.** Let $f\colon (X, \mathscr{M}_X) \to (Y, \mathscr{M}_Y)$ be a morphism of fine logarithmic schemes and $(N, \delta)$ a fine chart for $(Y, \mathscr{M}_Y)$ (II.5.13). By ([**50**] 3.5), $f$ is smooth (resp. étale) if and only if locally for the étale topology on $X$, $f$ admits a fine chart $((M, \gamma), (N, \delta), \vartheta\colon N \to M)$ (II.5.14) satisfying the following conditions:

(i) the kernel and the torsion subgroup of the cokernel (resp. the kernel and the cokernel) of the homomorphism $\vartheta^{\mathrm{gp}}\colon N^{\mathrm{gp}} \to M^{\mathrm{gp}}$ are finite of invertible order in $X$;

(ii) the induced morphism $X \to Y \times_{\mathbf{A}_N} \mathbf{A}_M$ (II.5.14.2) is étale in the classical sense.

## II.6. Faltings' almost purity theorem

**II.6.1.** We set $S = \mathrm{Spec}(\mathscr{O}_K)$ and denote by $\eta$ (resp. $s$) its generic (resp. closed) point and by $\overline{\eta}$ the generic geometric point corresponding to $\overline{K}$ (II.2.1). We endow $S$ with the logarithmic structure $\mathscr{M}_S$ defined by its closed point; in other words, $\mathscr{M}_S = j_*(\mathscr{O}_\eta^\times) \cap \mathscr{O}_S$, where $j\colon \eta \to S$ is the canonical injection. We fix a uniformizer $\pi$ of $\mathscr{O}_K$ and denote by $\iota\colon \mathbb{N} \to \Gamma(S, \mathscr{M}_S)$ the homomorphism defined by $\iota(1) = \pi$, which is a chart for $(S, \mathscr{M}_S)$.

**II.6.2.** For the remainder of this chapter, we fix a morphism of logarithmic schemes

(II.6.2.1) $$f\colon (X,\mathscr{M}_X)\to (S,\mathscr{M}_S),$$

a *toric* chart $(P,\gamma)$ for $(X,\mathscr{M}_X)$ (II.5.13), and a homomorphism $\vartheta\colon \mathbb{N}\to P$ such that the following conditions are satisfied:

($C_1$) The scheme $X=\mathrm{Spec}(R)$ is affine and connected.
($C_2$) The scheme $X_s=X\times_S s$ is nonempty.
($C_3$) The triple $((P,\gamma),(\mathbb{N},\iota),\vartheta)$ is a chart for $f$ (II.5.14); in other words, the diagram of homomorphisms

(II.6.2.2) $$\begin{array}{ccc} P & \xrightarrow{\gamma} & \Gamma(X,\mathscr{M}_X)\\ {\scriptstyle\vartheta}\uparrow & & \uparrow{\scriptstyle f^\flat}\\ \mathbb{N} & \xrightarrow{\iota} & \Gamma(S,\mathscr{M}_S)\end{array}$$

is commutative or, equivalently, the associated diagram of morphisms of logarithmic schemes

(II.6.2.3) $$\begin{array}{ccc} (X,\mathscr{M}_X) & \xrightarrow{\gamma^a} & \mathbf{A}_P\\ {\scriptstyle f}\downarrow & & \downarrow{\scriptstyle \mathbf{A}_\vartheta}\\ (S,\mathscr{M}_S) & \xrightarrow{\iota^a} & \mathbf{A}_{\mathbb{N}}\end{array}$$

is commutative.
($C_4$) The homomorphism $\vartheta$ is saturated (II.5.2).
($C_5$) The homomorphism $\vartheta^{\mathrm{gp}}\colon \mathbb{Z}\to P^{\mathrm{gp}}$ is injective, the order of the torsion subgroup of $\mathrm{coker}(\vartheta^{\mathrm{gp}})$ is prime to $p$, and the morphism of usual schemes

(II.6.2.4) $$X\to S\times_{\mathbf{A}_{\mathbb{N}}}\mathbf{A}_P$$

deduced from (II.6.2.3) is étale.
($C_6$) Let $\lambda=\vartheta(1)\in P$,

(II.6.2.5) $$L = \mathrm{Hom}_{\mathbb{Z}}(P^{\mathrm{gp}},\mathbb{Z}),$$
(II.6.2.6) $$\mathrm{H}(P) = \mathrm{Hom}(P,\mathbb{N}).$$

Note that $\mathrm{H}(P)$ is a fine, saturated, and sharp monoid and that the canonical homomorphism $\mathrm{H}(P)^{\mathrm{gp}}\to\mathrm{Hom}((P^\sharp)^{\mathrm{gp}},\mathbb{Z})$ is an isomorphism ([**58**] I 2.2.3). We assume that there exist $h_1,\dots,h_r\in\mathrm{H}(P)$ that are $\mathbb{Z}$-linearly independent in $L$, such that

(II.6.2.7) $$\ker(\lambda)\cap\mathrm{H}(P)=\{\sum_{i=1}^r a_ih_i|\ (a_1,\dots,a_r)\in\mathbb{N}^r\},$$

where we view $\lambda$ as a homomorphism $L\to\mathbb{Z}$.

Recall that $P^{\mathrm{gp}}$ is a free $\mathbb{Z}$-module of finite type. Note that we have

(II.6.2.8) $$S\times_{\mathbf{A}_{\mathbb{N}}}\mathbf{A}_P=\mathrm{Spec}(\mathscr{O}_K[P]/(\pi-e^\lambda)),$$

where $\lambda=\vartheta(1)$ (cf. II.5.12). We set

(II.6.2.9) $$L_\lambda=\mathrm{Hom}_{\mathbb{Z}}(P^{\mathrm{gp}}/\lambda\mathbb{Z},\mathbb{Z}),$$

which we identify with the kernel of the homomorphism $L\to\mathbb{Z}$ given by $y\mapsto\langle y,\lambda\rangle$ (II.6.2.5). Let $d$ be the rank of $L_\lambda$. We denote by $X^\circ$ the maximal open subscheme of

$X$ where the logarithmic structure $\mathscr{M}_X$ is trivial. We have $X^\circ = X \times_{\mathbf{A}_P} \mathbf{A}_{P^{\rm gp}}$, which is an affine open subscheme of $X_\eta$.

Let $\alpha\colon P \to R$ be the homomorphism induced by the chart $(P, \gamma)$. Note that for any $t \in P$, $\alpha(t)$ is invertible on $X^\circ$; in particular, $\alpha(t) \neq 0$.

**Proposition II.6.3.** (i) *The morphism $f$ is smooth and saturated.*

(ii) *The scheme $X$ is integral, normal, Cohen–Macaulay, and flat over $S$.*

(iii) *The scheme $X \otimes_{\mathscr{O}_K} \mathscr{O}_{\overline{K}}$ is normal.*

(iv) *The scheme $X \otimes_{\mathscr{O}_K} \overline{k}$ is reduced.*

(v) *The (usual) scheme $X \times_S \eta$ is smooth over $\eta$, $X^\circ \times_S \eta$ is the open complement in $X \times_S \eta$ of a divisor $D$ with strict normal crossings, and $\mathscr{M}_X|(X \times_S \eta)$ is the logarithmic structure on $X \times_S \eta$ defined by $D$.*

(i) This immediately follows from II.5.25 and ([**74**] Chap. II 3.5).

(ii) The last three properties follow from (i), ([**50**] 4.5), and ([**51**] 8.2 and 4.1); cf. also ([**73**] 1.5.1). Since $X$ is moreover noetherian and connected ($C_1$), it is then integral.

(iii) For every finite extension $K'$ of $K$, with the valuation ring $\mathscr{O}_{K'}$, endow $S' = \operatorname{Spec}(\mathscr{O}_{K'})$ with the logarithmic structure $\mathscr{M}_{S'}$ defined by its closed point, and let $f'\colon (X', \mathscr{M}_{X'}) \to (S', \mathscr{M}_{S'})$ be the morphism obtained from $f$ by base change in the category of logarithmic schemes by the canonical morphism $(S', \mathscr{M}_{S'}) \to (S, \mathscr{M}_S)$. Then $f'$ is smooth and saturated (II.5.20), and consequently $X'$ is normal by (ii). Since $X' = X \times_S S'$, the assertion follows by taking the direct limit over the finite extensions of $K$ contained in $\overline{K}$ ([**39**] 0.6.5.12(ii)).

(iv) This follows from (i) and ([**74**] Chap. II 4.2).

(v) Let $F$ be the face of $P$ generated by $\lambda$; that is, the set of elements $x \in P$ such that there exist $y \in P$ and $n \in \mathbb{N}$ such that $x + y = n\lambda$ ([**58**] I 1.4.2). We denote by $F^{-1}P$ the localization of $P$ by $F$ ([**58**] I 1.4.4). It immediately follows from the universal properties of localizations of monoids and rings that the canonical homomorphism $\mathbb{Z}[P] \to \mathbb{Z}[F^{-1}P]$ induces an isomorphism

$$\mathbb{Z}[P]_\lambda \xrightarrow{\sim} \mathbb{Z}[F^{-1}P]. \tag{II.6.3.1}$$

Let $P/F$ (resp. $\Lambda$) be the cokernel in the category of monoids of the canonical injection $F \to P$ (resp. of the homomorphism $\vartheta\colon \mathbb{N} \to P$) (cf. [**58**] I 1.1.5).

We have canonical isomorphisms

$$\Lambda^\sharp \xrightarrow{\sim} P/F \xrightarrow{\sim} (F^{-1}P)^\sharp. \tag{II.6.3.2}$$

The canonical homomorphism

$$\operatorname{Hom}(P/F, \mathbb{N}) \to \ker(\lambda) \cap \mathrm{H}(P) \tag{II.6.3.3}$$

is an isomorphism. As an amalgamated sum of the saturated homomorphism $\vartheta$ and the homomorphism $\mathbb{N} \to 0$, $\Lambda$ is saturated (II.5.2). Consequently, $P/F$ is saturated (II.6.3.2). Therefore, by ([**58**] I 2.2.3), we have a canonical isomorphism

$$P/F \xrightarrow{\sim} \operatorname{Hom}(\operatorname{Hom}(P/F, \mathbb{N}), \mathbb{N}). \tag{II.6.3.4}$$

Condition II.6.2($C_6$) then implies that $P/F$ is a free monoid of finite type. Consequently, there exists a homomorphism $P/F \to F^{-1}P$ that lifts the canonical isomorphism $P/F \xrightarrow{\sim} (F^{-1}P)^\sharp$ (II.6.3.2) so that the induced homomorphism $P/F \to \mathbb{Z}[F^{-1}P]$ is a chart for $\mathbf{A}_{F^{-1}P}$ ([**73**] 1.3.1). By (II.6.3.1), we deduce from this a chart

$$(X \times_S \eta, \mathscr{M}_X|(X \times_S \eta)) \to \mathbf{A}_{P/F}. \tag{II.6.3.5}$$

On the other hand, the logarithmic scheme $(X, \mathscr{M}_X)$ is regular by virtue of (i) and ([**51**] 8.2); cf. also ([**57**] 2.3) and the proof of ([**73**] 1.5.1). It then follows from ([**42**] 0.16.3.7

and 0.17.1.7) and the definitions ([**51**] 2.1 and [**57**] 2.2) that the scheme $X \times_S \eta$ is regular and therefore smooth over $\eta$, that $X^\circ \times_S \eta$ is the open complement of a divisor $D$ with strict normal crossings in $X \times_S \eta$, and that $\mathscr{M}_X|(X \times_S \eta)$ is the logarithmic structure on $X \times_S \eta$ defined by $D$.

**II.6.4.** For every integer $n \geq 1$, we set

(II.6.4.1) $$\mathscr{O}_{K_n} = \mathscr{O}_K[\zeta]/(\zeta^n - \pi),$$

which is a discrete valuation ring. Let $K_n$ be the field of fractions of $\mathscr{O}_{K_n}$ and $\pi_n$ the class of $\zeta$ in $\mathscr{O}_{K_n}$, which is a uniformizer of $\mathscr{O}_{K_n}$. We set $S_n = \operatorname{Spec}(\mathscr{O}_{K_n})$ which we endow with the logarithmic structure $\mathscr{M}_{S_n}$ defined by its closed point. We denote by $\tau_n \colon (S_n, \mathscr{M}_{S_n}) \to (S, \mathscr{M}_S)$ the canonical morphism and by $\iota_n \colon \mathbb{N} \to \Gamma(S_n, \mathscr{M}_{S_n})$ the homomorphism defined by $\iota_n(1) = \pi_n$. Note that $\iota_n$ is a chart for $(S_n, \mathscr{M}_{S_n})$ and that the diagram

(II.6.4.2) $$\begin{array}{ccc} (S_n, \mathscr{M}_{S_n}) & \xrightarrow{\iota_n^a} & \mathbf{A}_{\mathbb{N}} \\ {\scriptstyle \tau_n}\downarrow & & \downarrow{\scriptstyle \mathbf{A}_{\varpi_n}} \\ (S, \mathscr{M}_S) & \xrightarrow{\iota^a} & \mathbf{A}_{\mathbb{N}} \end{array}$$

is Cartesian (cf. II.5.12).

For all integers $m, n \geq 1$, we have $\varpi_{mn} = \varpi_m \circ \varpi_n$. We deduce from this a canonical morphism

(II.6.4.3) $$\tau_{m,n} \colon (S_{mn}, \mathscr{M}_{S_{mn}}) \to (S_n, \mathscr{M}_{S_n})$$

such that $\tau_{mn} = \tau_n \circ \tau_{m,n}$. For all integers $r, m, n \geq 1$, we have $\tau_{rm,n} = \tau_{m,n} \circ \tau_{r,mn}$. Hence the logarithmic schemes $(S_n, \mathscr{M}_{S_n})$ for $n \geq 1$ form a cofiltered inverse system indexed by the set $\mathbb{Z}_{\geq 1}$ ordered by the divisibility relation.

**II.6.5.** For every integer $n \geq 1$, we set

(II.6.5.1) $$(X_n, \mathscr{M}_{X_n}) = (X, \mathscr{M}_X) \times_{\mathbf{A}_P, \mathbf{A}_{\varpi_n}} \mathbf{A}_P$$

and we denote by $\rho_n \colon (X_n, \mathscr{M}_{X_n}) \to (X, \mathscr{M}_X)$ the canonical projection (cf. II.5.12). We also (abusively) denote by $\rho_n \colon X_n \to X$ the morphism of schemes underlying $\rho_n$. Then $\rho_n$ is finite, the scheme $X_n$ is affine with ring

(II.6.5.2) $$A_n = R \otimes_{\mathbb{Z}[P], \varpi_n} \mathbb{Z}[P],$$

and the canonical projection $(X_n, \mathscr{M}_{X_n}) \to \mathbf{A}_P$ is strict. Since the diagram (II.6.4.2) is Cartesian, there exists a unique morphism

(II.6.5.3) $$f_n \colon (X_n, \mathscr{M}_{X_n}) \to (S_n, \mathscr{M}_{S_n}),$$

that fits into the commutative diagram

(II.6.5.4)
$$\begin{array}{ccccc}
(X_n,\mathscr{M}_{X_n}) & & \longrightarrow & & \mathbf{A}_P \\
 & \overset{f_n}{\searrow} & & \overset{\mathbf{A}_\vartheta}{\swarrow} & \\
 & (S_n,\mathscr{M}_{S_n}) & \longrightarrow & \mathbf{A}_{\mathbb{N}} & \\
\Big\downarrow{\scriptstyle \rho_n} & \Big\downarrow & & \Big\downarrow{\scriptstyle \mathbf{A}_{\varpi_n}} & \Big\downarrow{\scriptstyle \mathbf{A}_{\varpi_n}} \\
 & (S,\mathscr{M}_S) & \longrightarrow & \mathbf{A}_{\mathbb{N}} & \\
 & \overset{f}{\nearrow} & & \overset{\mathbf{A}_\vartheta}{\nwarrow} & \\
(X,\mathscr{M}_X) & & \longrightarrow & & \mathbf{A}_P
\end{array}$$

We denote by $X_n^\circ$ the maximal open subscheme of $X_n$ where the logarithmic structure $\mathscr{M}_{X_n}$ is trivial. We have $X_n^\circ = X_n \times_{\mathbf{A}_P} \mathbf{A}_{P^{\mathrm{gp}}}$ and $f_n(X_n^\circ) = \mathrm{Spec}(K_n)$.

For all integers $m, n \geq 1$, we have $\varpi_{mn} = \varpi_m \circ \varpi_n$. We deduce from this a canonical morphism

(II.6.5.5)
$$\rho_{m,n} \colon (X_{mn}, \mathscr{M}_{X_{mn}}) \to (X_n, \mathscr{M}_{X_n})$$

such that $\rho_{mn} = \rho_n \circ \rho_{m,n}$. For all integers $r, m, n \geq 1$, we have $\rho_{rm,n} = \rho_{m,n} \circ \rho_{r,mn}$. Hence the logarithmic schemes $(X_n, \mathscr{M}_{X_n})$ form a cofiltered inverse system indexed by the set $\mathbb{Z}_{\geq 1}$ ordered by the divisibility relation.

**Proposition II.6.6.** *Let $n$ be an integer $\geq 1$.*

(i) *The morphism of usual schemes $X_n \to S_n \times_{\mathbf{A}_{\mathbb{N}}} \mathbf{A}_P$ deduced from* (II.6.5.4) *is étale and the morphism of logarithmic schemes $f_n$ is smooth and saturated.*

(ii) *The scheme $X_n$ is normal, Cohen–Macaulay, and flat over $S_n$.*

(iii) *The scheme $X_n \otimes_{\mathscr{O}_{K_n}} \mathscr{O}_{\overline{K}}$ is normal and has only a finite number of generic points; in particular, $X_n \otimes_{\mathscr{O}_{K_n}} \mathscr{O}_{\overline{K}}$ is a finite sum of normal integral schemes.*

(iv) *The morphism $X_n \otimes_{\mathscr{O}_{K_n}} K_n \to X \otimes_{\mathscr{O}_K} K$ deduced from $\rho_n$ is flat. If $X$ is moreover regular, then $\rho_n$ is flat.*

(v) *If $n$ is a power of $p$, then $X_n$ is integral, and the inverse image of any connected component of $X \otimes_{\mathscr{O}_K} \mathscr{O}_{\overline{K}}$ under the canonical morphism*
$$X_n \otimes_{\mathscr{O}_{K_n}} \mathscr{O}_{\overline{K}} \to X \otimes_{\mathscr{O}_K} \mathscr{O}_{\overline{K}}$$
*is integral.*

(vi) *The squares in the canonical commutative diagram*

(II.6.6.1)
$$\begin{array}{ccccc}
X_n & \longleftarrow & X_n^\circ & \longrightarrow & \mathbf{A}_{P^{\mathrm{gp}}} \\
\Big\downarrow{\scriptstyle \rho_n} & & \Big\downarrow & & \Big\downarrow{\scriptstyle \mathbf{A}_{\varpi_n}} \\
X & \longleftarrow & X^\circ & \longrightarrow & \mathbf{A}_{P^{\mathrm{gp}}}
\end{array}$$

*are Cartesian. In particular, $X_n^\circ$ is a principal homogeneous space over $X^\circ$ for the étale topology, under the group* $\mathrm{Hom}_{\mathbb{Z}}(P^{\mathrm{gp}}, \mu_n(\overline{K}))$.

(i) Indeed, the squares of the commutative diagram of morphisms of usual schemes

(II.6.6.2)
$$\begin{array}{ccccc}
X_n & \longrightarrow & S_n \times_{\mathbf{A}_{\mathbb{N}}} \mathbf{A}_P & \longrightarrow & \mathbf{A}_P \\
\Big\downarrow{\scriptstyle \rho_n} & & \Big\downarrow & & \Big\downarrow{\scriptstyle \mathbf{A}_{\varpi_n}} \\
X & \longrightarrow & S \times_{\mathbf{A}_{\mathbb{N}}} \mathbf{A}_P & \longrightarrow & \mathbf{A}_P
\end{array}$$

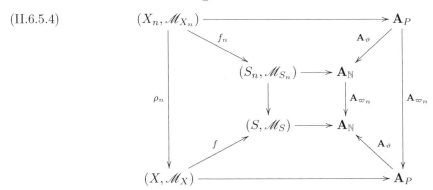

deduced from (II.6.5.4) are Cartesian.

(ii) This follows from (i) using the same proof as for II.6.3(ii).

(iii) It follows from (i) using the same argument as for II.6.3(iii) that $X_n \otimes_{\mathscr{O}_{K_n}} \mathscr{O}_{\overline{K}}$ is normal. On the other hand, since $X_n \otimes_{\mathscr{O}_{K_n}} \mathscr{O}_{\overline{K}}$ is flat over $\mathscr{O}_{\overline{K}}$, its generic points are the generic points of the scheme $X_n \otimes_{\mathscr{O}_{K_n}} \overline{K}$, which is noetherian. Consequently, the set of generic points of $X_n \otimes_{\mathscr{O}_{K_n}} \mathscr{O}_{\overline{K}}$ is finite. The last assertion follows in view of (**[39]** 0.2.1.6). Indeed, since $X_n \otimes_{\mathscr{O}_{K_n}} \mathscr{O}_{\overline{K}}$ is normal, two distinct irreducible components of $X_n \otimes_{\mathscr{O}_{K_n}} \mathscr{O}_{\overline{K}}$ do not meet.

(iv) This follows from (**[42]** 0.17.3.5) because $X \otimes_{\mathscr{O}_K} K$ is regular by virtue of II.6.3(v), $X_n$ is Cohen–Macaulay by (ii), and $\rho_n$ is finite.

(v) By (II.6.6.2), we have a Cartesian diagram of $S$-morphisms

(II.6.6.3)
$$\begin{array}{ccc} X_n & \longrightarrow & \mathrm{Spec}(\mathscr{O}_{K_n}[P]/(\pi_n - e^{\lambda})) \\ {\scriptstyle \rho_n}\downarrow & & \downarrow{\scriptstyle \gamma_n} \\ X & \longrightarrow & \mathrm{Spec}(\mathscr{O}_K[P]/(\pi - e^{\lambda})), \end{array}$$

where $\gamma_n$ is induced by the homomorphism $\varpi_n$. By II.6.2(C$_2$), there exists $x \in X_s$. Since $n$ is a power of $p$, the morphism $\rho_n \otimes_{\mathscr{O}_K} k$ is a universal homeomorphism. Hence $\rho_n^{-1}(x)$ contains a single point that we denote by $x_n$. On the other hand, since $X_n$ is flat over $S_n$, every generic point of $X_n$ is a generic point of $X_n \otimes_{\mathscr{O}_{K_n}} K_n$; consequently, its image by $\rho_n$ is the generic point of $X$ by virtue of (iv). Since $\rho_n$ is closed, it follows that $x_n$ is a specialization of all generic points of $X_n$. Since $X_n$ is noetherian and normal by (ii), it is integral.

The proof of the second assertion is similar to that of the first one. Indeed, one easily deduces from (II.6.6.3) a Cartesian diagram of $\mathscr{O}_{\overline{K}}$-morphisms

(II.6.6.4)
$$\begin{array}{ccc} X_n \otimes_{\mathscr{O}_{K_n}} \mathscr{O}_{\overline{K}} & \longrightarrow & \mathrm{Spec}(\mathscr{O}_{\overline{K}}[P]/(\pi_n - e^{\lambda})) \\ {\scriptstyle \beta_n}\downarrow & & \downarrow{\scriptstyle \alpha_n} \\ X \otimes_{\mathscr{O}_K} \mathscr{O}_{\overline{K}} & \longrightarrow & \mathrm{Spec}(\mathscr{O}_{\overline{K}}[P]/(\pi - e^{\lambda})) \end{array}$$

By (iii), $X \otimes_{\mathscr{O}_K} \mathscr{O}_{\overline{K}}$ is a finite sum of normal integral schemes; in particular, it has only a finite number of connected components, which are therefore open. Since $X$ is integral and flat over $S$ by II.6.3(ii), $G_K$ acts transitively on the set of connected components of $X \otimes_{\mathscr{O}_K} \mathscr{O}_{\overline{K}}$. Let $Y$ be a connected component of $X \otimes_{\mathscr{O}_K} \mathscr{O}_{\overline{K}}$ and $Y_n = \beta_n^{-1}(Y)$. In view of II.6.2(C$_2$), $Y \otimes_{\mathscr{O}_{\overline{K}}} \overline{k}$ is nonempty. Let $y \in Y \otimes_{\mathscr{O}_{\overline{K}}} \overline{k}$. Since $n$ is a power of $p$, the morphism $\alpha_n \otimes_{\mathscr{O}_{\overline{K}}} \overline{k}$ is a universal homeomorphism. Hence $\beta_n^{-1}(y)$ contains a single point that we denote by $y_n$. On the other hand, every generic point of $Y_n$ lies over the unique generic point of $X_n$ and therefore lies over the generic point of $X$. Consequently, the image of any generic point of $Y_n$ by $\beta_n$ is the generic point of $Y$. Since $\beta_n$ is closed, it follows that $y_n$ is a specialization of all the generic points of $Y_n$. On the other hand, $Y_n$ is a finite disjoint union of open and closed subschemes that are integral and normal by (iii). It is therefore integral.

(vi) Let us show that the canonical diagram

(II.6.6.5)
$$\begin{array}{ccc} P & \xrightarrow{\varpi_n} & P \\ \downarrow & & \downarrow \\ P^{\mathrm{gp}} & \xrightarrow{\varpi_n} & P^{\mathrm{gp}} \end{array}$$

is co-Cartesian. Indeed, the amalgamated sum $P^{\rm gp} \oplus_{P,\varpi_n} P$ is the quotient of $P^{\rm gp} \oplus P$ by the congruence relation $E$ defined by the set of pairs $((y,x),(y',x'))$ of elements of $P^{\rm gp} \oplus P$ for which there exist $z, z' \in P$ such that $y+z = y'+z' \in P^{\rm gp}$ and $x+nz = x'+nz' \in P$ (**[58]** I 1.1.5). It therefore suffices to show that $E$ is the congruence relation defined by the homomorphism

$$P^{\rm gp} \oplus P \to P^{\rm gp}, \quad (y,x) \to x - ny. \tag{II.6.6.6}$$

If $((y,x),(y',x')) \in E$, then $x - ny = x' - ny'$. Conversely, suppose $x = x' + n(y-y') \in P^{\rm gp}$. Since $P$ is integral, there exist $z, z' \in P$ such that $y+z = y'+z' \in P^{\rm gp}$; we therefore have $x + nz = x' + nz' \in P$, which proves the assertion. Consequently, the diagram

$$\begin{array}{ccc} \mathbf{A}_{P^{\rm gp}} & \longrightarrow & \mathbf{A}_P \\ \downarrow{\scriptstyle \mathbf{A}_{\varpi_n}} & & \downarrow{\scriptstyle \mathbf{A}_{\varpi_n}} \\ \mathbf{A}_{P^{\rm gp}} & \longrightarrow & \mathbf{A}_P \end{array} \tag{II.6.6.7}$$

induced by (II.6.6.5) is Cartesian. Hence the squares in the diagram (II.6.6.1) are Cartesian. The second assertion follows from the first one and the fact that the kernel of the étale isogeny $\mathbf{A}_{\varpi_n} \otimes_{\mathbb{Z}} K$ of $\mathbf{A}_{P^{\rm gp}} \otimes_{\mathbb{Z}} K$ corresponds to the $\mathbb{Z}[G_K]$-module $\mathrm{Hom}_{\mathbb{Z}}(P^{\rm gp}, \mu_n(\overline{K}))$.

**II.6.7.** Let $\kappa$ be the generic point of $X$ and $F$ the residue field of $X$ at $\kappa$ (that is, the field of fractions of $R$). For the remainder of this chapter, we fix a geometric point $\widetilde{\kappa}$ of $X \otimes_{\mathscr{O}_K} \mathscr{O}_{\overline{K}}$ over $\kappa$, in other words, the spectrum of a separably closed extension $F^a$ of $F$ containing $\overline{K}$. We denote by $\overline{F}$ the union of the finite extensions $N$ of $F$ contained in $F^a$ such that the integral closure of $R$ in $N$ is étale over $X^\circ$, and by $\overline{R}$ the integral closure of $R$ in $\overline{F}$.

For all integers $m, n \geq 1$, the morphism $\rho_{m,n}\colon X_{mn} \to X_n$ is finite and surjective. Consequently, by virtue of (**[42]** 8.3.8(i)), there exists an $X$-morphism

$$\widetilde{\kappa} \to \varprojlim_{n\geq 1} X_n, \tag{II.6.7.1}$$

where the inverse limit is indexed by the set $\mathbb{Z}_{\geq 1}$ ordered by the divisibility relation. *For the remainder of this chapter, we fix such a morphism.* Since the set of integers $n!$, for $n \geq 0$, is cofinal in $\mathbb{Z}_{\geq 1}$ for the divisibility relation, this corresponds to fixing a morphism

$$\widetilde{\kappa} \to \varprojlim_{n\geq 0} X_{n!}, \tag{II.6.7.2}$$

where the inverse limit is indexed by the set $\mathbb{N}$ with the usual ordering.

By II.6.6(vi), the morphism (II.6.7.1) factors through an $X$-morphism

$$\mathrm{Spec}(\overline{R}) \to \varprojlim_{n\geq 1} X_n. \tag{II.6.7.3}$$

We deduce from this a direct system of $R$-homomorphisms $u_n\colon A_n \to \overline{R}$, indexed by the set $\mathbb{Z}_{\geq 1}$ ordered by the divisibility relation. We denote by $B_n$ the image of $u_n$ and set

$$B_\infty = \varinjlim_{n\geq 1} B_n, \tag{II.6.7.4}$$

which we identify with a sub-$R$-algebra of $\overline{R}$. We denote by $H_n$ the field of fractions of $B_n$ and by $H_\infty$ the field of fractions of $B_\infty$.

On the other hand, the morphism (II.6.7.3) induces a morphism

$$\mathrm{Spec}(\mathscr{O}_{\overline{K}}) \to \varprojlim_{n\geq 1} S_n. \tag{II.6.7.5}$$

We can therefore extend the $u_n$'s to a direct system of $(R \otimes_{\mathscr{O}_K} \mathscr{O}_{\overline{K}})$-homomorphisms

(II.6.7.6) $$v_n \colon A_n \otimes_{\mathscr{O}_{K_n}} \mathscr{O}_{\overline{K}} \to \overline{R},$$

indexed by the set $\mathbb{Z}_{\geq 1}$ ordered by the divisibility relation. We denote by $R_n$ the image of $v_n$ and set

(II.6.7.7) $$R_\infty = \varinjlim_{n \geq 1} R_n,$$

which we identify with a sub-$(R \otimes_{\mathscr{O}_K} \mathscr{O}_{\overline{K}})$-algebra of $\overline{R}$. We denote by $F_n$ the field of fractions of $R_n$ and by $F_\infty$ the field of fractions of $R_\infty$. We set

(II.6.7.8) $$R_{p^\infty} = \varinjlim_{n \geq 0} R_{p^n},$$

where the direct limit is indexed by the set $\mathbb{N}$ with the usual ordering. We identify $R_{p^\infty}$ with a sub-$(R \otimes_{\mathscr{O}_K} \mathscr{O}_{\overline{K}})$-algebra of $R_\infty$ and denote the field of fractions of $R_{p^\infty}$ by $F_{p^\infty}$.

**Proposition II.6.8.** (i) *For every $n \geq 1$, $\mathrm{Spec}(B_n)$ is an open connected component of $X_n$ and $\mathrm{Spec}(R_n)$ is an open connected component of $X_n \otimes_{\mathscr{O}_{K_n}} \mathscr{O}_{\overline{K}}$.*

(ii) *The rings $B_n$, $R_n$ $(n \geq 1)$, $B_\infty$, $R_\infty$, and $R_{p^\infty}$ are normal integral domains.*

(iii) *For every $n \geq 0$, we have*

(II.6.8.1) $$\mathrm{Spec}(B_{p^n}) = X_{p^n},$$

(II.6.8.2) $$\mathrm{Spec}(R_{p^n}) = (X_{p^n} \otimes_{\mathscr{O}_{K_n}} \mathscr{O}_{\overline{K}}) \times_{(X \otimes_{\mathscr{O}_K} \mathscr{O}_{\overline{K}})} \mathrm{Spec}(R_1).$$

(iv) *The extensions $F_n$ $(n \geq 1)$, $F_\infty$, and $F_{p^\infty}$ of $F$ are Galois and we have canonical injective homomorphisms* (II.6.2.9)

(II.6.8.3) $$\mathrm{Gal}(F_\infty/F_1) \to L_\lambda \otimes \widehat{\mathbb{Z}}(1),$$

(II.6.8.4) $$\mathrm{Gal}(F_{p^\infty}/F_1) \xrightarrow{\sim} L_\lambda \otimes \mathbb{Z}_p(1),$$

*where the second is an isomorphism. Moreover, the diagram*

(II.6.8.5) $$\begin{array}{ccc} \mathrm{Gal}(F_\infty/F_1) & \longrightarrow & L_\lambda \otimes \widehat{\mathbb{Z}}(1) \\ \downarrow & & \downarrow \\ \mathrm{Gal}(F_{p^\infty}/F_1) & \longrightarrow & L_\lambda \otimes \mathbb{Z}_p(1), \end{array}$$

*where the vertical arrows are the canonical morphisms, is commutative.*

By II.6.6(vi), for any $n \geq 1$, the morphism (II.6.7.1) induces a morphism

(II.6.8.6) $$\widetilde{\kappa} \to X_n^\circ \otimes_{K_n} \overline{K}.$$

We denote by $\widetilde{\kappa}_n$ its image, which is a generic point of $X_n^\circ \otimes_{K_n} \overline{K}$. We denote by $\kappa_n$ the image of $\widetilde{\kappa}_n$ in $X_n^\circ$, which is a generic point of $X_n^\circ$. Note that $H_n$ is the residue field of $X_n$ at $\kappa_n$ and that $F_n$ is the residue field of $X_n \otimes_{\mathscr{O}_{K_n}} \mathscr{O}_{\overline{K}}$ at $\widetilde{\kappa}_n$.

(i) Since $X_n$ is noetherian and normal by II.6.6(ii), $\mathrm{Spec}(B_n)$ is the connected component of $X_n$ containing $\kappa_n$, which is obviously open in $X_n$. In view of II.6.6(iii), $\mathrm{Spec}(R_n)$ is the connected component of $X_n \otimes_{\mathscr{O}_{K_n}} \mathscr{O}_{\overline{K}}$ containing $\widetilde{\kappa}_n$, which is open in $X_n \otimes_{\mathscr{O}_{K_n}} \mathscr{O}_{\overline{K}}$.

(ii) These rings are clearly integral domains. For any $n \geq 1$, $B_n$ and $R_n$ are normal by virtue of (i) and II.6.6(ii)-(iii). The same therefore holds for $B_\infty$, $R_\infty$, and $R_{p^\infty}$.

(iii) This follows from (i) and II.6.6(v).

(iv) It follows from II.6.6(vi) that for every integer $n \geq 1$, $H_n$ is a Galois extension of $F$, whose Galois group is canonically isomorphic to a subgroup of $L \otimes_{\mathbb{Z}} \mu_n(\mathscr{O}_{\overline{K}})$; more precisely, $\mathrm{Gal}(H_n/F)$ is the decomposition subgroup of $\kappa_n$. By II.6.6(v), if $n$ is a power of $p$, we have

(II.6.8.7) $$\mathrm{Gal}(H_n/F) \simeq L \otimes_{\mathbb{Z}} \mu_n(\mathscr{O}_{\overline{K}}).$$

Let $M_n$ be the image of the canonical homomorphism $F \otimes_K K_n \to H_n$, so that $M_n$ is a Galois extension of $F$ whose Galois group is a subgroup of $\mathrm{Gal}(K_n/K)$. We then have a commutative diagram of field extensions

(II.6.8.8) $$\begin{array}{ccccc} & & H_n & \longrightarrow & F_n \\ & & \uparrow & & \uparrow \\ F & \longrightarrow & M_n & \longrightarrow & F_1 \\ \uparrow & & \uparrow & & \uparrow \\ K & \longrightarrow & K_n & \longrightarrow & \overline{K} \end{array}$$

where the homomorphisms $F \otimes_K K_n \to M_n$, $M_n \otimes_{K_n} \overline{K} \to F_1$, and $H_n \otimes_{M_n} F_1 \to F_n$ are surjective. It follows that $F_1$ is a Galois extension of $F$, with Galois group canonically isomorphic to a subgroup of $G_K$. Consequently, $F_n$ is a Galois extension of $F$, since it is the composition of the Galois extensions $H_n$ and $F_1$ of $F$. In particular, $F_n$ is a Galois extension of $F_1$, with Galois group canonically isomorphic to a subgroup of $\mathrm{Gal}(H_n/M_n)$. It follows from II.6.6(v) that if $n$ is a power of $p$, we have

(II.6.8.9) $$\mathrm{Gal}(F_n/F_1) \simeq \mathrm{Gal}(H_n/M_n).$$

On the other hand, we have a commutative diagram

(II.6.8.10) $$\begin{array}{ccccc} \mathrm{Gal}(H_n/F) & & \longrightarrow & & L \otimes_{\mathbb{Z}} \mu_n(\mathscr{O}_{\overline{K}}) \\ \downarrow & & & & \downarrow \lambda_n \\ \mathrm{Gal}(M_n/F) & \hookrightarrow & \mathrm{Gal}(K_n/K) & \xrightarrow{\sim} & \mu_n(\mathscr{O}_{\overline{K}}), \end{array}$$

where $\lambda_n$ is the morphism defined by $\lambda \in P$ (II.6.2.5) and the unlabeled arrows are the canonical morphisms. It follows that $\mathrm{Gal}(F_n/F_1)$ is canonically isomorphic to a subgroup of $\ker(\lambda_n)$. If $n$ is a power of $p$, then the isomorphisms (II.6.8.7) and (II.6.8.9) and a chase in the diagram (II.6.8.10) show that

(II.6.8.11) $$\mathrm{Gal}(F_n/F_1) \simeq \ker(\lambda_n).$$

The proposition follows by taking the inverse limit over $n$.

**Remark II.6.9.** Note that condition II.6.2($C_6$) is not used in the proofs of II.6.6 and II.6.8. In Proposition II.6.3, it is only used in the proof of (v).

**II.6.10.** We set $\Gamma = \mathrm{Gal}(\overline{F}/F)$, $\Gamma_\infty = \mathrm{Gal}(F_\infty/F)$, $\Gamma_{p^\infty} = \mathrm{Gal}(F_{p^\infty}/F)$, $\Delta = \mathrm{Gal}(\overline{F}/F_1)$, $\Delta_\infty = \mathrm{Gal}(F_\infty/F_1)$, $\Delta_{p^\infty} = \mathrm{Gal}(F_{p^\infty}/F_1)$, $\Sigma = \mathrm{Gal}(\overline{F}/F_\infty)$, and $\Sigma_0 = \mathrm{Gal}(F_\infty/F_{p^\infty})$.

(II.6.10.1)

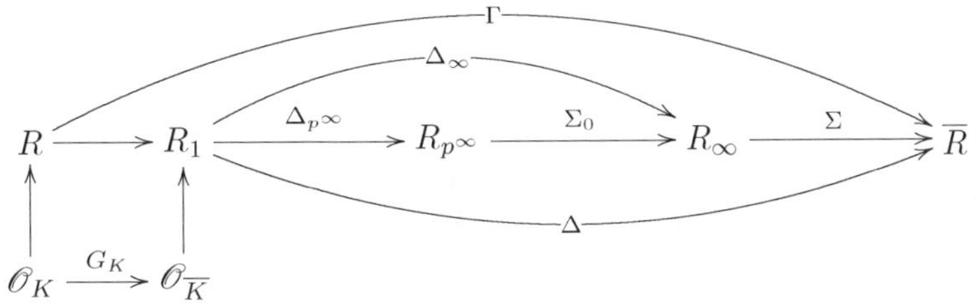

By II.6.8(iv), $\Delta_\infty$ is canonically isomorphic to a subgroup of $L_\lambda \otimes_{\mathbb{Z}} \widehat{\mathbb{Z}}(1)$, $\Delta_{p^\infty}$ is canonically isomorphic to $L_\lambda \otimes_{\mathbb{Z}} \mathbb{Z}_p(1)$, and $\Sigma_0$ is a profinite group of order prime to $p$.

We denote by $K^+$ the extension of $K$ contained in $\overline{K}$ such that $\mathrm{Gal}(\overline{K}/K^+)$ is the image of the canonical homomorphism $\mathrm{Gal}(F_1/F) \to G_K$. We have $K^+ = K$ if and only if $X \times_S \overline{\eta}$ is integral or, equivalently, connected.

**II.6.11.** Let $M$ be a $p$-primary torsion discrete $\mathbb{Z}$-$\Delta_\infty$-module. Since the $p$-cohomological dimension of $\Sigma_0$ is zero (**[65]** I Cor. 2 to Prop. 14), for any $q \geq 0$, the canonical morphism

(II.6.11.1) $$\mathrm{H}^q(\Delta_{p^\infty}, M^{\Sigma_0}) \to \mathrm{H}^q(\Delta_\infty, M)$$

is an isomorphism. Consequently, the $p$-cohomological dimension of $\Delta_\infty$ is equal to that of $\Delta_{p^\infty}$; that is, to the rank $d$ of $L_\lambda$ (II.3.24).

**Remarks II.6.12.** Let $A$ be a $\mathbb{Z}_p$-algebra that is complete and separated for the $p$-adic topology and $M$ an $A$-module that is complete and separated for the $p$-adic topology. Then:

(i) The canonical homomorphisms

(II.6.12.1) $$\mathrm{Hom}_{\mathbb{Z}_p}(\Delta_{p^\infty}, M) \to \mathrm{Hom}_{\mathbb{Z}}(\Delta_{p^\infty}, M) \to \mathrm{Hom}_{\mathbb{Z}}(\Delta_\infty, M)$$

are bijective. Indeed, since the multiplication by $p$ in $\Sigma_0$ is an isomorphism, for every homomorphism $\psi\colon \Sigma_0 \to M$, we have $\psi(\Sigma_0) \subset \cap_{n\geq 0} p^n M = 0$.

(ii) The canonical morphism

(II.6.12.2) $$\mathrm{Hom}_{\mathbb{Z}}(\Delta_\infty, A) \otimes_A M \to \mathrm{Hom}_{\mathbb{Z}}(\Delta_\infty, M)$$

is bijective. This follows from (i) because $\Delta_{p^\infty}$ is a free $\mathbb{Z}_p$-module of finite type.

(iii) Every homomorphism from $\Delta_\infty$ (resp. $\Delta_{p^\infty}$) to $M$ is continuous when we endow $\Delta_\infty$ (resp. $\Delta_{p^\infty}$) with the profinite topology and $M$ with the $p$-adic topology. Indeed, every homomorphism $\psi\colon \Delta_\infty \to M$ factors through a homomorphism $\varphi\colon \Delta_{p^\infty} \to M$ by (i), and $\varphi$ is clearly continuous for the $p$-adic topologies. In particular, when we endow $M$ with the trivial action of $\Delta_{p^\infty}$, the canonical morphisms

(II.6.12.3) $$\mathrm{Hom}_{\mathbb{Z}}(\Delta_{p^\infty}, M) \to \mathrm{H}^1_{\mathrm{cont}}(\Delta_{p^\infty}, M),$$

(II.6.12.4) $$\mathrm{Hom}_{\mathbb{Z}}(\Delta_\infty, M) \to \mathrm{H}^1_{\mathrm{cont}}(\Delta_\infty, M),$$

are isomorphisms.

**Lemma II.6.13.** *For every $a \in \mathscr{O}_{\overline{K}}$, the canonical homomorphism* (II.6.10)

(II.6.13.1) $$R_{p^\infty}/aR_{p^\infty} \to (R_\infty/aR_\infty)^{\Sigma_0}$$

*is an isomorphism.*

Let $N$ be a finite Galois extension of $F_{p^\infty}$ contained in $F_\infty$, $A$ the integral closure of $R_{p^\infty}$ in $N$, and $G = \mathrm{Gal}(N/F_{p^\infty})$. Since we have $R_{p^\infty} = A \cap F_{p^\infty}$ and $A = R_\infty \cap N$, the canonical homomorphisms $R_{p^\infty}/aR_{p^\infty} \to A/aA \to R_\infty/aR_\infty$ are injective. On the other hand, since the order of $G$ is prime to $p$ (and therefore invertible in $A$), the homomorphism

(II.6.13.2) $$R_{p^\infty} = A^G \to (A/aA)^G$$

is surjective. The assertion follows.

**Lemma II.6.14.** *The rings $\widehat{R_1}$, $\widehat{R_{p^\infty}}$, $\widehat{R_\infty}$, and $\widehat{\overline{R}}$ are $\mathscr{O}_C$-flat and the canonical homomorphisms $\widehat{R_1} \to \widehat{R_{p^\infty}}$, $\widehat{R_{p^\infty}} \to \widehat{R_\infty}$, and $\widehat{R_\infty} \to \widehat{\overline{R}}$ are injective.*

Since $R_1$, $R_{p^\infty}$, and $R_\infty$ are normal by II.6.8(ii), we have, for every $n \geq 0$,

$$p^n R_1 = (p^n R_{p^\infty}) \cap R_1, \quad p^n R_{p^\infty} = (p^n R_\infty) \cap R_{p^\infty}, \quad \text{and} \quad p^n R_\infty = (p^n \overline{R}) \cap R_\infty.$$

It follows that the canonical homomorphisms $\widehat{R_1} \to \widehat{R_{p^\infty}}$, $\widehat{R_{p^\infty}} \to \widehat{R_\infty}$, and $\widehat{R_\infty} \to \widehat{\overline{R}}$ are injective. On the other hand, by ([**11**] Chap. III § 2.11 Prop. 14 and Cor. 1), for every $n \geq 0$, we have

$$\widehat{R_1}/p^n\widehat{R_1} \simeq R_1/p^n R_1. \tag{II.6.14.1}$$

Let $x \in \widehat{R_1}$ be an element satisfying $px = 0$ and $\overline{x}$ the class of $x$ in $\widehat{R_1}/p^n\widehat{R_1}$ ($n \geq 1$). Since $R_1$ is $\mathscr{O}_{\overline{K}}$-flat, it follows from (II.6.14.1) that $\overline{x} \in p^{n-1}\widehat{R_1}/p^n\widehat{R_1}$. Consequently, $x \in \cap_{n\geq 0} p^n\widehat{R_1} = \{0\}$. Therefore $p$ is not a zero divisor in $\widehat{R_1}$ and hence $\widehat{R_1}$ is $\mathscr{O}_C$-flat ([**11**] Chap. VI § 3.6 Lem. 1). The same argument shows that $\widehat{R_\infty}$, $\widehat{R_{p^\infty}}$, and $\widehat{\overline{R}}$ are flat over $\mathscr{O}_C$.

**Proposition II.6.15.** *The ring $\widehat{R_1}$ is normal.*

First note that $\widehat{R_1}$ is an $\mathscr{O}_C$-algebra that is topologically of finite presentation ([**1**] 1.10.4), and then that $\widehat{R_1}[\frac{1}{p}]$ is an affinoid algebra over $C$.

Let us show that $\widehat{R_1}[\frac{1}{p}]$ is normal. We identify $\widehat{R_1}$ with the $p$-adic Hausdorff completion of $B = R_1 \otimes_{\mathscr{O}_{\overline{K}}} \mathscr{O}_C$ and let $\varphi\colon B \to \widehat{R_1}$ be the canonical homomorphism. Let $\mathfrak{q}$ be a maximal ideal of $\widehat{R_1}[\frac{1}{p}]$ and $\mathfrak{p} = \varphi^{-1}(\mathfrak{q})$. By virtue of ([**1**] 1.12.18), the canonical homomorphism $B_\mathfrak{p} \to (\widehat{R_1})_\mathfrak{q}$ induces an isomorphism between the Hausdorff completions of these local rings for the topologies defined by the respective maximal ideals. The scheme $\mathrm{Spec}(B[\frac{1}{p}])$ endowed with the inverse image of the logarithmic structure $\mathscr{M}_X$ is smooth over $\mathrm{Spec}(C)$ endowed with the trivial logarithmic structure. Consequently, $B[\frac{1}{p}]$ is normal by virtue of ([**51**] 4.1 and 8.2). Since $B[\frac{1}{p}]$ is an excellent ring, it follows from the above that the Hausdorff completions of $\widehat{R_1}[\frac{1}{p}]$ for each of its maximal ideals are normal ([**42**] 7.8.3(v)). Since $\widehat{R_1}[\frac{1}{p}]$ is an excellent ring ([**6**] 3.3.3), its localizations at each of its maximal ideals are normal ([**42**] 7.8.3(v)). Consequently, $\widehat{R_1}[\frac{1}{p}]$ is normal ([**42**] 7.8.3(iv)).

For every $h \in \widehat{R_1}[\frac{1}{p}]$, set

$$|h|_{\sup} = \sup_{x \in \mathrm{Max}(\widehat{R_1}[\frac{1}{p}])} |h(x)|, \tag{II.6.15.1}$$

where $\mathrm{Max}(\widehat{R_1}[\frac{1}{p}])$ is the maximal spectrum of $\widehat{R_1}[\frac{1}{p}]$ or, equivalently, the set of rigid points of $\mathrm{Spf}(\widehat{R_1})$ ([**1**] 3.3.2). It is a multiplicative seminorm on $\widehat{R_1}[\frac{1}{p}]$ ([**7**] 6.2.1/1). Let

$$B_{\sup} = \{h \in \widehat{R_1}[\frac{1}{p}] \mid |h|_{\sup} \leq 1\}. \tag{II.6.15.2}$$

Since $\widehat{R_1}$ is $\mathscr{O}_C$-flat (II.6.14) and $\widehat{R_1}[\frac{1}{p}]$ is reduced, we have $\widehat{R_1} \subset B_{\sup}$ and $B_{\sup}$ is the integral closure of $\widehat{R_1}$ in $\widehat{R_1}[\frac{1}{p}]$ ([**7**] 6.3.4/1 and 6.2.2/3). Since $R_1 \otimes_{\mathscr{O}_{\overline{K}}} \overline{k}$ is reduced by II.6.3(iv) and II.6.8(i), $\widehat{R_1} = B_{\sup}$ by ([**7**] 6.4.3/4; cf. also [**8**] 1.1). Hence $\widehat{R_1}$ is integrally closed in $\widehat{R_1}[\frac{1}{p}]$ and therefore normal.

**Theorem II.6.16** (Faltings' almost purity theorem, [**26**] § 2b)**.** *For every finite extension $N$ of $H_\infty$ contained in $\overline{F}$, the integral closure of $B_\infty$ in $N$ is almost étale over $B_\infty$* (II.6.7.4).

Let us note here that Scholze, in the setting of his theory of perfectoids, gives a generalization of this result ([**64**] 1.10 and 7.9).

**Corollary II.6.17.** *The extension $\overline{F}$ of $F_\infty$ is the union of a filtered direct system of finite Galois subextensions $E$ of $F_\infty$ such that the integral closure of $R_\infty$ in $E$ is almost étale over $R_\infty$* (II.6.7.7).

Let $N$ be a finite extension of $H_\infty$ contained in $\overline{F}$, $E$ the image of the canonical homomorphism $N \otimes_{H_\infty} F_\infty \to \overline{F}$, and $\mathscr{N}$ (resp. $\mathscr{E}$) the integral closure of $R$ in $N$ (resp. $E$). We know (II.6.16) that $\mathscr{N}$ is almost étale over $B_\infty$. Hence $\mathscr{N} \otimes_{B_\infty} R_\infty$ is almost étale over $R_\infty$ by V.7.4. Consequently, $\mathscr{E}$ is almost étale over $R_\infty$ by V.7.11 and V.7.4. If $N/H_\infty$ is a Galois extension, the same holds for $E/F_\infty$, giving the corollary.

**Corollary II.6.18.** *For every subring $A$ of $R_\infty$, the canonical morphism*

(II.6.18.1) $$\Omega^1_{R_\infty/A} \otimes_{R_\infty} \overline{R} \to \Omega^1_{\overline{R}/A}$$

*is an almost isomorphism.*

This follows from II.6.17 and ([**24**] I 2.4(i)).

**Corollary II.6.19.** *Let $M$ be an $\overline{R}$-module endowed with a continuous $\overline{R}$-semi-linear action of $\Sigma$ for the discrete topology on $M$. Then $\mathrm{H}^i(\Sigma, M)$ is almost zero for all $i \geq 1$, and the canonical morphism $M^\Sigma \otimes_{R_\infty} \overline{R} \to M$ is an almost isomorphism.*

Let $N$ be a finite Galois extension of $F_\infty$ contained in $\overline{F}$, $D$ the integral closure of $R_\infty$ in $N$, $G = \mathrm{Gal}(N/F_\infty)$, and $\Sigma_N = \mathrm{Gal}(\overline{F}/N)$. Suppose that $D$ is almost étale over $R_\infty$. Then $D$ is an almost $G$-torsor over $R_\infty$ by V.12.9. Using V.12.5 and V.12.8, we deduce that for every $i \geq 1$, $\mathrm{H}^i(G, M^{\Sigma_N})$ is almost zero, and the canonical morphism $M^\Sigma \otimes_{R_\infty} D \to M^{\Sigma_N}$ is an almost isomorphism. The corollary follows by taking the direct limit, by virtue of II.6.17.

**Corollary II.6.20.** *Let $M$ be an $\overline{R}$-module endowed with a continuous $\overline{R}$-semi-linear action of $\Delta$ for the discrete topology on $M$. Then the canonical morphism*

$$\mathrm{H}^i(\Delta_\infty, M^\Sigma) \to \mathrm{H}^i(\Delta, M)$$

*is an almost isomorphism for every $i \geq 0$.*

This follows from II.6.19 and from the spectral sequence

(II.6.20.1) $$E_1^{ij} = \mathrm{H}^i(\Delta_\infty, \mathrm{H}^j(\Sigma, M)) \Rightarrow \mathrm{H}^{i+j}(\Delta, M).$$

**Corollary II.6.21.** *Let $(M_n)_{n\in\mathbb{N}}$ be an inverse system of $\widehat{\overline{R}}$-representations of $\Sigma$* (II.3.1) *and $M$ its inverse limit. We suppose that for every $n \geq 0$, $M_n$ is annihilated by a power of $p$, that the action of $\Sigma$ on $M_n$ is continuous for the discrete topology, and that the morphism $M_{n+1} \to M_n$ is surjective. Then $\mathrm{H}^i_{\mathrm{cont}}(\Sigma, M)$ is almost zero for every integer $i \geq 1$.*

Indeed, by (II.3.10.4) and (II.3.10.5), we have

(II.6.21.1) $$0 \to \mathrm{R}^1\varprojlim_n \mathrm{H}^{i-1}(\Sigma, M_n) \to \mathrm{H}^i_{\mathrm{cont}}(\Sigma, M) \to \varprojlim_n \mathrm{H}^i(\Sigma, M_n) \to 0.$$

For every $q \geq 1$, $\varprojlim_n \mathrm{H}^q(\Sigma, M_n)$ and $\mathrm{R}^1\varprojlim_n \mathrm{H}^q(\Sigma, M_n)$ are almost zero by virtue of II.6.19 and ([**32**] 2.4.2(ii)). For every $n \geq 0$, let $C_n$ be the kernel of the surjective morphism $M_{n+1} \to M_n$, so that we have an exact sequence

(II.6.21.2) $$M^\Sigma_{n+1} \xrightarrow{\psi_n} M^\Sigma_n \longrightarrow \mathrm{H}^1(\Sigma, C_n).$$

Then $\operatorname{coker}(\psi_n)$ is almost zero by virtue of II.6.19. It follows that $\mathrm{R}^1\lim\limits_{\substack{\longleftarrow \\ n}} M_n^{\Sigma}$ is almost zero by ([**32**] 2.4.2(iii) and 2.4.3), giving the corollary.

**Corollary II.6.22.** *For every $a \in \mathscr{O}_{\overline{K}}$, the canonical homomorphism*

$$R_\infty/aR_\infty \to (\overline{R}/a\overline{R})^\Sigma \tag{II.6.22.1}$$

*is an almost isomorphism.*

Let $N$ be a finite Galois extension of $F_\infty$ contained in $\overline{F}$, $D$ the integral closure of $R_\infty$ in $N$, $G = \operatorname{Gal}(N/F_\infty)$, and $\operatorname{Tr}_G$ the $R_\infty$-linear endomorphism of $D$ (or of $D/aD$) induced by $\sum_{\sigma\in G}\sigma$. Since we have $D = \overline{R}\cap N$ and $R_\infty = D\cap F_\infty$, by II.6.8(ii), the homomorphisms $R_\infty/aR_\infty \to D/aD \to \overline{R}/a\overline{R}$ are injective. Suppose that $D$ is almost étale over $R_\infty$. Then $D$ is an almost $G$-torsor over $R_\infty$ by virtue of V.12.9. Consequently, the quotient

$$\frac{(D/aD)^G}{\operatorname{Tr}_G(D/aD)}$$

is almost zero by V.12.8. Since $\operatorname{Tr}_G(D) \subset R_\infty$, the homomorphism $R_\infty/aR_\infty \to (D/aD)^G$ is an almost isomorphism. The corollary follows by taking the direct limit, by II.6.17.

**Corollary II.6.23.** *The canonical homomorphism $\widehat{R_\infty} \to \widehat{\overline{R}}^\Sigma$ is an almost isomorphism.*

Indeed, this homomorphism is the inverse limit of the homomorphisms ($r \geq 0$)

$$R_\infty/p^rR_\infty \to (\overline{R}/p^r\overline{R})^\Sigma. \tag{II.6.23.1}$$

This can be easily verified or follows from (II.3.10.5). The corollary therefore follows from II.6.22 and ([**32**] 2.4.2(ii)).

**Corollary II.6.24.** *For every nonzero element $a$ of $\mathscr{O}_{\overline{K}}$ and every integer $i \geq 0$, the canonical morphisms*

$$\mathrm{H}^i(\Delta_{p^\infty}, R_{p^\infty}/aR_{p^\infty}) \to \mathrm{H}^i(\Delta, \overline{R}/a\overline{R}), \tag{II.6.24.1}$$

$$\mathrm{H}^i(\Delta_\infty, R_\infty/aR_\infty) \to \mathrm{H}^i(\Delta, \overline{R}/a\overline{R}), \tag{II.6.24.2}$$

*are almost isomorphisms.*

Indeed, (II.6.24.2) is an almost isomorphism by II.6.20 and II.6.22. On the other hand, the canonical morphism

$$\mathrm{H}^i(\Delta_{p^\infty}, R_{p^\infty}/aR_{p^\infty}) \to \mathrm{H}^i(\Delta_\infty, R_\infty/aR_\infty) \tag{II.6.24.3}$$

is an isomorphism by II.6.13 and (II.6.11.1).

**Corollary II.6.25.** *For every integer $i \geq 0$, the canonical morphisms*

$$\mathrm{H}^i_{\mathrm{cont}}(\Delta_{p^\infty}, \widehat{R_{p^\infty}}) \to \mathrm{H}^i_{\mathrm{cont}}(\Delta, \widehat{\overline{R}}), \tag{II.6.25.1}$$

$$\mathrm{H}^i_{\mathrm{cont}}(\Delta_\infty, \widehat{R_\infty}) \to \mathrm{H}^i_{\mathrm{cont}}(\Delta, \widehat{\overline{R}}), \tag{II.6.25.2}$$

*are almost isomorphisms.*

For every integer $r \geq 0$, let

$$\psi_r\colon \mathrm{H}^i(\Delta_\infty, R_\infty/p^rR_\infty) \to \mathrm{H}^i(\Delta, \overline{R}/p^r\overline{R}) \tag{II.6.25.3}$$

be the canonical homomorphism and $A_r$ (resp. $C_r$) its kernel (resp. cokernel). We know that the $\mathscr{O}_{\overline{K}}$-modules $A_r$ and $C_r$ are almost zero by II.6.24. Hence the $\mathscr{O}_{\overline{K}}$-modules

$$\lim_{\substack{\longleftarrow \\ r\geq 0}} A_r, \quad \lim_{\substack{\longleftarrow \\ r\geq 0}} C_r, \quad \mathrm{R}^1\lim_{\substack{\longleftarrow \\ r\geq 0}} A_r, \quad \mathrm{R}^1\lim_{\substack{\longleftarrow \\ r\geq 0}} C_r$$

are almost zero by ([**32**] 2.4.2(ii)). It follows that the morphisms

$$\lim_{\substack{\longleftarrow \\ r\geq 0}} \psi_r \quad \text{and} \quad \mathrm{R}^1\lim_{\substack{\longleftarrow \\ r\geq 0}} \psi_r$$

are almost isomorphisms, and the same holds for (II.6.25.2) in view of (II.3.10.4) and (II.3.10.5). The proof that (II.6.25.1) is an almost isomorphism is similar.

## II.7. Faltings extension

**II.7.1.** Let $K_0$ be the field of fractions of $\mathrm{W}(k)$ (II.2.3) and $\mathscr{D}_{K/K_0}$ the different of the extension $K/K_0$. By ([**30**] Thm. 1'), there exists a unique $G_K$-equivariant $\mathscr{O}_{\overline{K}}$-linear morphism

(II.7.1.1)
$$\phi\colon \overline{K}\otimes_{\mathbb{Z}_p}\mathbb{Z}_p(1)\to \Omega^1_{\mathscr{O}_{\overline{K}}/\mathscr{O}_K},$$

such that for all $\zeta\in\mathbb{Z}_p(1)$, $a\in\mathscr{O}_{\overline{K}}$, and $r\in\mathbb{N}$, if $\zeta_r\in\mu_{p^r}(\mathscr{O}_{\overline{K}})$ is the canonical image of $\zeta$, then

(II.7.1.2)
$$\phi(p^{-r}a\otimes\zeta)=a\cdot d\log(\zeta_r).$$

It is surjective with kernel $\rho^{-1}\mathscr{O}_{\overline{K}}(1)$, where $\rho$ is an element of $\mathscr{O}_{\overline{K}}$ with valuation $\frac{1}{p-1}+v(\mathscr{D}_{K/K_0})$.

**II.7.2.** The morphism (II.6.7.5) induces an $\mathscr{O}_K$-homomorphism

(II.7.2.1)
$$\lim_{\substack{\longrightarrow \\ n\geq 1}} \mathscr{O}_{K_n}\to\mathscr{O}_{\overline{K}},$$

where the direct limit is indexed by the set $\mathbb{Z}_{\geq 1}$ ordered by the divisibility relation. For every $n\geq 1$, we will from now on identify $\mathscr{O}_{K_n}$ with a sub-$\mathscr{O}_K$-algebra of $\mathscr{O}_{\overline{K}}$; in particular, we view $\pi_n$ as an element of $\mathscr{O}_{\overline{K}}$ (II.6.4). We have $\pi_1=\pi$ and $\pi_{mn}^m=\pi_n$ for all $m,n\geq 1$.

For every integer $n\geq 1$, there exists an element of $\Omega^1_{\mathscr{O}_{\overline{K}}/\mathscr{O}_K}$, which we denote by $d\log(\pi_n)$, such that for every integer $m\geq 1$ divisible by $p$, we have

(II.7.2.2)
$$d\log(\pi_n)=\frac{m}{\pi_{mn}}d\pi_{mn}\in\Omega^1_{\mathscr{O}_{\overline{K}}/\mathscr{O}_K}.$$

Indeed, $m\in\pi_{mn}\mathscr{O}_{K_{mn}}$, so that $m/\pi_{mn}\in\mathscr{O}_{K_{mn}}$, and for every $m'\geq 1$, we have

(II.7.2.3)
$$\frac{m}{\pi_{mn}}d\pi_{mn}=\frac{m}{\pi_{mn}}m'\pi_{mm'n}^{m'-1}d\pi_{mm'n}=\frac{mm'}{\pi_{mm'n}}d\pi_{mm'n}\in\Omega^1_{\mathscr{O}_{\overline{K}}/\mathscr{O}_K}.$$

Remember that the element $d\log(\pi_n)$ depends not only on $\pi_n$, but also on the homomorphism (II.7.2.1).

For all integers $m,n\geq 1$, we have

(II.7.2.4)
$$\pi_n d\log(\pi_n) = d\pi_n,$$

(II.7.2.5)
$$d\log(\pi_n) = m d\log(\pi_{mn}).$$

Since the canonical morphisms $\Omega^1_{\mathscr{O}_{K_n}/\mathscr{O}_K}\to\Omega^1_{\mathscr{O}_{\overline{K}}/\mathscr{O}_K}$ are injective ([**30**] 2.4 Lem. 4), for every integer $n\geq 1$, the annihilator of $d\log(\pi_n)$ in $\Omega^1_{\mathscr{O}_{\overline{K}}/\mathscr{O}_K}$ is $n\pi\mathscr{O}_{\overline{K}}$ ([**30**] 2.1 Lem. 1).

**II.7.3.** Consider the direct system of monoids $(\mathbb{N}^{(n)})_{n\geq 1}$, indexed by the set $\mathbb{Z}_{\geq 1}$ ordered by the divisibility relation, defined by $\mathbb{N}^{(n)} = \mathbb{N}$ for every $n \geq 1$ and whose transition homomorphism $\lambda_{n,mn}\colon \mathbb{N}^{(n)} \to \mathbb{N}^{(mn)}$ (for $m,n \geq 1$) is the Frobenius homomorphism $\varpi_m$ of order $m$ of $\mathbb{N}$ (II.5.12). We denote by $\mathbb{N}_\infty$ its direct limit,

$$\mathbb{N}_\infty = \varinjlim_{n\geq 1} \mathbb{N}^{(n)}. \tag{II.7.3.1}$$

For every $n \geq 1$, we denote by $a_n\colon \mathbb{N}^{(n)} \to \mathscr{O}_{\overline{K}}$ the homomorphism defined by $a_n(1) = \pi_n$. For all integers $m, n \geq 1$, the diagram

$$\begin{array}{ccc} \mathbb{N}^{(n)} & \xrightarrow{a_n} & \mathscr{O}_{\overline{K}} \\ {\scriptstyle \lambda_{n,mn}}\downarrow & & \| \\ \mathbb{N}^{(mn)} & \xrightarrow{a_{mn}} & \mathscr{O}_{\overline{K}} \end{array} \tag{II.7.3.2}$$

is commutative. By taking the direct limit, the $a_n$'s therefore define a homomorphism

$$a_\infty\colon \mathbb{N}_\infty \to \mathscr{O}_{\overline{K}}. \tag{II.7.3.3}$$

We will denote $\mathbb{N}^{(1)}$ (resp. $a_1$) simply by $\mathbb{N}$ (resp. $a$).

**Lemma II.7.4.** *Let $n$ be an integer $\geq 1$. Then:*

(i) *We have canonical isomorphisms*

$$\Omega^1_{(\mathscr{O}_{\overline{K}},\mathbb{N}^{(n)})/(\mathscr{O}_K,\mathbb{N})} \simeq \frac{\Omega^1_{\mathscr{O}_{\overline{K}}/\mathscr{O}_K} \oplus \mathscr{O}_{\overline{K}}/n\mathscr{O}_{\overline{K}}}{(d\pi_n - \pi_n)\mathscr{O}_{\overline{K}}}, \tag{II.7.4.1}$$

$$\Omega^1_{(\mathscr{O}_{\overline{K}},\mathbb{N}^{(n)})/(\mathscr{O}_{\overline{K}},\mathbb{N})} \simeq \mathscr{O}_{\overline{K}}/(n\mathscr{O}_{\overline{K}} + \pi_n\mathscr{O}_{\overline{K}}). \tag{II.7.4.2}$$

(ii) *If $p$ divides $n$, the kernel of the canonical morphism*

$$\Omega^1_{\mathscr{O}_{\overline{K}}/\mathscr{O}_K} \to \Omega^1_{(\mathscr{O}_{\overline{K}},\mathbb{N}^{(n)})/(\mathscr{O}_K,\mathbb{N})} \tag{II.7.4.3}$$

*is generated by $d\log(\pi)$ (II.7.2.2).*

(i) We have a commutative diagram with Cartesian square

$$\begin{array}{ccccc} \mathrm{Spec}(\mathscr{O}_{\overline{K}}) & \xrightarrow{j_n} & \mathrm{Spec}(\mathscr{O}_{\overline{K}}[\xi]/(\xi^n - \pi)) & \longrightarrow & \mathrm{Spec}(\mathscr{O}_{\overline{K}}[\mathbb{N}^{(n)}]) \\ & \searrow & \downarrow & \square & \downarrow \\ & & \mathrm{Spec}(\mathscr{O}_K) & \longrightarrow & \mathrm{Spec}(\mathscr{O}_K[\mathbb{N}]) \end{array} \tag{II.7.4.4}$$

where $j_n$ is the closed immersion defined by the equation $\xi - \pi_n$. On the other hand, we have a canonical isomorphism

$$\Omega^1_{(\mathscr{O}_{\overline{K}}[\mathbb{N}^{(n)}],\mathbb{N}^{(n)})/(\mathscr{O}_K[\mathbb{N}],\mathbb{N})} \simeq \Omega^1_{\mathscr{O}_{\overline{K}}/\mathscr{O}_K} \otimes_{\mathscr{O}_{\overline{K}}} \mathscr{O}_{\overline{K}}[\mathbb{N}^{(n)}] \oplus \mathscr{O}_{\overline{K}}[\mathbb{N}^{(n)}]/n\mathscr{O}_{\overline{K}}[\mathbb{N}^{(n)}]. \tag{II.7.4.5}$$

The isomorphism (II.7.4.1) follows directly from this. The isomorphism (II.7.4.2) is proved using a diagram analogous to (II.7.4.4).

(ii) Let $\omega \in \Omega^1_{\mathscr{O}_{\overline{K}}/\mathscr{O}_K}$ be such that its image in $\Omega^1_{(\mathscr{O}_{\overline{K}},\mathbb{N}^{(n)})/(\mathscr{O}_K,\mathbb{N})}$ is zero. By (II.7.4.1), there exists $x \in \mathscr{O}_{\overline{K}}$ such that $\omega = x(d\pi_n - \pi_n) \in \Omega^1_{\mathscr{O}_{\overline{K}}/\mathscr{O}_K} \oplus \mathscr{O}_{\overline{K}}/n\mathscr{O}_K$. Consequently, $x\pi_n \in n\mathscr{O}_{\overline{K}}$ and $\omega = xd\pi_n \in \mathscr{O}_{\overline{K}}d\log(\pi)$. Conversely, we have

$$d\log(\pi) = (n/\pi_n)(d\pi_n - \pi_n) \in \Omega^1_{\mathscr{O}_{\overline{K}}/\mathscr{O}_K} \oplus \mathscr{O}_{\overline{K}}/n\mathscr{O}_K. \tag{II.7.4.6}$$

Hence the image of $d\log(\pi)$ by the morphism (II.7.4.3) is zero by virtue of (II.7.4.1).

**Proposition II.7.5.** *The canonical morphism*

(II.7.5.1) $$\Omega^1_{\mathscr{O}_{\overline{K}}/\mathscr{O}_K} \to \Omega^1_{(\mathscr{O}_{\overline{K}},\mathbb{N}_\infty)/(\mathscr{O}_K,\mathbb{N})}$$

*is surjective and its kernel is generated by* $d\log(\pi)$. *In particular, the morphism* $\phi$ (II.7.1.1) *induces a surjective* $\mathscr{O}_{\overline{K}}$*-linear morphism*

(II.7.5.2) $$\overline{K} \otimes_{\mathbb{Z}_p} \mathbb{Z}_p(1) \to \Omega^1_{(\mathscr{O}_{\overline{K}},\mathbb{N}_\infty)/(\mathscr{O}_K,\mathbb{N})}$$

*with kernel* $(\pi\rho)^{-1}\mathscr{O}_{\overline{K}}(1)$.

Note that the second statement is an immediate consequence of the first and II.7.1. Let us prove the first statement. By the universal property of modules of logarithmic 1-differentials, the canonical morphism

(II.7.5.3) $$\varinjlim_{n\geq 1} \Omega^1_{(\mathscr{O}_{\overline{K}},\mathbb{N}^{(n)})/(\mathscr{O}_K,\mathbb{N})} \to \Omega^1_{(\mathscr{O}_{\overline{K}},\mathbb{N}_\infty)/(\mathscr{O}_K,\mathbb{N})},$$

where the direct limit is indexed by the set $\mathbb{Z}_{\geq 1}$ ordered by the divisibility relation, is an isomorphism. It therefore follows from II.7.4(ii) that the kernel of the morphism (II.7.5.1) is generated by $d\log(\pi)$. On the other hand, we have a canonical isomorphism

(II.7.5.4) $$\varinjlim_{n\geq 1} \Omega^1_{(\mathscr{O}_{\overline{K}},\mathbb{N}^{(n)})/(\mathscr{O}_{\overline{K}},\mathbb{N})} \xrightarrow{\sim} \Omega^1_{(\mathscr{O}_{\overline{K}},\mathbb{N}_\infty)/(\mathscr{O}_{\overline{K}},\mathbb{N})}.$$

By (II.7.4.2), for every integer $n \geq 1$, we have

(II.7.5.5) $$\Omega^1_{(\mathscr{O}_{\overline{K}},\mathbb{N}^{(n)})/(\mathscr{O}_{\overline{K}},\mathbb{N})} = \begin{cases} 0 & \text{if } (n,p)=1, \\ k_n \otimes_{\mathscr{O}_{K_n}} \mathscr{O}_{\overline{K}} & \text{if } p|n, \end{cases}$$

where $k_n$ is the residue field of $\mathscr{O}_{K_n}$. Observe that for all integers $m, n \geq 1$, the canonical morphism

(II.7.5.6) $$\Omega^1_{(\mathscr{O}_{\overline{K}},\mathbb{N}^{(n)})/(\mathscr{O}_{\overline{K}},\mathbb{N})} \to \Omega^1_{(\mathscr{O}_{\overline{K}},\mathbb{N}^{(nm)})/(\mathscr{O}_{\overline{K}},\mathbb{N})}$$

identifies with $m$ times the canonical morphism $k_n \otimes_{\mathscr{O}_{K_n}} \mathscr{O}_{\overline{K}} \to k_{mn} \otimes_{\mathscr{O}_{K_{mn}}} \mathscr{O}_{\overline{K}}$. It follows that $\Omega^1_{(\mathscr{O}_{\overline{K}},\mathbb{N}_\infty)/(\mathscr{O}_{\overline{K}},\mathbb{N})} = 0$ and consequently that the morphism (II.7.5.1) is surjective.

**Remark II.7.6.** For any integer $n \geq 1$, the canonical image of the element $d\log(\pi_n) \in \Omega^1_{\mathscr{O}_{\overline{K}}/\mathscr{O}_K}$ (II.7.2.2) in $\Omega^1_{(\mathscr{O}_{\overline{K}},\mathbb{N}_\infty)/(\mathscr{O}_K,\mathbb{N})}$ equals the element $d\log(1^{(n)}) \in \Omega^1_{(\mathscr{O}_{\overline{K}},\mathbb{N}^{(n)})/(\mathscr{O}_K,\mathbb{N})}$, where $1^{(n)}$ denotes the image of 1 by the canonical isomorphism $\mathbb{N} \xrightarrow{\sim} \mathbb{N}^{(n)}$. Indeed, in $\Omega^1_{(\mathscr{O}_{\overline{K}},\mathbb{N}^{(pn)})/(\mathscr{O}_K,\mathbb{N})}$, we have

(II.7.6.1) $$d\log(\pi_n) - d\log(1^{(n)}) = \frac{p}{\pi_{pn}}(d\pi_{pn} - \pi_{pn}d\log(1^{(pn)})) = 0.$$

Note that the equality $d\log(\pi_n) = d\log(1^{(n)})$ holds in $\Omega^1_{(\mathscr{O}_{\overline{K}},\mathbb{N}^{(mn)})/(\mathscr{O}_K,\mathbb{N})}$ for every integer $m \geq 1$ divisible by $p$ (but in general it does not hold when $m = 1$).

**Lemma II.7.7.** *For every flat* $\mathscr{O}_{\overline{K}}$*-algebra* $A$*, the* $\mathscr{O}_{\overline{K}}$*-modules*

(II.7.7.1) $$\Omega^1_{\mathscr{O}_{\overline{K}}/\mathscr{O}_K} \otimes_{\mathscr{O}_{\overline{K}}} A \quad \text{and} \quad \Omega^1_{(\mathscr{O}_{\overline{K}},\mathbb{N}_\infty)/(\mathscr{O}_K,\mathbb{N})} \otimes_{\mathscr{O}_{\overline{K}}} A$$

*do not have any nonzero* $\mathfrak{m}_{\overline{K}}$*-torsion.*

In view of (II.7.1.1) and (II.7.5.2), it suffices to show that the $\mathscr{O}_{\overline{K}}$-module $A[\frac{1}{p}]/A$ does not have any nonzero $\mathfrak{m}_{\overline{K}}$-torsion. Since the $\mathfrak{m}_{\overline{K}}$-torsion is contained in the $p$-torsion, it therefore suffices to show that the $\mathscr{O}_{\overline{K}}$-module $A/pA$ does not have any nonzero $\mathfrak{m}_{\overline{K}}$-torsion. A computation of valuations show that $\mathscr{O}_{\overline{K}}/p\mathscr{O}_{\overline{K}}$ does not have any nonzero $\mathfrak{m}_{\overline{K}}$-torsion; in other words, the morphism

(II.7.7.2) $$\mathscr{O}_{\overline{K}}/p\mathscr{O}_{\overline{K}} \to \oplus_{n\geq 1}\mathscr{O}_{\overline{K}}/p\mathscr{O}_{\overline{K}}, \quad x \mapsto (p^{1/n}x)_{n\geq 1}$$

is injective. The same then holds for the morphism obtained by extending the scalars from $\mathscr{O}_{\overline{K}}$ to $A$, giving the statement.

**II.7.8.** Consider the direct system of monoids $(P^{(n)})_{n\geq 1}$, indexed by the set $\mathbb{Z}_{\geq 1}$ ordered by the divisibility relation, defined by $P^{(n)} = P$ for every $n \geq 1$ and with transition homomorphism $i_{n,mn}\colon P^{(n)} \to P^{(mn)}$ (for $m, n \geq 1$) given by the Frobenius homomorphism $\varpi_m$ of order $m$ of $P$ (II.5.12). We denote by $P_\infty$ its direct limit,

(II.7.8.1) $$P_\infty = \lim_{\substack{\longrightarrow \\ n\geq 1}} P^{(n)}.$$

For every $n \geq 1$, we denote by

(II.7.8.2) $$P \xrightarrow{\sim} P^{(n)}, \quad t \mapsto t^{(n)},$$

the canonical isomorphism. For all $m, n \geq 1$ and $t \in P$, we have $i_{n,mn}(t^{(n)}) = (t^{(mn)})^m$ and consequently

(II.7.8.3) $$t^{(n)} = (t^{(mn)})^m \in P_\infty.$$

For every $n \geq 1$, we denote by $\alpha_n\colon P^{(n)} \to R_n$ the homomorphism induced by the canonical strict morphism $(X_n, \mathscr{M}_{X_n}) \to \mathbf{A}_P$ (II.6.5.1). The reader will remember that $R_n$ depends on the choice of the morphism (II.6.7.1). For all integers $m, n \geq 1$, the diagram

(II.7.8.4) $$\begin{array}{ccc} P^{(n)} & \xrightarrow{\alpha_n} & R_n \\ {\scriptstyle i_{n,mn}}\downarrow & & \downarrow \\ P^{(mn)} & \xrightarrow{\alpha_{mn}} & R_{mn} \end{array}$$

is commutative. By taking the direct limit, the $\alpha_n$'s therefore define a homomorphism

(II.7.8.5) $$\alpha_\infty\colon P_\infty \to R_\infty.$$

We also denote by $\alpha_\infty\colon P_\infty \to \overline{R}$ the composition of $\alpha_\infty$ and the canonical injection $R_\infty \to \overline{R}$. We will simply write $P$ for $P^{(1)}$. Note that $\alpha_1$ factors through $\alpha$ (II.6.2).

**Proposition II.7.9.** (i) *The canonical sequence*

(II.7.9.1) $$0 \to \Omega^1_{(R,P)/(\mathscr{O}_K,\mathbb{N})} \otimes_R R_\infty \to \Omega^1_{(R_\infty,P_\infty)/(\mathscr{O}_{\overline{K}},\mathbb{N})} \to \Omega^1_{(R_\infty,P_\infty)/(R_1,P)} \to 0$$

*is exact.*

(ii) *For every integer $m \geq 0$, the morphism*

(II.7.9.2) $$\mathrm{Hom}(p^{-m}\mathbb{Z}/\mathbb{Z}, \Omega^1_{(R_\infty,P_\infty)/(R_1,P)}) \to \Omega^1_{(R,P)/(\mathscr{O}_K,\mathbb{N})} \otimes_R (R_\infty/p^m R_\infty)$$

*deduced from* (II.7.9.1) *using the snake lemma is an isomorphism.*

(iii) *There exists a canonical $R_\infty$-linear isomorphism*

(II.7.9.3) $$(P^{\mathrm{gp}}/\lambda\mathbb{Z}) \otimes_{\mathbb{Z}} (R_\infty[\frac{1}{p}]/R_\infty) \xrightarrow{\sim} \Omega^1_{(R_\infty,P_\infty)/(R_1,P)}.$$

(i) By II.6.8(i), we have a canonical isomorphism

(II.7.9.4) $$\Omega^1_{(R_1,P)/(\mathscr{O}_{\overline{K}},\mathbb{N})} \xrightarrow{\sim} \Omega^1_{(R,P)/(\mathscr{O}_K,\mathbb{N})} \otimes_R R_1.$$

On the other hand, we have a canonical isomorphism

(II.7.9.5) $$\lim_{\substack{\longrightarrow \\ n\geq 1}} \Omega^1_{(R_n,P^{(n)})/(\mathscr{O}_{\overline{K}},\mathbb{N})} \xrightarrow{\sim} \Omega^1_{(R_\infty,P_\infty)/(\mathscr{O}_{\overline{K}},\mathbb{N})},$$

where the direct limit is indexed by the set $\mathbb{Z}_{\geq 1}$ ordered by the divisibility relation. It therefore suffices to show that for every $n \geq 1$, the canonical morphism

$$\Omega^1_{(R_1,P)/(\mathscr{O}_{\overline{K}},\mathbb{N})} \otimes_{R_1} R_n \to \Omega^1_{(R_n,P^{(n)})/(\mathscr{O}_{\overline{K}},\mathbb{N})} \tag{II.7.9.6}$$

is injective. Recall that $\mathrm{Spec}(R_n)$ is an open connected component of $X_n \otimes_{\mathscr{O}_{K_n}} \mathscr{O}_{\overline{K}}$ (II.6.8), that the canonical morphism $X \to \mathrm{Spec}(\mathscr{O}_K[P]/(\pi - e^{\lambda}))$ is étale (II.6.2.4), and that we have a Cartesian diagram of $\mathscr{O}_{\overline{K}}$-morphisms (II.6.6.4)

$$\begin{array}{ccc} X_n \otimes_{\mathscr{O}_{K_n}} \mathscr{O}_{\overline{K}} & \longrightarrow & \mathrm{Spec}(\mathscr{O}_{\overline{K}}[P^{(n)}]/(\pi_n - e^{\lambda^{(n)}})) \\ \downarrow & & \downarrow \\ X \otimes_{\mathscr{O}_K} \mathscr{O}_{\overline{K}} & \longrightarrow & \mathrm{Spec}(\mathscr{O}_{\overline{K}}[P]/(\pi - e^{\lambda})) \end{array} \tag{II.7.9.7}$$

where $\lambda^{(n)}$ is the image of $\lambda$ in $P^{(n)}$ by the isomorphism (II.7.8.2). Consider the commutative diagram with Cartesian squares

(II.7.9.8)

$$\begin{array}{ccccc} \mathrm{Spec}(\mathscr{O}_{\overline{K}}[P^{(n)}]/(\pi_n - e^{\lambda^{(n)}})) & \longrightarrow & \mathrm{Spec}(\mathscr{O}_{\overline{K}}[P^{(n)}]/(\pi - e^{n\lambda^{(n)}})) & \longrightarrow & \mathrm{Spec}(\mathscr{O}_{\overline{K}}[P^{(n)}]) \\ & \searrow & \downarrow & \square & \downarrow{\scriptstyle i_n^a} \\ & & \mathrm{Spec}(\mathscr{O}_{\overline{K}}[P]/(\pi - e^{\lambda})) & \longrightarrow & \mathrm{Spec}(\mathscr{O}_{\overline{K}}[P]) \\ & & \downarrow & \square & \downarrow \\ & & \mathrm{Spec}(\mathscr{O}_{\overline{K}}) & \longrightarrow & \mathrm{Spec}(\mathscr{O}_{\overline{K}}[\mathbb{N}]) \end{array}$$

where $i_n^a$ is the morphism induced by the canonical homomorphism $i_n \colon P \to P^{(n)}$. We have a canonical isomorphism

$$\Omega^1_{(\mathscr{O}_{\overline{K}}[P],P)/(\mathscr{O}_{\overline{K}}[\mathbb{N}],\mathbb{N})} \xrightarrow{\sim} (P^{\mathrm{gp}}/\lambda\mathbb{Z}) \otimes_{\mathbb{Z}} \mathscr{O}_{\overline{K}}[P] \tag{II.7.9.9}$$

such that for every $x \in P$, the image of $d\log(x)$ is the class of $x$ in $P^{\mathrm{gp}}/\lambda\mathbb{Z}$. We can therefore identify the morphism

$$\Omega^1_{(\mathscr{O}_{\overline{K}}[P],P)/(\mathscr{O}_{\overline{K}}[\mathbb{N}],\mathbb{N})} \otimes_{\mathscr{O}_{\overline{K}}[P]} \mathscr{O}_{\overline{K}}[P^{(n)}] \to \Omega^1_{(\mathscr{O}_{\overline{K}}[P^{(n)}],P^{(n)})/(\mathscr{O}_{\overline{K}}[\mathbb{N}],\mathbb{N})}$$

induced by $i_n^a$ with the $\mathscr{O}_{\overline{K}}[P^{(n)}]$-linear morphism

$$(P^{\mathrm{gp}}/\lambda\mathbb{Z}) \otimes_{\mathbb{Z}} \mathscr{O}_{\overline{K}}[P^{(n)}] \to (P^{\mathrm{gp}}/n\lambda\mathbb{Z}) \otimes_{\mathbb{Z}} \mathscr{O}_{\overline{K}}[P^{(n)}] \tag{II.7.9.10}$$

deduced from multiplication by $n$ in $P^{\mathrm{gp}}$.

It follows from the diagram (II.7.9.8) and what precedes it that we have a canonical isomorphism

$$\frac{(P^{\mathrm{gp}}/n\lambda\mathbb{Z}) \otimes_{\mathbb{Z}} R_n}{\overline{\lambda} \otimes \pi_n R_n} \xrightarrow{\sim} \Omega^1_{(R_n,P^{(n)})/(\mathscr{O}_{\overline{K}},\mathbb{N})}, \tag{II.7.9.11}$$

where $\overline{\lambda}$ denote the class of $\lambda$ in $P^{\mathrm{gp}}/n\lambda\mathbb{Z}$. We deduce from this a canonical surjective morphism

$$\Omega^1_{(R_n,P^{(n)})/(\mathscr{O}_{\overline{K}},\mathbb{N})} \to (P^{\mathrm{gp}}/\lambda\mathbb{Z}) \otimes_{\mathbb{Z}} R_n. \tag{II.7.9.12}$$

In view of (II.7.9.10), we identify the composition of the morphisms (II.7.9.6) and (II.7.9.12) with the multiplication by $n$ in $(P^{\mathrm{gp}}/\lambda\mathbb{Z}) \otimes_{\mathbb{Z}} R_n$, which is injective because the torsion subgroup of $P^{\mathrm{gp}}/\lambda\mathbb{Z}$ has order prime to $p$ and $R_n$ is flat over $\mathbb{Z}_p$. Consequently, the morphism (II.7.9.6) is injective.

(ii) It suffices to show that the multiplication by $p^m$ in $\Omega^1_{(R_\infty,P_\infty)/(\mathscr{O}_{\overline{K}},\mathbb{N})}$ is an isomorphism. For all integers $n,n'\geq 1$, we have a commutative diagram

(II.7.9.13)
$$\begin{array}{ccc} (P^{\rm gp}/n\lambda\mathbb{Z})\otimes_{\mathbb{Z}} R_n & \longrightarrow & (P^{\rm gp}/nn'\lambda\mathbb{Z})\otimes_{\mathbb{Z}} R_{nn'} \\ \downarrow & & \downarrow \\ \Omega^1_{(R_n,P^{(n)})/(\mathscr{O}_{\overline{K}},\mathbb{N})} & \xrightarrow{u_{n,nn'}} & \Omega^1_{(R_{nn'},P^{(nn')})/(\mathscr{O}_{\overline{K}},\mathbb{N})} \end{array}$$

where the vertical arrows are the surjective morphisms deduced from (II.7.9.11), $u_{n,nn'}$ is the canonical morphism, and the top horizontal arrow is induced by the multiplication by $n'$ in $P^{\rm gp}$ and the canonical homomorphism $R_n\to R_{nn'}$. It follows that

(II.7.9.14)
$$u_{n,nn'}(\Omega^1_{(R_n,P^{(n)})/(\mathscr{O}_{\overline{K}},\mathbb{N})})\subset n'\cdot\Omega^1_{(R_{nn'},P^{(nn')})/(\mathscr{O}_{\overline{K}},\mathbb{N})}.$$

Consequently, the multiplication by $p^m$ in $\Omega^1_{(R_\infty,P_\infty)/(\mathscr{O}_{\overline{K}},\mathbb{N})}$ is surjective.

Let $\omega\in\Omega^1_{(R_\infty,P_\infty)/(\mathscr{O}_{\overline{K}},\mathbb{N})}$ be such that $p^m\omega=0$. Then there exists $n\geq 1$ such that $\omega\in\Omega^1_{(R_n,P^{(n)})/(\mathscr{O}_{\overline{K}},\mathbb{N})}$ and $p^m\omega=0$ in $\Omega^1_{(R_n,P^{(n)})/(\mathscr{O}_{\overline{K}},\mathbb{N})}$. Consider the canonical exact sequence

(II.7.9.15)
$$\Omega^1_{(\mathscr{O}_{\overline{K}},\mathbb{N}^{(n)})/(\mathscr{O}_{\overline{K}},\mathbb{N})}\otimes_{\mathscr{O}_{\overline{K}}} R_n\xrightarrow{h_n}\Omega^1_{(R_n,P^{(n)})/(\mathscr{O}_{\overline{K}},\mathbb{N})}\longrightarrow\Omega^1_{(R_n,P^{(n)})/(\mathscr{O}_{\overline{K}},\mathbb{N}^{(n)})}\longrightarrow 0.$$

The $R_n$-module $\Omega^1_{(R_n,P^{(n)})/(\mathscr{O}_{\overline{K}},\mathbb{N}^{(n)})}$ is free of finite type by virtue of II.6.6(i) and II.6.8(i). Hence it is $\mathbb{Z}_p$-flat. Consequently, $\omega$ is contained in the image of $h_n$. It therefore follows from (II.7.5.5) and (II.7.5.6) that there exists $n'\geq 1$ such that the image of $\omega$ in $\Omega^1_{(R_{nn'},P^{(nn')})/(\mathscr{O}_{\overline{K}},\mathbb{N})}$ is zero. Hence $\omega$ is zero in $\Omega^1_{(R_\infty,P_\infty)/(\mathscr{O}_{\overline{K}},\mathbb{N})}$.

(iii) We have a canonical isomorphism

(II.7.9.16)
$$\lim_{\substack{\longrightarrow\\ n\geq 1}}\Omega^1_{(R_n,P^{(n)})/(R_1,P)}\stackrel{\sim}{\to}\Omega^1_{(R_\infty,P_\infty)/(R_1,P)},$$

where the direct limit is indexed by the set $\mathbb{Z}_{\geq 1}$ ordered by the divisibility relation. For every integer $n\geq 1$, it follows from the diagram (II.7.9.8) and what precedes it that we have a canonical isomorphism

(II.7.9.17)
$$M_n=\frac{(P^{\rm gp}/nP^{\rm gp})\otimes_{\mathbb{Z}} R_n}{\widetilde{\lambda}\otimes\pi_n R_n}\stackrel{\sim}{\to}\Omega^1_{(R_n,P^{(n)})/(R_1,P)},$$

where $\widetilde{\lambda}$ is the class of $\lambda$ in $P^{\rm gp}/nP^{\rm gp}$. We denote by $d\log(\lambda^{(n)})$ the image of $\widetilde{\lambda}\otimes 1$ in $M_n$, which can be justified by the isomorphism (II.7.9.17). For every integer $m\geq 1$, the multiplication by $m$ in $P^{\rm gp}$ and the canonical homomorphism $R_n\to R_{mn}$ induce a morphism $M_n\to M_{mn}$. The $(M_n)_{n\geq 1}$ form a direct system for the divisibility relation, and we have a canonical isomorphism

(II.7.9.18)
$$\lim_{\substack{\longrightarrow\\ n\geq 1}} M_n\stackrel{\sim}{\to}\Omega^1_{(R_\infty,P_\infty)/(R_1,P)}.$$

Let $n$ be an integer $\geq 1$. In $M_{pn}$, we have

(II.7.9.19)
$$d\log(\lambda^{(n)})=pd\log(\lambda^{(pn)})=\frac{p}{\pi_{pn}}\pi_{pn}d\log(\lambda^{(pn)})=0.$$

Consequently, for every integer $m\geq 1$ divisible by $p$, the composition of the canonical morphisms

(II.7.9.20)
$$(P^{\rm gp}/nP^{\rm gp})\otimes_{\mathbb{Z}} R_n\to M_n\to M_{mn}$$

factors through an $R_n$-linear morphism

(II.7.9.21) $$N_n = ((P^{\rm gp}/\lambda\mathbb{Z})/n(P^{\rm gp}/\lambda\mathbb{Z}))\otimes_{\mathbb{Z}} R_n \to M_{mn}.$$

On the other hand, we have a canonical surjective morphism

(II.7.9.22) $$M_{mn} \to ((P^{\rm gp}/\lambda\mathbb{Z})/mn(P^{\rm gp}/\lambda\mathbb{Z}))\otimes_{\mathbb{Z}} R_{mn}.$$

The composition of (II.7.9.21) and (II.7.9.22) is induced by the multiplication by $m$ in $P^{\rm gp}/\lambda\mathbb{Z}$ and the canonical homomorphism $R_n \to R_{mn}$. Since the torsion subgroup of $P^{\rm gp}/\lambda\mathbb{Z}$ is of order prime to $p$ and $R_n$ is $\mathbb{Z}_p$-flat, multiplication by $m$ induces an injective morphism

$$((P^{\rm gp}/\lambda\mathbb{Z})/n(P^{\rm gp}/\lambda\mathbb{Z}))\otimes_{\mathbb{Z}} R_n \to ((P^{\rm gp}/\lambda\mathbb{Z})/mn(P^{\rm gp}/\lambda\mathbb{Z}))\otimes_{\mathbb{Z}} R_n.$$

Since $R_n$ is a normal integral domain by II.6.8(ii), the canonical homomorphism

$$R_n/mnR_n \to R_{mn}/mnR_{mn}$$

is injective. It follows that the morphism (II.7.9.21) is injective.

For all integers $m, n \geq 1$, the multiplication by $m$ in $P^{\rm gp}/\lambda\mathbb{Z}$ and the canonical homomorphism $R_n \to R_{mn}$ induce a morphism $N_n \to N_{mn}$. The $(N_n)_{n\geq 1}$ form a direct system for the divisibility relation. Taking the direct limit of the morphisms (II.7.9.21), first with respect to the integers $m \geq 1$ that are multiples of $p$, and then with respect to the integers $n \geq 1$, gives an isomorphism

(II.7.9.23) $$\varinjlim_{n\geq 1} N_n \xrightarrow{\sim} \Omega^1_{(R_\infty,P_\infty)/(R_1,P)}.$$

It is clear that the canonical morphism

(II.7.9.24) $$\varinjlim_{s\geq 0} N_{p^s} \to \varinjlim_{n\geq 1} N_n,$$

where the first direct limit is indexed by the set $\mathbb{N}$ with the usual ordering, is an isomorphism. Statement (iii) follows.

**Corollary II.7.10.** *The $\mathscr{O}_{\overline{K}}$-modules $\Omega^1_{(R_\infty,P_\infty)/(R_1,P)}$ and $\Omega^1_{(R_\infty,P_\infty)/(R_1,P)}\otimes_{R_\infty}\overline{R}$ do not have any nonzero $\mathfrak{m}_{\overline{K}}$-torsion.*

It suffices to copy the proof of II.7.7, taking into account (II.7.9.3) and the fact that $R_\infty$ and $\overline{R}$ are $\mathscr{O}_{\overline{K}}$-flat ([**11**] Chap. VI § 3.6 Lem. 1).

**Corollary II.7.11.** *Every element of $\Omega^1_{(R_\infty,P_\infty)/(R,P)}$ (resp. $\Omega^1_{(\overline{R},P_\infty)/(R,P)}$) is annihilated by a power of $p$.*

Consider the exact sequence

(II.7.11.1) $$\Omega^1_{R_1/R}\otimes_{R_1} R_\infty \to \Omega^1_{(R_\infty,P_\infty)/(R,P)} \to \Omega^1_{(R_\infty,P_\infty)/(R_1,P)} \to 0.$$

By II.6.8(i), we have a canonical isomorphism

(II.7.11.2) $$\Omega^1_{R_1/R} \xrightarrow{\sim} \Omega^1_{\mathscr{O}_{\overline{K}}/\mathscr{O}_K}\otimes_{\mathscr{O}_{\overline{K}}} R_1.$$

On the other hand, by (II.7.9.3), every element of $\Omega^1_{(R_\infty,P_\infty)/(R_1,P)}$ is annihilated by a power of $p$, and the same holds for $\Omega^1_{\mathscr{O}_{\overline{K}}/\mathscr{O}_K}$ (II.7.1). Consequently, every element of $\Omega^1_{(R_\infty,P_\infty)/(R,P)}$ is annihilated by a power of $p$. Since the canonical morphism

(II.7.11.3) $$\Omega^1_{(R_\infty,P_\infty)/(R,P)}\otimes_{R_\infty}\overline{R} \to \Omega^1_{(\overline{R},P_\infty)/(R,P)}$$

is an almost isomorphism by virtue of II.6.18 and ([**50**] 1.7), it follows that every element of $\Omega^1_{(\overline{R},P_\infty)/(R,P)}$ is annihilated by a power of $p$.

**Remark II.7.12.** The isomorphism (II.7.9.2) can also be deduced from the isomorphism (II.7.9.3) as follows. First note that we have a canonical isomorphism

$$(P^{\mathrm{gp}}/\lambda\mathbb{Z}) \otimes_{\mathbb{Z}} R \xrightarrow{\sim} \Omega^1_{(R,P)/(\mathscr{O}_K,\mathbb{N})}. \tag{II.7.12.1}$$

For every integer $m \geq 0$, the $p^m$-torsion of $R_\infty[\frac{1}{p}]/R_\infty$ is canonically isomorphic to $R_\infty/p^m R_\infty$. Since the torsion subgroup of $P^{\mathrm{gp}}/\lambda\mathbb{Z}$ is of order prime to $p$, (II.7.9.3) induces an isomorphism

$$(P^{\mathrm{gp}}/\lambda\mathbb{Z}) \otimes_{\mathbb{Z}} (R_\infty/p^m R_\infty) \xrightarrow{\sim} \mathrm{Hom}(p^{-m}\mathbb{Z}/\mathbb{Z}, \Omega^1_{(R_\infty,P_\infty)/(R_1,P)}). \tag{II.7.12.2}$$

It follows from the proof of II.7.9 that this is the inverse of the isomorphism (II.7.9.2).

**Proposition II.7.13.** (i) *The kernel of the canonical morphism*

$$\Omega^1_{\mathscr{O}_{\overline{K}}/\mathscr{O}_K} \otimes_{\mathscr{O}_{\overline{K}}} R_\infty \to \Omega^1_{(R_\infty,P_\infty)/(R,P)} \tag{II.7.13.1}$$

*is generated by* $d\log(\pi)$ (II.7.2.2).

(ii) *The sequence of canonical morphisms*

$$0 \to \Omega^1_{(\mathscr{O}_{\overline{K}},\mathbb{N}_\infty)/(\mathscr{O}_K,\mathbb{N})} \otimes_{\mathscr{O}_{\overline{K}}} R_\infty \to \Omega^1_{(R_\infty,P_\infty)/(R,P)} \to \Omega^1_{(R_\infty,P_\infty)/(R_1,P)} \to 0 \tag{II.7.13.2}$$

*is exact and split.*

(i) We have a canonical isomorphism

$$\varinjlim_{n\geq 1} \Omega^1_{(R_n,P^{(n)})/(R,P)} \xrightarrow{\sim} \Omega^1_{(R_\infty,P_\infty)/(R,P)}, \tag{II.7.13.3}$$

where the direct limit is indexed by the set $\mathbb{Z}_{\geq 1}$ ordered by the divisibility relation. It therefore suffices to show that for every integer $n \geq 1$ such that $p$ divides $n$, the kernel of the canonical morphism

$$\Omega^1_{\mathscr{O}_{\overline{K}}/\mathscr{O}_K} \otimes_{\mathscr{O}_{\overline{K}}} R_n \to \Omega^1_{(R_n,P^{(n)})/(R,P)} \tag{II.7.13.4}$$

is generated by $d\log(\pi)$. By functoriality, this morphism factors through the canonical morphism

$$\Omega^1_{(\mathscr{O}_{\overline{K}},\mathbb{N}^{(n)})/(\mathscr{O}_K,\mathbb{N})} \otimes_{\mathscr{O}_{\overline{K}}} R_n \to \Omega^1_{(R_n,P^{(n)})/(R,P)}. \tag{II.7.13.5}$$

It therefore follows from II.7.4(ii) that $d\log(\pi)$ belongs to the kernel of the morphism (II.7.13.4).

Conversely, let us show that the kernel of the morphism (II.7.13.4) is contained in $R_n d\log(\pi)$. Recall that $\mathrm{Spec}(R_n)$ is an open connected component of $X_n \otimes_{\mathscr{O}_{K_n}} \mathscr{O}_{\overline{K}}$ (II.6.8) and that we have a Cartesian diagram of $\mathscr{O}_K$-morphisms (II.6.6.4)

$$\begin{array}{ccc} X_n \otimes_{\mathscr{O}_{K_n}} \mathscr{O}_{\overline{K}} & \longrightarrow & \mathrm{Spec}(\mathscr{O}_{\overline{K}}[P^{(n)}]/(\pi_n - e^{\lambda^{(n)}})) \\ \downarrow & & \downarrow \\ X & \longrightarrow & \mathrm{Spec}(\mathscr{O}_K[P]/(\pi - e^{\lambda})) \end{array} \tag{II.7.13.6}$$

where $\lambda^{(n)}$ is the image of $\lambda$ in $P^{(n)}$ by the isomorphism (II.7.8.2). Consider the commutative diagram with Cartesian square

(II.7.13.7)

$$\begin{array}{ccccc} \mathrm{Spec}(\mathscr{O}_{\overline{K}}[P^{(n)}]/(\pi_n - e^{\lambda^{(n)}})) & \longrightarrow & \mathrm{Spec}(\mathscr{O}_{\overline{K}}[P^{(n)}]/(\pi - e^{n\lambda^{(n)}})) & \longrightarrow & \mathrm{Spec}(\mathscr{O}_{\overline{K}}[P^{(n)}]) \\ & \searrow & \downarrow & \square & \downarrow \\ & & \mathrm{Spec}(\mathscr{O}_K[P]/(\pi - e^{\lambda})) & \longrightarrow & \mathrm{Spec}(\mathscr{O}_K[P]) \end{array}$$

We have a canonical isomorphism
(II.7.13.8)

$$\Omega^1_{(\mathscr{O}_{\overline{K}}[P^{(n)}],P^{(n)})/(\mathscr{O}_K[P],P)} \xrightarrow{\sim} (P^{\rm gp}/nP^{\rm gp})\otimes_{\mathbb{Z}}\mathscr{O}_{\overline{K}}[P^{(n)}]\oplus\Omega^1_{\mathscr{O}_{\overline{K}}/\mathscr{O}_K}\otimes_{\mathscr{O}_{\overline{K}}}\mathscr{O}_{\overline{K}}[P^{(n)}].$$

We deduce from this an isomorphism

$$\text{(II.7.13.9)}\qquad \Omega^1_{(R_n,P^{(n)})/(R,P)} \xrightarrow{\sim} \frac{(P^{\rm gp}/nP^{\rm gp})\otimes_{\mathbb{Z}} R_n\oplus\Omega^1_{\mathscr{O}_{\overline{K}}/\mathscr{O}_K}\otimes_{\mathscr{O}_{\overline{K}}} R_n}{(\overline{\lambda}\otimes\pi_n - d\pi_n)R_n},$$

where $\overline{\lambda}$ is the class of $\lambda$ in $P^{\rm gp}/nP^{\rm gp}$. Note for the proof of (ii) that for every integer $m\geq 1$, the canonical morphism

$$\text{(II.7.13.10)}\qquad \Omega^1_{(R_n,P^{(n)})/(R,P)}\to\Omega^1_{(R_{mn},P^{(mn)})/(R,P)}$$

is induced by the multiplication by $m$ in $P^{\rm gp}$, the identity of $\Omega^1_{\mathscr{O}_{\overline{K}}/\mathscr{O}_K}$, and the canonical homomorphism $R_n\to R_{mn}$.

Let $\omega\in\Omega^1_{\mathscr{O}_{\overline{K}}/\mathscr{O}_K}\otimes_{\mathscr{O}_{\overline{K}}} R_n$ be such that its image by the morphism (II.7.13.4) is zero. By (II.7.13.9), there exists $x\in R_n$ such that

$$\text{(II.7.13.11)}\qquad \omega = x(\overline{\lambda}\otimes\pi_n - d\pi_n)\in(P^{\rm gp}/nP^{\rm gp})\otimes_{\mathbb{Z}} R_n\oplus\Omega^1_{\mathscr{O}_{\overline{K}}/\mathscr{O}_K}\otimes_{\mathscr{O}_{\overline{K}}} R_n.$$

Consequently, $\overline{\lambda}\otimes\pi_n x=0$ in $(P^{\rm gp}/nP^{\rm gp})\otimes_{\mathbb{Z}} R_n$. Since the torsion subgroup of $P^{\rm gp}/\lambda\mathbb{Z}$ is of order prime to $p$ and $R_n$ is flat over $\mathbb{Z}_p$, the homomorphism

$$\text{(II.7.13.12)}\qquad (\lambda\mathbb{Z}/n\lambda\mathbb{Z})\otimes_{\mathbb{Z}} R_n\to(P^{\rm gp}/nP^{\rm gp})\otimes_{\mathbb{Z}} R_n$$

is injective. It follows that $\pi_n x\in nR_n$. Hence $x\in(n/\pi_n)R_n$, because $p$ divides $n$ and $R_n$ is flat over $\mathscr{O}_{\overline{K}}$ by virtue of II.6.6(ii) and II.6.8(i). It then follows from (II.7.13.11) that $\omega=-xd\pi_n\in R_n d\log(\pi)$.

(ii) By II.6.8(i), we have a canonical isomorphism

$$\text{(II.7.13.13)}\qquad \Omega^1_{R_1/R}\xrightarrow{\sim}\Omega^1_{\mathscr{O}_{\overline{K}}/\mathscr{O}_K}\otimes_{\mathscr{O}_{\overline{K}}} R_1.$$

Hence the sequence (II.7.13.2) is exact by (i), II.7.5, and by the canonical exact sequence

$$\text{(II.7.13.14)}\qquad \Omega^1_{R_1/R}\otimes_{R_1} R_\infty\to\Omega^1_{(R_\infty,P_\infty)/(R,P)}\to\Omega^1_{(R_\infty,P_\infty)/(R_1,P)}\to 0.$$

It remains to construct a splitting of (II.7.13.2). We have a canonical isomorphism

$$\text{(II.7.13.15)}\qquad \lim_{\substack{\longrightarrow\\ n\geq 1}}\Omega^1_{(R_n,P^{(n)})/(R_1,P)}\xrightarrow{\sim}\Omega^1_{(R_\infty,P_\infty)/(R_1,P)},$$

where the direct limit is indexed by the set $\mathbb{Z}_{\geq 1}$ ordered by the divisibility relation. Hence, in view of the isomorphism (II.7.13.3), it suffices to construct, for every integer $n\geq 1$ such that $p$ divides $n$, a right inverse of the canonical morphism

$$\text{(II.7.13.16)}\qquad \Omega^1_{(R_n,P^{(n)})/(R,P)}\to\Omega^1_{(R_n,P^{(n)})/(R_1,P)}$$

such that the family of these right inverses is a morphism of direct systems.

It follows from the diagram (II.7.9.8) and what precedes it that we have an isomorphism

$$\text{(II.7.13.17)}\qquad \Omega^1_{(R_n,P^{(n)})/(R_1,P)}\xrightarrow{\sim}\frac{(P^{\rm gp}/nP^{\rm gp})\otimes_{\mathbb{Z}} R_n}{\overline{\lambda}\otimes\pi_n R_n},$$

where $\overline{\lambda}$ is the class of $\lambda$ in $P^{\rm gp}/nP^{\rm gp}$. For every integer $m\geq 1$, the canonical morphism

$$\text{(II.7.13.18)}\qquad \Omega^1_{(R_n,P^{(n)})/(R,P)}\to\Omega^1_{(R_{mn},P^{(mn)})/(R,P)}$$

is induced by the multiplication by $m$ in $P^{\mathrm{gp}}$ and by the canonical homomorphism $R_n \to R_{mn}$. In view of (II.7.13.9) and (II.7.13.17), the morphism (II.7.13.16) is induced by the canonical projection

$$(P^{\mathrm{gp}}/nP^{\mathrm{gp}}) \otimes_{\mathbb{Z}} R_n \oplus \Omega^1_{\mathscr{O}_{\overline{K}}/\mathscr{O}_K} \otimes_{\mathscr{O}_{\overline{K}}} R_n \to (P^{\mathrm{gp}}/nP^{\mathrm{gp}}) \otimes_{\mathbb{Z}} R_n. \tag{II.7.13.19}$$

Since the torsion subgroup of $P^{\mathrm{gp}}/\lambda\mathbb{Z}$ is of order prime to $p$ and $R_n$ is flat over $\mathbb{Z}_p$, the $R_n$-linear morphism

$$R_n \to P^{\mathrm{gp}} \otimes_{\mathbb{Z}} R_n \tag{II.7.13.20}$$

defined by $\lambda$ admits an $R_n$-linear left inverse $u\colon P^{\mathrm{gp}} \otimes_{\mathbb{Z}} R_n \to R_n$. Consider the morphism

$$v_n\colon P^{\mathrm{gp}} \otimes_{\mathbb{Z}} R_n \to (P^{\mathrm{gp}}/nP^{\mathrm{gp}}) \otimes_{\mathbb{Z}} R_n \oplus \Omega^1_{\mathscr{O}_{\overline{K}}/\mathscr{O}_K} \otimes_{\mathscr{O}_{\overline{K}}} R_n \tag{II.7.13.21}$$

defined, for $x \in P^{\mathrm{gp}} \otimes_{\mathbb{Z}} R_n$ with class $\overline{x}$ in $(P^{\mathrm{gp}}/nP^{\mathrm{gp}}) \otimes_{\mathbb{Z}} R_n$, by

$$v_n(x) = \overline{x} - d\log(\pi_n) \otimes u(x), \tag{II.7.13.22}$$

where $d\log(\pi_n)$ is the element of $\Omega^1_{\mathscr{O}_{\overline{K}}/\mathscr{O}_K}$ defined in (II.7.2.2). In view of (II.7.13.9), $v_n$ induces an $R_n$-linear morphism

$$w_n\colon P^{\mathrm{gp}} \otimes_{\mathbb{Z}} R_n \to \Omega^1_{(R_n,P^{(n)})/(R,P)}. \tag{II.7.13.23}$$

For every $t \in P^{\mathrm{gp}}$, we have $w_n(nt) = -nu(t)d\log(\pi_n) = -u(t)d\log(\pi)$ (II.7.2.5). Since $p$ divides $n$, the image of $d\log(\pi)$ in $\Omega^1_{(R_n,P^{(n)})/(R,P)}$ is zero by the proof of (i); hence $w_n(nt) = 0$. On the other hand, by virtue of (II.7.2.4), we have

$$w_n(\lambda \otimes \pi_n) = \overline{\lambda} \otimes \pi_n - \pi_n d\log(\pi_n) = \overline{\lambda} \otimes \pi_n - d\pi_n = 0. \tag{II.7.13.24}$$

Consequently, in view of (II.7.13.17), $w_n$ induces an $R_n$-linear morphism

$$\omega_n\colon \Omega^1_{(R_n,P^{(n)})/(R_1,P)} \to \Omega^1_{(R_n,P^{(n)})/(R,P)} \tag{II.7.13.25}$$

that is the right inverse of the canonical morphism (II.7.13.16). Since $md\log(\pi_{mn}) = d\log(\pi_n)$ for every $m \geq 1$, the $\omega_n$'s form a morphism of direct systems for the divisibility relation, completing the proof.

**II.7.14.** There exists a unique map

$$\langle\ ,\ \rangle\colon \Gamma_\infty \times P_\infty \to \mu_\infty(\mathscr{O}_{\overline{K}}) = \varinjlim_{n\geq 1} \mu_n(\mathscr{O}_{\overline{K}}), \tag{II.7.14.1}$$

where the direct limit is indexed by the set $\mathbb{Z}_{\geq 1}$ ordered by the divisibility relation, such that for every $g \in \Gamma_\infty$ and every $x \in P^{(n)}$ $(n \geq 1)$, we have $\langle g, x\rangle \in \mu_n(\mathscr{O}_{\overline{K}})$ and

$$g(\alpha_\infty(x)) = \langle g, x\rangle \cdot \alpha_\infty(x), \tag{II.7.14.2}$$

where $\alpha_\infty$ is the homomorphism (II.7.8.5). Indeed, $R_\infty$ is an integral domain, and we have $\alpha_\infty(x)^n \in \alpha(P) \subset R$; hence $\alpha_\infty(x) \neq 0$ (II.7.8) and $\alpha_\infty(x)^n$ is invariant under $\Gamma_\infty$.

For every $g \in \Gamma_\infty$, the map $x \mapsto \langle g, x\rangle$ is a morphism of monoids from $P_\infty$ to $\mu_\infty(\mathscr{O}_{\overline{K}})$. For every $x \in P_\infty$, the map $g \mapsto \langle g, x\rangle$ is a 1-cocycle; in other words, for all $g, g' \in \Gamma_\infty$, we have

$$\langle gg', x\rangle = g(\langle g', x\rangle) \cdot \langle g, x\rangle. \tag{II.7.14.3}$$

Let $n$ be an integer $\geq 1$. Recall that we have a canonical isomorphism $P^{(n)} \xrightarrow{\sim} P$ (II.7.8.2) and that $\Delta_\infty$ is canonically isomorphic to a subgroup of $L_\lambda \otimes_{\mathbb{Z}} \widehat{\mathbb{Z}}(1)$ (II.6.8.3).

We therefore have a canonical homomorphism $\Delta_\infty \to L \otimes_{\mathbb{Z}} \mu_n(\mathscr{O}_{\overline{K}})$. By II.6.6(vi), the diagram

$$\begin{array}{ccc} \Delta_\infty \times P^{(n)} & \xrightarrow{\langle\ ,\ \rangle} & \mu_n(\mathscr{O}_{\overline{K}}) \\ \downarrow & \nearrow & \\ (L \otimes_{\mathbb{Z}} \mu_n(\mathscr{O}_{\overline{K}})) \times P & & \end{array} \tag{II.7.14.4}$$

where the slanted arrow is induced by the canonical pairing $L \otimes_{\mathbb{Z}} P^{\mathrm{gp}} \to \mathbb{Z}$, is commutative.

**II.7.15.** There exists a unique map

$$\langle\ ,\ \rangle \colon G_K \times \mathbb{N}_\infty \to \mu_\infty(\mathscr{O}_{\overline{K}}) \tag{II.7.15.1}$$

such that for every $g \in G_K$ and every $x \in \mathbb{N}^{(n)}$ $(n \geq 1)$, we have $\langle g, x\rangle \in \mu_n(\mathscr{O}_{\overline{K}})$ and

$$g(a_\infty(x)) = \langle g, x\rangle \cdot a_\infty(x), \tag{II.7.15.2}$$

where $a_\infty$ is the homomorphism (II.7.3.3). Identifying $\mathbb{N}^{(n)}$ with $\mathbb{N}$ through the canonical isomorphism $\mathbb{N}^{(n)} \xrightarrow{\sim} \mathbb{N}$, the relation (II.7.15.2) becomes $g(\pi_n^x) = \langle g, x\rangle \cdot \pi_n^x$. The map (II.7.15.1) has properties analogous to those of (II.7.14.1). Moreover, the diagram

$$\begin{array}{ccc} \Gamma_\infty \times \mathbb{N}_\infty & \longrightarrow & \Gamma_\infty \times P_\infty \\ \downarrow & & \downarrow{\scriptstyle \langle\ ,\ \rangle} \\ G_K \times \mathbb{N}_\infty & \xrightarrow{\langle\ ,\ \rangle} & \mu_\infty(\mathscr{O}_{\overline{K}}) \end{array} \tag{II.7.15.3}$$

where the unlabeled arrows are the canonical homomorphisms, is commutative, which justifies using the same notation.

**II.7.16.** Let $\mathscr{M}_\infty$ be the logarithmic structure on $\mathrm{Spec}(R_\infty)$ associated with the prelogarithmic structure defined by $(P_\infty, \alpha_\infty)$ (II.7.8). For every $g \in \Gamma_\infty$, denote by $\tau_g$ the automorphism of $\mathrm{Spec}(R_\infty)$ induced by $g$. The logarithmic structure $\tau_g^*(\mathscr{M}_\infty)$ on $\mathrm{Spec}(R_\infty)$ is associated with the prelogarithmic structure defined by $(P_\infty, g \circ \alpha_\infty)$. The homomorphism

$$P_\infty \to \Gamma(\mathrm{Spec}(R_\infty), \mathscr{M}_\infty), \quad x \mapsto \langle g, x\rangle \cdot x \tag{II.7.16.1}$$

induces a morphism of logarithmic structures on $\mathrm{Spec}(R_\infty)$

$$a_g \colon \tau_g^*(\mathscr{M}_\infty) \to \mathscr{M}_\infty. \tag{II.7.16.2}$$

Likewise, by virtue of (II.7.14.3), the homomorphism

$$P_\infty \to \Gamma(\mathrm{Spec}(R_\infty), \tau_g^*(\mathscr{M}_\infty)), \quad x \mapsto g(\langle g^{-1}, x\rangle) \cdot x \tag{II.7.16.3}$$

induces a morphism of logarithmic structures on $\mathrm{Spec}(R_\infty)$

$$b_g \colon \mathscr{M}_\infty \to \tau_g^*(\mathscr{M}_\infty). \tag{II.7.16.4}$$

We immediately see that $a_g$ and $b_g$ are isomorphisms inverse to each other (II.7.14.3), and that the map $g \mapsto (\tau_{g^{-1}}, a_{g^{-1}})$ is a left action of $\Gamma_\infty$ on $(\mathrm{Spec}(R_\infty), \mathscr{M}_\infty)$.

We denote by $\mathscr{L}$ the logarithmic structure on $\mathrm{Spec}(\overline{R})$ inverse image of $\mathscr{M}_\infty$. Then the action above lifts to a left action of $\Gamma$ on $(\mathrm{Spec}(\overline{R}), \mathscr{L})$.

Let $\mathscr{N}_\infty$ be the logarithmic structure on $\mathrm{Spec}(\mathscr{O}_{\overline{K}})$ associated with the prelogarithmic structure defined by $(\mathbb{N}_\infty, a_\infty)$ (II.7.3.3). Likewise, the map (II.7.15.1) defines a left action of $G_K$ on the logarithmic scheme $(\mathrm{Spec}(\mathscr{O}_{\overline{K}}), \mathscr{N}_\infty)$.

Let $\mathscr{M}$ be the logarithmic structure on $\mathrm{Spec}(R_1)$ associated with the prelogarithmic structure defined by $(P,\alpha)$ (II.7.8). For every $g\in\mathrm{Gal}(F_1/F)$, we denote by $u_g$ the automorphism of $\mathrm{Spec}(R_1)$ induced by $g$. The logarithmic structure $u_g^*(\mathscr{M})$ on $\mathrm{Spec}(R_1)$ is associated with the prelogarithmic structure defined by $(P,g\circ\alpha)$. Since $g\circ\alpha=\alpha$, we deduce from this a canonical isomorphism

$$c_g\colon u_g^*(\mathscr{M})\stackrel{\sim}{\to}\mathscr{M}. \tag{II.7.16.5}$$

The map $g\mapsto(u_{g^{-1}},c_{g^{-1}})$ is a left action of $\mathrm{Gal}(F_1/F)$ on the logarithmic scheme $(\mathrm{Spec}(R_1),\mathscr{M})$.

Let $\mathscr{N}$ be the logarithmic structure on $\mathrm{Spec}(\mathscr{O}_{\overline{K}})$ associated with the prelogarithmic structure defined by $(\mathbb{N},a)$ (II.7.3). Likewise, we define a left action of $G_K$ on the logarithmic scheme $(\mathrm{Spec}(\mathscr{O}_{\overline{K}}),\mathscr{N})$.

Note that all morphisms in the commutative diagram

$$\begin{array}{ccccc}(\mathrm{Spec}(\overline{R}),\mathscr{L}) & \longrightarrow & (\mathrm{Spec}(R_\infty),\mathscr{M}_\infty) & \longrightarrow & (\mathrm{Spec}(\mathscr{O}_{\overline{K}}),\mathscr{N}_\infty)\\ & & \downarrow & & \downarrow\\ & & (\mathrm{Spec}(R_1),\mathscr{M}) & \longrightarrow & (\mathrm{Spec}(\mathscr{O}_{\overline{K}}),\mathscr{N})\end{array} \tag{II.7.16.6}$$

are $\Gamma$-equivariant.

**II.7.17.** We denote by $\mathscr{E}_\infty$ the $\widehat{R_\infty}$-representation of $\Gamma_\infty$ defined by

$$\mathscr{E}_\infty=\mathrm{Hom}(\mathbb{Q}_p/\mathbb{Z}_p,\Omega^1_{(R_\infty,P_\infty)/(R,P)})\otimes_{\mathbb{Z}_p}\mathbb{Z}_p(-1), \tag{II.7.17.1}$$

where the action of $\Gamma_\infty$ comes from its action on $\Omega^1_{(R_\infty,P_\infty)/(R,P)}$ (II.7.16). In view of II.7.5 and II.7.9(ii), applying the functor $\mathrm{Hom}(\mathbb{Q}_p/\mathbb{Z}_p,-)\otimes_{\mathbb{Z}_p}\mathbb{Z}_p(-1)$ to the split exact sequence (II.7.13.2) gives an exact sequence of $\widehat{R_\infty}$-representations of $\Gamma_\infty$

$$0\to(\pi\rho)^{-1}\widehat{R_\infty}\to\mathscr{E}_\infty\to\Omega^1_{(R,P)/(\mathscr{O}_K,\mathbb{N})}\otimes_R\widehat{R_\infty}(-1)\to 0, \tag{II.7.17.2}$$

where we have written $(\pi\rho)^{-1}\widehat{R_\infty}$ instead of $(\pi\rho)^{-1}\mathscr{O}_C\otimes_{\mathscr{O}_C}\widehat{R_\infty}$, which is justified because $\widehat{R_\infty}$ is $\mathscr{O}_C$-flat (II.6.14). Note that $\Omega^1_{(R,P)/(\mathscr{O}_K,\mathbb{N})}$ is a free $R$-module of finite type. To simplify the notation, we set

$$\widetilde{\Omega}^1_{R/\mathscr{O}_K} = \Omega^1_{(R,P)/(\mathscr{O}_K,\mathbb{N})}, \tag{II.7.17.3}$$

$$\widetilde{\Omega}^i_{R/\mathscr{O}_K} = \wedge^i\widetilde{\Omega}^1_{R/\mathscr{O}_K}, \quad (i\geq 1). \tag{II.7.17.4}$$

The sequences (II.7.17.2) and (II.7.13.2) are split as sequences of $\widehat{R_\infty}$-modules (without $\Gamma_\infty$-action).

We denote by

$$\delta\colon\widetilde{\Omega}^1_{R/\mathscr{O}_K}\otimes_R\widehat{R_1}\to\mathrm{H}^1_{\mathrm{cont}}(\Delta_\infty,(\pi\rho)^{-1}\widehat{R_\infty})(1) \tag{II.7.17.5}$$

the boundary map of the long exact sequence of cohomology obtained by applying the functor $\Gamma(\Delta_\infty,-)(1)$ to the exact sequence (II.7.17.2). Later, we will show that $(\widehat{R_\infty})^{\Delta_\infty}=\widehat{R_1}$ (II.8.16).

**II.7.18.** For every $\zeta\in\mathbb{Z}_p(1)$, we denote by $d\log(\zeta)$ the element of $\mathscr{E}_\infty(1)$ defined by

$$d\log(\zeta)(p^{-n})=d\log(\zeta_n), \tag{II.7.18.1}$$

where $\zeta_n$ is the canonical image of $\zeta$ in $\mu_{p^n}(\mathscr{O}_{\overline{K}})$. It is clear that $d\log(\zeta)$ is the image of $1\otimes\zeta$ by the injection $(\pi\rho)^{-1}\widehat{R_\infty}(1)\to\mathscr{E}_\infty(1)$ (II.7.17.2).

For every $t \in P$, we denote by $d\log(\widetilde{t})$ the element of $\mathscr{E}_\infty(1)$ defined by

$$d\log(\widetilde{t})(p^{-n}) = d\log(t^{(p^n)}), \tag{II.7.18.2}$$

where $t^{(p^n)}$ is the image of $t$ in $P^{(p^n)}$ by the isomorphism (II.7.8.2). This element is well-defined by virtue of (II.7.8.3). It is clear that the image of $d\log(\widetilde{t})$ in $\widetilde{\Omega}^1_{R/\mathscr{O}_K} \otimes_R \widehat{R_\infty}$ is $d\log(t)$.

By (II.7.6), for every $n \geq 0$, $d\log(\widetilde{\lambda})(p^{-n})$ is the canonical image of the element $d\log(\pi_{p^n}) \in \Omega^1_{\mathscr{O}_{\overline{K}}/\mathscr{O}_K}$ (II.7.2.2) in $\widetilde{\Omega}^1_{(R_\infty,P_\infty)/(R,P)}$. In particular, we have $d\log(\widetilde{\lambda}) \in (\pi\rho)^{-1}\widehat{R_\infty}(1) \subset \mathscr{E}_\infty(1)$ (II.7.17.2).

The map $P \to \mathscr{E}_\infty(1)$ defined by $t \mapsto d\log(\widetilde{t})$ is a homomorphism; it therefore induces a homomorphism that we also denote by

$$P^{\mathrm{gp}} \to \mathscr{E}_\infty(1), \quad t \mapsto d\log(\widetilde{t}). \tag{II.7.18.3}$$

This fits into a commutative diagram

$$\begin{array}{ccccccccc} 0 & \longrightarrow & \mathbb{Z}\lambda & \longrightarrow & P^{\mathrm{gp}} & \longrightarrow & P^{\mathrm{gp}}/\mathbb{Z}\lambda & \longrightarrow & 0 \\ & & \downarrow & & \downarrow & & \downarrow & & \\ 0 & \longrightarrow & (\pi\rho)^{-1}\widehat{R_\infty}(1) & \longrightarrow & \mathscr{E}_\infty(1) & \longrightarrow & \widetilde{\Omega}^1_{R/\mathscr{O}_K} \otimes_R \widehat{R_\infty} & \longrightarrow & 0 \end{array} \tag{II.7.18.4}$$

where the vertical arrow on the right comes from the canonical isomorphism (II.7.12.1). It is useful to also denote by

$$d\log\colon P^{\mathrm{gp}} \to \widetilde{\Omega}^1_{R/\mathscr{O}_K} \tag{II.7.18.5}$$

the homomorphism induced by the logarithmic derivation $d\log\colon P \to \widetilde{\Omega}^1_{R/\mathscr{O}_K}$.

**II.7.19.** Let $t \in P$. We denote by

$$\widetilde{\chi}_t\colon \Gamma_\infty \to \mathbb{Z}_p(1) \tag{II.7.19.1}$$

the map that associates with each $g \in \Gamma_\infty$ the element

$$\widetilde{\chi}_t(g) = \lim_{\substack{\longleftarrow \\ n\geq 0}} \langle g, t^{(p^n)}\rangle, \tag{II.7.19.2}$$

where $t^{(p^n)}$ is the image of $t$ in $P^{(p^n)}$ by the isomorphism (II.7.8.2) and $\langle g, t^{(p^n)}\rangle \in \mu_{p^n}(\mathscr{O}_{\overline{K}})$ is defined in (II.7.14.1). By (II.7.14.3), for all $g, g' \in \Gamma_\infty$, we have

$$\widetilde{\chi}_t(gg') = g(\widetilde{\chi}_t(g'))\widetilde{\chi}_t(g). \tag{II.7.19.3}$$

Hence the restriction of $\widetilde{\chi}_t$ to $\Delta_\infty$ is a character with values in $\mathbb{Z}_p(1)$; we denote it by $\chi_t\colon \Delta_\infty \to \mathbb{Z}_p(1)$.

We clearly have $\widetilde{\chi}_0 = 1$, and for all $t, t' \in P$,

$$\widetilde{\chi}_{tt'} = \widetilde{\chi}_t \cdot \widetilde{\chi}_{t'}. \tag{II.7.19.4}$$

Consequently, the map $P \to \mathrm{Hom}(\Delta_\infty, \mathbb{Z}_p(1))$ defined by $t \mapsto \chi_t$ is a homomorphism. It therefore induces a homomorphism that we will also denote by

$$P^{\mathrm{gp}} \to \mathrm{Hom}(\Delta_\infty, \mathbb{Z}_p(1)), \quad t \mapsto \chi_t. \tag{II.7.19.5}$$

Since $\chi_\lambda = 1$, we deduce from this a homomorphism

$$P^{\mathrm{gp}}/\mathbb{Z}\lambda \to \mathrm{Hom}(\Delta_\infty, \mathbb{Z}_p(1)). \tag{II.7.19.6}$$

By (II.7.14.4), the latter is equal to the composition

$$P^{\mathrm{gp}}/\mathbb{Z}\lambda \to \mathrm{Hom}(L_\lambda \otimes_{\mathbb{Z}} \mathbb{Z}_p(1), \mathbb{Z}_p(1)) \to \mathrm{Hom}(\Delta_\infty, \mathbb{Z}_p(1)), \tag{II.7.19.7}$$

where the first arrow is induced by the canonical (biduality) morphism (II.6.2.9) and the second arrow by the canonical morphism $\nu\colon \Delta_\infty \to L_\lambda \otimes_{\mathbb{Z}} \mathbb{Z}_p(1)$ (II.6.8.3). Recall that $\nu$ induces an isomorphism $\Delta_{p^\infty} \xrightarrow{\sim} L_\lambda \otimes_{\mathbb{Z}} \mathbb{Z}_p(1)$ and that we have a canonical isomorphism (II.6.12.1)

$$\mathrm{Hom}_{\mathbb{Z}_p}(\Delta_{p^\infty}, \mathbb{Z}_p(1)) \xrightarrow{\sim} \mathrm{Hom}_{\mathbb{Z}}(\Delta_\infty, \mathbb{Z}_p(1)).$$

Since the torsion subgroup of $P^{\mathrm{gp}}/\mathbb{Z}\lambda$ is of order prime to $p$, the homomorphism (II.7.19.6) induces an isomorphism

(II.7.19.8) $$(P^{\mathrm{gp}}/\mathbb{Z}\lambda) \otimes_{\mathbb{Z}} \mathbb{Z}_p \xrightarrow{\sim} \mathrm{Hom}(\Delta_\infty, \mathbb{Z}_p(1)).$$

In view of (II.7.12.1) and (II.6.12.2), we deduce from this an $\widehat{R_1}$-linear isomorphism

(II.7.19.9) $$\widetilde{\delta}\colon \widetilde{\Omega}^1_{R/\mathscr{O}_K} \otimes_R \widehat{R_1} \xrightarrow{\sim} \mathrm{Hom}(\Delta_\infty, \widehat{R_1}(1)).$$

**II.7.20.** It immediately follows from the definitions that for all $t \in P$ and $g \in \Gamma_\infty$, we have

(II.7.20.1) $$g(d\log(\widetilde{t})) = d\log(\widetilde{t}) + d\log(\chi_t(g)).$$

Since both sides of the equation are homomorphisms from $P$ to $\mathscr{E}(1)$, we have equality for every $t \in P^{\mathrm{gp}}$. Consequently, the diagram

(II.7.20.2) $$\begin{array}{ccc} \widetilde{\Omega}^1_{R/\mathscr{O}_K} \otimes_R \widehat{R_1} & \xrightarrow{\ \widetilde{\delta}\ } & \mathrm{Hom}(\Delta_\infty, \widehat{R_1}(1)) \\ {\scriptstyle\delta}\downarrow & & \| \\ \mathrm{H}^1_{\mathrm{cont}}(\Delta_\infty, (\pi\rho)^{-1}\widehat{R_\infty}(1)) & \longleftarrow & \mathrm{H}^1_{\mathrm{cont}}(\Delta_\infty, \widehat{R_1}(1)) \end{array}$$

where $\delta$ is the morphism (II.7.17.5), $\widetilde{\delta}$ is the morphism (II.7.19.9), and the bottom horizontal arrow is induced by the canonical injection $\widehat{R_1} \to (\pi\rho)^{-1}\widehat{R_\infty}$, is commutative. Indeed, since $\widetilde{\Omega}^1_{R/\mathscr{O}_K}$ is generated over $R$ by the elements of the form $d\log(t)$ for $t \in P$, it suffices to show the commutativity of this diagram for these elements, which follows from (II.7.20.1).

**II.7.21.** Consider the commutative diagram of canonical morphisms (II.7.21.1)

$$\begin{array}{ccccc} \Omega^1_{(\mathscr{O}_{\overline{K}},\mathbb{N}_\infty)/(\mathscr{O}_K,\mathbb{N})} \otimes_{\mathscr{O}_{\overline{K}}} \overline{R} & \xrightarrow{\ u\ } & \Omega^1_{(R_\infty,P_\infty)/(R,P)} \otimes_{R_\infty} \overline{R} & \twoheadrightarrow & \Omega^1_{(R_\infty,P_\infty)/(R_1,P)} \otimes_{R_\infty} \overline{R} \\ \| & & {\scriptstyle a}\downarrow & & {\scriptstyle b}\downarrow \\ \Omega^1_{(\mathscr{O}_{\overline{K}},\mathbb{N}_\infty)/(\mathscr{O}_K,\mathbb{N})} \otimes_{\mathscr{O}_{\overline{K}}} \overline{R} & \xrightarrow{\ u'\ } & \Omega^1_{(\overline{R},P_\infty)/(R,P)} & \twoheadrightarrow & \Omega^1_{(\overline{R},P_\infty)/(R_1,P)} \end{array}$$

Since the sequence (II.7.13.2) is exact and split, $u$ is injective. On the other hand, the kernel of $a$ is annihilated by $\mathfrak{m}_{\overline{K}}$ by virtue of II.6.18 and (II.5.21.1). Consequently, the kernel of $u'$ is annihilated by $\mathfrak{m}_{\overline{K}}$. Since $\Omega^1_{(\mathscr{O}_{\overline{K}},\mathbb{N}_\infty)/(\mathscr{O}_K,\mathbb{N})} \otimes_{\mathscr{O}_{\overline{K}}} \overline{R}$ does not have any nonzero $\mathfrak{m}_{\overline{K}}$-torsion by II.7.7, $u'$ is injective.

Applying the "Tate module" functor $\mathrm{T}_p(-) = \mathrm{Hom}(\mathbb{Q}_p/\mathbb{Z}_p, -)$ to the diagram above, and setting

$$\Xi = \Omega^1_{(\mathscr{O}_{\overline{K}},\mathbb{N}_\infty)/(\mathscr{O}_K,\mathbb{N})} \otimes_{\mathscr{O}_{\overline{K}}} \overline{R},$$

we obtain a commutative diagram

(II.7.21.2)
$$\begin{array}{ccccc} \mathrm{T}_p(\Xi) & \hookrightarrow & \mathrm{T}_p(\Omega^1_{(R_\infty,P_\infty)/(R,P)}\otimes_{R_\infty}\overline{R}) & \xrightarrow{v} & \mathrm{T}_p(\Omega^1_{(R_\infty,P_\infty)/(R_1,P)}\otimes_{R_\infty}\overline{R}) \\ \| & & \downarrow{\scriptstyle \mathrm{T}_p(a)} & & \downarrow{\scriptstyle \mathrm{T}_p(b)} \\ \mathrm{T}_p(\Xi) & \hookrightarrow & \mathrm{T}_p(\Omega^1_{(\overline{R},P_\infty)/(R,P)}) & \xrightarrow{v'} & \mathrm{T}_p(\Omega^1_{(\overline{R},P_\infty)/(R_1,P)}) \end{array}$$

Since the sequence (II.7.13.2) is exact and split, $v$ is surjective. On the other hand, the kernel and cokernel of $b$ are annihilated by $\mathfrak{m}_{\overline{K}}$, and $\Omega^1_{(R_\infty,P_\infty)/(R_1,P)}\otimes_{R_\infty}\overline{R}$ does not have any nonzero $\mathfrak{m}_{\overline{K}}$-torsion by virtue of II.7.10. Consequently, $b$ is injective, and therefore $\mathrm{T}_p(b)$ is an isomorphism. It follows that $v'$ is surjective and that $\mathrm{T}_p(a)$ is an isomorphism.

It follows from (II.7.5.2) that the canonical morphism

(II.7.21.3)
$$\mathrm{T}_p(\Omega^1_{(\mathscr{O}_{\overline{K}},\mathbb{N}_\infty)/(\mathscr{O}_K,\mathbb{N})})\otimes_{\mathscr{O}_C}\widehat{\overline{R}}\to \mathrm{T}_p(\Omega^1_{(\mathscr{O}_{\overline{K}},\mathbb{N}_\infty)/(\mathscr{O}_K,\mathbb{N})}\otimes_{\mathscr{O}_{\overline{K}}}\overline{R})$$

is an isomorphism. Likewise, it follows from (II.7.9.3) that the canonical morphism

(II.7.21.4)
$$\mathrm{T}_p(\Omega^1_{(R_\infty,P_\infty)/(R_1,P)})\otimes_{\widehat{R_\infty}}\widehat{\overline{R}}\to \mathrm{T}_p(\Omega^1_{(R_\infty,P_\infty)/(R_1,P)}\otimes_{R_\infty}\overline{R})$$

is an isomorphism. Consequently, the canonical morphism

(II.7.21.5)
$$\mathrm{T}_p(\Omega^1_{(R_\infty,P_\infty)/(R,P)})\otimes_{\widehat{R_\infty}}\widehat{\overline{R}}\to \mathrm{T}_p(\Omega^1_{(R_\infty,P_\infty)/(R,P)}\otimes_{R_\infty}\overline{R})$$

is an isomorphism.

**II.7.22.** We denote by $\mathscr{E}$ the $\widehat{\overline{R}}$-representation of $\Gamma$ defined by

(II.7.22.1)
$$\mathscr{E}=\mathrm{Hom}(\mathbb{Q}_p/\mathbb{Z}_p,\Omega^1_{(\overline{R},P_\infty)/(R,P)})\otimes_{\mathbb{Z}_p}\mathbb{Z}_p(-1),$$

where the action of $\Gamma$ comes from its action on $\Omega^1_{(\overline{R},P_\infty)/(R,P)}$ (II.7.16). It follows from II.7.21 and (II.7.17.2) that we have a canonical exact sequence of $\widehat{\overline{R}}$-representations of $\Gamma$

(II.7.22.2)
$$0\to(\pi\rho)^{-1}\widehat{\overline{R}}\to\mathscr{E}\to\widetilde{\Omega}^1_{R/\mathscr{O}_K}\otimes_R\widehat{\overline{R}}(-1)\to 0,$$

called *Faltings extension.* We have an isomorphism of $\widehat{\overline{R}}$-representations of $\Gamma$

(II.7.22.3)
$$\mathscr{E}_\infty\otimes_{\widehat{R_\infty}}\widehat{\overline{R}}\xrightarrow{\sim}\mathscr{E},$$

inducing an isomorphism of the extensions (II.7.17.2) and (II.7.22.2). In particular, the sequence (II.7.22.2) is split as a sequence of $\widehat{\overline{R}}$-modules (without $\Gamma$-action).

## II.8. Galois cohomology

**Proposition II.8.1.** *Let $n$ be an integer $\geq 0$, $\nu\colon \Delta_{p^\infty}\to\mu_{p^n}(\mathscr{O}_{\overline{K}})$ a surjective homomorphism, $\zeta$ a generator of the group $\mu_{p^n}(\mathscr{O}_{\overline{K}})$, $a\in\mathscr{O}_{\overline{K}}$, and $\mathfrak{q}$ the ideal of $\mathscr{O}_{\overline{K}}$ generated by $a$ and $\zeta-1$. Let $A$ be an $\mathscr{O}_{\overline{K}}$-algebra that is complete and separated for the $p$-adic topology and $\mathscr{O}_{\overline{K}}$-flat. We denote by $A(\nu)$ the topological $A$-$\Delta_{p^\infty}$-module $A$, endowed with the $p$-adic topology and the action of $\Delta_{p^\infty}$ defined by the multiplication by $\nu$ (II.3.1).*

(i) *If $\nu=1$ (that is, $n=0$), then we have a canonical isomorphism of graded $A$-algebras*

(II.8.1.1)
$$\wedge\left(\mathrm{Hom}_{\mathbb{Z}_p}(\Delta_{p^\infty},A/aA)\right)\xrightarrow{\sim}\mathrm{H}^*_{\mathrm{cont}}(\Delta_{p^\infty},A(\nu)/aA(\nu)).$$

(ii) *If $\nu\neq 1$ (that is, $n\neq 0$), then $\mathrm{H}^i_{\mathrm{cont}}(\Delta_{p^\infty},A(\nu)/aA(\nu))$ is a free $A/\mathfrak{q}A$-module of finite type for every $i\geq 0$ and is zero for every $i\geq \mathrm{rg}(L)=d+1$ (II.6.2.5).*

(iii) *The inverse system* $(\mathrm{H}^*(\Delta_{p^\infty}, A(\nu)/p^r A(\nu)))_{r\geq 0}$ *satisfies the Mittag–Leffler condition uniformly in* $\nu$*; in other words, if for all integers* $r' \geq r \geq 0$*, we denote by*

$$h^\nu_{r,r'} \colon \mathrm{H}^*(\Delta_{p^\infty}, A(\nu)/p^{r'} A(\nu)) \to \mathrm{H}^*(\Delta_{p^\infty}, A(\nu)/p^r A(\nu)) \tag{II.8.1.2}$$

*the canonical morphism, then for every integer* $r \geq 1$*, there exists an integer* $r' \geq r$ *depending on* $d$ *but not on* $\nu$*, such that for every integer* $r'' \geq r'$*, the images of* $h^\nu_{r,r'}$ *and* $h^\nu_{r,r''}$ *are equal.*

Fix a $\mathbb{Z}_p$-basis $e_1, \dots, e_d$ of $\Delta_{p^\infty}$ and denote by $\mathrm{S}_A(\Delta_{p^\infty})$ the symmetric algebra of the $A$-module $\Delta_{p^\infty} \otimes_{\mathbb{Z}_p} A$. The ring $A$ is endowed with a structure of $\mathrm{S}_A(\Delta_{p^\infty})$-algebra defined by the homomorphism of $A$-algebras $\mathrm{S}_A(\Delta_{p^\infty}) \to A$ that sends $e_i$ to $\nu(e_i) - 1$ for $1 \leq i \leq d$; we denote it by $A(\nu)^\triangleright$. Note that the $\mathrm{S}_A(\Delta_{p^\infty})$-module underlying $A(\nu)^\triangleright$ is none other than the $\mathrm{S}_A(\Delta_{p^\infty})$-module associated with $A(\nu)$ and denoted by the same symbol in II.3.25. Consider the linear form

$$u \colon \Delta_{p^\infty} \otimes_{\mathbb{Z}_p} A \to A \tag{II.8.1.3}$$

that sends $e_i \otimes 1$ to $\nu(e_i) - 1$ for $1 \leq i \leq d$. It immediately follows from the definitions (II.2.6.3) and (II.2.7.4) that we have a canonical isomorphism

$$\mathbb{K}^\bullet_{\mathrm{S}_A(\Delta_{p^\infty})}((A(\nu)/aA(\nu))^\triangleright) \xrightarrow{\sim} \mathbb{K}^\bullet_A(u, A/aA). \tag{II.8.1.4}$$

By virtue of II.3.25, we deduce from this a canonical isomorphism

$$\mathrm{C}^\bullet_{\mathrm{cont}}(\Delta_{p^\infty}, A(\nu)/aA(\nu)) \xrightarrow{\sim} \mathbb{K}^\bullet_A(u, A/aA) \tag{II.8.1.5}$$

of $\mathbf{D}^+(\mathbf{Mod}(A))$, where the left-hand side is the complex of continuous cochains of $G$ with values in $A(\nu)/aA(\nu)$ (II.3.8).

(i) This follows from II.3.30 and II.6.12(iii).

Note that since the form $u$ (II.8.1.3) is zero, (II.8.1.5) provides an isomorphism of graded $A$-modules

$$\mathrm{H}^*_{\mathrm{cont}}(\Delta_{p^\infty}, A(\nu)/aA(\nu)) \xrightarrow{\sim} \wedge(\mathrm{Hom}_{\mathbb{Z}_p}(\Delta_{p^\infty}, A/aA)). \tag{II.8.1.6}$$

But it is not clear, a priori, that this is an isomorphism of graded $A$-algebras. We can, however, deduce it from II.3.28 (which essentially corresponds to the proof of II.3.30).

(ii) We denote also by $u$ the linear form $\Delta_{p^\infty} \otimes_{\mathbb{Z}_p} (A/aA) \to (A/aA)$ deduced from $u$ (II.8.1.3). By virtue of (II.8.1.5) and (II.2.6.8), it suffices to show that $\mathrm{H}_i(\mathbb{K}^{A/aA}_\bullet(u))$ is a free $(A/\mathfrak{q}A)$-module of finite type for every $i \geq 0$ and is zero for every $i \geq d+1$. The second statement is obvious. We also know that $\mathrm{H}_i(\mathbb{K}^{A/aA}_\bullet(u))$ is annihilated by $\mathfrak{q}$ (II.2.6). For $1 \leq j \leq d$, set $\zeta_j = \nu(e_j)$. We may assume that $\zeta_1 = \zeta \neq 1$ and that (II.2.1)

$$v(\zeta_1 - 1) \leq v(\zeta_2 - 1) \leq \cdots \leq v(\zeta_d - 1).$$

We proceed by induction on $d$. The statement for $d = 1$ is an immediate consequence of the flatness of $A$ over $\mathscr{O}_{\overline{K}}$. Suppose $d \geq 2$ and that the assertion holds for $d-1$. Denote by $G$ the sub-$\mathbb{Z}_p$-module of $\Delta_{p^\infty}$ generated by $e_1, \dots, e_{d-1}$ and by $u' \colon G \otimes_{\mathbb{Z}_p} (A/aA) \to A/aA$ the restriction of $u$ to $G \otimes_{\mathbb{Z}_p} (A/aA)$. By (II.2.6.4), we have a canonical isomorphism

$$\mathbb{K}^{A/aA}_\bullet(\zeta_d - 1) \otimes \mathbb{K}^{A/aA}_\bullet(u') \xrightarrow{\sim} \mathbb{K}^{A/aA}_\bullet(u), \tag{II.8.1.7}$$

where $\mathbb{K}^{A/aA}_\bullet(\zeta_d - 1)$ is the Koszul complex defined by the linear form $\zeta_d - 1$ in $A/aA$. By virtue of ([**41**] 1.1.4.1), for every integer $i$, we have an exact sequence

$$\begin{aligned} 0 &\to \mathrm{H}_0(\mathbb{K}^{A/aA}_\bullet(\zeta_d - 1) \otimes \mathrm{H}_i(\mathbb{K}^{A/aA}_\bullet(u'))) \\ &\to \mathrm{H}_i(\mathbb{K}^{A/aA}_\bullet(u)) \to \mathrm{H}_1(\mathbb{K}^{A/aA}_\bullet(\zeta_d - 1) \otimes \mathrm{H}_{i-1}(\mathbb{K}^{A/aA}_\bullet(u'))) \to 0. \end{aligned} \tag{II.8.1.8}$$

By the induction hypothesis, $\mathrm{H}_i(\mathbb{K}_\bullet^{A/aA}(u'))$ is a free $(A/\mathfrak{q}A)$-module of finite type for every $i \geq 0$. Since $(\zeta_d - 1) \in \mathfrak{q}$, we deduce from this an exact sequence

$$0 \to \mathrm{H}_i(\mathbb{K}_\bullet^{A/aA}(u')) \to \mathrm{H}_i(\mathbb{K}_\bullet^{A/aA}(u)) \to \mathrm{H}_{i-1}(\mathbb{K}_\bullet^{A/aA}(u')) \to 0. \tag{II.8.1.9}$$

Consequently, since $\mathrm{H}_i(\mathbb{K}_\bullet^{A/aA}(u))$ is annihilated by $\mathfrak{q}$, it is free of finite type over $A/\mathfrak{q}A$.

(iii) It follows from (i) that the inverse system $(\mathrm{H}^*(\Delta_{p^\infty}, A(\nu)/p^r A(\nu)))_{r\geq 0}$ satisfies the Mittag–Leffler condition when $\nu = 1$. We can therefore restrict ourselves to the characters $\nu \neq 1$. We set $A_r = A/p^r A$ and we denote also by $u$ the linear form $u \otimes_A \mathrm{id}_{A_r} : \Delta_{p^\infty} \otimes_{\mathbb{Z}_p} A_r \to A_r$ (II.8.1.3). By (II.8.1.5) and (II.2.6.8), it suffices to show the analogous statement for the inverse system $(\mathrm{H}_i(\mathbb{K}_\bullet^{A_r}(u)))_{r\geq 1}$. We proceed by induction on $d$. First take $d = 1$. The canonical homomorphism

$$\mathrm{H}_1(\mathbb{K}_\bullet^{A_{r+1}}(u)) \to \mathrm{H}_1(\mathbb{K}_\bullet^{A_r}(u)) \tag{II.8.1.10}$$

is clearly bijective, and since $v(\zeta - 1) \leq 1$, the canonical homomorphism

$$\mathrm{H}_0(\mathbb{K}_\bullet^{A_{r+1}}(u)) \to \mathrm{H}_0(\mathbb{K}_\bullet^{A_r}(u)) \tag{II.8.1.11}$$

is zero. The assertion therefore holds with $r' = r + 1$. Suppose $d \geq 2$ and that the assertion holds for $d - 1$. The assertion for $d$ then easily follows from the exact sequence (II.8.1.9) by taking $a = p^r$ for $r \geq 0$ (cf. the proof of **[41]** 0.13.2.1).

**Corollary II.8.2.** *Under the assumptions of* II.8.1, *the canonical homomorphism*

$$\mathrm{H}^*_{\mathrm{cont}}(\Delta_{p^\infty}, A(\nu)) \to \varprojlim_{r\geq 0} \mathrm{H}^*(\Delta_{p^\infty}, A(\nu)/p^r A(\nu)) \tag{II.8.2.1}$$

*is an isomorphism.*

Indeed, by (II.3.10.4) and (II.3.10.5), for every $i \geq 0$, we have an exact sequence

$$\begin{aligned} 0 \to \mathrm{R}^1\varprojlim_{r\geq 0} \mathrm{H}^{i-1}(\Delta_{p^\infty}, A(\nu)/p^r A(\nu)) \\ \to \mathrm{H}^i_{\mathrm{cont}}(\Delta_{p^\infty}, A(\nu)) \to \varprojlim_{r\geq 0} \mathrm{H}^i(\Delta_{p^\infty}, A(\nu)/p^r A(\nu)) \to 0 \end{aligned} \tag{II.8.2.2}$$

whose left term is zero by virtue of II.8.1(iii) and (II.3.10.2).

**Remark II.8.3.** Under the assumptions of II.8.1, if $\nu \neq 1$, the canonical homomorphism

$$\mathrm{H}^*_{\mathrm{cont}}(\Delta_{p^\infty}, A(\nu)) \otimes_A A/p^r A \to \mathrm{H}^*(\Delta_{p^\infty}, A(\nu)/p^r A(\nu)) \tag{II.8.3.1}$$

is not, in general, an isomorphism.

**Proposition II.8.4.** *Let $n$ be an integer $\geq 1$, $\nu\colon \Delta_{p^\infty} \to \mu_{p^n}(\mathscr{O}_{\overline{K}})$ a surjective homomorphism, $\zeta$ a generator of the group $\mu_{p^n}(\mathscr{O}_{\overline{K}})$, $a$ a nonzero element of $\mathscr{O}_{\overline{K}}$, $b = a(\zeta - 1)^{-1}$, and $\alpha$ a rational number. Let $A$ be an $\mathscr{O}_{\overline{K}}$-algebra that is complete and separated for the $p$-adic topology and $\mathscr{O}_{\overline{K}}$-flat, $N$ an $(A/aA)$-module, and $M$ a discrete $(A/aA)$-$\Delta_{p^\infty}$-module. We denote by $N(\nu)$ the discrete $A$-$\Delta_{p^\infty}$-module $N$ endowed with the action of $\Delta_{p^\infty}$ defined by the multiplication by $\nu$. Suppose that the following conditions are satisfied:*

(i) $\inf(v(a), \alpha) > v(\zeta - 1)$;
(ii) *$N$ is flat over $\mathscr{O}_{\overline{K}}/a\mathscr{O}_{\overline{K}}$;*
(iii) *$M$ is projective of finite type over $A/aA$, and is generated by a finite number of elements that are $\Delta_{p^\infty}$-invariant modulo $p^\alpha M$.*

*Then for every $i \geq 0$, we have*

$$(\zeta - 1) \cdot \mathrm{H}^i(\Delta_{p^\infty}, (M/bM) \otimes_A N(\nu)) = 0.$$

Since $M$ is a direct summand of a free $(A/aA)$-module of finite type, we can restrict to the case where it is free of finite type over $A/aA$. It therefore admits an $(A/aA)$-basis consisting of elements that are $\Delta_{p^\infty}$-invariant modulo $p^\alpha M$. Let $T = M \otimes_A N(\nu)$. Fix a $\mathbb{Z}_p$-base $e_1, \dots, e_d$ of $\Delta_{p^\infty}$ and denote by $\mathrm{S}_A(\Delta_{p^\infty})$ the symmetric algebra of the $A$-module $\Delta_{p^\infty} \otimes_{\mathbb{Z}_p} A$. By virtue of II.3.25, we have a canonical isomorphism

$$\text{(II.8.4.1)} \qquad \mathrm{C}^\bullet(\Delta_{p^\infty}, T/bT) \xrightarrow{\sim} \mathbb{K}^\bullet_{\mathrm{S}_A(\Delta_{p^\infty})}((T/bT)^\triangleright)$$

of $\mathbf{D}^+(\mathbf{Mod}(A))$. There exists an integer $i$ with $1 \le i \le d$ such that $\nu(e_i)$ is a generator of $\mu_{p^n}(\mathscr{O}_{\overline{K}})$; we may assume that $\nu(e_i) = \zeta$. Let $\varphi\colon \Delta_{p^\infty} \to \mathrm{Aut}_A(M)$ be the representation of $\Delta_{p^\infty}$ over $M$. Then there exists an $A$-linear endomorphism $U$ of $M$ such that $\varphi(e_i) = \mathrm{id}_M + p^\alpha U$. Let $c = p^\alpha(\zeta - 1)^{-1} \in \mathfrak{m}_{\overline{K}}$ and $V = \mathrm{id}_T + cU \otimes (\zeta \cdot \mathrm{id}_N)$, which is an $A$-linear automorphism of $T$, so that we have

$$\text{(II.8.4.2)} \qquad \varphi(e_i) \otimes (\nu(e_i) \cdot \mathrm{id}_N) - \mathrm{id}_T = (\zeta - 1)V \in \mathrm{End}_A(T).$$

For every integer $1 \le j \le d$, we have

$$\text{(II.8.4.3)} \qquad \begin{aligned} &(\varphi(e_j) \otimes (\nu(e_j) \cdot \mathrm{id}_N) - \mathrm{id}_T) \circ V^{-1} \\ &= \; V^{-1} \circ (\varphi(e_j) \otimes (\nu(e_j) \cdot \mathrm{id}_N) - \mathrm{id}_T) \in \mathrm{End}_A(T/bT). \end{aligned}$$

Indeed, since $\Delta_{p^\infty}$ is abelian, the products of these endomorphisms by $\zeta - 1$ are equal in $\mathrm{End}_A(T)$, which implies the relation (II.8.4.3) because $T$ is flat over $\mathscr{O}_{\overline{K}}/a\mathscr{O}_{\overline{K}}$. Hence $V^{-1}$ is an automorphism of the $\mathrm{S}_A(\Delta_{p^\infty})$-module $(T/bT)^\triangleright$ and it induces, for every $q \ge 0$, an $\mathrm{S}_A(\Delta_{p^\infty})$-linear automorphism of $\mathrm{H}^q(\mathbb{K}^\bullet_{\mathrm{S}_A(\Delta_{p^\infty})}((T/bT)^\triangleright))$. Now, this module is annihilated by $e_i$ (II.2.6). Consequently, for every $x \in \mathrm{H}^q(\mathbb{K}^\bullet_{\mathrm{S}_A(\Delta_{p^\infty})}((T/bT)^\triangleright))$, we have, by virtue of (II.8.4.2),

$$\text{(II.8.4.4)} \qquad e_i \cdot V^{-1}(x) = (\zeta - 1)x = 0.$$

**II.8.5.** We denote by $\Lambda$ the cokernel in the category of monoids of the homomorphism $\vartheta\colon \mathbb{N} \to P$ and by $q\colon P \to \Lambda$ the canonical homomorphism. By ([**58**] I 1.1.5), $\Lambda$ is the quotient of $P$ by the congruence relation $E$ consisting of the elements $(x, y) \in P \times P$ for which there exist $a, b \in \mathbb{N}$ such that $x + a\lambda = y + b\lambda$. That $E$ is a congruence relation means that it is an equivalence relation and that $E$ is a submonoid of $P \times P$. The group associated with $\Lambda$ is canonically identified with $P^{\mathrm{gp}}/\mathbb{Z}\lambda$. Since $P$ is integral, $\Lambda$ is integral; we can therefore identify it with the image of $P$ in $P^{\mathrm{gp}}/\mathbb{Z}\lambda$.

**Lemma II.8.6.** *We keep the notation of* II.8.5. *Then:*

(i) *The monoid $\Lambda$ is saturated.*

(ii) *For every $x \in \Lambda$, the set $q^{-1}(x)$ admits a unique minimal element $\widetilde{x}$ for the preordering on $P$ defined by the monoid structure.*

(iii) *For every $x \in \Lambda$ and every integer $n \ge 0$, we have $\widetilde{nx} = n\widetilde{x}$.*

(i) Indeed, $\Lambda$ is the amalgamated sum of the saturated homomorphism $\vartheta$ and the homomorphism $\mathbb{N} \to 0$. It is therefore saturated (II.5.2).

(ii) First note that two arbitrary elements of $q^{-1}(x)$ are necessarily comparable (II.8.5). Let us show that the set $q^{-1}(x)$ admits a minimal element $\widetilde{x}$. This corresponds to saying that for every $t \in P$, there exists $n \in \mathbb{N}$ such that the element $t - n\lambda$ of $P^{\mathrm{gp}}$ does not belong to $P$. Indeed, if this is not the case, then for every $n \ge 0$, the element $\alpha(t)/\pi^n$ of $R_K$ belongs to $R$, where $\alpha\colon P \to R$ is the homomorphism defined by the chart $(P, \gamma)$ (II.6.2). Since $\alpha(t) \ne 0$ and $R$ is a noetherian integral domain, it follows that $\pi$ is invertible in $R$, which contradicts condition II.6.2($\mathrm{C}_2$). It also follows that $-\lambda$ does not belong to $P$. Hence $\widetilde{x}$ is necessarily unique because $P$ is integral and $P^{\mathrm{gp}}$ is without torsion.

(iii) Since $-\lambda$ does not belong to $P$, we have $\widetilde{0} = 0$. We may therefore restrict to the case where $n$ is a prime number. First note that $q(\widetilde{nx}) = q(n\widetilde{x})$. If $\widetilde{nx} \neq n\widetilde{x}$, then there exists an $m \geq 1$ such that $n\widetilde{x} \geq m\lambda$. Since the homomorphism $\vartheta$ is saturated, there then exists $m' \in \mathbb{N}$ such that $\widetilde{x} \geq m'\lambda$ and $nm' \geq m$ by virtue of ([**74**] 4.1 p. 11; cf. also [**58**] I 4.8.13). Since $m' \geq 1$, the relation $\widetilde{x} \geq m'\lambda$ contradicts the fact that $\widetilde{x}$ is minimal.

**II.8.7.** We keep the notation of II.8.5. For every $n \geq 0$, we denote by $\Lambda^{(p^n)}$ the monoid over $\Lambda$ defined by the pair $(\Lambda, \varpi_{p^n})$; in other words, $\Lambda^{(p^n)}$ is the monoid $\Lambda$ and the structural homomorphism $\Lambda \to \Lambda^{(p^n)}$ is the Frobenius of order $p^n$ of $\Lambda$ (II.5.12). We naturally identify $\Lambda^{(p^n)}$ with the cokernel of the homomorphism $\vartheta\colon \mathbb{N}^{(p^n)} \to P^{(p^n)}$ and denote also by $q\colon P^{(p^n)} \to \Lambda^{(p^n)}$ the canonical homomorphism. Let

$$P_{p^\infty} = \varinjlim_{\mathbb{N}} P^{(p^n)} \quad \text{and} \quad \Lambda_{p^\infty} = \varinjlim_{\mathbb{N}} \Lambda^{(p^n)}.$$

We identify $P_{p^\infty}$ with a submonoid of $P_\infty$ (II.7.8.1). We denote also by $q\colon P_{p^\infty} \to \Lambda_{p^\infty}$ the direct limit of the canonical homomorphisms $q\colon P^{(p^n)} \to \Lambda^{(p^n)}$.

For every $t \in P_{p^\infty}$, the map

(II.8.7.1) $$\nu_t\colon \Delta_\infty \to \mu_{p^\infty}(\mathscr{O}_{\overline{K}}) = \varinjlim_{n\geq 0} \mu_{p^n}(\mathscr{O}_{\overline{K}}), \quad g \mapsto \langle g, t\rangle,$$

where $\langle g, t\rangle$ is defined in (II.7.14.1), induces, by virtue of (II.7.14.3), a homomorphism

(II.8.7.2) $$\nu_t\colon \Delta_{p^\infty} \to \mu_{p^\infty}(\mathscr{O}_{\overline{K}}).$$

It is clear that the map

(II.8.7.3) $$P_{p^\infty} \to \mathrm{Hom}(\Delta_{p^\infty}, \mu_{p^\infty}(\mathscr{O}_{\overline{K}})), \quad t \mapsto \nu_t$$

is a homomorphism. For every $n \geq 0$, denote by $\lambda^{(p^n)}$ the image of $\lambda$ in $P^{(p^n)}$ by the isomorphism (II.7.8.2). Since $\nu_{\lambda^{(p^n)}} = 1$, (II.8.7.3) induces a homomorphism that we denote also by

(II.8.7.4) $$\Lambda_{p^\infty} \to \mathrm{Hom}(\Delta_{p^\infty}, \mu_{p^\infty}(\mathscr{O}_{\overline{K}})), \quad x \mapsto \nu_x.$$

In fact, this induces, for every $n \geq 0$, a homomorphism

(II.8.7.5) $$\Lambda^{(p^n)} \to \mathrm{Hom}(\Delta_{p^\infty}, \mu_{p^n}(\mathscr{O}_{\overline{K}})).$$

Let

(II.8.7.6) $$\Xi_{p^\infty} = \mathrm{Hom}(\Delta_{p^\infty}, \mu_{p^\infty}(\mathscr{O}_{\overline{K}})),$$
(II.8.7.7) $$\Xi_{p^n} = \mathrm{Hom}(\Delta_{p^\infty}, \mu_{p^n}(\mathscr{O}_{\overline{K}})).$$

We identify $\Xi_{p^n}$ with a subgroup of $\Xi_{p^\infty}$.

**Lemma II.8.8.** *Under the assumptions of* (II.8.7), *let moreover $x \in \Lambda_{p^\infty}$, and $n$ be an integer $\geq 0$. Then we have $\nu_x(\Delta_{p^\infty}) \subset \mu_{p^n}(\mathscr{O}_{\overline{K}})$ if and only if $x \in \Lambda^{(p^n)} \subset \Lambda_{p^\infty}$; in particular, $\nu_x = 1$ if and only if $x \in \Lambda^{(1)} \subset \Lambda_{p^\infty}$.*

Suppose $x \in \Lambda^{(p^{n+m})}$ for an integer $m \geq 0$. By virtue of (II.7.14.4), we have a commutative diagram

(II.8.8.1)
$$\begin{array}{ccccc}
\Lambda^{(p^n)} & \xrightarrow{\varpi_{p^m}} & \Lambda^{(p^{n+m})} & & \\
\downarrow & & \downarrow & \searrow^{a} & \\
P^{\mathrm{gp}}/\mathbb{Z}\lambda & \xrightarrow{\cdot p^m} & P^{\mathrm{gp}}/\mathbb{Z}\lambda & & \\
\downarrow & & \downarrow & & \\
\mathrm{Hom}(\Delta_{p^\infty}, \mathbb{Z}_p(1)) & \xrightarrow{\cdot p^m} & \mathrm{Hom}(\Delta_{p^\infty}, \mathbb{Z}_p(1)) & \xrightarrow{b} & \mathrm{Hom}(\Delta_{p^\infty}, \mu_{p^{n+m}}(\mathscr{O}_{\overline{K}}))
\end{array}$$

where the lower vertical arrows come from the identification of $\Delta_{p^\infty}$ and $L_\lambda \otimes_{\mathbb{Z}} \mathbb{Z}_p(1)$ (II.6.10), $a$ is the homomorphism (II.8.7.5), and $b$ is the canonical morphism. The lower square is Cartesian because the torsion subgroup of $P^{\mathrm{gp}}/\mathbb{Z}\lambda$ is of order prime to $p$, and the upper square is Cartesian because $\Lambda$ is saturated by virtue of II.8.6(i). We have $\nu_x(\Delta_{p^\infty}) \subset \mu_{p^n}(\mathscr{O}_{\overline{K}})$ if and only if $a(x)$ is in the image of $p^m b$. The lemma then follows by a chase in the diagram (II.8.8.1).

**Lemma II.8.9.** *There exists a canonical decomposition of $R_{p^\infty}$ into a direct sum of $R_1$-modules of finite presentation that are stable under the action of $\Delta_{p^\infty}$,*

(II.8.9.1)
$$R_{p^\infty} = \bigoplus_{\nu \in \Xi_{p^\infty}} R_{p^\infty}^{(\nu)},$$

*such that the action of $\Delta_{p^\infty}$ on the factor $R_{p^\infty}^{(\nu)}$ is given by the character $\nu$. Moreover, for every $n \geq 0$, we have*

(II.8.9.2)
$$R_{p^n} = \bigoplus_{\nu \in \Xi_{p^n}} R_{p^\infty}^{(\nu)}.$$

For every $n \geq 1$, set $C_n = A_n \otimes_{\mathscr{O}_{K_n}} \mathscr{O}_{\overline{K}}$ (II.6.5.2) and denote by

(II.8.9.3)
$$\lambda_n \colon L \otimes_{\mathbb{Z}} \mu_n(\overline{K}) \to \mu_n(\overline{K})$$

the homomorphism defined by $\lambda$. We have a canonical isomorphism (II.6.6.4)

(II.8.9.4)
$$C_n \xrightarrow{\sim} \mathscr{O}_{\overline{K}}[P^{(n)}]/(\pi_n - e^{\lambda^{(n)}}) \otimes_{\mathscr{O}_{\overline{K}}[P]/(\pi - e^\lambda)} C_1,$$

where $\lambda^{(n)}$ denotes the image of $\lambda$ in $P^{(n)}$ by the canonical isomorphism (II.7.8.2). The group $L \otimes_{\mathbb{Z}} \mu_n(\overline{K})$ acts naturally on $\mathscr{O}_{\overline{K}}[P^{(n)}]$ by homomorphisms of $(\mathscr{O}_{\overline{K}}[P])$-algebras. In view of (II.8.9.4), we deduce from this an action of the group $\ker(\lambda_n)$ on $C_n$ by homomorphisms of $C_1$-algebras. Set

(II.8.9.5)
$$C_{p^\infty} = \varinjlim_{n \geq 0} C_{p^n},$$

where the direct limit is indexed by the set $\mathbb{N}$ with the usual ordering. We have canonical isomorphisms (II.6.8.4)

(II.8.9.6)
$$\Delta_{p^\infty} \xrightarrow{\sim} L_\lambda \otimes \mathbb{Z}_p(1) \xrightarrow{\sim} \varprojlim_{n \geq 0} \ker(\lambda_{p^n}).$$

Taking limits gives an action of $\Delta_{p^\infty}$ on $C_{p^\infty}$ by homomorphisms of $C_1$-algebras.

By virtue of II.6.8(iii), for every integer $n \geq 0$, we have a canonical $\ker(\lambda_{p^n})$-equivariant isomorphism

(II.8.9.7)
$$R_{p^n} \xrightarrow{\sim} C_{p^n} \otimes_{C_1} R_1.$$

From this we deduce a $\Delta_{p^\infty}$-equivariant isomorphism

$$R_{p^\infty} \xrightarrow{\sim} C_{p^\infty} \otimes_{C_1} R_1. \tag{II.8.9.8}$$

It therefore suffices to show that there exists a canonical decomposition of $C_{p^\infty}$ into a direct sum of $C_1$-modules of finite presentation that are stable under the action of $\Delta_{p^\infty}$,

$$C_{p^\infty} = \bigoplus_{\nu \in \Xi_{p^\infty}} C_{p^\infty}^{(\nu)}, \tag{II.8.9.9}$$

such that the action of $\Delta_{p^\infty}$ on the factor $C_{p^\infty}^{(\nu)}$ is given by the character $\nu$; moreover, for every $n \geq 0$, we have

$$C_{p^n} = \bigoplus_{\nu \in \Xi_{p^n}} C_{p^\infty}^{(\nu)}. \tag{II.8.9.10}$$

In view of (II.8.9.4), we can reduce to the case where $R = \mathscr{O}_K[P]/(\pi - e^\lambda)$. Note that $X$ is no longer necessarily connected after this reduction. Let $\Lambda$ be the cokernel in the category of monoids of the homomorphism $\vartheta \colon \mathbb{N} \to P$ and denote by $q \colon P \to \Lambda$ the canonical homomorphism (cf. II.8.5). We then have

$$R = \bigoplus_{x \in \Lambda} \mathscr{O}_K \cdot \alpha(\widetilde{x}), \tag{II.8.9.11}$$

where $\widetilde{x}$ is the minimal lift of $x$ in $P$ defined in II.8.6(ii) and $\alpha \colon P \to R$ is the homomorphism induced by the chart $(P, \gamma)$ (II.6.2). By (II.8.9.4), for every $n \geq 0$, we have

$$C_{p^n} = \mathscr{O}_{\overline{K}}[P^{(p^n)}]/(\pi_{p^n} - e^{\lambda^{(p^n)}}). \tag{II.8.9.12}$$

Denote by $\alpha_{p^n} \colon P^{(p^n)} \to C_{p^n}$ the homomorphism induced by the canonical strict morphism (II.6.5.1)

$$(X_n, \mathscr{M}_{X_n}) \to \mathbf{A}_P.$$

This notation is compatible with that introduced in II.7.8 and does not lead to any confusion. Taking the direct limit, we obtain a homomorphism

$$\alpha_{p^\infty} \colon P_{p^\infty} \to C_{p^\infty}. \tag{II.8.9.13}$$

We take again the notation of II.8.7 and for every $x \in \Lambda^{(p^n)}$, we denote by $\widetilde{x} \in P^{(p^n)}$ the unique minimal element of $q^{-1}(x)$ for the preordering on $P^{(p^n)}$ defined by its monoid structure (cf. II.8.6(ii)). We then have

$$C_{p^n} = \bigoplus_{x \in \Lambda^{(p^n)}} \mathscr{O}_{\overline{K}} \cdot \alpha_{p^n}(\widetilde{x}). \tag{II.8.9.14}$$

By virtue of II.8.6(iii), the maps $\Lambda^{(p^n)} \to P^{(p^n)}, x \mapsto \widetilde{x}$ are compatible. Therefore, by taking the direct limit, they define a map that we denote also by

$$\Lambda_{p^\infty} \to P_{p^\infty}, \quad x \mapsto \widetilde{x}. \tag{II.8.9.15}$$

We clearly have $q(\widetilde{x}) = x$. Since $\Lambda$ is integral and the torsion subgroup of $P^{\mathrm{gp}}/\mathbb{Z}\lambda$ is of order prime to $p$, the homomorphisms $\Lambda^{(p^n)} \to \Lambda^{(p^{n+1})}$ are injective. Taking the direct limit of the decomposition (II.8.9.14), we obtain

$$C_{p^\infty} = \bigoplus_{x \in \Lambda_{p^\infty}} \mathscr{O}_{\overline{K}} \cdot \alpha_{p^\infty}(\widetilde{x}). \tag{II.8.9.16}$$

Each element $\alpha_{p^\infty}(\widetilde{x}) \in C_{p^\infty}$ is an eigenvector for the action of $\Delta_{p^\infty}$. It immediately follows from (II.7.14.4) that the action of $\Delta_{p^\infty}$ on $\alpha_{p^\infty}(\widetilde{x})$ is given by the character $\nu_x$. For every $\nu \in \Xi_{p^\infty}$, we then set

(II.8.9.17) $$C_{p^\infty}^{(\nu)} = \bigoplus_{x \in \Lambda_{p^\infty} | \nu_x = \nu} \mathscr{O}_{\overline{K}} \cdot \alpha_{p^\infty}(\widetilde{x}),$$

so that we have

(II.8.9.18) $$C_{p^\infty} = \bigoplus_{\nu \in \Xi_{p^\infty}} C_{p^\infty}^{(\nu)}.$$

Since the map $\Lambda_{p^\infty} \to \Xi_{p^\infty}, x \mapsto \nu_x$ is a homomorphism, we see that $C_{p^\infty}^{(\nu)}$ is a sub-$C_1$-module of $C_{p^\infty}$. It follows from II.8.8 and (II.8.9.14) that for every $n \geq 0$, we have

(II.8.9.19) $$C_{p^n} = \bigoplus_{\nu \in \Xi_{p^n}} C_{p^\infty}^{(\nu)}.$$

Since $C_{p^n}$ is of finite presentation over $C_1$ for every $n \geq 0$, $C_{p^\infty}^{(\nu)}$ is of finite presentation over $C_1$ for every $\nu \in \Xi_{p^\infty}$.

**Theorem II.8.10.** *Let $a$ be a nonzero element of $\mathscr{O}_{\overline{K}}$ and $\zeta$ a primitive $p$th root of unity in $\mathscr{O}_{\overline{K}}$. Then:*

(i) *There exists a unique homomorphism of graded $R_1$-algebras*

(II.8.10.1) $$\wedge\left(\mathrm{Hom}_{\mathbb{Z}}(\Delta_{p^\infty}, R_1/aR_1)\right) \to \mathrm{H}^*(\Delta_{p^\infty}, R_{p^\infty}/aR_{p^\infty})$$

*whose degree one component is the composition of the canonical morphisms*

(II.8.10.2) $$\mathrm{Hom}_{\mathbb{Z}}(\Delta_{p^\infty}, R_1/aR_1) \xrightarrow{\sim} \mathrm{H}^1(\Delta_{p^\infty}, R_1/aR_1) \to \mathrm{H}^1(\Delta_{p^\infty}, R_{p^\infty}/aR_{p^\infty}).$$

*As a morphism of graded $R_1$-modules, this admits a canonical left inverse*

(II.8.10.3) $$\mathrm{H}^*(\Delta_{p^\infty}, R_{p^\infty}/aR_{p^\infty}) \to \wedge(\mathrm{Hom}_{\mathbb{Z}}(\Delta_{p^\infty}, R_1/aR_1)),$$

*whose kernel is annihilated by $\zeta - 1$.*

(ii) *The $R_1$-module $\mathrm{H}^i(\Delta_{p^\infty}, R_{p^\infty}/aR_{p^\infty})$ is almost of finite presentation for every $i \geq 0$ and is zero for every $i \geq \mathrm{rg}(L) = d+1$* (II.6.2.5).

(iii) *The inverse system $(\mathrm{H}^*(\Delta_{p^\infty}, R_{p^\infty}/p^r R_{p^\infty}))_{r \geq 0}$ satisfies the Mittag–Leffler condition; more precisely, if for all integers $r' \geq r \geq 0$, we denote by*

(II.8.10.4) $$h_{r,r'}\colon \mathrm{H}^*(\Delta_{p^\infty}, R_{p^\infty}/p^{r'} R_{p^\infty}) \to \mathrm{H}^*(\Delta_{p^\infty}, R_{p^\infty}/p^r R_{p^\infty})$$

*the canonical morphism, then for every integer $r \geq 1$, there exists an integer $r' \geq r$ depending only on $d$ and not on the other data in* II.6.2, *such that for every integer $r'' \geq r'$, the images of $h_{r,r'}$ and $h_{r,r''}$ are equal.*

Indeed, by II.8.9, we have a canonical decomposition of $R_{p^\infty}$ into a direct sum of $R_1[\Delta_{p^\infty}]$-modules

(II.8.10.5) $$R_{p^\infty} = \bigoplus_{\nu \in \Xi_{p^\infty}} R_{p^\infty}^{(\nu)} \otimes_{\mathscr{O}_{\overline{K}}} \mathscr{O}_{\overline{K}}(\nu),$$

where $\Delta_{p^\infty}$ acts trivially on $R_{p^\infty}^{(\nu)}$ and acts on $\mathscr{O}_{\overline{K}}(\nu) = \mathscr{O}_{\overline{K}}$ by the character $\nu$. Since the $R_{p^\infty}^{(\nu)}$'s are $\mathscr{O}_{\overline{K}}$-flat, by virtue of II.3.15, we have a canonical decomposition into a direct sum of $R_1$-modules

(II.8.10.6) $$\mathrm{H}^*(\Delta_{p^\infty}, R_{p^\infty}/aR_{p^\infty}) = \bigoplus_{\nu \in \Xi_{p^\infty}} \mathrm{H}^*(\Delta_{p^\infty}, \mathscr{O}_{\overline{K}}(\nu)/a\mathscr{O}_{\overline{K}}(\nu)) \otimes_{\mathscr{O}_{\overline{K}}} R_{p^\infty}^{(\nu)}.$$

(i) We have $R^{(1)}_{p^\infty} = R_1$ (II.8.9.2), so that the component for $\nu = 1$ of the decomposition (II.8.10.6) is the image of the canonical homomorphism of graded $R_1$-algebras

$$\text{(II.8.10.7)} \qquad \mathrm{H}^*(\Delta_{p^\infty}, \mathscr{O}_{\overline{K}}/a\mathscr{O}_{\overline{K}}) \otimes_{\mathscr{O}_{\overline{K}}} R_1 \to \mathrm{H}^*(\Delta_{p^\infty}, R_{p^\infty}/aR_{p^\infty}).$$

Moreover, the canonical homomorphism of graded $R_1$-algebras

$$\text{(II.8.10.8)} \qquad \wedge\,(\mathrm{Hom}_{\mathbb{Z}}(\Delta_{p^\infty}, \mathscr{O}_{\overline{K}}/a\mathscr{O}_{\overline{K}})) \otimes_{\mathscr{O}_{\overline{K}}} R_1 \to \wedge(\mathrm{Hom}_{\mathbb{Z}}(\Delta_{p^\infty}, R_1/aR_1))$$

is an isomorphism. The statement then follows from (II.8.10.6) and II.8.1 (applied with $A = \mathscr{O}_C$).

(ii) Let $i, n$ be two integers $\geq 0$ and $\zeta_n$ a primitive $p^n$th root of unity. It follows from (II.8.10.6), II.8.1, and II.8.9 that $\mathrm{H}^i(\Delta_{p^\infty}, R_{p^\infty}/aR_{p^\infty})$ is the direct sum of an $R_1$-module of finite presentation and an $R_1$-module annihilated by $\zeta_n - 1$. It is therefore almost of finite presentation over $R_1$. The second assertion is clear because the $p$-cohomological dimension of $\Delta_{p^\infty}$ is equal to $d$.

(iii) This follows from (II.8.10.6) and II.8.1(iii).

**Corollary II.8.11.** *For every nonzero element $a$ of $\mathscr{O}_{\overline{K}}$ and every integer $i \geq 0$, the kernel and cokernel of the canonical morphism*

$$\text{(II.8.11.1)} \qquad \mathrm{H}^i(\Delta_{p^\infty}, R_1/aR_1) \to \mathrm{H}^i(\Delta, \overline{R}/a\overline{R})$$

*are annihilated by $\mathfrak{m}_{\overline{K}}$ and $p^{\frac{1}{p-1}}\mathfrak{m}_{\overline{K}}$, respectively.*

Indeed, the canonical morphism

$$\text{(II.8.11.2)} \qquad \mathrm{H}^i(\Delta_{p^\infty}, R_{p^\infty}/aR_{p^\infty}) \to \mathrm{H}^i(\Delta, \overline{R}/a\overline{R})$$

is an almost isomorphism by II.6.24. On the other hand, it follows from (II.8.10.6), II.8.1 (applied with $A = \mathscr{O}_C$), II.3.15, and the fact that $R^{(1)}_{p^\infty} = R_1$ (II.8.9.2), that the canonical morphism

$$\text{(II.8.11.3)} \qquad \mathrm{H}^i(\Delta_{p^\infty}, R_1/aR_1) \to \mathrm{H}^i(\Delta_{p^\infty}, R_{p^\infty}/aR_{p^\infty})$$

is injective with cokernel annihilated by $\zeta - 1$, where $\zeta$ is a primitive $p$th root of unity. The statement follows because $v(\zeta - 1) = \frac{1}{p-1}$.

**Corollary II.8.12.** *The canonical homomorphism*

$$\text{(II.8.12.1)} \qquad \mathrm{H}^*_{\mathrm{cont}}(\Delta_{p^\infty}, \widehat{R_{p^\infty}}) \to \varprojlim_{r\geq 0} \mathrm{H}^*(\Delta_{p^\infty}, R_{p^\infty}/p^r R_{p^\infty})$$

*is an isomorphism.*

Indeed, by (II.3.10.4) and (II.3.10.5), for every $i \geq 0$, we have an exact sequence

$$\begin{aligned} 0 \to \mathrm{R}^1&\varprojlim_{r\geq 0} \mathrm{H}^{i-1}(\Delta_{p^\infty}, R_{p^\infty}/p^r R_{p^\infty}) \\ &\to \mathrm{H}^i_{\mathrm{cont}}(\Delta_{p^\infty}, \widehat{R_{p^\infty}}) \to \varprojlim_{r\geq 0} \mathrm{H}^i(\Delta_{p^\infty}, R_{p^\infty}/p^r R_{p^\infty}) \to 0 \end{aligned}$$

whose left term is zero by virtue of II.8.10(iii) and (II.3.10.2).

**II.8.13.** Let $a$ be a nonzero element of $\mathscr{O}_{\overline{K}}$. Since the torsion subgroup of $P^{\mathrm{gp}}/\mathbb{Z}\lambda$ is of order prime to $p$, we have a canonical isomorphism (II.6.10)

$$(P^{\mathrm{gp}}/\mathbb{Z}\lambda)\otimes_{\mathbb{Z}}\mathbb{Z}_p(-1)\xrightarrow{\sim}\mathrm{Hom}_{\mathbb{Z}_p}(\Delta_{p^\infty},\mathbb{Z}_p). \tag{II.8.13.1}$$

In view of (II.7.12.1), we deduce from this $\widehat{R_1}$-linear isomorphisms

$$\widetilde{\Omega}^1_{R/\mathscr{O}_K}\otimes_R(R_1/aR_1)(-1)\xrightarrow{\sim}\mathrm{Hom}_{\mathbb{Z}}(\Delta_{p^\infty},R_1/aR_1), \tag{II.8.13.2}$$

$$\widetilde{\Omega}^1_{R/\mathscr{O}_K}\otimes_R\widehat{R_1}(-1)\xrightarrow{\sim}\mathrm{Hom}_{\mathbb{Z}}(\Delta_{p^\infty},\widehat{R_1}). \tag{II.8.13.3}$$

We will, from now on, view the targets of these morphisms as cohomology groups (II.6.12). Moreover, the composition of (II.8.13.3) and the canonical isomorphism (II.6.12.1)

$$\mathrm{Hom}_{\mathbb{Z}}(\Delta_{p^\infty},\widehat{R_1})\xrightarrow{\sim}\mathrm{Hom}_{\mathbb{Z}}(\Delta_\infty,\widehat{R_1}) \tag{II.8.13.4}$$

is none other than the morphism $\widetilde{\delta}(-1)$ (II.7.19.9), by II.7.19.

**Proposition II.8.14.** *There exists a unique homomorphism of graded $\widehat{R_1}$-algebras*

$$\wedge\,(\widetilde{\Omega}^1_{R/\mathscr{O}_K}\otimes_R\widehat{R_1}(-1))\to\mathrm{H}^*_{\mathrm{cont}}(\Delta_{p^\infty},\widehat{R_{p^\infty}}) \tag{II.8.14.1}$$

*whose degree one component is induced by* (II.8.13.3)*. As a morphism of graded $\widehat{R_1}$-modules, it admits a canonical left inverse*

$$\mathrm{H}^*_{\mathrm{cont}}(\Delta_{p^\infty},\widehat{R_{p^\infty}})\to\wedge(\widetilde{\Omega}^1_{R/\mathscr{O}_K}\otimes_R\widehat{R_1}(-1)), \tag{II.8.14.2}$$

*whose kernel is annihilated by $p^{\frac{1}{p-1}}$.*

This follows from II.8.10(i) and II.8.12.

**Corollary II.8.15.** *We have $(\widehat{R_{p^\infty}})^{\Delta_{p^\infty}}=\widehat{R_1}$.*

Indeed, by II.8.14, the canonical homomorphism $\widehat{R_1}\to(\widehat{R_{p^\infty}})^{\Delta_{p^\infty}}$ admits a left inverse $(\widehat{R_{p^\infty}})^{\Delta_{p^\infty}}\to\widehat{R_1}$ whose kernel is annihilated by $p^{\frac{1}{p-1}}$. Since $\widehat{R_{p^\infty}}$ is $\mathscr{O}_C$-flat (II.6.14), the left inverse is injective, proving the statement.

**Remark II.8.16.** By II.6.13, (II.6.11.1), and (II.6.12.1), theorem II.8.10 and its corollaries II.8.11 and II.8.12 still hold when we replace $\Delta_{p^\infty}$ by $\Delta_\infty$ and $R_{p^\infty}$ by $R_\infty$. The same then holds for proposition II.8.14 and its corollary II.8.15. We therefore have $(\widehat{R_\infty})^{\Delta_\infty}=\widehat{R_1}$.

**Proposition II.8.17.** *Let $a$ be a nonzero element of $\mathscr{O}_{\overline{K}}$.*

(i) *There exists a unique homomorphism of graded $R_1$-algebras*

$$\wedge\,(\widetilde{\Omega}^1_{R/\mathscr{O}_K}\otimes_R(R_1/aR_1)(-1))\to\mathrm{H}^*(\Delta,\overline{R}/a\overline{R}) \tag{II.8.17.1}$$

*whose degree one component is induced by* (II.8.13.2)*. It is almost injective and its cokernel is annihilated by $p^{\frac{1}{p-1}}\mathfrak{m}_{\overline{K}}$.*

(ii) *The $R_1$-module $\mathrm{H}^i(\Delta,\overline{R}/a\overline{R})$ is almost of finite presentation for every $i\geq 0$ and almost zero for every $i\geq d+1$.*

(iii) *For all integers $r'\geq r\geq 0$, denote by*

$$\hbar_{r,r'}\colon\mathrm{H}^*(\Delta,\overline{R}/p^{r'}\overline{R})\to\mathrm{H}^*(\Delta,\overline{R}/p^{r}\overline{R}) \tag{II.8.17.2}$$

*the canonical morphism. Then for every integer $r\geq 1$, there exists an integer $r'\geq r$ depending only on $d$ and not on the other data in II.6.2, such that for every integer $r''\geq r'$, the images of $\hbar_{r,r'}$ and $\hbar_{r,r''}$ are almost isomorphic.*

This follows from II.6.24 and II.8.10.

**II.8.18.** For every $\mathscr{O}_{\overline{K}}$-module $M$, set

$$M^\flat = \mathrm{Hom}_{\mathscr{O}_{\overline{K}}}(\mathfrak{m}_{\overline{K}}, M). \tag{II.8.18.1}$$

The canonical morphism $M \to M^\flat$ is an almost isomorphism. A morphism of $\mathscr{O}_{\overline{K}}$-modules $u\colon M \to N$ is an almost isomorphism if and only if the associated morphism $u^\flat\colon M^\flat \to N^\flat$ is an isomorphism (V.2.5).

**Lemma II.8.19.** *The canonical morphism*

$$\jmath\colon \widehat{R_1} \to (\widehat{R_1})^\flat \tag{II.8.19.1}$$

*is an isomorphism.*

Since $\widehat{R_1}$ is flat over $\mathscr{O}_C$ (II.6.14), $\jmath$ is injective. Let us show that $\jmath$ is surjective. Let $u \in (\widehat{R_1})^\flat$. For every $\alpha \in \mathbb{Q}_{>0}$, set $x_\alpha = u(p^\alpha)$. Recall that $\widehat{R_1}[\frac{1}{p}]$ is an affinoid algebra over $C$ and that $\widehat{R_1}$ is the unit ball for the norm $|\ |_{\sup}$ on $\widehat{R_1}[\frac{1}{p}]$ (cf. the proof of II.6.15). The relations $x_{\alpha+\beta} = p^\alpha x_\beta = p^\beta x_\alpha$ ($\alpha, \beta \in \mathbb{Q}_{>0}$) then imply that

$$|p^{-\alpha} x_\alpha|_{\sup} \leq 1. \tag{II.8.19.2}$$

Consequently, $x = p^{-\alpha} x_\alpha \in \widehat{R_1}$ and $x$ is independent of $\alpha$. It is clear that $\jmath(x) = u$.

**Remark II.8.20.** The canonical morphism $\mathscr{O}_C \to (\mathscr{O}_C)^\flat$ is an isomorphism. The proof is a very simple variant of that of II.8.19.

**Proposition II.8.21.** *There exists a unique homomorphism of graded $\widehat{R_1}$-algebras*

$$\wedge\, (\widetilde{\Omega}^1_{R/\mathscr{O}_K} \otimes_R \widehat{R_1}(-1)) \to \mathrm{H}^*_{\mathrm{cont}}(\Delta, \widehat{\overline{R}}) \tag{II.8.21.1}$$

*whose degree one component is induced by* (II.8.13.3). *As a morphism of graded $\widehat{R_1}$-modules, it admits a canonical left inverse*

$$\mathrm{H}^*_{\mathrm{cont}}(\Delta, \widehat{\overline{R}}) \to \wedge(\widetilde{\Omega}^1_{R/\mathscr{O}_K} \otimes_R \widehat{R_1}(-1)), \tag{II.8.21.2}$$

*whose kernel is annihilated by $p^{\frac{1}{p-1}}$.*

Indeed, we have a commutative diagram of homomorphisms of graded $\widehat{R_1}$-algebras

$$\begin{array}{ccccc}
\wedge(\widetilde{\Omega}^1_{R/\mathscr{O}_K} \otimes_R \widehat{R_1}(-1)) & \underset{\overset{w}{\longleftarrow}}{\xrightarrow{\ v\ }} & \mathrm{H}^*_{\mathrm{cont}}(\Delta_{p^\infty}, \widehat{R_{p^\infty}}) & \xrightarrow{\ u\ } & \mathrm{H}^*_{\mathrm{cont}}(\Delta, \widehat{\overline{R}}) \\
\Big\downarrow{\scriptstyle \iota} & & \Big\downarrow & & \Big\downarrow \\
(\wedge(\widetilde{\Omega}^1_{R/\mathscr{O}_K} \otimes_R \widehat{R_1}(-1)))^\flat & \overset{\xrightarrow{\ v^\flat\ }}{\underset{w^\flat}{\longleftarrow}} & (\mathrm{H}^*_{\mathrm{cont}}(\Delta_{p^\infty}, \widehat{R_{p^\infty}}))^\flat & \xrightarrow{\ u^\flat\ } & (\mathrm{H}^*_{\mathrm{cont}}(\Delta, \widehat{\overline{R}}))^\flat
\end{array} \tag{II.8.21.3}$$

where the functor $(\ )^\flat$ is defined in (II.8.18.1), the vertical arrows and $u$ are the canonical morphisms, $v$ is the homomorphism (II.8.14.1), and $w$ is the section (II.8.14.2) of $v$. Then $\iota$ and $u^\flat$ are isomorphisms (II.8.19 and II.6.25). On the other hand, it follows from II.8.14 that the kernel of $w^\flat$ is annihilated by $p^{\frac{1}{p-1}}$. The proposition follows by a chase in the diagram (II.8.21.3) because $u \circ v$ is the homomorphism (II.8.21.1).

**Corollary II.8.22.** (i) *We have $(\widehat{\overline{R}})^\Delta = \widehat{R_1}$.*

(ii) *The $p$-primary torsion submodule $M^1$ of $\mathrm{H}^1_{\mathrm{cont}}(\Delta, \widehat{\overline{R}})$ is equal to the kernel of the morphism*

$$\iota\colon \mathrm{H}^1_{\mathrm{cont}}(\Delta, \widehat{\overline{R}}) \to \mathrm{H}^1_{\mathrm{cont}}(\Delta, (\pi\rho)^{-1}\widehat{\overline{R}}) \tag{II.8.22.1}$$

*deduced from the canonical injection* $\widehat{\overline{R}} \subset (\pi\rho)^{-1}\widehat{\overline{R}}$.

(iii) *The composition*

$$\widetilde{\Omega}^1_{R/\mathscr{O}_K} \otimes_R \widehat{R_1}(-1) \longrightarrow \mathrm{H}^1_{\mathrm{cont}}(\Delta, \widehat{\overline{R}}) \xrightarrow{\iota} \mathrm{H}^1_{\mathrm{cont}}(\Delta, (\pi\rho)^{-1}\widehat{\overline{R}}), \tag{II.8.22.2}$$

*where the first arrow is induced by* (II.8.13.3)*, is the boundary map of the long exact sequence of cohomology deduced from the short exact sequence* (II.7.22.2).

(iv) *For every integer* $i \geq 0$*, denote by* $M^i$ *the* $p$*-primary torsion submodule of* $\mathrm{H}^i_{\mathrm{cont}}(\Delta, \widehat{\overline{R}})$*. Then there exists a unique isomorphism of graded* $\widehat{R_1}$*-algebras*

$$\wedge\,(\widetilde{\Omega}^1_{R/\mathscr{O}_K} \otimes_R \widehat{R_1}(-1)) \xrightarrow{\sim} \oplus_{i\geq 0}(\mathrm{H}^i_{\mathrm{cont}}(\Delta, \widehat{\overline{R}})/M^i) \tag{II.8.22.3}$$

*such that the composition*

$$\widetilde{\Omega}^1_{R/\mathscr{O}_K} \otimes_R \widehat{R_1}(-1) \to \mathrm{H}^1_{\mathrm{cont}}(\Delta, \widehat{\overline{R}})/M^1 \to \mathrm{H}^1_{\mathrm{cont}}(\Delta, (\pi\rho)^{-1}\widehat{\overline{R}}), \tag{II.8.22.4}$$

*where the first arrow is the degree one component of* (II.8.22.3) *and the second arrow is induced by* $\iota$*, is the boundary map of the long exact sequence of cohomology deduced from the short exact sequence* (II.7.22.2).

(i) By II.8.21, the canonical homomorphism $\widehat{R_1} \to (\widehat{\overline{R}})^{\Delta}$ admits a left inverse $(\widehat{\overline{R}})^{\Delta} \to \widehat{R_1}$ whose kernel is annihilated by $p^{\frac{1}{p-1}}$. Since $\widehat{\overline{R}}$ is $\mathscr{O}_C$-flat (II.6.14), the left inverse is injective, proving the statement.

(ii) It follows from II.8.21 that $M^1$ is annihilated by $p^{\frac{1}{p-1}}$. On the other hand, the multiplication by $\pi\rho$ in $\widehat{\overline{R}}$ is the composition of the canonical injection $\widehat{\overline{R}} \to (\pi\rho)^{-1}\widehat{\overline{R}}$ and the isomorphism $(\pi\rho)^{-1}\widehat{\overline{R}} \xrightarrow{\sim} \widehat{\overline{R}}$ that, for every $x \in \widehat{\overline{R}}$, sends $(\pi\rho)^{-1}x$ to $x$. Since $v(\rho) \geq \frac{1}{p-1}$, we deduce from this that $\ker(\iota) = M^1$.

(iii) This follows from the definition of (II.8.13.3), (II.7.20.2), and (II.7.22.3).

(iv) This follows from (ii), (iii), and II.8.21.

**Proposition II.8.23.** *Let* $M$ *be a discrete* $R_1$*-*$\Delta_{p^\infty}$*-module,* $a$ *a nonzero element of* $\mathscr{O}_{\overline{K}}$*, and* $\alpha$ *a rational number. Suppose* $\inf(v(a), \alpha) > \frac{1}{p-1}$ *and that* $M$ *is a projective* $(R_1/aR_1)$*-module of finite type generated by a finite number of elements that are* $\Delta_{p^\infty}$*-invariant modulo* $p^\alpha M$*. Let* $b = ap^{-\frac{1}{p-1}}$*. Then for every* $i \geq 0$*, the kernel and cokernel of the canonical morphism*

$$\mathrm{H}^i(\Delta_{p^\infty}, M/bM) \to \mathrm{H}^i(\Delta, (M/bM) \otimes_{R_1} \overline{R}) \tag{II.8.23.1}$$

*are annihilated by* $\mathfrak{m}_{\overline{K}}$ *and* $p^{\frac{1}{p-1}}\mathfrak{m}_{\overline{K}}$*, respectively.*

Since $M$ is a direct summand of a free $(R_1/aR_1)$-module of finite type, we have, by II.6.13,

$$((M/bM) \otimes_{R_1} R_\infty)^{\Sigma_0} = (M/bM) \otimes_{R_1} R_{p^\infty}. \tag{II.8.23.2}$$

Therefore by virtue of (II.6.11.1), the canonical morphism

$$\mathrm{H}^i(\Delta_{p^\infty}, (M/bM) \otimes_{R_1} R_{p^\infty}) \to \mathrm{H}^i(\Delta_\infty, (M/bM) \otimes_{R_1} R_\infty) \tag{II.8.23.3}$$

is an isomorphism. On the other hand, by II.6.22, the canonical morphism

$$(M/bM) \otimes_{R_1} R_\infty \to ((M/bM) \otimes_{R_1} \overline{R})^{\Sigma} \tag{II.8.23.4}$$

is an almost isomorphism. Consequently, by virtue of II.6.20, the canonical morphism

$$\mathrm{H}^i(\Delta_\infty, (M/bM) \otimes_{R_1} R_\infty) \to \mathrm{H}^i(\Delta, (M/bM) \otimes_{R_1} \overline{R}) \tag{II.8.23.5}$$

is an almost isomorphism. It therefore suffices to show that the canonical morphism

$$\mathrm{H}^i(\Delta_{p^\infty}, M/bM) \to \mathrm{H}^i(\Delta_{p^\infty}, (M/bM) \otimes_{R_1} R_{p^\infty}) \tag{II.8.23.6}$$

is injective with cokernel annihilated by $p^{\frac{1}{p-1}}$.

By II.8.9, we have a canonical decomposition of $(M/bM)\otimes_{R_1} R_{p^\infty}$ into a direct sum of $R_1[\Delta_{p^\infty}]$-modules

$$(M/bM)\otimes_{R_1} R_{p^\infty} = \bigoplus_{\nu\in\Xi_{p^\infty}} (M/bM)\otimes_{R_1} R_{p^\infty}^{(\nu)}\otimes_{R_1} R_1(\nu), \tag{II.8.23.7}$$

where $\Delta_{p^\infty}$ acts trivially on $R_{p^\infty}^{(\nu)}$ and acts on $R_1(\nu)=R_1$ by the character $\nu$. Since $R_{p^\infty}^{(1)}=R_1$ (II.8.9.2), the statement follows from II.8.4 (applied to $A=R_1/aR_1$ and $N=R_{p^\infty}^{(\nu)}/aR_{p^\infty}^{(\nu)}$ for $\nu\neq 1$).

## II.9. Fontaine $p$-adic infinitesimal thickenings

**II.9.1.** Let us begin by recalling the following construction due to Grothendieck ([**36**] IV 3.3). Let $A$ be a commutative $\mathbb{Z}_{(p)}$-algebra and $n$ an integer $\geq 1$. The ring homomorphism (II.2.3.1)

$$\begin{array}{rrcc} \Phi_{n+1}\colon & \mathrm{W}_{n+1}(A/p^nA) & \to & A/p^nA \\ & (x_1,\dots,x_{n+1}) & \mapsto & x_1^{p^n}+px_2^{p^{n-1}}+\dots+p^nx_{n+1} \end{array} \tag{II.9.1.1}$$

vanishes on $\mathrm{V}^n(A/p^nA)$ and therefore induces, by taking the quotient, a ring homomorphism

$$\begin{array}{rrcc} \Phi'_{n+1}\colon & \mathrm{W}_{n}(A/p^nA) & \to & A/p^nA \\ & (x_1,\dots,x_{n}) & \mapsto & x_1^{p^n}+px_2^{p^{n-1}}+\dots+p^{n-1}x_n^p. \end{array} \tag{II.9.1.2}$$

The latter vanishes on

$$\mathrm{W}_n(pA/p^nA)=\ker(\mathrm{W}_n(A/p^nA)\to \mathrm{W}_n(A/pA)) \tag{II.9.1.3}$$

and in turn factors into a ring homomorphism

$$\theta_n\colon \mathrm{W}_n(A/pA)\to A/p^nA. \tag{II.9.1.4}$$

It immediately follows from the definition that the diagram

$$\begin{array}{ccc} \mathrm{W}_{n+1}(A/pA) & \xrightarrow{\theta_{n+1}} & A/p^{n+1}A \\ \Big\downarrow{\scriptstyle \mathrm{RF}} & & \Big\downarrow \\ \mathrm{W}_n(A/pA) & \xrightarrow{\theta_n} & A/p^nA \end{array} \tag{II.9.1.5}$$

where R is the restriction morphism (II.2.3.2), F is the Frobenius (II.2.3.4), and the unlabeled arrow is the canonical homomorphism, is commutative.

For every homomorphism of commutative $\mathbb{Z}_{(p)}$-algebras $\varphi\colon A\to B$, the diagram

$$\begin{array}{ccc} \mathrm{W}_n(A/pA) & \longrightarrow & \mathrm{W}_n(B/pB) \\ {\scriptstyle\theta_n}\Big\downarrow & & \Big\downarrow{\scriptstyle\theta_n} \\ A/p^nA & \longrightarrow & B/p^nB \end{array} \tag{II.9.1.6}$$

where the horizontal arrows are the morphisms induced by $\varphi$, is commutative.

**Proposition II.9.2.** *Let $A$ be a commutative $\mathbb{Z}_{(p)}$-algebra satisfying the following conditions:*

(i) *$A$ is $\mathbb{Z}_{(p)}$-flat.*
(ii) *$A$ is integrally closed in $A[\frac{1}{p}]$.*
(iii) *The absolute Frobenius of $A/pA$ is surjective.*

(iv) *There exist an integer $N \geq 1$ and a sequence $(p_n)_{0\leq n\leq N}$ of elements of $A$ such that $p_0 = p$ and $p_{n+1}^p = p_n$ for every $0 \leq n \leq N-1$.*

*For every integer $0 \leq n \leq N$, we set*

(II.9.2.1) $$\xi_n = [\overline{p}_n] - p \in \mathrm{W}_n(A/pA),$$

*where $\overline{p}_n$ is the class of $p_n$ in $A/pA$ and $[\overline{p}_n]$ is the multiplicative representative of $\overline{p}_n$. Then for all integers $n \geq 1$ and $i \geq 0$ such that $n+i \leq N$, the sequence*

(II.9.2.2) $$\mathrm{W}_n(A/pA) \xrightarrow{\cdot \mathrm{R}^i(\xi_{n+i})} \mathrm{W}_n(A/pA) \xrightarrow{\theta_n \circ \mathrm{F}^i} A/p^nA \longrightarrow 0$$

*is exact.*

Indeed, we have $\mathrm{FR}(\xi_{n+1}) = \xi_n$ and $\theta_n(\mathrm{F}^i(\mathrm{R}^i(\xi_{n+i}))) = \theta_n(\xi_n) = 0$. To establish the exactness of (II.9.2.2), we proceed by induction on $n$. Let $i$ be an integer with $0 \leq i \leq N-1$ and $\alpha, \beta \in A$ such that $\alpha^{p^{i+1}} = p\beta$. It follows from conditions (i) and (ii) that $\alpha \in p_{i+1}A$. Consequently, the sequence

(II.9.2.3) $$A/pA \xrightarrow{\cdot \overline{p}_{i+1}} A/pA \xrightarrow{\mathrm{F}^{i+1}} A/pA \longrightarrow 0$$

is exact and the statement is true for $n = 1$. Next, assume that the statement is true for $1 \leq n \leq N-1$ (and every $0 \leq i \leq N-n$) and let us prove it for $n+1$. Let $i$ be an integer with $0 \leq i \leq N-n-1$. We have a commutative diagram with exact lines (II.9.1.5)

(II.9.2.4) $$\begin{array}{ccccccccc}
0 & \longrightarrow & A/pA & \xrightarrow{\mathrm{V}^n} & \mathrm{W}_{n+1}(A/pA) & \xrightarrow{\mathrm{R}} & \mathrm{W}_n(A/pA) & \longrightarrow & 0 \\
 & & \downarrow{\scriptstyle \cdot \mathrm{R}^i(\xi_{i+1})} & & \downarrow{\scriptstyle \cdot \mathrm{R}^i(\xi_{n+i+1})} & & \downarrow{\scriptstyle \cdot \mathrm{R}^{i+1}(\xi_{n+i+1})} & & \\
0 & \longrightarrow & A/pA & \xrightarrow{\mathrm{V}^n} & \mathrm{W}_{n+1}(A/pA) & \xrightarrow{\mathrm{R}} & \mathrm{W}_n(A/pA) & \longrightarrow & 0 \\
 & & \downarrow{\scriptstyle \mathrm{F}^{i+1}} & & & & & & \\
 & & A/pA & & \downarrow{\scriptstyle \theta_{n+1}\circ \mathrm{F}^i} & & \downarrow{\scriptstyle \theta_n \circ \mathrm{F}^{i+1}} & & \\
 & & \downarrow{\scriptstyle \cdot p^n} & & & & & & \\
0 & \longrightarrow & p^nA/p^{n+1}A & \longrightarrow & A/p^{n+1}A & \longrightarrow & A/p^nA & \longrightarrow & 0
\end{array}$$

The induction hypothesis and (II.9.2.3) then imply the statement for $n+1$ using the snake lemma.

**II.9.3.** Let $A$ be a commutative $\mathbb{Z}_{(p)}$-algebra. We denote by $\mathscr{R}_A$ the inverse limit of the inverse system $(A/pA)_{\mathbb{N}}$ whose transition morphisms are the iterates of the Frobenius homomorphisms of $A/pA$:

(II.9.3.1) $$\mathscr{R}_A = \varprojlim_{x \mapsto x^p} A/pA.$$

This is a perfect ring of characteristic $p$. For every integer $n \geq 1$, the canonical projection $\mathscr{R}_A \to A/pA$ onto the $(n+1)$th component of the inverse system $(A/pA)_{\mathbb{N}}$ (that is, the component of index $n$) induces a homomorphism

(II.9.3.2) $$\nu_n \colon \mathrm{W}(\mathscr{R}_A) \to \mathrm{W}_n(A/pA).$$

Since $\nu_n = \mathrm{F} \circ \mathrm{R} \circ \nu_{n+1}$, taking the inverse limit gives a homomorphism

(II.9.3.3) $$\nu \colon \mathrm{W}(\mathscr{R}_A) \to \varprojlim_{n \geq 1} \mathrm{W}_n(A/pA),$$

where the transition morphisms of the inverse limit are the morphisms FR. One immediately verifies that it is bijective. In view of (II.9.1.5), the homomorphisms $\theta_n$ induce, by taking the inverse limit, a homomorphism

(II.9.3.4) $$\theta\colon \mathrm{W}(\mathscr{R}_A) \to \widehat{A},$$

where $\widehat{A}$ is the $p$-adic Hausdorff completion of $A$. We recover the homomorphism defined by Fontaine ([**29**] 2.2). We set

(II.9.3.5) $$\mathscr{A}_2(A) = \mathrm{W}(\mathscr{R}_A)/\ker(\theta)^2,$$

and write also $\theta\colon \mathscr{A}_2(A) \to \widehat{A}$ for the homomorphism induced by $\theta$ (cf. [**31**] 1.2.2).

For every homomorphism of commutative $\mathbb{Z}_{(p)}$-algebras $\varphi\colon A \to B$, the diagram

(II.9.3.6) $$\begin{array}{ccc} \mathrm{W}(\mathscr{R}_A) & \longrightarrow & \mathrm{W}(\mathscr{R}_B) \\ \Big\downarrow{\scriptstyle\theta} & & \Big\downarrow{\scriptstyle\theta} \\ \widehat{A} & \longrightarrow & \widehat{B} \end{array}$$

where the horizontal arrows are the morphisms induced by $\varphi$, is commutative (II.9.1.6). The correspondence $A \mapsto \mathscr{A}_2(A)$ is therefore functorial.

**Remark II.9.4.** The canonical projection $\mathscr{R}_{\mathrm{W}(k)} \to k$ onto the first component (that is, the component of index 0) is an isomorphism. It therefore induces an isomorphism $\mathrm{W}(\mathscr{R}_{\mathrm{W}(k)}) \xrightarrow{\sim} \mathrm{W}(k)$, which we use to identify these two rings. The homomorphism $\theta$ then identifies with the identity endomorphism of $\mathrm{W}(k)$.

**Proposition II.9.5** ([**73**] A.1.1 and A.2.2). *Let $A$ be a commutative $\mathbb{Z}_{(p)}$-algebra satisfying the following conditions:*

(i) *$A$ is $\mathbb{Z}_{(p)}$-flat.*
(ii) *$A$ is integrally closed in $A[\frac{1}{p}]$.*
(iii) *The absolute Frobenius of $A/pA$ is surjective.*
(iv) *There exists a sequence $(p_n)_{n\geq 0}$ of elements of $A$ such that $p_0 = p$ and $p_{n+1}^p = p_n$ for every $n \geq 0$.*

*We denote by $\underline{p}$ the element of $\mathscr{R}_A$ induced by the sequence $(p_n)_{n\geq 0}$ and set*

(II.9.5.1) $$\xi = [\underline{p}] - p \in \mathrm{W}(\mathscr{R}_A),$$

*where $[\underline{p}]$ is the multiplicative representative of $\underline{p}$. Then the sequence*

(II.9.5.2) $$0 \longrightarrow \mathrm{W}(\mathscr{R}_A) \xrightarrow{\cdot\xi} \mathrm{W}(\mathscr{R}_A) \xrightarrow{\theta} \widehat{A} \longrightarrow 0$$

*is exact.*

Indeed, for every $n \geq 1$, if we set

(II.9.5.3) $$\xi_n = [\overline{p}_n] - p \in \mathrm{W}_n(A/pA),$$

where $\overline{p}_n$ is the class of $p_n$ in $A/pA$, then the sequence

(II.9.5.4) $$\mathrm{W}_n(A/pA) \xrightarrow{\cdot\xi_n} \mathrm{W}_n(A/pA) \xrightarrow{\theta_n} A/p^nA \longrightarrow 0$$

is exact by virtue of II.9.2. Since the homomorphism $\mathrm{RF}\colon \mathrm{W}_{n+1}(A/pA) \to \mathrm{W}_n(A/pA)$ is surjective for every $n \geq 0$, the sequence (II.9.5.2) is exact in the middle and on the right ([**41**] 0.13.2.1(i) and 0.13.2.2).

If $a = (a_0, a_1, a_2, \dots) \in \mathrm{W}(\mathscr{R}_A)$ is such that $\xi a = 0$, then

(II.9.5.5) $$(\underline{p}a_0, \underline{p}^p a_1, \underline{p}^{p^2} a_2, \dots) = (0, a_0^p, a_1^p, \dots).$$

To show that $\xi$ is not a zero divisor in $\mathrm{W}(\mathscr{R}_A)$, it therefore suffices to show that $\underline{p}$ is not a zero divisor in $\mathscr{R}_A$. Let $y = (y_n)_{n\in\mathbb{N}} \in \mathscr{R}_A$ satisfy $\underline{p}y = 0$. For every $n \geq 0$, let $\widetilde{y}_n$ be a lift of $y_n$ in $A$. We have $p_n\widetilde{y}_n \in pA$. Consequently, $\widetilde{y}_n \in p_n^{p^n-1}A$ because $p$ is not a zero divisor in $A$. It follows that

$$y_n = y_{n+1}^p = (\widetilde{y}_{n+1}^p \mod pA) = 0 \tag{II.9.5.6}$$

because $p^{n+2} - p \geq p^{n+1}$.

**II.9.6.** Let $Y = (Y, \mathscr{M}_Y)$ be an affine logarithmic $\mathbb{Z}_{(p)}$-scheme with ring $A$, $M$ a monoid, and $u\colon M \to \Gamma(Y, \mathscr{M}_Y)$ a homomorphism. Consider the inverse system of multiplicative monoids $(A)_{n\in\mathbb{N}}$, where the transition morphisms are all equal to the $p$th power homomorphism. We denote by $Q$ the fibered product of the diagram of homomorphisms of monoids

$$\begin{array}{ccc} & & M \\ & & \downarrow \\ \varprojlim_{x\mapsto x^p} A & \longrightarrow & A \end{array} \tag{II.9.6.1}$$

where the horizontal arrow is the projection onto the first component (that is, the component of index 0) and the vertical arrow is the composition of $u$ and the canonical homomorphism $\Gamma(Y, \mathscr{M}_Y) \to A$. We denote by $q$ the composition

$$Q \longrightarrow \varprojlim_{x\mapsto x^p} A \longrightarrow \mathscr{R}_A \xrightarrow{[\ ]} \mathrm{W}(\mathscr{R}_A), \tag{II.9.6.2}$$

where the first and second arrows are the canonical homomorphisms (II.9.3.1) and $[\ ]$ is the multiplicative representative. It immediately follows from the definitions that the diagram

$$\begin{array}{ccc} Q & \longrightarrow & M \\ {\scriptstyle q}\downarrow & & \downarrow \\ \mathrm{W}(\mathscr{R}_A) & \xrightarrow{\theta} & \widehat{A} \end{array} \tag{II.9.6.3}$$

where the unlabeled arrows are the canonical morphisms, is commutative.

We endow $\widehat{Y} = \mathrm{Spec}(\widehat{A})$ with the logarithmic structure $\mathscr{M}_{\widehat{Y}}$ inverse image of $\mathscr{M}_Y$ and $\mathrm{Spec}(\mathrm{W}(\mathscr{R}_A))$ with the logarithmic structure $\mathscr{Q}$ associated with the prelogarithmic structure defined by $q$ (II.9.6.2). By (II.9.6.3), $\theta$ induces a morphism

$$(\widehat{Y}, \mathscr{M}_{\widehat{Y}}) \to (\mathrm{Spec}(\mathrm{W}(\mathscr{R}_A)), \mathscr{Q}). \tag{II.9.6.4}$$

The following statement is inspired by ([**73**] 1.4.2).

**Proposition II.9.7.** *We keep the assumptions of* II.9.6, *denote by* $Y^\circ$ *the maximal open subscheme of* $Y$ *where the logarithmic structure* $\mathscr{M}_Y$ *is trivial, and suppose that the following conditions are satisfied:*

(a) $A$ *is a normal integral domain.*
(b) $Y^\circ$ *is a nonempty simply connected* $\mathbb{Q}$*-scheme.*
(c) $M$ *is integral and there exist a fine and saturated monoid* $M'$ *and a homomorphism* $v\colon M' \to M$ *such that the induced homomorphism* $M' \to M/M^\times$ *is an isomorphism.*

*Then:*

(i) *The monoid* $Q$ *is integral and the group* $M'^{\mathrm{gp}}$ *is free.*

(ii) *We can complete the diagram* (II.9.6.1) *into a commutative diagram*

(II.9.7.1)
$$\begin{array}{ccc} M' & \xrightarrow{v} & M \\ {\scriptstyle w}\downarrow & & \downarrow \\ \varprojlim_{x\mapsto x^p} A & \longrightarrow & A \end{array}$$

*Denote by* $\beta\colon M' \to Q$ *the induced homomorphism.*

(iii) *The logarithmic structure* $\mathscr{Q}$ *on* $\mathrm{Spec}(\mathrm{W}(\mathscr{R}_A))$ *is associated with the prelogarithmic structure defined by the composition*

(II.9.7.2)
$$M' \xrightarrow{\beta} Q \xrightarrow{q} \mathrm{W}(\mathscr{R}_A).$$

*In particular, the logarithmic scheme* $(\mathrm{Spec}(\mathrm{W}(\mathscr{R}_A)), \mathscr{Q})$ *is fine and saturated.*

(iv) *If, moreover, the composition* $u \circ v\colon M' \to \Gamma(Y, \mathscr{M}_Y)$ *is a chart for* $Y$ (II.5.13), *then the morphism* (II.9.6.4) *is strict.*

(i) Since $Y^\circ$ is nonempty, the canonical image of $\Gamma(Y, \mathscr{M}_Y)$ in $A$ does not contain 0. It immediately follows that $Q$ is integral. On the other hand, since $M'$ is saturated, the torsion subgroup of $M'^{\mathrm{gp}}$ is contained in $M'$. But $M'$ is sharp; therefore $M'^{\mathrm{gp}}$ is without torsion.

(ii) Let $L$ be the field of fractions of $A$ and $\overline{L}$ an algebraic closure of $L$. Consider the inverse system of multiplicative monoids $(\overline{L})_{n\in\mathbb{N}}$, where the transition morphisms are all equal to the $p$th power homomorphism. By (i) and its proof, there exists a homomorphism

(II.9.7.3)
$$M'^{\mathrm{gp}} \to \varprojlim_{x\mapsto x^p} \overline{L}$$

that lifts the homomorphism $M'^{\mathrm{gp}} \to \overline{L}$ induced by the composition $M' \xrightarrow{v} M \to A$.

Let $t \in \Gamma(Y, \mathscr{M}_Y)$, $x$ its canonical image in $A$, and $y \in \overline{L}$ such that $y^p = x$. The extension $A[z]/z^p - x$ of $A$ is étale over $Y^\circ$ because $p$ is invertible in $Y^\circ$ and $x$ does not vanish at any points of $Y^\circ$. Since $Y^\circ$ is simply connected, it follows that $y \in L$ and consequently that $y \in A$ because $A$ is normal. It follows that the restriction of (II.9.7.3) to $M'$ induces a homomorphism

(II.9.7.4)
$$w\colon M' \to \varprojlim_{x\mapsto x^p} A$$

as desired.

(iii) Let $G$ be the inverse image of $M^\times$ under the canonical homomorphism $Q \to M$. It immediately follows from the definition (II.9.6.1) that $G$ is a subgroup of $Q$. On the other hand, the composition $M' \to Q/G \to M/M^\times$, where the first arrow is deduced from $\beta$, is an isomorphism. Consequently, $M' \to Q/G$ is an isomorphism. The statement then follows from ([**73**] 1.3.1).

(iv) This immediately follows from (iii).

**II.9.8.** For the remainder of this chapter, we fix a sequence $(p_n)_{n\geq 0}$ of elements of $\mathscr{O}_{\overline{K}}$ such that $p_0 = p$ and $p_{n+1}^p = p_n$ for every $n \geq 0$. We denote by $\underline{p}$ the element of $\mathscr{R}_{\mathscr{O}_{\overline{K}}}$ induced by the sequence $(p_n)_{n\geq 0}$ and set

(II.9.8.1)
$$\xi = [\underline{p}] - p \in \mathrm{W}(\mathscr{R}_{\mathscr{O}_{\overline{K}}}),$$

where $[\underline{p}]$ is the multiplicative representative of $\underline{p}$. By II.9.5, the sequence

(II.9.8.2)
$$0 \longrightarrow \mathrm{W}(\mathscr{R}_{\mathscr{O}_{\overline{K}}}) \xrightarrow{\cdot\xi} \mathrm{W}(\mathscr{R}_{\mathscr{O}_{\overline{K}}}) \xrightarrow{\theta} \mathscr{O}_C \longrightarrow 0$$

is exact. It induces an exact sequence

$$0 \longrightarrow \mathscr{O}_C \xrightarrow{\cdot\xi} \mathscr{A}_2(\mathscr{O}_{\overline{K}}) \xrightarrow{\theta} \mathscr{O}_C \longrightarrow 0, \tag{II.9.8.3}$$

where we have also denoted by $\cdot\xi$ the morphism induced by the multiplication by $\xi$ in $\mathscr{A}_2(\mathscr{O}_{\overline{K}})$ (II.9.3.5). The ideal $\ker(\theta)$ of $\mathscr{A}_2(\mathscr{O}_{\overline{K}})$ has square zero. It is a free $\mathscr{O}_C$-module with basis $\xi$. We will denote it by $\xi\mathscr{O}_C$. Note that, unlike $\xi$, this module does not depend on the choice of the sequence $(p_n)_{n\geq 0}$.

The Galois group $G_K$ acts naturally on $\mathrm{W}(\mathscr{R}_{\mathscr{O}_{\overline{K}}})$ by ring automorphisms, and the homomorphism $\theta$ (II.9.8.2) is $G_K$-equivariant. We deduce from this an action of $G_K$ on $\mathscr{A}_2(\mathscr{O}_{\overline{K}})$ by ring automorphisms such that the homomorphism $\theta$ (II.9.8.3) is $G_K$-equivariant.

We set

$$\overline{S} = \mathrm{Spec}(\mathscr{O}_{\overline{K}}) \quad \text{and} \quad \check{\overline{S}} = \mathrm{Spec}(\mathscr{O}_C), \tag{II.9.8.4}$$

which we endow with the logarithmic structures inverse images of $\mathscr{M}_S$ (II.6.1), denoted by $\mathscr{M}_{\overline{S}}$ and $\mathscr{M}_{\check{\overline{S}}}$, respectively. The action of $G_K$ on $\mathscr{O}_{\overline{K}}$ and $\mathscr{O}_C$ extends naturally to a left action on the logarithmic schemes $(\overline{S}, \mathscr{M}_{\overline{S}})$ and $(\check{\overline{S}}, \mathscr{M}_{\check{\overline{S}}})$, respectively.

We set

$$\mathscr{A}_2(\overline{S}) = \mathrm{Spec}(\mathscr{A}_2(\mathscr{O}_{\overline{K}})), \tag{II.9.8.5}$$

which we endow with the logarithmic structure $\mathscr{M}_{\mathscr{A}_2(\overline{S})}$ defined as follows. Let $Q_{\overline{S}}$ be the monoid and $q_{\overline{S}}\colon Q_{\overline{S}} \to \mathrm{W}(\mathscr{R}_{\mathscr{O}_{\overline{K}}})$ the homomorphism defined in II.9.6 (denoted by $Q$ and $q$) by taking for $(Y, \mathscr{M}_Y)$ the logarithmic scheme $(\overline{S}, \mathscr{M}_{\overline{S}})$ and for $u$ the canonical homomorphism $\Gamma(S, \mathscr{M}_S) \to \Gamma(\overline{S}, \mathscr{M}_{\overline{S}})$. We denote by $\mathscr{M}_{\mathscr{A}_2(\overline{S})}$ the logarithmic structure on $\mathscr{A}_2(\overline{S})$ associated with the prelogarithmic structure defined by the homomorphism $Q_{\overline{S}} \to \mathscr{A}_2(\mathscr{O}_{\overline{K}})$ induced by $q_{\overline{S}}$. By II.9.7, the logarithmic scheme $(\mathscr{A}_2(\overline{S}), \mathscr{M}_{\mathscr{A}_2(\overline{S})})$ is fine and saturated, and $\theta$ induces an exact closed immersion

$$i_{\overline{S}}\colon (\check{\overline{S}}, \mathscr{M}_{\check{\overline{S}}}) \to (\mathscr{A}_2(\overline{S}), \mathscr{M}_{\mathscr{A}_2(\overline{S})}). \tag{II.9.8.6}$$

The Galois group $G_K$ has a natural action on the monoid $Q_{\overline{S}}$ and the homomorphism $q_{\overline{S}}\colon Q_{\overline{S}} \to \mathrm{W}(\mathscr{R}_{\mathscr{O}_{\overline{K}}})$ is $G_K$-equivariant. From this, we deduce a left action of $G_K$ on the logarithmic scheme $(\mathscr{A}_2(\overline{S}), \mathscr{M}_{\mathscr{A}_2(\overline{S})})$. The morphism $i_{\overline{S}}$ is $G_K$-equivariant.

**Remark II.9.9.** We denote by $\xi^{-1}\mathscr{O}_C$ the dual $\mathscr{O}_C$-module of $\xi\mathscr{O}_C$. For every $\mathscr{O}_C$-module $M$, we denote the $\mathscr{O}_C$-modules $M \otimes_{\mathscr{O}_C} (\xi\mathscr{O}_C)$ and $M \otimes_{\mathscr{O}_C} (\xi^{-1}\mathscr{O}_C)$ simply by $\xi M$ and $\xi^{-1}M$, respectively. Note that, unlike $\xi$, these modules do not depend on the choice of the sequence $(p_n)_{n\geq 0}$. It is therefore important to not identify them with $M$.

For $\widehat{\overline{R_1}}$-algebras $A$, in the remainder of this chapter, we consider Higgs $A$-modules with coefficients in $\xi^{-1}\widetilde{\Omega}^1_{R/\mathscr{O}_K} \otimes_R A$ (cf. II.2.8). We will abusively say that they have coefficients in $\xi^{-1}\widetilde{\Omega}^1_{R/\mathscr{O}_K}$. We will denote the category of these modules by $\mathbf{HM}(A, \xi^{-1}\widetilde{\Omega}^1_{R/\mathscr{O}_K})$.

**Proposition II.9.10.** *The absolute Frobenius homomorphisms of $R_\infty/pR_\infty$ and $\overline{R}/p\overline{R}$ are surjective.*

Let us first show that the absolute Frobenius homomorphism of $R_\infty/pR_\infty$ is surjective. Recall that for every integer $n \geq 1$, $\mathrm{Spec}(R_n)$ is a connected component of

$X_n \otimes_{\mathscr{O}_{K_n}} \mathscr{O}_{\overline{K}}$ (II.6.8) and that we have a Cartesian diagram of $\mathscr{O}_{\overline{K}}$-morphisms (II.6.6.4)

(II.9.10.1)
$$\begin{array}{ccc} X_n \otimes_{\mathscr{O}_{K_n}} \mathscr{O}_{\overline{K}} & \longrightarrow & \operatorname{Spec}(\mathscr{O}_{\overline{K}}[P^{(n)}]/(\pi_n - e^{\lambda^{(n)}})) \\ \downarrow & & \downarrow \\ X \otimes_{\mathscr{O}_K} \mathscr{O}_{\overline{K}} & \longrightarrow & \operatorname{Spec}(\mathscr{O}_{\overline{K}}[P]/(\pi - e^{\lambda})) \end{array}$$

where for every $t \in P$, we have denoted by $t^{(n)}$ its image in $P^{(n)}$ by the canonical isomorphism (II.7.8.2). Since the canonical morphism $X \to S \times_{\mathbf{A}_{\mathbb{N}}} \mathbf{A}_P$ is étale (II.6.2.4), it suffices to show that the absolute Frobenius homomorphism of the direct limit of $\mathbb{F}_p$-algebras (with respect to the divisibility relation)

(II.9.10.2)
$$\lim_{\substack{\longrightarrow \\ n \geq 1}} (\mathscr{O}_{\overline{K}}/p\mathscr{O}_{\overline{K}})[P^{(n)}]/(\pi_n - e^{\lambda^{(n)}})$$

is surjective. This follows from the fact that the absolute Frobenius homomorphism of $\mathscr{O}_{\overline{K}}/p\mathscr{O}_{\overline{K}}$ is surjective and that for every $t \in P$, we have $e^{t^{(n)}} = e^{pt^{(pn)}}$ in (II.9.10.2).

Next, let us show that the absolute Frobenius homomorphism of $\overline{R}/p\overline{R}$ is surjective. Let $N$ be a finite extension of $F_\infty$ contained in $\overline{F}$ and $D$ the integral closure of $R$ in $N$. Suppose that $D$ is almost étale over $R_\infty$. Then $D/pD$ is almost étale over $R_\infty/pR_\infty$ (V.7.4(3)). Since the absolute Frobenius homomorphism of $R_\infty/pR_\infty$ is surjective, the absolute Frobenius homomorphism of $D/pD$ is almost surjective by virtue of V.7.9. For every $x \in D$, there exist $x', y \in D$ such that $p^{1/2}x = x'^p + py$. Then $x'' = p^{-1/(2p)}x' \in D$ and we have $x = x''^p + p^{1/2}y$. By the same argument, there exist $y', z \in D$ such that $y = y'^p + p^{1/2}z$. Consequently, we have $x \equiv (x'' + p^{1/(2p)}y')^p \mod pD$. Taking the direct limit, it follows, by virtue of II.6.17, that the absolute Frobenius homomorphism of $\overline{R}/p\overline{R}$ is surjective.

**II.9.11.** By II.9.5 and II.9.10, the sequence

(II.9.11.1)
$$0 \longrightarrow \mathrm{W}(\mathscr{R}_{\overline{R}}) \xrightarrow{\cdot\xi} \mathrm{W}(\mathscr{R}_{\overline{R}}) \xrightarrow{\theta} \widehat{\overline{R}} \longrightarrow 0,$$

where $\xi$ is the element (II.9.8.1), is exact. It induces an exact sequence

(II.9.11.2)
$$0 \longrightarrow \widehat{\overline{R}} \xrightarrow{\cdot\xi} \mathscr{A}_2(\overline{R}) \xrightarrow{\theta} \widehat{\overline{R}} \longrightarrow 0,$$

where we have also denoted by $\cdot\xi$ the morphism induced by the multiplication by $\xi$ in $\mathscr{A}_2(\overline{R})$. The ideal $\ker(\theta)$ of $\mathscr{A}_2(\overline{R})$ has square zero. It is a free $\widehat{\overline{R}}$-module with basis $\xi$, canonically isomorphic to $\xi\widehat{\overline{R}}$ (cf. II.9.9). The Galois group $\Gamma$ (II.6.10) acts naturally on $\mathrm{W}(\mathscr{R}_{\overline{R}})$ by ring automorphisms, and the homomorphism $\theta$ (II.9.11.1) is $\Gamma$-equivariant. We deduce from this an action of $\Gamma$ on $\mathscr{A}_2(\overline{R})$ by ring automorphisms such that the homomorphism $\theta$ (II.9.11.2) is $\Gamma$-equivariant.

We set

(II.9.11.3)
$$Y = \operatorname{Spec}(\overline{R}) \quad \text{and} \quad \widehat{Y} = \operatorname{Spec}(\widehat{\overline{R}})$$

and endow these with the logarithmic structures inverse images of $\mathscr{M}_X$ (II.6.2), denoted by $\mathscr{M}_Y$ and $\mathscr{M}_{\widehat{Y}}$, respectively. The actions of $\Gamma$ on $\overline{R}$ and $\widehat{\overline{R}}$ induce left actions on the logarithmic schemes $(Y, \mathscr{M}_Y)$ and $(\widehat{Y}, \mathscr{M}_{\widehat{Y}})$, respectively.

We set

(II.9.11.4)
$$\mathscr{A}_2(Y) = \operatorname{Spec}(\mathscr{A}_2(\overline{R}))$$

and endow this with the logarithmic structure $\mathscr{M}'_{\mathscr{A}_2(Y)}$ defined as follows. Let $Q_Y$ be the monoid and $q_Y \colon Q_Y \to \mathrm{W}(\mathscr{R}_{\overline{R}})$ the homomorphism defined in II.9.6 (denoted by $Q$ and

$q$) by taking for $u$ the canonical homomorphism $\Gamma(X, \mathscr{M}_X) \to \Gamma(Y, \mathscr{M}_Y)$. We denote by $\mathscr{M}'_{\mathscr{A}_2(Y)}$ the logarithmic structure on $\mathscr{A}_2(Y)$ associated with the prelogarithmic structure defined by the homomorphism $Q_Y \to \mathscr{A}_2(\overline{R})$ induced by $q_Y$. The homomorphism $\theta$ then induces a morphism (II.9.6.4)

$$i'_Y\colon (\widehat{Y}, \mathscr{M}_{\widehat{Y}}) \to (\mathscr{A}_2(Y), \mathscr{M}'_{\mathscr{A}_2(Y)}). \tag{II.9.11.5}$$

The Galois group $\Gamma$ acts naturally on the monoid $Q_Y$, and the homomorphism $q_Y\colon Q_Y \to \mathrm{W}(\mathscr{R}_{\overline{R}})$ is $\Gamma$-equivariant. We deduce from this a left action of $\Gamma$ on $(\mathscr{A}_2(Y), \mathscr{M}'_{\mathscr{A}_2(Y)})$. The morphism $i'_Y$ is $\Gamma$-equivariant.

Without the assumptions of II.9.7, we would not know whether the logarithmic scheme $(\mathscr{A}_2(Y), \mathscr{M}'_{\mathscr{A}_2(Y)})$ is fine and saturated and whether $i'_Y$ is an exact closed immersion. That is why we will endow $\mathscr{A}_2(Y)$ with a different logarithmic structure $\mathscr{M}_{\mathscr{A}_2(Y)}$ in II.9.12.

**II.9.12.** For every $t \in P$, we denote by $\widetilde{t}$ the element of $Q_Y$ defined by its projections (II.9.6.1)

$$(\alpha_\infty(t^{(p^n)}))_{n\in\mathbb{N}} \in \varprojlim_{\mathbb{N}} \overline{R} \quad \text{and} \quad \gamma(t) \in \Gamma(X, \mathscr{M}_X), \tag{II.9.12.1}$$

where $t^{(p^n)}$ is the image of $t$ in $P^{(p^n)}$ by the isomorphism (II.7.8.2) and $\alpha_\infty$ is the homomorphism (II.7.8.5). We will see that this notation is compatible with that introduced in (II.7.18.2) and does not cause any confusion. The reader will note that $\widetilde{t}$ depends on the choice of the morphism (II.6.7.1). The resulting map

$$P \to Q_Y, \quad t \mapsto \widetilde{t} \tag{II.9.12.2}$$

is a morphism of monoids. We denote by

$$\widetilde{q}_Y\colon P \to \mathscr{A}_2(\overline{R}) \tag{II.9.12.3}$$

the composition

$$P \longrightarrow Q_Y \xrightarrow{q_Y} \mathrm{W}(\mathscr{R}_{\overline{R}}) \longrightarrow \mathscr{A}_2(\overline{R}). \tag{II.9.12.4}$$

We endow $\mathscr{A}_2(Y)$ with the logarithmic structure $\mathscr{M}_{\mathscr{A}_2(Y)}$ associated with the prelogarithmic structure defined by $\widetilde{q}_Y$. The morphism (II.9.12.2) then induces a morphism of logarithmic structures on $\mathscr{A}_2(Y)$ (II.9.11)

$$\mathscr{M}_{\mathscr{A}_2(Y)} \to \mathscr{M}'_{\mathscr{A}_2(Y)}. \tag{II.9.12.5}$$

It is clear that the composition $\theta \circ \widetilde{q}_Y\colon P \to \widehat{\overline{R}}$ is induced by $\alpha$ (cf. II.7.8). Consequently, $\theta$ induces an exact closed immersion

$$i_Y\colon (\widehat{Y}, \mathscr{M}_{\widehat{Y}}) \to (\mathscr{A}_2(Y), \mathscr{M}_{\mathscr{A}_2(Y)}), \tag{II.9.12.6}$$

which factors through $i'_Y$ (II.9.11.5).

**Proposition II.9.13.** *Suppose that there exists a fine and saturated chart for $(X, \mathscr{M}_X)$, $M' \to \Gamma(X, \mathscr{M}_X)$ (II.5.13), such that the induced homomorphism*

$$M' \to \Gamma(X, \mathscr{M}_X)/\Gamma(X, \mathscr{O}_X^\times) \tag{II.9.13.1}$$

*is an isomorphism. Then the morphism $\mathscr{M}_{\mathscr{A}_2(Y)} \to \mathscr{M}'_{\mathscr{A}_2(Y)}$* (II.9.12.5) *is an isomorphism.*

Indeed, by II.9.7, the logarithmic scheme $(\mathscr{A}_2(Y), \mathscr{M}'_{\mathscr{A}_2(Y)})$ is fine and saturated, and the morphism $i'_Y$ (II.9.11.5) is an exact closed immersion. Consequently, for every geometric point $\overline{y}$ of $\widehat{Y}$, if we also denote by $\overline{y}$ the geometric point $i_Y(\overline{y}) = i'_Y(\overline{y})$ of $\mathscr{A}_2(Y)$, the homomorphism

(II.9.13.2) $$\mathscr{M}_{\mathscr{A}_2(Y),\overline{y}}/\mathscr{O}^{\times}_{\mathscr{A}_2(Y),\overline{y}} \to \mathscr{M}'_{\mathscr{A}_2(Y),\overline{y}}/\mathscr{O}^{\times}_{\mathscr{A}_2(Y),\overline{y}}$$

induced by (II.9.12.5) is an isomorphism. Since $\mathscr{M}'_{\mathscr{A}_2(Y),\overline{y}}$ is integral, it follows that the morphism $\mathscr{M}_{\mathscr{A}_2(Y),\overline{y}} \to \mathscr{M}'_{\mathscr{A}_2(Y),\overline{y}}$ (II.9.12.5) is an isomorphism; the proposition follows.

**Remark II.9.14.** Note that, unlike $\mathscr{M}'_{\mathscr{A}_2(Y)}$, the logarithmic structure $\mathscr{M}_{\mathscr{A}_2(Y)}$ depends on the chart $(P, \gamma)$ for $(X, \mathscr{M}_X)$ (II.6.2). Nevertheless, by II.5.17, after replacing $X$ by an affine open covering, if necessary, we may assume that the condition of II.9.13 is satisfied, in which case $\mathscr{M}_{\mathscr{A}_2(Y)}$ no longer depends on the chart $(P, \gamma)$.

**Remark II.9.15.** We denote by $\widetilde{\pi}$ the element of $Q_{\overline{S}}$ defined by its projections (II.9.6.1)

(II.9.15.1) $$(\pi_{p^n})_{n\in\mathbb{N}} \in \varprojlim_{\mathbb{N}} \mathscr{O}_{\overline{K}} \quad \text{and} \quad \pi \in \Gamma(S, \mathscr{M}_S)$$

(cf. II.6.4 and II.9.8) and by

(II.9.15.2) $$\widetilde{q}_{\overline{S}} \colon \mathbb{N} \to \mathscr{A}_2(\mathscr{O}_{\overline{K}})$$

the composition

(II.9.15.3) $$\mathbb{N} \longrightarrow Q_{\overline{S}} \xrightarrow{q_{\overline{S}}} \mathrm{W}(\mathscr{R}_{\mathscr{O}_{\overline{K}}}) \to \mathscr{A}_2(\mathscr{O}_{\overline{K}}),$$

where the first arrow sends 1 to $\widetilde{\pi}$. The logarithmic structure on $\mathscr{A}_2(\overline{S})$ associated with the prelogarithmic structure defined by $\widetilde{q}_{\overline{S}}$ is canonically isomorphic to $\mathscr{M}_{\mathscr{A}_2(\overline{S})}$ (II.9.8). This follows from II.9.7 as in the proof of II.9.13.

**II.9.16.** We have a canonical homomorphism $Q_{\overline{S}} \to Q_Y$ that fits into a commutative diagram

(II.9.16.1) $$\begin{array}{ccc} \mathbb{N} & \xrightarrow{\vartheta} & P \\ \downarrow & & \downarrow \\ Q_{\overline{S}} & \longrightarrow & Q_Y \\ {\scriptstyle q_{\overline{S}}}\downarrow & & \downarrow{\scriptstyle q_Y} \\ \mathrm{W}(\mathscr{R}_{\mathscr{O}_{\overline{K}}}) & \longrightarrow & \mathrm{W}(\mathscr{R}_{\overline{R}}) \end{array}$$

where $\vartheta$ is the homomorphism given in (II.6.2), the left upper vertical arrow sends 1 to $\widetilde{\pi}$ (II.9.15), and the one on the right is (II.9.12.2). We deduce from this a commutative diagram

(II.9.16.2) $$\begin{array}{ccc} (\widehat{Y}, \mathscr{M}_{\widehat{Y}}) & \xrightarrow{i_Y} & (\mathscr{A}_2(Y), \mathscr{M}_{\mathscr{A}_2(Y)}) \\ \downarrow & & \downarrow \\ (\check{\overline{S}}, \mathscr{M}_{\check{\overline{S}}}) & \xrightarrow{i_S} & (\mathscr{A}_2(\overline{S}), \mathscr{M}_{\mathscr{A}_2(\overline{S})}) \end{array}$$

**II.9.17.** We have a canonical homomorphism

$$\mathbb{Z}_p(1) \to \mathscr{R}^{\times}_{\overline{R}}. \tag{II.9.17.1}$$

For every $\zeta \in \mathbb{Z}_p(1)$, we denote also by $\zeta$ its image in $\mathscr{R}^{\times}_{\overline{R}}$. Since $\theta([\zeta]-1) = 0$, we obtain a group homomorphism

$$\mathbb{Z}_p(1) \to \mathscr{A}_2(\overline{R}), \quad \zeta \mapsto \log([\zeta]) = [\zeta] - 1, \tag{II.9.17.2}$$

whose image is contained in $\ker(\theta) = \xi\widehat{\overline{R}}$.

**Lemma II.9.18.** *The homomorphism $\mathbb{Z}_p(1) \to \mathscr{A}_2(\overline{R})$ (II.9.17.2) is $\Gamma$-equivariant and $\mathbb{Z}_p$-linear; its image generates the ideal $p^{\frac{1}{p-1}}\xi\widehat{\overline{R}}$ of $\mathscr{A}_2(\overline{R})$ and the $\widehat{\overline{R}}$-linear morphism*

$$\widehat{\overline{R}}(1) \to p^{\frac{1}{p-1}}\xi\widehat{\overline{R}} \tag{II.9.18.1}$$

*that sends $x \otimes \zeta$, where $x \in \widehat{\overline{R}}$ and $\zeta \in \mathbb{Z}_p(1)$, to $x \cdot \log([\zeta])$ is an isomorphism.*

The first statement is immediate because $\mathbb{Z}_p(1)$ and $\mathscr{A}_2(\overline{R})$ are complete and separated for the $p$-adic topologies. Let $\zeta = (\zeta_n)_{n\geq 0} \in \mathbb{Z}_p(1)$ such that $\zeta_1 \neq 1$ (we have $\zeta_0 = 1$). Denote by $\zeta'$ the canonical image of $(\zeta_{n+1})_{n\geq 0}$ in $\mathscr{R}_{\overline{R}}$ and set $\omega = \sum_{i=0}^{p-1}[\zeta']^i \in \mathrm{W}(\mathscr{R}_{\overline{R}})$. Then $\omega$ is a generator of $\ker(\theta)$ (**[73]** A.2.6), and in $\mathscr{A}_2(\overline{R})$, we have

$$[\zeta] - 1 = \omega([\zeta'] - 1) = \omega\theta([\zeta'] - 1) = \omega(\zeta_1 - 1). \tag{II.9.18.2}$$

Since $v(\zeta_1 - 1) = \frac{1}{p-1}$, the rest of the assertion follows.

**II.9.19.** The canonical homomorphism $\mathbb{Z}_p(1) \to (\mathscr{R}_{\overline{R}}, \times)$ (II.9.17.1) and the trivial homomorphism $\mathbb{Z}_p(1) \to \Gamma(X, \mathscr{M}_X)$ (with value 1) induce a homomorphism

$$\mathbb{Z}_p(1) \to Q_Y. \tag{II.9.19.1}$$

For all $g \in \Gamma$ and $t \in P$, we have, in $Q_Y$,

$$g(\widetilde{t}) = \widetilde{\chi}_t(g) \cdot \widetilde{t}, \tag{II.9.19.2}$$

where we have (abusively) denoted by $\widetilde{\chi}_t \colon \Gamma \to \mathbb{Z}_p(1)$ the map deduced from (II.7.19.1). We deduce from this the following relation in $\mathscr{A}_2(\overline{R})$:

$$g(\widetilde{q}_Y(t)) = [\widetilde{\chi}_t(g)] \cdot \widetilde{q}_Y(t), \tag{II.9.19.3}$$

where $[\widetilde{\chi}_t(g)]$ denotes the image of $\widetilde{\chi}_t(g)$ by the composition

$$\mathbb{Z}_p(1) \longrightarrow \mathscr{R}_{\overline{R}} \xrightarrow{[\ ]} \mathrm{W}(\mathscr{R}_{\overline{R}}) \longrightarrow \mathscr{A}_2(\overline{R}).$$

Denote by $\tau_g$ the automorphism of $\mathscr{A}_2(Y)$ induced by the action of $g$ on $\mathscr{A}_2(\overline{R})$. The logarithmic structure $\tau_g^*(\mathscr{M}_{\mathscr{A}_2(Y)})$ on $\mathscr{A}_2(Y)$ is associated with the prelogarithmic structure defined by the composition $g \circ \widetilde{q}_Y \colon P \to \mathscr{A}_2(\overline{R})$ (II.9.12.3). The map

$$P \to \Gamma(\mathscr{A}_2(Y), \mathscr{M}_{\mathscr{A}_2(Y)}), \quad t \mapsto [\widetilde{\chi}_t(g)] \cdot t \tag{II.9.19.4}$$

is a morphism of monoids (II.7.19.4). It therefore induces a morphism of logarithmic structures on $\mathscr{A}_2(Y)$

$$a_g \colon \tau_g^*(\mathscr{M}_{\mathscr{A}_2(Y)}) \to \mathscr{M}_{\mathscr{A}_2(Y)}. \tag{II.9.19.5}$$

Likewise, by virtue of (II.7.19.3), the morphism of monoids

$$P \to \Gamma(\mathscr{A}_2(Y), \tau_g^*(\mathscr{M}_{\mathscr{A}_2(Y)})), \quad t \mapsto [g(\widetilde{\chi}_t(g^{-1}))] \cdot t \tag{II.9.19.6}$$

induces a morphism of logarithmic structures on $\mathscr{A}_2(Y)$

$$b_g \colon \mathscr{M}_{\mathscr{A}_2(Y)} \to \tau_g^*(\mathscr{M}_{\mathscr{A}_2(Y)}). \tag{II.9.19.7}$$

We immediately see that $a_g$ and $b_g$ are isomorphisms inverse to each other (II.7.19.3), and that the map $g \mapsto (\tau_{g^{-1}}, a_{g^{-1}})$ is a left action of $\Gamma$ on the logarithmic scheme $(\mathscr{A}_2(Y), \mathscr{M}_{\mathscr{A}_2(Y)})$.

One immediately verifies that the morphism $i_Y$ (II.9.12.6) and the canonical morphism (II.9.16.2)

(II.9.19.8) $$(\mathscr{A}_2(Y), \mathscr{M}_{\mathscr{A}_2(Y)}) \to (\mathscr{A}_2(\overline{S}), \mathscr{M}_{\mathscr{A}_2(\overline{S})})$$

are $\Gamma$-equivariant. Moreover, for every $g \in \Gamma$, the diagram

(II.9.19.9) $$\begin{array}{ccc} \tau_g^*(\mathscr{M}_{\mathscr{A}_2(Y)}) & \longrightarrow & \tau_g^*(\mathscr{M}'_{\mathscr{A}_2(Y)}) \\ {\scriptstyle a_g}\downarrow & & \downarrow{\scriptstyle a'_g} \\ \mathscr{M}_{\mathscr{A}_2(Y)} & \longrightarrow & \mathscr{M}'_{\mathscr{A}_2(Y)} \end{array}$$

where the horizontal arrows are induced by the homomorphism (II.9.12.5) and $a'_g$ is the automorphism of logarithmic structures on $\mathscr{A}_2(Y)$ induced by the action of $g$ on $Q_Y$, is commutative.

## II.10. Higgs–Tate torsors and algebras

We will use the notation introduced in II.9.8, II.9.11, and II.9.12 for the remainder of this chapter.

**II.10.1.** We set $\overline{X} = X \times_S \overline{S}$ and $\check{\overline{X}} = X \times_S \check{\overline{S}}$, and endow these with the logarithmic structures inverse images of $\mathscr{M}_X$ (II.6.2), denoted by $\mathscr{M}_{\overline{X}}$ and $\mathscr{M}_{\check{\overline{X}}}$, respectively. We then have canonical isomorphisms

(II.10.1.1) $$(\overline{X}, \mathscr{M}_{\overline{X}}) \xrightarrow{\sim} (X, \mathscr{M}_X) \times_{(S, \mathscr{M}_S)} (\overline{S}, \mathscr{M}_{\overline{S}}),$$

(II.10.1.2) $$(\check{\overline{X}}, \mathscr{M}_{\check{\overline{X}}}) \xrightarrow{\sim} (X, \mathscr{M}_X) \times_{(S, \mathscr{M}_S)} (\check{\overline{S}}, \mathscr{M}_{\check{\overline{S}}}),$$

where the product is taken indifferently in the category of logarithmic schemes or in that of fine logarithmic schemes. We in particular deduce from this a left action of $G_K$ on the logarithmic schemes $(\overline{X}, \mathscr{M}_{\overline{X}})$ and $(\check{\overline{X}}, \mathscr{M}_{\check{\overline{X}}})$. We have a $\Gamma$-equivariant canonical morphism (II.9.11.3)

(II.10.1.3) $$(\widehat{Y}, \mathscr{M}_{\widehat{Y}}) \to (\check{\overline{X}}, \mathscr{M}_{\check{\overline{X}}}).$$

A *smooth* $(\mathscr{A}_2(\overline{S}), \mathscr{M}_{\mathscr{A}_2(\overline{S})})$*-deformation* of $(\check{\overline{X}}, \mathscr{M}_{\check{\overline{X}}})$ consists of a smooth morphism of fine logarithmic schemes $(\widetilde{X}, \mathscr{M}_{\widetilde{X}}) \to (\mathscr{A}_2(\overline{S}), \mathscr{M}_{\mathscr{A}_2(\overline{S})})$ and an $(\check{\overline{S}}, \mathscr{M}_{\check{\overline{S}}})$-isomorphism

(II.10.1.4) $$(\check{\overline{X}}, \mathscr{M}_{\check{\overline{X}}}) \xrightarrow{\sim} (\widetilde{X}, \mathscr{M}_{\widetilde{X}}) \times_{(\mathscr{A}_2(\overline{S}), \mathscr{M}_{\mathscr{A}_2(\overline{S})})} (\check{\overline{S}}, \mathscr{M}_{\check{\overline{S}}}).$$

Since $(\check{\overline{X}}, \mathscr{M}_{\check{\overline{X}}}) \to (\check{\overline{S}}, \mathscr{M}_{\check{\overline{S}}})$ is smooth and $\check{\overline{X}}$ is affine, such a deformation exists and is unique up to isomorphism by virtue of ([**50**], 3.14). Its automorphism group is isomorphic to

(II.10.1.5) $$\mathrm{Hom}_{\mathscr{O}_{\check{\overline{X}}}}(\widetilde{\Omega}^1_{R/\mathscr{O}_K} \otimes_R \mathscr{O}_{\check{\overline{X}}}, \xi \mathscr{O}_{\check{\overline{X}}}).$$

Recall that we have set $\widetilde{\Omega}^1_{R/\mathscr{O}_K} = \Omega^1_{(R,P)/(\mathscr{O}_K, \mathbb{N})}$ (II.7.17.3).

We fix such a deformation $(\widetilde{X}, \mathscr{M}_{\widetilde{X}})$ for the remainder of this section.

**Remark II.10.2.** By II.6.8(i), the scheme $\mathrm{Spec}(R_1)$ is an open connected component of $\overline{X}$. Consequently, $Z = \mathrm{Spec}(R_1 \otimes_{\mathscr{O}_{\overline{K}}} \mathscr{O}_C)$ is an open and closed subscheme of $\check{\overline{X}}$, through which the morphism (II.10.1.3) factors. An $(\mathscr{A}_2(\overline{S}), \mathscr{M}_{\mathscr{A}_2(\overline{S})})$-deformation of $(Z, \mathscr{M}_{\check{\overline{X}}}|Z)$ suffices for this chapter.

**II.10.3.** Set

$$\mathrm{T} = \mathrm{Hom}_{\widehat{\overline{R}}}(\widetilde{\Omega}^1_{R/\mathscr{O}_K} \otimes_R \widehat{\overline{R}}, \xi\widehat{\overline{R}}). \tag{II.10.3.1}$$

We identify the dual $\widehat{\overline{R}}$-module with $\xi^{-1}\widetilde{\Omega}^1_{R/\mathscr{O}_K} \otimes_R \widehat{\overline{R}}$ (II.9.9) and denote by $\mathscr{S}$ the associate symmetric $\widehat{\overline{R}}$-algebra (II.2.5)

$$\mathscr{S} = \oplus_{n\geq 0}\mathscr{S}^n = \mathrm{S}_{\widehat{\overline{R}}}(\xi^{-1}\widetilde{\Omega}^1_{R/\mathscr{O}_K} \otimes_R \widehat{\overline{R}}). \tag{II.10.3.2}$$

Denote by $\widehat{Y}_{\mathrm{zar}}$ the Zariski topos of $\widehat{Y} = \mathrm{Spec}(\widehat{\overline{R}})$ (II.9.11.3), by $\widetilde{\mathrm{T}}$ the $\mathscr{O}_{\widehat{Y}}$-module associated with T, and by $\mathbf{T}$ the $\widehat{Y}$-vector bundle associated with its dual; in other words,

$$\mathbf{T} = \mathrm{Spec}(\mathscr{S}). \tag{II.10.3.3}$$

Let $U$ be a Zariski open subscheme of $\widehat{Y}$ and $\widetilde{U}$ the corresponding open subscheme of $\mathscr{A}_2(Y)$ (cf. II.9.11). We denote by $\mathscr{L}(U)$ the set of morphisms represented by dotted arrows that complete the diagram

$$\begin{array}{ccc} (U, \mathscr{M}_{\widehat{Y}}|U) & \xrightarrow{i_Y|U} & (\widetilde{U}, \mathscr{M}_{\mathscr{A}_2(Y)}|\widetilde{U}) \\ \downarrow & & \vdots \\ (\check{\overline{X}}, \mathscr{M}_{\check{\overline{X}}}) & \longrightarrow & (\widetilde{X}, \mathscr{M}_{\widetilde{X}}) \\ \downarrow & & \downarrow \\ (\check{\overline{S}}, \mathscr{M}_{\check{\overline{S}}}) & \xrightarrow{i_S} & (\mathscr{A}_2(\overline{S}), \mathscr{M}_{\mathscr{A}_2(\overline{S})}) \end{array} \tag{II.10.3.4}$$

in such a way that it remains commutative. By II.5.23, the functor $U \mapsto \mathscr{L}(U)$ is a $\widetilde{\mathrm{T}}$-torsor of $\widehat{Y}_{\mathrm{zar}}$. We call it the *Higgs–Tate torsor* associated with $(\widetilde{X}, \mathscr{M}_{\widetilde{X}})$. We denote by $\mathscr{F}$ the $\widehat{\overline{R}}$-module of affine functions on $\mathscr{L}$ (cf. II.4.9). This fits into a canonical exact sequence (II.4.9.1)

$$0 \to \widehat{\overline{R}} \to \mathscr{F} \to \xi^{-1}\widetilde{\Omega}^1_{R/\mathscr{O}_K} \otimes_R \widehat{\overline{R}} \to 0. \tag{II.10.3.5}$$

By ([**45**] I 4.3.1.7), for every integer $n \geq 1$, this sequence induces an exact sequence (II.2.5)

$$0 \to \mathrm{S}^{n-1}_{\widehat{\overline{R}}}(\mathscr{F}) \to \mathrm{S}^n_{\widehat{\overline{R}}}(\mathscr{F}) \to \mathrm{S}^n_{\widehat{\overline{R}}}(\xi^{-1}\widetilde{\Omega}^1_{R/\mathscr{O}_K} \otimes_R \widehat{\overline{R}}) \to 0. \tag{II.10.3.6}$$

The $\widehat{\overline{R}}$-modules $(\mathrm{S}^n_{\widehat{\overline{R}}}(\mathscr{F}))_{n\in\mathbb{N}}$ therefore form a filtered direct system whose direct limit

$$\mathscr{C} = \varinjlim_{n\geq 0} \mathrm{S}^n_{\widehat{\overline{R}}}(\mathscr{F}) \tag{II.10.3.7}$$

is naturally endowed with a structure of $\widehat{\overline{R}}$-algebra. By II.4.10, the $\widehat{Y}$-scheme

$$\mathbf{L} = \mathrm{Spec}(\mathscr{C}) \tag{II.10.3.8}$$

is naturally a principal homogeneous $\mathbf{T}$-bundle over $\widehat{Y}$ that canonically represents $\mathscr{L}$. Remember that $\mathscr{L}$, $\mathscr{F}$, $\mathscr{C}$, and $\mathbf{L}$ depend on $(\widetilde{X}, \mathscr{M}_{\widetilde{X}})$.

**II.10.4.** We endow $\widehat{Y}$ with the natural left action of $\Delta$; for every $g \in \Delta$, the automorphism of $\widehat{Y}$ defined by $g$, which we denote also by $g$, is induced by the automorphism $g^{-1}$ of $\widehat{\overline{R}}$. We view $\widetilde{\mathrm{T}}$ as a $\Delta$-equivariant $\mathscr{O}_{\widehat{Y}}$-module using the descent datum corresponding to the $\widehat{R_1}$-module $\mathrm{Hom}_{\widehat{R_1}}(\widetilde{\Omega}^1_{R/\mathscr{O}_K} \otimes_R \widehat{R_1}, \xi\widehat{R_1})$ (cf. II.4.18). For every $g \in \Delta$, we therefore have a canonical isomorphism of $\mathscr{O}_{\widehat{Y}}$-modules

$$\tau_g^{\widetilde{\mathrm{T}}} \colon \widetilde{\mathrm{T}} \xrightarrow{\sim} g^*(\widetilde{\mathrm{T}}). \tag{II.10.4.1}$$

This induces an isomorphism of $\widehat{Y}$-group schemes

$$\tau_g^{\mathbf{T}} \colon \mathbf{T} \xrightarrow{\sim} g^\bullet(\mathbf{T}), \tag{II.10.4.2}$$

where $g^\bullet$ denotes the base change functor by the automorphism $g$ of $\widehat{Y}$ (II.4.1.1). We thus obtain a $\Delta$-equivariant structure on the $\widehat{Y}$-group scheme $\mathbf{T}$ (cf. II.4.17) and consequently a left action of $\Delta$ on $\mathbf{T}$ that is compatible with its action on $\widehat{Y}$; the automorphism of $\mathbf{T}$ defined by an element $g$ of $\Delta$ is the composition of $\tau_g^{\mathbf{T}}$ and the canonical projection $g^\bullet(\mathbf{T}) \to \mathbf{T}$. We deduce from this an action of $\Delta$ on $\mathscr{S}$ by ring automorphisms that is compatible with its action on $\widehat{\overline{R}}$; we call it the *canonical action*. Concretely, it is induced by the trivial action on $\mathrm{S}_{\widehat{R_1}}(\xi^{-1}\widetilde{\Omega}^1_{R/\mathscr{O}_K} \otimes_R \widehat{R_1})$.

The left action of $\Delta$ on the logarithmic scheme $(\mathscr{A}_2(Y), \mathscr{M}_{\mathscr{A}_2(Y)})$ defined in II.9.19 induces a $\Delta$-equivariant structure on the $\widetilde{\mathrm{T}}$-torsor $\mathscr{L}$ (cf. II.4.18). In other words, it induces, for every $g \in \Delta$, a $\tau_g^{\widetilde{\mathrm{T}}}$-equivariant isomorphism

$$\tau_g^{\mathscr{L}} \colon \mathscr{L} \xrightarrow{\sim} g^*(\mathscr{L}), \tag{II.10.4.3}$$

these isomorphisms being subject to compatibility relations (II.4.16.6). Indeed, for every Zariski open subscheme $U$ of $\widehat{Y}$, we take for

$$\tau_g^{\mathscr{L}}(U) \colon \mathscr{L}(U) \xrightarrow{\sim} \mathscr{L}(g(U)) \tag{II.10.4.4}$$

the isomorphism defined as follows. Let $\widetilde{U}$ be the open subscheme of $\mathscr{A}_2(Y)$ corresponding to $U$ and $\psi \in \mathscr{L}(U)$ that we will view as a morphism

$$\psi \colon (\widetilde{U}, \mathscr{M}_{\mathscr{A}_2(Y)}|\widetilde{U}) \to (\widetilde{X}, \mathscr{M}_{\widetilde{X}}). \tag{II.10.4.5}$$

The morphisms $i_Y$ (II.9.12.6) and $(\widehat{Y}, \mathscr{M}_{\widehat{Y}}) \to (\check{\overline{X}}, \mathscr{M}_{\check{\overline{X}}})$ (II.10.1.3) are $\Delta$-equivariant, so the composition

$$(g(\widetilde{U}), \mathscr{M}_{\mathscr{A}_2(Y)}|g(\widetilde{U})) \xrightarrow{g^{-1}} (\widetilde{U}, \mathscr{M}_{\mathscr{A}_2(Y)}|\widetilde{U}) \xrightarrow{\psi} (\widetilde{X}, \mathscr{M}_{\widetilde{X}}) \tag{II.10.4.6}$$

extends the canonical morphism $(g(U), \mathscr{M}_{\widehat{Y}}|g(U)) \to (\widetilde{X}, \mathscr{M}_{\widetilde{X}})$. It corresponds to the image of $\psi$ by $\tau_g^{\mathscr{L}}(U)$. One immediately verifies that the resulting morphism $\tau_g^{\mathscr{L}}$ is a $\tau_g^{\widetilde{\mathrm{T}}}$-equivariant isomorphism and that these isomorphisms satisfy the required compatibility relations (II.4.16.6).

By II.4.21, the $\Delta$-equivariant structures on $\widetilde{\mathrm{T}}$ and $\mathscr{L}$ induce a $\Delta$-equivariant structure on the $\mathscr{O}_{\widehat{Y}}$-module associated with $\mathscr{F}$, or, equivalently, an $\widehat{\overline{R}}$-semi-linear action of $\Delta$ on $\mathscr{F}$, such that the morphisms in the sequence (II.10.3.5) are $\Delta$-equivariant. We deduce from this a structure of $\Delta$-equivariant principal homogeneous $\mathbf{T}$-bundle over $\widehat{Y}$ on $\mathbf{L}$ (cf. II.4.20). For every $g \in \Delta$, we therefore have a $\tau_g^{\mathbf{T}}$-equivariant isomorphism

$$\tau_g^{\mathbf{L}} \colon \mathbf{L} \xrightarrow{\sim} g^\bullet(\mathbf{L}). \tag{II.10.4.7}$$

This structure determines a left action of $\Delta$ on $\mathbf{L}$ that is compatible with its action on $\widehat{Y}$; the automorphism of $\mathbf{L}$ defined by an element $g$ of $\Delta$ is the composition of $\tau_g^{\mathbf{L}}$ and the canonical projection $g^\bullet(\mathbf{L}) \to \mathbf{L}$. We thus obtain an action of $\Delta$ on $\mathscr{C}$ by ring automorphisms that is compatible with its action on $\widehat{\overline{R}}$; we call it the *canonical action*. Concretely, it is induced by the action of $\Delta$ on $\mathscr{F}$.

For every $g \in \Delta$, we denote by

$$\mathbf{L}(\widehat{Y}) \xrightarrow{\sim} \mathbf{L}(\widehat{Y}), \quad \psi \mapsto {}^g\psi \tag{II.10.4.8}$$

the composition of the isomorphisms

$$\tau_g^{\mathbf{L}} \colon \mathbf{L}(\widehat{Y}) \xrightarrow{\sim} g^\bullet(\mathbf{L})(\widehat{Y}), \tag{II.10.4.9}$$

$$g^\bullet(\mathbf{L})(\widehat{Y}) \xrightarrow{\sim} \mathbf{L}(\widehat{Y}), \quad \psi \mapsto \mathrm{pr} \circ \psi \circ g^{-1}, \tag{II.10.4.10}$$

where $g^{-1}$ acts on $\widehat{Y}$ and $\mathrm{pr} \colon g^\bullet(\mathbf{L}) \to \mathbf{L}$ is the canonical projection, so that the diagram

$$\begin{array}{ccc} \mathbf{L} & \xrightarrow{g} & \mathbf{L} \\ {\scriptstyle\psi}\uparrow & & \uparrow{\scriptstyle {}^g\psi} \\ \widehat{Y} & \xrightarrow{g} & \widehat{Y} \end{array} \tag{II.10.4.11}$$

is commutative.

**Definition II.10.5.** The $\widehat{\overline{R}}$-algebra $\mathscr{C}$ (II.10.3.7), endowed with the canonical action of $\Delta$ (II.10.4), is called the *Higgs–Tate algebra* associated with $(\widetilde{X}, \mathscr{M}_{\widetilde{X}})$. The $\widehat{\overline{R}}$-representation $\mathscr{F}$ (II.10.3.5) of $\Delta$ is called the *Higgs–Tate extension* associated with $(\widetilde{X}, \mathscr{M}_{\widetilde{X}})$.

Note that under the assumptions of II.9.13, these two $\widehat{\overline{R}}$-representations of $\Delta$ do not depend on the choice of the chart $(P, \gamma)$ (II.6.2), by II.9.13 and (II.9.19.9).

**II.10.6.** For every $u \in \mathrm{T} = \mathbf{T}(\widehat{Y})$, we denote by

$$\mathtt{t}_u \colon \mathbf{T} \to \mathbf{T} \tag{II.10.6.1}$$

the translation by $u$. For every $\psi \in \mathscr{L}(\widehat{Y})$, we denote by $\mathtt{t}_\psi$ the isomorphism of principal homogeneous $\mathbf{T}$-bundles over $\widehat{Y}$

$$\mathtt{t}_\psi \colon \mathbf{T} \xrightarrow{\sim} \mathbf{L}, \quad v \mapsto v + \psi. \tag{II.10.6.2}$$

The structure of $\Delta$-equivariant principal homogeneous $\mathbf{T}$-bundle over $\mathbf{L}$ transfers through $\mathtt{t}_\psi$ to a structure of $\Delta$-equivariant principal homogeneous $\mathbf{T}$-bundle over $\mathbf{T}$. For every $g \in \Delta$, we therefore have a $\tau_g^{\mathbf{T}}$-equivariant isomorphism

$$\tau_{g,\psi}^{\mathbf{T}} \colon \mathbf{T} \xrightarrow{\sim} g^\bullet(\mathbf{T}). \tag{II.10.6.3}$$

This structure determines a left action of $\Delta$ on $\mathbf{T}$ that is compatible with its action on $\widehat{Y}$; the automorphism of $\mathbf{T}$ defined by an element $g$ of $\Delta$ is the composition of $\tau_{g,\psi}^{\mathbf{T}}$ and the canonical projection $g^\bullet(\mathbf{T}) \to \mathbf{T}$. We deduce from this an action

$$\varphi_\psi \colon \Delta \to \mathrm{Aut}_{\widehat{R_1}}(\mathscr{S}) \tag{II.10.6.4}$$

of $\Delta$ on $\mathscr{S}$ (II.10.3.2) by ring automorphisms that is compatible with its action on $\widehat{\overline{R}}$; for every $g \in \Delta$, $\varphi_\psi(g)$ is induced by the automorphism of $\mathbf{T}$ defined by $g^{-1}$.

The isomorphism $\mathtt{t}_\psi^* \colon \mathscr{C} \xrightarrow{\sim} \mathscr{S}$ induced by $\mathtt{t}_\psi$ is $\Delta$-equivariant if we endow $\mathscr{C}$ with the canonical action of $\Delta$ and $\mathscr{S}$ with the action $\varphi_\psi$.

One immediately verifies that for every $g \in \Delta$, the diagram

(II.10.6.5)
$$\begin{array}{ccc} \mathbf{T} & \xrightarrow{\mathtt{t}_\psi} & \mathbf{L} \\ {\scriptstyle \tau_g^{\mathbf{T}}}\downarrow & & \downarrow{\scriptstyle \tau_g^{\mathbf{L}}} \\ g^\bullet(\mathbf{T}) & \xrightarrow{g^\bullet(\mathtt{t}_{{}^g\psi})} & g^\bullet(\mathbf{L}) \end{array}$$

is commutative (II.10.4.8). Consequently, the diagram

(II.10.6.6)
$$\begin{array}{ccc} \mathbf{T} & \xrightarrow{\tau_g^{\mathbf{T}}} & g^\bullet(\mathbf{T}) \\ & {\scriptstyle \tau_{g,\psi}^{\mathbf{T}}}\searrow & \downarrow{\scriptstyle g^\bullet(\mathtt{t}_{({}^g\psi-\psi)})} \\ & & g^\bullet(\mathbf{T}) \end{array}$$

is also commutative. It follows that

(II.10.6.7)
$$\varphi_\psi(g) = \mathtt{t}^*_{(\psi - {}^g\psi)} \circ g,$$

where $\mathtt{t}^*_{(\psi-{}^g\psi)}$ is the automorphism of $\mathscr{S}$ induced by $\mathtt{t}_{(\psi-{}^g\psi)}$ and $g$ acts on $\mathscr{S}$ by the canonical action.

**II.10.7.** The pairing $\mathrm{T}\otimes_{\widehat{\overline{R}}}(\xi^{-1}\widetilde{\Omega}^1_{R/\mathscr{O}_K}\otimes_R\widehat{\overline{R}}) \to \widehat{\overline{R}}$ extends into a pairing $\mathrm{T}\otimes_{\widehat{\overline{R}}}\mathscr{S} \to \mathscr{S}$, where the elements of T act as $\widehat{\overline{R}}$-derivations of $\mathscr{S}$. We thus define a morphism

(II.10.7.1)
$$\mathrm{T} \to \Gamma(\mathbf{T}, \mathbb{T}_{\mathbf{T}/\widehat{Y}}), \quad u \mapsto D_u,$$

where $\mathbb{T}_{\mathbf{T}/\widehat{Y}}$ is the tangent bundle of $\mathbf{T}$ over $\widehat{Y}$. This identifies T with the module of invariant vector fields of $\mathbf{T}$ over $\widehat{Y}$. It also induces an isomorphism

(II.10.7.2)
$$\mathrm{T}\otimes_{\widehat{\overline{R}}}\mathscr{O}_{\mathbf{T}} \xrightarrow{\sim} \mathbb{T}_{\mathbf{T}/\widehat{Y}}.$$

Denote by $\Pi = \oplus_{n\geq 0}\Pi_n$ the divided power algebra of T (II.2.5). We have a canonical pairing

(II.10.7.3)
$$\Pi_n \otimes_{\widehat{\overline{R}}} \mathscr{S}^{n+m} \to \mathscr{S}^m$$

that is perfect if $m = 0$ ([**5**] A.10). For every $u \in \mathrm{T} = \mathbf{T}(\widehat{Y})$, $D_u$ belongs to the divided power ideal of $\Pi$, and we can define $\exp(D_u)$ as a differential operator of $\mathscr{S}$ of infinite order. For every $x \in \mathscr{S}$, we have the Taylor formula

(II.10.7.4)
$$\mathtt{t}^*_u(x) = \exp(D_u)(x),$$

where $\mathtt{t}^*_u$ is the automorphism of $\mathscr{S}$ induced by $\mathtt{t}_u$ (II.10.6.1).

**II.10.8.** We have a canonical $\mathscr{S}$-linear isomorphism

(II.10.8.1)
$$\Omega^1_{\mathscr{S}/\widehat{\overline{R}}} \xrightarrow{\sim} \xi^{-1}\widetilde{\Omega}^1_{R/\mathscr{O}_K}\otimes_R\mathscr{S}.$$

We denote by

(II.10.8.2)
$$d_{\mathscr{S}}\colon \mathscr{S} \to \xi^{-1}\widetilde{\Omega}^1_{R/\mathscr{O}_K}\otimes_R\mathscr{S}$$

the universal $\widehat{\overline{R}}$-derivation of $\mathscr{S}$. Explicitly, the morphism (II.10.7.1) is defined as follows: for every $u \in \mathrm{T} = \mathrm{Hom}_{\widehat{\overline{R}}}(\widetilde{\Omega}^1_{R/\mathscr{O}_K}\otimes_R\widehat{\overline{R}}, \xi\widehat{\overline{R}})$, $D_u$ is the composed $\widehat{\overline{R}}$-derivation

(II.10.8.3)
$$\mathscr{S} \xrightarrow{d_{\mathscr{S}}} \xi^{-1}\widetilde{\Omega}^1_{R/\mathscr{O}_K}\otimes_R\mathscr{S} \xrightarrow{(\xi^{-1}\cdot u)\otimes \mathrm{id}} \mathscr{S}.$$

We can then prove again, directly, that the differential operator

$$\exp(D_u)\colon \mathscr{S} \to \mathscr{S} \tag{II.10.8.4}$$

is well-defined and that it is an isomorphism of $\widehat{\overline{R}}$-algebras (II.10.7.4). Indeed, for every $n \geq 0$, $D_u$ sends $\mathscr{S}^n$ to $\mathscr{S}^{n-1}$ (II.10.3.2); it is therefore nilpotent on $\oplus_{0 \leq i \leq n} \mathscr{S}^i$. Consequently, $\exp(D_u)$ is well-defined as an automorphism of the $\widehat{\overline{R}}[\frac{1}{p}]$-algebra $\mathscr{S}[\frac{1}{p}]$. On the other hand, for every $x \in \mathscr{S}^1 \subset \mathscr{S}$, we have

$$\exp(D_u)(x) = x + (\xi^{-1}u)(x). \tag{II.10.8.5}$$

It follows that $\exp(D_u)(\mathscr{S}) \subset \mathscr{S}$ and consequently that $\exp(D_u)(\mathscr{S}) = \mathscr{S}$ (because $D_{-u} = -D_u$).

**II.10.9.** The isomorphism (II.10.8.1) induces an isomorphism

$$\Omega^1_{\mathscr{C}/\widehat{\overline{R}}} \xrightarrow{\sim} \xi^{-1}\widetilde{\Omega}^1_{R/\mathscr{O}_K} \otimes_R \mathscr{C}. \tag{II.10.9.1}$$

We denote by

$$d_{\mathscr{C}}\colon \mathscr{C} \to \xi^{-1}\widetilde{\Omega}^1_{R/\mathscr{O}_K} \otimes_R \mathscr{C} \tag{II.10.9.2}$$

the universal $\widehat{\overline{R}}$-derivation of $\mathscr{C}$. For every $x \in \mathscr{F}$, $d_{\mathscr{C}}(x)$ is the canonical image of $x$ in $\xi^{-1}\widetilde{\Omega}^1_{R/\mathscr{O}_K} \otimes_R \widehat{\overline{R}}$ (II.10.3.5). Consequently, for every $g \in \Delta$, the diagram

$$\begin{array}{ccc} \mathscr{C} & \xrightarrow{\quad g \quad} & \mathscr{C} \\ {\scriptstyle d_{\mathscr{C}}}\downarrow & & \downarrow{\scriptstyle d_{\mathscr{C}}} \\ \xi^{-1}\widetilde{\Omega}^1_{R/\mathscr{O}_K} \otimes_R \mathscr{C} & \xrightarrow{\ \mathrm{id}\otimes g\ } & \xi^{-1}\widetilde{\Omega}^1_{R/\mathscr{O}_K} \otimes_R \mathscr{C} \end{array} \tag{II.10.9.3}$$

is commutative. For every $\psi \in \mathscr{L}(\widehat{Y})$, the diagram

$$\begin{array}{ccc} \mathscr{C} & \xrightarrow{\quad \mathtt{t}_\psi^* \quad} & \mathscr{S} \\ {\scriptstyle d_{\mathscr{C}}}\downarrow & & \downarrow{\scriptstyle d_{\mathscr{S}}} \\ \xi^{-1}\widetilde{\Omega}^1_{R/\mathscr{O}_K} \otimes_R \mathscr{C} & \xrightarrow{\ \mathrm{id}\otimes \mathtt{t}_\psi^*\ } & \xi^{-1}\widetilde{\Omega}^1_{R/\mathscr{O}_K} \otimes_R \mathscr{S} \end{array} \tag{II.10.9.4}$$

where $\mathtt{t}_\psi^*$ is the isomorphism induced by $\mathtt{t}_\psi$ (II.10.6.2), is clearly commutative. It follows that the diagram

$$\begin{array}{ccc} \mathscr{S} & \xrightarrow{\quad \varphi_\psi(g) \quad} & \mathscr{S} \\ {\scriptstyle d_{\mathscr{S}}}\downarrow & & \downarrow{\scriptstyle d_{\mathscr{S}}} \\ \xi^{-1}\widetilde{\Omega}^1_{R/\mathscr{O}_K} \otimes_R \mathscr{S} & \xrightarrow{\ \mathrm{id}\otimes \varphi_\psi(g)\ } & \xi^{-1}\widetilde{\Omega}^1_{R/\mathscr{O}_K} \otimes_R \mathscr{S} \end{array} \tag{II.10.9.5}$$

is commutative, where $\varphi_\psi$ is the action of $\Delta$ on $\mathscr{S}$ defined in (II.10.6.4).

**Remark II.10.10.** Let $(\widetilde{X}', \mathscr{M}_{\widetilde{X}'})$ be another smooth $(\mathscr{A}_2(\overline{S}), \mathscr{M}_{\mathscr{A}_2(\overline{S})})$-deformation of $(\check{\overline{X}}, \mathscr{M}_{\check{\overline{X}}})$, $\mathscr{L}'$ the Higgs–Tate torsor, $\mathscr{C}'$ the Higgs–Tate $\widehat{\overline{R}}$-algebra, and $\mathscr{F}'$ the Higgs–Tate $\widehat{\overline{R}}$-extension associated with $(\widetilde{X}', \mathscr{M}_{\widetilde{X}'})$. Then we have an isomorphism of $(\mathscr{A}_2(\overline{S}), \mathscr{M}_{\mathscr{A}_2(\overline{S})})$-deformations

$$u\colon (\widetilde{X}, \mathscr{M}_{\widetilde{X}}) \xrightarrow{\sim} (\widetilde{X}', \mathscr{M}_{\widetilde{X}'}). \tag{II.10.10.1}$$

The isomorphism of $\widetilde{\mathrm{T}}$-torsors $\mathscr{L} \xrightarrow{\sim} \mathscr{L}'$, $\psi \mapsto u \circ \psi$ (II.10.3.4) induces a $\Delta$-equivariant $\widehat{\overline{R}}$-linear isomorphism

(II.10.10.2) $$\mathscr{F}' \xrightarrow{\sim} \mathscr{F}$$

that fits into a commutative diagram (II.10.3.5)

(II.10.10.3)
$$\begin{array}{ccccccccc} 0 & \longrightarrow & \widehat{\overline{R}} & \longrightarrow & \mathscr{F}' & \longrightarrow & \xi^{-1}\widetilde{\Omega}^1_{R/\mathscr{O}_K} \otimes_R \widehat{\overline{R}} & \longrightarrow & 0 \\ & & \| & & \downarrow & & \| & & \\ 0 & \longrightarrow & \widehat{\overline{R}} & \longrightarrow & \mathscr{F} & \longrightarrow & \xi^{-1}\widetilde{\Omega}^1_{R/\mathscr{O}_K} \otimes_R \widehat{\overline{R}} & \longrightarrow & 0 \end{array}$$

We deduce from this a $\Delta$-equivariant $\widehat{\overline{R}}$-isomorphism

(II.10.10.4) $$\mathscr{C}' \xrightarrow{\sim} \mathscr{C}.$$

**II.10.11.** Every section $\psi \in \mathscr{L}(\widehat{Y})$ defines a splitting $v_\psi \colon \mathscr{F} \to \widehat{\overline{R}}$ of the exact sequence (II.10.3.5), namely the morphism that with any affine function $\ell$ on $\mathscr{L}$ associates $\ell(\psi)$. The morphism $\mathrm{id}_{\mathscr{F}} - v_\psi$ induces an $\widehat{\overline{R}}$-linear morphism

(II.10.11.1) $$u_\psi \colon \xi^{-1}\widetilde{\Omega}^1_{R/\mathscr{O}_K} \otimes_R \widehat{\overline{R}} \to \mathscr{F}$$

that is none other than the restriction of the isomorphism $(\mathtt{t}^*_\psi)^{-1} \colon \mathscr{S} \to \mathscr{C}$ (II.10.6.2) to $\mathscr{S}^1$.

For all $\psi, \psi' \in \mathscr{L}(\widehat{Y})$, the difference $\psi - \psi' \in \mathbf{T}(\widehat{Y}) = \mathrm{T}$ (II.10.3.1) determines an $\widehat{\overline{R}}$-linear morphism

(II.10.11.2) $$\sigma_{\psi,\psi'} \colon \xi^{-1}\widetilde{\Omega}^1_{R/\mathscr{O}_K} \otimes_R \widehat{\overline{R}} \to \widehat{\overline{R}}.$$

We have

(II.10.11.3) $$\sigma_{\psi,\psi'} = u_{\psi'} - u_\psi.$$

By (II.10.4.11), for all $g \in \Delta$ and $\psi \in \mathscr{L}(\widehat{Y})$, we have

(II.10.11.4) $$u_{{}^g\psi} = g \circ u_\psi \circ g^{-1}.$$

Consequently, for every $x \in \xi^{-1}\widetilde{\Omega}^1_{R/\mathscr{O}_K} \otimes_R \widehat{R_1}$, the map $g \mapsto \sigma_{\psi,{}^g\psi}(x)$ is a cocycle of $\Delta$ with values in $\widehat{\overline{R}}$. The induced map

(II.10.11.5) $$\xi^{-1}\widetilde{\Omega}^1_{R/\mathscr{O}_K} \otimes_R \widehat{R_1} \to \mathrm{H}^1(\Delta, \widehat{\overline{R}})$$

is clearly the boundary map of the long exact sequence of cohomology deduced from the exact sequence (II.10.3.5).

By (II.10.6.7) and (II.10.7.4), if we denote by $D_{\psi - {}^g\psi}$ the image of $\psi - {}^g\psi$ by the morphism (II.10.7.1), we have

(II.10.11.6) $$\varphi_\psi(g) = \exp(D_{\psi - {}^g\psi}) \circ g,$$

where $g$ denotes the canonical action of $g$ on $\mathscr{S}$ (II.10.4). In particular, by (II.10.8.5), for every $x \in \xi^{-1}\widetilde{\Omega}^1_{R/\mathscr{O}_K} \subset \mathscr{S}^1$, we have

(II.10.11.7) $$\varphi_\psi(g)(x) = x + \sigma_{\psi,{}^g\psi}(x).$$

**Remark II.10.12.** We say that an element $\psi$ of $\mathscr{L}(\widehat{Y})$ is *optimal* if for every $g \in \Delta$, the morphism $\psi - {}^g\psi \in \mathbf{T}(\widehat{Y}) = \mathrm{T}$ factors into

$$\widetilde{\Omega}^1_{R/\mathscr{O}_K} \otimes_R \widehat{\overline{R}} \to \widehat{\overline{R}}(1) \xrightarrow{\sim} p^{\frac{1}{p-1}}\xi\widehat{\overline{R}} \to \xi\widehat{\overline{R}}, \tag{II.10.12.1}$$

where the second arrow is the isomorphism (II.9.18.1) and the last arrow is the canonical injection.

Denote by $\mathscr{S}'$ the graded $\widehat{\overline{R}}$-algebra (II.2.5)

$$\mathscr{S}' = \mathrm{S}_{\widehat{\overline{R}}}(\widetilde{\Omega}^1_{R/\mathscr{O}_K} \otimes_R \widehat{\overline{R}}(-1)). \tag{II.10.12.2}$$

The canonical isomorphism (II.9.18.1)

$$\xi^{-1}\widetilde{\Omega}^1_{R/\mathscr{O}_K} \otimes_R \widehat{\overline{R}} \xrightarrow{\sim} p^{\frac{1}{p-1}}\widetilde{\Omega}^1_{R/\mathscr{O}_K} \otimes_R \widehat{\overline{R}}(-1) \tag{II.10.12.3}$$

induces an $\widehat{\overline{R}}$-homomorphism

$$\jmath \colon \mathscr{S} \to \mathscr{S}' \tag{II.10.12.4}$$

such that $\jmath \otimes_{\mathbb{Z}_p} \mathbb{Q}_p$ is an isomorphism. For every $\psi \in \mathscr{L}(\widehat{Y})$, we also denote by $\varphi_\psi$ the action of $\Delta$ on $\mathscr{S}'[\frac{1}{p}]$ deduced from its action on $\mathscr{S}$ defined in (II.10.6.4). This action preserves $\mathscr{S}'$ if and only if $\psi$ is optimal. Indeed, denote by

$$\varsigma_{\psi,{}^g\psi} \colon \widetilde{\Omega}^1_{R/\mathscr{O}_K} \otimes_R \widehat{\overline{R}}(-1) \to p^{-\frac{1}{p-1}}\widehat{\overline{R}} \tag{II.10.12.5}$$

the $\widehat{\overline{R}}$-linear morphism deduced from $\psi - {}^g\psi \in \mathbf{T}(\widehat{Y}) = \mathrm{T}$ and the isomorphism (II.9.18.1). For all $g \in \Delta$ and $x \in \widetilde{\Omega}^1_{R/\mathscr{O}_K}(-1) \subset \mathscr{S}'$, we then have

$$\varphi_\psi(x) = x + \varsigma_{\psi,{}^g\psi}(x). \tag{II.10.12.6}$$

Consequently, $\varphi_\psi$ preserves $\mathscr{S}'$ if and only if $\varsigma_{\psi,{}^g\psi}$ factors through $\widehat{\overline{R}} \subset p^{-\frac{1}{p-1}}\widehat{\overline{R}}$ for every $g \in \Delta$, which is equivalent to $\psi$ being optimal.

**II.10.13.** By II.6.2($\mathrm{C}_5$), there essentially exists a unique étale morphism

$$(\widetilde{X}_0, \mathscr{M}_{\widetilde{X}_0}) \to (\mathscr{A}_2(\overline{S}), \mathscr{M}_{\mathscr{A}_2(\overline{S})}) \times_{\mathbf{A}_{\mathbb{N}}} \mathbf{A}_P \tag{II.10.13.1}$$

that fits into a commutative diagram with Cartesian squares

$$\begin{array}{ccccc} (\check{\overline{X}}, \mathscr{M}_{\check{\overline{X}}}) & \longrightarrow & (\check{\overline{S}}, \mathscr{M}_{\check{\overline{S}}}) \times_{\mathbf{A}_{\mathbb{N}}} \mathbf{A}_P & \longrightarrow & (\check{\overline{S}}, \mathscr{M}_{\check{\overline{S}}}) \\ \downarrow & & \downarrow & & \downarrow{\scriptstyle i_{\overline{S}}} \\ (\widetilde{X}_0, \mathscr{M}_{\widetilde{X}_0}) & \longrightarrow & (\mathscr{A}_2(\overline{S}), \mathscr{M}_{\mathscr{A}_2(\overline{S})}) \times_{\mathbf{A}_{\mathbb{N}}} \mathbf{A}_P & \longrightarrow & (\mathscr{A}_2(\overline{S}), \mathscr{M}_{\mathscr{A}_2(\overline{S})}) \\ & & \downarrow & & \downarrow{\scriptstyle a} \\ & & \mathbf{A}_P & \longrightarrow & \mathbf{A}_{\mathbb{N}} \end{array} \tag{II.10.13.2}$$

where the morphism $a$ is defined by the chart $\mathbb{N} \to \Gamma(\mathscr{A}_2(\overline{S}), \mathscr{M}_{\mathscr{A}_2(\overline{S})}), 1 \mapsto \widetilde{\pi}$ (II.9.15). We say that $(\widetilde{X}_0, \mathscr{M}_{\widetilde{X}_0})$ is the smooth $(\mathscr{A}_2(\overline{S}), \mathscr{M}_{\mathscr{A}_2(\overline{S})})$-deformation of $(\check{\overline{X}}, \mathscr{M}_{\check{\overline{X}}})$ defined by the chart $(P, \gamma)$. We denote by $\mathscr{L}_0$ the Higgs–Tate torsor (II.10.3), by $\mathscr{C}_0$ the Higgs–Tate $\widehat{\overline{R}}$-algebra, and by $\mathscr{F}_0$ the Higgs–Tate $\widehat{\overline{R}}$-extension associated with $(\widetilde{X}_0, \mathscr{M}_{\widetilde{X}_0})$ (II.10.5).

The diagram

(II.10.13.3)
$$\begin{array}{ccccc}(\widehat{Y},\mathscr{M}_{\widehat{Y}}) & \xrightarrow{i_Y} & (\mathscr{A}_2(Y),\mathscr{M}_{\mathscr{A}_2(Y)}) & \xrightarrow{b} & \mathbf{A}_P \\ \downarrow & & \downarrow & & \downarrow \\ (\check{\overline{S}},\mathscr{M}_{\check{\overline{S}}}) & \xrightarrow{i_{\overline{S}}} & (\mathscr{A}_2(\overline{S}),\mathscr{M}_{\mathscr{A}_2(\overline{S})}) & \xrightarrow{a} & \mathbf{A}_{\mathbb{N}}\end{array}$$

where the morphism $b$ is defined by the canonical chart $P\to\Gamma(\mathscr{A}_2(Y),\mathscr{M}_{\mathscr{A}_2(Y)})$ (II.9.12), is commutative by (II.9.16.1). It is clear that the diagram

(II.10.13.4)
$$\begin{array}{ccccc}(\widehat{Y},\mathscr{M}_{\widehat{Y}}) & \longrightarrow & (\check{\overline{X}},\mathscr{M}_{\check{\overline{X}}}) & \longrightarrow & (\widetilde{X}_0,\mathscr{M}_{\widetilde{X}_0}) \\ {\scriptstyle i_Y}\downarrow & & {\scriptstyle \psi_0}\ \nearrow & & \downarrow \\ (\mathscr{A}_2(Y),\mathscr{M}_{\mathscr{A}_2(Y)}) & & \xrightarrow{\phi_0} & & (\mathscr{A}_2(\overline{S}),\mathscr{M}_{\mathscr{A}_2(\overline{S})})\times_{\mathbf{A}_{\mathbb{N}}}\mathbf{A}_P\end{array}$$

(without the dotted arrow), where $\phi_0$ is defined by (II.10.13.3), is commutative. We can complete it with a unique dotted arrow $\psi_0\in\mathscr{L}_0(\widehat{Y})$ in such a way that it remains commutative. We say that $\psi_0$ is the section of $\mathscr{L}_0(\widehat{Y})$ defined by the chart $(P,\gamma)$. The reader will note that $(\widetilde{X}_0,\mathscr{M}_{\widetilde{X}_0})$ and $\psi_0$ depend on the choice of the (II.6.7.1), because the charts $a$ and $b$ depend on it.

**Proposition II.10.14.** *We keep the notation of* II.10.13. *For all* $t\in P^{\mathrm{gp}}$ *and* $g\in\Delta$, *we have*

(II.10.14.1)
$$(\psi_0-{}^g\psi_0)(d\log(t))=-\log([\chi_t(g)]),$$

*where we view* $\psi_0-{}^g\psi_0$ *as an element of* $\mathbf{T}(\widehat{Y})=\mathrm{T}$ (II.10.3.1), $d\log(t)$ *is the canonical image of* $t$ *in* $\widetilde{\Omega}^1_{R/\mathscr{O}_K}$ (II.7.18.5), *and* $\log([\chi_t])$ *denotes the composition*

(II.10.14.2)
$$\Delta\longrightarrow\Delta_\infty\xrightarrow{\chi_t}\mathbb{Z}_p(1)\xrightarrow{\log([\ ])}p^{\frac{1}{p-1}}\xi\widehat{\overline{R}}\longrightarrow\xi\widehat{\overline{R}},$$

*where the first and last arrows are the canonical morphisms,* $\chi_t$ *is the homomorphism* (II.7.19.5), *and the third arrow is induced by the isomorphism* (II.9.18.1). *In particular,* $\psi_0$ *is optimal* (II.10.12).

Since the two sides of equation (II.10.14.1) are homomorphisms from $P^{\mathrm{gp}}$ to $\xi\widehat{\overline{R}}$, we can restrict to the case where $t\in P$. The morphisms $\phi_0$ and $\phi_0\circ g^{-1}$, where $\phi_0$ is the morphism defined in (II.10.13.4) and $g^{-1}$ acts on $(\mathscr{A}_2(Y),\mathscr{M}_{\mathscr{A}_2(Y)})$, extend the same morphism

$$(\widehat{Y},\mathscr{M}_{\widehat{Y}})\to(\mathscr{A}_2(\overline{S}),\mathscr{M}_{\mathscr{A}_2(\overline{S})})\times_{\mathbf{A}_{\mathbb{N}}}\mathbf{A}_P.$$

By the definitions and condition II.6.2($\mathrm{C}_5$), the difference $\phi_0-\phi_0\circ g^{-1}$ corresponds to the morphism $\psi_0-{}^g\psi_0\in\mathrm{T}$. On the other hand, we have $g(\widetilde{t})=[\chi_t(g)]\cdot\widetilde{t}$ in $\Gamma(\mathscr{A}_2(Y),\mathscr{M}_{\mathscr{A}_2(Y)})$ (II.9.19.4). The first statement follows from this in view of II.5.23 and (II.9.17.2). The second statement is an immediate consequence of the first.

**Lemma II.10.15.** *We have a commutative diagram*

(II.10.15.1)
$$\begin{array}{ccc}\xi^{-1}\widetilde{\Omega}^1_{R/\mathscr{O}_K}\otimes_R\widehat{R_1} & \xrightarrow{\partial} & \mathrm{H}^1(\Delta,\widehat{\overline{R}}) \\ {\scriptstyle u}\downarrow & & \uparrow{\scriptstyle w} \\ \mathrm{H}^1(\Delta,\xi^{-1}\widehat{\overline{R}}(1)) & \xrightarrow[\sim]{-v} & \mathrm{H}^1(\Delta,p^{\frac{1}{p-1}}\widehat{\overline{R}})\end{array}$$

*where $\partial$ is the boundary map of the long exact sequence of cohomology deduced from the short exact sequence* (II.10.3.5), *$u$ is induced by the morphism* (II.8.13.3), *$v$ is induced by the isomorphism* (II.9.18.1), *and $w$ is induced by the canonical injection $p^{\frac{1}{p-1}}\widehat{\overline{R}} \to \widehat{\overline{R}}$.*

Indeed, it suffices to show that the diagram

(II.10.15.2)
$$\begin{array}{ccc} \xi^{-1}\widetilde{\Omega}^1_{R/\mathscr{O}_K} \otimes_R \widehat{R_1} & \xrightarrow{\partial} & \mathrm{H}^1(\Delta, \widehat{\overline{R}}) \\ \downarrow{\scriptstyle \xi^{-1}\widetilde{\delta}} & & \uparrow{\scriptstyle w'} \\ \mathrm{Hom}_{\mathbb{Z}}(\Delta_\infty, \xi^{-1}\widehat{R_1}(1)) & \xrightarrow[\sim]{-v'} & \mathrm{Hom}_{\mathbb{Z}}(\Delta_\infty, p^{\frac{1}{p-1}}\widehat{R_1}) \end{array}$$

where $\widetilde{\delta}$ is the isomorphism (II.7.19.9), $v'$ comes from the isomorphism $\mathscr{O}_C(1) \xrightarrow{\sim} p^{\frac{1}{p-1}}\xi\mathscr{O}_C$ (II.9.18.1), and $w'$ is induced by the canonical injection $p^{\frac{1}{p-1}}\widehat{R_1} \to \widehat{\overline{R}}$, is commutative. By II.10.10, $\partial$ does not depend on the deformation $(\widetilde{X}, \mathscr{M}_{\widetilde{X}})$. We may therefore restrict to the case where $(\widetilde{X}, \mathscr{M}_{\widetilde{X}})$ is the deformation defined by the chart $(P, \gamma)$ (II.10.13). Let $\psi_0$ be the section of $\mathscr{L}_0(\widehat{Y})$ defined by the chart $(P, \gamma)$. By II.10.11, for every $x \in \xi^{-1}\widetilde{\Omega}^1_{R/\mathscr{O}_K} \otimes_R \widehat{R_1}$, $\partial(x)$ is the class of the cocycle $g \mapsto \sigma_{\psi_0, {}^g\psi_0}(x)$ of $\Delta$ with values in $\widehat{\overline{R}}$. The statement therefore follows from II.10.14.

**Remark II.10.16.** It immediately follows from II.10.15 that for every nonzero element $a$ of $\mathscr{O}_{\overline{K}}$, we have a commutative diagram

(II.10.16.1)
$$\begin{array}{ccc} \xi^{-1}\widetilde{\Omega}^1_{R/\mathscr{O}_K} \otimes_R (R_1/aR_1) & \xrightarrow{\partial_a} & \mathrm{H}^1(\Delta, \overline{R}/a\overline{R}) \\ \downarrow{\scriptstyle u_a} & & \uparrow{\scriptstyle w_a} \\ \mathrm{H}^1(\Delta, \xi^{-1}\overline{R}(1)/a\xi^{-1}\overline{R}(1)) & \xrightarrow[\sim]{-v_a} & \mathrm{H}^1(\Delta, p^{\frac{1}{p-1}}\overline{R}/ap^{\frac{1}{p-1}}\overline{R}) \end{array}$$

where $\partial_a$ is induced by the boundary map of the long exact sequence of cohomology deduced from the short exact sequence obtained by reducing (II.10.3.5) modulo $a$, $u_a$ is induced by the morphism (II.8.13.2), $v_a$ is induced by the isomorphism (II.9.18.1), and $w_a$ is induced by the canonical injection $p^{\frac{1}{p-1}}\overline{R} \to \overline{R}$.

**II.10.17.** We denote by $\varphi_0 = \varphi_{\psi_0}$ (II.10.6.4) the action of $\Delta$ on $\mathscr{S}$ induced by the section $\psi_0 \in \mathscr{L}_0(\widehat{Y})$ defined by the chart $(P, \gamma)$ (II.10.13). Let $t_1, \dots, t_d \in P^{\mathrm{gp}}$ be such that their images in $(P^{\mathrm{gp}}/\mathbb{Z}\lambda) \otimes_{\mathbb{Z}} \mathbb{Z}_p$ form a $\mathbb{Z}_p$-basis, so that $(d\log(t_i))_{1 \leq i \leq d}$ is an $R$-basis of $\widetilde{\Omega}^1_{R/\mathscr{O}_K}$ (II.7.18.5). For every $1 \leq i \leq d$, set $y_i = \xi^{-1} d\log(t_i) \in \xi^{-1}\widetilde{\Omega}^1_{R/\mathscr{O}_K} \subset \mathscr{S}$ and denote by $\chi_i$ the composition

(II.10.17.1)
$$\Delta \longrightarrow \Delta_\infty \xrightarrow{\chi_{t_i}} \mathbb{Z}_p(1) \xrightarrow{\log([\ ])} p^{\frac{1}{p-1}}\xi\mathscr{O}_C\ ,$$

where the first arrow is the canonical morphism, $\chi_{t_i}$ is the homomorphism (II.7.19.5), and the third arrow is induced by the isomorphism (II.9.18.1). It then follows from (II.10.11.6) and (II.10.14.1) that for every $g \in \Delta$, we have

(II.10.17.2)
$$\varphi_0(g) = \exp\Big(-\sum_{i=1}^{d} \xi^{-1}\frac{\partial}{\partial y_i} \otimes \chi_i(g)\Big) \circ g.$$

Consequently, $\varphi_0$ preserves the sub-$\widehat{R_1}$-algebra

(II.10.17.3)
$$\mathfrak{S} = \mathrm{S}_{\widehat{R_1}}(\xi^{-1}\widetilde{\Omega}^1_{R/\mathscr{O}_K} \otimes_R \widehat{R_1})$$

of $\mathscr{S}$, and the induced action of $\Delta$ on $\mathfrak{S}$ factors through $\Delta_{p^\infty}$.

**Theorem II.10.18.** *We keep the notation of* II.10.13 *and denote by* $(P^{\rm gp}/\mathbb{Z}\lambda)_{\rm lib}$ *the quotient of* $P^{\rm gp}/\mathbb{Z}\lambda$ *by its torsion submodule. Then giving a right inverse*

$$w\colon (P^{\rm gp}/\mathbb{Z}\lambda)_{\rm lib} \to P^{\rm gp} \tag{II.10.18.1}$$

*of the canonical morphism* $P^{\rm gp} \to (P^{\rm gp}/\mathbb{Z}\lambda)_{\rm lib}$ *uniquely determines an* $\widehat{\overline{R}}$*-linear and* $\Delta$*-equivariant morphism*

$$\beta_w\colon p^{-\frac{1}{p-1}}\mathscr{F}_0 \to \mathscr{E} \tag{II.10.18.2}$$

*that fits into a commutative diagram*

$$\begin{array}{ccccccccc} 0 & \longrightarrow & p^{-\frac{1}{p-1}}\widehat{\overline{R}} & \longrightarrow & p^{-\frac{1}{p-1}}\mathscr{F}_0 & \longrightarrow & p^{-\frac{1}{p-1}}\xi^{-1}\widetilde{\Omega}^1_{R/\mathscr{O}_K}\otimes_R\widehat{\overline{R}} & \longrightarrow & 0 \\ & & \big\downarrow & & \big\downarrow{\scriptstyle \beta_w} & & \big\downarrow{\scriptstyle -c} & & \\ 0 & \longrightarrow & (\pi\rho)^{-1}\widehat{\overline{R}} & \longrightarrow & \mathscr{E} & \longrightarrow & \widetilde{\Omega}^1_{R/\mathscr{O}_K}\otimes_R\widehat{\overline{R}}(-1) & \longrightarrow & 0 \end{array} \tag{II.10.18.3}$$

*where the lines come from the exact sequences* (II.7.22.2) *and* (II.10.3.5) *and* $c$ *is induced by the isomorphism* (II.9.18.1).

Since the torsion subgroup of $P^{\rm gp}/\mathbb{Z}\lambda$ is of order prime to $p$, the isomorphism (II.7.12.1) induces an isomorphism

$$(P^{\rm gp}/\mathbb{Z}\lambda)_{\rm lib}\otimes_{\mathbb{Z}} R \xrightarrow{\sim} \widetilde{\Omega}^1_{R/\mathscr{O}_K}.$$

Consequently, the composition of $w$ with the homomorphism $P^{\rm gp}\to\mathscr{E}(1), t\mapsto d\log(\widetilde{t})$ (II.7.18.3) induces an $\widehat{\overline{R}}$-linear morphism that we denote by

$$\sigma_{\psi_0}\colon \widetilde{\Omega}^1_{R/\mathscr{O}_K}\otimes_R\widehat{\overline{R}}\to\mathscr{E}(1). \tag{II.10.18.4}$$

By (II.7.18.4), $\sigma_{\psi_0}$ is a splitting of the exact sequence (II.7.22.2) twisted by $\mathbb{Z}_p(1)$. For every $x\in\mathscr{E}(1)$, set

$$\langle\psi_0, x\rangle = x - \sigma_{\psi_0}(\nu(x)) \in (\pi\rho)^{-1}\widehat{\overline{R}}(1), \tag{II.10.18.5}$$

where $\nu\colon\mathscr{E}(1)\to\widetilde{\Omega}^1_{R/\mathscr{O}_K}\otimes_R\widehat{\overline{R}}$ is the canonical morphism. We define a map

$$\langle\ ,\ \rangle\colon \mathscr{L}_0(\widehat{Y})\times\mathscr{E}(1)\to(\pi\rho)^{-1}\widehat{\overline{R}}(1) \tag{II.10.18.6}$$

as follows: for every $\phi\in\mathrm{T}$ (II.10.3.1), if we denote by $\overline{\phi}$ the composition

$$\widetilde{\Omega}^1_{R/\mathscr{O}_K}\otimes_R\widehat{\overline{R}}\xrightarrow{\phi}\xi\widehat{\overline{R}}\xrightarrow{\sim}p^{-\frac{1}{p-1}}\widehat{\overline{R}}(1)\longrightarrow(\pi\rho)^{-1}\widehat{\overline{R}}(1), \tag{II.10.18.7}$$

where the second arrow is induced by the isomorphism (II.9.18.1) and the third arrow is the canonical injection, we have

$$\langle\psi_0+\phi, x\rangle = \langle\psi_0, x\rangle - \overline{\phi}(\nu(x)). \tag{II.10.18.8}$$

For every $\psi\in\mathscr{L}_0(\widehat{Y})$, the morphism $\mathscr{E}(1)\to(\pi\rho)^{-1}\widehat{\overline{R}}(1), x\mapsto\langle\psi,x\rangle$ is a splitting of the exact sequence (II.7.22.2) twisted by $\mathbb{Z}_p(1)$.

Let us show that for all $\psi\in\mathscr{L}_0(\widehat{Y})$, $x\in\mathscr{E}(1)$, and $g\in\Delta$, we have

$$g(\langle\psi,x\rangle) = \langle{}^g\psi, g(x)\rangle. \tag{II.10.18.9}$$

First consider the case $\psi=\psi_0$. It then suffices to show that for every $x\in\widetilde{\Omega}^1_{R/\mathscr{O}_K}\otimes_R\widehat{\overline{R}}$, we have

$$\overline{\sigma}_{\psi_0,{}^g\psi_0}(x) = \sigma_{\psi_0}(x) - g(\sigma_{\psi_0}(g^{-1}x)), \tag{II.10.18.10}$$

where $\overline{\sigma}_{\psi_0, {}^g\psi_0}$ is the morphism defined in (II.10.18.7) from $\psi_0 - {}^g\psi_0 \in \mathrm{T}$ (II.10.11). Let $t \in P$, $y$ its canonical image in $(P^{\mathrm{gp}}/\mathbb{Z}\lambda)_{\mathrm{lib}}$, and $z = w(y) \in P^{\mathrm{gp}}$. Then we have

$$\sigma_{\psi_0}(d\log(t)) = d\log(\widetilde{z}).$$

On the other hand, by virtue of II.10.14, we have

$$\overline{\sigma}_{\psi_0, {}^g\psi_0}(d\log(t)) = (\chi_t(g))^{-1},$$

where $(\chi_t(g))^{-1} \in \mathbb{Z}_p(1) \subset \widehat{\overline{R}}(1) \subset (\pi\rho)^{-1}\widehat{\overline{R}}(1)$. Note that the character $\log([\ ])$ disappears from the formula because of the definition (II.10.18.7). Since the image of $\chi_t(g)$ by the canonical injection $(\pi\rho)^{-1}\widehat{\overline{R}}(1) \to \mathscr{E}(1)$ is $d\log(\chi_t(g))$ (II.7.18), it suffices to show the following relation in $\mathscr{E}(1)$:

(II.10.18.11) $$g(d\log(\widetilde{z})) = d\log(\widetilde{z}) + \log(\chi_t(g)).$$

We can obviously replace $t$ by a power, and therefore assume $t - z \in \mathbb{Z}\lambda$. Since $\Delta$ fixes $d\log(\widetilde{\lambda})$, the relation (II.10.18.11) follows from (II.7.20.1).

In the general case, for all $\phi \in \mathrm{T}$, $x \in \mathscr{E}(1)$, and $g \in \Delta$, we have

$$\begin{aligned} g(\langle \psi_0 + \phi, x\rangle) &= g(\langle \psi_0, x\rangle) - g \circ \overline{\phi}(\nu(x)) \\ &= \langle {}^g\psi_0, g(x)\rangle - g \circ \overline{\phi} \circ g^{-1}(\nu(g(x))) \\ &= \langle {}^g\psi_0 + g \circ \phi \circ g^{-1}, g(x)\rangle \\ &= \langle {}^g(\psi_0 + \phi), g(x)\rangle, \end{aligned}$$

which concludes the proof of (II.10.18.9).

For every $x \in \mathscr{E}(1)$, the map $\psi \mapsto \langle \psi, x\rangle$ is an affine function on $\mathscr{L}_0$ with values in $(\pi\rho)^{-1}\widehat{\overline{R}}(1)$ (cf. II.4.7). It is therefore naturally an element of $(\pi\rho)^{-1}\mathscr{F}_0(1)$. The map

(II.10.18.12) $$\mathscr{E}(1) \to (\pi\rho)^{-1}\mathscr{F}_0(1), \quad x \mapsto (\psi \mapsto \langle \psi, x\rangle)$$

is an $\widehat{\overline{R}}$-linear and $\Delta$-equivariant morphism by virtue of (II.10.18.9). The induced morphism

$$\mathscr{E} \to (\pi\rho)^{-1}\mathscr{F}_0$$

fits into a commutative diagram

$$\begin{array}{ccccccccc}
0 & \longrightarrow & (\pi\rho)^{-1}\widehat{\overline{R}} & \longrightarrow & \mathscr{E} & \longrightarrow & \widetilde{\Omega}^1_{R/\mathscr{O}_K} \otimes_R \widehat{\overline{R}}(-1) & \longrightarrow & 0 \\
 & & \| & & \Big\downarrow & & \Big\downarrow{\scriptstyle i_1} & & \\
 & & & & & & (\pi\rho)^{-1}p^{\frac{1}{p-1}}\widetilde{\Omega}^1_{R/\mathscr{O}_K} \otimes_R \widehat{\overline{R}}(-1) & \longrightarrow & 0 \\
 & & & & & & \Big\downarrow{\scriptstyle -c^{-1}} & & \\
0 & \longrightarrow & (\pi\rho)^{-1}\widehat{\overline{R}} & \longrightarrow & (\pi\rho)^{-1}\mathscr{F}_0 & \longrightarrow & (\pi\rho)^{-1}\xi^{-1}\widetilde{\Omega}^1_{R/\mathscr{O}_K} \otimes_R \widehat{\overline{R}} & \longrightarrow & 0
\end{array}$$

where $i_1$ is the canonical injection. Consider the commutative diagram

$$\begin{array}{ccccccccc}
0 & \longrightarrow & p^{-\frac{1}{p-1}}\widehat{\overline{R}} & \longrightarrow & p^{-\frac{1}{p-1}}\mathscr{F}_0 & \longrightarrow & p^{-\frac{1}{p-1}}\xi^{-1}\widetilde{\Omega}^1_{R/\mathscr{O}_K}\otimes_R\widehat{\overline{R}} & \longrightarrow & 0 \\
 & & \downarrow{\scriptstyle i_3} & & \downarrow & & \| & & \\
0 & \longrightarrow & (\pi\rho)^{-1}\widehat{\overline{R}} & \longrightarrow & \mathscr{H} & \longrightarrow & p^{-\frac{1}{p-1}}\xi^{-1}\widetilde{\Omega}^1_{R/\mathscr{O}_K}\otimes_R\widehat{\overline{R}} & \longrightarrow & 0 \\
 & & \| & & \downarrow & & \downarrow{\scriptstyle i_2} & & \\
0 & \longrightarrow & (\pi\rho)^{-1}\widehat{\overline{R}} & \longrightarrow & (\pi\rho)^{-1}\mathscr{F}_0 & \longrightarrow & (\pi\rho)^{-1}\xi^{-1}\widetilde{\Omega}^1_{R/\mathscr{O}_K}\otimes_R\widehat{\overline{R}} & \longrightarrow & 0
\end{array}$$

where $i_2$ and $i_3$ are the canonical injections and $\mathscr{H}$ is the inverse image of $(\pi\rho)^{-1}\mathscr{F}_0$ under $i_2$. Since $i_2\circ c^{-1} = c^{-1}\circ i_1$, we deduce from this an isomorphism $\alpha\colon \mathscr{H} \xrightarrow{\sim} \mathscr{E}$ that fits into a commutative diagram

$$\begin{array}{ccccccccc}
0 & \longrightarrow & (\pi\rho)^{-1}\widehat{\overline{R}} & \longrightarrow & \mathscr{H} & \longrightarrow & p^{-\frac{1}{p-1}}\xi^{-1}\widetilde{\Omega}^1_{R/\mathscr{O}_K}\otimes_R\widehat{\overline{R}} & \longrightarrow & 0 \\
 & & \| & & \downarrow{\scriptstyle \alpha} & & \downarrow{\scriptstyle -c} & & \\
0 & \longrightarrow & (\pi\rho)^{-1}\widehat{\overline{R}} & \longrightarrow & \mathscr{E} & \longrightarrow & \widetilde{\Omega}^1_{R/\mathscr{O}_K}\otimes_R\widehat{\overline{R}}(-1) & \longrightarrow & 0
\end{array}$$

The morphism $p^{-\frac{1}{p-1}}\mathscr{F}_0 \to \mathscr{E}$ composed of the canonical injection $p^{-\frac{1}{p-1}}\mathscr{F}_0 \to \mathscr{H}$ and $\alpha$ then has the desired property.

**Corollary II.10.19.** *There exists a* $\Delta$*-equivariant* $\widehat{\overline{R}}$*-linear morphism*

$$p^{-\frac{1}{p-1}}\mathscr{F} \to \mathscr{E} \tag{II.10.19.1}$$

*that fits into a commutative diagram*

$$\begin{array}{ccccccccc}
0 & \longrightarrow & p^{-\frac{1}{p-1}}\widehat{\overline{R}} & \longrightarrow & p^{-\frac{1}{p-1}}\mathscr{F} & \longrightarrow & p^{-\frac{1}{p-1}}\xi^{-1}\widetilde{\Omega}^1_{R/\mathscr{O}_K}\otimes_R\widehat{\overline{R}} & \longrightarrow & 0 \\
 & & \downarrow & & \downarrow & & \downarrow{\scriptstyle -c} & & \\
0 & \longrightarrow & (\pi\rho)^{-1}\widehat{\overline{R}} & \longrightarrow & \mathscr{E} & \longrightarrow & \widetilde{\Omega}^1_{R/\mathscr{O}_K}\otimes_R\widehat{\overline{R}}(-1) & \longrightarrow & 0
\end{array} \tag{II.10.19.2}$$

*where the lines come from the exact sequences* (II.7.22.2) *and* (II.10.3.5), *and* $c$ *is induced by the isomorphism* (II.9.18.1).

This follows from II.10.10 and II.10.18.

## II.11. Galois cohomology II

In this section, we slightly generalize the results of T. Tsuji from IV.

**II.11.1.** We denote by $\mathscr{S}$ the symmetric $\widehat{\overline{R}}$-algebra (II.10.3.2)

$$\mathscr{S} = \mathrm{S}_{\widehat{\overline{R}}}(\xi^{-1}\widetilde{\Omega}^1_{R/\mathscr{O}_K}\otimes_R\widehat{\overline{R}}) \tag{II.11.1.1}$$

and by $\widehat{\mathscr{S}}$ its $p$-adic Hausdorff completion. For every rational number $r \geq 0$, we denote by $\mathscr{S}^{(r)}$ the sub-$\widehat{\overline{R}}$-algebra of $\mathscr{S}$ defined by (II.2.5)

$$\mathscr{S}^{(r)} = \mathrm{S}_{\widehat{\overline{R}}}(p^r\xi^{-1}\widetilde{\Omega}^1_{R/\mathscr{O}_K}\otimes_R\widehat{\overline{R}}) \tag{II.11.1.2}$$

and by $\widehat{\mathscr{S}^{(r)}}$ its $p$-adic Hausdorff completion, which we always assume endowed with the $p$-adic topology. We endow $\widehat{\mathscr{S}^{(r)}} \otimes_{\mathbb{Z}_p} \mathbb{Q}_p$ with the $p$-adic topology (II.2.2). In view of II.6.14 and its proof, $\mathscr{S}^{(r)}$ and $\widehat{\mathscr{S}^{(r)}}$ are $\mathscr{O}_C$-flat. For all rational numbers $r' \geq r \geq 0$, we have a canonical injective homomorphism $\alpha^{r,r'} : \mathscr{S}^{(r')} \to \mathscr{S}^{(r)}$. One immediately verifies that the induced homomorphism $h_\alpha^{r,r'} : \widehat{\mathscr{S}^{(r')}} \to \widehat{\mathscr{S}^{(r)}}$ is injective. We set

$$\mathscr{S}^\dagger = \varinjlim_{r \in \mathbb{Q}_{>0}} \widehat{\mathscr{S}^{(r)}}, \tag{II.11.1.3}$$

which we identify with a sub-$\widehat{\overline{R}}$-algebra of $\widehat{\mathscr{S}}$ by the direct limit of the homomorphisms $h_\alpha^{0,r}$.

We have a canonical $\mathscr{S}^{(r)}$-isomorphism

$$\Omega^1_{\mathscr{S}^{(r)}/\widehat{\overline{R}}} \xrightarrow{\sim} \xi^{-1}\widetilde{\Omega}^1_{R/\mathscr{O}_K} \otimes_R \mathscr{S}^{(r)}. \tag{II.11.1.4}$$

We denote by

$$d_{\mathscr{S}^{(r)}} : \mathscr{S}^{(r)} \to \xi^{-1}\widetilde{\Omega}^1_{R/\mathscr{O}_K} \otimes_R \mathscr{S}^{(r)} \tag{II.11.1.5}$$

the universal $\widehat{\overline{R}}$-derivation of $\mathscr{S}^{(r)}$ and by

$$d_{\widehat{\mathscr{S}^{(r)}}} : \widehat{\mathscr{S}^{(r)}} \to \xi^{-1}\widetilde{\Omega}^1_{R/\mathscr{O}_K} \otimes_R \widehat{\mathscr{S}^{(r)}} \tag{II.11.1.6}$$

its extension to the completions (note that the $R$-module $\widetilde{\Omega}^1_{R/\mathscr{O}_K}$ is free of finite type). Since $\xi^{-1}\widetilde{\Omega}^1_{R/\mathscr{O}_K} \otimes_R \widehat{\overline{R}} \subset d_{\mathscr{S}^{(r)}}(\mathscr{S}^{(r)})$, $d_{\mathscr{S}^{(r)}}$ and $d_{\widehat{\mathscr{S}^{(r)}}}$ are also Higgs $\widehat{\overline{R}}$-fields with coefficients in $\xi^{-1}\widetilde{\Omega}^1_{R/\mathscr{O}_K}$ (cf. II.2.12, II.2.16, and II.9.9). We denote by $\mathbb{K}^\bullet(\widehat{\mathscr{S}^{(r)}}, p^r d_{\widehat{\mathscr{S}^{(r)}}})$ the Dolbeault complex of the Higgs module $(\widehat{\mathscr{S}^{(r)}}, p^r d_{\widehat{\mathscr{S}^{(r)}}})$ (II.2.8.2) and denote by $\widetilde{\mathbb{K}}^\bullet(\widehat{\mathscr{S}^{(r)}}, p^r d_{\widehat{\mathscr{S}^{(r)}}})$ the augmented Dolbeault complex
(II.11.1.7)

$$\widehat{\overline{R}} \to \mathbb{K}^0(\widehat{\mathscr{S}^{(r)}}, p^r d_{\widehat{\mathscr{S}^{(r)}}}) \to \mathbb{K}^1(\widehat{\mathscr{S}^{(r)}}, p^r d_{\widehat{\mathscr{S}^{(r)}}}) \to \cdots \to \mathbb{K}^n(\widehat{\mathscr{S}^{(r)}}, p^r d_{\widehat{\mathscr{S}^{(r)}}}) \to \dots,$$

where $\widehat{\overline{R}}$ is placed in degree $-1$ and the differential $\widehat{\overline{R}} \to \widehat{\mathscr{S}^{(r)}}$ is the canonical homomorphism.

For all rational numbers $r' \geq r \geq 0$, we have

$$p^{r'}(\mathrm{id} \times \alpha^{r,r'}) \circ d_{\mathscr{S}^{(r')}} = p^r d_{\mathscr{S}^{(r)}} \circ \alpha^{r,r'}. \tag{II.11.1.8}$$

The homomorphism $h_\alpha^{r,r'}$ therefore induces a morphism of complexes

$$\nu^{r,r'} : \widetilde{\mathbb{K}}^\bullet(\widehat{\mathscr{S}^{(r')}}, p^{r'} d_{\widehat{\mathscr{S}^{(r')}}}) \to \widetilde{\mathbb{K}}^\bullet(\widehat{\mathscr{S}^{(r)}}, p^r d_{\widehat{\mathscr{S}^{(r)}}}). \tag{II.11.1.9}$$

By (II.11.1.8), the derivations $p^r d_{\widehat{\mathscr{S}^{(r)}}}$ induce an $\widehat{\overline{R}}$-derivation

$$d_{\mathscr{S}^\dagger} : \mathscr{S}^\dagger \to \xi^{-1}\widetilde{\Omega}^1_{R/\mathscr{O}_K} \otimes_R \mathscr{S}^\dagger \tag{II.11.1.10}$$

that is none other than the restriction of $d_{\widehat{\mathscr{S}}}$ to $\mathscr{S}^\dagger$. We denote by $\mathbb{K}^\bullet(\mathscr{S}^\dagger, d_{\mathscr{S}^\dagger})$ the Dolbeault complex of the Higgs module $(\mathscr{S}^\dagger, d_{\mathscr{S}^\dagger})$. Since $\widehat{\overline{R}}$ is $\mathscr{O}_C$-flat (II.6.14), for every rational number $r \geq 0$, we have

$$\ker(d_{\mathscr{S}^\dagger}) = \ker(d_{\widehat{\mathscr{S}^{(r)}}}) = \widehat{\overline{R}}. \tag{II.11.1.11}$$

**Proposition II.11.2.** *For all rational numbers $r' > r > 0$, there exists a rational number $\alpha \geq 0$ depending on $r$ and $r'$, but not on the data in* II.6.2, *such that*

$$p^\alpha \nu^{r,r'} : \widetilde{\mathbb{K}}^\bullet(\widehat{\mathscr{S}^{(r')}}, p^{r'} d_{\widehat{\mathscr{S}^{(r')}}}) \to \widetilde{\mathbb{K}}^\bullet(\widehat{\mathscr{S}^{(r)}}, p^r d_{\widehat{\mathscr{S}^{(r)}}}) \tag{II.11.2.1}$$

*is homotopic to* 0 *by an* $\widehat{\overline{R}}$*-linear homotopy.*

Let $t_1, \dots, t_d \in P^{\rm gp}$ be such that their images in $(P^{\rm gp}/\mathbb{Z}\lambda) \otimes_{\mathbb{Z}} \mathbb{Z}_p$ form a $\mathbb{Z}_p$-basis, so that $(d\log(t_i))_{1 \leq i \leq d}$ form an $R$-basis of $\widetilde{\Omega}^1_{R/\mathscr{O}_K}$ (II.7.18.5). For every $1 \leq i \leq d$, set $y_i = \xi^{-1} d\log(t_i) \in \xi^{-1}\widetilde{\Omega}^1_{R/\mathscr{O}_K} \subset \mathscr{S}$. We denote by

$$h^{-1}\colon \widehat{\mathscr{S}^{(r')}} \otimes_{\mathbb{Z}_p} \mathbb{Q}_p \to \widehat{\overline{R}} \otimes_{\mathbb{Z}_p} \mathbb{Q}_p \tag{II.11.2.2}$$

the $\widehat{\overline{R}}$-linear morphism defined by

$$h^{-1}\Big(\sum_{\underline{n}=(n_1,\dots,n_d)\in\mathbb{N}^d} a_{\underline{n}} \prod_{1\leq i\leq d} y_i^{n_i}\Big) = a_0. \tag{II.11.2.3}$$

For every integer $m \geq 0$, there exists a unique $\widehat{\overline{R}}$-linear morphism

$$h^m\colon \xi^{-m-1}\widetilde{\Omega}^{m+1}_{R/\mathscr{O}_K} \otimes_R \widehat{\mathscr{S}^{(r')}} \otimes_{\mathbb{Z}_p} \mathbb{Q}_p \to \xi^{-m}\widetilde{\Omega}^m_{R/\mathscr{O}_K} \otimes_R \widehat{\mathscr{S}^{(r)}} \otimes_{\mathbb{Z}_p} \mathbb{Q}_p \tag{II.11.2.4}$$

such that for all $1 \leq i_1 < \dots < i_{m+1} \leq d$, we have

$$\begin{aligned} h^m\Big(&\sum_{\underline{n}=(n_1,\dots,n_d)\in\mathbb{N}^d} a_{\underline{n}} \prod_{1\leq i\leq d} y_i^{n_i} \otimes \xi^{-1}d\log(t_{i_1}) \wedge \dots \wedge \xi^{-1}d\log(t_{i_{m+1}})\Big) \\ &= \sum_{\underline{n}=(n_1,\dots,n_d)\in J_{i_1-1}} \frac{a_{\underline{n}}}{n_{i_1}+1} \prod_{1\leq i\leq d} y_i^{n_i+\delta_{ii_1}} \otimes \xi^{-1}d\log(t_{i_2}) \wedge \dots \wedge \xi^{-1}d\log(t_{i_{m+1}}), \end{aligned}$$

where $J_{i_1-1}$ is the subset of $\mathbb{N}^d$ consisting of the elements $\underline{n} = (n_1, \dots, n_d)$ such that $n_1 = \dots = n_{i_1-1} = 0$.

Let $\alpha$ be a rational number such that

$$\alpha \geq \sup_{x\in\mathbb{Q}_{\geq 0}} (\log_p(x+1) + (x+1)r - xr'), \tag{II.11.2.5}$$

where $\log_p$ is the logarithm to base $p$. For every integer $m \geq 0$, we clearly have

$$p^\alpha h^m(\xi^{-m-1}\widetilde{\Omega}^{m+1}_{R/\mathscr{O}_K} \otimes_R \widehat{\mathscr{S}^{(r')}}) \subset \xi^{-m}\widetilde{\Omega}^m_{R/\mathscr{O}_K} \otimes_R \widehat{\mathscr{S}^{(r)}}. \tag{II.11.2.6}$$

One immediately verifies that the morphisms $(p^\alpha h^m)_{m\geq -1}$ define a homotopy linking 0 to the morphism $p^\alpha \nu^{r,r'}$.

**Corollary II.11.3.** *For all rational numbers* $r' > r > 0$*, the canonical morphism* (II.11.1.9)

$$\widetilde{\mathbb{K}}^\bullet(\widehat{\mathscr{S}^{(r')}}, p^{r'} d_{\widehat{\mathscr{S}^{(r')}}}) \otimes_{\mathbb{Z}_p} \mathbb{Q}_p \to \widetilde{\mathbb{K}}^\bullet(\widehat{\mathscr{S}^{(r)}}, p^{r} d_{\widehat{\mathscr{S}^{(r)}}}) \otimes_{\mathbb{Z}_p} \mathbb{Q}_p \tag{II.11.3.1}$$

*is homotopic to* 0 *by a continuous homotopy.*

**Corollary II.11.4.** *The complex* $\mathbb{K}^\bullet(\mathscr{S}^\dagger, d_{\mathscr{S}^\dagger}) \otimes_{\mathbb{Z}_p} \mathbb{Q}_p$ *is a resolution of* $\widehat{\overline{R}}[\frac{1}{p}]$.

Indeed, we have a canonical isomorphism of complexes

$$\varinjlim_{r\in\mathbb{Q}_{>0}} \mathbb{K}^\bullet(\widehat{\mathscr{S}^{(r)}}, p^r d_{\widehat{\mathscr{S}^{(r)}}}) \otimes_{\mathbb{Z}_p} \mathbb{Q}_p \xrightarrow{\sim} \mathbb{K}^\bullet(\mathscr{S}^\dagger, d_{\mathscr{S}^\dagger}) \otimes_{\mathbb{Z}_p} \mathbb{Q}_p. \tag{II.11.4.1}$$

The corollary therefore follows from II.11.3.

**II.11.5.** We denote by $(\widetilde{X}_0, \mathscr{M}_{\widetilde{X}_0})$ the smooth $(\mathscr{A}_2(\overline{S}), \mathscr{M}_{\mathscr{A}_2(\overline{S})})$-deformation of $(\check{\overline{X}}, \mathscr{M}_{\check{\overline{X}}})$ defined by the chart $(P, \gamma)$ (II.10.13.1), by $\mathscr{L}_0$ the associated Higgs–Tate torsor (II.10.3), by $\psi_0 \in \mathscr{L}_0(\widehat{Y})$ the section defined by the same chart (II.10.13.4), and by $\varphi_0 = \varphi_{\psi_0}$ the action of $\Delta$ on $\mathscr{S}$ induced by $\psi_0$ (II.10.6.4). By II.11.6 below, for every rational number $r \geq 0$, the sub-$\widehat{\overline{R}}$-algebra $\mathscr{S}^{(r)}$ of $\mathscr{S}$ is stable under $\varphi_0$. Unless stated otherwise, we endow $\mathscr{S}^{(r)}$, $\widehat{\mathscr{S}^{(r)}}$, and $\mathscr{S}^{\dagger}$ with the action of $\Delta$ induced by $\varphi_0$. The derivations $d_{\mathscr{S}}$ and $d_{\widehat{\mathscr{S}}}$ are $\Delta$-equivariant by (II.10.9.5). The same therefore holds for the derivations $d_{\mathscr{S}^{(r)}}$, $d_{\widehat{\mathscr{S}^{(r)}}}$, and $d_{\mathscr{S}^{\dagger}}$, in view of (II.11.1.8).

**Proposition II.11.6.** *For every rational number $r \geq 0$, the sub-$\widehat{\overline{R}}$-algebra $\mathscr{S}^{(r)}$ of $\mathscr{S}$ is stable under the action $\varphi_0$ of $\Delta$ on $\mathscr{S}$, and the induced actions of $\Delta$ on $\mathscr{S}^{(r)}$ and $\widehat{\mathscr{S}^{(r)}}$ are continuous for the $p$-adic topologies.*

We take again the assumptions and notation of II.10.17. By (II.10.17.2), for every $g \in \Delta$ and every $1 \leq i \leq d$, we have

$$\varphi_0(g)(y_i) = y_i - \xi^{-1}\chi_i(g). \tag{II.11.6.1}$$

Since $\xi^{-1}\chi_i(g) \in p^{\frac{1}{p-1}}\mathscr{O}_C$ (II.9.18), $\mathscr{S}^{(r)}$ is stable under $\varphi_0(g)$. Let $\zeta$ be a generator of $\mathbb{Z}_p(1)$. There exists $a_g \in \mathbb{Z}_p$ such that $\chi_i(g) = [\zeta^{a_g}] - 1 \in \mathscr{A}_2(\mathscr{O}_{\overline{K}})$. By linearity, we have $\log([\zeta^{a_g}]) \in p^{v_p(a_g)}\xi\mathscr{O}_C$, and consequently $\varphi_0(g)(y_i) - y_i \in p^{v_p(a_g)}\mathscr{S}$. For every integer $n \geq 0$, since the set of $g \in \Delta$ such that $v_p(a_g) \geq n$ is an open subgroup of $\Delta$, it follows that the stabilizer of the class of $p^r y_i$ in $\mathscr{S}^{(r)}/p^n\mathscr{S}^{(r)}$ is open in $\Delta$. The second assertion follows because the action of $\Delta$ on $\overline{R}/p^n\overline{R}$ is continuous for the discrete topology.

**Theorem II.11.7** (IV.5.3.4). *Let $r$ be a rational number $> 0$. Then:*

(i) *The canonical morphism*

$$\widehat{R_1} \otimes_{\mathbb{Z}_p} \mathbb{Q}_p \to (\widehat{\mathscr{S}^{(r)}} \otimes_{\mathbb{Z}_p} \mathbb{Q}_p)^{\Delta} \tag{II.11.7.1}$$

*is an isomorphism.*

(ii) *For every integer $i \geq 1$,*

$$\varinjlim_{r \in \mathbb{Q}_{>0}} \mathrm{H}^i_{\mathrm{cont}}(\Delta, \widehat{\mathscr{S}^{(r)}} \otimes_{\mathbb{Z}_p} \mathbb{Q}_p) = 0. \tag{II.11.7.2}$$

The proof of this theorem will be given in II.11.17. We have pushed it to the end of this section because it requires rather heavy notation. Note that this statement is slightly more general than that of Tsuji (IV.5.3.4).

**Corollary II.11.8.** *For every rational number $r > 0$, we have $(\mathscr{S}^{\dagger})^{\Delta} = (\widehat{\mathscr{S}^{(r)}})^{\Delta} = \widehat{R_1}$.*

Indeed, $\widehat{\mathscr{S}^{(r)}}$ is $\mathscr{O}_C$-flat and the canonical homomorphism $\widehat{\overline{R}} \to \widehat{\mathscr{S}^{(r)}}$ is injective. We deduce from II.11.7 that $(\widehat{\mathscr{S}^{(r)}})^{\Delta} = \widehat{\mathscr{S}^{(r)}} \cap \widehat{R_1}[\frac{1}{p}]$. On the other hand, we have $\widehat{\mathscr{S}^{(r)}} \cap \widehat{\overline{R}}[\frac{1}{p}] = \widehat{\overline{R}}$ and $(\widehat{\overline{R}})^{\Delta} = \widehat{R_1}$ by II.8.22(i). It follows that $(\widehat{\mathscr{S}^{(r)}})^{\Delta} = \widehat{R_1}$ and consequently that $(\mathscr{S}^{\dagger})^{\Delta} = \widehat{R_1}$.

**II.11.9.** We denote by $\mathfrak{S}$ the sub-$\widehat{R_1}$-algebra of $\mathscr{S}$ defined by (II.2.5)

$$\mathfrak{S} = \mathrm{S}_{\widehat{R_1}}(\xi^{-1}\widetilde{\Omega}^1_{R/\mathscr{O}_K} \otimes_R \widehat{R_1}). \tag{II.11.9.1}$$

By II.10.17, the action $\varphi_0$ of $\Delta$ on $\mathscr{S}$ preserves $\mathfrak{S}$, and the induced action on $\mathfrak{S}$ factors through $\Delta_{p^\infty}$. We also denote by $\varphi_0$ the resulting action of $\Delta_{p^\infty}$ on $\mathfrak{S}$. Copying the proof of II.11.6, we prove that the latter is continuous for the $p$-adic topology on $\mathfrak{S}$. We can also deduce this property directly from II.11.6 by observing that for every integer $n \geq 0$, the canonical homomorphism $\mathfrak{S}/p^n\mathfrak{S} \to \mathscr{S}/p^n\mathscr{S}$ is injective (cf. the proof of II.6.14). We set

$$\mathscr{S}_\infty = \mathfrak{S} \otimes_{\widehat{R_1}} \widehat{R_\infty}, \tag{II.11.9.2}$$

$$\mathscr{S}_{p^\infty} = \mathfrak{S} \otimes_{\widehat{R_1}} \widehat{R_{p^\infty}}. \tag{II.11.9.3}$$

The action $\varphi_0$ of $\Delta_{p^\infty}$ on $\mathfrak{S}$ induces actions of $\Delta_{p^\infty}$ on $\mathscr{S}_{p^\infty}$ and of $\Delta_\infty$ on $\mathscr{S}_\infty$.

For every rational number $r \geq 0$, we denote by $\mathscr{S}_\infty^{(r)}$ the sub-$\widehat{R_\infty}$-algebra of $\mathscr{S}_\infty$ defined by

$$\mathscr{S}_\infty^{(r)} = \mathrm{S}_{\widehat{R_\infty}}(p^r\xi^{-1}\widetilde{\Omega}^1_{R/\mathscr{O}_K} \otimes_R \widehat{R_\infty}), \tag{II.11.9.4}$$

and by $\widehat{\mathscr{S}}_\infty^{(r)}$ its $p$-adic Hausdorff completion. In view of II.6.14 and its proof, $\mathscr{S}_\infty^{(r)}$ and $\widehat{\mathscr{S}}_\infty^{(r)}$ are $\mathscr{O}_C$-flat. For every rational number $r' \geq r$, we have a canonical injective homomorphism $\mathscr{S}_\infty^{(r')} \to \mathscr{S}_\infty^{(r)}$. One immediately verifies that the induced homomorphism $\widehat{\mathscr{S}}_\infty^{(r')} \to \widehat{\mathscr{S}}_\infty^{(r)}$ is injective. The proof of II.11.6 shows that $\mathscr{S}_\infty^{(r)}$ is stable under the action of $\Delta_\infty$ on $\mathscr{S}_\infty = \mathscr{S}_\infty^{(0)}$, and the induced actions of $\Delta_\infty$ on $\mathscr{S}_\infty^{(r)}$ and $\widehat{\mathscr{S}}_\infty^{(r)}$ are continuous for the $p$-adic topologies. Unless explicitly stated otherwise, we endow $\mathscr{S}_\infty^{(r)}$ and $\widehat{\mathscr{S}}_\infty^{(r)}$ with these actions and with $p$-adic topologies.

We denote by $\mathscr{S}_{p^\infty}^{(r)}$ the sub-$\widehat{R_{p^\infty}}$-algebra of $\mathscr{S}_{p^\infty}$ defined by

$$\mathscr{S}_{p^\infty}^{(r)} = \mathrm{S}_{\widehat{R_{p^\infty}}}(p^r\xi^{-1}\widetilde{\Omega}^1_{R/\mathscr{O}_K} \otimes_R \widehat{R_{p^\infty}}), \tag{II.11.9.5}$$

and by $\widehat{\mathscr{S}}_{p^\infty}^{(r)}$ its $p$-adic Hausdorff completion. The algebra $\mathscr{S}_{p^\infty}^{(r)}$ has properties analogous to those of $\mathscr{S}_\infty^{(r)}$. In particular, $\mathscr{S}_{p^\infty}^{(r)}$ is stable under the action of $\Delta_{p^\infty}$ on $\mathscr{S}_{p^\infty} = \mathscr{S}_{p^\infty}^{(0)}$, and the induced actions of $\Delta_{p^\infty}$ on $\mathscr{S}_{p^\infty}^{(r)}$ and $\widehat{\mathscr{S}}_{p^\infty}^{(r)}$ are continuous for the $p$-adic topologies. Unless explicitly stated otherwise, we endow $\mathscr{S}_{p^\infty}^{(r)}$ and $\widehat{\mathscr{S}}_{p^\infty}^{(r)}$ with these actions and with $p$-adic topologies.

**Lemma II.11.10.** *For every rational number $r > 0$ and every integer $i \geq 0$, the canonical morphism*

$$\mathrm{H}^i_{\mathrm{cont}}(\Delta_{p^\infty}, \widehat{\mathscr{S}}_{p^\infty}^{(r)}) \to \mathrm{H}^i_{\mathrm{cont}}(\Delta_\infty, \widehat{\mathscr{S}}_\infty^{(r)}) \tag{II.11.10.1}$$

*is an isomorphism, and the canonical morphism*

$$\mathrm{H}^i_{\mathrm{cont}}(\Delta_\infty, \widehat{\mathscr{S}}_\infty^{(r)}) \to \mathrm{H}^i_{\mathrm{cont}}(\Delta, \widehat{\mathscr{S}}^{(r)}) \tag{II.11.10.2}$$

*is an almost isomorphism.*

Let $n$ be an integer $\geq 0$. The canonical homomorphism

$$\mathscr{S}_{p^\infty}^{(r)}/p^n\mathscr{S}_{p^\infty}^{(r)} \to (\mathscr{S}_\infty^{(r)}/p^n\mathscr{S}_\infty^{(r)})^{\Sigma_0} \tag{II.11.10.3}$$

is an isomorphism by virtue of II.6.13. By (II.6.11.1), it follows that the canonical morphism

$$\mathrm{H}^i(\Delta_{p^\infty}, \mathscr{S}_{p^\infty}^{(r)}/p^n\mathscr{S}_{p^\infty}^{(r)}) \to \mathrm{H}^i(\Delta_\infty, \mathscr{S}_\infty^{(r)}/p^n\mathscr{S}_\infty^{(r)}) \tag{II.11.10.4}$$

is an isomorphism. We deduce from this, by (II.3.10.4) and (II.3.10.5), that the morphism (II.11.10.1) is an isomorphism.

On the other hand, the canonical homomorphism

$$\mathscr{S}_\infty^{(r)}/p^n\mathscr{S}_\infty^{(r)} \to (\mathscr{S}^{(r)}/p^n\mathscr{S}^{(r)})^{\Sigma} \tag{II.11.10.5}$$

is an almost isomorphism by virtue of II.6.22. By II.6.20, it follows that the canonical morphism

(II.11.10.6) $$\psi_n\colon \mathrm{H}^i(\Delta_\infty, \mathscr{S}_\infty^{(r)}/p^n\mathscr{S}_\infty^{(r)}) \to \mathrm{H}^i(\Delta, \mathscr{S}^{(r)}/p^n\mathscr{S}^{(r)})$$

is an almost isomorphism. We denote by $A_n$ (resp. $C_n$) the kernel (resp. cokernel) of $\psi_n$. Then the $\mathscr{O}_{\overline{K}}$-modules

$$\lim_{\substack{\longleftarrow \\ n\geq 0}} A_n, \quad \lim_{\substack{\longleftarrow \\ n\geq 0}} C_n, \quad \mathrm{R}^1\lim_{\substack{\longleftarrow \\ n\geq 0}} A_n, \quad \mathrm{R}^1\lim_{\substack{\longleftarrow \\ n\geq 0}} C_n$$

are almost zero by virtue of ([**32**] 2.4.2(ii)). Consequently, the morphisms

$$\lim_{\substack{\longleftarrow \\ n\geq 0}} \psi_n \quad \text{and} \quad \mathrm{R}^1\lim_{\substack{\longleftarrow \\ n\geq 0}} \psi_n$$

are almost isomorphisms. We deduce from this, by (II.3.10.4) and (II.3.10.5), that the morphism (II.11.10.2) is an isomorphism.

**II.11.11.** Let $t_1, \dots, t_d \in P^{\mathrm{gp}}$ be such that their images in $(P^{\mathrm{gp}}/\mathbb{Z}\lambda) \otimes_{\mathbb{Z}} \mathbb{Z}_p$ form a $\mathbb{Z}_p$-basis, $(\chi_{t_i})_{1\leq i\leq d}$ their images in $\mathrm{Hom}(\Delta_{p^\infty}, \mathbb{Z}_p(1))$ (II.7.19.5), and $\zeta$ a $\mathbb{Z}_p$-basis of $\mathbb{Z}_p(1)$. The $(\chi_{t_i})_{1\leq i\leq d}$ form a $\mathbb{Z}_p$-basis of $\mathrm{Hom}(\Delta_{p^\infty}, \mathbb{Z}_p(1))$ (II.7.19.8). Hence there exists a unique $\mathbb{Z}_p$-basis $(\gamma_i)_{1\leq i\leq d}$ of $\Delta_{p^\infty}$ such that $\chi_{t_i}(\gamma_j) = \delta_{ij}\zeta$ for all $1 \leq i, j \leq d$ (II.6.10). For every integer $0 \leq i \leq d$, we denote by ${}_i\Xi_{p^\infty}$ the subgroup of (II.8.7.7)

(II.11.11.1) $$\Xi_{p^\infty} = \mathrm{Hom}(\Delta_{p^\infty}, \mu_{p^\infty}(\mathscr{O}_{\overline{K}}))$$

consisting of the homomorphisms $\nu\colon \Delta_{p^\infty} \to \mu_{p^\infty}(\mathscr{O}_{\overline{K}})$ such that $\nu(\gamma_j) = 1$ for every $1 \leq j \leq i$.

The $(d\log(t_i))_{1\leq i\leq d}$ form an $R$-basis of $\widetilde{\Omega}^1_{R/\mathscr{O}_K}$ (II.7.18.5). For every $1 \leq i \leq d$ and every $\underline{n} = (n_1, \dots, n_d) \in \mathbb{N}^d$, set $y_i = \xi^{-1}d\log(t_i) \in \xi^{-1}\widetilde{\Omega}^1_{R/\mathscr{O}_K} \subset \mathscr{S}_{p^\infty}$, $|\underline{n}| = \sum_{i=1}^d n_i$, and $\underline{y}^{\underline{n}} = \prod_{i=1}^d y_i^{n_i} \in \mathscr{S}_{p^\infty}$. Note that $R_{p^\infty}$ is separated for the $p$-adic topology and therefore identifies with a subring of $\widehat{R_{p^\infty}}$; this follows, for example, from (II.8.9.8), (II.8.9.17), and the fact that $\mathrm{Spec}(R_1)$ is an open subscheme of $\mathrm{Spec}(C_1)$. For every rational number $r > 0$, every integer $0 \leq i \leq d$, and every $\nu \in {}_i\Xi_{p^\infty}$, in view of II.8.9 and using the same notation, we denote by ${}_i\mathscr{S}_{p^\infty}^{(r)}(\nu)$ and ${}_i\mathscr{S}_{p^\infty}^{(r)}$ the sub-$R_1$-modules of $\mathscr{S}_{p^\infty}^{(r)}$ defined by

(II.11.11.2) $${}_i\mathscr{S}_{p^\infty}^{(r)}(\nu) = \bigoplus_{\underline{n}\in J_i} p^{r|\underline{n}|} R_{p^\infty}^{(\nu)} \underline{y}^{\underline{n}},$$

(II.11.11.3) $${}_i\mathscr{S}_{p^\infty}^{(r)} = \bigoplus_{\nu'\in {}_i\Xi_{p^\infty}} {}_i\mathscr{S}_{p^\infty}^{(r)}(\nu'),$$

where $J_i$ is the subset of $\mathbb{N}^d$ consisting of the elements $\underline{n} = (n_1, \dots, n_d)$ such that $n_1 = \dots = n_i = 0$. We denote by ${}_i\widehat{\mathscr{S}}_{p^\infty}^{(r)}$ the $p$-adic Hausdorff completion of ${}_i\mathscr{S}_{p^\infty}^{(r)}$. Since the $p$-adic topology on $R_{p^\infty}$ is induced by the $p$-adic topology on $\widehat{R_{p^\infty}}$, it easily follows from (II.8.9.1) that the $p$-adic topology on ${}_i\mathscr{S}_{p^\infty}^{(r)}$ is induced by the $p$-adic topology on $\mathscr{S}_{p^\infty}^{(r)}$. Consequently, ${}_i\widehat{\mathscr{S}}_{p^\infty}^{(r)}$ is the closure of ${}_i\mathscr{S}_{p^\infty}^{(r)}$ in $\widehat{\mathscr{S}}_{p^\infty}^{(r)}$. It follows from II.8.9 and (II.11.6.1) that for every $1 \leq j \leq d$ and every $\nu \in {}_i\Xi_{p^\infty}$, $\gamma_j$ preserves ${}_i\mathscr{S}_{p^\infty}^{(r)}(\nu)$ and therefore also ${}_i\mathscr{S}_{p^\infty}^{(r)}$. If $1 \leq j \leq i$, $\gamma_j$ fixes ${}_i\mathscr{S}_{p^\infty}^{(r)}(\nu)$ and ${}_i\mathscr{S}_{p^\infty}^{(r)}$.

**Lemma II.11.12.** *Under the assumptions of* II.11.11, *let moreover $i$ be an integer such that $1 \leq i \leq d$, and $r, r'$ two rational numbers such that $r' > r > 0$. Then:*

(i) *We have* ${}_0\widehat{\mathscr{S}}^{(r)}_{p^\infty} = \widehat{\mathscr{S}}^{(r)}_{p^\infty}$ *and* ${}_d\widehat{\mathscr{S}}^{(r)}_{p^\infty} = \widehat{R_1}$.

(ii) *The kernel of the morphism*

$$\gamma_i - \mathrm{id}\colon {}_{(i-1)}\widehat{\mathscr{S}}^{(r)}_{p^\infty} \to {}_{(i-1)}\widehat{\mathscr{S}}^{(r)}_{p^\infty} \tag{II.11.12.1}$$

*is equal to* ${}_i\widehat{\mathscr{S}}^{(r)}_{p^\infty}$.

(iii) *There exists an integer* $\alpha \geq 0$ *depending on* $r$ *and* $r'$ *but not on the data* (II.6.2), *such that we have*

$$p^\alpha \cdot {}_{(i-1)}\widehat{\mathscr{S}}^{(r')}_{p^\infty} \subset (\gamma_i - \mathrm{id})({}_{(i-1)}\widehat{\mathscr{S}}^{(r)}_{p^\infty}). \tag{II.11.12.2}$$

(i) Since $R^{(1)}_{p^\infty} = R_1$ (II.8.9.2), we have ${}_d\widehat{\mathscr{S}}^{(r)}_{p^\infty} = \widehat{R_1}$. On the other hand, we have

$${}_0\mathscr{S}^{(r)}_{p^\infty} = \mathrm{S}_{R_{p^\infty}}(p^r\xi^{-1}\widetilde{\Omega}^1_{R/\mathscr{O}_K} \otimes_R R_{p^\infty}), \tag{II.11.12.3}$$

which implies that ${}_0\widehat{\mathscr{S}}^{(r)}_{p^\infty} = \widehat{\mathscr{S}}^{(r)}_{p^\infty}$.

(ii) For every integer $n \geq 0$, we clearly have

$$p^n \cdot {}_i\mathscr{S}^{(r)}_{p^\infty} = {}_i\mathscr{S}^{(r)}_{p^\infty} \cap (p^n \cdot {}_{(i-1)}\mathscr{S}^{(r)}_{p^\infty}). \tag{II.11.12.4}$$

It therefore suffices to show that the sequence

$$0 \longrightarrow {}_i\mathscr{S}^{(r)}_{p^\infty} \longrightarrow {}_{(i-1)}\mathscr{S}^{(r)}_{p^\infty} \xrightarrow{\gamma_i - \mathrm{id}} {}_{(i-1)}\mathscr{S}^{(r)}_{p^\infty} \tag{II.11.12.5}$$

is exact. Let $\nu \in {}_{(i-1)}\Xi_{p^\infty}$ and

$$z = \sum_{\underline{n} \in J_{i-1}} p^{r|\underline{n}|} a_{\underline{n}} y^{\underline{n}} \in {}_{(i-1)}\mathscr{S}^{(r)}_{p^\infty}(\nu). \tag{II.11.12.6}$$

We then have (II.11.6.1)

$$(\gamma_i - \mathrm{id})(z) = \sum_{\underline{n} \in J_{i-1}} p^{r|\underline{n}|} b_{\underline{n}} y^{\underline{n}} \in {}_{(i-1)}\mathscr{S}^{(r)}_{p^\infty}(\nu), \tag{II.11.12.7}$$

where for every $\underline{n} = (n_1, \dots, n_d) \in J_{i-1}$,

$$b_{\underline{n}} = (\nu(\gamma_i) - 1)a_{\underline{n}} + \sum_{\underline{m} = (m_1, \dots, m_d) \in J_{i-1}(\underline{n})} p^{r(m_i - n_i)} \binom{m_i}{n_i} \nu(\gamma_i) a_{\underline{m}} w^{m_i - n_i}, \tag{II.11.12.8}$$

$J_{i-1}(\underline{n})$ denotes the subset of $J_{i-1}$ consisting of the elements $\underline{m} = (m_1, \dots, m_d)$ such that $m_j = n_j$ for $j \neq i$ and $m_i > n_i$, and $w = \xi^{-1}\log([\zeta])$ is an element of valuation $\frac{1}{p-1}$ of $\mathscr{O}_C$ (II.9.18).

Suppose that $\gamma_i(z) = z$ and $\nu(\gamma_i) \neq 1$. Then $v(\nu(\gamma_i) - 1) \leq \frac{1}{p-1}$ and for every $\underline{n} = (n_1, \dots, n_d) \in J_{i-1}$, we have

$$a_{\underline{n}} = -(\nu(\gamma_i) - 1)^{-1} \sum_{\underline{m} = (m_1, \dots, m_d) \in J_{i-1}(\underline{n})} p^{r(m_i - n_i)} \binom{m_i}{n_i} \nu(\gamma_i) a_{\underline{m}} w^{m_i - n_i}.$$

It follows that for every $\alpha \in \mathbb{N}$ and every $\underline{n} \in J_{i-1}$, we have $a_{\underline{n}} \in p^{r\alpha} R^{(\nu)}_{p^\infty}$ (this is proved by induction on $\alpha$); hence $z = 0$ because $R^{(\nu)}_{p^\infty}$ is separated for the $p$-adic topology.

Suppose that $\gamma_i(z) = z$ and $\nu(\gamma_i) = 1$. Then for every $\underline{n} = (n_1, \dots, n_d) \in J_{i-1}$, if we set $\underline{n}' = (n'_1, \dots, n'_d) \in J_{i-1}(\underline{n})$ with $n'_i = n_i + 1$, we have

$$(n_i + 1)! a_{\underline{n}'} = -\sum_{\underline{m} = (m_1, \dots, m_d) \in J_{i-1}(\underline{n}')} p^{r(m_i - n_i - 1)} m_i! a_{\underline{m}} \frac{w^{m_i - n_i - 1}}{(m_i - n_i)!}.$$

We have $w^{m-1}/m! \in \mathscr{O}_C$ for every integer $m \geq 1$. It follows that for every $\alpha \in \mathbb{N}$ and every $\underline{n} = (n_1, \dots, n_d) \in J_{i-1}$ such that $n_i \geq 1$, we have $n_i! a_{\underline{n}} \in p^{r\alpha} R^{(\nu)}_{p^\infty}$ (this is proved by induction on $\alpha$); hence $a_{\underline{n}} = 0$. Consequently, $z \in {}_i\mathscr{S}^{(r)}_{p^\infty}(\nu)$, which concludes the proof of the exactness of the sequence (II.11.12.5).

(iii) It suffices to show that there exists an integer $\alpha \geq 0$ depending only on $r$ and $r'$ such that for every $\nu \in {}_{(i-1)}\Xi_{p^\infty}$, we have

$$p^\alpha({}_{(i-1)}\mathscr{S}^{(r')}_{p^\infty}(\nu)) \subset (\gamma_i - \mathrm{id})({}_{(i-1)}\widehat{\mathscr{S}}^{(r)}_{p^\infty}). \tag{II.11.12.9}$$

Indeed, if we set $M = (\gamma_i - \mathrm{id})({}_{(i-1)}\widehat{\mathscr{S}}^{(r)}_{p^\infty})$, then by $p$-adic completion, we will deduce from this a commutative diagram

$$\begin{array}{ccccc} p^\alpha({}_{(i-1)}\widehat{\mathscr{S}}^{(r')}_{p^\infty}) & \longrightarrow & \widehat{M} & \longrightarrow & {}_{(i-1)}\widehat{\mathscr{S}}^{(r)}_{p^\infty} \\ & & \big\uparrow & \nearrow_{\gamma_i - \mathrm{id}} & \\ & & {}_{(i-1)}\widehat{\mathscr{S}}^{(r)}_{p^\infty} & & \end{array} \tag{II.11.12.10}$$

where the vertical arrow is surjective by virtue of ([**1**] 1.8.5).

Suppose $\nu(\gamma_i) \neq 1$. By (II.11.12.7), we have

(II.11.12.11)
$$(\nu(\gamma_i) - 1)({}_{(i-1)}\mathscr{S}^{(r)}_{p^\infty}(\nu)) \subset (\gamma_i - \mathrm{id})({}_{(i-1)}\mathscr{S}^{(r)}_{p^\infty}(\nu)) + (\nu(\gamma_i) - 1)p^r({}_{(i-1)}\mathscr{S}^{(r)}_{p^\infty}(\nu)).$$

It follows that

$$(\nu(\gamma_i) - 1)({}_{(i-1)}\mathscr{S}^{(r)}_{p^\infty}(\nu)) \subset (\gamma_i - \mathrm{id})({}_{(i-1)}\widehat{\mathscr{S}}^{(r)}_{p^\infty}). \tag{II.11.12.12}$$

We may therefore take $\alpha = 1$ because $v(\nu(\gamma_i) - 1) \leq \frac{1}{p-1}$.

Suppose that $\nu(\gamma_i) = 1$, so that $\nu \in {}_i\Xi_{p^\infty}$. Denote by $(R^{(\nu)}_{p^\infty})^\wedge$ and $({}_{(i-1)}\mathscr{S}^{(r)}_{p^\infty}(\nu))^\wedge$ the $p$-adic Hausdorff completions of $R^{(\nu)}_{p^\infty}$ and ${}_{(i-1)}\mathscr{S}^{(r)}_{p^\infty}(\nu)$, respectively. Note that $({}_{(i-1)}\mathscr{S}^{(r)}_{p^\infty}(\nu))^\wedge$ can be identified with a sub-$\widehat{R_1}$-module of ${}_{(i-1)}\widehat{\mathscr{S}}^{(r)}_{p^\infty}$. Every element $z$ of $({}_{(i-1)}\mathscr{S}^{(r)}_{p^\infty}(\nu))^\wedge$ can be written as a series

$$z = \sum_{\underline{n} \in J_{i-1}} p^{r|\underline{n}|} a_{\underline{n}} y^{\underline{n}}, \tag{II.11.12.13}$$

where $a_{\underline{n}} \in (R^{(\nu)}_{p^\infty})^\wedge$ and $a_{\underline{n}}$ tends to 0 when $|\underline{n}|$ tends to infinity. Consequently, $(\gamma_i - \mathrm{id})(z)$ is also given by formula (II.11.12.7). It therefore suffices to show that there exists an integer $\alpha \geq 0$ depending only on $r$ and $r'$ such that for every

$$\sum_{\underline{n} \in J_{i-1}} p^{r'|\underline{n}|} b_{\underline{n}} y^{\underline{n}} \in {}_{(i-1)}\mathscr{S}^{(r')}_{p^\infty}(\nu), \tag{II.11.12.14}$$

the system of linear equations defined, for $\underline{n} = (n_1, \dots, n_d) \in J_{i-1}$, by

$$p^\alpha p^{(r'-r)|\underline{n}|} n_i! b_{\underline{n}} = \sum_{\underline{m} = (m_1, \dots, m_d) \in J_{i-1}(\underline{n})} p^{r(m_i - n_i)} m_i! a_{\underline{m}} \frac{w^{m_i - n_i}}{(m_i - n_i)!}, \tag{II.11.12.15}$$

admits a solution $a_{\underline{m}} \in (R^{(\nu)}_{p^\infty})^\wedge$ for $\underline{m} \in J_{i-1}$ such that $a_{\underline{m}}$ tends to 0 when $|\underline{m}|$ tends to infinity. For $\underline{n} = (n_1, \dots, n_d) \in J_{i-1}$, set

$$a'_{\underline{n}} = n_i! p^{-\frac{n_i}{p-1}} a_{\underline{n}}, \tag{II.11.12.16}$$

$$b'_{\underline{n}} = p^\alpha p^{(r'-r)|\underline{n}|} n_i! p^{-\frac{n_i}{p-1}} b_{\underline{n}}, \tag{II.11.12.17}$$

so that equation (II.11.12.15) becomes

$$b'_{\underline{n}} = p^{r+\frac{1}{p-1}} w \sum_{\underline{m}=(m_1,\dots,m_d)\in J_{i-1}(\underline{n})} p^{(r+\frac{1}{p-1})(m_i-n_i-1)} a'_{\underline{m}} \frac{w^{(m_i-n_i-1)}}{(m_i-n_i)!}. \tag{II.11.12.18}$$

Consider the $\widehat{R_1}$-linear endomorphism $\Phi$ of $(\oplus_{\underline{n}\in J_{i-1}} R^{(\nu)}_{p^\infty})^\wedge$ defined, for a sequence $(x_{\underline{n}})_{\underline{n}\in J_{i-1}}$ of elements of $(R^{(\nu)}_{p^\infty})^\wedge$ that tends to 0 when $|\underline{n}|$ tends to infinity, by

$$\Phi(\sum_{\underline{n}\in J_{i-1}} x_{\underline{n}}) = \sum_{\underline{n}\in J_{i-1}} z_{\underline{n}}, \tag{II.11.12.19}$$

where for $\underline{n} = (n_1, \dots, n_d) \in J_{i-1}$,

$$z_{\underline{n}} = \sum_{\underline{m}=(m_1,\dots,m_d)\in \{\underline{n}\}\cup J_{i-1}(\underline{n})} p^{(r+\frac{1}{p-1})(m_i-n_i)} x_{\underline{m}} \frac{w^{(m_i-n_i)}}{(m_i-n_i+1)!}. \tag{II.11.12.20}$$

Since $\Phi$ is congruent to the identity modulo $p^{r+\frac{1}{p-1}}$, it is surjective by virtue of ([**1**] 1.8.5). Consequently, for every sequence $b'_{\underline{n}} \in p^{r+\frac{1}{p-1}} w (R^{(\nu)}_{p^\infty})^\wedge$ for $\underline{n} \in J_{i-1}$ that tends to 0 when $|\underline{n}|$ tends to infinity, equation (II.11.12.18) admits a solution $a'_{\underline{m}} \in (R^{(\nu)}_{p^\infty})^\wedge$ for $\underline{m} \in J_{i-1}$ that tends to 0 when $|\underline{m}|$ tends to infinity. On the other hand, $v(w) = \frac{1}{p-1}$ and there exists an integer $\alpha \geq 0$ such that for every $n \in \mathbb{N}$, we have

$$(r'-r)n + v(n!) - \frac{n}{p-1} + \alpha \geq r + \frac{2}{p-1}. \tag{II.11.12.21}$$

The desired assertion follows by taking for $b'_{\underline{n}}$ for $\underline{n} \in J_{i-1}$ the elements defined by (II.11.12.17) (which are in fact zero except for finitely many).

**II.11.13.** We keep the assumptions of II.11.11. For every rational number $r > 0$, we define by induction, for every integer $0 \leq i \leq d$, a complex $\mathbb{K}_i^{(r),\bullet}$ of continuous $\widehat{R_{p^\infty}}$-representations of $\Delta_{p^\infty}$ by setting $\mathbb{K}_0^{(r),\bullet} = \widehat{\mathscr{S}^{(r)}_{p^\infty}}[0]$ and for every $1 \leq i \leq n$, $\mathbb{K}_i^{(r),\bullet}$ is the fiber of the morphism

$$\gamma_i - \mathrm{id}\colon \mathbb{K}_{i-1}^{(r),\bullet} \to \mathbb{K}_{i-1}^{(r),\bullet}. \tag{II.11.13.1}$$

It follows from II.3.25 and (II.2.7.9) that we have a canonical isomorphism

$$\mathrm{C}^\bullet_{\mathrm{cont}}(\Delta_{p^\infty}, \widehat{\mathscr{S}^{(r)}_{p^\infty}}) \xrightarrow{\sim} \mathbb{K}_d^{(r),\bullet} \tag{II.11.13.2}$$

in $\mathbf{D}^+(\mathbf{Mod}(\widehat{R_1}))$. For all rational numbers $r' > r > 0$, the canonical homomorphism $\widehat{\mathscr{S}^{(r')}_{p^\infty}} \to \widehat{\mathscr{S}^{(r)}_{p^\infty}}$ induces, for every integer $0 \leq i \leq d$, a morphism $\mathbb{K}_i^{(r'),\bullet} \to \mathbb{K}_i^{(r),\bullet}$ of complexes of continuous $\widehat{R_{p^\infty}}$-representations of $\Delta_{p^\infty}$.

**Proposition II.11.14.** *Under the assumptions of* II.11.11, *let moreover $i$ be an integer such that $0 \leq i \leq d$, and $r, r'$ two rational numbers such that $r' > r > 0$. Then:*

(i) *We have a canonical $\Delta_{p^\infty}$-equivariant $\widehat{R_1}$-linear isomorphism*

$${}_i\widehat{\mathscr{S}}^{(r)}_{p^\infty} \xrightarrow{\sim} \mathrm{H}^0(\mathbb{K}_i^{(r),\bullet}). \tag{II.11.14.1}$$

(ii) *There exists an integer $\alpha_i \geq 0$ depending on $r$, $r'$, and $i$ but not on the data in* II.6.2, *such that for every integer $j \geq 1$, the canonical morphism*

$$\mathrm{H}^j(\mathbb{K}_i^{(r'),\bullet}) \to \mathrm{H}^j(\mathbb{K}_i^{(r),\bullet}) \tag{II.11.14.2}$$

*is annihilated by $p^{\alpha_i}$.*

We proceed by induction on $i$. The statement is immediate for $i = 0$, by II.11.12(i). Suppose $i \geq 1$ and that the statement has been proved for $i - 1$. The distinguished triangle

$$\mathbb{K}_i^{(r),\bullet} \longrightarrow \mathbb{K}_{i-1}^{(r),\bullet} \xrightarrow{\gamma_i - \mathrm{id}} \mathbb{K}_{i-1}^{(r),\bullet} \xrightarrow{+1} \tag{II.11.14.3}$$

and the induction hypothesis induce an exact sequence

$$0 \longrightarrow \mathrm{H}^0(\mathbb{K}_i^{(r),\bullet}) \longrightarrow {}_{i-1}\widehat{\mathscr{F}}^{(r)}_{p^\infty} \xrightarrow{\gamma_i - \mathrm{id}} {}_{i-1}\widehat{\mathscr{F}}^{(r)}_{p^\infty}, \tag{II.11.14.4}$$

which implies statement (i) by virtue of II.11.12(ii).

For every integer $j \geq 1$, the distinguished triangle (II.11.14.3) induces an exact sequence of $\widehat{R_1}$-modules

$$0 \to C_j^{(r)} \to \mathrm{H}^j(\mathbb{K}_i^{(r),\bullet}) \to D_j^{(r)} \to 0, \tag{II.11.14.5}$$

where $C_j^{(r)}$ is a quotient of $\mathrm{H}^{j-1}(\mathbb{K}_{i-1}^{(r),\bullet})$ and $D_j^{(r)}$ is a submodule of $\mathrm{H}^j(\mathbb{K}_{i-1}^{(r),\bullet})$. Moreover, by the induction hypothesis, we have a canonical isomorphism

$$C_1^{(r)} \xrightarrow{\sim} {}_{i-1}\widehat{\mathscr{F}}^{(r)}_{p^\infty}/(\gamma_i - 1)({}_{i-1}\widehat{\mathscr{F}}^{(r)}_{p^\infty}). \tag{II.11.14.6}$$

The canonical morphisms $\mathbb{K}_{i-1}^{(r'),\bullet} \to \mathbb{K}_{i-1}^{(r),\bullet}$ and $\mathbb{K}_i^{(r'),\bullet} \to \mathbb{K}_i^{(r),\bullet}$ induce morphisms $C_j^{(r')} \to C_j^{(r)}$ and $D_j^{(r')} \to D_j^{(r)}$ that fit into a commutative diagram

$$\begin{array}{ccccccccc} 0 & \longrightarrow & C_j^{(r')} & \longrightarrow & \mathrm{H}^j(\mathbb{K}_i^{(r'),\bullet}) & \longrightarrow & D_j^{(r')} & \longrightarrow & 0 \\ & & \downarrow & & \downarrow & & \downarrow & & \\ 0 & \longrightarrow & C_j^{(r)} & \longrightarrow & \mathrm{H}^j(\mathbb{K}_i^{(r),\bullet}) & \longrightarrow & D_j^{(r)} & \longrightarrow & 0 \end{array} \tag{II.11.14.7}$$

where the vertical arrow in the middle is the canonical morphism.

Set $r'' = (r + r')/2$. By the induction hypothesis, there exists an integer $\alpha_{i-1} \geq 0$ depending only on $r$, $r'$, and $i - 1$, such that for every integer $j \geq 1$, the morphism $D_j^{(r')} \to D_j^{(r'')}$ is annihilated by $p^{\alpha_{i-1}}$. On the other hand, in view of the induction hypothesis and by virtue of (II.11.14.6) and II.11.12(ii), there exists an integer $\alpha'_{i-1} \geq 0$ depending only on $r$, $r'$, and $i - 1$, such that for every integer $j \geq 1$, the morphism $C_j^{(r'')} \to C_j^{(r)}$ is annihilated by $p^{\alpha'_{i-1}}$. Statement (ii) follows by taking $\alpha_i = \alpha_{i-1} + \alpha'_{i-1}$.

**Corollary II.11.15.** *Let $r, r'$ be two rational numbers such that $r' > r > 0$. Then:*

(i) *The canonical homomorphism*

$$\widehat{R_1} \to (\widehat{\mathscr{F}}^{(r)}_{p^\infty})^{\Delta_{p^\infty}} \tag{II.11.15.1}$$

*is an isomorphism.*

(ii) *There exists an integer $\alpha \geq 0$ depending on $r$, $r'$, and $d$, but not on the data in* II.6.2, *such that for every integer $j \geq 1$, the canonical morphism*

$$\mathrm{H}^j_{\mathrm{cont}}(\Delta_{p^\infty}, \widehat{\mathscr{F}}^{(r')}_{p^\infty}) \to \mathrm{H}^j_{\mathrm{cont}}(\Delta_{p^\infty}, \widehat{\mathscr{F}}^{(r)}_{p^\infty}) \tag{II.11.15.2}$$

*is annihilated by $p^\alpha$.*

This follows from (II.11.13.2) and II.11.14.

**Corollary II.11.16.** *Let $r, r'$ be two rational numbers such that $r' > r > 0$. Then:*

(i) *The canonical homomorphism*

$$\widehat{R_1} \to (\widehat{\mathscr{S}^{(r)}})^{\Delta} \tag{II.11.16.1}$$

*is an almost isomorphism.*

(ii) *There exists an integer $\alpha \geq 0$ depending on $r$, $r'$, and $d$ but not on the data in II.6.2, such that for every integer $j \geq 1$, the canonical morphism*

$$\mathrm{H}^j_{\mathrm{cont}}(\Delta, \widehat{\mathscr{S}^{(r')}}) \to \mathrm{H}^j_{\mathrm{cont}}(\Delta, \widehat{\mathscr{S}^{(r)}}) \tag{II.11.16.2}$$

*is annihilated by $p^{\alpha}$.*

This follows from II.11.10 and II.11.15.

**II.11.17.** We can now prove theorem II.11.7. Since $\Delta$ is compact, for every rational number $r > 0$, the canonical morphism (II.3.8)

$$\mathrm{C}^{\bullet}_{\mathrm{cont}}(\Delta, \widehat{\mathscr{S}^{(r)}}) \otimes_{\mathbb{Z}_p} \mathbb{Q}_p \to \mathrm{C}^{\bullet}_{\mathrm{cont}}(\Delta, \widehat{\mathscr{S}^{(r)}} \otimes_{\mathbb{Z}_p} \mathbb{Q}_p) \tag{II.11.17.1}$$

is an isomorphism. Theorem II.11.7 therefore follows from II.11.16.

**Proposition II.11.18.** *Let $r, r'$ be two rational numbers such that $r' > r > 0$. Then:*

(i) *For every integer $n \geq 1$, the canonical morphism*

$$R_1/p^n R_1 \to (\mathscr{S}^{(r)}/p^n \mathscr{S}^{(r)})^{\Delta} \tag{II.11.18.1}$$

*is almost injective. We denote by $\mathscr{H}_n^{(r)}$ its cokernel.*

(ii) *There exists an integer $\alpha \geq 0$ depending on $r$, $r'$, and $d$ but not on the data in II.6.2, such that for every integer $n \geq 1$, the canonical morphism $\mathscr{H}_n^{(r')} \to \mathscr{H}_n^{(r)}$ is annihilated by $p^{\alpha}$.*

(iii) *There exists an integer $\beta \geq 0$ depending on $r$, $r'$, and $d$ but not on the data in II.6.2, such that for all integers $n, q \geq 1$, the canonical morphism*

$$\mathrm{H}^q(\Delta, \mathscr{S}^{(r')}/p^n \mathscr{S}^{(r')}) \to \mathrm{H}^q(\Delta, \mathscr{S}^{(r)}/p^n \mathscr{S}^{(r)}) \tag{II.11.18.2}$$

*is annihilated by $p^{\beta}$.*

(i) This follows from II.11.16(i) and from the long exact sequence of cohomology associated with the short exact sequence of $\widehat{\overline{R}}$-representations of $\Delta$

$$0 \longrightarrow \widehat{\mathscr{S}^{(r)}} \xrightarrow{\cdot p^n} \widehat{\mathscr{S}^{(r)}} \longrightarrow \widehat{\mathscr{S}^{(r)}}/p^n \widehat{\mathscr{S}^{(r)}} \longrightarrow 0. \tag{II.11.18.3}$$

We also deduce from this an almost injective $\widehat{R_1}$-linear morphism

$$\mathscr{H}_n^{(r)} \to \mathrm{H}^1_{\mathrm{cont}}(\Delta, \widehat{\mathscr{S}^{(r)}}). \tag{II.11.18.4}$$

(ii) This follows from (II.11.18.4) and II.11.16(ii).

(iii) For all integers $n, q \geq 1$, the long exact sequence of cohomology deduced from (II.11.18.3) gives an exact sequence of $\widehat{R_1}$-modules

(II.11.18.5)

$$0 \to \mathrm{H}^q_{\mathrm{cont}}(\Delta, \widehat{\mathscr{S}^{(r)}})/p^n \mathrm{H}^q_{\mathrm{cont}}(\Delta, \widehat{\mathscr{S}^{(r)}}) \to \mathrm{H}^q(\Delta, \mathscr{S}^{(r)}/p^n \mathscr{S}^{(r)}) \to T_n^{(r),q} \to 0,$$

where $T_n^{(r),q}$ is a $p^n$-torsion submodule of $\mathrm{H}^{q+1}_{\mathrm{cont}}(\Delta, \widehat{\mathscr{S}^{(r)}})$. Let $r'' = (r + r')/2$. By II.11.16(ii), there exists an integer $\beta' > 0$ depending only on $r$, $r'$, and $d$, such that for every integer $q \geq 1$, the canonical morphisms

$$\mathrm{H}^q_{\mathrm{cont}}(\Delta, \widehat{\mathscr{S}^{(r')}}) \to \mathrm{H}^q_{\mathrm{cont}}(\Delta, \widehat{\mathscr{S}^{(r'')}}) \quad \text{and} \quad \mathrm{H}^q_{\mathrm{cont}}(\Delta, \widehat{\mathscr{S}^{(r'')}}) \to \mathrm{H}^q_{\mathrm{cont}}(\Delta, \widehat{\mathscr{S}^{(r)}})$$

are annihilated by $p^{\beta'}$. Statement (iii) follows by taking $\beta = 2\beta'$.

## II.12. Dolbeault representations

**II.12.1.** For the remainder of this chapter, we fix a smooth $(\mathscr{A}_2(\overline{S}), \mathscr{M}_{\mathscr{A}_2(\overline{S})})$-deformation $(\widetilde{X}, \mathscr{M}_{\widetilde{X}})$ of $(\check{\overline{X}}, \mathscr{M}_{\check{\overline{X}}})$ (II.10.1) and denote by $\mathscr{F}$ the associated Higgs–Tate $\widehat{\overline{R}}$-extension and by $\mathscr{C}$ the associated Higgs–Tate $\widehat{\overline{R}}$-algebra (II.10.5). We denote by $\widehat{\mathscr{C}}$ the $p$-adic Hausdorff completion of $\mathscr{C}$. For every rational number $r \geq 0$, we denote by $\mathscr{F}^{(r)}$ the $\widehat{\overline{R}}$-representation of $\Delta$ deduced from $\mathscr{F}$ by taking the inverse image under the multiplication by $p^r$ on $\xi^{-1}\widetilde{\Omega}^1_{R/\mathscr{O}_K} \otimes_R \widehat{\overline{R}}$, so that we have a locally split exact sequence of $\widehat{\overline{R}}$-modules

$$0 \longrightarrow \widehat{\overline{R}} \longrightarrow \mathscr{F}^{(r)} \xrightarrow{u^{(r)}} \xi^{-1}\widetilde{\Omega}^1_{R/\mathscr{O}_K} \otimes_R \widehat{\overline{R}} \longrightarrow 0. \tag{II.12.1.1}$$

By ([**45**] I 4.3.1.7), this sequence induces, for every integer $n \geq 1$, an exact sequence (II.2.5)

$$0 \to \mathrm{S}^{n-1}_{\widehat{\overline{R}}}(\mathscr{F}^{(r)}) \to \mathrm{S}^{n}_{\widehat{\overline{R}}}(\mathscr{F}^{(r)}) \to \mathrm{S}^{n}_{\widehat{\overline{R}}}(\xi^{-1}\widetilde{\Omega}^1_{R/\mathscr{O}_K} \otimes_R \widehat{\overline{R}}) \to 0. \tag{II.12.1.2}$$

The $\widehat{\overline{R}}$-modules $(\mathrm{S}^{n}_{\widehat{\overline{R}}}(\mathscr{F}^{(r)}))_{n\in\mathbb{N}}$ therefore form a filtered direct system whose direct limit

$$\mathscr{C}^{(r)} = \varinjlim_{n\geq 0} \mathrm{S}^{n}_{\widehat{\overline{R}}}(\mathscr{F}^{(r)}) \tag{II.12.1.3}$$

is naturally endowed with a structure of $\widehat{\overline{R}}$-algebra. There exists a unique homomorphism of $\widehat{\overline{R}}$-algebras

$$\mu^{(r)} \colon \mathscr{C}^{(r)} \to \mathscr{C}^{(r)} \otimes_{\widehat{\overline{R}}} \mathscr{S}^{(r)}, \tag{II.12.1.4}$$

where $\mathscr{S}^{(r)}$ is the $\widehat{\overline{R}}$-algebra defined in (II.11.1.2), such that for every $x \in \mathscr{F}^{(r)}$, we have

$$\mu^{(r)}(x) = x \otimes 1 + 1 \otimes (p^r \cdot u^{(r)}(x)). \tag{II.12.1.5}$$

This makes $\mathrm{Spec}(\mathscr{C}^{(r)})$ into a principal homogeneous $\mathrm{Spec}(\mathscr{S}^{(r)})$-bundle over $\widehat{Y}$ (cf. the proof of II.4.10).

The action of $\Delta$ on $\mathscr{F}^{(r)}$ induces an action on $\mathscr{C}^{(r)}$ by ring automorphisms that is compatible with its action on $\widehat{\overline{R}}$; we call it the *canonical action*. The $\widehat{\overline{R}}$-algebra $\mathscr{C}^{(r)}$ endowed with this action is called the *Higgs–Tate algebra* of thickness $r$ associated with $(\widetilde{X}, \mathscr{M}_{\widetilde{X}})$. We denote by $\widehat{\mathscr{C}^{(r)}}$ the $p$-adic Hausdorff completion of $\mathscr{C}^{(r)}$, which we always assume endowed with the $p$-adic topology. We endow $\widehat{\mathscr{C}^{(r)}} \otimes_{\mathbb{Z}_p} \mathbb{Q}_p$ with the $p$-adic topology (II.2.2). In view of II.6.14 and its proof, $\mathscr{C}^{(r)}$ and $\widehat{\mathscr{C}^{(r)}}$ are $\mathscr{O}_C$-flat. For all rational numbers $r' \geq r \geq 0$, we have an injective and $\Delta$-equivariant canonical $\widehat{\overline{R}}$-homomorphism $\alpha^{r,r'} \colon \mathscr{C}^{(r')} \to \mathscr{C}^{(r)}$. One immediately verifies that the induced homomorphism $h^{r,r'}_\alpha \colon \widehat{\mathscr{C}^{(r')}} \to \widehat{\mathscr{C}^{(r)}}$ is injective. Set

$$\mathscr{C}^{\dagger} = \varinjlim_{r\in\mathbb{Q}_{>0}} \widehat{\mathscr{C}^{(r)}}, \tag{II.12.1.6}$$

which we identify with a sub-$\widehat{\overline{R}}$-algebra of $\widehat{\mathscr{C}} = \widehat{\mathscr{C}^{(0)}}$ by the direct limit of the homomorphisms $(h^{0,r}_\alpha)_{r\in\mathbb{Q}_{>0}}$. The action of $\Delta$ on the rings $(\widehat{\mathscr{C}^{(r)}})_{r\in\mathbb{Q}_{>0}}$ induces an action on $\mathscr{C}^{\dagger}$ by ring automorphisms that is compatible with its actions on $\widehat{\overline{R}}$ and on $\widehat{\mathscr{C}}$.

We denote by

$$d_{\mathscr{C}^{(r)}} \colon \mathscr{C}^{(r)} \to \xi^{-1}\widetilde{\Omega}^1_{R/\mathscr{O}_K} \otimes_R \mathscr{C}^{(r)} \tag{II.12.1.7}$$

the universal $\widehat{\overline{R}}$-derivation of $\mathscr{C}^{(r)}$ and by

$$d_{\widehat{\mathscr{C}}^{(r)}}\colon \widehat{\mathscr{C}}^{(r)} \to \xi^{-1}\widetilde{\Omega}^1_{R/\mathscr{O}_K} \otimes_R \widehat{\mathscr{C}}^{(r)} \tag{II.12.1.8}$$

its extension to the completions (note that the $R$-module $\widetilde{\Omega}^1_{R/\mathscr{O}_K}$ is free of finite type). We immediately see that, as for $d_{\mathscr{C}}$ (II.10.9.3), the derivations $d_{\mathscr{C}^{(r)}}$ and $d_{\widehat{\mathscr{C}}^{(r)}}$ are $\Delta$-equivariant. Furthermore, $d_{\mathscr{C}^{(r)}}$ and $d_{\widehat{\mathscr{C}}^{(r)}}$ are also Higgs $\widehat{\overline{R}}$-fields with coefficients in $\xi^{-1}\widetilde{\Omega}^1_{R/\mathscr{O}_K}$, because $\xi^{-1}\widetilde{\Omega}^1_{R/\mathscr{O}_K} \otimes_R \overline{R} = d_{\mathscr{C}^{(r)}}(\mathscr{F}^{(r)}) \subset d_{\mathscr{C}^{(r)}}(\mathscr{C}^{(r)})$ (cf. II.2.12 and II.2.16). We denote by $\mathbb{K}^\bullet(\widehat{\mathscr{C}}^{(r)}, p^r d_{\widehat{\mathscr{C}}^{(r)}})$ the Dolbeault complex of $(\widehat{\mathscr{C}}^{(r)}, p^r d_{\widehat{\mathscr{C}}^{(r)}})$ (II.2.8.2) and by $\widetilde{\mathbb{K}}^\bullet(\widehat{\mathscr{C}}^{(r)}, p^r d_{\widehat{\mathscr{C}}^{(r)}})$ the augmented Dolbeault complex

$$\widehat{\overline{R}} \to \mathbb{K}^0(\widehat{\mathscr{C}}^{(r)}, p^r d_{\widehat{\mathscr{C}}^{(r)}}) \to \mathbb{K}^1(\widehat{\mathscr{C}}^{(r)}, p^r d_{\widehat{\mathscr{C}}^{(r)}}) \to \cdots \to \mathbb{K}^n(\widehat{\mathscr{C}}^{(r)}, p^r d_{\widehat{\mathscr{C}}^{(r)}}) \to \cdots, \tag{II.12.1.9}$$

where $\widehat{\overline{R}}$ is placed in degree $-1$ and the differential $\widehat{\overline{R}} \to \widehat{\mathscr{C}}^{(r)}$ is the canonical homomorphism.

For all rational numbers $r' \geq r \geq 0$, we have

$$p^{r'}(\mathrm{id} \times \alpha^{r,r'}) \circ d_{\mathscr{C}^{(r')}} = p^r d_{\mathscr{C}^{(r)}} \circ \alpha^{r,r'}. \tag{II.12.1.10}$$

Consequently, $h_\alpha^{r,r'}$ induces a morphism of complexes

$$\nu^{r,r'}\colon \widetilde{\mathbb{K}}^\bullet(\widehat{\mathscr{C}}^{(r')}, p^{r'} d_{\widehat{\mathscr{C}}^{(r')}}) \to \widetilde{\mathbb{K}}^\bullet(\widehat{\mathscr{C}}^{(r)}, p^r d_{\widehat{\mathscr{C}}^{(r)}}). \tag{II.12.1.11}$$

By (II.12.1.10), the derivations $p^r d_{\widehat{\mathscr{C}}^{(r)}}$ induce an $\widehat{\overline{R}}$-derivation

$$d_{\mathscr{C}^\dagger}\colon \mathscr{C}^\dagger \to \xi^{-1}\widetilde{\Omega}^1_{R/\mathscr{O}_K} \otimes_R \mathscr{C}^\dagger \tag{II.12.1.12}$$

that is none other than the restriction of $d_{\widehat{\mathscr{C}}}$ to $\mathscr{C}^\dagger$. This is also a Higgs $\widehat{\overline{R}}$-field with coefficients in $\xi^{-1}\widetilde{\Omega}^1_{R/\mathscr{O}_K}$. We denote by $\mathbb{K}^\bullet(\mathscr{C}^\dagger, d_{\mathscr{C}^\dagger})$ the Dolbeault complex of $(\mathscr{C}^\dagger, d_{\mathscr{C}^\dagger})$. Since $\widehat{\overline{R}}$ is $\mathscr{O}_C$-flat (II.6.14), for every rational number $r \geq 0$, we have

$$\ker(d_{\mathscr{C}^\dagger}) = \ker(d_{\widehat{\mathscr{C}}^{(r)}}) = \widehat{\overline{R}}. \tag{II.12.1.13}$$

**II.12.2.** When $(\widetilde{X}, \mathscr{M}_{\widetilde{X}})$ is the deformation $(\widetilde{X}_0, \mathscr{M}_{\widetilde{X}_0})$ of $(\check{\overline{X}}, \mathscr{M}_{\check{\overline{X}}})$ defined by the chart $(P, \gamma)$ (II.10.13.1), we add an index $_0$ to the objects defined in (II.12.1): $\mathscr{C}_0$, $\mathscr{C}_0^{(r)}$... The section $\psi_0 \in \mathscr{L}_0(\widehat{Y})$ defined by the chart $(P, \gamma)$ (II.10.13) then induces an isomorphism of $\widehat{\overline{R}}$-algebras

$$\mathscr{S} \xrightarrow{\sim} \mathscr{C}_0, \tag{II.12.2.1}$$

where $\mathscr{S}$ is the $\widehat{\overline{R}}$-algebra defined in (II.10.3.2). It is $\Delta$-equivariant when we endow $\mathscr{S}$ with the action $\varphi_0 = \varphi_{\psi_0}$ of $\Delta$ induced by $\psi_0$ (II.10.6.4). It is moreover compatible with the universal derivations, by (II.10.9.4). For every rational number $r \geq 0$, $\psi_0$ induces an $\widehat{\overline{R}}$-homomorphism $\mathscr{C}_0^{(r)} \to \widehat{\overline{R}}$ and consequently an isomorphism of $\widehat{\overline{R}}$-algebras

$$\mathscr{S}^{(r)} \xrightarrow{\sim} \mathscr{C}_0^{(r)}, \tag{II.12.2.2}$$

where $\mathscr{S}^{(r)}$ is the $\widehat{\overline{R}}$-algebra defined in (II.11.1.2). For all rational numbers $r' \geq r \geq 0$, the diagram

(II.12.2.3)
$$\begin{array}{ccc} \mathscr{S}^{(r')} & \xrightarrow{\sim} & \mathscr{C}_0^{(r')} \\ \downarrow & & \downarrow \\ \mathscr{S}^{(r)} & \xrightarrow{\sim} & \mathscr{C}_0^{(r)} \end{array}$$

where the vertical arrows are the canonical homomorphisms, is commutative. It follows that the isomorphism (II.12.2.2) is $\Delta$-equivariant when we endow $\mathscr{S}^{(r)}$ with the action of $\Delta$ induced by $\varphi_0$ (II.11.6). One also immediately verifies that the diagram (II.11.1.5)

(II.12.2.4)
$$\begin{array}{ccc} \mathscr{S}^{(r)} & \xrightarrow{\sim} & \mathscr{C}_0^{(r)} \\ {\scriptstyle d_{\mathscr{S}^{(r)}}}\downarrow & & \downarrow{\scriptstyle d_{\mathscr{C}_0^{(r)}}} \\ \xi^{-1}\widetilde{\Omega}^1_{R/\mathscr{O}_K} \otimes_R \mathscr{S}^{(r)} & = & \xi^{-1}\widetilde{\Omega}^1_{R/\mathscr{O}_K} \otimes_R \mathscr{C}_0^{(r)} \end{array}$$

is commutative.

For future reference, we rewrite in the following five propositions the main results of Section II.11 concerning the algebra $\mathscr{S}$ in terms of the algebra $\mathscr{C}$, for a general deformation $(\widetilde{X}, \mathscr{M}_{\widetilde{X}})$ of $(\check{\overline{X}}, \mathscr{M}_{\check{\overline{X}}})$, taking into account II.12.2 and II.10.10.

**Proposition II.12.3.** (i) *For all rational numbers $r' > r > 0$, there exists a rational number $\alpha \geq 0$ depending on $r$ and $r'$, but not on the data in* (II.6.2), *such that*

(II.12.3.1)
$$p^{\alpha}\nu^{r,r'} \colon \widetilde{\mathbb{K}}^{\bullet}(\widehat{\mathscr{C}}^{(r')}, p^{r'}d_{\widehat{\mathscr{C}}^{(r')}}) \to \widetilde{\mathbb{K}}^{\bullet}(\widehat{\mathscr{C}}^{(r)}, p^{r}d_{\widehat{\mathscr{C}}^{(r)}}),$$

*where $\nu^{r,r'}$ is the morphism* (II.12.1.11), *is homotopic to $0$ by an $\widehat{\overline{R}}$-linear homotopy.*

(ii) *For all rational numbers $r' > r > 0$, the canonical morphism*

(II.12.3.2)
$$\nu^{r,r'} \otimes_{\mathbb{Z}_p} \mathbb{Q}_p \colon \widetilde{\mathbb{K}}^{\bullet}(\widehat{\mathscr{C}}^{(r')}, p^{r'}d_{\widehat{\mathscr{C}}^{(r')}}) \otimes_{\mathbb{Z}_p} \mathbb{Q}_p \to \widetilde{\mathbb{K}}^{\bullet}(\widehat{\mathscr{C}}^{(r)}, p^{r}d_{\widehat{\mathscr{C}}^{(r)}}) \otimes_{\mathbb{Z}_p} \mathbb{Q}_p$$

*is homotopic to $0$ by a continuous homotopy.*

(iii) *The complex $\mathbb{K}^{\bullet}(\mathscr{C}^{\dagger}, d_{\mathscr{C}^{\dagger}}) \otimes_{\mathbb{Z}_p} \mathbb{Q}_p$ is a resolution of $\widehat{\overline{R}}[\frac{1}{p}]$.*

This follows from II.11.2, II.11.3, and II.11.4.

**Proposition II.12.4.** *For every rational number $r \geq 0$, the actions of $\Delta$ on $\mathscr{C}^{(r)}$ and $\widehat{\mathscr{C}}^{(r)}$ are continuous for the $p$-adic topologies.*

This follows from II.11.6.

**Proposition II.12.5.** *Let $r$ be a rational number $> 0$. Then:*

(i) *The canonical morphism*

(II.12.5.1)
$$\widehat{R_1} \otimes_{\mathbb{Z}_p} \mathbb{Q}_p \to (\widehat{\mathscr{C}}^{(r)} \otimes_{\mathbb{Z}_p} \mathbb{Q}_p)^{\Delta}$$

*is an isomorphism.*

(ii) *For every integer $i \geq 1$, we have*

(II.12.5.2)
$$\lim_{\substack{\longrightarrow \\ r \in \mathbb{Q}_{>0}}} \mathrm{H}^i_{\mathrm{cont}}(\Delta, \widehat{\mathscr{C}}^{(r)} \otimes_{\mathbb{Z}_p} \mathbb{Q}_p) = 0.$$

This follows from II.11.7.

**Corollary II.12.6.** *For every rational number $r > 0$, we have* $(\mathscr{C}^\dagger)^\Delta = (\widehat{\mathscr{C}^{(r)}})^\Delta = \widehat{R_1}$.

This follows from II.11.8 (or from II.12.5).

**Proposition II.12.7.** *Let $r, r'$ be two rational numbers such that $r' > r > 0$. Then:*

(i) *For every integer $n \geq 1$, the canonical homomorphism*

$$R_1/p^nR_1 \to (\mathscr{C}^{(r)}/p^n\mathscr{C}^{(r)})^\Delta \tag{II.12.7.1}$$

*is almost injective. We denote by $\mathscr{H}_n^{(r)}$ its cokernel.*

(ii) *There exists an integer $\alpha \geq 0$ depending on $r$, $r'$, and $d$ but not on the data in* II.6.2, *such that for every integer $n \geq 1$, the canonical morphism $\mathscr{H}_n^{(r')} \to \mathscr{H}_n^{(r)}$ is annihilated by $p^\alpha$.*

(iii) *There exists an integer $\beta \geq 0$ depending on $r$, $r'$, and $d$ but not on the data in* II.6.2, *such that for all integers $n, q \geq 1$, the canonical morphism*

$$\mathrm{H}^q(\Delta, \mathscr{C}^{(r')}/p^n\mathscr{C}^{(r')}) \to \mathrm{H}^q(\Delta, \mathscr{C}^{(r)}/p^n\mathscr{C}^{(r)}) \tag{II.12.7.2}$$

*is annihilated by $p^\beta$.*

This follows from II.11.18.

**II.12.8.** For every $\widehat{\overline{R}}$-representation $M$ of $\Delta$, we denote by $\mathbb{H}(M)$ the $\widehat{R_1}$-module defined by

$$\mathbb{H}(M) = (M \otimes_{\widehat{\overline{R}}} \mathscr{C}^\dagger)^\Delta. \tag{II.12.8.1}$$

We endow it with the Higgs $\widehat{R_1}$-field with coefficients in $\xi^{-1}\widetilde{\Omega}^1_{R/\mathscr{O}_K}$ induced by $d_{\mathscr{C}^\dagger}$ (II.12.1.12) (cf. II.9.9). We thus define a functor

$$\mathbb{H}\colon \mathbf{Rep}_{\widehat{\overline{R}}}(\Delta) \to \mathbf{HM}(\widehat{R_1}, \xi^{-1}\widetilde{\Omega}^1_{R/\mathscr{O}_K}). \tag{II.12.8.2}$$

**II.12.9.** For every Higgs $\widehat{R_1}$-module $(N, \theta)$ with coefficients in $\xi^{-1}\widetilde{\Omega}^1_{R/\mathscr{O}_K}$ (II.9.9), we denote by $\mathbb{V}(N)$ the $\widehat{\overline{R}}$-module defined by

$$\mathbb{V}(N) = (N \otimes_{\widehat{R_1}} \mathscr{C}^\dagger)^{\theta_{\rm tot}=0}, \tag{II.12.9.1}$$

where $\theta_{\rm tot} = \theta \otimes \mathrm{id} + \mathrm{id} \otimes d_{\mathscr{C}^\dagger}$ is the total Higgs $\widehat{R_1}$-field on $N \otimes_{\widehat{R_1}} \mathscr{C}^\dagger$ (II.2.8.8). We endow it with the $\widehat{\overline{R}}$-semi-linear action of $\Delta$ induced by its natural action on $\mathscr{C}^\dagger$. We thus define a functor

$$\mathbb{V}\colon \mathbf{HM}(\widehat{R_1}, \xi^{-1}\widetilde{\Omega}^1_{R/\mathscr{O}_K}) \to \mathbf{Rep}_{\widehat{\overline{R}}}(\Delta). \tag{II.12.9.2}$$

**Remarks II.12.10.** (i) It follows from II.10.10 that the functors $\mathbb{H}$ and $\mathbb{V}$ do not depend on the choice of the deformation $(\widetilde{X}, \mathscr{M}_{\widetilde{X}})$, up to noncanonical isomorphism.

(ii) For every $\widehat{\overline{R}}$-representation $M$ of $\Delta$, the canonical morphism

$$\mathbb{H}(M) \otimes_{\widehat{R_1}} \widehat{R_1}[\frac{1}{p}] \to \mathbb{H}(M \otimes_{\widehat{\overline{R}}} \widehat{\overline{R}}[\frac{1}{p}]) \tag{II.12.10.1}$$

is an isomorphism.

(iii) For every Higgs $\widehat{R_1}$-module $(N, \theta)$ with coefficients in $\xi^{-1}\widetilde{\Omega}^1_{R/\mathscr{O}_K}$, the canonical morphism

$$\mathbb{V}(N) \otimes_{\widehat{\overline{R}}} \widehat{\overline{R}}[\frac{1}{p}] \to \mathbb{V}(N \otimes_{\widehat{R_1}} \widehat{R_1}[\frac{1}{p}]) \tag{II.12.10.2}$$

is an isomorphism.

**Definition II.12.11.** We say that a continuous $\widehat{\overline{R}}$-representation $M$ of $\Delta$ is *Dolbeault* if the following conditions are satisfied:

(i) $M$ is a projective $\widehat{\overline{R}}$-module of finite type, endowed with the $p$-adic topology.
(ii) $\mathbb{H}(M)$ is a projective $\widehat{R_1}$-module of finite type.
(iii) The canonical $\mathscr{C}^\dagger$-linear morphism

$$\mathbb{H}(M)\otimes_{\widehat{R_1}}\mathscr{C}^\dagger\to M\otimes_{\widehat{\overline{R}}}\mathscr{C}^\dagger \tag{II.12.11.1}$$

is an isomorphism.

This notion does not depend on the choice of $(\widetilde{X},\mathscr{M}_{\widetilde{X}})$ (II.10.10 and II.12.10(i)).

**Definition II.12.12.** A Higgs $\widehat{R_1}$-module $(N,\theta)$ with coefficients in $\xi^{-1}\widetilde{\Omega}^1_{R/\mathscr{O}_K}$ is called *solvable* if the following conditions are satisfied:

(i) $N$ is a projective $\widehat{R_1}$-module of finite type.
(ii) $\mathbb{V}(N)$ is a projective $\widehat{\overline{R}}$-module of finite type.
(iii) The canonical $\mathscr{C}^\dagger$-linear morphism

$$\mathbb{V}(N)\otimes_{\widehat{\overline{R}}}\mathscr{C}^\dagger\to N\otimes_{\widehat{R_1}}\mathscr{C}^\dagger \tag{II.12.12.1}$$

is an isomorphism.

This notion does not depend on the choice of $(\widetilde{X},\mathscr{M}_{\widetilde{X}})$ (II.10.10 and II.12.10(i)).

**Lemma II.12.13.** *Let $M$ be a Dolbeault $\widehat{\overline{R}}$-representation of $\Delta$. Then the Higgs $\widehat{R_1}$-module $\mathbb{H}(M)$ is solvable, and we have a functorial and $\Delta$-equivariant canonical $\widehat{\overline{R}}$-isomorphism*

$$\mathbb{V}(\mathbb{H}(M))\xrightarrow{\sim}M. \tag{II.12.13.1}$$

Indeed, the canonical $\mathscr{C}^\dagger$-linear morphism

$$\mathbb{H}(M)\otimes_{\widehat{R_1}}\mathscr{C}^\dagger\to M\otimes_{\widehat{\overline{R}}}\mathscr{C}^\dagger \tag{II.12.13.2}$$

is a $\Delta$-equivariant isomorphism of Higgs $\widehat{\overline{R}}$-modules with coefficients in $\xi^{-1}\widetilde{\Omega}^1_{R/\mathscr{O}_K}$, where $\mathscr{C}^\dagger$ is endowed with the Higgs field $d_{\mathscr{C}^\dagger}$ (II.12.1.12), $\mathbb{H}(M)$ is endowed with the trivial action of $\Delta$, and $M$ is endowed with the zero Higgs field (cf. II.2.13). Since $M$ is $\widehat{\overline{R}}$-flat and $\ker(d_{\mathscr{C}^\dagger})=\widehat{\overline{R}}$, we deduce from this a $\Delta$-equivariant $\widehat{\overline{R}}$-isomorphism $\mathbb{V}(\mathbb{H}(M))\xrightarrow{\sim}M$. The canonical $\mathscr{C}^\dagger$-linear morphism

$$\mathbb{V}(\mathbb{H}(M))\otimes_{\widehat{\overline{R}}}\mathscr{C}^\dagger\to\mathbb{H}(M)\otimes_{\widehat{\overline{R}}}\mathscr{C}^\dagger \tag{II.12.13.3}$$

can then be identified with the inverse of the isomorphism (II.12.13.2), which shows that $\mathbb{H}(M)$ is solvable.

**Lemma II.12.14.** *Let $(N,\theta)$ be a solvable Higgs $\widehat{R_1}$-module with coefficients in $\xi^{-1}\widetilde{\Omega}^1_{R/\mathscr{O}_K}$. Then the $\widehat{\overline{R}}$-representation $\mathbb{V}(N)$ of $\Delta$ is Dolbeault, and we have a functorial canonical isomorphism of Higgs $\widehat{R_1}$-modules*

$$\mathbb{H}(\mathbb{V}(N))\xrightarrow{\sim}N. \tag{II.12.14.1}$$

For all rational numbers $r'\geq r\geq 0$, the canonical morphism

$$N\otimes_{\widehat{R_1}}\widehat{\mathscr{C}}^{(r')}\to N\otimes_{\widehat{R_1}}\widehat{\mathscr{C}}^{(r)}$$

is injective because $N$ is $\widehat{R_1}$-flat. Since $\mathbb{V}(N)$ is an $\widehat{\overline{R}}$-module of finite type, there exists a rational number $r > 0$ such that we have

$$\mathbb{V}(N) = (N \otimes_{\widehat{R_1}} \widehat{\mathscr{C}}^{(r)})^{\theta_{\rm tot}^{(r)}=0}, \tag{II.12.14.2}$$

where $\theta_{\rm tot}^{(r)} = \theta \otimes {\rm id} + p^r {\rm id} \otimes d_{\widehat{\mathscr{C}}^{(r)}}$ is the total Higgs $\widehat{R_1}$-field on $N \otimes_{\widehat{R_1}} \widehat{\mathscr{C}}^{(r)}$. On the other hand, since $\widehat{\mathscr{C}}^{(r)}$ is $\mathscr{O}_C$-flat and $N$ is $\widehat{R_1}$-flat, $\xi^{-1}\widetilde{\Omega}^1_{R/\mathscr{O}_K} \otimes_R N \otimes_{\widehat{R_1}} \widehat{\mathscr{C}}^{(r)}$ is $\mathscr{O}_C$-flat. Consequently, for every $n \geq 0$ and every $x \in N \otimes_{\widehat{R_1}} \widehat{\mathscr{C}}^{(r)}$ such that $\theta_{\rm tot}^{(r)}(p^n x) = 0$, we have $\theta_{\rm tot}^{(r)}(x) = 0$. It follows that

$$(p^n N \otimes_{\widehat{R_1}} \widehat{\mathscr{C}}^{(r)}) \cap \mathbb{V}(N) = p^n \mathbb{V}(N). \tag{II.12.14.3}$$

Hence the $p$-adic topology on $\mathbb{V}(N)$ is induced by that on $N \otimes_{\widehat{R_1}} \widehat{\mathscr{C}}^{(r)}$. It then follows from II.12.4 that the action of $\Delta$ on $\mathbb{V}(N)$ is continuous for the $p$-adic topology.

The canonical $\mathscr{C}^\dagger$-linear morphism

$$\mathbb{V}(N) \otimes_{\widehat{\overline{R}}} \mathscr{C}^\dagger \to N \otimes_{\widehat{R_1}} \mathscr{C}^\dagger \tag{II.12.14.4}$$

is a $\Delta$-equivariant isomorphism of Higgs $\widehat{\overline{R}}$-modules with coefficients in $\xi^{-1}\widetilde{\Omega}^1_{R/\mathscr{O}_K}$, where $\mathscr{C}^\dagger$ is endowed with the Higgs field $d_{\mathscr{C}^\dagger}$, $\mathbb{V}(N)$ is endowed with the zero Higgs field, and $N$ is endowed with the trivial action of $\Delta$ (cf. II.2.13). Since $N$ is a direct summand of a free $\widehat{R_1}$-module of finite type, we have$(N \otimes_{\widehat{R_1}} \mathscr{C}^\dagger)^\Delta = N$ (II.12.6). We deduce from this an isomorphism of Higgs $\widehat{R_1}$-modules $\mathbb{H}(\mathbb{V}(N)) \stackrel{\sim}{\to} N$. The canonical $\mathscr{C}^\dagger$-linear morphism

$$\mathbb{H}(\mathbb{V}(M)) \otimes_{\widehat{R_1}} \mathscr{C}^\dagger \to \mathbb{V}(M) \otimes_{\widehat{R_1}} \mathscr{C}^\dagger \tag{II.12.14.5}$$

then identifies with the inverse of the isomorphism (II.12.14.4), which shows that $\mathbb{V}(N)$ is Dolbeault.

**Proposition II.12.15.** *The functors $\mathbb{V}$ and $\mathbb{H}$ induce equivalences of categories quasi-inverse to each other between the category of Dolbeault $\widehat{\overline{R}}$-representations of $\Delta$ and that of solvable Higgs $\widehat{R_1}$-modules with coefficients in $\xi^{-1}\widetilde{\Omega}^1_{R/\mathscr{O}_K}$.*

This follows from II.12.13 and II.12.14.

**Definition II.12.16.** We say that a continuous $\widehat{\overline{R}}[\frac{1}{p}]$-representation $M$ of $\Delta$ is *Dolbeault* if the following conditions are satisfied:

(i) $M$ is a projective $\widehat{\overline{R}}[\frac{1}{p}]$-module of finite type, endowed with the $p$-adic topology (II.2.2).
(ii) $\mathbb{H}(M)$ is a projective $\widehat{R_1}[\frac{1}{p}]$-module of finite type (II.12.8.1).
(iii) The canonical morphism

$$\mathbb{H}(M) \otimes_{\widehat{R_1}} \mathscr{C}^\dagger \to M \otimes_{\widehat{\overline{R}}} \mathscr{C}^\dagger \tag{II.12.16.1}$$

is an isomorphism.

This notion does not depend on the choice of $(\widetilde{X}, \mathscr{M}_{\widetilde{X}})$ (II.10.10 and II.12.10(i)).

**Remark II.12.17.** A continuous $\widehat{\overline{R}}[\frac{1}{p}]$-representation $M$ of $\Delta$ is Dolbeault if and only if it satisfies conditions (i) and (ii) of II.12.16, as well as the following condition:

(iii') There exists a rational number $r > 0$ such that $\mathbb{H}(M)$ is contained in $M \otimes_{\widehat{\overline{R}}} \widehat{\mathscr{C}}^{(r)}$ (II.12.1.3) and the canonical morphism

$$\mathbb{H}(M) \otimes_{\widehat{R_1}} \widehat{\mathscr{C}}^{(r)} \to M \otimes_{\widehat{\overline{R}}} \widehat{\mathscr{C}}^{(r)} \tag{II.12.17.1}$$

is an isomorphism.

Indeed, for any rational number $r > 0$, the canonical morphism $M \otimes_{\widehat{\overline{R}}} \widehat{\mathscr{C}^{(r)}} \to M \otimes_{\widehat{\overline{R}}} \mathscr{C}^{\dagger}$ is injective because $M$ is $\widehat{\overline{R}}$-flat. Condition (iii') clearly implies condition II.12.16(iii). Conversely, suppose that condition II.12.16(iii) holds. Since $\mathbb{H}(M)$ is an $\widehat{R_1}[\frac{1}{p}]$-module of finite type, there exists a rational number $r > 0$ such that $\mathbb{H}(M)$ is contained in $M \otimes_{\widehat{\overline{R}}} \widehat{\mathscr{C}^{(r)}}$. Since $M$ is of finite type over $\widehat{\overline{R}}[\frac{1}{p}]$, after taking a smaller $r$, if necessary, we may assume that the morphism (II.12.17.1) is surjective. Moreover, since $\mathbb{H}(M)$ is $\widehat{R_1}$-flat, for every rational number $r > 0$, the canonical morphism

$$\mathbb{H}(M) \otimes_{\widehat{R_1}} \widehat{\mathscr{C}^{(r)}} \to \mathbb{H}(M) \otimes_{\widehat{R_1}} \mathscr{C}^{\dagger} \tag{II.12.17.2}$$

is injective. The morphism (II.12.17.1) is therefore injective by virtue of II.12.16(iii).

**Definition II.12.18.** We say that a Higgs $\widehat{R_1}[\frac{1}{p}]$-module $(N, \theta)$ with coefficients in $\xi^{-1}\widetilde{\Omega}^1_{R/\mathscr{O}_K}$ is *solvable* if the following conditions are satisfied:

(i) $N$ is a projective $\widehat{R_1}[\frac{1}{p}]$-module of finite type.

(ii) $\mathbb{V}(N)$ is a projective $\widehat{\overline{R}}[\frac{1}{p}]$-module of finite type.

(iii) The canonical morphism

$$\mathbb{V}(N) \otimes_{\widehat{\overline{R}}} \mathscr{C}^{\dagger} \to N \otimes_{\widehat{R_1}} \mathscr{C}^{\dagger} \tag{II.12.18.1}$$

is an isomorphism.

This notion does not depend on the choice of $(\widetilde{X}, \mathscr{M}_{\widetilde{X}})$ (II.10.10 and II.12.10(i)).

**Remark II.12.19.** A Higgs $\widehat{R_1}[\frac{1}{p}]$-module $(N, \theta)$ with coefficients in $\xi^{-1}\widetilde{\Omega}^1_{R/\mathscr{O}_K}$ is solvable if and only if it satisfies conditions (i) and (ii) of II.12.18, as well as the following condition:

(iii') There exists a rational number $r > 0$ such that $\mathbb{V}(N)$ is contained in $N \otimes_{\widehat{R_1}} \widehat{\mathscr{C}^{(r)}}$ and the canonical morphism

$$\mathbb{V}(N) \otimes_{\widehat{\overline{R}}} \widehat{\mathscr{C}^{(r)}} \to N \otimes_{\widehat{R_1}} \widehat{\mathscr{C}^{(r)}} \tag{II.12.19.1}$$

is an isomorphism.

The proof, similar to that of II.12.17, is left to the reader.

**Lemma II.12.20.** *For every Dolbeault $\widehat{\overline{R}}[\frac{1}{p}]$-representation $M$ of $\Delta$, the Higgs $\widehat{R_1}[\frac{1}{p}]$-module $\mathbb{H}(M)$ is solvable, and we have a functorial and $\Delta$-equivariant canonical $\widehat{\overline{R}}[\frac{1}{p}]$-isomorphism*

$$\mathbb{V}(\mathbb{H}(M)) \xrightarrow{\sim} M. \tag{II.12.20.1}$$

Indeed, the canonical $\mathscr{C}^{\dagger}$-linear morphism

$$\mathbb{H}(M) \otimes_{\widehat{R_1}} \mathscr{C}^{\dagger} \to M \otimes_{\widehat{\overline{R}}} \mathscr{C}^{\dagger} \tag{II.12.20.2}$$

is a $\Delta$-equivariant $\mathscr{C}^{\dagger}$-isomorphism of Higgs $\widehat{\overline{R}}$-modules with coefficients in $\xi^{-1}\widetilde{\Omega}^1_{R/\mathscr{O}_K}$, where $\mathscr{C}^{\dagger}$ is endowed with the Higgs field $d_{\mathscr{C}^{\dagger}}$ (II.12.1.12), $\mathbb{H}(M)$ is endowed with the trivial action of $\Delta$, and $M$ is endowed with the zero Higgs field (cf. II.2.13). Since $M$ is $\widehat{\overline{R}}$-flat and $\ker(d_{\mathscr{C}^{\dagger}}) = \widehat{\overline{R}}$, we deduce from this a $\Delta$-equivariant $\widehat{\overline{R}}[\frac{1}{p}]$-isomorphism $\mathbb{V}(\mathbb{H}(M)) \xrightarrow{\sim} M$. The canonical $\mathscr{C}^{\dagger}$-linear morphism

$$\mathbb{V}(\mathbb{H}(M)) \otimes_{\widehat{\overline{R}}} \mathscr{C}^{\dagger} \to \mathbb{H}(M) \otimes_{\widehat{\overline{R}}} \mathscr{C}^{\dagger} \tag{II.12.20.3}$$

can then be identified with the inverse of the isomorphism (II.12.20.2), which shows that $\mathbb{H}(M)$ is solvable.

**Lemma II.12.21.** *Let $V$ be a projective $\widehat{\overline{R}}[\frac{1}{p}]$-module of finite type, $T$ a sub-$\widehat{\overline{R}}$-module of finite type of $V$ such that $V = T \otimes_{\mathbb{Z}_p} \mathbb{Q}_p$, and $r$ a rational number $\geq 0$. Let $\mathscr{M} = V \otimes_{\widehat{\overline{R}}} \widehat{\mathscr{C}^{(r)}}$ and denote by $\mathscr{M}^\circ$ the canonical image of $T \otimes_{\widehat{\overline{R}}} \widehat{\mathscr{C}^{(r)}}$ in $\mathscr{M}$. Then the canonical morphism $V \to \mathscr{M}$ is injective and there exists an integer $m \geq 0$ such that $T \subset V \cap \mathscr{M}^\circ \subset p^{-m}T$.*

We choose a projective $\widehat{\overline{R}}[\frac{1}{p}]$-module of finite type $V'$ and an $\widehat{\overline{R}}[\frac{1}{p}]$-isomorphism $\varphi\colon V \oplus V' \xrightarrow{\sim} (\widehat{\overline{R}}[\frac{1}{p}])^n$, where $n$ is an integer $\geq 1$. Let $T'$ be a sub-$\widehat{\overline{R}}$-module of finite type of $V'$ such that $V' = T' \otimes_{\mathbb{Z}_p} \mathbb{Q}_p$. Let $\mathscr{M}' = V' \otimes_{\widehat{\overline{R}}} \widehat{\mathscr{C}^{(r)}}$ and denote by $\mathscr{M}'^\circ$ the canonical image of $T' \otimes_{\widehat{\overline{R}}} \widehat{\mathscr{C}^{(r)}}$ in $\mathscr{M}'$. The isomorphism $\varphi$ induces a $(\widehat{\mathscr{C}^{(r)}} \otimes_{\mathbb{Z}_p} \mathbb{Q}_p)$-isomorphism $\phi\colon \mathscr{M} \oplus \mathscr{M}' \xrightarrow{\sim} (\widehat{\mathscr{C}^{(r)}} \otimes_{\mathbb{Z}_p} \mathbb{Q}_p)^n$. Since the structural homomorphism $\widehat{\overline{R}} \to \widehat{\mathscr{C}^{(r)}}$ is injective (II.12.1.13), the canonical morphism $V \to \mathscr{M}$ is injective. On the other hand, there exists an integer $j \geq 0$ such that we have $p^j \widehat{\overline{R}}^n \subset \varphi(T \oplus T') \subset p^{-j}\widehat{\overline{R}}^n$ and therefore $p^j(\widehat{\mathscr{C}^{(r)}})^n \subset \phi(\mathscr{M}^\circ \oplus \mathscr{M}'^\circ) \subset p^{-j}(\widehat{\mathscr{C}^{(r)}})^n$. It immediately follows from II.10.10 and (II.12.2.2) that $\widehat{\overline{R}}[\frac{1}{p}] \cap \widehat{\mathscr{C}^{(r)}} = \widehat{\overline{R}}$. Consequently,

$$\varphi(T \oplus T') \subset \varphi((V \cap \mathscr{M}^\circ) \oplus (V' \cap \mathscr{M}'^\circ)) \subset p^{-j}\widehat{\overline{R}}^n \subset p^{-2j}\varphi(T \oplus T'), \tag{II.12.21.1}$$

giving the statement of the lemma.

**Lemma II.12.22.** *For every solvable Higgs $\widehat{R_1}[\frac{1}{p}]$-module $(N, \theta)$ with coefficients in $\xi^{-1}\widetilde{\Omega}^1_{R/\mathscr{O}_K}$, the $\widehat{\overline{R}}[\frac{1}{p}]$-representation $\mathbb{V}(N)$ of $\Delta$ is Dolbeault, and we have a functorial canonical isomorphism of Higgs $\widehat{R_1}[\frac{1}{p}]$-modules*

$$\mathbb{H}(\mathbb{V}(N)) \xrightarrow{\sim} N. \tag{II.12.22.1}$$

Let us first show that $\mathbb{V}(N)$ is a continuous representation of $\Delta$ for the $p$-adic topology (II.2.2). By II.12.19, there exists a rational number $r > 0$ such that the canonical $\widehat{\mathscr{C}^{(r)}}$-linear morphism

$$\phi\colon \mathbb{V}(N) \otimes_{\widehat{\overline{R}}} \widehat{\mathscr{C}^{(r)}} \to N \otimes_{\widehat{R_1}} \widehat{\mathscr{C}^{(r)}} \tag{II.12.22.2}$$

is a $\Delta$-equivariant isomorphism. Let $T$ be a sub-$\widehat{\overline{R}}$-module of finite type of $\mathbb{V}(N)$ such that $\mathbb{V}(N) = T \otimes_{\mathbb{Z}_p} \mathbb{Q}_p$. Let $\mathscr{M} = \mathbb{V}(N) \otimes_{\widehat{\overline{R}}} \widehat{\mathscr{C}^{(r)}}$ and denote by $\mathscr{M}^\circ$ the canonical image of $T \otimes_{\widehat{\overline{R}}} \widehat{\mathscr{C}^{(r)}}$ in $\mathscr{M}$. Let $N^\circ$ be a sub-$\widehat{R_1}$-module of finite type of $N$ such that $N = N^\circ \otimes_{\mathbb{Z}_p} \mathbb{Q}_p$. Let $\mathscr{N} = N \otimes_{\widehat{R_1}} \widehat{\mathscr{C}^{(r)}}$ and denote by $\mathscr{N}^\circ$ the canonical image of $N^\circ \otimes_{\widehat{\overline{R}}} \widehat{\mathscr{C}^{(r)}}$ in $\mathscr{N}$. Since $\mathscr{M}^\circ$ and $\mathscr{N}^\circ$ are $\widehat{\overline{R}}$-modules of finite type, there exists an integer $n \geq 0$ such that $p^n\mathscr{N}^\circ \subset \phi(\mathscr{M}^\circ) \subset p^{-n}\mathscr{N}^\circ$. On the other hand, by II.12.21, the canonical morphism $\mathbb{V}(N) \to \mathscr{M}$ is injective and there exists an integer $m \geq 0$ such that $T \subset \mathbb{V}(N) \cap \mathscr{M}^\circ \subset p^{-m}T$. It follows that

$$p^nT \subset \mathbb{V}(N) \cap \phi^{-1}(\mathscr{N}^\circ) \subset p^{-m-n}T. \tag{II.12.22.3}$$

Let $x \in T$ and $\nu$ be an integer $\geq 0$. By (II.12.22.3), there exist an integer $b \geq 1$, $y_1, \dots, y_b \in N^\circ$, and $\alpha_1, \dots, \alpha_b \in \widehat{\mathscr{C}^{(r)}}$ such that $p^n\phi(x) = \sum_{1 \leq i \leq b} \alpha_i y_i$. By virtue of

II.12.4, there exists an open subgroup $\Delta_{x,\nu}$ of $\Delta$ such that for every $g \in \Delta_{x,\nu}$ and every $1 \leq i \leq b$, we have $g(\alpha_i) - \alpha_i \in p^{m+2n+\nu}\widehat{\mathscr{C}}^{(r)}$. It follows that

$$g(x) - x \in \mathbb{V}(N) \cap (p^{m+n+\nu}\phi^{-1}(\mathscr{N}^\circ)) \subset p^\nu T. \tag{II.12.22.4}$$

Since $T$ is of finite type over $\widehat{\overline{R}}$, we deduce from this that there exists an open subgroup $\Delta'$ of $\Delta$ such that for every $g \in \Delta'$, we have $g(T) \subset T$. Consequently, $T' = \sum_{g\in\Delta} g(T)$ is a sub-$\widehat{\overline{R}}$-module of finite type of $\mathbb{V}(N)$, stable under $\Delta$. Replacing $T$ by $T'$, we reduce to the case where $T$ is stable under the action of $\Delta$. It then follows from (II.12.22.4) that the action of $\Delta$ on $T$ is continuous for the $p$-adic topology, and the same therefore holds for the action of $\Delta$ on $\mathbb{V}(N)$.

The canonical $\mathscr{C}^\dagger$-linear morphism

$$\mathbb{V}(N) \otimes_{\widehat{\overline{R}}} \mathscr{C}^\dagger \to N \otimes_{\widehat{R_1}} \mathscr{C}^\dagger \tag{II.12.22.5}$$

is a $\Delta$-equivariant $\mathscr{C}^\dagger$-isomorphism of Higgs $\widehat{\overline{R}}$-modules with coefficients in $\xi^{-1}\widetilde{\Omega}^1_{R/\mathscr{O}_K}$, where $\mathscr{C}^\dagger$ is endowed with the Higgs $\widehat{\overline{R}}$-field $d_{\mathscr{C}^\dagger}$, $N$ is endowed with the trivial action of $\Delta$, and $\mathbb{V}(N)$ is endowed with the zero Higgs $\widehat{\overline{R}}$-field (cf. II.2.13). Since $N$ is a direct summand of a free $\widehat{R_1}[\frac{1}{p}]$-module of finite type, we have $(N \otimes_{\widehat{R_1}} \mathscr{C}^\dagger)^\Delta = N$ (II.12.6). We deduce from this an isomorphism of Higgs $\widehat{R_1}[\frac{1}{p}]$-modules $\mathbb{H}(\mathbb{V}(N)) \xrightarrow{\sim} N$. The canonical $\mathscr{C}^\dagger$-linear morphism

$$\mathbb{H}(\mathbb{V}(N)) \otimes_{\widehat{R_1}} \mathscr{C}^\dagger \to \mathbb{V}(N) \otimes_{\widehat{R_1}} \mathscr{C}^\dagger \tag{II.12.22.6}$$

is then identified with the inverse of the isomorphism (II.12.22.5), which shows that $\mathbb{V}(N)$ is a Dolbeault $\widehat{\overline{R}}[\frac{1}{p}]$-representation of $\Delta$.

**Remark II.12.23.** The proof of II.12.22 given above is due to Tsuji. It is simpler and more elegant than the proof we had given in an earlier version of this text.

**Proposition II.12.24.** *The functors $\mathbb{H}$ and $\mathbb{V}$ induce equivalences of categories quasi-inverse to each other between the category of Dolbeault $\widehat{\overline{R}}[\frac{1}{p}]$-representations of $\Delta$ and that of the solvable Higgs $\widehat{R_1}[\frac{1}{p}]$-modules with coefficients in $\xi^{-1}\widetilde{\Omega}^1_{R/\mathscr{O}_K}$.*

This follows from II.12.20 and II.12.22.

**II.12.25.** Let $M$ be a Dolbeault $\widehat{\overline{R}}[\frac{1}{p}]$-representation of $\Delta$, $(\mathbb{H}(M), \theta)$ the associated Higgs $\widehat{R_1}[\frac{1}{p}]$-module with coefficients in $\xi^{-1}\widetilde{\Omega}^1_{R/\mathscr{O}_K}$ (II.12.8.2), and $\theta_{\text{tot}} = \theta \otimes \mathrm{id} + \mathrm{id} \otimes d_{\mathscr{C}^\dagger}$ the total Higgs $\widehat{R_1}$-field over $\mathbb{H}(M) \otimes_{\widehat{R_1}} \mathscr{C}^\dagger$. It immediately follows from (II.12.16.1) that we have a functorial canonical isomorphism of complexes of $\widehat{\overline{R}}$-representations

$$\mathbb{K}^\bullet(\mathbb{H}(M) \otimes_{\widehat{R_1}} \mathscr{C}^\dagger, \theta_{\text{tot}}) \xrightarrow{\sim} M \otimes_{\widehat{\overline{R}}} \mathbb{K}^\bullet(\mathscr{C}^\dagger, d_{\mathscr{C}^\dagger}), \tag{II.12.25.1}$$

where $\mathbb{K}^\bullet(-,-)$ denotes the Dolbeault complex (II.2.8.4). By II.12.6, for every $i \geq 0$, since $\mathbb{H}(M) \otimes_R \widetilde{\Omega}^i_{R/\mathscr{O}_K}$ is a direct summand of a free $\widehat{R_1}[\frac{1}{p}]$-module of finite type, we have

$$(\xi^{-i}\mathbb{H}(M) \otimes_{\widehat{R_1}} \mathscr{C}^\dagger \otimes_R \widetilde{\Omega}^i_{R/\mathscr{O}_K})^\Delta = \xi^{-i}\mathbb{H}(M) \otimes_R \widetilde{\Omega}^i_{R/\mathscr{O}_K}. \tag{II.12.25.2}$$

We deduce from this a functorial canonical isomorphism of complexes of $\widehat{R_1}[\frac{1}{p}]$-modules

$$\mathbb{K}^\bullet(\mathbb{H}(M), \theta) \xrightarrow{\sim} (M \otimes_{\widehat{\overline{R}}} \mathbb{K}^\bullet(\mathscr{C}^\dagger, d_{\mathscr{C}^\dagger}))^\Delta, \tag{II.12.25.3}$$

where the functor $(-)^\Delta$ on the right-hand side is defined component-wise. This result can be refined as follows.

**Proposition II.12.26** (**[27]** § 3, IV.5.3.2). *Let $M$ be a Dolbeault $\widehat{\overline{R}}[\frac{1}{p}]$-representation of $\Delta$ and $(\mathbb{H}(M),\theta)$ the associated Higgs $\widehat{R_1}[\frac{1}{p}]$-module with coefficients in $\xi^{-1}\widetilde{\Omega}^1_{R/\mathscr{O}_K}$* (II.12.8.2). *Then we have a functorial canonical isomorphism in $\mathbf{D}^+(\mathbf{Mod}(\widehat{R_1}[\frac{1}{p}]))$*

$$\text{(II.12.26.1)} \qquad \mathrm{C}^\bullet_{\mathrm{cont}}(\Delta, M) \xrightarrow{\sim} \mathbb{K}^\bullet(\mathbb{H}(M),\theta),$$

*where $\mathrm{C}^\bullet_{\mathrm{cont}}(\Delta, M)$ is the complex of continuous cochains of $\Delta$ with values in $M$* (II.3.8) *and $\mathbb{K}^\bullet(\mathbb{H}(M),\theta)$ is the Dolbeault complex* (II.2.8.4).

Indeed, for every integer $i \geq 0$ and all rational numbers $r' > r > 0$, the canonical morphism

$$\text{(II.12.26.2)} \quad \mathrm{C}^i_{\mathrm{cont}}(\Delta, M \otimes_{\widehat{\overline{R}}} \widetilde{\mathbb{K}}^\bullet(\widehat{\mathscr{C}}^{(r')}, p^{r'} d_{\widehat{\mathscr{C}}^{(r')}})) \to \mathrm{C}^i_{\mathrm{cont}}(\Delta, M \otimes_{\widehat{\overline{R}}} \widetilde{\mathbb{K}}^\bullet(\widehat{\mathscr{C}}^{(r)}, p^{r} d_{\widehat{\mathscr{C}}^{(r)}}))$$

is homotopic to 0, by virtue of II.12.3(ii). Consequently, the complex

$$\text{(II.12.26.3)} \qquad \varinjlim_{r\in\mathbb{Q}_{>0}} \mathrm{C}^i_{\mathrm{cont}}(\Delta, M \otimes_{\widehat{\overline{R}}} \widetilde{\mathbb{K}}^\bullet(\widehat{\mathscr{C}}^{(r)}, p^{r} d_{\widehat{\mathscr{C}}^{(r)}}))$$

is acyclic. The first spectral sequence of the bicomplex (**[41]** § 0 (11.3.2.2))

$$\text{(II.12.26.4)} \qquad \varinjlim_{r\in\mathbb{Q}_{>0}} \mathrm{C}^\bullet_{\mathrm{cont}}(\Delta, M \otimes_{\widehat{\overline{R}}} \widetilde{\mathbb{K}}^\bullet(\widehat{\mathscr{C}}^{(r)}, p^{r} d_{\widehat{\mathscr{C}}^{(r)}}))$$

then implies that the associated simple complex (II.3.6.2)

$$\text{(II.12.26.5)} \qquad \varinjlim_{r\in\mathbb{Q}_{>0}} \int \mathrm{C}^\bullet_{\mathrm{cont}}(\Delta, M \otimes_{\widehat{\overline{R}}} \widetilde{\mathbb{K}}^\bullet(\widehat{\mathscr{C}}^{(r)}, p^{r} d_{\widehat{\mathscr{C}}^{(r)}}))$$

is acyclic. The canonical morphisms $\widehat{\overline{R}}[0] \to \mathbb{K}^\bullet(\widehat{\mathscr{C}}^{(r)}, p^{r} d_{\widehat{\mathscr{C}}^{(r)}})$ therefore induce a quasi-isomorphism

$$\text{(II.12.26.6)} \qquad \mathrm{C}^\bullet_{\mathrm{cont}}(\Delta, M) \to \varinjlim_{r\in\mathbb{Q}_{>0}} \int \mathrm{C}^\bullet_{\mathrm{cont}}(\Delta, M \otimes_{\widehat{\overline{R}}} \mathbb{K}^\bullet(\widehat{\mathscr{C}}^{(r)}, p^{r} d_{\widehat{\mathscr{C}}^{(r)}})).$$

By II.12.17, there exists a rational number $r_0 > 0$ such that $\mathbb{H}(M)$ is contained in $M \otimes_{\widehat{\overline{R}}} \widehat{\mathscr{C}}^{(r_0)}$ and that for every rational number $0 < r \leq r_0$, the canonical morphism

$$\text{(II.12.26.7)} \qquad \mathbb{H}(M) \otimes_{\widehat{R_1}} \widehat{\mathscr{C}}^{(r)} \to M \otimes_{\widehat{\overline{R}}} \widehat{\mathscr{C}}^{(r)}$$

is bijective. If we write $\theta^{(r)}_{\mathrm{tot}} = \theta \otimes \mathrm{id} + p^r \mathrm{id} \otimes d_{\widehat{\mathscr{C}}^{(r)}}$ for the total Higgs $\widehat{R_1}$-field over $\mathbb{H}(M) \otimes_{\widehat{R_1}} \widehat{\mathscr{C}}^{(r)}$, then we deduce from this an isomorphism

$$\text{(II.12.26.8)} \qquad \mathbb{K}^\bullet(\mathbb{H}(M) \otimes_{\widehat{R_1}} \widehat{\mathscr{C}}^{(r)}, \theta^{(r)}_{\mathrm{tot}}) \xrightarrow{\sim} M \otimes_{\widehat{\overline{R}}} \mathbb{K}^\bullet(\widehat{\mathscr{C}}^{(r)}, p^{r} d_{\widehat{\mathscr{C}}^{(r)}}).$$

For every integer $i \geq 0$, the canonical morphism

$$\text{(II.12.26.9)} \qquad \mathbb{H}(M) \otimes_{\widehat{R_1}} \mathrm{C}^i_{\mathrm{cont}}(\Delta, \widehat{\mathscr{C}}^{(r)}) \to \mathrm{C}^i_{\mathrm{cont}}(\Delta, \mathbb{H}(M) \otimes_{\widehat{R_1}} \widehat{\mathscr{C}}^{(r)})$$

is an isomorphism. Indeed, we can reduce to the case where $\mathbb{H}(M)$ is a free $\widehat{R_1}[\frac{1}{p}]$-module of finite type, in which case the assertion follows from the compactness of $\Delta$. On the other hand, the derivation $d_{\widehat{\mathscr{C}}^{(r)}}$ induces a Higgs $\widehat{R_1}$-field $\delta^{i,(r)}$ on $\mathrm{C}^i_{\mathrm{cont}}(\Delta, \widehat{\mathscr{C}}^{(r)})$ with coefficients in $\xi^{-1}\widetilde{\Omega}^1_{R/\mathscr{O}_K}$. The differentials of the complex $\mathrm{C}^\bullet_{\mathrm{cont}}(\Delta, \widehat{\mathscr{C}}^{(r)})$ are morphisms of Higgs modules. If we denote by $\vartheta^{i,(r)}_{\mathrm{tot}} = \theta \otimes \mathrm{id} + p^r \mathrm{id} \otimes \delta^{i,(r)}$ the total Higgs $\widehat{R_1}$-field on $\mathbb{H}(M) \otimes_{\widehat{R_1}} \mathrm{C}^i_{\mathrm{cont}}(\Delta, \widehat{\mathscr{C}}^{(r)})$, then the canonical morphism

$$\text{(II.12.26.10)} \quad \mathbb{K}^\bullet(\mathbb{H}(M) \otimes_{\widehat{R_1}} \mathrm{C}^i_{\mathrm{cont}}(\Delta, \widehat{\mathscr{C}}^{(r)}), \vartheta^{i,(r)}_{\mathrm{tot}}) \to \mathrm{C}^i_{\mathrm{cont}}(\Delta, \mathbb{K}^\bullet(\mathbb{H}(M) \otimes_{\widehat{R_1}} \widehat{\mathscr{C}}^{(r)}, \theta^{(r)}_{\mathrm{tot}}))$$

is an isomorphism (II.12.26.9). It follows from II.12.5, using the first spectral sequence of bicomplexes, that the canonical morphism

$$\text{(II.12.26.11)} \qquad \mathbb{K}^\bullet(\mathbb{H}(M),\theta) \to \lim_{\substack{\longrightarrow \\ r\in\mathbb{Q}_{>0}}} \int \mathbb{K}^\bullet(\mathbb{H}(M)\otimes_{\widehat{R_1}} \mathrm{C}^\bullet_{\mathrm{cont}}(\Delta,\widehat{\mathscr{C}^{(r)}}),\vartheta^{\bullet,(r)}_{\mathrm{tot}})$$

is a quasi-isomorphism. The proposition follows.

## II.13. Small representations

We keep the assumptions and notation of II.12 in this section.

**Definition II.13.1.** Let $G$ be a topological group, $A$ an $\mathscr{O}_C$-algebra that is complete and separated for the $p$-adic topology, endowed with a continuous action of $G$ (by homomorphisms of $\mathscr{O}_C$-algebras), $\alpha$ a rational number $> 0$, and $M$ a continuous $A$-representation of $G$, endowed with the $p$-adic topology.

(i) We say that $M$ is *$\alpha$-quasi-small* if the $A$-module $M$ is complete and separated for the $p$-adic topology, and is generated by a finite number of elements that are $G$-invariant modulo $p^\alpha M$.

(ii) We say that $M$ is *$\alpha$-small* if $M$ is a free $A$-module of finite type having a basis over $A$ consisting of elements that are $G$-invariant modulo $p^\alpha M$.

(iii) We say that $M$ is *quasi-small* (resp. *small*) if it is $\alpha'$-quasi-small (resp. $\alpha'$-small) for a rational number $\alpha' > \frac{2}{p-1}$.

We denote by $\mathbf{Rep}_A^{\alpha\text{-qsf}}(G)$ (resp. $\mathbf{Rep}_A^{\mathrm{qsf}}(G)$) the full subcategory of $\mathbf{Rep}_A^{\mathrm{cont}}(G)$ made up of the $\alpha$-quasi-small (resp. quasi-small) $A$-representations of $G$ whose underlying $A$-module is $\mathscr{O}_C$-flat, and by $\mathbf{Rep}_A^{\alpha\text{-small}}(G)$ (resp. $\mathbf{Rep}_A^{\mathrm{small}}(G)$) the full subcategory of $\mathbf{Rep}_A^{\mathrm{cont}}(G)$ made up of the $\alpha$-small (resp. small) $A$-representations of $G$.

If the action of $G$ on $A$ is trivial, then an $A$-representation $M$ of $G$ is $\alpha$-small if and only if it is $\alpha$-quasi-small and $M$ is a free $A$-module of finite type.

**Definition II.13.2.** Let $G$ be a topological group, $A$ an $\mathscr{O}_C$-algebra that is complete and separated for the $p$-adic topology, endowed with a continuous action of $G$ (by homomorphisms of $\mathscr{O}_C$-algebras). We endow $A[\frac{1}{p}]$ with the $p$-adic topology (II.2.2). We say that a continuous $A[\frac{1}{p}]$-representation $M$ of $G$ is *small* if the following conditions are satisfied:

(i) $M$ is a projective $A[\frac{1}{p}]$-module of finite type, endowed with the $p$-adic topology (II.2.2).

(ii) There exist a rational number $\alpha > \frac{2}{p-1}$ and a sub-$A$-module of finite type $M^\circ$ of $M$ that is stable under $G$, generated by a finite number of elements that are $G$-invariant modulo $p^\alpha M^\circ$, and that generates $M$ over $A[\frac{1}{p}]$.

We denote by $\mathbf{Rep}_{A[\frac{1}{p}]}^{\mathrm{small}}(G)$ the full subcategory of $\mathbf{Rep}_{A[\frac{1}{p}]}^{\mathrm{cont}}(G)$ made up of the small $A$-representations of $G$.

Note that unlike in the integral case (II.13.1), we do not require $M$ to be free over $A[\frac{1}{p}]$.

**Remarks II.13.3.** Let $G$ be a topological group and $A$ an $\mathscr{O}_C$-algebra that is separated and complete for the $p$-adic topology, endowed with a continuous action of $G$ (by homomorphisms of $\mathscr{O}_C$-algebras).

(i) Let $M$ be a projective $A[\frac{1}{p}]$-module of finite type and $M^\circ$ a sub-$A$-module of finite type of $M$. Then $M^\circ$ is complete and separated for the $p$-adic topology.

Indeed, $M^\circ$ is complete by virtue of ([**11**] Chap. III § 2.12 Cor. 1 to Prop. 16). On the other hand, after adding a direct summand to $M$, if necessary, we may assume that it is free of finite type over $A[\frac{1}{p}]$. Consequently, there exists an integer $m \geq 0$ such that $p^m M^\circ$ is contained in a free $A$-module of finite type $N$. Hence $\cap_{n\geq 0} p^n M^\circ \subset \cap_{n \geq 0} p^n N = 0$.

(ii) Let $M$ be a small $A[\frac{1}{p}]$-representation of $G$ and $M^\circ$ a sub-$A$-module of finite type of $M$ satisfying condition II.13.2(ii). It then follows from (i) that $M^\circ$ is a quasi-small $A$-representation of $G$.

**Definition II.13.4.** Let $\beta$ be a rational number $> 0$ and $(N, \theta)$ a Higgs $\widehat{R_1}$-module with coefficients in $\xi^{-1}\widetilde{\Omega}^1_{R/\mathscr{O}_K}$ (II.9.9).

(i) We say that $(N, \theta)$ is *$\beta$-quasi-small* if $N$ is of finite type over $\widehat{R_1}$ and if $\theta$ is a multiple of $p^\beta$ in $\xi^{-1}\mathrm{End}_{\widehat{R_1}}(N) \otimes_R \widetilde{\Omega}^1_{R/\mathscr{O}_K}$ (II.2.8.10). We then also say that the Higgs $\widehat{R_1}$-field $\theta$ is *$\beta$-quasi-small.*

(ii) We say that $(N, \theta)$ is *$\beta$-small* if it is $\beta$-quasi-small and if $N$ is free of finite type over $\widehat{R_1}$. We then also say that the Higgs $\widehat{R_1}$-field $\theta$ is *$\beta$-small.*

(iii) We say that $(N, \theta)$ is *quasi-small* (resp. *small*) if it is $\beta'$-quasi-small (resp. $\beta'$-small) for a rational number $\beta' > \frac{1}{p-1}$. We then also say that the Higgs $\widehat{R_1}$-field $\theta$ is *quasi-small* (resp. *small*).

We denote by $\mathbf{HM}^{\beta\text{-qsf}}(\widehat{R_1}, \xi^{-1}\widetilde{\Omega}^1_{R/\mathscr{O}_K})$ the full subcategory of $\mathbf{HM}(\widehat{R_1}, \xi^{-1}\widetilde{\Omega}^1_{R/\mathscr{O}_K})$ (II.9.9) made up of the $\beta$-quasi-small Higgs $\widehat{R_1}$-modules whose underlying $\widehat{R_1}$-module is $\mathscr{O}_C$-flat and by $\mathbf{HM}^{\beta\text{-small}}(\widehat{R_1}, \xi^{-1}\widetilde{\Omega}^1_{R/\mathscr{O}_K})$ the full subcategory of $\mathbf{HM}(\widehat{R_1}, \xi^{-1}\widetilde{\Omega}^1_{R/\mathscr{O}_K})$ made up of the $\beta$-small Higgs $\widehat{R_1}$-modules.

**Definition II.13.5.** A Higgs $\widehat{R_1}[\frac{1}{p}]$-module $(N, \theta)$ with coefficients in $\xi^{-1}\widetilde{\Omega}^1_{R/\mathscr{O}_K}$ is said to be *small* if the following conditions are satisfied:

(i) $N$ is a projective $\widehat{R_1}[\frac{1}{p}]$-module of finite type.

(ii) There exist a rational number $\beta > \frac{1}{p-1}$ and a sub-$\widehat{R_1}$-module of finite type $N^\circ$ of $N$ that generates it over $\widehat{R_1}[\frac{1}{p}]$, such that we have

(II.13.5.1)
$$\theta(N^\circ) \subset p^\beta \xi^{-1} N^\circ \otimes_R \widetilde{\Omega}^1_{R/\mathscr{O}_K}.$$

We denote by $\mathbf{HM}^{\mathrm{small}}(\widehat{R_1}[\frac{1}{p}], \xi^{-1}\widetilde{\Omega}^1_{R/\mathscr{O}_K})$ the full subcategory of $\mathbf{HM}(\widehat{R_1}[\frac{1}{p}], \xi^{-1}\widetilde{\Omega}^1_{R/\mathscr{O}_K})$ made up of the small Higgs $\widehat{R_1}[\frac{1}{p}]$-modules (cf. II.9.9).

Note that unlike in the integral case (II.13.4), we do not require $N$ to be free over $\widehat{R_1}[\frac{1}{p}]$.

**Lemma II.13.6.** *Let $A$ be an integral domain with field of fractions $L$, $B$ a subring of $L$ containing $A$, $N$ a flat $B$-module of finite type, $u_1, \dots, u_\ell$ $B$-linear endomorphisms of $N$ that commute pairwise and such that for every $1 \leq i \leq \ell$, the characteristic polynomial of the endomorphism $u_i \otimes \mathrm{id}$ of $N \otimes_B L$ has coefficients in $A$. Then $N$ is generated over $B$ by a sub-$A$-module of finite type $M$ such that $u_i(M) \subset M$ for every $1 \leq i \leq \ell$.*

We proceed by induction on the number of endomorphisms. Suppose that the assertion has been proved for $\ell - 1$ and let us show it for $\ell$. Let $M'$ be a sub-$A$-module of finite type of $N$ that generates it over $B$ and satisfies $u_i(M') \subset M'$ for every $1 \leq i \leq \ell - 1$. Let $P(X) = X^n + a_1 X^{n-1} + \cdots + a_n \in A[X]$ be the characteristic polynomial of the endomorphism $u_\ell \otimes \mathrm{id}$ of $N \otimes_B L$. Since $N$ is $B$-flat, we can identify it with a sub-$B$-module of $N \otimes_B L$. Consequently, the endomorphism $P(u_\ell)$ of $N$ is zero and the sub-$A$-module $M = \sum_{i=0}^{n-1} u_\ell^i(M')$ of $N$ has the desired property.

**Lemma II.13.7.** *Let $(N,\theta)$ be a Higgs $\widehat{R_1}[\frac{1}{p}]$-module with coefficients in $\xi^{-1}\widetilde{\Omega}^1_{R/\mathscr{O}_K}$ such that the following conditions are satisfied:*

(i) *$N$ is a projective $\widehat{R_1}[\frac{1}{p}]$-module of finite type.*

(ii) *There exists a rational number $\beta > \frac{1}{p-1}$ such that for every $i \geq 1$, the $i$th characteristic invariant of $\theta$ belongs to $p^{i\beta}\xi^{-i}\mathrm{S}^i_R(\widetilde{\Omega}^1_{R/\mathscr{O}_K}) \otimes_R \widehat{R_1}$ (II.2.8.5).*

*Then $(N,\theta)$ is small.*

Since $\mathrm{Spec}(\widehat{R}_1)$ is $\mathscr{O}_C$-flat, its generic points are the generic points of $\mathrm{Spec}(\widehat{R}_1[\frac{1}{p}])$. Since the latter is noetherian, $\mathrm{Spec}(\widehat{R}_1)$ has only a finite number of generic points. On the other hand, $\widehat{R}_1$ is normal by virtue of II.6.15. It is therefore isomorphic to a finite product $\prod_{j=1}^n A_j$ of normal integral domains. By II.13.6, for every $1 \leq j \leq n$, there exists a sub-$A_j$-module of finite type $N_j^\circ$ of $N \otimes_{\widehat{R}_1} A_j$ that generates it over $A_j[\frac{1}{p}]$, such that if we denote by $\theta_j$ the Higgs $A_j[\frac{1}{p}]$-field over $N \otimes_{\widehat{R}_1} A_j$ with coefficients in $\xi^{-1}\widetilde{\Omega}^1_{R/\mathscr{O}_K}$ induced by $\theta$, we have

$$\theta_j(N_j^\circ) \subset p^\beta \xi^{-1} N_j^\circ \otimes_R \widetilde{\Omega}^1_{R/\mathscr{O}_K}. \tag{II.13.7.1}$$

The lemma follows.

**II.13.8.** Recall that on $C$, the logarithmic function $\log(x)$ converges when $x \in 1 + \mathfrak{m}_C$, and the exponential function $\exp(x)$ converges when $v(x) > \frac{1}{p-1}$. For every $x \in C$ such that $v(x) > \frac{1}{p-1}$, we have

$$\begin{aligned} \exp(x) &\equiv 1 + x \mod (x\mathfrak{m}_C), \\ \log(1+x) &\equiv x \mod (x\mathfrak{m}_C), \end{aligned}$$

$\exp(\log(1+x)) = 1 + x$ and $\log(\exp(x)) = x$.

**II.13.9.** Let $M$ be an $\widehat{R_1}$-module of finite type. By ([**1**] 1.10.2), $M$ is complete and separated for the $p$-adic topology, and the same therefore holds for $\mathrm{End}_{\widehat{R_1}}(M)$. Suppose, moreover, that $M$ is $\mathscr{O}_C$-flat. Then $\mathrm{End}_{\widehat{R_1}}(M)$ is $\mathscr{O}_C$-flat and for every rational number $\alpha \geq 0$, the canonical homomorphism $p^\alpha \mathrm{End}_{\widehat{R_1}}(M) \to \mathrm{Hom}_{\widehat{R_1}}(M, p^\alpha M)$ is an isomorphism. Let $u$ be an $\widehat{R_1}$-linear endomorphism of $M$ and $\alpha$ a rational number $> 0$. If $u$ induces an automorphism of $M/p^\alpha M$, then $u$ is an automorphism of $M$. Indeed, $u$ is surjective by Nakayama's lemma. If $x \in M$ is such that $u(x) = 0$, then there exists $y \in M$ such that $x = p^\alpha y$; since $M$ is $\mathscr{O}_C$-flat, we have $u(y) = 0$; it follows that $x \in \cap_{n\geq 0} p^{n\alpha} M = 0$, giving the assertion. We can therefore identify $\mathrm{id} + p^\alpha \mathrm{End}_{\widehat{R_1}}(M)$ with a subgroup of $\mathrm{Aut}_{\widehat{R_1}}(M)$. Suppose $\alpha > \frac{1}{p-1}$ and $v = u - \mathrm{id} \in p^\alpha \mathrm{End}_{\widehat{R_1}}(M)$. For every $x \in M$, the series

$$\exp(v)(x) = \sum_{n\geq 0} \frac{1}{n!} v^n(x), \tag{II.13.9.1}$$

$$\log(u)(x) = \sum_{n\geq 0} \frac{(-1)^n}{n+1} v^{n+1}(x), \tag{II.13.9.2}$$

then converge in $M$, and define two $\widehat{R_1}$-endomorphisms $\exp(v)$ and $\log(u)$ of $M$. Moreover, we have $\exp(v) \in \mathrm{id} + p^\alpha \mathrm{End}_{\widehat{R_1}}(M)$ and $\log(u) \in p^\alpha \mathrm{End}_{\widehat{R_1}}(M)$, and $\exp(\log(u)) = u$ and $\log(\exp(v)) = v$.

**II.13.10.** Let $M$ be an $\mathscr{O}_C$-flat $\widehat{R_1}$-module of finite type, $\alpha$ a rational number $> \frac{1}{p-1}$, and $\beta = \alpha - \frac{1}{p-1}$. We denote by $\Psi_M$ the composed isomorphism

$$\begin{array}{ccc} \mathrm{Hom}_{\mathbb{Z}}(\Delta_\infty, p^\alpha \mathrm{End}_{\widehat{R_1}}(M)) & \xrightarrow{\sim} & p^\beta \xi^{-1}\mathrm{End}_{\widehat{R_1}}(M) \otimes_{\widehat{R_1}} \mathrm{Hom}_{\mathbb{Z}}(\Delta_\infty, \widehat{R_1}(1)) \\ & \searrow^{\Psi_M} & \downarrow{\scriptstyle \mathrm{id}\otimes\widetilde{\delta}} \\ & & p^\beta\xi^{-1}\mathrm{End}_{\widehat{R_1}}(M) \otimes_R \widetilde{\Omega}^1_{R/\mathscr{O}_K} \end{array} \tag{II.13.10.1}$$

where $\widetilde{\delta}$ is the isomorphism (II.7.19.9) and the horizontal isomorphism comes from (II.6.12.2) and from the canonical isomorphism $\mathscr{O}_C(1) \xrightarrow{\sim} p^{\frac{1}{p-1}}\xi\mathscr{O}_C$ (II.9.18.1). Note that the $\widehat{R_1}$-module $\mathrm{End}_{\widehat{R_1}}(M)$ is complete and separated for the $p$-adic topology and $\mathscr{O}_C$-flat (II.13.9). We can write $\Psi_M$ explicitly as follows. Let $t_1, \dots, t_d \in P^{\mathrm{gp}}$ be such that their images in $(P^{\mathrm{gp}}/\mathbb{Z}\lambda) \otimes_{\mathbb{Z}} \mathbb{Z}_p$ form a $\mathbb{Z}_p$-basis, $(\chi_{t_i})_{1\leq i\leq d}$ their images in $\mathrm{Hom}_{\mathbb{Z}}(\Delta_\infty, \mathbb{Z}_p(1))$ (II.7.19.5), and $(d\log(t_i))_{1\leq i\leq d}$ their images in $\widetilde{\Omega}^1_{R/\mathscr{O}_K}$ (II.7.18.5). The $(d\log(t_i))_{1\leq i\leq d}$ then form an $R$-basis of $\widetilde{\Omega}^1_{R/\mathscr{O}_K}$ (II.7.12.1) and the $(\chi_{t_i})_{1\leq i\leq d}$ form a $\mathbb{Z}_p$-basis of $\mathrm{Hom}_{\mathbb{Z}}(\Delta_\infty, \mathbb{Z}_p(1))$ (II.7.19.8). For every $1 \leq i \leq d$, denote by $\chi_i$ the composition

$$\Delta_\infty \xrightarrow{\chi_{t_i}} \mathbb{Z}_p(1) \xrightarrow{\log([\ ])} p^{\frac{1}{p-1}}\xi\mathscr{O}_C\ , \tag{II.13.10.2}$$

where the second arrow is induced by the isomorphism (II.9.18.1). In view of (II.6.12.2), for every homomorphism $\phi\colon \Delta_\infty \to p^\alpha\mathrm{End}_{\widehat{R_1}}(M)$, there exist $\phi_i \in p^\beta\mathrm{End}_{\widehat{R_1}}(M)$ ($1 \leq i \leq d$) such that

$$\phi = \sum_{i=1}^{d} \xi^{-1}\phi_i \otimes \chi_i. \tag{II.13.10.3}$$

Concretely, if $\zeta$ is a $\mathbb{Z}_p$-basis of $\mathbb{Z}_p(1)$, there exist elements $\gamma_1, \dots, \gamma_d \in \Delta_\infty$ such that for all $1 \leq i, j \leq d$, we have $\chi_{t_i}(\gamma_j) = \delta_{ij}\zeta$. For every $1 \leq i \leq d$, we therefore have

$$\xi^{-1}\log([\zeta])\phi_i = \phi(\gamma_i). \tag{II.13.10.4}$$

Recall that $\xi^{-1}\log([\zeta])$ is an element of $\mathscr{O}_C$ with valuation $\frac{1}{p-1}$ (II.9.18). Since $\widetilde{\delta}(\chi_{t_i}) = d\log(t_i)$ for every $1 \leq i \leq d$, we have

$$\Psi_M(\phi) = \sum_{i=1}^{d} \xi^{-1}\phi_i \otimes d\log(t_i). \tag{II.13.10.5}$$

Let $\varphi$ be an $\alpha$-quasi-small $\widehat{R_1}$-representation of $\Delta_\infty$ over $M$. Since $\Delta_\infty$ acts trivially on $\widehat{R_1}$, $\varphi$ is a homomorphism

$$\varphi\colon \Delta_\infty \to \mathrm{Aut}_{\widehat{R_1}}(M) \tag{II.13.10.6}$$

whose image is contained in the subgroup $\mathrm{id} + p^\alpha\mathrm{End}_{\widehat{R_1}}(M)$ of $\mathrm{Aut}_{\widehat{R_1}}(M)$. Since $\Delta_\infty$ is abelian, we can define the homomorphism (II.13.9)

$$\log(\varphi)\colon \Delta_\infty \to p^\alpha\mathrm{End}_{\widehat{R_1}}(M). \tag{II.13.10.7}$$

Moreover, it follows from (II.13.10.4) and (II.13.10.5) that $\Psi_M(\log(\varphi)) \wedge \Psi_M(\log(\varphi)) = 0$; in other words, $\Psi_M(\log(\varphi))$ is a $\beta$-quasi-small Higgs $\widehat{R_1}$-field on $M$ with coefficients in $\xi^{-1}\widetilde{\Omega}^1_{R/\mathscr{O}_K}$. We thus obtain a functor

$$\mathbf{Rep}^{\alpha\text{-qsf}}_{\widehat{R_1}}(\Delta_\infty) \to \mathbf{HM}^{\beta\text{-qsf}}(\widehat{R_1}, \xi^{-1}\widetilde{\Omega}^1_{R/\mathscr{O}_K}), \quad (M, \varphi) \mapsto (M, \Psi_M(\log(\varphi))). \tag{II.13.10.8}$$

Let $\theta$ be a $\beta$-quasi-small Higgs $\widehat{R_1}$-field over $M$ with coefficients in $\xi^{-1}\widetilde{\Omega}^1_{R/\mathscr{O}_K}$. In view of (II.13.10.3) and (II.13.10.4), since $\theta \wedge \theta = 0$, the image of the homomorphism $\Psi_M^{-1}(\theta)\colon \Delta_\infty \to p^\alpha \mathrm{End}_{\widehat{R_1}}(M)$ consists of endomorphisms of $M$ that commute pairwise. We can therefore define the homomorphism (II.13.9)

$$\exp(\Psi_M^{-1}(\theta))\colon \Delta_\infty \to \mathrm{Aut}_{\widehat{R_1}}(M), \tag{II.13.10.9}$$

which is clearly an $\alpha$-quasi-small $\widehat{R_1}$-representation of $\Delta_\infty$ over $M$. We thus define a functor

(II.13.10.10)
$$\mathbf{HM}^{\beta\text{-qsf}}(\widehat{R_1}, \xi^{-1}\widetilde{\Omega}^1_{R/\mathscr{O}_K}) \to \mathbf{Rep}^{\alpha\text{-qsf}}_{\widehat{R_1}}(\Delta_\infty), \quad (M,\theta) \mapsto (M, \exp(\Psi_M^{-1}(\theta))).$$

**Proposition II.13.11** (Faltings, [**27**]). *For all rational numbers $\alpha > \frac{1}{p-1}$ and $\beta = \alpha - \frac{1}{p-1}$, the functors* (II.13.10.8) *and* (II.13.10.10) *are equivalences of categories quasi-inverse to each other. They induce equivalences of categories*

$$\mathbf{Rep}^{\alpha\text{-small}}_{\widehat{R_1}}(\Delta_\infty) \overset{\sim}{\to} \mathbf{HM}^{\beta\text{-small}}(\widehat{R_1}, \xi^{-1}\widetilde{\Omega}^1_{R/\mathscr{O}_K}), \tag{II.13.11.1}$$

$$\mathbf{HM}^{\beta\text{-small}}(\widehat{R_1}, \xi^{-1}\widetilde{\Omega}^1_{R/\mathscr{O}_K}) \overset{\sim}{\to} \mathbf{Rep}^{\alpha\text{-small}}_{\widehat{R_1}}(\Delta_\infty) \tag{II.13.11.2}$$

*quasi-inverse to each other.*

The first assertion is an immediate consequence of the definition of the functors (II.13.10.8) and (II.13.10.10). The second assertion follows because $\widehat{R_1}$ is $\mathscr{O}_C$-flat (II.6.14).

**Remark II.13.12.** Under the assumptions of II.13.10, if $(M, \varphi)$ is an object of the category $\mathbf{Rep}^{\alpha\text{-qsf}}_{\widehat{R_1}}(\Delta_\infty)$ and $(M,\theta)$ is an object of the category $\mathbf{HM}^{\beta\text{-qsf}}(\widehat{R_1}, \xi^{-1}\widetilde{\Omega}^1_{R/\mathscr{O}_K})$ that correspond via the functors (II.13.10.8) and (II.13.10.10), then $\varphi$ and $\theta$ are linked by the following formulas:

$$\varphi = \exp\left(\sum_{i=1}^{d} \xi^{-1}\theta_i \otimes \chi_i\right), \tag{II.13.12.1}$$

$$\theta = \sum_{i=1}^{d} \xi^{-1}\theta_i \otimes d\log(t_i). \tag{II.13.12.2}$$

Since the $\chi_i$ factor through $\Delta_{p^\infty}$ (II.7.19), $\varphi$ factors through $\Delta_{p^\infty}$.

**II.13.13.** For every rational number $r \geq 0$, we denote by $\mathfrak{S}^{(r)}$ the sub-$\widehat{R_1}$-algebra of $\mathscr{S}^{(r)}$ (II.11.1.2) defined by

$$\mathfrak{S}^{(r)} = \mathrm{S}_{\widehat{R_1}}(p^r\xi^{-1}\widetilde{\Omega}^1_{R/\mathscr{O}_K} \otimes_R \widehat{R_1}), \tag{II.13.13.1}$$

and by $\widehat{\mathfrak{S}}^{(r)}$ its $p$-adic Hausdorff completion, which we endow with the $p$-adic topology. Note that $\widehat{\mathfrak{S}}^{(r)}$ is $\widehat{R_1}$-flat by virtue of ([**1**] 1.12.4) and therefore $\mathscr{O}_C$-flat (II.6.14). We set $\mathfrak{S} = \mathfrak{S}^{(0)}$ and $\widehat{\mathfrak{S}} = \widehat{\mathfrak{S}}^{(0)}$. For all rational numbers $r' \geq r \geq 0$, we have a canonical injective homomorphism $\mathrm{a}^{r,r'}\colon \mathfrak{S}^{(r')} \to \mathfrak{S}^{(r)}$. One immediately verifies that the induced homomorphism $\widehat{\mathrm{a}}^{r,r'}\colon \widehat{\mathfrak{S}}^{(r')} \to \widehat{\mathfrak{S}}^{(r)}$ is injective. We have a canonical $\mathfrak{S}^{(r)}$-isomorphism

$$\Omega^1_{\mathfrak{S}^{(r)}/\widehat{R_1}} \overset{\sim}{\to} \xi^{-1}\widetilde{\Omega}^1_{R/\mathscr{O}_K} \otimes_R \mathfrak{S}^{(r)}. \tag{II.13.13.2}$$

We denote by

$$d_{\mathfrak{S}^{(r)}}\colon \mathfrak{S}^{(r)} \to \xi^{-1}\widetilde{\Omega}^1_{R/\mathscr{O}_K} \otimes_R \mathfrak{S}^{(r)} \tag{II.13.13.3}$$

the universal $\widehat{R_1}$-derivation of $\mathfrak{S}^{(r)}$, which is also a Higgs $\widehat{R_1}$-field with coefficients in $\xi^{-1}\widetilde{\Omega}^1_{R/\mathscr{O}_K}$ because $\xi^{-1}\widetilde{\Omega}^1_{R/\mathscr{O}_K}\otimes_R\widehat{R_1}\subset d_{\mathfrak{S}^{(r)}}(\mathfrak{S}^{(r)})$. We denote by

$$d_{\widehat{\mathfrak{S}}^{(r)}}\colon \widehat{\mathfrak{S}}^{(r)}\to \xi^{-1}\widetilde{\Omega}^1_{R/\mathscr{O}_K}\otimes_R\widehat{\mathfrak{S}}^{(r)} \tag{II.13.13.4}$$

its extension to the completions. For all rational numbers $r'\geq r\geq 0$, we have

$$p^{r'-r}(\mathrm{id}\times \mathsf{a}^{r,r'})\circ d_{\mathfrak{S}^{(r')}} = d_{\mathfrak{S}^{(r)}}\circ \mathsf{a}^{r,r'}. \tag{II.13.13.5}$$

We denote by $(\widetilde{X}_0,\mathscr{M}_{\widetilde{X}_0})$ the smooth $(\mathscr{A}_2(\overline{S}),\mathscr{M}_{\mathscr{A}_2(\overline{S})})$-deformation of $(\check{\overline{X}},\mathscr{M}_{\check{\overline{X}}})$ defined by the chart $(P,\gamma)$ (II.10.13.1), by $\mathscr{L}_0$ the associated Higgs–Tate torsor (II.10.3), by $\psi_0\in\mathscr{L}_0(\widehat{Y})$ the section defined by the same chart (II.10.13.4), and by $\varphi_0=\varphi_{\psi_0}$ the action of $\Delta$ on $\mathscr{S}$ induced by $\psi_0$ (II.10.6.4). By II.10.17, $\varphi_0$ preserves $\mathfrak{S}$ and the induced action of $\Delta$ on $\mathfrak{S}$ factors through $\Delta_{p^\infty}$. We also denote the resulting action of $\Delta_{p^\infty}$ on $\mathfrak{S}$ by $\varphi_0$.

Let $t_1,\dots,t_d\in P^{\mathrm{gp}}$ be such that their images in $(P^{\mathrm{gp}}/\mathbb{Z}\lambda)\otimes_{\mathbb{Z}}\mathbb{Z}_p$ form a $\mathbb{Z}_p$-basis, $(\chi_{t_i})_{1\leq i\leq d}$ their images in $\mathrm{Hom}_{\mathbb{Z}}(\Delta_{p^\infty},\mathbb{Z}_p(1))$ (II.7.19.5), and $(d\log(t_i))_{1\leq i\leq d}$ their images in $\widetilde{\Omega}^1_{R/\mathscr{O}_K}$ (II.7.18.5). For every $1\leq i\leq d$, we set $y_i=\xi^{-1}d\log(t_i)\in\xi^{-1}\widetilde{\Omega}^1_{R/\mathscr{O}_K}\subset\mathfrak{S}$, and denote by $\chi_i$ the composition

$$\Delta_{p^\infty}\xrightarrow{\chi_{t_i}}\mathbb{Z}_p(1)\xrightarrow{\log([\ ])}p^{\frac{1}{p-1}}\xi\mathscr{O}_C, \tag{II.13.13.6}$$

where the second arrow is induced by the isomorphism (II.9.18.1). By (II.10.17.2), for every $g\in\Delta_{p^\infty}$, we have

$$\varphi_0(g)=\exp\Big(-\sum_{i=1}^d\xi^{-1}\frac{\partial}{\partial y_i}\otimes\chi_i(g)\Big). \tag{II.13.13.7}$$

It follows that $\varphi_0$ is continuous for the $p$-adic topology on $\mathfrak{S}$ (cf. II.11.9). For every rational number $r\geq 0$, $\varphi_0$ preserves $\mathfrak{S}^{(r)}$, and the induced actions of $\Delta_{p^\infty}$ on $\mathfrak{S}^{(r)}$ and $\widehat{\mathfrak{S}}^{(r)}$ are continuous for the $p$-adic topologies (cf. II.11.6). Unless explicitly stated otherwise, we endow $\mathfrak{S}^{(r)}$ and $\widehat{\mathfrak{S}}^{(r)}$ with these actions. It immediately follows from (II.13.13.7) that $d_{\mathfrak{S}^{(r)}}$ and $d_{\widehat{\mathfrak{S}}^{(r)}}$ are $\Delta_{p^\infty}$-equivariant.

**II.13.14.** We keep the notation of II.13.13 and let moreover $\beta$, $r$ be two rational numbers such that $\beta>r+\frac{1}{p-1}$ and $r\geq 0$, and $(N,\theta)$ a $\beta$-quasi-small Higgs $\widehat{R_1}$-module with coefficients in $\xi^{-1}\widetilde{\Omega}^1_{R/\mathscr{O}_K}$ (II.13.4) such that $N$ is $\mathscr{O}_C$-flat. There is a unique way to write

$$\theta=\sum_{i=1}^d\theta_i\otimes y_i, \tag{II.13.14.1}$$

where the $\theta_i$ are endomorphisms of $N$ that belong to $p^\beta\mathrm{End}_{\widehat{R_1}}(N)$ and commute pairwise. For every $\underline{n}=(n_1,\dots,n_d)\in\mathbb{N}^d$, let $|\underline{n}|=\sum_{i=1}^d n_i$, $\underline{n}!=\prod_{i=1}^d n_i!$, $\theta^{\underline{n}}=\prod_{i=1}^d\theta_i^{n_i}\in\mathrm{End}_{\widehat{R_1}}(N)$, and $y^{\underline{n}}=\prod_{i=1}^d y_i^{n_i}\in\mathfrak{S}$. Note that $N\otimes_{\widehat{R_1}}\widehat{\mathfrak{S}}^{(r)}$ is complete and separated for the $p$-adic topology ([**1**] 1.10.2), and that it is $\mathscr{O}_C$-flat because $\widehat{\mathfrak{S}}^{(r)}$ is $\widehat{R_1}$-flat ([**1**] 1.12.4). Consequently, for every $z\in N\otimes_{\widehat{R_1}}\widehat{\mathfrak{S}}^{(r)}$, the series

$$\sum_{\underline{n}\in\mathbb{N}^d}\frac{1}{\underline{n}!}(\theta^{\underline{n}}\otimes y^{\underline{n}})(z) \tag{II.13.14.2}$$

converges in $N\otimes_{\widehat{R_1}}\widehat{\mathscr{C}}^{(r)}$, and defines an $\widehat{\mathscr{C}}^{(r)}$-linear endomorphism of $N\otimes_{\widehat{R_1}}\widehat{\mathscr{C}}^{(r)}$, which we denote by

$$\exp_r(\theta)\colon N\otimes_{\widehat{R_1}}\widehat{\mathscr{C}}^{(r)}\to N\otimes_{\widehat{R_1}}\widehat{\mathscr{C}}^{(r)}. \tag{II.13.14.3}$$

For every rational number $r'$ such that $0\leq r'\leq r$, the diagram

$$\begin{array}{ccc} N\otimes_{\widehat{R_1}}\widehat{\mathscr{C}}^{(r)} & \xrightarrow{\exp_r(\theta)} & N\otimes_{\widehat{R_1}}\widehat{\mathscr{C}}^{(r)} \\ {\scriptstyle \mathrm{id}\otimes \mathrm{a}^{r',r}}\downarrow & & \downarrow{\scriptstyle \mathrm{id}\otimes \mathrm{a}^{r',r}} \\ N\otimes_{\widehat{R_1}}\widehat{\mathscr{C}}^{(r')} & \xrightarrow{\exp_{r'}(\theta)} & N\otimes_{\widehat{R_1}}\widehat{\mathscr{C}}^{(r')} \end{array} \tag{II.13.14.4}$$

is commutative. We may therefore leave the index $r$ out of the notation $\exp_r(\theta)$ without risk of confusion.

**Proposition II.13.15.** *Let $\beta$, $r$ be two rational numbers such that $\beta > r+\frac{1}{p-1}$ and $r\geq 0$, $N$ an $\widehat{R_1}$-module of finite type that is $\mathscr{O}_C$-flat, $\theta$ a $\beta$-quasi-small Higgs $\widehat{R_1}$-field over $N$ with coefficients in $\xi^{-1}\widetilde{\Omega}^1_{R/\mathscr{O}_K}$, and $\varphi$ the quasi-small $\widehat{R_1}$-representation of $\Delta_\infty$ over $N$ associated with $\theta$ by the functor* (II.13.10.10). *Then the $\widehat{\mathscr{C}}^{(r)}$-module $N\otimes_{\widehat{R_1}}\widehat{\mathscr{C}}^{(r)}$ is complete and separated for the p-adic topology, and the endomorphism* (II.13.14.3)

$$\exp_r(\theta)\colon N\otimes_{\widehat{R_1}}\widehat{\mathscr{C}}^{(r)}\to N\otimes_{\widehat{R_1}}\widehat{\mathscr{C}}^{(r)} \tag{II.13.15.1}$$

*is a $\Delta_\infty$-equivariant isomorphism of modules with integrable p-adic $p^r$-connections with respect to the extension $\widehat{\mathscr{C}}^{(r)}/\widehat{R_1}$* (II.2.14), *where $\widehat{\mathscr{C}}^{(r)}$ is endowed with the action of $\Delta_\infty$ induced by $\varphi_0$* (II.13.13.7) *and with the p-adic $p^r$-connection $p^r d_{\widehat{\mathscr{C}}^{(r)}}$* (II.13.13.4), *the module $N$ on the left-hand side is endowed with the trivial action of $\Delta_\infty$ and the Higgs $\widehat{R_1}$-field $\theta$, and the module $N$ on the right-hand side is endowed with the action $\varphi$ of $\Delta_\infty$ and the zero Higgs $\widehat{R_1}$-field (cf.* II.2.17*). In particular, $\exp_r(\theta)$ is an isomorphism of Higgs $\widehat{R_1}$-modules with coefficients in $\xi^{-1}\widetilde{\Omega}^1_{R/\mathscr{O}_K}$* (II.2.16).

First note that $N\otimes_{\widehat{R_1}}\widehat{\mathscr{C}}^{(r)}$ is complete and separated for the $p$-adic topology by virtue of ([**1**] 1.10.2(ii)). Let $x_1,\dots,x_n$ be $\widehat{R_1}$-generators of $N$, and for every $1\leq\ell\leq d$, let $A_\ell=(m^\ell_{ij})_{1\leq i,j\leq n}$ be an $n\times n$ matrix with coefficients in $\widehat{R_1}$ such that for every $1\leq j\leq n$, we have

$$\theta_\ell(x_j)=p^\beta\sum_{i=1}^n m^\ell_{ij}x_i. \tag{II.13.15.2}$$

Note that the matrices $A_\ell$ do not commute pairwise in general (but they do commute if $N$ is free over $\widehat{R_1}$ with basis $x_1,\dots,x_n$). For every $\underline{n}=(n_1,\dots,n_d)\in\mathbb{N}^d$, we set

$$\underline{A}^{\underline{n}}=A_1^{n_1}\cdot A_2^{n_2}\cdots A_d^{n_d}\in \mathrm{Mat}_n(\widehat{R_1}). \tag{II.13.15.3}$$

The series

$$E=\sum_{\underline{n}\in\mathbb{N}^d}\frac{p^{\beta|\underline{n}|}}{\underline{n}!}\underline{A}^{\underline{n}}\otimes \underline{y}^{\underline{n}} \tag{II.13.15.4}$$

then defines an $n\times n$ matrix with coefficients in $\widehat{\mathscr{C}}^{(r)}$. For all $(a_1,\dots,a_n)\in(\widehat{\mathscr{C}}^{(r)})^n$, we have

$$\exp_r(\theta)(\sum_{i=1}^n x_i\otimes a_i)=\sum_{i=1}^n x_i\otimes b_i\in N\otimes_{\widehat{R_1}}\widehat{\mathscr{C}}^{(r)},$$

where $(b_1, \ldots, b_n) = E \cdot (a_1, \ldots, a_n)$.

Since the matrix $E - \mathrm{id}$ has coefficients in $p^{\beta-r}\widehat{\mathfrak{S}}^{(r)}$, the determinant of $E$ is invertible in $\widehat{\mathfrak{S}}^{(r)}$. If $E^{-1} = (f_{ij})_{1 \leq i,j \leq n}$ is the inverse matrix of $E$ in $\mathrm{GL}_n(\widehat{\mathfrak{S}}^{(r)})$, then for every $1 \leq i \leq d$, we have

$$(\text{II.13.15.5}) \qquad \exp_r(\theta)(\sum_{j=1}^{d} x_j \otimes f_{ji}) = x_i \otimes 1.$$

Hence $\exp_r(\theta)$ is surjective.

Let $x \in N \otimes_{\widehat{R_1}} \widehat{\mathfrak{S}}^{(r)}$ be such that $\exp_r(\theta)(x) = 0$. Since $E \equiv \mathrm{id} \mod (p^{\beta-r})$, there exists $y \in N \otimes_{\widehat{R_1}} \widehat{\mathfrak{S}}^{(r)}$ such that $x = p^{\beta-r}y$. Since $N \otimes_{\widehat{R_1}} \widehat{\mathfrak{S}}^{(r)}$ is $\mathscr{O}_C$-flat (II.13.14), we have $\exp_r(\theta)(y) = 0$. It follows that $x \in \cap_{n \geq 0} p^{n(\beta-r)}(N \otimes_{\widehat{R_1}} \widehat{\mathfrak{S}}^{(r)})$ and consequently that $x = 0$ because $N \otimes_{\widehat{R_1}} \widehat{\mathfrak{S}}^{(r)}$ is separated for the $p$-adic topology ([1] 1.10.2). Hence $\exp_r(\theta)$ is bijective.

It immediately follows from the definition (II.13.14.2) that the diagram

$$(\text{II.13.15.6})$$

$$\begin{array}{ccc} N \otimes_{\widehat{R_1}} \widehat{\mathfrak{S}}^{(r)} & \xrightarrow{\ \exp_r(\theta)\ } & N \otimes_{\widehat{R_1}} \widehat{\mathfrak{S}}^{(r)} \\ {\scriptstyle \nabla^{(r)}}\big\downarrow & & \big\downarrow{\scriptstyle p^r \mathrm{id} \otimes d_{\widehat{\mathfrak{S}}^{(r)}}} \\ \xi^{-1}\widetilde{\Omega}^1_{R/\mathscr{O}_K} \otimes_R N \otimes_{\widehat{R_1}} \widehat{\mathfrak{S}}^{(r)} & \xrightarrow{\exp_r(\theta) \otimes \mathrm{id}} & \xi^{-1}\widetilde{\Omega}^1_{R/\mathscr{O}_K} \otimes_R N \otimes_{\widehat{R_1}} \widehat{\mathfrak{S}}^{(r)} \end{array}$$

where $\nabla^{(r)} = \theta \otimes \mathrm{id} + p^r \mathrm{id} \otimes d_{\widehat{\mathfrak{S}}^{(r)}}$ (II.2.17.2), is commutative modulo $p^n$ for every $n \geq 1$. It is therefore commutative; in other words, $\exp_r(\theta)$ is a morphism of modules with $p$-adic $p^r$-connections with respect to the extension $\widehat{\mathfrak{S}}^{(r)}/\widehat{R_1}$.

By II.13.12 and (II.13.13.7), for every $g \in \Delta_\infty$, we have

$$(\text{II.13.15.7}) \qquad k\varphi(g) = \exp(\sum_{i=1}^{d} \xi^{-1}\theta_i \otimes \chi_i(g)),$$

$$(\text{II.13.15.8}) \qquad \varphi_0(g) = \exp(-\sum_{i=1}^{d} \xi^{-1}\frac{\partial}{\partial y_i} \otimes \chi_i(g)).$$

On the other hand, if we view $\exp(-\xi^{-1}\frac{\partial}{\partial y_i} \otimes \chi_i(g))$ as an $\widehat{R_1}$-automorphism of $\widehat{\mathfrak{S}}^{(r)}$, we have

$$(\mathrm{id}_N \otimes \exp(-\xi^{-1}\frac{\partial}{\partial y_i} \otimes \chi_i(g))) \circ \exp_r(\theta)$$

$$= (\exp(-\xi^{-1}\theta_i \otimes \chi_i(g)) \otimes \mathrm{id}_{\widehat{\mathfrak{S}}^{(r)}}) \circ \exp_r(\theta) \circ (\mathrm{id}_N \otimes \exp(-\xi^{-1}\frac{\partial}{\partial y_i} \otimes \chi_i(g))).$$

Indeed, it suffices to verify this equation modulo $p^n$ for every $n \geq 1$, which follows formally from the definition (II.13.14.2) (we use first that $\exp(-\xi^{-1}\frac{\partial}{\partial y_i} \otimes \chi_i(g))$ is a homomorphism of the $\widehat{R_1}$-algebra $\mathfrak{S}^{(r)}$). Since the endomorphisms $\theta_i$ commute pairwise, it follows that

$$(\text{II.13.15.9}) \qquad (\exp(\sum_{i=1}^{d} \xi^{-1}\theta_i \otimes \chi_i(g)) \otimes \exp(-\sum_{i=1}^{d} \xi^{-1}\frac{\partial}{\partial y_i} \otimes \chi_i(g))) \circ \exp_r(\theta)$$

$$= \exp_r(\theta) \circ (\mathrm{id}_N \otimes \exp(-\sum_{i=1}^{d} \xi^{-1}\frac{\partial}{\partial y_i} \otimes \chi_i(g))).$$

Consequently, the morphism $\exp_r(\theta)$ is $\Delta_\infty$-equivariant.

**Corollary II.13.16.** *Under the assumptions of* II.13.15, *we have a functorial and* $\Delta$-*equivariant* $\widehat{\mathscr{C}}^{(r)}$-*isomorphism of Higgs* $\widehat{\overline{R}}$-*modules with coefficients in* $\xi^{-1}\widetilde{\Omega}^1_{R/\mathscr{O}_K}$

$$N \otimes_{\widehat{R_1}} \widehat{\mathscr{C}}^{(r)} \xrightarrow{\sim} N \otimes_{\widehat{R_1}} \widehat{\mathscr{C}}^{(r)}, \tag{II.13.16.1}$$

*where* $\widehat{\mathscr{C}}^{(r)}$ *is endowed with the canonical action of* $\Delta$ *and the Higgs* $\widehat{\overline{R}}$-*field* $p^r d_{\widehat{\mathscr{C}}^{(r)}}$ (II.12.1.8), *the module* $N$ *on the left-hand side is endowed with the trivial action of* $\Delta_\infty$ *and the Higgs* $\widehat{R_1}$-*field* $\theta$, *and the module* $N$ *on the right-hand side is endowed with the action* $\varphi$ *of* $\Delta_\infty$ *and the zero Higgs* $\widehat{R_1}$-*field. If, moreover, the deformation* $(\widetilde{X}, \mathscr{M}_{\widetilde{X}})$ *is defined by the chart* $(P, \gamma)$ (II.10.13.1), *then the isomorphism is canonical.*

Indeed, by II.13.15, $\exp_r(\theta)$ induces a functorial and $\Delta$-equivariant $\widehat{\mathscr{F}}^{(r)}$-isomorphism of Higgs $\widehat{\overline{R}}$-modules with coefficients in $\xi^{-1}\widetilde{\Omega}^1_{R/\mathscr{O}_K}$ (II.11.1)

$$N \otimes_{\widehat{R_1}} \widehat{\mathscr{F}}^{(r)} \xrightarrow{\sim} N \otimes_{\widehat{R_1}} \widehat{\mathscr{F}}^{(r)}, \tag{II.13.16.2}$$

where $\widehat{\mathscr{F}}^{(r)}$ is endowed with the action of $\Delta$ induced by $\varphi_0$ (II.11.5) and the Higgs $\widehat{\overline{R}}$-field $p^r d_{\widehat{\mathscr{F}}^{(r)}}$ (II.11.1.6), the module $N$ on the left-hand side is endowed with the trivial action of $\Delta_\infty$ and the Higgs $\widehat{R_1}$-field $\theta$, and the module $N$ on the right-hand side is endowed with the action $\varphi$ of $\Delta_\infty$ and the zero Higgs $\widehat{R_1}$-field. The corollary follows in view of II.10.10 and II.12.2.

**Remark II.13.17.** Under the assumptions of II.13.15, by virtue of II.2.15, $\exp_r(\theta)$ induces a functorial and $\Delta$-equivariant $\widehat{\mathscr{C}}^{(r)}$-isomorphism of modules with integrable $p$-adic $p^r$-connections with respect to the extension $\widehat{\mathscr{C}}^{(r)}/\widehat{\overline{R}}$,

$$N \widehat{\otimes}_{\widehat{R_1}} \widehat{\mathscr{C}}^{(r)} \xrightarrow{\sim} N \widehat{\otimes}_{\widehat{R_1}} \widehat{\mathscr{C}}^{(r)}, \tag{II.13.17.1}$$

where $\widehat{\mathscr{C}}^{(r)}$ is endowed with the canonical action of $\Delta$ and the $p$-adic $p^r$-connection $p^r d_{\widehat{\mathscr{C}}^{(r)}}$ (II.12.1.8), the module $N$ on the left-hand side is endowed with the trivial action of $\Delta_\infty$ and the Higgs $\widehat{R_1}$-field $\theta$, and the module $N$ on the right-hand side is endowed with the action $\varphi$ of $\Delta_\infty$ and the zero Higgs $\widehat{R_1}$-field (cf. II.2.17).

**Corollary II.13.18.** *Under the assumptions of* II.13.15, *we have a functorial and* $\Delta$-*equivariant* $\mathscr{C}^\dagger$-*isomorphism of Higgs* $\widehat{\overline{R}}$-*modules with coefficients in* $\xi^{-1}\widetilde{\Omega}^1_{R/\mathscr{O}_K}$

$$N \otimes_{\widehat{R_1}} \mathscr{C}^\dagger \xrightarrow{\sim} N \otimes_{\widehat{R_1}} \mathscr{C}^\dagger, \tag{II.13.18.1}$$

*where* $\mathscr{C}^\dagger$ *is endowed with the canonical action of* $\Delta$ *and the Higgs* $\widehat{\overline{R}}$-*field Higgs* $d_{\mathscr{C}^\dagger}$ (II.12.1.12), *the module* $N$ *on the left-hand side is endowed with the trivial action of* $\Delta_\infty$ *and the* $\widehat{R_1}$-*field* $\theta$, *and the module* $N$ *on the right-hand side is endowed with the action* $\varphi$ *of* $\Delta_\infty$ *and the zero Higgs* $\widehat{R_1}$-*field* (II.2.8.7). *If, moreover, the deformation* $(\widetilde{X}, \mathscr{M}_{\widetilde{X}})$ *is defined by the chart* $(P, \gamma)$ (II.10.13.1), *then the isomorphism is canonical.*

This follows from II.13.16.

**Corollary II.13.19.** *Let* $N$ *be a projective* $\widehat{R_1}$-*module of finite type,* $\theta$ *a quasi-small Higgs* $\widehat{R_1}$-*field over* $N$ *with coefficients in* $\xi^{-1}\widetilde{\Omega}^1_{R/\mathscr{O}_K}$, *and* $\varphi$ *the quasi-small* $\widehat{R_1}$-*representation of* $\Delta_\infty$ *over* $N$ *associated with* $\theta$ *by the functor* (II.13.10.10). *Then we have a functorial isomorphism of Higgs* $\widehat{R_1}$-*modules with coefficients in* $\xi^{-1}\widetilde{\Omega}^1_{R/\mathscr{O}_K}$

$$\mathbb{H}((N, \varphi) \otimes_{\widehat{R_1}} \widehat{\overline{R}}) \xrightarrow{\sim} (N, \theta), \tag{II.13.19.1}$$

*where* $\mathbb{H}$ *is the functor* (II.12.8.2), *and a functorial isomorphism of* $\widehat{\overline{R}}$*-representations of* $\Delta$

(II.13.19.2) $$\mathbb{V}(N,\theta) \xrightarrow{\sim} (N,\varphi) \otimes_{\widehat{R_1}} \widehat{\overline{R}},$$

*where* $\mathbb{V}$ *is the functor* (II.12.9.2). *If, moreover, the deformation* $(\widetilde{X}, \mathscr{M}_{\widetilde{X}})$ *is defined by the chart* $(P,\gamma)$ (II.10.13.1), *then the isomorphisms are canonical.*

This follows from II.13.18 and the fact that $\ker(d_{\mathscr{C}^\dagger}) = \widehat{\overline{R}}$ and $(\mathscr{C}^\dagger)^\Delta = \widehat{R_1}$ (II.12.6).

**Corollary II.13.20.** *Every small Higgs* $\widehat{R_1}$*-module with coefficients in* $\xi^{-1}\widetilde{\Omega}^1_{R/\mathscr{O}_K}$ *is solvable, and its image by* $\mathbb{V}$ *is a small* $\widehat{\overline{R}}$*-representation of* $\Delta$.

This follows from II.13.18 and II.13.19.

This statement will be strengthened in II.14.7.

**II.13.21.** Let $(N,\theta)$ be a small Higgs $\widehat{R_1}[\frac{1}{p}]$-module with coefficients in $\xi^{-1}\widetilde{\Omega}^1_{R/\mathscr{O}_K}$ (II.13.5), $\beta$ a rational number $> \frac{1}{p-1}$, $N^\circ$ a sub-$\widehat{R_1}$-module of finite type of $N$ that generates it over $\widehat{R_1}[\frac{1}{p}]$, such that we have

(II.13.21.1) $$\theta(N^\circ) \subset p^\beta \xi^{-1} N^\circ \otimes_R \widetilde{\Omega}^1_{R/\mathscr{O}_K},$$

and $\theta^\circ$ the Higgs $\widehat{R_1}$-field on $N^\circ$ with coefficients in $\xi^{-1}\widetilde{\Omega}^1_{R/\mathscr{O}_K}$ induced by $\theta$. Then $(N^\circ, \theta^\circ)$ is a quasi-small Higgs $\widehat{R_1}$-module (II.13.4). We denote by $\varphi^\circ$ the quasi-small $\widehat{R_1}$-representation of $\Delta_\infty$ over $N^\circ$ associated with $\theta^\circ$ by the functor (II.13.10.10) and by $\varphi$ the small $\widehat{R_1}[\frac{1}{p}]$-representation of $\Delta_\infty$ over $N$ deduced from $\varphi^\circ$. Let us show that $\varphi$ does not depend on the choice of $N^\circ$, and that the correspondence $(N,\theta) \mapsto (N,\varphi)$ defines a functor (II.13.2)

(II.13.21.2) $$\mathbf{HM}^{\rm small}(\widehat{R_1}[\frac{1}{p}], \xi^{-1}\widetilde{\Omega}^1_{R/\mathscr{O}_K}) \to \mathbf{Rep}^{\rm small}_{\widehat{R_1}[\frac{1}{p}]}(\Delta_\infty).$$

Indeed, let $u\colon (N,\theta) \to (N_1,\theta_1)$ be a morphism of small Higgs $\widehat{R_1}[\frac{1}{p}]$-modules with coefficients in $\xi^{-1}\widetilde{\Omega}^1_{R/\mathscr{O}_K}$ and $N_1^\circ$ a sub-$\widehat{R_1}$-module of finite type of $N_1$ that generates it over $\widehat{R_1}[\frac{1}{p}]$, such that we have $u(N^\circ) \subset N_1^\circ$ and

(II.13.21.3) $$\theta_1(N_1^\circ) \subset p^\beta \xi^{-1} N_1^\circ \otimes_R \widetilde{\Omega}^1_{R/\mathscr{O}_K}.$$

We denote by $\theta_1^\circ$ the Higgs $\widehat{R_1}$-field on $N_1^\circ$ with coefficients in $\xi^{-1}\widetilde{\Omega}^1_{R/\mathscr{O}_K}$ induced by $\theta_1$, by $\varphi_1^\circ$ the $\widehat{R_1}$-representation of $\Delta_\infty$ over $N_1^\circ$ associated with $\theta_1^\circ$ by the functor (II.13.10.10), and by $\varphi_1$ the $\widehat{R_1}[\frac{1}{p}]$-representation of $\Delta_\infty$ over $N_1$ induced by $\varphi_1^\circ$. Since the morphism $(N^\circ,\varphi^\circ) \to (N_1^\circ,\varphi_1^\circ)$ induced by $u$ is clearly $\Delta_\infty$-equivariant, the same holds for $u\colon (N,\varphi) \to (N_1,\varphi_1)$.

Let us show that $\varphi$ does not depend on the choice of $N^\circ$. Let $\gamma$ be a rational number $> \frac{1}{p-1}$ and $N^\star$ a sub-$\widehat{R_1}$-module of finite type of $N$ that generates it over $\widehat{R_1}[\frac{1}{p}]$, such that we have

(II.13.21.4) $$\theta(N^\star) \subset p^\gamma \xi^{-1} N^\star \otimes_R \widetilde{\Omega}^1_{R/\mathscr{O}_K}.$$

Replacing $\beta$ and $\gamma$ by the least of the two and $N^\star$ by $N^\circ + N^\star$, we may assume that $\beta = \gamma$ and $N^\circ \subset N^\star$. Applying the above to the identity of $N$, we deduce that $\varphi$ does not depend on the choice of $N^\circ$.

Let us show that for every morphism $v\colon (N,\theta) \to (N',\theta')$ of small Higgs $\widehat{R_1}[\frac{1}{p}]$-modules with coefficients in $\xi^{-1}\widetilde{\Omega}^1_{R/\mathscr{O}_K}$, if $\varphi'$ is the small $\widehat{R_1}[\frac{1}{p}]$-representation of $\Delta_\infty$ over $N'$ associated with $\theta'$, then $v\colon (N,\varphi) \to (N',\varphi')$ is $\Delta_\infty$-equivariant. Let $\beta'$ be a rational number $> \frac{1}{p-1}$ and $N'^\circ$ a sub-$\widehat{R_1}$-module of finite type of $N'$ that generates it over $\widehat{R_1}[\frac{1}{p}]$, such that we have

$$\theta'(N'^\circ) \subset p^{\beta'}\xi^{-1}N'^\circ \otimes_R \widetilde{\Omega}^1_{R/\mathscr{O}_K}. \tag{II.13.21.5}$$

Replacing $\beta$ and $\beta'$ by the least of the two and $N'^\circ$ by $v(N^\circ)+N'^\circ$, we may assume that $\beta=\beta'$ and $v(N^\circ)\subset N'^\circ$. We then conclude as above that $v\colon (N,\varphi)\to(N',\varphi')$ is $\Delta_\infty$-equivariant.

**II.13.22.** Let $(N,\varphi)$ be a small $\widehat{R_1}[\frac{1}{p}]$-representation of $\Delta_\infty$ (II.13.2), $\alpha$ a rational number $> \frac{2}{p-1}$, and $N^\circ$ a sub-$\widehat{R_1}$-module of finite type of $N$ that is stable under $\Delta_\infty$ and generated by a finite number of elements that are $\Delta_\infty$-invariant modulo $p^\alpha N^\circ$, and that generates $N$ over $\widehat{R_1}[\frac{1}{p}]$. By II.13.3(ii), the $\widehat{R_1}$-representation $\varphi^\circ$ of $\Delta_\infty$ over $N^\circ$ induced by $\varphi$ is quasi-small. Denote by $\theta^\circ$ the quasi-small Higgs $\widehat{R_1}$-field on $N^\circ$ with coefficients in $\xi^{-1}\widetilde{\Omega}^1_{R/\mathscr{O}_K}$ associated with $\varphi^\circ$ by the functor (II.13.10.8) and by $\theta$ the small Higgs $\widehat{R_1}[\frac{1}{p}]$-field on $N$ induced by $\theta^\circ$. Proceeding as in II.13.21, one shows that $(N,\theta)$ does not depend on the choice of $N^\circ$ and that the correspondence $(N,\varphi)\mapsto(N,\theta)$ defines a functor (II.13.5)

$$\mathbf{Rep}^{\rm small}_{\widehat{R_1}[\frac{1}{p}]}(\Delta_\infty) \to \mathbf{HM}^{\rm small}(\widehat{R_1}[\frac{1}{p}], \xi^{-1}\widetilde{\Omega}^1_{R/\mathscr{O}_K}). \tag{II.13.22.1}$$

It immediately follows from II.13.10 that the functors (II.13.21.2) and (II.13.22.1) are quasi-inverse to each other.

**Proposition II.13.23.** *Let $(N,\theta)$ be a small Higgs $\widehat{R_1}[\frac{1}{p}]$-module with coefficients in $\xi^{-1}\widetilde{\Omega}^1_{R/\mathscr{O}_K}$* (II.13.5)*, and $\varphi$ the small $\widehat{R_1}[\frac{1}{p}]$-representation of $\Delta_\infty$ over $N$ associated with $\theta$ by the functor* (II.13.21.2)*. Then:*

(i) *We have a functorial $\Delta$-equivariant $\mathscr{C}^\dagger$-isomorphism of Higgs $\widehat{\overline{R}}$-modules with coefficients in $\xi^{-1}\widetilde{\Omega}^1_{R/\mathscr{O}_K}$,*

$$N\otimes_{\widehat{R_1}}\mathscr{C}^\dagger \xrightarrow{\sim} N\otimes_{\widehat{R_1}}\mathscr{C}^\dagger, \tag{II.13.23.1}$$

*where $\mathscr{C}^\dagger$ is endowed with the canonical action of $\Delta$ and the Higgs $\widehat{\overline{R}}$-field $d_{\mathscr{C}^\dagger}$* (II.12.1.12)*, the module $N$ on the left-hand side is endowed with the trivial action of $\Delta_\infty$ and the Higgs $\widehat{R_1}$-field $\theta$, and the module $N$ on the right-hand side is endowed with the action $\varphi$ of $\Delta_\infty$ and the zero Higgs $\widehat{R_1}$-field. If, moreover, the deformation $(\widetilde{X},\mathscr{M}_{\widetilde{X}})$ is defined by the chart $(P,\gamma)$* (II.10.13.1)*, then the isomorphism is canonical.*

(ii) *The Higgs $\widehat{R_1}[\frac{1}{p}]$-module $(N,\theta)$ is solvable and we have a functorial $\Delta$-equivariant $\widehat{\overline{R}}[\frac{1}{p}]$-isomorphism*

$$\mathbb{V}(N) \xrightarrow{\sim} (N,\varphi)\otimes_{\widehat{R_1}}\widehat{\overline{R}}, \tag{II.13.23.2}$$

*where $\mathbb{V}$ is the functor* (II.12.9.2)*. If, moreover, the deformation $(\widetilde{X},\mathscr{M}_{\widetilde{X}})$ is defined by the chart $(P,\gamma)$* (II.10.13.1)*, then the isomorphism is canonical.*

(iii) *The* $\widehat{\overline{R}}[\frac{1}{p}]$*-representation* $\mathbb{V}(N)$ *of* $\Delta$ *is small and Dolbeault, and we have a functorial isomorphism of Higgs* $\widehat{R_1}[\frac{1}{p}]$*-modules*

$$\mathbb{H}(\mathbb{V}(N)) \xrightarrow{\sim} (N, \theta), \tag{II.13.23.3}$$

*where* $\mathbb{H}$ *is the functor* (II.12.8.2). *If, moreover, the deformation* $(\widetilde{X}, \mathscr{M}_{\widetilde{X}})$ *is defined by the chart* $(P, \gamma)$ (II.10.13.1), *then the isomorphism is canonical.*

The isomorphism (II.13.23.1) follows from II.13.18; the functoriality is proved as in II.13.21. The other assertions follow in view of the fact that $\ker(d_{\mathscr{C}^\dagger}) = \widehat{\overline{R}}$ and $(\mathscr{C}^\dagger)^\Delta = \widehat{R_1}$ (II.12.6).

**Proposition II.13.24** (IV.5.3.10)**.** *Let* $N$ *be a projective* $\widehat{R_1}[\frac{1}{p}]$*-module of finite type,* $\theta$ *a Higgs* $\widehat{R_1}[\frac{1}{p}]$*-field over* $N$ *with coefficients in* $\xi^{-1}\widetilde{\Omega}^1_{R/\mathscr{O}_K}$, $M$ *an* $\widehat{\overline{R}}[\frac{1}{p}]$*-module,* $r$ *a rational number* $> 0$*, and*

$$N \otimes_{\widehat{R_1}} \widehat{\mathscr{C}^{(r)}} \xrightarrow{\sim} M \otimes_{\widehat{\overline{R}}} \widehat{\mathscr{C}^{(r)}} \tag{II.13.24.1}$$

*a* $\widehat{\mathscr{C}^{(r)}}$*-linear isomorphism of Higgs* $\widehat{\overline{R}}$*-modules with coefficients in* $\xi^{-1}\widetilde{\Omega}^1_{R/\mathscr{O}_K}$*, where* $\widehat{\mathscr{C}^{(r)}}$ *is endowed with the Higgs field* $p^r d_{\widehat{\mathscr{C}^{(r)}}}$ (II.12.1.8) *and* $M$ *is endowed with the zero Higgs field. Then* $(N, \theta)$ *is a small Higgs* $\widehat{R_1}[\frac{1}{p}]$*-module with coefficients in* $\xi^{-1}\widetilde{\Omega}^1_{R/\mathscr{O}_K}$ (II.13.5).

Indeed, in view of II.10.10 and II.12.2, the isomorphism (II.13.24.1) induces an $\widehat{\mathscr{S}^{(r)}}$-linear isomorphism of Higgs $\widehat{\overline{R}}$-modules with coefficients in $\xi^{-1}\widetilde{\Omega}^1_{R/\mathscr{O}_K}$,

$$N \otimes_{\widehat{R_1}} \widehat{\mathscr{S}^{(r)}} \xrightarrow{\sim} M \otimes_{\widehat{\overline{R}}} \widehat{\mathscr{S}^{(r)}}, \tag{II.13.24.2}$$

where $\widehat{\mathscr{S}^{(r)}}$ is endowed with the Higgs field $p^r d_{\widehat{\mathscr{S}^{(r)}}}$ (II.11.1.6) and $M$ is endowed with the zero Higgs field. This induces an $\widehat{\overline{R}}$-linear isomorphism

$$N \otimes_{\widehat{R_1}} \widehat{\overline{R}} \xrightarrow{\sim} M. \tag{II.13.24.3}$$

Let $t_1, \dots, t_d \in P^{\rm gp}$ be such that their images in $(P^{\rm gp}/\mathbb{Z}\lambda) \otimes_{\mathbb{Z}} \mathbb{Z}_p$ form a $\mathbb{Z}_p$-basis, so that $(d\log(t_i))_{1 \le i \le d}$ is an $R$-basis of $\widetilde{\Omega}^1_{R/\mathscr{O}_K}$ (II.7.12.1). For every $1 \le i \le d$, we set $y_i = \xi^{-1} d\log(t_i) \in \xi^{-1}\widetilde{\Omega}^1_{R/\mathscr{O}_K} \subset \mathscr{S}$. For every $\underline{n} = (n_1, \dots, n_d) \in \mathbb{N}^d$, we set $|\underline{n}| = \sum_{i=1}^d n_i$ and $y^{\underline{n}} = \prod_{i=1}^d y_i^{n_i} \in \mathscr{S}$. If we endow $\widehat{\overline{R}}[\frac{1}{p}]$ with the $p$-adic topology (II.2.2), the $\widehat{\overline{R}}[\frac{1}{p}]$-algebra $\widehat{\mathscr{S}^{(r)}}[\frac{1}{p}]$ is canonically identified with

$$\{\sum_{\underline{n} \in \mathbb{N}^d} a_{\underline{n}} y^{\underline{n}} \in \widehat{\overline{R}}[\frac{1}{p}][[y_1, \dots, y_d]] \mid \lim_{|\underline{n}| \to +\infty} p^{-r|\underline{n}|} a_{\underline{n}} = 0\}. \tag{II.13.24.4}$$

Note that under the conventions of (II.11.1.6), we have $d_{\widehat{\mathscr{S}^{(r)}}}(p^r y_i) = 1 \otimes y_i$ for every $1 \le i \le d$.

The $\widehat{\overline{R}}[\frac{1}{p}][[y_1, \dots, y_d]]$-module $N \otimes_{\widehat{R_1}[\frac{1}{p}]} \widehat{\overline{R}}[\frac{1}{p}][[y_1, \dots, y_d]]$ is complete and separated for the $(y_1, \dots, y_d)$-adic topology. Indeed, it is complete by virtue of (**[11]** Chap. III § 2.12 Cor. 1 to Prop. 16) and is separated because it is a direct summand of a free module of finite type. Consequently, every formal series $\sum_{\underline{n} \in \mathbb{N}^d} x_{\underline{n}} \otimes y^{\underline{n}}$, where $x_{\underline{n}} \in N \otimes_{\widehat{R_1}} \widehat{\overline{R}}$, converges in $N \otimes_{\widehat{R_1}[\frac{1}{p}]} \widehat{\overline{R}}[\frac{1}{p}][[y_1, \dots, y_d]]$. Conversely, every element $x$ of

$$N \otimes_{\widehat{R_1}[\frac{1}{p}]} \widehat{\overline{R}}[\frac{1}{p}][[y_1, \dots, y_d]]$$

can be written in a unique way as

$$x = \sum_{\underline{n} \in \mathbb{N}^d} x_{\underline{n}} \otimes \underline{y}^{\underline{n}}, \tag{II.13.24.5}$$

where $x_{\underline{n}} \in N \otimes_{\widehat{R_1}} \widehat{\overline{R}}$. If we endow the $\widehat{\overline{R}}[\frac{1}{p}]$-module $N \otimes_{\widehat{R_1}} \widehat{\overline{R}}$ with the $p$-adic topology (II.2.2), then the $\widehat{\mathscr{S}^{(r)}}[\frac{1}{p}]$-module $N \otimes_{\widehat{R_1}} \widehat{\mathscr{S}^{(r)}}$ is identified with the $\widehat{\overline{R}}[\frac{1}{p}]$-module

$$\{\sum_{\underline{n} \in \mathbb{N}^d} x_{\underline{n}} \otimes \underline{y}^{\underline{n}} \mid x_{\underline{n}} \in N \otimes_{\widehat{R_1}} \widehat{\overline{R}}, \lim_{|\underline{n}| \to +\infty} p^{-r|\underline{n}|} x_{\underline{n}} = 0\}. \tag{II.13.24.6}$$

Indeed, the latter clearly contains $N \otimes_{\widehat{R_1}} \widehat{\mathscr{S}^{(r)}}$. To prove equality, we can reduce to the case where $N$ is free of finite type over $\widehat{R_1}[\frac{1}{p}]$, in which case the assertion is obvious (II.13.24.4).

Let $x \in N$, $z \in M$ the image of $x \otimes 1$ by the isomorphism (II.13.24.3), and

$$\sum_{\underline{n} \in \mathbb{N}^d} x_{\underline{n}} \otimes \underline{y}^{\underline{n}} \in N \otimes_{\widehat{R_1}} \widehat{\mathscr{S}^{(r)}} \tag{II.13.24.7}$$

the image of $z \otimes 1 \in M \otimes_{\widehat{\overline{R}}} \widehat{\mathscr{S}^{(r)}}$ by the inverse of the isomorphism (II.13.24.2), where $x_{\underline{n}} \in N \otimes_{\widehat{R_1}} \widehat{\overline{R}}$. We clearly have $x_{\underline{0}} = x$. Let us write

$$\theta = \sum_{i=1}^{d} \theta_i \otimes y_i, \tag{II.13.24.8}$$

where the $\theta_i$ are $\widehat{R_1}[\frac{1}{p}]$-endomorphisms of $N$ that commute pairwise. Since the section (II.13.24.7) is annihilated by $\theta \otimes 1 + 1 \otimes p^r d_{\widehat{\mathscr{S}^{(r)}}}$, for every $\underline{n} = (n_1, \dots, n_d) \in \mathbb{N}^d$ and every $1 \leq i \leq d$, we have

$$\theta_i(x_{\underline{n}}) + (n_i + 1) x_{\underline{n} + \underline{1}_i} = 0, \tag{II.13.24.9}$$

where $\underline{1}_i$ is the element of $\mathbb{N}^d$ whose components are all zero except for the $i$th, which is 1. It follows that

$$x_{\underline{n}} = (-1)^{|\underline{n}|} (\prod_{1 \leq i \leq d} \frac{1}{n_i!} \theta_i^{n_i})(x). \tag{II.13.24.10}$$

Consequently, $p^{-r|\underline{n}|}(\prod_{1 \leq i \leq d} \frac{1}{n_i!} \theta_i^{n_i})(x)$ tends to 0 in $N \otimes_{\widehat{R_1}} \widehat{\overline{R}}$ when $|\underline{n}|$ tends to infinity (II.13.24.6).

The $p$-adic topology on $N$ is induced by the $p$-adic topology on $N \otimes_{\widehat{R_1}} \widehat{\overline{R}}$. Indeed, since $N$ is projective of finite type over $\widehat{R_1}[\frac{1}{p}]$, we can reduce to the case where $N$ is free of finite type, or even to the case where $N = \widehat{R_1}[\frac{1}{p}]$. For every $n \geq 1$, since the canonical homomorphism $\widehat{R_1}/p^n \widehat{R_1} \to \widehat{\overline{R}}/p^n \widehat{\overline{R}}$ is injective (cf. the proof of II.6.14), we have $p^n \widehat{\overline{R}} \cap \widehat{R_1}[\frac{1}{p}] = p^n \widehat{R_1}$, giving the assertion.

It follows from the above that for every $x \in N$, $p^{-r|\underline{n}|}(\prod_{1 \leq i \leq d} \frac{1}{n_i!} \theta_i^{n_i})(x)$ tends to 0 in $N$ when $|\underline{n}|$ tends to infinity.

Let $N_0$ be a sub-$\widehat{R_1}$-module of finite type of $N$ that generates it over $\widehat{R_1}[\frac{1}{p}]$, and $\varepsilon$ a rational number such that $\frac{1}{p-1} < \varepsilon < r + \frac{1}{p-1}$. Since the sequence $p^{(r-\varepsilon)n} n!$ tends to 0 in $\mathscr{O}_C$ when $n$ tends to infinity, for every $x \in N$, $p^{-\varepsilon|\underline{n}|}(\prod_{1\leq i\leq d}\theta_i^{n_i})(x)$ tends to 0 in $N$ when $|\underline{n}|$ tends to infinity. We may therefore consider the sub-$\widehat{R_1}$-module

$$N^\circ = \sum_{\underline{n}\in\mathbb{N}^d} p^{-\varepsilon|\underline{n}|}(\prod_{1\leq i\leq d}\theta_i^{n_i})(N_0) \tag{II.13.24.11}$$

of $N$. It is of finite type over $\widehat{R_1}$ and generates $N$ over $\widehat{R_1}[\frac{1}{p}]$. Since we have

$$\theta(N^\circ) \subset p^\varepsilon\xi^{-1}N^\circ \otimes_R \widetilde{\Omega}^1_{R/\mathscr{O}_K}, \tag{II.13.24.12}$$

$(N,\theta)$ is a small Higgs $\widehat{R_1}[\frac{1}{p}]$-module with coefficients in $\xi^{-1}\widetilde{\Omega}^1_{R/\mathscr{O}_K}$.

**Corollary II.13.25.** *A Higgs $\widehat{R_1}[\frac{1}{p}]$-module with coefficients in $\xi^{-1}\widetilde{\Omega}^1_{R/\mathscr{O}_K}$ is solvable* (II.12.18) *if and only if it is small* (II.13.5).

This follows from II.13.23(ii) and II.13.24.

**Corollary II.13.26.** *Any Dolbeault $\widehat{\overline{R}}[\frac{1}{p}]$-representation of $\Delta$* (II.12.16) *is small* (II.13.2).

This follows from II.12.24, II.13.25, and II.13.23(iii).

We will strengthen this statement, under certain assumptions, in II.14.8.

## II.14. Descent of small representations and applications

We keep the assumptions and notation of II.12 in this section.

**Proposition II.14.1.** *Let $a$ be a nonzero element of $\mathscr{O}_{\overline{K}}$, $\alpha$ a rational number $> \frac{1}{p-1}$, $M_1$ and $M_2$ two $\alpha$-small $(R_1/aR_1)$-representations of $\Delta_{p^\infty}$* (II.13.1), *and*

$$\overline{u}\colon M_1\otimes_{R_1}\overline{R}\to M_2\otimes_{R_1}\overline{R} \tag{II.14.1.1}$$

*a $\Delta$-equivariant $\overline{R}$-linear morphism. Suppose that $v(a) > \frac{1}{p-1}+\alpha$ and set $b = ap^{-\alpha-\frac{1}{p-1}}$. Then there exists a unique $\Delta_{p^\infty}$-equivariant $R_1$-linear morphism*

$$u\colon M_1/bM_1 \to M_2/bM_2 \tag{II.14.1.2}$$

*such that $u\otimes_{R_1}\overline{R}\equiv\overline{u} \mod bM_2\otimes_{R_1}\overline{R}$.*

Denote by $M$ the discrete $R_1$-$\Delta_{p^\infty}$-module $\mathrm{Hom}_{R_1}(M_1,M_2)$, which is an $\alpha$-small $(R_1/aR_1)$-representation of $\Delta_{p^\infty}$. Since $R_1$ is normal by II.6.8(ii), we have $p^\alpha bR_1 = (p^\alpha b\overline{R})\cap R_1$. Consequently, the canonical morphism

$$\mathrm{H}^0(\Delta_{p^\infty}, M/p^\alpha bM)\to \mathrm{H}^0(\Delta,(M/p^\alpha bM)\otimes_{R_1}\overline{R}) \tag{II.14.1.3}$$

is injective and its cokernel is annihilated by $p^{\frac{1}{p-1}}\mathfrak{m}_{\overline{K}}$ by virtue of II.8.23. Hence there exists a $\Delta_{p^\infty}$-equivariant $R_1$-linear morphism

$$v\colon M_1/p^\alpha bM_1\to M_2/p^\alpha bM_2 \tag{II.14.1.4}$$

such that $v\otimes_{R_1}\overline{R}\equiv p^\alpha\overline{u} \mod p^\alpha bM_2\otimes_{R_1}\overline{R}$. Since $p^\alpha R_1 = (p^\alpha\overline{R})\cap R_1$ and since $R_1$ is $\mathscr{O}_{\overline{K}}$-flat, representing $v$ by a matrix with coefficients in $R_1/p^\alpha bR_1$, we see that there exists a unique $\Delta_{p^\infty}$-equivariant $R_1$-linear morphism

$$u\colon M_1/bM_1\to M_2/bM_2 \tag{II.14.1.5}$$

such that $v = p^\alpha u$. Using once again that $\overline{R}$ is $\mathscr{O}_{\overline{K}}$-flat, we deduce that $u\otimes_{R_1}\overline{R}\equiv\overline{u}$ mod $bM_2\otimes_{R_1}\overline{R}$. The uniqueness of $u$ follows from the fact that the canonical homomorphism $R_1/bR_1\to\overline{R}/b\overline{R}$ is injective (cf. the proof of II.6.14).

**Proposition II.14.2.** *Let $\alpha$ be rational number $> \frac{1}{p-1}$, $a$ a nonzero element of $\mathscr{O}_{\overline{K}}$ such that $v(a) > \alpha$, and $M$ a free $(\overline{R}/a\overline{R})$-module of rank $r \geq 1$, endowed with the discrete topology and a continuous $\overline{R}$-semi-linear action of $\Delta$ such that $M$ admits a basis $e_1, \dots, e_r$ consisting of elements that are $\Delta$-invariant modulo $p^{2\alpha}M$. Then there exist a discrete $R_1$-$\Delta_{p^\infty}$-module $N$ whose underlying module is free of rank $r$ over $R_1/ap^{-\alpha}R_1$, with a basis $f_1, \dots, f_r$ consisting of elements that are $\Delta_{p^\infty}$-invariant modulo $p^\alpha N$, and a $\Delta$-equivariant $\overline{R}$-linear isomorphism*

$$N \otimes_{R_1} \overline{R} \xrightarrow{\sim} M/ap^{-\alpha}M \tag{II.14.2.1}$$

*that sends $f_j \otimes 1 \mod p^\alpha N \otimes_{R_1} \overline{R}$ to $e_j \mod p^\alpha M$ for every $1 \leq j \leq r$.*

The proposition is obvious if $v(a) \leq 3\alpha$, in which case we take $N = (R_1/ap^{-\alpha}R_1)^r$ endowed with the trivial representation of $\Delta_{p^\infty}$ and the canonical basis. Suppose, therefore, $v(a) > 3\alpha$. Let $n$ be an integer $\geq 1$ such that

$$\varepsilon = \frac{v(a) - 3\alpha}{n} < \frac{1}{3}(\alpha - \frac{1}{p-1}). \tag{II.14.2.2}$$

Let us show by a finite induction that for every $0 \leq i \leq n$, the proposition holds for the $\overline{R}$-representation $M/p^{3\alpha+i\varepsilon}M$; in other words, that there exist a discrete $R_1$-$\Delta_{p^\infty}$-module $N_i$ whose underlying module is free of rank $r$ over $R_1/p^{2\alpha+i\varepsilon}R_1$, with a basis $f_1^{(i)}, \dots, f_r^{(i)}$ consisting of elements that are $\Delta_{p^\infty}$-invariant modulo $p^\alpha N_i$, and a $\Delta$-equivariant $\overline{R}$-linear isomorphism

$$N_i \otimes_{R_1} \overline{R} \xrightarrow{\sim} M/p^{2\alpha+i\varepsilon}M \tag{II.14.2.3}$$

that sends $f_j^{(i)} \otimes 1 \mod p^\alpha N_i \otimes_{R_1} \overline{R}$ to $e_j \mod p^\alpha M$ for every $1 \leq j \leq r$. The representation $N = N_n$ will then answer the question because $2\alpha + n\varepsilon = v(a) - \alpha$.

We take $N_0 = (R_1/p^{2\alpha}R_1)^r$ endowed with the trivial representation of $\Delta_{p^\infty}$ and the canonical basis. Suppose that $N_i$ has been constructed for $0 \leq i < n$ and let us construct $N_{i+1}$. By II.3.35, the obstruction to lifting $N_i$ to a discrete $(R_1/p^{3\alpha+i\varepsilon}R_1)$-$\Delta_{p^\infty}$-module whose underlying module is free of finite type over $R_1/p^{3\alpha+i\varepsilon}R_1$ is an element $\mathfrak{o}$ of $\mathrm{H}^2(\Delta_{p^\infty}, \mathrm{Mat}_r(R_1/p^\alpha R_1))$. On the other hand, the morphism

$$\mathrm{H}^2(\Delta_{p^\infty}, \mathrm{Mat}_r(R_1/p^\alpha R_1)) \to \mathrm{H}^2(\Delta, \mathrm{Mat}_r(\overline{R}/p^\alpha \overline{R})) \tag{II.14.2.4}$$

is almost injective by virtue of II.8.11. The image of $\mathfrak{o}$ in $\mathrm{H}^2(\Delta, \mathrm{Mat}_r(\overline{R}/p^\alpha\overline{R}))$ is zero because the representation $M/p^{2\alpha+i\varepsilon}M$ lifts to $M/p^{3\alpha+i\varepsilon}M$; hence $p^\varepsilon \mathfrak{o} = 0$. Consequently, by virtue of (II.3.35.4), $N_i/p^{2\alpha+(i-1)\varepsilon}N_i$ lifts to a discrete $(R_1/p^{3\alpha+(i-1)\varepsilon}R_1)$-$\Delta_{p^\infty}$-module $N'_{i+1}$ whose underlying module is free of finite type over $R_1/p^{3\alpha+(i-1)\varepsilon}R_1$. By II.3.34, the lift $N'_{i+1} \otimes_{R_1} \overline{R}$ of $M/p^{2\alpha+(i-1)\varepsilon}M$ is deduced from $M/p^{3\alpha+(i-1)\varepsilon}M$ by twisting by an element $\overline{\mathfrak{c}}$ of $\mathrm{H}^1(\Delta, \mathrm{Mat}_r(\overline{R}/p^\alpha\overline{R}))$. By virtue of II.8.11, the cokernel of the canonical morphism

$$\mathrm{H}^1(\Delta_{p^\infty}, \mathrm{Mat}_r(R_1/p^\alpha R_1)) \to \mathrm{H}^1(\Delta, \mathrm{Mat}_r(\overline{R}/p^\alpha\overline{R})) \tag{II.14.2.5}$$

is annihilated by $p^{\frac{1}{p-1}+\varepsilon}$. Since $\alpha > \frac{1}{p-1} + 3\varepsilon$, $p^{\alpha-2\varepsilon}\overline{\mathfrak{c}}$ is the image of an element

$$\mathfrak{c}' \in \mathrm{H}^1(\Delta_{p^\infty}, \mathrm{Mat}_r(R_1/p^\alpha R_1)).$$

Twisting $N'_{i+1}/p^{2\alpha+(i+1)\varepsilon}N'_{i+1}$ by $-\mathfrak{c}'$, we obtain a discrete $(R_1/p^{2\alpha+(i+1)\varepsilon}R_1)$-$\Delta_{p^\infty}$-module $N_{i+1}$ whose underlying module is free of finite type over $R_1/p^{2\alpha+(i+1)\varepsilon}R_1$, which lifts $N_i/p^{\alpha+(i+1)\varepsilon}N_i$. Note that $2\alpha + (i-1)\varepsilon > \alpha + (i+1)\varepsilon$. Then $N_{i+1}$ has the required properties by the last assertion of II.3.34.

**Proposition II.14.3.** *Let $a$ be a nonzero element of $\mathscr{O}_{\overline{K}}$, $\alpha$ and $\beta$ two rational numbers such that $v(a) > \alpha > \beta > \frac{1}{p-1}$, and $M$ a free $(\overline{R}/a\overline{R})$-module of rank $r \geq 1$, endowed with the discrete topology and a continuous $\overline{R}$-semi-linear action of $\Delta$ such that $M$ admits a basis consisting of elements that are $\Delta$-invariant modulo $p^{2\alpha}M$. Then there exist a discrete $R_1$-$\Delta_{p^\infty}$-module $N$ whose underlying module is free of rank $r$ over $R_1/ap^{-\alpha}R_1$, having a basis consisting of elements that are $\Delta_{p^\infty}$-invariant modulo $p^{2\beta}N$, and a $\Delta$-equivariant $\overline{R}$-linear isomorphism*

$$N \otimes_{R_1} \overline{R} \xrightarrow{\sim} M/ap^{-\alpha}M. \tag{II.14.3.1}$$

The proposition is obvious if $v(a) \leq 3\alpha$, in which case we take $N = (R_1/ap^{-\alpha}R_1)^r$ endowed with the trivial representation of $\Delta_{p^\infty}$. Suppose, therefore, $v(a) > 3\alpha$. Let $n$ be an integer $\geq 1$ such that

$$\varepsilon = \frac{v(a) - 3\alpha}{n} < \inf(\frac{1}{3}(\alpha - \frac{1}{p-1}), \alpha - \beta). \tag{II.14.3.2}$$

Let us show by a finite induction that for every $0 \leq i \leq n$, the proposition holds for the $\overline{R}$-representation $M/p^{3\alpha+i\varepsilon}M$; in other words, that there exists a discrete $R_1$-$\Delta_{p^\infty}$-module $N_i$ whose underlying module is free of rank $r$ over $R_1/p^{2\alpha+i\varepsilon}R_1$, having a basis consisting of elements that are $\Delta_{p^\infty}$-invariant modulo $p^{2(\alpha-\varepsilon)}N_i$ and a $\Delta$-equivariant $\overline{R}$-linear isomorphism

$$N_i \otimes_{R_1} \overline{R} \xrightarrow{\sim} M/p^{2\alpha+i\varepsilon}M. \tag{II.14.3.3}$$

The representation $N = N_n$ will then answer the question because $2\alpha + n\varepsilon = v(a) - \alpha$ and $\alpha - \varepsilon > \beta$.

We take $N_0 = (R_1/p^{2\alpha}R_1)^r$ endowed with the trivial representation of $\Delta_{p^\infty}$ and the canonical basis. Suppose that $N_i$ has been constructed for $0 \leq i < n$ and let us construct $N_{i+1}$. By II.3.35 (applied with $a^q = p^{3\alpha+i\varepsilon}$, $a^n = p^{2\alpha+i\varepsilon}$, and $a^m = p^{2(\alpha-\varepsilon)}$), the obstruction to lifting $N_i$ to a discrete $(R_1/p^{3\alpha+i\varepsilon}R_1)$-$\Delta_{p^\infty}$-module whose underlying module is free of finite type over $R_1/p^{3\alpha+i\varepsilon}R_1$ is an element $\mathfrak{o}$ of $\mathrm{H}^2(\Delta_{p^\infty}, \mathrm{Mat}_r(R_1/p^\alpha R_1))$. On the other hand, the morphism

$$\mathrm{H}^2(\Delta_{p^\infty}, \mathrm{Mat}_r(R_1/p^\alpha R_1)) \to \mathrm{H}^2(\Delta, \mathrm{Mat}_r(\overline{R}/p^\alpha \overline{R})) \tag{II.14.3.4}$$

is almost injective by virtue of II.8.11. The image of $\mathfrak{o}$ in $\mathrm{H}^2(\Delta, \mathrm{Mat}_r(\overline{R}/p^\alpha\overline{R}))$ is zero because the representation $M/p^{2\alpha+i\varepsilon}M$ lifts to $M/p^{3\alpha+i\varepsilon}M$; hence $p^\varepsilon\mathfrak{o} = 0$. Consequently, by (II.3.35.4), $N_i/p^{2\alpha+(i-1)\varepsilon}N_i$ lifts to a discrete $(R_1/p^{3\alpha+(i-1)\varepsilon}R_1)$-$\Delta_{p^\infty}$-module $N'_{i+1}$ whose underlying module is free of finite type over $R_1/p^{3\alpha+(i-1)\varepsilon}R_1$. By II.3.34, the lift $N'_{i+1} \otimes_{R_1} \overline{R}$ of $M/p^{2\alpha+(i-1)\varepsilon}M$ is deduced from the lift $M/p^{3\alpha+(i-1)\varepsilon}M$ by twisting by an element $\bar{\mathfrak{c}}$ of $\mathrm{H}^1(\Delta, \mathrm{Mat}_r(\overline{R}/p^\alpha\overline{R}))$. By virtue of II.6.24, the cokernel of the canonical morphism

$$\mathrm{H}^1(\Delta_{p^\infty}, \mathrm{Mat}_r(R_{p^\infty}/p^\alpha R_{p^\infty})) \to \mathrm{H}^1(\Delta, \mathrm{Mat}_r(\overline{R}/p^\alpha\overline{R})) \tag{II.14.3.5}$$

is annihilated by $p^\varepsilon$. Hence $p^\varepsilon\bar{\mathfrak{c}}$ is the image of an element

$$\mathfrak{c}' \in \mathrm{H}^1(\Delta_{p^\infty}, \mathrm{Mat}_r(R_{p^\infty}/p^\alpha R_{p^\infty})).$$

By the proof of II.8.10(i), we have $\mathfrak{c}' = \mathfrak{c}'_1 + \mathfrak{c}'_2$, where $p^{\frac{1}{p-1}}\mathfrak{c}'_2 = 0$ and $\mathfrak{c}'_1$ is the image of an element

$$\mathfrak{c}_1 \in \mathrm{H}^1(\Delta_{p^\infty}, \mathrm{Mat}_r(R_1/p^\alpha R_1)).$$

Twisting $N'_{i+1}/p^{3\alpha+(i-2)\varepsilon}N'_{i+1}$ by $-\mathfrak{c}_1$, we obtain a discrete $(R_1/p^{3\alpha+(i-2)\varepsilon}R_1)$-$\Delta_{p^\infty}$-module $N''_{i+1}$ whose underlying module is free of finite type over $R_1/p^{3\alpha+(i-2)\varepsilon}R_1$ and that lifts $N_i/p^{2\alpha+(i-2)\varepsilon}N_i$. By the last assertion of II.3.34, the lift $N''_{i+1} \otimes_{R_1} \overline{R}$ of

$M/p^{2\alpha+(i-2)\varepsilon}M$ is deduced from the lift $M/p^{3\alpha+(i-2)\varepsilon}M$ by twisting by the image $\overline{\mathfrak{c}}_2$ of $\mathfrak{c}'_2$ in $\mathrm{H}^1(\Delta, \mathrm{Mat}_r(\overline{R}/p^\alpha \overline{R}))$. Since $p^{\frac{1}{p-1}}\overline{\mathfrak{c}}_2 = 0$, we deduce from this, again by the last assertion of II.3.34, that there exists a $\Delta$-equivariant $\overline{R}$-linear isomorphism

$$(N''_{i+1}/p^{3\alpha-\frac{1}{p-1}+(i-2)\varepsilon}N''_{i+1}) \otimes_{R_1} \overline{R} \xrightarrow{\sim} M/p^{3\alpha-\frac{1}{p-1}+(i-2)\varepsilon}M. \tag{II.14.3.6}$$

Since $\alpha > \frac{1}{p-1} + 3\varepsilon$, the representation $N_{i+1} = N''_{i+1}/p^{2\alpha+(i+1)\varepsilon}N''_{i+1}$ has the desired properties.

**Proposition II.14.4** (Faltings, **[27]**). *The functor* (II.13.1)

$$\mathbf{Rep}^{\mathrm{small}}_{\widehat{R_1}}(\Delta_\infty) \to \mathbf{Rep}^{\mathrm{small}}_{\widehat{\overline{R}}}(\Delta), \quad M \mapsto M \otimes_{\widehat{R_1}} \widehat{\overline{R}} \tag{II.14.4.1}$$

*is an equivalence of categories.*

Indeed, since the canonical functor

$$\mathbf{Rep}^{\mathrm{small}}_{\widehat{R_1}}(\Delta_{p^\infty}) \to \mathbf{Rep}^{\mathrm{small}}_{\widehat{R_1}}(\Delta_\infty) \tag{II.14.4.2}$$

is an equivalence of categories (II.13.12), it suffices to show that the functor

$$\mathbf{Rep}^{\mathrm{small}}_{\widehat{R_1}}(\Delta_{p^\infty}) \to \mathbf{Rep}^{\mathrm{small}}_{\widehat{\overline{R}}}(\Delta), \quad M \mapsto M \otimes_{\widehat{R_1}} \widehat{\overline{R}} \tag{II.14.4.3}$$

is an equivalence of categories. This functor is fully faithful by virtue of II.14.1. Let us show that it is essentially surjective. Let $\alpha, \beta$ be two rational numbers such that $\alpha > \beta > \frac{1}{p-1}$ and $M$ a $(2\alpha)$-small $\widehat{\overline{R}}$-representation of rank $r \geq 1$. By II.14.3, for every integer $n > \alpha$, there exist a discrete $R_1$-$\Delta_{p^\infty}$-module $N_n$ whose underlying module is free of rank $r$ over $R_1/p^{n-\alpha}R_1$, with a basis consisting of elements that are $\Delta_{p^\infty}$-invariant modulo $p^{2\beta}N$, and a $\Delta_{p^\infty}$-equivariant $\overline{R}$-linear isomorphism

$$N_n \otimes_{R_1} \overline{R} \xrightarrow{\sim} M/p^{n-\alpha}M. \tag{II.14.4.4}$$

By virtue of II.14.1, for all integers $n \geq m > \alpha$, there exists a unique $\Delta_{p^\infty}$-equivariant $R_1$-linear isomorphism

$$N_n/p^{m-\alpha}N_n \xrightarrow{\sim} N_m \tag{II.14.4.5}$$

that is compatible with the isomorphisms (II.14.4.4). Consequently, the $R_1$-modules $(N_n)_{n>\alpha}$ form an inverse system, and if $N$ is its inverse limit, then we have a $\Delta$-equivariant $\widehat{\overline{R}}$-linear isomorphism

$$N \otimes_{\widehat{R_1}} \widehat{\overline{R}} \xrightarrow{\sim} M. \tag{II.14.4.6}$$

**Remarks II.14.5.** (i) Though similar, the proofs of Propositions II.14.2 and II.14.3 differ significantly in their last steps that consist of defining $N_{i+1}$ from $N'_{i+1}$. Note that as far as the descent of the basis is concerned, the proof of II.14.3 does not give better results than that of II.14.2, which is why we left this part out in statement II.14.3. Proposition II.14.2 is due to Faltings (**[27]** Lemma 1).

(ii) The descent statement established by Faltings in **[27]** is in fact slightly weaker than II.14.4.

**Corollary II.14.6.** *Every small $\widehat{\overline{R}}$-representation of $\Delta$ is Dolbeault, and its image by $\mathbb{H}$* (II.12.8.2) *is a small Higgs $\widehat{R_1}$-module with coefficients in $\xi^{-1}\widetilde{\Omega}^1_{R/\mathscr{O}_K}$.*

This follows from II.14.4, II.13.11, II.13.18, and II.13.19.

**Corollary II.14.7.** *The functors $\mathbb{H}$* (II.12.8.2) *and $\mathbb{V}$* (II.12.9.2) *induce equivalences of categories quasi-inverse to each other between the category of small $\widehat{\overline{R}}$-representations of $\Delta$ and that of small Higgs $\widehat{R_1}$-modules with coefficients in $\xi^{-1}\widetilde{\Omega}^1_{R/\mathscr{O}_K}$.*

This follows from II.12.15, II.13.20 and II.14.6.

**Proposition II.14.8.** *The following statements are equivalent:*

(i) *For every small* $\widehat{\overline{R}}[\frac{1}{p}]$*-representation* $M$ *of* $\Delta$ (II.13.2), *there exist a small Higgs* $\widehat{R_1}[\frac{1}{p}]$*-module* $N$ *with coefficients in* $\xi^{-1}\widetilde{\Omega}^1_{R/\mathscr{O}_K}$ (II.13.5) *and a* $\Delta$*-equivariant* $\widehat{\overline{R}}[\frac{1}{p}]$*-isomorphism*

$$M \xrightarrow{\sim} \mathbb{V}(N). \tag{II.14.8.1}$$

(ii) *Every small* $\widehat{\overline{R}}[\frac{1}{p}]$*-representation of* $\Delta$ *is Dolbeault* (II.12.16).

(iii) *An* $\widehat{\overline{R}}[\frac{1}{p}]$*-representation of* $\Delta$ *is Dolbeault if and only if it is small.*

(iv) *The functor* (II.13.2)

$$\mathbf{Rep}^{\mathrm{small}}_{\widehat{R_1}[\frac{1}{p}]}(\Delta_\infty) \to \mathbf{Rep}^{\mathrm{small}}_{\widehat{\overline{R}}[\frac{1}{p}]}(\Delta), \quad M \mapsto M \otimes_{\widehat{R_1}} \widehat{\overline{R}} \tag{II.14.8.2}$$

*is an equivalence of categories.*

Indeed, the implication (i)⇒(ii) follows from II.13.23(iii), the implication (ii)⇒(iii) is a consequence of II.13.26, and the implication (iii)⇒(i) follows from II.12.15 and II.13.25.

Next, let us show (i)⇒(iv). For every small $\widehat{\overline{R}}[\frac{1}{p}]$-representation $M$ of $\Delta$, the Higgs $\widehat{R_1}[\frac{1}{p}]$-module $\mathbb{H}(M)$ is small by (i) and II.13.23(iii). The functor (II.13.21.2) therefore associates with it a small $\widehat{R_1}[\frac{1}{p}]$-representation $\varphi$ of $\Delta_\infty$ on $\mathbb{H}(M)$. The correspondence $M \mapsto (\mathbb{H}(M), \varphi)$ defines a functor

$$\mathbf{Rep}^{\mathrm{small}}_{\widehat{\overline{R}}[\frac{1}{p}]}(\Delta) \to \mathbf{Rep}^{\mathrm{small}}_{\widehat{R_1}[\frac{1}{p}]}(\Delta_\infty). \tag{II.14.8.3}$$

Let us show that the functors (II.14.8.2) and (II.14.8.3) are quasi-inverse to each other. For every small $\widehat{\overline{R}}[\frac{1}{p}]$-representation $M$ of $\Delta$, we have functorial $\Delta$-equivariant $\widehat{\overline{R}}$-isomorphisms (II.13.23.2)

$$\mathbb{H}(M) \otimes_{\widehat{R_1}} \widehat{\overline{R}} \xrightarrow{\sim} \mathbb{V}(\mathbb{H}(M)) \xrightarrow{\sim} M. \tag{II.14.8.4}$$

Moreover, let $(N, \varphi)$ be a small $\widehat{R_1}[\frac{1}{p}]$-representation of $\Delta_\infty$, and $\theta$ the small Higgs $\widehat{R_1}[\frac{1}{p}]$-field over $N$ with coefficients in $\xi^{-1}\widetilde{\Omega}^1_{R/\mathscr{O}_K}$ associated with $\varphi$ by the functor (II.13.22.1). By II.13.23, we have a functorial isomorphism of Higgs $\widehat{R_1}$-modules

$$\mathbb{H}((N, \varphi) \otimes_{\widehat{R_1}} \widehat{\overline{R}}) \xrightarrow{\sim} (N, \theta). \tag{II.14.8.5}$$

Applying the functor (II.13.21.2), quasi-inverse of the functor (II.13.22.1), we obtain a functorial $\Delta_\infty$-equivariant $\widehat{R_1}$-isomorphism

$$\mathbb{H}((N, \varphi) \otimes_{\widehat{R_1}} \widehat{\overline{R}}) \xrightarrow{\sim} (N, \varphi). \tag{II.14.8.6}$$

The functors (II.14.8.2) and (II.14.8.3) are therefore quasi-inverse to each other.

Finally, let us show (iv)⇒(i). Let $M$ be a small $\widehat{\overline{R}}[\frac{1}{p}]$-representation of $\Delta$. By (iv), there exist a small $\widehat{R_1}[\frac{1}{p}]$-representation $(N, \varphi)$ of $\Delta_\infty$ and a $\Delta$-equivariant $\widehat{\overline{R}}$-isomorphism

$$M \xrightarrow{\sim} (N, \varphi) \otimes_{\widehat{R_1}} \widehat{\overline{R}}. \tag{II.14.8.7}$$

Let $\theta$ be the small Higgs $\widehat{R_1}[\frac{1}{p}]$-field on $N$ with coefficients in $\xi^{-1}\widetilde{\Omega}^1_{R/\mathscr{O}_K}$ associated with $\varphi$ by the functor (II.13.22.1). By virtue of II.13.23(ii), since the functors (II.13.21.2) and

(II.13.22.1) are quasi-inverse to each other, we have a $\Delta$-equivariant $\widehat{\overline{R}}[\frac{1}{p}]$-isomorphism $M \xrightarrow{\sim} \mathbb{V}(N, \theta)$, proving the proposition.

**Corollary II.14.9.** *Assume that the equivalent statements of* II.14.8 *hold. Then for every small $\widehat{\overline{R}}[\frac{1}{p}]$-representation $M$ of $\Delta$, the $\widehat{\overline{R}}[\frac{1}{p}]$-representation* $\mathrm{Hom}_{\widehat{\overline{R}}[\frac{1}{p}]}(M, \widehat{\overline{R}}[\frac{1}{p}])$ *of $\Delta$ is small.*

Indeed, there exist a small $\widehat{R_1}[\frac{1}{p}]$-representation $N$ of $\Delta_\infty$ and a $\Delta$-equivariant $\widehat{\overline{R}}$-isomorphism $M \xrightarrow{\sim} N \otimes_{\widehat{R_1}} \widehat{\overline{R}}$. We deduce from this a $\Delta$-equivariant $\widehat{\overline{R}}$-isomorphism

$$\mathrm{Hom}_{\widehat{\overline{R}}[\frac{1}{p}]}(M, \widehat{\overline{R}}[\frac{1}{p}]) \xrightarrow{\sim} \mathrm{Hom}_{\widehat{R_1}[\frac{1}{p}]}(N, \widehat{R_1}[\frac{1}{p}]) \otimes_{\widehat{R_1}} \widehat{\overline{R}}. \tag{II.14.9.1}$$

Let $\alpha$ be a rational number $> \frac{2}{p-1}$ and $N^\circ$ a sub-$\widehat{R_1}$-module of finite type of $N$ stable under $\Delta_\infty$ and generated by a finite number of elements that are $\Delta_\infty$-invariant modulo $p^\alpha N^\circ$, and that generates $N$ over $\widehat{R_1}[\frac{1}{p}]$. By virtue of (**[1]** 10.10.2(iii)), $N^\circ$ is a coherent $\widehat{R_1}$-module. The same therefore holds for $\mathrm{Hom}_{\widehat{R_1}}(N^\circ, \widehat{R_1})$, and the canonical morphism

$$\mathrm{Hom}_{\widehat{R_1}}(N^\circ, \widehat{R_1}) \otimes_{\widehat{R_1}} \widehat{R_1}[\frac{1}{p}] \to \mathrm{Hom}_{\widehat{R_1}[\frac{1}{p}]}(N, \widehat{R_1}[\frac{1}{p}]) \tag{II.14.9.2}$$

is an isomorphism. On the other hand, since $\widehat{R_1}$ is $\mathscr{O}_{\overline{K}}$-flat, for every rational number $\beta > 0$, the canonical morphism

$$\mathrm{Hom}_{\widehat{R_1}}(N^\circ, \widehat{R_1}) \otimes_{\mathscr{O}_{\overline{K}}} \mathscr{O}_{\overline{K}}/p^\beta \mathscr{O}_{\overline{K}} \to \mathrm{Hom}_{R_1}(N^\circ/p^\beta N^\circ, R_1/p^\beta R_1) \tag{II.14.9.3}$$

is injective. It follows that the representation of $\Delta_\infty$ over $\mathrm{Hom}_{\widehat{R_1}}(N^\circ, \widehat{R_1})$ is continuous for the $p$-adic topology, and is quasi-small, giving the corollary.

**Corollary II.14.10.** *Assume that the equivalent statements of* II.14.8 *hold. Then for all small $\widehat{\overline{R}}[\frac{1}{p}]$-representations $M$ and $M'$ of $\Delta$, and every $\Delta$-equivariant surjective $\widehat{\overline{R}}[\frac{1}{p}]$-linear morphism $u \colon M' \to M$, the $\widehat{\overline{R}}[\frac{1}{p}]$-representation of $\Delta$ over the kernel of $u$ is small.*

This follows from II.14.9.

**Lemma II.14.11.** *Let $\alpha, \varepsilon$ be two rational numbers such that $0 < \varepsilon < \alpha$ and $M$ an $\alpha$-quasi-small $\widehat{\overline{R}}$-representation of $\Delta$. Then:*

(i) *The $R_\infty$-module $(M/p^\alpha M)^\Sigma$ is almost of finite type.*
(ii) *The $\widehat{\overline{R}}$-module $p^\varepsilon M$ is generated by a finite number of elements that are, on the one hand, $\Delta$-invariant modulo $p^\alpha M$ and, on the other hand, $\Sigma$-invariant.*
(iii) *The canonical morphism $M^\Sigma \otimes_{\widehat{R_\infty}} \widehat{\overline{R}} \to M$ is almost surjective.*

Let $x_1, \dots, x_d$ be generators of $M$ over $\widehat{\overline{R}}$ that are $\Delta$-invariant modulo $p^\alpha M$.

(i) We denote by $u \colon (\overline{R}/p^\alpha \overline{R})^d \to M/p^\alpha M$ the $\Delta$-equivariant surjective $\overline{R}$-linear morphism defined by the $(x_i)_{1 \leq i \leq d}$, and by $C$ its kernel. Then $\mathrm{H}^1(\Sigma, C)$ is almost zero by virtue of II.6.19. On the other hand, the canonical morphism

$$R_\infty/p^\alpha R_\infty \to (\overline{R}/p^\alpha \overline{R})^\Sigma$$

is an almost isomorphism (II.6.22). Consequently, the $R_\infty$-linear morphism

$$(R_\infty/p^\alpha R_\infty)^d \to (M/p^\alpha M)^\Sigma \tag{II.14.11.1}$$

defined by the $(x_i)_{1 \leq i \leq d}$ is almost surjective.

(ii) The $\widehat{\overline{R}}$-module $p^\alpha M$ is complete and separated for the $p$-adic topology; it is complete by virtue of ([**11**] Chap. III § 2.12 Cor. 1 to Prop. 16) and is separated as a submodule of $M$. Therefore $\mathrm{H}^1_{\mathrm{cont}}(\Sigma, p^\alpha M)$ is almost zero by virtue of II.6.21. Consequently, the canonical morphism

$$M^\Sigma \to (M/p^\alpha M)^\Sigma \tag{II.14.11.2}$$

is almost surjective. The classes of the elements $p^\varepsilon x_1, \dots, p^\varepsilon x_n$ in $(M/p^\alpha M)^\Sigma$ lift to elements $x'_1, \dots, x'_d \in M^\Sigma$. For every $1 \leq i \leq d$, we have

$$x'_i \in M^\Sigma \cap (p^\varepsilon M) = (p^\varepsilon M)^\Sigma. \tag{II.14.11.3}$$

On the other hand, by Nakayama's lemma, the $(x'_i)_{1 \leq i \leq d}$ generate $p^\varepsilon M$ over $\widehat{\overline{R}}$.

(iii) This follows from (ii).

**Lemma II.14.12.** *Let $M$ be an $\widehat{\overline{R}}$-module that is complete and separated for the $p$-adic topology, endowed with a continuous $\widehat{\overline{R}}$-semi-linear action of $\Sigma$, and $x_1, \dots, x_d$ elements of $M^\Sigma$ that generate $M$ over $\widehat{\overline{R}}$. Then the $\widehat{R_\infty}$-linear morphism $\widehat{R_\infty}^d \to M^\Sigma$ defined by the $(x_i)_{1 \leq i \leq d}$, is almost surjective.*

Denote by $C$ the kernel of the $\Sigma$-equivariant surjective $\widehat{\overline{R}}$-linear morphism $\widehat{\overline{R}}^d \to M$ defined by the $(x_i)_{1 \leq i \leq d}$. Since $M$ is separated, $C$ is closed in $\widehat{\overline{R}}^d$ for the $p$-adic topology. It is therefore complete and separated for the topology induced by the $p$-adic topology on $\widehat{\overline{R}}^d$; in other words, $C$ is isomorphic to the inverse limit of the inverse system of $\widehat{\overline{R}}$-modules $(C/(C \cap p^n \widehat{\overline{R}}^d))_{n \in \mathbb{N}}$. Consequently, $\mathrm{H}^1_{\mathrm{cont}}(\Sigma, C)$ is almost zero by virtue of II.6.21. On the other hand, the canonical homomorphism $\widehat{R_\infty} \to \widehat{\overline{R}}^\Sigma$ is an almost isomorphism by II.6.23, giving the lemma.

**Lemma II.14.13.** *Let $\alpha$ be a rational number $> 0$ and $M$ an $\alpha$-quasi-small $\widehat{R_\infty}$-representation of $\Delta_\infty$ such that $M$ is $\mathbb{Z}_p$-flat. Then the $\widehat{R_{p^\infty}}$-module $M^{\Sigma_0}$ endowed with the induced action of $\Delta_{p^\infty}$, is an $\alpha$-quasi-small $\widehat{R_{p^\infty}}$-representation of $\Delta_{p^\infty}$, and the canonical morphism*

$$M^{\Sigma_0} \otimes_{\widehat{R_{p^\infty}}} \widehat{R_\infty} \to M \tag{II.14.13.1}$$

*is surjective.*

By (II.3.10.4) and (II.3.10.5), we have

$$0 \to \mathrm{R}^1 \lim_{\substack{\longleftarrow \\ n}} (M/p^n M)^{\Sigma_0} \to \mathrm{H}^1_{\mathrm{cont}}(\Sigma_0, M) \to \lim_{\substack{\longleftarrow \\ n}} \mathrm{H}^1(\Sigma_0, M/p^n M) \to 0. \tag{II.14.13.2}$$

On the other hand, since $\Sigma_0$ is a profinite group of order prime to $p$, $\mathrm{H}^1(\Sigma_0, M/p^n M) = 0$ for every $n \geq 0$, and the inverse system $((M/p^n M)^{\Sigma_0})_{n \geq 0}$ satisfies the Mittag–Leffler condition. It follows that $\mathrm{H}^1_{\mathrm{cont}}(\Sigma_0, M) = 0$. Likewise, we have $\mathrm{H}^1_{\mathrm{cont}}(\Sigma_0, p^\alpha M) = 0$ because $p^\alpha M$ is complete and separated for the $p$-adic topology. Consequently, the canonical morphism

$$M^{\Sigma_0} \to (M/p^\alpha M)^{\Sigma_0} \tag{II.14.13.3}$$

is surjective.

Let $x_1, \dots, x_d$ be generators of $M$ over $R_\infty$ that are $\Delta_\infty$-invariant modulo $p^\alpha M$. Their classes in $(M/p^\alpha M)^{\Sigma_0}$ lift to elements $x'_1, \dots, x'_d$ of $M^{\Sigma_0}$ that generate $M$ over

$\widehat{R_\infty}$. Therefore the canonical morphism $M^{\Sigma_0} \otimes_{\widehat{R_{p^\infty}}} \widehat{R_\infty} \to M$ is surjective. Copying the proof of II.14.12, we show that the morphism

$$\widehat{R_{p^\infty}}^d \to M^{\Sigma_0} \tag{II.14.13.4}$$

defined by the $(x'_i)_{1\leq i\leq d}$ is surjective. Consequently, $M^{\Sigma_0}$ is an $\widehat{R_{p^\infty}}$-module of finite type. It is therefore complete and separated for the $p$-adic topology; it is complete by virtue of ([**11**] Chap. III § 2.12 Cor. 1 to Prop. 16) and is separated as a submodule of $M$. On the other hand, since $M$ is $\mathbb{Z}_p$-flat, the $p$-adic topology on $M^{\Sigma_0}$ is clearly induced by the $p$-adic topology on $M$. Hence the representation of $\Delta_{p^\infty}$ over $M^{\Sigma_0}$ induced by that of $\Delta$ is continuous for the $p$-adic topology. For every $1 \leq i \leq d$ and every $g \in \Delta_{p^\infty}$, we have $g(x'_i) - x'_i \in M^{\Sigma_0} \cap p^\alpha M = p^\alpha M^{\Sigma_0}$. It follows that $M^{\Sigma_0}$ is an $\alpha$-quasi-small $\widehat{R_{p^\infty}}$-representation of $\Delta_{p^\infty}$, giving the lemma.

**Proposition II.14.14.** *Let $\alpha, \varepsilon$ be two rational numbers such that $0 < 2\varepsilon < \alpha$ and $M$ an $\alpha$-quasi-small $\widehat{\overline{R}}$-representation of $\Delta$ such that $M$ is $\mathbb{Z}_p$-flat. Then there exist an $(\alpha - 2\varepsilon)$-quasi-small $\widehat{R_{p^\infty}}$-representation $M'$ of $\Delta_{p^\infty}$ and a $\Delta$-equivariant surjective $\widehat{\overline{R}}$-linear morphism*

$$M' \otimes_{\widehat{R_{p^\infty}}} \widehat{\overline{R}} \to p^\varepsilon M. \tag{II.14.14.1}$$

By II.14.11(ii), there exist generators $x_1, \ldots, x_d$ of $p^\varepsilon M$ over $\widehat{\overline{R}}$ that are, on the one hand, $\Delta$-invariant modulo $p^\alpha M$ and, on the other hand, $\Sigma$-invariant. For every $1 \leq i \leq d$ and every $g \in \Delta$, $g(x_i) - x_i \in p^\alpha M \cap M^\Sigma = p^\alpha M^\Sigma$. Hence by virtue of II.14.12 (applied to the $(\alpha - \varepsilon)$-quasi-small $\widehat{\overline{R}}$-representation $p^\varepsilon M$ of $\Sigma$), there exist $(a_{ij})_{1\leq j\leq d} \in \widehat{R_\infty}^d$ such that

$$g(x_i) - x_i = \sum_{j=1}^{d} p^{\alpha-2\varepsilon} a_{ij} x_j. \tag{II.14.14.2}$$

Denote by $M_1$ the sub-$\widehat{R_\infty}$-module of $p^\varepsilon M$ generated by $x_1, \ldots, x_d$. By (II.14.14.2), $M_1$ is stable under the action of $\Delta$. Since $M_1 \subset M^\Sigma$, the induced action of $\Delta$ on $M_1$ factors through $\Delta_\infty$.

The $\widehat{R_\infty}$-module $M_1$ is complete and separated for the $p$-adic topology; it is complete by virtue of ([**11**] Chap. III § 2.12 Cor. 1 to Prop. 16) and is separated as a submodule of $M$. For every integer $n \geq 2\varepsilon$, we have

$$p^n M_1 \subset M_1 \cap p^n M \subset p^n M^\Sigma \subset p^{n-2\varepsilon} M_1, \tag{II.14.14.3}$$

where the last inclusion is a consequence of II.14.12. Therefore the $p$-adic topology on $M_1$ is induced by the $p$-adic topology on $M$. In view of (II.14.14.2), it follows that $M_1$ is an $(\alpha - 2\varepsilon)$-quasi-small $\widehat{R_\infty}$-representation of $\Delta_\infty$. Since the canonical morphism

$$M_1 \otimes_{\widehat{R_\infty}} \widehat{\overline{R}} \to p^\varepsilon M \tag{II.14.14.4}$$

is clearly surjective, the proposition follows from II.14.13 applied to $M_1$.

**Proposition II.14.15.** *Suppose $d = 1$, and let $M$ be a small $\widehat{\overline{R}}[\frac{1}{p}]$-representation of $\Delta$. Then there exist a small $\widehat{R_1}$-representation $M'$ of $\Delta_\infty$ and a $\Delta$-equivariant surjective $\widehat{\overline{R}}[\frac{1}{p}]$-linear morphism*

$$M' \otimes_{\widehat{R_1}} \widehat{\overline{R}}[\frac{1}{p}] \to M. \tag{II.14.15.1}$$

Indeed, by II.13.3, there exist a quasi-small $\widehat{\overline{R}}$-representation $M^\circ$ of $\Delta$ such that $M^\circ$ is $\mathbb{Z}_p$-flat and a $\Delta$-equivariant $\widehat{\overline{R}}[\frac{1}{p}]$-linear isomorphism

$$M^\circ \otimes_{\widehat{\overline{R}}} \widehat{\overline{R}}[\frac{1}{p}] \xrightarrow{\sim} M. \tag{II.14.15.2}$$

By virtue of II.14.14, after replacing $M^\circ$ by $p^\varepsilon M^\circ$, if necessary, for a rational number $\varepsilon > 0$, we may assume that there exist a quasi-small $\widehat{R_{p^\infty}}$-representation $N$ of $\Delta_{p^\infty}$ and a $\Delta$-equivariant surjective $\widehat{\overline{R}}$-linear morphism

$$N \otimes_{\widehat{R_{p^\infty}}} \widehat{\overline{R}} \to M^\circ. \tag{II.14.15.3}$$

There exist a rational number $\alpha > \frac{1}{p-1}$ and generators $x_1, \dots, x_d$ of $N$ over $\widehat{R_{p^\infty}}$ that are $\Delta_{p^\infty}$-invariant modulo $p^{2\alpha}N$. We denote by $N'$ the free $\widehat{R_{p^\infty}}$-module with basis $e_1, \dots, e_d$ and by $\sigma\colon N' \to N$ the $\widehat{R_{p^\infty}}$-linear morphism that sends $e_i$ to $x_i$ for every $1 \leq i \leq d$. Denote by

$$\varphi_0\colon \Delta_{p^\infty} \to \mathrm{Aut}_{\widehat{R_1}}(N') \tag{II.14.15.4}$$

the trivial $\widehat{R_{p^\infty}}$-representation of $\Delta_{p^\infty}$ with respect to the basis $e_1, \dots, e_d$. We choose a $\mathbb{Z}_p$-basis $\gamma$ of $\Delta_{p^\infty}$. There exists an $\widehat{R_{p^\infty}}$-linear automorphism $u$ of $N'$ such that for every $1 \leq i \leq d$, we have $u(e_i) - e_i \in p^{2\alpha}N'$ and

$$\sigma(u(e_i)) = \gamma(x_i). \tag{II.14.15.5}$$

For every $g \in \Delta_{p^\infty}$, denote by ${}^g u$ the $\widehat{R_{p^\infty}}$-linear automorphism $\varphi_0(g) \circ u \circ \varphi_0(g^{-1})$ of $N'$. If $A \in \mathrm{GL}_d(\widehat{R_{p^\infty}})$ is the matrix of $u$ with respect to the basis $e_1, \dots, e_d$ of $N'$, then $g(A)$ is the matrix of ${}^g u$ with respect to the same basis. Let $r$ be an integer $\geq 0$ and $A_r$ the class of $A$ in $\mathrm{GL}_d(R_{p^\infty}/p^r R_{p^\infty})$. There exists an integer $m$ such that $A_r$ belongs to $\mathrm{GL}_d(R_{p^m}/p^r R_{p^m})$, so that $\gamma^{p^m}(A_r) = A_r$. Hence for every $n \geq 0$, we have

$$A_r\gamma(A_r)\gamma^2(A_r)\dots\gamma^{p^{m+n}-1}(A_r) = (A_r\gamma(A_r)\gamma^2(A_r)\dots\gamma^{p^m-1}(A_r))^{p^n}. \tag{II.14.15.6}$$

Since $A \equiv \mathrm{id} \mod (p^{2\alpha}\widehat{R_{p^\infty}})$, for $n$ sufficiently large, the product (II.14.15.6) is equal to the identity of $\mathrm{GL}_d(R_{p^\infty}/p^r R_{p^\infty})$. Consequently, for every $y \in N'$, the sequence of elements of $N'$

$$n \mapsto (u \circ \varphi_0(\gamma))^{p^n}(y) = u \circ ({}^\gamma u) \circ \dots \circ ({}^{\gamma^{p^n-1}} u) \circ \varphi_0(\gamma^{p^n})(y) \tag{II.14.15.7}$$

converges to $y$ for the $p$-adic topology. It follows that the homomorphism

$$\mathbb{Z} \to \mathrm{Aut}_{\widehat{R_1}}(N'), \quad n \mapsto (u \circ \varphi_0(\gamma))^n \tag{II.14.15.8}$$

extends to an $\widehat{R_{p^\infty}}$-representation $\varphi$ of $\Delta_{p^\infty}$ over $N'$, where we identify $\mathbb{Z}$ with a subgroup of $\Delta_{p^\infty}$ by the injection $n \mapsto \gamma^n$. It is clear that $\varphi$ is a continuous representation for the $p$-adic topology of $N'$, and is even small. Moreover, since $\varphi(\gamma) = u \circ \varphi_0(\gamma)$, the morphism $\sigma$ is $\Delta_{p^\infty}$-equivariant (II.14.15.5). The proposition then follows from II.14.4.

**Proposition II.14.16.** *Suppose $d = 1$ and that $\widehat{\overline{R}}[\frac{1}{p}]$ is faithfully flat over $\widehat{R_1}[\frac{1}{p}]$* (cf. II.14.17(i)). *Then the statements of* II.14.8 *are equivalent to the following statement:*

- ($\star$) *For all small $\widehat{\overline{R}}[\frac{1}{p}]$-representations $M$ and $M'$ of $\Delta$, and every $\Delta$-equivariant surjective $\widehat{\overline{R}}[\frac{1}{p}]$-linear morphism $u\colon M' \to M$, the $\widehat{\overline{R}}[\frac{1}{p}]$-representation of $\Delta$ on the kernel of $u$ is small.*

By II.14.8 and II.14.10, it suffices to show that ($\star$) implies II.14.8(i). Let $M$ be a small $\widehat{\overline{R}}[\frac{1}{p}]$-representation of $\Delta$. By II.14.15, there exist a small $\widehat{\overline{R}}$-representation $M'$ of $\Delta$ and a $\Delta$-equivariant surjective $\widehat{\overline{R}}[\frac{1}{p}]$-linear morphism

$$u\colon M'\otimes_{\widehat{\overline{R}}}\widehat{\overline{R}}[\frac{1}{p}]\to M. \tag{II.14.16.1}$$

By assumption, the $\widehat{\overline{R}}[\frac{1}{p}]$-representation of $\Delta$ on the kernel of $u$ is small. Applying II.14.15 once again, we obtain a small $\widehat{\overline{R}}$-representation $M''$ of $\Delta$ and an exact sequence of $\Delta$-equivariant $\widehat{\overline{R}}[\frac{1}{p}]$-linear morphisms

$$M''\otimes_{\widehat{\overline{R}}}\widehat{\overline{R}}[\frac{1}{p}]\xrightarrow{v}M'\otimes_{\widehat{\overline{R}}}\widehat{\overline{R}}[\frac{1}{p}]\xrightarrow{u}M\longrightarrow 0. \tag{II.14.16.2}$$

Replacing $M''$ by $p^nM''$ for an integer $n\geq 0$, we may assume that $v(M'')\subset M'$. We denote by $(N,\theta)$ the Higgs $\widehat{R_1}[\frac{1}{p}]$-module with coefficients in $\xi^{-1}\widetilde{\Omega}^1_{R/\mathscr{O}_K}$ that is the cokernel of the morphism

$$\mathbb{H}(v)\colon \mathbb{H}(M''\otimes_{\widehat{\overline{R}}}\widehat{\overline{R}}[\frac{1}{p}])\to\mathbb{H}(M'\otimes_{\widehat{\overline{R}}}\widehat{\overline{R}}[\frac{1}{p}]). \tag{II.14.16.3}$$

By virtue of II.14.6, the $\widehat{\overline{R}}$-representation $M'$ of $\Delta$ is Dolbeault, the Higgs $\widehat{R_1}$-module $\mathbb{H}(M')$ is small and solvable, and we have a functorial and $\Delta$-equivariant canonical $\widehat{\overline{R}}$-isomorphism (II.12.13.1)

$$\mathbb{V}(\mathbb{H}(M'))\xrightarrow{\sim}M'. \tag{II.14.16.4}$$

Consequently, by (II.13.19.2), we have a functorial $\widehat{\overline{R}}$-isomorphism

$$\mathbb{H}(M')\otimes_{\widehat{R_1}}\widehat{\overline{R}}\xrightarrow{\sim}M'. \tag{II.14.16.5}$$

We deduce from this an isomorphism

$$N\otimes_{\widehat{R_1}}\widehat{\overline{R}}\xrightarrow{\sim}M. \tag{II.14.16.6}$$

Since $\widehat{\overline{R}}[\frac{1}{p}]$ is faithfully flat over $\widehat{R_1}[\frac{1}{p}]$, $N$ is projective of finite type over $\widehat{R_1}[\frac{1}{p}]$. Consequently, as a quotient of $\mathbb{H}(M'\otimes_{\widehat{\overline{R}}}\widehat{\overline{R}}[\frac{1}{p}])$, the Higgs $\widehat{R_1}[\frac{1}{p}]$-module $(N,\theta)$ is small. Hence by virtue of II.13.23(ii) and (II.14.16.4), we have a $\Delta$-equivariant $\widehat{\overline{R}}[\frac{1}{p}]$-isomorphism $M\xrightarrow{\sim}\mathbb{V}(N)$, giving statement II.14.8(i).

**Remarks II.14.17.** (i) We expect $\widehat{\overline{R}}[\frac{1}{p}]$ to be always faithfully flat over $\widehat{R_1}[\frac{1}{p}]$. This statement has been established by Tsuji if $\mathscr{M}_X$ is defined by a divisor with strict normal crossings on $X$.

(ii) Statements II.14.15 and II.14.16 are directly inspired by the approach of Faltings ([**27**] Theorem 3, p. 852).

(iii) We expect the equivalent statements of II.14.8 to hold for every $d$.

## II.15. Hodge–Tate representations

**II.15.1.** Consider the exact sequence (II.7.22.2)

$$0\to(\pi\rho)^{-1}\widehat{\overline{R}}\to\mathscr{E}\to\widetilde{\Omega}^1_{R/\mathscr{O}_K}\otimes_R\widehat{\overline{R}}(-1)\to 0 \tag{II.15.1.1}$$

and denote also by $1 \in \mathscr{E}$ the image of $1 \in (\pi\rho)^{-1}\widehat{\overline{R}}$. For every $n \geq 0$, let (II.2.5)

$$\iota_n \colon \mathrm{S}^n(\mathscr{E}) \to \mathrm{S}^{n+1}(\mathscr{E}) \tag{II.15.1.2}$$

be the $\widehat{\overline{R}}$-linear morphism defined as follows. For $n = 0$, $\iota_0$ is the composition of the canonical injections $\widehat{\overline{R}} \to (\pi\rho)^{-1}\widehat{\overline{R}} \to \mathscr{E}$. For $n \geq 1$ and for all $x_1, \dots, x_n \in \mathscr{E}$, we have

$$\iota_n([x_1 \otimes \cdots \otimes x_n]) = [1 \otimes x_1 \otimes \cdots \otimes x_n]. \tag{II.15.1.3}$$

The $\mathrm{S}^n(\mathscr{E})$'s then form a direct system. We set

$$\mathscr{C}_{\mathrm{HT}} = \varinjlim_{n\geq 0} \mathrm{S}^n(\mathscr{E}). \tag{II.15.1.4}$$

The multiplication morphisms $\mathrm{S}^m(\mathscr{E}) \otimes_{\widehat{\overline{R}}} \mathrm{S}^n(\mathscr{E}) \to \mathrm{S}^{m+n}(\mathscr{E})$ $(m, n \geq 0)$ induce a structure of $\widehat{\overline{R}}$-algebra on $\mathscr{C}_{\mathrm{HT}}$. Note that $\mathscr{C}_{\mathrm{HT}}$ is naturally an $\widehat{\overline{R}}$-representation of $\Gamma$. The resulting representation is called the *Hyodo ring* ([**43**] § 1).

**Remark II.15.2.** Note that $p$ is invertible in $\mathscr{C}_{\mathrm{HT}}$, so that $\mathscr{C}_{\mathrm{HT}} \simeq \mathscr{C}_{\mathrm{HT}}[\frac{1}{p}]$. Indeed, we have $1 = \pi\rho \cdot (\pi\rho)^{-1}$, where $(\pi\rho)^{-1} \in (\pi\rho)^{-1}\widehat{\overline{R}} \subset \mathscr{E} \subset \mathscr{C}_{\mathrm{HT}}$.

**II.15.3.** For all integers $i \geq 0$ and $n \geq 0$, we define a $\Gamma$-equivariant $\widehat{\overline{R}}$-linear morphism

$$\varkappa_{i,n} \colon \mathrm{S}^n(\mathscr{E}) \otimes_R \widetilde{\Omega}^i_{R/\mathscr{O}_K}(-i) \to \mathrm{S}^{n-1}(\mathscr{E}) \otimes_R \widetilde{\Omega}^{i+1}_{R/\mathscr{O}_K}(-i-1) \tag{II.15.3.1}$$

by setting $\varkappa_{i,0} = 0$, and if $n \geq 1$,

$$\varkappa_{i,n}([x_1 \otimes \cdots \otimes x_n] \otimes \omega) = \sum_{1 \leq i \leq n} [x_1 \otimes \cdots \otimes x_{i-1} \otimes x_{i+1} \otimes \cdots \otimes x_n] \otimes (u(x_i) \wedge \omega), \tag{II.15.3.2}$$

where $x_1, \dots, x_n \in \mathscr{E}$, $\omega \in \widetilde{\Omega}^n_{R/\mathscr{O}_K}(-n)$, and $u \colon \mathscr{E} \to \widetilde{\Omega}^1_{R/\mathscr{O}_K} \otimes_R \widehat{\overline{R}}(-1)$ is the canonical morphism (II.15.1.1). One immediately verifies that we have $\varkappa_{i+1,n-1} \circ \varkappa_{i,n} = 0$ and that the diagram

$$\begin{array}{ccc} \mathrm{S}^n(\mathscr{E}) \otimes_R \widetilde{\Omega}^i_{R/\mathscr{O}_K}(-i) & \xrightarrow{\varkappa_{i,n}} & \mathrm{S}^{n-1}(\mathscr{E}) \otimes_R \widetilde{\Omega}^{i+1}_{R/\mathscr{O}_K}(-i-1) \\ \Big\downarrow {\scriptstyle \iota_n \otimes \mathrm{id}} & & \Big\downarrow {\scriptstyle \iota_{n-1} \otimes \mathrm{id}} \\ \mathrm{S}^{n+1}(\mathscr{E}) \otimes_R \widetilde{\Omega}^i_{R/\mathscr{O}_K}(-i) & \xrightarrow{\varkappa_{i,n+1}} & \mathrm{S}^n(\mathscr{E}) \otimes_R \widetilde{\Omega}^{i+1}_{R/\mathscr{O}_K}(-i-1) \end{array}$$

is commutative. Taking the direct limit, we obtain a $\Gamma$-equivariant $\widehat{\overline{R}}[\frac{1}{p}]$-linear morphism

$$\varkappa_i \colon \mathscr{C}_{\mathrm{HT}} \otimes_R \widetilde{\Omega}^i_{R/\mathscr{O}_K}(-i) \to \mathscr{C}_{\mathrm{HT}} \otimes_R \widetilde{\Omega}^{i+1}_{R/\mathscr{O}_K}(-i-1). \tag{II.15.3.3}$$

These morphisms define a complex $\mathscr{K}^\bullet$ of $\widehat{\overline{R}}[\frac{1}{p}]$-modules by setting

$$\mathscr{K}^i = \begin{cases} \mathscr{C}_{\mathrm{HT}} \otimes_R \widetilde{\Omega}^i_{R/\mathscr{O}_K}(-i) & \text{if } i \geq 0, \\ 0 & \text{if } i < 0. \end{cases} \tag{II.15.3.4}$$

**II.15.4.** Let $V$ be a $\mathbb{Q}_p$-representation of $\Gamma$. For every integer $i$, we denote by $\mathrm{D}^i(V)$ the $\widehat{R}[\frac{1}{p}]$-module defined by

$$\mathrm{D}^i(V) = (V \otimes_{\mathbb{Q}_p} \mathscr{C}_{\mathrm{HT}}(i))^{\Gamma}. \tag{II.15.4.1}$$

Taking the $\Gamma$-invariants of the complex $V \otimes_{\mathbb{Q}_p} \mathscr{K}^\bullet(i)$ (II.15.3.4), we obtain an $\widehat{R}[\frac{1}{p}]$-linear complex, denoted by $\mathbb{D}^i(V)$,

$$\mathrm{D}^i(V) \to \mathrm{D}^{i-1}(V) \otimes_R \widetilde{\Omega}^1_{R/\mathscr{O}_K} \to \mathrm{D}^{i-2}(V) \otimes_R \widetilde{\Omega}^2_{R/\mathscr{O}_K} \to \dots, \tag{II.15.4.2}$$

where $\mathrm{D}^i(V)$ is placed in degree zero. We endow the $\widehat{R}[\frac{1}{p}]$-module $\oplus_{i \in \mathbb{Z}} \mathrm{D}^i(V)(-i)$ with the Higgs field with coefficients in $\widetilde{\Omega}^1_{R/\mathscr{O}_K}(-1)$ induced by $\varkappa_0$ (II.15.3.3). The Dolbeault complex of the resulting Higgs module is $\oplus_{i \in \mathbb{Z}} \mathbb{D}^i(V)(-i)$ (II.2.8.4).

**Definition II.15.5** ([**43**] 2.1). We say that a continuous $\mathbb{Q}_p$-representation $V$ of $\Gamma$ is *Hodge–Tate* if the following conditions are satisfied:

(i) $V$ is a $\mathbb{Q}_p$-vector space of finite dimension, endowed with the $p$-adic topology (II.2.2).

(ii) The canonical morphism

$$\oplus_{i \in \mathbb{Z}} \mathrm{D}^i(V) \otimes_{\widehat{R}[\frac{1}{p}]} \mathscr{C}_{\mathrm{HT}}(-i) \to V \otimes_{\mathbb{Q}_p} \mathscr{C}_{\mathrm{HT}} \tag{II.15.5.1}$$

is an isomorphism.

We can make the following remarks.

II.15.5.1. Hyodo shows in loc. cit. that for every continuous $\mathbb{Q}_p$-representation $V$ of $\Gamma$ of finite dimension, the morphism (II.15.5.1) is injective.

II.15.5.2. For every Hodge–Tate $\mathbb{Q}_p$-representation $V$ of $\Gamma$, the $\widehat{R}[\frac{1}{p}]$-modules $\mathrm{D}^i(V)$ ($i \in \mathbb{Z}$) are locally free of finite type and are zero except for a finite number of integers $i$, called the *Hodge–Tate weights* of $V$ (cf. [**14**] 4.2.7).

**Proposition II.15.6.** *Let $(\widetilde{X}_0, \mathscr{M}_{\widetilde{X}_0})$ be the smooth $(\mathscr{A}_2(\overline{S}), \mathscr{M}_{\mathscr{A}_2(\overline{S})})$-deformation of $(\check{\overline{X}}, \mathscr{M}_{\check{\overline{X}}})$ defined by the chart $(P, \gamma)$ (II.10.13.1), $(P^{\mathrm{gp}}/\mathbb{Z}\lambda)_{\mathrm{lib}}$ the quotient of $P^{\mathrm{gp}}/\mathbb{Z}\lambda$ by its torsion submodule, and*

$$w \colon (P^{\mathrm{gp}}/\mathbb{Z}\lambda)_{\mathrm{lib}} \to P^{\mathrm{gp}} \tag{II.15.6.1}$$

*a right inverse of the canonical morphism $P^{\mathrm{gp}} \to (P^{\mathrm{gp}}/\mathbb{Z}\lambda)_{\mathrm{lib}}$. We denote by $\mathscr{C}_0$ the Higgs–Tate $\widehat{\overline{R}}$-algebra (II.10.5) and by $\mathscr{F}_0$ the Higgs–Tate $\widehat{\overline{R}}$-extension of (II.10.5) associated with $(\widetilde{X}_0, \mathscr{M}_{\widetilde{X}_0})$. Then there exists a canonical isomorphism of $\widehat{\overline{R}}[\frac{1}{p}]$-algebras*

$$\widetilde{\beta}_w \colon \mathscr{C}_0[\frac{1}{p}] \to \mathscr{C}_{\mathrm{HT}} \tag{II.15.6.2}$$

*such that for every $x \in p^{-\frac{1}{p-1}}\mathscr{F}_0 \subset \mathscr{C}_0[\frac{1}{p}]$, we have*

$$\widetilde{\beta}_w(x) = \beta_w(x) \in \mathscr{E} \subset \mathscr{C}_{\mathrm{HT}}, \tag{II.15.6.3}$$

*where $\beta_w$ is the morphism (II.10.18.2). Moreover, $\widetilde{\beta}_w$ is $\Delta$-equivariant and the diagram* (II.15.6.4)

$$\begin{array}{ccccc}
\mathscr{C}_0 & & \xrightarrow{\quad\quad\widetilde{\beta}_w\quad\quad} & & \mathscr{C}_{\mathrm{HT}} \\
\Big\downarrow{\scriptstyle d_{\mathscr{C}_0}} & & & & \Big\downarrow{\scriptstyle \varkappa_0} \\
\xi^{-1}\mathscr{C}_0 \otimes_R \widetilde{\Omega}^1_{R/\mathscr{O}_K} & \xrightarrow[\sim]{v} & p^{-\frac{1}{p-1}}\mathscr{C}_0 \otimes_R \widetilde{\Omega}^1_{R/\mathscr{O}_K}(-1) & \xrightarrow{\widetilde{\beta}_w \otimes \mathrm{id}} & \mathscr{C}_{\mathrm{HT}} \otimes_R \widetilde{\Omega}^1_{R/\mathscr{O}_K}(-1)
\end{array}$$

*where $v$ is the isomorphism induced by* (II.9.18.1), *is commutative.*

This follows from (II.10.3.7) and II.10.18.

**Proposition II.15.7.** *We keep the assumptions of* II.15.6 *and let $V$ be a Hodge–Tate $\mathbb{Q}_p$-representation of $\Gamma$, $(\mathbb{H}_0(V\otimes_{\mathbb{Z}_p}\widehat{\overline{R}}),\theta)$ the Higgs module with coefficients in $\xi^{-1}\widetilde{\Omega}^1_{R/\mathscr{O}_K}$ associated with the $\widehat{\overline{R}}[\frac{1}{p}]$-representation $V\otimes_{\mathbb{Z}_p}\widehat{\overline{R}}$ of $\Delta$ by the functor* (II.12.8.2) *with respect to the deformation $(\widetilde{X}_0,\mathscr{M}_{\widetilde{X}_0})$, and $\theta'$ the Higgs field on $\mathbb{H}_0(V\otimes_{\mathbb{Z}_p}\widehat{\overline{R}})$ with coefficients in $\widetilde{\Omega}^1_{R/\mathscr{O}_K}(-1)$ deduced from $\theta$ and the isomorphism $\widehat{\overline{R}}(1)\xrightarrow{\sim}p^{\frac{1}{p-1}}\xi\widehat{\overline{R}}$* (II.9.18.1). *Then $V\otimes_{\mathbb{Z}_p}\widehat{\overline{R}}$ is a Dolbeault $\widehat{\overline{R}}[\frac{1}{p}]$-representation of $\Delta$, and we have a functorial canonical $\widehat{R_1}[\frac{1}{p}]$-isomorphism of Higgs $\widehat{R_1}[\frac{1}{p}]$-modules with coefficients in $\widetilde{\Omega}^1_{R/\mathscr{O}_K}(-1)$*

$$\oplus_{i\in\mathbb{Z}}\,\mathrm{D}^i(V)\otimes_{\widehat{R}}\widehat{R_1}(-i)\xrightarrow{\sim}\mathbb{H}_0(V\otimes_{\mathbb{Z}_p}\widehat{\overline{R}}),\tag{II.15.7.1}$$

*where the left-hand side is endowed with the Higgs field induced by $\varkappa_0$* (II.15.3.3) *and the right-hand side is endowed with $\theta'$.*

First, note that $V\otimes_{\mathbb{Z}_p}\widehat{\overline{R}}$ is a continuous $\widehat{\overline{R}}[\frac{1}{p}]$-representation of $\Delta$. If we endow $\mathrm{D}^i(V)$ with the trivial action of $\Gamma$, $V\otimes_{\mathbb{Z}_p}\widehat{\overline{R}}$ with the trivial Higgs field, and $\oplus_{i\in\mathbb{Z}}\mathrm{D}^i(V)\otimes_{\widehat{R}}\widehat{\overline{R}}(-i)$ and $\mathscr{C}_{\mathrm{HT}}$ with the Higgs fields induced by $\varkappa_0$, then the canonical morphism

$$\oplus_{i\in\mathbb{Z}}\,\mathrm{D}^i(V)\otimes_{\widehat{R}[\frac{1}{p}]}\mathscr{C}_{\mathrm{HT}}(-i)\to V\otimes_{\mathbb{Q}_p}\mathscr{C}_{\mathrm{HT}}\tag{II.15.7.2}$$

is a $\Gamma$-equivariant $\mathscr{C}_{\mathrm{HT}}$-linear isomorphism of Higgs $\widehat{\overline{R}}[\frac{1}{p}]$-modules with coefficients in $\widetilde{\Omega}^1_{R/\mathscr{O}_K}(-1)$. In view of II.15.6, we deduce from this a $\Delta$-equivariant $\mathscr{C}_0^\dagger$-linear isomorphism of Higgs $\widehat{\overline{R}}[\frac{1}{p}]$-modules with coefficients in $\widetilde{\Omega}^1_{R/\mathscr{O}_K}(-1)$

$$\oplus_{i\in\mathbb{Z}}\,\mathrm{D}^i(V)\otimes_{\widehat{R}}\mathscr{C}_0^\dagger(-i)\xrightarrow{\sim}V\otimes_{\mathbb{Z}_p}\mathscr{C}_0^\dagger,\tag{II.15.7.3}$$

where $\mathscr{C}_0^\dagger[\frac{1}{p}]$ is endowed with the Higgs field induced by $d_{\mathscr{C}_0^\dagger}$ and by the isomorphism $\widehat{\overline{R}}(1)\xrightarrow{\sim}p^{\frac{1}{p-1}}\xi\widehat{\overline{R}}$ (II.9.18.1). Since $\mathrm{D}^i(V)$ is a direct summand of a free $\widehat{R}[\frac{1}{p}]$-module of finite type (II.15.5.2), we have $(\mathrm{D}^i(V)\otimes_{\widehat{R}}\mathscr{C}_0^\dagger)^\Delta=\mathrm{D}^i(V)\otimes_{\widehat{R}}\widehat{R_1}$ by virtue of II.12.6. Consequently, by taking, in (II.15.7.3), the invariants under $\Delta$, we obtain an isomorphism of Higgs $\widehat{R_1}[\frac{1}{p}]$-modules with coefficients in $\widetilde{\Omega}^1_{R/\mathscr{O}_K}(-1)$

$$\oplus_{i\in\mathbb{Z}}\,\mathrm{D}^i(V)\otimes_{\widehat{R}}\widehat{R_1}(-i)\xrightarrow{\sim}(V\otimes_{\mathbb{Z}_p}\mathscr{C}_0^\dagger)^\Delta=\mathbb{H}_0(V\otimes_{\mathbb{Z}_p}\widehat{\overline{R}}).\tag{II.15.7.4}$$

Moreover, the canonical morphism

$$\mathbb{H}_0(V\otimes_{\mathbb{Z}_p}\widehat{\overline{R}})\otimes_{\widehat{R_1}}\mathscr{C}_0^\dagger\to V\otimes_{\mathbb{Z}_p}\mathscr{C}_0^\dagger\tag{II.15.7.5}$$

identifies with the isomorphism (II.15.7.3). Hence $V\otimes_{\mathbb{Z}_p}\widehat{\overline{R}}$ is a Dolbeault $\widehat{\overline{R}}[\frac{1}{p}]$-representation of $\Delta$.

CHAPTER III

# Representations of the fundamental group and the torsor of deformations. Global aspects

AHMED ABBES AND MICHEL GROS

## III.1. Introduction

In this chapter we continue the construction and study of the $p$-adic Simpson correspondence started in Chapter II, following the general approach summarized in Chapter I. After fixing the notation and general conventions in III.2, we develop, in Sections III.3 to III.7, preliminaries that are useful for later on. Section III.3 contains results and complements on the notion of locally irreducible schemes. In III.4, we fix the logarithmic geometry setting of our constructions. The interlude III.5 contains a number of results on the Koszul complex used in different places in this book. Section III.6 develops the formalism of additive categories up to isogeny. Section III.7 is devoted to the study of the inverse systems of a ringed topos, in particular to the notion of adic modules and to the finiteness conditions adapted to this setting.

Let $K$ be a complete discrete valuation ring of characteristic 0, with algebraically closed residue field of characteristic $p$ and let $\overline{K}$ be an algebraic closure of $K$. We denote by $\mathscr{O}_K$ the valuation ring of $K$, by $\mathscr{O}_{\overline{K}}$ the integral closure of $\mathscr{O}_K$ in $\overline{K}$, and by $\mathscr{O}_C$ the $p$-adic Hausdorff completion of $\mathscr{O}_{\overline{K}}$. We set $S = \operatorname{Spec}(\mathscr{O}_K)$ and endow it with the logarithmic structure $\mathscr{M}_S$ defined by its closed point. We consider in this chapter a logarithmic scheme $(X, \mathscr{M}_X)$ smooth and saturated over $(S, \mathscr{M}_S)$, satisfying a local condition (III.4.7) corresponding to the assumptions made in the first part (II.6.2). We denote by $X^\circ$ the maximal open subscheme of $X$ where the logarithmic structure $\mathscr{M}_X$ is trivial. The topological setting in which the $p$-adic Simpson correspondence takes place is that of the Faltings topos $\widetilde{E}$ associated with the canonical morphism $X^\circ \otimes_K \overline{K} \to X$, whose detailed study has been undertaken independently in Chapter VI. In III.8 we endow it with a ring $\overline{\mathscr{B}}$. We then introduce, in III.9, the topos $\widetilde{E}_s$ special fiber of $\widetilde{E}$ and the ringed topos $(\widetilde{E}_s^{\mathbb{N}^\circ}, \breve{\overline{\mathscr{B}}})$ $p$-adic formal completion of $(\widetilde{E}, \overline{\mathscr{B}})$, where our main constructions will take place.

As sketched in the general introduction in Chapter I, the $p$-adic Simpson correspondence depends on a smooth (logarithmic) deformation of $X \otimes_{\mathscr{O}_K} \mathscr{O}_C$ over the universal $p$-adic infinitesimal thickening of $\mathscr{O}_C$ of order $\leq 1$, introduced by Fontaine (II.9.8). In the remainder of this introduction, we assume that there exists such a deformation, which we fix. In Section III.10, we define, for every rational number $r \geq 0$, the *Higgs–Tate* $\breve{\overline{\mathscr{B}}}$*-algebra of thickness* $r$, denoted by $\breve{\mathscr{C}}^{(r)}$. These algebras form a direct system: for all rational numbers $r \geq r' \geq 0$, we have a canonical homomorphism $\breve{\mathscr{C}}^{(r)} \to \breve{\mathscr{C}}^{(r')}$. These are sheaf-theoretic analogues of the Higgs–Tate algebras introduced in the first part (II.12.1). They are naturally endowed with Higgs fields. Section III.11 contains two acyclicity results that are fundamental for what comes after. We prove (III.11.24) that the direct limit of the Dolbeault complexes of $\breve{\mathscr{C}}^{(r)}$, for $r \in \mathbb{Q}_{>0}$, is a resolution of $\breve{\overline{\mathscr{B}}}$, up to isogeny. On the other hand, if we denote by $\mathfrak{X}$ the formal scheme $p$-adic completion of $X \otimes_{\mathscr{O}_K} \mathscr{O}_{\overline{K}}$, then we have a canonical morphism of ringed topos

$$\top\colon (\widetilde{E}_s^{\mathbb{N}^\circ}, \breve{\overline{\mathscr{B}}}) \to (\mathfrak{X}_{\text{zar}}, \mathscr{O}_{\mathfrak{X}}).$$

We prove (III.11.18) that the canonical homomorphism

$$\mathscr{O}_{\mathfrak{X}}[\frac{1}{p}] \to \lim_{\substack{\longrightarrow \\ r\in\mathbb{Q}_{>0}}} \top_*(\breve{\mathscr{C}}^{(r)})[\frac{1}{p}]$$

is an isomorphism and that for every integer $q \geq 1$,

$$\lim_{\substack{\longrightarrow \\ r\in\mathbb{Q}_{>0}}} \mathrm{R}^q\top_*(\breve{\mathscr{C}}^{(r)})[\frac{1}{p}] = 0.$$

Section III.12 is devoted to the construction of the $p$-adic Simpson correspondence. We introduce the notions of Dolbeault modules and solvable Higgs bundles (III.12.11) and we show (III.12.26) that they lead to two equivalent categories. We in fact construct two explicit equivalences of categories that are quasi-inverse to each other. We also prove the compatibility of this correspondence with the relevant cohomologies on each side (III.12.34). In III.13 we establish a link with the constructions developed in the first part of Chapter II for affine schemes of a certain type, called *small* by Faltings. In III.14 we study the functoriality of the $p$-adic Simpson correspondence for étale morphisms. Section III.15 is devoted to studying the fibered category of Dolbeault modules over the restricted étale site of $X$. We show the local character for the étale topology on $X$ of the Dolbeault property for modules (III.15.4). The analogous assertion for the solvability property of Higgs bundles is equivalent to the fact that the fibered category of Dolbeault modules is a stack (III.15.5). We say that a Higgs bundle is *small* if its Higgs field satisfies a certain divisibility condition with respect to a lattice, and that it is *locally small* if this condition holds locally (III.15.6). We show that every solvable Higgs bundle is locally small (III.15.8). Conversely, if the Dolbeault modules form a stack, then every locally small Higgs bundle is solvable (III.15.10). For a small affine scheme, we show unconditionally that every small Higgs bundle is solvable (III.15.9).

## III.2. Notation and conventions

*All rings in this chapter have an identity element; all ring homomorphisms map the identity element to the identity element. We mostly consider commutative rings, and rings are assumed to be commutative unless stated otherwise; in particular, when we take a ringed topos $(X, A)$, the ring $A$ is assumed to be commutative unless stated otherwise.*

**III.2.1.** In this chapter, $p$ denotes a prime number, $K$ a complete discrete valuation ring of characteristic 0, with *algebraically closed* residue field $k$ of characteristic $p$, and $\overline{K}$ an algebraic closure of $K$. We denote by $\mathscr{O}_K$ the valuation ring of $K$, by $\mathscr{O}_{\overline{K}}$ the integral closure of $\mathscr{O}_K$ in $\overline{K}$, by $\mathfrak{m}_{\overline{K}}$ the maximal ideal of $\mathscr{O}_{\overline{K}}$, and by $v$ the valuation of $\overline{K}$ normalized so that $v(p) = 1$. We denote by $\mathscr{O}_C$ the $p$-adic Hausdorff completion of $\mathscr{O}_{\overline{K}}$, by $C$ its field of fractions, and by $\mathfrak{m}_C$ its maximal ideal. Unless stated otherwise, we view $\mathscr{O}_C$ as an adic ring, endowed with the $p$-adic topology ([**1**] 1.8.7); it is a 1-valuative ring ([**1**] 1.9.9).

We choose a compatible system $(\beta_n)_{n>0}$ of $n$th roots of $p$ in $\mathscr{O}_{\overline{K}}$. For any rational number $\varepsilon > 0$, we set $p^\varepsilon = (\beta_n)^{\varepsilon n}$, where $n$ is an integer $> 0$ such that $\varepsilon n$ is an integer.

For any abelian group $A$, we denote by $\widehat{A}$ its $p$-adic Hausdorff completion.

We set $S = \mathrm{Spec}(\mathscr{O}_K)$, $\overline{S} = \mathrm{Spec}(\mathscr{O}_{\overline{K}})$, and $\check{\overline{S}} = \mathrm{Spec}(\mathscr{O}_C)$. We denote by $s$ (resp. $\eta$, resp. $\overline{\eta}$) the closed point of $S$ (resp. generic point of $S$, resp. generic point of $\overline{S}$). For any integer $n \geq 1$, we set $S_n = \mathrm{Spec}(\mathscr{O}_K/p^n\mathscr{O}_K)$. For any $S$-scheme $X$, we set

$$\overline{X} = X \times_S \overline{S}, \quad \check{\overline{X}} = X \times_S \check{\overline{S}}, \quad \text{and} \quad X_n = X \times_S S_n. \tag{III.2.1.1}$$

We endow $S$ with the logarithmic structure $\mathscr{M}_S$ defined by its closed point; in other words, $\mathscr{M}_S = j_*(\mathscr{O}_\eta^\times) \cap \mathscr{O}_S$, where $j\colon \eta \to S$ is the canonical injection (cf. II.5.9). Note that a homomorphism of monoids $\iota\colon \mathbb{N} \to \Gamma(S, \mathscr{M}_S)$ is a chart for $(S, \mathscr{M}_S)$ (II.5.13) if and only if $\iota(1)$ is a uniformizer of $\mathscr{O}_K$.

We endow $\overline{S}$ and $\check{\overline{S}}$ with the logarithmic structures $\mathscr{M}_{\overline{S}}$ and $\mathscr{M}_{\check{\overline{S}}}$ inverse images of $\mathscr{M}_S$ (cf. II.5.10).

We denote by $\mathscr{S} = \mathrm{Spf}(\mathscr{O}_C)$ the formal scheme $p$-adic completion of $\overline{S}$ or, equivalently, of $\check{\overline{S}}$.

**III.2.2.** Recall (II.9.3) that Fontaine associates functorially with every $\mathbb{Z}_{(p)}$-algebra $A$, the ring

$$\mathscr{R}_A = \varprojlim_{x \mapsto x^p} A/pA \tag{III.2.2.1}$$

and a homomorphism $\theta$ from the ring $\mathrm{W}(\mathscr{R}_A)$ of Witt vectors of $\mathscr{R}_A$ to the $p$-adic Hausdorff completion $\widehat{A}$ of $A$. We set

$$\mathscr{A}_2(A) = \mathrm{W}(\mathscr{R}_A)/\ker(\theta)^2 \tag{III.2.2.2}$$

and denote also by $\theta\colon \mathscr{A}_2(A) \to \widehat{A}$ the homomorphism induced by $\theta$.

**III.2.3.** In this chapter, we fix a sequence $(p_n)_{n\geq 0}$ of elements of $\mathscr{O}_{\overline{K}}$ such that $p_0 = p$ and $p_{n+1}^p = p_n$ for every $n \geq 0$. We denote by $\underline{p}$ the element of $\mathscr{R}_{\mathscr{O}_{\overline{K}}}$ (III.2.2.1) induced by the sequence $(p_n)_{n\geq 0}$ and set

$$\xi = [\underline{p}] - p \in \mathrm{W}(\mathscr{R}_{\mathscr{O}_{\overline{K}}}), \tag{III.2.3.1}$$

where $[\underline{p}]$ is the multiplicative representative of $\underline{p}$. The homomorphism $\theta\colon \mathrm{W}(\mathscr{R}_{\mathscr{O}_{\overline{K}}}) \to \mathscr{O}_C$ is surjective and its kernel is generated by $\xi$, which is not a zero divisor in $\mathrm{W}(\mathscr{R}_{\mathscr{O}_{\overline{K}}})$ (II.9.5). We therefore have an exact sequence

$$0 \longrightarrow \mathscr{O}_C \xrightarrow{\cdot\xi} \mathscr{A}_2(\mathscr{O}_{\overline{K}}) \xrightarrow{\theta} \mathscr{O}_{\overline{K}} \longrightarrow 0, \tag{III.2.3.2}$$

where we have denoted also by $\cdot\xi$ the morphism induced by the multiplication by $\xi$ in $\mathscr{A}_2(\overline{R})$. The ideal $\ker(\theta)$ of $\mathscr{A}_2(\mathscr{O}_{\overline{K}})$ has square zero. It is a free $\mathscr{O}_C$-module with basis $\xi$. We will denote it by $\xi\mathscr{O}_C$. Note that, unlike $\xi$, this module does not depend on the choice of the sequence $(p_n)_{n\geq 0}$.

We denote by $\xi^{-1}\mathscr{O}_C$ the dual $\mathscr{O}_C$-module of $\xi\mathscr{O}_C$. For any $\mathscr{O}_C$-module $M$, we denote the $\mathscr{O}_C$-modules $M \otimes_{\mathscr{O}_C} (\xi\mathscr{O}_C)$ and $M \otimes_{\mathscr{O}_C} (\xi^{-1}\mathscr{O}_C)$ simply by $\xi M$ and $\xi^{-1}M$, respectively. Note that, unlike $\xi$, these modules do not depend on the choice of the sequence $(p_n)_{n\geq 0}$. It is therefore important to not identify them with $M$.

We set

$$\mathscr{A}_2(\overline{S}) = \mathrm{Spec}(\mathscr{A}_2(\mathscr{O}_{\overline{K}})) \tag{III.2.3.3}$$

which we endow with the logarithmic structure $\mathscr{M}_{\mathscr{A}_2(\overline{S})}$ defined in (II.9.8). The logarithmic scheme $(\mathscr{A}_2(\overline{S}), \mathscr{M}_{\mathscr{A}_2(\overline{S})})$ is then fine and saturated, and $\theta$ induces an exact closed immersion

$$i_{\overline{S}}\colon (\check{\overline{S}}, \mathscr{M}_{\check{\overline{S}}}) \to (\mathscr{A}_2(\overline{S}), \mathscr{M}_{\mathscr{A}_2(\overline{S})}). \tag{III.2.3.4}$$

**III.2.4.** For any abelian category $\mathbf{A}$, we denote by $\mathbf{D}(\mathbf{A})$ its derived category and by $\mathbf{D}^-(\mathbf{A})$, $\mathbf{D}^+(\mathbf{A})$, and $\mathbf{D}^b(\mathbf{A})$ the full subcategories of $\mathbf{D}(\mathbf{A})$ of complexes with cohomology bounded from above, from below, and from both sides, respectively. Unless mentioned otherwise, complexes in $\mathbf{A}$ have a differential of degree $+1$, the degree being written as an exponent.

**III.2.5.** For this entire chapter, we fix a universe $\mathbb{U}$ with an element of infinite cardinality. We call category of $\mathbb{U}$-sets, and denote by **Ens**, the category of sets that are in $\mathbb{U}$. Unless stated otherwise, the schemes in this chapter are assumed to be elements of the universe $\mathbb{U}$. We denote by **Sch** the category of schemes that are elements of $\mathbb{U}$.

**III.2.6.** Following the conventions of ([**2**] VI), we use the adjective *coherent* as a synonym for quasi-compact and quasi-separated.

**III.2.7.** Let $(X, A)$ be a ringed topos. We denote the category of $A$-modules of $X$ by $\mathbf{Mod}(A)$ or $\mathbf{Mod}(A, X)$. If $M$ is an $A$-module, then $\mathrm{S}_A(M)$ (resp. $\wedge_A(M)$, resp. $\Gamma_A(M)$) is the symmetric algebra (resp. the exterior algebra, resp. the divided power algebra) of $M$ ([**45**] I 4.2.2.6). For any integer $n \geq 0$, we denote the homogeneous part of degree $n$ by $\mathrm{S}^n_A(M)$ (resp. $\wedge^n_A(M)$, resp. $\Gamma^n_A(M)$). We will leave the ring $A$ out of the notation when there is no risk of confusion. Forming these algebras commutes with localizing over an object of $X$.

**Definition III.2.8** ([**4**] I 1.3.1)**.** Let $(X, A)$ be a ringed topos. We say that an $A$-module $M$ of $X$ is *locally projective of finite type* if the following equivalent conditions are satisfied:

(i) $M$ is of finite type and the functor $\mathscr{H}om_A(M, \cdot)$ is exact;
(ii) $M$ is of finite type and every epimorphism of $A$-modules $N \to M$ admits locally a section;
(iii) $M$ is locally a direct summand of a free $A$-module of finite type.

When $X$ has enough points, and for every point $x$ of $X$, the stalk of $A$ at $x$ is a local ring, the locally projective $A$-modules of finite type are locally free $A$-modules of finite type ([**4**] I 2.15.1).

**III.2.9.** For any scheme $X$, we denote by $\mathbf{\acute{E}t}_{/X}$ (resp. $X_{\text{ét}}$) the étale site (resp. topos) of $X$. We denote by $\mathbf{\acute{E}t}_{\mathrm{f}/X}$ the finite étale site of $X$, that is, the full subcategory of $\mathbf{\acute{E}t}_{/X}$ made up of the finite étale schemes over $X$, endowed with the topology induced by that of $\mathbf{\acute{E}t}_{/X}$, and we denote by $X_{\text{fét}}$ the finite étale topos of $X$, that is, the topos of sheaves of $\mathbb{U}$-sets on $\mathbf{\acute{E}t}_{\mathrm{f}/X}$ (cf. VI.9.2). The canonical injection $\mathbf{\acute{E}t}_{\mathrm{f}/X} \to \mathbf{\acute{E}t}_{/X}$ induces a morphism of topos

(III.2.9.1)
$$\rho_X \colon X_{\text{ét}} \to X_{\text{fét}}.$$

We denote by $X_{\text{zar}}$ the Zariski topos of $X$ and by

(III.2.9.2)
$$u_X \colon X_{\text{ét}} \to X_{\text{zar}}$$

the canonical morphism ([**2**] VII 4.2.2). If $F$ is a quasi-coherent $\mathscr{O}_X$-module of $X_{\text{zar}}$, we denote also by $F$ the sheaf on $X_{\text{ét}}$ defined, for every étale $X$-scheme $X'$, by ([**2**] VII 2 c))

(III.2.9.3)
$$F(X') = \Gamma(X', F \otimes_{\mathscr{O}_X} \mathscr{O}_{X'}).$$

This abuse of notation does not lead to any confusion. We have a canonical isomorphism

(III.2.9.4)
$$u_{X*}(F) \xrightarrow{\sim} F.$$

We therefore view $u_X$ as a morphism from the ringed topos $(X_{\text{ét}}, \mathscr{O}_X)$ to the ringed topos $(X_{\text{zar}}, \mathscr{O}_X)$. For modules, we use the notation $u_X^{-1}$ to denote the inverse image in the sense of abelian sheaves, and we keep the notation $u_X^*$ for the inverse image in the sense of modules. The isomorphism (III.2.9.4) induces by adjunction a morphism

$$u_X^*(F) \to F. \tag{III.2.9.5}$$

This is an isomorphism if $F$ is an $\mathscr{O}_X$-module of finite presentation. Indeed, since the question is local, we may restrict to the case where there exists an exact sequence of $\mathscr{O}_X$-modules $\mathscr{O}_X^m \to \mathscr{O}_X^n \to F \to 0$ of $X_{\text{zar}}$. This induces an exact sequence of $\mathscr{O}_X$-modules $\mathscr{O}_X^m \to \mathscr{O}_X^n \to F \to 0$ of $X_{\text{ét}}$. The assertion follows in view of the right exactness of the functor $u_X^*$.

**III.2.10.** Let $X$ be a connected scheme and $\overline{x}$ a geometric point of $X$. We denote by

$$\omega_{\overline{x}} \colon \mathbf{\acute{E}t}_{\mathrm{f}/X} \to \mathbf{Ens} \tag{III.2.10.1}$$

the fiber functor at $\overline{x}$, which associates with each finite étale cover $Y$ of $X$ the set of geometric points of $Y$ over $\overline{x}$, by $\pi_1(X, \overline{x})$ the fundamental group of $X$ at $\overline{x}$ (that is, the automorphism group of the functor $\omega_{\overline{x}}$), and by $\mathbf{B}_{\pi_1(X,\overline{x})}$ the classifying topos of the profinite group $\pi_1(X, \overline{x})$, that is, the category of discrete $\mathbb{U}$-sets endowed with a continuous left action of $\pi_1(X, \overline{x})$ ([**2**] IV 2.7). Then $\omega_{\overline{x}}$ induces a fully faithful functor

$$\mu_{\overline{x}}^+ \colon \mathbf{\acute{E}t}_{\mathrm{f}/X} \to \mathbf{B}_{\pi_1(X,\overline{x})} \tag{III.2.10.2}$$

whose essential image is the full subcategory of $\mathbf{B}_{\pi_1(X,\overline{x})}$ made up of finite sets ([**37**] V § 4 and § 7). Let $(X_i)_{i\in I}$ be an inverse system on a filtered ordered set $I$ in $\mathbf{\acute{E}t}_{\mathrm{f}/X}$ that prorepresents $\omega_{\overline{x}}$, normalized by the fact that the transition morphisms $X_i \to X_j$ $(i \geq j)$ are epimorphisms and that every epimorphism $X_i \to X'$ of $\mathbf{\acute{E}t}_{\mathrm{f}/X}$ is equivalent to a suitable epimorphism $X_i \to X_j$ $(j \leq i)$. Such a pro-object is essentially unique. It is called the *normalized universal cover of* $X$ *at* $\overline{x}$ or the *normalized fundamental pro-object of* $\mathbf{\acute{E}t}_{\mathrm{f}/X}$ *at* $\overline{x}$. Note that the set $I$ is $\mathbb{U}$-small. The functor

$$\nu_{\overline{x}} \colon X_{\text{fét}} \to \mathbf{B}_{\pi_1(X,\overline{x})}, \quad F \mapsto \varinjlim_{i\in I} F(X_i) \tag{III.2.10.3}$$

is an equivalence of categories that extends the functor $\mu_{\overline{x}}^+$ (cf. VI.9.8). We call it the fiber functor of $X_{\text{fét}}$ at $\overline{x}$.

**III.2.11.** We keep the assumptions of III.2.10 and let moreover $R$ be a ring of $X_{\text{fét}}$. Set $R_{\overline{x}} = \nu_{\overline{x}}(R)$, which is a ring endowed with the discrete topology and a continuous action of $\pi_1(X, \overline{x})$ by ring homomorphisms. We denote by $\mathbf{Rep}^{\text{disc}}_{R_{\overline{y}}}(\pi_1(Y, \overline{y}))$ the category of continuous $R_{\overline{y}}$-representations of $\pi_1(Y, \overline{y})$ for which the topology is discrete (II.3.1). By restricting the functor $\nu_{\overline{x}}$ to $R$-modules, we obtain an equivalence of categories that we denote also by

$$\nu_{\overline{x}} \colon \mathbf{Mod}(R) \xrightarrow{\sim} \mathbf{Rep}^{\text{disc}}_{R_{\overline{x}}}(\pi_1(X, \overline{x})). \tag{III.2.11.1}$$

An $R$-module $M$ of $X_{\text{fét}}$ is of finite type (resp. locally projective of finite type (III.2.8)) if and only if the $R_{\overline{x}}$-module underlying $\nu_{\overline{x}}(M)$ is of finite type (resp. projective of finite type). Indeed, the condition is necessary by virtue of VI.9.9. Let us show that it is sufficient. First suppose that the $R_{\overline{x}}$-module $\nu_{\overline{x}}(M)$ is of finite type. Let $N$ be a free $R_{\overline{x}}$-module with basis $e_1, \dots, e_d$ and $u \colon N \to \nu_{\overline{x}}(M)$ an $R_{\overline{x}}$-linear epimorphism. Since the desired assertion is local for $X_{\text{fét}}$, after replacing $X$ by a finite étale cover, if necessary, we may assume that $\pi_1(X, \overline{x})$ fixes the elements $u(e_1), \dots, u(e_d)$ of $\nu_{\overline{x}}(M)$. Endowing $N$ with the unique $R_{\overline{x}}$-representation of $\pi_1(X, \overline{x})$ that fixes $e_1, \dots, e_d$, the homomorphism

$u$ is then $\pi_1(X, \overline{x})$-equivariant. Consequently, $M$ is an $R$-module of finite type. Suppose, moreover, that the $R_{\overline{x}}$-module $\nu_{\overline{x}}(M)$ is projective of finite type. Let $v\colon \nu_{\overline{x}}(M) \to N$ be an $R_{\overline{x}}$-linear splitting of $u$. After once more replacing $X$ by an étale cover, if necessary, we may assume that $\pi_1(X, \overline{x})$ fixes the elements $v(u(e_1)), \dots, v(u(e_d))$ of $N$. Consequently, $v$ is $\pi_1(X, \overline{x})$-equivariant. It follows that $M$ is the direct summand of a free $R$-module of finite type, giving the assertion.

## III.3. Locally irreducible schemes

**III.3.1.** Let $X$ be a scheme whose set of irreducible components is locally finite. Recall that the following conditions are equivalent ([**39**] 0.2.1.6):

(i) The irreducible components of $X$ are open.
(ii) The irreducible components of $X$ are identical to its connected components.
(iii) The connected components of $X$ are irreducible.
(iv) Two distinct irreducible components of $X$ do not meet.

The scheme $X$ is then the sum of the schemes induced on its irreducible components. When these conditions are satisfied, we say that $X$ is *locally irreducible*. This notion is clearly local on $X$; that is, if $(X_i)_{i \in I}$ is an open covering of $X$, then $X$ is locally irreducible if and only if for every $i \in I$, the same holds for $X_i$.

**Remarks III.3.2.** (i) The set of irreducible components of a locally noetherian scheme is locally finite ([**39**] 0.2.2.2).

(ii) A normal scheme is locally irreducible if and only if the set of its irreducible components is locally finite. Indeed, every normal scheme clearly satisfies condition III.3.1(iv).

(iii) A locally irreducible scheme $X$ is étale-locally connected; that is, for every étale morphism $X' \to X$, every connected component of $X'$ is an open set in $X'$ (VI.9.7.3).

**Lemma III.3.3.** *Let $X$ be a normal and locally irreducible scheme and $f\colon Y \to X$ an étale morphism. Then $Y$ is normal and locally irreducible.*

Indeed, $Y$ is normal by virtue of ([**62**] VII Proposition 2). It therefore suffices to show that the set of its irreducible components is locally finite, by III.3.2(ii). Since the question is local on $X$ and $Y$, we may restrict ourselves to the case where they are affine, so that $f$ is quasi-compact and consequently quasi-finite. The desired assertion then follows from ([**42**] 2.3.6(iii)).

**Lemma III.3.4.** *Let $X$ be a normal scheme and $j\colon U \to X$ a dense and quasi-compact open immersion. Then $X$ is locally irreducible if and only if the same holds for $U$.*

Indeed, if $X$ is locally irreducible, then so is $U$. Conversely, suppose that $U$ is locally irreducible and let us show that the same holds for $X$. We may clearly restrict to the case where $X$ is quasi-compact. Consequently, $U$ is quasi-compact and therefore has only a finite number of irreducible components, by III.3.1(i). The same therefore holds for $X$ because $X$ and $U$ have the same generic points. Since $X$ is normal, it is locally irreducible by virtue of III.3.2(ii).

**Lemma III.3.5.** *Let $A$ be a henselian local ring and $B$ an $A$-algebra that is an integral domain and integral over $A$. Then $B$ is a henselian local ring. If, moreover, $A$ is strictly local, then the same holds for $B$.*

Let us view $B$ as a filtered direct limit of sub-$A$-algebras of finite type $(B_i)_{i \in I}$. Since for every $i \in I$, $B_i$ is an integral domain and is finite over $A$, it is local and henselian. For all $(i, j) \in I^2$ such that $i \leq j$, since the transition morphism $B_i \to B_j$ is finite, it is local. Consequently, $B$ is local and henselian ([**62**] I § 3 Proposition 1). Suppose that $A$ is strictly local. Since the homomorphism $A \to B$ is local, the residue field of $B$ is an algebraic extension of that of $A$. It is therefore separably closed. Consequently, $B$ is strictly local.

**Lemma III.3.6.** *Let $X$ be a scheme, $\overline{x}$ a geometric point of $X$, $X'$ the strict localization of $X$ at $\overline{x}$, $Y$ an $X$-scheme, $Y' = Y \times_X X'$, and $f\colon Y' \to Y$ the canonical projection. Then the canonical homomorphism $f^{-1}(\mathscr{O}_Y) \to \mathscr{O}_{Y'}$ is an isomorphism of $Y'_{\text{ét}}$.*

We view $X'$ as a cofiltered inverse limit of affine étale neighborhoods $(X_i)_{\in I}$ of $\overline{x}$ in $X$ (cf. [**2**] VIII 4.5) and set $Y_i = Y \times_X X_i$ for every $i \in I$. The scheme $Y'$ is then the inverse limit of the schemes $(Y_i)_{\in I}$ ([**42**] 8.2.5). Let $\overline{y}$ be a geometric point of $Y'$. For every $i \in I$, the canonical projection $Y_i \to Y$ induces an isomorphism between the strict localizations of $Y_i$ and $Y$ at the canonical images of $\overline{y}$. On the other hand, it follows from ([**39**] 0.6.1.6) and ([**62**] I § 3 Proposition 1) that the strict localization of $Y'$ at $\overline{y}$ is the inverse limit of the strict localizations of the $Y_i$'s at the canonical images of $\overline{y}$. Consequently, the canonical projection $Y' \to Y$ induces an isomorphism between the strict localizations of $Y'$ at $\overline{y}$ and $Y$ at $f(\overline{y})$; in other words, the canonical homomorphism $\mathscr{O}_{Y,f(\overline{y})} \to \mathscr{O}_{Y',\overline{y}}$ is an isomorphism, giving the lemma.

**Lemma III.3.7.** *Let $f\colon Y \to X$ be an integral morphism of schemes, $\overline{x}$ a geometric point of $X$, and $X'$ the strict localization of $X$ at $\overline{x}$. Suppose that $Y$ is normal and that $f_{\overline{x}}\colon Y_{\overline{x}} \to \overline{x}$ is a universal homeomorphism. Then $X' \times_X Y$ is normal and strictly local.*

Viewing $X'$ as a cofiltered inverse limit of affine étale neighborhoods $(X_i)_{\in I}$ of $\overline{x}$ in $X$ ([**2**] VIII 4.5), $X' \times_X Y$ is canonically isomorphic to the inverse limit of the schemes $(X_i \times_X Y)_{i\in I}$ ([**42**] 8.2.5). For every $i \in I$, $X_i \times_X Y$ is normal ([**62**] VII prop. 2). Since for every $(i,j) \in I^2$ such that $i \leq j$, the morphism $X_j \to X_i$ is étale, every irreducible component of $X_j \times_X Y$ dominates an irreducible component of $X_i \times_X Y$ ([**42**] 2.3.5(ii)). It follows that $X' \times_X Y$ is normal ([**39**] 0.6.5.12(ii)). On the other hand, since $X' \times_X Y$ is integral over $X'$ and $f_{\overline{x}}$ is a universal homeomorphism, $X' \times_X Y$ is strictly local ([**62**] I § 3 Proposition 2).

## III.4. Adequate logarithmic schemes

**III.4.1.** We will use in this chapter the conventions and notation of logarithmic geometry introduced in II.5. We denote by $(S, \mathscr{M}_S)$ the logarithmic trait fixed in (III.2.1).

**Proposition III.4.2.** *Let $f\colon (X, \mathscr{M}_X) \to (S, \mathscr{M}_S)$ be a smooth and saturated morphism* (II.5.18) *of fine logarithmic schemes, $X^\circ$ the maximal open subscheme of $X$ where the logarithmic structure $\mathscr{M}_X$ is trivial, and $j\colon X^\circ \to X$ the canonical injection. Then:*

(i) *The scheme $X$ is $S$-flat and the scheme $X_{\overline{s}}$ is reduced.*
(ii) *The logarithmic scheme $(X, \mathscr{M}_X)$ is regular* ([**51**] 2.1 and [**57**] 2.2).
(iii) *The schemes $X$ and $X \times_S \overline{S}$ are normal and locally irreducible* (III.3.1).
(iv) *The immersion $j\colon X^\circ \to X$ is schematically dominant, and we have canonical isomorphisms*

$$\mathscr{M}_X^{\text{gp}} \xrightarrow{\sim} j_*(\mathscr{O}_{X^\circ}^\times), \tag{III.4.2.1}$$

$$\mathscr{M}_X \xrightarrow{\sim} j_*(\mathscr{O}_{X^\circ}^\times) \cap \mathscr{O}_X. \tag{III.4.2.2}$$

*In particular, the canonical homomorphism $\mathscr{M}_X \to \mathscr{O}_X$ is a monomorphism.*

(i) This follows from ([**50**] 4.5) and ([**74**] II 4.2).

(ii) Since $(X, \mathscr{M}_X)$ is saturated by ([**74**] II 2.12), it is regular by virtue of ([**51**] 8.2); cf. also ([**57**] 2.3) and the proof of ([**73**] 1.5.1).

(iii) The scheme $X$ is normal by virtue of (ii) and ([**51**] 4.1); cf. also ([**73**] 1.5.1). By taking a direct limit, we deduce that the scheme $X \times_S \overline{S}$ is normal (cf. the proof of II.6.3(iii)). Since $X$ is locally noetherian, it is locally irreducible by III.3.2(ii). On the other hand, since $X \times_S \overline{S}$ is $\overline{S}$-flat, its generic points are the generic points of the scheme $X \times_S \overline{\eta}$, which is locally noetherian. Consequently, the set of generic points of $X \times_S \overline{S}$ is locally finite, and $X \times_S \overline{S}$ is locally irreducible by virtue of III.3.2(ii).

(iv) This follows from (ii) and ([**51**] 11.6); cf. also ([**57**] 2.6).

**Lemma III.4.3.** *Let $f\colon (X, \mathscr{M}_X) \to (S, \mathscr{M}_S)$ be a smooth and saturated morphism of fine logarithmic schemes, $(\mathbb{N}, \iota)$ a chart for $(S, \mathscr{M}_S)$, and $\overline{x}$ a geometric point of $X$ over $\overline{s}$. Then there exist an étale neighborhood $U$ of $\overline{x}$ in $X$, a chart $(P, \gamma)$ for $(U, \mathscr{M}_X|U)$, and a homomorphism of monoids $\vartheta\colon \mathbb{N} \to P$ such that the following conditions are satisfied:*

(i) *$((P, \gamma), (\mathbb{N}, \iota), \vartheta)$ is a chart for the restriction $f_U\colon (U, \mathscr{M}|U) \to (S, \mathscr{M}_S)$ of $f$ to $U$* (II.5.14)*; in other words, the diagram of homomorphisms of monoids*

(III.4.3.1)
$$\begin{array}{ccc} P & \xrightarrow{\gamma} & \Gamma(U, \mathscr{M}_X) \\ \uparrow{\scriptstyle\vartheta} & & \uparrow{\scriptstyle f_U^\flat} \\ \mathbb{N} & \xrightarrow{\iota} & \Gamma(S, \mathscr{M}_S) \end{array}$$

*is commutative or, equivalently, the associated diagram of morphisms of logarithmic schemes*

(III.4.3.2)
$$\begin{array}{ccc} (U, \mathscr{M}_X|U) & \xrightarrow{\gamma^a} & \mathbf{A}_P \\ \downarrow{\scriptstyle f_U} & & \downarrow{\scriptstyle \mathbf{A}_\vartheta} \\ (S, \mathscr{M}_S) & \xrightarrow{\iota^a} & \mathbf{A}_{\mathbb{N}} \end{array}$$

*is commutative.*

(ii) *$P$ is toric; that is, $P$ is fine and saturated and $P^{\mathrm{gp}}$ is free over $\mathbb{Z}$* (II.5.1).

(iii) *The homomorphism $\vartheta$ is saturated* (II.5.2).

(iv) *The homomorphism $\vartheta^{\mathrm{gp}}\colon \mathbb{Z} \to P^{\mathrm{gp}}$ is injective, the torsion subgroup of the cokernel of $\vartheta^{\mathrm{gp}}$ is of order prime to $p$, and the morphism of usual schemes*

(III.4.3.3)
$$U \to S \times_{\mathbf{A}_{\mathbb{N}}} \mathbf{A}_P$$

*deduced from* (III.4.3.2) *is étale.*

(v) *There exists a subgroup $A$ of $P$ such that $\gamma$ induces an isomorphism*

(III.4.3.4)
$$P/A \xrightarrow{\sim} \mathscr{M}_{X,\overline{x}}/\mathscr{O}^\times_{X,\overline{x}}.$$

We adapt the proof of ([**49**] 4.1; cf. § 6). Set

(III.4.3.5)
$$\widetilde{\Omega}^1_{\widetilde{X}/\widetilde{S}} = \Omega^1_{(X,\mathscr{M}_X)/(S,\mathscr{M}_S)},$$

and denote by $\lambda \in \Gamma(X, \mathscr{M}_X)$ the canonical image of $\iota(1) = \pi \in \Gamma(S, \mathscr{M}_S) = \mathscr{O}_K - \{0\}$. Let $t_1, \dots, t_r \in \mathscr{M}_{X,\overline{x}}$ be such that $d\log(t_1), \dots, d\log(t_r)$ form a basis of the $\mathscr{O}_{X,\overline{x}}$-module $\widetilde{\Omega}^1_{\widetilde{X}/\widetilde{S},\overline{x}}$. Set $H = \mathbb{N}^{r+1}$ and consider the homomorphism

(III.4.3.6)
$$\varphi\colon H \to \mathscr{M}_{X,\overline{x}}, \quad (n_1, \dots, n_{r+1}) \mapsto \prod_{i=1}^r t_i^{n_i} \cdot \lambda^{n_{r+1}}.$$

Denote by $\alpha\colon \mathscr{M}^{\mathrm{gp}}_{X,\overline{x}} \to \mathscr{M}^{\mathrm{gp}}_{X,\overline{x}}/\mathscr{O}^\times_{X,\overline{x}}$ the canonical projection and by $L$ the image of the homomorphism

(III.4.3.7)
$$\alpha \circ \varphi^{\mathrm{gp}}\colon H^{\mathrm{gp}} \to \mathscr{M}^{\mathrm{gp}}_{X,\overline{x}}/\mathscr{O}^\times_{X,\overline{x}}.$$

Since $\mathscr{M}_X$ is fine and saturated ([**74**] II 2.12), $\mathscr{M}^{\rm gp}_{X,\overline{x}}/\mathscr{O}^{\times}_{X,\overline{x}}$ is a free $\mathbb{Z}$-module of finite type. By step 2 of ([**49**] page 331), we see that the cokernel of $\alpha \circ \varphi^{\rm gp}$ is annihilated by an integer invertible in $\mathscr{O}_{X,\overline{x}}$. Hence there exist a $\mathbb{Z}$-basis $e_1,\dots,e_d$ of $\mathscr{M}^{\rm gp}_{X,\overline{x}}/\mathscr{O}^{\times}_{X,\overline{x}}$ and integers $f_1,\dots,f_d$ such that the $e_1^{f_1},\dots,e_d^{f_d}$ form a $\mathbb{Z}$-basis of $L$, that $f_i$ divides $f_{i+1}$ for every $1 \leq i \leq d-1$, and that $f_d$ is invertible in $\mathscr{O}_{X,\overline{x}}$. Let $F_1,\dots,F_d \in H^{\rm gp}$ and $\widetilde{e}_1,\dots,\widetilde{e}_d \in \mathscr{M}^{\rm gp}_{X,\overline{x}}$ be such that $\alpha(\varphi^{\rm gp}(F_i)) = e_i^{f_i}$ and $\alpha(\widetilde{e}_i) = e_i$ for every $1 \leq i \leq d$. Then there exists $u_i \in \mathscr{O}^{\times}_{X,\overline{x}}$ such that $\varphi^{\rm gp}(F_i) = u_i\widetilde{e}_i^{f_i}$. Since $\mathscr{O}^{\times}_{X,\overline{x}}$ is $f_i$-divisible, there exists $v_i \in \mathscr{O}^{\times}_{X,\overline{x}}$ such that $v_i^{f_i} = u_i$. Replacing $\widetilde{e}_i$ by $v_i\widetilde{e}_i$, we may assume that $\varphi^{\rm gp}(F_i) = \widetilde{e}_i^{f_i}$. We denote by $\beta\colon \mathscr{M}^{\rm gp}_{X,\overline{x}}/\mathscr{O}^{\times}_{X,\overline{x}} \to \mathscr{M}^{\rm gp}_{X,\overline{x}}$ the splitting of $\alpha$ defined by $\beta(e_i) = \widetilde{e}_i$ for every $1 \leq i \leq d$, by $\rho\colon H^{\rm gp} \to L$ the surjective homomorphism induced by $\alpha \circ \varphi^{\rm gp}$, by $\sigma\colon L \to H^{\rm gp}$ the splitting of $\rho$ defined by $\sigma(e_i^{f_i}) = F_i$ for every $1 \leq i \leq d$, by $M$ the kernel of $\rho$, and by $\tau\colon H^{\rm gp} \to M$ the homomorphism that to each $h \in H^{\rm gp}$ associates $h - \sigma(\rho(h))$. We set $G = M \oplus \mathscr{M}^{\rm gp}_{X,\overline{x}}/\mathscr{O}^{\times}_{X,\overline{x}}$ and

(III.4.3.8) $$\phi\colon G = M \oplus \mathscr{M}^{\rm gp}_{X,\overline{x}}/\mathscr{O}^{\times}_{X,\overline{x}} \to \mathscr{M}^{\rm gp}_{X,\overline{x}}, \quad (m,t) \mapsto \phi(m,t) = \varphi^{\rm gp}(m)\cdot\beta(t).$$

We have $\phi \circ (\tau \oplus \alpha \circ \varphi^{\rm gp}) = \varphi^{\rm gp}\colon H^{\rm gp} \to \mathscr{M}^{\rm gp}_{X,\overline{x}}$. One immediately verifies this for the elements of $M$ and for the elements $(F_i)_{1\leq i\leq d}$. Set $P = \phi^{-1}(\mathscr{M}_{X,\overline{x}})$. By ([**50**] 2.10), there exist an étale neighborhood $U$ of $\overline{x}$ in $X$ and a homomorphism $\gamma\colon P \to \Gamma(U,\mathscr{M}_X)$ that is a chart for $(U,\mathscr{M}_X|U)$ and whose stalk $\gamma_{\overline{x}}\colon P \to \mathscr{M}_{X,\overline{x}}$ at $\overline{x}$ is induced by $\phi$. The homomorphism $\tau\oplus\alpha\circ\varphi^{\rm gp}\colon H^{\rm gp} \to G$ induces a homomorphism $H \to P$ and consequently a homomorphism $\vartheta\colon \mathbb{N} \to P$ that makes diagram (III.4.3.1) commutative.

Since the homomorphism $G \to \mathscr{M}^{\rm gp}_{X,\overline{x}}/\mathscr{O}^{\times}_{X,\overline{x}}$ induced by $\phi$ is surjective, $P^{\rm gp} = G$ and $P$ is integral. It immediately follows from the definition that there exists a subgroup $A$ of $P$ such that $\gamma_{\overline{x}}$ induces an isomorphism

(III.4.3.9) $$P/A \xrightarrow{\sim} \mathscr{M}_{X,\overline{x}}/\mathscr{O}^{\times}_{X,\overline{x}}.$$

Consequently, $A = P^{\times}$. Since $\mathscr{M}_{X,\overline{x}}$ is saturated, $\mathscr{M}_{X,\overline{x}}/\mathscr{O}^{\times}_{X,\overline{x}}$ is saturated and hence $P$ is saturated (II.5.1). It follows that $P$ is toric. The homomorphism $\vartheta$ is saturated by virtue of (III.4.3.9) and ([**74**] I 3.16).

The homomorphism $\tau\oplus\alpha\circ\varphi^{\rm gp}\colon H^{\rm gp} \to G$ is injective and its cokernel is isomorphic to that of $\alpha\circ\varphi^{\rm gp}$. It follows that the homomorphism $\vartheta^{\rm gp}\colon \mathbb{Z} \to P^{\rm gp}$ is injective and that the torsion subgroup of $\operatorname{coker}(\vartheta^{\rm gp})$ is of order prime to $p$. Step 4 of ([**49**] § 6 page 332) shows that the morphism of usual schemes $U \to S \times_{\mathbf{A}_{\mathbb{N}}} \mathbf{A}_P$ deduced from (III.4.3.2) is étale.

**Definition III.4.4.** Let $f\colon (X,\mathscr{M}_X) \to (S,\mathscr{M}_S)$ be a morphism of logarithmic schemes, $(P,\gamma)$ a chart for $(X,\mathscr{M}_X)$, $(\mathbb{N},\iota)$ a chart for $(S,\mathscr{M}_S)$, and $\vartheta\colon \mathbb{N} \to P$ a homomorphism of monoids such that $((P,\gamma),(\mathbb{N},\iota),\vartheta)$ is a chart for $f$ (II.5.14); in other words, such that the diagram of homomorphisms of monoids

(III.4.4.1) $$\begin{array}{ccc} P & \xrightarrow{\gamma} & \Gamma(X,\mathscr{M}_X) \\ \uparrow{\scriptstyle\vartheta} & & \uparrow{\scriptstyle f^{\flat}} \\ \mathbb{N} & \xrightarrow{\iota} & \Gamma(S,\mathscr{M}_S) \end{array}$$

is commutative or, equivalently, that the associated diagram of morphisms of logarithmic schemes

$$\begin{array}{ccc} (X, \mathscr{M}_X) & \xrightarrow{\gamma^a} & \mathbf{A}_P \\ {\scriptstyle f}\downarrow & & \downarrow{\scriptstyle \mathbf{A}_\vartheta} \\ (S, \mathscr{M}_S) & \xrightarrow{\iota^a} & \mathbf{A}_{\mathbb{N}} \end{array} \tag{III.4.4.2}$$

is commutative. We say that the chart $((P, \gamma), (\mathbb{N}, \iota), \vartheta)$ is *adequate* if the following conditions are satisfied:

(i) The monoid $P$ is toric; that is, $P$ is fine and saturated and $P^{\mathrm{gp}}$ is free over $\mathbb{Z}$.
(ii) The homomorphism $\vartheta$ is saturated.
(iii) The homomorphism $\vartheta^{\mathrm{gp}}\colon \mathbb{Z} \to P^{\mathrm{gp}}$ is injective, the torsion subgroup of the cokernel $\vartheta^{\mathrm{gp}}$ is of order prime to $p$, and the morphism of usual schemes

$$X \to S \times_{\mathbf{A}_{\mathbb{N}}} \mathbf{A}_P \tag{III.4.4.3}$$

deduced from (III.4.4.2) is étale.
(iv) Set $\lambda = \vartheta(1) \in P$ and

$$L = \mathrm{Hom}_{\mathbb{Z}}(P^{\mathrm{gp}}, \mathbb{Z}), \tag{III.4.4.4}$$

$$\mathrm{H}(P) = \mathrm{Hom}(P, \mathbb{N}). \tag{III.4.4.5}$$

Note that $\mathrm{H}(P)$ is a fine, saturated, and sharp monoid, and that the canonical homomorphism $\mathrm{H}(P)^{\mathrm{gp}} \to \mathrm{Hom}((P^\sharp)^{\mathrm{gp}}, \mathbb{Z})$ is an isomorphism ([**58**] I 2.2.3), where $P^\sharp$ denotes the quotient $P/P^\times$ (II.5.1). We suppose that there exist $h_1, \dots, h_r \in \mathrm{H}(P)$ that are $\mathbb{Z}$-linearly independent in $L$, such that

$$\ker(\lambda) \cap \mathrm{H}(P) = \{\sum_{i=1}^{r} a_i h_i |\ (a_1, \dots, a_r) \in \mathbb{N}^r\}, \tag{III.4.4.6}$$

where we view $\lambda$ as a homomorphism $L \to \mathbb{Z}$.

Note that every morphism of logarithmic schemes that admits an adequate chart is smooth and saturated (II.5.25 and [**74**] Chapter II 3.5).

**Lemma III.4.5.** *Let $X$ be an affine scheme, $U$ a schematically dense open subscheme of $X$, $j\colon U \to X$ the canonical injection, $\mathscr{M}$ the multiplicative submonoid $j_*(\mathscr{O}_U^\times) \cap \mathscr{O}_X$ of $j_*(\mathscr{O}_U)$, and $\lambda \in \Gamma(X, \mathscr{M})$. Denote by $\Gamma(X, \mathscr{M})_\lambda$ the localization of the monoid $\Gamma(X, \mathscr{M})$ by $\lambda$ ([**58**] I 1.4.4) and by $X_\lambda$ the open subscheme of $X$ where the canonical image of $\lambda$ in $\Gamma(X, \mathscr{O}_X)$ is invertible. Then the canonical homomorphism*

$$\Gamma(X, \mathscr{M})_\lambda \to \Gamma(X_\lambda, \mathscr{M}) \tag{III.4.5.1}$$

*is an isomorphism.*

Indeed, set $A = \Gamma(X, \mathscr{O}_X)$ and let us identify $\lambda$ with an element of $A$ and $\Gamma(X, \mathscr{M})$ (resp. $\Gamma(X_\lambda, \mathscr{M})$) with the multiplicative submonoid of $A$ (resp. $A_\lambda$) formed by the elements $f$ such that $f|U$ is a unit. One immediately verifies that for every monoid $P$, every homomorphism $u\colon \Gamma(X, \mathscr{M}) \to P$ such that $u(\lambda)$ is invertible factors uniquely through $\Gamma(X_\lambda, \mathscr{M})$, giving the lemma.

**Proposition III.4.6.** *Let $f\colon (X, \mathscr{M}_X) \to (S, \mathscr{M}_S)$ be a smooth and saturated morphism of fine logarithmic schemes, $\overline{x}$ a geometric point of $X$ over $s$, $X'$ the strict localization of $X$ at $\overline{x}$, and $\mathscr{M}_{X'}$ the logarithmic structure inverse image of $\mathscr{M}_X$ on $X'$ (II.5.10). Then the following conditions are equivalent:*

(a) *There exists an étale neighborhood $U$ of $\overline{x}$ in $X$ such that the restriction $f_U\colon (U, \mathscr{M}_X|U) \to (S, \mathscr{M}_S)$ of $f$ to $U$ admits an adequate chart* (III.4.4).

(b) *There exists an étale neighborhood $U$ of $\overline{x}$ in $X$ such that the scheme $U_\eta$ is smooth over $\eta$ and that the logarithmic structure $\mathscr{M}_X|U_\eta$ on $U_\eta$ is defined by a divisor with strict normal crossings on $U_\eta$.*

(c) *The scheme $X'_\eta$ is regular and the logarithmic structure $\mathscr{M}_{X'}|X'_\eta$ is defined by a divisor with strict normal crossings on $X'_\eta$.*

The implication (a)⇒(b) follows from II.6.3(v); note that conditions ($C_1$) and ($C_2$) of II.6.2 do not play any role in the proof of II.6.3(v). We denote by $\nu\colon X' \to X$ the canonical morphism and by $X^\circ$ the maximal open subscheme of $X$ where the logarithmic structure $\mathscr{M}_X$ is trivial. We set $X'^\circ = X' \times_X X^\circ$ and denote by $j\colon X^\circ \to X$ and $j'\colon X'^\circ \to X'$ the canonical injections. The canonical homomorphism $\nu^{-1}(\mathscr{O}_X) \to \mathscr{O}_{X'}$ is an isomorphism of $X'_{\text{ét}}$ (III.3.6). Consequently, the canonical homomorphism

$$\nu^{-1}(\mathscr{M}_X) \to \mathscr{M}_{X'} \tag{III.4.6.1}$$

is also an isomorphism. We have $\mathscr{M}_X = j_*(\mathscr{O}^\times_{X^\circ}) \cap \mathscr{O}_X$ by virtue of III.4.2(iv). Since $\nu$ is universally locally acyclic, it follows that

$$\mathscr{M}_{X'} = j'_*(\mathscr{O}^\times_{X'^\circ}) \cap \mathscr{O}_{X'}. \tag{III.4.6.2}$$

The implication (b)⇒(c) follows. Let us show the implication (c)⇒(a). By III.4.3, after replacing $X$ by an étale neighborhood of $\overline{x}$ in $X$, if necessary, we may assume that $f$ admits a chart $((P,\gamma),(\mathbb{N},\iota),\vartheta)$ satisfying conditions (i), (ii), and (iii) of III.4.4 and that there exists a subgroup $A$ of $P$ such that $\gamma$ induces an isomorphism

$$P/A \xrightarrow{\sim} \mathscr{M}_{X,\overline{x}}/\mathscr{O}^\times_{X,\overline{x}}. \tag{III.4.6.3}$$

Set $\lambda = \vartheta(1) \in P$ and

$$L = \operatorname{Hom}_{\mathbb{Z}}(P^{\text{gp}}, \mathbb{Z}), \tag{III.4.6.4}$$

$$\text{H}(P) = \operatorname{Hom}(P, \mathbb{N}). \tag{III.4.6.5}$$

Note that $\text{H}(P)$ is a fine, saturated, and sharp monoid, and that the canonical homomorphism $\text{H}(P)^{\text{gp}} \to \operatorname{Hom}((P^\sharp)^{\text{gp}}, \mathbb{Z})$ is an isomorphism (**[58]** I 2.2.3), where $P^\sharp$ denotes the quotient $P/P^\times$. Let $F$ be the face of $P$ generated by $\lambda$, that is, the set of elements $\alpha \in P$ such that there exist $\beta \in P$ and $n \in \mathbb{N}$ such that $\alpha + \beta = n\lambda$ (**[58]** I 1.4.2). Denote by $P/F$ the quotient of $P$ by $F$ (cf. **[58]** I 1.1.5). The canonical homomorphism

$$\operatorname{Hom}(P/F, \mathbb{N}) \to \ker(\lambda) \cap \text{H}(P), \tag{III.4.6.6}$$

where we view $\lambda$ as a homomorphism $L \to \mathbb{Z}$, is an isomorphism. It therefore suffices to show that the monoid $P/F$ is free of finite type. Let $G$ be the face of $\mathscr{M}_{X,\overline{x}}$ generated by $\gamma(\lambda)$. In view of (III.4.6.3), it also suffices to show that the monoid $\mathscr{M}_{X,\overline{x}}/G$ is free of finite type. We have a canonical isomorphism

$$\mathscr{M}_{X,\overline{x}}/G \xrightarrow{\sim} (G^{-1}\mathscr{M}_{X,\overline{x}})^\sharp. \tag{III.4.6.7}$$

On the other hand, we have $\mathscr{M}_{X,\overline{x}} = \Gamma(X', \mathscr{M}_{X'})$ (III.4.6.1), and the canonical homomorphism

$$G^{-1}\Gamma(X', \mathscr{M}_{X'}) \to \Gamma(X'_\eta, \mathscr{M}_{X'}) \tag{III.4.6.8}$$

is an isomorphism by III.4.5 and (III.4.6.2). The desired implication then follows from the fact that the monoid $\Gamma(X'_\eta, \mathscr{M}^\sharp_{X'})$ is free of finite type in view of (c).

**Definition III.4.7.** We say that a morphism $f\colon (X, \mathscr{M}_X) \to (S, \mathscr{M}_S)$ of fine logarithmic schemes is *adequate* if it is smooth and saturated, if the morphism of underlying schemes $X \to S$ is of finite type, and if for every geometric point $\overline{x}$ of $X$ over $s$, the equivalent conditions of III.4.6 are satisfied.

The notion of adequate morphism of logarithmic schemes corresponds to Faltings' notion of morphism *with toroidal singularities.* In his terminology, the morphism $f$ is said to be *small* if it admits an adequate chart, if $X$ is affine and connected, and if $X_s$ is nonempty, in other words, if the conditions of II.6.2 are satisfied.

**Proposition III.4.8.** *Let $f\colon (X, \mathscr{M}_X) \to (S, \mathscr{M}_S)$ be an adequate morphism of fine logarithmic schemes, $K'$ a finite extension of $K$, $\mathscr{O}_{K'}$ the integral closure of $\mathscr{O}_K$ in $K'$, and $S' = \mathrm{Spec}(\mathscr{O}_{K'})$. We endow $S'$ with the logarithmic structure $\mathscr{M}_{S'}$ defined by its closed point. We have a canonical morphism $(S', \mathscr{M}_{S'}) \to (S, \mathscr{M}_S)$. Set*

(III.4.8.1) $$(X', \mathscr{M}_{X'}) = (X, \mathscr{M}_X) \times_{(S, \mathscr{M}_S)} (S', \mathscr{M}_{S'}),$$

*where the product is taken in the category of logarithmic schemes. Then the canonical projection $f'\colon (X', \mathscr{M}_{X'}) \to (S', \mathscr{M}_{S'})$ is adequate.*

Indeed, $f'$ is smooth and saturated ([**74**] II 2.11). Since $X' = X \times_S S'$, it is of finite type over $S'$. On the other hand, condition III.4.6(b) is clearly satisfied at every geometric point of $X'$ over the closed point of $S'$.

## III.5. Variations on the Koszul complex

**III.5.1.** In this section, $(X, A)$ denotes a ringed topos. For every morphism of $A$-modules $u\colon E \to F$, there exists on the bigraded algebra $\mathrm{S}(E) \otimes_A \Lambda(F)$ (III.2.7) (anti-commutative in the second degree; [**45**] I 4.3.1.1) a unique $A$-derivation

(III.5.1.1) $$d_u\colon \mathrm{S}(E) \otimes_A \Lambda(F) \to \mathrm{S}(E) \otimes_A \Lambda(F)$$

of bidegree $(-1, 1)$ such that for all local sections $x_1, \dots, x_n$ of $E$ ($n \geq 1$) and $y$ of $\wedge(F)$, we have

(III.5.1.2) $$d_u([x_1 \otimes \cdots \otimes x_n] \otimes y) = \sum_{i=1}^{n} [x_1 \otimes \cdots \otimes x_{i-1} \otimes x_{i+1} \otimes \cdots \otimes x_n] \otimes (u(x_i) \wedge y),$$

(III.5.1.3) $$d_u(1 \otimes y) = 0.$$

It moreover satisfies $d_u \circ d_u = 0$. For every integer $n \geq 0$, the homogeneous part of degree $n$ of $\mathrm{S}(E) \otimes_A \Lambda(F)$ gives a complex

(III.5.1.4) $$0 \to \mathrm{S}^n(E) \to \mathrm{S}^{n-1}(E) \otimes_A F \to \cdots \to E \otimes_A \wedge^{n-1}(F) \to \wedge^n(F) \to 0.$$

The algebra $\mathrm{S}(E) \otimes_A \Lambda(F)$ endowed with the derivation $d_u$ depends functorially on $u$.

If $A$ is a $\mathbb{Q}$-algebra, identifying $\mathrm{S}(E)$ with the divided power algebra $\Gamma(E)$ of $E$, $d_u$ identifies with the $A$-derivation of $\Gamma(E) \otimes_A \Lambda(F)$ defined in ([**45**] I 4.3.1.2(b)). It then follows from ([**45**] I 4.3.1.6) that if $u$ is an isomorphism of flat $A$-modules and if $n > 0$, the sequence (III.5.1.4) is exact.

**III.5.2.** Let

(III.5.2.1) $$0 \to A \to E \to F \to 0$$

be an exact sequence of flat $A$-modules. By ([**45**] I 4.3.1.7), it induces, for every integer $n \geq 1$, an exact sequence (III.2.7)

(III.5.2.2) $$0 \to \mathrm{S}^{n-1}(E) \to \mathrm{S}^n(E) \to \mathrm{S}^n(F) \to 0.$$

We deduce from this an exact sequence

(III.5.2.3) $$0 \to \mathrm{S}^{n-1}(F) \to \mathrm{S}^n(E)/\mathrm{S}^{n-2}(E) \to \mathrm{S}^n(F) \to 0.$$

We thus define a functor $\mathrm{S}^n$ from the category $\mathbf{Ext}(F, A)$ of extensions of $F$ by $A$ to the category $\mathbf{Ext}(\mathrm{S}^n(F), \mathrm{S}^{n-1}(F))$ of extensions of $\mathrm{S}^n(F)$ by $\mathrm{S}^{n-1}(F)$. Taking the groups of isomorphism classes of the objects of these categories, we obtain a homomorphism, which we denote also by

(III.5.2.4) $$\mathrm{S}^n \colon \mathrm{Ext}^1_A(F, A) \to \mathrm{Ext}^1_A(\mathrm{S}^n(F), \mathrm{S}^{n-1}(F)).$$

**III.5.3.** Let $F$ be a locally projective $A$-module of finite type (III.2.8) and $n$ an integer $\geq 1$. The local-to-global spectral sequence of Ext (**[2]** V 6.1) gives an isomorphism

(III.5.3.1) $$\mathrm{Ext}^1_A(F, A) \xrightarrow{\sim} \mathrm{H}^1(X, \mathscr{H}om_A(F, A)).$$

Likewise, since the $A$-module $\mathrm{S}^n(F)$ is locally free of finite type, we have a canonical isomorphism

(III.5.3.2) $$\mathrm{Ext}^1_A(\mathrm{S}^n(F), \mathrm{S}^{n-1}(F)) \xrightarrow{\sim} \mathrm{H}^1(X, \mathscr{H}om_A(\mathrm{S}^n(F), \mathrm{S}^{n-1}(F))).$$

On the other hand, we denote by

(III.5.3.3) $$J_n \colon \mathscr{H}om_A(F, A) \to \mathscr{H}om_A(\mathrm{S}^n(F), \mathrm{S}^{n-1}(F))$$

the morphism that for every $U \in \mathrm{Ob}(X)$, associates with each morphism $u \colon F|U \to A|U$ the restriction to $\mathrm{S}^n(F)|U$ of the derivation $d_u$ of $\mathrm{S}(F|U)$ defined in (III.5.1.1). This induces a pairing

(III.5.3.4) $$\mathscr{H}om_A(F, A) \otimes_A \mathrm{S}^n(F) \to \mathrm{S}^{n-1}(F).$$

**Proposition III.5.4.** *For every locally projective $A$-module of finite type $F$ and every integer $n \geq 1$, the diagram*

(III.5.4.1) $$\begin{array}{ccc} \mathrm{Ext}^1_A(F, A) & \xrightarrow{\mathrm{S}^n} & \mathrm{Ext}^1_A(\mathrm{S}^n(F), \mathrm{S}^{n-1}(F)) \\ \downarrow & & \downarrow \\ \mathrm{H}^1(X, \mathscr{H}om_A(F, A)) & \xrightarrow{J_n} & \mathrm{H}^1(X, \mathscr{H}om_A(\mathrm{S}^n(F), \mathrm{S}^{n-1}(F))) \end{array}$$

*where $\mathrm{S}^n$ is the morphism* (III.5.2.4), *$J_n$ is the morphism* (III.5.3.3), *and the vertical arrows are the isomorphisms* (III.5.3.1) *and* (III.5.3.2), *is commutative.*

We first recall (**[2]** V 3.4) that the Cartan–Leray spectral sequence associated with the coverings of the final object of $X$ induce an isomorphism

(III.5.4.2) $$\check{\mathrm{H}}^1(X, \mathscr{H}om_A(F, A)) \xrightarrow{\sim} \mathrm{H}^1(X, \mathscr{H}om_A(F, A)),$$

where the source is the Čech cohomology group (**[2]** V (2.4.5.4)). We can explicitly describe the isomorphism

(III.5.4.3) $$\mathrm{Ext}^1_A(F, A) \xrightarrow{\sim} \check{\mathrm{H}}^1(X, \mathscr{H}om_A(F, A))$$

composed of (III.5.3.1) and the inverse of (III.5.4.2), as follows. Let

(III.5.4.4) $$0 \to A \to E \xrightarrow{\nu} F \to 0$$

be an exact sequence of $A$-modules. Since $F$ is locally projective of finite type, there exists a family $\mathscr{U} = (U_i)_{i \in I}$ of objects of $X$, epimorphic over the final object, such that for every $i \in I$, there exists an $(A|U_i)$-linear section $\varphi_i \colon F|U_i \to E|U_i$ of $\nu|U_i$. For every $(i, j) \in I^2$, setting $U_{i,j} = U_i \times U_j$, the difference $\varphi_{i,j} = \varphi_i|U_{i,j} - \varphi_j|U_{i,j}$ defines a morphism from $F|U_{i,j}$ to $A|U_{i,j}$. The collection $(\varphi_{i,j})$ is a 1-cocycle for the covering $\mathscr{U}$ with coefficients in $\mathscr{H}om_A(F, A)$ whose class in $\check{\mathrm{H}}^1(X, \mathscr{H}om_A(F, A))$ is the canonical image of the extension (III.5.4.4) (that is, its image by the isomorphism (III.5.4.3)). For any $i \in I$, we denote by $\psi_i^n$ the composition

(III.5.4.5) $$\mathrm{S}^n(F)|U_i \xrightarrow{\mathrm{S}^n(\varphi_i)} \mathrm{S}^n(E)|U_i \longrightarrow (\mathrm{S}^n(E)/\mathrm{S}^{n-2}(E))|U_i\ ,$$

where the second arrow is the canonical projection. This is clearly a splitting over $U_i$ of the exact sequence (III.5.2.3)

(III.5.4.6) $$0 \to \mathrm{S}^{n-1}(F) \to \mathrm{S}^n(E)/\mathrm{S}^{n-2}(E) \to \mathrm{S}^n(F) \to 0$$

deduced from (III.5.4.4). For every $(i,j) \in I^2$, the difference $\psi_{i,j}^n = \psi_i^n|U_{i,j} - \psi_j^n|U_{i,j}$ defines a morphism from $\mathrm{S}^n(F)|U_{i,j}$ to $\mathrm{S}^{n-1}(F)|U_{i,j}$. The collection $(\psi_{i,j}^n)$ is a 1-cocycle for the covering $\mathscr{U}$ with coefficients in $\mathscr{H}om_A(\mathrm{S}^n(F), \mathrm{S}^{n-1}(F))$ whose class in the group $\check{\mathrm{H}}^1(X, \mathscr{H}om_A(\mathrm{S}^n(F), \mathrm{S}^{n-1}(F)))$ is the canonical image of the extension (III.5.4.6).

For every $(i,j) \in I^2$ and all local sections $x_1, \dots, x_n$ of $F|U_{i,j}$,

(III.5.4.7)
$$\begin{aligned}
&\psi_{i,j}^n([x_1 \otimes \cdots \otimes x_n]) \\
&= \psi_i^n([x_1 \otimes \cdots \otimes x_n]) - \psi_j^n([x_1 \otimes \cdots \otimes x_n]) \\
&= [(\varphi_j(x_1) + \varphi_{i,j}(x_1)) \otimes \cdots \otimes (\varphi_j(x_n) + \varphi_{i,j}(x_n))] \\
&\quad -[\varphi_j(x_1) \otimes \cdots \otimes \varphi_j(x_n)] \mod (\mathrm{S}^{n-2}(E)) \\
&= \sum_{\alpha=1}^{n} \varphi_{i,j}(x_\alpha)[x_1 \otimes \cdots \otimes x_{\alpha-1} \otimes x_{\alpha+1} \otimes \cdots \otimes x_n] \\
&= J_n(\varphi_{i,j})([x_1 \otimes \cdots \otimes x_n]).
\end{aligned}$$

The proposition follows.

**III.5.5.** Let

(III.5.5.1) $$0 \to A \to E \to F \to 0$$

be an exact sequence of locally projective $A$-modules of finite type and $n, q$ integers $\geq 0$. By III.5.2, the exact sequence (III.5.5.1) induces an exact sequence

(III.5.5.2) $$0 \to \mathrm{S}^n(F) \to \mathrm{S}^{n+1}(E)/\mathrm{S}^{n-1}(E) \to \mathrm{S}^{n+1}(F) \to 0.$$

On the other hand, the pairing (III.5.3.4) induces a pairing

(III.5.5.3) $$\mathrm{H}^1(X, \mathscr{H}om_A(F, A)) \otimes_{A(X)} \mathrm{H}^q(X, \mathrm{S}^{n+1}(F)) \to \mathrm{H}^{q+1}(X, \mathrm{S}^n(F)).$$

It immediately follows from III.5.4 that the boundary map

(III.5.5.4) $$\mathrm{H}^q(X, \mathrm{S}^{n+1}(F)) \to \mathrm{H}^{q+1}(X, \mathrm{S}^n(F))$$

of the long exact sequence of cohomology deduced from the short exact sequence (III.5.5.2) is induced by the cup product with the class of the extension (III.5.5.1) by the pairing (III.5.5.3).

Denote by

(III.5.5.5) $$\partial\colon \Gamma(X, F) \to \mathrm{H}^1(X, A)$$

the boundary map of the long exact sequence of cohomology deduced from the short exact sequence (III.5.5.1). It again follows from III.5.4 (more precisely, from (III.5.4.7)) that we have a commutative diagram

(III.5.5.6) $$\begin{array}{ccc}
\mathrm{S}^{n+1}(\Gamma(X,F)) & \longrightarrow & \Gamma(X, \mathrm{S}^{n+1}(F)) \\
\downarrow{\scriptstyle \alpha} & & \downarrow \\
\mathrm{S}^n(\Gamma(X,F)) \otimes_{A(X)} \mathrm{H}^1(X,A) & \longrightarrow & \mathrm{H}^1(X, \mathrm{S}^n(F))
\end{array}$$

where $\alpha$ is the restriction to $\mathrm{S}^{n+1}(\Gamma(X,F))$ of the $A(X)$-derivation $d_\partial$ of

$$\mathrm{S}(\Gamma(X,F)) \otimes_{A(X)} \wedge(\mathrm{H}^1(X,A))$$

defined in (III.5.1.1) with respect to the morphism $\partial$, the top (resp. bottom) horizontal arrow is the canonical morphism (resp. the morphism induced by the cup product) and the right vertical arrow is the morphism (III.5.5.4) for $q=0$. Using the associativity of the cup product, we deduce that the diagram

(III.5.5.7)
$$\begin{array}{ccc}
\mathrm{S}^{n+1}(\Gamma(X,F)) \otimes_{A(X)} \mathrm{H}^q(X,A) & \longrightarrow & \mathrm{H}^q(X,\mathrm{S}^{n+1}(F)) \\
\Big\downarrow{\scriptstyle \alpha\otimes \mathrm{id}} & & \Big\downarrow \\
\mathrm{S}^n(\Gamma(X,F)) \otimes_{A(X)} \mathrm{H}^1(X,A) \otimes_{A(X)} \mathrm{H}^q(X,A) & & \\
\Big\downarrow{\scriptstyle \mathrm{id}\otimes\cup} & & \Big\downarrow \\
\mathrm{S}^n(\Gamma(X,F)) \otimes_{A(X)} \mathrm{H}^{q+1}(X,A) & \longrightarrow & \mathrm{H}^{q+1}(X,\mathrm{S}^n(F))
\end{array}$$

where $\cup$ is the cup product of the $A(X)$-algebra $\oplus_{i\geq 0}\mathrm{H}^i(X,A)$, the horizontal morphisms are induced by the cup product, and the right vertical arrow is the morphism (III.5.5.4), is commutative.

**III.5.6.** Let $f\colon (X,A)\to(Y,B)$ be a morphism of ringed topos and

(III.5.6.1)
$$0\to A\to E\to F\to 0$$

an exact sequence of locally projective $A$-modules of finite type. We denote by

(III.5.6.2)
$$u\colon f_*(F)\to \mathrm{R}^1 f_*(A)$$

the boundary map of the long exact sequence of cohomology deduced from the short exact sequence (III.5.6.1). By III.5.2, the exact sequence (III.5.6.1) induces, for every integer $n\geq 0$, an exact sequence

(III.5.6.3)
$$0\to \mathrm{S}^n(F)\to \mathrm{S}^{n+1}(E)/\mathrm{S}^{n-1}(E)\to \mathrm{S}^{n+1}(F)\to 0.$$

**Proposition III.5.7.** *Under the assumptions of* III.5.6, *for all integers* $n,q\geq 0$, *we have a commutative diagram*

(III.5.7.1)
$$\begin{array}{ccc}
\mathrm{S}^{n+1}(f_*(F)) \otimes_B \mathrm{R}^q f_*(A) & \longrightarrow & \mathrm{R}^q f_*(\mathrm{S}^{n+1}(F)) \\
\Big\downarrow{\scriptstyle \alpha\otimes \mathrm{id}} & & \\
\mathrm{S}^n(f_*(F)) \otimes_B \mathrm{R}^1 f_*(A) \otimes_B \mathrm{R}^q f_*(A) & & \Big\downarrow{\scriptstyle \partial} \\
\Big\downarrow{\scriptstyle \mathrm{id}\otimes\cup} & & \\
\mathrm{S}^n(f_*(F)) \otimes_B \mathrm{R}^{q+1} f_*(A) & \longrightarrow & \mathrm{R}^{q+1} f_*(\mathrm{S}^n(F))
\end{array}$$

*where $\partial$ is the boundary map of the long exact sequence of cohomology deduced from the short exact sequence* (III.5.6.1), *$\alpha$ is the restriction to $\mathrm{S}^{n+1}(f_*(F))$ of the $B$-derivation $d_u$ of $\mathrm{S}(f_*(F))\otimes_B \wedge(\mathrm{R}^1 f_*(A))$ defined in* (III.5.1.1) *with respect to the morphism $u$* (III.5.6.2), *$\cup$ is the cup product of the $B$-algebra $\oplus_{i\geq 0}\mathrm{R}^i f_*(A)$, and the horizontal morphisms are induced by the cup product.*

Indeed, $\mathrm{R}^q f_*(\mathrm{S}^{n+1}(F))$ is the sheaf on $X$ (for the canonical topology) associated with the presheaf that with each $V\in \mathrm{Ob}(Y)$ associates $\mathrm{H}^q(f^*(V),\mathrm{S}^{n+1}(F))$, and $d$ is induced by the boundary map

(III.5.7.2)
$$\mathrm{H}^q(f^*(V),\mathrm{S}^{n+1}(F))\to \mathrm{H}^{q+1}(f^*(V),\mathrm{S}^n(F))$$

of the long exact sequence of cohomology deduced from the short exact sequence (III.5.6.3). The proposition then follows from (III.5.5.7).

## III.6. Additive categories up to isogeny

**Definition III.6.1.** Let $\mathbf{C}$ be an additive category.

(i) A morphism $u\colon M \to N$ of $\mathbf{C}$ is called an *isogeny* if there exist an integer $n \neq 0$ and a morphism $v\colon N \to M$ in $\mathbf{C}$ such that $v \circ u = n \cdot \mathrm{id}_M$ and $u \circ v = n \cdot \mathrm{id}_N$.

(ii) An object $M$ of $\mathbf{C}$ is said to be of *finite exponent* if there exists an integer $n \neq 0$ such that $n \cdot \mathrm{id}_M = 0$.

Let us complete the terminology and make a few remarks.

III.6.1.1. The family of isogenies of $\mathbf{C}$ allows a bilateral fractional computation ([**45**] I 1.4.2). We call *category of objects of* $\mathbf{C}$ *up to isogeny*, and denote by $\mathbf{C}_{\mathbb{Q}}$, the localized category of $\mathbf{C}$ with respect to isogenies. We denote by

(III.6.1.2) $$F\colon \mathbf{C} \to \mathbf{C}_{\mathbb{Q}}, \quad M \mapsto M_{\mathbb{Q}}$$

the localization functor. One easily verifies that for all $M, N \in \mathrm{Ob}(\mathbf{C})$, we have

(III.6.1.3) $$\mathrm{Hom}_{\mathbf{C}_{\mathbb{Q}}}(M_{\mathbb{Q}}, N_{\mathbb{Q}}) = \mathrm{Hom}_{\mathbf{C}}(M, N) \otimes_{\mathbb{Z}} \mathbb{Q}.$$

In particular, the category $\mathbf{C}_{\mathbb{Q}}$ is additive and the localization functor is additive. An object $M$ of $\mathbf{C}$ is of finite exponent if and only if $M_{\mathbb{Q}}$ is zero.

III.6.1.4. If $\mathbf{C}$ is an abelian category, then the category $\mathbf{C}_{\mathbb{Q}}$ is abelian and the localization functor $F\colon \mathbf{C} \to \mathbf{C}_{\mathbb{Q}}$ is exact. In fact, $\mathbf{C}_{\mathbb{Q}}$ identifies canonically with the quotient category of $\mathbf{C}$ by the thick subcategory $\mathbf{E}$ of objects of finite exponent. Indeed, denote by $\mathbf{C}/\mathbf{E}$ the quotient category of $\mathbf{C}$ by $\mathbf{E}$ and by $T\colon \mathbf{C} \to \mathbf{C}/\mathbf{E}$ the canonical functor ([**34**] III §1). For every $M \in \mathrm{Ob}(\mathbf{E})$, we have $F(M) = 0$. Consequently, there exists a unique functor $F'\colon \mathbf{C}/\mathbf{E} \to \mathbf{C}_{\mathbb{Q}}$ such that $F = F' \circ T$. On the other hand, for every $M \in \mathrm{Ob}(\mathbf{C})$ and every integer $n \neq 0$, $T(n \cdot \mathrm{id}_M)$ is an isomorphism. Hence there exists a unique functor $T'\colon \mathbf{C}_{\mathbb{Q}} \to \mathbf{C}/\mathbf{E}$ such that $T = T' \circ F$. We immediately see that $T'$ and $F'$ are equivalences of categories quasi-inverse to each other.

III.6.1.5. Every additive (resp. exact) functor between additive (resp. abelian) categories $\mathbf{C} \to \mathbf{C}'$ extends uniquely to an additive (resp. exact) functor $\mathbf{C}_{\mathbb{Q}} \to \mathbf{C}'_{\mathbb{Q}}$ compatible with the localization functors.

III.6.1.6. If $\mathbf{C}$ is an abelian category, the localization functor $\mathbf{C} \to \mathbf{C}_{\mathbb{Q}}$ transforms injective objects into injective objects ([**34**] III Corollary 1 to Proposition 1). In particular, if $\mathbf{C}$ has enough injectives, the same holds for $\mathbf{C}_{\mathbb{Q}}$.

**III.6.2.** Let $(X, A)$ be a ringed topos. We denote by $\mathbf{Mod}(A)$ the category of $A$-modules of $X$ and by $\mathbf{Mod}_{\mathbb{Q}}(A)$, instead of $\mathbf{Mod}(A)_{\mathbb{Q}}$, the category of $A$-modules of $X$ up to isogeny (III.6.1.1). The tensor product of $A$-modules induces a bifunctor

(III.6.2.1) $$\mathbf{Mod}_{\mathbb{Q}}(A) \times \mathbf{Mod}_{\mathbb{Q}}(A) \to \mathbf{Mod}_{\mathbb{Q}}(A), \quad (M, N) \mapsto M \otimes_{A_{\mathbb{Q}}} N,$$

making $\mathbf{Mod}_{\mathbb{Q}}(A)$ into a symmetric monoidal category with $A_{\mathbb{Q}}$ as identity element. The objects of $\mathbf{Mod}_{\mathbb{Q}}(A)$ will also be called $A_{\mathbb{Q}}$-*modules*. This terminology is justified by viewing $A_{\mathbb{Q}}$ as a monoid of $\mathbf{Mod}_{\mathbb{Q}}(A)$. If $M$ and $N$ are two $A_{\mathbb{Q}}$-modules, we denote by $\mathrm{Hom}_{A_{\mathbb{Q}}}(M, N)$ the group of morphisms from $M$ to $N$ in $\mathbf{Mod}_{\mathbb{Q}}(A)$. The bifunctor "sheaf of morphisms" on the category of $A$-modules of $X$ induces a bifunctor

(III.6.2.2) $$\mathbf{Mod}_{\mathbb{Q}}(A) \times \mathbf{Mod}_{\mathbb{Q}}(A) \to \mathbf{Mod}_{\mathbb{Q}}(A), \quad (M, N) \mapsto \mathscr{H}om_{A_{\mathbb{Q}}}(M, N).$$

The bifunctors (III.6.2.1) and (III.6.2.2) inherit the same exactness properties as the bifunctors on the category $\mathbf{Mod}(A)$ that gave birth to them.

**III.6.3.** For any morphism of ringed topos $f\colon (Y,B)\to(X,A)$, we denote also by

(III.6.3.1) $$f^*\colon \mathbf{Mod}_{\mathbb{Q}}(A) \to \mathbf{Mod}_{\mathbb{Q}}(B),$$

(III.6.3.2) $$f_*\colon \mathbf{Mod}_{\mathbb{Q}}(B) \to \mathbf{Mod}_{\mathbb{Q}}(A),$$

the functors induced by the functors inverse image under $f$ and direct image by $f$, so that the first is a left adjoint of the second. The first functor is exact and the second is left exact. We denote by

(III.6.3.3) $$\mathrm{R}f_*\colon \mathbf{D}^+(\mathbf{Mod}_{\mathbb{Q}}(B)) \to \mathbf{D}^+(\mathbf{Mod}_{\mathbb{Q}}(A)),$$

(III.6.3.4) $$\mathrm{R}^q f_*\colon \mathbf{Mod}_{\mathbb{Q}}(B) \to \mathbf{Mod}_{\mathbb{Q}}(A), \quad (q\in\mathbb{N}),$$

the right derived functors of $f_*$ (III.6.3.2). This notation does not lead to any confusion with that of the right derived functors of the functor $f_*\colon \mathbf{Mod}(B)\to\mathbf{Mod}(A)$, because the localization functor $\mathbf{Mod}(B)\to\mathbf{Mod}_{\mathbb{Q}}(B)$ is exact and transforms injective objects into injective objects.

**Definition III.6.4.** Let $(X,A)$ be a ringed topos. We say that an $A_{\mathbb{Q}}$-module $M$ is *flat* (or $A_{\mathbb{Q}}$*-flat*) if the functor $N\mapsto M\otimes_{A_{\mathbb{Q}}} N$ from the category $\mathbf{Mod}_{\mathbb{Q}}(A)$ to itself is exact.

We can make the following remarks.

III.6.4.1. Let $M$ be an $A$-module. Then $M_{\mathbb{Q}}$ is $A_{\mathbb{Q}}$-flat if and only if for every injective morphism of $A$-modules $u\colon N\to N'$, the kernel of $u\otimes \mathrm{id}_M$ is of finite exponent (III.6.1.4).

III.6.4.2. If $M$ is a flat $A$-module, then $M_{\mathbb{Q}}$ is $A_{\mathbb{Q}}$-flat.

III.6.4.3. Let $B$ be an $A$-algebra and $M$ a flat $A_{\mathbb{Q}}$-module. Then $M\otimes_{A_{\mathbb{Q}}} B_{\mathbb{Q}}$ is $B_{\mathbb{Q}}$-flat.

III.6.4.4. Let $B$ be an $A$-algebra such that the functor

$$\mathbf{Mod}(A)\to\mathbf{Mod}(B), \quad N\mapsto N\otimes_A B$$

is exact and faithful. Then an $A_{\mathbb{Q}}$-module $M$ is flat if and only if the $B_{\mathbb{Q}}$-module $M\otimes_{A_{\mathbb{Q}}} B_{\mathbb{Q}}$ is flat.

**III.6.5.** Let $(X,A)$ be a ringed topos and $U$ an object of $X$. We denote by $j_U\colon X_{/U}\to X$ the localization of $X$ at $U$. For any $F\in\mathrm{Ob}(X)$, we will denote the sheaf $j_U^*(F)$ also by $F|U$. The topos $X_{/U}$ will be ringed by $A|U$. Since the extension by zero functor $j_{U!}\colon \mathbf{Mod}(A|U)\to\mathbf{Mod}(A)$ is exact and faithful ([**2**] IV 11.3.1), it induces an exact and faithful functor that we denote also by

(III.6.5.1) $$j_{U!}\colon \mathbf{Mod}_{\mathbb{Q}}(A|U)\to\mathbf{Mod}_{\mathbb{Q}}(A), \quad P\mapsto j_{U!}(P),$$

and that we also call the *extension by zero*. It is a left adjoint of the functor (III.6.3.1)

(III.6.5.2) $$j_U^*\colon \mathbf{Mod}_{\mathbb{Q}}(A)\to\mathbf{Mod}_{\mathbb{Q}}(A|U).$$

For any $A_{\mathbb{Q}}$-module $M$, we will denote the $(A|U)_{\mathbb{Q}}$-module $j_U^*(M)$ also by $M|U$. For every $A$-module $N$, we have, by definition, $(N|U)_{\mathbb{Q}} = N_{\mathbb{Q}}|U$.

**Lemma III.6.6.** *Let $(X,A)$ be a ringed topos and $U$ an object of $X$. Then:*

(i) *For every flat $(A_{\mathbb{Q}}|U)$-module $P$, $j_{U!}(P)$ is $A_{\mathbb{Q}}$-flat.*

(ii) *For every flat $A_{\mathbb{Q}}$-module $M$, $j_U^*(M)$ is $(A_{\mathbb{Q}}|U)$-flat.*

Indeed, it immediately follows from ([**2**] IV 12.11) that for every $A_{\mathbb{Q}}$-module $M$ and every $(A_{\mathbb{Q}}|U)$-module $P$, we have a functorial canonical isomorphism

(III.6.6.1) $$j_{U!}(P\otimes_{(A_{\mathbb{Q}}|U)} j_U^*(M)) \xrightarrow{\sim} j_{U!}(P)\otimes_{A_{\mathbb{Q}}} M.$$

(i) This follows from (III.6.6.1) and the fact that the functors $j_U^*$ and $j_{U!}$ are exact.

(ii) It follows from (III.6.6.1) and the fact that the functor $j_{U!}$ is exact that the functor

$$\text{(III.6.6.2)} \qquad \mathbf{Mod}_{\mathbb{Q}}(A|U) \to \mathbf{Mod}_{\mathbb{Q}}(A), \quad P \mapsto j_{U!}(P \otimes_{(A_{\mathbb{Q}}|U)} j_U^*(M))$$

is exact. Since the functor $j_{U!}$ is moreover faithful (III.6.5), it follows that the functor $P \mapsto P \otimes_{(A_{\mathbb{Q}}|U)} j_U^*(M)$ on the category $\mathbf{Mod}_{\mathbb{Q}}(A|U)$ is exact; the statement follows.

**Lemma III.6.7.** *Let $(X, A)$ be a ringed topos, $(U_i)_{1 \leq i \leq n}$ a finite covering of the final object of $X$, and $M, N$ two $A_{\mathbb{Q}}$-modules. For all $1 \leq i, j \leq n$, we set $U_{ij} = U_i \times U_j$. Then:*

(i) *The diagram of maps of sets*

(III.6.7.1)

$$\mathrm{Hom}_{A_{\mathbb{Q}}}(M, N) \to \prod_{1 \leq i \leq n} \mathrm{Hom}_{(A_{\mathbb{Q}}|U_i)}(M|U_i, N|U_i) \rightrightarrows \prod_{1 \leq i,j \leq n} \mathrm{Hom}_{(A_{\mathbb{Q}}|U_{i,j})}(M|U_{ij}, N|U_{ij})$$

*is exact.*

(ii) *The module $M$ is zero if and only if for every $1 \leq i \leq n$, $M|U_i$ is zero.*

(iii) *The module $M$ is $A_{\mathbb{Q}}$-flat if and only if for every $1 \leq i \leq n$, $M|U_i$ is $(A_{\mathbb{Q}}|U_i)$-flat.*

(i) Let $M^\circ, N^\circ$ be two $A$-modules such that $M = M^\circ_{\mathbb{Q}}$ and $N = N^\circ_{\mathbb{Q}}$, and $u, v \colon M^\circ \to N^\circ$ two $A$-linear morphisms. Suppose that for every $1 \leq i \leq n$, we have $u_{\mathbb{Q}}|U_i = v_{\mathbb{Q}}|U_i$. Then there exists an integer $m \neq 0$ such that $m \cdot u|U_i = m \cdot v|U_i$. It follows that $m \cdot u = m \cdot v$, giving the left exactness of (III.6.7.1). On the other hand, let, for every $1 \leq i \leq n$, $u_i \colon M^\circ|U_i \to N^\circ|U_i$ be an $(A|U_i)$-linear morphism such that $(u_{i,\mathbb{Q}})_{1 \leq i \leq n}$ is in the kernel of the double arrow of (III.6.7.1). Then there exists an integer $m' \neq 0$ such that for every $1 \leq i, j \leq n$, we have $m' \cdot u_i|U_{ij} = m' \cdot u_j|U_{ij}$. Consequently, the morphisms $(m' \cdot u_i)_{1 \leq i \leq n}$ glue to an $A$-linear morphism $w \colon M^\circ \to N^\circ$. It is clear that $(u_{i,\mathbb{Q}})_{1 \leq i \leq n}$ is the canonical image of $m'^{-1} w_{\mathbb{Q}}$, giving the exactness in the middle of (III.6.7.1).

(ii) Indeed, $M$ is zero if and only if $\mathrm{id}_M = 0$. The statement therefore follows from (i).

(iii) Indeed, the condition is necessary by virtue of III.6.6(ii), and it is sufficient in view of (ii) and ([**2**] IV 12.11).

**III.6.8.** Let $(X, A)$ be a ringed topos and $E$ an $A$-module. A *Higgs $A$-isogeny with coefficients in $E$* is a quadruple

$$\text{(III.6.8.1)} \qquad (M, N, u \colon M \to N, \theta \colon M \to N \otimes_A E)$$

consisting of two $A$-modules $M$ and $N$ and two $A$-linear morphisms $u$ and $\theta$ satisfying the following property: there exist an integer $n \neq 0$ and an $A$-linear morphism $v \colon N \to M$ such that $v \circ u = n \cdot \mathrm{id}_M$, $u \circ v = n \cdot \mathrm{id}_N$ and that $(M, (v \otimes \mathrm{id}_E) \circ \theta)$ and $(N, \theta \circ v)$ are Higgs $A$-modules with coefficients in $E$ (II.2.8). Note that $u$ induces an isogeny of Higgs modules from $(M, (v \otimes \mathrm{id}_E) \circ \theta)$ to $(N, \theta \circ v)$ (III.6.1), whence the terminology. Let $(M, N, u, \theta)$, $(M', N', u', \theta')$ be two Higgs $A$-isogenies with coefficients in $E$. A morphism from $(M, N, u, \theta)$ to $(M', N', u', \theta')$ consists of two $A$-linear morphisms $\alpha \colon M \to M'$ and $\beta \colon N \to N'$ such that $\beta \circ u = u' \circ \alpha$ and $(\beta \otimes \mathrm{id}_E) \circ \theta = \theta' \circ \alpha$. We denote by $\mathbf{HI}(A, E)$ the category of Higgs $A$-isogenies with coefficients in $E$. It is an additive category. We denote by $\mathbf{HI}_{\mathbb{Q}}(A, E)$ the category of objects of $\mathbf{HI}(A, E)$ up to isogeny.

**III.6.9.** Let $(X, A)$ be a ringed topos, $E$ an $A$-module, and $(M, N, u, \theta)$ a Higgs $A$-isogeny with coefficients in $E$. For every $i \geq 1$, we denote by

$$\text{(III.6.9.1)} \qquad \theta_i \colon M \otimes_A \wedge^i E \to N \otimes_A \wedge^{i+1} E$$

the $A$-linear morphism defined for all local sections $m$ of $M$ and $\omega$ of $\wedge^i E$ by $\theta_i(m\otimes\omega) = \theta(m)\wedge\omega$. We denote by

$$\overline{\theta}_i \colon M_{\mathbb{Q}} \otimes_{A_{\mathbb{Q}}} (\wedge^i E)_{\mathbb{Q}} \to M_{\mathbb{Q}} \otimes_{A_{\mathbb{Q}}} (\wedge^{i+1} E)_{\mathbb{Q}} \tag{III.6.9.2}$$

the morphism of $\mathbf{Mod}_{\mathbb{Q}}(A)$ composed of the image of $\theta_i$ and the inverse of the image of $u\otimes \mathrm{id}_{\wedge^{i+1}E}$. Let $v\colon N\to M$ be an $A$-linear morphism and $n$ a nonzero integer such that $v\circ u = n\cdot \mathrm{id}_M$ and that $(M,(v\otimes\mathrm{id}_E)\circ\theta)$ is a Higgs $A$-module with coefficients in $E$. We denote by

$$\vartheta_i \colon M\otimes_A \wedge^i E \to M\otimes_A \wedge^{i+1}E \tag{III.6.9.3}$$

the $A$-linear morphism induced by $(v\otimes \mathrm{id}_E)\circ\theta$ (II.2.8.3). The canonical image of $\vartheta_i$ in $\mathbf{Mod}_{\mathbb{Q}}(A)$ is then equal to $n\cdot\overline{\theta}_i$. It follows that $\overline{\theta}_{i+1}\circ\overline{\theta}_i = 0$ (cf. II.2.8.2). The *Dolbeault* complex of $(M,N,u,\theta)$, denoted by $\mathbb{K}^\bullet(M,N,u,\theta)$, is the cochain complex of $\mathbf{Mod}_{\mathbb{Q}}(A)$

$$M_{\mathbb{Q}} \xrightarrow{\overline{\theta}_0} M_{\mathbb{Q}}\otimes_{A_{\mathbb{Q}}} E_{\mathbb{Q}} \xrightarrow{\overline{\theta}_1} M_{\mathbb{Q}}\otimes_{A_{\mathbb{Q}}} (\wedge^2 E)_{\mathbb{Q}} \to \dots, \tag{III.6.9.4}$$

where $M_{\mathbb{Q}}$ is placed in degree zero and the differentials are of degree one. We thus obtain a functor from the category $\mathbf{HI}(A,E)$ to the category of complexes of $\mathbf{Mod}_{\mathbb{Q}}(A)$. Every isogeny of $\mathbf{HI}(A,E)$ induces an isomorphism of the associated Dolbeault complexes. The functor "Dolbeault complex" therefore induces a functor from $\mathbf{HI}_{\mathbb{Q}}(A,E)$ to the category of complexes of $\mathbf{Mod}_{\mathbb{Q}}(A)$.

**III.6.10.** Let $(X,A)$ be a ringed topos, $B$ an $A$-algebra, and $\lambda\in\Gamma(X,A)$. A *$\lambda$-isoconnection with respect to the extension $B/A$* (or simply *$\lambda$-isoconnection* when there is no risk of confusion) is a quadruple

$$(M,N,u\colon M\to N,\nabla\colon M\to \Omega^1_{B/A}\otimes_B N) \tag{III.6.10.1}$$

where $M$ and $N$ are $B$-modules, $u$ is an isogeny of $B$-modules (III.6.1), and $\nabla$ is an $A$-linear morphism such that for all local sections $x$ of $B$ and $t$ of $M$, we have

$$\nabla(xt) = \lambda d(x)\otimes u(t) + x\nabla(t). \tag{III.6.10.2}$$

For every $B$-linear morphism $v\colon N\to M$ for which there exists an integer $n$ such that $u\circ v = n\cdot\mathrm{id}_N$ and $v\circ u = n\cdot \mathrm{id}_M$, the pairs $(M,(\mathrm{id}\otimes v)\circ\nabla)$ and $(N,\nabla\circ v)$ are modules with $(n\lambda)$-connections (II.2.10), and $u$ is a morphism from $(M,(\mathrm{id}\otimes v)\circ\nabla)$ to $(N,\nabla\circ v)$. We say that the $\lambda$-isoconnection $(M,N,u,\nabla)$ is *integrable* if there exist a $B$-linear morphism $v\colon N\to M$ and an integer $n\neq 0$ such that $u\circ v = n\cdot\mathrm{id}_N$, $v\circ u = n\cdot\mathrm{id}_M$, and that the $(n\lambda)$-connections $(\mathrm{id}\otimes v)\circ\nabla$ on $M$ and $\nabla\circ v$ on $N$ are integrable.

Let $(M,N,u,\nabla)$, $(M',N',u',\nabla')$ be two $\lambda$-isoconnections. A morphism from $(M,N,u,\nabla)$ to $(M',N',u',\nabla')$ consists of two $B$–linear morphisms $\alpha\colon M\to M'$ and $\beta\colon N\to N'$ such that $\beta\circ u = u'\circ\alpha$ and $(\mathrm{id}\otimes\beta)\circ\nabla = \nabla'\circ\alpha$.

**III.6.11.** Let $f\colon (X',A')\to(X,A)$ be a morphism of ringed topos, $B$ an $A$-algebra, $B'$ an $A'$-algebra, $\alpha\colon f^*(B)\to B'$ a homomorphism of $A'$-algebras, $\lambda\in\Gamma(X,A)$, and $(M,N,u,\nabla)$ a $\lambda$-isoconnection with respect to the extension $B/A$. Denote by $\lambda'$ the canonical image of $\lambda$ in $\Gamma(X',A')$, by $d'\colon B'\to\Omega^1_{B'/A'}$ the universal $A'$-derivation of $B'$, and by

$$\gamma\colon f^*(\Omega^1_{B/A})\to\Omega^1_{B'/A'} \tag{III.6.11.1}$$

the canonical $\alpha$-linear morphism. We immediately see that $(f^*(M),f^*(N),f^*(u),f^*(\nabla))$ is a $\lambda'$-isoconnection with respect to the extension $f^*(B)/A'$, which is integrable if $(M,N,u,\nabla)$ is.

There exists a unique $A'$-linear morphism

$$\nabla'\colon B'\otimes_{f^*(B)} f^*(M)\to \Omega^1_{B'/A'}\otimes_{f^*(B)} f^*(N) \tag{III.6.11.2}$$

such that for all local sections $x'$ of $B'$ and $t$ of $f^*(M)$, we have

$$\nabla'(x' \otimes t) = \lambda' d'(x') \otimes f^*(u)(t) + x'(\gamma \otimes \mathrm{id}_{f^*(N)})(f^*(\nabla)(t)). \tag{III.6.11.3}$$

The quadruple $(B' \otimes_{f^*(B)} f^*(M), B' \otimes_{f^*(B)} f^*(N), \mathrm{id}_{B'} \otimes_{f^*(B)} f^*(u), \nabla')$ is a $\lambda'$-isoconnection with respect to the extension $B'/A'$, which is integrable if $(M, N, u, \nabla)$ is.

**III.6.12.** Let $(X, A)$ be a ringed topos, $B$ an $A$-algebra, $\lambda \in \Gamma(X, A)$, and $(M, N, u, \nabla)$ an integrable $\lambda$-isoconnection with respect to the extension $B/A$. Suppose that there exist an $A$-module $E$ and a $B$-isomorphism $\gamma\colon E \otimes_A B \xrightarrow{\sim} \Omega^1_{B/A}$ such that for every local section $\omega$ of $E$, we have $d(\gamma(\omega \otimes 1)) = 0$. Denote by $\theta\colon M \to E \otimes_A N$ the morphism induced by $\nabla$ and $\gamma$. Then $(M, N, u, \theta)$ is a Higgs $A$-isogeny with coefficients in $E$ (cf. II.2.12).

Let $(M', N', u', \theta')$ be a Higgs $A$-isogeny with coefficients in $E$. There exists a unique $A$-linear morphism

$$\nabla'\colon M \otimes_A M' \to \Omega^1_{B/A} \otimes_B N \otimes_A N' \tag{III.6.12.1}$$

such that for all local sections $t$ of $M$ and $t'$ of $M'$, we have

$$\nabla'(t \otimes t') = \nabla(t) \otimes_A u'(t') + (\gamma \otimes_B \mathrm{id}_{N \otimes_A N'})(u(t) \otimes_A \theta'(t')). \tag{III.6.12.2}$$

The quadruple $(M \otimes_A M', N \otimes_A N', u \otimes u', \nabla')$ is an integrable $\lambda$-isoconnection.

**III.6.13.** Let $A$ be an adic ring, $I$ an ideal of definition of $A$, $\lambda \in A$, and $B$ an adic $A$-algebra, that is, $B$ is an $A$-algebra that is complete and separated for the $(IB)$-adic topology. Recall that the canonical topology on the $B$-module $\Omega^1_{B/A}$ is deduced from that on $B$ (**[42]** 0.20.4.5). We denote by $\widehat{\Omega}^1_{B/A}$ its Hausdorff completion and denote also by

$$d\colon B \to \widehat{\Omega}^1_{B/A} \tag{III.6.13.1}$$

the universal continuous $A$-derivation of $B$. An *adic $\lambda$-isoconnection* (or *$I$-adic $\lambda$-isoconnection*) with respect to the extension $B/A$ is a quadruple

$$(M, N, u\colon M \to N, \nabla\colon M \to \widehat{\Omega}^1_{B/A} \widehat{\otimes}_B N) \tag{III.6.13.2}$$

where $M$ and $N$ are $B$-modules that are complete and separated for the $(IB)$-adic topologies, $u$ is an isogeny of $B$-modules (III.6.1), and $\nabla$ is an $A$-linear morphism such that for all $x \in B$ and $t \in M$, we have

$$\nabla(xt) = \lambda d(x) \widehat{\otimes} u(t) + x\nabla(t). \tag{III.6.13.3}$$

For every $B$-linear morphism $v\colon N \to M$ for which there exists an integer $n$ such that $u \circ v = n \cdot \mathrm{id}_N$ and $v \circ u = n \cdot \mathrm{id}_M$, the pairs $(M, (\mathrm{id} \widehat{\otimes} v) \circ \nabla)$ and $(N, \nabla \circ v)$ are modules with adic $(n\lambda)$-connections (II.2.14), and $u$ is a morphism from $(M, (\mathrm{id} \widehat{\otimes} v) \circ \nabla)$ to $(N, \nabla \circ v)$. We say that the adic $\lambda$-isoconnection $(M, N, u, \nabla)$ is *integrable* if there exist a $B$-linear morphism $v\colon N \to M$ and an integer $n \neq 0$ such that $u \circ v = n \cdot \mathrm{id}_N$, $v \circ u = n \cdot \mathrm{id}_M$, and that the adic $(n\lambda)$-connections $(\mathrm{id} \widehat{\otimes} v) \circ \nabla$ on $M$ and $\nabla \circ v$ on $N$ are integrable (cf. II.2.14).

Let $(M, N, u, \nabla)$, $(M', N', u', \nabla')$ be two adic $\lambda$-isoconnections with respect to the extension $B/A$. A morphism from $(M, N, u, \nabla)$ to $(M', N', u', \nabla')$ consists of two $B$-linear morphisms $\alpha\colon M \to M'$ and $\beta\colon N \to N'$ such that $\beta \circ u = u' \circ \alpha$ and $(\mathrm{id} \widehat{\otimes} \beta) \circ \nabla = \nabla' \circ \alpha$.

**III.6.14.** Let $A$ be an adic ring, $\lambda \in A$, $B$ an adic $A$-algebra, and $(M, N, u, \nabla)$ an integrable adic $\lambda$-isoconnection with respect to the extension $B/A$. Suppose that the following conditions are satisfied:

(i) $A$ admits an ideal of definition of finite type $I$, and if we set $A_1 = A/I$ and $B_1 = B \otimes_A A_1$, the $B_1$-module $\Omega^1_{B_1/A_1}$ is of finite type.

(ii) There exist a *free* $A$-module *of finite type* $E$ and a $B$-linear isomorphism $\gamma \colon E \otimes_A B \xrightarrow{\sim} \widehat{\Omega}^1_{B/A}$ such that $\gamma(E) \subset d(B)$.

Note that $\widehat{\Omega}^1_{B/A} \widehat{\otimes}_B N = \widehat{\Omega}^1_{B/A} \otimes_B N = E \otimes_A N$. We denote by $\theta \colon M \to E \otimes_A N$ the morphism induced by $\nabla$ and $\gamma$. It then follows from (II.2.16) that $(M, N, u, \theta)$ is a Higgs $A$-isogeny with coefficients in $E$.

**III.6.15.** Recall that $\mathscr{S}$ denotes the formal scheme $\mathrm{Spf}(\mathscr{O}_C)$ (III.2.1). Let $\mathfrak{X}$ be a formal $\mathscr{S}$-scheme locally of finite presentation (cf. [**1**] 2.3.15), $\mathscr{J}$ a coherent ideal of definition of $\mathfrak{X}$, and $\mathscr{F}$ an $\mathscr{O}_{\mathfrak{X}}$-module. The formal scheme $\mathfrak{X}$ is therefore idyllic ([**1**] 2.6.13). Following ([**1**] 2.10.1), we call *rigid closure* of $\mathscr{F}$ and denote by $\mathscr{H}^0_{\mathrm{rig}}(\mathscr{F})$, the $\mathscr{O}_{\mathfrak{X}}$-module

(III.6.15.1)
$$\mathscr{H}^0_{\mathrm{rig}}(\mathscr{F}) = \varinjlim_{n \geq 0} \mathscr{H}om_{\mathscr{O}_{\mathfrak{X}}}(\mathscr{J}^n, \mathscr{F}).$$

This notion does not depend on the ideal $\mathscr{J}$. On the other hand, we set

(III.6.15.2)
$$\mathscr{F}[\frac{1}{p}] = \mathscr{F}_{\mathbb{Q}_p} = \mathscr{F} \otimes_{\mathbb{Z}_p} \mathbb{Q}_p.$$

Since $p\mathscr{O}_{\mathfrak{X}}$ is an ideal of definition for $\mathfrak{X}$, the canonical morphism $\mathscr{F}_{\mathbb{Q}_p} \to \mathscr{H}^0_{\mathrm{rig}}(\mathscr{F})$ is an isomorphism by ([**1**] 2.10.5). We say that $\mathscr{F}$ is *rig-null* if the canonical morphism $\mathscr{F} \to \mathscr{F}_{\mathbb{Q}_p}$ is zero (cf. [**1**] 2.10.1.4).

Consider the following conditions:

(i) $\mathscr{F}_{\mathbb{Q}_p} = 0$.

(ii) $\mathscr{F}$ is rig-null.

(iii) There exists an integer $n \geq 1$ such that $p^n \mathscr{F} = 0$.

By ([**1**] 2.10.10), we then have (iii)$\Rightarrow$(i)$\Leftrightarrow$(ii). Moreover, if $\mathfrak{X}$ is quasi-compact and if $\mathscr{F}$ is of finite type, the three conditions are equivalent. It follows that if $\mathfrak{X}$ is quasi-compact and if $\mathscr{F}$ is of finite type, for every $\mathscr{O}_{\mathfrak{X}}$-module $\mathscr{G}$, the canonical homomorphism

(III.6.15.3)
$$\mathrm{Hom}_{\mathscr{O}_{\mathfrak{X}}}(\mathscr{F}, \mathscr{G}) \otimes_{\mathbb{Z}_p} \mathbb{Q}_p \to \mathrm{Hom}_{\mathscr{O}_{\mathfrak{X}}[\frac{1}{p}]}(\mathscr{F}_{\mathbb{Q}_p}, \mathscr{G}_{\mathbb{Q}_p})$$

is injective.

**Lemma III.6.16.** *Let $\mathfrak{X}$ be a formal $\mathscr{S}$-scheme of finite presentation. We denote by* $\mathbf{Mod}^{\mathrm{coh}}(\mathscr{O}_{\mathfrak{X}})$ *(resp.* $\mathbf{Mod}^{\mathrm{coh}}(\mathscr{O}_{\mathfrak{X}}[\frac{1}{p}])$*) the category of coherent $\mathscr{O}_{\mathfrak{X}}$-modules (resp. $\mathscr{O}_{\mathfrak{X}}[\frac{1}{p}]$-modules) and by* $\mathbf{Mod}^{\mathrm{coh}}_{\mathbb{Q}}(\mathscr{O}_{\mathfrak{X}})$ *the category of coherent $\mathscr{O}_{\mathfrak{X}}$-modules up to isogeny. Then the canonical functor*

(III.6.16.1)
$$\mathbf{Mod}^{\mathrm{coh}}(\mathscr{O}_{\mathfrak{X}}) \to \mathbf{Mod}^{\mathrm{coh}}(\mathscr{O}_{\mathfrak{X}}[\frac{1}{p}]), \quad \mathscr{F} \mapsto \mathscr{F}_{\mathbb{Q}_p},$$

*induces an equivalence of abelian categories*

(III.6.16.2)
$$\mathbf{Mod}^{\mathrm{coh}}_{\mathbb{Q}}(\mathscr{O}_{\mathfrak{X}}) \xrightarrow{\sim} \mathbf{Mod}^{\mathrm{coh}}(\mathscr{O}_{\mathfrak{X}}[\frac{1}{p}]).$$

First note that the functor (III.6.16.1) is well-defined by virtue of ([**1**] 2.10.24(i)) and that it induces an exact functor

$$\mathbf{Mod}^{\mathrm{coh}}_{\mathbb{Q}}(\mathscr{O}_{\mathfrak{X}}) \to \mathbf{Mod}^{\mathrm{coh}}(\mathscr{O}_{\mathfrak{X}}[\frac{1}{p}]). \tag{III.6.16.3}$$

This is essentially surjective by virtue of ([**1**] 2.10.24(ii)). Let us show that it is fully faithful. Let $\mathscr{F}$, $\mathscr{G}$ be two coherent $\mathscr{O}_{\mathfrak{X}}$-modules. The canonical homomorphism

$$\mathrm{Hom}_{\mathscr{O}_{\mathfrak{X}}}(\mathscr{F},\mathscr{G}) \otimes_{\mathbb{Z}_p} \mathbb{Q}_p \to \mathrm{Hom}_{\mathscr{O}_{\mathfrak{X}}[\frac{1}{p}]}(\mathscr{F}_{\mathbb{Q}_p},\mathscr{G}_{\mathbb{Q}_p}) \tag{III.6.16.4}$$

is injective by (III.6.15.3). On the other hand, for every $\mathscr{O}_{\mathfrak{X}}$-linear morphism $v\colon \mathscr{F} \to \mathscr{G}_{\mathbb{Q}_p}$, there exists an integer $n \geq 0$ such that $v(p^n\mathscr{F})$ is contained in the image of the canonical morphism $c_{\mathscr{G}}\colon \mathscr{G} \to \mathscr{G}_{\mathbb{Q}_p}$. Since $\mathscr{G}_{\mathrm{tor}} = \ker(c_{\mathscr{G}})$ is coherent ([**1**] 2.10.14), there exists an integer $m \geq 0$ such that $p^m\mathscr{G}_{\mathrm{tor}} = 0$. It follows that there exists an $\mathscr{O}_{\mathfrak{X}}$-linear morphism $w\colon \mathscr{F} \to \mathscr{G}$ such that $c_{\mathscr{G}} \circ w = p^{n+m}v$. The homomorphism (III.6.16.4) is therefore surjective; the lemma follows.

**Lemma III.6.17.** *Let $\mathfrak{X} = \mathrm{Spf}(R)$ be an affine formal $\mathscr{S}$-scheme of finite presentation and $\mathscr{F}$ a coherent $\mathscr{O}_{\mathfrak{X}}[\frac{1}{p}]$-module. Then $\mathscr{F}$ is a locally projective $\mathscr{O}_{\mathfrak{X}}[\frac{1}{p}]$-module of finite type* (III.2.8) *if and only if $\Gamma(\mathfrak{X},\mathscr{F})$ is a projective $R[\frac{1}{p}]$-module of finite type.*

Recall that the ring $R[\frac{1}{p}]$ is noetherian ([**1**] 1.10.2(i)). By III.6.16 and ([**1**] 2.7.2), there exists a coherent $R$-module $M$ such that $\mathscr{F} = (M^{\Delta})_{\mathbb{Q}_p}$. We have $\Gamma(\mathfrak{X},\mathscr{F}) = M_{\mathbb{Q}_p}$ by virtue of ([**1**] (2.10.5.1)).

Let us first suppose that $\mathscr{F}$ is a locally projective $\mathscr{O}_{\mathfrak{X}}[\frac{1}{p}]$-module of finite type and let us show that $\Gamma(\mathfrak{X},\mathscr{F})$ is a projective $R[\frac{1}{p}]$-module of finite type. By ([**1**] 2.7.4, (2.10.5.1) and 5.1.11), there exists a faithfully flat $R$-algebra $R'$ that is topologically of finite presentation such that the $R'$-module $\Gamma(\mathfrak{X},\mathscr{F}) \otimes_R R'$ is projective of finite type. We deduce from this by faithfully flat descent that $\Gamma(\mathfrak{X},\mathscr{F})$ is a projective $R[\frac{1}{p}]$-module of finite type.

Next, let us suppose that $\Gamma(\mathfrak{X},\mathscr{F})$ is a projective $R[\frac{1}{p}]$-module of finite type and let us show that $\mathscr{F}$ is a locally projective $\mathscr{O}_{\mathfrak{X}}[\frac{1}{p}]$-module of finite type. In view of ([**1**] 1.10.2(iii) and (2.10.5.1)), we may assume that the $R[\frac{1}{p}]$-module $\Gamma(\mathfrak{X},\mathscr{F})$ is free of finite type. Hence there exist an integer $n \geq 0$ and an $R$-linear morphism $R^n \to M$ whose kernel and cokernel are torsion. Consequently, $\mathscr{F}$ is a free $\mathscr{O}_{\mathfrak{X}}[\frac{1}{p}]$-module of finite type.

**Lemma III.6.18.** *Let $A$ be a ring, $t \in A$, and $M$ an $A$-module of finite type. We assume that $A$ is complete and separated for the $(tA)$-adic topology, that $t$ is not a zero divisor in $A$, and that $M_t$ is a projective $A_t$-module. We denote by $\widehat{M}$ the Hausdorff completion of $M$ for the $(tA)$-adic topology. Then the canonical morphism $M \to \widehat{M}$ induces an isomorphism $M_t \xrightarrow{\sim} \widehat{M}_t$.*

Indeed, denote by $M'$ the image of the canonical morphism $M \to M_t$. There exist an integer $n \geq 1$ and an injective $A_t$-linear morphism $u\colon M_t \to A_t^n$. Since $t$ is not a zero divisor in $A$, there exists an integer $i \geq 0$ such that $M' \subset u^{-1}(t^{-i}A)^n$. Consequently, $M'$ is contained in a free $A$-module of finite type, and is therefore separated for the $(tA)$-adic topology. On the other hand, $M$ and $M'$ are complete for the $(tA)$-adic topologies by ([**11**] Chapter III § 2.12 Corollary 1 to Proposition 16). It follows that the canonical morphism $M \to \widehat{M}$ is surjective and that $M'$ is complete and separated for the $(tA)$-adic topology. Consequently, the canonical surjective morphism $M \to M'$ factors into $M \to \widehat{M} \to M'$. Since $M_t \to M'_t$ is an isomorphism, $M_t \to \widehat{M}_t$ is also an isomorphism.

**III.6.19.** Let $\mathfrak{X}$ be a formal $\mathscr{S}$-scheme of finite presentation and $\mathscr{E}$ an $\mathscr{O}_{\mathfrak{X}}$-module. We denote by $\mathbf{HM}(\mathscr{O}_{\mathfrak{X}}[\frac{1}{p}], \mathscr{E}_{\mathbb{Q}_p})$ the category of Higgs $\mathscr{O}_{\mathfrak{X}}[\frac{1}{p}]$-modules with coefficients in $\mathscr{E}_{\mathbb{Q}_p}$ (II.2.8), by $\mathbf{HM}^{\mathrm{coh}}(\mathscr{O}_{\mathfrak{X}}[\frac{1}{p}], \mathscr{E}_{\mathbb{Q}_p})$ the full subcategory made up of the Higgs modules whose underlying $\mathscr{O}_{\mathfrak{X}}[\frac{1}{p}]$-module is coherent, by $\mathbf{HI}(\mathscr{O}_{\mathfrak{X}}, \mathscr{E})$ the category of Higgs $\mathscr{O}_{\mathfrak{X}}$-isogenies with coefficients in $\mathscr{E}$ (III.6.8), and by $\mathbf{HI}^{\mathrm{coh}}(\mathscr{O}_{\mathfrak{X}}, \mathscr{E})$ the full subcategory made up of the quadruples $(\mathscr{M}, \mathscr{N}, u, \theta)$ such that the $\mathscr{O}_{\mathfrak{X}}$-modules $\mathscr{M}$ and $\mathscr{N}$ are coherent. These are additive categories. We denote by $\mathbf{HI}_{\mathbb{Q}}(\mathscr{O}_{\mathfrak{X}}, \mathscr{E})$ (resp. $\mathbf{HI}^{\mathrm{coh}}_{\mathbb{Q}}(\mathscr{O}_{\mathfrak{X}}, \mathscr{E})$) the category of objects of $\mathbf{HI}(\mathscr{O}_{\mathfrak{X}}, \mathscr{E})$ (resp. $\mathbf{HI}^{\mathrm{coh}}(\mathscr{O}_{\mathfrak{X}}, \mathscr{E})$) up to isogeny (III.6.1.1). We have a functor

(III.6.19.1)
$$\mathbf{HI}(\mathscr{O}_{\mathfrak{X}}, \mathscr{E}) \to \mathbf{HM}(\mathscr{O}_{\mathfrak{X}}[\frac{1}{p}], \mathscr{E}_{\mathbb{Q}_p}), \quad (\mathscr{M}, \mathscr{N}, u, \theta) \mapsto (\mathscr{M}_{\mathbb{Q}_p}, (u^{-1}_{\mathbb{Q}_p} \otimes \mathrm{id}_{\mathscr{E}_{\mathbb{Q}_p}}) \circ \theta_{\mathbb{Q}_p}).$$

**Lemma III.6.20.** *Under the assumptions of* III.6.19, *let moreover* $(\mathscr{M}, \mathscr{N}, u, \theta)$ *and* $(\mathscr{M}', \mathscr{N}', u', \theta')$ *be two objects of* $\mathbf{HI}(\mathscr{O}_{\mathfrak{X}}, \mathscr{E})$ *such that* $\mathscr{M}$ *and* $\mathscr{N}$ *are* $\mathscr{O}_{\mathfrak{X}}$*-modules of finite type, and* $(\mathscr{M}_{\mathbb{Q}_p}, \widetilde{\theta})$ *and* $(\mathscr{M}'_{\mathbb{Q}_p}, \widetilde{\theta}')$ *their respective images by the functor* (III.6.19.1). *Then the canonical homomorphism*

(III.6.20.1)
$$\mathrm{Hom}_{\mathbf{HI}(\mathscr{O}_{\mathfrak{X}}, \mathscr{E})}((\mathscr{M}, \mathscr{N}, u, \theta), (\mathscr{M}', \mathscr{N}', u', \theta')) \otimes_{\mathbb{Z}_p} \mathbb{Q}_p \to \mathrm{Hom}_{\mathbf{HM}(\mathscr{O}_{\mathfrak{X}}[\frac{1}{p}], \mathscr{E}_{\mathbb{Q}_p})}((\mathscr{M}_{\mathbb{Q}_p}, \widetilde{\theta}), (\mathscr{M}_{\mathbb{Q}_p}, \widetilde{\theta}'))$$

*is injective.*

Indeed, let $\alpha\colon \mathscr{M} \to \mathscr{M}'$ and $\beta\colon \mathscr{N} \to \mathscr{N}'$ be two $\mathscr{O}_{\mathfrak{X}}$-linear morphisms defining a morphism from $(\mathscr{M}, \mathscr{N}, u, \theta)$ to $(\mathscr{M}', \mathscr{N}', u', \theta')$ in $\mathbf{HI}(\mathscr{O}_{\mathfrak{X}}, \mathscr{E})$, whose image $\alpha_{\mathbb{Q}_p}$ by the homomorphism (III.6.20.1) is zero. Since $(\alpha(\mathscr{M}))_{\mathbb{Q}_p} = 0$ and $\mathscr{M}$ is of finite type over $\mathscr{O}_{\mathfrak{X}}$, there exists an integer $n \geq 0$ such that $p^n \alpha = 0$ (III.6.15). Likewise, since $\beta_{\mathbb{Q}_p} = 0$, there exists an integer $m \geq 0$ such that $p^m \beta = 0$. The lemma follows.

**Lemma III.6.21.** *Under the assumptions of* III.6.19, *suppose moreover that* $\mathscr{E}$ *is coherent. Then the functor*

(III.6.21.1)
$$\mathbf{HI}^{\mathrm{coh}}_{\mathbb{Q}}(\mathscr{O}_{\mathfrak{X}}, \mathscr{E}) \to \mathbf{HM}^{\mathrm{coh}}(\mathscr{O}_{\mathfrak{X}}[\frac{1}{p}], \mathscr{E}_{\mathbb{Q}_p})$$

*induced by* (III.6.19.1) *is an equivalence of categories.*

Let $N$ be a coherent $\mathscr{O}_{\mathfrak{X}}[\frac{1}{p}]$-module, $\theta$ a Higgs $\mathscr{O}_{\mathfrak{X}}[\frac{1}{p}]$-field on $N$ with coefficients in $\mathscr{E}_{\mathbb{Q}_p}$. By ([**1**] 2.10.24(ii)), there exist a coherent $\mathscr{O}_{\mathfrak{X}}$-module $\mathscr{N}$ and an isomorphism $u\colon \mathscr{N}_{\mathbb{Q}_p} \xrightarrow{\sim} N$. We may assume that $\mathscr{N}$ has no $p$-torsion ([**1**] 2.10.14). By the proof of III.6.16, there exist an integer $n \geq 0$ and an $\mathscr{O}_{\mathfrak{X}}$-linear morphism $\vartheta\colon \mathscr{N} \to \mathscr{N} \otimes_{\mathscr{O}_{\mathfrak{X}}} \mathscr{E}$ such that the diagram

(III.6.21.2)
$$\begin{array}{ccc} \mathscr{N} & \longrightarrow & N \\ \downarrow{\scriptstyle \vartheta} & & \downarrow{\scriptstyle p^n\theta} \\ \mathscr{N} \otimes_{\mathscr{O}_{\mathfrak{X}}} \mathscr{E} & \longrightarrow & N \otimes_{\mathscr{O}_{\mathfrak{X}}[\frac{1}{p}]} \mathscr{E}_{\mathbb{Q}_p} \end{array}$$

where the horizontal arrows are induced by $u$, is commutative. After multiplying $\vartheta$ by a power of $p$, if necessary, we may assume that $\vartheta \wedge \vartheta = 0$ (III.6.15.3) (that is, that $\vartheta$ is a Higgs field). Since $\mathscr{N}$ has no $p$-torsion, the morphism $\vartheta$ factors into two $\mathscr{O}_{\mathfrak{X}}$-linear morphisms

(III.6.21.3)
$$\mathscr{N} \xrightarrow{\nu_n} p^n \mathscr{N} \xrightarrow{\vartheta_n} \mathscr{N} \otimes_{\mathscr{O}_{\mathfrak{X}}} \mathscr{E}\ ,$$

where $\nu_n$ is the isomorphism induced by the multiplication by $p^n$ on $\mathscr{N}$. The composition

$$p^n\mathscr{N} \xrightarrow{\vartheta_n} \mathscr{N} \otimes_{\mathscr{O}_{\mathfrak{X}}} \mathscr{E} \xrightarrow{\nu_n \otimes \mathrm{id}_{\mathscr{E}}} (p^n\mathscr{N}) \otimes_{\mathscr{O}_{\mathfrak{X}}} \mathscr{E} \tag{III.6.21.4}$$

is then also a Higgs field. Denote by $\iota_n \colon p^n\mathscr{N} \to \mathscr{N}$ the canonical injection. We have $\vartheta \circ \iota_n = p^n\vartheta_n$, so that the diagram

$$\begin{array}{ccccc} p^n\mathscr{N} & \xrightarrow{\iota_n} & \mathscr{N} & \longrightarrow & N \\ {\scriptstyle\vartheta_n}\downarrow & & & & \downarrow{\scriptstyle\theta} \\ \mathscr{N} \otimes_{\mathscr{O}_{\mathfrak{X}}} \mathscr{E} & & \longrightarrow & & N \otimes_{\mathscr{O}_{\mathfrak{X}}[\frac{1}{p}]} \mathscr{E}_{\mathbb{Q}_p} \end{array} \tag{III.6.21.5}$$

is commutative. Consequently, $(p^n\mathscr{N}, \mathscr{N}, \iota_n, \vartheta_n)$ is an object of $\mathbf{HI}^{\mathrm{coh}}(\mathscr{O}_{\mathfrak{X}}, \mathscr{E})$ whose image by the functor (III.6.19.1) is isomorphic to $(N, \theta)$. The functor (III.6.21.1) is therefore essentially surjective. We know by III.6.20 that it is faithful. Let us show that it is full. Let $(\mathscr{M}, \mathscr{N}, u, \theta)$ and $(\mathscr{M}', \mathscr{N}', u', \theta')$ be two objects of $\mathbf{HI}^{\mathrm{coh}}(\mathscr{O}_{\mathfrak{X}}, \mathscr{E})$, $(\mathscr{M}_{\mathbb{Q}_p}, \widetilde{\theta})$ and $(\mathscr{M}'_{\mathbb{Q}_p}, \widetilde{\theta}')$ their respective images by the functor (III.6.21.1), and $\lambda \colon \mathscr{M}_{\mathbb{Q}_p} \to \mathscr{M}'_{\mathbb{Q}_p}$ an $\mathscr{O}_{\mathfrak{X}}[\frac{1}{p}]$-linear morphism such that $(\lambda \otimes \mathrm{id}_{\mathscr{E}_{\mathbb{Q}_p}}) \circ \widetilde{\theta} = \widetilde{\theta}' \circ \lambda$. By the proof of III.6.16, there exist an integer $n \geq 0$ and an $\mathscr{O}_{\mathfrak{X}}$-linear morphism $\alpha \colon \mathscr{M} \to \mathscr{M}'$ such that the diagram

$$\begin{array}{ccc} \mathscr{M} & \longrightarrow & \mathscr{M}_{\mathbb{Q}_p} \\ {\scriptstyle\alpha}\downarrow & & \downarrow{\scriptstyle p^n\lambda} \\ \mathscr{M}' & \longrightarrow & \mathscr{M}'_{\mathbb{Q}_p} \end{array} \tag{III.6.21.6}$$

where the horizontal arrows are the canonical morphisms, is commutative. By ([**1**] 2.10.22(i) and 2.10.10), there exists an integer $m \geq 0$ such that $p^m$ annihilates the kernel and cokernel of $u$. Hence there exists an $\mathscr{O}_{\mathfrak{X}}$-linear morphism $\beta \colon \mathscr{N} \to \mathscr{N}'$ such that $\beta \circ u = u' \circ (p^{2m}\alpha)$. Denote by

$$c \colon \mathscr{N}' \otimes_{\mathscr{O}_{\mathfrak{X}}} \mathscr{E} \to \mathscr{N}'_{\mathbb{Q}_p} \otimes_{\mathscr{O}_{\mathfrak{X}}[\frac{1}{p}]} \mathscr{E}_{\mathbb{Q}_p}$$

the canonical morphism. Since $(\mathscr{N}' \otimes_{\mathscr{O}_{\mathfrak{X}}} \mathscr{E})_{\mathrm{tor}} = \ker(c)$ is coherent ([**1**] 2.10.14), there exists an integer $q \geq 0$ such that $p^q \ker(c) = 0$. The relation $(\lambda \otimes \mathrm{id}_{\mathscr{E}_{\mathbb{Q}_p}}) \circ \widetilde{\theta} = \widetilde{\theta}' \circ \lambda$ then implies that $((p^q\beta) \otimes \mathrm{id}_{\mathscr{E}}) \circ \theta = \theta' \circ (p^{q+2m}\alpha)$. The functor (III.6.21.1) is therefore full.

**Proposition III.6.22.** *Let $\mathfrak{X}$ be a formal $\mathscr{S}$-scheme of finite presentation, $f \colon \mathfrak{X}' \to \mathfrak{X}$ a faithfully flat morphism of finite presentation* ([**1**] 5.1.7), *$\mathfrak{X}'' = \mathfrak{X}' \times_{\mathfrak{X}} \mathfrak{X}'$, and $\mathrm{p}_1, \mathrm{p}_2 \colon \mathfrak{X}'' \to \mathfrak{X}'$ the canonical projections.*

(i) *Let $\mathscr{F}$ and $\mathscr{G}$ be two coherent $\mathscr{O}_{\mathfrak{X}}[\frac{1}{p}]$-modules, $\mathscr{F}'$ and $\mathscr{G}'$ their inverse images on $\mathfrak{X}'$, and $\mathscr{F}''$ and $\mathscr{G}''$ their inverse images on $\mathfrak{X}''$. Then the diagram of maps of sets*

$$\mathrm{Hom}_{\mathscr{O}_{\mathfrak{X}}[\frac{1}{p}]}(\mathscr{F}, \mathscr{G}) \to \mathrm{Hom}_{\mathscr{O}_{\mathfrak{X}'}[\frac{1}{p}]}(\mathscr{F}', \mathscr{G}') \rightrightarrows \mathrm{Hom}_{\mathscr{O}_{\mathfrak{X}''}[\frac{1}{p}]}(\mathscr{F}'', \mathscr{G}'') \tag{III.6.22.1}$$

*defined by the base change functors by $f$, $\mathrm{p}_1$, and $\mathrm{p}_2$ is exact.*

(ii) *For every coherent $\mathscr{O}_{\mathfrak{X}}[\frac{1}{p}]$-module $\mathscr{F}'$, every descent datum on $\mathscr{F}'$ with respect to $f$ is effective.*

This follows from the proof of ([**1**] 5.11.12).

**Remark III.6.23.** Let $\mathfrak{X}$ be a formal $\mathscr{S}$-scheme of finite presentation and $\mathscr{F}$, $\mathscr{G}$ two coherent $\mathscr{O}_{\mathfrak{X}}[\frac{1}{p}]$-modules. Denote by $\mathfrak{X}^{\mathrm{rig}}$ the rigid space associated with $\mathfrak{X}$ ([**1**] 4.1.6), by $\mathfrak{X}^{\mathrm{rig}}_{\mathrm{ad}}$ the admissible topos of $\mathfrak{X}^{\mathrm{rig}}$ ([**1**] 4.4.1), and by

$$\varrho_{\mathfrak{X}}\colon (\mathfrak{X}^{\mathrm{rig}}_{\mathrm{ad}}, \mathscr{O}_{\mathfrak{X}^{\mathrm{rig}}}) \to (\mathfrak{X}_{\mathrm{zar}}, \mathscr{O}_{\mathfrak{X}}[\frac{1}{p}]) \tag{III.6.23.1}$$

the canonical morphism of ringed topos ([**1**] (4.7.5.1)). By ([**1**] 2.10.24(ii), 4.7.8, and 4.7.28), the adjunction morphism $\mathscr{F} \to \varrho_{\mathfrak{X}*}(\varrho^*_{\mathfrak{X}}(\mathscr{F}))$ is an isomorphism. Consequently, the map

$$\mathrm{Hom}_{\mathscr{O}_{\mathfrak{X}^{\mathrm{rig}}}}(\varrho^*_{\mathfrak{X}}(\mathscr{F}), \varrho^*_{\mathfrak{X}}(\mathscr{G})) \to \mathrm{Hom}_{\mathscr{O}_{\mathfrak{X}}[\frac{1}{p}]}(\mathscr{F}, \mathscr{G}), \quad u \mapsto \varrho_{\mathfrak{X}*}(u) \tag{III.6.23.2}$$

is bijective; it is the adjunction isomorphism. Statement III.6.22(i) then immediately follows from ([**1**] 5.11.12(i)). Statement III.6.22(ii) does not formally result from the statement of ([**1**] 5.11.12(ii)), but does result from the same proof.

## III.7. Inverse systems of a topos

**III.7.1.** In this section, $X$ denotes a $\mathbb{U}$-topos and $I$ a $\mathbb{U}$-small category (III.2.5). We always consider $X$ as endowed with its canonical topology, which makes it into a $\mathbb{U}$-site. We endow $X \times I$ with the total topology associated with the constant fibered site $X \times I \to I$ with fiber $X$ (cf. VI.7.1 and [**2**] VI 7.4.1), which makes it into a $\mathbb{U}$-site. Recall ([**2**] VI 7.4.7) that the topos of sheaves of $\mathbb{U}$-sets on $X \times I$ is canonically equivalent to the category $\mathbf{Hom}(I^\circ, X)$ of functors from $I^\circ$ to $X$, which we also denote by $X^{I^\circ}$. In particular, $X^{I^\circ}$ is a $\mathbb{U}$-topos. This last fact can be seen directly ([**2**] IV 1.2). We refer to VI.7.4 for the description of the rings and modules of $X^{I^\circ}$.

For any $i \in \mathrm{Ob}(I)$, we denote by

$$\alpha_{i!}\colon X \to X \times I \tag{III.7.1.1}$$

the functor that associates with each object $F$ of $X$ the pair $(F, i)$. Since this is cocontinuous ([**2**] VI 7.4.2), it defines a morphism of topos ([**2**] IV 4.7)

$$\alpha_i\colon X \to X^{I^\circ}. \tag{III.7.1.2}$$

By ([**2**] VI 7.4.7), for all $F \in \mathrm{Ob}(X^{I^\circ})$ and $i \in \mathrm{Ob}(I)$, we have

$$\alpha_i^*(F) = F(i). \tag{III.7.1.3}$$

We denote also by

$$\alpha_{i!}\colon X \to X^{I^\circ} \tag{III.7.1.4}$$

the composition of $\alpha_{i!}$ (III.7.1.1) and the canonical functor $X \times I \to X^{I^\circ}$. For all $j \in \mathrm{Ob}(I)$ and $F \in \mathrm{Ob}(X)$, we have

$$\alpha_{i!}(F)(j) = F \times (\mathrm{Hom}_I(j, i))_X, \tag{III.7.1.5}$$

where $(\mathrm{Hom}_I(j, i))_X$ is the constant sheaf on $X$ with value $\mathrm{Hom}_I(j, i)$. By ([**2**] VI 7.4.3(4)), the functor $\alpha_{i!}$ (III.7.1.4) is a left adjoint of $\alpha_i^*$.

For every morphism $f\colon i \to j$ of $I$, we have a morphism

$$\rho_f\colon \alpha_i \to \alpha_j, \tag{III.7.1.6}$$

defined at the level of the inverse images, for each $F \in \mathrm{Ob}(X^{I^\circ})$, by the morphism $F(j) \to F(i)$ induced by $f$ ([**2**] VI (7.4.5.2)). If $f$ and $g$ are two composable morphisms of $I$, then we have $\rho_{gf} = \rho_g \rho_f$.

**Remarks III.7.2.** (i) Let $U \in \mathrm{Ob}(X)$, $i \in \mathrm{Ob}(I)$. A family $((U_n, i_n) \to (U, i))_{n \in N}$ is covering for the total topology on $X \times I$ if and only if it is refined by a family $((V_m, i) \to (U, i))_{m \in M}$, where $(V_m \to U)_{m \in M}$ is a covering family of $X$ ([**2**] VI 7.4.2(1)).

(ii) It follows from (i) that a sieve $\mathscr{R}$ of $X^{I^\circ}$ is universal strict epimorphic, that is, covers the final object of $X^{I^\circ}$ for the canonical topology, if and only if for every $i \in \mathrm{Ob}(I)$, there exists a refinement $(U_{i,n})_{n \in N_i}$ of the final object of $X$ such that for every $n \in N_i$, $\alpha_{i!}(U_{i,n})$ is an object of $\mathscr{R}$ (III.7.1.4).

(iii) Suppose that fibered products are representable in $I$. The total topology on $X \times I$ then coincides with the covanishing topology with respect to the constant fibered site $X \times I \to I$ with fiber $X$, when we endow $I$ with the chaotic topology (VI.5.4). In other words, the total topology on $X \times I$ is generated by the *pretopology* formed by the vertical coverings (VI.5.3 and VI.5.7).

**Remarks III.7.3.** (i) The direct $\mathbb{U}$-limits (resp. the finite inverse limits) in $X^{I^\circ}$ can be computed term-wise; in other words, for every functor $\varphi \colon N \to X^{I^\circ}$ such that the category $N$ is $\mathbb{U}$-small (resp. finite), if $F$ is an object of $X^{I^\circ}$ that represents the direct (resp. inverse) limit of $\varphi$, then for every $i \in \mathrm{Ob}(I)$, the direct (resp. inverse) limit of $\alpha_i^* \circ \varphi$ is representable by $\alpha_i^*(F)$.

(ii) Let $\iota$ be a final object of $I$. By (III.7.1.5), for every $j \in \mathrm{Ob}(I)$, $\alpha_j^* \alpha_{\iota !}$ is the identity functor of $X$. Hence by virtue of (i), $\alpha_{\iota !}$ is left exact, and the pair $(\alpha_{\iota !}, \alpha_\iota^*)$ forms a morphism of topos

(III.7.3.1) $$\beta_\iota \colon X^{I^\circ} \to X.$$

The morphism $\beta_\iota \alpha_\iota \colon X \to X$ is isomorphic to the identity morphism (cf. [**2**] VI 7.4.12).

(iii) Let $i$ be an object of $I$ that is not final. Then there exists $j \in \mathrm{Ob}(I)$ such that $\mathrm{Hom}_I(j, i)$ is not a singleton. In particular, $\alpha_j^* \alpha_{i!}$ does not transform the final object into the final object (III.7.1.5). Consequently, $\alpha_{i!}$ is not left exact, and the pair $(\alpha_{i!}, \alpha_i^*)$ does not form a morphism of topos, unlike what was stated in ([**22**] line 20 page 59). Nevertheless, the isomorphism (4.4) and the spectral sequence (4.5) of loc. cit. are correct by virtue of (VI.7.7 and VI.7.8).

**III.7.4.** The functor

(III.7.4.1) $$\lambda^* \colon X \to X^{I^\circ}$$

that associates with each object $F$ of $X$ the constant functor $I^\circ \to X$ with value $F$ is left exact by virtue of III.7.3(i). It admits as a right adjoint the functor

(III.7.4.2) $$\lambda_* \colon X^{I^\circ} \to X$$

that associates with a functor $I^\circ \to X$ its inverse limit ([**2**] II 4.1(3)). The pair $(\lambda^*, \lambda_*)$ therefore defines a morphism of topos

(III.7.4.3) $$\lambda \colon X^{I^\circ} \to X.$$

**III.7.5.** Every morphism of $\mathbb{U}$-topos $f \colon X \to Y$ induces a Cartesian morphism of constant fibered topos over $I$

(III.7.5.1) $$f \times \mathrm{id}_I \colon X \times I \to Y \times I.$$

By ([**2**] VI 7.4.10), this induces a morphism of topos

(III.7.5.2) $$f^{I^\circ} \colon X^{I^\circ} \to Y^{I^\circ}$$

such that for every $F \in X^{I^\circ}$, we have ([**2**] VI (7.4.9.2))

(III.7.5.3) $$(f^{I^\circ})_*(F) = f_* \circ F.$$

We immediately see that for every $G \in Y^{I^\circ}$, we have

$$(f^{I^\circ})^*(G) = f^* \circ G. \tag{III.7.5.4}$$

Denote by $\mathrm{R}^q(f^{I^\circ})_*$ ($q \in \mathbb{N}$) the right derived functors of the functor $(f^{I^\circ})_*$ for abelian groups. By VI.7.7, for every abelian group $F$ of $X^{I^\circ}$ and every $i \in \mathrm{Ob}(I)$, we have a functorial canonical isomorphism

$$\mathrm{R}^q(f^{I^\circ})_*(F)(i) \xrightarrow{\sim} \mathrm{R}^q f_*(F(i)). \tag{III.7.5.5}$$

**Proposition III.7.6.** *Assume that fibered products are representable in $I$; let moreover $U$ be an object of $X$, $j_{U!}\colon X_{/U} \to X$ the canonical functor, $j_U\colon X_{/U} \to X$ the localization morphism of $X$ at $U$, and $j_{\lambda^*(U)}\colon (X^{I^\circ})_{/\lambda^*(U)} \to X^{I^\circ}$ the localization morphism of $X^{I^\circ}$ at $\lambda^*(U)$. Then:*

(i) *The total topology on $X_{/U} \times I$ is induced by the total topology on $X \times I$ by the functor $j_{U!} \times \mathrm{id}_I$.*

(ii) *There exists a canonical equivalence of topos*

$$h\colon (X_{/U})^{I^\circ} \xrightarrow{\sim} (X^{I^\circ})_{/\lambda^*(U)}, \tag{III.7.6.1}$$

*such that $(j_U)^{I^\circ} = j_{\lambda^*(U)} \circ h$ (III.7.5.2). In particular, for every $F \in \mathrm{Ob}(X^{I^\circ})$, $F \times \lambda^*(U)$ identifies with the functor $I^\circ \to X_{/U}, i \mapsto F(i) \times U$.*

First note that the canonical topology on $X_{/U}$ is induced by the canonical topology on $X$ by the functor $j_{U!}$, and that in this case the "extension by the empty set" functor identifies with the functor $j_{U!}$ (**[2]** IV 1.2, III 3.5 and 5.4), whence the notation.

(i) The total topology on $X \times I$ is generated by the pretopology formed by the vertical coverings by III.7.2(iii), and likewise for $X_{/U} \times I$. The statement therefore follows from (**[2]** III 3.3 and II 1.4).

(ii) For all $V \in \mathrm{Ob}(X)$ and $i \in \mathrm{Ob}(I)$, we have a canonical isomorphism (III.7.1.3)

$$\lambda^*(U)(V \times i) = \mathrm{Hom}_{X^{I^\circ}}(\alpha_{i!}(V), \lambda^*(U)) \xrightarrow{\sim} \mathrm{Hom}_X(V, \alpha_i^*(\lambda^*(U))) = U(V). \tag{III.7.6.2}$$

Consequently, the functor $j_{U!} \times \mathrm{id}_I\colon X_{/U} \times I \to X \times I$ factors canonically into

$$X_{/U} \times I \xrightarrow[\sim]{e} (X \times I)_{/\lambda^*(U)} \xrightarrow{j'_{\lambda^*(U)}} X \times I, \tag{III.7.6.3}$$

where $e$ is an equivalence of categories and $j'_{\lambda^*(U)}$ is the canonical functor. By (i) and (**[2]** III 5.4), $e$ induces an equivalence of topos

$$h\colon (X_{/U})^{I^\circ} \xrightarrow{\sim} (X^{I^\circ})_{/\lambda^*(U)}. \tag{III.7.6.4}$$

For every $F \in \mathrm{Ob}(X^{I^\circ})$, we have $j^*_{\lambda^*(U)}(F) = F \circ j'_{\lambda^*(U)}$ by (**[2]** III 2.3 and 5.2(2)). Hence $h^*(j^*_{\lambda^*(U)}(F)) = F \circ (j_{U!} \times \mathrm{id}_I)$. Since $F \circ (j_{U!} \times \mathrm{id}_I) = ((j_U)^{I^\circ})^*(F)$ (III.7.5.4), it follows that $(j_U)^{I^\circ} = j_{\lambda^*(U)} \circ h$.

**III.7.7.** Let $J$ be a $\mathbb{U}$-small ordered set. We denote also by $J$ the category defined by $J$, that is, the category made up of the elements of $J$, with at most one arrow with given source and target, and for all $i, j \in J$, the set $\mathrm{Hom}_J(i,j)$ is nonempty if and only if $i \leq j$. It is often convenient to use for the objects $F\colon J^\circ \to X$ of $X^{J^\circ}$ the index notation $(F_j)_{j \in J}$ or even $(F_j)$, where $F_j = F(j)$ for every $j \in J$. We say that an object $(F_j)$ of $X^{J^\circ}$ is *strict* if for all elements $i \geq j$ of $J$, the transition morphism $F_i \to F_j$ is an epimorphism.

In this work, we will limit ourselves to the case where $J$ is either the ordered set of natural numbers $\mathbb{N}$, or one of the ordered subsets $[n] = \{0, 1, \ldots, n\}$. Note that in each of these cases, fibered products are representable in $J$.

**Lemma III.7.8.** *Let $n$ be an integer $\geq 0$ and $\iota_n\colon X\times[n]\to X\times\mathbb{N}$ the canonical injection functor. Then:*

(i) *The total topology on $X\times[n]$ is induced by the total topology on $X\times\mathbb{N}$ by the functor $\iota_n$.*

(ii) *The functor $\iota_n$ is continuous and cocontinuous for the total topologies. Denote by*

(III.7.8.1) $$\varphi_n\colon X^{[n]^\circ}\to X^{\mathbb{N}^\circ}$$

*the associated morphism of topos. For every $F=(F_i)_{i\in\mathbb{N}}\in\mathrm{Ob}(X^{\mathbb{N}^\circ})$, we have*

(III.7.8.2) $$\varphi_n^*(F)=(F_i)_{i\in[n]}.$$

(iii) *The functor $\varphi_n^*\colon : X^{\mathbb{N}^\circ}\to X^{[n]^\circ}$ admits as left adjoint the functor*

(III.7.8.3) $$\varphi_{n!}\colon X^{[n]^\circ}\to X^{\mathbb{N}^\circ},$$

*defined for every $F=(F_i)_{i\in[n]}\in\mathrm{Ob}(X^{[n]^\circ})$ by*

(III.7.8.4) $$\varphi_{n!}(F)=(F_i)_{i\in\mathbb{N}},$$

*where for every $i\geq n+1$, $F_i=\emptyset$ is the initial object of $X$.*

(i) The total topology on $X\times\mathbb{N}$ is generated by the pretopology formed by the vertical coverings, by III.7.2(iii), and likewise for $X\times[n]$. The statement therefore follows from ([**2**] III 3.3 and II 1.4).

(ii) Denote by $\widehat{X}$ (resp. $(X\times\mathbb{N})^\wedge$, resp. $(X\times[n])^\wedge$) the category of presheaves of $\mathbb{U}$-sets on $X$ (resp. $X\times\mathbb{N}$, resp. $X\times[n]$). We then have an equivalence of categories (III.7.1.1)

(III.7.8.5) $$(X\times\mathbb{N})^\wedge\xrightarrow{\sim}\widehat{X}^{\mathbb{N}^\circ}=\mathbf{Hom}(\mathbb{N}^\circ,\widehat{X}),\quad F\mapsto(F\circ\alpha_{i!})_{i\in\mathbb{N}},$$

and likewise for $(X\times[n])^\wedge$. By composition, the functor $\iota_n$ induces the functor

(III.7.8.6) $$\widehat{\iota}_n^*\colon\widehat{X}^{\mathbb{N}^\circ}\to\widehat{X}^{[n]^\circ},\quad(F_i)_{i\in\mathbb{N}}\mapsto(F_i)_{i\in[n]}.$$

This admits as right adjoint the functor

(III.7.8.7) $$\widehat{\iota}_{n*}\colon\widehat{X}^{[n]^\circ}\to\widehat{X}^{\mathbb{N}^\circ},\quad(F_i)_{i\in[n]}\mapsto(F_i)_{i\in\mathbb{N}},$$

where for every $i\geq n+1$, $F_i=F_n$ and the morphism $F_i\to F_{i-1}$ is the identity of $F_n$. The functor $\widehat{\iota}_n^*$ admits as left adjoint the functor

(III.7.8.8) $$\widehat{\iota}_{n!}\colon\widehat{X}^{[n]^\circ}\to\widehat{X}^{\mathbb{N}^\circ},\quad(F_i)_{i\in[n]}\mapsto(F_i)_{i\in\mathbb{N}},$$

where for every $i\geq n+1$, $F_i=\emptyset$ is the initial object of $\widehat{X}$. It is clear that $\widehat{\iota}_n^*$ transforms sheaves on $X\times\mathbb{N}$ into sheaves on $X\times[n]$ and that $\widehat{\iota}_{n*}$ transforms sheaves on $X\times[n]$ into sheaves on $X\times\mathbb{N}$. Consequently, $i_n$ is continuous and cocontinuous. The continuity also follows from (i). The formula (III.7.8.2) is a consequence of (III.7.8.6) and ([**2**] III 2.3).

(iii) This follows from (III.7.8.8) and ([**2**] III 1.3).

**Lemma III.7.9.** *Let $U$ be an object of $X$, $j_U\colon X_{/U}\to X$ the localization morphism of $X$ at $U$, $n\in\mathbb{N}$, and $j_{(U,n)}\colon(X^{\mathbb{N}^\circ})_{/\alpha_{n!}(U)}\to X^{\mathbb{N}^\circ}$ the localization morphism of $X^{\mathbb{N}^\circ}$ at $\alpha_{n!}(U)$ (III.7.1.4). We then have a canonical equivalence of topos*

(III.7.9.1) $$h\colon(X_{/U})^{[n]^\circ}\xrightarrow{\sim}(X^{\mathbb{N}^\circ})_{/\alpha_{n!}(U)}$$

*such that $j_{(U,n)}\circ h$ is the composition*

(III.7.9.2) $$(X_{/U})^{[n]^\circ}\xrightarrow{\varphi_n}(X_{/U})^{\mathbb{N}^\circ}\xrightarrow{(j_U)^{\mathbb{N}^\circ}}X^{\mathbb{N}^\circ},$$

*where the first arrow is the morphism* (III.7.8.1) *and the second arrow is the morphism* (III.7.5.2).

Denote by $u\colon \alpha_{n!}(U) \to \lambda^*(U)$ the adjoint of the canonical isomorphism $U \xrightarrow{\sim} \alpha_n^*(\lambda^*(U))$ and by $\widetilde{\alpha}_{n!}(U)$ the image of $(U,n)$ by the canonical functor $X_{/U} \times \mathbb{N} \to (X_{/U})^{\mathbb{N}^\circ}$ (III.7.1.4). By III.7.6(ii), we have a canonical equivalence of topos

(III.7.9.3) $$g\colon (X_{/U})^{\mathbb{N}^\circ} \xrightarrow{\sim} (X^{\mathbb{N}^\circ})_{/\lambda^*(U)}.$$

In view of (III.7.6.3) and ([**2**] III 5.4), we have $g(\widetilde{\alpha}_{n!}(U)) = u$. We may therefore restrict ourselves to the case where $U$ is the final object $e_X$ of $X$ by virtue of III.7.6(ii) and ([**2**] IV 5.6).

The canonical injection functor $\iota_n\colon X \times [n] \to X \times \mathbb{N}$ factors canonically into

(III.7.9.4) $$X \times [n] \xrightarrow[\sim]{\nu} (X \times \mathbb{N})_{/(e_X,n)} \xrightarrow{j'_n} X \times \mathbb{N}\,,$$

where $\nu$ is an equivalence of categories and $j'_n$ is the canonical functor. By III.7.8(i) and ([**2**] III 5.4), $\nu$ induces an equivalence of topos

(III.7.9.5) $$h\colon X^{[n]^\circ} \xrightarrow{\sim} (X^{\mathbb{N}^\circ})_{/\alpha_{n!}(e_X)}.$$

For every $F = (F_i)_{i\in\mathbb{N}} \in \mathrm{Ob}(X^{\mathbb{N}^\circ})$, we have $j^*_{(e_X,n)}(F) = F \circ j'_n$ ([**2**] III 2.3 and 5.2(2)). Hence $h^*(j^*_{(e_X,n)}(F)) = F \circ \iota_n$. Since $F \circ \iota_n = (F_i)_{i\in[n]} = \varphi_n^*(F)$ (III.7.8.2), it follows that $j_{(e_X,n)} \circ h = \varphi_n$.

**Proposition III.7.10** (cf. VI.7.9). *Let $n$ be an integer $\geq 0$, $A = (A_i)_{i\in[n]}$ a ring of $X^{[n]^\circ}$, and $M = (M_i)_{i\in[n]}$ an $A$-module of $X^{[n]^\circ}$. For every integer $q \geq 0$, we then have a canonical isomorphism*

(III.7.10.1) $$\mathrm{H}^q(X^{[n]^\circ}, M) \xrightarrow{\sim} \mathrm{H}^q(X, M_n).$$

**Proposition III.7.11** (cf. VI.7.10). *Let $A = (A_n)_{n\in\mathbb{N}}$ be a ring of $X^{\mathbb{N}^\circ}$, $U$ an object of $X$, and $M = (M_n)_{n\in\mathbb{N}}$ an $A$-module of $X^{\mathbb{N}^\circ}$. For every integer $q \geq 0$, we then have a canonical and functorial exact sequence*

(III.7.11.1) $$0 \to \mathrm{R}^1 \varprojlim_{n\in\mathbb{N}^\circ} \mathrm{H}^{q-1}(U, M_n) \to \mathrm{H}^q(\lambda^*(U), M) \to \varprojlim_{n\in\mathbb{N}^\circ} \mathrm{H}^q(U, M_n) \to 0,$$

*where we have set $\mathrm{H}^{-1}(U, M_n) = 0$ for every $n \in \mathbb{N}$.*

Indeed, we may restrict ourselves to the case where $U$ is the final object of $X$ (III.7.6), in which case the proposition is a special case of VI.7.10.

**III.7.12.** Let $A = (A_n)_{n\in\mathbb{N}}$ be a ring of $X^{\mathbb{N}^\circ}$, and $M = (M_n)_{n\in\mathbb{N}}$ and $N = (N_n)_{n\in\mathbb{N}}$ two $A$-modules of $X^{\mathbb{N}^\circ}$. We then have a bifunctorial canonical isomorphism

(III.7.12.1) $$M \otimes_A N \xrightarrow{\sim} (M_n \otimes_{A_n} N_n)_{n\in\mathbb{N}}.$$

Indeed, for every $n \in \mathbb{N}$, we have a canonical isomorphism ([**2**] IV 13.4)

(III.7.12.2) $$\alpha_n^*(M \otimes_A N) \xrightarrow{\sim} \alpha_n^*(M) \otimes_{\alpha_n^*(A)} \alpha_n^*(N).$$

Likewise, for every integer $q \geq 0$, we have functorial canonical isomorphisms (III.2.7)

(III.7.12.3) $$\mathrm{S}^q_A(M) \xrightarrow{\sim} (\mathrm{S}^q_{A_n}(M_n))_{n\in\mathbb{N}},$$

(III.7.12.4) $$\wedge^q_A(M) \xrightarrow{\sim} (\wedge^q_{A_n}(M_n))_{n\in\mathbb{N}}.$$

**Lemma III.7.13.** *Let $A = (A_n)_{n\in\mathbb{N}}$ be a ring of $X^{\mathbb{N}^\circ}$ and $M = (M_n)_{n\in\mathbb{N}}$ an $A$-module of $X^{\mathbb{N}^\circ}$. Then $M$ is $A$-flat if and only if for every integer $n \geq 0$, $M_n$ is $A_n$-flat.*

Indeed, the condition is necessary by virtue of (III.7.1.3) and ([**2**] V 1.7.1), and it is sufficient by III.7.3(i) and (III.7.12.1).

**Proposition III.7.14.** *Let $A = (A_n)_{n\in\mathbb{N}}$ be a ring of $X^{\mathbb{N}^\circ}$ and $M = (M_n)_{n\in\mathbb{N}}$ a strict $A$-module of $X^{\mathbb{N}^\circ}$ (III.7.7). Then $M$ is of finite type over $A$ if and only if for every integer $n \geq 0$, $M_n$ is of finite type over $A_n$.*

We clearly only have to show that the condition is sufficient (III.7.1.3). Suppose that for every integer $n \geq 0$, the $A_n$-module $M_n$ is of finite type. Let $n$ be an integer $\geq 0$, and $U \in \mathrm{Ob}(X)$ such that $M_n|U$ is generated over $A_n|U$ by a finite number of sections $s_1, \dots, s_\ell \in \Gamma(U, M_n)$. Since $M$ is strict, for every integer $0 \leq i \leq n$, $M_i|U$ is generated over $A_i|U$ by the canonical images of the sections $s_1, \dots, s_\ell$ in $\Gamma(U, M_i)$. By virtue of III.7.9, we have a canonical equivalence of categories

$$h\colon (X_{/U})^{[n]^\circ} \xrightarrow{\sim} (X^{\mathbb{N}^\circ})_{/\alpha_{n!}(U)} \tag{III.7.14.1}$$

such that the inverse image $h^*(A|\alpha_{n!}(U))$ is isomorphic to the ring $(A_i|U)_{i\in[n]}$ of $(X_{/U})^{[n]^\circ}$ and the inverse image $h^*(M|\alpha_{n!}(U))$ is isomorphic to the module $(M_i|U)_{i\in[n]}$ of $(X_{/U})^{[n]^\circ}$. By III.7.10, we have a canonical isomorphism

$$\Gamma((X_{/U})^{[n]^\circ}, (M_i|U)_{i\in[n]}) \xrightarrow{\sim} \Gamma(U, M_n). \tag{III.7.14.2}$$

In view of III.7.3(i), the module $(M_i|U)_{i\in[n]}$ is then generated over $(A_i|U)_{i\in[n]}$ by the sections $s_1, \dots, s_\ell \in \Gamma(U, M_n)$.

For every integer $m \geq 0$, let $(U_{m,j})_{j\in J_m}$ be a refinement of the final object of $X$ such that for every $j \in J_m$, $M_m|U_{m,j}$ is generated over $A_m|U_{m,j}$ by a finite number of sections of $\Gamma(U_{m,j}, M_m)$. By the above, for every $m \in \mathbb{N}$ and every $j \in J_m$, the module $M|\alpha_{m!}(U_{m,j})$ is generated over $A|\alpha_{m!}(U_{m,j})$ by a finite number of sections of $\Gamma(U_{m,j}, M_m)$. On the other hand, the family $(\alpha_{m!}(U_{m,j}))_{m\in\mathbb{N}, j\in J_m}$ is a refinement of the final object of $X^{\mathbb{N}^\circ}$ by virtue of III.7.2(ii). Consequently, $M$ is of finite type over $A$.

**Definition III.7.15.** Let $n$ be an integer $\geq 0$ and $A = (A_i)_{i\in[n]}$ a ring of $X^{[n]^\circ}$. We say that an $A$-module $(M_i)_{i\in[n]}$ of $X^{[n]^\circ}$ is *adic* if for all integers $i$ and $j$ such that $0 \leq i \leq j \leq n$, the morphism $M_j \otimes_{A_j} A_i \to M_i$ deduced from the transition morphism $M_j \to M_i$ is an isomorphism.

**Definition III.7.16.** Let $A = (A_i)_{i\in\mathbb{N}}$ be a ring of $X^{\mathbb{N}^\circ}$. We say that an $A$-module $(M_i)_{i\in\mathbb{N}}$ of $X^{\mathbb{N}^\circ}$ is *adic* if for all integers $i$ and $j$ such that $0 \leq i \leq j$, the morphism $M_j \otimes_{A_j} A_i \to M_i$ deduced from the transition morphism $M_j \to M_i$ is an isomorphism.

The notions of adic modules defined here are special cases of the notion of co-Cartesian modules introduced in VI.7.11.

**Lemma III.7.17.** *Suppose that $X$ has enough points. Let moreover $R$ be a ring of $X$ and $J$ an ideal of $R$. For any integer $n \geq 0$, we set $R_n = R/J^{n+1}$. We denote by $\breve{R}$ the ring $(R_n)_{n\in\mathbb{N}}$ of $X^{\mathbb{N}^\circ}$. Then an adic $\breve{R}$-module $(M_n)_{n\in\mathbb{N}}$ of $X^{\mathbb{N}^\circ}$ is of finite type if and only if the $R_0$-module $M_0$ is of finite type.*

We clearly only have to show that the condition is sufficient. Suppose that the $R_0$-module $M_0$ is of finite type. Let $\varphi\colon X \to \mathbf{Ens}$ be a fiber functor and $n$ an integer $\geq 1$. Since we have $\varphi(R_n) = \varphi(R)/(\varphi(J))^{n+1}$ and $\varphi(M_0) = \varphi(M_n)/(\varphi(J)\varphi(M_n))$, $\varphi(M_n)$ is of finite type over $\varphi(R_n)$ by ([**1**] 1.8.5). Consequently, $M_n$ is of finite type over $R_n$, giving the lemma by virtue of III.7.14.

**III.7.18.** Let $f\colon Y \to X$ be a morphism of $\mathbb{U}$-topos, $A = (A_n)_{n\in\mathbb{N}}$ a ring of $X^{\mathbb{N}^\circ}$, $B = (B_n)_{n\in\mathbb{N}}$ a ring of $Y^{\mathbb{N}^\circ}$, and $u\colon A \to (f^{\mathbb{N}^\circ})_*(B)$ a ring homomorphism. We view $f^{\mathbb{N}^\circ}\colon Y^{\mathbb{N}^\circ} \to X^{\mathbb{N}^\circ}$ as a morphism of ringed topos (by $A$ and $B$, respectively). For modules, we use the notation $(f^{\mathbb{N}^\circ})^{-1}$ to denote the inverse image in the sense of abelian sheaves, and we keep the notation $(f^{\mathbb{N}^\circ})^*$ for the inverse image in the sense of modules. Giving $u$ is equivalent to giving, for every $n \in \mathbb{N}$, a ring homomorphism $u_n\colon A_n \to f_*(B_n)$ such that these homomorphisms are compatible with the transition morphisms of $A$ and $B$ (III.7.5.3). The homomorphism $(f^{\mathbb{N}^\circ})^{-1}(A) \to B$ adjoint to $u$ corresponds to the system of homomorphisms $f^*(A_n) \to B_n$ adjoint to $u_n$ ($n \in \mathbb{N}$) (III.7.5.4). For any $n \in \mathbb{N}$, we denote by

(III.7.18.1) $$f_n\colon (Y, B_n) \to (X, A_n)$$

the morphism of ringed topos defined by $f$ and $u_n$. By (III.7.5.4) and (III.7.12.1), for every $A$-module $M = (M_n)_{n\in\mathbb{N}}$ of $X^{\mathbb{N}^\circ}$, we have a functorial canonical isomorphism

(III.7.18.2) $$(f^{\mathbb{N}^\circ})^*(M) \xrightarrow{\sim} (f_n^*(M_n)).$$

Consequently, if $M$ is adic, the same holds for $(f^{\mathbb{N}^\circ})^*(M)$.

**Lemma III.7.19.** *Let $n$ be an integer $\geq 0$, $A = (A_i)_{i\in[n]}$ a ring of $X^{[n]^\circ}$, and $(M_i)_{i\in[n]}$ an $A$-module of $X^{[n]^\circ}$. The $A$-module $M$ is locally projective of finite type* (III.2.8) *if and only if $M$ is adic and the $A_n$-module $M_n$ is locally projective of finite type.*

First, suppose that $M$ is locally projective of finite type over $A$. The $A_n$-module $M_n$ is then locally projective of finite type (III.7.1.3). Let us show that $M$ is adic. By III.7.2(ii), there exists a refinement $(U_j)_{j\in J}$ of the final object of $X$ such that for every $j \in J$, $M|\alpha_{n!}(U_j)$ is a direct summand of a free $A|\alpha_{n!}(U_j)$-module of finite type. We have $\lambda^*(U_j) = \alpha_{n!}(U_j)$ (III.7.1.5). In view of III.7.6(ii), we may then restrict ourselves to the case where $M$ is a direct summand of a free $A$-module of finite type. Hence there exist an integer $d \geq 1$, an $A$-module $N = (N_i)_{i\in[n]}$, and an $A$-linear isomorphism $A^d \xrightarrow{\sim} M \oplus N$. It follows that for all integers $i$ and $j$ such that $0 \leq i \leq j \leq n$, the canonical morphisms $M_j \otimes_{A_j} A_i \to M_i$ and $N_j \otimes_{A_j} A_i \to N_i$ are isomorphisms; in other words, $M$ and $N$ are adic.

Next, suppose that $M$ is adic and that the $A_n$-module $M_n$ is locally projective of finite type. Let us show that the $A$-module $M$ is locally projective of finite type. In view of III.7.6(ii), we may restrict ourselves to the case where $M_n$ is a direct summand of a free $A_n$-module of finite type. Hence there exist an integer $d \geq 1$, an $A_n$-module $N_n$, and an $A_n$-linear isomorphism $A_n^d \xrightarrow{\sim} M_n \oplus N_n$. This induces an $A$-linear isomorphism

(III.7.19.1) $$A^d \xrightarrow{\sim} M \oplus (N_n \otimes_{A_n} A_i)_{i\in[n]}.$$

**Proposition III.7.20.** *Let $A = (A_n)_{n\in\mathbb{N}}$ be a ring of $X^{\mathbb{N}^\circ}$ and $M = (M_n)_{n\in\mathbb{N}}$ an $A$-module of $X^{\mathbb{N}^\circ}$. The $A$-module $M$ is locally projective of finite type* (III.2.8) *if and only if $M$ is adic and for every integer $n \geq 0$, the $A_n$-module $M_n$ is locally projective of finite type.*

First, suppose that $M$ is locally projective of finite type over $A$. For every integer $n \geq 0$, the $(A_i)_{i\in[n]}$-module $(M_i)_{i\in[n]}$ is locally projective of finite type by III.7.8(ii). It follows that $M$ is adic and that for every integer $n \geq 0$, the $A_n$-module $M_n$ is locally projective of finite type, by virtue of III.7.19. Next, suppose that $M$ is adic and that for every integer $n \geq 0$, the $A_n$-module $M_n$ is locally projective of finite type. Let $n$ be an integer $\geq 0$ and $U \in \mathrm{Ob}(X)$. By III.7.9, we have a canonical equivalence of topos

(III.7.20.1) $$h\colon (X_{/U})^{[n]^\circ} \xrightarrow{\sim} (X^{\mathbb{N}^\circ})_{/\alpha_{n!}(U)}$$

such that the inverse image $h^*(A|\alpha_{n!}(U))$ is isomorphic to the ring $(A_i|U)_{i\in[n]}$ of $(X_{/U})^{[n]^\circ}$ and the inverse image $h^*(M|\alpha_{n!}(U))$ is isomorphic to the module $(M_i|U)_{i\in[n]}$ of $(X_{/U})^{[n]^\circ}$. Consequently, the $A|\alpha_{n!}(U)$-module $M|\alpha_{n!}(U)$ is locally projective of finite type by virtue of III.7.19. In view of III.7.2(ii), we deduce from this that the $A$-module $M$ is locally projective of finite type.

**III.7.21.** Let $Y$ be a connected scheme, $\overline{y}$ a geometric point of $Y$, $Y_{\text{fét}}$ the finite étale topos of $Y$ (III.2.9), $\mathbf{B}_{\pi_1(Y,\overline{y})}$ the classifying topos of the profinite group $\pi_1(Y,\overline{y})$, and

(III.7.21.1) $$\nu_{\overline{y}}\colon Y_{\text{fét}} \to \mathbf{B}_{\pi_1(Y,\overline{y})}$$

the fiber functor of $Y_{\text{fét}}$ at $\overline{y}$ (III.2.10). Let $R$ be a ring of $Y_{\text{fét}}$. Set $R_{\overline{y}} = \nu_{\overline{y}}(R)$, which is a ring endowed with the discrete topology and with a continuous action of $\pi_1(Y,\overline{y})$ by ring homomorphisms. We denote by $\widehat{R}_{\overline{y}}$ the $p$-adic Hausdorff completion of $R_{\overline{y}}$ that we endow with the $p$-adic topology and with the action of $\pi_1(Y,\overline{y})$ induced by that on $R_{\overline{y}}$. By ([**11**] III § 2.11 Proposition 14 and Corollary 1; cf. also [**1**] 1.8.7), for every integer $n\geq 1$, we have

(III.7.21.2) $$\widehat{R}_{\overline{y}}/p^n\widehat{R}_{\overline{y}} \simeq R_{\overline{y}}/p^nR_{\overline{y}}.$$

The action of $\pi_1(Y,\overline{y})$ on $\widehat{R}_{\overline{y}}$ is therefore continuous.

For any topological ring $A$ endowed with a continuous action of $\pi_1(Y,\overline{y})$ by ring homomorphisms, we denote by $\mathbf{Rep}^{\text{cont}}_A(\pi_1(Y,\overline{y}))$ the category of continuous $A$-representations of $\pi_1(Y,\overline{y})$ (II.3.1). Consider the following categories:

(a) $\mathbf{Rep}^{\text{disc}}_{R_{\overline{y}}}(\pi_1(Y,\overline{y}))$: the full subcategory of $\mathbf{Rep}^{\text{cont}}_{R_{\overline{y}}}(\pi_1(Y,\overline{y}))$ made up of the continuous $R_{\overline{y}}$-representations of $\pi_1(Y,\overline{y})$ for which the topology is discrete;

(b) $p$-$\mathbf{ad}(\mathbf{Rep}^{\text{disc}}_{R_{\overline{y}}}(\pi_1(Y,\overline{y})))$: the category of $p$-adic inverse systems of the category $\mathbf{Rep}^{\text{disc}}_{R_{\overline{y}}}(\pi_1(Y,\overline{y}))$—an inverse system $(M_n)_{n\in\mathbb{N}}$ of $\mathbf{Rep}^{\text{disc}}_{R_{\overline{y}}}(\pi_1(Y,\overline{y}))$ is said to be *$p$-adic* if for every integer $n\geq 0$, $p^{n+1}M_n=0$ and for all integers $m\geq n\geq 0$, the morphism $M_m/p^{n+1}M_m\to M_n$ induced by the transition morphism $M_m\to M_n$ is an isomorphism ([**38**] V 3.1.1);

(c) $p$-$\mathbf{ad}_{\text{ft}}(\mathbf{Rep}^{\text{disc}}_{R_{\overline{y}}}(\pi_1(Y,\overline{y})))$: the full subcategory of $p$-$\mathbf{ad}(\mathbf{Rep}^{\text{disc}}_{R_{\overline{y}}}(\pi_1(Y,\overline{y})))$ made up of the $p$-adic inverse systems of finite type—we say that a $p$-adic inverse system $(M_n)_{n\in\mathbb{N}}$ of $\mathbf{Rep}^{\text{disc}}_{R_{\overline{y}}}(\pi_1(Y,\overline{y}))$ is *of finite type* if for every integer $n\geq 0$, $M_n$ is an $R_{\overline{y}}$-module of finite type, or, equivalently, if $M_0$ is an $R_{\overline{y}}$-module of finite type ([**1**] 1.8.5);

(d) $\mathbf{Rep}^{p\text{-aft}}_{\widehat{R}_{\overline{y}}}(\pi_1(Y,\overline{y}))$: the full subcategory of $\mathbf{Rep}^{\text{cont}}_{\widehat{R}_{\overline{y}}}(\pi_1(Y,\overline{y}))$ made up of the $p$-adic $\widehat{R}_{\overline{y}}$-representations of finite type of $\pi_1(Y,\overline{y})$—an $\widehat{R}_{\overline{y}}$-representation of $\pi_1(Y,\overline{y})$ is said to be *$p$-adic of finite type* if it is continuous for the $p$-adic topology and if the underlying $\widehat{R}_{\overline{y}}$-module is separated of finite type. Note that every $\widehat{R}_{\overline{y}}$-module of finite type is complete for the $p$-adic topology ([**11**] Chapter III § 2.11 Corollary 1 to Proposition 16).

The inverse limit defines a functor

(III.7.21.3) $$p\text{-}\mathbf{ad}(\mathbf{Rep}^{\text{disc}}_{R_{\overline{y}}}(\pi_1(Y,\overline{y}))) \to \mathbf{Rep}^{\text{cont}}_{\widehat{R}_{\overline{y}}}(\pi_1(Y,\overline{y}))$$

that induces an equivalence of categories

(III.7.21.4) $$p\text{-}\mathbf{ad}_{\text{ft}}(\mathbf{Rep}^{\text{disc}}_{R_{\overline{y}}}(\pi_1(Y,\overline{y}))) \xrightarrow{\sim} \mathbf{Rep}^{p\text{-aft}}_{\widehat{R}_{\overline{y}}}(\pi_1(Y,\overline{y})).$$

Indeed, for every object $(M_n)$ of $p\text{-}\mathbf{ad}_{\mathrm{ft}}(\mathbf{Rep}^{\mathrm{disc}}_{R_{\overline{y}}}(\pi_1(Y,\overline{y})))$, the $\widehat{R}_{\overline{y}}$-module

$$\widehat{M}=\lim_{\substack{\longleftarrow \\ n\in\mathbb{N}}} M_n$$

is of finite type and for every $n\in\mathbb{N}$, we have $\widehat{M}/p^{n+1}\widehat{M}\simeq M_n$ by ([**11**] III § 2.11 Proposition 14 and Corollary 1).

By restricting the functor $\nu_{\overline{y}}$ (III.7.21.1) to $\breve{R}$-modules, we obtain an equivalence of categories that we denote also by

(III.7.21.5) $$\nu_{\overline{y}}\colon \mathbf{Mod}(R)\xrightarrow{\sim}\mathbf{Rep}^{\mathrm{disc}}_{R_{\overline{y}}}(\pi_1(Y,\overline{y})).$$

We denote by $\breve{R}$ the ring $(R/p^{n+1}R)_{n\in\mathbb{N}}$ of $Y^{\mathbb{N}^\circ}_{\text{fét}}$ and by $\mathbf{Mod}^{\mathrm{ad}}(\breve{R})$ (resp. $\mathbf{Mod}^{\mathrm{aft}}(\breve{R})$) the category of adic $\breve{R}$-modules (resp. adic $\breve{R}$-module of finite type) of $Y^{\mathbb{N}^\circ}_{\text{fét}}$. An $\breve{R}$-module $(M_n)_{n\in\mathbb{N}}$ of $Y^{\mathbb{N}^\circ}_{\text{fét}}$ is adic (resp. adic of finite type) if and only if the inverse system $(\nu_{\overline{y}}(M_n))$ of $\mathbf{Rep}^{\mathrm{disc}}_{R_{\overline{y}}}(\pi_1(Y,\overline{y}))$ is $p$-adic (resp. $p$-adic of finite type by virtue of III.7.14). The functors (III.7.21.5) and (III.7.21.3) therefore induce a functor

(III.7.21.6) $$\mathbf{Mod}^{\mathrm{ad}}(\breve{R})\to\mathbf{Rep}^{\mathrm{cont}}_{\widehat{R}_{\overline{y}}}(\pi_1(Y,\overline{y}))$$

and an equivalence of categories

(III.7.21.7) $$\mathbf{Mod}^{\mathrm{aft}}(\breve{R})\xrightarrow{\sim}\mathbf{Rep}^{p\text{-aft}}_{\widehat{R}_{\overline{y}}}(\pi_1(Y,\overline{y})).$$

## III.8. Faltings ringed topos

**III.8.1.** For this section, we fix a commutative diagram of morphisms of schemes

(III.8.1.1) $$\begin{array}{ccc} Y & \xrightarrow{\;j\;} & \overline{X} \\ & {\scriptstyle h}\searrow & \downarrow{\scriptstyle \hbar} \\ & & X \end{array}$$

such that $\overline{X}$ is normal and locally irreducible (III.3.1) and that $j$ is a quasi-compact open immersion. Note that $\overline{X}$ and therefore $Y$ are étale-locally connected by III.3.2(iii). For any $X$-scheme $U$, we set

(III.8.1.2) $$\overline{U}=U\times_X\overline{X}\quad\text{and}\quad U_Y=U\times_X Y.$$

We denote by

(III.8.1.3) $$\pi\colon E\to\mathbf{\acute{E}t}_{/X}$$

the Faltings fibered $\mathbb{U}$-site associated with the morphism $h$ (VI.10.1). Recall that the objects of $E$ are the morphisms of schemes $V\to U$ over $h$ such that the morphism $U\to X$ is étale and that the morphism $V\to U_Y$ is finite étale. Let $(V'\to U')$ and $(V\to U)$ be two objects of $E$. A morphism from $(V'\to U')$ to $(V\to U)$ consists of an $X$-morphism $U'\to U$ and a $Y$-morphism $V'\to V$ such that the diagram

(III.8.1.4) $$\begin{array}{ccc} V' & \longrightarrow & U' \\ \downarrow & & \downarrow \\ V & \longrightarrow & U \end{array}$$

is commutative. The functor $\pi$ is then defined for every $(V\to U)\in\mathrm{Ob}(E)$ by

(III.8.1.5) $$\pi(V\to U)=U.$$

For every $U \in \mathrm{Ob}(\mathbf{\acute{E}t}_{/X})$, the fiber of $E$ over $U$ can be canonically identified with the finite étale site of $U_Y$ (III.2.9). We denote by

$$\alpha_{U!}\colon \mathbf{\acute{E}t}_{\mathrm{f}/U_Y} \to E, \quad V \mapsto (V \to U) \tag{III.8.1.6}$$

the canonical functor (VI.5.1.2).

We denote by

$$\mathfrak{F} \to \mathbf{\acute{E}t}_{/X} \tag{III.8.1.7}$$

the fibered $\mathbb{U}$-topos associated with $\pi$. The fiber of $\mathfrak{F}$ over every $U \in \mathrm{Ob}(\mathbf{\acute{E}t}_{/X})$ is canonically equivalent to the finite étale topos $(U_Y)_{\text{fét}}$ of $U_Y$ (III.2.9) and the inverse image functor for every morphism $f\colon U' \to U$ of $\mathbf{\acute{E}t}_{/X}$ identifies with the functor $(f_Y)^*_{\text{fét}}\colon (U_Y)_{\text{fét}} \to (U'_Y)_{\text{fét}}$ inverse image under the morphism of topos $(f_Y)_{\text{fét}}\colon (U'_Y)_{\text{fét}} \to (U_Y)_{\text{fét}}$ (VI.9.3). We denote by

$$\mathfrak{F}^{\vee} \to (\mathbf{\acute{E}t}_{/X})^{\circ} \tag{III.8.1.8}$$

the fibered category obtained by associating with each $U \in \mathrm{Ob}(\mathbf{\acute{E}t}_{/X})$ the category $(U_Y)_{\text{fét}}$, and with each morphism $f\colon U' \to U$ of $\mathbf{\acute{E}t}_{/X}$ the functor $(f_Y)_{\text{fét}*}\colon (U'_Y)_{\text{fét}} \to (U_Y)_{\text{fét}}$ direct image by the morphism of topos $(f_Y)_{\text{fét}}$. We denote by

$$\mathscr{P}^{\vee} \to (\mathbf{\acute{E}t}_{/X})^{\circ} \tag{III.8.1.9}$$

the fibered category obtained by associating with each $U \in \mathrm{Ob}(\mathbf{\acute{E}t}_{/X})$ the category $(\mathbf{\acute{E}t}_{\mathrm{f}/U_Y})^{\wedge}$ of presheaves of $\mathbb{U}$-sets on $\mathbf{\acute{E}t}_{\mathrm{f}/U_Y}$, and with each morphism $f\colon U' \to U$ of $\mathbf{\acute{E}t}_{/X}$ the functor

$$(f_Y)_{\text{fét}*}\colon (\mathbf{\acute{E}t}_{\mathrm{f}/U'_Y})^{\wedge} \to (\mathbf{\acute{E}t}_{\mathrm{f}/U_Y})^{\wedge} \tag{III.8.1.10}$$

obtained by composing with the inverse image functor $f_Y^+\colon \mathbf{\acute{E}t}_{\mathrm{f}/U_Y} \to \mathbf{\acute{E}t}_{\mathrm{f}/U'_Y}$.

**III.8.2.** We denote by $\widehat{E}$ the category of presheaves of $\mathbb{U}$-sets on $E$. We then have an equivalence of categories (VI.5.2)

$$\begin{array}{ccc} \widehat{E} & \to & \mathbf{Hom}_{(\mathbf{\acute{E}t}_{/X})^{\circ}}((\mathbf{\acute{E}t}_{/X})^{\circ}, \mathscr{P}^{\vee}) \\ F & \mapsto & \{U \mapsto F \circ \alpha_{U!}\}. \end{array} \tag{III.8.2.1}$$

From now on, we will identify $F$ with the section $\{U \mapsto F \circ \alpha_{U!}\}$ that is associated with it by this equivalence.

We endow $E$ with the covanishing topology defined by $\pi$ (VI.5.3) and denote by $\widetilde{E}$ the topos of sheaves of $\mathbb{U}$-sets on $E$. The resulting site and topos are called the *Faltings site and topos* associated with $h$ (VI.10.1). If $F$ is a presheaf on $E$, we denote by $F^a$ the associated sheaf. By VI.5.11, the functor (III.8.2.1) induces a fully faithful functor

$$\widetilde{E} \to \mathbf{Hom}_{(\mathbf{\acute{E}t}_{/X})^{\circ}}((\mathbf{\acute{E}t}_{/X})^{\circ}, \mathfrak{F}^{\vee}) \tag{III.8.2.2}$$

whose essential images consists of the sections $\{U \mapsto F_U\}$ satisfying a gluing condition.

**III.8.3.** The functor $\alpha_{X!}\colon \mathbf{\acute{E}t}_{\mathrm{f}/Y} \to E$ is continuous and left exact (VI.5.32). It therefore defines a morphism of topos (VI.10.6.3)

$$\beta\colon \widetilde{E} \to Y_{\text{fét}}. \tag{III.8.3.1}$$

The functor

$$\sigma^+\colon \mathbf{\acute{E}t}_{/X} \to E, \quad U \mapsto (U_Y \to U) \tag{III.8.3.2}$$

is continuous and left exact (VI.5.32). It therefore defines a morphism of topos (VI.10.6.4)

$$\sigma\colon \widetilde{E} \to X_{\text{ét}}. \tag{III.8.3.3}$$

On the other hand, the functor

$$\Psi^+\colon E \to \mathbf{\acute{E}t}_{/Y}, \quad (V \to U) \mapsto V \tag{III.8.3.4}$$

is continuous and left exact (VI.10.7). It therefore defines a morphism of topos

$$\Psi\colon Y_{\text{ét}} \to \widetilde{E}. \tag{III.8.3.5}$$

We have canonical morphisms ((VI.10.8.3) and (VI.10.8.4))

$$\sigma^* \quad \to \quad \Psi_* h_{\text{ét}}^*, \tag{III.8.3.6}$$

$$\beta^* \quad \to \quad \Psi_* \rho_Y^*, \tag{III.8.3.7}$$

where $\rho_Y\colon Y_{\text{ét}} \to Y_{\text{fét}}$ is the canonical morphism (III.2.9.1). If $X$ is quasi-separated and $Y$ is coherent, (III.8.3.7) is an isomorphism by virtue of VI.10.9(iii).

**Remark III.8.4.** It immediately follows from VI.5.10 that $\{U \mapsto U_Y\}$ is a sheaf on $E$. It is therefore a final object of $\widetilde{E}$ and we have canonical isomorphisms

$$\sigma^*(X) \xrightarrow{\sim} \{U \mapsto U_Y\} \xleftarrow{\sim} \beta^*(Y). \tag{III.8.4.1}$$

**III.8.5.** Consider a commutative diagram

$$\begin{array}{ccc} Y' & \xrightarrow{h'} & X' \\ \downarrow & & \downarrow \\ Y & \xrightarrow{h} & X \end{array} \tag{III.8.5.1}$$

and denote by $E'$ the Faltings topos associated with the morphism $h'$ and by $\widetilde{E}'$ the topos of sheaves of $\mathbb{U}$-sets on $E'$ (VI.10.1). Then we have a continuous left exact functor (VI.10.12)

$$\Phi^+\colon E \to E', \quad (V \to U) \mapsto (V \times_Y Y' \to U \times_X X'). \tag{III.8.5.2}$$

It defines a morphism of topos

$$\Phi\colon \widetilde{E}' \to \widetilde{E}. \tag{III.8.5.3}$$

**III.8.6.** We denote by $D$ the covanishing site associated with the functor $h^+\colon \mathbf{\acute{E}t}_{/X} \to \mathbf{\acute{E}t}_{/Y}$ induced by $h$ (VI.4.1). The topos of sheaves of $\mathbb{U}$-sets on $D$ is the covanishing topos $X_{\text{ét}} \overleftarrow{\times}_{X_{\text{ét}}} Y_{\text{ét}}$ of the morphism $h_{\text{ét}}\colon Y_{\text{ét}} \to X_{\text{ét}}$ induced by $h$ (VI.3.12 and VI.4.10). Every object of $E$ is naturally an object of $D$. We thus define a fully faithful functor

$$\rho^+\colon E \to D. \tag{III.8.6.1}$$

It is continuous and left exact (VI.10.15). It therefore defines a morphism of topos

$$\rho\colon X_{\text{ét}} \overleftarrow{\times}_{X_{\text{ét}}} Y_{\text{ét}} \to \widetilde{E}. \tag{III.8.6.2}$$

Giving a point $X_{\text{ét}} \overleftarrow{\times}_{X_{\text{ét}}} Y_{\text{ét}}$ is equivalent to giving a pair of geometric points $\overline{x}$ of $X$ and $\overline{y}$ of $Y$ and a specialization arrow $u$ from $h(\overline{y})$ to $\overline{x}$, that is, an $X$-morphism $u\colon \overline{y} \to X_{(\overline{x})}$, where $X_{(\overline{x})}$ denotes the strict localization of $X$ at $\overline{x}$ (VI.10.18). Such a point will be denoted by $(\overline{y} \rightsquigarrow \overline{x})$ or by $(u\colon \overline{y} \rightsquigarrow \overline{x})$. We denote by $\rho(\overline{y} \rightsquigarrow \overline{x})$ its image by $\rho$, which is therefore a point of $\widetilde{E}$.

If $X$ and $Y$ are coherent, then as $(\overline{y} \rightsquigarrow \overline{x})$ goes through the family of points of $X_{\text{ét}} \overleftarrow{\times}_{X_{\text{ét}}} Y_{\text{ét}}$, the family of fiber functors of $\widetilde{E}$ associated with the points $\rho(\overline{y} \rightsquigarrow \overline{x})$ is conservative by virtue of VI.10.21.

**III.8.7.** Suppose that $X$ is strictly local, with closed point $x$. We denote by $E_{\text{scoh}}$ the full subcategory of $E$ made up of the $(V \to U)$'s such that the morphism $U \to X$ is separated and coherent, which we endow with the topology induced by that on $E$. The canonical injection functor $E_{\text{scoh}} \to E$ then induces by restriction an equivalence of categories between $\widetilde{E}$ and the topos of sheaves of $\mathbb{U}$-sets on $E_{\text{scoh}}$ (VI.10.4).

For any étale, separated, and coherent morphism $U \to X$, we denote by $U^{\text{f}}$ the disjoint union of the strict localizations of $U$ at the points of $U_x$; it is an open and closed subscheme of $U$ that is finite over $X$ ([**42**] 18.5.11). By VI.10.23, the functor

(III.8.7.1) $$\theta^+ : E_{\text{scoh}} \to \mathbf{\acute{E}t}_{\text{f}/Y}, \quad (V \to U) \mapsto V \times_U U^{\text{f}}$$

is continuous and left exact. It therefore defines a morphism of topos

(III.8.7.2) $$\theta \colon Y_{\text{fét}} \to \widetilde{E}.$$

We have a canonical isomorphism (VI.10.24.3)

(III.8.7.3) $$\beta\theta \xrightarrow{\sim} \text{id}_{Y_{\text{fét}}}.$$

We deduce from this a base change morphism (VI.10.24.4)

(III.8.7.4) $$\beta_* \to \theta^*.$$

This is an isomorphism by virtue of VI.10.27; in particular, the functor $\beta_*$ is exact.

**III.8.8.** Let $\overline{x}$ be a geometric point of $X$, $X'$ the strict localization of $X$ at $\overline{x}$, $Y' = Y \times_X X'$, and $h' \colon Y' \to X'$ the canonical projection. We denote by $E'$ the Faltings site associated with the morphism $h'$, by $\widetilde{E}'$ the topos of sheaves of $\mathbb{U}$-sets on $E'$, by

(III.8.8.1) $$\beta' \colon \widetilde{E}' \to Y'_{\text{fét}}$$

the canonical morphism (III.8.3.1), by

(III.8.8.2) $$\Phi \colon \widetilde{E}' \to \widetilde{E}$$

the functoriality morphism (III.8.5.3), and by

(III.8.8.3) $$\theta \colon Y'_{\text{fét}} \to \widetilde{E}'$$

the morphism (III.8.7.2). We denote by

(III.8.8.4) $$\varphi_{\overline{x}} \colon \widetilde{E} \to Y'_{\text{fét}}$$

the composed functor $\theta^* \circ \Phi^*$.

If $X$ and $Y$ are coherent, then as $\overline{x}$ goes through the set of geometric points of $X$, the family of functors $\varphi_{\overline{x}}$ is conservative by virtue of VI.10.32.

We denote by $\mathfrak{V}_{\overline{x}}$ the category of $\overline{x}$-pointed étale $X$-schemes ([**2**] VIII 3.9) or, equivalently, the category of neighborhoods of the point of $X_{\text{ét}}$ associated with $\overline{x}$ in the site $\mathbf{\acute{E}t}_{/X}$ ([**2**] IV 6.8.2). For any object $(U, \mathfrak{p} \colon \overline{x} \to U)$ of $\mathfrak{V}_{\overline{x}}$, we denote also by $\mathfrak{p} \colon X' \to U$ the morphism deduced from $\mathfrak{p}$ ([**2**] VIII 7.3) and by

(III.8.8.5) $$\mathfrak{p}_Y : Y' \to U_Y$$

its base change by $h$. By VI.10.37, for every sheaf $F = \{U \mapsto F_U\}$ of $\widetilde{E}$, we have a functorial canonical isomorphism

(III.8.8.6) $$\lim_{\substack{\longrightarrow \\ (U,\mathfrak{p}) \in \mathfrak{V}_{\overline{x}}^{\circ}}} (\mathfrak{p}_Y)^*_{\text{fét}}(F_U) \xrightarrow{\sim} \varphi_{\overline{x}}(F).$$

**III.8.9.** We denote by $\mathscr{B}$ the presheaf on $E$ defined for every $(V \to U) \in \mathrm{Ob}(E)$, by

$$\mathscr{B}((V \to U)) = \Gamma(\overline{U}, \mathscr{O}_{\overline{U}}) \tag{III.8.9.1}$$

and by $\mathscr{B}^a$ the associated sheaf. By VI.5.34(ii) and with the notation conventions of III.2.9, we have a canonical isomorphism

$$\sigma^*(\hbar_*(\mathscr{O}_{\overline{X}})) \xrightarrow{\sim} \mathscr{B}^a. \tag{III.8.9.2}$$

**III.8.10.** For any $(V \to U) \in \mathrm{Ob}(E)$, we denote by $\overline{U}^V$ the integral closure of $\overline{U} = U \times_X \overline{X}$ in $V$. For every morphism $(V' \to U') \to (V \to U)$ of $E$, we have a canonical morphism $\overline{U}'^{V'} \to \overline{U}^V$ that fits into a commutative diagram

$$\begin{array}{ccccccc} V' & \longrightarrow & \overline{U}'^{V'} & \longrightarrow & \overline{U}' & \longrightarrow & U' \\ \downarrow & & \downarrow & & \downarrow & & \downarrow \\ V & \longrightarrow & \overline{U}^{V} & \longrightarrow & \overline{U} & \longrightarrow & U \end{array} \tag{III.8.10.1}$$

We denote by $\overline{\mathscr{B}}$ the presheaf on $E$ defined for every $(V \to U) \in \mathrm{Ob}(E)$, by

$$\overline{\mathscr{B}}((V \to U)) = \Gamma(\overline{U}^V, \mathscr{O}_{\overline{U}^V}). \tag{III.8.10.2}$$

For any $U \in \mathrm{Ob}(\mathbf{\acute{E}t}_{/X})$, we set (III.8.1.6)

$$\overline{\mathscr{B}}_U = \overline{\mathscr{B}} \circ \alpha_{U!}. \tag{III.8.10.3}$$

**Remarks III.8.11.** Let $(V \to U)$ be an object of $E$. Then:

(i) Since $V$ is integral over $U_Y$, the canonical morphism $V \to U_Y \times_{\overline{U}} \overline{U}^V$ is an isomorphism. In particular, the canonical morphism $V \to \overline{U}^V$ is a schematically dominant open immersion.

(ii) The scheme $\overline{U}^V$ is normal and locally irreducible (III.3.1). Indeed, $\overline{U}$ and $V$ are normal and locally irreducible by III.3.3. Let $U_0$ be an open subscheme of $\overline{U}$ having only finitely many irreducible components. Then $U_0 \times_{\overline{U}} \overline{U}^V$ is the finite sum of the integral closures of $U_0$ in the generic points of $V$ that lie over $U_0$, each of which is an integral normal scheme by virtue of ([**40**] 6.3.7).

(iii) For every étale $U$-scheme $U'$, setting $V' = V \times_U U'$, the canonical morphism (III.8.10.1)

$$\overline{U}'^{V'} \to \overline{U}^V \times_U U' \tag{III.8.11.1}$$

is an isomorphism. Indeed, $\overline{U}^V \times_U U'$ is normal and locally irreducible by (ii) and III.3.3, and the canonical morphism $V' \to \overline{U}^V \times_U U'$ is a schematically dominant open immersion by virtue of (i) and ([**42**] 11.10.5). The assertion follows because $\overline{U}'^{V'}$ identifies with the integral closure of $\overline{U}^V \times_U U'$ in $V'$.

**Lemma III.8.12.** *Let $(V \to U)$ be an object of $E$, $f\colon W \to V$ a torsor for the étale topology on $V$ under a finite constant group $G$. We denote by $\overline{f}\colon \overline{U}^W \to \overline{U}^V$ the morphism induced by $f$. Then the natural action of $G$ on $\overline{U}^W$ is admissible* ([**37**] V Definition 1.7) *and $(\overline{U}^V, \overline{f})$ is a quotient of $\overline{U}^W$ by $G$; in other words, the canonical morphism*

$$\mathscr{O}_{\overline{U}^V} \to \overline{f}_*(\mathscr{O}_{\overline{U}^W})^G \tag{III.8.12.1}$$

*is an isomorphism. In particular, the canonical morphism*

$$\Gamma(\overline{U}^V, \mathscr{O}_{\overline{U}^V}) \to \Gamma(\overline{U}^W, \mathscr{O}_{\overline{U}^W})^G \tag{III.8.12.2}$$

*is an isomorphism.*

Indeed, the action of $G$ on $\overline{U}^W$ is admissible by virtue of ([**37**] V Corollary 1.8). Denote by $Z$ the quotient of $\overline{U}^W$ by $G$. By III.8.11(i) and ([**37**] Proposition 1.9), $f$ induces an isomorphism $V \xrightarrow{\sim} U_Y \times_{\overline{U}} Z$, and the morphism $V \to Z$ is a schematically dominant open immersion. Since $Z$ is integral over $\overline{U}$, we deduce from this that the morphism $Z \to \overline{U}^V$ induced by $\overline{f}$ is an isomorphism, giving the isomorphism (III.8.12.1). The isomorphism (III.8.12.2) follows because the functor $\Gamma(\overline{U}^V, -)$ on the Zariski topos of $\overline{U}^V$ is left exact.

**III.8.13.** Let $U$ be an object of $\mathbf{\acute{E}t}_{/X}$ and $\overline{y}$ a geometric point of $U_Y$. Since the scheme $\overline{U}$ is locally irreducible (III.3.3), it is the sum of the schemes induced on its irreducible components. We denote by $\overline{U}^\star$ the irreducible component of $\overline{U}$ (or, equivalently, its connected component) containing $\overline{y}$. Likewise, $U_Y$ is the sum of the induced schemes on its irreducible components. We set $V = \overline{U}^\star \times_{\overline{X}} Y$, which is the irreducible component of $U_Y$ containing $\overline{y}$. We denote by $\mathbf{B}_{\pi_1(V,\overline{y})}$ the classifying topos of the profinite group $\pi_1(V,\overline{y})$, by $(V_i)_{i\in I}$ the normalized universal cover of $V$ at $\overline{y}$ (III.2.10), and by

$$\nu_{\overline{y}} \colon V_{\text{fét}} \xrightarrow{\sim} \mathbf{B}_{\pi_1(V,\overline{y})}, \quad F \mapsto \varinjlim_{i\in I} F(V_i) \tag{III.8.13.1}$$

the fiber functor of $V_{\text{fét}}$ at $\overline{y}$ (III.2.10.3). For each $i \in I$, $(V_i \to U)$ is naturally an object of $E$. We can therefore consider the filtered inverse system of schemes $(\overline{U}^{V_i})_{i\in I}$. We set

$$\overline{R}^{\overline{y}}_U = \varinjlim_{i\in I} \Gamma(\overline{U}^{V_i}, \mathscr{O}_{\overline{U}^{V_i}}), \tag{III.8.13.2}$$

which is a continuous discrete representation of $\pi_1(V,\overline{y})$ by virtue of III.8.12; in other words, it is an object of $\mathbf{B}_{\pi_1(V,\overline{y})}$.

**Remark III.8.14.** We keep the assumptions of III.8.13 and denote by $\overline{U}^{\overline{y}}$ the inverse limit in the category of $\overline{U}$-schemes of the filtered inverse system $(\overline{U}^{V_i})_{i\in I}$, which exists by virtue of ([**42**] 8.2.3). By ([**2**] VI 5.3), if $\overline{U}$ is coherent, we have a canonical isomorphism

$$\Gamma(\overline{U}^{\overline{y}}, \mathscr{O}_{\overline{U}^{\overline{y}}}) \xrightarrow{\sim} \overline{R}^{\overline{y}}_U. \tag{III.8.14.1}$$

**Lemma III.8.15.** *Under the assumptions of* III.8.13, $\overline{\mathscr{B}}_U$ (III.8.10.3) *is a sheaf for the étale topology of* $\mathbf{\acute{E}t}_{\text{f}/U_Y}$, *and we have a canonical isomorphism*

$$\nu_{\overline{y}}(\overline{\mathscr{B}}_U|V) \xrightarrow{\sim} \overline{R}^{\overline{y}}_U. \tag{III.8.15.1}$$

It suffices to show that the restriction of the presheaf $\overline{\mathscr{B}}_U$ to the site $\mathbf{\acute{E}t}_{\text{f}/V}$ is a sheaf. The isomorphism (III.8.15.1) will then immediately follow from the definition. Denote by

$$\mu^+_{\overline{y}} \colon \mathbf{\acute{E}t}_{\text{f}/V} \to \mathbf{B}_{\pi_1(V,\overline{y})} \tag{III.8.15.2}$$

the fiber functor at $\overline{y}$ (III.2.10.2). For every $W \in \mathrm{Ob}(\mathbf{\acute{E}t}_{\text{f}/V})$, we have a functorial isomorphism

$$\mu^+_{\overline{y}}(W) \xrightarrow{\sim} \varinjlim_{i\in I} \mathrm{Hom}_V(V_i, W). \tag{III.8.15.3}$$

We deduce from this a morphism that is functorial in $W$:

$$\Gamma(\overline{U}^W, \mathscr{O}_{\overline{U}^W}) \to \mathrm{Hom}_{\mathbf{B}_{\pi_1(V,\overline{y})}}(\mu^+_{\overline{y}}(W), \overline{R}^{\overline{y}}_U). \tag{III.8.15.4}$$

This is an isomorphism by virtue of III.8.12, giving the desired statement.

**Proposition III.8.16.** *The presheaf $\overline{\mathscr{B}}$ on $E$ is a sheaf for the covanishing topology.*

Let $(V \to U) \in \mathrm{Ob}(E)$ and $(U_i \to U)_{i\in I}$ be a covering of $\mathbf{\acute{E}t}_{/X}$. For any $(i,j) \in I^2$, set $V_i = V \times_U U_i$, $U_{ij} = U_i \times_U U_j$, and $V_{ij} = U_{ij} \times_U V$. We have $\overline{U}_i^{V_i} \simeq \overline{U}^V \times_U U_i$ and $\overline{U}_{ij}^{V_{ij}} \simeq \overline{U}^V \times_U U_{ij}$ by III.8.11(iii). Viewing $\mathscr{O}_{\overline{U}^V}$ as a sheaf on the étale topos of $\overline{U}^V$, the étale covering $(\overline{U}_i^{V_i} \to \overline{U}^V)_{i\in I}$ induces an exact sequence of maps of sets

$$\overline{\mathscr{B}}((V \to U)) \to \prod_{i\in I} \overline{\mathscr{B}}((V_i \to U_i)) \rightrightarrows \prod_{(i,j)\in I^2} \overline{\mathscr{B}}((V_{ij} \to U_{ij})). \tag{III.8.16.1}$$

The proposition follows in view of III.8.15 and VI.5.10.

**III.8.17.** From now on, we will say that $\overline{\mathscr{B}}$ is the ring of $\widetilde{E}$ associated with $\overline{X}$. By III.8.9, the canonical homomorphism $\mathscr{B} \to \overline{\mathscr{B}}$ induces a homomorphism $\sigma^*(\hbar_*(\mathscr{O}_{\overline{X}})) \to \overline{\mathscr{B}}$ (cf. III.2.9). Unless mentioned otherwise, we view $\sigma\colon \widetilde{E} \to X_{\text{ét}}$ (III.8.3.3) as a morphism of ringed topos (by $\overline{\mathscr{B}}$ and $\hbar_*(\mathscr{O}_{\overline{X}})$, respectively). For modules, we use the notation $\sigma^{-1}$ to denote the inverse image in the sense of abelian sheaves, and we keep the notation $\sigma^*$ for the inverse image in the sense of modules.

**Remark III.8.18.** The presheaf $\overline{\mathscr{B}}$ is not in general a sheaf for the topology on $E$ originally defined by Faltings in ([**26**] page 214). Indeed, suppose that $\hbar$ is integral, that $j$ is not closed, and that there exists an affine open immersion $Z \to X$ such that $Y = Z \times_X \overline{X}$, where $j\colon Y \to \overline{X}$ is the canonical projection. Consider two affine schemes $U$ and $U'$ of $\mathbf{\acute{E}t}_{/X}$ and a surjective morphism $U' \to U$ such that $U'_Y \to U_Y$ is finite and $\overline{U}' \to \overline{U}$ is not integral. We could, for example, take for $U$ an affine open subscheme of $X$ such that the open immersion $j_U\colon U_Y \to \overline{U}$ is not closed and for $U'$ the sum of $U$ and $U_Z$. Set $V = U'_Y$. It is clear that $(V \to U)$ is an object of $E$ and that $(V \to U') \to (V \to U)$ is a covering for the topology on $E$ defined by Faltings in ([**26**] page 214). But the sequence

$$\overline{\mathscr{B}}((V \to U)) \to \overline{\mathscr{B}}((V \to U')) \rightrightarrows \overline{\mathscr{B}}((V \to U' \times_U U')) \tag{III.8.18.1}$$

is not exact. Indeed, let us suppose that it is. The diagram

$$V \to U' \overset{\Delta}{\to} U' \times_U U' \rightrightarrows U', \tag{III.8.18.2}$$

where $\Delta$ is the diagonal morphism and the double arrow represents the two canonical projections, is commutative. Consequently, the double arrow in (III.8.18.1) is made up of two copies of the same morphism. Hence the canonical map

$$\overline{\mathscr{B}}((V \to U)) \to \overline{\mathscr{B}}((V \to U')) \tag{III.8.18.3}$$

is an isomorphism. Since $\Gamma(\overline{U}', \mathscr{O}_{\overline{U}'}) \subset \overline{\mathscr{B}}((V \to U'))$, it follows that $\Gamma(\overline{U}', \mathscr{O}_{\overline{U}'})$ is integral over $\Gamma(\overline{U}, \mathscr{O}_{\overline{U}})$, and consequently that $\overline{U}' \to \overline{U}$ is integral because $\overline{U}$ and $\overline{U}'$ are affine, which contradicts the assumptions.

**Proposition III.8.19.** *Suppose that $X$ and $\overline{X}$ are strictly local and let $\overline{y}$ be a geometric point of $Y$. Denote by $x$ the closed point of $X$ and by $(\overline{y} \rightsquigarrow x)$ the point of $X_{\text{ét}} \overleftarrow{\times}_{X_{\text{ét}}} Y_{\text{ét}}$ defined by the unique specialization arrow from $h(\overline{y})$ to $x$ (III.8.6). Then:*

(i) *The stalk $\overline{\mathscr{B}}_{\rho(\overline{y}\rightsquigarrow x)}$ of $\overline{\mathscr{B}}$ at the point $\rho(\overline{y} \rightsquigarrow x)$ of $\widetilde{E}$ (III.8.6.2) is a normal strictly local ring.*

(ii) *We have a canonical isomorphism*

(III.8.19.1)
$$(\hbar_*(\mathscr{O}_{\overline{X}}))_x \xrightarrow{\sim} \Gamma(\overline{X}, \mathscr{O}_{\overline{X}}).$$

(iii) *The homomorphism*

(III.8.19.2)
$$(\hbar_*(\mathscr{O}_{\overline{X}}))_x \to \overline{\mathscr{B}}_{\rho(\overline{y} \rightsquigarrow x)}$$

*induced by the canonical homomorphism* $\sigma^{-1}(\hbar_*(\mathscr{O}_{\overline{X}})) \to \overline{\mathscr{B}}$ (III.8.17) *is injective and local.*

(i) First note that $Y$ is integral. Let $(V_i)_{i\in I}$ be the normalized universal cover of $Y$ at the point $\overline{y}$ (III.2.10). By VI.10.36, we have a canonical isomorphism

(III.8.19.3)
$$\lim_{\substack{\longrightarrow \\ i\in I}} \overline{\mathscr{B}}((V_i \to X)) \xrightarrow{\sim} \overline{\mathscr{B}}_{\rho(\overline{y} \rightsquigarrow x)}.$$

For every $i \in I$, the scheme $\overline{X}^{V_i}$ is normal, integral, and integral over $\overline{X}$ (III.8.11). It is therefore strictly local by virtue of III.3.5. On the other hand, for all $(i,j) \in I^2$ with $j \geq i$, the transition morphism $\overline{X}^{V_j} \to \overline{X}^{V_i}$ is integral and dominant. In particular, the transition homomorphism $\overline{\mathscr{B}}((V_i \to X)) \to \overline{\mathscr{B}}((V_j \to X))$ is local. It follows that the ring $\overline{\mathscr{B}}_{\rho(\overline{y} \rightsquigarrow x)}$ is local, normal, and henselian ([**39**] 0.6.5.12(ii) and [**62**] I § 3 Proposition 1). Since the homomorphism $\Gamma(\overline{X}, \mathscr{O}_{\overline{X}}) \to \overline{\mathscr{B}}_{\rho(\overline{y} \rightsquigarrow x)}$ is integral and therefore local, the residue field of $\overline{\mathscr{B}}_{\rho(\overline{y} \rightsquigarrow x)}$ is an algebraic extension of that of $\Gamma(\overline{X}, \mathscr{O}_{\overline{X}})$. It is therefore separably closed.

(ii) This immediately follows from the fact that $X$ is strictly local.

(iii) Recall that we have a canonical isomorphism (VI.10.18.1)

(III.8.19.4)
$$(\sigma^{-1}(\hbar_*(\mathscr{O}_{\overline{X}})))_{\rho(\overline{y} \rightsquigarrow x)} \xrightarrow{\sim} \hbar_*(\mathscr{O}_{\overline{X}})_x.$$

In view of (ii) and (III.8.19.3), the stalk of the canonical homomorphism $\sigma^{-1}(\hbar_*(\mathscr{O}_{\overline{X}})) \to \overline{\mathscr{B}}$ at $\rho(\overline{y} \rightsquigarrow x)$ identifies with the canonical homomorphism

(III.8.19.5)
$$\Gamma(\overline{X}, \mathscr{O}_{\overline{X}}) \to \overline{\mathscr{B}}_{\rho(\overline{y} \rightsquigarrow x)} = \lim_{\substack{\longrightarrow \\ i\in I}} \overline{\mathscr{B}}((V_i \to X)),$$

which is clearly injective and integral, and therefore local.

**III.8.20.** Consider a commutative diagram

(III.8.20.1)
$$\begin{array}{ccccc} Y' & \xrightarrow{j'} & \overline{X}' & \xrightarrow{\hbar'} & X' \\ \downarrow{\scriptstyle g'} & & \downarrow{\scriptstyle \overline{g}} & & \downarrow{\scriptstyle g} \\ Y & \xrightarrow{j} & \overline{X} & \xrightarrow{\hbar} & X \end{array}$$

and set $h' = \hbar' \circ j'$. Suppose that $\overline{X}'$ is normal and locally irreducible and that $j'$ is a quasi-compact open immersion. We denote by $E'$ (resp. $\widetilde{E}'$) the Faltings site (resp. topos) associated with the morphism $h'$ (III.8.2), by $\overline{\mathscr{B}}'$ the ring of $\widetilde{E}'$ associated with $\overline{X}'$ (III.8.17), and by

(III.8.20.2)
$$\Phi\colon \widetilde{E}' \to \widetilde{E}$$

the functoriality morphism (III.8.5.3). For any $(V' \to U') \in \mathrm{Ob}(E')$, we set $\overline{U}' = U' \times_{X'} \overline{X}'$ and denote by $\overline{U}'^{V'}$ the integral closure of $\overline{U}'$ in $V'$, so that

(III.8.20.3)
$$\overline{\mathscr{B}}'((V' \to U')) = \Gamma(\overline{U}'^{V'}, \mathscr{O}_{\overline{U}'^{V'}}).$$

For any $(V \to U) \in \mathrm{Ob}(E)$, set $V' = V \times_Y Y'$ and $U' = U \times_X X'$, so that $(V' \to U')$ is an object of $E'$ and that we have a commutative diagram

(III.8.20.4)
$$\begin{array}{ccccccc} Y' & \longleftarrow & V' & \longrightarrow & \overline{U}' & \longrightarrow & \overline{X}' \\ \downarrow & \square & \downarrow & & \downarrow & \square & \downarrow \\ Y & \longleftarrow & V & \longrightarrow & \overline{U} & \longrightarrow & \overline{X} \end{array}$$

We deduce from this a morphism

(III.8.20.5)
$$\overline{U}'^{V'} \to \overline{U}^{V},$$

and consequently a ring homomorphism of $\widetilde{E}$

(III.8.20.6)
$$\overline{\mathscr{B}} \to \Phi_*(\overline{\mathscr{B}}').$$

From now on, we will consider $\Phi$ as a morphism of ringed topos (by $\overline{\mathscr{B}}'$ and $\overline{\mathscr{B}}$, respectively). For modules, we use the notation $\Phi^{-1}$ to denote the inverse image in the sense of abelian sheaves, and we keep the notation $\Phi^*$ for the inverse image in the sense of modules.

**Lemma III.8.21.** *We keep the assumptions of* III.8.20 *and moreover suppose that $g$ is étale and that the two squares in diagram* (III.8.20.1) *are Cartesian, so that $(Y' \to X')$ is an object of $E$. Then:*

(i) *The morphism*

(III.8.21.1)
$$\Phi^{-1}(\overline{\mathscr{B}}) \to \overline{\mathscr{B}}'$$

*adjoint to the morphism* (III.8.20.6) *is an isomorphism.*

(ii) *The morphism of ringed topos $\Phi\colon (\widetilde{E}', \overline{\mathscr{B}}') \to (\widetilde{E}, \overline{\mathscr{B}})$ identifies with the localization morphism of $(\widetilde{E}, \overline{\mathscr{B}})$ at $\sigma^*(X')$.*

Indeed, the morphism of topos $\Phi\colon \widetilde{E}' \to \widetilde{E}$ identifies with the localization morphism of $\widetilde{E}$ at $\sigma^*(X') = (Y' \to X')^a$ by virtue of VI.10.14. It therefore suffices to show the first statement. The functor $\Phi^*\colon \widetilde{E} \to \widetilde{E}'$ identifies with the restriction functor by the canonical functor $E' \to E$. For every $(V \to U) \in \mathrm{Ob}(E')$, we have a canonical isomorphism

(III.8.21.2)
$$U \times_{X'} \overline{X}' \xrightarrow{\sim} U \times_X \overline{X} = \overline{U}.$$

We deduce from this an isomorphism (that is functorial in $(V \to U)$)

(III.8.21.3)
$$\Phi^{-1}(\overline{\mathscr{B}})((V \to U)) \xrightarrow{\sim} \overline{\mathscr{B}}'((V \to U)).$$

It remains to show that this is adjoint to the morphism (III.8.20.6). Set $U' = U \times_X X'$ and $V' = V \times_Y Y'$. The structural morphisms $U \to X'$ and $V \to Y'$ induce sections $U \to U'$ and $V \to V'$ of the canonical projections $U' \to U$ and $V' \to V$. We deduce from this a commutative diagram

(III.8.21.4)
$$\begin{array}{ccccc} & V & \longrightarrow & \overline{U} & \\ & \downarrow & & \downarrow & \\ \mathrm{id}_V & V' & \longrightarrow & \overline{U}' & \mathrm{id}_{\overline{U}} \\ & \downarrow & & \downarrow & \\ & V & \longrightarrow & \overline{U} & \end{array}$$

and consequently morphisms $\overline{U}^V \to \overline{U}'^{V'} \to \overline{U}^V$ whose composition is the identity of $\overline{U}^V$. The composition

(III.8.21.5)
$$\Phi^{-1}(\overline{\mathscr{B}})((V \to U)) \to \Phi^{-1}(\Phi_*(\overline{\mathscr{B}}'))((V \to U)) = \overline{\mathscr{B}}'((V' \to U')) \to \overline{\mathscr{B}}'((V \to U)),$$

where the first arrow is induced by (III.8.20.6) and the last arrow is the adjunction morphism, is therefore the isomorphism (III.8.21.3); the first statement follows.

**Lemma III.8.22.** *We keep the assumptions of* III.8.20 *and moreover suppose that $X'$ is the strict localization of $X$ at a geometric point $\overline{x}$ and that the two squares in diagram* (III.8.20.1) *are Cartesian. Let $(V \to U)$ be an object of $E$ and $U' = U \times_X X'$, $V' = V \times_Y Y'$. Then the canonical morphism $\overline{U}'^{V'} \to \overline{U}^V \times_X X'$* (III.8.20.5) *is an isomorphism.*

Indeed, we have a commutative diagram with Cartesian squares

(III.8.22.1)
$$\begin{array}{ccccccc} V' & \longrightarrow & \overline{U}' & \longrightarrow & U' & \longrightarrow & X' \\ \downarrow & & \downarrow & & \downarrow & & \downarrow \\ V & \longrightarrow & \overline{U} & \longrightarrow & U & \longrightarrow & X \end{array}$$

We can therefore identify $\overline{U}'^{V'}$ with the integral closure of $\overline{U}^V \times_X X'$ in $V'$. It then suffices to show that the canonical morphism $V' \to \overline{U}^V \times_X X'$ is a quasi-compact and schematically dominant open immersion and that $\overline{U}^V \times_X X'$ is normal and locally irreducible. The first statement follows from III.8.11(i) and ([**42**] 11.10.5). Since the second statement is local, to prove it we may restrict to the case where $X$ is affine. We consider $X'$ as a cofiltered inverse limit of affine étale neighborhoods $(X_i)_{\in I}$ of $\overline{x}$ in $X$ (cf. [**2**] VIII 4.5). Then $\overline{U}^V \times_X X'$ is canonically isomorphic to the inverse limit of the schemes $(\overline{U}^V \times_X X_i)_{i \in I}$ ([**42**] 8.2.5). Since each $\overline{U}^V \times_X X_i$ is normal by III.8.11(ii) and ([**62**] VII Proposition 2), $\overline{U}^V \times_X X'$ is normal by ([**39**] 0.6.5.12(ii)). On the other hand, since $\overline{X}'$ is normal and locally irreducible by assumption, the same holds for $V'$ by III.3.3. It follows that $\overline{U}^V \times_X X'$ is locally irreducible by virtue of III.3.4, giving the lemma.

**Proposition III.8.23.** *We keep the assumptions of* III.8.20 *and moreover suppose that $\hbar$ is coherent, that $X'$ is the strict localization of $X$ at a geometric point $\overline{x}$, and that the two squares in diagram* (III.8.20.1) *are Cartesian. Then the morphism*

(III.8.23.1)
$$\Phi^{-1}(\overline{\mathscr{B}}) \to \overline{\mathscr{B}}'$$

*adjoint to the morphism* (III.8.20.6) *is an isomorphism.*

We choose an affine open neighborhood $X_0$ of the image of $\overline{x}$ in $X$ and denote by $I$ the category of $\overline{x}$-pointed étale $X_0$-schemes that are affine over $X_0$ (cf. [**2**] VIII 3.9 and 4.5). We denote by

(III.8.23.2)
$$\mathscr{E} \to I$$

the fibered $\mathbb{U}$-site defined in (VI.11.2.2): the fiber of $\mathscr{E}$ over an object $U$ of $I$ is the Faltings site associated with the canonical projection $h_U : U_Y \to U$, and the inverse image functor associated with a morphism $f : U' \to U$ in $I$ is the functor $\Phi_f^+ : \mathscr{E}_U \to \mathscr{E}_{U'}$ defined in (III.8.5.2). We denote by

(III.8.23.3)
$$\mathscr{F} \to I$$

the fibered $\mathbb{U}$-topos associated with $\mathscr{E}/I$. The fiber of $\mathscr{F}$ over an object $U$ of $I$ is the topos $\widetilde{\mathscr{E}}_U$ of sheaves of $\mathbb{U}$-sets on the covanishing site $\mathscr{E}_U$, and the inverse image functor with respect to a morphism $f\colon U' \to U$ of $I$ is the inverse image functor under the morphism of topos $\Phi_f\colon \widetilde{\mathscr{E}}_{U'} \to \widetilde{\mathscr{E}}_U$ defined in (III.8.5.3). We denote by

(III.8.23.4) $$\mathscr{F}^\vee \to I^\circ$$

the fibered category obtained by associating with each object $U$ of $I$ the category $\mathscr{F}_U = \widetilde{\mathscr{E}}_U$ and with each morphism $f\colon U' \to U$ of $I$ the direct image functor by the morphism of topos $\Phi_f$. Recall (VI.10.14) that for every $U \in \mathrm{Ob}(I)$, $\widetilde{\mathscr{E}}_U$ is canonically equivalent to the topos $\widetilde{E}_{/(U_{\overline{Y}} \to U)^a}$, where $(U_{\overline{Y}} \to U)^a$ denotes the sheaf associated with $(U_{\overline{Y}} \to U)$.

By virtue of VI.11.3, the topos $\widetilde{E}'$ is canonically equivalent to the inverse limit of the fibered topos $\mathscr{F}/I$. We endow $\mathscr{E}$ with the total topology ([**2**] VI 7.4.1) and denote by $\mathbf{Top}(\mathscr{E})$ the topos of sheaves of $\mathbb{U}$-sets on $\mathscr{E}$. Then we have a canonical morphism (VI.11.4.2)

(III.8.23.5) $$\varpi\colon \widetilde{E}' \to \mathbf{Top}(\mathscr{E})$$

and a canonical commutative diagram of functors (VI.11.4.3)

(III.8.23.6) $$\begin{array}{ccc} \widetilde{E}' & \xrightarrow{\sim} & \mathbf{Hom}_{\mathrm{cart}/I^\circ}(I^\circ, \mathscr{F}^\vee) \\ {\scriptstyle \varpi_*}\downarrow & & \downarrow \\ \mathbf{Top}(\mathscr{E}) & \xrightarrow{\sim} & \mathbf{Hom}_{I^\circ}(I^\circ, \mathscr{F}^\vee) \end{array}$$

where the horizontal arrows are equivalences of categories and the right vertical arrow is the canonical injection (cf. [**2**] VI 8.2.9).

For any object $U$ of $I$, we denote by $\overline{\mathscr{B}}_{/U}$ the ring of $\widetilde{\mathscr{E}}_U$ associated with $\overline{U}$ (III.8.17). For every morphism $f\colon U' \to U$ of $I$, we have a canonical homomorphism $\overline{\mathscr{B}}_{/U} \to \Phi_{f*}(\overline{\mathscr{B}}_{/U'})$ (III.8.20.6). These homomorphisms satisfy a compatibility relation for the composition of morphisms of $I$ of type ([**1**] (1.1.2.2)). They therefore define a ring of $\mathbf{Top}(\mathscr{E})$ that we denote by $\{U \mapsto \overline{\mathscr{B}}_{/U}\}$ (not to be confused with the ring $\overline{\mathscr{B}} = \{U \mapsto \overline{\mathscr{B}}_U\}$ of $\widetilde{E}$ (III.8.10.3)). For every $U \in \mathrm{Ob}(I)$, we have a canonical morphism $g_U : X' \to U$ that induces a morphism of ringed topos (III.8.20.6)

(III.8.23.7) $$\Phi_U\colon (\widetilde{E}', \overline{\mathscr{B}}') \to (\widetilde{\mathscr{E}}_U, \overline{\mathscr{B}}_{/U}).$$

The collection of homomorphisms $\overline{\mathscr{B}}_{/U} \to \Phi_{U*}(\overline{\mathscr{B}}')$ defines a ring homomorphism of $\mathbf{Top}(\mathscr{E})$

(III.8.23.8) $$\{U \mapsto \overline{\mathscr{B}}_{/U}\} \to \varpi_*(\overline{\mathscr{B}}').$$

Let us show that the adjoint homomorphism

(III.8.23.9) $$\varpi^*(\{U \mapsto \overline{\mathscr{B}}_{/U}\}) \to \overline{\mathscr{B}}'$$

is an isomorphism. Since the functor $\varpi_*$ is fully faithful (III.8.23.6), it suffices to show that the induced homomorphism

(III.8.23.10) $$\varpi_*(\varpi^*(\{U \mapsto \overline{\mathscr{B}}_{/U}\})) \to \varpi_*(\overline{\mathscr{B}}')$$

is an isomorphism. In view of ([**2**] VI 8.5.3), this is equivalent to showing that for every $U \in \mathrm{Ob}(I)$, the canonical homomorphism of $\widetilde{\mathscr{E}}_U$

(III.8.23.11) $$\varinjlim_{f\colon U' \to U} \Phi_{f*}(\overline{\mathscr{B}}_{/U'}) \to \Phi_{U*}(\overline{\mathscr{B}}'),$$

where the limit is taken over the morphisms $f\colon U' \to U$ of $I$, is an isomorphism.

Let $(V_1 \to U_1)$ be an object of $E_{/(U_Y \to U)}$ such that the morphism $U_1 \to U$ is coherent. The canonical morphism

$$\overline{U}_1^{V_1} \times_U X' \to \varprojlim_{U' \in \mathrm{Ob}(I_{/U})} \overline{U}_1^{V_1} \times_U U' \tag{III.8.23.12}$$

is an isomorphism by ([**42**] 8.2.5). Since $\hbar$ is coherent, the scheme $\overline{U}_1^{V_1}$ is coherent. By ([**2**] VI 5.2), III.8.11(iii), and III.8.22, it follows that the canonical homomorphism

$$\varinjlim_{U' \in \mathrm{Ob}(I_{/U})} \overline{\mathscr{B}}_{/U'}((V_1 \times_U U' \to U_1 \times_U U')) \to \overline{\mathscr{B}}'((V_1 \times_U X' \to U_1 \times_U X')) \tag{III.8.23.13}$$

is an isomorphism. On the other hand, since $h$ is coherent, the sheaf $(V_1 \to U_1)^a$ of $\widetilde{\mathscr{E}}_U$ associated with $(V_1 \to U_1)$ is coherent by VI.10.5(i). Hence by virtue of VI.10.4, VI.10.5(ii), and ([**2**] VI 5.3), the isomorphism (III.8.23.13) shows that (III.8.23.11) is an isomorphism.

We deduce from the isomorphism (III.8.23.9) and from (VI.11.4.4) that the canonical homomorphism

$$\varinjlim_{i \in I^\circ} \Phi_U^{-1}(\overline{\mathscr{B}}_{/U}) \to \overline{\mathscr{B}}' \tag{III.8.23.14}$$

is an isomorphism. For every $U \in \mathrm{Ob}(I)$, the canonical morphism $U \to X$ induces a morphism of ringed topos

$$\rho_U \colon (\widetilde{\mathscr{E}}_U, \overline{\mathscr{B}}_{/U}) \to (\widetilde{E}, \overline{\mathscr{B}}). \tag{III.8.23.15}$$

Since the homomorphism $\rho_U^{-1}(\overline{\mathscr{B}}) \to \overline{\mathscr{B}}_{/U}$ is an isomorphism by virtue of III.8.21(i), the proposition follows from the isomorphism (III.8.23.14).

## III.9. Faltings topos over a trait

**III.9.1.** In this section, $S$ denotes the trait fixed in III.2.1. We fix a coherent $S$-scheme $X$ and an open subscheme $X^\circ$ of $X_\eta$. For any $X$-scheme $U$, we set

$$U^\circ = U \times_X X^\circ. \tag{III.9.1.1}$$

Recall that for every $S$-scheme $Y$, we have set $\overline{Y} = Y \times_S \overline{S}$ and for every integer $n \geq 1$, $Y_n = Y \times_S S_n$ (III.2.1.1). We denote by $j \colon X^\circ \to X$ and $\hbar \colon \overline{X} \to X$ the canonical morphisms. We suppose that $j$ is quasi-compact and that $\overline{X}$ is normal and locally irreducible (III.3.1). Note that $\overline{X}$ and $\overline{X}^\circ$ are coherent and étale-locally connected by III.3.2(iii). We propose to apply the constructions of III.8 to the morphisms in the top row of the following commutative diagram:

$$\begin{array}{ccccc} \overline{X}^\circ & \xrightarrow{j_{\overline{X}}} & \overline{X} & \xrightarrow{\hbar} & X \\ \downarrow & & \downarrow & \Box & \downarrow \\ \overline{\eta} & \longrightarrow & \overline{S} & \longrightarrow & S \end{array} \tag{III.9.1.2}$$

We denote by $E$ (resp. $\widetilde{E}$) the Faltings site (resp. topos) associated with the morphism $\hbar \circ j_{\overline{X}} \colon \overline{X}^\circ \to X$ (III.8.2), and by $\overline{\mathscr{B}}$ the ring of $\widetilde{E}$ associated with $\overline{X}$ (III.8.17), which is then an $\mathscr{O}_{\overline{K}}$-algebra. We denote by

$$\sigma \colon \widetilde{E} \to X_{\text{ét}}, \tag{III.9.1.3}$$

$$\rho \colon X_{\text{ét}} \overleftarrow{\times}_{X_{\text{ét}}} \overline{X}^\circ_{\text{ét}} \to \widetilde{E}, \tag{III.9.1.4}$$

the canonical morphisms (III.8.3.3) and (III.8.6.2), respectively.

For any integer $n \geq 0$, we set

$$\overline{\mathscr{B}}_n = \overline{\mathscr{B}}/p^n\overline{\mathscr{B}}. \tag{III.9.1.5}$$

For any $U \in \mathrm{Ob}(\mathbf{\acute{E}t}_{/X})$, we set $\overline{\mathscr{B}}_U = \overline{\mathscr{B}} \circ \alpha_{U!}$ (III.8.1.6) and

$$\overline{\mathscr{B}}_{U,n} = \overline{\mathscr{B}}_U/p^n\overline{\mathscr{B}}_U. \tag{III.9.1.6}$$

Note that the canonical homomorphism $\overline{\mathscr{B}}_{U,n} \to \overline{\mathscr{B}}_n \circ \alpha_{U!}$ is not in general an isomorphism; this is why we will not use the notation $\overline{\mathscr{B}}_{n,U}$. Nevertheless, the correspondences $\{U \mapsto p^n\overline{\mathscr{B}}_U\}$ and $\{U \mapsto \overline{\mathscr{B}}_{U,n}\}$ naturally form presheaves on $E$ (III.8.2.1), and the canonical morphisms

$$\{U \mapsto p^n\overline{\mathscr{B}}_U\}^a \to p^n\overline{\mathscr{B}}, \tag{III.9.1.7}$$

$$\{U \mapsto \overline{\mathscr{B}}_{U,n}\}^a \to \overline{\mathscr{B}}_n, \tag{III.9.1.8}$$

where the terms on the left-hand side denote the associated sheaves in $\widetilde{E}$, are isomorphisms by VI.8.2 and VI.8.9.

**Lemma III.9.2.** *The ring $\overline{\mathscr{B}}$ is $\mathscr{O}_{\overline{K}}$-flat.*

For every $(V \to U) \in \mathrm{Ob}(E)$, since $V$ is an $\overline{\eta}$-scheme, $\overline{U}^V$ is $\overline{S}$-flat by virtue of III.8.11(i). Consequently, $\overline{\mathscr{B}}$ does not have any $\mathscr{O}_{\overline{K}}$-torsion and is therefore $\mathscr{O}_{\overline{K}}$-flat ([**11**] Chapter VI § 3.6 Lemma 1).

**III.9.3.** Since $X_\eta$ is a subobject of the final object $X$ of $X_{\text{ét}}$ ([**2**] IV 8.3), $\sigma^*(X_\eta) = (\overline{X}^\circ \to X_\eta)^a$ is subobject of the final object of $\widetilde{E}$ (III.8.3.3). Recall that the topos $\widetilde{E}_{/\sigma^*(X_\eta)}$ is canonically equivalent to the Faltings topos associated with the morphism $\overline{X}^\circ \to X_\eta$ (VI.10.14). Denote by

$$\gamma \colon \widetilde{E}_{/\sigma^*(X_\eta)} \to \widetilde{E} \tag{III.9.3.1}$$

the localization morphism of $\widetilde{E}$ at $\sigma^*(X_\eta)$, which we identify with the functoriality morphism induced by the canonical injection $X_\eta \to X$ (III.8.5). We then have a sequence of three adjoint functors

$$\gamma_! \colon \widetilde{E}_{/\sigma^*(X_\eta)} \to \widetilde{E}, \quad \gamma^* \colon \widetilde{E} \to \widetilde{E}_{/\sigma^*(X_\eta)}, \quad \gamma_* \colon \widetilde{E}_{/\sigma^*(X_\eta)} \to \widetilde{E}, \tag{III.9.3.2}$$

in the sense that for any two consecutive functors in the sequence, the one on the right is the right adjoint of the other. The functors $\gamma_!$ and $\gamma_*$ are fully faithful ([**2**] IV 9.2.4).

We denote by $\widetilde{E}_s$ the closed subtopos of $\widetilde{E}$ complement of the open subtopos $\sigma^*(X_\eta)$, that is, the full subcategory of $\widetilde{E}$ made up of the sheaves $F$ such that $\gamma^*(F)$ is a final object of $\widetilde{E}_{/\sigma^*(X_\eta)}$ ([**2**] IV 9.3.5), and by

$$\delta \colon \widetilde{E}_s \to \widetilde{E} \tag{III.9.3.3}$$

the canonical embedding, that is, the morphism of topos such that $\delta_* \colon \widetilde{E}_s \to \widetilde{E}$ is the canonical injection functor. For any $F \in \mathrm{Ob}(\widetilde{E})$, we set $F_s = \delta^*(F)$.

We denote by $\mathbf{Pt}(\widetilde{E})$, $\mathbf{Pt}(\widetilde{E}_{/\sigma^*(X_\eta)})$, and $\mathbf{Pt}(\widetilde{E}_s)$ the categories of points of $\widetilde{E}$, $\widetilde{E}_{/\sigma^*(X_\eta)}$, and $\widetilde{E}_s$, respectively, and by

$$u \colon \mathbf{Pt}(\widetilde{E}_{/\sigma^*(X_\eta)}) \to \mathbf{Pt}(\widetilde{E}) \quad \text{and} \quad v \colon \mathbf{Pt}(\widetilde{E}_s) \to \mathbf{Pt}(\widetilde{E}) \tag{III.9.3.4}$$

the functors induced by $\gamma$ and $\delta$, respectively. These functors are fully faithful and every point of $\widetilde{E}$ belongs to the essential image of exactly one of these functors ([**2**] IV 9.7.2).

**Remark III.9.4.** By definition, for every $F \in \mathrm{Ob}(\widetilde{E}_s)$, the canonical projection

(III.9.4.1) $$\sigma^*(X_\eta) \times \delta_*(F) \to \sigma^*(X_\eta)$$

is an isomorphism. Hence there exists a unique morphism $\sigma^*(X_\eta) \to \delta_*(F)$ of $\widetilde{E}$; in other words, $\delta^*(\sigma^*(X_\eta))$ is an initial object of $\widetilde{E}_s$.

**Lemma III.9.5.** (i) *Let* $(\overline{y} \rightsquigarrow \overline{x})$ *be a point of* $X_{\text{ét}} \overleftarrow{\times}_{X_{\text{ét}}} \overline{X}^{\circ}_{\text{ét}}$ (III.8.6). *Then* $\rho(\overline{y} \rightsquigarrow \overline{x})$ *belongs to the essential image of* $u$ *(resp.* $v$*)* (III.9.3.4) *if and only if* $\overline{x}$ *lies over* $\eta$ *(resp.* $s$*).*

(ii) *The family of points of* $\widetilde{E}_{/\sigma^*(X_\eta)}$ *(resp.* $\widetilde{E}_s$*) defined by the family of points* $\rho(\overline{y} \rightsquigarrow \overline{x})$ *of* $\widetilde{E}$ *such that* $\overline{x}$ *lies over* $\eta$ *(resp.* $s$*) is conservative.*

(i) Indeed, $\rho(\overline{y} \rightsquigarrow \overline{x})$ belongs to the essential image of $u$ (resp. $v$) if and only if $(\sigma^*(X_\eta))_{\rho(\overline{y} \rightsquigarrow \overline{x})}$ is a singleton (resp. empty). On the other hand, we have a canonical isomorphism (VI.10.18.1)

(III.9.5.1) $$(\sigma^*(X_\eta))_{\rho(\overline{y} \rightsquigarrow \overline{x})} \stackrel{\sim}{\to} (X_\eta)_{\overline{x}},$$

giving the desired statement.

(ii) This follows from (i), VI.10.21, and ([**2**] IV 9.7.3).

**Lemma III.9.6.** *For every sheaf* $F = \{U \mapsto F_U\}$ *of* $\widetilde{E}$*, the following properties are equivalent:*

(i) *$F$ is an object of* $\widetilde{E}_s$*.*
(ii) *For every* $U \in \mathbf{\acute{E}t}_{/X_\eta}$*,* $F_U$ *is a final object of* $\overline{U}^{\circ}_{\text{fét}}$*, that is, is representable by* $\overline{U}^{\circ}$*.*
(iii) *For every point* $(\overline{y} \rightsquigarrow \overline{x})$ *of* $X_{\text{ét}} \overleftarrow{\times}_{X_{\text{ét}}} \overline{X}^{\circ}_{\text{ét}}$ (III.8.6) *such that* $\overline{x}$ *lies over* $\eta$*, the stalk* $F_{\rho(\overline{y} \rightsquigarrow \overline{x})}$ *of* $F$ *at* $\rho(\overline{y} \rightsquigarrow \overline{x})$ *is a singleton.*

Indeed, by VI.5.38, we have a canonical isomorphism

(III.9.6.1) $$\gamma^*(F) \stackrel{\sim}{\to} \{U' \mapsto F_{U'}\}, \quad (U' \in \mathrm{Ob}(\mathbf{\acute{E}t}_{/X_\eta})).$$

Since $\{U' \mapsto \overline{U}'^{\circ}\}$, for $U' \in \mathrm{Ob}(\mathbf{\acute{E}t}_{/X_\eta})$, is a final object of $\widetilde{E}_{/\sigma^*(X_\eta)}$ (III.8.4), conditions (i) and (ii) are equivalent. On the other hand, conditions (i) and (iii) are equivalent by virtue of III.9.5(ii).

**Lemma III.9.7.** *For every* $n \geq 0$*, the ring* $\overline{\mathscr{B}}_n$ (III.9.1.5) *is an object of* $\widetilde{E}_s$*.*

Indeed, for every point $(\overline{y} \rightsquigarrow \overline{x})$ of $X_{\text{ét}} \overleftarrow{\times}_{X_{\text{ét}}} \overline{X}^{\circ}_{\text{ét}}$ (III.8.6) such that $\overline{x}$ lies over $\eta$, the canonical homomorphism $\sigma^{-1}(\mathscr{O}_X) \to \overline{\mathscr{B}}$ (III.8.17) induces a homomorphism $\mathscr{O}_{X,\overline{x}} \to \overline{\mathscr{B}}_{\rho(\overline{y} \rightsquigarrow \overline{x})}$ (VI.10.18.1). Consequently, $p$ is invertible in $\overline{\mathscr{B}}_{\rho(\overline{y} \rightsquigarrow \overline{x})}$, giving the lemma by virtue of III.9.6.

**III.9.8.** We denote by $a\colon X_s \to X$ and $b\colon X_\eta \to X$ the canonical injections. The topos $\widetilde{E}_{/\sigma^*(X_\eta)}$ being canonically equivalent to the Faltings topos associated with the morphism $\overline{X}^{\circ} \to X_\eta$, we denote by

(III.9.8.1) $$\sigma_\eta\colon \widetilde{E}_{/\sigma^*(X_\eta)} \to X_{\eta,\text{ét}}$$

the canonical morphism (III.8.3.3). By (VI.10.12.6), the diagram

(III.9.8.2) $$\begin{array}{ccc} \widetilde{E}_{/\sigma^*(X_\eta)} & \xrightarrow{\sigma_\eta} & X_{\eta,\text{ét}} \\ \Big\downarrow{\scriptstyle \gamma} & & \Big\downarrow{\scriptstyle b} \\ \widetilde{E} & \xrightarrow{\sigma} & X_{\text{ét}} \end{array}$$

is commutative up to canonical isomorphism. By virtue of ([**2**] IV 9.4.3), there exists a morphism

(III.9.8.3) $$\sigma_s \colon \widetilde{E}_s \to X_{s,\text{ét}},$$

unique up to isomorphism, such that the diagram

(III.9.8.4) $$\begin{array}{ccc} \widetilde{E}_s & \xrightarrow{\sigma_s} & X_{s,\text{ét}} \\ {\scriptstyle \delta}\big\downarrow & & \big\downarrow{\scriptstyle a} \\ \widetilde{E} & \xrightarrow{\sigma} & X_{\text{ét}} \end{array}$$

is commutative up to isomorphism. By definition, we have a canonical isomorphism

(III.9.8.5) $$\sigma^* \circ a_* \xrightarrow{\sim} \delta_* \circ \sigma_s^*.$$

Since the functors $a_*$ and $\delta_*$ are exact, for every abelian group $F$ of $\widetilde{E}_s$ and every integer $i \geq 0$, we have a canonical isomorphism

(III.9.8.6) $$a_*(\mathrm{R}^i\sigma_{s*}(F)) \xrightarrow{\sim} \mathrm{R}^i\sigma_*(\delta_* F).$$

**III.9.9.** For any integer $\geq 1$, denote by $a \colon X_s \to X$, $a_n \colon X_s \to X_n$, $\iota_n \colon X_n \to X$, and $\overline{\iota}_n \colon \overline{X}_n \to \overline{X}$ the canonical injections. Since the residue field of $\mathscr{O}_K$ is algebraically closed, there exists a unique $S$-morphism $s \to \overline{S}$. This induces closed immersions $\overline{a} \colon X_s \to \overline{X}$ and $\overline{a}_n \colon X_s \to \overline{X}_n$ that lift $a$ and $a_n$, respectively.

(III.9.9.1) $$\begin{array}{ccccc} & & \overset{\overline{a}}{\frown\!\!\!\longrightarrow} & & \\ X_s & \xrightarrow[\overline{a}_n]{} & \overline{X}_n & \xrightarrow[\overline{\iota}_n]{} & \overline{X} \\ \Big\| & & \big\downarrow{\scriptstyle \hbar_n} & & \big\downarrow{\scriptstyle \hbar} \\ X_s & \xrightarrow{a_n} & X_n & \xrightarrow{\iota_n} & X \\ & & \underset{a}{\smile\!\!\!\longrightarrow} & & \end{array}$$

The canonical homomorphism $\sigma^{-1}(\hbar_*(\mathscr{O}_{\overline{X}})) \to \overline{\mathscr{B}}$ (III.8.17) induces a homomorphism (III.9.1.5)

(III.9.9.2) $$\sigma^{-1}(\iota_{n*}(\hbar_{n*}(\mathscr{O}_{\overline{X}_n}))) \to \overline{\mathscr{B}}_n.$$

Since $\overline{a}_n$ is a universal homeomorphism, we may view $\mathscr{O}_{\overline{X}_n}$ as a sheaf of $X_{s,\text{ét}}$ (cf. III.2.9). We can then identify the rings $\iota_{n*}(\hbar_{n*}(\mathscr{O}_{\overline{X}_n}))$ and $a_*(\mathscr{O}_{\overline{X}_n})$ of $X_{\text{ét}}$ and consequently the rings $\sigma^{-1}(\iota_{n*}(\hbar_{n*}(\mathscr{O}_{\overline{X}_n})))$ and $\delta_*(\sigma_s^*(\mathscr{O}_{\overline{X}_n}))$ of $\widetilde{E}_s$ (III.9.8.5). Since $\overline{\mathscr{B}}_n$ is an object of $\widetilde{E}_s$ (III.9.7), we may view (III.9.9.2) as a homomorphism of $\widetilde{E}_s$

(III.9.9.3) $$\sigma_s^*(\mathscr{O}_{\overline{X}_n}) \to \overline{\mathscr{B}}_n.$$

The morphism $\sigma_s$ (III.9.8.3) is therefore underlying a morphism of ringed topos, which we denote by

(III.9.9.4) $$\sigma_n \colon (\widetilde{E}_s, \overline{\mathscr{B}}_n) \to (X_{s,\text{ét}}, \mathscr{O}_{\overline{X}_n}).$$

**III.9.10.** We denote by $\breve{\overline{\mathscr{B}}}$ the ring $(\overline{\mathscr{B}}_{n+1})_{n\in\mathbb{N}}$ of $\widetilde{E}_s^{\mathbb{N}^\circ}$ (III.7.7) and by $\mathscr{O}_{\breve{\overline{X}}}$ the ring $(\mathscr{O}_{\overline{X}_{n+1}})_{n\in\mathbb{N}}$ of $X_{s,\mathrm{zar}}^{\mathbb{N}^\circ}$ or of $X_{s,\mathrm{\acute{e}t}}^{\mathbb{N}^\circ}$, depending on the context. This abuse of notation does not lead to any confusion (cf. III.2.9). By III.7.5, the morphisms $(\sigma_{n+1})_{n\in\mathbb{N}}$ (III.9.9.4) induce a morphism of ringed topos

$$\breve{\sigma}\colon (\widetilde{E}_s^{\mathbb{N}^\circ}, \breve{\overline{\mathscr{B}}}) \to (X_{s,\mathrm{\acute{e}t}}^{\mathbb{N}^\circ}, \mathscr{O}_{\breve{\overline{X}}}). \tag{III.9.10.1}$$

For every integer $n \geq 1$, we denote by

$$u_n\colon (X_{s,\mathrm{\acute{e}t}}, \mathscr{O}_{\overline{X}_n}) \to (X_{s,\mathrm{zar}}, \mathscr{O}_{\overline{X}_n}) \tag{III.9.10.2}$$

the canonical morphism of ringed topos (cf. III.2.9 and III.9.9). Then the morphisms $(u_{n+1})_{n\in\mathbb{N}}$ induce a morphism of ringed topos

$$\breve{u}\colon (X_{s,\mathrm{\acute{e}t}}^{\mathbb{N}^\circ}, \mathscr{O}_{\breve{\overline{X}}}) \to (X_{s,\mathrm{zar}}^{\mathbb{N}^\circ}, \mathscr{O}_{\breve{\overline{X}}}). \tag{III.9.10.3}$$

We denote by $\mathfrak{X}$ the formal scheme $p$-adic completion of $\overline{X}$. The topological space underlying $\mathfrak{X}$ is canonically isomorphic to $X_s$. It is ringed by the sheaf of topological rings $\mathscr{O}_{\mathfrak{X}}$ inverse limit of the sheaves of pseudo-discrete rings $(\mathscr{O}_{\overline{X}_{n+1}})_{n\in\mathbb{N}}$ (**[39]** 0.3.9.1). We denote by

$$\lambda\colon X_{s,\mathrm{zar}}^{\mathbb{N}^\circ} \to X_{s,\mathrm{zar}} \tag{III.9.10.4}$$

the morphism defined in (III.7.4.3). The sheaf of rings $\lambda_*(\mathscr{O}_{\breve{\overline{X}}})$ is canonically isomorphic to the sheaf of rings (without topologies) underlying $\mathscr{O}_{\mathfrak{X}}$ (**[39]** 0.3.9.1 and 0.3.2.6). We then view $\lambda$ as a morphism of ringed topos with structure sheaves $\mathscr{O}_{\breve{\overline{X}}}$ and $\mathscr{O}_{\mathfrak{X}}$, respectively. We denote by

$$\top\colon (\widetilde{E}_s^{\mathbb{N}^\circ}, \breve{\overline{\mathscr{B}}}) \to (X_{s,\mathrm{zar}}, \mathscr{O}_{\mathfrak{X}}) \tag{III.9.10.5}$$

the composed morphism of ringed topos $\lambda \circ \breve{u} \circ \breve{\sigma}$.

**III.9.11.** Let $g\colon X' \to X$ be a coherent morphism and $X'^{\rhd}$ an open subscheme of $X'^{\circ} = X^\circ \times_X X'$. For any $X'$-scheme $U'$, we set

$$U'^{\rhd} = U' \times_{X'} X'^{\rhd}. \tag{III.9.11.1}$$

We denote by $j'\colon X'^{\rhd} \to X'$ the canonical injection and by $\hbar'\colon \overline{X}' \to X'$ the canonical morphism (III.9.1.1). Suppose that $\overline{X}'$ is normal and locally irreducible. We denote by $E'$ (resp. $\widetilde{E}'$) the Faltings site (resp. topos) associated with the morphism $\hbar' \circ j'_{\overline{X}'}\colon \overline{X}'^{\rhd} \to X'$ (III.8.2), by $\overline{\mathscr{B}}'$ the ring of $\widetilde{E}'$ associated with $\overline{X}'$ (III.8.17), and by

$$\sigma'\colon (\widetilde{E}', \overline{\mathscr{B}}') \to (X'_{\mathrm{\acute{e}t}}, \hbar'_*(\mathscr{O}_{\overline{X}'})) \tag{III.9.11.2}$$

the canonical morphism of ringed topos (III.8.17). We denote by $\widetilde{E}'_s$ the closed subtopos of $\widetilde{E}'$ complement of the open subtopos $\sigma'^*(X'_\eta)$, by

$$\delta'\colon \widetilde{E}'_s \to \widetilde{E}' \tag{III.9.11.3}$$

the canonical embedding, and by

$$\sigma'_s\colon \widetilde{E}'_s \to X'_{s,\mathrm{\acute{e}t}} \tag{III.9.11.4}$$

the canonical morphism of topos (III.9.8.3). For any integer $n \geq 1$, we set $\overline{\mathscr{B}}'_n = \overline{\mathscr{B}}'/p^n\overline{\mathscr{B}}'$, and denote by

$$\sigma'_n\colon (\widetilde{E}'_s, \overline{\mathscr{B}}'_n) \to (X'_{s,\mathrm{\acute{e}t}}, \mathscr{O}_{\overline{X}'_n}) \tag{III.9.11.5}$$

the morphism of ringed topos induced by $\sigma'$ (III.9.9.4).

We denote by

$$\Phi\colon (\widetilde{E}', \overline{\mathscr{B}}') \to (\widetilde{E}, \overline{\mathscr{B}}) \tag{III.9.11.6}$$

the morphism of ringed topos deduced from $g$ by functoriality (III.8.20). By (VI.10.12.6) and the definitions III.8.17 and III.8.20, the diagram of morphisms of ringed topos

$$\begin{array}{ccc} (\widetilde{E}', \overline{\mathscr{B}}') & \xrightarrow{\ \Phi\ } & (\widetilde{E}, \overline{\mathscr{B}}) \\ {\scriptstyle \sigma'}\downarrow & & \downarrow{\scriptstyle \sigma} \\ (X'_{\text{ét}}, \hbar'_*(\mathscr{O}_{\overline{X}'})) & \xrightarrow{\ \overline{g}\ } & (X_{\text{ét}}, \hbar_*(\mathscr{O}_{\overline{X}})) \end{array} \tag{III.9.11.7}$$

where $\overline{g}$ is the morphism induced by $g$, is commutative up to canonical isomorphism, in the sense of ([**1**] 1.2.3). We have a canonical isomorphism $\Phi^*(\sigma^*(X_\eta)) \simeq \sigma'^*(X'_\eta)$ (III.8.5.2). By virtue of ([**2**] IV 9.4.3), there consequently exists a morphism of topos

$$\Phi_s\colon \widetilde{E}'_s \to \widetilde{E}_s, \tag{III.9.11.8}$$

unique up to isomorphism, such that the diagram

$$\begin{array}{ccc} \widetilde{E}'_s & \xrightarrow{\ \Phi_s\ } & \widetilde{E}_s \\ {\scriptstyle \delta'}\downarrow & & \downarrow{\scriptstyle \delta} \\ \widetilde{E}' & \xrightarrow{\ \Phi\ } & \widetilde{E} \end{array} \tag{III.9.11.9}$$

is commutative up to isomorphism, and even 2-Cartesian. It follows from (VI.10.12.6) and ([**2**] IV 9.4.3) that the diagram of morphisms of topos

$$\begin{array}{ccc} \widetilde{E}'_s & \xrightarrow{\ \Phi_s\ } & \widetilde{E}_s \\ {\scriptstyle \sigma'_s}\downarrow & & \downarrow{\scriptstyle \sigma_s} \\ X'_{s,\text{ét}} & \xrightarrow{\ g_s\ } & X_{s,\text{ét}} \end{array} \tag{III.9.11.10}$$

is commutative up to canonical isomorphism.

The canonical homomorphism $\Phi^{-1}(\overline{\mathscr{B}}) \to \overline{\mathscr{B}}'$ induces a homomorphism $\Phi_s^*(\overline{\mathscr{B}}_n) \to \overline{\mathscr{B}}'_n$. The morphism $\Phi_s$ is therefore underlying a morphism of ringed topos, which we denote by

$$\Phi_n\colon (\widetilde{E}'_s, \overline{\mathscr{B}}'_n) \to (\widetilde{E}_s, \overline{\mathscr{B}}_n). \tag{III.9.11.11}$$

It follows from (III.9.11.7) and (III.9.11.10) that the diagram of morphisms of ringed topos

$$\begin{array}{ccc} (\widetilde{E}'_s, \overline{\mathscr{B}}'_n) & \xrightarrow{\ \Phi_n\ } & (\widetilde{E}_s, \overline{\mathscr{B}}_n) \\ {\scriptstyle \sigma'_n}\downarrow & & \downarrow{\scriptstyle \sigma_n} \\ (X'_{s,\text{ét}}, \mathscr{O}_{\overline{X}'_n}) & \xrightarrow{\ \overline{g}_n\ } & (X_{s,\text{ét}}, \mathscr{O}_{\overline{X}_n}) \end{array} \tag{III.9.11.12}$$

where $\overline{g}_n$ is the morphism induced by $g$, is commutative up to canonical isomorphism, in the sense of ([**1**] 1.2.3).

Set $\breve{\overline{\mathscr{B}}}' = (\overline{\mathscr{B}}'_{n+1})_{n\in\mathbb{N}}$, which is a ring of $\widetilde{E}'^{\mathbb{N}^\circ}_s$. By III.7.5, the morphisms $(\Phi_{n+1})_{n\in\mathbb{N}}$ define a morphism of ringed topos

$$\breve{\Phi}\colon (\widetilde{E}'^{\mathbb{N}^\circ}_s, \breve{\overline{\mathscr{B}}}') \to (\widetilde{E}^{\mathbb{N}^\circ}_s, \breve{\overline{\mathscr{B}}}). \tag{III.9.11.13}$$

We denote by $\mathfrak{X}'$ the formal scheme $p$-adic completion of $\overline{X}'$ and by

$$\top' \colon (\widetilde{E}'^{\mathbb{N}^\circ}_s, \breve{\overline{\mathscr{B}}}') \to (\mathfrak{X}'_{\mathrm{zar}}, \mathscr{O}_{\mathfrak{X}'}) \tag{III.9.11.14}$$

the morphism of ringed topos defined in (III.9.10.5) with respect to $(X', X'^{\rhd})$. It immediately follows from (III.9.11.12) and from the functorial character of the morphisms $\breve{u}$ (III.9.10.3) and $\lambda$ (III.9.10.4) that the diagram of morphisms of ringed topos

$$\begin{array}{ccc} (\widetilde{E}'^{\mathbb{N}^\circ}_s, \breve{\overline{\mathscr{B}}}') & \xrightarrow{\breve{\Phi}} & (\widetilde{E}^{\mathbb{N}^\circ}_s, \breve{\overline{\mathscr{B}}}) \\ {\scriptstyle \top'}\downarrow & & \downarrow{\scriptstyle \top} \\ (X'_{s,\mathrm{zar}}, \mathscr{O}_{\mathfrak{X}'}) & \xrightarrow{\mathfrak{g}} & (X_{s,\mathrm{zar}}, \mathscr{O}_{\mathfrak{X}}) \end{array} \tag{III.9.11.15}$$

where $\mathfrak{g}$ is the morphism induced by $g$, is commutative up to canonical isomorphism, in the sense of ([**1**] 1.2.3). We deduce from this, for every $\breve{\overline{\mathscr{B}}}$-module $\mathscr{F}$ and every integer $q \geq 0$, a base change morphism

$$\mathfrak{g}^*(\mathrm{R}^q\top_*(\mathscr{F})) \to \mathrm{R}^q\top'_*(\breve{\Phi}^*(\mathscr{F})). \tag{III.9.11.16}$$

**Lemma III.9.12.** *We keep the assumptions of* III.9.11 *and moreover suppose that $g$ is étale and $X'^{\rhd} = X'^{\circ}$. Then $\Phi_s$* (III.9.11.8) *is canonically isomorphic to the localization morphism of $\widetilde{E}_s$ at $\sigma_s^*(X'_s)$.*

First note that $(\overline{X}'^{\rhd} \to X')$ is an object of $E$ and that the associated sheaf is none other than $\sigma^*(X')$. On the other hand, we have a canonical isomorphism $\delta^*(\sigma^*(X')) \xrightarrow{\sim} \sigma_s^*(X'_s)$ (III.9.8.4). Denote by $j\colon \widetilde{E}_{/\sigma^*(X')} \to \widetilde{E}$ (resp. $j_s\colon (\widetilde{E}_s)_{/\sigma_s^*(X'_s)} \to \widetilde{E}_s$) the localization morphism of $\widetilde{E}$ at $\sigma^*(X')$ (resp. of $\widetilde{E}_s$ at $\sigma_s^*(X'_s)$). By ([**2**] IV 5.10), the morphism $\delta$ induces a morphism

$$\delta_{/\sigma^*(X')}\colon (\widetilde{E}_s)_{/\sigma_s^*(X'_s)} \to \widetilde{E}_{/\sigma^*(X')} \tag{III.9.12.1}$$

that fits into a commutative diagram up to canonical isomorphism

$$\begin{array}{ccc} (\widetilde{E}_s)_{/\sigma_s^*(X'_s)} & \xrightarrow{j_s} & \widetilde{E}_s \\ {\scriptstyle \delta_{/\sigma^*(X')}}\downarrow & & \downarrow{\scriptstyle \delta} \\ \widetilde{E}_{/\sigma^*(X')} & \xrightarrow{j} & \widetilde{E} \end{array} \tag{III.9.12.2}$$

This diagram is in fact 2-Cartesian by ([**2**] IV 5.11). On the other hand, the topos $\widetilde{E}'$ and $\widetilde{E}_{/\sigma^*(X')}$ are canonically equivalent and the morphism $\Phi$ identifies with $j$ by virtue of VI.10.14. Since the diagram (III.9.11.9) is also 2-Cartesian, the lemma follows.

**Lemma III.9.13.** *We keep the assumptions of* III.9.11 *and moreover suppose that $X'^{\rhd} = X'^{\circ}$ and that one of the following two conditions is satisfied:*

(i) *$g$ is étale;*

(ii) *$X'$ is the strict localization of $X$ at a geometric point $\overline{x}$.*

*Then for every integer $n \geq 1$, the homomorphism $\Phi_s^*(\overline{\mathscr{B}}_n) \to \overline{\mathscr{B}}'_n$ is an isomorphism.*

This follows from III.8.21(i) and III.8.23.

**Proposition III.9.14.** *We keep the assumptions of* III.9.11 *and moreover suppose that $g$ is étale and $X'^{\rhd} = X'^{\circ}$, and denote also by $\lambda\colon \widetilde{E}^{\mathbb{N}^\circ}_s \to \widetilde{E}_s$ the morphism of topos defined in* (III.7.4.3). *Then $\breve{\Phi}$* (III.9.11.13) *is canonically isomorphic to the localization morphism of the ringed topos $(\widetilde{E}^{\mathbb{N}^\circ}_s, \breve{\overline{\mathscr{B}}})$ at $\lambda^*(\sigma_s^*(X'_s))$.*

Indeed, the morphism of topos $\Phi_s^{\mathbb{N}^\circ}\colon \widetilde{E}_s'^{\mathbb{N}^\circ} \to \widetilde{E}_s^{\mathbb{N}^\circ}$ identifies with the localization morphism of $\widetilde{E}_s^{\mathbb{N}^\circ}$ at $\lambda^*(\sigma_s^*(X_s'))$, by III.9.12 and III.7.6(ii). On the other hand, the canonical homomorphism $(\Phi_s^{\mathbb{N}^\circ})^*(\breve{\overline{\mathscr{B}}}) \to \breve{\overline{\mathscr{B}}}$ is an isomorphism by virtue of III.9.13 and (III.7.5.4); the proposition follows.

**Corollary III.9.15.** *We keep the assumptions of* III.9.11 *and moreover suppose that $g$ is an open immersion and $X'^{\rhd} = X'^{\circ}$. Then for every $\breve{\overline{\mathscr{B}}}$-module $\mathscr{F}$ and every integer $q \geq 0$, the base change* (III.9.11.16)

(III.9.15.1) $$\mathfrak{g}^*(\mathrm{R}^q\top_*(\mathscr{F})) \to \mathrm{R}^q\top'_*(\breve{\Phi}^*(\mathscr{F}))$$

*is an isomorphism.*

Indeed, the squares of the diagram of morphisms of topos

(III.9.15.2) $$\begin{array}{ccc} \widetilde{E}_s^{\mathbb{N}^\circ} & \xrightarrow{\lambda} & \widetilde{E}_s \\ \Big\downarrow{\scriptstyle \sigma_s^{\mathbb{N}^\circ}} & & \Big\downarrow{\scriptstyle \sigma_s} \\ X_{s,\text{ét}}^{\mathbb{N}^\circ} & \xrightarrow{\lambda} & X_{s,\text{ét}} \\ \Big\downarrow{\scriptstyle u_s^{\mathbb{N}^\circ}} & & \Big\downarrow{\scriptstyle u_s} \\ X_{s,\text{zar}}^{\mathbb{N}^\circ} & \xrightarrow{\lambda} & X_{s,\text{zar}} \end{array}$$

where $u_s$ is the canonical morphism (III.2.9.2) and $\lambda$ (abusively) denotes the morphisms defined in (III.7.4.3), are commutative up to canonical isomorphisms (III.7.5.4). Writing $\mathtt{T} = \lambda \circ u_s^{\mathbb{N}^\circ} \circ \sigma_s^{\mathbb{N}^\circ}$, which is the morphism of topos underlying $\top$ (III.9.10.5), we deduce from this an isomorphism

(III.9.15.3) $$\mathtt{T}^*(X_s') \xrightarrow{\sim} \lambda^*(\sigma_s^*(X_s')).$$

It then follows from (III.9.11.15) and III.9.14 that $\top'$ identifies with the morphism $\top_{/X_s'}$ (cf. **[2]** IV 5.10); the corollary follows.

## III.10. Higgs–Tate algebras

**III.10.1.** In this section, $(S, \mathscr{M}_S)$ denotes the logarithmic trait fixed in III.2.1 and

(III.10.1.1) $$f\colon (X, \mathscr{M}_X) \to (S, \mathscr{M}_S)$$

an adequate morphism of logarithmic schemes (III.4.7). We denote by $X^\circ$ the maximal open subscheme of $X$ where the logarithmic structure $\mathscr{M}_X$ is trivial; it is an open subscheme of $X_\eta$. For any usual $X$-scheme $U$, we set

(III.10.1.2) $$U^\circ = U \times_X X^\circ.$$

Recall that for every $S$-scheme $Y$, we have set $\overline{Y} = Y \times_S \overline{S}$, $\breve{\overline{Y}} = Y \times_S \breve{\overline{S}}$ and for every integer $n \geq 1$, $Y_n = Y \times_S S_n$ (III.2.1.1). We denote by $j\colon X^\circ \to X$ and $\hbar\colon \overline{X} \to X$ the canonical morphisms. To alleviate the notation, we set

(III.10.1.3) $$\widetilde{\Omega}^1_{X/S} = \Omega^1_{(X,\mathscr{M}_X)/(S,\mathscr{M}_S)},$$

which we view as a sheaf of $X_{\text{zar}}$ or $X_{\text{ét}}$, depending on the context (cf. III.2.9). For any integer $n \geq 1$, we set

(III.10.1.4) $$\widetilde{\Omega}^1_{\overline{X}_n/\overline{S}_n} = \widetilde{\Omega}^1_{X/S} \otimes_{\mathscr{O}_X} \mathscr{O}_{\overline{X}_n},$$

which we view as a sheaf of $X_{s,\text{zar}}$ or $X_{s,\text{ét}}$, depending on the context (cf. III.2.9 and III.9.9). We endow $\overline{X}$ and $\check{\overline{X}}$ with the logarithmic structures $\mathscr{M}_{\overline{X}}$ and $\mathscr{M}_{\check{\overline{X}}}$ inverse images of $\mathscr{M}_X$. We then have canonical isomorphisms (III.2.1)

$$(\overline{X}, \mathscr{M}_{\overline{X}}) \xrightarrow{\sim} (X, \mathscr{M}_X) \times_{(S,\mathscr{M}_S)} (\overline{S}, \mathscr{M}_{\overline{S}}), \tag{III.10.1.5}$$

$$(\check{\overline{X}}, \mathscr{M}_{\check{\overline{X}}}) \xrightarrow{\sim} (X, \mathscr{M}_X) \times_{(S,\mathscr{M}_S)} (\check{\overline{S}}, \mathscr{M}_{\check{\overline{S}}}), \tag{III.10.1.6}$$

where the product is taken indifferently in the category of logarithmic schemes or in that of fine logarithmic schemes.

A *smooth* $(\mathscr{A}_2(\overline{S}), \mathscr{M}_{\mathscr{A}_2(\overline{S})})$*-deformation* of $(\check{\overline{X}}, \mathscr{M}_{\check{\overline{X}}})$ (III.2.3) consists of a smooth morphism of fine logarithmic schemes

$$\widetilde{f}\colon (\widetilde{X}, \mathscr{M}_{\widetilde{X}}) \to (\mathscr{A}_2(\overline{S}), \mathscr{M}_{\mathscr{A}_2(\overline{S})}) \tag{III.10.1.7}$$

and an $(\check{\overline{S}}, \mathscr{M}_{\check{\overline{S}}})$-isomorphism

$$(\check{\overline{X}}, \mathscr{M}_{\check{\overline{X}}}) \xrightarrow{\sim} (\widetilde{X}, \mathscr{M}_{\widetilde{X}}) \times_{(\mathscr{A}_2(\overline{S}),\mathscr{M}_{\mathscr{A}_2(\overline{S})})} (\check{\overline{S}}, \mathscr{M}_{\check{\overline{S}}}), \tag{III.10.1.8}$$

where the product is taken indifferently in the category of logarithmic schemes or in that of fine logarithmic schemes (cf. **[50]** 3.14). For the remainder of this section, *we assume that there exists such a deformation* $(\widetilde{X}, \mathscr{M}_{\widetilde{X}})$*, which we fix.*

**III.10.2.** By III.4.2(iii), the schemes $X$ and $\overline{X}$ are normal and locally irreducible. On the other hand, since $X$ is noetherian, $j$ is quasi-compact. We can therefore apply the constructions of III.9 to the morphisms in the top row of the following commutative diagram:

$$\begin{array}{ccccc} \overline{X}^{\circ} & \xrightarrow{j_{\overline{X}}} & \overline{X} & \xrightarrow{\hbar} & X \\ \downarrow & & \downarrow & \square & \downarrow \\ \overline{\eta} & \longrightarrow & \overline{S} & \longrightarrow & S \end{array} \tag{III.10.2.1}$$

We denote by

$$\pi\colon E \to \mathbf{\acute{E}t}_{/X} \tag{III.10.2.2}$$

the Faltings fibered $\mathbb{U}$-site associated with the morphism $h = \hbar \circ j_{\overline{X}}\colon \overline{X}^{\circ} \to X$ (III.8.1). We endow $E$ with the covanishing topology and denote by $\widetilde{E}$ the topos of sheaves of $\mathbb{U}$-sets on $E$ (III.8.2), by $\overline{\mathscr{B}}$ the ring of $\widetilde{E}$ associated with $\overline{X}$ (III.8.17), and by

$$\sigma\colon \widetilde{E} \to X_{\text{ét}}, \tag{III.10.2.3}$$

$$\rho\colon X_{\text{ét}} \overleftarrow{\times}_{X_{\text{ét}}} \overline{X}^{\circ}_{\text{ét}} \to \widetilde{E}, \tag{III.10.2.4}$$

the canonical morphisms (III.8.3.3) and (III.8.6.2). We denote by $\widetilde{E}_s$ the closed subtopos of $\widetilde{E}$ complement of the open subtopos $\sigma^*(X_\eta)$ (III.9.3), by

$$\delta\colon \widetilde{E}_s \to \widetilde{E} \tag{III.10.2.5}$$

the canonical embedding, and by

$$\sigma_s\colon \widetilde{E}_s \to X_{s,\text{ét}} \tag{III.10.2.6}$$

the morphism of topos induced by $\sigma$ (III.9.8.3). For any integer $n \geq 0$, we set

$$\overline{\mathscr{B}}_n = \overline{\mathscr{B}}/p^n\overline{\mathscr{B}}. \tag{III.10.2.7}$$

For any $U \in \mathrm{Ob}(\mathbf{\acute{E}t}_{/X})$, we set $\overline{\mathscr{B}}_U = \overline{\mathscr{B}} \circ \alpha_{U!}$ (III.8.1.6) and

$$\overline{\mathscr{B}}_{U,n} = \overline{\mathscr{B}}_U / p^n \overline{\mathscr{B}}_U. \tag{III.10.2.8}$$

Note that the canonical homomorphism $\overline{\mathscr{B}}_{U,n} \to \overline{\mathscr{B}}_n \circ \alpha_{U!}$ is not in general an isomorphism (cf. (III.9.1.8)). Recall (III.9.7) that $\overline{\mathscr{B}}_n$ is a ring of $\widetilde{E}_s$. If $n \geq 1$, we denote by

$$\sigma_n \colon (\widetilde{E}_s, \overline{\mathscr{B}}_n) \to (X_{s,\text{ét}}, \mathscr{O}_{\overline{X}_n}) \tag{III.10.2.9}$$

the canonical morphism of ringed topos (III.9.9.4).

**III.10.3.** We denote by $\mathbf{P}$ the full subcategory of $\mathbf{\acute{E}t}_{/X}$ made up of the affine schemes $U$ such that the morphism $(U, \mathscr{M}_X|U) \to (S, \mathscr{M}_S)$ induced by $f$ admits an adequate chart (III.4.4). We endow $\mathbf{P}$ with the topology induced by that of $\mathbf{\acute{E}t}_{/X}$. Since $X$ is noetherian and therefore quasi-separated, any object of $\mathbf{P}$ is coherent over $X$. Consequently, $\mathbf{P}$ is a $\mathbb{U}$-small family that topologically generates the site $\mathbf{\acute{E}t}_{/X}$ and is stable under fibered products. We denote by

$$\pi_{\mathbf{P}} \colon E_{\mathbf{P}} \to \mathbf{P} \tag{III.10.3.1}$$

the fibered site deduced from $\pi$ (III.10.2.2) by base change by the canonical injection functor $\mathbf{P} \to \mathbf{\acute{E}t}_{/X}$. We endow $E_{\mathbf{P}}$ with the covanishing topology defined by $\pi_{\mathbf{P}}$ and we denote by $\widetilde{E}_{\mathbf{P}}$ the topos of sheaves of $\mathbb{U}$-sets on $E_{\mathbf{P}}$. By VI.5.21 and VI.5.22, the topology on $E_{\mathbf{P}}$ is induced by that on $E$ through the canonical projection functor $E_{\mathbf{P}} \to E$, and the latter induces by restriction an equivalence of categories

$$\widetilde{E} \xrightarrow{\sim} \widetilde{E}_{\mathbf{P}}. \tag{III.10.3.2}$$

**Remark III.10.4.** Let $U$ be an object of $\mathbf{P}$, $\overline{y}$ a generic geometric point of $\overline{U}^{\circ}$, and $\overline{R}^{\overline{y}}_U$ the ring defined in (III.8.13.2). Since the schemes $U$ and $\overline{U}$ are locally irreducible (III.3.3), they are the sums of the schemes induced on their irreducible components. Denote by $U^c$ (resp. $\overline{U}^{\star}$) the irreducible component of $U$ (resp. $\overline{U}$) containing $\overline{y}$. Likewise, $\overline{U}^{\circ}$ is the sum of the schemes induced on its irreducible components and $\overline{U}^{\star\circ} = \overline{U}^{\star} \times_X X^{\circ}$ is the irreducible component of $\overline{U}^{\circ}$ containing $\overline{y}$. Then $U^c$ is naturally an object of $\mathbf{P}$ over $U$, and the canonical homomorphism $\overline{R}^{\overline{y}}_U \to \overline{R}^{\overline{y}}_{U^c}$ is a $\pi_1(\overline{U}^{\star\circ}, \overline{y})$-equivariant isomorphism. If $U^c_s = \emptyset$, then $\overline{R}^{\overline{y}}_U$ is an $\overline{K}$-algebra. Suppose $U^c_s \neq \emptyset$, so that the morphism $(U^c, \mathscr{M}_X|U^c) \to (S, \mathscr{M}_S)$ induced by $f$ satisfies the conditions of II.6.2. The algebra $\overline{R}^{\overline{y}}_U$ endowed with the action of $\pi_1(\overline{U}^{\star\circ}, \overline{y})$ then corresponds to the algebra $\overline{R}$ endowed with the action of $\Delta$ introduced in II.6.7 and II.6.10; whence the notation. The coordinate ring of the affine scheme $\overline{U}^{\star}$ corresponds to the algebra $R_1$ in loc. cit. by II.6.8(i).

**III.10.5.** We denote by $\mathbf{Q}$ the full subcategory of $\mathbf{P}$ (III.10.3) made up of the connected affine schemes $U$ such that there exists a fine and saturated chart $M \to \Gamma(U, \mathscr{M}_X)$ for $(U, \mathscr{M}_X|U)$ that induces an isomorphism

$$M \xrightarrow{\sim} \Gamma(U, \mathscr{M}_X)/\Gamma(U, \mathscr{O}_X^{\times}). \tag{III.10.5.1}$$

This chart is a priori independent of the adequate chart required in the definition of the objects of $\mathbf{P}$. We endow $\mathbf{Q}$ with the topology induced by that of $\mathbf{\acute{E}t}_{/X}$. It follows from II.5.17 that $\mathbf{Q}$ is a topologically generating subcategory of $\mathbf{\acute{E}t}_{/X}$. We denote by

$$\pi_{\mathbf{Q}} \colon E_{\mathbf{Q}} \to \mathbf{Q} \tag{III.10.5.2}$$

the fibered site deduced from $\pi$ (III.10.2.2) by base change by the canonical injection functor $\mathbf{Q} \to \mathbf{\acute{E}t}_{/X}$. The canonical projection functor $E_{\mathbf{Q}} \to E$ is fully faithful and the category $E_{\mathbf{Q}}$ is $\mathbb{U}$-small and topologically generates the site $E$. We endow $E_{\mathbf{Q}}$ with the topology induced by that on $E$. By restriction, the topos $\widetilde{E}$ is then equivalent to the category of sheaves of $\mathbb{U}$-sets on $E_{\mathbf{Q}}$ ([**2**] III 4.1). Notice that in general, since $\mathbf{Q}$ is not stable under fibered products, we cannot speak of the covanishing topology on $E_{\mathbf{Q}}$ associated with $\pi_{\mathbf{Q}}$, and even less apply VI.5.21 and VI.5.22.

**III.10.6.** We denote by $\widehat{E}_{\mathbf{Q}}$ the category of presheaves of $\mathbb{U}$-sets on $E_{\mathbf{Q}}$ and by

(III.10.6.1) $$\mathscr{P}^{\vee}_{\mathbf{Q}} \to \mathbf{Q}^{\circ}$$

the fibered category obtained by associating with each $U \in \mathrm{Ob}(\mathbf{Q})$ the category $(\mathbf{\acute{E}t}_{\mathrm{f}/\overline{U}^{\circ}})^{\wedge}$ of presheaves of $\mathbb{U}$-sets on $\mathbf{\acute{E}t}_{\mathrm{f}/\overline{U}^{\circ}}$, and with each morphism $f\colon U' \to U$ of $\mathbf{Q}$ the functor

(III.10.6.2) $$\overline{f}^{\circ}_{\mathrm{f\acute{e}t}*}\colon (\mathbf{\acute{E}t}_{\mathrm{f}/\overline{U}'^{\circ}})^{\wedge} \to (\mathbf{\acute{E}t}_{\mathrm{f}/\overline{U}^{\circ}})^{\wedge}$$

obtained by composing with the inverse image functor $\mathbf{\acute{E}t}_{\mathrm{f}/\overline{U}^{\circ}} \to \mathbf{\acute{E}t}_{\mathrm{f}/\overline{U}'^{\circ}}$ under the morphism $\overline{f}^{\circ}: \overline{U}'^{\circ} \to \overline{U}^{\circ}$ deduced from $f$; in other words, $\mathscr{P}^{\vee}_{\mathbf{Q}}$ is the fibered category on $\mathbf{Q}^{\circ}$ deduced from the fibered category (III.8.1.9) by base change by the canonical injection functor $\mathbf{Q} \to \mathbf{\acute{E}t}_{/X}$. For any $U \in \mathrm{Ob}(\mathbf{Q})$, we denote by $\alpha_{U!}\colon \mathbf{\acute{E}t}_{\mathrm{f}/\overline{U}^{\circ}} \to E_{\mathbf{Q}}$ the canonical functor (III.8.1.6). By ([**37**] VI 12; cf. also [**1**] 1.1.2), we have an equivalence of categories

(III.10.6.3) $$\begin{array}{rcl} \widehat{E}_{\mathbf{Q}} & \xrightarrow{\sim} & \mathbf{Hom}_{\mathbf{Q}^{\circ}}(\mathbf{Q}^{\circ}, \mathscr{P}^{\vee}_{\mathbf{Q}}) \\ F & \mapsto & \{U \mapsto F \circ \alpha_{U!}\}. \end{array}$$

We will, from now on, identify $F$ with the section $\{U \mapsto F \circ \alpha_{U!}\}$ that is associated with it by this equivalence.

Since $E_{\mathbf{Q}}$ is a topologically generating subcategory of $E$, the "associated sheaf" functor on $E_{\mathbf{Q}}$ induces a functor that we denote also by

(III.10.6.4) $$\widehat{E}_{\mathbf{Q}} \to \widetilde{E}, \quad F \mapsto F^{a}.$$

Let $F = \{W \mapsto G_W\}$ ($W \in \mathrm{Ob}(\mathbf{\acute{E}t}_{/X})$) be an object of $\widehat{E}$ (III.8.2.1) and $F_{\mathbf{Q}} = \{U \mapsto G_U\}$ ($U \in \mathrm{Ob}(\mathbf{Q})$) the object of $\widehat{E}_{\mathbf{Q}}$ obtained by restricting $F$ to $E_{\mathbf{Q}}$. It immediately follows from ([**2**] II 3.0.4) and from the definition of the "associated sheaf" functor ([**2**] II 3.4) that we have a canonical isomorphism of $\widetilde{E}$

(III.10.6.5) $$(F_{\mathbf{Q}})^{a} \xrightarrow{\sim} F^{a}.$$

**Remark III.10.7.** Let $F = \{U \mapsto F_U\}$ be a presheaf on $E_{\mathbf{Q}}$. For each $U \in \mathrm{Ob}(\mathbf{Q})$, denote by $F^{a}_{U}$ the sheaf of $\overline{U}^{\circ}_{\mathrm{f\acute{e}t}}$ associated with $F_U$. Then $\{U \mapsto F^{a}_{U}\}$ is a presheaf on $E_{\mathbf{Q}}$ and we have a canonical morphism $\{U \mapsto F_U\} \to \{U \mapsto F^{a}_{U}\}$ of $\widehat{E}_{\mathbf{Q}}$, inducing an isomorphism between the associated sheaves. This assertion does not follow directly from VI.5.17 because $\mathbf{Q}$ is not stable under fibered products. However, the proof is similar. The only point that needs to be verified is the isomorphism (VI.5.17.3). Let $G = \{W \mapsto G_W\}$ ($W \in \mathrm{Ob}(\mathbf{\acute{E}t}_{/X})$) be an object of $\widetilde{E}$ and $G_{\mathbf{Q}} = \{U \mapsto G_U\}$ ($U \in \mathrm{Ob}(\mathbf{Q})$) the object of $\widehat{E}_{\mathbf{Q}}$ obtained by restricting $G$ to $E_{\mathbf{Q}}$. For every $U \in \mathrm{Ob}(\mathbf{Q})$, $G_U$ is a sheaf of $\overline{U}^{\circ}_{\mathrm{f\acute{e}t}}$ (III.8.2.2). Consequently, the map

(III.10.7.1) $$\mathrm{Hom}_{\widehat{E}_{\mathbf{Q}}}(\{U \mapsto F^{a}_{U}\}, \{U \mapsto G_U\}) \to \mathrm{Hom}_{\widehat{E}_{\mathbf{Q}}}(\{U \mapsto F_U\}, \{U \mapsto G_U\})$$

induced by the canonical morphism $\{U \mapsto F_U\} \to \{U \mapsto F^{a}_{U}\}$ is an isomorphism. The assertion follows.

**III.10.8.** Let $(\overline{y} \rightsquigarrow \overline{x})$ be a point of $X_{\text{ét}} \overleftarrow{\times}_{X_{\text{ét}}} \overline{X}^{\circ}_{\text{ét}}$ (III.8.6) such that $\overline{x}$ lies over $s$ and $X'$ the strict localization of $X$ at $\overline{x}$. Recall that giving a neighborhood of the point of $X_{\text{ét}}$ associated with $\overline{x}$ in the site $\mathbf{\acute{E}t}_{/X}$ (resp. $\mathbf{P}$ (III.10.3), resp. $\mathbf{Q}$ (III.10.5)) is equivalent to giving an $\overline{x}$-pointed étale $X$-scheme (resp. of $\mathbf{P}$, resp. of $\mathbf{Q}$) ([**2**] IV 6.8.2). These objects naturally form a cofiltered category, which we denote by $\mathfrak{V}_{\overline{x}}$ (resp. $\mathfrak{V}_{\overline{x}}(\mathbf{P})$, resp. $\mathfrak{V}_{\overline{x}}(\mathbf{Q})$). The categories $\mathfrak{V}_{\overline{x}}(\mathbf{P})$ and $\mathfrak{V}_{\overline{x}}(\mathbf{Q})$ are $\mathbb{U}$-small, and the canonical injection functors $\mathbf{Q} \to \mathbf{P} \to \mathbf{\acute{E}t}_{/X}$ induce fully faithful cofinal functors $\mathfrak{V}_{\overline{x}}(\mathbf{Q}) \to \mathfrak{V}_{\overline{x}}(\mathbf{P}) \to \mathfrak{V}_{\overline{x}}$. For any object $(U, \mathfrak{p} \colon \overline{x} \to U)$ of $\mathfrak{V}_{\overline{x}}$, we denote also by $\mathfrak{p} \colon X' \to U$ the $X$-morphism deduced by $\mathfrak{p}$ ([**2**] VIII 7.3) and we set

$$\overline{\mathfrak{p}}^{\circ} = \mathfrak{p} \times_X \overline{X}^{\circ} \colon \overline{X}'^{\circ} \to \overline{U}^{\circ}. \tag{III.10.8.1}$$

By virtue of III.3.7, $\overline{X}'$ is normal and strictly local (and in particular integral). The $X$-morphism $u \colon \overline{y} \to X'$ defining $(\overline{y} \rightsquigarrow \overline{x})$ lifts to a $\overline{X}^{\circ}$-morphism $v \colon \overline{y} \to \overline{X}'^{\circ}$ and therefore induces a geometric point of $\overline{X}'^{\circ}$ that we (abusively) denote also by $\overline{y}$. For any $(U, \mathfrak{p}) \in \mathrm{Ob}(\mathfrak{V}_{\overline{x}})$, we (abusively) denote also by $\overline{y}$ the geometric point $\overline{\mathfrak{p}}^{\circ}(v(\overline{y}))$ of $\overline{U}^{\circ}$. Since $\overline{U}$ is locally irreducible (III.3.3), it is the sum of the schemes induced on its irreducible components. Denote by $\overline{U}^{\star}$ the irreducible component of $\overline{U}$ containing $\overline{y}$. Likewise, $\overline{U}^{\circ}$ is the sum of the schemes induced on its irreducible component and $\overline{U}^{\star\circ} = \overline{U}^{\star} \times_X X^{\circ}$ is the irreducible component on $\overline{U}^{\circ}$ containing $\overline{y}$. The morphism $\overline{\mathfrak{p}}^{\circ} \colon \overline{X}'^{\circ} \to \overline{U}^{\circ}$ therefore factors through $\overline{U}^{\star\circ}$.

We denote by

$$\varphi_{\overline{x}} \colon \widetilde{E} \to \overline{X}'^{\circ}_{\text{fét}} \tag{III.10.8.2}$$

the canonical functor (III.8.8.4) and by

$$\nu_{\overline{y}} \colon \overline{X}'^{\circ}_{\text{fét}} \xrightarrow{\sim} \mathbf{B}_{\pi_1(\overline{X}'^{\circ}, \overline{y})} \tag{III.10.8.3}$$

the fiber functor of $\overline{X}'^{\circ}_{\text{fét}}$ at $\overline{y}$ (III.2.10.3). By (III.8.8.6), we have a canonical isomorphism

$$\varphi_{\overline{x}}(\overline{\mathscr{B}}) \xrightarrow{\sim} \lim_{\substack{\longrightarrow \\ (U,\mathfrak{p}) \in \mathfrak{V}_{\overline{x}}^{\circ}}} (\overline{\mathfrak{p}}^{\circ})^{*}_{\text{fét}}(\overline{\mathscr{B}}_U), \tag{III.10.8.4}$$

where $\overline{\mathscr{B}}_U$ is the sheaf of $\overline{U}^{\circ}_{\text{fét}}$ defined in (III.8.10.3). In view of (III.8.15.1), we deduce from this an isomorphism of $\mathscr{O}_{\overline{K}}$-algebras of $\mathbf{B}_{\pi_1(\overline{X}'^{\circ}, \overline{y})}$

$$\nu_{\overline{y}}(\varphi_{\overline{x}}(\overline{\mathscr{B}})) \xrightarrow{\sim} \lim_{\substack{\longrightarrow \\ (U,\mathfrak{p}) \in \mathfrak{V}_{\overline{x}}^{\circ}}} \overline{R}^{\overline{y}}_U, \tag{III.10.8.5}$$

where $\overline{R}^{\overline{y}}_U$ is the $\mathscr{O}_{\overline{K}}$-algebra of $\mathbf{B}_{\pi_1(\overline{U}^{\star\circ}, \overline{y})}$ defined in (III.8.13.2). By VI.10.31 and VI.9.9, the ring underlying $\nu_{\overline{y}}(\varphi_{\overline{x}}(\overline{\mathscr{B}}))$ is canonically isomorphic to the stalk $\overline{\mathscr{B}}_{\rho(\overline{y} \rightsquigarrow \overline{x})}$.

**Remark III.10.9.** For every geometric point $\overline{x}$ of $X$ over $s$, there exists a point $(\overline{y} \rightsquigarrow \overline{x})$ of $X_{\text{ét}} \overleftarrow{\times}_{X_{\text{ét}}} \overline{X}^{\circ}_{\text{ét}}$ (III.8.6). Indeed, denote by $X'$ the strict localization of $X$ at $\overline{x}$. By III.3.7, $\overline{X}'$ is normal and strictly local (and in particular integral). Since $X^{\circ}$ is schematically dense in $X$ by III.4.2(iv), $\overline{X}'^{\circ}$ is integral and nonempty ([**42**] 11.10.5). Let $v \colon \overline{y} \to \overline{X}'^{\circ}$ be a geometric point of $\overline{X}'^{\circ}$. We denote also by $\overline{y}$ the geometric point of $\overline{X}^{\circ}$ and by $u \colon \overline{y} \to X'$ the $X$-morphism induced by $v$. We thus obtain a point $(\overline{y} \rightsquigarrow \overline{x})$ of $X_{\text{ét}} \overleftarrow{\times}_{X_{\text{ét}}} \overline{X}^{\circ}_{\text{ét}}$.

**Proposition III.10.10.** *Let $(\overline{y} \rightsquigarrow \overline{x})$ be a point of $X_{\text{ét}} \overleftarrow{\times}_{X_{\text{ét}}} \overline{X}^{\circ}_{\text{ét}}$ (III.8.6) such that $\overline{x}$ lies over $s$ and $X'$ the strict localization of $X$ at $\overline{x}$. Then:*

(i) *The stalk* $\overline{\mathscr{B}}_{\rho(\overline{y}\rightsquigarrow\overline{x})}$ *of* $\overline{\mathscr{B}}$ *at* $\rho(\overline{y}\rightsquigarrow\overline{x})$ *is a normal and strictly local ring.*
(ii) *The stalk* $\hbar_*(\mathscr{O}_{\overline{X}})_{\overline{x}}$ *of* $\hbar_*(\mathscr{O}_{\overline{X}})$ *at* $\overline{x}$ *is a normal and strictly local ring.*
(iii) *The homomorphism*

$$\hbar_*(\mathscr{O}_{\overline{X}})_{\overline{x}} \to \overline{\mathscr{B}}_{\rho(\overline{y}\rightsquigarrow\overline{x})} \tag{III.10.10.1}$$

*induced by the canonical homomorphism* $\sigma^{-1}(\hbar_*(\mathscr{O}_{\overline{X}})) \to \overline{\mathscr{B}}$ (III.8.17) *is injective and local.*

Denote by $j'\colon X'^{\circ} \to X'$ the canonical injection and by $g$, $\overline{g}$, and $\hbar'$ the canonical arrows of the following Cartesian diagram:

$$\begin{array}{ccc} \overline{X}' & \xrightarrow{\ \overline{g}\ } & \overline{X} \\ {\scriptstyle \hbar'}\downarrow & \Box & \downarrow{\scriptstyle \hbar} \\ X' & \xrightarrow{\ g\ } & X \end{array} \tag{III.10.10.2}$$

By III.3.7, $\overline{X}'$ is normal and strictly local (and in particular integral). We denote by $E'$ (resp. $\widetilde{E}'$) the Faltings site (resp. topos) associated with the morphism $h' = \hbar' \circ j'_{\overline{X}'}\colon \overline{X}'^{\circ} \to X'$ (III.8.2) and by $\overline{\mathscr{B}}'$ the ring of $\widetilde{E}'$ associated with $\overline{X}'$ (III.8.17). We denote by

$$\sigma'\colon \widetilde{E}' \to X'_{\text{ét}}, \tag{III.10.10.3}$$

$$\rho'\colon X'_{\text{ét}} \overleftarrow{\times}_{X'_{\text{ét}}} \overline{X}'^{\circ}_{\text{ét}} \to \widetilde{E}', \tag{III.10.10.4}$$

the canonical morphisms (III.8.3.3) and (III.8.6.2), respectively, and by

$$\Phi\colon (\widetilde{E}', \overline{\mathscr{B}}') \to (\widetilde{E}, \overline{\mathscr{B}}) \tag{III.10.10.5}$$

the morphism of ringed topos deduced from $g$ by functoriality (III.8.20). The $X$-morphism $u\colon \overline{y} \to X'$ defining $(\overline{y} \rightsquigarrow \overline{x})$ induces an $X'$-morphism $v\colon \overline{y} \to \overline{X}'^{\circ}$. We (abusively) denote also by $\overline{x}$ the closed point of $X'$, by $\overline{y}$ the geometric point of $\overline{X}'^{\circ}$ defined by $v$, and by $(\overline{y} \rightsquigarrow \overline{x})$ the point of $X'_{\text{ét}} \overleftarrow{\times}_{X'_{\text{ét}}} \overline{X}'^{\circ}_{\text{fét}}$ defined by $u$. The points $\rho(\overline{y} \rightsquigarrow \overline{x})$ and $\Phi(\rho'(\overline{y} \rightsquigarrow \overline{x}))$ of $\widetilde{E}$ are then canonically isomorphic (VI.10.17).

(i) Since the canonical homomorphism $\Phi^{-1}(\overline{\mathscr{B}}) \to \overline{\mathscr{B}}'$ is an isomorphism by III.8.23, it induces an isomorphism

$$\overline{\mathscr{B}}_{\rho(\overline{y}\rightsquigarrow\overline{x})} \xrightarrow{\sim} \overline{\mathscr{B}}'_{\rho'(\overline{y}\rightsquigarrow\overline{x})}. \tag{III.10.10.6}$$

The statement then follows from III.8.19(i).

(ii) By III.3.6, the canonical morphism $\overline{g}^{-1}(\mathscr{O}_{\overline{X}}) \to \mathscr{O}_{\overline{X}'}$ is an isomorphism of $\overline{X}'_{\text{ét}}$. By ([**2**] VIII 5.2), we deduce from this a canonical isomorphism

$$\hbar_*(\mathscr{O}_{\overline{X}})_{\overline{x}} \xrightarrow{\sim} \Gamma(\overline{X}', \mathscr{O}_{\overline{X}'}). \tag{III.10.10.7}$$

The statement follows by virtue of III.3.7.

(iii) The diagram of morphisms of topos

$$\begin{array}{ccc} \widetilde{E}' & \xrightarrow{\ \Phi\ } & \widetilde{E} \\ {\scriptstyle \sigma'}\downarrow & & \downarrow{\scriptstyle \sigma} \\ X'_{\text{ét}} & \xrightarrow{\ g\ } & X_{\text{ét}} \end{array} \tag{III.10.10.8}$$

is commutative up to canonical isomorphism (VI.10.12.6). Moreover, the diagram

$$\begin{array}{ccccc} \sigma'^{-1}(g^{-1}(\hbar_*(\mathscr{O}_{\overline{X}}))) & = & \Phi^{-1}(\sigma^{-1}(\hbar_*\mathscr{O}_{\overline{X}})) & \longrightarrow & \Phi^{-1}(\overline{\mathscr{B}}) \\ \downarrow{\scriptstyle c} & & & & \downarrow{\scriptstyle b} \\ \sigma'^{-1}(\hbar'_*(\overline{g}^{-1}(\mathscr{O}_{\overline{X}}))) & \xrightarrow{a} & \sigma'^{-1}(\hbar'_*(\mathscr{O}_{\overline{X}'})) & \longrightarrow & \overline{\mathscr{B}}' \end{array} \tag{III.10.10.9}$$

where $c$ is the base change morphism with respect to the diagram (III.10.10.2) and the other arrows are the canonical morphisms, is commutative. Since $\hbar$ is integral, $c$ is an isomorphism ([2] VIII 5.6). On the other hand, $a$ is an isomorphism (III.3.6) and $b$ is an isomorphism (III.8.23). The statement then follows from III.8.19(iii).

**Corollary III.10.11.** (i) *The topos $\widetilde{E}_s$ is locally ringed by $\overline{\mathscr{B}}_s = \overline{\mathscr{B}}|\widetilde{E}_s$.*

(ii) *For every integer $n \geq 1$, $\sigma_n : (\widetilde{E}_s, \overline{\mathscr{B}}_n) \to (X_{s,\text{ét}}, \mathscr{O}_{\overline{X}_n})$ (III.10.2.9) is a morphism of locally ringed topos.*

This follows from III.9.5, III.10.10, and ([2] IV 13.9).

**Proposition III.10.12.** *The absolute Frobenius endomorphism of $\overline{\mathscr{B}}_1$ is surjective.*

Let $(\overline{y} \rightsquigarrow \overline{x})$ be a point of $X_{\text{ét}} \overleftarrow{\times}_{X_{\text{ét}}} \overline{X}^{\circ}_{\text{ét}}$ (III.8.6) such that $\overline{x}$ lies over $s$ and that $\overline{y}$ is a generic geometric point of $\overline{X}^{\circ}$, and $X'$ the strict localization of $X$ at $\overline{x}$. Using the notation of III.10.8, we have a canonical isomorphism (III.10.8.5)

$$\nu_{\overline{y}}(\varphi_{\overline{x}}(\overline{\mathscr{B}}_1)) \xrightarrow{\sim} \lim_{\substack{\longrightarrow \\ (U,\mathfrak{p}) \in \mathfrak{V}_{\overline{x}}(\mathbf{P})^{\circ}}} \overline{R}^{\overline{y}}_U / p\overline{R}^{\overline{y}}_U. \tag{III.10.12.1}$$

By the functoriality of the isomorphism (III.8.8.6), the latter is compatible with the absolute Frobenius endomorphisms of $\overline{\mathscr{B}}_1$ and $\overline{R}^{\overline{y}}_U / p\overline{R}^{\overline{y}}_U$. For every $(U, \mathfrak{p}) \in \text{Ob}(\mathfrak{V}_{\overline{x}}(\mathbf{P}))$, the absolute Frobenius endomorphism of $\overline{R}^{\overline{y}}_U / p\overline{R}^{\overline{y}}_U$ is surjective by virtue of III.10.4 and II.9.10. The proposition follows by virtue of III.9.5 and III.9.7.

**III.10.13.** Let $Y$ be an object of $\mathbf{Q}$ (III.10.5) such that $Y_s \neq \emptyset$ and $\overline{y}$ a geometric point of $\overline{Y}^{\circ}$. Since $\overline{Y}$ is locally irreducible (III.3.3), it is the sum of the schemes induced on its irreducible components. We denote by $\overline{Y}^{\star}$ the irreducible component of $\overline{Y}$ containing $\overline{y}$. Likewise, $\overline{Y}^{\circ}$ is the sum of the schemes induced on its irreducible components, and $\overline{Y}^{\star\circ} = \overline{Y}^{\star} \times_X X^{\circ}$ is the irreducible component of $\overline{Y}^{\circ}$ containing $\overline{y}$. We denote by $\overline{R}^{\overline{y}}_Y$ the ring defined in (III.8.13.2) and by $\widehat{\overline{R}}^{\overline{y}}_Y$ its $p$-adic Hausdorff completion. We set (III.2.2.1)

$$\mathscr{R}_{\overline{R}^{\overline{y}}_Y} = \lim_{\substack{\longleftarrow \\ x \mapsto x^p}} \overline{R}^{\overline{y}}_Y / p\overline{R}^{\overline{y}}_Y, \tag{III.10.13.1}$$

and we denote by $\theta_Y : \mathrm{W}(\mathscr{R}_{\overline{R}^{\overline{y}}_Y}) \to \widehat{\overline{R}}^{\overline{y}}_Y$ Fontaine's homomorphism defined in (II.9.3.4). We set

$$\mathscr{A}_2(\overline{R}^{\overline{y}}_Y) = \mathrm{W}(\mathscr{R}_{\overline{R}^{\overline{y}}_Y}) / \ker(\theta_Y)^2, \tag{III.10.13.2}$$

and denote also by $\theta_Y : \mathscr{A}_2(\overline{R}^{\overline{y}}_Y) \to \widehat{\overline{R}}^{\overline{y}}_Y$ the homomorphism induced by $\theta_Y$ (II.9.3.5). Finally, we set

$$\overline{Y}^{\overline{y}} = \text{Spec}(\overline{R}^{\overline{y}}_Y), \tag{III.10.13.3}$$

$$\widehat{\overline{Y}}^{\overline{y}} = \text{Spec}(\widehat{\overline{R}}^{\overline{y}}_Y), \tag{III.10.13.4}$$

$$\mathscr{A}_2(\overline{Y}^{\overline{y}}) = \text{Spec}(\mathscr{A}_2(\overline{R}^{\overline{y}}_Y)). \tag{III.10.13.5}$$

Note that since $\overline{Y}$ is affine, $\overline{Y}^{\overline{y}}$ is none other than the scheme defined in III.8.14. We endow $\overline{Y}^{\overline{y}}$ (resp. $\widehat{\overline{Y}}^{\overline{y}}$) with the logarithmic structure $\mathscr{M}_{\overline{Y}^{\overline{y}}}$ (resp. $\mathscr{M}_{\widehat{\overline{Y}}^{\overline{y}}}$) inverse image of $\mathscr{M}_X$ (II.5.10) and $\mathscr{A}_2(\overline{Y}^{\overline{y}})$ with the logarithmic structure $\mathscr{M}_{\mathscr{A}_2(\overline{Y}^{\overline{y}})}$ defined as follows. Let $Q_Y$ be the monoid and $q_Y : Q_Y \to \mathrm{W}(\mathscr{R}_{\overline{R}^{\overline{y}}_Y})$ the homomorphism defined in II.9.6 (denoted by $Q$ and $q$ in loc. cit.) by taking for $u$ the canonical homomorphism $\Gamma(Y, \mathscr{M}_X) \to \Gamma(\overline{Y}^{\overline{y}}, \mathscr{M}_{\overline{Y}^{\overline{y}}})$. We denote by $\mathscr{M}_{\mathscr{A}_2(\overline{Y}^{\overline{y}})}$ the logarithmic structure on $\mathscr{A}_2(\overline{Y}^{\overline{y}})$ associated with the prelogarithmic structure defined by the homomorphism $Q_Y \to \mathscr{A}_2(\overline{R}^{\overline{y}}_Y)$ induced by $q_Y$. The homomorphism $\theta_Y$ then induces a morphism (II.9.6.4)

$$i_Y : (\widehat{\overline{Y}}^{\overline{y}}, \mathscr{M}_{\widehat{\overline{Y}}^{\overline{y}}}) \to (\mathscr{A}_2(\overline{Y}^{\overline{y}}), \mathscr{M}_{\mathscr{A}_2(\overline{Y}^{\overline{y}})}). \tag{III.10.13.6}$$

The logarithmic scheme $(\mathscr{A}_2(\overline{Y}^{\overline{y}}), \mathscr{M}_{\mathscr{A}_2(\overline{Y}^{\overline{y}})})$ is fine and saturated and $i_Y$ is an exact closed immersion. Indeed, all fiber functors of $\overline{Y}^{\star\circ}_{\text{fét}}$ being isomorphic, it suffices to show this assertion in the case where $\overline{y}$ is localized at a generic point of $\overline{Y}$. In view of III.10.4, the notation above then corresponds to that introduced in II.9.11, with the exception of $\mathscr{M}_{\mathscr{A}_2(\overline{Y}^{\overline{y}})}$, which rather corresponds to the logarithmic structure $\mathscr{M}'_{\mathscr{A}_2(\overline{Y}^{\overline{y}})}$ in loc. cit. But since $Y$ is an object of $\mathbf{Q}$, the latter is canonically isomorphic to the logarithmic structure $\mathscr{M}_{\mathscr{A}_2(\overline{Y}^{\overline{y}})}$ introduced in II.9.12 by virtue of II.9.13; whence the assertion (and notation).

We set

$$\mathrm{T}^{\overline{y}}_Y = \mathrm{Hom}_{\widehat{\overline{R}}^{\overline{y}}_Y}(\widetilde{\Omega}^1_{X/S}(Y) \otimes_{\mathscr{O}_X(Y)} \widehat{\overline{R}}^{\overline{y}}_Y, \xi \widehat{\overline{R}}^{\overline{y}}_Y) \tag{III.10.13.7}$$

and identify the dual $\widehat{\overline{R}}^{\overline{y}}_Y$-module with $\xi^{-1}\widetilde{\Omega}^1_{X/S}(Y) \otimes_{\mathscr{O}_X(Y)} \widehat{\overline{R}}^{\overline{y}}_Y$ (cf. III.2.3). We denote by $\widehat{\overline{Y}}^{\overline{y}}_{\text{zar}}$ the Zariski topos of $\widehat{\overline{Y}}^{\overline{y}}$, by $\widetilde{\mathrm{T}}^{\overline{y}}_Y$ the $\mathscr{O}_{\widehat{\overline{Y}}^{\overline{y}}}$-module associated with $\mathrm{T}^{\overline{y}}_Y$, and by $\mathbf{T}^{\overline{y}}_Y$ the $\widehat{\overline{Y}}^{\overline{y}}$-bundle associated with its dual, in other words, (III.2.7)

$$\mathbf{T}^{\overline{y}}_Y = \mathrm{Spec}(\mathrm{S}_{\widehat{\overline{R}}^{\overline{y}}_Y}(\xi^{-1}\widetilde{\Omega}^1_{X/S}(Y) \otimes_{\mathscr{O}_X(Y)} \widehat{\overline{R}}^{\overline{y}}_Y)). \tag{III.10.13.8}$$

Let $U$ be a Zariski open subscheme of $\widehat{\overline{Y}}^{\overline{y}}$ and $\widetilde{U}$ the open subscheme of $\mathscr{A}_2(\overline{Y}^{\overline{y}})$ defined by $U$. We denote by $\mathscr{L}^{\overline{y}}_Y(U)$ the set of morphisms represented by dotted arrows that complete the diagram

$$\begin{array}{ccc}
(U, \mathscr{M}_{\widehat{\overline{Y}}^{\overline{y}}}|U) & \xrightarrow{i_Y|U} & (\widetilde{U}, \mathscr{M}_{\mathscr{A}_2(\overline{Y}^{\overline{y}})}|\widetilde{U}) \\
\downarrow & & \vdots \\
(\check{\overline{X}}, \mathscr{M}_{\check{\overline{X}}}) & \longrightarrow & (\widetilde{\overline{X}}, \mathscr{M}_{\widetilde{\overline{X}}}) \\
\downarrow & & \downarrow \\
(\check{\overline{S}}, \mathscr{M}_{\check{\overline{S}}}) & \xrightarrow{i_{\overline{S}}} & (\mathscr{A}_2(\overline{S}), \mathscr{M}_{\mathscr{A}_2(\overline{S})})
\end{array} \tag{III.10.13.9}$$

in such a way that it remains commutative. By II.5.23, the functor $U \mapsto \mathscr{L}^{\overline{y}}_Y(U)$ is a $\widetilde{\mathrm{T}}^{\overline{y}}_Y$-torsor of $\widehat{\overline{Y}}^{\overline{y}}_{\text{zar}}$. We denote by $\mathscr{F}^{\overline{y}}_Y$ the $\widehat{\overline{R}}^{\overline{y}}_Y$-module of affine functions on $\mathscr{L}^{\overline{y}}_Y$ (cf.

II.4.9). It fits into a canonical exact sequence (II.4.9.1)

(III.10.13.10) $$0 \to \widehat{\overline{R}}^{\overline{y}}_Y \to \mathscr{F}^{\overline{y}}_Y \to \xi^{-1}\widetilde{\Omega}^1_{X/S}(Y) \otimes_{\mathscr{O}_X(Y)} \widehat{\overline{R}}^{\overline{y}}_Y \to 0.$$

This sequence induces, for every integer $m \geq 1$, an exact sequence (III.5.2.2)

(III.10.13.11) $$0 \to \mathrm{S}^{m-1}_{\widehat{\overline{R}}^{\overline{y}}_Y}(\mathscr{F}^{\overline{y}}_Y) \to \mathrm{S}^{m}_{\widehat{\overline{R}}^{\overline{y}}_Y}(\mathscr{F}^{\overline{y}}_Y) \to \mathrm{S}^{m}_{\widehat{\overline{R}}^{\overline{y}}_Y}(\xi^{-1}\widetilde{\Omega}^1_{X/S}(Y) \otimes_{\mathscr{O}_X(Y)} \widehat{\overline{R}}^{\overline{y}}_Y) \to 0.$$

The $\widehat{\overline{R}}^{\overline{y}}_Y$-modules $(\mathrm{S}^{m}_{\widehat{\overline{R}}^{\overline{y}}_Y}(\mathscr{F}^{\overline{y}}_Y))_{m\in\mathbb{N}}$ therefore form a filtered direct system, whose direct limit

(III.10.13.12) $$\mathscr{C}^{\overline{y}}_Y = \lim_{\substack{\longrightarrow \\ m\geq 0}} \mathrm{S}^{m}_{\widehat{\overline{R}}^{\overline{y}}_Y}(\mathscr{F}^{\overline{y}}_Y)$$

is naturally endowed with a structure of $\widehat{\overline{R}}^{\overline{y}}_Y$-algebra. By II.4.10, the $\widehat{\overline{Y}}^{\overline{y}}$-scheme

(III.10.13.13) $$\mathbf{L}^{\overline{y}}_Y = \mathrm{Spec}(\mathscr{C}^{\overline{y}}_Y)$$

is naturally a principal homogeneous $\mathbf{T}^{\overline{y}}_Y$-bundle on $\widehat{\overline{Y}}^{\overline{y}}$ that canonically represents $\mathscr{L}^{\overline{y}}_Y$. Note that $\mathscr{L}^{\overline{y}}_Y$, $\mathscr{F}^{\overline{y}}_Y$, $\mathscr{C}^{\overline{y}}_Y$, and $\mathbf{L}^{\overline{y}}_Y$ depend on the choice of the deformation $(\widetilde{X}, \mathscr{M}_{\widetilde{X}})$ fixed in III.10.1.

The group $\pi_1(\overline{Y}^{\star\circ}, \overline{y})$ has a natural left action on the logarithmic schemes $(\widehat{\overline{Y}}^{\overline{y}}, \mathscr{M}_{\widehat{\overline{Y}}^{\overline{y}}})$ and $(\mathscr{A}_2(\overline{Y}^{\overline{y}}), \mathscr{M}_{\mathscr{A}_2(\overline{Y}^{\overline{y}})})$, and the morphism $i_Y$ is $\pi_1(\overline{Y}^{\star\circ}, \overline{y})$-equivariant (cf. II.9.11). Proceeding as in II.10.4, we endow $\widetilde{\mathrm{T}}^{\overline{y}}_Y$ with a canonical structure of $\pi_1(\overline{Y}^{\star\circ}, \overline{y})$-equivariant $\mathscr{O}_{\widehat{\overline{Y}}^{\overline{y}}}$-module and $\mathscr{L}^{\overline{y}}_Y$ with a canonical structure of $\pi_1(\overline{Y}^{\star\circ}, \overline{y})$-equivariant $\widetilde{\mathrm{T}}^{\overline{y}}_Y$-torsor (cf. II.4.18). By II.4.21, these two structures induce a $\pi_1(\overline{Y}^{\star\circ}, \overline{y})$-equivariant structure on the $\mathscr{O}_{\widehat{\overline{Y}}^{\overline{y}}}$-module associated with $\mathscr{F}^{\overline{y}}_Y$, or, equivalently, an $\widehat{\overline{R}}^{\overline{y}}_Y$-semi-linear action of $\pi_1(\overline{Y}^{\star\circ}, \overline{y})$ on $\mathscr{F}^{\overline{y}}_Y$ such that the morphisms in the sequence (III.10.13.10) are $\pi_1(\overline{Y}^{\star\circ}, \overline{y})$-equivariant. We deduce from this an action of $\pi_1(\overline{Y}^{\star\circ}, \overline{y})$ on $\mathscr{C}^{\overline{y}}_Y$ by ring automorphisms that is compatible with its action on $\widehat{\overline{R}}^{\overline{y}}_Y$.

**Lemma III.10.14.** *Under the assumptions of* III.10.13, *the actions of* $\pi_1(\overline{Y}^{\star\circ}, \overline{y})$ *on* $\mathscr{F}^{\overline{y}}_Y$ *and on* $\mathscr{C}^{\overline{y}}_Y$ *are continuous for the p-adic topologies.*

Indeed, in view of III.8.15 and the fact that all fiber functors of $\overline{Y}^{\star\circ}_{\text{fét}}$ are isomorphic (III.2.10.3), we may assume that $\overline{y}$ is localized at a generic point of $\overline{Y}$. We denote by $\widetilde{Y} \to \widetilde{X}$ the unique étale morphism that lifts $\check{\overline{Y}} \to \check{\overline{X}}$ ([42] 18.1.2) and by $\mathscr{M}_{\check{\overline{Y}}}$ (resp. $\mathscr{M}_{\widetilde{Y}}$) the logarithmic structure on $\check{\overline{Y}}$ (resp. $\widetilde{Y}$) inverse image of $\mathscr{M}_{\check{\overline{X}}}$ (resp. $\mathscr{M}_{\widetilde{X}}$). Let $U$ be a Zariski open subscheme of $\widehat{\overline{Y}}^{\overline{y}}$ and $\widetilde{U}$ the open subscheme of $\mathscr{A}_2(\overline{Y}^{\overline{y}})$ defined by $U$. The set $\mathscr{L}^{\overline{y}}_Y(U)$ is then canonically isomorphic to the set of morphisms represented by dotted arrows that complete the diagram

(III.10.14.1)
$$\begin{array}{ccc} (U, \mathscr{M}_{\widehat{\overline{Y}}^{\overline{y}}}|U) & \xrightarrow{i_Y|U} & (\widetilde{U}, \mathscr{M}_{\mathscr{A}_2(\overline{Y}^{\overline{y}})}|\widetilde{U}) \\ \downarrow & & \vdots \\ (\check{\overline{Y}}, \mathscr{M}_{\check{\overline{Y}}}) & \longrightarrow & (\widetilde{Y}, \mathscr{M}_{\widetilde{Y}}) \\ \downarrow & & \downarrow \\ (\check{\overline{S}}, \mathscr{M}_{\check{\overline{S}}}) & \xrightarrow{i_{\overline{S}}} & (\mathscr{A}_2(\overline{S}), \mathscr{M}_{\mathscr{A}_2(\overline{S})}) \end{array}$$

in such a way that it remains commutative. The $\widehat{\overline{R}}^{\overline{y}}_Y$-algebra $\mathscr{C}^{\overline{y}}_Y$ endowed with the action of $\pi_1(\overline{Y}^{\star\circ}, \overline{y})$ therefore identifies with the Higgs–Tate algebra associated with $(Y, \mathscr{M}_Y, \widetilde{Y}, \mathscr{M}_{\widetilde{Y}})$ defined in II.10.5 (cf. III.10.4). The lemma follows by virtue of II.10.4.

**III.10.15.** Let $Y$ be an object of $\mathbf{Q}$ (III.10.3) such that $Y_s \neq \emptyset$ and $n$ an integer $\geq 0$. If $A$ is a ring and $M$ an $A$-module, we denote also by $A$ (resp. $M$) the constant sheaf with value $A$ (resp. $M$) of $\overline{Y}^{\circ}_{\text{fét}}$. Since the scheme $\overline{Y}^{\circ}$ is locally irreducible, it is the sum of the schemes induced on its irreducible components. Let $W$ be an irreducible component of $\overline{Y}^{\circ}$ and $\Pi(W)$ its fundamental groupoid (VI.9.10). In view of III.8.15 and VI.9.11, the sheaf $\overline{\mathscr{B}}_Y|W$ of $W_{\text{fét}}$ defines a functor

(III.10.15.1)
$$\Pi(W) \to \mathbf{Ens}, \quad \overline{y} \mapsto \overline{R}^{\overline{y}}_Y.$$

We deduce from this a functor

(III.10.15.2)
$$\Pi(W) \to \mathbf{Ens}, \quad \overline{y} \mapsto \mathscr{F}^{\overline{y}}_Y / p^n \mathscr{F}^{\overline{y}}_Y.$$

By III.10.14, for every geometric point $\overline{y}$ of $W$, $\mathscr{F}^{\overline{y}}_Y / p^n \mathscr{F}^{\overline{y}}_Y$ is a discrete continuous representation of $\pi_1(W, \overline{y})$. Consequently, by virtue of VI.9.11, the functor (III.10.15.2) defines a $(\overline{\mathscr{B}}_{Y,n}|W)$-module $\mathscr{F}_{W,n}$ of $W_{\text{fét}}$, unique up to canonical isomorphism, where $\overline{\mathscr{B}}_{Y,n} = \overline{\mathscr{B}}_Y / p^n \overline{\mathscr{B}}_Y$ (III.10.2.8). By descent ([**35**] II 3.4.4), there exists a $\overline{\mathscr{B}}_{Y,n}$-module $\mathscr{F}_{Y,n}$ of $\overline{Y}^{\circ}_{\text{fét}}$, unique up to canonical isomorphism, such that for every irreducible component $W$ of $\overline{Y}^{\circ}$, we have $\mathscr{F}_{Y,n}|W = \mathscr{F}_{W,n}$.

The exact sequence (III.10.13.10) induces an exact sequence of $\overline{\mathscr{B}}_{Y,n}$-modules

(III.10.15.3)
$$0 \to \overline{\mathscr{B}}_{Y,n} \to \mathscr{F}_{Y,n} \to \xi^{-1}\widetilde{\Omega}^1_{X/S}(Y) \otimes_{\mathscr{O}_X(Y)} \overline{\mathscr{B}}_{Y,n} \to 0.$$

This induces, for every integer $m \geq 1$, an exact sequence (III.5.2.2)

$$0 \to \mathrm{S}^{m-1}_{\overline{\mathscr{B}}_{Y,n}}(\mathscr{F}_{Y,n}) \to \mathrm{S}^{m}_{\overline{\mathscr{B}}_{Y,n}}(\mathscr{F}_{Y,n}) \to \mathrm{S}^{m}_{\overline{\mathscr{B}}_{Y,n}}(\xi^{-1}\widetilde{\Omega}^1_{X/S}(Y) \otimes_{\mathscr{O}_X(Y)} \overline{\mathscr{B}}_{Y,n}) \to 0.$$

The $\overline{\mathscr{B}}_{Y,n}$-modules $(\mathrm{S}^{m}_{\overline{\mathscr{B}}_{Y,n}}(\mathscr{F}_{Y,n}))_{m \in \mathbb{N}}$ therefore form a direct system whose direct limit

(III.10.15.4)
$$\mathscr{C}_{Y,n} = \varinjlim_{m \geq 0} \mathrm{S}^{m}_{\overline{\mathscr{B}}_{Y,n}}(\mathscr{F}_{Y,n})$$

is naturally endowed with a structure of $\overline{\mathscr{B}}_{Y,n}$-algebra of $\overline{Y}^{\circ}_{\text{fét}}$. Note that $\mathscr{F}_{Y,n}$ and $\mathscr{C}_{Y,n}$ depend on the choice of the deformation $(\widetilde{X}, \mathscr{M}_{\widetilde{X}})$ fixed in III.10.1.

**III.10.16.** Let $g\colon Y \to Z$ be a morphism of $\mathbf{Q}$ such that $Y_s \neq \emptyset$, $\overline{y}$ a geometric point of $\overline{Y}^{\circ}$, and $\overline{z} = \overline{g}(\overline{y})$. Recall that $\overline{Y}$ and $\overline{Z}$ are sums of the schemes induced on their irreducible components (III.3.3). We denote by $\overline{Y}^{\star}$ the irreducible component of $\overline{Y}$ containing $\overline{y}$ and by $\overline{Z}^{\star}$ the irreducible component of $\overline{Z}$ containing $\overline{z}$, so that $\overline{g}(\overline{Y}^{\star}) \subset \overline{Z}^{\star}$. We take again the notation of III.10.13 for $Y$ and for $Z$. The morphism $\overline{g}^{\circ}\colon \overline{Y}^{\circ} \to \overline{Z}^{\circ}$ induces a group homomorphism $\pi_1(\overline{Y}^{\star\circ}, \overline{y}) \to \pi_1(\overline{Z}^{\star\circ}, \overline{z})$. The canonical morphism $(\overline{g}^{\circ})^*_{\text{fét}}(\overline{\mathscr{B}}_Z) \to \overline{\mathscr{B}}_Y$ induces a $\pi_1(\overline{Y}^{\star\circ}, \overline{y})$-equivariant ring homomorphism (III.8.15.1)

$$\overline{R}^{\overline{z}}_Z \to \overline{R}^{\overline{y}}_Y \tag{III.10.16.1}$$

and consequently a $\pi_1(\overline{Y}^{\star\circ}, \overline{y})$-equivariant morphism of schemes $h\colon \widehat{\overline{Y}}^{\overline{y}} \to \widehat{\overline{Z}}^{\overline{z}}$. Since $g$ is étale, we have a canonical $\pi_1(\overline{Y}^{\star\circ}, \overline{y})$-equivariant $\mathscr{O}_{\widehat{\overline{Z}}^{\overline{z}}}$-linear morphism $u\colon \widetilde{\mathrm{T}}^{\overline{z}}_Z \to h_*(\widetilde{\mathrm{T}}^{\overline{y}}_Y)$ such that the adjoint morphism $u^{\sharp}\colon h^*(\widetilde{\mathrm{T}}^{\overline{z}}_Z) \to \widetilde{\mathrm{T}}^{\overline{y}}_Y$ is an isomorphism. It immediately follows from the definitions (III.10.13.9) that we have a canonical $u$-equivariant and $\pi_1(\overline{Y}^{\star\circ}, \overline{y})$-equivariant morphism

$$v\colon \mathscr{L}^{\overline{z}}_Z \to h_*(\mathscr{L}^{\overline{y}}_Y). \tag{III.10.16.2}$$

By II.4.22, the pair $(u, v)$ induces a $\pi_1(\overline{Y}^{\star\circ}, \overline{y})$-equivariant and $\widehat{\overline{R}}^{\overline{z}}_Y$-linear isomorphism

$$\mathscr{F}^{\overline{y}}_Y \xrightarrow{\sim} \mathscr{F}^{\overline{z}}_Z \otimes_{\widehat{\overline{R}}^{\overline{z}}_Z} \widehat{\overline{R}}^{\overline{y}}_Y \tag{III.10.16.3}$$

and, consequently, a $\pi_1(\overline{Y}^{\star\circ}, \overline{y})$-equivariant and $\widehat{\overline{R}}^{\overline{z}}_Z$-linear morphism

$$\mathscr{F}^{\overline{z}}_Z \to \mathscr{F}^{\overline{y}}_Y \tag{III.10.16.4}$$

that fits into a commutative diagram

$$\begin{array}{ccccccccc} 0 & \longrightarrow & \widehat{\overline{R}}^{\overline{z}}_Z & \longrightarrow & \mathscr{F}^{\overline{z}}_Z & \longrightarrow & \xi^{-1}\widetilde{\Omega}^1_{X/S}(Z) \otimes_{\mathscr{O}_X(Z)} \widehat{\overline{R}}^{\overline{z}}_Z & \longrightarrow & 0 \\ & & \downarrow & & \downarrow & & \downarrow & & \\ 0 & \longrightarrow & \widehat{\overline{R}}^{\overline{y}}_Y & \longrightarrow & \mathscr{F}^{\overline{y}}_Y & \longrightarrow & \xi^{-1}\widetilde{\Omega}^1_{X/S}(Y) \otimes_{\mathscr{O}_X(Y)} \widehat{\overline{R}}^{\overline{y}}_Y & \longrightarrow & 0 \end{array} \tag{III.10.16.5}$$

We deduce from this a $\pi_1(\overline{Y}^{\star\circ}, \overline{y})$-equivariant homomorphism of $\widehat{\overline{R}}^{\overline{z}}_Z$-algebras

$$\mathscr{C}^{\overline{z}}_Z \to \mathscr{C}^{\overline{y}}_Y. \tag{III.10.16.6}$$

We denote by $\Pi(\overline{Y}^{\star\circ})$ and $\Pi(\overline{Z}^{\star\circ})$ the fundamental groupoids of $\overline{Y}^{\star\circ}$ and $\overline{Z}^{\star\circ}$, respectively, and by

$$\gamma\colon \Pi(\overline{Y}^{\star\circ}) \to \Pi(\overline{Z}^{\star\circ}) \tag{III.10.16.7}$$

the functor induced by the inverse image functor $\mathbf{\acute{E}t}_{\mathrm{f}/\overline{Z}^{\star\circ}} \to \mathbf{\acute{E}t}_{\mathrm{f}/\overline{Y}^{\star\circ}}$. For any integer $n \geq 0$, we denote by $F_{Y,n}\colon \Pi(\overline{Y}^{\star\circ}) \to \mathbf{Ens}$ and $F_{Z,n}\colon \Pi(\overline{Z}^{\star\circ}) \to \mathbf{Ens}$ the functors associated by VI.9.11 with the objects $\mathscr{F}_{Y,n}|\overline{Y}^{\star\circ}$ of $\overline{Y}^{\star\circ}_{\text{fét}}$ and $\mathscr{F}_{Z,n}|\overline{Z}^{\star\circ}$ of $\overline{Z}^{\star\circ}_{\text{fét}}$, respectively. The morphism (III.10.16.4) clearly induces a morphism of functors

$$F_{Z,n} \circ \gamma \to F_{Y,n}. \tag{III.10.16.8}$$

By VI.9.11, we deduce from this a $(\overline{g}^{\circ})^*_{\text{fét}}(\overline{\mathscr{B}}_{Z,n})$-linear morphism

$$(\overline{g}^{\circ})^*_{\text{fét}}(\mathscr{F}_{Z,n}) \to \mathscr{F}_{Y,n} \tag{III.10.16.9}$$

and therefore by adjunction a $\overline{\mathscr{B}}_{Z,n}$-linear morphism

(III.10.16.10) $$\mathscr{F}_{Z,n} \to \overline{g}^{\circ}_{\text{fét}*}(\mathscr{F}_{Y,n}).$$

It follows from (III.10.16.5) that the diagram

(III.10.16.11)
$$\begin{array}{ccccccccc} 0 & \longrightarrow & (\overline{g}^{\circ})^*_{\text{fét}}(\overline{\mathscr{B}}_{Z,n}) & \longrightarrow & (\overline{g}^{\circ})^*_{\text{fét}}(\mathscr{F}_{Z,n}) & \longrightarrow & \xi^{-1}\widetilde{\Omega}^1_{X/S}(Z) \times_{\mathscr{O}_X(Z)} (\overline{g}^{\circ})^*_{\text{fét}}(\overline{\mathscr{B}}_{Z,n}) & \longrightarrow & 0 \\ & & \downarrow & & \downarrow & & \downarrow & & \\ 0 & \longrightarrow & \overline{\mathscr{B}}_{Y,n} & \longrightarrow & \mathscr{F}_{Y,n} & \longrightarrow & \xi^{-1}\widetilde{\Omega}^1_{X/S}(Y) \times_{\mathscr{O}_X(Y)} \overline{\mathscr{B}}_{Y,n} & \longrightarrow & 0 \end{array}$$

is commutative. We deduce from this a homomorphism of $(\overline{g}^{\circ})^*_{\text{fét}}(\overline{\mathscr{B}}_{Z,n})$-algebras

(III.10.16.12) $$(\overline{g}^{\circ})^*_{\text{fét}}(\mathscr{C}_{Z,n}) \to \mathscr{C}_{Y,n},$$

and therefore by adjunction a homomorphism of $\overline{\mathscr{B}}_{Z,n}$-algebras

(III.10.16.13) $$\mathscr{C}_{Z,n} \to \overline{g}^{\circ}_{\text{fét}*}(\mathscr{C}_{Y,n}).$$

**III.10.17.** For any integer $n \geq 0$ and any object $Y$ of $\mathbf{Q}$ such that $Y_s = \emptyset$, we set $\mathscr{C}_{Y,n} = \mathscr{F}_{Y,n} = 0$. The exact sequence (III.10.15.3) still holds in this case, because $\overline{\mathscr{B}}_Y$ is a $\overline{K}$-algebra. The morphisms (III.10.16.10) and (III.10.16.13) are then defined for every morphism of $\mathbf{Q}$, and they verify cocycle relations of the type ([**1**] (1.1.2.2)).

**III.10.18.** Let $r$ be a rational number $\geq 0$, $n$ an integer $\geq 0$, and $Y$ an object of $\mathbf{Q}$. We denote by $\mathscr{F}^{(r)}_{Y,n}$ the extension of $\overline{\mathscr{B}}_{Y,n}$-modules of $\overline{Y}^{\circ}_{\text{fét}}$ deduced from $\mathscr{F}_{Y,n}$ (III.10.15.3) by inverse image under the morphism of multiplication by $p^r$ on $\xi^{-1}\widetilde{\Omega}^1_{X/S}(Y) \otimes_{\mathscr{O}_X(Y)} \overline{\mathscr{B}}_{Y,n}$, so that we have a canonical exact sequence of $\overline{\mathscr{B}}_{Y,n}$-modules

(III.10.18.1) $$0 \to \overline{\mathscr{B}}_{Y,n} \to \mathscr{F}^{(r)}_{Y,n} \to \xi^{-1}\widetilde{\Omega}^1_{X/S}(Y) \otimes_{\mathscr{O}_X(Y)} \overline{\mathscr{B}}_{Y,n} \to 0.$$

For every integer $m \geq 1$, this induces an exact sequence of $\overline{\mathscr{B}}_{Y,n}$-modules (III.5.2.2)

$$0 \to \mathrm{S}^{m-1}_{\overline{\mathscr{B}}_{Y,n}}(\mathscr{F}^{(r)}_{Y,n}) \to \mathrm{S}^{m}_{\overline{\mathscr{B}}_{Y,n}}(\mathscr{F}^{(r)}_{Y,n}) \to \mathrm{S}^{m}_{\overline{\mathscr{B}}_{Y,n}}(\xi^{-1}\widetilde{\Omega}^1_{X/S}(Y) \otimes_{\mathscr{O}_X(Y)} \overline{\mathscr{B}}_{Y,n}) \to 0.$$

The $\overline{\mathscr{B}}_{Y,n}$-modules $(\mathrm{S}^{m}_{\overline{\mathscr{B}}_{Y,n}}(\mathscr{F}^{(r)}_{Y,n}))_{m \in \mathbb{N}}$ therefore form a direct system whose direct limit

(III.10.18.2) $$\mathscr{C}^{(r)}_{Y,n} = \varinjlim_{m \geq 0} \mathrm{S}^{m}_{\overline{\mathscr{B}}_{Y,n}}(\mathscr{F}^{(r)}_{Y,n})$$

is naturally endowed with a structure of $\overline{\mathscr{B}}_{Y,n}$-algebra of $\overline{Y}^{\circ}_{\text{fét}}$.

For all rational numbers $r \geq r' \geq 0$, we have a canonical $\overline{\mathscr{B}}_{Y,n}$-linear morphism

(III.10.18.3) $$\mathsf{a}^{r,r'}_{Y,n} \colon \mathscr{F}^{(r)}_{Y,n} \to \mathscr{F}^{(r')}_{Y,n}$$

that lifts the multiplication by $p^{r-r'}$ on $\xi^{-1}\widetilde{\Omega}^1_{X/S}(Y) \otimes_{\mathscr{O}_X(Y)} \overline{\mathscr{B}}_{Y,n}$ and that extends the identity of $\overline{\mathscr{B}}_{Y,n}$ (III.10.18.1). It induces a homomorphism of $\overline{\mathscr{B}}_{Y,n}$-algebras

(III.10.18.4) $$\alpha^{r,r'}_{Y,n} \colon \mathscr{C}^{(r)}_{Y,n} \to \mathscr{C}^{(r')}_{Y,n}.$$

**III.10.19.** Let $r$ be a rational number $\geq 0$, $n$ an integer $\geq 0$, and $g\colon Y \to Z$ a morphism of $\mathbf{Q}$. The diagram (III.10.16.11) induces a $(\overline{g}^{\circ})^{*}_{\text{fét}}(\overline{\mathscr{B}}_{Z,n})$-linear morphism

$$(\overline{g}^{\circ})^{*}_{\text{fét}}(\mathscr{F}^{(r)}_{Z,n}) \to \mathscr{F}^{(r)}_{Y,n} \tag{III.10.19.1}$$

that fits into a commutative diagram
(III.10.19.2)

$$\begin{array}{ccccccccc} 0 & \to & (\overline{g}^{\circ})^{*}_{\text{fét}}(\overline{\mathscr{B}}_{Z,n}) & \to & (\overline{g}^{\circ})^{*}_{\text{fét}}(\mathscr{F}^{(r)}_{Z,n}) & \to & \xi^{-1}\widetilde{\Omega}^{1}_{X/S}(Z) \times_{\mathscr{O}_X(Z)} (\overline{g}^{\circ})^{*}_{\text{fét}}(\overline{\mathscr{B}}_{Z,n}) & \to & 0 \\ & & \downarrow & & \downarrow & & \downarrow & & \\ 0 & \to & \overline{\mathscr{B}}_{Y,n} & \to & \mathscr{F}^{(r)}_{Y,n} & \to & \xi^{-1}\widetilde{\Omega}^{1}_{X/S}(Y) \times_{\mathscr{O}_X(Y)} \overline{\mathscr{B}}_{Y,n} & \to & 0 \end{array}$$

We deduce from this by adjunction a $\overline{\mathscr{B}}_{Z,n}$-linear morphism

$$\mathscr{F}^{(r)}_{Z,n} \to (\overline{g}^{\circ})_{\text{fét}*}(\mathscr{F}^{(r)}_{Y,n}). \tag{III.10.19.3}$$

We also deduce from this a morphism of $(\overline{g}^{\circ})^{*}_{\text{fét}}(\overline{\mathscr{B}}_{Z,n})$-algebras

$$(\overline{g}^{\circ})^{*}_{\text{fét}}(\mathscr{C}^{(r)}_{Z,n}) \to \mathscr{C}^{(r)}_{Y,n}, \tag{III.10.19.4}$$

and therefore by adjunction a morphism of $\overline{\mathscr{B}}_{Z,n}$-algebras

$$\mathscr{C}^{(r)}_{Z,n} \to (\overline{g}^{\circ})_{\text{fét}*}(\mathscr{C}^{(r)}_{Y,n}). \tag{III.10.19.5}$$

The morphisms (III.10.19.3) and (III.10.19.5) satisfy cocycles relations of the type ([**1**] (1.1.2.2)).

**Lemma III.10.20.** *For every rational number $r \geq 0$, every integer $n \geq 0$, and every morphism $g\colon Y \to Z$ of $\mathbf{Q}$, the morphisms* (III.10.19.1) *and* (III.10.19.4) *induce isomorphisms*

$$(\overline{g}^{\circ})^{*}_{\text{fét}}(\mathscr{F}^{(r)}_{Z,n}) \otimes_{(\overline{g}^{\circ})^{*}_{\text{fét}}(\overline{\mathscr{B}}_{Z,n})} \overline{\mathscr{B}}_{Y,n} \overset{\sim}{\to} \mathscr{F}^{(r)}_{Y,n}, \tag{III.10.20.1}$$

$$(\overline{g}^{\circ})^{*}_{\text{fét}}(\mathscr{C}^{(r)}_{Z,n}) \otimes_{(\overline{g}^{\circ})^{*}_{\text{fét}}(\overline{\mathscr{B}}_{Z,n})} \overline{\mathscr{B}}_{Y,n} \overset{\sim}{\to} \mathscr{C}^{(r)}_{Y,n}. \tag{III.10.20.2}$$

Indeed, the first isomorphism results from the diagram (III.10.19.2) and the fact that the canonical morphism

$$\widetilde{\Omega}^{1}_{X/S}(Z) \otimes_{\mathscr{O}_X(Z)} \mathscr{O}_X(Y) \to \widetilde{\Omega}^{1}_{X/S}(Y) \tag{III.10.20.3}$$

is an isomorphism. The second isomorphism follows from the first.

**III.10.21.** Let $r$ be a rational number $\geq 0$ and $n$ an integer $\geq 0$. By III.10.19, the correspondences

$$\{Y \mapsto \mathscr{F}^{(r)}_{Y,n}\} \quad \text{and} \quad \{Y \mapsto \mathscr{C}^{(r)}_{Y,n}\}, \quad (Y \in \mathrm{Ob}(\mathbf{Q})), \tag{III.10.21.1}$$

define presheaves on $E_{\mathbf{Q}}$ (III.10.5.2) of modules and algebras, respectively, with respect to the ring $\{Y \mapsto \overline{\mathscr{B}}_{Y,n}\}$. Let

$$\mathscr{F}^{(r)}_{n} = \{Y \mapsto \mathscr{F}^{(r)}_{Y,n}\}^{a}, \tag{III.10.21.2}$$

$$\mathscr{C}^{(r)}_{n} = \{Y \mapsto \mathscr{C}^{(r)}_{Y,n}\}^{a}, \tag{III.10.21.3}$$

be the associated sheaves in $\widetilde{E}$ (III.10.6.4). By (III.9.1.8) and (III.10.6.5), $\mathscr{F}^{(r)}_{n}$ is a $\overline{\mathscr{B}}_{n}$-module; we call it the *Higgs–Tate $\overline{\mathscr{B}}_{n}$-extension of thickness* $r$ associated with $(f, \widetilde{X}, \mathscr{M}_{\widetilde{X}})$. Likewise, $\mathscr{C}^{(r)}_{n}$ is a $\overline{\mathscr{B}}_{n}$-algebra; we call it the *Higgs–Tate $\overline{\mathscr{B}}_{n}$-algebra of thickness* $r$ associated with $(f, \widetilde{X}, \mathscr{M}_{\widetilde{X}})$. We set $\mathscr{F}_{n} = \mathscr{F}^{(0)}_{n}$ and $\mathscr{C}_{n} = \mathscr{C}^{(0)}_{n}$, and call these the *Higgs–Tate* $\overline{\mathscr{B}}_n$*-extension* and the *Higgs–Tate* $\overline{\mathscr{B}}_n$*-algebra*, respectively, associated with $(f, \widetilde{X}, \mathscr{M}_{\widetilde{X}})$.

For all rational numbers $r \geq r' \geq 0$, the morphisms (III.10.18.3) induce a $\overline{\mathscr{B}}_n$-linear morphism

(III.10.21.4) $$\mathsf{a}_n^{r,r'} : \mathscr{F}_n^{(r)} \to \mathscr{F}_n^{(r')}.$$

The homomorphisms (III.10.18.4) induce a homomorphism of $\overline{\mathscr{B}}_n$-algebras

(III.10.21.5) $$\alpha_n^{r,r'} : \mathscr{C}_n^{(r)} \to \mathscr{C}_n^{(r')}.$$

For all rational numbers $r \geq r' \geq r'' \geq 0$, we have

(III.10.21.6) $$\mathsf{a}_n^{r,r''} = \mathsf{a}_n^{r',r''} \circ \mathsf{a}_n^{r,r'} \quad \text{and} \quad \alpha_n^{r,r''} = \alpha_n^{r',r''} \circ \alpha_n^{r,r'}.$$

**Proposition III.10.22.** *Let $r$ be a rational number $\geq 0$ and $n$ an integer $\geq 1$. Then:*

(i) *The sheaves $\mathscr{F}_n^{(r)}$ and $\mathscr{C}_n^{(r)}$ are objects of $\widetilde{E}_s$.*

(ii) *We have a canonical locally split exact sequence of $\overline{\mathscr{B}}_n$-modules ((III.10.1.4) and (III.10.2.9))*

(III.10.22.1) $$0 \to \overline{\mathscr{B}}_n \to \mathscr{F}_n^{(r)} \to \sigma_n^*(\xi^{-1}\widetilde{\Omega}^1_{\overline{X}_n/\overline{S}_n}) \to 0.$$

*It induces, for every integer $m \geq 1$, an exact sequence of $\overline{\mathscr{B}}_n$-modules* (III.2.7)

(III.10.22.2) $$0 \to \mathrm{S}^{m-1}_{\overline{\mathscr{B}}_n}(\mathscr{F}_n^{(r)}) \to \mathrm{S}^m_{\overline{\mathscr{B}}_n}(\mathscr{F}_n^{(r)}) \to \sigma_n^*(\mathrm{S}^m_{\mathscr{O}_{\overline{X}_n}}(\xi^{-1}\widetilde{\Omega}^1_{\overline{X}_n/\overline{S}_n})) \to 0.$$

*In particular, the $\overline{\mathscr{B}}_n$-modules $(\mathrm{S}^m_{\overline{\mathscr{B}}_n}(\mathscr{F}_n^{(r)}))_{m\in\mathbb{N}}$ form a filtered direct system.*

(iii) *We have a canonical isomorphism of $\overline{\mathscr{B}}_n$-algebras*

(III.10.22.3) $$\mathscr{C}_n^{(r)} \xrightarrow{\sim} \varinjlim_{m\geq 0} \mathrm{S}^m_{\overline{\mathscr{B}}_n}(\mathscr{F}_n^{(r)}).$$

(iv) *For all rational numbers $r \geq r' \geq 0$, the diagram*

(III.10.22.4) $$\begin{array}{ccccccccc} 0 & \longrightarrow & \overline{\mathscr{B}}_n & \longrightarrow & \mathscr{F}_n^{(r)} & \longrightarrow & \sigma_n^*(\xi^{-1}\widetilde{\Omega}^1_{\overline{X}_n/\overline{S}_n}) & \longrightarrow & 0 \\ & & \| & & \downarrow{\scriptstyle \mathsf{a}_n^{r,r'}} & & \downarrow{\scriptstyle \cdot p^{r-r'}} & & \\ 0 & \longrightarrow & \overline{\mathscr{B}}_n & \longrightarrow & \mathscr{F}_n^{(r')} & \longrightarrow & \sigma_n^*(\xi^{-1}\widetilde{\Omega}^1_{\overline{X}_n/\overline{S}_n}) & \longrightarrow & 0 \end{array}$$

*where the horizontal arrows are the exact sequences* (III.10.22.1) *and the right vertical arrow denotes the multiplication by $p^{r-r'}$, is commutative. Moreover, the morphisms $\mathsf{a}_n^{r,r'}$ and $\alpha_n^{r,r'}$ are compatible with the isomorphisms* (III.10.22.3) *for $r$ and $r'$.*

(i) Indeed, since $\mathscr{F}_{Y,n}^{(r)} = \mathscr{C}_{Y,n}^{(r)} = 0$ for every $Y \in \mathrm{Ob}(\mathbf{Q})$ such that $Y_s = \emptyset$, we have $\mathscr{F}_n^{(r)}|\sigma^*(X_\eta) = \mathscr{C}_n^{(r)}|\sigma^*(X_\eta) = 0$ by virtue of ([**2**] III 5.5).

(ii) We have a canonical isomorphism (III.9.8.4)

(III.10.22.5) $$\sigma_n^*(\xi^{-1}\widetilde{\Omega}^1_{\overline{X}_n/\overline{S}_n}) \xrightarrow{\sim} \sigma^{-1}(\xi^{-1}\widetilde{\Omega}^1_{X/S}) \otimes_{\sigma^{-1}(\mathscr{O}_X)} \overline{\mathscr{B}}_n.$$

Hence by virtue of VI.5.34(ii), VI.8.9, VI.5.17, and (III.10.6.5), $\sigma_n^*(\xi^{-1}\widetilde{\Omega}^1_{\overline{X}_n/\overline{S}_n})$ is the sheaf of $\widetilde{E}$ associated with the presheaf on $E_{\mathbf{Q}}$ defined by the correspondence

(III.10.22.6) $$\{Y \mapsto \xi^{-1}\widetilde{\Omega}^1_{X/S}(Y) \otimes_{\mathscr{O}_X(Y)} \overline{\mathscr{B}}_{Y,n}\}, \quad (Y \in \mathrm{Ob}(\mathbf{Q})).$$

The exact sequence (III.10.22.1) then follows from (III.10.18.1) and (III.10.19.2) because the "associated sheaf" functor is exact ([2] II 4.1). It is locally split because the $\overline{\mathscr{B}}_n$-module $\sigma_n^*(\xi^{-1}\widetilde{\Omega}^1_{\overline{X}_n/\overline{S}_n})$ is locally free of finite type. The exact sequence (III.10.22.2) follows by (III.5.2.2).

(iii) We easily deduce from III.10.7 and ([2] IV 12.10) that for every integer $m \geq 0$, $\mathrm{S}^m_{\overline{\mathscr{B}}_n}(\mathscr{F}_n^{(r)})$ is the sheaf associated with the presheaf

(III.10.22.7) $$\{Y \mapsto \mathrm{S}^m_{\overline{\mathscr{B}}_{Y,n}}(\mathscr{F}_{Y,n}^{(r)})\}, \quad (Y \in \mathrm{Ob}(\mathbf{Q})).$$

The proposition follows in view of III.10.7 and the fact that the "associated sheaf" functor commutes with direct limits ([2] II 4.1).

(iv) This immediately follows from the proofs of (ii) and (iii).

**III.10.23.** Let $r$ be a rational number $\geq 0$ and $n$ an integer $\geq 1$. By III.10.22, we have a canonical $\mathscr{C}_n^{(r)}$-linear isomorphism

(III.10.23.1) $$\Omega^1_{\mathscr{C}_n^{(r)}/\overline{\mathscr{B}}_n} \xrightarrow{\sim} \sigma_n^*(\xi^{-1}\widetilde{\Omega}^1_{\overline{X}_n/\overline{S}_n}) \otimes_{\overline{\mathscr{B}}_n} \mathscr{C}_n^{(r)}.$$

The universal $\overline{\mathscr{B}}_n$-derivation of $\mathscr{C}_n^{(r)}$ corresponds via this isomorphism to the unique $\overline{\mathscr{B}}_n$-derivation

(III.10.23.2) $$d_n^{(r)} : \mathscr{C}_n^{(r)} \to \sigma_n^*(\xi^{-1}\widetilde{\Omega}^1_{\overline{X}_n/\overline{S}_n}) \otimes_{\overline{\mathscr{B}}_n} \mathscr{C}_n^{(r)}$$

that extends the canonical morphism $\mathscr{F}_n^{(r)} \to \sigma_n^*(\xi^{-1}\widetilde{\Omega}^1_{\overline{X}_n/\overline{S}_n})$ (III.10.22.1). It follows from III.10.22(iv) that for all rational numbers $r \geq r' \geq 0$, we have

(III.10.23.3) $$p^{r-r'}(\mathrm{id} \otimes \alpha_n^{r,r'}) \circ d_n^{(r)} = d_n^{(r')} \circ \alpha_n^{r,r'}.$$

**III.10.24.** Let $Y$ be an object of $\mathbf{Q}$ (III.10.5) such that $Y_s \neq \emptyset$ and $\overline{y}$ a geometric point of $\overline{Y}^\circ$. We take again the notation of (III.10.13). For any rational number $r \geq 0$, we denote by $\mathscr{F}_Y^{\overline{y},(r)}$ the extension of $\widehat{\overline{R}}_Y^{\overline{y}}$-modules deduced from $\mathscr{F}_Y^{\overline{y}}$ (III.10.13.10) by inverse image under the morphism of multiplication by $p^r$ on $\xi^{-1}\widetilde{\Omega}^1_{X/S}(Y) \otimes_{\mathscr{O}_X(Y)} \widehat{\overline{R}}_Y^{\overline{y}}$, so that we have an exact sequence of $\widehat{\overline{R}}_Y^{\overline{y}}$-modules

(III.10.24.1) $$0 \to \widehat{\overline{R}}_Y^{\overline{y}} \to \mathscr{F}_Y^{\overline{y},(r)} \to \xi^{-1}\widetilde{\Omega}^1_{X/S}(Y) \otimes_{\mathscr{O}_X(Y)} \widehat{\overline{R}}_Y^{\overline{y}} \to 0.$$

This induces for every integer $m \geq 1$ an exact sequence of $\widehat{\overline{R}}_Y^{\overline{y}}$-modules (III.5.2.2)

(III.10.24.2) $$0 \to \mathrm{S}^{m-1}_{\widehat{\overline{R}}_Y^{\overline{y}}}(\mathscr{F}_Y^{\overline{y},(r)}) \to \mathrm{S}^m_{\widehat{\overline{R}}_Y^{\overline{y}}}(\mathscr{F}_Y^{\overline{y},(r)}) \to \mathrm{S}^m_{\widehat{\overline{R}}_Y^{\overline{y}}}(\xi^{-1}\widetilde{\Omega}^1_{X/S}(Y) \otimes_{\mathscr{O}_X(Y)} \widehat{\overline{R}}_Y^{\overline{y}}) \to 0.$$

In particular, the $\widehat{\overline{R}}_Y^{\overline{y}}$-modules $(\mathrm{S}^m_{\widehat{\overline{R}}_Y^{\overline{y}}}(\mathscr{F}_Y^{\overline{y},(r)}))_{m \in \mathbb{N}}$ form a filtered direct system whose direct limit

(III.10.24.3) $$\mathscr{C}_Y^{\overline{y},(r)} = \lim_{\substack{\longrightarrow \\ m \geq 0}} \mathrm{S}^m_{\widehat{\overline{R}}_Y^{\overline{y}}}(\mathscr{F}_Y^{\overline{y},(r)}),$$

is naturally endowed with a structure of $\widehat{\overline{R}}_Y^{\overline{y}}$-algebra.

It follows from III.10.16 that the formation of $\mathscr{F}_Y^{\overline{y},(r)}$ and $\mathscr{C}_Y^{\overline{y},(r)}$ is functorial in the pair $(Y, \overline{y})$. More precisely, let $g : Y \to Z$ be a morphism of $\mathbf{Q}$ and $\overline{z}$ the image of $\overline{y}$ by the morphism $\overline{g}^\circ\colon \overline{Y}^\circ \to \overline{Z}^\circ$. It immediately follows from III.10.16 that the canonical diagram

(III.10.24.4)
$$\begin{array}{ccccccccc} 0 & \longrightarrow & \widehat{\overline{R}}_Z^{\overline{z}} & \longrightarrow & \mathscr{F}_Z^{\overline{z},(r)} & \longrightarrow & \xi^{-1}\widetilde{\Omega}^1_{X/S}(Z)\otimes_{\mathscr{O}_X(Z)}\widehat{\overline{R}}_Z^{\overline{z}} & \longrightarrow & 0 \\ & & \downarrow & & \downarrow & & \downarrow & & \\ 0 & \longrightarrow & \widehat{\overline{R}}_Y^{\overline{y}} & \longrightarrow & \mathscr{F}_Y^{\overline{y},(r)} & \longrightarrow & \xi^{-1}\widetilde{\Omega}^1_{X/S}(Y)\otimes_{\mathscr{O}_X(Y)}\widehat{\overline{R}}_Y^{\overline{y}} & \longrightarrow & 0 \end{array}$$

is commutative. Since the canonical morphism

(III.10.24.5)
$$\widetilde{\Omega}^1_{X/S}(Z)\otimes_{\mathscr{O}_X(Z)}\mathscr{O}_X(Y)\to\widetilde{\Omega}^1_{X/S}(Y)$$

is an isomorphism, we deduce from this that the canonical morphisms

(III.10.24.6)
$$\mathscr{F}_Z^{\overline{z},(r)}\otimes_{\widehat{\overline{R}}_Z^{\overline{z}}}\widehat{\overline{R}}_Y^{\overline{y}} \to \mathscr{F}_Y^{\overline{y},(r)},$$

(III.10.24.7)
$$\mathscr{C}_Z^{\overline{z},(r)}\otimes_{\widehat{\overline{R}}_Z^{\overline{z}}}\widehat{\overline{R}}_Y^{\overline{y}} \to \mathscr{C}_Y^{\overline{y},(r)},$$

are isomorphisms.

Since $\overline{Y}$ is locally irreducible (III.3.3), it is the sum of the schemes induced on its irreducible components. We denote by $\overline{Y}^\star$ the irreducible component of $\overline{Y}$ containing $\overline{y}$. Likewise, $\overline{Y}^\circ$ is the sum of the schemes induced on its irreducible components, and $\overline{Y}^{\star\circ} = \overline{Y}^\star \times_X X^\circ$ is the irreducible component of $\overline{Y}^\circ$ containing $\overline{y}$. We denote by $\mathbf{B}_{\pi_1(\overline{Y}^{\star\circ},\overline{y})}$ the classifying topos of the profinite group $\pi_1(\overline{Y}^{\star\circ},\overline{y})$ and by

(III.10.24.8)
$$\nu_{\overline{y}}\colon \overline{Y}^{\star\circ}_{\text{fét}} \xrightarrow{\sim} \mathbf{B}_{\pi_1(\overline{Y}^{\star\circ},\overline{y})}$$

the fiber functor at $\overline{y}$ (III.2.10.3). We then have a canonical ring isomorphism (III.8.15.1)

(III.10.24.9)
$$\nu_{\overline{y}}(\overline{\mathscr{B}}_Y|\overline{Y}^{\star\circ}) \xrightarrow{\sim} \overline{R}_Y^{\overline{y}}.$$

Since $\nu_{\overline{y}}$ is exact and commutes with direct limits, for every integer $n\geq 0$, we have canonical isomorphisms of $\overline{R}_Y^{\overline{y}}$-modules and of $\overline{R}_Y^{\overline{y}}$-algebras, respectively,

(III.10.24.10)
$$\nu_{\overline{y}}(\mathscr{F}_{Y,n}^{(r)}|\overline{Y}^{\star\circ}) \xrightarrow{\sim} \mathscr{F}_Y^{\overline{y},(r)}/p^n\mathscr{F}_Y^{\overline{y},(r)},$$

(III.10.24.11)
$$\nu_{\overline{y}}(\mathscr{C}_{Y,n}^{(r)}|\overline{Y}^{\star\circ}) \xrightarrow{\sim} \mathscr{C}_Y^{\overline{y},(r)}/p^n\mathscr{C}_Y^{\overline{y},(r)}.$$

**III.10.25.** Let $(\overline{y}\rightsquigarrow\overline{x})$ be a point of $X_{\text{ét}}\overleftarrow{\times}_{X_{\text{ét}}}\overline{X}^\circ_{\text{ét}}$ (III.8.6) such that $\overline{x}$ lies over $s$ and $X'$ the strict localization of $X$ at $\overline{x}$. We take again the notation of III.10.8. Set

(III.10.25.1)
$$\overline{R}_{X'}^{\overline{y}} = \lim_{\substack{\longrightarrow \\ (U,\mathfrak{p})\in\mathfrak{V}_{\overline{x}}^{\circ}}} \overline{R}_U^{\overline{y}},$$

where $\overline{R}_U^{\overline{y}}$ is the ring defined in (III.8.13.2). We denote by $\widehat{\overline{R}}_{X'}^{\overline{y}}$ the $p$-adic Hausdorff completion of $\overline{R}_{X'}^{\overline{y}}$. We have a canonical isomorphism (III.10.8.5)

(III.10.25.2)
$$\nu_{\overline{y}}(\varphi_{\overline{x}}(\overline{\mathscr{B}})) \xrightarrow{\sim} \overline{R}_{X'}^{\overline{y}}.$$

For any rational number $r \geq 0$, we set

$$\mathscr{F}_{X'}^{\overline{y},(r)} = \lim_{\substack{\longrightarrow \\ (U,\mathfrak{p})\in \mathfrak{V}_{\overline{x}}(\mathbf{Q})^{\circ}}} \mathscr{F}_U^{\overline{y},(r)} \otimes_{\widehat{\overline{R}}{}_U^{y}} \widehat{\overline{R}}{}_{X'}^{y}, \tag{III.10.25.3}$$

$$\mathscr{C}_{X'}^{\overline{y},(r)} = \lim_{\substack{\longrightarrow \\ (U,\mathfrak{p})\in \mathfrak{V}_{\overline{x}}(\mathbf{Q})^{\circ}}} \mathscr{C}_U^{\overline{y},(r)} \otimes_{\widehat{\overline{R}}{}_U^{y}} \widehat{\overline{R}}{}_{X'}^{y}, \tag{III.10.25.4}$$

where $\mathscr{F}_U^{\overline{y},(r)}$ is the $\widehat{\overline{R}}{}_U^{y}$-module defined in (III.10.24.1) and $\mathscr{C}_U^{\overline{y},(r)}$ is the $\widehat{\overline{R}}{}_U^{y}$-algebra defined in (III.10.24.3). We denote by $\widehat{\mathscr{C}}_{X'}^{\overline{y},(r)}$ the $p$-adic Hausdorff completion of $\mathscr{C}_{X'}^{\overline{y},(r)}$. By (III.10.24.6) and (III.10.24.7), for every object $(U,\mathfrak{p})$ of $\mathfrak{V}_{\overline{x}}(\mathbf{Q})$, the canonical homomorphisms

$$\mathscr{F}_U^{\overline{y},(r)} \otimes_{\widehat{\overline{R}}{}_U^{y}} \widehat{\overline{R}}{}_{X'}^{y} \to \mathscr{F}_{X'}^{\overline{y},(r)}, \tag{III.10.25.5}$$

$$\mathscr{C}_U^{\overline{y},(r)} \otimes_{\widehat{\overline{R}}{}_U^{y}} \widehat{\overline{R}}{}_{X'}^{y} \to \mathscr{C}_{X'}^{\overline{y},(r)}, \tag{III.10.25.6}$$

are isomorphisms.

**Remark III.10.26.** Under the assumptions of III.10.25, for every integer $n \geq 1$, the canonical morphisms

$$\lim_{\substack{\longrightarrow \\ (U,\mathfrak{p})\in \mathfrak{V}_{\overline{x}}(\mathbf{Q})^{\circ}}} \mathscr{F}_U^{\overline{y},(r)}/p^n\mathscr{F}_U^{\overline{y},(r)} \to \mathscr{F}_{X'}^{\overline{y},(r)}/p^n\mathscr{F}_{X'}^{\overline{y},(r)}, \tag{III.10.26.1}$$

$$\lim_{\substack{\longrightarrow \\ (U,\mathfrak{p})\in \mathfrak{V}_{\overline{x}}(\mathbf{Q})^{\circ}}} \mathscr{C}_U^{\overline{y},(r)}/p^n\mathscr{C}_U^{\overline{y},(r)} \to \mathscr{C}_{X'}^{\overline{y},(r)}/p^n\mathscr{C}_{X'}^{\overline{y},(r)}, \tag{III.10.26.2}$$

are isomorphisms.

**Lemma III.10.27.** *Let $(\overline{y} \rightsquigarrow \overline{x})$ be a point of $X_{\text{ét}} \overleftarrow{\times}_{X_{\text{ét}}} \overline{X}^{\circ}_{\text{ét}}$ (III.8.6) such that $\overline{x}$ lies over $s$, $X'$ the strict localization of $X$ at $\overline{x}$, and $r$ a rational number $\geq 0$. We take again the notation of III.10.8 and III.10.25; moreover, we set $R_1' = \Gamma(\overline{X}', \mathscr{O}_{\overline{X}'})$ and denote by $\widehat{R}_1'$ its $p$-adic Hausdorff completion. Then:*

(i) *The rings $R_1'$ and $\overline{R}{}_{X'}^{y}$ are $\mathscr{O}_{\overline{K}}$-flat, normal, and integral domains, and the canonical homomorphism $R_1' \to \overline{R}{}_{X'}^{y}$ is injective and integral.*

(ii) *The rings $\widehat{R}_1'$, $\widehat{\overline{R}}{}_{X'}^{y}$, $\mathscr{C}_{X'}^{\overline{y},(r)}$, and $\widehat{\mathscr{C}}_{X'}^{\overline{y},(r)}$ are $\mathscr{O}_C$-flat.*

(iii) *For every integer $n \geq 1$, the canonical homomorphism $R_1'/p^nR_1' \to \overline{R}{}_{X'}^{y}/p^n\overline{R}{}_{X'}^{y}$ is injective.*

(iv) *The canonical homomorphism $\widehat{R}_1' \to \widehat{\overline{R}}{}_{X'}^{y}$ is injective, and the $p$-adic topology on $\widehat{R}_1'[\frac{1}{p}]$ is induced by the $p$-adic topology on $\widehat{\overline{R}}{}_{X'}^{y}[\frac{1}{p}]$ (II.2.2).*

(i) The ring $R_1'$ is normal and an integral domain by virtue of III.3.7, and it is clearly $\mathscr{O}_{\overline{K}}$-flat. For every object $(U,\mathfrak{p})$ of $\mathfrak{V}_{\overline{x}}(\mathbf{P})$, the ring $\overline{R}{}_U^{y}$ (III.8.13.2) is $\mathscr{O}_{\overline{K}}$-flat, normal, and an integral domain by III.8.11 and ([**39**] 0.6.5.12(ii)). Moreover, denoting by $\overline{U}^{\star}$ the irreducible component of $\overline{U}$ containing $\overline{y}$, the canonical homomorphism $\Gamma(\overline{U}^{\star}, \mathscr{O}_{\overline{U}^{\star}}) \to \overline{R}{}_U^{y}$ is injective and integral. One immediately verifies that for every morphism $(U',\mathfrak{p}') \to (U,\mathfrak{p})$ of $\mathfrak{V}_{\overline{x}}(\mathbf{P})$, the canonical homomorphism $\overline{R}{}_U^{y} \to \overline{R}{}_{U'}^{y}$ is injective. Consequently, $\overline{R}{}_{X'}^{y}$ is $\mathscr{O}_{\overline{K}}$-flat, normal, and an integral domain. Since $\overline{X}'$ is integral by III.3.7, we have

$$R_1' \simeq \lim_{\substack{\longrightarrow \\ (U,\mathfrak{p})\in \mathfrak{V}_{\overline{x}}(\mathbf{P})^{\circ}}} \Gamma(\overline{U}^{\star}, \mathscr{O}_{\overline{U}^{\star}}). \tag{III.10.27.1}$$

The canonical homomorphism $R'_1 \to \overline{R}^{\overline{y}}_{X'}$ is therefore injective and integral.

(ii) By ([**11**] Chapter III §2.11 Proposition 14 and Corollary 1), for every $n \geq 1$, we have

$$\widehat{R}'_1/p^n\widehat{R}'_1 \simeq R'_1/p^nR'_1. \tag{III.10.27.2}$$

Let $\alpha \in \widehat{R}'_1$ be such that $p\alpha = 0$ and $\overline{\alpha}$ its class in $R'_1/p^nR'_1$ $(n \geq 1)$. Since $R'_1$ is $\mathscr{O}_{\overline{K}}$-flat, it follows from (III.10.27.2) that $\overline{\alpha} \in p^{n-1}\widehat{R}'_1/p^n\widehat{R}'_1$. We deduce from this that $\alpha \in \cap_{n\geq 0}p^n\widehat{R}'_1 = \{0\}$. Consequently, $p$ is not a zero divisor in $\widehat{R}'_1$, and hence $\widehat{R}'_1$ is $\mathscr{O}_C$-flat ([**11**] Chapter VI §3.6 Lemma 1). We prove similarly that $\widehat{\overline{R}}^{\overline{y}}_{X'}$ is $\mathscr{O}_C$-flat. Consequently, $\mathscr{C}^{\overline{y},(r)}_{X'}$ is $\mathscr{O}_C$-flat (III.10.25.6). As above, we deduce from this that $\widehat{\mathscr{C}}^{\overline{y},(r)}_{X'}$ is $\mathscr{O}_C$-flat.

(iii) This immediately follows from (i).

(iv) The first assertion immediately follows from (iii). By (iii) and (III.10.27.2), for every $n \geq 1$, we have $\widehat{R}'_1 \cap p^n\widehat{\overline{R}}^{\overline{y}}_{X'} = p^n\widehat{R}'_1$. Since $\widehat{R}'_1$ is $\mathscr{O}_C$-flat, it follows that $\widehat{R}'_1[\frac{1}{p}] \cap p^n\widehat{\overline{R}}^{\overline{y}}_{X'} = p^n\widehat{R}'_1$, giving the second assertion.

**Proposition III.10.28.** *Let $\overline{x}$ be a point of $X$ over $s$, $X'$ the strict localization of $X$ at $\overline{x}$, $r$ a rational number $\geq 0$, and $n$ an integer $\geq 0$. Then, with the notation of* (III.10.8), *we have canonical isomorphisms of $\overline{X}'^{\circ}_{\text{fét}}$*

$$\lim_{\substack{\longrightarrow \\ (U,\mathfrak{p})\in\mathfrak{V}_{\overline{x}}(\mathbf{Q})^{\circ}}} (\overline{\mathfrak{p}}^{\circ})^*_{\text{fét}}(\mathscr{C}^{(r)}_{U,n}) \xrightarrow{\sim} \varphi_{\overline{x}}(\mathscr{C}^{(r)}_n), \tag{III.10.28.1}$$

$$\lim_{\substack{\longrightarrow \\ (U,\mathfrak{p})\in\mathfrak{V}_{\overline{x}}(\mathbf{Q})^{\circ}}} (\overline{\mathfrak{p}}^{\circ})^*_{\text{fét}}(\mathscr{F}^{(r)}_{U,n}) \xrightarrow{\sim} \varphi_{\overline{x}}(\mathscr{F}^{(r)}_n). \tag{III.10.28.2}$$

Note that the proposition does not follow directly from VI.10.37 because $\mathbf{Q}$ is not stable under fibered products. Set (III.8.2.1)

$$\mathscr{C}^{(r)}_n = \{U \mapsto C_U\}, \quad U \in \text{Ob}(\mathbf{\acute{E}t}_{/X}). \tag{III.10.28.3}$$

By virtue of VI.10.37, we have a functorial canonical isomorphism

$$\varphi_{\overline{x}}(\mathscr{C}^{(r)}_n) \xrightarrow{\sim} \lim_{\substack{\longrightarrow \\ (U,\mathfrak{p})\in\mathfrak{V}^{\circ}_{\overline{x}}}} (\overline{\mathfrak{p}}^{\circ})^*_{\text{fét}}(C_U). \tag{III.10.28.4}$$

We may obviously replace $\mathfrak{V}_{\overline{x}}$ by the category $\mathfrak{V}_{\overline{x}}(\mathbf{Q})$ (III.10.8). To prove (III.10.28.1), it therefore suffices to prove that the canonical morphism

$$\gamma\colon \lim_{\substack{\longrightarrow \\ (U,\mathfrak{p})\in\mathfrak{V}_{\overline{x}}(\mathbf{Q})^{\circ}}} (\overline{\mathfrak{p}}^{\circ})^*_{\text{fét}}(\mathscr{C}^{(r)}_{U,n}) \to \lim_{\substack{\longrightarrow \\ (U,\mathfrak{p})\in\mathfrak{V}_{\overline{x}}(\mathbf{Q})^{\circ}}} (\overline{\mathfrak{p}}^{\circ})^*_{\text{fét}}(C_U) \tag{III.10.28.5}$$

is an isomorphism. Let $\overline{y}$ be a geometric point of $\overline{X}'^{\circ}$ and $\phi_{\overline{y}}\colon \overline{X}'^{\circ}_{\text{fét}} \to \mathbf{Ens}$ the fiber functor associated with the point $\rho_{\overline{X}'^{\circ}}(\overline{y})$ of $\overline{X}'^{\circ}_{\text{fét}}$ (III.2.9.1). It also suffices to show that $\phi_{\overline{y}}(\gamma)$ is an isomorphism (VI.9.6).

We denote also by $\overline{y}$ the geometric point of $\overline{X}^{\circ}$ induced by $\overline{y}$, by $u\colon \overline{y} \to X'$ the canonical $X$-morphism, and by $(\overline{y} \rightsquigarrow \overline{x})$ the point of $X_{\text{ét}} \overleftarrow{\times}_{X_{\text{ét}}} \overline{X}^{\circ}_{\text{ét}}$ defined by $u$. Let $\mathscr{Q}_{\rho(\overline{y}\rightsquigarrow\overline{x})}$ be the category of $\rho(\overline{y} \rightsquigarrow \overline{x})$-pointed objects of $E_{\mathbf{Q}}$ (III.10.5); in other words, the category of triples $((V \to U), \mathfrak{p}, \mathfrak{q})$ consisting of an object $(V \to U)$ of $E_{\mathbf{Q}}$, an $X$-morphism $\mathfrak{p}\colon \overline{x} \to U$, and an $\overline{X}^{\circ}$-morphism $\mathfrak{q}\colon \overline{y} \to V$ such that if we denote also by

$\mathfrak{p}\colon X' \to U$ the $X$-morphism induced by $\mathfrak{p}$, the diagram

$$\begin{array}{ccc} \overline{y} & \xrightarrow{u} & X' \\ {\scriptstyle \mathfrak{q}}\downarrow & & \downarrow{\scriptstyle \mathfrak{p}} \\ V & \longrightarrow & U \end{array} \qquad \text{(III.10.28.6)}$$

is commutative. By VI.10.20, $\mathscr{Q}_{\rho(\overline{y}\rightsquigarrow\overline{x})}$ is canonically equivalent to the category of neighborhoods of $\rho(\overline{y}\rightsquigarrow\overline{x})$ in $E_{\mathbf{Q}}$ ([2] IV 6.8.2). It is therefore cofiltered and for every presheaf $F=\{U\mapsto F_U\}$ on $E_{\mathbf{Q}}$, we have a functorial canonical isomorphism ([2] IV (6.8.4))

$$(F^a)_{\rho(\overline{y}\rightsquigarrow\overline{x})} \xrightarrow{\sim} \lim_{\substack{\longrightarrow \\ ((V\to U),\mathfrak{p},\mathfrak{q})\in\mathscr{Q}^{\circ}_{\rho(\overline{y}\rightsquigarrow\overline{x})}}} F_U(V). \qquad \text{(III.10.28.7)}$$

We have a functor

$$\alpha\colon \mathscr{Q}_{\rho(\overline{y}\rightsquigarrow\overline{x})} \to \mathfrak{V}_{\overline{x}}(\mathbf{Q}), \quad ((V\to U),\mathfrak{p},\mathfrak{q})\mapsto (U,\mathfrak{p}). \qquad \text{(III.10.28.8)}$$

For every $(U,\mathfrak{p})\in \mathrm{Ob}(\mathfrak{V}_{\overline{x}}(\mathbf{Q}))$, the fiber of $\alpha$ over $(U,\mathfrak{p})$ is canonically equivalent to the category $\mathscr{D}^{\overline{y}}_{(U,\mathfrak{p})}$ of $\overline{\mathfrak{p}}^{\circ}(\overline{y})$-pointed, finite étale $\overline{U}^{\circ}$-schemes (III.10.8.1).

In view of (VI.9.3.4) and ([2] IV (6.8.4)), $\phi_{\overline{y}}(\gamma)$ identifies with the canonical map (III.10.28.9)

$$\phi_{\overline{y}}(\gamma)\colon \lim_{\substack{\longrightarrow \\ (U,\mathfrak{p})\in\mathfrak{V}_{\overline{x}}(\mathbf{Q})^{\circ}}} \lim_{\substack{\longrightarrow \\ (V,\mathfrak{q})\in(\mathscr{D}^{\overline{y}}_{(U,\mathfrak{p})})^{\circ}}} \mathscr{C}^{(r)}_{U,n}(V) \to \lim_{\substack{\longrightarrow \\ (U,\mathfrak{p})\in\mathfrak{V}_{\overline{x}}(\mathbf{Q})^{\circ}}} \lim_{\substack{\longrightarrow \\ (V,\mathfrak{q})\in(\mathscr{D}^{\overline{y}}_{(U,\mathfrak{p})})^{\circ}}} C_U(V).$$

We may clearly replace each of the double direct limits above with a direct limit on the category $\mathscr{Q}^{\circ}_{\rho(\overline{y}\rightsquigarrow\overline{x})}$. It then follows from (III.10.28.7) that $\phi_{\overline{y}}(\gamma)$ is bijective; this gives the isomorphism (III.10.28.1). The proof of (III.10.28.2) is similar.

**Corollary III.10.29.** *Let $(\overline{y}\rightsquigarrow\overline{x})$ be a point of $X_{\text{ét}} \overleftarrow{\times}_{X_{\text{ét}}} \overline{X}^{\circ}_{\text{ét}}$ such that $\overline{x}$ lies over $s$, $X'$ the strict localization of $X$ at $\overline{x}$, $n$ an integer $\geq 0$, and $r$ a rational number $\geq 0$. Then, using the notation of* III.10.8 *and* III.10.24, *we have canonical isomorphisms of* $\mathbf{B}_{\pi_1(\overline{X}'^{\circ},\overline{y})}$

$$\nu_{\overline{y}}(\varphi_{\overline{x}}(\mathscr{C}^{(r)}_n)) \xrightarrow{\sim} \lim_{\substack{\longrightarrow \\ (U,\mathfrak{p})\in\mathfrak{V}_{\overline{x}}(\mathbf{Q})^{\circ}}} \mathscr{C}^{\overline{y},(r)}_U/p^n\mathscr{C}^{\overline{y},(r)}_U, \qquad \text{(III.10.29.1)}$$

$$\nu_{\overline{y}}(\varphi_{\overline{x}}(\mathscr{F}^{(r)}_n)) \xrightarrow{\sim} \lim_{\substack{\longrightarrow \\ (U,\mathfrak{p})\in\mathfrak{V}_{\overline{x}}(\mathbf{Q})^{\circ}}} \mathscr{F}^{\overline{y},(r)}_U/p^n\mathscr{F}^{\overline{y},(r)}_U. \qquad \text{(III.10.29.2)}$$

This follows from III.10.28, (III.10.24.10), and (III.10.24.11).

**Remarks III.10.30.** We keep the assumptions of III.10.29 and moreover suppose that $n\geq 1$.

(i) The isomorphisms (III.10.25.2), (III.10.29.1), and (III.10.29.2) are compatible with each other.

(ii) We have a canonical isomorphism of $\mathbf{B}_{\pi_1(\overline{X}'^{\circ},\overline{y})}$

$$\nu_{\overline{y}}(\varphi_{\overline{x}}(\sigma^*_n(\xi^{-1}\widetilde{\Omega}^1_{\overline{X}_n/\overline{S}_n}))) \xrightarrow{\sim} \lim_{\substack{\longrightarrow \\ (U,\mathfrak{p})\in\mathfrak{V}^{\circ}_{\overline{x}}}} \xi^{-1}\widetilde{\Omega}^1_{X/S}(U)\otimes_{\mathscr{O}_X(U)}(\overline{R}^{\overline{y}}_U/p^n\overline{R}^{\overline{y}}_U). \qquad \text{(III.10.30.1)}$$

Indeed, by (III.10.22.5), VI.5.34(ii), VI.8.9, and VI.5.17, $\sigma^*_n(\xi^{-1}\widetilde{\Omega}^1_{\overline{X}_n/\overline{S}_n})$ is the sheaf of $\widetilde{E}$ associated with the presheaf on $E_{\mathbf{Q}}$ defined by the correspondence

$$\{U\mapsto \xi^{-1}\widetilde{\Omega}^1_{X/S}(U)\otimes_{\mathscr{O}_X(U)}\overline{\mathscr{B}}_{U,n}\}, \quad (U\in\mathrm{Ob}(\mathbf{\acute{E}t}_{/X})). \qquad \text{(III.10.30.2)}$$

The assertion therefore follows from VI.10.37 and (III.8.15.1).

(iii) The image of the exact sequence (III.10.22.1) by the composed functor $\nu_{\overline{y}} \circ \varphi_{\overline{x}}$ identifies with the inverse limit on the category $\mathfrak{V}_{\overline{x}}(\mathbf{Q})^\circ$ of the exact sequences

(III.10.30.3)
$$0 \to \overline{R}^{\overline{y}}_U/p^n\overline{R}^{\overline{y}}_U \to \mathscr{F}^{\overline{y},(r)}_U/p^n\mathscr{F}^{\overline{y},(r)}_U \to \xi^{-1}\widetilde{\Omega}^1_{X/S}(U)\otimes_{\mathscr{O}_X(U)}(\overline{R}^{\overline{y}}_U/p^n\overline{R}^{\overline{y}}_U) \to 0$$

deduced from (III.10.24.1).

(iv) The image of the isomorphism (III.10.22.3) by the composed functor $\nu_{\overline{y}} \circ \varphi_{\overline{x}}$ identifies with the isomorphism

(III.10.30.4)
$$\lim_{\substack{\longrightarrow \\ (U,\mathfrak{p})\in\mathfrak{V}_{\overline{x}}(\mathbf{Q})^\circ}} \mathscr{C}^{\overline{y},(r)}_U/p^n\mathscr{C}^{\overline{y},(r)}_U \xrightarrow{\sim} \lim_{\substack{\longrightarrow \\ (U,\mathfrak{p})\in\mathfrak{V}_{\overline{x}}(\mathbf{Q})^\circ}} \lim_{\substack{\longrightarrow \\ m\geq 0}} \mathrm{S}^m_{\overline{R}^{\overline{y}}_U}(\mathscr{F}^{\overline{y},(r)}_U/p^n\mathscr{F}^{\overline{y},(r)}_U)$$

deduced from (III.10.24.3). The proof is similar to that of III.10.28 and is left to the reader.

**III.10.31.** We denote by $\breve{\overline{\mathscr{B}}}$ the ring $(\overline{\mathscr{B}}_{n+1})_{n\in\mathbb{N}}$ of $\widetilde{E}_s^{\mathbb{N}^\circ}$ (III.7.7), by $\mathscr{O}_{\breve{\overline{X}}}$ the ring $(\mathscr{O}_{\overline{X}_{n+1}})_{n\in\mathbb{N}}$ of $X_{s,\text{ét}}^{\mathbb{N}^\circ}$ (not to be confused with $\mathscr{O}_{\breve{X}}$ (III.10.1.2)), and by $\xi^{-1}\widetilde{\Omega}^1_{\breve{\overline{X}}/\breve{\overline{S}}}$ the $\mathscr{O}_{\breve{\overline{X}}}$-module $(\xi^{-1}\widetilde{\Omega}^1_{\overline{X}_{n+1}/\overline{S}_{n+1}})_{n\in\mathbb{N}}$ of $X_{s,\text{ét}}^{\mathbb{N}^\circ}$ (III.10.1.4). We denote by

(III.10.31.1)
$$\breve{\sigma}\colon (\widetilde{E}_s^{\mathbb{N}^\circ}, \breve{\overline{\mathscr{B}}}) \to (X_{s,\text{ét}}^{\mathbb{N}^\circ}, \mathscr{O}_{\breve{\overline{X}}})$$

the morphism of ringed topos induced by the $(\sigma_{n+1})_{n\in\mathbb{N}}$ (III.10.2.9).

Let $r$ be a rational number $\geq 0$. For all integers $m \geq n \geq 1$, we have a canonical $\overline{\mathscr{B}}_m$-linear morphism $\mathscr{F}^{(r)}_m \to \mathscr{F}^{(r)}_n$ and a canonical homomorphism of $\overline{\mathscr{B}}_m$-algebras $\mathscr{C}^{(r)}_m \to \mathscr{C}^{(r)}_n$, compatible with the exact sequence (III.10.22.1) and the isomorphism (III.10.22.3) and such that the induced morphisms

(III.10.31.2)
$$\mathscr{F}^{(r)}_m \otimes_{\overline{\mathscr{B}}_m} \overline{\mathscr{B}}_n \to \mathscr{F}^{(r)}_n \quad\text{and}\quad \mathscr{C}^{(r)}_m \otimes_{\overline{\mathscr{B}}_m} \overline{\mathscr{B}}_n \to \mathscr{C}^{(r)}_n$$

are isomorphisms. These morphisms form compatible systems when $m$ and $n$ vary, so that $(\mathscr{F}^{(r)}_{n+1})_{n\in\mathbb{N}}$ and $(\mathscr{C}^{(r)}_{n+1})_{n\in\mathbb{N}}$ are inverse systems. We call *Higgs–Tate $\breve{\overline{\mathscr{B}}}$-extension of thickness* $r$ associated with $(f, \widetilde{X}, \mathscr{M}_{\widetilde{X}})$, and denote by $\breve{\mathscr{F}}^{(r)}$, the $\breve{\overline{\mathscr{B}}}$-module $(\mathscr{F}^{(r)}_{n+1})_{n\in\mathbb{N}}$ of $\widetilde{E}_s^{\mathbb{N}^\circ}$. We call *Higgs–Tate $\breve{\overline{\mathscr{B}}}$-algebra of thickness* $r$ associated with $(f, \widetilde{X}, \mathscr{M}_{\widetilde{X}})$, and denote by $\breve{\mathscr{C}}^{(r)}$, the $\breve{\overline{\mathscr{B}}}$-algebra $(\mathscr{C}^{(r)}_{n+1})_{n\in\mathbb{N}}$ of $\widetilde{E}_s^{\mathbb{N}^\circ}$. These are adic $\breve{\overline{\mathscr{B}}}$-modules (III.7.16). By III.7.3(i), (III.7.5.4), and (III.7.12.1), the exact sequence (III.10.22.1) induces an exact sequence of $\breve{\overline{\mathscr{B}}}$-modules

(III.10.31.3)
$$0 \to \breve{\overline{\mathscr{B}}} \to \breve{\mathscr{F}}^{(r)} \to \breve{\sigma}^*(\xi^{-1}\widetilde{\Omega}^1_{\breve{\overline{X}}/\breve{\overline{S}}}) \to 0.$$

Since the $\mathscr{O}_X$-module $\widetilde{\Omega}^1_{X/S}$ is locally free of finite type, the $\breve{\overline{\mathscr{B}}}$-module $\breve{\sigma}^*(\xi^{-1}\widetilde{\Omega}^1_{\breve{\overline{X}}/\breve{\overline{S}}})$ is locally free of finite type and the sequence (III.10.31.3) is locally split. By (III.5.2.2), it induces, for every integer $m \geq 1$, an exact sequence of $\breve{\overline{\mathscr{B}}}$-modules

(III.10.31.4)
$$0 \to \mathrm{S}^{m-1}_{\breve{\overline{\mathscr{B}}}}(\breve{\mathscr{F}}^{(r)}) \to \mathrm{S}^m_{\breve{\overline{\mathscr{B}}}}(\breve{\mathscr{F}}^{(r)}) \to \breve{\sigma}^*(\mathrm{S}^m_{\mathscr{O}_{\breve{\overline{X}}}}(\xi^{-1}\widetilde{\Omega}^1_{\breve{\overline{X}}/\breve{\overline{S}}})) \to 0.$$

In particular, the $\breve{\overline{\mathscr{B}}}$-modules $(\mathrm{S}^m_{\breve{\overline{\mathscr{B}}}}(\breve{\mathscr{F}}^{(r)}))_{m\in\mathbb{N}}$ form a filtered direct system. By III.7.3(i), (III.7.12.3), and (III.10.22.3), we have a canonical isomorphism of $\breve{\overline{\mathscr{B}}}$-algebras

(III.10.31.5)
$$\breve{\mathscr{C}}^{(r)} \xrightarrow{\sim} \lim_{\substack{\longrightarrow \\ m\geq 0}} \mathrm{S}^m_{\breve{\overline{\mathscr{B}}}}(\breve{\mathscr{F}}^{(r)}).$$

Set $\breve{\mathscr{F}} = \breve{\mathscr{F}}^{(0)}$ and $\breve{\mathscr{C}} = \breve{\mathscr{C}}^{(0)}$. We call these the *Higgs–Tate $\overline{\breve{\mathscr{B}}}$-extension* and the *Higgs–Tate $\overline{\breve{\mathscr{B}}}$-algebra*, respectively, associated with $(f, \widetilde{X}, \mathscr{M}_{\widetilde{X}})$. For all rational numbers $r \geq r' \geq 0$, the morphisms $(\mathsf{a}_n^{r,r'})_{n\in\mathbb{N}}$ (III.10.21.4) induce a $\overline{\breve{\mathscr{B}}}$-linear morphism

$$\breve{\mathsf{a}}^{r,r'}\colon \breve{\mathscr{F}}^{(r)} \to \breve{\mathscr{F}}^{(r')}. \tag{III.10.31.6}$$

The homomorphisms $(\alpha_n^{r,r'})_{n\in\mathbb{N}}$ (III.10.21.5) induce a homomorphism of $\overline{\breve{\mathscr{B}}}$-algebras

$$\breve{\alpha}^{r,r'}\colon \breve{\mathscr{C}}^{(r)} \to \breve{\mathscr{C}}^{(r')}. \tag{III.10.31.7}$$

For all rational numbers $r \geq r' \geq r'' \geq 0$, we have

$$\breve{\mathsf{a}}^{r,r''} = \breve{\mathsf{a}}^{r',r''} \circ \breve{\mathsf{a}}^{r,r'} \quad \text{and} \quad \breve{\alpha}^{r,r''} = \breve{\alpha}^{r',r''} \circ \breve{\alpha}^{r,r'}. \tag{III.10.31.8}$$

The derivations $(d_{n+1}^{(r)})_{n\in\mathbb{N}}$ (III.10.23.2) define a morphism

$$\breve{d}^{(r)}\colon \breve{\mathscr{C}}^{(r)} \to \breve{\sigma}^*(\xi^{-1}\widetilde{\Omega}^1_{\breve{\overline{X}}/\breve{\overline{S}}}) \otimes_{\overline{\breve{\mathscr{B}}}} \breve{\mathscr{C}}^{(r)}, \tag{III.10.31.9}$$

that is none other than the universal $\overline{\breve{\mathscr{B}}}$-derivation of $\breve{\mathscr{C}}^{(r)}$. It extends the canonical morphism $\breve{\mathscr{F}}^{(r)} \to \breve{\sigma}^*(\xi^{-1}\widetilde{\Omega}^1_{\breve{\overline{X}}/\breve{\overline{S}}})$. For all rational numbers $r \geq r' \geq 0$, we have

$$p^{r-r'}(\mathrm{id} \otimes \breve{\alpha}^{r,r'}) \circ \breve{d}^{(r)} = \breve{d}^{(r')} \circ \breve{\alpha}^{r,r'}. \tag{III.10.31.10}$$

**Remarks III.10.32.** Let $r$ be a rational number $\geq 0$ and $n$ an integer $\geq 1$.

(i) For every integer $m \geq 0$, the canonical morphisms $\mathrm{S}^m_{\overline{\mathscr{B}}_n}(\mathscr{F}_n^{(r)}) \to \mathscr{C}_n^{(r)}$ and $\mathrm{S}^m_{\overline{\breve{\mathscr{B}}}}(\breve{\mathscr{F}}^{(r)}) \to \breve{\mathscr{C}}^{(r)}$ are injective. Indeed, for every integer $m' \geq m$, the canonical morphism $\mathrm{S}^m_{\overline{\mathscr{B}}_n}(\mathscr{F}_n^{(r)}) \to \mathrm{S}^{m'}_{\overline{\mathscr{B}}_n}(\mathscr{F}_n^{(r)})$ is injective (III.10.22.2). Since filtered direct limits commute with finite inverse limits in $\widetilde{E}_s^{\mathbb{N}^\circ}$, $\mathrm{S}^m_{\overline{\mathscr{B}}_n}(\mathscr{F}_n^{(r)}) \to \mathscr{C}_n^{(r)}$ is injective. The second assertion follows from the first by III.7.3(i).

(ii) We have $\sigma_n^*(\xi^{-1}\widetilde{\Omega}^1_{\overline{X}_n/\overline{S}_n}) = d_n^{(r)}(\mathscr{F}_n^{(r)}) \subset d_n^{(r)}(\mathscr{C}_n^{(r)})$ (III.10.23.2). Consequently, the derivation $d_n^{(r)}$ is a Higgs $\overline{\mathscr{B}}_n$-field with coefficients in $\sigma_n^*(\xi^{-1}\widetilde{\Omega}^1_{\overline{X}_n/\overline{S}_n})$ by II.2.12.

(iii) We have $\breve{\sigma}^*(\xi^{-1}\widetilde{\Omega}^1_{\breve{\overline{X}}/\breve{\overline{S}}}) = \breve{d}^{(r)}(\breve{\mathscr{F}}^{(r)}) \subset \breve{d}^{(r)}(\breve{\mathscr{C}}^{(r)})$ (III.10.31.9). Consequently, the derivation $\breve{d}^{(r)}$ is a Higgs $\overline{\breve{\mathscr{B}}}$-field with coefficients in $\breve{\sigma}^*(\xi^{-1}\widetilde{\Omega}^1_{\breve{\overline{X}}/\breve{\overline{S}}})$.

**Proposition III.10.33.** *For every rational number $r \geq 0$, the functor*

$$\mathbf{Mod}(\overline{\breve{\mathscr{B}}}) \to \mathbf{Mod}(\breve{\mathscr{C}}^{(r)}), \quad M \mapsto M \otimes_{\overline{\breve{\mathscr{B}}}} \breve{\mathscr{C}}^{(r)} \tag{III.10.33.1}$$

*is exact and faithful; in particular, $\breve{\mathscr{C}}^{(r)}$ is $\overline{\breve{\mathscr{B}}}$-flat.*

Since the $\mathscr{O}_X$-module $\widetilde{\Omega}^1_{X/S}$ is locally free of finite type, the exact sequence (III.10.31.3) is locally split. A local splitting of this sequence induces, for every integer $m \geq 0$, a local splitting of the exact sequence (III.10.31.4). We deduce from this that $\mathrm{S}^m_{\overline{\breve{\mathscr{B}}}}(\breve{\mathscr{F}}^{(r)})$ is $\overline{\breve{\mathscr{B}}}$-flat and that the canonical homomorphism $\overline{\breve{\mathscr{B}}} \to \breve{\mathscr{C}}^{(r)}$ admits locally sections. The proposition follows in view of (III.10.31.5).

## III.11. Cohomological computations

**III.11.1.** We keep the assumptions and general notation of III.10 in this section. We moreover denote by $d = \dim(X/S)$ the relative dimension of $X$ over $S$, by $\breve{\overline{\mathscr{B}}}$ the ring $(\overline{\mathscr{B}}_{n+1})_{n\in\mathbb{N}}$ of $\widetilde{E}_s^{\mathbb{N}^\circ}$ (III.7.7), and by $\mathscr{O}_{\breve{\overline{X}}}$ the ring $(\mathscr{O}_{\overline{X}_{n+1}})_{n\in\mathbb{N}}$ of $X_{s,\mathrm{zar}}^{\mathbb{N}^\circ}$ or of $X_{s,\mathrm{\acute{e}t}}^{\mathbb{N}^\circ}$, depending on the context (cf. III.2.9 and III.9.9). For all integers $i, n \geq 1$, we set

$$\widetilde{\Omega}^i_{X/S} = \wedge^i(\widetilde{\Omega}^1_{X/S}) \quad \text{and} \quad \widetilde{\Omega}^i_{\overline{X}_n/\overline{S}_n} = \wedge^i(\widetilde{\Omega}^1_{\overline{X}_n/\overline{S}_n}). \tag{III.11.1.1}$$

We denote by $\xi^{-i}\widetilde{\Omega}^i_{\breve{\overline{X}}/\breve{\overline{S}}}$ the $\mathscr{O}_{\breve{\overline{X}}}$-module $(\xi^{-i}\widetilde{\Omega}^i_{\overline{X}_{n+1}/\overline{S}_{n+1}})_{n\in\mathbb{N}}$. We have a canonical isomorphism (III.7.12.4)

$$\xi^{-i}\widetilde{\Omega}^i_{\breve{\overline{X}}/\breve{\overline{S}}} \xrightarrow{\sim} \wedge^i(\xi^{-1}\widetilde{\Omega}^1_{\breve{\overline{X}}/\breve{\overline{S}}}). \tag{III.11.1.2}$$

For any integer $n \geq 1$, we denote by

$$\sigma_n \colon (\widetilde{E}_s, \overline{\mathscr{B}}_n) \to (X_{s,\mathrm{\acute{e}t}}, \mathscr{O}_{\overline{X}_n}) \tag{III.11.1.3}$$

the canonical morphism of ringed topos (III.10.2.9), by

$$u_n \colon (X_{s,\mathrm{\acute{e}t}}, \mathscr{O}_{\overline{X}_n}) \to (X_{s,\mathrm{zar}}, \mathscr{O}_{\overline{X}_n}) \tag{III.11.1.4}$$

the canonical morphism of ringed topos (III.2.9), and by

$$\tau_n \colon (\widetilde{E}_s, \overline{\mathscr{B}}_n) \to (X_{s,\mathrm{zar}}, \mathscr{O}_{\overline{X}_n}) \tag{III.11.1.5}$$

the composition $u_n \circ \sigma_n$. We denote by

$$\breve{\sigma} \colon (\widetilde{E}_s^{\mathbb{N}^\circ}, \breve{\overline{\mathscr{B}}}) \to (X_{s,\mathrm{\acute{e}t}}^{\mathbb{N}^\circ}, \mathscr{O}_{\breve{\overline{X}}}), \tag{III.11.1.6}$$

$$\breve{u} \colon (X_{s,\mathrm{\acute{e}t}}^{\mathbb{N}^\circ}, \mathscr{O}_{\breve{\overline{X}}}) \to (X_{s,\mathrm{zar}}^{\mathbb{N}^\circ}, \mathscr{O}_{\breve{\overline{X}}}), \tag{III.11.1.7}$$

$$\breve{\tau} \colon (\widetilde{E}_s^{\mathbb{N}^\circ}, \breve{\overline{\mathscr{B}}}) \to (X_{s,\mathrm{zar}}^{\mathbb{N}^\circ}, \mathscr{O}_{\breve{\overline{X}}}), \tag{III.11.1.8}$$

the morphisms of ringed topos induced by the $(\sigma_{n+1})_{n\in\mathbb{N}}$, $(u_{n+1})_{n\in\mathbb{N}}$, and $(\tau_{n+1})_{n\in\mathbb{N}}$, respectively, so that $\breve{\tau} = \breve{u} \circ \breve{\sigma}$.

Recall that we have set $\mathscr{S} = \mathrm{Spf}(\mathscr{O}_C)$ (III.2.1). We denote by $\mathfrak{X}$ the formal scheme $p$-adic completion of $\overline{X}$. It is a formal $\mathscr{S}$-scheme of finite presentation ([**1**] 2.3.15). It is therefore idyllic ([**1**] 2.6.13). We denote by $\xi^{-i}\widetilde{\Omega}^i_{\mathfrak{X}/\mathscr{S}}$ the $p$-adic completion of the $\mathscr{O}_{\overline{X}}$-module $\xi^{-i}\widetilde{\Omega}^i_{\overline{X}/\overline{S}} = \xi^{-i}\widetilde{\Omega}^i_{X/S} \otimes_{\mathscr{O}_X} \mathscr{O}_{\overline{X}}$ ([**1**] 2.5.1). We have a canonical isomorphism ([**1**] 2.5.5(ii))

$$\xi^{-i}\widetilde{\Omega}^i_{\mathfrak{X}/\mathscr{S}} \xrightarrow{\sim} \wedge^i(\xi^{-1}\widetilde{\Omega}^1_{\mathfrak{X}/\mathscr{S}}). \tag{III.11.1.9}$$

We denote by

$$\lambda \colon (X_{s,\mathrm{zar}}^{\mathbb{N}^\circ}, \mathscr{O}_{\breve{\overline{X}}}) \to (X_{s,\mathrm{zar}}, \mathscr{O}_{\mathfrak{X}}) \tag{III.11.1.10}$$

the morphism defined in (III.9.10.5) and by

$$\top \colon (\widetilde{E}_s^{\mathbb{N}^\circ}, \breve{\overline{\mathscr{B}}}) \to (X_{s,\mathrm{zar}}, \mathscr{O}_{\mathfrak{X}}) \tag{III.11.1.11}$$

the composition $\lambda \circ \breve{\tau}$. For modules, we use the notation $\top^{-1}$ to denote the inverse image in the sense of abelian sheaves, and we keep the notation $\top^*$ for the inverse image in the sense of modules. We do likewise for $\breve{\sigma}$ and $\breve{\tau}$. For every $\mathscr{O}_{\mathfrak{X}}$-module $\mathscr{N}$ of $X_{s,\mathrm{zar}}$, we have a canonical isomorphism

$$\top^*(\mathscr{N}) \xrightarrow{\sim} \breve{\tau}^*((\mathscr{N}/p^{n+1}\mathscr{N})_{n\in\mathbb{N}}). \tag{III.11.1.12}$$

In particular, $\top^*(\mathscr{N})$ is adic (III.7.18).

**III.11.2.** For every integer $n \geq 1$, the adjunction morphism of $X_{s,\text{ét}}$

$$\xi^{-1}\widetilde{\Omega}^1_{\overline{X}_n/\overline{S}_n} \to \sigma_{n*}(\sigma_n^*(\xi^{-1}\widetilde{\Omega}^1_{\overline{X}_n/\overline{S}_n})) \tag{III.11.2.1}$$

induces an $\mathscr{O}_{\overline{X}_n}$-linear morphism

$$\xi^{-1}\widetilde{\Omega}^1_{\overline{X}_n/\overline{S}_n} \to \tau_{n*}(\sigma_n^*(\xi^{-1}\widetilde{\Omega}^1_{\overline{X}_n/\overline{S}_n})). \tag{III.11.2.2}$$

Likewise, in view of (III.7.5.3), the adjunction morphism

$$\xi^{-1}\widetilde{\Omega}^1_{\breve{\overline{X}}/\breve{\overline{S}}} \to \breve{\sigma}_*(\breve{\sigma}^*(\xi^{-1}\widetilde{\Omega}^1_{\breve{\overline{X}}/\breve{\overline{S}}})) \tag{III.11.2.3}$$

induces an $\mathscr{O}_{\breve{\overline{X}}}$-linear morphism

$$\xi^{-1}\widetilde{\Omega}^1_{\breve{\overline{X}}/\breve{\overline{S}}} \to \breve{\tau}_*(\breve{\sigma}^*(\xi^{-1}\widetilde{\Omega}^1_{\breve{\overline{X}}/\breve{\overline{S}}})). \tag{III.11.2.4}$$

The latter induces an $\mathscr{O}_{\mathfrak{X}}$-linear morphism

$$\xi^{-1}\widetilde{\Omega}^1_{\mathfrak{X}/\mathscr{S}} \to \top_*(\breve{\sigma}^*(\xi^{-1}\widetilde{\Omega}^1_{\breve{\overline{X}}/\breve{\overline{S}}})). \tag{III.11.2.5}$$

Note that the adjoint morphism

$$\top^*(\xi^{-1}\widetilde{\Omega}^1_{\mathfrak{X}/\mathscr{S}}) \to \breve{\sigma}^*(\xi^{-1}\widetilde{\Omega}^1_{\breve{\overline{X}}/\breve{\overline{S}}}) \tag{III.11.2.6}$$

is an isomorphism by (III.11.1.12) and the remark following (III.2.9.5).

We denote by

$$\partial_n\colon \xi^{-1}\widetilde{\Omega}^1_{\overline{X}_n/\overline{S}_n} \to \mathrm{R}^1\sigma_{n*}(\overline{\mathscr{B}}_n) \tag{III.11.2.7}$$

the $\mathscr{O}_{\overline{X}_n}$-linear morphism of $X_{s,\text{ét}}$ composed of the morphism (III.11.2.1) and the boundary map of the long exact sequence of cohomology deduced from the canonical exact sequence (III.10.22.1)

$$0 \to \overline{\mathscr{B}}_n \to \mathscr{F}_n \to \sigma_n^*(\xi^{-1}\widetilde{\Omega}^1_{\overline{X}_n/\overline{S}_n}) \to 0. \tag{III.11.2.8}$$

We denote by

$$\breve{\partial}\colon \xi^{-1}\widetilde{\Omega}^1_{\breve{\overline{X}}/\breve{\overline{S}}} \to \mathrm{R}^1\breve{\sigma}_*(\breve{\overline{\mathscr{B}}}) \tag{III.11.2.9}$$

the $\mathscr{O}_{\breve{\overline{X}}}$-linear morphism of $X_{s,\text{ét}}^{\mathbb{N}^\circ}$ composed of the morphism (III.11.2.3) and the boundary map of the long exact sequence of cohomology deduced from the canonical exact sequence (III.10.31.3)

$$0 \to \breve{\overline{\mathscr{B}}} \to \breve{\mathscr{F}} \to \breve{\sigma}^*(\xi^{-1}\widetilde{\Omega}^1_{\breve{\overline{X}}/\breve{\overline{S}}}) \to 0. \tag{III.11.2.10}$$

In view of (III.7.5.4), (III.7.5.5), and (III.7.12.1), we can identify $\breve{\partial}$ with the morphism $(\partial_n)_{n\geq 1}$.

We denote by

$$\delta\colon \xi^{-1}\widetilde{\Omega}^1_{\mathfrak{X}/\mathscr{S}} \to \mathrm{R}^1\top_*(\breve{\overline{\mathscr{B}}}) \tag{III.11.2.11}$$

the $\mathscr{O}_{\mathfrak{X}}$-linear morphism of $X_{s,\text{zar}}$ composed of the morphism (III.11.2.5) and the boundary map of the long exact sequence of cohomology deduced from the canonical exact sequence (III.10.31.3)

$$0 \to \breve{\overline{\mathscr{B}}} \to \breve{\mathscr{F}} \to \breve{\sigma}^*(\xi^{-1}\widetilde{\Omega}^1_{\breve{\overline{X}}/\breve{\overline{S}}}) \to 0. \tag{III.11.2.12}$$

**Proposition III.11.3.** *Let $n$ be an integer $\geq 1$.*

(i) *There exists a unique homomorphism of graded $\mathscr{O}_{\overline{X}_n}$-algebras of $\overline{X}_{n,\text{ét}}$*

$$\wedge\,(\xi^{-1}\widetilde{\Omega}^1_{\overline{X}_n/\overline{S}_n}) \to \oplus_{i\geq 0}\mathrm{R}^i\sigma_{n*}(\overline{\mathscr{B}}_n) \tag{III.11.3.1}$$

*whose degree one component is the morphism* $\partial_n$ (III.11.2.7). *Moreover, its kernel is annihilated by* $p^{\frac{2d}{p-1}}\mathfrak{m}_{\overline{K}}$ *and its cokernel is annihilated by* $p^{\frac{2d+1}{p-1}}\mathfrak{m}_{\overline{K}}$ (III.11.1).

(ii) *For every integer* $i \geq d+1$, $\mathrm{R}^i\sigma_{n*}(\overline{\mathscr{B}}_n)$ *is almost zero.*

Let $\overline{x}$ be a geometric point of $X$ over $s$, $X'$ the strict localization of $X$ of $\overline{x}$, and

$$\varphi_{\overline{x}}\colon \widetilde{E} \to \overline{X}'^{\circ}_{\text{fét}} \tag{III.11.3.2}$$

the functor (III.8.8.4). By VI.10.30 and (III.9.8.6), for every integer $i \geq 0$, we have a canonical isomorphism

$$(\mathrm{R}^i\sigma_{n*}(\overline{\mathscr{B}}_n))_{\overline{x}} \xrightarrow{\sim} \mathrm{H}^i(\overline{X}'^{\circ}_{\text{fét}}, \varphi_{\overline{x}}(\overline{\mathscr{B}}_n)). \tag{III.11.3.3}$$

By virtue of III.3.7, $\overline{X}'$ is normal and strictly local (and in particular integral). Since $\overline{X}'$ is $\overline{S}$-flat, $\overline{X}'^{\circ}$ is integral and nonempty. Let $v\colon \overline{y} \to \overline{X}'^{\circ}$ be a generic geometric point of $\overline{X}'^{\circ}$ and

$$\nu_{\overline{y}}\colon \overline{X}'^{\circ}_{\text{fét}} \xrightarrow{\sim} \mathbf{B}_{\pi_1(\overline{X}'^{\circ},\overline{y})} \tag{III.11.3.4}$$

the associated fiber functor (III.2.10.3). We denote also by $\overline{y}$ the geometric point of $\overline{X}^{\circ}$ and by $u\colon \overline{y} \to X'$ the $X$-morphism induced by $v$. We thus obtain a point $(\overline{y} \rightsquigarrow \overline{x})$ of $X_{\text{ét}} \overleftarrow{\times}_{X_{\text{ét}}} \overline{X}^{\circ}_{\text{ét}}$.

We denote by $\mathfrak{V}_{\overline{x}}$ (resp. $\mathfrak{V}_{\overline{x}}(\mathbf{Q})$) the category of neighborhoods of the point of $X_{\text{ét}}$ associated with $\overline{x}$ in the site $\mathbf{\acute{E}t}_{/X}$ (resp. $\mathbf{Q}$ (III.10.5)). For any $(U, \mathfrak{p}\colon \overline{x} \to U) \in \mathrm{Ob}(\mathfrak{V}_{\overline{x}})$, we denote also by $\mathfrak{p}\colon X' \to U$ the morphism deduced from $\mathfrak{p}$ and we set

$$\overline{\mathfrak{p}}^{\circ} = \mathfrak{p} \times_X \overline{X}^{\circ}\colon \overline{X}'^{\circ} \to \overline{U}^{\circ}. \tag{III.11.3.5}$$

We (abusively) denote also by $\overline{y}$ the geometric point $\overline{\mathfrak{p}}^{\circ}(v(\overline{y}))$ of $\overline{U}^{\circ}$. Note that $\overline{y}$ is localized at a generic point of $\overline{U}^{\circ}$ because $\overline{\mathfrak{p}}^{\circ}$ is plat. Since $\overline{U}$ is locally irreducible (III.3.3), it is the sum of the schemes induced on its irreducible components. Denote by $\overline{U}^{\star}$ the irreducible component of $\overline{U}$ containing $\overline{y}$. Likewise, $\overline{U}^{\circ}$ is the sum of the schemes induced on its irreducible components, and $\overline{U}^{\star\circ} = \overline{U}^{\star} \times_X X^{\circ}$ is the irreducible component of $\overline{U}^{\circ}$ containing $\overline{y}$. The morphism $\overline{\mathfrak{p}}^{\circ}$ factors through $\overline{U}^{\star\circ}$. We have a canonical isomorphism (III.10.8.4)

$$\varphi_{\overline{x}}(\overline{\mathscr{B}}) \xrightarrow{\sim} \varinjlim_{(U,\mathfrak{p})\in\mathfrak{V}^{\circ}_{\overline{x}}} (\overline{\mathfrak{p}}^{\circ})^*_{\text{fét}}(\overline{\mathscr{B}}_U). \tag{III.11.3.6}$$

It induces, for every integer $i \geq 0$, an isomorphism

$$\mathrm{H}^i(\overline{X}'^{\circ}_{\text{fét}}, \varphi_{\overline{x}}(\overline{\mathscr{B}}_n)) \xrightarrow{\sim} \varinjlim_{(U,\mathfrak{p})\in\mathfrak{V}^{\circ}_{\overline{x}}} \mathrm{H}^i((\overline{U}^{\circ}_{\text{fét}})_{/\overline{U}^{\star\circ}}, \overline{\mathscr{B}}_{U,n}). \tag{III.11.3.7}$$

Indeed, if $(Z, \mathfrak{q})$ is an object of $\mathfrak{V}_{\overline{x}}$ such that $Z$ is affine, then it suffices to apply VI.11.10 to the full subcategory of $(\mathfrak{V}_{\overline{x}})_{/(Z,\mathfrak{q})}$ made up of the $(U,\mathfrak{p})$'s such that $U$ is affine. In view of (III.8.15.1) and (VI.9.8.6), we deduce from (III.11.3.3) and (III.11.3.7) an isomorphism

$$(\mathrm{R}^i\sigma_{n*}(\overline{\mathscr{B}}_n))_{\overline{x}} \xrightarrow{\sim} \varinjlim_{(U,\mathfrak{p})\in\mathfrak{V}^{\circ}_{\overline{x}}} \mathrm{H}^i(\pi_1(\overline{U}^{\star\circ},\overline{y}), \overline{R}^{\overline{y}}_U/p^n\overline{R}^{\overline{y}}_U). \tag{III.11.3.8}$$

The isomorphisms (III.11.3.3) and (III.11.3.7) are clearly compatible with cup products. The same therefore holds for (III.11.3.8).

On the other hand, since $\overline{X}'$ is strictly local (III.3.7), it identifies with the strict localization of $\overline{X}$ at $\overline{x}$. We therefore have a canonical isomorphism

$$\widetilde{\Omega}^1_{\overline{X}_n/\overline{S}_n,\overline{x}} \xrightarrow{\sim} \lim_{\substack{\longrightarrow \\ (U,\mathfrak{p})\in\mathfrak{V}^{\circ}_{\overline{x}}}} \widetilde{\Omega}^1_{\overline{X}_n/\overline{S}_n}(\overline{U}^{\star}), \tag{III.11.3.9}$$

where we view $\widetilde{\Omega}^1_{\overline{X}_n/\overline{S}_n}$ as a sheaf of $\overline{X}_{\text{ét}}$.

(i) Let $(U,\mathfrak{p})$ be an object of $\mathfrak{V}_{\overline{x}}(\mathbf{Q})$. Note that the schemes $U$, $\overline{U}$, and $\overline{U}^{\star}$ are affine. The exact sequence of $\overline{R}^{\overline{y}}_U$-representations of $\pi_1(\overline{U}^{\star\circ},\overline{y})$

$$0 \to \overline{R}^{\overline{y}}_U/p^n\overline{R}^{\overline{y}}_U \to \mathscr{F}^{\overline{y}}_U/p^n\mathscr{F}^{\overline{y}}_U \to \xi^{-1}\widetilde{\Omega}^1_{X/S}(U)\otimes_{\mathscr{O}_X(U)}(\overline{R}^{\overline{y}}_U/p^n\overline{R}^{\overline{y}}_U) \to 0 \tag{III.11.3.10}$$

deduced from (III.10.13.10), induces an $\mathscr{O}_{\overline{X}_n}(\overline{U}^{\star})$-linear morphism

$$\alpha_{(U,\mathfrak{p})}\colon \xi^{-1}\widetilde{\Omega}^1_{\overline{X}_n/\overline{S}_n}(\overline{U}^{\star}) \to \mathrm{H}^1(\pi_1(\overline{U}^{\star\circ},\overline{y}),\overline{R}^{\overline{y}}_U/p^n\overline{R}^{\overline{y}}_U). \tag{III.11.3.11}$$

We can describe this morphism explicitly in view of III.10.4. Indeed, by II.10.16, we have a commutative diagram
(III.11.3.12)

$$\begin{array}{ccc}
\xi^{-1}\widetilde{\Omega}^1_{\overline{X}_n/\overline{S}_n}(\overline{U}^{\star}) & \xrightarrow{\alpha_{(U,\mathfrak{p})}} & \mathrm{H}^1(\pi_1(\overline{U}^{\star\circ},\overline{y}),\overline{R}^{\overline{y}}_U/p^n\overline{R}^{\overline{y}}_U) \\
\Big\downarrow{\scriptstyle a} & & \Big\uparrow{\scriptstyle c} \\
\mathrm{H}^1(\pi_1(\overline{U}^{\star\circ},\overline{y}),\xi^{-1}\overline{R}^{\overline{y}}_U(1)/p^n\xi^{-1}\overline{R}^{\overline{y}}_U(1)) & \xrightarrow[\sim]{-b} & \mathrm{H}^1(\pi_1(\overline{U}^{\star\circ},\overline{y}),p^{\frac{1}{p-1}}\overline{R}^{\overline{y}}_U/p^{n+\frac{1}{p-1}}\overline{R}^{\overline{y}}_U)
\end{array}$$

where $a$ is induced by the morphism defined in (II.8.13.2), $b$ is induced by the isomorphism

$$\widehat{\overline{R}}^{\overline{y}}_U(1) \xrightarrow{\sim} p^{\frac{1}{p-1}}\xi\widehat{\overline{R}}^{\overline{y}}_U \tag{III.11.3.13}$$

defined in II.9.18, and $c$ is induced by the canonical injection $p^{\frac{1}{p-1}}\overline{R}^{\overline{y}}_U \to \overline{R}^{\overline{y}}_U$.

By virtue of II.8.17(i), there exists a unique homomorphism of graded $\mathscr{O}_{\overline{X}_n}(\overline{U}^{\star})$-algebras

$$\wedge(\xi^{-1}\widetilde{\Omega}^1_{\overline{X}_n/\overline{S}_n}(\overline{U}^{\star})) \to \oplus_{i\geq 0}\mathrm{H}^i(\pi_1(\overline{U}^{\star\circ},\overline{y}),\xi^{-i}\overline{R}^{\overline{y}}_U(i)/p^n\xi^{-i}\overline{R}^{\overline{y}}_U(i)) \tag{III.11.3.14}$$

whose component in degree one is the morphism $a$. This is almost injective and its cokernel is annihilated by $p^{\frac{1}{p-1}}\mathfrak{m}_{\overline{K}}$. We deduce from this that there exists a unique homomorphism of graded $\mathscr{O}_{\overline{X}_n}(\overline{U}^{\star})$-algebras

$$\wedge(\xi^{-1}\widetilde{\Omega}^1_{\overline{X}_n/\overline{S}_n}(\overline{U}^{\star})) \to \oplus_{i\geq 0}\mathrm{H}^i(\pi_1(\overline{U}^{\star\circ},\overline{y}),\overline{R}^{\overline{y}}_U/p^n\overline{R}^{\overline{y}}_U) \tag{III.11.3.15}$$

whose component in degree one is $\alpha_{(U,\mathfrak{p})}$. A chase on the diagram (III.11.3.12) shows that the kernel of (III.11.3.15) is annihilated by $p^{\frac{2d}{p-1}}\mathfrak{m}_{\overline{K}}$. Since $\mathrm{H}^i(\pi_1(\overline{U}^{\star\circ},\overline{y}),\overline{R}^{\overline{y}}_U/p^n\overline{R}^{\overline{y}}_U)$ is almost zero for every $i\geq d+1$ by virtue of II.8.17(ii), the cokernel of (III.11.3.15) is annihilated by $p^{\frac{2d+1}{p-1}}\mathfrak{m}_{\overline{K}}$.

On the other hand, by III.10.30(iii), the image of the exact sequence (III.11.2.8) by the functor $\nu_{\overline{y}}\circ\varphi_{\overline{x}}$ identifies with the direct limit on the category $\mathfrak{V}_{\overline{x}}(\mathbf{Q})^{\circ}$ of the exact sequences (III.11.3.10). Consequently, by VI.10.30(iii), the stalk of the morphism $\partial_n$ (III.11.2.7) at $\overline{x}$ identifies with the direct limit on the category $\mathfrak{V}_{\overline{x}}(\mathbf{Q})^{\circ}$ of the morphisms $\alpha_{(U,\mathfrak{p})}$. We deduce from this the existence (and uniqueness) of the homomorphism (III.11.3.1) in view of III.9.5(ii). The stalk of the latter at $\overline{x}$ identifies with the direct limit on the category $\mathfrak{V}_{\overline{x}}(\mathbf{Q})^{\circ}$ of the homomorphisms (III.11.3.15). The statement follows because filtered direct limits are exact.

(ii) This follows from (III.11.3.8), III.9.5(ii), and II.8.17(ii).

**Corollary III.11.4.** (i) *There exists a unique homomorphism of graded $\mathscr{O}_{\breve{\overline{X}}}$-algebras of $X_{s,\text{ét}}^{\mathbb{N}^\circ}$*

$$\wedge(\xi^{-1}\widetilde{\Omega}^1_{\breve{\overline{X}}/\breve{\overline{S}}}) \to \oplus_{i\geq 0}\mathrm{R}^i\breve{\sigma}_*(\breve{\overline{\mathscr{B}}}) \tag{III.11.4.1}$$

*whose component in degree one is induced by the morphism $\breve{\partial}$* (III.11.2.9). *Moreover, its kernel is annihilated by $p^{\frac{2d}{p-1}}\mathfrak{m}_{\overline{K}}$ and its cokernel is annihilated by $p^{\frac{2d+1}{p-1}}\mathfrak{m}_{\overline{K}}$.*

(ii) *For every integer $i \geq d+1$, $\mathrm{R}^i\breve{\sigma}_*(\breve{\overline{\mathscr{B}}})$ is almost zero.*

This follows from III.11.3, III.7.3(i), and (III.7.5.5).

**Proposition III.11.5.** (i) *For all integers $i \geq 1$ and $j \geq 1$, the $\mathscr{O}_{\breve{\overline{X}}}$-module*

$$\mathrm{R}^i\breve{u}_*(\mathrm{R}^j\breve{\sigma}_*(\breve{\overline{\mathscr{B}}}))$$

*is annihilated by $p^{\frac{4d+1}{p-1}}\mathfrak{m}_{\overline{K}}$.*

(ii) *For every integer $q \geq 0$, the kernel of the morphism*

$$\mathrm{R}^q\breve{\tau}_*(\breve{\overline{\mathscr{B}}}) \to \breve{u}_*(\mathrm{R}^q\breve{\sigma}_*(\breve{\overline{\mathscr{B}}})) \tag{III.11.5.1}$$

*induced by the Cartan–Leray spectral sequence is annihilated by $p^{\frac{(d+1)(4d+1)}{p-1}}\mathfrak{m}_{\overline{K}}$ and its cokernel is annihilated by $p^{\frac{q(4d+1)}{p-1}}\mathfrak{m}_{\overline{K}}$.*

(i) Indeed, it follows from III.11.3(i) and ([**2**] VII 4.3) that for every integer $n \geq 1$, the $\mathscr{O}_{\overline{X}_n}$-modules $\mathrm{R}^iu_{n*}(\mathrm{R}^j\sigma_{n*}(\overline{\mathscr{B}}_n))$ are annihilated by $p^{\frac{4d+1}{p-1}}\mathfrak{m}_{\overline{K}}$. The statement follows in view of (III.7.5.5).

(ii) We may clearly restrict ourselves to the case where $q \geq 1$. Consider the Cartan–Leray spectral sequence ([**2**] V 5.4)

$$\mathrm{E}_2^{i,j} = \mathrm{R}^i\breve{u}_*(\mathrm{R}^j\breve{\sigma}_*(\breve{\overline{\mathscr{B}}})) \Rightarrow \mathrm{R}^{i+j}\breve{\tau}_*(\breve{\overline{\mathscr{B}}}), \tag{III.11.5.2}$$

and denote by $(\mathrm{E}_i^q)_{0\leq i\leq q}$ the abutment filtration on $\mathrm{R}^q\breve{\tau}_*(\breve{\overline{\mathscr{B}}})$. Note that $\mathrm{E}_1^q$ is the kernel of the morphism (III.11.5.1). We know that $\mathrm{E}_2^{i,j}$ is annihilated by $p^{\frac{4d+1}{p-1}}\mathfrak{m}_{\overline{K}}$ for all $i \geq 1$ and $j \geq 0$ by (i) and that it is almost zero for all $i \geq 0$ and $j \geq d+1$ by virtue of III.11.4(ii). We deduce from this that

$$\mathrm{E}_i^q/\mathrm{E}_{i+1}^q = \mathrm{E}_\infty^{i,q-i} \tag{III.11.5.3}$$

is annihilated by $p^{\frac{4d+1}{p-1}}\mathfrak{m}_{\overline{K}}$ for every $i \geq 1$ and that it is almost zero for every $i \leq q-d-1$. Consequently, $\mathrm{E}_1^q$ is annihilated by $p^{\frac{(4d+1)(d+1)}{p-1}}\mathfrak{m}_{\overline{K}}$; the first assertion follows. On the other hand, we have $\mathrm{E}_\infty^{0,q} = \mathrm{E}_{q+2}^{0,q}$, and the cokernel of the morphism (III.11.5.1) identifies with the cokernel of the composition of the canonical injections

$$\mathrm{E}_{q+2}^{0,q} \to \mathrm{E}_{q+1}^{0,q} \to \cdots \to \mathrm{E}_2^{0,q}. \tag{III.11.5.4}$$

The cokernel of each of these injections is annihilated by $p^{\frac{4d+1}{p-1}}\mathfrak{m}_{\overline{K}}$ by (i); the second assertion follows.

**Lemma III.11.6.** *Let $\breve{\mathscr{M}} = (\mathscr{M}_{n+1})_{n\in\mathbb{N}}$ be an $\mathscr{O}_{\breve{\overline{X}}}$-module of $X_{s,\text{zar}}^{\mathbb{N}^\circ}$ such that for all integers $m \geq n \geq 1$, $\mathscr{M}_n$ is a quasi-coherent $\mathscr{O}_{\overline{X}_n}$-module on $\overline{X}_n$ and the canonical morphism $\mathscr{M}_m \to \mathscr{M}_n$ is surjective. Then $\mathrm{R}^i\lambda_*(\breve{\mathscr{M}})$* (III.11.1.10) *is zero for every integer $i \geq 1$.*

Indeed, by ([**2**] V 5.1), $\mathrm{R}^i\lambda_*(\breve{\mathscr{M}})$ is the sheaf associated with the presheaf on $X_{s,\mathrm{zar}}$ that to a Zariski open subscheme $U$ of $X_s$ associates the $\mathscr{O}_{\breve{\overline{X}}}(U)$-module $\mathrm{H}^i(\lambda^*(U),\breve{\mathscr{M}})$. By virtue of III.7.11, we have a canonical exact sequence

$$0 \to \mathrm{R}^1\varprojlim_{n\geq 1} \mathrm{H}^{i-1}(U,\mathscr{M}_n) \to \mathrm{H}^i(\lambda^*(U),\breve{\mathscr{M}}) \to \varprojlim_{n\geq 1} \mathrm{H}^i(U,\mathscr{M}_n) \to 0. \tag{III.11.6.1}$$

From now on, we identify the Zariski sites of $X_s$ and $\overline{X}_1$ (III.9.9). Let $U$ be an affine open subscheme of $\overline{X}_1$. For every integer $n\geq 1$, $U$ determines an affine open subscheme of $\overline{X}_n$ ([**39**] 2.3.5). Consequently, $\mathrm{H}^i(U,\mathscr{M}_n)=0$ and for all integers $m\geq n\geq 1$, the canonical morphism

$$\mathrm{H}^0(U,\mathscr{M}_m)\to \mathrm{H}^0(U,\mathscr{M}_n) \tag{III.11.6.2}$$

is surjective, so that the inverse system of abelian groups $(\mathrm{H}^0(U,\mathscr{M}_{n+1}))_{n\in\mathbb{N}}$ satisfies the Mittag–Leffler condition. We deduce from this that $\mathrm{H}^i(\lambda^*(U),\breve{\mathscr{M}})=0$ ([**47**] 1.15 and [**63**] 3.1). The lemma follows because the affine open subschemes of $\overline{X}_1$ form a topologically generating family for the Zariski site of $\overline{X}_1$.

**Proposition III.11.7.** *For every integer $j\geq 0$, set $\alpha_j=\frac{(d+1+j)(4d+1)+6d+1}{p-1}$. Then:*

(i) *For all integers $i\geq 1$ and $j\geq 0$, the $\mathscr{O}_{\breve{\overline{X}}}$-module $\mathrm{R}^i\lambda_*(\mathrm{R}^j\breve{\tau}_*(\breve{\overline{\mathscr{B}}}))$ is annihilated by $p^{\alpha_j}\mathfrak{m}_{\overline{K}}$.*

(ii) *For every integer $q\geq 0$, the kernel and cokernel of the morphism*

$$\mathrm{R}^q\top_*(\breve{\overline{\mathscr{B}}})\to \lambda_*(\mathrm{R}^q\breve{\tau}_*(\breve{\overline{\mathscr{B}}})) \tag{III.11.7.1}$$

*induced by the Cartan–Leray spectral sequence* (III.11.1.11), *are annihilated by $p^{\sum_{j=0}^{q-1}\alpha_j}\mathfrak{m}_{\overline{K}}$.*

(i) Indeed, by virtue of III.11.4(i), we have an $\mathscr{O}_{\breve{\overline{X}}}$-linear morphism of $X_{s,\mathrm{zar}}^{\mathbb{N}^\circ}$

$$\xi^{-j}\widetilde{\Omega}^j_{\breve{\overline{X}}/\breve{\overline{S}}}\to \breve{u}_*(\mathrm{R}^j\breve{\sigma}_*(\breve{\overline{\mathscr{B}}})), \tag{III.11.7.2}$$

whose kernel is annihilated by $p^{\frac{2d}{p-1}}\mathfrak{m}_{\overline{K}}$ and whose cokernel is annihilated by $p^{\frac{4d+1}{p-1}}\mathfrak{m}_{\overline{K}}$. In view of III.11.6, we deduce from this that $\mathrm{R}^i\lambda_*(\breve{u}_*(\mathrm{R}^j\breve{\sigma}_*(\breve{\overline{\mathscr{B}}})))$ is annihilated by $p^{\frac{6d+1}{p-1}}\mathfrak{m}_{\overline{K}}$. On the other hand, by III.11.5(ii), we have an $\mathscr{O}_{\breve{\overline{X}}}$-linear morphism

$$\mathrm{R}^j\breve{\tau}(\breve{\overline{\mathscr{B}}})\to \breve{u}_*(\mathrm{R}^j\breve{\sigma}_*(\breve{\overline{\mathscr{B}}})) \tag{III.11.7.3}$$

whose kernel is annihilated by $p^{\frac{(d+1)(4d+1)}{p-1}}\mathfrak{m}_{\overline{K}}$ and whose cokernel is annihilated by $p^{\frac{j(4d+1)}{p-1}}\mathfrak{m}_{\overline{K}}$. The statement follows.

(ii) The proof is similar to that of III.11.5(ii). We may clearly assume $q\geq 1$. Consider the Cartan–Leray spectral sequence

$$\mathrm{E}_2^{i,j}=\mathrm{R}^i\lambda_*(\mathrm{R}^j\breve{\tau}_*(\breve{\overline{\mathscr{B}}}))\Rightarrow \mathrm{R}^{i+j}\top_*(\breve{\overline{\mathscr{B}}}), \tag{III.11.7.4}$$

and denote by $(\mathrm{E}^q_i)_{0\leq i\leq q}$ the abutment filtration on $\mathrm{R}^q\top_*(\breve{\overline{\mathscr{B}}})$. Note that $\mathrm{E}^q_1$ is the kernel of the morphism (III.11.7.1). By (i), $\mathrm{E}_2^{i,j}$ is annihilated by $p^{\alpha_j}\mathfrak{m}_{\overline{K}}$ for all $i\geq 1$ and $j\geq 0$. The same then holds for

$$\mathrm{E}^q_i/\mathrm{E}^q_{i+1}=\mathrm{E}_\infty^{i,q-i} \tag{III.11.7.5}$$

for every $i \geq 1$. Hence $\mathrm{E}_1^q$ is annihilated by $p^{\sum_{j=0}^{q-1}\alpha_j}\mathfrak{m}_{\overline{K}}$. On the other hand, we have $\mathrm{E}_\infty^{0,q} = \mathrm{E}_{q+2}^{0,q}$, and the cokernel of the morphism (III.11.7.1) identifies with the cokernel of the composition of the canonical injections

$$\mathrm{E}_{q+2}^{0,q} \to \mathrm{E}_{q+1}^{0,q} \to \cdots \to \mathrm{E}_2^{0,q}. \tag{III.11.7.6}$$

For every integer $m$ such that $2 \leq m \leq q+1$, the cokernel of the canonical injection $\mathrm{E}_{m+1}^{0,q} \to \mathrm{E}_m^{0,q}$ is annihilated by $p^{\alpha_{q-m+1}}\mathfrak{m}_{\overline{K}}$ by (i); the desired statement follows.

**Corollary III.11.8.** *There exists a unique isomorphism of graded $\mathscr{O}_{\mathfrak{X}}[\frac{1}{p}]$-algebras*

$$\wedge\,(\xi^{-1}\widetilde{\Omega}^1_{\mathfrak{X}/\mathscr{S}}[\frac{1}{p}]) \xrightarrow{\sim} \oplus_{i\geq 0}\mathrm{R}^i\top_*(\breve{\overline{\mathscr{B}}})[\frac{1}{p}] \tag{III.11.8.1}$$

*whose component in degree one is the morphism $\delta \otimes_{\mathbb{Z}_p} \mathbb{Q}_p$* (III.11.2.11).

Indeed, the homomorphism (III.11.4.1) induces a homomorphism of graded $\mathscr{O}_{\mathfrak{X}}$-algebras

$$\wedge\,(\xi^{-1}\widetilde{\Omega}^1_{\mathfrak{X}/\mathscr{S}}) \to \oplus_{i\geq 0}\lambda_*(\breve{u}_*(\mathrm{R}^i\breve{\sigma}_*(\breve{\overline{\mathscr{B}}}))) \tag{III.11.8.2}$$

whose kernel and cokernel are rig-null by III.11.4(i) (cf. III.6.15). On the other hand, the morphisms

$$\oplus_{i\geq 0}\,\mathrm{R}^i\top_*(\breve{\overline{\mathscr{B}}}) \to \oplus_{i\geq 0}\lambda_*(\mathrm{R}^i\breve{\tau}_*(\breve{\overline{\mathscr{B}}})) \to \oplus_{i\geq 0}\lambda_*(\breve{u}_*(\mathrm{R}^i\breve{\sigma}_*(\breve{\overline{\mathscr{B}}}))) \tag{III.11.8.3}$$

induced by the Cartan–Leray spectral sequences are homomorphisms of graded $\mathscr{O}_{\mathfrak{X}}$-algebras (cf. [**41**] 0.12.2.6) whose kernels and cokernels are rig-null by virtue of III.11.5(ii) and III.11.7(ii). We obtain the desired isomorphism (III.11.8.1) by applying the functor $-\otimes_{\mathbb{Z}_p}\mathbb{Q}_p$.

**Lemma III.11.9.** *Let $\mathscr{M}$ be a locally free $\mathscr{O}_{\mathfrak{X}}$-module of finite type and $q$ an integer $\geq 0$. For every integer $n \geq 1$, set $\mathscr{M}_n = \mathscr{M} \otimes_{\mathscr{O}_{\mathfrak{X}}} \mathscr{O}_{\overline{X}_n}$, and let $\breve{\mathscr{M}} = (\mathscr{M}_{n+1})_{n\in\mathbb{N}}$, which we view as a sheaf of $X_{s,\text{ét}}^{\mathbb{N}^\circ}$* (III.2.9). *We then have a canonical isomorphism*

$$\mathscr{M} \otimes_{\mathscr{O}_{\mathfrak{X}}} \mathrm{R}^q\top_*(\breve{\overline{\mathscr{B}}}) \xrightarrow{\sim} \mathrm{R}^q\top_*(\breve{\sigma}^*(\breve{\mathscr{M}})). \tag{III.11.9.1}$$

Indeed, we have a canonical isomorphism $\breve{u}_*(\breve{\mathscr{M}}) \xrightarrow{\sim} \breve{\mathscr{M}}$. On the other hand, by virtue of ([**1**] 2.8.5), we have a canonical isomorphism $\mathscr{M} \xrightarrow{\sim} \lambda_*(\breve{\mathscr{M}})$. The adjunction morphism $\breve{\mathscr{M}} \to \breve{\sigma}_*(\breve{\sigma}^*(\breve{\mathscr{M}}))$ then induces an $\mathscr{O}_{\mathfrak{X}}$-linear morphism

$$\mathscr{M} \to \top_*(\breve{\sigma}^*(\breve{\mathscr{M}})). \tag{III.11.9.2}$$

We deduce from this, by cup product, an $\mathscr{O}_{\mathfrak{X}}$-linear morphism

$$\mathscr{M} \otimes_{\mathscr{O}_{\mathfrak{X}}} \mathrm{R}^q\top_*(\breve{\overline{\mathscr{B}}}) \to \mathrm{R}^q\top_*(\breve{\sigma}^*(\breve{\mathscr{M}})). \tag{III.11.9.3}$$

To see that this is an isomorphism, we can reduce to the case where $\mathscr{M} = \mathscr{O}_{\mathfrak{X}}$ (since the question is local for the Zariski topology on $\mathfrak{X}$), in which case the statement is immediate.

**III.11.10.** By ([**45**] I 4.3.1.7), the canonical exact sequence (III.10.31.3)

$$0 \to \breve{\overline{\mathscr{B}}} \to \breve{\mathscr{F}} \to \breve{\sigma}^*(\xi^{-1}\widetilde{\Omega}^1_{\breve{\overline{X}}/\breve{\overline{S}}}) \to 0 \tag{III.11.10.1}$$

induces, for every integer $m \geq 1$, an exact sequence (III.2.7)

$$0 \to \mathrm{S}^{m-1}_{\breve{\overline{\mathscr{B}}}}(\breve{\mathscr{F}}) \to \mathrm{S}^m_{\breve{\overline{\mathscr{B}}}}(\breve{\mathscr{F}}) \to \breve{\sigma}^*(\mathrm{S}^m_{\mathscr{O}_{\breve{\overline{X}}}}(\xi^{-1}\widetilde{\Omega}^1_{\breve{\overline{X}}/\breve{\overline{S}}})) \to 0. \tag{III.11.10.2}$$

We endow $\mathrm{S}^m_{\breve{\mathscr{B}}}(\breve{\mathscr{F}})$ with the exhaustive decreasing filtration $(\mathrm{S}^{m-i}_{\breve{\mathscr{B}}}(\breve{\mathscr{F}}))_{i\in\mathbb{N}}$. We then have a canonical exact sequence

(III.11.10.3)
$$0\to\breve{\sigma}^*(\mathrm{S}^{m-1}_{\mathscr{O}_{\breve{X}}}(\xi^{-1}\widetilde{\Omega}^1_{\breve{X}/\breve{S}}))\to\mathrm{S}^m_{\breve{\mathscr{B}}}(\breve{\mathscr{F}})/\mathrm{S}^{m-2}_{\breve{\mathscr{B}}}(\breve{\mathscr{F}})\to\breve{\sigma}^*(\mathrm{S}^m_{\mathscr{O}_{\breve{X}}}(\xi^{-1}\widetilde{\Omega}^1_{\breve{X}/\breve{S}}))\to 0.$$

For all integers $i$ and $j$, set

(III.11.10.4)
$$\mathrm{E}_1^{i,j}=\mathrm{R}^{i+j}\top_*(\breve{\sigma}^*(\mathrm{S}^{-i}_{\mathscr{O}_{\breve{X}}}(\xi^{-1}\widetilde{\Omega}^1_{\breve{X}/\breve{S}}))),$$

and denote by

(III.11.10.5)
$$d_1^{i,j}\colon \mathrm{E}_1^{i,j}\to\mathrm{E}_1^{i+1,j}$$

the morphism induced by the exact sequence (III.11.10.3). By III.11.9, we have a canonical $\mathscr{O}_{\mathfrak{X}}$-linear isomorphism

(III.11.10.6)
$$\mathrm{S}^{-i}(\xi^{-1}\widetilde{\Omega}^1_{\mathfrak{X}/\mathscr{S}})\otimes_{\mathscr{O}_{\mathfrak{X}}}\mathrm{R}^{i+j}\top_*(\breve{\overline{\mathscr{B}}})\xrightarrow{\sim}\mathrm{E}_1^{i,j}.$$

In view of III.11.8, we deduce from this an isomorphism

(III.11.10.7)
$$\mathrm{S}^{-i}(\xi^{-1}\widetilde{\Omega}^1_{\mathfrak{X}/\mathscr{S}}[\frac{1}{p}])\otimes_{\mathscr{O}_{\mathfrak{X}}[\frac{1}{p}]}\wedge^{i+j}(\xi^{-1}\widetilde{\Omega}^1_{\mathfrak{X}/\mathscr{S}}[\frac{1}{p}])\xrightarrow{\sim}\mathrm{E}_1^{i,j}[\frac{1}{p}].$$

**Proposition III.11.11.** *We have a commutative diagram*

(III.11.11.1)
$$\begin{array}{ccc}\mathrm{S}^{-i}(\xi^{-1}\widetilde{\Omega}^1_{\mathfrak{X}/\mathscr{S}}[\frac{1}{p}])\otimes_{\mathscr{O}_{\mathfrak{X}}[\frac{1}{p}]}\wedge^{i+j}(\xi^{-1}\widetilde{\Omega}^1_{\mathfrak{X}/\mathscr{S}}[\frac{1}{p}]) & \xrightarrow{\sim} & \mathrm{E}_1^{i,j}[\frac{1}{p}]\\ \downarrow{\scriptstyle \phi^{i,j}} & & \downarrow{\scriptstyle d_1^{i,j}\otimes_{\mathbb{Z}_p}\mathbb{Q}_p}\\ \mathrm{S}^{-i-1}(\xi^{-1}\widetilde{\Omega}^1_{\mathfrak{X}/\mathscr{S}}[\frac{1}{p}])\otimes_{\mathscr{O}_{\mathfrak{X}}[\frac{1}{p}]}\wedge^{i+j+1}(\xi^{-1}\widetilde{\Omega}^1_{\mathfrak{X}/\mathscr{S}}[\frac{1}{p}]) & \xrightarrow{\sim} & \mathrm{E}_1^{i+1,j}[\frac{1}{p}]\end{array}$$

*where $\phi^{i,j}$ is the restriction of the $\mathscr{O}_{\mathfrak{X}}[\frac{1}{p}]$-derivation of*

(III.11.11.2)
$$\mathrm{S}(\xi^{-1}\widetilde{\Omega}^1_{\mathfrak{X}/\mathscr{S}}[\frac{1}{p}])\otimes_{\mathscr{O}_{\mathfrak{X}}[\frac{1}{p}]}\wedge(\xi^{-1}\widetilde{\Omega}^1_{\mathfrak{X}/\mathscr{S}}[\frac{1}{p}])$$

*defined in* (III.5.1.1) *relative to the identity morphism of $\xi^{-1}\widetilde{\Omega}^1_{\mathfrak{X}/\mathscr{S}}[\frac{1}{p}]$, and the horizontal arrows are the isomorphisms* (III.11.10.7).

Indeed, by virtue of III.5.7 and in view of the morphism (III.11.2.5), we have a commutative diagram

(III.11.11.3)
$$\begin{array}{ccc}\mathrm{S}^{-i}(\xi^{-1}\widetilde{\Omega}^1_{\mathfrak{X}/\mathscr{S}})\otimes_{\mathscr{O}_{\mathfrak{X}}}\mathrm{R}^{i+j}\top_*(\breve{\overline{\mathscr{B}}}) & \xrightarrow{\sim} & \mathrm{E}_1^{i,j}\\ \downarrow{\scriptstyle \alpha\otimes\mathrm{id}} & & \\ \mathrm{S}^{-i-1}(\xi^{-1}\widetilde{\Omega}^1_{\mathfrak{X}/\mathscr{S}})\otimes_{\mathscr{O}_{\mathfrak{X}}}\mathrm{R}^1\top_*(\breve{\overline{\mathscr{B}}})\otimes_{\mathscr{O}_{\mathfrak{X}}}\mathrm{R}^{i+j}\top_*(\breve{\overline{\mathscr{B}}}) & & \downarrow{\scriptstyle d_1^{i,j}}\\ \downarrow{\scriptstyle \mathrm{id}\otimes\cup} & & \\ \mathrm{S}^{-i-1}(\xi^{-1}\widetilde{\Omega}^1_{\mathfrak{X}/\mathscr{S}})\otimes_{\mathscr{O}_{\mathfrak{X}}}\mathrm{R}^{i+j+1}\top_*(\breve{\overline{\mathscr{B}}}) & \xrightarrow{\sim} & \mathrm{E}_1^{i+1,j}\end{array}$$

where the morphism $\alpha$ is the restriction to $\mathrm{S}^{-i}(\xi^{-1}\widetilde{\Omega}^1_{\mathfrak{X}/\mathscr{S}})$ of the $\mathscr{O}_{\mathfrak{X}}$-derivation $d_\delta$ of the algebra $\mathrm{S}(\xi^{-1}\widetilde{\Omega}^1_{\mathfrak{X}/\mathscr{S}})\otimes_{\mathscr{O}_{\mathfrak{X}}}\wedge(\mathrm{R}^1\top_*(\breve{\overline{\mathscr{B}}}))$ defined in (III.5.1.1) relative to the morphism $\delta$ (III.11.2.11), $\cup$ is the cup product of the $\mathscr{O}_{\mathfrak{X}}$-algebra $\oplus_{i\geq 0}\mathrm{R}^i\top_*(\breve{\overline{\mathscr{B}}})$, and the horizontal arrows are the isomorphisms (III.11.10.6). The proposition follows.

**Proposition III.11.12.** *Let $m$ be an integer $\geq 1$. Then:*

(i) *The morphism*

$$\mathrm{T}_*(\mathrm{S}^{m-1}_{\breve{\mathscr{B}}}(\breve{\mathscr{F}}))[\frac{1}{p}] \to \mathrm{T}_*(\mathrm{S}^{m}_{\breve{\mathscr{B}}}(\breve{\mathscr{F}}))[\frac{1}{p}] \tag{III.11.12.1}$$

*induced by* (III.11.10.2) *is an isomorphism.*

(ii) *For every integer $q \geq 1$, the morphism*

$$\mathrm{R}^q\mathrm{T}_*(\mathrm{S}^{m-1}_{\breve{\mathscr{B}}}(\breve{\mathscr{F}}))[\frac{1}{p}] \to \mathrm{R}^q\mathrm{T}_*(\mathrm{S}^{m}_{\breve{\mathscr{B}}}(\breve{\mathscr{F}}))[\frac{1}{p}] \tag{III.11.12.2}$$

*induced by* (III.11.10.2) *is zero.*

For all integers $i$ and $j$, we set (III.11.10.4)

$$_m\mathrm{E}_1^{i,j} = \begin{cases} \mathrm{E}_1^{i-m,j+m} & \text{if } i \geq 0, \\ 0 & \text{if } i < 0. \end{cases} \tag{III.11.12.3}$$

We denote by

$$_m\mathrm{E}_1^{i,j} \Rightarrow \mathrm{R}^{i+j}\mathrm{T}_*(\mathrm{S}^{m}_{\breve{\mathscr{B}}}(\breve{\mathscr{F}})) \tag{III.11.12.4}$$

the spectral sequence of hypercohomology of the filtered $\breve{\overline{\mathscr{B}}}$-module $\mathrm{S}^{m}_{\breve{\mathscr{B}}}(\breve{\mathscr{F}})$ (III.11.10.2) whose differentials $_md_1^{i,j}$ are given by (III.11.10.5)

$$_md_1^{i,j} = \begin{cases} d_1^{i-m,j+m} & \text{if } i \geq 0, \\ 0 & \text{of } i < 0. \end{cases} \tag{III.11.12.5}$$

For any integer $q \geq 0$, denote by $(_m\mathrm{E}_i^q)_{i\in\mathbb{Z}}$ the abutment filtration on $\mathrm{R}^q\mathrm{T}_*(\mathrm{S}^{m}_{\breve{\mathscr{B}}}(\breve{\mathscr{F}}))$, so that we have

$$_m\mathrm{E}_\infty^{i,q-i} = {_m\mathrm{E}_i^q}/{_m\mathrm{E}_{i+1}^q}. \tag{III.11.12.6}$$

We then have

$$_m\mathrm{E}_i^q = \begin{cases} \mathrm{R}^q\mathrm{T}_*(\mathrm{S}^{m}_{\breve{\mathscr{B}}}(\breve{\mathscr{F}})) & \text{if } i \leq 0, \\ 0 & \text{if } i \geq m+1, \end{cases} \tag{III.11.12.7}$$

and $_m\mathrm{E}_\infty^{0,q} \subset {_m\mathrm{E}_1^{0,q}}$. We deduce from this that the image of the canonical morphism

$$\mathrm{R}^q\mathrm{T}_*(\mathrm{S}^{m-1}_{\breve{\mathscr{B}}}(\breve{\mathscr{F}})) \to \mathrm{R}^q\mathrm{T}_*(\mathrm{S}^{m}_{\breve{\mathscr{B}}}(\breve{\mathscr{F}})) \tag{III.11.12.8}$$

is $_m\mathrm{E}_1^q$ and that its cokernel is $_m\mathrm{E}_\infty^{0,q}$.

On the other hand, it follows from III.11.11 and III.5.1 that for all integers $i$ and $q$ satisfying one of the following conditions:

(i) $q = 0$ and $i < m$,

(ii) $q \geq 1$ and $i \geq 1$,

we have

$$_m\mathrm{E}_\infty^{i,q-i}[\frac{1}{p}] = {_m\mathrm{E}_2^{i,q-i}}[\frac{1}{p}] = 0. \tag{III.11.12.9}$$

The proposition follows.

**Proposition III.11.13.** *Let $r, r'$ be two rational numbers such that $r > r' > 0$. Then:*

(i) *For every integer $n \geq 1$, the canonical homomorphism* (III.10.21.3)

$$\mathscr{O}_{\overline{X}_n} \to \sigma_{n*}(\mathscr{C}_n^{(r)}) \tag{III.11.13.1}$$

*is almost injective. We denote its cokernel by $\mathscr{H}_n^{(r)}$.*

(ii) *There exists a rational number* $\alpha > 0$ *such that for every integer* $n \geq 1$, *the morphism*

$$\mathscr{H}_n^{(r)} \to \mathscr{H}_n^{(r')} \tag{III.11.13.2}$$

*induced by the homomorphism* $\alpha_n^{r,r'} : \mathscr{C}_n^{(r)} \to \mathscr{C}_n^{(r')}$ (III.10.21.5) *is annihilated by* $p^\alpha$.

(iii) *There exists a rational number* $\beta > 0$ *such that for all integers* $n, q \geq 1$, *the canonical morphism*

$$\mathrm{R}^q\sigma_{n*}(\mathscr{C}_n^{(r)}) \to \mathrm{R}^q\sigma_{n*}(\mathscr{C}_n^{(r')}) \tag{III.11.13.3}$$

*is annihilated by* $p^\beta$.

Let $n$ and $q$ be two integers such that $n \geq 1$ and $q \geq 0$, $\overline{x}$ a geometric point of $X$ over $s$, $X'$ the strict localization of $X$ at $\overline{x}$, and

$$\varphi_{\overline{x}} \colon \widetilde{E} \to \overline{X}'^{\circ}_{\text{fét}} \tag{III.11.13.4}$$

the functor (III.8.8.4). By VI.10.30 and (III.9.8.6), we have a canonical isomorphism

$$(\mathrm{R}^q\sigma_{n*}(\mathscr{C}_n^{(r)}))_{\overline{x}} \xrightarrow{\sim} \mathrm{H}^q(\overline{X}'^{\circ}_{\text{fét}}, \varphi_{\overline{x}}(\mathscr{C}_n^{(r)})). \tag{III.11.13.5}$$

By virtue of III.3.7, $\overline{X}'$ is normal and strictly local (and in particular integral). Since $\overline{X}'$ is $\overline{S}$-flat, $\overline{X}'^{\circ}$ is integral and nonempty. Let $v \colon \overline{y} \to \overline{X}'^{\circ}$ be a generic geometric point of $\overline{X}'^{\circ}$ and

$$\nu_{\overline{y}} \colon \overline{X}'^{\circ}_{\text{fét}} \xrightarrow{\sim} \mathbf{B}_{\pi_1(\overline{X}'^{\circ}, \overline{y})} \tag{III.11.13.6}$$

the associated fiber functor (III.2.10.3). We denote also by $\overline{y}$ the geometric point of $\overline{X}^{\circ}$ and by $u \colon \overline{y} \to X'$ the morphism induced by $v$. We thus obtain a point $(\overline{y} \rightsquigarrow \overline{x})$ of $X_{\text{ét}} \overleftarrow{\times}_{X_{\text{ét}}} \overline{X}^{\circ}_{\text{ét}}$.

We denote by $\mathfrak{V}_{\overline{x}}$ (resp. $\mathfrak{V}_{\overline{x}}(\mathbf{Q})$) the category of neighborhoods of the point of $X_{\text{ét}}$ associated with $\overline{x}$ in the site $\mathbf{\acute{E}t}_{/X}$ (resp. $\mathbf{Q}$ (III.10.5)). For any $(U, \mathfrak{p} \colon \overline{x} \to U) \in \mathrm{Ob}(\mathfrak{V}_{\overline{x}})$, we denote also by $\mathfrak{p} \colon X' \to U$ the morphism deduced from $\mathfrak{p}$, and we set

$$\overline{\mathfrak{p}}^{\circ} = \mathfrak{p} \times_X \overline{X}^{\circ} \colon \overline{X}'^{\circ} \to \overline{U}^{\circ}. \tag{III.11.13.7}$$

We (abusively) denote also by $\overline{y}$ the geometric point $\overline{\mathfrak{p}}^{\circ}(v(\overline{y}))$ of $\overline{U}^{\circ}$. Note that $\overline{y}$ is localized at a generic point of $\overline{U}^{\circ}$ because $\overline{\mathfrak{p}}^{\circ}$ is flat. Since $\overline{U}$ is locally irreducible (III.3.3), it is the sum of the schemes induced on its irreducible components. Denote by $\overline{U}^{\star}$ the irreducible component of $\overline{U}$ containing $\overline{y}$. Likewise, $\overline{U}^{\circ}$ is the sum of the schemes induced on its irreducible components, and $\overline{U}^{\star\circ} = \overline{U}^{\star} \times_X X^{\circ}$ is the irreducible component of $\overline{U}^{\circ}$ containing $\overline{y}$. The morphism $\overline{\mathfrak{p}}^{\circ}$ factors through $\overline{U}^{\star\circ}$. By (III.10.29.1), we have a canonical isomorphism of $\mathbf{B}_{\pi_1(\overline{X}'^{\circ}, \overline{y})}$

$$\nu_{\overline{y}}(\varphi_{\overline{x}}(\mathscr{C}_n^{(r)})) \xrightarrow{\sim} \lim_{\substack{\longrightarrow \\ (U,\mathfrak{p}) \in \mathfrak{V}_{\overline{x}}(\mathbf{Q})^{\circ}}} \mathscr{C}_U^{\overline{y},(r)}/p^n\mathscr{C}_U^{\overline{y},(r)}. \tag{III.11.13.8}$$

By virtue of VI.11.10 and (VI.9.8.6), this induces an isomorphism

$$\mathrm{H}^q(\overline{X}'^{\circ}_{\text{fét}}, \varphi_{\overline{x}}(\mathscr{C}_n^{(r)})) \xrightarrow{\sim} \lim_{\substack{\longrightarrow \\ (U,\mathfrak{p}) \in \mathfrak{V}_{\overline{x}}(\mathbf{Q})^{\circ}}} \mathrm{H}^q(\pi_1(\overline{U}^{\star\circ}, \overline{y}), \mathscr{C}_U^{\overline{y},(r)}/p^n\mathscr{C}_U^{\overline{y},(r)}). \tag{III.11.13.9}$$

We deduce from (III.11.13.5) and (III.11.13.9) an isomorphism

$$(\mathrm{R}^q\sigma_{n*}(\mathscr{C}_n^{(r)}))_{\overline{x}} \xrightarrow{\sim} \lim_{\substack{\longrightarrow \\ (U,\mathfrak{p}) \in \mathfrak{V}_{\overline{x}}(\mathbf{Q})^{\circ}}} \mathrm{H}^q(\pi_1(\overline{U}^{\star\circ}, \overline{y}), \mathscr{C}_U^{\overline{y},(r)}/p^n\mathscr{C}_U^{\overline{y},(r)}). \tag{III.11.13.10}$$

On the other hand, we prove, as in (III.11.3.9), that we have a canonical isomorphism

$$(\mathscr{O}_{\overline{X}_n})_{\overline{x}} \xrightarrow{\sim} \lim_{\substack{\longrightarrow \\ (U,\mathfrak{p})\in\mathfrak{V}_{\overline{x}}^{\circ}}} \mathscr{O}_{\overline{X}_n}(\overline{U}^{\star}), \tag{III.11.13.11}$$

where we view $\mathscr{O}_{\overline{X}_n}$ on the left as a sheaf of $X_{s,\text{ét}}$ and on the right as a sheaf of $\overline{X}_{\text{ét}}$.

The stalk of the morphism (III.11.13.1) at $\overline{x}$ identifies with the direct limit on the category $\mathfrak{V}_{\overline{x}}(\mathbf{Q})^{\circ}$ of the canonical morphism

$$\mathscr{O}_{\overline{X}_n}(\overline{U}^{\star}) \to (\mathscr{C}_U^{\overline{y},(r)}/p^n\mathscr{C}_U^{\overline{y},(r)})^{\pi_1(\overline{U}^{\star\circ},\overline{y})}. \tag{III.11.13.12}$$

Since filtered direct limits are exact, the proposition then follows from II.12.7, in view of III.10.4 and the proof of III.10.14.

**Corollary III.11.14.** *Let $r, r'$ be two rational numbers such that $r > r' > 0$. Then:*

(i) *The canonical homomorphism of $X_{s,\text{ét}}^{\mathbb{N}^{\circ}}$ (III.10.31)*

$$\mathscr{O}_{\breve{\overline{X}}} \to \breve{\sigma}_*(\breve{\mathscr{C}}^{(r)}) \tag{III.11.14.1}$$

*is almost injective. We denote its cokernel by $\breve{\mathscr{H}}^{(r)}$.*

(ii) *There exists a rational number $\alpha > 0$ such that the morphism*

$$\breve{\mathscr{H}}^{(r)} \to \breve{\mathscr{H}}^{(r')} \tag{III.11.14.2}$$

*induced by the canonical homomorphism $\breve{\alpha}^{r,r'}\colon \breve{\mathscr{C}}^{(r)} \to \breve{\mathscr{C}}^{(r')}$ (III.10.31.7) is annihilated by $p^{\alpha}$.*

(iii) *There exists a rational number $\beta > 0$ such that for every integer $q \geq 1$, the canonical morphism of $X_{s,\text{ét}}^{\mathbb{N}^{\circ}}$*

$$\mathrm{R}^q\breve{\sigma}_*(\breve{\mathscr{C}}^{(r)}) \to \mathrm{R}^q\breve{\sigma}_*(\breve{\mathscr{C}}^{(r')}) \tag{III.11.14.3}$$

*is annihilated by $p^{\beta}$.*

This follows from III.11.13, III.7.3(i), and (III.7.5.5).

**Lemma III.11.15.** *Let $r, r'$ be two rational numbers such that $r > r' > 0$. Then:*

(i) *The canonical homomorphism of $X_{s,\text{zar}}^{\mathbb{N}^{\circ}}$ (III.10.31.1)*

$$\mathscr{O}_{\breve{\overline{X}}} \to \breve{\tau}_*(\breve{\mathscr{C}}^{(r)}) \tag{III.11.15.1}$$

*is almost injective. We denote its cokernel by $\breve{\mathscr{K}}^{(r)}$.*

(ii) *There exists a rational number $\alpha > 0$ such that the morphism*

$$\breve{\mathscr{K}}^{(r)} \to \breve{\mathscr{K}}^{(r')} \tag{III.11.15.2}$$

*induced by the canonical homomorphism $\breve{\alpha}^{r,r'}\colon \breve{\mathscr{C}}^{(r)} \to \breve{\mathscr{C}}^{(r')}$ (III.10.31.7) is annihilated by $p^{\alpha}$.*

(iii) *For every integer $q \geq 1$, there exists a rational number $\beta > 0$ such that the canonical morphism of $X_{s,\text{zar}}^{\mathbb{N}^{\circ}}$*

$$\mathrm{R}^q\breve{\tau}_*(\breve{\mathscr{C}}^{(r)}) \to \mathrm{R}^q\breve{\tau}_*(\breve{\mathscr{C}}^{(r')}) \tag{III.11.15.3}$$

*is annihilated by $p^{\beta}$.*

Let us use the notation of III.11.14 and moreover denote by $\breve{\mathscr{N}}^{(r)}$ and $\breve{\mathscr{M}}^{(r)}$ the kernels of the morphisms (III.11.14.1) and (III.11.15.1), respectively.

(i) Since $\breve{\mathscr{M}}^{(r)} = \breve{u}_*(\breve{\mathscr{N}}^{(r)})$, the statement follows from III.11.14(i).

(ii) Since $\mathrm{R}^1\breve{u}_*(\mathscr{O}_{\breve{\overline{X}}}) = 0$ by (III.7.5.5) and ([**2**] VII 4.3), we have an exact sequence

$$0 \to \mathrm{R}^1\breve{u}_*(\breve{\mathscr{N}}^{(r)}) \to \breve{\mathscr{K}}^{(r)} \to \breve{u}_*(\breve{\mathscr{H}}^{(r)}). \tag{III.11.15.4}$$

The statement then follows from III.11.14(i)-(ii).

(iii) Consider the Cartan–Leray spectral sequence

(III.11.15.5) $${}^r\mathrm{E}_2^{i,j} = \mathrm{R}^i\breve{u}_*(\mathrm{R}^j\breve{\sigma}_*(\breve{\mathscr{C}}^{(r)})) \Rightarrow \mathrm{R}^{i+j}\breve{\tau}_*(\breve{\mathscr{C}}^{(r)})$$

and denote by $({}^r\mathrm{E}_i^q)_{0\leq i\leq q}$ the abutment filtration on $\mathrm{R}^q\breve{\tau}_*(\breve{\mathscr{C}}^{r)})$, so that we have

(III.11.15.6) $${}^r\mathrm{E}_i^q/{}^r\mathrm{E}_{i+1}^q = {}^r\mathrm{E}_\infty^{i,q-i}.$$

For every integer $0 \leq i \leq q+1$, set $r_i = r' + (q+1-i)(r-r')/(q+1)$. By III.11.14(iii), for every integer $0 \leq i \leq q-1$, there exists a rational number $\beta_i > 0$ such that the canonical morphism

(III.11.15.7) $${}^{r_i}\mathrm{E}_2^{i,q-i} \to {}^{r_{i+1}}\mathrm{E}_2^{i,q-i}$$

is annihilated by $p^{\beta_i}$. The same then holds for the morphism ${}^{r_i}\mathrm{E}_\infty^{i,q-i} \to {}^{r_{i+1}}\mathrm{E}_\infty^{i,q-i}$. On the other hand, $\mathrm{R}^q\breve{u}_*(\mathscr{O}_{\breve{\overline{X}}}) = 0$ by (III.7.5.5) and ([**2**] VII 4.3). We deduce from this that the canonical morphism

(III.11.15.8) $$\mathrm{R}^q\breve{u}_*(\breve{\sigma}_*(\breve{\mathscr{C}}^{(r')})) \to \mathrm{R}^q\breve{u}_*(\breve{\mathscr{H}}^{(r')})$$

is almost injective by III.11.14(i). Consequently, by virtue of III.11.14(ii), there exists a rational number $\beta_q > 0$ such that the canonical morphism

(III.11.15.9) $${}^{r_q}\mathrm{E}_2^{q,0} \to {}^{r_{q+1}}\mathrm{E}_2^{q,0}$$

is annihilated by $p^{\beta_q}$. The same then holds for the morphism ${}^{r_q}\mathrm{E}_\infty^{q,0} \to {}^{r_{q+1}}\mathrm{E}_\infty^{q,0}$. The desired statement follows by taking $\beta = \sum_{i=0}^q \beta_i$.

**Proposition III.11.16.** *Let $r, r'$ be two rational numbers such that $r > r' > 0$. Then:*

(i) *The canonical homomorphism* (III.10.31.1)

(III.11.16.1) $$\mathscr{O}_{\mathfrak{X}} \to \mathsf{T}_*(\breve{\mathscr{C}}^{(r)})$$

*is injective. We denote its cokernel by $\mathscr{L}^{(r)}$.*

(ii) *There exists a rational number $\alpha > 0$ such that the morphism*

(III.11.16.2) $$\mathscr{L}^{(r)} \to \mathscr{L}^{(r')}$$

*induced by the canonical homomorphism $\breve{\alpha}^{r,r'}: \breve{\mathscr{C}}^{(r)} \to \breve{\mathscr{C}}^{(r')}$* (III.10.31.7) *is annihilated by $p^\alpha$.*

(iii) *For every integer $q \geq 1$, there exists a rational number $\beta > 0$ such that the canonical morphism*

(III.11.16.3) $$\mathrm{R}^q\mathsf{T}_*(\breve{\mathscr{C}}^{(r)}) \to \mathrm{R}^q\mathsf{T}_*(\breve{\mathscr{C}}^{(r')})$$

*is annihilated by $p^\beta$.*

Let us use the notation of III.11.15 and moreover denote by $\breve{\mathscr{M}}^{(r)}$ the kernel of the morphism (III.11.15.1).

(i) The kernel of the morphism (III.11.16.1) is canonically isomorphic to $\lambda_*(\breve{\mathscr{M}}^{(r)})$ (III.11.1.10). It is therefore almost zero by virtue of III.11.15(i). Since $\mathscr{O}_{\mathfrak{X}}$ is $\mathscr{O}_C$-flat by III.4.2(i) (rig-pure in the terminology of [**1**] 2.10.1.4), the morphism (III.11.16.1) is injective.

(ii) Since $\mathrm{R}^1\lambda_*(\mathscr{O}_{\breve{\overline{X}}}) = 0$ by III.11.6, we have an exact sequence

(III.11.16.4) $$0 \to \mathrm{R}^1\lambda_*(\breve{\mathscr{M}}^{(r)}) \to \mathscr{L}^{(r)} \to \lambda_*(\breve{\mathscr{K}}^{(r)}).$$

The statement then follows from III.11.15(i)-(ii).

(iii) The proof is similar to that of III.11.15(iii). Consider the Cartan–Leray spectral sequence

(III.11.16.5) $${}^r\mathrm{E}_2^{i,j} = \mathrm{R}^i\lambda_*(\mathrm{R}^j\breve{\tau}_*(\breve{\mathscr{C}}^{(r)})) \Rightarrow \mathrm{R}^{i+j}\top_*(\breve{\mathscr{C}}^{(r)}),$$

and denote by $({}^r\mathrm{E}_i^q)_{0\leq i\leq q}$ the abutment filtration on $\mathrm{R}^q\top_*(\breve{\mathscr{C}}^{r)})$, so that we have

(III.11.16.6) $${}^r\mathrm{E}_i^q/{}^r\mathrm{E}_{i+1}^q = {}^r\mathrm{E}_\infty^{i,q-i}.$$

For every integer $0 \leq i \leq q+1$, set $r_i = r' + (q+1-i)(r-r')/(q+1)$. By III.11.15(iii), for every integer $0 \leq i \leq q-1$, there exists a rational number $\beta_i > 0$ such that the canonical morphism

(III.11.16.7) $${}^{r_i}\mathrm{E}_2^{i,q-i} \to {}^{r_{i+1}}\mathrm{E}_2^{i,q-i}$$

is annihilated by $p^{\beta_i}$. The same then holds for the morphism ${}^{r_i}\mathrm{E}_\infty^{i,q-i} \to {}^{r_{i+1}}\mathrm{E}_\infty^{i,q-i}$. On the other hand, $\mathrm{R}^q\breve{u}_*(\mathscr{O}_{\breve{\overline{X}}}) = 0$ by III.11.6. We deduce from this that the canonical morphism

(III.11.16.8) $$\mathrm{R}^q\lambda_*(\breve{\tau}_*(\breve{\mathscr{C}}^{(r')})) \to \mathrm{R}^q\lambda_*(\breve{\mathscr{K}}^{(r')})$$

is almost injective by III.11.15(i). Consequently, by virtue of III.11.15(ii), there exists a rational number $\beta_q > 0$ such that the canonical morphism

(III.11.16.9) $${}^{r_q}\mathrm{E}_2^{q,0} \to {}^{r_{q+1}}\mathrm{E}_2^{q,0}$$

is annihilated by $p^{\beta_q}$. The same then holds for the morphism ${}^{r_q}\mathrm{E}_\infty^{q,0} \to {}^{r_{q+1}}\mathrm{E}_\infty^{q,0}$. The desired statement follows by taking $\beta = \sum_{i=0}^q \beta_i$.

**Corollary III.11.17.** *Let $r$, $r'$ be two rational numbers such that $r > r' > 0$.*

(i) *The canonical homomorphism*

(III.11.17.1) $$u^r\colon \mathscr{O}_{\mathfrak{X}}[\frac{1}{p}] \to \top_*(\breve{\mathscr{C}}^{(r)})[\frac{1}{p}]$$

*admits (as an $\mathscr{O}_{\mathfrak{X}}[\frac{1}{p}]$-linear morphism) a canonical left inverse*

(III.11.17.2) $$v^r\colon \top_*(\breve{\mathscr{C}}^{(r)})[\frac{1}{p}] \to \mathscr{O}_{\mathfrak{X}}[\frac{1}{p}].$$

(ii) *The composition*

(III.11.17.3) $$\top_*(\breve{\mathscr{C}}^{(r)})[\frac{1}{p}] \xrightarrow{v^r} \mathscr{O}_{\mathfrak{X}}[\frac{1}{p}] \xrightarrow{u^{r'}} \top_*(\breve{\mathscr{C}}^{(r')})[\frac{1}{p}]$$

*is the canonical homomorphism.*

(iii) *For every integer $q \geq 1$, the canonical morphism*

(III.11.17.4) $$\mathrm{R}^q\top_*(\breve{\mathscr{C}}^{(r)})[\frac{1}{p}] \to \mathrm{R}^q\top_*(\breve{\mathscr{C}}^{(r')})[\frac{1}{p}]$$

*is zero.*

Indeed, by III.11.16(i)-(ii), $u^r$ is injective and there exists a unique $\mathscr{O}_{\mathfrak{X}}[\frac{1}{p}]$-linear morphism

(III.11.17.5) $$v^{r,r'}\colon \top_*(\breve{\mathscr{C}}^{(r)})[\frac{1}{p}] \to \mathscr{O}_{\mathfrak{X}}[\frac{1}{p}]$$

such that $u^{r'} \circ v^{r,r'}$ is the canonical homomorphism $\top_*(\breve{\mathscr{C}}^{(r)})[\frac{1}{p}] \to \top_*(\breve{\mathscr{C}}^{(r')})[\frac{1}{p}]$. Since we have $u^{r'} \circ v^{r,r'} \circ u^r = u^{r'}$, we deduce from this that $v^{r,r'}$ is a left inverse of $u^r$. One immediately verifies that it does not depend on $r'$, giving statements (i) and (ii). Statement (iii) immediately follows from III.11.16(iii).

**Corollary III.11.18.** *The canonical homomorphism*

$$\mathscr{O}_{\mathfrak{X}}[\frac{1}{p}] \to \varinjlim_{r\in\mathbb{Q}_{>0}} \top_*(\breve{\mathscr{C}}^{(r)})[\frac{1}{p}] \tag{III.11.18.1}$$

*is an isomorphism and for every integer* $q\geq 1$,

$$\varinjlim_{r\in\mathbb{Q}_{>0}} \mathrm{R}^q\top_*(\breve{\mathscr{C}}^{(r)})[\frac{1}{p}] = 0. \tag{III.11.18.2}$$

**III.11.19.** Recall that for every rational number $r\geq 0$ and every integer $n\geq 1$, the universal $\overline{\mathscr{B}}_n$-derivation of $\mathscr{C}_n^{(r)}$ (III.10.23.2)

$$d_n^{(r)}\colon \mathscr{C}_n^{(r)} \to \sigma_n^*(\xi^{-1}\widetilde{\Omega}^1_{\overline{X}_n/\overline{S}_n}) \otimes_{\overline{\mathscr{B}}_n} \mathscr{C}_n^{(r)} \tag{III.11.19.1}$$

is a Higgs $\overline{\mathscr{B}}_n$-field with coefficients in $\sigma_n^*(\xi^{-1}\widetilde{\Omega}^1_{\overline{X}_n/\overline{S}_n})$ (III.10.32). We denote the Dolbeault complex of the Higgs $\overline{\mathscr{B}}_n$-module $(\mathscr{C}_n^{(r)}, p^r d_n^{(r)})$ (II.2.8.2) by $\mathbb{K}^\bullet(\mathscr{C}_n^{(r)}, p^r d_n^{(r)})$ and the augmented Dolbeault complex

$$\overline{\mathscr{B}}_n \to \mathbb{K}^0(\mathscr{C}_n^{(r)}, p^r d_n^{(r)}) \to \mathbb{K}^1(\mathscr{C}_n^{(r)}, p^r d_n^{(r)}) \to \mathbb{K}^2(\mathscr{C}_n^{(r)}, p^r d_n^{(r)}) \to \dots, \tag{III.11.19.2}$$

where $\overline{\mathscr{B}}_n$ is placed in degree $-1$ and the differential $\overline{\mathscr{B}}_n \to \mathscr{C}_n^{(r)}$ is the canonical homomorphism, by $\widetilde{\mathbb{K}}^\bullet(\mathscr{C}_n^{(r)}, p^r d_n^{(r)})$.

For all rational numbers $r\geq r'\geq 0$, we have (III.10.23.3)

$$p^r(\mathrm{id}\otimes \alpha_n^{r,r'})\circ d_n^{(r)} = p^{r'} d_n^{(r')} \circ \alpha_n^{r,r'}, \tag{III.11.19.3}$$

where $\alpha_n^{r,r'}\colon \mathscr{C}_n^{(r)} \to \mathscr{C}_n^{(r')}$ is the homomorphism (III.10.21.5). Consequently, $\alpha_n^{r,r'}$ induces a morphism

$$\nu_n^{r,r'}\colon \widetilde{\mathbb{K}}^\bullet(\mathscr{C}_n^{(r)}, p^r d_n^{(r)}) \to \widetilde{\mathbb{K}}^\bullet(\mathscr{C}_n^{(r')}, p^{r'} d_n^{(r')}). \tag{III.11.19.4}$$

**Proposition III.11.20.** *For all rational numbers* $r>r'>0$, *there exists a rational number* $\alpha\geq 0$ *such that for all integers* $n$ *and* $q$ *with* $n\geq 1$, *the morphism*

$$\mathrm{H}^q(\nu_n^{r,r'})\colon \mathrm{H}^q(\widetilde{\mathbb{K}}^\bullet(\mathscr{C}_n^{(r)}, p^r d_n^{(r)})) \to \mathrm{H}^q(\widetilde{\mathbb{K}}^\bullet(\mathscr{C}_n^{(r')}, p^{r'} d_n^{(r')})) \tag{III.11.20.1}$$

*is annihilated by* $p^\alpha$.

Let $n$ be an integer $\geq 1$, $\overline{x}$ a geometric point of $X$ over $s$, $X'$ the strict localization of $X$ at $\overline{x}$, and

$$\varphi_{\overline{x}}\colon \widetilde{E} \to \overline{X}'^{\circ}_{\text{fét}} \tag{III.11.20.2}$$

the functor (III.8.8.4). By virtue of III.3.7, $\overline{X}'$ is normal and strictly local (and in particular integral). Since $\overline{X}'$ is $\overline{S}$-flat, $\overline{X}'^{\circ}$ is integral and nonempty. Let $v\colon \overline{y} \to \overline{X}'^{\circ}$ be a generic geometric point, and

$$\nu_{\overline{y}}\colon \overline{X}'^{\circ}_{\text{fét}} \xrightarrow{\sim} \mathbf{B}_{\pi_1(\overline{X}'^{\circ},\overline{y})} \tag{III.11.20.3}$$

the associated fiber functor (III.2.10.3). We denote also by $\overline{y}$ the geometric point of $\overline{X}^{\circ}$ and by $u\colon \overline{y}\to X'$ the morphism induced by $v$. We thus obtain a point $(\overline{y}\rightsquigarrow\overline{x})$ of $X_{\text{ét}} \overleftarrow{\times}_{X_{\text{ét}}} \overline{X}^{\circ}_{\text{ét}}$.

We denote by $\mathfrak{V}_{\overline{x}}$ (resp. $\mathfrak{V}_{\overline{x}}(\mathbf{Q})$) the category of neighborhoods of the point of $X_{\text{ét}}$ associated with $\overline{x}$ in the site $\mathbf{\acute{E}t}_{/X}$ (resp. $\mathbf{Q}$ (III.10.5)). For any $(U, \mathfrak{p}\colon \overline{x}\to U) \in \mathrm{Ob}(\mathfrak{V}_{\overline{x}})$, we denote also by $\mathfrak{p}\colon X'\to U$ the morphism deduced from $\mathfrak{p}$, and we set

$$\overline{\mathfrak{p}}^{\circ} = \mathfrak{p}\times_X \overline{X}^{\circ}\colon \overline{X}'^{\circ} \to \overline{U}^{\circ}. \tag{III.11.20.4}$$

We (abusively) denote also by $\overline{y}$ the geometric point $\overline{\mathfrak{p}}^\circ(v(\overline{y}))$ of $\overline{U}^\circ$. Note that $\overline{y}$ is localized at a generic point of $\overline{U}^\circ$ because $\overline{\mathfrak{p}}^\circ$ is flat. Since $\overline{U}$ is locally irreducible (III.3.3), it is the sum of the schemes induced on its irreducible components. Denote by $\overline{U}^\star$ the irreducible component of $\overline{U}$ containing $\overline{y}$. Likewise, $\overline{U}^\circ$ is the sum of the schemes induced on its irreducible components, and $\overline{U}^{\star\circ} = \overline{U}^\star \times_X X^\circ$ is the irreducible component of $\overline{U}^\circ$ containing $\overline{y}$. The morphism $\overline{\mathfrak{p}}^\circ$ factors through $\overline{U}^{\star\circ}$.

By (III.10.8.5) and (III.10.29.1), we have canonical isomorphisms of $\mathbf{B}_{\pi_1(\overline{X}'^\circ,\overline{y})}$

$$\nu_{\overline{y}}(\varphi_{\overline{x}}(\overline{\mathscr{B}}_n)) \xrightarrow{\sim} \varinjlim_{(U,\mathfrak{p})\in\mathfrak{V}_{\overline{x}}^\circ} \overline{R}_U^{\overline{y}}/p^n\overline{R}_U^{\overline{y}}, \tag{III.11.20.5}$$

$$\nu_{\overline{y}}(\varphi_{\overline{x}}(\mathscr{C}_n^{(r)})) \xrightarrow{\sim} \varinjlim_{(U,\mathfrak{p})\in\mathfrak{V}_{\overline{x}}(\mathbf{Q})^\circ} \mathscr{C}_U^{\overline{y},(r)}/p^n\mathscr{C}_U^{\overline{y},(r)}. \tag{III.11.20.6}$$

The rings underlying these representations are canonically isomorphic to the stalks of $\overline{\mathscr{B}}_n$ and $\mathscr{C}_n^{(r)}$ at $\rho(\overline{y} \rightsquigarrow \overline{x})$ (VI.10.31 and VI.9.9). On the other hand, we have canonical isomorphisms (III.11.13.11) and (III.11.3.9)

$$(\mathscr{O}_{\overline{X}_n})_{\overline{x}} \xrightarrow{\sim} \varinjlim_{(U,\mathfrak{p})\in\mathfrak{V}_{\overline{x}}^\circ} \mathscr{O}_{\overline{X}_n}(\overline{U}^\star), \tag{III.11.20.7}$$

$$(\widetilde{\Omega}^1_{\overline{X}_n/\overline{S}_n})_{\overline{x}} \xrightarrow{\sim} \varinjlim_{(U,\mathfrak{p})\in\mathfrak{V}_{\overline{x}}^\circ} \widetilde{\Omega}^1_{\overline{X}_n/\overline{S}_n}(\overline{U}^\star), \tag{III.11.20.8}$$

where we view $\mathscr{O}_{\overline{X}_n}$ and $\widetilde{\Omega}^1_{\overline{X}_n/\overline{S}_n}$ on the left as sheaves of $X_{s,\text{ét}}$ and on the right as sheaves of $\overline{X}_{\text{ét}}$. These modules are canonically isomorphic to the stalks of $\sigma_s^*(\mathscr{O}_{\overline{X}_n})$ and $\sigma_s^*(\widetilde{\Omega}^1_{\overline{X}_n/\overline{S}_n})$ at $\rho(\overline{y} \rightsquigarrow \overline{x})$ (VI.10.18.1). It follows from III.10.30 that the stalk of the derivation $d_n^{(r)}$ (III.10.23.2) at $\rho(\overline{y} \rightsquigarrow \overline{x})$ identifies with the direct limit on the category $\mathfrak{V}_{\overline{x}}(\mathbf{Q})^\circ$ of the universal $(\overline{R}_U^{\overline{y}}/p^n\overline{R}_U^{\overline{y}})$-derivations

$$\mathscr{C}_U^{\overline{y},(r)}/p^n\mathscr{C}_U^{\overline{y},(r)} \to \xi^{-1}\widetilde{\Omega}^1_{X/S}(U) \otimes_{\mathscr{O}_X(U)} (\mathscr{C}_U^{\overline{y},(r)}/p^n\mathscr{C}_U^{\overline{y},(r)}). \tag{III.11.20.9}$$

By III.9.5, the family of points $\rho(\overline{y} \rightsquigarrow \overline{x})$ of $\widetilde{E}_s$ is conservative. Since filtered direct limits are exact, the proposition then follows from II.12.3(i), in view of III.10.4 and the proof of III.10.14.

**III.11.21.** For any rational number $r \geq 0$, we denote also by

$$\breve{d}^{(r)}: \breve{\mathscr{C}}^{(r)} \to \mathrm{T}^*(\xi^{-1}\widetilde{\Omega}^1_{\mathfrak{X}/\mathscr{S}}) \otimes_{\breve{\overline{\mathscr{B}}}} \breve{\mathscr{C}}^{(r)} \tag{III.11.21.1}$$

the $\breve{\overline{\mathscr{B}}}$-derivation induced by $\breve{d}^{(r)}$ (III.10.31.9) and the isomorphism (III.11.2.6), which we identify with the universal $\breve{\overline{\mathscr{B}}}$-derivation of $\breve{\mathscr{C}}^{(r)}$. It is a Higgs $\breve{\overline{\mathscr{B}}}$-field with coefficients in $\mathrm{T}^*(\xi^{-1}\widetilde{\Omega}^1_{\mathfrak{X}/\mathscr{S}})$ (III.10.32). We denote by $\mathbb{K}^\bullet(\breve{\mathscr{C}}^{(r)}, p^r\breve{d}^{(r)})$ the Dolbeault complex of the Higgs $\breve{\overline{\mathscr{B}}}$-module $(\breve{\mathscr{C}}^{(r)}, p^r\breve{d}^{(r)})$ and by $\widetilde{\mathbb{K}}^\bullet(\breve{\mathscr{C}}^{(r)}, p^r\breve{d}^{(r)})$ the augmented Dolbeault complex

$$\breve{\overline{\mathscr{B}}} \to \mathbb{K}^0(\breve{\mathscr{C}}^{(r)}, p^r\breve{d}^{(r)}) \to \mathbb{K}^1(\breve{\mathscr{C}}^{(r)}, p^r\breve{d}^{(r)}) \to \cdots \to \mathbb{K}^n(\breve{\mathscr{C}}^{(r)}, p^r\breve{d}^{(r)}) \to \dots, \tag{III.11.21.2}$$

where $\breve{\overline{\mathscr{B}}}$ is placed in degree $-1$ and the differential $\breve{\overline{\mathscr{B}}} \to \breve{\mathscr{C}}^{(r)}$ is the canonical homomorphism.

For all rational numbers $r \geq r' \geq 0$, we have (III.10.31.10)

$$p^r(\mathrm{id} \otimes \breve{\alpha}^{r,r'}) \circ \breve{d}^{(r)} = p^{r'}\breve{d}^{(r')} \circ \breve{\alpha}^{r,r'}, \tag{III.11.21.3}$$

where $\breve{\alpha}^{r,r'}\colon \breve{\mathscr{C}}^{(r)} \to \breve{\mathscr{C}}^{(r')}$ is the homomorphism (III.10.31.7). Consequently, $\breve{\alpha}^{r,r'}$ induces a morphism of complexes

$$\breve{\nu}^{r,r'}\colon \widetilde{\mathbb{K}}^{\bullet}(\breve{\mathscr{C}}^{(r)}, p^r\breve{d}^{(r)}) \to \widetilde{\mathbb{K}}^{\bullet}(\breve{\mathscr{C}}^{(r')}, p^{r'}\breve{d}^{(r')}). \tag{III.11.21.4}$$

We denote by $\mathbf{Mod}_{\mathbb{Q}}(\breve{\overline{\mathscr{B}}})$ the category of $\breve{\overline{\mathscr{B}}}$-modules of $\widetilde{E}_s^{\mathbb{N}^\circ}$ up to isogeny (III.6.2), and by $\mathbb{K}^{\bullet}_{\mathbb{Q}}(\breve{\mathscr{C}}^{(r)}, p^r\breve{d}^{(r)})$ and $\widetilde{\mathbb{K}}^{\bullet}_{\mathbb{Q}}(\breve{\mathscr{C}}^{(r)}, p^r\breve{d}^{(r)})$ the images of the complexes $\mathbb{K}^{\bullet}(\breve{\mathscr{C}}^{(r)}, p^r\breve{d}^{(r)})$ and $\widetilde{\mathbb{K}}^{\bullet}(\breve{\mathscr{C}}^{(r)}, p^r\breve{d}^{(r)})$ in $\mathbf{Mod}_{\mathbb{Q}}(\breve{\overline{\mathscr{B}}})$.

**Proposition III.11.22.** *For all rational numbers $r > r' > 0$ and every integer $q$, the canonical morphism* (III.11.21.4)

$$\mathrm{H}^q(\breve{\nu}^{r,r'}_{\mathbb{Q}})\colon \mathrm{H}^q(\widetilde{\mathbb{K}}^{\bullet}_{\mathbb{Q}}(\breve{\mathscr{C}}^{(r)}, p^r\breve{d}^{(r)})) \to \mathrm{H}^q(\widetilde{\mathbb{K}}^{\bullet}_{\mathbb{Q}}(\breve{\mathscr{C}}^{(r')}, p^{r'}\breve{d}^{(r')})) \tag{III.11.22.1}$$

*is zero.*

This follows from III.11.20 and III.7.3(i).

**Corollary III.11.23.** *Let $r$, $r'$ be two rational numbers such that $r > r' > 0$.*

(i) *The canonical morphism*

$$u^r\colon \breve{\overline{\mathscr{B}}}_{\mathbb{Q}} \to \mathrm{H}^0(\mathbb{K}^{\bullet}_{\mathbb{Q}}(\breve{\mathscr{C}}^{(r)}, p^r\breve{d}^{(r)})) \tag{III.11.23.1}$$

*admits a canonical left inverse*

$$v^r\colon \mathrm{H}^0(\mathbb{K}^{\bullet}_{\mathbb{Q}}(\breve{\mathscr{C}}^{(r)}, p^r\breve{d}^{(r)})) \to \breve{\overline{\mathscr{B}}}_{\mathbb{Q}}. \tag{III.11.23.2}$$

(ii) *The composition*

$$\mathrm{H}^0(\mathbb{K}^{\bullet}_{\mathbb{Q}}(\breve{\mathscr{C}}^{(r)}, p^r\breve{d}^{(r)})) \xrightarrow{v^r} \breve{\overline{\mathscr{B}}}_{\mathbb{Q}} \xrightarrow{u^{r'}} \mathrm{H}^0(\mathbb{K}^{\bullet}_{\mathbb{Q}}(\breve{\mathscr{C}}^{(r')}, p^{r'}\breve{d}^{(r')})) \tag{III.11.23.3}$$

*is the canonical morphism.*

(iii) *For every integer $q \geq 1$, the canonical morphism*

$$\mathrm{H}^q(\mathbb{K}^{\bullet}_{\mathbb{Q}}(\breve{\mathscr{C}}^{(r)}, p^r\breve{d}^{(r)})) \to \mathrm{H}^q(\mathbb{K}^{\bullet}_{\mathbb{Q}}(\breve{\mathscr{C}}^{(r')}, p^{r'}\breve{d}^{(r')})) \tag{III.11.23.4}$$

*is zero.*

Indeed, consider the canonical commutative diagram (without the dotted arrow)

$$\begin{array}{ccccc} \breve{\overline{\mathscr{B}}}_{\mathbb{Q}} & \xrightarrow{u^r} & \mathrm{H}^0(\mathbb{K}^{\bullet}_{\mathbb{Q}}(\breve{\mathscr{C}}^{(r)}, p^r\breve{d}^{(r)})) & \twoheadrightarrow & \mathrm{H}^0(\widetilde{\mathbb{K}}^{\bullet}_{\mathbb{Q}}(\breve{\mathscr{C}}^{(r)}, p^r\breve{d}^{(r)})) \\ \| & \swarrow{\scriptstyle v^{r,r'}} & \downarrow{\scriptstyle \varpi^{r,r'}} & & \downarrow{\scriptstyle \mathrm{H}^0(\breve{\nu}^{r,r'}_{\mathbb{Q}})} \\ \breve{\overline{\mathscr{B}}}_{\mathbb{Q}} & \xrightarrow[u^{r'}]{} & \mathrm{H}^0(\mathbb{K}^{\bullet}_{\mathbb{Q}}(\breve{\mathscr{C}}^{(r')}, p^{r'}\breve{d}^{(r')})) & \twoheadrightarrow & \mathrm{H}^0(\widetilde{\mathbb{K}}^{\bullet}_{\mathbb{Q}}(\breve{\mathscr{C}}^{(r')}, p^{r'}\breve{d}^{(r')})) \end{array} \tag{III.11.23.5}$$

It follows from III.11.22 that $u^r$ and consequently $u^{r'}$ are injective, and that there exists a unique morphism $v^{r,r'}$ as above such that $\varpi^{r,r'} = u^{r'} \circ v^{r,r'}$. Since we have $u^{r'} \circ v^{r,r'} \circ u^r = u^{r'}$, it follows that $v^{r,r'}$ is a left inverse of $u^r$. One immediately verifies that it does not depend on $r'$, giving statements (i) and (ii). Statement (iii) immediately follows from III.11.22.

**Corollary III.11.24.** *The canonical morphism*

$$\breve{\overline{\mathscr{B}}}_{\mathbb{Q}} \to \varinjlim_{r \in \mathbb{Q}_{>0}} \mathrm{H}^0(\mathbb{K}^{\bullet}_{\mathbb{Q}}(\breve{\mathscr{C}}^{(r)}, p^r\breve{d}^{(r)})) \tag{III.11.24.1}$$

*is an isomorphism, and for every integer $q \geq 1$,*

$$\varinjlim_{r \in \mathbb{Q}_{>0}} \mathrm{H}^q(\mathbb{K}^{\bullet}_{\mathbb{Q}}(\breve{\mathscr{C}}^{(r)}, p^r\breve{d}^{(r)})) = 0. \tag{III.11.24.2}$$

This follows from III.11.23.

**Remark III.11.25.** Filtered direct limits are not a priori representable in the category $\mathbf{Mod}_{\mathbb{Q}}(\breve{\overline{\mathscr{B}}})$.

## III.12. Dolbeault modules

**III.12.1.** We keep the assumptions and general notation of III.10 in this section. We moreover denote by $\mathbf{Mod}(\breve{\overline{\mathscr{B}}})$ the category of $\breve{\overline{\mathscr{B}}}$-modules of $\widetilde{E}_s^{\mathbb{N}^\circ}$ (III.10.31), by $\mathbf{Mod}^{\rm ad}(\breve{\overline{\mathscr{B}}})$ (resp. $\mathbf{Mod}^{\rm aft}(\breve{\overline{\mathscr{B}}})$) the full subcategory made up of adic $\breve{\overline{\mathscr{B}}}$-modules (resp. adic $\breve{\overline{\mathscr{B}}}$-modules of finite type) (III.7.16), and by $\mathbf{Mod}_{\mathbb{Q}}(\breve{\overline{\mathscr{B}}})$ (resp. $\mathbf{Mod}_{\mathbb{Q}}^{\rm ad}(\breve{\overline{\mathscr{B}}})$, resp. $\mathbf{Mod}_{\mathbb{Q}}^{\rm aft}(\breve{\overline{\mathscr{B}}})$) the category of objects of $\mathbf{Mod}(\breve{\overline{\mathscr{B}}})$ (resp. $\mathbf{Mod}^{\rm ad}(\breve{\overline{\mathscr{B}}})$, resp. $\mathbf{Mod}^{\rm aft}(\breve{\overline{\mathscr{B}}})$) up to isogeny (III.6.1.1). The category $\mathbf{Mod}_{\mathbb{Q}}(\breve{\overline{\mathscr{B}}})$ is then abelian and the canonical functors

$$\mathbf{Mod}_{\mathbb{Q}}^{\rm aft}(\breve{\overline{\mathscr{B}}}) \to \mathbf{Mod}_{\mathbb{Q}}^{\rm ad}(\breve{\overline{\mathscr{B}}}) \to \mathbf{Mod}_{\mathbb{Q}}(\breve{\overline{\mathscr{B}}}) \tag{III.12.1.1}$$

are fully faithful.

We denote by $\mathfrak{X}$ the formal $\mathscr{S}$-scheme $p$-adic completion of $\overline{X}$ and by $\xi^{-1}\widetilde{\Omega}^1_{\mathfrak{X}/\mathscr{S}}$ the $p$-adic completion of the $\mathscr{O}_{\overline{X}}$-module $\xi^{-1}\widetilde{\Omega}^1_{\overline{X}/\overline{S}} = \xi^{-1}\widetilde{\Omega}^1_{X/S} \otimes_{\mathscr{O}_X} \mathscr{O}_{\overline{X}}$ (cf. III.2.1). We denote by $\mathbf{Mod}^{\rm coh}(\mathscr{O}_{\mathfrak{X}})$ (resp. $\mathbf{Mod}^{\rm coh}(\mathscr{O}_{\mathfrak{X}}[\frac{1}{p}])$) the category of coherent $\mathscr{O}_{\mathfrak{X}}$-modules (resp. $\mathscr{O}_{\mathfrak{X}}[\frac{1}{p}]$-modules) of $X_{s,\rm zar}$ (III.6.15).

Let

$$\top\colon (\widetilde{E}_s^{\mathbb{N}^\circ}, \breve{\overline{\mathscr{B}}}) \to (X_{s,\rm zar}, \mathscr{O}_{\mathfrak{X}}) \tag{III.12.1.2}$$

be the morphism of ringed topos defined in (III.11.1.11). The functor $\top_*$ induces an additive left exact functor that we denote also by

$$\top_*\colon \mathbf{Mod}_{\mathbb{Q}}(\breve{\overline{\mathscr{B}}}) \to \mathbf{Mod}(\mathscr{O}_{\mathfrak{X}}[\frac{1}{p}]). \tag{III.12.1.3}$$

By III.6.16, the functor $\top^*$ induces an additive functor that we denote also by

$$\top^*\colon \mathbf{Mod}^{\rm coh}(\mathscr{O}_{\mathfrak{X}}[\frac{1}{p}]) \to \mathbf{Mod}_{\mathbb{Q}}^{\rm aft}(\breve{\overline{\mathscr{B}}}). \tag{III.12.1.4}$$

For every coherent $\mathscr{O}_{\mathfrak{X}}[\frac{1}{p}]$-module $\mathscr{F}$ and every $\breve{\overline{\mathscr{B}}}_{\mathbb{Q}}$-module $\mathscr{G}$, we have a bifunctorial canonical homomorphism

$$\mathrm{Hom}_{\breve{\overline{\mathscr{B}}}_{\mathbb{Q}}}(\top^*(\mathscr{F}), \mathscr{G}) \to \mathrm{Hom}_{\mathscr{O}_{\mathfrak{X}}[\frac{1}{p}]}(\mathscr{F}, \top_*(\mathscr{G})), \tag{III.12.1.5}$$

that is injective by (III.6.15.3) and III.6.16. We abusively call *adjoint* of a $\breve{\overline{\mathscr{B}}}_{\mathbb{Q}}$-linear morphism $\top^*(\mathscr{F}) \to \mathscr{G}$ its image by the homomorphism (III.12.1.5).

We denote by

$$\mathrm{R}\top_*\colon \mathbf{D}^+(\mathbf{Mod}_{\mathbb{Q}}(\breve{\overline{\mathscr{B}}})) \to \mathbf{D}^+(\mathbf{Mod}(\mathscr{O}_{\mathfrak{X}}[\frac{1}{p}])), \tag{III.12.1.6}$$

$$\mathrm{R}^q\top_*\colon \mathbf{Mod}_{\mathbb{Q}}(\breve{\overline{\mathscr{B}}}) \to \mathbf{Mod}(\mathscr{O}_{\mathfrak{X}}[\frac{1}{p}]), \quad (q \in \mathbb{N}), \tag{III.12.1.7}$$

the right derived functors of the functor $\top_*$ (III.12.1.3). This notation does not lead to any confusion with that of the right derived functors of the functor $\top_*\colon \mathbf{Mod}(\breve{\overline{\mathscr{B}}}) \to \mathbf{Mod}(\mathscr{O}_{\mathfrak{X}})$, because the localization functor $\mathbf{Mod}(\breve{\overline{\mathscr{B}}}) \to \mathbf{Mod}_{\mathbb{Q}}(\breve{\overline{\mathscr{B}}})$ is exact and transforms injective objects into injective objects.

**III.12.2.** Let $\mathscr{M}$ be an $\mathscr{O}_{\mathfrak{X}}$-module, $\mathscr{N}$ a $\breve{\overline{\mathscr{B}}}$-module, and $q$ an integer $\geq 0$. The adjunction morphism $\mathscr{M} \to \mathsf{T}_*(\mathsf{T}^*(\mathscr{M}))$ and the cup product induce a bifunctorial morphism

$$\mathscr{M} \otimes_{\mathscr{O}_{\mathfrak{X}}} \mathrm{R}^q\mathsf{T}_*(\mathscr{N}) \to \mathrm{R}^q\mathsf{T}_*(\mathsf{T}^*(\mathscr{M}) \otimes_{\breve{\overline{\mathscr{B}}}} \mathscr{N}). \tag{III.12.2.1}$$

We can make the following remarks:

(i) For every $\mathscr{O}_{\mathfrak{X}}$-module $\mathscr{M}'$, the composition

$$\begin{array}{ccc} \mathscr{M} \otimes_{\mathscr{O}_{\mathfrak{X}}} \mathscr{M}' \otimes_{\mathscr{O}_{\mathfrak{X}}} \mathrm{R}^q\mathsf{T}_*(\mathscr{N}) & \longrightarrow & \mathscr{M} \otimes_{\mathscr{O}_{\mathfrak{X}}} \mathrm{R}^q\mathsf{T}_*(\mathsf{T}^*(\mathscr{M}') \otimes_{\breve{\overline{\mathscr{B}}}} \mathscr{N}) \\ & \searrow & \downarrow \\ & & \mathrm{R}^q\mathsf{T}_*(\mathsf{T}^*(\mathscr{M} \otimes_{\mathscr{O}_{\mathfrak{X}}} \mathscr{M}') \otimes_{\breve{\overline{\mathscr{B}}}} \mathscr{N}) \end{array} \tag{III.12.2.2}$$

of the morphisms induced by the morphisms (III.12.2.1) with respect to $\mathscr{M}$ and $\mathscr{M}'$ is none other than the morphism (III.12.2.1) with respect to $\mathscr{M} \otimes_{\mathscr{O}_{\mathfrak{X}}} \mathscr{M}'$.

(ii) When $q = 0$, the morphism (III.12.2.1) is the composition

$$\mathscr{M} \otimes_{\mathscr{O}_{\mathfrak{X}}} \mathsf{T}_*(\mathscr{N}) \to \mathsf{T}_*(\mathsf{T}^*(\mathscr{M} \otimes_{\mathscr{O}_{\mathfrak{X}}} \mathsf{T}_*(\mathscr{N}))) \to \mathsf{T}_*(\mathsf{T}^*(\mathscr{M}) \otimes_{\breve{\overline{\mathscr{B}}}} \mathscr{N}), \tag{III.12.2.3}$$

where the first arrow is the adjunction morphism and the second arrow is induced by the canonical morphism $\mathsf{T}^*(\mathsf{T}_*(\mathscr{N})) \to \mathscr{N}$. Its adjoint

$$\mathsf{T}^*(\mathscr{M} \otimes_{\mathscr{O}_{\mathfrak{X}}} \mathsf{T}_*(\mathscr{N})) \to \mathsf{T}^*(\mathscr{M}) \otimes_{\breve{\overline{\mathscr{B}}}} \mathscr{N} \tag{III.12.2.4}$$

is therefore induced by the canonical morphism $\mathsf{T}^*(\mathsf{T}_*(\mathscr{N})) \to \mathscr{N}$.

**III.12.3.** Let $\mathscr{F}$ be a coherent $\mathscr{O}_{\mathfrak{X}}[\frac{1}{p}]$-module, $\mathscr{G}$ a $\breve{\overline{\mathscr{B}}}_{\mathbb{Q}}$-module, and $q$ an integer $\geq 0$. In view of III.6.16, the morphism (III.12.2.1) induces a bifunctorial morphism

$$\mathscr{F} \otimes_{\mathscr{O}_{\mathfrak{X}}[\frac{1}{p}]} \mathrm{R}^q\mathsf{T}_*(\mathscr{G}) \to \mathrm{R}^q\mathsf{T}_*(\mathsf{T}^*(\mathscr{F}) \otimes_{\breve{\overline{\mathscr{B}}}_{\mathbb{Q}}} \mathscr{G}). \tag{III.12.3.1}$$

We can make the following remarks:

(i) Let $\mathscr{F}'$ be a coherent $\mathscr{O}_{\mathfrak{X}}[\frac{1}{p}]$-module. It follows from III.12.2(i) that the composition

$$\begin{array}{ccc} \mathscr{F} \otimes_{\mathscr{O}_{\mathfrak{X}}[\frac{1}{p}]} \mathscr{F}' \otimes_{\mathscr{O}_{\mathfrak{X}}[\frac{1}{p}]} \mathrm{R}^q\mathsf{T}_*(\mathscr{G}) & \longrightarrow & \mathscr{F} \otimes_{\mathscr{O}_{\mathfrak{X}}[\frac{1}{p}]} \mathrm{R}^q\mathsf{T}_*(\mathsf{T}^*(\mathscr{F}') \otimes_{\breve{\overline{\mathscr{B}}}_{\mathbb{Q}}} \mathscr{G}) \\ & \searrow & \downarrow \\ & & \mathrm{R}^q\mathsf{T}_*(\mathsf{T}^*(\mathscr{F} \otimes_{\mathscr{O}_{\mathfrak{X}}[\frac{1}{p}]} \mathscr{F}') \otimes_{\breve{\overline{\mathscr{B}}}_{\mathbb{Q}}} \mathscr{G}) \end{array} \tag{III.12.3.2}$$

of the morphisms induced by the morphisms (III.12.3.1) with respect to $\mathscr{F}$ and $\mathscr{F}'$ is none other than the morphism (III.12.3.1) with respect to $\mathscr{F} \otimes_{\mathscr{O}_{\mathfrak{X}}[\frac{1}{p}]} \mathscr{F}'$.

(ii) Let $\mathscr{L}$ be a coherent $\mathscr{O}_{\mathfrak{X}}[\frac{1}{p}]$-module, $u\colon \mathsf{T}^*(\mathscr{L}) \to \mathscr{G}$ a $\breve{\overline{\mathscr{B}}}_{\mathbb{Q}}$-linear morphism, and $v\colon \mathscr{L} \to \mathsf{T}_*(\mathscr{G})$ the adjoint morphism (III.12.1.5). It then follows from III.12.2(ii) and III.6.16 that the morphism

$$\mathscr{F} \otimes_{\mathscr{O}_{\mathfrak{X}}[\frac{1}{p}]} \mathscr{L} \to \mathsf{T}_*(\mathsf{T}^*(\mathscr{F}) \otimes_{\breve{\overline{\mathscr{B}}}_{\mathbb{Q}}} \mathscr{G}) \tag{III.12.3.3}$$

induced by (III.12.3.1) and $v$ is the adjoint of the morphism

$$\mathsf{T}^*(\mathscr{F} \otimes_{\mathscr{O}_{\mathfrak{X}}[\frac{1}{p}]} \mathscr{L}) \to \mathsf{T}^*(\mathscr{F}) \otimes_{\breve{\overline{\mathscr{B}}}_{\mathbb{Q}}} \mathscr{G} \tag{III.12.3.4}$$

induced by $u$.

**Lemma III.12.4.** (i) *Let $\mathscr{M}$ be a locally free $\mathscr{O}_{\mathfrak{X}}$-module of finite type, $\mathscr{N}$ a $\breve{\overline{\mathscr{B}}}$-module, and $q$ an integer $\geq 0$. Then the canonical morphism* (III.12.2.1)

(III.12.4.1) $$\mathscr{M} \otimes_{\mathscr{O}_{\mathfrak{X}}} \mathrm{R}^q\top_*(\mathscr{N}) \to \mathrm{R}^q\top_*(\top^*(\mathscr{M}) \otimes_{\breve{\overline{\mathscr{B}}}} \mathscr{N})$$

*is an isomorphism.*

(ii) *Let $\mathscr{F}$ be a locally projective $\mathscr{O}_{\mathfrak{X}}[\frac{1}{p}]$-module of finite type* (III.2.8), *$\mathscr{G}$ a $\breve{\overline{\mathscr{B}}}_{\mathbb{Q}}$-module, and $q$ an integer $\geq 0$. Then the canonical morphism* (III.12.3.1)

(III.12.4.2) $$\mathscr{F} \otimes_{\mathscr{O}_{\mathfrak{X}}[\frac{1}{p}]} \mathrm{R}^q\top_*(\mathscr{G}) \to \mathrm{R}^q\top_*(\top^*(\mathscr{F}) \otimes_{\breve{\overline{\mathscr{B}}}_{\mathbb{Q}}} \mathscr{G})$$

*is an isomorphism.*

We only prove (ii); the proof of (i) is similar and simpler. There exists a Zariski open covering $(U_i)_{i\in I}$ of $X$ such that for every $i \in I$, the restriction of $\mathscr{F}$ to $(U_i)_s$ is a direct summand of a free $(\mathscr{O}_{\mathfrak{X}}|U_i)[\frac{1}{p}]$-module of finite type. In view of III.9.15, we may then restrict to the case where $\mathscr{F}$ is a direct summand of a free $\mathscr{O}_{\mathfrak{X}}[\frac{1}{p}]$-module of finite type, and even to the case where $\mathscr{F}$ is a free $\mathscr{O}_{\mathfrak{X}}[\frac{1}{p}]$-module of finite type, in which case the assertion is obvious.

**III.12.5.** We denote by $\mathbf{HI}(\mathscr{O}_{\mathfrak{X}}, \xi^{-1}\widetilde{\Omega}^1_{\mathfrak{X}/\mathscr{S}})$ the category of Higgs $\mathscr{O}_{\mathfrak{X}}$-isogenies with coefficients in $\xi^{-1}\widetilde{\Omega}^1_{\mathfrak{X}/\mathscr{S}}$ (III.6.8) and by $\mathbf{HI}^{\mathrm{coh}}(\mathscr{O}_{\mathfrak{X}}, \xi^{-1}\widetilde{\Omega}^1_{\mathfrak{X}/\mathscr{S}})$ the full subcategory made up of the quadruples $(\mathscr{M}, \mathscr{N}, u, \theta)$ such that $\mathscr{M}$ and $\mathscr{N}$ are coherent $\mathscr{O}_{\mathfrak{X}}$-modules. These are additive categories. We denote by $\mathbf{HI}_{\mathbb{Q}}(\mathscr{O}_{\mathfrak{X}}, \xi^{-1}\widetilde{\Omega}^1_{\mathfrak{X}/\mathscr{S}})$ (resp. $\mathbf{HI}^{\mathrm{coh}}_{\mathbb{Q}}(\mathscr{O}_{\mathfrak{X}}, \xi^{-1}\widetilde{\Omega}^1_{\mathfrak{X}/\mathscr{S}})$) the category of objects of $\mathbf{HI}(\mathscr{O}_{\mathfrak{X}}, \xi^{-1}\widetilde{\Omega}^1_{\mathfrak{X}/\mathscr{S}})$ (resp. $\mathbf{HI}^{\mathrm{coh}}(\mathscr{O}_{\mathfrak{X}}, \xi^{-1}\widetilde{\Omega}^1_{\mathfrak{X}/\mathscr{S}})$) up to isogeny (III.6.1.1).

By a Higgs $\mathscr{O}_{\mathfrak{X}}[\frac{1}{p}]$-module with coefficients in $\xi^{-1}\widetilde{\Omega}^1_{\mathfrak{X}/\mathscr{S}}$, we mean a Higgs $\mathscr{O}_{\mathfrak{X}}[\frac{1}{p}]$-module with coefficients in $\xi^{-1}\widetilde{\Omega}^1_{\mathfrak{X}/\mathscr{S}}[\frac{1}{p}]$ (II.2.8). From now on, we will leave the Higgs field out of the notation of a Higgs module when we do not need it explicitly. We denote by $\mathbf{HM}(\mathscr{O}_{\mathfrak{X}}[\frac{1}{p}], \xi^{-1}\widetilde{\Omega}^1_{\mathfrak{X}/\mathscr{S}})$ the category of Higgs $\mathscr{O}_{\mathfrak{X}}[\frac{1}{p}]$-modules with coefficients in $\xi^{-1}\widetilde{\Omega}^1_{\mathfrak{X}/\mathscr{S}}$ and by $\mathbf{HM}^{\mathrm{coh}}(\mathscr{O}_{\mathfrak{X}}[\frac{1}{p}], \xi^{-1}\widetilde{\Omega}^1_{\mathfrak{X}/\mathscr{S}})$ the full subcategory made up of Higgs modules whose underlying $\mathscr{O}_{\mathfrak{X}}[\frac{1}{p}]$-module is coherent. The functor (III.6.19.1)

(III.12.5.1) $$\begin{array}{rcl}\mathbf{HI}(\mathscr{O}_{\mathfrak{X}}, \xi^{-1}\widetilde{\Omega}^1_{\mathfrak{X}/\mathscr{S}}) & \to & \mathbf{HM}(\mathscr{O}_{\mathfrak{X}}[\frac{1}{p}], \xi^{-1}\widetilde{\Omega}^1_{\mathfrak{X}/\mathscr{S}}) \\ (\mathscr{M}, \mathscr{N}, u, \theta) & \mapsto & (\mathscr{M}_{\mathbb{Q}_p}, (\mathrm{id} \otimes u^{-1}_{\mathbb{Q}_p}) \circ \theta_{\mathbb{Q}_p})\end{array}$$

induces a functor

(III.12.5.2) $$\mathbf{HI}_{\mathbb{Q}}(\mathscr{O}_{\mathfrak{X}}, \xi^{-1}\widetilde{\Omega}^1_{\mathfrak{X}/\mathscr{S}}) \to \mathbf{HM}(\mathscr{O}_{\mathfrak{X}}[\frac{1}{p}], \xi^{-1}\widetilde{\Omega}^1_{\mathfrak{X}/\mathscr{S}}).$$

By III.6.21, this induces an equivalence of categories

(III.12.5.3) $$\mathbf{HI}^{\mathrm{coh}}_{\mathbb{Q}}(\mathscr{O}_{\mathfrak{X}}, \xi^{-1}\widetilde{\Omega}^1_{\mathfrak{X}/\mathscr{S}}) \xrightarrow{\sim} \mathbf{HM}^{\mathrm{coh}}(\mathscr{O}_{\mathfrak{X}}[\frac{1}{p}], \xi^{-1}\widetilde{\Omega}^1_{\mathfrak{X}/\mathscr{S}}).$$

**Definition III.12.6.** A *Higgs $\mathscr{O}_{\mathfrak{X}}[\frac{1}{p}]$-bundle with coefficients in $\xi^{-1}\widetilde{\Omega}^1_{\mathfrak{X}/\mathscr{S}}$* is a Higgs $\mathscr{O}_{\mathfrak{X}}[\frac{1}{p}]$-module with coefficients in $\xi^{-1}\widetilde{\Omega}^1_{\mathfrak{X}/\mathscr{S}}$ whose underlying $\mathscr{O}_{\mathfrak{X}}[\frac{1}{p}]$-module is locally projective of finite type (III.2.8).

**III.12.7.** Let $r$ be a rational number $\geq 0$. We denote also by

$$\breve{d}^{(r)}\colon \breve{\mathscr{C}}^{(r)} \to \mathrm{T}^*(\xi^{-1}\widetilde{\Omega}^1_{\mathfrak{X}/\mathscr{S}}) \otimes_{\breve{\overline{\mathscr{B}}}} \breve{\mathscr{C}}^{(r)} \tag{III.12.7.1}$$

the $\breve{\overline{\mathscr{B}}}$-derivation induced by $\breve{d}^{(r)}$ (III.10.31.9) and the isomorphism (III.11.2.6), which we identify with the universal $\breve{\overline{\mathscr{B}}}$-derivation of $\breve{\mathscr{C}}^{(r)}$. It is a Higgs $\breve{\overline{\mathscr{B}}}$-field with coefficients in $\mathrm{T}^*(\xi^{-1}\widetilde{\Omega}^1_{\mathfrak{X}/\mathscr{S}})$ (III.10.32). We denote by $\Xi^r$ the category of integrable $p^r$-isoconnections with respect to the extension $\breve{\mathscr{C}}^{(r)}/\breve{\overline{\mathscr{B}}}$ (III.6.10). It is an additive category. We denote by $\Xi^r_{\mathbb{Q}}$ the category of objects of $\Xi^r$ up to isogeny (III.6.1.1). By III.6.12 and III.10.32(iii), every object of $\Xi^r$ is a Higgs $\breve{\overline{\mathscr{B}}}$-isogeny with coefficients in $\mathrm{T}^*(\xi^{-1}\widetilde{\Omega}^1_{\mathfrak{X}/\mathscr{S}})$ (III.6.8). In particular, we can associate, functorially, with every object of $\Xi^r_{\mathbb{Q}}$ a Dolbeault complex in $\mathbf{Mod}_{\mathbb{Q}}(\breve{\overline{\mathscr{B}}})$ (cf. III.6.9).

Consider the functor

$$\mathfrak{S}^r\colon \mathbf{Mod}(\breve{\overline{\mathscr{B}}}) \to \Xi^r, \quad \mathscr{M} \mapsto (\breve{\mathscr{C}}^{(r)} \otimes_{\breve{\overline{\mathscr{B}}}} \mathscr{M}, \breve{\mathscr{C}}^{(r)} \otimes_{\breve{\overline{\mathscr{B}}}} \mathscr{M}, \mathrm{id}, p^r\breve{d}^{(r)} \otimes \mathrm{id}), \tag{III.12.7.2}$$

and denote also by

$$\mathfrak{S}^r\colon \mathbf{Mod}_{\mathbb{Q}}(\breve{\overline{\mathscr{B}}}) \to \Xi^r_{\mathbb{Q}} \tag{III.12.7.3}$$

the induced functor. Consider, on the other hand, the functor

$$\mathscr{K}^r\colon \Xi^r \to \mathbf{Mod}(\breve{\overline{\mathscr{B}}}), \quad (\mathscr{F}, \mathscr{G}, u, \nabla) \mapsto \ker(\nabla), \tag{III.12.7.4}$$

and denote also by

$$\mathscr{K}^r\colon \Xi^r_{\mathbb{Q}} \to \mathbf{Mod}_{\mathbb{Q}}(\breve{\overline{\mathscr{B}}}) \tag{III.12.7.5}$$

the induced functor. It is clear that the functor (III.12.7.2) is a left adjoint of the functor (III.12.7.4). Consequently, the functor (III.12.7.3) is a left adjoint of the functor (III.12.7.5).

By III.6.12, if $(\mathscr{N}, \mathscr{N}', v, \theta)$ is a Higgs $\mathscr{O}_{\mathfrak{X}}$-isogeny with coefficients in $\xi^{-1}\widetilde{\Omega}^1_{\mathfrak{X}/\mathscr{S}}$,

$$(\breve{\mathscr{C}}^{(r)} \otimes_{\breve{\overline{\mathscr{B}}}} \mathrm{T}^*(\mathscr{N}), \breve{\mathscr{C}}^{(r)} \otimes_{\breve{\overline{\mathscr{B}}}} \mathrm{T}^*(\mathscr{N}'), \mathrm{id} \otimes_{\breve{\overline{\mathscr{B}}}} \mathrm{T}^*(v), p^r\breve{d}^{(r)} \otimes \mathrm{T}^*(v) + \mathrm{id} \otimes \mathrm{T}^*(\theta)) \tag{III.12.7.6}$$

is an object of $\Xi^r$. We thus obtain a functor

$$\mathrm{T}^{r+}\colon \mathbf{HI}(\mathscr{O}_{\mathfrak{X}}, \xi^{-1}\widetilde{\Omega}^1_{\mathfrak{X}/\mathscr{S}}) \to \Xi^r. \tag{III.12.7.7}$$

By (III.12.5.3), this induces a functor that we denote also by

$$\mathrm{T}^{r+}\colon \mathbf{HM}^{\mathrm{coh}}(\mathscr{O}_{\mathfrak{X}}[\frac{1}{p}], \xi^{-1}\widetilde{\Omega}^1_{\mathfrak{X}/\mathscr{S}}) \to \Xi^r_{\mathbb{Q}}. \tag{III.12.7.8}$$

Let $(\mathscr{F}, \mathscr{G}, u, \nabla)$ be an object of $\Xi^r$. In view of III.12.4(i), $\nabla$ induces an $\mathscr{O}_{\mathfrak{X}}$-linear morphism

$$\mathrm{T}_*(\nabla)\colon \mathrm{T}_*(\mathscr{F}) \to \xi^{-1}\widetilde{\Omega}^1_{\mathfrak{X}/\mathscr{S}} \otimes_{\mathscr{O}_{\mathfrak{X}}} \mathrm{T}_*(\mathscr{G}). \tag{III.12.7.9}$$

We easily deduce from III.12.3(i) that $(\mathrm{T}_*(\mathscr{F}), \mathrm{T}_*(\mathscr{G}), \mathrm{T}_*(u), \mathrm{T}_*(\nabla))$ is a Higgs $\mathscr{O}_{\mathfrak{X}}$-isogeny with coefficients in $\xi^{-1}\widetilde{\Omega}^1_{\mathfrak{X}/\mathscr{S}}$. We thus obtain a functor

$$\mathrm{T}^r_+\colon \Xi^r \to \mathbf{HI}(\mathscr{O}_{\mathfrak{X}}, \xi^{-1}\widetilde{\Omega}^1_{\mathfrak{X}/\mathscr{S}}). \tag{III.12.7.10}$$

The composition of the functors (III.12.7.10) and (III.12.5.1) induces a functor that we denote also by

$$\mathrm{T}^r_+\colon \Xi^r_{\mathbb{Q}} \to \mathbf{HM}(\mathscr{O}_{\mathfrak{X}}[\frac{1}{p}], \xi^{-1}\widetilde{\Omega}^1_{\mathfrak{X}/\mathscr{S}}). \tag{III.12.7.11}$$

It is clear that the functor (III.12.7.7) is a left adjoint of the functor (III.12.7.10). We deduce from this that for all $\mathscr{N} \in \mathrm{Ob}(\mathbf{HM}^{\mathrm{coh}}(\mathscr{O}_{\mathfrak{X}}[\frac{1}{p}], \xi^{-1}\widetilde{\Omega}^1_{\mathfrak{X}/\mathscr{S}}))$ and $\mathscr{A} \in \mathrm{Ob}(\Xi^r_{\mathbb{Q}})$, we have a bifunctorial canonical homomorphism

$$\mathrm{Hom}_{\Xi^r_{\mathbb{Q}}}(\mathrm{T}^{r+}(\mathscr{N}), \mathscr{A}) \to \mathrm{Hom}_{\mathbf{HM}(\mathscr{O}_{\mathfrak{X}}[\frac{1}{p}], \xi^{-1}\widetilde{\Omega}^1_{\mathfrak{X}/\mathscr{S}})}(\mathscr{N}, \mathrm{T}^r_+(\mathscr{A})), \tag{III.12.7.12}$$

which is injective by III.6.20 and III.6.21. We abusively call *adjoint* of a morphism $\mathrm{T}^{r+}(\mathscr{N}) \to \mathscr{A}$ of $\Xi^r_{\mathbb{Q}}$ its image by the homomorphism (III.12.7.12).

**III.12.8.** Let $r$, $r'$ be two rational numbers such that $r \geq r' \geq 0$ and $(\mathscr{F}, \mathscr{G}, u, \nabla)$ an integrable $p^r$-isoconnection with respect to the extension $\breve{\mathscr{C}}^{(r)}/\breve{\overline{\mathscr{B}}}$. By (III.11.21.3), there exists a unique $\breve{\overline{\mathscr{B}}}$-linear morphism

$$\nabla' \colon \breve{\mathscr{C}}^{(r')} \otimes_{\breve{\mathscr{C}}^{(r)}} \mathscr{F} \to \mathrm{T}^*(\xi^{-1}\widetilde{\Omega}^1_{\mathfrak{X}/\mathscr{S}}) \otimes_{\breve{\overline{\mathscr{B}}}} \breve{\mathscr{C}}^{(r')} \otimes_{\breve{\mathscr{C}}^{(r)}} \mathscr{G} \tag{III.12.8.1}$$

such that for all local sections $x'$ of $\breve{\mathscr{C}}^{(r')}$ and $s$ of $\mathscr{F}$, we have

$$\nabla'(x' \otimes_{\breve{\mathscr{C}}^{(r)}} s) = p^{r'} \breve{d}^{(r')}(x') \otimes_{\breve{\mathscr{C}}^{(r)}} u(s) + x' \otimes_{\breve{\mathscr{C}}^{(r)}} \nabla(s). \tag{III.12.8.2}$$

The quadruple $(\breve{\mathscr{C}}^{(r')} \otimes_{\breve{\mathscr{C}}^{(r)}} \mathscr{F}, \breve{\mathscr{C}}^{(r')} \otimes_{\breve{\mathscr{C}}^{(r)}} \mathscr{G}, \mathrm{id} \otimes_{\breve{\mathscr{C}}^{(r)}} u, \nabla')$ is an integrable $p^{r'}$-isoconnection with respect to the extension $\breve{\mathscr{C}}^{(r')}/\breve{\overline{\mathscr{B}}}$. We thus obtain a functor

$$\epsilon^{r,r'} : \Xi^r \to \Xi^{r'}. \tag{III.12.8.3}$$

This induces a functor that we denote also by

$$\epsilon^{r,r'} : \Xi^r_{\mathbb{Q}} \to \Xi^{r'}_{\mathbb{Q}}. \tag{III.12.8.4}$$

We have a canonical isomorphism of functors from $\mathbf{Mod}(\breve{\overline{\mathscr{B}}})$ to $\Xi^{r'}$ (resp. from $\mathbf{Mod}_{\mathbb{Q}}(\breve{\overline{\mathscr{B}}})$ to $\Xi^{r'}_{\mathbb{Q}}$)

$$\epsilon^{r,r'} \circ \mathfrak{S}^r \xrightarrow{\sim} \mathfrak{S}^{r'}. \tag{III.12.8.5}$$

On the other hand, we have a canonical isomorphism of functors from $\mathbf{HI}(\mathscr{O}_{\mathfrak{X}}, \xi^{-1}\widetilde{\Omega}^1_{\mathfrak{X}/\mathscr{S}})$ to $\Xi^{r'}$ (resp. from $\mathbf{HM}^{\mathrm{coh}}(\mathscr{O}_{\mathfrak{X}}[\frac{1}{p}], \xi^{-1}\widetilde{\Omega}^1_{\mathfrak{X}/\mathscr{S}})$ to $\Xi^{r'}_{\mathbb{Q}}$ )

$$\epsilon^{r,r'} \circ \mathrm{T}^{r+} \xrightarrow{\sim} \mathrm{T}^{r'+}. \tag{III.12.8.6}$$

The diagram

$$\begin{array}{ccc} \mathscr{F} & \xrightarrow{\quad\nabla\quad} & \mathrm{T}^*(\xi^{-1}\widetilde{\Omega}^1_{\mathfrak{X}/\mathscr{S}}) \otimes_{\breve{\overline{\mathscr{B}}}} \mathscr{G} \\ {\scriptstyle \breve{\alpha}^{r,r'} \otimes_{\breve{\mathscr{C}}^{(r)}} \mathrm{id}} \downarrow & & \downarrow {\scriptstyle \mathrm{id} \otimes_{\breve{\overline{\mathscr{B}}}} \breve{\alpha}^{r,r'} \otimes_{\breve{\mathscr{C}}^{(r)}} \mathrm{id}} \\ \breve{\mathscr{C}}^{(r')} \otimes_{\breve{\mathscr{C}}^{(r)}} \mathscr{F} & \xrightarrow{\quad\nabla'\quad} & \mathrm{T}^*(\xi^{-1}\widetilde{\Omega}^1_{\mathfrak{X}/\mathscr{S}}) \otimes_{\breve{\overline{\mathscr{B}}}} \breve{\mathscr{C}}^{(r')} \otimes_{\breve{\mathscr{C}}^{(r)}} \mathscr{G} \end{array} \tag{III.12.8.7}$$

is clearly commutative. We deduce from this a canonical morphism of functors from $\Xi^r$ to $\mathbf{Mod}(\breve{\overline{\mathscr{B}}})$ (resp. from $\Xi^r_{\mathbb{Q}}$ to $\mathbf{Mod}_{\mathbb{Q}}(\breve{\overline{\mathscr{B}}})$)

$$\mathscr{K}^r \to \mathscr{K}^{r'} \circ \epsilon^{r,r'}. \tag{III.12.8.8}$$

We also deduce a canonical morphism of functors from $\Xi^r$ to $\mathbf{HI}(\mathscr{O}_{\mathfrak{X}}, \xi^{-1}\widetilde{\Omega}^1_{\mathfrak{X}/\mathscr{S}})$ (resp. from $\Xi^r_{\mathbb{Q}}$ to $\mathbf{HM}(\mathscr{O}_{\mathfrak{X}}[\frac{1}{p}], \xi^{-1}\widetilde{\Omega}^1_{\mathfrak{X}/\mathscr{S}})$)

$$\mathrm{T}^r_+ \longrightarrow \mathrm{T}^{r'}_+ \circ \epsilon^{r,r'}. \tag{III.12.8.9}$$

For every rational number $r''$ such that $r' \geq r'' \geq 0$, we have a canonical isomorphism of functors from $\Xi^r$ to $\Xi^{r''}$ (resp. from $\Xi^r_{\mathbb{Q}}$ to $\Xi^{r''}_{\mathbb{Q}}$)

$$\epsilon^{r',r''} \circ \epsilon^{r,r'} \xrightarrow{\sim} \epsilon^{r,r''}. \tag{III.12.8.10}$$

**Remark III.12.9.** Under the assumptions of III.12.8,

$$(\breve{\mathscr{C}}^{(r')} \otimes_{\breve{\mathscr{C}}^{(r)}} \mathscr{F}, \breve{\mathscr{C}}^{(r')} \otimes_{\breve{\mathscr{C}}^{(r)}} \mathscr{G}, \mathrm{id} \otimes_{\breve{\mathscr{C}}^{(r)}} u, p^{r-r'}\nabla')$$

is the integrable $p^r$-isoconnection with respect to the extension $\breve{\mathscr{C}}^{(r')}/\breve{\overline{\mathscr{B}}}$ deduced from $(\mathscr{F}, \mathscr{G}, u, \nabla)$ by extension of the scalars by $\breve{\alpha}^{r,r'}$, defined in III.6.11. This shift can be explained by the fact that the canonical homomorphism $\Omega^1_{\breve{\mathscr{C}}^{(r)}/\breve{\overline{\mathscr{B}}}} \to \Omega^1_{\breve{\mathscr{C}}^{(r')}/\breve{\overline{\mathscr{B}}}}$ identifies with

$$p^{r-r'}\mathrm{id} \otimes \breve{\alpha}^{r,r'} : \mathrm{T}^*(\xi^{-1}\widetilde{\Omega}^1_{\mathfrak{X}/\mathscr{S}}) \otimes_{\breve{\overline{\mathscr{B}}}} \breve{\mathscr{C}}^{(r)} \to \mathrm{T}^*(\xi^{-1}\widetilde{\Omega}^1_{\mathfrak{X}/\mathscr{S}}) \otimes_{\breve{\overline{\mathscr{B}}}} \breve{\mathscr{C}}^{(r')}. \tag{III.12.9.1}$$

**Definition III.12.10.** Let $\mathscr{M}$ be an object of $\mathbf{Mod}^{\mathrm{aft}}_{\mathbb{Q}}(\breve{\overline{\mathscr{B}}})$ (III.12.1) and $\mathscr{N}$ a Higgs $\mathscr{O}_{\mathfrak{X}}[\frac{1}{p}]$-bundle with coefficients in $\xi^{-1}\widetilde{\Omega}^1_{\mathfrak{X}/\mathscr{S}}$ (III.12.6).

(i) Let $r$ be a rational number $> 0$. We say that $\mathscr{M}$ and $\mathscr{N}$ are *r-associated* if there exists an isomorphism of $\Xi^r_{\mathbb{Q}}$

$$\alpha\colon \mathrm{T}^{r+}(\mathscr{N}) \xrightarrow{\sim} \mathfrak{S}^r(\mathscr{M}). \tag{III.12.10.1}$$

We then also say that the triple $(\mathscr{M}, \mathscr{N}, \alpha)$ is *r-admissible.*

(ii) We say that $\mathscr{M}$ and $\mathscr{N}$ are *associated* if there exists a rational number $r > 0$ such that $\mathscr{M}$ and $\mathscr{N}$ are $r$-associated.

Note that for all rational numbers $r \geq r' > 0$, if $\mathscr{M}$ and $\mathscr{N}$ are $r$-associated, they are $r'$-associated, in view of (III.12.8.5) and (III.12.8.6).

**Definition III.12.11.** (i) A *Dolbeault* $\breve{\overline{\mathscr{B}}}_{\mathbb{Q}}$*-module* is an object of $\mathbf{Mod}^{\mathrm{aft}}_{\mathbb{Q}}(\breve{\overline{\mathscr{B}}})$ for which there exists an associated Higgs $\mathscr{O}_{\mathfrak{X}}[\frac{1}{p}]$-bundle with coefficients in $\xi^{-1}\widetilde{\Omega}^1_{\mathfrak{X}/\mathscr{S}}$.

(ii) We say that a Higgs $\mathscr{O}_{\mathfrak{X}}[\frac{1}{p}]$-bundle with coefficients in $\xi^{-1}\widetilde{\Omega}^1_{\mathfrak{X}/\mathscr{S}}$ is *solvable* if it admits an associated Dolbeault module.

We denote by $\mathbf{Mod}^{\mathrm{Dolb}}_{\mathbb{Q}}(\breve{\overline{\mathscr{B}}})$ the full subcategory of $\mathbf{Mod}^{\mathrm{aft}}_{\mathbb{Q}}(\breve{\overline{\mathscr{B}}})$ made up of Dolbeault $\breve{\overline{\mathscr{B}}}_{\mathbb{Q}}$-modules, and by $\mathbf{HM}^{\mathrm{sol}}(\mathscr{O}_{\mathfrak{X}}[\frac{1}{p}], \xi^{-1}\widetilde{\Omega}^1_{\mathfrak{X}/\mathscr{S}})$ the full subcategory of $\mathbf{HM}(\mathscr{O}_{\mathfrak{X}}[\frac{1}{p}], \xi^{-1}\widetilde{\Omega}^1_{\mathfrak{X}/\mathscr{S}})$ made up of solvable Higgs $\mathscr{O}_{\mathfrak{X}}[\frac{1}{p}]$-bundles with coefficients in $\xi^{-1}\widetilde{\Omega}^1_{\mathfrak{X}/\mathscr{S}}$.

**Proposition III.12.12.** *Every Dolbeault* $\breve{\overline{\mathscr{B}}}_{\mathbb{Q}}$*-module is* $\breve{\overline{\mathscr{B}}}_{\mathbb{Q}}$*-flat* (III.6.4).

Let $\mathscr{M}$ be a Dolbeault $\breve{\overline{\mathscr{B}}}_{\mathbb{Q}}$-module, $\mathscr{N}$ a Higgs $\mathscr{O}_{\mathfrak{X}}[\frac{1}{p}]$-bundle with coefficients in $\xi^{-1}\widetilde{\Omega}^1_{\mathfrak{X}/\mathscr{S}}$, $r$ a rational number $> 0$, and $\alpha\colon \mathrm{T}^{r+}(\mathscr{N}) \xrightarrow{\sim} \mathfrak{S}^r(\mathscr{M})$ an isomorphism of $\Xi^r_{\mathbb{Q}}$. Since the $\mathscr{O}_{\mathfrak{X}}[\frac{1}{p}]$-module $\mathscr{N}$ is locally free of finite type, the $\breve{\mathscr{C}}^{(r)}_{\mathbb{Q}}$-module $\mathrm{T}^*(\mathscr{N}) \otimes_{\breve{\overline{\mathscr{B}}}_{\mathbb{Q}}} \breve{\mathscr{C}}^{(r)}_{\mathbb{Q}}$ is flat by III.6.7(iii). It follows that $\mathscr{M} \otimes_{\breve{\overline{\mathscr{B}}}_{\mathbb{Q}}} \breve{\mathscr{C}}^{(r)}_{\mathbb{Q}}$ is $\breve{\mathscr{C}}^{(r)}_{\mathbb{Q}}$-flat. Consequently, $\mathscr{M}$ is $\breve{\overline{\mathscr{B}}}_{\mathbb{Q}}$-flat by virtue of III.6.4.4 and III.10.33.

**III.12.13.** For every $\breve{\overline{\mathscr{B}}}_{\mathbb{Q}}$-module $\mathscr{M}$ and all rational numbers $r \geq r' \geq 0$, the morphism (III.12.8.9) and the isomorphism (III.12.8.5) induce a morphism

$$\text{(III.12.13.1)} \qquad \mathrm{T}^r_+(\mathfrak{S}^r(\mathscr{M})) \to \mathrm{T}^{r'}_+(\mathfrak{S}^{r'}(\mathscr{M}))$$

of $\mathbf{HM}(\mathscr{O}_{\mathfrak{X}}[\frac{1}{p}], \xi^{-1}\widetilde{\Omega}^1_{\mathfrak{X}/\mathscr{S}})$. We thus obtain a filtered direct system $(\mathrm{T}^r_+(\mathfrak{S}^r(\mathscr{M})))_{r\in\mathbb{Q}_{\geq 0}}$. We denote by $\mathscr{H}$ the functor

$$\text{(III.12.13.2)} \qquad \mathscr{H}\colon \mathbf{Mod}_{\mathbb{Q}}(\breve{\overline{\mathscr{B}}}) \to \mathbf{HM}(\mathscr{O}_{\mathfrak{X}}[\frac{1}{p}], \xi^{-1}\widetilde{\Omega}^1_{\mathfrak{X}/\mathscr{S}}), \quad \mathscr{M} \mapsto \varinjlim_{r\in\mathbb{Q}_{>0}} \mathrm{T}^r_+(\mathfrak{S}^r(\mathscr{M})).$$

For every object $\mathscr{N}$ of $\mathbf{HM}(\mathscr{O}_{\mathfrak{X}}[\frac{1}{p}], \xi^{-1}\widetilde{\Omega}^1_{\mathfrak{X}/\mathscr{S}})$ and all rational numbers $r \geq r' \geq 0$, the morphism (III.12.8.8) and the isomorphism (III.12.8.6) induce a morphism

$$\text{(III.12.13.3)} \qquad \mathscr{K}^r(\mathrm{T}^{r+}(\mathscr{N})) \to \mathscr{K}^{r'}(\mathrm{T}^{r'+}(\mathscr{N}))$$

of $\mathbf{Mod}_{\mathbb{Q}}(\breve{\overline{\mathscr{B}}})$. We thus obtain a filtered direct system $(\mathscr{K}^r(\mathrm{T}^{r+}(\mathscr{N})))_{r\geq 0}$. Recall (III.11.25) that filtered direct limits are not a priori representable in the category $\mathbf{Mod}_{\mathbb{Q}}(\breve{\overline{\mathscr{B}}})$.

**Lemma III.12.14.** *We have a canonical isomorphism of* $\mathbf{HM}(\mathscr{O}_{\mathfrak{X}}[\frac{1}{p}], \xi^{-1}\widetilde{\Omega}^1_{\mathfrak{X}/\mathscr{S}})$

$$\text{(III.12.14.1)} \qquad (\mathscr{O}_{\mathfrak{X}}[\frac{1}{p}], 0) \xrightarrow{\sim} \mathscr{H}(\breve{\overline{\mathscr{B}}}_{\mathbb{Q}}).$$

This follows from III.11.18.

**Lemma III.12.15.** *Let $r$ be a rational number $\geq 0$ and $\mathscr{N}$ a Higgs $\mathscr{O}_{\mathfrak{X}}[\frac{1}{p}]$-bundle with coefficients in $\xi^{-1}\widetilde{\Omega}^1_{\mathfrak{X}/\mathscr{S}}$* (III.12.6). *We have a canonical isomorphism*

$$\text{(III.12.15.1)} \qquad \gamma^r\colon \mathscr{N} \otimes_{\mathscr{O}_{\mathfrak{X}}[\frac{1}{p}]} \mathrm{T}^r_+(\mathfrak{S}^r(\breve{\overline{\mathscr{B}}}_{\mathbb{Q}})) \xrightarrow{\sim} \mathrm{T}^r_+(\mathrm{T}^{r+}(\mathscr{N}))$$

*of* $\mathbf{HM}(\mathscr{O}_{\mathfrak{X}}[\frac{1}{p}], \xi^{-1}\widetilde{\Omega}^1_{\mathfrak{X}/\mathscr{S}})$, *where the left-hand side is the tensor product of Higgs modules* (II.2.8.8). *Moreover, we have the following properties:*

(i) *The morphism*

$$\text{(III.12.15.2)} \qquad \mathscr{N} \to \mathrm{T}^r_+(\mathrm{T}^{r+}(\mathscr{N}))$$

*induced by $\gamma^r$ and the canonical morphism $\mathscr{O}_{\mathfrak{X}}[\frac{1}{p}] \to \mathrm{T}^r_+(\mathfrak{S}^r(\breve{\overline{\mathscr{B}}}_{\mathbb{Q}}))$ is the adjoint of the identity of $\mathrm{T}^{r+}(\mathscr{N})$* (III.12.7.12).

(ii) *For every rational number $r'$ such that $r \geq r' \geq 0$, the diagram*

$$\text{(III.12.15.3)} \qquad \begin{array}{ccc} \mathscr{N} \otimes_{\mathscr{O}_{\mathfrak{X}}[\frac{1}{p}]} \mathrm{T}^r_+(\mathfrak{S}^r(\breve{\overline{\mathscr{B}}}_{\mathbb{Q}})) & \xrightarrow{\gamma^r} & \mathrm{T}^r_+(\mathrm{T}^{r+}(\mathscr{N})) \\ \downarrow & & \downarrow \\ \mathscr{N} \otimes_{\mathscr{O}_{\mathfrak{X}}[\frac{1}{p}]} \mathrm{T}^{r'}_+(\mathfrak{S}^{r'}(\breve{\overline{\mathscr{B}}}_{\mathbb{Q}})) & \xrightarrow{\gamma^{r'}} & \mathrm{T}^{r'}_+(\mathrm{T}^{r'+}(\mathscr{N})) \end{array}$$

*where the vertical arrows are induced by the morphism* (III.12.8.9) *and the isomorphisms* (III.12.8.5) *and* (III.12.8.6), *is commutative.*

Indeed, by III.12.4(ii), we have canonical isomorphisms of $\mathscr{O}_{\mathfrak{X}}[\frac{1}{p}]$-modules

$$\mathscr{N} \otimes_{\mathscr{O}_{\mathfrak{X}}[\frac{1}{p}]} \mathsf{T}_*(\breve{\mathscr{C}}^{(r)}_{\mathbb{Q}}) \xrightarrow{\sim} \mathsf{T}_*(\mathsf{T}^*(\mathscr{N}) \otimes_{\breve{\overline{\mathscr{B}}}_{\mathbb{Q}}} \breve{\mathscr{C}}^{(r)}_{\mathbb{Q}}), \tag{III.12.15.4}$$

$$\begin{aligned} &\xi^{-1}\widetilde{\Omega}^1_{\mathfrak{X}/\mathscr{S}} \otimes_{\mathscr{O}_{\mathfrak{X}}} \mathsf{T}_*(\mathsf{T}^*(\mathscr{N}) \otimes_{\breve{\overline{\mathscr{B}}}_{\mathbb{Q}}} \breve{\mathscr{C}}^{(r)}_{\mathbb{Q}}) \\ &\qquad \xrightarrow{\sim} \mathsf{T}_*(\mathsf{T}^*(\xi^{-1}\widetilde{\Omega}^1_{\mathfrak{X}/\mathscr{S}} \otimes_{\mathscr{O}_{\mathfrak{X}}} \mathscr{N}) \otimes_{\breve{\overline{\mathscr{B}}}_{\mathbb{Q}}} \breve{\mathscr{C}}^{(r)}_{\mathbb{Q}}), \end{aligned} \tag{III.12.15.5}$$

$$\begin{aligned} &\xi^{-1}\widetilde{\Omega}^1_{\mathfrak{X}/\mathscr{S}} \otimes_{\mathscr{O}_{\mathfrak{X}}} \mathscr{N} \otimes_{\mathscr{O}_{\mathfrak{X}}[\frac{1}{p}]} \mathsf{T}_*(\breve{\mathscr{C}}^{(r)}_{\mathbb{Q}}) \\ &\qquad \xrightarrow{\sim} \mathsf{T}_*(\mathsf{T}^*(\xi^{-1}\widetilde{\Omega}^1_{\mathfrak{X}/\mathscr{S}} \otimes_{\mathscr{O}_{\mathfrak{X}}} \mathscr{N}) \otimes_{\breve{\overline{\mathscr{B}}}_{\mathbb{Q}}} \breve{\mathscr{C}}^{(r)}_{\mathbb{Q}}). \end{aligned} \tag{III.12.15.6}$$

The third isomorphism is induced by the first two by III.12.3(i). Moreover, in view of the bifunctoriality of the isomorphism (III.12.4.2), the diagram
(III.12.15.7)

$$\begin{array}{ccc} \mathscr{N} \otimes_{\mathscr{O}_{\mathfrak{X}}[\frac{1}{p}]} \mathsf{T}_*(\breve{\mathscr{C}}^{(r)}_{\mathbb{Q}}) & \longrightarrow & \mathsf{T}_*(\mathsf{T}^*(\mathscr{N}) \otimes_{\breve{\overline{\mathscr{B}}}_{\mathbb{Q}}} \breve{\mathscr{C}}^{(r)}_{\mathbb{Q}}) \\ {\scriptstyle \theta\otimes\mathrm{id}+p^r\mathrm{id}\otimes\mathsf{T}_*(\breve{d}^{(r)}_{\mathbb{Q}})}\downarrow & & \downarrow{\scriptstyle \mathsf{T}_*(\mathsf{T}^*(\theta)\otimes\mathrm{id}+p^r\mathrm{id}\otimes\breve{d}^{(r)})} \\ \xi^{-1}\widetilde{\Omega}^1_{\mathfrak{X}/\mathscr{S}} \otimes_{\mathscr{O}_{\mathfrak{X}}} \mathscr{N} \otimes_{\mathscr{O}_{\mathfrak{X}}[\frac{1}{p}]} \mathsf{T}_*(\breve{\mathscr{C}}^{(r)}_{\mathbb{Q}}) & \longrightarrow & \mathsf{T}_*(\mathsf{T}^*(\xi^{-1}\widetilde{\Omega}^1_{\mathfrak{X}/\mathscr{S}} \otimes_{\mathscr{O}_{\mathfrak{X}}} \mathscr{N}) \otimes_{\breve{\overline{\mathscr{B}}}_{\mathbb{Q}}} \breve{\mathscr{C}}^{(r)}_{\mathbb{Q}}) \end{array}$$

where $\theta$ is the Higgs field of $\mathscr{N}$, is commutative. We then take for $\gamma^r$ (III.12.15.1) the isomorphism (III.12.15.4). Statement (i) follows from III.12.2(ii) and III.6.21. Statement (ii) is a consequence of the bifunctoriality of the isomorphism (III.12.4.2).

**III.12.16.** Let $r$ be a rational number $> 0$ and $(\mathscr{M}, \mathscr{N}, \alpha)$ an $r$-admissible triple. For any rational number $r'$ such that $0 < r' \leq r$, we denote by

$$\alpha^{r'} : \mathsf{T}^{r'+}(\mathscr{N}) \xrightarrow{\sim} \mathfrak{S}^{r'}(\mathscr{M}) \tag{III.12.16.1}$$

the isomorphism of $\Xi^{r'}_{\mathbb{Q}}$ induced by $\epsilon^{r,r'}(\alpha)$ and the isomorphisms (III.12.8.5) and (III.12.8.6), and by

$$\beta^{r'} : \mathscr{N} \to \mathsf{T}^{r'}_{+}(\mathfrak{S}^{r'}(\mathscr{M})) \tag{III.12.16.2}$$

its adjoint (III.12.7.12).

**Proposition III.12.17.** *Under the assumptions of* III.12.16, *let moreover $r'$, $r''$ be two rational numbers such that $0 < r'' < r' \leq r$. Then:*

(i) *The composition*

$$\mathscr{N} \xrightarrow{\beta^{r'}} \mathsf{T}^{r'}_{+}(\mathfrak{S}^{r'}(\mathscr{M})) \longrightarrow \mathscr{H}(\mathscr{M}), \tag{III.12.17.1}$$

*where the second arrow is the canonical morphism* (III.12.13.2), *is an isomorphism that does not depend on $r'$.*

(ii) *The composition*

$$\mathsf{T}^{r'}_{+}(\mathfrak{S}^{r'}(\mathscr{M})) \longrightarrow \mathscr{H}(\mathscr{M}) \xrightarrow{\sim} \mathscr{N} \xrightarrow{\beta^{r''}} \mathsf{T}^{r''}_{+}(\mathfrak{S}^{r''}(\mathscr{M})), \tag{III.12.17.2}$$

*where the first arrow is the canonical morphism* (III.12.13.2) *and the second arrow is the inverse of the isomorphism* (III.12.17.1), *is the canonical morphism* (III.12.13.1).

(i) For any rational number $0 < t \leq r$, we denote by

$$\gamma^t : \mathscr{N} \otimes_{\mathscr{O}_{\mathfrak{X}}[\frac{1}{p}]} \mathsf{T}^t_{+}(\mathfrak{S}^t(\breve{\overline{\mathscr{B}}}_{\mathbb{Q}})) \xrightarrow{\sim} \mathsf{T}^t_{+}(\mathsf{T}^{t+}(\mathscr{N})) \tag{III.12.17.3}$$

the isomorphism (III.12.15.1) of $\mathbf{HM}(\mathscr{O}_{\mathfrak{X}}[\frac{1}{p}], \xi^{-1}\widetilde{\Omega}^1_{\mathfrak{X}/\mathscr{S}})$, and by

(III.12.17.4) $$\delta^t \colon \mathscr{N} \otimes_{\mathscr{O}_{\mathfrak{X}}[\frac{1}{p}]} \mathsf{T}^t_+(\mathfrak{S}^t(\breve{\overline{\mathscr{B}}}_{\mathbb{Q}})) \xrightarrow{\sim} \mathsf{T}^t_+(\mathfrak{S}^t(\mathscr{M}))$$

the composition $\mathsf{T}^t_+(\alpha^t) \circ \gamma^t$. The diagram

(III.12.17.5) $$\begin{array}{ccc} \mathscr{N} \otimes_{\mathscr{O}_{\mathfrak{X}}[\frac{1}{p}]} \mathsf{T}^{r'}_+(\mathfrak{S}^{r'}(\breve{\overline{\mathscr{B}}}_{\mathbb{Q}})) & \xrightarrow{\delta^{r'}} & \mathsf{T}^{r'}_+(\mathfrak{S}^{r'}(\mathscr{M})) \\ \downarrow & & \downarrow \\ \mathscr{N} \otimes_{\mathscr{O}_{\mathfrak{X}}[\frac{1}{p}]} \mathsf{T}^{r''}_+(\mathfrak{S}^{r''}(\breve{\overline{\mathscr{B}}}_{\mathbb{Q}})) & \xrightarrow{\delta^{r''}} & \mathsf{T}^{r''}_+(\mathfrak{S}^{r''}(\mathscr{M})) \end{array}$$

where the vertical arrows are the canonical morphisms (III.12.13.1), is commutative by virtue of III.12.15(ii). The isomorphisms $(\delta^t)_{0<t\leq r}$ induce, by direct limit, an isomorphism of $\mathbf{HM}(\mathscr{O}_{\mathfrak{X}}[\frac{1}{p}], \xi^{-1}\widetilde{\Omega}^1_{\mathfrak{X}/\mathscr{S}})$

(III.12.17.6) $$\delta \colon \mathscr{N} \otimes_{\mathscr{O}_{\mathfrak{X}}[\frac{1}{p}]} \mathscr{H}(\breve{\overline{\mathscr{B}}}_{\mathbb{Q}}) \xrightarrow{\sim} \mathscr{H}(\mathscr{M}).$$

Consider the commutative diagram

(III.12.17.7) $$\begin{array}{ccccc} \mathscr{N} & \xrightarrow{\iota^{r'}} & \mathscr{N} \otimes_{\mathscr{O}_{\mathfrak{X}}[\frac{1}{p}]} \mathsf{T}^{r'}_+(\mathfrak{S}^{r'}(\breve{\overline{\mathscr{B}}}_{\mathbb{Q}})) & \xrightarrow{\delta^{r'}} & \mathsf{T}^{r'}_+(\mathfrak{S}^{r'}(\mathscr{M})) \\ & \searrow & \downarrow & & \downarrow \\ & & \mathscr{N} \otimes_{\mathscr{O}_{\mathfrak{X}}[\frac{1}{p}]} \mathscr{H}(\breve{\overline{\mathscr{B}}}_{\mathbb{Q}}) & \xrightarrow{\delta} & \mathscr{H}(\mathscr{M}) \end{array}$$

where $\iota^{r'}$ is induced by the canonical morphism $\mathscr{O}_{\mathfrak{X}}[\frac{1}{p}] \to \mathsf{T}^{r'}_+(\mathfrak{S}^{r'}(\breve{\overline{\mathscr{B}}}_{\mathbb{Q}}))$ and the vertical arrows are the canonical morphisms. By III.12.15(i), we have

(III.12.17.8) $$\delta^{r'} \circ \iota^{r'} = \mathsf{T}^{r'}_+(\alpha^{r'}) \circ \gamma^{r'} \circ \iota^{r'} = \beta^{r'}.$$

The statement follows by virtue of III.12.14.

(ii) This follows from (III.12.17.5), (III.12.17.7), and III.11.17(ii).

**Corollary III.12.18.** *For every Dolbeault $\breve{\overline{\mathscr{B}}}_{\mathbb{Q}}$-module $\mathscr{M}$, $\mathscr{H}(\mathscr{M})$* (III.12.13.2) *is a solvable Higgs $\mathscr{O}_{\mathfrak{X}}[\frac{1}{p}]$-bundle associated with $\mathscr{M}$. In particular, $\mathscr{H}$ induces a functor that we denote also by*

(III.12.18.1) $$\mathscr{H} \colon \mathbf{Mod}^{\mathrm{Dolb}}_{\mathbb{Q}}(\breve{\overline{\mathscr{B}}}) \to \mathbf{HM}^{\mathrm{sol}}(\mathscr{O}_{\mathfrak{X}}[\frac{1}{p}], \xi^{-1}\widetilde{\Omega}^1_{\mathfrak{X}/\mathscr{S}}), \quad \mathscr{M} \mapsto \mathscr{H}(\mathscr{M}).$$

**Corollary III.12.19.** *For every Dolbeault $\breve{\overline{\mathscr{B}}}_{\mathbb{Q}}$-module $\mathscr{M}$, there exist a rational number $r > 0$ and an isomorphism of $\Xi^r_{\mathbb{Q}}$*

(III.12.19.1) $$\alpha \colon \mathsf{T}^{r+}(\mathscr{H}(\mathscr{M})) \xrightarrow{\sim} \mathfrak{S}^r(\mathscr{M})$$

*satisfying the following properties. For any rational number $r'$ such that $0 < r' \leq r$, denote by*

(III.12.19.2) $$\alpha^{r'} \colon \mathsf{T}^{r'+}(\mathscr{H}(\mathscr{M})) \xrightarrow{\sim} \mathfrak{S}^{r'}(\mathscr{M})$$

*the isomorphism of $\Xi^{r'}_{\mathbb{Q}}$ induced by $\epsilon^{r,r'}(\alpha)$ and the isomorphisms* (III.12.8.5) *and* (III.12.8.6)*, and by*

(III.12.19.3) $$\beta^{r'} \colon \mathscr{H}(\mathscr{M}) \to \mathsf{T}^{r'}_+(\mathfrak{S}^{r'}(\mathscr{M}))$$

*its adjoint* (III.12.7.12)*. Then:*

(i) *For every rational number $r'$ such that $0 < r' \leq r$, the morphism $\beta^{r'}$ is a left inverse of the canonical morphism* $\varpi^{r'}\colon \mathrm{T}_{+}^{r'}(\mathfrak{S}^{r'}(\mathscr{M})) \to \mathscr{H}(\mathscr{M})$.
(ii) *For all rational numbers $r'$ and $r''$ such that $0 < r'' < r' \leq r$, the composition*

$$\mathrm{T}_{+}^{r'}(\mathfrak{S}^{r'}(\mathscr{M})) \xrightarrow{\varpi^{r'}} \mathscr{H}(\mathscr{M}) \xrightarrow{\beta^{r''}} \mathrm{T}_{+}^{r''}(\mathfrak{S}^{r''}(\mathscr{M})) \tag{III.12.19.4}$$

*is the canonical morphism.*

**Remark III.12.20.** Under the assumptions of III.12.19, the isomorphism $\alpha$ is not a priori uniquely determined by $(\mathscr{M}, r)$, but for every rational number $0 < r' < r$, the morphism $\alpha^{r'}$ (III.12.19.2) depends only on $\mathscr{M}$, on which it depends functorially (cf. the proof of III.12.26).

**III.12.21.** Let $r$ be a rational number $> 0$ and $(\mathscr{M}, \mathscr{N}, \alpha)$ an $r$-admissible triple. To avoid any ambiguity with (III.12.16.1), we denote by

$$\breve{\alpha}\colon \mathfrak{S}^{r}(\mathscr{M}) \to \mathrm{T}^{r+}(\mathscr{N}) \tag{III.12.21.1}$$

the inverse of $\alpha$ in $\Xi^{r}_{\mathbb{Q}}$. For any rational number $r'$ such that $0 < r' \leq r$, we denote by

$$\breve{\alpha}^{r'}\colon \mathfrak{S}^{r'}(\mathscr{M}) \xrightarrow{\sim} \mathrm{T}^{r'+}(\mathscr{N}) \tag{III.12.21.2}$$

the isomorphism of $\Xi^{r'}_{\mathbb{Q}}$ induced by $\epsilon^{r,r'}(\breve{\alpha})$ and the isomorphisms (III.12.8.5) and (III.12.8.6), and by

$$\breve{\beta}^{r'}\colon \mathscr{M} \to \mathscr{K}^{r'}(\mathrm{T}^{r'+}(\mathscr{N})) \tag{III.12.21.3}$$

the adjoint morphism.

**Proposition III.12.22.** *Under the assumptions of* III.12.21, *let moreover $r'$, $r''$ be two rational numbers such that $0 < r'' < r' \leq r$. Then:*

(i) *The direct limit $\mathscr{V}(\mathscr{N})$ of the direct system $(\mathscr{K}^{t}(\mathrm{T}^{t+}(\mathscr{N})))_{t\in\mathbb{Q}_{>0}}$* (III.12.13.3) *is representable in* $\mathbf{Mod}_{\mathbb{Q}}(\breve{\overline{\mathscr{B}}})$.
(ii) *The composition*

$$\mathscr{M} \xrightarrow{\breve{\beta}^{r'}} \mathscr{K}^{r'}(\mathrm{T}^{r'+}(\mathscr{N})) \longrightarrow \mathscr{V}(\mathscr{N}), \tag{III.12.22.1}$$

*where the second arrow is the canonical morphism, is an isomorphism, which does not depend on $r'$.*
(iii) *The composition*

$$\mathscr{K}^{r'}(\mathrm{T}^{r'+}(\mathscr{N})) \longrightarrow \mathscr{V}(\mathscr{N}) \xrightarrow{\sim} \mathscr{M} \xrightarrow{\breve{\beta}^{r''}} \mathscr{K}^{r''}(\mathrm{T}^{r''+}(\mathscr{N})), \tag{III.12.22.2}$$

*where the first arrow is the canonical morphism and the second arrow is the inverse of the isomorphism* (III.12.22.1), *is the canonical morphism* (III.12.13.3).

(i) Since $\mathscr{M}$ is $\breve{\overline{\mathscr{B}}}_{\mathbb{Q}}$-flat by III.12.12, for every rational number $t \geq 0$, we have a canonical isomorphism of $\mathbf{Mod}_{\mathbb{Q}}(\breve{\overline{\mathscr{B}}})$

$$\gamma^{t}\colon \mathscr{M} \otimes_{\breve{\overline{\mathscr{B}}}_{\mathbb{Q}}} \mathscr{K}^{t}(\mathfrak{S}^{t}(\breve{\overline{\mathscr{B}}}_{\mathbb{Q}})) \xrightarrow{\sim} \mathscr{K}^{t}(\mathfrak{S}^{t}(\mathscr{M})). \tag{III.12.22.3}$$

We denote by

$$\delta^{t}\colon \mathscr{M} \otimes_{\breve{\overline{\mathscr{B}}}_{\mathbb{Q}}} \mathscr{K}^{t}(\mathfrak{S}^{t}(\breve{\overline{\mathscr{B}}}_{\mathbb{Q}})) \xrightarrow{\sim} \mathscr{K}^{t}(\mathrm{T}^{t+}(\mathscr{N})) \tag{III.12.22.4}$$

the composition $\mathscr{K}^t(\check{\alpha}^t) \circ \gamma^t$. The diagram

(III.12.22.5)
$$\begin{array}{ccc} \mathscr{M} \otimes_{\breve{\overline{\mathscr{B}}}_{\mathbb{Q}}} \mathscr{K}^{r'}(\mathfrak{S}^{r'}(\breve{\overline{\mathscr{B}}}_{\mathbb{Q}})) & \xrightarrow{\delta^{r'}} & \mathscr{K}^{r'}(\mathrm{T}^{r'+}(\mathscr{N})) \\ \downarrow & & \downarrow \\ \mathscr{M} \otimes_{\breve{\overline{\mathscr{B}}}_{\mathbb{Q}}} \mathscr{K}^{r''}(\mathfrak{S}^{r''}(\breve{\overline{\mathscr{B}}}_{\mathbb{Q}})) & \xrightarrow{\delta^{r''}} & \mathscr{K}^{r''}(\mathrm{T}^{r''+}(\mathscr{N})) \end{array}$$

where the vertical arrows are induced by the morphism (III.12.8.8) and the isomorphisms (III.12.8.5) and (III.12.8.6), is clearly commutative. The statement then follows from III.11.23.

(ii) By III.11.23, the canonical morphism

(III.12.22.6)
$$\mathscr{M} \to \varinjlim_{t\in\mathbb{Q}_{>0}} \mathscr{M} \otimes_{\breve{\overline{\mathscr{B}}}_{\mathbb{Q}}} \mathscr{K}^t(\mathfrak{S}^t(\breve{\overline{\mathscr{B}}}_{\mathbb{Q}}))$$

is an isomorphism. The isomorphisms $(\delta^t)_{0<t\leq r}$ then induce, by direct limit, an isomorphism

(III.12.22.7)
$$\delta\colon \mathscr{M} \xrightarrow{\sim} \mathscr{V}(\mathscr{N}).$$

It immediately follows from the definitions that the diagram

(III.12.22.8)
$$\begin{array}{ccc} \mathscr{M} \otimes_{\breve{\overline{\mathscr{B}}}_{\mathbb{Q}}} \mathscr{K}^{r'}(\mathfrak{S}^{r'}(\breve{\overline{\mathscr{B}}}_{\mathbb{Q}})) & \xrightarrow{\delta^{r'}} & \mathscr{K}^{r'}(\mathrm{T}^{r'+}(\mathscr{N})) \\ \uparrow{\scriptstyle \iota^{r'}} & & \downarrow \\ \mathscr{M} & \xrightarrow{\delta} & \mathscr{V}(\mathscr{N}) \end{array}$$

where $\iota^{r'}$ is induced by the canonical morphism $\breve{\overline{\mathscr{B}}}_{\mathbb{Q}} \to \mathscr{K}^{r'}(\mathfrak{S}^{r'}(\breve{\overline{\mathscr{B}}}_{\mathbb{Q}}))$ and the unlabeled arrow is the canonical morphism, is commutative. One immediately verifies that

(III.12.22.9)
$$\delta^{r'} \circ \iota^{r'} = \mathscr{K}^{r'}(\check{\alpha}^{r'}) \circ \gamma^{r'} \circ \iota^{r'} = \check{\beta}^{r'}.$$

The statement follows.

(iii) This follows from (III.12.22.5) and III.11.23(ii).

**Corollary III.12.23.** *We have a functor*

(III.12.23.1)
$$\mathscr{V}\colon \mathbf{HM}^{\mathrm{sol}}(\mathscr{O}_{\mathfrak{X}}[\frac{1}{p}], \xi^{-1}\widetilde{\Omega}^1_{\mathfrak{X}/\mathscr{S}}) \to \mathbf{Mod}^{\mathrm{Dolb}}_{\mathbb{Q}}(\breve{\overline{\mathscr{B}}}), \quad \mathscr{N} \mapsto \varinjlim_{r\in\mathbb{Q}_{>0}} \mathscr{K}^r(\mathrm{T}^{r+}(\mathscr{N})).$$

*Moreover, for every object* $\mathscr{N}$ *of* $\mathbf{HM}^{\mathrm{sol}}(\mathscr{O}_{\mathfrak{X}}[\frac{1}{p}], \xi^{-1}\widetilde{\Omega}^1_{\mathfrak{X}/\mathscr{S}})$, $\mathscr{V}(\mathscr{N})$ *is associated with* $\mathscr{N}$.

**Corollary III.12.24.** *For every solvable Higgs* $\mathscr{O}_{\mathfrak{X}}[\frac{1}{p}]$*-bundle* $\mathscr{N}$ *with coefficients in* $\xi^{-1}\widetilde{\Omega}^1_{\mathfrak{X}/\mathscr{S}}$, *there exist a rational number* $r > 0$ *and an isomorphism of* $\Xi^r_{\mathbb{Q}}$

(III.12.24.1)
$$\check{\alpha}\colon \mathfrak{S}^r(\mathscr{V}(\mathscr{N})) \xrightarrow{\sim} \mathrm{T}^{r+}(\mathscr{N})$$

*satisfying the following properties. For any rational number* $r'$ *such that* $0 < r' \leq r$, *denote by*

(III.12.24.2)
$$\check{\alpha}^{r'}\colon \mathfrak{S}^{r'}(\mathscr{V}(\mathscr{N})) \xrightarrow{\sim} \mathrm{T}^{r'+}(\mathscr{N})$$

*the isomorphism of* $\Xi^{r'}_{\mathbb{Q}}$ *induced by* $\epsilon^{r,r'}(\check{\alpha})$ *and the isomorphisms* (III.12.8.5) *and* (III.12.8.6), *and by*

(III.12.24.3)
$$\check{\beta}^{r'}\colon \mathscr{V}(\mathscr{N}) \to \mathscr{K}^{r'}(\mathrm{T}^{r'+}(\mathscr{N}))$$

*its adjoint. Then:*

(i) *For every rational number $r'$ such that $0 < r' \leq r$, the morphism $\breve{\beta}^{r'}$ is a right inverse of the canonical morphism $\varpi^{r'}\colon \mathscr{K}^{r'}(\mathrm{T}^{r'+}(\mathscr{N})) \to \mathscr{V}(\mathscr{N})$.*

(ii) *For all rational numbers $r'$ and $r''$ such that $0 < r'' < r' \leq r$, the composition*

$$\mathscr{K}^{r'}(\mathrm{T}^{r'+}(\mathscr{N})) \xrightarrow{\varpi^{r'}} \mathscr{V}(\mathscr{N}) \xrightarrow{\breve{\beta}^{r''}} \mathscr{K}^{r''}(\mathrm{T}^{r''+}(\mathscr{N})) \tag{III.12.24.4}$$

*is the canonical morphism.*

**Remark III.12.25.** Under the assumptions of III.12.24, the isomorphism $\breve{\alpha}$ is not a priori uniquely determined by $(\mathscr{N}, r)$, but for every rational number $0 < r' < r$, the morphism $\breve{\alpha}^{r'}$ (III.12.24.2) depends only on $\mathscr{N}$, on which it depends functorially (cf. the proof of III.12.26).

**Theorem III.12.26.** *The functors* (III.12.18.1) *and* (III.12.23.1)

$$\mathbf{Mod}^{\mathrm{Dolb}}_{\mathbb{Q}}(\breve{\overline{\mathscr{B}}}) \underset{\mathscr{V}}{\overset{\mathscr{H}}{\rightleftarrows}} \mathbf{HM}^{\mathrm{sol}}(\mathscr{O}_{\mathfrak{X}}[\tfrac{1}{p}], \xi^{-1}\widetilde{\Omega}^1_{\mathfrak{X}/\mathscr{S}}) \tag{III.12.26.1}$$

*are equivalences of categories quasi-inverse to each other.*

For every object $\mathscr{M}$ of $\mathbf{Mod}^{\mathrm{Dolb}}_{\mathbb{Q}}(\breve{\overline{\mathscr{B}}})$, $\mathscr{H}(\mathscr{M})$ is a solvable Higgs $\mathscr{O}_{\mathfrak{X}}[\frac{1}{p}]$-bundle associated with $\mathscr{M}$, by virtue of III.12.18. We choose a rational number $r_{\mathscr{M}} > 0$ and an isomorphism of $\Xi^{r_{\mathscr{M}}}_{\mathbb{Q}}$

$$\alpha_{\mathscr{M}}\colon \mathrm{T}^{r_{\mathscr{M}}+}(\mathscr{H}(\mathscr{M})) \xrightarrow{\sim} \mathfrak{S}^{r_{\mathscr{M}}}(\mathscr{M}) \tag{III.12.26.2}$$

satisfying the properties of III.12.19. For any rational number $r$ such that $0 < r \leq r_{\mathscr{M}}$, we denote by

$$\alpha^r_{\mathscr{M}}\colon \mathrm{T}^{r+}(\mathscr{H}(\mathscr{M})) \xrightarrow{\sim} \mathfrak{S}^{r}(\mathscr{M}) \tag{III.12.26.3}$$

the isomorphism of $\Xi^r_{\mathbb{Q}}$ induced by $\epsilon^{r_{\mathscr{M}},r}(\alpha_{\mathscr{M}})$ and the isomorphisms (III.12.8.5) and (III.12.8.6), by

$$\breve{\alpha}_{\mathscr{M}}\colon \mathfrak{S}^{r_{\mathscr{M}}}(\mathscr{M}) \xrightarrow{\sim} \mathrm{T}^{r_{\mathscr{M}}+}(\mathscr{H}(\mathscr{M})), \tag{III.12.26.4}$$

$$\breve{\alpha}^r_{\mathscr{M}}\colon \mathfrak{S}^{r}(\mathscr{M}) \xrightarrow{\sim} \mathrm{T}^{r+}(\mathscr{H}(\mathscr{M})), \tag{III.12.26.5}$$

the inverses of $\alpha_{\mathscr{M}}$ and $\alpha^r_{\mathscr{M}}$, respectively, and by

$$\beta^r_{\mathscr{M}}\colon \mathscr{H}(\mathscr{M}) \to \mathrm{T}^r_+(\mathfrak{S}^r(\mathscr{M})), \tag{III.12.26.6}$$

$$\breve{\beta}^r_{\mathscr{M}}\colon \mathscr{M} \to \mathscr{K}^r(\mathrm{T}^{r+}(\mathscr{H}(\mathscr{M}))), \tag{III.12.26.7}$$

the adjoint morphisms of $\alpha^r_{\mathscr{M}}$ and $\breve{\alpha}^r_{\mathscr{M}}$, respectively. Note that $\breve{\alpha}^r_{\mathscr{M}}$ is induced by $\epsilon^{r_{\mathscr{M}},r}(\breve{\alpha}_{\mathscr{M}})$ and the isomorphisms (III.12.8.5) and (III.12.8.6). By III.12.22(ii), the composition

$$\mathscr{M} \xrightarrow{\breve{\beta}^r_{\mathscr{M}}} \mathscr{K}^r(\mathrm{T}^{r+}(\mathscr{H}(\mathscr{M}))) \longrightarrow \mathscr{V}(\mathscr{H}(\mathscr{M})), \tag{III.12.26.8}$$

where the second arrow is the canonical morphism, is an isomorphism, which a priori depends on $\alpha_{\mathscr{M}}$ but not on $r$. Let us show that this isomorphism depends only on $\mathscr{M}$ (but not on the choice of $\alpha_{\mathscr{M}}$) and that it depends on it functorially. It suffices to show that for every morphism $u\colon \mathscr{M} \to \mathscr{M}'$ of $\mathbf{Mod}^{\mathrm{Dolb}}_{\mathbb{Q}}(\breve{\overline{\mathscr{B}}})$ and every rational number

$0 < r < \inf(r_{\mathscr{M}}, r_{\mathscr{M}'})$, the diagram of $\Xi^r_{\mathbb{Q}}$

(III.12.26.9)
$$\begin{array}{ccc} \mathrm{T}^{r+}(\mathscr{H}(\mathscr{M})) & \xrightarrow{\alpha^r_{\mathscr{M}}} & \mathfrak{S}^r(\mathscr{M}) \\ {\scriptstyle \mathrm{T}^{r+}(\mathscr{H}(u))}\downarrow & & \downarrow{\scriptstyle \mathfrak{S}^r(u)} \\ \mathrm{T}^{r+}(\mathscr{H}(\mathscr{M}')) & \xrightarrow{\alpha^r_{\mathscr{M}'}} & \mathfrak{S}^r(\mathscr{M}') \end{array}$$

is commutative. Let $r$, $r'$ be two rational numbers such that $0 < r < r' < \inf(r_{\mathscr{M}}, r_{\mathscr{M}'})$. Consider the diagram

(III.12.26.10)
$$\begin{array}{ccccc} \mathrm{T}^{r'}_{+}(\mathfrak{S}^{r'}(\mathscr{M})) & \xrightarrow{\varpi^{r'}_{\mathscr{M}}} & \mathscr{H}(\mathscr{M}) & \xrightarrow{\beta^r_{\mathscr{M}}} & \mathrm{T}^{r}_{+}(\mathfrak{S}^{r}(\mathscr{M})) \\ {\scriptstyle \mathrm{T}^{r'}_{+}(\mathfrak{S}^{r'}(u))}\downarrow & & \downarrow{\scriptstyle \mathscr{H}(u)} & & \downarrow{\scriptstyle \mathrm{T}^{r}_{+}(\mathfrak{S}^{r}(u))} \\ \mathrm{T}^{r'}_{+}(\mathfrak{S}^{r'}(\mathscr{M}')) & \xrightarrow{\varpi^{r'}_{\mathscr{M}'}} & \mathscr{H}(\mathscr{M}') & \xrightarrow{\beta^r_{\mathscr{M}'}} & \mathrm{T}^{r}_{+}(\mathfrak{S}^{r}(\mathscr{M}')) \end{array}$$

where $\varpi^{r'}_{\mathscr{M}}$ and $\varpi^{r'}_{\mathscr{M}'}$ are the canonical morphisms. It follows from III.12.19(ii) that the large rectangle is commutative. Since the left square is commutative and $\varpi^{r'}_{\mathscr{M}'}$ is surjective by III.12.19(i), the right square is also commutative. The desired assertion follows in view of the injectivity of (III.12.7.12).

Likewise, for every object $\mathscr{N}$ of $\mathbf{HM}^{\mathrm{sol}}(\mathscr{O}_{\mathfrak{X}}[\frac{1}{p}], \xi^{-1}\widetilde{\Omega}^1_{\mathfrak{X}/\mathscr{S}})$, $\mathscr{V}(\mathscr{N})$ is a Dolbeault $\breve{\overline{\mathscr{B}}}_{\mathbb{Q}}$-module associated with $\mathscr{N}$, by virtue of III.12.23. We choose a rational number $r_{\mathscr{N}} > 0$ and an isomorphism of $\Xi^{r_{\mathscr{N}}}_{\mathbb{Q}}$

$$\breve{\alpha}_{\mathscr{N}} \colon \mathfrak{S}^{r_{\mathscr{N}}}(\mathscr{V}(\mathscr{N})) \xrightarrow{\sim} \mathrm{T}^{r_{\mathscr{N}}+}(\mathscr{N}) \tag{III.12.26.11}$$

satisfying the properties of III.12.24. For any rational number $r$ such that $0 < r \leq r_{\mathscr{N}}$, we denote by

$$\breve{\alpha}^r_{\mathscr{N}} \colon \mathfrak{S}^{r}(\mathscr{V}(\mathscr{N})) \xrightarrow{\sim} \mathrm{T}^{r+}(\mathscr{N}) \tag{III.12.26.12}$$

the isomorphism of $\Xi^r_{\mathbb{Q}}$ induced by $\epsilon^{r_{\mathscr{N}}, r}(\breve{\alpha}_{\mathscr{N}})$ and the isomorphisms (III.12.8.5) and (III.12.8.6), by

$$\alpha_{\mathscr{N}} \colon \mathrm{T}^{r_{\mathscr{N}}+}(\mathscr{N}) \xrightarrow{\sim} \mathfrak{S}^{r_{\mathscr{N}}}(\mathscr{V}(\mathscr{N})), \tag{III.12.26.13}$$

$$\alpha^r_{\mathscr{N}} \colon \mathrm{T}^{r+}(\mathscr{N}) \xrightarrow{\sim} \mathfrak{S}^{r}(\mathscr{V}(\mathscr{N})), \tag{III.12.26.14}$$

the inverses of $\breve{\alpha}_{\mathscr{M}}$ and $\breve{\alpha}^r_{\mathscr{N}}$, respectively, and by

$$\breve{\beta}^r_{\mathscr{N}} \colon \mathscr{V}(\mathscr{N}) \to \mathscr{K}^r(\mathrm{T}^{r+}(\mathscr{N})), \tag{III.12.26.15}$$

$$\beta^r_{\mathscr{N}} \colon \mathscr{N} \to \mathrm{T}^r_{+}(\mathfrak{S}^r(\mathscr{V}(\mathscr{N}))), \tag{III.12.26.16}$$

the adjoint morphisms of $\breve{\alpha}^r_{\mathscr{N}}$ and $\alpha^r_{\mathscr{N}}$, respectively. By III.12.17(i), the composition

$$\mathscr{N} \xrightarrow{\beta^r} \mathrm{T}^r_{+}(\mathfrak{S}^r(\mathscr{V}(\mathscr{N}))) \longrightarrow \mathscr{H}(\mathscr{V}(\mathscr{N})), \tag{III.12.26.17}$$

where the second arrow is the canonical morphism, is an isomorphism, which a priori depends on $\breve{\alpha}_{\mathscr{N}}$ but not on $r$. Let us show that this isomorphism depends only on $\mathscr{N}$ (but not on the choice of $\breve{\alpha}_{\mathscr{N}}$) and that it depends on it functorially. It suffices to show that for every morphism $v \colon \mathscr{N} \to \mathscr{N}'$ of $\mathbf{HM}^{\mathrm{sol}}(\mathscr{O}_{\mathfrak{X}}[\frac{1}{p}], \xi^{-1}\widetilde{\Omega}^1_{\mathfrak{X}/\mathscr{S}})$ and every rational number $0 < r < \inf(r_{\mathscr{N}}, r_{\mathscr{N}'})$, the diagram of $\Xi^r_{\mathbb{Q}}$

(III.12.26.18)
$$\begin{array}{ccc}\mathfrak{S}^r(\mathscr{V}(\mathscr{N})) & \xrightarrow{\check{\alpha}^r_{\mathscr{N}}} & \mathrm{T}^{r+}(\mathscr{N}) \\ \downarrow{\scriptstyle \mathfrak{S}^r(\mathscr{V}(v))} & & \downarrow{\scriptstyle \mathrm{T}^{r+}(v)} \\ \mathfrak{S}^r(\mathscr{V}(\mathscr{N}')) & \xrightarrow{\check{\alpha}^r_{\mathscr{N}'}} & \mathrm{T}^{r+}(\mathscr{N}')\end{array}$$

is commutative. Let $r$, $r'$ be two rational numbers such that $0 < r < r' < \inf(r_{\mathscr{N}}, r_{\mathscr{N}'})$. Consider the diagram of $\mathbf{Mod}_{\mathbb{Q}}(\breve{\overline{\mathscr{B}}})$

(III.12.26.19)
$$\begin{array}{ccccc}\mathscr{K}^{r'}(\mathrm{T}^{r'+}(\mathscr{N})) & \xrightarrow{\varpi^{r'}_{\mathscr{N}}} & \mathscr{V}(\mathscr{N}) & \xrightarrow{\check{\beta}^r_{\mathscr{M}}} & \mathscr{K}^r(\mathrm{T}^{r+}(\mathscr{N})) \\ \downarrow{\scriptstyle \mathscr{K}^{r'}(\mathrm{T}^{r'+}(v))} & & \downarrow{\scriptstyle \mathscr{V}(v)} & & \downarrow{\scriptstyle \mathscr{K}^r(\mathrm{T}^{r+}(v))} \\ \mathscr{K}^{r'}(\mathrm{T}^{r'+}(\mathscr{N}')) & \xrightarrow{\varpi^{r'}_{\mathscr{N}'}} & \mathscr{V}(\mathscr{N}') & \xrightarrow{\check{\beta}^r_{\mathscr{N}'}} & \mathscr{K}^r(\mathrm{T}^{r+}(\mathscr{N}'))\end{array}$$

where $\varpi^{r'}_{\mathscr{N}}$ and $\varpi^{r'}_{\mathscr{N}'}$ are the canonical morphisms. It follows from III.12.24(ii) that the large rectangle is commutative. Since the left square is commutative and $\varpi^{r'}_{\mathscr{N}}$ is invertible on the right by III.12.24(i), the right square is also commutative; the desired assertion follows.

**III.12.27.** Let $(\overline{y} \rightsquigarrow \overline{x})$ be a point of $X_{\text{ét}} \overleftarrow{\times}_{X_{\text{ét}}} \overline{X}^{\circ}_{\text{ét}}$ (III.8.6) such that $\overline{x}$ lies over $s$, $X'$ the strict localization of $X$ at $\overline{x}$, $R'_1 = \Gamma(\overline{X}', \mathscr{O}_{\overline{X}'})$, and $\widehat{R}'_1$ its $p$-adic Hausdorff completion. For any coherent $\mathscr{O}_{\mathfrak{X}}$-module $\mathscr{F}$, we (abusively) set

(III.12.27.1)
$$\mathscr{F}_{\overline{x}} = \lim_{\substack{\longleftarrow \\ n \in \mathbb{N}^{\circ}}} \Gamma(\overline{X}'_n, \mathscr{F} \otimes_{\mathscr{O}_{\mathfrak{X}}} \mathscr{O}_{\overline{X}'_n}).$$

By III.6.16, the resulting functor $\mathbf{Mod}^{\mathrm{coh}}(\mathscr{O}_{\mathfrak{X}}) \to \mathbf{Mod}(\widehat{R}'_1)$ induces a functor that we denote also by

(III.12.27.2)
$$\mathbf{Mod}^{\mathrm{coh}}(\mathscr{O}_{\mathfrak{X}}[\tfrac{1}{p}]) \to \mathbf{Mod}(\widehat{R}'_1[\tfrac{1}{p}]), \quad \mathscr{F} \mapsto \mathscr{F}_{\overline{x}}.$$

By III.9.5(i), $\rho(\overline{y} \rightsquigarrow \overline{x})$ is a point of $\widetilde{E}_s$. For any object $\mathfrak{F} = (\mathfrak{F}_n)_{n \in \mathbb{N}}$ of $\widetilde{E}_s^{\mathbb{N}^{\circ}}$, we (abusively) set

(III.12.27.3)
$$\mathfrak{F}_{\rho(\overline{y} \rightsquigarrow \overline{x})} = \lim_{\substack{\longleftarrow \\ n \in \mathbb{N}^{\circ}}} (\mathfrak{F}_n)_{\rho(\overline{y} \rightsquigarrow \overline{x})}.$$

Take care that the resulting functor $\widetilde{E}_s^{\mathbb{N}^{\circ}} \to \mathbf{Ens}$ is not a priori a fiber functor. Let us take again the notation of III.10.25. By (III.10.25.2), VI.10.31, and VI.9.9, we have a canonical isomorphism

(III.12.27.4)
$$\breve{\overline{\mathscr{B}}}_{\rho(\overline{y} \rightsquigarrow \overline{x})} \xrightarrow{\sim} \widehat{\overline{R}}^{\overline{y}}_{X'}.$$

The functor (III.12.27.3) induces functors that we denote also by

(III.12.27.5)
$$\mathbf{Mod}(\breve{\overline{\mathscr{B}}}) \to \mathbf{Mod}(\widehat{\overline{R}}^{\overline{y}}_{X'}), \quad \mathfrak{M} \mapsto \mathfrak{M}_{\rho(\overline{y} \rightsquigarrow \overline{x})},$$

(III.12.27.6)
$$\mathbf{Mod}_{\mathbb{Q}}(\breve{\overline{\mathscr{B}}}) \to \mathbf{Mod}(\widehat{\overline{R}}^{\overline{y}}_{X'}[\tfrac{1}{p}]), \quad \mathscr{M} \mapsto \mathscr{M}_{\rho(\overline{y} \rightsquigarrow \overline{x})}.$$

By III.10.26, III.10.29, VI.10.31, and VI.9.9, for every rational number $r \geq 0$, we have a canonical isomorphism

(III.12.27.7)
$$\breve{\mathscr{C}}^{(r)}_{\rho(\overline{y} \rightsquigarrow \overline{x})} \xrightarrow{\sim} \widehat{\mathscr{C}}^{\overline{y},(r)}_{X'}.$$

**Lemma III.12.28.** *Under the assumptions of* III.12.27, *let moreover* $\mathscr{F}$ *be a coherent* $\mathscr{O}_{\mathfrak{X}}$*-module,* $(U, \mathfrak{p}\colon \overline{x} \to U)$ *an object of* $\mathfrak{V}_{\overline{x}}$ (III.10.8) *such that* $U$ *is affine, and* $\mathscr{U}$ *the formal scheme p-adic completion of* $\overline{U}$. *Then:*

(i) *The* $\widehat{R}'_1$*-module* $\mathscr{F}_{\overline{x}}$ *is of finite type, and is complete and separated for the p-adic topology.*

(ii) *We have a canonical and functorial isomorphism*

$$\mathscr{F}_{\overline{x}} \xrightarrow{\sim} \Gamma(\mathscr{U}, \mathscr{F} \otimes_{\mathscr{O}_{\mathfrak{X}}} \mathscr{O}_{\mathscr{U}}) \widehat{\otimes}_{\mathscr{O}_{\mathscr{U}}(\mathscr{U})} \widehat{R}'_1. \tag{III.12.28.1}$$

(iii) *We have a canonical and functorial isomorphism*

$$(\top^*(\mathscr{F}))_{\rho(\overline{y} \rightsquigarrow \overline{x})} \xrightarrow{\sim} \mathscr{F}_{\overline{x}} \widehat{\otimes}_{\widehat{R}'_1} \widehat{\overline{R}}^{\overline{y}}_{X'}. \tag{III.12.28.2}$$

(i) This follows from ([**11**] Chapter III § 2.11 Proposition 14).
(ii) This immediately follows from the definition.
(iii) By ([**2**] VIII 5.2 and VII 5.8), we have a canonical isomorphism

$$(\mathscr{F} \otimes_{\mathscr{O}_{\mathfrak{X}}} \mathscr{O}_{\overline{X}_n})_{\overline{x}} \xrightarrow{\sim} \Gamma(\overline{X}'_n, \mathscr{F} \otimes_{\mathscr{O}_{\mathfrak{X}}} \mathscr{O}_{\overline{X}'_n}), \tag{III.12.28.3}$$

where we view $\mathscr{F} \otimes_{\mathscr{O}_{\mathfrak{X}}} \mathscr{O}_{\overline{X}_n}$ on the left-hand side as a sheaf of $X_{s,\text{ét}}$ (III.9.9). In view of (III.2.9.5), (III.11.1.12), and (VI.10.18.1), we deduce from this a canonical and functorial isomorphism

$$(\top^*(\mathscr{F}))_{\rho(\overline{y} \rightsquigarrow \overline{x})} \xrightarrow{\sim} \varprojlim_{n \in \mathbb{N}^{\circ}} (\Gamma(\overline{X}'_n, \mathscr{F} \otimes_{\mathscr{O}_{\mathfrak{X}}} \mathscr{O}_{\overline{X}'_n}) \otimes_{R'_1} \overline{R}^{\overline{y}}_{X'}). \tag{III.12.28.4}$$

The statement follows in view of ([**11**] Chapter III § 2.11 Corollary 1 to Proposition 14).

**III.12.29.** We keep the assumptions and notation of III.12.27 and moreover set $R' = \Gamma(X', \mathscr{O}_{X'})$ and

$$\widetilde{\Omega}^1_{X/S}(X') = \Gamma(X', \widetilde{\Omega}^1_{X/S} \otimes_{\mathscr{O}_X} \mathscr{O}_{X'}). \tag{III.12.29.1}$$

For any $R'$-algebra $A$, by a Higgs $A$-module with coefficients in $\xi^{-1}\widetilde{\Omega}^1_{X/S}(X')$ we mean a Higgs $A$-module with coefficients in $\xi^{-1}\widetilde{\Omega}^1_{X/S}(X') \otimes_{R'} A$ (II.2.8). In view of (III.12.5.3) and the fact that the $R'$-module $\widetilde{\Omega}^1_{X/S}(X')$ is free of finite type, the functor (III.12.27.2) induces a functor
(III.12.29.2)

$$\mathbf{HM}^{\text{coh}}(\mathscr{O}_{\mathfrak{X}}[\tfrac{1}{p}], \xi^{-1}\widetilde{\Omega}^1_{\mathfrak{X}/\mathscr{S}}) \to \mathbf{HM}(\widehat{R}'_1[\tfrac{1}{p}], \xi^{-1}\widetilde{\Omega}^1_{X/S}(X')), \quad (\mathscr{N}, \theta) \mapsto (\mathscr{N}_{\overline{x}}, \theta_{\overline{x}}).$$

Let $r$ be a rational number $\geq 0$. By (III.10.25.6), we have a canonical isomorphism

$$\Omega^1_{\mathscr{C}^{\overline{y},(r)}_{X'}/\widehat{\overline{R}}^{\overline{y}}_{X'}} \xrightarrow{\sim} \xi^{-1}\widetilde{\Omega}^1_{X/S}(X') \otimes_{R'} \mathscr{C}^{\overline{y},(r)}_{X'}. \tag{III.12.29.3}$$

We denote by

$$d_{\mathscr{C}^{\overline{y},(r)}_{X'}} \colon \mathscr{C}^{\overline{y},(r)}_{X'} \to \xi^{-1}\widetilde{\Omega}^1_{X/S}(X') \otimes_{R'} \mathscr{C}^{\overline{y},(r)}_{X'} \tag{III.12.29.4}$$

the universal $\widehat{\overline{R}}^{\overline{y}}_{X'}$-derivation of $\mathscr{C}^{\overline{y},(r)}_{X'}$ (cf. III.10.25) and by

$$d_{\widehat{\mathscr{C}}^{\overline{y},(r)}_{X'}} \colon \widehat{\mathscr{C}}^{\overline{y},(r)}_{X'} \to \xi^{-1}\widetilde{\Omega}^1_{X/S}(X') \otimes_{R'} \widehat{\mathscr{C}}^{\overline{y},(r)}_{X'} \tag{III.12.29.5}$$

its extension to the $p$-adic completions (the $R'$-module $\widetilde{\Omega}^1_{X/S}(X')$ being free of finite type). These are Higgs $\widehat{\overline{R}}^{\overline{y}}_{X'}$-fields with coefficients in $\xi^{-1}\widetilde{\Omega}^1_{X/S}(X')$ by II.2.12 and II.2.16.

We denote by $\Xi^{\mathrm{aft},r}$ the full subcategory of $\Xi^r$ (III.12.7) made up of integrable $p^r$-isoconnections $(\mathscr{F},\mathscr{G},u,\nabla)$ with respect to the extension $\breve{\mathscr{C}}^{(r)}/\breve{\mathscr{B}}$ such that the $\breve{\mathscr{C}}^{(r)}$-modules $\mathscr{F}$ and $\mathscr{G}$ are adic of finite type (III.7.16), and by $\Xi'^r$ the category of integrable $p$-adic $p^r$-isoconnections with respect to the extension $\widehat{\mathscr{C}}_{X'}^{\overline{y},(r)}/\widehat{\overline{R}}_{X'}^{\overline{y}}$ (III.6.13). These are additive categories. We denote by $\Xi^{\mathrm{aft},r}_{\mathbb{Q}}$ and $\Xi'^r_{\mathbb{Q}}$ the categories of objects of $\Xi^{\mathrm{aft},r}$ and $\Xi'^r$ up to isogeny (III.6.1.1). In view of ([**11**] Chapter III § 2.11 Proposition 14 and Corollary 1), the functor (III.12.27.3) induces an additive functor

(III.12.29.6) $$\Xi^{\mathrm{aft},r}\to\Xi'^r.$$

By III.6.14, every object of $\Xi'^r$ is a Higgs $\widehat{\overline{R}}_{X'}^{\overline{y}}$-isogeny with coefficients in $\xi^{-1}\widetilde{\Omega}^1_{X/S}(X')$. We therefore have a functor

(III.12.29.7)

$$\Xi'^r_{\mathbb{Q}}\to\mathbf{HM}(\widehat{\overline{R}}_{X'}^{\overline{y}}[\frac{1}{p}],\xi^{-1}\widetilde{\Omega}^1_{X/S}(X')),\quad (M,N,u,\nabla)\mapsto(M_{\mathbb{Q}_p},(\mathrm{id}\otimes u_{\mathbb{Q}_p}^{-1})\circ\nabla_{\mathbb{Q}_p}).$$

We deduce from this a functor

(III.12.29.8) $$\phi_{\rho(\overline{y}\rightsquigarrow\overline{x})}\colon\Xi^{\mathrm{aft},r}_{\mathbb{Q}}\to\mathbf{HM}(\widehat{\overline{R}}_{X'}^{\overline{y}}[\frac{1}{p}],\xi^{-1}\widetilde{\Omega}^1_{X/S}(X')).$$

Let $\mathfrak{M}$ be an adic $\breve{\mathscr{B}}$-module of finite type. By ([**11**] Chapter III § 2.11 Proposition 14 and Corollary 1), the $\widehat{\overline{R}}_{X'}^{\overline{y}}$-module $\mathfrak{M}_{\rho(\overline{y}\rightsquigarrow\overline{x})}$ is of finite type, and is complete and separated for the $p$-adic topology. We have a canonical and functorial isomorphism (III.12.7.3)

(III.12.29.9) $$\phi_{\rho(\overline{y}\rightsquigarrow\overline{x})}(\mathfrak{S}^r(\mathfrak{M}))\xrightarrow{\sim}((\widehat{\mathscr{C}}_{X'}^{(r),\overline{y}}\widehat{\otimes}_{\widehat{\overline{R}}_{X'}^{\overline{y}}}\mathfrak{M}_{\rho(\overline{y}\rightsquigarrow\overline{x})})_{\mathbb{Q}_p},(p^rd_{\widehat{\mathscr{C}}_{X'}^{\overline{y},(r)}}\widehat{\otimes}\mathrm{id})_{\mathbb{Q}_p}).$$

**Lemma III.12.30.** *Under the assumptions of* III.12.27 *and* III.12.29*, let moreover* $(\mathscr{N},\theta)$ *be a Higgs* $\mathscr{O}_{\mathfrak{X}}[\frac{1}{p}]$*-bundle with coefficients in* $\xi^{-1}\widetilde{\Omega}^1_{\mathfrak{X}/\mathscr{S}}$*. Then:*

(i) *The* $\widehat{R}'_1[\frac{1}{p}]$*-module* $\mathscr{N}_{\overline{x}}$ *is projective of finite type.*

(ii) *For every rational number* $r\geq 0$*, we have a canonical and functorial isomorphism*

(III.12.30.1) $$\phi_{\rho(\overline{y}\rightsquigarrow\overline{x})}(\top^{r+}(\mathscr{N},\theta))\xrightarrow{\sim}(\widehat{\mathscr{C}}_{X'}^{(r),\overline{y}}\otimes_{\widehat{R}'_1}\mathscr{N}_{\overline{x}},p^rd_{\widehat{\mathscr{C}}_{X'}^{\overline{y},(r)}}\otimes\mathrm{id}+\mathrm{id}\otimes\theta_{\overline{x}}).$$

Indeed, let $\mathfrak{N}$ be a coherent $\mathscr{O}_{\mathfrak{X}}$-module such that $\mathfrak{N}_{\mathbb{Q}_p}=\mathscr{N}$ (III.6.16).

(i) Let $(U,\mathfrak{p}\colon\overline{x}\to U)$ be an object of $\mathfrak{V}_{\overline{x}}(\mathbf{P})$ and $\mathscr{U}$ the formal scheme $p$-adic completion of $\overline{U}$. By III.12.28(ii), we have a canonical isomorphism

(III.12.30.2) $$\Gamma(\mathscr{U},\mathfrak{N}\otimes_{\mathscr{O}_{\mathfrak{X}}}\mathscr{O}_{\mathscr{U}})\widehat{\otimes}_{\mathscr{O}_{\mathscr{U}}(\mathscr{U})}\widehat{R}'_1\xrightarrow{\sim}\mathfrak{N}_{\overline{x}}.$$

We have $\Gamma(\mathscr{U},\mathfrak{N}\otimes_{\mathscr{O}_{\mathfrak{X}}}\mathscr{O}_{\mathscr{U}})\otimes_{\mathbb{Z}_p}\mathbb{Q}_p=\Gamma(\mathscr{U},\mathscr{N}\otimes_{\mathscr{O}_{\mathfrak{X}}}\mathscr{O}_{\mathscr{U}})$ ([**1**] (2.10.5.1)). By III.6.17, the $\mathscr{O}_{\mathscr{U}}(\mathscr{U})[\frac{1}{p}]$-module $\Gamma(\mathscr{U},\mathscr{N}\otimes_{\mathscr{O}_{\mathfrak{X}}}\mathscr{O}_{\mathscr{U}})$ is projective of finite type. We deduce from this, by virtue of III.6.18 and III.10.27(ii), that the morphism

(III.12.30.3) $$\Gamma(\mathscr{U},\mathfrak{N}\otimes_{\mathscr{O}_{\mathfrak{X}}}\mathscr{O}_{\mathscr{U}})\otimes_{\mathscr{O}_{\mathscr{U}}(\mathscr{U})}\widehat{R}'_1[\frac{1}{p}]\to\mathscr{N}_{\overline{x}}$$

induced by (III.12.30.2) is an isomorphism; the desired statement follows.

(ii) By III.12.28(iii), we have a canonical and functorial isomorphism

(III.12.30.4) $$(\top^*(\mathfrak{N})\otimes_{\breve{\mathscr{B}}}\breve{\mathscr{C}}^{(r)})_{\rho(\overline{y}\rightsquigarrow\overline{x})}\xrightarrow{\sim}\mathfrak{N}_{\overline{x}}\widehat{\otimes}_{\widehat{R}'_1}\widehat{\mathscr{C}}_{X'}^{(r),\overline{y}}.$$

The statement then follows from (i), III.6.18, and III.10.27(ii).

**Lemma III.12.31.** *Under the assumptions of* III.12.27 *and* III.12.29, *let moreover* $\mathscr{M}$ *be an object of* $\mathbf{Mod}^{\mathrm{aft}}_{\mathbb{Q}}(\breve{\overline{\mathscr{B}}})$, $(\mathscr{N},\theta)$ *a Higgs* $\mathscr{O}_{\mathfrak{X}}[\frac{1}{p}]$*-bundle with coefficients in* $\xi^{-1}\widetilde{\Omega}^1_{\mathfrak{X}/\mathscr{S}}$, $r$ *a rational number* $> 0$, *and*

$$\alpha\colon \top^{r+}(\mathscr{N},\theta) \xrightarrow{\sim} \mathfrak{S}^r(\mathscr{M}) \tag{III.12.31.1}$$

*an isomorphism of* $\Xi^r_{\mathbb{Q}}$ (III.12.7). *Then:*

(i) *The* $\widehat{\overline{R}}{}^{\overline{y}}_{X'}[\frac{1}{p}]$*-module* $\mathscr{M}_{\rho(\overline{y}\rightsquigarrow\overline{x})}$ *is projective of finite type.*

(ii) *The isomorphism* $\phi_{\rho(\overline{y}\rightsquigarrow\overline{x})}(\alpha)$ (III.12.29.8) *induces a* $\widehat{\mathscr{C}}{}^{\overline{y},(r)}_{X'}$*-linear isomorphism*

$$\alpha_{\rho(\overline{y}\rightsquigarrow\overline{x})}\colon \widehat{\mathscr{C}}{}^{\overline{y},(r)}_{X'} \otimes_{\widehat{R}'_1} \mathscr{N}_{\overline{x}} \xrightarrow{\sim} \widehat{\mathscr{C}}{}^{\overline{y},(r)}_{X'} \otimes_{\widehat{\overline{R}}{}^{\overline{y}}_{X'}} \mathscr{M}_{\rho(\overline{y}\rightsquigarrow\overline{x})} \tag{III.12.31.2}$$

*of Higgs* $\widehat{\overline{R}}{}^{\overline{y}}_{X'}$*-modules with coefficients in* $\xi^{-1}\widetilde{\Omega}^1_{X/S}(X')$, *where* $\mathscr{N}_{\overline{x}}$ *is endowed with the Higgs field* $\theta_{\overline{x}}$ (III.12.29.2), $\widehat{\mathscr{C}}{}^{\overline{y},(r)}_{X'}$ *is endowed with the Higgs field* $p^r d_{\widehat{\mathscr{C}}{}^{\overline{y},(r)}_{X'}}$ (III.12.29.5), *and* $\mathscr{M}_{\rho(\overline{y}\rightsquigarrow\overline{x})}$ *is endowed with the zero Higgs field.*

Indeed, let $\mathfrak{M}$ be an object of $\mathbf{Mod}^{\mathrm{aft}}(\breve{\overline{\mathscr{B}}})$ such that $\mathscr{M} = \mathfrak{M}_{\mathbb{Q}_p}$. By (III.12.29.9) and III.12.30, $\phi_{\rho(\overline{y}\rightsquigarrow\overline{x})}(\alpha)$ is a $\widehat{\mathscr{C}}{}^{\overline{y},(r)}_{X'}$-linear isomorphism

$$\widehat{\mathscr{C}}{}^{\overline{y},(r)}_{X'} \otimes_{\widehat{R}'_1} \mathscr{N}_{\overline{x}} \xrightarrow{\sim} (\widehat{\mathscr{C}}{}^{\overline{y},(r)}_{X'} \widehat{\otimes}_{\widehat{\overline{R}}{}^{\overline{y}}_{X'}} \mathfrak{M}_{\rho(\overline{y}\rightsquigarrow\overline{x})}) \otimes_{\mathbb{Z}_p} \mathbb{Q}_p \tag{III.12.31.3}$$

of Higgs $\widehat{\overline{R}}{}^{\overline{y}}_{X'}$-modules with coefficients in $\xi^{-1}\widetilde{\Omega}^1_{X/S}(X')$, where $\mathscr{N}_{\overline{x}}$ is endowed with the Higgs field $\theta_{\overline{x}}$, $\widehat{\mathscr{C}}{}^{\overline{y},(r)}_{X'}$ is endowed with the Higgs field $p^r d_{\widehat{\mathscr{C}}{}^{\overline{y},(r)}_{X'}}$, and $\mathfrak{M}_{\rho(\overline{y}\rightsquigarrow\overline{x})}$ is endowed with the zero Higgs field. Consider an $\widehat{\overline{R}}{}^{\overline{y}}_{X'}$-augmentation $u\colon \widehat{\mathscr{C}}{}^{\overline{y},(r)}_{X'} \to \widehat{\overline{R}}{}^{\overline{y}}_{X'}$, which exists by (III.10.25.6). By virtue of III.6.18, III.10.27(ii), III.12.30(i), and (III.12.31.3), the canonical morphism

$$(\widehat{\overline{R}}{}^{\overline{y}}_{X'} \otimes_{\widehat{\mathscr{C}}{}^{\overline{y},(r)}_{X'}} (\widehat{\mathscr{C}}{}^{\overline{y},(r)}_{X'} \widehat{\otimes}_{\widehat{\overline{R}}{}^{\overline{y}}_{X'}} \mathfrak{M}_{\rho(\overline{y}\rightsquigarrow\overline{x})})) \otimes_{\mathbb{Z}_p} \mathbb{Q}_p \to (\widehat{\overline{R}}{}^{\overline{y}}_{X'} \widehat{\otimes}_{\widehat{\mathscr{C}}{}^{\overline{y},(r)}_{X'}} (\widehat{\mathscr{C}}{}^{\overline{y},(r)}_{X'} \widehat{\otimes}_{\widehat{\overline{R}}{}^{\overline{y}}_{X'}} \mathfrak{M}_{\rho(\overline{y}\rightsquigarrow\overline{x})})) \otimes_{\mathbb{Z}_p} \mathbb{Q}_p$$

is an isomorphism. The right-hand side identifies canonically with $\mathfrak{M}_{\rho(\overline{y}\rightsquigarrow\overline{x})} \otimes_{\mathbb{Z}_p} \mathbb{Q}_p$. The isomorphism (III.12.31.3) therefore induces, by base change by $u$, an isomorphism

$$\widehat{\overline{R}}{}^{\overline{y}}_{X'} \otimes_{\widehat{R}'_1} \mathscr{N}_{\overline{x}} \xrightarrow{\sim} \mathfrak{M}_{\rho(\overline{y}\rightsquigarrow\overline{x})} \otimes_{\mathbb{Z}_p} \mathbb{Q}_p = \mathscr{M}_{\rho(\overline{y}\rightsquigarrow\overline{x})}. \tag{III.12.31.4}$$

Consequently, the $\widehat{\overline{R}}{}^{\overline{y}}_{X'}[\frac{1}{p}]$-module $\mathscr{M}_{\rho(\overline{y}\rightsquigarrow\overline{x})}$ is projective of finite type by virtue of III.12.30(i). By III.6.18 and III.10.27(ii), the canonical morphism

$$(\widehat{\mathscr{C}}{}^{\overline{y},(r)}_{X'} \otimes_{\widehat{\overline{R}}{}^{\overline{y}}_{X'}} \mathfrak{M}_{\rho(\overline{y}\rightsquigarrow\overline{x})}) \otimes_{\mathbb{Z}_p} \mathbb{Q}_p \to (\widehat{\mathscr{C}}{}^{\overline{y},(r)}_{X'} \widehat{\otimes}_{\widehat{\overline{R}}{}^{\overline{y}}_{X'}} \mathfrak{M}_{\rho(\overline{y}\rightsquigarrow\overline{x})}) \otimes_{\mathbb{Z}_p} \mathbb{Q}_p \tag{III.12.31.5}$$

is therefore an isomorphism, giving the lemma.

**Lemma III.12.32.** *For every flat* $\breve{\overline{\mathscr{B}}}_{\mathbb{Q}}$*-module* $\mathscr{M}$ *and every integer* $q \geq 0$, *we have a functorial canonical isomorphism* (III.11.21)

$$\varinjlim_{r\in\mathbb{Q}_{>0}} \mathrm{R}^q\top_*(\mathscr{M} \otimes_{\breve{\overline{\mathscr{B}}}_{\mathbb{Q}}} \mathbb{K}^\bullet_{\mathbb{Q}}(\breve{\mathscr{C}}^{(r)}, p^r\breve{d}^{(r)})) \xrightarrow{\sim} \mathrm{R}^q\top_*(\mathscr{M}), \tag{III.12.32.1}$$

*where* $\mathrm{R}^q\top_*(\mathscr{M} \otimes_{\breve{\overline{\mathscr{B}}}_{\mathbb{Q}}} \mathbb{K}^\bullet_{\mathbb{Q}}(\breve{\mathscr{C}}^{(r)}, p^r\breve{d}^{(r)}))$ *denotes the hypercohomology of the functor* $\top_*$ (III.12.1.3) *with respect to the complex* $\mathscr{M} \otimes_{\breve{\overline{\mathscr{B}}}_{\mathbb{Q}}} \mathbb{K}^\bullet_{\mathbb{Q}}(\breve{\mathscr{C}}^{(r)}, p^r\breve{d}^{(r)})$.

Indeed, the spectral sequence of hypercohomology of the functor $\top_*$ induces, for every rational number $r \geq 0$, a functorial spectral sequence

(III.12.32.2)
$$ {}^r\mathrm{E}_2^{i,j} = \mathrm{R}^i\top_*(\mathscr{M} \otimes_{\breve{\overline{\mathscr{B}}}_{\mathbb{Q}}} \mathrm{H}^j(\mathbb{K}^\bullet_{\mathbb{Q}}(\breve{\mathscr{C}}^{(r)}, p^r\breve{d}^{(r)}))) \Rightarrow \mathrm{R}^{i+j}\top_*(\mathscr{M} \otimes_{\breve{\overline{\mathscr{B}}}_{\mathbb{Q}}} \mathbb{K}^\bullet_{\mathbb{Q}}(\breve{\mathscr{C}}^{(r)}, p^r\breve{d}^{(r)})).$$

By III.11.23(iii), for all integers $i \geq 0$ and $j \geq 1$ and all rational numbers $r > r' > 0$, the canonical morphism

$$ {}^r\mathrm{E}_2^{i,j} \to {}^{r'}\mathrm{E}_2^{i,j} \qquad \text{(III.12.32.3)}$$

is zero. We therefore have

$$ \lim_{\substack{\longrightarrow \\ r\in\mathbb{Q}_{>0}}} {}^r\mathrm{E}_2^{i,j} = 0. \qquad \text{(III.12.32.4)}$$

On the other hand, it follows from III.11.23(ii) that the canonical morphisms

$$ \breve{\overline{\mathscr{B}}}_{\mathbb{Q}} \to \mathrm{H}^0(\mathbb{K}^\bullet_{\mathbb{Q}}(\breve{\mathscr{C}}^{(r)}, p^r\breve{d}^{(r)})), \qquad \text{(III.12.32.5)}$$

for $r \in \mathbb{Q}_{>0}$, induce an isomorphism

$$ \mathrm{R}^i\top_*(\mathscr{M}) \xrightarrow{\sim} \lim_{\substack{\longrightarrow \\ r\in\mathbb{Q}_{>0}}} {}^r\mathrm{E}_2^{i,0}. \qquad \text{(III.12.32.6)}$$

Since filtered direct limits are representable in $\mathbf{Mod}(\mathscr{O}_{\mathfrak{X}}[\frac{1}{p}])$ and commute with finite inverse limits ([**2**] II 4.3), the lemma follows.

**Lemma III.12.33.** *Let $\mathscr{N}$ be a Higgs $\mathscr{O}_{\mathfrak{X}}[\frac{1}{p}]$-bundle with coefficients in $\xi^{-1}\widetilde{\Omega}_{\mathfrak{X}/\mathscr{S}}$ and $q$ an integer $\geq 0$. Denote by $\mathbb{K}^\bullet(\mathscr{N})$ the Dolbeault complex of $\mathscr{N}$* (II.2.8.2) *and for any rational number $r \geq 0$, by $\mathbb{K}^\bullet(\top^{r+}(\mathscr{N}))$ the Dolbeault complex of $\top^{r+}(\mathscr{N})$* (III.12.7). *We then have a functorial canonical isomorphism*

$$ \lim_{\substack{\longrightarrow \\ r\in\mathbb{Q}_{>0}}} \mathrm{R}^q\top_*(\mathbb{K}^\bullet(\top^{r+}(\mathscr{N}))) \xrightarrow{\sim} \mathrm{H}^q(\mathbb{K}^\bullet(\mathscr{N})), \qquad \text{(III.12.33.1)}$$

*where $\mathrm{R}^q\top_*(\mathbb{K}^\bullet(\top^{r+}(\mathscr{N})))$ denotes the hypercohomology of the functor $\top_*$* (III.12.1.3) *with respect to the complex $\mathbb{K}^\bullet(\top^{r+}(\mathscr{N}))$.*

Let $r$ be a rational number $> 0$, and $i, j$ two integers $\geq 0$. In view of III.12.4(i), $\breve{d}^{(r)}$ (III.11.21.1) induces an $\mathscr{O}_{\mathfrak{X}}$-linear morphism

$$ \delta^{j,(r)}\colon \mathrm{R}^j\top_*(\breve{\mathscr{C}}^{(r)}) \to \xi^{-1}\widetilde{\Omega}^1_{\mathfrak{X}/\mathscr{S}} \otimes_{\mathscr{O}_{\mathfrak{X}}} \mathrm{R}^j\top_*(\breve{\mathscr{C}}^{(r)}), \qquad \text{(III.12.33.2)}$$

which is clearly a Higgs $\mathscr{O}_{\mathfrak{X}}$-field on $\mathrm{R}^j\top_*(\breve{\mathscr{C}}^{(r)})$ with coefficients in $\xi^{-1}\widetilde{\Omega}^1_{\mathfrak{X}/\mathscr{S}}$. We denote by $\theta$ the Higgs $\mathscr{O}_{\mathfrak{X}}[\frac{1}{p}]$-field on $\mathscr{N}$ and by $\vartheta^{j,(r)}_{\mathrm{tot}} = \theta\otimes\mathrm{id}+p^r\mathrm{id}\otimes\delta^{j,(r)}$ the total Higgs $\mathscr{O}_{\mathfrak{X}}[\frac{1}{p}]$-field on $\mathscr{N} \otimes_{\mathscr{O}_{\mathfrak{X}}} \mathrm{R}^j\top_*(\widehat{\mathscr{C}}^{(r)})$ (II.2.8.8). By III.12.4(ii), we have a canonical $\mathscr{O}_{\mathfrak{X}}[\frac{1}{p}]$-linear isomorphism

$$ \mathrm{R}^j\top_*(\mathbb{K}^i(\top^{r+}(\mathscr{N}))) \xrightarrow{\sim} \mathbb{K}^i(\mathscr{N} \otimes_{\mathscr{O}_{\mathfrak{X}}} \mathrm{R}^j\top_*(\breve{\mathscr{C}}^{(r)}), \vartheta^{j,(r)}_{\mathrm{tot}}), \qquad \text{(III.12.33.3)}$$

which is compatible with the differentials of the two Dolbeault complexes.

On the other hand, we have a functorial canonical spectral sequence

$$ {}^r\mathrm{E}_1^{i,j} = \mathrm{R}^j\top_*(\mathbb{K}^i(\top^{r+}(\mathscr{N}))) \Rightarrow \mathrm{R}^{i+j}\top_*(\mathbb{K}^\bullet(\top^{r+}(\mathscr{N}))). \qquad \text{(III.12.33.4)}$$

By III.11.18 and (III.12.33.3), for every $i \geq 0$, we have a canonical isomorphism

$$ \lim_{\substack{\longrightarrow \\ r\in\mathbb{Q}_{>0}}} {}^r\mathrm{E}_1^{i,0} \xrightarrow{\sim} \mathbb{K}^i(\mathscr{N}, \theta), \qquad \text{(III.12.33.5)}$$

and for every $j \geq 1$, we have

$$\lim_{\substack{\longrightarrow \\ r\in\mathbb{Q}_{>0}}} {}^r\mathrm{E}_1^{i,j} = 0. \tag{III.12.33.6}$$

Moreover, the isomorphisms (III.12.33.5) (for $i \in \mathbb{N}$) form an isomorphism of complexes. The lemma follows ([**2**] II 4.3).

**Theorem III.12.34.** *Let $\mathscr{M}$ be a Dolbeault $\breve{\overline{\mathscr{B}}}_{\mathbb{Q}}$-module and $q$ an integer $\geq 0$. Denote by $\mathbb{K}^\bullet(\mathscr{H}(\mathscr{M}))$ the Dolbeault complex of the Higgs $\mathscr{O}_{\mathfrak{X}}[\frac{1}{p}]$-bundle $\mathscr{H}(\mathscr{M})$* (II.2.8.2). *We then have a functorial canonical isomorphism of $\mathscr{O}_{\mathfrak{X}}[\frac{1}{p}]$-modules*

$$\mathrm{R}^q\top_*(\mathscr{M}) \xrightarrow{\sim} \mathrm{H}^q(\mathbb{K}^\bullet(\mathscr{H}(\mathscr{M}))). \tag{III.12.34.1}$$

Indeed, $\mathscr{H}(\mathscr{M})$ is a solvable Higgs $\mathscr{O}_{\mathfrak{X}}[\frac{1}{p}]$-bundle associated with $\mathscr{M}$ by virtue of III.12.18. We choose a rational number $r_{\mathscr{M}} > 0$ and an isomorphism of $\Xi_{\mathbb{Q}}^{r_{\mathscr{M}}}$

$$\alpha_{\mathscr{M}} \colon \top^{r_{\mathscr{M}}+}(\mathscr{H}(\mathscr{M})) \xrightarrow{\sim} \mathfrak{S}^{r_{\mathscr{M}}}(\mathscr{M}) \tag{III.12.34.2}$$

satisfying the properties of III.12.19. For any rational number $r$ such that $0 < r < r_{\mathscr{M}}$, we denote by

$$\alpha_{\mathscr{M}}^r \colon \top^{r+}(\mathscr{H}(\mathscr{M})) \xrightarrow{\sim} \mathfrak{S}^r(\mathscr{M}) \tag{III.12.34.3}$$

the isomorphism of $\Xi_{\mathbb{Q}}^r$ induced by $\epsilon^{r_{\mathscr{M}},r}(\alpha_{\mathscr{M}})$ and the isomorphisms (III.12.8.5) and (III.12.8.6). By the proof of III.12.26, $\alpha_{\mathscr{M}}^r$ depends only on $\mathscr{M}$ (but not on $\alpha_{\mathscr{M}}$), and depends on it functorially. We denote by $\mathbb{K}^\bullet(\top^{r+}(\mathscr{H}(\mathscr{M})))$ the Dolbeault complex of $\top^{r+}(\mathscr{H}(\mathscr{M}))$ in $\mathbf{Mod}_{\mathbb{Q}}(\breve{\overline{\mathscr{B}}})$ (cf. III.12.7). Since $\mathscr{M}$ is $\breve{\overline{\mathscr{B}}}_{\mathbb{Q}}$-flat by III.12.12, $\alpha_{\mathscr{M}}^r$ induces an isomorphism (III.11.21)

$$\mathbb{K}^\bullet(\top^{r+}(\mathscr{H}(\mathscr{M}))) \xrightarrow{\sim} \mathscr{M} \otimes_{\breve{\overline{\mathscr{B}}}_{\mathbb{Q}}} \mathbb{K}_{\mathbb{Q}}^\bullet(\breve{\mathscr{C}}^{(r)}, p^r \breve{d}^{(r)}). \tag{III.12.34.4}$$

We deduce from this a functorial canonical isomorphism of $\mathscr{O}_{\mathfrak{X}}[\frac{1}{p}]$-modules

$$\lim_{\substack{\longrightarrow \\ r\in\mathbb{Q}_{>0}}} \mathrm{R}^q\top_*(\mathbb{K}^\bullet(\top^{r+}(\mathscr{H}(\mathscr{M})))) \xrightarrow{\sim} \lim_{\substack{\longrightarrow \\ r\in\mathbb{Q}_{>0}}} \mathrm{R}^q\top_*(\mathscr{M} \otimes_{\breve{\overline{\mathscr{B}}}_{\mathbb{Q}}} \mathbb{K}_{\mathbb{Q}}^\bullet(\breve{\mathscr{C}}^{(r)}, p^r \breve{d}^{(r)})), \tag{III.12.34.5}$$

where $\mathrm{R}^q\top_*(-)$ denotes the hypercohomology of the functor $\top_*$. The theorem follows in view of III.12.32 and III.12.33.

## III.13. Dolbeault modules on a small affine scheme

**III.13.1.** We keep the assumptions and general notation of III.10 and III.12 in this section. We moreover suppose that $X$ is an object of $\mathbf{Q}$ (III.10.5), in other words, that the following conditions are satisfied:

(i) $X$ is affine and connected;
(ii) $f\colon (X, \mathscr{M}_X) \to (S, \mathscr{M}_S)$ admits an adequate chart (III.4.4);
(iii) there exists a fine and saturated chart $M \to \Gamma(X, \mathscr{M}_X)$ for $(X, \mathscr{M}_X)$ inducing an isomorphism

$$M \xrightarrow{\sim} \Gamma(X, \mathscr{M}_X)/(X, \mathscr{O}_X^\times). \tag{III.13.1.1}$$

We set $R = \Gamma(X, \mathscr{O}_X)$, $R_1 = R \otimes_{\mathscr{O}_K} \mathscr{O}_{\overline{K}}$, and

$$\widetilde{\Omega}^1_{R/\mathscr{O}_K} = \Gamma(X, \widetilde{\Omega}^1_{X/S}). \tag{III.13.1.2}$$

We denote by $\widehat{R_1}$ the $p$-adic Hausdorff completion of $R_1$, by $\delta\colon \widetilde{E}_s \to \widetilde{E}$ the canonical embedding (III.10.2.5), and by

$$\beta\colon \widetilde{E} \to \overline{X}^\circ_{\text{fét}} \tag{III.13.1.3}$$

the canonical morphism (III.8.3.1). For any integer $n \geq 1$, we denote by

$$\beta_n \colon (\widetilde{E}_s, \overline{\mathscr{B}}_n) \to (\overline{X}^{\circ}_{\text{fét}}, \overline{\mathscr{B}}_{X,n}) \tag{III.13.1.4}$$

the morphism of ringed topos defined by the morphism of topos $\beta \circ \delta$ and by the canonical homomorphism $\overline{\mathscr{B}}_{X,n} \to \beta_*(\overline{\mathscr{B}}_n)$ (cf. III.10.2). Recall that the latter is not in general an isomorphism (III.9.1.8). We denote by $\breve{\overline{\mathscr{B}}}_X$ the ring $(\overline{\mathscr{B}}_{X,n+1})_{n\in\mathbb{N}}$ of $(\overline{X}^{\circ}_{\text{fét}})^{\mathbb{N}^{\circ}}$ and by

$$\breve{\beta} \colon (\widetilde{E}_s^{\mathbb{N}^{\circ}}, \breve{\overline{\mathscr{B}}}) \to ((\overline{X}^{\circ}_{\text{fét}})^{\mathbb{N}^{\circ}}, \breve{\overline{\mathscr{B}}}_X) \tag{III.13.1.5}$$

the morphism of ringed topos induced by the morphisms $(\beta_{n+1})_{n\in\mathbb{N}}$.

If $A$ is a ring and $M$ an $A$-module, we denote also by $A$ (resp. $M$) the constant sheaf with value $A$ (resp. $M$) of $\overline{X}^{\circ}_{\text{fét}}$ or $(\overline{X}^{\circ}_{\text{fét}})^{\mathbb{N}^{\circ}}$, depending on the context.

**Proposition III.13.2.** *For every coherent $\mathscr{O}_{\mathfrak{X}}$-module $\mathscr{N}$, we have a canonical and functorial $\breve{\overline{\mathscr{B}}}$-linear isomorphism*

$$\breve{\beta}^*(\mathscr{N}(\mathfrak{X}) \otimes_{\widehat{R_1}} \breve{\overline{\mathscr{B}}}_X) \xrightarrow{\sim} \mathrm{T}^*(\mathscr{N}). \tag{III.13.2.1}$$

For any integer $n \geq 1$, we set $\mathscr{N}_n = \mathscr{N}/p^n\mathscr{N}$, which we view as an $\mathscr{O}_{\overline{X}_n}$-module of $X_{s,\text{ét}}$ or $X_{\text{ét}}$, depending on the context (cf. III.2.9 and III.9.9). By (III.11.1.12) and the remark following (III.2.9.5), we have a canonical isomorphism

$$\mathrm{T}^*(\mathscr{N}) \xrightarrow{\sim} (\sigma^*_{n+1}(\mathscr{N}_{n+1}))_{n\in\mathbb{N}}. \tag{III.13.2.2}$$

For every integer $n \geq 1$, we have a canonical isomorphism (III.9.8.4)

$$\sigma^*_n(\mathscr{N}_n) \xrightarrow{\sim} \sigma^{-1}(\mathscr{N}_n) \otimes_{\sigma^{-1}(\mathscr{O}_{\overline{X}_n})} \overline{\mathscr{B}}_n. \tag{III.13.2.3}$$

Hence by virtue of VI.5.34(ii), VI.8.9, and VI.5.17, the $\overline{\mathscr{B}}_n$-module $\sigma^*_n(\mathscr{N}_n)$ is the sheaf of $\widetilde{E}$ associated with the presheaf

$$\{U \mapsto \mathscr{N}_n(U_s) \otimes_{\mathscr{O}_{\overline{X}_n}(U_s)} \overline{\mathscr{B}}_{U,n}\}, \quad (U \in \mathrm{Ob}(\mathbf{\acute{E}t}_{/X})). \tag{III.13.2.4}$$

For every affine object $U$ of $\mathbf{\acute{E}t}_{/X}$, we have canonical isomorphisms

$$\mathscr{N}(U_s) \xrightarrow{\sim} \mathscr{N}(\mathfrak{X}) \otimes_{\widehat{R_1}} \mathscr{O}_{\mathfrak{X}}(U_s), \tag{III.13.2.5}$$

$$\mathscr{N}_n(U_s) \xrightarrow{\sim} \mathscr{N}(U_s)/p^n\mathscr{N}(U_s), \tag{III.13.2.6}$$

$$\mathscr{O}_{\overline{X}_n}(U_s) \xrightarrow{\sim} \mathscr{O}_{\mathfrak{X}}(U_s)/p^n\mathscr{O}_{\mathfrak{X}}(U_s). \tag{III.13.2.7}$$

On the other hand, by virtue of VI.5.34(i), VI.8.9, and VI.5.17, the $\overline{\mathscr{B}}_n$-module $\beta^*_n(\mathscr{N}(\mathfrak{X}) \otimes_{\widehat{R_1}} \breve{\overline{\mathscr{B}}}_{X,n})$ is the sheaf of $\widetilde{E}$ associated with the presheaf

$$\{U \mapsto \mathscr{N}(\mathfrak{X}) \otimes_{\widehat{R_1}} \overline{\mathscr{B}}_{U,n}\}, \quad (U \in \mathrm{Ob}(\mathbf{\acute{E}t}_{/X})). \tag{III.13.2.8}$$

By (III.10.6.5), we deduce from this a canonical and functorial $\overline{\mathscr{B}}_n$-linear isomorphism

$$\beta^*_n(\mathscr{N}(\mathfrak{X}) \otimes_{\widehat{R_1}} \overline{\mathscr{B}}_{X,n}) \xrightarrow{\sim} \sigma^*_n(\mathscr{N}_n). \tag{III.13.2.9}$$

The proposition follows in view of (III.7.5.4) and (III.13.2.2).

**III.13.3.** Let $r$ be a rational number $\geq 0$. We denote by $\breve{\mathscr{F}}_X^{(r)}$ the $\breve{\overline{\mathscr{B}}}_X$-module $(\mathscr{F}_{X,n+1}^{(r)})_{n\in\mathbb{N}}$ and by $\breve{\mathscr{C}}_X^{(r)}$ the $\breve{\overline{\mathscr{B}}}_X$-algebra $(\mathscr{C}_{X,n+1}^{(r)})_{n\in\mathbb{N}}$ (cf. III.10.18). By III.7.3(i) and (III.7.12.1), we have an exact sequence of $\breve{\overline{\mathscr{B}}}_X$-modules

$$0 \to \breve{\overline{\mathscr{B}}}_X \to \breve{\mathscr{F}}_X^{(r)} \to \xi^{-1}\widetilde{\Omega}^1_{R/\mathscr{O}_K} \otimes_R \breve{\overline{\mathscr{B}}}_X \to 0. \tag{III.13.3.1}$$

In view of III.7.3(i) and (III.7.12.3), we have a canonical isomorphism of $\breve{\overline{\mathscr{B}}}_X$-algebras

$$\breve{\mathscr{C}}_X^{(r)} \xrightarrow{\sim} \varinjlim_{m\geq 0} \mathrm{S}^m_{\breve{\overline{\mathscr{B}}}_X}(\breve{\mathscr{F}}_X^{(r)}). \tag{III.13.3.2}$$

For all rational numbers $r \geq r' \geq 0$, the morphisms $(\mathsf{a}_{X,n+1}^{r,r'})_{n\in\mathbb{N}}$ (III.10.18.3) induce a $\breve{\overline{\mathscr{B}}}_X$-linear morphism

$$\breve{\mathsf{a}}_X^{r,r'} : \breve{\mathscr{F}}_X^{(r)} \to \breve{\mathscr{F}}_X^{(r')}. \tag{III.13.3.3}$$

The homomorphisms $(\alpha_{X,n+1}^{r,r'})_{n\in\mathbb{N}}$ (III.10.18.4) induce a homomorphism of $\breve{\overline{\mathscr{B}}}_X$-algebras

$$\breve{\alpha}_X^{r,r'} : \breve{\mathscr{C}}_X^{(r)} \to \breve{\mathscr{C}}_X^{(r')}. \tag{III.13.3.4}$$

For all rational numbers $r \geq r' \geq r'' \geq 0$, we have

$$\breve{\mathsf{a}}_X^{r,r''} = \breve{\mathsf{a}}_X^{r',r''} \circ \breve{\mathsf{a}}_X^{r,r'} \quad \text{and} \quad \breve{\alpha}_X^{r,r''} = \breve{\alpha}_X^{r',r''} \circ \breve{\alpha}_X^{r,r'}. \tag{III.13.3.5}$$

We have a canonical $\breve{\mathscr{C}}_X^{(r)}$-linear isomorphism

$$\Omega^1_{\breve{\mathscr{C}}_X^{(r)}/\breve{\overline{\mathscr{B}}}_X} \xrightarrow{\sim} \xi^{-1}\widetilde{\Omega}^1_{R/\mathscr{O}_K} \otimes_R \breve{\mathscr{C}}_X^{(r)}. \tag{III.13.3.6}$$

The universal $\breve{\overline{\mathscr{B}}}_X$-derivation of $\breve{\mathscr{C}}_X^{(r)}$ corresponds via this isomorphism to the unique $\breve{\overline{\mathscr{B}}}_X$-derivation

$$\breve{d}_X^{(r)} : \breve{\mathscr{C}}_X^{(r)} \to \xi^{-1}\widetilde{\Omega}^1_{R/\mathscr{O}_K} \otimes_R \breve{\mathscr{C}}_X^{(r)} \tag{III.13.3.7}$$

that extends the canonical morphism $\breve{\mathscr{F}}_X^{(r)} \to \xi^{-1}\widetilde{\Omega}^1_{R/\mathscr{O}_K} \otimes_R \breve{\overline{\mathscr{B}}}_X$ (III.13.3.1). Since

$$\xi^{-1}\widetilde{\Omega}^1_{R/\mathscr{O}_K} \otimes_R \breve{\overline{\mathscr{B}}}_X = \breve{d}_X^{(r)}(\breve{\mathscr{F}}_X^{(r)}) \subset \breve{d}_X^{(r)}(\breve{\mathscr{C}}_X^{(r)}), \tag{III.13.3.8}$$

the derivation $\breve{d}_X^{(r)}$ is a Higgs $\breve{\overline{\mathscr{B}}}_X$-field with coefficients in $\xi^{-1}\widetilde{\Omega}^1_{R/\mathscr{O}_K}$ by II.2.12. For all rational numbers $r \geq r' \geq 0$, we have

$$p^{r-r'}(\mathrm{id} \otimes \breve{\alpha}_X^{r,r'}) \circ \breve{d}_X^{(r)} = \breve{d}_X^{(r')} \circ \breve{\alpha}_X^{r,r'}. \tag{III.13.3.9}$$

**Proposition III.13.4.** *For every rational number $r \geq 0$, the canonical morphisms*

$$\breve{\beta}^*(\breve{\mathscr{F}}_X^{(r)}) \xrightarrow{\sim} \breve{\mathscr{F}}^{(r)}, \tag{III.13.4.1}$$

$$\breve{\beta}^*(\breve{\mathscr{C}}_X^{(r)}) \xrightarrow{\sim} \breve{\mathscr{C}}^{(r)}, \tag{III.13.4.2}$$

*are isomorphisms. Moreover, for all rational numbers $r \geq r' \geq 0$, the morphisms $\breve{\beta}^*(\breve{\mathsf{a}}_X^{r,r'})$ and $\breve{\beta}^*(\breve{\alpha}_X^{r,r'})$ identify with the morphisms $\breve{\mathsf{a}}^{r,r'}$ (III.10.31.6) and $\breve{\alpha}^{r,r'}$ (III.10.31.7), respectively.*

For any $U \in \mathrm{Ob}(\mathbf{\acute{E}t}_{/X})$, we denote by $g_U : U \to X$ the canonical morphism. By virtue of VI.5.34(i), VI.8.9, and VI.5.17, for every integer $n \geq 1$, the $\overline{\mathscr{B}}_n$-module $\beta_n^*(\mathscr{F}_{X,n}^{(r)})$ is canonically isomorphic to the sheaf associated with the presheaf on $E$ defined by the correspondence

$$\{U \mapsto (\overline{g}^{\circ}_U)^*_{\text{fét}}(\mathscr{F}^{(r)}_{X,n}) \otimes_{(\overline{g}^{\circ}_U)^*_{\text{fét}}(\overline{\mathscr{B}}_{X,n})} \overline{\mathscr{B}}_{U,n}\}. \tag{III.13.4.3}$$

For every $Y \in \mathrm{Ob}(\mathbf{Q})$, the canonical homomorphism

$$(\overline{g}^{\circ}_Y)^*_{\text{fét}}(\mathscr{F}^{(r)}_{X,n}) \otimes_{(\overline{g}^{\circ}_Y)^*_{\text{fét}}(\overline{\mathscr{B}}_{X,n})} \overline{\mathscr{B}}_{Y,n} \to \mathscr{F}^{(r)}_{Y,n} \tag{III.13.4.4}$$

is an isomorphism by virtue of III.10.20. Consequently, the morphism

$$\beta^*_n(\mathscr{F}^{(r)}_{X,n}) \to \mathscr{F}^{(r)}_n \tag{III.13.4.5}$$

adjoint to the canonical morphism $\mathscr{F}^{(r)}_{X,n} \to \beta_{n*}(\mathscr{F}^{(r)}_n)$, is an isomorphism by (III.10.6.5). We deduce from this, in view of (III.7.5.4) and (III.7.12.1), that the canonical morphism (III.13.4.1) is an isomorphism. We prove likewise that the canonical homomorphism (III.13.4.2) is an isomorphism; we can also deduce this from (III.13.4.1).The last assertion is obvious by adjunction.

**Remark III.13.5.** It follows from III.13.4 that for every rational number $r \geq 0$, $\breve{\beta}^*(\breve{d}^{(r)}_X)$ identifies with the derivation $\breve{d}^{(r)}$ (III.10.31.9). We can construct the identification explicitly as follows. By the proof of III.10.22(ii), for every integer $n \geq 1$, the diagram

$$\begin{array}{ccc} \mathscr{F}^{(r)}_{X,n} & \longrightarrow & \xi^{-1}\widetilde{\Omega}^1_{R/\mathscr{O}_K} \otimes_R \overline{\mathscr{B}}_{X,n} \\ \downarrow & & \downarrow \\ \beta_*(\mathscr{F}^{(r)}_n) & \longrightarrow & \beta_*(\sigma^*_n(\xi^{-1}\widetilde{\Omega}^1_{\overline{X}_n/\overline{S}_n})) \end{array} \tag{III.13.5.1}$$

where the vertical arrows are the canonical morphisms and the horizontal arrows come from the exact sequences (III.10.18.1) and (III.10.22.1), is commutative. We deduce from this by adjunction a commutative diagram

$$\begin{array}{ccc} \beta^*_n(\mathscr{F}^{(r)}_{X,n}) & \longrightarrow & \beta^*_n(\xi^{-1}\widetilde{\Omega}^1_{R/\mathscr{O}_K} \otimes_R \overline{\mathscr{B}}_{X,n}) \\ \downarrow & & \downarrow \\ \mathscr{F}^{(r)}_n & \longrightarrow & \sigma^*_n(\xi^{-1}\widetilde{\Omega}^1_{\overline{X}_n/\overline{S}_n}) \end{array} \tag{III.13.5.2}$$

whose vertical arrows are isomorphisms, by the proofs of III.13.2 and III.13.4. Consequently, the diagram

$$\begin{array}{ccc} \breve{\beta}^*(\breve{\mathscr{C}}^{(r)}_X) & \xrightarrow{\breve{\beta}^*(\breve{d}^{(r)}_X)} & \breve{\beta}^*(\xi^{-1}\widetilde{\Omega}^1_{R/\mathscr{O}_K} \otimes_R \breve{\mathscr{C}}^{(r)}_X) \\ \downarrow & & \downarrow \\ \breve{\mathscr{C}}^{(r)} & \xrightarrow{\breve{d}^{(r)}} & \breve{\sigma}^*(\xi^{-1}\widetilde{\Omega}^1_{\breve{\overline{X}}/\breve{\overline{S}}}) \otimes_{\breve{\overline{\mathscr{B}}}} \breve{\mathscr{C}}^{(r)} \end{array} \tag{III.13.5.3}$$

where the vertical arrows are the isomorphisms induced by (III.13.2.1) and (III.13.4.2) is commutative.

**III.13.6.** Let $r$ be a rational number $\geq 0$. We denote by $\mathbf{Mod}(\breve{\overline{\mathscr{B}}}_X)$ the category of $\breve{\overline{\mathscr{B}}}_X$-modules, by $\Theta^r$ the category of integrable $p^r$-isoconnections with respect to the extension $\breve{\mathscr{C}}_X^{(r)}/\breve{\overline{\mathscr{B}}}_X$ (III.6.10), and by $\mathfrak{S}_X^r$ the functor

(III.13.6.1) $\mathfrak{S}_X^r\colon \mathbf{Mod}(\breve{\overline{\mathscr{B}}}_X) \to \Theta^r, \quad \mathscr{M} \mapsto (\breve{\mathscr{C}}_X^{(r)} \otimes_{\breve{\overline{\mathscr{B}}}_X} \mathscr{M}, \breve{\mathscr{C}}_X^{(r)} \otimes_{\breve{\overline{\mathscr{B}}}_X} \mathscr{M}, \mathrm{id}, p^r \breve{d}_X^{(r)} \otimes \mathrm{id}).$

By III.6.12, if $(\mathscr{N}, \mathscr{N}', v, \theta)$ is a Higgs $\mathscr{O}_{\mathfrak{X}}$-isogeny with coefficients in $\xi^{-1}\widetilde{\Omega}^1_{\mathfrak{X}/\mathscr{S}}$ (III.12.5),

(III.13.6.2) $$(\breve{\mathscr{C}}_X^{(r)} \otimes_{\widehat{\overline{R_1}}} \mathscr{N}(\mathfrak{X}), \breve{\mathscr{C}}_X^{(r)} \otimes_{\widehat{\overline{R_1}}} \mathscr{N}'(\mathfrak{X}), \mathrm{id} \otimes_{\widehat{\overline{R_1}}} v, p^r \breve{d}_X^{(r)} \otimes v + \mathrm{id} \otimes \theta)$$

is an object of $\Theta^r$. We thus obtain a functor

(III.13.6.3) $$\mathsf{T}_X^{r+}\colon \mathbf{HI}(\mathscr{O}_{\mathfrak{X}}, \xi^{-1}\widetilde{\Omega}^1_{\mathfrak{X}/\mathscr{S}}) \to \Theta^r.$$

**Proposition III.13.7.** *For every rational number $r \geq 0$, the diagrams of functors*

(III.13.7.1) $$\begin{array}{ccc} \mathbf{Mod}(\breve{\overline{\mathscr{B}}}_X) & \xrightarrow{\mathfrak{S}_X^r} & \Theta^r \\ \downarrow{\scriptstyle \breve{\beta}^*} & & \downarrow{\scriptstyle \breve{\beta}^*} \\ \mathbf{Mod}(\breve{\overline{\mathscr{B}}}) & \xrightarrow{\mathfrak{S}^r} & \Xi^r \end{array}$$

(III.13.7.2) $$\begin{array}{ccc} \mathbf{HI}^{\mathrm{coh}}(\mathscr{O}_{\mathfrak{X}}, \xi^{-1}\widetilde{\Omega}^1_{\mathfrak{X}/\mathscr{S}}) & \xrightarrow{\mathsf{T}_X^{r+}} & \Theta^r \\ & \searrow^{\mathsf{T}^{r+}} & \downarrow{\scriptstyle \breve{\beta}^*} \\ & & \Xi^r \end{array}$$

*where the inverse image functor under $\breve{\beta}$ for $p^r$-isoconnections is defined in* (III.6.11), *are commutative up to canonical isomorphisms.*

This follows from III.13.2, III.13.4, and III.13.5.

**Proposition III.13.8.** *Let $\epsilon, r$ be two rational numbers such that $\epsilon > r + \frac{1}{p-1}$ and $r > 0$, $\mathscr{N}$ an $\mathscr{S}$-flat coherent $\mathscr{O}_{\mathfrak{X}}$-module, and $\theta$ a Higgs $\mathscr{O}_{\mathfrak{X}}$-field on $\mathscr{N}$ with coefficients in $\xi^{-1}\widetilde{\Omega}^1_{\mathfrak{X}/\mathscr{S}}$ such that*

(III.13.8.1) $$\theta(\mathscr{N}) \subset p^{\epsilon}\xi^{-1}\widetilde{\Omega}^1_{\mathfrak{X}/\mathscr{S}} \otimes_{\mathscr{O}_{\mathfrak{X}}} \mathscr{N}.$$

*We denote also by $\mathscr{N}$ the Higgs $\mathscr{O}_{\mathfrak{X}}$-isogeny $(\mathscr{N}, \mathscr{N}, \mathrm{id}, \theta)$ with coefficients in $\xi^{-1}\widetilde{\Omega}^1_{\mathfrak{X}/\mathscr{S}}$. Then there exist an adic $\breve{\overline{\mathscr{B}}}_X$-module of finite type $\mathscr{M}$ of $(\overline{X}^{\circ}_{\mathrm{f\acute{e}t}})^{\mathbb{N}^{\circ}}$ and an isomorphism of $\Theta^r$*

(III.13.8.2) $$\mathsf{T}_X^{r+}(\mathscr{N}) \xrightarrow{\sim} \mathfrak{S}_X^r(\mathscr{M}).$$

We may clearly assume that $X_s$ is nonempty, so that $(X, \mathscr{M}_X)$ satisfies the assumptions of II.6.2. Let $\overline{y}$ be a generic geometric point of $\overline{X}^{\circ}$. Since $\overline{X}$ is locally irreducible (III.3.1), it is the sum of the schemes induced on its irreducible components. We denote by $\overline{X}_{\langle \overline{y} \rangle}$ the irreducible component of $\overline{X}$ containing $\overline{y}$. Likewise, $\overline{X}^{\circ}$ is the sum of the schemes induced on its irreducible components, and $\overline{X}^{\circ}_{\langle \overline{y} \rangle} = \overline{X}_{\langle \overline{y} \rangle} \times_X X^{\circ}$ is the irreducible component of $\overline{X}^{\circ}$ containing $\overline{y}$. We set $R_1^{\overline{y}} = \Gamma(\overline{X}_{\langle \overline{y} \rangle}, \mathscr{O}_{\overline{X}})$ and $\Delta_{\overline{y}} = \pi_1(\overline{X}^{\circ}_{\langle \overline{y} \rangle}, \overline{y})$. We denote by $\widehat{R}_1^{\overline{y}}$ the $p$-adic Hausdorff completion of $R_1^{\overline{y}}$, by $\mathbf{B}_{\Delta_{\overline{y}}}$ the classifying topos of $\Delta_{\overline{y}}$, by

(III.13.8.3) $$\nu_{\overline{y}}\colon \overline{X}^{\circ}_{\langle \overline{y} \rangle, \mathrm{f\acute{e}t}} \xrightarrow{\sim} \mathbf{B}_{\Delta_{\overline{y}}}$$

the fiber functor at $\overline{y}$ (III.2.10.3), by $\overline{R}^{\overline{y}}_X$ the ring defined in (III.8.13.2), and by $\widehat{\overline{R}}^{\overline{y}}_X$ its $p$-adic Hausdorff completion. We denote by $\mathscr{C}^{\overline{y},(r)}_X$ the $\widehat{\overline{R}}^{\overline{y}}_X$-algebra defined in (III.10.24.3), by $\widehat{\mathscr{C}}^{\overline{y},(r)}_X$ its $p$-adic Hausdorff completion, by

(III.13.8.4) $$d_{\mathscr{C}^{\overline{y},(r)}_X} : \mathscr{C}^{\overline{y},(r)}_X \to \xi^{-1}\widetilde{\Omega}^1_{R/\mathscr{O}_K} \otimes_R \mathscr{C}^{\overline{y},(r)}_X$$

the universal $\widehat{\overline{R}}^{\overline{y}}_X$-derivation of $\mathscr{C}^{\overline{y},(r)}_X$, and by

(III.13.8.5) $$d_{\widehat{\mathscr{C}}^{\overline{y},(r)}_X} : \widehat{\mathscr{C}}^{\overline{y},(r)}_X \to \xi^{-1}\widetilde{\Omega}^1_{R/\mathscr{O}_K} \otimes_R \widehat{\mathscr{C}}^{\overline{y},(r)}_X$$

its extension to the completions (note that the $R$-module $\widetilde{\Omega}^1_{R/\mathscr{O}_K}$ is free of finite type). It follows from III.8.15 and the definitions that we have canonical isomorphisms

(III.13.8.6) $$\nu_{\overline{y}}(\overline{\mathscr{B}}_X|\overline{X}^{\circ}_{\langle\overline{y}\rangle}) \overset{\sim}{\to} \overline{R}^{\overline{y}}_X,$$

(III.13.8.7) $$\nu_{\overline{y}}(\mathscr{C}^{(r)}_{X,n}|\overline{X}^{\circ}_{\langle\overline{y}\rangle}) \overset{\sim}{\to} \mathscr{C}^{\overline{y},(r)}_X/p^n\mathscr{C}^{\overline{y},(r)}_X.$$

The objects $\Delta_{\overline{y}}$, $R^{\overline{y}}_1$, $\overline{R}^{\overline{y}}_X$, and $\mathscr{C}^{\overline{y},(r)}_X$ correspond to the objects $\Delta$, $R_1$, $\overline{R}$, and $\mathscr{C}^{(r)}$ defined in II, by taking $\widetilde{\kappa} = \overline{y}$ in II.6.7. From now on, we will use the constructions of II.13. Set $N_{\overline{y}} = \Gamma(\overline{X}_{\langle\overline{y}\rangle,s}, \mathscr{N})$ and denote by

(III.13.8.8) $$\theta_{\overline{y}} : N_{\overline{y}} \to \xi^{-1}\widetilde{\Omega}^1_{R/\mathscr{O}_K} \otimes_R N_{\overline{y}}$$

the Higgs $\widehat{R}^{\overline{y}}_1$-field with coefficients in $\xi^{-1}\widetilde{\Omega}^1_{R/\mathscr{O}_K}$ induced by $\theta$, which is $\epsilon$-quasi-small in the sense of II.13.4. We associate with it, by the functor (II.13.10.10), a quasi-small $\widehat{R}^{\overline{y}}_1$-representation $\varphi_{\overline{y}}$ of $\Delta_{\overline{y}}$ on $N_{\overline{y}}$. By II.13.17, we have a $\Delta_{\overline{y}}$-equivariant $\widehat{\mathscr{C}}^{\overline{y},(r)}_X$-isomorphism of modules with $p$-adic $p^r$-connections with respect to the extension $\widehat{\mathscr{C}}^{\overline{y},(r)}_X/\widehat{\overline{R}}^{\overline{y}}$,

(III.13.8.9) $$u_{\overline{y}} : N_{\overline{y}} \widehat{\otimes}_{\widehat{R}^{\overline{y}}_1} \widehat{\mathscr{C}}^{\overline{y},(r)}_X \overset{\sim}{\to} N_{\overline{y}} \widehat{\otimes}_{\widehat{R}^{\overline{y}}_1} \widehat{\mathscr{C}}^{\overline{y},(r)}_X,$$

where $\widehat{\mathscr{C}}^{\overline{y},(r)}_X$ is endowed with the canonical action of $\Delta_{\overline{y}}$ and with the $p$-adic $p^r$-connection $p^r d_{\widehat{\mathscr{C}}^{\overline{y},(r)}_X}$, the module $N_{\overline{y}}$ in the source is endowed with the trivial action of $\Delta_{\overline{y}}$ and with the Higgs $\widehat{R}^{\overline{y}}_1$-field $\theta_{\overline{y}}$, and the module $N_{\overline{y}}$ in the target is endowed with the action $\varphi_{\overline{y}}$ of $\Delta_{\overline{y}}$ and with the zero Higgs $\widehat{R}^{\overline{y}}_1$-field.

There exists an inverse system $(\mathscr{L}_{n+1})_{n\in\mathbb{N}}$ of $R_1$-modules of $\overline{X}^{\circ}_{\text{fét}}$ such that for every generic geometric point $\overline{y}$ of $\overline{X}^{\circ}$, we have an isomorphism of inverse systems of $R^{\overline{y}}_1$-representations of $\Delta_{\overline{y}}$,

(III.13.8.10) $$(\nu_{\overline{y}}(\mathscr{L}_{n+1}|\overline{X}^{\circ}_{\langle\overline{y}\rangle}))_{n\in\mathbb{N}} \overset{\sim}{\to} (N_{\overline{y}}/p^{n+1}N_{\overline{y}}, \varphi_{\overline{y}})_{n\in\mathbb{N}}.$$

This immediately follows from the definition of the functor (II.13.10.10) (cf. VI.9.8). By III.2.11, $\mathscr{L}_n$ is of finite type on $R_1$. For all integers $m \geq n \geq 1$, the morphism $\mathscr{L}_m/p^n\mathscr{L}_m \to \mathscr{L}_n$ induced by the transition morphism $\mathscr{L}_m \to \mathscr{L}_n$ is an isomorphism. We set

(III.13.8.11) $$\mathscr{M} = (\mathscr{L}_{n+1} \otimes_{R_1} \overline{\mathscr{B}}_{X,n+1})_{n\in\mathbb{N}}.$$

It is an adic $\breve{\overline{\mathscr{B}}}_X$-module of finite type of $(\overline{X}^{\circ}_{\text{fét}})^{\mathbb{N}^{\circ}}$ by virtue of III.7.14. The isomorphisms (III.13.8.9) induce a $\breve{\mathscr{C}}^{(r)}_X$-linear isomorphism

(III.13.8.12) $$u : \mathscr{N}(\mathfrak{X}) \otimes_{\widehat{R_1}} \breve{\mathscr{C}}^{(r)}_X \overset{\sim}{\to} \mathscr{M} \otimes_{\breve{\overline{\mathscr{B}}}_X} \breve{\mathscr{C}}^{(r)}_X$$

such that the diagram

(III.13.8.13)
$$\begin{array}{ccc} \mathscr{N}(\mathfrak{X}) \otimes_{\widehat{R_1}} \breve{\mathscr{C}}_X^{(r)} & \xrightarrow{u} & \mathscr{M} \otimes_{\breve{\overline{\mathscr{B}}}_X} \breve{\mathscr{C}}_X^{(r)} \\ \Big\downarrow{\scriptstyle \theta\otimes\mathrm{id}+p^r\mathrm{id}\otimes\breve{d}_X^{(r)}} & & \Big\downarrow{\scriptstyle p^r\mathrm{id}\otimes\breve{d}_X^{(r)}} \\ \xi^{-1}\widetilde{\Omega}^1_{R/\mathscr{O}_K}\otimes_R \mathscr{N}(\mathfrak{X}) \otimes_{\widehat{R_1}} \breve{\mathscr{C}}_X^{(r)} & \xrightarrow{\mathrm{id}\otimes u} & \xi^{-1}\widetilde{\Omega}^1_{R/\mathscr{O}_K}\otimes_R \mathscr{M} \otimes_{\breve{\overline{\mathscr{B}}}_X} \breve{\mathscr{C}}_X^{(r)} \end{array}$$

is commutative, giving the proposition.

**Corollary III.13.9.** *Under the assumptions of* III.13.8, *if* $\mathscr{N}_{\mathbb{Q}_p}$ *is a locally projective* $\mathscr{O}_{\mathfrak{X}}[\frac{1}{p}]$*-module of finite type,* $(\mathscr{N}_{\mathbb{Q}_p}, \theta_{\mathbb{Q}_p})$ *is a solvable Higgs* $\mathscr{O}_{\mathfrak{X}}[\frac{1}{p}]$*-bundle.*

This follows from III.13.7 and III.13.8.

## III.14. Inverse image of a Dolbeault module under an étale morphism

**III.14.1.** We keep the assumptions and general notation of III.10 and III.12 in this section. Let, moreover, $g\colon X' \to X$ be an étale morphism of finite type. We endow $X'$ with the logarithmic structure $\mathscr{M}_{X'}$ inverse image of $\mathscr{M}_X$ and we denote by $f'\colon (X', \mathscr{M}_{X'}) \to (S, \mathscr{M}_S)$ the morphism induced by $f$ and $g$. Note that $f'$ is adequate (III.4.7) and that $X'^{\circ} = X^{\circ} \times_X X'$ is the maximal open subscheme of $X'$ where the logarithmic structure $\mathscr{M}_{X'}$ is trivial. We endow $\overline{X}'$ and $\check{\overline{X}}'$ with the logarithmic structures $\mathscr{M}_{\overline{X}'}$ and $\mathscr{M}_{\check{\overline{X}}'}$ inverse images of $\mathscr{M}_{X'}$. There exists essentially a unique étale morphism $\widetilde{g}\colon \widetilde{X}' \to \widetilde{X}$ that fits into a Cartesian diagram (III.10.1)

(III.14.1.1)
$$\begin{array}{ccc} \check{\overline{X}}' & \longrightarrow & \widetilde{X}' \\ \Big\downarrow{\scriptstyle \check{\overline{g}}} & & \Big\downarrow{\scriptstyle \widetilde{g}} \\ \check{\overline{X}} & \longrightarrow & \widetilde{X} \end{array}$$

We endow $\widetilde{X}'$ with the logarithmic structure $\mathscr{M}_{\widetilde{X}'}$ inverse image of $\mathscr{M}_{\widetilde{X}}$, so that $(\widetilde{X}', \mathscr{M}_{\widetilde{X}'})$ is a smooth $(\mathscr{A}_2(\overline{S}), \mathscr{M}_{\mathscr{A}_2(\overline{S})})$-deformation of $(\check{\overline{X}}', \mathscr{M}_{\check{\overline{X}}'})$.

We associate with $(f', \widetilde{X}', \mathscr{M}_{\widetilde{X}'})$ objects analogous to those defined in III.10 and III.12 for $(f, \widetilde{X}, \mathscr{M}_{\widetilde{X}})$, which we denote by the same symbols equipped with an exponent $'$. We denote by

(III.14.1.2)
$$\Phi\colon \widetilde{E}' \to \widetilde{E},$$

(III.14.1.3)
$$\Phi_s\colon \widetilde{E}'_s \to \widetilde{E}_s,$$

the morphisms of topos (III.8.5.3) and (III.9.11.8) induced by functoriality by $g$. By VI.10.14, $\Phi$ identifies with the localization of $\widetilde{E}$ at $\sigma^*(X')$. Moreover, the canonical homomorphism $\Phi^{-1}(\overline{\mathscr{B}}) \to \overline{\mathscr{B}}'$ is an isomorphism by virtue of III.8.21(i). For every integer $n \geq 1$, $\Phi_s$ is underlying a canonical morphism of ringed topos (III.9.11.11)

(III.14.1.4)
$$\Phi_n\colon (\widetilde{E}'_s, \overline{\mathscr{B}}'_n) \to (\widetilde{E}_s, \overline{\mathscr{B}}_n).$$

Since the homomorphism $\Phi_s^*(\overline{\mathscr{B}}_n) \to \overline{\mathscr{B}}'_n$ is an isomorphism (III.9.13), there is no difference for $\overline{\mathscr{B}}_n$-modules between the inverse image under $\Phi_s$ in the sense of abelian sheaves and the inverse image under $\Phi_n$ in the sense of modules. The diagram of morphisms of ringed topos (III.9.11.12)

(III.14.1.5)
$$\begin{array}{ccc}(\widetilde{E}'_s, \overline{\mathscr{B}}'_n) & \xrightarrow{\Phi_n} & (\widetilde{E}_s, \overline{\mathscr{B}}_n) \\ \big\downarrow{\scriptstyle \sigma'_n} & & \big\downarrow{\scriptstyle \sigma_n} \\ (X'_{s,\text{ét}}, \mathscr{O}_{\overline{X}'_n}) & \xrightarrow{\overline{g}_n} & (X_{s,\text{ét}}, \mathscr{O}_{\overline{X}_n})\end{array}$$

where $\overline{g}_n$ is the morphism induced by $g$, is commutative up to canonical isomorphism.

**III.14.2.** Since every object of $E'$ is naturally an object of $E$, we denote by $\jmath\colon E' \to E$ the canonical functor. This factors through an equivalence of categories

(III.14.2.1)
$$E' \xrightarrow{\sim} E_{/(\overline{X}'^{\circ} \to X')},$$

that is even an equivalence of categories over $\mathbf{\acute{E}t}_{/X'}$, where we view $E_{/(\overline{X}'^{\circ} \to X')}$ as an $(\mathbf{\acute{E}t}_{/X'})$-category by base change of the canonical fibration $\pi\colon E \to \mathbf{\acute{E}t}_{/X}$ (III.10.2.2). By VI.5.38, the covanishing topology on $E'$ is induced by that on $E$ through $\jmath$. Consequently, $\jmath$ is continuous and cocontinuous ([**2**] III 5.2). Moreover, $\Phi$ identifies with the localization morphism of $\widetilde{E}$ at $\sigma^*(X') = (\overline{X}'^{\circ} \to X')^a$ (VI.10.14). In particular, $\Phi^*$ is none other than the restriction functor by $\jmath$. We denote by $\mathbf{Q}'$ the full subcategory of $\mathbf{\acute{E}t}_{/X'}$ made up of the objects that are in $\mathbf{Q}$ (III.10.5) and by

(III.14.2.2)
$$\pi'_{\mathbf{Q}'}\colon E'_{\mathbf{Q}'} \to \mathbf{Q}'$$

the fibered category deduced by base change from the canonical fibration $\pi'\colon E' \to \mathbf{\acute{E}t}_{/X'}$. The functor $\jmath$ therefore induces a functor $\jmath_{\mathbf{Q}}\colon E'_{\mathbf{Q}'} \to E_{\mathbf{Q}}$ that fits into a commutative diagram up to canonical isomorphism

(III.14.2.3)
$$\begin{array}{ccc}E'_{\mathbf{Q}'} & \xrightarrow{\jmath_{\mathbf{Q}}} & E_{\mathbf{Q}} \\ \big\downarrow{\scriptstyle u'} & & \big\downarrow{\scriptstyle u} \\ E' & \xrightarrow{\jmath} & E\end{array}$$

where $u$ and $u'$ are the canonical projection functors. The functors $u$ and $u'$ are fully faithful, and the category $E_{\mathbf{Q}}$ (resp. $E'_{\mathbf{Q}'}$) is $\mathbb{U}$-small and topologically generates the site $E$ (resp. $E'$) by III.10.5. It immediately follows from (III.14.2.1) that $\jmath_{\mathbf{Q}}$ factors through an equivalence of categories

(III.14.2.4)
$$E'_{\mathbf{Q}'} \xrightarrow{\sim} (E_{\mathbf{Q}})_{/\widehat{u}^*(\overline{X}'^{\circ} \to X')},$$

where $\widehat{u}^*(\overline{X}'^{\circ} \to X')$ is the presheaf on $E_{\mathbf{Q}}$ deduced from $(\overline{X}'^{\circ} \to X')$ by restriction by $u$. We endow $E_{\mathbf{Q}}$ (resp. $E'_{\mathbf{Q}'}$) with the topology induced by that on $E$ (resp. $E'$). Since the functor $\Phi^*\colon \widetilde{E} \to \widetilde{E}'$ is essentially surjective, the topology on $E'_{\mathbf{Q}'}$ is induced by that on $E_{\mathbf{Q}}$ by $\jmath_{\mathbf{Q}}$ ([**2**] II 2.2). Consequently, $\jmath_{\mathbf{Q}}$ is continuous and cocontinuous ([**2**] III 5.2).

**III.14.3.** Let $Y$ be an object of $\mathbf{Q}'$ (III.14.2) such that $Y_s \neq \emptyset$, $\overline{y}$ a geometric point of $\overline{Y}^{\circ}$, and $\overline{Y}^{\star}$ the irreducible component of $\overline{Y}$ containing $\overline{y}$ (III.3.3). Consider the objects associated with $Y$ in III.10.13 and III.10.15 relative to $(f, \widetilde{X}, \mathscr{M}_{\widetilde{X}})$. Note that the ring $\overline{R}^{\overline{y}}_Y$ (III.8.13.2), the logarithmic structure $\mathscr{M}_{\mathscr{A}_2(\overline{Y}^{\overline{y}})}$ on $\mathscr{A}_2(\overline{Y}^{\overline{y}})$ (III.10.13.5), and the $\widehat{\overline{R}}^{\overline{y}}_Y$-module $\mathrm{T}^{\overline{y}}_Y$ (III.10.13.7) do not change whether we use $f$ or $f'$. Replacing $(f, \widetilde{X}, \mathscr{M}_{\widetilde{X}})$ by $(f', \widetilde{X}', \mathscr{M}_{\widetilde{X}'})$, we denote by $\mathscr{L}'^{\overline{y}}_Y$ the $\widetilde{\mathrm{T}}^{\overline{y}}_Y$-torsor of $\widehat{\overline{Y}}^{\overline{y}}_{\text{zar}}$ defined in (III.10.13.9), by

$\mathscr{F}'^{\overline{y}}_Y$ the $\widehat{\overline{R}}^{\overline{y}}_Y$-module defined in (III.10.13.10), and by $\mathscr{C}'^{\overline{y}}_Y$ the $\widehat{\overline{R}}^{\overline{y}}_Y$-algebra defined in (III.10.13.12). Let $U$ be a Zariski open subscheme of $\widehat{\overline{Y}}^{\overline{y}}$ and $\widetilde{U}$ the open subscheme of $\mathscr{A}_2(\overline{Y}^{\overline{y}})$ defined by $U$. Consider the commutative diagram (without the dotted arrow)

(III.14.3.1)
$$\begin{array}{ccccc}
(U, \mathscr{M}_{\widehat{\overline{Y}}^{\overline{y}}}|U) & \longrightarrow & (\check{\overline{X}}', \mathscr{M}_{\check{\overline{X}}'}) & \xrightarrow{\check{\overline{g}}} & (\check{\overline{X}}, \mathscr{M}_{\check{\overline{X}}}) \\
\Big\downarrow{\scriptstyle i_Y|U} & & \Big\downarrow & & \Big\downarrow \\
(\widetilde{U}, \mathscr{M}_{\mathscr{A}_2(\overline{Y}^{\overline{y}})}|\widetilde{U}) & \overset{\psi}{\dashrightarrow} & (\widetilde{X}', \mathscr{M}_{\widetilde{X}'}) & \xrightarrow{\widetilde{g}} & (\widetilde{X}, \mathscr{M}_{\widetilde{X}})
\end{array}$$

Since $\widetilde{g}$ is étale, the map

(III.14.3.2)
$$\mathscr{L}'^{\overline{y}}_Y(U) \to \mathscr{L}^{\overline{y}}_Y(U), \quad \psi \mapsto \widetilde{g} \circ \psi,$$

is bijective. We deduce from this a $\pi_1(\overline{Y}^{\star\circ}, \overline{y})$-equivariant and $\widehat{\overline{R}}^{\overline{y}}_Y$-linear isomorphism

(III.14.3.3)
$$\mathscr{F}^{\overline{y}}_Y \xrightarrow{\sim} \mathscr{F}'^{\overline{y}}_Y,$$

which fits into a commutative diagram

(III.14.3.4)
$$\begin{array}{ccccccccc}
0 & \longrightarrow & \widehat{\overline{R}}^{\overline{y}}_Y & \longrightarrow & \mathscr{F}^{\overline{y}}_Y & \longrightarrow & \xi^{-1}\widetilde{\Omega}^1_{X/S}(Y) \otimes_{\mathscr{O}_X(Y)} \widehat{\overline{R}}^{\overline{y}}_Y & \longrightarrow & 0 \\
 & & \Big\| & & \Big\downarrow & & \Big\downarrow & & \\
0 & \longrightarrow & \widehat{\overline{R}}^{\overline{y}}_Y & \longrightarrow & \mathscr{F}'^{\overline{y}}_Y & \longrightarrow & \xi^{-1}\widetilde{\Omega}^1_{X'/S}(Y) \otimes_{\mathscr{O}_{X'}(Y)} \widehat{\overline{R}}^{\overline{y}}_Y & \longrightarrow & 0
\end{array}$$

where the horizontal lines are the exact sequences (III.10.13.10). The isomorphism (III.14.3.3) induces a $\pi_1(\overline{Y}^{\star\circ}, \overline{y})$-equivariant isomorphism of $\widehat{\overline{R}}^{\overline{y}}_Y$-algebras

(III.14.3.5)
$$\mathscr{C}^{\overline{y}}_Y \xrightarrow{\sim} \mathscr{C}'^{\overline{y}}_Y.$$

Let $n$ be an integer $\geq 1$. Replacing $(f, \widetilde{X}, \mathscr{M}_{\widetilde{X}})$ by $(f', \widetilde{X}', \mathscr{M}_{\widetilde{X}'})$, we denote by $\mathscr{F}'_{Y,n}$ the $\overline{\mathscr{B}}'_Y$-module of $\overline{Y}^{\circ}_{\text{fét}}$ defined in (III.10.15.3) and by $\mathscr{C}'_{Y,n}$ the $\overline{\mathscr{B}}'_Y$-algebra of $\overline{Y}^{\circ}_{\text{fét}}$ defined in (III.10.15.4). By III.8.21(ii), we have a canonical ring isomorphism of $\overline{Y}^{\circ}_{\text{fét}}$

(III.14.3.6)
$$\overline{\mathscr{B}}_Y \xrightarrow{\sim} \overline{\mathscr{B}}'_Y.$$

In view of the above, we have a canonical $\overline{\mathscr{B}}_Y$-linear isomorphism

(III.14.3.7)
$$\mathscr{F}_{Y,n} \xrightarrow{\sim} \mathscr{F}'_{Y,n}.$$

We deduce from this an isomorphism of $\overline{\mathscr{B}}_Y$-algebras

(III.14.3.8)
$$\mathscr{C}_{Y,n} \xrightarrow{\sim} \mathscr{C}'_{Y,n}.$$

**III.14.4.** Let $n$ be an integer $\geq 1$ and $r$ a rational number $\geq 0$. By ([**2**] III 2.3(2)), since the canonical functor $\jmath_{\mathbf{Q}} \colon E'_{\mathbf{Q}'} \to E_{\mathbf{Q}}$ is cocontinuous (III.14.2), the isomorphisms (III.14.3.7) induce an isomorphism of $\overline{\mathscr{B}}'_n$-modules

(III.14.4.1)
$$\rho_n \colon \Phi^*_n(\mathscr{F}_n) \xrightarrow{\sim} \mathscr{F}'_n$$

that fits into a commutative diagram

$$\begin{array}{ccccccccc}0 & \longrightarrow & \Phi_n^*(\overline{\mathscr{B}}_n) & \longrightarrow & \Phi_n^*(\mathscr{F}_n) & \longrightarrow & \Phi_n^*(\sigma_n^*(\xi^{-1}\widetilde{\Omega}^1_{\overline{X}_n/\overline{S}_n})) & \longrightarrow & 0 \\ & & \| & & \downarrow{\scriptstyle \rho_n} & & \downarrow & & \\ 0 & \longrightarrow & \overline{\mathscr{B}}'_n & \longrightarrow & \mathscr{F}'_n & \longrightarrow & \sigma_n'^*(\xi^{-1}\widetilde{\Omega}^1_{\overline{X}'_n/\overline{S}_n}) & \longrightarrow & 0\end{array} \qquad \text{(III.14.4.2)}$$

where the horizontal lines are the exact sequences deduced from (III.10.22.1) (cf. the proof of III.10.22(ii)). Likewise, the isomorphisms (III.14.3.8) induce an isomorphism of $\overline{\mathscr{B}}'_n$-algebras

$$\gamma_n\colon \Phi_n^*(\mathscr{C}_n) \xrightarrow{\sim} \mathscr{C}'_n, \qquad \text{(III.14.4.3)}$$

which is compatible with $\rho_n$ via the isomorphisms (III.10.22.3).

By (III.14.4.2), $\rho_n$ induces a $\overline{\mathscr{B}}'_n$-linear isomorphism

$$\rho_n^{(r)}\colon \Phi_n^*(\mathscr{F}_n^{(r)}) \to \mathscr{F}_n'^{(r)} \qquad \text{(III.14.4.4)}$$

that fits into a commutative diagram

$$\begin{array}{ccccccccc}0 & \longrightarrow & \Phi_n^*(\overline{\mathscr{B}}_n) & \longrightarrow & \Phi_n^*(\mathscr{F}_n^{(r)}) & \longrightarrow & \Phi_n^*(\sigma_n^*(\xi^{-1}\widetilde{\Omega}^1_{\overline{X}_n/\overline{S}_n})) & \longrightarrow & 0 \\ & & \| & & \downarrow{\scriptstyle \rho_n^{(r)}} & & \downarrow & & \\ 0 & \longrightarrow & \overline{\mathscr{B}}'_n & \longrightarrow & \mathscr{F}_n'^{(r)} & \longrightarrow & \sigma_n'^*(\xi^{-1}\widetilde{\Omega}^1_{\overline{X}'_n/\overline{S}_n}) & \longrightarrow & 0\end{array} \qquad \text{(III.14.4.5)}$$

where the horizontal lines are the exact sequences deduced from (III.10.22.1). We deduce from this an isomorphism of $\overline{\mathscr{B}}'_n$-algebras

$$\gamma_n^{(r)}\colon \Phi_n^*(\mathscr{C}_n^{(r)}) \xrightarrow{\sim} \mathscr{C}_n'^{(r)}. \qquad \text{(III.14.4.6)}$$

**III.14.5.** We denote by $\mathfrak{X}$ (resp. $\mathfrak{X}'$) the formal scheme $p$-adic completion of $\overline{X}$ (resp. $\overline{X}'$), by

$$\mathfrak{g}\colon \mathfrak{X}' \to \mathfrak{X} \qquad \text{(III.14.5.1)}$$

the extension of $\overline{g}\colon \overline{X}' \to \overline{X}$ to the completions, and by

$$\breve{\Phi}\colon (\widetilde{E}_s'^{\mathbb{N}^\circ}, \breve{\overline{\mathscr{B}}}') \to (\widetilde{E}_s^{\mathbb{N}^\circ}, \breve{\overline{\mathscr{B}}}) \qquad \text{(III.14.5.2)}$$

the morphism of ringed topos induced by the morphisms $(\Phi_n)_{n\geq 1}$ (III.14.1.4) (cf. III.7.5). We denote also by

$$\breve{\Phi}^*\colon \mathbf{Mod}_{\mathbb{Q}}(\breve{\overline{\mathscr{B}}}) \to \mathbf{Mod}_{\mathbb{Q}}(\breve{\overline{\mathscr{B}}}') \qquad \text{(III.14.5.3)}$$

the functor induced by the inverse image under $\breve{\Phi}$. By III.9.14, $\breve{\Phi}$ is canonically isomorphic to the localization morphism of the ringed topos $(\widetilde{E}_s^{\mathbb{N}^\circ}, \breve{\overline{\mathscr{B}}})$ at $\lambda^*(\sigma_s^*(X'_s))$, where $\lambda\colon \widetilde{E}_s^{\mathbb{N}^\circ} \to \widetilde{E}_s$ is the morphism of topos defined in (III.7.4.3). Consequently, there is no difference for $\breve{\overline{\mathscr{B}}}$-modules between the inverse image under $\breve{\Phi}$ in the sense of abelian sheaves and the inverse image in the sense of modules. The diagram of morphisms of ringed topos

(III.14.5.4)
$$\begin{array}{ccc}(\widetilde{E}'^{\mathbb{N}^\circ}_s, \breve{\overline{\mathscr{B}}}') & \xrightarrow{\breve{\Phi}} & (\widetilde{E}^{\mathbb{N}^\circ}_s, \breve{\overline{\mathscr{B}}}) \\ {\scriptstyle \top'}\downarrow & & \downarrow{\scriptstyle \top} \\ (X'_{s,\mathrm{zar}}, \mathscr{O}_{\mathfrak{X}'}) & \xrightarrow{\mathfrak{g}} & (X_{s,\mathrm{zar}}, \mathscr{O}_{\mathfrak{X}})\end{array}$$

where $\top$ and $\top'$ are the morphisms of ringed topos defined in (III.11.1.11), is commutative up to canonical isomorphism (III.9.11.15). Since the canonical morphism

(III.14.5.5)
$$\mathfrak{g}^*(\xi^{-1}\widetilde{\Omega}^1_{\mathfrak{X}/\mathscr{S}}) \to \xi^{-1}\widetilde{\Omega}^1_{\mathfrak{X}'/\mathscr{S}}$$

is an isomorphism, it induces by (III.14.5.4) an isomorphism

(III.14.5.6)
$$\delta\colon \breve{\Phi}^*(\top^*(\xi^{-1}\widetilde{\Omega}^1_{\mathfrak{X}/\mathscr{S}})) \xrightarrow{\sim} \top'^*(\xi^{-1}\widetilde{\Omega}^1_{\mathfrak{X}'/\mathscr{S}}).$$

Let $r$ be a rational number $\geq 0$. In view of (III.7.5.4) and (III.7.12.1), the isomorphisms $(\rho^{(r)}_n)_{n\geq 1}$ (III.14.4.4) induce a $\breve{\overline{\mathscr{B}}}'$-linear isomorphism

(III.14.5.7)
$$\breve{\rho}^{(r)}\colon \breve{\Phi}^*(\breve{\mathscr{F}}^{(r)}) \xrightarrow{\sim} \breve{\mathscr{F}}'^{(r)}.$$

Likewise, the isomorphisms $(\gamma^{(r)}_n)_{n\geq 1}$ (III.14.4.6) induce an isomorphism of $\breve{\overline{\mathscr{B}}}'$-algebras

(III.14.5.8)
$$\breve{\gamma}^{(r)}\colon \breve{\Phi}^*(\breve{\mathscr{C}}^{(r)}) \xrightarrow{\sim} \breve{\mathscr{C}}'^{(r)}.$$

It immediately follows from (III.14.4.5) that the diagram

(III.14.5.9)
$$\begin{array}{ccc}\breve{\Phi}^*(\breve{\mathscr{C}}^{(r)}) & \xrightarrow{\breve{\gamma}^{(r)}} & \breve{\mathscr{C}}'^{(r)} \\ {\scriptstyle \breve{\Phi}^*(\breve{d}^{(r)})}\downarrow & & \downarrow{\scriptstyle \breve{d}'^{(r)}} \\ \breve{\Phi}^*(\top^*(\xi^{-1}\widetilde{\Omega}^1_{\mathfrak{X}/\mathscr{S}})\otimes_{\breve{\overline{\mathscr{B}}}}\breve{\mathscr{C}}^{(r)}) & \xrightarrow{\delta\otimes\breve{\gamma}^{(r)}} & \top'^*(\xi^{-1}\widetilde{\Omega}^1_{\mathfrak{X}'/\mathscr{S}})\otimes_{\breve{\overline{\mathscr{B}}}'}\breve{\mathscr{C}}'^{(r)}\end{array}$$

where $\breve{d}^{(r)}$ and $\breve{d}'^{(r)}$ are the derivations (III.12.7.1), is commutative. For all rational numbers $r \geq r' \geq 0$, the diagram

(III.14.5.10)
$$\begin{array}{ccc}\breve{\Phi}^*(\breve{\mathscr{C}}^{(r)}) & \xrightarrow{\breve{\gamma}^{(r)}} & \breve{\mathscr{C}}'^{(r)} \\ {\scriptstyle \breve{\Phi}^*(\breve{\alpha}^{r,r'})}\downarrow & & \downarrow{\scriptstyle \breve{\alpha}'^{r,r'}} \\ \breve{\Phi}^*(\breve{\mathscr{C}}^{(r')}) & \xrightarrow{\breve{\gamma}^{(r')}} & \breve{\mathscr{C}}'^{(r')}\end{array}$$

where $\breve{\alpha}^{r,r'}$ and $\breve{\alpha}'^{r,r'}$ are the canonical homomorphisms (III.10.31.7), is commutative.

**III.14.6.** We (abusively) denote by

(III.14.6.1)
$$\mathfrak{g}^*\colon \mathbf{HM}(\mathscr{O}_{\mathfrak{X}}[\tfrac{1}{p}], \xi^{-1}\widetilde{\Omega}^1_{\mathfrak{X}/\mathscr{S}}) \to \mathbf{HM}(\mathscr{O}_{\mathfrak{X}'}[\tfrac{1}{p}], \xi^{-1}\widetilde{\Omega}^1_{\mathfrak{X}'/\mathscr{S}})$$

the inverse image functor for the Higgs modules (II.2.9) induced by $\mathfrak{g}$ and the canonical morphism (III.14.5.5). We define likewise an inverse image functor (III.12.5)

(III.14.6.2)
$$\mathfrak{g}^*\colon \mathbf{HI}(\mathscr{O}_{\mathfrak{X}}, \xi^{-1}\widetilde{\Omega}^1_{\mathfrak{X}/\mathscr{S}}) \to \mathbf{HI}(\mathscr{O}_{\mathfrak{X}'}, \xi^{-1}\widetilde{\Omega}^1_{\mathfrak{X}'/\mathscr{S}}).$$

This induces a functor that we denote also by

(III.14.6.3)
$$\mathfrak{g}^*\colon \mathbf{HI}_{\mathbb{Q}}(\mathscr{O}_{\mathfrak{X}}, \xi^{-1}\widetilde{\Omega}^1_{\mathfrak{X}/\mathscr{S}}) \to \mathbf{HI}_{\mathbb{Q}}(\mathscr{O}_{\mathfrak{X}'}, \xi^{-1}\widetilde{\Omega}^1_{\mathfrak{X}'/\mathscr{S}}).$$

The diagram of functors

$$\begin{array}{ccc} \mathbf{HI}_{\mathbb{Q}}(\mathscr{O}_{\mathfrak{X}}, \xi^{-1}\widetilde{\Omega}^1_{\mathfrak{X}/\mathscr{S}}) & \longrightarrow & \mathbf{HM}(\mathscr{O}_{\mathfrak{X}}[\frac{1}{p}], \xi^{-1}\widetilde{\Omega}^1_{\mathfrak{X}/\mathscr{S}}) \\ {\scriptstyle \mathfrak{g}^*}\downarrow & & \downarrow{\scriptstyle \mathfrak{g}^*} \\ \mathbf{HI}_{\mathbb{Q}}(\mathscr{O}_{\mathfrak{X}'}, \xi^{-1}\widetilde{\Omega}^1_{\mathfrak{X}'/\mathscr{S}}) & \longrightarrow & \mathbf{HM}(\mathscr{O}_{\mathfrak{X}'}[\frac{1}{p}], \xi^{-1}\widetilde{\Omega}^1_{\mathfrak{X}'/\mathscr{S}}) \end{array} \tag{III.14.6.4}$$

where the horizontal arrows are the functors (III.12.5.2) is commutative up to canonical isomorphism.

**III.14.7.** Let $r$ be a rational number $\geq 0$. By III.6.11 and (III.14.5.9), for every object $(\mathscr{F}, \mathscr{G}, u, \nabla)$ of $\Xi^r$ (III.12.7), $(\breve{\Phi}^*(\mathscr{F}), \breve{\Phi}^*(\mathscr{G}), \breve{\Phi}^*(u), \breve{\Phi}^*(\nabla))$ identifies with an object of $\Xi'^r$ through the isomorphisms $\breve{\gamma}^{(r)}$ (III.14.5.8) and $\delta$ (III.14.5.6). We deduce from this a functor that we denote also by

$$\breve{\Phi}^* \colon \Xi^r \to \Xi'^r. \tag{III.14.7.1}$$

This induces a functor that we denote also by

$$\breve{\Phi}^* \colon \Xi^r_{\mathbb{Q}} \to \Xi'^r_{\mathbb{Q}}. \tag{III.14.7.2}$$

The diagrams of functors

$$\begin{array}{ccc} \mathbf{Mod}(\breve{\overline{\mathscr{B}}}) & \xrightarrow{\mathfrak{S}^r} & \Xi^r \\ {\scriptstyle \breve{\Phi}^*}\downarrow & & \downarrow{\scriptstyle \breve{\Phi}^*} \\ \mathbf{Mod}(\breve{\overline{\mathscr{B}}}') & \xrightarrow{\mathfrak{S}'^r} & \Xi'^r \end{array} \tag{III.14.7.3}$$

where the horizontal arrows are the functors (III.12.7.2), and

$$\begin{array}{ccc} \mathbf{HI}(\mathscr{O}_{\mathfrak{X}}, \xi^{-1}\widetilde{\Omega}^1_{\mathfrak{X}/\mathscr{S}}) & \xrightarrow{\top^{r+}} & \Xi^r \\ {\scriptstyle \mathfrak{g}^*}\downarrow & & \downarrow{\scriptstyle \breve{\Phi}^*} \\ \mathbf{HI}(\mathscr{O}_{\mathfrak{X}'}, \xi^{-1}\widetilde{\Omega}^1_{\mathfrak{X}'/\mathscr{S}}) & \xrightarrow{\top'^{r+}} & \Xi'^r \end{array} \tag{III.14.7.4}$$

where the horizontal arrows are the functors (III.12.7.7), are clearly commutative up to canonical isomorphisms. By III.9.14, the diagram of functors

$$\begin{array}{ccc} \Xi^r & \xrightarrow{\mathscr{K}^r} & \mathbf{Mod}(\breve{\overline{\mathscr{B}}}) \\ {\scriptstyle \Phi^*}\downarrow & & \downarrow{\scriptstyle \breve{\Phi}^*} \\ \Xi'^r & \xrightarrow{\mathscr{K}'^r} & \mathbf{Mod}(\breve{\overline{\mathscr{B}}}') \end{array} \tag{III.14.7.5}$$

where $\mathscr{K}^r$ and $\mathscr{K}'^r$ are the functors (III.12.7.4), is commutative up to canonical isomorphism.

The base change morphism relative to the diagram (III.14.5.4) induces a morphism of functors from $\Xi^r$ to $\mathbf{HI}(\mathscr{O}_{\mathfrak{X}'}, \xi^{-1}\widetilde{\Omega}^1_{\mathfrak{X}'/\mathscr{S}})$

$$\mathfrak{g}^* \circ \top^r_+ \to \top'^r_+ \circ \breve{\Phi}^*, \tag{III.14.7.6}$$

where $\top^r_+$ and $\top'^r_+$ are the functors (III.12.7.10). By (**[2]** XVII 2.1.3), this is the adjoint of the morphism

$$\top'^{r+} \circ \mathfrak{g}^* \circ \top^r_+ \xrightarrow{\sim} \breve{\Phi}^* \circ \top^{r+} \circ \top^r_+ \to \breve{\Phi}^*, \tag{III.14.7.7}$$

where the first arrow is the isomorphism underlying the diagram (III.14.7.4) and the second arrow is the adjunction map. Consequently, for every object $\mathscr{N}$ of $\mathbf{HI}(\mathscr{O}_{\mathfrak{X}}, \xi^{-1}\widetilde{\Omega}^1_{\mathfrak{X}/\mathscr{S}})$ and every object $\mathscr{F}$ of $\Xi^r$, the diagram of maps of sets

(III.14.7.8)
$$\begin{array}{ccc}
\operatorname{Hom}_{\Xi^r}(\mathsf{T}^{r+}(\mathscr{N}), \mathscr{F}) & \longrightarrow & \operatorname{Hom}_{\mathbf{HI}(\mathscr{O}_{\mathfrak{X}}, \xi^{-1}\widetilde{\Omega}^1_{\mathfrak{X}/\mathscr{S}})}(\mathscr{N}, \mathsf{T}^r_+(\mathscr{F})) \\
\downarrow a & & \downarrow b \\
\operatorname{Hom}_{\Xi'^r}(\mathsf{T}'^{r+}(\mathfrak{g}^*(\mathscr{N})), \breve{\Phi}^*(\mathscr{F})) & \longrightarrow & \operatorname{Hom}_{\mathbf{HI}(\mathscr{O}_{\mathfrak{X}'}, \xi^{-1}\widetilde{\Omega}^1_{\mathfrak{X}'/\mathscr{S}})}(\mathfrak{g}^*(\mathscr{N}), \mathsf{T}'^r_+(\breve{\Phi}^*(\mathscr{F})))
\end{array}$$

where the horizontal arrows are the adjunction isomorphisms, $a$ is induced by the functor $\breve{\Phi}^*$ and the isomorphism underlying the diagram (III.14.7.4), and $b$ is induced by the functor $\mathfrak{g}^*$ and the morphism (III.14.7.6), is commutative.

For all rational numbers $r \geq r' \geq 0$, the diagram of functors

(III.14.7.9)
$$\begin{array}{ccc}
\Xi^r & \xrightarrow{\epsilon^{r,r'}} & \Xi^{r'} \\
{\scriptstyle \breve{\Phi}^*}\downarrow & & \downarrow{\scriptstyle \breve{\Phi}^*} \\
\Xi'^r & \xrightarrow{\epsilon'^{r,r'}} & \Xi'^{r'}
\end{array}$$

where the horizontal arrows are the functors (III.12.8.3), is commutative up to canonical isomorphism. It immediately follows from (III.14.5.10) that the diagram of morphisms of functors

(III.14.7.10)
$$\begin{array}{ccccc}
\mathfrak{g}^* \circ \mathsf{T}^r_+ & & \longrightarrow & & \mathfrak{g}^* \circ \mathsf{T}^{r'}_+ \circ \epsilon^{r,r'} \\
\downarrow & & & & \downarrow \\
\mathsf{T}'^r_+ \circ \breve{\Phi}^* & \longrightarrow & \mathsf{T}'^{r'}_+ \circ \epsilon^{r,r'} \circ \breve{\Phi}^* & = & \mathsf{T}'^{r'}_+ \circ \breve{\Phi}^* \circ \epsilon^{r,r'}
\end{array}$$

where the horizontal arrows are induced by the morphism (III.12.8.9), the vertical arrows are induced by the morphism (III.14.7.6), and the identification indicated by the symbol $=$ comes from the diagram (III.14.7.9), is commutative. Consequently, the composition

(III.14.7.11)
$$\mathfrak{g}^* \circ \mathsf{T}^r_+ \circ \mathfrak{S}^r \to \mathsf{T}'^r_+ \circ \breve{\Phi}^* \circ \mathfrak{S} \xrightarrow{\sim} \mathsf{T}'^r_+ \circ \mathfrak{S}'^r \circ \breve{\Phi}^*,$$

where the first arrow is induced by (III.14.7.6) and the second arrow is the isomorphism underlying the diagram (III.14.7.3), induces by direct limit, for $r \in \mathbb{Q}_{>0}$, a morphism of functors from $\mathbf{Mod}_{\mathbb{Q}}(\breve{\overline{\mathscr{B}}})$ to $\mathbf{HM}(\mathscr{O}_{\mathfrak{X}'}[\frac{1}{p}], \xi^{-1}\widetilde{\Omega}^1_{\mathfrak{X}'/\mathscr{S}})$

(III.14.7.12)
$$\mathfrak{g}^* \circ \mathscr{H} \to \mathscr{H}' \circ \breve{\Phi}^*,$$

where $\mathscr{H}$ and $\mathscr{H}'$ are the functors (III.12.13.2).

**Proposition III.14.8.** *Suppose that $g$ is an open immersion. Then:*

(i) *For every rational number $r \geq 0$, the morphism* (III.14.7.6) *is an isomorphism. It makes commutative the diagram of functors*

(III.14.8.1)
$$\begin{array}{ccc}
\Xi^r & \xrightarrow{\mathsf{T}^r_+} & \mathbf{HI}(\mathscr{O}_{\mathfrak{X}}, \xi^{-1}\widetilde{\Omega}^1_{\mathfrak{X}/\mathscr{S}}) \\
{\scriptstyle \breve{\Phi}^*}\downarrow & & \downarrow{\scriptstyle \mathfrak{g}^*} \\
\Xi'^r & \xrightarrow{\mathsf{T}'^r_+} & \mathbf{HI}(\mathscr{O}_{\mathfrak{X}'}, \xi^{-1}\widetilde{\Omega}^1_{\mathfrak{X}'/\mathscr{S}})
\end{array}$$

(ii) *The morphism* (III.14.7.12) *is an isomorphism. It makes commutative the diagram of functors*

(III.14.8.2)
$$\begin{array}{ccc} \mathbf{Mod}_{\mathbb{Q}}(\breve{\overline{\mathscr{B}}}) & \xrightarrow{\mathscr{H}} & \mathbf{HM}(\mathscr{O}_{\mathfrak{X}}[\frac{1}{p}], \xi^{-1}\widetilde{\Omega}^1_{\mathfrak{X}/\mathscr{S}}) \\ \downarrow{\scriptstyle \breve{\Phi}^*} & & \downarrow{\scriptstyle \mathfrak{g}^*} \\ \mathbf{Mod}_{\mathbb{Q}}(\breve{\overline{\mathscr{B}}}') & \xrightarrow{\mathscr{H}'} & \mathbf{HM}(\mathscr{O}_{\mathfrak{X}'}[\frac{1}{p}], \xi^{-1}\widetilde{\Omega}^1_{\mathfrak{X}'/\mathscr{S}}) \end{array}$$

(i) This follows from III.9.15.
(ii) This follows from (i) and the definitions.

**Proposition III.14.9.** *Let $\mathscr{M}$ be a Dolbeault $\breve{\overline{\mathscr{B}}}_{\mathbb{Q}}$-module and $\mathscr{N}$ a solvable Higgs $\mathscr{O}_{\mathfrak{X}}[\frac{1}{p}]$-bundle with coefficients in $\xi^{-1}\widetilde{\Omega}^1_{\mathfrak{X}/\mathscr{S}}$. Then $\breve{\Phi}^*(\mathscr{M})$ is a Dolbeault $\breve{\overline{\mathscr{B}}}'_{\mathbb{Q}}$-module and $\mathfrak{g}^*(\mathscr{N})$ is a solvable Higgs $\mathscr{O}_{\mathfrak{X}'}[\frac{1}{p}]$-bundle with coefficients in $\xi^{-1}\widetilde{\Omega}^1_{\mathfrak{X}'/\mathscr{S}}$. If, moreover, $\mathscr{M}$ and $\mathscr{N}$ are associated, then $\breve{\Phi}^*(\mathscr{M})$ and $\mathfrak{g}^*(\mathscr{N})$ are associated*

Indeed, $\mathfrak{g}^*(\mathscr{N})$ is a Higgs $\mathscr{O}_{\mathfrak{X}'}[\frac{1}{p}]$-bundle with coefficients in $\xi^{-1}\widetilde{\Omega}^1_{\mathfrak{X}'/\mathscr{S}}$ and $\breve{\Phi}^*(\mathscr{M})$ is an object of $\mathbf{Mod}^{\mathrm{aft}}_{\mathbb{Q}}(\breve{\overline{\mathscr{B}}}')$. Suppose that there exist a rational number $r > 0$ and an isomorphism of $\Xi^r_{\mathbb{Q}}$

(III.14.9.1)
$$\alpha\colon \mathrm{T}^{r+}(\mathscr{N}) \xrightarrow{\sim} \mathfrak{S}^r(\mathscr{M}).$$

In view of (III.14.7.3) and (III.14.7.4), $\breve{\Phi}^*(\alpha)$ induces an isomorphism of $\Xi'^r_{\mathbb{Q}}$

(III.14.9.2)
$$\alpha'\colon \mathrm{T}'^{r+}(\mathfrak{g}^*(\mathscr{N})) \xrightarrow{\sim} \mathfrak{S}'^r(\breve{\Phi}^*(\mathscr{M})),$$

giving the proposition.

**III.14.10.** By III.14.9, $\breve{\Phi}^*$ induces a functor

(III.14.10.1)
$$\breve{\Phi}^*\colon \mathbf{Mod}^{\mathrm{Dolb}}_{\mathbb{Q}}(\breve{\overline{\mathscr{B}}}) \to \mathbf{Mod}^{\mathrm{Dolb}}_{\mathbb{Q}}(\breve{\overline{\mathscr{B}}}'),$$

and $\mathfrak{g}^*$ induces a functor

(III.14.10.2)
$$\mathfrak{g}^*\colon \mathbf{HM}^{\mathrm{sol}}(\mathscr{O}_{\mathfrak{X}}[\frac{1}{p}], \xi^{-1}\widetilde{\Omega}^1_{\mathfrak{X}/\mathscr{S}}) \to \mathbf{HM}^{\mathrm{sol}}(\mathscr{O}_{\mathfrak{X}'}[\frac{1}{p}], \xi^{-1}\widetilde{\Omega}^1_{\mathfrak{X}'/\mathscr{S}}).$$

**Proposition III.14.11.** (i) *The diagram of functors*

(III.14.11.1)
$$\begin{array}{ccc} \mathbf{Mod}^{\mathrm{Dolb}}_{\mathbb{Q}}(\breve{\overline{\mathscr{B}}}) & \xrightarrow{\mathscr{H}} & \mathbf{HM}^{\mathrm{sol}}(\mathscr{O}_{\mathfrak{X}}[\frac{1}{p}], \xi^{-1}\widetilde{\Omega}^1_{\mathfrak{X}/\mathscr{S}}) \\ \downarrow{\scriptstyle \breve{\Phi}^*} & & \downarrow{\scriptstyle \mathfrak{g}^*} \\ \mathbf{Mod}^{\mathrm{Dolb}}_{\mathbb{Q}}(\breve{\overline{\mathscr{B}}}') & \xrightarrow{\mathscr{H}'} & \mathbf{HM}^{\mathrm{sol}}(\mathscr{O}_{\mathfrak{X}'}[\frac{1}{p}], \xi^{-1}\widetilde{\Omega}^1_{\mathfrak{X}'/\mathscr{S}}) \end{array}$$

*where $\mathscr{H}$ and $\mathscr{H}'$ are the functors* (III.12.18.1) *is commutative up to canonical isomorphism.*

(ii) *The diagram of functors*

(III.14.11.2)
$$\begin{array}{ccc} \mathbf{HM}^{\mathrm{sol}}(\mathscr{O}_{\mathfrak{X}}[\frac{1}{p}], \xi^{-1}\widetilde{\Omega}^1_{\mathfrak{X}/\mathscr{S}}) & \xrightarrow{\mathscr{V}} & \mathbf{Mod}^{\mathrm{Dolb}}_{\mathbb{Q}}(\breve{\overline{\mathscr{B}}}) \\ \downarrow{\scriptstyle \mathfrak{g}^*} & & \downarrow{\scriptstyle \breve{\Phi}^*} \\ \mathbf{HM}^{\mathrm{sol}}(\mathscr{O}_{\mathfrak{X}'}[\frac{1}{p}], \xi^{-1}\widetilde{\Omega}^1_{\mathfrak{X}'/\mathscr{S}}) & \xrightarrow{\mathscr{V}'} & \mathbf{Mod}^{\mathrm{Dolb}}_{\mathbb{Q}}(\breve{\overline{\mathscr{B}}}') \end{array}$$

*where $\mathscr{V}$ and $\mathscr{V}'$ are the functors* (III.12.23.1) *is commutative up to canonical isomorphism.*

(i) For every object $\mathscr{M}$ of $\mathbf{Mod}_{\mathbb{Q}}^{\mathrm{Dolb}}(\breve{\overline{\mathscr{B}}})$, $\mathscr{M}$ and $\mathscr{H}(\mathscr{M})$ are associated by virtue of III.12.18. We choose a rational number $r_{\mathscr{M}} > 0$ and an isomorphism of $\Xi_{\mathbb{Q}}^{r_{\mathscr{M}}}$

$$\alpha_{\mathscr{M}} \colon \mathsf{T}^{r_{\mathscr{M}}+}(\mathscr{H}(\mathscr{M})) \xrightarrow{\sim} \mathfrak{S}^{r_{\mathscr{M}}}(\mathscr{M}) \tag{III.14.11.3}$$

satisfying the properties of III.12.19. For any rational number $r$ such that $0 < r \leq r_{\mathscr{M}}$, we denote by

$$\alpha^r_{\mathscr{M}} \colon \mathsf{T}^{r+}(\mathscr{H}(\mathscr{M})) \xrightarrow{\sim} \mathfrak{S}^{r}(\mathscr{M}) \tag{III.14.11.4}$$

the isomorphism of $\Xi_{\mathbb{Q}}^{r}$ induced by $\epsilon^{r_{\mathscr{M}},r}(\alpha_{\mathscr{M}})$ (III.12.8.4) and the isomorphisms (III.12.8.5) and (III.12.8.6). In view of (III.14.7.3) and (III.14.7.4), $\breve{\Phi}^*(\alpha_{\mathscr{M}})$ induces an isomorphism of $\Xi_{\mathbb{Q}}^{\prime r_{\mathscr{M}}}$

$$\alpha'_{\mathscr{M}} \colon \mathsf{T}^{\prime r_{\mathscr{M}}+}(\mathfrak{g}^*(\mathscr{H}(\mathscr{M}))) \xrightarrow{\sim} \mathfrak{S}^{\prime r_{\mathscr{M}}}(\breve{\Phi}^*(\mathscr{M})). \tag{III.14.11.5}$$

Likewise, $\breve{\Phi}^*(\alpha^r_{\mathscr{M}})$ induces an isomorphism of $\Xi_{\mathbb{Q}}^{\prime r}$

$$\alpha'^r_{\mathscr{M}} \colon \mathsf{T}^{\prime r+}(\mathfrak{g}^*(\mathscr{H}(\mathscr{M}))) \xrightarrow{\sim} \mathfrak{S}^{\prime r}(\breve{\Phi}^*(\mathscr{M})), \tag{III.14.11.6}$$

that we can also deduce from $\epsilon'^{r_{\mathscr{M}},r}(\alpha'_{\mathscr{M}})$ by (III.14.7.9). We denote by

$$\beta'^r_{\mathscr{M}} \colon \mathfrak{g}^*(\mathscr{H}(\mathscr{M})) \to \mathsf{T}'^r_+(\mathfrak{S}'^r(\breve{\Phi}^*(\mathscr{M}))) \tag{III.14.11.7}$$

its adjoint (III.12.7.12). By III.14.9, $\breve{\Phi}^*(\mathscr{M})$ is a Dolbeault $\breve{\overline{\mathscr{B}}}'_{\mathbb{Q}}$-module and $\mathfrak{g}^*(\mathscr{H}(\mathscr{M}))$ is a solvable Higgs $\mathscr{O}_{\mathfrak{X}'}[\frac{1}{p}]$-bundle with coefficients in $\xi^{-1}\widetilde{\Omega}^1_{\mathfrak{X}'/\mathscr{S}}$, associated with $\breve{\Phi}^*(\mathscr{M})$. Consequently, by virtue of III.12.17(i), the composition

$$\mathfrak{g}^*(\mathscr{H}(\mathscr{M})) \xrightarrow{\beta'^r_{\mathscr{M}}} \mathsf{T}'^r_+(\mathfrak{S}'^r(\breve{\Phi}^*(\mathscr{M}))) \longrightarrow \mathscr{H}'(\breve{\Phi}^*(\mathscr{M})), \tag{III.14.11.8}$$

where the second arrow is the canonical morphism (III.12.13.2), is an isomorphism that depends a priori on $\alpha_{\mathscr{M}}$ but not on $r$. By the proof of III.12.26, for every morphism $u \colon \mathscr{M} \to \mathscr{M}'$ of $\mathbf{Mod}_{\mathbb{Q}}^{\mathrm{Dolb}}(\breve{\overline{\mathscr{B}}})$ and every rational number $r$ such that $0 < r < \inf(r_{\mathscr{M}}, r_{\mathscr{M}'})$, the diagram of $\Xi_{\mathbb{Q}}^{r}$

$$\begin{array}{ccc} \mathsf{T}^{r+}(\mathscr{H}(\mathscr{M})) & \xrightarrow{\alpha^r_{\mathscr{M}}} & \mathfrak{S}^r(\mathscr{M}) \\ \downarrow{\scriptstyle \mathsf{T}^{r+}(\mathscr{H}(u))} & & \downarrow{\scriptstyle \mathfrak{S}^r(u)} \\ \mathsf{T}^{r+}(\mathscr{H}(\mathscr{M}')) & \xrightarrow{\alpha^r_{\mathscr{M}'}} & \mathfrak{S}^r(\mathscr{M}') \end{array} \tag{III.14.11.9}$$

is commutative. We deduce from this that the composed isomorphism (III.14.11.8)

$$\mathfrak{g}^*(\mathscr{H}(\mathscr{M})) \xrightarrow{\sim} \mathscr{H}'(\breve{\Phi}^*(\mathscr{M})) \tag{III.14.11.10}$$

depends only on $\mathscr{M}$ (but not on the choice of $\alpha_{\mathscr{M}}$) and that it depends on it functorially; whence the statement.

(ii) The proof is similar to that of (i) and is left to the reader.

**Remarks III.14.12.** Let $\mathscr{M}$ be a Dolbeault $\breve{\overline{\mathscr{B}}}_{\mathbb{Q}}$-module.

(i) The canonical morphism (III.14.7.12)

$$\mathfrak{g}^*(\mathscr{H}(\mathscr{M})) \to \mathscr{H}'(\breve{\Phi}^*(\mathscr{M})) \tag{III.14.12.1}$$

is an isomorphism; it is the isomorphism underlying the commutative diagram (III.14.11.1). Indeed, let us use the notation of the proof of III.14.11(i). We moreover denote by

(III.14.12.2) $$\beta^r_{\mathscr{M}} \colon \mathscr{H}(\mathscr{M}) \to \mathrm{T}^r_+(\mathfrak{S}^r(\mathscr{M}))$$

the adjoint morphism of $\alpha^r_{\mathscr{M}}$. It follows from (III.14.7.8) that the morphism $\beta'^r_{\mathscr{M}}$ (III.14.11.7) is equal to the composition

(III.14.12.3)
$$\mathfrak{g}^*(\mathscr{H}(\mathscr{M})) \xrightarrow{\mathfrak{g}^*(\beta^r_{\mathscr{M}})} \mathfrak{g}^*(\mathrm{T}^r_+(\mathfrak{S}^r(\mathscr{M}))) \longrightarrow \mathrm{T}'^r_+(\breve{\Phi}^*(\mathfrak{S}^r(\mathscr{M}))) \xrightarrow{\sim} \mathrm{T}'^r_+(\mathfrak{S}'^r(\breve{\Phi}^*(\mathscr{M}))),$$

where the second arrow is the morphism (III.14.7.6) and the last arrow is the isomorphism underlying the diagram (III.14.7.3). On the other hand, the direct limit of the morphisms $\beta^r_{\mathscr{M}}$, for $r \in \mathbb{Q}_{>0}$, is the identity, and the direct limit of the morphisms $\beta'^r_{\mathscr{M}}$, for $r \in \mathbb{Q}_{>0}$, is equal to the composed isomorphism (III.14.11.8) underlying the commutative diagram (III.14.11.1).

(ii) Let $r$ be a rational number $> 0$ and

(III.14.12.4) $$\alpha \colon \mathrm{T}^{r+}(\mathscr{H}(\mathscr{M})) \xrightarrow{\sim} \mathfrak{S}^r(\mathscr{M})$$

an isomorphism of $\Xi^r_{\mathbb{Q}}$ satisfying the properties of III.12.19. By (i), (III.14.7.3), and (III.14.7.4), we can identify $\breve{\Phi}^*(\alpha)$ with an isomorphism

(III.14.12.5) $$\alpha' \colon \mathrm{T}'^{r+}(\mathscr{H}'(\breve{\Phi}^*(\mathscr{M}))) \xrightarrow{\sim} \mathfrak{S}'^r(\breve{\Phi}^*(\mathscr{M})).$$

On the other hand, $\breve{\Phi}^*(\mathscr{M})$ is a Dolbeault $\breve{\overline{\mathscr{B}}}'_{\mathbb{Q}}$-module by III.14.9. It immediately follows from III.12.17 that $\alpha'$ satisfies the properties of III.12.19.

## III.15. Fibered category of Dolbeault modules

**III.15.1.** We keep the assumptions and general notation of III.10 and III.12 in this section. We denote by $\psi$ the composition

(III.15.1.1) $$\psi \colon \widetilde{E}_s^{\mathbb{N}^\circ} \xrightarrow{\lambda} \widetilde{E}_s \xrightarrow{\sigma_s} X_{s,\text{ét}} \xrightarrow{\iota_{\text{ét}}} X_{\text{ét}},$$

where $\lambda$ is the morphism of topos defined in (III.7.4.3), $\sigma_s$ is the canonical morphism of topos (III.10.2.6), and $\iota \colon X_s \to X$ is the canonical injection. For any object $U$ of $\mathbf{\acute{E}t}_{/X}$, we denote by $f_U \colon (U, \mathscr{M}_X|U) \to (S, \mathscr{M}_S)$ the morphism induced by $f$, and by $\widetilde{U} \to \widetilde{X}$ the unique étale morphism that lifts $\check{\overline{U}} \to \check{\overline{X}}$, so that $(\widetilde{U}, \mathscr{M}_{\widetilde{X}}|\widetilde{U})$ is a smooth $(\mathscr{A}_2(\overline{S}), \mathscr{M}_{\mathscr{A}_2(\overline{S})})$-deformation of $(\check{\overline{U}}, \mathscr{M}_{\check{\overline{X}}}|\check{\overline{U}})$. The localization of the ringed topos $(\widetilde{E}_s^{\mathbb{N}^\circ}, \breve{\overline{\mathscr{B}}})$ at $\psi^*(U)$ is canonically equivalent to the analogous ringed topos associated with $f_U$ by virtue of III.9.14. For every rational number $r \geq 0$, the restriction of the $\breve{\overline{\mathscr{B}}}$-algebra $\breve{\mathscr{C}}^{(r)}$ over $\psi^*(U)$ is canonically isomorphic to the analogous $(\breve{\overline{\mathscr{B}}}|\psi^*(U))$-algebra associated with the deformation $(\widetilde{U}, \mathscr{M}_{\widetilde{X}}|\widetilde{U})$, by (III.14.5.8). We denote by $\mathbf{Mod}(\breve{\overline{\mathscr{B}}}|\psi^*(U))$ the category of $(\breve{\overline{\mathscr{B}}}|\psi^*(U))$-modules of $(\widetilde{E}_s^{\mathbb{N}^\circ})_{/\psi^*(U)}$, by $\mathbf{Mod}_{\mathbb{Q}}(\breve{\overline{\mathscr{B}}}|\psi^*(U))$ the category of $(\breve{\overline{\mathscr{B}}}|\psi^*(U))$-modules up to isogeny, by $\Xi^r_U$ the category of integrable $p^r$-isoconnections with respect to the extension $(\breve{\mathscr{C}}^{(r)}|\psi^*(U))/(\breve{\overline{\mathscr{B}}}|\psi^*(U))$ (cf. III.12.7), by $\Xi^r_{U,\mathbb{Q}}$ the category of objects of $\Xi^r_U$ up to isogeny, and by $\mathbf{Mod}^{\text{Dolb}}_{\mathbb{Q}}(\breve{\overline{\mathscr{B}}}|\psi^*(U))$ the category of Dolbeault $(\breve{\overline{\mathscr{B}}}|\psi^*(U))_{\mathbb{Q}}$-modules with respect to the deformation $(\widetilde{U}, \mathscr{M}_{\widetilde{X}}|\widetilde{U})$ (cf. III.12.11). By

III.14.9, for every morphism $g\colon U' \to U$ of $\mathbf{\acute{E}t}_{/X}$, the restriction functor

$$\mathbf{Mod}_{\mathbb{Q}}(\breve{\overline{\mathscr{B}}}|\psi^*(U)) \to \mathbf{Mod}_{\mathbb{Q}}(\breve{\overline{\mathscr{B}}}|\psi^*(U')), \quad \mathscr{M} \mapsto \mathscr{M}|\psi^*(U'), \tag{III.15.1.2}$$

induces a functor

$$\mathbf{Mod}_{\mathbb{Q}}^{\mathrm{Dolb}}(\breve{\overline{\mathscr{B}}}|\psi^*(U)) \to \mathbf{Mod}_{\mathbb{Q}}^{\mathrm{Dolb}}(\breve{\overline{\mathscr{B}}}|\psi^*(U')). \tag{III.15.1.3}$$

We denote by

$$\Xi^r_U \to \Xi^r_{U'}, \qquad A \mapsto A|\psi^*(U'), \tag{III.15.1.4}$$

$$\Xi^r_{U,\mathbb{Q}} \to \Xi^r_{U',\mathbb{Q}}, \qquad B \mapsto B|\psi^*(U'), \tag{III.15.1.5}$$

the restriction functors defined in (III.14.7.1) and (III.14.7.2), respectively.

**Lemma III.15.2.** *Let $r$ be a rational number $\geq 0$, $A, B$ two objects of $\Xi^r_{X,\mathbb{Q}}$, and $(U_i)_{i\in I}$ an étale covering of $X$. For all $(i,j) \in I^2$, we set $U_{ij} = U_i \times_X U_j$. Then the diagram of maps of sets*

$$\begin{aligned}\mathrm{Hom}_{\Xi^r_{X,\mathbb{Q}}}(A,B) \to &\prod_{i\in I} \mathrm{Hom}_{\Xi^r_{U_i,\mathbb{Q}}}(A|\psi^*(U_i), B|\psi^*(U_i)) \\ &\rightrightarrows \prod_{(i,j)\in I^2} \mathrm{Hom}_{\Xi^r_{U_{ij},\mathbb{Q}}}(A|\psi^*(U_{ij}), B|\psi^*(U_{ij}))\end{aligned} \tag{III.15.2.1}$$

*is exact.*

Indeed, since $X$ is quasi-compact, we may assume that $I$ is finite, in which case the assertion easily follows from III.6.7.

**III.15.3.** We denote by $\mathrm{MOD}(\breve{\overline{\mathscr{B}}})$ the fibered (and even split **[37]** VI § 9) $(\widetilde{E}_s^{\mathbb{N}^\circ})$-category of $\breve{\overline{\mathscr{B}}}$-modules over $\widetilde{E}_s^{\mathbb{N}^\circ}$ (**[35]** II 3.4.1). It is a stack over $\widetilde{E}_s^{\mathbb{N}^\circ}$ by (**[35]** II 3.4.4). We denote by $\mathbf{\acute{E}t}_{\mathrm{coh}/X}$ the full subcategory of $\mathbf{\acute{E}t}_{/X}$ made up of étale schemes of finite presentation over $X$ and by

$$\mathrm{MOD}'(\breve{\overline{\mathscr{B}}}) \to \mathbf{\acute{E}t}_{\mathrm{coh}/X} \tag{III.15.3.1}$$

the base change of $\mathrm{MOD}(\breve{\overline{\mathscr{B}}})$ (**[37]** VI § 3) by $\psi^* \circ \varepsilon$, where $\psi$ is the morphism (III.15.1.1) and $\varepsilon\colon \mathbf{\acute{E}t}_{\mathrm{coh}/X} \to X_{\mathrm{\acute{e}t}}$ is the canonical functor. This is also a stack by (**[35]** II 3.1.1). We deduce from this a fibered category

$$\mathrm{MOD}'_{\mathbb{Q}}(\breve{\overline{\mathscr{B}}}) \to \mathbf{\acute{E}t}_{\mathrm{coh}/X}, \tag{III.15.3.2}$$

whose fiber over an object $U$ of $\mathbf{\acute{E}t}_{\mathrm{coh}/X}$ is the category $\mathbf{Mod}_{\mathbb{Q}}(\breve{\overline{\mathscr{B}}}|\psi^*(U))$ and the inverse image functor under a morphism $U' \to U$ of $\mathbf{\acute{E}t}_{\mathrm{coh}/X}$ is the restriction functor (III.15.1.2). Note that this is not a priori a stack. It induces a fibered category

$$\mathrm{MOD}^{\mathrm{Dolb}}_{\mathbb{Q}}(\breve{\overline{\mathscr{B}}}) \to \mathbf{\acute{E}t}_{\mathrm{coh}/X} \tag{III.15.3.3}$$

whose fiber over an object $U$ of $\mathbf{\acute{E}t}_{\mathrm{coh}/X}$ is the category $\mathbf{Mod}_{\mathbb{Q}}^{\mathrm{Dolb}}(\breve{\overline{\mathscr{B}}}|\psi^*(U))$ and the inverse image functor under a morphism $U' \to U$ of $\mathbf{\acute{E}t}_{\mathrm{coh}/X}$ is the restriction functor (III.15.1.3).

**Proposition III.15.4.** *Let $\mathscr{M}$ be an object of $\mathbf{Mod}_{\mathbb{Q}}^{\mathrm{aft}}(\breve{\overline{\mathscr{B}}})$ and $(U_i)_{i\in I}$ a covering of $\mathbf{\acute{E}t}_{\mathrm{coh}/X}$. Then $\mathscr{M}$ is Dolbeault if and only if for every $i \in I$, the $(\breve{\overline{\mathscr{B}}}|\psi^*(U_i))_{\mathbb{Q}}$-module $\mathscr{M}|\psi^*(U_i)$ is Dolbeault.*

Indeed, the condition is necessary by virtue of III.14.9. Suppose that for every $i \in I$, $\mathscr{M}|\psi^*(U_i)$ is Dolbeault and let us show that $\mathscr{M}$ is Dolbeault. Since $X$ is quasi-compact, we may assume that $I$ is finite. For any $i \in I$, denote by $\mathfrak{X}_i$ the formal scheme $p$-adic completion of $\overline{U}_i$. For any $(i,j) \in I^2$, set $U_{ij} = U_i \times_X U_j$ and denote by $\mathfrak{X}_{ij}$ the formal scheme $p$-adic completion of $\overline{U}_{ij}$. By virtue of III.14.11(i), we have a canonical isomorphism of Higgs $\mathscr{O}_{\mathfrak{X}_{ij}}[\frac{1}{p}]$-modules with coefficients in $\xi^{-1}\widetilde{\Omega}^1_{\mathfrak{X}_{ij}/\mathscr{S}}$

$$\mathscr{H}_i(\mathscr{M}|\psi^*(U_i)) \otimes_{\mathscr{O}_{\mathfrak{X}_i}} \mathscr{O}_{\mathfrak{X}_{ij}} \xrightarrow{\sim} \mathscr{H}_{ij}(\mathscr{M}|\psi^*(U_{ij})), \tag{III.15.4.1}$$

where $\mathscr{H}_i$ and $\mathscr{H}_{ij}$ are the functors (III.12.18.1) associated with $(f_{U_i}, \widetilde{U}_i, \mathscr{M}_{\widetilde{X}}|\widetilde{U}_i)$ and $(f_{U_{ij}}, \widetilde{U}_{ij}, \mathscr{M}_{\widetilde{X}}|\widetilde{U}_{ij})$, respectively. We deduce from this a descent datum $\delta$ on the Higgs modules $(\mathscr{H}_i(\mathscr{M}|\psi^*(U_i)))_{i\in I}$ relative to the étale covering $(\mathfrak{X}_i \to \mathfrak{X})_{i\in I}$. Since the latter is effective by III.6.22, there exist a Higgs $\mathscr{O}_{\mathfrak{X}}[\frac{1}{p}]$-bundle $\mathscr{N}$ and for every $i \in I$, an isomorphism of Higgs $\mathscr{O}_{\mathfrak{X}_i}[\frac{1}{p}]$-modules with coefficients in $\xi^{-1}\widetilde{\Omega}^1_{\mathfrak{X}_i/\mathscr{S}}$

$$\mathscr{N} \otimes_{\mathscr{O}_{\mathfrak{X}}} \mathscr{O}_{\mathfrak{X}_i} \xrightarrow{\sim} \mathscr{H}_i(\mathscr{M}|\psi^*(U_i)), \tag{III.15.4.2}$$

that induce the descent datum $\delta$.

For any $(i,j) \in I^2$ and any rational number $r > 0$, we denote by $\top_i^{r+}$ and $\mathfrak{S}_i^r$ (resp. $\top_{ij}^{r+}$ and $\mathfrak{S}_{ij}^r$) the functors (III.12.7.10) and (III.12.7.2) associated with $(f_{U_i}, \widetilde{U}_i, \mathscr{M}_{\widetilde{X}}|\widetilde{U}_i)$ (resp. $(f_{U_{ij}}, \widetilde{U}_{ij}, \mathscr{M}_{\widetilde{X}}|\widetilde{U}_{ij})$). For every $i \in I$, we choose a rational number $r_i > 0$ and an isomorphism of $\Xi^{r_i}_{U_i,\mathbb{Q}}$

$$\alpha_i \colon \top_i^{r_i+}(\mathscr{H}_i(\mathscr{M}|\psi^*(U_i))) \xrightarrow{\sim} \mathfrak{S}_i^{r_i}(\mathscr{M}|\psi^*(U_i)) \tag{III.15.4.3}$$

satisfying the properties of III.12.19. For every $(i,j) \in I^2$, $\mathscr{M}|\psi^*(U_{ij})$ is Dolbeault by virtue of III.14.9. By (III.14.7.3), (III.14.7.4), and III.14.11(i), $\alpha_i|\psi^*(U_{ij})$ identifies with an isomorphism

$$\alpha_i|\psi^*(U_{ij}) \colon \top_{ij}^{r_i+}(\mathscr{H}_{ij}(\mathscr{M}|\psi^*(U_{ij}))) \xrightarrow{\sim} \mathfrak{S}_{ij}^{r_i}(\mathscr{M}|\psi^*(U_{ij})). \tag{III.15.4.4}$$

This satisfies the properties of III.12.19, in view of III.14.12. For any rational number $r$ such that $0 < r \leq r_i$, we denote by $\epsilon_i^{r_i,r} \colon \Xi^{r_i}_{U_i,\mathbb{Q}} \to \Xi^r_{U_i,\mathbb{Q}}$ the functor (III.12.8.4) associated with $(f_{U_i}, \widetilde{U}_i, \mathscr{M}_{\widetilde{X}}|\widetilde{U}_i)$ and by

$$\alpha_i^r \colon \top_i^{r+}(\mathscr{H}_i(\mathscr{M}|\psi^*(U_i))) \xrightarrow{\sim} \mathfrak{S}_i^r(\mathscr{M}|\psi^*(U_i)) \tag{III.15.4.5}$$

the isomorphism of $\Xi^r_{U_i,\mathbb{Q}}$ induced by $\epsilon_i^{r_i,r}(\alpha_i)$ and the isomorphisms (III.12.8.5) and (III.12.8.6). By (III.14.7.3) and (III.14.7.4), we can identify $\alpha_i^r$ with an isomorphism

$$\alpha_i^r \colon \top^{r+}(\mathscr{N})|\psi^*(U_i) \xrightarrow{\sim} \mathfrak{S}^r(\mathscr{M})|\psi^*(U_i). \tag{III.15.4.6}$$

It follows from the proof of III.12.26 that for every rational number $0 < r < \inf(r_i, r_j)$, we have in $\Xi^r_{U_{ij},\mathbb{Q}}$

$$\alpha_i^r|\psi^*(U_{ij}) = \alpha_j^r|\psi^*(U_{ij}). \tag{III.15.4.7}$$

By virtue of III.15.2, for every rational number $0 < r < \inf(r_i, i \in I)$, the isomorphisms $(\alpha_i^r)_{i\in I}$ glue to an isomorphism of $\Xi^r_{\mathbb{Q}}$

$$\alpha^r \colon \top^{r+}(\mathscr{N}) \xrightarrow{\sim} \mathfrak{S}^r(\mathscr{M}). \tag{III.15.4.8}$$

Consequently, $\mathscr{M}$ is Dolbeault.

**Proposition III.15.5.** *The following conditions are equivalent:*

(i) *The fibered category* (III.15.3.3)

(III.15.5.1) $$\mathbf{MOD}^{\mathrm{Dolb}}_{\mathbb{Q}}(\breve{\overline{\mathscr{B}}}) \to \mathbf{\acute{E}t}_{\mathrm{coh}/X}$$

*is a stack* (**[35]** II 1.2.1).

(ii) *For every covering* $(U_i \to U)_{i\in I}$ *of* $\mathbf{\acute{E}t}_{\mathrm{coh}/X}$, *denoting by* $\mathscr{U}$ *(resp. for every* $i \in I$, *by* $\mathscr{U}_i$*) the formal scheme p-adic completion of* $\overline{U}$ *(resp.* $\overline{U}_i$*), a Higgs* $\mathscr{O}_{\mathscr{U}}[\frac{1}{p}]$*-bundle* $\mathscr{N}$ *with coefficients in* $\xi^{-1}\widetilde{\Omega}^1_{\mathscr{U}/\mathscr{S}}$ *is solvable if and only if for every* $i \in I$, *the Higgs* $\mathscr{O}_{\mathscr{U}_i}[\frac{1}{p}]$*-bundle* $\mathscr{N} \otimes_{\mathscr{O}_{\mathscr{U}}} \mathscr{O}_{\mathscr{U}_i}$ *with coefficients in* $\xi^{-1}\widetilde{\Omega}^1_{\mathscr{U}_i/\mathscr{S}}$ *is solvable.*

Let $(U_i \to U)_{i\in I}$ be a covering of $\mathbf{\acute{E}t}_{\mathrm{coh}/X}$. For every $(i,j) \in I^2$, set $U_{ij} = U_i \times_X U_j$. Denote by $\mathscr{U}$ the formal scheme $p$-adic completion of $\overline{U}$ and by $\mathscr{H}^\star$ and $\mathscr{V}^\star$ the functors (III.12.18.1) and (III.12.23.1) associated with $(f_U, \widetilde{U}, \mathscr{M}_{\widetilde{X}}|\widetilde{U})$. For any $i \in I$, denote by $\mathscr{U}_i$ the formal scheme $p$-adic completion of $\overline{U}_i$ and by $\mathscr{H}_i$ and $\mathscr{V}_i$ the functors (III.12.18.1) and (III.12.23.1) associated with $(f_{U_i}, \widetilde{U}_i, \mathscr{M}_{\widetilde{X}}|\widetilde{U}_i)$.

Let us first show (i)⇒(ii). Let $\mathscr{N}$ be a Higgs $\mathscr{O}_{\mathscr{U}}[\frac{1}{p}]$-bundle with coefficients in $\xi^{-1}\widetilde{\Omega}^1_{\mathscr{U}/\mathscr{S}}$. If $\mathscr{N}$ is solvable, then for every $i \in I$, $\mathscr{N} \otimes_{\mathscr{O}_{\mathscr{U}}} \mathscr{O}_{\mathscr{U}_i}$ is solvable by III.14.9. Conversely, suppose that for every $i \in I$, $\mathscr{N} \otimes_{\mathscr{O}_{\mathscr{U}}} \mathscr{O}_{\mathscr{U}_i}$ is solvable, and let us show that $\mathscr{N}$ is solvable. For every $i \in I$, $\mathscr{M}_i = \mathscr{V}_i(\mathscr{N} \otimes_{\mathscr{O}_{\mathscr{U}}} \mathscr{O}_{\mathscr{U}_i})$ is a Dolbeault $\breve{\overline{\mathscr{B}}}_{\mathbb{Q}}|\psi^*(U_i)$-module. By III.14.11(ii), the canonical descent datum on the Higgs bundles $(\mathscr{N} \otimes_{\mathscr{O}_{\mathscr{U}}} \mathscr{O}_{\mathscr{U}_i})_{i\in I}$ relative to the étale covering $(\mathscr{U}_i \to \mathscr{U})_{i\in I}$ induces a descent datum $\delta$ on the Dolbeault modules $(\mathscr{M}_i)_{i\in I}$ relative to the covering $(U_i \to U)_{i\in I}$. Since the latter is effective by (i), there exists a Dolbeault $(\breve{\overline{\mathscr{B}}}_{\mathbb{Q}}|\psi^*(U))$-module $\mathscr{M}$ and for every $i \in I$, an isomorphism of $\breve{\overline{\mathscr{B}}}_{\mathbb{Q}}|\psi^*(U_i)$-modules

(III.15.5.2) $$\mathscr{M}|\psi^*(U_i) \xrightarrow{\sim} \mathscr{M}_i$$

that induce the descent datum $\delta$. By virtue of III.12.26 and III.14.11(i), we have a canonical isomorphism of Higgs $\mathscr{O}_{\mathscr{U}}[\frac{1}{p}]$-bundles $\mathscr{H}^\star(\mathscr{M}) \xrightarrow{\sim} \mathscr{N}$. Consequently, $\mathscr{N}$ is solvable.

Next, let us show (ii)⇒(i). For all $(\breve{\overline{\mathscr{B}}}_{\mathbb{Q}}|\psi^*(U))$-modules $\mathscr{M}$ and $\mathscr{M}'$, the diagram of maps of sets

(III.15.5.3) $$\begin{aligned}\mathrm{Hom}_{\breve{\overline{\mathscr{B}}}_{\mathbb{Q}}|\psi^*(U)}(\mathscr{M},\mathscr{M}') &\to \prod_{i\in I}\mathrm{Hom}_{\breve{\overline{\mathscr{B}}}_{\mathbb{Q}}|\psi^*(U_i)}(\mathscr{M}|\psi^*(U_i),\mathscr{M}'|\psi^*(U_i))\\ &\rightrightarrows \prod_{(i,j)\in I^2}\mathrm{Hom}_{\breve{\overline{\mathscr{B}}}_{\mathbb{Q}}|\psi^*(U_{ij})}(\mathscr{M}|\psi^*(U_{ij}),\mathscr{M}'|\psi^*(U_{ij}))\end{aligned}$$

is exact. Indeed, since $U$ is quasi-compact, we may assume that $I$ is finite, in which case the assertion follows from III.6.7.

For every $i \in I$, let $\mathscr{M}_i$ be a Dolbeault $(\breve{\overline{\mathscr{B}}}_{\mathbb{Q}}|\psi^*(U_i))$-module and let $\delta$ be a descent datum on $(\mathscr{M}_i)_{i\in I}$ relative to the covering $(U_i \to U)_{i\in I}$. Let us show that $\delta$ is effective. By assumption, for every $i \in I$, $\mathscr{N}_i = \mathscr{H}_i(\mathscr{M}_i)$ is a solvable Higgs $\mathscr{O}_{\mathscr{U}_i}[\frac{1}{p}]$-bundle with coefficients in $\xi^{-1}\widetilde{\Omega}^1_{\mathscr{U}_i/\mathscr{S}}$. In view of III.14.11(i), $\delta$ induces a descent datum $\gamma$ on the Higgs bundles $(\mathscr{N}_i)_{i\in I}$ relative to the étale covering $(\mathscr{U}_i \to \mathscr{U})_{\in I}$. Since the latter is effective by III.6.22, there exist a Higgs $\mathscr{O}_{\mathscr{U}}[\frac{1}{p}]$-bundle $\mathscr{N}$ and for every $i \in I$, an isomorphism of Higgs $\mathscr{O}_{\mathscr{U}_i}[\frac{1}{p}]$-modules

(III.15.5.4) $$\mathscr{N} \otimes_{\mathscr{O}_{\mathscr{U}}} \mathscr{O}_{\mathscr{U}_i} \xrightarrow{\sim} \mathscr{N}_i,$$

that induce the descent datum $\gamma$. By (ii), $\mathscr{N}$ is solvable. Consequently, $\mathscr{M} = \mathscr{V}^{\star}(\mathscr{N})$ is a Dolbeault $(\breve{\overline{\mathscr{B}}}_{\mathbb{Q}}|\psi^*(U))$-module. By III.12.26 and III.14.11(ii), for every $i \in I$, we have a canonical isomorphism of $(\breve{\overline{\mathscr{B}}}_{\mathbb{Q}}|\psi^*(U_i))$-modules $\mathscr{M}|\psi^*(U_i) \xrightarrow{\sim} \mathscr{M}_i$, that induce the descent datum $\delta$, proving the assertion.

**Definition III.15.6.** Let $(\mathscr{N}, \theta)$ be a Higgs $\mathscr{O}_{\mathfrak{X}}[\frac{1}{p}]$-bundle with coefficients in $\xi^{-1}\widetilde{\Omega}^1_{\mathfrak{X}/\mathscr{S}}$.

(i) We say that $(\mathscr{N}, \theta)$ is *small* if there exist a coherent sub-$\mathscr{O}_{\mathfrak{X}}$-module $\mathfrak{N}$ of $\mathscr{N}$ that generates it over $\mathscr{O}_{\mathfrak{X}}[\frac{1}{p}]$ and a rational number $\varepsilon > \frac{1}{p-1}$ such that

$$\theta(\mathfrak{N}) \subset p^{\varepsilon}\xi^{-1}\widetilde{\Omega}^1_{\mathfrak{X}/\mathscr{S}} \otimes_{\mathscr{O}_{\mathfrak{X}}} \mathfrak{N}. \tag{III.15.6.1}$$

(ii) We say that $(\mathscr{N}, \theta)$ is *locally small* if there exists an open covering $(U_i)_{i\in I}$ of $X_s$ such that for every $i \in I$, $(\mathscr{N}|U_i, \theta|U_i)$ is small.

**Remark III.15.7.** Suppose that $X$ is affine and that the $\mathscr{O}_X$-module $\widetilde{\Omega}^1_{X/S}$ is free of finite type. Set $R = \Gamma(X, \mathscr{O}_X)$ and $R_1 = R \otimes_{\mathscr{O}_K} \mathscr{O}_{\overline{K}}$ and denote by $\widehat{R}$ and $\widehat{R}_1$ their $p$-adic Hausdorff completions. Let $(\mathscr{N}, \theta)$ be a Higgs $\mathscr{O}_{\mathfrak{X}}[\frac{1}{p}]$-bundle with coefficients in $\xi^{-1}\widetilde{\Omega}^1_{\mathfrak{X}/\mathscr{S}}$. Set $N = \Gamma(\mathfrak{X}, \mathscr{N})$, which is a projective $\widehat{R}_1[\frac{1}{p}]$-module of finite type by III.6.17, and denote also by

$$\theta\colon N \to \xi^{-1}\widetilde{\Omega}^1_{X/S}(X) \otimes_R N \tag{III.15.7.1}$$

the Higgs $\widehat{R}_1[\frac{1}{p}]$-field induced by $\theta$. Then $(\mathscr{N}, \theta)$ is small if and only if there exist a sub-$\widehat{R}_1$-module of finite type $N^\circ$ of $N$ that generates it over $\widehat{R}_1[\frac{1}{p}]$ and a rational number $\varepsilon > \frac{1}{p-1}$ such that

$$\theta(N^\circ) \subset p^{\varepsilon}\xi^{-1}\widetilde{\Omega}^1_{X/S}(X) \otimes_R N^\circ. \tag{III.15.7.2}$$

Indeed, the condition is necessary by virtue of ([**1**] (2.10.5.1)) and it is sufficient in view of (III.12.5.3) and ([**1**] 1.10.2).

**Proposition III.15.8.** *Every solvable Higgs $\mathscr{O}_{\mathfrak{X}}[\frac{1}{p}]$-bundle $(\mathscr{N}, \theta)$ with coefficients in $\xi^{-1}\widetilde{\Omega}^1_{\mathfrak{X}/\mathscr{S}}$ is locally small.*

Indeed, we may restrict to the case where $X$ is affine and the $\mathscr{O}_X$-module $\widetilde{\Omega}^1_{X/S}$ is free of rank $d$ (III.14.9). Let us then show that $(\mathscr{N}, \theta)$ is small. Set $R = \Gamma(X, \mathscr{O}_X)$ and $R_1 = R \otimes_{\mathscr{O}_K} \mathscr{O}_{\overline{K}}$ and denote by $\widehat{R}$ and $\widehat{R}_1$ their $p$-adic Hausdorff completions. Let $\omega_1, \dots, \omega_d \in \Gamma(X, \xi^{-1}\widetilde{\Omega}^1_{X/S})$ be an $\mathscr{O}_X$-basis of $\xi^{-1}\widetilde{\Omega}^1_{X/S}$. Set $N = \Gamma(\mathfrak{X}, \mathscr{N})$, which is a projective $\widehat{R}_1[\frac{1}{p}]$-module of finite type by III.6.17, and denote also by

$$\theta\colon N \to \xi^{-1}\widetilde{\Omega}^1_{X/S}(X) \otimes_R N \tag{III.15.8.1}$$

the Higgs $\widehat{R}_1[\frac{1}{p}]$-field induced by $\theta$. We write

$$\theta = \sum_{i=1}^{d} \theta_i \otimes \omega_i, \tag{III.15.8.2}$$

where the $\theta_i$'s are $\widehat{R}_1[\frac{1}{p}]$-endomorphisms of $N$ that commute pairwise.

Let $r$ be a rational number $> 0$, $\mathscr{M}$ an object of $\mathbf{Mod}^{\mathrm{aft}}_{\mathbb{Q}}(\breve{\overline{\mathscr{B}}})$, and

$$\alpha\colon \top^{r+}(\mathscr{N}) \xrightarrow{\sim} \mathfrak{S}^r(\mathscr{M}) \tag{III.15.8.3}$$

an isomorphism of $\Xi^r_{\mathbb{Q}}$ (III.12.10). Let $(\overline{y} \rightsquigarrow \overline{x})$ be a point of $X_{\text{ét}} \overleftarrow{\times}_{X_{\text{ét}}} \overline{X}^{\circ}_{\text{ét}}$ (III.8.6) such that $\overline{x}$ lies over $s$ and $X'$ the strict localization of $X$ at $\overline{x}$. We take again the notation of III.12.27 and III.12.29; however, to emphasize the dependence on $\overline{x}$, we set $R_{\overline{x}} = \Gamma(X', \mathscr{O}_{X'})$ and $R_{\overline{x},1} = \Gamma(\overline{X}', \mathscr{O}_{\overline{X}'})$ and denote by $\widehat{R}_{\overline{x},1}$ the $p$-adic Hausdorff completion of $R_{\overline{x},1}$ (instead of $R'$, $R'_1$, and $\widehat{R}'_1$, respectively). By III.12.31, $\alpha$ induces a $\mathscr{C}^{\overline{y},(r)}_{X'}$-linear isomorphism

$$\mathscr{C}^{\overline{y},(r)}_{X'} \otimes_{\widehat{R}_{\overline{x},1}} \mathscr{N}_{\overline{x}} \stackrel{\sim}{\to} \mathscr{C}^{\overline{y},(r)}_{X'} \otimes_{\widehat{\overline{R}}^{\overline{y}}_{X'}} \mathscr{M}_{\rho(\overline{y} \rightsquigarrow \overline{x})} \tag{III.15.8.4}$$

of Higgs $\widehat{\overline{R}}^{\overline{y}}_{X'}$-modules with coefficients in $\xi^{-1}\widetilde{\Omega}^1_{X/S}(X')$ (III.12.29.1), where $\mathscr{N}_{\overline{x}}$ is endowed with the Higgs field $\theta_{\overline{x}}$ (III.12.29.2), $\mathscr{C}^{\overline{y},(r)}_{X'}$ is endowed with the Higgs field $p^r d_{\mathscr{C}^{\overline{y},(r)}_{X'}}$ (III.12.29.5), and $\mathscr{M}_{\rho(\overline{y} \rightsquigarrow \overline{x})}$ is endowed with the zero Higgs field (III.12.27.6).

Let $\mathscr{F}$ be a coherent $\mathscr{O}_{\mathfrak{X}}$-module such that $\mathscr{N} = \mathscr{F}_{\mathbb{Q}_p}$ (III.6.16), so that $N = \Gamma(\mathfrak{X}, \mathscr{F}) \otimes_{\mathbb{Z}_p} \mathbb{Q}_p$ ([**1**] (2.10.5.1)). By virtue of III.6.18, III.10.27(ii), and III.12.28(ii), we have a canonical isomorphism

$$\mathscr{N}_{\overline{x}} \stackrel{\sim}{\to} N \otimes_{\widehat{R}_1} \widehat{R}_{\overline{x},1}. \tag{III.15.8.5}$$

We will from now on identify these two modules. Endow the $\widehat{R}_1[\frac{1}{p}]$-module $N$ (resp. the $\widehat{R}_{\overline{x},1}[\frac{1}{p}]$-module $\mathscr{N}_{\overline{x}}$, resp. the $\widehat{\overline{R}}^{\overline{y}}_{X'}[\frac{1}{p}]$-module $\mathscr{N}_{\overline{x}} \otimes_{\widehat{R}_{\overline{x},1}} \widehat{\overline{R}}^{\overline{y}}_{X'}$) with the $p$-adic topology (II.2.2). Note that the $p$-adic topology on $\mathscr{N}_{\overline{x}}$ is induced by that on $\mathscr{N}_{\overline{x}} \otimes_{\widehat{R}_{\overline{x},1}} \widehat{\overline{R}}^{\overline{y}}_{X'}$. Indeed, since $\mathscr{N}_{\overline{x}}$ is projective of finite type over $\widehat{R}_{\overline{x},1}[\frac{1}{p}]$ (III.12.30), we may restrict ourselves to the case where $\mathscr{N}_{\overline{x}}$ is free of finite type, or even to the case where $\mathscr{N}_{\overline{x}} = \widehat{R}_{\overline{x},1}[\frac{1}{p}]$, for which the assertion has been established in III.10.27(iv).

For every $z \in N$ and every $\underline{n} = (n_1, \dots, n_d) \in \mathbb{N}^d$, $p^{-r|\underline{n}|}(\prod_{1 \le i \le d} \frac{1}{n_i!}\theta_i^{n_i})(z)$ tends to 0 in $\mathscr{N}_{\overline{x}} = N \otimes_{\widehat{R}_1} \widehat{R}_{\overline{x},1}$ when $|\underline{n}|$ tends to infinity. This follows from the isomorphism (III.15.8.4) by the same proof as II.13.24, in view of (III.10.25.6).

Let $\pi$ be a uniformizer of $\mathscr{O}_K$, $(x_i)_{i \in I}$ the generic points of $X_s$, and for every $i \in I$, let $\overline{x}_i$ be a generic point of $X$ localized at $x_i$. Since $X$ is $S$-flat and $X_s$ is reduced (III.4.2), for every integer $n \geq 1$, the canonical homomorphism

$$R/\pi^n R \to \prod_{i \in I} R_{\overline{x}_i}/\pi^n R_{\overline{x}_i} \tag{III.15.8.6}$$

is injective. We deduce from this that the canonical homomorphism

$$R_1/p^n R_1 \to \prod_{i \in I} R_{\overline{x}_i,1}/p^n R_{\overline{x}_i,1} \tag{III.15.8.7}$$

is injective. The homomorphism $\widehat{R}_1 \to \prod_{i \in I} \widehat{R}_{\overline{x}_i,1}$ is therefore injective. On the other hand, by ([**11**] Chapter III §2.11 Proposition 14 and Corollary 1), we have

$$\widehat{R}_1/p^n \widehat{R}_1 \simeq R_1/p^n R_1, \tag{III.15.8.8}$$

and likewise for the $R_{\overline{x}_i,1}$. We deduce from this that $\widehat{R}_1 \cap p^n(\oplus_{i \in I} \widehat{R}_{\overline{x}_i,1}) = p^n \widehat{R}_1$ and consequently that

$$\widehat{R}_1[\frac{1}{p}] \cap p^n(\oplus_{i \in I} \widehat{R}_{\overline{x}_i,1}) = p^n \widehat{R}_1. \tag{III.15.8.9}$$

Since the $\widehat{R}_1[\frac{1}{p}]$-module $N$ is projective of finite type, it follows that the $p$-adic topology on $N$ is induced by the product of the $p$-adic topologies on $\prod_{i \in I} N \otimes_{\widehat{R}_1} \widehat{R}_{\overline{x}_i,1}$.

It follows from the above that for every $z \in N$ and every $\underline{n} = (n_1, \dots, n_d) \in \mathbb{N}^d$,

$$p^{-r|\underline{n}|}\Big(\prod_{1 \leq i \leq d} \frac{1}{n_i!}\theta_i^{n_i}\Big)(z) \tag{III.15.8.10}$$

tends to 0 in $N$ when $|\underline{n}|$ tends to infinity. Let $N_0$ be a sub-$\widehat{R}_1$-module of finite type of $N$ that generates it over $\widehat{R}_1[\frac{1}{p}]$ and $\varepsilon$ a rational number such that $\frac{1}{p-1} < \varepsilon < r + \frac{1}{p-1}$. Since the sequence $p^{(r-\varepsilon)n} n!$ tends to 0 in $\mathscr{O}_C$ when $n$ tends to infinity, for every $z \in N$, $p^{-\varepsilon|\underline{n}|}(\prod_{1 \leq i \leq d} \theta_i^{n_i})(z)$ tends to 0 in $N$ when $|\underline{n}|$ tends to infinity. We can therefore consider the sub-$\widehat{R}_1$-module

$$N^\circ = \sum_{\underline{n} \in \mathbb{N}^d} p^{-\varepsilon|\underline{n}|}\Big(\prod_{1 \leq i \leq d} \theta_i^{n_i}\Big)(N_0) \tag{III.15.8.11}$$

of $N$. It is of finite type over $\widehat{R}_1$ and it generates $N$ over $\widehat{R}_1[\frac{1}{p}]$. Since we have

$$\theta(N^\circ) \subset p^\varepsilon \xi^{-1} \widetilde{\Omega}^1_{X/S}(X) \otimes_R N^\circ, \tag{III.15.8.12}$$

$(\mathscr{N}, \theta)$ is a small Higgs $\mathscr{O}_{\mathfrak{X}}[\frac{1}{p}]$-bundle with coefficients in $\xi^{-1}\widetilde{\Omega}^1_{\mathfrak{X}/\mathscr{S}}$ (III.15.7).

**Proposition III.15.9.** *Suppose that $X$ is an object of* **Q** (III.10.5), *in other words, that the following conditions are satisfied:*

(i) *$X$ is affine and connected;*
(ii) *$f\colon (X, \mathscr{M}_X) \to (S, \mathscr{M}_S)$ admits an adequate chart* (III.4.4);
(iii) *there exists a fine and saturated chart $M \to \Gamma(X, \mathscr{M}_X)$ for $(X, \mathscr{M}_X)$ inducing an isomorphism*

$$M \xrightarrow{\sim} \Gamma(X, \mathscr{M}_X)/\Gamma(X, \mathscr{O}_X^\times). \tag{III.15.9.1}$$

*Then every small Higgs $\mathscr{O}_{\mathfrak{X}}[\frac{1}{p}]$-bundle with coefficients in $\xi^{-1}\widetilde{\Omega}^1_{\mathfrak{X}/\mathscr{S}}$ is solvable.*

This statement is mentioned here as a reminder (III.13.9).

**Corollary III.15.10.** *If the conditions of* III.15.5 *are satisfied, every locally small Higgs $\mathscr{O}_{\mathfrak{X}}[\frac{1}{p}]$-bundle with coefficients in $\xi^{-1}\widetilde{\Omega}^1_{\mathfrak{X}/\mathscr{S}}$ is solvable.*

CHAPTER IV

# Cohomology of Higgs isocrystals

TAKESHI TSUJI

## IV.1. Introduction

In **[27]**, G. Faltings established a $p$-adic analogue of the Simpson correspondence between small Higgs bundles and small generalized representations or vector bundles on Faltings topos. In his theory and also in another approach by A. Abbes and M. Gros (cf. Chapters II and III), we need to assume the existence of a certain first infinitesimal smooth deformation of the relevant smooth variety and to choose and fix one such deformation to develop the theories. The purpose of this chapter is to remove this restriction by giving an interpretation of small Higgs bundles in a way similar to the interpretation of modules with integrable connections in terms of stratifications and crystals on crystalline sites (cf. **[3]**).

Let us recall the basic settings in the $p$-adic Simpson correspondence. Let $V$ be a complete discrete valuation ring of mixed characteristic $(0, p)$ whose residue field $k$ is algebraically closed, let $K$ be its field of fractions, and let $\overline{V}$ be the integral closure of $V$ in a fixed algebraic closure $\overline{K}$ of $K$. Following **[29]** and **[31]**, we define $R_{\overline{V}}$ to be the perfection $\varprojlim_{\mathrm{Frob}} \overline{V}/p\overline{V}$ and consider the ring of Witt vectors $A(\overline{V}) := W(R_{\overline{V}})$, which is endowed with a natural surjective homomorphism $\theta\colon W(R_{\overline{V}}) \to \widehat{\overline{V}}(:= \varprojlim_m \overline{V}/p^m\overline{V})$ characterized by $\theta([a]) = \lim_{m\to\infty} \widetilde{a}_m^{p^m}$. Here $a = (a_n)_{n\in\mathbb{N}} \in R_{\overline{V}}$, $\widetilde{a}_m$ denotes a lifting of $a_m \in \overline{V}/p\overline{V}$ in $\overline{V}$ and $[a]$ denotes the Teichmüller representative of $a$. It is known that the kernel of $\theta$ is generated by a nonzero divisor $\xi$. We define $A_N(\overline{V})$ to be $W(R_{\overline{V}})/\xi^N W(R_{\overline{V}})$.

Let $X$ be a smooth scheme separated of finite type over $V$, and we consider the $p$-adic formal scheme $X_1 := X \times_{\mathrm{Spec}(V)} \mathrm{Spf}(\widehat{\overline{V}})$. A (rational) Higgs bundle on $X_1$ which we consider is a locally finitely generated projective $\mathcal{O}_{X_1,\mathbb{Q}_p}(= \mathcal{O}_{X_1} \otimes_{\mathbb{Z}_p} \mathbb{Q}_p)$-module $\mathcal{M}$ endowed with an $\mathcal{O}_{X_1,\mathbb{Q}_p}$-linear homomorphism (called a Higgs field) $\theta\colon \mathcal{M} \to \mathcal{M} \otimes_{\mathcal{O}_{X_1}} \xi^{-1}\Omega^1_{X_1/\widehat{\overline{V}}}$ such that $\theta\wedge\theta = 0$, where $\xi^{-1}$ denotes $(\xi A_2(\overline{V}))^{-1}\otimes_{\widehat{\overline{V}}}$. In **[27]** and in Chapters II and III of this volume, a smooth lifting of $X_1$ over $A_2(\overline{V})$ is fixed. In fact, we will consider, more generally, a regular scheme $X$ flat separated of finite type over $V$ and a horizontal divisor $D$ on $X$ such that the union of $D$ and the special fiber $X_k$ of $X$ is a reduced divisor with normal crossings, and always work in the category of log schemes and log formal schemes giving $X$ the log structure defined by $D + X_k$. A Higgs field will take its values in the tensor product with the differential module with log poles. To simplify the exposition, we assume that $X$ is smooth over $V$ and ignore log structures in the introduction.[1]

[1]Note that the definition of the site $(X_1/A(\overline{V}))^r_{\mathrm{HIGGS}}$ without log structures is different from that with log structures because we work with big sites. We need to restrict to an object with a strict morphism $z\colon T_1 \to X_1$ to have the same one. It causes no problem except that the sites are functorial only for strict morphisms.

We start by explaining the idea of interpreting Higgs fields in terms of stratifications. Let $\mathscr{A}$ denote the category of $A(\overline{V})$-algebras $A$ endowed with an ideal $I$ containing $\xi A$, and let $\mathscr{A}'$ denote the full subcategory of $\mathscr{A}$ consisting of $(A, I)$ such that $I = \xi A$ and $A$ is $\xi$-torsion free. Then the inclusion functor $\mathscr{A}' \to \mathscr{A}$ has a left adjoint functor $D\colon \mathscr{A} \to \mathscr{A}'$, which is simply given by associating to $(A, I)$ the $A$-subalgebra $A[\frac{I}{\xi}]$ of $A[\frac{1}{\xi}]$ generated by $x\xi^{-1}$ $(x \in I)$ (a partial blow up along $I$). This construction, which we call Higgs envelope, plays the role of PD-envelopes for nilpotent Higgs fields as follows. Let $B$ be a smooth ring over $\overline{V}$ with coordinates $t_1, \dots, t_d \in B^\times$, and assume that we are given a smooth lifting $\widetilde{B}$ of $B$ over $A(\overline{V})$ and coordinates $\widetilde{t}_1, \dots, \widetilde{t}_d \in \widetilde{B}^\times$ which are liftings of $t_1, \dots, t_d$. Let $\widetilde{B}(\nu)$ $(\nu \in \mathbb{N})$ be the tensor product of $\nu + 1$ copies of $\widetilde{B}$ over $A(\overline{V})$, let $I_{\widetilde{B}(\nu)}$ be the kernel of $\widetilde{B}(\nu) \to B$, let $\widetilde{C}(\nu)$ be the image of $(\widetilde{B}(\nu), I_{\widetilde{B}(\nu)})$ under the functor $D$ above, and put $C(\nu) = \widetilde{C}(\nu)/\xi\widetilde{C}(\nu)$. Then, for either of the two $B$-algebra structures, we have an isomorphism $B[x_1, \dots, x_d] \cong C(1)$ defined by $x_i \mapsto ((\widetilde{t}_i \otimes 1 - 1 \otimes \widetilde{t}_i)\xi^{-1} \bmod \xi)$. We have a similar description of $C(2)$. This allows us to show that, defining a Higgs stratification on a $B$-module $M$ similarly as an HPD stratification using

$$B \rightleftarrows C(1) \mathrel{\substack{\rightarrow\\\leftarrow\\\rightarrow\\\leftarrow\\\rightarrow}} C(2),$$

the data of a Higgs stratification is equivalent to the data of a nilpotent Higgs field on $M$, i.e., a $B$-linear homomorphism $\theta\colon M \to M \otimes_B \xi^{-1}\Omega^1_{B/\overline{V}}$ such that $\theta \wedge \theta = 0$ and the endomorphisms $\theta_i$ $(1 \le i \le d)$ of $M$ defined by $\theta = \sum_{1 \le i \le d} \theta_i \otimes \xi^{-1} d\log(t_i)$ are nilpotent.

Working with $p$-adic formal schemes over $\mathrm{Spf}(A_N(\overline{V}))$, we can give a similar interpretation of a Higgs field on a locally finitely generated projective $\mathcal{O}_{X_1,\mathbb{Q}_p}$-module $\mathcal{M}$ satisfying the following convergence condition: For local coordinates $t_1, \dots, t_d$ and the endomorphism $\theta_i$ defined in the same way as above, we have

$$\prod_{1 \le i \le d} \frac{1}{n_i!} \theta_i^{n_i}(x) \to 0$$

as $\sum_{1 \le i \le d} n_i \to \infty$ for a local section $x$ of $\mathcal{M}$ (cf. the proof of Theorem IV.3.4.16). This condition is weaker than the condition "small" by Faltings, which will be necessary when we consider the cohomology later. Having the interpretation above, we are naturally lead to introducing an analogue $(X_1/A(\overline{V}))^\infty_{\mathrm{HIGGS}}$ of (big) crystalline sites, whose object is simply a pair $(T_\bullet, z)$ of a direct system consisting of $p$-adic formal schemes $T_N$ over $\mathrm{Spf}(A_N(\overline{V}))$ and closed immersions $T_N \to T_{N+1}$ over $\mathrm{Spf}(A_{N+1}(\overline{V}))$, and a morphism $z\colon T_1 \to X_1$ over $\mathrm{Spf}(\widehat{\overline{V}})$ such that $T_N$ are flat over $\mathrm{Spf}(\mathbb{Z}_p)$, the kernel of $\mathcal{O}_{T_N} \to \mathcal{O}_{T_n}$ $(N > n)$ is generated by $\xi^n$ and the multiplication by $\xi$ on $\mathcal{O}_{T_{N+1}}$ induces an injective morphism $\mathcal{O}_{T_N} \to \mathcal{O}_{T_{N+1}}$ (cf. Definition IV.3.1.1). A Higgs bundle satisfying the convergence condition above is interpreted as an "isocrystal" on the site $(X_1/A(\overline{V}))^\infty_{\mathrm{HIGGS}}$, which we call a Higgs isocrystal (cf. Definition IV.3.3.1), when we are given a compatible system of smooth liftings $X_N \to \mathrm{Spf}(A_N(\overline{V}))$ of $X_1$, which exists if $X$ is affine (cf. Theorem IV.3.4.16).

Next we explain the idea behind the interpretation of the $p$-adic Simpson correspondence by Faltings in terms of Higgs isocrystals. Suppose first that $X$ is connected and affine, has nonempty special fiber, and admits coordinates consisting of invertible functions. Let $A = \Gamma(X, \mathcal{O}_X)$, let $\overline{\mathcal{K}}$ be an algebraic closure containing $\overline{K}$ of the field of fractions $\mathcal{K} := \mathrm{Frac}(A)$ of $A$, let $\mathcal{K}^{\mathrm{ur}}$ denote the union of finite extensions $\mathcal{L}$ of $\mathcal{K}$ contained in $\overline{\mathcal{K}}$ such that the integral closure of $A \otimes_V K$ in $\mathcal{L}$ is étale, and let $\overline{A}$ denote the integral closure of $A$ in $\mathcal{K}^{\mathrm{ur}}$. Put $\overline{s} = \mathrm{Frac}(\overline{\mathcal{K}})$. Then $\overline{A}$ and its $p$-adic completion

$\widehat{\overline{A}}$ are naturally endowed with continuous actions of $\Delta_X := \pi_1(X_{\overline{K}}, \overline{s}) \cong \mathrm{Gal}(\mathcal{K}^{\mathrm{ur}}/\mathcal{K}\overline{K})$, where $X_{\overline{K}} = X \times_{\mathrm{Spec}(V)} \mathrm{Spec}(\overline{K})$. By an $\widehat{\overline{A}}_{\mathbb{Q}_p}$-representation of $\Delta_X$, we mean a finitely generated projective $\widehat{\overline{A}}_{\mathbb{Q}_p}$-module endowed with a continuous semi-linear action of $\Delta_X$. It is known, by the theory of almost étale extensions, that the absolute Frobenius of $\overline{A}/p\overline{A}$ is surjective (cf. Theorem IV.5.3.6 (2)), and one can apply the same construction as $A_N(\overline{V})$ to $\overline{A}$, obtaining $p$-adically complete $A_N(\overline{V})$-algebras $A_N(\overline{A})$ such that $A_1(\overline{A}) \cong \widehat{\overline{A}}$ (cf. IV.5.1). The direct system $\mathrm{Spf}(A_N(\overline{A}))$ with the natural morphism $\mathrm{Spf}(A_1(\overline{A})) = \mathrm{Spf}(\widehat{\overline{A}}) \to X_1$ becomes an object of the site $(X_1/A(\overline{V}))^{\infty}_{\mathrm{HIGGS}}$ endowed with an action of $\Delta_X$. Now, simply by evaluating Higgs isocrystals on this object, we obtain a functor from the category of Higgs isocrystals on $(X_1/A(\overline{V}))^{\infty}_{\mathrm{HIGGS}}$ (finite on $X_1$) to that of $\widehat{\overline{A}}_{\mathbb{Q}_p}$-representations of $\Delta_X$. We prove that this functor is fully faithful (cf. Theorem IV.5.3.3).

The continuity of the action of $\Delta_X$ on the above evaluation and the fully faithfulness of the functor follow from the description of the functor in terms of a "period ring" explained below (cf. Corollary IV.5.2.13). Choose a compatible system of smooth liftings $X_N$ of $X_1$ over $A_N(\overline{V})$, and let $\mathscr{D} = (\mathscr{D}_N)$ be the Higgs envelope of $X_1$ in $(X_N \times_{\mathrm{Spf}(A_N(\overline{V}))} \mathrm{Spf}(A_N(\overline{A})))_{N\in\mathbb{N}_{>0}}$, which is an object of $(X_1/A(\overline{V}))^{\infty}_{\mathrm{HIGGS}}$. Then the ring $\mathscr{A}_1(\overline{A}) = \Gamma(\mathscr{D}_1, \mathscr{O}_{\mathscr{D}_1})$ has both a continuous action of $\Delta_X$ (Proposition IV.5.2.6) and a Higgs field, and plays the role of "period ring" for the local $p$-adic Higgs correspondence as follows. For a Higgs isocrystal $\mathcal{F}$ (finite on $X_1$) and the Higgs bundle $(\mathcal{M}, \theta)$ corresponding to $\mathcal{F}$ via the lifting $(X_N)_N$, the $\widehat{\overline{A}}_{\mathbb{Q}_p}$-representation of $\Delta_X$ associated to $\mathcal{F}$ is given by $(\mathscr{A}_1(\overline{A}) \otimes_{\Gamma(X_1, \mathscr{O}_{X_1})} \Gamma(X_1, \mathcal{M}))^{\theta=0}$ (cf. Proposition IV.5.2.12). We also see that a Higgs bundle $(\mathcal{M}, \theta)$ is "admissible" with respect to the period ring $\mathscr{A}_1(\overline{A})$ if and only if $\theta$ satisfies the convergence condition above (cf. Proposition IV.5.3.10).

If we start with a general $X$ and a Higgs isocrystal on $(X_1/A(\overline{V}))^{\infty}_{\mathrm{HIGGS}}$, we obtain a compatible system of representations on étale affine connected $X$-schemes $U$ satisfying the conditions on $X$ in the previous paragraph. One can interpret the compatible system, as Faltings did in **[27]**, in terms of "vector bundles" on the Faltings site of $X$ when the Higgs isocrystal admits a suitable lattice globally (cf. IV.6.4). We prove that the functor thus obtained is fully faithful (Theorem IV.6.4.9).

For a small Higgs vector bundle $(\mathcal{M}, \theta)$ on $X_1$, Faltings also proved, in **[27]**, that the hypercohomology of the complex $(\mathcal{M} \otimes_{\mathscr{O}_{X_1}} \xi^{-\bullet}\Omega^{\bullet}_{X_1/\widehat{\overline{V}}}, \theta^{\bullet})$ is canonically isomorphic to the Galois cohomology of the corresponding representation on $X$ in the local case, and to the cohomology of the corresponding sheaf on the Faltings site of $X$ in the global case. It is natural to ask whether one can also interpret the cohomology of the above complex in terms of the site $(X_1/A(\overline{V}))^{\infty}_{\mathrm{HIGGS}}$ and prove a comparison theorem of cohomologies for our Higgs isocrystals. We give positive answers to them as follows.

Following an analogy with the comparison between crystalline cohomology and de Rham cohomology, one can construct a "linearization" of the complex above (when a compatible system of smooth liftings $X_N$ of $X_1$ over $A_N(\overline{V})$ is given). However the complex does not give a resolution of the corresponding Higgs isocrystal, i.e., an analogue of Poincaré lemma does not hold; a complex of the form

$$(B_1[x_1, \ldots, x_d]^{\wedge}_{\mathbb{Q}_p} \otimes_{\widehat{\overline{V}}} \xi^{-\bullet}\Omega^{\bullet}, \theta^{\bullet})$$

$$\Omega = \oplus_{1\leq i\leq d} \widehat{\overline{V}} d\log(t_i), \quad \Omega^q = \wedge^q\Omega, \quad \theta(x_i) = \xi^{-1} d\log(t_i)$$

appears for an affine object $(\mathrm{Spf}(B_N))_N$ of $(X_1/A(\overline{V}))^{\infty}_{\mathrm{HIGGS}}$ and it does not give a resolution of $B_1$ unless $d = 0$. Here $^\wedge$ denotes the $p$-adic completion. The $B_1$-algebra $B_1[x_1,\ldots,x_d]^\wedge$ in the above complex is obtained as the ring of the reduction mod $\xi$ of the Higgs envelope of $\mathrm{Spf}(B_1)$ in $(\mathrm{Spf}(B_N)\times_{\mathrm{Spf}(A_N(\overline{V}))} X_N)_N$. As it is well-known (cf. **[55]**), we obtain a resolution of $B_1$ if we replace $B_1[x_1,\ldots,x_d]^\wedge$ by the weak completion $B[x_1,\ldots,x_d]^\dagger$. With this fact in mind, we introduce the notion of Higgs envelope of level $r$ (cf. Definition IV.2.2.10) for $r \in \mathbb{N}_{>0}$, which gives a subring

$$\left\{\sum_{\underline{n}\in\mathbb{N}^d} a_{\underline{n}} \prod_{1\le i\le d} x_i^{n_i} \,\middle|\, a_{\underline{n}} \in B_{1,\mathbb{Q}_p},\ p^{-[\frac{\sum n_i}{r}]} a_{\underline{n}} \to 0\ \left(\sum n_i \to \infty\right)\right\}$$

of $B_1[x_1,\ldots,x_d]^\wedge_{\mathbb{Q}_p}$ in the above setting (cf. Proposition IV.2.3.17 and Lemma IV.2.3.14), whose union over $r$ is the weak completion tensored with $\mathbb{Q}_p$. Based on this notion, we further introduce an inverse system of sites $(X_1/A(\overline{V}))^r_{\mathrm{HIGGS}}$ $(r \in \mathbb{N})$ (cf. Definition IV.3.1.1) and the notion of a Higgs isocrystal on $(X_1/A(\overline{V}))^r_{\mathrm{HIGGS}}$ (cf. Definition IV.3.3.1), and give an interpretation of the above cohomology in terms of the cohomology of the inverse limit $(X_1/A(\overline{V}))_{\mathrm{HIGGS}}$ of the sites $(X_1/A(\overline{V}))^r_{\mathrm{HIGGS}}$ (cf. IV.4.5). If we are given a compatible system of smooth liftings $X_N$ of $X_1$ over $A_N(\overline{V})$, then the category of Higgs isocrystals on $(X_1/A(\overline{V}))^r_{\mathrm{HIGGS}}$ is equivalent to the category of Higgs bundles on $X_1$ satisfying the convergence condition

$$p^{-[\frac{\sum_i n_i}{r}]} \prod_{1\le i\le d} \frac{1}{n_i!}\theta_i^{n_i}(x) \to 0$$

(cf. Theorem IV.3.4.16). A Higgs bundle $(\mathcal{M},\theta)$ on $X_1$ is small in the sense of Faltings (cf. Definition IV.3.6.5) if and only if $\mathcal{M}$ admits a "lattice" globally (as an $\mathcal{O}_{X_1,\mathbb{Q}_p}$-module) and satisfies the convergence condition above for some $r \in \mathbb{N}_{>0}$; the existence of a "lattice" always holds if $X$ is affine and $\mathcal{M}$ is finitely generated and projective (cf. Proposition IV.3.6.6 and Corollary IV.3.6.4). For a Higgs isocrystal $\mathcal{F}$ on $(X_1/A(\overline{V}))^r_{\mathrm{HIGGS}}$, the corresponding Higgs bundle $(\mathcal{M},\theta)$ on $X_1$ and the inverse image $\mathcal{F}^\dagger$ of $\mathcal{F}$ on $(X_1/A(\overline{V}))_{\mathrm{HIGGS}}$, we prove that the analogue of linearization gives a resolution of $\mathcal{F}^\dagger$ and that the derived direct image of $\mathcal{F}^\dagger$ under the canonical morphism of topos to $(X_1)^\sim_{\text{ét}}$ is canonically isomorphic to the complex $\mathcal{M}\otimes_{\mathcal{O}_{X_1}} \xi^{-\bullet}\Omega^\bullet_{X_1/\widehat{\overline{V}}}$ (cf. Corollary IV.4.5.8).

We also prove that, for a Higgs isocrystal $\mathcal{F}$ on $(X_1/A(\overline{V}))^r_{\mathrm{HIGGS}}$ admitting a global lattice of its pull-back $\mathcal{F}^\infty$ on $(X_1/A(\overline{V}))^\infty_{\mathrm{HIGGS}}$ (cf. Definition IV.3.5.4), the cohomology of the pull-back $\mathcal{F}^\dagger$ of $\mathcal{F}$ on $(X_1/A(\overline{V}))_{\mathrm{HIGGS}}$ is canonically isomorphic to the cohomology of the sheaf on the Faltings site of $X$ corresponding to $\mathcal{F}$. The key local ingredients in the proof are as follows. One can define a period ring $\mathscr{A}_1^s(\overline{A})$ of level $s$ similarly as $\mathscr{A}_1(\overline{A})$ using the Higgs envelope of level $s$. Then, for the $\widehat{\overline{A}}_{\mathbb{Q}_p}$-representation $V$ of $\Delta_X$ corresponding to $\mathcal{F}$, $M = \Gamma(X_1,\mathcal{M})$ and $A_1 = \Gamma(X_1,\mathcal{O}_{X_1})$, we obtain a direct system of complexes of $\Delta_X$-modules

$$(V \to \xi^{-\bullet} M \otimes_{A_1} \mathscr{A}_1^s(\overline{A}) \otimes_A \Omega^\bullet_A)_{s\ge r}$$

such that the transition morphisms are homotopic to zero as morphism of complexes of topological modules (cf. the proof of Proposition IV.5.2.15). On the other hand, one can show, by using the theory of almost étale extensions by Faltings (cf. Theorem IV.5.3.4), that

$$\varinjlim_s H^q_{\mathrm{cont}}(\Delta_X, \mathscr{A}_1^s(\overline{A})_{\mathbb{Q}_p}) = \begin{cases} (A_1)_{\mathbb{Q}_p} & (\text{if } q = 0),\\ 0 & (\text{if } q > 0),\end{cases}$$

which is an analogue of the computation of Galois cohomology of the period ring for Hodge-Tate representations by O. Hyodo in **[43]** (1.2).

In IV.2, we develop a general theory of Higgs envelopes. We define Higgs envelopes of level $r$ ($r \in \mathbb{N}_{>0} \cup \{\infty\}$) and prove their existence for $p$-adically complete algebras in IV.2.1 and then for $p$-adic fine log formal schemes in IV.2.2. In IV.2.3, we give an explicit description of Higgs envelopes for power series rings and sections of smooth morphisms. In IV.2.4, we study the relation between the differential module and the Higgs envelope of the diagonal immersion for a smooth morphism of $p$-adic fine log formal schemes. In IV.2.5, we study the relation between Higgs envelopes and torsors of deformations (which play a key role in the approach by Abbes and Gros in Chapters II and III) under a certain general setting.

In IV.3, we introduce the notion of Higgs isocrystals of level $r$ ($r \in \mathbb{N}_{>0} \cup \{\infty\}$) and the sites on which they live. We define the site $(X/B)^r_{\mathrm{HIGGS}}$ in IV.3.1 for $r \in \mathbb{N}_{>0} \cup \{\infty\}$. After preliminaries on finitely generated and projective $\mathcal{O}_{X,\mathbb{Q}_p}$-modules and their lattices on a $p$-adic formal scheme $X$ in IV.3.2, we define in IV.3.3 Higgs isocrystals of level $r$ and also Higgs crystals, which provide us the notion of lattices on Higgs isocrystals in IV.3.5. In IV.3.4, we give a description of a Higgs isocrystal and also of a Higgs crystal in terms of a (generalized) Higgs vector bundle on the Higgs envelope of an embedding of $X$ into a smooth $p$-adic fine log formal scheme (similarly as the case of crystalline sites). In IV.3.6, we compare our convergence conditions on Higgs fields with the condition "small" (divisibility by a certain rational power of $p$) by Faltings.

In IV.4, we study the relation between the cohomology of Higgs isocrystals of finite level and that of the corresponding (generalized) Higgs vector bundles. We first define the projection morphism of topos to the étale topos in IV.4.1 in the same way as crystalline sites. In IV.4.2, we introduce the notion of linearizations for our site, and prove that the linearization of a finitely generated projective module is acyclic under the projection to the étale site. We prove a Poincaré lemma for Higgs isocrystals of finite level in IV.4.3, and after preliminaries on the coherence of relevant topos and morphisms of topos in IV.4.4, which are necessary for the computation of the cohomology of the inverse limit of the sites $(X/B)^r_{\mathrm{HIGGS}}$ ($r \in \mathbb{N}$), we prove a comparison theorem of cohomologies in IV.4.5.

We discuss the local $p$-adic Simpson correspondence in IV.5. After preliminaries on the log structure and the continuity of the action of Galois group for the ring $A_N(\overline{A}) = W(R_{\overline{A}})/\xi^N W(R_{\overline{A}})$ associated to a certain kind of affine fs (=fine and saturated) log scheme in IV.5.1, we construct a functor from the category of Higgs isocrystals to that of $\widehat{\overline{A}}_{\mathbb{Q}_p}$-representations for a sufficiently small affine fs log scheme $U$ whose underlying scheme is normal and of finite type over $V$ and has reduced special fiber. We also construct a canonical homomorphism from the cohomology of a Higgs isocrystal of finite level to the continuous Galois cohomology of the corresponding $\widehat{\overline{A}}_{\mathbb{Q}_p}$-representation. In IV.5.3, we prove that the functor is fully faithful and the comparison homomorphism of cohomologies is an isomorphism in the case where $U$ has semi-stable reduction. We also study the admissibility of Higgs bundles with respect to the period ring associated to a compatible system of smooth liftings $(U_N)$ of $U_1$. In IV.5.4, we compare our "period rings" with the Higgs-Tate algebras introduced in II.10.3 and II.12.1.

In the final section IV.6, we study the global $p$-adic Simpson correspondence. A sheaf on our Faltings site of an fs log scheme $X$ of finite type over $V$ is equivalent to the data assigning to each strict étale affine log scheme $U$ over $X$ a sheaf on the finite étale site of the log trivial locus of the geometric generic fiber $U_{\overline{K}} = U \times_{\mathrm{Spec}(K)} \mathrm{Spec}(\overline{K})$ of $U$ satisfying a certain gluing condition for strict étale coverings $(U_\alpha \to U)_\alpha$. We restrict ourselves to affine $U$ because we also work with two variants simultaneously where $U_{\overline{K}}$ is replaced by the geometric generic fiber of the Spec of the $p$-adic henselization or $p$-adic completion of $\Gamma(U, \mathcal{O}_U)$. We then assume that the underlying scheme of $X$ is normal, separated of finite type over $V$ and has reduced special fiber. After preliminaries on Faltings sites in IV.6.1, IV.6.2, and IV.6.3, we construct a functor from the category of Higgs crystals to that of sheaves on the Faltings site of $X$ and prove its fully faithfulness (up to isogeny) in the semi-stable reduction case in IV.6.4. After some preliminaries on the cohomology of the projection to the étale site in IV.6.5 and inverse systems and direct systems of sheaves in IV.6.6, we construct a comparison homomorphism from the cohomology of a Higgs isocrystal of finite level (admitting a lattice) to that of the corresponding sheaf on the Faltings site in IV.6.7 and prove that it is an isomorphism in the semi-stable reduction case in IV.6.8.

**Acknowledgments.** The author learned the work on a $p$-adic Simpson correspondence by G. Faltings **[27]** and obtained the idea of Higgs (iso)crystals and the proof of the Galois acyclicity of the "overconvergent period ring" **[75]** through his partial participation (in January 2009) in the workshop on the paper **[27]** held at the University of Rennes I in 2008-2009. He was away from the topic for a few years after the workshop, and then started to work on it again in the occasion of the preparation of his talk at the summer school: Higgs bundles on $p$-adic curves and representation theory, held at Mainz University in 2012; he introduced the notion of Higgs envelopes and Higgs isocrystals of finite level and established a theory interpreting the cohomology of Higgs bundles in terms of a certain site, which is the main theme of this chapter. The author would like to thank the organizers of the workshop and the summer school and, in particular, Ahmed Abbes, who kindly invited me to both of them. Without these two occasions, the works in this chapter would never have been accomplished. The project of this book was initiated by Ahmed Abbes and Michel Gros, and later they kindly invited the author to join them by writing the aforementioned results. Some of the results in this chapter were obtained through his discussions with them during his stay at IHES (Institut Hautes Études Scientifiques) in September 2013 and March 2014. The author would like to express his sincere gratitude to Ahmed Abbes and Michel Gros for their kindness and constant support for writing this article, and to IHES for its hospitality. The work of this chapter was financially supported by JSPS Grants-in-Aid for Scientific Research, Grant Number 24540009.

**Notation.** Throughout this chapter, we fix a universe $\mathscr{U}$ and a universe $\mathscr{V}$ such that $\mathscr{U} \in \mathscr{V}$, and we only consider groups, rings, schemes, formal schemes, and log schemes which belong to $\mathscr{U}$. By a topos (resp. a fibered topos), we mean a $\mathscr{V}$-topos (resp. a fibered $\mathscr{V}$-topos). For a sheaf of rings $\mathcal{O}$ (resp. an inverse system of sheaves of rings $(\mathcal{O}_m)_{m \in \mathbb{N}}$ on a site $\mathcal{C}$), we write $\mathcal{M}od(\mathcal{C}, \mathcal{O})$ (resp. $\mathcal{M}od(\mathcal{C}, \mathcal{O}_\bullet)$) for the category of sheaves of $\mathcal{O}$-modules (resp. inverse systems of sheaves of $\mathcal{O}_m$-modules).

For abelian groups, rings, sheaves of abelian groups, and sheaves of rings, the subscript $\mathbb{Q}$ means $\otimes_{\mathbb{Z}} \mathbb{Q}$. Similarly the subscript $\mathbb{Q}_p$ means $\otimes_{\mathbb{Z}_p} \mathbb{Q}_p$ for $\mathbb{Z}_p$-modules, $\mathbb{Z}_p$-algebras, etc. We say that a log scheme is affine if its underlying scheme is affine. We say that a morphism of log schemes is affine (resp. of finite type) if its underlying morphism of schemes is affine (resp. of finite type).

In IV.2–IV.4, we fix a $\mathbb{Z}_p$-algebra $R$ with a principal ideal $I$ generated by a regular element such that $R/I$ is $p$-torsion free and $p$-adically complete and separated, which implies that, for every $N \in \mathbb{N}_{>0}$, $R/I^N$ is $p$-torsion free and $p$-adically complete and separated. For $N \in \mathbb{N}_{>0}$, let $R_N$ denote $R/I^N$ and let $S_N$ denote the affine formal scheme associated to $R_N$ with the $p$-adic topology. Let $\xi$ be a generator of $I$.

## IV.2. Higgs envelopes

**IV.2.1. $p$-adically complete rings.** In this subsection, we prove the existence of the Higgs envelope of level $r$ ($r \in \mathbb{N}_{>0} \cup \{\infty\}$) for an inverse system of $p$-adically complete and separated $R_N$-algebras with surjective transition maps (Proposition IV.2.1.8, Definition IV.2.1.9).

**Definition IV.2.1.1.** (1) We define $\mathscr{A}$ to be the category of $R$-algebras with a decreasing filtration $F^nA \subset A$ ($n \in \mathbb{N}$) by ideals such that $F^0A = A$ and $\xi^n A \subset F^n A$ for every $n \in \mathbb{N}$. A morphism in $\mathscr{A}$ is a homomorphism of $R$-algebras compatible with the filtrations.

(2) For $r \in \mathbb{N}_{>0}$ (resp. $r = \infty$), we define $\mathscr{A}^r_{\mathrm{alg}}$ to be the full subcategory of $\mathscr{A}$ consisting of $(A, F^nA) \in \mathrm{Ob}\,\mathscr{A}$ satisfying the following conditions:

(i) The rings $A$ and $A/F^nA$ ($n \in \mathbb{N}_{>0}$) are $p$-torsion free.

(ii) The ring $A$ is $\xi$-torsion free.

(iii) For every $s \in \mathbb{N} \cap [0, r]$ (resp. $s \in \mathbb{N}$), we have $pF^sA \subset \xi^s A$ (resp. $F^sA = \xi^s A$).

(iv) The inverse image of $F^nA$ under $\xi\colon A \to A$ is $F^{n-1}A$ for every $n \in \mathbb{N}_{>0}$.

(3) For $r \in \mathbb{N}_{>0} \cup \{\infty\}$, we define $\mathscr{A}^r_p$ to be the full subcategory of $\mathscr{A}^r_{\mathrm{alg}}$ consisting of $(A, F^nA) \in \mathrm{Ob}\,\mathscr{A}^r_{\mathrm{alg}}$ satisfying the following condition:

(i) The rings $A$ and $A/F^nA$ ($n \in \mathbb{N}_{>0}$) are $p$-adically complete and separated.

(4) For $r \in \mathbb{N}_{>0} \cup \{\infty\}$, we define $\mathscr{A}^r$ to be the full subcategory of $\mathscr{A}^r_p$ consisting of $(A, F^nA) \in \mathrm{Ob}\,\mathscr{A}^r_p$ satisfying the following condition:

(i) The natural homomorphism $A \to \varprojlim_n A/F^nA$ is an isomorphism.

For an object $(A, F^nA)$ of $\mathscr{A}^r_{\mathrm{alg}}$ and $n \in \mathbb{N}_{>0}$, the multiplication by $\xi$ induces an injective homomorphism $A/F^{n-1}A \to A/F^nA$, which is denoted by $[\xi]$ in the following.

**Proposition IV.2.1.2.** *For every $r \in \mathbb{N}_{>0} \cup \{\infty\}$, the inclusion functors $\mathscr{A}^r \to \mathscr{A}^r_p \to \mathscr{A}^r_{\mathrm{alg}} \to \mathscr{A}$ have left adjoint functors.*

**Lemma IV.2.1.3.** *Let $M$ be a module and let $\widehat{M}$ be $\varprojlim_m M/p^mM$.*

(1) *Assume that $M$ is $p$-torsion free. Then $\widehat{M}$ is $p$-torsion free, the projection $\widehat{M} \to M/p^mM$ induces an isomorphism $\widehat{M}/p^m\widehat{M} \xrightarrow{\cong} M/p^mM$ for every $m \in \mathbb{N}$, and the natural homomorphism $\widehat{M} \to \varprojlim_m \widehat{M}/p^m\widehat{M}$ is an isomorphism.*

(2) *Let $N \subset M$ be a submodule such that the quotient $L := M/N$ is $p$-torsion free, and let $\widehat{N}$ (resp. $\widehat{L}$) be $\varprojlim_m N/p^mN$ (resp. $\varprojlim_m L/p^mL$). Then the sequence $0 \to \widehat{N} \to \widehat{M} \to \widehat{L} \to 0$ is exact.*

Proof. Since $M$ is $p$-torsion free, the homomorphism $p^m\colon M \to M$ induces injective homomorphisms $M/p^NM \to M/p^{m+N}M$ whose cokernel is isomorphic to $M/p^mM$. By taking $\varprojlim_N$, we obtain an exact sequence $0 \to \widehat{M} \xrightarrow{p^m} \widehat{M} \to M/p^mM \to 0$, which implies (1). The claim (2) is obtained by taking $\varprojlim_m$ of the exact sequences $0 \to N/p^mN \to M/p^mM \to L/p^mL \to 0$. $\square$

**Lemma IV.2.1.4.** *Let $r \in \mathbb{N}_{>0}$. Any object $(A, F^nA)$ of $\mathscr{A}^r_{\mathrm{alg}}$ has the following properties:*

(1) *Let $n \in \mathbb{N}$ and let $m$ be the smallest integer satisfying $nr^{-1} \leq m$. Then we have $p^mF^nA \subset \xi^nA \subset F^nA$. In particular, we have $F^nA_{\mathbb{Q}} = \xi^nA_{\mathbb{Q}}$.*

(2) *For any $n, n' \in \mathbb{N}$, we have $F^nA \cdot F^{n'}A \subset F^{n+n'}A$.*

Proof. (1) It is enough to prove $pF^{n+s}A \subset \xi^sF^nA$ for $n \in \mathbb{N}$ and $s \in \mathbb{N} \cap [0, r]$. Let $a \in pF^{n+s}A$. Then $a \in pF^sA \subset \xi^sA$. Since the inverse image of $F^{n+s}A$ by $\xi^s\colon A \to A$ is $F^nA$, we obtain $a \in \xi^sF^nA$.

(2) Since $A$ and $A/F^nA$ are $p$-torsion free, the last equality of (1) implies $A\cap\xi^nA_{\mathbb{Q}} = F^nA$. This immediately implies the claim. □

PROOF OF PROPOSITION IV.2.1.2. Left adjoint of $\mathscr{A}^r_{\mathrm{alg}} \to \mathscr{A}$: Let $(A, F^nA)$ be an object of $\mathscr{A}$. Let $B$ be the $A$-subalgebra of $A[\frac{1}{p\xi}]$ generated by $p^{-1}$ and $a\xi^{-n}$ ($n \in \mathbb{N}_{>0}, a \in F^nA$), and define the decreasing filtration $F^nB$ ($n \in \mathbb{N}$) to be $\xi^nB$. The natural homomorphism $A \to B$ is compatible with the filtrations. For an $A$-subalgebra $A'$ of $B$, we define the decreasing filtration $F^nA'$ by ideals to be $A' \cap F^nB$. Then $(A', F^nA')$ is an object of $\mathscr{A}$ and satisfies the conditions (i), (ii), (iv) in the definition of $\mathscr{A}^r_{\mathrm{alg}}$ because $p \in B^\times$, $\xi$ is regular in $B$, and the homomorphism $A'/F^nA' \to B/F^nB$ is injective. Let $\mathcal{S}^r$ be the set of all $A$-subalgebras such that $pF^sA' \subset \xi^sA'$ for every $s \in \mathbb{N}\cap[0,r]$ if $r \in \mathbb{N}_{>0}$, and $F^sA' = \xi^sA'$ for every $s \in \mathbb{N}$ if $r = \infty$. The set $\mathcal{S}^r$ contains $B$, and any $A' \in \mathcal{S}^r$ with the filtration $F^nA'$ is an object of $\mathscr{A}^r_{\mathrm{alg}}$. Let $A_0$ be the intersection $\cap_{A'\in\mathcal{S}^r} A'$. Since $p$ and $\xi$ are regular in $B$, we have $\xi^sA_0 = \cap\xi^sA'$, $F^sA_0 = \cap F^sA'$, and $pF^sA_0 = \cap pF^sA'$. Hence $A_0 \in \mathcal{S}^r$. We assert that $(A_0, F^nA_0) \in \mathrm{Ob}\,\mathscr{A}^r_{\mathrm{alg}}$ with the morphism $(A, F^nA) \to (A_0, F^nA_0)$ in $\mathscr{A}$ satisfies the desired universal property, i.e., any morphism $f\colon (A, F^nA) \to (C, F^nC)$ in $\mathscr{A}$ with $(C, F^nC) \in \mathrm{Ob}\,\mathscr{A}^r_{\mathrm{alg}}$ uniquely factors as $(A, F^nA) \to (A_0, F^nA_0) \xrightarrow{g} (C, F^nC)$. Let us prove the existence of $g$. We note that the natural homomorphisms $C \to C[\frac{1}{p}] \to C[\frac{1}{p\xi}]$ are injective. Let $\widetilde{f}\colon A[\frac{1}{p\xi}] \to C[\frac{1}{p\xi}]$ be the homomorphism induced by $f$. Then $F^nC[\frac{1}{p}] = \xi^nC[\frac{1}{p}]$ (cf. Lemma IV.2.1.4 (1)) implies $\widetilde{f}(B) \subset C[\frac{1}{p}]$. Let $f_p$ be the homomorphism $B \to C[\frac{1}{p}]$ induced by $\widetilde{f}$. Let $A'$ be the $A$-subalgebra $f_p^{-1}(C)$ of $B$. For $n \in \mathbb{N}$, we have $f_p(F^nA') \subset f_p(A') \cap f_p(\xi^nB) \subset C \cap \xi^nC[\frac{1}{p}] = C \cap F^nC[\frac{1}{p}] = F^nC$. We claim that $A'$ is contained in $\mathcal{S}^r$, which implies that $f_p(A_0) \subset C$ and $f_p$ induces the desired morphism $g$. If $r \in \mathbb{N}_{>0}$ (resp. $r = \infty$), we have $f_p(pF^sA') \subset pF^sC \subset \xi^sC$ for all $s \in \mathbb{N}\cap[0,r]$ (resp. $f_p(F^sA') \subset F^sC = \xi^sC$ for all $s \in \mathbb{N}$). For any $x \in F^sB = \xi^sB$, $f_p(x) \in \xi^sC$ implies $x \in \xi^sA'$ because $\xi$ is regular in $C[\frac{1}{p}]$. Hence $(A', F^nA') \in \mathcal{S}^r$. The uniqueness of $g$ follows from $A[\frac{1}{p\xi}] \xrightarrow{\cong} A_0[\frac{1}{p\xi}]$ and the injectivity of $C \to C[\frac{1}{p\xi}]$.

Left adjoint of $\mathscr{A}^r_p \to \mathscr{A}^r_{\mathrm{alg}}$: For a module $M$, let $M[p^m]$ (resp. $\widehat{M}$) denote the kernel of $p^m\colon M \to M$ (resp. $\varprojlim_m M/p^mM$). Let $(A, F^nA)$ be an object of $\mathscr{A}^r_{\mathrm{alg}}$. We give the $R$-algebra $\widehat{A}$ the decreasing filtration $F^n\widehat{A}$ ($n \in \mathbb{N}$) by ideals defined by the image of the natural homomorphism $\widehat{F^nA} \to \widehat{A}$. We assert that $(\widehat{A}, F^n\widehat{A})$ is an object of $\mathscr{A}^r_p$ and has the desired universal property. By Lemma IV.2.1.3 (2), the homomorphism $\widehat{F^nA} \to F^n\widehat{A}$ is an isomorphism and the quotient $\widehat{A}/F^n\widehat{A}$ is isomorphic to $\widehat{A/F^nA}$. The latter implies $\xi^n\widehat{A} \subset F^n\widehat{A}$. Applying Lemma IV.2.1.3 (1) to $A$ and $A/F^nA$, we see that $\widehat{A}$ and $\widehat{A}/F^n\widehat{A}$ are $p$-torsion free and $p$-adically complete and separated. Since $A/F^1A$ is $p$-torsion free and $p(F^1A/\xi A) = 0$, we have $(A/\xi A)[p^m] = (A/\xi A)[p]$ and $(A/(F^nA + \xi A))[p^m] = (A/(F^nA + \xi A))[p]$ for $m \in \mathbb{N}_{>0}$. This implies that the image of the kernel of $\xi\colon A/p^{m+1}A \to A/p^{m+1}A$ (resp. $[\xi] \mod p^{m+1} : (A/F^{n-1}A)/p^{m+1} \to (A/F^nA)/p^{m+1}$) in $A/p^mA$ (resp. $(A/F^{n-1}A)/p^m$) is 0. By taking the inverse limit over $m$, we see that the homomorphism $\xi\colon \widehat{A} \to \widehat{A}$ is injective and induces an injective homomorphism $\widehat{A}/F^{n-1}\widehat{A} \to \widehat{A}/F^n\widehat{A}$. If $r \in \mathbb{N}_{>0}$ (resp. $r = \infty$), taking the $p$-adic completion of $F^sA \xrightarrow[\cong]{p} pF^sA \subset \xi^sA \xleftarrow[\cong]{\xi^s} A$ ($s \in \mathbb{N}\cap[0,r]$) (resp. $A \xrightarrow[\cong]{\xi^s} F^sA$ ($s \in \mathbb{N}$)), we obtain $pF^s\widehat{A} \subset \xi^s\widehat{A}$ (resp. $F^s\widehat{A} = \xi^s\widehat{A}$). For any morphism $f\colon (A, F^nA) \to (B, F^nB)$ in $\mathscr{A}^r_{\mathrm{alg}}$ with $(B, F^nB) \in \mathrm{Ob}\,\mathscr{A}^r_p$, a factorization through $(\widehat{A}, F^n\widehat{A})$ is obtained by taking the

$p$-adic completion of $f$ and using $B \cong \varprojlim_m B/p^m B$ and $F^n B \cong \varprojlim_m F^n B/p^m F^n B$ ($n \in \mathbb{N}$) (cf. Lemma IV.2.1.3 (2) for the last isomorphism). The uniqueness of the factorization follows from $A/p^m A \xrightarrow{\cong} \widehat{A}/p^m \widehat{A}$ (cf. Lemma IV.2.1.3 (1)) and $B \xrightarrow{\cong} \varprojlim_m B/p^m B$.

Left adjoint of $\mathscr{A}^r \to \mathscr{A}_p^r$. Let $(A, F^n A)$ be an object of $\mathscr{A}_p^r$. Let $B$ be the $A$-algebra $\varprojlim_N A/F^N A$ endowed with the decreasing filtration by ideals $F^n B = \varprojlim_N F^n A/F^N A$. By taking the inverse limit of the exact sequence $0 \to F^n A/F^N A \to A/F^N A \to A/F^n A \to 0$, we obtain an isomorphism $B/F^n B \xrightarrow{\cong} A/F^n A$. This implies $(B, F^n B) \in \operatorname{Ob} \mathscr{A}$. We assert that $(B, F^n B) \in \operatorname{Ob} \mathscr{A}^r$ and the morphism $(A, F^n A) \to (B, F^n B)$ in $\mathscr{A}_p^r$ satisfies the desired universal property. Since $A/F^N A$ ($N \in \mathbb{N}$) are $p$-torsion free, $B$ and $B/F^n B$ are $p$-torsion free. By taking the inverse limit of the injective homomorphisms $[\xi]\colon A/F^N A \to A/F^{N+1} A$ ($N \in \mathbb{N}_{>0}$), we see that $\xi$ is regular in $B$. If $r \in \mathbb{N}_{>0}$ (resp. $r = \infty$), taking the inverse limit of $F^s A/F^N A \xrightarrow[\cong]{p} p(F^s A/F^N A) \subset \xi^s(A/F^N A) \xleftarrow[\cong]{[\xi]^s} A/F^{N-s} A$ for $s \in \mathbb{N} \cap [0, r]$ (resp. $F^s A/F^N A = \xi^s(A/F^N A) \xleftarrow[\cong]{[\xi]^s} A/F^{N-s} A$ for $s \in \mathbb{N}$), we obtain $pF^s B \subset \xi^s B$ (resp. $F^s B = \xi^s B$). The exact sequences $0 \to A/F^N A \xrightarrow{p^m} A/F^N A \to (A/F^N A)/p^m \to 0$ induce an isomorphism $B/p^m B \xrightarrow{\cong} \varprojlim_N (A/F^N A)/p^m$. Taking $\varprojlim_m$, we obtain $\varprojlim_m B/p^m B \cong \varprojlim_m \varprojlim_N (A/F^N A)/p^m \cong \varprojlim_N \varprojlim_m (A/F^N A)/p^m \cong \varprojlim_N A/F^N A = B$. The remaining conditions are verified using $B/F^n B \cong A/F^n A$. For any morphism $f\colon (A, F^n A) \to (C, F^n C)$ in $\mathscr{A}_p^r$ with $(C, F^n C) \in \operatorname{Ob} \mathscr{A}^r$, a factorization through $(B, F^n B)$ is obtained by taking the inverse limit of the homomorphisms $A/F^N A \to C/F^N C$ induced by $f$ and using $C \cong \varprojlim_N C/F^N C$. The uniqueness of the factorization follows from $C \xrightarrow{\cong} \varprojlim_N C/F^N C$ and $A/F^N A \cong B/F^N B$. □

**Definition IV.2.1.5.** (1) We define $\mathscr{A}_{p,\bullet}$ to be the category of inverse systems of $R$-algebras $(A_N)_{N \in \mathbb{N}_{>0}}$ such that $A_N$ ($N \in \mathbb{N}_{>0}$) are $p$-adically complete and separated, $\xi^N A_N = 0$ for $N \in \mathbb{N}_{>0}$, and the transition maps $\pi\colon A_{N+1} \to A_N$ ($N \in \mathbb{N}_{>0}$) are surjective.

(2) For $r \in \mathbb{N}_{>0}$ (resp. $r = \infty$), we define $\mathscr{A}_\bullet^r$ to be the full subcategory of $\mathscr{A}_{p,\bullet}$ consisting of $(A_N) \in \operatorname{Ob} \mathscr{A}_{p,\bullet}$ satisfying the following conditions.

(i) The rings $A_N$ ($N \in \mathbb{N}_{>0}$) are $p$-torsion free.

(ii) For every $s \in \mathbb{N} \cap [0, r]$ (resp. $s \in \mathbb{N}$) and $N \in \mathbb{N}_{>0}$, we have $pF^s A_N \subset \xi^s A_N$ (resp. $F^s A_N = \xi^s A_N$), where $F^n A_N$ ($n \in \mathbb{N}$) is the decreasing filtration of $A_N$ by ideals defined by $F^0 A_N = A_N$, $F^n A_N = \operatorname{Ker}(\pi\colon A_N \to A_n)$ ($1 \le n \le N$) and $F^n A_N = 0$ ($n > N$).

(iii) The kernel of the homomorphism $A_N \to A_N; x \mapsto \xi x$ is $F^{N-1} A_N$ for every $N \in \mathbb{N}_{>0}$.

The inverse system $R_1 \leftarrow R_2 \leftarrow R_3 \leftarrow \cdots$ is an object of $\mathscr{A}_\bullet^r$ for every $r \in \mathbb{N}_{>0} \cup \{\infty\}$.

For an object $(A_N)$ of $\mathscr{A}_\bullet^r$, the multiplication by $\xi$ on $A_N$ induces an injective homomorphism $A_{N-1} \to A_N$, which is denoted by $[\xi]$ in the following.

**Lemma IV.2.1.6.** *Let $r \in \mathbb{N}_{>0} \cup \{\infty\}$. Any object $(A_N)$ of $\mathscr{A}_\bullet^r$ have the following properties.*

(1) *The kernel of $A_{N+M} \to A_{N+M}; x \mapsto \xi^N x$ is $F^M A_{N+M}$ for $N, M \in \mathbb{N}_{>0}$.*

(2) *Assume $r \in \mathbb{N}_{>0}$, let $n \in \mathbb{N}$ and let $m$ be the smallest integer such that $nr^{-1} \le m$. Then, for $N \in \mathbb{N}_{>0}$, we have $p^m F^n A_N \subset \xi^n A_N \subset F^n A_N$. In particular, $(F^n A_N)_{\mathbb{Q}} = \xi^n A_{N,\mathbb{Q}}$.*

(3) *For $n, n' \in \mathbb{N}$ and $N \in \mathbb{N}_{>0}$, we have $F^n A_N \cdot F^{n'} A_N \subset F^{n+n'} A_N$.*

PROOF. (1) For positive integers $n \leq N$, the inverse image of $F^{n-1}A_n$ by $A_N \to A_n$ is $F^{n-1}A_N$. Hence, by the condition (iii) in the definition of $\mathscr{A}^r_\bullet$, the inverse image of $F^nA_N$ under $\xi\colon A_N \to A_N$ is $F^{n-1}A_N$. This implies the claim.

(2) It suffices to prove $pF^{n+s}A_N \subset \xi^s F^n A_N$ for $s \in \mathbb{N} \cap [0, r]$ and $n, N \in \mathbb{N}$ such that $n + s < N$. For $a \in pF^{n+s}A_N$, we have $a \in pF^sA_N \subset \xi^s A_N$. Since the inverse image of $F^{n+s}A_N$ under $\xi^s\colon A_N \to A_N$ is $F^nA_N$ by the proof of (1), we have $a \in \xi^s F^n A_N$.

(3) The claim is obvious if $r = \infty$. Assume $r \in \mathbb{N}_{>0}$. For $n, N \in \mathbb{N}_{>0}$, $A_N$ and $A_N/F^nA_N \cong A_n$ are $p$-torsion free. Hence $F^nA_N = F^nA_{N,\mathbb{Q}} \cap A_N$, and the claim follows from the last equality of (2). □

**Lemma IV.2.1.7.** *(1) For $(A_N) \in \mathrm{Ob}\,\mathscr{A}^r_\bullet$, the $R$-algebra $A = \varprojlim_N A_N$ with the decreasing filtration $F^nA = \varprojlim_N F^nA_N$ by ideals is an object of $\mathscr{A}^r$. This construction gives an equivalence of categories $\mathscr{A}^r_\bullet \to \mathscr{A}^r$, whose quasi-inverse is given by $(A, F^nA) \mapsto (A/F^NA)$.*

*(2) The functor $\mathscr{A}_{p,\bullet} \to \mathscr{A}$ defined by $(A_N) \mapsto (\varprojlim_N A_N, \varprojlim_N F^nA_N)$ is fully faithful.*

PROOF. (1) Let $(A_N) \in \mathrm{Ob}\,\mathscr{A}^r_\bullet$ and let $(A, F^nA)$ be as in the claim. We prove $(A, F^nA) \in \mathrm{Ob}\,\mathscr{A}^r$. By taking $\varprojlim_N$ of the exact sequences $0 \to F^nA_N \to A_N \to A_n \to 0$, we obtain an isomorphism $A/F^nA \cong A_n$. This implies $(A, F^nA) \in \mathrm{Ob}\,\mathscr{A}$. Since $A_N$ $(N \in \mathbb{N}_{>0})$ are $p$-torsion free, $A$ and $A/F^nA$ are $p$-torsion free. By taking $\varprojlim_N$ of the injective homomorphisms $[\xi]\colon A_N \to A_{N+1}$, we see that $\xi$ is regular in $A$. If $r \in \mathbb{N}_{>0}$ (resp. $r = \infty$), by taking $\varprojlim_N$ of $F^sA_N \xrightarrow[p]{\cong} pF^sA_N \subset \xi^sA_N \xleftarrow[{[\xi]^s}]{\cong} A_{N-s}$ for $s \in \mathbb{N} \cap [0, r]$ (resp. $F^sA_N = \xi^sA_N \xleftarrow[{[\xi]^s}]{\cong} A_{N-s}$ for $s \in \mathbb{N}$), we obtain $pF^sA \subset \xi^sA$ (resp. $F^sA = \xi^sA$). By taking $\varprojlim_N$ of the exact sequences $0 \to A_N \xrightarrow{p^m} A_N \to A_N/p^mA_N \to 0$, we obtain $A/p^mA \xrightarrow{\cong} \varprojlim_N A_N/p^mA_N$. Taking $\varprojlim_m$ and using $A_N \cong \varprojlim_m A_N/p^mA_N$, we obtain $\varprojlim_m A/p^mA \cong \varprojlim_N A_N = A$. The remaining conditions immediately follow from $A/F^nA \cong A_n$. For an object $(A, F^nA) \in \mathrm{Ob}\,\mathscr{A}^r$, it is immediate to see that $(A/F^NA)_{N\in\mathbb{N}_{>0}} \in \mathrm{Ob}\,\mathscr{A}^r_\bullet$ just noting that $F^n(A/F^NA)$ is the image of $F^nA$ in $A/F^NA$. It is straightforward to verify that the two functors are quasi-inverses of each other.

(2) Let $(A_N)$ be an object of $\mathscr{A}_{p,\bullet}$, and put $(A, F^nA) = (\varprojlim_N A_N, \varprojlim_N F^nA_N)$. Then, by the same argument as in the beginning of the proof of (1), we have $A_n \cong A/F^nA$ compatible with $n$, and $(A, F^nA) \in \mathrm{Ob}\,\mathscr{A}$. For another object $(B_N)$ of $\mathscr{A}_{p,\bullet}$ and the object $(B, F^nB) = (\varprojlim_N B_N, \varprojlim_N F^nB_N)$ of $\mathscr{A}$, we see that the map

$$\mathrm{Hom}_{\mathscr{A}}((A, F^nA), (B, F^nB)) \longrightarrow \mathrm{Hom}_{\mathscr{A}_{p,\bullet}}((A_N), (B_N))$$

sending $f$ to $(A_N \cong A/F^NA \xrightarrow{f \bmod F^N} B/F^NB \cong B_N)_N$ is the inverse of the map $\mathrm{Hom}_{\mathscr{A}_{p,\bullet}}((A_N), (B_N)) \to \mathrm{Hom}_{\mathscr{A}}((A, F^nA), (B, F^nB))$ defined by the functor in question. □

Combining Lemma IV.2.1.7 and Proposition IV.2.1.2, we obtain the following.

**Proposition IV.2.1.8.** *For every $r \in \mathbb{N}_{>0} \cup \{\infty\}$, the inclusion functor $\mathscr{A}^r_\bullet \to \mathscr{A}_{p,\bullet}$ has a left adjoint functor $D^r_{\mathrm{Higgs}}\colon \mathscr{A}_{p,\bullet} \to \mathscr{A}^r_\bullet$.*

**Definition IV.2.1.9.** For $r \in \mathbb{N}_{>0} \cup \{\infty\}$ and $(A_N) \in \mathrm{Ob}\,\mathscr{A}_{p,\bullet}$, we call $D^r_{\mathrm{Higgs}}((A_N))$ the *Higgs envelope of $(A_N)$ of level $r$*.

**Lemma IV.2.1.10.** *Let $(f_N)\colon (A_N) \to (A'_N)$ be a morphism in $\mathscr{A}_{p,\bullet}$. Let $A_{N,m}$, $A'_{N,m}$ and $f_{N,m}$ be the reduction mod $p^m$ of $A_N$, $A'_N$ and $f_N$, respectively, and assume that* *the natural homomorphism* $A'_{N+1,m}\otimes_{A_{N+1,m}}A_{N,m}\to A'_{N,m}$ *is an isomorphism for every* $m,N\in\mathbb{N}_{>0}$. *Let* $r\in\mathbb{N}_{>0}\cup\{\infty\}$.

(1) *If* $f_{N,m}$ *is flat for every* $m,N\in\mathbb{N}_{>0}$ *and* $(A_N)\in\mathrm{Ob}\,\mathscr{A}^r_\bullet$, *then* $(A'_N)\in\mathrm{Ob}\,\mathscr{A}^r_\bullet$.

(2) *If* $f_{N,m}$ *is faithfully flat for every* $m,N\in\mathbb{N}_{>0}$ *and* $(A'_N)\in\mathrm{Ob}\,\mathscr{A}^r_\bullet$, *then* $(A_N)\in\mathrm{Ob}\,\mathscr{A}^r_\bullet$.

PROOF. Since $A_N$ (resp. $A'_N$) is $p$-adically complete and separated, $A_N$ (resp. $A'_N$) is flat over $\mathbb{Z}_p$ if and only if $A_{N,m}$ (resp. $A'_{N,m}$) is flat over $\mathbb{Z}/p^m\mathbb{Z}$ for every $m\in\mathbb{N}_{>0}$. If $f_{N,m}$ is flat (resp. faithfully flat) and $A_{N,m}$ (resp. $A'_{N,m}$) is flat over $\mathbb{Z}/p^m\mathbb{Z}$, then $A'_{N,m}$ (resp. $A_{N,m}$) is flat over $\mathbb{Z}/p^m\mathbb{Z}$. Hence we may assume that $A_N$ and $A'_N$ are flat over $\mathbb{Z}_p$. Assume that $f_{N,m}$ is flat for every $m,N\in\mathbb{N}_{>0}$. Let $n,N,m$ be positive integers such that $n<N$. Then since $A'_{N,m}\otimes_{A_{N,m}}A_{n,m}\xrightarrow{\cong}A'_{n,m}$, we have exact sequences

$$0\to F^nA_N/p^mF^nA_N\otimes_{A_{N,m}}A'_{N,m}\to A'_{N,m}\to A'_{n,m}\to 0,$$
$$0\to F^nA'_N/p^mF^nA'_N\to A'_{N,m}\to A'_{n,m}\to 0.$$

Hence we have an isomorphism $F^nA_N/p^mF^nA_N\otimes_{A_{N,m}}A'_{N,m}\xrightarrow{\cong}F^nA'_N/p^mF^nA'_N$. Now the claim follows from Sublemma IV.2.1.11 below. □

**Sublemma IV.2.1.11.** *Let* $(A_N)$ *be an object of* $\mathscr{A}_{p,\bullet}$ *and assume that* $A_N$ *is* $p$*-torsion free for every* $N\in\mathbb{N}_{>0}$. *Let* $r\in\mathbb{N}_{>0}$ *(resp.* $r=\infty$*). Then* $(A_N)$ *is an object of* $\mathscr{A}^r_\bullet$ *if and only if the following two conditions hold.*

(i) *For* $N,m,s\in\mathbb{N}_{>0}$ *such that* $s<N$, *the composition of* $F^{N-s}A_N/p^mF^{N-s}A_N\to A_N/p^mA_N\xrightarrow{\xi^s}A_N/p^mA_N$ *is* 0, *and hence* $\xi^s\colon A_N/p^mA_N\to A_N/p^mA_N$ *induces a homomorphism* $g_{N,m,s}\colon A_{N-s}/p^mA_{N-s}\to F^sA_N/p^mF^sA_N$.

(ii) *For* $N,m\in\mathbb{N}_{>0}$ *and* $s\in\mathbb{N}\cap[0,r]$ *(resp.* $s\in\mathbb{N}$*) such that* $s<N$, $p\cdot\mathrm{Cok}(g_{N,m,s})=0$ *and the homomorphism* $\mathrm{Ker}(g_{N,m+1,s})\to\mathrm{Ker}(g_{N,m,s})$ *induced by the projection* $A_{N-s}/p^{m+1}A_{N-s}\to A_{N-s}/p^mA_{N-s}$ *is* 0 *(resp. the homomorphism* $g_{N,m,s}$ *is an isomorphism).*

PROOF. The necessity follows from Lemma IV.2.1.6 (1) and $p\cdot(F^sA_N/\xi^sA_N)=0$ (resp. $F^sA_N=\xi^sA_N$) for $s\in\mathbb{N}\cap[0,r]$ (resp. $s\in\mathbb{N}$). Let us prove the sufficiency. By the condition (i), the composition of $F^{N-s}A_N\to A_N\xrightarrow{\xi^s}F^sA_N$ is 0. Hence $\xi^s\colon A_N\to F^sA_N$ induces a homomorphism $g_{N,s}\colon A_{N-s}\to F^sA_N$. Since $g_{N,s}=\varprojlim_m g_{N,m,s}$, the condition (ii) implies that $\mathrm{Cok}(g_{N,s})=F^sA_N/\xi^sA_N$ is annihilated by $p$ and $g_{N,s}$ is injective (resp. $g_{N,s}$ is an isomorphism) for $N\in\mathbb{N}_{>0}$ and $s\in\mathbb{N}\cap[0,r]$ (resp. $s\in\mathbb{N}$) such that $s<N$. □

**IV.2.2. $p$-adic fine log formal schemes.** In this subsection, we prove the existence of the Higgs envelope of level $r$ ($r\in\mathbb{N}_{>0}\cup\{\infty\}$) and its compatibility with flat morphisms for a direct system of $p$-adic fine log formal schemes over $S_N$ satisfying certain conditions (Proposition IV.2.2.9, Definition IV.2.2.10).

By a *$p$-adic formal scheme*, we mean an adic formal scheme $X$ such that $p\mathcal{O}_X$ is an ideal of definition. Let $X$ be a $p$-adic formal scheme. We have a canonical morphism $X\to\mathrm{Spf}(\mathbb{Z}_p)$. Let $X_m$ denote $X\times_{\mathrm{Spf}(\mathbb{Z}_p)}\mathrm{Spec}(\mathbb{Z}/p^m\mathbb{Z})$. We define a *fine log structure $M$ on $X$* to be a family of fine log structures $M_m$ on $X_m$ ($m\in\mathbb{N}_{>0}$) and exact closed immersions $(X_m,M_m)\to(X_{m+1},M_{m+1})$ ($m\in\mathbb{N}_{>0}$) extending the natural closed immersions $X_m\to X_{m+1}$. We define a *$p$-adic fine log formal scheme* to be a $p$-adic formal scheme endowed with a fine log structure. We say that a $p$-adic fine log formal scheme is *affine* if its underlying formal scheme is affine. A *morphism of $p$-adic fine log formal schemes* $f\colon(X,M)\to(Y,N)$ is a family of morphisms of fine log schemes

$f_m\colon (X_m, M_m) \to (Y_m, N_m)$ $(m \in \mathbb{N}_{>0})$ compatible with the exact closed immersions $(X_m, M_m) \to (X_{m+1}, M_{m+1})$ and $(Y_m, N_m) \to (Y_{m+1}, N_{m+1})$. Note that giving a morphism of $p$-adic formal schemes $X \to Y$ is equivalent to giving a family of morphisms of schemes $X_m \to Y_m$ $(m \in \mathbb{N}_{>0})$ compatible with the closed immersions $X_m \hookrightarrow X_{m+1}$ and $Y_m \hookrightarrow Y_{m+1}$ (cf. **[1]** Proposition 2.2.2). Let $f\colon (X, M) \to (Y, N)$ be a morphism of $p$-adic fine log formal schemes. We say that $f$ is *smooth* (resp. *étale*, resp. an *exact closed immersion*, resp. a *closed immersion*, resp. *strict*, resp. *integral*) if $f_m$ is smooth (resp. étale, resp. ...) for every $m \in \mathbb{N}_{>0}$. We say that $f$ is an *open immersion* if $f_m$ is strict and the morphism of schemes underlying $f_m$ is an open immersion for every $m \in \mathbb{N}_{>0}$. We say that $f$ is an *immersion* if it is the composition $g \circ h$ of an open immersion $g$ and a closed immersion $h$. We say that $f$ is *affine* if the morphism of $p$-adic formal schemes underlying $f$ is affine. Finite inverse limits are representable in the category of $p$-adic fine log formal schemes. In the following, a $p$-adic fine log formal scheme is denoted by a single letter such as $X$, $Y$, $T$. For a $p$-adic fine log formal scheme $X$, we write $(M_{X_m})$ for the fine log structure of $X$.

For a closed immersion $i\colon X \to Y$ of $p$-adic fine log formal schemes and $n \in \mathbb{N}_{>0}$, the $n$th infinitesimal neighborhood $D_m$ of the reduction mod $p^m$ of $i$ for $m \in \mathbb{N}_{>0}$ gives a $p$-adic fine log formal scheme $D$ with a factorization $X \to D \to Y$ of $i$. We call $D$ the $n$th *infinitesimal neighborhood* of $i$.

For a $p$-adic fine log formal scheme $X$, we define the site $X_{\text{ét}}$ (resp. $X_{\text{ÉT}}$) to be the category of $p$-adic fine log formal schemes strict étale over $X$ (resp. $p$-adic fine log formal schemes over $X$) endowed with the topology associated to the pretopology defined by

$$\mathrm{Cov}(U) = \{(u_\alpha\colon U_\alpha \to U)_{\alpha\in A} | u_\alpha \text{ is strict étale for every } \alpha \in A \text{ and } U = \cup_{\alpha\in A} u_\alpha(U_\alpha)\}$$

for $U \in \mathrm{Ob}\, X_{\text{ét}}$ (resp. $\mathrm{Ob}\, X_{\text{ÉT}}$). For any morphism of $p$-adic fine log formal schemes $f\colon X' \to X$, the functor $X_{\text{ét}} \to X'_{\text{ét}}$ (resp. $X_{\text{ÉT}} \to X'_{\text{ÉT}}$); $U \mapsto U \times_X X'$ defines a morphism of sites, and hence a morphism of topos $f_{\text{ét}}\colon X'^{\sim}_{\text{ét}} \to X^{\sim}_{\text{ét}}$ (resp. $f_{\text{ÉT}}\colon X'^{\sim}_{\text{ÉT}} \to X^{\sim}_{\text{ÉT}}$) (cf. **[2]** I Proposition 5.4 4), III Proposition 1.3 5)). The functor $X'_{\text{ÉT}} \to X_{\text{ÉT}}$; $(u\colon U' \to X') \mapsto (f \circ u\colon U' \to X)$ is cocontinuous (cf. **[2]** III Définition 2.1) and is a left adjoint of the functor $X_{\text{ÉT}} \to X'_{\text{ÉT}}$ considered above. Hence we have a canonical functorial isomorphism $(f^*_{\text{ÉT}}\mathcal{F})(u\colon U' \to X') \cong \mathcal{F}(f \circ u\colon U' \to X)$ for $\mathcal{F} \in \mathrm{Ob}\, X_{\text{ÉT}}$ and $u\colon U' \to X' \in \mathrm{Ob}\, X'_{\text{ÉT}}$ (cf. **[2]** III Proposition 2.5).

For a $p$-adic fine log formal scheme $X$, we define the sheaf of rings $\mathcal{O}_X$ (resp. the sheaf of monoids $M_X$) on $X_{\text{ét}}$ by $\Gamma(U, \mathcal{O}_X) = \varprojlim_m \Gamma(U_m, \mathcal{O}_{U_m})$ (resp. $\Gamma(U, M_X) = \varprojlim_m \Gamma(U_m, M_{X_m})$). Let $f\colon X \to Y$ be a morphism of $p$-adic fine log formal schemes. The morphism of topos $f_{\text{ét}}$ naturally extends to a morphism of ringed topos $(X_{\text{ét}}, \mathcal{O}_X) \to (Y_{\text{ét}}, \mathcal{O}_Y)$, which is also denoted by $f_{\text{ét}}$. For a sheaf of $\mathcal{O}_Y$-modules $\mathcal{F}$ on $Y_{\text{ét}}$, we write $f^{-1}(\mathcal{F})$ (resp. $f^*(\mathcal{F})$) for the inverse image of $\mathcal{F}$ as a sheaf of abelian groups (resp. $f^{-1}(\mathcal{F}) \otimes_{f^{-1}(\mathcal{O}_Y)} \mathcal{O}_X$). For a sheaf of $\mathcal{O}_X$-modules $\mathcal{F}$ on $X_{\text{ét}}$ and a sheaf of $\mathcal{O}_Y$-modules $\mathcal{G}$ on $Y_{\text{ét}}$, we abbreviate $\mathcal{F} \otimes_{f^{-1}(\mathcal{O}_Y)} f^{-1}(\mathcal{G})$ to $\mathcal{F} \otimes_{\mathcal{O}_Y} \mathcal{G}$. We define the sheaf of $\mathcal{O}_X$-modules $\Omega^q_{X/Y}$ on $X_{\text{ét}}$ by $\Gamma(U, \Omega^q_{X/Y}) = \varprojlim_m \Gamma(U_m, \Omega^q_{X_m/Y_m})$. For $t = (t_m) \in \Gamma(U, M_X) = \varprojlim_m \Gamma(U_m, M_{X_m})$, we define $d\log t \in \Gamma(U, \Omega^1_{X/Y})$ to be $(d\log t_m)_{m\in\mathbb{N}_{>0}}$. If $f$ is smooth, $\Omega^q_{X/Y}$ is a locally free $\mathcal{O}_X$-module of finite type, and strict étale locally, there exist $t_1, \ldots, t_d \in \Gamma(X, M_X)$ such that $d\log t_i$ $(1 \leq i \leq d)$ form a basis of $\Omega^1_{X/Y}$. We call such $t_1, \ldots, t_d$ *log coordinates of* $X$ *over* $Y$.

We endow $S_N = \mathrm{Spf}(R_N)$ with the trivial log structure.

**Definition IV.2.2.1.** (1) We define $\mathscr{C}$ to be the category of sequences of morphisms $Y_1 \to Y_2 \to Y_3 \to \cdots \to Y_N \to Y_{N+1} \to \cdots$ of $p$-adic fine log formal schemes $Y_N$

over $S_N$ compatible with the closed immersions $S_N \to S_{N+1}$ and satisfying the following conditions.

(i) The morphism $Y_1 \to Y_2$ is an immersion, and for each $N \geq 2$, the morphism $Y_N \to Y_{N+1}$ is an exact closed immersion.

(ii) For each $N \geq 2$, the morphism of schemes underlying the reduction mod $p$ of $Y_N \to Y_{N+1}$ is a nilpotent immersion.

A morphism $(Y_N) \to (Y'_N)$ in $\mathscr{C}$ is a family of morphisms $f_N\colon Y_N \to Y'_N$ over $S_N$ for $N \in \mathbb{N}_{>0}$ compatible with the immersions $Y_N \to Y_{N+1}$ and $Y'_N \to Y'_{N+1}$.

(2) For $r \in \mathbb{N}_{>0}$ (resp. $r = \infty$), we define $\mathscr{C}^r$ to be the full subcategory of $\mathscr{C}$ consisting of $(Y_N) \in \mathrm{Ob}\,\mathscr{C}$ satisfying the following conditions.

(i) The morphism $Y_1 \to Y_2$ is an exact closed immersion.

(ii) The morphism $p\colon \mathcal{O}_{Y_N} \to \mathcal{O}_{Y_N}$ is injective for every $N \in \mathbb{N}_{>0}$.

(iii) For every $s \in \mathbb{N} \cap [0, r]$ (resp. $s \in \mathbb{N}$) and $N \in \mathbb{N}_{>0}$, we have $pF^s\mathcal{O}_{Y_N} \subset \xi^s\mathcal{O}_{Y_N}$ (resp. $F^s\mathcal{O}_{Y_N} = \xi^s\mathcal{O}_{Y_N}$), where $F^n\mathcal{O}_{Y_N}$ is the decreasing filtration of $\mathcal{O}_{Y_N}$ by ideals defined by $F^0\mathcal{O}_{Y_N} = \mathcal{O}_{Y_N}$, $F^n\mathcal{O}_{Y_N} = \mathrm{Ker}(\mathcal{O}_{Y_N} \to \mathcal{O}_{Y_n})$ $(1 \leq n \leq N)$, and $F^n\mathcal{O}_{Y_N} = 0$ $(n > N)$.

(iv) For every $N \in \mathbb{N}_{>0}$, the kernel of $\xi\colon \mathcal{O}_{Y_N} \to \mathcal{O}_{Y_N}$ is $F^{N-1}\mathcal{O}_{Y_N}$.

(3) Let $Y = (Y_N)$ be an object of $\mathscr{C}$, let $X$ be a $p$-adic fine log formal scheme over $S_1$, and let $i\colon X \to Y_1$ be an immersion over $S_1$. Then we write $(i\colon X \hookrightarrow Y)$ for the object of $\mathscr{C}$ defined by the composition of $X \xrightarrow{i} Y_1 \to Y_2$ and the exact closed immersions $Y_N \to Y_{N+1}$ $(N \geq 2)$.

For an object $Y = (Y_N)$ of $\mathscr{C}^r$, the multiplication by $\xi$ on $\mathcal{O}_{Y_N}$ induces an injective morphism $\mathcal{O}_{Y_{N-1}} \to \mathcal{O}_{Y_N}$, which is denoted by $[\xi]$ in the following.

**Definition IV.2.2.2.** Let $f = (f_N)\colon Y = (Y_N) \to Y' = (Y'_N)$ be a morphism in $\mathscr{C}$.

(1) We say that $f$ is *Cartesian* if the morphism $Y'_N \to Y'_{N+1} \times_{Y_{N+1}} Y_N$ induced by $f$ is an isomorphism for every $N \in \mathbb{N}_{>0}$.

(2) We say that $f$ is *smooth* (resp. *étale*, resp. *strict*, resp. *integral*, resp. *affine*, resp. an *exact closed immersion*, resp. an *open immersion*) if $f_N$ is smooth (resp. étale, resp. strict, resp. integral, resp. affine, resp. an exact closed immersion, resp. an open immersion) for every $N \in \mathbb{N}_{>0}$.

(3) We say that a family of morphisms $(u_\alpha\colon Y_\alpha \to Y)$ in $\mathscr{C}$ is a *strict étale covering* (resp. a *Zariski covering*) if $u_\alpha$ is strict étale (resp. an open immersion) and Cartesian for every $\alpha \in A$ and $\cup_{\alpha\in A} u_{\alpha,N}(Y_{\alpha,N}) = Y_N$ for every $N \in \mathbb{N}_{>0}$ (or equivalently, for $N = 2$).

**Remark IV.2.2.3.** For $Y = (Y_N) \in \mathscr{C}$ and a strict étale morphism $V_2 \to Y_2$, there exists a strict étale Cartesian morphism $U = (U_N) \to Y = (Y_N)$ in $\mathscr{C}$ with $U_2 = V_2$, and it is unique up to a unique isomorphism. By **[39]** Proposition (5.1.9), we see that, if $U_2$ is affine and $Y_1 \to Y_2$ is a closed immersion, then $U_N$ is affine for every $N \in \mathbb{N}_{>0}$.

Let $r$ denote a positive integer or $\infty$ in the following. Let $Y = (Y_N)$ be an object of $\mathscr{C}$ such that $Y_2$ is affine and $Y_1 \to Y_2$ is a closed immersion. Then, by Remark IV.2.2.3, $Y_N$ is affine, and $A_N = \Gamma(Y_N, \mathcal{O}_{Y_N})$ defines an object $A = (A_N)$ of $\mathscr{A}_{p,\bullet}$.

**Lemma IV.2.2.4.** *Let the notation and assumption be as above. If $Y$ is an object of $\mathscr{C}^r$, then $A$ is an object of $\mathrm{Ob}\,\mathscr{A}^r_\bullet$.*

Proof. By taking $\Gamma(Y_N, -)$ of the factorization $\mathcal{O}_{Y_N} \twoheadrightarrow \mathcal{O}_{Y_{N-s}} \xrightarrow[\cong]{[\xi]^s} \xi^s\mathcal{O}_{Y_N} \hookrightarrow \mathcal{O}_{Y_N}$ of the multiplication by $\xi^s$ on $\mathcal{O}_{Y_N}$, we obtain $\Gamma(Y_N, \xi^s\mathcal{O}_{Y_N}) = \xi^s A_N$. Similarly, since the multiplication by $p$ on $F^s\mathcal{O}_{Y_N}$ induces an isomorphism $F^s\mathcal{O}_{Y_N} \xrightarrow{\cong} pF^s\mathcal{O}_{Y_N}$, we have $\Gamma(Y_N, pF^s\mathcal{O}_{Y_N}) = p\Gamma(Y_N, F^s\mathcal{O}_{Y_N}) = pF^s A_N$. Hence $A$ satisfies the condition (ii) in the definition of $\mathscr{A}^r_\bullet$. The other two conditions immediately follow from the definition of $\mathscr{C}^r$.

□

**Corollary IV.2.2.5.** *Let $Y$ be an object of $\mathscr{C}^r$, then the morphism of schemes underlying the reduction mod $p$ of the exact closed immersion $Y_1 \to Y_2$ is nilpotent.*

PROOF. If a morphism $U \to Y$ in $\mathscr{C}$ is an open immersion and Cartesian, then $U$ is an object of $\mathscr{C}^r$. Hence, if $U_2$ is affine, $(\Gamma(U_N, \mathcal{O}_{U_N}))_N$ is an object of $\mathscr{A}^r_\bullet$ by Lemma IV.2.2.4. By Lemma IV.2.1.6 (3), we have $\Gamma(U_2, F^1\mathcal{O}_{U_2}) \cdot \Gamma(U_2, F^1\mathcal{O}_{U_2}) = 0$. □

By Corollary IV.2.2.5, for $Y = (Y_N) \in \mathscr{C}^r$ and a strict étale morphism $V_1 \to Y_1$, there exists a strict Cartesian morphism $U = (U_N) \to Y = (Y_N)$ in $\mathscr{C}^r$ with $U_1 = V_1$, which is unique up to a unique isomorphism. If $U_1$ is affine, then $U_N$ is affine for every $N \in \mathbb{N}_{>0}$.

**Lemma IV.2.2.6.** *Let $Y$ and $A$ be as before Lemma* IV.2.2.4. *Then $Y$ is an object of $\mathscr{C}^r$ if and only if the morphism $Y_1 \to Y_2$ is an exact closed immersion and $A \in \mathrm{Ob}\,\mathscr{A}^r_\bullet$.*

PROOF. The necessity follows from Lemma IV.2.2.4. Let us prove the sufficiency. Let $V_2 \to Y_2$ be a strict étale morphism such that $V_2$ is affine. As in Remark IV.2.2.3, there exists a strict étale Cartesian morphism $U = (U_N) \to Y = (Y_N)$ in $\mathscr{C}$ such that $U_2 = V_2$. By applying Lemma IV.2.1.10 (1) to the morphism $(A_N) \to (\Gamma(U_N, \mathcal{O}_{Y_N}))$ in $\mathscr{A}_{p,\bullet}$, we obtain $(\Gamma(U_N, \mathcal{O}_{Y_N})) \in \mathrm{Ob}\,\mathscr{A}^r_\bullet$. Varying $V_2$, we see that $(Y_N)$ satisfies the conditions (ii), (iii), and (iv) in the definition of the category $\mathscr{C}^r$. □

By Lemma IV.2.2.6, the sequence $S = (S_1 \hookrightarrow S_2 \hookrightarrow S_3 \hookrightarrow \cdots)$ is an object of $\mathscr{C}^r$ for every $r \in \mathbb{N}_{>0} \cup \{\infty\}$.

**Lemma IV.2.2.7.** (1) *Let $f\colon Y' \to Y$ be a Cartesian morphism in $\mathscr{C}$ such that the morphism of schemes underlying the reduction mod $p^m$ of $f_N\colon Y'_N \to Y_N$ is flat for every $N, m \in \mathbb{N}_{>0}$. If $Y$ is an object of $\mathscr{C}^r$, then $Y'$ is also an object of $\mathscr{C}^r$.*

(2) *Let $(u_\alpha\colon Y_\alpha \to Y)_{\alpha \in A}$ be a family of strict Cartesian morphisms in $\mathscr{C}$ such that $A$ is finite, the morphism of schemes underlying the reduction mod $p^m$ of $u_{\alpha,N}\colon Y_{\alpha,N} \to Y_N$ is flat and quasi-compact for every $N, m \in \mathbb{N}_{>0}$ and $\alpha \in A$, and $\cup_{\alpha \in A} u_{\alpha,N}(Y_{\alpha,N}) = Y_N$ for every $N \in \mathbb{N}_{>0}$. If $Y_\alpha$ is an object of $\mathscr{C}^r$ for every $\alpha \in A$, then $Y$ is an object of $\mathscr{C}^r$.*

PROOF. (1) If $Y \in \mathrm{Ob}\,\mathscr{C}^r$, $Y'_1 \hookrightarrow Y'_2$ is an exact closed immersion since $f$ is Cartesian. Since the question is Zariski local on $Y_2$ and $Y'_2$, we may assume that $Y_2$ and $Y'_2$ are affine. By Lemma IV.2.2.4, the inverse system $(\Gamma(Y_N, \mathcal{O}_{Y_N}))_N$ is an object of $\mathscr{A}^r_\bullet$. By the assumption on $f$ and $Y$, we see that the morphism $(\Gamma(Y_N, \mathcal{O}_{Y_N}))_N \to (\Gamma(Y'_N, \mathcal{O}_{Y'_N}))_N$ in $\mathscr{A}_{p,\bullet}$ satisfies the assumption in Lemma IV.2.1.10 (1). Hence $(\Gamma(Y'_N, \mathcal{O}_{Y'_N}))_N$ is an object of $\mathscr{A}^r_\bullet$. By Lemma IV.2.2.6, $Y'$ is an object of $\mathscr{C}^r$.

(2) By fpqc descent of closed immersions of schemes (cf. **[42]** Proposition (2.7.1) (xii)), the morphism of schemes underlying the reduction mod $p^m$ of $Y_1 \to Y_2$ is a closed immersion. Then we see that $Y_1 \to Y_2$ is an exact closed immersion since $u_\alpha$ is strict, $Y_{\alpha,1} \to Y_{\alpha,2}$ is an exact closed immersion, and $\cup_{\alpha \in A} u_{\alpha,1}(Y_{\alpha,1}) = Y_1$. Since the question is Zariski local on $Y_2$ and $Y_{\alpha,2}$, we may assume that $Y_2$ and $Y_{\alpha,2}$ are affine. By Lemma IV.2.2.4, the inverse system $(\prod_{\alpha \in A} \Gamma(Y_{\alpha,N}, \mathcal{O}_{Y_{\alpha,N}}))_N$ is an object of $\mathscr{A}^r_\bullet$. By the assumption on $(u_\alpha)_\alpha$, the morphism $(\Gamma(Y_N, \mathcal{O}_{Y_N}))_N \to (\prod_{\alpha \in A} \Gamma(Y_{\alpha,N}, \mathcal{O}_{Y_{\alpha,N}}))_N$ in $\mathscr{A}_{p,\bullet}$ satisfies the assumption in Lemma IV.2.1.10 (2). Hence $(\Gamma(Y_N, \mathcal{O}_{Y_N}))_N \in \mathrm{Ob}\,\mathscr{A}^r_\bullet$, and Lemma IV.2.2.6 implies $Y \in \mathrm{Ob}\,\mathscr{C}^r$. □

**Lemma IV.2.2.8.** *Let $f\colon Y' \to Y$ and $g\colon Y'' \to Y$ be two morphisms in $\mathscr{C}$. Then $Y'''_N = Y''_N \times_{Y_N} Y'_N$ ($N \in \mathbb{N}_{>0}$) with the induced morphisms $Y'''_N \to Y'''_{N+1}$ define an object $Y'''$ of $\mathscr{C}$ and $Y'''$ with the canonical morphisms $Y''' \to Y'$ and $Y''' \to Y$ represents the fiber product of $Y' \to Y \leftarrow Y''$ in $\mathscr{C}$.*

PROOF. In the category of fine log schemes, closed immersions, open immersions, exact closed immersions, and nilpotent immersions in underlying schemes are stable under base changes. This implies the first claim. The second claim is obvious. □

By Lemma IV.2.2.8, we see that Cartesian morphisms in $\mathscr{C}$ are stable under base changes.

**Proposition IV.2.2.9.** (1) *The inclusion functor $\mathscr{C}^r \to \mathscr{C}$ has a right adjoint functor $D^r_{\mathrm{Higgs}} \colon \mathscr{C} \to \mathscr{C}^r$.*

(2) *For $Y \in \mathscr{C}$, the morphism $D^r_{\mathrm{Higgs}}(Y)_1 \to Y_1$ defined by the adjunction morphism $D^r_{\mathrm{Higgs}}(Y) \to Y$ is affine and strict. If $Y_1 \to Y_2$ is a closed immersion, then $D^r_{\mathrm{Higgs}}(Y) \to Y$ is affine.*

(3) *Let $f \colon Y' \to Y$ be a strict Cartesian morphism in $\mathscr{C}$ such that the morphism of schemes underlying the reduction mod $p^m$ of $f_N \colon Y'_N \to Y_N$ is flat for every $N, m \in \mathbb{N}_{>0}$. Then the natural morphism $D^r_{\mathrm{Higgs}}(Y') \to D^r_{\mathrm{Higgs}}(Y) \times_Y Y'$ in $\mathscr{C}$ is an isomorphism.*

**Definition IV.2.2.10.** For an object $Y$ of $\mathscr{C}$, we call $D^r_{\mathrm{Higgs}}(Y)$ the *Higgs envelope of $Y$ of level $r$.*

**Corollary IV.2.2.11.** *Finite fiber products and finite products are representable in $\mathscr{C}^r$.*

PROOF. Since $S$ is a final object of $\mathscr{C}^r$, it suffices to prove the representability of finite fiber products. Let $T' \to T \leftarrow T''$ be morphisms in $\mathscr{C}^r$, and let $T'''$ be the fiber product of this diagram in $\mathscr{C}$ (cf. Lemma IV.2.2.8). Then $D^r_{\mathrm{Higgs}}(T''')$ gives the fiber product in $\mathscr{C}^r$. □

In the rest of this subsection, we prove Proposition IV.2.2.9.

**Definition IV.2.2.12.** Let $Y$ be an object of $\mathscr{C}$. We say that a pair $(T, u)$ of an object $T$ of $\mathscr{C}^r$ and a morphism $u \colon T \to Y$ in $\mathscr{C}$ is an *affine Higgs envelope of $Y$ of level $r$* if it satisfies the following two conditions.

(a) The morphism $u$ is affine.

(b) For every $T' \in \mathscr{C}^r$, the map $\mathrm{Hom}_{\mathscr{C}^r}(T', T) \to \mathrm{Hom}_{\mathscr{C}}(T, Y); f \mapsto u \circ f$ is bijective.

**Lemma IV.2.2.13.** *Let $Y$ be an object of $\mathscr{C}$ and let $u \colon T \to Y$ be an affine Higgs envelope of $Y$ of level $r$. Let $\widetilde{Y} \to Y$ be a strict Cartesian morphism in $\mathscr{C}$ such that the morphism of schemes underlying the reduction mod $p^m$ of $\widetilde{Y}_N \to Y_N$ is flat for every $m, N \in \mathbb{N}_{>0}$. Then the base change $\widetilde{u} \colon \widetilde{T} \to \widetilde{Y}$ of $u$ by $\widetilde{Y} \to Y$ is an affine Higgs envelope of $\widetilde{Y}$ of level $r$.*

PROOF. Since $u$ is affine, $\widetilde{u}$ is affine. By the assumption on $\widetilde{Y} \to Y$, the morphism $\widetilde{T} \to T$ is strict Cartesian and the underlying morphism of schemes of the reduction mod $p^m$ of $\widetilde{T}_N \to T_N$ is flat for every $m, N \in \mathbb{N}_{>0}$. By Lemma IV.2.2.7 (1), we see that $\widetilde{T}$ is an object of $\mathscr{C}^r$. For any $T' \in \mathrm{Ob}\,\mathscr{C}^r$, we have bijections

$$\mathrm{Hom}_{\mathscr{C}}(T', \widetilde{T}) \xrightarrow{\sim} \mathrm{Hom}_{\mathscr{C}}(T', T) \times_{\mathrm{Hom}_{\mathscr{C}}(T', Y)} \mathrm{Hom}_{\mathscr{C}}(T', \widetilde{Y}) \xrightarrow{\sim} \mathrm{Hom}_{\mathscr{C}}(T', \widetilde{Y})$$

because $\mathrm{Hom}_{\mathscr{C}}(T', T) \to \mathrm{Hom}_{\mathscr{C}}(T', Y)$ is bijective by the assumption on $u$. □

**Proposition IV.2.2.14.** *Let $Y$ be an object of $\mathscr{C}$ such that $Y_2$ is affine and $Y_1 \to Y_2$ is an exact closed immersion. Then $Y$ has an affine Higgs envelope of level $r$.*

PROOF. Let $A = (A_N)$ be the object of $\mathscr{A}_{p,\bullet}$ defined by $A_N = \Gamma(Y_N, \mathcal{O}_{Y_N})$, and let $B = (B_N)$ be the object $D^r_{\mathrm{Higgs}}(A)$ of $\mathscr{A}^r_{\bullet}$ (cf. Proposition IV.2.1.8). We have the adjunction morphism $A \to B$. By Lemma IV.2.1.6 (3), the square of $F^N B_{N+1} = \mathrm{Ker}(B_{N+1} \to B_N)$ is 0. Hence $\mathrm{Spf}(B_N)$ endowed with the inverse image of $M_{Y_N}$ defines an object

$T = (T_N)$ of $\mathscr{C}$. Since $Y_1 \to Y_2$ is an exact closed immersion, $T$ is an object of $\mathscr{C}^r$ by Lemma IV.2.2.6. We prove that the affine morphism $u\colon T \to Y$ is an affine Higgs envelope of level $r$. Let $T'$ be an object of $\mathscr{C}^r$ and let $f\colon T' \to Y$ be a morphism in $\mathscr{C}$. It suffices to prove that there exists a unique morphism $g\colon T' \to T$ such that $u \circ g = f$. Since $u$ is a strict morphism, we see that giving a morphism $g$ as above is equivalent to giving a compatible system of $A_N$-algebra homomorphisms $B_N \to \Gamma(T'_N, \mathcal{O}_{T'_N})$. If $T'_N$ are affine, the inverse system $(\Gamma(T'_N, \mathcal{O}_{T'_N}))_N$ is an object of $\mathscr{A}^r_\bullet$ by Lemma IV.2.2.6, and there exists a unique such compatible system by the definition of $D^r_{\mathrm{Higgs}}$. In the general case, choose Zariski coverings $(T'_\alpha \to T')_{\alpha\in A}$ and $(T'_{\alpha\beta;\gamma} \to T'_\alpha \times_{T'} T'_\beta)_{\gamma\in\Gamma_{\alpha\beta}}$ $((\alpha,\beta) \in A^2)$ in $\mathscr{C}$ (cf. Definition IV.2.2.2 (3)) such that $T'_{\alpha,N}$ and $T'_{\alpha\beta;\gamma,N}$ are affine. Then $T'_\alpha$ and $T'_{\alpha\beta;\gamma}$ are objects of $\mathscr{C}^r$. Hence we are reduced to the affine case by using the inverse system of exact sequences of $A_N$-algebras:

$$\left(\Gamma(T'_N, \mathcal{O}_{T'_N}) \to \prod_\alpha \Gamma(T'_{\alpha,N}, \mathcal{O}_{T'_{\alpha,N}}) \rightrightarrows \prod_{\alpha\beta}\prod_\gamma \Gamma(T'_{\alpha\beta;\gamma,N}, \mathcal{O}_{T'_{\alpha\beta;\gamma,N}})\right)_N. \qquad \square$$

**Lemma IV.2.2.15.** *Let $f\colon Y' \to Y$ be an étale morphism in $\mathscr{C}$ such that $f_1\colon Y'_1 \to Y_1$ is an isomorphism. Then for any $T \in \mathscr{C}^r$, the morphism $\mathrm{Hom}_{\mathscr{C}}(T, Y') \to \mathrm{Hom}_{\mathscr{C}}(T, Y); g \mapsto f \circ g$ is bijective.*

PROOF. Let $h\colon T \to Y$ be a morphism in $\mathscr{C}$. For $N \in \mathbb{N}_{>0}$, the morphism $T_N \to T_{N+1}$ is an exact closed immersion and the underlying morphism of schemes of its reduction mod $p^m$ is a nilpotent immersion for every $m \in \mathbb{N}_{>0}$ (cf. Corollary IV.2.2.5). Since $Y'_{N+1} \to Y_{N+1}$ is étale, this implies that any morphism $T_N \to Y'_N$ over $Y_N$ has a unique lifting $T_{N+1} \to Y'_{N+1}$ over $Y_{N+1}$. Since $f_1$ is an isomorphism, we see that there exists a unique morphism $g\colon T \to Y'$ such that $f \circ g = h$. $\square$

**Lemma IV.2.2.16.** *Let $Y$ be an object of $\mathscr{C}$ such that $Y_1 \to Y_2$ is a closed immersion, and let $y$ be a point on $Y_1$. Then there exist morphisms $Y'' \xrightarrow{g} Y' \xrightarrow{f} Y$ in $\mathscr{C}$ such that $y$ is contained in the image of $Y''_1$, $Y'_N$ and $Y''_N$ are affine for every $N \in \mathbb{N}_{>0}$, $f$ is strict étale and Cartesian, $g$ is étale, $g_1\colon Y''_1 \to Y'_1$ is an isomorphism, and $Y''_1 \to Y''_2$ is an exact closed immersion.*

PROOF. For $N = 1, 2$, let $Y_{N,1}$ denote the reduction mod $p$ of $Y_N$. Then there exists a strict étale morphism of fine log schemes $f_{2,1}\colon Y'_{2,1} \to Y_{2,1}$ such that $y$ is contained in the image of $Y'_{1,1} := Y_{1,1} \times_{Y_{2,1}} Y'_{2,1} \to Y_{1,1}$, $Y'_{2,1}$ is affine, and the closed immersion $Y'_{1,1} \to Y'_{2,1}$ has a factorization $Y'_{1,1} \xrightarrow{i''_{1,1}} Y''_{2,1} \xrightarrow{g_{2,1}} Y'_{2,1}$, where $Y''_{2,1}$ is affine, $g_{2,1}$ is étale, and $i''_{1,1}$ is an exact closed immersion. Since $Y'_{1,1} \to Y'_{2,1}$ is a closed immersion, $Y'_{1,1}$ is also affine. By **[50]** Proposition (3.14), there exist a unique strict étale lifting $f_2\colon Y'_2 \to Y_2$ of $f_{2,1}$ and a unique étale lifting $g_2\colon Y''_2 \to Y'_2$ of $g_{2,1}$. Put $Y'_1 := Y_1 \times_{Y_2} Y'_2$. Then, since $g_2$ is étale, the factorization $g_{2,1} \circ i''_{1,1}$ has a unique lifting $Y'_1 \xrightarrow{i''_1} Y''_2 \xrightarrow{g_2} Y'_2$. The morphism $i''_1$ is an exact closed immersion because $Y'_1 \to Y'_2$ is a closed immersion and $i''_{1,1}$ is an exact closed immersion. Put $Y''_1 = Y'_1$. By **[50]** Proposition (3.14), the morphism $f_2\colon Y'_2 \to Y_2$ extends uniquely to a strict étale Cartesian morphism $Y' \to Y$ in $\mathscr{C}$, and the morphisms $\mathrm{id}\colon Y''_1 \to Y'_1$ and $g_2\colon Y''_2 \to Y'_2$ extend uniquely to an étale morphism $g\colon Y'' \to Y'$ in $\mathscr{C}$ such that the morphism $Y''_N \to Y''_{N+1} \times_{Y'_{N+1}} Y'_N$ is an isomorphism for every integer $N \geq 2$. $\square$

By combining Proposition IV.2.2.14 with Lemmas IV.2.2.15 and IV.2.2.16, we obtain the following proposition.

**Proposition IV.2.2.17.** *Let $Y$ be an object of $\mathscr{C}$ such that $Y_1 \to Y_2$ is a closed immersion. Then, for every point $y$ of $Y_1$, there exists a strict étale Cartesian morphism $Y' \to Y$ in $\mathscr{C}$ such that $y$ is contained in the image of $Y_1'$ and there exists an affine Higgs envelope of $Y'$ of level $r$.*

PROOF. Let $Y'' \xrightarrow{g} Y' \xrightarrow{f} Y$ be as in Lemma IV.2.2.16. Then $Y''$ has an affine Higgs envelope of level $r$ by Proposition IV.2.2.14, and it is also an affine Higgs envelope of $Y'$ of level $r$ by Lemma IV.2.2.15. □

**Lemma IV.2.2.18.** *Let $(u_\alpha\colon Y_\alpha \to Y)_{\alpha\in A}$ be a family of morphisms in $\mathscr{C}$ satisfying one of the following two conditions.*

(a) *The family $(u_\alpha)$ is a Zariski covering.*

(b) *The family $(u_\alpha)$ is a strict étale covering, the set $A$ is finite, and the morphism of schemes underlying the reduction mod $p^m$ of $u_{\alpha,N}\colon Y_{\alpha,N} \to Y_N$ is quasi-compact for every $\alpha \in A$ and $m, N \in \mathbb{N}_{>0}$.*

*Put $Y_{\alpha\beta} = Y_\alpha \times_Y Y_\beta$ for $\alpha, \beta \in A$ and $Y_{\alpha\beta\gamma} = Y_\alpha \times_Y Y_\beta \times_Y Y_\gamma$ for $\alpha, \beta, \gamma \in A$.*

(1) *Let $Z \to Y$ and $W \to Y$ be morphisms in $\mathscr{C}$, and let $Z_\alpha$ and $W_\alpha$ (resp. $Z_{\alpha\beta}$ and $W_{\alpha\beta}$) be the base changes of $Z$ and $W$ by $Y_\alpha \to Y$ (resp. $Y_{\alpha\beta} \to Y$). Then the following sequence is exact.*

$$\mathrm{Hom}_Y(Z, W) \to \prod_{\alpha\in A} \mathrm{Hom}_{Y_\alpha}(Z_\alpha, W_\alpha) \rightrightarrows \prod_{(\alpha,\beta)\in A^2} \mathrm{Hom}_{Y_{\alpha\beta}}(Z_{\alpha\beta}, W_{\alpha\beta})$$

(2) *Assume that we are given an affine morphism $Z_\alpha \to Y_\alpha$ in $\mathscr{C}$ for each $\alpha \in A$ and an isomorphism $\iota_{\alpha\beta}\colon Z_\beta \times_{Y_\beta} Y_{\alpha\beta} \xrightarrow{\cong} Z_\alpha \times_{Y_\alpha} Y_{\alpha\beta}$ over $Y_{\alpha\beta}$ for each $\alpha, \beta \in A$ such that $(\iota_{\alpha\beta} \times_{Y_{\alpha\beta}} Y_{\alpha\beta\gamma}) \circ (\iota_{\beta\gamma} \times_{Y_{\beta\gamma}} Y_{\alpha\beta\gamma}) = \iota_{\alpha\gamma} \times_{Y_{\alpha\gamma}} Y_{\alpha\beta\gamma}$ for every $(\alpha, \beta, \gamma) \in A^3$. Then there exist an affine morphism $Z \to Y$ in $\mathscr{C}$ and an isomorphism $\iota_\alpha\colon Z \times_Y Y_\alpha \xrightarrow{\cong} Z_\alpha$ over $Y_\alpha$ for each $\alpha \in A$ such that $\iota_{\alpha\beta} \circ (\iota_\beta \times_{Y_\beta} Y_{\alpha\beta}) = \iota_\alpha \times_{Y_\alpha} Y_{\alpha\beta}$. Furthermore the above $(Z \to Y, \iota_\alpha)$ is unique up to a unique isomorphism.*

**Lemma IV.2.2.19.** *Let $(u_\alpha\colon X_\alpha \to X)_{\alpha\in A}$ be a family of strict morphisms of log schemes satisfying one of the following two conditions.*

(a) *The family of morphisms of schemes underlying $(u_\alpha)_{\alpha\in A}$ is an open covering.*

(b) *The set $A$ is finite. The family of morphisms of schemes underlying $(u_\alpha)_{\alpha\in A}$ is a covering by quasi-compact étale morphisms.*

*Put $X_{\alpha\beta} = X_\alpha \times_X X_\beta$ for $\alpha, \beta \in A$, and $X_{\alpha\beta\gamma} = X_\alpha \times_X X_\beta \times_X X_\gamma$ for $\alpha, \beta, \gamma \in A$.*

(1) *Let $U \to X$ and $V \to X$ be morphisms of log schemes. Let $U_\alpha$ and $V_\alpha$ (resp. $U_{\alpha\beta}$ and $V_{\alpha\beta}$) be the base changes of $U$ and $V$ by $X_\alpha \to X$ (resp. $X_{\alpha\beta} \to X$). Then the following sequence is exact.*

$$\mathrm{Hom}_X(U, V) \to \prod_{\alpha\in A} \mathrm{Hom}_{X_\alpha}(U_\alpha, V_\alpha) \rightrightarrows \prod_{(\alpha,\beta)\in A^2} \mathrm{Hom}_{X_{\alpha\beta}}(U_{\alpha\beta}, V_{\alpha\beta}).$$

(2) *Assume that we are give a morphism of log schemes $U_\alpha \to X_\alpha$ whose underlying morphism of schemes is affine for each $\alpha \in A$, and an isomorphism $\iota_{\alpha\beta}\colon U_\beta \times_{X_\beta} X_{\alpha\beta} \xrightarrow{\cong} U_\alpha \times_{X_\alpha} X_{\alpha\beta}$ over $X_{\alpha\beta}$ for each $(\alpha, \beta) \in A^2$ such that $(\iota_{\alpha\beta} \times_{X_{\alpha\beta}} X_{\alpha\beta\gamma}) \circ (\iota_{\beta\gamma} \times_{X_{\beta\gamma}} X_{\alpha\beta\gamma}) = \iota_{\alpha\gamma} \times_{X_{\alpha\gamma}} X_{\alpha\beta\gamma}$. Then there exists a morphism of log schemes $U \to X$ and an isomorphism $\iota_\alpha\colon U \times_X X_\alpha \cong U_\alpha$ over $X_\alpha$ for each $\alpha$ such that the underlying morphism of schemes of $U \to X$ is affine, and $\iota_{\alpha\beta} \circ (\iota_\beta \times_{X_\beta} X_{\alpha\beta}) = \iota_\alpha \times_{X_\alpha} X_{\alpha\beta}$. Furthermore the above $(U \to X, \iota_\alpha)$ is unique up to unique isomorphisms.*

PROOF. For a log scheme $Z$ (resp. a morphism of log schemes $h\colon Z' \to Z$), let $\mathring{Z}$ (resp. $\mathring{h}$) denote the underlying scheme of $Z$ (resp. the morphism of schemes underlying $h$).

(1) The morphism $V \to X$ has a factorization $V \xrightarrow{f} \overline{V} \xrightarrow{g} X$ such that the underlying morphism of $f$ is the identity map and the morphism $g$ is strict. Let $V_\alpha \to \overline{V}_\alpha \to X_\alpha$ and $V_{\alpha\beta} \to \overline{V}_{\alpha\beta} \to X_{\alpha\beta}$ denote the base changes of the above factorization by $X_\alpha \to X$ and $X_{\alpha\beta} \to X$, respectively. Let $\star$ denote $\emptyset$, $\alpha$ or $\alpha\beta$. Then we have $\mathrm{Hom}_{X_\star}(U_\star, \overline{V}_\star) = \mathrm{Hom}_{\mathring{X}_\star}(\mathring{U}_\star, \mathring{\overline{V}}_\star)$. By **[37]** VIII Théorème 5.2, this implies that the claim holds for $\mathrm{Hom}_{X_\star}(U_\star, \overline{V}_\star)$. Hence it suffices to prove that, for any morphism $h\colon U \to \overline{V}$ over $X$ and its base changes $h_\alpha$ and $h_{\alpha\beta}$ by $X_\alpha \to X$ and $X_{\alpha\beta} \to X$, the claim also holds for $\mathrm{Hom}_{\overline{V}_\star}(U_\star, V_\star) = \mathrm{Hom}_{M_{\overline{V}_\star}}(M_{V_\star}, h_{\star *}(M_{U_\star}))$. This is an immediate consequence of the étale descent of morphisms of sheaves of monoids on $\mathring{V}_{\text{ét}}$ with respect to the covering $(\mathring{V}_\alpha \to \mathring{V})_{\alpha \in A}$.

(2) The uniqueness follows from (1). Let us prove the existence. By fpqc descent of affine schemes (cf. **[37]** VIII Théorème 2.1), there exists an affine morphism of schemes $\mathring{f}\colon \mathring{U} \to \mathring{X}$ and an isomorphism $\mathring{\iota}_\alpha\colon \mathring{U} \times_{\mathring{X}} \mathring{X}_\alpha \xrightarrow{\cong} \mathring{U}_\alpha$ over $\mathring{X}_\alpha$ for each $\alpha \in A$ such that $\mathring{\iota}_{\alpha\beta} \circ (\mathring{\iota}_\beta \times_{\mathring{X}_\beta} \mathring{X}_{\alpha\beta}) = \mathring{\iota}_\alpha \times_{\mathring{X}_\alpha} \mathring{X}_{\alpha\beta}$. Put $M_\alpha = \mathring{\iota}_\alpha^*(M_{U_\alpha})$. Then the descent data $(\iota_{\alpha\beta})_{\alpha\beta}$ is equivalent to isomorphisms of log structures $\tau_{\alpha\beta}\colon M_\beta|_{\mathring{U}\times_{\mathring{X}}\mathring{X}_{\alpha\beta}} \xrightarrow{\cong} M_\alpha|_{\mathring{U}\times_{\mathring{X}}\mathring{X}_{\alpha\beta}}$ $(\alpha, \beta \in A)$ satisfying the cocycle condition and the compatibility with the morphisms $\mathring{f}_\alpha^{-1}(M_{X_\alpha}) \to M_\alpha$ $(\alpha \in A)$ induced by $U_\alpha \to X_\alpha$. Here $\mathring{f}_\alpha$ denotes the morphism $\mathring{U} \times_{\mathring{X}} \mathring{X}_\alpha \to \mathring{X}_\alpha$. By gluing of sheaves on the étale site $\mathring{U}_{\text{ét}}$ with respect to an étale covering, we obtain a log structure $M$ on $\mathring{U}$, an isomorphism $\tau_\alpha\colon M|_{\mathring{U}\times_{\mathring{X}}\mathring{X}_\alpha} \xrightarrow{\cong} M_\alpha$ of log structures for each $\alpha \in A$, and a morphism of log structures $\mathring{f}^{-1}(M_X) \to M$ compatible with $\tau_{\alpha\beta}$ and $\mathring{f}_\alpha^{-1}(M_{X_\alpha}) \to M_\alpha$ $(\alpha \in A)$. These data give a morphism of log schemes $f\colon U = (\mathring{U}, M) \to X$ and an isomorphism $\iota_\alpha\colon U \times_X X_\alpha \xrightarrow{\cong} U_\alpha$ over $X_\alpha$ for each $\alpha \in A$ satisfying the desired properties. □

Proof of Lemma IV.2.2.18. We obtain the claim (1) just by applying Lemma IV.2.2.19 (1) to the reduction mod $p^m$ of $X_N \to Y_N$ and $W_N \to Y_N$ for each $m, N \in \mathbb{N}_{>0}$. Similarly we can apply Lemma IV.2.2.19 (2) to the reduction mod $p^m$ of $\iota_{\beta,N}$ for $m, N \in \mathbb{N}_{>0}$, obtaining the claim (2) by fpqc descent of immersions of schemes (**[42]** Proposition (2.7.1)). □

**Lemma IV.2.2.20.** *Let $(u_\alpha\colon Y_\alpha \to Y)_{\alpha\in A}$ be as in Lemma* IV.2.2.18. *If $Y_\alpha$ has an affine Higgs envelope of level $r$ for every $\alpha \in A$, then $Y$ also has an affine Higgs envelope of level $r$.*

Proof. Let $Y_\alpha$ and $Y_{\alpha\beta\gamma}$ be as in Lemma IV.2.2.18. Let $T_\alpha \to Y_\alpha$ be an affine Higgs envelope of $Y_\alpha$ of level $r$. By Lemma IV.2.2.13, $T_\delta \times_{Y_\delta} Y_{\alpha\beta}$ for $\delta \in \{\alpha, \beta\}$ (resp. $T_\delta \times_{Y_\delta} Y_{\alpha\beta\gamma}$ for $\delta \in \{\alpha, \beta, \gamma\}$) is an affine Higgs envelope of $Y_{\alpha\beta}$ (resp. $Y_{\alpha\beta\gamma}$) of level $r$. Hence there exists a unique $Y_{\alpha\beta}$-isomorphism $\iota_{\alpha\beta}\colon T_\beta \times_{Y_\beta} Y_{\alpha\beta} \xrightarrow{\cong} T_\alpha \times_{Y_\alpha} Y_{\alpha\beta}$ for each $(\alpha, \beta) \in A^2$ satisfying the cocycle condition in Lemma IV.2.2.18 (2). Therefore there exists an affine morphism $T \to Y$ in $\mathscr{C}$ and a $Y_\alpha$-isomorphism $\iota_\alpha\colon Y_\alpha \times_Y T \xrightarrow{\cong} T_\alpha$ for each $\alpha \in A$ compatible with $\iota_{\alpha\beta}$. By Lemma IV.2.2.7 (2), $T$ is an object of $\mathscr{C}^r$. Let $f\colon T' \to Y$ be a morphism in $\mathscr{C}$ such that $T' \in \mathrm{Ob}\,\mathscr{C}^r$. Then, for each $\alpha \in A$ (resp. $(\alpha, \beta) \in A^2$), there exists a unique $Y_\alpha$-morphism (resp. $Y_{\alpha\beta}$-morphism) $T' \times_Y Y_\alpha \to T \times_Y Y_\alpha (\cong T_\alpha)$ (resp. $T' \times_Y Y_{\alpha\beta} \to T \times_Y Y_{\alpha\beta} (\cong T_\alpha \times_{Y_\alpha} Y_{\alpha\beta})$). By Lemma IV.2.2.18 (1), we see that there exists a unique $Y$-morphism $T' \to T$. Hence $T \to Y$ is an affine Higgs envelope of level $r$. □

Proof of Proposition IV.2.2.9. Let $Y \in \mathrm{Ob}\,\mathscr{C}$. There exists a Cartesian open immersion $f\colon Y' \to Y$ in $\mathscr{C}$ such that $f_1 = \mathrm{id}$ and $Y'_1 \to Y'_2$ is a closed immersion. For any

$T' \in \mathscr{C}^r$, the morphism of topological spaces underlying $T'_1 \to T'_N$ is a homeomorphism for every $N \in \mathbb{N}_{>0}$ (cf. Corollary IV.2.2.5). This implies that the map $\mathrm{Hom}_{\mathscr{C}}(T', Y') \to \mathrm{Hom}_{\mathscr{C}}(T', Y)$ induced by $f$ is bijective. Hence we may assume that $Y_1 \to Y_2$ is a closed immersion. Let $U \to Y$ be an open immersion such that $U_2$ is the complement of the image of $Y_1$ in $Y_2$. Then $U_1 = \emptyset$, and $\emptyset \to U$ is an affine Higgs envelope of level $r$. Combining this with Proposition IV.2.2.17, we obtain a strict étale covering $(Y_\alpha \to Y)_{\alpha \in A}$ such that $Y_\alpha$ has an affine Higgs envelope of level $r$. If $Y_N$ are affine, then by Lemma IV.2.2.13, we may assume that $Y_{\alpha,N}$ are affine and $A$ is finite. By Lemma IV.2.2.20, we see that $Y$ has an affine Higgs envelope of $r$. The general case is reduced to the affine case by choosing a Zariski covering $(U_\beta \to Y)_{\beta \in B}$ such that $U_{\beta,N}$ are affine and applying Lemma IV.2.2.20. This completes the proof of (1) and (2). The claim (3) follows from Lemma IV.2.2.13. □

**IV.2.3. Rings of power series.** For $r \in \mathbb{N}_{>0} \cup \{\infty\}$, we give an explicit description of the Higgs envelope of level $r$ of a ring of power series with coefficients in an object $A = (A_N)$ of $\mathscr{A}^r_\bullet$ (Proposition IV.2.3.15). We also prove an analogous result for $p$-adic fine log formal schemes as follows. Let $f\colon Y' \to Y$ be a smooth Cartesian morphism in $\mathscr{C}$ (Definitions IV.2.2.1, IV.2.2.2) such that $Y_1$ is affine, and let $i\colon Y_1 \to Y'_1$ be a section of $f_1$ over $S_1$. Under the assumption on the existence of certain log coordinates of $f$, we give an explicit description of the Higgs envelope of level $r$ of $(i\colon Y_1 \hookrightarrow Y')$ (Definition IV.2.2.1 (3)) in terms of the Higgs envelope of level $r$ of $Y$ and the log coordinates (Proposition IV.2.3.17).

**Definition IV.2.3.1.** For $r \in \mathbb{N}_{>0}$ (resp. $r = \infty$), let $\mathscr{M}^r_{\mathrm{alg},\bullet}$ denote the category of inverse systems $(M_N)_{N \in \mathbb{N}_{>0}}$ of $R$-modules satisfying the following conditions:

(0) For each $N \in \mathbb{N}_{>0}$, $\xi^N M_N = 0$ and the transition map $M_{N+1} \to M_N$ is surjective.

(i) The module $M_N$ is $p$-torsion free for every $N \in \mathbb{N}_{>0}$.

(ii) For every $s \in \mathbb{N} \cap [0, r]$ (resp. $s \in \mathbb{N}$) and $N \in \mathbb{N}_{>0}$, we have $pF^s M_N \subset \xi^s M_N$ (resp. $F^s M_N = \xi^s M_N$), where $F^n M_N$ ($n \in \mathbb{N}$) is the decreasing filtration of $M_N$ by $R_N$-submodules defined by $F^0 M_N = M_N$, $F^n M_N = \mathrm{Ker}(M_N \to M_n)$ ($1 \le n \le N$), and $F^n M_N = 0$ ($n > N$).

(iii) For every $N \in \mathbb{N}_{>0}$, the kernel of $\xi\colon M_N \to M_N$ is $F^{N-1} M_N$.

Let $M = (M_N)$ be an object of $\mathscr{M}^r_{\mathrm{alg},\bullet}$. Since $M_N$ is $p$-torsion free, the natural homomorphism $M_N \to M_{N,\mathbb{Q}}$ is injective. We regard $M_N$ as a submodule of $M_{N,\mathbb{Q}}$. Since $M_N / F^n M_N \cong M_n$ is $p$-torsion free for $n \in \mathbb{N} \cap [1, N]$, we have $F^n M_N = M_N \cap F^n M_{N,\mathbb{Q}}$ in $M_{N,\mathbb{Q}}$ for $n \in \mathbb{N}$.

**Lemma IV.2.3.2.** *Let $r \in \mathbb{N}_{>0} \cup \{\infty\}$ and let $M = (M_N)$ be an object of $\mathscr{M}^r_{\mathrm{alg},\bullet}$.*

*(1) For $s, n, N \in \mathbb{N}_{>0}$ such that $s \le n \le N$, the inverse image of $F^n M_N$ under the morphism $\xi^s\colon M_N \to M_N$ is $F^{n-s} M_N$.*

*(2) Assume $r \in \mathbb{N}_{>0}$. For $s \in \mathbb{N} \cap [0, r]$, $n \in \mathbb{N}$ and $N \in \mathbb{N}_{>0}$, we have $pF^{n+s} M_N \subset \xi^s F^n M_N$.*

PROOF. The same as the proof of Lemma IV.2.1.6. □

**Lemma IV.2.3.3.** *Let $r \in \mathbb{N}_{>0} \cup \{\infty\}$ and let $M = (M_N)$ be an object of $\mathscr{M}^r_{\mathrm{alg},\bullet}$.*

*(1) For $N, M \in \mathbb{N}_{>0}$, the homomorphism $\xi^M\colon M_{N+M} \to M_{N+M}$ decomposes as $M_{N+M} \to M_N \xrightarrow{[\xi^M]} M_{N+M}$ with $[\xi^M]$ an injective homomorphism.*

*(2) Assume $r \in \mathbb{N}_{>0}$. For $n \in \mathbb{N}$, $N \in \mathbb{N}_{>0}$, and the smallest integer $m$ such that $nr^{-1} \le m$, we have $p^m F^n M_N \subset \xi^n M_N \subset F^n M_N$. In particular, we have $(F^n M_N)_{\mathbb{Q}} = \xi^n (M_N)_{\mathbb{Q}}$.*

(3) *For* $n, m \in \mathbb{N}$ *and* $N \in \mathbb{N}_{>0}$ *such that* $m \leq n$, *we have* $\xi^m M_N \cap F^n M_N = \xi^m F^{n-m} M_N$.

PROOF. (1) and (3) (resp. (2)) follow(s) from Lemma IV.2.3.2 (1) (resp. (2)). □

Let $r \in \mathbb{N}_{>0} \cup \{\infty\}$ and let $M = (M_N)$ be an object of $\mathscr{M}^r_{\mathrm{alg},\bullet}$. For $n \in \mathbb{N}_{>0}$ and a submodule $P$ (resp. an element $x$) of $\xi^n (M_{N+n})_{\mathbb{Q}} = (F^n M_{N+n})_{\mathbb{Q}}$, we define $\xi^{-n} P$ (resp. $\xi^{-n} x$) to be the inverse image of $P$ (resp. $x$) by the isomorphism $[\xi^n]_{\mathbb{Q}} \colon (M_N)_{\mathbb{Q}} \xrightarrow{\cong} \xi^n (M_{N+n})_{\mathbb{Q}}$.

We assume $r \in \mathbb{N}_{>0}$ until Lemma IV.2.3.13. For $m, n \in \mathbb{N}$ and $N \in \mathbb{N}_{>0}$, we define the $R_N$-submodule $M_N^{(m)_r,n}$ of $M_{N,\mathbb{Q}}$ by

$$M_N^{(m)_r,n} := p^{-n_1} \xi^{n-n_2} F^{n_2} M_{N+\max\{0,n_2-n\}},$$

where $n_1 = \min\{[\frac{n-m}{r}], 0\}$ and $n_2 = n - m - n_1 r$. If $n \leq m + r - 1$, then $n_1 = [\frac{n-m}{r}] \leq 0$ and $0 \leq n_2 \leq r-1$. If $n \geq m$, we have $n_1 = 0$ and $n_2 = n - m$, which imply $M_N^{(m)_r,n} = \xi^m F^{n-m} M_N$. The homomorphism $(M_{N+1})_{\mathbb{Q}_p} \to (M_N)_{\mathbb{Q}_p}$ induces a surjective homomorphism $M_{N+1}^{(m)_r,n} \to M_N^{(m)_r,n}$.

**Lemma IV.2.3.4.** *For* $m, n \in \mathbb{N}$ *and* $N \in \mathbb{N}_{>0}$ *such that* $m \geq n$, *we have*

$$p^{[\frac{m-n}{r}]+1} \xi^n M_N \subset M_N^{(m)_r,n} \subset p^{[\frac{m-n}{r}]} \xi^n M_N.$$

PROOF. Put $n - m = n_1 r + n_2$ ($n_1 \in \mathbb{Z}, n_2 \in \mathbb{Z} \cap [0, r-1]$). We have $[\frac{m-n}{r}] = -n_1$ (if $n_2 = 0$), $-n_1 - 1$ (if $n_2 \neq 0$). If $n_2 = 0$, then we have $M_N^{(m)_r,n} = p^{-n_1} \xi^n M_N$. If $n_2 \neq 0$, then we have $\xi^{n_2} M_{N'} \subset F^{n_2} M_{N'} \subset p^{-1} \xi^{n_2} M_{N'}$ ($N' = N + \max\{0, n_2 - n\}$), which implies $p^{-n_1} \xi^n M_N \subset M_N^{(m)_r,n} \subset p^{-n_1-1} \xi^n M_N$. □

**Lemma IV.2.3.5.** *Let* $m, n \in \mathbb{N}$ *and define* $m_1, m_2, n_2 \in \mathbb{N}$, *and* $n_1 \in \mathbb{Z}$ *by* $m = m_1 r + m_2$, $0 \leq m_2 \leq r-1$, $n_1 = \min\{[\frac{n-m}{r}], 0\}$ *and* $n_2 = n - m - n_1 r$. *Then* $n - n_2 < 0$ *if and only if* $0 \leq n \leq m_2 - 1$. *In this case, we have* $n_2 - n = r - m_2$ *and* $n_1 = -1 - m_1$. *In particular,* $1 \leq n_2 - n \leq r - 1$ *and* $-n_1 \geq 1$.

PROOF. If $n \geq m$, then $n_1 = 0$ and $n - n_2 = m \geq 0$. Assume $n < m$. Put $n = m_2 + l_1 r + l_2$ ($l_1, l_2 \in \mathbb{Z}, 0 \leq l_2 \leq r-1$). Then $n - m = (l_1 - m_1) r + l_2$, which implies $n_1 = l_1 - m_1$, $n_2 = l_2$ and $n - n_2 = m_2 + l_1 r$. On the other hand, $0 \leq n = m_2 + l_1 r + l_2 \leq r - 1 + l_1 r + r - 1$ implies $l_1 \geq -1$. Hence $n - n_2 < 0$ if and only if $l_1 = -1$, which is equivalent to $n \leq m_2 - r + r - 1 = m_2 - 1$. Now the remaining assertions are obvious. □

**Lemma IV.2.3.6.** *For* $M = (M_N) \in \mathrm{Ob}\, \mathscr{M}^r_{\mathrm{alg},\bullet}$ *and* $t, s, N \in \mathbb{Z}$ *such that* $N > 0$, $t \geq 0$, *and* $t + s \geq 0$, *we have* $[\xi^t](\xi^s F^n M_{N+\max\{0,-s\}}) = \xi^{t+s} F^n M_{N+t}$.

PROOF. If $s \geq 0$, the claim follows from the surjectivity of $\xi^s F^n M_{N+t} \to \xi^s F^n M_N$. If $s < 0$, using $[\xi^t] = [\xi^{t+s}][\xi^{-s}]$, we see that the left-hand side coincides with $[\xi^{t+s}](F^n M_{N-s}) = \xi^{t+s} F^n M_{N+t}$. □

**Lemma IV.2.3.7.** *Let* $M = (M_N)$ *be an object of* $\mathscr{M}^r_{\mathrm{alg},\bullet}$. *Let* $N$ *be a positive integer.*

(1) *We have* $M_N^{(m)_r,n} \subset F^n M_N$ *for* $n, m \in \mathbb{N}$.

(2) *Let* $n, m \in \mathbb{N}$, *and put* $n_1 = \min\{[\frac{n-m}{r}], 0\}$ *and* $n_2 = n - m - n_1 r$. *Then we have* $M_N^{(m)_r,n} \cap F^{n+1} M_N \subset M_N^{(m)_r,n+1}$. *The equality holds if* $n \geq m$ *or* $0 \leq n_2 \leq r - 2$.

(3) *For* $n, m \in \mathbb{N}$, *we have* $M_N^{(m+1)_r,n} \subset M_N^{(m)_r,n}$.

PROOF. (1) Put $n_1 = \min\{[\frac{n-m}{r}], 0\}(\leq 0)$ and $n_2 = n-m-n_1r(\geq 0)$. If $n-n_2 \geq 0$, then $M_N^{(m)_r,n} \subset \xi^{n-n_2}F^{n_2}M_N \subset F^nM_N$ by Lemma IV.2.3.3 (3). If $n-n_2 < 0$, then $1 \leq n_2-n \leq r-1$ and $-n_1 \geq 1$ by Lemma IV.2.3.5. Hence, by Lemma IV.2.3.2 (2), we have

$$[\xi^{n_2-n}](M_N^{(m)_r,n}) \subset pF^{n_2}M_{N+n_2-n} \subset \xi^{n_2-n}F^nM_{N+n_2-n} = [\xi^{n_2-n}](F^nM_N).$$

(2) Put $n_1' = \min\{[\frac{(n+1)-m}{r}], 0\}$ and $n_2' = (n+1)-m-n_1'r$. We have $n_1' = n_1$ or $n_1+1$, and $n_1' = n_1+1$ holds if and only if $[\frac{(n+1)-m}{r}] = [\frac{n-m}{r}]+1 \leq 0$, which is equivalent to $n < m$ and $n_2 = r-1$. Choose $N' \in \mathbb{N}$ such that $N' \geq n_2 - n$. By (1) and $F^{n+1}M_N = F^{n+1}M_{N,\mathbb{Q}} \cap M_N$, it suffices to prove that $[\xi^{N'}](M_N^{(m)_r,n}) \cap [\xi^{N'}](F^{n+1}M_{N,\mathbb{Q}})$ is equal to (resp. contained in) $[\xi^{N'}](M_N^{(m)_r,n+1})$ if $n_1' = n_1$ (resp. $n_1' = n_1+1$). By Lemma IV.2.3.6 and Lemma IV.2.3.3 (2)(3), the former is equal to

$$p^{-n_1}\xi^{N'+n-n_2}F^{n_2}M_{N+N'} \cap F^{N'+n+1}M_{N+N',\mathbb{Q}} = p^{-n_1}\xi^{N'+n-n_2}F^{n_2+1}M_{N+N'},$$

which coincides with the latter if $n_1' = n_1$. If $n_1' = n_1+1$, we have

$$[\xi^{N'}](M_N^{(m)_r,n+1}) = p^{-n_1-1}\xi^{N'+n-n_2+r}F^{n_2+1-r}M_{N+N'}$$

by Lemma IV.2.3.6. Hence the claim follows from Lemma IV.2.3.2 (2).

(3) Put $n_1 = \min\{[\frac{n-m}{r}], 0\}$, $n_2 = n-m-n_1r$, $n_1' = \min\{[\frac{n-(m+1)}{r}], 0\}$, $n_2' = n-(m+1)-n_1'r$. Then $(n_1', n_2') = (n_1, n_2-1)$ or $(n_1-1, n_2+(r-1))$. Choose $N' \in \mathbb{N}$ such that $N'+n-n_2 \geq 0$ and $N'+n-n_2' \geq 0$. Then, by Lemma IV.2.3.6, we have

$$[\xi^{N'}](M_N^{(m)_r,n}) = p^{-n_1}\xi^{N'+n-n_2}F^{n_2}M_{N+N'},$$

$$[\xi^{N'}](M_N^{(m+1)_r,n}) = \begin{cases} p^{-n_1}\xi^{N'+n-n_2+1}F^{n_2-1}M_{N+N'} & (\text{if } n_1' = n_1), \\ p^{-n_1+1}\xi^{N'+n-n_2-(r-1)}F^{n_2+(r-1)}M_{N+N'} & (\text{if } n_1' = n_1-1). \end{cases}$$

By Lemma IV.2.3.3 (3) and Lemma IV.2.3.2 (2), we obtain

$$[\xi^{N'}](M_N^{(m+1)_r,n}) \subset [\xi^{N'}](M_N^{(m)_r,n}).$$ □

Let $M = (M_N) \in \mathscr{M}^r_{\mathrm{alg},\bullet}$. For $m \in \mathbb{N}$ and $N \in \mathbb{N}_{>0}$, we define the $R_N$-submodule $M_N^{(m)_r}$ of $M_N$ to be $\sum_{n\geq 0} M_N^{(m)_r,n}$. The homomorphism $M_{N+1,\mathbb{Q}} \to M_{N,\mathbb{Q}}$ induces a surjective homomorphism $M_{N+1}^{(m)_r} \to M_N^{(m)_r}$. We write $M^{(m)_r}$ for the inverse system $(M_N^{(m)_r})$. We define the decreasing filtration $F^nM_N^{(m)_r}$ $(n \in \mathbb{N})$ on $M_N^{(m)_r}$ to be $M_N^{(m)_r} \cap F^nM_{N,\mathbb{Q}}$.

**Corollary IV.2.3.8.** *Let $M = (M_N)$ be an object of $\mathscr{M}^r_{\mathrm{alg},\bullet}$.*

(1) *For $n, m \in \mathbb{N}$ and $N \in \mathbb{N}_{>0}$, we have $F^nM_N^{(m)_r} = \sum_{n'\geq n} M_N^{(m)_r,n'}$.*

(2) *For $m \in \mathbb{N}$ and $N \in \mathbb{N}_{>0}$, we have $M_N^{(m)_r} = \sum_{m\geq n\geq 0} M_N^{(m)_r,n}$.*

**Proposition IV.2.3.9.** *Let $M = (M_N)$ be an object of $\mathscr{M}^r_{\mathrm{alg},\bullet}$.*

(1) *For $m \in \mathbb{N}$, the inverse system $M^{(m)_r}$ is an object of $\mathscr{M}^r_{\mathrm{alg},\bullet}$.*

(2) *For $m \in \mathbb{N}$, we have $(M^{(m)_r})^{(1)_r} = M^{(m+1)_r}$.*

**Lemma IV.2.3.10.** *For an object $M = (M_N)$ of $\mathscr{M}^r_{\mathrm{alg},\bullet}$, $M^{(1)_r}$ is again an object of $\mathscr{M}^r_{\mathrm{alg},\bullet}$.*

PROOF. By Corollary IV.2.3.8 (2), we have

$$M_N^{(1)_r} = M_N^{(1)_r,0} + M_N^{(1)_r,1} = p\xi^{-(r-1)}F^{r-1}M_{N+r-1} + \xi M_N.$$

Since $M_N^{(1)_r} \subset M_N$, it remains to prove $pF^s M_N^{(1)_r} \subset \xi^s M_N^{(1)_r}$ for $s \in \mathbb{N} \cap [1, r]$. By Corollary IV.2.3.8 (1) and Lemma IV.2.3.7 (2), we have $F^s M_N^{(1)_r} = \xi F^{s-1} M_N$. By Lemma IV.2.3.3 (3), we obtain

$$[\xi^{r-1}](pF^s M_N^{(1)_r}) = p\xi^r F^{s-1} M_{N+r-1} \subset p\xi^s F^{r-1} M_{N+r-1} = [\xi^{r-1}](\xi^s M_N^{(1)_r,0}). \quad \square$$

**Lemma IV.2.3.11.** *For $M = (M_N) \in \mathrm{Ob}\,\mathscr{M}^r_{\mathrm{alg},\bullet}$ and $m, n \in \mathbb{N}$, we have $\xi M_N^{(m)_r,n} = M_N^{(m+1)_r,n+1}$.*

Proof. Put $n_1 = \min\{[\frac{n-m}{r}], 0\}$ and $n_2 = n - m - n_1 r$. Since $(n+1) - (m+1) = n - m$, we have

$$M_N^{(m)_r,n} = p^{-n_1}\xi^{n-n_2} F^{n_2} M_{N+\max\{0,n_2-n\}},$$
$$M_N^{(m+1)_r,n+1} = p^{-n_1}\xi^{n+1-n_2} F^{n_2} M_{N+\max\{0,n_2-(n+1)\}}.$$

Choose $N' \in \mathbb{N}$ such that $N' + n - n_2 \geq 0$. Then, by Lemma IV.2.3.6, we have

$$[\xi^{N'}](\xi M_N^{(m)_r,n}) = p^{-n_1}\xi^{N'+n+1-n_2} F^{n_2} M_{N+N'} = [\xi^{N'}](M_N^{(m+1)_r,n+1}),$$

which implies the claim. $\square$

**Lemma IV.2.3.12.** *For $M = (M_N) \in \mathrm{Ob}\,\mathscr{M}^r_{\mathrm{alg},\bullet}$ and $n \in \mathbb{N}$, $n \geq r-1$, we have $p\xi^{-(r-1)} M_{N+(r-1)}^{(m)_r,n} \subset M_N^{(m+1)_r,n-(r-1)}$. Furthermore the equality holds if $n < m + r$.*

Proof. Put $n_1 = \min\{[\frac{n-m}{r}], 0\}$, $n_2 = n - m - n_1 r$, $n_1' = \min\{[\frac{n-(r-1)-(m+1)}{r}], 0\} = \min\{[\frac{n-m}{r}] - 1, 0\}$ and $n_2' = n - (r-1) - (m+1) - n_1' r$. Then we have $(n_1', n_2') = (n_1 - 1, n_2)$ if $[\frac{n-m}{r}] \leq 0$ ($\Leftrightarrow n < m + r$), and $(n_1', n_2') = (n_1, n_2 - r)$ otherwise. Choose $N' \in \mathbb{N}$ such that $N' + n - n_2 - (r-1) \geq 0$ and $N' - (r-1) \geq 0$. It suffices to prove that $[\xi^{N'}](p\xi^{-(r-1)} M_{N+(r-1)}^{(m)_r,n})$ is equal to (resp. contained in) $[\xi^{N'}](M_N^{(m+1)_r,n-(r-1)})$ if $n_1' = n_1 - 1$ (resp. $n_1' = n_1$). By Lemma IV.2.3.6, the former is

$$p[\xi^{N'-(r-1)}](M_N^{(m)_r,n}) = p^{-n_1+1}\xi^{N'-(r-1)+n-n_2} F^{n_2} M_{N+N'},$$

which coincides with the latter if $n_1' = n_1 - 1$. If $n_1' = n_1$, we have

$$[\xi^{N'}](M_N^{(m+1)_r,n-(r-1)}) = p^{-n_1}\xi^{N'+n-n_2+1} F^{n_2-r} M_{N+N'}$$

by Lemma IV.2.3.6, which implies the claim by Lemma IV.2.3.2 (2). $\square$

Proof of Proposition IV.2.3.9. Let $(1)_m$ (resp. $(2)_m$) denote the claim (1) (resp. (2)) for an $m \in \mathbb{N}$. Then $(1)_0$ is trivial by $M^{(0)_r} = M$. $(1)_m$ and $(2)_m$ imply $(1)_{m+1}$ by Lemma IV.2.3.10. Hence it suffices to prove that $(1)_m$ implies $(2)_m$. Assume that $(1)_m$ holds. By Corollary IV.2.3.8, we have

$$\begin{aligned}(M_N^{(m)_r})^{(1)_r} &= p\xi^{-(r-1)} F^{r-1} M_{N+(r-1)}^{(m)_r} + \xi M_N^{(m)_r} \\ &= \sum_{n \geq r-1} p\xi^{-(r-1)} M_{N+r-1}^{(m)_r,n} + \sum_{n \geq 0} \xi M_N^{(m)_r,n}.\end{aligned}$$

Hence the claim $(2)_m$ follows from Lemmas IV.2.3.11 and IV.2.3.12. Note $r - 1 < m + r$. $\square$

Let $A = (A_N)$ be an object of $\mathscr{A}^r_\bullet$ (Definition IV.2.1.5 (2)). Then the underlying inverse system of $R$-modules is an object of $\mathscr{M}^r_{\mathrm{alg},\bullet}$. Hence, by applying the above construction, we obtain objects $A^{(m)_r} = (A_N^{(m)_r})$ ($m \in \mathbb{N}$) of $\mathscr{M}^r_{\mathrm{alg},\bullet}$. Since $A_N \supset A_N^{(m)_r} \supset A_N^{(m)_r,0} \supset p^{-n_1} A_N$, where $-m = n_1 r + n_2$ ($n_1 \in \mathbb{Z}, n_2 \in \mathbb{Z} \cap [0, r-1]$), we see that $A_N^{(m)_r}$ is $p$-adically complete and separated.

**Lemma IV.2.3.13.** *Let the notation and assumption be as above.*

(1) *We have* $A_N^{(m)_r} \cdot A_N^{(m')_r} \subset A_N^{(m+m')_r}$ *for* $m, m' \in \mathbb{N}$ *and* $N \in \mathbb{N}_{>0}$.

(2) *We have* $A_N^{(1)_r} \cdot A_N^{(m)_r} \supset pA_N^{(m+1)_r}$ *for* $m \in \mathbb{N}$ *and* $N \in \mathbb{N}_{>0}$.

PROOF. (1) It suffices to prove $A_N^{(m)_r,n} \cdot A_N^{(m')_r,n'} \subset A_N^{(m+m')_r,n+n'}$ for $n, n', m, m' \in \mathbb{N}$ and $N \in \mathbb{N}_{>0}$. Put $n_1 = \min\{[\frac{n-m}{r}], 0\}$ and $n_2 = n - m - n_1 r$. We define $(n_1', n_2')$ and $(n_1'', n_2'')$ similarly using $(n', m')$ and $(n+n', m+m')$, respectively. Put $l = n_1'' - (n_1 + n_1')$. Then we have $l \geq 0$ and $n_2 + n_2' = n_2'' + lr$. Choose $N' \in \mathbb{N}$ such that $N' + n - n_2 \geq 0$ and $N' + n' - n_2' \geq 0$. Since the images of $A_{N+N'}^{(m)_r,n}$ and $A_{N+N'}^{(m')_r,n'}$ under the projection $A_{N+N'} \to A_N$ are $A_N^{(m)_r,n}$ and $A_N^{(m')_r,n'}$, respectively, we have

$$\begin{aligned}
[\xi^{2N'}](A_N^{(m)_r,n} \cdot A_N^{(m')_r,n'}) &= [\xi^{N'}](A_{N+N'}^{(m)_r,n}) \cdot [\xi^{N'}](A_{N+N'}^{(m')_r,n}) \\
&= p^{-n_1}\xi^{N'+n-n_2}F^{n_2}A_{N+2N'} \cdot p^{-n_1'}\xi^{N'+n'-n_2'}F^{n_2'}A_{N+2N'} \\
&\subset p^{-n_1''+l}\xi^{2N'+(n+n')-n_2''-lr}F^{n_2''+lr}A_{N+2N'} \\
&\subset p^{-n_1''}\xi^{2N'+(n+n')-n_2''}F^{n_2''}A_{N+2N'} \\
&= [\xi^{2N'}](A_N^{(m+m')_r,n+n'})
\end{aligned}$$

by Lemma IV.2.1.6 (3), Lemma IV.2.3.6, and Lemma IV.2.3.2 (2).

(2) Since $A_N^{(1)_r} \supset A_N^{(1)_r,0} = p\xi^{-(r-1)}F^{r-1}A_{N+r-1} \supset pA_N$, we obtain $A_N^{(m)_r} \cdot A_N^{(1)_r} \supset pA_N^{(m)_r} \supset pA_N^{(m+1)_r}$ from Lemma IV.2.3.7 (3). □

Let $r \in \mathbb{N}_{>0} \cup \{\infty\}$ and let $A = (A_N)$ be an object of $\mathscr{A}_{\bullet}^r$. If $r = \infty$, we define $A_N^{(m)_r}$ to be $A_N$ for $m \in \mathbb{N}$ and $N \in \mathbb{N}_{>0}$. Let $d \in \mathbb{N}_{>0}$, and define the object $B = (B_N)$ of $\mathscr{A}_{p,\bullet}$ over $A = (A_N)$ by $B_1 = A_1$, $B_N = A_N\{T_1, \ldots, T_d\}(= \varprojlim_m A_N/p^m A_N[T_1, \ldots, T_d])$ $(N \geq 2)$, $B_2 \to B_1; T_i \mapsto 0$, and $B_{N+1} \to B_N; T_i \mapsto T_i$ $(N \geq 2)$. We define the $R_N$-submodule $A_N[W_1, \ldots, W_d]_r$ of $A_N[W_1, \ldots, W_d]$ to be

$$\bigoplus_{\underline{m} \in \mathbb{N}^d} A_N^{(|\underline{m}|)_r} \underline{W}^{\underline{m}},$$

which is an $A_N$-subalgebra by Lemma IV.2.3.13 (1). Here $\underline{W}^{\underline{m}} = \prod_{1 \leq i \leq d} W_i^{m_i}$ and $|\underline{m}| = m_1 + \cdots + m_d$ for $\underline{m} = (m_1, \ldots, m_d) \in \mathbb{N}^d$. We define $A_N\{W_1, \ldots, W_d\}_r$ to be the $p$-adic completion of $A_N[W_1, \ldots, W_d]_r$. The natural homomorphism $A_{N+1}\{W_1, \ldots, W_d\}_r \to A_N\{W_1, \ldots, W_d\}_r$ is surjective, and hence we obtain an object $(A_N\{W_1, \ldots, W_d\}_r)$ of $\mathscr{A}_{p,\bullet}$, which is denoted by $A\{W_1, \ldots, W_d\}_r$ in the following. We define a morphism $u\colon B \to A\{W_1, \ldots, W_d\}_r$ in $\mathscr{A}_{p,\bullet}$ over $A$ by

$$B_N = A_N\{T_1, \ldots, T_d\} \to A_N\{W_1, \ldots, W_d\}_r; T_i \mapsto \xi W_i, \qquad N \geq 2.$$

Note that we have $\xi \in \xi A_N = A_N^{(1)_r,1} \subset A_N^{(1)_r}$.

**Lemma IV.2.3.14.** *Assume* $r \in \mathbb{N}_{>0}$. *Then, for* $N \in \mathbb{N}_{>0}$, *we have*

$$(A_N\{W_1, \ldots, W_d\}_r)_{\mathbb{Q}} = \left\{ \sum_{\underline{m} \in \mathbb{N}^d} a_{\underline{m}} \underline{W}^{\underline{m}} \in A_{N,\mathbb{Q}}[[W_1, \ldots, W_d]] \,\middle|\, \lim_{|\underline{m}| \to \infty} p^{-[\frac{|\underline{m}|}{r}]} a_{\underline{m}} = 0 \right\}.$$

PROOF. By Lemma IV.2.3.4 and Lemma IV.2.3.7 (1), we have $p^{[\frac{m}{r}]+1}A_N \subset A_N^{(m)_r} \subset p^{[\frac{m-N+1}{r}]}A_N$ for $m \geq N - 1$. □

**Proposition IV.2.3.15.** *Under the notation and assumption as above, $A\{W_1,\ldots,W_d\}_r$ is an object of $\mathscr{A}^r_\bullet$ and $D^r_{\mathrm{Higgs}}(B)$ with the adjunction morphism $B \to D^r_{\mathrm{Higgs}}(B)$ (Proposition IV.2.1.8) is isomorphic to $u\colon B \to A\{W_1,\ldots,W_d\}_r$.*

PROOF. To simplify the notation, we abbreviate $W_1,\ldots,W_d$ to $\underline{W}$ in the notation $A_N\{W_1,\ldots,W_d\}_r$, $A\{W_1,\ldots,W_d\}$, etc. Since $A_N^{(|\underline{m}|)_r}$ (resp. $F^nA_N^{(|\underline{m}|)_r}$) is $p$-adically complete and separated, an element of $A_N\{\underline{W}\}_r$ (resp. $F^n(A_N\{\underline{W}\}_r)$) is written uniquely in the form $\sum_{\underline{m}\in\mathbb{N}^d} a_{\underline{m}}\underline{W}^{\underline{m}}$ where $a_{\underline{m}} \in p^{l(|\underline{m}|)}A_N^{(|\underline{m}|)_r}$ (resp. $p^{l(|\underline{m}|)}F^nA_N^{(|\underline{m}|)_r}$) for some sequence $l(m)\in\mathbb{N}$ ($m\in\mathbb{N}$) tending to $\infty$ as $m\to\infty$. Hence, by applying Proposition IV.2.3.9 (1) to $A$ in the case $r\in\mathbb{N}_{>0}$, we see that $A\{\underline{W}\}_r$ is an object of $\mathscr{A}^r_\bullet$.

Let $C=(C_N)$ be an object of $\mathscr{A}^r_\bullet$ and let $f=(f_N)\colon B\to C$ be a morphism in $\mathscr{A}^r_\bullet$. It suffices to prove that there exists a unique morphism $g=(g_N)\colon A\{\underline{W}\}_r\to C$ such that $f=g\circ u$. Since $f_{N+1}(T_i)\in F^1C_{N+1}\subset \xi\cdot C_{N+1,\mathbb{Q}}$ by Lemma IV.2.3.3 (2) if $r\in\mathbb{N}_{>0}$, there exists a unique $w_{i,N}\in C_{N,\mathbb{Q}}$ such that $[\xi](w_{i,N})=f_{N+1}(T_i)$. These elements form an inverse system $(w_{i,N})\in\varprojlim_N C_{N,\mathbb{Q}}$. We define the compatible system of $A_N$-algebra homomorphisms $h_N\colon A_N[\underline{W}]_{\mathbb{Q}}\to C_{N,\mathbb{Q}}$ by $h_N(W_i)=w_{i,N}$. We assert $h_N(A_N[\underline{W}]_r)\subset C_N$. If $r=\infty$, $F^1C_{N+1}=\xi C_{N+1}$ implies $w_{i,N}\in C_N$. Hence the claim is trivial. Assume $r\in\mathbb{N}_{>0}$. For $\underline{m}=(m_i)\in\mathbb{N}^d$, let $\underline{w}_N^{\underline{m}}$ (resp. $\underline{T}^{\underline{m}}$) denote $\prod_{1\le i\le d}w_{i,N}^{m_i}$ (resp. $\prod_{1\le i\le d}T_i^{m_i}$). For $\underline{m}\in\mathbb{N}^d$ such that $0\le|\underline{m}|\le r$, we have $p[\xi^{|\underline{m}|}]\underline{w}_N^{\underline{m}}=pf_{N+|\underline{m}|}(\underline{T}^{\underline{m}})\in pF^{|\underline{m}|}C_{N+|\underline{m}|}\subset \xi^{|\underline{m}|}C_{N+|\underline{m}|}$ by Lemma IV.2.1.6 (3). Hence $p\underline{w}_N^{\underline{m}}\in C_N$. We also have $\xi w_{i,N}=f_N(T_i)\in F^1C_N$. Since $f_N(A_N^{(m)_r,n})\subset C_N^{(m)_r,n}$ by definition, the proof of the above claim is reduced to showing that $C_N^{(m)_r,n}\underline{w}_N^{\underline{m}}\subset C_N$ for $m,n\in\mathbb{N}$, $N\in\mathbb{N}_{>0}$, and $\underline{m}\in\mathbb{N}^d$ such that $|\underline{m}|=m$. Put $n_1=\min\{[\frac{n-m}{r}],0\}$ and $n_2=n-m-n_1r$. Recall $C_N^{(m)_r,n}=p^{-n_1}\xi^{n-n_2}F^{n_2}C_{N+\max\{0,n_2-n\}}$. If $n-n_2\ge|\underline{m}|=m$, the claim follows from $\xi^m\underline{w}_N^{\underline{m}}\in F^mC_N$ and $-n_1\ge 0$. If $0\le n-n_2\le|\underline{m}|=m$, choosing a decomposition $\underline{m}=\underline{m}'+\underline{m}''$ ($\underline{m}',\underline{m}''\in\mathbb{N}^d$) such that $|\underline{m}''|=n-n_2$, we obtain $\xi^{n-n_2}\underline{w}_N^{\underline{m}''}\in F^{n-n_2}C_N$ and $p^{-n_1}\underline{w}_N^{\underline{m}'}\in C_N$ because $|\underline{m}'|=m-(n-n_2)=-n_1r$. This implies the claim. Suppose $n-n_2<0$. Then we have $1\le n_2-n\le r-1$ and $-n_1\ge 1$ by Lemma IV.2.3.5. By Lemma IV.2.3.6, we have

$$[\xi^r](C_N^{(m)_r,n}\underline{w}_N^{\underline{m}})=p^{-n_1}\xi^{r+n-n_2}F^{n_2}C_{N+r}\cdot\underline{w}_{N+r}^{\underline{m}}.$$

Put $l=r+n-n_2(\ge 0)$. Then $m-l=(-n_1-1)r\ge 0$. Choosing a decomposition $\underline{m}=\underline{m}'+\underline{m}''$ ($\underline{m}',\underline{m}''\in\mathbb{N}^d$) such that $|\underline{m}''|=l$, we obtain $p^{-n_1-1}\underline{w}_N^{\underline{m}'}\in C_N$ and

$$p\xi^lF^{n_2}C_{N+r}\cdot\underline{w}_{N+r}^{\underline{m}''}\subset pF^lC_{N+r}F^{n_2}C_{N+r}\subset pF^{l+n_2}C_{N+r}\subset\xi^rF^nC_{N+r}$$

by Lemma IV.2.1.6 (3) and Lemma IV.2.3.2 (2). Hence $[\xi^r](C_N^{(m)_r,n}\underline{w}_N^{\underline{m}})\subset[\xi^r](F^nC_N)$, which implies the claim.

Now, by taking the $p$-adic completion of $h_N|_{A_N[\underline{W}]_r}\colon A_N[\underline{W}]_r\to C_N$, we obtain a morphism $g\colon A\{\underline{W}\}_r\to C$ such that $f=g\circ u$. It remains to prove the uniqueness. Let $g'=(g'_N)\colon A\{\underline{W}\}_r\to C$ be a morphism in $\mathscr{A}^r_\bullet$ such that $f=g'\circ u$. Then, letting $g'_{N,\mathbb{Q}}$ denote the morphism $A_N\{\underline{W}\}_{r,\mathbb{Q}}\to C_{N,\mathbb{Q}}$ induced by $g'_N$, we have $f_{N+1}(T_i)=\xi g'_{N+1,\mathbb{Q}}(W_i)=[\xi]g'_{N,\mathbb{Q}}(W_i)$, which implies $g'_{N,\mathbb{Q}}(W_i)=w_{i,N}$. Hence $g'_N|_{A_N[\underline{W}]_r}=g_N|_{A_N[\underline{W}]_r}$. By taking the $p$-adic completion, we obtain $g'_N=g_N$. □

**Lemma IV.2.3.16.** *For $N,l\in\mathbb{N}_{>0}$, we have the following equality, where $^\wedge$ denotes the $p$-adic completion.*

$$(A_N\{W_1,\ldots,W_d\}_r)^{(l)_r}=\left(\bigoplus_{\underline{m}\in\mathbb{N}^d}A_N^{(|\underline{m}|+l)_r}\underline{W}^{\underline{m}}\right)^\wedge.$$

PROOF. The claim is trivial if $r = \infty$. Assume $r \in \mathbb{N}_{>0}$. For $M \in \mathrm{Ob}\,\mathscr{M}^r_{\mathrm{alg},\bullet}$ and $m, l \in \mathbb{N}$, the submodule $M_N^{(m)_r}$ of $M_N$ is the sum of finite number of submodules $M_N^{(m)_r,n}$ $(0 \leq n \leq m)$ by Corollary IV.2.3.8 (2), and we have $(M^{(m)_r})^{(l)_r} = M^{(m+l)_r}$ by Proposition IV.2.3.9. Hence the claim follows from the fact that an element of $F^n(A_N\{W_1, \ldots, W_d\}_r)$ is written uniquely in the form $\sum_{\underline{m}\in\mathbb{N}^d} a_{\underline{m}} \underline{W}^{\underline{m}}$ where $a_{\underline{m}} \in p^{l(|\underline{m}|)} F^n A_N^{(|\underline{m}|)_r}$ for some sequence $l(m) \in \mathbb{N}$ $(m \in \mathbb{N})$ tending to $\infty$ as $m \to \infty$. □

**Proposition IV.2.3.17.** *Let $f\colon Y' \to Y$ be a morphism in the category $\mathscr{C}$ (Definition IV.2.2.1 (1)) such that $f_1\colon Y'_1 \to Y_1$ is the identity, $Y_1$ is affine, $f_N\colon Y'_N \to Y_N$ $(N \geq 2)$ are smooth, and the morphisms $Y'_N \to Y'_{N+1} \times_{Y_{N+1}} Y_N$ $(N \geq 2)$ induced by $f$ are isomorphisms. We further assume that there exist $t_i = (t_{i,N}) \in \varprojlim_N \Gamma(Y'_N, M_{Y'_N})$ and $s_i = (s_{i,N}) \in \varprojlim_N \Gamma(Y_N, M_{Y_N})$ $(1 \leq i \leq d)$ such that $\{d\log(t_{i,N})\}_{1\leq i\leq d}$ is a basis of $\Omega^1_{Y'_N/Y_N}$ for $N \geq 2$ and $s_{i,1} = t_{i,1}$. Let $r \in \mathbb{N}_{>0} \cup \{\infty\}$, and let $D$ and $D'$ be $D^r_{\mathrm{Higgs}}(Y)$ and $D^r_{\mathrm{Higgs}}(Y')$, respectively, for which $D_N$ and $D'_N$ $(N \in \mathbb{N}_{>0})$ are affine by Proposition IV.2.2.9 (2) and the remark after Corollary IV.2.2.5. Put $A_N = \Gamma(D_N, \mathcal{O}_{D_N})$ and $A'_N = \Gamma(D'_N, \mathcal{O}_{D'_N})$, which define objects $A = (A_N)$ and $A' = (A'_N)$ of $\mathscr{A}^r_\bullet$ (cf. Lemma IV.2.2.4). For $i \in \{1, 2, \ldots, d\}$ and $N \in \mathbb{N}$, $N \geq 2$, let $u_{i,N}$ denote the unique element of $1 + F^1 A'_N$ such that $p^*_{Y',N}(t_{i,N}) = p^*_{Y',N} f^*_N(s_{i,N}) u_{i,N}$ in $\Gamma(D'_N, M_{D'_N})$, where $p_{Y'}$ denotes the canonical morphism $D' \to Y'$. Then we have a canonical isomorphism*

$$A\{W_1, \ldots, W_d\}_r \xrightarrow{\cong} A'$$

*in $\mathscr{A}^r_\bullet$ over $A$ such that the image of $W_i$ in $(A'_N)_{\mathbb{Q}}$ is $\xi^{-1}(u_{i,N+1} - 1)$.*

PROOF. For any $T \in \mathrm{Ob}\,\mathscr{C}^r$, the natural map $\mathrm{Hom}_{\mathscr{C}}(T, D \times_Y Y') \to \mathrm{Hom}_{\mathscr{C}}(T, Y')$ is bijective because $\mathrm{Hom}_{\mathscr{C}}(T, D) \to \mathrm{Hom}_{\mathscr{C}}(T, Y)$ is bijective. Hence we may replace $Y' \to Y$ with $D \times_Y Y' \to D$ and assume $D = Y$. For an integral monoid $P$, let $\mathbb{Z}_p\{P\}$ denote the $p$-adic completion of $\mathbb{Z}_p[P]$, and let $\mathrm{Spf}(\mathbb{Z}_p)\{P\}$ denote the $p$-adic formal scheme $\mathrm{Spf}(\mathbb{Z}_p\{P\})$ endowed with the log structure associated to the inclusion map $P \hookrightarrow \mathbb{Z}_p\{P\}$. Let $\underline{1}_i$ $(1 \leq i \leq d)$ denote the element of $\mathbb{N}^d$ whose $i$-th component is 1 and other components are 0. Then, for $N \in \mathbb{N}$ such that $N \geq 2$, we have a commutative diagram of $p$-adic fine log formal schemes

$$\begin{array}{ccc} Y'_N & \longrightarrow & \mathrm{Spf}(\mathbb{Z}_p)\{\mathbb{N}^d \oplus \mathbb{N}^d\} \\ {\scriptstyle f_N}\downarrow & & \downarrow \\ Y_N & \longrightarrow & \mathrm{Spf}(\mathbb{Z}_p)\{\mathbb{N}^d\}, \end{array}$$

where the lower (resp. upper) horizontal morphism is defined by $\mathbb{N}^d \to \Gamma(Y_N, M_{Y_N}); \underline{1}_i \mapsto s_{i,N}$ (resp. $\mathbb{N}^d \oplus \mathbb{N}^d \to \Gamma(Y'_N, M_{Y'_N}); (\underline{1}_i, 0), (0, \underline{1}_i) \mapsto f^*_N(s_{i,N}), t_{i,N}$) and the right vertical morphism is defined by $\mathbb{N}^d \to \mathbb{N}^d \oplus \mathbb{N}^d; a \mapsto (a, 0)$. Let $Y''_N$ be the fiber product of $\mathrm{Spf}(\mathbb{Z}_p)\{\mathbb{N}^d \oplus \mathbb{N}^d\} \to \mathrm{Spf}(\mathbb{Z}_p)\{\mathbb{N}^d\} \leftarrow Y_N$ for $N \in \mathbb{N}$, $N \geq 2$. Then the above diagram induces an étale morphism $Y'_N \to Y''_N$. Hence, by Lemma IV.2.2.15, we may replace $Y'_N$ with $Y''_N$ for $N \geq 2$ and assume that the above diagram is Cartesian. Let $Q$ be the inverse image of $\mathbb{N}^d$ by $\mathbb{Z}^d \oplus \mathbb{Z}^d \to \mathbb{Z}^d; (a, b) \mapsto a + b$. For $N \in \mathbb{N}$, $N \geq 2$, let $\widetilde{Y}'_N$ be the fiber product of $Y'_N \to \mathrm{Spf}(\mathbb{Z}_p)\{\mathbb{N}^d \oplus \mathbb{N}^d\} \leftarrow \mathrm{Spf}(\mathbb{Z}_p)\{Q\}$. Then the natural morphism $\widetilde{Y}'_N \to Y'_N$ is étale, and the morphism $Q \to \Gamma(Y'_1, M_{Y'_1}); (-\underline{1}_i, \underline{1}_i), (0, \underline{1}_i) \mapsto 1, t_{i,1}$ defines a factorization $Y'_1 \to \widetilde{Y}''_2 \to Y'_2$ because $t_{i,1} = s_{i,1}$ on $Y'_1 = Y_1$. Putting $\widetilde{Y}'_1 = Y'_1$, we obtain an object $\widetilde{Y}' = (\widetilde{Y}'_N)$ of $\mathscr{C}$ and the étale morphism $\widetilde{Y}' \to Y'$ induces an isomorphism $D^r_{\mathrm{Higgs}}(\widetilde{Y}') \cong D'$ by Lemma IV.2.2.15. Since $\widetilde{Y}' \to Y$ is strict, the closed immersion

$\widetilde{Y}'_1 \to \widetilde{Y}'_2$ is exact. Let $v_{i,N}$ denote the image of $(-\underline{1}_i, \underline{1}_i) \in Q^\times$ in $\Gamma(\widetilde{Y}'_N, \mathcal{O}^\times_{\widetilde{Y}'_N})$. Then the pull-back of $v_{i,N}$ on $D^r_{\mathrm{Higgs}}(\widetilde{Y}')_N \cong D'_N$ is $u_{i,N}$. On the other hand, for $N \geq 2$, $\widetilde{Y}'_N$ is affine and we have an isomorphism

$$A_N[V_1, V_1^{-1}, \ldots, V_d, V_d^{-1}]^\wedge \cong \Gamma(\widetilde{Y}'_N, \mathcal{O}_{\widetilde{Y}'_N})$$

sending $V_i$ to $v_{i,N}$, where $^\wedge$ denotes the $p$-adic completion. Replacing $\widetilde{Y}'_N$ $(N \geq 2)$ with the open $p$-adic fine log formal scheme of $\widetilde{Y}'_N$ defined by the open formal subscheme $\mathrm{Spf}(A_N\{V_1 - 1, \ldots, V_d - 1\})$ of the underlying $p$-adic formal scheme of $\widetilde{Y}'_N$, we obtain an object $\widetilde{Y}''$ of $\mathscr{C}$ with $D^r_{\mathrm{Higgs}}(\widetilde{Y}'') \xrightarrow{\cong} D^r_{\mathrm{Higgs}}(\widetilde{Y}')$. Now the claim follows from the proof of Proposition IV.2.2.14 and Proposition IV.2.3.15. □

**IV.2.4. Differential modules.** For $r \in \mathbb{N}_{>0} \cup \{\infty\}$, $B \in \mathrm{Ob}\,\mathscr{C}^r$ (Definition IV.2.2.1 (2)), a smooth Cartesian morphism $Y \to B$ in $\mathscr{C}$ (Definition IV.2.2.2 (1), (2)), and an immersion $i\colon X \to Y_1$ of $p$-adic fine log formal schemes, we construct a canonical derivation of the structure sheaf of the Higgs envelope of level $r$ of $(i\colon X \to Y)$ (Definitions IV.2.2.1 (3), IV.2.2.10) with values in the differential module of $Y/B$ (see (IV.2.4.8)), and study its properties (Lemma IV.2.4.10, Proposition IV.2.4.13).

**Lemma IV.2.4.1.** *Let $r \in \mathbb{N}_{>0} \cup \{\infty\}$ and let $T = (T_N)$ be an object of $\mathscr{C}^r$.*

*(1) For $N, M \in \mathbb{N}_{>0}$, the morphism $\xi^N\colon \mathcal{O}_{T_{N+M}} \to \mathcal{O}_{T_{N+M}}$ factors as $\mathcal{O}_{T_{N+M}} \to \mathcal{O}_{T_M} \xrightarrow{[\xi^N]} \mathcal{O}_{T_{N+M}}$ with $[\xi^N]$ injective.*

*(2) Assume $r \in \mathbb{N}_{>0}$. For $n, l \in \mathbb{N}$ and the smallest integer $m$ such that $nr^{-1} \leq m$, we have $p^m F^{l+n}\mathcal{O}_{T_N} \subset \xi^n F^l \mathcal{O}_{T_N}$. In particular, we have $F^n\mathcal{O}_{T_N,\mathbb{Q}} = \xi^n \mathcal{O}_{T_N,\mathbb{Q}}$.*

Proof. (1) By the same argument as the proof of Lemma IV.2.1.6 (1), we see that the inverse image of $F^n\mathcal{O}_{T_N}$ under $\xi\colon \mathcal{O}_{T_N} \to \mathcal{O}_{T_N}$ is $F^{n-1}\mathcal{O}_{T_N}$ for positive integers $n \leq N$.

(2) As in the proof of Lemma IV.2.1.6 (2), we have $pF^{n+s}\mathcal{O}_{T_N} \subset \xi^s\mathcal{O}_{T_N} \cap F^{n+s}\mathcal{O}_{T_N} = \xi^s F^n \mathcal{O}_{T_N}$ for $s \in \mathbb{N} \cap [0, r]$ and $n, N \in \mathbb{N}$ such that $n + s < N$. □

Let $r \in \mathbb{N}_{>0} \cup \{\infty\}$ and let $T$ be an object of $\mathscr{C}^r$. For $n \in \mathbb{N}_{>0}$ and a submodule $\mathcal{P}$ of $\xi^n\mathcal{O}_{T_{N+n},\mathbb{Q}} = F^n\mathcal{O}_{T_{N+n},\mathbb{Q}}$, we define $\xi^{-n}\mathcal{P}$ to be the inverse image of $\mathcal{P}$ under the isomorphism $[\xi^n]_{\mathbb{Q}}\colon \mathcal{O}_{T_N,\mathbb{Q}} \xrightarrow{\cong} \xi^n\mathcal{O}_{T_{N+n},\mathbb{Q}}$. Let $m \in \mathbb{N}$ and $N \in \mathbb{N}_{>0}$. If $r \in \mathbb{N}_{>0}$, we define the subsheaf $\mathcal{O}^{(m)_r}_{T_N}$ of $\mathcal{O}_{T_N,\mathbb{Q}}$ to be

$$\sum_{n \in \mathbb{N}} p^{-n_1}\xi^{n-n_2}F^{n_2}\mathcal{O}_{T_{N+\max\{0,n_2-n\}}},$$

where $n_1 = \min\{[\frac{n-m}{r}], 0\}$ and $n_2 = n - m - n_1 r$. We have $\mathcal{O}^{(m)_r}_{T_N} \supset p^{-m_1}\mathcal{O}_{T_N}$, where $-m = m_1 r + m_2$ $(m_1 \in \mathbb{Z}, m_2 \in \mathbb{Z} \cap [0, r-1])$. If $r = \infty$, we define $\mathcal{O}^{(m)_r}_{T_N} = \mathcal{O}_{T_N}$.

**Lemma IV.2.4.2.** *Let $m \in \mathbb{N}$ and $N \in \mathbb{N}_{>0}$.*

*(1) The sheaf $\mathcal{O}^{(m)_r}_{T_N}$ is an ideal of $\mathcal{O}_{T_N}$.*

*(2) Assume that $T_1$ is affine, and let $A = (A_N)$ be the object of $\mathscr{A}^r_\bullet$ defined by $A_N = \Gamma(T_N, \mathcal{O}_{T_N})$ (cf. Lemma IV.2.2.4 (1)). Then the ideal $A^{(m)_r}_N$ of $A_N$ coincides with $\Gamma(T_N, \mathcal{O}^{(m)_r}_{T_N})$.*

*(3) We have $p\mathcal{O}^{(m+1)_r}_{T_N} \subset \mathcal{O}^{(1)_r}_{T_N} \cdot \mathcal{O}^{(m)_r}_{T_N} \subset \mathcal{O}^{(m+1)_r}_{T_N}$.*

**Sublemma IV.2.4.3.** *Assume that $r \in \mathbb{N}_{>0}$ and $T_1$ is affine, and let $A = (A_N)$ be the object of $\mathscr{A}^r_\bullet$ defined by $A_N = \Gamma(T_N, \mathcal{O}_{T_N})$.*

(1) *For* $n \in \mathbb{N}$ *and* $m, N \in \mathbb{N}_{>0}$, *the* $\mathcal{O}_{T_N}/p^m\mathcal{O}_{T_N}$-*module* $F^n\mathcal{O}_{T_N}/p^mF^n\mathcal{O}_{T_N}$ *is quasi-coherent and the natural morphism* $F^nA_N/p^mF^nA_N \to \Gamma(T_N, F^n\mathcal{O}_{T_N}/p^mF^n\mathcal{O}_{T_N})$ *is an isomorphism.*

(2) *For* $m, N \in \mathbb{N}_{>0}$ *and* $s \in \mathbb{N} \cap [0, r]$, *the* $\mathcal{O}_{T_N}/p^m\mathcal{O}_{T_N}$-*module*

$$p\xi^{-s}F^{n+s}\mathcal{O}_{T_{N+s}}/p^mF^{n+s}\mathcal{O}_{T_N}$$

*is quasi-coherent and the natural morphism*

$$p\xi^{-s}F^{n+s}A_{N+s}/p^mF^{n+s}A_N \to \Gamma(T_N, p\xi^{-s}F^{n+s}\mathcal{O}_{T_{N+s}}/p^mF^{n+s}\mathcal{O}_{T_N})$$

*is an isomorphism.*

PROOF. (1) Since $\mathcal{O}_{T_N}$ and $\mathcal{O}_{T_n}$ are $p$-torsion free, we have an exact sequence

$$0 \to F^n\mathcal{O}_{T_N}/p^mF^n\mathcal{O}_{T_N} \to \mathcal{O}_{T_N}/p^m\mathcal{O}_{T_N} \to \mathcal{O}_{T_n}/p^m\mathcal{O}_{T_n} \to 0.$$

We also have an exact sequence with $\mathcal{O}_{T_N}$ replaced by $A_N$. Hence the claim follows from the fact that $\mathcal{O}_{T_N}/p^m\mathcal{O}_{T_N}$ and $\mathcal{O}_{T_n}/p^m\mathcal{O}_{T_n}$ are quasi-coherent $\mathcal{O}_{T_N}/p^m\mathcal{O}_{T_N}$-modules.

(2) Note that we have $pF^{n+s}\mathcal{O}_{T_N} \subset p\xi^{-s}F^{n+s}\mathcal{O}_{T_{N+s}} \subset F^n\mathcal{O}_{T_N}$ by Lemma IV.2.4.1 (2), and similar inclusions for $A$. We have a commutative diagram

$$\begin{array}{ccccc}
\dfrac{F^n\mathcal{O}_{T_N}}{p^mF^n\mathcal{O}_{T_N}} & \xrightarrow[{[\xi^s]}]{\sim} & \dfrac{\xi^sF^n\mathcal{O}_{T_{N+s}}}{p^m\xi^sF^n\mathcal{O}_{T_{N+s}}} & & \\
\uparrow & & \uparrow & & \\
\dfrac{p\xi^{-s}F^{n+s}\mathcal{O}_{T_{N+s}}}{p^mF^n\mathcal{O}_{T_N}} & \xrightarrow[{[\xi^s]}]{\sim} & \dfrac{pF^{n+s}\mathcal{O}_{T_{N+s}}}{p^m\xi^sF^n\mathcal{O}_{T_{N+s}}} & \xleftarrow{p} & \dfrac{F^{n+s}\mathcal{O}_{T_{N+s}}}{p^mF^{n+s}\mathcal{O}_{T_{N+s}}}
\end{array}$$

and a similar diagram for $A$. We obtain the claim by comparing the $\Gamma(T_N, -)$ of the diagram for $\mathcal{O}_{T_\bullet}$ with the diagram for $A$ and noting that $F^n\mathcal{O}_{T_N}/p^mF^N\mathcal{O}_{T_N}$ and $F^{n+s}\mathcal{O}_{T_{N+s}}/p^mF^{n+s}\mathcal{O}_{T_{N+s}}$ are quasi-coherent by (1). □

PROOF OF LEMMA IV.2.4.2. The claim is obvious if $r = \infty$. We assume $r \in \mathbb{N}_{>0}$.

(1) The same as the proof of Lemma IV.2.3.7 (1) using Lemma IV.2.4.1 (2) and $\xi^mF^n\mathcal{O}_{T_N} \subset F^{n+m}\mathcal{O}_{T_N}$ which follows from Lemma IV.2.4.1 (1).

(2) Let $l$ be a positive integer such that $-l \leq [\frac{-m}{r}]$, and let $\mathcal{O}_{T_N,l}^{(m)_r}$ (resp. $A_{N,l}^{(m)_r}$) be the image of $\mathcal{O}_{T_N}^{(m)_r}$ (resp. $A_N^{(m)_r}$) in $\mathcal{O}_{T_N}/p^l\mathcal{O}_{T_N}$ (resp. $A_N/p^lA_N$). We have $p^l\mathcal{O}_{T_N} \subset \mathcal{O}_{T_N}^{(m)_r}$, $p^lA_{T_N} \subset A_{T_N}^{(m)_r}$ and the morphism $p^lA_N \to \Gamma(T_N, p^l\mathcal{O}_{T_N})$ is an isomorphism. Hence it suffices to prove that $A_{N,l}^{(m)_r} \to \Gamma(T_N, \mathcal{O}_{T_N,l}^{(m)_r})$ is an isomorphism. For $n \in \mathbb{N}$, put $n_1 = \min\{[\frac{n-m}{r}], 0\}$ and $n_2 = n - m - n_1r$. Then $\mathcal{O}_{T_N,l}^{(m)_r}$ is the sum of the images of morphisms of $\mathcal{O}_{T_N}/p^l\mathcal{O}_{T_N}$-modules

$$p\xi^{n-n_2}F^{n_2}\mathcal{O}_{T_{N+n-n_2}}/p^lF^n\mathcal{O}_{T_N} \xrightarrow{p^{-n_1-1}} \mathcal{O}_{T_N}/p^l\mathcal{O}_{T_N} \qquad (\text{if } n - n_2 < 0),$$

$$F^{n_2}\mathcal{O}_{T_N}/p^lF^{n_2}\mathcal{O}_{T_N} \xrightarrow{p^{-n_1}\xi^{n-n_2}} \mathcal{O}_{T_N}/p^l\mathcal{O}_{T_N} \qquad (\text{if } n - n_2 \geq 0).$$

Hence the claim follows from Sublemma IV.2.4.3.

(3) By applying (2) to a strict étale Cartesian morphism $U \to T$ in $\mathscr{C}$ such that $U_1$ is affine, we are reduced to Lemma IV.2.3.13. □

Let $r \in \mathbb{N}_{>0} \cup \{\infty\}$ and let $B \in \mathrm{Ob}\,\mathscr{C}^r$. Let $X$ be a $p$-adic fine log formal scheme over $B_1$, let $Y \to B$ be a smooth Cartesian morphism in $\mathscr{C}$ (Definition IV.2.2.2), and let $X \to Y_1$ be an immersion over $B_1$. For $\nu \in \mathbb{N}$, let $Y(\nu)$ denote the fiber product over $B$ of $\nu + 1$ copies of $Y$ (cf. Lemma IV.2.2.8). The immersion $X \to Y(\nu)_2$ and the exact closed immersion $Y(\nu)_N \to Y(\nu)_{N+1}$ ($N \geq 2$) define an object of $\mathscr{C}$. Let $D(\nu)$ denote its image under the functor $D^r_{\mathrm{Higgs}}$ in Proposition IV.2.2.9 (1). We write $D$ for $D(0)$. Let $\Delta_D\colon D \to D(1)$ be the exact closed immersion induced by the diagonal morphisms $Y_N \to Y(1)_N$ ($N \geq 2$), and let $p_i$ ($i = 1, 2$) denote the morphism $D(1) \to D$ induced by the $i$-th projection $Y(1)_N \to Y_N$ ($N \geq 2$). Let $p_{ij}$ ($(i,j) = (1,2), (2,3), (1,3)$) denote the morphism $D(2) \to D(1)$ induced by the morphism defined by $i$-th and $j$-th projections $Y(2)_N \to Y(1)_N$ ($N \geq 2$), and let $q_i$ ($i = 1, 2, 3$) denote the morphism $D(2) \to D$ induced by the $i$-th projection $Y(2)_N \to Y_N$ ($N \geq 2$).

**Remark IV.2.4.4.** Let $Y' \to Y$ be a strict étale Cartesian morphism, let $X'$ be $X \times_{Y_1} Y'_1$, and define $Y'(\nu)$ and $D'(\nu)$ in the same way as $Y(\nu)$ and $D(\nu)$ using $Y'$ and $X' \to Y'_1$. Then, by applying Lemma IV.2.2.15 to $Y'(\nu) \to Y' \times_B Y(\nu-1)$ and Proposition IV.2.2.9 (3) to $Y' \times_B Y(\nu - 1) \to Y(\nu)$ and $X' \to X$, we see that the natural morphism $D'(\nu)_1 \to D(\nu)_1 \times_X X'$ is an isomorphism and the morphism $D'(\nu) \to D(\nu)$ is strict étale and Cartesian. This will allow us to discuss properties about $D(\nu)$ étale locally on $Y_1$.

Let $D(1)^1_N$ denote the first infinitesimal neighborhood of $\Delta_{D,N}\colon D_N \to D(1)_N$ (cf. the beginning of IV.2.2). Since the reduction mod $p$ of $D_N \to D(1)^1_N$ is a nilpotent immersion, we may regard a sheaf on $D(1)^1_{N,\text{ét}}$ as a sheaf on $D_{1,\text{ét}}$. We define the sheaf $\Omega^1_{D_N/B_N}$ on $D_{1,\text{ét}}$ by

$$\Omega^1_{D_N/B_N} := \mathrm{Ker}(\mathcal{O}_{D(1)^1_N} \to \mathcal{O}_{D_N})/(p\text{-tor}).$$

The two $\mathcal{O}_{D_N}$-module structures on $\Omega^1_{D_N/B_N}$ defined by $p_1^*$ and $p_2^*$ coincide. Let $Y(1)^1_N$ be the first infinitesimal neighborhood of the diagonal immersion $Y_N \to Y(1)_N$. We have a canonical isomorphism

$$\Omega^1_{Y_N/B_N} \cong \mathrm{Ker}(\mathcal{O}_{Y(1)^1_N} \to \mathcal{O}_{Y_N}).$$

Hence the morphism $D(1)^1_N \to Y(1)^1_N$ defined by the natural morphisms $D_N \to Y_N$ and $D(1)_N \to Y(1)_N$ induces an $\mathcal{O}_{D_N}$-linear morphism

$$\mathcal{O}_{D_N} \otimes_{\mathcal{O}_{Y_N}} \Omega^1_{Y_N/B_N} \longrightarrow \Omega^1_{D_N/B_N}. \tag{IV.2.4.5}$$

**Proposition IV.2.4.6.** *Let the notation and assumption be as above. Let $N$ be an integer $\geq 2$.*

(1) *The morphism* (IV.2.4.5) *induces an injective morphism*

$$\mathcal{O}_{D_{N-1}} \otimes_{\mathcal{O}_{Y_{N-1}}} \Omega^1_{Y_{N-1}/B_{N-1}} \longrightarrow \Omega^1_{D_N/B_N}.$$

*Furthermore, there exists a unique isomorphism*

$$\xi\mathcal{O}_{S_N} \otimes_{\mathcal{O}_{S_{N-1}}} \Omega^1_{D_{N-1}/B_{N-1}} \otimes \mathbb{Q} \xrightarrow{\cong} \mathcal{O}_{D_{N-1}} \otimes_{\mathcal{O}_{Y_{N-1}}} \Omega^1_{Y_{N-1}/B_{N-1}} \otimes \mathbb{Q}$$

*such that the following diagram is commutative.*

$$\begin{array}{ccc}
\xi\mathcal{O}_{S_N} \otimes_{\mathcal{O}_{S_N}} \Omega^1_{D_N/B_N} \otimes \mathbb{Q} & \longrightarrow & \Omega^1_{D_N/B_N} \otimes \mathbb{Q} \\
\downarrow & & \uparrow \\
\xi\mathcal{O}_{S_N} \otimes_{\mathcal{O}_{S_{N-1}}} \Omega^1_{D_{N-1}/B_{N-1}} \otimes \mathbb{Q} & \xrightarrow{\cong} & \mathcal{O}_{D_{N-1}} \otimes_{\mathcal{O}_{Y_{N-1}}} \Omega^1_{Y_{N-1}/B_{N-1}} \otimes \mathbb{Q}.
\end{array}$$

(2) *The isomorphism in* (1) *induces an isomorphism*

$$\xi\mathcal{O}_{S_N} \otimes_{\mathcal{O}_{S_{N-1}}} \Omega^1_{D_{N-1}/B_{N-1}} \cong \mathcal{O}^{(1)_r}_{D_{N-1}} \otimes_{\mathcal{O}_{Y_{N-1}}} \Omega^1_{Y_{N-1}/B_{N-1}}.$$

PROOF. By Remark IV.2.4.4, the question is strict étale local on $Y_1$ and we may assume that $Y_1$ is affine, $X \to Y_1$ is a closed immersion, and there exist $(t_{i,N})_N \in \varprojlim_N \Gamma(Y_N, M_{Y_N})$ ($1 \leq i \leq d$) such that $d\log t_{i,N}$ ($1 \leq i \leq d$) is a basis of $\Omega^1_{Y_N/B_N}$ for every $N$. By Proposition IV.2.2.9 (2), $D_N$ and $D_N(1)$ are affine. Let $A$ and $A(1)$ be the objects of $\mathscr{A}^r_\bullet$ defined by their coordinate rings (cf. Lemma IV.2.2.4). The $p$-adic fine log formal scheme $D(1)^1_N$ is also affine. Let $A(1)^1_N$ denote $\Gamma(D(1)^1_N, \mathcal{O}_{D(1)^1_N})$. Let $p_{Y(1)} = (p_{Y(1),N})_N$ denote the natural morphism $D(1) \to Y(1)$. Since $p^*_{Y(1),1}p^*_{2,1}(t_{i,1}) = p^*_{Y(1),1}p^*_{1,1}(t_{i,1})$ on $D(1)_1$ and $D(1)_1 \to D(1)_N$ is an exact closed immersion, there exists a unique $u_{i,N} \in 1 + F^1A(1)_N$ such that $p^*_{Y(1),N}p^*_{2,N}(t_{i,N}) = u_{i,N} \cdot p^*_{Y(1),N}p^*_{1,N}(t_{i,N})$. Put $w_{i,N} := \xi^{-1}(u_{i,N+1} - 1) \in A(1)_{N,\mathbb{Q}}$. By Proposition IV.2.3.17, we have an isomorphism $A\{W_1, \dots, W_d\}_r \xrightarrow{\cong} A(1)$ over $A$ with respect to $p^*_1\colon A \to A(1)$ such that the image of $W_i$ in $A(1)_{N,\mathbb{Q}}$ is $w_{i,N}$. By Lemma IV.2.3.13, we have

$$p(\oplus_{\underline{m}\in\mathbb{N}^d, |\underline{m}|=2} A_N^{(|\underline{m}|)_r} \underline{w}_N^{\underline{m}})^\wedge \subset \operatorname{Ker}(A_N(1) \to A(1)^1_N) \subset (\oplus_{\underline{m}\in\mathbb{N}^d, |\underline{m}|=2} A_N^{(|\underline{m}|)_r} \underline{w}_N^{\underline{m}})^\wedge,$$

where $\underline{w}_N^{\underline{m}} = \prod_{1\le i\le d} w_{i,N}^{m_i}$ and $^\wedge$ denotes the $p$-adic completion. Hence we have an isomorphism

$$\operatorname{Ker}(A(1)^1_N \to A_N)/(p\text{-tor}) \xrightarrow{\cong} \oplus_{1\le i\le d} A_N^{(1)_r} \overline{w}_{i,N},$$

where $\overline{w}_{i,N}$ denotes the image of $w_{i,N}$ in $A(1)^1_N$. Applying this to all strict étale Cartesian morphisms $Y' \to Y$ and $X \times_{Y_1} Y'_1$ such that $Y'_1$ is affine, we obtain an isomorphism

$$\Omega^1_{D_N/B_N} \xrightarrow{\cong} \oplus_{1\le i\le d} \mathcal{O}_{Y_N}^{(1)_r} \overline{w}_{i,N}$$

(cf. Lemma IV.2.4.2 (2)). Now the claim follows from the fact that the image of $d\log t_{i,N} \in \Omega^1_{Y_N/B_N}$ in $\Omega^1_{D_N/B_N}$ is $\xi \cdot \overline{w}_{i,N}$. □

For $N \in \mathbb{N}_{>0}$, we have an isomorphism $[\xi]\colon \mathcal{O}_{S_N} \xrightarrow{\cong} \xi\mathcal{O}_{S_{N+1}}$. Hence $\xi\mathcal{O}_{S_{N+1}}$ is a free $\mathcal{O}_{S_N}$-module of rank 1. For $n \in \mathbb{N}_{>0}$ and an $\mathcal{O}_{Y_N}$-module $\mathcal{M}$, we define $\xi^{-n}\mathcal{M}$ to be $(\xi\mathcal{O}_{S_{N+1}})^{\otimes(-n)} \otimes_{\mathcal{O}_{S_N}} \mathcal{M}$. For $x \in \mathcal{M}$, we define $\xi^{-n}x$ to be $\xi^{\otimes(-n)} \otimes x$. By Proposition IV.2.4.6 (2), we have a canonical isomorphism

(IV.2.4.7)
$$\Omega^1_{D_N/B_N} \cong \xi^{-1}\mathcal{O}_{D_N}^{(1)_r} \otimes_{\mathcal{O}_{Y_N}} \Omega^1_{Y_N/B_N}.$$

Hence the $\mathcal{O}_{B_N}$-linear derivation $\mathcal{O}_{D_N} \to \Omega^1_{D_N/B_N}; x \mapsto p^*_{2,N}(x) - p^*_{1,N}(x)$ induces an $\mathcal{O}_{B_N}$-linear derivation

(IV.2.4.8)
$$\theta\colon \mathcal{O}_{D_N} \longrightarrow \xi^{-1}\mathcal{O}_{D_N}^{(1)_r} \otimes_{\mathcal{O}_{Y_N}} \Omega^1_{Y_N/B_N}.$$

Since the derivation $d\colon \mathcal{O}_{Y_N} \to \Omega^1_{Y_N/B_N}$ is defined by $d(x) = p^*_2(x) - p^*_1(x)$, we see that the following diagram is commutative.

(IV.2.4.9)
$$\begin{array}{ccc} \mathcal{O}_{D_N} & \xrightarrow{\theta} & \xi^{-1}\mathcal{O}_{D_N}^{(1)_r} \otimes_{\mathcal{O}_{Y_N}} \Omega^1_{Y_N/B_N} \\ \uparrow & & \uparrow \\ p^{-1}_{Y,N}(\mathcal{O}_{Y_N}) & \xrightarrow{d} & p^{-1}_{Y,N}(\Omega^1_{Y_N/B_N}), \end{array}$$

where $p_Y = (p_{Y,N})_N$ denotes the natural morphism $D \to Y$ and the right vertical morphism is defined by $\omega \mapsto \xi^{-1}(\xi) \otimes \omega$.

This implies that one can define a morphism

$$\theta^q\colon \xi^{-q}\mathcal{O}_{D_N} \otimes_{\mathcal{O}_{Y_N}} \Omega^q_{Y_N/B_N} \longrightarrow \xi^{-q-1}\mathcal{O}_{D_N}^{(1)_r} \otimes_{\mathcal{O}_{Y_N}} \Omega^{q+1}_{Y_N/B_N}$$

by

$$\theta^q(\xi^{-q}a \otimes \omega) = \xi^{-q}\theta(a) \wedge \omega + \xi^{-q}a \otimes d^q\omega.$$

**Lemma IV.2.4.10.** (1) *For* $m, q \in \mathbb{N}$, *we have*

$$\theta^q(\xi^{-q}\mathcal{O}_{D_N}^{(m)_r} \otimes_{\mathcal{O}_{Y_N}} \Omega^q_{Y_N/B_N}) \subset \xi^{-q-1}\mathcal{O}_{D_N}^{(m+1)_r} \otimes_{\mathcal{O}_{Y_N}} \Omega^{q+1}_{Y_N/B_N}.$$

(2) *For* $q \in \mathbb{N}$, *we have* $\theta^{q+1} \circ \theta^q = 0$.

We will give a proof of Lemma IV.2.4.10 after the proof of Proposition IV.2.4.13.

Now assume that $Y_1$ is affine, $X \to Y_1$ is a closed immersion, and there exist $t_i = (t_{i,N}) \in \varprojlim_N \Gamma(Y_N, M_{Y_N})$ $(1 \leq i \leq d)$ such that $\{d\log(t_{i,N})\}_{1 \leq i \leq d}$ is a basis of $\Omega^1_{Y_N/B_N}$ for every $N \in \mathbb{N}_{>0}$. Let $A_D(\nu) = (A_D(\nu)_N)$ be the object of $\mathscr{A}^r_\bullet$ defined by $A_D(\nu)_N = \Gamma(D(\nu)_N, \mathcal{O}_{D(\nu)_N})$ (cf. Lemma IV.2.2.4). We write $A_D$ for $A_D(0)$. Since the images of $p_{1,N}^*p_{Y,N}^*(t_{i,N})$, $p_{2,N}^*p_{Y,N}^*(t_{i,N}) \in \Gamma(D(1)_N, M_{D(1)_N})$ in $\Gamma(D(1)_1, M_{D(1)_1})$ coincide and $D(1)_1 \to D(1)_N$ is an exact closed immersion, there exists a unique $u_{i,N} \in 1 + F^1A_D(1)_N$ such that $p_{1,N}^*p_{Y,N}^*(t_{i,N})u_{i,N} = p_{2,N}^*p_{Y,N}^*(t_{i,N})$ in $\Gamma(D(1)_N, M_{D(1)_N})$. Put $w_{i,N} = \xi^{-1}(u_{i,N+1} - 1) \in A_D(1)_{N,\mathbb{Q}}$.

By Proposition IV.2.3.17, either of the two homomorphisms $p_{1,N}^*, p_{2,N}^*\colon A_{D,N} \to A_D(1)_N$ induces an isomorphism

$$A_{D,N}\{W_1, \ldots, W_d\}_r \xrightarrow{\cong} A_D(1)_N \qquad \text{(IV.2.4.11)}$$

sending $W_i$ to $w_{i,N}$ (after $\otimes\mathbb{Q}$), and any of the three homomorphisms $q_i^*\colon A_{D,N} \to A_D(2)_N$ $(i = 1, 2, 3)$ induces an isomorphism

$$A_{D,N}\{W_1, \ldots, W_d, W_1', \ldots, W_d'\}_r \xrightarrow{\cong} A_D(2)_N \qquad \text{(IV.2.4.12)}$$

sending $W_i$ and $W_i'$ to $p_{12}^*(w_{i,N})$ and $p_{23}^*(w_{i,N})$.

We define the endomorphisms $\theta_i\colon A_{D,N} \to A_{D,N}$ $(1 \leq i \leq d)$ by the formula $\theta(x) = \sum_{1 \leq i \leq d} \xi^{-1}\theta_i(x) \otimes d\log(t_{i,N})$. For $\underline{m} = (m_1, \ldots, m_d) \in \mathbb{N}^d$, we set $|\underline{m}| = \sum_{1 \leq i \leq d} m_i$ and $\underline{m}! = \prod_{1 \leq i \leq d}(m_i)!$.

**Proposition IV.2.4.13.** *The endomorphisms* $\theta_i$ *of* $A_{D,N}$ $(1 \leq i \leq d)$ *have the following properties.*

(1) $\theta_i \circ \theta_j = \theta_j \circ \theta_i$ $(i \neq j)$.

*Put* $\theta_{\underline{m}} = \prod_{1 \leq i \leq d} \prod_{0 \leq j \leq m_i - 1}(\theta_i - j\xi)$ *for* $\underline{m} = (m_1, \ldots, m_d) \in \mathbb{N}^d$.

(2) *For* $x \in A_{D,N}^{(m)_r}$ *and* $\underline{m} \in \mathbb{N}^d$, $\frac{1}{\underline{m}!}\theta_{\underline{m}}(x)$ *is contained in* $A_{D,N}^{(|\underline{m}|+m)_r}$, *and there exists a sequence of non-negative integers* $l(n)$ $(n \in \mathbb{N})$ *such that* $\lim_{n \to \infty} l(n) = \infty$ *and* $\frac{1}{\underline{m}!}\theta_{\underline{m}}(x) \in p^{l(|\underline{m}|)}A_{D,N}^{(|\underline{m}|+m)_r}$.

(3) *For* $x \in A_{D,N}^{(m)_r}$, *we have the following equality in* $A_D(1)_N^{(m)_r}$.

$$p_{2,N}^*(x) = \sum_{\underline{m} \in \mathbb{N}^d} \frac{1}{\underline{m}!} p_{1,N}^*(\theta_{\underline{m}}(x)) \prod_{1 \leq i \leq d} w_{i,N}^{m_i}.$$

Proof. We abbreviate $p_{i,N}$, $p_{i,j,N}$ and $q_{i,N}$ to $p_i$, $p_{ij}$ and $q_i$, respectively. We define the homomorphisms $\Theta_{\underline{m}}\colon A_{D,N} \to A_{D,N}^{(|\underline{m}|)_r}$ $(\underline{m} \in \mathbb{N}^d)$ by the equality

$$p_2^*(x) = \sum_{\underline{m} \in \mathbb{N}^d} p_1^*(\Theta_{\underline{m}}(x))\underline{w}_N^{\underline{m}} \qquad (x \in A_{D,N})$$

in $A_D(1)_N$, where $\underline{w}_N^{\underline{m}} = \prod_{1 \leq i \leq d} w_{i,N}^{m_i}$ for $\underline{m} = (m_1, \ldots, m_d) \in \mathbb{N}^d$. Since $p_2^*(A_{D,N}^{(l)_r}) \subset A_D(1)_N^{(l)_r}$, we have $\Theta_{\underline{m}}(A_{D,N}^{(l)_r}) \subset A_{D,N}^{(|\underline{m}|+l)_r}$ by Lemma IV.2.3.16. We have $\theta_i = \Theta_{\underline{1}_i}$ $(1 \leq i \leq d)$ by definition. For $x \in A_{D,N}$, we have the following two expansions of $q_3^*(x)$

in $A_D(2)_N$.

$$q_3^*(x) = \sum_{\underline{l}\in\mathbb{N}^d} q_2^*(\Theta_{\underline{l}}(x))p_{23}^*(\underline{w}_N^{\underline{l}}) = \sum_{\underline{l},\underline{n}\in\mathbb{N}^d} q_1^*(\Theta_{\underline{n}}\circ\Theta_{\underline{l}})p_{23}^*(\underline{w}_N^{\underline{l}})p_{12}^*(\underline{w}_N^{\underline{n}}),$$

$$q_3^*(x) = \sum_{\underline{m}\in\mathbb{N}^d} q_1^*(\Theta_{\underline{m}}(x))p_{13}^*(\underline{w}_N^{\underline{m}}).$$

Since $p_{13}^*(u_{i,N+1}) = p_{23}^*(u_{i,N+1})p_{12}^*(u_{i,N+1})$, we have $p_{13}^*(w_{i,N}) = \xi p_{23}^*(w_{i,N})p_{12}^*(w_{i,N}) + p_{23}^*(w_{i,N}) + p_{12}^*(w_{i,N})$. Comparing the coefficients of $p_{23}(\underline{w}_N^{\underline{l}})p_{12}(\underline{w}_N^{\underline{n}})$ for $\underline{n},\underline{l}\in\mathbb{N}^d$, we obtain

$$\Theta_{\underline{n}}\circ\Theta_{\underline{l}} = \sum_{\underline{n}=\underline{m}_1+\underline{m}_3,\,\underline{l}=\underline{m}_2+\underline{m}_3} \frac{(\underline{m}_1+\underline{m}_2+\underline{m}_3)!}{\underline{m}_1!\underline{m}_2!\underline{m}_3!}\xi^{|\underline{m}_3|}\Theta_{\underline{m}_1+\underline{m}_2+\underline{m}_3}.$$

Hence we have $\theta_i\circ\theta_j = \Theta_{\underline{1}_i+\underline{1}_j} = \theta_j\circ\theta_i$ $(i\neq j)$ and

$$\theta_i\circ\Theta_{\underline{m}} = (m_i+1)\Theta_{\underline{m}+\underline{1}_i} + m_i\xi\Theta_{\underline{m}} \Longleftrightarrow \Theta_{\underline{m}+\underline{1}_i} = \frac{1}{m_i+1}(\theta_i - m_i\xi)\circ\Theta_{\underline{m}}$$

for $\underline{m}\in\mathbb{N}^d$. By induction on $|\underline{m}|$, we see $\Theta_{\underline{m}} = \frac{1}{\underline{m}!}\theta_{\underline{m}}$ for $\underline{m}\in\mathbb{N}^d$, which implies (2) and (3). □

PROOF OF LEMMA IV.2.4.10. By Remark IV.2.4.4, the question is étale local on $Y_1$. Hence we see $\theta(\mathcal{O}_{D_N}^{(m)_r}) \subset \xi^{-1}\mathcal{O}_{D_N}^{(m+1)_r}\otimes_{\mathcal{O}_{Y_N}}\Omega^1_{Y_N/B_N}$ by Lemma IV.2.4.2 (2) and Proposition IV.2.4.13 (2). Therefore the claim (1) follows from $\xi\mathcal{O}_{D_N}^{(m)_r} \subset \mathcal{O}_{D_N}^{(1)_r}\cdot\mathcal{O}_{D_N}^{(m)_r} \subset \mathcal{O}_{D_N}^{(m+1)_r}$ (Lemma IV.2.4.2 (3)). The claim (2) follows from Proposition IV.2.4.13 (1). □

**IV.2.5. Torsors of deformations and Higgs envelopes.** Let $B$ be an object of $\mathscr{C}^\infty$, let $Y$ be an object of $\mathscr{C}^\infty$ over $B$, let $X\to B$ be a smooth Cartesian morphism in $\mathscr{C}$, and let $z\colon Y_1\to X_1$ be a morphism of $p$-adic fine log formal schemes over $B_1$. Assume that $Y_1$ is affine. Let $\mathcal{T}_{\mathrm{Zar}}$ be the sheaf of $\mathcal{O}_{Y_1}$-modules $\mathcal{H}om_{\mathcal{O}_{Y_1}}(z^*(\xi^{-1}\Omega^1_{X_1/B_1}),\mathcal{O}_{Y_1})$ on the Zariski site $(Y_1)_{\mathrm{Zar}}$. Then, following II.10.3, one can construct a $\mathcal{T}_{\mathrm{Zar}}$-torsor $\mathcal{L}_{\mathrm{Zar}}$ from $Y\to B\leftarrow X$ and $z$. In this subsection, we prove that the principal homogeneous space $L$ associated to $\mathcal{L}_{\mathrm{Zar}}$ (cf. Proposition II.4.10) is canonically isomorphic to the first component $D_1$ of the Higgs envelope $D$ of level $\infty$ of $Y_1\hookrightarrow X\times_B Y$ (Theorem IV.2.5.2). We also give a description (Proposition IV.2.5.16) of the first components of the Higgs envelopes of finite level in terms of the symmetric tensor product of the module of affine functions on $\mathcal{L}_{\mathrm{Zar}}$ (cf. II.4.7), which will allow us, in IV.5.4, to compare the period ring of finite level $r$ defined in IV.5.2 with the $p$-adic completion of the Higgs-Tate algebra of depth $1/r$ defined in II.12.1. We prove the compatibility with the canonical derivations (Proposition IV.2.5.19) and functoriality (Proposition IV.2.5.25) for the isomorphism $L\cong D_1$.

For a $p$-adic fine log formal scheme $Z$ (resp. a morphism $f$ of $p$-adic fine log formal schemes), let ${}_nZ$ (resp. ${}_nf$) denote the reduction mod $p^n$ of $Z$ (resp. $f$) in this subsection.

As in II.10.3, we consider the following torsor. Let $U$ be an open $p$-adic fine log formal subscheme of $Y_1$, which means an open formal subscheme of the formal scheme underlying $Y_1$ endowed with the restriction of the log structure of $Y_1$, and let $\widetilde{U}$ be the open $p$-adic fine log formal subscheme of $Y_2$ defined by $U$. Let $\mathcal{L}_{\mathrm{Zar}}(U)$ denote the set of morphisms $\widetilde{U}\to X_2$ over $B_2$ whose composition with the natural closed immersion $U\hookrightarrow\widetilde{U}$ coincides with the composition $U\subset Y_1\xrightarrow{z} X_1\hookrightarrow X_2$. This construction is obviously functorial on $U$, and the presheaf of sets $\mathcal{L}_{\mathrm{Zar}}\colon (Y_1)_{\mathrm{Zar}}\to(\mathrm{Sets})$ on the Zariski site $(Y_1)_{\mathrm{Zar}}$ of $Y_1$ admits a natural structure of a torsor under $\mathcal{T}_{\mathrm{Zar}} := \mathcal{H}om_{\mathcal{O}_{Y_1}}(z^*(\xi^{-1}\Omega^1_{X_1/B_1}),\mathcal{O}_{Y_1})$, where $z^*(-) = \mathcal{O}_{Y_1}\otimes_{z^{-1}(\mathcal{O}_{X_1})} z^{-1}(-)$ by **[50]** Proposition (3.9) (cf. II.5.23). Note that the multiplication by $\xi$ on $\mathcal{O}_{Y_1}$ induces an isomorphism $\mathcal{O}_{Y_1} \xrightarrow{\cong} \mathrm{Ker}(\mathcal{O}_{Y_2} \to \mathcal{O}_{Y_1})$ by Definition IV.2.2.1 (2) (iii), (iv), $\mathcal{O}_{Y_1}$ and $\mathcal{O}_{Y_2}$ are $p$-torsion free by Definition IV.2.2.1 (2) (ii), and the natural homomorphism $z^*(\xi^{-1}\Omega^1_{X_1/B_1}) \to \varprojlim_n ({}_nz^*(\xi^{-1}\Omega^1_{{}_nX_1/{}_nB_1}))$ is an isomorphism.

For each $n \in \mathbb{N}_{>0}$, we put ${}_n\mathcal{T}_{\mathrm{Zar}} := \mathcal{H}om_{\mathcal{O}_{{}_nY_1}}({}_nz^*(\xi^{-1}\Omega^1_{{}_nX_1/{}_nB_1}), \mathcal{O}_{{}_nY_1})$ and define the ${}_n\mathcal{T}_{\mathrm{Zar}}$-torsor ${}_n\mathcal{L}_{\mathrm{Zar}}$ on $({}_nY_1)_{\mathrm{Zar}} = (Y_1)_{\mathrm{Zar}}$ as follows. For an open fine log subscheme $U$ of ${}_nY_1$ and the open fine log subscheme $\widetilde{U}$ of ${}_nY_2$ defined by $U$, we define ${}_n\mathcal{L}_{\mathrm{Zar}}(U)$ to be the set of morphisms $\widetilde{U} \to {}_nX_2$ over ${}_nB_2$ whose composition with the natural closed immersion $U \hookrightarrow \widetilde{U}$ coincides with the composition of $U \subset {}_nY_1 \xrightarrow{{}_nz} {}_nX_1 \to {}_nX_2$. For $n \in \mathbb{N}_{>0}$, we have a natural morphism ${}_{n+1}\mathcal{L}_{\mathrm{Zar}} \to {}_n\mathcal{L}_{\mathrm{Zar}}$ equivariant with respect to the natural homomorphism ${}_{n+1}\mathcal{T}_{\mathrm{Zar}} \to {}_n\mathcal{T}_{\mathrm{Zar}}$. Let ${}_nT$ be the vector bundle $\mathrm{Spec}(\mathrm{Sym}^\bullet_{\mathcal{O}_{{}_nY_1}}({}_nz^*(\xi^{-1}\Omega^1_{{}_nX_1/{}_nB_1})))$ associated to ${}_n\mathcal{T}_{\mathrm{Zar}}$ endowed with the inverse image of the log structure of ${}_nY_1$. Then, by applying the construction in II.4.9, we obtain an ${}_nT$-principal homogeneous space ${}_nL$ in the category of strict fine log schemes over ${}_nY_1$ locally trivial with respect to the Zariski topology on ${}_nY_1$, and a canonical isomorphism

$$ {}_n\mathcal{L}_{\mathrm{Zar}} \xrightarrow{\cong} \varphi_{{}_nY_1}({}_nL) $$

of ${}_n\mathcal{T}_{\mathrm{Zar}}$-torsors, where $\varphi_{{}_nY_1}$ is the functor defined as (II.4.3.1). By II.4.13, we have a natural isomorphism ${}_{n+1}T \times_{{}_{n+1}Y_1} {}_nY_1 \cong {}_nT$ of group fine log schemes and an isomorphism of ${}_nT$-principal homogeneous spaces ${}_{n+1}L \times_{{}_{n+1}Y_1} {}_nY_1 \cong {}_nL$. By taking the inverse limit with respect to $n$, we obtain a group $p$-adic fine log formal scheme $T$ over $Y_1$ and a $T$-principal homogeneous space $L$ in the category of strict $p$-adic fine log formal schemes over $Y_1$ locally trivial with respect to the Zariski topology on $Y_1$.

Let $\mathrm{PHS}(T/Y_1)$ denote the category of $T$-principal homogeneous spaces in the category of strict $p$-adic fine log formal schemes over $Y_1$ locally trivial with respect to the Zariski topology on $Y_1$. Let $\mathrm{Tors}(\mathcal{T}_{\mathrm{Zar}}, (Y_1)_{\mathrm{Zar}})$ denote the category of $\mathcal{T}_{\mathrm{Zar}}$-torsors on $(Y_1)_{\mathrm{Zar}}$. Similarly as (II.4.3.1), we have a canonical functor

$$ \varphi_{Y_1} \colon \mathrm{PHS}(T/Y_1) \longrightarrow \mathrm{Tors}(\mathcal{T}_{\mathrm{Zar}}, (Y_1)_{\mathrm{Zar}}) $$

defined by $Z \mapsto \mathrm{Hom}_{Y_1}(-, Z)$. By the above construction, we have a canonical isomorphism

$$ \mathcal{L}_{\mathrm{Zar}} \xrightarrow{\cong} \varphi_{Y_1}(L). $$

Note that for $p$-adic fine log formal schemes $Z$ and $Z'$, the natural map $\mathrm{Hom}(Z, Z') \to \varprojlim_n \mathrm{Hom}({}_nZ, {}_nZ')$ is bijective by the definition of fine log structures on $p$-adic formal schemes in the beginning of IV.2.2 and **[1]** Proposition 2.2.2. By applying Proposition II.4.4 to ${}_nT$ and ${}_n\mathcal{T}_{\mathrm{Zar}}$ on ${}_nY_1$, and then II.4.13 to the morphisms ${}_nY_1 \to {}_{n+1}Y_1$, we obtain the following proposition.

**Proposition IV.2.5.1.** *The functor $\varphi_{Y_1}$ above is fully faithful.*

The morphism $z\colon Y_1 \to X_1$ and the identity morphism of $Y_1$ induce an immersion $\widetilde{z}\colon Y_1 \to Y_1 \times_{B_1} X_1$. We define the object $D \in \mathrm{Ob}\,\mathscr{C}^\infty$ to be $D^\infty_{\mathrm{Higgs}}(\widetilde{z}\colon Y_1 \hookrightarrow X \times_B Y)$ (cf. Definition IV.2.2.1 (3) and Proposition IV.2.2.9 (1)). The purpose of this subsection is to prove the following isomorphism and study its properties (cf. Propositions IV.2.5.16, IV.2.5.19, and IV.2.5.25).

**Theorem IV.2.5.2.** *There exists a canonical isomorphism of $p$-adic fine log formal schemes $D_1 \xrightarrow{\cong} L$ over $Y_1$.*

We will regard $D_1$ as a $T$-principal homogeneous space by the action of $T$ on $D_1$ induced by that on $L$ via the canonical isomorphism in Theorem IV.2.5.2.

**Lemma IV.2.5.3.** *The canonical morphism $D \to Y$ in $\mathscr{C}^\infty$ is strict smooth Cartesian and $D_1$ is affine.*

PROOF. We may replace $X$ with $X \times_B Y$ and $z$ with $\widetilde{z}\colon Y_1 \to X_1 \times_{B_1} Y_1$, and assume that $Y = B$. By Proposition IV.2.2.9 (3), the claim is strict étale local on $Y_1$. Hence the claim follows from Proposition IV.2.2.9 (2) and Proposition IV.2.3.17. □

Let $(\mathrm{AffSmStr}/Y_1)_{\mathrm{Zar}}$ denote the category of strict smooth affine $p$-adic fine log formal schemes over $Y_1$ endowed with the Zariski topology, and let $(Y_1)_{\mathrm{ZarAff}}$ denote the category of open affine $p$-adic fine log formal schemes of $Y_1$ endowed with the Zariski topology. The restriction functor induces an equivalence of categories of topos $(Y_1)^{\sim}_{\mathrm{Zar}} \xrightarrow{\cong} (Y_1)^{\sim}_{\mathrm{ZarAff}}$. By Lemma IV.2.5.3, $D_1$ is an object of $(\mathrm{AffSmStr}/Y_1)_{\mathrm{Zar}}$. Let $\mathcal{T}$ denote the sheaf on $(\mathrm{AffSmStr}/Y_1)_{\mathrm{Zar}}$ defined by $\mathcal{T}(U \xrightarrow{u} Y_1) = \Gamma(U, \mathcal{H}om_{\mathcal{O}_U}(u^*z^*(\xi^{-1}\Omega^1_{X_1/B_1}), \mathcal{O}_U))$. We construct a natural extension of the $\mathcal{T}_{\mathrm{Zar}}$-torsor $\mathcal{L}_{\mathrm{Zar}}$ to a $\mathcal{T}$-torsor $\mathcal{L}$ on $(\mathrm{AffSmStr}/Y_1)_{\mathrm{Zar}}$ and prove that $\mathcal{L}$ is canonically isomorphic to the sheaf represented by $D_1$.

First let us define the torsor $\mathcal{L}$. For $(u\colon U \to Y_1) \in \mathrm{Ob}\,(\mathrm{AffSmStr}/Y_1)$, let $\mathrm{Lift}(U/Y_2)$ denote the category defined as follows. An object is a smooth strict morphism of $p$-adic fine log formal schemes $\widetilde{u}\colon \widetilde{U} \to Y_2$ with an isomorphism $U \xrightarrow{\cong} \widetilde{U} \times_{Y_2} Y_1$ over $Y_1$ (i.e., a smooth strict lifting of $U \to Y_1$). A morphism is a morphism of $p$-adic fine log formal schemes over $Y_2$ compatible with the isomorphisms from $U$ in the obvious sense.

For an object $\widetilde{U}$ of $\mathrm{Lift}(U/Y_2)$, we write $F^1\mathcal{O}_{{}_n\widetilde{U}}$ for the kernel of $\mathcal{O}_{{}_n\widetilde{U}} \to \mathcal{O}_{{}_nU}$. By Definition IV.2.2.1 (2), the multiplication by $\xi$ on $\mathcal{O}_{{}_nY_2}$ induces an isomorphism $\mathcal{O}_{{}_nY_1} \xrightarrow{\cong} F^1\mathcal{O}_{{}_nY_2}$ for each $n \in \mathbb{N}_{>0}$. Let $\widetilde{U}$ be an object of $\mathrm{Lift}(U/Y_2)$. Then, since the morphism of schemes underling ${}_nU \to {}_nY_1$ is flat, we see that the multiplication $\xi$ on $\mathcal{O}_{{}_nU_2}$ induces an isomorphism

$$\mathcal{O}_{{}_nU} \xrightarrow{\cong} F^1\mathcal{O}_{{}_n\widetilde{U}}. \tag{IV.2.5.4}$$

**Lemma IV.2.5.5.** *The category* $\mathrm{Lift}(U/Y_2)$ *is nonempty. Any two objects of* $\mathrm{Lift}(U/Y_2)$ *are isomorphic to each other. Furthermore every morphism in* $\mathrm{Lift}(U/Y_2)$ *is an isomorphism.*

PROOF. The first two claims follow from **[50]** Proposition (3.14). By (IV.2.5.4), a morphism $v\colon \widetilde{U}' \to \widetilde{U}$ in $\mathrm{Lift}(U/Y_2)$ induces an isomorphism $v^*\colon F^1\mathcal{O}_{{}_n\widetilde{U}} \xrightarrow{\cong} F^1\mathcal{O}_{{}_n\widetilde{U}'}$ for each $n \in \mathbb{N}_{>0}$. This implies that ${}_nv$ $(n \in \mathbb{N}_{>0})$ and hence $v$ are isomorphisms. □

For each object $\widetilde{u}\colon \widetilde{U} \to Y_2$ of $\mathrm{Lift}(U/Y_2)$, we define $\mathcal{L}_U(\widetilde{U})$ to be the set of morphisms $\widetilde{U} \to X_2$ over $B_2$ such that the composition with $U \hookrightarrow \widetilde{U}$ coincides with the composition of $U \xrightarrow{u} Y_1 \xrightarrow{z} X_1 \hookrightarrow X_2$. By Sublemma IV.3.4.2 (2) and Corollary IV.2.2.5, the set $\mathcal{L}_U(\widetilde{U})$ is nonempty. A morphism $v\colon \widetilde{U}' \to \widetilde{U}$ in $\mathrm{Lift}(U/Y_2)$, which is an isomorphism by Lemma IV.2.5.5, induces a bijection

$$v^*\colon \mathcal{L}_U(\widetilde{U}) \to \mathcal{L}_U(\widetilde{U}'); g \mapsto g \circ v.$$

We have $\mathrm{id}^* = \mathrm{id}$ and $(w \circ v)^* = v^* \circ w^*$ for two composable morphisms $v$ and $w$ in $\mathrm{Lift}(U/Y_2)$. Thus we obtain a functor

$$\mathcal{L}_U\colon \mathrm{Lift}(U/Y_2) \to (\mathrm{Sets}).$$

**Lemma IV.2.5.6.** *Let $v, v'\colon \widetilde{U}' \to \widetilde{U}$ be two morphisms in* $\mathrm{Lift}(U/Y_2)$ *with the same source and target. Then we have $v^* = v'^*\colon \mathcal{L}_U(\widetilde{U}) \to \mathcal{L}_U(\widetilde{U}')$.*

PROOF. By (IV.2.5.4), the two homomorphisms $F^1\mathcal{O}_{{}_n\widetilde{U}} \to F^1\mathcal{O}_{{}_n\widetilde{U}'}$ induced by $v$ and $v'$ coincide. On the other hand, for an element $\widetilde{z}$ of $\mathcal{L}_Y(Y_2)$, which exists by Sublemma

IV.3.4.2 (2), the image of $\widetilde{z}\circ\widetilde{u}\in\mathcal{L}_U(\widetilde{U})$ under the two maps in the lemma coincide. Hence the claim follows from **[50]** Proposition (3.9) (cf. II.5.23). □

We define the set $\mathcal{L}(U)$ by

$$\mathcal{L}(U) := \Gamma(\mathrm{Lift}(U/Y_2), \mathcal{L}_U).$$

By Lemmas IV.2.5.5 and IV.2.5.6, the projection map $\mathcal{L}(U)\to\mathcal{L}_U(\widetilde{U})$ is bijective for any object $\widetilde{U}$ of $\mathrm{Lift}(U/Y_2)$. By **[50]** Proposition (3.9) (cf. II.5.23) and (IV.2.5.4), the set $\mathcal{L}_U(\widetilde{U})$ for an object $\widetilde{U}$ of $\mathrm{Lift}(U/Y_2)$ is naturally regarded as a $\Gamma(U,\mathcal{T})$-torsor. For a morphism $v\colon\widetilde{U}'\to\widetilde{U}$ in $\mathrm{Lift}(U/Y_2)$, the bijection $v^*\colon\mathcal{L}_U(\widetilde{U})\xrightarrow{\cong}\mathcal{L}_U(\widetilde{U}')$ is an isomorphism of $\Gamma(U,\mathcal{T})$-torsors. Hence $\mathcal{L}(U)$ has a natural structure of a $\Gamma(U,\mathcal{T})$-torsor.

The $\Gamma(U,\mathcal{T})$-torsor $\mathcal{L}(U)$ is functorial on $U$ as follows. Let $v\colon U'\to U$ be a morphism in $(\mathrm{AffSmStr}/Y_1)_{\mathrm{Zar}}$, and let $\widetilde{U}$ (resp. $\widetilde{U}'$) be an object of $\mathrm{Lift}(U/Y_2)$ (resp. $\mathrm{Lift}(U'/Y_2)$). Then, by the smoothness of the morphism $\widetilde{U}\to Y_2$, there exists a $Y_2$-morphism $\widetilde{v}\colon\widetilde{U}'\to\widetilde{U}$ which lifts $v\colon U'\to U$. By composing with $\widetilde{v}$, we obtain a map $\widetilde{v}^*\colon\mathcal{L}_U(\widetilde{U})\to\mathcal{L}_U(\widetilde{U}'); g\mapsto g\circ\widetilde{v}$.

**Lemma IV.2.5.7.** *For two liftings $\widetilde{v}$ and $\widetilde{v}'$ of $v$, the maps $\widetilde{v}^*,\widetilde{v}'^*\colon\mathcal{L}_U(\widetilde{U})\to\mathcal{L}_U(\widetilde{U}')$ coincide.*

PROOF. By (IV.2.5.4), we have the following commutative diagrams.

$$\begin{array}{ccc}
\mathcal{O}_{{}_nU} & \xrightarrow{\cong} & F^1\mathcal{O}_{{}_n\widetilde{U}} \\
{\scriptstyle v^*}\downarrow & & \downarrow\downarrow{\scriptstyle \widetilde{v}^*,\ \widetilde{v}'^*} \\
\mathcal{O}_{{}_nU'} & \xrightarrow{\cong} & F^1\mathcal{O}_{{}_n\widetilde{U}'}.
\end{array}$$

Hence the right two vertical homomorphisms coincide and the same argument as the proof of Lemma IV.2.5.6 shows the claim. □

By Lemmas IV.2.5.5, IV.2.5.6, and IV.2.5.7, we see that the composition of

$$\mathcal{L}(U)\xrightarrow{\cong}\mathcal{L}_U(\widetilde{U})\xrightarrow{\widetilde{v}^*}\mathcal{L}_U(\widetilde{U}')\xleftarrow{\cong}\mathcal{L}(U')$$

does not depend on the choice of $\widetilde{U}$, $\widetilde{U}'$, and $\widetilde{v}$. We denote the composition by $\mathcal{L}(v)$. By the construction of the bijection in **[50]** Proposition (3.9), we see that $\mathcal{L}(v)$ is compatible with the torsor structures via $v^*\colon\Gamma(U,\mathcal{T})\to\Gamma(U',\mathcal{T})$. Thus we obtain a $\mathcal{T}$-torsor $\mathcal{L}$ on $(\mathrm{AffSmStr}/Y_1)_{\mathrm{Zar}}$.

Let $D$ be an object of $\mathscr{C}^\infty$ defined before Theorem IV.2.5.2, and let $\mathcal{D}$ denote the sheaf on $(\mathrm{AffSmStr}/Y_1)_{\mathrm{Zar}}$ represented by the object $D_1$ of $(\mathrm{AffSmStr}/Y_1)_{\mathrm{Zar}}$.

**Theorem IV.2.5.8.** *There exists a canonical isomorphism of sheaves $\mathcal{L}\cong\mathcal{D}$ on the site $(\mathrm{AffSmStr}/Y_1)_{\mathrm{Zar}}$.*

Theorem IV.2.5.8 and Proposition IV.2.5.1 immediately imply Theorem IV.2.5.2 as follows.

PROOF OF THEOREM IV.2.5.2. By Theorem IV.2.5.8, $D_1$ has naturally a structure of an object of $\mathrm{PHS}(T/Y_1)$, and there exist canonical isomorphisms $\varphi_{Y_1}(L)|_{(Y_1)_{\mathrm{ZarAff}}}\cong\mathcal{L}_{\mathrm{Zar}}|_{(Y_1)_{\mathrm{ZarAff}}}\cong\varphi_{Y_1}(D_1)|_{(Y_1)_{\mathrm{ZarAff}}}$ of $\mathcal{T}_{\mathrm{Zar}}|_{(Y_1)_{\mathrm{ZarAff}}}$-torsors. Since the restriction functor $(Y_1)^{\sim}_{\mathrm{Zar}}\to(Y_1)^{\sim}_{\mathrm{ZarAff}}$ is an equivalence of categories, Proposition IV.2.5.1 implies that there exists a unique isomorphism $L\cong D_1$ in $\mathrm{PHS}(T/Y_1)$ inducing the composition of the above isomorphisms. □

By the transport of structures via the canonical isomorphism in Theorem IV.2.5.8, the sheaf $\mathcal{D}$ acquires a $\mathcal{T}$-torsor structure. By the proof of Theorem IV.2.5.2, it induces the same $T$-action on $D_1$ as the one defined after Theorem IV.2.5.2.

Let $U_1$ be an object of $(\mathrm{AffSmStr}/Y_1)_{\mathrm{Zar}}$ and let $f\colon U \to Y$ be a smooth strict Cartesian morphism in $\mathscr{C}$ with the given $U_1 \to Y_1$ over $S_1$. We have $U \in \mathrm{Ob}\,\mathscr{C}^\infty$ by Lemma IV.2.2.7 (1). Then we have natural maps

$$\text{(IV.2.5.9)} \qquad \mathrm{Hom}_Y(U, (\widetilde{z}\colon Y_1 \hookrightarrow X \times_B Y)) \longrightarrow \mathcal{L}_{U_1}(U_2),$$

$$\text{(IV.2.5.10)} \qquad \mathrm{Hom}_Y(U, D) \longrightarrow \mathrm{Hom}_{Y_1}(U_1, D_1).$$

**Proposition IV.2.5.11.** (1) *The two maps* (IV.2.5.9) *and* (IV.2.5.10) *are surjective.*

(2) *There exists a unique bijective map*

$$\iota_U\colon \mathcal{L}_{U_1}(U_2) \xrightarrow{\cong} \mathrm{Hom}_{Y_1}(U_1, D_1)$$

*such that the following diagram is commutative, where the bottom horizontal bijection is induced by the adjunction morphism $D \to (\widetilde{z}\colon Y_1 \hookrightarrow X \times_B Y)$ in $\mathscr{C}$.*

$$\begin{array}{ccc} \mathcal{L}_{U_1}(U_2) & \xrightarrow{\cong} & \mathrm{Hom}_{Y_1}(U_1, D_1) \\ \uparrow & & \uparrow \\ \mathrm{Hom}_{\mathscr{C}_{/Y}}(U, (\widetilde{z}\colon Y_1 \hookrightarrow X \times_B Y)) & \xleftarrow{\cong} & \mathrm{Hom}_{\mathscr{C}^\infty_{/Y}}(U, D). \end{array}$$

Proof. (1) The morphism (IV.2.5.10) is surjective by Lemma IV.2.5.3 and Sublemma IV.3.4.2. By definition, giving an element of $\mathcal{L}_{U_1}(U_2)$ is equivalent to giving a morphism $U_2 \to X_2 \times_{B_2} Y_2$ over $Y_2$ such that the composition with $U_1 \to U_2$ coincides with the composition of $U_1 \to Y_1 \xrightarrow{\widetilde{z}} X_2 \times_{B_2} Y_2$. Hence the morphism (IV.2.5.9) is also surjective by Sublemma IV.3.4.2.

(2) By the description of the element of $\mathcal{L}_{U_1}(U_2)$ in the proof of (1), we may replace $X$ by $X \times_B Y$ and $z\colon Y_1 \to X_1$ by the section $Y_1 \to X_1 \times_{B_1} Y_1$ induced by $z$, and assume $Y = B$. Let $(Y_\alpha \to Y)_{\alpha\in A}$ be a strict étale covering such that $Y_{\alpha,1}$ is affine and $A$ is a finite set, and let $Y_{\alpha\beta}$ denote $Y_\alpha \times_Y Y_\beta$ for $(\alpha,\beta) \in A^2$. Let $U_\alpha$ and $X_\alpha$ (resp. $U_{\alpha\beta}$ and $X_{\alpha\beta}$) denote the base change of $U$ and $X$ under $Y_\alpha \to Y$ (resp. $Y_{\alpha\beta} \to Y$), and define the set $\mathcal{L}_{U_{\alpha,1}}(U_{\alpha,2})$ (resp. $\mathcal{L}_{U_{\alpha\beta,1}}(U_{\alpha\beta,2})$) in the same way as $\mathcal{L}_{U_1}(U_2)$ by using $U_\alpha \to Y_\alpha \leftarrow X_\alpha$ and $Y_{\alpha,1} \to X_{\alpha,1}$ (resp. $U_{\alpha\beta} \to Y_{\alpha\beta} \leftarrow X_{\alpha\beta}$ and $Y_{\alpha\beta,1} \to X_{\alpha\beta,1}$). Then, by Lemma IV.2.2.19 (1), we see that the sequence $\mathcal{L}_{U_1}(U_2) \to \prod_\alpha \mathcal{L}_{U_{\alpha,1}}(U_{\alpha,2}) \rightrightarrows \prod_{\alpha\beta} \mathcal{L}_{U_{\alpha\beta,1}}(U_{\alpha\beta,2})$ is exact. Proposition IV.2.2.9 (3), Lemma IV.2.2.18, and Lemma IV.2.2.19 imply similar exact sequences for the other three terms of the diagram in question. Hence the claim holds for $(U, Y, X)$ if it holds for $(U_\alpha, Y_\alpha, X_\alpha)$ and $(U_{\alpha\beta}, Y_{\alpha\beta}, X_{\alpha\beta})$. By Corollary IV.2.2.5 and Lemma IV.2.2.15, we also see that every term in the diagram does not change if we replace $X$ with an open $p$-adic fine log formal scheme containing the image of $Y_1$. Hence it suffices to prove the claim in the case where $X_N$ ($N \geq 1$) are affine, there exist $(t_{i,N})_N \in \varprojlim_N \Gamma(X_N, M_{X_N})$ ($1 \leq i \leq d$) such that $d\log(t_{i,N})$ ($1 \leq i \leq d$) is a basis of $\Omega^1_{X_N/Y_N}$ for every $N$.

Let $\mathrm{Spf}(\mathbb{Z}_p)\{\mathbb{N}^d\}$ be the $p$-adic fine log formal scheme defined as in the proof of Proposition IV.2.3.17, and let $Y_N\{\mathbb{N}^d\}$ denote $Y_N \times_{\mathrm{Spf}(\mathbb{Z}_p)} \mathrm{Spf}(\mathbb{Z}_p)\{\mathbb{N}^d\}$. Then the $Y_N$-morphism $X_N \to Y_N\{\mathbb{N}^d\}$ induced by $\mathbb{N}^d \to \Gamma(X_N, M_{X_N}); \underline{1}_i \mapsto t_{i,N}$ is étale and defines a morphism $X \to Y\{\mathbb{N}^d\}$ in $\mathscr{C}$. Hence we have the following bijections, where $i_{U,1}$ denotes the morphism $U_1 \to U_2$ and $\Gamma(U, M_U)$ denotes the inverse limit $\varprojlim_N \Gamma(U_N, M_{U_N})$.

$$\mathcal{L}_{U_1}(U_2) \to \{(a_i)_i \in \Gamma(U_2, M_{U_2})^d \mid i_{U,1}^*(a_i) = f_1^* z^*(t_{i,1})\}; g \mapsto (g^*(t_{i,2}))_i,$$

$$\begin{aligned}\mathrm{Hom}_Y(U, (z\colon Y_1 \hookrightarrow X)) &\to \{((a_{i,N})_N)_i \in \Gamma(U, M_U)^d \mid a_{i,1} = f_1^* z^*(t_{i,1})\}\\ h &\mapsto ((h_N^*(t_{i,N}))_{N\geq 2})_i.\end{aligned}$$

Choose a lifting $(s_{i,N})_N \in \varprojlim_N \Gamma(Y_N, M_{Y_N})$ of $s_{i,1} := z^*(t_{i,1})$. Then, by Proposition IV.2.3.17, we obtain isomorphisms $\Gamma(Y_N, \mathcal{O}_{Y_N})\{W_{1,N}, \dots, W_{d,N}\} \cong \Gamma(D_N, \mathcal{O}_{D_N})$ compatible with $N$. By Proposition IV.2.2.9 (2), we also see that the morphism $D \to Y$ is strict. Hence, we have the following bijections, where $\Gamma(U, \mathcal{O}_U)$ denotes $\varprojlim_N \Gamma(U_N, \mathcal{O}_{U_N})$.

$$\mathrm{Hom}_{Y_1}(U_1, D_1) \to \Gamma(U_1, \mathcal{O}_{U_1})^d; \varphi \mapsto (\varphi^*(W_{i,1})),$$
$$\mathrm{Hom}_Y(U, D) \to \Gamma(U, \mathcal{O}_U)^d; \psi \mapsto (\psi_N^*(W_{i,N}))_N.$$

Let $h \in \mathrm{Hom}_Y(U, (z\colon Y_1 \hookrightarrow X))$, let $\psi$ be the corresponding $Y$-morphism $U \to D$, and let $((a_{i,N})_N)_i \in \Gamma(U, M_U)^d$ (resp. $((b_{i,N})_N)_i \in \Gamma(U, \mathcal{O}_U)^d$) be the image of $h$ (resp. $\psi$) under the bijections above. Then, by the construction of the isomorphism in Proposition IV.2.3.17, we have an equality

$$a_{i,2} = f_2^*(s_{i,2}) \cdot (1 + [\xi] b_{i,1})$$

in $\Gamma(U_2, M_{U_2})$, where $[\xi]$ denotes the isomorphism $\Gamma(U_1, \mathcal{O}_{U_1}) \xrightarrow{\cong} \mathrm{Ker}(\Gamma(U_2, \mathcal{O}_{U_2}) \to \Gamma(U_1, \mathcal{O}_{U_1}))$ induced by the multiplication by $\xi$ on $\mathcal{O}_{U_2}$ (cf. Definition IV.2.2.1 (2) (iii) (iv)). Hence $a_{i,2}$ and $b_{i,1}$ determine each other. This completes the proof by (1) and the descriptions of $\mathcal{L}_{U_1}(U_2)$ and $\mathrm{Hom}_{Y_1}(U_1, D_1)$ above. □

**Lemma IV.2.5.12.** *The composition of the morphism $\iota_U$ with the projection $\mathcal{L}(U_1) \xrightarrow{\cong} \mathcal{L}_{U_1}(U_2)$ is independent of the choice of $U$.*

Proof. Let $U' \to Y$ be another strict smooth Cartesian morphism in $\mathscr{C}^\infty$ with $U'_1 = U_1$. Then, by the smoothness of $U \to Y$ and Sublemma IV.3.4.2 (2), there exists a morphism $v\colon U' \to U$ over $Y$ such that $v_1$ is the identity of $U'_1 = U_1$. Put $C_N = \Gamma(U_N, \mathcal{O}_{U_N})$ and $C'_N = \Gamma(U'_N, \mathcal{O}_{U'_N})$. Then, by Lemma IV.2.2.6, we have an exact sequence $0 \to C^{(\prime)}_{N-1} \xrightarrow{[\xi]} C^{(\prime)}_N \to C^{(\prime)}_1 \to 0$. Hence we see that $v_N$ is an isomorphism by induction on $N$. Now the claim follows from the commutative diagram below.

$$\begin{array}{ccccccc}
\mathcal{L}_{U_1}(U_2) & \longleftarrow & \mathrm{Hom}_Y(U, (Y_1 \hookrightarrow X \times_B Y)) & \xleftarrow{\cong} & \mathrm{Hom}_Y(U, D) & \longrightarrow & \mathrm{Hom}_{Y_1}(U_1, D_1) \\
\cong \Big\downarrow v_2^* & & -\circ v \Big\downarrow \cong & & -\circ v \Big\downarrow \cong & & \Big\| \\
\mathcal{L}_{U_1}(U'_2) & \longleftarrow & \mathrm{Hom}_Y(U', (Y_1 \hookrightarrow X \times_B Y)) & \xleftarrow{\cong} & \mathrm{Hom}_Y(U', D) & \longrightarrow & \mathrm{Hom}_{Y_1}(U'_1, D_1).
\end{array}$$

□

Let $\iota_{U_1}$ denote the canonical bijection

$$\mathcal{L}(U_1) \xrightarrow{\cong} \mathrm{Hom}_{Y_1}(U_1, D_1)$$

obtained by the composition in Lemma IV.2.5.12. Now Theorem IV.2.5.8 follows from the following lemma.

**Lemma IV.2.5.13.** *For a morphism* $v_1\colon U_1' \to U_1$ *in* $(\mathrm{AffSmStr}/Y_1)_{\mathrm{Zar}}$*, the following diagram is commutative.*

$$\begin{array}{ccc} \mathcal{L}(U_1) & \xrightarrow[\iota_{U_1}]{\cong} & \mathrm{Hom}_{Y_1}(U_1, D_1) \\ {\scriptstyle \mathcal{L}(v_1)}\downarrow & & \downarrow{\scriptstyle -\circ v_1} \\ \mathcal{L}(U_1') & \xrightarrow[\iota_{U_1'}]{\cong} & \mathrm{Hom}_{Y_1}(U_1', D_1) \end{array}$$

PROOF. Let $U \to Y$ and $U' \to Y$ be strict smooth Cartesian liftings of $U_1 \to Y_1$ and $U_1' \to Y_1$, respectively. By the smoothness of $U \to Y$ and Sublemma IV.3.4.2 (2), there exists a lifting $v\colon U' \to U$ of $v_1$. Then the claim follows from the same kind of diagram as in the proof of Lemma IV.2.5.12 induced by $v$. □

Next let us consider the object $D^r$ ($r \in \mathbb{N}_{>0}$) of $\mathscr{C}^r$ defined by $D^r := D^r_{\mathrm{Higgs}}(Y_1 \overset{\tilde{z}}{\hookrightarrow} X \times_B Y)$ (cf. Proposition IV.2.2.9 (1)). By Proposition IV.2.2.9 (2), $D_1^r$ is affine. Let $A_{Y_1}$, $A_L$, $A_{D_1}$, and $A_{D_1^r}$ denote the rings of coordinates of the underlying affine $p$-adic formal schemes of $Y_1$, $L$, $D_1$, and $D_1^r$, respectively. Let ${}_n\Omega_{\mathrm{Zar}}$ and $\Omega_{\mathrm{Zar}}$ denote the sheaves $\mathcal{H}om_{\mathcal{O}_{{}_nY_1}}({}_n\mathcal{T}_{\mathrm{Zar}}, \mathcal{O}_{{}_nY_1})$ and $\mathcal{H}om_{\mathcal{O}_{Y_1}}(\mathcal{T}_{\mathrm{Zar}}, \mathcal{O}_{Y_1})$. Let ${}_n\mathcal{F}$ be the sheaf of affine functions on the ${}_n\mathcal{T}_{\mathrm{Zar}}$-torsor ${}_n\mathcal{L}_{\mathrm{Zar}}$ (II.4.9). Then we have an exact sequence of $\mathcal{O}_{{}_nY_1}$-modules $0 \to \mathcal{O}_{{}_nY_1} \to {}_n\mathcal{F} \to {}_n\Omega_{\mathrm{Zar}} \to 0$ compatible with $n$. By taking the global sections and the inverse limit over $n$, we obtain an exact sequence of $A_{Y_1}$-modules

$$0 \longrightarrow A_{Y_1} \xrightarrow{c} M \longrightarrow \Gamma(Y_1, \Omega_{\mathrm{Zar}}) \longrightarrow 0.$$

**Lemma IV.2.5.14.** *Under the notation and the assumption above, the* $A_{Y_1}$*-module* $\Gamma(Y_1, \Omega_{\mathrm{Zar}})$ *is finitely generated and projective.*

PROOF. Put ${}_nP := \Gamma(Y_1, {}_n\Omega_{\mathrm{Zar}})$, ${}_nA_{Y_1} := \Gamma({}_nY_1, \mathcal{O}_{{}_nY_1})$ and $P := \Gamma(Y_1, \Omega_{\mathrm{Zar}})$. We have $P = \varprojlim_n ({}_nP)$ and ${}_{n+1}P/p^n \xrightarrow{\cong} {}_nP$. Since ${}_n\Omega_{\mathrm{Zar}}$ is a locally free $\mathcal{O}_{{}_nY_1}$-module of finite type, ${}_nP$ is a finitely generated projective ${}_nA_{Y_1}$-module. Choose a surjective ${}_1A_{Y_1}$-linear homomorphism ${}_1\varphi\colon {}_1A_{Y_1}^{\oplus r} \to {}_1P$ and then a compatible system of ${}_nA_{Y_1}$-linear liftings ${}_n\varphi\colon {}_nA_{Y_1}^{\oplus r} \to {}_nP$. By Nakayama's lemma, ${}_n\varphi$ is surjective. Since the natural homomorphisms $\mathrm{Ker}({}_{n+1}\varphi) \to \mathrm{Ker}({}_n\varphi)$ are surjective, we see that there exists a compatible system of ${}_nA_{Y_1}$-linear homomorphisms ${}_ns\colon {}_nP \to {}_nA_{Y_1}^{\oplus r}$ such that ${}_n\varphi \circ {}_ns = \mathrm{id}_{{}_nP}$. Letting $s$ and $\varphi$ denote the inverse limits of ${}_ns$ and ${}_n\varphi$, we obtain an $A_{Y_1}$-linear isomorphism $P \oplus \mathrm{Ker}(\varphi) \xrightarrow{\cong} A_{Y_1}^{\oplus r}; (a, b) \mapsto s(a) + b$. □

For $m \in \mathbb{N}$, let $S^m_{A_{Y_1}}(M)$ denote the $m$-th symmetric tensor product of the $A_{Y_1}$-module $M$. Then the homomorphism $c$ induces an injective $A_{Y_1}$-linear homomorphism $c^m\colon S^m_{A_{Y_1}}(M) \to S^{m+1}_{A_{Y_1}}(M)$. Let $C_{A_{Y_1}}(M)$ denote the direct limit $\varinjlim_m (S^m_{A_{Y_1}}(M))$, and let $\widehat{C}_{A_{Y_1}}(M)$ denote the $p$-adic completion of $C_{A_{Y_1}}(M)$. Then, by the construction of the $T$-torsor $L$, we have a canonical isomorphism

$$\widehat{C}_{A_{Y_1}}(M) \xrightarrow{\cong} A_L. \tag{IV.2.5.15}$$

Identifying $S^m_{A_{Y_1}}(M)$ with its image in $C_{A_{Y_1}}(M)$ under the canonical injection, we define the $A_{Y_1}$-subalgebra $C^{(r)}_{A_{Y_1}}(M)$ by

$$C^{(r)}_{A_{Y_1}}(M) := \sum_{m \in \mathbb{N}} p^{\lceil \frac{m}{r} \rceil} S^m_{A_{Y_1}}(M).$$

Let $\widehat{C}^{(r)}_{A_{Y_1}}(M)$ denote the $p$-adic completion of $C^{(r)}_{A_{Y_1}}(M)$.

**Proposition IV.2.5.16.** *For $r \in \mathbb{N}_{>0}$, the natural homomorphisms $A_{D_1^r} \to A_{D_1}$ and $\widehat{C}^{(r)}_{A_{Y_1}}(M) \to \widehat{C}_{A_{Y_1}}(M)$ are injective, and the composition of the isomorphism* (IV.2.5.15) *and the isomorphism $A_L \xrightarrow{\cong} A_{D_1}$ (Theorem* IV.2.5.2*) induces an isomorphism*

$$\widehat{C}^{(r)}_{A_{Y_1}}(M) \xrightarrow{\cong} A_{D_1^r}.$$

PROOF. Replacing $X$ with $X \times_B Y$ and $z$ with the morphism $\widetilde{z}\colon Y_1 \to X_1 \times_{B_1} Y_1$, we may assume that $Y = B$. Let ${}_n\mathcal{F}'$ be the sheaf of affine functions on the ${}_n\mathcal{T}_{\text{Zar}}$-torsor ${}_n\mathcal{D}_{\text{Zar}} := \varphi_{{}_nY_1}({}_nD_1)$, and let $M'$ denote the $A_{Y_1}$-module $\varprojlim_n \Gamma(Y_1, {}_n\mathcal{F}')$, which is an extension of $\Gamma(Y_1, \Omega_{\text{Zar}})$ by $A_{Y_1}$. Let ${}_nL'$ be the ${}_nT$-principal homogeneous space associated to the ${}_n\mathcal{T}_{\text{Zar}}$-torsor ${}_n\mathcal{D}_{\text{Zar}}$ (II.4.9) and put $L' = ({}_nL')_{n\geq 1}$, which is a $T$-principal homogeneous space. The isomorphism in Theorem IV.2.5.2 induces an isomorphism of extensions $M \xrightarrow{\cong} M'$, and it suffices to prove the claim for $M'$, $L'$, and the canonical isomorphism $A_{L'} \cong A_{D_1}$ instead of $M$, $L$, and $A_L \cong A_{D_1}$. By Lemma IV.2.5.14, we have an exact sequence

$$0 \longrightarrow \bigoplus_{m\in\mathbb{N}} p^{\lceil\frac{m+1}{r}\rceil} S^m_{A_{Y_1}}(M') \xrightarrow{(*)} \bigoplus_{m\in\mathbb{N}} p^{\lceil\frac{m}{r}\rceil} S^m_{A_{Y_1}}(M') \longrightarrow C^{(r)}_{A_{Y_1}}(M') \longrightarrow 0,$$

where $(*)$ is induced by $(\mathrm{id}, -c^m)\colon S^m_{A_{Y_1}}(M') \to S^m_{A_{Y_1}}(M') \oplus S^{m+1}_{A_{Y_1}}(M')$. It remains exact after $\otimes_{\mathbb{Z}}\mathbb{Z}/p^N\mathbb{Z}$. For any strict étale covering $\mathfrak{Y} = (u_\alpha\colon Y_\alpha \to Y)_{\alpha\in A}$ such that $\sharp A < \infty$ and $Y_{\alpha,1}$ are affine, and any quasi-coherent $\mathcal{O}_{{}_nY_1}$-module $\mathcal{G}$ on $Y_{1,\text{Zar}}$, the Čech cohomology $\check{H}^q(\mathfrak{Y}, \mathcal{G})$ vanishes if $q > 0$ and $\check{H}^0(\mathfrak{Y}, \mathcal{G}) = \Gamma(Y, \mathcal{G})$. Hence, by Propositions IV.2.5.22 and IV.2.5.24 below and Lemma IV.2.2.15, we see that the question is strict étale local on $Y_1$ and we may assume that there exist $(t_{i,N}) \in \varprojlim_N \Gamma(X_N, M_{X_N})$ $(1 \leq i \leq d)$ such that $d\log(t_{i,N})$ $(1 \leq i \leq d)$ is a basis of $\Omega^1_{X_N/Y_N}$ for every $N \in \mathbb{N}_{>0}$. Choosing a lifting $(s_{i,N})_N \in \varprojlim_N \Gamma(Y_N, M_{Y_N})$ of $z^*(t_{i,1})$ and applying Proposition IV.2.3.17, we obtain an isomorphism $A_{D_1} \cong A_{Y_1}\{W_1, \ldots, W_d\}_r$, and an isomorphism

$$A^r_{D_1} \cong \left(\bigoplus_{\underline{m}\in\mathbb{N}^d} p^{\lceil\frac{|\underline{m}|}{r}\rceil} A_{Y_1} \underline{W}^{\underline{m}}\right)^{\wedge}.$$

Note that we have $A_{Y_1}^{(m)_r} = p^{\lceil\frac{m}{r}\rceil} A_{Y_1}$ by $F^n A_{Y_N} = \xi^n A_{Y_N}$. Hence it suffices to prove that the image of $M'$ under the natural isomorphism $\widehat{C}_{A_{Y_1}}(M') \xrightarrow{\cong} A_{D_1}$ is $A_{Y_1} \oplus (\oplus_{1\leq i\leq d} A_{Y_1} W_i)$.

We identify $\mathcal{D}$ with $\mathcal{O}_{Y_1}^{\oplus d}$ by the coordinates $W_i$ $(1 \leq i \leq d)$. Then, by the proof of Proposition IV.2.5.11, we see that the action of $f \in \mathcal{T}(U)$ on $\mathcal{D}(U) = \mathcal{O}_U(U)^{\oplus d}$ is given by $(a_i)_{1\leq i\leq d} \mapsto (a_i + f(\xi^{-1} d\log(t_{i,1})))_{1\leq i\leq d}$. This implies that the action of $T$ on $D_1$ is given by

$$A_{D_1} \longrightarrow S_{A_{Y_1}}(\Omega)\widehat{\otimes}_{A_{Y_1}} A_{D_1}; W_i \mapsto \xi^{-1} d\log(t_{i,1}) \otimes 1 + 1 \otimes W_i,$$

where $\Omega = \Gamma(Y_1, \Omega_{\text{Zar}})$. We identify ${}_n\mathcal{D}_{\text{Zar}}$ with $\mathcal{O}_{{}_nY_1}^{\oplus d}$ by the coordinates $W_i$ $(1 \leq i \leq d)$. Then it induces isomorphisms ${}_n\mathcal{F}' \xrightarrow{\cong} \mathcal{O}_{{}_nY_1} \oplus (\mathcal{O}_{{}_nY_1})^{\oplus d}; f \mapsto (f(0), (f(\underline{1}_i) - f(0))_{1\leq i\leq d})$ and ${}_nL' \cong {}_nY_1[V_1, \ldots, V_d]$. We identify $\varphi_{{}_nY_1}({}_nL')$ with $\mathcal{O}_{{}_nY_1}^{\oplus d}$ by the coordinates $V_i$. Then we see that the canonical isomorphism of ${}_n\mathcal{T}_{\text{Zar}}$-torsors ${}_n\mathcal{D}_{\text{Zar}} \cong \varphi_{{}_nY_1}({}_nL')$ (cf. (II.4.9.7)) is simply given by the identity map of $\mathcal{O}_{Y_1}^{\oplus r}$. On the other hand, the action of ${}_nT$ on ${}_nL'$ is given by

$$\mathcal{O}_{{}_nY_1}[V_1, \ldots, V_d] \to S_{\mathcal{O}_{{}_nY_1}}({}_n\Omega_{\text{Zar}}) \otimes_{\mathcal{O}_{{}_nY_1}} \mathcal{O}_{{}_nY_1}[V_1, \ldots, V_d]; V_i \mapsto \xi^{-1} d\log t_{i,1} \otimes 1 + 1 \otimes V_i$$

(cf. (II.4.9.4)). Hence the unique ${}_nT$-equivariant isomorphism ${}_nD_1 \xrightarrow{\cong} {}_nL'$ inducing the above isomorphism ${}_n\mathcal{D}_{\mathrm{Zar}} \cong \varphi_{Y_1}({}_nL')$ (cf. II.4.4) is given by $V_i \mapsto W_i$. This completes the proof. □

For $n \in \mathbb{N}$, we have a canonical $\mathcal{O}_{{}_nY_1}$-linear isomorphism $\mathcal{O}_{{}_nT} \otimes_{\mathcal{O}_{{}_nY_1}} {}_n\Omega_{\mathrm{Zar}} \xrightarrow{\cong} \Omega^1_{{}_nT/{}_nY_1}$ on $(T_1)_{\mathrm{Zar}}$ equivariant with respect to the translation by any Zariski local section ${}_nY_1 \supset U \to {}_nT$ of ${}_nT \to {}_nY_1$. This induces a canonical $\mathcal{O}_{{}_nY_1}$-linear isomorphism $\mathcal{O}_{{}_nL} \otimes_{\mathcal{O}_{{}_nY_1}} {}_n\Omega_{\mathrm{Zar}} \xrightarrow{\cong} \Omega^1_{{}_nL/{}_nY_1}$ on $(L_1)_{\mathrm{Zar}}$ equivariant with respect to the translation by any Zariski local section of ${}_nT \to {}_nY_1$ (cf. II.10.9). By Lemma IV.2.5.14, we see that $\Omega_{\mathrm{Zar}}$ is a direct factor of a free $\mathcal{O}_{Y_1}$-module of finite type. Hence the above isomorphism induces an $A_L$-linear isomorphism $A_L \otimes_{A_{Y_1}} \Gamma(Y_1, \Omega_{\mathrm{Zar}}) \xrightarrow{\cong} \Gamma(L, \Omega^1_{L/Y_1})$ and then an $A_{Y_1}$-linear derivation

(IV.2.5.17) $$\theta_L \colon A_L \to A_L \otimes_{A_{Y_1}} \Gamma(Y_1, \Omega_{\mathrm{Zar}}).$$

We obtain the following $A_{Y_1}$-linear derivation from (IV.2.4.8).

(IV.2.5.18) $$\theta_{D_1} \colon A_{D_1} \to A_{D_1} \otimes_{A_{Y_1}} \Gamma(Y_1, \Omega_{\mathrm{Zar}}).$$

**Proposition IV.2.5.19.** *The isomorphism $A_L \cong A_{D_1}$ induced by that in Theorem IV.2.5.2 is compatible with the derivations* (IV.2.5.17) *and* (IV.2.5.18).

PROOF. Replacing $X$ and $z$ with $X \times_B Y$ and $\tilde{z} \colon Y_1 \to X_1 \times_{B_1} Y_1$, we may assume that $Y = B$. Let ${}_n\mathcal{F}'$, ${}_n\mathcal{D}_{\mathrm{Zar}}$, $M'$, and $L'$ be as in the beginning of the proof of Proposition IV.2.5.16. It suffices to prove the claim for $L'$ instead of $L$ and for the canonical isomorphism $A_{L'} \cong A_{D_1}$. By Propositions IV.2.5.22 and IV.2.5.24 below, Lemma IV.2.2.15, and a similar argument as the proof of Proposition IV.2.5.16, we see that the question is strict étale local on $Y_1$ and we may assume that there exist $(t_{i,N}) \in \varprojlim_N \Gamma(X_N, M_{X_N})$ $(1 \leq i \leq d)$ as in the proof of Proposition IV.2.5.16. We use the notation in loc. cit. Since $\theta_{L'}|_{M'}$ is induced by the canonical projection $M' \to \Gamma(Y_1, \Omega_{\mathrm{Zar}})$, we see $\theta_{L'}(V_i) = 1 \otimes \xi^{-1} d\log(t_{i,1})$ $(1 \leq i \leq d)$ by the explicit description of the action of $T$ on $D_1$ given in loc. cit. Now the claim follows from $\theta_{D_1}(W_i) = 1 \otimes \xi^{-1} d\log(t_{i,1})$ $(1 \leq i \leq d)$. □

We discuss the functoriality of the isomorphisms in Theorems IV.2.5.2 and IV.2.5.8.

We define the category $\mathscr{C}_{\mathrm{Tors}}$ as follows: An object $\boldsymbol{X} = (Y \to B \leftarrow X, z)$ consists of morphisms $Y \to B \leftarrow X$ in $\mathscr{C}$ and a $B_1$-morphism $z \colon Y_1 \to X_1$ of $p$-adic fine log formal schemes satisfying the conditions in the beginning of this subsection. A morphism $\boldsymbol{\alpha} \colon (Y' \to B' \leftarrow X', z') \to (Y \to B \leftarrow X, z)$ is a triple $(\alpha_Y, \alpha_B, \alpha_X)$ consisting of morphisms $\alpha_Y \colon Y' \to Y$, $\alpha_B \colon B' \to B$, and $\alpha_X \colon X' \to X$ in $\mathscr{C}$ making the following diagrams commutative.

(IV.2.5.20) $$\begin{array}{ccccc} Y' & \longrightarrow & B' & \longleftarrow & X' \\ \downarrow{\scriptstyle \alpha_Y} & & \downarrow{\scriptstyle \alpha_B} & & \downarrow{\scriptstyle \alpha_X} \\ Y & \longrightarrow & B & \longleftarrow & X, \end{array} \qquad \begin{array}{ccc} Y'_1 & \xrightarrow{z'} & X'_1 \\ \downarrow{\scriptstyle \alpha_{Y,1}} & & \downarrow{\scriptstyle \alpha_{X,1}} \\ Y_1 & \xrightarrow{z} & X_1. \end{array}$$

We often abbreviate $\alpha_Y$ to $\alpha$ in the following. For an object $\boldsymbol{X}$ of $\mathscr{C}_{\mathrm{Tors}}$, we write $\mathcal{L}_{\boldsymbol{X},\mathrm{Zar}}$, $\mathcal{T}_{\boldsymbol{X},\mathrm{Zar}}$, $T_{\boldsymbol{X}}$, $\mathcal{L}_{\boldsymbol{X},U}$, $\mathcal{L}_{\boldsymbol{X}}$, $\mathcal{T}_{\boldsymbol{X}}$, $D_{\boldsymbol{X}}$, and $\mathcal{D}_{\boldsymbol{X}}$ for $\mathcal{L}_{\mathrm{Zar}}$, $\mathcal{T}_{\mathrm{Zar}}$, $T$, $\mathcal{L}_U$, $\mathcal{L}$, etc. constructed from $\boldsymbol{X}$ as above.

We begin by studying the functoriality for a morphism $\boldsymbol{\alpha} = (\alpha_Y, \alpha_B, \alpha_X) \colon \boldsymbol{X}' = (Y' \to B' \leftarrow X') \to \boldsymbol{X} = (Y \to B \leftarrow X)$ satisfying one of the following conditions.

(Cond 1) The morphism $X' \to X \times_B B'$ induced by $\alpha_X$ and $\alpha_B$ is an isomorphism.

(Cond 2) $B = B'$, $Y = Y'$, and $\alpha_B$ and $\alpha_Y$ are the identity morphisms.

We first consider the case where $\boldsymbol{\alpha}$ satisfies (Cond 1). Fiber products are representable in $(\mathrm{AffSmStr}/Y_1)_{\mathrm{Zar}}$, and the functor $(\mathrm{AffSmStr}/Y_1)_{\mathrm{Zar}} \to (\mathrm{AffSmStr}/Y_1')_{\mathrm{Zar}}$ defined by the base change under the morphism $\alpha_1 \colon Y_1' \to Y_1$ preserves finite fiber products, final objects, and Zariski open coverings. Hence the functor above is a morphism of sites and defines a morphism of topos, whose direct image functor will be simply denoted by $\alpha_{1*}$.

Let $u \colon U \to Y_1$ be an object of $(\mathrm{AffSmStr}/Y_1)_{\mathrm{Zar}}$, let $u' \colon U' \to Y_1'$ be the base change of $u$ under the morphism $\alpha_1 \colon Y_1' \to Y_1$, and let $\alpha_U \colon U' \to U$ denote the natural morphism. By (Cond 1), the $\mathcal{O}_{X_1'}$-linear homomorphism $\alpha_{X,1}^*\Omega^1_{X_1/B_1} \to \Omega^1_{X_1'/B_1'}$ on $(X_1')_{\mathrm{Zar}}$ is an isomorphism, which induces an isomorphism $\alpha_U^*(\mathcal{H}om_{\mathcal{O}_U}(u^*z^*(\xi^{-1}\Omega^1_{X_1/B_1}), \mathcal{O}_U)) \cong \mathcal{H}om_{\mathcal{O}_{U'}}(\alpha_U^*u^*z^*(\xi^{-1}\Omega^1_{X_1/B_1}), \mathcal{O}_{U'}) \cong \mathcal{H}om_{\mathcal{O}_{U'}}(u'^*z'^*(\xi^{-1}\Omega^1_{X_1'/B_1'}), \mathcal{O}_{U'})$. By taking $\Gamma(U', -)$, we obtain a homomorphism $\mathcal{T}_{\boldsymbol{X}}(U) \to \mathcal{T}_{\boldsymbol{X}'}(U') = \alpha_{1*}\mathcal{T}_{\boldsymbol{X}'}(U)$. Varying $U$, we obtain a morphism

$$\alpha_{\mathcal{T}}^* \colon \mathcal{T}_{\boldsymbol{X}} \to \alpha_{1*}(\mathcal{T}_{\boldsymbol{X}'})$$

compatible with the actions of $\mathcal{O}_{Y_1}$ and $\mathcal{O}_{Y_1'}$ via the morphism $\mathcal{O}_{Y_1} \to \alpha_{1*}\mathcal{O}_{Y_1'}$.

Let $u \colon U \to Y_1$ be an object of $(\mathrm{AffSmStr}/Y_1)_{\mathrm{Zar}}$, let $\widetilde{u} \colon \widetilde{U} \to Y_2$ be an object of $\mathrm{Lift}(U/Y_2)$, and let $u' \colon U' \to Y_1'$ (resp. $\widetilde{u}' \colon \widetilde{U}' \to Y_2'$) be the base change of $u$ (resp. $\widetilde{u}$) by $\alpha_1$ (resp. $\alpha_2$). By (Cond 1), for any $g \in \mathcal{L}_{\boldsymbol{X},U}(\widetilde{U})$, there exists a unique $B_2'$-morphism $g' \colon \widetilde{U}' \to X_2'$ such that $\alpha_{X,2} \circ g' \colon \widetilde{U}' \to X_2$ coincides with the composition of $\widetilde{U}' \to \widetilde{U} \xrightarrow{g} X_2$. It is straightforward to verify that $g'$ belongs to $\mathcal{L}_{\boldsymbol{X}',U'}(\widetilde{U}')$. Thus we obtain a map $\alpha^*_{\mathcal{L},\widetilde{U},\widetilde{U}'} \colon \mathcal{L}_{\boldsymbol{X},U}(\widetilde{U}) \to \mathcal{L}_{\boldsymbol{X},U'}(\widetilde{U}')$. By the definition of the torsor structures, we see that this morphism is $\Gamma(U, \alpha_{\mathcal{T}}^*)$-equivariant. We also see that this construction is functorial on the pair $(u, \widetilde{u})$ in the obvious sense. Hence the composition of $\mathcal{L}_{\boldsymbol{X}}(U) \cong \mathcal{L}_{\boldsymbol{X},U}(\widetilde{U}) \xrightarrow{\alpha^*_{\mathcal{L},\widetilde{U},\widetilde{U}'}} \mathcal{L}_{\boldsymbol{X}',U'}(\widetilde{U}') \cong \mathcal{L}_{\boldsymbol{X}'}(U') = \alpha_{1*}\mathcal{L}_{\boldsymbol{X}}(U)$ does not depend on the choice of $\widetilde{U}$ and induces an $\alpha_{\mathcal{T}}^*$-equivariant morphism

$$\alpha_{\mathcal{L}}^* \colon \mathcal{L}_{\boldsymbol{X}} \longrightarrow \alpha_{1*}\mathcal{L}_{\boldsymbol{X}'}.$$

The morphisms $\alpha_{Y,1} \colon Y_1' \to Y_1$ and $\alpha_X \times_{\alpha_B} \alpha_Y \colon X' \times_{B'} Y' \to X \times_B Y$ induce a morphism $D_{\boldsymbol{X}'} \to D_{\boldsymbol{X}}$ compatible with the morphism $\alpha_Y \colon Y' \to Y$. By (Cond 1), Proposition IV.2.2.9 (3), Lemma IV.2.2.18 (1), and Proposition IV.2.3.17, we see that the morphism $\alpha_D \colon D_{\boldsymbol{X}'} \to D_{\boldsymbol{X}} \times_Y Y'$ induced by $D_{\boldsymbol{X}'} \to D_{\boldsymbol{X}}$ is an isomorphism, and we obtain a morphism

$$\alpha_{\mathcal{D}}^* \colon \mathcal{D}_{\boldsymbol{X}} \longrightarrow \alpha_{1*}\mathcal{D}_{\boldsymbol{X}'}.$$

**Proposition IV.2.5.21.** *For a morphism $\boldsymbol{\alpha} \colon \boldsymbol{X}' \to \boldsymbol{X}$ in $\mathscr{C}_{\mathrm{Tors}}$ satisfying* (Cond 1), *the following diagram is commutative. In particular, $\alpha_{\mathcal{D}}^*$ is $\alpha_{\mathcal{T}}^*$-equivariant.*

$$\begin{array}{ccc}
\mathcal{L}_{\boldsymbol{X}} & \xrightarrow{\alpha_{\mathcal{L}}^*} & \alpha_{1*}\mathcal{L}_{\boldsymbol{X}'} \\
{\scriptstyle \text{Thm. IV.2.5.8}}\downarrow{\scriptstyle \cong} & & {\scriptstyle \cong}\downarrow{\scriptstyle \text{Thm. IV.2.5.8}} \\
\mathcal{D}_{\boldsymbol{X}} & \xrightarrow{\alpha_{\mathcal{D}}^*} & \alpha_{1*}\mathcal{D}_{\boldsymbol{X}'}.
\end{array}$$

PROOF. Let $Z$ (resp. $Z'$) denote $X \times_B Y$ (resp. $X' \times_{B'} Y'$). By (Cond 1), we have isomorphisms $(Y_1' \hookrightarrow Z') \xrightarrow{\cong} (Y_1 \hookrightarrow Z) \times_Y Y'$ and $D_{\boldsymbol{X}'} \xrightarrow{\cong} D_{\boldsymbol{X}} \times_Y Y'$ in the category $\mathscr{C}$. Hence, for a strict smooth affine Cartesian morphism $U \to Y$ in $\mathscr{C}$ and its base change $U' \to Y'$ under the morphism $\alpha_Y \colon Y' \to Y$, we have the following commutative diagram, where the vertical morphisms except $\alpha^*_{\mathcal{L},U_2,U_2'}$ are defined by the base change under $\alpha_Y$.

$$\begin{array}{ccccccc}
\mathcal{L}_{\boldsymbol{X},U_1}(U_2) & \longleftarrow & \mathrm{Hom}_Y(U,(Y_1 \hookrightarrow Z)) & \xleftarrow{\cong} & \mathrm{Hom}_Y(U,D_{\boldsymbol{X}}) & \longrightarrow & \mathrm{Hom}_{Y_1}(U_1,D_{\boldsymbol{X},1}) \\
{\scriptstyle \alpha^*_{\mathcal{L},U_2,U_2'}}\downarrow & & \downarrow & & \downarrow & & \downarrow \\
\mathcal{L}_{\boldsymbol{X}',U_1'}(U_2') & \longleftarrow & \mathrm{Hom}_{Y'}(U',(Y_1' \hookrightarrow Z')) & \xleftarrow{\cong} & \mathrm{Hom}_{Y'}(U',D_{\boldsymbol{X}'}) & \longrightarrow & \mathrm{Hom}_{Y_1'}(U_1',D_{\boldsymbol{X}',1}).
\end{array}$$

This implies that the diagram in the proposition commutes for the sections on $U_1$. $\square$

The constructions of $\alpha^*_{\mathcal{T}}$ and $\alpha^*_{\mathcal{L}}$ still work after taking the reduction mod $p^n$, and we obtain a morphism ${}_n\alpha^*_{\mathcal{T}} \colon {}_n\mathcal{T}_{\boldsymbol{X},\mathrm{Zar}} \to \alpha_{1*}({}_n\mathcal{T}_{\boldsymbol{X}',\mathrm{Zar}})$ on $(Y_1)_{\mathrm{Zar}}$ compatible with the actions of $\mathcal{O}_{{}_nY_1}$ and $\mathcal{O}_{{}_nY_1'}$, and an ${}_n\alpha^*_{\mathcal{T}}$-equivariant morphism ${}_n\alpha^*_{\mathcal{L}} \colon {}_n\mathcal{L}_{\boldsymbol{X},\mathrm{Zar}} \to \alpha_{1*}({}_n\mathcal{L}_{\boldsymbol{X}',\mathrm{Zar}})$. The morphism ${}_n\alpha^*_{\mathcal{T}}$ induces an isomorphism $\mathcal{O}_{{}_nY_1'} \otimes_{\alpha_1^{-1}(\mathcal{O}_{{}_nY_1})} \alpha_1^{-1}({}_n\mathcal{T}_{\boldsymbol{X},\mathrm{Zar}}) \xrightarrow{\cong} {}_n\mathcal{T}_{\boldsymbol{X}',\mathrm{Zar}}$. By II.4.13, we obtain an isomorphism ${}_n\alpha_L \colon {}_nL_{\boldsymbol{X}'} \xrightarrow{\cong} {}_nL_{\boldsymbol{X}} \times_{{}_nY_1} {}_nY_1'$ of ${}_nT_{\boldsymbol{X}'}$-principal homogeneous spaces, and then an isomorphism of $T_{\boldsymbol{X}'}$-principal homogeneous spaces

$$\alpha_L \colon L_{\boldsymbol{X}'} \xrightarrow{\cong} L_{\boldsymbol{X}} \times_{Y_1} Y_1'.$$

**Proposition IV.2.5.22.** *For a morphism $\boldsymbol{\alpha} \colon \boldsymbol{X}' \to \boldsymbol{X}$ in $\mathscr{C}_{\mathrm{Tors}}$ satisfying* (Cond 1), *the following diagram is commutative. In particular, the isomorphism $\alpha_{D,1} \colon D_{\boldsymbol{X}',1} \xrightarrow{\cong} D_{\boldsymbol{X},1} \times_{Y_1} Y_1'$ is $T_{\boldsymbol{X}'}$-equivariant.*

$$\begin{array}{ccc}
L_{\boldsymbol{X}'} & \xrightarrow[\alpha_L]{\cong} & L_{\boldsymbol{X}} \times_{Y_1} Y_1' \\
{\scriptstyle \text{Thm. IV.2.5.2}}\downarrow{\scriptstyle \cong} & & {\scriptstyle \cong}\downarrow{\scriptstyle \text{Thm. IV.2.5.2}} \\
D_{\boldsymbol{X}',1} & \xrightarrow[\alpha_{D,1}]{\cong} & D_{\boldsymbol{X},1} \times_{Y_1} Y_1'.
\end{array}$$

Proof. By (II.4.13.8) and Proposition IV.2.5.21, we have the following commutative diagram on $(Y_1)_{\mathrm{Zar}}$.

$$\begin{array}{ccccc}
\varphi_{Y_1}(L_{\boldsymbol{X}}) & \xleftarrow{\cong} & \mathcal{L}_{\boldsymbol{X},\mathrm{Zar}} & \xleftarrow[\text{Thm. IV.2.5.8}]{\cong} & \varphi_{Y_1}(D_{\boldsymbol{X},1}) \\
\downarrow & & \downarrow{\scriptstyle \alpha^*_{\mathcal{L}}} & & \downarrow \\
\alpha_{1*}(\varphi_{Y_1'}(L_{\boldsymbol{X}'})) & \xleftarrow{\cong} & \alpha_{1*}(\mathcal{L}_{\boldsymbol{X}',\mathrm{Zar}}) & \xleftarrow[\text{Thm. IV.2.5.8}]{\cong} & \alpha_{1*}(\varphi_{Y_1'}(D_{\boldsymbol{X}',1})).
\end{array}$$

Here the left (resp. right) vertical morphism is naturally induced by $\alpha_L$ (resp. $\alpha_{D,1}$). By taking the adjoints of the vertical morphisms, we obtain a diagram, whose outer square is commutative.

$$\begin{array}{ccccc}
\alpha_1^{-1}(\varphi_{Y_1}(D_{\boldsymbol{X},1})) & \longrightarrow & \varphi_{Y_1}(D_{\boldsymbol{X},1} \times_{Y_1} Y_1') & \xrightarrow{\cong} & \varphi_{Y_1}(D_{\boldsymbol{X}',1}) \\
{\scriptstyle \text{Thm. IV.2.5.2}}\downarrow{\scriptstyle \cong} & & {\scriptstyle \text{Thm. IV.2.5.2}}\downarrow{\scriptstyle \cong} & & {\scriptstyle \text{Thm. IV.2.5.2}}\downarrow{\scriptstyle \cong} \\
\alpha_1^{-1}(\varphi_{Y_1}(L_{\boldsymbol{X}})) & \longrightarrow & \varphi_{Y_1}(L_{\boldsymbol{X}} \times_{Y_1} Y_1') & \xrightarrow{\cong} & \varphi_{Y_1}(L_{\boldsymbol{X}'}).
\end{array}$$

By using the action of $\mathcal{T}_{\boldsymbol{X}',\mathrm{Zar}}$, we see that the right square is also commutative. Hence the claim follows from Proposition IV.2.5.1. $\square$

Next we consider a morphism $\boldsymbol{\alpha} \colon \boldsymbol{X}' = (Y \to X' \leftarrow B, z') \to \boldsymbol{X} = (Y \to X \leftarrow B, z)$ satisfying (Cond 2).

The $\mathcal{O}_{X_1'}$-linear morphism $\alpha^*_{X,1}(\Omega^1_{X_1/B_1}) \to \Omega^1_{X_1'/B_1}$ induces an $\mathcal{O}_{Y_1}$-linear morphism

$$\alpha_{\mathcal{T}*} \colon \mathcal{T}_{\boldsymbol{X}'} \to \mathcal{T}_{\boldsymbol{X}}.$$

For an object $U$ of $(\mathrm{AffSmStr}/Y_1)_{\mathrm{Zar}}$ and an object $\widetilde{U}$ of $\mathrm{Lift}(U/Y_2)$, the composition with $\alpha_{X,2}\colon X_2' \to X_2$ induces an $\alpha_{\mathcal{T}*}(U)$-equivariant map $\mathcal{L}_{\boldsymbol{X}',U}(\widetilde{U}) \to \mathcal{L}_{\boldsymbol{X},U}(\widetilde{U})$, which is functorial on the pair $(u, \widetilde{u})$. Hence it defines an $\alpha_{\mathcal{T}*}$-equivariant morphism

$$\alpha_{\mathcal{L}*}\colon \mathcal{L}_{\boldsymbol{X}'} \to \mathcal{L}_{\boldsymbol{X}}.$$

The morphisms $\alpha_{Y_1}$ and $\alpha_X \times \alpha_Y$ induce a morphism $\alpha_D\colon D_{\boldsymbol{X}'} \to D_{\boldsymbol{X}}$. The composition with $\alpha_{D,1}$ gives a morphism

$$\alpha_{\mathcal{D}*}\colon \mathcal{D}_{\boldsymbol{X}'} \longrightarrow \mathcal{D}_{\boldsymbol{X}}.$$

**Proposition IV.2.5.23.** *For a morphism $\boldsymbol{\alpha}\colon \boldsymbol{X}' \to \boldsymbol{X}$ in $\mathscr{C}_{\mathrm{Tors}}$ satisfying* (Cond 2), *the following diagram is commutative. In particular, $\alpha_{\mathcal{D}*}$ is $\alpha_{\mathcal{T}*}$-equivariant.*

$$\begin{array}{ccc}
\mathcal{L}_{\boldsymbol{X}'} & \xrightarrow{\alpha_{\mathcal{L}*}} & \mathcal{L}_{\boldsymbol{X}} \\
\text{Thm. IV.2.5.8}\ \downarrow\cong & & \cong\downarrow\ \text{Thm. IV.2.5.8} \\
\mathcal{D}_{\boldsymbol{X}'} & \xrightarrow{\alpha_{\mathcal{D}*}} & \mathcal{D}_{\boldsymbol{X}}.
\end{array}$$

PROOF. Let $Z$ (resp. $Z'$) denote $Y \times_B X$ (resp. $Y \times_B X'$). Let $U \to Y$ be a strict smooth affine Cartesian morphism in $\mathscr{C}$. Then the compositions with $X_2' \to X_2$, $(Y_1 \hookrightarrow Z') \to (Y_1 \hookrightarrow Z)$, $D_{\boldsymbol{X}'} \to D_{\boldsymbol{X}}$, and $D_{\boldsymbol{X}',1} \to D_{\boldsymbol{X},1}$ give the following commutative diagram.

$$\begin{array}{ccccccc}
\mathcal{L}_{\boldsymbol{X}',U_1}(U_2) & \longleftarrow & \mathrm{Hom}_Y(U, (Y_1 \hookrightarrow Z')) & \xleftarrow{\cong} & \mathrm{Hom}_Y(U, D_{\boldsymbol{X}'}) & \longrightarrow & \mathrm{Hom}_{Y_1}(U_1, D_{\boldsymbol{X}',1}) \\
\downarrow & & \downarrow & & \downarrow & & \downarrow \\
\mathcal{L}_{\boldsymbol{X},U_1}(U_2) & \longleftarrow & \mathrm{Hom}_Y(U, (Y_1 \hookrightarrow Z)) & \xleftarrow{\cong} & \mathrm{Hom}_Y(U, D_{\boldsymbol{X}}) & \longrightarrow & \mathrm{Hom}_{Y_1}(U_1, D_{\boldsymbol{X},1}).
\end{array}$$

This implies that the diagram in the proposition is commutative for sections on $U_1$. □

We continue to assume that $\boldsymbol{\alpha}$ satisfies (Cond 2). The constructions of $\alpha_{\mathcal{T}*}$ and $\alpha_{\mathcal{L}*}$ above still work after taking the reduction mod $p^n$, and we obtain an $\mathcal{O}_{{}_nY_1}$-linear morphism ${}_n\alpha_{\mathcal{T}*}\colon {}_n\mathcal{T}_{\boldsymbol{X}',\mathrm{Zar}} \to {}_n\mathcal{T}_{\boldsymbol{X},\mathrm{Zar}}$ and an ${}_n\alpha_{\mathcal{T}*}$-equivariant morphism ${}_n\alpha_{\mathcal{L}*}\colon {}_n\mathcal{L}_{\boldsymbol{X}',\mathrm{Zar}} \to {}_n\mathcal{L}_{\boldsymbol{X},\mathrm{Zar}}$. By II.4.12, we obtain a morphism ${}_n\alpha_T\colon {}_nT_{\boldsymbol{X}'} \to {}_nT_{\boldsymbol{X}}$ and an ${}_n\alpha_T$-equivariant morphism ${}_n\alpha_L\colon {}_nL_{\boldsymbol{X}'} \to {}_nL_{\boldsymbol{X}}$. These form a compatible system with respect to $n$ and define a morphism

$$\alpha_T\colon T_{\boldsymbol{X}'} \to T_{\boldsymbol{X}}$$

and an $\alpha_T$-equivariant morphism

$$\alpha_L\colon L_{\boldsymbol{X}'} \to L_{\boldsymbol{X}}.$$

**Proposition IV.2.5.24.** *For a morphism $\boldsymbol{\alpha}\colon \boldsymbol{X}' \to \boldsymbol{X}$ in $\mathscr{C}_{\mathrm{Tors}}$ satisfying* (Cond 2), *the morphism $\alpha_{D,1}\colon D_{\boldsymbol{X}',1} \to D_{\boldsymbol{X},1}$ is $\alpha_T$-equivariant, and the following diagram is commutative.*

$$\begin{array}{ccc}
L_{\boldsymbol{X}'} & \xrightarrow{\alpha_L} & L_{\boldsymbol{X}} \\
\text{Thm. IV.2.5.2}\ \downarrow\cong & & \cong\downarrow\ \text{Thm. IV.2.5.2} \\
D_{\boldsymbol{X}',1} & \xrightarrow{\alpha_{D,1}} & D_{\boldsymbol{X},1}.
\end{array}$$

PROOF. The first claim follows from Proposition IV.2.5.23. By (II.4.12.4) and Proposition IV.2.5.23, the following diagram is commutative.

$$\begin{array}{ccccc}
\varphi_{Y_1}(L_{\boldsymbol{X}'}) & \xleftarrow{\ \cong\ } & \mathcal{L}_{\boldsymbol{X}',\mathrm{Zar}} & \xleftarrow[\text{Thm. IV.2.5.8}]{\cong} & \varphi_{Y_1}(D_{\boldsymbol{X}',1}) \\
{\scriptstyle \varphi_{Y_1}(\alpha_L)}\downarrow & & {\scriptstyle \alpha_{\mathcal{L}*}}\downarrow & & \downarrow{\scriptstyle \varphi_{Y_1}(\alpha_{D,1})} \\
\varphi_{Y_1}(L_{\boldsymbol{X}}) & \xleftarrow{\ \cong\ } & \mathcal{L}_{\boldsymbol{X},\mathrm{Zar}} & \xleftarrow[\text{Thm. IV.2.5.8}]{\cong} & \varphi_{Y_1}(D_{\boldsymbol{X},1}).
\end{array}$$

Hence it suffices to prove that, for two $\alpha_T$-equivariant morphisms $\sigma, \sigma'\colon L_{\boldsymbol{X}'} \to L_{\boldsymbol{X}}$, $\varphi_{Y_1}(\sigma) = \varphi_{Y_1}(\sigma')$ implies $\sigma = \sigma'$. By considering the morphisms of sheaves on $(\mathrm{AffSmStr}/Y_1)_{\mathrm{Zar}}$ associated to $\sigma$ and $\sigma'$, we see that there exists a unique $T_{\boldsymbol{X}}$-equivariant automorphism $\tau\colon L_{\boldsymbol{X}} \to L_{\boldsymbol{X}}$ such that $\sigma' = \tau \circ \sigma$. The equality $\varphi_{Y_1}(\sigma) = \varphi_{Y_1}(\sigma')$ implies $\varphi_{Y_1}(\tau) = \mathrm{id}$. By Proposition IV.2.5.1, we have $\tau = \mathrm{id}$, which implies $\sigma = \sigma'$. □

Let $\boldsymbol{\alpha}\colon \boldsymbol{X}' = (Y' \to B' \leftarrow X', z') \to (Y \to B \leftarrow X, z)$ be a morphism in $\mathscr{C}_{\mathrm{Tors}}$. Let $X'' \to B'$ denote the base change of $X \to B$ by $\alpha_B\colon B' \to B$, let $z''\colon Y_1' \to X_1''$ denote the $B_1'$-morphism induced by $Y_1' \xrightarrow{\alpha_{Y,1}} Y_1 \to X_1$, and let $\boldsymbol{X}''$ denote the object $(Y' \to B' \leftarrow X'', z'')$ of $\mathscr{C}_{\mathrm{Tors}}$. Then the morphism $\alpha$ decomposes naturally as $\boldsymbol{X}' \xrightarrow{\boldsymbol{\alpha}^{II}} \boldsymbol{X}'' \xrightarrow{\boldsymbol{\alpha}^{I}} \boldsymbol{X}$, and $\boldsymbol{\alpha}^I$ (resp. $\boldsymbol{\alpha}^{II}$) satisfies (Cond 1) (resp. (Cond 2)). We define the group morphism

$$\alpha_T\colon T_{\boldsymbol{X}'} \to T_{\boldsymbol{X}} \times_{Y_1} Y_1'$$

to be the composition of $\alpha_T^{II}\colon T_{\boldsymbol{X}'} \to T_{\boldsymbol{X}''}$ with the natural isomorphism $T_{\boldsymbol{X}''} \cong T_{\boldsymbol{X}} \times_{Y_1} Y_1'$. Composing $\alpha_L^{II}\colon L_{\boldsymbol{X}'} \to L_{\boldsymbol{X}''}$ with $\alpha_L^I\colon L_{\boldsymbol{X}''} \xrightarrow{\cong} L_{\boldsymbol{X}} \times_{Y_1} Y_1'$, we obtain an $\alpha_T$-equivariant morphism

$$\alpha_L\colon L_{\boldsymbol{X}'} \to L_{\boldsymbol{X}} \times_{Y_1} Y_1'.$$

On the other hand, the morphisms $\alpha_{Y,1}\colon Y_1' \to Y_1$ and $\alpha_X \times_{\alpha_B} \alpha_Y\colon X' \times_{B'} Y' \to X \times_B Y$ induce a morphism $D_{\boldsymbol{X}'} \to D_{\boldsymbol{X}}$ and then a $Y'$-morphism

$$\alpha_D\colon D_{\boldsymbol{X}'} \to D_{\boldsymbol{X}} \times_Y Y'.$$

**Proposition IV.2.5.25.** *For a morphism $\boldsymbol{\alpha}\colon \boldsymbol{X}' \to \boldsymbol{X}$ in $\mathscr{C}_{\mathrm{Tors}}$, the following diagram is commutative. In particular, the morphism $\alpha_{D,1}$ is $\alpha_T$-equivariant.*

$$\begin{array}{ccc}
L_{\boldsymbol{X}'} & \xrightarrow{\ \alpha_L\ } & L_{\boldsymbol{X}} \times_{Y_1} Y_1' \\
{\scriptstyle \text{Thm. IV.2.5.2}}\ \downarrow{\scriptstyle \cong} & & {\scriptstyle \cong}\downarrow\ {\scriptstyle \text{Thm. IV.2.5.2}} \\
D_{\boldsymbol{X}',1} & \xrightarrow{\ \alpha_{D,1}\ } & D_{\boldsymbol{X},1} \times_{Y_1} Y_1'.
\end{array}$$

PROOF. The morphism $\alpha_D$ coincides with the composition of $\alpha_D^{II}\colon D_{\boldsymbol{X}'} \to D_{\boldsymbol{X}''}$ and $\alpha_D^I\colon D_{\boldsymbol{X}''} \to D_{\boldsymbol{X}} \times_Y Y'$. Hence the claim immediately follows from Propositions IV.2.5.22 and IV.2.5.24. □

**Corollary IV.2.5.26.** *For morphisms $\boldsymbol{\alpha}\colon \boldsymbol{X}' \to \boldsymbol{X}$, $\boldsymbol{\alpha}'\colon \boldsymbol{X}'' \to \boldsymbol{X}'$ in $\mathscr{C}_{\mathrm{Tors}}$, and their composition $\boldsymbol{\alpha}'' := \boldsymbol{\alpha} \circ \boldsymbol{\alpha}'$, we have $\alpha_L'' = (\alpha_L \times 1_{Y_1''}) \circ \alpha_L'$ and $\alpha_D'' = (\alpha_D \times 1_{Y''}) \circ \alpha_D'$.*

PROOF. The second equality is obvious by the construction of morphisms, and it implies the first equality by Proposition IV.2.5.25. □

## IV.3. Higgs isocrystals and Higgs crystals

**IV.3.1. Sites.** For $r \in \mathbb{N}_{>0} \cup \{\infty\}$, $B \in \mathrm{Ob}\,\mathscr{C}^r$ and a $p$-adic fine log formal scheme $X$ over $B_1$, we define a site $(X/B)^r_{\mathrm{HIGGS}}$ (Definitions IV.3.1.1, IV.3.1.4), an analogue of the crystalline site for Higgs fields, on which Higgs isocrystals and Higgs crystals will be defined in IV.3.3. We have a description of sheaves on the site $(X/B)^r_{\mathrm{HIGGS}}$ in terms of certain compatible systems of étale sheaves on objects of $(X/B)^r_{\mathrm{HIGGS}}$ in a way similar to the crystalline site (see after Definition IV.3.1.4). We then study functoriality and two kinds of localizations: one with respect to a morphism of $p$-adic fine log formal schemes $U \to X$ and the other with respect to an object of the site.

**Definition IV.3.1.1.** Let $r \in \mathbb{N}_{>0} \cup \{\infty\}$, let $B \in \mathrm{Ob}\,\mathscr{C}^r$, and let $X$ be a $p$-adic fine log formal scheme over $B_1$. We define the category $(X/B)^r_{\mathrm{HIGGS}}$ as follows. An object is a pair $(T, z)$ of an object $T$ of $\mathscr{C}^r_{/B}$ and a $B_1$-morphism $z\colon T_1 \to X$. A morphism $(T', z') \to (T, z)$ is a morphism $u\colon T' \to T$ in $\mathscr{C}^r_{/B}$ such that $z \circ u_1 = z'$.

We say that a morphism $u\colon (T', z') \to (T, z)$ in $(X/B)^r_{\mathrm{HIGGS}}$ is étale (resp. strict, resp. Cartesian) if the underlying morphism $u\colon T' \to T$ in $\mathscr{C}^r$ is étale (resp. strict, resp. Cartesian).

**Proposition IV.3.1.2.** *Let $r$, $B$, and $X$ be as in Definition* IV.3.1.1.

(1) *Finite fiber products and nonempty finite products are representable in the category $(X/B)^r_{\mathrm{HIGGS}}$.*

(2) *Finite fiber products are compatible with the forgetful functor $(X/B)^r_{\mathrm{HIGGS}} \to \mathscr{C}^r/B$.*

Proof. The fiber product of $(T', z') \to (T, z) \leftarrow (T'', z'')$ in $(X/B)^r_{\mathrm{HIGGS}}$ is represented by the fiber product $T'''$ of $T' \to T \leftarrow T''$ in $\mathscr{C}^r_{/B}$ (cf. Corollary IV.2.2.11) endowed with the composition of $T'''_1 \to T'_1 \xrightarrow{z'} X$, which coincides with that of $T'''_1 \to T''_1 \xrightarrow{z''} X$. The product of two objects $(T, z)$ and $(T', z')$ of $(X/B)^r_{\mathrm{HIGGS}}$ is represented by

$$T'' := D^r_{\mathrm{Higgs}}(T_1 \times_X T'_1 \hookrightarrow T \times_B T')$$

endowed with the composition of $T''_1 \to T_1 \xrightarrow{z} X$, which coincides with that of $T''_1 \to T'_1 \xrightarrow{z'} X$. □

For an object $(T, z)$ of $(X/B)^r_{\mathrm{HIGGS}}$, we define the set $\mathrm{Cov}((T, z))$ to be

$$\left\{ (u_\alpha\colon (T_\alpha, z_\alpha) \to (T, z))_{\alpha \in A} \;\middle|\; \begin{array}{l} \text{(i) } u_\alpha \text{ is strict étale and Cartesian for all } \alpha \in A \\ \text{(ii) } \bigcup_{\alpha \in A} u_{\alpha,1}(T_{\alpha,1}) = T_1 \end{array} \right\}.$$

**Lemma IV.3.1.3.** *Let $r$, $B$, and $X$ be as in Definition* IV.3.1.1. *The sets* $\mathrm{Cov}((T, z))$ *for $(T, z) \in \mathrm{Ob}\,(X/B)^r_{\mathrm{HIGGS}}$ satisfy the axiom of pretopology* ([**2**] II *Définition* 1.3).

Proof. By Lemma IV.2.2.7 (1) and the construction of fiber products in $\mathscr{C}^r$ given in the proof of Corollary IV.2.2.11, we see that $\mathrm{Cov}((T, z))$ is stable under base change. The stability under composition and $\mathrm{id}_{(T,z)} \in \mathrm{Cov}((T, z))$ are obvious. □

**Definition IV.3.1.4.** Let $r$, $B$, and $X$ be as in Definition IV.3.1.1. We endow the category $(X/B)^r_{\mathrm{HIGGS}}$ with the topology associated to the pretopology $\mathrm{Cov}((T, z))$, $(T, z) \in \mathrm{Ob}\,(X/B)^r_{\mathrm{HIGGS}}$. We define the sheaf of rings $\mathscr{O}_{X/B,1}$ on $(X/B)^r_{\mathrm{HIGGS}}$ by $\mathscr{O}_{X/B,1}(T, z) = \Gamma(T_1, \mathscr{O}_{T_1})$.

Let $r$, $B$, and $X$ be as in Definition IV.3.1.1. Let $\mathscr{F}$ be a sheaf of sets (resp. $\mathscr{O}_{X/B,1}$-modules) on the site $(X/B)^r_{\mathrm{HIGGS}}$. Then, for each object $(T, z)$ of $(X/B)^r_{\mathrm{HIGGS}}$, one can define a sheaf of sets (resp. $\mathscr{O}_{T_1}$-modules) $\mathscr{F}_{(T,z)}$ on $T_{1,\text{ét}}$ by $\mathscr{F}_{(T,z)}(v_1\colon U_1 \to T_1) =$

$\mathcal{F}((U, z \circ v_1))$, where $v\colon U \to T$ is the unique strict étale Cartesian morphism in $\mathscr{C}^r$ such that $U_1 \to T_1$ is the given morphism $v_1$. For a morphism $u\colon (T', z') \to (T, z)$ in $(X/B)^r_{\mathrm{HIGGS}}$, we can define a morphism of sheaves of sets (resp. $\mathcal{O}_{T'_1}$-modules)

$$\tau_u\colon u_1^{-1}(\mathcal{F}_{(T,z)}) \to \mathcal{F}_{(T',z')} \quad (\text{resp. } \tau_u\colon u_1^*(\mathcal{F}_{(T,z)}) \to \mathcal{F}_{(T',z')})$$

satisfying the following conditions:

(i) For any morphism $u\colon (T', z') \to (T, z)$ such that the underlying morphism $u\colon T' \to T$ in $\mathscr{C}$ is strict étale and Cartesian, the morphism $\tau_u$ is an isomorphism.

(ii) For any composable morphisms $(T'', z'') \xrightarrow{v} (T', z') \xrightarrow{u} (T, z)$, we have $\tau_v \circ v_1^{-1}(\tau_u) = \tau_{uv}$ (resp. $\tau_v \circ v_1^*(\tau_u) = \tau_{uv}$).

The category of sheaves of sets (resp. $\mathcal{O}_{X/B,1}$-modules) on $(X/B)^r_{\mathrm{HIGGS}}$ is equivalent to the category of $(\mathcal{F}_{(T,z)}, \tau_u)$ satisfying the conditions (i) and (ii) above.

**Proposition IV.3.1.5.** *Let $r, r' \in \mathbb{N}_{>0} \cup \{\infty\}$ such that $r' \geq r$. Let $B$ (resp. $B'$) be an object of $\mathscr{C}^r$ (resp. $\mathscr{C}^{r'}$) and let $X$ (resp. $X'$) be a p-adic fine log formal scheme over $B_1$ (resp. $B'_1$). Let $g\colon B' \to B$ be a morphism in $\mathscr{C}$, and let $f\colon X' \to X$ be a morphism of p-adic fine log formal schemes compatible with the morphism $g_1\colon B'_1 \to B_1$. Then the functor*

$$(X'/B')^{r'}_{\mathrm{HIGGS}} \longrightarrow (X/B)^r_{\mathrm{HIGGS}}$$

*defined by $(T \xrightarrow{h} B', z) \mapsto (T \xrightarrow{g\circ h} B, f \circ z)$ is cocontinuous (cf.* **[2]** III *Définition* 2.1), *and induces a morphism of topos (cf.* **[2]** III *Proposition* 2.3)

$$f_{\mathrm{HIGGS}}\colon (X'/B')^{r'\sim}_{\mathrm{HIGGS}} \longrightarrow (X/B)^{r\sim}_{\mathrm{HIGGS}}.$$

*We have $f^*_{\mathrm{HIGGS}}(\mathcal{O}_{X/B,1}) = \mathcal{O}_{X'/B',1}$.*

Proof. By **[2]** III Définition 2.1 and **[2]** II Proposition 1.4, the cocontinuity follows from the following fact: Let $(T \xrightarrow{h} B', z) \in \mathrm{Ob}\,(X'/B')^{r'}_{\mathrm{HIGGS}}$ and let $(u_\alpha\colon T_\alpha \to T)_{\alpha\in A}$ be a strict étale covering in $\mathscr{C}$. Note that $T_\alpha \in \mathrm{Ob}\,\mathscr{C}^{r'}$ by Lemma IV.2.2.7 (1). Let $h_\alpha$ denote $h \circ u_\alpha\colon T_\alpha \to B'$ and let $z_\alpha$ denote $z \circ u_{\alpha,1}\colon T_{\alpha,1} \to X'$. Then a morphism $\psi\colon (\widetilde{T} \xrightarrow{\widetilde{h}} B', \widetilde{z}) \to (T \xrightarrow{h} B', z)$ in $(X'/B')^{r'}_{\mathrm{HIGGS}}$ factors thorough a morphism $u_\alpha\colon (T_\alpha \xrightarrow{h_\alpha} B', z_\alpha) \to (T \xrightarrow{h} B', z)$ for some $\alpha \in A$ if and only if the morphism $\psi\colon (\widetilde{T} \xrightarrow{g\circ\widetilde{h}} B, f\circ\widetilde{z}) \to (T \xrightarrow{g\circ h} B, f \circ z)$ in $(X/B)^r_{\mathrm{HIGGS}}$ factors through a morphism $u_\alpha\colon (T_\alpha \xrightarrow{g\circ h_\alpha} B, f \circ z_\alpha) \to (T \xrightarrow{g\circ h} B, f \circ z)$ for some $\alpha \in A$. □

**Proposition IV.3.1.6.** *Let $r$, $r'$, $B$, $B'$, $X$, $X'$, $f$, and $g$ be the same as in Proposition* IV.3.1.5 *and assume that the morphism $X' \to X \times_{B_1} B'_1$ induced by $f$ and $g_1$ is an isomorphism. Then the functor*

$$f^*\colon (X/B)^r_{\mathrm{HIGGS}} \longrightarrow (X'/B')^{r'}_{\mathrm{HIGGS}}$$

*defined by $(T, z) \mapsto (D^{r'}_{\mathrm{Higgs}}(T \times_B B'), z')$ is continuous (cf.* **[2]** III *Définition* 1.1) *and is a right adjoint of the functor in Proposition* IV.3.1.5, *where $z'$ is the composition of $D^{r'}_{\mathrm{Higgs}}(T \times_B B')_1 \to T_1 \times_{B_1} B'_1 \xrightarrow{z\times\mathrm{id}} X \times_{B_1} B'_1 \cong X'$. Hence the functor $f^*$ defines a morphism of sites, and it induces the morphism of topos $f_{\mathrm{HIGGS}}$ in Proposition* IV.3.1.5 *(cf.* **[2]** III *Proposition* 2.5).

Proof. For $T \in \mathrm{Ob}\,(X/B)^r_{\mathrm{HIGGS}}$ and $(T_\alpha \to T)_{\alpha\in A} \in \mathrm{Cov}(T)$, we have $(f^*(T_\alpha) \to f^*(T))_{\alpha\in A} \in \mathrm{Cov}(f^*(T))$ and $f^*(T_\alpha \times_T T_\beta) \to f^*(T_\alpha) \times_{f^*(T)} f^*(T_\beta)$ is an isomorphism by Proposition IV.2.2.9 (3). Hence $f^*$ is continuous. Let $j$ be the functor $(X'/B')^{r'}_{\mathrm{HIGGS}} \to (X/B)^r_{\mathrm{HIGGS}}$ in Proposition IV.3.1.5. Let $(T, z) \in \mathrm{Ob}\,(X/B)^r_{\mathrm{HIGGS}}$, let $(T', z')$ be $f^*(T, z)$

and let $u$ be the composition of $T' = D^{r'}_{\mathrm{HIGGS}}(T \times_B B') \to T \times_B B' \to T$, which induces a morphism $j(T', z') \to (T, z)$ functorial in $(T, z)$. It suffices to prove that, for any $(\widetilde{T}', \widetilde{z}') \in \mathrm{Ob}\,(X'/B')^{r'}_{\mathrm{HIGGS}}$, the map

$$\mathrm{Hom}_{(X'/B')^{r'}_{\mathrm{HIGGS}}}((\widetilde{T}', \widetilde{z}'), (T', z')) \longrightarrow \mathrm{Hom}_{(X/B)^{r}_{\mathrm{HIGGS}}}(j(\widetilde{T}', \widetilde{z}'), (T, z))$$

induced by $u$ is bijective. By the adjoint property of $D^{r'}_{\mathrm{HIGGS}}$, we see that the map

$$\mathrm{Hom}_{\mathscr{C}^{r'}_{/B'}}(\widetilde{T}', T') \longrightarrow \mathrm{Hom}_{\mathscr{C}^{r}_{/B}}(\widetilde{T}', T)$$

induced by $u$ is bijective. For a morphism $\psi\colon \widetilde{T}' \to T'$ over $B'$, we see that $\psi_1\colon \widetilde{T}'_1 \to T'_1$ is compatible with $\widetilde{z}'$ and $z'$ if and only if $u_1 \circ \psi_1\colon \widetilde{T}'_1 \to T_1$ is compatible with $\widetilde{z}'$ and $z$ noting that the morphism $X' \to X \times_{B_1} B'_1$ is an isomorphism. □

**Proposition IV.3.1.7.** *Let $r \in \mathbb{N}_{>0} \cup \{\infty\}$, let $B$ be an object of $\mathscr{C}^r$, and let $X' \to X$ be a strict étale morphism of $p$-adic fine log formal schemes over $B_1$. Let $f^*\colon (X/B)^r_{\mathrm{HIGGS}} \to (X'/B)^r_{\mathrm{HIGGS}}$ be the functor associating to $(T, z)$ the strict étale lifting $T' \to T$ of $T_1 \times_X X' \to T_1$ endowed with the morphism $T'_1 = T_1 \times_X X' \to X'$. Then $f^*$ is continuous and a right adjoint of the functor in Proposition* IV.3.1.5. *Hence the functor $f^*$ defines a morphism of sites and it induces the morphism of topos $f_{\mathrm{HIGGS}}$ in Proposition* IV.3.1.5.

Proof. For $T \in \mathrm{Ob}\,(X/B)^r_{\mathrm{HIGGS}}$ and $(T_\alpha \to T)_{\alpha \in A} \in \mathrm{Cov}(T)$, we have $(f^*(T_\alpha) \to f^*(T))_{\alpha \in A} \in \mathrm{Cov}(f^*(T))$ and $f^*(T_\alpha \times_T T_\beta) \to f^*(T_\alpha) \times_{f^*(T)} f^*(T_\beta)$ is an isomorphism. Hence $f^*$ is continuous. For $(T, z) \in \mathrm{Ob}\,(X/B)^r_{\mathrm{HIGGS}}$ and $(T', z') := f^*(T, z), (T'', z'') \in \mathrm{Ob}\,(X'/B)^r_{\mathrm{HIGGS}}$, the composition with $f\colon X' \to X$ induces a bijection $\mathrm{Hom}_{X'}(T''_1, T'_1) \to \mathrm{Hom}_X(T''_1, T_1)$. Let $f_T$ be the canonical morphism $T' \to T$. Then, for an $X'$-morphism $v_1\colon T''_1 \to T'_1$ and a lifting $u\colon T'' \to T$ of $f_{T,1} \circ v_1\colon T''_1 \to T_1$, there exists a unique lifting $v\colon T'' \to T'$ of $v_1$ such that $u = f_T \circ v$. Hence the composition with $f_T$ induces a bijection $\mathrm{Hom}_{(X'/B)^r_{\mathrm{HIGGS}}}((T'', z''), (T', z')) \xrightarrow{\cong} \mathrm{Hom}_{(X/B)^r_{\mathrm{HIGGS}}}((T'', f \circ z''), (T, z))$. □

**Corollary IV.3.1.8.** *Let $r \in \mathbb{N}_{>0} \cup \{\infty\}$, let $B$ be an object of $\mathscr{C}^r$, let $X$ be a $p$-adic fine log formal scheme over $B_1$, and let $(u_\alpha\colon X_\alpha \to X)_{\alpha \in A}$ be a strict étale covering. Let $X_{\alpha\beta}$ denote $X_\alpha \times_X X_\beta$ and let $u_{\alpha\beta}\colon X_{\alpha\beta} \to X$ be the canonical morphism. Let $\mathcal{F}_i$ $(i = 1, 2)$ be sheaves of $\mathcal{O}_{X/B,1}$-modules on $(X/B)^r_{\mathrm{HIGGS}}$, and put $\mathcal{F}_{i,\alpha} := u^*_{\alpha,\mathrm{HIGGS}}(\mathcal{F}_i)$ and $\mathcal{F}_{i,\alpha\beta} := u^*_{\alpha\beta,\mathrm{HIGGS}}(\mathcal{F}_i)$. Then the following sequence is exact.*

$$\mathrm{Hom}_{(X/B)^r_{\mathrm{HIGGS}}}(\mathcal{F}_1, \mathcal{F}_2) \to \prod_{\alpha \in A} \mathrm{Hom}_{(X_\alpha/B)^r_{\mathrm{HIGGS}}}(\mathcal{F}_{1,\alpha}, \mathcal{F}_{2,\alpha})$$
$$\rightrightarrows \prod_{(\alpha,\beta) \in A^2} \mathrm{Hom}_{(X_{\alpha\beta}/B)^r_{\mathrm{HIGGS}}}(\mathcal{F}_{1,\alpha\beta}, \mathcal{F}_{2,\alpha\beta}).$$

Proof. For a strict étale morphism $f\colon X' \to X$, $(T, z) \in \mathrm{Ob}\,(X/B)^r_{\mathrm{HIGGS}}$ and $(T', z') := f^*(T, z) \in \mathrm{Ob}\,(X'/B)^r_{\mathrm{HIGGS}}$, the morphism

$$\mathcal{F}_i(T, z) \to f_{\mathrm{HIGGS}*} f^*_{\mathrm{HIGGS}} \mathcal{F}_i(T, z) = \mathcal{F}_i(T', f \circ z')$$

is induced by the natural morphism $T' \to T$. On the other hand, for an object $(T, z)$ of $(X/B)^r_{\mathrm{HIGGS}}$, letting $(T_\alpha, z_\alpha) := u^*_\alpha(T, z)$ and $(T_{\alpha\beta}, z_{\alpha\beta}) := u^*_{\alpha\beta}(T, z)$, we see that the family of canonical morphisms $(T_{\alpha,1} \to T_1)_{\alpha \in A}$ is a strict étale covering and the natural morphism $T_{\alpha\beta,1} \to T_{\alpha,1} \times_{T_1} T_{\beta,1}$ is an isomorphism. Hence the sequence

$$\mathcal{F}_i \to \prod_{\alpha \in A} u_{\alpha,\mathrm{HIGGS}*} u^*_{\alpha,\mathrm{HIGGS}} \mathcal{F}_i \rightrightarrows \prod_{(\alpha,\beta) \in A^2} u_{\alpha\beta,\mathrm{HIGGS}*} u^*_{\alpha\beta,\mathrm{HIGGS}} \mathcal{F}_i$$

is exact. This implies the claim. □

**Proposition IV.3.1.9.** *Let $r \in \mathbb{N}_{>0} \cup \{\infty\}$, let $B$ be an object of $\mathscr{C}^r$, and let $X$ be a $p$-adic fine log formal scheme over $B_1$.*

(1) *Let $u\colon U \to X$ be a morphism of $p$-adic fine log formal schemes, and define the presheaf $U^\wedge$ on $(X/B)^r_{\mathrm{HIGGS}}$ by*

$$U^\wedge((T,z)) = \left\{ f\colon T_1 \to U \;\middle|\; \begin{array}{l} f \text{ is a morphism of } p\text{-adic fine log formal schemes} \\ \text{such that } u \circ f = z \end{array} \right\}.$$

*Then the functor $((X/B)^r_{\mathrm{HIGGS}})_{/U^\wedge} \to (U/B)^r_{\mathrm{HIGGS}}$ defined by $((T,z),f) \mapsto (T,f)$ for $(T,z) \in \mathrm{Ob}\,(X/B)^r_{\mathrm{HIGGS}}$ and $f \in U^\wedge((T,z))$, is an isomorphism of sites.*

(2) *Let $(D,z_D) \in \mathrm{Ob}\,(X/B)^r_{\mathrm{HIGGS}}$. Then the topology on $((X/B)^r_{\mathrm{HIGGS}})_{/(D,z_D)}$ induced by that of $(X/B)^r_{\mathrm{HIGGS}}$ coincides with the topology associated to the pretopology defined by strict étale coverings of $T$ for $((T,z) \to (D,z_D)) \in \mathrm{Ob}\,((X/B)^r_{\mathrm{HIGGS}})_{/(D,z_D)}$.*

This follows from the following lemma.

**Lemma IV.3.1.10.** *Let $C$ be a $\mathscr{U}$-site defined by a pretopology $\mathrm{Cov}(X)$ $(X \in \mathrm{Ob}\,C)$ and let $U$ be a presheaf on $C$ with values in the category of $\mathscr{U}$-sets. For an object $X \to U$ of $C_{/U}$, we define the set $\mathrm{Cov}(X \to U)$ of families of morphisms in $C_{/U}$ with target $X \to U$ by*

$$\mathrm{Cov}(X \to U) = \{(f_\alpha\colon (X_\alpha \to U) \to (X \to U))_{\alpha \in A} | (f_\alpha\colon X_\alpha \to X)_{\alpha\in A} \in \mathrm{Cov}(X)\}.$$

(1) *The data $\mathrm{Cov}(X \to U)$ $((X \to U) \in \mathrm{Ob}\,C_{/U})$ is a pretopology on $C_{/U}$.*

(2) *The topology $\mathcal{T}$ on $C_{/U}$ induced by that of $C$ coincides with the topology $\mathcal{T}'$ on $C_{/U}$ defined by the pretopology $\mathrm{Cov}(X \to U)$ $((X \to U) \in \mathrm{Ob}\,C_{/U})$.*

Proof. The proof of (1) is straightforward. Let us prove (2). Let $R$ be a sieve of an object $u\colon X \to U$ of $C_{/U}$. Then, by **[2]** III Proposition 5.2 (1), $R$ is a covering for the topology $\mathcal{T}$ if and only if the sieve $j_!R$ of $X (= j_!(X \to U))$ is a covering, where $j$ denotes the functor $C_{/U} \to C$ forgetting morphisms to $U$. For an object $X'$ of $C$, we have the following commutative diagram.

$$\begin{array}{ccc} \mathrm{Hom}_C(X',X) & =\!=\!= & \coprod_{u'\colon X' \to U} \mathrm{Hom}_{C_{/U}}((u'\colon X' \to U),(u\colon X \to U)) \\ \uparrow & & \uparrow \\ j_!R(X) & =\!=\!=\!=\!=\!= & \coprod_{u'\colon X' \to U} R(u'\colon X' \to U). \end{array}$$

By **[2]** II Proposition 1.4, $j_!R$ is a covering if and only if there exists $(f_\alpha\colon X_\alpha \to X)_{\alpha\in A} \in \mathrm{Cov}(X)$ such that $f_\alpha \in (j_!R)(X_\alpha)$, which is equivalent to $(f_\alpha\colon (u \circ f_\alpha\colon X_\alpha \to U) \to (u\colon X \to U))_{\alpha \in A} \in R(u \circ f_\alpha\colon X \to U)$ by the commutative diagram above. By using **[2]** II Proposition 1.4 again, we see that $j_!R$ is a covering if and only if $R$ is a covering for the topology $\mathcal{T}'$. □

Let $r$, $B$, and $X$ be the same as in Proposition IV.3.1.9 and let $(D,z_D)$ be an object of $(X/B)^r_{\mathrm{HIGGS}}$. Then, by Proposition IV.3.1.9, we see that the category of sheaves of sets on $(((X/B)^r_{\mathrm{HIGGS}})_{/(D,z_D)})^\sim$ is naturally equivalent to the category of the following data: a sheaf of sets $\mathcal{F}_{((T,z),v)}$ on $T_{1,\text{ét}}$ for each object $v\colon (T,z) \to (D,z_D)$ of $((X/B)^r_{\mathrm{HIGGS}})_{/(D,z_D)}$ and a morphism $\tau_f\colon f_1^{-1}(\mathcal{F}_{(T,z),v}) \to \mathcal{F}_{(T',z'),v'}$ for each morphism $f\colon ((T',z'),v') \to ((T,z),v)$ in $((X/B)^r_{\mathrm{HIGGS}})_{/(D,z_D)}$ which satisfy the following two conditions:

(i) The morphism $\tau_f$ is an isomorphism if the underlying morphism $f\colon T' \to T$ is strict étale and Cartesian.

(ii) We have $\tau_g \circ g_1^{-1}(\tau_f) = \tau_{f\circ g}$ for any two composable morphisms $((T'',z''),v'') \xrightarrow{g} ((T',z'),v') \xrightarrow{f} ((T,z),v)$.

With the description above, one can define a morphism of topos

$$\pi_D\colon (((X/B)^r_{\mathrm{HIGGS}})_{/(D,z_D)})^\sim \to D^\sim_{1,\text{ét}}$$

as follows. The direct image functor $\pi_{D*}$ is defined by $\pi_{D*}\mathcal{F} = \mathcal{F}_{(D,z_D),\mathrm{id}}$. The inverse image $\pi_D^*\mathcal{G}$ of $\mathcal{G} \in \mathrm{Ob}\, D^\sim_{1,\text{ét}}$ is defined by the data $v_1^{-1}(\mathcal{G})$ for each $v\colon (T,z) \to (D,z_D)$ and the canonical isomorphism $f_1^{-1}v_1^{-1}(\mathcal{G}) \xrightarrow{\cong} (v'_1)^{-1}(\mathcal{G})$ for each morphism $f\colon ((T',z'),v') \to ((T,z),v)$. It is obvious that the functor $\pi_D^*$ preserves finite inverse limits, and the pair $(\pi_D^*, \pi_{D*})$ defines the desired morphism of topos $\pi_D$ by the following lemma.

**Lemma IV.3.1.11.** *The functor $\pi_D^*$ is a left adjoint of the functor $\pi_{D*}$.*

PROOF. It suffices to construct a functorial isomorphism

$$\mathrm{Hom}(\mathcal{G}, \pi_{D*}\mathcal{F}) \cong \mathrm{Hom}(\pi_D^*\mathcal{G}, \mathcal{F})$$

for $\mathcal{G} \in \mathrm{Ob}\, D^\sim_{1,\text{ét}}$ and $\mathcal{F} \in \mathrm{Ob}\,(((X/B)^r_{\mathrm{HIGGS}})_{/(D,z_D)})^\sim$. One can construct a map $F$ from the left to the right by associating to $\varphi\colon \mathcal{G} \to \pi_{D*}\mathcal{F} = \mathcal{F}_{(D,z_D),\mathrm{id}}$, the morphism $\pi_D^*\mathcal{G} \to \mathcal{F}$ defined by $\tau_v \circ v^{-1}(\varphi)\colon v^{-1}(\mathcal{G}) \to v^{-1}(\mathcal{F}_{(D,z_D),\mathrm{id}}) \to \mathcal{F}_{(T,z),v}$ for $v\colon (T,z) \to (D,z_D)$. One can define the map $G$ in the opposite direction by sending $\psi\colon \pi_D^*\mathcal{G} \to \mathcal{F}$ to $\psi_{(D,z_D),\mathrm{id}}\colon \mathcal{G} \to \mathcal{F}_{(D,z_D),\mathrm{id}} = \pi_{D*}\mathcal{F}$. It is straightforward to check that $F \circ G = \mathrm{id}$ and $G \circ F = \mathrm{id}$. □

Let $j_D$ be the canonical morphism of topos $(((X/B)^r_{\mathrm{HIGGS}})_{/(D,z_D)})^\sim \to (X/B)^{r\sim}_{\mathrm{HIGGS}}$. By [2] V Proposition 4.11 1), the functor $j_D^*$ preserves injective sheaves of abelian groups. On the other hand, the direct image functor $\pi_{D*}$ is obviously exact by its construction. Hence, for any sheaf of abelian groups $\mathcal{F}$ on $(((X/B)^r_{\mathrm{HIGGS}})_{/(D,z_D)})^\sim$, we have an isomorphism

(IV.3.1.12) $$R\Gamma((D,z_D),\mathcal{F}) \cong R\Gamma(((D,z_D),\mathrm{id}), j_D^*\mathcal{F}) \cong R\Gamma(D_{1,\text{ét}}, \mathcal{F}_{(D,z_D)}),$$

where we use $\pi_{D*}j_D^*\mathcal{F} = \mathcal{F}_{(D,z_D)}$ for the second isomorphism.

**IV.3.2. Lattices of projective modules.** Let $X$ be a $p$-adic formal scheme flat over $\mathbb{Z}_p$. In this subsection, we study integral structures of finitely generated projective $\mathcal{O}_{X,\mathbb{Q}_p}$-modules.

Recall that we always work on $X_{\text{ét}}$. Let $X_{\text{étaff}}$ be the full subcategory of $X_{\text{ét}}$ consisting of affine $p$-adic formal schemes étale over $X$ endowed with the topology associated to the pretopology defined by étale coverings. Then the topology of $X_{\text{étaff}}$ coincides with the one induced by the topology of $X_{\text{ét}}$, and the inclusion functor $X_{\text{étaff}} \to X_{\text{ét}}$ induces an equivalence of categories $X^\sim_{\text{ét}} \xrightarrow{\sim} X^\sim_{\text{étaff}}$ (cf. [2] III Corollaire 3.3, Théorème 4.1). For any $U \in \mathrm{Ob}\, X_{\text{étaff}}$, the functor $\Gamma(U,-)$ from the category of sheaves of abelian groups on $X_{\text{étaff}}$ (or on $X_{\text{ét}}$) to that of abelian groups commutes with filtered direct limits and direct sums.

Let $A$ be $\Gamma(X,\mathcal{O}_X)$, which is $p$-adically complete and separated and $p$-torsion free. Let $\mathcal{PM}(\mathcal{O}_{X,\mathbb{Q}_p})$ (resp. $\mathcal{PM}(A_{\mathbb{Q}_p})$) denote the category of finitely generated projective $\mathcal{O}_{X,\mathbb{Q}_p}$ (resp. $A_{\mathbb{Q}_p}$)-modules.

**Lemma IV.3.2.1.** (1) *For $\mathcal{M} \in \mathrm{Ob}\,\mathcal{PM}(\mathcal{O}_{X,\mathbb{Q}_p})$, we have $\Gamma(X,\mathcal{M}) \in \mathrm{Ob}\,\mathcal{PM}(A_{\mathbb{Q}_p})$.*

(2) *For $M \in \mathrm{Ob}\,\mathcal{PM}(A_{\mathbb{Q}_p})$, the presheaf $\mathcal{M}$ of $\mathcal{O}_{X,\mathbb{Q}_p}$-modules on $X_{\text{étaff}}$ defined by $\Gamma(U,\mathcal{M}) = M \otimes_A \Gamma(U,\mathcal{O}_U)$ ($U \in \mathrm{Ob}\, X_{\text{étaff}}$) is a finitely generated and projective sheaf of $\mathcal{O}_{X,\mathbb{Q}_p}$-modules on $X_{\text{étaff}}$.*

(3) *The constructions of (1) and (2) define equivalences of categories between $\mathcal{PM}(\mathcal{O}_{X,\mathbb{Q}_p})$ and $\mathcal{PM}(A_{\mathbb{Q}_p})$ inverse to each other.*

PROOF. (1) There exist an object $\mathcal{M}'$ of $\mathcal{PM}(\mathcal{O}_{X,\mathbb{Q}_p})$ and an isomorphism of $\mathcal{O}_{X,\mathbb{Q}_p}$-modules $\mathcal{M}\oplus\mathcal{M}'\cong\mathcal{O}_{X,\mathbb{Q}_p}^{\oplus r}$. By taking $\Gamma(X,-)$ of the isomorphism, we obtain the claim.

(2) There exist an $A$-module $M'$ and an isomorphism of $A$-modules $M\oplus M'\cong A_{\mathbb{Q}_p}^{\oplus r}$. For $U\in\operatorname{Ob}X_{\text{étaff}}$, we have an isomorphism $\Gamma(U,\mathcal{M})\oplus(M'\otimes_A\Gamma(U,\mathcal{O}_U))\cong\Gamma(U,\mathcal{O}_U)_{\mathbb{Q}_p}^{\oplus r}\cong\Gamma(U,\mathcal{O}_{X,\mathbb{Q}_p}^{\oplus r})$. Hence $\mathcal{M}$ is a sheaf and isomorphic to a direct summand of $\mathcal{O}_{X,\mathbb{Q}_p}^{\oplus r}$.

(3) For $M$ and $\mathcal{M}$ as in (2), we have $\Gamma(X,\mathcal{M})=M$ by definition. Hence it remains to prove that, for $\mathcal{M}\in\operatorname{Ob}\mathcal{PM}(\mathcal{O}_{X,\mathbb{Q}_p})$ and $U\in\operatorname{Ob}X_{\text{étaff}}$, the natural homomorphism $\Gamma(X,\mathcal{M})\otimes_A\Gamma(U,\mathcal{O}_U)\to\Gamma(U,\mathcal{M})$ is an isomorphism. This is reduced to the case $\mathcal{M}=\mathcal{O}_{X,\mathbb{Q}_p}$ by taking $\mathcal{M}'\in\operatorname{Ob}\mathcal{PM}(\mathcal{O}_{X,\mathbb{Q}_p})$ as in (1). □

It is natural to consider finitely generated $A$-modules $M$ with $M_{\mathbb{Q}_p}$ projective over $A_{\mathbb{Q}_p}$ as integral structures of objects of $\mathcal{PM}(A_{\mathbb{Q}_p})$. This is justified by Lemma IV.3.2.3 below. Let $\mathcal{LPM}(A)$ denote the category of such $A$-modules. For $M\in\operatorname{Ob}\mathcal{LPM}(A)$, let $M_{\text{tor}}$ denote the kernel of the natural homomorphism $M\to M_{\mathbb{Q}_p}$.

**Lemma IV.3.2.2.** *Let $M\in\operatorname{Ob}\mathcal{LPM}(A)$.*
(1) *There exists an $N\in\mathbb{N}$ such that $p^NM_{\text{tor}}=0$.*
(2) *The $A$-module $M$ is $p$-adically complete and separated.*

PROOF. (1) Choose a surjective homomorphism $f\colon A^{\oplus r}\to M$ of $A$-modules. Since $M_{\mathbb{Q}_p}$ is projective, there exists a decomposition $A_{\mathbb{Q}_p}^{\oplus r}=M_1\oplus M_2$ as $A_{\mathbb{Q}_p}$-modules such that $M_2=\operatorname{Ker}f_{\mathbb{Q}_p}$. Since $M_1$ and $M_2$ are finitely generated $A_{\mathbb{Q}_p}$-modules, there exist finitely generated $A$-submodules $M_i^\circ\subset M_{i,\mathbb{Q}_p}$ $(i=1,2)$ such that $M_1^\circ\oplus M_2^\circ\subset A^{\oplus r}$ and $M_{i,\mathbb{Q}_p}^\circ=M_i$ $(i=1,2)$. There exist $N_1,N_2\in\mathbb{N}$ such that $p^{N_1}A^{\oplus r}\subset M_1^\circ\oplus M_2^\circ$ and $p^{N_2}f(M_2^\circ)=0$. For any $x\in A^{\oplus r}$ such that $f_{\mathbb{Q}_p}(x)=0$, we have $p^{N_1}x\in M_2^\circ$ and $p^{N_1+N_2}f(x)\in p^{N_2}f(M_2^\circ)=0$.

(2) Put $\overline{M}=\operatorname{Im}(M\to M_{\mathbb{Q}_p})$, which is an object of $\mathcal{LPM}(A)$. By (1), there exists an exact sequence $0\to M_{\text{tor}}\to\varprojlim_m M/p^mM\to\varprojlim_m\overline{M}/p^m\overline{M}\to0$. Hence we are reduced to the case where $M$ is $p$-torsion free. Then there exists a $p$-torsion free $M'\in\operatorname{Ob}\mathcal{LPM}(A)$ and an isomorphism $M_{\mathbb{Q}_p}\oplus M'_{\mathbb{Q}_p}\cong A_{\mathbb{Q}_p}^{\oplus r}$. Choosing $N\in\mathbb{N}$ such that $p^NA^{\oplus r}\subset M\oplus M'\subset p^{-N}A^{\oplus r}$, we see that $M$ is $p$-adically complete and separated. □

For an additive category $\mathscr{A}$, we define the additive category $\mathscr{A}_{\mathbb{Q}}$ by $\operatorname{Ob}\mathscr{A}_{\mathbb{Q}}=\operatorname{Ob}\mathscr{A}$ and $\operatorname{Hom}_{\mathscr{A}_{\mathbb{Q}}}(x,y)=\operatorname{Hom}(x,y)\otimes_{\mathbb{Z}}\mathbb{Q}$ for $x,y\in\operatorname{Ob}\mathscr{A}$.

**Lemma IV.3.2.3.** *The natural functor $\mathcal{LPM}(A)\to\mathcal{PM}(A_{\mathbb{Q}_p})$ defined by $M\mapsto M_{\mathbb{Q}_p}$ induces an equivalence of categories $\mathcal{LPM}(A)_{\mathbb{Q}}\xrightarrow{\sim}\mathcal{PM}(A_{\mathbb{Q}_p})$.*

PROOF. Since the functor $\mathcal{LPM}(A)\to\mathcal{PM}(A_{\mathbb{Q}_p})$ is obviously essentially surjective, it remains to prove that the natural homomorphism

$$\operatorname{Hom}_A(M,M')\otimes_{\mathbb{Z}}\mathbb{Q}\longrightarrow\operatorname{Hom}_{A_{\mathbb{Q}_p}}(M_{\mathbb{Q}_p},M'_{\mathbb{Q}_p})$$

is an isomorphism for $M,M'\in\operatorname{Ob}\mathcal{LPM}(A)$. This follows from Lemma IV.3.2.2 (1) for $M'$. For the surjectivity, note that $p^N\colon M'\to M'$ factors through $\operatorname{Im}(M'\to M'_{\mathbb{Q}_p})$ for a sufficiently large $N\in\mathbb{N}$ by the lemma. □

For a general $p$-adic affine formal scheme $X$, we don't have a general theory relating coherent (or finitely generated) $\mathcal{O}_X$-modules and certain $A$-modules. Therefore we will interpret an object $M$ of $\mathcal{LPM}(A)$ in terms of $\mathcal{O}_X$-modules by taking the inverse limit of the quasi-coherent $\mathcal{O}_{X_m}$-modules associated to $M/p^mM$.

First we interpret objects of $\mathcal{LPM}(A)$ in terms of inverse systems of $A_m$-modules. We consider an inverse system of $A_m$-modules $(M_m)_{m\in\mathbb{N}_{>0}}$ such that $M_1$ is a finitely generated $A_1$-module, the transition maps induce isomorphisms $M_{m+1}\otimes_{A_{m+1}}A_m \xrightarrow{\cong} M_m$ ($m \in \mathbb{N}_{>0}$), and $(\varprojlim_m M_m)_{\mathbb{Q}_p}$ is a projective $A_{\mathbb{Q}_p}$-module. Let $\mathcal{LPM}(A_\bullet)$ be the category of inverse systems of $A_m$-modules as above. A morphism is a compatible system of $A_m$-linear homomorphisms.

**Proposition IV.3.2.4** (cf. **[39]** Chap. 0 Proposition 7.2.9). *The functor $\mathcal{LPM}(A) \to \mathcal{LPM}(A_\bullet)$ defined by $M \mapsto (M \otimes_A A_m)_{m\in\mathbb{N}_{>0}}$ is an equivalence of categories. A quasi-inverse is given by $(M_m) \mapsto \varprojlim_m M_m$.*

PROOF. For an object $M$ of $\mathcal{LPM}(A)$, we have $M \xrightarrow{\cong} \varprojlim_m (M \otimes_A A_m)$ by Lemma IV.3.2.2 (2). Hence $(M \otimes_A A_m)_m$ is an object of $\mathcal{LPM}(A_\bullet)$. Conversely, for an object $(M_m)$ of $\mathcal{LPM}(A_\bullet)$, the inverse limit $M := \varprojlim_m M_m$ is a finitely generated $A$-module and the natural homomorphism $M/p^mM \to M_m$ ($m \in \mathbb{N}_{>0}$) is an isomorphism by **[39]** Chap. 0 Proposition (7.2.9). Hence $M$ is an object of $\mathcal{LPM}(A)$ and the functor $(M_m) \mapsto M$ is a quasi-inverse. □

We define the category $\mathcal{LPM}(\mathcal{O}_{X_\bullet})$ as follows. An object is an inverse system $(\mathcal{M}_m)_{m\in\mathbb{N}_{>0}}$ of quasi-coherent $\mathcal{O}_{X_m}$-modules such that $\Gamma(X_1, \mathcal{M}_1)$ is a finitely generated $A_1$-module, the transition maps induce isomorphisms $\mathcal{M}_{m+1} \otimes_{\mathcal{O}_{X_{m+1}}} \mathcal{O}_{X_m} \xrightarrow{\cong} \mathcal{M}_m$, and $(\varprojlim_m \Gamma(X, \mathcal{M}_m))_{\mathbb{Q}_p}$ is a projective $A_{\mathbb{Q}_p}$-module. A morphism is a compatible system of $\mathcal{O}_{X_m}$-linear homomorphisms. Then we have a canonical equivalence of categories

$$\mathcal{LPM}(A_\bullet) \xrightarrow{\sim} \mathcal{LPM}(\mathcal{O}_{X_\bullet}) \tag{IV.3.2.5}$$

defined by $(M_m) \mapsto (\widetilde{M_m})$ and $(\mathcal{M}_m) \mapsto (\Gamma(X, \mathcal{M}_m))$.

Let $\mathcal{LPM}(\mathcal{O}_X)$ be the full subcategory of the category of $\mathcal{O}_X$-modules consisting of $\mathcal{O}_X$-modules $\mathcal{M}$ such that $\mathcal{M}_m := \mathcal{M} \otimes_{\mathcal{O}_X} \mathcal{O}_{X_m}$ is a quasi-coherent $\mathcal{O}_{X_m}$-module for every $m \in \mathbb{N}_{>0}$, $\Gamma(X, \mathcal{M}_1)$ is a finitely generated $A_1$-module, the natural morphism $\mathcal{M} \to \varprojlim_m \mathcal{M}_m$ is an isomorphism, and $\Gamma(X, \mathcal{M})_{\mathbb{Q}_p}$ is a finitely generated and projective $A_{\mathbb{Q}_p}$-module.

**Lemma IV.3.2.6.** (1) *The functor $\mathcal{LPM}(\mathcal{O}_X) \to \mathcal{LPM}(\mathcal{O}_{X_\bullet}); \mathcal{M} \mapsto (\mathcal{M} \otimes_{\mathcal{O}_X} \mathcal{O}_{X_m})_m$ is an equivalence of categories. Its quasi-inverse is given by $(\mathcal{M}_m) \mapsto (\varprojlim_m \mathcal{M}_m)$.*

(2) *For $\mathcal{M} \in \mathrm{Ob}\,\mathcal{LPM}(\mathcal{O}_X)$ and $U \in \mathrm{Ob}\,X_{\text{étaff}}$, the natural $\Gamma(U, \mathcal{O}_U)$-linear homomorphism $\Gamma(X, \mathcal{M}) \otimes_A \Gamma(U, \mathcal{O}_U) \xrightarrow{\cong} \Gamma(U, \mathcal{M})$ is an isomorphism and $\Gamma(U, \mathcal{M})$ is $p$-adically complete and separated.*

PROOF. (1) It is obvious that $(\mathcal{M} \otimes_{\mathcal{O}_X} \mathcal{O}_{X_m})_m \in \mathcal{LPM}(\mathcal{O}_{X_\bullet})$. To prove that the functor in the opposite direction is well-defined and gives a quasi-inverse, it suffices to prove that, for $(\mathcal{M}_m) \in \mathcal{LPM}(\mathcal{O}_{X_\bullet})$, the natural morphism $(\varprojlim_n \mathcal{M}_n) \otimes_{\mathcal{O}_X} \mathcal{O}_{X_m} \to \mathcal{M}_m$ is an isomorphism. Let $U \in \mathrm{Ob}\,X_{\text{étaff}}$ and put $A' = \Gamma(U, \mathcal{O}_X)$. Put $\mathcal{M} := \varprojlim_m \mathcal{M}_m$, $M_m = \Gamma(X, \mathcal{M}_m)$ and $M := \varprojlim_m M_m$. Then by (IV.3.2.5), Proposition IV.3.2.4, and Lemma IV.3.2.2 (2), we see that $M \otimes_A A'$ is $p$-adically complete and separated. Hence we have $\Gamma(U, \mathcal{M}) = \varprojlim_m M_m \otimes_{A_m} A'_m \cong M \otimes_A A'$, which implies $\Gamma(U, \mathcal{M})/p^m\Gamma(U, \mathcal{M}) \xrightarrow{\cong} M_m \otimes_{A_m} A'_m \cong \Gamma(U, \mathcal{M}_m)$. Varying $U$, we obtain $\mathcal{M}/p^m\mathcal{M} \xrightarrow{\cong} \mathcal{M}_m$.

(2) We have $(\mathcal{M} \otimes_{\mathcal{O}_X} \mathcal{O}_{X_m}) \in \mathrm{Ob}\,\mathcal{LPM}(\mathcal{O}_{X_\bullet})$ and $\mathcal{M} \cong \varprojlim_m (\mathcal{M} \otimes_{\mathcal{O}_X} \mathcal{O}_{X_m})$ by (1). Hence we obtain (2) from the above proof of (1). □

**Proposition IV.3.2.7.** (1) *For $\mathcal{M} \in \mathcal{LPM}(\mathcal{O}_X)$, we have $\Gamma(X, \mathcal{M}) \in \mathcal{LPM}(A)$. The functor $\mathcal{LPM}(\mathcal{O}_X) \to \mathcal{LPM}(A)$ defined by $\mathcal{M} \mapsto \Gamma(X, \mathcal{M})$ is an equivalence of categories.*

(2) *For* $\mathcal{M} \in \mathcal{LPM}(\mathcal{O}_X)$, *we have* $\mathcal{M}_{\mathbb{Q}_p} \in \mathcal{PM}(\mathcal{O}_{X,\mathbb{Q}_p})$. *The functor* $\mathcal{LPM}(\mathcal{O}_X) \to \mathcal{PM}(\mathcal{O}_{X,\mathbb{Q}_p})$ *defined by* $\mathcal{M} \mapsto \mathcal{M}_{\mathbb{Q}_p}$ *induces an equivalence of categories* $\mathcal{LPM}(\mathcal{O}_X)_{\mathbb{Q}} \xrightarrow{\sim} \mathcal{PM}(\mathcal{O}_{X,\mathbb{Q}_p})$.

PROOF. (1) The equivalence of categories $\mathcal{LPM}(\mathcal{O}_X) \xrightarrow{\sim} \mathcal{LPM}(A)$ obtained by the composition of the equivalences of categories in Proposition IV.3.2.4, (IV.3.2.5), and Lemma IV.3.2.6 is given by $\mathcal{M} \mapsto \varprojlim_m \Gamma(X, \mathcal{M}/p^m\mathcal{M})$, and the latter $A$-module is isomorphic to $\Gamma(X, \varprojlim_m \mathcal{M}/p^m\mathcal{M}) \cong \Gamma(X, \mathcal{M})$.

(2) By combining Lemma IV.3.2.1, Lemma IV.3.2.3, and the equivalence of categories in (1), we obtain a functor $\mathcal{LPM}(\mathcal{O}_X) \to \mathcal{PM}(\mathcal{O}_{X,\mathbb{Q}_p})$ inducing an equivalence of categories $\mathcal{LPM}(\mathcal{O}_X)_{\mathbb{Q}} \xrightarrow{\sim} \mathcal{PM}(\mathcal{O}_{X,\mathbb{Q}_p})$. By Lemma IV.3.2.1, we see that the image of $\mathcal{M} \in \operatorname{Ob} \mathcal{LPM}(\mathcal{O}_X)$ under this functor is given by $U \mapsto \Gamma(X, \mathcal{M}) \otimes_A \Gamma(U, \mathcal{O}_U)_{\mathbb{Q}_p}$ ($U \in \operatorname{Ob} X_{\text{étaff}}$). Hence the claim follows from Lemma IV.3.2.6 (2). □

**Lemma IV.3.2.8.** *Let* $f\colon X' \to X$ *be a morphism of affine p-adic formal schemes flat over* $\mathbb{Z}_p$, *and put* $A = \Gamma(X, \mathcal{O}_X)$ *and* $A' = \Gamma(X', \mathcal{O}_{X'})$. *Then the inverse image functors* $\mathcal{PM}(\mathcal{O}_{X,\mathbb{Q}_p}) \to \mathcal{PM}(\mathcal{O}_{X',\mathbb{Q}_p})$, $\mathcal{PM}(A_{\mathbb{Q}_p}) \to \mathcal{PM}(A'_{\mathbb{Q}_p})$, $\mathcal{LPM}(A) \to \mathcal{LPM}(A')$, $\mathcal{LPM}(A_\bullet) \to \mathcal{LPM}(A'_\bullet)$, $\mathcal{LPM}(\mathcal{O}_{X_\bullet}) \to \mathcal{LPM}(\mathcal{O}_{X'_\bullet})$, *and* $\mathcal{LPM}(\mathcal{O}_X) \to \mathcal{LPM}(\mathcal{O}_{X'})$ *are defined by* $\otimes_{f^{-1}(\mathcal{O}_{X,\mathbb{Q}_p})}\mathcal{O}_{X',\mathbb{Q}_p}$, $\otimes_{A_{\mathbb{Q}_p}} A'_{\mathbb{Q}_p}$, $\otimes_A A'$, $\otimes_{A_m} A'_m$, $\otimes_{f^{-1}(\mathcal{O}_{X_m})}\mathcal{O}_{X'_m}$, *and* $\mathcal{F} \mapsto \varprojlim_m (\mathcal{F} \otimes_{f^{-1}(\mathcal{O}_X)} \mathcal{O}_{X'_m})$. *Furthermore these functors are compatible with the equivalences of categories in Lemma* IV.3.2.1, *Lemma* IV.3.2.3, *Proposition* IV.3.2.4, (IV.3.2.5), *Lemma* IV.3.2.6, *and Proposition* IV.3.2.7

PROOF. For $\mathcal{M} \in \operatorname{Ob}\mathcal{PM}(\mathcal{O}_{X,\mathbb{Q}_p})$ (resp. $M \in \operatorname{Ob}\mathcal{PM}(A_{\mathbb{Q}_p})$), there exist $\mathcal{M}' \in \operatorname{Ob}\mathcal{PM}(\mathcal{O}_{X,\mathbb{Q}_p})$ (resp. $M' \in \operatorname{Ob}\mathcal{PM}(A'_{\mathbb{Q}_p})$) and an isomorphism $\mathcal{M} \oplus \mathcal{M}' \cong \mathcal{O}_{X,\mathbb{Q}_p}^{\oplus r}$ (resp. $M \oplus M' \cong A_{\mathbb{Q}_p}^{\oplus r}$). By taking $\otimes_{f^{-1}(\mathcal{O}_{X,\mathbb{Q}_p})}\mathcal{O}_{X',\mathbb{Q}_p}$ (resp. $\otimes_{A_{\mathbb{Q}_p}} A'_{\mathbb{Q}_p}$), we see that $\mathcal{M} \otimes_{\mathcal{O}_{X,\mathbb{Q}_p}} \mathcal{O}_{X',\mathbb{Q}_p}$ (resp. $M \otimes_{A_{\mathbb{Q}_p}} A'_{\mathbb{Q}_p}$) is an object of $\mathcal{PM}(\mathcal{O}_{X',\mathbb{Q}_p})$ (resp. $\mathcal{PM}(A'_{\mathbb{Q}_p})$). Hence the first three functors are well-defined and compatible with the equivalences of categories in Lemma IV.3.2.1 and Lemma IV.3.2.3. For $(M_m) \in \operatorname{Ob}\mathcal{LPM}(A_\bullet)$ and $M = \varprojlim_m M_m \in \operatorname{Ob}\mathcal{LPM}(A)$, $M \otimes_A A'$ is $p$-adically complete and separated by Lemma IV.3.2.2 (2). Hence we have an isomorphism $M \otimes_A A' \cong \varprojlim_m (M_m \otimes_{A_m} A'_m)$. This implies the well-definedness of the fourth and fifth functors and the compatibility with the equivalences of categories in Proposition IV.3.2.4 and (IV.3.2.5). Now the claims for the last functor and the equivalence of categories in Lemma IV.3.2.6 are obvious. □

**Definition IV.3.2.9.** (1) For a $p$-adic formal scheme $X$ flat over $\mathbb{Z}_p$, we define the category $\mathcal{LPM}^{\mathrm{loc}}(\mathcal{O}_X)$ to be the category of $\mathcal{O}_X$-modules which belong to $\mathcal{LPM}(\mathcal{O}_X)$ strict étale locally on $X$.

(2) For a morphism $f\colon X' \to X$ of $p$-adic log formal schemes flat over $\mathbb{Z}_p$, we define the functor $\widehat{f}^*\colon \mathcal{LPM}^{\mathrm{loc}}(\mathcal{O}_X) \to \mathcal{LPM}^{\mathrm{loc}}(\mathcal{O}_{X'})$ by $\widehat{f}^*(\mathcal{M}) = \varprojlim_m f_m^*(\mathcal{M} \otimes_{\mathcal{O}_X} \mathcal{O}_{X_m})$. (This is well-defined by Lemma IV.3.2.8.)

For an object $\mathcal{M}$ of $\mathcal{LPM}^{\mathrm{loc}}(\mathcal{O}_X)$, the sheaf of $\mathcal{O}_{X_m}$-modules $\mathcal{M}_m := \mathcal{M} \otimes_{\mathcal{O}_X} \mathcal{O}_{X_m}$ is quasi-coherent and the natural morphism $\mathcal{M} \to \varprojlim_m \mathcal{M}_m$ is an isomorphism. For two composable morphisms $g\colon X'' \to X'$ and $f\colon X' \to X$ of $p$-adic fine log formal schemes flat over $\mathbb{Z}_p$, we have a canonical isomorphism of functors $\widehat{g}^* \circ \widehat{f}^* \cong \widehat{(f \circ g)}^*$.

**Lemma IV.3.2.10.** *Let* $\mathcal{M}$ *be an object of* $\mathcal{LPM}(\mathcal{O}_X)$ *and let* $\overline{\mathcal{M}}$ *be the image of* $\mathcal{M} \to \mathcal{M}_{\mathbb{Q}_p}$.

(1) *The* $\mathcal{O}_X$*-module* $\overline{\mathcal{M}}$ *is an object of* $\mathcal{LPM}(\mathcal{O}_X)$.

(2) *There exists* $N \in \mathbb{N}$ *such that* $p^N \cdot \operatorname{Ker}(\mathcal{M} \to \overline{\mathcal{M}}) = 0$.

(3) *Let $M$ and $\overline{M}$ be the images $\Gamma(X,\mathcal{M})$ and $\Gamma(X,\overline{\mathcal{M}})$ of $\mathcal{M}$ and $\overline{\mathcal{M}}$ under the equivalence of categories in Proposition* IV.3.2.7 (1). *Then the natural homomorphism $M/M_{\mathrm{tor}}\to\overline{M}$ is an isomorphism.*

PROOF. Put $M=\Gamma(X,\mathcal{M})$ and $\overline{M}=M/M_{\mathrm{tor}}$, which is an object of $\mathcal{LPM}(A)$ by Proposition IV.3.2.7 (1), and let $\overline{\mathcal{M}}'$ be the object of $\mathcal{LPM}(\mathcal{O}_X)$ corresponding to $\overline{M}$ by loc. cit. By Lemma IV.3.2.2 (1), there exists $N\in\mathbb{N}$ such that $p^N M_{\mathrm{tor}}=0$. Under this notation, it suffices to prove that, for every $U\in\mathrm{Ob}\,X_{\text{étaff}}$, the homomorphism $\Gamma(U,\mathcal{M})\to\Gamma(U,\overline{\mathcal{M}}')$ is surjective, its kernel is annihilated by $p^N$, and $\Gamma(U,\overline{\mathcal{M}}')$ is $p$-torsion free. Put $A_U=\Gamma(U,\mathcal{O}_U)$. By Lemma IV.3.2.6 (2), we have $\Gamma(U,\mathcal{M})\cong M\otimes_A A_U$ and $\Gamma(U,\overline{\mathcal{M}}')\cong\overline{M}\otimes_A A_U$, and these modules are $p$-adically complete and separated. Hence the first and second claims hold. Since $\overline{M}$ is $p$-torsion free, the multiplication by $p$ on $\overline{M}$ induces an injective homomorphism $\overline{M}/p^{m-1}\overline{M}\to\overline{M}/p^m\overline{M}$. By taking the extension of scalars by the flat homomorphism $A/p^mA\to A_U/p^mA_U$ and then $\varprojlim_m$, we see that $\overline{M}\otimes_A A_U$ is $p$-torsion free. □

**Lemma IV.3.2.11.** *Let $f\colon X'\to X$ be a morphism of affine $p$-adic formal schemes flat over $\mathbb{Z}_p$, let $\mathcal{M}$ be an object of $\mathcal{LPM}(\mathcal{O}_X)$ and let $\overline{\mathcal{M}}$ be the image of $\mathcal{M}\to\mathcal{M}_{\mathbb{Q}_p}$, which is an object of $\mathcal{LPM}(\mathcal{O}_X)$ by Lemma* IV.3.2.10. *If the reduction mod $p^m$ of $f$ is flat for every $m\in\mathbb{N}_{>0}$, then, via the identification $(\widehat{f}^*(\mathcal{M}))_{\mathbb{Q}}=f^*(\mathcal{M}_{\mathbb{Q}})=f^*(\overline{\mathcal{M}}_{\mathbb{Q}})=\widehat{f}^*(\overline{\mathcal{M}})_{\mathbb{Q}}$ (cf. Lemma* IV.3.2.8), *we have $\widehat{f}^*(\overline{\mathcal{M}})=\mathrm{Im}(\widehat{f}^*(\mathcal{M})\to\widehat{f}(\mathcal{M})_{\mathbb{Q}})$. In particular, if $\mathcal{M}$ is $p$-torsion free, then $\widehat{f}^*(\mathcal{M})$ is $p$-torsion free.*

PROOF. Put $A=\Gamma(X,\mathcal{O}_X)$ and $A'=\Gamma(X',\mathcal{O}_{X'})$. By Lemma IV.3.2.10 and Lemma IV.3.2.8, it suffices to prove the corresponding claim for the functor $\mathcal{LPM}(A)\to\mathcal{LPM}(A')\colon M\mapsto M\otimes_A A'$. For $M\in\mathrm{Ob}\,\mathcal{LPM}(A)$ and $\overline{M}=\mathrm{Im}(M\to M_{\mathbb{Q}})$, the homomorphism $M\otimes_A A'\to\overline{M}\otimes_A A'$ is surjective and the kernel is annihilated by a power of $p$. Since $\overline{M}$ is $p$-torsion free, the multiplication by $p$ on $\overline{M}/p^m\overline{M}$ factors through an injective homomorphism $\overline{M}/p^{m-1}\overline{M}\to\overline{M}/p^m\overline{M}$. Since $A/p^mA\to A'/p^mA'$ is flat, it induces an injective homomorphism $\overline{M}\otimes_A(A'/p^{m-1}A')\to\overline{M}\otimes_A(A'/p^mA')$. By taking the inverse limit over $m$, we see that $\overline{M}\otimes_A A'$ is $p$-torsion free. This completes the proof. □

**IV.3.3. Higgs isocrystals and Higgs crystals.** For $r\in\mathbb{N}_{>0}\cup\{\infty\}$ (resp. $r=\infty$), $B\in\mathrm{Ob}\,\mathscr{C}^r$, and a $p$-adic fine log formal scheme $X$ over $B_1$, we define Higgs isocrystals of level $r$ (resp. Higgs crystals) on $X/B$.

**Definition IV.3.3.1.** Let $r\in\mathbb{N}_{>0}\cup\{\infty\}$, let $B\in\mathscr{C}^r$, and let $X$ be a $p$-adic fine log formal scheme over $B_1$. A *Higgs isocrystal on $X/B$ of lever $r$* (or a *Higgs isocrystal* on $(X/B)^r_{\mathrm{HIGGS}}$) is a sheaf $\mathcal{F}$ of $\mathcal{O}_{X/B,1,\mathbb{Q}}$-modules on $(X/B)^r_{\mathrm{HIGGS}}$ such that the corresponding system of sheaves of $\mathcal{O}_{T_1,\mathbb{Q}}$-modules $\mathcal{F}_{(T,z)}$ on $T_{1,\text{ét}}$ for $(T,z)\in\mathrm{Ob}\,(X/B)^r_{\mathrm{HIGGS}}$ satisfies the following conditions.

(i) For any morphism $u\colon(T',z')\to(T,z)$ in $(X/B)^r_{\mathrm{HIGGS}}$, $\tau_u\colon u_1^*(\mathcal{F}_{(T,z)})\to\mathcal{F}_{(T',z')}$ is an isomorphism.

(ii) There exists a strict étale covering $(X_\alpha\to X)$ satisfying the following condition: For any object $(T,z)$ of $(X/B)^r_{\mathrm{HIGGS}}$ such that $T_1$ is affine and $z$ factors through $X_\alpha\to X$ for some $\alpha$, the $\mathcal{O}_{T_1,\mathbb{Q}}$-module $\mathcal{F}_{(T,z)}$ is finitely generated and projective.

We define a morphism of Higgs isocrystals on $X/B$ of level $r$ to be a morphism of sheaves of $\mathcal{O}_{X/B,1,\mathbb{Q}}$-modules. Let $\mathrm{HC}^r_{\mathbb{Q}_p}(X/B)$ denote the category of Higgs isocrystals on $X/B$ of level $r$. For a $p$-adic fine log formal scheme $U$ over $X$, we say that a Higgs isocrystal $\mathcal{F}$ on $X/B$ of level $r$ is *finite on $U$* if there exists a strict étale covering $(X_\alpha\to$

$X$) for $\mathcal{F}$ as in condition (ii) such that the morphism $U \to X$ factors through $X_\alpha \to X$ for some $\alpha$. We write $\mathrm{HC}^r_{\mathbb{Q}_p,U\text{-fin}}(X/B)$ for the full subcategory of $\mathrm{HC}^r_{\mathbb{Q}_p}(X/B)$ consisting of objects finite on $U$.

**Definition IV.3.3.2.** Let $B$ be an object of $\mathscr{C}^\infty$ and let $X$ be a $p$-adic fine log formal scheme over $B_1$. A *Higgs crystal on* $X/B$ is a sheaf $\mathcal{F}$ of $\mathcal{O}_{X/B,1}$-modules on $(X/B)^\infty_{\mathrm{HIGGS}}$ such that the corresponding system of sheaves of $\mathcal{O}_{T_1}$-modules $\mathcal{F}_{(T,z)}$ on $T_{1,\text{ét}}$ for $(T,z) \in \mathrm{Ob}\,(X/B)^\infty_{\mathrm{HIGGS}}$ satisfies the following conditions.

(i) For every $(T,z) \in \mathrm{Ob}\,(X/B)^\infty_{\mathrm{HIGGS}}$, the $\mathcal{O}_{T_1}$-module $\mathcal{F}_{(T,z)}$ is an object of $\mathcal{LPM}^{\mathrm{loc}}(\mathcal{O}_{T_1})$ (cf. Definition IV.3.2.9).

(ii) For every morphism $u\colon (T',z') \to (T,z)$ in $(X/B)^\infty_{\mathrm{HIGGS}}$, the natural morphism $\widehat{u}^*(\mathcal{F}_{(T,z)}) \to \mathcal{F}_{(T',z')}$ is an isomorphism.

(iii) There exists a strict étale covering $(X_\alpha \to X)$ satisfying the following condition: For any $(T,z) \in \mathrm{Ob}\,(X/B)^\infty_{\mathrm{HIGGS}}$ such that $T_1$ is affine and $z$ factors through $X_\alpha \to X$ for some $\alpha$, the $\mathcal{O}_{T_1}$-module $\mathcal{F}_{(T,z)}$ is an object of $\mathcal{LPM}(\mathcal{O}_{T_1})$.

We define a morphism of Higgs crystals on $X/B$ to be a morphism as sheaves of $\mathcal{O}_{X/B,1}$-modules. Let $\mathrm{HC}_{\mathbb{Z}_p}(X/B)$ denote the category of Higgs crystals on $X/B$. For a $p$-adic fine log formal scheme $U$ over $X$, we say that a Higgs crystal $\mathcal{F}$ on $X/B$ is *finite on* $U$ if there exists a strict étale covering $(X_\alpha \to X)$ for $\mathcal{F}$ as in the condition (iii) such that the morphism $U \to X$ factors through $X_\alpha \to X$ for some $\alpha$. We write $\mathrm{HC}_{\mathbb{Z}_p,U\text{-fin}}(X/B)$ for the full subcategory of $\mathrm{HC}_{\mathbb{Z}_p}(X/B)$ consisting of objects finite on $U$.

**Remark IV.3.3.3.** We can define a Higgs crystal on $(X/B)^r_{\mathrm{HIGGS}}$ for $r \in \mathbb{N}_{>0}$ in the same way as Definition IV.3.3.2. However it is not useful as an integral structure of a Higgs crystal of level $r$ as we will see in Remark IV.3.4.13.

**IV.3.4. Local Description.** Let $r \in \mathbb{N}_{>0} \cup \{\infty\}$, let $B$ be an object of $\mathscr{C}^r$, let $X$ be a $p$-adic fine log formal scheme over $B_1$, let $Y \to B$ be a smooth Cartesian morphism in $\mathscr{C}$, and let $X \to Y_1$ be an immersion over $B_1$. Let $D$ denote the Higgs envelope of level $r$ of $X \hookrightarrow Y$. In this subsection, we give a description of a Higgs isocrystal of level $r$ on $X/B$ in terms of an $\mathcal{O}_{D_1,\mathbb{Q}_p}$-module on $D_1$ with a derivation taking its values in $\xi^{-1}\Omega_{Y_1/B_1}$ compatible with the canonical derivation on $\mathcal{O}_{D_1}$ (IV.2.4.8). We also give a similar description of a Higgs crystal when $r = \infty$. See Theorem IV.3.4.16. Similarly to the crystalline site, the descriptions are given via a kind of stratifications on modules on $D_1$ defined in terms of the Higgs envelopes of the diagonal immersions of $X$ into the fiber products of copies of $Y$.

**Lemma IV.3.4.1.** *For any object* $(T,z)$ *of* $(X/B)^r_{\mathrm{HIGGS}}$ *such that* $T_1$ *is affine, a morphism* $z\colon T_1 \to X$ *extends to a morphism in* $\mathscr{C}_{/B}$ *from* $T$ *to the object* $(X \hookrightarrow Y)$ *of* $\mathscr{C}$ *(cf. Definition* IV.2.2.1 (3)*).*

Proof. This immediately follows from Corollary IV.2.2.5 and Sublemma IV.3.4.2 (2) below. □

**Sublemma IV.3.4.2.** (1) *Let* $j\colon \overline{U} \to U$ *be an exact closed immersion of fine log schemes whose underlying morphism of schemes is a nilpotent immersion. Let*

$$\begin{array}{ccc} W' & \xrightarrow{s'} & V \\ {\scriptstyle i}\downarrow & & \downarrow{\scriptstyle f} \\ W & \xrightarrow{s} & U \end{array} \qquad\qquad \begin{array}{ccc} \overline{W}' & \xrightarrow{\overline{s}'} & \overline{V} \\ {\scriptstyle \overline{i}}\downarrow & & \downarrow{\scriptstyle \overline{f}} \\ \overline{W} & \xrightarrow{\overline{s}} & \overline{U} \end{array}$$

*be commutative diagrams of fine log schemes such that the right diagram is the base change of the left one by the morphism* $j$, $f$ *is smooth,* $i$ *is an exact closed immersion whose underlying morphism of schemes is a nilpotent immersion, and the underlying scheme of* $\overline{W}$ *is affine. Then, for any morphism* $\overline{g}\colon \overline{W} \to \overline{V}$ *such that* $\overline{f} \circ \overline{g} = \overline{s}$ *and* $\overline{g} \circ \overline{i} = \overline{s}'$*, there exists a lifting* $g\colon W \to V$ *of* $\overline{g}$ *such that* $f \circ g = s$ *and* $g \circ i = s'$*.*

(2) *Consider a commutative diagram of* $p$*-adic fine log formal schemes*

$$\begin{array}{ccc} W' & \xrightarrow{s'} & V \\ {\scriptstyle i}\downarrow & & \downarrow{\scriptstyle f} \\ W & \xrightarrow{s} & U \end{array}$$

*such that* $f$ *is smooth,* $i$ *is an exact closed immersion, the morphism of schemes underlying the reduction mod* $p$ *of* $i$ *is a nilpotent immersion, and the underlying formal scheme of* $W$ *is affine. Then there exists a morphism* $g\colon W \to V$ *such that* $f \circ g = s$ *and* $g \circ i = s'$*.*

PROOF. (1) Let $\mathcal{I}$ (resp. $\overline{\mathcal{I}}$) be an ideal of $\mathcal{O}_W$ (resp. $\mathcal{O}_{\overline{W}}$) defining $W'$ in $W$ (resp. $\overline{W}'$ in $\overline{W}$). By factorizing the exact closed immersion $i\colon W' \to W$ into a sequence of exact closed immersions defined by $\mathcal{I}^n$ for $n \in \mathbb{N}_{>0}$, we are reduced to the case $\mathcal{I}^2 = 0$. By the assumptions on $i$ and $f$, there exists a morphism $g\colon W \to V$ such that $f \circ g = s$ and $g \circ i = s'$. Since $\Omega^1_{V/U}$ is locally free of finite type and $W$ is affine, the homomorphism $\mathrm{Hom}_{\mathcal{O}_{W'}}(\mathcal{O}_{W'} \otimes_{\mathcal{O}_V} \Omega^1_{V/U}, \mathcal{I}) \to \mathrm{Hom}_{\mathcal{O}_{\overline{W}}}(\mathcal{O}_{\overline{W}'} \otimes_{\mathcal{O}_{\overline{V}}} \Omega^1_{\overline{V}/\overline{U}}, \overline{\mathcal{I}})$ is surjective. By **[50]** Proposition (3.9), we can modify $g$ in such a way that the base change of $g$ by $j$ coincides with $\overline{g}$.

(2) We obtain the claim by applying (1) repeatedly to the reduction mod $p^n$ of the diagram in question. □

We define $Y(\nu)$, $D(\nu)$ ($\nu \in \mathbb{N}$), $D$, $\Delta_D$, $p_i$, $p_{ij}$, and $q_i$ in the same way as after the proof of Lemma IV.2.4.2.

**Definition IV.3.4.3.** (1) We define the category $\mathrm{HS}^r_{\mathbb{Q}_p}(X, Y/B)$ as follows. An object is a pair $(\mathcal{M}, \varepsilon)$ of an $\mathcal{O}_{D_1,\mathbb{Q}}$-module and an $\mathcal{O}_{D(1)_1,\mathbb{Q}}$-linear isomorphism $\varepsilon\colon p_2^*(\mathcal{M}) \xrightarrow{\cong} p_1^*(\mathcal{M})$ satisfying the following conditions.

(i) There exists a strict étale covering $(X_\alpha \to X)_{\alpha \in A}$ such that $\mathcal{M}|_{D_{1,\alpha}}$ ($D_{1,\alpha} = D_1 \times_X X_\alpha$) is a finitely generated projective $\mathcal{O}_{D_{1,\alpha},\mathbb{Q}}$-module for every $\alpha \in A$.

(ii) (a) $\Delta_D^*(\varepsilon) = \mathrm{id}_{\mathcal{M}}$. (b) $p_{12}^*(\varepsilon) \circ p_{23}^*(\varepsilon) = p_{13}^*(\varepsilon)\colon q_3^*(\mathcal{M}) \xrightarrow{\cong} q_1^*(\mathcal{M})$.

A morphism $f\colon (\mathcal{M}, \varepsilon) \to (\mathcal{M}', \varepsilon')$ in $\mathrm{HS}^r_{\mathbb{Q}_p}(X, Y/B)$ is an $\mathcal{O}_{D_1,\mathbb{Q}}$-linear morphism $f\colon \mathcal{M} \to \mathcal{M}'$ such that $\varepsilon' \circ p_2^*(f) = p_1^*(f) \circ \varepsilon$.

(2) Suppose that $r = \infty$. We define the category $\mathrm{HS}_{\mathbb{Z}_p}(X, Y/B)$ as follows. An object is a pair $(\mathcal{M}, \varepsilon)$ of an object $\mathcal{M}$ of $\mathcal{LPM}^{\mathrm{loc}}(\mathcal{O}_{D_1})$ and an isomorphism $\varepsilon\colon \widehat{p}_2^*(\mathcal{M}) \xrightarrow{\cong} \widehat{p}_1^*(\mathcal{M})$ in $\mathcal{LPM}^{\mathrm{loc}}(\mathcal{O}_{D(1)_1})$ satisfying the following conditions.

(i) There exists a strict étale covering $(X_\alpha \to X)_{\alpha \in A}$ such that $X_\alpha$ is affine and $\mathcal{M}|_{D_{1,\alpha}}$ ($D_{1,\alpha} = D_1 \times_X X_\alpha$) is an object of $\mathcal{LPM}(\mathcal{O}_{D_{1,\alpha}})$ for every $\alpha \in A$.

(ii) (a) $\widehat{\Delta}_D^*(\varepsilon) = \mathrm{id}_{\mathcal{M}}$. (b) $\widehat{p}_{12}^*(\varepsilon) \circ \widehat{p}_{23}^*(\varepsilon) = \widehat{p}_{13}^*(\varepsilon)\colon \widehat{q}_3^*(\mathcal{M}) \xrightarrow{\cong} \widehat{q}_1^*(\mathcal{M})$.

A morphism $f\colon (\mathcal{M}, \varepsilon) \to (\mathcal{M}', \varepsilon')$ in $\mathrm{HS}_{\mathbb{Z}_p}(X, Y/B)$ is a morphism $f\colon \mathcal{M} \to \mathcal{M}'$ in $\mathcal{LPM}^{\mathrm{loc}}(\mathcal{O}_{D_1})$ such that $\varepsilon' \circ \widehat{p}_2^*(f) = \widehat{p}_1^*(f) \circ \varepsilon$.

**Proposition IV.3.4.4.** (1) *There exists a canonical equivalence of categories between* $\mathrm{HC}^r_{\mathbb{Q}_p}(X/B)$ *and* $\mathrm{HS}^r_{\mathbb{Q}_p}(X, Y/B)$*.*

(2) *Suppose* $r = \infty$*. Then there exists a canonical equivalence of categories between* $\mathrm{HC}_{\mathbb{Z}_p}(X/B)$ *and* $\mathrm{HS}_{\mathbb{Z}_p}(X, Y/B)$*.*

PROOF. Standard argument using Lemma IV.3.4.1 and the adjoint property of the functor $D^r_{\mathrm{Higgs}}$. □

Let $\mathcal{M}$ be an $\mathcal{O}_{D_1}$-module endowed with an additive morphism $\theta\colon \mathcal{M} \to \xi^{-1}\mathcal{M}\otimes_{\mathcal{O}_{Y_1}} \Omega^1_{Y_1/B_1}$ such that $\theta(ax) = a\theta(x) + x \otimes \theta(a)$ for local sections $a \in \mathcal{O}_{D_1}$ and $x \in \mathcal{M}$. (See (IV.2.4.8) for $\theta$ on $\mathcal{O}_{D_1}$.) Since the right vertical morphism in (IV.2.4.9) is 0 if $N = 1$, the morphism $\theta$ is $p^{-1}_{Y,1}(\mathcal{O}_{Y_1})$-linear, where $p_Y = (p_{Y,N})$ denotes the natural morphism $D \to Y$. Hence we can define an additive morphism

$$\theta^q\colon \xi^{-q}\mathcal{M}\otimes_{\mathcal{O}_{Y_1}} \Omega^q_{Y_1/B_1} \to \xi^{-q-1}\mathcal{M}\otimes_{\mathcal{O}_{Y_1}} \Omega^{q+1}_{Y_1/B_1}$$

by

$$\theta^q(\xi^{-q}x\otimes\omega) = \xi^{-q}\theta(x)\wedge\omega.$$

**Lemma IV.3.4.5.** *If $\theta^1\circ\theta = 0$, then $\theta^{q+1}\circ\theta^q = 0$ for every $q \in \mathbb{N}$.*

PROOF. For $q, q' \in \mathbb{N}$, $y \in \xi^{-q}\mathcal{M}\otimes_{\mathcal{O}_{Y_1}}\Omega^q_{Y_1/B_1}$, and $\eta \in \Omega^{q'}_{Y_1/B_1}$, we have $\theta^{q+q'}(\xi^{-q'}y\wedge \eta) = \xi^{-q'}(\theta^q(y))\wedge\eta$. Hence, for $x \in \mathcal{M}$ and $\omega \in \Omega^q_{Y_1/B_1}$, we have $\theta^{q+1}\circ\theta^q(\xi^{-q}x\otimes\omega) = \theta^{q+1}(\xi^{-q}\theta(x)\wedge\omega) = \xi^{-q}\theta^1(\theta(x))\wedge\omega = 0$. □

**Lemma IV.3.4.6.** *Let $\mathcal{M}$ be as above, and suppose that $Y_1$ is affine and we are given $(t_{i,N}) \in \varprojlim_N \Gamma(Y_N, M_{Y_N})$ $(1 \le i \le d)$ such that $\{d\log(t_{i,N})\}_{1\le i\le d}$ is a basis of $\Omega^1_{Y_N/B_N}$ for every $N$. Let $\theta_i$ $(1 \le i \le d)$ be the endomorphisms of $\mathcal{M}$ defined by $\theta(x) = \sum_{1\le i\le d}\xi^{-1}\theta_i(x)\otimes d\log(t_{i,1})$ for $x \in \mathcal{M}$.*

*(1) We have $\theta^1\circ\theta = 0$ if and only if $\theta_i\circ\theta_j = \theta_j\circ\theta_i$ for every $i, j \in \{1, 2, \ldots, d\}$.*

*(2) If $\theta^1\circ\theta = 0$, then we have*

$$\theta^{\underline{m}}(ax) = \sum_{\underline{m}_1+\underline{m}_2=\underline{m}} \frac{\underline{m}!}{\underline{m}_1!\underline{m}_2!}\theta^{\underline{m}_1}(a)\theta^{\underline{m}_2}(x)$$

*for $a \in \mathcal{O}_{D_1}$ and $x \in \mathcal{M}$. Here we define the endomorphisms $\theta_i$ of $\mathcal{O}_{D_1}$ in the same way as $\mathcal{M}$ and $\theta^{\underline{m}} = \prod_{1\le i\le d}\theta_i^{m_i}$ for $\underline{m} = (m_i) \in \mathbb{N}^d$.*

PROOF. (1) follows from the formula

$$\theta^1\circ\theta(x) = \sum_{1\le i<j\le d}\xi^{-2}(\theta_i\circ\theta_j - \theta_j\circ\theta_i)(x)\otimes(d\log t_{i,1}\wedge d\log t_{j,1})$$

for $x \in \mathcal{M}$.

(2) follows from $\theta_i(ax) = a\theta_i(x) + \theta_i(a)x$ $(a \in \mathcal{O}_{D_1}, x \in \mathcal{M})$ by induction on $|\underline{m}|$. □

**Definition IV.3.4.7.** (1) We define the category $'\mathrm{HM}^r_{\mathbb{Q}_p}(X, Y/B)$ (resp. $'\mathrm{HM}_{\mathbb{Z}_p}(X, Y/B)$ when $r = \infty$) as follows. An object is a pair $(\mathcal{M}, \theta)$ of an $\mathcal{O}_{D_1,\mathbb{Q}}$ (resp. $\mathcal{O}_{D_1}$)-module and an additive morphism $\theta\colon \mathcal{M} \to \xi^{-1}\mathcal{M}\otimes_{\mathcal{O}_{Y_1}}\Omega^1_{Y_1/B_1}$ such that $\theta(ax) = a\theta(x) + x\otimes\theta(a)$ for $a \in \mathcal{O}_{D_1,\mathbb{Q}}$ (resp. $\mathcal{O}_{D_1}$) and $x \in \mathcal{M}$. A morphism $f\colon (\mathcal{M}, \theta) \to (\mathcal{M}', \theta')$ is an $\mathcal{O}_{D_1,\mathbb{Q}}$ (resp. $\mathcal{O}_{D_1}$)-linear morphism $f\colon \mathcal{M} \to \mathcal{M}'$ such that $(\xi^{-1}f\otimes\mathrm{id})\circ\theta = \theta'\circ f$.

(2) We define $\mathrm{HM}^r_{\mathbb{Q}_p}(X, Y/B)$ (resp. $\mathrm{HM}_{\mathbb{Z}_p}(X, Y/B)$ when $r = \infty$) to be the full subcategory of $'\mathrm{HM}^r_{\mathbb{Q}_p}(X, Y/B)$ (resp. $'\mathrm{HM}_{\mathbb{Z}_p}(X, Y/B)$) consisting of $(\mathcal{M}, \theta)$ such that $\theta^1\circ\theta = 0$.

Let $(\mathcal{M}, \varepsilon)$ be one of the following.

(Case I) $\mathcal{M}$ is an $\mathcal{O}_{D_1,\mathbb{Q}}$-module and $\varepsilon$ is an $\mathcal{O}_{D(1)_1,\mathbb{Q}}$-linear isomorphism $p_2^*(\mathcal{M}) \xrightarrow{\cong} p_1^*(\mathcal{M})$ satisfying the conditions (i) and (ii) (a) in Definition IV.3.4.3 (1).

(Case II) $r=\infty$, $\mathcal{M}$ is an object of $\mathcal{LPM}^{\mathrm{loc}}(\mathcal{O}_{D_1})$, and $\varepsilon$ is an isomorphism $\widehat{p}_2^*(\mathcal{M})\xrightarrow{\cong}\widehat{p}_1^*(\mathcal{M})$ in $\mathcal{LPM}^{\mathrm{loc}}(\mathcal{O}_{D(1)_1})$ satisfying the conditions (i) and (ii) (a) in Definition IV.3.4.3 (2).

Let $D(1)^1_N$ and $\Omega^1_{D_N/B_N}$ be as before Proposition IV.2.4.6. If $r=\infty$, (IV.2.4.11) and Remark IV.2.4.4 imply that $\mathcal{O}_{D(1)^1_1}$ is locally free of finite rank over $\mathcal{O}_{D_1}$ via $p_i^*$ $(i=1,2)$. Hence, in both cases, the isomorphism $\varepsilon$ induces an isomorphism

$$\varepsilon^1\colon \mathcal{O}_{D(1)^1_1}\otimes_{\mathcal{O}_{D_1}}\mathcal{M}\xrightarrow{\cong}\mathcal{M}\otimes_{\mathcal{O}_{D_1}}\mathcal{O}_{D(1)^1_1}$$

and an additive morphism

$$\begin{aligned}\theta\colon \mathcal{M}\to\mathcal{M}\otimes_{\mathcal{O}_{D_1}}\Omega^1_{D_1/B_1}&\cong\xi^{-1}\mathcal{M}\otimes_{\mathcal{O}_{Y_1}}\Omega^1_{Y_1/B_1}\\ x\mapsto\varepsilon^1(1\otimes x)-x\otimes 1.&\end{aligned}$$

For local sections $a\in\mathcal{O}_{D_1}$ and $x\in\mathcal{M}$, we have $\theta(ax)=a\theta(x)+x\otimes\theta(a)$, i.e., the pair $(\mathcal{M},\theta)$ is an object of $'\mathrm{HM}^r_{\mathbb{Q}_p}(X,Y/B)$ in the case I (resp. $'\mathrm{HM}_{\mathbb{Z}_p}(X,Y/B)$ in the case II).

Now let us study the condition (ii) (b) in Definition IV.3.4.3 strict étale locally on $Y_1$. Let $Y'\to Y$ be a strict étale Cartesian morphism, let $X'$ be $X\times_{Y_1}Y'_1$, and define $Y'(\nu)$, $D'(\nu)$ $(\nu\in\mathbb{N})$, $D'$, $p'_i$, $p'_{ij}$, and $q'_i$ in the same way as $Y(\nu)$, $D(\nu)$, etc. using $Y'$ and $X'\to Y'_1$. Then $D'(\nu)\to D(\nu)$ is a strict étale Cartesian morphism and $D'(\nu)_1\cong D(\nu)_1\times_X X'$ (cf. Remark IV.2.4.4). Hence $\mathcal{M}'=\mathcal{M}|_{D'_1}$ with $\theta'=\theta|_{D'_1}$ is obtained from $\mathcal{M}'$ with $\varepsilon'=\varepsilon|_{D'(1)_1}$ by the above construction applied to $Y'$ and $X'\to Y'_1$. Assume that $X'$ and $Y'_1$ are affine and there exist $t_i=(t_{i,N})\in\varprojlim_N\Gamma(Y'_N,M_{Y'_N})$ $(1\le i\le d)$ such that $\{d\log(t_{i,N})\}_{1\le i\le d}$ is a basis of $\Omega^1_{Y'_N/B'_N}$ for every $N\in\mathbb{N}_{>0}$. In the case I (resp. II), we also assume that $\mathcal{M}'$ is a finitely generated and projective $\mathcal{O}_{D'_1,\mathbb{Q}}$-module (resp. an object of $\mathcal{LPM}(\mathcal{O}_{D'_1})$). Let $A_{D'}(\nu)=(A_{D'}(\nu)_N)$ be the object of $\mathscr{A}^r_\bullet$ defined by $A_{D'}(\nu)_N=\Gamma(D'(\nu)_N,\mathcal{O}_{D'(\nu)_N})$. We write $A_{D'}$ for $A_{D'}(0)$. Put $M'=\Gamma(D'_1,\mathcal{M})$. For $i\in\{1,2\}$ (resp. $i\in\{1,2,3\}$), let $p'^*_iM'$ (resp. $q'^*_iM'$) denote the scalar extension of $M'$ by $p'^*_{i,1}\colon A_{D',1}\to A_{D'}(1)_1$ (resp. $q'^*_{i,1}\colon A_{D',1}\to A_{D'}(2)_1$), and let $p'^*_i$ (resp. $q'^*_i$) also denote the natural homomorphism $M'\to p'^*_iM'$ (resp. $M'\to q'^*_iM'$). Note that, in the case II, we have $p'^*_iM'=\Gamma(D'(1)_1,\widehat{p}'^*_{i,1}\mathcal{M}')$ and $q'^*_iM'=\Gamma(D'(2)_1,\widehat{q}'^*_{i,1}\mathcal{M}')$ and these are $p$-adically complete and separated by Lemma IV.3.2.8 and Lemma IV.3.2.2 (2).

In the case I, choose a finitely generated $A_{D',1}$-submodule $M'^{\circ}$ of $M'$ such that $M'^{\circ}_{\mathbb{Q}}=M'$. Since $M'$ is a finite projective $A_{D',1,\mathbb{Q}}$-module and hence a direct summand of a finite free $A_{D',1,\mathbb{Q}}$-module, the isomorphisms (IV.2.4.11) and (IV.2.4.12) induce isomorphisms

$$p'^*_iM'\cong\left(\bigoplus_{\underline{m}\in\mathbb{N}^d}A^{(|\underline{m}|)_r}_{D',1}M'^{\circ}\underline{W}^{\underline{m}}\right)^{\wedge}\otimes\mathbb{Q}\qquad(i=1,2),\tag{IV.3.4.8}$$

$$q'^*_jM'\cong\left(\bigoplus_{\underline{l},\underline{n}\in\mathbb{N}^d}A^{(|\underline{l}+\underline{n}|)_r}_{D',1}M'^{\circ}\underline{W}^{\underline{n}}\underline{W}'^{\underline{l}}\right)^{\wedge}\otimes\mathbb{Q}\qquad(j=1,2,3),\tag{IV.3.4.9}$$

where $^\wedge$ denotes the $p$-adic completion. Similarly, in the case II, we obtain isomorphisms

$$p_i'^*M' \cong \left(\bigoplus_{\underline{m}\in\mathbb{N}^d} M'\underline{W}^{\underline{m}}\right)^{\wedge} \qquad (i=1,2), \tag{IV.3.4.10}$$

$$q_j'^*M' \cong \left(\bigoplus_{\underline{l},\underline{n}\in\mathbb{N}^d} M'\underline{W}^{\underline{n}}\underline{W}'^{\underline{l}}\right)^{\wedge} \qquad (j=1,2,3). \tag{IV.3.4.11}$$

We define endomorphisms $\Theta'_{\underline{m}}$ ($\underline{m}\in\mathbb{N}^d$) of $M'$ by

$$\varepsilon'(p_2'^*(x)) = \sum_{\underline{m}\in\mathbb{N}^d} p_1'^*(\Theta'_{\underline{m}}(x)) \otimes \prod_{1\le i\le d} w_{i,1}^{m_i} \qquad (x\in M').$$

We define endomorphisms $\theta'_i$ ($1\le i\le d$) of $M'$ by the formula $\theta'(x)=\sum_{1\le i\le d}\xi^{-1}\theta'_i(x)\otimes d\log(t_{i,1})$ for $x\in M'$.

**Lemma IV.3.4.12.** *Let notation and assumption be as above. In the case* I (*resp.* II), *we endow $M'$ with the $p$-adic topology defined by $M'^{\circ}$ (resp. the $p$-adic topology). Assume that $\mathcal{M}'$ is $p$-torsion free in the case* II.

(1) *The isomorphism $\varepsilon'$ satisfies the condition* (ii) (b) *in Definition* IV.3.4.3 *if and only if $\theta'_i\circ\theta'_j=\theta'_j\circ\theta'_i$ ($i\neq j$) and $\Theta'_{\underline{m}}=\frac{1}{\underline{m}!}\prod_{1\le i\le d}\theta_i'^{m_i}$ ($\underline{m}\in\mathbb{N}^d$).*

(2) *If the equivalent conditions in* (1) *hold, then $(M',\theta')$ satisfies the following convergence:*

(Conv) *For any $x\in M'$, $p^{-[|\underline{m}|r^{-1}]}\frac{1}{\underline{m}!}\prod_{1\le i\le d}\theta_i'^{m_i}(x)$ converges to $0$ as $|\underline{m}|\to\infty$, where $|\underline{m}|r^{-1}=0$ if $r=\infty$.*

(3) *In the case* II, *if the equivalent conditions in* (1) *hold, then $(M',\theta')$ satisfies the following integrality:*

(Int) *For any $x\in M'$ and $\underline{m}\in\mathbb{N}^d$, we have $\frac{1}{\underline{m}!}\prod_{1\le i\le d}\theta_i'^{m_i}(x)\in M'$.*

Proof. The claim (3) immediately follows from (1). We prove (1) and (2).

(1) To simplify the notation, we omit $'$ from the notation $M'$, $\Theta'_{\underline{m}}$, $\theta'_i$, $\varepsilon'$, $p'_i$, $p'_{ij}$, and $q'_i$. We also abbreviate $p'_{i,N}$, $p'_{ij,N}$, and $q'_{i,N}$ to $p_i$, $p_{ij}$, and $q_i$, respectively. We have $\theta_i=\Theta_{\underline{1}_i}$ by definition. For $x\in M$, we have the following equalities in $q_3^*(M)$:

$$\begin{aligned} p_{12}^*(\varepsilon)\circ p_{23}^*(\varepsilon)(q_3^*(x)) &= p_{12}^*(\varepsilon)\left(\sum_{\underline{l}\in\mathbb{N}^d} q_2^*(\Theta_{\underline{l}}(x))\otimes p_{23}^*(\underline{w}^{\underline{l}})\right)\\ &= \sum_{\underline{l},\underline{n}\in\mathbb{N}^d} q_1^*(\Theta_{\underline{n}}\circ\Theta_{\underline{l}}(x))\otimes p_{23}^*(\underline{w}^{\underline{l}})p_{12}^*(\underline{w}^{\underline{n}})\\ p_{13}^*(\varepsilon)(q_3^*(x)) &= \sum_{\underline{m}\in\mathbb{N}^d} q_1^*(\Theta_{\underline{m}}(x))\otimes q_{13}^*(\underline{w}^{\underline{m}}),\end{aligned}$$

where $\underline{w}^{\underline{m}}=\prod_{1\le i\le d}w_{i,1}^{m_i}$ for $\underline{m}=(m_i)\in\mathbb{N}^d$. Since $p_{13}^*(u_{i,2})=p_{23}^*(u_{i,2})p_{12}^*(u_{i,2})$ and $\xi A_{D'}(2)_1=0$, we have $p_{13}^*(w_{i,1})=p_{23}^*(w_{i,1})+p_{12}^*(w_{i,1})$. Hence, by (IV.3.4.9) and (IV.3.4.11), the condition (ii) (b) in Definition IV.3.4.3 is equivalent to

$$\Theta_{\underline{n}}\circ\Theta_{\underline{l}}=\frac{(\underline{n}+\underline{l})!}{\underline{n}!\underline{l}!}\Theta_{\underline{n}+\underline{l}} \qquad (\underline{n},\underline{l}\in\mathbb{N}^d).$$

These equalities imply $\theta_i\circ\theta_j=\Theta_{\underline{1}_i+\underline{1}_j}=\theta_j\circ\theta_i$ ($i\neq j$) and

$$\theta_i\circ\Theta_{\underline{m}}=(m_i+1)\Theta_{\underline{m}+\underline{1}_i} \iff \Theta_{\underline{m}+\underline{1}_i}=\frac{1}{m_i+1}\theta_i\circ\Theta_{\underline{m}}$$

for $\underline{m} \in \mathbb{N}^d$ and $i \in \{1, 2, \ldots, d\}$. This implies the necessity. The sufficiency is obvious.

(2) If $r \in \mathbb{N}_{>0}$, we have $p^{[\frac{m}{r}]+1}A_{D',1} \subset A_{D',1}^{(m)_r} = A_{D',1}^{(m)_r,0} \subset p^{[\frac{m}{r}]}A_{D',1}$ by Lemma IV.2.3.4. If $r = \infty$, we have $A_{D',1}^{(m)_r} = A_{D',1}$ by definition. Hence the claim follows from (IV.3.4.8) and (IV.3.4.10). □

**Remark IV.3.4.13.** If we define a Higgs crystal on $(X/B)^r_{\mathrm{HIGGS}}$ similarly as Definition IV.3.3.2, then the integrality in Lemma IV.3.4.12 (3) becomes $\frac{1}{\underline{m}!}\prod_{1 \leq i \leq d} \theta_i'^{m_i}(x) \in A_{D',1}^{(|\underline{m}|)_r} M'$ by (IV.2.4.11). In the case $X \to B_1$ is smooth and integral, $X = Y_1$ and $B \in \mathscr{C}^\infty$, which is typical, we have $A_{D',1}^{(1)_r} = p\xi^{-(r-1)}F^{r-1}A_{Y'_r} = pA_{X'}$ and hence $\theta_i(M') \subset pM'$, where $A_{Y'_N} = \Gamma(Y'_N, \mathcal{O}_{Y'_N})$ and $A_{X'} = \Gamma(X', \mathcal{O}_{X'})$. This implies that, if $p \geq 5$ (resp. $p = 3$), the pair $(M'_{\mathbb{Q}}, \theta'_{\mathbb{Q}})$ automatically satisfies the convergence condition in Lemma IV.3.4.12 (2) for every level $\geq 2$ (resp. $\geq 3$).

**Definition IV.3.4.14.** We define $\mathrm{HB}^r_{\mathbb{Q}_p,\mathrm{conv}}(X, Y/B)$ (resp. $\mathrm{HB}_{\mathbb{Z}_p,\mathrm{conv}}(X, Y/B)$ when $r = \infty$) to be the full subcategory of $\mathrm{HM}^r_{\mathbb{Q}_p}(X, Y/B)$ (resp. $\mathrm{HM}_{\mathbb{Z}_p}(X, Y/B)$) (cf. Definition IV.3.4.7) consisting of $(\mathcal{M}, \theta)$ satisfying the following condition: There exists a strict étale covering $(Y_\alpha \to Y)_{\alpha \in A}$ in $\mathscr{C}$ and $(t_{i,N,\alpha})_N \in \varprojlim_N \Gamma(Y_{\alpha,N}, M_{Y_{\alpha,N}})$ $(1 \leq i \leq d_\alpha)$ for each $\alpha \in A$ such that

(i) $d\log t_{i,N,\alpha}$ $(1 \leq i \leq d_\alpha)$ is a basis of $\Omega^1_{Y_{\alpha,N}/B_N}$ for every $N \in \mathbb{N}_{>0}$,

(ii) $Y_{\alpha,1}$ is affine,

(iii) the restriction of $\mathcal{M}$ to $D_{\alpha,1} = Y_{\alpha,1} \times_{Y_1} D_1$ is a finitely generated projective $\mathcal{O}_{D_{\alpha,1},\mathbb{Q}}$-module (resp. a $p$-torsion free object of $\mathcal{LPM}(\mathcal{O}_{D_{\alpha,1}})$),

(iv) the pair $(\Gamma(D_{\alpha,1}, \mathcal{M}), \Gamma(D_{\alpha,1}, \theta))$ satisfies (Conv) (resp. and (Int)) in Lemma IV.3.4.12 with respect to $(t_{i,N,\alpha})_N$ $(1 \leq i \leq d_\alpha)$.

**Remark IV.3.4.15.** Let $(\mathcal{M}, \theta)$ be an object of the category $\mathrm{HB}^r_{\mathbb{Q}_p,\mathrm{conv}}(X, Y/B)$ (resp. $\mathrm{HB}_{\mathbb{Z}_p,\mathrm{conv}}(X, Y/B)$) and let $(Y_\alpha \to Y)$ be as in Definition IV.3.4.14. Then, by Lemma IV.3.4.6 (2) and Proposition IV.2.4.13 (2), we see that, for any strict étale Cartesian morphism $Y' \to Y_\alpha$ such that $Y'_1$ is affine and the pull-back of $(t_{i,N,\alpha})$ on $Y'$, the pair $(\Gamma(D'_1, \mathcal{M}), \Gamma(D'_1, \theta))$ $(D'_1 = Y'_1 \times_{Y_1} D_1)$ also satisfies (Conv) (resp. and (Int)) in Lemma IV.3.4.12.

**Theorem IV.3.4.16.** *In the case* I (*resp.* II), *Proposition* IV.3.4.4 *and the construction of $\theta$ after Definition* IV.3.4.7 *give an equivalence of categories from the category* $\mathrm{HC}^r_{\mathbb{Q}_p}(X/B)$ (*resp. the full subcategory of* $\mathrm{HC}_{\mathbb{Z}_p}(X/B)$ *consisting of $\mathcal{F}$ such that $\mathcal{F}_D$ is p-torsion free*) *to the category* $\mathrm{HB}^r_{\mathbb{Q}_p,\mathrm{conv}}(X, Y/B)$ (*resp.* $\mathrm{HB}_{\mathbb{Z}_p,\mathrm{conv}}(X, Y/B)$).

**Remark IV.3.4.17.** Suppose $r = \infty$, let $(\mathcal{M}, \varepsilon)$ be an object of $\mathrm{HS}_{\mathbb{Z}_p}(X, Y/B)$, and let $\overline{\mathcal{M}}$ be the image of $\mathcal{M} \to \mathcal{M}_{\mathbb{Q}_p}$, which is an object of $\mathcal{LPM}^{\mathrm{loc}}(\mathcal{O}_{D_1})$ and satisfies the condition (2) (i) in Definition IV.3.4.3 by Lemma IV.3.2.10. By (IV.2.4.11), Remark IV.2.4.4, and $r = \infty$, we see that $p_{i,1} \colon D(1)_1 \to D_1$ and $p_{ij,1} \colon D(2)_1 \to D(1)_1$ are flat after taking the reduction mod $p^m$ for every $m \in \mathbb{N}_{>0}$. Hence, by Lemma IV.3.2.11, the isomorphism $\varepsilon$ induces an isomorphism $\overline{\varepsilon} \colon \widehat{p}_2^*\overline{\mathcal{M}} \xrightarrow{\cong} \widehat{p}_1^*\overline{\mathcal{M}}$ and the pair $(\overline{\mathcal{M}}, \overline{\varepsilon})$ becomes a $p$-torsion free object of $\mathrm{HS}_{\mathbb{Z}_p}(X, Y/B)$.

Proof of Theorem IV.3.4.16. By Proposition IV.3.4.4, it suffices to prove that the category $\mathrm{HS}^r_{\mathbb{Q}_p}(X, Y/B)$ (resp. the full subcategory of $\mathrm{HS}_{\mathbb{Z}_p}(X, Y/B)$ consisting of $(\mathcal{M}, \varepsilon)$ such that $\mathcal{M}$ is $p$-torsion free) is equivalent to the category $\mathrm{HB}^r_{\mathbb{Q}_p,\mathrm{conv}}(X, Y/B)$ (resp. $\mathrm{HB}_{\mathbb{Z}_p,\mathrm{conv}}(X, Y/B)$) via the construction of $\theta$ after Definition IV.3.4.7. For an object $(\mathcal{M}, \varepsilon)$ of $\mathrm{HS}^r_{\mathbb{Q}_p}(X, Y/B)$ (resp. $\mathrm{HS}_{\mathbb{Z}_p}(X, Y/B)$ such that $\mathcal{M}$ is $p$-torsion free), we see that the associated pair $(\mathcal{M}, \theta)$ satisfies the conditions in Definitions IV.3.4.7 and

IV.3.4.14 by Lemmas IV.3.4.12 and IV.3.4.6 (1). Let us construct a quasi-inverse functor. Let $(\mathcal{M},\theta)$ be an object of $\mathrm{HB}^r_{\mathbb{Q}_p,\mathrm{conv}}(X,Y/B)$ (resp. $\mathrm{HB}^r_{\mathbb{Z}_p,\mathrm{conv}}(X,Y/B)$).

First let us consider the case where $Y_1$ is affine, $\mathcal{M}$ is a finitely generated projective $\mathcal{O}_{D_1,\mathbb{Q}}$-module (resp. an object of $\mathcal{LPM}(\mathcal{O}_{D_1})$), and there exist $(t_{i,N})\in\varprojlim_N\Gamma(Y_N,M_{Y_N})$ $(1\leq i\leq d)$ such that $\{d\log t_{i,N}\}$ is a basis of $\Omega^1_{Y_N/B_N}$ and $(M,\theta):=(\Gamma(D_1,\mathcal{M}),\Gamma(D_1,\theta))$ satisfies (Conv) (resp. and (Int)) in Lemma IV.3.4.12. Put $A_N=\Gamma(D_N,\mathcal{O}_{D_N})$ and $A(\nu)_N=\Gamma(D(\nu)_N,\mathcal{O}_{D(\nu)_N})$ $(\nu\in\mathbb{N})$. We define $w_{i,N}\in A(1)_N$ as before (IV.2.4.11). Let $p_i^*$, $q_j^*$, $p_{ij}^*$, and $\Delta^*$ denote the ring homomorphisms among $A_1$, $A(1)_1$, and $A(2)_1$ induced by the morphisms $p_{i,1}$, $q_{j,1}$, $p_{ij,1}$, and $\Delta_{D,1}$ among $D_1$, $D(1)_1$, and $D(2)_1$. Let $p_i^*M$ (resp. $q_j^*M$) denote the scalar extension of $M$ by $p_i$ (resp. $q_j$) and let $p_i^*$ (resp. $q_j^*$) also denote the natural homomorphism $M\to p_i^*M$ (resp. $M\to p_j^*M$). Then we have isomorphisms similar to (IV.3.4.8) and (IV.3.4.9) for a finitely generated $A_1$-submodule $M^\circ$ of $M$ such that $M^\circ_{\mathbb{Q}}=M$ (resp. (IV.3.4.10) and (IV.3.4.11)). We define the endomorphisms $\theta_i$ $(1\leq i\leq d)$ of $M$ by $\theta(x)=\sum_{1\leq i\leq d}\xi^{-1}\theta_i(x)\otimes d\log(t_{i,1})$ as usual. Since $p^{[\frac{m}{r}]+1}A_1\subset A_1^{(m)_r}\subset p^{[\frac{m}{r}]}A_1$ if $r\in\mathbb{N}_{>0}$ by Lemma IV.2.3.4, one can define an additive map $\alpha\colon M\to p_1^*M$ by

$$\alpha(x)=\sum_{\underline{m}\in\mathbb{N}^d}\underline{w}^{\underline{m}}\cdot p_1^*\left(\frac{1}{\underline{m}!}\theta^{\underline{m}}(x)\right),$$

where $\underline{w}^{\underline{m}}=\prod_{1\leq i\leq d}w_{i,1}^{m_i}$ and $\theta^{\underline{m}}=\prod_{1\leq i\leq d}\theta_i^{m_i}$ for $\underline{m}=(m_i)\in\mathbb{N}^d$. Since $p_2^*(a)=\sum_{\underline{m}\in\mathbb{N}^d}\underline{w}^{\underline{m}}\cdot p_1^*(\frac{1}{\underline{m}!}\theta^{\underline{m}}(a))$ for $a\in A_1$ by Proposition IV.2.4.13 (3), we obtain the equality $p_2^*(a)\alpha(x)=\alpha(ax)$ for $a\in A_1$ and $x\in M$ from Lemma IV.3.4.6 (2). Hence the morphism $\alpha$ induces an $A(1)_1$-linear homomorphism $\varepsilon\colon p_2^*M\to p_1^*M$. It is clear that the scalar extension of $\varepsilon$ by $\Delta^*\colon A(1)_1\to A$ is the identity. By the same argument as the proof of Lemma IV.3.4.12, we see that $\varepsilon$ satisfies the cocycle condition $p_{12}^*(\varepsilon)\circ p_{23}^*(\varepsilon)=p_{13}^*(\varepsilon)\colon q_3^*M\to q_1^*M$. Let $\tau$ (resp. $\iota$) be the morphism $D(1)\to D(2)$ (resp. the isomorphism $D(1)\xrightarrow{\cong}D(1)$) induced by the morphism $\tau_Y\colon Y(1)\to Y(2)$ (resp. the isomorphism $\iota_Y\colon Y(1)\xrightarrow{\cong}Y(1)$) defined by $q_1\circ\tau_Y=q_3\circ\tau_Y=p_1$ and $q_2\circ\tau_Y=p_2$ (resp. $p_1\circ\iota_Y=p_2$ and $p_2\circ\iota_Y=p_1$). Let $\tau^*$ (resp. $\iota^*$) denote the homomorphism $A(2)_1\to A(1)_1$ (resp. the isomorphism $A(1)_1\xrightarrow{\cong}A(1)_1$) induced by $\tau$ (resp. $\iota$). Then, by taking the extension of scalars of the above cocycle condition, we see that the composition of $p_1^*M\xrightarrow{\iota^*(\varepsilon)}p_2^*M\xrightarrow{\varepsilon}p_1^*M$ is the identity. This implies that $\varepsilon$ is an isomorphism. By Lemma IV.3.4.12 (1), we see that $\varepsilon$ constructed above is independent of the choice of $(t_{i,N})$.

In the general case, by applying the above argument to $Y_\alpha\to Y$ in the condition of Definition IV.3.4.14 and using Lemma IV.3.2.1 (3), Proposition IV.3.2.7 (1), and Lemma IV.3.2.8, we obtain isomorphisms $\varepsilon_\alpha\colon p_2^*\mathcal{M}|_{D(1)_1\times_YY_\alpha}\xrightarrow{\cong}p_1^*\mathcal{M}|_{D(1)_1\times_YY_\alpha}$, which glue and give an isomorphism $\varepsilon\colon p_2^*\mathcal{M}\xrightarrow{\cong}p_1^*\mathcal{M}$ satisfying the conditions (ii) (a) and (b) in Definition IV.3.4.3 (cf. Remark IV.3.4.15). Thus we obtain a functor in the opposite direction and Lemma IV.3.4.12 (1) implies that this gives a quasi-inverse. □

**IV.3.5. Lattices of Higgs isocrystals.** For $B\in\mathrm{Ob}\,\mathscr{C}^\infty$ and a morphism of $p$-adic fine log formal schemes $X\to B_1$, we define a kind of an integral structure on a Higgs isocrystal of level $r$ $(r\in\mathbb{N}_{>0})$ on $X/B$ (Definition IV.3.5.4), which is justified by the two preliminary propositions: IV.3.5.1 and IV.3.5.3 when $X\to B_1$ is smooth and integral.

**Proposition IV.3.5.1.** *Let $B\in\mathrm{Ob}\,\mathscr{C}^\infty$ and let $X$ be a $p$-adic fine log formal scheme over $B_1$ such that the morphism of schemes underlying the reduction mod $p$ of $X\to B_1$*

*is of finite type. Then the functor*

$$\mathrm{HC}_{\mathbb{Z}_p}(X/B)_{\mathbb{Q}} \longrightarrow \mathrm{HC}^{\infty}_{\mathbb{Q}_p}(X/B); \mathcal{F} \mapsto \mathcal{F}_{\mathbb{Q}_p}$$

*is fully faithful.*

**Proposition IV.3.5.2.** *Let $B \in \mathrm{Ob}\,\mathscr{C}^{\infty}$ and let $X$ be an affine $p$-adic fine log formal scheme over $B_1$. Assume that there exists a smooth Cartesian morphism $Y \to B$ and an immersion $X \to Y_1$ over $B_1$ satisfying the following condition: There exist $(t_{i,N}) \in \varprojlim_N \Gamma(Y_N, M_{Y_N})$ $(1 \leq i \leq d)$ such that $\{d\log(t_{i,N})\}_{1\leq i\leq d}$ is a basis of $\Omega^1_{Y_N/B_N}$ for every $N \in \mathbb{N}_{>0}$. Then the functor*

$$\mathrm{HC}_{\mathbb{Z}_p,X\text{-fin}}(X/B)_{\mathbb{Q}} \longrightarrow \mathrm{HC}^{\infty}_{\mathbb{Q}_p,X\text{-fin}}(X/B); \mathcal{F} \mapsto \mathcal{F}_{\mathbb{Q}_p}$$

*is an equivalence of categories.*

Proof. We apply Theorem IV.3.4.16 to $Y$ and $X \hookrightarrow Y_1$ in the assumption. Then the essential image of $\mathrm{HC}^{\infty}_{\mathbb{Q}_p,X\text{-fin}}(X/B)$ (resp. the full subcategory of $\mathrm{HC}_{\mathbb{Z}_p,X\text{-fin}}(X/B)$ consisting of $\mathcal{F}$ such that $\mathcal{F}_D$ is $p$-torsion free) under the equivalence of categories in Theorem IV.3.4.16 is the full subcategory consisting of $(\mathcal{M},\theta)$ such that $\mathcal{M} \in \mathcal{PM}(\mathcal{O}_{D_1,\mathbb{Q}})$ (resp. $\mathcal{LPM}(\mathcal{O}_{D_1})$). By the proof of the theorem, we also see that such $(\mathcal{M},\theta)$ satisfies (Conv) (resp. (Conv) and (Int)) in Lemma IV.3.4.12 for $r = \infty$ with respect to $(t_{i,N})_N$ $(1 \leq i \leq d)$. By Proposition IV.3.4.4, Lemma IV.3.2.8, and Proposition IV.3.2.7 (2), we see that the functor in the proposition is fully faithful. Let $(\mathcal{M},\theta)$ be an object of $\mathrm{HB}_{\mathbb{Q}_p,\mathrm{conv}}(X, Y/B)$ such that $\mathcal{M} \in \mathcal{PM}(\mathcal{O}_{D_1,\mathbb{Q}_p})$, put $M = \Gamma(D_1,\mathcal{M})$, let $\theta$ also denote $\Gamma(D_1,\theta)$, and define the endomorphisms $\theta_i$ $(1 \leq i \leq d)$ of $M$ by $\theta(x) = \sum_{1\leq i\leq d} \xi^{-1}\theta_i(x)d\log(t_{i,1})$. Choose a finitely generated $A_{D,1}$-submodule $M^{\circ\prime}$ such that $M^{\circ\prime}_{\mathbb{Q}_p} = M$. Since $\theta_i$ satisfy (Conv) in Lemma IV.3.4.12 for $r = \infty$, Lemma IV.3.4.6 (2) and Proposition IV.2.4.13 (2) imply that there exists a finite subset $S \subset \mathbb{N}^d$ such that $\frac{1}{\underline{m}!}\prod_{1\leq i\leq d}\theta_i^{m_i}(M^{\circ\prime}) \subset M^{\circ\prime}$ for all $\underline{m} \in \mathbb{N}^d \backslash S$. Hence $M^{\circ} = \sum_{\underline{m}\in\mathbb{N}^d}\frac{1}{\underline{m}!}\prod_{1\leq i\leq d}\theta_i^{m_i}(M^{\circ\prime})$ is a finitely generated $A_{D,1}$-module and satisfies $\frac{1}{\underline{m}!}\prod_{1\leq i\leq d}\theta_i^{m_i}(M^{\circ}) \subset M^{\circ}$ for every $\underline{m} \in \mathbb{N}^d$. Let $\mathcal{M}^{\circ}$ be the object of $\mathcal{LPM}(\mathcal{O}_{D_1})$ corresponding to $M^{\circ}$ (cf. Proposition IV.3.2.7 (1)). Then the natural morphism $\mathcal{M}^{\circ} \to \mathcal{M}$ is injective (cf. Lemma IV.3.2.10), $\mathcal{M}^{\circ}_{\mathbb{Q}_p} \cong \mathcal{M}$, the image of $\mathcal{M}^{\circ}$ in $\mathcal{M}$ is stable under $\theta$ (cf. Lemma IV.3.2.6 (2), Lemma IV.3.4.6 (2), and Proposition IV.2.4.13 (2)), and $\mathcal{M}^{\circ}$ with the induced $\theta$ satisfies (Conv) in Lemma IV.3.4.12. This completes the proof of the essential surjectivity. □

Proof of Proposition IV.3.5.1. By the assumption on $X$, there exists a strict étale covering $(X_\alpha \to X)_{\alpha\in A}$ such that $\mathcal{F}$ is $X_\alpha$-finite and $X_\alpha$ satisfies the assumption of Proposition IV.3.5.2 for each $\alpha \in A$. Furthermore, for each $\alpha, \beta \in A$, there exists an affine open covering $X_{\alpha\beta} = \cup_{\gamma\in\Gamma_{\alpha\beta}} X_{\alpha\beta;\gamma}$. For $X_\alpha \hookrightarrow Y_\alpha$ satisfying the condition in Proposition IV.3.5.2, $X_{\alpha\beta;\gamma} \hookrightarrow Y_\alpha \times_B Y_\beta$ also satisfies the condition. Hence we obtain the proposition by combining Corollary IV.3.1.8 and Proposition IV.3.5.2. □

**Proposition IV.3.5.3.** *Let $B \in \mathscr{C}^{\infty}$ and let $X \to B_1$ be a smooth integral morphism of $p$-adic fine log formal schemes. Then, for every $r \in \mathbb{N}_{>0}$, the functor*

$$\mathrm{HC}^{r}_{\mathbb{Q}_p}(X/B) \longrightarrow \mathrm{HC}^{\infty}_{\mathbb{Q}_p}(X/B)$$

*induced by the inclusion functor $(X/B)^{\infty}_{\mathrm{HIGGS}} \hookrightarrow (X/B)^{r}_{\mathrm{HIGGS}}$ is fully faithful.*

Proof. By Corollary IV.3.1.8, we are reduced to the case where $X$ is affine. Choose a smooth integral Cartesian morphism $Y \to B$ in $\mathscr{C}$ such that $Y_1 = X$. Then we can apply Theorem IV.3.4.16 to $X$ and $\mathrm{id}\colon X \to Y_1$. Since the reduction mod $p^m$ of the underlying morphism of formal schemes of $Y_N \to B_N$ is flat, we have $Y \in \mathscr{C}^{\infty}$ by Lemma IV.2.2.7

(1) and hence $D^r_{\mathrm{Higgs}}(X_1 \hookrightarrow Y) = Y$ for every $r \in \mathbb{N}_{>0} \cup \{\infty\}$. Now the proposition follows from Theorem IV.3.4.16. □

**Definition IV.3.5.4.** Let $B \in \mathrm{Ob}\,\mathscr{C}^\infty$ and let $X \to B_1$ be a morphism of $p$-adic fine log formal schemes. For $r \in \mathbb{N}_{>0}$, we define the category $\mathrm{HC}^r_{\mathbb{Z}_p}(X/B)$ as follows. An object is a triple consisting of an object $\mathcal{F}$ of $\mathrm{HC}^r_{\mathbb{Q}_p}(X/B)$, an object $\mathcal{F}^\circ$ of $\mathrm{HC}_{\mathbb{Z}_p}(X/B)$, and an isomorphism $\iota_{\mathcal{F}} \colon \mathcal{F}^\circ_{\mathbb{Q}_p} \xrightarrow{\cong} \mathcal{F}|_{(X/B)^\infty_{\mathrm{HIGGS}}}$. A morphism in $(\mathcal{F}, \mathcal{F}^\circ, \iota_{\mathcal{F}}) \to (\mathcal{G}, \mathcal{G}^\circ, \iota_{\mathcal{G}})$ is a pair of morphisms $f \colon \mathcal{F} \to \mathcal{G}$ and $f^\circ \colon \mathcal{F}^\circ \to \mathcal{G}^\circ$ such that $f|_{(X/B)^\infty_{\mathrm{HIGGS}}} \circ \iota_{\mathcal{F}} = \iota_{\mathcal{G}} \circ f^\circ_{\mathbb{Q}_p}$.

Under the notation in Definition IV.3.5.4, we have a natural functor

$$\mathrm{HC}^r_{\mathbb{Z}_p}(X/B) \longrightarrow \mathrm{HC}_{\mathbb{Z}_p}(X/B)$$

defined by $(\mathcal{F}, \mathcal{F}^\circ, \iota_{\mathcal{F}}) \mapsto \mathcal{F}^\circ$ and $(f, f^\circ) \mapsto f^\circ$. This is an equivalence of categories if $r = \infty$. If $X \to B_1$ is smooth and integral, this functor is fully faithful by Proposition IV.3.5.3. The following diagram is commutative up to canonical isomorphism.

$$\text{(IV.3.5.5)} \qquad \begin{array}{ccc} \mathrm{HC}^r_{\mathbb{Z}_p}(X/B) & \longrightarrow & \mathrm{HC}_{\mathbb{Z}_p}(X/B) \\ \downarrow & & \downarrow \\ \mathrm{HC}^r_{\mathbb{Q}_p}(X/B) & \longrightarrow & \mathrm{HC}^\infty_{\mathbb{Q}_p}(X/B). \end{array}$$

Here the left vertical functor is defined by $(\mathcal{F}, \mathcal{F}^\circ, \iota_{\mathcal{F}}) \mapsto \mathcal{F}$ and $(f, f^\circ) \mapsto f$.

If the morphism of schemes underlying the reduction mod $p$ of $X \to B_1$ is of finite type, then we see that the functor $\mathrm{HC}^r_{\mathbb{Z}_p}(X/B)_{\mathbb{Q}} \to \mathrm{HC}^r_{\mathbb{Q}_p}(X/B)$ defined by $(\mathcal{F}, \mathcal{F}^\circ, \iota_{\mathcal{F}}) \mapsto \mathcal{F}$ is fully faithful by Proposition IV.3.5.1.

**IV.3.6. Convergence and divisibility of Higgs fields.** We discuss the relation between the convergence condition on a Higgs field in Lemma IV.3.4.12 (2) and the divisibility of a Higgs field by a positive rational power of $p$ considered in **[27]**.

Let $V$ be a complete discrete valuation ring of mixed characteristic $(0, p)$, let $\overline{K}$ be an algebraic closure of the field of fractions $K$ of $V$, let $\overline{V}$ be the integral closure of $V$ in $\overline{K}$, and let $\widehat{\overline{V}}$ be the $p$-adic completion of $\overline{V}$. Let $v$ be a valuation of $\overline{V}$, and for $\alpha \in \mathbb{Q}_{>0}$, let $p^\alpha$ denote an element of $\overline{V}$ such that $v(p^\alpha) = \alpha v(p)$.

Let $U \to B_1$ be a smooth integral morphism of $p$-adic fine log formal schemes over $\mathrm{Spf}(\widehat{\overline{V}})$ such that the morphism $p \colon \mathcal{O}_{B_1} \to \mathcal{O}_{B_1}$ is injective, the underling formal scheme of $U$ is affine, and there exist $t_1, \ldots, t_d \in \Gamma(U, M_U)$ such that $\{d\log(t_i) | 1 \leq i \leq d\}$ is a basis of $\Omega^1_{U/B_1}$. Choose $t_i$ as above, and put $A = \Gamma(U, \mathcal{O}_U)$ and $\Omega^1 = \Gamma(U, \Omega^1_{U/B_1})$. The assumption implies that $A$ is $p$-torsion free.

Let $M$ be a finitely generated projective $A_{\mathbb{Q}_p}$-module and let $\theta \colon M \to M \otimes_A \xi^{-1}\Omega^1$ be a Higgs field of $M$, i.e., an $A$-linear homomorphism such that $\theta^1 \circ \theta = 0$, where $\theta^1$ is defined as before Lemma IV.3.4.5. We define the endomorphisms $\theta_i$ of $M$ by $\theta(x) = \sum_{1 \leq i \leq d} \theta_i(x) \otimes \xi^{-1} d\log(t_i)$ $(x \in M)$. For $\underline{n} = (n_1, \ldots, n_d) \in \mathbb{N}^d$, let $|\underline{n}|$, $\underline{n}!$ and $\theta^{\underline{n}}$ denote $\sum_{1 \leq i \leq d} n_i$, $\prod_{1 \leq i \leq d} n_i!$ and $\prod_{1 \leq i \leq d} \theta_i^{n_i}$, respectively.

As in Lemma IV.3.4.12 (2), we consider the following convergence of $\theta$ for $r \in \mathbb{N}_{>0} \cup \{\infty\}$.

$$(\mathrm{Conv})_r \qquad \text{For any } x \in M,\ p^{-r^{-1}|\underline{n}|} \tfrac{1}{\underline{n}!} \theta^{\underline{n}}(x)\ (\underline{n} \in \mathbb{N}^d) \text{ converges to } 0 \text{ as } |\underline{n}| \to \infty.$$

Here we understand $r^{-1}|\underline{n}|$ to be 0 if $r = \infty$. If we are given an object $B \in \mathscr{C}^\infty$ such that $B_1$ is the $p$-adic fine log formal scheme above and a smooth Cartesian morphism $Y \to B$ such that $Y_1 \to B_1$ is the morphism $U \to B_1$ above, then the category of Higgs isocrystals on $(U/B)^r_{\mathrm{HIGGS}}$ finite on $U$ is equivalent to the category of pairs $(M, \theta)$ satisfying $(\mathrm{Conv})_r$

(cf. Theorem IV.3.4.16 together with its proof, and Lemma IV.3.2.1). Note that we have $D^r_{\mathrm{HIGGS}}(Y) = Y$ because $Y$ is an object of $\mathscr{C}^\infty$ by Lemma IV.2.2.7 (1).

For $\alpha \in \mathbb{Q}_{>\frac{1}{p-1}}$, we first consider a similar convergence condition as follows.

$(\mathrm{Conv})'_\alpha$ For any $x \in M$, $p^{-\alpha|\underline{n}|}\underline{\theta}^{\underline{n}}(x)$ ($\underline{n} \in \mathbb{N}^d$) converges to 0 as $|\underline{n}| \to \infty$.

Then since $\frac{p^{\alpha n}}{n!}$ (resp. $\frac{n!}{p^{\alpha n}}$) converges to 0 as $n \to \infty$ if $\alpha > \frac{1}{p-1}$ (resp. $\alpha < \frac{1}{p-1}$), we obtain the following implications between the two conditions.

**Lemma IV.3.6.1.** *Let $r \in \mathbb{N}_{>0}$ and $\alpha \in \mathbb{Q}_{>\frac{1}{p-1}}$.*

(1) $(\mathrm{Conv})_r$ *implies* $(\mathrm{Conv})'_\alpha$ *if* $\alpha < \frac{1}{p-1} + \frac{1}{r}$.

(2) $(\mathrm{Conv})'_\alpha$ *implies* $(\mathrm{Conv})_r$ *if* $\frac{1}{p-1} + \frac{1}{r} < \alpha$.

For $\alpha \in \mathbb{Q}_{>\frac{1}{p-1}}$, we consider the following condition as in **[27]** Definition 2.

$(\mathrm{Div})_\alpha$ There exists a finitely generated $A$-submodule $M^\circ$ such that $M^\circ_{\mathbb{Q}_p} = M$ and $p^{-\alpha}\theta(M^\circ) \subset M^\circ \otimes_A \xi^{-1}\Omega^1$.

**Proposition IV.3.6.2.** *Let $\alpha, \beta \in \mathbb{Q}_{>\frac{1}{p-1}}$.*

(1) $(\mathrm{Div})_\alpha$ *implies* $(\mathrm{Conv})'_\beta$ *if* $\beta < \alpha$.

(2) $(\mathrm{Conv})'_\alpha$ *implies* $(\mathrm{Div})_\alpha$.

PROOF. (1) follows from $(p^{-\beta}\theta_i)^n(M^\circ) \subset p^{(\alpha-\beta)n}M^\circ$. Let us prove (2). Suppose that $(M, \theta)$ satisfies $(\mathrm{Conv})'_\alpha$. Choose a finitely generated $A$-submodule $M^{\circ\prime}$ of $M$ such that $M^{\circ\prime} \otimes_{\mathbb{Z}_p} \mathbb{Q}_p = M$ and let $x_\lambda$, $\lambda \in \Lambda$, $\sharp\Lambda < \infty$ be generators of $M^{\circ\prime}$ over $A$. Then $(\mathrm{Conv})'_\alpha$ implies that there exists $N \in \mathbb{N}$ such that for every $\underline{n} \in \mathbb{N}^d$ satisfying $|\underline{n}| \geq N$, we have $p^{-\alpha|\underline{n}|}\underline{\theta}^{\underline{n}}(x_\lambda) \in M^{\circ\prime}, \lambda \in \Lambda$, which implies, since $\theta_i$ are $A$-linear, that $p^{-\alpha|\underline{n}|}\underline{\theta}^{\underline{n}}(M^{\circ\prime}) \subset M^{\circ\prime}$. Hence the $A$-submodule $M^\circ := \sum_{\underline{n}\in\mathbb{N}^d} p^{-\alpha|\underline{n}|}\underline{\theta}^{\underline{n}}(M^{\circ\prime})$ is finitely generated, $M^\circ$ generates $M$ over $A_{\mathbb{Q}_p}$, and for any $1 \leq i \leq d$, we have

$$p^{-\alpha}\theta_i(M^\circ) = \sum_{\underline{n}\in\mathbb{N}^d} p^{-\alpha}\theta_i p^{-\alpha|\underline{n}|}\underline{\theta}^{\underline{n}}(M^{\circ\prime}) \subset M^\circ.$$

Thus we see that $(M, \theta)$ satisfies $(\mathrm{Div})_\alpha$. □

Following **[27]** p.852, we define a small Higgs field as follows.

**Definition IV.3.6.3.** Let $U \to B_1$, $A$ and $\Omega^1$ be as above, and let $M$ be a finitely generated projective $A_{\mathbb{Q}_p}$-module. We say that a Higgs field $\theta\colon M \to M \otimes_A \xi^{-1}\Omega^1$ is *small* if it satisfies $(\mathrm{Div})_\alpha$ for some $\alpha \in \mathbb{Q}_{>\frac{1}{p-1}}$.

**Corollary IV.3.6.4.** *Let $M$ be a finitely generated projective $A_{\mathbb{Q}_p}$-module and let $\theta$ be a Higgs field $M \to M \otimes_A \xi^{-1}\Omega^1$ on $M$. Then $\theta$ is small if and only if it satisfies* $(\mathrm{Conv})_r$ *for some $r \in \mathbb{N}_{>0}$.*

PROOF. This immediately follows from Lemma IV.3.6.1 and Proposition IV.3.6.2. □

In the global case, we have a similar equivalence assuming the existence of a "lattice" of the underlying module as follows.

Let $X \to B_1$ be a smooth integral morphism of $p$-adic fine log formal schemes over $\mathrm{Spf}(\widehat{\overline{V}})$ such that the morphism $p\colon \mathcal{O}_{B_1} \to \mathcal{O}_{B_1}$ is injective. Let $\mathcal{M}$ be an $\mathcal{O}_{X,\mathbb{Q}_p}$-module locally finitely generated and projective, and let $\theta\colon \mathcal{M} \to \mathcal{M} \otimes_{\mathcal{O}_X} \xi^{-1}\Omega^1_{X/B_1}$ be a Higgs field, i.e., an $\mathcal{O}_{X,\mathbb{Q}_p}$-linear morphism such that $\theta^1 \circ \theta = 0$, where $\theta^1$ is defined as before Lemma IV.3.4.5. Following **[27]** p.852 and p.855, we define a small Higgs field as follows. (See also II.13.1.)

**Definition IV.3.6.5.** Let $\mathcal{M}$ and $\theta$ be as above. We say that $\theta$ is *small* if there exist $\alpha \in \mathbb{Q}_{>\frac{1}{p-1}}$ and an $\mathcal{O}_X$-submodule $\mathcal{M}^\circ \subset \mathcal{M}$ such that $\mathcal{M}^\circ \in \mathrm{Ob}\,\mathcal{LPM}_{\mathrm{loc}}(\mathcal{O}_X)$ (cf. Definition IV.3.2.9 (1)), $\mathcal{M}^\circ_{\mathbb{Q}_p} = \mathcal{M}$, and $p^{-\alpha}\theta(\mathcal{M}^\circ) \subset \mathcal{M}^\circ \otimes_{\mathcal{O}_X} \xi^{-1}\Omega^1_{X/B_1}$.

For $r \in \mathbb{N}_{>0} \cup \{\infty\}$, let $\mathrm{HB}^r_{\mathbb{Q}_p,\mathrm{conv}}(X/B_1)$ denote the category whose object is an $\mathcal{O}_{X,\mathbb{Q}_p}$-module locally finitely generated and projective $\mathcal{M}$ with a Higgs field $\theta\colon \mathcal{M} \to \mathcal{M} \otimes_{\mathcal{O}_X} \xi^{-1}\Omega^1_{X/B_1}$ satisfying $(\mathrm{Conv})_r$ strict étale locally on $X$, i.e., there exists a strict étale covering $(U_\alpha \to X)_{\alpha\in A}$ and $t_{\alpha,i} \in \Gamma(U_\alpha, M_X)$ $(1 \leq i \leq d_\alpha)$ such that $\{d\log(t_{\alpha,i}) | 1 \leq i \leq d_\alpha\}$ is a basis of $\Omega^1_{U_\alpha/B_1}$, $U_\alpha$ is affine, $\mathcal{M}|_{U_\alpha}$ is a finitely generated projective $\mathcal{O}_{U_\alpha,\mathbb{Q}_p}$-module, and the pair $(\Gamma(U_\alpha,\mathcal{M}), \Gamma(U_\alpha,\theta))$ satisfies $(\mathrm{Conv})_r$ with respect to $t_{\alpha,i}$.

If we are given an object $B \in \mathscr{C}^\infty$ such that $B_1$ is the $p$-adic fine log formal scheme above and a smooth Cartesian morphism $Y \to B$ such that $Y_1 \to B_1$ is the morphism $X \to B_1$ above, then the category $\mathrm{HB}^r_{\mathbb{Q}_p,\mathrm{conv}}(X/B_1)$ is equivalent to the category of Higgs isocrystals $\mathrm{HC}^r_{\mathbb{Q}_p}(X/B)$ on $(X/B)^r_{\mathrm{HIGGS}}$ by Theorem IV.3.4.16.

**Proposition IV.3.6.6.** *Let $\mathcal{M}$ be an $\mathcal{O}_{X,\mathbb{Q}_p}$-module locally finitely generated and projective, and let $\theta\colon \mathcal{M} \to \mathcal{M} \otimes_{\mathcal{O}_X} \xi^{-1}\Omega^1_{X/B_1}$ be a Higgs field on $\mathcal{M}$. We further assume that there exists an $\mathcal{O}_X$-submodule $\mathcal{M}^{\circ\prime} \subset \mathcal{M}$ such that $\mathcal{M}^{\circ\prime} \in \mathrm{Ob}\,\mathcal{LPM}^{\mathrm{loc}}(\mathcal{O}_X)$ and $\mathcal{M}^{\circ\prime}_{\mathbb{Q}_p} = \mathcal{M}$. Then $\theta$ is small if and only if $(\mathcal{M},\theta) \in \mathrm{Ob}\,(\mathrm{HB}^r_{\mathbb{Q}_p,\mathrm{conv}}(X/B_1))$ for some $r \in \mathbb{N}_{>0}$.*

PROOF. The necessity follows from Corollary IV.3.6.4. Let us prove the sufficiency. Let $(\mathcal{M},\theta)$ be an object of $\mathrm{HB}^r_{\mathbb{Q}_p,\mathrm{conv}}(X/B_1)$. Let $\alpha \in \mathbb{Q}$ such that $\frac{1}{p-1} < \alpha < \frac{1}{p-1} + \frac{1}{r}$ and let $U \to X$ be a strict étale morphism satisfying the conditions on $U_\alpha$ for $(\mathcal{M},\theta)$ in the definition of $\mathrm{HB}^r_{\mathbb{Q}_p,\mathrm{conv}}(X/B_1)$. We further assume that $\mathcal{M}^{\circ\prime}|_U \in \mathcal{LPM}(\mathcal{O}_U)$ (cf. Remark IV.3.4.15). Then choosing $t_i \in \Gamma(U, M_U)$ $(1 \leq i \leq d)$ such that $\{d\log(t_i)\}$ is a basis of $\Omega^1_{U/B_1}$ and applying Lemma IV.3.6.1 (1) and the proof of Proposition IV.3.6.2 (2) to the pair $(\Gamma(U,\mathcal{M}), \Gamma(U,\theta))$ and $\Gamma(U,\mathcal{M}^{\prime\circ})$, we obtain an $\mathcal{O}_U$-submodule $\mathcal{M}^\circ_U$ of $\mathcal{M}|_U$ such that $\mathcal{M}^\circ_U \in \mathcal{LPM}(\mathcal{O}_U)$, $\mathcal{M}^\circ_{U,\mathbb{Q}_p} = \mathcal{M}|_U$, and $p^{-\alpha}\theta|_U(\mathcal{M}^\circ_U) \subset \mathcal{M}^\circ_U \otimes_{\mathcal{O}_U} \xi^{-1}\Omega^1_{U/B_1}$ (cf. Lemma IV.3.2.1 (3) and Proposition IV.3.2.7 (1)). We see that the sheaf $\mathcal{M}^\circ_U$ does not depend on the choice of $t_i$ as follows. If you make another choice $t'_i$ and define $\theta'_i$ using $t'_i$, then $\theta'_i$ is a linear combination of $\theta_j$ with coefficients in $\Gamma(U,\mathcal{O}_U)$ and hence $\prod_{1\leq i\leq d}(\theta'_i)^{n_i}$ $(\underline{n} = (n_i) \in \mathbb{N}^d)$ is a linear combination of $\prod_{1\leq i\leq d}\theta_i^{m_i}$ for $\underline{m} = (m_i) \in \mathbb{N}^d$ such that $|\underline{m}| = |\underline{n}|$. This implies that the $\mathcal{O}_U$-submodule of $\mathcal{M}|_U$ defined by using $t'_i$ is contained in $\mathcal{M}^\circ_U$. By exchanging $t_i$ and $t'_i$ in this argument, we see that the two submodules coincide. Thus the above $\mathcal{M}^\circ_U$'s glue together and give an $\mathcal{O}_X$-submodule $\mathcal{M}^\circ$ of $\mathcal{M}$ such that $\mathcal{M}^\circ_{\mathbb{Q}_p} = \mathcal{M}$, $\mathcal{M}^\circ \in \mathcal{LPM}^{\mathrm{loc}}(\mathcal{O}_X)$, and $p^{-\alpha}\theta(\mathcal{M}^\circ) \subset \mathcal{M}^\circ \otimes_{\mathcal{O}_X} \xi^{-1}\Omega^1_{X/B_1}$. □

## IV.4. Cohomology of Higgs isocrystals

**IV.4.1. Projections to étale sites.** Let $r \in \mathbb{N}_{>0} \cup \{\infty\}$, let $B \in \mathrm{Ob}\,\mathscr{C}^r$, and let $X$ be a $p$-adic fine log formal scheme over $B_1$. We construct a morphism of topos $U_{X/B}\colon (X/B)^{r\sim}_{\mathrm{HIGGS}} \to X^\sim_{\mathrm{\acute{E}T}}$ (Proposition IV.4.1.2) in a way similar to the crystalline site (cf. **[3]** III 4.4) and prove its functoriality (Proposition IV.4.1.3).

For a sheaf of sets $\mathcal{F}$ on $(X/B)^r_{\mathrm{HIGGS}}$, we define the presheaf $U_{X/B*}(\mathcal{F})$ on $X_{\mathrm{\acute{E}T}}$ by

$$U_{X/B*}(\mathcal{F})(u\colon U \to X) = \Gamma((U/B)^r_{\mathrm{HIGGS}}, u^*_{\mathrm{HIGGS}}(\mathcal{F})).$$

**Lemma IV.4.1.1.** *Under the notation and assumption above, $U_{X/B*}(\mathcal{F})$ is a sheaf on $X_{\mathrm{\acute{E}T}}$.*

PROOF. Put $\mathcal{G} = U_{X/B*}(\mathcal{F})$. For an object $u\colon U \to X$ of $X_{\text{ÉT}}$, giving a section $x$ of $\mathcal{G}(U)$ is equivalent to giving a section $x_{(T,z)}$ of $\Gamma((T, u \circ z), \mathcal{F})$ for each $(T, z) \in \operatorname{Ob}(U/B)^r_{\text{HIGGS}}$ such that for every morphism $f\colon (\widetilde{T}, \widetilde{z}) \to (T, z)$ in $(U/B)^r_{\text{HIGGS}}$, we have $f^*(z_{(T,z)}) = z_{(\widetilde{T},\widetilde{z})}$. With this description, the pull-back $v^*\colon \mathcal{G}(U) \to \mathcal{G}(U')$ by a morphism $v\colon U' \to U$ in $X_{\text{ÉT}}$ is given by $v^*(x)_{(T',z')} = x_{(T',v\circ z')}$.

Let $(u_\alpha\colon U_\alpha \to U)_{\alpha\in A}$ be a strict étale covering of $(u\colon U \to X) \in \operatorname{Ob} X_{\text{ÉT}}$. Put $U_{\alpha\beta} = U_\alpha \times_U U_\beta$ and let $p_\alpha, p_\beta$ denote the projections $U_{\alpha\beta} \to U_\alpha, U_\beta$. It suffices to prove that the sequence $\mathcal{G}(U) \to \prod_{\alpha\in A} \mathcal{G}(U_\alpha) \rightrightarrows \prod_{(\alpha,\beta)\in A^2} \mathcal{G}(U_{\alpha\beta})$ is exact. Let $(T, z)$ be an object of $(U/B)^r_{\text{HIGGS}}$. Let $T_{\alpha,1}$ be $T_1 \times_U U_\alpha$, let $v_{\alpha,1}$, $z_\alpha$ denote the projections $T_{\alpha,1} \to T_1, U_\alpha$, and let $v_\alpha\colon T_\alpha \to T$ be the unique strict étale Cartesian lifting of $v_{\alpha,1}$. Put $U_{\alpha\beta} = U_\alpha \times_U U_\beta$ (resp. $T_{\alpha\beta} = T_\alpha \times_T T_\beta$), and let $q_\alpha$, $q_\beta$ denote the projections $T_{\alpha\beta} \to T_\alpha$, $T_\beta$. Let $z_{\alpha\beta}\colon T_{\alpha\beta,1} \to U_{\alpha\beta}$ be the morphism induced by $z_\alpha$ and $z_\beta$. Then $(T_\alpha, z_\alpha)$ (resp. $(T_{\alpha\beta}, z_{\alpha\beta})$) is an object of $(U_\alpha/B)^r_{\text{HIGGS}}$ (resp. $(U_{\alpha\beta}/B)^r_{\text{HIGGS}}$), and the morphism $v_\alpha$ (resp. $q_\alpha$, resp. $q_\beta$) defines a morphism $(T_\alpha, u_\alpha \circ z_\alpha) \to (T, z)$ (resp. $(T_{\alpha\beta}, p_\alpha \circ z_{\alpha\beta}) \to (T_\alpha, z_\alpha)$, resp. $(T_{\alpha\beta}, p_\beta \circ z_{\alpha\beta}) \to (T_\beta, z_\beta)$). Since $(v_\alpha\colon T_\alpha \to T)_{\alpha\in A}$ is a strict étale covering, we have an exact sequence

$$(*)\quad \Gamma((T, u\circ z), \mathcal{F}) \to \prod_{\alpha\in A} \Gamma((T_\alpha, u\circ u_\alpha \circ z_\alpha), \mathcal{F}) \rightrightarrows \prod_{(\alpha,\beta)\in A^2} \Gamma((T_{\alpha\beta}, u\circ u_{\alpha\beta}\circ z_{\alpha\beta}), \mathcal{F}),$$

where $u_{\alpha\beta} = u_\alpha \circ p_\alpha = u_\beta \circ p_\beta$. This implies that $\mathcal{G}(U) \to \prod_{\alpha\in A} \mathcal{G}(U_\alpha)$ is injective. Suppose that $(x_\alpha) \in \prod_{\alpha\in A} \Gamma(U_\alpha, \mathcal{G})$ satisfies $p_\alpha^*(x_\alpha) = p_\beta^*(x_\beta)$ for every $(\alpha, \beta) \in A^2$. Put $x_{\alpha\beta} = p_\alpha^*(x_\alpha) = p_\beta^*(x_\beta)$. Then the image of the element $(x_{\alpha,(T_\alpha,z_\alpha)})_\alpha$ in the middle term of $(*)$ under the two maps are both $(x_{\alpha\beta,(T_{\alpha\beta},z_{\alpha\beta})})_{\alpha\beta}$. Hence there exists a unique $x_{(T,z)} \in \Gamma((T, u\circ z), \mathcal{F})$ such that $v_\alpha^*(x_{(T,z)}) = x_{\alpha,(T_\alpha,z_\alpha)}$. Since this construction is functorial in $(T, z)$, we see that $x_{(T,z)}$ defines a section $x$ of $\Gamma(U, \mathcal{G})$. Now it remains to show that the image of $x$ in $\Gamma(U_\gamma, \mathcal{G})$ $(\gamma \in A)$ is $x_\gamma$, i.e., for any object $(T, z')$ of $(U_\gamma/B)^r_{\text{HIGGS}}$, we have $x_{(T,u_\gamma\circ z')} = x_{\gamma,(T,z')}$ in $\Gamma((T, u\circ u_\gamma \circ z'), \mathcal{F})$. Put $z = u_\gamma \circ z'\colon T_1 \to U$. By the construction of $x$, it suffices to prove that the image of $x_{\gamma,(T,z')}$ under the morphism $v_\alpha^*\colon \Gamma((T, u\circ z), \mathcal{F}) \to \Gamma((T_\alpha, u\circ u_\alpha \circ z_\alpha), \mathcal{F})$ is $x_{\alpha,(T_\alpha,z_\alpha)}$. The factorization $z = u_\gamma \circ z'\colon T_1 \to U_\gamma \to U$ induces a factorization $z_\alpha = p_\alpha \circ z'_\alpha\colon T_{\alpha,1} \to U_{\alpha\gamma} \to U_\alpha$, and $v_\alpha$ defines a morphism $(T_\alpha, p_\gamma \circ z'_\alpha) \to (T, z')$. Hence we have $x_{\alpha,(T_\alpha,z_\alpha)} = x_{\alpha\gamma,(T_\alpha,z'_\alpha)} = v_\alpha^*(x_{\gamma,(T,z')})$. □

For a sheaf of sets $\mathcal{G}$ on $X_{\text{ÉT}}$, we define the sheaf of sets $U^*_{X/B}(\mathcal{G})$ on $(X/B)^r_{\text{HIGGS}}$ by $U^*_{X/B}(\mathcal{G})((T, z)) = \mathcal{G}(z\colon T_1 \to X)$.

**Proposition IV.4.1.2.** *The functor $U^*_{X/B}$ is a left adjoint of $U_{X/B*}$ and left exact, so that the pair $(U^*_{X/B}, U_{X/B*})$ defines a morphism of topos*

$$U_{X/B}\colon (X/B)^{r\sim}_{\text{HIGGS}} \longrightarrow X^\sim_{\text{ÉT}}.$$

PROOF. Let $\mathcal{F} \in \operatorname{Ob}(X/B)^{r\sim}_{\text{HIGGS}}$ and $\mathcal{G} \in \operatorname{Ob} X^\sim_{\text{ÉT}}$. We can define a map

$$F\colon \operatorname{Hom}(U^*_{X/B}(\mathcal{G}), \mathcal{F}) \to \operatorname{Hom}(\mathcal{G}, U_{X/B*}\mathcal{F})$$

as follows. Let $\varphi$ be a morphism $U^*_{X/B}(\mathcal{G}) \to \mathcal{F}$. For $(u\colon U \to X) \in \operatorname{Ob}(X_{\text{ÉT}})$ and $(T, z) \in \operatorname{Ob}(U/B)^r_{\text{HIGGS}}$, the morphism $\varphi$ defines a map

$$\psi_{U,(T,z)}\colon \Gamma(u\colon U \to X, \mathcal{G}) \xrightarrow{z^*} \Gamma(u\circ z\colon T_1 \to X, \mathcal{G}) = \Gamma((T, u\circ z), U^*_{X/B}\mathcal{G})$$
$$\xrightarrow{\Gamma((T,u\circ z),\varphi)} \Gamma((T, u\circ z), \mathcal{F}).$$

For a morphism $f\colon (T',z')\to(T,z)$ in $(U/B)^r_{\mathrm{HIGGS}}$, we have $f^*\circ\psi_{U,(T,z)}=\psi_{U,(T',z')}$. Hence, by the description of sections of $U_{X/B*}\mathcal{F}$ in the proof of Lemma IV.4.1.1, we see that $(\psi_{U,(T,z)})_{(T,z)}$ defines a morphism $\psi_U\colon \Gamma(U,\mathcal{G})\to\Gamma(U,U_{X/T*}\mathcal{F})$. For a morphism $v\colon U'\to U$ in $X_{\text{ÉT}}$ and $(T',z')\in\mathrm{Ob}\,(U'/B)^r_{\mathrm{HIGGS}}$, we also have $\psi_{U',(T',z')}\circ v^*=\psi_{U,(T',v\circ z')}$. This implies that $v^*\circ\psi_U=\psi_{U'}\circ v^*$ and $(\psi_U)_U$ defines a morphism $\psi\colon\mathcal{G}\to U_{X/B*}\mathcal{F}$. We define $F(\varphi)$ to be $\psi$. This construction is obviously functorial in $\mathcal{F}$ and $\mathcal{G}$. Conversely we can define a map $G\colon \mathrm{Hom}(\mathcal{G},U_{X/B*}\mathcal{F})\to\mathrm{Hom}(U^*_{X/B}\mathcal{G},\mathcal{F})$ as follows. Let $\psi$ be a morphism $\mathcal{G}\to U_{X/B*}\mathcal{F}$. Then, for $(T,z)\in\mathrm{Ob}\,(X/B)^r_{\mathrm{HIGGS}}$, $\psi$ defines a map

$$\begin{aligned}\varphi_{(T,z)}\colon U^*_{X/B}\mathcal{G}((T,z))=\mathcal{G}(z\colon T_1\to X)&\xrightarrow{\Gamma(z\colon T_1\to X,\psi)}\Gamma((T_1/B)^r_{\mathrm{HIGGS}},z^*_{\mathrm{HIGGS}}\mathcal{F})\\ &\longrightarrow\Gamma((T,\mathrm{id}_{T_1}),z^*_{\mathrm{HIGGS}}\mathcal{F})=\Gamma((T,z),\mathcal{F}).\end{aligned}$$

For a morphism $f\colon(T',z')\to(T,z)$, we see that $f^*\circ\varphi_{(T,z)}=\varphi_{(T',z')}\circ f^*$. Hence $(\varphi_{(T,z)})_{(T,z)}$ defines a morphism $\varphi\colon U^*_{X/B}\mathcal{G}\to\mathcal{F}$. We define $G(\psi)$ to be $\varphi$. Now it is straightforward to verify $F\circ G=\mathrm{id}$ and $G\circ F=\mathrm{id}$. □

We define the morphism of topos

$$u_{X/B}\colon (X/B)^{r\sim}_{\mathrm{HIGGS}}\longrightarrow X^\sim_{\text{ét}}$$

to be the composition of $U_{X/B}$ with the canonical morphism of topos $X^\sim_{\text{ÉT}}\to X^\sim_{\text{ét}}$ (defined by the inclusion functor $X_{\text{ét}}\to X_{\text{ÉT}}$, which is a morphism of sites).

**Proposition IV.4.1.3.** *Let $r$, $r'$, $B$, $B'$, $X$, $X'$, $g$, and $f$ be the same as in Proposition* IV.3.1.5. *Then the following diagram of topos is commutative up to canonical isomorphism.*

(IV.4.1.4)
$$\begin{array}{ccc}(X'/B')^{r'\sim}_{\mathrm{HIGGS}} & \xrightarrow{f_{\mathrm{HIGGS}}} & (X/B)^{r\sim}_{\mathrm{HIGGS}}\\ \downarrow{\scriptstyle U_{X'/B'}} & & \downarrow{\scriptstyle U_{X/B}}\\ X'^\sim_{\text{ÉT}} & \xrightarrow{f_{\text{ÉT}}} & X^\sim_{\text{ÉT}}.\end{array}$$

Proof. It suffices to prove that there exists a canonical isomorphism $f^*_{\mathrm{HIGGS}}\circ U^*_{X/B}\cong U^*_{X'/B'}\circ f^*_{\text{ÉT}}$, which is verified as follows: For $\mathcal{F}\in\mathrm{Ob}\,X^\sim_{\text{ÉT}}$ and $(T',z')\in\mathrm{Ob}\,(X'/B')^{r'}_{\mathrm{HIGGS}}$, we have

$$\begin{aligned}f^*_{\mathrm{HIGGS}}(U^*_{X/B}\mathcal{F})((T',z'))&=U^*_{X/B}\mathcal{F}((T',f\circ z'))=\mathcal{F}(f\circ z'\colon T'_1\to X),\\ U^*_{X'/B'}(f^*_{\text{ÉT}}\mathcal{F})((T',z'))&=f^*_{\text{ÉT}}\mathcal{F}(z'\colon T'_1\to X')\cong\mathcal{F}(f\circ z'\colon T'_1\to X).\end{aligned}$$

□

**IV.4.2. Linearizations.** Let $r\in\mathbb{N}_{>0}$, let $B\in\mathrm{Ob}\,\mathscr{C}^r$, let $X$ be a $p$-adic fine log formal scheme over $B_1$, let $Y\to B$ be a smooth Cartesian morphism in $\mathscr{C}$ (Definition IV.2.2.2), and let $i\colon X\to Y_1$ be an immersion over $B_1$. Let $D$ denote $D^r_{\mathrm{Higgs}}(i\colon X\hookrightarrow Y)$ (cf. Definition IV.2.2.1 (3) and Proposition IV.2.2.9 (1)). The object $D$ of $\mathscr{C}^r$ with the canonical morphism $z_D\colon D_1\to X$ is an object of $(X/B)^r_{\mathrm{HIGGS}}$.

For a sheaf of $\mathcal{O}_{D_1,\mathbb{Q}}$-modules $\mathcal{F}$ on $D_{1,\text{ét}}$, we define a sheaf of $\mathcal{O}_{X/B,1,\mathbb{Q}}$-modules $L_Y(\mathcal{F})$ on $(X/B)^r_{\mathrm{HIGGS}}$, which we call the *linearization* of $\mathcal{F}$. We follow the construction of the linearization on the crystalline site given in **[5]** 6.10.1 Remark. We then prove that the derived direct image of $L_Y(\mathcal{F})$ under the projection to the small étale site $X_{\text{ét}}$ gives the direct image of $\mathcal{F}$ by $z_D\colon D_1\to X$ when $\mathcal{F}$ is finitely generated and projective strict étale locally on $X$ (Proposition IV.4.2.1). The necessity of the last condition on $\mathcal{F}$ stems from the fact that the morphism of topological spaces underlying

$D^r_{\mathrm{HIGGS}}(Z)_1 \to Z_1$ for $Z \in \mathscr{C}$ is not a homeomorphism in general, which is different from the case of divided power envelopes. Except this point, one can adapt the argument via Čech-Alexander complex in **[3]** V. Théorème 1.2.5 and Proposition 2.2.2. i) to our context without difficulty.

Let $\mathcal{F}$ be a sheaf of $\mathcal{O}_{D_1,\mathbb{Q}}$-modules on $D_{1,\text{ét}}$. Let $(T,z)$ be an object of $(X/B)^r_{\mathrm{HIGGS}}$. The identity morphism of $T_1$ and the morphism $T_1 \xrightarrow{z} X \xrightarrow{i} Y_1$ induce an immersion $i_T\colon T_1 \to T_1 \times_{B_1} Y_1$. Let $D_T$ denote $D^r_{\mathrm{Higgs}}(i_T\colon T_1 \hookrightarrow T \times_B Y)$, and let $p_D\colon D_T \to D$ be the morphism in $\mathscr{C}^r$ induced by $z\colon T_1 \to X$ and the projection $T \times_B Y \to Y$. Let $p_T\colon D_T \to T$ be the morphism induced by $\mathrm{id}\colon T_1 \to T_1$ and the projection $T \times_B Y \to T$. We define $L_Y(\mathcal{F})((T,z))$ to be $\Gamma(D_{T,1}, p^*_{D_1}(\mathcal{F}))$ regarded as an $\Gamma(T_1, \mathcal{O}_{T_1,\mathbb{Q}})$-module via the ring homomorphism $p^*_{T_1}\colon \Gamma(T_1,\mathcal{O}_{T_1,\mathbb{Q}}) \to \Gamma(D_{T,1}, \mathcal{O}_{D_{T,1},\mathbb{Q}})$. This construction is functorial in $(T,z)$, and, by Proposition IV.2.2.9 (3), defines a sheaf of $\mathcal{O}_{X/B,1.\mathbb{Q}}$-modules $L_Y(\mathcal{F})$ on $(X/B)^r_{\mathrm{HIGGS}}$; for an object $(T,z)$ of $(X/B)^r_{\mathrm{HIGGS}}$, the sheaf of $\mathcal{O}_{T_1,\mathbb{Q}}$-modules $L_Y(\mathcal{F})_T$ on $(T_1)_{\text{ét}}$ is given by $p_{T,1*}p^*_{D,1}(\mathcal{F})$.

**Proposition IV.4.2.1.** *Let $r$, $B$, $X$, $Y$, $X \to Y_1$, $D$, and $z_D$ be as above. Let $\mathcal{F}$ be a sheaf of $\mathcal{O}_{D_1,\mathbb{Q}}$-modules on $D_{1,\text{ét}}$. Assume that there exists a strict étale covering $(X_\alpha \to X)_{\alpha\in A}$ such that $\mathcal{F}|_{D_{1,\alpha}}$ ($D_{1,\alpha} = X_\alpha \times_X D_1$) is a finitely generated and projective $\mathcal{O}_{D_{1,\alpha},\mathbb{Q}}$-module for every $\alpha \in A$. Then we have a canonical isomorphism*

$$Ru_{X/B*}(L_Y(\mathcal{F})) \cong z_{D*}(\mathcal{F}).$$

**Lemma IV.4.2.2.** *Let $X = \mathrm{Spf}(A)$ be a $p$-adic fine log formal scheme and let $\mathcal{F}$ be a finitely generated projective $\mathcal{O}_{X,\mathbb{Q}}$-module on $X_{\text{ét}}$. Then we have $H^q(X_{\text{ét}}, \mathcal{F}) = 0$ for $q > 0$.*

For an abelian category $\mathscr{A}$, let $\mathscr{A}^{\mathbb{N}^\circ}$ denote the category of inverse system of objects of $\mathscr{A}$ indexed by $\mathbb{N}$ (cf. III.7, IV.6.6). If $\mathscr{A}$ has enough injectives, then $\mathscr{A}^{\mathbb{N}^\circ}$ has enough injectives (cf. **[47]** (1.1) Proposition a)). An object $(A_n)$ of $\mathscr{A}^{\mathbb{N}^\circ}$ is injective if and only if $A_n$ ($n \in \mathbb{N}$) are injective and the transition maps $A_{n+1} \to A_n$ ($n \in \mathbb{N}$) are split surjections (cf. **[47]** (1.1) Proposition b)). In particular, if an additive functor $f\colon \mathscr{A} \to \mathscr{B}$ between abelian categories preserves injectives, then the induced functor $f^{\mathbb{N}^\circ}\colon \mathscr{A}^{\mathbb{N}^\circ} \to \mathscr{B}^{\mathbb{N}^\circ}$ preserves injectives.

**Lemma IV.4.2.3.** *Let $C$ be a site and let $S(C,\mathbb{Z})$ be the category of sheaves of abelian groups on $C$. Let $(\mathcal{F}_n)$ be an object of $S(C,\mathbb{Z})^{\mathbb{N}^\circ}$ and assume that there exists a subset $\mathscr{S} \subset \mathrm{Ob}\, C$ satisfying the following conditions.*

(i) *For every $X \in \mathscr{S}$, $H^q(X,\mathcal{F}_n) = 0$ ($q > 0$) and $R^1 \varprojlim_n H^0(X, \mathcal{F}_n) = 0$.*

(ii) *For every $X \in \mathrm{Ob}\, C$, there exists a covering $(X_\alpha \to X)_{\alpha \in A}$ such that $X_\alpha \in \mathscr{S}$ for all $\alpha \in A$.*

*Then we have $R^q \varprojlim_n (\mathcal{F}_n) = 0$ ($q > 0$).*

Proof. Let $(\mathcal{F}_n) \to (\mathcal{I}^\bullet_n)$ be an injective resolution in $S(C,\mathbb{Z})^{\mathbb{N}^\circ}$. Let $X \in \mathscr{S}$. Since $\mathcal{I}^q_n$ ($q, n \in \mathbb{N}$) are injective, the assumption (i) implies that $\Gamma(X,\mathcal{F}_n) \to \Gamma(X, \mathcal{I}^\bullet_n)$ is a resolution. Since $\mathcal{I}^q_{n+1} \to \mathcal{I}^q_n$ ($q, n \in \mathbb{N}$) are split surjections, $\Gamma(X,\mathcal{I}^q_{n+1}) \to \Gamma(X,\mathcal{I}^q_n)$ are surjective. Hence we have $H^q(\varprojlim_n \Gamma(X,\mathcal{I}^\bullet_n)) = 0$ ($q \geq 2$) and $H^1(\varprojlim_n \Gamma(X,\mathcal{I}^\bullet_n)) = R^1 \varprojlim_n \Gamma(X,\mathcal{F}_n) = 0$. Now the assumption (ii) implies the claim because $R^q \varprojlim_n \mathcal{F}_n = \mathcal{H}^q(\varprojlim_n (\mathcal{I}^\bullet_n))$. □

Proof of Lemma IV.4.2.2. Since $\mathcal{F}$ is a direct summand of a finitely generated free $\mathcal{O}_{X,\mathbb{Q}}$-module, it suffices to prove $H^q(X_{\text{ét}}, \mathcal{O}_{X,\mathbb{Q}}) = 0$ for $q > 0$. By Lemma IV.4.4.6 (4) and **[2]** VI Corollaire 5.2, we have $H^q(X_{\text{ét}}, \mathcal{O}_{X,\mathbb{Q}}) = H^q(X_{\text{ét}}, \mathcal{O}_X) \otimes \mathbb{Q}$. The presheaf $\mathcal{G}_n$ ($n \in \mathbb{N}_{>0}$) on $X_{\text{ét}}$ defined by $\Gamma(Y,\mathcal{G}) = \Gamma(Y_n, \mathcal{O}_{Y_n})$, $Y_n = Y \times_{\mathrm{Spf}(\mathbb{Z}_p)} \mathrm{Spec}(\mathbb{Z}/p^n\mathbb{Z})$

is a sheaf and we have $\mathcal{O}_X = \varprojlim \mathcal{G}_n$. If $Y$ is affine, then we have $H^q(Y_{\text{ét}}, \mathcal{G}_n) \cong H^q(Y_{n,\text{ét}}, \mathcal{O}_{Y_n}) \cong H^q(Y_{n,\text{Zar}}, \mathcal{O}_{Y_n}) = 0$ for $q > 0$ (cf. [2] VII Proposition 4.3 for the second isomorphism) and $H^0(Y_{\text{ét}}, \mathcal{G}_{n+1}) \to H^0(Y_{\text{ét}}, \mathcal{G}_n)$ is surjective. Hence, by Lemma IV.4.2.3, we have $R\Gamma(X_{\text{ét}}, \mathcal{O}_X) \cong R\Gamma(X_{\text{ét}}, -) \circ R\varprojlim_n((\mathcal{G}_n)_n) \cong R\varprojlim_n \circ R(\Gamma(X_{\text{ét}}, -)^{\mathbb{N}^\circ})((\mathcal{G}_n)_n) \cong R\varprojlim_n(H^0(X_{\text{ét}}, \mathcal{G}_n))_n \cong \varprojlim_n H^0(X_{\text{ét}}, \mathcal{G}_n)$. □

Proof of Proposition IV.4.2.1. For a strict étale morphism $u\colon U \to X$, we can define a natural morphism

$$\Phi_U : \Gamma(U, z_{D*}\mathcal{F}) = \Gamma(D_1 \times_X U, \mathcal{F}) \longrightarrow \Gamma((U/B)^r_{\text{HIGGS}}, u^*_{\text{HIGGS}}(L_Y(\mathcal{F})))$$

as follows. For $(T, z) \in \text{Ob}\,(U/B)^r_{\text{HIGGS}}$, we have

$$\Gamma((T, z), u^*_{\text{HIGGS}}(L_Y(\mathcal{F}))) = \Gamma(D_{T,1}, p^*_{D,1}\mathcal{F}),$$

where $p_D\colon D_T \to D$ is the morphism in $\mathscr{C}^r$ associated to $(T, u \circ z)$ as in the definition of $L_Y(\mathcal{F})$. The morphism $z\colon T_1 \to U$ gives rise to a factorization $D_{T,1} \to D_1 \times_X U \to D_1$ of $p_{D,1}$, which induces a natural morphism $\Gamma(D_1 \times_X U, \mathcal{F}) \to \Gamma((T, z), u^*_{\text{HIGGS}} L_Y(\mathcal{F}))$. Observing that this morphism is functorial in $(T, z)$, we obtain the desired morphism. Varying $U$, we also obtain a morphism $z_{D*}\mathcal{F} \to u_{X/B*}L_Y(\mathcal{F})$.

For a strict étale morphism $u\colon U \to X$, the functor

$$u^*_{\text{HIGGS}} \colon (X/B)^{r,\sim}_{\text{HIGGS}} \to (U/B)^{r,\sim}_{\text{HIGGS}}$$

preserves injective sheaves of abelian groups by Proposition IV.3.1.9 (1) and [2] V Proposition 4.11 1). Hence it suffices to prove that $H^q((U/B)^r_{\text{HIGGS}}, u^*_{\text{HIGGS}} L_Y(\mathcal{F})) = 0$ $(q > 0)$ and that the morphism $\Phi_U$ constructed above is an isomorphism for $u$ satisfying the following conditions. There exists an $X$-morphism $U \to X_\alpha$ for some $\alpha$, there exists a strict étale Cartesian lifting $Y_U \to Y$ of $u\colon U \to X$, and $U$ is affine. For an object $(T, z)$ of $(U/B)^r_{\text{HIGGS}}$, the natural morphism $D^r_{\text{Higgs}}(T_1 \hookrightarrow T \times_B Y_U) \to D^r_{\text{Higgs}}(T_1 \hookrightarrow T \times_B Y)$ is an isomorphism by Lemma IV.2.2.15. Hence we have a natural isomorphism $u^*_{\text{HIGGS}} L_Y(\mathcal{F}) \cong L_{Y_U}(v_1^*\mathcal{F})$, where $v_1$ denotes the morphism $D^r_{\text{Higgs}}(U \hookrightarrow Y_U)_1 \cong D_1 \times_{X_1} U_1 \to D_1$. By replacing $X$ with $U$, we may assume that $X = U$ and $\mathcal{F}$ is a finitely generated projective $\mathcal{O}_{D_1,\mathbb{Q}}$-module.

Let $\varepsilon$ denote the natural functor $(X/B)^r_{\text{HIGGS}} \to (X/B)^{r\sim}_{\text{HIGGS}}$. By Lemma IV.3.4.1 (and [2] II Proposition 4.3 2)), we see that $\varepsilon(D, z_D)$ is a covering of the final object $e$ of $(X/B)^{r\sim}_{\text{HIGGS}}$. We define $D(\nu)$ $(\nu \in \mathbb{Z})$ in the same way as before Remark IV.2.4.4 and let $z_{D(\nu)}$ denote the natural morphism $D(\nu)_1 \to X$. Then $\varepsilon(D(\nu), z_{D(\nu)})$ represents the product of $\nu + 1$ copies of $\varepsilon(D, z_D)$. By Proposition IV.2.2.9 (2), $D(\nu)_1$ is affine. Now, by applying [2] V Corollaire 3.3 to the covering $\varepsilon(D, z_D) \to e$ and using (IV.3.1.12), we obtain a spectral sequence

$$E_2^{a,b} = H^a(\nu \mapsto H^b(D(\nu)_{1,\text{ét}}, L_Y(\mathcal{F})_{(D(\nu), z_{D(\nu)})})) \Longrightarrow H^{a+b}((X/B)^r_{\text{HIGGS}}, L_Y(\mathcal{F})).$$

For any object $(T, z)$ of $(X/B)^r_{\text{HIGGS}}$, the morphism $p_{T,1}\colon D_{T,1} = D^r_{\text{Higgs}}(T_1 \hookrightarrow T \times_B Y)_1 \to T_1$ is affine by Proposition IV.2.2.9 (2). Hence, if $T_1$ is affine, we obtain $H^q(T_{1,\text{ét}}, L_Y(\mathcal{F})_{(T,z)}) = 0$ for $q > 0$ from Lemma IV.4.2.2. Hence we have $E_2^{a,b} = 0$ for $b > 0$ and obtain

$$H^a((X/B)^r_{\text{HIGGS}}, L_X(\mathcal{F})) \cong H^a(\nu \mapsto \Gamma(D(\nu)_{1,\text{ét}}, L_Y(\mathcal{F})_{(D(\nu), z_{D(\nu)})})).$$

We see that the morphism $D(\nu) \times_D Y \to Y(\nu) \times_B Y = Y(\nu + 1)$ induces an isomorphism $D^r_{\text{Higgs}}(D(\nu)_1 \hookrightarrow D(\nu) \times_B Y) \xrightarrow{\cong} D(\nu + 1)$ by verifying that the source satisfies the universal property defining the target. Hence we have an isomorphism $\Gamma(D(\nu)_{1,\text{ét}}, L_Y(\mathcal{F})_{(D(\nu), z_{D(\nu)})}) \cong \Gamma(D(\nu + 1)_{1,\text{ét}}, p^*_{\nu+2,1}\mathcal{F})$, where $p_{\nu+2}$ denotes the morphism $D(\nu+1) \to D$ defined by the projection to the $(\nu+2)$-th component $Y(\nu+1) \to Y$.

With this description, the differential morphism from the degree $\nu$ term to the degree $(\nu+1)$ term is given by the alternating sum of the pull-backs by the morphisms $D(\nu+2) \to D(\nu+1)$ defined by forgetting one of the components except the last one. We also see that the composition $\Gamma(D_{1,\text{ét}}, \mathcal{F}) \to \Gamma((X/B)^r_{\text{HIGGS}}, L_Y(\mathcal{F})) \to \Gamma(D_{1,\text{ét}}, L_Y(\mathcal{F})_{(D,z_D)}) \cong \Gamma(D(1)_{1,\text{ét}}, p_2^*\mathcal{F})$ is given by the pull-back by the projection to the second component. Hence the claim follows from **[3]** V Lemme 2.2.1. □

**IV.4.3. Poincaré lemma.** Let $r$, $B$, $X$, $Y$, $i\colon X \to Y_1$, and $(D, z_D)$ be the same as the beginning of IV.4.2. For a Higgs isocrystal $\mathcal{F}$ of level $r$ on $X/B$, we construct a complex on $(X/B)^r_{\text{HIGGS}}$

(IV.4.3.1)
$$\mathcal{F} \to L_Y(\mathcal{F}_D) \to L_Y(\xi^{-1}\mathcal{F}_D \otimes_{\mathcal{O}_{Y_1}} \Omega^1_{Y_1/B_1}) \to \cdots \to L_Y(\xi^{-q}\mathcal{F}_D \otimes_{\mathcal{O}_{Y_1}} \Omega^q_{Y_1/B_1}) \to \cdots$$

and study its properties: a compatibility with the complex $(\xi^{-\bullet}\mathcal{F}_D \otimes_{\mathcal{O}_{Y_1}} \Omega^\bullet_{Y_1/B_1}, \theta^\bullet)$ on $(D_1)_{\text{ét}}$ (Lemma IV.4.3.2) and the vanishing of $\mathcal{H}^\bullet$ of the complex (IV.4.3.1) after raising the level from $r$ to $r+1$ (Proposition IV.4.3.4), which may be regarded as a Poincaré lemma.

Let $(T, z)$ be an object of $(X/B)^r_{\text{HIGGS}}$, and define $D_T$, $p_D$, and $p_T$ as in IV.4.2. Then by applying Lemma IV.2.4.10 to a smooth Cartesian morphism $Y \times_B T \to T$ and the immersion $T_1 \to T_1 \times_{B_1} Y_1$, we obtain a complex

$$p_{T,1}^{-1}(\mathcal{O}_{T_1}) \to \mathcal{O}_{D_{T,1}} \to \xi^{-1}\mathcal{O}^{(1)_r}_{D_{T,1}} \otimes_{\mathcal{O}_{Y_1}} \Omega^1_{Y_1/B_1} \to \cdots \to \xi^{-q}\mathcal{O}^{(q)_r}_{D_{T_1}} \otimes_{\mathcal{O}_{Y_1}} \Omega^q_{Y_1/B_1} \to \cdots$$

whose differential maps are $p_{T,1}^{-1}(\mathcal{O}_{T_1})$-linear. Hence, by taking $p_{T,1}^{-1}(\mathcal{F}_T)\otimes_{p_{T,1}^{-1}(\mathcal{O}_{T_1})}$, and using the isomorphisms $p_{T,1}^*(\mathcal{F}_T) \xrightarrow{\cong} \mathcal{F}_{D_T} \xleftarrow{\cong} p_{D,1}^*(\mathcal{F}_D)$, we obtain a complex

$$p_{T,1}^{-1}(\mathcal{F}_T) \to p_{D,1}^*(\mathcal{F}_D) \xrightarrow{\theta^0} p_{D,1}^*(\xi^{-1}\mathcal{F}_D \otimes_{\mathcal{O}_{Y_1}} \Omega^1_{Y_1/B_1}) \xrightarrow{\theta^1} \cdots$$
$$\cdots \xrightarrow{\theta^{q-1}} p_{D,1}^*(\xi^{-q}\mathcal{F}_D \otimes_{\mathcal{O}_{Y_1}} \Omega^q_{Y_1/B_1}) \xrightarrow{\theta^q} \cdots.$$

By taking $\Gamma(D_T, -)$ and varying $(T, z)$, we obtain the desired complex (IV.4.3.1).

We have a complex $(\xi^{-q}\mathcal{F}_D \otimes_{\mathcal{O}_{Y_1}} \Omega^q_{Y_1/B_1}, \theta^q)_{q\in\mathbb{N}}$ by Theorem IV.3.4.16 and Lemma IV.3.4.5.

**Lemma IV.4.3.2.** *Under the notation and assumption as above, the following diagram is commutative for every $q \in \mathbb{N}$.*

$$\begin{array}{ccc}
u_{X/B*}(L_Y(\xi^{-q}\mathcal{F}_D \otimes_{\mathcal{O}_{Y_1}} \Omega^q_{Y_1/B_1})) & \xrightarrow{u_{X/S*}(\theta^q)} & u_{X/B*}(L_Y(\xi^{-q-1}\mathcal{F}_D \otimes_{\mathcal{O}_{Y_1}} \Omega^{q+1}_{Y_1/B_1})) \\
\uparrow{\scriptstyle \textit{Prop. IV.4.2.1}\ \cong} & & \uparrow{\scriptstyle \textit{Prop. IV.4.2.1}\ \cong} \\
z_{D*}(\xi^{-q}\mathcal{F}_D \otimes_{\mathcal{O}_{Y_1}} \Omega^q_{Y_1/B_1}) & \xrightarrow{z_{D*}(\theta^q)} & z_{D*}(\xi^{-q-1}\mathcal{F}_D \otimes_{\mathcal{O}_{Y_1}} \Omega^{q+1}_{Y_1/B_1})
\end{array}$$

PROOF. It suffices to prove that $\Gamma(U, -)$ of the diagram is commutative for any strict étale morphism $u\colon U \to X$ which has a strict étale Cartesian lifting $Y_U \to Y$. Let $(T, z)$ be an object of $(U/B)^r_{\text{HIGGS}}$. It is enough to prove that the following diagrams are commutative, where the vertical morphisms are defined as in the first paragraph of the proof of Proposition IV.4.2.1.

$$\begin{array}{ccc}\Gamma(U_1\times_{X_1}D_1,\xi^{-q}\mathcal{F}_D\otimes\Omega^q_{Y_1/B_1}) & \xrightarrow{\Gamma(U_1\times_{X_1}D_1,\theta^q)} & \Gamma(U_1\times_{X_1}D_1,\xi^{-q-1}\mathcal{F}_D\otimes\Omega^{q+1}_{Y_1/B_1})\\ \downarrow & & \downarrow\\ \Gamma((T,u\circ z),L_Y(\xi^{-q}\mathcal{F}_D\otimes\Omega^q_{Y_1/B_1})) & \xrightarrow{\Gamma((T,u\circ z),\theta^q)} & \Gamma((T,u\circ z),L_Y(\xi^{-q-1}\mathcal{F}_D\otimes\Omega^{q+1}_{Y_1/B_1})),\end{array}$$

where $\otimes$ denotes $\otimes_{\mathcal{O}_{Y_1}}$. By the definition of $\theta^q$ ($q\geq 2$) before Lemma IV.2.4.10 and Lemma IV.3.4.5, the claim is reduced to the case $q=0$.

We can construct $D(\nu)$ (resp. $D_U(\nu)$, resp. $D_T(\nu)$, resp. $D_{U,T}(\nu)$) from $Y/B$ and $X\hookrightarrow Y_1$ (resp. $Y_U/B$ and $U\hookrightarrow Y_{U,1}$, resp. $(T\times_B Y)/T$ and $T_1\hookrightarrow T_1\times_{B_1}Y_1$, resp. $(T\times_B Y_U)/T$ and $T_1\hookrightarrow T_1\times_{B_1}Y_{U,1}$) as before Remark IV.2.4.4. We omit $(\nu)$ if $\nu=0$. Let $\overline{D}_U(1)$ (resp. $\overline{D}_{U,T}(1)$) be the first infinitesimal neighborhood of $D_{U,1}\hookrightarrow D_U(1)_1$ (resp. $D_{U,T,1}\hookrightarrow D_{U,T}(1)_1$). The natural morphism $D_U(\nu)\to D(\nu)$ is strict étale and Cartesian, and it induces an isomorphism $D_U(\nu)_1\xrightarrow{\cong} D(\nu)_1\times_X U$ (cf. Remark IV.2.4.4). Hence the upper horizontal morphism in the diagram above for $q=0$ is induced by the morphism $p_2^*-p_1^*\colon \mathcal{F}_{D_U}\to\mathcal{F}_{D_U(1)}\otimes_{\mathcal{O}_{D_U(1)_1}}\mathcal{O}_{\overline{D}_U(1)}/(p\text{-tor})$ on $D_1\times_X U\cong D_{U,1}$. The natural morphisms $D_{U,T}(\nu)\to D_T(\nu)$ are isomorphisms by Lemma IV.2.2.15. Hence the lower horizontal morphism for $q=0$ is induced by $p_2^*-p_1^*\colon\mathcal{F}_{D_{U,T}}\to\mathcal{F}_{D_{U,T}(1)}\otimes_{\mathcal{O}_{D_{U,T}(1)}}\mathcal{O}_{\overline{D}_{U,T}(1)}/(p\text{-tor})$ and the isomorphisms $\mathcal{F}_{D_{U,T}}\cong\mathcal{F}_{D_T}\cong L_Y(\mathcal{F}_D)_T$. Hence the commutativity of the diagram for $q=0$ follows from the compatibility of the natural morphisms $D_{U,T}(1)\to D_U(1)$ and $D_{U,T}\to D_U$ with the projections $p_i\colon D_{U,T}(1)\to D_{U,T}$ and $D_U(1)\to D_U$ for $i=1,2$. □

Next we prove a kind of Poincaré lemma. Put $B'=D^{r+1}_{\mathrm{Higgs}}(B)$, $X'=X\times_{B_1}B'_1$, and $Y'=Y\times_B B'$. We have a $B'_1$-immersion $X'\to Y'_1$ induced by the $B_1$-immersion $X\to Y_1$. Let $\mu\colon (X'/B')^{r+1}_{\mathrm{HIGGS}}\to(X/B)^r_{\mathrm{HIGGS}}$ denote the morphism of topos defined by the natural morphisms $X'\to X$ and $B'\to B$. Let $D'$ be $D^{r+1}_{\mathrm{Higgs}}(X'\hookrightarrow Y')$ and let $\mu_D\colon D'\to D$ be the morphism induced by $X'\to X$ and $Y'\to Y$.

For any object $(T,z)$ of $(X'/B')^{r+1}_{\mathrm{HIGGS}}$, define $D'_T$ to be $D^{r+1}_{\mathrm{Higgs}}(T_1\hookrightarrow T\times_{B'}Y')$ as in the definition of linearization. Regarding $(T,z)$ as an object of $(X/B)^r_{\mathrm{HIGGS}}$, we may also define $D_T$, and we have a natural morphism $\mu_T\colon D'_T\to D_T$ compatible with $\mu_D\colon D'\to D$. Hence for a sheaf of $\mathcal{O}_{D_1,\mathbb{Q}}$-modules $\mathcal{G}$, the morphism $\mu_T$ induces a morphism $\Gamma(D_{T,1},p^*_{D,1}(\mathcal{G}))\to\Gamma(D'_{T,1},p^*_{D',1}(\mu_D^*(\mathcal{G})))$. This construction is functorial in $(T,z)$. Hence varying $(T,z)$, we obtain a canonical morphism

$$\mu^*(L_Y(\mathcal{G}))\longrightarrow L_{Y'}(\mu_D^*(\mathcal{G})).\qquad\text{(IV.4.3.3)}$$

**Proposition IV.4.3.4.** *Let $\mathcal{F}$ be a Higgs isocrystal on $(X/B)^r_{\mathrm{HIGGS}}$, and let $\mathcal{F}'$ be the pull-back $\mu^*(\mathcal{F})$ on $(X'/B')^{r+1}_{\mathrm{HIGGS}}$. Then the morphisms* (IV.4.3.3) *for $\xi^{-q}\mathcal{F}_D\otimes_{\mathcal{O}_{Y_1}}\Omega^q_{Y_1/B_1}$ ($q\in\mathbb{N}$) define a morphism of complexes*

$$\begin{array}{ccccccccc}\mu^*\mathcal{F} & \to & \mu^*(L_Y(\mathcal{F}_D)) & \to & \mu^*(L_Y(\xi^{-1}\mathcal{F}_D\otimes\Omega^1_{Y_1/B_1})) & \to & \mu^*(L_Y(\xi^{-2}\mathcal{F}_D\otimes\Omega^2_{Y_1/B_1})) & \to & \cdots\\ \| & & \downarrow & & \downarrow & & \downarrow & & \\ \mathcal{F}' & \to & L_{Y'}(\mathcal{F}'_{D'}) & \to & L_{Y'}(\xi^{-1}\mathcal{F}'_{D'}\otimes\Omega^1_{Y'_1/B'_1}) & \to & L_{Y'}(\xi^{-2}\mathcal{F}'_{D'}\otimes\Omega^2_{Y'_1/B'_1}) & \to & \cdots,\end{array}$$

*where $\otimes$ in the upper (resp. lower) complex denotes $\otimes_{\mathcal{O}_{Y_1}}$ (resp. $\otimes_{\mathcal{O}_{Y'_1}}$). Furthermore the morphism induces a zero morphism on each $\mathcal{H}^q$ ($q\in\mathbb{Z}$).*

PROOF. Let $z_D$, $z_{D'}$, and $\mu_X$ denote the natural morphisms $D_1 \to X$, $D'_1 \to X'$, and $X' \to X$, respectively. Then we have isomorphisms $\mu_D^*(\mathcal{F}_{(D,z_D)}) \cong \mathcal{F}_{(D',\mu_X\circ z_{D'})} = \mathcal{F}'_{(D',z_{D'})}$ and $\mathcal{O}_{Y'_1}\otimes_{\mathcal{O}_{Y_1}}\Omega^q_{Y_1/B_1} \xrightarrow{\cong} \Omega^q_{Y'_1/B'_1}$, which induce an isomorphism $\mathcal{F}'_{(D',z_{D'})}\otimes_{\mathcal{O}_{Y'_1}}\Omega^q_{Y'_1/B'_1} \cong \mu_D^*(\mathcal{F}_{(D,z_D)}\otimes_{\mathcal{O}_{Y_1}}\Omega^q_{Y_1/B_1})$. Hence (IV.4.3.3) defines the vertical morphisms.

Let $(T,z)$ be an object of $(X'/B')^{r+1}_{\text{HIGGS}}$, let $Y_T(\nu)$ ($\nu\in\mathbb{N}$) be the fiber product of $(\nu+1)$ copies of $Y\times_B T = Y'\times_{B'} T$ over $T$, and let $D_T(\nu)$ (resp. $D'_T(\nu)$) be $D^s_{\text{Higgs}}(T_1\hookrightarrow Y_T(\nu))$ for $s=r$ (resp. $s=r+1$). Then we have a morphism $D'_T(\bullet)\to D_T(\bullet)$ of simplicial objects of $\mathscr{C}^r$ compatible with the morphisms to $Y_T(\bullet)$. Hence, by the definition of $\theta$ and $\theta^q$, the following diagrams are commutative.

$$\begin{array}{ccc}
\mu_{T,1}^{-1}(\xi^{-q}\mathcal{O}_{D_{T,1},\mathbb{Q}}\otimes_{\mathcal{O}_{Y_1}}\Omega^q_{Y_1/B_1}) & \xrightarrow{\theta^q} & \mu_{T,1}^{-1}(\xi^{-q-1}\mathcal{O}_{D_{T,1},\mathbb{Q}}\otimes_{\mathcal{O}_{Y_1}}\Omega^{q+1}_{Y_1/B_1}) \\
\downarrow & & \downarrow \\
\xi^{-q}\mathcal{O}_{D'_{T,1},\mathbb{Q}}\otimes_{\mathcal{O}_{Y'_1}}\Omega^q_{Y'_1/B'_1} & \xrightarrow{\theta^q} & \xi^{-q-1}\mathcal{O}_{D'_{T,1},\mathbb{Q}}\otimes_{\mathcal{O}_{Y'_1}}\Omega^{q+1}_{Y'_1/B'_1}
\end{array}$$

By taking $p'^{-1}_{T,1}(\mathcal{F}_T)\otimes_{p'^{-1}_{T,1}(\mathcal{O}_T)}$, where $p'_T$ denotes the morphism $D'_T\to T$, we see that $\Gamma((T,z),-)$ of the diagram in the proposition is commutative.

Let us prove the second claim. Choose a strict étale covering $(u_\alpha\colon X_\alpha\to X)_{\alpha\in A}$ for $\mathcal{F}$ as in Definition IV.3.3.1 (ii). Let $u\colon U\to X$ be a strict étale morphism satisfying the following conditions: The morphism $u$ factors through $u_\alpha$ for some $\alpha\in A$ and there exist a strict étale Cartesian lifting $Y_U\to Y$ of $u$ and $t_i=(t_{i,N})\in\varprojlim_N\Gamma(Y_{U,N},M_{Y_{U,N}})$ $(1\le i\le d)$ such that $d\log t_{i,N}$ $(1\le i\le d)$ form a basis of $\Omega^1_{Y_{U,N}/B_N}$ for every $N\in\mathbb{N}_{>0}$. Then, for $(T,z)\in\text{Ob}\,(U'/B')^{r+1}_{\text{HIGGS}}$, $U'=U\times_B B'$, the natural morphisms $D^s_{\text{Higgs}}(T_1\hookrightarrow Y_U(\nu)\times_B T)\to D^s_{\text{Higgs}}(T_1\hookrightarrow Y(\nu)\times_B T)$ for $s=r,r+1$ are isomorphisms by Lemma IV.2.2.15. Hence we may replace $X$, $Y$, and $\mathcal{F}$ by $U$, $Y_U$, and $u^*_{\text{HIGGS}}\mathcal{F}$. We will prove that $\Gamma(T,-)$ of the morphism in the proposition is homotopic to 0 by a $\Gamma(T_1,\mathcal{O}_{T_1})$-linear homotopy when $T_1$ is affine. Suppose that $T_1$ is affine, which implies that $D_{T,1}$ and $D'_{T,1}$ are affine by Proposition IV.2.2.9 (2). Put $A_1=\Gamma(T_1,\mathcal{O}_{T_1})$, $C=\Gamma(D_{T,1},\mathcal{O}_{D_{T,1}})$, $C'=\Gamma(D'_{T,1},\mathcal{O}_{D'_{T,1}})$, and $M=\Gamma(T,\mathcal{F}_T)$. Then, we have $\Gamma(D_{T,1},p^*_{T,1}(\mathcal{F}_T))=C\otimes_{A_1}M$ and $\Gamma(D'_{T,1},p'^*_{T,1}(\mathcal{F}_T))=C'\otimes_{A_1}M$ since $\mathcal{F}_T$ is a finitely generated projective $\mathcal{O}_{T_1,\mathbb{Q}}$-module. Hence the claim is reduced to the case $\mathcal{F}=\mathcal{O}_{X/B,1,\mathbb{Q}}$. Choose a lifting $(s_{i,N})\in\varprojlim_N\Gamma(T_N,M_{T_N})$ of the image of $t_{i,1}$ under $\Gamma(Y_1,M_{Y_1})\to\Gamma(X,M_X)\to\Gamma(T_1,M_{T_1})$. Then, by Proposition IV.2.3.17, we obtain isomorphisms $C\cong A_1\{W_1,\ldots,W_d\}_r$ and $C'\cong A_1\{W_1,\ldots,W_d\}_{r+1}$ compatible with the natural homomorphism $C\to C'$. With this description, $\theta$ is characterized by $\theta(W_i)=\xi^{-1}d\log t_{i,1}$ $(1\le i\le d)$. Put $\omega_i=\xi^{-1}d\log t_{i,1}$. Then, by Lemma IV.2.3.14, we can define the desired $A_{1,\mathbb{Q}}$-linear homotopy by

$$k^0\left(\sum_{\underline{m}\in\mathbb{N}^d}a_{\underline{m}}\underline{W}^{\underline{m}}\right)=a_0$$

for $a_{\underline{m}}\in A_{1,\mathbb{Q}}$ such that $p^{-[\frac{|\underline{m}|}{r}]}a_{\underline{m}}\to 0$ as $|\underline{m}|\to 0$, and

$$k^q\left(\sum_{\underline{m}\in\mathbb{N}^d}a_{\underline{m}}\underline{W}^{\underline{m}}\omega_{i_1}\wedge\cdots\wedge\omega_{i_q}\right)=\sum_{\substack{\underline{m}\in\mathbb{N}^d\\ m_i=0\text{ for }i<i_1}}\frac{1}{m_{i_1}+1}a_{\underline{m}}\underline{W}^{\underline{m}+\underline{1}_{i_1}}\omega_{i_2}\wedge\cdots\wedge\omega_{i_q}$$

for $q>0$, $1\le i_1<\ldots<i_q\le d$, and $a_{\underline{m}}\in A_{1,\mathbb{Q}}$ such that $p^{-[\frac{|\underline{m}|}{r}]}a_{\underline{m}}\to 0$. Note $p^{-[\frac{|\underline{m}|+1}{r+1}]}\frac{1}{m_{i_1}+1}a_{\underline{m}}\to 0$ as $|\underline{m}|\to 0$. $\square$

**IV.4.4. Coherence.** In this subsection, we study coherence of topos $(X/B)^{r\sim}_{\mathrm{HIGGS}}$, $X^{\sim}_{\text{ét}}$, $X^{\sim}_{\text{ÉT}}$ and morphisms between them, which will be used to describe the cohomology of the inverse limit of the sites $(X/B)^r_{\mathrm{HIGGS}}$ ($r \in \mathbb{N}_{>0}$) as the direct limit of the cohomology of each $(X/B)^r_{\mathrm{HIGGS}}$ in IV.4.5.

We say that a morphism of $p$-adic fine log formal schemes $f\colon X \to Y$ is quasi-compact (resp. quasi-separated) if the morphism of schemes underlying the reduction mod $p^m$ of $f$ is quasi-compact (resp. quasi-separated) for every $m \in \mathbb{N}_{>0}$. We say that $f$ is coherent if it is quasi-compact and quasi-separated. Quasi-compact (resp. quasi-separated) morphisms of $p$-adic fine log formal schemes are stable under compositions and base changes (cf. **[42]** Proposition (1.1.2) (ii), (iii), Proposition (1.2.2) (ii), (iii)). Note that for a scheme with a coherent log structure $X$, the morphism of schemes underlying $X^{\mathrm{int}} \to X$ is a closed immersion, and hence quasi-compact and quasi-separated. Let $f\colon X \to Y$ and $g\colon Y \to Z$ be two composable morphisms of $p$-adic fine log formal schemes. If $g \circ f$ is quasi-compact and $g$ is quasi-separated, then $f$ is quasi-compact (cf. **[42]** Proposition (1.2.4)). If $g \circ f$ is quasi-separated, then $f$ is quasi-separated (cf. **[42]** Proposition (1.2.2) (v)). In particular, if $X$ is an affine $p$-adic fine log formal scheme, then any morphism $X \to Y$ of $p$-adic fine log formal schemes is quasi-separated.

We say that a morphism $f\colon Y' \to Y$ in the category $\mathscr{C}$ is quasi-compact (resp. quasi-separated, resp. coherent) if the morphism of $p$-adic fine log formal schemes $f_N\colon Y'_N \to Y_N$ is quasi-compact (resp. quasi-separated, resp. coherent) for every $N \in \mathbb{N}_{>0}$. Since the morphisms of schemes underlying the reduction mod $p$ of $Y_2 \to Y_N$ and $Y'_2 \to Y'_N$ are nilpotent immersions, the condition for $N = 2$ implies that for every $N \geq 2$. Similarly, if $f$ is a morphism in the subcategory $\mathscr{C}^r$, then the condition for $N = 1$ implies the condition for every $N \in \mathbb{N}_{>0}$ (cf. Corollary IV.2.2.5).

Let $r \in \mathbb{N}_{>0} \cup \{\infty\}$, let $B \in \mathrm{Ob}\,\mathscr{C}^r$, and let $X$ be a $p$-adic fine log formal scheme over $B_1$. We define $(X/B)^r_{\mathrm{HIGGS,coh}}$ to be the full subcategory of $(X/B)^r_{\mathrm{HIGGS}}$ consisting of $(T, z)$ such that the structure morphism $T \to B$ is coherent. We endow $(X/B)^r_{\mathrm{HIGGS,coh}}$ with the topology induced by that of $(X/B)^r_{\mathrm{HIGGS}}$.

**Lemma IV.4.4.1.** *The full subcategory $(X/B)^r_{\mathrm{HIGGS,coh}}$ of $(X/B)^r_{\mathrm{HIGGS}}$ is stable under finite fiber products. If the morphism of $p$-adic fine log formal schemes $X \to B_1$ is quasi-separated, then it is also stable under finite products.*

PROOF. By the proofs of Proposition IV.3.1.2 (1) and Corollary IV.2.2.11, the fiber product of $(T', z') \to (T, z) \leftarrow (T'', z'')$ in $(X/B)^r_{\mathrm{HIGGS}}$ is represented by $T''' = D^r_{\mathrm{Higgs}}(T' \times_T T'')$. Suppose that the morphisms $T, T', T'' \to B$ are coherent. Then the morphism $T' \to T$ and its base change $T' \times_T T'' \to T''$ are coherent. Since $T''' \to T' \times_T T''$ is affine by Proposition IV.2.2.9 (2), the composition $T''' \to T' \times_T T'' \to T'' \to B$ is coherent. By the proof of Proposition IV.3.1.2 (1), the product of $(T, z)$ and $(T', z') \in \mathrm{Ob}\,(X/B)^r_{\mathrm{HIGGS}}$ in $(X/B)^r_{\mathrm{HIGGS}}$ is represented by $T'' = D^r_{\mathrm{Higgs}}(T_1 \times_X T'_1 \hookrightarrow T \times_B T')$. Suppose that $T, T' \to B$ are coherent and $X \to B_1$ is quasi-separated. Then $z\colon T_1 \to X$ and its base change $T_1 \times_X T'_1 \to T'_1$ are coherent. Since $T''_1 \to T_1 \times_X T'_1$ is affine by Proposition IV.2.2.9 (2), the composition of $T''_1 \to T_1 \times_X T'_1 \to T'_1 \to B_1$ is coherent. □

For an object $(T, z)$ of $(X/B)^r_{\mathrm{HIGGS,coh}}$, we define $\mathrm{Cov}_{\mathrm{coh}}((T, z))$ to be

$$\{(u_\alpha\colon (T_\alpha, z_\alpha) \to (T, z))_{\alpha \in A} \in \mathrm{Cov}((T, z)) | (T_\alpha, z_\alpha) \in \mathrm{Ob}\,(X/B)^r_{\mathrm{HIGGS,coh}} \text{ for all } \alpha \in A\}.$$

**Lemma IV.4.4.2.** *Assume that the morphism $B \to S$ is quasi-separated.*

(1) *An object $(T, z)$ of $(X/B)^r_{\mathrm{HIGGS}}$ such that $T_1$ is affine belongs to $(X/B)^r_{\mathrm{HIGGS,coh}}$.*

(2) *The sets $\mathrm{Cov}_{\mathrm{coh}}((T, z))$ for $(T, z) \in \mathrm{Ob}\,(X/B)^r_{\mathrm{HIGGS,coh}}$ satisfy the axiom of pretopology, and the topology of $(X/B)^r_{\mathrm{HIGGS,coh}}$ is induced by this pretopology.*

(3) *The continuous functor* $(X/B)^r_{\mathrm{HIGGS,coh}} \to (X/B)^r_{\mathrm{HIGGS}}$ *induces an equivalence of categories* $(X/B)^{r\sim}_{\mathrm{HIGGS}} \xrightarrow{\sim} (X/B)^{r\sim}_{\mathrm{HIGGS,coh}}$.

PROOF. (1) If $T_1$ is affine, then $T_N$ are affine and the morphism $T \to S$ is coherent. Hence, if $B \to S$ is quasi-separated, the morphism $T \to B$ is coherent.

(2) The first claim follows from the fact that the subcategory $(X/B)^r_{\mathrm{HIGGS,coh}}$ is stable under fiber products (Lemma IV.4.4.1). By **[2]** III Corollaire 3.3, it also implies that a family of morphisms $\mathcal{U} = (u_\alpha\colon (T_\alpha, z_\alpha) \to (T, z))_{\alpha\in A}$ in $(X/B)^r_{\mathrm{HIGGS,coh}}$ is a covering if and only if there exists a refinement $\mathcal{U}' \in \mathrm{Cov}((T, z))$ of $\mathcal{U}$. By (1), every $\mathcal{V} \in \mathrm{Cov}((T, z))$ admits a refinement by a $\mathcal{V}' \in \mathrm{Cov}_{\mathrm{coh}}((T, z))$. Hence, we may replace $\mathrm{Cov}((T, z))$ with $\mathrm{Cov}_{\mathrm{coh}}((T, z))$ in the above equivalence.

(3) This follows from (1) and **[2]** III Théorème 4.1. □

**Proposition IV.4.4.3.** *Assume that the morphism* $B \to S$ *is coherent, and the morphism* $X \to B_1$ *is quasi-separated.*

(1) *The topos* $(X/B)^{r\sim}_{\mathrm{HIGGS}}$ *is algebraic* (**[2]** VI *Définition* 2.3).

(2) *If the morphism* $X \to B_1$ *is quasi-compact, then the topos* $(X/B)^{r\sim}_{\mathrm{HIGGS}}$ *is coherent.*

PROOF. Put $C = (X/B)^r_{\mathrm{HIGGS,coh}}$ to simplify the notation. By Lemma IV.4.4.2 (3) and the assumption on $B \to S$, it suffices to prove the claims for $C^\sim$. Let $\varepsilon$ denote the canonical functor $C \to C^\sim$.

(1) For any $(T, z) \in \mathrm{Ob}\, C$, the composition of $T_1 \to B_1 \to S_1$ is quasi-compact by the assumption on $B \to S$. Since $S_1$ is quasi-compact, $T_1$ is quasi-compact, which implies that $(T, z)$ is a quasi-compact object of the site $C$ (**[2]** VI Définition 1.1) by Lemma IV.4.4.2 (2). By Lemma IV.4.4.1 and Proposition IV.3.1.2 (1), fiber products are representable in $C$. Hence, by **[2]** VI Corollaire 2.1.1, $\varepsilon(T)$ is a coherent object of $C^\sim$ for any $T \in \mathrm{Ob}\, C$. By Lemma IV.4.4.1, Proposition IV.3.1.2 (1) and the assumption on $X \to B_1$, we see that $T \times T'$ are representable in $C$ for any $T, T' \in \mathrm{Ob}\, C$, which implies that $\varepsilon(T)\times\varepsilon(T') = \varepsilon(T\times T')$ is coherent. Now the claim follows from one of the defining properties **[2]** VI Proposition 2.2 (ii bis) of an algebraic topos.

(2) By (1), it suffices to prove that the final object $e$ of $C^\sim$ is coherent (cf. **[2]** VI Définition 2.3). By the proof of (1), $\varepsilon(T)$ and $\varepsilon(T) \times \varepsilon(T')$ are coherent for any $T$, $T' \in \mathrm{Ob}\, C$. Hence, by **[2]** VI Corollaire 1.17, it suffices to prove that there exists a covering of $e$ of the form $(\varepsilon(T_\alpha) \to e)_{\alpha\in A}$ such that $T_\alpha \in \mathrm{Ob}\, C$ and $A$ is a finite set.

Since $X \to B_1$, $B_1 \to S_1$, and $S_1$ are quasi-compact, $X$ is quasi-compact. Hence there exists a strict étale covering $(X_\alpha \to X)_{\alpha\in A}$ such that $A$ is a finite set, $X_\alpha$ is affine, and there exists a chart $P_\alpha \to \Gamma(\overline{X}_\alpha, M_{\overline{X}_\alpha})$ of $M_{\overline{X}_\alpha}$, where $\overline{X}_\alpha = X_\alpha \times_{\mathrm{Spf}(\mathbb{Z}_p)} \mathrm{Spec}(\mathbb{Z}/p\mathbb{Z})$. Choose a surjective morphism of monoids $h_\alpha\colon \mathbb{N}^{d_\alpha} \to P_\alpha$ for some $d_\alpha \in \mathbb{N}$, and a surjective homomorphism of $R_1/pR_1$-algebras $f_\alpha\colon R_1/pR_1[T_\lambda; \lambda \in \Lambda_\alpha] \to \Gamma(\overline{X}_\alpha, \mathcal{O}_{\overline{X}_\alpha})$. Let $Y_{\alpha,N}$ (resp. $Z_{\alpha,N}$) be $\mathrm{Spf}(R_N[\mathbb{N}^{d_\alpha}]^\wedge)$ (resp. $\mathrm{Spf}(R_N[\mathbb{N}^{d_\alpha}][T_\lambda; \lambda \in \Lambda_\alpha]^\wedge)$) endowed with the log structure naturally defined by $\mathbb{N}^{d_\alpha}$. Here $^\wedge$ denotes the $p$-adic completion. Then $Y_{\alpha,N}$ and $Z_{\alpha,N}$ naturally define objects $Y$ and $Z$ of $\mathscr{C}$ with morphisms $Z \to Y \to S$ in $\mathscr{C}$. The morphism $Y \to S$ is smooth and Cartesian. The morphisms $h_\alpha$ and $f_\alpha$ induce an $S_1$-closed immersion $\overline{X}_\alpha \to Z_{\alpha,1}$ over $S_1$, which is lifted to an $S_1$-closed immersion $i_\alpha\colon X_\alpha \to Z_\alpha$ because $Y_{\alpha,1} \to S_1$ is smooth. Let $(T, z)$ be an object of $C$ such that $T_1$ is affine (and hence $T_N$ are affine), and suppose that there exist $\alpha \in A$ and an $X$-morphism $v\colon T_1 \to X_\alpha$. Then, by Sublemma IV.3.4.2 (2), we see that the morphism $T_1 \to X_\alpha \to Y_{1,\alpha}$ over $S_1$ has a lifting $T \to Y_\alpha$ over $S$, which obviously admits a factorization $T \to Z_\alpha \to Y_\alpha$ lifting $T_1 \xrightarrow{i_\alpha\circ v} Z_{\alpha,1} \to Y_{\alpha,1}$. Thus we obtain a morphism $T \to (X_\alpha \hookrightarrow Z_\alpha \times_S B)$ over $B$ compatible with $v\colon T_1 \to X_\alpha$. Hence

$T_\alpha = D^r_{\mathrm{Higgs}}(X_\alpha \hookrightarrow Z_\alpha \times_S B)$ with the natural morphism $z_\alpha\colon T_{\alpha,1} \to X$ gives the desired covering of $e$. Note that $T_{\alpha,1}$ is affine since $T_{\alpha,1} \to X_\alpha$ is affine by Proposition IV.2.2.9 (2), which implies that $(T_\alpha, z_\alpha)$ is an object of $C$ by Lemma IV.4.4.2 (1). □

**Proposition IV.4.4.4.** *Let the notation and assumption be the same as in Proposition* IV.3.1.6. *If the morphisms $B, B' \to S$ are coherent, then the morphism of topos $f_{\mathrm{HIGGS}}\colon (X'/B')^{r'\sim}_{\mathrm{HIGGS}} \to (X/B)^{r\sim}_{\mathrm{HIGGS}}$ is coherent.*

PROOF. For an object $(T,z)$ of $(X/B)^r_{\mathrm{HIGGS,coh}}$, the morphism $D^{r'}_{\mathrm{Higgs}}(T \times_B B') \to T \times_B B'$ is affine, in particular, coherent by Proposition IV.2.2.9 (2). Hence the functor $f^*$ in Proposition IV.3.1.6 induces a functor $f^*_{\mathrm{coh}}\colon (X/B)^r_{\mathrm{HIGGS,coh}} \to (X'/B')^{r'}_{\mathrm{HIGGS,coh}}$. For a morphism $u\colon (T',z') \to (T,z)$ in $(X/B)^r_{\mathrm{HIGGS,coh}}$ such that $u$ is strict étale, the natural morphism $D^{r'}_{\mathrm{HIGGS}}(T' \times_B B') \to D^{r'}_{\mathrm{HIGGS}}(T \times_B B') \times_T T'$ is an isomorphism by Proposition IV.2.2.9 (3). Hence $f^*_{\mathrm{coh}}$ is continuous by Lemma IV.4.4.2 (2). By Lemma IV.4.4.2 (3) and Proposition IV.3.1.6, the functor $f^*_{\mathrm{coh}}$ is a morphism of site, and it suffices to prove that the morphism of topos associated to $f^*_{\mathrm{coh}}$ is coherent. By Lemma IV.4.4.1 and Proposition IV.3.1.2 (1), fiber products are representable in $(X/B)^r_{\mathrm{HIGGS,coh}}$ and $(X'/B')^{r'}_{\mathrm{HIGGS,coh}}$. On the other hand, for every object $(T,z)$ of $(X/B)^r_{\mathrm{HIGGS,coh}}$ (resp. $(X'/B')^{r'}_{\mathrm{HIGGS,coh}}$), $T_1$ is quasi-compact because $B_1 \to S_1$ (resp. $B'_1 \to S_1$) and $S_1$ are quasi-compact. Hence $T$ is a quasi-compact object. Now the claim follows from **[2]** VI Corollaire 3.3. □

Next we discuss coherence of projections to étale sites.

Let $X$ be a $p$-adic fine log formal scheme. We define $X_{\text{ét,coh}}$ (resp. $X_{\text{ÉT,coh}}$) to be the full subcategory of $X_{\text{ét}}$ (resp. $X_{\text{ÉT}}$) consisting of $u\colon U \to X$ such that $U$ is quasi-compact and $u$ is coherent. We endow $X_{\text{ét,coh}}$ (resp. $X_{\text{ÉT,coh}}$) with the topology induced by that of $X_{\text{ét}}$ (resp. $X_{\text{ÉT}}$).

**Lemma IV.4.4.5.** *The full subcategory $X_{\text{ét,coh}}$ of $X_{\text{ét}}$ is stable under finite fiber products and nonempty finite products, and the full subcategory $X_{\text{ÉT,coh}}$ of $X_{\text{ÉT}}$ is stable under finite fiber products.*

PROOF. For morphisms $U' \to U \leftarrow U''$ in $X_{\text{ét,coh}}$ (resp. $X_{\text{ÉT,coh}}$), the morphism $U' \to U$ and its base change $U' \times_U U'' \to U''$ are coherent because $U \to X$ is quasi-separated and $U' \to X$ is coherent. Since $U''$ is quasi-compact and $U'' \to X$ is coherent, this implies that $U' \times_U U''$ is quasi-compact and $U' \times_U U'' \to X$ is coherent. For two objects $U$ and $U'$ of $X_{\text{ét,coh}}$, the morphism $U \times_X U' \to U$ is coherent since $U' \to X$ is coherent. Hence $U \times_X U'$ is quasi-compact and $U \times_X U' \to X$ is coherent because $U$ is quasi-compact and $U \to X$ is coherent. □

**Lemma IV.4.4.6.** *Assume that $X \to \mathrm{Spf}(\mathbb{Z}_p)$ is quasi-separated. Let $\star$ denote* ét *or* ÉT.

(1) *The sets*

$$\mathrm{Cov}_{\mathrm{coh}}(U) = \left\{ (U_\alpha \to U)_{\alpha\in A} \;\middle|\; \begin{array}{l} \text{(i) } (U_\alpha \to U)_{\alpha \in A} \textit{ is a strict étale covering.} \\ \text{(ii) } U_\alpha \in \mathrm{Ob}\, X_{\star,\mathrm{coh}}. \end{array} \right\}$$

*for $U \in \mathrm{Ob}\, X_{\star,\mathrm{coh}}$ satisfy the axiom of pretopology, and it induces the topology of $X_{\star,\mathrm{coh}}$.*

(2) *The natural morphism of topos $X^\sim_\star \to X^\sim_{\star,\mathrm{coh}}$ is an equivalence of categories.*

(3) *Any object of $X^\sim_{\star,\mathrm{coh}}$ represented by an object of $X_{\star,\mathrm{coh}}$ is coherent.*

(4) *The topos $X^\sim_\star$ is algebraic. If $X$ is quasi-compact, then the topos $X^\sim_\star$ is coherent.*

(5) *The canonical morphism of topos $X^\sim_{\text{ÉT}} \to X^\sim_{\text{ét}}$ is coherent.*

PROOF. (1) (2) An object $U \to X$ of $X_\star$ with $U$ affine is an object of $X_{\star,\mathrm{coh}}$ by the assumption on $X \to \mathrm{Spf}(\mathbb{Z}_p)$. Hence one can prove the claims in the same way as the proof of Lemma IV.4.4.2 (2) and (3) by using Lemma IV.4.4.5 and **[2]** III Corollaire 3.3, Théorème 4.1.

(3) By (1) and the definition of $X_{\star,\mathrm{coh}}$, every object of $X_{\star,\mathrm{coh}}$ is quasi-compact. Hence the claim follows from Lemma IV.4.4.5 and **[2]** VI Corollaire 2.1.1.

(4) By (2), it suffices to prove the claims for $X^{\sim}_{\star,\mathrm{coh}}$. Let $\varepsilon$ denote the canonical functor $X_{\star,\mathrm{coh}} \to X^{\sim}_{\star,\mathrm{coh}}$. Then, by (3) and Lemma IV.4.4.5, $\varepsilon(U)$ and $\varepsilon(U) \times \varepsilon(U') = \varepsilon(U \times_X U')$ are coherent for any $U, U' \in \mathrm{Ob}\, X_{\star,\mathrm{coh}}$. Hence $X^{\sim}_{\star,\mathrm{coh}}$ is algebraic by the defining property **[2]** VI Proposition 2.2 (ii bis) of algebraic topos. If $X$ is quasi-compact, then $X \in \mathrm{Ob}\, X_{\star,\mathrm{coh}}$, which implies that the final object $\varepsilon(X)$ of $X^{\sim}_{\star,\mathrm{coh}}$ is coherent.

(5) The inclusion functor $\iota\colon X_{\text{ét}} \to X_{\text{ÉT}}$ induces a functor $\iota_{\mathrm{coh}}\colon X_{\text{ét},\mathrm{coh}} \to X_{\text{ÉT},\mathrm{coh}}$. By (1), the functor $\iota_{\mathrm{coh}}$ is continuous. By Lemma IV.4.4.5, finite inverse limits are representable in $X_{\text{ét},\mathrm{coh}}$ and the functor $\iota_{\mathrm{coh}}$ is left exact. Hence $\iota_{\mathrm{coh}}$ is a morphism of site (cf. **[2]** I Proposition 5.4 4), III Proposition 1.3 5)). By (1), every object of $X_{\text{ét},\mathrm{coh}}$ and $X_{\text{ÉT},\mathrm{coh}}$ is quasi-compact. Hence the claim follows from (2), Lemma IV.4.4.5, and **[2]** VI Corollaire 3.3. □

**Lemma IV.4.4.7.** *Let $f\colon X' \to X$ be a morphism of $p$-adic fine log formal schemes. If $X, X' \to \mathrm{Spf}(\mathbb{Z}_p)$ are quasi-separated and $f$ is quasi-compact, then the morphisms of topos $f_{\text{ét}}\colon X'^{\sim}_{\text{ét}} \to X^{\sim}_{\text{ét}}$ and $f_{\text{ÉT}}\colon X'^{\sim}_{\text{ÉT}} \to X^{\sim}_{\text{ÉT}}$ are coherent.*

PROOF. Let $\star$ denote ét or ÉT. Since $f$ is quasi-compact, the functor $f^*\colon X_\star \to X'_\star; U \mapsto U \times_X X'$ induces a functor $f^*_{\mathrm{coh}}\colon X_{\star,\mathrm{coh}} \to X'_{\star,\mathrm{coh}}$. By Lemma IV.4.4.6 (1), the functor $f^*_{\mathrm{coh}}$ is continuous. Hence it is a morphism of sites by Lemma IV.4.4.6 (2). Since every object of $X_{\star,\mathrm{coh}}$ and $X'_{\star,\mathrm{coh}}$ is quasi-compact by Lemma IV.4.4.6 (1), the claim follows from Lemma IV.4.4.6 (2), Lemma IV.4.4.5, and **[2]** VI Corollaire 3.3. □

**Proposition IV.4.4.8.** *Let $r \in \mathbb{N}_{>0} \cup \{\infty\}$, let $B \in \mathrm{Ob}\,\mathscr{C}^r$, and let $X$ be a $p$-adic fine log formal scheme over $B_1$. Assume that $X \to B_1$ is quasi-separated and $B \to S$ is coherent. Then the morphism of topos $U_{X/B}\colon (X/B)^{r\sim}_{\mathrm{HIGGS}} \to X^{\sim}_{\text{ÉT}}$ is coherent.*

PROOF. Let $\varepsilon_{X/B}$ and $\varepsilon_X$ denote the canonical functors $(X/B)^r_{\mathrm{HIGGS}} \to (X/B)^{r\sim}_{\mathrm{HIGGS}}$ and $X_{\text{ÉT}} \to X^{\sim}_{\text{ÉT}}$, respectively. By Lemma IV.4.4.2 (3) and the proof of Proposition IV.4.4.3 (1), the objects $\varepsilon_{X/B}(T,z)$, $(T,z) \in \mathrm{Ob}\,(X/B)^r_{\mathrm{HIGGS,coh}}$ are coherent and generate $(X/B)^{r\sim}_{\mathrm{HIGGS}}$. Similarly, by Lemma IV.4.4.6 (2) and (3), the objects $\varepsilon_X(U)$, $U \in \mathrm{Ob}\, X_{\text{ÉT},\mathrm{coh}}$ are coherent and generate $X^{\sim}_{\text{ÉT}}$. Hence, by **[2]** VI Proposition 3.2, it suffices to prove that $U^*_{X/B}(\varepsilon_X(U))$ is coherent for $U \in \mathrm{Ob}\, X_{\text{ÉT},\mathrm{coh}}$.

We have $\Gamma((T,z), U^*_{X/B}(\varepsilon_X(U))) = \Gamma(T_1, \varepsilon_X(U)) = \mathrm{Hom}_X(T_1, U)$ for an object $(T,z)$ of $(X/B)^r_{\mathrm{HIGGS}}$ (cf. **[17]** IV Proposition 6.3.1 (iii)). Since $U$ is quasi-compact, there exists a strict étale covering $(u_\alpha\colon U_\alpha \to U)_{\alpha\in A}$ such that $U_\alpha$ is affine, $A$ is a finite set, and there exists a chart $P_\alpha \to \Gamma(\overline{U}_\alpha, M_{\overline{U}_\alpha})$, where $\overline{U}_\alpha = U_\alpha \times_{\mathrm{Spf}(\mathbb{Z}_p)} \mathrm{Spec}(\mathbb{Z}/p\mathbb{Z})$ for each $\alpha \in A$. As in the proof of Proposition IV.4.4.3 (2), we can construct $(T_\alpha, z_\alpha) \in \mathrm{Ob}\,((X/B)^r_{\mathrm{HIGGS,coh}})$ with a morphism $w_\alpha\colon T_{\alpha,1} \to U_\alpha$ over $X$ satisfying the following property: For any $(T,z) \in \mathrm{Ob}\,(X/B)^r_{\mathrm{HIGGS}}$ such that $T_1$ is affine and any morphism $w\colon T_1 \to U_\alpha$ over $X$, there exists a morphism $f\colon (T,z) \to (T_\alpha, z_\alpha)$ such that $z_\alpha \circ f_1 = w$. This property implies that the family of morphisms $(\varepsilon_{X/B}(T_\alpha, z_\alpha) \to U^*_{X/B}(\varepsilon_X(U)))_{\alpha\in A}$ defined by the composition $v_\alpha := u_\alpha \circ z_\alpha\colon T_{\alpha,1} \to U_\alpha \to U$ is a covering (cf. **[2]** II Proposition 4.3 2)). Let $(T_{\alpha\beta}, v_{\alpha\beta})$ be the product of $(T_\alpha, v_\alpha)$ and $(T_\beta, v_\beta)$ in the category $(U/B)^r_{\mathrm{HIGGS,coh}}$ (cf. Lemma IV.4.4.1, Proposition IV.3.1.2 (1)). Note that the composition of $U \to X \to B$ is quasi-separated because $U \to X$ is coherent. Let $z_{\alpha\beta}$ denote the composition of

$T_{\alpha\beta,1} \xrightarrow{v_{\alpha\beta}} U \to X$. Then we see that the object $(T_{\alpha\beta}, z_{\alpha\beta})$ represents the fiber product of $(T_\alpha, z_\alpha) \to U^*_{X/B}(\varepsilon_X(U)) \leftarrow (T_\beta, z_\beta)$ in the category of presheaves on $(X/B)^r_{\mathrm{HIGGS}}$ (cf. [2] I Proposition 5.11 and Proposition IV.3.1.9 (1)). Hence $\varepsilon_{X/B}(T_\alpha, z_\alpha) \times_{U^*_{X/B}(\varepsilon_X(U))} \varepsilon_{X/B}(T_\beta, z_\beta) = \varepsilon(T_{\alpha\beta}, z_{\alpha\beta})$ is coherent. Now the claim follows from [2] VI Corollaire 1.17. □

**IV.4.5. Cohomology.** Let $r_0 \in \mathbb{N}_{>0}$ and let $B \in \mathrm{Ob}\,\mathscr{C}^{r_0}$. For each integer $r \geq r_0$, let $B^r$ denote $D^r_{\mathrm{Higgs}}(B^{r_0})$. We have a sequence of morphisms

$$\cdots \to B^{r+1} \to B^r \to \cdots \to B^{r_0+1} \to B^{r_0} = B.$$

We assume that $B \to S$ is coherent. By Proposition IV.2.2.9 (2), the morphism $B^r \to B^{r_0}$ for $r \geq r_0$ is affine, in particular, coherent. Let $X \to B_1$ be a coherent morphism of $p$-adic fine log formal schemes. For an integer $r \geq r_0$, let $X^r$ be the base change $X \times_{B_1} B^r_1$.

Suppose that we are given a smooth Cartesian morphism $Y \to B$ in $\mathscr{C}$ and an immersion $X \to Y_1$ over $B_1$, and let $D^r$ $(r \geq r_0)$ denote the Higgs envelope of level $r$ of $X^r \to Y^r := Y \times_B B^r$. Then, for $r \in \mathbb{N}$, $r \geq r_0$ and a Higgs isocrystal $\mathcal{F}$ of level $r$ on $X^r/B^r$, one obtains an object $(\mathcal{F}^s_{D^s}, \theta)$ of $\mathrm{HB}^s_{\mathbb{Q}_p,\mathrm{conv}}(X^s, Y^s/B^s)$ associated to the pull-back $\mathcal{F}^s$ of $\mathcal{F}$ on $(X^s/B^s)^s_{\mathrm{HIGGS}}$ by the equivalence of categories in Theorem IV.3.4.16 for each $s \in \mathbb{N}$, $s \geq r$. In this subsection, we prove a comparison theorem (Theorem IV.4.5.6) between the cohomology of the pull-back of $\mathcal{F}$ on the inverse limit of the sites $(X^s/B^s)^s_{\mathrm{HIGGS}}$ $(s \geq r_0)$ and the complexes $\xi^{-\bullet}\mathcal{F}^s_{D^s} \otimes_{\mathcal{O}_{Y^s_1}} \Omega^\bullet_{Y^s_1/B^s_1}$ $(s \geq r)$ associated to $(\mathcal{F}^s_{D^s}, \theta)$ by Lemma IV.3.4.5

We define the category $(X^\bullet/B^\bullet)_{\mathrm{HIGGS}}$ as follows. An object is a pair $(r, (T, z))$ of an integer $r \geq r_0$ and an object $(T, z)$ of $(X^r/B^r)^r_{\mathrm{HIGGS}}$. For two objects $(r, (T, z))$ and $(r', (T', z'))$, there is no morphism $(r', (T', z')) \to (r, (T, z))$ if $r' < r$. If $r' \geq r$, a morphism $(r', (T', z')) \to (r, (T, z))$ is a morphism $f\colon T' \to T$ in $\mathscr{C}^r$ compatible with the morphisms $X^{r'} \to X^r$ and $B^{r'} \to B^r$.

Let $I^{r_0}$ denote the opposite category of the category associated to the ordered set $\{r \in \mathbb{N} | r \geq r_0\}$ (cf. IV.6.6). Then we have a natural functor $\pi\colon (X^\bullet/B^\bullet)_{\mathrm{HIGGS}} \to I^{r_0}$ defined by $((T, z), r) \mapsto r$. The fiber of $\pi$ over $r \in I^{r_0}$ can be identified with the category $(X^r/B^r)^r_{\mathrm{HIGGS}}$.

**Lemma IV.4.5.1.** *The pair $((X^\bullet/B^\bullet)_{\mathrm{HIGGS}}, \pi)$ is a fibered site over $I^{r_0}$.*

Proof. Let $r, r' \in I^{r_0}$ such that $r' \geq r$ and let $m\colon r' \to r$ be the unique morphism. Then, by Proposition IV.3.1.6, the inverse image functor $m^*\colon (X^r/B^r)^r_{\mathrm{HIGGS}} \to (X^{r'}/B^{r'})^{r'}_{\mathrm{HIGGS}}$ is given by the morphism of sites induced by the natural morphisms $X^{r'} \to X^r$ and $B^{r'} \to B^r$ as in loc. cit. □

Let $(X^\infty/B^\infty)_{\mathrm{HIGGS}}$ be the inverse limit of the fibered site $\pi\colon (X^\bullet/B^\bullet)_{\mathrm{HIGGS}} \to I^{r_0}$ ([2] VI Définition 8.2.5). The associated topos $(X^\infty/B^\infty)^\sim_{\mathrm{HIGGS}}$ is an inverse limit of the fibered topos $(X^\bullet/B^\bullet)^{\sim/I^{r_0}} \to I^{r_0}$ ([2] VI Théorème 8.2.3 2)). The composition of $(X^r/B^r)^r_{\mathrm{HIGGS}} \to (X^\bullet/B^\bullet)_{\mathrm{HIGGS}} \to (X^\infty/B^\infty)_{\mathrm{HIGGS}}$ is a morphism of site ([2] VI Théorème 8.2.3 1)). Let $\mu_r$ denote the associated morphism of topos $(X^\infty/B^\infty)^\sim_{\mathrm{HIGGS}} \to (X^r/B^r)^{r\sim}_{\mathrm{HIGGS}}$. For integers $r' \geq r \geq r_0$, let $\mu_{rr'}\colon (X^{r'}/B^{r'})^{r'\sim}_{\mathrm{HIGGS}} \to (X^r/B^r)^{r\sim}_{\mathrm{HIGGS}}$ be the morphism of topos defined by the inverse image functor between the fibers of $(X^\bullet/B^\bullet)_{\mathrm{HIGGS}}$ over $r$ and $r'$, which coincides with the morphism of topos induced by the morphisms $X^{r'} \to X^r$ and $B^{r'} \to B^r$ (cf. the proof of Lemma IV.4.5.1). Then we have a canonical isomorphism $\mu_{rr'} \circ \mu_{r'} \cong \mu_r$. We have a natural isomorphism $\mu^*_{rr'}(\mathcal{O}_{X^r/B^r,1}) \xrightarrow{\cong} \mathcal{O}_{X^{r'}/B^{r'},1}$ of sheaves of rings. We define the sheaf of rings $\mathcal{O}_{X^\infty/B^\infty,1}$ on $(X^\infty/B^\infty)_{\mathrm{HIGGS}}$

to be $\varinjlim_{r\geq r_0}\mu_r^*(\mathcal{O}_{X^r/B^r,1})$, which is canonically isomorphic to $\mu_r^*(\mathcal{O}_{X^r/B^r,1})$ for any $r\geq r_0$.

By Proposition IV.4.4.3, $(X^r/B^r)^{r\sim}_{\text{HIGGS}}$ is coherent for every $r\geq r_0$. By Proposition IV.4.4.4, the morphism $\mu_{rr'}$ is coherent. Hence, by **[2]** VI Corollaire 8.7.7, we have a canonical isomorphism

$$H^q((X^\infty/B^\infty)_{\text{HIGGS}},\mu_r^*\mathcal{F})\cong\varinjlim_{s\geq r}H^q((X^s/B^s)_{\text{HIGGS}},\mu_{rs}^*(\mathcal{F})) \tag{IV.4.5.2}$$

for an integer $r\geq r_0$ and a sheaf of $\mathcal{O}_{X^r/B^r,1,\mathbb{Q}}$-modules $\mathcal{F}$ on $(X^r/B^r)^r_{\text{HIGGS}}$.

Let $X^\bullet_{\text{ét}}\to I^{r_0}$ be the fibered site defined by the sequence of morphisms of $p$-adic fine log formal schemes

$$\cdots\to X_{r+1}\to X_r\to\cdots\to X_{r_0+1}\to X_{r_0}=X.$$

Let $X^\infty_{\text{ét}}$ denote the inverse limit of the fibered site $X^\bullet_{\text{ét}}\to I^{r_0}$. The associated topos $(X^\infty)^\sim_{\text{ét}}$ is an inverse limit of the fibered topos $(X^\bullet)^{\sim/I^{r_0}}_{\text{ét}}\to I^{r_0}$. Let $\mu_r\colon(X^\infty)^\sim_{\text{ét}}\to(X^r)^\sim_{\text{ét}}$ $(r\geq r_0)$ and $\mu_{rs}\colon(X^s)^\sim_{\text{ét}}\to(X^r)^\sim_{\text{ét}}$ $(s\geq r\geq r_0)$ be the natural morphisms of topos.

By Proposition IV.4.1.3, the morphisms of topos $u_{X^r/B^r}$ $(r\geq r_0)$ define a Cartesian morphism of fibered topos over $I^{r_0}$

$$u_{X^\bullet/B^\bullet}\colon(X^\bullet/B^\bullet)^{\sim/I^{r_0}}_{\text{HIGGS}}\longrightarrow(X^\bullet_{\text{ét}})^{\sim/I^{r_0}}.$$

Let $u_{X^\infty/B^\infty}\colon(X^\infty/B^\infty)^\sim_{\text{HIGGS}}\to(X^\infty_{\text{ét}})^\sim$ denote the morphism of topos induced by $u_{X^\bullet/B^\bullet}$. By Propositions IV.4.4.3, IV.4.4.4, IV.4.4.8, and Lemmas IV.4.4.6, IV.4.4.7, the topos $(X^r/B^r)^{r\sim}_{\text{HIGGS}}$ and $(X^r_{\text{ét}})^\sim$ are coherent for $r\geq r_0$, and the morphisms of topos $(X^s/B^s)^{s\sim}_{\text{HIGGS}}\to(X^r/B^r)^{r\sim}_{\text{HIGGS}}$, $(X^s_{\text{ét}})^\sim\to(X^r_{\text{ét}})^\sim$, and $u_{X^r/B^r}\colon(X^r/B^r)^{r\sim}_{\text{HIGGS}}\to(X^r_{\text{ét}})^\sim$ are coherent for $s\geq r\geq r_0$. Hence, by **[2]** VI Corollaire 8.7.5, we have a canonical isomorphism

$$R^qu_{X^\infty/B^\infty*}(\mu_r^*(\mathcal{F}))\cong\varinjlim_{s\geq r}\mu_s^*(R^qu_{X^r/B^r*}(\mu_{rs}^*(\mathcal{F})) \tag{IV.4.5.3}$$

for an integer $r\geq r_0$ and a sheaf of $\mathcal{O}_{X^r/B^r,1,\mathbb{Q}}$- modules $\mathcal{F}$ on $(X^r/B^r)^r_{\text{HIGGS}}$.

Let $Y\to B$ be a smooth Cartesian morphism in the category $\mathscr{C}$ and let $i\colon X\to Y_1$ be an immersion over $B_1$. For an integer $r\geq r_0$, let $Y^r\to B^r$ denote the base change of $Y\to B$ by the morphism $B^r\to B$ in the category $\mathscr{C}$, and let $i^r\colon X^r\to Y_1^r$ be the base change of $i$ by $B_1^r\to B_1$. Let $D^r$ be $D^r_{\text{Higgs}}(i^r\colon X^r\hookrightarrow Y^r)$ and let $z_{D^r}$ be the natural morphism $D_1^r\to X^r$ over $B_1^r$.

Let $r$ be an integer $\geq r_0$ and let $\mathcal{F}$ be a Higgs isocrystal on $(X^r/B^r)^r_{\text{HIGGS}}$. We will give a description of $Ru_{X^\infty/B^\infty*}(\mu_r^*(\mathcal{F}))$ in terms of the complexes $(\xi^{-\bullet}\mu_{rs}^*(\mathcal{F})_{D^s}\otimes_{\mathcal{O}_{Y_1^s}}\Omega^\bullet_{Y_1^s/B_1},\theta^\bullet)$ $(s\geq r)$.

Let $(X^\bullet/B^\bullet)^{\geq r}_{\text{HIGGS}}\to I^r$ be the base change of $(X^\bullet/B^\bullet)_{\text{HIGGS}}\to I^{r_0}$ by the inclusion functor $I^r\to I^{r_0}$. The site $(X^\infty/B^\infty)_{\text{HIGGS}}$ is also the direct limit of the fibered site $(X^\bullet/B^\bullet)^{\geq r}_{\text{HIGGS}}\to I^r$. Let $(X^\bullet/B^\bullet)^{\geq r\sim}_{\text{HIGGS}}$ be the topos associated to the total site of the fibered site $(X^\bullet/B^\bullet)^{\geq r}_{\text{HIGGS}}\to I^r$. Let $Q\colon(X^\infty/B^\infty)^\sim_{\text{HIGGS}}\to(X^\bullet/B^\bullet)^{\geq r\sim}_{\text{HIGGS}}$ be the morphism of topos induced by the functor $(X^\bullet/B^\bullet)^{\geq r}_{\text{HIGGS}}\to(X^\infty/B^\infty)_{\text{HIGGS}}$. For $s\geq r$, let $\mathcal{F}^s$ denote the Higgs isocrystal $\mu_{rs}^*\mathcal{F}$ on $(X^s/B^s)^s_{\text{HIGGS}}$, and let $\mathcal{F}^\bullet$ denote the sheaf in $(X^\bullet/B^\bullet)^{\geq r\sim}_{\text{HIGGS}}$ defined by $\mathcal{F}^s$ and the natural morphisms $\mathcal{F}^s\to\mu_{ss'*}\mathcal{F}^{s'}$ $(s'\geq s\geq r)$ (cf. **[2]** VI Proposition 7.4.7). Let $\mathcal{F}^\dagger$ be $Q^*(\mathcal{F}^\bullet)$, which is canonical isomorphic to $\varinjlim_{s\geq r}\mu_s^*(\mathcal{F}^s)$ and also to $\mu_r^*(\mathcal{F}^r)$ (cf. **[2]** VI Proposition 8.5.2).

By Proposition IV.4.3.4, we obtain a complex

$$(\mathcal{F}^s\longrightarrow L_{Y^s}(\xi^{-\bullet}(\mathcal{F}^s_{D^s})\otimes_{\mathcal{O}_{Y_1^s}}\Omega^\bullet_{Y_1^s/B_1^s}))_{s\geq r} \tag{IV.4.5.4}$$

in $(X^\bullet/B^\bullet)^{\geq r\sim}_{\mathrm{HIGGS}}$. By the second claim of Proposition IV.4.3.4 and [2] VI Proposition 8.5.2, we obtain the following.

**Proposition IV.4.5.5.** *The pull-back of* (IV.4.5.4) *by the morphism of topos* $Q$ *gives a resolution of* $\mathcal{F}^\dagger (\cong \mu_r^*\mathcal{F})$

$$\mathcal{F}^\dagger \longrightarrow Q^*((L_{Y^s}(\xi^{-\bullet}(\mathcal{F}^s_{D^s}) \otimes_{\mathcal{O}_{Y_1^s}} \Omega^\bullet_{Y_1^s/B_1^s}))_{s\geq r}).$$

By [2] VI Théorème 8.7.3, we have a canonical isomorphism

$$R^q u_{X^\infty/B^\infty *} Q^*(\mathcal{G}) \cong \varinjlim_{s\geq r} \mu_s^* R^q u_{X^s/B^s *}\mathcal{G}^s$$

for a sheaf of abelian groups $\mathcal{G} = (\mathcal{G}^s)_{s\geq r}$ on $(X^\bullet/B^\bullet)^{\geq r\sim}_{\mathrm{HIGGS}}$. Hence, by Proposition IV.4.2.1 and Lemma IV.4.3.2, we obtain the following.

**Theorem IV.4.5.6.** *Let* $r$ *be an integer* $\geq r_0$, *let* $\mathcal{F}$ *be a Higgs isocrystal on the site* $(X^r/B^r)^r_{\mathrm{HIGGS}}$, *and let* $\mathcal{F}^s$ $(s \geq r)$ *denote the pull-back of* $\mathcal{F}$ *on* $(X^s/B^s)^s_{\mathrm{HIGGS}}$, *which is a Higgs isocrystal. Then there exists a canonical isomorphism*

$$Ru_{X^\infty/B^\infty *}\mu_r^*(\mathcal{F}) \cong \varinjlim_{s\geq r}(\mu_s^*(z_{D^s *}(\xi^{-\bullet}\mathcal{F}^s_{D^s} \otimes_{\mathcal{O}_{Y_1^s}} \Omega^\bullet_{Y_1^s/B_1^s}, \theta^\bullet))).$$

If $B$ is an object of $\mathscr{C}^\infty$, then $B$ is contained in $\mathscr{C}^r$ for every $r \in \mathbb{N}_{>0}$. Hence, we may assume $r_0 = 1$ and we have $B^r = B$ and $X^r = X$ for every $r \in \mathbb{N}_{>0}$. The functor $\mu_{r*}\colon (X^\infty)^\sim_{\text{ét}} \to X^r_{\text{ét}}$ is an equivalence of categories for $r \in \mathbb{N}_{>0}$. Hence we may regard $u_{X_\infty/B_\infty}$ as a morphism of topos $(X^\infty/B^\infty)^\sim_{\mathrm{HIGGS}} \to X^\sim_{\text{ét}}$.

**Corollary IV.4.5.7.** *Assume that* $r_0 = 1$ *and* $B$ *is an object of* $\mathscr{C}^\infty$. *Let* $r$ *be a positive integer, let* $\mathcal{F}$ *be a Higgs isocrystal on* $(X/B)^r_{\mathrm{HIGGS}}$, *and let* $\mathcal{F}^s$ $(s \geq r)$ *denote the pull-back of* $\mathcal{F}$ *on* $(X/B)^s_{\mathrm{HIGGS}}$. *Then there exists a canonical isomorphism*

$$Ru_{X^\infty/B^\infty *}\mu_r^*(\mathcal{F}) \cong \varinjlim_{s\geq r}(z_{D^s *}(\xi^{-\bullet}\mathcal{F}^s_{D^s} \otimes_{\mathcal{O}_{Y_1}} \Omega^\bullet_{Y_1/B_1}, \theta^\bullet)).$$

If $X \to B_1$ is smooth and integral and $i\colon X \to Y_1$ is an isomorphism, then $Y_N \to B_N$ is smooth and integral. In particular, the morphism of schemes underlying the reduction mod $p^m$ of $Y_N \to B_N$ is flat for every $N, m \in \mathbb{N}_{>0}$. Hence, if $B$ is an object of $\mathscr{C}^\infty$, then for every $r \in \mathbb{N}_{>0}$, $Y$ is an object of $\mathscr{C}^r$ by Lemma IV.2.2.7 (1), and therefore we have $Y = D^r$ and $X \cong Y_1 = D^r_1$.

**Corollary IV.4.5.8.** *Assume that* $r_0 = 1$, $B$ *is an object of* $\mathscr{C}^\infty$, $X \to B_1$ *is smooth and integral, and* $i\colon X \to Y_1$ *is an isomorphism. Let* $r$ *be a positive integer, let* $\mathcal{F}$ *be a Higgs isocrystal on* $(X/B)^r_{\mathrm{HIGGS}}$, *and let* $(\mathcal{M}, \theta)$ *be the Higgs vector bundle on* $X$ *associated to* $\mathcal{F}$ *via* $Y$ *and* $i$ *(cf. Theorem* IV.3.4.16*). Then there exists a canonical isomorphism*

$$Ru_{X^\infty/B^\infty *}\mu_r^*(\mathcal{F}) \cong (\xi^{-\bullet}\mathcal{M} \otimes_{\mathcal{O}_X} \Omega^\bullet_{X/B_1}, \theta^\bullet).$$

## IV.5. Representations of the fundamental group

**IV.5.1. The $p$-adic fine log formal schemes $D_N(\overline{U})$ and $\overline{U}$.** Let $U$ be an affine fs log scheme over $\mathbb{Z}_p$ satisfying the following conditions: The ring $A = \Gamma(U, \mathcal{O}_U)$ is a noetherian normal ring flat over $\mathbb{Z}_p$, and $U$ satisfies the conditions in [73] Lemma 1.3.2, i.e., there exists a chart $\alpha\colon P_U \to M_U$ such that $P$ is a finitely generated and saturated monoid and the morphism $P \to \Gamma(U, M_U)/\Gamma(U, \mathcal{O}_U^\times)$ induced by $\alpha$ is an isomorphism.

Let $\mathcal{A}$ be one of the following three $A$-algebras:

(Case I) The algebra $A$ itself.

(Case II) The henselization of $A$ with respect to the ideal $pA$.

(Case III) The $p$-adic completion of $A$.

Then $\mathcal{A}$ is noetherian and, in the cases I and II, it is normal. In the case III, we also assume that $\mathcal{A}$ is normal, which holds if $A$ is of finite type over a complete noetherian local ring (cf. **[42]** Scholie (7.8.3) (ii), (iii), (v)).

Let $\mathcal{U}$ be Spec($\mathcal{A}$) endowed with the inverse image of $M_U$. Let $\mathcal{U}_{\mathrm{triv}}$ be the open subscheme of Spec($\mathcal{A}$) defined by $\{x \in \mathrm{Spec}(\mathcal{A}) | M_{\mathcal{U},\overline{x}} = \mathcal{O}^{\times}_{\mathcal{U},\overline{x}}\}$. If $P$ is generated by $a_1, \ldots, a_r \in P$, we have $\mathcal{U}_{\mathrm{triv}} = \mathrm{Spec}(\mathcal{A}[\alpha(a_1a_2\cdots a_r)^{-1}])$. Put $\mathcal{A}_{\mathrm{triv}} = \Gamma(\mathcal{U}_{\mathrm{triv}}, \mathcal{O}_{\mathcal{U}_{\mathrm{triv}}})$.

Let $\overline{s} \to \mathcal{U}_{\mathrm{triv}}$ be a geometric point of $\mathcal{U}_{\mathrm{triv}}$ such that the image $s$ of $\overline{s}$ in $\mathcal{U}_{\mathrm{triv}}$ is of codimension 0 and $\overline{\mathcal{K}} := \Gamma(\overline{s}, \mathcal{O}_{\overline{s}})$ is an algebraic closure of the residue field $\mathcal{K} := \kappa(s)$ of $\mathcal{U}_{\mathrm{triv}}$ at $s$. We further assume that $p$ is not invertible on the connected component of $\mathcal{U}$ containing $s$, which always holds in the cases II and III. We define $\mathcal{K}^{\mathrm{ur}}$ to be the union of all finite extensions $\mathcal{L}$ of $\mathcal{K}$ contained in $\overline{\mathcal{K}}$ such that the integral closures of $\mathcal{A}_{\mathrm{triv},\mathbb{Q}_p}$ in $\mathcal{L}$ are étale over $\mathcal{A}_{\mathrm{triv},\mathbb{Q}_p}$. Let $\overline{\mathcal{A}}$ denote the integral closure of $\mathcal{A}$ in $\mathcal{K}^{\mathrm{ur}}$ and let $\widehat{\overline{\mathcal{A}}}$ denote the $p$-adic completion of $\overline{\mathcal{A}}$. We have $\overline{\mathcal{A}}/p\overline{\mathcal{A}} \neq 0$ and hence $\widehat{\overline{\mathcal{A}}} \neq 0$ by assumption. Let $G_{(U,\overline{s})}$ denote the Galois group $\mathrm{Gal}(\mathcal{K}^{\mathrm{ur}}/\mathcal{K})$, which is canonically isomorphic to the fundamental group of the connected component of $\mathcal{U}_{\mathrm{triv},\mathbb{Q}_p}$ containing $s$ with respect to the base point $\overline{s} \to \mathcal{U}_{\mathrm{triv},\mathbb{Q}_p}$.

Let $R_{\overline{\mathcal{A}}}$ be the inverse limit of

$$\overline{\mathcal{A}}/p\overline{\mathcal{A}} \xleftarrow{F} \overline{\mathcal{A}}/p\overline{\mathcal{A}} \xleftarrow{F} \overline{\mathcal{A}}/p\overline{\mathcal{A}} \xleftarrow{F} \overline{\mathcal{A}}/p\overline{\mathcal{A}} \xleftarrow{F} \cdots,$$

where $F$ denotes the absolute Frobenius of $\overline{\mathcal{A}}/p\overline{\mathcal{A}}$. We have a canonical ring homomorphism $\theta\colon W(R_{\overline{\mathcal{A}}}) \to \widehat{\overline{\mathcal{A}}}$ characterized by $\theta([x]) = \lim_{m\to\infty} \widetilde{x_m}^{p^m}$ for $x = (x_0, x_1, \ldots) \in R_{\overline{\mathcal{A}}}$, $x_n \in \overline{\mathcal{A}}/p\overline{\mathcal{A}}$, where $[\ ]\colon \overline{\mathcal{A}} \to W(R_{\overline{\mathcal{A}}})$ denotes the Teichmüller representative and $\tilde{}$ denotes a lifting of an element of $\overline{\mathcal{A}}/p\overline{\mathcal{A}}$ to $\overline{\mathcal{A}}$. The kernel of $\theta$ is generated by a nonzero divisor $\xi$ (cf. **[29]** 2.4. Proposition, **[25]** II b), **[73]** Corollary A2.2). One can construct a generator as follows. Let $\nu_n \in \overline{\mathcal{A}}$ be a compatible system of $p^n$-th roots of $-p$, i.e., $\nu_0 = -p$ and $(\nu_{n+1})^p = \nu_n$, and let $\underline{-p}$ be $(\nu_n \mod p)_{n\geq 0} \in R_{\overline{\mathcal{A}}}$. Then the element $\xi_p := [\underline{-p}] + p$ of $W(R_{\overline{\mathcal{A}}})$ generates the ideal $\mathrm{Ker}(\theta)$. Put $\widetilde{\overline{\mathcal{A}}} = \theta(W(R_{\overline{\mathcal{A}}}))$, which is $p$-adically complete and separated (cf. **[73]** Sublemma A2.12). For $N \in \mathbb{N}_{>0}$, let $A_N(\overline{\mathcal{A}})$ denote $W(R_{\overline{\mathcal{A}}})/\xi^N W(R_{\overline{\mathcal{A}}})$, which is also $p$-adically complete and separated.

**Lemma IV.5.1.1.** *In the cases* II *and* III, *we have* $\widehat{\overline{\mathcal{A}}} = \widetilde{\overline{\mathcal{A}}}$.

PROOF. It suffices to prove that the absolute Frobenius of $\overline{\mathcal{A}}/p\overline{\mathcal{A}}$ is surjective (cf. **[31]** 1.2.2, **[25]** II b), **[73]** Lemma A 1.1). We prove that, for $a \in \overline{\mathcal{A}}$, all solutions of $x^{p^2} - px = a$ in $\overline{\mathcal{K}}$ are contained in $\overline{\mathcal{A}}$. Choose a finite extension $\mathcal{L}$ of $\mathcal{K}$ contained in $\mathcal{K}^{\mathrm{ur}}$ such that $a \in \mathcal{L}$, and let $\mathcal{B}$ be the integral closure of $\mathcal{A}$ in $\mathcal{L}$. Then the ring $\mathcal{B}' := \mathcal{B}[X]/(X^{p^2} - pX - a)$ is integral over $\mathcal{A}$, which implies that the pair $(\mathcal{B}', p\mathcal{B}')$ is henselian. Hence the image of $(X^{p^2} - pX - a)' = p(-1 + pX^{p^2-1})$ in $\mathcal{B}'_{\mathbb{Q}_p}$ is invertible, $\mathcal{B}'_{\mathbb{Q}_p}/\mathcal{B}_{\mathbb{Q}_p}$ is étale, and hence all solutions of $x^{p^2} - px - a = 0$ in $\overline{\mathcal{K}}$ are contained in $\overline{\mathcal{A}}$. □

**Lemma IV.5.1.2.** *The action of* $G_{(U,\overline{s})}$ *on* $A_N(\overline{\mathcal{A}})$ *is continuous with respect to the* $p$*-adic topology of* $A_N(\overline{\mathcal{A}})$.

PROOF (cf. the proof of **[73]** Lemma 1.4.4). It suffices to prove that the action of $G_{(U,\overline{s})}$ on $A_N(\overline{\mathcal{A}})/p^m A_N(\overline{\mathcal{A}}) = W_m(R_{\overline{\mathcal{A}}})/\xi^N W_m(R_{\overline{\mathcal{A}}})$ is continuous with respect to the discrete topology, which is reduced to proving that the stabilizer of the image of $[x]$ in $A_N(\overline{\mathcal{A}})/p^m$ is open for any $x \in R_{\overline{\mathcal{A}}}$. For the polynomials $S_n(X_0, \ldots, X_n, Y_0, \ldots, Y_n) \in \mathbb{Z}[X_0, \ldots, X_n, Y_0, \ldots, Y_n]$ $(n \in \mathbb{N})$ defining the addition of Witt vectors, let $\varphi_n(X_0, Y_0)$

denote $S_n(X_0, 0, \ldots, 0, Y_0, 0, \ldots, 0)$. Then we have

$$(X_0)^{p^n} + (Y_0)^{p^n} = \varphi_0(X_0, Y_0)^{p^n} + p\varphi_1(X_0, Y_0)^{p^{n-1}} + \cdots + p^n\varphi_n(X_0, Y_0),$$

which implies $\varphi_0(X_0, Y_0) = X_0 + Y_0$ and $\varphi_n(X_0, 0) = \varphi_n(0, Y_0) = 0$ for $n \geq 1$. Hence, for $y \in R_{\overline{\mathcal{A}}}$ and $M \in \mathbb{N}_{>0}$, the sum $[x] + [(\underline{-p})^M y]$ in $W_m(R_{\overline{\mathcal{A}}})$ is written in the form

$$\begin{aligned}&(x + (\underline{-p})^M y, (\underline{-p})^M y_1, \ldots, (\underline{-p})^M y_{m-1})\\ =&[x + (\underline{-p})^M y] + p[(\underline{-p})^{p^{-1}M}(y_1)^{p^{-1}}] + \cdots + p^{m-1}[(\underline{-p})^{p^{-(m-1)}M}(y_{m-1})^{p^{-(m-1)}}]\end{aligned}$$

$(y_1, \ldots, y_{m-1} \in R_{\overline{\mathcal{A}}})$. Since $[\underline{-p}] = \xi_p - p$, we have $[\underline{-p}]^{N+m} = 0$ in $A_N(\overline{\mathcal{A}})/p^m$. Hence, if $M \geq p^{m-1}(N+m)$, the morphism $[\;]\colon R_{\overline{\mathcal{A}}} \to A_N(\overline{\mathcal{A}})/p^m$ factors through the quotient $R_{\overline{\mathcal{A}}}/(\underline{-p})^M R_{\overline{\mathcal{A}}}$. For any $l \in \mathbb{N}$, the $l$-th power of the absolute Frobenius induces an isomorphism $R_{\overline{\mathcal{A}}}/(\underline{-p})^{p^l} R_{\overline{\mathcal{A}}} \xrightarrow{\cong} R_{\overline{\mathcal{A}}}/(\underline{-p})R_{\overline{\mathcal{A}}}$, and the projection to the first component induces an injection $R_{\overline{\mathcal{A}}}/(\underline{-p})R_{\overline{\mathcal{A}}} \hookrightarrow \overline{\mathcal{A}}/p\overline{\mathcal{A}}$ (cf. **[73]** Lemma A2.1). Hence the action of $G_{(U,\overline{s})}$ on $R_{\overline{\mathcal{A}}}/(\underline{-p})^M R_{\overline{\mathcal{A}}}$ is continuous with respect to the discrete topology for any $M \in \mathbb{N}_{>0}$. This completes the proof. □

**Lemma IV.5.1.3** (cf. **[73]** Lemma 1.4.1). (1) *Let $\overline{\mathcal{A}_{\text{triv}}}$ be the integral closure of $\mathcal{A}_{\text{triv}}$ in $\mathcal{K}^{\text{ur}}$. Then, for any $a \in \overline{\mathcal{A}_{\text{triv}}}^{\times}$, all solutions of $x^p = a$ in $\overline{\mathcal{K}}$ are contained in $\overline{\mathcal{A}_{\text{triv}}}^{\times}$.*

(2) *The composition of the morphisms $\Gamma(U, M_U) \to \mathcal{A} \to \widehat{\overline{\mathcal{A}}}$ factors through $\widetilde{\overline{\mathcal{A}}}$.*

PROOF. (1) Choose a finite extension $\mathcal{L}$ of $\mathcal{K}$ contained in $\mathcal{K}^{\text{ur}}$ such that $a \in \mathcal{L}$, let $\mathcal{B}$ be the integral closure of $\mathcal{A}$ in $\mathcal{L}$, and put $\mathcal{B}_{\text{triv}} = \mathcal{A}_{\text{triv}} \otimes_{\mathcal{A}} \mathcal{B}$. Since $a$ and $a^{-1}$ are integral over $\mathcal{A}_{\text{triv}}$, we have $a \in \mathcal{B}_{\text{triv}}^{\times}$, which implies that $\mathcal{B}_{\text{triv},\mathbb{Q}_p}[X]/(X^p - a)$ is étale over $\mathcal{B}_{\text{triv},\mathbb{Q}}$. Hence every $p$-th root $b \in \overline{\mathcal{K}}$ of $a$ is contained in $\overline{\mathcal{A}}_{\text{triv}}^{\times}$. Note that $b$ and $b^{-1}$ are integral over $\mathcal{A}_{\text{triv}}$.

(2) The image of $\Gamma(U, M_U)$ in $\overline{\mathcal{K}}$ is contained in $\overline{\mathcal{A}_{\text{triv}}}^{\times} \cap \overline{\mathcal{A}}$. For any $a \in \overline{\mathcal{A}_{\text{triv}}}^{\times} \cap \overline{\mathcal{A}}$, choose a compatible system of $p^n$-th roots $a_n \in \overline{\mathcal{K}}$ of $a$. Then $a_n \in \overline{\mathcal{A}}$ by (1), and we have $a = \theta((a_n \mod p)_{n \geq 0}) \in \widetilde{\overline{\mathcal{A}}}$. □

We define $\overline{U}$ to be the $p$-adic formal scheme $\mathrm{Spf}(\widetilde{\overline{\mathcal{A}}})$ endowed with the log structure associated to $\Gamma(U, M_U) \to \widetilde{\overline{\mathcal{A}}}$. The action of $G_{(U,\overline{s})}$ on $\widetilde{\overline{\mathcal{A}}}$ induces its action on $\overline{U}$. Let $Q$ be the fiber product of the diagram of integral monoids

$$\varprojlim(\overline{\mathcal{A}}\backslash\{0\} \xleftarrow{f} \overline{\mathcal{A}}\backslash\{0\} \xleftarrow{f} \overline{\mathcal{A}}\backslash\{0\} \xleftarrow{f} \cdots) \longrightarrow \overline{\mathcal{A}}\backslash\{0\} \longleftarrow \Gamma(U, M_U),$$

where $f$ is the morphism defined by $f(x) = x^p$ and the left map is the projection to the first component. For $N \in \mathbb{N}_{>0}$, we define $D_N(\overline{U})$ to be the $p$-adic formal scheme $\mathrm{Spf}(A_N(\overline{\mathcal{A}}))$ endowed with the log structure associated to the composition of

$$Q \longrightarrow \varprojlim_f(\overline{\mathcal{A}}\backslash\{0\}) \longrightarrow R_{\overline{\mathcal{A}}} \xrightarrow{[\,]} A_N(\overline{\mathcal{A}}).$$

The natural actions of $G_{(U,\overline{s})}$ on $Q$ and $A_N(\overline{\mathcal{A}})$ define its action on $D_N(\overline{U})$. For $N \in \mathbb{N}_{>0}$, the surjective homomorphism $A_{N+1}(\overline{\mathcal{A}}) \to A_N(\overline{\mathcal{A}})$ and the identity map of $Q$ induces a morphism $i_{\overline{U},N}\colon D_N(\overline{U}) \to D_{N+1}(\overline{U})$ compatible with the actions of $G_{(U,\overline{s})}$. The isomorphism $A_1(\overline{\mathcal{A}}) \to \widetilde{\overline{\mathcal{A}}}$ induced by $\theta$ and the projection $Q \to \Gamma(U, M_U)$ induce a morphism $i_{\overline{U}}\colon \overline{U} \to D_1(\overline{U})$ compatible with the actions of $G_{(U,\overline{s})}$.

**Lemma IV.5.1.4** (cf. **[73]** Lemma 1.4.2). *The log structures of $\overline{U}$ and $D_N(\overline{U})$ $(N \in \mathbb{N}_{>0})$ are fine and saturated, the morphism $i_{\overline{U}}$ is an isomorphism, and the morphisms $i_{\overline{U},N}$ $(N \in \mathbb{N}_{>0})$ are exact closed immersions.*

PROOF. Let $\overline{U}_n$ and $D_N(\overline{U})_n$ denote the reduction mod $p^n$ of $\overline{U}$ and $D_N(\overline{U})$. Since the chart $\alpha$ in the condition on $U$ induces an isomorphism $P \times \Gamma(U, \mathcal{O}_U^\times) \xrightarrow{\cong} \Gamma(U, M_U)$, the log structure $M_{\overline{U}}$ is fine and saturated and the composition $P \to \Gamma(U, M_U) \to \Gamma(\overline{U}_n, M_{\overline{U}_n})$ gives the chart of $M_{\overline{U}_n}$. Since $P^\times = \{1\}$ and $P$ is fine and saturated, $P^{\mathrm{gp}}$ is torsion free. Hence by Lemma IV.5.1.3 (1), there exists a lifting $P^{\mathrm{gp}} \to \varprojlim_n \overline{\mathcal{A}}_{\mathrm{triv}}^\times$ of the composition of $P^{\mathrm{gp}} \to \Gamma(U, M_U)^{\mathrm{gp}} \to \mathcal{A}_{\mathrm{triv}}^\times \to \overline{\mathcal{A}}_{\mathrm{triv}}^\times$, which induces a lifting $g\colon P \to \varprojlim_f (\overline{\mathcal{A}} \backslash \{0\})$ and then a morphism $h := (g, \Gamma(U, \alpha))\colon P \to Q$. Let $G$ be the fiber product of the diagram of groups $\varprojlim_f \overline{\mathcal{A}}^\times \to \overline{\mathcal{A}}^\times \leftarrow \Gamma(U, \mathcal{O}_U^\times)$. Then $h$ and the natural inclusion $G \hookrightarrow Q$ induce an isomorphism $P \times G \cong Q$. Indeed, for any $((x_n), x) \in Q$, there exists a unique decomposition $x = au$ $(a \in P, u \in \Gamma(U, \mathcal{O}_U^\times))$, which implies the injectivity. Furthermore $(x_n g(a)_n^{-1})^{p^n}$ =(the image of $u$) implies $(x_n g(a)_n^{-1}) \in \overline{\mathcal{A}}^\times$, and we obtain the desired decomposition $((x_n), x) = h(a) \cdot ((x_n g(a)_n^{-1}), u)$. Now we see that the log structure of $D_N(\overline{U})$ is fine and saturated and $P \to Q \to \Gamma(D_N(\overline{U})_n, M_{D_N(\overline{U})_n})$ gives the chart of $M_{D_N(\overline{U})_n}$. The claims on $i_{\overline{U}}$ and $i_{\overline{U},N}$ are obvious because the projection $Q \to \Gamma(U, M_U)$ is compatible with $h$ and $\Gamma(U, \alpha)$. □

We identify $\overline{U}$ with $D_1(\overline{U})$ by the isomorphism $i_{\overline{U}}$ in the following.

Let $\widehat{U}$ be the $p$-adic formal completion of $U$. Here and hereafter, the $p$-adic formal completion of a fine log scheme $X$ means the $p$-adic formal completion of the underlying scheme endowed with the inverse image of $M_X$ on the reduction mod $p^n$ of the completion for each $n \in \mathbb{N}_{>0}$. If the homomorphism $A \to \widehat{\overline{\mathcal{A}}}$ factors through $\widetilde{\overline{\mathcal{A}}}$, which is always true in the cases II and III by Lemma IV.5.1.1, the identity map of $\Gamma(U, M_U)$ and the homomorphism $\varprojlim_n A/p^n A \to \widetilde{\overline{\mathcal{A}}}$ induce a morphism of $p$-adic fine log formal schemes $\overline{U} \to \widehat{U}$.

The above constructions are functorial in $(U, \overline{s})$ as follows. Let $U'$ be another fs log scheme satisfying the same conditions as $U$. We define $A'$, $\mathcal{A}'$, $\mathcal{U}'$, and $\mathcal{U}'_{\mathrm{triv}}$ using $U'$ in the same way as $A$, $\mathcal{A}$, etc. Let $\overline{s}' \to \mathcal{U}'_{\mathrm{triv}}$ be a geometric point satisfying the same conditions as $\overline{s}$, and define $\overline{\mathcal{A}}'$, $G_{(U', \overline{s}')}$, $\overline{U}'$, $D_N(\overline{U}')$, $i_{\overline{U}',N}$ using $U'$ in the same way as $\overline{\mathcal{A}}$, $G_{(U,\overline{s})}$, etc. Let $f\colon U' \to U$ be a morphism over $\mathbb{Z}_p$ and let $h$ be a path from $\overline{s}' \to \mathcal{U}'_{\mathrm{triv}} \xrightarrow{\boldsymbol{f}} \mathcal{U}_{\mathrm{triv}}$ to $\overline{s} \to \mathcal{U}_{\mathrm{triv}}$, where $\boldsymbol{f}$ denotes the morphism induced by $f$. Then the morphism $f$ induces a homomorphism $u\colon A \to A'$ and $\boldsymbol{u}\colon \mathcal{A} \to \mathcal{A}'$, and the path $h$ induces a homomorphism $\overline{\boldsymbol{u}}\colon \overline{\mathcal{A}} \to \overline{\mathcal{A}}'$ compatible with $u$. For any $\sigma \in G_{(U', \overline{s}')}$, there exists $\rho_h(\sigma) \in G_{(U,\overline{s})}$ uniquely such that $\sigma \circ \overline{\boldsymbol{u}} = \overline{\boldsymbol{u}} \circ \rho_h(\sigma)$ (cf. **[37]** I Corollaire 5.4). This correspondence defines a continuous homomorphism $\rho_h\colon G_{(\overline{U}', \overline{s}')} \to G_{(U,\overline{s})}$. The morphisms $\Gamma(U, M_U) \to \Gamma(U', M_{U'})$ and $R_{\overline{\mathcal{A}}} \to R_{\overline{\mathcal{A}}'}$ induced by $f$ and $\overline{\boldsymbol{u}}$ naturally give a morphism $\overline{f}\colon \overline{U}' \to \overline{U}$ and $\overline{f}_N\colon D_N(\overline{U}') \to D_N(\overline{U})$ compatible with the actions of $G_{(U,\overline{s})}$ and $G_{(U',\overline{s}')}$ via $\rho_h$, and with the exact closed immersions $i_{\overline{U},N}$ and $i_{\overline{U}',N}$. We have $\overline{f} = \overline{f}_1$.

**IV.5.2. Representations associated to Higgs crystals and Higgs isocrystals.** Let $V$ be a complete discrete valuation ring of mixed characteristic $(0, p)$ such that the residue field $k$ of $V$ is algebraically closed, and let $K$ be the field of fractions of $V$. Choose and fix an algebraic closure $\overline{K}$ of $K$, and let $\overline{V}$ denote the integral closure of $V$ in $\overline{K}$.

We take the ring $W(R_{\overline{V}})$ and the ideal $\mathrm{Ker}(\theta\colon W(R_{\overline{V}}) \to \widehat{\overline{V}})$ as the base $(R, I)$ of our theory of Higgs isocrystals and Higgs crystals. See IV.5.1 for $R_{\overline{V}}$ and $\theta$. Let $\Sigma$ be

$\mathrm{Spec}(V)$ endowed with the trivial log structure or the canonical log structure (i.e., the log structure defined by the closed point). The $p$-adic fine log formal schemes $D_N(\overline{\Sigma})$ over $S_N = \mathrm{Spf}(A_N(\overline{V}))$ and the exact closed immersions $i_{\overline{\Sigma},N}\colon D_N(\overline{\Sigma}) \to D_{N+1}(\overline{\Sigma})$ compatible with the closed immersions $S_N \to S_{N+1}$ define an object of $\mathscr{C}^\infty$ (cf. Lemma IV.2.2.6), which is denoted by $D(\overline{\Sigma})$ in the following. Let $\widehat{\Sigma}$ be the $p$-adic completion of $\Sigma$. We have a natural morphism $\overline{\Sigma} \to \widehat{\Sigma}$ of $p$-adic fine log formal schemes.

**Lemma IV.5.2.1.** *Let $A$ be a flat finitely generated $V$-algebra such that $A\otimes_V k$ is reduced and $A$ is normal. Let $\mathcal{A}$ be one of $A$* (Case I)*, the henselization of $A$ with respect to the ideal $pA$* (Case II)*, and the $p$-adic completion of $A$* (Case III).

(1) *The ring $\mathcal{A}$ is noetherian and normal.*

(2) *Let $\prod_{i\in I} \mathcal{A}_i$ be the decomposition of $\mathcal{A}$ into the product of finite number of normal domains. For $i \in I$ such that $\mathcal{A}_i/p\mathcal{A}_i \neq 0$ (which always holds in the cases* II *and* III*), $\mathcal{A}_i \otimes_V \overline{V}$ is a normal domain.*

Proof. (1) This is trivial in the case I and follows from **[62]** XI §2 (resp. **[42]** Scholie (7.8.3) (ii), (iii), (v)) in the case II (resp. III).

(2) It suffices to prove that $\mathcal{A}_i \otimes_V V'$ is a normal domain for the integer ring $V'$ of any finite extension $K'$ of $K$. We write $\mathcal{B}$, $\mathcal{B}_{V'}$, and $\mathcal{B}_k$ for $\mathcal{A}_i$, $\mathcal{A}_i \otimes_V V'$, and $\mathcal{A}_i \otimes_V k$. Note that $\mathcal{B}_{V'} \otimes_{V'} k \cong \mathcal{B}_k$ is reduced. We first prove that $\mathcal{B}_{V'}$ is normal. Since $\mathcal{B}_{V'}$ is noetherian, it suffices to prove that $\mathcal{B}_{V'}$ satisfies $(R_1)$ and $(S_2)$. Since $\mathcal{B}_{V'} \otimes_{V'} K' = \mathcal{B} \otimes_V K'$ is étale over $\mathcal{B} \otimes_V K$, $\mathcal{B}_{V'} \otimes_{V'} K'$ is normal. Hence it remains to check the conditions for a prime ideal $\mathfrak{p}$ of $\mathcal{B}_{V'}$ containing $p$. Since $\mathcal{B}_k$ is reduced, $\mathcal{B}_k$ satisfies $(R_0)$ and $(S_1)$. Put $\overline{\mathfrak{p}} = \mathfrak{p}\mathcal{B}_k$. Since $V' \to \mathcal{B}_{V'}$ is flat, we have $\mathrm{ht}\,\overline{\mathfrak{p}} = \mathrm{ht}\,\mathfrak{p} - 1$. If $\mathrm{ht}\,\mathfrak{p} = 1$, then $\mathrm{ht}\,\overline{\mathfrak{p}} = 0$, which implies that $(\mathcal{B}_k)_{\overline{\mathfrak{p}}}$ is regular, i.e., a field. Hence the maximal ideal of $(\mathcal{B}_{V'})_{\mathfrak{p}}$ is generated by a uniformizer of $V'$, and $(\mathcal{B}_{V'})_{\mathfrak{p}}$ is regular. If $\mathrm{ht}\,\mathfrak{p} \geq 2$, then $\mathrm{ht}\,\overline{\mathfrak{p}} \geq 1$, which implies $\mathrm{depth}(\mathcal{B}_k)_{\overline{\mathfrak{p}}} \geq 1$. Since a uniformizer of $V'$ is a nonzero divisor in $(\mathcal{B}_{V'})_{\mathfrak{p}}$, this implies $\mathrm{depth}(\mathcal{B}_{V'})_{\mathfrak{p}} \geq 2$. Thus we see that $\mathcal{B}_{V'}$ is normal.

In the cases II and III, the pairs $(\mathcal{B}, p\mathcal{B})$ and $(\mathcal{B}_{V'}, p\mathcal{B}_{V'})$ are henselian because $\mathcal{B}$ and $\mathcal{B}_{V'}$ are finite over $\mathcal{A}$ (**[62]** XI §2 Proposition 2). Hence we have $\pi_0(\mathrm{Spec}(\mathcal{B}\otimes_V V')) \xrightarrow{\cong} \pi_0(\mathrm{Spec}(\mathcal{B}_k)) \xleftarrow{\cong} \pi_0(\mathrm{Spec}(\mathcal{B}))$. In the case I, since $\mathcal{B}_{V'} \to \mathcal{B}\otimes_V K'$ is injective, it suffices to prove that $K$ is algebraically closed in $\mathrm{Frac}(\mathcal{B})$. Let $L$ be a finite extension of $K$ contained in $\mathrm{Frac}(\mathcal{B})$. Then the ring of integers $V_L$ of $L$ is contained in $\mathcal{B}$ because $\mathcal{B}$ is normal. Since $V_L \to \mathcal{B}$ is faithfully flat by the assumption $\mathcal{A}_i/p\mathcal{A}_i \neq 0$, the homomorphism $V_L \otimes_V k \to \mathcal{B}_k$ is injective. Hence $V_L \otimes_V k$ is reduced, i.e., a field, and therefore $L = K$ because $k$ is algebraically closed. □

In this subsection, we consider an fs log scheme $U$ over $\Sigma$ satisfying the conditions in the first paragraph in IV.5.1. We further assume that the underlying scheme of $U$ is of finite type over $\mathrm{Spec}(V)$ and its special fiber is reduced. Let $A$, $\mathcal{A}$, $\mathcal{U}$, and $\mathcal{U}_{\mathrm{triv}}$ be as in the beginning of IV.5.1, let $\mathcal{U}_{\overline{K},\mathrm{triv}}$ be $\mathcal{U}_{\mathrm{triv}} \times_{\mathrm{Spec}(K)} \mathrm{Spec}(\overline{K})$, and let $\overline{s} \to \mathcal{U}_{\overline{K},\mathrm{triv}}$ be a geometric point such that the composition with $\mathcal{U}_{\overline{K},\mathrm{triv}} \to \mathcal{U}_{\mathrm{triv}}$ satisfies the condition in the beginning of IV.5.1. We keep the notation in IV.5.1 such as $\mathcal{K}$, $\overline{\mathcal{K}}$, $\mathcal{K}^{\mathrm{ur}}$, $G_{(U,\overline{s})}$, $\overline{\widetilde{\mathcal{A}}}$, $\overline{U}$, and $D_N(\overline{U})$. In the case I, we further assume that the composition of $A \to \mathcal{A} \to \widehat{\overline{\mathcal{A}}}$ factors through $\widetilde{\overline{\mathcal{A}}}$. Recall that we have $\widehat{\overline{\mathcal{A}}} = \widetilde{\overline{\mathcal{A}}}$ in the cases II and III (Lemma IV.5.1.1). Hence in all cases, we have a canonical morphism $\overline{U} \to \widehat{U}$.

The morphism $\overline{s} \to \mathcal{U}_{\overline{K},\mathrm{triv}}$ induces an extension $\overline{K} \to \overline{\mathcal{K}}$ of the inclusion map $K \to \mathcal{K}$. Hence, by the last paragraph of IV.5.1, we have a continuous homomorphism $G_{(U,\overline{s})} = \mathrm{Gal}(\mathcal{K}^{\mathrm{ur}}/\mathcal{K}) \to G_\Sigma = \mathrm{Gal}(\overline{K}/K)$, morphisms $D_N(\overline{U}) \to D_N(\overline{\Sigma})$ compatible with the actions of $G_{(U,\overline{s})}$ and $G_\Sigma$, and a commutative diagram

$$\text{(IV.5.2.2)} \qquad \begin{array}{ccc} \overline{U} = D_1(\overline{U}) & \longrightarrow & \widehat{U} \\ \downarrow & & \downarrow \\ \overline{\Sigma} = D_1(\overline{\Sigma}) & \longrightarrow & \widehat{\Sigma}. \end{array}$$

The $p$-adic fine log formal schemes $D_N(\overline{U})$ over $S_N$ and the exact closed immersions $i_{\overline{U},N}\colon D_N(\overline{U}) \to D_{N+1}(\overline{U})$ define an object of $\mathscr{C}^\infty$ (cf. Lemma IV.2.2.6), which is denoted by $D(\overline{U})$. The morphisms $D_N(\overline{U}) \to D_N(\overline{\Sigma})$ define a morphism $D(\overline{U}) \to D(\overline{\Sigma})$ in $\mathscr{C}^\infty$. Let $U_1$ be $\widehat{U} \times_{\widehat{\Sigma}} \overline{\Sigma}$. Then the commutative diagram (IV.5.2.2) induces a morphism $z_{\overline{U}}\colon \overline{U} \to U_1$ over $\overline{\Sigma}$, and the pair $(D(\overline{U}), z_{\overline{U}})$ becomes an object of $(U_1/D(\overline{\Sigma}))^\infty_{\mathrm{HIGGS}}$. Let $\Delta_{(U,\overline{s})}$ be the kernel of the continuous homomorphism $G_{(U,\overline{s})} \to G_\Sigma$, which coincides with $\mathrm{Gal}(\mathcal{K}^{\mathrm{ur}}/\mathcal{K}\overline{K})$ and is canonically isomorphic to the fundamental group of the connected component of $\mathcal{U}_{\overline{K},\mathrm{triv}}$ containing the image of $\overline{s}$ with respect to the base point $\overline{s} \to \mathcal{U}_{\overline{K},\mathrm{triv}}$. Then $\Delta_{(U,\overline{s})}$ acts on the object $(D(\overline{U}), z_{\overline{U}})$.

Let $X \to \Sigma$ be a morphism of fine log schemes whose underlying morphism of schemes is of finite type, and suppose that we are given a morphism $h\colon U \to X$ over $\Sigma$. Let $\widehat{h}\colon \widehat{U} \to \widehat{X}$ be the $p$-adic completion of $h$, and let $h_1\colon U_1 \to X_1$ be the base change of $\widehat{h}$ by the morphism $\overline{\Sigma} \to \widehat{\Sigma}$. Let $r \in \mathbb{N}_{>0} \cup \{\infty\}$ and let $\mathcal{F}$ be a Higgs isocrystal on $(X_1/D(\overline{\Sigma}))^r_{\mathrm{HIGGS}}$ finite on $U_1$ (cf. Definition IV.3.3.1). We define $V^r_{(U,\overline{s}),\mathrm{HIGGS}}(\mathcal{F})$ by

$$V^r_{(U,\overline{s}),\mathrm{HIGGS}}(\mathcal{F}) = \Gamma((D(\overline{U}), h_1 \circ z_{\overline{U}}), \mathcal{F}),$$

which is a finitely generated projective $\widetilde{\overline{\mathcal{A}}}_{\mathbb{Q}_p}$-module. The action of $\Delta_{(U,\overline{s})}$ on $(D(\overline{U}), z_{\overline{U}})$ induces a semi-linear action of $\Delta_{(U,\overline{s})}$ on $V^r_{(U,\overline{s}),\mathrm{HIGGS}}(\mathcal{F})$. We have an obvious identity $V^r_{(U,\overline{s}),\mathrm{HIGGS}}(\mathcal{F}) = V^r_{(U,\overline{s}),\mathrm{HIGGS}}(h^*_{1,\mathrm{HIGGS}}(\mathcal{F}))$. For $r' \in \mathbb{N}_{>0} \cup \{\infty\}$ such that $r' > r$ and the morphism of topos $\mu_{r,r'}\colon (X_1/D(\overline{\Sigma}))^{r'}_{\mathrm{HIGGS}} \to (X_1/D(\overline{\Sigma}))^r_{\mathrm{HIGGS}}$, we have

$$V^{r'}_{(U,\overline{s}),\mathrm{HIGGS}}(\mu^*_{r,r'}(\mathcal{F})) = V^r_{(U,\overline{s}),\mathrm{HIGGS}}(\mathcal{F}).$$

Similarly for a Higgs crystal $\mathcal{F}$ on $X/D(\overline{\Sigma})$ finite on $U_1$ (cf. Definition IV.3.3.2), we can define an $\widetilde{\overline{\mathcal{A}}}$-module $T_{(U,\overline{s}),\mathrm{HIGGS}}(\mathcal{F})$ belonging to $\mathcal{LPM}(\widetilde{\overline{\mathcal{A}}})$ (cf. IV.3.2) endowed with a semi-linear action of $\Delta_{(U,\overline{s})}$ by

$$T_{(U,\overline{s}),\mathrm{HIGGS}}(\mathcal{F}) := \Gamma((D(\overline{U}), h_1 \circ z_{\overline{U}}), \mathcal{F}).$$

We will construct a "period ring" for the functors $V^r_{(U,\overline{s}),\mathrm{HIGGS}}$ and $T_{(U,\overline{s}),\mathrm{HIGGS}}$. Let $Y$ be a smooth fine log scheme over $\Sigma$ and let $i$ be an immersion $X \to Y$ over $\Sigma$. Let $\widehat{Y}$ be the $p$-adic completion of $Y$ and let $Y_1$ be $\widehat{Y} \times_{\widehat{\Sigma}} \overline{\Sigma}$. Then the morphism $Y_1 \to \overline{\Sigma}$ is smooth and the immersion $i$ induces an immersion $i_1\colon X_1 \to Y_1$ over $\overline{\Sigma}$. We further assume that we are given a compatible system of smooth liftings $Y_N \to D_N(\overline{\Sigma})$ ($N \in \mathbb{N}_{>0}$) of $Y_1 \to \overline{\Sigma}$. The smooth $p$-adic fine log formal schemes $Y_N$ over $D_N(\overline{\Sigma})$ and the closed immersions $Y_N \to Y_{N+1}$ define an object $Y_\bullet$ of $\mathscr{C}$ with a smooth Cartesian morphism $Y \to D(\overline{\Sigma})$. The morphism $\overline{U} \xrightarrow{z_{\overline{U}}} U_1 \xrightarrow{h_1} X_1 \xrightarrow{i_1} Y_1$ and the identity morphism of $\overline{U}$ define an immersion $i_{\overline{U},Y}\colon \overline{U} \to Y_1 \times_{\overline{\Sigma}} \overline{U}$. For $r \in \mathbb{N}_{>0} \cup \{\infty\}$, we define the object $\mathscr{D}^r_{X,Y}(\overline{U}) = (\mathscr{D}^r_{X,Y,N}(\overline{U}))$ of $\mathscr{C}^r$ by

$$\mathscr{D}^r_{X,Y}(\overline{U}) = D^r_{\mathrm{Higgs}}(i_{\overline{U},Y}\colon \overline{U} \hookrightarrow Y_\bullet \times_{D(\overline{\Sigma})} D(\overline{U})).$$

The action of $\Delta_{(U,\overline{s})}$ on $(D(\overline{U}), z_{\overline{U}})$ induces its action on $\mathscr{D}^r_{X,Y}(\overline{U})$. By Proposition IV.2.2.9 (2), $\mathscr{D}^r_{X,Y,N}(\overline{U})$ is affine. We define the object $\mathscr{A}^r_{X,Y}(\overline{\mathcal{A}}) = (\mathscr{A}^r_{X,Y,N}(\overline{\mathcal{A}}))$ of $\mathscr{A}^r_\bullet$ by

$$\mathscr{A}^r_{X,Y,N}(\overline{\mathcal{A}}) = \Gamma(\mathscr{D}^r_{X,Y,N}(\overline{U}), \mathcal{O}_{\mathscr{D}^r_{X,Y,N}(\overline{U})})$$

(cf. Lemma IV.2.2.6), which is naturally endowed with the action of $\Delta_{(U,\overline{s})}$. We abbreviate $\mathscr{A}^r_{X,Y}(-)$ and $\mathscr{D}^r_{X,Y}(-)$ to $\mathscr{A}^r(-)$ and $\mathscr{D}^r(-)$ if there is no risk of confusion.

In order to study some properties of $\mathscr{A}^r_N(\overline{\mathcal{A}})$, we assume that there exist a strict étale fine log scheme $Y'$ over $Y$ and a factorization $U \xrightarrow{j} Y' \to Y$ of $i \circ h \colon U \to Y$ such that $Y'$ satisfies the following conditions.

**Condition IV.5.2.3.** (i) The underlying scheme of $Y'$ is affine.

(ii) There exist $t_i \in \Gamma(Y', M_{Y'})$ $(1 \leq i \leq d)$ such that $d\log t_i$ $(1 \leq i \leq d)$ form a basis of $\Omega^1_{Y'/\Sigma}$.

Then there exists a compatible system of strict étale liftings $Y'_N \to Y_N$ of $Y'_1 := \widehat{Y}' \times_{\widehat{\Sigma}} \overline{\Sigma} \to Y_1$ uniquely up to a unique isomorphism. We see that $Y'_N$ is affine, there exist $(t_{i,N})_N \in \varprojlim_N \Gamma(Y'_N, M_{Y'_N})$ $(1 \leq i \leq d)$ such that $t_{i,1}$ coincides with the image of $t_i$, and $d\log(t_{i,N})$ $(1 \leq i \leq d)$ also form a basis of $\Omega^1_{Y'_N/D_N(\overline{\Sigma})}$ for each $N \in \mathbb{N}_{>0}$.

**Remark IV.5.2.4.** If a smooth $p$-adic fine log formal scheme $Y'$ over $\overline{\Sigma}$ satisfies the condition (i), then there always exists a compatible system of smooth liftings $Y'_N \to D_N(\overline{\Sigma})$ ([**50**] Proposition (3.14) (1)).

Let $i_{\overline{U},Y'}$ be the immersion $\overline{U} \to Y'_1 \times_{\overline{\Sigma}} \overline{U}$ induced by $j$. Then the morphism $Y'_\bullet \to Y_\bullet$ induces an isomorphism $\mathscr{D}^r_{X,Y}(\overline{U}) \cong D^r_{\mathrm{Higgs}}(i_{\overline{U},Y'} \colon \overline{U} \to Y'_\bullet \times_{D(\overline{\Sigma})} D(\overline{U}))$ by Lemma IV.2.2.15. We will first give an explicit description of $\mathscr{A}^r_N(\overline{\mathcal{A}})$. Let $s_i$ be the image of $t_i$ under $j^* \colon \Gamma(Y', M_{Y'}) \to \Gamma(U, M_U)$ and let $Q$ be the monoid $\varprojlim_f \overline{A}\backslash\{0\} \times_{\overline{A}\backslash\{0\}} \Gamma(U, M_U)$ used in the definition of the log structure of $D_N(\overline{U})$ in IV.5.1. By Lemma IV.5.1.3 (1), there exists $(s_{i,n})_{n\in\mathbb{N}} \in \varprojlim_f \overline{A}\backslash\{0\}$ such that the pair $((s_{i,n}), s_i)$ becomes an element of $Q$. Choose such a $(s_{i,n})$ and let $\underline{s}_{i,N}$ be the image of $(s_i, (s_{i,n}))$ in $\Gamma(D_N(\overline{U}), M_{D_N(\overline{U})})$. Since $\underline{s}_{i,1}$ coincides with the image of $s_i$, we can apply Proposition IV.2.3.17 to $\mathrm{id}_{\overline{U}}$, $i_{\overline{U},Y'}$, the inverse image of $t_{i,N}$ to $Y'_N \times_{D_N(\overline{\Sigma})} D_N(\overline{U})$ and $\underline{s}_{i,N}$, and obtain an isomorphism

$$A(\overline{\mathcal{A}})\{W_1, \ldots, W_d\}_r \xrightarrow{\cong} \mathscr{A}^r(\overline{\mathcal{A}}) \tag{IV.5.2.5}$$

for $r \in \mathbb{N}_{>0} \cup \{\infty\}$, where the image of $W_i$ $(1 \leq i \leq d)$ in $\mathscr{A}^r_N(\overline{\mathcal{A}})_{\mathbb{Q}_p}$ is given as follows: Let $\pi_{Y'}$ and $\pi_{D(\overline{U})}$ denote the natural morphisms $\mathscr{D}^r(\overline{U}) \to Y'_\bullet, D(\overline{U})$. Then $\pi^*_{Y',N}(t_{i,N})$ and $\pi^*_{D(\overline{U}),N}(\underline{s}_{i,N})$ have the same inverse image in $\Gamma(\mathscr{D}^r_1(\overline{U}), M_{\mathscr{D}^r_1(\overline{U})})$, and there exists a unique $u_{i,N} \in 1 + F^1\mathscr{A}^r_N(\overline{\mathcal{A}})$ such that $\pi^*_{Y',N}(t_{i,N}) = \pi^*_{D(\overline{U}),N}(\underline{s}_{i,N})u_{i,N}$. Let $w_{i,N}$ be the unique element of $\mathscr{A}^r_N(\overline{\mathcal{A}})_{\mathbb{Q}_p}$ such that $[\xi](w_{i,N}) = u_{i,N+1} - 1$. Then the image of $W_i$ in $\mathscr{A}^r_N(\overline{\mathcal{A}})_{\mathbb{Q}_p}$ is $w_{i,N}$.

**Proposition IV.5.2.6.** *Suppose that $h \colon U \to X$ satisfies the assumption before Condition IV.5.2.3. Then, for $r \in \mathbb{N}_{>0} \cup \{\infty\}$ and $N \in \mathbb{N}_{>0}$, the action of $\Delta_{(U,\overline{s})}$ on $\mathscr{A}^r_N(\overline{\mathcal{A}})$ is continuous with respect to the $p$-adic topology.*

Proof. It suffices to prove that the action of $\Delta_{(U,\overline{s})}$ on $\mathscr{A}^r_N(\overline{\mathcal{A}})/p^m$ is continuous with respect to the discrete topology. By (IV.5.2.5), this is further reduced to proving that, for $\underline{m} \in \mathbb{N}^d$ and $a \in A_N(\overline{\mathcal{A}})^{(|\underline{m}|)_r}$, the stabilizer of the image of $a\underline{w}^{\underline{m}}_N$ in $\mathscr{A}^r_N(\overline{\mathcal{A}})/p^m$ is open. Here $\underline{w}^{\underline{m}}_N = \prod_{1 \leq i \leq d} w^{m_i}_{i,N}$. Choose $c \in \mathbb{N}$ such that $p^c A_N(\overline{\mathcal{A}}) \subset A_N(\overline{\mathcal{A}})^{(\nu)_r}$ for every $\nu \in \mathbb{N}$, $0 \leq \nu \leq |\underline{m}|$, and a generator $\varepsilon = (\varepsilon_n)$ of $\mathbb{Z}_p(1) = \varprojlim_n \mu_{p^n}(\overline{\mathcal{A}})$. Let $\underline{\varepsilon}$ be $(\varepsilon_n \mod p) \in R_{\overline{\mathcal{A}}}$. For $g \in \Delta_{(U,\overline{s})}$, define $\eta_i(g) = (\eta_{i,n}(g)) \in \mathbb{Z}_p = \varprojlim_n \mathbb{Z}/p^n$ by $g(s_{i,n}) = s_{i,n}\varepsilon_n^{-\eta_{i,n}(g)}$. Then we have

(IV.5.2.7) $$g(w_{i,N}) = [\underline{\varepsilon}^{\eta_i(g)}]w_{i,N} + \xi^{-1}([\underline{\varepsilon}^{\eta_i(g)}] - 1).$$

Choose $l \in \mathbb{N}$ such that $p^l \geq N+1$. Then, since $[\underline{\varepsilon}] - 1 \in F^1 A_{N+1}(\overline{\mathcal{A}})$, we have $[\underline{\varepsilon}]^{p^l} = 1$ in $A_{N+1}(\overline{\mathcal{A}})/p$. Hence $[\underline{\varepsilon}]^{p^{l+m+c-1}} = 1$ in $A_{N+1}(\overline{\mathcal{A}})/p^{m+c}$, which implies $\xi^{-1}([\underline{\varepsilon}^{\eta}] - 1) \in p^{m+c}A_N(\overline{\mathcal{A}})$ for $\eta \in p^{l+m+c-1}\mathbb{Z}_p$. By Lemma IV.5.1.2, there exists an open subgroup $\Delta' \subset \Delta_{(U,\overline{s})}$ such that $g(a) - a \in p^{c+m}A_N(\overline{\mathcal{A}})(\subset p^m A_N(\overline{\mathcal{A}})^{(|\underline{m}|)_r})$ for every $g \in \Delta'$ and $\eta_i(\Delta') \in p^{l+m+c-1}\mathbb{Z}_p$ for $1 \leq i \leq d$. For $g \in \Delta'$, the above description of $g(w_{i,N})$ implies

$$g(a)(g(\underline{w}_N^{\underline{m}}) - \underline{w}_N^{\underline{m}}) \in p^{c+m} \oplus_{|\underline{n}|\leq|\underline{m}|} A_N(\overline{\mathcal{A}})\underline{w}_N^{\underline{m}} \subset p^m \mathscr{A}_N^r(\overline{\mathcal{A}}).$$

Hence $g(a\underline{w}_N^{\underline{m}}) - a\underline{w}_N^{\underline{m}} = g(a)(g(\underline{w}_N^{\underline{m}}) - \underline{w}_N^{\underline{m}}) + (g(a) - a)\underline{w}_N^{\underline{m}} \in p^m \mathscr{A}_N^r(\overline{\mathcal{A}})$. □

For $q \in \mathbb{N}$, put $\Omega^q = \Gamma(U, (i \circ h)^*\Omega^q_{Y/\Sigma})$ which is isomorphic to $\Gamma(U, j^*\Omega^q_{Y'/\Sigma})$ and hence finite free over $A$ by Condition IV.5.2.3. By applying (IV.2.4.8) and Lemma IV.2.4.10 to the smooth Cartesian morphism $Y_\bullet \times_{D(\overline{\Sigma})} D(\overline{U}) \to D(\overline{U})$ and the immersion $i_{\overline{U},Y} \colon \overline{U} \to Y_1 \times_{\overline{\Sigma}} \overline{U}$, we obtain a derivation

(IV.5.2.8) $$\theta \colon \mathscr{A}_N^r(\overline{\mathcal{A}}) \longrightarrow \xi^{-1}\mathscr{A}_N^r(\overline{\mathcal{A}}) \otimes_A \Omega^1$$

and a complex

(IV.5.2.9) $$A_N(\overline{\mathcal{A}}) \to (\xi^{-q}\mathscr{A}_N^r(\overline{\mathcal{A}}) \otimes_A \Omega^q, \theta^q)_{q\in\mathbb{N}}.$$

**Proposition IV.5.2.10.** *Suppose that $h \colon U \to X$ satisfies the assumption before Condition* IV.5.2.3.

(1) *For $r \in \mathbb{N}_{>0} \cup \{\infty\}$ and $N \in \mathbb{N}_{>0}$, we have $\mathscr{A}_N^r(\overline{\mathcal{A}})^{\theta=0} = A_N(\overline{\mathcal{A}})$.*

(2) *For $r \in \mathbb{N}_{>0}$, the natural homomorphism $\mathscr{A}_N^r(\overline{\mathcal{A}}) \to \mathscr{A}_N^{r+1}(\overline{\mathcal{A}})$ induces a morphism of complexes from the complex* (IV.5.2.9) *for $\mathscr{A}_N^r(\overline{\mathcal{A}})$ to the complex* (IV.5.2.9) *for $\mathscr{A}_N^{r+1}(\overline{\mathcal{A}})$. Furthermore, it becomes homotopic to zero after taking $\otimes_{\mathbb{Z}_p}\mathbb{Q}_p$, and a homotopy is given by $A_N(\overline{\mathcal{A}})_{\mathbb{Q}_p}$-linear continuous homomorphisms.*

PROOF. (1) Under the description (IV.5.2.5) of $\mathscr{A}_N(\overline{\mathcal{A}})$, $\theta$ of $\mathscr{A}_N(\overline{\mathcal{A}})$ is given by $\theta(w_{i,N}) = \xi^{-1}u_{i,N}d\log t_{i,N}$ because $\theta(u_{i,N+1}) = u_{i,N+1}d\log(t_{i,N+1})$. This immediately implies the claim.

(2) Put $\overline{Y} = Y_\bullet \times_{D(\overline{\Sigma})} D(\overline{U})$, let $\overline{Y}(\nu)$ ($\nu \in \mathbb{N}$) be the fiber product over $D(\overline{U})$ of $(\nu+1)$ copies of $\overline{Y}$, and let $\overline{D}^s(\nu)$ ($s \in \{r, r+1\}, \nu \in \mathbb{N}$) be $D^s_{\mathrm{HIGGS}}(\overline{U} \hookrightarrow \overline{Y}(\nu))$. Then we have a natural morphism $\overline{D}^{r+1}(\bullet) \to \overline{D}^r(\bullet)$ of simplicial objects of $\mathscr{C}^r/D(\overline{\Sigma})$ compatible with the natural morphisms to $\overline{Y}(\bullet)$. Hence the construction of $\theta$ and $\theta^q$ implies the first claim. Put $\omega_i = \theta(W_i) = \xi^{-1}u_{i,N}d\log t_{i,N}$. Then, since $\theta(\omega_i) = \xi^{-1}u_{i,N}d\log t_{i,N} \wedge d\log t_{i,N} = 0$, we can construct the desired homotopy in the same way as the proof of Proposition IV.4.3.4 using Lemma IV.2.3.14. □

Let $D^r$ be $D^r_{\mathrm{Higgs}}(X_1 \hookrightarrow Y_\bullet)$. By applying (IV.2.4.8) to $Y_\bullet \to D(\Sigma)$ and $X_1 \hookrightarrow Y_1$, we obtain a derivation $\theta \colon \mathcal{O}_{D_1^r} \to \xi^{-1}\mathcal{O}_{D_1^r} \otimes_{\mathcal{O}_Y} \Omega^1_{Y/\Sigma}$. From the commutative diagram

(IV.5.2.11) $$\begin{array}{ccc} (\overline{U} \hookrightarrow Y_\bullet \times_{D(\overline{\Sigma})} D(\overline{U})) & \longrightarrow & D(\overline{U}) \\ \downarrow & & \downarrow \\ (X_1 \hookrightarrow Y_\bullet) & \longrightarrow & D(\overline{\Sigma}) \end{array}$$

in $\mathscr{C}$, we obtain a morphism $\pi_D\colon \mathscr{D}^r(\overline{U}) \to D^r$ in $\mathscr{C}^r$ and a homomorphism $\pi_{D,1}^{-1}(\mathcal{O}_{D_1^r}) \to \mathcal{O}_{\mathscr{D}_1^r(\overline{U})}$ compatible with $\theta$.

Assume that $h\colon U \to X$ is strict étale and put $A_{D,1}^r = \Gamma(D_1^r \times_{X_1} U_1, \mathcal{O}_{D_1^r})$. Then since $\mathscr{D}_1^r(\overline{U}) \xrightarrow{\pi_{D,1}} D_1^r \to X_1$ canonically factors through $U_1$ as $\mathscr{D}_1^r(\overline{U}) \to \overline{U} \xrightarrow{z_{\overline{U}}} U_1 \to X_1$, we obtain a canonical homomorphism $A_{D_1}^r \to \mathscr{A}_1^r(\overline{\mathcal{A}})$ compatible with $\theta$.

**Proposition IV.5.2.12.** *Assume that $h\colon U \to X$ is strict étale and satisfies the condition before Condition* IV.5.2.3. *Let $r \in \mathbb{N}_{>0} \cup \{\infty\}$ (resp. $r = \infty$), let $\mathcal{F}$ be a Higgs isocrystal on $(X_1/D(\overline{\Sigma}))_{\mathrm{HIGGS}}^r$ finite on $U_1$ (resp. a Higgs crystal on $X_1/D(\overline{\Sigma})$ finite on $U_1$ such that $\mathcal{F}_{D^\infty}$ is $p$-torsion free), and let $(\mathcal{M}, \theta)$ be the object of* $\mathrm{HB}_{\mathbb{Q}_p,\mathrm{conv}}^r(X_1, Y_\bullet/D(\overline{\Sigma}))$ *(resp.* $\mathrm{HB}_{\mathbb{Z}_p,\mathrm{conv}}(X_1, Y_\bullet/D(\overline{\Sigma}))$*) associated to $\mathcal{F}$ by Theorem* IV.3.4.16. *Let $M$ be the $A_{D,1}^r$-module $\Gamma(D_1^r \times_{X_1} U_1, \mathcal{M})$, let $\theta$ also denote the morphism $\Gamma(D_1^r \times_{X_1} U_1, \theta)\colon M \to M \otimes_A \Omega^1$, and let $V(\mathcal{F})$ denote $V_{(U,\overline{s}),\mathrm{HIGGS}}^r(\mathcal{F})$ (resp. $T_{(U,\overline{s}),\mathrm{HIGGS}}(\mathcal{F})$). We define the endomorphism $\theta$ on $\mathscr{A}_1^r(\overline{\mathcal{A}}) \otimes_{\widetilde{\mathcal{A}}} V(\mathcal{F})$ by $\theta \otimes \mathrm{id}_{V(\mathcal{F})}$. Then one can define a $\Delta_{(U,\overline{s})}$-equivariant homomorphism*

$$\theta\colon \mathscr{A}_1^r(\overline{\mathcal{A}}) \otimes_{A_{D,1}^r} M \longrightarrow \xi^{-1}\mathscr{A}_1^r(\overline{\mathcal{A}}) \otimes_{A_{D,1}^r} M \otimes_A \Omega^1$$

*by $\theta(a \otimes m) = a \otimes \theta(m) + \theta(a) \otimes m$ ($a \in \mathscr{A}_1^r(\overline{\mathcal{A}})$, $m \in M$) and there exists a canonical $\Delta_{(U,\overline{s})}$-equivariant $\mathscr{A}_1^r(\overline{\mathcal{A}})$-linear isomorphism*

$$\mathscr{A}_1^r(\overline{\mathcal{A}}) \otimes_{\widetilde{\mathcal{A}}} V(\mathcal{F}) \xrightarrow{\cong} \mathscr{A}_1^r(\overline{\mathcal{A}}) \otimes_{A_{D,1}^r} M$$

*compatible with $\theta$ and functorial in $\mathcal{F}$. In the case of isocrystal, it induces an $\widetilde{\mathcal{A}}_{\mathbb{Q}_p}$-linear isomorphism*

$$V(\mathcal{F}) \cong (M \otimes_{A_{D,1}^r} \mathscr{A}_1^r(\overline{\mathcal{A}}))^{\theta=0}.$$

Proof. The first claim follows from the compatibility of the homomorphism $A_{D,1}^r \to \mathscr{A}_1^r(\overline{\mathcal{A}})$ with $\theta$. Let $\overline{D}^r$ be $D_{\mathrm{Higgs}}^r(\overline{U} \hookrightarrow Y_\bullet \times_{D(\overline{\Sigma})} D(\overline{U}))$, let $z_{\overline{D}^r}$ be the composition of $\overline{D}_1^r \to \overline{U} \to U_1 \to X_1$, and let $z_{D^r}$ be the canonical morphism $D_1^r \to X_1$. Then we have morphisms

$$(D(\overline{U}), h_1 \circ z_{\overline{U}}) \longleftarrow (\overline{D}^r, z_{\overline{D}^r}) \longrightarrow (D^r, z_{D^r})$$

in $(X_1/D(\overline{\Sigma}))_{\mathrm{HIGGS}}^r$ compatible with the actions of $\Delta_{(U,\overline{s})}$. Hence we obtain the following $\Delta_{(U,\overline{s})}$-equivariant $\mathscr{A}_1^r(\overline{\mathcal{A}})$-linear isomorphisms (cf. Lemma IV.3.2.8), where $\overline{M} = \Gamma((\overline{D}^r, z_{\overline{D}^r}), \mathcal{F})$.

$$\mathscr{A}_1^r(\overline{\mathcal{A}}) \otimes_{\widetilde{\mathcal{A}}} V(\mathcal{F}) \xrightarrow{\cong} \overline{M} \xleftarrow{\cong} \mathscr{A}_1^r(\overline{\mathcal{A}}) \otimes_{A_{D,1}^r} M.$$

Note that $\mathcal{F}_{D^r}|_{D_1^r \times_{X_1} U_1}$, $\mathcal{F}_{\overline{D}^r}$ and $\mathcal{F}_{D(\overline{U})}$ belong to $\mathcal{PM}(-)$ (resp. $\mathcal{LPM}(-)$) by the finiteness of $\mathcal{F}$ on $U_1$. Let $\overline{\mathcal{F}}$ be the inverse image of $\mathcal{F}$ on $(\overline{U}/D(\overline{U}))_{\mathrm{HIGGS}}^r$ and let $(\overline{\mathcal{M}}, \overline{\theta})$ be the object of $\mathrm{HM}_{\mathbb{Q}_p}^r(\overline{U}, Y_\bullet \times_{D(\overline{\Sigma})} D(\overline{U})/D(\overline{U}))$ (resp. $\mathrm{HM}_{\mathbb{Z}_p}(\overline{U}, Y_\bullet \times_{D(\overline{\Sigma})} D(\overline{U})/D(\overline{U}))$) associated to $\overline{\mathcal{F}}$. (See after Definition IV.3.4.7.) Then we have $\overline{M} = \Gamma(\overline{D}_1^r, \overline{\mathcal{M}})$ and obtain a homomorphism $\Gamma(\overline{D}_1^r, \overline{\theta})\colon \overline{M} \to \overline{M} \otimes_A \Omega^1$, which is also denoted by $\overline{\theta}$. By the commutative diagram (IV.5.2.11), we see that the above isomorphisms are compatible with $\theta \otimes \mathrm{id}_{V(\mathcal{F})}$, $\overline{\theta}$ and $\theta$. The last claim follows from Proposition IV.5.2.10 (1). □

For a $p$-adically complete and separated algebra $C$ and a finitely generated $C_{\mathbb{Q}_p}$-module $M$, we define the $p$-adic topology of $M$ to be the topology defined by $p^m M^\circ$ ($m \in \mathbb{N}$), where $M^\circ$ is a finitely generated $C$-submodule of $M$ such that $M^\circ_{\mathbb{Q}_p} = M$. The subset $W$ of $M$ is open if and only if, for any $w \in W$, there exists an $m \in \mathbb{N}$ such that $w + p^m M^\circ \subset W$. The topology does not depend on the choice of $M^\circ$ and $M$ becomes a topological $C_{\mathbb{Q}_p}$-module.

**Corollary IV.5.2.13.** *Let $\mathcal{F}$ be a Higgs isocrystal on $(X_1/D(\overline{\Sigma}))^r_{\mathrm{HIGGS}}$ (resp. a Higgs crystal on $X_1/D(\overline{\Sigma})$) finite on $U_1$. Then the action of $\Delta_{(U,\overline{s})}$ on $V^r_{(U,\overline{s}),\mathrm{HIGGS}}(\mathcal{F})$ (resp. $T_{(U,\overline{s}),\mathrm{HIGGS}}(\mathcal{F})$) is continuous with respect to the $p$-adic topology.*

PROOF. Put $r=\infty$ in the second case. By replacing $\mathcal{F}$ with $h^*_{1,\mathrm{HIGGS}}\mathcal{F}$, we may assume $X=U$. Since $U$ is affine and we have a chart $P_U\to M_U$ globally, there exists a $\Sigma$-closed immersion $X=U\to Y$ into a smooth $p$-adic fine log scheme $Y$ over $\Sigma$ satisfying Condition IV.5.2.3. Choose a compatible system of smooth liftings $Y_N\to D_N(\overline{\Sigma})$ (cf. Remark IV.5.2.4) and define the period ring $\mathscr{A}^r_1(\overline{\mathcal{A}})$. Put $W:=V^r_{(U,\overline{s}),\mathrm{HIGGS}}(\mathcal{F})$ (resp. $T_{(U,\overline{s}),\mathrm{HIGGS}}(\mathcal{F})$). By Proposition IV.5.2.12, we have a $\Delta_{(U,\overline{s})}$-equivariant $\mathscr{A}^r_1(\overline{\mathcal{A}})$-linear isomorphism $\mathscr{A}^r_1(\overline{\mathcal{A}})\otimes_{\tilde{\overline{\mathcal{A}}}}W\xrightarrow{\cong}\mathscr{A}^r_1(\overline{\mathcal{A}})\otimes_{A^r_{D,1}}M$. Since $W$ is a finitely generated projective $\tilde{\overline{\mathcal{A}}}_{\mathbb{Q}_p}$-module (i.e., a direct summand of a finite free module) (resp. $W$ is an object of $\mathcal{LPM}(\tilde{\overline{\mathcal{A}}})$ and $\tilde{\overline{\mathcal{A}}}/p^m\tilde{\overline{\mathcal{A}}}\to\mathscr{A}^\infty_1(\overline{\mathcal{A}})/p^m\mathscr{A}^\infty_1(\overline{\mathcal{A}})$ is faithfully flat), we see that the natural homomorphism $W\to\mathscr{A}^r_1(\overline{\mathcal{A}})\otimes_{\tilde{\overline{\mathcal{A}}}}W$ is injective and the $p$-adic topology of $W$ is induced by the $p$-adic topology of $\mathscr{A}^r_1(\overline{\mathcal{A}})\otimes_{\tilde{\overline{\mathcal{A}}}}W$. On the other hand, in the case of isocrystals, the $p$-adic topology of $\mathscr{A}^r_1(\overline{\mathcal{A}})\otimes_{A^r_{D,1}}M$ is defined by $p^m(\mathscr{A}^r_1(\overline{\mathcal{A}})\cdot M^\circ)$ ($m\in\mathbb{N}$), where $M^\circ$ is a finitely generated $A^r_{D,1}$-submodule of $M$ such that $M^\circ_{\mathbb{Q}_p}=M$. Hence the claim follows from Proposition IV.5.2.6. □

Let $\mathrm{Rep}^{\mathcal{PM}}_{\mathrm{cont}}(\Delta_{(U,\overline{s})},\tilde{\overline{\mathcal{A}}}_{\mathbb{Q}_p})$ (resp. $\mathrm{Rep}^{\mathcal{LPM}}_{\mathrm{cont}}(\Delta_{(U,\overline{s})},\tilde{\overline{\mathcal{A}}})$) denote the category of finitely generated projective $\tilde{\overline{\mathcal{A}}}_{\mathbb{Q}_p}$-modules (resp. objects of $\mathcal{LPM}(\tilde{\overline{\mathcal{A}}})$) $W$ endowed with a semi-linear action of $\Delta_{(U,\overline{s})}$ continuous with respect to the $p$-adic topology of $W$. By Corollary IV.5.2.13, we obtain a functor

$$V^r_{(U,\overline{s}),\mathrm{HIGGS}}\colon \mathrm{HC}^r_{\mathbb{Q}_p,U_1\text{-fin}}(X_1/D(\overline{\Sigma}))\longrightarrow \mathrm{Rep}^{\mathcal{PM}}_{\mathrm{cont}}(\Delta_{(U,\overline{s})},\tilde{\overline{\mathcal{A}}}_{\mathbb{Q}_p}),$$

$$T_{(U,\overline{s}),\mathrm{HIGGS}}\colon \mathrm{HC}_{\mathbb{Z}_p,U_1\text{-fin}}(X_1/D(\overline{\Sigma}))\longrightarrow \mathrm{Rep}^{\mathcal{LPM}}_{\mathrm{cont}}(\Delta_{(U,\overline{s})},\tilde{\overline{\mathcal{A}}}).$$

We will need the following lemma in IV.6.4. We define the category $\mathrm{Rep}^{\mathcal{LPM}}_{\mathrm{cont}}(\tilde{\overline{\mathcal{A}}}_\bullet)$ as follows. An object is an inverse system of $\tilde{\overline{\mathcal{A}}}_m:=\tilde{\overline{\mathcal{A}}}/p^m\tilde{\overline{\mathcal{A}}}$-modules endowed with semi-linear actions of $\Delta_{(U,\overline{s})}$ for $m\in\mathbb{N}$ such that the underlying inverse system of $\tilde{\overline{\mathcal{A}}}_m$-modules is contained in $\mathcal{LPM}(\tilde{\overline{\mathcal{A}}}_\bullet)$. A morphism is a morphism in $\mathcal{LPM}(\tilde{\overline{\mathcal{A}}}_\bullet)$ compatible with the actions of $\Delta_{(U,\overline{s})}$.

**Lemma IV.5.2.14.** (1) *The functor $\mathrm{Rep}^{\mathcal{LPM}}_{\mathrm{cont}}(\tilde{\overline{\mathcal{A}}})\to\mathrm{Rep}^{\mathcal{LPM}}_{\mathrm{cont}}(\tilde{\overline{\mathcal{A}}}_\bullet); T\mapsto(T\otimes_{\tilde{\overline{\mathcal{A}}}}\tilde{\overline{\mathcal{A}}}_m)_m$ is an equivalence of categories. A quasi-inverse is given by $(M_m)\mapsto\varprojlim_m M_m$.*

(2) *The functor $\mathrm{Rep}^{\mathcal{LPM}}_{\mathrm{cont}}(\tilde{\overline{\mathcal{A}}})_{\mathbb{Q}}\to\mathrm{Rep}^{\mathcal{LPM}}_{\mathrm{cont}}(\tilde{\overline{\mathcal{A}}}_{\mathbb{Q}_p})$ is an equivalence of categories.*

PROOF. The claim (1) follows from Proposition IV.3.2.4. By Lemma IV.3.2.2 (1), we see that the functor in (2) is fully faithful. It remains to prove that it is essentially surjective. Let $W$ be an object of $\mathrm{Rep}^{\mathcal{PM}}_{\mathrm{cont}}(\Delta_{(U,\overline{s})},\tilde{\overline{\mathcal{A}}}_{\mathbb{Q}_p})$ and choose a generator $x_1,\ldots,x_r\in W$ as an $\tilde{\overline{\mathcal{A}}}_{\mathbb{Q}_p}$-module and let $T$ be the $\tilde{\overline{\mathcal{A}}}$-module generated by $x_i$ ($1\le i\le r$). Then, for each $i\in\{1,\ldots,r\}$, the map $\Delta_{(U,\overline{s})}\to W/T; g\mapsto(g(x_i)\mod T)$ is continuous, $\Delta_{(U,\overline{s})}$ is compact, and $W/T$ is discrete. Hence the image of the map is finite and the $\tilde{\overline{\mathcal{A}}}$-submodule generated by $g(x_i)$ ($g\in\Delta_{(U,\overline{s})}, i\in\{1,\ldots,r\}$) is finitely generated and stable under the action of $\Delta_{(U,\overline{s})}$. This completes the proof of (2). □

**Proposition IV.5.2.15.** *Let $r \in \mathbb{N}_{>0}$ and let $\mathcal{F}$ be a Higgs isocrystal on $(U_1/D(\overline{\Sigma}))^r_{\mathrm{HIGGS}}$ finite on $U_1$. Let $(U_1/D(\overline{\Sigma}))_{\mathrm{HIGGS}}$ denote the inverse limit of the fibered site $(s \mapsto (U_1/D(\overline{\Sigma}))^s_{\mathrm{HIGGS}})$ defined as in IV.4.5, and let $\mathcal{F}^\dagger$ be the pull-back of $\mathcal{F}$ on $(U_1/D(\overline{\Sigma}))_{\mathrm{HIGGS}}$. Then there exists a canonical morphism*

$$R\Gamma((U_1/D(\overline{\Sigma}))_{\mathrm{HIGGS}}, \mathcal{F}^\dagger) \longrightarrow C^\bullet_{\mathrm{cont}}(\Delta_{(U,\overline{s})}, V^r_{U,\mathrm{HIGGS}}(\mathcal{F})) \tag{IV.5.2.16}$$

*in $D^+(C\text{-Vect})$ functorial in $\mathcal{F}$. Here $C^\bullet_{\mathrm{cont}}(\Delta_{(U,\overline{s})}, -)$ denotes the inhomogeneous continuous cochain complex and $C$ denotes the completion of $\overline{K}$ with respect to the valuation.*

Proof. Put $X = U$ and choose $X \hookrightarrow Y$ and $Y_N$ $(N \in \mathbb{N}_{>0})$ as in the proof of Corollary IV.5.2.13. For an integer $s \geq r$, let $\mathcal{F}^s$ denote the inverse image of $\mathcal{F}^r$ on $(U_1/D(\overline{\Sigma}))^s_{\mathrm{HIGGS}}$. Put $W = V^r_{(U,\overline{s}),\mathrm{HIGGS}}(\mathcal{F}) = V^s_{(U,\overline{s}),\mathrm{HIGGS}}(\mathcal{F}^s)$ $(s \geq r)$, $A_{D^s_1} = \Gamma(D^s_1, \mathcal{O}_{D^s_1})$, and $M^s = \Gamma(D^s_1, \mathcal{F}^s_{D^s})$ $(s \geq r)$, where $D^s = D^s_{\mathrm{HIGGS}}(X \hookrightarrow Y_\bullet)$ as before Proposition IV.5.2.12. From Proposition IV.5.2.12 and (IV.5.2.9), we obtain morphisms of complexes

$$W \longrightarrow \xi^{-\bullet}\mathscr{A}^s_1(\overline{A}) \otimes_{\widetilde{\mathcal{A}}} W \otimes_A \Omega^\bullet \xrightarrow{\cong} \xi^{-\bullet}\mathscr{A}^s_1(\overline{A}) \otimes_{A_{D^s_1}} M^s \otimes_A \Omega^\bullet$$

compatible with $s$ in the obvious sense. By Proposition IV.5.2.10 (2), we obtain a quasi-isomorphism

$$C^\bullet_{\mathrm{cont}}(\Delta_{(U,\overline{s})}, W) \longrightarrow \varinjlim_{s \geq r} C^\bullet_{\mathrm{cont}}(\Delta_{(U,\overline{s})}, \xi^{-\bullet}\mathscr{A}^s_1(\overline{A}) \otimes_{A_{D^s_1}} M^s \otimes_A \Omega^\bullet). \tag{IV.5.2.17}$$

Let $z_{D^s}$ be the canonical morphism $D^s_1 \to U_1$. Since $U_1$ and $z_{D^s}$ are affine, Lemma IV.4.2.2 implies that the natural morphism

$$\xi^{-\bullet}M^s \otimes_A \Omega^\bullet \longrightarrow R\Gamma(U_{1,\text{ét}}, z_{D^s *}(\xi^{-\bullet}\mathcal{F}^s_{D^s} \otimes_{\mathcal{O}_{Y_1}} \Omega^\bullet_{Y_1/\overline{\Sigma}}))$$

is an isomorphism in $D^+(C\text{-Vect})$. By Lemma IV.4.4.6 (4), **[2]** VI Corollaire 5.2 and Corollary IV.4.5.7, we obtain an isomorphism

(IV.5.2.18)
$$R\Gamma((U_1/D(\overline{\Sigma}))_{\mathrm{HIGGS}}, \mathcal{F}^\dagger) \cong R\Gamma(U_{1,\text{ét}}, Ru_{U_1/D(\overline{\Sigma})*}\mathcal{F}^\dagger) \cong \varinjlim_{s \geq r}(\xi^{-\bullet}M^s \otimes_A \Omega^\bullet).$$

We have a natural morphism of complexes from the last term of (IV.5.2.18) to the target of (IV.5.2.17), which induces the desired morphism.

We need to verify the independence of the choice of $(i\colon X \to Y, Y_\bullet)$. Choose another $(i'\colon X \to Y', Y'_\bullet)$. By considering $(X \to Y \times_\Sigma Y', Y_\bullet \times_{D(\overline{\Sigma})} Y'_\bullet)$, we are reduced to the case where we are given morphisms $f\colon Y' \to Y$ and $f_\bullet\colon Y'_\bullet \to Y_\bullet$ compatible with $i$, $i'$, $Y_1 = \widehat{Y} \times_{\widehat{\Sigma}} \overline{\Sigma}$, and $Y'_1 = \widehat{Y}' \times_{\widehat{\Sigma}} \overline{\Sigma}$. Let $\overline{D}^s$ be $D^s_{\mathrm{HIGGS}}(\overline{U} \hookrightarrow Y_\bullet \times_{D(\overline{\Sigma})} D(\overline{U}))$ and let $\overline{M}^s$ be $\Gamma(\overline{D}^s, \mathcal{F}^s)$. We define $\mathscr{A}'^s_1(\overline{A})$, $\overline{D}'^s$, $M'^s$, etc. in the same way as those without $'$ using $(i', Y'_\bullet)$. Then the independence follows from the commutative diagram

$$\begin{array}{ccccc}
\mathscr{A}^s_1(\overline{A}) \otimes_{\widetilde{\mathcal{A}}} W & \xrightarrow{\cong} & \overline{M}^s & \xleftarrow{\cong} & \mathscr{A}^s_1(\overline{A}) \otimes_{A_{D^s_1}} M^s \\
\downarrow & & \downarrow & & \downarrow \\
\mathscr{A}'^s_1(\overline{A}) \otimes_{\widetilde{\mathcal{A}}} W & \xrightarrow{\cong} & \overline{M}'^s & \xleftarrow{\cong} & \mathscr{A}'^s(\overline{A}) \otimes_{A_{D'^s_1}} M'^s
\end{array}$$

compatible with $\theta$'s and the commutative diagram

$$\begin{array}{ccc} & & \varinjlim_{s\geq r} z_{D^s*}(\xi^{-\bullet}\mathcal{F}^s_{D^s}\otimes_{\mathcal{O}_{Y_1}}\Omega^{\bullet}_{Y_1/\overline{\Sigma}}) \\ & \nearrow & \\ Ru_{U_1/D(\overline{\Sigma})*}(\mathcal{F}^{\dagger}) & & \downarrow \\ & \searrow & \\ & & \varinjlim_{s\geq r} z_{D'^s*}(\xi^{-\bullet}\mathcal{F}^s_{D'^s}\otimes_{\mathcal{O}_{Y'_1}}\Omega^{\bullet}_{Y'_1/\overline{\Sigma}}). \end{array}$$

Here the vertical morphisms are the ones naturally induced from $f$ and $f_\bullet$, and the second commutative diagram is derived from the following commutative diagram.

$$\begin{array}{ccc} & & Q^*((L_{Y_\star}(\xi^{-\bullet}\mathcal{F}^s_{D^s}\otimes_{\mathcal{O}_{Y_1}}\Omega^{\bullet}_{Y_1/\overline{\Sigma}}))_{s\geq r}) \\ & \nearrow^{\text{Prop.IV.4.5.5}} & \\ \mathcal{F}^{\dagger} & & \downarrow \\ & \searrow_{\text{Prop.IV.4.5.5}} & \\ & & Q^*((L_{Y'_\star}(\xi^{-\bullet}\mathcal{F}^s_{D'^s}\otimes_{\mathcal{O}_{Y'_1}}\Omega^{\bullet}_{Y'_1/\overline{\Sigma}}))_{s\geq r}). \end{array}$$

□

**IV.5.3. Semi-stable reduction case.** We keep the notation and the assumption in IV.5.2 and assume that $U\times_{\mathrm{Spec}(V)}\mathrm{Spec}(k)$ is nonempty. In the case I (resp. the cases II and III), we further assume that $U$ (resp. $U\times_{\mathrm{Spec}(V)}\mathrm{Spec}(k)$) is connected. Then $\mathcal{A}$ and $\mathcal{A}\otimes_V\overline{V}$ are normal domains and the homomorphism $\mathcal{A}\otimes_V\overline{V}\to\overline{\mathcal{A}}$ induced by $\bar{s}\to\mathcal{U}_{\overline{K},\mathrm{triv}}$ is injective (cf. Lemma IV.5.2.1). We further assume that the log structure of $\Sigma$ is defined by the closed point and $U\to\Sigma$ satisfies the following condition.

**Condition IV.5.3.1.** The morphism $U\to\Sigma$ admits a chart $(\mathbb{N}_\Sigma\to M_\Sigma,\mathbb{N}^{d+1}_U\to M_U, h\colon\mathbb{N}\to\mathbb{N}^{d+1})$ such that the morphism $U\to\mathrm{Spec}(\mathbb{Z})[\mathbb{N}^{d+1}]\times_{\mathrm{Spec}(\mathbb{Z})[\mathbb{N}]}\Sigma$ induced by the chart is strict étale and the morphism $h$ is given by $\mathbb{N}\to\mathbb{N}^{d+1}; 1\mapsto\sum_{1\leq i\leq e}\underline{1}_i$ for some integer $e\geq 1$. Here, for a monoid $P$, $\mathrm{Spec}(\mathbb{Z})[P]$ denotes $\mathrm{Spec}(\mathbb{Z}[P])$ endowed with the log structure associated to $P\hookrightarrow\mathbb{Z}[P]$, and $\underline{1}_i$ denotes the element of $\mathbb{N}^{d+1}$ whose $i$-th component is 1 and other components are 0.

We choose and fix such a chart as in Condition IV.5.3.1 in the following. Under this assumption, we will prove the following theorems.

**Theorem IV.5.3.2** (cf. [27] §3). *Let $r\in\mathbb{N}_{>0}$ and let $\mathcal{F}$ be a Higgs isocrystal on $(U_1/D(\overline{\Sigma}))^r_{\mathrm{HIGGS}}$ finite on $U_1$. Let $(U_1/D(\overline{\Sigma}))^{\sim}_{\mathrm{HIGGS}}$ denote the inverse limit of the fibered site $(s\mapsto(U_1/D(\overline{\Sigma}))^s_{\mathrm{HIGGS}})$ defined as in* IV.4.5*, and let $\mathcal{F}^{\dagger}$ be the pull-back of $\mathcal{F}$ on $(U_1/D(\overline{\Sigma}))_{\mathrm{HIGGS}}$. Then the morphism* (IV.5.2.16) *in Proposition* IV.5.2.15 *is an isomorphism in $D^+(C\text{-Vect})$.*

**Theorem IV.5.3.3** (cf. [27] Theorem 3). *For $r\in\mathbb{N}_{>0}\cup\{\infty\}$, the functor*

$$V^r_{(U,\bar{s}),\mathrm{HIGGS}}\colon \mathrm{HC}^r_{\mathbb{Q}_p,U_1\text{-fin}}(U_1/D(\overline{\Sigma}))\to\mathrm{Rep}^{\mathcal{PM}}_{\mathrm{cont}}(\Delta_{(U,\bar{s})},\widetilde{\overline{\mathcal{A}}}_{\mathbb{Q}_p})$$

*is fully faithful.*

Put $X=U$. Since $U$ is smooth over $\Sigma$, we can use $Y=U$ and a compatible system of smooth liftings $Y_N$ ($N\in\mathbb{N}_{>0}$) of $Y_1$ to define the rings $\mathscr{A}^r_N(\overline{\mathcal{A}})$ ($r\in\mathbb{N}_{>0}\cup\{\infty\}$, $N\in\mathbb{N}_{>0}$). We will also discuss the admissibility of Higgs vector bundles with respect to the period ring $\mathscr{A}^r_1(\overline{\mathcal{A}})$ in Proposition IV.5.3.10.

Let $t_i\in\Gamma(Y,M_Y)=\Gamma(U,M_U)$ ($0\leq i\leq d$) be the image of $\underline{1}_{i+1}\in\mathbb{N}^{d+1}$ under the chart $\mathbb{N}^{d+1}_U\to M_U$. Then $t_i$ ($1\leq i\leq d$) satisfy Condition IV.5.2.3 (ii). Choosing a lifting

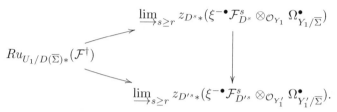

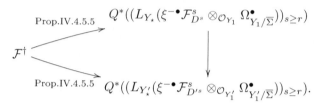

$(t_{i,N}) \in \varprojlim_N \Gamma(Y_N, M_{Y_N})$ of the image of $t_i$ in $\Gamma(Y_1, M_{Y_1})$, we obtain the isomorphism (IV.5.2.5). We will derive Theorems IV.5.3.2 and IV.5.3.3 from the following theorem.

**Theorem IV.5.3.4.** *With the above notation, we have:*

(1) $\mathscr{A}_1^{\infty}(\overline{\mathcal{A}})^{\Delta_{(U,\overline{s})}} = (\mathcal{A} \otimes_V \overline{V})^{\wedge}$,

(2) $\varinjlim_{r\in\mathbb{N}_{>0}} H^i_{\mathrm{cont}}(\Delta_{(U,\overline{s})}, \mathscr{A}_1^r(\overline{\mathcal{A}})_{\mathbb{Q}_p}) = \begin{cases} 0 & \text{if } i>0, \\ (\mathcal{A}\otimes_V \overline{V})^{\wedge}_{\mathbb{Q}_p} & \text{if } i=0, \end{cases}$

*where* $^\wedge$ *denotes the p-adic completion.*

Let $\pi$ denote the image of $1 \in \mathbb{N}$ in $\Gamma(\Sigma, M_\Sigma) \subset \Gamma(\Sigma, \mathcal{O}_\Sigma) = V$ by the chart of $M_\Sigma$, which is a uniformizer of $V$. We have an étale homomorphism $V[T_0, \ldots, T_d]/(T_0 \cdots T_{e-1} - \pi) \to A; T_i \mapsto t_i$. Choose a compatible system $(\pi_n)$ of $p^n$-th roots of $\pi$ in $\overline{V}$, let $V_n$ be $V[\pi_n]$, and define $A_n$ to be the cofiber product of

$$V_n[T_{0,n}, \ldots, T_{d,n}]/(T_{0,n} \cdots T_{e-1,n} - \pi_n) \longleftarrow V[T_0, \ldots, T_d]/(T_0 \cdots T_{e-1} - \pi) \longrightarrow A,$$

where the left homomorphism is defined by $T_i \mapsto (T_{i,n})^{p^n}$. The ring $A_n$ is normal and flat over $V_n$, $\mathrm{Spec}(A_n/\pi_n A_n)$ is reduced, and the morphism $\mathrm{Spec}(A_n/\pi_n A_n) \to \mathrm{Spec}(A/\pi A)$ is a homeomorphism. Observing that the extension of residue fields for $A \otimes_V V_n \to A_n$ at height one primes containing $p$ are purely inseparable of degree $p^{nd}$, we see that $A_n$ is a normal domain. Note that $A \otimes_V V_n$ is a normal domain by Lemma IV.5.2.1. We put $\mathcal{A}_n = \mathcal{A} \otimes_A A_n$, which is the henselization of $(A_n, pA_n)$ (resp. the $p$-adic completion of $A_n$) in the case II (resp. III). We have a natural injective homomorphism $\mathcal{A}_n \otimes_{V_n} \overline{V} \to \mathcal{A}_{n+1} \otimes_{V_{n+1}} \overline{V}$. By Lemma IV.5.2.1, we see that $\mathcal{A}_n \otimes_{V_n} \overline{V}$ is a normal domain. Since $\mathcal{A}_{\mathbb{Q}_p} \to \mathcal{A}_{n,\mathbb{Q}_p}$ is finite étale, there exists an injective homomorphism $\mathcal{A}_n \otimes_{V_n} \overline{V} \hookrightarrow \overline{\mathcal{A}}$ compatible with $n$. We choose and fix such an embedding, regard $\mathcal{A}_n \otimes_{V_n} \overline{V}$ as a subring of $\overline{\mathcal{A}}$, and define $\mathcal{A}_\infty$ to be the union of $\mathcal{A}_n \otimes_{V_n} \overline{V}$. We write $t_{i,n}$ for the image of $T_{i,n}$ in $\mathcal{A}_\infty$.

For $n \in \mathbb{N}$, let $\mathcal{K}_n$ denote the field of fractions of $\mathcal{A}_n \otimes_{V_n} \overline{V}$, and put $\mathcal{K}_\infty := \cup_{n\in\mathbb{N}} \mathcal{K}_n$. Then, for $n \in \mathbb{N} \cup \{\infty\}$, $\mathcal{K}_n$ is a Galois extension of $\mathcal{K}\overline{K}$. Let $\Delta_n$ denote its Galois group. For $n \in \mathbb{N}$, we have an isomorphism $\Delta_n \cong \mu_{p^n}^{\oplus d}$ defined by $g \mapsto (\varepsilon_{i,n})$, $g(t_{i,n}) = t_{i,n}\varepsilon_{i,n}$ $(1 \le i \le d)$. This isomorphism is compatible with $n$ and induces an isomorphism

$$\text{(IV.5.3.5)} \qquad \Delta_\infty \xrightarrow{\cong} \mathbb{Z}_p(1)^{\oplus d}.$$

Let $H$ be the kernel of the homomorphism $\Delta_{(U,\overline{s})} \to \Delta_\infty$.

We use the following theorem of Faltings which is a consequence of his almost purity theorem for almost étale extensions. Let $\overline{\mathfrak{m}}$ denote the maximal ideal of $\overline{V}$.

**Theorem IV.5.3.6** (G. Faltings **[26]**). (1) *For* $n \in \mathbb{N}_{>0}$, $H^q(H, \overline{\mathcal{A}}/p^n\overline{\mathcal{A}})$ $(q > 0)$ *is annihilated by* $\overline{\mathfrak{m}}$ *and the cokernel of the injective homomorphism* $\mathcal{A}_\infty/p^n\mathcal{A}_\infty \to H^0(H, \overline{\mathcal{A}}/p^n\overline{\mathcal{A}})$ *is annihilated by* $\overline{\mathfrak{m}}$.

(2) *The absolute Frobenius of* $\overline{\mathcal{A}}/p\overline{\mathcal{A}}$ *is surjective in the case* I *(i.e.,* $\mathcal{A} = A$*). In particular, we have* $\widetilde{\overline{\mathcal{A}}} = \widehat{\overline{\mathcal{A}}}$ *(cf. Lemma* IV.5.1.1*).*

We define $w_{i,1} \in \mathscr{A}_1^r(\overline{\mathcal{A}})$ appearing in the isomorphism (IV.5.2.5) by using $t_i$ and $((t_{i,n}), t_i) \in Q$. We abbreviate $w_{i,1}$ to $w_i$ in the following. Since $F^n A_N(\overline{\mathcal{A}}) = \xi^n A_N(\overline{\mathcal{A}})$, we have $A_1(\overline{\mathcal{A}})^{(m)_r} = p^{\lceil \frac{m}{r} \rceil}\widehat{\overline{\mathcal{A}}}$ (cf. IV.2.3) for $r \in \mathbb{N}_{>0}$ and $m \in \mathbb{N}$, where $\lceil x \rceil$ $(x \in \mathbb{R})$ denotes the smallest integer $l$ satisfying $x \le l$. We understand $\frac{m}{\infty}$ to be $0$ for $m \in \mathbb{N}$ in the following. For $\underline{m} = (m_1, m_2, \ldots, m_d) \in \mathbb{N}^d$, we write $w^{\underline{m}}$ for $\prod_{1 \le i \le d} w_i^{m_i}$. Then, for

$r \in \mathbb{N}_{>0} \cup \{\infty\}$, we have

$$\mathscr{A}_1^r(\overline{\mathcal{A}}) = \left(\bigoplus_{\underline{m}\in\mathbb{N}^d} p^{\lceil \frac{|\underline{m}|}{r} \rceil}\widehat{\overline{\mathcal{A}}} w^{\underline{m}}\right)^{\wedge}. \tag{IV.5.3.7}$$

For $r \in \mathbb{N}_{>0} \cup \{\infty\}$, we define the subalgebras $\mathscr{A}_1^{r,\circ}(\overline{\mathcal{A}})$ and $\mathscr{A}_1^{r,\circ}(\mathcal{A}_\infty)$ of $\mathscr{A}_1^r(\overline{\mathcal{A}})_{\mathbb{Q}_p}$ as follows.

$$\mathscr{A}_1^{r,\circ}(\overline{\mathcal{A}}) = \left\{ \sum_{\underline{m}\in\mathbb{N}^d} a_{\underline{m}} w^{\underline{m}} \in \mathscr{A}_1^r(\overline{\mathcal{A}})_{\mathbb{Q}_p} \middle| a_{\underline{m}} \in p^{\frac{|\underline{m}|}{r}} \widehat{\overline{\mathcal{A}}} \text{ for all } \underline{m} \in \mathbb{N}^d \right\},$$

$$\mathscr{A}_1^{r,\circ}(\mathcal{A}_\infty) = \left\{ \sum_{\underline{m}\in\mathbb{N}^d} a_{\underline{m}} w^{\underline{m}} \in \mathscr{A}_1^r(\overline{\mathcal{A}})_{\mathbb{Q}_p} \middle| a_{\underline{m}} \in p^{\frac{|\underline{m}|}{r}} \widehat{\mathcal{A}}_\infty \text{ for all } \underline{m} \in \mathbb{N}^d \right\}.$$

By (IV.5.2.7), we see that $\mathscr{A}_1^{r,\circ}(\overline{\mathcal{A}})$ and $\mathscr{A}_1^{r,\circ}(\mathcal{A}_\infty)$ are stable under the action of $\Delta_{(U,\overline{s})}$, and the action on the latter factors through $\Delta_\infty$. We have $\mathscr{A}_1^{\infty}(\overline{\mathcal{A}}) = \mathscr{A}_1^{\infty,\circ}(\overline{\mathcal{A}})$ and $\mathscr{A}_1^r(\overline{\mathcal{A}}) \subset \mathscr{A}_1^{r,\circ}(\overline{\mathcal{A}}) \subset p^{-1}\mathscr{A}_1^r(\overline{\mathcal{A}})$ if $r \in \mathbb{N}_{>0}$. We endow $\mathscr{A}_1^{r,\circ}(\mathcal{A}_\infty)_{\mathbb{Q}_p}$ with the $p$-adic topology induced by $\mathscr{A}_1^{r,\circ}(\mathcal{A}_\infty)$. Since the intersection of $\mathscr{A}_1^{r,\circ}(\mathcal{A}_\infty)_{\mathbb{Q}_p}$ and $\mathscr{A}_1^{r,\circ}(\overline{\mathcal{A}})$ is $\mathscr{A}_1^{r,\circ}(\mathcal{A}_\infty)$, it is the same as the topology induced by that of $\mathscr{A}_1^r(\overline{\mathcal{A}})_{\mathbb{Q}_p}$.

Theorem IV.5.3.6 (1) above implies the following.

**Corollary IV.5.3.8.** *For $r \in \mathbb{N}_{>0} \cup \{\infty\}$ and $i \in \mathbb{N}$, the kernel and the cokernel of the homomorphism*

$$H^i_{\text{cont}}(\Delta_\infty, \mathscr{A}_1^{r,\circ}(\mathcal{A}_\infty)) \longrightarrow H^i_{\text{cont}}(\Delta_{(U,\overline{s})}, \mathscr{A}_1^r(\overline{\mathcal{A}}))$$

*are annihilated by $\overline{\mathfrak{m}}$.*

Proof. The action of $H$ on $w_i$ is trivial. Hence, by Theorem IV.5.3.6 and the Hochschild-Serre spectral sequence for $H \subset \Delta_{(U,\overline{s})}$, we see that the kernel and the cokernel of the homomorphism

$$H^i(\Delta_\infty, \mathscr{A}_1^{r,\circ}(\mathcal{A}_\infty)/p^n) \longrightarrow H^i(\Delta_{(U,\overline{s})}, \mathscr{A}_1^{r,\circ}(\overline{\mathcal{A}})/p^n)$$

are annihilated by $\overline{\mathfrak{m}}$. Since $\mathscr{A}_1^{r,\circ}(\overline{\mathcal{A}})$ is $p$-adically complete and separated, we have the following exact sequence (cf. **[47]** §2):

$$0 \longrightarrow R^1 \varprojlim_n H^{i-1}(\Delta_{(U,\overline{s})}, \mathscr{A}_1^{r,\circ}(\overline{\mathcal{A}})/p^n) \to H^i_{\text{cont}}(\Delta_{(U,\overline{s})}, \mathscr{A}_1^{r,\circ}(\overline{\mathcal{A}}))$$
$$\longrightarrow \varprojlim_n H^i(\Delta_{(U,\overline{s})}, \mathscr{A}_1^{r,\circ}(\overline{\mathcal{A}})/p^n) \to 0.$$

We have the same exact sequence for $\mathcal{A}_\infty$ and $\Delta_\infty$. Hence the kernel and the cokernel of the homomorphism

$$H^i_{\text{cont}}(\Delta_\infty, \mathscr{A}_1^{r,\circ}(\mathcal{A}_\infty)) \longrightarrow H^i_{\text{cont}}(\Delta_{(U,\overline{s})}, \mathscr{A}_1^{r,\circ}(\overline{\mathcal{A}}))$$

are annihilated by $\overline{\mathfrak{m}}$. □

The isomorphism (IV.5.3.5) induces an isomorphism

$$\text{Hom}_{\text{cont}}(\Delta_\infty, \mu_{p^\infty}(\overline{V})) \cong (\mathbb{Q}_p/\mathbb{Z}_p)^{\oplus d}.$$

For $\underline{\alpha} \in (\mathbb{Q}_p/\mathbb{Z}_p)^{\oplus d}$, we denote by $\chi_{\underline{\alpha}}$ the corresponding character of $\Delta_\infty$.

The ring $\overline{V}[T_{0,n}, \ldots, T_{d,n}]/(\prod_{0\leq i\leq e-1} T_{i,n} - \pi_n)$ is a free $\overline{V}$-module with basis $T^{\underline{m}}_{\bullet,n} := \prod_{0\leq i\leq d} T_{i,n}^{m_i}$ ($\underline{m} = (m_0, m_1, \ldots, m_d) \in \mathscr{S}$), where $\mathscr{S} = \mathbb{N}^{d+1} - ((\mathbb{N}_{>0})^e \times \mathbb{N}^{d+1-e})$. For $\underline{l} = (l_1, l_2, \ldots, l_d) \in (\mathbb{N} \cap [0, p^n[)^d$, put

$$\mathscr{S}^n_{\underline{l}} := \left\{ (m_0, m_1, \ldots, m_d) \in \mathscr{S} \,\middle|\, \begin{matrix} m_i - m_0 \equiv l_i \bmod p^n (1 \leq i \leq e-1), \\ m_i \equiv l_i \bmod p^n (e \leq i \leq d). \end{matrix} \right\}.$$

Then $\mathscr{S}$ is a disjoint union of $\mathscr{S}_{\underline{l}}^n$ ($\underline{l} \in (\mathbb{N} \cap [0, p^n[)^d$) and $\oplus_{\underline{m} \in \mathscr{S}_{\underline{l}}^n} \overline{V} \cdot T_{\bullet,n}^{\underline{m}}$ is stable under the action of $\overline{V}[T_0, \ldots, T_d]/(T_0 \cdots T_{e-1} - \pi)$. This implies that $\mathcal{A}_\infty$ has a decomposition as an $\mathcal{A} \otimes_V \overline{V}$-module

$$\mathcal{A}_\infty = \bigoplus_{\underline{\alpha} \in (\mathbb{Q}_p/\mathbb{Z}_p)^{\oplus d}} \mathcal{A}_{\infty, \underline{\alpha}},$$

where $\mathcal{A}_{\infty,\underline{\alpha}}$ is $\Delta_\infty$-stable and $g(x) = \chi_{\underline{\alpha}}(g)x$ for $x \in \mathcal{A}_{\infty,\underline{\alpha}}$ and $g \in \Delta_\infty$. We have $\mathcal{A}_{\infty,0} = \mathcal{A} \otimes_V \overline{V}$.

For $a \in \mathbb{N}$, $0 \leq a \leq d$, let $I_a$ (resp. $J_a$) denote the submodule of $(\mathbb{Q}_p/\mathbb{Z}_p)^{\oplus d}$ (resp. the subset of $\mathbb{N}^d$) consisting of elements whose first $a$ components are 0. For $r \in \mathbb{N}_{>0} \cup \{\infty\}$ and $a \in \mathbb{N} \cap [0, d]$, we define ${}_a\mathscr{A}^r$ to be the $\Delta_\infty$-stable submodule

$$\Bigg(\bigoplus_{(\underline{m},\underline{\alpha}) \in J_a \times I_a} p^{\frac{|\underline{m}|}{r}} \mathcal{A}_{\infty,\underline{\alpha}} w^{\underline{m}}\Bigg)^\wedge$$

of $\mathscr{A}_1^{r,\circ}(\mathcal{A}_\infty)$, where $^\wedge$ denotes the $p$-adic completion. We have

$${}_0\mathscr{A}^r = \mathscr{A}_1^{r,\circ}(\mathcal{A}_\infty), \qquad {}_d\mathscr{A}^r = (\mathcal{A} \otimes_V \overline{V})^\wedge.$$

Choose a generator $\varepsilon = (\varepsilon_n)$ of $\mathbb{Z}_p(1)(\overline{V})$ and let $\gamma_i \in \Delta_\infty$ be the element corresponding to the element of $\mathbb{Z}_p(1)^{\oplus d}$ whose $i$-th component is $\varepsilon$ and other components are 0.

**Proposition IV.5.3.9.** *Let $a$ be an integer such that $1 \leq a \leq d$.*

(1) *For $r \in \mathbb{N}_{>0} \cup \{\infty\}$, the $\gamma_a$-invariant part of ${}_{a-1}\mathscr{A}^r$ is ${}_a\mathscr{A}^r$.*

(2) *For $r \in \mathbb{N}_{>1}$, there exists an integer $l$ such that $p^l \cdot {}_{a-1}\mathscr{A}^{r-1}$ is contained in the image of $\gamma_a - 1 \colon {}_{a-1}\mathscr{A}^r \to {}_{a-1}\mathscr{A}^r$.*

PROOF. The action of $\gamma_a$ on $w_i$ ($i \neq a$) is trivial. Hence it suffices to prove the following two claims. Let $\mathscr{A}_{a,\underline{\alpha}}^r$ be the submodule

$$\Bigg(\bigoplus_{\nu \in \mathbb{N}} p^{\frac{\nu}{r}} \mathcal{A}_{\infty,\underline{\alpha}} w_a^\nu\Bigg)^\wedge$$

of $\mathscr{A}_1^{r,\circ}(\mathcal{A}_\infty)$, which is stable under the action of $\gamma_a$.

(i) For $r \in \mathbb{N}_{>0} \cup \{\infty\}$, the $\gamma_a$-invariant part of $\mathscr{A}_{a,\underline{\alpha}}^r$ is 0 if $\chi_{\underline{\alpha}}(\gamma_a) \neq 1$ and $\widehat{\mathcal{A}}_{\infty,\underline{\alpha}}$ if $\chi_{\underline{\alpha}}(\gamma_a) = 1$.

(ii) For $r \in \mathbb{N}_{>1}$, there exists an integer $l$ independent on $\underline{\alpha}$ such that $p^l \mathscr{A}_{a,\underline{\alpha}}^{r-1}$ is contained in the image of $\gamma_a - 1 \colon \mathscr{A}_{a,\underline{\alpha}}^r \to \mathscr{A}_{a,\underline{\alpha}}^r$.

For $a_\nu \in \widehat{\mathcal{A}}_{\infty,\underline{\alpha}}$ ($\nu \in \mathbb{N}$), $a_\nu \to 0$ ($\nu \to \infty$), we have

$$(\gamma_a - 1)\Big(\sum_{\nu \in \mathbb{N}} p^{\frac{\nu}{r}} a_\nu w_a^\nu\Big) = \sum_{\nu \in \mathbb{N}} p^{\frac{\nu}{r}} b_\nu w_a^\nu,$$

$$b_\nu = (\chi_{\underline{\alpha}}(\gamma_a) - 1)a_\nu + \sum_{\mu > \nu} p^{\frac{1}{r}(\mu-\nu)} \chi_{\underline{\alpha}}(\gamma_a) a_\mu \binom{\mu}{\nu} \eta^{\mu-\nu}$$

by (IV.5.2.7), where $\eta$ is the element $(\xi^{-1}([\underline{\varepsilon}^{-1}] - 1) \mod \xi) \in \overline{V}$. We have $v_p(\eta) = \frac{1}{p-1}$, where the valuation $v_p$ of $\overline{V}$ is normalized by $v_p(p) = 1$.

Proof of (i). Suppose $b_\nu = 0$ for all $\nu \in \mathbb{N}$. If $\chi_{\underline{\alpha}}(\gamma_a) \neq 1$, then $v_p(\chi_{\underline{\alpha}}(\gamma_a) - 1) \leq \frac{1}{p-1}$, and we have

$$a_\nu = -(\chi_{\underline{\alpha}}(\gamma_a) - 1)^{-1} \sum_{\mu > \nu} p^{\frac{1}{r}(\mu-\nu)} \chi_{\underline{\alpha}}(\gamma_a) a_\mu \binom{\mu}{\nu} \eta^{\mu-\nu}.$$

For any $l \in \mathbb{N}$, there exists $\nu_l \in \mathbb{N}$ such that $a_\nu \in p^l \widehat{\mathcal{A}}_{\infty,\underline{\alpha}}$ for all $\nu \geq \nu_l$. Then, by the above equality for $0 \leq \nu \leq \nu_l - 1$, we see $a_\nu \in p^l \widehat{\mathcal{A}}_{\infty,\underline{\alpha}}$ for all $\nu \in \mathbb{N}$. Hence $a_\nu \in \cap_{l \in \mathbb{N}} p^l \widehat{\mathcal{A}}_{\infty,\underline{\alpha}} = 0$ for all $\nu \in \mathbb{N}$. If $\chi_{\underline{\alpha}}(\gamma_a) = 1$, then we have

$$(\nu+1)! a_{\nu+1} = -\sum_{\mu \geq \nu+2} p^{\frac{1}{r}(\mu-\nu-1)} \mu! a_\mu \frac{\eta^{\mu-\nu-1}}{(\mu-\nu)!}.$$

Since $v_p(\eta) = \frac{1}{p-1}$, we have $\eta^{m-1}/m! \in \overline{V}$ for $m \in \mathbb{N}_{>0}$. For any $l \in \mathbb{N}$, there exists $\nu_l \in \mathbb{N}$ such that $(\nu+1)! a_{\nu+1} \in p^l \widehat{\mathcal{A}}_{\infty,\underline{\alpha}}$ for all $\nu \geq \nu_l$. By the above equality for $0 \leq \nu \leq \nu_l - 1$, we see $(\nu+1)! a_{\nu+1} \in p^l \widehat{\mathcal{A}}_{\infty,\underline{\alpha}}$ for all $\nu \in \mathbb{N}$. Hence $(\nu+1)! a_{\nu+1} \in \cap_{l \in \mathbb{N}} p^l \widehat{\mathcal{A}}_{\infty,\underline{\alpha}} = 0$ for all $\nu \in \mathbb{N}$.

Proof of (ii). For the case $\chi_{\underline{\alpha}}(\gamma_a) \neq 1$, $\gamma_a - 1$ on $\mathscr{A}^r_{a,\underline{\alpha}}$ is divisible by $\chi_{\underline{\alpha}}(\gamma_a) - 1$ and it is congruent to the identity modulo $p^{\frac{1}{r}}$. Hence $(\chi_{\underline{\alpha}}(\gamma_a) - 1)^{-1}(\gamma_a - 1)$ is bijective and we may take $l = 1$ since $v_p(\chi_{\underline{\alpha}}(\gamma_a) - 1) \leq \frac{1}{p-1}$. Suppose $\chi_{\underline{\alpha}}(\gamma_a) = 1$. Let $\sum_\nu p^{\frac{\nu}{r-1}} c_\nu w_a^\nu$, $c_\nu \in \widehat{\mathcal{A}}_{\infty,\underline{\alpha}}$ be an element of $\mathscr{A}^{r-1}_{a,\underline{\alpha}}$. It suffices to show that there exists an integer $l \geq 1$ depending only on $r$ such that the equality

$$p^l p^{\frac{\nu}{r}} \nu! p^{(\frac{1}{r-1} - \frac{1}{r})\nu} c_\nu = \sum_{\mu > \nu} p^{\frac{\mu}{r}} \mu! a_\mu \frac{\eta^{\mu-\nu}}{(\mu-\nu)!}$$

has a solution $a_\nu \in \widehat{\mathcal{A}}_{\infty,\underline{\alpha}}$ such that $a_\nu \to 0$ as $\nu \to \infty$. Set $c'_\nu = p^l \nu! p^{(\frac{1}{r-1} - \frac{1}{r})\nu} p^{-\frac{\nu}{p-1}} c_\nu$ and $a'_\nu = \nu! p^{-\frac{\nu}{p-1}} a_\nu$. Then the above equation is written as

$$c'_\nu = p^{\frac{1}{r} + \frac{1}{p-1}} \eta \sum_{\mu \geq \nu+1} p^{(\frac{1}{r} + \frac{1}{p-1})(\mu-\nu-1)} a'_\mu \frac{\eta^{\mu-\nu-1}}{(\mu-\nu)!}.$$

The endomorphism of $(\oplus_{\nu \in \mathbb{N}_{>0}} \mathcal{A}_{\infty,\underline{\alpha}})^\wedge$ defined by

$$(x_\nu) \mapsto (y_\nu), \quad y_\nu = \sum_{\mu \geq \nu} p^{(\frac{1}{r} + \frac{1}{p-1})(\mu-\nu)} x_\mu \frac{\eta^{\mu-\nu}}{(\mu-\nu+1)!}$$

is congruent to the identity map modulo $p^{\frac{1}{r} + \frac{1}{p-1}}$. Hence the above equation has a solution $a'_\nu \in \widehat{\mathcal{A}}_{\infty,\underline{\alpha}}$, $\nu \in \mathbb{N}_{>0}$ such that $a'_\nu \to 0$ as $\nu \to \infty$ if $c'_\nu \in p^{\frac{1}{r} + \frac{1}{p-1}} \eta \widehat{\mathcal{A}}_{\infty,\underline{\alpha}}$ and $c'_\nu \to 0$ as $\nu \to \infty$. Since $\frac{p^{\frac{\nu}{p-1}}}{\nu!} \in \overline{V}$, we have $a_\nu \in \widehat{\mathcal{A}}_{\infty,\underline{\alpha}}$ and $a_\nu \to 0$ as $\nu \to \infty$ for such a solution $a'_\nu$. Since $v_p(\eta) = \frac{1}{p-1}$ and $\mathrm{ord}_p(\nu! p^{(\frac{1}{r-1} - \frac{1}{r})\nu} p^{-\frac{\nu}{p-1}}) \to \infty$ as $\nu \to \infty$, it is enough to take an integer $l$ such that $\inf\{\mathrm{ord}_p(\nu! p^{(\frac{1}{r-1} - \frac{1}{r})\nu} p^{-\frac{\nu}{p-1}}) | \nu \in \mathbb{N}\} - \frac{1}{r} - \frac{2}{p-1} \geq -l$. □

Proof of Theorem IV.5.3.4. The equality (1) follows from Proposition IV.5.3.9 (1) and Corollary IV.5.3.8. Let us prove (2). By Corollary IV.5.3.8, it suffices to prove

$$\varinjlim_r H^i_{\mathrm{cont}}(\Delta_\infty, \mathscr{A}^{r,\circ}_1(\mathcal{A}_\infty)_{\mathbb{Q}_p}) = \begin{cases} 0 & \text{if } i > 0, \\ (\mathcal{A} \otimes_V \overline{V})^\wedge_{\mathbb{Q}_p} & \text{if } i = 0. \end{cases}$$

For $r \in \mathbb{N}_{>0}$, define the complex $C^r_a$ $(0 \leq a \leq d)$ inductively by $C^r_0 = \mathscr{A}^{r,\circ}_1(\mathcal{A}_\infty)$ and $C^r_a = \mathrm{fiber}(C^r_{a-1} \xrightarrow{\gamma_a - 1} C^r_{a-1})$. Then the complex $C^\bullet_{\mathrm{cont}}(\Delta_\infty, \mathscr{A}^{r,\circ}_1(\mathcal{A}_\infty))$ is isomorphic to $C^r_d$ in the derived category $D^+(\mathbb{Z}_p\text{-Mod})$. By using Proposition IV.5.3.9, one can show that the natural homomorphism

$$\varinjlim_r {}_a\mathscr{A}^r_{\mathbb{Q}_p} \longrightarrow \varinjlim_r C^r_{a,\mathbb{Q}_p}$$

is a quasi-isomorphism by induction on $a$. The quasi-isomorphism for $a = d$ implies the desired claim. □

PROOF OF THEOREM IV.5.3.2. We apply the construction of (IV.5.2.16) in the proof of Proposition IV.5.2.15 to $Y = X = U$ and $Y_N$ ($N \in \mathbb{N}_{>0}$). We follow the notation in loc. cit. Then we have $A_{D_1^s} = (\mathcal{A} \otimes_V \overline{V})^\wedge$ and $M^s = M^r$ for all $s \geq r$, and $M^r$ is a finitely generated projective $(\mathcal{A} \otimes_V \overline{V})^\wedge_{\mathbb{Q}_p}$-module. Hence Theorem IV.5.3.4 implies that the following natural morphism of complexes is a quasi-isomorphism.

$$\xi^{-\bullet} M^r \otimes_A \Omega^\bullet = \varinjlim_{s \geq r} \xi^{-\bullet} M^s \otimes_A \Omega^\bullet \longrightarrow \varinjlim_{s \geq r} C^\bullet_{\mathrm{cont}}(\Delta_{(U,\overline{s})}, \xi^{-\bullet} \mathscr{A}_1^s(\overline{\mathcal{A}}) \otimes_{(\mathcal{A} \otimes_V \overline{V})^\wedge} M^s \otimes_A \Omega^\bullet).$$

Combining with the quasi-isomorphisms (IV.5.2.17) and (IV.5.2.18), we obtain the theorem. □

PROOF OF THEOREM IV.5.3.3. Let $\mathcal{F}$ be a Higgs isocrystal on $(U_1/A(\Sigma))^r_{\mathrm{HIGGS}}$ finite on $U_1$. By Proposition IV.5.2.12, we have a functorial $\Delta_{(U,\overline{s})}$-equivariant $\mathscr{A}_1^r(\overline{\mathcal{A}})$-linear isomorphism compatible with $\theta$

$$\mathscr{A}_1^r(\overline{\mathcal{A}})_{\mathbb{Q}_p} \otimes_{\widehat{\overline{\mathcal{A}}}_{\mathbb{Q}_p}} V^r_{(U,\overline{s}),\mathrm{HIGGS}}(\mathcal{F}) \cong \mathscr{A}_1^r(\overline{\mathcal{A}})_{\mathbb{Q}_p} \otimes_{(\mathcal{A} \otimes_V \overline{V})^\wedge_{\mathbb{Q}_p}} M.$$

By Theorem IV.5.3.4, we have $(\mathscr{A}_1^r(\overline{\mathcal{A}})_{\mathbb{Q}_p})^{\Delta_{(U,\overline{s})}} = (\mathcal{A} \otimes_V \overline{V})^\wedge_{\mathbb{Q}_p}$. Since $M$ is a finitely generated projective $(\mathcal{A} \otimes_V \overline{V})^\wedge_{\mathbb{Q}_p}$-module, we obtain an $(\mathcal{A} \otimes_V \overline{V})^\wedge_{\mathbb{Q}_p}$-linear isomorphism

$$M \cong (\mathscr{A}_1^r(\overline{\mathcal{A}})_{\mathbb{Q}_p} \otimes_{\widehat{\overline{\mathcal{A}}}_{\mathbb{Q}_p}} V^r_{(U,\overline{s}),\mathrm{HIGGS}}(\mathcal{F}))^{\Delta_{(U,\overline{s})}}$$

compatible with $\theta$. □

We discuss the admissibility of Higgs vector bundles with respect to the period ring $\mathscr{A}_1^r(\overline{\mathcal{A}})$. Let $A_1$ denote the $p$-adic completion of $A \otimes_V \overline{V}$ and let $\Omega^1$ denote $\Gamma(U, \Omega^1_{U/\Sigma})$. Let $M$ be a finitely generated projective $A_{1,\mathbb{Q}_p}$-module and let $\theta\colon M \to M \otimes_A \Omega^1$ be a Higgs field, i.e., an $A_{1,\mathbb{Q}_p}$-linear homomorphism such that $\theta^1 \circ \theta = 0$, where $\theta^1$ is defined as before Lemma IV.3.4.5.

Let $r \in \mathbb{N}_{>0} \cup \{\infty\}$. If $\theta$ satisfies the condition $(\mathrm{Conv})_r$ in IV.3.6, then it comes from a Higgs isocrystal on $(U_1/D(\Sigma))^r_{\mathrm{HIGGS}}$ finite on $U_1$ by Theorem IV.3.4.16. Hence $V^r_{\mathrm{HIGGS}}(M) := (\mathscr{A}_1^r(\overline{\mathcal{A}}) \otimes_{A_1} M)^{\theta=0}$ belongs to $\mathrm{Rep}^{\mathcal{PM}}_{\mathrm{cont}}(\Delta_{(U,\overline{s})}, \widehat{\overline{\mathcal{A}}}_{\mathbb{Q}_p})$ and the natural $\mathscr{A}_1^r(\overline{\mathcal{A}})$-linear homomorphism

$$\mathscr{A}_1^r(\overline{\mathcal{A}}) \otimes_{\widehat{\overline{A}}} V^r_{\mathrm{HIGGS}}(M) \longrightarrow \mathscr{A}_1^r(\overline{\mathcal{A}}) \otimes_{A_1} M$$

is an isomorphism by Proposition IV.5.2.12 and Corollary IV.5.2.13. One can prove the converse as follows.

**Proposition IV.5.3.10.** *Let $r \in \mathbb{N}_{>0} \cup \{\infty\}$, let $M$ be a finitely generated projective $A_{1,\mathbb{Q}_p}$-module, and let $\theta\colon M \to M \otimes_A \Omega^1$ be a Higgs field. Suppose that there exists an $\widehat{\overline{\mathcal{A}}}_{\mathbb{Q}_p}$-module $W$ and an $\mathscr{A}_1^r(\overline{\mathcal{A}})$-linear isomorphism*

$$\mathscr{A}_1^r(\overline{\mathcal{A}}) \otimes_{\widehat{\overline{\mathcal{A}}}} W \xrightarrow{\cong} \mathscr{A}_1^r(\overline{\mathcal{A}}) \otimes_{A_1} M$$

*compatible with the Higgs fields. Then $W$ is a finitely generated projective $\widehat{\overline{\mathcal{A}}}_{\mathbb{Q}_p}$-module, the above homomorphism induces an isomorphism $W \xrightarrow{\cong} (\mathscr{A}_1^r(\overline{\mathcal{A}}) \otimes_{A_1} M)^{\theta=0}$, and the Higgs field $\theta$ of $M$ satisfies the condition $(\mathrm{Conv})_r$ in IV.3.6.*

PROOF. We use the description (IV.5.3.7) of $\mathscr{A}_1^r(\overline{\mathcal{A}})$. By taking the reduction mod $(w_1, \ldots, w_d)$ of the isomorphism in the assumption, we obtain an $\widehat{\overline{\mathcal{A}}}_{\mathbb{Q}_p}$-linear isomorphism

$$W \cong \widehat{\overline{\mathcal{A}}} \otimes_{A_1} M, \tag{IV.5.3.11}$$

which implies that $W$ is a finitely generated projective $\widehat{\mathcal{A}}_{\mathbb{Q}_p}$-module. From Proposition IV.5.2.10 (1), we obtain the second claim. It remains to prove the last claim. Since $M$ is a finitely generated projective $A_{1,\mathbb{Q}_p}$-module, we have

$$\mathscr{A}_1^r(\overline{A}) \otimes_{A_1} M = \Big\{ \sum_{\underline{m}\in\mathbb{N}^d} w^{\underline{m}} \otimes x_{\underline{m}} \Big| x_{\underline{m}} \in \widehat{\mathcal{A}} \otimes_{A_1} M, p^{-\lceil r^{-1}|\underline{m}|\rceil} x_{\underline{m}} \to 0 \ (|\underline{m}| \to \infty) \Big\}.$$

Let $x \in M$ and let $v \in W$ be the image of $1 \otimes x$ under the isomorphism (IV.5.3.11). Then the image of $1 \otimes v$ under the isomorphism in the assumption of the proposition is written as

$$\sum_{\underline{m}\in\mathbb{N}^d} w^{\underline{m}} \otimes x_{\underline{m}}, \quad x_0 = x, \ x_{\underline{m}} \in \widehat{\mathcal{A}} \otimes_{A_1} M, \ p^{-\lceil r^{-1}|\underline{m}|\rceil} x_{\underline{m}} \to 0 \ (|\underline{m}| \to \infty).$$

Since this section is annihilated by $\theta_i$ $(1 \leq i \leq d)$, $\theta_i(w_i) = \xi^{-1} d\log(t_i)$, and $\theta_i(w_j) = 0$ $(j \neq i)$, we have $\theta_i(x_{\underline{m}}) = -(m_i + 1)x_{\underline{m}+\underline{1}_i}$ for $1 \leq i \leq d$ and $\underline{m} \in \mathbb{N}^d$, where $\underline{1}_i$ denotes the element of $\mathbb{N}^d$ whose $i$-th component is 1 and other components are 0. This implies that $x_{\underline{m}} = (-1)^{|\underline{m}|} \frac{1}{\underline{m}!} \prod_{1\leq i\leq d} \theta_i^{m_i}(x)$ and hence $p^{-\lceil r^{-1}|\underline{m}|\rceil} \frac{1}{\underline{m}!} \prod_{1\leq i\leq d} \theta_i^{m_i}(x)$ converges to 0 in $\widehat{\mathcal{A}} \otimes_{A_1} M$ as $|\underline{m}| \to \infty$. Since $M$ is a finitely generated projective module over $A_{1,\mathbb{Q}_p}$ and the homomorphism $A_1/pA_1 \to \widehat{\mathcal{A}}/p\widehat{\mathcal{A}}$ is injective, this implies that the sequence also converges in $M$. This completes the proof. □

**IV.5.4. Comparison with the approach via Higgs-Tate torsors.** In this subsection, we compare the period ring $\mathscr{A}^r_{X,Y,1}(\overline{\mathcal{A}})$ defined before Condition IV.5.2.3 with the algebras $\mathscr{C}$ and $\mathscr{C}^{(r)}$ $(r \in \mathbb{Q}_{>0})$ introduced in II.10.3 and II.12.1.

Let $\Sigma$, $\overline{\Sigma}$, $\widehat{\Sigma}$, and $D(\overline{\Sigma})$ be the same as the beginning of IV.5.2. We consider the case where the log structure of $\Sigma$ is defined by its closed point. Let $X \to \Sigma$ be a smooth morphism of fine and saturated log schemes satisfying the conditions in II.6.2 strict étale locally on $X$, and let $h\colon U \to X$ be a strict étale morphism such that the morphism $U \to \Sigma$ satisfies the conditions mentioned above. We further assume that $U$ satisfies the conditions in **[73]** Lemma 1.3.2 (cf. the beginning of IV.5.1). Then $U$ satisfies the conditions in the beginning of IV.5.1. We consider (Case I), i.e., $\mathcal{A} = A$ and $\mathcal{U} = U$ (cf. IV.5.1). As in IV.5.2, choose a geometric point $\overline{s} \to \mathcal{U}_{\overline{K},\mathrm{triv}}$ whose image is of codimension 0. Then, with the notation in IV.5.1, we have $\widehat{\overline{\mathcal{A}}} = \widetilde{\overline{\mathcal{A}}}$ by Proposition II.9.10, which is a consequence of the almost purity theorem by G. Faltings (cf. the proof of Lemma IV.5.1.1). Hence, as in IV.5.2, we have a morphism $D(\overline{U}) \to D(\overline{\Sigma})$ in $\mathscr{C}^\infty$ compatible with the actions of $G_{(U,\overline{s})}$ and $G_\Sigma$ via the natural morphism $G_{(U,\overline{s})} \to G_\Sigma$ and a $\overline{\Sigma}$-morphism $z_{\overline{U}}\colon \overline{U} = D_1(\overline{U}) \to U_1 = \widehat{U} \times_{\widehat{\Sigma}} \overline{\Sigma}$.

Suppose that we are given a smooth Cartesian lifting $X_\bullet = (X_N)_{N\in\mathbb{N}_{>0}} \to D(\overline{\Sigma})$ of $X_1 \to \overline{\Sigma} = D_1(\overline{\Sigma})$. Then we may apply the construction of the period ring before Condition IV.5.2.3; the object $\mathscr{D}^r_{X,X_\bullet}(\overline{U}) = (\mathscr{D}^r_{X,X_\bullet,N}(\overline{U}))$ of $\mathscr{C}^r$ for $r \in \mathbb{N}_{>0} \cup \{\infty\}$ is defined to be $D^r_{\mathrm{Higgs}}(\overline{U} \hookrightarrow X_\bullet \times_{D(\overline{\Sigma})} D(\overline{U}))$, and the period ring $\mathscr{A}_{X,X_\bullet,1}(\overline{U})$ is defined to be the coordinate ring of $\mathscr{D}^r_{X,X_\bullet,1}(\overline{U})$.

We may also apply IV.2.5 to $D(\overline{U}) \to D(\overline{\Sigma}) \leftarrow X_\bullet$ and the composition of $z_{\overline{U}}\colon \overline{U} \to U_1$ and $h_1\colon U_1 \to X_1$. We obtain a group $p$-adic fine log formal scheme $T^{\overline{U}}_{X_2}$ and a $T^{\overline{U}}_{X_2}$-principal homogenous space $L^{\overline{U}}_{X_2}$ over $\overline{U}$. This principal homogeneous space is a natural formal scheme analogue of that associated to the Higgs-Tate torsor defined in II.10.3. The action of $\Delta_{(U,\overline{s})}(= \mathrm{Ker}(G_{(U,\overline{s})} \to G_\Sigma))$ on $\overline{U}$ induces an action on $T^{\overline{U}}_{X_2}$ and an equivariant action on the $T^{\overline{U}}_{X_2}$-principal homogeneous space $L^{\overline{U}}_{X_2}$. Put $C^{\overline{U}}_{X_2} =$

$\Gamma(L^{\overline{U}}_{X_2}, \mathcal{O}_{L^{\overline{U}}_{X_2}})$ and define the $\widehat{\overline{\mathcal{A}}}$-submodule of "affine functions" $M^{\overline{U}}_{X_2}$ of $C^{\overline{U}}_{X_2}$ as before Lemma IV.2.5.14, which is a finitely generated projective $\widehat{\overline{\mathcal{A}}}$-module by Lemma IV.2.5.14. As before Proposition IV.2.5.16, we define $C_{\widehat{\overline{\mathcal{A}}}}(M^{\overline{U}}_{X_2})$ to be the direct limit $\varinjlim_m S^m_{\widehat{\overline{\mathcal{A}}}}(M^{\overline{U}}_{X_2})$ whose transition maps are induced by $\widehat{\overline{\mathcal{A}}} \hookrightarrow M^{\overline{U}}_{X_2}$, $C^{(r)}_{\widehat{\overline{\mathcal{A}}}}(M^{\overline{U}}_{X_2})$ ($r \in \mathbb{N}_{>0}$) to be its subring $\sum_{m\in\mathbb{N}} p^{\lceil \frac{m}{r} \rceil} S^m_{\widehat{\overline{\mathcal{A}}}}(M^{\overline{U}}_{X_2})$, and $\widehat{C}_{\widehat{\overline{\mathcal{A}}}}(M^{\overline{U}}_{X_2})$ and $\widehat{C}^{(r)}_{\widehat{\overline{\mathcal{A}}}}(M^{\overline{U}}_{X_2})$ to be their $p$-adic completions. These are naturally endowed with the actions of $\Delta_{(U,\overline{s})}$ and we have $\widehat{C}_{\widehat{\overline{\mathcal{A}}}}(M^{\overline{U}}_{X_2}) = C^{\overline{U}}_{X_2}$ by the construction of $L^{\overline{U}}_{X_2}$. The $\widehat{\overline{\mathcal{A}}}$-algebra $C^{\overline{U}}_{X_2}$ is naturally endowed with the $\widehat{\overline{\mathcal{A}}}$-linear derivation $\theta\colon C^{\overline{U}}_{X_2} \to C^{\overline{U}}_{X_2} \otimes_A \Omega^1$ (cf. (IV.2.5.17)), where $\Omega^1 = \Gamma(U, h^*(\Omega^1_{X/\Sigma}))$.

By Theorem IV.2.5.2 and Proposition IV.2.5.22, we have a natural $\Delta_{(U,\overline{s})}$-equivariant isomorphism

$$T^{\overline{U}}_{X_2} \cong \mathscr{D}^{\infty}_{X,X_{\bullet},1}(\overline{U}).$$

It induces a $\Delta_{(U,\overline{s})}$-equivariant isomorphism

$$C^{\overline{U}}_{X_2} \xrightarrow{\cong} \mathscr{A}^{\infty}_{X,X_{\bullet},1}(\overline{U}), \tag{IV.5.4.1}$$

which is compatible with the derivations $\theta$'s by Proposition IV.2.5.19. By Proposition IV.2.5.16, it further induces a $\Delta_{(U,\overline{s})}$-equivariant isomorphism

$$\widehat{C}^{(r)}_{\widehat{\overline{\mathcal{A}}}}(M^{\overline{U}}_{X_2}) \xrightarrow{\cong} \mathscr{A}^{r}_{X,X_{\bullet},1}(\overline{U}) \tag{IV.5.4.2}$$

for $r \in \mathbb{N}_{>0}$.

Now let us compare the algebras $\widehat{C}_{\widehat{\overline{\mathcal{A}}}}(M^{\overline{U}}_{X_2})$ and $\widehat{C}^{(r)}_{\widehat{\overline{\mathcal{A}}}}(M^{\overline{U}}_{X_2})$ with the algebras $\mathscr{C}$ and $\mathscr{C}^{(r)}$ in II.10.3 and II.12.1.

Under the notation and the assumption in IV.5.1, we can define fine and saturated log schemes $\check{\overline{U}}$ and $D_N(\check{\overline{U}})$ by replacing $\mathrm{Spf}(\widetilde{\overline{\mathcal{A}}})$ and $\mathrm{Spf}(A_N(\overline{\mathcal{A}}))$ with $\mathrm{Spec}(\widetilde{\overline{\mathcal{A}}})$ and $\mathrm{Spec}(A_N(\overline{\mathcal{A}}))$ in the definition of $\overline{U}$ and $D_N(\overline{U})$ in IV.5.1. Note that the proof of Lemma IV.5.1.4 still works after this modification. We have natural actions of $G_{(U,\overline{s})}$ on these log schemes, a $G_{(U,\overline{s})}$-equivariant isomorphism $i_{\check{\overline{U}}}\colon \check{\overline{U}} \to D_1(\check{\overline{U}})$, and $G_{(U,\overline{s})}$-equivariant exact closed immersions $i_{\check{\overline{U}},N}\colon D_N(\check{\overline{U}}) \to D_{N+1}(\check{\overline{U}})$. We identify $\check{\overline{U}}$ with $D_1(\check{\overline{U}})$ via $i_{\check{\overline{U}}}$. If $\widehat{\overline{\mathcal{A}}} = \widetilde{\overline{\mathcal{A}}}$, then we have a natural strict morphism $\check{\overline{U}} \to U$.

Let us return to the settings in this subsection. By applying the above construction to $(\Sigma, \mathrm{Spec}(\overline{K}))$ (resp. $(U,\overline{s})$), we obtain $\check{\overline{\Sigma}}$, $D_N(\check{\overline{\Sigma}})$ and $i_{\check{\overline{\Sigma}},N}$ (resp. $\check{\overline{U}}$, $D_N(\check{\overline{U}})$ and $i_{\check{\overline{U}},N}$). Since $\overline{s}$ is a geometric point of $\mathcal{U}_{\overline{K},\mathrm{triv}}$, we have natural morphisms $D_N(\check{\overline{U}}) \to D_N(\check{\overline{\Sigma}})$ compatible with $i_{\check{\overline{\Sigma}},N}$ and $i_{\check{\overline{U}},N}$ and also with the actions of $G_{(U,\overline{s})}$ and $G_{\Sigma}$. Let $\check{U}_1 \xrightarrow{h} \check{X}_1 \to \check{\overline{\Sigma}}$ be the base change of $U \xrightarrow{h} X \to \Sigma$ by the morphism $\check{\overline{\Sigma}} \to \Sigma$. Since $\widetilde{\overline{\mathcal{A}}} = \widehat{\overline{\mathcal{A}}}$ as mentioned above, we have a natural strict morphism $\check{\overline{U}} \to \check{U}_1$ over $\check{\overline{\Sigma}}$.

Choose a smooth morphism $\check{X}_2 \to D_2(\check{\overline{\Sigma}})$ endowed with an isomorphism $\check{X}_2 \times_{D_2(\check{\overline{\Sigma}})} \check{\overline{\Sigma}} \cong \check{X}_1$ over $\check{\overline{\Sigma}}$. Then we can apply II.10.3 to the commutative diagram

$$\begin{array}{ccc} \check{\overline{U}} & \longrightarrow & \check{X}_2 \\ \downarrow & & \downarrow \\ \check{\overline{\Sigma}} & \xrightarrow{\;i_{\check{\overline{\Sigma}},1}\;} & D_2(\check{\overline{\Sigma}}), \end{array}$$

obtaining a vector bundle $T_{\check{X}_2}^{\check{\overline{U}}}$ and a $T_{\check{X}_2}^{\check{\overline{U}}}$-principal homogeneous space $L_{\check{X}_2}^{\check{\overline{U}}}$ over the underlying scheme of $\check{\overline{U}}$. Precisely speaking, we have to work with the diagram in which $\check{X}_2$ is replaced by the unique strict étale lifting $\check{U}_2 \to \check{X}_2$ of $\check{U}_1 \to \check{X}_1$, but the resulting principal vector bundle and principal homogeneous space are canonically identified with $T_{\check{X}_2}^{\check{\overline{U}}}$ and $L_{\check{X}_2}^{\check{\overline{U}}}$ above (cf. the proof of Lemma III.10.14). Put $\mathscr{C}_{\check{X}_2}^{\check{\overline{U}}} := \Gamma(L_{\check{X}_2}^{\check{\overline{U}}}, \mathcal{O}_{L_{\check{X}_2}^{\check{\overline{U}}}})$, and let $\mathscr{F}_{\check{X}_2}^{\check{\overline{U}}}$ be the submodule of $\mathscr{C}_{\check{X}_2}^{\check{\overline{U}}}$ consisting of affine functions. The $\widehat{\overline{\mathcal{A}}}$-module $\mathscr{C}_{\check{X}_2}^{\check{\overline{U}}}$ is naturally endowed with an $\widehat{\overline{\mathcal{A}}}$-linear derivation $\theta \colon \mathscr{C}_{\check{X}_2}^{\check{\overline{U}}} \to \mathscr{C}_{\check{X}_2}^{\check{\overline{U}}} \otimes_A \Omega^1$ (cf. II.10.9). Let $\widehat{\mathscr{C}_{\check{X}_2}^{\check{\overline{U}}}}$ denote the $p$-adic completion of $\mathscr{C}_{\check{X}_2}^{\check{\overline{U}}}$.

The morphisms $\overline{U} \to U_1 \to X_1 \to \overline{\Sigma}$ and $D(\overline{U}) \to D(\overline{\Sigma})$ are naturally identified with the $p$-adic formal completion of the morphisms $\check{\overline{U}} \to \check{U}_1 \to \check{X}_1 \to \check{\overline{\Sigma}}$ and the compatible system of morphisms $D_N(\check{\overline{U}}) \to D_N(\check{\overline{\Sigma}})$ ($N \in \mathbb{N}_{>0}$). Under this identification, the $p$-adic formal completion of $\check{X}_2$ is regarded as a smooth lifting of $X_1$ over $D_2(\overline{\Sigma})$. Suppose that we are given an isomorphism between the above completion and the lifting $X_2$ of $X_1$ which we have chosen. Then by the construction of $T_{X_2}^{\overline{U}}$, $L_{X_2}^{\overline{U}}$, and $M_{X_2}^{\overline{U}}$, we see that $M_{X_2}^{\overline{U}}$ is canonically isomorphic to $\mathscr{F}_{\check{X}_2}^{\check{\overline{U}}}$ as extensions of $\xi^{-1}\widehat{\overline{\mathcal{A}}} \otimes_A \Omega^1$ by $\widehat{\overline{\mathcal{A}}}$ endowed with actions of $\Delta_{(U,\overline{s})}$. Combined with (IV.5.4.1), this isomorphism induces $\Delta_{(U,\overline{s})}$-equivariant isomorphisms

$$\text{(IV.5.4.3)} \qquad \widehat{\mathscr{C}_{\check{X}_2}^{\check{\overline{U}}}} \xrightarrow{\cong} \widehat{C}_{\widehat{\overline{\mathcal{A}}}}(M_{X_2}^{\overline{U}}) = C_{X_2}^{\overline{U}} \xrightarrow{\cong} \mathscr{A}_{X,X,1}^{\infty}(\overline{U})$$

compatible with the derivations $\theta$.

As in II.12.1, we define $\mathscr{F}_{\check{X}_2}^{\check{\overline{U}},(r)}$ ($r \in \mathbb{Q}_{>0}$) to be the pull-back of the extension $0 \to \widehat{\overline{\mathcal{A}}} \to \mathscr{F}_{\check{X}_2}^{\check{\overline{U}}} \to \xi^{-1}\widehat{\overline{\mathcal{A}}} \otimes_A \Omega^1 \to 0$ by $p^r(\xi^{-1}\widehat{\overline{\mathcal{A}}} \otimes_A \Omega) \subset \xi^{-1}\widehat{\overline{\mathcal{A}}} \otimes_A \Omega^1$, and $\mathscr{C}_{\check{X}_2}^{\check{\overline{U}},(r)}$ to be the $\widehat{\overline{\mathcal{A}}}$-subalgebra $\varinjlim_m S_{\widehat{\overline{\mathcal{A}}}}^m(\mathscr{F}_{\check{X}_2}^{\check{\overline{U}},(r)})$ of $\mathscr{C}_{\check{X}_2}^{\check{\overline{U}}}$. Let $\widehat{\mathscr{C}_{\check{X}_2}^{\check{\overline{U}},(r)}}$ denote the $p$-adic completion of $\mathscr{C}_{\check{X}_2}^{\check{\overline{U}},(r)}$. If we identify $\mathscr{C}_{\check{X}_2}^{\check{\overline{U}}}$ with $C_{\widehat{\overline{\mathcal{A}}}}(M_{X_2}^{\overline{U}})$ by the isomorphism induced by $\mathscr{F}_{\check{X}_2}^{\check{\overline{U}}} \cong M_{X_2}^{\overline{U}}$, then we have $p\mathscr{C}_{\check{X}_2}^{\check{\overline{U}},(1/r)} \subset C_{\widehat{\overline{\mathcal{A}}}}^{(r)}(M_{X_2}^{\overline{U}}) \subset \mathscr{C}_{\check{X}_2}^{\check{\overline{U}},(1/r)}$ for $r \in \mathbb{N}_{>0}$. By (IV.5.4.2), we see that the isomorphism (IV.5.4.3) induces an injective homomorphism

$$\text{(IV.5.4.4)} \qquad \mathscr{A}_{X,X,1}^r(\overline{U}) \hookrightarrow \widehat{\mathscr{C}_{\check{X}_2}^{\check{\overline{U}},(1/r)}}$$

whose cokernel as a homomorphism of $\widehat{\overline{\mathcal{A}}}$-modules is annihilated by $p$.

By Proposition IV.2.5.22, we see that the homomorphisms (IV.5.4.3) and (IV.5.4.4) are functorial with respect to $X$-morphisms between $U$'s.

## IV.6. Comparison with Faltings cohomology

**IV.6.1. Faltings site.** Let $V$, $k$, $K$, $\overline{K}$, $\overline{V}$, and $\Sigma$ be the same as in the beginning of IV.5.2. Let $U$ be an fs log scheme over $\Sigma$ whose underlying scheme is affine and of finite type over $\operatorname{Spec}(V)$. Let $A_U$ denote $\Gamma(U, \mathcal{O}_U)$. Similarly as in IV.5.1, let $\mathcal{A}_U$ denote one of the following three $A_U$-algebras:

(Case I) The algebra $A_U$ itself.
(Case II) The henselization of $A_U$ with respect to the ideal $pA_U$.
(Case III) The $p$-adic completion of $A_U$.

For an extension $L$ of $K$ contained in $\overline{K}$, put $\mathcal{A}_{U,L} := \mathcal{A}_U \otimes_V L$, let $\mathcal{U}_L$ denote $\mathrm{Spec}(\mathcal{A}_{U,L})$ endowed with the inverse image of $M_U$, and let $\mathcal{U}_{L,\mathrm{triv}}$ denote the open subscheme of $\mathrm{Spec}(\mathcal{A}_{U,L})$ defined by $\{x \in \mathrm{Spec}(\mathcal{A}_{U,L}) | M_{\mathcal{U}_L,\overline{x}} = \mathcal{O}^{\times}_{\mathcal{U}_L,\overline{x}}\}$. By using charts, we see that the morphism $\mathcal{U}_{L,\mathrm{triv}} \to \mathcal{U}_L$ is affine strict étale locally on $\mathcal{U}_L$. Hence $\mathcal{U}_{L,\mathrm{triv}} \to \mathcal{U}_L$ is affine, which implies that $\mathcal{U}_{L,\mathrm{triv}}$ is affine (cf. **[42]** Proposition (2.7.1)). Let $\mathcal{A}_{U,L,\mathrm{triv}}$ denote $\Gamma(\mathcal{U}_{L,\mathrm{triv}}, \mathcal{O}_{\mathcal{U}_{L,\mathrm{triv}}})$.

We consider the category $\mathcal{P}$ defined as follows. An object of $\mathcal{P}$ is a triple $\mathfrak{U} = (U, \mathcal{V}, v)$ consisting of an fs log scheme $U$ over $\Sigma$ whose underlying scheme is affine and of finite type over $\mathrm{Spec}(V)$ and a finite étale morphism $v\colon \mathcal{V} \to \mathcal{U}_{\overline{K},\mathrm{triv}}$. A morphism $\mathfrak{f} = (f, g)\colon \mathfrak{U}' = (U', \mathcal{V}', v') \to \mathfrak{U} = (U, \mathcal{V}, v)$ is a pair of a morphism $f\colon U' \to U$ over $\Sigma$ and a morphism $g\colon \mathcal{V}' \to \mathcal{V}$ such that $v \circ g = \boldsymbol{f}_{\overline{K},\mathrm{triv}} \circ v'$, where $\boldsymbol{f}_{\overline{K},\mathrm{triv}}$ denotes the morphism $\mathcal{U}'_{\overline{K},\mathrm{triv}} \to \mathcal{U}_{\overline{K},\mathrm{triv}}$ induced by $f$. The composition of two morphisms is defined in the obvious way. We often abbreviate $(U, \mathcal{V}, v)$ to $(U, \mathcal{V})$ in the following.

The fiber product of $(U', \mathcal{V}') \to (U, \mathcal{V}) \leftarrow (U'', \mathcal{V}'')$ in $\mathcal{P}$ is representable as follows: Let $U'''$ be the fiber product of $U' \to U \leftarrow U''$, and let $\mathcal{V}'''$ be the fiber product of $\mathcal{V}' \times_{\mathcal{U}'_{\overline{K},\mathrm{triv}}} \mathcal{U}'''_{\overline{K},\mathrm{triv}} \to \mathcal{V} \times_{\mathcal{U}_{\overline{K},\mathrm{triv}}} \mathcal{U}'''_{\overline{K},\mathrm{triv}} \leftarrow \mathcal{V}'' \times_{\mathcal{U}''_{\overline{K},\mathrm{triv}}} \mathcal{U}'''_{\overline{K},\mathrm{triv}}$, which is finite étale over $\mathcal{U}'''_{\overline{K},\mathrm{triv}}$. Then the object $(U''', \mathcal{V}''')$ of $\mathcal{P}$ with the natural morphisms $(U''', \mathcal{V}''') \to (U', \mathcal{V}'), (U'', \mathcal{V}'')$ represents the fiber product.

We say that a morphism $\mathfrak{f}\colon (U', \mathcal{V}') \to (U, \mathcal{V})$ in $\mathcal{P}$ is *horizontal* (resp. *vertical*) if the morphism $\mathcal{V}' \to \mathcal{V} \times_{\mathcal{U}_{\mathrm{triv},\overline{K}}} \mathcal{U}'_{\mathrm{triv},\overline{K}}$ (resp. $U' \to U$) induced by $\mathfrak{f}$ is an isomorphism. By the above explicit construction of fiber products, we see that $\mathfrak{f}$ is horizontal if and only if the morphism $(U', \mathcal{V}') \to (U, \mathcal{V}) \times_{(U, \mathcal{U}_{\mathrm{triv},\overline{K}})} (U', \mathcal{U}'_{\mathrm{triv},\overline{K}})$ induced by $\mathfrak{f}$ is an isomorphism.

For an object $\mathfrak{U} = (U, \mathcal{V})$ of $\mathrm{Ob}\,\mathcal{P}$, we define $\mathrm{Cov}^h(\mathfrak{U})$ (resp. $\mathrm{Cov}^v(\mathfrak{U})$) to be the set of families of horizontal (resp. vertical) morphisms $((f_\alpha, g_\alpha)\colon (U_\alpha, \mathcal{V}_\alpha) \to (U, \mathcal{V}))_{\alpha \in A}$ such that $(f_\alpha\colon U_\alpha \to U)_{\alpha \in A}$ (resp. $(g_\alpha\colon \mathcal{V}_\alpha \to \mathcal{V})_{\alpha \in A}$) is a strict étale covering (resp. a finite étale covering). We call a family of morphisms belonging to $\mathrm{Cov}^h(\mathfrak{U})$ (resp. $\mathrm{Cov}^v(\mathfrak{U})$) a *horizontal strict étale covering* (resp. a *vertical finite étale covering*) of $\mathfrak{U}$. Horizontal strict étale coverings and vertical finite étale coverings are stable under base changes and compositions. The latter means that, for $(\mathfrak{U}_\alpha \to \mathfrak{U})_{\alpha \in A} \in \mathrm{Cov}^h(\mathfrak{U})$ (resp. $\mathrm{Cov}^v(\mathfrak{U})$) and $(\mathfrak{U}_{\alpha\beta} \to \mathfrak{U}_\alpha)_{\beta \in B_\alpha} \in \mathrm{Cov}^h(\mathfrak{U}_\alpha)$ (resp. $\mathrm{Cov}^v(\mathfrak{U}_\alpha)$), we have $(\mathfrak{U}_{\alpha\beta} \to \mathfrak{U})_{\alpha \in A, \beta \in B_\alpha} \in \mathrm{Cov}^h(\mathfrak{U})$ (resp. $\mathrm{Cov}^v(\mathfrak{U})$). Put $\mathrm{Cov}(\mathfrak{U}) = \mathrm{Cov}^h(\mathfrak{U}) \cup \mathrm{Cov}^v(\mathfrak{U})$.

Let $X$ be an fs log scheme over $\Sigma$ whose underlying scheme is separated of finite type over $\mathrm{Spec}(V)$. We define the sites $(\mathcal{P}/X)_{\text{ét-fét}}$ and $(\mathcal{P}/X)_{\text{fét}}$ associated to $X$ as follows. We first define the category $\mathcal{P}/X$ as follows. An object is a pair $(\mathfrak{U}, u)$ of an object $\mathfrak{U}$ of $\mathcal{P}$ and a strict étale morphism $u\colon U \to X$ over $\Sigma$. A morphism $\mathfrak{f} = (f, g)\colon (\mathfrak{U}', u') \to (\mathfrak{U}, u)$ is a morphism $\mathfrak{U}' \to \mathfrak{U}$ in $\mathcal{P}$ such that $u \circ f = u'$. The fiber product of $(\mathfrak{U}', u') \to (\mathfrak{U}, u) \leftarrow (\mathfrak{U}'', u'')$ in $\mathcal{P}/X$ is represented by $\mathfrak{U}''' = (U''', \mathcal{V}''') := \mathfrak{U}' \times_{\mathfrak{U}} \mathfrak{U}''$ with the natural morphism $u'''\colon U''' \to X$. Similarly the product of $(\mathfrak{U}, u)$ and $(\mathfrak{U}', u')$ in $\mathcal{P}/X$ is representable as follows. Put $\mathfrak{U} = (U, \mathcal{V})$ and $\mathfrak{U}' = (U', \mathcal{V}')$. Let $U'' = U \times_X U'$, which is affine because the underlying scheme of $X$ is separated by assumption, and let $\mathcal{V}''$ be the fiber product of $\mathcal{V} \times_{\mathcal{U}_{\overline{K},\mathrm{triv}}} \mathcal{U}''_{\overline{K},\mathrm{triv}}$ and $\mathcal{V}' \times_{\mathcal{U}'_{\overline{K},\mathrm{triv}}} \mathcal{U}''_{\overline{K},\mathrm{triv}}$ over $\mathcal{U}''_{\overline{K},\mathrm{triv}}$. Then the pair $\mathfrak{U}'' = (U'', \mathcal{V}'')$ is an object of $\mathcal{P}$ and $\mathfrak{U}''$ with the natural morphism $U'' \to X$ represents the product of $(\mathfrak{U}, u)$ and $(\mathfrak{U}', u')$. If $X$ is affine, then $\mathcal{P}/X$ is naturally regarded as a full subcategory of the category $\mathcal{P}_{/(X, \mathcal{X}_{\overline{K},\mathrm{triv}})}$ of objects of $\mathcal{P}$ over $(X, \mathcal{X}_{\overline{K},\mathrm{triv}})$. We often omit $u$ in the notation $(\mathfrak{U}, u)$ and write simply $\mathfrak{U}$.

For an object $(\mathfrak{U}, u)$ of $\mathcal{P}/X$, a family of morphisms $(\mathfrak{U}_\alpha \to \mathfrak{U})_{\alpha \in A} \in \mathrm{Cov}(\mathfrak{U})$ is naturally regarded as a family of morphisms in $\mathcal{P}/X$. We define $(\mathcal{P}/X)_{\text{ét-fét}}$ (resp. $(\mathcal{P}/X)_{\text{fét}}$) to be the category $\mathcal{P}/X$ endowed with the topology generated by $\mathrm{Cov}(\mathfrak{U})$ (resp. $\mathrm{Cov}^v(\mathfrak{U})$).

Let $X_{\text{étaff}}$ be the category of affine fs log schemes strict étale over $X$ endowed with the topology associated to the pretopology defined by strict étale covering. The category $\mathcal{P}/X$ with the functor $(\mathcal{P}/X) \to X_{\text{étaff}}; (U, \mathcal{V}) \mapsto U$ is a fibered category such that the fiber over the object $U$ of $X_{\text{étaff}}$ is canonically identified with $(\mathcal{U}_{\overline{K},\text{triv}})_{\text{fét}}$. For a morphism $f: U' \to U$ in $X_{\text{étaff}}$, the inverse image functor between the fibers is given by the base change under the morphism $f_{\overline{K},\text{triv}}: \mathcal{U}'_{\overline{K},\text{triv}} \to \mathcal{U}_{\overline{K},\text{triv}}$, which is continuous and defines a morphism of sites. Thus $\mathcal{P}/X$ with the finite étale topology on each fiber becomes a fibered site and the site $(\mathcal{P}/X)_{\text{fét}}$ defined above is the category $\mathcal{P}/X$ with the total topology (cf. **[2]** VI Définition 7.4.1, Proposition 7.4.2). The fibered site $\mathcal{P}/X$ and the strict étale site $X_{\text{étaff}}$ satisfies the conditions (i), (ii), and (iii) in VI.5.1 and $(\mathcal{P}/X)_{\text{ét-fét}}$ is the category $\mathcal{P}/X$ with the covanishing topology associated to the fibered site and the strict étale topology of $X_{\text{étaff}}$ defined in VI.5.3.

The sets $\text{Cov}(\mathfrak{U})$ (resp. $\text{Cov}^v(\mathfrak{U})$) for $(\mathfrak{U}, u) \in \mathcal{P}/X$ are stable under base changes in $\mathcal{P}/X$. Hence, by **[2]** II Corollaire 2.3, sheaves on $(\mathcal{P}/X)_{\text{ét-fét}}$ (resp. $(\mathcal{P}/X)_{\text{fét}}$) are characterized as follows.

**Lemma IV.6.1.1** (cf. Proposition VI.5.10). *A presheaf $\mathcal{F}$ on $\mathcal{P}/X$ is a sheaf on the site $(\mathcal{P}/X)_{\text{ét-fét}}$ (resp. $(\mathcal{P}/X)_{\text{fét}}$) if and only if for every $\mathfrak{U} \in \text{Ob}(\mathcal{P}/X)$ and $(\mathfrak{U}_\alpha \to \mathfrak{U})_{\alpha\in A} \in \text{Cov}(\mathfrak{U})$ (resp. $\text{Cov}^v(\mathfrak{U})$), the following sequence is exact*

$$\mathcal{F}(\mathfrak{U}) \to \prod_{\alpha\in A} \mathcal{F}(\mathfrak{U}_\alpha) \rightrightarrows \prod_{(\alpha,\beta)\in A^2} \mathcal{F}(\mathfrak{U}_\alpha \times_{\mathfrak{U}} \mathfrak{U}_\beta).$$

The identity functor $(\mathcal{P}/X)_{\text{fét}} \to (\mathcal{P}/X)_{\text{ét-fét}}$ is continuous. Hence the identity functor $(\mathcal{P}/X)_{\text{ét-fét}} \to (\mathcal{P}/X)_{\text{fét}}$ is cocontinuous and these functors induce a morphism of topos (cf. **[2]** III Proposition 2.5)

$$v_{\mathcal{P}/X}: (\mathcal{P}/X)^{\sim}_{\text{ét-fét}} \to (\mathcal{P}/X)^{\sim}_{\text{fét}}. \tag{IV.6.1.2}$$

Under the interpretation in terms of total topology and covanishing topology above, this coincides with the morphism $\delta$ in (VI.5.16.1). For $\mathcal{F} \in \text{Ob}((\mathcal{P}/X)^{\sim}_{\text{fét}})$, the inverse image $v^*_{\mathcal{P}/X}\mathcal{F}$ is the sheaf on $(\mathcal{P}/X)^{\sim}_{\text{ét-fét}}$ associated to $\mathcal{F}$ regarded as a presheaf on $(\mathcal{P}/X)_{\text{ét-fét}}$.

For $\mathfrak{U} \in \text{Ob}\,\mathcal{P}$, let $\text{Cov}^h_f(\mathfrak{U})$ (resp. $\text{Cov}^v_f(\mathfrak{U})$) be the subset of $\text{Cov}^h(\mathfrak{U})$ (resp. $\text{Cov}^v(\mathfrak{U})$) consisting of $(\mathfrak{U}_\alpha \to \mathfrak{U})_{\alpha\in A}$ such that $A$ is a finite set. Put $\text{Cov}_f(\mathfrak{U}) = \text{Cov}^h_f(\mathfrak{U}) \cup \text{Cov}^v_f(\mathfrak{U})$. Since $U$ and $\mathcal{U}_{\overline{K},\text{triv}}$ is affine, for any $(\mathfrak{U}_\alpha \to \mathfrak{U})_{\alpha\in A} \in \text{Cov}(\mathfrak{U})$, there exists a finite subset $A'$ of $A$ such that $(\mathfrak{U}_\alpha \to \mathfrak{U})_{\alpha\in A'}$ belongs to $\text{Cov}_f(\mathfrak{U})$. Hence the topology of $(\mathcal{P}/X)_{\text{ét-fét}}$ (resp. $(\mathcal{P}/X)_{\text{fét}}$) is generated by $\text{Cov}_f(\mathfrak{U})$ (resp. $\text{Cov}^v_f(\mathfrak{U})$), and Lemma IV.6.1.1 with $\text{Cov}(\mathfrak{U})$ (resp. $\text{Cov}^v(\mathfrak{U})$) replaced by $\text{Cov}_f(\mathfrak{U})$ (resp. $\text{Cov}^v_f(\mathfrak{U})$) still holds.

**Proposition IV.6.1.3.** *Let $\mathfrak{U}$ be an object of $\mathcal{P}/X$ and let $R$ be a sieve of $\mathfrak{U}$. Then $R$ is a covering sieve for the topology of $(\mathcal{P}/X)_{\text{ét-fét}}$ if and only if there exist $(\mathfrak{f}_\alpha: \mathfrak{U}_\alpha \to \mathfrak{U})_{\alpha\in A} \in \text{Cov}^h_f(\mathfrak{U})$ and $(\mathfrak{g}_{\alpha\beta}: \mathfrak{U}_{\alpha\beta} \to \mathfrak{U}_\alpha)_{\beta\in B_\alpha} \in \text{Cov}^v_f(\mathfrak{U}_\alpha)$ for each $\alpha \in A$ such that $\mathfrak{f}_\alpha \circ \mathfrak{g}_{\alpha\beta} \in R(\mathfrak{U}_{\alpha\beta})$ for every $\alpha \in A$ and $\beta \in B_\alpha$.*

Proof. Let $J(\mathfrak{U})$ be the set of sieves $R$ of $\mathfrak{U}$ such that there exist $\mathfrak{f}_\alpha$ and $\mathfrak{g}_{\alpha\beta}$ as in the proposition satisfying $\mathfrak{f}_\alpha \circ \mathfrak{g}_{\alpha\beta} \in R(\mathfrak{U}_{\alpha\beta})$. It is sufficient to prove that $J(\mathfrak{U})$ satisfies the axiom of topology in **[2]** II Définition 1.1. The condition T3) follows from $(\text{id}_{\mathfrak{U}}: \mathfrak{U} \to \mathfrak{U}) \in \text{Cov}^h_f(\mathfrak{U})$. The condition T1) follows from the stability of horizontal strict étale coverings (resp. vertical finite étale coverings) by base change. It remains to prove T2), i.e., for $R \in J(\mathfrak{U})$ and a sieve $R'$ of $\mathfrak{U}$, if for any $\mathfrak{U}' \in \text{Ob}(\mathcal{P}/X)$ and any morphism $\mathfrak{U}' \to R$ the sieve $R' \times_{\mathfrak{U}} \mathfrak{U}'$ belongs to $J(\mathfrak{U}')$, then $R'$ belongs to $J(\mathfrak{U})$. The assumption $R \in J(\mathfrak{U})$ means that there exist $(\mathfrak{f}_\alpha: \mathfrak{U}_\alpha \to \mathfrak{U})_{\alpha\in A} \in \text{Cov}^h_f(\mathfrak{U})$ and $(\mathfrak{g}_{\alpha\beta}: \mathfrak{U}_{\alpha\beta} \to \mathfrak{U}_\alpha)_{\beta\in B_\alpha} \in \text{Cov}^v_f(\mathfrak{U}_\alpha)$ for each $\alpha \in A$ such that the composition $\mathfrak{f}_\alpha \circ$

$\mathfrak{g}_{\alpha\beta}\colon \mathfrak{U}_{\alpha\beta} \to \mathfrak{U}$ factors through $R$. By assumption, we have $R' \times_{\mathfrak{U}} \mathfrak{U}_{\alpha\beta} \in J(\mathfrak{U}_{\alpha\beta})$, which implies that there exist $(\mathfrak{f}_{\alpha\beta;\gamma}\colon \mathfrak{U}_{\alpha\beta;\gamma} \to \mathfrak{U}_{\alpha\beta})_{\gamma\in\Gamma_{\alpha\beta}} \in \mathrm{Cov}^h_f(\mathfrak{U}_{\alpha\beta})$ and $(\mathfrak{g}_{\alpha\beta;\gamma\delta}\colon \mathfrak{U}_{\alpha\beta;\gamma\delta} \to \mathfrak{U}_{\alpha\beta;\gamma})_{\delta\in\Delta_{\alpha\beta;\gamma}} \in \mathrm{Cov}^v_f(\mathfrak{U}_{\alpha\beta;\gamma})$ for each $\gamma \in \Gamma_{\alpha\beta}$ such that $\mathfrak{U}_{\alpha\beta;\gamma\delta} \to \mathfrak{U}_{\alpha\beta}$ factors through $R' \times_{\mathfrak{U}} \mathfrak{U}_{\alpha\beta}$. Then we see that the composition $\mathfrak{U}_{\alpha\beta;\gamma\delta} \to \mathfrak{U}_{\alpha\beta;\gamma} \to \mathfrak{U}_{\alpha\beta} \to \mathfrak{U}_{\alpha} \to \mathfrak{U}$ factors through $R'$. We will exchange the order of the vertical finite étale coverings and the horizontal strict étale coverings in the middle of the composition above. Let $U_\alpha$, $U_{\alpha\beta}$, and $U_{\alpha\beta;\gamma}$ denote the first component of $\mathfrak{U}_\alpha$, $\mathfrak{U}_{\alpha\beta}$, and $\mathfrak{U}_{\alpha\beta;\gamma}$, respectively. Note that the morphism $U_{\alpha\beta} \to U_\alpha$ induced by $\mathfrak{g}_{\alpha\beta}$ is an isomorphism. Since $B_\alpha$ is a finite set, there exists a strict étale covering $(U'_{\alpha\kappa} \to U_\alpha)_{\kappa\in K_\alpha}$, $\sharp K_\alpha < \infty$ such that for every $\kappa \in K_\alpha$ and $\beta \in B_\alpha$, there exist $\gamma = \gamma(\kappa,\beta) \in \Gamma_{\alpha\beta}$ and a morphism $U'_{\alpha\kappa} \to U_{\alpha\beta;\gamma}$ over $U_\alpha$. For $\kappa \in K_\alpha$, let $(\mathfrak{U}'_{\alpha\kappa;\beta} \to \mathfrak{U}'_{\alpha\kappa})_{\beta\in B_\alpha} \in \mathrm{Cov}^v_f(\mathfrak{U}'_{\alpha\kappa})$ be the base change of $(\mathfrak{U}_{\alpha\beta} \to \mathfrak{U}_\alpha)_{\beta\in B_\alpha} \in \mathrm{Cov}^v_f(\mathfrak{U}_\alpha)$ by the morphism $(U'_{\alpha\kappa}, \mathcal{U}'_{\alpha\kappa,\mathrm{triv},\overline{K}}) \to (U_\alpha, \mathcal{U}_{\alpha,\mathrm{triv},\overline{K}})$. Then $(\mathfrak{U}'_{\alpha\kappa} \to \mathfrak{U}_\alpha)_{\kappa\in K_\alpha}$ belongs to $\mathrm{Cov}^h_f(\mathfrak{U}_\alpha)$ and we have the following commutative diagram whose three squares are Cartesian:

$$\begin{array}{ccccc}
\mathfrak{U}'_{\alpha\kappa;\beta} & \longrightarrow & \mathfrak{U}_{\alpha\beta} & \longleftarrow & \mathfrak{U}_{\alpha\beta;\gamma} \\
\downarrow & & \downarrow & & \\
\mathfrak{U}'_{\alpha\kappa} & \longrightarrow & \mathfrak{U}_{\alpha} & & \downarrow \\
\downarrow & & \downarrow & & \\
(U'_{\alpha\kappa}, \mathcal{U}'_{\alpha\kappa,\mathrm{triv},\overline{K}}) & \longrightarrow & (U_\alpha, \mathcal{U}_{\alpha,\mathrm{triv},\overline{K}}) & \longleftarrow & (U_{\alpha\beta;\gamma}, \mathcal{U}_{\alpha\beta;\gamma,\mathrm{triv},\overline{K}})
\end{array}$$

Hence for $\kappa \in K_\alpha$, $\beta \in B_\alpha$, and $\gamma = \gamma(\kappa,\beta) \in \Gamma_{\alpha\beta}$, a morphism $U'_{\alpha\kappa} \to U_{\alpha\beta;\gamma}$ over $U_\alpha$ induces a morphism $\mathfrak{U}'_{\alpha\kappa;\beta} \to \mathfrak{U}_{\alpha\beta;\gamma}$ over $\mathfrak{U}_{\alpha\beta}$. Let $(\mathfrak{U}'_{\alpha\kappa;\beta\delta} \to \mathfrak{U}'_{\alpha\kappa;\beta})_{\delta\in\Delta_{\alpha\beta;\gamma}} \in \mathrm{Cov}^v_f(\mathfrak{U}'_{\alpha\kappa;\beta})$ be the base change of $(\mathfrak{U}_{\alpha\beta;\gamma\delta} \to \mathfrak{U}_{\alpha\beta;\gamma})_{\delta\in\Delta_{\alpha\beta;\gamma}} \in \mathrm{Cov}^v_f(\mathfrak{U}_{\alpha\beta;\gamma})$ by a $\mathfrak{U}_{\alpha\beta}$-morphism $\mathfrak{U}'_{\alpha\kappa;\beta} \to \mathfrak{U}_{\alpha\beta;\gamma}$ as above. Then we have $(\mathfrak{U}'_{\alpha\kappa;\beta\delta} \to \mathfrak{U}'_{\alpha\kappa})_{\beta,\delta} \in \mathrm{Cov}^v_f(\mathfrak{U}'_{\alpha\kappa})$ and $(\mathfrak{U}'_{\alpha\kappa} \to \mathfrak{U})_{\alpha,\kappa} \in \mathrm{Cov}^h_f(\mathfrak{U})$, and their composition $(\mathfrak{U}'_{\alpha\kappa;\beta\delta} \to \mathfrak{U})$ refines $(\mathfrak{U}_{\alpha\beta;\gamma\delta} \to \mathfrak{U})$. Hence $R' \in J(\mathfrak{U})$. □

**Corollary IV.6.1.4** (cf. Proposition VI.5.38). *Let $f\colon X' \to X$ be a strict étale morphism of fs log schemes over $\Sigma$ such that $X'$ satisfies the same condition as $X$. Let $X'^{\wedge}$ denote the presheaf on $(\mathcal{P}/X)$ defined by $X'^{\wedge}((U,\mathcal{V}),u) = \mathrm{Hom}_X(U,X')$. Then the canonical isomorphism of categories $(\mathcal{P}/X')_{\text{ét-fét}} \xrightarrow{\cong} (\mathcal{P}/X)_{\text{ét-fét}}/X'^{\wedge}$ (resp. $(\mathcal{P}/X')_{\text{fét}} \xrightarrow{\cong} (\mathcal{P}/X)_{\text{fét}}/X'^{\wedge}$) defines an isomorphism of sites, where the target is endowed with the topology induced by that of $(\mathcal{P}/X)_{\text{ét-fét}}$ (resp. $(\mathcal{P}/X)_{\text{fét}}$).*

Proof. It suffices to prove that the topology of $(\mathcal{P}/X')_{\text{ét-fét}}$ (resp. $(\mathcal{P}/X')_{\text{fét}}$) is induced by the functor $j\colon (\mathcal{P}/X') \to (\mathcal{P}/X); (\mathfrak{U}',u') \mapsto (\mathfrak{U}', f\circ u')$ and the topology of $(\mathcal{P}/X)_{\text{ét-fét}}$ (resp. $(\mathcal{P}/X)_{\text{fét}}$). Let $(\mathfrak{U}',u') \in \mathrm{Ob}\,(\mathcal{P}/X')$ and let $R$ be a sieve of $(\mathfrak{U}',u')$. By **[2]** III Proposition 5.2 1), it suffices to prove that $R$ is a covering sieve for the topology of $(\mathcal{P}/X')_{\text{ét-fét}}$ (resp. $(\mathcal{P}/X')_{\text{fét}}$) if and only if $j_!R$ is a covering sieve of $(\mathfrak{U}', f\circ u')$ for the topology of $(\mathcal{P}/X)_{\text{ét-fét}}$ (resp. $(\mathcal{P}/X)_{\text{fét}}$). For $(\mathfrak{U},u) \in \mathrm{Ob}\,(\mathcal{P}/X)$, we have

$$(j_!(\mathfrak{U}',u'))(\mathfrak{U},u) = \mathrm{Hom}((\mathfrak{U},u),(\mathfrak{U}',f\circ u')) = \coprod_{v\in\mathrm{Hom}_X(U,X')} \mathrm{Hom}((\mathfrak{U},v),(\mathfrak{U}',u')),$$

$$(j_!R)(\mathfrak{U},u) = \coprod_{v\in\mathrm{Hom}_X(U,X')} R(\mathfrak{U},v).$$

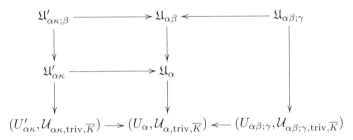

By Proposition IV.6.1.3 (resp. By definition), the sieve $R$ is a covering if and only if there exist $(\mathfrak{f}_\alpha\colon \mathfrak{U}'_\alpha \to \mathfrak{U}') \in \mathrm{Cov}^h_f(\mathfrak{U}')$ and $(\mathfrak{g}_{\alpha\beta}\colon \mathfrak{U}'_{\alpha\beta} \to \mathfrak{U}'_\alpha) \in \mathrm{Cov}^v_f(\mathfrak{U}'_\alpha)$ (resp. there exists $(\mathfrak{f}_\alpha\colon \mathfrak{U}'_\alpha \to \mathfrak{U}') \in \mathrm{Cov}^v(\mathfrak{U}'))$ such that $\mathfrak{f}_\alpha \circ \mathfrak{g}_{\alpha\beta}$ (resp. $\mathfrak{f}_\alpha$) factors through $R$ in $(\mathcal{P}/X')^\wedge$. By the above description of $j_!R$, the last condition is equivalent to saying that $\mathfrak{f}_\alpha \circ \mathfrak{g}_{\alpha\beta}$ (resp. $\mathfrak{f}_\alpha$) factors through $j_!R$ in $(\mathcal{P}/X)^\wedge$. By applying Proposition IV.6.1.3 to $(\mathcal{P}/X)_{\text{ét-fét}}$ (resp. By the definition of $(\mathcal{P}/X)_{\text{fét}}$), we obtain the claim. □

By Corollary IV.6.1.4, the functor $(\mathcal{P}/X')_{\text{ét-fét}} \to (\mathcal{P}/X)_{\text{ét-fét}}; (\mathfrak{U}', u') \to (\mathfrak{U}', f \circ u')$ is continuous and cocontinuous (cf. [2] III Proposition 5.2 2) and induces a morphism of topos

(IV.6.1.5)
$$f_{\mathcal{P},\text{ét-fét}}\colon (\mathcal{P}/X')^{\sim}_{\text{ét-fét}} \longrightarrow (\mathcal{P}/X)^{\sim}_{\text{ét-fét}}.$$

Similarly, we obtain a morphism of topos

(IV.6.1.6)
$$f_{\mathcal{P},\text{fét}}\colon (\mathcal{P}/X')^{\sim}_{\text{fét}} \longrightarrow (\mathcal{P}/X)^{\sim}_{\text{fét}}.$$

**Proposition IV.6.1.7.** *Let $f\colon X' \to X$ be as in Corollary* IV.6.1.4 *and assume that $X'$ is affine. We define the functor $f^*\colon (\mathcal{P}/X) \to (\mathcal{P}/X')$ by $((U, \mathcal{V}), u) \mapsto ((U', \mathcal{V}'), u')$, where $U' = X' \times_X U$, $\mathcal{V}' = \mathcal{V} \times_{\mathcal{U}_{\overline{K},\mathrm{triv}}} \mathcal{U}'_{\overline{K},\mathrm{triv}}$, and $u'$ is the natural morphism $U' \to X'$. Then the functor $f^*$ is continuous with respect to the topologies of $(\mathcal{P}/X)_{\text{ét-fét}}$ and $(\mathcal{P}/X')_{\text{ét-fét}}$ (resp. $(\mathcal{P}/X)_{\text{fét}}$ and $(\mathcal{P}/X')_{\text{fét}}$) and it is a right adjoint of the functor $j\colon (\mathcal{P}/X') \to (\mathcal{P}/X); (\mathfrak{U}', u') \mapsto (\mathfrak{U}', f \circ u')$. Hence $f^*$ defines a morphism of sites $(\mathcal{P}/X')_{\text{ét-fét}} \to (\mathcal{P}/X)_{\text{ét-fét}}$ (resp. $(\mathcal{P}/X')_{\text{fét}} \to (\mathcal{P}/X)_{\text{fét}}$) and induces the morphism of topos $f_{\mathcal{P},\text{ét-fét}}$ (resp. $f_{\mathcal{P},\text{fét}}$) above (cf.* [2] III *Proposition* 2.5).

PROOF. The functor $f^*$ preserves finite fiber products and induces $\mathrm{Cov}(\mathfrak{U}, u) \to \mathrm{Cov}(f^*(\mathfrak{U}, u))$ (resp. $\mathrm{Cov}^v(\mathfrak{U}, u) \to \mathrm{Cov}^v(f^*(\mathfrak{U}, u)))$ for any $(\mathfrak{U}, u) \in \mathrm{Ob}\,(\mathcal{P}/X)$. Hence $f^*$ is continuous. Let $(\mathfrak{U}, u) \in \mathrm{Ob}\,(\mathcal{P}/X)$, $(\mathfrak{U}', u') := f^*(\mathfrak{U}, u)$, and $(\mathfrak{U}'', u'') \in \mathrm{Ob}\,(\mathcal{P}/X')$. Put $\mathfrak{U} = (U, \mathcal{V})$, $\mathfrak{U}' = (U', \mathcal{V}')$, and $\mathfrak{U}'' = (U'', \mathcal{V}'')$. Then the composition with the natural morphism $f_U\colon U' \to U$ induces a bijection $\mathrm{Hom}_{X'}(U'', U') \to \mathrm{Hom}_X(U'', U)$. Given a morphism $U'' \to U'$ over $X'$, the composition with the natural morphism $f_{\mathcal{V}}\colon \mathcal{V}' \to \mathcal{V}$ induces a bijection $\mathrm{Hom}_{\mathcal{U}'_{\mathrm{triv},\overline{K}}}(\mathcal{V}'', \mathcal{V}') \to \mathrm{Hom}_{\mathcal{U}_{\mathrm{triv},\overline{K}}}(\mathcal{V}'', \mathcal{V})$. Hence the composition with $f_U$ and $f_{\mathcal{V}}$ induces a bijection $\mathrm{Hom}((\mathfrak{U}'', u''), (\mathfrak{U}', u')) \xrightarrow{\cong} \mathrm{Hom}((\mathfrak{U}'', f \circ u''), (\mathfrak{U}, u))$. □

We will later interpret the representations constructed in IV.5.2 in terms of sheaves on $(\mathcal{P}/X)_{\text{ét-fét}}$. Since the representations are constructed only for sufficiently small affine fs log schemes, we need to work with a full subcategory of $\mathcal{P}/X$ of the following type. Let $\mathcal{C}$ be a full subcategory of $X_{\text{étaff}}$ such that for any $U \in \mathrm{Ob}\,X_{\text{étaff}}$, there exists a strict étale covering $(U_\alpha \to U)_{\alpha \in A}$ such that $U_\alpha \in \mathrm{Ob}\,\mathcal{C}$ for every $\alpha \in A$. We define $\mathcal{P}/\mathcal{C}$ to be the full subcategory of $\mathcal{P}/X$ consisting of $(U, \mathcal{V})$ such that $U \in \mathrm{Ob}\,\mathcal{C}$ and define the site $(\mathcal{P}/\mathcal{C})_{\text{ét-fét}}$ (resp. $(\mathcal{P}/\mathcal{C})_{\text{fét}}$) to be the category $\mathcal{P}/\mathcal{C}$ endowed with the topology induced by that of $(\mathcal{P}/X)_{\text{ét-fét}}$ (cf. [2] III §3) (resp. the pretopology $\mathrm{Cov}^v(\mathfrak{U})$ ($\mathfrak{U} \in \mathrm{Ob}\,(\mathcal{P}/\mathcal{C})_{\text{fét}}$)).

For any $(U, \mathcal{V}) \in (\mathcal{P}/X)_{\text{ét-fét}}$, a strict étale covering $(U_\alpha \to U)_{\alpha \in A}$ by objects of $\mathcal{C}$ induces a horizontal strict étale covering $((U, \mathcal{V}) \times_{(U, \mathcal{U}_{\overline{K},\mathrm{triv}})} (U_\alpha, \mathcal{U}_{\alpha,\overline{K},\mathrm{triv}}) \to (U, \mathcal{V}))_{\alpha \in A}$ by objects of $\mathcal{P}/\mathcal{C}$. Hence the inclusion functor $w\colon (\mathcal{P}/\mathcal{C}) \to (\mathcal{P}/X)$ induces an equivalence of categories (cf. [2] III Théorème 4.1)

(IV.6.1.8)
$$w_s\colon (\mathcal{P}/X)^{\sim}_{\text{ét-fét}} \xrightarrow{\sim} (\mathcal{P}/\mathcal{C})^{\sim}_{\text{ét-fét}}.$$

The inclusion functor $w\colon (\mathcal{P}/\mathcal{C})_{\text{fét}} \to (\mathcal{P}/X)_{\text{fét}}$ is continuous. Using (IV.6.1.8), we also see that the identity functor $(\mathcal{P}/\mathcal{C})_{\text{fét}} \to (\mathcal{P}/\mathcal{C})_{\text{ét-fét}}$ is continuous. Hence the identity functor $(\mathcal{P}/\mathcal{C})_{\text{ét-fét}} \to (\mathcal{P}/\mathcal{C})_{\text{fét}}$ is cocontinuous and these two functors induce a morphism of topos (cf. [2] III Proposition 2.5)

(IV.6.1.9)
$$v_{\mathcal{P}/\mathcal{C}}\colon (\mathcal{P}/\mathcal{C})^{\sim}_{\text{ét-fét}} \to (\mathcal{P}/\mathcal{C})^{\sim}_{\text{fét}}.$$

The inverse image functor is given by the sheafification.

Let $f\colon X' \to X$ be a strict étale morphism of fs log schemes over $\Sigma$ such that $X'$ satisfies the same condition as $X$. Let $\mathcal{C}'$ be the full subcategory of $X'_{\text{étaff}}$ such that every object of $X'_{\text{étaff}}$ admits a strict étale covering by objects of $\mathcal{C}'$ and, for every object $u'\colon U' \to X'$ of $\mathcal{C}'$, the composite $f \circ u'$ is an object of $\mathcal{C}$.

**Lemma IV.6.1.10.** *Let the notation and the assumption be as above. Then the functor* $(\mathcal{P}/\mathcal{C}')_{\text{ét-fét}} \to (\mathcal{P}/\mathcal{C})_{\text{ét-fét}}$ *(resp.* $(\mathcal{P}/\mathcal{C}')_{\text{fét}} \to (\mathcal{P}/\mathcal{C})_{\text{ét-fét}}$*) defined by* $(u'\colon U' \to X', \mathcal{V}) \mapsto (f \circ u', \mathcal{V})$ *is continuous and cocontinuous.*

Proof. We first prove the claim for $(\mathcal{P}/\mathcal{C})_{\text{ét-fét}}$ and $(\mathcal{P}/\mathcal{C}')_{\text{ét-fét}}$. We have a commutative diagram of categories

$$\begin{array}{ccc} (\mathcal{P}/\mathcal{C}')_{\text{ét-fét}} & \xrightarrow{w'} & (\mathcal{P}/X')_{\text{ét-fét}} \\ {\scriptstyle j'}\downarrow & & \downarrow{\scriptstyle j} \\ (\mathcal{P}/\mathcal{C})_{\text{ét-fét}} & \xrightarrow{w} & (\mathcal{P}/X)_{\text{ét-fét}}, \end{array}$$

where $w$ and $w'$ are inclusion functors and $j$ and $j'$ are defined by the composition with $f$. By Corollary IV.6.1.4 and **[2]** III Proposition 5.2 2), the functor $j$ is continuous and cocontinuous. Hence, by (IV.6.1.8) applied to $w$ and $w'$, we see that $j'$ is continuous. By the proof of **[2]** III Théorème 4.1, the functor $w'$ is cocontinuous. Hence the composition $w \circ j' = j \circ w'$ is cocontinuous. Let $\widehat{j}'^{*}$ (resp. $\widehat{w}^{*}$) denote the functor $(\mathcal{P}/\mathcal{C})^{\wedge}_{\text{ét-fét}} \to (\mathcal{P}/\mathcal{C}')^{\wedge}_{\text{ét-fét}}$ (resp. $(\mathcal{P}/X)^{\wedge}_{\text{ét-fét}} \to (\mathcal{P}/\mathcal{C})^{\wedge}_{\text{ét-fét}}$) defined by the composition with $j'$ (resp. $w$), and let $\widehat{j}'_{*}$ (resp. $\widehat{w}_{*}$) denote its right adjoint. Then, for a sheaf $\mathcal{F}$ on $(\mathcal{P}/\mathcal{C}')_{\text{ét-fét}}$, $\widehat{w}_{*}\widehat{j}'_{*}\mathcal{F}$ is a sheaf by **[2]** III Proposition 2.2. Since $w$ is fully faithful, we see that the adjunction morphism $\widehat{w}^{*}\widehat{w}_{*}\widehat{j}'_{*}\mathcal{F} \to \widehat{j}'_{*}\mathcal{F}$ is an isomorphism (cf. **[2]** I Proposition 5.6). Hence $\widehat{j}'_{*}\mathcal{F}$ is a sheaf on $(\mathcal{P}/\mathcal{C})_{\text{ét-fét}}$ because $w$ is continuous. By **[2]** III Proposition 2.2, $j'$ is cocontinuous.

Next we prove the second case. The topologies are induced by the pretopologies defined by vertical finite étale coverings and the functor in question is compatible with the fiber products of $\mathfrak{U}' \xrightarrow{\mathfrak{f}} \mathfrak{U} \leftarrow \mathfrak{U}''$ with $\mathfrak{f}$ vertical. This implies the continuity. For an object $(\mathfrak{U}, u)$ of $\mathcal{P}/\mathcal{C}'$, the functor $(\mathcal{P}/\mathcal{C}')/(\mathfrak{U}, u) \to (\mathcal{P}/\mathcal{C})/(\mathfrak{U}, f \circ u)$ is fully faithful. Hence, by **[2]** II Proposition 1.4, we see that the functor $(\mathcal{P}/\mathcal{C}')_{\text{fét}} \to (\mathcal{P}/\mathcal{C})_{\text{fét}}$ satisfies the property defining cocontinuity (cf. **[2]** III Définition 2.1). □

We obtain from Lemma IV.6.1.10 a sequence of three adjoint functors between $(\mathcal{P}/\mathcal{C}')^{\sim}_{\text{ét-fét}}$ and $(\mathcal{P}/\mathcal{C})^{\sim}_{\text{ét-fét}}$ (resp. $(\mathcal{P}/\mathcal{C}')^{\sim}_{\text{fét}}$ and $(\mathcal{P}/\mathcal{C})^{\sim}_{\text{fét}}$)

$$f^{\mathcal{C},\mathcal{C}'}_{\mathcal{P},\text{ét-fét}!},\ f^{\mathcal{C},\mathcal{C}'*}_{\mathcal{P},\text{ét-fét}},\ f^{\mathcal{C},\mathcal{C}'}_{\mathcal{P},\text{ét-fét}*} \quad (\text{resp. } f^{\mathcal{C},\mathcal{C}'}_{\mathcal{P},\text{fét}!},\ f^{\mathcal{C},\mathcal{C}'*}_{\mathcal{P},\text{fét}},\ f^{\mathcal{C},\mathcal{C}'}_{\mathcal{P},\text{fét}*}). \tag{IV.6.1.11}$$

These are equivalences of categories if $X' = X$ and $f = \mathrm{id}_X$ in the first case. The pairs $(f^{\mathcal{C},\mathcal{C}'*}_{\mathcal{P},\text{ét-fét}}, f^{\mathcal{C},\mathcal{C}'}_{\mathcal{P},\text{ét-fét}*})$ and $(f^{\mathcal{C},\mathcal{C}'*}_{\mathcal{P},\text{fét}}, f^{\mathcal{C},\mathcal{C}'}_{\mathcal{P},\text{fét}*})$ define morphisms of topos

$$f^{\mathcal{C},\mathcal{C}'}_{\mathcal{P},\text{ét-fét}}\colon (\mathcal{P}/\mathcal{C}')^{\sim}_{\text{ét-fét}} \longrightarrow (\mathcal{P}/\mathcal{C})^{\sim}_{\text{ét-fét}}, \tag{IV.6.1.12}$$

$$f^{\mathcal{C},\mathcal{C}'}_{\mathcal{P},\text{fét}}\colon (\mathcal{P}/\mathcal{C}')^{\sim}_{\text{fét}} \longrightarrow (\mathcal{P}/\mathcal{C})^{\sim}_{\text{fét}}, \tag{IV.6.1.13}$$

which coincide with $f_{\mathcal{P},\text{ét-fét}}$ and $f_{\mathcal{P},\text{fét}}$ if $\mathcal{C} = X_{\text{étaff}}$ and $\mathcal{C}' = X'_{\text{étaff}}$. We see that the following diagram of topos is commutative up to canonical isomorphism by looking at the corresponding diagram of sites and cocontinuous morphisms.

(IV.6.1.14)
$$\begin{array}{ccc} (\mathcal{P}/\mathcal{C}')^{\sim}_{\text{ét-fét}} & \xrightarrow{f^{\mathcal{C},\mathcal{C}'}_{\mathcal{P},\text{ét-fét}}} & (\mathcal{P}/\mathcal{C})^{\sim}_{\text{ét-fét}} \\ \downarrow v_{\mathcal{P}/\mathcal{C}'} & & \downarrow v_{\mathcal{P}/\mathcal{C}} \\ (\mathcal{P}/\mathcal{C}')^{\sim}_{\text{fét}} & \xrightarrow{f^{\mathcal{C},\mathcal{C}'}_{\mathcal{P},\text{fét}}} & (\mathcal{P}/\mathcal{C})^{\sim}_{\text{fét}}. \end{array}$$

We write $\iota^{\mathcal{C},\mathcal{C}'}_{\mathcal{P},\text{fét}}$ and $\iota^{\mathcal{C},\mathcal{C}'}_{\mathcal{P},\text{ét-fét}}$ for $f^{\mathcal{C},\mathcal{C}'}_{\mathcal{P},\text{fét}}$ and $f^{\mathcal{C},\mathcal{C}'}_{\mathcal{P},\text{ét-fét}}$ if $X = X'$ and $f = \mathrm{id}_X$. We abbreviate the superscript $\mathcal{C}, \mathcal{C}'$ if there is no risk of confusion in the following.

For an object $\mathfrak{U} = (U, \mathcal{V})$ of $\mathcal{P}/X$, let $\mathcal{A}_{\mathcal{V}}$ denote the integral closure of $\mathcal{A}_U$ in $\Gamma(\mathcal{V}, \mathcal{O}_{\mathcal{V}})$.

**Lemma IV.6.1.15.** *The presheaf $\mathcal{V} \mapsto \mathcal{A}_{\mathcal{V}}$ on $(\mathcal{U}_{\overline{K},\text{triv}})_{\text{fét}}$ is a sheaf.*

PROOF. Since $\mathcal{V} \mapsto \Gamma(\mathcal{V}, \mathcal{O}_{\mathcal{V}})$ is a sheaf on $(\mathcal{U}_{\overline{K},\text{triv}})_{\text{fét}}$, it suffices to prove the following claim. For a finite étale covering $(\mathcal{V}_\alpha \to \mathcal{V})_{\alpha \in A}$, a section $x \in \Gamma(\mathcal{V}, \mathcal{O}_{\mathcal{V}})$ is integral over $\mathcal{A}_U$ if its image $x_\alpha$ in $\Gamma(\mathcal{V}_\alpha, \mathcal{O}_{\mathcal{V}_\alpha})$ is integral over $\mathcal{A}_U$ for every $\alpha \in A$. Since $\mathcal{V}$ is affine, there exists a finite subset $A'$ of $A$ such that $(\mathcal{V}_\alpha \to \mathcal{V})_{\alpha \in A'}$ is still a covering. For each $\alpha \in A'$, choose a monic polynomial $f_\alpha \in \mathcal{A}_U[X]$ such that $f_\alpha(x_\alpha) = 0$. Then, we have $f(x) = 0$ for $f = \prod_{\alpha \in A'} f_\alpha$ because $\Gamma(\mathcal{V}, \mathcal{O}_{\mathcal{V}}) \to \prod_{\alpha \in A'} \Gamma(\mathcal{V}_\alpha, \mathcal{O}_{\mathcal{V}_\alpha})$ is injective. □

We define the sheaf $\mathcal{O}^{\text{int}}_{\mathcal{U}_{\overline{K},\text{triv}}}$ on $(\mathcal{U}_{\overline{K},\text{triv}})_{\text{fét}}$ by $\mathcal{O}^{\text{int}}_{\mathcal{U}_{\overline{K},\text{triv}}}(\mathcal{V}) = \mathcal{A}_{\mathcal{V}}$, and define the sheaf $\mathcal{O}^v_{\mathcal{P}/\mathcal{C}}$ on $(\mathcal{P}/\mathcal{C})_{\text{fét}}$ by $\mathcal{O}^v_{\mathcal{P}/\mathcal{C}}(U, \mathcal{V}) = \mathcal{O}^{\text{int}}_{\mathcal{U}_{\overline{K},\text{triv}}}(\mathcal{V}) = \mathcal{A}_{\mathcal{V}}$. Let $\mathcal{O}_{\mathcal{P}/\mathcal{C}}$ denote the sheaf on $(\mathcal{P}/\mathcal{C})_{\text{ét-fét}}$ associated to $\mathcal{O}^v_{\mathcal{P}/\mathcal{C}}$. We have $\mathcal{O}_{\mathcal{P}/\mathcal{C}} = v^*_{\mathcal{P}/\mathcal{C}}(\mathcal{O}^v_{\mathcal{P}/\mathcal{C}})$.

For $m \in \mathbb{N}$, we define $\mathcal{O}_{\mathcal{P}/\mathcal{C},m}$ (resp. $\mathcal{O}^v_{\mathcal{P}/\mathcal{C},m}$) to be the quotient $\mathcal{O}_{\mathcal{P}/\mathcal{C}}/p^m\mathcal{O}_{\mathcal{P}/\mathcal{C}}$ (resp. $\mathcal{O}^v_{\mathcal{P}/\mathcal{C}}/p^m\mathcal{O}^v_{\mathcal{P}/\mathcal{C}}$) as a sheaf on $(\mathcal{P}/\mathcal{C})_{\text{ét-fét}}$ (resp. $(\mathcal{P}/\mathcal{C})_{\text{fét}}$). We have a canonical isomorphism

$$v^*_{\mathcal{P}/\mathcal{C}}(\mathcal{O}^v_{\mathcal{P}/\mathcal{C},m}) \cong \mathcal{O}_{\mathcal{P}/\mathcal{C},m}.$$

We write $\mathcal{O}^\star_{\mathcal{P}/X}$ and $\mathcal{O}^\star_{\mathcal{P}/X,m}$ ($\star = v, \emptyset$) for $\mathcal{O}^\star_{\mathcal{P}/\mathcal{C}}$ and $\mathcal{O}^\star_{\mathcal{P}/\mathcal{C},m}$ if $\mathcal{C} = X_{\text{étaff}}$.

**Lemma IV.6.1.16.** *Assume that $A_U$ is flat over $V$, $A_U$ is normal, and $A_U \otimes_V k$ is reduced. Then, in the cases* I *and* III, *we have $\mathcal{O}^v_{\mathcal{P}/\mathcal{C}} = \mathcal{O}_{\mathcal{P}/\mathcal{C}}$, i.e., $\mathcal{O}^v_{\mathcal{P}/\mathcal{C}}$ is a sheaf on $(\mathcal{P}/\mathcal{C})_{\text{ét-fét}}$.*

PROOF. By Lemma IV.6.1.15 and the remark before Proposition IV.6.1.3, it suffices to prove the following. For $\mathfrak{U} = (U, \mathcal{V}) \in \mathrm{Ob}\,(\mathcal{P}/\mathcal{C})$, $(\mathfrak{U}_\alpha = (U_\alpha, \mathcal{V}_\alpha) \to \mathfrak{U})_{\alpha \in A} \in \mathrm{Cov}^h_f(\mathfrak{U})$, and $\mathfrak{U}_{\alpha\beta} = (U_{\alpha\beta}, \mathcal{V}_{\alpha\beta}) := \mathfrak{U}_\alpha \times_{\mathfrak{U}} \mathfrak{U}_\beta$, the sequence $\mathcal{A}_{\mathcal{V}} \to \prod_{\alpha \in A} \mathcal{A}_{\mathcal{V}_\alpha} \rightrightarrows \prod_{(\alpha,\beta) \in A^2} \mathcal{A}_{\mathcal{V}_{\alpha\beta}}$ is exact. Put $\widetilde{U} = \mathcal{U}_{\overline{K},\text{triv}}$, $\widetilde{U}_\alpha = \mathcal{U}_{\alpha,\overline{K},\text{triv}}$, and $\widetilde{U}_{\alpha\beta} = \mathcal{U}_{\alpha\beta,\overline{K},\text{triv}}$ to simplify the notation. In the cases I and III, the sequence $\mathcal{A}_U \to \prod_{\alpha \in A} \mathcal{A}_{U_\alpha} \rightrightarrows \prod_{(\alpha,\beta) \in A^2} \mathcal{A}_{U_{\alpha\beta}}$ is exact, and it remains exact after $\otimes_V \overline{V}$. By Lemma IV.5.2.1, we see that $\mathcal{A}_U \otimes_V \overline{V}$, $\mathcal{A}_{U_\alpha} \otimes_V \overline{V}$, and $\mathcal{A}_{U_{\alpha\beta}} \otimes_V \overline{V}$ are products of finite numbers of normal domains. Hence, $\mathcal{A}_{\widetilde{U}}$, $\mathcal{A}_{\widetilde{U}_\alpha}$, and $\mathcal{A}_{\widetilde{U}_{\alpha\beta}}$ are also products of finite numbers of normal domains, and the following sequence is exact.

$$(*) \qquad \mathcal{A}_{\widetilde{U}} \to \prod_{\alpha \in A} \mathcal{A}_{\widetilde{U}_\alpha} \rightrightarrows \prod_{(\alpha,\beta) \in A^2} \mathcal{A}_{\widetilde{U}_{\alpha\beta}}.$$

We also see that the sequence $\Gamma(\widetilde{U}, \mathcal{O}_{\widetilde{U}}) \to \prod_{\alpha \in A} \Gamma(\widetilde{U}_\alpha, \mathcal{O}_{\widetilde{U}_\alpha}) \rightrightarrows \prod_{(\alpha,\beta) \in A^2} \Gamma(\widetilde{U}_{\alpha\beta}, \mathcal{O}_{\widetilde{U}_{\alpha\beta}})$ is exact. Since the morphism $\mathfrak{U}_\alpha \to \mathfrak{U}$ is horizontal, the natural morphisms $\mathcal{V}_\alpha \to \widetilde{U}_\alpha \times_{\widetilde{U}} \mathcal{V}$ and $\mathcal{V}_{\alpha\beta} \to \widetilde{U}_{\alpha\beta} \times_{\widetilde{U}} \mathcal{V}$ are isomorphisms. Since $\mathcal{V}_\alpha \to \widetilde{U}_\alpha$ is flat, this implies that the sequence $\Gamma(\mathcal{V},\mathcal{O}_{\mathcal{V}}) \to \prod_{\alpha\in A}\Gamma(\mathcal{V}_\alpha,\mathcal{O}_{\mathcal{V}_\alpha}) \rightrightarrows \prod_{(\alpha,\beta)\in A^2}\Gamma(\mathcal{V}_{\alpha\beta},\mathcal{O}_{\mathcal{V}_{\alpha\beta}})$ is also exact. Hence it remains to prove that an element $x\in\Gamma(\mathcal{V},\mathcal{O}_{\mathcal{V}})$ is integral over $\mathcal{A}_{\widetilde{U}}$ if its pull-back in $\Gamma(\mathcal{V}_\alpha,\mathcal{O}_{\mathcal{V}_\alpha})$ is integral over $\mathcal{A}_{\widetilde{U}_\alpha}$ for every $\alpha\in A$. Let $x_\alpha$ be the image of $x$ in $\Gamma(\mathcal{V}_\alpha,\mathcal{O}_{\mathcal{V}_\alpha})$, and let $\Phi_\alpha(T)\in\Gamma(\widetilde{U}_\alpha,\mathcal{O}_{\widetilde{U}_\alpha})[T]$ be the characteristic polynomial of $x_\alpha$ over $\widetilde{U}_\alpha$. Since $\mathcal{A}_{\widetilde{U}_\alpha}$ is a product of finite number of normal domains, the coefficients of $\Phi_\alpha(T)$ lie in $\mathcal{A}_{\widetilde{U}_\alpha}$. The inverse images of $\Phi_\alpha(T)$ and $\Phi_\beta(T)$ in $\mathcal{A}_{\widetilde{U}_{\alpha\beta}}[T]$ coincide because they are both the characteristic polynomial of the inverse image of $x$ in $\Gamma(\mathcal{V}_{\alpha\beta},\mathcal{O}_{\mathcal{V}_{\alpha\beta}})$ over $\widetilde{U}_{\alpha\beta}$. By the exact sequence $(*)$, there exists $\Phi(T)\in\mathcal{A}_{\widetilde{U}}[T]$ whose pull-back in $\mathcal{A}_{\widetilde{U}_\alpha}[T]$ is $\Phi_\alpha(T)$. The vanishing $\Phi_\alpha(x_\alpha)=0$ for every $\alpha$ implies $\Phi(x)=0$ and hence $x$ is integral over $\mathcal{A}_{\widetilde{U}}$. □

For $f\colon X'\to X$, $\mathcal{C}$, and $\mathcal{C}'$ as before Lemma IV.6.1.10, we have

$$f^{\mathcal{C},\mathcal{C}'*}_{\mathcal{P},\text{ét-fét}}(\mathcal{O}_{\mathcal{P}/\mathcal{C},m})\cong\mathcal{O}_{\mathcal{P}/\mathcal{C}',m}$$

by (IV.6.1.14). Hence we have a diagram of ringed topos commutative up to canonical isomorphism

(IV.6.1.17)
$$\begin{array}{ccc}((\mathcal{P}/\mathcal{C}')^\sim_{\text{ét-fét}},\mathcal{O}_{\mathcal{P}/\mathcal{C}',m}) & \xrightarrow{f^{\mathcal{C},\mathcal{C}'}_{\mathcal{P},\text{ét-fét}}} & ((\mathcal{P}/\mathcal{C})^\sim_{\text{ét-fét}},\mathcal{O}_{\mathcal{P}/\mathcal{C},m})\\ \downarrow{\scriptstyle v_{\mathcal{P}/\mathcal{C}'}} & & \downarrow{\scriptstyle v_{\mathcal{P}/\mathcal{C}}}\\ ((\mathcal{P}/\mathcal{C}')^\sim_{\text{fét}},\mathcal{O}^v_{\mathcal{P}/\mathcal{C}',m}) & \xrightarrow{f^{\mathcal{C},\mathcal{C}'}_{\mathcal{P},\text{fét}}} & ((\mathcal{P}/\mathcal{C})^\sim_{\text{fét}},\mathcal{O}^v_{\mathcal{P}/\mathcal{C}',m}).\end{array}$$

If $X'=X$ and $f=\mathrm{id}$, we have the following equivalences of categories which are quasi-inverse of each other, where $\mathcal{M}\mathrm{od}((\mathcal{P}/\mathcal{C}^{(\prime)})_{\text{ét-fét}},\mathcal{O}_{\mathcal{P}/\mathcal{C}^{(\prime)},\bullet})$ denotes the category of inverse systems of $\mathcal{O}_{\mathcal{P}/\mathcal{C}^{(\prime)},m}$-modules for $m\in\mathbb{N}_{>0}$.

(IV.6.1.18)
$$\mathcal{M}\mathrm{od}((\mathcal{P}/\mathcal{C}')_{\text{ét-fét}},\mathcal{O}_{\mathcal{P}/\mathcal{C}',\bullet}) \underset{\iota^{\mathcal{C},\mathcal{C}'*}_{\mathcal{P},\text{ét-fét}}}{\overset{\iota^{\mathcal{C},\mathcal{C}'}_{\mathcal{P},\text{ét-fét}*}}{\rightleftarrows}} \mathcal{M}\mathrm{od}((\mathcal{P}/\mathcal{C})_{\text{ét-fét}},\mathcal{O}_{\mathcal{P}/\mathcal{C},\bullet}).$$

**Proposition IV.6.1.19.** *Let $(f_\alpha\colon X_\alpha\to X)_{\alpha\in A}$ be a strict étale covering such that $X_\alpha$ is affine. Let $X_{\alpha\beta}$ denote $X_\alpha\times_X X_\beta$ and let $f_{\alpha\beta}\colon X_{\alpha\beta}\to X$ denote the natural morphism. Then, for a sheaf of $\mathcal{O}_{\mathcal{P}/X,m}$-modules $\mathcal{F}_i$ $(i=1,2)$, the following sequence is exact.*

$$\begin{aligned}\mathrm{Hom}_{\mathcal{O}_{\mathcal{P}/X,m}}(\mathcal{F}_1,\mathcal{F}_2)&\to\prod_{\alpha\in A}\mathrm{Hom}_{\mathcal{O}_{\mathcal{P}/X_\alpha,m}}(f^*_{\alpha,\text{ét-fét}}\mathcal{F}_1,f^*_{\alpha,\text{ét-fét}}\mathcal{F}_2)\\ &\rightrightarrows\prod_{(\alpha,\beta)\in A^2}\mathrm{Hom}_{\mathcal{O}_{\mathcal{P}/X_{\alpha\beta},m}}(f^*_{\alpha\beta,\text{ét-fét}}\mathcal{F}_1,f^*_{\alpha\beta,\text{ét-fét}}\mathcal{F}_2).\end{aligned}$$

Proof. For a strict étale morphism $f\colon X'\to X$ with $X'$ affine, an object $(\mathfrak{U},u)$ of $(X/\mathcal{P})_{\text{ét-fét}}$, and $(\mathfrak{U}',u'):=f^*(\mathfrak{U},u)$ (cf. Proposition IV.6.1.7), the morphism $\mathcal{F}_i(\mathfrak{U},u)\to f_{\text{ét-fét}*}f^*_{\text{ét-fét}}\mathcal{F}_i(\mathfrak{U},u)=\mathcal{F}_i(\mathfrak{U}',f\circ u')$ is induced by the natural morphism $(\mathfrak{U}',f\circ u')\to(\mathfrak{U},u)$. On the other hand, for $(\mathfrak{U},u)\in\mathrm{Ob}\,(\mathcal{P}/X)_{\text{ét-fét}}$, letting $\mathfrak{U}_\alpha$ and $\mathfrak{U}_{\alpha\beta}$ denote the objects of $\mathcal{P}$ underlying $f^*_{\alpha,\text{ét-fét}}(\mathfrak{U},u)$ and $f^*_{\alpha\beta,\text{ét-fét}}(\mathfrak{U},u)$, the family of natural morphisms $(\mathfrak{U}_\alpha\to\mathfrak{U})_{\alpha\in A}$ is a horizontal strict étale covering and the natural morphism $\mathfrak{U}_{\alpha\beta}\to\mathfrak{U}_\alpha\times_{\mathfrak{U}}\mathfrak{U}_\beta$ is an isomorphism. Hence the sequence

$$\mathcal{F}_i\to\prod_{\alpha\in A}f_{\alpha,\text{ét-fét}*}f^*_{\alpha,\text{ét-fét}}\mathcal{F}_i\rightrightarrows\prod_{(\alpha,\beta)\in A^2}f_{\alpha\beta,\text{ét-fét}*}f^*_{\alpha\beta,\text{ét-fét}}\mathcal{F}_i$$

is exact. This implies the claim. □

**Definition IV.6.1.20.** Let $C$ be a $\overline{V}$-linear abelian category.

(1) We say that an object $x$ of $C$ is *almost zero* if $a \cdot \mathrm{id}_x = 0$ for all $a \in \overline{\mathfrak{m}}$.

(2) We say that a morphism $f\colon x \to y$ in $C$ is an *almost isomorphism* if $\mathrm{Ker}(f)$ and $\mathrm{Cok}(f)$ is almost zero.

**Lemma IV.6.1.21.** *Let $C$ be a $\overline{V}$-linear abelian category.*

*(1) Let $x \to y \to z$ be an exact sequence in $C$. If $x$ and $z$ are almost zero, then $y$ is almost zero.*

*(2) If morphisms $f\colon x \to y$ and $g\colon y \to z$ in $C$ are almost isomorphisms, then $g \circ f$ is also an almost isomorphism.*

*(3) If $f\colon x \to y$ is an almost isomorphism in $C$, then for every $a \in \overline{\mathfrak{m}}$, there exists a morphism $g\colon y \to x$ such that $g \circ f = a \cdot \mathrm{id}_x$ and $f \circ g = a \cdot \mathrm{id}_y$.*

PROOF. It is straightforward to prove (1) and (2). Let us prove (3). Let $f\colon x \to y$ be an almost isomorphism in $C$ and let $a \in \overline{\mathfrak{m}}$. Choose $b, c \in \overline{\mathfrak{m}}$ such that $a = bc$. Since $b \cdot \mathrm{id}_{\mathrm{Cok} f} = 0$, $b \cdot \mathrm{id}_y$ factors through $\varphi_b\colon y \to \mathrm{Im}(f)$. Since $c \cdot \mathrm{id}_{\mathrm{Ker} f} = 0$, $c \cdot \mathrm{id}_x$ factors through $\psi_c\colon \mathrm{Im} f \to x$. The composition $\psi_c \circ \varphi_b\colon y \to x$ satisfies the desired property. □

**Corollary IV.6.1.22.** *Let $C$ be a site and let $f\colon \mathcal{F} \to \mathcal{G}$ be a morphism of sheaves of $\overline{V}$-modules on $C$. Then $f$ is an almost isomorphism of sheaves of $\overline{V}$-modules if and only if it is an almost isomorphism of presheaves of $\overline{V}$-modules.*

**Proposition IV.6.1.23.** *Let $X$ and $\mathcal{C}$ be as above. Let $\mathcal{F}$ be a presheaf of $\overline{V}$-modules on $(\mathcal{P}/X)_{\text{ét-fét}}$ satisfying the following two conditions:*

*(a) For any $\mathfrak{U} \in \mathrm{Ob}(\mathcal{P}/\mathcal{C})$ and $(\mathfrak{U}_\alpha \to \mathfrak{U}) \in \mathrm{Cov}_f^v(\mathfrak{U})$ such that $\mathfrak{U}_\alpha \in \mathrm{Ob}(\mathcal{P}/\mathcal{C})$, the morphism*

$$\mathcal{F}(\mathfrak{U}) \to \mathrm{Ker}\Big(\prod_{\alpha \in A} \mathcal{F}(\mathfrak{U}_\alpha) \rightrightarrows \prod_{(\alpha,\beta)\in A^2} \mathcal{F}(\mathfrak{U}_\alpha \times_{\mathfrak{U}} \mathfrak{U}_\beta)\Big)$$

*is an almost isomorphism.*

*(b) For any $\mathfrak{U} \in \mathrm{Ob}(\mathcal{P}/\mathcal{C})$, $(\mathfrak{U}_\alpha \to \mathfrak{U})_{\alpha \in A} \in \mathrm{Cov}_f^h(\mathfrak{U})$, and $(\mathfrak{U}_{\alpha\beta;\gamma} \to \mathfrak{U}_\alpha \times_{\mathfrak{U}} \mathfrak{U}_\beta)_{\gamma \in \Gamma_{\alpha\beta}} \in \mathrm{Cov}_f^h(\mathfrak{U}_\alpha \times_{\mathfrak{U}} \mathfrak{U}_\beta)$ for $(\alpha, \beta) \in A^2$ such that $\mathfrak{U}_\alpha, \mathfrak{U}_{\alpha\beta;\gamma} \in \mathrm{Ob}(\mathcal{P}/\mathcal{C})$, the morphism*

$$\mathcal{F}(\mathfrak{U}) \to \mathrm{Ker}\Big(\prod_{\alpha \in A} \mathcal{F}(\mathfrak{U}_\alpha) \rightrightarrows \prod_{(\alpha,\beta)\in A^2} \prod_{\gamma \in \Gamma_{\alpha\beta}} \mathcal{F}(\mathfrak{U}_{\alpha\beta;\gamma})\Big)$$

*is an almost isomorphism.*

*Then for the sheaf $\mathcal{F}^a$ on the site $(\mathcal{P}/\mathcal{C})_{\text{ét-fét}}$ associated to $\mathcal{F}$, the canonical morphism $\mathcal{F}(\mathfrak{U}) \to \mathcal{F}^a(\mathfrak{U})$ is an almost isomorphism for every $\mathfrak{U} \in \mathrm{Ob}(\mathcal{P}/\mathcal{C})$.*

PROOF. By Proposition IV.6.1.3 and the construction of the sheafification in **[2]** II §3, it suffices to prove that for $(\mathfrak{f}_\alpha)$ and $(\mathfrak{g}_{\alpha\beta})$ as in Proposition IV.6.1.3, the morphism

$$\mathcal{F}(\mathfrak{U}) \longrightarrow \mathrm{Ker}\Big(\prod_{\alpha\beta} \mathcal{F}(\mathfrak{U}_{\alpha\beta}) \rightrightarrows \prod_{\alpha\beta,\alpha'\beta'} \mathcal{F}(\mathfrak{U}_{\alpha\beta} \times_{\mathfrak{U}} \mathfrak{U}_{\alpha'\beta'})\Big)$$

is an almost isomorphism. (If $L$ denotes the functor defined in **[2]** II 3.0.5, then the above claim implies that $\mathcal{F} \to L\mathcal{F}$ is an almost isomorphism. Hence $L\mathcal{F}$ also satisfies the conditions (a) and (b). Applying the above claim again, we see that $L\mathcal{F} \to LL\mathcal{F} = \mathcal{F}^a$ is an almost isomorphism.) The conditions (a) and (b) on $\mathcal{F}$ immediately imply that the above homomorphism is almost injective. Assume that $(x_{\alpha\beta}) \in \prod_{\alpha\beta} \mathcal{F}(\mathfrak{U}_{\alpha\beta})$ is contained in the kernel. Let $\varepsilon \in \mathbb{Q}_{>0}$. Since the restrictions of $x_{\alpha\beta}$ and $x_{\alpha\beta'}$ on $\mathfrak{U}_{\alpha\beta} \times_{\mathfrak{U}_\alpha} \mathfrak{U}_{\alpha\beta'}$ coincide, we see that there exists $x_\alpha \in \mathcal{F}(\mathfrak{U}_\alpha)$ such that $x_\alpha|_{\mathfrak{U}_{\alpha\beta}} = p^\varepsilon x_{\alpha\beta}$ for every $\beta \in B_\alpha$ by the condition (a). For each $(\alpha, \alpha') \in A^2$, choose $(\mathfrak{U}_{\alpha\alpha';\gamma} \to \mathfrak{U}_\alpha \times_{\mathfrak{U}} \mathfrak{U}_{\alpha'})_{\gamma \in B_{\alpha\alpha'}} \in \mathrm{Cov}_f^h(\mathfrak{U}_\alpha \times_{\mathfrak{U}} \mathfrak{U}_{\alpha'})$ such that $\mathfrak{U}_{\alpha\alpha';\gamma} \in \mathrm{Ob}\, \mathcal{C}$. For $(\beta, \beta') \in B_\alpha \times B_{\alpha'}$, let $(\mathfrak{U}_{\alpha\beta,\alpha'\beta';\gamma} \to \mathfrak{U}_{\alpha\beta} \times_{\mathfrak{U}} \mathfrak{U}_{\alpha'\beta'})_{\gamma \in B_{\alpha\alpha'}}$ be the base change of the above covering. Since $(\mathfrak{U}_{\alpha\beta}\times_{\mathfrak{U}}\mathfrak{U}_{\alpha'\beta'}\to\mathfrak{U}_\alpha\times_{\mathfrak{U}}\mathfrak{U}_{\alpha'})_{(\beta,\beta')\in B_\alpha\times B_{\alpha'}}$ belongs to $\mathrm{Cov}^v_f(\mathfrak{U}_\alpha\times_{\mathfrak{U}}\mathfrak{U}_{\alpha'})$, we see that $\mathfrak{U}_{\alpha\beta,\alpha'\beta';\gamma}\in\mathrm{Ob}\,\mathcal{C}$ and the family of natural morphisms $(\mathfrak{U}_{\alpha\beta,\alpha'\beta';\gamma}\to\mathfrak{U}_{\alpha\alpha';\gamma})_{(\beta,\beta')\in B_\alpha\times B_{\alpha'}}$ belongs to $\mathrm{Cov}^v_f(\mathfrak{U}_{\alpha\alpha';\gamma})$. Since the restrictions of $x_{\alpha\beta}$ and $x_{\alpha'\beta'}$ on $\mathfrak{U}_{\alpha\beta,\alpha'\beta';\gamma}$ coincide by assumption, we see that the restrictions of $p^\varepsilon x_\alpha$ and $p^\varepsilon x_{\alpha'}$ on $\mathfrak{U}_{\alpha\alpha';\gamma}$ coincide by the condition (a). Hence there exists $x\in\mathcal{F}(\mathfrak{U})$ whose restriction on $\mathfrak{U}_\alpha$ is $p^{2\varepsilon}x_\alpha$ by the condition (b). □

**Corollary IV.6.1.24.** *Let $X$ and $\mathcal{C}$ be as above, let $\mathcal{C}'$ be a full subcategory of $X_{\text{étaff}}$ containing $\mathcal{C}$, let $\mathcal{F}$ be a presheaf of $\overline{V}$-modules on $(\mathcal{P}/\mathcal{C}')_{\text{ét-fét}}$ satisfying the conditions* (a) *and* (b) *in Proposition* IV.6.1.23, *and let $\mathcal{F}^a$ be the sheaf on $(\mathcal{P}/\mathcal{C}')_{\text{ét-fét}}$ associated to $\mathcal{F}$. Then the homomorphism $\mathcal{F}(\mathfrak{U})\to\mathcal{F}^a(\mathfrak{U})$ is an almost isomorphism for every $\mathfrak{U}\in\mathrm{Ob}\,(\mathcal{P}/\mathcal{C})$.*

Proof. Give $\mathcal{P}/\mathcal{C}'$ and $\mathcal{P}/X$ the topologies defined by the identity coverings. Then the inclusion functor $(\mathcal{P}/\mathcal{C}')\to(\mathcal{P}/X)$ induces a diagram of categories and adjoint pairs of functors, commutative up to canonical isomorphisms (cf. Lemma IV.6.1.10 and **[2]** I Proposition 5.1). Here $a$ denotes the sheafification functors.

$$\begin{array}{ccc}
\mathcal{M}od((\mathcal{P}/\mathcal{C}'),\overline{V}) & \underset{u^{-1}}{\overset{u_*}{\leftrightarrows}} & \mathcal{M}od((\mathcal{P}/X),\overline{V})\\
i\uparrow\downarrow a & & i\uparrow\downarrow a\\
\mathcal{M}od((\mathcal{P}/\mathcal{C}')_{\text{ét-fét}},\overline{V}) & \underset{u^{-1}_{\text{ét-fét}}}{\overset{u_{\text{ét-fét}*}}{\leftrightarrows}} & \mathcal{M}od((\mathcal{P}/X)_{\text{ét-fét}},\overline{V}).
\end{array}$$

For $\mathfrak{U}\in\mathrm{Ob}\,(\mathcal{P}/\mathcal{C}')$, we have $(u^{-1}\mathcal{F})(\mathfrak{U})=\varinjlim_{\mathfrak{U}\to\mathfrak{U}',\mathfrak{U}'\in\mathrm{Ob}\,(\mathcal{P}/\mathcal{C}')}\mathcal{F}(\mathfrak{U}')=\mathcal{F}(\mathfrak{U})$. Hence $u^{-1}\mathcal{F}$ satisfies the conditions in Proposition IV.6.1.23 and the adjunction map $\mathcal{F}\to u_*u^{-1}\mathcal{F}$ is an isomorphism. From the above diagram, we obtain the following commutative diagram.

$$\begin{array}{ccccc}
\mathcal{F} & \longrightarrow & ia\mathcal{F} & \longrightarrow & iu_{\text{ét-fét}*}u^{-1}_{\text{ét-fét}}a\mathcal{F}\\
 & \searrow\cong & & & \downarrow\cong\\
 & & u_*u^{-1}\mathcal{F} & \longrightarrow & u_*iau^{-1}\mathcal{F}.
\end{array}$$

The upper right morphism is an isomorphism because $u_{\text{ét-fét}*}$ is an equivalence of categories. The restriction of the bottom arrow on $\mathcal{P}/\mathcal{C}$ is an almost isomorphism by Proposition IV.6.1.23. Hence it also holds for $\mathcal{F}\to ia\mathcal{F}$. □

**IV.6.2. A description of sheaves on $(\mathcal{P}/\mathcal{C})_{\text{fét}}$ in terms of representations.** In order to interpret the representations constructed in IV.5.2 in terms of sheaves on $(\mathcal{P}/\mathcal{C})_{\text{ét-fét}}$, we give a description of sheaves on $(\mathcal{P}/\mathcal{C})_{\text{fét}}$ in terms of compatible system of modules on "universal coverings" of $\mathcal{U}_{\text{triv},\overline{K}}$ under the assumption that the underlying scheme of $X$ is normal and has reduced special fiber.

We first start by reviewing a description of sheaves on $(\mathcal{U}_{\text{triv},\overline{K}})_{\text{fét}}$. To avoid the dependence on the choice of geometric points, we work with the category of all geometric points whose images are of codimension 0 instead of the fundamental group at one geometric point (cf. Lemma VI.9.11, **[37]** V Proposition 5.8).

Let $U$ be an fs log scheme over $\Sigma$ satisfying the following conditions: The underlying scheme of $U$ is affine and normal. The morphism of schemes underlying $U\to\Sigma$ is flat of finite type and its special fiber is reduced. We define $A_U$, $\mathcal{A}_U$, and $\mathcal{U}_{L,\text{triv}}$ as in the beginning of IV.6.1.

Let $\overline{s}\to\mathcal{U}_{\overline{K},\text{triv}}$ be a geometric point of $\mathcal{U}_{\overline{K},\text{triv}}$ such that the image $s^g$ of $\overline{s}$ is of codimension 0 and $\Gamma(\overline{s},\mathcal{O}_{\overline{s}})$ is an algebraic closure of the residue field $\kappa(s^g)$ at $s^g$. Let

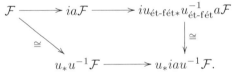

$s$ be the image of $\overline{s}$ in $\mathcal{U}_{K,\mathrm{triv}}$ and let $\kappa(s)$ be the residue field at $s$. Then we have an isomorphism $\kappa(s)\otimes_K \overline{K} \xrightarrow{\cong} \kappa(s^g)$ in the cases II and III by Lemma IV.5.2.1, and an algebraic closure of $K$ in $\kappa(s)$ is finite over $K$ in the case I. Let $\mathscr{S}_{K,\overline{s}}$ be the set of all finite Galois extensions $\mathcal{L}$ of $\kappa(s)$ contained in $\kappa(\overline{s})$ such that the normalization of $\mathcal{U}_{K,\mathrm{triv}}$ in $\mathcal{L}$ is étale over $\mathcal{U}_{K,\mathrm{triv}}$. We define the set $\mathscr{S}_{\overline{s}}$ to be $\{\mathcal{L}\kappa(s^g)|\mathcal{L}\in\mathscr{S}_{K,\overline{s}}\}$. For $\mathcal{L}\in\mathscr{S}_{\overline{s}}$, let $U_{\mathcal{L}}$ denote the normalization of $\mathcal{U}_{\overline{K},\mathrm{triv}}$ in $\mathcal{L}$ (cf. **[40]** Proposition (6.3.4)). Choose $\mathcal{L}_0\in\mathscr{S}_{K,\overline{s}}$ such that $\mathcal{L}=\mathcal{L}_0\kappa(s^g)$ and let $L$ be an algebraic closure of $K$ in $\mathcal{L}$, which is finite over $K$ because an algebraic closure of $K$ in $\kappa(s)$ is finite over $K$. Let $U_{L,\mathcal{L}_0}$ be the normalization of $\mathcal{U}_{L,\mathrm{triv}}$ in $\mathcal{L}_0$. Then $U_{L,\mathcal{L}_0}$ is finite étale over $\mathcal{U}_{L,\mathrm{triv}}$ and the natural morphism $U_{L,\mathcal{L}_0}\times_{\mathcal{U}_{L,\mathrm{triv}}}\mathcal{U}_{\overline{K},\mathrm{triv}}\to U_{\mathcal{L}}$ is an isomorphism. This implies that $U_{\mathcal{L}}\to U_{\overline{K},\mathrm{triv}}$ is finite étale for $\mathcal{L}\in\mathscr{S}_{\overline{s}}$ and we have an isomorphism

$$\text{(IV.6.2.1)}\qquad \bigsqcup_{\sigma}\sigma\times 1\colon \bigsqcup_{\sigma\in\mathrm{Gal}(\mathcal{L}'/\mathcal{L})} U_{\mathcal{L}'}\xrightarrow{\cong} U_{\mathcal{L}'}\times_{U_{\mathcal{L}}}U_{\mathcal{L}'}$$

for $\mathcal{L},\mathcal{L}'\in\mathscr{S}_{\overline{s}}$ such that $\mathcal{L}\subset\mathcal{L}'$. Let $\mathcal{V}\to\mathcal{U}_{\overline{K},\mathrm{triv}}$ be a finite étale morphism. Then it comes from a finite étale morphism $\mathcal{V}_L\to\mathcal{U}_{L,\mathrm{triv}}$ for some finite extension $L\subset\overline{K}$ of $K$. Hence we see that there exists $\mathcal{L}\in\mathscr{S}_{\overline{s}}$ such that $\mathcal{V}\times_{\mathcal{U}_{\overline{K},\mathrm{triv}}}U_{\mathcal{L}}$ is isomorphic to the disjoint union of a finite number of copies of $U_{\mathcal{L}}$ as a $U_{\mathcal{L}}$-scheme. This property will allow us to use the inverse system of finite étale coverings $(U_{\mathcal{L}})_{\mathcal{L}\in\mathscr{S}_{\overline{s}}}$ as a "universal covering" of the connected component of $\mathcal{U}_{\mathrm{triv},\overline{K}}$ containing $s^g$. We define $\kappa(s)^{\mathrm{ur}}$ to be the union of $\mathcal{L}\in\mathscr{S}_{\overline{s}}$ (which coincides with the union of $\mathcal{L}\in\mathscr{S}_{K,\overline{s}}$) and $s^{\mathrm{ur}}$ to be $\mathrm{Spec}(\kappa(s)^{\mathrm{ur}})$. Let $\Delta_{\overline{s}}$ denote the Galois group $\mathrm{Gal}(\kappa(s)^{\mathrm{ur}}/\kappa(s^g))$.

For another geometric point $\overline{s}'\to\mathcal{U}_{\overline{K},\mathrm{triv}}$ satisfying the same conditions, we define a morphism from $\overline{s}'$ to $\overline{s}$ to be a morphism $h\colon s'^{\mathrm{ur}}\to s^{\mathrm{ur}}$ compatible with the morphisms $s'^{\mathrm{ur}}$, $s^{\mathrm{ur}}\to\mathcal{U}_{\mathrm{triv},\overline{K}}$. Let $(\mathcal{U}_{\overline{K},\mathrm{triv}})_{\mathrm{gpt}}$ denote the category of $\overline{s}\to\mathcal{U}_{\mathrm{triv},\overline{K}}$ and morphisms defined above. By definition, we have a canonical isomorphism $\Delta_{\overline{s}}\cong\mathrm{Aut}_{(\mathcal{U}_{\overline{K},\mathrm{triv}})_{\mathrm{gpt}}}(\overline{s})^{\circ}$ for $\overline{s}\in(\mathcal{U}_{\overline{K},\mathrm{triv}})_{\mathrm{gpt}}$.

Now we are ready to give a description of sheaves on $(\mathcal{U}_{\overline{K},\mathrm{triv}})_{\mathrm{f\acute{e}t}}$. Associated to a sheaf of sets $\mathcal{F}$ on $(\mathcal{U}_{\overline{K},\mathrm{triv}})_{\mathrm{f\acute{e}t}}$, one can define a presheaf of sets $s_U(\mathcal{F})$ on $(\mathcal{U}_{\overline{K},\mathrm{triv}})_{\mathrm{gpt}}$ by

$$s_U(\mathcal{F})(\overline{s})=\varinjlim_{\mathcal{L}\in\mathscr{S}_{\overline{s}}}\mathcal{F}(U_{\mathcal{L}});$$

a morphism $h\colon\overline{s}'\to\overline{s}$ in $(\mathcal{U}_{\overline{K},\mathrm{triv}})_{\mathrm{gpt}}$ induces an isomorphism $h^*\colon\kappa(s)^{\mathrm{ur}}\xrightarrow{\cong}\kappa(s')^{\mathrm{ur}}$ over $\kappa(s)=\kappa(s')$, and then an isomorphism $U_{h^*(\mathcal{L})}\xrightarrow{\cong}U_{\mathcal{L}}$ for each $\mathcal{L}\in\mathscr{S}_{\overline{s}}$, which induces an isomorphism $s_U(\mathcal{F})(\overline{s})\xrightarrow{\cong}s_U(\mathcal{F})(\overline{s}')$. By construction, the natural left action of $\Delta_{\overline{s}}(\cong\mathrm{Aut}(\overline{s})^{\circ})$ on $s_U(\mathcal{F})(\overline{s})$ is continuous. Having this observation, we define $(\mathcal{U}_{\overline{K},\mathrm{triv}})^{\wedge,\mathrm{cont}}_{\mathrm{gpt}}$ to be the full subcategory of $(\mathcal{U}_{\overline{K},\mathrm{triv}})^{\wedge}_{\mathrm{gpt}}$ consisting of presheaves $\mathcal{G}$ such that the action of $\Delta_{\overline{s}}$ on $\mathcal{G}(\overline{s})$ is continuous for every $\overline{s}\in\mathrm{Ob}\,(\mathcal{U}_{\overline{K},\mathrm{triv}})_{\mathrm{gpt}}$. Then the above construction gives a functor

$$s_U\colon(\mathcal{U}_{\mathrm{triv},\overline{K}})^{\sim}_{\mathrm{f\acute{e}t}}\longrightarrow(\mathcal{U}_{\mathrm{triv},\overline{K}})^{\wedge,\mathrm{cont}}_{\mathrm{gpt}}.$$

One can also construct a functor in the opposite direction as follows. Let $\mathcal{G}\in\mathrm{Ob}\,(\mathcal{U}_{\overline{K},\mathrm{triv}})^{\wedge,\mathrm{cont}}_{\mathrm{gpt}}$. For $\mathcal{V}\in\mathrm{Ob}\,(\mathcal{U}_{\overline{K},\mathrm{triv}})_{\mathrm{f\acute{e}t}}$, we define the presheaf $\mathcal{G}_{\mathcal{V}}$ on $(\mathcal{U}_{\overline{K},\mathrm{triv}})_{\mathrm{gpt}}$ by

$$\mathcal{G}_{\mathcal{V}}(\overline{s})=\mathrm{Map}(\mathrm{Hom}_{\mathcal{U}_{\overline{K},\mathrm{triv}}}(s^{\mathrm{ur}},\mathcal{V}),\mathcal{G}(\overline{s})).$$

For a morphism $h\colon\overline{s}'\to\overline{s}$, we define the morphism $\mathcal{G}_{\mathcal{V}}(h)\colon\mathcal{G}_{\mathcal{V}}(\overline{s})\to\mathcal{G}_{\mathcal{V}}(\overline{s}')$ by composing with the bijection $\mathrm{Hom}(s'^{\mathrm{ur}},\mathcal{V})\to\mathrm{Hom}(s^{\mathrm{ur}},\mathcal{V})$ induced by $h$ and $\mathcal{G}(h)\colon\mathcal{G}(\overline{s})\to\mathcal{G}(\overline{s}')$. For a morphism $g\colon\mathcal{V}'\to\mathcal{V}$ in $(\mathcal{U}_{\overline{K},\mathrm{triv}})_{\mathrm{f\acute{e}t}}$, the map $g\circ-\colon\mathrm{Hom}(s^{\mathrm{ur}},\mathcal{V}')\to\mathrm{Hom}(s^{\mathrm{ur}},\mathcal{V})$

induces a morphism of presheaves $\mathcal{G}_{\mathcal{V}} \to \mathcal{G}_{\mathcal{V}'}$. We define the presheaf $r_U(\mathcal{G})$ on $(\mathcal{U}_{\overline{K},\mathrm{triv}})_{\text{fét}}$ associated to $\mathcal{G}$ by

$$r_U(\mathcal{G})(\mathcal{V}) = \Gamma((\mathcal{U}_{\overline{K},\mathrm{triv}})_{\mathrm{gpt}}, \mathcal{G}_{\mathcal{V}}) = \varprojlim_{(\mathcal{U}_{\overline{K},\mathrm{triv}})_{\mathrm{gpt}}} \mathcal{G}_{\mathcal{V}}.$$

This construction is obviously functorial in $\mathcal{G}$. Let $P$ be a finite subset of $\mathrm{Ob}\,(\mathcal{U}_{\overline{K},\mathrm{triv}})_{\mathrm{gpt}}$ such that the natural map $P \to \pi_0(\mathcal{U}_{\overline{K},\mathrm{triv}})$ is bijective. Then the evaluation at $\overline{s} \in P$ induces an isomorphism

$$r_U(\mathcal{G})(\mathcal{V}) \xrightarrow{\cong} \prod_{\overline{s}\in P} \mathrm{Map}_{\Delta_{\overline{s}}}(\mathrm{Hom}_{\mathcal{U}_{\overline{K},\mathrm{triv}}}(s^{\mathrm{ur}}, \mathcal{V}), \mathcal{G}(\overline{s})). \tag{IV.6.2.2}$$

**Lemma IV.6.2.3.** *The presheaf $r_U(\mathcal{G})$ is a sheaf on $(\mathcal{U}_{\overline{K},\mathrm{triv}})_{\text{fét}}$.*

PROOF. For any finite number of finite étale schemes $\mathcal{V}_\lambda$ $(\lambda \in \Lambda)$ over $\mathcal{U}_{\overline{K},\mathrm{triv}}$, we have $r_U(\mathcal{G})(\sqcup_{\lambda\in\Lambda}\mathcal{V}_\lambda) = \prod_{\lambda\in\Lambda} r_U(\mathcal{G})(\mathcal{V}_\lambda)$. Hence it suffices to prove the following claim. For any finite étale surjective morphism $\mathcal{V}' \to \mathcal{V}$ of finite étale schemes over $\mathcal{U}_{\overline{K},\mathrm{triv}}$ and $\overline{s} \in \mathrm{Ob}\,(\mathcal{U}_{\overline{K},\mathrm{triv}})_{\mathrm{gpt}}$, the sequence $\mathcal{G}_{\mathcal{V}}(\overline{s}) \to \mathcal{G}_{\mathcal{V}'}(\overline{s}) \rightrightarrows \mathcal{G}_{\mathcal{V}'\times_{\mathcal{V}}\mathcal{V}'}(\overline{s})$ is exact. This follows from $\mathrm{Hom}(s^{\mathrm{ur}}, \mathcal{V}' \times_{\mathcal{V}} \mathcal{V}') = \mathrm{Hom}(s^{\mathrm{ur}}, \mathcal{V}') \times_{\mathrm{Hom}(s^{\mathrm{ur}},\mathcal{V})} \mathrm{Hom}(s^{\mathrm{ur}}, \mathcal{V}')$ and the surjectivity of the map $\mathrm{Hom}(s^{\mathrm{ur}}, \mathcal{V}') \to \mathrm{Hom}(s^{\mathrm{ur}}, \mathcal{V})$. □

By Lemma IV.6.2.3, we obtain a functor

$$r_U\colon (\mathcal{U}_{\overline{K},\mathrm{triv}})_{\mathrm{gpt}}^{\wedge,\mathrm{cont}} \to (\mathcal{U}_{\overline{K},\mathrm{triv}})_{\text{fét}}^{\sim}.$$

**Proposition IV.6.2.4.** *The functors $r_U$ and $s_U$ defined above are equivalences of categories which are canonically quasi-inverse of each other.*

PROOF. We write $\widetilde{U}$ for $\mathcal{U}_{\overline{K},\mathrm{triv}}$ to simplify the notation. We prove that there exist canonical and functorial isomorphisms $\mathcal{F} \cong r_U \circ s_U(\mathcal{F})$ and $s_U \circ r_U(\mathcal{G}) \cong \mathcal{G}$ for $\mathcal{F} \in \mathrm{Ob}\,(\widetilde{U})_{\text{fét}}^{\sim}$ and $\mathcal{G} \in \mathrm{Ob}\,(\widetilde{U})_{\mathrm{gpt}}^{\wedge,\mathrm{cont}}$. For $\mathcal{V} \in \mathrm{Ob}\,\widetilde{U}_{\text{fét}}$, $\overline{s} \in \mathrm{Ob}\,\widetilde{U}_{\mathrm{gpt}}$, and a morphism $u\colon s^{\mathrm{ur}} \to \mathcal{V}$ over $\widetilde{U}$, there exists an $\mathcal{L} \in \mathscr{S}_{\overline{s}}$ such that $u$ factors through a $\widetilde{U}$-morphism $u_{\mathcal{L}}\colon U_{\mathcal{L}} \to \mathcal{V}$ uniquely, and the morphism $\mathcal{F}(\mathcal{V}) \to s_U(\mathcal{F})(\overline{s}) = \varinjlim_{\mathcal{L}\in\mathscr{S}_{\overline{s}}} \mathcal{F}(U_{\mathcal{L}})$ defined by $\mathcal{F}(u_{\mathcal{L}})$ is independent of the choice of $\mathcal{L}$. Let $\mathcal{F}(u)$ denote the morphism. Then we obtain a canonical morphism $\mathcal{F}(\mathcal{V}) \to s_U(\mathcal{F})_{\mathcal{V}}(\overline{s}) = \mathrm{Map}(\mathrm{Hom}_{\widetilde{U}}(s^{\mathrm{ur}}, \mathcal{V}), s_U(\mathcal{F})(\overline{s}))$ by $x \mapsto (u \mapsto \mathcal{F}(u)(x))$. It is straightforward to check that this map is compatible with the morphism $s_U(\mathcal{F})_{\mathcal{V}}(\overline{s}) \to s_U(\mathcal{F})_{\mathcal{V}}(\overline{s}')$ induced by a morphism $\overline{s}' \to \overline{s}$ in $\widetilde{U}_{\mathrm{gpt}}$ and defines a morphism $\Phi(\mathcal{V})\colon \mathcal{F}(\mathcal{V}) \to r_U \circ s_U(\mathcal{F})(\mathcal{V}) = \Gamma(\widetilde{U}_{\mathrm{gpt}}, s_U(\mathcal{F})_{\mathcal{V}})$. One can also verify that this defines a morphism of sheaves $\Phi\colon \mathcal{F} \to r_U \circ s_U(\mathcal{F})$. We show that $\Phi(\mathcal{V})$ for $\mathcal{V} \in \mathrm{Ob}\,\widetilde{U}_{\text{fét}}$ is an isomorphism. Choose $P$ as before (IV.6.2.2). Then for each $\overline{s}$ and sufficiently large $\mathcal{L} \in \mathscr{S}_{\overline{s}}$, $U_{\mathcal{L}} \times_{\widetilde{U}} \mathcal{V}$ is isomorphic to the disjoint union of a finite number of copies of $\mathcal{U}_{\mathcal{L}}$ and hence we have a canonical isomorphism

$$\mathcal{F}(U_{\mathcal{L}} \times_{\widetilde{U}} \mathcal{V})^{\Delta_{\mathcal{L}}} \xrightarrow{\cong} \mathrm{Map}_{\Delta_{\mathcal{L}}}(\mathrm{Hom}_{\widetilde{U}}(U_{\mathcal{L}}, \mathcal{V}), \mathcal{F}(U_{\mathcal{L}})); x \mapsto (f \mapsto \mathcal{F}((\mathrm{id}_{U_{\mathcal{L}}}, f))(x)),$$

where $\Delta_{\mathcal{L}} = \mathrm{Gal}(\mathcal{L}/\kappa(s^g))$. By taking $\prod_{\overline{s}\in P} \varinjlim_{\mathcal{L}\in\mathscr{S}_{\overline{s}}}$ and using (IV.6.2.1), we obtain

$$\mathcal{F}(\mathcal{V}) \cong \prod_{\overline{s}\in P} \varinjlim_{\mathcal{L}\in\mathscr{S}_{\overline{s}}} \mathcal{F}(U_{\mathcal{L}} \times_{\widetilde{U}} \mathcal{V})^{\Delta_{\mathcal{L}}} \cong \prod_{\overline{s}\in P} \mathrm{Map}_{\Delta_{\overline{s}}}(\mathrm{Hom}_{\widetilde{U}}(s^{\mathrm{ur}}, \mathcal{V}), s_U(\mathcal{F})(\overline{s})).$$

We see that this coincides with $\Phi(\mathcal{V})$ via the description (IV.6.2.2) of $r_U \circ s_U(\mathcal{F})(\mathcal{V})$.

For $\mathcal{G} \in (\widetilde{U})^{\wedge,\mathrm{cont}}_{\mathrm{gpt}}$ and $\overline{s} \in \widetilde{U}_{\mathrm{gpt}}$, we have canonical isomorphisms

$$s_U \circ r_U(\mathcal{G})(\overline{s}) = \varinjlim_{\mathcal{L} \in \mathscr{S}_{\overline{s}}} r_U(\mathcal{G})(U_{\mathcal{L}}) \xrightarrow{\cong} \varinjlim_{\mathcal{L} \in \mathscr{S}_{\overline{s}}} \mathrm{Map}_{\Delta_{\overline{s}}}(\mathrm{Hom}_{\widetilde{U}}(s^{\mathrm{ur}}, U_{\mathcal{L}}), \mathcal{G}(\overline{s}))$$
$$\xrightarrow{\cong} \varinjlim_{\mathcal{L} \in \mathscr{S}_{\overline{s}}} \mathrm{Map}_{\Delta_{\mathcal{L}}}(\mathrm{Hom}_{\widetilde{U}}(U_{\mathcal{L}}, U_{\mathcal{L}}), \mathcal{G}(\overline{s})^{\mathrm{Gal}(\kappa(\overline{s})/\mathcal{L})}) \xrightarrow{\cong} \varinjlim_{\mathcal{L} \in \mathscr{S}_{\overline{s}}} \mathcal{G}(\overline{s})^{\mathrm{Gal}(\kappa(\overline{s})/\mathcal{L})} = \mathcal{G}(\overline{s}),$$

where the second (resp. fourth) isomorphism is defined by evaluating at $\overline{s}$ (resp. $\mathrm{id}_{U_{\mathcal{L}}}$) and the last equality is a consequence of the continuity of the action of $\Delta_{\overline{s}}$ on $\mathcal{G}(\overline{s})$. These isomorphisms are obviously functorial in $\mathcal{G}$. One can verify the compatibility with the morphisms $s_U \circ r_U(\mathcal{G})(\overline{s}) \to s_U \circ r_U(\mathcal{G})(\overline{s}')$ and $\mathcal{G}(\overline{s}) \to \mathcal{G}(\overline{s}')$ induced by a morphism $\overline{s}' \to \overline{s}$ simply going back to the definition of the functors $s_U$ and $r_U$. □

Let $X$ be an fs log scheme over $\Sigma$ satisfying the following conditions: The underlying scheme is normal, the morphisms of schemes underlying $X \to \Sigma$ is flat, separated, and of finite type, and its special fiber is reduced. Let $\mathcal{C}$ be a full subcategory of $X_{\mathrm{étaff}}$ such that any object of $X_{\mathrm{étaff}}$ admits a strict étale covering by objects of $\mathcal{C}$. We describe sheaves on $(\mathcal{P}/\mathcal{C})_{\mathrm{fét}}$ in terms of presheaves on the category $\mathcal{C}_{\mathrm{gpt}}$ defined as follows. An object of $\mathcal{C}_{\mathrm{gpt}}$ is a pair $(U, \overline{s})$ of $U \in \mathrm{Ob}\,\mathcal{C}$ and $\overline{s} \in \mathrm{Ob}\,(\mathcal{U}_{\overline{K},\mathrm{triv}})_{\mathrm{gpt}}$. A morphism $(f, h)\colon (U', \overline{s}') \to (U, \overline{s})$ is a pair of a morphism $f\colon U' \to U$ in $\mathcal{C}$ and a morphism $h\colon s'^{\mathrm{ur}} \to s^{\mathrm{ur}}$ compatible with the morphism $\mathcal{U}'_{\overline{K},\mathrm{triv}} \to \mathcal{U}_{\overline{K},\mathrm{triv}}$ induced by $f$. A morphism $(f, h)\colon (U', \overline{s}') \to (U, \overline{s})$ induces a homomorphism $h^*\colon \kappa(s)^{\mathrm{ur}} \to \kappa(s')^{\mathrm{ur}}$ and then a compatible system of morphisms $h_{\mathcal{L}}\colon U'_{h^*(\mathcal{L})\cdot\kappa(s'^g)} \to U_{\mathcal{L}}$ for $\mathcal{L} \in \mathscr{S}_{\overline{s}}$.

Let $\mathcal{C}^{\wedge,\mathrm{cont}}_{\mathrm{gpt}}$ be the full subcategory of $\mathcal{C}^{\wedge}_{\mathrm{gpt}}$ consisting of $\mathcal{G}$ such that for each $(U, \overline{s}) \in \mathrm{Ob}\,\mathcal{C}_{\mathrm{gpt}}$, the action of $\Delta_{(U,\overline{s})} := \mathrm{Gal}(\kappa(\overline{s})/\kappa(s^g))(\cong \mathrm{Aut}_{\mathcal{C}_{\mathrm{gpt}}}((U, \overline{s}))^{\circ})$ on $\mathcal{G}(U, \overline{s})$ is continuous. Then one can define a functor

$$s_{\mathcal{C}}\colon (\mathcal{P}/\mathcal{C})^{\sim}_{\mathrm{fét}} \longrightarrow (\mathcal{C}_{\mathrm{gpt}})^{\wedge,\mathrm{cont}}$$

by

$$s_{\mathcal{C}}(\mathcal{F})(U, \overline{s}) = \varinjlim_{\mathcal{L} \in \mathscr{S}_{\overline{s}}} \mathcal{F}(U, U_{\mathcal{L}}) \,(= s_U(\mathcal{F}|_{(\mathcal{U}_{\mathrm{triv},\overline{K}})_{\mathrm{fét}}})(\overline{s}));$$

for a morphism $(f, h)\colon (U', \overline{s}') \to (U, \overline{s})$, the compatible system of morphisms $h_{\mathcal{L}}$ ($\mathcal{L} \in \mathscr{S}_{\overline{s}}$) induces a morphism $s_{\mathcal{C}}(\mathcal{F})(U, \overline{s}) \to s_{\mathcal{C}}(\mathcal{F})(U', \overline{s}')$.

For $\mathcal{G} \in \mathrm{Ob}\,(\mathcal{C}^{\wedge,\mathrm{cont}}_{\mathrm{gpt}})$ and $(U, \mathcal{V}) \in \mathrm{Ob}\,(\mathcal{P}/\mathcal{C})$, we define the presheaf $\mathcal{G}_{(U,\mathcal{V})}$ on $(\mathcal{C}_{/U})_{\mathrm{gpt}}$ by

$$\mathcal{G}_{(U,\mathcal{V})}(U', \overline{s}) = \mathrm{Map}(\mathrm{Hom}_{\mathcal{U}_{\overline{K},\mathrm{triv}}}(s^{\mathrm{ur}}, \mathcal{V}), \mathcal{G}(U', \overline{s}));$$

for a morphism $(f, h)\colon (U'', \overline{s}') \to (U', \overline{s})$ in $(\mathcal{C}_{/U})_{\mathrm{gpt}}$, we define the map $\mathcal{G}_{(U,\mathcal{V})}(U', \overline{s}) \to \mathcal{G}_{(U,\mathcal{V})}(U'', \overline{s}')$ by the composition with $\mathcal{G}(f, h)\colon \mathcal{G}(U', \overline{s}) \to \mathcal{G}(U'', \overline{s}')$ and the inverse of the bijection $- \circ h\colon \mathrm{Hom}_{\mathcal{U}_{\overline{K},\mathrm{triv}}}(s^{\mathrm{ur}}, \mathcal{V}) \to \mathrm{Hom}_{\mathcal{U}_{\overline{K},\mathrm{triv}}}(s'^{\mathrm{ur}}, \mathcal{V})$. For a morphism $\mathfrak{f} = (f, g)\colon (U', \mathcal{V}') \to (U, \mathcal{V})$ in $\mathcal{P}/\mathcal{C}$, let $f^*(\mathcal{G}_{(U,\mathcal{V})})$ denote the composition of $\mathcal{G}_{(U,\mathcal{V})}$ with the functor $(\mathcal{C}_{/U'})_{\mathrm{gpt}} \to (\mathcal{C}_{/U})_{\mathrm{gpt}}$ induced by $f$. Then we obtain a morphism $\mathcal{G}_{\mathfrak{f}}\colon f^*(\mathcal{G}_{(U,\mathcal{V})}) \to \mathcal{G}_{(U',\mathcal{V}')}$ by the composition with $g \circ -\colon \mathrm{Hom}_{\mathcal{U}'_{\overline{K},\mathrm{triv}}}(s^{\mathrm{ur}}, \mathcal{V}') \to \mathrm{Hom}_{\mathcal{U}_{\overline{K},\mathrm{triv}}}(s^{\mathrm{ur}}, \mathcal{V})$. Now one can define a presheaf $r_{\mathcal{C}}(\mathcal{G})$ on $(\mathcal{P}/\mathcal{C})_{\mathrm{fét}}$ by

$$r_{\mathcal{C}}(\mathcal{G})(U, \mathcal{V}) := \Gamma((\mathcal{C}_{/U})_{\mathrm{gpt}}, \mathcal{G}_{(U,\mathcal{V})}),$$
$$r_{\mathcal{C}}(\mathcal{G})(\mathfrak{f}) := \Gamma((\mathcal{C}_{/U'})_{\mathrm{gpt}}, \mathcal{G}_{\mathfrak{f}}) \circ f^*\colon \Gamma((\mathcal{C}_{/U})_{\mathrm{gpt}}, \mathcal{G}_{(U,\mathcal{V})}) \longrightarrow \Gamma((\mathcal{C}_{/U'})_{\mathrm{gpt}}, f^*(\mathcal{G}_{(U,\mathcal{V})}))$$
$$\longrightarrow \Gamma((\mathcal{C}_{/U'})_{\mathrm{gpt}}, \mathcal{G}_{(U',\mathcal{V}')}).$$

**Lemma IV.6.2.5.** *Let* $\mathcal{G} \in \mathrm{Ob}\,\mathcal{C}^{\wedge,\mathrm{cont}}_{\mathrm{gpt}}$.

(1) *For* $(U,\mathcal{V})\in\mathrm{Ob}\,(\mathcal{P}/\mathcal{C})$*, the following restriction map is bijective.*

$$r_{\mathcal{C}}(\mathcal{G})(U,\mathcal{V})=\mathcal{G}_{(U,\mathcal{V})}((\mathcal{C}_{/U})_{\mathrm{gpt}})\to\mathcal{G}_{(U,\mathcal{V})}((\mathcal{U}_{\mathrm{triv},\overline{K}})_{\mathrm{gpt}})=r_U(\mathcal{G}|_{(\mathcal{U}_{\overline{K},\mathrm{triv}})_{\mathrm{gpt}}}).$$

(2) *The presheaf* $r_{\mathcal{C}}(\mathcal{G})$ *is a sheaf on* $(\mathcal{P}/\mathcal{C})_{\mathrm{fét}}$.

PROOF. (2) follows from (1) and Lemma IV.6.2.3. Let us prove (1). The injectivity follows from the fact that for any $(U',\overline{s}')\in\mathrm{Ob}\,(\mathcal{C}_{/U})_{\mathrm{gpt}}$, there exists $\overline{s}\in(\mathcal{U}_{\overline{K},\mathrm{triv}})_{\mathrm{gpt}}$ and a morphism $(U',\overline{s}')\to(U,\overline{s})$ in $(\mathcal{C}_{/U})_{\mathrm{gpt}}$. For any two morphisms $h,h_1\colon(U',\overline{s}')\to(U,\overline{s}),(U,\overline{s}_1)$, there exists a unique isomorphism $\sigma\colon s_1^{\mathrm{ur}}\xrightarrow{\cong}s^{\mathrm{ur}}$ such that $h=\sigma\circ h_1$. This implies the surjectivity. □

By Lemma IV.6.2.5, we obtain a functor

$$r_{\mathcal{C}}\colon\mathcal{C}_{\mathrm{gpt}}^{\wedge,\mathrm{cont}}\to(\mathcal{P}/\mathcal{C})^{\sim}_{\mathrm{fét}}.$$

**Proposition IV.6.2.6.** *The functors* $r_{\mathcal{C}}$ *and* $s_{\mathcal{C}}$ *defined above are equivalences of categories which are canonically quasi-inverse of each other.*

PROOF. We prove that there exist canonical and functorial isomorphisms $r_{\mathcal{C}}s_{\mathcal{C}}(\mathcal{F})\cong\mathcal{F}$ and $s_{\mathcal{C}}r_{\mathcal{C}}(\mathcal{G})\cong\mathcal{G}$ for $\mathcal{F}\in\mathrm{Ob}\,(\mathcal{P}/\mathcal{C})^{\sim}_{\mathrm{fét}}$ and $\mathcal{G}\in\mathrm{Ob}\,(\mathcal{C}_{\mathrm{gpt}}^{\wedge,\mathrm{cont}})$. For $U\in\mathrm{Ob}\,\mathcal{C}$ and $\widetilde{U}=\mathcal{U}_{\overline{K},\mathrm{triv}}$, we have canonical isomorphisms

$$(r_{\mathcal{C}}s_{\mathcal{C}}(\mathcal{F}))|_{\widetilde{U}_{\mathrm{fét}}}=r_U(s_{\mathcal{C}}(\mathcal{F})|_{\widetilde{U}_{\mathrm{gp}}})\cong r_Us_U(\mathcal{F}|_{\widetilde{U}_{\mathrm{fét}}})\cong\mathcal{F}|_{\widetilde{U}_{\mathrm{fét}}}$$

by the definition of $s_{\mathcal{C}}$, Lemma IV.6.2.5 (1), and Proposition IV.6.2.4. For $(f\colon U'\to U,\overline{s})\in\mathrm{Ob}\,(\mathcal{C}_{/U})_{\mathrm{gpt}}$ and $\mathcal{V}\in\mathrm{Ob}\,\widetilde{U}_{\mathrm{fét}}$, any morphism $u\colon s^{\mathrm{ur}}\to\mathcal{V}$ over $\widetilde{U}$ factors through a morphism $u_{\mathcal{L}}\colon U'_{\mathcal{L}}\to\mathcal{V}$ for a sufficiently large $\mathcal{L}\in\mathscr{S}_{\overline{s}}$, and the morphism $\mathcal{F}(U,\mathcal{V})\to\varinjlim_{\mathcal{L}\in\mathscr{S}_{\overline{s}}}\mathcal{F}(U',U'_{\mathcal{L}})=:\mathcal{F}(U',\overline{s})$ induced by $u_{\mathcal{L}}$ depends only on $u$, which we denote by $\mathcal{F}(f,u)$. Then, for $(U,\mathcal{V})\in\mathrm{Ob}\,(\mathcal{P}/\mathcal{C})$, the above isomorphism

$$\mathcal{F}(U,\mathcal{V})\xrightarrow{\cong}r_{\mathcal{C}}s_{\mathcal{C}}(\mathcal{F})(U,\mathcal{V})=\varprojlim_{(f\colon U'\to U,\overline{s})\in\mathrm{Ob}\,(\mathcal{C}_{/U})_{\mathrm{gpt}}}\mathrm{Map}(\mathrm{Hom}_{\widetilde{U}}(s^{\mathrm{ur}},\mathcal{V}),\mathcal{F}(U',\overline{s}))$$

is given by $x\mapsto(u\mapsto\mathcal{F}(f,u)(x))$. For a morphism $(f,g)\colon(U',\mathcal{V}')\to(U,\mathcal{V})$ in $(\mathcal{P}/\mathcal{C})_{\mathrm{fét}}$, a morphism $f'\colon U''\to U'$ in $\mathcal{C}$, $\overline{s}\in\mathrm{Ob}\,(\mathcal{U}''_{\overline{K},\mathrm{triv}})_{\mathrm{gpt}}$, and a morphism $h\colon\overline{s}\to\mathcal{V}'$ compatible with $f'$, we have $\mathcal{F}(f',h)\circ\mathcal{F}(f,g)=\mathcal{F}(f\circ f',g\circ h)$. This implies that the above isomorphism is functorial in $(U,\mathcal{V})$ and defines the desired isomorphism.

Let $\mathcal{G}\in\mathrm{Ob}\,(\mathcal{C}_{\mathrm{gpt}}^{\wedge,\mathrm{cont}})$. By the proof of Proposition IV.6.2.4 and Lemma IV.6.2.5, we have a canonical isomorphism

$$s_{\mathcal{C}}r_{\mathcal{C}}(\mathcal{G})(U,\overline{s})\cong\varinjlim_{\mathcal{L}\in\mathscr{S}_{\overline{s}}}\mathrm{Map}_{\Delta_{\overline{s}}}(\mathrm{Hom}_{\widetilde{U}}(s^{\mathrm{ur}},U_{\mathcal{L}}),\mathcal{G}(U,\overline{s}))\xrightarrow{\cong}\mathcal{G}(U,\overline{s})$$

defined by the evaluation at the canonical morphism $s^{\mathrm{ur}}\to U_{\mathcal{L}}$. Via the first isomorphism, the restriction map by $(f,h)\colon(U',\overline{s}')\to(U,\overline{s})$ of the source is given by the composition with the morphisms $\mathrm{Hom}_{\widetilde{U}'}(s'^{\mathrm{ur}},U'_{h^*(\mathcal{L})\kappa(s'^g)})\to\mathrm{Hom}_{\widetilde{U}}(s^{\mathrm{ur}},U_{\mathcal{L}})$ and $\mathcal{G}(U,\overline{s})\to\mathcal{G}(U',\overline{s}')$ induced by $(f,h)$, and the first morphism preserves the canonical morphisms. Hence the above isomorphism is functorial in $(U,\overline{s})$ and gives the desired isomorphism. □

**Definition IV.6.2.7.** Let $X$ be an fs log scheme over $\Sigma$ satisfying the following conditions: The underlying scheme is normal, the morphism of schemes underlying $X\to\Sigma$ is flat, separated, and of finite type, and its special fiber is reduced. Let $\mathcal{C}$ be a full subcategory of $X_{\mathrm{étaff}}$ such that any object of $X_{\mathrm{étaff}}$ admits a strict étale covering by objects of $\mathcal{C}$.

(1) We define $\mathcal{O}_{\mathcal{C}_{\mathrm{gpt}}}$ and $\mathcal{O}_{\mathcal{C}_{\mathrm{gpt}},m}$ ($m\in\mathbb{N}$) to be the ring objects $s_{\mathcal{C}}(\mathcal{O}^v_{\mathcal{P}/\mathcal{C}})$ and $s_{\mathcal{C}}(\mathcal{O}^v_{\mathcal{P}/\mathcal{C},m})$ of $(\mathcal{C}_{\mathrm{gpt}})^{\wedge,\mathrm{cont}}$.

(2) For $m \in \mathbb{N}$, we define $\mathcal{M}od(\mathcal{C}_{\mathrm{gpt}}, \mathcal{O}_{\mathcal{C}_{\mathrm{gpt}},m})$ to be the category of $\mathcal{O}_{\mathcal{C}_{\mathrm{gpt}},m}$-modules in $(\mathcal{C}_{\mathrm{gpt}})^{\wedge,\mathrm{cont}}$. We define $\mathcal{M}od(\mathcal{C}_{\mathrm{gpt}}, \mathcal{O}_{\mathcal{C}_{\mathrm{gpt}},\bullet})$ to be the category of inverse systems of $\mathcal{O}_{\mathcal{C}_{\mathrm{gpt}},m}$-modules for $m \in \mathbb{N}$.

(3) For $m \in \mathbb{N}$, we define $\mathcal{M}od_{\mathrm{cocart}}(\mathcal{C}_{\mathrm{gpt}}, \mathcal{O}_{\mathcal{C}_{\mathrm{gpt}},m})$ to be the full subcategory of $\mathcal{M}od(\mathcal{C}_{\mathrm{gpt}}, \mathcal{O}_{\mathcal{C}_{\mathrm{gpt}},m})$ consisting of $\mathcal{G}$ such that for every morphism $(f,h)\colon (U',\overline{s}') \to (U,\overline{s})$ in $\mathcal{C}_{\mathrm{gpt}}$, the natural homomorphism

$$\mathcal{O}_{\mathcal{C}_{\mathrm{gpt}},m}(U',\overline{s}') \otimes_{\mathcal{O}_{\mathcal{C}_{\mathrm{gpt}},m}(U,\overline{s})} \mathcal{G}(U,\overline{s}) \longrightarrow \mathcal{G}(U',\overline{s}')$$

is an isomorphism. We define $\mathcal{M}od_{\mathrm{cocart}}(\mathcal{C}_{\mathrm{gpt}}, \mathcal{O}_{\mathcal{C}_{\mathrm{gpt}},\bullet})$ to be the full subcategory of $\mathcal{M}od(\mathcal{C}_{\mathrm{gpt}}, \mathcal{O}_{\mathcal{C}_{\mathrm{gpt}},\bullet})$ consisting of $(\mathcal{G}_m)_{m\in\mathbb{N}}$ with $\mathcal{G}_m$ an object of $\mathcal{M}od_{\mathrm{cart}}(\mathcal{C}_{\mathrm{gpt}}, \mathcal{O}_{\mathcal{C}_{\mathrm{gpt}},m})$ for every $m \in \mathbb{N}$.

(4) For $m \in \mathbb{N}$, we define $\mathcal{M}od_{\mathrm{cocart}}((\mathcal{P}/\mathcal{C})_{\mathrm{f\acute{e}t}}, \mathcal{O}^v_{\mathcal{P}/\mathcal{C},m})$ to be the full subcategory of $\mathcal{M}od((\mathcal{P}/\mathcal{C})_{\mathrm{f\acute{e}t}}, \mathcal{O}^v_{\mathcal{P}/\mathcal{C},m})$ consisting of $\mathcal{F}$ such that $r_{\mathcal{C}}(\mathcal{F})$ belongs to the category $\mathcal{M}od_{\mathrm{cocart}}(\mathcal{C}_{\mathrm{gpt}}, \mathcal{O}_{\mathcal{C}_{\mathrm{gpt}},m})$. We define $\mathcal{M}od_{\mathrm{cocart}}((\mathcal{P}/\mathcal{C})_{\mathrm{f\acute{e}t}}, \mathcal{O}^v_{\mathcal{P}/\mathcal{C},\bullet})$ to be the full subcategory of $\mathcal{M}od((\mathcal{P}/\mathcal{C})_{\mathrm{f\acute{e}t}}, \mathcal{O}^v_{\mathcal{P}/\mathcal{C},\bullet})$ consisting of $(\mathcal{F}_m)_{m\in\mathbb{N}}$ such that $\mathcal{F}_m$ is an object of $\mathcal{M}od_{\mathrm{cocart}}((\mathcal{P}/\mathcal{C})_{\mathrm{f\acute{e}t}}, \mathcal{O}^v_{\mathcal{P}/\mathcal{C},m})$ for every $m \in \mathbb{N}$.

We have $\mathcal{O}_{\mathcal{C}_{\mathrm{gpt}},m} = \mathcal{O}_{\mathcal{C}_{\mathrm{gpt}}}/p^m\mathcal{O}_{\mathcal{C}_{\mathrm{gpt}}}$ and $\mathcal{O}_{\mathcal{C}_{\mathrm{gpt}}}(U,\overline{s})$ is the integral closure of $\mathcal{A}_U$ in $\kappa(s)^{\mathrm{ur}}$.

**Proposition IV.6.2.8.** *Let $X$ and $\mathcal{C}$ be as in Definition* IV.6.2.7 *and assume that $X \in \mathrm{Ob}\,\mathcal{C}$, $X$ is connected, $X \times_{\mathrm{Spec}(\mathbb{Z}_p)} \mathrm{Spec}(\mathbb{Z}/p\mathbb{Z})$ is connected (resp. nonempty) in the cases* II *and* III *(resp. the case* I*), and $\mathcal{X}_{\overline{K},\mathrm{triv}}$ is nonempty. Let $\overline{t} \in \mathrm{Ob}\,(\mathcal{X}_{\overline{K},\mathrm{triv}})_{\mathrm{gpt}}$, and let $\mathrm{Rep}_{\mathrm{cont}}(\Delta_{\overline{t}}, \mathcal{O}_{\mathcal{C}_{\mathrm{gpt}},m}(X,\overline{t}))$ denote the category of $\mathcal{O}_{\mathcal{C}_{\mathrm{gpt}},m}(X,\overline{t})$-modules with the discrete topology endowed with a continuous semi-linear action of $G_{\overline{t}}$. Then the functor*

$$\mathrm{ev}_{\overline{t}}\colon \mathcal{M}od_{\mathrm{cocart}}(\mathcal{C}_{\mathrm{gpt}}, \mathcal{O}_{\mathcal{C}_{\mathrm{gpt}},m}) \longrightarrow \mathrm{Rep}_{\mathrm{cont}}(\Delta_{\overline{t}}, \mathcal{O}_{\mathcal{C}_{\mathrm{gpt}},m}(X,\overline{t})); \mathcal{G} \mapsto \mathcal{G}(X,\overline{t})$$

*is an equivalence of categories.*

PROOF. We write $\mathcal{O}$ for $\mathcal{O}_{\mathcal{C}_{\mathrm{gpt}},m}$ to simplify the notation. For any $(U,\overline{s}) \in \mathrm{Ob}\,\mathcal{C}_{\mathrm{gpt}}$, there exists a morphism $(U,\overline{s}) \to (X,\overline{t})$ because $\mathcal{X}_{\overline{K},\mathrm{triv}}$ is connected by Lemma IV.5.2.1. Hence the functor $\mathrm{ev}_{\overline{t}}$ is faithful. Let $\mathcal{G}_1$ and $\mathcal{G}_2$ be two objects of $\mathcal{M}od_{\mathrm{cocart}}(\mathcal{C}_{\mathrm{gpt}}, \mathcal{O})$ and let $\varphi\colon \mathcal{G}_1(X,\overline{t}) \to \mathcal{G}_2(X,\overline{t})$ be an $\mathcal{O}(X,\overline{t})$-linear $\Delta_{\overline{t}}$-equivariant homomorphism. Then, for each $(U,\overline{s}) \in \mathrm{Ob}\,\mathcal{C}_{\mathrm{gpt}}$, choosing a morphism $(f,h)\colon (U,\overline{s}) \to (X,\overline{t})$, one obtains an $\mathcal{O}(U,\overline{s})$-linear homomorphism $\psi(U,\overline{s})\colon \mathcal{G}_1(U,\overline{s}) \to \mathcal{G}_2(U,\overline{s})$ by the composition of

$$\mathcal{G}_1(U,\overline{s}) \xleftarrow[\mathcal{G}_1(f,h)]{\cong} \mathcal{O}(U,\overline{s}) \otimes_{\mathcal{O}(X,\overline{t})} \mathcal{G}_1(X,\overline{t}) \xrightarrow{\mathrm{id}\otimes\varphi} \mathcal{O}(U,\overline{s}) \otimes_{\mathcal{O}(X,\overline{t})} \mathcal{G}_2(X,\overline{t}) \xrightarrow[\mathcal{G}_2(f,h)]{\cong} \mathcal{G}_2(U,\overline{s}).$$

Let $(f,h')\colon (U,\overline{s}) \to (X,\overline{t})$ be another morphism in $\mathcal{C}_{\mathrm{gpt}}$. (Note that $f$ is unchanged because $X$ is the final object of $\mathcal{C}$.) Then there exists a unique $\sigma \in \Delta_{\overline{t}}$ such that $h' = \sigma \circ h$ and we have $\mathcal{G}_i(f,h') = \mathcal{G}_i(f,h) \circ \sigma\colon \mathcal{G}_i(X,\overline{t}) \to \mathcal{G}_i(U,\overline{s})$ $(i = 1,2)$. Hence, the $\Delta_{\overline{t}}$-equivariance of $\varphi$ implies that $\psi(U,\overline{s})$ is independent of the choice of $(f,h)$. Now it is straightforward to verify that $\psi(U,\overline{s})$ defines a morphism $\psi\colon \mathcal{G}_1 \to \mathcal{G}_2$ and $\psi(X,\overline{t}) = \varphi$. Thus we see that $\mathrm{ev}_{\overline{t}}$ is full.

It remains to prove that $\mathrm{ev}_{\overline{t}}$ is essentially surjective. Let $G \in \mathrm{Ob}\,\mathrm{Rep}_{\mathrm{cont}}(\Delta_{\overline{t}}, \mathcal{O}(X,\overline{t}))$. For $(U,\overline{s}) \in \mathrm{Ob}\,\mathcal{C}_{\mathrm{gpt}}$, let $I_{(U,\overline{s})}$ denote the set of morphisms $h\colon s^{\mathrm{ur}} \to t^{\mathrm{ur}}$ compatible with the structure morphism $f\colon U \to X$. The set $I_{(U,\overline{s})}$ is a $\Delta_{\overline{t}}$-torsor under the right action $h \mapsto \sigma \circ h$. We define the $\mathcal{O}(U,\overline{s})$-module $\mathcal{G}(U,\overline{s})$ to be

$$\left\{(x_h) \in \prod_{h\in I_{(U,\overline{s})}} h^*(G) \,\middle|\, (\mathrm{id}\otimes\sigma)(x_{\sigma\circ h}) = x_h \text{ for all } \sigma \in \Delta_{\overline{t}}\right\},$$

where $h^*(G) = \mathcal{O}(U,\overline{s})\ {}_{\mathcal{O}(f,h)}\otimes_{\mathcal{O}(X,\overline{t})} G$. For a morphism $(u,w)\colon (U',\overline{s}') \to (U,\overline{s})$ in $\mathcal{C}_{\mathrm{gpt}}$, one can define a morphism $\mathcal{G}(u,w)\colon \mathcal{G}(U,\overline{s}) \to \mathcal{G}(U',\overline{s}')$ by sending $(x_h)$ to $(y_{h'})$ defined by $y_{h\circ w} = (w^*\otimes \mathrm{id})(x_h)$ $(h \in I_{(U,\overline{s})})$. (Note that $I_{(U,\overline{s})} \to I_{(U',\overline{s}')}; h \mapsto h\circ w$ is an isomorphism of $\Delta_{\overline{t}}$-torsors.) It is straightforward to verify that $\mathcal{G}(\mathrm{id}) = \mathrm{id}$ and $\mathcal{G}((u,w)\circ(u',w')) = \mathcal{G}(u',w')\circ \mathcal{G}(u,w)$. For $(U,\overline{s}) \in \mathrm{Ob}\,\mathcal{C}_{\mathrm{gpt}}$ and a morphism $(f,h)\colon (U,\overline{s}) \to (X,\overline{t})$, we have a continuous homomorphism $r_h\colon \Delta_{\overline{s}} \to \Delta_{\overline{t}}$ defined by $r_h(\tau)\circ h = h\circ \tau$ and the projection to the $h$-component induces an $\mathcal{O}(U,\overline{s})$-linear isomorphism $\mathcal{G}(U,\overline{s}) \xrightarrow{\cong} h^*(G)$, through which the action of $\Delta_{\overline{s}}$ on $\mathcal{G}(U,\overline{s})$ is transferred to the action of $\Delta_{\overline{s}}$ on $h^*(G)$ defined by $\tau\otimes r_h(\tau)$ $(\tau \in \Delta_{\overline{s}})$. The latter action is continuous. Thus we obtain an object $\mathcal{G}$ of $\mathcal{M}\mathrm{od}_{\mathrm{cocart}}(\mathcal{C}_{\mathrm{gpt}},\mathcal{O})$ and a $\Delta_{\overline{t}}$-equivariant isomorphism $\mathcal{G}(X,\overline{t}) \xrightarrow{\cong} \mathrm{id}^*(G) = G$. □

Let $X$ and $\mathcal{C}$ be as in Definition IV.6.2.7, let $X' \to X$ be a strict étale morphism of fs log schemes over $\Sigma$, and let $\mathcal{C}'$ be a full subcategory of $X'_{\mathrm{\acute{e}taff}}$ such that $X'$ and $\mathcal{C}'$ satisfy the same conditions as $X$ and $\mathcal{C}$ and, for every $(u\colon U \to X') \in \mathrm{Ob}\,\mathcal{C}'$, $f\circ u\colon U \to X$ is an object of $\mathcal{C}$. Then the composition with the functor $f_{\mathrm{gpt}}\colon \mathcal{C}'_{\mathrm{gpt}} \to \mathcal{C}_{\mathrm{gpt}}; (u\colon U \to X',\overline{s}) \mapsto (f\circ u\colon U \to X,\overline{s})$ induces the following functor.

$$f_{\mathrm{gpt}}^{\mathcal{C},\mathcal{C}'*}\colon \mathcal{C}_{\mathrm{gpt}}^{\wedge,\mathrm{cont}} \longrightarrow (\mathcal{C}'_{\mathrm{gpt}})^{\wedge,\mathrm{cont}}.$$

We see that the diagram

$$\begin{array}{ccc}
(\mathcal{C}_{\mathrm{gpt}})^{\wedge,\mathrm{cont}} & \xrightarrow{f_{\mathrm{gpt}}^{\mathcal{C},\mathcal{C}'*}} & (\mathcal{C}'_{\mathrm{gpt}})^{\wedge,\mathrm{cont}} \\
\sim\big\downarrow r_{\mathcal{C}} & & \sim\big\downarrow r_{\mathcal{C}'} \\
(\mathcal{P}/\mathcal{C})^{\sim}_{\mathrm{f\acute{e}t}} & \xrightarrow{f_{\mathcal{P},\mathrm{f\acute{e}t}}^{\mathcal{C},\mathcal{C}'*}} & (\mathcal{P}/\mathcal{C}')^{\sim}_{\mathrm{f\acute{e}t}}
\end{array}$$

is commutative up to canonical isomorphism by going back to the constructions of $r_{\mathcal{C}}$ and $r_{\mathcal{C}'}$ and using Lemma IV.6.2.5 (1). See (IV.6.1.13) for the definition of $f_{\mathcal{P},\mathrm{f\acute{e}t}}^{\mathcal{C},\mathcal{C}'*}$. We abbreviate $f_{\mathrm{gpt}}^{\mathcal{C},\mathcal{C}'*}$ to $f^*_{\mathrm{gpt}}$ if there is no risk of confusion in the following.

We have $f^*_{\mathrm{gpt}}(\mathcal{O}_{\mathcal{C}_{\mathrm{gpt}},m}) \cong \mathcal{O}_{\mathcal{C}'_{\mathrm{gpt}},m}$ and $f^*_{\mathcal{P},\mathrm{f\acute{e}t}}(\mathcal{O}^v_{\mathcal{P}/\mathcal{C},m}) \cong \mathcal{O}^v_{\mathcal{P}/\mathcal{C}',m}$ and the above diagram induces the following diagrams commutative up to canonical isomorphism.

(IV.6.2.9)
$$\begin{array}{ccc}
\mathcal{M}\mathrm{od}_{\mathrm{cocart}}(\mathcal{C}_{\mathrm{gpt}},\mathcal{O}_{\mathcal{C}_{\mathrm{gpt}},m}) & \xrightarrow{f^*_{\mathrm{gpt}}} & \mathcal{M}\mathrm{od}_{\mathrm{cocart}}(\mathcal{C}'_{\mathrm{gpt}},\mathcal{O}_{\mathcal{C}'_{\mathrm{gpt}},m}) \\
\sim\big\downarrow r_{\mathcal{C}} & & \sim\big\downarrow r_{\mathcal{C}'} \\
\mathcal{M}\mathrm{od}_{\mathrm{cocart}}((\mathcal{P}/\mathcal{C})_{\mathrm{f\acute{e}t}},\mathcal{O}^v_{\mathcal{P}/\mathcal{C},m}) & \xrightarrow{f^*_{\mathcal{P},\mathrm{f\acute{e}t}}} & \mathcal{M}\mathrm{od}_{\mathrm{cocart}}((\mathcal{P}/\mathcal{C}')_{\mathrm{f\acute{e}t}},\mathcal{O}^v_{\mathcal{P}/\mathcal{C}',m}),
\end{array}$$

(IV.6.2.10)
$$\begin{array}{ccc}
\mathcal{M}\mathrm{od}_{\mathrm{cocart}}(\mathcal{C}_{\mathrm{gpt}},\mathcal{O}_{\mathcal{C}_{\mathrm{gpt}},\bullet}) & \xrightarrow{f^*_{\mathrm{gpt}}} & \mathcal{M}\mathrm{od}_{\mathrm{cocart}}(\mathcal{C}'_{\mathrm{gpt}},\mathcal{O}_{\mathcal{C}'_{\mathrm{gpt}},\bullet}) \\
\sim\big\downarrow r_{\mathcal{C}} & & \sim\big\downarrow r_{\mathcal{C}'} \\
\mathcal{M}\mathrm{od}_{\mathrm{cocart}}((\mathcal{P}/\mathcal{C})_{\mathrm{f\acute{e}t}},\mathcal{O}^v_{\mathcal{P}/\mathcal{C},\bullet}) & \xrightarrow{f^*_{\mathcal{P},\mathrm{f\acute{e}t}}} & \mathcal{M}\mathrm{od}_{\mathrm{cocart}}((\mathcal{P}/\mathcal{C}')_{\mathrm{f\acute{e}t}},\mathcal{O}^v_{\mathcal{P}/\mathcal{C}',\bullet}).
\end{array}$$

**IV.6.3. Sheaves of $\mathcal{O}_{\mathcal{P}/\mathcal{C}}$-modules in the semi-stable reduction case.** In this subsection, we study sheaves of $\mathcal{O}_{\mathcal{P}/\mathcal{C}}$-modules when $X$ has semi-stable reduction.

Assume that the log structure of $\Sigma$ is defined by the closed point, and let $U$ be an affine fs log scheme over $\Sigma$ satisfying Condition IV.5.3.1. We keep the notation in IV.6.1 and IV.6.2. Note that $U$ satisfies the conditions in IV.6.2. We abbreviate $A_U$ and $\mathcal{A}_U$

to $A$ and $\mathcal{A}$ and write $\widetilde{U}$ for $\mathcal{U}_{\overline{K},\mathrm{triv}}$. Let $\widetilde{A}$ denote $\Gamma(\widetilde{U}, \mathcal{O}_{\widetilde{U}})$. Choose and fix a chart as in Condition IV.5.3.1, let $\pi$ denote the image of $1 \in \mathbb{N}$ in $\Gamma(\Sigma, M_\Sigma) \subset \Gamma(\Sigma, \mathcal{O}_\Sigma) = V$ by the chart of $M_\Sigma$, and let $t_i$ denote the image of $\underline{1}_{i+1} \in \mathbb{N}^{d+1}$ by the chart of $M_U$. Then we have an étale homomorphism $V[T_0, \ldots, T_d]/(T_0 \cdots T_{e-1} - \pi) \to A; T_i \mapsto t_i$ and $\widetilde{A} = \mathcal{A} \otimes_V \overline{K}[(T_0 \cdots T_d)^{-1}]$. Choose a compatible system $(\pi_n)$ of $n!$-th roots of $\pi$ in $\overline{V}$, let $V_n$ be $V[\pi_n]$, and define $A_n$ to be the cofiber product of

$$V_n[T_{0,n}, \ldots, T_{d,n}]/(T_{0,n} \cdots T_{e-1,n} - \pi_n) \longleftarrow V[T_0, \ldots, T_d]/(T_0 \cdots T_{e-1} - \pi) \longrightarrow A,$$

where the left homomorphism is defined by $T_i \mapsto (T_{i,n})^{n!}$. Then the ring $A_n$ is normal and flat over $V_n$ and $A_n/\pi_n A_n$ is reduced. We put $\mathcal{A}_n = \mathcal{A} \otimes_A A_n$, which is the henselization of $(A_n, pA_n)$ (resp. the $p$-adic completion of $A_n$) in the case II (resp. III). In particular, $\mathcal{A}_n$ is noetherian and normal. The natural homomorphism $\mathcal{A}_n \otimes_{V_n} \overline{V} \to \mathcal{A}_{n+1} \otimes_{V_{n+1}} \overline{V}$ is injective. Put $\widetilde{A}_n = (\mathcal{A}_n \otimes_{V_n} \overline{K})[(T_0 \cdots T_d)^{-1}]$ and $\widetilde{U}_n = \mathrm{Spec}(\widetilde{A}_n)$. We have $\widetilde{U} = \widetilde{U}_0$ and

$$\widetilde{A}_n = \widetilde{A} \otimes_{\overline{K}[T_i^{\pm 1}; 1 \leq i \leq d]} \overline{K}[T_{i,n}^{\pm 1}; 1 \leq i \leq d].$$

Hence $\widetilde{U}_n \to \widetilde{U}$ is finite étale. If we put $\Delta_n = \mathrm{Gal}(\overline{K}(T_{i,n}; 1 \leq i \leq d)/\overline{K}(T_i; 1 \leq i \leq d))$, we have a natural action of $\Delta_n$ on $\widetilde{U}_n/\widetilde{U}$ and it induces an isomorphism $\sqcup \sigma \times 1\colon \sqcup_{\sigma \in \Delta_n} \widetilde{U}_n \xrightarrow{\cong} \widetilde{U}_n \times_{\widetilde{U}} \widetilde{U}_n$. Hence, for any sheaf $\mathcal{F}$ on $\widetilde{U}_{\text{fét}}$ and any $\mathcal{V} \in \mathrm{Ob}(\widetilde{U}_{\text{fét}})$, we have an isomorphism

$$\mathcal{F}(\mathcal{V}) \xrightarrow{\cong} \mathcal{F}(\mathcal{V} \times_{\widetilde{U}} \widetilde{U}_n)^{\Delta_n}. \tag{IV.6.3.1}$$

By applying Lemma IV.5.2.1 to $A_n/V_n$, we see that $\mathcal{A}_n \otimes_{V_n} \overline{V}$ is a product of finite number of normal domains, which implies

$$\mathcal{O}_{\widetilde{U}}^{\mathrm{int}}(\widetilde{U}_n) = \mathcal{A}_n \otimes_{V_n} \overline{V}. \tag{IV.6.3.2}$$

The number of the connected components of $\widetilde{U}_n$ is bounded when $n$ varies. Hence it is stable when $n \to \infty$, i.e., there exists $n_0 \in \mathbb{N}$ such that $\pi_0(\widetilde{U}_n) \to \pi_0(\widetilde{U}_{n_0})$ is bijective for every $n \geq n_0$. In the case I, this follows from the fact that $\overline{K}[T_i^{\pm 1}; 1 \leq i \leq d] \to \widetilde{A}$ is étale. In the cases II and III, the number of the connected components of $\widetilde{U}_n$ coincides with that of $\mathrm{Spec}(\mathcal{A}_n/\pi_n\mathcal{A}_n) = \mathrm{Spec}(A_n/\pi_n A_n)$ by Lemma IV.5.2.1. Hence the claim follows from the facts that $k[T_i; 0 \leq i \leq d]/(T_0 \cdots T_{e-1}) \to A/\pi A$ is étale and that the number of irreducible components of the special fiber of $k[T_{i,n}; 0 \leq i \leq d]/(T_{0,n} \cdots T_{e-1,n})$ is $e$ for every $n$.

For $\overline{s} \in \widetilde{U}_{\mathrm{gpt}}$, there exists an extension $\widetilde{A}_n \to \kappa(s)^{\mathrm{ur}}$ of $\widetilde{A} \to \kappa(s^g)$ and its image is independent on the choice of extension. Let $\kappa(s_n)$ denote the field generated by the image. Let $P$ be a finite set of liftings $v\colon \overline{s} \to \varprojlim_n \widetilde{U}_n$ of $\overline{s} \to \widetilde{U}$ such that the natural map $P \to \pi_0(\widetilde{U}_n)$, sending $v$ to the connected component containing its image in $\widetilde{U}_n$, is bijective for $n \geq n_0$. For $n \geq n_0$, each $v \in P$ induces an open and closed immersion $U_{\kappa(s_n)} \to \widetilde{U}_n$. By taking the union of these immersions, we obtain an isomorphism

$$\bigsqcup_P U_{\kappa(s_n)} \xrightarrow{\cong} \widetilde{U}_n \times_{\widetilde{U}} U_{\kappa(s^g)} \tag{IV.6.3.3}$$

for $n \geq n_0$. Recall that, for $\mathcal{L} \in \mathscr{S}_{\overline{s}}$, $U_{\mathcal{L}}$ denotes the integral closure of $\widetilde{U}$ in $\mathcal{L}$ (cf. IV.6.2).

We use the following theorem of Faltings.

**Theorem IV.6.3.4** (cf. **[26]** 2b). *Let notation and assumption be as above. Then for any* $n_1 \in \mathbb{N}$ *and* $\mathcal{V} \in \mathrm{Ob}(\widetilde{U}_{n_1})_{\text{fét}}$, *the homomorphism* $\varinjlim_n \mathcal{O}_{\widetilde{U}}^{\mathrm{int}}(\widetilde{U}_n) \to \varinjlim_n \mathcal{O}_{\widetilde{U}}^{\mathrm{int}}(\widetilde{U}_n \times_{\widetilde{U}_{n_1}} \mathcal{V})$ *is an almost étale covering* (*cf. Definition* V.7.1).

**Corollary IV.6.3.5.** *Let notation and assumption be as above, and let* $\overline{s} \in \mathrm{Ob}\,\widetilde{U}_{\mathrm{gpt}}$ *and* $\mathcal{V} \in \mathrm{Ob}\,\widetilde{U}_{\mathrm{f\acute{e}t}}$.

(1) *Let* $\mathcal{L} \in \mathscr{S}_{\overline{s}}$, *and put* $\kappa(s_\infty) = \cup_n \kappa(s_n)$ *and* $\mathcal{L}_n = \kappa(s_n)\mathcal{L}$ $(n \in \mathbb{N} \cup \{\infty\})$. *Then the homomorphism* $\varinjlim_n \mathcal{O}^{\mathrm{int}}_{\widetilde{U}}(\mathcal{V} \times_{\widetilde{U}} U_{\kappa(s_n)}) \to \varinjlim_n \mathcal{O}^{\mathrm{int}}_{\widetilde{U}}(\mathcal{V} \times_{\widetilde{U}} U_{\mathcal{L}_n})$ *is an almost étale* $\mathrm{Gal}(\mathcal{L}_\infty/\kappa(s_\infty))$*-covering (cf. Definition* V.12.2, *Lemma* V.12.3).

(2) *Let* $\mathcal{F}$ *be a sheaf of* $\mathcal{O}^{\mathrm{int}}_{\widetilde{U}}$*-modules on* $\widetilde{U}_{\mathrm{f\acute{e}t}}$. *Then the following homomorphism is an almost isomorphism*

$$\varinjlim_n \mathcal{F}(\mathcal{V} \times_{\widetilde{U}} U_{\kappa(s_n)}) \otimes_{\varinjlim_n \mathcal{O}^{\mathrm{int}}_{\widetilde{U}}(\mathcal{V} \times_{\widetilde{U}} U_{\kappa(s_n)})} \varinjlim_{\mathcal{L} \in \mathscr{S}_{\overline{s}}} \mathcal{O}^{\mathrm{int}}_{\widetilde{U}}(\mathcal{V} \times_{\widetilde{U}} U_{\mathcal{L}}) \longrightarrow \varinjlim_{\mathcal{L} \in \mathscr{S}_{\overline{s}}} \mathcal{F}(\mathcal{V} \times_{\widetilde{U}} U_{\mathcal{L}}).$$

PROOF. (1) Choose $n_1 \in \mathbb{N}$ such that $n_1 \geq n_0$ and $\mathcal{L}_\infty \cong \mathcal{L}_{n_1} \otimes_{\kappa(s_{n_1})} \kappa(s_\infty)$. Then for any $n \in \mathbb{N}$, $n \geq n_1$, we have $U_{\kappa(s_n)} \times_{U_{\kappa(n_1)}} U_{\mathcal{L}_{n_1}} \cong U_{\mathcal{L}_n}$. Choose a lifting $v \colon \overline{s} \to \varprojlim_n \widetilde{U}_n$ of $\overline{s} \to \widetilde{U}$. Then it induces a compatible system of open and closed immersions $U_{\kappa(s_n)} \to \widetilde{U}_n$ and hence a compatible system of isomorphisms $U_{\kappa(s_n)} \cong \widetilde{U}_n \times_{\widetilde{U}_{n_1}} U_{\kappa(s_{n_1})}$ by (IV.6.3.3). By Theorem IV.6.3.4, we see that $\varinjlim_n \mathcal{O}^{\mathrm{int}}_{\widetilde{U}}(U_{\kappa(s_n)}) \to \varinjlim_n \mathcal{O}^{\mathrm{int}}_{\widetilde{U}}(U_{\mathcal{L}_n})$ is an almost étale $\mathrm{Gal}(\mathcal{L}_\infty/\kappa(s_\infty))$-covering (cf. Proposition V.12.9). Hence its base change by $\varinjlim_n \mathcal{O}^{\mathrm{int}}_{\widetilde{U}}(U_{\kappa(s_n)}) \to \varinjlim_n \mathcal{O}^{\mathrm{int}}_{\widetilde{U}}(\mathcal{V} \times_{\widetilde{U}} U_{\kappa(s_n)})$ is an almost étale $\mathrm{Gal}(\mathcal{L}_\infty/\kappa(s_\infty))$-covering and is almost isomorphic to the homomorphism in the claim.

(2) Let $\mathcal{L} \in \mathscr{S}_{\overline{s}}$ and let $\mathcal{L}_n$ be as in (1). Then by (IV.6.2.1), we have an isomorphism

$$\varinjlim_n \mathcal{F}(\mathcal{V} \times_{\widetilde{U}} U_{\kappa(s_n)}) \xrightarrow{\cong} (\varinjlim_n \mathcal{F}(\mathcal{V} \times_{\widetilde{U}} U_{\mathcal{L}_n}))^{\mathrm{Gal}(\mathcal{L}_\infty/\kappa(s_\infty))}.$$

By (1) and almost Galois descent (cf. Proposition V.12.5), we obtain an almost isomorphism

$$\varinjlim_n \mathcal{F}(\mathcal{V} \times_{\widetilde{U}} U_{\kappa(s_n)}) \otimes_{\varinjlim_n \mathcal{O}^{\mathrm{int}}_{\widetilde{U}}(\mathcal{V} \times_{\widetilde{U}} U_{\kappa(s_n)})} \varinjlim_n \mathcal{O}^{\mathrm{int}}_{\widetilde{U}}(\mathcal{V} \times_{\widetilde{U}} U_{\mathcal{L}_n}) \xrightarrow{\approx} \varinjlim_n \mathcal{F}(\mathcal{V} \times_{\widetilde{U}} U_{\mathcal{L}_n}).$$

Varying $\mathcal{L}$, we obtain the claim. □

Recall that $\mathcal{O}_{\mathcal{C}_{\mathrm{gpt}}}$ denotes the ring object $s_{\mathcal{C}}(\mathcal{O}^v_{\mathcal{P}/\mathcal{C}})$ of $\mathcal{C}^{\wedge,\mathrm{cont}}_{\mathrm{gpt}}$ corresponding to $\mathcal{O}^v_{\mathcal{P}/\mathcal{C}}$ (cf. Definition IV.6.2.7).

**Proposition IV.6.3.6.** *Let* $X$ *be an fs log scheme satisfying Condition* IV.5.3.1 *strict étale locally and the underlying scheme of* $X$ *is separated. Let* $\mathcal{C}$ *be a full subcategory of* $X_{\mathrm{\acute{e}taff}}$ *satisfying the following two conditions:*

(a) *Every object of* $X_{\mathrm{\acute{e}taff}}$ *admits a strict étale covering by objects of* $\mathcal{C}$.

(b) *Every object* $U$ *of* $\mathcal{C}$ *satisfies Condition* IV.5.3.1.

*Let* $\mathcal{F}$ *be an object of* $\mathcal{M}\mathrm{od}_{\mathrm{cocart}}((\mathcal{P}/\mathcal{C})_{\mathrm{f\acute{e}t}}, \mathcal{O}^v_{\mathcal{P}/\mathcal{C},m})$ *(cf. Definition* IV.6.2.7). *Then* $\mathcal{F}$ *satisfies the condition* (b) *of Proposition* IV.6.1.23.

**Lemma IV.6.3.7.** *Let* $X$ *and* $\mathcal{C}$ *be as in Proposition* IV.6.3.6 *and let* $\mathcal{F}$ *be an object of* $\mathcal{M}\mathrm{od}_{\mathrm{cocart}}((\mathcal{P}/\mathcal{C})_{\mathrm{f\acute{e}t}}, \mathcal{O}^v_{\mathcal{P}/\mathcal{C},m})$. *Let* $(f,g) \colon (U', \mathcal{V}') \to (U, \mathcal{V})$ *be a horizontal morphism in* $\mathcal{P}/\mathcal{C}$, *and let* $\widetilde{U}$ *and* $\widetilde{U}'$ *(resp.* $\mathcal{A}$ *and* $\mathcal{A}'$*) denote the schemes* $\mathcal{U}_{\overline{K},\mathrm{triv}}$ *and* $\mathcal{U}'_{\overline{K},\mathrm{triv}}$ *(resp. the* $\overline{V}$*-algebras* $\mathcal{A}_U$ *and* $\mathcal{A}_{U'}$*) associated to* $U$ *and* $U'$ *(cf. the beginning of* IV.6.1). *Choose a chart of* $U \to \Sigma$ *as in Condition* IV.5.3.1 *and define a compatible system of finite étale morphisms* $\widetilde{U}_n \to \widetilde{U}$ $(n \in \mathbb{N})$ *as in the beginning of this subsection. Put* $\widetilde{U}'_n = \widetilde{U}_n \times_{\widetilde{U}} \widetilde{U}'$. *Then the following natural homomorphism is an almost isomorphism.*

$$\varinjlim_n \mathcal{F}(U, \mathcal{V} \times_{\widetilde{U}} \widetilde{U}_n) \otimes_{\mathcal{A}} \mathcal{A}' \longrightarrow \varinjlim_n \mathcal{F}(U', \mathcal{V}' \times_{\widetilde{U}'} \widetilde{U}'_n).$$

PROOF. For $\overline{s} \in \mathrm{Ob}\, \widetilde{U}_{\mathrm{gpt}}$ (resp. $\overline{s}' \in \mathrm{Ob}\, \widetilde{U}'_{\mathrm{gpt}}$), we define $\kappa(s_n)$ (resp. $\kappa(s'_n)$) as before Theorem IV.6.3.4. Let $\mathcal{F}(U, \mathcal{V}, \overline{s})$ (resp. $\mathcal{F}(U, \mathcal{V}, s_\infty)$) denote the direct limit $\varinjlim_{\mathcal{L} \in \mathscr{S}_{\overline{s}}} \mathcal{F}(U, \mathcal{V} \times_{\widetilde{U}} U_{\mathcal{L}})$ (resp. $\varinjlim_n \mathcal{F}(U, \mathcal{V} \times_{\widetilde{U}} U_{\kappa(s_n)})$) and define similarly for $\mathcal{O}^v_{\mathcal{P}/\mathcal{C}}$ and for $(U', \mathcal{V}')$. Since we have an isomorphism $\sqcup(f, \mathrm{id}_{U_{\mathcal{L}}})\colon \sqcup_{f \in \mathrm{Hom}_{\widetilde{U}}(U_{\mathcal{L}}, \mathcal{V})} U_{\mathcal{L}} \xrightarrow{\cong} \mathcal{V} \times_{\widetilde{U}} U_{\mathcal{L}}$ for a sufficiently large $\mathcal{L} \in \mathscr{S}_{\overline{s}}$, we have a canonical isomorphism

$$\mathcal{F}(U, \mathcal{V}, \overline{s}) \cong \bigoplus_{\mathrm{Hom}_{\widetilde{U}}(s^{\mathrm{ur}}, \mathcal{V})} s_{\mathcal{C}} \mathcal{F}(U, \overline{s}).$$

We have the same type of descriptions of $\mathcal{O}^v_{\mathcal{P}/\mathcal{C}}(U, \mathcal{V}, \overline{s})$, $\mathcal{F}(U', \mathcal{V}', \overline{s}')$ and $\mathcal{O}^v_{\mathcal{P}/\mathcal{C}}(U', \mathcal{V}', \overline{s}')$. Choose a morphism $(f, h)\colon (U', \overline{s}') \to (U, \overline{s})$ over $f$. Then it induces homomorphisms $\mathcal{F}(U, \mathcal{V}, \overline{s}) \to \mathcal{F}(U', \mathcal{V}', \overline{s}')$ and $\mathcal{O}^v_{\mathcal{P}/\mathcal{C}}(U, \mathcal{V}, \overline{s}) \to \mathcal{O}^v_{\mathcal{P}/\mathcal{C}}(U', \mathcal{V}', \overline{s}')$ and the above descriptions are compatible via the map $\mathrm{Hom}_{\widetilde{U}'}(\overline{s}', \mathcal{V}') \to \mathrm{Hom}_{\widetilde{U}}(\overline{s}, \mathcal{V})$ induced by $f$, $g$ and $h$ (cf. the definition of the functor $s_{\mathcal{C}}$ in IV.6.2). Hence the condition on $\mathcal{F}$ implies that the homomorphism

$$\mathcal{F}(U, \mathcal{V}, \overline{s}) \otimes_{\mathcal{O}^v_{\mathcal{P}/\mathcal{C}}(U, \mathcal{V}, \overline{s})} \mathcal{O}^v_{\mathcal{P}/\mathcal{C}}(U', \mathcal{V}', \overline{s}') \xrightarrow{\cong} \mathcal{F}(U', \mathcal{V}', \overline{s}')$$

is an isomorphism. By Corollary IV.6.3.5 (2), we obtain an almost isomorphism

$$\begin{aligned} &\mathcal{F}(U, \mathcal{V}, s_\infty) \otimes_{\mathcal{O}^v_{\mathcal{P}/\mathcal{C}}(U, \mathcal{V}, s_\infty)} \mathcal{O}^v_{\mathcal{P}/\mathcal{C}}(U', \mathcal{V}', \overline{s}') \\ &\qquad \xrightarrow{\approx} \mathcal{F}(U', \mathcal{V}', s'_\infty) \otimes_{\mathcal{O}^v_{\mathcal{P}/\mathcal{C}}(U', \mathcal{V}', s'_\infty)} \mathcal{O}^v_{\mathcal{P}/\mathcal{C}}(U', \mathcal{V}', \overline{s}'). \end{aligned}$$

Since $\mathcal{O}^v_{\mathcal{P}/\mathcal{C}}(U', \mathcal{V}', s'_\infty) \to \mathcal{O}^v_{\mathcal{P}/\mathcal{C}}(U', \mathcal{V}', \overline{s}')$ is almost faithfully flat by Corollary IV.6.3.5 (1) (cf. Proposition V.8.11 and Definition V.12.2), we may replace $\mathcal{O}^v_{\mathcal{P}/\mathcal{C}}(U', \mathcal{V}', \overline{s}')$ by $\mathcal{O}^v_{\mathcal{P}/\mathcal{C}}(U', \mathcal{V}', s'_\infty)$ in the above almost isomorphism. Varying $\overline{s}$ and $\overline{s}'$ and using (IV.6.3.3), we obtain an almost isomorphism:

$$\mathcal{F}(U, \mathcal{V}_\infty) \otimes_{\mathcal{O}^v_{\mathcal{P}/\mathcal{C}}(U, \mathcal{V}_\infty)} \mathcal{O}^v_{\mathcal{P}/\mathcal{C}}(U', \mathcal{V}'_\infty) \xrightarrow{\approx} \mathcal{F}(U', \mathcal{V}'_\infty),$$

where $\mathcal{F}(U, \mathcal{V}_\infty)$ denotes $\varinjlim_n \mathcal{F}(U, \mathcal{V} \times_{\widetilde{U}} \widetilde{U}_n)$ and similarly for the other three terms. On the other hand, since $(f, g)$ is horizontal, we see that the natural morphism $(\mathcal{V} \times_{\widetilde{U}} \widetilde{U}_n) \times_{\widetilde{U}_n} \widetilde{U}'_n \xleftarrow{\cong} \mathcal{V}' \times_{\widetilde{U}'} \widetilde{U}'_n$ is an isomorphism, which implies, by Theorem IV.6.3.4, that the homomorphism

$$\mathcal{O}^v_{\mathcal{P}/\mathcal{C}}(U, \mathcal{V}_\infty) \otimes_{\varinjlim_n \mathcal{O}^v_{\mathcal{P}/\mathcal{C}}(U, \widetilde{U}_n)} \varinjlim_n \mathcal{O}^v_{\mathcal{P}/\mathcal{C}}(U', \widetilde{U}'_n) \longrightarrow \mathcal{O}^v_{\mathcal{P}/\mathcal{C}}(U', \mathcal{V}'_\infty)$$

is an almost isomorphism (cf. Proposition V.7.11). Furthermore, from (IV.6.3.2) for $U$ and $U'$, we obtain an isomorphism

$$\varinjlim_n \mathcal{O}^v_{\mathcal{P}/\mathcal{C}}(U, \widetilde{U}_n) \otimes_{\mathcal{A}} \mathcal{A}' \xrightarrow{\cong} \varinjlim_n \mathcal{O}^v_{\mathcal{P}/\mathcal{C}}(U', \widetilde{U}'_n).$$

Combining with the above two almost isomorphisms, we obtain the lemma. □

PROOF OF PROPOSITION IV.6.3.6. Let $\mathfrak{U} = (U, \mathcal{V})$, $\mathfrak{U}_\alpha = (U_\alpha, \mathcal{V}_\alpha)$ ($\alpha \in A$), and $\mathfrak{U}_{\alpha\beta;\gamma} = (U_{\alpha\beta;\gamma}, \mathcal{V}_{\alpha\beta;\gamma})_{\gamma \in \Gamma_{\alpha\beta}}$ be as in Proposition IV.6.1.23 (b). Let $\widetilde{U}$ denote $\mathcal{U}_{\overline{K},\mathrm{triv}}$ associated to $U$ and define $\widetilde{U}_\alpha$ and $\widetilde{U}_{\alpha\beta;\gamma}$ similarly. Choose a chart as in Condition IV.5.3.1 and define $\widetilde{U}_n \to \widetilde{U}$. Put $\widetilde{U}_{\alpha,n} = \widetilde{U}_n \times_{\widetilde{U}} \widetilde{U}_\alpha$ and $\widetilde{U}_{\alpha\beta;\gamma,n} := \widetilde{U}_n \times_{\widetilde{U}} \widetilde{U}_{\alpha\beta;\gamma}$. By (IV.6.3.1), it suffices to prove that the following homomorphism is an almost isomorphism

$$\mathcal{F}(U, \mathcal{V}_\infty) \longrightarrow \mathrm{Ker}\left( \prod_{\alpha \in A} \mathcal{F}(U_\alpha, \mathcal{V}_{\alpha,\infty}) \rightrightarrows \prod_{(\alpha,\beta) \in A^2} \prod_{\gamma \in \Gamma_{\alpha\beta}} \mathcal{F}(U_{\alpha\beta;\gamma}, \mathcal{V}_{\alpha\beta;\gamma,\infty}) \right),$$

where $\mathcal{F}(U_\star, \mathcal{V}_{\star,\infty}) = \varinjlim_n \mathcal{F}(U_\star, \mathcal{V}_\star \times_{\widetilde{U}_\star} \widetilde{U}_{\star,n})$ for $\star = \emptyset, \alpha, \alpha\beta; \gamma$. Let $\mathcal{A}$, $\mathcal{A}_\alpha$, $\mathcal{A}_{\alpha\beta}$, and $\mathcal{A}_{\alpha\beta;\gamma}$ denote $\mathcal{A}_U$, $\mathcal{A}_{U_\alpha}$, $\mathcal{A}_{U_\alpha \times_U U_\beta}$, and $\mathcal{A}_{U_{\alpha\beta;\gamma}}$. Then noting $\mathcal{A}/p^m\mathcal{A} = A_U/p^m A_U$, etc., we see that $\mathcal{A}/p^m \to \prod_{\alpha\in A} \mathcal{A}_\alpha/p^m$ and $\mathcal{A}_{\alpha\beta}/p^m \to \prod_{\gamma\in\Gamma_{\alpha\beta}} \mathcal{A}_{\alpha\beta;\gamma}/p^m$ are faithfully flat and $\mathcal{A}_{\alpha\beta}/p^m \cong \mathcal{A}_\alpha/p^m \otimes_{\mathcal{A}_\alpha/p^m} \mathcal{A}_\beta/p^m$. Hence we obtain the claim from Lemma IV.6.3.7 applied to $\mathfrak{U}_\alpha \to \mathfrak{U}$ and $\mathfrak{U}_{\alpha\beta;\gamma} \to \mathfrak{U}$. □

**Corollary IV.6.3.8.** *Let $X$ and $\mathcal{C}$ be the same as in Proposition* IV.6.3.6. *Then the natural homomorphism $\mathcal{O}^v_{\mathcal{P}/\mathcal{C},m} \to \mathcal{O}_{\mathcal{P}/\mathcal{C},m}$ is an almost isomorphism.*

**Proposition IV.6.3.9.** *Let $X$ and $\mathcal{C}$ be as in Proposition* IV.6.3.6. *Then the functor*

$$(v^*_{\mathcal{P}/\mathcal{C}})_{\mathbb{Q}} \colon \mathcal{M}\mathrm{od}_{\mathrm{cocart}}((\mathcal{P}/\mathcal{C})_{\mathrm{f\acute{e}t}}, \mathcal{O}^v_{\mathcal{P}/\mathcal{C},\bullet})_{\mathbb{Q}} \to \mathcal{M}\mathrm{od}((\mathcal{P}/\mathcal{C})_{\text{ét-fét}}, \mathcal{O}_{\mathcal{P}/\mathcal{C},\bullet})_{\mathbb{Q}}$$

*is fully faithful.*

Proof. Let $\mathcal{F}_\bullet$ and $\mathcal{G}_\bullet$ be objects of $\mathcal{M}\mathrm{od}_{\mathrm{cocart}}((\mathcal{P}/\mathcal{C})_{\mathrm{f\acute{e}t}}, \mathcal{O}^v_{\mathcal{P}/\mathcal{C},\bullet})$. It suffices to prove that the homomorphism

$$\mathrm{Hom}_{\mathcal{O}^v_{\mathcal{P}/\mathcal{C},\bullet}}(\mathcal{F}_\bullet, \mathcal{G}_\bullet) \longrightarrow \mathrm{Hom}_{\mathcal{O}_{\mathcal{P}/\mathcal{C},\bullet}}(v^*_{\mathcal{P}/\mathcal{C}}(\mathcal{F}_\bullet), v^*_{\mathcal{P}/\mathcal{C}}(\mathcal{G}_\bullet))$$

is an almost isomorphism. By Proposition IV.6.3.6 and Corollary IV.6.1.24, the adjunction morphism $\mathcal{G}_\bullet \to v_{\mathcal{P}/\mathcal{C}*} v^*_{\mathcal{P}/\mathcal{C}} \mathcal{G}_\bullet$ is an almost isomorphism. This implies the claim. □

**IV.6.4. Higgs crystals and modules on Faltings sites.** Let $V$, $k$, $K$, $\overline{K}$, $\overline{V}$, $\Sigma$, $\widehat{\Sigma}$, $\overline{\Sigma}$, and $D(\overline{\Sigma})$ be the same as in the beginning of IV.5.2. We also take the base ring $R$ of our theory of Higgs crystals as in loc. cit.

Throughout this subsection, let $X$ denote an fs log scheme over $\Sigma$ satisfying the following conditions: the underlying scheme is normal, the morphism of schemes underlying $X \to \Sigma$ is flat, separated, and of finite type, its special fiber is reduced, and the log structure of $X$ is trivial at every codimension zero point. In the case I (cf. IV.5.1 and IV.6.1), we further assume that the log structure of $\Sigma$ is defined by the closed point and $X$ satisfies Condition IV.5.3.1 strict étale locally. Let $\widehat{X}$ denote the $p$-adic completion of $X$ and let $X_1$ denote $\widehat{X} \times_{\widehat{\Sigma}} \overline{\Sigma}$ as in IV.5.2.

**Definition IV.6.4.1.** Let $X$ be as above.

(1) We define $\mathcal{C}_X$ to be the full subcategory of $X_{\mathrm{\acute{e}taff}}$ consisting of $U$ satisfying the following conditions:

(a) The underlying scheme of $U$ is affine.

(b) There exists a chart $\alpha\colon P_U \to M_U$ such that $P$ is a finitely generated and saturated monoid, and the morphism $P \to \Gamma(U, M_U)/\Gamma(U, \mathcal{O}_U^\times)$ induced by $\alpha$ is an isomorphism (cf. **[73]** Lemma 1.3.2).

(c) In the case I, $U$ satisfies Condition IV.5.3.1.

(2) For a covering sieve $R$ of $X$ with respect to the site $X_{\mathrm{\acute{e}taff}}$, we define $\mathcal{C}_R$ to be the full subcategory of $\mathcal{C}_X$ consisting of $u\colon U \to X$ such that $u \in R(U)$.

By **[73]** Lemma 1.3.3, any object of $X_{\mathrm{\acute{e}taff}}$ admits a strict étale covering by objects of $\mathcal{C}_R$ for any $R$.

Let $(h\colon U \to X, \overline{s})$ be an object of $(\mathcal{C}_X)_{\mathrm{gpt}}$ and let $\mathcal{A}_U$ be as in the beginning of IV.6.1. If $p$ is not invertible on the connected component of $\mathrm{Spec}(\mathcal{A}_U \otimes_V \overline{V})$ (cf. Lemma IV.5.2.1) containing the image $s_g$ of $\overline{s}$, which always holds in the cases II and III, then $(U, \overline{s})$ satisfies the conditions after Lemma IV.5.2.1 (cf. Theorem IV.5.3.6 (2)). In this case, we write $(D(U,\overline{s}), z_{(U,\overline{s})})$ for the object $(D(\overline{U}), z_{\overline{U}})$ of $(U_1/D(\overline{\Sigma}))^{\infty}_{\mathrm{HIGGS}}$ constructed in IV.5.2. If $p$ is invertible on the connected component of $\mathrm{Spec}(\mathcal{A}_U \otimes_V \overline{V})$ containing $s_g$, then we define $(D(U,\overline{s}), z_{(U,\overline{s})})$ to be $\emptyset$. By the functoriality mentioned in the last paragraph of IV.5.1, we see that the construction of $(D(U,\overline{s}), z_{(U,\overline{s})})$ is functorial in $(U,\overline{s})$ and obtain a functor

$$D_X\colon (\mathcal{C}_X)_{\mathrm{gpt}} \longrightarrow (X_1/D(\overline{\Sigma}))^{\infty}_{\mathrm{HIGGS}}$$

by associating $(D(U,\overline{s}), h_1\circ z_{(U,\overline{s})})$ to $(h\colon U\to X,\overline{s})$.

**Definition IV.6.4.2.** Let $R$ be a covering sieve of $X$ with respect to the site $X_{\text{étaff}}$. We say that $\mathcal{F}\in \mathrm{Ob}\,(\mathrm{HC}_{\mathbb{Z}_p}(X_1/D(\overline{\Sigma})))$ (cf. Definition IV.3.3.2) is *$R$-finite* if $\mathcal{F}_{(T,z)}\in \mathrm{Ob}\,\mathcal{LPM}(\mathcal{O}_{T_1})$ for every $(T,z)\in \mathrm{Ob}\,(X_1/D(\overline{\Sigma}))^{\infty}_{\mathrm{HIGGS}}$ (cf. IV.3.2) such that $T_1$ is affine and $z$ factors through $u_1\colon U_1\to X_1$ for some $u\colon U\to X\in \mathrm{Ob}\,X_{\text{étaff}}$ satisfying $u\in R(U)$. We write $\mathrm{HC}_{\mathbb{Z}_p,R\text{-fin}}(X_1/D(\overline{\Sigma}))$ for the full subcategory of $\mathrm{HC}_{\mathbb{Z}_p}(X_1/D(\overline{\Sigma}))$ consisting of $R$-finite objects.

**Remark IV.6.4.3.** Since $X_1\times_{\overline{\Sigma}}\mathrm{Spec}(k) = X\times_{\mathrm{Spec}(V)}\mathrm{Spec}(k)$, every object of $(X_1)_{\text{ét}}$ admits a strict étale covering by objects coming from $X_{\text{étaff}}$ by base change. Hence, for every $\mathcal{F}\in \mathrm{Ob}\,(\mathrm{HC}_{\mathbb{Z}_p}(X_1/D(\overline{\Sigma})))$, there exists a covering sieve $R$ of $X$ such that $\mathcal{F}$ is $R$-finite.

Let $R$ be a covering sieve of $X$ with respect to the site $X_{\text{étaff}}$, and let $\mathcal{F}$ be an $R$-finite Higgs crystal on $X_1/D(\overline{\Sigma})$. Then, for each $(U,\overline{s})\in \mathrm{Ob}\,(\mathcal{C}_R)_{\mathrm{gpt}}$, the evaluation of $\mathcal{F}$ on $D_X(U,\overline{s})$, denoted $\mathcal{T}^R_{X,\mathrm{gpt}}(\mathcal{F})(U,\overline{s})$, is an object of $\mathcal{LPM}(\mathcal{O}_{\mathcal{C}_{R,\mathrm{gpt}}}(U,\overline{s}))$ (cf. Proposition IV.3.2.7 (1)). Put $\mathcal{T}^R_{X,\mathrm{gpt},m}(\mathcal{F})(U,\overline{s}) := \mathcal{T}^R_{X,\mathrm{gpt}}(\mathcal{F})(U,\overline{s})\otimes_{\mathbb{Z}}\mathbb{Z}/p^m\mathbb{Z}$ for $m\in\mathbb{N}$. Then the left action of $\Delta_{\overline{s}}$ on $\mathcal{T}^R_{X,\mathrm{gpt},m}(\mathcal{F})(U,\overline{s})$ induced by the right action of $\Delta_{\overline{s}}$ on $(U,\overline{s})$ is continuous by Corollary IV.5.2.13. The morphism $(U',\overline{s}')\to(U,\overline{s})$ in $\mathcal{C}_R$ induces an isomorphism

$$\mathcal{O}_{\mathcal{C}_{R,\mathrm{gpt}},m}(U',\overline{s}')\otimes_{\mathcal{O}_{\mathcal{C}_{R,\mathrm{gpt}},m}(U,\overline{s})}\mathcal{T}^R_{X,\mathrm{gpt},m}(\mathcal{F})(U,\overline{s})\xrightarrow{\cong}\mathcal{T}^R_{X,\mathrm{gpt},m}(\mathcal{F})(U',\overline{s}')$$

compatible with $m$ (cf. Lemma IV.3.2.8 and Definition IV.3.3.2). Thus we obtain an object $\mathcal{T}^R_{X,\mathrm{gpt},\bullet}(\mathcal{F}) = (\mathcal{T}^R_{X,\mathrm{gpt},m}(\mathcal{F}))_m$ of $\mathcal{M}\mathrm{od}_{\mathrm{cocart}}(\mathcal{C}_{R,\mathrm{gpt}},\mathcal{O}_{R,\mathrm{gpt},\bullet})$ (cf. Definition IV.6.2.7 (3)). This construction is obviously functorial in $\mathcal{F}$ and we obtain a functor

$$\text{(IV.6.4.4)}\qquad \mathcal{T}^R_{X,\mathrm{gpt},\bullet}\colon \mathrm{HC}_{\mathbb{Z}_p,R\text{-fin}}(X_1/D(\overline{\Sigma}))\longrightarrow \mathcal{M}\mathrm{od}_{\mathrm{cocart}}(\mathcal{C}_{R,\mathrm{gpt}},\mathcal{O}_{\mathcal{C}_{R,\mathrm{gpt}},\bullet}).$$

Composing with the equivalence of categories $r_{\mathcal{C}_R}$ (cf. Proposition IV.6.2.6), the functor

$$v^*_{\mathcal{P}/\mathcal{C}_R}\colon \mathcal{M}\mathrm{od}_{\mathrm{cocart}}((\mathcal{P}/\mathcal{C}_R)_{\text{fét}},\mathcal{O}^v_{\mathcal{P}/\mathcal{C}_R,\bullet})\longrightarrow \mathcal{M}\mathrm{od}((\mathcal{P}/\mathcal{C}_R)_{\text{ét-fét}},\mathcal{O}_{\mathcal{P}/\mathcal{C}_R,\bullet})$$

and the equivalence of categories

$$\iota^{X,\mathcal{C}_R}_{\mathcal{P},\text{ét-fét}*}\colon \mathcal{M}\mathrm{od}((\mathcal{P}/\mathcal{C}_R)_{\text{ét-fét}},\mathcal{O}_{\mathcal{P}/\mathcal{C}_R,\bullet})\xrightarrow{\sim}\mathcal{M}\mathrm{od}((\mathcal{P}/X)_{\text{ét-fét}},\mathcal{O}_{\mathcal{P}/X,\bullet})$$

(cf. (IV.6.1.18)), we obtain a functor

$$\text{(IV.6.4.5)}\qquad \mathcal{T}^R_{X,\bullet}\colon \mathrm{HC}_{\mathbb{Z}_p,R\text{-fin}}(X_1/D(\overline{\Sigma}))\longrightarrow \mathcal{M}\mathrm{od}((\mathcal{P}/X)_{\text{ét-fét}},\mathcal{O}_{\mathcal{P}/X,\bullet}).$$

**Lemma IV.6.4.6.** *Let $f\colon X'\to X$ be a strict étale morphism of fs log schemes over $\Sigma$ such that $X'$ satisfies the same conditions as $X$. Let $R$ (resp. $R'$) be a covering sieve of $X$ (resp. $X'$) with respect to the site $X_{\text{étaff}}$ (resp. $X'_{\text{étaff}}$) such that for every $u'\colon U'\to X'\in R'(U')$, the composition $f\circ u'$ is contained in $R(U')$. Then the following diagram is commutative up to canonical isomorphism.*

$$\begin{array}{ccc}
\mathrm{HC}_{\mathbb{Z}_p,R\text{-}\mathit{fin}}(X_1/D(\overline{\Sigma})) & \xrightarrow{\mathcal{T}^R_{X,\bullet}} & \mathcal{M}\mathrm{od}((\mathcal{P}/X)_{\text{ét-fét}},\mathcal{O}_{\mathcal{P}/X,\bullet}) \\
\Big\downarrow{\scriptstyle f^*_{1,\mathrm{HIGGS}}} & & \Big\downarrow{\scriptstyle f^*_{\mathcal{P},\text{ét-fét}}} \\
\mathrm{HC}_{\mathbb{Z}_p,R'\text{-}\mathit{fin}}(X'_1/D(\overline{\Sigma})) & \xrightarrow{\mathcal{T}^{R'}_{X',\bullet}} & \mathcal{M}\mathrm{od}((\mathcal{P}/X')_{\text{ét-fét}},\mathcal{O}_{\mathcal{P}/X',\bullet}).
\end{array}$$

Proof. By assumption, $u' \in \mathcal{C}_{R'}$ implies $f \circ u' \in \mathcal{C}_R$. Define $f_{\mathrm{gpt}}$ and $f^*_{\mathrm{gpt}}$ as before (IV.6.2.9). Then the diagram

$$\begin{array}{ccc} \mathcal{C}_{R,\mathrm{gpt}} & \xrightarrow{D_X} & (X_1/D(\overline{\Sigma}))^{\infty}_{\mathrm{HIGGS}} \\ {\scriptstyle f_{\mathrm{gpt}}}\uparrow & & \uparrow \\ \mathcal{C}_{R',\mathrm{gpt}} & \xrightarrow{D_{X'}} & (X'_1/D(\overline{\Sigma}))^{\infty}_{\mathrm{HIGGS}} \end{array}$$

is commutative, which implies that the diagram

$$\begin{array}{ccc} \mathrm{HC}_{\mathbb{Z}_p}(X_1/D(\overline{\Sigma})) & \xrightarrow{\mathcal{T}^R_{X,\mathrm{gpt},\bullet}} & \mathcal{M}\mathrm{od}_{\mathrm{cocart}}(\mathcal{C}_{R,\mathrm{gpt}}, \mathcal{O}_{\mathcal{C}_{R,\mathrm{gpt}},\bullet}) \\ {\scriptstyle f^*_{1,\mathrm{HIGGS}}}\downarrow & & \downarrow{\scriptstyle f^*_{\mathrm{gpt}}} \\ \mathrm{HC}_{\mathbb{Z}_p}(X'_1/D(\overline{\Sigma})) & \xrightarrow{\mathcal{T}^{R'}_{X',\mathrm{gpt},\bullet}} & \mathcal{M}\mathrm{od}_{\mathrm{cocart}}(\mathcal{C}_{R',\mathrm{gpt}}, \mathcal{O}_{R',\mathrm{gpt},\bullet}) \end{array}$$

is commutative. Hence the claim follows from (IV.6.2.10) and (IV.6.1.17). □

**Corollary IV.6.4.7.** *For two covering sieves $R' \subset R$ of $X$ with respect to the site $X_{\text{étaff}}$, we have a canonical isomorphism of functors:*

$$\mathcal{T}^{R'}_{X,\bullet}|_{\mathrm{HC}_{\mathbb{Z}_p,R\text{-fin}}(X_1/D(\overline{\Sigma}))} \cong \mathcal{T}^R_{X,\bullet} \colon \mathrm{HC}_{\mathbb{Z}_p,R\text{-fin}}(X_1/D(\overline{\Sigma})) \longrightarrow \mathcal{M}\mathrm{od}((\mathcal{P}/X)_{\text{ét-fét}}, \mathcal{O}_{\mathcal{P}/X,\bullet}).$$

By Corollary IV.6.4.7 and Remark IV.6.4.3, we obtain a functor

$$\mathcal{T}_{X,\bullet} \colon \mathrm{HC}_{\mathbb{Z}_p}(X_1/D(\overline{\Sigma})) \longrightarrow \mathcal{M}\mathrm{od}((\mathcal{P}/X)_{\text{ét-fét}}, \mathcal{O}_{\mathcal{P}/X,\bullet}). \tag{IV.6.4.8}$$

**Theorem IV.6.4.9.** *Assume that the log structure of $\Sigma$ is defined by the closed point and $X$ satisfies Condition* IV.5.3.1 *strict étale locally. Then the functor*

$$(\mathcal{T}_{X,\bullet})_{\mathbb{Q}} \colon \mathrm{HC}_{\mathbb{Z}_p}(X_1/D(\overline{\Sigma}))_{\mathbb{Q}} \to \mathcal{M}\mathrm{od}((\mathcal{P}/X)_{\text{ét-fét}}, \mathcal{O}_{\mathcal{P}/X,\bullet})_{\mathbb{Q}}$$

*is fully faithful.*

Proof. First let us prove that the functor restricted to $\mathrm{HC}_{\mathbb{Z}_p,X\text{-fin}}(X_1/D(\overline{\Sigma}))$ is fully faithful when $X$ is an object of $\mathcal{C}_X$ and satisfies the conditions on $U$ in the beginning of IV.5.3. By Propositions IV.6.3.9 and IV.6.2.8, it suffices to prove that the functor

$$(\mathcal{T}^X_{X,\mathrm{gpt},\bullet})(-)(X,\overline{t})_{\mathbb{Q}} \colon \mathrm{HC}_{\mathbb{Z}_p,X\text{-fin}}(X_1/D(\overline{\Sigma}))_{\mathbb{Q}} \to \mathrm{Rep}_{\mathrm{cont}}(\Delta_{\overline{t}}, \mathcal{O}_{\mathcal{C}_X,\mathrm{gpt},\bullet}(X,\overline{t}))_{\mathbb{Q}}$$

is fully faithful for $\overline{t} \in \mathrm{Ob}\,(\mathcal{X}_{\overline{K},\mathrm{triv}})_{\mathrm{gpt}}$. This follows from Lemma IV.5.2.14, Proposition IV.3.5.1, and Theorem IV.5.3.3.

Let $R$ be a covering sieve of $X$ with respect to the site $X_{\text{étaff}}$. Choose a strict étale covering $(X_\alpha \to X)_{\alpha\in A}$ and a strict étale covering $(X_{\alpha\beta;\gamma} \to X_{\alpha\beta})_{\gamma\in\Gamma_{\alpha\beta}}$ of $X_{\alpha\beta} := X_\alpha \times_X X_\beta$ such that $A$ and $\Gamma_{\alpha\beta}$ are finite sets, and $X_\alpha$ and $X_{\alpha\beta;\gamma}$ are objects of $\mathcal{C}_R$ and satisfy the conditions on $X$ in the previous paragraph. Let $\mathcal{F}_i$ $(i = 1,2)$ be objects of $\mathrm{HC}_{\mathbb{Z}_p,R\text{-fin}}(X/D(\overline{\Sigma}))$ and let $\mathcal{F}_{i,\alpha}$ (resp. $\mathcal{F}_{i,\alpha\beta;\gamma}$) be their pull-backs on $X_\alpha$ (resp. $X_{\alpha\beta;\gamma}$), which are $X_\alpha$ (resp. $X_{\alpha\beta;\gamma}$)-finite. We may apply the above special case of the theorem to $\mathcal{F}_{i,\alpha}$ and $\mathcal{F}_{i,\alpha\beta;\gamma}$. Hence, by Corollary IV.3.1.8, Proposition IV.6.1.19, and Lemma IV.6.4.6, we see that the homomorphism $\mathrm{Hom}(\mathcal{F}_1,\mathcal{F}_2)_{\mathbb{Q}} \to \mathrm{Hom}((\mathcal{T}_{X,\bullet})_{\mathbb{Q}}(\mathcal{F}_1), (\mathcal{T}_{X,\bullet})_{\mathbb{Q}}(\mathcal{F}_2))$ is an isomorphism. □

**IV.6.5. Projections to étale sites.** Let $X$ be an fs log scheme over $\Sigma$ whose underlying scheme is separated and of finite type over $V$. In this subsection, we give a way to describe the higher direct images by a natural projection from the Faltings site $(\mathcal{P}/X)_{\text{ét-fét}}$ to the étale site $X_{\text{étaff}}$ in terms of the cohomology on $(\mathcal{U}_{\overline{K},\text{triv}})_{\text{fét}}$ for each $U \in X_{\text{étaff}}$ (cf. Proposition IV.6.5.22).

Let $X_{\text{traff}}$ denote the category $X_{\text{étaff}}$ endowed with the topology defined by the identity covering. Every presheaf is a sheaf on $X_{\text{traff}}$, i.e., $(X_{\text{traff}})^\sim = (X_{\text{étaff}})^\wedge$. The identity functor $X_{\text{traff}} \to X_{\text{étaff}}$ is continuous and the identity functor $X_{\text{étaff}} \to X_{\text{traff}}$ is cocontinuous. By **[2]** III Proposition 2.5, these two functors induce a morphism of topos

(IV.6.5.1)
$$v_X \colon X^\sim_{\text{étaff}} \to X^\sim_{\text{traff}} (= X^\wedge_{\text{étaff}}).$$

The functor $v_{X*}$ is the inclusion and the functor $v_X^*$ is the sheafification.

The functor $X_{\text{étaff}} \to (\mathcal{P}/X)_{\text{ét-fét}}$ (resp. $X_{\text{traff}} \to (\mathcal{P}/X)_{\text{fét}}$); $U \mapsto (U, \mathcal{U}_{\overline{K},\text{triv}})$ is continuous because it preserves finite fiber products and the image of a strict étale covering is a horizontal strict étale covering. The functor $(\mathcal{P}/X)_{\text{ét-fét}} \to X_{\text{étaff}}$ (resp. $(\mathcal{P}/X)_{\text{fét}} \to X_{\text{traff}}$); $(U, \mathcal{V}) \to U$ is a left adjoint of the above functor. Hence it is cocontinuous and these two functors induce a morphism of topos (cf. **[2]** III Proposition 2.5)

(IV.6.5.2)
$$\pi^h_{X,\text{ét}} \colon (\mathcal{P}/X)^\sim_{\text{ét-fét}} \longrightarrow X^\sim_{\text{étaff}} \quad (\text{resp. } \pi^h_X \colon (\mathcal{P}/X)^\sim_{\text{fét}} \to X^\wedge_{\text{étaff}}).$$

Under the interpretation of the topology of $(\mathcal{P}/X)_{\text{ét-fét}}$ in terms of covanishing topology, the above functor coincides with the morphism $\sigma$ in (VI.5.32.7) if $X$ is affine. We see that the diagram of topos

(IV.6.5.3)
$$\begin{array}{ccc} (\mathcal{P}/X)^\sim_{\text{ét-fét}} & \xrightarrow{v_{\mathcal{P}/X}} & (\mathcal{P}/X)^\sim_{\text{fét}} \\ {\scriptstyle \pi^h_{X,\text{ét}}}\downarrow & & \downarrow{\scriptstyle \pi^h_X} \\ X^\sim_{\text{étaff}} & \xrightarrow{v_X} & X^\wedge_{\text{étaff}} \end{array}$$

is commutative up to canonical isomorphism by looking at the corresponding diagram of sites and continuous (or cocontinuous) morphisms.

Similarly, if $X$ is affine, then the functor $\mathcal{X}_{\overline{K},\text{triv}} \to (\mathcal{P}/X)_{\text{ét-fét}}$ (resp. $(\mathcal{P}/X)_{\text{fét}}$) defined by $\mathcal{V} \mapsto (\mathcal{X}_{\overline{K},\text{triv}}, \mathcal{V})$ is continuous and defines a morphism of sites because it preserves finite inverse limits and the image of a finite étale covering is a vertical finite étale covering. Thus we obtain a diagram of topos commutative up to canonical isomorphism

(IV.6.5.4)
$$\begin{array}{ccc} (\mathcal{P}/X)^\sim_{\text{ét-fét}} & \xrightarrow{v_{\mathcal{P}/X}} & (\mathcal{P}/X)^\sim_{\text{fét}} \\ & {\scriptstyle \pi^v_{X,\text{ét}}}\searrow & \downarrow{\scriptstyle \pi^v_X} \\ & & (\mathcal{X}_{\overline{K},\text{triv}})^\sim_{\text{fét}}. \end{array}$$

Under the interpretation in terms of covanishing topology, the morphism $\pi^v_{X,\text{ét}}$ coincides with the morphism $\beta$ in (VI.5.32.2).

**Lemma IV.6.5.5.** *Let $X$ be as above.*

(1) *The topos $(\mathcal{P}/X)_{\text{ét-fét}}$ is coherent and the topos $(\mathcal{P}/X)_{\text{fét}}$ is algebraic. If $X$ is affine, then the topos $(\mathcal{P}/X)_{\text{fét}}$ is coherent.*

(2) *The topos $X^\sim_{\text{étaff}}$ is coherent and the topos $X^\wedge_{\text{étaff}}$ is algebraic. If $X$ is affine, then the topos $X^\wedge_{\text{étaff}}$ and $(\mathcal{X}_{\overline{K},\text{triv}})_{\text{fét}}$ are coherent.*

(3) *The morphisms of topos $\pi^h_{X,\text{ét}}$ and $\pi^h_X$ are coherent.*

(4) *If $X$ is affine, then the morphisms of topos $\pi^v_{X,\text{ét}}$ and $\pi^v_X$ are coherent.*

PROOF. For a site $C$, let $\varepsilon$ denote the canonical functor $C^\wedge \to C^\sim$.

(1) By Proposition IV.6.1.3 (resp. Since the topology of $(\mathcal{P}/X)_{\text{fét}}$ is defined by $\mathrm{Cov}^v_f(-)$), every object of the site $(\mathcal{P}/X)_{\text{ét-fét}}$ (resp. $(\mathcal{P}/X)_{\text{fét}}$) is quasi-compact (cf. **[2]** VI Définition 1.1). Since finite fiber products and finite products are representable in $\mathcal{P}/X$, we see that $\varepsilon(\mathfrak{U})$ and $\varepsilon(\mathfrak{U}) \times \varepsilon(\mathfrak{U}')(= \varepsilon(\mathfrak{U} \times \mathfrak{U}'))$ are coherent for any objects $\mathfrak{U}$ and $\mathfrak{U}'$ of $\mathcal{P}/X$ by **[2]** VI Corollaire 2.1.1. Hence the topos $(\mathcal{P}/X)_{\text{ét-fét}}$ (resp. $(\mathcal{P}/X)_{\text{fét}}$) satisfies the condition **[2]** VI Proposition 2.2 (ii) bis defining algebraic topos. If $X$ is affine, then the final object $\varepsilon((X, \mathcal{X}_{\overline{K},\text{triv}}))$ is coherent, and hence the topos in question is coherent. In the general case, there exists a strict étale covering $(U_\alpha \to X)_{\alpha \in A}$ such that $U_\alpha$ is affine and $A$ is a finite set. Letting $\mathfrak{U}_\alpha = (U_\alpha, \mathcal{U}_{\alpha,\overline{K},\text{triv}}) \in \mathrm{Ob}\,(\mathcal{P}/X)$, we see that the finite set $(\varepsilon(\mathfrak{U}_\alpha))_{\alpha \in A}$ of objects of $(\mathcal{P}/X)^\sim_{\text{ét-fét}}$ is a covering of the final object and the product $\varepsilon(\mathfrak{U}_\alpha) \times \varepsilon(\mathfrak{U}_\beta)$ is coherent as we have seen above. Hence the final object of $(\mathcal{P}/X)^\sim_{\text{ét-fét}}$ is coherent by **[2]** VI Corollaire 1.17.

(2) Every object of the site $X_{\text{étaff}}$ (resp. $(\mathcal{X}_{\overline{K},\text{triv}})_{\text{fét}}$, resp. $X_{\text{traff}}$) is quasi-compact. Since finite fiber products and finite products are representable, the same argument as (1) shows that the associated topos is algebraic and the coherence follows from the existence of finite family of strict étale covering of $X$ by objects of the site (resp. $\mathcal{X}_{\overline{K},\text{triv}} \in \mathrm{Ob}\,(\mathcal{X}_{\overline{K},\text{triv}})_{\text{fét}}$, resp. $X \in \mathrm{Ob}\, X_{\text{étaff}}$).

(3) (4) Since finite fiber products are representable and every object is quasi-compact in every relevant category, the claim follows from **[2]** VI Corollaire 3.3. □

Let $\mathcal{C}$ be a full subcategory of $X_{\text{étaff}}$ such that every object of $X_{\text{étaff}}$ admits a strict étale covering by objects of $\mathcal{C}$. Let $\mathcal{C}_{\text{ét}}$ (resp. $\mathcal{C}_{\text{tr}}$) denote the category $\mathcal{C}$ endowed with the topology induced by that of $X_{\text{étaff}}$ via the inclusion functor (resp. the trivial topology). By **[2]** III Théorème 4.1, the restriction functor induces an equivalence of categories

(IV.6.5.6) $$X^\sim_{\text{étaff}} \xrightarrow{\sim} \mathcal{C}^\sim_{\text{ét}}.$$

The identity functor $\mathcal{C}_{\text{tr}} \to \mathcal{C}_{\text{ét}}$ is continuous and the identity functor $\mathcal{C}_{\text{ét}} \to \mathcal{C}_{\text{tr}}$ is cocontinuous. By **[2]** III Proposition 2.5, these two functors induce a morphism of topos

(IV.6.5.7) $$v_{\mathcal{C}} \colon \mathcal{C}^\sim_{\text{ét}} \longrightarrow \mathcal{C}^\wedge.$$

The functor $\mathcal{C}_{\text{ét}} \to (\mathcal{P}/\mathcal{C})_{\text{ét-fét}}$ (resp. $\mathcal{C}_{\text{tr}} \to (\mathcal{P}/\mathcal{C})_{\text{fét}}$); $U \mapsto (U, \mathcal{U}_{\overline{K},\text{triv}})$ is continuous by (IV.6.1.8) (resp. obviously) and the functor $(\mathcal{P}/\mathcal{C})_{\text{ét-fét}} \to \mathcal{C}_{\text{ét}}$ (resp. $(\mathcal{P}/\mathcal{C})_{\text{fét}} \to \mathcal{C}_{\text{tr}}$); $(U, \mathcal{V}) \mapsto U$ is its left adjoint. Hence, by **[2]** III Proposition 2.5, the latter functor is cocontinuous and this adjoint pair of functors induces a morphism of topos

(IV.6.5.8) $$\pi^h_{\mathcal{C},\text{ét}} \colon (\mathcal{P}/\mathcal{C})^\sim_{\text{ét-fét}} \longrightarrow \mathcal{C}^\sim_{\text{ét}} \quad (\text{resp. } \pi^h_{\mathcal{C}} \colon (\mathcal{P}/\mathcal{C})^\sim_{\text{fét}} \longrightarrow \mathcal{C}^\wedge).$$

We also have a canonical isomorphism of morphisms of topos

(IV.6.5.9) $$\pi^h_{\mathcal{C}} \circ v_{\mathcal{P}/\mathcal{C}} \cong v_{\mathcal{C}} \circ \pi^h_{\mathcal{C},\text{ét}},$$

which follows from the corresponding isomorphism for continuous (or cocontinuous) morphisms of sites.

Since $X_1 \times_{\mathrm{Spf}(\overline{V})} \mathrm{Spec}(k) \cong X \times_{\mathrm{Spec}(V)} \mathrm{Spec}(k)$, we may regard sheaves on $(X_1)_{\text{ét}}$ as sheaves on $X_{\text{ét}}$ with supports in $X \times_{\mathrm{Spec}(V)} \mathrm{Spec}(k)$. Put $\mathcal{O}_{X_1,m} = \mathcal{O}_{X_1}/p^m\mathcal{O}_{X_1}$. We define $\mathcal{O}_{\mathcal{C}_1,m}$ to be the restriction of $\mathcal{O}_{X_1,m}$ on $\mathcal{C}_{\text{ét}}$. Let $\mathcal{O}_{\mathcal{C}_1,m}$ also denote $v_{\mathcal{C}*}\mathcal{O}_{\mathcal{C}_1,m}$. Then, for $U \in \mathrm{Ob}\,\mathcal{C}$, we have a canonical homomorphism of $\overline{V}$-algebras

$$\Gamma(U, \mathcal{O}_X) \otimes_V \overline{V} = A_U \otimes_V \overline{V} \longrightarrow \Gamma((U, \mathcal{U}_{\overline{K},\text{triv}}), \mathcal{O}^v_{\mathcal{P}/\mathcal{C}}) = \mathcal{A}_U \otimes_V \overline{V}$$

functorial in $U$ (cf. Lemma IV.5.2.1 for the last equality). Since $\Gamma(U, \mathcal{O}_{\mathcal{C}_1,m}) = (A_U \otimes_V \overline{V})/p^m$, this induces

$$\mathcal{O}_{\mathcal{C}_1,m} \longrightarrow \pi^h_{\mathcal{C}*}(\mathcal{O}^v_{\mathcal{P}/\mathcal{C},m})$$

and then

$$\mathcal{O}_{\mathcal{C}_1,m} = v_{\mathcal{C}}^*\mathcal{O}_{\mathcal{C}_1,m} \longrightarrow v_{\mathcal{C}}^*\pi^h_{\mathcal{C}*}\mathcal{O}^v_{\mathcal{P}/\mathcal{C},m} \longrightarrow \pi^h_{\mathcal{C},\text{ét}*}v^*_{\mathcal{P}/\mathcal{C}}(\mathcal{O}^v_{\mathcal{P}/\mathcal{C},m}) \cong \pi^h_{\mathcal{C},\text{ét}*}(\mathcal{O}_{\mathcal{P}/\mathcal{C},m}).$$

Thus we obtain a commutative diagram of ringed topos

(IV.6.5.10)
$$\begin{array}{ccc}
((\mathcal{P}/\mathcal{C})^{\sim}_{\text{ét-fét}}, \mathcal{O}_{\mathcal{P}/\mathcal{C},m}) & \xrightarrow{v_{\mathcal{P}/\mathcal{C}}} & ((\mathcal{P}/\mathcal{C})^{\sim}_{\text{fét}}, \mathcal{O}^v_{\mathcal{P}/\mathcal{C},m}) \\
\downarrow{\scriptstyle \pi^h_{\mathcal{C},\text{ét}}} & & \downarrow{\scriptstyle \pi^h_{\mathcal{C}}} \\
(\mathcal{C}^{\sim}_{\text{ét}}, \mathcal{O}_{\mathcal{C}_1,m}) & \xrightarrow{v_{\mathcal{C}}} & (\mathcal{C}^{\wedge}, \mathcal{O}_{\mathcal{C}_1,m}).
\end{array}$$

The functors $\pi^h_{\mathcal{C}*}$ and $\pi^{h*}_{\mathcal{C}}$ are described in terms of $\mathcal{C}^{\wedge,\text{cont}}_{\text{gpt}}$ via $r_{\mathcal{C}}$ and $s_{\mathcal{C}}$ (cf. Proposition IV.6.2.6) as follows. We define a functor $\pi^h_{\mathcal{C},\text{gpt}*}\colon \mathcal{C}^{\wedge,\text{cont}}_{\text{gpt}} \longrightarrow \mathcal{C}^{\wedge}$ by

$$(\pi^h_{\mathcal{C},\text{gpt}*}\mathcal{G})(U) = \Gamma((\mathcal{C}_{/U})_{\text{gpt}}, \mathcal{G}|_U),$$

where $\mathcal{G}|_U$ denotes the composition of $\mathcal{G}$ with the forgetful functor $(\mathcal{C}_{/U})_{\text{gpt}} \to \mathcal{C}_{\text{gpt}}$. For a morphism $f\colon U' \to U$, $\mathcal{G}|_{U'}$ coincides with the composition of $\mathcal{G}|_U$ with the functor $f_{\text{gpt}}\colon (\mathcal{C}_{/U'})_{\text{gpt}} \to (\mathcal{C}_{/U})_{\text{gpt}}$ induced by $f$. Therefore we obtain a map $\Gamma((\mathcal{C}_{/U})_{\text{gpt}}, \mathcal{G}|_U) \to \Gamma((\mathcal{C}_{/U'})_{\text{gpt}}, \mathcal{G}|_{U'})$ by composition with $f_{\text{gpt}}$. By the construction of $r_{\mathcal{C}}$, we have a canonical isomorphism

(IV.6.5.11)
$$\pi^h_{\mathcal{C}*} \circ r_{\mathcal{C}} \cong \pi^h_{\mathcal{C},\text{gpt}*}.$$

We define a functor $\pi^{h*}_{\mathcal{C},\text{gpt}}\colon \mathcal{C}^{\wedge} \longrightarrow \mathcal{C}^{\wedge,\text{cont}}_{\text{gpt}}$ by

$$(\pi^{h*}_{\mathcal{C},\text{gpt}}\mathcal{F})(U,\bar{s}) = \mathcal{F}(U), \quad (\pi^{h*}_{\mathcal{C},\text{gpt}}\mathcal{F})(f,h) = \mathcal{F}(f).$$

Then we have a canonical isomorphism

(IV.6.5.12)
$$s_{\mathcal{C}} \circ \pi^{h*}_{\mathcal{C}} \cong \pi^{h*}_{\mathcal{C},\text{gpt}}.$$

Let $f\colon X' \to X$ be a strict étale morphism of fs log schemes over $\Sigma$ such that $X'$ satisfies the same condition as $X$. Then the functor $X'_{\text{étaff}} \to X_{\text{étaff}}$ (resp. $X'_{\text{traff}} \to X_{\text{traff}}$); $(u'\colon U' \to X) \mapsto (f \circ u'\colon U' \to X)$ is continuous and cocontinuous and induces a morphism of topos $f_{\text{étaff}}\colon X'^{\sim}_{\text{étaff}} \to X^{\sim}_{\text{étaff}}$ (resp. $f_{\text{traff}}\colon X'^{\wedge}_{\text{étaff}} \to X^{\wedge}_{\text{étaff}}$). We see that the following diagrams of topos are commutative up to canonical isomorphisms by looking at the corresponding diagrams of sites and cocontinuous morphisms.

(IV.6.5.13)
$$\begin{array}{ccc}
(\mathcal{P}/X')^{\sim}_{\text{ét-fét}} & \xrightarrow{f_{\mathcal{P},\text{ét-fét}}} & (\mathcal{P}/X)^{\sim}_{\text{ét-fét}} \\
\downarrow{\scriptstyle \pi^h_{X',\text{ét}}} & & \downarrow{\scriptstyle \pi^h_{X,\text{ét}}} \\
(X'_{\text{étaff}})^{\sim} & \xrightarrow{f_{\text{étaff}}} & X^{\sim}_{\text{étaff}},
\end{array}
\qquad
\begin{array}{ccc}
(\mathcal{P}/X')^{\sim}_{\text{fét}} & \xrightarrow{f_{\mathcal{P},\text{fét}}} & (\mathcal{P}/X)^{\sim}_{\text{fét}} \\
\downarrow{\scriptstyle \pi^h_{X'}} & & \downarrow{\scriptstyle \pi^h_{X}} \\
(X'_{\text{étaff}})^{\wedge} & \xrightarrow{f_{\text{traff}}} & X^{\wedge}_{\text{étaff}}.
\end{array}$$

We also have a canonical isomorphism of morphisms of topos

(IV.6.5.14)
$$v_X \circ f_{\text{étaff}} \cong f_{\text{traff}} \circ v_{X'}.$$

If $X'$ is affine, then the functor $X_{\text{étaff}} \to X'_{\text{étaff}}$ (resp. $X_{\text{traff}} \to X'_{\text{traff}}$); $U \mapsto U \times_X X'$ is continuous and a right adjoint of the cocontinuous functor $u' \mapsto f \circ u'$ above. Hence it defines a morphism of sites and induces the morphism of topos $f_{\text{étaff}}$ (resp. $f_{\text{traff}}$) (cf. **[2]** III Proposition 2.5). We can also verify the commutativity of the diagram (IV.6.5.13) by looking at the corresponding diagrams of sites and continuous morphisms (cf. Proposition IV.6.1.7).

Similarly, if $X$ and $X'$ are affine, then we have the following diagrams of topos commutative up to canonical isomorphisms, where $f$ denotes the morphism $\mathcal{X}'_{\overline{K},\mathrm{triv}} \to \mathcal{X}_{\overline{K},\mathrm{triv}}$ induced by $f$.

(IV.6.5.15)

$$\begin{array}{ccc} (\mathcal{P}/X')^{\sim}_{\text{ét-fét}} & \xrightarrow{f_{\mathcal{P},\text{ét-fét}}} & (\mathcal{P}/X)^{\sim}_{\text{ét-fét}} \\ \downarrow{\scriptstyle \pi^v_{X',\text{ét}}} & & \downarrow{\scriptstyle \pi^v_{X,\text{ét}}} \\ (\mathcal{X}'_{\overline{K},\mathrm{triv}})^{\sim}_{\text{fét}} & \xrightarrow{f_{\text{fét}}} & (\mathcal{X}_{\overline{K},\mathrm{triv}})^{\sim}_{\text{fét}}, \end{array} \qquad \begin{array}{ccc} (\mathcal{P}/X')^{\sim}_{\text{fét}} & \xrightarrow{f_{\mathcal{P},\text{fét}}} & (\mathcal{P}/X)^{\sim}_{\text{fét}} \\ \downarrow{\scriptstyle \pi^v_{X'}} & & \downarrow{\scriptstyle \pi^v_{X}} \\ (\mathcal{X}'_{\overline{K},\mathrm{triv}})^{\sim}_{\text{fét}} & \xrightarrow{f_{\text{fét}}} & (\mathcal{X}_{\overline{K},\mathrm{triv}})^{\sim}_{\text{fét}}. \end{array}$$

**Lemma IV.6.5.16.** *Let $f\colon X' \to X$ be as above and assume that $X'$ is affine. Then the morphisms of topos $f_{\mathcal{P},\text{ét-fét}}\colon (\mathcal{P}/X')^{\sim}_{\text{ét-fét}} \to (\mathcal{P}/X)^{\sim}_{\text{ét-fét}}$ and $f_{\mathcal{P},\text{fét}}\colon (\mathcal{P}/X')^{\sim}_{\text{fét}} \to (\mathcal{P}/X)^{\sim}_{\text{fét}}$ are coherent.*

PROOF. The same as the proof of Proposition IV.6.5.5 (3) using Proposition IV.6.1.7. □

Let $\mathcal{C}$ and $\mathcal{C}'$ be full subcategories of $X_{\text{étaff}}$ and $X'_{\text{étaff}}$ such that every object of $X_{\text{étaff}}$ (resp. $X'_{\text{étaff}}$) admits a strict étale covering by objects of $\mathcal{C}$ (resp. $\mathcal{C}'$) and $u'\colon U' \to X' \in \mathrm{Ob}\,\mathcal{C}'$ implies $f \circ u' \in \mathrm{Ob}\,\mathcal{C}$. Then we see that the functor $\mathcal{C}'_{\text{ét}} \to \mathcal{C}_{\text{ét}}; u' \mapsto f \circ u'$ is continuous and cocontinuous by the same argument as the proof of Lemma IV.6.1.10 and the functor $\mathcal{C}'_{\mathrm{tr}} \to \mathcal{C}_{\mathrm{tr}}; u' \mapsto f \circ u'$ is obviously continuous and cocontinuous. Hence similarly as $f_{\text{étaff}}$, $f_{\text{traff}}$, (IV.6.5.13), and (IV.6.5.14), we obtain morphisms of topos $f^{\mathcal{C},\mathcal{C}'}_{\text{ét}}\colon \mathcal{C}'^{\sim}_{\text{ét}} \to \mathcal{C}^{\sim}_{\text{ét}}$ and $f^{\mathcal{C},\mathcal{C}'}_{\mathrm{tr}}\colon \mathcal{C}'^{\wedge} \to \mathcal{C}^{\wedge}$ and canonical isomorphisms of morphisms of topos

$$\pi^h_{\mathcal{C},\text{ét}} \circ f^{\mathcal{C},\mathcal{C}'}_{\mathcal{P},\text{ét-fét}} \cong f^{\mathcal{C},\mathcal{C}'}_{\text{ét}} \circ \pi^h_{\mathcal{C}',\text{ét}}, \quad \pi^h_{\mathcal{C}} \circ f^{\mathcal{C},\mathcal{C}'}_{\mathcal{P},\text{fét}} \cong f^{\mathcal{C},\mathcal{C}'}_{\mathrm{tr}} \circ \pi^h_{\mathcal{C}'} \tag{IV.6.5.17}$$

$$v_{\mathcal{C}} \circ f^{\mathcal{C},\mathcal{C}'}_{\text{ét}} \cong f^{\mathcal{C},\mathcal{C}'}_{\mathrm{tr}} \circ v_{\mathcal{C}'}. \tag{IV.6.5.18}$$

We have $f^{\mathcal{C},\mathcal{C}'*}_{\bullet}\mathcal{O}_{\mathcal{C}_1,m} = \mathcal{O}_{\mathcal{C}'_1,m}$ ($\bullet = \text{ét}, \mathrm{tr}$) and the above isomorphisms (IV.6.5.17) and (IV.6.5.18) are naturally extended to isomorphisms of morphisms of ringed topos (cf. (IV.6.1.17) and (IV.6.5.10)).

There are also canonical isomorphism of functors

$$f^{\mathcal{C},\mathcal{C}'*}_{\text{ét}} \circ \pi^h_{\mathcal{C},\text{ét}*} \xrightarrow{\cong} \pi^h_{\mathcal{C}',\text{ét}*} \circ f^{\mathcal{C},\mathcal{C}'*}_{\mathcal{P},\text{ét-fét}}, \quad f^{\mathcal{C},\mathcal{C}'*}_{\mathrm{tr}} \circ \pi^h_{\mathcal{C}*} \xrightarrow{\cong} \pi^h_{\mathcal{C}'*} \circ f^{\mathcal{C},\mathcal{C}'*}_{\mathcal{P},\text{fét}} \tag{IV.6.5.19}$$
$$f^{\mathcal{C},\mathcal{C}'*}_{\mathrm{tr}} \circ \pi^h_{\mathcal{C},\mathrm{gpt}*} \xrightarrow{\cong} \pi^h_{\mathcal{C}',\mathrm{gpt}*} \circ f^{\mathcal{C},\mathcal{C}'*}_{\mathrm{gpt}}.$$

The first and the second isomorphisms are derived from the corresponding isomorphisms of continuous morphisms of sites. The third isomorphism follows from Lemma IV.6.2.5 (1) for $\mathcal{V} = \mathcal{U}_{\overline{K},\mathrm{triv}}$.

**Lemma IV.6.5.20.** *Assume that $X$ is affine. The functor $(\mathcal{X}_{\overline{K},\mathrm{triv}})_{\text{fét}} \to (\mathcal{P}/X)_{\text{fét}}; \mathcal{V} \mapsto (X, \mathcal{V})$ is cocontinuous. In particular, the functor $\pi^v_{X*}\colon (\mathcal{P}/X)^{\sim}_{\text{fét}} \to (\mathcal{X}_{\overline{K},\mathrm{triv}})^{\sim}_{\text{fét}}$ is exact (cf.* **[2]** III *Proposition* 2.3 1)).

PROOF. Since the functor $\mathcal{V} \mapsto (X, \mathcal{V})$ is fully faithful, the claim immediately follows from the definition of the topologies of the source and target. □

**Corollary IV.6.5.21.** *Let $\mathcal{F}$ be a sheaf of abelian groups on $(\mathcal{P}/X)_{\text{fét}}$. Then the natural homomorphism $H^q((\mathcal{X}_{\overline{K},\mathrm{triv}})_{\text{fét}}, \pi^v_{X*}\mathcal{F})) \to H^q((\mathcal{P}/X)_{\text{fét}}, \mathcal{F})$ is an isomorphism for every $q \in \mathbb{N}$.*

In the rest of this subsection, we prove the following proposition.

**Proposition IV.6.5.22** (cf. Théorème VI.10.40). *Let $X$ be as above. For an object $\mathcal{F}$ of $D^+((\mathcal{P}/X)_{\text{fét}},\mathbb{Z})$, the following base change morphism with respect to the diagram* (IV.6.5.3) *is an isomorphism.*

$$v_X^* R\pi^h_{X*}\mathcal{F} \longrightarrow R\pi^h_{X,\text{ét}*}v^*_{\mathcal{P}/X}\mathcal{F}.$$

We prove Proposition IV.6.5.22 by looking at the stalks at every geometric point of $X$. Let $\nu\colon \overline{x}\to X$ be a geometric point and choose a filtered inverse system $(X_\lambda,\nu_\lambda)_{\lambda\in\Lambda}$ of affine strict étale fs log schemes over $X$ with liftings $\nu_\lambda\colon\overline{x}\to X_\lambda$ of $\nu$ such that $\varinjlim_\lambda \Gamma(X_\lambda,\mathcal{O}_\lambda)=\mathcal{O}_{X,\overline{x}}$. Let $I$ be the category defined by $\operatorname{Ob} I=\Lambda$ and $\sharp\operatorname{Hom}_I(\lambda,\mu)=1$ if $\lambda\geq\mu$ and 0 otherwise. We define the category $I_e$ by adding a final object $e$ to $I$, i.e., $I_e=I\sqcup\{e\}$ and $e$ is the final object of $I_e$. We set $X_e=X$. Then the inverse system defines a functor $\underline{X}_e\colon I_e\to X_{\text{étaff}}$. We define the category $\mathcal{P}/\underline{X}_e$ as follows. An object is a pair $(i,(\mathfrak{U},u))\in\operatorname{Ob} I_e\times\operatorname{Ob}(\mathcal{P}/X_i)$ and a morphism $(i',(\mathfrak{U}',u'))\to(i,(\mathfrak{U},u))$ is a pair of a morphism $m\colon i'\to i$ (which is unique if it exists) and a morphism $\mathfrak{U}'\to\mathfrak{U}$ in $\mathcal{P}$ compatible with the morphism $\underline{X}(m)\colon X_{i'}\to X_i$. Then we have a natural functor $(\mathcal{P}/\underline{X}_e)\to I_e$, whose fiber over $i\in\operatorname{Ob} I_e$ is naturally identified with $\mathcal{P}/X_i$, and $\mathcal{P}/\underline{X}_e$ becomes a fibered category over $I_e$. For a morphism $m\colon i'\to i$ in $I_e$, the inverse image functor is given by the functor $\underline{X}(m)^*\colon(\mathcal{P}/X_i)\to(\mathcal{P}/X_{i'})$ (cf. Proposition IV.6.1.7). Hence giving the topology defined by $\operatorname{Cov}(-)$ (resp. $\operatorname{Cov}^v(-)$), we obtain a fibered site $(\mathcal{P}/\underline{X}_e)_{\text{ét-fét}}$ (resp. $(\mathcal{P}/\underline{X}_e)_{\text{fét}}$) over $I_e$. We define the fibered site $(\mathcal{P}/\underline{X})_{\text{ét-fét}}$ (resp. $(\mathcal{P}/\underline{X})_{\text{fét}}$) over $I$ using $\underline{X}:=\underline{X}_e|_I\colon I\to X_{\text{étaff}}$, which coincides with the base change of $(\mathcal{P}/\underline{X}_e)_{\text{ét-fét}}$ (resp. $(\mathcal{P}/\underline{X}_e)_{\text{fét}}$) by the inclusion functor $I\to I_e$. We can define similarly the fibered sites $(\underline{X}_e)_{\text{étaff}}$ and $(\underline{X}_e)_{\text{traff}}$ over $I_e$ and the fibered site $\underline{\widetilde{X}}_{\text{fét}}$ over $I$ whose fibers are $(X_i)_{\text{étaff}}$, $(X_i)_{\text{traff}}$, and $(\widetilde{X}_i)_{\text{fét}}$, where $\widetilde{X}_i$ denotes the scheme $\mathcal{X}_{i,\overline{K},\text{triv}}$ associated to $X_i$. From (IV.6.5.3), (IV.6.5.4), (IV.6.5.13), (IV.6.1.14), and (IV.6.5.15), we obtain the following diagrams of fibered sites over $I_e$ and $I$ which are commutative up to canonical isomorphisms. Every functor is Cartesian.

(IV.6.5.23)
$$\begin{array}{ccc}
(\mathcal{P}/\underline{X}_e)_{\text{ét-fét}} & \xrightarrow{v_{\mathcal{P}/\underline{X}_e}} & (\mathcal{P}/\underline{X}_e)_{\text{fét}}\\
\downarrow{\scriptstyle \pi^h_{\underline{X}_e,\text{ét}}} & & \downarrow{\scriptstyle \pi^h_{\underline{X}_e}}\\
(\underline{X}_e)_{\text{étaff}} & \xrightarrow{v_{\underline{X}_e}} & (\underline{X}_e)_{\text{traff}},
\end{array}
\qquad
\begin{array}{ccc}
(\mathcal{P}/\underline{X})_{\text{ét-fét}} & \xrightarrow{v_{\mathcal{P}/\underline{X}}} & (\mathcal{P}/\underline{X})_{\text{fét}}\\
 & {\scriptstyle \pi^v_{\underline{X},\text{ét}}}\searrow & \downarrow{\scriptstyle \pi^v_{\underline{X}}}\\
 & & \underline{\widetilde{X}}_{\text{fét}}.
\end{array}$$

Let $X_\infty$ be $\operatorname{Spec}(\mathcal{O}_{X,\overline{x}})$ endowed with the inverse image of $M_X$ and let $\widetilde{X}_\infty$ be the inverse limit of $\mathcal{X}_{\lambda,\overline{K},\text{triv}}$. Then, by **[2]** VI Propositions 6.4, 6.5, 8.3.3 and **[42]** Théorème (8.8.2), Théorème (8.10.5), Proposition (17.7.8), we see that the inverse limits $\varprojlim_{I_e}(\underline{X}_e)_{\text{étaff}}$, $\varprojlim_{I_e}(\underline{X}_e)_{\text{traff}}$, and $\varprojlim_I\underline{\widetilde{X}}_{\text{fét}}$ are canonically equivalent to the sites $X_{\infty,\text{étaff}}$, $X_{\infty,\text{traff}}$, and $\widetilde{X}_{\infty,\text{fét}}$. Hence by taking the inverse limit of (IV.6.5.23) (cf. **[2]** VI Définition 8.2.5), we obtain the diagram of sites commutative up to canonical isomorphisms (cf. **[2]** VI Lemme 8.2.2, 8.2.7).

(IV.6.5.24)
$$\begin{array}{ccc}
\varprojlim_{I_e}(\mathcal{P}/\underline{X}_e)_{\text{ét-fét}} & \xrightarrow{\underleftarrow{v}_{\mathcal{P}/\underline{X}_e}} & \varprojlim_{I_e}(\mathcal{P}/\underline{X}_e)_{\text{fét}}\\
\downarrow{\scriptstyle \underleftarrow{\pi}^h_{\underline{X}_e,\text{ét}}} & & \downarrow{\scriptstyle \underleftarrow{\pi}^h_{\underline{X}_e}}\\
X_{\infty,\text{étaff}} & \xrightarrow{v_{X_\infty}} & X_{\infty,\text{traff}},
\end{array}
\qquad
\begin{array}{ccc}
\varprojlim_{I}(\mathcal{P}/\underline{X})_{\text{ét-fét}} & \xrightarrow{\underleftarrow{v}_{\mathcal{P}/\underline{X}}} & \varprojlim_{I}(\mathcal{P}/\underline{X})_{\text{fét}}\\
 & {\scriptstyle \underleftarrow{\pi}^v_{\underline{X},\text{ét}}}\searrow & \downarrow{\scriptstyle \underleftarrow{\pi}^v_{\underline{X}}}\\
 & & \widetilde{X}_{\infty,\text{fét}}.
\end{array}$$

For the fibered site $\pi: C \to J$ and an object $x$ of $C$, we write $[x]$ for the image of $x$ in $\varinjlim_J C$. For a morphism $m: j \to \pi(x)$, the canonical morphism $m^*(x) \to x$ induces an isomorphism $[m^*(x)] \xrightarrow{\cong} [x]$.

**Lemma IV.6.5.25.** (1) *The functor* $\Gamma(X_\infty, -) : (X_\infty)^\sim_{\text{étaff}} \to \mathscr{V}$-Sets *is exact.*

(2) *The canonical morphism of functors* $\Gamma(X_\infty, -) \circ v^*_{X_\infty} \to \Gamma(X_\infty, -)$ *is an isomorphism.*

PROOF. (1) follows from $\Gamma(X_\infty, \mathcal{F}) \cong \mathcal{F}_{\overline{x}}$. (2) follows from the fact that $v^*_{X_\infty}$ is the sheafification with respect to the strict étale topology and that every strict étale covering of $X_\infty$ is refined by the trivial covering. □

**Lemma IV.6.5.26.** *The functors* $\widetilde{X}_{\infty,\text{fét}} \to \varinjlim_I (\mathcal{P}/\underline{X})_{\text{ét-fét}}$ *and* $\widetilde{X}_{\infty,\text{fét}} \to \varinjlim_I (\mathcal{P}/\underline{X})_{\text{fét}}$ *defining the morphisms of sites* $\underleftarrow{\pi}^{v}_{\underline{X},\text{ét}}$ *and* $\underleftarrow{\pi}^{v}_{\underline{X}}$ *in* (IV.6.5.24) *are cocontinuous. In particular, the functors* $\underleftarrow{\pi}^{v}_{\underline{X},\text{ét}*}: \varinjlim_I (\mathcal{P}/\underline{X})^\sim_{\text{ét-fét}} \to \widetilde{X}^\sim_{\infty,\text{fét}}$ *and* $\underleftarrow{\pi}^{v}_{\underline{X},*}: \varinjlim_I (\mathcal{P}/\underline{X})^\sim_{\text{fét}} \to \widetilde{X}^\sim_{\infty,\text{fét}}$ *are exact (cf.* **[2]** III *Proposition* 2.3 1)).

PROOF. We prove the cocontinuity of the first functor and then explain how one can simplify the argument to prove the claim for the second functor. Since the functor $(\widetilde{X}_i)_{\text{fét}} \to (\mathcal{P}/X_i)_{\text{ét-fét}}; \mathcal{V} \mapsto (X_i, \mathcal{V})$ is fully faithful, we see that the functor $\varinjlim_I \underline{\widetilde{X}}_{\text{fét}} \to \varinjlim_I (\mathcal{P}/\underline{X})_{\text{ét-fét}}$ is fully faithful by the explicit construction of the direct limit of a fibered site in **[2]** VI Propositions 6.4, 6.5. Let $i \in \operatorname{Ob} I$ and $\mathcal{V}_i \in \operatorname{Ob}(\widetilde{X}_{i,\text{fét}})$. For $j \in \operatorname{Ob} I$ such that $j \geq i$, put $\mathcal{V}_j = \mathcal{V}_i \times_{\widetilde{X}_i} \widetilde{X}_j$ and $\mathfrak{U}_j = (X_j, \mathcal{V}_j) \in \operatorname{Ob}(\mathcal{P}/X_j)$. Let $R$ be a covering sieve of $[i, \mathfrak{U}_i]$ on $(\varinjlim_I \mathcal{P}/\underline{X})_{\text{ét-fét}}$. Since finite fiber products are representable in $\mathcal{P}/X_i$ and the inverse image functors between the fibers of $\mathcal{P}/\underline{X}$ preserve finite fiber products, we may apply **[2]** VI Proposition 8.3.3 to $(\mathcal{P}/\underline{X})_{\text{ét-fét}}$. Hence, by Proposition IV.6.1.3, we see that there exist $j \geq i$, $(\mathfrak{U}_\alpha \to \mathfrak{U}_j)_{\alpha \in A} \in \operatorname{Cov}^h_f(\mathfrak{U}_j)$, and $(\mathfrak{U}_{\alpha\beta} \to \mathfrak{U}_\alpha)_{\beta \in B_\alpha} \in \operatorname{Cov}^v_f(\mathfrak{U}_\alpha)$ such that the composition $[j, \mathfrak{U}_{\alpha\beta}] \to [j, \mathfrak{U}_j] \xrightarrow{\sim} [i, \mathfrak{U}_i]$ is a section of $R$. Put $\mathfrak{U}_\alpha = (U_\alpha, \mathcal{V}_\alpha)$. Since $(U_\alpha \times_{X_i} X_\infty \to X_\infty)_\alpha$ is a strict étale covering and $\Gamma(X_\infty, \mathcal{O}_{X_\infty})$ is strictly henselian local, there exist $\alpha$ and a section $X_\infty \to U_\alpha \times_{X_j} X_\infty$, which comes from a section $X_{j'} \to U_\alpha \times_{X_j} X_{j'}$ for some $j' \geq j$, $j' \in \operatorname{Ob} I$. This implies that the inverse image of $(\mathfrak{U}_{\alpha\beta} \to \mathfrak{U}_j)_{\alpha\beta}$ by the unique morphism $j' \to j$ is refined by a vertical étale covering $((X_{j'}, \mathcal{V}_\gamma) \to \mathfrak{U}_{j'})_\gamma$. Thus we see that $R$ is refined by a sieve generated by $[j', (X_{j'}, \mathcal{V}_\gamma)] \to [j', \mathfrak{U}_{j'}] \xrightarrow{\cong} [i, \mathfrak{U}_i]$. Let $\mathcal{V}'_\infty$ be an object of $\varinjlim_I \underline{\widetilde{X}}_{\text{fét}}$ and let $\mathfrak{U}'_\infty$ be its image in $\varinjlim_I (\mathcal{P}/\underline{X})_{\text{fét}}$. Then, for a morphism $\mathcal{V}'_\infty \to [i, \mathcal{V}_i]$ and its image $\mathfrak{U}'_\infty \to [i, \mathfrak{U}_i]$, the latter factors through $[j', (X_{j'}, \mathcal{V}_\gamma)]$ if and only if the former factors through $[j', \mathcal{V}_\gamma]$ by the fully faithfulness mentioned above. Furthermore $([j', \mathcal{V}_\gamma] \to [j', \mathcal{V}_{j'}] \xrightarrow{\cong} [i, \mathcal{V}_i])_\gamma$ is a covering by **[2]** VI Proposition 8.3.3. Thus we see that the functor $\varinjlim_I \underline{\widetilde{X}}_{\text{fét}} \to \varinjlim_I (\mathcal{P}/\underline{X})_{\text{ét-fét}}$ satisfies the defining property of cocontinuous functor.

In the case of the second functor, we can take a vertical finite étale covering $(\mathfrak{U}_\alpha \to \mathfrak{U}_j)$ for some $j \geq i$ from the beginning, and the refinement is not necessary. □

**Lemma IV.6.5.27.** *For a sheaf of sets* $\mathcal{F}$ *on* $\varinjlim_I (\mathcal{P}/\underline{X})_{\text{fét}}$, *the morphism*

$$\underleftarrow{\pi}^{v}_{\underline{X},\text{ét}*}(\text{adj}): \underleftarrow{\pi}^{v}_{\underline{X},*}\mathcal{F} \longrightarrow \underleftarrow{\pi}^{v}_{\underline{X},*}\underleftarrow{v}_{\mathcal{P}/\underline{X}*}\underleftarrow{v}^*_{\mathcal{P}/X}\mathcal{F} \cong \underleftarrow{\pi}^{v}_{\underline{X},\text{ét}*}\underleftarrow{v}^*_{\mathcal{P}/X}\mathcal{F}$$

*is an isomorphism.*

PROOF. The morphism $\mathcal{F} \to \underleftarrow{\nu}_{\mathcal{P}/\underline{X}*}\underleftarrow{\nu}^*_{\mathcal{P}/X}\mathcal{F}$ is a sheafification of $\mathcal{F}$ for the topology of $\varprojlim_I(\mathcal{P}/\underline{X})_{\text{ét-fét}}$. Since the functor $\varprojlim_I \widetilde{X}_{\text{fét}} \to \varprojlim_I(\mathcal{P}/\underline{X})_{\text{ét-fét}}$ is continuous and cocontinuous by Lemma IV.6.5.26, the morphism $\underleftarrow{\pi}^v_{\underline{X},*}\mathcal{F} \to \underleftarrow{\pi}^v_{\underline{X},*}\underleftarrow{\nu}_{\mathcal{P}/\underline{X}*}\underleftarrow{\nu}^*_{\mathcal{P}/X}\mathcal{F}$ is the sheafification of $\underleftarrow{\pi}^v_{\underline{X},*}\mathcal{F}$ (cf. **[2]** III Proposition 2.3 2)), which is an isomorphism because $\underleftarrow{\pi}^v_{\underline{X},*}\mathcal{F}$ is a sheaf. □

**Corollary IV.6.5.28.** *For $\mathcal{F} \in D^+(\varprojlim_I(\mathcal{P}/\underline{X})_{\text{fét}}, \mathbb{Z})$, the following natural morphism is an isomorphism.*

$$R\Gamma(\varprojlim_I(\mathcal{P}/\underline{X})_{\text{fét}}, \mathcal{F}) \longrightarrow R\Gamma(\varprojlim_I(\mathcal{P}/\underline{X})_{\text{ét-fét}}, \underleftarrow{\nu}^*_{\mathcal{P}/\underline{X}}\mathcal{F}).$$

PROOF. By Lemma IV.6.5.26 and Lemma IV.6.5.27, we obtain isomorphisms

$$R\Gamma(\varprojlim_I(\mathcal{P}/\underline{X})_{\text{ét-fét}}, \underleftarrow{\nu}^*_{\mathcal{P}/\underline{X}}\mathcal{F}) \xleftarrow{\cong} R\Gamma((\widetilde{X}_\infty)_{\text{fét}}, \underleftarrow{\pi}^v_{\underline{X},\text{ét}*}\underleftarrow{\nu}^*_{\mathcal{P}/\underline{X}}\mathcal{F})$$

$$\xleftarrow{\cong} R\Gamma((\widetilde{X}_\infty)_{\text{fét}}, \underleftarrow{\pi}^v_{\underline{X},*}\mathcal{F}) \xrightarrow{\cong} R\Gamma(\varprojlim_I(\mathcal{P}/\underline{X})_{\text{fét}}, \mathcal{F})$$

and it is straightforward to verify that the composition coincides with the natural pull-back morphism. □

For a fibered site $\pi\colon C \to J$ where $J$ belongs to the universe $\mathscr{U}$ and associated to a filtered ordered set, we use the following notation. For $i, j \in \operatorname{Ob} J$ such that $j \geq i$, let $\mu_{ij}$ denote the morphism of topos $C_j^\sim \to C_i^\sim$ induced by the inverse image functor by the unique morphism $j \to i$. For $i \in \operatorname{Ob} J$, let $\mu_i$ denote the morphism of topos $(\varprojlim_J C)^\sim \to C_i^\sim$ induced by the natural functor $C_i \to \varprojlim_J C$ which defines a morphism of sites. We have a canonical isomorphism $\mu_{ij} \circ \mu_j \cong \mu_i$ for $j \geq i$. (See **[2]** VI Théorème 8.2.3.)

By (IV.6.5.23), we see that the functors $\mu_i$ and $\mu_{ij}$ are compatible with the morphisms of topos induced by those of sites in the first (resp. second) diagram in (IV.6.5.24) and the morphisms of topos in (IV.6.5.3) (resp. (IV.6.5.4)) for $X_i$, $i \in I_e$ (resp. $i \in I$).

In particular, we have the following diagram commutative up to canonical isomorphisms.

(IV.6.5.29)

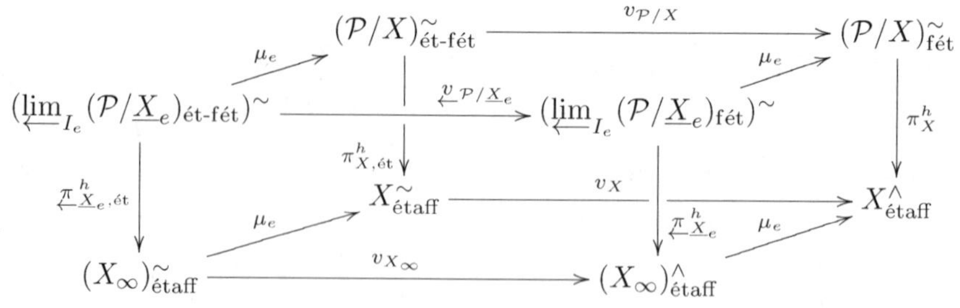


**Lemma IV.6.5.30.** *For $\mathcal{F} \in D^+((\mathcal{P}/X)_{\text{ét-fét}}, \mathbb{Z})$ and $\mathcal{G} \in D^+((\mathcal{P}/X)_{\text{fét}}, \mathbb{Z})$, the following base change morphisms are isomorphisms.*

$$\mu_e^* R\pi^h_{X,\text{ét}*}\mathcal{F} \longrightarrow R\underleftarrow{\pi}^h_{\underline{X}_e,\text{ét}*}\mu_e^*\mathcal{F}, \qquad \mu_e^* R\pi^h_{X*}\mathcal{G} \longrightarrow R\underleftarrow{\pi}^h_{\underline{X}_e*}\mu_e^*\mathcal{G}.$$

PROOF. By Corollary IV.6.1.4, the pull-back morphisms

$$\mu_{ei}^* R\pi^h_{X,\text{ét}*}(\mathcal{F}) \longrightarrow R\pi^h_{X_i,\text{ét}*}\mu_{ei}^*(\mathcal{F}), \qquad \mu_{ei}^* R\pi^h_{X*}(\mathcal{G}) \longrightarrow R\pi^h_{X_i*}\mu_{ei}^*(\mathcal{G})$$

are isomorphisms for every $i \in \operatorname{Ob} I$. By Lemma IV.6.5.5 (1), (2), (3) and Lemma IV.6.5.16, we may apply **[2]** VI Corollaire 8.7.5 to the morphisms of fibered sites $\pi^h_{\underline{X},\text{ét}}$ and $\pi^h_{\underline{X}}$, and obtain isomorphisms in the lemma. □

Proof of Proposition IV.6.5.22. It suffices to prove that the stalk at every geometric point $\overline{x} \to X$ is an isomorphism. Since

$$\mathcal{G}_{\overline{x}} = \Gamma(X_{\infty,\text{étaff}}, \mathcal{G}) = R\Gamma(X_{\infty,\text{étaff}}, \mathcal{G})$$

for a sheaf of abelian groups $\mathcal{G}$ on $X_{\infty,\text{étaff}}$ (cf. Lemma IV.6.5.25), it suffices to prove that the morphism

$$R\Gamma(X_{\infty,\text{étaff}}, \mu_e^* v_X^* R\pi^h_{X*}\mathcal{F}) \longrightarrow R\Gamma(X_{\infty,\text{étaff}}, \mu_e^* R\pi^h_{X,\text{ét}*} v^*_{\mathcal{P}/X}\mathcal{F})$$

is an isomorphism. We write $\mu$, $v$, $\pi$, $\underleftarrow{v}$, and $\underleftarrow{\pi}$ for $\mu_e$, $v_\bullet$, $\pi^h_\bullet$, $\underleftarrow{v}_\bullet$, and $\underleftarrow{\pi}^h_\bullet$ with $\bullet = X$, etc. to simplify the notation. Then, by considering the compositions of the base change morphisms along the vertical faces in the diagram (IV.6.5.29), we obtain the following commutative diagram.

$$\begin{array}{ccccc}
\mu^* v^* R\pi_*\mathcal{F} & \xrightarrow{\varphi} & \mu^* R\pi_* v^*\mathcal{F} & \longrightarrow & R\underleftarrow{\pi}_*\mu^* v^*\mathcal{F} \\
\Vert & & & & \Vert \\
\underleftarrow{v}^*\mu^* R\pi_*\mathcal{F} & \longrightarrow & \underleftarrow{v}^* R\underleftarrow{\pi}_*\mu^*\mathcal{F} & \xrightarrow{\psi} & R\underleftarrow{\pi}_*\underleftarrow{v}^*\mu^*\mathcal{F}
\end{array}$$

The upper right and the lower left morphisms are isomorphisms by Lemma IV.6.5.30. After taking $R\Gamma((X_\infty)_{\text{étaff}}, -)$, the lower right morphisms are put in the following commutative diagram.

$$\begin{array}{ccc}
R\Gamma(X_{\infty,\text{étaff}}, \underleftarrow{v}^* R\underleftarrow{\pi}_*\mu^*\mathcal{F}) & \xrightarrow{R\Gamma(X_{\infty,\text{étaff}},\psi)} & R\Gamma(X_{\infty,\text{étaff}}, R\underleftarrow{\pi}_*\underleftarrow{v}^*\mu^*\mathcal{F}) \\
\uparrow & \nearrow & \\
R\Gamma(X_{\infty,\text{traff}}, R\underleftarrow{\pi}_*\mu^*\mathcal{F}) & &
\end{array}$$

The vertical morphism is an isomorphism by Lemma IV.6.5.25 and the right slanted morphism is an isomorphism by Corollary IV.6.5.28. Hence $R\Gamma(X_{\infty,\text{étaff}}, \psi)$ is an isomorphism, which implies that $R\Gamma(X_{\infty,\text{étaff}}, \varphi)$ is an isomorphism. □

**IV.6.6. Topos of inverse systems of sheaves and direct systems of sheaves.** In this subsection, we review basic facts on the topos of inverse systems of sheaves and the topos of direct systems of sheaves (cf. III.7).

Let $C$ be a site, and let $\Lambda$ be a $\mathscr{U}$-small ordered set. We regard $\Lambda$ as a category as follows. The set of objects is $\Lambda$ and $\text{Hom}_\Lambda(\lambda, \mu)$ is empty if $\lambda > \mu$ and consists of one element if $\lambda \leq \mu$. Let $\Lambda^\circ$ denote the opposite category of $\Lambda$.

The product category $C \times \Lambda$ with the projection $C \times \Lambda \to \Lambda$ is a fibered site (cf. **[2]** VI 7.2.1); every fiber is identified with the site $C$ and the inverse image functor is given by the identity. We write $C^\Lambda$ for the fibered site $C \times \Lambda$ endowed with the total topology (cf. **[2]** VI 7.4.1). Then the presheaf $\mathcal{F}$ on $C^\Lambda$ is a sheaf if and only if its restriction $\mathcal{F}_\lambda$ on the fiber over $\lambda \in \Lambda$ is a sheaf on $C$ for every $\lambda \in \Lambda$ (cf. **[2]** VI Proposition 7.4.4). Hence we have isomorphisms of categories.

$$\text{(IV.6.6.1)} \qquad (C^\Lambda)^\sim \xrightarrow{\cong} \underline{\text{Hom}}_{\Lambda^\circ}(\Lambda^\circ, C^\sim \times \Lambda^\circ),$$

$$\text{(IV.6.6.2)} \qquad (C^\Lambda)^\wedge \xrightarrow{\cong} \underline{\text{Hom}}_{\Lambda^\circ}(\Lambda^\circ, C^\wedge \times \Lambda^\circ)$$

defined by $\mathcal{F} \mapsto (\mathcal{F}_\lambda)$ (cf. **[2]** VI Proposition 7.4.7). The target of the first (resp. the second) functor is the category of inverse systems of sheaves (resp. presheaves) of sets on $C$ indexed by $\Lambda$, which is denoted by $C^{\sim\Lambda^\circ}$ (resp. $C^{\wedge\Lambda^\circ}$) following III.7.1.

Let $u\colon C^{\Lambda} \to C$ be the projection functor. Then the right adjoint $\widehat{u}_*\colon (C^{\Lambda})^{\wedge} \to C^{\wedge}$ of the functor $\widehat{u}^*\colon C^{\wedge} \to (C^{\Lambda})^{\wedge}; \mathcal{G} \mapsto \mathcal{G}\circ u$ is given by

$$\widehat{u}_*\mathcal{F} = \varprojlim_{\lambda\in\Lambda} \mathcal{F}_{\lambda}.$$

Indeed, for $\mathcal{F} \in \mathrm{Ob}\,(C^{\Lambda})^{\wedge}$ and $\mathcal{G} \in \mathrm{Ob}\,C^{\wedge}$, we have $\mathrm{Hom}(\mathcal{G}\circ u, \mathcal{F}) \cong \mathrm{Hom}((\mathcal{G})_{\lambda}, (\mathcal{F}_{\lambda})_{\lambda}) \cong \mathrm{Hom}(\mathcal{G}, \varprojlim_{\Lambda} \mathcal{F}_{\lambda})$. Since the inverse limit preserves sheaves, we see that the functor $u$ is cocontinuous (cf. [2] III Proposition 2.2) and defines a morphism of topos

(IV.6.6.3) $$\underleftarrow{l}\colon (C^{\Lambda})^{\sim} \to C^{\sim}, \quad \underleftarrow{l}_*(\mathcal{F}) = \varprojlim_{\Lambda} \mathcal{F}_{\lambda}, \quad \underleftarrow{l}^*(\mathcal{G}) = \mathcal{G}\circ u.$$

Let $f\colon C^{\sim} \to D^{\sim}$ be a morphism of topos. Then the compositions with $f_*\times \mathrm{id}_{\Lambda^{\circ}}$ and $f^*\times \mathrm{id}_{\Lambda^{\circ}}$ define a morphism of topos $f^{\Lambda^{\circ}}\colon C^{\sim\Lambda^{\circ}} \to D^{\sim\Lambda^{\circ}}$; the left exactness of the inverse image functor follows from the fact that the natural morphism $(\varprojlim_I \mathcal{F})(\lambda) \to \varprojlim_I(\mathcal{F}(\lambda))$ $(\lambda\in\Lambda)$ is an isomorphism for an inverse system $\mathcal{F}\colon I^{\circ} \to C^{\sim\Lambda^{\circ}}$ (or $D^{\sim\Lambda^{\circ}}$) over a $\mathscr{U}$-small category $I$. Combining with the isomorphisms (IV.6.6.1) for $C$ and $D$, we obtain a morphism of topos $(C^{\Lambda})^{\sim} \to (D^{\Lambda})^{\sim}$, which is also denoted by $f^{\Lambda^{\circ}}$ in the following.

By looking at the compositions of inverse image functors, we see that the following diagram of topos is commutative up to canonical isomorphism.

(IV.6.6.4) $$\begin{array}{ccc} (C^{\Lambda})^{\sim} & \xrightarrow{\underleftarrow{l}} & C^{\sim} \\ {\scriptstyle f^{\Lambda^{\circ}}}\downarrow & & \downarrow{\scriptstyle f} \\ (D^{\Lambda})^{\sim} & \xrightarrow{\underleftarrow{l}} & D^{\sim}. \end{array}$$

If we replace $\Lambda$ with $\Lambda^{\circ}$, we obtain a topos of direct systems of sheaves on $C$ indexed by $\Lambda$ as follows. The product $C\times\Lambda^{\circ}$ with the projection $C\times\Lambda^{\circ} \to \Lambda^{\circ}$ is a fibered site, and we write $C^{\Lambda^{\circ}}$ for the fibered site $C\times\Lambda^{\circ}$ with the total topology. Then the presheaf $\mathcal{F}$ of sets on $C^{\Lambda^{\circ}}$ is a sheaf if and only if its restriction on the fiber over $\lambda$ is a sheaf on $C$ for every $\lambda\in\Lambda$, and we have an isomorphism f categories defined by $\mathcal{F} \mapsto (\mathcal{F}_{\lambda})$

(IV.6.6.5) $$(C^{\Lambda^{\circ}})^{\sim} \xrightarrow{\cong} \underline{\mathrm{Hom}}_{\Lambda}(\Lambda, C^{\sim}\times\Lambda),$$

(IV.6.6.6) $$(C^{\Lambda^{\circ}})^{\wedge} \xrightarrow{\cong} \underline{\mathrm{Hom}}_{\Lambda}(\Lambda, C^{\wedge}\times\Lambda).$$

The target of the first (resp. the second) functor is the category of direct systems of sheaves (resp. presheaves) on $C$ indexed by $\Lambda$, which is denoted by $C^{\sim\Lambda}$ (resp. $C^{\wedge\Lambda}$).

Let $v\colon C^{\Lambda^{\circ}} \to C$ be the projection functor, which is continuous by the above characterization of sheaves on $C^{\Lambda^{\circ}}$. The left adjoint $\widehat{v}_!\colon (C^{\Lambda^{\circ}})^{\wedge} \to C^{\wedge}$ of the functor $\widehat{v}^*\colon C^{\wedge} \to (C^{\Lambda^{\circ}})^{\wedge}; \mathcal{G} \mapsto \mathcal{G}\circ v$ is given by

$$\widehat{v}_!\mathcal{F} = \varinjlim_{\lambda\in\Lambda} \mathcal{F}_{\lambda}.$$

Indeed, for $\mathcal{F} \in \mathrm{Ob}\,(C^{\Lambda^{\circ}})^{\wedge}$ and $\mathcal{G} \in \mathrm{Ob}\,C^{\wedge}$, we have bijections $\mathrm{Hom}(\mathcal{F}, \mathcal{G}\circ v) \cong \mathrm{Hom}((\mathcal{F}_{\lambda})_{\lambda}, (\mathcal{G})_{\lambda}) \cong \mathrm{Hom}(\varinjlim_{\Lambda} \mathcal{F}_{\lambda}, \mathcal{G})$.

Suppose, from now on, that $\Lambda$ is a filtered ordered set. Then the functor $\widehat{v}_!$ is exact. Therefore the functor $v$ defines a morphism of sites and hence a morphism of topos

(IV.6.6.7) $$\underrightarrow{l}\colon C^{\sim} \to (C^{\Lambda^{\circ}})^{\sim}, \quad \underrightarrow{l}_*\mathcal{G} = \mathcal{G}\circ v, \quad \underrightarrow{l}^*\mathcal{F} = \varinjlim_{\lambda\in\Lambda} \mathcal{F}_{\lambda}.$$

In the description of $\underrightarrow{l}^*\mathcal{F}$, we consider the direct limit as sheaves of sets. Note that the inverse image functor $\underrightarrow{l}^*$ gives the direct limit of sheaves, while the direct image functor

$\underset{\leftarrow}{l}_*$ gives the inverse limit of sheaves. We have

$$\underset{\rightarrow}{l}^* \circ \underset{\rightarrow}{l}_* = \mathrm{id}_{C^\sim}. \tag{IV.6.6.8}$$

A morphism of topos $f\colon C^\sim \to D^\sim$ induces a morphism of topos $f^\Lambda\colon C^{\sim\Lambda} \to D^{\sim\Lambda}$ whose direct (resp. inverse) image functor is defined by the composition with $f_* \times \mathrm{id}_\Lambda$ (resp. $f^* \times \mathrm{id}_\Lambda$). Combining with the isomorphisms (IV.6.6.5) for $C$ and $D$, we obtain a morphism of topos $(C^{\Lambda^\circ})^\sim \to (D^{\Lambda^\circ})^\sim$, which is also denoted by $f^\Lambda$. By looking at the compositions of the direct image functors, we see that the following diagram of topos is commutative up to canonical isomorphism.

$$\begin{array}{ccc} C^\sim & \xrightarrow{\underset{\rightarrow}{l}} & (C^{\Lambda^\circ})^\sim \\ {\scriptstyle f}\downarrow & & \downarrow{\scriptstyle f^\Lambda} \\ D^\sim & \xrightarrow{\underset{\rightarrow}{l}} & (D^{\Lambda^\circ})^\sim. \end{array} \tag{IV.6.6.9}$$

**IV.6.7. Comparison morphism.** Let $V$, $k$, $K$, $\overline{K}$, $\overline{V}$, $\Sigma$, $\widehat{\Sigma}$, $\overline{\Sigma}$, and $D(\overline{\Sigma})$ be the same as in the beginning of IV.5.2 and take the ring $W(R_{\overline{V}})$ as the base ring $R$ of our theory of Higgs crystals as in loc. cit. Let $X$ be an fs log scheme over $\Sigma$ satisfying the conditions in the beginning of IV.6.4, and define $\widehat{X}$ and $X_1$ as in loc. cit.

Let $(X_1/D(\overline{\Sigma}))_{\mathrm{HIGGS}}$ be the inverse limit of the fibered site $(r \mapsto (X_1/D(\overline{\Sigma}))^r_{\mathrm{HIGGS}})$ defined as in IV.4.5. Let $\mu_r\colon (X_1/D(\overline{\Sigma}))^\sim_{\mathrm{HIGGS}} \to (X_1/D(\overline{\Sigma}))^{r\sim}_{\mathrm{HIGGS}}$ ($r \in \mathbb{N}_{>0}$) be the canonical morphism of topos and let $\mathcal{O}_{X_1/D(\overline{\Sigma}),1}$ be the sheaf of rings $\varinjlim_r \mu_r^*(\mathcal{O}_{X_1/D(\overline{\Sigma}),1})$ ($\cong \mu_s^*(\mathcal{O}_{X_1/D(\overline{\Sigma}),1})$, $s \in \mathbb{N}_{>0}$). The compatible system of morphisms of ringed topos $u_{X_1/D(\overline{\Sigma})}\colon ((X_1/D(\overline{\Sigma}))^{r\sim}_{\mathrm{HIGGS}}, \mathcal{O}_{X_1/D(\overline{\Sigma}),1}) \to ((X_1)^\sim_{\text{ét}}, \mathcal{O}_{X_1})$ ($r \in \mathbb{N}_{>0}$) induces a morphism of ringed topos

$$u_{X_1/D(\overline{\Sigma})}\colon ((X_1/D(\overline{\Sigma}))^\sim_{\mathrm{HIGGS}}, \mathcal{O}_{X_1/D(\overline{\Sigma}),1}) \longrightarrow ((X_1)^\sim_{\text{ét}}, \mathcal{O}_{X_1}).$$

Under the notation in IV.6.6, we regard the inverse system of sheaves of rings $\mathcal{O}_{\mathcal{P}/X,\bullet} = (\mathcal{O}_{\mathcal{P}/X,m})_{m\in\mathbb{N}}$ as a sheaf of rings on $(\mathcal{P}/X)^{\mathbb{N}}_{\text{ét-fét}}$ by the isomorphism (IV.6.6.1). Similarly, we regard the inverse system of sheaves of rings $\mathcal{O}_{X_1,\bullet} = (\mathcal{O}_{X_1,m})_{m\in\mathbb{N}}$ as a sheaf of rings on $(X_1)^{\mathbb{N}}_{\text{ét}}$ and also on $X^{\mathbb{N}}_{\text{étaff}}$. Then we have morphisms of ringed topos (cf. (IV.6.5.10))

$$((\mathcal{P}/X)^{\mathbb{N}\sim}_{\text{ét-fét}}, \mathcal{O}_{\mathcal{P}/X,\bullet}) \xrightarrow{\pi^{h,\mathbb{N}^\circ}_{X,\text{ét}}} (X^{\mathbb{N}\sim}_{\text{étaff}}, \mathcal{O}_{X_1,\bullet}) \xrightarrow{\underset{\leftarrow}{l}} (X^\sim_{\text{étaff}}, \mathcal{O}_{X_1}).$$

Let $(\mathcal{F}, \mathcal{F}^\circ)$ be an object of $\mathrm{HC}^r_{\mathbb{Z}_p}(X_1/D(\overline{\Sigma}))$ (cf. Definition IV.3.5.4). In this subsection, we will construct a canonical morphism

$$Ru_{X_1/D(\overline{\Sigma})*}(\mu_r^*\mathcal{F}) \longrightarrow \mathbb{Q} \otimes R\underset{\leftarrow}{l}_* R\pi^{h,\mathbb{N}^\circ}_{X,\text{ét}*}\mathcal{T}_{X,\bullet}(\mathcal{F}^\circ) \tag{IV.6.7.1}$$

in $D^+(X_{\text{étaff}}, \mathcal{O}_{X_1,\mathbb{Q}})$. See IV.6.4 for the definition of the functor $\mathcal{T}_{X,\bullet}$. By taking $R\Gamma(X_{\text{étaff}}, -)$, we obtain a canonical morphism in $D^+(C\text{-Mod})$

$$R\Gamma((X_1/D(\overline{\Sigma}))_{\mathrm{HIGGS}}, \mu_r^*\mathcal{F}) \longrightarrow \mathbb{Q} \otimes R\Gamma((\mathcal{P}/X)^{\mathbb{N}}_{\text{ét-fét}}, \mathcal{T}_{X,\bullet}(\mathcal{F}^\circ)). \tag{IV.6.7.2}$$

Here $C$ denotes the completion of $\overline{K}$ with respect to the valuation.

We begin with some preliminaries. Let $\mathcal{C}$ be a full subcategory of $X_{\text{étaff}}$ such that every object of $X_{\text{étaff}}$ admits a strict étale covering by objects of $\mathcal{C}$. We define the functors

$\eta_{\mathcal{P}/\mathcal{C}}$ and $\eta_{\mathcal{C}}$ by the following compositions:

(IV.6.7.3)

$$\eta_{\mathcal{P}/\mathcal{C}}\colon \mathcal{M}\mathrm{od}(\mathcal{C}_{\mathrm{gpt}}, \mathcal{O}_{\mathcal{C}_{\mathrm{gpt}},\bullet}) \xrightarrow[r_{\mathcal{C}}]{\sim} \mathcal{M}\mathrm{od}((\mathcal{P}/\mathcal{C})_{\mathrm{f\acute{e}t}}, \mathcal{O}^{v}_{\mathcal{P}/\mathcal{C},\bullet})$$
$$\xrightarrow{v^*_{\mathcal{P}/\mathcal{C}}} \mathcal{M}\mathrm{od}((\mathcal{P}/\mathcal{C})_{\text{ét-fét}}, \mathcal{O}_{\mathcal{P}/\mathcal{C},\bullet}) \xrightarrow[\mathrm{id}^{X,\mathcal{C}}_{\mathcal{P},\text{ét-fét}*}]{\sim} \mathcal{M}\mathrm{od}((\mathcal{P}/X)_{\text{ét-fét}}, \mathcal{O}_{\mathcal{P}/X,\bullet}),$$

(IV.6.7.4) $$\eta_{\mathcal{C}}\colon \mathcal{M}\mathrm{od}(\mathcal{C}_{\mathrm{tr}}, \mathcal{O}_{\mathcal{C}_1,\bullet}) \xrightarrow{v^*_{\mathcal{C}}} \mathcal{M}\mathrm{od}(\mathcal{C}_{\mathrm{\acute{e}t}}, \mathcal{O}_{\mathcal{C}_1,\bullet}) \xrightarrow[\mathrm{id}^{X,\mathcal{C}}_{\mathrm{\acute{e}t}*}]{\sim} \mathcal{M}\mathrm{od}(X_{\mathrm{\acute{e}taff}}, \mathcal{O}_{X_1,\bullet}).$$

Then we have a natural morphism of functors

(IV.6.7.5) $$\eta_{\mathcal{C}} \circ \pi^h_{\mathcal{C},\mathrm{gpt}*} \longrightarrow \pi^h_{X,\mathrm{\acute{e}t}*} \circ \eta_{\mathcal{P}/\mathcal{C}}$$

defined by the composition

$$\mathrm{id}^{X,\mathcal{C}}_{\mathrm{\acute{e}t}*} \circ v^*_{\mathcal{C}} \circ \pi^h_{\mathcal{C},\mathrm{gpt}*} \xrightarrow[\text{(IV.6.5.11)}]{\cong} \mathrm{id}^{X,\mathcal{C}}_{\mathrm{\acute{e}t}*} \circ v^*_{\mathcal{C}} \circ \pi^h_{\mathcal{C}*} \circ r_{\mathcal{C}} \xrightarrow[\text{(IV.6.5.10)}]{} \mathrm{id}^{X,\mathcal{C}}_{\mathrm{\acute{e}t}*} \circ \pi^h_{\mathcal{C},\mathrm{\acute{e}t}*} \circ v^*_{\mathcal{P}/\mathcal{C}} \circ r_{\mathcal{C}}$$
$$\xrightarrow[\text{(IV.6.5.17)}]{\cong} \pi^h_{X,\mathrm{\acute{e}t}*} \circ \mathrm{id}^{X,\mathcal{C}}_{\mathcal{P},\text{ét-fét}*} \circ v^*_{\mathcal{P}/\mathcal{C}} \circ r_{\mathcal{C}}.$$

We define the functor $\eta_{\mathcal{C},\mathbb{Q}}$ by

(IV.6.7.6) $$\eta_{\mathcal{C},\mathbb{Q}}\colon \mathcal{M}\mathrm{od}(\mathcal{C}_{\mathrm{tr}}, \mathcal{O}_{\mathcal{C}_1,\mathbb{Q}}) \xrightarrow{v^*_{\mathcal{C}}} \mathcal{M}\mathrm{od}(\mathcal{C}_{\mathrm{\acute{e}t}}, \mathcal{O}_{\mathcal{C}_1,\mathbb{Q}}) \xrightarrow[\mathrm{id}^{X,\mathcal{C}}_{\mathrm{\acute{e}t}*}]{\sim} \mathcal{M}\mathrm{od}(X_{\mathrm{\acute{e}taff}}, \mathcal{O}_{X_1,\mathbb{Q}}).$$

Then we have a morphism of functors

(IV.6.7.7) $$\eta_{\mathcal{C},\mathbb{Q}} \circ (\mathbb{Q} \otimes \varprojlim) \longrightarrow (\mathbb{Q} \otimes \varprojlim) \circ \eta_{\mathcal{C}}$$

defined by the composition

$$\mathrm{id}^{X,\mathcal{C}}_{\mathrm{\acute{e}t}*} \circ v^*_{\mathcal{C}} \circ (\mathbb{Q} \otimes \varprojlim) \longrightarrow \mathrm{id}^{X,\mathcal{C}}_{\mathrm{\acute{e}t}*} \circ (\mathbb{Q} \otimes \varprojlim) \circ v^*_{\mathcal{C}} \xrightarrow{\sim} (\mathbb{Q} \otimes \varprojlim) \circ \mathrm{id}^{\mathcal{C},X}_{\mathrm{\acute{e}t}*} \circ v^*_{\mathcal{C}}$$

(cf. (IV.6.6.4) for the first morphism). By combining (IV.6.7.5) and (IV.6.7.7), we obtain a morphism of functors

(IV.6.7.8) $$\eta_{\mathcal{C},\mathbb{Q}} \circ (\mathbb{Q} \otimes \varprojlim) \circ \pi^h_{\mathcal{C},\mathrm{gpt}*} \longrightarrow (\mathbb{Q} \otimes \varprojlim) \circ \pi^h_{X,\mathrm{\acute{e}t}*} \circ \eta_{\mathcal{P}/\mathcal{C}}.$$

The morphisms (IV.6.7.5) and (IV.6.7.7) are compatible with the pull-backs by a strict étale morphism as follows. Let $X$ and $\mathcal{C}$ be as above, let $f\colon X' \to X$ be a strict étale morphism, and let $\mathcal{C}'$ be a full subcategory of $X'_{\mathrm{\acute{e}taff}}$ such that every object of $X'_{\mathrm{\acute{e}taff}}$ admits a strict étale covering by objects of $\mathcal{C}'$ and that $u\colon U' \to X' \in \mathrm{Ob}\,(\mathcal{C}')$ implies $f \circ u \in \mathrm{Ob}\,(\mathcal{C})$. We obtain the following isomorphism from (IV.6.2.10), (IV.6.1.17), and (IV.6.1.18).

(IV.6.7.9) $$f^*_{\mathcal{P},\text{ét-fét}} \circ \eta_{\mathcal{P}/\mathcal{C}} \xrightarrow{\cong} \eta_{\mathcal{P}/\mathcal{C}'} \circ f^{\mathcal{C},\mathcal{C}'*}_{\mathrm{gpt}}.$$

**Lemma IV.6.7.10.** *Let $f$, $\mathcal{C}$, and $\mathcal{C}'$ be as above. Then the following diagrams of functors are commutative, where the isomorphisms* (A) *are induced by* (IV.6.5.18) *and* (IV.6.5.6).

$$\begin{array}{ccccc}
f^*_{\mathrm{\acute{e}t}} \circ \eta_{\mathcal{C}} \circ \pi^h_{\mathcal{C},\mathrm{gpt}*} & \xrightarrow[\text{(A)}]{\cong} & \eta_{\mathcal{C}'} \circ f^{\mathcal{C},\mathcal{C}'*}_{\mathrm{tr}} \circ \pi^h_{\mathcal{C},\mathrm{gpt}*} & \xrightarrow[\text{(IV.6.5.19)}]{\cong} & \eta_{\mathcal{C}'} \circ \pi^h_{\mathcal{C}',\mathrm{gpt}*} \circ f^{\mathcal{C},\mathcal{C}'*}_{\mathrm{gpt}} \\
\downarrow{\scriptstyle \text{(IV.6.7.5)}} & & & & \downarrow{\scriptstyle \text{(IV.6.7.5)}} \\
f^*_{\mathrm{\acute{e}t}} \circ \pi^h_{X,\mathrm{\acute{e}t}*} \circ \eta_{\mathcal{P}/\mathcal{C}} & \xrightarrow[\text{(IV.6.5.19)}]{\cong} & \pi^h_{X',\mathrm{\acute{e}t}*} \circ f^*_{\mathcal{P},\text{ét-fét}} \circ \eta_{\mathcal{P}/\mathcal{C}} & \xrightarrow[\text{(IV.6.7.9)}]{\cong} & \pi^h_{X',\mathrm{\acute{e}t}*} \circ \eta_{\mathcal{P}/\mathcal{C}'} \circ f^{\mathcal{C},\mathcal{C}'*}_{\mathrm{gpt}}
\end{array}$$

$$\begin{array}{ccccc} f_{\text{ét}}^* \circ \eta_{\mathcal{C},\mathbb{Q}} \circ (\mathbb{Q}\otimes \varprojlim) & \xrightarrow[\text{(A)}]{\cong} & \eta_{\mathcal{C}',\mathbb{Q}} \circ f_{\text{tr}}^{\mathcal{C},\mathcal{C}'*} \circ (\mathbb{Q}\otimes \varprojlim) & \xrightarrow{\cong} & \eta_{\mathcal{C}',\mathbb{Q}} \circ (\mathbb{Q}\otimes \varprojlim) \circ f_{\text{tr}}^{\mathcal{C},\mathcal{C}'*} \\ \downarrow{\scriptstyle (\text{IV.6.7.7})} & & & & \downarrow{\scriptstyle (\text{IV.6.7.7})} \\ f_{\text{ét}}^* \circ (\mathbb{Q}\otimes \varprojlim) \circ \eta_{\mathcal{C}} & \xrightarrow{\cong} & (\mathbb{Q}\otimes \varprojlim) \circ f_{\text{ét}}^* \circ \eta_{\mathcal{C}} & \xrightarrow[\text{(A)}]{\cong} & (\mathbb{Q}\otimes \varprojlim) \circ \eta_{\mathcal{C}'} \circ f_{\text{tr}}^{\mathcal{C},\mathcal{C}'*} \end{array}$$

PROOF. For the first diagram, it suffices to prove three similar diagrams obtained by replacing $(\eta_{\mathcal{P}/\mathcal{C}^{(\prime)}}, \eta_{\mathcal{C}^{(\prime)}})$ with $(r_{\mathcal{C}^{(\prime)}}, \mathrm{id})$, $(v^*_{\mathcal{P}/\mathcal{C}^{(\prime)}}, v^*_{\mathcal{C}^{(\prime)}})$, and $(\mathrm{id}_{\mathcal{P},\text{ét-fét}}^{X^{(\prime)},\mathcal{C}^{(\prime)}*}, \mathrm{id}_{\text{ét}}^{X^{(\prime)},\mathcal{C}^{(\prime)}*})$ are commutative. The diagram for $(r_{\mathcal{C}^{(\prime)}}, \mathrm{id})$ commutes by the constructions of (IV.6.2.10), (IV.6.5.19), and (IV.6.5.11). The commutativity of the other two diagrams follows from the compatibility of base change morphisms with compositions applied to $v_{\dots} \circ f_{\dots}^{\mathcal{C},\mathcal{C}'} = f_{\dots}^{\mathcal{C},\mathcal{C}'} \circ v_{\dots}$ and $\mathrm{id}_{\dots}^{X,\mathcal{C}} \circ f_{\dots}^{\mathcal{C},\mathcal{C}'} = f_{\dots} \circ \mathrm{id}_{\dots}^{X',\mathcal{C}'}$. For the second diagram, it suffices to prove that the two similar diagrams obtained by replacing $\eta_{\mathcal{C}^{(\prime)}}$ with $v^*_{\mathcal{C}^{(\prime)}}$ and $\mathrm{id}_{\text{ét}}^{X^{(\prime)},\mathcal{C}^{(\prime)}*}$ are commutative, which also follows from the compatibility of base change morphisms with compositions. □

Now let us construct the morphism (IV.6.7.1). We first consider the case where we are given a smooth fine log scheme $Y$ over $\Sigma$ and an immersion $X \to Y$ over $\Sigma$. Let $\widehat{Y}$ be the $p$-adic completion of $Y$, let $Y_1$ be $\widehat{Y} \times_{\widehat{\Sigma}} \overline{\Sigma}$ and let $i_1$ denote the immersion $X_1 \to Y_1$ over $\overline{\Sigma}$ induced by $i$. We further assume that we are given a smooth Cartesian morphism $Y_\bullet \to D(\overline{\Sigma})$ in the category $\mathscr{C}$ (cf. Definitions IV.2.2.1 and IV.2.2.2) which lifts $Y_1 \to \overline{\Sigma}$.

Let $D^r$ ($r \in \mathbb{N}_{>0}$) be $D^r_{\text{Higgs}}(i\colon X_1 \hookrightarrow Y_\bullet)$ and let $z_{D^r}$ be the natural morphism $D^r_1 \to X_1$. let $(\mathcal{M}^s, \theta)$ ($s \geq r$) denote the object of $\mathrm{HB}^s_{\mathbb{Q}_p,\text{conv}}(X_1, Y_\bullet/D(\overline{\Sigma}))$ corresponding to $\mu^*_{rs}(\mathcal{F}) \in \mathrm{Ob}\,(\mathrm{HC}^s_{\mathbb{Q}_p}(X_1/D(\overline{\Sigma})))$ by the equivalence of categories in Theorem IV.3.4.16. Then, by Corollary IV.4.5.7, we have a canonical isomorphism

$$Ru_{X_1/D(\overline{\Sigma})*}\mu^*_r(\mathcal{F}) \cong \varinjlim_{s\geq r} z_{D^s*}(\xi^{-\bullet}\mathcal{M}^s \otimes_{\mathcal{O}_X} i^*(\Omega^\bullet_{Y/\Sigma})) \tag{IV.6.7.11}$$

in $D^+((X_1)_{\text{ét}}, \mathcal{O}_{X_1,\mathbb{Q}})$. Using (IV.6.7.11), we construct the morphism (IV.6.7.1) along the following lines. Let $(\mathcal{O}_{\mathcal{P}/X,\bullet})^{\mathbb{N}}$ denote the sheaf of rings $\underrightarrow{l}_*(\mathcal{O}_{\mathcal{P}/X,\bullet})$ on $((\mathcal{P}/X)^{\mathbb{N}}_{\text{ét,fét}})^{\mathbb{N}^\circ}$ (cf. (IV.6.6.7)), which corresponds to the constant direct system defined by $\mathcal{O}_{\mathcal{P}/X,\bullet}$ via the equivalence of categories (IV.6.6.5). Similarly we define the sheaves of rings $\mathcal{O}^{\mathbb{N}}_{X_1,\bullet}$ and $\mathcal{O}^{\mathbb{N}}_{X_1}$ on $(X^{\mathbb{N}}_{\text{étaff}})^{\mathbb{N}^\circ}$ and $X^{\mathbb{N}^\circ}_{\text{étaff}}$, respectively. We first construct a morphism of complexes of $\mathcal{O}^{\mathbb{N}}_{\mathcal{P}/X,\bullet}$-modules on $((\mathcal{P}/X)^{\mathbb{N}}_{\text{ét-fét}})^{\mathbb{N}^\circ}$

$$\underrightarrow{l}_*\mathcal{T}_{X,\bullet}(\mathcal{F}^\circ) \longrightarrow \mathcal{T}^{\star+1}_{X,\bullet}\Omega_Y(\mathcal{F}^\circ) \tag{IV.6.7.12}$$

by a method similar to the construction of $V \to \xi^{-\bullet}\mathscr{A}^s_1(\overline{A}) \otimes_{\widetilde{\overline{A}}} V \otimes_A \Omega^\bullet$ ($s \in \mathbb{N}$) in the proof of Proposition IV.5.2.15. Then we prove that (IV.6.7.12) induces an isomorphism

$$\mathbb{Q} \otimes R\underleftarrow{l}_* R\pi^{h,\mathbb{N}^\circ}_{X,\text{ét}*}\mathcal{T}_{X,\bullet}(\mathcal{F}^\circ) \xrightarrow{\cong} \underrightarrow{l}^*\mathbb{Q} \otimes R\underleftarrow{l}^{\mathbb{N}}_* R(\pi^{h,\mathbb{N}^\circ}_{X,\text{ét}})^{\mathbb{N}}_*(\mathcal{T}^{\star+1}_{X,\bullet}\Omega_Y(\mathcal{F}^\circ)) \tag{IV.6.7.13}$$

in $D^+(X_{\text{étaff}}, \mathcal{O}_{X_1,\mathbb{Q}})$. Finally we construct a canonical morphism of direct systems of complexes of $\mathcal{O}_{X_1,\mathbb{Q}}$-modules

$$(z_{D^s*}(\xi^{-\bullet}\mathcal{M}^s \otimes_{\mathcal{O}_X} i^*\Omega^\bullet_{Y/\Sigma}))_{s\geq r} \longrightarrow (\mathbb{Q} \otimes \underleftarrow{l}_*\pi^{h,\mathbb{N}^\circ}_{X,\text{ét}*}\mathcal{T}^s_{X,\bullet}\Omega_Y(\mathcal{F}^\circ))_{s\geq r}. \tag{IV.6.7.14}$$

By taking the direct limit of (IV.6.7.14) and combining with (IV.6.7.11) and (IV.6.7.13), we obtain (IV.6.7.1).

Let $\mathcal{C}^Y$ be the full subcategory of $\mathcal{C}_X$ (cf. Definition IV.6.4.1 (1)) consisting of $U \in \mathrm{Ob}\,\mathcal{C}_X$ satisfying the following condition: There exists a strict étale morphism $Y' \to Y$

such that $Y'$ satisfies Condition IV.5.2.3 and the morphism $U \to X$ factors through $X \times_Y Y' \to X$. For $(h\colon U \to X, \overline{s}) \in \mathcal{C}^Y_{\mathrm{gpt}}$ and $r \in \mathbb{N}_{>0}$, by using the object $D_X(U,\overline{s}) = (D(U,\overline{s}), h_1 \circ z_{(U,\overline{s})})$ of $(X_1/D(\Sigma))^{\infty}_{\mathrm{HIGGS}}$ (cf. IV.6.4), we define the object $\mathscr{D}^r_{X,Y}(U,\overline{s}) = (\mathscr{D}^r_{X,Y,N}(U,\overline{s}))$ of $\mathscr{C}^r$ by

$$\mathscr{D}^r_{X,Y}(U,\overline{s}) := D^r_{\mathrm{Higgs}}(D(U,\overline{s})_1 \hookrightarrow Y_\bullet \times_{D(\Sigma)} D(U,\overline{s}))$$

(cf. the construction of $\mathscr{D}^r_{X,Y}(\overline{U})$ before Condition IV.5.2.3). This construction is functorial in $(U,\overline{s})$ and we can define a ring object $\mathscr{A}^r_{X,Y,m}$ of $(\mathcal{C}^Y_{\mathrm{gpt}})^{\wedge,\mathrm{cont}}$ by

$$\mathscr{A}^r_{X,Y,m}(U,\overline{s}) := \Gamma(\mathscr{D}^r_{X,Y,1}(U,\overline{s}), \mathcal{O}_{\mathscr{D}^r_{X,Y,1}(U,\overline{s})})/p^m,$$

which is an $\mathcal{O}_{\mathcal{C}^Y_{\mathrm{gpt}},m}$-algebra (cf. Proposition IV.5.2.6). For a sheaf $\mathcal{H}$ on $X_{\text{ét}}$, we also write $\mathcal{H}$ for the object of $\mathcal{C}^Y_{\mathrm{gpt}}$ defined by $(U,\overline{s}) \mapsto \mathcal{H}(U)$. Then $\mathscr{A}^r_{X,Y,m}$ is also an $\mathcal{O}_{X_1}$-algebra. The construction of the complex (IV.5.2.9) is functorial in $(U,\overline{s})$. By Proposition IV.5.2.10 (2), we obtain a direct system of complexes in $\mathcal{M}od(\mathcal{C}^Y_{\mathrm{gpt}}, \mathcal{O}_{\mathcal{C}^Y_{\mathrm{gpt}},\bullet})$

$$(\mathcal{O}_{\mathcal{C}^Y_{\mathrm{gpt}},\bullet} \longrightarrow (\xi^{-q}\mathscr{A}^{r+1}_{X,Y,\bullet} \otimes_{\mathcal{O}_X} i^*(\Omega^q_{Y/\Sigma}), \theta^q)_{q\in\mathbb{N}})_{r\in\mathbb{N}}. \tag{IV.6.7.15}$$

Choose a covering sieve $R$ of $X$ in $X_{\text{étaff}}$ such that $\mathcal{F}^\circ$ is $R$-finite (cf. Definition IV.6.4.2) and $\mathcal{F}_{(T,z)} \in \mathrm{Ob}\,\mathcal{PM}(\mathcal{O}_{T_1,\mathbb{Q}_p})$ for every $(T,z) \in \mathrm{Ob}\,(X_1/D(\Sigma))^r_{\mathrm{HIGGS}}$ such that $T_1$ is affine and $z$ factors through $u_1\colon U_1 \to X_1$ for some $u\colon U \to X \in \mathrm{Ob}\,X_{\text{étaff}}$ satisfying $u \in R(U)$. See Remark IV.6.4.3 for the existence of such an $R$. Let $\mathcal{C}$ be a full subcategory of $X_{\text{étaff}}$ contained in both $\mathcal{C}^Y$ and $\mathcal{C}_R$ (cf. Definition IV.6.4.1 (2)) such that every object of $X_{\text{étaff}}$ admits a strict étale covering by objects of $\mathcal{C}$ and that, for a morphism $U' \to U$ in $X_{\text{étaff}}$, $U \in \mathrm{Ob}\,\mathcal{C}$ and $U' \in \mathrm{Ob}\,\mathcal{C}_R$ imply $U' \in \mathrm{Ob}\,\mathcal{C}$. By taking the tensor product of $\mathcal{T}^R_{X,\mathrm{gpt},\bullet}(\mathcal{F}^\circ)|_{\mathcal{C}_{\mathrm{gpt}}}$ (cf. (IV.6.4.4)) and (IV.6.7.15)$|_{\mathcal{C}_{\mathrm{gpt}}}$ over $\mathcal{O}_{\mathcal{C}^Y_{\mathrm{gpt}},\bullet}|_{\mathcal{C}_{\mathrm{gpt}}}$, we obtain a direct system of complexes in $\mathcal{M}od(\mathcal{C}_{\mathrm{gpt}}, \mathcal{O}_{\mathcal{C}_{\mathrm{gpt}},\bullet})$

$$(\mathcal{T}^R_{X,\mathrm{gpt},\bullet}(\mathcal{F}^\circ)|_{\mathcal{C}_{\mathrm{gpt}}} \longrightarrow \mathcal{T}^{s+1}_{\mathcal{C},\mathrm{gpt},\bullet}\Omega_Y(\mathcal{F}^\circ))_{s\in\mathbb{N}}. \tag{IV.6.7.16}$$

**Lemma IV.6.7.17.** *For $r \in \mathbb{N}_{>0}$, there exists $N_r \in \mathbb{N}_{>0}$ such that the homomorphism from $\mathcal{H}^q$ of the $r$-th term of* (IV.6.7.16) *to $\mathcal{H}^q$ of the $(r+1)$-th term is annihilated by $p^{N_r}$ for every $q \in \mathbb{Z}$.*

PROOF. It suffices to prove that there exists an integer $N_r$ determined only by $p$ and $r$ such that the natural morphism from the complex (IV.5.2.9) for $\mathscr{A}^r_1(\overline{\mathcal{A}})$ to the complex (IV.5.2.9) for $\mathscr{A}^{r+1}_1(\overline{\mathcal{A}})$ is homotopic to 0 by an $A_1(\overline{\mathcal{A}})$-linear homotopy after multiplied by $p^{N_r}$. Since $F^nA_N(\overline{\mathcal{A}}) = \xi^nA_N(\overline{\mathcal{A}})$, we have $A_1(\overline{\mathcal{A}})^{(m)_r} = p^{\lceil \frac{m}{r}\rceil}A_1(\overline{\mathcal{A}})$. Hence, by the construction of the homotopy in the proof of Proposition IV.5.2.10 (2), it suffices to show that the $p$-adic valuation of $\frac{1}{m'+1}p^{\lceil\frac{m}{r}\rceil}p^{-\lceil\frac{m+1}{r+1}\rceil}$ is bounded for $0 \le m' \le m$. This follows from $\lim_{m\to\infty}(-\log_p(m+1) + \lceil\frac{m}{r}\rceil - \lceil\frac{m+1}{r+1}\rceil) = \infty$. □

By applying $\eta_{\mathcal{P}/\mathcal{C}}$ (IV.6.7.3) to (IV.6.7.16), we obtain (IV.6.7.12). Note that the image of $\mathcal{T}^R_{X,\mathrm{gpt},\bullet}(\mathcal{F}^\circ)|_{\mathcal{C}_{\mathrm{gpt}}}$ under $\eta_{\mathcal{P}/\mathcal{C}}$ is canonically isomorphic to $\mathcal{T}_{X,\bullet}(\mathcal{F}^\circ)$ by (IV.6.7.9) applied to $f = \mathrm{id}_X$ and $\mathcal{C} \subset \mathcal{C}_R$. We also see that the above complex does not depend on the choice of $R$ and $\mathcal{C}$ up to canonical isomorphisms. Note that if we choose another $\mathcal{C}'$, then $\mathcal{C} \cap \mathcal{C}'$ also satisfies the conditions on $\mathcal{C}$.

**Lemma IV.6.7.18.** *Regard* (IV.6.7.12) *as a complex of* $\mathscr{O}^{\mathbb{N}}_{\mathcal{P}/X,\bullet}$*-modules. Then its image under the composition of the following functors are* 0.

$$D^+(((\mathcal{P}/X)^{\mathbb{N}}_{\text{ét-fét}})^{\mathbb{N}^\circ},\mathscr{O}^{\mathbb{N}}_{\mathcal{P}/X,\bullet})\xrightarrow{R(\pi^{h,\mathbb{N}^\circ}_{X,\text{ét}})^{\mathbb{N}}_*}D^+((X^{\mathbb{N}}_{\text{étaff}})^{\mathbb{N}^\circ},\mathscr{O}^{\mathbb{N}}_{X_1,\bullet})\xrightarrow{R\underset{\leftarrow}{l}^{\mathbb{N}}_*}D^+(X^{\mathbb{N}^\circ}_{\text{étaff}},\mathscr{O}^{\mathbb{N}}_{X_1})$$
$$\xrightarrow{\mathbb{Q}\otimes}D^+(X^{\mathbb{N}^\circ}_{\text{étaff}},\mathscr{O}^{\mathbb{N}}_{X_1,\mathbb{Q}})\xrightarrow{\underset{\to}{l}^*}D^+(X_{\text{étaff}},\mathscr{O}_{X_1,\mathbb{Q}}).$$

PROOF. Let $\mathcal{K}=(\mathcal{K}_s)_{s\in\mathbb{N}}$ denote (IV.6.7.12) regarded as a complex of $\mathscr{O}^{\mathbb{N}}_{\mathcal{P}/X,\bullet}$-modules. Since the functor $\eta_{\mathcal{P}/\mathcal{C}}$ is exact, Lemma IV.6.7.17 implies that the transition map $\mathcal{H}^q(\mathcal{K}_s)\to\mathcal{H}^q(\mathcal{K}_{s+1})$ is annihilated by $p^{N_s}$ for every $s\in\mathbb{N}$. We have a spectral sequence

$$E_2^{a,b}=R^a(\underset{\leftarrow}{l}\circ\pi^{h,\mathbb{N}^\circ}_{X,\text{ét}})^{\mathbb{N}}_*\mathcal{H}^b(\mathcal{K})\Longrightarrow R^{a+b}(\underset{\leftarrow}{l}\circ\pi^{h,\mathbb{N}^\circ}_{X,\text{ét}})^{\mathbb{N}}_*(\mathcal{K})$$

in $\mathcal{M}\mathrm{od}(X^{\mathbb{N}^\circ}_{\text{étaff}},\underset{\to}{l}_*\mathscr{O}_{X_1})$, and the $s$-th term of $E_2^{a,b}$ is given by $R^a(\underset{\leftarrow}{l}\circ\pi^{h,\mathbb{N}^\circ}_{X,\text{ét}})_*\mathcal{H}^b(\mathcal{K}_s)$ (cf. **[2]** VI the first paragraph of the proof of Lemme 8.7.2). Hence we have $\underset{\to}{l}^*\mathbb{Q}\otimes R^{a+b}(\underset{\leftarrow}{l}\circ\pi^{h,\mathbb{N}^\circ}_{X,\text{ét}})^{\mathbb{N}}_*(\mathcal{K})=0.$ □

The upper (resp. lower) square of the diagram below is commutative up to canonical isomorphisms by (IV.6.6.9) (resp. (IV.6.6.4)).

$$\begin{array}{ccc}
D^+((\mathcal{P}/X)^{\mathbb{N}}_{\text{ét-fét}},\mathscr{O}_{\mathcal{P}/X,\bullet}) & \xrightarrow{\underset{\to}{l}_*} & D^+(((\mathcal{P}/X)^{\mathbb{N}}_{\text{ét-fét}})^{\mathbb{N}^\circ},\mathscr{O}^{\mathbb{N}}_{\mathcal{P}/X,\bullet})\\
\downarrow{\scriptstyle R\pi^{h,\mathbb{N}^\circ}_{X,\text{ét}*}} & & \downarrow{\scriptstyle R(\pi^{h,\mathbb{N}^\circ}_{X,\text{ét}})^{\mathbb{N}}_*}\\
D^+(X^{\mathbb{N}}_{\text{étaff}},\mathscr{O}_{X_1,\bullet}) & \xrightarrow{\underset{\to}{l}_*} & D^+((X^{\mathbb{N}}_{\text{étaff}})^{\mathbb{N}^\circ},\mathscr{O}^{\mathbb{N}}_{X_1,\bullet})\\
\downarrow{\scriptstyle R\underset{\leftarrow}{l}_*} & & \downarrow{\scriptstyle R\underset{\leftarrow}{l}^{\mathbb{N}}_*}\\
D^+(X_{\text{étaff}},\mathscr{O}_{X_1}) & \xrightarrow{\underset{\to}{l}_*} & D^+(X^{\mathbb{N}^\circ}_{\text{étaff}},\mathscr{O}^{\mathbb{N}}_{X_1}).
\end{array}$$

Hence by (IV.6.6.8), we see that the image of $\underset{\to}{l}_*\mathcal{T}_{X,\bullet}(\mathcal{F}^\circ)$ under the composition of the functors in Lemma IV.6.7.18 is canonically isomorphic to $\mathbb{Q}\otimes R\underset{\leftarrow}{l}_*R\pi^{h,\mathbb{N}^\circ}_{X,\text{ét}*}\mathcal{T}_{X,\bullet}(\mathcal{F}^\circ)$. By Lemma IV.6.7.18, we obtain the isomorphism (IV.6.7.13).

It remains to construct (IV.6.7.14). Let $(U,\overline{s})\in\mathrm{Ob}\,\mathcal{C}_{\text{gpt}}$. By Proposition IV.5.2.12 and the choice of $\mathcal{C}$, we have a canonical $\Delta_{(U,\overline{s})}$-equivariant $\varprojlim_m\mathscr{A}^s_{X,Y,m}(U,\overline{s})$-linear isomorphism

(IV.6.7.19)
$$\mathcal{M}^s(D_1^s\times_{X_1}U_1)\otimes_{\mathscr{O}_{D_1^s}(D_1^s\times_{X_1}U_1)}\varprojlim_m\mathscr{A}^s_{X,Y,m}(U,\overline{s})$$
$$\xrightarrow{\cong}\mathbb{Q}\otimes\varprojlim_m(\mathcal{T}^R_{X,\text{gpt},m}(\mathcal{F}^\circ)(U,\overline{s})\otimes_{\mathscr{O}_{\mathcal{C}_{\text{gpt},m}}(U,\overline{s})}\mathscr{A}^s_{X,Y,m}(U,\overline{s}))$$

compatible with $\theta$. By construction, this isomorphism is compatible with $s$ and functorial with respect to $(U,\overline{s})$. Hence, the above isomorphism (IV.6.7.19) induces a morphism of direct systems of complexes of $\mathscr{O}_{\mathcal{C}_1,\mathbb{Q}}(:=\mathscr{O}_{X_1,\mathbb{Q}}|_{\mathcal{C}})$-modules on $\mathcal{C}_{\text{tr}}$

(IV.6.7.20) $$(z_{D^s*}(\xi^{-\bullet}\mathcal{M}^s\otimes_{\mathscr{O}_X}i^*\Omega^\bullet_{Y/\Sigma})|_{\mathcal{C}})_{s\geq r}\longrightarrow(\mathbb{Q}\otimes\varprojlim_m\pi^h_{\mathcal{C},\text{gpt}*}\mathcal{T}^s_{\mathcal{C},\text{gpt},m}\Omega_Y(\mathcal{F}^\circ))_{s\geq r}.$$

By taking $\eta_{\mathcal{C},\mathbb{Q}}$ of (IV.6.7.20) and using (IV.6.7.8), we obtain (IV.6.7.14). By Lemma IV.6.7.10 for $f=\mathrm{id}_X$, we see that the morphism (IV.6.7.14) obtained above is independent of the choice of $\mathcal{C}$. Thus we obtain the morphism (IV.6.7.1).

We can verify the independence of the choice of $(i\colon X\to Y,Y_\bullet)$ as follows. Choose another $(i'\colon X\to Y',Y'_\bullet)$. By considering $(X\to Y\times_\Sigma Y',Y_\bullet\times_{D(\overline{\Sigma})}Y'_\bullet)$, we are reduced to the case where there exist morphisms $g\colon Y' \to Y$ and $g_\bullet\colon Y'_\bullet \to Y_\bullet$ compatible with $i$, $i'$, $Y_1 \cong \widehat{Y} \times_{\widehat{\Sigma}} \overline{\Sigma}$, and $Y'_1 = \widehat{Y}' \times_{\widehat{\Sigma}} \overline{\Sigma}$. Then the independence follows from the functoriality of (IV.6.7.12) and (IV.6.7.14) below applied to $f = \mathrm{id}_X$, $g$, $g_\bullet$ and $\mathcal{C} = \mathcal{C}' \subset \mathcal{C}^Y \cap \mathcal{C}^{Y'} \cap \mathcal{C}_R$.

Let $i'\colon X' \to Y'$, $Y'_\bullet$ be another data satisfying the same assumptions as $i\colon X \to Y, Y_\bullet$, and let $f\colon X' \to X$ and $g\colon Y' \to Y$ be morphisms over $\Sigma$ such that $g \circ i' = i \circ f$ and $f$ is strict étale. Let $g_\bullet\colon Y'_\bullet \to Y_\bullet$ be a morphism over $D(\overline{\Sigma})$ such that $g_1 = \widehat{g} \times_{\widehat{\Sigma}} \overline{\Sigma}$. Let $R'$ be a covering sieve of $X'$ in $X'_{\text{étaff}}$ defined by $R'(u\colon U \to X') = R(f \circ u\colon U \to X)$. Let $(\mathcal{F}', \mathcal{F}'^\circ)$ be $f^*_{1,\mathrm{HIGGS}}(\mathcal{F}, \mathcal{F}^\circ)$. Assume that there exist full subcategories $\mathcal{C} \subset \mathcal{C}_R \cap \mathcal{C}^Y$ and $\mathcal{C}' \subset \mathcal{C}_{R'} \cap \mathcal{C}^{Y'}$ satisfying the condition before Lemma IV.6.7.10.

By the proof of Lemma IV.6.4.6, we have a canonical isomorphism

$$f_{\mathrm{gp}}^{\mathcal{C},\mathcal{C}'*}(\mathcal{T}^R_{X,\mathrm{gp},\bullet}(\mathcal{F}^\circ)|_{\mathcal{C}_{\mathrm{gpt}}}) \xrightarrow{\cong} \mathcal{T}^{R'}_{X',\mathrm{gp},\bullet}(\mathcal{F}'^\circ)|_{\mathcal{C}'_{\mathrm{gpt}}}. \tag{IV.6.7.21}$$

Hence by the construction of (IV.6.7.16), we see that the morphisms $g$ and $g_\bullet$ naturally induce a direct system of morphisms in $\mathcal{M}\mathrm{od}(\mathcal{C}'_{\mathrm{gpt}}, \mathcal{O}_{\mathcal{C}'_{\mathrm{gpt}},\bullet})$.

$$(f_{\mathrm{gpt}}^{\mathcal{C},\mathcal{C}'*}(\mathcal{T}^{s+1}_{\mathcal{C},\mathrm{gpt},\bullet}\Omega_Y(\mathcal{F}^\circ)) \longrightarrow \mathcal{T}^{s+1}_{\mathcal{C}',\mathrm{gpt},\bullet}\Omega_{Y'}(\mathcal{F}'^\circ))_{s\in\mathbb{N}} \tag{IV.6.7.22}$$

such that (IV.6.7.21) and (IV.6.7.22) are compatible with (IV.6.7.16). By taking $\eta_{\mathcal{P}/\mathcal{C}}$ of (IV.6.7.22) and using (IV.6.7.9), we obtain a morphism on $((\mathcal{P}/X')^{\mathbb{N}}_{\text{ét-fét}})^{\mathbb{N}^\circ}$

$$f^*_{\mathcal{P},\text{ét-fét}}(\mathcal{T}^{\star+1}_{X,\bullet}\Omega_Y(\mathcal{F}^\circ)) \longrightarrow \mathcal{T}^{\star+1}_{X',\bullet}\Omega_{Y'}(\mathcal{F}'^\circ) \tag{IV.6.7.23}$$

such that the following diagram is commutative.

(IV.6.7.24)

$$\begin{array}{ccccc}
f^*_{\mathcal{P},\text{ét-fét}}\underrightarrow{l}_*(\mathcal{T}_{X,\bullet}(\mathcal{F}^\circ)) & \xrightarrow{\cong} & \underrightarrow{l}_* f^*_{\mathcal{P},\text{ét-fét}}(\mathcal{T}_{X,\bullet}(\mathcal{F}^\circ)) & \xrightarrow[\text{Lemma IV.6.4.6}]{\cong} & \underrightarrow{l}_*(\mathcal{T}_{X',\bullet}(\mathcal{F}'^\circ)) \\
\Big\downarrow{\scriptstyle(\mathrm{IV.6.7.12})} & & & & \Big\downarrow{\scriptstyle(\mathrm{IV.6.7.12})} \\
f^*_{\mathcal{P},\text{ét-fét}}(\mathcal{T}^{\star+1}_{X,\bullet}\Omega_Y(\mathcal{F}^\circ)) & & \xrightarrow{(\mathrm{IV.6.7.23})} & & \mathcal{T}^{\star+1}_{X',\bullet}\Omega_{Y'}(\mathcal{F}'^\circ).
\end{array}$$

We define $D'^s$ and $\mathcal{M}'^s$ similarly as $D^s$ and $\mathcal{M}^s$ using $i'\colon X' \to Y'$, $Y'_\bullet$ and $(\mathcal{F}', \mathcal{F}'^\circ)$. Then, by the construction of (IV.6.7.20), we see that the composition of

$$\begin{aligned}
f_{\mathrm{tr}}^{\mathcal{C},\mathcal{C}'*}(\mathbb{Q}\otimes \varprojlim_m \pi^h_{\mathcal{C},\mathrm{gpt}*}\mathcal{T}^s_{\mathcal{C},\mathrm{gpt},m}\Omega_Y(\mathcal{F}^\circ)) &\xrightarrow[(\mathrm{IV.6.5.19})]{\cong} \mathbb{Q}\otimes \varprojlim_m \pi^h_{\mathcal{C}',\mathrm{gpt}*} f_{\mathrm{gpt}}^{\mathcal{C},\mathcal{C}'*}\mathcal{T}^s_{\mathcal{C},\mathrm{gpt},m}\Omega_Y(\mathcal{F}^\circ) \\
&\xrightarrow{(\mathrm{IV.6.7.22})} \mathbb{Q}\otimes \varprojlim_m \pi^h_{\mathcal{C}',\mathrm{gpt}*}\mathcal{T}^s_{\mathcal{C}',\mathrm{gpt},m}\Omega_{Y'}(\mathcal{F}'^\circ)
\end{aligned}$$

and the natural morphism

$$f_{\mathrm{tr}}^{\mathcal{C},\mathcal{C}'*}(z_{D^s*}(\xi^{-\bullet}\mathcal{M}^s \otimes_{\mathcal{O}_X} i^*\Omega^\bullet_{Y/\Sigma})|_{\mathcal{C}}) \longrightarrow (z_{D'^s*}(\xi^{-\bullet}\mathcal{M}'^s \otimes_{\mathcal{O}_{X'}} i'^*\Omega^\bullet_{Y'/\Sigma}))|_{\mathcal{C}'}$$

are compatible with (IV.6.7.20). By taking $\eta_{\mathcal{C}',\mathbb{Q}}$ and using Lemma IV.6.7.10, we see that the following diagram is commutative, where $\underleftarrow{\pi}^h_{X^{(\prime)},\text{ét}*,\mathbb{Q}}$ denotes $\mathbb{Q}\otimes\varprojlim_m \pi^h_{X^{(\prime)},\text{ét}*}$.

(IV.6.7.25)

$$\begin{array}{ccc}
f^*_{\text{ét}}(z_{D^s*}(\xi^{-\bullet}\mathcal{M}^s \otimes_{\mathcal{O}_X} i^*\Omega^\bullet_{Y/\Sigma})) & \xrightarrow{(\mathrm{IV.6.7.14})} & f^*_{\text{ét}}\underleftarrow{\pi}^h_{X,\text{ét}*,\mathbb{Q}}(\mathcal{T}^s_{X,m}\Omega_Y(\mathcal{F}^\circ)) \\
\Big\downarrow & & \cong\Big\downarrow{\scriptstyle(\mathrm{IV.6.5.19})} \\
& & \underleftarrow{\pi}^h_{X',\text{ét}*,\mathbb{Q}} f^*_{\mathcal{P},\text{ét-fét}}(\mathcal{T}^s_{X,m}\Omega_Y(\mathcal{F}^\circ)) \\
& & \Big\downarrow{\scriptstyle(\mathrm{IV.6.7.23})} \\
z_{D'^s*}(\xi^{-\bullet}\mathcal{M}'^s \otimes_{\mathcal{O}_{X'}} i'^*\Omega^\bullet_{Y'/\Sigma}) & \xrightarrow{(\mathrm{IV.6.7.14})} & \underleftarrow{\pi}^h_{X',\text{ét}*,\mathbb{Q}}(\mathcal{T}^s_{X',m}\Omega_{Y'}(\mathcal{F}'^\circ)).
\end{array}$$

Now let us construct the morphism (IV.6.7.1) for a general $X$. We reduce it to the special case above by using étale cohomological descent.

Let $\Delta$ be the category whose objects are ordered sets $[\nu] = \{0, 1, \ldots, \nu\}$ ($\nu \in \mathbb{N}$) and whose morphisms are non-decreasing maps. Recall that a simplicial object of a category $C$ means a functor $\Delta^\circ \to C$. Choose a strict étale hypercovering $\underline{X} = (X^{[\nu]})_{\nu\in\mathbb{N}}$ of $X$ such that $X^{[\nu]}$ are quasi-compact, an immersion of simplicial $p$-adic fine log formal schemes $\underline{i} = (i^{[\nu]})\colon \underline{X} \hookrightarrow \underline{Y} = (Y^{[\nu]})$ over $\Sigma$ such that each $Y^{[\nu]}$ is smooth over $\Sigma$, and a simplicial object $\underline{Y}_\bullet = (Y_\bullet^{[\nu]})$ of $\mathscr{C}$ over $D(\overline{\Sigma})$ such that each $Y_\bullet^{[\nu]}$ is smooth and Cartesian over $D(\overline{\Sigma})$ and $\underline{Y}_1$ coincides with $(\widehat{Y}^{[\nu]} \times_{\widehat{\Sigma}} \overline{\Sigma})_{\nu\in\mathbb{N}}$. Let $\underline{X}_1$ denote the simplicial $p$-adic fine log formal scheme $(\widehat{X}^{[\nu]} \times_{\widehat{\Sigma}} \overline{\Sigma})_{\nu\in\mathbb{N}}$ over $\overline{\Sigma}$ and let $\underline{i}_1 = (i_1^{[\nu]})$ denote the immersion $\underline{X}_1 \to \underline{Y}_1$ induced by $\underline{i}$.

One can define a fibered ringed topos over $X_{\text{étaff}}$ whose fiber over $U \in \operatorname{Ob} X_{\text{étaff}}$ is $((\mathcal{P}/U)^{\mathbb{N}\sim}_{\text{ét-fét}}, \mathcal{O}_{\mathcal{P}/U,\bullet})$ and whose inverse image functor by a morphism $U' \to U$ in $X_{\text{étaff}}$ is the inverse image functor of ringed topos (cf. (IV.6.1.12), (IV.6.1.17)). By taking the base change of the fibered ringed topos by the functor $\underline{X}\colon \Delta^\circ \to X_{\text{étaff}}$ and considering the category of sections of the associated ringed $\Delta$-topos, we obtain a ringed topos $((\mathcal{P}/\underline{X})^{\mathbb{N}\sim}_{\text{ét-fét}}, \mathcal{O}_{\mathcal{P}/\underline{X},\bullet})$ (cf. **[2]** $\mathrm{V}^{\mathrm{bis}}$ Définition (1.2.1), (1.2.5), Proposition (1.2.12), Définition (1.3.1), (1.3.4)). Note that a fibered topos over a category $D$ is the same as "une catégorie bifibrée en duaux de topos au-dessus de $D$" in **[2]** $\mathrm{V}^{\mathrm{bis}}$ Définition (1.2.2). An object of $(\mathcal{P}/\underline{X})^{\mathbb{N}\sim}_{\text{ét-fét}}$ is a data $(\mathcal{G}^{[\nu]}, \rho_s)_{[\nu]\in \operatorname{Ob}\Delta, s\in \operatorname{Mor}\Delta}$ consisting of a sheaf $\mathcal{G}^{[\nu]}$ on $(\mathcal{P}/X^{[\nu]})^{\mathbb{N}}_{\text{ét-fét}}$ for each $\nu \in \mathbb{N}$ and a morphism $\rho_s\colon (\underline{s}^{\mathbb{N}}_{\mathcal{P},\text{ét-fét}})^{-1}(\mathcal{G}^{[\nu]}) \to \mathcal{G}^{[\mu]}$ for each $(s\colon [\nu] \to [\mu]) \in \operatorname{Mor}(\Delta)$, where $\underline{s}$ denotes the morphism $X^{[\mu]} \to X^{[\nu]}$ corresponding to $s$, such that $\rho_{\mathrm{id}} = \mathrm{id}$ and $\rho_{st} = \rho_s \circ (\underline{s}^{\mathbb{N}}_{\mathcal{P},\text{ét-fét}})^{-1}(\rho_t)$ for two composable morphisms $[\kappa] \xrightarrow{t} [\nu] \xrightarrow{s} [\mu]$. Similarly, one can define ringed topos $(((\mathcal{P}/\underline{X})^{\mathbb{N}}_{\text{ét-fét}})^{\mathbb{N}^\circ\sim}, \mathcal{O}^{\mathbb{N}}_{\mathcal{P}/\underline{X},\bullet})$, $(\underline{X}^{\sim}_{\text{étaff}}, \mathcal{O}_{\underline{X}_1})$, $(\underline{X}^{\mathbb{N}\sim}_{\text{étaff}}, \mathcal{O}_{\underline{X}_1,\bullet})$, $((\underline{X}^{\mathbb{N}}_{\text{étaff}})^{\mathbb{N}^\circ\sim}, \mathcal{O}^{\mathbb{N}}_{\underline{X}_1,\bullet})$, $(\underline{X}^{\sim}_{1,\text{ét}}, \mathcal{O}_{\underline{X}_1})$, $((\underline{X}_1/D(\overline{\Sigma}))^{r,\sim}_{\mathrm{HIGGS}}, \mathcal{O}_{\underline{X}_1/D(\overline{\Sigma}),1})$ and $((\underline{X}_1/D(\overline{\Sigma}))^{\sim}_{\mathrm{HIGGS}}, \mathcal{O}_{\underline{X}_1/D(\overline{\Sigma}),1})$ by using the fibered ringed topos over $X_{\text{étaff}}$ defined by $(((\mathcal{P}/U)^{\mathbb{N}}_{\text{ét-fét}})^{\mathbb{N}^\circ\sim}, \mathcal{O}^{\mathbb{N}}_{\mathcal{P}/U,\bullet})$, $(U^{\sim}_{\text{étaff}}, \mathcal{O}_{U_1})$, etc. for $U \in \operatorname{Ob} X_{\text{étaff}}$. We may naturally regard an object of $\underline{X}^{\sim}_{1,\text{ét}}$ as an object of $\underline{X}^{\sim}_{\text{étaff}}$.

Let $C^{\Delta}_X$ be the constant simplicial object of $X_{\text{étaff}}$ defined by $X$. Then the morphism of functors $\underline{X} \to C^{\Delta}_X$ induces morphisms of ringed topos $\overline{\theta}\colon ((\mathcal{P}/\underline{X})^{\mathbb{N}\sim}_{\text{ét-fét}}, \mathcal{O}_{\mathcal{P}/\underline{X},\bullet}) \to ((\mathcal{P}/X)^{\mathbb{N}\sim}_{\text{ét-fét}}, \mathcal{O}_{\mathcal{P}/X,\bullet})$, $\overline{\theta}\colon (\underline{X}^{\sim}_{\text{étaff}}, \mathcal{O}_{\underline{X}_1}) \to (X^{\sim}_{\text{étaff}}, \mathcal{O}_{X_1})$, etc. (cf. **[2]** $\mathrm{V}^{\mathrm{bis}}$ (2.1.3), (2.2.1)). For $(X^{\sim}_{\text{étaff}}, \mathcal{O}_{X_1})$, we have an isomorphism of functors

$$\mathrm{id}_{D^+(X_{\text{étaff}},\mathcal{O}_{X_1})} \xrightarrow{\cong} R\overline{\theta}_*\overline{\theta}^*. \tag{IV.6.7.26}$$

By (IV.6.5.17) and the remark following it, we see that the morphisms of ringed topos $\pi^{h,\mathbb{N}^\circ}_{U,\text{ét}}\colon ((\mathcal{P}/U)^{\mathbb{N}\sim}_{\text{ét-fét}}, \mathcal{O}_{\mathcal{P}/U,\bullet}) \to (U^{\mathbb{N}\sim}_{\text{étaff}}, \mathcal{O}_{U_1,\bullet})$ for $U \in \operatorname{Ob} X_{\text{étaff}}$ induce a Cartesian morphism between the corresponding fibered ringed topos and then a morphism of ringed topos

$$\pi^{h,\mathbb{N}^\circ}_{\underline{X},\text{ét}}\colon ((\mathcal{P}/\underline{X})^{\mathbb{N}\sim}_{\text{ét-fét}}, \mathcal{O}_{\mathcal{P}/\underline{X},\bullet}) \longrightarrow (\underline{X}^{\mathbb{N}\sim}_{\text{étaff}}, \mathcal{O}_{\underline{X}_1,\bullet})$$

compatible with $\overline{\theta}$'s (cf. **[2]** $\mathrm{V}^{\mathrm{bis}}$ Proposition (1.2.15), Lemme (1.2.16)). Similarly one can construct morphisms of ringed topos

$$(\pi^{h,\mathbb{N}^\circ}_{\underline{X},\text{ét}})^{\mathbb{N}}\colon (((\mathcal{P}/\underline{X})^{\mathbb{N}}_{\mathrm{HIGGS}})^{\mathbb{N}^\circ\sim}, \mathcal{O}^{\mathbb{N}}_{\mathcal{P}/\underline{X},\bullet}) \longrightarrow ((\underline{X}^{\mathbb{N}}_{\text{étaff}})^{\mathbb{N}^\circ\sim}, \mathcal{O}^{\mathbb{N}}_{\underline{X}_1,\bullet}),$$
$$u_{\underline{X}/D(\overline{\Sigma})}\colon ((\underline{X}/D(\overline{\Sigma}))^{\star\sim}_{\mathrm{HIGGS}}, \mathcal{O}_{\underline{X}/D(\overline{\Sigma}),1}) \longrightarrow (\underline{X}^{\sim}_{1,\text{ét}}, \mathcal{O}_{\underline{X}_1}),$$
$$\mu_{r\star}\colon ((\underline{X}/D(\overline{\Sigma}))^{\star\sim}_{\mathrm{HIGGS}}, \mathcal{O}_{\underline{X}/D(\overline{\Sigma}),1}) \longrightarrow ((\underline{X}/D(\overline{\Sigma}))^{r\sim}_{\mathrm{HIGGS}}, \mathcal{O}_{\underline{X}/D(\overline{\Sigma}),1})$$

compatible with $\overline{\theta}$'s. Here $\star = s \in \mathbb{N}$ or $\emptyset$ and assume $s \geq r$ for the last morphism. By (IV.6.6.4) the morphisms of topos $\underleftarrow{l}$ (resp. $\underleftarrow{l}^{\mathbb{N}}$) induce a morphism of ringed topos $(\underline{X}^{\mathbb{N}\sim}_{\text{étaff}}, \mathcal{O}_{\underline{X}_1,\bullet}) \to (\underline{X}^{\sim}_{\text{étaff}}, \mathcal{O}_{\underline{X}_1})$ (resp. $((\underline{X}^{\mathbb{N}}_{\text{étaff}})^{\mathbb{N}^\circ\sim}, \mathcal{O}^{\mathbb{N}}_{\underline{X}_1,\bullet}) \to (\underline{X}^{\mathbb{N}^\circ\sim}_{\text{étaff}}, \mathcal{O}^{\mathbb{N}}_{\underline{X}_1})$), which is also denoted by $\underleftarrow{l}$ (resp. $\underleftarrow{l}^{\mathbb{N}}$). Similarly, by (IV.6.6.9), the morphisms of topos $\underrightarrow{l}$ induce morphisms of ringed topos $((\mathcal{P}/\underline{X})^{\mathbb{N}\sim}_{\text{ét-fét}}, \mathcal{O}_{\mathcal{P}/\underline{X},\bullet}) \to (((\mathcal{P}/\underline{X})^{\mathbb{N}}_{\text{ét-fét}})^{\mathbb{N}^\circ\sim}, \mathcal{O}^{\mathbb{N}}_{\mathcal{P}/\underline{X},\bullet})$, $(\underline{X}^{\sim}_{\text{étaff}}, \mathcal{O}_{\underline{X}_1}) \to (\underline{X}^{\mathbb{N}^\circ\sim}_{\text{étaff}}, \mathcal{O}^{\mathbb{N}}_{\underline{X}_1})$, etc., which are also denoted by $\underrightarrow{l}$. They are compatible with $\overline{\theta}$'s.

Let $r \in \mathbb{N}_{>0}$ and let $(\mathcal{F}, \mathcal{F}^\circ)$ be an object of $\mathrm{HC}^r_{\mathbb{Z}_p}(X_1/D(\overline{\Sigma}))$. Let $(\mathcal{F}^{[\nu]}, \mathcal{F}^{[\nu],\circ})$ be the pull-back of $(\mathcal{F}, \mathcal{F}^\circ)$ by the morphism $X_1^{[\nu]} \to X_1$. Then the sheaves of $\mathcal{O}_{\mathcal{P}/X^{[\nu]},\bullet}$-modules $\mathcal{T}_{X^{[\nu]},\bullet}(\mathcal{F}^{[\nu],\circ})$ and the isomorphisms $\underline{s}^*_{\text{HIGGS}}(\mathcal{T}_{X^{[\nu]},\bullet}(\mathcal{F}^{[\nu],\circ})) \to \mathcal{T}_{X^{[\mu]},\bullet}(\mathcal{F}^{[\mu],\circ})$ (cf. Lemma IV.6.4.6) for morphisms $s\colon [\nu] \to [\mu]$ in $\Delta$ define a sheaf of $\mathcal{O}_{\mathcal{P}/\underline{X},\bullet}$-modules on $(\mathcal{P}/\underline{X})^{\mathbb{N},\sim}_{\text{ét-fét}}$, which is denoted by $\mathcal{T}_{\underline{X},\bullet}(\underline{\mathcal{F}}^\circ)$. By Lemma IV.6.4.6, we have a canonical isomorphism

$$\overline{\theta}^*(\mathcal{T}_{X,\bullet}(\mathcal{F}^\circ)) \cong \mathcal{T}_{\underline{X},\bullet}(\underline{\mathcal{F}}^\circ). \tag{IV.6.7.27}$$

Similarly we obtain a complex of $\mathcal{O}^{\mathbb{N}}_{\mathcal{P}/\underline{X},\bullet}$-modules $\mathcal{T}^{\star+1}_{\underline{X},\bullet}\Omega_{\underline{Y}}(\underline{\mathcal{F}}^\circ)$ from $\mathcal{T}^{\star+1}_{X^{[\nu]},\bullet}\Omega_{Y^{[\nu]}}(\mathcal{F}^{[\nu],\circ})$ and (IV.6.7.23) applied to the morphism $X^{[\mu]} \to X^{[\nu]}$ associated to a morphism $s\colon [\nu] \to [\mu]$ in $\Delta$. By (IV.6.7.24), we see that the morphisms (IV.6.7.12) for $X^{[\nu]} \hookrightarrow Y^{[\nu]}$ and $\mathcal{F}^{[\nu],\circ}$ define a morphism of complexes of $\mathcal{O}^{\mathbb{N}}_{\mathcal{P}/\underline{X},\bullet}$-modules on $((\mathcal{P}/\underline{X})^{\mathbb{N}}_{\text{ét-fét}})^{\mathbb{N}^\circ\sim}$

$$\underrightarrow{l}_*\mathcal{T}_{\underline{X},\bullet}(\underline{\mathcal{F}}^\circ) \longrightarrow \mathcal{T}^{\star+1}_{\underline{X},\bullet}\Omega_{\underline{Y}}(\underline{\mathcal{F}}^\circ). \tag{IV.6.7.28}$$

Let $R\pi^h_{\underline{X}*}$ denote the composition (cf. Lemma IV.6.7.18).

$$\underrightarrow{l}^* \circ \mathbb{Q} \otimes \circ R\underleftarrow{l}^{\mathbb{N}}_* \circ R(\pi^{h,\mathbb{N}^\circ}_{\underline{X},\text{ét}})^{\mathbb{N}}_*\colon D^+(((\mathcal{P}/\underline{X})^{\mathbb{N}}_{\text{ét-fét}})^{\mathbb{N}^\circ\sim}, \mathcal{O}^{\mathbb{N}}_{\mathcal{P}/\underline{X},\bullet}) \to D^+(\underline{X}^{\sim}_{\text{étaff}}, \mathcal{O}_{\underline{X}_1,\mathbb{Q}}).$$

Then, since the four functors appearing above can be computed fiber by fiber (by **[2]** $\mathrm{V}^{\mathrm{bis}}$ Corollaire (1.3.12) for the right derived functors), Lemma IV.6.7.18 implies that (IV.6.7.28) induces an isomorphism

$$R\pi^h_{\underline{X}*}(\underrightarrow{l}_*\mathcal{T}_{\underline{X},\bullet}(\underline{\mathcal{F}}^\circ)) \xrightarrow{\cong} R\pi^h_{\underline{X}*}(\mathcal{T}^{\star+1}_{\underline{X},\bullet}\Omega_{\underline{Y}}(\underline{\mathcal{F}}^\circ))$$

in $D^+(\underline{X}^{\sim}_{1,\text{étaff}}, \mathcal{O}_{\underline{X}_1,\mathbb{Q}})$. By the same argument as after Lemma IV.6.7.18, we see $R\pi^h_{\underline{X}*} \circ \underrightarrow{l}_* \cong \mathbb{Q} \otimes \circ R\underleftarrow{l}_* \circ R\pi^{h,\mathbb{N}^\circ}_{\underline{X},\text{ét}*}$ and obtain an isomorphism

$$\mathbb{Q} \otimes R\underleftarrow{l}_* R\pi^{h,\mathbb{N}^\circ}_{\underline{X},\text{ét}*}\mathcal{T}_{\underline{X},\bullet}(\underline{\mathcal{F}}^\circ) \xrightarrow{\cong} R\pi^h_{\underline{X}*}(\mathcal{T}^{\star+1}_{\underline{X},\bullet}\Omega_{\underline{Y}}(\underline{\mathcal{F}}^\circ)). \tag{IV.6.7.29}$$

**Lemma IV.6.7.30.** *The base change morphism $\overline{\theta}^* R\underleftarrow{l}_* R\pi^{h,\mathbb{N}^\circ}_{X,\text{ét}*} \to R\underleftarrow{l}_* R\pi^{h,\mathbb{N}^\circ}_{\underline{X},\text{ét}*}\overline{\theta}^*$ between the functors from $D^+((\mathcal{P}/X)^{\mathbb{N}}_{\text{ét-fét}}, \mathcal{O}_{\mathcal{P}/X,\bullet})$ to $D^+(\underline{X}^{\sim}_{\text{étaff}}, \mathcal{O}_{\underline{X}_1})$ is an isomorphism.*

Proof. Since the right derived functors $R\underleftarrow{l}_*$ and $R\pi^{h,\mathbb{N}^\circ}_{\underline{X},\text{ét}*}$ appearing in the target are computed fiber by fiber (cf. **[2]** $\mathrm{V}^{\mathrm{bis}}$ Corollaire (1.3.12)), it suffices to prove the base change theorems for $R\underleftarrow{l}_*$ and $R\pi^{h,\mathbb{N}^\circ}_{-,\text{ét}*}$ with respect to a strict étale morphism $f\colon X' \to X$. By Corollary IV.6.1.4 (resp. Lemma IV.3.1.10 applied to $X_{\text{étaff}}$ and $X'$), **[47]** (1.1) Proposition b), and **[2]** V Proposition 4.11 1), the inverse image of an injective sheaf of $\mathcal{O}_{X/\mathcal{P},\bullet}$ (resp. $\mathcal{O}_{X_1,\bullet}$)-modules by $f$ is again injective. Hence it suffices to verify the base change theorems for the functors $\underleftarrow{l}_*$ and $\pi^{h,\mathbb{N}^\circ}_{-,\text{ét}*}$, which are obvious. □

From (IV.6.7.27) and (IV.6.7.29) and Lemma IV.6.7.30, we obtain an isomorphism

$$\overline{\theta}^*(\mathbb{Q} \otimes R\underleftarrow{l}_* R\pi^{h,\mathbb{N}^\circ}_{X,\text{ét}*}\mathcal{T}_{X,\bullet}(\mathcal{F}^\circ)) \xrightarrow{\cong} R\pi^h_{\underline{X}*}(\mathcal{T}^{\star+1}_{\underline{X},\bullet}\Omega_{\underline{Y}}(\underline{\mathcal{F}}^\circ)). \tag{IV.6.7.31}$$

Let $\underline{\mathcal{F}}$ denote the sheaf of $\mathcal{O}_{\underline{X}/D(\overline{\Sigma}),1,\mathbb{Q}}$-modules $\underline{\theta}^*(\mathcal{F})$ on $(\underline{X}/D(\overline{\Sigma}))^{r\sim}_{\mathrm{HIGGS}}$. Let $\mathcal{F}^s$, $\mathcal{F}^{[\nu],s}$, and $\underline{\mathcal{F}}^s$ for $s\in\mathbb{N}_{\geq r}$ (resp. $\mathcal{F}^\dagger$, $\mathcal{F}^{[\nu],\dagger}$, and $\underline{\mathcal{F}}^\dagger$) be the inverse image of $\mathcal{F}$, $\mathcal{F}^{[\nu]}$, and $\underline{\mathcal{F}}$ by the inverse image functors $\mu^*_{rs}$ (resp. $\mu^*_r$). The sheaf $\mathcal{F}^{[\nu],s}$ (resp. $\mathcal{F}^{[\nu],\dagger}$) is canonically isomorphic to the inverse image of $\mathcal{F}^s$ (resp. $\mathcal{F}^\dagger$) by the morphism $X_1^{[\nu]}\to X_1$, and we have canonical isomorphisms $\underline{\mathcal{F}}^s\cong\overline{\theta}^*(\mathcal{F}^s)$ and $\underline{\mathcal{F}}^\dagger\cong\overline{\theta}^*(\mathcal{F}^\dagger)$. Hence the sheaf $\underline{\mathcal{F}}^s$ (resp. $\underline{\mathcal{F}}^\dagger$) is canonically isomorphic to the sheaf defined by the system $(\mathcal{F}^{[\nu],s})_\nu$ (resp. $(\mathcal{F}^{[\nu],\dagger})_\nu$).

Let $D^{[\nu],s}$ ($s\in\mathbb{N}_{>0}$) be $D^s_{\mathrm{Higgs}}(X_1^{[\nu]}\hookrightarrow Y_\bullet^{[\nu]})$, let $z^{[\nu],s}$ be the natural morphism $D_1^{[\nu],s}\to X_1^{[\nu]}$, and let $(\mathcal{M}^{[\nu],s},\theta)$ ($s\geq r$) denote the object of $\mathrm{HB}^s_{\mathbb{Q}_p,\mathrm{conv}}(X_1^{[\nu]},Y_\bullet^{[\nu]}/D(\overline{\Sigma}))$ corresponding to $\mathcal{F}^{[\nu],s}$ by Theorem IV.3.4.16. We define the complex $\mathcal{C}_{Y^{[\nu]}}(\mathcal{M}^{[\nu],s})$ of $\mathcal{O}_{X_1^{[\nu]},\mathbb{Q}}$-modules to be $z_*^{[\nu],s}(\mathcal{M}^{[\nu],s}\otimes_{\mathcal{O}_{Y_1^{[\nu]}}}\Omega^\bullet_{Y_1^{[\nu]}/\overline{\Sigma}})$. For a morphism $\tau\colon[\nu]\to[\mu]$, we have a natural morphism $\underline{\tau}^*_{1,\mathrm{\acute{e}t}}(\mathcal{C}_{Y^{[\nu]}}(\mathcal{M}^{[\nu],s}))\to\mathcal{C}_{Y^{[\mu]}}(\mathcal{M}^{[\mu],s})$ compatible with $s$ and with compositions. Here $\underline{\tau}_1$ denotes the morphism $X_1^{[\mu]}\to X_1^{[\nu]}$ corresponding to $\tau$. These data define a direct system of complexes of $\mathcal{O}_{\underline{X}_1,\bullet}$-modules $(\mathcal{C}_{\underline{Y}}(\underline{\mathcal{M}}^s))_{s\geq r}$ on $\underline{X}^\sim_{1,\mathrm{\acute{e}t}}$, which we also regard as that on $\underline{X}^\sim_{\mathrm{\acute{e}taff}}$. Now, by (IV.6.7.25), the morphisms (IV.6.7.14) for $X^{[\nu]}\hookrightarrow Y^{[\nu]}$, $Y_\bullet^{[\nu]}$ and $(\mathcal{F}^{[\nu],s},\mathcal{F}^{[\nu]\circ})$ induce a morphism of complexes of $\mathcal{O}_{\underline{X}_1,\mathbb{Q}}$-modules

(IV.6.7.32)
$$\varinjlim_s\mathcal{C}_{\underline{Y}}(\underline{\mathcal{M}}^s)\longrightarrow\boldsymbol{\pi}^h_{\underline{X}*}(\mathcal{T}^{\star+1}_{\underline{X},\bullet}\Omega_{\underline{Y}}(\underline{\mathcal{F}}^\circ))$$

on $\underline{X}^\sim_{\mathrm{\acute{e}taff}}$, where $\boldsymbol{\pi}^h_{\underline{X}*}$ denotes the composition

$$\underset{\rightarrow}{l}^*\circ\mathbb{Q}\otimes\circ\underset{\leftarrow}{l}^{\mathbb{N}}_*\circ(\pi^{h,\mathbb{N}^\circ}_{\underline{X},\mathrm{\acute{e}t}*})^{\mathbb{N}}\colon(((\mathcal{P}/\underline{X})^{\mathbb{N}}_{\mathrm{\acute{e}t\text{-}f\acute{e}t}})^{\mathbb{N}^\circ\sim},\mathcal{O}^{\mathbb{N}}_{\mathcal{P}/\underline{X},\bullet})\to(\underline{X}^\sim_{\mathrm{\acute{e}taff}},\mathcal{O}_{\underline{X}_1,\mathbb{Q}}).$$

**Proposition IV.6.7.33.** *There exists a canonical isomorphism in* $D^+(\underline{X}^\sim_{\mathrm{\acute{e}taff}},\mathcal{O}_{\underline{X}_1,\mathbb{Q}})$

$$Ru_{\underline{X}_1/D(\overline{\Sigma})*}(\underline{\mathcal{F}}^\dagger)\cong\varinjlim_s\mathcal{C}_{\underline{Y}}(\underline{\mathcal{M}}^s).$$

PROOF. By Proposition IV.4.5.5, we have a resolution

$$\mathcal{F}^{[\nu]\dagger}\longrightarrow\varinjlim_s\mu_s^*(L_{Y_\bullet^{[\nu]}}(\xi^{-\bullet}\mathcal{M}^{[\nu],s}\otimes_{\mathcal{O}_{Y_1^{[\nu]}}}\Omega^\bullet_{Y_1^{[\nu]}/\overline{\Sigma}}))$$

for each $\nu\in\mathbb{N}$. By construction, we have a canonical morphism

$$\underline{\tau}^*_{1,\mathrm{HIGGS}}(L_{Y_\bullet^{[\nu]}}(\xi^{-\bullet}\mathcal{M}^{[\nu],s}\otimes_{\mathcal{O}_{Y_1^{[\nu]}}}\Omega^\bullet_{Y_1^{[\nu]}/\overline{\Sigma}}))\longrightarrow L_{Y_\bullet^{[\mu]}}(\xi^{-\bullet}\mathcal{M}^{[\mu],s}\otimes_{\mathcal{O}_{Y_1^{[\mu]}}}\Omega^\bullet_{Y_1^{[\mu]}/\overline{\Sigma}})$$

for the morphism $\underline{\tau}\colon X_1^{[\mu]}\to X_1^{[\nu]}$ corresponding to a morphism $\tau\colon[\nu]\to[\mu]$. This morphism is compatible with $s$, the compositions of $\tau$'s, and the resolution of $\mathcal{F}^{[\nu]\dagger}$ above. Thus we obtain a resolution $\underline{\mathcal{F}}^\dagger\to\overline{\mathcal{C}}_{Y_\bullet}(\underline{\mathcal{M}})$ of $\underline{\mathcal{F}}^\dagger$ as an $\mathcal{O}_{\underline{X}_1/D(\overline{\Sigma}),1,\mathbb{Q}}$-module on $(\underline{X}_1/D(\overline{\Sigma}))^\sim_{\mathrm{HIGGS}}$. Since the right derived functor $Ru_{\underline{X}_1/D(\overline{\Sigma})*}$ can be computed fiber by fiber ([2] V$^{\mathrm{bis}}$ Corollaire (1.3.12)), applying the argument before Theorem IV.4.5.6 to each $X_{[\nu]}$, we obtain

$$Ru_{\underline{X}_1/D(\overline{\Sigma})*}(\overline{\mathcal{C}}_{Y_\bullet}(\underline{\mathcal{M}}))\cong u_{\underline{X}_1/D(\overline{\Sigma})*}(\overline{\mathcal{C}}_{Y_\bullet}(\underline{\mathcal{M}}))\cong\varinjlim_s\mathcal{C}_{\underline{Y}}(\underline{\mathcal{M}}^s).$$ □

**Proposition IV.6.7.34.** *The following base change morphism in* $D^+(\underline{X}^\sim_{\mathrm{\acute{e}taff}},\mathcal{O}_{\underline{X}_1,\mathbb{Q}})$ *is an isomorphism*

$$\overline{\theta}^*Ru_{X_1/D(\overline{\Sigma})*}(\mathcal{F}^\dagger)\longrightarrow Ru_{\underline{X}_1/D(\overline{\Sigma})*}\overline{\theta}^*(\mathcal{F}^\dagger)\cong Ru_{\underline{X}_1/D(\overline{\Sigma})*}(\underline{\mathcal{F}}^\dagger).$$

PROOF. Since $Ru_{\underline{X}_1/D(\overline{\Sigma})*}$ can be computed fiber by fiber (cf. **[2]** $\mathrm{V}^{\mathrm{bis}}$ Corollaire (1.3.12)), it suffices to prove that the base change morphism

$$f^*_{1,\text{ét}}(R^q u_{X_1/D(\overline{\Sigma})*}\mathcal{F}^\dagger) \to R^q u_{X'_1/D(\overline{\Sigma})*}(f^*_{1,\mathrm{HIGGS}}(\mathcal{F}^\dagger))$$

is an isomorphism for a strict étale morphism $f\colon X' \to X$ such that $X'$ is affine. As before Theorem IV.4.5.6, we have an isomorphism

$$\varinjlim_{s\geq r} R^q u_{X_1/D(\overline{\Sigma})*}\mathcal{F}^s \xrightarrow{\cong} R^q u_{X_1/D(\overline{\Sigma})*}\mathcal{F}^\dagger$$

and a similar isomorphism for $X'$ and $f^*_{1,\mathrm{HIGGS}}(\mathcal{F}^s)$. Thus we are reduced to proving the base change theorem for $Ru_{X_1/D(\overline{\Sigma})*}\colon D^+((X_1/D(\overline{\Sigma}))^r_{\mathrm{HIGGS}}, \mathcal{O}_{X_1/D(\overline{\Sigma}),1,\mathbb{Q}}) \to D^+(X_{1,\text{ét}}, \mathcal{O}_{X_1,\mathbb{Q}})$ with respect to $f_1\colon X'_1 \to X_1$. By Proposition IV.3.1.9 (1) and **[2]** V Proposition 4.11 1), the inverse image of an injective sheaf of $\mathcal{O}_{X_1/D(\overline{\Sigma}),1,\mathbb{Q}}$-modules by $f_1$ is injective. Hence it suffices to prove the base change theorem for $u_{X_1/D(\overline{\Sigma})*}$, which is obvious. □

From (IV.6.7.32), Proposition IV.6.7.33, and Proposition IV.6.7.34, we obtain a morphism

$$\overline{\theta}^* Ru_{X_1/D(\overline{\Sigma})*}(\mathcal{F}^\dagger) \longrightarrow R\pi^h_{\underline{X},\text{ét}*}(\mathcal{T}^{\star+1}_{\underline{X},\bullet}\Omega_{\underline{Y}}(\underline{\mathcal{F}}^\circ)) \tag{IV.6.7.35}$$

in $D^+(\underline{X}^\sim_{\text{étaff}}, \mathcal{O}_{\underline{X}_1,\mathbb{Q}})$. Composing (IV.6.7.35) with the inverse of (IV.6.7.31), we obtain a morphism

$$\overline{\theta}^* Ru_{X_1/D(\overline{\Sigma})*}(\mathcal{F}^\dagger) \longrightarrow \overline{\theta}^*(\mathbb{Q}\otimes R\underleftarrow{l}_* R\pi^{h,\mathbb{N}^\circ}_{X,\text{ét}*}\mathcal{T}_{X,\bullet}(\mathcal{F}^\circ)). \tag{IV.6.7.36}$$

By taking $R\overline{\theta}_*$ and using (IV.6.7.26), we obtain the morphism (IV.6.7.1).

**IV.6.8. Comparison theorem.** We keep the notation in the previous subsection. In this subsection, we will prove the following theorem.

**Theorem IV.6.8.1.** *Assume that the log structure of* $\Sigma$ *is defined by the closed point and* $X$ *satisfies Condition* IV.5.3.1 *strict étale locally. Then, for any* $r \in \mathbb{N}_{>0}$ *and an object* $(\mathcal{F}, \mathcal{F}^\circ)$ *of* $\mathrm{HC}^r_{\mathbb{Z}_p}(X_1/D(\overline{\Sigma}))$*, the morphisms* (IV.6.7.1) *and* (IV.6.7.2) *are isomorphisms.*

The claim for (IV.6.7.2) obviously follows from that of (IV.6.7.1). By the construction of (IV.6.7.1), we are reduced to the case where $X$ is affine and satisfies Condition IV.5.3.1, $\mathcal{F}$ and $\mathcal{F}^\circ$ are finite on $X_1$ (cf. Definitions IV.3.3.1, IV.3.3.2), and $X \in \mathrm{Ob}\,\mathcal{C}_X$ (cf. Definition IV.6.4.1 (1)), which we assume in the following. Then $X_1 = \widehat{X}\times_{\widehat{\Sigma}}\overline{\Sigma}$ has a smooth Cartesian lifting $X_\bullet \to D(\overline{\Sigma})$ and we can construct the morphism (IV.6.7.1) by using $\mathrm{id}\colon X \to X$ and $X_\bullet$. Let $\mathcal{C}$ denote $\mathcal{C}^X = \mathcal{C}_X$ in the following (cf. the construction of $\mathscr{A}^r_{X,Y,m}$ in IV.6.7). By the construction of $\mathscr{A}^r_{X,X,\bullet}$ before (IV.6.7.15), the morphisms $z_{(U,\overline{s})}\colon D(U,\overline{s})_1 \to U_1$ for $(U,\overline{s}) \in \mathcal{C}_{\mathrm{gpt}}$ induce a morphism of $\mathcal{O}_{\mathcal{C}_1,\bullet}$-modules on $\mathcal{C}^{\mathbb{N}}_{\mathrm{tr}}$

$$\mathcal{O}_{\mathcal{C}_1,\bullet} \longrightarrow \pi^h_{\mathcal{C},\mathrm{gpt}*}\mathscr{A}^r_{X,X,\bullet} \tag{IV.6.8.2}$$

compatible with $r \in \mathbb{N}_{>0}$. By applying $\eta_{\mathcal{C}}$ (IV.6.7.4) and using (IV.6.7.5), we obtain a morphism of $\mathcal{O}_{X_1,\bullet}$-modules on $X^{\mathbb{N}}_{\text{étaff}}$

$$\mathcal{O}_{X_1,\bullet} \longrightarrow \pi^{h,\mathbb{N}^\circ}_{X,\text{ét}*}\eta_{\mathcal{P}/\mathcal{C}}(\mathscr{A}^r_{X,X,\bullet}) \tag{IV.6.8.3}$$

compatible with $r \in \mathbb{N}_{>0}$.

We will derive Theorem IV.6.8.1 from the isomorphism (IV.6.7.19) and the following theorem, which is an analogue of Theorem IV.5.3.4.

**Theorem IV.6.8.4.** *The following morphism in* $D^+(X_{\text{étaff}}, \mathscr{O}_{X_1,\mathbb{Q}})$ *induced by* (IV.6.8.3) *is an isomorphism.*

$$\mathscr{O}_{X_1,\mathbb{Q}} \cong \underset{\to}{l}^*\mathbb{Q} \otimes \underset{\to}{l}_*\mathscr{O}_{X_1} \longrightarrow \underset{\to}{l}^*\mathbb{Q} \otimes R\underset{\leftarrow}{l}^{\mathbb{N}}_* R(\pi^{h,\mathbb{N}^\circ}_{X,\text{ét}})^{\mathbb{N}}_* \eta_{\mathcal{P}/\mathcal{C}}(\mathscr{A}^{\star+1}_{X,X,\bullet}).$$

**Proposition IV.6.8.5.** *Let* $U$, $d$, $\overline{s}$, $\mathcal{A}$, $\mathscr{A}_1^r(\overline{\mathcal{A}})$ *be as in* IV.5.3. *Let* $\mathcal{A}_{\overline{V}}$ *be* $\mathcal{A}\otimes_V \overline{V}$. *Then there exist a positive integer* $N$ *depending only on* $d$ *and positive integers* $M_r$ $(r \in \mathbb{N}_{>0})$ *depending only on* $d$ *and* $r$ *such that, for any* $r, m, i \in \mathbb{N}_{>0}$, *the following natural homomorphisms are annihilated by* $p^{M_r}$.

$$H^i(\Delta_{(U,\overline{s})}, \mathscr{A}_1^r(\overline{\mathcal{A}})/p^m) \longrightarrow H^i(\Delta_{(U,\overline{s})}, \mathscr{A}_1^{r+N}(\overline{\mathcal{A}})/p^m),$$

$$\text{Ker}(\mathcal{A}_{\overline{V}}/p^m \to H^0(\Delta_{(U,\overline{s})}, \mathscr{A}_1^r(\overline{\mathcal{A}})/p^m)) \to \text{Ker}(\mathcal{A}_{\overline{V}}/p^m \to H^0(\Delta_{(U,\overline{s})}, \mathscr{A}_1^{r+N}(\overline{\mathcal{A}})/p^m)),$$

$$\text{Cok}(\mathcal{A}_{\overline{V}}/p^m \to H^0(\Delta_{(U,\overline{s})}, \mathscr{A}_1^r(\overline{\mathcal{A}})/p^m)) \to \text{Cok}(\mathcal{A}_{\overline{V}}/p^m \to H^0(\Delta_{(U,\overline{s})}, \mathscr{A}_1^{r+N}(\overline{\mathcal{A}})/p^m)).$$

PROOF. Let $\mathscr{A}_1^{r,\circ}(\overline{\mathcal{A}})$ and $\mathscr{A}_1^{r\circ}(\mathcal{A}_\infty)$ be as before Corollary IV.5.3.8. Then as in the proof of Corollary IV.5.3.8, the kernel and cokernel of $H^q(\Delta_\infty, \mathscr{A}_1^{r,\circ}(\mathcal{A}_\infty)/p^m) \to H^q(\Delta_{(U,\overline{s})}, \mathscr{A}_1^{r,\circ}(\overline{\mathcal{A}})/p^m)$ are annihilated by $\overline{\mathfrak{m}}$ for every $q \in \mathbb{N}$. We define ${}_a\mathscr{A}^r$ $(a \in \mathbb{N}\cap[0,d]$, $r \in \mathbb{N}_{>0})$ and $\gamma_i \in \Delta_\infty$ $(i \in \mathbb{N}\cap[1,d])$ as before Proposition IV.5.3.9. We define the complexes $C_a^r$ $(a \in \mathbb{N}\cap[0,d]$, $r \in \mathbb{N}_{>0})$ as in the proof of Theorem IV.5.3.4. Then $R\Gamma(\Delta_\infty, \mathscr{A}_1^{r,\circ}(\mathcal{A}_\infty)/p^m)$ is canonically isomorphic to $C_d^r/p^mC_d^r$ in $D^+(\mathbb{Z}/p^m\text{-Mod})$. Hence it suffices to prove the following claim; the claim for $a = d$ implies the proposition. For $a \in \mathbb{N}\cap[0,d]$ and $r \in \mathbb{N}_{>0}$, let $\overline{C}_a^r$ be the mapping fiber of ${}_a\mathscr{A}^r \to C_a^r$. Then there exist $N_a$ and $M_{r,a}$ for $a \in \mathbb{N}\cap[0,d]$ and $r \in \mathbb{N}_{>0}$ such that the following holds: The natural homomorphism $H^q(\overline{C}_a^r) \to H^q(\overline{C}_a^{r+N_a})$ is annihilated by $p^{M_{r,a}}$ for $a \in \mathbb{N}\cap[0,d]$, $r \in \mathbb{N}_{>0}$, and $q \in \mathbb{Z}$.

We prove the claim by induction on $a$. The claim is trivial if $a = 0$ because ${}_0\mathscr{A}^r = C_0^r$. Let $a \in \mathbb{N}\cap[1,d]$ and assume that the claim holds for $\overline{C}_{a-1}^r$. By the construction of $C_a^r$, $\overline{C}_a^r$ is canonically isomorphic to the mapping fiber of $({}_a\mathscr{A}^r \to C_{a-1}^r) \longrightarrow (0 \to C_{a-1}^r)$. Hence, letting $\overline{D}_a^r$ (resp. $\overline{E}_a^r$) denote the mapping fiber of $\gamma_a - 1\colon \overline{C}_{a-1}^r \to \overline{C}_{a-1}^r$ (resp. the complex ${}_{a-1}\mathscr{A}^r/{}_a\mathscr{A}^r \xrightarrow{\gamma_a-1} {}_a\mathscr{A}^r \to 0 \to \cdots$), we obtain a short exact sequence

$$0 \longrightarrow \overline{C}_a^r \longrightarrow \overline{D}_a^r \longrightarrow \overline{E}_a^r \longrightarrow 0.$$

By the induction hypothesis, we see that the homomorphism $H^q(\overline{D}_a^r) \to H^q(\overline{D}_a^{r+2N_{a-1}})$ is annihilated by $p^{2M_{r,a-1}}$ for every $q \in \mathbb{Z}$. By Proposition IV.5.3.9, we have $H^q(\overline{E}_a^r) = 0$ $(q \neq 1)$ and the homomorphism $H^1(\overline{E}_a^r) \to H^1(\overline{E}_a^{r+1})$ is annihilated by $p^{l_{a,r}}$ for some $l_{a,r} \in \mathbb{N}$. Hence the homomorphism $H^q(\overline{C}_a^r) \to H^q(\overline{C}_a^{r+2N_{a-1}+1})$ is annihilated by $p^{2M_{r,a-1}+l_{a,r}}$. □

Let us consider the following commutative diagram of ringed topos.

(IV.6.8.6)

$$\begin{array}{ccccccc}
 & & ((\mathcal{P}/\mathcal{C})^{\mathbb{N}\sim}_{\text{ét-fét}}, \mathscr{O}_{\mathcal{P}/\mathcal{C},\bullet}) & \xrightarrow{\text{id}^{X,\mathcal{C}}_{\mathcal{P},\text{ét-fét}}} & & & ((\mathcal{P}/X)^{\mathbb{N}\sim}_{\text{ét-fét}}, \mathscr{O}_{\mathcal{P}/X,\bullet}) \\
 & \swarrow^{\pi^h_{\mathcal{C},\text{ét}}} & & & & \swarrow^{\pi^h_{X,\text{ét}}} & \\
(\mathcal{C}^{\mathbb{N}\sim}_{\text{ét}}, \mathscr{O}_{\mathcal{C}_1,\bullet}) & & \xrightarrow{\text{id}^{X,\mathcal{C}}_{\text{ét}}} & & (X^{\mathbb{N}\sim}_{\text{étaff}}, \mathscr{O}_{X_1,\bullet}) & & \\
 & & \downarrow{\scriptstyle v_{\mathcal{P}/\mathcal{C}}} & & & & \downarrow{\scriptstyle v_{\mathcal{P}/X}} \\
\downarrow{\scriptstyle v_{\mathcal{C}}} & & ((\mathcal{P}/\mathcal{C})^{\mathbb{N}\sim}_{\text{fét}}, \mathscr{O}^v_{\mathcal{P}/\mathcal{C},\bullet}) & \xrightarrow{\text{id}^{X,\mathcal{C}}_{\mathcal{P},\text{fét}}} & \downarrow{\scriptstyle v_X} & & ((\mathcal{P}/X)^{\mathbb{N}\sim}_{\text{fét}}, \mathscr{O}^v_{\mathcal{P}/X,\bullet}) \\
 & \swarrow^{\pi^h_{\mathcal{C}}} & & & & \swarrow^{\pi^h_X} & \\
(\mathcal{C}^{\mathbb{N}\sim}_{\text{tr}}, \mathscr{O}_{\mathcal{C}_1,\bullet}) & & \xrightarrow{\text{id}^{X,\mathcal{C}}_{\text{tr}}} & & (X^{\mathbb{N}\sim}_{\text{tr}}, \mathscr{O}_{X_1,\bullet}) & &
\end{array}$$

**Lemma IV.6.8.7.** (1) *The base change morphisms* $v^*_{\mathcal{P}/X}\mathrm{id}^{X,\mathcal{C}}_{\mathcal{P},\mathrm{fét}*} \longrightarrow \mathrm{id}^{X,\mathcal{C}}_{\mathcal{P},\mathrm{ét}\text{-}\mathrm{fét}*}v^*_{\mathcal{P}/\mathcal{C}}$ *and* $v^*_X\mathrm{id}^{X,\mathcal{C}}_{\mathrm{tr}*} \longrightarrow \mathrm{id}^{X,\mathcal{C}}_{\mathrm{ét}*}v^*_{\mathcal{C}} = \eta_{\mathcal{C}}$ *are isomorphisms.*

(2) *The morphism* (IV.6.7.5) *coincides with the composition of the following morphisms, where* $(*)$ *is the base change morphism for the right vertical face in* (IV.6.8.6).

$$\eta_{\mathcal{C}}\pi^h_{\mathcal{C},\mathrm{gpt}*} \xleftarrow[(1)]{\cong} v^*_X\mathrm{id}^{X,\mathcal{C}}_{\mathrm{tr}*}\pi^h_{\mathcal{C},\mathrm{gpt}*} \xrightarrow[(\mathrm{IV}.6.5.11)]{\cong} v^*_X\mathrm{id}^{X,\mathcal{C}}_{\mathrm{tr}*}\pi^h_{\mathcal{C}*}r_{\mathcal{C}} \cong v^*_X\pi^h_{X*}\mathrm{id}^{X,\mathcal{C}}_{\mathcal{P},\mathrm{fét}*}r_{\mathcal{C}}$$
$$\xrightarrow{(*)} \pi^h_{X,\mathrm{ét}*}v^*_{\mathcal{P}/X}\mathrm{id}^{X,\mathcal{C}}_{\mathcal{P},\mathrm{fét}*}r_{\mathcal{C}} \xrightarrow[(1)]{\cong} \pi^h_{X,\mathrm{ét}*}\eta_{\mathcal{P}/\mathcal{C}}.$$

PROOF. (1) We prove that the first morphism is an isomorphism. Since $\mathrm{id}^{X,\mathcal{C}*}_{\mathcal{P},\mathrm{ét}\text{-}\mathrm{fét}}$ is an equivalence of categories, it suffices to prove that it is an isomorphism after taking $\mathrm{id}^{X,\mathcal{C}*}_{\mathcal{P},\mathrm{ét}\text{-}\mathrm{fét}}\circ$ and composing with $\mathrm{id}^{X,\mathcal{C}*}_{\mathcal{P},\mathrm{ét}\text{-}\mathrm{fét}}\mathrm{id}^{X,\mathcal{C}}_{\mathcal{P},\mathrm{ét}\text{-}\mathrm{fét}*}v^*_{\mathcal{P}/\mathcal{C}} \xrightarrow{\cong} v^*_{\mathcal{P}/\mathcal{C}}$. The morphism thus obtained coincides with $\mathrm{id}^{X,\mathcal{C}*}_{\mathcal{P},\mathrm{ét}\text{-}\mathrm{fét}}v^*_{\mathcal{P}/X}\mathrm{id}^{X,\mathcal{C}}_{\mathcal{P},\mathrm{fét}*} \cong v^*_{\mathcal{P}/\mathcal{C}}\mathrm{id}^{X,\mathcal{C}*}_{\mathcal{P},\mathrm{fét}}\mathrm{id}^{X,\mathcal{C}}_{\mathcal{P},\mathrm{fét}*} \xrightarrow{\mathrm{adj}} v^*_{\mathcal{P}/\mathcal{C}}$. Hence it suffices to prove that adj: $\mathrm{id}^{X,\mathcal{C}*}_{\mathcal{P},\mathrm{fét}}\mathrm{id}^{X,\mathcal{C}}_{\mathcal{P},\mathrm{fét}*} \to \mathrm{id}$ is an isomorphism, which is reduced to $\widehat{\mathrm{id}^{X,\mathcal{C}}_{\mathcal{P},\mathrm{fét}}}^{\,*}\widehat{\mathrm{id}^{X,\mathcal{C}}_{\mathcal{P},\mathrm{fét}*}} \xrightarrow{\cong} \mathrm{id}$, i.e., the fully faithfulness of the functor $(\mathcal{P}/\mathcal{C}) \to (\mathcal{P}/X)$ (cf. **[2]** I Proposition 5.6). One can prove the second isomorphism by the same argument.

(2) We write $v$, $\pi$, and $\iota$ for $v_\bullet$, $\pi^h_\bullet$, and $\mathrm{id}^{X,\mathcal{C}}_\star$, where $\bullet = X, \mathcal{C}$, etc. and $\star = \mathrm{ét}, \mathrm{tr}$, etc. to simplify the notation. Then by considering the compositions of the base change morphisms for vertical faces in the diagram (IV.6.8.6), we obtain the following commutative diagram of the functors from $\mathcal{M}od((\mathcal{P}/\mathcal{C})_{\mathrm{fét}}, \mathcal{O}^v_{\mathcal{P}/\mathcal{C},\bullet})$ to $\mathcal{M}od(X_{\mathrm{étaff}}, \mathcal{O}_{X_1,\bullet})$.

$$\begin{array}{ccccc}
\iota_*v^*\pi_* & \longrightarrow & \iota_*\pi_*v^* & \xrightarrow{\cong} & \pi_*\iota_*v^* \\
\cong\Big\uparrow(1) & & & & \cong\Big\uparrow(1) \\
v^*\iota_*\pi_* & \xrightarrow{\cong} & v^*\pi_*\iota_* & \longrightarrow & \pi_*v^*\iota_*.
\end{array}$$

By composing with $r_{\mathcal{C}}$ from the left, we obtain the claim. □

**Proposition IV.6.8.8.** *Let* $N$ *and* $M_r$ $(r \in \mathbb{N}_{>0})$ *be as in Proposition* IV.6.8.5. *Then, for any* $r, m, i \in \mathbb{N}_{>0}$, *the following natural morphisms are annihilated by* $p^{M_r}$.

$$R^i\pi^{h,\mathbb{N}^\circ}_{X,\mathrm{ét}*}(\eta_{\mathcal{P}/\mathcal{C}}(\mathscr{A}^r_{X,X,\bullet})) \longrightarrow R^i\pi^{h,\mathbb{N}^\circ}_{X,\mathrm{ét}*}(\eta_{\mathcal{P}/\mathcal{C}}(\mathscr{A}^{r+N}_{X,X,\bullet})),$$
$$\mathrm{Ker}(\mathcal{O}_{X_1,\bullet} \to \pi^{h,\mathbb{N}^\circ}_{X,\mathrm{ét}*}(\eta_{\mathcal{P}/\mathcal{C}}(\mathscr{A}^r_{X,X,\bullet}))) \longrightarrow \mathrm{Ker}(\mathcal{O}_{X_1,\bullet} \to \pi^{h,\mathbb{N}^\circ}_{X,\mathrm{ét}*}(\eta_{\mathcal{P}/\mathcal{C}}(\mathscr{A}^{r+N}_{X,X,\bullet}))),$$
$$\mathrm{Cok}(\mathcal{O}_{X_1,\bullet} \to \pi^{h,\mathbb{N}^\circ}_{X,\mathrm{ét}*}(\eta_{\mathcal{P}/\mathcal{C}}(\mathscr{A}^r_{X,X,\bullet}))) \longrightarrow \mathrm{Cok}(\mathcal{O}_{X_1,\bullet} \to \pi^{h,\mathbb{N}^\circ}_{X,\mathrm{ét}*}(\eta_{\mathcal{P}/\mathcal{C}}(\mathscr{A}^{r+N}_{X,X,\bullet}))).$$

PROOF. Let $\mathscr{A}^r_{\bullet,\mathrm{fét}}$ be the object $\mathrm{id}^{X,\mathcal{C}}_{\mathcal{P},\mathrm{fét}*}r_{\mathcal{C}}\mathscr{A}^r_{X,X,\bullet}$ of $\mathcal{M}od((\mathcal{P}/X)_{\mathrm{fét}}, \mathcal{O}^v_{\mathcal{P}/X,\bullet})$. By Lemma IV.6.8.7 (1), we have isomorphisms $\mathcal{O}_{X_1,\bullet} \cong v^*_X\mathrm{id}^{X,\mathcal{C}}_{\mathrm{tr}*}\mathcal{O}_{\mathcal{C}_1,\bullet}$ and $\eta_{\mathcal{P}/\mathcal{C}}\mathscr{A}^r_{X,X,\bullet} \cong v^*_{\mathcal{P}/X}\mathscr{A}^r_{\bullet,\mathrm{fét}}$. On the other hand, the morphism (IV.6.8.2) induces a morphism

$$\mathrm{id}^{X,\mathcal{C}}_{\mathrm{tr}*}\mathcal{O}_{\mathcal{C}_1,\bullet} \longrightarrow \mathrm{id}^{X,\mathcal{C}}_{\mathrm{tr}*}\pi^h_{\mathcal{C},\mathrm{gpt}*}\mathscr{A}^r_{X,X,\bullet} \xrightarrow{\cong} \mathrm{id}^{X,\mathcal{C}}_{\mathrm{tr}*}\pi^h_{\mathcal{C}*}r_{\mathcal{C}}\mathscr{A}^r_{X,X,\bullet} \cong \pi^h_{X*}\mathscr{A}^r_{\bullet,\mathrm{fét}}.$$

Lemma IV.6.8.7 (2) implies that the morphism (IV.6.8.3) is obtained by taking $v^*_X$ of the above morphism and composing with the base change morphism $v^*_X\pi^h_{X*}\mathscr{A}^r_{\bullet,\mathrm{fét}} \to \pi^h_{X,\mathrm{ét}*}v^*_{\mathcal{P}/X}\mathscr{A}^r_{\bullet,\mathrm{fét}}$.

We have a canonical isomorphism $\Gamma(U, R^i\pi^h_{X*}\mathcal{G}) \cong R^i\Gamma((\mathcal{U}_{\overline{K},\mathrm{triv}})_{\mathrm{fét}}, \mathcal{G}|_{\mathcal{U}_{\overline{K},\mathrm{triv}}})$ for $U \in \mathrm{Ob}\,X_{\mathrm{étaff}}$ and a sheaf of abelian groups $\mathcal{G}$ on $(\mathcal{P}/X)_{\mathrm{fét}}$ (cf. **[2]** VI the first paragraph of the proof of Lemme 8.7.2). Hence Proposition IV.6.8.5 implies that the following morphisms are annihilated by $p^{M_r}$ after restricting to $\mathcal{C}'_{\rm tr}$, where $\mathcal{C}'$ is the full subcategory of $\mathcal{C}$ consisting of $U$ satisfying the condition in the beginning of IV.5.3.

$$R^i\pi^{h,\mathbb{N}^\circ}_{X*}(\mathscr{A}^r_{\bullet,\text{fét}}) \longrightarrow R^i\pi^{h,\mathbb{N}^\circ}_{X*}(\mathscr{A}^{r+N}_{\bullet,\text{fét}}),$$
$$\text{Ker}(\text{id}^{X,\mathcal{C}}_{\text{tr}*}\mathcal{O}_{\mathcal{C}_1,\bullet} \to \pi^{h,\mathbb{N}^\circ}_{X*}(\mathscr{A}^r_{\bullet,\text{fét}})) \longrightarrow \text{Ker}(\text{id}^{X,\mathcal{C}}_{\text{tr}*}\mathcal{O}_{\mathcal{C}_1,\bullet} \to \pi^{h,\mathbb{N}^\circ}_{X*}(\mathscr{A}^{r+N}_{\bullet,\text{fét}})),$$
$$\text{Cok}(\text{id}^{X,\mathcal{C}}_{\text{tr}*}\mathcal{O}_{\mathcal{C}_1,\bullet} \to \pi^{h,\mathbb{N}^\circ}_{X*}(\mathscr{A}^r_{\bullet,\text{fét}})) \longrightarrow \text{Cok}(\text{id}^{X,\mathcal{C}}_{\text{tr}*}\mathcal{O}_{\mathcal{C}_1,\bullet} \to \pi^{h,\mathbb{N}^\circ}_{X*}(\mathscr{A}^{r+N}_{\bullet,\text{fét}})).$$

Note that the adjunction morphisms $\text{id}^{X,\mathcal{C}*}_{\text{tr}}\text{id}^{X,\mathcal{C}}_{\text{tr}*} \to \text{id}$ and $\text{id}^{X,\mathcal{C}*}_{\mathcal{P},\text{fét}}\text{id}^{X,\mathcal{C}}_{\mathcal{P},\text{fét}*} \to \text{id}$ are isomorphisms (cf. the proof of Lemma IV.6.8.7). Now, by taking $v^*_X$ and using Proposition IV.6.5.22, we obtain the claim. Note that every object of $X_{\text{étaff}}$ admits a strict étale covering by objects of $\mathcal{C}'$. □

**Lemma IV.6.8.9.** *Let $C$ be a site, let $\mathbb{Z}/p^\bullet$ be the sheaf of rings on $C^{\mathbb{N}}$ defined by the inverse system of sheaves of rings $(\mathbb{Z}/p^m)_{m\in\mathbb{N}}$ on $C$, and let $(\mathbb{Z}/p^\bullet)^{\mathbb{N}}$ denote the sheaf of rings $\underrightarrow{l}_*(\mathbb{Z}/p^\bullet)$ on $(C^{\mathbb{N}})^{\mathbb{N}^\circ}$. Let $\mathcal{K}$ be the complex of $(\mathbb{Z}/p^\bullet)^{\mathbb{N}}$-modules bounded below on $(C^{\mathbb{N}})^{\mathbb{N}^\circ}$, which corresponds to a direct system of complexes of $\mathbb{Z}/p^\bullet$-modules $(\mathcal{K}^r)_{r\in\mathbb{N}}$ on $C^{\mathbb{N}}$. Assume that there exist $N\in\mathbb{N}$ and $M_r\in\mathbb{N}$ $(r\in\mathbb{N})$ such that the morphism $\mathcal{H}^i(\mathcal{K}^r)\to\mathcal{H}^i(\mathcal{K}^{r+N})$ is annihilated by $p^{M_r}$ for every $r\in\mathbb{N}$ and $i\in\mathbb{Z}$. Then we have $\underrightarrow{l}^*\mathbb{Q}\otimes R\underleftarrow{l}^{\mathbb{N}}_*(\mathcal{K})=0$.*

Proof. We have a spectral sequence

$$E_2^{a,b}=R^a\underleftarrow{l}^{\mathbb{N}}_*\mathcal{H}^b(\mathcal{K}) \Longrightarrow R^{a+b}\underleftarrow{l}^{\mathbb{N}}_*(\mathcal{K})$$

in $\mathcal{M}od(C^{\mathbb{N}^\circ},\underrightarrow{l}_*\mathbb{Z}_p)$, and the $r$-th component of $E_2^{a,b}$ is given by $R^a\underleftarrow{l}_*\mathcal{H}^b(\mathcal{K}^r)$. By the assumption on $\mathcal{K}^r$, the morphism $\mathbb{Q}\otimes R^a\underleftarrow{l}_*\mathcal{H}^b(\mathcal{K}^r)\to\mathbb{Q}\otimes R^a\underleftarrow{l}_*\mathcal{H}^b(\mathcal{K}^{r+N})$ vanishes. Hence $\underrightarrow{l}^*\mathbb{Q}\otimes R^{a+b}\underleftarrow{l}^{\mathbb{N}}(\mathcal{K})=0$. □

Proof of Theorem IV.6.8.4. Consider a distinguished triangle

$$\mathcal{O}^{\mathbb{N}}_{X_1,\bullet} \longrightarrow R(\pi^{h,\mathbb{N}^\circ}_{X,\text{ét}})^{\mathbb{N}}_*(\eta_{\mathcal{P}/\mathcal{C}}(\mathscr{A}^{\star+1}_{X,X,\bullet})) \longrightarrow \mathcal{K} \longrightarrow$$

in $D^+((X^{\mathbb{N}}_{\text{étaff}})^{\mathbb{N}^\circ},\mathcal{O}^{\mathbb{N}}_{X_1,\bullet})$. Then Proposition IV.6.8.8 implies that $\mathcal{K}$ satisfies the condition in Lemma IV.6.8.9. Hence $\underrightarrow{l}^*\mathbb{Q}\otimes R\underleftarrow{l}^{\mathbb{N}}_*(\mathcal{K})=0$. □

Proof of Theorem IV.6.8.1. Let $(\mathcal{M},\varepsilon)$ be the object of $\text{HS}_{\mathbb{Z}_p}(X_1,X_\bullet/D(\overline{\Sigma}))$ corresponding to $\mathcal{F}^\circ$ by the equivalence of categories in Proposition IV.3.4.4 (2), and let $\overline{\mathcal{M}}$ be the image of $\mathcal{M}\to\mathcal{M}_{\mathbb{Q}_p}$, which is naturally endowed with the structure $\overline{\varepsilon}$ of an object of $\text{HS}_{\mathbb{Z}_p}(X_1,X_\bullet/D(\overline{\Sigma}))$ induced by $\varepsilon$ (cf. Remark IV.3.4.17). Since we assume that $\mathcal{F}^\circ$ is finite on $X_1$, $\mathcal{M}$ and $\overline{\mathcal{M}}$ are objects of $\mathcal{LPM}(\mathcal{O}_{X_1})$. Let $(\overline{\mathcal{M}},\theta)$ denote the object of $\text{HB}_{\mathbb{Z}_p}(X_1,X_\bullet/D(\overline{\Sigma}))$ corresponding to $(\overline{\mathcal{M}},\overline{\varepsilon})$ by Theorem IV.3.4.16. For an integer $s\geq r$, the object of $\text{HB}^s_{\mathbb{Q}_p}(X_1,X_\bullet/D(\overline{\Sigma}))$ corresponding to the object $\mu^*_{rs}(\mathcal{F})$ of $\text{HC}^s_{\mathbb{Q}_p}(X_1/D(\overline{\Sigma}))$ is canonically isomorphic to $(\overline{\mathcal{M}}_{\mathbb{Q}_p},\theta_{\mathbb{Q}_p})$.

Let $\mathcal{C}^\wedge_{\text{gpt}}$ be the category of presheaves on $\mathcal{C}_{\text{gpt}}$, and define the functors $\pi^h_{\text{gpt}*}\colon\mathcal{C}^\wedge_{\text{gpt}}\to\mathcal{C}^\wedge$ and $\pi^{h*}_{\text{gpt}}\colon\mathcal{C}^\wedge\to\mathcal{C}^\wedge_{\text{gpt}}$ in the same way as before (IV.6.5.11) and (IV.6.5.12), respectively. The functor $\pi^{h*}_{\text{gpt}}$ naturally becomes a left adjoint of $\pi^h_{\text{gpt}*}$. We define the presheaves of rings $\mathscr{A}^s_{X,X}$ on $\mathcal{C}_{\text{gpt}}$ by $\mathscr{A}^s_{X,X}(U,\overline{s})=\Gamma(\mathscr{D}^s_{X,X,1}(U,\overline{s}),\mathcal{O}_{\mathscr{D}^s_{X,X,1}(U,\overline{s})})$ (cf. the definition of $\mathscr{A}^r_{X,Y,m}$ before (IV.6.7.15)). Put $\mathscr{A}\mathcal{T}^s=\mathscr{A}^s_{X,X}\otimes_{\widehat{\mathcal{O}}_{\mathcal{C}_{\text{gpt}}}}\mathcal{T}^X_{X,\text{gpt}}(\mathcal{F}^\circ)$ and $\mathscr{A}\mathcal{M}^s=\mathscr{A}^s_{X,X}\otimes_{\pi^{h*}_{\text{gpt}}\mathcal{O}_{\mathcal{C}_1}}\pi^{h*}_{\text{gpt}}(\overline{\mathcal{M}}|_{\mathcal{C}})$ for $s\in\mathbb{N}$, $s\geq r$. Here $\widehat{\mathcal{O}}_{\mathcal{C}_{\text{gpt}}}$ denotes the presheaf of rings $\varprojlim_m \mathcal{O}_{\mathcal{C}_{\mathrm{gpt},m}}$ on $\mathcal{C}_{\mathrm{gpt}}$. Then, from Proposition IV.5.2.12, we obtain an isomorphism of complexes of $\mathscr{A}^s_{X,X,\mathbb{Q}_p}$-modules on $\mathcal{C}_{\mathrm{gpt}}$

$$c\colon \mathscr{A}\mathcal{T}^s_{\mathbb{Q}_p} \otimes_{\pi^{h*}_{\mathrm{gpt}}(\mathcal{O}_X|_{\mathcal{C}})} \pi^{h*}_{\mathrm{gpt}}(\Omega^\bullet_{X/\Sigma}|_{\mathcal{C}}) \xrightarrow{\cong} \mathscr{A}\mathcal{M}^s_{\mathbb{Q}_p} \otimes_{\pi^{h*}_{\mathrm{gpt}}(\mathcal{O}_X|_{\mathcal{C}})} \pi^{h*}_{\mathrm{gpt}}(\Omega^\bullet_{X/\Sigma}|_{\mathcal{C}})$$

compatible with $s$. We assert that there exists $N \in \mathbb{N}$ such that $p^N c((\mathscr{A}\mathcal{T}^s)/p\text{-tor}) \subset (\mathscr{A}\mathcal{M}^s)/p\text{-tor}$ and $p^N c^{-1}((\mathscr{A}\mathcal{M}^s)/p\text{-tor}) \subset (\mathscr{A}\mathcal{T}^s)/p\text{-tor}$ for every $s$. This is reduced to the claim for $s = r$ and the section on $(X, \bar{t}) \in \mathcal{C}_{\mathrm{gpt}}$ because we have $\overline{\mathcal{M}}(U) = \overline{\mathcal{M}}(X) \otimes_{\mathcal{O}_{\mathcal{C}_1}(X)} \mathcal{O}_{\mathcal{C}_1}(U)$ and $\mathcal{T}^X_{X,\mathrm{gpt}}(\mathcal{F}^\circ)(U,\bar{s}) = \mathcal{T}^X_{X,\mathrm{gpt}}(\mathcal{F}^\circ)(X,\bar{t}) \otimes_{\widehat{\mathcal{O}}_{\mathcal{C}_{\mathrm{gpt}}}(X)} \widehat{\mathcal{O}}_{\mathcal{C}_{\mathrm{gpt}}}(U)$ for an object $(U, \bar{s})$ of $\mathcal{C}_{\mathrm{gpt}}$ over $(X, \bar{t})$ (cf. Lemma IV.3.2.8). This claim is obvious because $\overline{\mathcal{M}}(X)$ (resp. $\mathcal{T}^X_{X,\mathrm{gpt}}(X,\bar{t})$) is an object of $\mathcal{LPM}(\mathcal{O}_{\mathcal{C}_1}(X))$ (resp. $\mathcal{LPM}(\widehat{\mathcal{O}}_{\mathcal{C}_{\mathrm{gpt}}}(X))$). The last fact also implies that there exists $N' \in \mathbb{N}$ such that the $p$-torsion parts of $\mathscr{A}\mathcal{M}^s$ and $\mathscr{A}\mathcal{T}^s$ are annihilated by $p^{N'}$ for every $s \geq r$ by Lemma IV.6.8.10. Now, by using Lemma IV.6.8.11, we obtain morphisms

$$\mathscr{A}\mathcal{T}^s \otimes_{\pi^{h*}_{\mathrm{gpt}}(\mathcal{O}_X|_{\mathcal{C}})} \pi^{h*}_{\mathrm{gpt}}(\Omega^\bullet_{X/\Sigma}|_{\mathcal{C}}) \underset{\psi}{\overset{\varphi}{\rightleftarrows}} \mathscr{A}\mathcal{M}^s \otimes_{\pi^{h*}_{\mathrm{gpt}}(\mathcal{O}_X|_{\mathcal{C}})} \pi^{h*}_{\mathrm{gpt}}(\Omega^\bullet_{X/\Sigma}|_{\mathcal{C}})$$

such that $\varphi_{\mathbb{Q}} = p^{N+N'}c$, $\psi_{\mathbb{Q}} = p^{N+N'}c^{-1}$, $\varphi \circ \psi = p^{2(N+N')}\mathrm{id}$, and $\psi \circ \varphi = p^{2(N+N')}\mathrm{id}$. Let $\mathscr{A}\mathcal{T}^s_\bullet\Omega$ (resp. $\mathscr{A}\mathcal{M}^s_\bullet\Omega$) denote the complex in $\mathcal{M}\mathrm{od}(\mathcal{C}^{\wedge,\mathrm{cont}}_{\mathrm{gpt}}, \mathcal{O}_{\mathcal{C}_{\mathrm{gpt}},\bullet})$ obtained by taking the reduction mod $p^m$ of the source (resp. target) of the morphism $\varphi$ above. Let $\overline{\mathcal{M}}_\bullet$, $\varphi_\bullet$, and $\psi_\bullet$ denote the inverse systems obtained from $\overline{\mathcal{M}}$, $\varphi$, and $\psi$ by taking the reduction mod $p^m$ ($m \in \mathbb{Z}$). Since $(\overline{\mathcal{M}}/p^m\overline{\mathcal{M}})|_{\mathcal{C}}$ coincides with the reduction mod $p^m$ of $\overline{\mathcal{M}}|_{\mathcal{C}}$ as a presheaf, we obtain morphisms of complexes in $\mathcal{M}\mathrm{od}(\mathcal{C}_{\mathrm{tr}}, \mathcal{O}_{\mathcal{C}_1,\bullet})$

$$(\overline{\mathcal{M}}_\bullet \otimes_{\mathcal{O}_X} \Omega^\bullet_{X/\Sigma})|_{\mathcal{C}} \longrightarrow \pi^h_{\mathcal{C},\mathrm{gpt}*}(\mathscr{A}\mathcal{M}^s_\bullet\Omega) \xrightarrow{\pi^h_{\mathcal{C},\mathrm{gpt}*}(\varphi_\bullet)} \pi^h_{\mathcal{C},\mathrm{gpt}*}(\mathscr{A}\mathcal{T}^s_\bullet\Omega)$$

compatible with $s$. The morphism (IV.6.7.20) is obtained by applying $p^{-N-N'} \cdot \mathbb{Q} \otimes \underleftarrow{l}_*$ to the composition of the above morphisms. By taking $\eta_{\mathcal{C}}$ and composing with (IV.6.7.5), we obtain morphisms of complexes in $\mathcal{M}\mathrm{od}(X_{\text{étaff}}, \mathcal{O}_{X_1,\bullet})$

$$\overline{\mathcal{M}}_\bullet \otimes_{\mathcal{O}_X} \Omega^\bullet_{X/\Sigma} \longrightarrow \pi^h_{X,\text{ét}*}\eta_{\mathcal{P}/\mathcal{C}}(\mathscr{A}\mathcal{M}^s_\bullet\Omega) \xrightarrow{\pi^h_{X,\text{ét}*}\eta_{\mathcal{P}/\mathcal{C}}(\varphi_\bullet)} \pi^h_{X,\text{ét}*}\eta_{\mathcal{P}/\mathcal{C}}(\mathscr{A}\mathcal{T}^s_\bullet\Omega).$$

The morphism (IV.6.7.14) is obtained by taking $p^{-N-N'} \cdot \mathbb{Q} \otimes \underleftarrow{l}_*$ of the composition of the above morphisms. Hence it suffices to prove that the following morphism is an isomorphism for each $q \in \mathbb{N}$.

$$\underrightarrow{l}^*\mathbb{Q} \otimes \underleftarrow{l}_*^{\mathbb{N}_{\geq r}}(\mathcal{M}_\bullet \otimes_{\mathcal{O}_X} \Omega^q_{X/\Sigma})_{s \geq r} \longrightarrow \underrightarrow{l}^*\mathbb{Q} \otimes R\underleftarrow{l}_*^{\mathbb{N}_{\geq r}} R(\pi^{h,\mathbb{N}^\circ}_{X,\text{ét}})_*^{\mathbb{N}_{\geq r}}(\eta_{\mathcal{P}/\mathcal{C}}(\mathscr{A}\mathcal{M}^s_\bullet\Omega^q))_{s\geq r}.$$

The last claim follows from Theorem IV.6.8.4 since $\mathcal{M}_\bullet \otimes_{\mathcal{O}_X} \Omega^q_{X/\Sigma}$ is a direct factor of $\mathcal{O}^{\oplus n}_{X_1,\bullet}$ for some $n \in \mathbb{N}$ in the category $\mathcal{LPM}(\mathcal{O}_{X_1,\bullet})_{\mathbb{Q}}$ by Lemma IV.3.2.6 (1) and Proposition IV.3.2.7 (2). □

**Lemma IV.6.8.10.** *Let $A$ be a $p$-adically complete and separated algebra flat over $\mathbb{Z}_p$. Then for any object $M$ of $\mathcal{LPM}(A)$, there exists an integer $N$ satisfying the following condition. For any $A$-algebra $A'$ flat over $\mathbb{Z}_p$, the kernel of $M \otimes_A A' \to (M \otimes_A A')_{\mathbb{Q}_p}$ is annihilated by $p^N$.*

Proof. By Lemma IV.3.2.2 (1), there exists $N_1 \in \mathbb{N}$ such that $p^{N_1}(\mathrm{Ker}(M \to M_{\mathbb{Q}_p})) = 0$. Choose a surjective $A$-linear homomorphism $f\colon A^{\oplus r} \to M$. Since $M_{\mathbb{Q}_p}$ is a projective $A_{\mathbb{Q}_p}$-module, there exists an $A_{\mathbb{Q}_p}$-linear homomorphism $g\colon M_{\mathbb{Q}_p} \to A^{\oplus r}_{\mathbb{Q}_p}$ such that $f_{\mathbb{Q}_p} \circ g = \mathrm{id}_{M_{\mathbb{Q}_p}}$. Let $h$ denote the composition $M \to M_{\mathbb{Q}_p} \xrightarrow{g} A^{\oplus r}_{\mathbb{Q}_p}$. Then there exists $N_2 \in \mathbb{N}$ such that $p^{N_2}h(M) \subset A^{\oplus r}$. The composition $p^{N_1}f \circ p^{N_2}h$ coincides with $p^{N_1+N_2}\mathrm{id}_M$. By taking the scalar extension by $A \to A'$, we obtain a factorization $M \otimes_A A' \to (A')^{\oplus r} \to M \otimes_A A'$ of the multiplication by $p^{N_1+N_2}$ on $M \otimes_A A'$. □

**Lemma IV.6.8.11.** *Let $M$ be a $\mathbb{Z}_p$-module, let $N \in \mathbb{N}$, let $\overline{M}$ be the image of $M \to M_{\mathbb{Q}_p}$, and assume that* $\mathrm{Ker}(M \to M_{\mathbb{Q}_p})$ *is annihilated by $p^N$. Then the morphism $p^N : M \to M$ uniquely factors through $\overline{M} \to M$.*

PROOF. Obvious. □

CHAPTER V

# Almost étale coverings

Takeshi Tsuji

## V.1. Introduction

In this chapter, we explain the theory of almost étale coverings introduced by G. Faltings in **[24]** I 2 and **[26]** 1, 2a. We can find a more systematic and complete argument in the book **[32]**, where they work with the category of modules "modulo almost isomorphisms." In this chapter, we try to follow faithfully the original approach by G. Faltings, and to give detailed proofs of the results in **[26]** 1, 2a and the beginning of 2c, adding some preliminaries if necessary. We do not discuss the so-called "almost purity theorem" for a variety with a certain type of log smooth reduction proved in **[26]** 2b; we refer the reader to II.6 for a detailed explanation of the theorem.

In the introduction, we give a history of almost étale extensions briefly. The idea of "almost étaleness" dates back to the work **[71]** of Tate. Let $K$ be a complete discrete valuation field of mixed characteristic $(0,p)$ with a perfect residue field, let $\overline{K}$ be an algebraic closure of $K$, let $C$ be the completion of $\overline{K}$, let $K_n$ be the subfield of $\overline{K}$ generated by $p^n$-th roots of unity over $K$, and let $K_\infty$ denote the union $\cup_{n\in\mathbb{N}}K_n$. Let $O_K$ and $O_{K_\infty}$ denote the rings of integers of $K$ and $O_{K_\infty}$, respectively. Let $v$ denote a valuation of $\overline{K}$. Then, for any finite extension $L$ of $K$ contained in $\overline{K}$, letting $L_n := LK_n$, he proved that the valuation of a generator of the relative different $\mathcal{D}_{L_n/K_n}$ $(n \in \mathbb{N})$ converges to 0 as $n \to \infty$, i.e., the extension $L_\infty/K_\infty$ is "almost étale." (Precisely speaking, he considered an arbitrary ramified $\mathbb{Z}_p$-extension, and also determined the behavior of the valuations more precisely.) Using this almost étaleness, he computed the continuous Galois cohomology $H^q(\mathrm{Gal}(\overline{K}/K), C(r))$ $(q = 0, 1, r \in \mathbb{Z})$ and applied it to a $p$-divisible group over $O_K$; in particular, he proved the Hodge-Tate decomposition for a $p$-divisible group over $O_K$. The above "almost étaleness" implies the vanishing of the higher continuous Galois cohomology $H^q(\mathrm{Gal}(\overline{K}/K_\infty), C(r))$ $(q > 0)$ (cf. Proposition V.12.8) and thus the computation of $H^q(\mathrm{Gal}(\overline{K}/K), C(r))$ is reduced to that of $H^q(\mathrm{Gal}(K_\infty/K), C(r))$ which is easier.

The above results on the almost étaleness and the computation of the Galois cohomology were generalized to the imperfect residue field case by O. Hyodo **[44]**, to an affine smooth scheme over $O_K$ having invertible coordinate functions and then to a certain affine log smooth scheme over $O_K$ by G. Faltings in **[24]** and **[26]**, respectively. See also **[23]** for the case of curves. For a finite homomorphism $A \to B$ of flat algebras over $O_{K_\infty}$ such that $A[\frac{1}{p}] \to B[\frac{1}{p}]$ is étale, an idempotent $e_{B/A}$ of $(B \otimes_A B)[\frac{1}{p}]$ is defined by the diagonal immersion, which is open and closed. In **[24]**, Faltings defined the almost étaleness by the condition that $\mathfrak{m}_{K_\infty}e_{B/A}$ is contained in the image of $B \otimes_A B$ and the trace map $B[\frac{1}{p}] \to A[\frac{1}{p}]$ maps $B$ into $A$. Here $\mathfrak{m}_{K_\infty}$ denotes the maximal ideal of $O_{K_\infty}$. With this definition, he proved that, for a smooth algebra $A$ over $O_K$ with invertible coordinates $t_1, \dots, t_d$, $A_\infty := A_{O_K} \otimes_{O_K} O_{K_\infty}[t_1^{\pm p^{-\infty}}, \dots, t_d^{\pm p^{-\infty}}]$ and a finite étale algebra $C$ over $A_\infty[\frac{1}{p}]$, the normalization of $A_\infty$ in $C$ is almost étale over $A_\infty$. This theorem is called the "almost purity theorem" because the proof is based on the almost étaleness over the completion of the localization at a codimension one point lying on the special fiber.

Using this generalization, he proved the Hodge-Tate decomposition of the étale cohomology of a proper smooth variety, and also gave, in **[25]**, a proof of the crystalline conjecture: $C_{\mathrm{crys}}$ by Fontaine.

Actually the proof of the above almost purity theorem for a smooth ring in **[24]** has a gap in the Lefschetz argument. An alternative proof was given by Faltings in **[26]** being generalized to a log smooth setting. In the proof, he needed the notion of almost étaleness for algebras over $W(\varprojlim_{x^p \leftarrow x} O_{\overline{K}}/pO_{\overline{K}})$ on which $p$ is nilpotent and the theory of almost étale extension in **[24]** was generalized to a more general setting applicable to such algebras. Finally we should mention that the almost purity theorem also plays a key role in the theory of $p$-adic Simpson correspondence by G. Faltings **[27]**: the main theme of this book and that the almost purity theorem by Faltings was generalized to more general rings by P. Scholze in **[64]**

**Notation.** As in **[26]** 1, let $\overline{V}$ denote a commutative ring endowed with a sequence of principal ideals $\mathfrak{m}_\alpha \subsetneq \overline{V}$ indexed by $\alpha \in \Lambda^+$, where $\Lambda$ is a subgroup of $\mathbb{Q}$ dense in $\mathbb{R}$ and $\Lambda^+ = \{\alpha \in \Lambda | \alpha > 0\}$. Choose a generator $\pi^\alpha$ of $\mathfrak{m}_\alpha$ for each $\alpha \in \Lambda^+$. We further assume that $\pi^\alpha$ $(\alpha \in \Lambda^+)$ are nonzero divisors in $\overline{V}$ and, for $\alpha, \beta \in \Lambda^+$, we have $\pi^\alpha \cdot \pi^\beta = u_{\alpha,\beta} \cdot \pi^{\alpha+\beta}$, $u_{\alpha,\beta} \in \overline{V}^*$. Set $\mathfrak{m} := \cup_{\alpha \in \Lambda^+} \mathfrak{m}_\alpha$.

## V.2. Almost isomorphisms

**Definition V.2.1.** (1) We say that a homomorphism of $\overline{V}$-modules $M \to N$ is an *almost isomorphism* if the kernel and the cokernel are annihilated by $\mathfrak{m}$. We denote it by $M \overset{\approx}{\to} N$.

(2) We say that a $\overline{V}$-module $M$ is *almost zero* if it is annihilated by $\mathfrak{m}$. We denote it by $M \approx 0$.

For homomorphisms of $\overline{V}$-modules $L \xrightarrow{f} M \xrightarrow{g} N$, if two of $f$, $g$, and $g \circ f$ are almost isomorphisms, then so is the rest. This follows from the fact that, for any $\alpha \in \Lambda^+$, there exist $\alpha_1, \alpha_2 \in \Lambda^+$ such that $\alpha = \alpha_1 + \alpha_2$.

**Lemma V.2.2.** *Let $R$ be a $\overline{V}$-algebra and let $f\colon M \to N$ be a homomorphism of $R$-modules such that the kernel and the cokernel are annihilated by $\mathfrak{m}_\alpha$ for an $\alpha \in \Lambda^+$. Then, there exists an $R$-homomorphism $g\colon N \to M$ such that $f \circ g = \pi^{3\alpha} \cdot \mathrm{id}_N$ and $g \circ f = \pi^{3\alpha} \cdot \mathrm{id}_M$.*

PROOF. Since $\pi^\alpha \cdot \mathrm{Ext}^1_R(N, \mathrm{Ker}(f)) = 0$ and $\pi^\alpha \cdot N \subset f(M)$, there exists $g'\colon N \to M$ such that $f \circ g' = (\pi^\alpha)^2 \cdot \mathrm{id}_N$. Then $f \circ g' \circ f = (\pi^\alpha)^2 \cdot f$ and, since $\pi^\alpha \cdot \mathrm{Ker}(f) = 0$, $(\pi^\alpha g') \circ f = (\pi^\alpha)^3 \cdot \mathrm{id}_M$. Hence, for $g = \pi^\alpha \cdot g'$, we have $g \circ f = (\pi^\alpha)^3 \cdot \mathrm{id}_M$ and $f \circ g = (\pi^\alpha)^3 \cdot \mathrm{id}_N$. □

**Proposition V.2.3.** *Let $R$ be a $\overline{V}$-algebra and let $f\colon M \to M'$, $g\colon N \to N'$ be two almost isomorphisms of $R$-modules. Then, the homomorphisms $f \oplus g$, $f \otimes_R g$, $\wedge^r_R f$ $(r \in \mathbb{Z}, r \geq 0)$, and $\mathrm{Hom}_R(M', N) \to \mathrm{Hom}_R(M, N')$; $\varphi \mapsto g \circ \varphi \circ f$ are almost isomorphisms. Similarly for $\mathrm{Tor}_i^R$ and $\mathrm{Ext}^i_R$.*

By Proposition V.2.3, for a $\overline{V}$-module $M$, the canonical homomorphisms

$$
\begin{aligned}
&M = \mathrm{Hom}_{\overline{V}}(\overline{V}, M) \to \mathrm{Hom}_{\overline{V}}(\mathfrak{m}, M) \\
&\mathfrak{m} \otimes_{\overline{V}} M \to \overline{V} \otimes_{\overline{V}} M = M
\end{aligned}
\tag{V.2.4}
$$

are almost isomorphisms.

**Proposition V.2.5.** *Let $f\colon M \to N$ be a homomorphism of $\overline{V}$-modules.*

(1) *$f$ is an almost isomorphism if and only if the homomorphism* $\mathrm{Hom}_{\overline{V}}(\mathfrak{m}, M) \to \mathrm{Hom}_{\overline{V}}(\mathfrak{m}, N)$ *induced by $f$ is an isomorphism.*

(2) *$f$ is an almost isomorphism if and only if the homomorphism $\mathfrak{m}\otimes_{\overline{V}} M \to \mathfrak{m}\otimes_{\overline{V}} N$ induced by $f$ is an isomorphism.*

PROOF. The sufficiency is trivial by the remark before the proposition. The necessity follows from Lemma V.2.6 below. □

**Lemma V.2.6.** *Let $M$ be a $\overline{V}$-module almost zero. Then we have* $\mathrm{Tor}_i^{\overline{V}}(\mathfrak{m}, M) = 0$ *and* $\mathrm{Ext}^i_{\overline{V}}(\mathfrak{m}, M) = 0$ *for all integers $i \geq 0$.*

PROOF. For any $\alpha \in \Lambda^+$ and $m \in M$, we have $\pi^\alpha \otimes m = u_{\alpha_1,\alpha_2}^{-1}\pi^{\alpha_1} \otimes \pi^{\alpha_2}\cdot m = 0$ in $\mathfrak{m}\otimes_{\overline{V}} M$, where $\alpha_1, \alpha_2 \in \Lambda^+$ such that $\alpha = \alpha_1+\alpha_2$. Hence $\mathfrak{m}\otimes_{\overline{V}} M = 0$. Similarly, for $f \in \mathrm{Hom}_{\overline{V}}(\mathfrak{m}, M)$ and $\alpha \in \Lambda^+$, we have $f(\pi^\alpha) = \pi^{\alpha_1}\cdot f(u_{\alpha_1,\alpha_2}^{-1}\pi^{\alpha_2}) = 0$. Hence $\mathrm{Hom}_{\overline{V}}(\mathfrak{m}, M) = 0$. Since $\mathfrak{m}_\alpha$ is a free $\overline{V}$-module of rank 1 and $\mathfrak{m} = \varinjlim_{\alpha\in\Lambda^+}\mathfrak{m}_\alpha$, $\mathfrak{m}$ is flat over $\overline{V}$. Hence $\mathrm{Tor}_i^{\overline{V}}(\mathfrak{m}, M) = 0$ for $i > 0$. Choose a strictly decreasing sequence $\alpha_n \in \Lambda^+$ $(n = 0, 1, \dots)$ converging to 0. Then, we can construct a free resolution of $\mathfrak{m}$:

$$0 \longleftarrow \mathfrak{m} \overset{\Phi}{\longleftarrow} \oplus_{n\geq 0}\overline{V} \overset{\Psi}{\longleftarrow} \oplus_{n\geq 0}\overline{V} \longleftarrow 0,$$

where $\Phi((x_n)) = \sum_{n\geq 0}\pi^{\alpha_n}x_n$, $\Psi((x_n)) = (y_n)$, $y_0 = x_0$, $y_n = x_n - a_{n-1}x_{n-1}$ $(n \geq 1)$, and $a_n \in \mathfrak{m}$ $(n \geq 0)$ is defined by $\pi^{\alpha_n} = a_n\cdot\pi^{\alpha_{n+1}}$. Since the endomorphism $\Psi^*$ on $\mathrm{Hom}_{\overline{V}}(\oplus_{n\geq 0}\overline{V}, M)$ is the identity if $M \approx 0$, this implies $\mathrm{Ext}^i_{\overline{V}}(\mathfrak{m}, M) = 0$ for $i \geq 0$. □

Proposition V.2.5 implies that, for any $\overline{V}$-module $M$, the canonical homomorphisms (see (V.2.4))

$$\mathrm{Hom}_{\overline{V}}(\mathfrak{m}, M) \longrightarrow \mathrm{Hom}_{\overline{V}}(\mathfrak{m}, \mathrm{Hom}_{\overline{V}}(\mathfrak{m}, M))$$
$$\mathfrak{m}\otimes_{\overline{V}}(\mathfrak{m}\otimes_{\overline{V}} M) \longrightarrow \mathfrak{m}\otimes_{\overline{V}} M$$

are isomorphisms.

For a $\overline{V}$-algebra $R$, we can define a product on $\mathrm{Hom}_{\overline{V}}(\mathfrak{m}, R)$ by

$$\mathrm{Hom}_{\overline{V}}(\mathfrak{m}, R)\times\mathrm{Hom}_{\overline{V}}(\mathfrak{m}, R) \to \mathrm{Hom}_{\overline{V}}(\mathfrak{m}\otimes_{\overline{V}}\mathfrak{m}, R\otimes_{\overline{V}} R) \to \mathrm{Hom}_{\overline{V}}(\mathfrak{m}, R),$$

where the first map is $(f, g) \mapsto f\otimes g$ and the second one is induced by $\mathfrak{m}\otimes_{\overline{V}}\mathfrak{m} \xrightarrow{\sim} \mathfrak{m}$ and the multiplication $R\otimes_{\overline{V}} R \to R$. We have the formula $(f\cdot g)(xy) = f(x)g(y)$ for $f, g \in \mathrm{Hom}_{\overline{V}}(\mathfrak{m}, R)$ and $x, y \in \mathfrak{m}$. This product makes $\mathrm{Hom}_{\overline{V}}(\mathfrak{m}, R)$ an $R$-algebra; the structure homomorphism is the canonical homomorphism $R \to \mathrm{Hom}_{\overline{V}}(\mathfrak{m}, R); a \mapsto (x \mapsto x\cdot a)$. For a $\overline{V}$-algebra homomorphism $R \to S$, the induced homomorphism $\mathrm{Hom}_{\overline{V}}(\mathfrak{m}, R) \to \mathrm{Hom}_{\overline{V}}(\mathfrak{m}, S)$ is a ring homomorphism.

**Remark V.2.7.** Let $R$ be a $\overline{V}$-algebra such that the canonical homomorphism $R \to \mathrm{Hom}_{\overline{V}}(\mathfrak{m}, R)$ is an isomorphism. Then, for two free $R$-modules $L$ and $M$ of finite rank, if we are given an $R$-module $N$, an almost isomorphism $f\colon N \xrightarrow{\approx} L$, and a homomorphism $g\colon N \to M$, then they induce a homomorphism from $L$ to $M$

$$L \xrightarrow{\sim} \mathrm{Hom}_{\overline{V}}(\mathfrak{m}, L) \xleftarrow{\sim} \mathrm{Hom}_{\overline{V}}(\mathfrak{m}, N) \to \mathrm{Hom}_{\overline{V}}(\mathfrak{m}, M) \xleftarrow{\sim} M$$

by Proposition V.2.5.

## V.3. Almost finitely generated projective modules

**Proposition V.3.1.** *Let $R$ be a $\overline{V}$-algebra and let $M$ be an $R$-module. Then the following conditions are equivalent:*

(i) $\mathrm{Ext}^i_R(M,N) \approx 0$ *for all $R$-modules $N$ and all integers $i > 0$.*

(ii) $\mathrm{Ext}^1_R(M,N) \approx 0$ *for all $R$-modules $N$.*

(iii) *For any surjective homomorphism $f\colon L \to N$ and homomorphism $g\colon M \to N$ of $R$-modules, and any $\alpha \in \Lambda^+$, there exists a homomorphism of $R$-modules $h\colon M \to L$ such that $f \circ h = \pi^\alpha \cdot g$.*

(iv) *For any $\alpha \in \Lambda^+$, there exist a free $R$-module $L$ and homomorphisms of $R$-modules $f\colon M \to L$, $g\colon L \to M$ such that $g \circ f = \pi^\alpha \cdot \mathrm{id}_M$.*

PROOF. (i)⇒(ii)⇒(iii) are trivial. (iii)⇒(iv): Choose a free $R$-module $L$ and a surjection $g\colon L \to M$. Then (iii) implies that, for any $\alpha \in \Lambda^+$, there exists $f\colon M \to L$ such that $g \circ f = \pi^\alpha \cdot \mathrm{id}_M$. (iv)⇒(i): The composite of the homomorphisms $\mathrm{Ext}^i_R(M,N) \to \mathrm{Ext}^i_R(L,N) \to \mathrm{Ext}^i_R(M,N)$ induced by $f$ and $g$ in the condition (iv) is the multiplication by $\pi^\alpha$ and the middle term vanishes for $i > 0$. □

**Definition V.3.2.** Let $R$ be a $\overline{V}$-algebra and let $M$ be an $R$-module. We say that $M$ is *almost projective* if $M$ satisfies the equivalent conditions of Proposition V.3.1.

**Proposition V.3.3.** (1) *Let $R$ be a $\overline{V}$-algebra and let $M \to M'$ be an almost isomorphism of $R$-modules. Then, $M$ is almost projective over $R$ if and only if $M'$ is almost projective.*

(2) *Let $R \to R'$ be a homomorphism of $\overline{V}$-algebras and let $M$ be an almost projective $R$-module. Then $M \otimes_R R'$ is an almost projective $R'$-module.*

PROOF. If $M \xrightarrow{\approx} M'$, then $\mathrm{Ext}^i_R(M,N) \xrightarrow{\approx} \mathrm{Ext}^i_R(M',N)$ $(i > 0)$ for any $R$-module $N$ (Proposition V.2.3). This implies (1). The claim (2) is trivial by the condition (iv) of Proposition V.3.1. □

**Definition V.3.4.** Let $R$ be a $\overline{V}$-algebra and let $M$ be an $R$-module. We say that $M$ is *almost finitely generated* if, for any $\alpha \in \Lambda^+$, there exist a finitely generated $R$-module $N$ and homomorphisms of $R$-modules $\varphi_\alpha\colon M \to N$, $\psi_\alpha\colon N \to M$ such that $\psi_\alpha \circ \varphi_\alpha = \pi^\alpha \cdot \mathrm{id}_M$ and $\varphi_\alpha \circ \psi_\alpha = \pi^\alpha \cdot \mathrm{id}_N$. We say that $M$ is *almost finitely generated projective* if it is almost finitely generated and almost projective.

**Lemma V.3.5.** *Let $R$ be a $\overline{V}$-algebra and let $M$ be an $R$-module. Then $M$ is almost finitely generated if and only if, for any $\alpha \in \Lambda^+$, there exist a free $R$-module $L$ of finite rank and a homomorphism of $R$-modules $f\colon L \to M$ such that $\pi^\alpha \cdot \mathrm{Cok}(f) = 0$.*

PROOF. The necessity is trivial. For $f\colon L \to M$ in the condition in Lemma V.3.5, $N = f(L)$ is finitely generated and the multiplication by $\pi^\alpha$ on $M$ factors through $N$. Hence the inclusion $N \hookrightarrow M$ and $\pi^\alpha\colon M \to N (\subset M)$ satisfies the condition of Definition V.3.4. □

**Proposition V.3.6.** (1) *Let $R$ be a $\overline{V}$-algebra and let $f\colon M \to M'$ be an almost isomorphism of $R$-modules. Then $M$ is almost finitely generated if and only if $M'$ is almost finitely generated.*

(2) *Let $R \to R'$ be a homomorphism of $\overline{V}$-algebras and let $M$ be an $R$-module. If $M$ is almost finitely generated over $R$, then $M \otimes_R R'$ is almost finitely generated over $R'$.*

PROOF. (1) follows from Lemma V.2.2 and (2) is trivial by definition. □

**Proposition V.3.7.** *Let $R$ be a $\overline{V}$-algebra and let $M$ be an $R$-module. Then, $M$ is almost finitely generated projective if and only if, for any $\alpha \in \Lambda^+$, there exist a free $R$-module $L$ of finite rank and homomorphisms of $R$-modules $f\colon M \to L$, $g\colon L \to M$ such that $g \circ f = \pi^\alpha \cdot \mathrm{id}_M$.*

PROOF. The sufficiency follows from the condition (iv) of Proposition V.3.1 and Lemma V.3.5. Let us prove the necessity. By Lemma V.3.5, for any $\alpha \in \Lambda^+$, there exist a free $R$-module $L$ of finite rank and a homomorphism of $R$-modules $g\colon L \to M$ such that $\pi^\alpha \cdot \mathrm{Cok}(g) = 0$. Since the image of $\pi^\alpha \mathrm{id}_M\colon M \to M$ is contained in $\mathrm{Im}(g)$ and $M$ is almost finitely generated and projective, there exists $f\colon M \to L$ such that $g \circ f = (\pi^\alpha)^2 \cdot \mathrm{id}_M$. □

**Proposition V.3.8.** *Let $R$ be a $\overline{V}$-algebra and let $M$, $N$ be almost finitely generated projective $R$-modules. Then $\mathrm{Hom}_R(M, N)$, $M \otimes_R N$, and $\wedge^i_R M$ $(i \geq 1)$ are also almost finitely generated projective.*

PROOF. For $\alpha \in \Lambda^+$, by Proposition V.3.7, there exist free $R$-modules of finite rank $L_1$, $L_2$ and homomorphisms of $R$-modules $M \xrightarrow{f_1} L_1 \xrightarrow{g_1} M$, $N \xrightarrow{f_2} L_2 \xrightarrow{g_2} N$ such that $g_1 \circ f_1 = \pi^\alpha \cdot \mathrm{id}_M$, $g_2 \circ f_2 = \pi^\alpha \cdot \mathrm{id}_N$. Then the composites of the homomorphisms

$$\mathrm{Hom}_R(M, N) \to \mathrm{Hom}_R(L_1, L_2) \to \mathrm{Hom}_R(M, N)$$
$$M \otimes_R N \to L_1 \otimes_R L_2 \to M \otimes_R N$$
$$\wedge^i_R M \to \wedge^i_R L_1 \to \wedge^i_R M$$

induced by $f_1$, $g_1$, $f_2$, and $g_2$ are the multiplication by $(\pi^\alpha)^2$, $(\pi^\alpha)^2$, and $(\pi^\alpha)^i$, respectively, and the middle modules are free $R$-modules of finite rank. By Proposition V.3.7, this completes the proof. □

**Proposition V.3.9.** *Let $R \to R'$ be an almost isomorphism of $\overline{V}$-algebras and let $M$ be an $R$-module. Then $R$ is almost projective (resp. almost finitely generated) over $R$ if and only if $M' := M \otimes_R R'$ is almost projective (resp. almost finitely generated) over $R'$.*

PROOF. The necessity is already proved (Proposition V.3.3 and Proposition V.3.6). Let us prove the sufficiency. Since the canonical homomorphism $M \to M'$ is an almost isomorphism (Proposition V.2.3), it suffices to prove that $M'$ is almost projective (resp. almost finitely generated) over $R$ by Proposition V.3.3 (resp. Proposition V.3.6). First note that $L \to L \otimes_R R'$ is an almost isomorphism for any free $R$-module $L$. By Lemma V.3.5, this implies the claim in the second case. In the first case, for $\alpha \in \Lambda^+$, choose a free $R$-module $L$ and $R'$-linear homomorphisms $f\colon M' \to L \otimes_R R'$ and $g'\colon L \otimes_R R' \to M'$ such that $g' \circ f' = \pi^\alpha \cdot \mathrm{id}_{M'}$. Then, by the above remark and Proposition V.2.3, $\mathrm{Ext}^i_R(L \otimes_R R', N) \xrightarrow{\approx} \mathrm{Ext}^i_R(L, N) = 0$ $(i > 0)$ and hence $(\pi^\alpha)^2 \cdot \mathrm{Ext}^i_R(M', N) = 0$ $(i > 0)$ for any $R$-module $N$. Varying $\alpha$, we see that $M'$ is almost projective over $R$. □

## V.4. Trace

**Lemma V.4.1.** *Let $R$ be a $\overline{V}$-algebra and let $M$ and $N$ be almost finitely generated projective $R$-modules. Then, the natural homomorphism $M \otimes_R \mathrm{Hom}_R(N, R) \to \mathrm{Hom}_R(N, M)$ is an almost isomorphism.*

Proof. For $\alpha \in \Lambda^+$, choose $L_i$, $f_i$, $g_i$ ($i = 1, 2$) as in the proof of Proposition V.3.8. Then, we have a commutative diagram

$$\begin{array}{ccc} M \otimes_R \mathrm{Hom}_R(N, R) & \longrightarrow & \mathrm{Hom}_R(N, M) \\ \downarrow & & \downarrow \\ L_1 \otimes_R \mathrm{Hom}_R(L_2, R) & \longrightarrow & \mathrm{Hom}_R(L_2, L_1) \\ \downarrow & & \downarrow \\ M \otimes_R \mathrm{Hom}_R(N, R) & \longrightarrow & \mathrm{Hom}_R(N, M), \end{array}$$

where the vertical homomorphisms are induced by $f_1$, $g_1$, $f_2$, $g_2$. The composites of the vertical homomorphisms are the multiplication by $(\pi^\alpha)^2$ and the middle homomorphism is an isomorphism. Hence the kernel and the cokernel of the homomorphism in question are annihilated by $(\pi^\alpha)^2$. □

Let $R$ be a $\overline{V}$-algebra and let $M$ be an almost finitely generated projective $R$-module. We define the trace map

$$\mathrm{tr}_R \colon \mathrm{End}_R(M) \longrightarrow \mathrm{Hom}_{\overline{V}}(\mathfrak{m}, R)$$

functorial on $M$ as the composite:

$$\begin{aligned} \mathrm{End}_R(M) &\to \mathrm{Hom}_{\overline{V}}(\mathfrak{m}, \mathrm{End}_R(M)) \\ &\xleftarrow{\sim} \mathrm{Hom}_{\overline{V}}(\mathfrak{m}, M \otimes_R \mathrm{Hom}_R(M, R)) \to \mathrm{Hom}_{\overline{V}}(\mathfrak{m}, R). \end{aligned}$$

Here, the second homomorphism is an isomorphism by Lemma V.4.1 and Proposition V.2.5 (1), and the last homomorphism is induced by $M \otimes_R \mathrm{Hom}_R(M, R) \to R; m \otimes f \mapsto f(m)$.

**Proposition V.4.2.** (1) *Let $R$ be a $\overline{V}$-algebra and let $M$ be a finitely generated projective $R$-module. Then, the trace map defined above coincides with the usual trace map $\mathrm{End}_R(M) \to R$ followed by the canonical homomorphism $R \to \mathrm{Hom}_{\overline{V}}(\mathfrak{m}, R)$.*

(2) *Let $R \to R'$ be a homomorphism of $\overline{V}$-algebra and let $M$ be an almost finitely generated projective $R$-module. Let $f \in \mathrm{End}_R(M)$ and let $f' \in \mathrm{End}_{R'}(M \otimes_R R')$ be the base change of $f$. Then $\mathrm{tr}_{R'}(f')$ is the image of $\mathrm{tr}_R(f)$ under the homomorphism $\mathrm{Hom}_{\overline{V}}(\mathfrak{m}, R) \to \mathrm{Hom}_{\overline{V}}(\mathfrak{m}, R')$.*

Proof. Trivial by definition. □

**Proposition V.4.3.** *Let $R$ be a $\overline{V}$-algebra, let $f \colon M \to M'$ be an almost isomorphism of almost finitely generated projective $R$-modules, and let $g$ and $g'$ be $R$-linear endomorphisms of $M$ and $M'$, respectively, such that $f \circ g = g' \circ f$. Then $\mathrm{tr}_R(g) = \mathrm{tr}_R(g')$. Especially, if $M \approx 0$, then the trace map $\mathrm{tr}_R \colon \mathrm{End}_R(M) \to \mathrm{Hom}_{\overline{V}}(\mathfrak{m}, R)$ is $0$.*

Proof. This follows from the commutative diagrams

$$\begin{array}{ccc} \mathrm{Hom}_{\overline{V}}(\mathfrak{m}, M \otimes_R M^*) & \xleftarrow{\ \sim\ } & \mathrm{Hom}_{\overline{V}}(\mathfrak{m}, M \otimes_R M'^*) \\ \downarrow & & \downarrow \wr \\ \mathrm{Hom}_{\overline{V}}(\mathfrak{m}, R) & \longleftarrow & \mathrm{Hom}_{\overline{V}}(\mathfrak{m}, M' \otimes_R M'^*) \end{array}$$

and

$$\begin{array}{ccc} \mathrm{Hom}_{\overline{V}}(\mathfrak{m}, M \otimes_R M^*) & \xleftarrow{\ \sim\ } & \mathrm{Hom}_{\overline{V}}(\mathfrak{m}, M \otimes_R M'^*) \\ \wr\downarrow & & \downarrow \wr \\ \mathrm{Hom}_{\overline{V}}(\mathfrak{m}, M' \otimes_R M^*) & \xleftarrow{\ \sim\ } & \mathrm{Hom}_{\overline{V}}(\mathfrak{m}, M' \otimes_R M'^*), \end{array}$$

where $M^* = \mathrm{Hom}_{\overline{V}}(M, R)$ and $M'^* = \mathrm{Hom}_{\overline{V}}(M', R)$. □

**Proposition V.4.4.** *Let $R$ be a $\overline{V}$-algebra and let $M_1$, $M_2$ be two almost finitely generated projective $R$-modules. Then, for any $R$-linear endomorphisms $f_1$ and $f_2$ of $M_1$ and $M_2$, respectively, we have*

$$\mathrm{tr}_R(f \otimes g) = \mathrm{tr}_R(f) \cdot \mathrm{tr}_R(g) \quad (\textit{resp. } \mathrm{tr}_R(f \oplus g) = \mathrm{tr}_R(f) + \mathrm{tr}_R(g)).$$

PROOF. The first case follows from the commutative diagram:

$$\begin{array}{ccc}
\mathrm{End}_R(M_1) \otimes_R \mathrm{End}_R(M_2) & \longrightarrow & \mathrm{End}_R(M_1 \otimes_R M_2) \\
\uparrow & & \uparrow \\
(M_1 \otimes_R M_1^*) \otimes_R (M_2 \otimes_R M_2^*) & \longrightarrow & (M_1 \otimes_R M_2) \otimes_R (M_1 \otimes_R M_2)^* \\
\cong (M_1 \otimes_R M_2) \otimes_R (M_1^* \otimes_R M_2^*) & & \\
\downarrow & & \downarrow \\
R \otimes_R R & \xrightarrow{\sim} & R.
\end{array}$$

The second case is easy. □

## V.5. Rank and determinant

To avoid the confusion, we will denote by $\mathrm{tr}^{\mathrm{cl}}$ (resp. $\det^{\mathrm{cl}}$) the usual trace (resp. determinant) map for a projective (resp. free) module of finite rank. The following relation between usual determinant and trace will play a key role.

**Proposition V.5.1.** *Let $R$ be a commutative ring and let $M$ be a free $R$-module of finite rank. Then, for any $R$-linear endomorphism $f$ of $M$, we have*

$$\det{}^{\mathrm{cl}}_R(1 + f) = \textstyle\sum_{s=0}^{\infty} \mathrm{tr}^{\mathrm{cl}}_R(\wedge^s f).$$

*Note $\wedge^s f = 0$ if $s \geq \mathrm{rank}_R M + 1$ and hence the right-hand sum is finite.*

PROOF. Explicit computation. Choose a basis $\{e_i | 1 \leq i \leq r\}$ of $M$, and describe $(\wedge^r(1+f))(e_1 \wedge \cdots \wedge e_r)$ $(= (\det^{\mathrm{cl}}_R(1+f)) \cdot e_1 \wedge \cdots \wedge e_r)$ in terms of the matrix associated to $\wedge^s f$ with respect to the base $\{e_{i_1} \wedge \cdots \wedge e_{i_s} | 1 \leq i_1 < \cdots < i_s \leq r\}$ of $\wedge^s M$. □

Let $R$ be a $\overline{V}$-algebra. The canonical injection $\mathrm{Hom}_{\overline{V}}(\mathfrak{m}, R[T]) \hookrightarrow \mathrm{Hom}_{\overline{V}}(\mathfrak{m}, R)[[T]]$ is a ring homomorphism and we will regard the former as a subring of the latter in the following. For an almost finitely generated projective $R[T]$-module $\mathcal{M}$ and an $R[T]$-linear endomorphism $F$ of $\mathcal{M}$, we set

(V.5.2)
$$\det(1 + T \cdot F) := \sum_{s=0}^{\infty} T^s \cdot \mathrm{tr}_{R[T]}(\wedge^s F) \in \mathrm{Hom}_{\overline{V}}(\mathfrak{m}, R)[[T]]$$
$$\left( = \sum_{s=0}^{\infty} \mathrm{tr}_{R[T]}(\wedge^s (T \cdot F)) \right).$$

If $\mathcal{M} = M \otimes_R R[T]$ for an almost finitely generated projective $R$-module $M$ and $F$ arises from an $R$-linear endomorphism $f$ of $M$ by the base change $R \to R[T]$, then, by Proposition V.4.2 (2), we have

(V.5.3)
$$\det(1 + T \cdot F) = \sum_{i=0}^{\infty} T^i \cdot \mathrm{tr}_R(\wedge^i f).$$

**Proposition V.5.4.** *Let $R$ be a $\overline{V}$-algebra, let $M$ be an almost finitely generated projective $R$-module, let $f$ and $g$ be $R$-linear endomorphisms of $M$, and let $F$ and $G$ denote the $R[T]$-linear endomorphisms $f \otimes \mathrm{id}_{R[T]}$ and $g \otimes \mathrm{id}_{R[T]}$ of $M \otimes_R R[T]$. Then we have*

$$\det(1 + T \cdot F) \cdot \det(1 + T \cdot G) = \det(1 + T \cdot (F + G + T \cdot (F \circ G))).$$

**Lemma V.5.5.** *Let $R$ be a $\overline{V}$-algebra and let $\varphi\colon M \to N$ and $\psi\colon N \to M$ be homomorphisms of almost finitely generated projective $R$-modules such that $\psi \circ \varphi = a \cdot \mathrm{id}_M$ for an $a \in R$. Then, for an $R$-linear endomorphism $f$ of $M$, we have*

$$\mathrm{tr}_R(\wedge^s(\varphi \circ f \circ \psi)) = a^s \cdot \mathrm{tr}_R(\wedge^s f)$$

*for any integer $s \geq 0$.*

Proof. Since $\wedge^s\psi \circ \wedge^s\varphi = a^s \cdot \mathrm{id}_{\wedge^s M}$ and $\wedge^s(\varphi \circ f \circ \psi) = \wedge^s\varphi \circ \wedge^s f \circ \wedge^s\psi$, it suffices to prove the lemma for $s = 1$, in which case, it follows from the commutative diagram:

$$\begin{array}{ccccc}
\mathrm{End}_R(M) & \longleftarrow & M \otimes_R \mathrm{Hom}_R(M, R) & \longrightarrow & R \\
\downarrow{\scriptstyle \varphi\circ - \circ\psi} & & \downarrow{\scriptstyle \varphi\otimes(-\circ\psi)} & & \downarrow{\scriptstyle a\cdot} \\
\mathrm{End}_R(N) & \longleftarrow & N \otimes_R \mathrm{Hom}_R(N, R) & \longrightarrow & R.
\end{array}$$

□

Proof of Proposition V.5.4. Let $\alpha \in \Lambda^+$ and choose a free $R$-module $L$ of finite rank and $R$-linear homomorphisms $\varphi\colon M \to L$, $\psi\colon L \to M$ such that $\psi \circ \varphi = \pi^\alpha \cdot \mathrm{id}_M$ (Proposition V.3.7). Set $f' := \varphi \circ f \circ \psi$, $g' := \varphi \circ g \circ \psi \in \mathrm{End}_R(L)$ and let $F'$, $G'$ be the base changes of $f'$, $g'$ by $R \to R[T]$. Let $\kappa_\alpha$ be the ring homomorphism $\mathrm{Hom}_{\overline{V}}(\mathfrak{m}, R)[[T]] \to \mathrm{Hom}_{\overline{V}}(\mathfrak{m}, R)[[T]]$ defined by $\kappa_\alpha(\sum_n a_n T^n) = \sum_n a_n(\pi^\alpha T)^n$. Then, applying Lemma V.5.5 to $\varphi$, $\psi$, $f$ (resp. $g$) and using (V.5.3), we obtain $\kappa_\alpha(\det(1+T\cdot F)) = \det(1+T\cdot F')$ (resp. $\kappa_\alpha(\det(1 + T \cdot G)) = \det(1 + T \cdot G')$). On the other hand, by (V.5.2), Proposition V.5.1, and Proposition V.4.2 (1), we have

$$\det(1 + T \cdot F') \cdot \det(1 + T \cdot G') = \det(1 + T \cdot (F' + G' + T \cdot F' \circ G')).$$

We assert that the right-hand side coincides with $\kappa_\alpha(\det(1+T\cdot(F+G+T\cdot F\circ G)))$. The base change of $F + G + T \cdot F \circ G$ under $R[T] \to R[T]; T \mapsto \pi^\alpha \cdot T$ is $F + G + \pi^\alpha T \cdot F \circ G$. Hence, by Proposition V.4.2 (2), we obtain

$$\kappa_\alpha(\det(1 + T \cdot (F + G + T \cdot F \circ G))) = \sum_{s=0}^{\infty} (\pi^\alpha T)^s \mathrm{tr}_{R[T]}(\wedge^s(F + G + \pi^\alpha T \cdot F \circ G)).$$

Let $\Phi$ and $\Psi$ be the base changes of $\varphi$ and $\psi$ by $R \to R[T]$. Then $\Psi \circ \Phi = \pi^\alpha \cdot \mathrm{id}_{M\otimes_R R[T]}$ and $\Phi \circ (F + G + \pi^\alpha T \cdot F \circ G) \circ \Psi = F' + G' + T \cdot F \circ G'$. Now the assertion follows from Lemma V.5.5.

Thus we obtain the required equality modulo $\mathrm{Ker}(\kappa_\alpha)$. Since $\mathrm{Hom}_{\overline{V}}(\mathfrak{m}, R)$ is $\mathfrak{m}$-torsion free, varying $\alpha$, we obtain the exact equality. □

**Definition V.5.6.** Let $R$ be a $\overline{V}$-algebra and let $M$ be an almost finitely generated projective $R$-module. For an integer $r \geq 0$, we say that $M$ has *rank* $\leq r$ over $R$ if $\wedge_R^{r+1} M \approx 0$, and denote it by $\mathrm{rank}_R M \leq r$.

By Proposition V.2.3, if $M \xrightarrow{\approx} M'$, then $\mathrm{rank}_R M \leq r$ if and only if $\mathrm{rank}_R M' \leq r$. For a ring homomorphism $R \to R'$ and $M' := M \otimes_R R'$, if $\mathrm{rank}_R M \leq r$, then $\mathrm{rank}_R M' \leq r$. The converse is also true if $R \xrightarrow{\approx} R'$. For the last claim, we use $\wedge_R^{r+1} M \xrightarrow{\approx} (\wedge_R^{r+1} M) \otimes_R R' \cong \wedge_{R'}^{r+1} M'$.

We will only use the following consequence of Proposition V.5.4.

**Proposition V.5.7.** *Let $R$ be a $\overline{V}$-algebra, let $M$ be an almost finitely generated projective $R$-module of rank $\leq r$, and let $f$ and $g$ be two $R$-linear endomorphisms of $M$. Then, we have*

$$\mathrm{tr}_R(\wedge^r f \circ g) = \mathrm{tr}_R(\wedge^r f) \cdot \mathrm{tr}_R(\wedge^r g).$$

By setting $f = g = \mathrm{id}_M$, we obtain the following corollary.

**Corollary V.5.8.** *Let $R$, $M$ be as in Proposition* V.5.7. *Then,* $\mathrm{tr}_R(\mathrm{id}_{\wedge^r M})$ *is an idempotent of* $\mathrm{Hom}_{\overline{V}}(\mathfrak{m}, R)$.

PROOF OF PROPOSITION V.5.7. Let $F$, $G \in \mathrm{End}_{R[T]}(M \otimes_R R[T])$ denote the base changes of $f$ and $g$. Since $\wedge^s M \approx 0$ $(s \geq r+1)$, by Proposition V.4.3, $\mathrm{tr}_R(\wedge^s f)$, $\mathrm{tr}_R(\wedge^s g)$, and $\mathrm{tr}_{R[T]}(\wedge^s(F + G + T \cdot F \circ G))$ vanish for $s \geq r+1$. Hence, by Proposition V.5.4 and (V.5.3), we see that $\mathrm{tr}_R(\wedge^s f) \cdot \mathrm{tr}_R(\wedge^s g)$ is the coefficient of $T^{2r}$ in $\sum_{s=0}^{r} T^s \cdot \mathrm{tr}_{R[T]}(\wedge^s(F + G + T \cdot F \circ G))$. For any $f_0$, $f_1 \in \mathrm{End}_R(M)$ and their base changes $F_0$, $F_1$ by $R \to R[T]$, we have $\wedge^s(F_0 + T \cdot F_1) = \sum_{t=0}^{s} T^t \cdot G_{t,s}$, where $G_{t,s}$ denotes the base change of the $R$-linear endomorphism $g_{t,s} \colon \wedge^s M \to \wedge^s M$ defined by

$$g_{t,s}(x_1 \wedge \cdots \wedge x_s) = \sum_{\substack{\varepsilon \in \{0,1\}^s \\ \sharp\{i \mid \varepsilon_i = 1\} = t}} f_{\varepsilon_1}(x_1) \wedge \cdots \wedge f_{\varepsilon_s}(x_s).$$

Hence $\mathrm{tr}_{R[T]}(\wedge^s(F_0 + T \cdot F_1)) = \sum_{t=0}^{s} T^t \cdot \mathrm{tr}_R(g_{t,s})$. Now it is easy to see that the coefficient of $T^{2r}$ in question is $\mathrm{tr}_R(\wedge^r f \circ g)$. □

**Proposition V.5.9.** *Let $R$ be a $\overline{V}$-algebra and let $M$ be an almost finitely generated projective $R$-module of rank $\leq r$. Set $L := \wedge_R^r M$ and $L^* := \mathrm{Hom}_R(L, R)$. Let $t$ denote the homomorphism $\mathrm{Hom}_{\overline{V}}(\mathfrak{m}, \mathrm{End}_R(L)) \to \mathrm{Hom}_{\overline{V}}(\mathfrak{m}, R)$ appearing in the definition of* $\mathrm{tr}_R$ (V.4)*, and let $h$ denote the homomorphism $\mathrm{Hom}_{\overline{V}}(\mathfrak{m}, R) \to \mathrm{Hom}_{\overline{V}}(\mathfrak{m}, \mathrm{End}_R(L))$ induced by $R \to \mathrm{End}_R(L); a \mapsto a \cdot \mathrm{id}_L$. Then $t \circ h = \mathrm{tr}_R(\mathrm{id}_L) \cdot \mathrm{id}$ and $h \circ t = \mathrm{id}$.*

**Lemma V.5.10.** *Let $R$ be a commutative ring, let $M$ be an $R$-module, and let $M^*$ denote $\mathrm{Hom}_R(M, R)$. Then, for any integer $s \geq 0$, there exists a canonical $R$-linear homomorphism functorial on $M$:*

$$\Phi \colon \wedge^s M^* \longrightarrow \mathrm{Hom}_R(\wedge^s M, R)$$

*characterized by*

$$\Phi(f_1 \wedge \cdots \wedge f_s)(x_1 \wedge \cdots \wedge x_s) = \sum_{\sigma \in \mathfrak{S}_s} \mathrm{sgn}(\sigma) f_{\sigma(1)}(x_1) \wedge \cdots \wedge f_{\sigma(s)}(x_s).$$

*Furthermore, if $M$ is free of finite rank, $\Phi$ is an isomorphism.*

PROOF. Straightforward. □

**Lemma V.5.11.** *Let $R$ be a $\overline{V}$-algebra and let $M$ be an almost finitely generated projective $R$-module. Then, for any integer $s \geq 0$, the homomorphism $\Phi \colon \wedge^s M^* \to \mathrm{Hom}_R(\wedge^s M, R)$ in Lemma* V.5.10 *is an almost isomorphism.*

PROOF. The same argument as the proof of Lemma V.4.1. □

**Lemma V.5.12.** *Let $R$ be a $\overline{V}$-algebra and let $M$ be an almost finitely generated projective $R$-module. Then the canonical homomorphism*

$$M \to \mathrm{Hom}_R(\mathrm{Hom}_R(M, R), R)$$

*is an almost isomorphism.*

PROOF. The same argument as the proof of Lemma V.4.1. □

**Lemma V.5.13.** *Let $M$ and $N$ be two $\overline{V}$-modules, let $f,\ g: M \to N$ be two $\overline{V}$-linear homomorphisms, and let $h\colon L \to M$ be an almost isomorphism. Suppose that $N$ is $\mathfrak{m}$-torsion free. Then, $f = g$ if and only if $f \circ h = g \circ h$.*

PROOF. Trivial. □

PROOF OF PROPOSITION V.5.9. For the first equality, by Lemma V.5.13, it suffices to prove that the composite $R \overset{\cdot \mathrm{id}_L}{\to} \mathrm{End}_R(L) \overset{\mathrm{tr}_R}{\to} \mathrm{Hom}_{\overline{V}}(\mathfrak{m}, R)$ is the canonical homomorphism $\iota_R\colon R \to \mathrm{Hom}_{\overline{V}}(\mathfrak{m}, R)$ (V.2.4) multiplied by $\mathrm{tr}_R(\mathrm{id}_L)$, which is trivial. Let us prove the second equality. By Lemma V.5.13 and Lemma V.5.11, it suffices to prove that $h \circ tr_R$ and the canonical homomorphism $\iota_{\mathrm{End}_R(L)}\colon \mathrm{End}_R(L) \to \mathrm{Hom}_{\overline{V}}(\mathfrak{m}, \mathrm{End}_R(L))$ (V.2.4) coincides after composing with

$$\kappa\colon \wedge^r M \otimes_R \wedge^r M^* \xrightarrow[\mathrm{id}_L \otimes \Phi]{\approx} L \otimes_R L^* \xrightarrow[\text{Lemma V.4.1}]{\approx} \mathrm{End}_R(L),$$

where $\Phi$ is as in Lemma V.5.10. Take $\mu_i \in M$, $\lambda_i \in M^*$ $(1 \leq i \leq r)$ and define $f \in \mathrm{End}_R(M)$ by $f(x) = \sum_{1 \leq i \leq r} \lambda_i(x) \cdot \mu_i$. Then we have

$$\kappa((\mu_1 \wedge \cdots \wedge \mu_r) \otimes (\lambda_1 \wedge \cdots \wedge \lambda_r)) = \wedge^r f,$$
$$\mathrm{tr}_R \circ \kappa((\mu_1 \wedge \cdots \wedge \mu_r) \otimes (\lambda_1 \wedge \cdots \wedge \lambda_r)) = \iota_R(\Phi(\lambda_1 \wedge \cdots \wedge \lambda_r)(\mu_1 \wedge \cdots \wedge \mu_r)).$$

Hence it suffices to prove $(\wedge^r f)(\phi_1 \wedge \cdots \wedge \phi_r) = \Phi(\lambda_1 \wedge \cdots \wedge \lambda_r)(\mu_1 \wedge \cdots \wedge \mu_r) \cdot \phi_1 \wedge \cdots \wedge \phi_r$ modulo $\mathfrak{m}$-torsion for any $\phi_i \in M$ $(1 \leq i \leq r)$.

Take $\nu_i \in M^*$ $(1 \leq i \leq r)$ and define $g \in \mathrm{End}_R(M)$ by $g(x) = \sum_{1 \leq i \leq r} \nu_i(x)\phi_i$. Then, $\mathrm{tr}_R(\wedge^r f) = \iota_R(\Phi(\lambda_1 \wedge \cdots \wedge \lambda_r)(\mu_1 \wedge \cdots \wedge \mu_r))$, $\mathrm{tr}_R(\wedge^r g) = \iota(\Phi(\nu_1 \wedge \cdots \wedge \nu_r)(\phi_1 \wedge \cdots \wedge \phi_r))$ and $\mathrm{tr}_R(\wedge^r (f \circ g)) = \iota_R(\Phi(\nu_1 \wedge \cdots \wedge \nu_r)(\wedge^r f(\phi_1 \wedge \cdots \wedge \phi_r)))$. By Proposition V.5.7, we obtain

$$\begin{aligned}&\iota_R(\Phi(\nu_1 \wedge \cdots \wedge \nu_r)(\wedge^r f(\phi_1 \wedge \cdots \wedge \phi_r)))\\ =&\iota_R(\Phi(\lambda_1 \wedge \cdots \wedge \lambda_r)(\mu_1 \wedge \cdots \wedge \mu_r) \cdot \Phi(\nu_1 \wedge \cdots \wedge \nu_r)(\phi_1 \wedge \cdots \wedge \phi_r)).\end{aligned}$$

Since we have almost isomorphisms

$$\begin{aligned}L &\xrightarrow{\approx} \mathrm{Hom}_R(L^*, R) \quad \text{(Lemma V.5.12)}\\ &\xrightarrow[\Phi^*]{\approx} \mathrm{Hom}_R(\wedge^r M^*, R) \quad \text{(Lemma V.5.11, Proposition V.2.3)}\\ &\xrightarrow[\iota_R \circ -]{\approx} \mathrm{Hom}_R(\wedge^r M^*, \mathrm{Hom}_{\overline{V}}(\mathfrak{m}, R)) \quad \text{(Proposition V.2.3)},\end{aligned}$$

varying $\nu_i$ $(1 \leq i \leq r)$, we obtain the required congruence. □

**Definition V.5.14.** Let $R$ be a $\overline{V}$-algebra and let $M$ be an almost finitely generated projective $R$-module. For an integer $r \geq 0$, we say that $M$ has *rank* $r$ over $R$ if $\wedge_R^{r+1} M \approx 0$ and $\mathrm{tr}_R(\mathrm{id}_{\wedge^r M}) = 1$, and denote it by $\mathrm{rank}_R M = r$.

By Proposition V.4.3, if $M \overset{\approx}{\to} M'$, $\mathrm{rank}_R M = r$ if and only if $\mathrm{rank}_R M' = r$. By Proposition V.4.2 (2), for a ring homomorphism $R \to R'$ and $M' := M \otimes_R R'$, if $\mathrm{rank}_R M = r$, then $\mathrm{rank}_{R'} M' = r$. By Proposition V.2.5, the converse is also true if $R \overset{\approx}{\to} R'$.

**Corollary V.5.15.** *Let $R$ be a $\overline{V}$-algebra, let $M$ be an almost finitely generated projective $R$-module of rank $\leq r$, and set $L := \wedge^r M$, $L^* := \mathrm{Hom}_R(L, R)$. If $\mathrm{tr}_R(\mathrm{id}_{\wedge^r M}) = 1$ (resp. $0$), then the canonical homomorphism $L \otimes_R L^* \to R$ is an almost isomorphism (resp. $\mathrm{rank}_R M \leq r - 1$, that is, $L \approx 0$).*

Let $R$ be a $\overline{V}$-algebra and let $M$ be an almost finitely generated projective $R$-module of rank $\leq r$. Set $e = \mathrm{tr}_R(\mathrm{id}_{\wedge^r M})$ and define $R_r := \mathrm{Hom}_{\overline{V}}(\mathfrak{m}, R)/(e-1)$, $R' := \mathrm{Hom}_{\overline{V}}(\mathfrak{m}, R)/(e)$. Then, by Corollary V.5.8, $\mathrm{Hom}_{\overline{V}}(\mathfrak{m}, R) \xrightarrow{\sim} R_r \times R'$. Set $M_r := M \otimes_R R_r$ and $M' := M \otimes_R R'$. Then, by Proposition V.4.2 (2), $\mathrm{tr}_{R_r}(\mathrm{id}_{\wedge^r M_r}) = 1$ and $\mathrm{tr}_{R'}(\mathrm{id}_{\wedge^r_{M'}}) = 0$. By Corollary V.5.15, $\mathrm{rank}_{R_r}(M_r) = r$ and $\mathrm{rank}_{R'} M' \leq r-1$. Repeating this procedure and using $\mathrm{Hom}_{\overline{V}}(\mathfrak{m}, R) \xrightarrow{\sim} \mathrm{Hom}_{\overline{V}}(\mathfrak{m}, \mathrm{Hom}_{\overline{V}}(\mathfrak{m}, R))$, we obtain the following proposition.

**Proposition V.5.16.** *Let $R$ be a $\overline{V}$-algebra and let $M$ be an almost finitely generated projective $R$-module of rank $\leq r$. Then, there exists a unique decomposition* $\mathrm{Hom}_{\overline{V}}(\mathfrak{m}, R) = R_r \times \cdots \times R_0$ *such that* $\mathrm{rank}_{R_i}(M \otimes_R R_i) = i$ $(0 \leq i \leq r)$.

PROOF. We have already proven the existence. Suppose that we have another decomposition $R'_r \times \cdots \times R'_0$. Then the base change $M_{i,j}$ $(0 \leq i, j \leq r)$ of $M$ to $R_{i,j} := R_i \otimes_{\mathrm{Hom}_{\overline{V}}(\mathfrak{m}, R)} R'_j$ has rank $i$ and $j$ at the same time. Hence, by Corollary V.5.15, $R_{i,j} \approx 0$ if $i \neq j$. On the other hand $R_{i,j}$ is a direct factor of $\mathrm{Hom}_{\overline{V}}(\mathfrak{m}, R)$. Hence $R_{i,j}$ is $\mathfrak{m}$-torsion free and $R_{i,j} = 0$ if $i \neq j$, which implies $R_i = R'_i$ $(0 \leq i \leq r)$. □

By the uniqueness, the decomposition is compatible with base changes. For $R$, $M$ and the decomposition in Proposition V.5.16, we define the determinant map

$$\det_R \colon \mathrm{End}_R(M) \longrightarrow \mathrm{Hom}_{\overline{V}}(\mathfrak{m}, R)$$

by $\det_R(f) = (\mathrm{tr}_{R_i}(\wedge^i f_i))_{0 \leq i \leq r}$, where $f_i$ denotes the base change of $f$ to $R_i$. Note $R_i \xrightarrow{\sim} \mathrm{Hom}_{\overline{V}}(\mathfrak{m}, R_i)$. We have $\det_R(f \circ g) = \det_R(f) \cdot \det_R(g)$ (Proposition V.5.7) and $\det_R(\mathrm{id}_M) = 1$. The map $\det_R$ is compatible with base changes.

## V.6. Almost flat modules and almost faithfully flat modules

**Definition V.6.1.** Let $R$ be a $\overline{V}$-algebra and let $M$ be an $R$-module. We say that $M$ is *almost flat* if it satisfies the following equivalent conditions:

(i) $\mathrm{Tor}_i^R(M, N) \approx 0$ for all $R$-modules $N$ and all integers $i \geq 1$.

(ii) $\mathrm{Tor}_1^R(M, N) \approx 0$ for all $R$-modules $N$.

(iii) For any injective homomorphism $f : N_1 \to N_2$ of $R$-modules, the kernel of $\mathrm{id}_M \otimes f \colon M \otimes_R N_1 \to M \otimes_R N_2$ is almost zero.

We say that an $R$-algebra is *almost flat* if it is almost flat as an $R$-module.

**Proposition V.6.2.** (1) *For a $\overline{V}$-algebra $R$ and an almost isomorphism $M \xrightarrow{\approx} M'$ of $R$-modules, $M$ is almost flat if and only if $M'$ is almost flat.*

(2) *For a homomorphism $R \to R'$ of $\overline{V}$-algebra and an $R$-module $M$, if $M$ is almost flat over $R$, then $M' := M \otimes_R R'$ is almost flat over $R'$. If $R \xrightarrow{\approx} R'$, then the converse is also true.*

PROOF. (1) follows from Proposition V.2.3. For (2), use the condition (iii) in Definition V.6.1 and $M \otimes_R N \xrightarrow{\approx} M' \otimes_R N \cong M' \otimes_{R'} (N \otimes_R R')$ for an $R$-module $N$ if $R \xrightarrow{\approx} R'$. □

**Proposition V.6.3.** *Let $R$ be a $\overline{V}$-algebra and let $M$ be an $R$-module. If $M$ is almost projective, then $M$ is almost flat.*

PROOF. This follows from the condition (iv) in Proposition V.3.1 similarly as the proof of (iv)⇒(i) of Proposition V.3.1. □

**Definition V.6.4.** Let $R$ be a $\overline{V}$-algebra and let $M$ be an $R$-module. We say that $M$ is *almost faithfully flat* if it is almost flat and, for any $R$-module $N$, $N \otimes_R M \approx 0$ implies $N \approx 0$. We say that an $R$-algebra is almost faithfully flat if it is *almost faithfully flat* as an $R$-module.

**Proposition V.6.5.** (1) *For a $\overline{V}$-algebra $R$ and an almost isomorphism $M \stackrel{\approx}{\to} M'$ of $R$-modules, $M$ is almost faithfully flat if and only if $M'$ is almost faithfully flat.*

(2) *For a homomorphism of $\overline{V}$-algebras $R \to R'$ and an $R$-module $M$, if $M$ is almost faithfully flat over $R$, then $M' := M \otimes_R R'$ is almost faithfully flat over $R'$. If $R \stackrel{\approx}{\to} R'$, the converse is also true.*

PROOF. (1) follows from Proposition V.6.2 (1) and Proposition V.2.3. The first assertion of (2) is trivial. For the second assertion, we use almost isomorphisms $M \otimes_R N \stackrel{\approx}{\to} R' \otimes_R (M \otimes_R N) \cong M' \otimes_{R'} (N \otimes_R R')$ and $N \stackrel{\approx}{\to} N \otimes_R R'$. □

**Proposition V.6.6.** *Let $R \stackrel{f}{\to} R' \stackrel{g}{\to} R''$ be homomorphisms of $\overline{V}$-algebras. If $f$ and $g$ are almost flat (resp. almost faithfully flat), then $g \circ f$ is also almost flat (resp. almost faithfully flat).*

PROOF. The first case follows from the condition (iii) of Definition V.6.1 and Proposition V.2.3, and then the second case is trivial by definition. □

**Proposition V.6.7.** *Let $R$ be a $\overline{V}$-algebra and let $M$ be an almost finitely generated projective $R$-module of rank $r \geq 1$. Then $M$ is almost faithfully flat over $R$.*

**Lemma V.6.8.** *Let $R$ be a $\overline{V}$-algebra and let $M$ be an almost flat $R$-module. Then $M$ is almost faithfully flat if and only if, for any ideal $I$ of $R$, $M/IM \approx 0$ implies $R/IR \approx 0$.*

PROOF. The necessity is trivial. Let us prove the sufficiency. For an $R$-module $N$, suppose $M \otimes_R N \approx 0$. Then, for any $x \in N$, we have an injection $R/I \hookrightarrow N; \overline{a} \mapsto a \cdot x$, where $I = \{a \in R | a \cdot x = 0\}$. The kernel of the induced homomorphisms $M/IM \to N \otimes_R M$ is almost zero. Hence $M/IM \approx 0$, which implies $R/I \approx 0$, i.e., $\mathfrak{m} \cdot x = 0$ by assumption. □

PROOF OF PROPOSITION V.6.7. Let $I$ be an ideal of $R$ and suppose $M/IM \approx 0$. Then, since $\mathrm{rank}_{R/I}(M/IM) = r$, we have $\mathrm{tr}_R(\mathrm{id}_{\wedge^r M/IM}) = 1$. On the other hand, since $M/IM \approx 0$, we have $\mathrm{tr}_R(\mathrm{id}_{\wedge^r M/IM}) = 0$ by Proposition V.2.3 and Proposition V.4.3. Hence $\mathrm{Hom}_{\overline{V}}(\mathfrak{m}, R/I) = 0$ and $R/I \approx 0$. By Lemma V.6.8 and Proposition V.6.3, $M$ is almost faithfully flat. □

The following lemma will be useful in V.8 and V.9.

**Lemma V.6.9.** *Let $R$ be a $\overline{V}$-algebra, let $M$ be an almost flat $R$-module, and let $C^\bullet$ be a complex of $R$-modules. Then, the canonical homomorphism $H^q(C^\bullet) \otimes_R M \to H^q(C^\bullet \otimes_R M)$ is an almost isomorphism for $q \in \mathbb{Z}$. Especially, if $H^q(C^\bullet) \approx 0$, then $H^q(C^\bullet \otimes_R M) \approx 0$, and the converse is also true if $M$ is almost faithfully flat over $R$.*

PROOF. Put $Z^q := \mathrm{Ker}(d^q \colon C^q \to C^{q+1})$ and $B^q := \mathrm{Im}(d^{q-1} \colon C^{q-1} \to C^q)$. Let $C'^\bullet$ denote the complex $C^\bullet \otimes_R M$ and define $Z'^q$ and $B'^q$ similarly. Since the canonical homomorphism $B^q \otimes_R M \to B'^q$ is surjective and the kernel of $B^q \otimes_R M \to C'^q (= C^q \otimes_R M)$ is almost zero, we have $B^q \otimes_R M \stackrel{\approx}{\to} B'^q$. Comparing the exact sequence $0 \to Z'^q \to C'^q \to B'^{q+1} \to 0$ with the exact sequence $Z^q \otimes_R M \to C'^q \to B^{q+1} \otimes_R M \to 0$ in which the kernel of the left homomorphism is almost zero, we see $Z^q \otimes_R M \stackrel{\approx}{\to} Z'^q$. This completes the proof. □

## V.7. Almost étale coverings

**Definition V.7.1.** Let $f\colon R \to S$ be a homomorphism of $\overline{V}$-algebras. We say that $f$ is an *almost étale covering* if it satisfies the following conditions:

(i) $S$ is almost finitely generated projective of finite rank as an $R$-module.

(ii) $S$ is almost projective as an $S \otimes_R S$-module.

**Lemma V.7.2.** *Let $f\colon R \to S$ be a surjective homomorphism of $\overline{V}$-algebras and let $S'$ denote the image of the homomorphism $g\colon \mathrm{Hom}_{\overline{V}}(\mathfrak{m}, R) \to \mathrm{Hom}_{\overline{V}}(\mathfrak{m}, S)$ induced by $f$. Then $S$ is almost projective as an $R$-module if and only if the closed immersion $\mathrm{Spec}(S') \hookrightarrow \mathrm{Spec}(\mathrm{Hom}_{\overline{V}}(\mathfrak{m}, R))$ induced by $g$ is an open immersion.*

Proof. First let us prove the sufficiency. The condition implies that $S'$ is projective as a $\mathrm{Hom}_{\overline{V}}(\mathfrak{m}, R)$-module. Since $S \otimes_R \mathrm{Hom}_{\overline{V}}(\mathfrak{m}, R) \xrightarrow{\approx} S'$ and $R \xrightarrow{\approx} \mathrm{Hom}_{\overline{V}}(\mathfrak{m}, R)$, $S$ is almost projective as an $R$-module by Proposition V.3.3 (1) and Proposition V.3.9. Conversely suppose that $S$ is almost projective as an $R$-module. Then, $e = \mathrm{tr}_R(\mathrm{id}_S)$ is an idempotent (because $\mathrm{rank}_R S \leq 1$) and we have $\mathrm{Hom}_{\overline{V}}(\mathfrak{m}, R) = R_1 \times R_0$, where $R_1 = \mathrm{Hom}_{\overline{V}}(\mathfrak{m}, R)/(e-1)$, $R_0 = \mathrm{Hom}_{\overline{V}}(\mathfrak{m}, R)/(e)$. We will prove $\mathrm{Hom}_{\overline{V}}(\mathfrak{m}, I) = \{0\} \times R_0$ for $I = \mathrm{Ker}(f)$. By Proposition V.4.2 (2), the image of $e$ in $\mathrm{Hom}_{\overline{V}}(\mathfrak{m}, S)$ is $\mathrm{tr}_S(\mathrm{id}_S) = 1$. Hence $\mathrm{Hom}_{\overline{V}}(\mathfrak{m}, I) \supset \{0\} \times R_0$. For $\varphi \in \mathrm{Hom}_{\overline{V}}(\mathfrak{m}, I)$, $\varphi(\pi^\alpha) \cdot e = \mathrm{tr}_R(\varphi(\pi^\alpha) \cdot \mathrm{id}_S) = 0$ for any $\alpha \in \Lambda^+$, which implies $\varphi \cdot e = 0$. Hence $\mathrm{Hom}_{\overline{V}}(\mathfrak{m}, I) \subset \{0\} \times R_0$. □

By Lemma V.7.2, the condition (ii) in Definition V.7.1 can be replaced by

(ii)′ The closed immersion $\mathrm{Spec}(\mathrm{Hom}_{\overline{V}}(\mathfrak{m}, S)) \hookrightarrow \mathrm{Spec}(\mathrm{Hom}_{\overline{V}}(\mathfrak{m}, S \otimes_R S))$ is an open immersion.

This implies that the above definition is compatible with the usual étale coverings as follows.

**Proposition V.7.3.** *Let $f\colon R \to S$ be a homomorphism of $\overline{V}$-algebras and suppose that $\pi^\alpha$ is invertible on $R$ for some (or, equivalently, all) $\alpha \in \Lambda^+$. Then $f$ is an almost étale covering if and only if $f$ is finite étale.*

Proof. The conditions (i) and (ii)′ are equivalent to saying that $S$ is flat of finite presentation as an $R$-module (**[11]** II §5 n° 2 Corollaire 2) and the closed immersion $\mathrm{Spec}(S) \hookrightarrow \mathrm{Spec}(S \otimes_R S)$ is an open immersion. Hence the proposition follows from **[42]** Proposition (1.4.7), Corollaire (17.4.2) and Corollaire (17.6.2). □

**Proposition V.7.4.** *Let $R$ be a $\overline{V}$-algebra.*

(1) *For an almost isomorphism of $R$-algebras $S \xrightarrow{\approx} S'$, $S$ is an almost étale covering over $R$ if and only if $S'$ is an almost étale covering over $R$.*

(2) *For $R$-algebras $S_1$ and $S_2$, $S_1 \times S_2$ is an almost étale covering over $R$ if and only if $S_1$ and $S_2$ are almost étale coverings over $R$.*

(3) *For $R$-algebras $S$ and $R'$, if $S$ is an almost étale covering over $R$, then the base change $S' := S \otimes_R R'$ is an almost étale covering over $R'$. The converse is also true if $R \xrightarrow{\approx} R'$.*

Proof. For (1), the equivalence in the condition (i) follows from Proposition V.3.6 (1), Proposition V.3.3 (1), and the remark after Definition V.5.6. By Proposition V.2.3, we have $S \otimes_R S \xrightarrow{\approx} S' \otimes_R S'$ and $S \otimes_{S \otimes_R S} (S' \otimes_R S') \xrightarrow{\approx} S'$. Hence we obtain the equivalence in the condition (ii) from Proposition V.3.9 and Proposition V.3.3 (1). The proof of (2) is straightforward using $\wedge^s_R(S_1 \oplus S_2) \cong \oplus_{s=s_1+s_2}(\wedge^{s_1}_R S_1) \otimes_R (\wedge^{s_2}_R S_2)$ and $\mathrm{Ext}^i_R(S_1 \oplus S_2, N) \cong \mathrm{Ext}^i_R(S_1, N) \oplus \mathrm{Ext}^i_R(S_2, N)$. The claim (3) immediately follows from Proposition V.3.3 (2), Proposition V.3.6 (2), Proposition V.3.9, and the remark after Definition V.5.6. □

**Proposition V.7.5.** *Let $f\colon R \to S$ be a homomorphism of $\overline{V}$-algebra and suppose that $f$ is an almost étale covering and* $\mathrm{rank}_R S = r$. *Then, there exist an almost faithfully flat $R$-algebra $R'$ and an almost isomorphism of $R'$-algebras $S \otimes_R R' \overset{\approx}{\to} (R')^r$.*

PROOF. We prove by induction on $r$. The case $r = 0$ is trivial. Let $r \geq 1$ and assume that the proposition is true for $r-1$. The homomorphism $R \to \mathrm{Hom}_{\overline{V}}(\mathfrak{m}, S)$ is almost faithfully flat (Proposition V.6.7) and we have an almost isomorphism of $\mathrm{Hom}_{\overline{V}}(\mathfrak{m}, S)$-algebras $\mathrm{Hom}_{\overline{V}}(\mathfrak{m}, S) \otimes_R S \overset{\approx}{\to} \mathrm{Hom}_{\overline{V}}(\mathfrak{m}, S \otimes_R S)$. The last algebra is of the form $\mathrm{Hom}_{\overline{V}}(\mathfrak{m}, S) \times R'$ by the remark after Lemma V.7.2. By Proposition V.7.4 (2) and Lemma V.7.6 below, $\mathrm{Hom}_{\overline{V}}(\mathfrak{m}, S) \to R'$ is an almost étale covering and $\mathrm{rank}_{\mathrm{Hom}_{\overline{V}}(\mathfrak{m},S)}(R') = r-1$. By Proposition V.6.6, we are reduced to the case $r-1$. □

**Lemma V.7.6.** *Let $R$ be a $\overline{V}$-algebra and let $M_1$, $M_2$ be two almost finitely generated projective $R$-modules. For integers $r \geq r_1 \geq 0$, if* $\mathrm{rank}_R(M_1 \oplus M_2) = r$ *and* $\mathrm{rank}_R(M_1) = r_1$, *then* $\mathrm{rank}_R(M_2) = r - r_1$.

PROOF. Set $M = M_1 \oplus M_2$ and $r_2 = r - r_1$. Since $\wedge^{r+1} M \supset \wedge^{r_1} M_1 \otimes_R \wedge^{r_2+1} M_2$, we have $\wedge^{r_1} M_1 \otimes_R \wedge^{r_2+1} M_2 \approx 0$. Taking $(\wedge^{r_1} M_1)^* \otimes -$ and using $(\wedge^{r_1} M_1)^* \otimes_R \wedge^{r_1} M_1 \overset{\approx}{\to} R$ (Corollary V.5.15), we obtain $\wedge^{r_2+1} M_2 \approx 0$. Then, we have $\wedge^r M \cong \oplus_{r=s_1+s_2} \wedge^{s_1} M_1 \otimes_R \wedge^{s_2} M_2 \overset{\approx}{\leftarrow} \wedge^{r_1} M_1 \otimes_R \wedge^{r_2} M_2$ and $1 = \mathrm{tr}_R(\mathrm{id}_{\wedge^r M}) = \mathrm{tr}_R(\mathrm{id}_{\wedge^{r_1} M_1} \otimes \mathrm{id}_{\wedge^{r_2} M_2}) = \mathrm{tr}_R(\mathrm{id}_{\wedge^{r_1} M_1}) \cdot \mathrm{tr}_R(\mathrm{id}_{\wedge^{r_2} M_2}) = \mathrm{tr}_R(\mathrm{id}_{\wedge^{r_2} M_2})$ by Proposition V.4.3 and Proposition V.4.4. □

**Proposition V.7.7.** *Let $f\colon R \to S$ and $g\colon S \to T$ be homomorphisms of $\overline{V}$-algebras. If $f$ and $g$ are almost étale coverings, then $g \circ f$ is also an almost étale covering.*

PROOF. For $\alpha \in \Lambda^+$, choose integers $r, s \geq 0$ and homomorphisms $S \overset{\varphi}{\to} R^{\oplus r} \overset{\psi}{\to} S$ and $T \overset{\eta}{\to} S^{\oplus s} \overset{\theta}{\to} T$ as $R$-modules and as $S$-modules, respectively, such that $\psi \circ \varphi = \pi^\alpha \cdot \mathrm{id}_S$ and $\theta \circ \eta = \pi^\alpha \cdot \mathrm{id}_T$ (Proposition V.3.7). Then the $R$-linear homomorphisms $\varphi^{\oplus s} \circ \eta\colon T \to R^{\oplus rs}$ and $\theta \circ \psi^{\oplus s}\colon R^{\oplus rs} \to T$ satisfy $(\theta \circ \psi^{\oplus s}) \circ (\varphi^{\oplus s} \circ \eta) = (\pi^\alpha)^2 \cdot \mathrm{id}_T$. Hence, $T$ is almost finitely generated projective as an $R$-module. The homomorphism $T \otimes_R T \to T$ factors as $T \otimes_R T \to T \otimes_S T \to T$ and the first homomorphism is isomorphic to the base change of $S \otimes_R S \to S$ by $S \otimes_R S \to T \otimes_R T$. Hence, by Proposition V.3.3 (2) and the same argument as above, we see that $T$ is almost projective as a $T \otimes_R T$-module. It remains to prove that $T$ is of finite rank over $R$. By Proposition V.5.16 and Proposition V.7.5, there exist an almost faithfully flat $R$-algebra $R_r \times \cdots \times R_0$ and an almost isomorphism $S \otimes_R R_i \overset{\approx}{\to} (R_i)^i$ of $R_i$-algebras for each $0 \leq i \leq r$. Then, for an integer $t \geq 0$, $\mathrm{rank}_R(T) \leq t$ if and only if $\mathrm{rank}_{R_i}((T \otimes_R R_i) \otimes_{S \otimes_R R_i} (R_i)^i) \leq t$ for every $0 \leq i \leq r$. Thus, we are reduced to the case $S = R^r$, in which case the assertion is trivial. □

**Proposition V.7.8.** *Let $f$ and $g$ be the same as in Proposition V.7.7. If $f$ and $g \circ f$ are almost étale coverings, then $g$ is also an almost étale covering.*

PROOF. We consider the following commutative diagram, in which every square is co-Cartesian:

$$\begin{array}{ccccc}
T & \xleftarrow{\ g\ } & S & & \\
{\scriptstyle h_T}\uparrow & & {\scriptstyle h}\uparrow & & \\
T \otimes_R S & \xleftarrow{\ g\otimes 1\ } & S \otimes_R S & \xleftarrow{\ i_2\ } & S \\
{\scriptstyle i_T}\uparrow & & {\scriptstyle i_1}\uparrow & & {\scriptstyle f}\uparrow \\
T & \xleftarrow{\ g\ } & S & \xleftarrow{\ f\ } & R
\end{array}$$

Since $g \circ f$ is an almost étale covering, so is $g \otimes 1 \circ i_2$ by Proposition V.7.4 (3). Since $f$ is an almost étale covering, $S$ is almost finitely generated projective as an $S \otimes_R S$-module. Since $h$ is surjective, this implies that $h$ and hence $h_T$ are almost étale coverings. By Proposition V.7.7, $g = h_T \circ (g \otimes 1) \circ i_2$ is an almost étale covering. $\square$

**Proposition V.7.9.** *Let $p$ be a prime and let $f\colon R \to S$ be a homomorphism of $\overline{V}/p\overline{V}$-algebra. Let $S^{(p)}$ be the base change of $S$ by the absolute Frobenius $F_R\colon R \to R$ of $R$ and let $F_{S/R}\colon S^{(p)} \to S$ be the unique homomorphism of $R$-algebras such that the composite with $S \to S^{(p)}$ is the absolute Frobenius $F_S$ of $S$. If $f$ is an almost étale covering, then $F_{S/R}$ is an almost isomorphism.*

PROOF. By Proposition V.5.16 and Proposition V.7.5, there exists an almost faithfully flat $R$-algebra $R_r \times \cdots \times R_0$ and an almost isomorphism $S \otimes_R R_i \xrightarrow{\approx} (R_i)^i$ of $R_i$-algebras for each $0 \le i \le r$. Since the relative Frobenius $F_{S/R}$ is compatible with base changes and functorial on $S$, we are reduced to the case $S = R^r$, in which case the proposition is trivial. $\square$

For $\pi^\alpha$-torsion free algebras, we have the following criterion for almost étale coverings, which was adopted as a definition in **[24]** 2.1 Definition.

**Proposition V.7.10.** *Let $f\colon R \to S$ be a homomorphism of $\overline{V}$-algebras and assume that $R$ and $S$ are $\pi^\alpha$-torsion free for some (or equivalently all) $\alpha \in \Lambda^+$. Let $R_\pi$ and $S_\pi$ denote the rings $R[\frac{1}{\pi^{\alpha_0}}] = R[\frac{1}{\pi^\alpha}(\alpha \in \Lambda^+)]$ and $S[\frac{1}{\pi^{\alpha_0}}] = S[\frac{1}{\pi^\alpha}(\alpha \in \Lambda^+)]$ $(\alpha_0 \in \Lambda^+)$. Then $f$ is an almost étale covering if and only if it satisfies the following conditions:*

(i) *$R_\pi \to S_\pi$ is finite étale.*

(ii) *$\pi^\alpha \cdot \mathrm{tr}_{S_\pi/R_\pi}(S) \subset R$ for all $\alpha \in \Lambda^+$.*

(iii) *Let $e_{S_\pi/R_\pi} \in S_\pi \otimes_{R_\pi} S_\pi$ be the idempotent corresponding to the image of the open and closed diagonal immersion $\mathrm{Spec}(S_\pi) \hookrightarrow \mathrm{Spec}(S_\pi \otimes_{R_\pi} S_\pi)$. Then $\pi^\alpha \cdot e_{S_\pi/R_\pi}$ is contained in the image of $S \otimes_R S$ for all $\alpha \in \Lambda^+$.*

Note that the condition (i) implies that $S_\pi$ is projective of finite rank as an $R_\pi$-module (**[42]** Proposition (1.4.7) and **[11]** II §5 n° 2 Corollaire 2) and hence the trace map $\mathrm{tr}_{S_\pi/R_\pi}$ is well-defined.

PROOF. For the necessity, the condition (i) (resp. (ii), resp. (iii)) follows from Proposition V.7.3 and Proposition V.7.4 (3) (resp. Proposition V.4.2, resp. the condition (ii)$'$ after Lemma V.7.2). Let us prove the sufficiency. Set $e := e_{S_\pi/R_\pi}$ for simplicity. Since $e$ corresponds to $(1,0)$ under the decomposition $S_\pi \otimes_{R_\pi} S_\pi \cong S_\pi \times (S_\pi \otimes_{R_\pi} S_\pi)/(e)$ defined by $e$, for $b \in S$, the trace of $(b \otimes 1) \cdot e$ with respect to $S_\pi \to S_\pi \otimes_{R_\pi} S_\pi; y \mapsto 1 \otimes y$ is $b$. Hence, for $\alpha \in \Lambda^+$, if we write $\pi^\alpha \cdot e$ in the form $\sum_{1 \le i \le n} x_i \otimes y_i$, $x_i, y_i \in S$ (the condition (i)), then $\pi^\alpha \cdot b = \sum_{1 \le i \le n} \mathrm{tr}_{S_\pi/R_\pi}(bx_i) \cdot y_i$. By the condition (ii), we can construct $R$-linear homomorphisms $\varphi\colon S \to R^n; b \mapsto (\pi^\alpha \cdot \mathrm{tr}_{S_\pi/R_\pi}(bx_i))_i$ and $\psi\colon R^n \to S; (a_i) \mapsto \sum_{1 \le i \le n} a_i \cdot y_i$, and $\psi \circ \varphi = (\pi^\alpha)^2 \cdot \mathrm{id}_S$. By Proposition V.3.7, $S$ is almost finitely generated projective as an $R$-module. By the condition (i), $\wedge_{R_\pi}^{r+1} S_\pi = 0$ for some integer $r \ge 0$. Since $\wedge_R^{r+1} S$ is almost flat over $R$ by Proposition V.3.8 and Proposition V.6.3, the kernel of $\wedge_R^{r+1} S \to \wedge_R^{r+1} S \otimes_R R_\pi \cong \wedge_{R_\pi}^{r+1} S_\pi = 0$ is almost zero. Hence $\mathrm{rank}_R S \le r$. For the $S_\pi \otimes_{R_\pi} S_\pi$-linear section $i\colon S_\pi \to S_\pi \otimes_{R_\pi} S_\pi; a \mapsto \tilde{a} \cdot e$ of the surjective homomorphism $m : S_\pi \otimes_{R_\pi} S_\pi \to S_\pi$, $\pi^\alpha \cdot i(S)$ is contained in the image $\overline{S \otimes_R S}$ of $S \otimes_R S$ in $S_\pi \otimes_{R_\pi} S_\pi$ for any $\alpha \in \Lambda^+$ by the condition (iii). Hence we have homomorphisms $S \xrightarrow{\pi^\alpha \cdot i} \overline{S \otimes_R S} \xrightarrow{m} S$ of $S \otimes_R S$-modules whose composite is $\pi^\alpha \cdot \mathrm{id}_S$. On the other hand, by Proposition V.3.8 and Proposition V.6.3, $S \otimes_R S$ is almost flat over $R$ and hence we have $S \otimes_R S \xrightarrow{\approx} \overline{S \otimes_R S}$ and $\mathrm{Ext}^i_{S \otimes_R S}(\overline{S \otimes_R S}, N) \xleftarrow{\approx} \mathrm{Ext}^i_{S \otimes_R S}(S \otimes_R S, N) = 0$ $(i > 0)$ for any $S \otimes_R S$-modules

$N$. Therefore $(\pi^\alpha)^2 \cdot \mathrm{Ext}^i_{S\otimes_R S}(S,N) = 0$ $(i>0)$ for any $S\otimes_R S$-module $N$ and any $\alpha\in\Lambda^+$. □

**Proposition V.7.11.** *Let $f\colon R\to S$ be a homomorphism of $\overline{V}$-algebras and assume that $f$ is an almost étale covering and that $R$ is a normal domain and $\pi^\alpha$-torsion free for some (or equivalently all) $\alpha\in\Lambda^+$. Let $R_\pi$ and $S_\pi$ denote the rings $R[\frac{1}{\pi^{\alpha_0}}] = R[\frac{1}{\pi^\alpha}(\alpha\in\Lambda)]$ and $S[\frac{1}{\pi^{\alpha_0}}] = S[\frac{1}{\pi^\alpha}(\alpha\in\Lambda)]$ $(\alpha_0\in\Lambda^+)$, and let $S'$ denote the normalization of $R$ in $S_\pi$. Then, the natural homomorphism $S\to S_\pi$ induces an almost isomorphism $S\overset{\approx}{\to}\mathrm{Hom}_{\overline{V}}(\mathfrak{m},S')$. Here we regard $\mathrm{Hom}_{\overline{V}}(\mathfrak{m},S')$ as a subring of $S_\pi$ by the natural injection $\mathrm{Hom}_{\overline{V}}(\mathfrak{m},S')\hookrightarrow\mathrm{Hom}_{\overline{V}}(\mathfrak{m},S_\pi)\cong S_\pi$. Especially $R\to S'$ is an almost étale covering (Proposition* V.7.4 (1)).

Note that $R_\pi\to S_\pi$ is finite étale by Proposition V.7.3 and Proposition V.7.4 (3). By **[42]** Proposition (17.5.7), $S_\pi$ and hence $S'$ are normal.

Proof. Since $R\to S$ is almost flat by Proposition V.6.3, the kernel of the homomorphism $S\to S\otimes_R R_\pi\cong S_\pi$ is almost zero. Hence we may assume that $S$ is $\pi^\alpha$-torsion free for all $\alpha\in\Lambda^+$. It suffices to prove $\mathfrak{m}\cdot S\subset S'$ and $\mathfrak{m}\cdot S'\subset S$. For any $\alpha\in\Lambda^+$, choose a homomorphism $f\colon R^{\oplus r}\to S$ $(r\geq 1)$ of $R$-modules whose cokernel is annihilated by $\pi^\alpha$ (Lemma V.3.5). For any $y\in S$, there exists $g\in\mathrm{End}_R(R^{\oplus r})$ such that $f\circ g = (\pi^\alpha y\cdot \mathrm{id}_S)\circ f$. For the monic polynomial $F(T):=\det(T-g)\in R[T]$, we have $F(g)=0$ and hence $F(\pi^\alpha y)=0$, which implies $\pi^\alpha y\in S'$. Varying $\alpha$ and $y$, we obtain $\mathfrak{m}\cdot S\subset S'$. Next let us prove $\mathfrak{m}\cdot S'\subset S$. Let $e\in S_\pi\otimes_{R_\pi}S_\pi$ be the idempotent corresponding to the open and closed immersion $\mathrm{Spec}(S_\pi)\to\mathrm{Spec}(S_\pi\otimes_{R_\pi}S_\pi)$. By Proposition V.7.10, for any $\alpha\in\Lambda^+$, $\pi^\alpha e$ is written in the form $\sum_{1\leq i\leq n}x_i\otimes y_i$, $x_i,y_i\in\mathfrak{m}\cdot S\subset S'$. As in the proof of Proposition V.7.10, we have $\pi^\alpha\cdot y = \sum_{1\leq i\leq n}\mathrm{tr}_{S_\pi/R_\pi}(yx_i)\cdot y_i$ for any $y\in S'$. Since $R$ is a normal domain, $\mathrm{tr}_{S_\pi/R_\pi}(S')\subset R$. Hence, we have $\pi^\alpha y\in S$. Varying $\alpha$ and $y$, we obtain $\mathfrak{m}\cdot S'\subset S$ □

We need the following Proposition in V.12.

**Proposition V.7.12.** *Let $f\colon R\to S$ be a homomorphism of $\overline{V}$-algebras. If $f$ is an almost étale covering and almost faithfully flat (e.g., $\mathrm{rk}_R S = r\geq 1$ (Proposition* V.6.7)), *then* $\mathrm{tr}_{S/R}(S)\supset\mathfrak{m}\cdot\mathrm{Hom}_{\overline{V}}(\mathfrak{m},R)$.

Proof. By Proposition V.4.2 (2), $\mathrm{tr}_{S/R}\otimes 1\colon S\otimes_R S\to\mathrm{Hom}_{\overline{V}}(\mathfrak{m},R)\otimes_R S$ followed by the almost isomorphism $\mathrm{Hom}_{\overline{V}}(\mathfrak{m},R)\otimes_R S\overset{\approx}{\to}\mathrm{Hom}_{\overline{V}}(\mathfrak{m},S)$ becomes $\mathrm{tr}_{S\otimes_R S/S}$, where we regard $S\otimes_R S$ as an $S$-algebra by $y\mapsto 1\otimes y$. Since $f$ is almost faithfully flat, it suffices to prove $\mathrm{tr}_{S\otimes_R S/S}(S\otimes_R S)\supset\mathfrak{m}\cdot\mathrm{Hom}_{\overline{V}}(\mathfrak{m},S)$. Let $e$ be the idempotent of $\mathrm{Hom}_{\overline{V}}(\mathfrak{m},S\otimes_R S)$ corresponding to the open and closed immersion $\mathrm{Spec}(\mathrm{Hom}_{\overline{V}}(\mathfrak{m},S))\hookrightarrow\mathrm{Spec}(\mathrm{Hom}_{\overline{V}}(\mathfrak{m},S\otimes_R S))$ (cf. the remark after Lemma V.7.2). Then, for any $\alpha\in\Lambda^+$, there exists $x\in S\otimes_R S$ whose image in $\mathrm{Hom}_{\overline{V}}(\mathfrak{m},S\otimes_R S)$ is $\pi^\alpha\cdot e$, and by Propositions V.4.2 (1), V.4.3, and V.4.4, we have $\mathrm{tr}_{S\otimes_R S/S}(x) = \mathrm{tr}_{\mathrm{Hom}_{\overline{V}}(\mathfrak{m},S\otimes_R S)/S}(\pi^\alpha e) = \mathrm{tr}_{\mathrm{Hom}_{\overline{V}}(\mathfrak{m},S)/S}(\pi^\alpha) = \mathrm{tr}_{S/S}(\pi^\alpha) = \pi^\alpha$. Varying $\alpha$, we obtain the claim. □

## V.8. Almost faithfully flat descent I

We discuss almost faithfully flat descent of some properties of modules and algebras.

**Proposition V.8.1.** *Let $R\to R'$ be an almost faithfully flat homomorphism of $\overline{V}$-algebras, and let $M$ be an $R$-module. If $M' := M\otimes_R R'$ is almost finitely generated over $R'$, then $M$ is almost finitely generated over $R$.*

PROOF (cf. [37]VIII Corollaire 1.11). We have $M = \varinjlim_{i \in I} M_i$, where $M_i$ are finitely generated $R$-submodules of $M$ and $I$ is a filtered ordered set. Then $M' = \varinjlim_{i \in I} M_i'$, where $M_i' = M_i \otimes_R R'$. By assumption, for any $\alpha \in \Lambda^+$, there exists a finitely generated $R'$-submodule $N'$ of $M'$ such that $\pi^\alpha \cdot M'/N' = 0$. Choose $i \in I$ such that the image of $M_i'$ in $M'$ contains $N$. Then $M'/($ the image of $M_i') \cong (M/M_i) \otimes_R S$ is annihilated by $\pi^\alpha$. By Lemma V.8.2 below, $\pi^{2\alpha} \cdot M/M_i = 0$. □

**Lemma V.8.2.** *Let $R \to R'$ and $M$ be as in Proposition V.8.1. Then the kernel of $M \to M \otimes_R R'$ is almost zero.*

PROOF. Let $N$ denote the kernel. Then the homomorphism $N \otimes_R R' \to M \otimes_R R'$ is zero and its kernel is almost zero by the condition (iii) in Definition V.6.1. Hence $N \otimes_R R' \approx 0$, which implies $N \approx 0$. □

**Definition V.8.3.** Let $R$ be a $\overline{V}$-algebra and let $M$ be an $R$-module. We say that $M$ is *almost finitely presented* if, for any $\alpha \in \Lambda^+$, there exist a finitely presented $R$-module $N$ and homomorphisms of $R$-modules $\varphi_\alpha \colon M \to N$, $\psi_\alpha \colon N \to M$ such that $\psi_\alpha \circ \varphi_\alpha = \pi^\alpha \cdot \mathrm{id}_M$ and $\varphi_\alpha \circ \psi_\alpha = \pi^\alpha \cdot \mathrm{id}_N$.

By Lemma V.2.2, we have the following criterion.

**Lemma V.8.4.** *Let $R$ be a $\overline{V}$-algebra and let $M$ be an $R$-module. Then $M$ is almost finitely presented if and only if, for any $\alpha \in \Lambda^+$, there exist a free $R$-module $L$ of finite rank, a homomorphism of $R$-modules $f \colon L \to M$, and a finitely generated $R$-submodule $N$ of* $\mathrm{Ker}(f)$ *such that* $\mathrm{Cok}(f)$ *and* $\mathrm{Ker}(f)/N$ *are annihilated by $\pi^\alpha$.*

**Proposition V.8.5.** *Let $R \to R'$ be an almost faithfully flat homomorphism of $\overline{V}$-algebras and let $M$ be an $R$-module. If $M \otimes_R R'$ is almost finitely presented over $R'$, then $M$ is almost finitely presented over $R$.*

**Lemma V.8.6.** *Let $R$ be a $\overline{V}$-algebra, let $M$ be an $R$-module, let $L_1$ and $L_2$ be two free $R$-modules of finite rank, and let $f_i \colon L_i \to M$ $(i = 1, 2)$ and $g \colon L_1 \to L_2$ be homomorphisms of $R$-modules such that $f_2 \circ g = f_1$. Let $\alpha \in \Lambda^+$ and assume $\pi^\alpha \cdot \mathrm{Cok}(g) = 0$.*

(1) *If there exists a finitely generated $R$-submodule $N_1$ of* $\mathrm{Ker}(f_1)$ *and $\beta \in \Lambda^+$ such that $\pi^\beta \cdot \mathrm{Ker}(f_1)/N_1 = 0$, then there exists a finitely generated $R$-submodule $N_2$ of* $\mathrm{Ker}(f_2)$ *such that $\pi^{\alpha+\beta} \cdot \mathrm{Ker}(f_2)/N_2 = 0$.*

(2) *If there exists a finitely generated $R$-submodule $N_2$ of* $\mathrm{Ker}(f_2)$ *and $\beta \in \Lambda^+$ such that $\pi^\beta \cdot \mathrm{Ker}(f_2)/N_2 = 0$, then there exists a finitely generated $R$-submodule $N_1$ of* $\mathrm{Ker}(f_1)$ *such that $\pi^{2\alpha+\beta} \cdot \mathrm{Ker}(f_1)/N_1 = 0$.*

PROOF. By assumption, there exists $s \colon L_2 \to L_1$ such that $g \circ s = \pi^\alpha \cdot \mathrm{id}_{L_2}$. We define the homomorphism $h \colon L_1 \to \mathrm{Ker}(g)$ by $h(x) = \pi^\alpha \cdot x - s \circ g(x)$. Since the restriction of $h$ to $\mathrm{Ker}(g)$ is the multiplication by $\pi^\alpha$, the cokernel of $h$ is annihilated by $\pi^\alpha$. On the other hand, we have an exact sequence $0 \to \mathrm{Ker}(g) \to \mathrm{Ker}(f_1) \xrightarrow{g} \mathrm{Ker}(f_2)$ and the cokernel of the last homomorphism is killed by $\pi^\alpha$. The lemma follows easily from these facts. □

PROOF OF PROPOSITION V.8.5. Set $M' := M \otimes_R R'$. By Proposition V.8.1, $M$ is almost finitely generated over $R$. For $\alpha \in \Lambda^+$, choose a free $R$-module $L$ of finite rank and a homomorphism of $R$-modules $f \colon L \to M$ such that $\pi^\alpha \cdot \mathrm{Cok}(f) = 0$. Let $L'$ and $f'$ denote the base change of $L$ and $f$ by $R \to R'$. We have $\pi^\alpha \cdot \mathrm{Cok}(f') = 0$ and $\mathrm{Ker}(f) \otimes_R R' \xrightarrow{\approx} \mathrm{Ker}(f')$ (Lemma V.6.9). Choose a free $R'$-module $L_1'$ of finite rank and a homomorphism of $R'$-modules $f_1' \colon L_1' \to M'$ such that $\pi^\alpha \cdot \mathrm{Cok}(f_1') = 0$ and there exists a finitely generated $R'$-submodule $N_1'$ of $\mathrm{Ker}(f_1')$ such that $\pi^\alpha \cdot \mathrm{Ker}(f_1')/N_1' = 0$

(Lemma V.8.4). Then there are homomorphisms $g\colon L' \to L'_1$ and $h\colon L'_1 \to L'$ such that $f'_1 \circ g = \pi^\alpha \cdot f'$ and $f' \circ h = \pi^\alpha \cdot f'_1$. Choose such $g$ and $h$. Then the composite of $g + \pi^\alpha \cdot \mathrm{id}_{L'_1}\colon L' \oplus L'_1 \to L'_1$ (resp. $\pi^\alpha \cdot \mathrm{id}_{L'} + h\colon L' \oplus L'_1 \to L'$) with $f'_1$ (resp. $f'$) becomes $\pi^\alpha \cdot (f' + f'_1)$. Using Lemma V.8.6 twice, we see that there exists a finitely generated $R'$-submodule $N'$ of $\mathrm{Ker}(f')$ such that $\pi^{4\alpha} \cdot \mathrm{Ker}(f')/N' = 0$. Hence, there exists a finitely generated $R'$-submodule $N''$ of $(\mathrm{Ker}(f)) \otimes_R R'$ such that $\pi^{5\alpha} \cdot (\mathrm{Ker}(f) \otimes_R R')/N'' = 0$. By the same argument as the proof of Proposition V.8.1, there exists a finitely generated $R$-submodule $N$ of $\mathrm{Ker}(f)$ such that $\pi^{6\alpha} \cdot \mathrm{Ker}(f)/N = 0$. By Lemma V.8.4, $M$ is almost finitely presented. □

**Proposition V.8.7.** *Let $R \to R'$ be an almost faithfully flat homomorphism of $\overline{V}$-algebras and let $M$ be an $R$-module. If $M \otimes_R R'$ is an almost finitely generated projective $R$-module, then $M$ is an almost finitely generated projective $R$-module.*

To prove this proposition, we need the following lemmas.

**Lemma V.8.8.** *Let $R$ be a $\overline{V}$-algebra and let $M$ be an $R$-module. If $M$ is almost finitely generated projective over $R$, then $M$ is almost finitely presented over $R$.*

PROOF. For $\alpha \in \Lambda^+$, choose $L$, $f$, and $g$ as in Proposition V.3.7. Then we have an $R$-linear homomorphism $L \to \mathrm{Ker}(g)$ defined by $x \mapsto \pi^\alpha \cdot x - f \circ g(x)$, and its cokernel is annihilated by $\pi^\alpha$ because its restriction to $\mathrm{Ker}(g)$ is the multiplication by $\pi^\alpha$. Hence, by Lemma V.8.4, $M$ is almost finitely presented. □

**Lemma V.8.9.** *Let $R \to R'$ be an almost flat homomorphism of $\overline{V}$-algebras and let $M$ and $N$ be two $R$-modules. If $M$ is finitely presented, then the canonical homomorphism $\mathrm{Hom}_R(M, N) \otimes_R R' \to \mathrm{Hom}_{R'}(M \otimes_R R', N \otimes_R R')$ is an almost isomorphism.*

PROOF. Choose a presentation $L_2 \to L_1 \to M \to 0$ with $L_1$ and $L_2$ free $R$-modules of finite rank. Then we have a commutative diagram

$$\begin{array}{ccc} \mathrm{Hom}_R(M,N)\otimes_R R' & \longrightarrow & \mathrm{Ker}\{\mathrm{Hom}_R(L_1,N)\otimes_R R' \to \mathrm{Hom}_R(L_2,N)\otimes_R R'\} \\ \downarrow & & \downarrow\wr \\ \mathrm{Hom}_{R'}(M',N') & \xrightarrow{\sim} & \mathrm{Ker}\{\mathrm{Hom}_{R'}(L'_1,N') \to \mathrm{Hom}_{R'}(L'_2,N')\}, \end{array}$$

where $M'$ denotes $M \otimes_R R'$ and similarly for $N'$, $L'_1$, and $L'_2$. By Lemma V.6.9, the upper horizontal homomorphism is an almost isomorphism and it implies the lemma. □

PROOF OF PROPOSITION V.8.7. By Lemma V.8.8 and Proposition V.8.5, $M$ is almost finitely presented. For $\alpha \in \Lambda^+$, choose a finitely presented $R$-module $M$ and homomorphisms $f\colon M_1 \to M$ and $g\colon M \to M_1$ such that $f \circ g = \pi^\alpha \cdot \mathrm{id}_M$ and $g \circ f = \pi^\alpha \cdot \mathrm{id}_{M_1}$. Set $M' := M \otimes_R R'$ and $M'_1 := M_1 \otimes_R R'$. By assumption, for any $R'$-module $N'$, we have $\mathrm{Ext}^i_{R'}(M', N') \approx 0$ $(i > 0)$ and hence $\pi^{2\alpha} \cdot \mathrm{Ext}^i_{R'}(M'_1, N') = 0$ $(i > 0)$. Choose a free $R$-module $L$ of finite rank and a surjective homomorphism $h\colon L \to M_1$. Let $C$ denote the cokernel of the homomorphism $\mathrm{Hom}_R(M_1, L) \to \mathrm{Hom}_R(M_1, M_1)$ induced by $h$. Then, we have a commutative diagram whose two lines are exact:

$$\begin{array}{ccccccc} \mathrm{Hom}_R(M_1,L)\otimes_R R' & \longrightarrow & \mathrm{Hom}_R(M_1,M_1)\otimes_R R' & \longrightarrow & C\otimes_R R' & \longrightarrow & 0 \\ \downarrow & & \downarrow & & \downarrow & & \\ \mathrm{Hom}_{R'}(M'_1,L') & \longrightarrow & \mathrm{Hom}_{R'}(M'_1,M'_1) & \longrightarrow & \mathrm{Ext}^1_{R'}(M'_1,\mathrm{Ker}(h')), & & \end{array}$$

where $L' = L \otimes_R R'$ and $h' = h \otimes \mathrm{id}_{R'}$. By Lemma V.8.9, the left and middle vertical homomorphisms are almost isomorphisms. Hence, the right one is almost injective. By

Lemma V.8.2, $\pi^{3\alpha} \cdot C = 0$ and there exists $s\colon M_1 \to L$ such that $h \circ s = \pi^{3\alpha} \cdot \mathrm{id}_{M_1}$. This implies that, for any $R$-module $N$, $\pi^{3\alpha} \cdot \mathrm{Ext}^i_R(M_1, N) = 0$ $(i > 0)$ and hence $\pi^{4\alpha} \cdot \mathrm{Ext}^i_R(M, N) = 0$ $(i > 0)$. Varying $\alpha$, we obtain the proposition. □

**Corollary V.8.10.** *Let $R \to R'$ be an almost faithfully flat homomorphism of $\overline{V}$-algebras, and let $S$ be an $R$-algebra. If $S \otimes_R R'$ is an almost étale covering of $R'$, then $S$ is an almost étale covering of $R$.*

**Proposition V.8.11.** *Let $R \to R'$ be an almost faithfully flat homomorphism of $\overline{V}$-algebras and let $M$ be an $R$-module. If $M \otimes_R R'$ is almost flat (resp. almost faithfully flat) over $R'$, then $M$ is almost flat (resp. almost faithfully flat) over $R$.*

PROOF. Straightforward. □

## V.9. Almost faithfully flat descent II

We will discuss almost faithfully flat descent of modules and algebras.

**Proposition V.9.1.** *Let $f\colon R \to S$ be an almost faithfully flat homomorphism of $\overline{V}$-algebras and let $M$ be an $R$-module. Then the sequence*

$$0 \longrightarrow M \xrightarrow{\varepsilon} M \otimes_R S \xrightarrow{1\otimes d^0} M \otimes_R (S \otimes_R S) \xrightarrow{1\otimes d^1}$$
$$\cdots \longrightarrow M \otimes_R (S^{\otimes q}) \xrightarrow{1\otimes d^{q-1}} M \otimes_R (S^{\otimes (q+1)}) \longrightarrow \cdots$$

*is almost exact, that is, the cohomology groups are almost zero, where*

$$d^{q-1}(y_0 \otimes \cdots \otimes y_{q-1}) = \sum_{i=0}^{q} (-1)^i y_0 \otimes \cdots \otimes y_{i-1} \otimes 1 \otimes y_i \otimes \cdots \otimes y_{q-1}.$$

PROOF. By taking the base change by $f$ and using Lemma V.6.9, we are reduced to the case that there exists a homomorphism $s\colon S \to R$ such that $s \circ f = \mathrm{id}_R$. Then, the $R$-linear homomorphism $k^q\colon S^{\otimes(q+1)} \to S^{\otimes q}$ defined by $k^q(y_0 \otimes \cdots \otimes y_q) = s(y_0) \cdot y_1 \otimes \cdots \otimes y_q$ satisfy $k^{q+1} \circ d^q + d^{q-1} \circ k^q = 1$ $(q \geq 0)$ and $k^0 \circ f = 1$. Hence $\{1 \otimes k^q\}_{q\geq 0}$ is a contracting homotopy. □

In the following until the end of V.9, we write $\widetilde{M}$ for $\mathrm{Hom}_{\overline{V}}(\mathfrak{m}, M)$ to simplify the notation. Let us recall its fundamental properties briefly. We have a canonical almost isomorphism of $\overline{V}$-modules $M \xrightarrow{\approx} \widetilde{M}$ functorial on $M$. If $M \to N$ is an almost isomorphism, the induced homomorphism $\widetilde{M} \to \widetilde{N}$ is an isomorphism (Proposition V.2.5 (1)). We can construct a canonical homomorphism $\widetilde{M} \to \widetilde{\widetilde{M}}$ in two ways: one by taking $\widetilde{\ }$ of $M \to \widetilde{M}$ and the other by putting $N = \widetilde{M}$ in $N \to \widetilde{N}$. The coincidence is trivial and, by the first construction, this becomes an isomorphism. If $A$ is a $\overline{V}$-algebra, $\widetilde{A}$ has a canonical ring structure and $A \to \widetilde{A}$ is a ring homomorphism. For a $\overline{V}$-algebra homomorphism $A \to B$, $\widetilde{A} \to \widetilde{B}$ is a ring homomorphism. For a $\overline{V}$-algebra $R$ and an $R$-module $M$, $\widetilde{M}$ has a natural $R$-module structure and $M \to \widetilde{M}$ is $R$-linear. For an $R$-linear homomorphism $M \to N$, $\widetilde{M} \to \widetilde{N}$ is also $R$-linear.

For a $\overline{V}$-algebra $R$, let $R$-$\underline{\mathrm{Mod}}$ (resp. $R$-$\underline{\mathrm{Alg}}$) denote the category of $R$-modules (resp. $R$-algebras) and let $R$-$\widetilde{\underline{\mathrm{Mod}}}$ (resp. $R$-$\widetilde{\underline{\mathrm{Alg}}}$) denote the full subcategory of $R$-$\underline{\mathrm{Mod}}$ (resp. $R$-$\underline{\mathrm{Alg}}$) consisting of $R$-modules $M$ (resp. $R$-algebras $A$) such that $M \to \widetilde{M}$ (resp. $A \to \widetilde{A}$) are isomorphisms.

**Proposition V.9.2.** *With the above notation, the functor* $R\text{-}\underline{\mathrm{Mod}} \to R\text{-}\widetilde{\underline{\mathrm{Mod}}}$ *(resp.* $R\text{-}\underline{\mathrm{Alg}} \to R\text{-}\widetilde{\underline{\mathrm{Alg}}}$*) associating* $\widetilde{M}$ *(resp.* $\widetilde{A}$*) to an* $R$*-module* $M$ *(resp. an* $R$*-algebra* $A$*) is a left adjoint of the forgetful functor.*

PROOF. For $M \in R\text{-}\underline{\mathrm{Mod}}$ and $N \in R\text{-}\widetilde{\underline{\mathrm{Mod}}}$, the homomorphism $\mathrm{Hom}_R(\widetilde{M}, N) \to \mathrm{Hom}_R(M, N)$ induced by the canonical homomorphism $M \to \widetilde{M}$ is bijective. Indeed, it is injective since $N(\cong \widetilde{N})$ is $\mathfrak{m}$-torsion free, and any $R$-linear homomorphism $\varphi\colon M \to N$ is the image of the composite $\widetilde{M} \overset{\widetilde{\varphi}}{\to} \widetilde{N} \overset{\cong}{\leftarrow} N$. We can prove the second case similarly. □

For a homomorphism of $\overline{V}$-algebras $f\colon R \to S$, the functor $S\text{-}\underline{\mathrm{Mod}} \to R\text{-}\underline{\mathrm{Mod}}$ (resp. $S\text{-}\underline{\mathrm{Alg}} \to R\text{-}\underline{\mathrm{Alg}}$) restricting the scalars by $f$ induces a functor $S\text{-}\widetilde{\underline{\mathrm{Mod}}} \to R\text{-}\widetilde{\underline{\mathrm{Mod}}}$ (resp. $S\text{-}\widetilde{\underline{\mathrm{Alg}}} \to R\text{-}\widetilde{\underline{\mathrm{Alg}}}$). Let $\widetilde{f}_*$ denote the functor $R\text{-}\widetilde{\underline{\mathrm{Mod}}} \to S\text{-}\widetilde{\underline{\mathrm{Mod}}}$ (resp. $R\text{-}\widetilde{\underline{\mathrm{Alg}}} \to S\text{-}\widetilde{\underline{\mathrm{Alg}}}$) associating $\widetilde{M \otimes_R S}$ to $M$ (resp. $\widetilde{A \otimes_R S}$ to $A$).

**Proposition V.9.3.** *With the above notation, the functor* $\widetilde{f}_*$ *is a left adjoint of the functor restricting the scalars by* $f$*.*

PROOF. For any $M \in R\text{-}\widetilde{\underline{\mathrm{Mod}}}$ and $N \in S\text{-}\widetilde{\underline{\mathrm{Mod}}}$, we have

$$\mathrm{Hom}_S(\widetilde{M \otimes_R S}, N) \xrightarrow{\sim} \mathrm{Hom}_S(M \otimes_R S, N) \xrightarrow{\sim} \mathrm{Hom}_R(M, N).$$

Similarly for the second case. □

**Corollary V.9.4.** *For any two composable homomorphisms of* $\overline{V}$*-algebras* $f\colon R \to S$*,* $g\colon S \to T$*, there exists a canonical isomorphism* $\widetilde{g}_* \circ \widetilde{f}_* \cong \widetilde{g \circ f}_*$ *satisfying the obvious cocycle condition.*

If we denote by $f_*$ the functor $R\text{-}\underline{\mathrm{Mod}} \to S\text{-}\underline{\mathrm{Mod}}$; $M \mapsto M \otimes_R S$, we have $\widetilde{f}_* M = \widetilde{f_* M}$, the adjunction $M \to \widetilde{f}_* M$ is given by $M \overset{\mathrm{adj}}{\to} f_* M \to \widetilde{f_* M}$ or $M \to \widetilde{M} \overset{\widetilde{\mathrm{adj}}}{\to} \widetilde{f_* M}$, and the following diagram is commutative

$$\begin{array}{ccc} \widetilde{g \circ f_* M} & = & \widetilde{(g \circ f)_* M} \\ \downarrow\wr & & \downarrow\wr \\ \widetilde{g}_* \widetilde{f}_* M = \widetilde{g_* \widetilde{f_* M}} & \xleftarrow{\sim} & \widetilde{g_* f_* M}. \end{array}$$

Now we will prove almost faithfully flat descent for $R\text{-}\widetilde{\underline{\mathrm{Mod}}}$, $R\text{-}\widetilde{\underline{\mathrm{Alg}}}$, and $\widetilde{f}_*$.

**Proposition V.9.5.** *Let* $f\colon R \to S$ *be an almost faithfully flat homomorphism of* $\overline{V}$*-algebras, let* $f_0$ *and* $f_1$ *be the homomorphisms* $S \to S \otimes_R S$ *sending* $a$ *to* $a \otimes 1$ *and* $1 \otimes a$*, respectively, and let* $g$ *be* $f_0 \circ f = f_1 \circ f$*. Then, for any* $M \in R\text{-}\widetilde{\underline{\mathrm{Mod}}}$*, the sequence*

$$M \longrightarrow \widetilde{f}_* M \underset{i_1}{\overset{i_0}{\rightrightarrows}} \widetilde{g}_* M$$

*is exact, where* $i_\nu$ $(\nu = 0, 1)$ *denotes the adjunction map with respect to* $f_\nu$*:*

$$\widetilde{f}_* M \longrightarrow \widetilde{f}_{\nu *} \widetilde{f}_* M \cong \widetilde{g}_* M.$$

PROOF. By Proposition V.9.1, we have

$$M \xrightarrow{\approx} \mathrm{Ker}(f_* M \xrightarrow{i_0 - i_1} g_* M),$$

where $i_\nu$ is defined similarly. Since the functor $\widetilde{\ }$ is left exact, by taking $\widetilde{\ }$ of this almost isomorphism and using Proposition V.2.5 (1), we obtain the proposition. □

**Corollary V.9.6.** *Let the notation and the assumption be the same as Proposition* V.9.5. *For any* $M, N \in R\text{-}\widetilde{\underline{\mathrm{Mod}}}$ *(resp.* $A, B \in R\text{-}\widetilde{\underline{\mathrm{Alg}}}$*), the following sequence is exact:*

$$\mathrm{Hom}_R(M,N) \longrightarrow \mathrm{Hom}_S(\widetilde{f}_*M, \widetilde{f}_*N) \overset{\widetilde{f}_{0*}}{\underset{\widetilde{f}_{1*}}{\rightrightarrows}} \mathrm{Hom}_{S\otimes_R S}(\widetilde{g}_*M, \widetilde{g}_*N)$$

$$(\textit{resp. } \mathrm{Hom}_{R\text{-alg}}(A,B) \longrightarrow \mathrm{Hom}_{S\text{-alg}}(\widetilde{f}_*A, \widetilde{f}_*B) \overset{\widetilde{f}_{0*}}{\underset{\widetilde{f}_{1*}}{\rightrightarrows}} \mathrm{Hom}_{S\otimes_R S\text{-alg}}(\widetilde{g}_*A, \widetilde{g}_*B))$$

PROOF. The second case is reduced to the first one, which follows from Proposition V.9.5 since $\mathrm{Hom}_S(\widetilde{f}_*M, \widetilde{f}_*N) \cong \mathrm{Hom}_R(M, \widetilde{f}_*N)$ and $\mathrm{Hom}_{S\otimes_R S}(\widetilde{g}_*M, \widetilde{g}_*N) \cong \mathrm{Hom}_R(M, \widetilde{g}_*N)$ by Proposition V.9.2. □

**Proposition V.9.7.** *Let* $f$, $f_0$, $f_1$, *and* $g$ *be as in Proposition* V.9.5 *and let* $f_{01}$, $f_{12}$, *and* $f_{02}$ *denote the homomorphism* $S \otimes_R S \to S \otimes_R S \otimes_R S$ *sending* $a_0 \otimes a_1$ *to* $a_0 \otimes a_1 \otimes 1$, $1 \otimes a_0 \otimes a_1$, *and* $a_0 \otimes 1 \otimes a_1$, *respectively. Let* $M \in S\text{-}\widetilde{\underline{\mathrm{Mod}}}$ *and suppose that we are given an* $S \otimes_R S$*-linear isomorphism* $\phi\colon \widetilde{f}_{0*}M \xrightarrow{\sim} \widetilde{f}_{1*}M$ *such that* $\widetilde{f}_{12*}(\phi) \circ \widetilde{f}_{01*}(\phi) = \widetilde{f}_{02*}(\phi)$. *Then, up to unique isomorphisms, there exists a unique* $N \in R\text{-}\widetilde{\underline{\mathrm{Mod}}}$ *endowed with an isomorphism* $\iota\colon \widetilde{f}_*N \xrightarrow{\sim} M$ *which makes the following diagram commute:*

$$\begin{array}{ccccc}
\widetilde{f}_{0*}\widetilde{f}_*N & \xleftarrow{\sim} & \widetilde{g}_*N & \xrightarrow{\sim} & \widetilde{f}_{1*}\widetilde{f}_*N \\
{\scriptstyle \widetilde{f}_{0*}(\iota)}\downarrow & & & & \downarrow{\scriptstyle \widetilde{f}_{1*}(\iota)} \\
\widetilde{f}_{0*}M & & \xrightarrow[\phi]{\sim} & & \widetilde{f}_{1*}M
\end{array}$$

**Lemma V.9.8.** *Let* $f\colon R \to S$ *be an almost flat homomorphism of* $\overline{V}$*-algebras. Then, for any exact sequence* $0 \to M_1 \to M_2 \to M_3$ *with* $M_i \in R\text{-}\widetilde{\underline{\mathrm{Mod}}}$ $(i = 1, 2, 3)$, *the sequence* $0 \to \widetilde{f}_*M_1 \to \widetilde{f}_*M_2 \to \widetilde{f}_*M_3$ *is exact.*

PROOF. By Lemma V.6.9, we have

$$f_*M_1 \xrightarrow{\approx} \mathrm{Ker}(f_*M_2 \to f_*M_3).$$

By Proposition V.2.5 (1) and the left exactness of $\widetilde{\ }$, we obtain the lemma. □

PROOF OF PROPOSITION V.9.7. Let $\alpha_0$ (resp. $\alpha_1$) be the homomorphism $M \xrightarrow{\mathrm{adj}} \widetilde{f}_{0*}M \underset{\phi}{\xrightarrow{\sim}} \widetilde{f}_{1*}M$ (resp. $M \xrightarrow{\mathrm{adj}} \widetilde{f}_{1*}M$), which is compatible with $f_0$ (resp. $f_1$), and define the $R$-module $N$ to be the kernel of $\alpha_0 - \alpha_1$. Since the functor $\widetilde{\ }$ is left exact, we have $N \xrightarrow{\sim} \widetilde{N}$, that is, $N \in R\text{-}\widetilde{\underline{\mathrm{Mod}}}$. We can verify easily that the $S$-linear homomorphism $\iota\colon \widetilde{f}_*N \to M$ induced by the inclusion makes the diagram in the proposition commute. (Show the commutativity after composing with $N \xrightarrow{\mathrm{adj}} \widetilde{g}_*N$.) We will prove that $\iota$ is an isomorphism. Let $g_0$, $g_1$, and $g_2$ be the homomorphisms $S \to S \otimes_R S \otimes_R S$ sending $a \in S$ to $a \otimes 1 \otimes 1$, $1 \otimes a \otimes 1$, and $1 \otimes 1 \otimes a$, respectively. First, note that we have a commutative diagram of $\overline{V}$-algebras whose squares are co-Cartesian:

$$(*) \qquad \begin{array}{ccccc}
R & \xrightarrow{f} & S & \overset{f_0}{\underset{f_1}{\rightrightarrows}} & S \otimes_R S \\
{\scriptstyle f}\downarrow & & \downarrow{\scriptstyle f_0} & & \downarrow{\scriptstyle f_{01}} \\
S & \xrightarrow{f_1} & S \otimes_R S & \overset{f_{02}}{\underset{f_{12}}{\rightrightarrows}} & S \otimes_R S \otimes_R S.
\end{array}$$

Hence, by taking the base change by $f$ of the exact sequence

$$N \longrightarrow M \underset{\alpha_1}{\overset{\alpha_0}{\rightrightarrows}} \widetilde{f}_{1*}M$$

compatible with the first line, we obtain an exact sequence

$$\widetilde{f}_*N \longrightarrow \widetilde{f}_{0*}M \underset{\beta_1}{\overset{\beta_0}{\rightrightarrows}} \widetilde{f}_{01*}\widetilde{f}_{1*}M \cong \widetilde{g}_{1*}M$$

compatible with the second line by Lemma V.9.8. By the following commutative diagram, $\beta_1$ is $\widetilde{f}_{0*}M \xrightarrow{\text{adj}} \widetilde{f}_{12*}\widetilde{f}_{0*}M \cong \widetilde{g}_{1*}M$.

$$\begin{array}{ccc}
M & \xrightarrow{\quad\text{adj}\quad} & \widetilde{f}_{1*}M \\
{\scriptstyle\text{adj}}\downarrow & & \downarrow{\scriptstyle\text{adj}} \\
\widetilde{f}_{0*}M & \xrightarrow{\text{adj}} & \widetilde{f}_{12*}\widetilde{f}_{0*}M \cong \widetilde{g}_{1*}M \cong \widetilde{f}_{01*}\widetilde{f}_{1*}M.
\end{array}$$

Similarly, we see that $\beta_0$ is the composite $\widetilde{f}_{0*}M \xrightarrow{\text{adj}} \widetilde{f}_{02*}\widetilde{f}_{0*}M \cong \widetilde{g}_{0*}M \underset{\widetilde{f}_{01*}\phi}{\xrightarrow{\sim}} \widetilde{g}_{1*}M$. On the other hand, since $f_1$ is almost faithfully flat by Proposition V.6.5 (2), by applying Proposition V.9.5 to $f_1$, we obtain another exact sequence

$$M \longrightarrow \widetilde{f}_{1*}M \underset{\gamma_1}{\overset{\gamma_0}{\rightrightarrows}} \widetilde{g}_{2*}M$$

compatible with the second line of $(*)$, where $\gamma_i$ $(i = 1, 2)$ is the adjunction $\widetilde{f}_{1*}M \to \widetilde{f}_{i2*}\widetilde{f}_{1*}M \cong \widetilde{g}_{2*}M$. Now we can check that the following diagram is commutative and hence $\iota$ is an isomorphism.

$$\begin{array}{ccccc}
\widetilde{f}_*N & \longrightarrow & \widetilde{f}_{0*}M & \underset{\beta_1}{\overset{\beta_0}{\rightrightarrows}} & \widetilde{g}_{1*}M \cong \widetilde{f}_{12*}\widetilde{f}_{0*}M \\
{\scriptstyle\iota}\downarrow & & \wr\downarrow{\scriptstyle\phi} & & \wr\downarrow{\scriptstyle\widetilde{f}_{12*}\phi} \\
M & \longrightarrow & \widetilde{f}_{1*}M & \underset{\gamma_1}{\overset{\gamma_0}{\rightrightarrows}} & \widetilde{g}_{2*}M \cong \widetilde{f}_{12*}\widetilde{f}_{1*}M
\end{array}$$

(For the commutativity of the left square, we consider the composite with $\text{adj}\colon N \to \widetilde{f}_*N$. For $\widetilde{f}_{12*}\phi \circ \beta_0 = \gamma_0 \circ \phi$, we use the cocycle condition on $\phi$.)

It remains to prove the uniqueness. Suppose that we have $(N_1, \iota_1)$, $(N_2, \iota_2)$ satisfying the condition. Then, the following diagram is commutative for $\nu = 1, 2$, where $\alpha_i$ is as above and $\delta_i$ is $\widetilde{f}_*N_\nu \overset{\text{adj}}{\to} \widetilde{f}_{i*}\widetilde{f}_*N_\nu \cong \widetilde{g}_*N_\nu$.

$$\begin{array}{ccc}
\widetilde{f}_*N_\nu & \underset{\delta_1}{\overset{\delta_0}{\rightrightarrows}} & \widetilde{g}_*N_\nu \\
\wr\downarrow{\scriptstyle\iota_\nu} & & \wr\downarrow{\scriptstyle\widetilde{f}_{1*}\iota_\nu} \\
M & \underset{\alpha_1}{\overset{\alpha_0}{\rightrightarrows}} & \widetilde{f}_{1*}M.
\end{array}$$

By Proposition V.9.5, $N_\nu \xrightarrow{\sim} \operatorname{Ker}(\delta_0 - \delta_1)$. Hence there exists a unique $R$-linear isomorphism $N_1 \xrightarrow{\sim} N_2$ compatible with $\iota_1$ and $\iota_2$. □

Let $R$-$\widetilde{\underline{\text{Et}}}$ be the full subcategory of $R$-$\widetilde{\underline{\text{Alg}}}$ consisting of $A \in R$-$\widetilde{\underline{\text{Alg}}}$ which is an almost étale covering over $R$ (Definition V.7.1). For a homomorphism of $\overline{V}$-algebras $f\colon R \to S$, the canonical homomorphism $A \otimes_R S \to \widetilde{f}_*A$ $(A \in S$-$\widetilde{\underline{\text{Alg}}})$ is an almost isomorphism.

Hence, by Proposition V.7.4 (1), (3) and Corollary V.8.10, if $A \in R\text{-}\widetilde{\underline{\mathrm{Et}}}$, then $\widetilde{f}_*A \in S\text{-}\widetilde{\underline{\mathrm{Et}}}$ and the converse is also true if $f$ is almost faithfully flat. If we denote by $R\text{-}\underline{\mathrm{Et}}$, the full subcategory of $R\text{-}\underline{\mathrm{Alg}}$ consisting of almost étale coverings over $R$, we have a functor $\widetilde{\ } : R\text{-}\underline{\mathrm{Et}} \to R\text{-}\widetilde{\underline{\mathrm{Et}}}; A \to \widetilde{A}$, which is a left adjoint of the forgetful functor.

**Corollary V.9.9.** *Let $f$, $f_0$, $f_1$, $g$, $f_{01}$, $f_{12}$, and $f_{02}$ be the same as Proposition* V.9.7. *Let $A \in S\text{-}\widetilde{\underline{\mathrm{Alg}}}$ (resp. $S\text{-}\widetilde{\underline{\mathrm{Et}}}$) and suppose that we are given an $S$-algebra isomorphism $\phi\colon \widetilde{f}_{0*}A \xrightarrow{\sim} \widetilde{f}_{1*}A$ such that $\widetilde{f}_{12*}(\phi)\circ\widetilde{f}_{01*}(\phi) = \widetilde{f}_{02*}(\phi)$. Then, up to unique isomorphisms, there exists a unique $B \in R\text{-}\widetilde{\underline{\mathrm{Alg}}}$ (resp. $R\text{-}\widetilde{\underline{\mathrm{Et}}}$) endowed with an $S$-algebra isomorphism $\iota\colon \widetilde{f}_*B \xrightarrow{\sim} A$, which makes the following diagram commute.*

$$\begin{array}{ccccc} \widetilde{f}_{0*}\widetilde{f}_*B & \xleftarrow{\sim} & \widetilde{g}_*B & \xrightarrow{\sim} & \widetilde{f}_{1*}\widetilde{f}_*B \\ \Big\downarrow{\scriptstyle \widetilde{f}_{0*}(\iota)} & & & & \Big\downarrow{\scriptstyle \widetilde{f}_{1*}(\iota)} \\ \widetilde{f}_{0*}A & & \xrightarrow[\phi]{\sim} & & \widetilde{f}_{1*}A \end{array}$$

PROOF. Immediate from the construction of $N$ and the proof of the uniqueness in the proof of Proposition V.9.7. □

## V.10. Liftings

We study liftings of almost étale coverings with respect to nilpotent immersions. To do it, we first review the definition and some basic properties of Hochschild cohomology. Let $A$ be a commutative ring and let $B$ be any commutative $A$-algebra. Recall that giving a $B \otimes_A B$-module $M$ is equivalent to giving a module $M$ endowed with left and right $B$-module structures which induce the same $A$-module structure and commute with each other (i.e., $(x \cdot m) \cdot y = x \cdot (m \cdot y)$ for $x, y \in B$, $m \in M$); the correspondence is given by the formula: $(x \otimes y) \cdot m = (x \cdot m) \cdot y = x \cdot (m \cdot y)$. We define the chain complex $C_\bullet(B/A)$ of $B \otimes_A B$-modules as follows: The degree $n$ part ($n \geq 0$) is the $(n+2)$-fold tensor product $B^{\otimes n}_{/A}$ of $B$ over $A$ endowed with the $B\otimes_A B$-module structure: $(x\otimes y)\cdot(b_0\otimes b_1\otimes\cdots\otimes b_{n+1}) = x \cdot b_0 \otimes b_1 \otimes \cdots \otimes y \cdot b_{n+1}$. The differential $d_n\colon C_n(B/A) \to C_{n-1}(B/A)$ $(n \geq 1)$ is defined by

$$d_n(b_0 \otimes b_1 \otimes \cdots \otimes b_{n+1}) = \sum_{i=0}^{n} (-1)^i b_0 \otimes \cdots \otimes b_i b_{i+1} \otimes \cdots \otimes b_{n+1}.$$

If we regard $B$ as a $B \otimes_A B$-module by $(x \otimes y) \cdot b = x \cdot b \cdot y$, we have a $B \otimes_A B$-linear augmentation $\varepsilon\colon C_0(B/A) \to B; b_0 \otimes b_1 \mapsto b_0 \cdot b_1$.

**Lemma V.10.1.** *The complex $C_\bullet(B/A) \xrightarrow{\varepsilon} B$ is homotopically equivalent to zero as a complex of left (resp. right) $B$-modules.*

PROOF. A homotopy is explicitly given by $B \to C_0(B/A); b \mapsto b \otimes 1$ (resp. $1 \otimes b$) and $C_n(B/A) \to C_{n+1}(B/A); b_0 \otimes b_1 \otimes \cdots \otimes b_{n+1} \mapsto (-1)^{n+1} b_0 \otimes b_1 \otimes \cdots \otimes b_{n+1} \otimes 1$ (resp. $1 \otimes b_0 \otimes b_1 \otimes \cdots \otimes b_{n+1}$). □

For a $B \otimes_A B$-module $M$, we define the Hochschild cohomology $H^*(B/A, M)$ to be the cohomology of the (cochain) complex

$$\mathrm{Hom}_{B\otimes_A B}(C_\bullet(B/A), M).$$

The degree $n$-part of this complex is naturally identified with the module

$$\mathrm{Hom}_{A\text{-multilin}}(B^n, M)$$

of $A$-multilinear maps $B^n \to M$ and the differential is given by

(V.10.2)
$$\begin{aligned}(d^n g)(b_1, \dots, b_{n+1}) &= b_1 \cdot g(b_2, \dots, b_{n+1}) \\ &+ \sum_{i=1}^{n}(-1)^i g(b_1, \dots, b_i b_{i+1}, \dots, b_{n+1}) \\ &+ (-1)^{n+1} g(b_1, \dots, b_n) \cdot b_{n+1}.\end{aligned}$$

Hence, we have

(V.10.3)
$$H^0(B/A, M) = \{m \in M |\ x \cdot m = m \cdot x \text{ for all } x \in B\},$$

(V.10.4)
$$H^1(B/A, M) = \{f \in \mathrm{Hom}_A(B, M)|\ f(xy) = x{\cdot}f(y){+}f(x){\cdot}y\}/\{x \mapsto x{\cdot}m{-}m{\cdot}x|\ m \in M\}.$$

**Proposition V.10.5.** *For any $B$-module $N$, if we regard $\mathrm{Hom}_A(B, N)$ as a $B \otimes_A B$-module by $((x \otimes y) \cdot f)(b) = x \cdot f(y \cdot b)$ ($f \in \mathrm{Hom}_A(B, N)$, $x, y, b \in B$), then we have*

$$H^i(B/A, \mathrm{Hom}_A(B, N)) = \begin{cases} 0 & (i > 0), \\ N & (i = 0). \end{cases}$$

PROOF. This follows from Lemma V.10.1 and Lemma V.10.6 below. □

**Lemma V.10.6.** *Let $M$ be a $B \otimes_A B$-module and let $N$ be a $B$-module. Then there exists a canonical isomorphism functorial on $M$ and $N$:*

$$\begin{gathered}\mathrm{Hom}_{B\otimes_A B}(M, \mathrm{Hom}_A(B, N)) \cong \mathrm{Hom}_{B\text{-}left}(M, N) \\ \varphi \mapsto \psi, \quad \psi(m) = (\varphi(m))(1) \\ \varphi, \quad (\varphi(m))(b) = \psi(m \cdot b) \leftarrow\!\shortmid \psi\end{gathered}$$

**Corollary V.10.7.** *If $B$ is a projective $B \otimes_A B$-module, then for any $B \otimes_A B$-module $M$, we have $H^i(B/A, M) = 0$ $(i > 0)$.*

PROOF. For any two $B \otimes_A B$-modules $M$ and $N$, we define the $B \otimes_A B$-module structure on $\mathrm{Hom}_{B\text{-right}}(M, N)$ by $((x \otimes y) \cdot f)(m) = x \cdot f(y \cdot m)$, which is functorial on $M$ and $N$. Then the natural isomorphisms

$$M \cong \mathrm{Hom}_{B\text{-right}}(B, M), \quad \mathrm{Hom}_{B\text{-right}}(B \otimes_A B, M) \cong \mathrm{Hom}_A(B, M)$$

are $B \otimes_A B$-linear. Here we regard the right-hand side of the second isomorphism as a $B \otimes_A B$-module in the same way as Proposition V.10.5 using the left $B$-module structure on $M$. By the assumption, $B$ is a direct summand of $B \otimes_A B$ as a $B \otimes_A B$-module. Hence the claim follows from Proposition V.10.5. □

**Corollary V.10.8.** *Let $R \to S$ be a homomorphism of $\overline{V}$-algebras such that $S$ is an almost projective $S \otimes_R S$-module. Then, for any $S \otimes_R S$-module $M$, the natural almost isomorphisms*

$$H^i(S/R, M) \to H^i(S/R, \mathrm{Hom}_{\overline{V}}(\mathfrak{m}, M)) \quad (i > 0)$$

*vanish. Especially, $H^i(S/R, M) \approx 0$ $(i > 0)$ and, if $M \to \mathrm{Hom}_{\overline{V}}(\mathfrak{m}, M)$ is an isomorphism, $H^i(S/R, M) = 0$ $(i > 0)$.*

Proof. By Lemma V.7.2, we have $\mathrm{Hom}_{\overline{V}}(\mathfrak{m}, S \otimes_R S) = \mathrm{Hom}_{\overline{V}}(\mathfrak{m}, S) \times R_1$. Let $S'$ be the fiber product of $\mathrm{Hom}_{\overline{V}}(\mathfrak{m}, S) \to \mathrm{Hom}_{\overline{V}}(\mathfrak{m}, S \otimes_R S) \leftarrow S \otimes_R S$ in the category of $S \otimes_R S$-modules. Then, we have homomorphisms $S' \to S \otimes_R S \to S$ of $S \otimes_R S$-modules whose composite are almost isomorphisms. By a similar argument as the proof of Corollary V.10.7, we see that the homomorphisms

$$H^i(S/R, M) \to H^i(S/R, \mathrm{Hom}_{S\text{-right}}(S', M))$$

vanish. Since $\mathrm{Hom}_{\overline{V}}(\mathfrak{m}, M) \xrightarrow{\sim} \mathrm{Hom}_{\overline{V}}(\mathfrak{m}, \mathrm{Hom}_{S\text{-right}}(S', M))$ by Propositions V.2.3 and V.2.5 (1), this implies the claim. □

**Corollary V.10.9.** *Let $R \to S$ be as in Corollary V.10.8. Then, for any $S$-module $M$, the natural almost isomorphism*

$$\mathrm{Der}_R(S, M) \to \mathrm{Der}_R(S, \mathrm{Hom}_{\overline{V}}(\mathfrak{m}, M))$$

*vanishes. Especially, $\mathrm{Der}_R(S, M) \approx 0$ and, if $M \to \mathrm{Hom}_{\overline{V}}(\mathfrak{m}, M)$ is an isomorphism, $\mathrm{Der}_R(S, M) = 0$.*

Proof. Apply Corollary V.10.8 to $M$ with the $S \otimes_R S$-module structure induced by $S \otimes_R S \to S; x \otimes y \mapsto xy$ and use (V.10.4). □

**Proposition V.10.10.** *Let*

$$\begin{array}{ccc} S & \xrightarrow{t} & \overline{T} \\ {\scriptstyle f}\uparrow & & \uparrow{\scriptstyle g} \\ R & \xrightarrow{s} & T \end{array}$$

*be a commutative diagram of $\overline{V}$-algebras such that $g$ is almost surjective (i.e., $g(T) \supset \pi^\alpha \overline{T}$ for all $\alpha \in \Lambda^+$), $(\mathrm{Ker}(g))^2 = 0$, and $T \xrightarrow{\sim} \mathrm{Hom}_{\overline{V}}(\mathfrak{m}, T)$, $\overline{T} \xrightarrow{\sim} \mathrm{Hom}_{\overline{V}}(\mathfrak{m}, \overline{T})$. If $S$ is almost projective as an $S \otimes_R S$-module (resp. $f$ is an almost étale covering), then there exists at most one (resp. unique) homomorphism $u\colon S \to T$ such that $g \circ u = t$ and $u \circ f = s$.*

Proof. Set $I := \mathrm{Ker}(g)$. First let us prove the uniqueness. Suppose that we have two homomorphisms $u_1$, $u_2$ satisfying the conditions. Then their difference $u_1 - u_2 \colon S \to I$ is an $R$-derivation, which vanishes by Corollary V.10.9 because $I \xrightarrow{\sim} \mathrm{Hom}_{\overline{V}}(\mathfrak{m}, I)$ by assumption. Here we regard $I$ as an $S$-module via $t$. Next let us prove the existence when $f$ is an almost étale covering. For any $\alpha \in \Lambda^+$, since $S$ is an almost projective $R$-module and $g$ is almost surjective, there exists an $R$-linear lifting $u_\alpha \colon S \to T$ of $\pi^\alpha t$. Define the $R$-linear map $v_\alpha \colon S \times S \to I$ by $v_\alpha(x, y) = (\pi^\alpha)^2 u_\alpha(x \cdot y) - \pi^\alpha \cdot u_\alpha(x) \cdot u_\alpha(y)$. Then, using the fact that the action of $\pi^\alpha \cdot x$ ($x \in S$) on $I$ coincides with that of $u_\alpha(x) \in T$, we see that the coboundary (V.10.2) of $v_\alpha$:

$$dv_\alpha(x, y, z) = x \cdot v_\alpha(y, z) - v_\alpha(xy, z) + v_\alpha(x, yz) - z \cdot v_\alpha(x, y)$$

vanishes. By Corollary V.10.8, there exists an $R$-linear map $h_\alpha \colon S \to I$ such that

$$v_\alpha(x, y) = dh_\alpha(x, y) = x \cdot h_\alpha(y) - h_\alpha(xy) + h_\alpha(x) \cdot y,$$

which implies

$$(\pi^\alpha)^3 u_\alpha(xy) - (\pi^\alpha)^2 u_\alpha(x) \cdot u_\alpha(y) = u_\alpha(x) \cdot h_\alpha(y) - \pi^\alpha \cdot h_\alpha(x \cdot y) + h_\alpha(x) \cdot u_\alpha(y).$$

If we set $u'_\alpha := (\pi^\alpha)^2 \cdot u_\alpha + h_\alpha$, which is a lifting of $(\pi^\alpha)^3 t$, we obtain $(\pi^\alpha)^3 u'_\alpha(xy) = u'_\alpha(x) \cdot u'_\alpha(y)$. For each $\alpha \in 3\Lambda^+$, by multiplying $u'_{\alpha/3}$ by $\pi^\alpha/(\pi^{\alpha/3})^3$, we obtain an $R$-linear lifting $w_\alpha \colon S \to T$ of $\pi^\alpha t$ such that

(V.10.11) $$\pi^\alpha \cdot w_\alpha(y) = w_\alpha(x) \cdot w_\alpha(y).$$

Suppose that we have another such lifting $\widetilde{w}_\alpha$. Then the $R$-linear map $\pi^\alpha\cdot(w_\alpha-\widetilde{w}_\alpha)\colon S\to I$ is an $R$-derivation, which vanishes by Corollary V.10.9. By replacing $w_\alpha$ with $w_{\alpha/2}$ multiplied by $\pi^\alpha(\pi^{\alpha/2})^{-1}$ for $\alpha\in 6\Lambda_+$, we obtain a system of $R$-linear liftings $(w_\alpha\colon S\to T)_{\alpha\in 6\Lambda^+}$ of $\pi^\alpha t$ satisfying (V.10.11) and $w_\beta=\frac{\pi^\beta}{\pi^\alpha}\cdot w_\alpha$ for any $\alpha,\beta\in 6\Lambda^+$, $\beta\geq\alpha$. Define the homomorphism $u\colon S\to \mathrm{Hom}_{\overline{V}}(\mathfrak{m},T)=T$ by $u(x)(a\cdot\pi^\alpha)=a\cdot w_\alpha(x)$ $(x\in S,\alpha\in 6\Lambda^+,a\in\overline{V})$. We see easily that this is well-defined, $R$-linear, and multiplicative and satisfies $g\circ u=t$. Since $u(1)$ is an idempotent lifting of $1\in\overline{T}$ and $I^2=0$, we have $u(1)=1$. Hence $u$ is the required homomorphism. □

To prove the existence of liftings of almost étale coverings, we first prove that of almost projective modules.

**Proposition V.10.12.** *Let $R$ be a $\overline{V}$-algebra, let $I$ be an ideal of $R$ such that $I^2=0$, and set $\overline{R}:=R/I$. Then, for any almost projective (resp. almost finitely generated projective) $\overline{R}$-module $\overline{M}$, there exists an almost projective (resp. almost finitely generated projective) $R$-module $M$ with an almost isomorphism of $\overline{R}$-modules $M\otimes_R\overline{R}\overset{\approx}{\to}\overline{M}$. Furthermore, if $\wedge^{r+1}_{\overline{R}}\overline{M}\approx 0$, then $\wedge^{r+1}_R M\approx 0$ for any $R$-module $M$ as above.*

Proof. For each $\alpha\in\Lambda^+$, choose a free (resp. finite free) $R$-module $L_\alpha$ and $\overline{R}$-linear homomorphisms $\overline{f_\alpha}\colon\overline{M}\to\overline{L_\alpha}$, $\overline{g_\alpha}\colon\overline{L_\alpha}\to\overline{M}$ $(\overline{L_\alpha}:=L_\alpha\otimes_R\overline{R})$ such that $\overline{g_\alpha}\circ\overline{f_\alpha}=\pi^\alpha\cdot\mathrm{id}_{\overline{M}}$. Set $\overline{e_\alpha}:=\overline{f_\alpha}\circ\overline{g_\alpha}$, which satisfies $\overline{e_\alpha}^2=\pi^\alpha\cdot\overline{e_\alpha}$. Choose a lifting $e_\alpha\in\mathrm{End}_R(L_\alpha)$ of $\overline{e_\alpha}$. Then we have $(e_\alpha^2-\pi^\alpha\cdot e_\alpha)^2=0$ and $e'_{3\alpha}:=3\pi^\alpha\cdot(e_\alpha)^2-2(e_\alpha)^3$ is a lifting of $(\pi^\alpha)^2\overline{e_\alpha}$ satisfying $(e'_{3\alpha})^2=(\pi^\alpha)^3\cdot e'_{3\alpha}$. For each $\alpha\in\Lambda^+$, by choosing $\beta\in\Lambda^+$ such that $3\beta\leq\alpha$ and replacing $\overline{f_\alpha}$ by $\pi^\alpha(\pi^\beta)^{-1}\cdot\overline{f}_\beta$, we may assume that $\overline{e_\alpha}$ has a lifting $e_\alpha\in\mathrm{End}_R(L_\alpha)$ satisfying $e_\alpha^2=\pi^\alpha\cdot e_\alpha$.

Set $M_\alpha:=L_\alpha/(e_\alpha-\pi^\alpha)L_\alpha$. Then the endomorphism $e_\alpha$ induces an $R$-linear map $M_\alpha\to L_\alpha$ such that the composite $M_\alpha\to L_\alpha\to M_\alpha$ is $\pi^\alpha\cdot\mathrm{id}_{M_\alpha}$. Hence $\pi^\alpha\cdot\mathrm{Ext}^i_R(M_\alpha,-)=0$ and $\pi^\alpha\cdot\mathrm{Tor}^i_R(M_\alpha,-)=0$ $(i>0)$. If we set $\overline{M_\alpha}:=M_\alpha\otimes_R\overline{R}(\cong\overline{L_\alpha}/(\overline{e_\alpha}-\pi^\alpha)\overline{L_\alpha})$, then $\overline{f_\alpha}$ and $\overline{g_\alpha}$ induce $R$-linear maps $\overline{\varphi_\alpha}\colon\overline{M}\to\overline{M_\alpha}$ and $\overline{\psi_\alpha}\colon\overline{M_\alpha}\to\overline{M}$ satisfying $\overline{\varphi_\alpha}\circ\overline{\psi_\alpha}=\pi^\alpha\cdot\mathrm{id}_{\overline{M_\alpha}}$ and $\overline{\psi_\alpha}\circ\overline{\varphi_\alpha}=\pi^\alpha\cdot\mathrm{id}_{\overline{M}}$.

For $\alpha,\beta\in\Lambda^+$, the composite $M_\beta\to\overline{M_\beta}\overset{\overline{\psi_\beta}}{\to}\overline{M}\overset{\overline{\varphi_\alpha}}{\to}\overline{M_\alpha}$ has a lifting $\tau_{\alpha,\beta}\colon M_\beta\to M_\alpha$ after multiplication by $\pi^\beta$ because $\pi^\beta\cdot\mathrm{Ext}^1_R(M_\beta,-)=0$, and we have $\overline{\psi_\alpha}\circ\overline{\tau_{\alpha,\beta}}=\pi^\alpha\cdot\pi^\beta\cdot\overline{\psi_\beta}$, where $\overline{\tau_{\alpha,\beta}}$ denotes the reduction of $\tau_{\alpha,\beta}$ mod $I$. When $\alpha\leq\beta/3$, if we set $\tau'_{\alpha,\beta}:=\frac{\pi^\beta}{(\pi^\alpha)^3}\tau_{\alpha,\beta}$, then we have $(\pi^\alpha)^2\overline{\psi_\alpha}\circ\overline{\tau'_{\alpha,\beta}}=(\pi^\beta)^2\overline{\psi_\beta}$. Choose a sequence $\alpha_n\in\Lambda^+$ $(n\geq 0)$ such that $\alpha_{n+1}\leq\alpha_n/3$ and define $M_0$ to be the direct limit of $\{M_{\alpha_n},\tau'_{\alpha_{n+1},\alpha_n}\}_{n\geq 0}$. The homomorphisms $(\pi^{\alpha_n})^2\overline{\psi_{\alpha_n}}$ induce $\overline{\psi}\colon\overline{M_0}(:=M_0\otimes_R\overline{R})\to\overline{M}$, which is an almost isomorphism.

Consider the morphism of exact sequences:

$$\begin{array}{ccccccc}
I\otimes_{\overline{R}}\overline{M_{\alpha_n}} & \xrightarrow{h_n} & M_{\alpha_n} & \longrightarrow & \overline{M_{\alpha_n}} & \longrightarrow & 0\\
\downarrow & & \downarrow & & \downarrow & & \\
I\otimes_{\overline{R}}\overline{M_0} & \xrightarrow{h_0} & M_0 & \longrightarrow & \overline{M_0} & \longrightarrow & 0.
\end{array}$$

Since $\pi^{\alpha_n}\cdot\mathrm{Tor}_1^R(\overline{R},M_{\alpha_n})=0$, $\pi^{\alpha_n}\cdot\mathrm{Ker}(h_n)=0$ and $\mathrm{Ker}(h_0)\approx 0$. Using the almost isomorphism $\overline{\psi}$, we see that the kernels (resp. the cokernels) of the left and the right vertical maps are killed by $\pi^{3\alpha_n}$ (resp. $\pi^{4\alpha_n}$). Hence the kernel and the cokernel of the middle map are killed by $\pi^{8\alpha_n}$. Since $\pi^{\alpha_n}\cdot\mathrm{Ext}^i_R(M_{\alpha_n},-)=0$ $(i>0)$, varying $n$, we see that $M_0$ is almost projective. If $\overline{M}$ is almost finitely generated $M_{\alpha_n}$ is finitely generated, and hence $M_0$ is almost finitely generated. The last claim follows from the exact sequence $I \otimes_{\overline{R}} \wedge_{\overline{R}}^{r+1}\overline{M_0} \to \wedge_R^{r+1}M_0 \to \wedge_{\overline{R}}^{r+1}\overline{M_0} \to 0$ and $\wedge_{\overline{R}}^{r+1}\overline{M_0} \overset{\approx}{\to} \wedge_{\overline{R}}^{r+1}\overline{M}$. □

**Theorem V.10.13.** *Let $R$ be a $\overline{V}$-algebra, let $I$ be an ideal of $R$ such that $I^2 = 0$, and set $\overline{R} := R/I$. Then, for any almost étale coverings $\overline{S}$ of $\overline{R}$, there exists an almost étale covering $S$ of $R$ with an almost isomorphism $S \otimes_R \overline{R} \overset{\approx}{\to} \overline{S}$ of $\overline{R}$-algebras.*

PROOF. We may assume $\overline{S} \overset{\sim}{\to} \mathrm{Hom}_{\overline{V}}(\mathfrak{m}, \overline{S})$. (The general case is reduced to this case by taking the fiber product of $\overline{S} \overset{\approx}{\to} \mathrm{Hom}_{\overline{V}}(\mathfrak{m}, \overline{S}) \leftarrow S'$ for an almost étale lifting $S'$ of $\mathrm{Hom}_{\overline{V}}(\mathfrak{m}, \overline{S})$.) By Proposition V.10.12, there exists an almost finitely generated projective $R$-module $M$ of finite rank with an almost isomorphism $M \otimes_R \overline{R} \overset{\approx}{\to} \overline{S}$ of $\overline{R}$-modules. By replacing $M$ with $\mathrm{Hom}_{\overline{V}}(\mathfrak{m}, M)$, we may assume $M \overset{\sim}{\to} \mathrm{Hom}_{\overline{V}}(\mathfrak{m}, M)$. We denote by $\overline{m}$ the multiplication of $\overline{S}$.

Let $N$ denote the kernel of $M \to \overline{S}$. Since $I \cdot M \overset{\approx}{\to} N$ and $N$ does not have $\mathfrak{m}$-torsion, $N$ is an $\overline{R}$-module. By taking $\mathrm{Hom}_{\overline{V}}(\mathfrak{m}, -)$ of the almost isomorphisms

$$N \overset{\approx}{\leftarrow} I \cdot M \overset{\approx}{\leftarrow} I \otimes_R M \cong I \otimes_{\overline{R}} (M \otimes_R \overline{R}) \overset{\approx}{\to} I \otimes_{\overline{R}} \overline{S},$$

we obtain an isomorphism $N \cong \mathrm{Hom}_{\overline{V}}(\mathfrak{m}, I \otimes_{\overline{R}} \overline{S})$ and we can naturally regard $N$ as an $\overline{S}$-module. For $\alpha \in \Lambda^+ \cup \{0\}$, if we are given an $R$-bilinear lifting $m_\alpha \colon M \times M \to M$ of $\pi^\alpha \overline{m}$ ($\pi^0 = 1$), then, for $x \in M$, $y \in N$, and the image $\overline{x} \in \overline{S}$ of $x$, we have

$$(\pi^\alpha \cdot \overline{x}) \cdot y = m_\alpha(x, y) = m_\alpha(y, x). \tag{V.10.14}$$

By Proposition V.3.8, $M \otimes_R M$ is an almost projective $R$-module. Hence, for each $\alpha \in \Lambda^+$, there exists an $R$-bilinear lifting $m_\alpha \colon M \times M \to M$ of $\pi^\alpha \cdot \overline{m}$. Consider the $R$-multilinear map $g_\alpha \colon M \times M \times M \to N$ defined by $g_\alpha(x, y, z) = m_\alpha(m_\alpha(x, y), z) - m_\alpha(x, m_\alpha(y, z))$. $g_\alpha(x, y, z) = 0$ if one of $x$, $y$, $z$ is contained in $I \cdot M$. Since $I \cdot M \overset{\approx}{\to} N$ and $N$ does not have $\mathfrak{m}$-torsion, this also holds for $N$. Hence, if we denote by $\overline{S}'$ the image of $M$ in $\overline{S}$, $g_\alpha$ induces an $\overline{R}$-multilinear map $\overline{g_\alpha} \colon \overline{S}' \times \overline{S}' \times \overline{S}' \to N$. This extends to an $\overline{R}$-multilinear map $\overline{S} \times \overline{S} \times \overline{S} \to \mathrm{Hom}_{\overline{V}}(\mathfrak{m}, N) = N$ which sends $(x, y, z)$ to the $n$ defined by $n(a(\pi^\beta)^3) = a \cdot \overline{g_\alpha}(\pi^\beta x, \pi^\beta y, \pi^\beta z)$ ($\beta \in \Lambda^+, a \in \mathfrak{m}$). We denote this extension also by $\overline{g_\alpha}$. The coboundary (V.10.2) of $\pi^\alpha \overline{g_\alpha}$ vanishes. Indeed, for each $\beta \in \Lambda^+$ and $x, y, z, w \in \overline{S}$, if we choose a lifting $x_\beta$, $y_\beta$, $z_\beta$, $w_\beta \in M$ of $\pi^\beta x$, $\pi^\beta y$, $\pi^\beta z$, $\pi^\beta w \in \overline{S}'$, we have

$$\begin{aligned}(d(\pi^\alpha \overline{g_\alpha})(x, y, z, w))((\pi^\beta)^4) =& m_\alpha(x_\beta, g_\alpha(y_\beta, z_\beta, w_\beta)) - g_\alpha(m_\alpha(x_\beta, y_\beta), z_\beta, w_\beta) \\ &+ g_\alpha(x_\beta, m_\alpha(y_\beta, z_\beta), w_\beta) - g_\alpha(x_\beta, y_\beta, m_\alpha(z_\beta, w_\beta)) \\ &+ m_\alpha(g_\alpha(x_\beta, y_\beta, z_\beta), w_\beta) \\ =& 0\end{aligned}$$

by (V.10.14). By Corollary V.10.8, there exists an $\overline{R}$-bilinear map $\overline{h_\alpha} \colon \overline{S} \times \overline{S} \to N$ such that $\pi^\alpha \cdot \overline{g_\alpha} = d\overline{h_\alpha}$, which implies:

$$\begin{aligned}&(\pi^\alpha)^2 m_\alpha(m_\alpha(x, y), z) - (\pi^\alpha)^2 m_\alpha(x, m_\alpha(y, z)) \\ =& m_\alpha(x, h_\alpha(y, z)) - h_\alpha(m_\alpha(x, y), z) + h_\alpha(x, m_\alpha(y, z)) - m_\alpha(h_\alpha(x, y), z)\end{aligned}$$

for $x, y, z \in M$ by (V.10.14), where $h_\alpha$ denotes the $R$-bilinear map $M \times M \to \overline{S} \times \overline{S} \overset{\overline{h_\alpha}}{\to} N$. Now the lifting $(\pi^\alpha)^2 m_\alpha + h_\alpha$ of $(\pi^\alpha)^3 \overline{m}$ is associative. For each $\alpha \in \Lambda^+$, choosing $\beta \in \Lambda^+$ such that $3\beta \leq \alpha$ and applying the above argument to $\beta$, we see that $\pi^\alpha \cdot \overline{m}$ has an associative lifting. We choose such a lifting $m_\alpha$.

For $a \in M$, the $R$-linear map $g_a\colon M \to N; x \mapsto m_\alpha(a,x) - m_\alpha(x,a)$ induces an $\overline{R}$-linear map $\overline{g_a}\colon \overline{S} \to \mathrm{Hom}_{\overline{V}}(\mathfrak{m}, N) = N$ similarly as above. Then $\pi^\alpha \cdot \overline{g_a}$ is an $\overline{R}$-derivation. Indeed, for $x, y \in \overline{S}$, $\beta \in \Lambda^+$, and liftings $x_\beta, y_\beta \in M$ of $\pi^\beta x$, $\pi^\beta y$, we have

$$\begin{aligned}&\pi^\alpha\{(\overline{g_a}(xy) - x\cdot \overline{g_a}(y) - y \cdot \overline{g_a}(x))((\pi^\beta)^2)\}\\ =&g_a(m_\alpha(x_\beta, y_\beta)) - m_\alpha(x_\beta, g_a(y_\beta)) - m_\alpha(g_a(x_\beta), y_\beta) = 0\end{aligned}$$

by (V.10.14) and the associativity of $m_\alpha$. By Corollary V.10.9, $\pi^\alpha \cdot \overline{g_a} = 0$ and hence $\pi^\alpha \cdot m_\alpha$ is commutative. For each $\alpha \in \Lambda^+$, choosing $\beta \in \Lambda^+$ such that $2\beta \le \alpha$ and applying the above argument to $\beta$, we see that $\pi^\alpha \overline{m}_\alpha$ has an $R$-bilinear associative commutative lifting $M \times M \to M$.

Suppose that we have two $R$-bilinear associative commutative liftings $m_\alpha, m'_\alpha\colon M \times M \to M$ of $\pi^\alpha \overline{m}$. Then the $R$-bilinear map $g = m'_\alpha - m_\alpha\colon M \times M \to N$ induces an $\overline{R}$-bilinear map $\overline{g}\colon \overline{S} \times \overline{S} \to N$ similarly as above. For $x, y, z \in \overline{S}$, $\beta \in \Lambda^+$, and liftings $x_\beta, y_\beta, z_\beta \in M$ of $\pi^\beta x$, $\pi^\beta y$, $\pi^\beta z$, by (V.10.14), we have

$$\begin{aligned}&(d(\pi^\alpha \overline{g})(x,y,z))((\pi^\beta)^3)\\ =&m'_\alpha(x_\beta, g(y_\beta, z_\beta)) - g(m_\alpha(x_\beta, y_\beta), z_\beta) + g(x_\beta, m_\alpha(y_\beta, z_\beta)) - m'_\alpha(g(x_\beta, y_\beta), z_\beta)\\ =&0.\end{aligned}$$

By Corollary V.10.8, there exists an $R$-linear map $\overline{h}\colon \overline{S} \to N$ such that $\pi^\alpha \overline{g} = d\overline{h}$. By (V.10.14), this implies

$$(\pi^\alpha)^2 m'_\alpha(x,y) - (\pi^\alpha)^2 m_\alpha(x,y) = m_\alpha(x, h(y)) - h(m'_\alpha(x,y)) + m_\alpha(h(x), y),$$

where $h$ denotes $M \to \overline{S} \xrightarrow{\overline{h}} N$, and hence the homomorphism

$$(\pi^\alpha)^2 \cdot \mathrm{id}_M + h\colon (M, (\pi^\alpha)^2 m'_\alpha) \to (M, m_\alpha) \tag{V.10.15}$$

is multiplicative.

For each $\alpha \in \Lambda^+$, choose a commutative associative $R$-bilinear lifting $m_\alpha$ to $M$ of $\pi^\alpha \overline{m}$ and define

$$n_\alpha\colon \pi^{3\alpha}\overline{V} \otimes_{\overline{V}} M \times \pi^{3\alpha}\overline{V} \otimes_{\overline{V}} M \to \pi^{3\alpha}\overline{V} \otimes_{\overline{V}} M$$

by transporting $(\pi^\alpha)^2 m_\alpha$ via $(\pi^\alpha)^3\colon M \xrightarrow{\sim} \pi^{3\alpha}\overline{V} \otimes_{\overline{V}} M$, which is compatible with $\overline{m}$. For $\alpha, \beta \in \Lambda^+$ such that $\alpha \le \beta/3$, by applying the above claim to the two liftings $\frac{\pi^\beta}{\pi^\alpha} m_\alpha$ and $m_\beta$ of $\pi^\beta \overline{m}$, we obtain an $R$-linear map $h_{\alpha,\beta}\colon M \to N$ such that $(\pi^\beta)^2 \cdot \mathrm{id}_M + h_{\alpha,\beta}\colon (M, (\pi^\beta)^2 m_\beta) \to (M, \frac{\pi^\beta}{\pi^\alpha} m_\alpha)$ is multiplicative. Composing with the multiplicative map $\frac{\pi^\beta}{(\pi^\alpha)^3}\colon (M, \frac{\pi^\beta}{\pi^\alpha} m_\alpha) \to (M, (\pi^\alpha)^2 m_\alpha)$ and transporting via $(\pi^\gamma)^3\colon (M, (\pi^\gamma)^2 m_\gamma) \xrightarrow{\sim} (\pi^{3\gamma}\overline{V} \otimes_{\overline{V}} M, n_\gamma)$ $(\gamma = \alpha, \beta)$, we obtain an $R$-linear map $k_{\alpha,\beta}\colon \pi^{3\beta}\overline{V} \otimes_{\overline{V}} M \to \pi^{3\alpha}\overline{V} \otimes_{\overline{V}} N$ such that $1 + k_{\alpha,\beta}\colon (\pi^{3\beta}\overline{V} \otimes_{\overline{V}} M, n_\beta) \to (\pi^{3\alpha}\overline{V} \otimes_{\overline{V}} M, n_\alpha)$ is multiplicative. Choose a sequence $\alpha_n \in \Lambda^+$ $(n \ge 0)$ satisfying $\alpha_{n+1} \le \alpha_n/3$ and define the $R$-module $M'$ with the associative commutative $R$-bilinear map $m'\colon M' \times M' \to M'$ to be the direct limit of

$$\{(\pi^{3\alpha_n}\overline{V} \otimes_{\overline{V}} M, n_{\alpha_n}), 1 + k_{\alpha_{n+1},\alpha_n}\}_n.$$

We have a natural $R$-linear multiplicative map $M' \to \overline{S}$ inducing an almost isomorphism $M' \otimes_R \overline{R} \xrightarrow{\approx} \overline{S}$. Note $(\pi^{3\alpha_n}\overline{V} \otimes_{\overline{V}} M) \otimes_R \overline{R} \xrightarrow{\approx} \pi^{3\alpha_n}\overline{V} \otimes_{\overline{V}} \overline{S}$ and $1 + k_{\alpha_{n+1},\alpha_n}$ is a lifting of $1\colon \pi^{3\alpha_n}\overline{V} \otimes_{\overline{V}} \overline{S} \to \pi^{3\alpha_{n+1}}\overline{V} \otimes_{\overline{V}} \overline{S}$.

We assert that $M'$ is an almost finitely generated projective $R$-module of finite rank. Set $M_n := \pi^{3\alpha_n}\overline{V} \otimes_{\overline{V}} M$. If we denote by $\overline{\phantom{x}}$ the reduction mod $I$ of an $R$-module, we have a commutative diagram with exact lines:

$$\begin{array}{ccccccc} I \otimes_{\overline{R}} \overline{M}_n & \longrightarrow & M_n & \longrightarrow & \overline{M}_n & \longrightarrow & 0 \\ \downarrow & & \downarrow & & \downarrow & & \\ I \otimes_{\overline{R}} \overline{M}' & \longrightarrow & M' & \longrightarrow & \overline{M}' & \longrightarrow & 0. \end{array}$$

Since $M_n$ is almost projective and hence almost flat (Proposition V.6.3) over $R$, the two left horizontal homomorphisms are almost injective. Since the composite $\pi^{3\alpha_n}\overline{V} \otimes_{\overline{V}} \overline{M} \cong \overline{M}_n \to \overline{M}' \overset{\approx}{\to} \overline{S}$ is induced by $\overline{M} \overset{\approx}{\to} \overline{S}$, the kernel and the cokernel of the left and the right vertical maps are annihilated by $\pi^{4\alpha_n}$. Hence the kernel and the cokernel of the middle map are annihilated by $\pi^{9\alpha_n}$. Since $M_n$ is an almost finitely generated projective $R$-module, varying $n$, we see that so is $M'$. $M'$ is of finite rank by $\overline{M}' \overset{\approx}{\to} \overline{S}$ and the last claim of Proposition V.10.12.

We set $S := \mathrm{Hom}_{\overline{V}}(\mathfrak{m}, M')$ and define the $R$-bilinear map $m : S \times S \to S$ by the composite of

$$\mathrm{Hom}_{\overline{V}}(\mathfrak{m}, M) \times \mathrm{Hom}_{\overline{V}}(\mathfrak{m}, M') \to \mathrm{Hom}_{\overline{V}}(\mathfrak{m} \otimes_{\overline{V}} \mathfrak{m}, M' \otimes_{\overline{V}} M') \to \mathrm{Hom}_{\overline{V}}(\mathfrak{m}, M'),$$

where the left map is defined by $(f, g) \mapsto f \otimes g$ and the right one is induced by $m' \colon M' \otimes_{\overline{V}} M' \to M'$ and $\mathfrak{m} \otimes_{\overline{V}} \mathfrak{m} \overset{\sim}{\to} \mathfrak{m}; a \otimes b \mapsto a \cdot b$. We have $m(f, g)(a \cdot b) = m'(f(a), g(b))$ $(a, b \in \mathfrak{m})$. We see easily that $\mathfrak{m}$ is associative commutative and the homomorphism $S \to \mathrm{Hom}_{\overline{V}}(\mathfrak{m}, \overline{S}) = \overline{S}$ induced by $M' \to \overline{S}$ is multiplicative. We prove that $1_{\overline{S}} \in \overline{S}$ has an idempotent lifting $e \in S$. Since $S \to \overline{S}$ is almost surjective, for each $\alpha \in \Lambda^+$, there exists a lifting $f_\alpha \in S$ of $\pi^\alpha \cdot 1_{\overline{S}}$. Set $J := \mathrm{Ker}(S \to \overline{S})$. Then $I \cdot S \overset{\approx}{\to} J$ and $J$ does not have $\mathfrak{m}$-torsion, which implies $m(J, J) = 0$. Hence $(f_\alpha^2 - \pi^\alpha \cdot f_\alpha)^2 = 0$ and $f'_{3\alpha} := 3\pi^\alpha \cdot f_\alpha^2 - 2f_\alpha^3$ is a lifting of $(\pi^\alpha)^3 \cdot 1_{\overline{S}}$ satisfying $(f'_{3\alpha})^2 = (\pi^\alpha)^3 \cdot f'_{3\alpha}$. Choosing $\beta \in \Lambda^+$ such that $3\beta \leq \alpha$ and applying the above argument to $\beta$, we see that $\pi^\alpha \cdot 1_{\overline{S}}$ has a lifting $e_\alpha$ such that $e_\alpha^2 = \pi^\alpha \cdot e_\alpha$. If we have two liftings $e_\alpha$ and $e'_\alpha = e_\alpha + x$ $(x \in J)$ of $\pi^\alpha \cdot 1_{\overline{S}}$ satisfying $e_\alpha^2 = \pi^\alpha \cdot e_\alpha$, $e'^2_\alpha = \pi^\alpha \cdot e'_\alpha$, by taking the difference of the two equations, we obtain $(2e_\alpha - \pi^\alpha) \cdot x = 0$. Multiplying $2e_\alpha - \pi^\alpha$, we obtain $(\pi^\alpha)^2 \cdot x = 0$, i.e., $(\pi^\alpha)^2 e_\alpha = (\pi^\alpha)^2 e'_\alpha$. Choose a lifting $e_\alpha \in S$ of $\pi^\alpha \cdot 1_{\overline{S}}$ for each $\alpha \in \Lambda^+$ and define $e \in \mathrm{Hom}_{\overline{V}}(\mathfrak{m}, S) = S$ by $e(a \cdot (\pi^\alpha)^3) = a \cdot (\pi^\alpha)^2 \cdot e_\alpha$ $(a \in \overline{V}, \alpha \in \Lambda^+)$, which is well-defined by the above uniqueness. We can verify easily that $e$ is an idempotent lifting of $1_{\overline{S}}$. By (V.10.14) with $\alpha = 0$, the multiplication by $e$ on $J$ is the identity. Hence the multiplication by $e$ on $S$ is bijective, and for each $x \in S$, by choosing $y \in S$ such that $x = e \cdot y$, we see $e \cdot x = e \cdot (e \cdot y) = (e \cdot e) \cdot y = e \cdot y = x$. Thus $S$ becomes an $R$-algebra with an almost isomorphism $S \otimes_R \overline{R} \overset{\approx}{\to} \overline{S}$.

To prove that $S$ is the required lifting, it is enough to prove that $S$ is an almost projective $S \otimes_R S$-module. Let $\overline{e}$ be the idempotent of $\mathrm{Hom}_{\overline{V}}(\mathfrak{m}, \overline{S} \otimes_{\overline{R}} \overline{S})$ corresponding to the "diagonal immersion" $\mathrm{Spec}(\overline{S}) \to \mathrm{Spec}(\mathrm{Hom}_{\overline{V}}(\mathfrak{m}, \overline{S} \otimes_{\overline{R}} \overline{S}))$ (Lemma V.7.2). Then, by Lemma V.10.16 below, $\overline{e}$ has a unique idempotent lifting $e \in \mathrm{Hom}_{\overline{V}}(\mathfrak{m}, S \otimes_R S)$ and its image in $S$ is the unique idempotent lifting $1_S$ of $1_{\overline{S}}$. Let $J_1$ (resp. $\overline{J}_1$) be the kernel of $\mathrm{Hom}_{\overline{V}}(\mathfrak{m}, S \otimes_R S) \to S$ (resp. $\mathrm{Hom}_{\overline{V}}(\mathfrak{m}, \overline{S} \otimes_{\overline{R}} \overline{S}) \to \overline{S}$). Since $S \otimes_R \overline{R} \overset{\approx}{\to} \overline{S}$ and $S$ is almost projective and hence almost flat (Proposition V.6.3) over $R$, we have $J_1 \otimes_R \overline{R} \overset{\approx}{\to} \overline{J}_1$. Since $I \cdot J_1 \overset{\approx}{\to} \mathrm{Ker}(J_1 \to \overline{J}_1)$ and $\overline{e} \cdot \overline{J}_1 = 0$, we have $\pi^\alpha \cdot e \cdot J_1 \subset I \cdot J_1$ and hence $(\pi^\alpha)^2 e \cdot J_1 = (\pi^\alpha e)^2 \cdot J_1 = 0$ for each $\alpha \in \Lambda^+$. Since $J_1$ does not have $\mathfrak{m}$-torsion, $e \cdot J_1 = 0$. □

**Lemma V.10.16.** . *Let $R$, $I$, and $\overline{R}$ be the same as in Theorem* V.10.13. *Let $\overline{S}$ be an $\overline{R}$-algebra such that $\overline{S} \overset{\sim}{\to} \mathrm{Hom}_{\overline{V}}(\mathfrak{m}, \overline{S})$ and let $S$ be an $R$-algebra with an almost isomorphism* *of $\overline{R}$-algebras $S \otimes_R \overline{R} \overset{\approx}{\to} \overline{S}$ such that $S \overset{\sim}{\to} \mathrm{Hom}_{\overline{V}}(\mathfrak{m}, S)$. Then, any idempotent $\overline{e}$ of $\overline{S}$ has a unique idempotent lifting $e \in S$.*

PROOF. Set $J := \mathrm{Ker}(S \to \overline{S})$. Then $I \cdot S \overset{\approx}{\to} J$ and $J$ does not have $\mathfrak{m}$-torsion. Hence $J^2 = 0$. The proof of the existence is the same as the construction of an idempotent lifting $1_{\overline{S}}$ to $S$ in the proof Theorem V.10.13. If we have two idempotent liftings $e$ and $e' = e + x$ $(x \in J)$ of $\overline{e}$. Then, by taking the difference of $e^2 = e$ and $e'^2 = e'$, we obtain $(2e - 1) \cdot x = 0$. Multiplying $2e - 1$, we obtain $x = 0$. □

As in V.9, we denote by $R$-$\widetilde{\underline{\mathrm{Et}}}$, the category of almost étale coverings $S$ of $R$ such that $S \overset{\sim}{\to} \mathrm{Hom}_{\overline{V}}(\mathfrak{m}, S)$.

**Corollary V.10.17.** *Let $R$, $I$, and $\overline{R}$ be the same as Theorem V.10.13. Then the functor $R$-$\widetilde{\underline{Et}} \to \overline{R}$-$\widetilde{\underline{Et}}$; $S \mapsto \mathrm{Hom}_{\overline{V}}(\mathfrak{m}, S \otimes_R \overline{R})$ is an equivalence of categories.*

PROOF. Let $S, T \in R$-$\widetilde{\underline{\mathrm{Et}}}$, and put $\overline{S} = \mathrm{Hom}_{\overline{V}}(\mathfrak{m}, S \otimes_R \overline{R})$ and $\overline{T} = \mathrm{Hom}_{\overline{V}}(\mathfrak{m}, T \otimes_R \overline{R})$. Then, for any homomorphism of $\overline{R}$-algebras $\overline{h}\colon \overline{S} \to \overline{T}$, there exists a unique homomorphism $h\colon S \to T$ of $R$-algebras inducing $\overline{h}$ by Proposition V.10.10, i.e., the functor in question is fully faithful. Note that $T \to \overline{T}$ is almost surjective with a square-zero kernel. It is essentially surjective by Theorem V.10.13. □

## V.11. Group cohomology of discrete $A$-$G$-modules

Let $G$ be a profinite group and let $A$ be a commutative ring. A discrete $A$-$G$-module is an $A$-module with an $A$-linear action of $G$ continuous with respect to the discrete topology on $M$. Discrete $A$-$G$-modules form an abelian category, which we denote by $A$-$G$-$\underline{\mathrm{discMod}}$.

Let $M$ be an $A$-module endowed with the discrete topology. The induced module $\mathrm{Ind}_{A,G}(M)$ is defined to be the $A$-module of continuous maps of $G$ to $M$. The group $G$ acts on $\mathrm{Ind}_{A,G}(M)$ by the formula

$$(g \cdot f)(x) = f(x \cdot g) \quad (f \in \mathrm{Ind}_{A,G}(M),\ g \in G,\ x \in G).$$

The $A$-module $\mathrm{Ind}_{A,G}(M)$ with the above action of $G$ is a discrete $A$-$G$-module and functorial on $M$. Thus we obtain an exact functor

$$\mathrm{Ind}_{A,G}\colon A\text{-}\underline{\mathrm{Mod}} \to A\text{-}G\text{-}\underline{\mathrm{discMod}}.$$

We say that a discrete $A$-$G$-module is *induced* if it is isomorphic to the induced module of an $A$-module.

**Proposition V.11.1.** *The functor $\mathrm{Ind}_{A,G}$ is a right adjoint of the forgetful functor. Especially $\mathrm{Ind}_{A,G}$ preserves injectives.*

PROOF. For a discrete $A$-$G$-module $M$, we have a canonical and functorial morphism $\mathrm{adj}_M\colon M \to \mathrm{Ind}_{A,G}(M)$ of discrete $A$-$G$-modules which sends $m$ to the map $x \mapsto x \cdot m$. For an $A$-module $N$, we can verify that the homomorphism

$$\mathrm{Hom}_A(M, N) \to \mathrm{Hom}_{A,G}(M, \mathrm{Ind}_{A,G}(N)); f \mapsto \mathrm{Ind}_{A,G}(f) \circ \mathrm{adj}_M$$

is an isomorphism. Indeed, the inverse is given by $\varphi \mapsto \mathrm{adj}'_N \circ \varphi$, where $\mathrm{adj}'_N$ denotes the $A$-linear homomorphism $\mathrm{Ind}_{A,G}(N) \to N$ defined by $\psi \mapsto \psi(1)$. □

**Corollary V.11.2.** (1) *The category $A$-$G$-$\underline{\mathrm{discMod}}$ has enough injectives.*

(2) *An object of $A$-$G$-$\underline{\mathrm{discMod}}$ is injective if and only if it is isomorphic to a direct factor of $\mathrm{Ind}_{A,G}(I)$ for an injective $A$-module $I$.*

PROOF. Let $M$ be a discrete $A$-$G$-module and choose an $A$-linear injection $f$ of $M$ into an injective $A$-module $I$. Then, the morphism in $A$-$G$-$\underline{\text{discMod}}$ $\text{Ind}_{A,G}(f) \circ \text{adj}_M \colon M \to \text{Ind}_{A,G}(I)$ is injective, where $\text{adj}_M$ is as in the proof of Proposition V.11.1. Furthermore, if $M$ is injective, this admits a section. Now the corollary follows easily from Proposition V.11.1. □

Let

$$H^q_A(G,-)\colon A\text{-}G\text{-}\underline{\text{discMod}} \to A\text{-}\underline{\text{Mod}} \quad (q \geq 0),$$
$$R\Gamma_A(G,-)\colon D^+(A\text{-}G\text{-}\underline{\text{discMod}}) \to D^+(A\text{-}\underline{\text{Mod}})$$

be the right derived functors of the left exact functor

$$\Gamma_A(G,-)\colon A\text{-}G\text{-}\underline{\text{discMod}} \to A\text{-}\underline{\text{Mod}}; M \mapsto M^G,$$

where $M^G$ denotes the $G$-invariant part.

**Lemma V.11.3.** *For an $A$-module $M$, we have a canonical and functorial isomorphism*

$$R\Gamma_A(G, \text{Ind}_{A,G}(M)) \cong M.$$

PROOF. We have a canonical and functorial isomorphism $M \xrightarrow{\sim} \Gamma_A(G, \text{Ind}_{A,G}(M))$ which sends $m$ to the constant map $x \mapsto m$ $(x \in G)$. Hence the claim follows from Proposition V.11.1 and the exactness of $\text{Ind}_{A,G}$. □

**Corollary V.11.4.** *Let $A \to B$ be a homomorphism of commutative rings. Then we have canonical isomorphisms of functors*

$$F \circ H^q_B(G,-) \cong H^q_A(G,-) \circ F \quad (q \geq 0),$$
$$F \circ R\Gamma_B(G,-) \cong R\Gamma_A(G,-) \circ F,$$

*where $F$ denote the exact functors $B$-$G$-$\underline{\text{discMod}} \to A$-$G$-$\underline{\text{discMod}}$, $B$-$\underline{\text{Mod}} \to A$-$\underline{\text{Mod}}$ restricting the action of $B$ to $A$ and the functors $D^+(B\text{-}G\text{-}\underline{\text{discMod}}) \to D^+(A\text{-}G\text{-}\underline{\text{discMod}})$, $D^+(B\text{-}\underline{\text{Mod}}) \to D^+(A\text{-}\underline{\text{Mod}})$ induced by them.*

PROOF. For any $B$-module $M$, we have $F(\text{Ind}_{B,G}(M)) \cong \text{Ind}_{A,G}(F(M))$. Hence, by Corollary V.11.2 (2) and Lemma V.11.3, the functor $F\colon B$-$G$-$\underline{\text{discMod}} \to A$-$G$-$\underline{\text{discMod}}$ sends injectives to objects acyclic with respect to $\Gamma_A(G,-)$. □

In the following, we abbreviate $H^q_A(G,-)$, $R\Gamma_A(G,-)$ to $H^q(G,-)$, $R\Gamma(G,-)$ if there is no risk of confusion.

Let $H$ be a closed normal subgroup of $G$. Then the groups $H$ and $G/H$ endowed with the induced topology are again profinite groups. We also denote by $\Gamma(H,-)$ the left exact functor

$$A\text{-}G\text{-}\underline{\text{discMod}} \to A\text{-}G/H\text{-}\underline{\text{discMod}}; M \mapsto M^H,$$

and by $H^q(H,-)$ $(q \geq 0)$, $R\Gamma(H,-)$ its right derived functors. This notation is justified by the following proposition.

**Proposition V.11.5.** *The following diagrams commute up to canonical isomorphisms of functors:*

$$\begin{array}{ccc} A\text{-}G\text{-}\underline{\text{discMod}} & \xrightarrow{H^q(H,-)} & A\text{-}G/H\text{-}\underline{\text{discMod}} \\ \downarrow & & \downarrow \\ A\text{-}H\text{-}\underline{\text{discMod}} & \xrightarrow{H^q(H,-)} & A\text{-}\underline{\text{Mod}}, \end{array}$$

$$\begin{array}{ccc} D^+(A\text{-}G\text{-}\underline{\mathrm{discMod}}) & \xrightarrow{R\Gamma(H,-)} & D^+(A\text{-}G/H\text{-}\underline{\mathrm{discMod}}) \\ \downarrow & & \downarrow \\ D^+(A\text{-}H\text{-}\underline{\mathrm{discMod}}) & \xrightarrow{R\Gamma(H,-)} & D^+(A\text{-}\underline{\mathrm{Mod}}). \end{array}$$

PROOF. This follows from Corollary V.11.2 (2), Lemma V.11.3, and the following Lemma V.11.6. □

**Lemma V.11.6.** *Induced discrete $A$-$G$-modules are induced discrete $A$-$H$-modules.*

PROOF. Choose a continuous section $s\colon G/H \to G$ (**[65]** I Proposition 1). Then, we have a homeomorphism $G \cong H \times G/H$; $g \mapsto (s(gH)^{-1}g, gH)$, $s(\overline{g})h \leftarrow (h, \overline{g})$ compatible with the right action of $H$, where $H$ acts on $H \times G/H$ by $(h, \overline{g}) \mapsto (hk, \overline{g})$ $(k \in H)$. Hence, there are isomorphisms of discrete $A$-$H$-modules

$$\begin{aligned} \mathrm{Ind}_{A,G}(M) &\cong \mathrm{Map}_{\mathrm{cont}}(H \times G/H, M) \\ &\cong \mathrm{Map}_{\mathrm{cont}}(H, \mathrm{Map}_{\mathrm{cont}}(G/H, M)) \\ &= \mathrm{Ind}_{A,H}(\mathrm{Map}_{\mathrm{cont}}(G/H, M)) \end{aligned}$$

□

**Proposition V.11.7.** *The functor $\Gamma(H, -)\colon A\text{-}G\text{-}\underline{\mathrm{Mod}} \to A\text{-}G/H\text{-}\underline{\mathrm{Mod}}$ preserves injectives (resp. induced modules). Especially, we have a spectral sequence*

$$E_1^{a,b} = H^a(G/H, H^b(H, M)) \Longrightarrow H^{a+b}(G, M)$$

*for a discrete $A$-$G$-module $M$.*

PROOF. This follows from $\Gamma(H, -) \circ \mathrm{Ind}_{A,G} \cong \mathrm{Ind}_{A,G/H}$ and Corollary V.11.2 (2). □

Finally, we will recall the explicit description of $R\Gamma(G, -)$ in terms of the inhomogeneous cochain complex. Let $M$ be a discrete $A$-$G$-module. For each integer $q \geq 0$, we denote by $K^q(G, M)$ the $A$-module of continuous maps of the $(q+1)$-fold products $G^{q+1}$ to $M$ endowed with the action of $G$ defined by

$$(g \cdot f)(g_0, \dots, g_q) = g \cdot f(g^{-1}g_0, \dots, g^{-1}g_q).$$

Then $K^q(G, M)$ are discrete $A$-$G$-modules. We define the homomorphisms of discrete $A$-$G$-modules $d^q\colon K^q(G, M) \to K^{q+1}(G, M)$ $(q \geq 0)$ and $\varepsilon\colon M \to K^0(G, M)$ by the formulae

$$\begin{aligned} (d^q f)(g_0, \dots, g_{q+1}) &= \sum_{i=0}^{q+1} (-1)^i f(g_0, \dots, \hat{g}_i, \dots, g_{q+1}), \\ \varepsilon(m)(g_0) &= m. \end{aligned}$$

We have $d^{q+1} \circ d^q = 0$ $(q \geq 0)$ and $d^0 \circ \varepsilon = 0$.

**Lemma V.11.8.** *The complex $M \xrightarrow{\varepsilon} K^\bullet(G, M)$ is homotopy equivalent to zero as a complex of $A$-modules.*

PROOF. The homotopy between the identity and the zero endomorphisms of the complex in question is given by $k^q\colon K^q(G, M) \to K^{q-1}(G, M)$; $(k^q f)(g_0, \dots, g_{q-1}) = f(1, g_0, \dots, g_{q-1})$ $(q \geq 1)$ and $k^0\colon K^0(G, M) \to M$; $k^0 f = f(1)$. □

**Lemma V.11.9.** *The discrete $A$-$G$-modules $K^q(G, M)$ $(q \geq 0)$ are induced.*

PROOF. By the change of variables

$$G^{q+1} \xrightarrow{\sim} G \times G^q; (g_0, \ldots, g_q) \mapsto (g_0, g_0^{-1}g_1, g_0^{-1}g_2, \ldots, g_0^{-1}g_q),$$

we may replace the action of $G$ on $K^q(G, M)$ by

$$(g \cdot f)(g_0, g_1, \ldots, g_q) = g \cdot f(g^{-1}g_0, g_1, \ldots, g_q).$$

Then we have isomorphisms of discrete $A$-$G$-modules

$$K^q(G, M) \cong \mathrm{Map}_{\mathrm{cont}}(G \times G^q, M) \cong \mathrm{Ind}_{A,G}(\mathrm{Map}_{\mathrm{cont}}(G^q, M)),$$

where the first one is defined by $f \mapsto F(g_0, g_1, \ldots, g_q) = g_0 \cdot f(g_0^{-1}, g_1, \cdots g_q)$ and the action of $G$ on the middle module is defined by $(g \cdot \varphi)(g_0, g_1, \ldots, g_q) = \varphi(g_0g, g_1, \ldots, g_q)$. □

The inhomogeneous cochain complex $C^\bullet(G, M)$ is defined as follows: The $A$-module $C^q(G, M)$ $(q \geq 0)$ is an $A$-module of continuous maps of $G^q$ to $M$ and the differential $d^q\colon C^q(G, M) \to C^{q+1}(G, M)$ $(q \geq 0)$ is defined by

$$\begin{aligned}(d^q f)(g_1, \ldots, g_{q+1}) &= g_1 \cdot f(g_2, \ldots, g_{q+1}) \\ &+ \sum_{i=1}^{q}(-1)^i f(g_1, \ldots, g_i \cdot g_{i+1}, \ldots, g_{q+1}) \\ &+ (-1)^{q+1} f(g_1, \ldots, g_q).\end{aligned} \tag{V.11.10}$$

By the restriction with respect to the continuous map

$$G^q \to G^{q+1}; (g_1, \ldots, g_q) \mapsto (1, g_1, g_1g_2, \ldots, g_1g_2 \cdots g_q),$$

we obtain an isomorphism

$$K^\bullet(G, M)^G \xleftarrow{\sim} C^\bullet(G, M). \tag{V.11.11}$$

By Lemmas V.11.3, V.11.8, and V.11.9, we obtain a canonical and functorial isomorphism in $D^+(A\text{-}\underline{\mathrm{Mod}})$:

$$R\Gamma(G, M) \cong C^\bullet(G, M). \tag{V.11.12}$$

## V.12. Galois cohomology

Throughout this section, let $G$ denote a finite group.

**Lemma V.12.1.** *Let $A$ be a commutative ring and let $B$ be an $A$-algebra endowed with an action of $G$. Let $M$ be an arbitrary $B$-module endowed with a semi-linear action of $G$ (i.e., $g(b \cdot m) = g(b) \cdot m$ for $b \in B$, $m \in M$, $g \in G$). Let $\mathrm{tr}_G$ denote the $A$-linear homomorphisms $B \to B^G$ and $M \to M^G$ defined by $\mathrm{tr}_G(x) = \sum_{g \in G} g(x)$ $(x \in B$ or $M)$.*

*(1) For any $a \in A$ whose image in $B^G$ is contained in $tr_G(B)$, $a \cdot H^q(G, M) = 0$ $(q > 0)$ and $a(M^G/\mathrm{tr}_G(M)) = 0$.*

*(2) Let $a \in A$ and suppose that there exist $b_i$, $c_i \in B$ $(1 \leq i \leq r)$ such that $\sum_{i=1}^r b_i \cdot c_i = a \cdot 1_B$ and that $\sum_{i=1}^r b_i \cdot g(c_i \cdot b) = 0$ for all $b \in B$, $g \in G\backslash\{1\}$. Then, the kernel and the cokernel of $\psi\colon B \otimes_A M^G \to M$; $b \otimes x \mapsto b \cdot x$ are killed by $a$.*

PROOF. (1) As in V.11, there exists a canonical resolution $M \xrightarrow{\varepsilon} K^\bullet(G, M)$ as $A$-$G$-modules (Lemma V.11.8) such that $H^i(G, K^q(G, M)) = 0$ $(i > 0,\ q \geq 0)$ (Lemma V.11.9 and Lemma V.11.3). By the $B$-module structure of $M$, the $A$-module structure on $K^q(G, M)$ naturally extends to a $B$-module structure, the action of $G$ on $K^q(G, M)$

becomes $B$-semi-linear, and the above resolution becomes $B$-linear. For any $b \in B$, the composite of the $A$-linear homomorphisms $K^\bullet(G,M)^G \to K^\bullet(G,M); x \mapsto b\cdot x$ and $\mathrm{tr}_G\colon K^\bullet(G,M) \to K^\bullet(G,M)^G$ is the multiplication by $a$. Since $H^q(K^\bullet(G,M)^G) = H^q(G,M)$ and $H^q(K^\bullet(G,M)) = 0$ (if $q>0$), $M$ (if $q=0$), this implies (1).

(2) Define the $A$-linear homomorphism $\varphi\colon M \to B\otimes_A M^G$ by $\varphi(m) = \sum_{i=1}^r b_i \otimes \mathrm{tr}_G(c_i m)$. Then we have $\varphi\circ\psi = a\cdot 1_{B\otimes_A M^G}$ and $\psi\circ\varphi = a\cdot 1_M$. □

**Definition V.12.2.** Let $R$ be a $\overline{V}$-algebra. An $R$-algebra $S$ endowed with an action of $G$ is a *$G$-covering* (of $R$) if there exists an almost faithfully flat homomorphism $R\to R'$ and an almost isomorphism $S' = S\otimes_R R' \overset{\approx}{\to} \prod_G R'$ over $R'$ compatible with the actions of $G$. Here the action of $G$ on $\prod_G R'$ is defined by $g((x_h)_{h\in G}) = (x_{hg})_{h\in G}$ $(g\in G, x_h \in R')$.

**Lemma V.12.3.** *Let $R$ be a $\overline{V}$-algebra, and let $f\colon R\to S$ be a $G$-covering of $R$. Then $f$ is an almost étale covering and* $\mathrm{rk}_R S = |G|$.

PROOF. The first claim follows from Proposition V.7.4 (1) and Corollary V.8.10. Let $R\to R'$ be an almost faithfully flat homomorphism as in Definition V.12.2. Then, by Proposition V.4.2 (1) and the remark after Definition V.5.14, $\mathrm{rk}_{R'}S' = |G|$ where $S' = S\otimes_R R'$. Since $(\wedge_R^{|G|+1}S)\otimes_R R' \cong \wedge_{R'}^{|G|+1}S' \approx 0$, we have $\wedge_R^{|G|+1}S\approx 0$. On the other hand, the kernel of $R\to R'$ is almost zero by Lemma V.8.2, and hence $\mathrm{Hom}_{\overline{V}}(\mathfrak{m},R)\to \mathrm{Hom}_{\overline{V}}(\mathfrak{m},R')$ is injective. By Proposition V.4.2 (2), we obtain $\mathrm{tr}_{S/R}(\mathrm{id}_{\wedge_R^{|G|}S}) = 1$. □

**Lemma V.12.4.** *Let $R$ be a $\overline{V}$-algebra and let $S$ be an $R$-algebra endowed with an action of $G$. Then $S$ is a $G$-covering of $R$ if and only if the homomorphism $S\otimes_R S\to\prod_G S$ defined by $x\otimes y\mapsto (x\cdot g(y))_{g\in G}$ is an almost isomorphism and $S$ is an almost faithfully flat $R$-algebra.*

If we regard $S\otimes_R S$ as an $S$-algebra by $S\to S\otimes_R S; y\mapsto y\otimes 1$ and define the action of $G$ on $S\otimes_R S$ (resp. $\prod_G S$) by $x\otimes y\mapsto x\otimes g(y)$ $(x,y\in S, g\in G)$ (resp. as in Definition V.12.2), then the homomorphism $S\otimes_R S\to\prod_G S$ in Lemma V.12.4 is an $S$-algebra homomorphism compatible with the actions of $G$.

PROOF. By the above remark, the sufficiency is trivial. Conversely suppose that $S$ is a $G$-covering of $R$ and choose a faithfully flat homomorphism $R\to R'$ as in Definition V.12.2. Then, one sees easily that $S'\otimes_{R'}S'\to\prod_G S'$ for the base change $S' := S\otimes_R R'$ is an almost isomorphism and hence it also holds for $S/R$. By Proposition V.6.5 (1), $R'\to S'$ is almost faithfully flat. By Proposition V.8.1, $R\to S$ is almost faithfully flat. □

**Proposition V.12.5.** *Let $R$ be a $\overline{V}$-algebra and let $S$ be a $G$-covering of $R$. Then, for any $S$-module $M$ endowed with a semi-linear action of $G$, the homomorphism $S\otimes_R M^G\to M; y\otimes m\mapsto y\cdot m$ is an almost isomorphism.*

PROOF. The almost isomorphism $S\otimes_R S\overset{\approx}{\to}\prod_G S; x\otimes y\mapsto (x\cdot g(y))_{g\in G}$ (Lemma V.12.4) induces an isomorphism $\mathrm{Hom}_{\overline{V}}(\mathfrak{m},S\otimes_R S)\cong\prod_G\mathrm{Hom}_{\overline{V}}(\mathfrak{m},S)$ by Proposition V.2.5 (1). Let $e$ be the idempotent of $\mathrm{Hom}_{\overline{V}}(\mathfrak{m},S\otimes_R S)$ corresponding to $(\delta_{1,g})_{g\in G}$ in the right-hand side, where $\delta_{1,g}=1$ (if $g=1$), $0$ (otherwise). For any $\alpha\in\Lambda^+$, choose $x_i', y_i'\in S$ $(1\le i\le r)$ such that $\pi^\alpha e = \sum_{1\le i\le r} x_i'\otimes y_i'$ in $\mathrm{Hom}_{\overline{V}}(\mathfrak{m},S\otimes_R S)$ and set $x_i := \pi^\alpha x_i'$, $y_i := \pi^\alpha y_i'$. Then we have $\sum_{1\le i\le r} x_i\cdot y_i = (\pi^\alpha)^3$ and $\sum_{1\le i\le r} x_i\cdot g(y_i y) = 0$ for $g\in G\backslash\{1\}$ and $y\in S$. By applying Lemma V.12.1 (2), we obtain the proposition. □

**Corollary V.12.6.** *Let $R$ be a $\overline{V}$-algebra and let $S$ be a $G$-covering of $R$. Then the natural homomorphism $R\to S^G$ is an almost isomorphism.*

PROOF. By Proposition V.12.5, the homomorphism $S^G \otimes_R S \to S; x \otimes y \mapsto x \cdot y$ is an almost isomorphism, and hence so is $S \to S^G \otimes_R S$. Since $R \to S$ is almost faithfully flat by Lemma V.12.4, we have $R \overset{\approx}{\to} S^G$. □

**Lemma V.12.7.** *Let $R$ be a $\overline{V}$-algebra, let $S$ be a $G$-covering of $R$, and let $\mathrm{tr}_G$ denote the $R$-linear homomorphism $S \to S^G; y \mapsto \sum_{g \in G} g(y)$. Then $\mathrm{tr}_G$ followed by the natural almost isomorphism $S^G \overset{\approx}{\to} \mathrm{Hom}_{\overline{V}}(\mathfrak{m}, S^G)$ coincides with the composition of $\mathrm{tr}_{S/R}\colon S \to \mathrm{Hom}_{\overline{V}}(\mathfrak{m}, R)$ and the isomorphism $\mathrm{Hom}_{\overline{V}}(\mathfrak{m}, R) \overset{\sim}{\to} \mathrm{Hom}_{\overline{V}}(\mathfrak{m}, S^G)$ (Corollary V.12.6, Proposition V.2.5 (1)).*

PROOF. If we regard $S \otimes_R S$ as an $S$-algebra by the homomorphism $S \to S \otimes_R S; y \mapsto y \otimes 1$, there exists an almost isomorphism of $S$-algebras $S \otimes_R S \overset{\approx}{\to} \prod_G S; x \otimes y \mapsto (x \cdot g(y))_{g \in G}$ by Lemma V.12.4. By Propositions V.4.2 and V.4.3, for $y \in S$, the image of $\mathrm{tr}_{S/R}(y)$ in $\mathrm{Hom}_{\overline{V}}(\mathfrak{m}, S)$ is $\mathrm{tr}_{S \otimes_R S/S}(1 \otimes y) = \mathrm{tr}_{\prod_G S/S}((g(y))_{g \in G})$ = the image of $\sum_{g \in G} g(y)$. □

**Proposition V.12.8.** *Let $R$ be a $\overline{V}$-algebra and let $S$ be a $G$-covering of $R$. Then, for any $S$-module $M$ with a semi-linear action of $G$, we have $\mathfrak{m} \cdot H^q(G, M) = 0$ $(q > 0)$ and $\mathfrak{m} \cdot (M^G/\mathrm{tr}_G(M)) = 0$, where $\mathrm{tr}_G$ denotes the homomorphism $M \to M^G; m \mapsto \sum_{g \in G} g(m)$.*

PROOF. This follows from Lemma V.12.1(1), Lemma V.12.7, and Proposition V.7.12. Note that $R \to S$ is almost faithfully flat by Lemma V.12.4. □

**Proposition V.12.9.** *Let $R$ be a normal domain over $\overline{V}$ which is $\pi^\alpha$-torsion free for all $\alpha \in \Lambda^+$, let $L$ be a finite Galois extension of $K := \mathrm{Frac}(R)$, and let $S$ be the normalization of $R$ in $L$. If $R \to S$ is an almost étale covering, then it is a $\mathrm{Gal}(L/K)$-covering.*

PROOF. Set $G := \mathrm{Gal}(L/K)$ and let $R_\pi$ and $S_\pi$ be as in Proposition V.7.11. First let us prove $\mathrm{rank}_R S = |G|$ and hence $S$ is almost faithfully flat over $R$ (Proposition V.6.7). By Proposition V.5.16, we have a decomposition $\mathrm{Hom}_{\overline{V}}(\mathfrak{m}, R) = R_r \times \cdots \times R_0$, $R_r \neq 0$ such that $\mathrm{rank}_{R_i}(S \otimes_R R_i) = i$ $(0 \leq i \leq r)$. By taking $\otimes_R R_\pi$, we obtain a decomposition $R_\pi \cong \mathrm{Hom}_{\overline{V}}(\mathfrak{m}, R) \otimes_R R_\pi = R_{r,\pi} \times \cdots \times R_{0,\pi}$, where $R_{i,\pi} = R_i[\frac{1}{\pi^\alpha}(\alpha \in \Lambda^+)]$. Since $R_i$ are $\pi^\alpha$-torsion free and $R_\pi$ is a domain, we have $R_i = 0$ $(0 \leq i < r)$ and $\mathrm{rank}_R S = r$. Since $R_\pi \to S_\pi$ is finite étale (Propositions V.7.3, V.7.4 (3)), we have $S \otimes_R K = L$ and $\mathrm{rank}_R S = \mathrm{rank}_K L = |G|$. Now, by Lemma V.12.4, it suffices to prove that the homomorphism $\varphi\colon S \otimes_R S \to \prod_G S; x \otimes y \mapsto x \cdot g(y)$ of almost étale coverings of $S$ is an almost isomorphism. After inverting $\pi^\alpha$ $(\alpha \in \Lambda^+)$, this becomes an isomorphism because both sides become normal rings finite étale over $S_\pi$ (**[42]** Proposition (17.5.7)) and the homomorphism becomes the isomorphism $L \otimes_K L \overset{\sim}{\to} \prod_G L; x \to x \cdot g(y)$ after $\otimes_R K$. By Proposition V.7.11, $\varphi$ is an almost isomorphism. □

CHAPTER VI

# Covanishing topos and generalizations

Ahmed Abbes and Michel Gros

## VI.1. Introduction

**VI.1.1.** The aim of this chapter is to consolidate the topogical foundation necessary for Faltings' approach in $p$-adic Hodge theory **[24, 26, 27]**. Its genesis has been motivated by our work on the $p$-adic Simpson correspondence (cf. I–III). Schematically, Faltings' approach consists of two steps. The first, of a local nature, is a generalization of the Galois techniques of Tate, Sen, and Fontaine to certain affine schemes over $p$-adic local fields. It uses, in an essential manner, his theory of almost étale extensions (cf. V). The second step, of a more global nature, links the $p$-adic étale cohomology of certain schemes over $p$-adic local fields with the Galois cohomology studied in the first step. To do this, Faltings uses, on the one hand, the notion of $K(\pi, 1)$-schemes **[24]** and, on the other hand, a new topos **[26]**. The latter has been explicitly defined rather late compared to the rest of the theory (**[26]** page 214), and in our opinion has not received the attention it deserves.

**VI.1.2.** We have benefited from a letter from Deligne to Illusie **[16]** (prior to **[26]**). In this letter, Deligne suggests that the topos that Faltings needs should in reality be a *covanishing* topos, in other words, a special case of the oriented product of topos that he had introduced to develop the formalism of vanishing cycles in greater dimensions **[46]**. The first author of the present work noted in 2008 that the covanishing topology of Deligne differs in general from that considered by Faltings in (**[26]** page 214); however, the latter uses a characterization of sheaves that holds for covanishing topos (cf. VI.1.5).[1] It turns out that this characterization is not in general satisfied by the topos initially considered by Faltings (cf. VI.10.2 for a counterexample). In fact, one of the main sheaves used by Faltings, namely the structural sheaf, is not in general a sheaf for the topology he defined in (**[26]** page 214); but it is a sheaf for the covanishing topology (cf. III.8.16 and III.8.18). In this chapter, we propose to correct the definition of the topos introduced by Faltings and to develop it following Deligne's suggestion. It is this new topos that we will call *Faltings topos*; the one introduced in (**[26]** page 214) seems of little interest.

**VI.1.3.** Since the results of this chapter are rather technical, we will now give a detailed summary of them. Let us begin by recalling the definition of oriented products of topos due to Deligne. Let $f\colon \widetilde{X} \to \widetilde{S}$ and $g\colon \widetilde{Y} \to \widetilde{S}$ be two morphisms of $\mathbb{U}$-topos, where $\mathbb{U}$ is a fixed universe. The *oriented product* $\widetilde{X} \overleftarrow{\times}_{\widetilde{S}} \widetilde{Y}$ is a $\mathbb{U}$-topos endowed with two morphisms

$$\mathrm{p}_1\colon \widetilde{X} \overleftarrow{\times}_{\widetilde{S}} \widetilde{Y} \to \widetilde{X} \quad \text{and} \quad \mathrm{p}_2\colon \widetilde{X} \overleftarrow{\times}_{\widetilde{S}} \widetilde{Y} \to \widetilde{Y} \tag{VI.1.3.1}$$

[1]This problem does not seem to have been observed beforehand.

and a 2-morphism

$$\tau\colon g\mathrm{p}_2 \to f\mathrm{p}_1, \tag{VI.1.3.2}$$

such that the quadruple $(\widetilde{X} \overleftarrow{\times}_{\widetilde{S}} \widetilde{Y}, \mathrm{p}_1, \mathrm{p}_2, \tau)$ is universal in the 2-category of $\mathbb{U}$-topos. We can explicitly construct for it an underlying site $C$ from the data of $\mathbb{U}$-sites $X$, $Y$, and $S$ underlying $\widetilde{X}$, $\widetilde{Y}$, and $\widetilde{S}$, respectively, in which finite inverse limits are representable, and of two continuous left exact functors $f^+\colon S \to X$ and $g^+\colon S \to Y$ defining $f$ and $g$ (cf. VI.3.1 and VI.3.7). Following Illusie, we call $\widetilde{X} \overleftarrow{\times}_{\widetilde{S}} \widetilde{S}$ the *vanishing topos* of $f$, and $\widetilde{S} \overleftarrow{\times}_{\widetilde{S}} \widetilde{Y}$ the *covanishing topos* of $g$. The first topos has been used by Deligne to develop the formalism of vanishing cycles of $f$, and has been studied by Gabber, Illusie **[46]**, Laumon **[53]**, and Orgogozo **[60]**. The second topos is the prototype of the Faltings topos. We can explicitly construct for it another underlying site $D$, simpler than $C$, that we call the *covanishing* site of $g^+$ (cf. VI.4.1 and VI.4.10).

**VI.1.4.** Strictly speaking, the Faltings topos is not in fact a covanishing topos, but a special case of a more general notion that we develop in this chapter, and that we call *generalized covanishing topos*. Let $I$ be a $\mathbb{U}$-site, $\widetilde{I}$ the topos of sheaves of $\mathbb{U}$-sets on $I$,

$$\pi\colon E \to I \tag{VI.1.4.1}$$

a cleaved and normalized fibered category over the category underlying $I$. We suppose that the following conditions are satisfied:

(i) Fibered products are representable in $I$.
(ii) For every $i \in \mathrm{Ob}(I)$, the fiber $E_i$ of $E$ over $i$ is endowed with a topology making it into an $\mathbb{U}$-site, and finite inverse limits are representable in $E_i$. We denote by $\widetilde{E}_i$ the topos of sheaves of $\mathbb{U}$-sets on $E_i$.
(iii) For every morphism $f\colon i \to j$ of $I$, the inverse image functor $f^+\colon E_j \to E_i$ is continuous and left exact. It therefore defines a morphism of topos that we will (abusively) denote also by $f\colon \widetilde{E}_i \to \widetilde{E}_j$.

The functor $\pi$ is in fact a fibered $\mathbb{U}$-site (cf. VI.5.1). We call *covanishing* topology on $E$ the topology generated by the families of coverings $(V_n \to V)_{n\in\Sigma}$ of the following two types:

(v) There exists $i \in \mathrm{Ob}(I)$ such that $(V_n \to V)_{n\in\Sigma}$ is a covering family of $E_i$.
(c) There exists a covering family of morphisms $(f_n\colon i_n \to i)_{n\in\Sigma}$ of $I$ such that $V_n$ is isomorphic to $f_n^+(V)$ for every $n \in \Sigma$.

The coverings of type (v) are called *vertical*, and those of type (c) are called *Cartesian*. The resulting site is called the *covanishing* site associated with the fibered site $\pi$; it is a $\mathbb{U}$-site. We call *covanishing* topos associated with the fibered site $\pi$, and denote by $\widetilde{E}$, the topos of sheaves of $\mathbb{U}$-sets on $E$ (cf. VI.5.3). When $I$ is endowed with the chaotic topology, that is, with the coarsest topology on $I$, we recover the total topology on $E$ associated with the fibered site $\pi$ (cf. VI.5.4).

**VI.1.5.** We show (VI.5.10) that giving a sheaf $F$ on $E$ is equivalent to giving, for every object $i$ of $I$, a sheaf $F_i$ on $E_i$ and, for every morphism $f\colon i \to j$ of $I$, a morphism $F_j \to f_*(F_i)$, these morphisms being subject to compatibility relations, such that for every covering family $(f_n\colon i_n \to i)_{n\in\Sigma}$ of $I$, if for every $(m,n) \in \Sigma^2$, we set $i_{mn} = i_m \times_i i_n$ and we denote by $f_{mn}\colon i_{mn} \to i$ the canonical morphism, the sequence of morphisms of sheaves on $E_i$

$$F_i \to \prod_{n\in\Sigma} (f_n)_*(F_{i_n}) \rightrightarrows \prod_{(m,n)\in\Sigma^2} (f_{mn})_*(F_{i_{mn}}) \tag{VI.1.5.1}$$

is exact. We will, from now on, identify $F$ with the functor $\{i \mapsto F_i\}$ associated with it.

We study the functoriality of the covanishing sites and topos with respect to the fibration $\pi$ (VI.5.18 and VI.5.36) and by base change (VI.5.20). We also establish a number of coherence properties. We show (VI.5.28), among other things, that if there exists a full subcategory $I'$ of $I$, $\mathbb{U}$-small and topologically generating, made up of quasi-compact objects stable under fibered products, such that for every $i \in \mathrm{Ob}(I')$, the topos $\widetilde{E}_i$ is coherent and that for every morphism $f\colon i \to j$ of $I'$, the morphism of topos $f\colon \widetilde{E}_i \to \widetilde{E}_j$ is coherent, then the topos $\widetilde{E}$ is locally coherent. If, moreover, the category $I$ admits a final object that belongs to $I'$, then the topos $\widetilde{E}$ is coherent.

**VI.1.6.** The link with Deligne's covanishing topos is as follows. Let $X$ and $Y$ be two $\mathbb{U}$-sites in which finite inverse limits are representable, and $f^+\colon X \to Y$ a continuous left exact functor. We denote by $\widetilde{X}$ and $\widetilde{Y}$ the topos of sheaves of $\mathbb{U}$-sets on $X$ and $Y$, respectively, and by $f\colon \widetilde{Y} \to \widetilde{X}$ the morphism of topos defined by $f^+$. Consider the category $\mathrm{Fl}(Y)$ of morphisms of $Y$, and the "target functor"

$$\mathrm{Fl}(Y) \to Y, \tag{VI.1.6.1}$$

that makes $\mathrm{Fl}(Y)$ into a cleaved and normalized fibered category over $Y$; the fiber over any $V \in \mathrm{Ob}(Y)$ is canonically equivalent to the category $Y_{/V}$. Endowing each fiber $Y_{/V}$ with the topology induced by that of $Y$, $\mathrm{Fl}(Y)/Y$ becomes a fibered $\mathbb{U}$-site, satisfying the conditions of VI.1.4. Let

$$\pi\colon E \to X \tag{VI.1.6.2}$$

be the fibered site deduced from $\mathrm{Fl}(Y)/Y$ by base change by the functor $f^+$. The covanishing site $E$ associated with $\pi$ (VI.1.4) is canonically equivalent to the covanishing site $D$ associated with the functor $f^+$ (VI.1.3); whence the terminology. The covanishing topos $\widetilde{E}$ associated with $\pi$ is therefore canonically equivalent to the covanishing topos $\widetilde{X} \overleftarrow{\times}_{\widetilde{X}} \widetilde{Y}$ associated with $f$ (cf. VI.5.5).

**VI.1.7.** We keep the assumptions of VI.1.4 and moreover suppose that finite inverse limits are representable in $I$; in view of VI.1.4(i), it is equivalent to requiring $I$ to admit a final object $\iota$, which we assume fixed from now on. We can then define for $\widetilde{E}$ analogues of the canonical projections of the oriented product (VI.1.3.1). On the one hand, the canonical injection functor $\alpha_{\iota !}\colon E_\iota \to E$ is continuous and left exact (cf. VI.5.32). It therefore defines a morphism of topos

$$\beta\colon \widetilde{E} \to \widetilde{E}_\iota, \tag{VI.1.7.1}$$

analogous to the second projection $\mathrm{p}_2$ (VI.1.3.1). On the other hand, we take a final object $e$ of $E_\iota$, which exists by VI.1.4(ii), and which we assume fixed from now on. There then exists essentially a unique Cartesian section of $\pi$ (VI.1.4.1)

$$\sigma^+\colon I \to E \tag{VI.1.7.2}$$

such that $\sigma^+(\iota) = e$. For every $i \in \mathrm{Ob}(I)$, $\sigma^+(i)$ is a final object of $E_i$. One easily verifies that $\sigma^+$ is continuous and left exact (cf. VI.5.32). It therefore defines a morphism of topos

$$\sigma\colon \widetilde{E} \to \widetilde{I}, \tag{VI.1.7.3}$$

analogous to the first projection $\mathrm{p}_1$ (VI.1.3.1).

**VI.1.8.** We keep the assumptions of VI.1.7 and moreover let $V$ be an object of $E$, $c = \pi(V)$, and

$$\varpi\colon E_{/V} \to I_{/c} \qquad \text{(VI.1.8.1)}$$

the functor induced by $\pi$. For every morphism $f\colon i \to c$ of $I$, the fiber of $\varpi$ over the object $(i, f)$ of $I_{/c}$ is canonically equivalent to the category $(E_i)_{/f^+(V)}$. Endowing $I_{/c}$ with the topology induced by that of $I$, and each fiber $(E_i)_{/f^+(V)}$ with the topology induced by that on $E_i$, $\varpi$ becomes a fibered site satisfying the conditions of VI.1.4. We show (VI.5.38) that the covanishing topology on $E_{/V}$ is induced by that on $E$ through the canonical functor $E_{/V} \to E$. In particular, the topos of sheaves of $\mathbb{U}$-sets on the covanishing site $E_{/V}$ is canonically equivalent to $\widetilde{E}_{/\varepsilon(V)}$, where $\varepsilon\colon E \to \widetilde{E}$ is the canonical functor.

**VI.1.9.** One of the main characteristics of Deligne's covanishing topos is the existence of a morphism of *co-nearby cycles* (cf. VI.4.13). This also extends to the generalized covanishing topos introduced in VI.1.4. We keep the assumptions of VI.1.7 and moreover consider a $\mathbb{U}$-site $X$ and a continuous and left exact functor $\Psi^+\colon E \to X$. We denote by $\widetilde{X}$ the topos of sheaves of $\mathbb{U}$-sets on $X$ and by

$$\Psi\colon \widetilde{X} \to \widetilde{E} \qquad \text{(VI.1.9.1)}$$

the morphism of topos associated with $\Psi^+$. We set $u^+ = \Psi^+ \circ \sigma^+\colon I \to X$ and

$$u = \sigma\Psi\colon \widetilde{X} \to \widetilde{I}. \qquad \text{(VI.1.9.2)}$$

For every $i \in \mathrm{Ob}(I)$, the functor $\Psi^+$ induces a functor $\Psi_i^+\colon E_i \to X_{/u^+(i)}$. When we endow $X_{/u^+(i)}$ with the topology induced by that of $X$, $\Psi_i^+$ is exact and left continuous (cf. VI.6.1). It therefore defines a morphism of topos

$$\Psi_i\colon \widetilde{X}_{/u^*(i)} \to \widetilde{E}_i. \qquad \text{(VI.1.9.3)}$$

The morphism $\Psi_\iota$ is none other than the composition $\beta\Psi\colon \widetilde{X} \to \widetilde{E}_\iota$. From the relation $\Psi^*\beta^* = \Psi_\iota^*$, we deduce by adjunction a morphism

$$\beta^* \to \Psi_*\Psi_\iota^*. \qquad \text{(VI.1.9.4)}$$

Generalizing an important property of the covanishing topos of Deligne, we show (VI.6.3) that if for every $i \in \mathrm{Ob}(I)$, the adjunction morphism $\mathrm{id} \to \Psi_{i*}\Psi_i^*$ is an isomorphism, then the adjunction morphisms $\mathrm{id} \to \beta_*\beta^*$ and $\beta^* \to \Psi_*\Psi_\iota^*$ are isomorphisms.

**VI.1.10.** We keep the assumptions of VI.1.9 and moreover suppose that finite inverse limits are representable in $X$. We denote by $\varpi\colon D \to I$ the fibered sited associated with the functor $u^+\colon I \to X$ defined in (VI.1.6), and we endow $D$ with the covanishing topology associated with $\varpi$. We thus obtain the covanishing topos associated with the the functor $u^+$ (VI.1.3), whose topos of sheaves of $\mathbb{U}$-sets is $\widetilde{I} \overleftarrow{\times}_{\widetilde{I}} \widetilde{X}$. By the universal property of oriented products, the morphisms $u\colon \widetilde{X} \to \widetilde{I}$ and $\mathrm{id}_{\widetilde{X}}$ and the 2-morphism $\mathrm{id}_u$ define a morphism of topos, called morphism of *co-nearby cycles* (cf. VI.4.13)

$$\Psi_D\colon \widetilde{X} \to \widetilde{I} \overleftarrow{\times}_{\widetilde{I}} \widetilde{X}. \qquad \text{(VI.1.10.1)}$$

On the other hand, the functors $\Psi_i^+$ for every $i \in \mathrm{Ob}(I)$ define a Cartesian $I$-functor $\rho^+\colon E \to D$. The latter is continuous and left exact (cf. VI.6.4). It therefore defines a morphism of topos

$$\rho\colon \widetilde{I} \overleftarrow{\times}_{\widetilde{I}} \widetilde{X} \to \widetilde{E}. \qquad \text{(VI.1.10.2)}$$

One immediately verifies that the squares of the diagram

(VI.1.10.3)
$$\begin{array}{ccccc}
\widetilde{I} & \xleftarrow{\mathrm{p}_1} & \widetilde{I} \stackrel{\leftarrow}{\times}_{\widetilde{I}} \widetilde{X} & \xrightarrow{\mathrm{p}_2} & \widetilde{X} \\
\| & & \downarrow{\scriptstyle \rho} & & \downarrow{\scriptstyle \Psi_\iota} \\
\widetilde{I} & \xleftarrow{\sigma} & \widetilde{E} & \xrightarrow{\beta} & \widetilde{E}_\iota
\end{array}$$

and the diagram

(VI.1.10.4)
$$\begin{array}{ccc}
\widetilde{X} & \xrightarrow{\Psi_D} & \widetilde{I} \stackrel{\leftarrow}{\times}_{\widetilde{I}} \widetilde{X} \\
 & {\scriptstyle \Psi}\searrow & \downarrow{\scriptstyle \rho} \\
 & & \widetilde{E}
\end{array}$$

are commutative up to canonical isomorphisms.

**VI.1.11.** We keep the assumptions of VI.1.7 and let $R$ be a ring of $\widetilde{E}$; it amounts to giving, for every $i \in \mathrm{Ob}(I)$, a ring $R_i$ of $\widetilde{E}_i$, and for every morphism $f : i \to j$ of $I$, a ring homomorphism $R_j \to f_*(R_i)$, these homomorphisms being subject to compatibility and gluing relations (VI.1.5.1). We develop in VI.8 a number of results on the category of $R$-modules of $\widetilde{E}$, in particular, on the tensor product (VI.8.10) and the sheaf of morphisms (VI.8.12). We also study elementary cohomological invariants of the ringed topos $(\widetilde{E}, R)$. As an intermediary step, we revisit in VI.7 the formalism of ringed total topos ([**2**] VI 8.6). For the sake of simplification, suppose that the category $I$ is $\mathbb{U}$-small (cf. VI.8.4 for the general case). We denote by $\mathbf{Top}(E)$ the total topos associated with the fibered site $\pi$, that is, the topos of sheaves of $\mathbb{U}$-sets on the total site $E$ (cf. VI.7.1). We then have a canonical embedding of topos (VI.5.16.1)

(VI.1.11.1) $$\delta : \widetilde{E} \to \mathbf{Top}(E)$$

such that the functor $\delta_*$ is the canonical inclusion functor of $\widetilde{E}$ in $\mathbf{Top}(E)$. We consider it as a morphism of ringed topos (by $R$ and $\delta_*(R)$, respectively), and we compute in VI.8.5 the right derived functors of $\delta_*$. The Cartan–Leray spectral sequence associated with $\delta$ then gives a spectral sequence that computes the higher direct images of a morphism of ringed generalized covanishing topos (VI.8.8). Note that the Cartan–Leray spectral sequence associated with $\delta$ that computes the cohomology of an $R$-module of $\widetilde{E}$ is the same as that associated with $\beta$ (VI.8.6).

**VI.1.12.** The remainder of the chapter is devoted to studying a special case of the generalized covanishing topos, namely the Faltings topos. As a prelude, we develop in VI.9 a few results on the finite étale topos that we have not been able to find in the literature. For any scheme $X$, we denote by $\mathbf{\acute{E}t}_{/X}$ its étale site, that is, the category of étale schemes over $X$ (elements of $\mathbb{U}$), endowed with the étale topology, and by $X_{\text{ét}}$ the topos of sheaves of $\mathbb{U}$-sets on $\mathbf{\acute{E}t}_{/X}$. We call *finite étale* site of $X$, and denote by $\mathbf{\acute{E}t}_{\mathrm{f}/X}$, the full subcategory of $\mathbf{\acute{E}t}_{/X}$ made up of finite étale covers of $X$, endowed with the topology induced by that of $\mathbf{\acute{E}t}_{/X}$. We call *finite étale* topos of $X$, and denote by $X_{\text{fét}}$, the topos of sheaves of $\mathbb{U}$-sets on $\mathbf{\acute{E}t}_{\mathrm{f}/X}$. The canonical injection functor $\mathbf{\acute{E}t}_{\mathrm{f}/X} \to \mathbf{\acute{E}t}_{/X}$ induces a morphism of topos

(VI.1.12.1) $$\rho_X : X_{\text{ét}} \to X_{\text{fét}}.$$

We show (VI.9.18) that if $X$ is a coherent scheme having a finite number of connected components, then the adjunction morphism $\mathrm{id} \to \rho_{X*}\rho_X^*$ is an isomorphism; in particular, the functor $\rho_X^*\colon X_{\text{fét}} \to X_{\text{ét}}$ is fully faithful. Its essential image consists of the filtered direct limits of locally constant constructible sheaves (VI.9.20); we also establish a variant for abelian torsion sheaves.

We denote by **Sch** the category of schemes elements of $\mathbb{U}$, by $\mathscr{R}$ the category of finite étale covers (that is, the full subcategory of the category of morphisms of **Sch** made up of finite étale covers), and by

(VI.1.12.2) $$\mathscr{R} \to \mathbf{Sch}$$

the "target functor," which makes $\mathscr{R}$ into a cleaved and normalized fibered category over **Sch**; the fiber over a scheme $X$ is canonically equivalent to the category $\mathbf{\acute{E}t}_{\text{f}/X}$. We consider $\mathscr{R}/\mathbf{Sch}$ as a fibered $\mathbb{U}$-site by endowing each fiber with the étale topology.

**VI.1.13.** Let $f\colon Y \to X$ be a morphism of schemes. We call *Faltings fibered site* associated with $f$ the fibered $\mathbb{U}$-site

(VI.1.13.1) $$\pi\colon E \to \mathbf{\acute{E}t}_{/X}$$

deduced from $\mathscr{R}/\mathbf{Sch}$ (VI.1.12.2) by base change by the functor

(VI.1.13.2) $$\mathbf{\acute{E}t}_{/X} \to \mathbf{Sch}, \quad U \mapsto U \times_X Y.$$

We can describe the category $E$ explicitly as follows (cf. VI.10.1). The objects of $E$ are the morphisms of schemes $V \to U$ over $f\colon Y \to X$ such that the morphism $U \to X$ is étale and that the morphism $V \to U_Y = U \times_X Y$ is finite étale. Let $(V' \to U')$, $(V \to U)$ be two objects of $E$. A morphism from $(V' \to U')$ to $(V \to U)$ consists of an $X$-morphism $U' \to U$ and a $Y$-morphism $V' \to V$ such that the diagram

(VI.1.13.3) $$\begin{array}{ccc} V' & \longrightarrow & U' \\ \downarrow & & \downarrow \\ V & \longrightarrow & U \end{array}$$

is commutative. The functor $\pi$ is then defined for every object $(V \to U)$ of $E$ by $\pi(V \to U) = U$. This clearly satisfies the conditions of VI.1.4 and VI.1.7. We can therefore apply to it the constructions developed above. We endow $E$ with the covanishing topology associated with $\pi$. The resulting covanishing site is called the *Faltings site* associated with $f$; it is a $\mathbb{U}$-site. We call *Faltings topos* associated with $f$, and denote by $\widetilde{E}$, the topos of sheaves of $\mathbb{U}$-sets on $E$. In fact, Faltings limits himself to the following case. Let $K$ be a complete discrete valuation ring of characteristic 0, with perfect residue field of characteristic $p > 0$, $\mathscr{O}_K$ the valuation ring of $K$, $\overline{K}$ an algebraic closure of $K$, $X$ a separated $\mathscr{O}_K$-scheme of finite type, and $X^\circ$ an open subscheme of $X$. Faltings only considers the case where $f$ is the canonical morphism $X^\circ_{\overline{K}} \to X$.

Many results on the Faltings site and topos in VI.10 are direct applications of those established in VI.5 and VI.6 for generalized covanishing sites and topos. In particular, we find in this paragraph a study of the functoriality with respect to $f$ (VI.10.12), and of the localization with respect to an object of $E$ (VI.10.14). In the remainder of this introduction, we summarize other more specific properties.

**VI.1.14.** We keep the assumptions of VI.1.13 and moreover suppose that $X$ is quasi-separated. We denote by $\mathbf{\acute{E}t}_{\text{coh}/X}$ (resp. $\mathbf{\acute{E}t}_{\text{scoh}/X}$) the full subcategory of $\mathbf{\acute{E}t}_{/X}$ made up of étale schemes of finite presentation over $X$ (resp. étale separated schemes of finite presentation over $X$), endowed with the topology induced by that on $\mathbf{\acute{E}t}_{/X}$; they are $\mathbb{U}$-small sites. We denote by $\star$ the symbol "coh" or "scoh." Recall that the restriction functor from $X_{\text{ét}}$ to the topos of sheaves of $\mathbb{U}$-sets on $\mathbf{\acute{E}t}_{\star/X}$ is an equivalence of categories. We denote by

$$\pi_\star\colon E_\star \to \mathbf{\acute{E}t}_{\star/X} \tag{VI.1.14.1}$$

the fibered site deduced from $\pi$ by base change by the canonical injection functor

$$\mathbf{\acute{E}t}_{\star/X} \to \mathbf{\acute{E}t}_{/X}, \tag{VI.1.14.2}$$

and by $\Phi\colon E_\star \to E$ the canonical projection. We show (VI.5.21) that if we endow $E_\star$ with the covanishing topology defined by $\pi_\star$ and denote by $\widetilde{E}_\star$ the topos of sheaves of $\mathbb{U}$-sets on $E_\star$, the functor $\Phi$ induces by restriction an equivalence of categories $\widetilde{E} \xrightarrow{\sim} \widetilde{E}_\star$. Moreover, the covanishing topology on $E_\star$ is induced by that on $E$ through the functor $\Phi$ (VI.5.22).

The subcategory $E_{\text{coh}}$ allows us to apply the coherence results established earlier to the Faltings topos. We show (VI.10.5), for example, that if $X$ and $Y$ are coherent, then the topos $\widetilde{E}$ is coherent; in particular, it has enough points. We will see further on the interest of introducing the subcategory $E_{\text{scoh}}$ (VI.1.17).

**VI.1.15.** We keep the assumptions of VI.1.13 and endow $\mathbf{\acute{E}t}_{/X}$ with the final object $X$ and $E$ with the final object $(Y \to X)$. The functors $\alpha_{X!}$ and $\sigma^+$ introduced in VI.1.7 are explicitly defined by

$$\alpha_{X!}\colon \mathbf{\acute{E}t}_{\text{f}/Y} \to E, \qquad V \mapsto (V \to X), \tag{VI.1.15.1}$$

$$\sigma^+\colon \mathbf{\acute{E}t}_{/X} \to E, \qquad U \mapsto (U_Y \to U). \tag{VI.1.15.2}$$

They are left exact and continuous. They therefore define two morphisms of topos

$$\beta\colon \widetilde{E} \to Y_{\text{fét}}, \tag{VI.1.15.3}$$

$$\sigma\colon \widetilde{E} \to X_{\text{ét}}. \tag{VI.1.15.4}$$

On the other hand, the functor

$$\Psi^+\colon E \to \mathbf{\acute{E}t}_{/Y}, \quad (V \to U) \mapsto V \tag{VI.1.15.5}$$

is continuous and left exact (VI.10.7); it therefore defines a morphism of topos

$$\Psi\colon Y_{\text{ét}} \to \widetilde{E}. \tag{VI.1.15.6}$$

We have $f_{\text{ét}} = \sigma\Psi$. For every object $U$ of $\mathbf{\acute{E}t}_{/X}$, we can identify the morphism $\Psi_U$ defined in (VI.1.9.3) with the canonical morphism $\rho_{U_Y}\colon (U_Y)_{\text{ét}} \to (U_Y)_{\text{fét}}$ (VI.1.12.1); in particular, we have $\beta\Psi = \rho_Y$. From the isomorphism $\Psi^*\beta^* = \rho_Y^*$, we deduce by adjunction a morphism

$$\beta^* \to \Psi_*\rho_Y^*. \tag{VI.1.15.7}$$

Suppose, moreover, that $X$ is quasi-separated and that $Y$ is coherent and étale-locally connected (that is, for every étale morphism $X' \to X$, every connected component of $X'$ is an open set of $X'$). We show (VI.10.9) that the adjunction morphisms $\mathrm{id} \to \beta_*\beta^*$ and $\beta^* \to \Psi_*\rho_Y^*$ are isomorphisms.

**VI.1.16.** We keep the assumptions of VI.1.13. We denote by $\varpi\colon D \to \mathbf{\acute{E}t}_{/X}$ the fibered site associated with the inverse image functor $f^+\colon \mathbf{\acute{E}t}_{/X} \to \mathbf{\acute{E}t}_{/Y}$ defined in VI.1.6, and we endow $D$ with the covanishing topology associated with $\varpi$. We thus obtain the covanishing site associated with the functor $f^+$, whose topos of sheaves of $\mathbb{U}$-sets is $X_{\text{ét}} \overleftarrow{\times}_{X_{\text{ét}}} Y_{\text{ét}}$. Every object of $E$ is naturally an object of $D$. We therefore have a fully faithful left exact functor $\rho^+\colon E \to D$ that is none other than the functor of the same name defined in the more general setting of VI.1.10. Since the latter is continuous and left exact, it defines a morphism of topos

(VI.1.16.1) $$\rho\colon X_{\text{ét}} \overset{\leftarrow}{\times}_{X_{\text{ét}}} Y_{\text{ét}} \to \widetilde{E}.$$

We refer to VI.10.15 for more details. We show (VI.10.21) that if $X$ and $Y$ are coherent, then the family of points of $\widetilde{E}$ images by $\rho$ of the points of $X_{\text{ét}} \overset{\leftarrow}{\times}_{X_{\text{ét}}} Y_{\text{ét}}$ is conservative. Note that giving a point of $X_{\text{ét}} \overset{\leftarrow}{\times}_{X_{\text{ét}}} Y_{\text{ét}}$ is equivalent to giving a pair of geometric points $\overline{x}$ of $X$ and $\overline{y}$ of $Y$ and a specialization arrow from $f(\overline{y})$ to $\overline{x}$, that is, an $X$-morphism $\overline{y} \to X_{(\overline{x})}$, where $X_{(\overline{x})}$ denotes the strict localization of $X$ at $\overline{x}$ (cf. VI.10.18).

**VI.1.17.** We keep the assumptions of VI.1.13 and moreover suppose that $X$ is strictly local, with closed point $x$. For any separated étale $X$-scheme of finite presentation $U$, we denote by $U^{\text{f}}$ the disjoint sum of the strict localizations of $U$ at the points of $U_x$; it is an open and closed subscheme of $U$, which is finite over $X$ (cf. VI.10.22). Consider the fibered site

(VI.1.17.1) $$\pi_{\text{scoh}}\colon E_{\text{scoh}} \to \mathbf{\acute{E}t}_{\text{scoh}/X}$$

defined in (VI.1.14.1), and endow $E_{\text{scoh}}$ with the covanishing topology associated with $\pi_{\text{scoh}}$. For every object $(V \to U)$ of $E_{\text{scoh}}$, $V \times_U U^{\text{f}} = V \times_{U_Y} U^{\text{f}}_Y$ is a finite étale cover of $Y$. We thus obtain a functor

(VI.1.17.2) $$\theta^+\colon E_{\text{scoh}} \to \mathbf{\acute{E}t}_{\text{f}/Y}, \quad (V \to U) \mapsto V \times_U U^{\text{f}}.$$

It is continuous and left exact (VI.10.23). It therefore defines a morphism of topos

(VI.1.17.3) $$\theta\colon Y_{\text{fét}} \to \widetilde{E},$$

that we study following the approach introduced in VI.1.9 (cf. VI.10.24). We have a canonical isomorphism $\beta\theta \overset{\sim}{\to} \mathrm{id}_{Y_{\text{fét}}}$, which induces a base change morphism

(VI.1.17.4) $$\beta_* \to \theta^*.$$

We show (VI.10.27) that the base change morphism $\beta_* \to \theta^*$ is an isomorphism; in particular, the functor $\beta_*$ is exact. We deduce from this (VI.10.28) that for every sheaf $F$ of $\widetilde{E}$, the canonical map

(VI.1.17.5) $$\Gamma(\widetilde{E}, F) \to \Gamma(Y_{\text{fét}}, \theta^* F)$$

is bijective, and that for every abelian sheaf $F$ of $\widetilde{E}$, the canonical map

(VI.1.17.6) $$\mathrm{H}^i(\widetilde{E}, F) \to \mathrm{H}^i(Y_{\text{fét}}, \theta^* F)$$

is bijective for every $i \geq 0$.

**VI.1.18.** We keep the assumptions of VI.1.13 and moreover suppose that $f$ is coherent. Let $\overline{x}$ be a geometric point of $X$, $\underline{X}$ the strict localization of $X$ at $\overline{x}$, $\underline{Y} = Y \times_X \underline{X}$, and $\underline{f}\colon \underline{Y} \to \underline{X}$ the canonical projection. We denote by $\underline{E}$ the Faltings site associated with $\underline{f}$, by $\underline{\widetilde{E}}$ the topos of sheaves of $\mathbb{U}$-sets on $\underline{E}$, and by

(VI.1.18.1) $$\theta\colon \underline{Y}_{\text{fét}} \to \underline{\widetilde{E}}$$

the morphism of topos defined in (VI.1.17.3). Since the functor

(VI.1.18.2) $$\Phi^+\colon E \to \underline{E}, \quad (V \to U) \mapsto (V \times_Y \underline{Y} \to U \times_X \underline{X})$$

is continuous and left exact (cf. VI.10.12), it defines a morphism of topos

(VI.1.18.3) $$\Phi\colon \underline{\widetilde{E}} \to \widetilde{E}.$$

We show (VI.10.30) that for every sheaf $F$ of $\widetilde{E}$, we have a functorial canonical isomorphism

$$\sigma_*(F)_{\overline{x}} \xrightarrow{\sim} \Gamma(\underline{Y}_{\text{fét}}, \theta^*(\Phi^* F)); \tag{VI.1.18.4}$$

and for every abelian sheaf $F$ of $\widetilde{E}$ and every integer $i \geq 0$, we have a functorial canonical isomorphism

$$\mathrm{R}^i\sigma_*(F)_{\overline{x}} \xrightarrow{\sim} \mathrm{H}^i(\underline{Y}_{\text{fét}}, \theta^*(\Phi^* F)). \tag{VI.1.18.5}$$

The proof of this result depends on the computation of an inverse limit of Faltings topos (VI.11.3) and its cohomological consequences (VI.11.6).

In VI.10.40 we give a global variant of the isomorphism (VI.1.18.5).

**VI.1.19.** Finally, let us point out that the generalized covanishing topos may be suitable for other applications, including rigid and henselian variants of the Faltings topos. We want to thank L. Illusie for having passed on Deligne's letter **[16]** and his article **[46]**, which have been the main source of inspiration for this work. After we sent him a first version of this work, O. Gabber sent us a copy of an email he had sent to L. Illusie in 2006 in which he defines the covanishing topology for a fibered site over a site and sketches statements that overlap some of our results. We are very grateful for this exchange and the thus offered prospects for further development. A large part of this work was written during the stay of the first author (A.A.) at the University of Tokyo during the autumn of 2010 and the winter of 2011. He wishes to thank this university for its hospitality.

## VI.2. Notation and conventions

*All rings in this chapter have an identity element; all ring homomorphisms map the identity element to the identity element.*

**VI.2.1.** For this entire chapter, we fix a universe $\mathbb{U}$ with an element of infinite cardinality. We call category of $\mathbb{U}$-sets, and denote by **Ens**, the category of sets that are in $\mathbb{U}$. It is a punctual $\mathbb{U}$-topos that we also denote by **Pt** ([**2**] IV 2.2). Unless stated otherwise, the schemes in this chapter are assumed to be elements of the universe $\mathbb{U}$. We denote by **Sch** the category of schemes elements of $\mathbb{U}$.

**VI.2.2.** For a category $\mathscr{C}$, we denote by $\mathrm{Ob}(\mathscr{C})$ the set of its objects, by $\mathscr{C}^\circ$ the opposite category, and for $X, Y \in \mathrm{Ob}(\mathscr{C})$, by $\mathrm{Hom}_{\mathscr{C}}(X, Y)$ (or $\mathrm{Hom}(X, Y)$ when there is no ambiguity) the set of morphisms from $X$ to $Y$.

If $\mathscr{C}$ and $\mathscr{C}'$ are two categories, we denote by $\mathrm{Hom}(\mathscr{C}, \mathscr{C}')$ the set of functors from $\mathscr{C}$ to $\mathscr{C}'$, and by $\mathbf{Hom}(\mathscr{C}, \mathscr{C}')$ the category of functors from $\mathscr{C}$ to $\mathscr{C}'$.

Let $I$ be a category and $\mathscr{C}, \mathscr{C}'$ two categories over $I$ ([**37**] VI 2). We denote by $\mathrm{Hom}_I(\mathscr{C}, \mathscr{C}')$ the set of $I$-functors from $\mathscr{C}$ to $\mathscr{C}'$ and by $\mathrm{Hom}_{\mathrm{cart}/I}(\mathscr{C}, \mathscr{C}')$ the set of Cartesian functors ([**37**] VI 5.2). We denote by $\mathbf{Hom}_I(\mathscr{C}, \mathscr{C}')$ the category of $I$-functors from $\mathscr{C}$ to $\mathscr{C}'$ and by $\mathbf{Hom}_{\mathrm{cart}/I}(\mathscr{C}, \mathscr{C}')$ the full subcategory made up of Cartesian functors.

**VI.2.3.** Let $\mathscr{C}$ be a category. We denote by $\widehat{\mathscr{C}}$ the category of presheaves of $\mathbb{U}$-sets on $\mathscr{C}$, that is, the category of contravariant functors on $\mathscr{C}$ with values in **Ens** ([**2**] I 1.2). If $\mathscr{C}$ is endowed with a topology ([**2**] II 1.1), we denote by $\widetilde{\mathscr{C}}$ the topos of sheaves of $\mathbb{U}$-sets on $\mathscr{C}$ ([**2**] II 2.1).

For an object $F$ of $\widehat{\mathscr{C}}$, we denote by $\mathscr{C}_{/F}$ the following category ([**2**] I 3.4.0). The objects of $\mathscr{C}_{/F}$ are the pairs consisting of an object $X$ of $\mathscr{C}$ and a morphism $u$ from $X$ to

$F$. If $(X, u)$ and $(Y, v)$ are two objects, a morphism from $(X, u)$ to $(Y, v)$ is a morphism $g\colon X \to Y$ such that $u = v \circ g$.

**VI.2.4.** For a ringed topos $(\mathscr{E}, R)$, we denote by $\mathbf{Mod}(R)$ or $\mathbf{Mod}(R, \mathscr{E})$ the category of (left) $R$-modules of $\mathscr{E}$, by $\mathbf{D}(R)$ its derived category, and by $\mathbf{D}^-(R)$, $\mathbf{D}^+(R)$, and $\mathbf{D}^{\mathrm{b}}(R)$ the full subcategories of $\mathbf{D}(R)$ made up of complexes with cohomology bounded from above, from below, and from both sides, respectively.

**VI.2.5.** Following the conventions of ([**2**] VI), we use the adjective *coherent* as a synonym for quasi-compact and quasi-separated.

## VI.3. Oriented products of topos

The notion of oriented products of topos, recalled below, is due to Deligne. It has been studied by Gabber, Illusie, Laumon, and Orgogozo [**46, 53, 60**].

**VI.3.1.** In this section, $X$, $Y$, and $S$ denote $\mathbb{U}$-sites ([**2**] II 3.0.2) in which finite inverse limits are representable, and

(VI.3.1.1) $$f^+\colon S \to X \quad \text{and} \quad g^+\colon S \to Y$$

two continuous left exact functors. We denote by $\widetilde{X}$, $\widetilde{Y}$, and $\widetilde{S}$ the topos of sheaves of $\mathbb{U}$-sets on $X$, $Y$, and $S$, respectively, by

(VI.3.1.2) $$f\colon \widetilde{X} \to \widetilde{S} \quad \text{and} \quad g\colon \widetilde{Y} \to \widetilde{S}$$

the morphisms of topos defined by $f^+$ and $g^+$ ([**2**] IV 4.9.2), respectively, and by $\varepsilon_X\colon X \to \widetilde{X}$, $\varepsilon_Y\colon Y \to \widetilde{Y}$, and $\varepsilon_S\colon S \to \widetilde{S}$ the canonical functors. Let $e_X$, $e_Y$, and $e_S$ be final objects of $X$, $Y$, and $S$, respectively, which exist by assumption. Since the canonical functors are left exact, $\varepsilon_X(e_X)$, $\varepsilon_Y(e_Y)$, and $\varepsilon_S(e_S)$ are final objects of $\widetilde{X}$, $\widetilde{Y}$, and $\widetilde{S}$, respectively.

We denote by $C$ the category of triples

$$(W, U \to f^+(W), V \to g^+(W)),$$

where $W$ is an object of $S$, $U \to f^+(W)$ is a morphism of $X$, and $V \to g^+(W)$ is a morphism of $Y$; such an object will be denoted by $(U \to W \leftarrow V)$. Let $(U \to W \leftarrow V)$ and $(U' \to W' \leftarrow V')$ be two objects of $C$. A morphism from $(U' \to W' \leftarrow V')$ to $(U \to W \leftarrow V)$ consists of three morphisms $U \to U'$, $V \to V'$, and $W \to W'$ of $X$, $Y$, and $S$, respectively, such that the diagrams

(VI.3.1.3) $$\begin{array}{ccc} U' & \longrightarrow & f^+(W') \\ \downarrow & & \downarrow \\ U & \longrightarrow & f^+(W) \end{array} \qquad \begin{array}{ccc} V' & \longrightarrow & g^+(W') \\ \downarrow & & \downarrow \\ V & \longrightarrow & g^+(W) \end{array}$$

are commutative.

It immediately follows from the definition and the fact that the functors $f^+$ and $g^+$ are left exact that finite inverse limits in $C$ are representable.

We endow $C$ with the topology generated by the coverings

$$\{(U_i \to W_i \leftarrow V_i) \to (U \to W \leftarrow V)\}_{i \in I}$$

of the following three types:

(a) $V_i = V$, $W_i = W$ for every $i \in I$, and $(U_i \to U)_{i \in I}$ is a covering family.
(b) $U_i = U$, $W_i = W$ for every $i \in I$, and $(V_i \to V)_{i \in I}$ is a covering family.
(c) $I = \{'\}$, $U' = U$, and the morphism $V' \to V \times_{g^+(W)} g^+(W')$ is an isomorphism (there is no condition on the morphism $W' \to W$).

Note that each of these families is stable under base change. We denote by $\widetilde{C}$ the topos of sheaves of $\mathbb{U}$-sets on $C$. For a presheaf $F$ on $C$, we denote by $F^a$ the associated sheaf.

**Lemma VI.3.2.** *A presheaf $F$ on $C$ is a sheaf if and only if the following conditions are satisfied:*

(i) *For every covering family $(Z_i \to Z)_{i\in I}$ of $C$ of type* (a) *or* (b)*, the sequence*

$$F(Z) \to \prod_{i\in I} F(Z_i) \rightrightarrows \prod_{(i,j)\in I\times J} F(Z_i \times_Z Z_j) \tag{VI.3.2.1}$$

*is exact.*

(ii) *For every covering $(U \to W' \leftarrow V') \to (U \to W \leftarrow V)$ of type* (c)*, the map*

$$F(U \to W \leftarrow V) \to F(U \to W' \leftarrow V') \tag{VI.3.2.2}$$

*is bijective.*

Indeed, after extending the universe $\mathbb{U}$, if necessary, we may assume that the categories $X$ and $Y$ are small ([**2**] II 2.7(2)). For every covering $(U \to W' \leftarrow V') \to (U \to W \leftarrow V)$ of type (c), the diagonal morphism

$$(U \to W' \leftarrow V') \to (U \to W' \times_W W' \leftarrow V' \times_V V')$$

is a covering of type (c) that equalizes the two canonical projections

$$(U \to W' \times_W W' \leftarrow V' \times_V V') \rightrightarrows (U \to W' \leftarrow V').$$

The proposition therefore follows from ([**2**] II 2.3, I 3.5, and I 2.12).

**Remark VI.3.3.** It follows from VI.3.2 that the morphism of sheaves associated with a covering of type (c) is an isomorphism; in particular, the topology on $C$ is not always coarser than the canonical topology.

**VI.3.4.** The functors

$$\mathrm{p}_1^+ \colon X \to C, \quad U \mapsto (U \to e_Z \leftarrow e_Y), \tag{VI.3.4.1}$$

$$\mathrm{p}_2^+ \colon Y \to C, \quad V \mapsto (e_X \to e_Z \leftarrow V), \tag{VI.3.4.2}$$

are left exact and continuous ([**2**] III 1.6). They therefore define two morphisms of topos ([**2**] IV 4.9.2)

$$\mathrm{p}_1 \colon \widetilde{C} \to \widetilde{X}, \tag{VI.3.4.3}$$

$$\mathrm{p}_2 \colon \widetilde{C} \to \widetilde{Y}. \tag{VI.3.4.4}$$

On the other hand, we have a 2-morphism

$$\tau \colon g\mathrm{p}_2 \to f\mathrm{p}_1, \tag{VI.3.4.5}$$

given by the following morphism of functors $(g\mathrm{p}_2)_* \to (f\mathrm{p}_1)_*$: for every sheaf $F$ on $C$ and every $W \in \mathrm{Ob}(S)$,

$$g_*(\mathrm{p}_{2*}(F))(W) \to f_*(\mathrm{p}_{1*}(F))(W) \tag{VI.3.4.6}$$

is the composition

$$F(e_X \to e_Z \leftarrow g^+(W)) \to F(f^+(W) \to W \leftarrow g^+(W)) \to F(f^+(W) \to e_Z \leftarrow e_Y),$$

where the first arrow is the canonical morphism and the second arrow is the inverse of the isomorphism (VI.3.2.2).

**Remark VI.3.5.** For every $W \in \mathrm{Ob}(S)$, the morphism $\tau\colon (f\mathrm{p}_1)^*(W) \to (g\mathrm{p}_2)^*(W)$ (VI.3.4.5) is the morphism

(VI.3.5.1) $$(\mathrm{p}_1^+(f^+W))^a \to (\mathrm{p}_2^+(g^+W))^a,$$

composed of

$$(f^+W \to e_Z \leftarrow e_Y)^a \to (f^+W \to W \leftarrow g^+W)^a \to (e_X \to e_Z \leftarrow g^+W)^a,$$

where the first arrow is the inverse of the canonical isomorphism (VI.3.3) and the second arrow is the canonical morphism.

**Lemma VI.3.6.** *For every object $Z = (U \to W \leftarrow V)$ of $C$, we have a Cartesian diagram*

(VI.3.6.1) $$\begin{array}{ccc} Z^a & \longrightarrow & \mathrm{p}_2^*(V) \\ \downarrow & & \downarrow \\ \mathrm{p}_1^*(U) & \xrightarrow{\ i\ } & (g\mathrm{p}_2)^*(W) \end{array}$$

*where the unlabeled arrows are the canonical morphisms and $i$ is the composition of the morphism $\mathrm{p}_1^*(U) \to (f\mathrm{p}_1)^*(W)$ and the morphism $\tau\colon (f\mathrm{p}_1)^*(W) \to (g\mathrm{p}_2)^*(W)$ (VI.3.4.5).*

We have a commutative canonical diagram of $C$ with Cartesian squares

(VI.3.6.2) $$\begin{array}{ccccc} Z & \longrightarrow & (f^+W \to W \leftarrow V) & \longrightarrow & \mathrm{p}_2^+V \\ \downarrow & & \downarrow & & \downarrow \\ (U \to W \leftarrow g^+W) & \longrightarrow & (f^+W \to W \leftarrow g^+W) & \xrightarrow{\ v\ } & \mathrm{p}_2^+(g^+W) \\ \downarrow & & \downarrow{\scriptstyle u} & & \\ \mathrm{p}_1^+U & \longrightarrow & \mathrm{p}_1^+(f^+W) & & \end{array}$$

On the other hand, $u^a$ is an isomorphism (VI.3.3) and $\tau\colon (f\mathrm{p}_1)^*W \to (g\mathrm{p}_2)^*W$ is equal to $v^a \circ (u^a)^{-1}$ (VI.3.5). Since the canonical functor $C \to \widetilde{C}$ is left exact, we deduce from this that the diagram (VI.3.6.1) is Cartesian.

**Theorem VI.3.7.** *Let $T$ be a $\mathbb{U}$-topos, $a\colon T \to \widetilde{X}$, $b\colon T \to \widetilde{Y}$ two morphisms of topos, and $t\colon gb \to fa$ a 2-morphism. Then there exists a triple*

$$(h\colon T \to \widetilde{C}, \alpha\colon \mathrm{p}_1 h \xrightarrow{\sim} a, \beta\colon \mathrm{p}_2 h \xrightarrow{\sim} b),$$

*unique up to unique isomorphism, consisting of a morphism of topos $h$ and two isomorphisms of morphisms of topos $\alpha$ and $\beta$, such that the diagram*

(VI.3.7.1) $$\begin{array}{ccc} g\mathrm{p}_2 h & \xrightarrow{\ \tau * h\ } & f\mathrm{p}_1 h \\ {\scriptstyle g*\beta}\downarrow & & \downarrow{\scriptstyle f*\alpha} \\ gb & \xrightarrow{\ t\ } & fa \end{array}$$

*is commutative.*

The uniqueness of $(h, \alpha, \beta)$ is clear. Indeed, by VI.3.6, the "restriction" $h^+\colon C \to T$ of the functor $h^*$ to $C$ is necessarily given, for any object $Z = (U \to W \leftarrow V)$ of $C$, by

(VI.3.7.2) $$h^+(Z) = a^*(U) \times_{(gb)^*(W)} b^*(V),$$

where the morphism $a^*(U) \to (gb)^*(W)$ is the composition

$$a^*(U) \to (fa)^*(W) \xrightarrow{t} (gb)^*(W).$$

Let us show that the resulting functor $h^+$ is a morphism of sites and that the associated morphism of topos answers the question. The functor $h^+$ is clearly left exact and transforms covering families of $C$ of type (a) or (b) into covering families of $T$. On the other hand, if

$$(U \to W' \leftarrow V') \to (U \to W \leftarrow V)$$

is a covering of $C$ of type (c), the square

$$\text{(VI.3.7.3)} \qquad \begin{array}{ccc} b^*V' & \longrightarrow & b^*V \\ \downarrow & & \downarrow \\ b^*(g^+W') & \longrightarrow & b^*(g^+W) \end{array}$$

is Cartesian, and consequently the morphism

$$\text{(VI.3.7.4)} \qquad h^+((U \to W' \leftarrow V')) \to h^+((U \to W \leftarrow V))$$

is an isomorphism. Hence the functor $h^+$ is continuous by virtue of VI.3.2. We deduce from this that $h^+$ is a morphism of sites ([**2**] IV 4.9.4); it therefore defines a morphism of topos $h\colon T \to \widetilde{C}$.

We have canonical isomorphisms $\alpha\colon a^* \xrightarrow{\sim} h^*\mathrm{p}_1^*$ and $\beta\colon b^* \xrightarrow{\sim} h^*\mathrm{p}_2^*$ whose "restrictions" to $X$ and $Y$, respectively, are the tautological isomorphisms. To verify that the diagram (VI.3.7.1) is commutative, it suffices to show that its "restriction" to $S$ is. For every $W \in \mathrm{Ob}(S)$, consider the diagram

$$\text{(VI.3.7.5)} \qquad \begin{array}{ccccc} h^+((f^+W \to W \leftarrow g^+W)) & \xrightarrow{u} & (fa)^*(W) & \xrightarrow{t} & (gb)^*(W) \\ \| & & \downarrow{\scriptstyle \alpha(f^*W)} & & \downarrow{\scriptstyle \beta(g^*W)} \\ h^+((f^+W \to W \leftarrow g^+W)) & \xrightarrow{v} & h^*((f\mathrm{p}_1)^*W) & \xrightarrow{h^*(\tau)} & h^*((g\mathrm{p}_2)^*W) \end{array}$$

where $u$ is the projection deduced from the formula (VI.3.7.2) and $v$ is the canonical morphism. By definition, we have $v = \alpha(f^*W) \circ u$. On the other hand, $u' = t \circ u$ is the projection deduced from the formula (VI.3.7.2) and $v' = h^*(\tau) \circ v$ is the canonical morphism. By definition, we have $v' = \beta(g^*W) \circ u'$. Since $u$ and $v$ are isomorphisms, we deduce from this that the right square in (VI.3.7.5) is commutative. Hence the diagram (VI.3.7.1) is commutative.

**VI.3.8.** Let $X'$, $Y'$, $S'$ be three $\mathbb{U}$-sites in which finite inverse limits are representable, and $f'^+\colon S' \to X'$, $g'^+\colon S' \to Y'$ two continuous left exact functors. We denote by $\widetilde{X}'$, $\widetilde{Y}'$, and $\widetilde{S}'$ the topos of sheaves of $\mathbb{U}$-sets on $X'$, $Y'$, and $S'$, respectively, and by

$$\text{(VI.3.8.1)} \qquad f'\colon \widetilde{X}' \to \widetilde{S}' \quad \text{and} \quad g'\colon \widetilde{Y}' \to \widetilde{S}'$$

the morphisms of topos defined by $f'^+$ and $g'^+$. We denote by $C'$ the site associated with the functors $(f'^+, g'^+)$ defined in (VI.3.1), by $\widetilde{C}'$ the topos of sheaves of $\mathbb{U}$-sets on $C'$, by $\mathrm{p}'_1\colon \widetilde{C}' \to \widetilde{X}'$ and $\mathrm{p}'_2\colon \widetilde{C}' \to \widetilde{Y}'$ the canonical projections, and by $\tau'\colon g'\mathrm{p}'_2 \to f'\mathrm{p}'_1$

the canonical 2-morphism (VI.3.4). Consider a diagram of morphisms of topos

(VI.3.8.2)
$$\begin{array}{ccccc} \widetilde{X}' & \xrightarrow{f'} & \widetilde{S}' & \xleftarrow{g'} & \widetilde{Y}' \\ {\scriptstyle u}\downarrow & & \downarrow{\scriptstyle w} & & \downarrow{\scriptstyle v} \\ \widetilde{X} & \xrightarrow{f} & \widetilde{S} & \xleftarrow{g} & \widetilde{Y} \end{array}$$

and two 2-morphisms

(VI.3.8.3)
$$a\colon wf' \to fu \quad \text{and} \quad b\colon gv \to wg'.$$

By VI.3.7, the morphisms of topos $u\mathrm{p}'_1\colon \widetilde{C}' \to \widetilde{X}$ and $v\mathrm{p}'_2\colon \widetilde{C}' \to \widetilde{Y}$ and the composed 2-morphism $t$

(VI.3.8.4)
$$gv\mathrm{p}'_2 \xrightarrow{b*\mathrm{p}'_2} wg'\mathrm{p}'_2 \xrightarrow{h*\tau'} wf'\mathrm{p}'_1 \xrightarrow{a*\mathrm{p}'_1} fu\mathrm{p}'_1\,,$$

define a morphism of topos

(VI.3.8.5)
$$h\colon \widetilde{C}' \to \widetilde{C}$$

and 2-isomorphisms $\alpha\colon \mathrm{p}_1 h \xrightarrow{\sim} u\mathrm{p}'_1$ and $\beta\colon \mathrm{p}_2 h \xrightarrow{\sim} v\mathrm{p}'_2$ making the following diagram commutative:

(VI.3.8.6)
$$\begin{array}{ccc} g\mathrm{p}_2 h & \xrightarrow{\tau*h} & f\mathrm{p}_1 h \\ {\scriptstyle g*\beta}\downarrow & & \downarrow{\scriptstyle f*\alpha} \\ gv\mathrm{p}'_2 & \xrightarrow{t} & fu\mathrm{p}'_1 \end{array}$$

**Corollary VI.3.9.** *Under the assumptions of* VI.3.8, *if $u$, $v$, and $w$ are equivalences of topos, and $a$ and $b$ are 2-isomorphisms, then $h$ is an equivalence of topos.*

This follows from VI.3.7.

It follows from VI.3.9 that the topos $\widetilde{C}$ depends only on the pair of morphisms of topos $(f, g)$, up to canonical equivalence. This justifies the following terminology and notation.

**Definition VI.3.10.** The topos $\widetilde{C}$ is called the *oriented product* of $\widetilde{X}$ and $\widetilde{Y}$ over $\widetilde{S}$, and denoted by $\widetilde{X} \overleftarrow{\times}_{\widetilde{S}} \widetilde{Y}$.

Under the assumptions of VI.3.8, we denote the morphism $h$ (VI.3.8.5) by

(VI.3.10.1)
$$u \overleftarrow{\times}_w v\colon \widetilde{X}' \overleftarrow{\times}_{\widetilde{S}'} \widetilde{Y}' \to \widetilde{X} \overleftarrow{\times}_{\widetilde{S}} \widetilde{Y}.$$

**Corollary VI.3.11.** *Giving a point of $\widetilde{X} \overleftarrow{\times}_{\widetilde{S}} \widetilde{Y}$ is equivalent to giving a pair of points $x\colon \mathbf{Pt} \to \widetilde{X}$ and $y\colon \mathbf{Pt} \to \widetilde{Y}$ and a 2-morphism $u\colon gy \to fx$.*

This follows from VI.3.7.

**Definition VI.3.12** ([**46**] 4.1). The topos $\widetilde{X} \overleftarrow{\times}_{\widetilde{S}} \widetilde{S}$ is called the *vanishing* topos of $f$, and the topos $\widetilde{S} \overleftarrow{\times}_{\widetilde{S}} \widetilde{Y}$ is called the *covanishing* topos of $g$.

We will give in VI.4.10 a simpler description of $\widetilde{S} \overleftarrow{\times}_{\widetilde{S}} \widetilde{Y}$ due to Deligne ([**46**] 4.6).

**VI.3.13.** We consider $\widetilde{X}$, $\widetilde{Y}$, and $\widetilde{S}$ as $\mathbb{U}$-sites endowed with the canonical topologies; the associated $\mathbb{U}$-topos then identify with $\widetilde{X}$, $\widetilde{Y}$, and $\widetilde{S}$, respectively ([**2**] IV 1.2). On the other hand, the functors $f^*\colon \widetilde{S} \to \widetilde{X}$ and $g^*\colon \widetilde{S} \to \widetilde{Y}$ are clearly continuous and left exact. We can therefore consider the site $C^\dagger$ associated with $(f^*, g^*)$ defined in VI.3.1. For the moment, denote by $\widetilde{C}^\dagger$ the topos of sheaves of $\mathbb{U}$-sets on $C^\dagger$, by $\pi_1^+\colon \widetilde{X} \to C^\dagger$ and $\pi_2^+\colon \widetilde{Y} \to C^\dagger$ the functors defined in (VI.3.4.1) and (VI.3.4.2), by

$$\pi_1\colon \widetilde{C}^\dagger \to \widetilde{X}, \tag{VI.3.13.1}$$

$$\pi_2\colon \widetilde{C}^\dagger \to \widetilde{Y}, \tag{VI.3.13.2}$$

the associated morphisms of topos, and by $\nu\colon g\pi_2 \to f\pi_1$ the 2-morphism defined in (VI.3.4.5).

The canonical functors $\varepsilon_X$, $\varepsilon_Y$, and $\varepsilon_S$ induce a left exact functor

$$\varphi^+\colon C \to C^\dagger. \tag{VI.3.13.3}$$

This transforms coverings of $C$ of type (a) (resp. (b), resp. (c)) into coverings of $C^\dagger$ of the same type. It then follows from VI.3.2 that for every sheaf $F$ on $C^\dagger$, $F \circ \varphi^+$ is a sheaf on $C$. Consequently, $\varphi^+$ is continuous. It therefore defines a morphism of topos

$$\varphi\colon \widetilde{C}^\dagger \to \widetilde{C}. \tag{VI.3.13.4}$$

We have canonical isomorphisms

$$\varphi^+ \circ \mathrm{p}_1^+ \xrightarrow{\sim} \pi_1^+ \circ \varepsilon_X \quad \text{and} \quad \varphi^+ \circ \mathrm{p}_2^+ \xrightarrow{\sim} \pi_2^+ \circ \varepsilon_Y.$$

We deduce from this isomorphisms

$$\pi_1 \xrightarrow{\sim} \mathrm{p}_1\varphi \quad \text{and} \quad \pi_2 \xrightarrow{\sim} \mathrm{p}_2\varphi. \tag{VI.3.13.5}$$

Moreover, it immediately follows from the definition (VI.3.4.5) that $\tau * \varphi$ identifies with $\nu$. Consequently, $\varphi$ is an equivalence of topos by virtue of VI.3.7. Indeed, $\varphi$ is the morphism of topos (VI.3.8.5) defined in VI.3.8 by taking for $u$, $v$, and $w$ the identity morphisms of $\widetilde{X}$, $\widetilde{Y}$, and $\widetilde{S}$, respectively. From now on, we identify $\widetilde{C}^\dagger$ with the topos $\widetilde{X} \overleftarrow{\times}_{\widetilde{S}} \widetilde{Y}$ through the equivalence $\varphi$, the morphism $\pi_1$ (resp. $\pi_2$) with $\mathrm{p}_1$ (resp. $\mathrm{p}_2$), and the 2-morphism $\nu$ with $\tau$.

**VI.3.14.** Let $(F \to H \leftarrow G)$ be an object of $C^\dagger$ (VI.3.13). Recall ([**2**] IV 5.1) that the category $\widetilde{X}_{/F}$ is a $\mathbb{U}$-topos, called the topos induced on $F$ by $\widetilde{X}$, and that we have a canonical morphism $j_F\colon \widetilde{X}_{/F} \to \widetilde{X}$, called the localization morphism of $\widetilde{X}$ at $F$. Likewise, we have localization morphisms $j_G\colon \widetilde{Y}_{/G} \to \widetilde{Y}$ and $j_H\colon \widetilde{S}_{/H} \to \widetilde{S}$. Denote by $f'$ the composition

$$\widetilde{X}_{/F} \longrightarrow \widetilde{X}_{/f^*(H)} \xrightarrow{f_{/H}} \widetilde{S}_{/H}, \tag{VI.3.14.1}$$

where the first arrow is the localization morphism of $\widetilde{X}_{/f^*(H)}$ at $F \to f^*(H)$ ([**2**] IV 5.5), and the second arrow is the morphism deduced from $f$ ([**2**] IV 5.10). We define the morphism $g'\colon \widetilde{Y}_{/G} \to \widetilde{S}_{/H}$ likewise. The squares of the diagram

$$\begin{array}{ccccc} \widetilde{X}_{/F} & \xrightarrow{f'} & \widetilde{S}_{/H} & \xleftarrow{g'} & \widetilde{Y}_{/G} \\ {\scriptstyle j_F}\downarrow & & \downarrow{\scriptstyle j_H} & & \downarrow{\scriptstyle j_G} \\ \widetilde{X} & \xrightarrow{f} & \widetilde{S} & \xleftarrow{g} & \widetilde{Y} \end{array} \tag{VI.3.14.2}$$

are commutative up to canonical isomorphisms. We denote by $\widetilde{X}_{/F} \overleftarrow{\times}_{\widetilde{S}_{/H}} \widetilde{Y}_{/G}$ the oriented product of $\widetilde{X}_{/F}$ and $\widetilde{Y}_{/G}$ over $\widetilde{S}_{/H}$. The diagram (VI.3.14.2) then induces a canonical morphism (VI.3.10.1)

(VI.3.14.3) $$j_F \overleftarrow{\times}_{j_H} j_G \colon \widetilde{X}_{/F} \overleftarrow{\times}_{\widetilde{S}_{/H}} \widetilde{Y}_{/G} \to \widetilde{X} \overleftarrow{\times}_{\widetilde{S}} \widetilde{Y}.$$

We denote by $F \overleftarrow{\times}_H G$ the sheaf of $\widetilde{X} \overleftarrow{\times}_{\widetilde{S}} \widetilde{Y}$ associated with $(F \to H \leftarrow G)$ (cf. VI.3.13). By VI.3.6, we have a canonical isomorphism

(VI.3.14.4) $$F \overleftarrow{\times}_H G \xrightarrow{\sim} \mathrm{p}_1^*(F) \times_{(g\mathrm{p}_2)^*(H)} \mathrm{p}_2^*(G),$$

where the morphism $\mathrm{p}_1^*(F) \to (g\mathrm{p}_2)^*(H)$ is the composition of the morphism $\mathrm{p}_1^*(F) \to (f\mathrm{p}_1)^*(H)$ and the morphism $\tau \colon (f\mathrm{p}_1)^*(H) \to (g\mathrm{p}_2)^*(H)$ (VI.3.4.5). We denote by

(VI.3.14.5) $$j_{F \overleftarrow{\times}_H G} \colon (\widetilde{X} \overleftarrow{\times}_{\widetilde{S}} \widetilde{Y})_{/(F \overleftarrow{\times}_H G)} \to \widetilde{X} \overleftarrow{\times}_S \widetilde{Y}$$

the localization morphism of $\widetilde{X} \overleftarrow{\times}_{\widetilde{S}} \widetilde{Y}$ at $F \overleftarrow{\times}_H G$.

By constructions analogous to (VI.3.14.1), the morphisms $\mathrm{p}_1$ and $\mathrm{p}_2$ induce morphisms

(VI.3.14.6) $$q_1 \colon (\widetilde{X} \overleftarrow{\times}_{\widetilde{S}} \widetilde{Y})_{/(F \overleftarrow{\times}_H G)} \to \widetilde{X}_{/F},$$

(VI.3.14.7) $$q_2 \colon (\widetilde{X} \overleftarrow{\times}_{\widetilde{S}} \widetilde{Y})_{/(F \overleftarrow{\times}_H G)} \to \widetilde{Y}_{/G},$$

that fit into a diagram with commutative squares up to canonical isomorphisms

(VI.3.14.8) $$\begin{array}{ccccc} \widetilde{X}_{/F} & \xleftarrow{q_1} & (\widetilde{X} \overleftarrow{\times}_{\widetilde{S}} \widetilde{Y})_{/(F \overleftarrow{\times}_H G)} & \xrightarrow{q_2} & \widetilde{Y}_{/G} \\ \Big\downarrow{\scriptstyle j_F} & & \Big\downarrow{\scriptstyle j_{F \overleftarrow{\times}_H G}} & & \Big\downarrow{\scriptstyle j_G} \\ \widetilde{X} & \xleftarrow{\mathrm{p}_1} & \widetilde{X} \overleftarrow{\times}_{\widetilde{S}} \widetilde{Y} & \xrightarrow{\mathrm{p}_2} & \widetilde{Y} \end{array}$$

The 2-morphism $\tau \colon g\mathrm{p}_2 \to f\mathrm{p}_1$ (VI.3.4.5) and the commutative diagram

(VI.3.14.9) $$\begin{array}{ccccc} F \overleftarrow{\times}_H G & & \longrightarrow & & \mathrm{p}_2^* G \\ \Big\downarrow & & & & \Big\downarrow \\ \mathrm{p}_1^* F & \longrightarrow & \mathrm{p}_1^*(f^* H) & \xrightarrow{\tau} & \mathrm{p}_2^*(g^* H) \end{array}$$

induce, for every $L \in \mathrm{Ob}(\widetilde{S}_{/H})$, a functorial morphism

(VI.3.14.10) $$\mathrm{p}_1^*(f^* L) \times_{\mathrm{p}_1^*(f^* H)} (F \overleftarrow{\times}_H G) \to \mathrm{p}_2^*(g^* L) \times_{\mathrm{p}_2^*(g^* H)} (F \overleftarrow{\times}_H G).$$

Since $\mathrm{p}_1^*$ and $\mathrm{p}_2^*$ are left exact, we obtain a 2-morphism

(VI.3.14.11) $$\tau' \colon g' q_2 \to f' q_1$$

such that $j_H * \tau' = \tau * j_{F \overleftarrow{\times}_H G}$. By VI.3.7, $q_1$, $q_2$, and $\tau'$ define a morphism

(VI.3.14.12) $$m \colon (\widetilde{X} \overleftarrow{\times}_{\widetilde{S}} \widetilde{Y})_{/(F \overleftarrow{\times}_H G)} \to \widetilde{X}_{/F} \overleftarrow{\times}_{\widetilde{S}_{/H}} \widetilde{Y}_{/G}.$$

**Proposition VI.3.15.** *The morphism $m$ is an equivalence of topos, and we have a canonical isomorphism*

(VI.3.15.1) $$(j_F \overleftarrow{\times}_{j_H} j_G) \circ m \xrightarrow{\sim} j_{F \overleftarrow{\times}_H G}.$$

First note that the isomorphism (VI.3.15.1) follows from VI.3.7, in view of (VI.3.14.8) and the relation $j_H * \tau' = \tau * j_{F \overleftarrow{\times}_H G}$. Denote by

$$j_{(F \to H \leftarrow G)} \colon C^\dagger_{/(F \to H \leftarrow G)} \to C^\dagger \tag{VI.3.15.2}$$

the canonical functor, by $\mathscr{T}$ the topology of $C^\dagger$ (VI.3.13), and by $\mathscr{T}_1$ the topology of $C^\dagger_{/(F \to H \leftarrow G)}$ induced by $\mathscr{T}$ via $j_{(F \to H \leftarrow G)}$. A family $(L_i \to L)_{i \in I}$ of morphisms of $C^\dagger_{/(F \to H \leftarrow G)}$ is covering for $\mathscr{T}_1$ if and only if its image by $j_{(F \to H \leftarrow G)}$ is covering in $C^\dagger$ for $\mathscr{T}$ ([**2**] III 5.2(1)). By virtue of ([**2**] III 5.4), $(\widetilde{X} \overleftarrow{\times}_{\widetilde{S}} \widetilde{Y})_{/(F \overleftarrow{\times}_H G)}$ is canonically equivalent to the topos of sheaves of $\mathbb{U}$-sets on the site $(C^\dagger_{/(F \to H \leftarrow G)}, \mathscr{T}_1)$.

To define a site underlying the topos $\widetilde{X}_{/F} \overleftarrow{\times}_{\widetilde{S}_{/H}} \widetilde{Y}_{/G}$, we consider $\widetilde{X}_{/F}$, $\widetilde{Y}_{/G}$, and $\widetilde{S}_{/H}$ as $\mathbb{U}$-sites endowed with the canonical topologies (cf. VI.3.13). For all $F' \in \mathrm{Ob}(\widetilde{X}_{/F})$ and $H' \in \mathrm{Ob}(\widetilde{S}_{/H})$, taking a morphism $F' \to f'^*(H')$ of $\widetilde{X}_{/F}$ corresponds to taking a morphism $F' \to f^*(H')$ of $\widetilde{X}$ over $F \to f^*(H)$. Consequently, the site associated with the pair of functors $f'^* \colon \widetilde{S}_{/H} \to \widetilde{X}_{/F}$ and $g'^* \colon \widetilde{S}_{/H} \to \widetilde{Y}_{/G}$ defined in VI.3.1 identifies canonically with the category $C^\dagger_{/(F \to H \leftarrow G)}$, endowed with a topology $\mathscr{T}_2$, which is a priori coarser than the topology $\mathscr{T}_1$. The identity functor of $C^\dagger_{/(F \to H \leftarrow G)}$ then defines a morphism of sites

$$\mathrm{id} \colon (C^\dagger_{/(F \to H \leftarrow G)}, \mathscr{T}_1) \to (C^\dagger_{/(F \to H \leftarrow G)}, \mathscr{T}_2). \tag{VI.3.15.3}$$

Let us show that $m$ is the morphism of topos associated with (VI.3.15.3).

It follows from the proof of VI.3.7, in particular from (VI.3.7.2), that the restriction

$$m^+ \colon C^\dagger_{/(F \to H \leftarrow G)} \to (\widetilde{X} \overleftarrow{\times}_{\widetilde{Z}} \widetilde{Y})_{/(F \overleftarrow{\times}_H G)} \tag{VI.3.15.4}$$

of the functor $m^*$ is given, for every object $(F' \to H' \leftarrow G')$ of $C^\dagger_{/(F \to H \leftarrow G)}$, by

$$m^+((F' \to H' \leftarrow G')) = q_1^*(F') \times_{(g'q_2)^*(H')} q_2^*(G'), \tag{VI.3.15.5}$$

where the morphism $q_1^*(F') \to (g'q_2)^*(H')$ is the composition of the morphism $q_1^*(F') \to (f'q_1)^*(H')$ and the morphism $\tau' \colon (f'q_1')^*(H') \to (g'q_2)^*(H')$ (VI.3.14.11). We have canonical isomorphisms

$$q_1^*(F') \simeq \mathrm{p}_1^*(F') \times_{\mathrm{p}_1^*(F)} (F \overleftarrow{\times}_H G), \tag{VI.3.15.6}$$

$$q_2^*(G') \simeq \mathrm{p}_2^*(G') \times_{\mathrm{p}_2^*(G)} (F \overleftarrow{\times}_H G), \tag{VI.3.15.7}$$

$$(g'q_2)^*(H') \simeq (g\mathrm{p}_2)^*(H') \times_{(g\mathrm{p}_2)^*(H)} (F \overleftarrow{\times}_H G). \tag{VI.3.15.8}$$

Moreover, in view of (VI.3.14.9), the morphism $q_1^*(F') \to (g'q_2)^*(H')$ comes from the composition

$$\mathrm{p}_1^*(F') \to (f\mathrm{p}_1)^*(H') \xrightarrow{\tau} (g\mathrm{p}_2)^*(H').$$

We deduce from this an isomorphism ([**2**] I 2.5.0)

$$m^+((F' \to H' \leftarrow G')) \simeq \mathrm{p}_1^*(F') \times_{(g\mathrm{p}_2)^*(H')} \mathrm{p}_2^*(G'). \tag{VI.3.15.9}$$

Consequently, by virtue of VI.3.6 and ([**2**] III 5.5), the diagram

(VI.3.15.10)
$$\begin{array}{ccc} \widetilde{X}_{/F} \overleftarrow{\times}_{\widetilde{S}_{/H}} \widetilde{Y}_{/G} & \xrightarrow{m^*} & (\widetilde{X} \overleftarrow{\times}_{\widetilde{S}} \widetilde{Y})_{/(F \overleftarrow{\times}_H G)} \\ \uparrow & & \uparrow \\ C^{\dagger}_{/(F\to H\leftarrow G)} & = & C^{\dagger}_{/(F\to H\leftarrow G)} \end{array}$$

where the vertical arrows are the canonical functors, is commutative. Hence $m$ is the morphism of topos associated with (VI.3.15.3).

To show that $m$ is an equivalence of topos, it suffices to show that $\mathscr{T}_1 = \mathscr{T}_2$, or that $\mathscr{T}_1$ is coarser than $\mathscr{T}_2$, or that the canonical functor

(VI.3.15.11)
$$j_{(F\to H\leftarrow G)}\colon (C^{\dagger}_{/(F\to H\leftarrow G)}, \mathscr{T}_2) \to (C^{\dagger}, \mathscr{T})$$

is cocontinuous ([**2**] III 2.1). The functor $j_{(F\to H\leftarrow G)}$ is a left adjoint of the functor

(VI.3.15.12)
$$\begin{array}{rccc} j^+_{(F\to H\leftarrow G)}\colon & (C^{\dagger}, \mathscr{T}) & \to & (C^{\dagger}_{/(F\to H\leftarrow G)}, \mathscr{T}_2), \\ & L & \mapsto & L \times (F \to H \leftarrow G). \end{array}$$

Recall that finite inverse limits are representable in $C^{\dagger}$. We show as above that the diagram

(VI.3.15.13)
$$\begin{array}{ccc} \widetilde{X} \overleftarrow{\times}_{\widetilde{S}} \widetilde{Y} & \xrightarrow{(j_F \overleftarrow{\times}_{j_H} j_G)^*} & \widetilde{X}_{/F} \overleftarrow{\times}_{\widetilde{S}_{/H}} \widetilde{Y}_{/G} \\ \uparrow & & \uparrow \\ C^{\dagger} & \xrightarrow{j^+_{(F\to H\leftarrow G)}} & C^{\dagger}_{/(F\to H\leftarrow G)} \end{array}$$

where the vertical arrows are the canonical functors, is commutative. Consequently, $j^+_{(F\to H\leftarrow G)}$ is continuous by virtue of ([**2**] III 1.6), and therefore $j_{(F\to H\leftarrow G)}$ is cocontinuous by ([**2**] III 2.5).

## VI.4. Covanishing topos

**VI.4.1.** In this section, $X$ and $Y$ denote two $\mathbb{U}$-sites in which finite inverse limits are representable, and $f^+\colon X \to Y$ a continuous left exact functor. We denote by $\widetilde{X}$ and $\widetilde{Y}$ the topos of sheaves of $\mathbb{U}$-sets on $X$ and $Y$, respectively, by $f\colon \widetilde{Y} \to \widetilde{X}$ the morphism of topos associated with $f^+$, and by $\varepsilon_X\colon X \to \widetilde{X}$ and $\varepsilon_Y\colon Y \to \widetilde{Y}$ the canonical functors. Let $e_X$ and $e_Y$ be final objects of $X$ and $Y$, respectively, which exist by assumption. Since the canonical functors are left exact, $\varepsilon_X(e_X)$ and $\varepsilon_Y(e_Y)$ are final objects of $\widetilde{X}$ and $\widetilde{Y}$, respectively.

We denote by $D$ the category of pairs $(U, V \to f^+(U))$, where $U$ is an object of $X$ and $V \to f^+(U)$ is a morphism of $Y$; such an object will be denoted by $(V \to U)$. Let $(V \to U)$, $(V' \to U')$ be two objects of $D$. A morphism from $(V' \to U')$ to $(V \to U)$ consists of two morphisms $V' \to V$ of $Y$ and $U' \to U$ of $X$, such that the diagram

$$\begin{array}{ccc} V' & \longrightarrow & f^+(U') \\ \downarrow & & \downarrow \\ V & \longrightarrow & f^+(U) \end{array}$$

is commutative. It immediately follows from the definition and the fact that the functor $f^+$ is left exact that finite inverse limits in $D$ are representable.

We call *covanishing* topology on $D$ the topology generated by the coverings

$$\{(V_i \to U_i) \to (V \to U)\}_{i \in I}$$

of the following two types:

($\alpha$) $U_i = U$ for every $i \in I$, and $(V_i \to V)_{i\in I}$ is a covering family.

($\beta$) $(U_i \to U)_{i\in I}$ is a covering family, and for every $i \in I$, the canonical morphism $V_i \to V \times_{f^+(U)} f^+(U_i)$ is an isomorphism.

Note that each of these families is stable under base change. The resulting site is called the *covanishing* site associated with the functor $f^+$; it is a $\mathbb{U}$-site. We denote by $\widehat{D}$ (resp. $\widetilde{D}$) the category of presheaves (resp. the topos of sheaves) of $\mathbb{U}$-sets on $D$. We say that $\widetilde{D}$ is the *covanishing* topos associated with the functor $f^+$. We will show in VI.4.10 that the terminology does not lead to any confusion with that introduced in VI.3.12. For a presheaf $F$ on $D$, we denote by $F^a$ the associated sheaf.

**Remark VI.4.2.** The topology on $D$ is generated by the coverings

$$\{(V_{ij} \to U_i) \to (V \to U)\}_{(i,j)}$$

satisfying the following conditions:

(i) The family $(U_i \to U)_i$ is covering.

(ii) For every $i$, the family $(V_{ij} \to V \times_{f^+(U)} f^+(U_i))_j$ is covering.

Notice that in general, the family of these coverings is not stable under composition and therefore does not form a pretopology.

**VI.4.3.** We denote by $\widehat{Y}$ the category of presheaves of $\mathbb{U}$-sets on $Y$, and by $\mathscr{Q}$ the split category of presheaves of $\mathbb{U}$-sets on $Y$ ([**35**] I 2.6.1), that is, the fibered category on $Y$ obtained by associating with each $V \in \mathrm{Ob}(Y)$ the category $(Y_{/V})^\wedge = \widehat{Y}_{/V}$, and with each morphism $h\colon V' \to V$ of $Y$ the functor $h^*\colon \widehat{Y}_{/V} \to \widehat{Y}_{/V'}$ defined by composition with the functor $Y_{/h}\colon Y_{/V'} \to Y_{/V}$. Note that $h^*$ is also the base change in $\widehat{Y}$ by $h$. Since fibered products are representable in $Y$, $h^*$ admits a right adjoint, namely the "Weil restriction" functor $h_*\colon \widehat{Y}_{/V'} \to \widehat{Y}_{/V}$, defined, for all $F \in \mathrm{Ob}(\widehat{Y}_{/V'})$ and $W \in \mathrm{Ob}(Y_{/V})$, by

$$h_*(F)(W) = F(W \times_V V'). \tag{VI.4.3.1}$$

We denote by $\mathscr{Q}^\vee$ the cleaved and normalized fibered category over $Y^\circ$ obtained by associating with each $V \in \mathrm{Ob}(Y)$ the category $\widehat{Y}_{/V}$, and with each morphism $h\colon V' \to V$ of $Y$ the functor $h_*$ ([**1**] 1.1.2), and by

$$\mathscr{P} \to X \tag{VI.4.3.2}$$

$$\mathscr{P}^\vee \to X^\circ \tag{VI.4.3.3}$$

the fibered categories deduced from $\mathscr{Q}$ and $\mathscr{Q}^\vee$, respectively, by base change by the functor $f^+\colon X \to Y$. By ([**37**] VI 12; cf. also [**1**] 1.1.2), we have an equivalence of categories

$$\begin{array}{rcl} \widehat{D} & \xrightarrow{\sim} & \mathbf{Hom}_{X^\circ}(X^\circ, \mathscr{P}^\vee) \\ F & \mapsto & \{U \mapsto F_U\}, \end{array} \tag{VI.4.3.4}$$

defined, for every $(V \to U) \in \mathrm{Ob}(D)$, by the relation

$$F_U(V) = F(V \to U). \tag{VI.4.3.5}$$

From now on, we identify $F$ with the section $\{U \mapsto F_U\}$ that is associated with it by this equivalence.

**Proposition VI.4.4.** *A presheaf $F = \{U \mapsto F_U\}$ on $D$ is a sheaf if and only if the following conditions are satisfied:*

(i) *For every $U \in \mathrm{Ob}(X)$, $F_U$ is a sheaf on $Y_{/f^+(U)}$.*

(ii) *For every covering family $(U_i \to U)_{i\in I}$ of $X$, if for $(i,j) \in I^2$, we set $U_{ij} = U_i \times_U U_j$ and we denote by $h_i \colon f^+(U_i) \to f^+(U)$ and $h_{ij} \colon f^+(U_{ij}) \to f^+(U)$ the structural morphisms, then the sequence of morphisms of sheaves on $Y_{/f^+(U)}$*

(VI.4.4.1)
$$F_U \to \prod_{i\in I} h_{i*}(F_{U_i}) \rightrightarrows \prod_{(i,j)\in I^2} h_{ij*}(F_{U_{ij}})$$

*is exact.*

Indeed, after extending the universe $\mathbb{U}$, if necessary, we may assume that the category $X$ is small ([**2**] II 2.7(2)). The proposition then follows from ([**2**] II 2.3, I 3.5, I 2.12, and II 4.1(3)).

**Remark VI.4.5.** Condition VI.4.4(ii) corresponds to saying that, for every $(V \to U) \in \mathrm{Ob}(D)$, if we set $V_i = V \times_{f^+(U)} f^+(U_i)$ and $V_{ij} = V \times_{f^+(U)} f^+(U_{ij})$, the sequence of maps of sets

(VI.4.5.1)
$$F_U(V) \to \prod_{i\in I} F_{U_i}(V_i) \rightrightarrows \prod_{(i,j)\in I^2} F_{U_{ij}}(V_{ij})$$

is exact.

**Remarks VI.4.6.** (i) For every object $(V \to U)$ of $D$, the diagram

(VI.4.6.1)
$$\begin{array}{ccc} (V \to U)^a & \longrightarrow & (V \to e_X)^a \\ \downarrow & & \downarrow \\ (f^+(U) \to U)^a & \longrightarrow & (f^+(U) \to e_X)^a \end{array}$$

is Cartesian in $\widetilde{D}$. Indeed, the canonical functor $D \to \widetilde{D}$ is left exact.

(ii) Let $W$ be an object of $X$ and $F = \{U \mapsto F_U\}$ the presheaf on $D$ defined by $(f^+(W) \to W)$. For every $U \in \mathrm{Ob}(X)$, $F_U$ is the constant presheaf on $Y_{/f^+(U)}$ with value $\mathrm{Hom}_X(U,W)$. In particular, the topology on $D$ is not in general coarser than the canonical topology.

(iii) Let $V$ be an object of $Y$, $F = \{U \mapsto F_U\}$ the presheaf on $D$ defined by $(V \to e_X)$. For every $U \in \mathrm{Ob}(X)$, $F_U$ is the presheaf $V \times f^+(U)$ on $Y_{/f^+(U)}$. If the topologies on $X$ and $Y$ are coarser than the canonical topologies, $F$ is a sheaf on $D$ by virtue of VI.4.4.

**VI.4.7.** Denote by $C$ the site associated with the pair of functors $(\mathrm{id}_X, f^+)$ defined in VI.3.1, and consider the functors

(VI.4.7.1) $\iota^+ \colon D \to C, \quad (V \to U) \mapsto (U \to U \leftarrow V),$

(VI.4.7.2) $\jmath^+ \colon C \to D, \quad (U \to W \leftarrow V) \mapsto (V \times_{f^+(W)} f^+(U) \to U).$

It is clear that $\iota^+$ is a left adjoint of $\jmath^+$, that the adjunction morphism $\mathrm{id} \to \jmath^+ \circ \iota^+$ is an isomorphism (that is, $\iota^+$ is fully faithful), and that $\iota^+$ and $\jmath^+$ are left exact.

**Proposition VI.4.8.** (i) *The functors $\iota^+$ and $\jmath^+$ are continuous.*

(ii) *The topology on $D$ is induced by that on $C$ through the functor $\iota^+$.*

(i) The functor $\iota^+$ transforms covering families of $D$ of type $(\alpha)$ into covering families of $C$ of type (b), and covering families of $D$ of type $(\beta)$ into covering families of $C$:

$$\begin{array}{ccc} V_i & \longrightarrow & U_i \\ \downarrow & \square & \downarrow \\ V & \longrightarrow & U \end{array} \quad \mapsto \quad \begin{array}{ccccc} U_i & \longrightarrow & U_i & \longleftarrow & V_i \\ \downarrow & & \downarrow & \square & \downarrow \\ U_i & \longrightarrow & U & \longleftarrow & V \\ \downarrow & & \downarrow & & \downarrow \\ U & \longrightarrow & U & \longleftarrow & V \end{array} \tag{VI.4.8.1}$$

Let $G$ be a presheaf on $C$ and $F = \{U \mapsto F_U\} = G \circ \iota^+$. For every $(V \to U) \in \mathrm{Ob}(D)$, we have

$$F_U(V) = G(U \to U \leftarrow V). \tag{VI.4.8.2}$$

Consequently, if $G$ is a sheaf on $C$, $F$ is a sheaf on $D$ by virtue of VI.3.2, VI.4.4, and (VI.4.8.1); hence $\iota^+$ is continuous.

The functor $\jmath^+$ transforms coverings of $C$ of type $(a)$ (resp. $(b)$) into coverings of $D$ of type $(\beta)$ (resp. $(\alpha)$), and coverings of $C$ of type $(c)$ into isomorphisms. Consequently, for every sheaf $F$ on $D$, $F \circ \jmath^+$ is a sheaf on $C$ by virtue of VI.3.2; hence $\jmath^+$ is continuous.

(ii) We know that the topology on $D$ is induced by the canonical topology on $\widetilde{D}$ ([**2**] III 3.5); in other words, the topology on $D$ is the finest such that every $F \in \mathrm{Ob}(\widetilde{D})$ is a sheaf. By (i), we can consider the functors

$$\iota_s \colon \widetilde{C} \to \widetilde{D}, \quad G \mapsto G \circ \iota^+, \tag{VI.4.8.3}$$

$$\jmath_s \colon \widetilde{D} \to \widetilde{C}, \quad F \mapsto F \circ \jmath^+. \tag{VI.4.8.4}$$

The adjunction isomorphism $\mathrm{id} \xrightarrow{\sim} \jmath^+ \circ \iota^+$ induces an isomorphism $\iota_s \circ \jmath_s \xrightarrow{\sim} \mathrm{id}$. The functor $\iota_s$ is therefore essentially surjective. Consequently, the topology on $D$ is the finest such that, for every $G \in \mathrm{Ob}(\widetilde{C})$, $\iota_s(G)$ is a sheaf on $D$; whence the proposition.

**VI.4.9.** Since the functors $\iota^+$ and $\jmath^+$ are continuous and left exact (VI.4.8), they define morphisms of topos ([**2**] IV 4.9.2)

$$\iota \colon \widetilde{C} \to \widetilde{D}, \tag{VI.4.9.1}$$

$$\jmath \colon \widetilde{D} \to \widetilde{C}. \tag{VI.4.9.2}$$

The adjunction morphisms $\mathrm{id} \to \jmath^+ \circ \iota^+$ and $\iota^+ \circ \jmath^+ \to \mathrm{id}$ induce morphisms $\iota_* \circ \jmath_* \to \mathrm{id}$ and $\mathrm{id} \to \jmath_* \circ \iota_*$ that make $\iota_*$ into a right adjoint of $\jmath_*$.

**Proposition VI.4.10.** *The adjunction morphisms* $\iota_* \circ \jmath_* \to \mathrm{id}$ *and* $\mathrm{id} \to \jmath_* \circ \iota_*$ *are isomorphisms. In particular,* $\iota$ (VI.4.9.1) *and* $\jmath$ (VI.4.9.2) *are equivalences of topos quasi-inverse to each other.*

Indeed, since the adjunction morphism $\mathrm{id} \to \jmath^+ \circ \iota^+$ is an isomorphism, $\iota_* \circ \jmath_* \to \mathrm{id}$ is an isomorphism. On the other hand, the adjunction morphism $\iota^+ \circ \jmath^+ \to \mathrm{id}$ is defined, for every object $(U \to W \leftarrow V)$ of $C$, by the canonical morphism

$$(U \to U \leftarrow V \times_{f^+(W)} f^+(U)) \to (U \to W \leftarrow V), \tag{VI.4.10.1}$$

which is a covering of type (c). Since the morphism of sheaves associated with (VI.4.10.1) is an isomorphism in $\widetilde{C}$ (VI.3.3), $\mathrm{id} \to \jmath_* \circ \iota_*$ is an isomorphism.

**VI.4.11.** The functors

(VI.4.11.1) $$\mathrm{p}_1^+ \colon X \to D, \quad U \mapsto (f^+(U) \to U),$$

(VI.4.11.2) $$\mathrm{p}_2^+ \colon Y \to D, \quad V \mapsto (V \to e_X),$$

are left exact and continuous ([**2**] III 1.6). They therefore define two morphisms of topos ([**2**] IV 4.9.2)

(VI.4.11.3) $$\mathrm{p}_1 \colon \widetilde{D} \to \widetilde{X},$$

(VI.4.11.4) $$\mathrm{p}_2 \colon \widetilde{D} \to \widetilde{Y}.$$

For every $U \in \mathrm{Ob}(X)$, the morphism $(U \to U \leftarrow f^+(U))^a \to (U \to e_X \leftarrow e_Y)^a$ is an isomorphism of $\widetilde{C}$ (VI.3.3). The morphisms $\mathrm{p}_1 \circ \iota$ and $\mathrm{p}_2 \circ \iota$ therefore identify with morphisms $\mathrm{p}_1 \colon \widetilde{C} \to \widetilde{X}$ and $\mathrm{p}_2 \colon \widetilde{C} \to \widetilde{Y}$ defined in (VI.3.4); whence the terminology. The 2-morphism (VI.3.4.5)

(VI.4.11.5) $$\tau \colon f\mathrm{p}_2 \to \mathrm{p}_1$$

is then defined by the following morphism of functors $(f\mathrm{p}_2)_* \to \mathrm{p}_{1*}$: for every sheaf $F$ on $D$ and every $U \in \mathrm{Ob}(X)$,

(VI.4.11.6) $$f_*(\mathrm{p}_{2*}(F))(U) \to \mathrm{p}_{1*}(F)(U)$$

is the canonical map

$$F(f^+(U) \to e_X) \to F(f^+(U) \to U).$$

The 2-morphism $\tau$ induces a base change morphism

(VI.4.11.7) $$f_* \to \mathrm{p}_{1*}\mathrm{p}_2^*$$

composed of

$$f_* \longrightarrow f_*\mathrm{p}_{2*}\mathrm{p}_2^* \xrightarrow{\tau * \mathrm{p}_2^*} \mathrm{p}_{1*}\mathrm{p}_2^*,$$

where the first morphism is deduced from the adjunction morphism. For every ring $\Lambda$, the morphism (VI.4.11.7) induces a morphism of functors from $\mathbf{D}^+(\widetilde{Y}, \Lambda)$ to $\mathbf{D}^+(\widetilde{X}, \Lambda)$

(VI.4.11.8) $$\mathrm{R}f_* \to \mathrm{R}\mathrm{p}_{1*}\mathrm{p}_2^*.$$

**Proposition VI.4.12.** (i) *For every sheaf $F$ on $X$, $\mathrm{p}_1^*(F)$ is the sheaf associated with the presheaf $\{U \mapsto F(U)\}$ on $D$* (VI.4.3.4), *where for every $U \in \mathrm{Ob}(X)$, $F(U)$ is the constant presheaf on $Y_{/f^+(U)}$ with value $F(U)$.*

(ii) *For every sheaf $F$ on $Y$, $\mathrm{p}_2^*(F)$ is the sheaf $\{U \mapsto F \times f^*(U)\}$.*

(iii) *For every sheaf $F$ on $X$, the morphism $\tau \colon \mathrm{p}_1^*(F) \to (f\mathrm{p}_2)^*(F)$* (VI.4.11.5) *is induced by the morphism of presheaves on $D$ defined, for every $U \in \mathrm{Ob}(X)$, by the morphism of presheaves on $f^+(U)$*

(VI.4.12.1) $$F(U) \to f^*(F \times U)$$

*given, for every $(V \to U) \in \mathrm{Ob}(D)$, by the composition*

(VI.4.12.2) $$F(U) \to (f^*F)(f^+U) \to (f^*F)(V).$$

(iv) *The adjunction morphism* $\mathrm{id} \to \mathrm{p}_{2*}\mathrm{p}_2^*$ *is an isomorphism.*

Indeed, after extending the universe $\mathbb{U}$, if necessary, we may assume that the categories $X$ and $Y$ are $\mathbb{U}$-small ([**2**] II 3.6, and III 1.5).

(i) By ([**2**] I 5.1 and III 1.3), the sheaf $\mathrm{p}_1^*(F)$ is the sheaf on $D$ associated with the presheaf $G$ defined for $(V \to U) \in \mathrm{Ob}(D)$ by

(VI.4.12.3) $$G(V \to U) = \varinjlim_{(P,u) \in I^{\circ}_{(V \to U)}} F(P),$$

where $I_{(V\to U)}$ is the category of pairs $(P,u)$ consisting of an object $P$ of $X$ and a morphism $u\colon (V\to U)\to(f^+(P)\to P)$ of $D$. This category admits as initial object the pair consisting of $U$ and the canonical morphism $(V\to U)\to(f^+(U)\to U)$. We therefore have $G(V\to U)=F(U)$.

(ii) The sheaf $\mathrm{p}_2^*(F)$ is the sheaf on $D$ associated with the presheaf $H$ defined for $(V\to U)\in \mathrm{Ob}(D)$ by

(VI.4.12.4) $$H(V\to U)=\varinjlim_{(Q,v)\in J^{\circ}_{(V\to U)}} F(Q),$$

where $J_{(V\to U)}$ is the category of pairs $(Q,v)$ consisting of an object $Q$ of $Y$ and a morphism $v\colon (V\to U)\to(Q\to e_X)$ of $D$. This category admits as initial object the pair consisting of $V$ and the canonical morphism $(V\to U)\to(V\to e_X)$. We therefore have $H(V\to U)=F(V)$. Consequently, $H=\{U\mapsto F\times f^*(U)\}$, which is indeed a sheaf on $D$ by virtue of VI.4.4.

(iii) Denote by $G$ the presheaf on $D$ associated with $F$ defined in (VI.4.12.3) and by $H$ the sheaf on $D$ associated with $f^*(F)$ defined in (VI.4.12.4). For every object $(V\to U)$ of $D$, we have a functor

(VI.4.12.5) $$I_{(V\to U)}\to J_{(V\to U)},$$

defined by $(P,u)\mapsto(f^+(P),v)$, where $v$ is the composition

$$(V\to U)\xrightarrow{u}(f^+(P)\to P)\to(f^+(P)\to e_X).$$

Then the composition

$$\varinjlim_{(P,u)\in I^{\circ}_{(V\to U)}} F(P)\to \varinjlim_{(P,u)\in I^{\circ}_{(V\to U)}} (f^*F)(f^+P)\to \varinjlim_{(Q,v)\in J^{\circ}_{(V\to U)}} (f^*F)(Q),$$

where the first arrow is the canonical map and the second arrow is induced by the functor (VI.4.12.5), is equal to the map

(VI.4.12.6) $$G(V\to U)\to H(V\to U)$$

defined in (VI.4.12.2). Consequently, the morphism of sheaves $\mathrm{p}_1^*(F)\to(f\mathrm{p}_2)^*(F)$ associated with (VI.4.12.6) is the adjoint of the morphism (VI.4.11.6), giving the proposition.

(iv) Indeed, for every sheaf $F$ on $Y$ and every $V\in\mathrm{Ob}(Y)$, the adjunction morphism $F(V)\to(\mathrm{p}_2^*F)(V\to e_X)$ identifies with the identity morphism of $F(V)$ by virtue of (ii).

**VI.4.13.** The functor

(VI.4.13.1) $$\Psi^+\colon D\to Y,\quad (V\to U)\mapsto V$$

is clearly left exact. For every sheaf $F$ on $Y$, we have

(VI.4.13.2) $$F\circ\Psi^+=\{U\mapsto F|f^+(U)\},$$

where for every morphism $g\colon U'\to U$ of $X$, if we set $h=f^+(g)$, the transition morphism

(VI.4.13.3) $$F|f^+(U)\to h_*(F|f^+(U'))$$

is the adjoint of the canonical isomorphism $h^*(F|f^+(U))\xrightarrow{\sim}F|f^+(U')$. It follows from VI.4.4 that $F\circ\Psi^+$ is a sheaf on $D$. Consequently, $\Psi^+$ is continuous. It therefore defines a morphism of topos

(VI.4.13.4) $$\Psi\colon\widetilde{Y}\to\widetilde{D}$$

such that $\mathrm{p}_1\Psi = f$, $\mathrm{p}_2\Psi = \mathrm{id}_{\widetilde{Y}}$, and $\tau * \Psi = \mathrm{id}_f$, where $\tau$ is the 2-morphism (VI.4.11.5). Consequently, $\jmath\Psi$ is the morphism defined by the morphisms $f\colon \widetilde{Y} \to \widetilde{X}$ and $\mathrm{id}_{\widetilde{Y}}$ and the 2-morphism $\mathrm{id}_f$, in view of the universal property of oriented products (VI.3.7):

(VI.4.13.5)
$$\begin{array}{ccccc} & & \widetilde{Y} & & \\ & \swarrow^{f} & \downarrow{\scriptstyle \jmath\Psi} & \searrow^{\mathrm{id}_{\widetilde{Y}}} & \\ \widetilde{X} & \xleftarrow{\mathrm{p}_1} & \widetilde{X} \overset{\leftarrow}{\times}_{\widetilde{X}} \widetilde{Y} & \xrightarrow{\mathrm{p}_2} & \widetilde{Y} \\ & \searrow & & \swarrow_{f} & \\ & & \widetilde{X} & & \end{array}$$

The morphism $\Psi$ (or $\jmath\Psi$) is called the morphism of *co-nearby cycles*. From the relation $\mathrm{p}_{2*}\Psi_* = \mathrm{id}_{\widetilde{Y}}$, we obtain by adjunction a morphism

(VI.4.13.6)
$$\mathrm{p}_2^* \to \Psi_*.$$

**Proposition VI.4.14.** *The morphism* $\mathrm{p}_2^* \to \Psi_*$ (VI.4.13.6) *is an isomorphism; in particular, the functor* $\Psi_*$ *is exact.*

Indeed, for every sheaf $F$ on $Y$ and every $(V \to U) \in \mathrm{Ob}(D)$, we have a commutative diagram

(VI.4.14.1)
$$\begin{array}{ccc} \mathrm{p}_2^*(F)(V \to e_X) & \longrightarrow & \Psi_*(F)(V \to e_X) \\ \downarrow & & \downarrow \\ \mathrm{p}_2^*(F)(V \to U) & \longrightarrow & \Psi_*(F)(V \to U) \end{array}$$

where the horizontal arrows are the maps (VI.4.13.6) and the vertical arrows are the canonical maps. The latter are isomorphisms by virtue of VI.4.12(ii). On the other hand, the top horizontal arrow is induced by the morphism

(VI.4.14.2)
$$\mathrm{p}_{2*}\mathrm{p}_2^* \to \mathrm{p}_{2*}\Psi_*$$

deduced from (VI.4.13.6). The composition of (VI.4.14.2) and the adjunction morphism $\mathrm{id} \to \mathrm{p}_{2*}\mathrm{p}_2^*$ is the canonical isomorphism $\mathrm{id} \xrightarrow{\sim} \mathrm{p}_{2*}\Psi_*$. It then follows from VI.4.12(iv) that (VI.4.14.2) is an isomorphism. Consequently, the bottom horizontal arrow of (VI.4.14.1) is an isomorphism, giving the proposition.

**Proposition VI.4.15.** (i) *For every sheaf of sets $F$ on $Y$, the morphism* (VI.4.11.7)

(VI.4.15.1)
$$f_*(F) \to \mathrm{p}_{1*}\mathrm{p}_2^*(F)$$

*is an isomorphism.*

(ii) *Let $\Lambda$ be a ring. For every complex $F$ of* $\mathbf{D}^+(\widetilde{Y}, \Lambda)$, *the morphism* (VI.4.11.8)

(VI.4.15.2)
$$\mathrm{R}f_*(F) \to \mathrm{R}\mathrm{p}_{1*}\mathrm{p}_2^*(F)$$

*is an isomorphism.*

(i) Consider the commutative diagram

(VI.4.15.3)
$$\begin{array}{ccccc} f_* & \xrightarrow{a} & f_*\mathrm{p}_{2*}\mathrm{p}_2^* & \xrightarrow{\tau * \mathrm{p}_2^*} & \mathrm{p}_{1*}\mathrm{p}_2^* \\ & \searrow^{d} & \downarrow{\scriptstyle b} & & \downarrow{\scriptstyle c} \\ & & f_*\mathrm{p}_{2*}\Psi_* & \xrightarrow{\tau * \Psi_*} & \mathrm{p}_{1*}\Psi_* \end{array}$$

where $a$ is induced by the adjunction morphism and $b$ and $c$ are induced by (VI.4.13.6). Since $d = b \circ a$ is the isomorphism deduced from the relation $\mathrm{p}_2\Psi = \mathrm{id}_{\widetilde{Y}}$, $(\tau * \Psi_*) \circ d$ identifies with the isomorphism deduced from the relation $\mathrm{p}_1\Psi = f$ (VI.4.13). On the other hand, $c$ is an isomorphism by virtue of VI.4.14, giving the statement.

(ii) Let $\widetilde{\tau}\colon \mathrm{R}f_*\mathrm{R}\mathrm{p}_{2*} \to \mathrm{R}\mathrm{p}_{1*}$ be the morphism induced by $\tau$ (VI.4.11.5). Since $\Psi_*$ is exact by virtue of VI.4.14, the diagram (VI.4.15.3) induces a commutative diagram

(VI.4.15.4)
$$\begin{array}{ccccc} \mathrm{R}f_* & \xrightarrow{\alpha} & \mathrm{R}f_*\mathrm{R}\mathrm{p}_{2*}\mathrm{p}_2^* & \xrightarrow{\widetilde{\tau}*\mathrm{p}_2^*} & \mathrm{R}\mathrm{p}_{1*}\mathrm{p}_2^* \\ & \searrow^{\delta} & \downarrow{\scriptstyle \beta} & & \downarrow{\scriptstyle \gamma} \\ & & \mathrm{R}f_*\mathrm{R}\mathrm{p}_{2*}\Psi_* & \xrightarrow{\widetilde{\tau}*\Psi_*} & \mathrm{R}\mathrm{p}_{1*}\Psi_* \end{array}$$

On the other hand, since $\Psi_*$ is exact, $\delta$ is the isomorphism deduced from the relation $\mathrm{p}_2\Psi = \mathrm{id}_{\widetilde{Y}}$, and consequently $(\widetilde{\tau} * \Psi_*) \circ \delta$ identifies with the isomorphism deduced from the relation $\mathrm{p}_1\Psi = f$ (VI.4.13). Since $\gamma$ is an isomorphism by virtue of VI.4.14, the statement follows.

**Remark VI.4.16.** Proposition VI.4.15 and its proof are due to Gabber ([**46**] 4.9). It is a special case of a base change theorem for oriented topos ([**46**] 2.4), which, nevertheless, requires more restrictive coherence assumptions.

**VI.4.17.** Let $(B \to A)$ be an object of $D$. We denote by $j_A\colon X_{/A} \to X$ and $j_B\colon Y_{/B} \to Y$ the canonical functors, and endow $X_{/A}$ and $Y_{/B}$ with the topologies induced by those on $X$ and $Y$ by the functors $j_A$ and $j_B$, respectively. Denote by $(X_{/A})^\sim$ the topos of sheaves of $\mathbb{U}$-sets on $X_{/A}$. By ([**2**] III 5.2), the functor $j_A$ is continuous and cocontinuous. It therefore induces a sequence of three adjoint functors:

(VI.4.17.1)
$$j_{A!}\colon (X_{/A})^\sim \to \widetilde{X}, \quad j_A^*\colon \widetilde{X} \to (X_{/A})^\sim, \quad j_{A*}\colon (X_{/A})^\sim \to \widetilde{X}$$

in the sense that for any two consecutive functors in the sequence, the one on the right is right adjoint to the other. The functor $j_{A!}$ factors through an equivalence of categories $(X_{/A})^\sim \xrightarrow{\sim} \widetilde{X}_{/A^a}$, where $A^a = \varepsilon_X(A)$ ([**2**] III 5.4), and the pair $(j_A^*, j_{A*})$ defines a morphism of topos $\widetilde{X}_{/A^a} \to \widetilde{X}$, called localization morphism of $\widetilde{X}$ at $A^a$, and likewise for $j_B$.

Finite inverse limits are representable in $X_{/A}$ and $Y_{/B}$. On the other hand, the functor

(VI.4.17.2)
$$f'^+\colon X_{/A} \to Y_{/B}, \quad U \mapsto f^+(j_A(U)) \times_{f^+(A)} B$$

is left exact and continuous by virtue of ([**2**] III 1.6 and 3.3). The morphism of topos

(VI.4.17.3)
$$f'\colon \widetilde{Y}_{/B^a} \to \widetilde{X}_{/A^a}$$

associated with $f'^+$ identifies with the composition

$$\widetilde{Y}_{/B^a} \longrightarrow \widetilde{Y}_{/f^*(A)} \xrightarrow{f_{/A^a}} \widetilde{X}_{/A^a},$$

where the first arrow is the localization morphism associated with $B^a \to f^*(A^a)$ ([**2**] IV 5.5) and the second arrow is the morphism deduced from $f$ ([**2**] IV 5.10).

We denote by $D'$ (resp. $\widetilde{D}'$) the covanishing site (resp. topos) associated with the functor $f'^+$. The functors $j_A$ and $j_B$ induce a functor

(VI.4.17.4)
$$j_{(B\to A)}\colon D' \to D,$$

that factors through an equivalence of categories

(VI.4.17.5)
$$n\colon D' \xrightarrow{\sim} D_{/(B\to A)}.$$

**Proposition VI.4.18.** *Under the assumptions of* VI.4.17, *the covanishing topology on* $D'$ *is induced by the covanishing topology on* $D$ *through the functor* $j_{(B\to A)}$ (VI.4.17.4); *in particular,* $n$ (VI.4.17.5) *induces an equivalence of topos*

$$m\colon \widetilde{D}_{/(B\to A)^a} \stackrel{\sim}{\to} \widetilde{D}'. \tag{VI.4.18.1}$$

We identify the topos $\widetilde{D}$ and $\widetilde{X} \overleftarrow{\times}_{\widetilde{X}} \widetilde{Y}$ and the topos $\widetilde{D}'$ and $\widetilde{X}_{/A^a} \overleftarrow{\times}_{\widetilde{X}_{/A^a}} \widetilde{Y}_{/B^a}$ using the equivalences (VI.4.9.1). By VI.3.15, we have an equivalence of topos

$$m\colon (\widetilde{X} \overleftarrow{\times}_{\widetilde{X}} \widetilde{Y})_{/(B\to A)^a} \stackrel{\sim}{\to} \widetilde{X}_{/A^a} \overleftarrow{\times}_{\widetilde{X}_{/A^a}} \widetilde{Y}_{/B^a}. \tag{VI.4.18.2}$$

A priori, the covanishing topology on $D'$ is coarser than the topology induced by the covanishing topology on $D$ by the functor $j_{(B\to A)}$. But it follows from the proof of VI.3.15, in particular from (VI.3.15.10), that the diagram

$$\begin{array}{ccc} \widetilde{X}_{/A^a} \overleftarrow{\times}_{\widetilde{X}_{/A^a}} \widetilde{Y}_{/B^a} & \xrightarrow{m^*} & (\widetilde{X} \overleftarrow{\times}_{\widetilde{X}} \widetilde{Y})_{/(B\to A)^a} \\ \uparrow & & \uparrow \\ D' & \xrightarrow{\quad n \quad} & D_{/(B\to A)} \end{array} \tag{VI.4.18.3}$$

where the vertical arrows are the canonical functors, is commutative. We deduce from this, by ([**2**] III 3.5), that the covanishing topology on $D'$ is induced by the covanishing topology on $D$ by the functor $j_{(B\to A)}$. Note that the equivalence (VI.4.18.1) induced by $n$ identifies with the equivalence (VI.4.18.2) by virtue of (VI.4.18.3); whence the notation.

**Remark VI.4.19.** We can give a direct proof of VI.4.18 that does not pass through VI.3.15, but that uses the same arguments. The proof becomes particularly simple when $B = f^+(A)$. We will treat this case directly in a more generalized setting (VI.5.38).

**VI.4.20.** Recall (VI.3.11) that giving a point of $\widetilde{X} \overleftarrow{\times}_{\widetilde{X}} \widetilde{Y}$ is equivalent to giving a pair of points $x\colon \mathbf{Pt} \to \widetilde{X}$ and $y\colon \mathbf{Pt} \to \widetilde{Y}$ and a 2-morphism $u\colon f(y) \to x$. Such a point will be denoted by $(y \to x)$, or by $(u\colon y \to x)$. For all $F \in \mathrm{Ob}(\widetilde{X})$ and $G \in \mathrm{Ob}(\widetilde{Y})$, we have functorial canonical isomorphisms

$$(\mathrm{p}_1^* F)_{(y\to x)} \stackrel{\sim}{\to} F_x, \tag{VI.4.20.1}$$

$$(\mathrm{p}_2^* G)_{(y\to x)} \stackrel{\sim}{\to} G_y. \tag{VI.4.20.2}$$

By VI.3.7, the map

$$(\mathrm{p}_1^* F)_{(y\to x)} \to (\mathrm{p}_2^*(f^*F))_{(y\to x)} \tag{VI.4.20.3}$$

induced by $\tau$ (VI.4.11.5), identifies canonically with the specialization morphism $F_x \to F_{f(y)}$ defined by $u$. We identify the topos $\widetilde{D}$ and $\widetilde{X} \overleftarrow{\times}_{\widetilde{X}} \widetilde{Y}$ by the equivalence (VI.4.9.2). The isomorphisms (VI.4.13.6) and (VI.4.20.2) induce a functorial canonical isomorphism

$$(\Psi_* G)_{(y\to x)} \stackrel{\sim}{\to} G_y. \tag{VI.4.20.4}$$

It follows from VI.3.7 and (VI.4.13.5) that we have a functorial canonical isomorphism of points of $\widetilde{X} \overleftarrow{\times}_{\widetilde{X}} \widetilde{Y}$

$$\Psi(y) \stackrel{\sim}{\to} (y \to f(y)). \tag{VI.4.20.5}$$

By VI.4.6(i), for every $(V \to U) \in \mathrm{Ob}(D)$, we have a functorial canonical isomorphism

$$(V \to U)^a_{(y\to x)} \stackrel{\sim}{\to} U^a_x \times_{U^a_{f(y)}} V^a_y, \tag{VI.4.20.6}$$

where the exponent $^a$ denotes the associated sheaves, the map $V_y^a \to U_{f(y)}^a$ is induced by the structural morphism $V \to f^+(U)$, and the map $U_x^a \to U_{f(y)}^a$ is the specialization morphism defined by $u$.

## VI.5. Generalized covanishing topos

**VI.5.1.** In this section, $I$ denotes a $\mathbb{U}$-site, $\widetilde{I}$ the topos of sheaves of $\mathbb{U}$-sets on $I$, and

$$\pi\colon E \to I \tag{VI.5.1.1}$$

a cleaved and normalized fibered category over the category underlying $I$ ([**37**] VI 7.1). We suppose that the following conditions are satisfied:

(i) Fibered products are representable in $I$.
(ii) For every $i \in \mathrm{Ob}(I)$, the fiber $E_i$ of $E$ over $i$ is endowed with a topology making it into a $\mathbb{U}$-site, and finite inverse limits are representable in $E_i$. We denote by $\widetilde{E}_i$ the topos of sheaves of $\mathbb{U}$-sets on $E_i$.
(iii) For every morphism $f\colon i \to j$ of $I$, the inverse image functor $f^+\colon E_j \to E_i$ is continuous and left exact. It therefore defines a morphism of topos that we (abusively) denote also by $f\colon \widetilde{E}_i \to \widetilde{E}_j$ ([**2**] IV 4.9.2).

Condition (i) will be strengthened from VI.5.32 onward. For every $i \in \mathrm{Ob}(I)$, we denote by

$$\alpha_{i!}\colon E_i \to E \tag{VI.5.1.2}$$

the canonical inclusion functor.

The functor $\pi$ is in fact a fibered $\mathbb{U}$-site ([**2**] VI 7.2.1 and 7.2.4). We denote by

$$\mathscr{F} \to I \tag{VI.5.1.3}$$

the fibered $\mathbb{U}$-topos associated with $\pi$ ([**2**] VI 7.2.6). The fiber of $\mathscr{F}$ over any $i \in \mathrm{Ob}(I)$ is canonically equivalent to the topos $\widetilde{E}_i$, and the inverse image functor under any morphism $f\colon i \to j$ of $I$ identifies with the inverse image functor under the morphism of topos $f\colon \widetilde{E}_i \to \widetilde{E}_j$. We denote by

$$\mathscr{F}^\vee \to I^\circ \tag{VI.5.1.4}$$

the fibered category obtained by associating with each $i \in \mathrm{Ob}(I)$ the category $\widetilde{E}_i$, and with each morphism $f\colon i \to j$ of $I$ the functor $f_*\colon \widetilde{E}_i \to \widetilde{E}_j$ direct image by the morphism of topos $f\colon \widetilde{E}_i \to \widetilde{E}_j$. We denote by

$$\mathscr{P}^\vee \to I^\circ \tag{VI.5.1.5}$$

the fibered category obtained by associating with each $i \in \mathrm{Ob}(I)$ the category $\widehat{E}_i$ of presheaves of $\mathbb{U}$-sets on $E_i$, and with each morphism $f\colon i \to j$ of $I$ the functor $f_*\colon \widehat{E}_i \to \widehat{E}_j$ obtained by composing with the inverse image functor $f^+\colon E_j \to E_i$. This notation convention does not follow that of ([**2**] I 5.0); we make it so that the canonical $I^\circ$-functor $\mathscr{F}^\vee \to \mathscr{P}^\vee$ becomes compatible with inverse image functors.

**VI.5.2.** Note that $E$ is a $\mathbb{U}$-category. We denote by $\widehat{E}$ the category of presheaves of $\mathbb{U}$-sets on $E$. Observe that since the category $\widehat{E}$ is not naturally fibered over $I$, the notation $\widehat{E}_i$ for the fibers of $\mathscr{P}^\vee$ over $I^\circ$ does not lead to any confusion.

By ([**37**] VI 12; cf. also [**1**] 1.1.2) and with the notation of (VI.2.2), we have an equivalence of categories

(VI.5.2.1)
$$\begin{array}{rcl} \widehat{E} & \stackrel{\sim}{\to} & \mathbf{Hom}_{I^\circ}(I^\circ, \mathscr{P}^\vee) \\ F & \mapsto & \{i \mapsto F \circ \alpha_{i!}\}, \end{array}$$

where $\alpha_{i!}$ is the functor (VI.5.1.2). From now on, we identify $F$ with the section $\{i \mapsto F \circ \alpha_{i!}\}$ that is associated with it by this equivalence.

**VI.5.3.** We call *covanishing* topology on $E$ the topology generated by the families of coverings $(V_n \to V)_{n\in\Sigma}$ of the following two types:

(v) There exists $i \in \mathrm{Ob}(I)$ such that $(V_n \to V)_{n\in\Sigma}$ is a covering family of $E_i$.

(c) There exists a covering family of morphisms $(f_n : i_n \to i)_{n\in\Sigma}$ of $I$ such that $V_n$ is isomorphic to $f_n^+(V)$ for every $n \in \Sigma$.

The coverings of type (v) are called *vertical*, and those of type (c) are called *Cartesian*. The resulting site is called *covanishing* site associated with the fibered site $\pi$ (VI.5.1.1); it is a $\mathbb{U}$-site. We call *covanishing* topos associated with the fibered site $\pi$, and denote by $\widetilde{E}$, the topos of sheaves of $\mathbb{U}$-sets on $E$. We denote by

(VI.5.3.1)
$$\varepsilon : E \to \widetilde{E}$$

the canonical functor.

**Example VI.5.4.** Suppose that $I$ is endowed with the trivial or chaotic topology, that is, with the coarsest of all topologies on $I$ ([**2**] II 1.1.4). Note that under this assumption, requiring that $I$ be a $\mathbb{U}$-site amounts to requiring that $I$ be equivalent to a $\mathbb{U}$-small category. The total topology on $E$ associated with the fibered site $\pi$ ([**2**] VI 7.4.1) is generated by the vertical coverings, by virtue of ([**2**] VI 7.4.2(1)). It is therefore equal to the covanishing topology on $E$ .

**Example VI.5.5.** Let $X$ and $Y$ be two $\mathbb{U}$-sites in which finite inverse limits are representable and $f^+ : X \to Y$ a continuous left exact functor. We associate with $f^+$ a fibered $\mathbb{U}$-site

(VI.5.5.1)
$$\pi : E \to X$$

satisfying the conditions of VI.5.1, as follows. Consider the category $\mathrm{Fl}(Y)$ of morphisms of $Y$, and the "target functor"

(VI.5.5.2)
$$\mathrm{Fl}(Y) \to Y,$$

which makes $\mathrm{Fl}(Y)$ into a cleaved and normalized fibered category over $Y$: the fiber over any object $V$ of $Y$ is canonically equivalent to the category $Y_{/V}$, and for every morphism $h : V' \to V$ of $Y$, the inverse image functor $h^* : Y_{/V} \to Y_{/V'}$ is none other than the base change functor by $h$. Endowing each fiber $Y_{/V}$ with the topology induced by that of $Y$, $\mathrm{Fl}(Y)/Y$ becomes a fibered $\mathbb{U}$-site satisfying the conditions of VI.5.1. We then take for $\pi$ the fibered site deduced from $\mathrm{Fl}(Y)/Y$ by base change by the functor $f^+$. The covanishing site $E$ associated with the fibered site $\pi$ (VI.5.3) is canonically equivalent to the covanishing site $D$ associated with the functor $f^+$ (VI.4.1) by virtue of ([**2**] III 5.2(1)), whence the terminology.

**Lemma VI.5.6.** (i) *Fibered products are representable in $E$.*

(ii) *The functors $\pi$ and $\alpha_{i!}$ (VI.5.1.2), for every $i \in \mathrm{Ob}(I)$, commute with fibered products.*

(iii) *The family of vertical (resp. Cartesian) coverings of $E$ is stable under base change.*

(i) Consider a commutative diagram of $E$

(VI.5.6.1)
$$\begin{array}{ccc} X & \longrightarrow & V \\ \downarrow & & \downarrow \\ U & \longrightarrow & W \end{array}$$

over a commutative diagram of $I$

(VI.5.6.2)
$$\begin{array}{ccc} x & \longrightarrow & v \\ \downarrow & & \downarrow \\ u & \longrightarrow & w \end{array}$$

Then $u \times_w v$ is representable in $I$. Denote by $U'$, $V'$, and $W'$ the inverse images of $U$, $V$, and $W$ over $u \times_w v$ under the canonical morphisms from $u \times_w v$ to $u$, $v$, and $w$, respectively. Then $U' \times_{W'} V'$ is representable in $E_{u\times_w v}$. Since the inverse image functors of $\pi$ are left exact (VI.5.1(iii)), the diagram (VI.5.6.1) uniquely determines a morphism $X \to U' \times_{W'} V'$ over the canonical morphism $x \to u \times_w v$. Consequently, $U' \times_{W'} V'$ represents the fibered product $U \times_W V$ in $E$.

(ii) & (iii) These immediately follow from the proof of (i).

**Remark VI.5.7.** One easily verifies that the family of vertical (resp. horizontal) coverings forms a pretopology (VI.5.6). It is not the case for their union, which causes many difficulties.

**Lemma VI.5.8.** *Let $(V_m \to V)_{m\in M}$ be a* finite *vertical covering of $E$ and for every $m \in M$, let $(V_{m,n} \to V_m)_{n\in N_m}$ be a Cartesian covering of $E$. Then, there exists a Cartesian covering $(W_\ell \to V)_{\ell\in L}$ such that for all $m \in M$ and $\ell \in L$, there exist $n_\ell \in N_m$ and a $V_m$-morphism $V_m \times_V W_\ell \to V_{m,n_\ell}$; in particular, the covering $(V_m \times_V W_\ell \to V)_{m\in M,\ell\in L}$ refines the covering $(V_{m,n} \to V)_{m\in M,n\in N_m}$.*

Indeed, for every $m \in M$, $(\pi(V_{m,n}) \to \pi(V))_{n\in N_m}$ is a covering of $I$. Since $M$ is finite and $I$ is stable under fibered products, there exists a covering $(f_\ell \colon i_\ell \to \pi(V))_{\ell\in L}$ of $I$ such that for all $m \in M$ and $\ell \in L$, there exist $n_\ell \in N_m$ and a $\pi(V)$-morphism $g_{m,\ell} \colon i_\ell \to \pi(V_{m,n_\ell})$ of $I$. For any $\ell \in L$, set $W_\ell = f_\ell^+(V)$. For every $m \in M$, the morphism $g_{m,\ell}$ then induces a $V_m$-morphism $V_m \times_V W_\ell \to V_{m,n_\ell}$, giving the lemma.

**Proposition VI.5.9** (IV.6.1.3). *Suppose that for every $i \in \mathrm{Ob}(I)$, every object of $E_i$ is quasi-compact. Then a sieve $R$ of an object $V$ of $E$ is covering if and only if there exist a Cartesian covering $(V_n \to V)_{n\in N}$ and for every $n \in N$, a finite vertical covering $(V_{n,m} \to V_n)_{n\in M_n}$ such that for all $n \in N$ and $m \in M_n$, we have a $V$-morphism $V_{n,m} \to R$.*

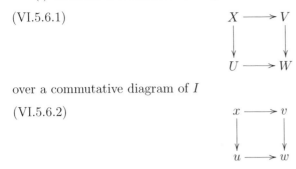

For every object $V$ of $E$, denote by $J(V)$ the set of sieves $R$ of $V$ in $E$ satisfying the required property. By definition, every sieve of $J(V)$ is covering for the covanishing topology, and every sieve generated by a Cartesian (resp. vertical) covering of $V$ belongs to $J(V)$. It therefore suffices to show that the $J(V)$, for $V \in \mathrm{Ob}(E)$, define a topology ([**2**] II 1.1). It is clear that $V$ belongs to $J(V)$ (axiom (T3) of loc. cit.). The stability under base change (axiom (T1) of loc. cit.) follows from VI.5.6(iii). It remains to establish the local character (axiom (T2) of loc. cit.). Let $V \in \mathrm{Ob}(E)$, $R$ and $R'$ two sieves of $V$ such that $R \in J(V)$ and that for every $W \in \mathrm{Ob}(E)$ and every morphism $W \to R$, the sieve $R' \times_V W$ belongs to $J(W)$. Let us show that $R'$ belongs to $J(V)$. By assumption, there exist a Cartesian covering $(V_n \to V)_{n\in N}$ and for every $n \in N$, a finite vertical covering $(V_{n,m} \to V_n)_{n\in M_n}$ such that for all $n \in N$ and $m \in M_n$, the composition $V_{n,m} \to V$ belongs to $R$. Moreover, for all $n \in N$ and $m \in M_n$, there exist a Cartesian covering $(V^{\alpha}_{n,m} \to V_{n,m})_{\alpha \in A_{n,m}}$ and for every $\alpha \in A_{n,m}$, a finite vertical covering $(V^{\alpha,\beta}_{n,m} \to V^{\alpha}_{n,m})_{\beta \in B^{\alpha}_{n,m}}$ such that for all $\alpha \in A_{n,m}$ and $\beta \in B^{\alpha}_{n,m}$, the composition $V^{\alpha,\beta}_{m,n} \to V$ belongs to $R'$. By VI.5.8, for every $n \in N$, there exists a Cartesian covering $(W_{n,\ell} \to V_n)_{\ell \in L_n}$ such that for all $m \in M_n$ and $\ell \in L_n$, there exist $\alpha_{n,m,\ell} \in A_{n,m}$ and a $V_{n,m}$-morphism $p_{n,m,\ell}\colon V_{n,m} \times_{V_n} W_{n,\ell} \to V^{\alpha_{n,m,\ell}}_{n,m}$. For any $n \in N$, $m \in M_n$, and $\ell \in L_n$, we denote by $(W^{m,\beta}_{n,\ell} \to V_{n,m} \times_{V_n} W_{n,\ell})_{\beta \in B^{\alpha_{n,m,\ell}}_{n,m}}$ the inverse image under $p_{n,m,\ell}$ of the vertical covering $(V^{\alpha_\ell,\beta}_{n,m} \to V^{\alpha_{n,m,\ell}}_{n,m})_{\beta \in B^{\alpha_{n,m,\ell}}_{n,m}}$. It is clear that $(W_{n,\ell} \to V)_{n \in N, \ell \in L_n}$ is a Cartesian covering, and that for all $n \in N$ and $\ell \in L_n$, $(W^{m,\beta}_{n,\ell} \to W_{n,\ell})_{m \in M_n, \beta \in B^{\alpha_{n,m,\ell}}_{n,m}}$ is a finite vertical covering. For all $n \in N$, $\ell \in L_n$, $m \in M_n$, and $\beta \in B^{\alpha_{n,m,\ell}}_{n,m}$, the composition $W^{\beta,m}_{n,\ell} \to V$ belongs to $R'$, giving the required property.

**Proposition VI.5.10.** *A presheaf $F = \{i \mapsto F_i\}$ on $E$ is a sheaf, if and only if the following conditions are satisfied:*

(i) *For every $i \in \mathrm{Ob}(I)$, $F_i$ is a sheaf on $E_i$.*

(ii) *For every covering family $(f_n\colon i_n \to i)_{n \in \Sigma}$ of $I$, if for every $(m,n) \in \Sigma^2$, we set $i_{mn} = i_m \times_i i_n$ and we denote by $f_{mn}\colon i_{mn} \to i$ the canonical morphism, then the sequence of morphisms of sheaves on $E_i$*

(VI.5.10.1)
$$F_i \to \prod_{n \in \Sigma} (f_n)_*(F_{i_n}) \rightrightarrows \prod_{(m,n) \in \Sigma^2} (f_{mn})_*(F_{i_{mn}})$$

*is exact.*

Indeed, after extending the universe $\mathbb{U}$, if necessary, we may assume that the category $I$ is small ([**2**] II 2.7(2)). The proposition then follows from VI.5.6 and ([**2**] II 2.3, I 3.5, I 2.12, and II 4.1(3)).

**Corollary VI.5.11.** *The functor* (VI.5.2.1) *induces an equivalence of categories between $\widetilde{E}$ and the full subcategory of $\mathbf{Hom}_{I^\circ}(I^\circ, \mathscr{F}^\vee)$ made up of sections $\{i \mapsto F_i\}$ such that for every covering family $(f_n\colon i_n \to i)_{n \in \Sigma}$ of $I$, if for every $(m,n) \in \Sigma^2$, we set $i_{mn} = i_m \times_i i_n$ and we denote by $f_{mn}\colon i_{mn} \to i$ the canonical morphism, the sequence of sheaves on $E_i$*

(VI.5.11.1)
$$F_i \to \prod_{n \in \Sigma} (f_n)_*(F_{i_n}) \rightrightarrows \prod_{(m,n) \in \Sigma^2} (f_{mn})_*(F_{i_{mn}})$$

*is exact.*

**Corollary VI.5.12.** *For every $i \in \mathrm{Ob}(I)$, the functor $\alpha_{i!}\colon E_i \to E$* (VI.5.1.2) *is continuous.*

**Remark VI.5.13.** Condition VI.5.10(ii) is equivalent to the following condition:

(ii') For every $i \in \mathrm{Ob}(I)$ and every covering sieve $\mathfrak{R}$ of $i$ in $I$, the canonical morphism

(VI.5.13.1)
$$F_i \to \lim_{\substack{\longleftarrow \\ (i',u) \in \mathfrak{R}^\circ}} u_*(F_{i'}),$$

where $u\colon i' \to i$ denotes the structural morphism, is an isomorphism.

Indeed, after extending the universe $\mathbb{U}$, if necessary, we may assume that the category $I$ is small. The assertion then follows from ([**2**] I 2.12) applied to the functor $\mathfrak{R}^\circ \to \widetilde{E}_i$, $(i',u) \mapsto u_*(F_{i'})$.

**Remark VI.5.14.** Suppose that every object of $I$ is quasi-compact. Then Proposition VI.5.10 remains true if in (ii) we restrict to *finite* covering families $(f_n\colon i_n \to i)_{n \in \Sigma}$

of $I$. Indeed, every Cartesian covering of $E$ admits a finite covering family. Consequently, the topology on $E$ is generated by the *finite* Cartesian coverings and the vertical coverings, giving the assertion.

**Remark VI.5.15.** The functors $\alpha_{i!}$ are cocontinuous for every $i \in \mathrm{Ob}(I)$ if and only if the covanishing topology of $E$ is equal to its total topology associated with $\pi$ (VI.5.4). Indeed, the latter is by definition the coarsest topology that makes the functors $\alpha_{i!}$ continuous for every $i \in \mathrm{Ob}(I)$ ([**2**] VI 7.4.1), and also the finest topology that makes the functors $\alpha_{i!}$ cocontinuous for every $i \in \mathrm{Ob}(I)$ ([**2**] VI 7.4.3(2)). The assertion therefore follows from VI.5.12. Note that the covanishing and total topologies on $E$ are not in general equal, in view of VI.5.10.

**VI.5.16.** Suppose that $I$ is equivalent to a $\mathbb{U}$-small category. It follows from VI.5.10 and ([**2**] VI 7.4.7) that the identity functor $\mathrm{id}_E\colon E \to E$ is continuous when we endow the source with the total topology associated with the fibered site $\pi$ (VI.5.4) and the target with the covanishing topology. Denote by $\mathbf{Top}(E)$ the *total topos* associated with the fibered site $\pi$, that is, the topos of sheaves of $\mathbb{U}$-sets on the total site $E$. We therefore have a canonical morphism of topos ([**2**] IV 4.9.2)

(VI.5.16.1) $$\delta\colon \widetilde{E} \to \mathbf{Top}(E)$$

such that the functor $\delta_*$ is the canonical inclusion functor of $\widetilde{E}$ in $\mathbf{Top}(E)$; it is an embedding ([**2**] IV 9.1.1). Note that the diagram

(VI.5.16.2) $$\begin{array}{ccc} \widetilde{E} & \searrow & \\ {\scriptstyle \delta_*}\downarrow & & \\ \mathbf{Top}(E) & \xrightarrow{\sim} & \mathbf{Hom}_{I^\circ}(I^\circ, \mathscr{F}^\vee) \end{array}$$

where the horizontal arrow is the canonical equivalence of categories ([**2**] VI 7.4.7) and the slanted arrow is induced by the functor (VI.5.2.1), is commutative up to canonical isomorphism. On the other hand, for every object $F$ of $\mathbf{Top}(E)$, $\delta^*(F)$ is canonically isomorphic to the sheaf associated with the presheaf $F$ on the covanishing site $E$. Indeed, after extending $\mathbb{U}$, if necessary, we may assume that the category $E$ is $\mathbb{U}$-small ([**2**] II 3.6 and III 1.5), in which case the assertion follows from ([**2**] I 5.1 and III 1.3).

**Lemma VI.5.17.** *Let $F = \{i \mapsto F_i\}$ be a presheaf on $E$. For each $i \in \mathrm{Ob}(I)$, denote by $F_i^a$ the sheaf of $\widetilde{E}_i$ associated with $F_i$. Then $\{i \mapsto F_i^a\}$ is a presheaf on $E$ and we have a canonical morphism $\{i \mapsto F_i\} \to \{i \mapsto F_i^a\}$ of $\widehat{E}$ inducing an isomorphism between the associated sheaves.*

For any morphism $f\colon i' \to i$ of $I$, denote by $\gamma_f\colon F_i \to f_*(F_{i'})$ the transition morphism of $F$ associated with $f$ (VI.5.2.1). The canonical morphism $F_{i'} \to F_{i'}^a$ induces a morphism $f_*(F_{i'}) \to f_*(F_{i'}^a)$. Since $f_*(F_{i'}^a)$ is a sheaf of $\widetilde{E}_i$, $\gamma_f$ induces a morphism

(VI.5.17.1) $$\gamma_f^a\colon F_i^a \to f_*(F_{i'}^a).$$

The morphisms $\gamma_f$ satisfy cocycle relations of type ([**1**] (1.1.2.2)), deduced from the composition of the morphisms in $I$. These induce analogous cocycle relations for the morphisms $\gamma_f^a$. Consequently, $\{i \mapsto F_i^a\}$ is a section of the fibered category $\mathscr{P}^\vee$, and is therefore a presheaf on $E$ (VI.5.2.1) (cf. [**37**] VI 12 or [**1**] 1.1.2). Moreover, we have a canonical morphism

(VI.5.17.2) $$\{i \mapsto F_i\} \to \{i \mapsto F_i^a\}.$$

Let $G = \{i \mapsto G_i\}$ be a sheaf on $E$. For every $i \in \mathrm{Ob}(I)$, $G_i$ is a sheaf of $\widetilde{E}_i$ by virtue of VI.5.10. We immediately deduce from this that the map

(VI.5.17.3) $$\mathrm{Hom}_{\widehat{E}}(\{i \mapsto F_i^a\}, \{i \mapsto G_i\}) \to \mathrm{Hom}_{\widehat{E}}(\{i \mapsto F_i\}, \{i \mapsto G_i\})$$

induced by (VI.5.17.2) is an isomorphism. Consequently, the morphism (VI.5.17.2) induces an isomorphism between the associated sheaves.

**VI.5.18.** Let $\pi' \colon E' \to I$ be a cleaved and normalized fibered $\mathbb{U}$-site satisfying the conditions of VI.5.1, and

(VI.5.18.1) $$\Phi \colon E' \to E$$

a Cartesian $I$-functor (VI.2.2). We endow $E'$ with the covanishing topology defined by $\pi'$, and we denote by $\widetilde{E}'$ the topos of sheaves of $\mathbb{U}$-sets on $E'$. We associate with $\pi'$ objects analogous to those associated with $\pi$, and we denote them by the same letters equipped with an exponent $'$. For any $i \in \mathrm{Ob}(I)$, we denote by $\Phi_i \colon E'_i \to E_i$ the functor induced by $\Phi$ on the fibers at $i$, and by $\widehat{\Phi}_i^* \colon \widehat{E}_i \to \widehat{E}'_i$ the functor obtained by composing with $\Phi_i$. For every morphism $f \colon j \to i$ of $I$, we have an isomorphism of functors

(VI.5.18.2) $$\Phi_j \circ f^+_{E'} \xrightarrow{\sim} f^+_E \circ \Phi_i,$$

where $f^+_E$ and $f^+_{E'}$ are the inverse image functors of $E$ and $E'$, respectively. It induces an isomorphism of functors

(VI.5.18.3) $$\widehat{\Phi}_i^* \circ f_{E*} \xrightarrow{\sim} f_{E'*} \circ \widehat{\Phi}_j^*,$$

where $f_{E*}$ and $f_{E'*}$ are the inverse image functors of $\mathscr{P}^\vee$ and $\mathscr{P}'^\vee$, respectively (VI.5.1.5). The isomorphisms (VI.5.18.2) satisfy a cocycle relation of type ([**1**] (1.1.2.2)), which induces an analogous relation for the isomorphisms (VI.5.18.3). By ([**37**] VI 12; cf. also [**1**] 1.1.2), the functors $\widehat{\Phi}_i^*$ therefore define a Cartesian $I^\circ$-functor

(VI.5.18.4) $$\mathscr{P}^\vee \to \mathscr{P}'^\vee.$$

One immediately verifies that the diagram of functors

(VI.5.18.5) $$\begin{array}{ccc} \widehat{E} & \xrightarrow{\sim} & \mathbf{Hom}_{I^\circ}(I^\circ, \mathscr{P}^\vee) \\ {\scriptstyle \widehat{\Phi}^*}\downarrow & & \downarrow \\ \widehat{E}' & \xrightarrow{\sim} & \mathbf{Hom}_{I^\circ}(I^\circ, \mathscr{P}'^\vee) \end{array}$$

where $\widehat{\Phi}^*$ is the functor defined by the composition with $\Phi$, the right vertical arrow is defined by the composition with (VI.5.18.4), and the horizontal arrows are the equivalences of categories (VI.5.2.1), is commutative up to canonical isomorphism. Consequently, for every presheaf $F = \{i \mapsto F_i\}$ on $E$, we have

(VI.5.18.6) $$\widehat{\Phi}^*(F) = \{i \mapsto \widehat{\Phi}_i^*(F_i)\}.$$

Suppose that for every $i \in \mathrm{Ob}(I)$, the functor $\Phi_i$ is continuous. Then the functor $\widehat{\Phi}_i^*$ induces a functor $\Phi_{i,s} \colon \widetilde{E}_i \to \widetilde{E}'_i$, that commutes with inverse limits ([**2**] III 1.2). It follows from VI.5.10 and (VI.5.18.5) that for every sheaf $F$ on $E$, $\widehat{\Phi}^*(F)$ is a sheaf on $E'$, and, consequently, that $\Phi$ is continuous for the covanishing topologies of $E$ and $E'$. It therefore induces a functor ([**2**] III 1.1.1)

(VI.5.18.7) $$\Phi_s \colon \widetilde{E} \to \widetilde{E}'.$$

If, moreover, for every $i \in \mathrm{Ob}(I)$, the functor $\Phi_{i,s} \colon \widetilde{E}_i \to \widetilde{E}'_i$ is an equivalence of categories, then $\Phi_s$ is an equivalence of categories by virtue of VI.5.11.

**Remark VI.5.19.** We consider the fibered topos $\mathscr{F}/I$ (VI.5.1.3) as a fibered $\mathbb{U}$-site by endowing each fiber with the canonical topology. This clearly satisfies the conditions of VI.5.1. Denote by

$$\varepsilon_I \colon E \to \mathscr{F} \tag{VI.5.19.1}$$

the canonical Cartesian $I$-functor, which induces on the fibers the canonical functors $\varepsilon_i \colon E_i \to \widetilde{E}_i$ (**[2]** VI (7.2.6.7)). It follows from VI.5.18 that $\varepsilon_I$ induces an equivalence between the covanishing topos associated with $E/I$ and $\mathscr{F}/I$.

**VI.5.20.** Let $I'$ be a $\mathbb{U}$-site in which fibered products are representable and $\varphi \colon I' \to I$ a continuous functor that commutes with fibered products. We denote by

$$\pi' \colon E' \to I' \tag{VI.5.20.1}$$

the base change of $\pi$ (VI.5.1.1) by $\varphi$, and by

$$\Phi \colon E' \to E \tag{VI.5.20.2}$$

the canonical projection (**[37]** VI § 3). Then $E'/I'$ is a fibered site satisfying the conditions of VI.5.1. The fibered $\mathbb{U}$-topos $\mathscr{F}' \to I'$ associated with $\pi'$ is canonically $I'$-equivalent to the fibered topos deduced from $\mathscr{F}/I$ by base change by $\varphi$. We denote by

$$\mathscr{F}'^{\vee} \to I'^{\circ}, \tag{VI.5.20.3}$$

$$\mathscr{P}'^{\vee} \to I'^{\circ}, \tag{VI.5.20.4}$$

the fibered categories associated with $\pi'$, defined in (VI.5.1.4) and (VI.5.1.5), respectively. They are canonically $I'^{\circ}$-equivalent to the fibered categories deduced from $\mathscr{F}^{\vee}/I^{\circ}$ and $\mathscr{P}^{\vee}/I^{\circ}$ by base change by $\varphi^{\circ}$.

We denote by $\widehat{E}'$ the category of presheaves of $\mathbb{U}$-sets on $E'$. One immediately verifies that the diagram of functors

$$\begin{array}{ccc} \widehat{E} & \xrightarrow{\sim} & \mathbf{Hom}_{I^{\circ}}(I^{\circ}, \mathscr{P}^{\vee}) \\ {\scriptstyle \widehat{\Phi}^*}\downarrow & & \downarrow \\ \widehat{E}' & \xrightarrow{\sim} & \mathbf{Hom}_{I'^{\circ}}(I'^{\circ}, \mathscr{P}'^{\vee}) \end{array} \tag{VI.5.20.5}$$

where the horizontal arrows are the equivalences of categories (VI.5.2.1), $\widehat{\Phi}^*$ is the functor defined by the composition with $\Phi$, and the right vertical arrow is the canonical functor (**[37]** VI § 3), is commutative up to canonical isomorphism. Consequently, for every presheaf $F = \{i \mapsto F_i\}$ on $E$, we have

$$\widehat{\Phi}^*(F) = \{i' \mapsto F_{\varphi(i')}\}, \tag{VI.5.20.6}$$

where for every $i' \in \mathrm{Ob}(I')$, we have identified the fibers $E'_{i'}$ and $E_{\varphi(i')}$.

We endow $E'$ with the covanishing topology associated with the fibered site $\pi'$, and we denote by $\widetilde{E}'$ the topos of sheaves of $\mathbb{U}$-sets on $E'$. It immediately follows from VI.5.10 and (VI.5.20.5) that for every sheaf $F$ on $E$, $\widehat{\Phi}^*(F)$ is a sheaf on $E'$, and consequently, that the functor $\Phi$ is continuous. It therefore induces a functor (**[2]** III 1.1.1)

$$\Phi_s \colon \widetilde{E} \to \widetilde{E}'. \tag{VI.5.20.7}$$

**Proposition VI.5.21.** *We keep the assumptions of* VI.5.20 *and moreover suppose that the following conditions are satisfied:*

(i) *The category $I'$ is $\mathbb{U}$-small and the functor $\varphi$ is fully faithful.*
(ii) *The topology on $I'$ is induced by that on $I$ through the functor $\varphi$.*
(iii) *Every object of $I$ can be covered by objects coming from $I'$.*

*Then the functor* $\Phi_s\colon \widetilde{E} \to \widetilde{E}'$ (VI.5.20.7) *is an equivalence of categories.*

For every object $i$ of $I$, we denote by $I'_{/i}$ the category of pairs $(i', u)$, where $i' \in \mathrm{Ob}(I')$ and $u\colon \varphi(i') \to i$ is a morphism of $I$. Let $(i', u)$, $(i'', v)$ be two objects of $I'_{/i}$. A morphism from $(i', u)$ to $(i'', v)$ is a morphism $w\colon i' \to i''$ of $I'$ such that $u = v \circ \varphi(w)$. Note that the functor $I'_{/i} \to I_{/i}$, $(i', u) \mapsto (\varphi(i'), u)$ is fully faithful.

Let us first show that for every object $F = \{i \mapsto F_i\}$ of $\widetilde{E}$ and every object $i$ of $I$, the morphism of sheaves on $E_i$

$$F_i \to \lim_{\substack{\longleftarrow \\ (i',u) \in I'^{\circ}_{/i}}} u_*(F_{\varphi(i')}) \tag{VI.5.21.1}$$

is an isomorphism. Indeed, it follows from VI.5.10 and the assumptions that the sequence

$$F_i \to \prod_{(i',u)\in \mathrm{Ob}(I'_{/i})} u_*(F_{\varphi(i')}) \rightrightarrows \prod_{((i',u),(i'',v))\in \mathrm{Ob}(I'_{/i})^2} \ \prod_{(j,w)\in \mathrm{Ob}(I'_{/\varphi(i')\times_i \varphi(i'')})} w_*(F_{\varphi(j)})$$

is exact. On the other hand, the canonical morphism

$$\lim_{\substack{\longleftarrow \\ (i',u) \in I'^{\circ}_{/i}}} u_*(F_{\varphi(i')}) \to$$

$$\ker\left(\prod_{(i',u)\in \mathrm{Ob}(I'_{/i})} u_*(F_{\varphi(i')}) \rightrightarrows \prod_{((i',u),(i'',v))\in \mathrm{Ob}(I'_{/i})^2} \ \prod_{(j,w)\in \mathrm{Ob}(I'_{/\varphi(i')\times_i \varphi(i'')})} w_*(F_{\varphi(j)})\right)$$

is a monomorphism by ([**2**] II 4.1(3)), giving the assertion.

The isomorphism (VI.5.21.1) shows that the functor $\Phi_s$ is fully faithful. Let us show that $\Phi_s$ is essentially surjective. Let $F' = \{i' \mapsto F'_{i'}\}$ be a sheaf of $\widetilde{E}'$. Since for every $i \in \mathrm{Ob}(I)$, the category $I'_{/i}$ is $\mathbb{U}$-small, we define the sheaf $F_i$ of $\widetilde{E}_i$ by the formula ([**2**] II 4.1)

$$F_i = \lim_{\substack{\longleftarrow \\ (i',u) \in I'^{\circ}_{/i}}} u_*(F'_{i'}). \tag{VI.5.21.2}$$

For every morphism $f\colon j \to i$ of $I$, the functor $f_*\colon \widetilde{E}_j \to \widetilde{E}_i$ commutes with inverse limits. We deduce from this that $\{i \mapsto F_i\}$ is a section of the fibered category $\mathscr{P}^\vee/I^\circ$ (VI.5.1.5) and is therefore a presheaf on $E$ (VI.5.2.1). We clearly have a canonical isomorphism $\Phi^*(F) \xrightarrow{\sim} F'$ (VI.5.20.6). It therefore suffices to show that $F$ is a sheaf on $E$, or, equivalently, by VI.5.10 and VI.5.13, that for every $i \in \mathrm{Ob}(I)$ and every covering sieve $\mathfrak{R}$ of $i$ in $I$, the canonical morphism

$$F_i \to \lim_{\substack{\longleftarrow \\ (j,v) \in \mathfrak{R}^{\circ}}} v_*(F_j), \tag{VI.5.21.3}$$

where $v\colon j \to i$ denotes the structural morphism, is an isomorphism. For any $j \in \mathrm{Ob}(I)$, denote by $J_{/j}$ the full subcategory of $I'_{/j}$ made up of pairs $(j', v)$ such that $\varphi(j')$ is an object of $\mathfrak{R}$. We have $J_{/j} = I'_{/j}$ if $j \in \mathrm{Ob}(\mathfrak{R})$. Since the functors $u_*$ commute with inverse limits, it suffices to show that the canonical morphism

$$F_i \to \lim_{\substack{\longleftarrow \\ (i',u) \in J^{\circ}_{/i}}} u_*(F'_{i'}) \tag{VI.5.21.4}$$

is an isomorphism. For every $(i',u) \in \mathrm{Ob}(I'_{/i})$, $J_{/\varphi(i')}$ is a covering sieve of $i'$ in $I'$ ([**2**] III 1.6). Consequently, the canonical morphism

$$F'_{i'} \to \lim_{\substack{\longleftarrow \\ (j',v)\in J^{\circ}_{/\varphi(i')}}} v_*(F'_{j'}) \tag{VI.5.21.5}$$

is an isomorphism (VI.5.13.1). Applying the functor $u_*$ and taking the inverse limit over the category $I'^{\circ}_{/i}$, we deduce from this that the canonical morphism

$$\lim_{\substack{\longleftarrow \\ (i',u)\in I'^{\circ}_{/i}}} u_*(F'_{i'}) \to \lim_{\substack{\longleftarrow \\ (j',v)\in J^{\circ}_{/i}}} v_*(F'_{j'}) \tag{VI.5.21.6}$$

is an isomorphism, and consequently that (VI.5.21.4) is an isomorphism.

**Proposition VI.5.22.** *We keep the assumptions of* VI.5.21 *and moreover suppose that for every $i' \in \mathrm{Ob}(I')$, the category $E'_{i'}$ is $\mathbb{U}$-small. Then the covanishing topology of $E'$ is induced by that on $E$ by the functor $\Phi$* (VI.5.20.2).

The functor $\Phi_s$ (VI.5.20.7) admits a left adjoint $\Phi^s$ that extends $\Phi$, that is, that fits into a diagram commutative up to isomorphism

$$\begin{array}{ccc} E & \xrightarrow{\varepsilon} & \widetilde{E} \\ {\scriptstyle\Phi}\uparrow & & \uparrow{\scriptstyle\Phi^s} \\ E' & \xrightarrow{\varepsilon'} & \widetilde{E}' \end{array} \tag{VI.5.22.1}$$

where the horizontal arrows are the canonical functors ([**2**] III 1.2). Since $\Phi_s$ is an equivalence of categories by virtue of VI.5.21, $\Phi^s$ is a quasi-inverse of $\Phi_s$. We deduce from this an isomorphism $\varepsilon' \xrightarrow{\sim} \Phi_s \circ \varepsilon \circ \Phi$. Consequently, the covanishing topology of $E'$ is induced by the canonical topology of $\widetilde{E}$ by the functor $\varepsilon \circ \Phi$ ([**2**] III 3.5).

On the other hand, it follows from the assumptions that the category $E'$ is $\mathbb{U}$-small and that the functor $\Phi$ is fully faithful. Hence by virtue of the comparison lemma ([**2**] III 4.1), if we endow $E'$ with the topology induced by that on $E$, the restriction functor from $\widetilde{E}$ to the category of sheaves of $\mathbb{U}$-sets on $E'$ is an equivalence of categories. The reasoning above then shows that the topology on $E'$ induced by that on $E$ is also induced by the canonical topology of $\widetilde{E}$ by the functor $\varepsilon \circ \Phi$, giving the proposition.

**Remark VI.5.23.** We keep the assumption of VI.5.21 and moreover suppose that for every $i' \in \mathrm{Ob}(I')$, the category $E'_{i'}$ is $\mathbb{U}$-small. We can then deduce VI.5.21 from VI.5.22 and ([**2**] III 4.1), even though we proceeded in the other direction.

**Lemma VI.5.24.** *With every $\mathbb{U}$-small category $J$ and every functor*

$$\phi\colon J \to \widetilde{E}, \quad j \mapsto F_j = \{i \mapsto F_{j,i}\} \tag{VI.5.24.1}$$

*are canonically associated the following data:*

(i) *a sheaf $\{i \mapsto \lim\limits_{\substack{\longleftarrow \\ j\in J}} F_{j,i}\}$ on $E$ and a canonical isomorphism*

$$\{i \mapsto \lim_{\substack{\longleftarrow \\ j\in J}} F_{j,i}\} \xrightarrow{\sim} \lim_{\substack{\longleftarrow \\ J}} \phi, \tag{VI.5.24.2}$$

(ii) *a presheaf $\{i \mapsto \lim\limits_{\substack{\longrightarrow \\ j\in J}} F_{j,i}\}$ on $E$, where the limits are taken in $\widetilde{E}_i$, and a canonical isomorphism*

$$\lim_{\substack{\longrightarrow \\ J}} \phi \xrightarrow{\sim} \{i \mapsto \lim_{\substack{\longrightarrow \\ j\in J}} F_{j,i}\}^a, \tag{VI.5.24.3}$$

*where the right-hand side is the sheaf on* $E$ *associated with the presheaf* $\{i \mapsto \varinjlim_{j\in J} F_{j,i}\}$.

*In particular, for every* $i \in I$, *the canonical functor*

$$\widetilde{E} \to \widetilde{E}_i, \quad \{i \mapsto F_i\} \mapsto F_i \tag{VI.5.24.4}$$

*commutes with inverse* $\mathbb{U}$*-limits.*

It immediately follows from (VI.5.2.1) and ([**2**] I 3.1) that the functor

$$\widehat{E} \to \widehat{E}_i, \quad \{i \mapsto G_i\} \mapsto G_i \tag{VI.5.24.5}$$

commutes with inverse and direct $\mathbb{U}$-limits. Moreover, with every $\mathbb{U}$-small category $J$ and every functor

$$\psi\colon J \to \widehat{E}, \quad j \mapsto G_j = \{i \mapsto G_{j,i}\} \tag{VI.5.24.6}$$

are canonically associated two presheaves $\{i \mapsto \varinjlim_{j\in J} G_{j,i}\}$ and $\{i \mapsto \varprojlim_{j\in J} G_{j,i}\}$ on $E$, and two canonical morphisms

$$\{i \mapsto \varprojlim_{j\in J} G_{j,i}\} \to \varprojlim_{J} \psi, \tag{VI.5.24.7}$$

$$\varinjlim_{J} \psi \to \{i \mapsto \varinjlim_{j\in J} G_{j,i}\}. \tag{VI.5.24.8}$$

The latter are therefore isomorphisms.

Suppose that $\psi$ is induced by a functor $\phi$ as in (VI.5.24.1). By ([**2**] II 4.1(3)), the isomorphism (VI.5.24.7) induces the isomorphism (VI.5.24.2); in particular, $\{i \mapsto \varprojlim_{j\in J} G_{j,i}\}$ is a sheaf on $E$. On the other hand, by virtue of VI.5.17 and ([**2**] II 4.1), $\{i \mapsto \varinjlim_{j\in J} F_{j,i}\}$ is a presheaf on $E$, and the isomorphism (VI.5.24.8) induces the isomorphism (VI.5.24.3).

**Proposition VI.5.25.** *Suppose that the following conditions are satisfied:*

(i) *Every object of* $I$ *is quasi-compact.*
(ii) *For every object* $i$ *of* $I$, *the topos* $\widetilde{E}_i$ *is algebraic* ([**2**] VI 2.3).
(iii) *For every morphism* $f\colon i \to j$ *of* $I$, *the morphism of topos* $f\colon \widetilde{E}_i \to \widetilde{E}_j$ *is coherent* ([**2**] VI 3.1).

*Then for every filtered* $\mathbb{U}$*-small category* $J$ *and every functor*

$$\phi\colon J \to \widetilde{E}, \quad j \mapsto F_j = \{i \mapsto F_{j,i}\}, \tag{VI.5.25.1}$$

$\{i \mapsto \varinjlim_{j\in J} F_{j,i}\}$ *is a sheaf on* $E$, *and we have a canonical isomorphism*

$$\varinjlim_{J} \phi \xrightarrow{\sim} \{i \mapsto \varinjlim_{j\in J} F_{j,i}\}. \tag{VI.5.25.2}$$

*In particular, for every* $i \in I$, *the canonical functor*

$$\widetilde{E} \to \widetilde{E}_i, \quad \{i \mapsto F_i\} \mapsto F_i \tag{VI.5.25.3}$$

*commutes with filtered direct* $\mathbb{U}$*-limits.*

In view of VI.5.24(ii), it suffices to show that $\{i \mapsto \varinjlim_{j\in J} F_{j,i}\}$ is a sheaf on $E$. Let $(f_n\colon i_n \to i)_{n\in\Sigma}$ be a *finite* covering family of $I$. For any $(m,n) \in \Sigma^2$, we set $i_{mn} = i_m \times_i i_n$ and we denote by $f_{mn}\colon i_{mn} \to i$ the canonical morphism. For every $j \in J$, the sequence

$$F_{j,i} \to \prod_{n\in\Sigma} (f_n)_*(F_{j,i_n}) \rightrightarrows \prod_{(m,n)\in\Sigma^2} (f_{mn})_*(F_{j,i_{mn}}) \tag{VI.5.25.4}$$

is exact. Since $\Sigma$ is finite, we deduce from this by taking the direct limit over $J$ that the sequence

$$\varinjlim_{j\in J} F_{j,i} \to \prod_{n\in\Sigma} \varinjlim_{j\in J} (f_n)_*(F_{j,i_n}) \rightrightarrows \prod_{(m,n)\in\Sigma^2} \varinjlim_{j\in J} (f_{mn})_*(F_{j,i_{mn}}) \tag{VI.5.25.5}$$

is exact ([**2**] II 4.3(4)). Since the functors $(f_n)_*$ and $(f_{mn})_*$ commute with filtered direct limits of sheaves of sets by virtue of ([**2**] VI 5.1 and VII 5.14), we deduce from this that the sequence

$$\varinjlim_{j\in J} F_{j,i} \to \prod_{n\in\Sigma} (f_n)_*(\varinjlim_{j\in J} F_{j,i_n}) \rightrightarrows \prod_{(m,n)\in\Sigma^2} (f_{mn})_*(\varinjlim_{j\in J} F_{j,i_{mn}}) \tag{VI.5.25.6}$$

is exact. The desired assertion follows in view of VI.5.14.

**Corollary VI.5.26.** *We keep the assumptions of* VI.5.25 *and moreover let $V$ be an object of $E$ and $v$ its image in $I$. Then $V$ is quasi-compact in $E$ if and only if it is quasi-compact in $E_v$.*

It immediately follows from the definition of the covanishing topology (VI.5.3) that if $V$ is quasi-compact in $E$, it is also quasi-compact in $E_v$. Let us show the converse implication. Denote by $\varepsilon_v\colon E_v \to \widetilde{E}_v$ the canonical functor. By VI.5.25, for every filtered $\mathbb{U}$-small category $J$ and every functor

$$J \to \widetilde{E}, \quad j \mapsto F_j = \{i \mapsto F_{j,i}\}, \tag{VI.5.26.1}$$

we can identify the canonical maps

$$\varinjlim_{j\in J} \mathrm{Hom}_{\widetilde{E}}(\varepsilon(V), F_j) \quad \to \quad \mathrm{Hom}_{\widetilde{E}}(\varepsilon(V), \varinjlim_{j\in J} F_j), \tag{VI.5.26.2}$$

$$\varinjlim_{j\in J} \mathrm{Hom}_{\widetilde{E}_v}(\varepsilon_v(V), F_{j,v}) \quad \to \quad \mathrm{Hom}_{\widetilde{E}_v}(\varepsilon_v(V), \varinjlim_{j\in J} F_{j,v}). \tag{VI.5.26.3}$$

If $V$ is quasi-compact in $E_v$, the map (VI.5.26.3) is injective by ([**2**] VI 1.2 and 1.23(i)). The same therefore holds for the map (VI.5.26.2). Consequently, $V$ is quasi-compact in $E$ by virtue of ([**2**] VI 1.23(i)).

**Proposition VI.5.27.** *Suppose that every object of $I$ is quasi-compact and that for every $i \in \mathrm{Ob}(I)$, every object of $E_i$ is quasi-compact. Then:*

(i) *For every object $i$ of $I$, the topos $\widetilde{E}_i$ is coherent.*
(ii) *For every morphism $f\colon i \to j$ of $I$, the morphism of topos $f\colon \widetilde{E}_i \to \widetilde{E}_j$ is coherent.*
(iii) *For every object $V$ of $E$, $\varepsilon(V)$ is a coherent object of $\widetilde{E}$; in particular, the topos $\widetilde{E}$ is locally coherent.*
(iv) *If, moreover, the category $E$ admits a final object $e$, then the topos $\widetilde{E}$ is coherent.*

(i) Denote by $\varepsilon_i\colon E_i \to \widetilde{E}_i$ the canonical functor. For every $U \in \mathrm{Ob}(E_i)$, $\varepsilon_i(U)$ is a coherent object of $\widetilde{E}_i$ by virtue of VI.5.1(ii) and ([**2**] VI 2.1). On the other hand, $E_i$ admits a final object $e_i$ (VI.5.1(ii)). Since $\varepsilon_i$ is left exact, $\varepsilon_i(e_i)$ is the final object of

$\widetilde{E}_i$. For every $U \in \mathrm{Ob}(E_i)$, the diagonal morphism $\delta\colon \varepsilon_i(U) \to \varepsilon_i(U) \times_{\varepsilon_i(e_i)} \varepsilon_i(U)$ is the image by $\varepsilon_i$ of the diagonal morphism $U \to U \times_{e_i} U$. Hence $\delta$ is quasi-compact because its source and target are coherent; in other words, $\varepsilon_i(U)$ is quasi-separated over $\varepsilon_i(e_i)$. Consequently, the topos $\widetilde{E}_i$ is coherent ([**2**] VI 2.3).

(ii) This follows from ([**2**] VI 3.3).

(iii) Every object of $E$ is quasi-compact by VI.5.9 (or by (i), (ii), and VI.5.26). Since fibered products are representable in $E$ (VI.5.6), the proposition follows from ([**2**] VI 2.1).

(iv) Since the canonical functor $\varepsilon\colon E \to \widetilde{E}$ is left exact, $\varepsilon(e)$ is the final object of $\widetilde{E}$. For every $V \in \mathrm{Ob}(E)$, the diagonal morphism $\delta\colon \varepsilon(V) \to \varepsilon(V) \times_{\varepsilon(e)} \varepsilon(V)$ is the image by $\varepsilon$ of the diagonal morphism $V \to V \times_e V$ (VI.5.6). Hence $\delta$ is quasi-compact because its source and target are coherent by (iii); in other words, $\varepsilon(V)$ is quasi-separated over $\varepsilon(e)$. The proposition follows in view of (iii).

**Proposition VI.5.28.** *Suppose that the following conditions are satisfied:*

(i) *There exists a $\mathbb{U}$-small, topologically generating, full subcategory $I'$ of $I$, made up of quasi-compact objects, and stable under fibered products.*
(ii) *For every object $i$ of $I'$, the topos $\widetilde{E}_i$ is coherent.*
(iii) *For every morphism $f\colon i \to j$ of $I'$, the morphism of topos $f\colon \widetilde{E}_i \to \widetilde{E}_j$ is coherent.*

*Then the topos $\widetilde{\mathrm{E}}$ is locally coherent. If, moreover, the category $I$ admits a final object that belongs to $I'$, then the topos $\widetilde{\mathrm{E}}$ is coherent.*

By VI.5.21, we may restrict to the case where $I = I'$. By virtue of VI.5.19, it suffices to show the analogous proposition for the fibered site $\mathscr{F}/I$. Denote by $\mathscr{C}$ the full subcategory of $\mathscr{F}$ made up of the objects that are coherent in their fibers. By condition (iii), the structural functor $\mathscr{C} \to I$ makes $\mathscr{C}/I$ into a cleaved and normalized fibered category. The fiber $\mathscr{C}_i$ of $\mathscr{C}$ over an object $i$ of $I$ identifies with the full subcategory of $\widetilde{E}_i$ made up of the coherent objects of $\widetilde{E}_i$. It is generating in $\widetilde{E}_i$, and is stable under fibered products by ([**2**] VI 2.2). Since the final object of $\widetilde{E}_i$ is an object of $\mathscr{C}_i$ by assumption, finite inverse limits are representable in $\mathscr{C}_i$ ([**2**] I 2.3.1). Endowing each fiber $\mathscr{C}_i$ with the topology induced by the canonical topology on $\widetilde{E}_i$, $\mathscr{C}/I$ becomes a fibered site satisfying the conditions of VI.5.1. The covanishing topos associated with $\mathscr{C}/I$ and $\mathscr{F}/I$ are equivalent by VI.5.18. On the other hand, if $I$ admits a final object $\iota$, then the final object $e$ of $\mathscr{C}_\iota$, which exists by VI.5.1(ii), is a final object of $\mathscr{C}$ (this easily follows from VI.5.1(iii); cf. VI.5.32 below). We thus reduce to the statement of Proposition VI.5.27.

**Corollary VI.5.29.** *Under the assumptions of* VI.5.28, *the topos $\widetilde{\mathrm{E}}$ has enough points.*

This follows from ([**2**] VI § 9).

**Corollary VI.5.30.** *Let $\widetilde{X}$, $\widetilde{Y}$ be two coherent $\mathbb{U}$-topos and $f\colon \widetilde{Y} \to \widetilde{X}$ a coherent morphism. Then the oriented product $\widetilde{X} \overleftarrow{\times}_{\widetilde{X}} \widetilde{Y}$ is coherent. In particular, it has enough points.*

By VI.5.31 below, there exists a $\mathbb{U}$-small and generating full subcategory $X$ (resp. $Y$) of $\widetilde{X}$ (resp. $\widetilde{Y}$), made up of coherent objects of $\widetilde{X}$ (resp. $\widetilde{Y}$) and stable under finite inverse limits, such that for every $U \in \mathrm{Ob}(X)$, $f^*(U)$ belongs to $Y$. The oriented product $\widetilde{X} \overleftarrow{\times}_{\widetilde{X}} \widetilde{Y}$ is equivalent to the covanishing topos associated with the functor $f^*\colon X \to Y$ (VI.4.10). On the other hand, for every object $V$ of $Y$, the topos $\widetilde{Y}_{/V}$ is coherent ([**2**] VI 2.4.2), and for every morphism $V' \to V$ of $Y$, the localization morphism $\widetilde{Y}_{/V'} \to \widetilde{Y}_{/V}$ is coherent ([**2**] VI 3.3). The corollary therefore follows from VI.5.5 and VI.5.28.

**Lemma VI.5.31.** *For every morphism of $\mathbb{U}$-topos $f\colon \widetilde{Y} \to \widetilde{X}$, there exists a $\mathbb{U}$-small and generating full subcategory $X$ (resp. $Y$) of $\widetilde{X}$ (resp. $\widetilde{Y}$), stable under finite inverse limits such that for every $U \in \mathrm{Ob}(X)$, $f^*(U)$ belongs to $Y$. If, moreover, $\widetilde{X}$ and $\widetilde{Y}$ are coherent and $f$ is coherent, then we may assume that $X$ (resp. $Y$) is made up of coherent objects.*

Indeed, there exists a $\mathbb{U}$-small and generating full subcategory $X$ (resp. $Y$) of $\widetilde{X}$ (resp. $\widetilde{Y}$) such that for every $U \in \mathrm{Ob}(X)$, $f^*(U)$ belongs to $Y$. By the first argument of ([**2**] IV 1.2.3), after extending $X$, and then $Y$, if necessary, we may assume that they are stable under finite inverse limits.

Next, suppose that $\widetilde{X}$ and $\widetilde{Y}$ are coherent and that $f$ is coherent. By ([**2**] VI 2.1), there exists a $\mathbb{U}$-small and generating full subcategory $X$ (resp. $Y$) of $\widetilde{X}$ (resp. $\widetilde{Y}$), made up of coherent objects, such that for every $U \in \mathrm{Ob}(X)$, $f^*(U)$ belongs to $Y$. On the other hand, the full subcategory of $\widetilde{X}$ (resp. $\widetilde{Y}$) made up of the coherent objects of $\widetilde{X}$ (resp. $\widetilde{Y}$) is stable under finite inverse limits ([**2**] VI 2.4.4). The argument of ([**2**] IV 1.2.3) then shows that after extending $X$, and then $Y$, if necessary, we may assume that they are stable under finite inverse limits.

**VI.5.32.** For the remainder of this section, in addition to the assumptions made in VI.5.1, we suppose that the following condition is satisfied:

(i') Finite inverse limits are representable in $I$.

In view of VI.5.1(i), this is equivalent to requiring that $I$ admits a final object ([**2**] I 2.3.1). We fix from now on a final object $\iota$ of $I$ and a final object $e$ of $E_\iota$ (which exists by VI.5.1(ii)). For every $i \in \mathrm{Ob}(I)$, we denote by

$$f_i\colon i \to \iota \tag{VI.5.32.1}$$

the canonical morphism. By VI.5.1(iii), $f_i^+(e)$ is a final object of $E_i$. Consequently, $e$ is a final object of $E$, and finite inverse limits are representable in $E$ by VI.5.6(i) and ([**2**] I 2.3.1).

The canonical injection functor $\alpha_{\iota!}\colon E_\iota \to E$ commutes with fibered products (VI.5.6) and transforms a final object into a final object; it is therefore left exact. On the other hand, it is continuous (VI.5.12). It then defines a morphism of topos ([**2**] IV 4.9.2)

$$\beta\colon \widetilde{E} \to \widetilde{E}_\iota. \tag{VI.5.32.2}$$

For every sheaf $F = \{i \mapsto F_i\}$ on $E$, we have a canonical isomorphism

$$\beta_*(F) \xrightarrow{\sim} F_\iota. \tag{VI.5.32.3}$$

There exists essentially a unique Cartesian section of $\pi$ (VI.5.1.1)

$$\sigma^+\colon I \to E \tag{VI.5.32.4}$$

such that $\sigma^+(\iota) = e$. It is defined, in view of ([**1**] 1.1.2), for every object $i$ of $I$, by $\sigma^+(i) = f_i^+(e)$ and for every morphism $f\colon j \to i$ of $I$ by the canonical isomorphism

$$\sigma^+(j) \xrightarrow{\sim} f^+(\sigma^+(i)). \tag{VI.5.32.5}$$

For every $i \in \mathrm{Ob}(I)$, $\sigma^+(i)$ is a final object of $E_i$, and we have a canonical morphism of functors

$$\mathrm{id}_E \to \sigma^+ \circ \pi. \tag{VI.5.32.6}$$

It immediately follows from the proof of VI.5.6 that $\sigma^+$ commutes with fibered products, and is therefore left exact. On the other hand, $\sigma^+$ is continuous by virtue of ([**2**] III 1.6).

It therefore defines a morphism of topos

$$\sigma\colon \widetilde{E} \to \widetilde{I}. \tag{VI.5.32.7}$$

**Remark VI.5.33.** The morphisms $\beta$ (VI.5.32.2) and $\sigma$ (VI.5.32.7) identify canonically with those defined from the fibered site $\mathscr{F}/I$ (VI.5.19).

**Lemma VI.5.34.** (i) *For every sheaf $F$ of $\widetilde{E}_\iota$, $\beta^*(F)$ is the sheaf associated with the presheaf $\{i \mapsto f_i^*F\}$ on $E$, where for every morphism $f\colon j \to i$ of $I$, the transition morphism*

$$f_i^*F \to f_*(f_j^*F) \tag{VI.5.34.1}$$

*is the adjoint of the canonical isomorphism $f^*(f_i^*F) \xrightarrow{\sim} f_j^*F$. Moreover, the adjunction morphism $F \to \beta_*(\beta^*F)$ factors through the morphism of presheaves on $E_\iota$*

$$F \to \{i \mapsto f_i^*F\} \circ \alpha_{\iota !} \tag{VI.5.34.2}$$

*defined by the identity of $F$.*

(ii) *For every sheaf $F$ of $\widetilde{I}$, $\sigma^*(F)$ is the sheaf associated with the presheaf $\{i \mapsto F(i)\}$ on $E$, where for any $i \in \mathrm{Ob}(I)$, we have denoted by $F(i)$ the constant presheaf on $E_i$ with value $F(i)$.*

Indeed, after extending $\mathbb{U}$, if necessary, we may assume that the categories $I$ and $E_\iota$ are $\mathbb{U}$-small ([**2**] II 3.6 and III 1.5).

(i) By ([**2**] I 5.1 and III 1.3), $\beta^*(F)$ is the sheaf associated with the presheaf $G$ on $E$ defined, for every $V \in \mathrm{Ob}(E)$, by

$$G(V) = \varinjlim_{(U,u)\in A_V^\circ} F(U), \tag{VI.5.34.3}$$

where $A_V$ is the category of pairs $(U,u)$ consisting of an object $U$ of $E_\iota$ and a morphism $u\colon V \to U$ of $E$. If $i$ denotes the image of $V$ in $I$, $A_V$ identifies with the category of pairs $(U,u)$ consisting of an object $U$ of $E_\iota$ and a morphism $u\colon V \to f_i^+(U)$ of $E_i$. It then follows from VI.5.17 that $\beta^*(F)$ is the sheaf associated with the presheaf $\{i \mapsto f_i^*F\}$ on $E$ defined by the transition morphisms (VI.5.34.1). The last assertion immediately follows from the above.

(ii) Indeed, $\sigma^*(F)$ is the sheaf associated with the presheaf $H$ on $E$ defined, for every $V \in \mathrm{Ob}(E)$, by

$$H(V) = \varinjlim_{(i,u)\in B_V^\circ} F(i), \tag{VI.5.34.4}$$

where $B_V$ is the category of pairs $(i,u)$ consisting of an object $i$ of $I$ and a morphism $u\colon V \to \sigma^+(i)$ of $E$. This category admits as initial object the pair consisting of $\pi(V)$ and the canonical morphism $V \to \sigma^+(\pi(V))$ (VI.5.32.6). We therefore have $H(V) = F(\pi(V))$.

**Proposition VI.5.35.** *Suppose that the following conditions are satisfied:*

(i) *There exists a $\mathbb{U}$-small and generating full subcategory $I'$ of $I$, made up of quasi-compact objects, and stable under finite inverse limits.*
(ii) *For every object $i$ of $I'$, the topos $\widetilde{E}_i$ is coherent.*
(iii) *For every morphism $f\colon i \to j$ of $I'$, the morphism of topos $f\colon \widetilde{E}_i \to \widetilde{E}_j$ is coherent.*

*Then the morphisms $\beta$ (VI.5.32.2) and $\sigma$ (VI.5.32.7) are coherent.*

In view of VI.5.33, proceeding as in the proof of VI.5.28, we can reduce to the case where every object of $I$ is quasi-compact and for every $i \in \mathrm{Ob}(I)$, every object of $E_i$ is quasi-compact. Consequently, every object of $E$ is quasi-compact by virtue of VI.5.27. The proposition then follows from ([**2**] VI 3.3).

**VI.5.36.** Let $\pi'\colon E' \to I$ be a cleaved and normalized fibered $\mathbb{U}$-site satisfying the conditions of VI.5.1, and

(VI.5.36.1) $$\Phi^+\colon E' \to E$$

a Cartesian $I$-functor. We endow $E'$ with the covanishing topology defined by $\pi'$, and we denote by $\widetilde{E}'$ the topos of sheaves of $\mathbb{U}$-sets on $E'$. We associate with $\pi'$ objects analogous to those associated with $\pi$, and we denote them by the same letters equipped with an exponent $'$. Suppose that for every $i \in \mathrm{Ob}(I)$, the functor $\Phi_i^+\colon E'_i \to E_i$, induced by $\Phi^+$ on the fibers at $i$, is continuous and left exact. This then defines a morphism of topos

(VI.5.36.2) $$\Phi_i\colon \widetilde{E}_i \to \widetilde{E}'_i,$$

characterized by $\Phi_{i*}(F) = F \circ \Phi_i$ and $\Phi_i^*$ extends $\Phi_i^+$ ([**2**] IV 4.9.4). We know (VI.5.18) that the functor $\Phi^+$ is continuous for the covanishing topologies of $E$ and $E'$. On the other hand, it commutes with fibered products (cf. the proof of VI.5.6(i)), and $\Phi^+(e) = \Phi_\iota^+(e)$ is a final object of $E'_\iota$ and therefore of $E'$ (VI.5.32). Consequently, $\Phi^+$ is left exact. It therefore defines a morphism of topos

(VI.5.36.3) $$\Phi\colon \widetilde{E} \to \widetilde{E}'$$

characterized by $\Phi_*(F) = F \circ \Phi^+$ and $\Phi^*$ extends $\Phi^+$. By (VI.5.18.6), for every sheaf $F = \{i \mapsto F_i\}$ on $E$, we have

(VI.5.36.4) $$\Phi_*(F) = \{i \mapsto \Phi_{i*}(F_i)\}.$$

**VI.5.37.** Let $V$ be an object of $E$ and $c = \pi(V)$. We denote by

(VI.5.37.1) $$\varpi\colon E_{/V} \to I_{/c}$$

the functor induced by $\pi$, and by

(VI.5.37.2) $$\gamma_V\colon E_{/V} \to E$$

the canonical functor. For every morphism $f\colon i \to c$ of $I$, the fiber of $\varpi$ over the object $(i, f)$ of $I_{/c}$ is canonically equivalent to the category $(E_i)_{/f^+(V)}$. Endowing $I_{/c}$ with the topology induced by that of $I$, and each fiber $(E_i)_{/f^+(V)}$ with the topology induced by that of $E_i$, $\varpi$ becomes a fibered site satisfying the conditions of VI.5.1. We then endow $E_{/V}$ with the covanishing topology associated with $\varpi$.

**Proposition VI.5.38.** *Under the assumptions of* VI.5.37, *the covanishing topology of $E_{/V}$ is induced by that of $E$ through the functor $\gamma_V$. In particular, the topos of sheaves of $\mathbb{U}$-sets on $E_{/V}$ is canonically equivalent to $\widetilde{E}_{/\varepsilon(V)}$. The restriction functor from $\widetilde{E}$ to $\widetilde{E}_{/\varepsilon(V)}$ is isomorphic to the functor*

(VI.5.38.1) $$\widetilde{E} \to \widetilde{E}_{/\varepsilon(V)}, \quad \{i \mapsto F_i\} \mapsto \{(i,f) \mapsto F_i \times f^*(V)\},$$

*where $(i, f)$ denotes an object of $I_{/c}$; in other words, $f\colon i \to c$ is a morphism of $I$.*

Let $F = \{i \mapsto F_i\}$ be a sheaf on $E$. By ([**2**] III 5.4), we have a canonical isomorphism of presheaves on $E_{/V}$

(VI.5.38.2) $$F \circ \gamma_V \xrightarrow{\sim} \{(i,f) \mapsto F_i \times f^*(V)\}.$$

For every morphism $g\colon (i,f)\to(j,h)$ of $I_{/c}$, the diagram of morphisms of topos

(VI.5.38.3)
$$\begin{array}{ccc} (\widetilde{E}_i)_{/f^*(V)} & \xrightarrow{g_\varpi} & (\widetilde{E}_j)_{/h^*(V)} \\ \downarrow & & \downarrow \\ \widetilde{E}_i & \xrightarrow{g_\pi} & \widetilde{E}_j \end{array}$$

where $g_\varpi$ and $g_\pi$ denote the morphisms of topos associated with $g$ by $\varpi$ and $\pi$, respectively, and the vertical arrows are the localization morphisms, is commutative up to canonical isomorphism ([2] IV 5.10). The transition morphism

(VI.5.38.4)
$$F_j\times h^*(V)\to g_{\varpi*}(F_i\times f^*(V))$$

of the presheaf $F\circ\gamma_V$ is the composition

(VI.5.38.5)
$$F_j\times h^*(V)\to g_{\pi*}(F_i)\times h^*(V)\xrightarrow{\sim} g_{\varpi*}(F_i\times f^*(V)),$$

where the first arrow is induced by the transition morphism of $F$, and the second arrow is the base change morphism associated with (VI.5.38.3) ([1] (1.2.2.2)), which is in fact an isomorphism.

Let $(i,f)$ be an object of $I_{/c}$ and $(g_n\colon i_n\to i)_{n\in\Sigma}$ a covering family of $I$. For any $(m,n)\in\Sigma^2$, we set $i_{mn}=i_m\times_i i_n$ and we denote by $g_{mn}\colon i_{mn}\to i$ the canonical morphism. Set $f_n=f\circ g_n$ and $f_{mn}=f\circ g_{mn}$. Since the restriction functor $\widetilde{E}_i\to(\widetilde{E}_i)_{/f^*(V)}$ admits a left adjoint, it commutes with inverse limits. Consequently, in view of (VI.5.38.5), the exact sequence of morphisms of sheaves on $E_i$

(VI.5.38.6)
$$F_i\to\prod_{n\in\Sigma}(g_n)_{\pi*}(F_{i_n})\rightrightarrows\prod_{(m,n)\in\Sigma^2}(g_{mn})_{\pi*}(F_{i_{mn}})$$

induces an exact sequence of morphisms of sheaves on $(\widetilde{E}_i)_{/f^*(V)}$

(VI.5.38.7)
$$F_i\times f^*(V)\to\prod_{n\in\Sigma}(g_n)_{\varpi*}(F_{i_n}\times f_n^*(V))\rightrightarrows\prod_{(m,n)\in\Sigma^2}(g_{mn})_{\varpi*}(F_{i_{mn}}\times f_{mn}^*(V)).$$

We deduce from this that $F\circ\gamma_V$ is a sheaf for the covanishing topology of $E_{/V}$. Hence $\gamma_V$ is continuous; in other words, the covanishing topology of $E_{/V}$ is coarser than the topology induced by that of $E$. To show the first assertion of the proposition, it suffices to show that $\gamma_V$ is cocontinuous ([2] III 2.1).

The functor $\gamma_V$ is a left adjoint of the functor

(VI.5.38.8)
$$\Phi^+\colon E\to E_{/V},\quad W\mapsto W\times V.$$

Let $G=\{(i,f)\mapsto G_{(i,f)}\}$ be a sheaf on $E_{/V}$ (where $(i,f)\in \mathrm{Ob}(I_{/c})$). For any $i\in\mathrm{Ob}(I)$, we denote by $p_i\colon i\times c\to c$ and $q_i\colon i\times c\to i$ the canonical projections, and by

(VI.5.38.9)
$$q_i'\colon(\widetilde{E}_{i\times c})_{/p_i^*(V)}\to\widetilde{E}_i$$

the composition

$$(\widetilde{E}_{i\times c})_{/p_i^*(V)}\longrightarrow\widetilde{E}_{i\times c}\xrightarrow{q_i}\widetilde{E}_i,$$

where the first arrow is the localization morphism. We then have a canonical isomorphism

(VI.5.38.10)
$$G\circ\Phi^+\simeq\{i\mapsto q'_{i*}(G_{i\times c})\}.$$

For every morphism $f\colon j\to i$ of $I$, the transition morphism

(VI.5.38.11)
$$q'_{i*}(G_{i\times c})\to f_{\pi*}(q'_{j*}(G_{j\times c}))$$

is the composition

$$q'_{i*}(G_{i\times c})\to q'_{i*}((f\times\mathrm{id}_c)_{\varpi*}(G_{j\times c}))\xrightarrow{\sim} f_{\pi*}(q'_{j*}(G_{j\times c})),$$

where the first arrow comes from the transition morphism of $F$ and the second arrow is the canonical isomorphism (VI.5.38.3). Let $(i_n \to i)_{n\in\Sigma}$ be a covering family of $I$. Since the functor $q'_{i*}$ commutes with inverse limits, the gluing relation (VI.5.10.1) for $G$ relative to the covering $(i_n \times c \to i \times c)_{n\in\Sigma}$ of $I_{/c}$ implies the analogous relation for $G \circ \Phi^+$ relative to the covering $(i_n \to i)_{n\in\Sigma}$. We deduce from this that $G \circ \Phi^+$ is a sheaf on $E$ by VI.5.10; hence $\Phi^+$ is continuous. Consequently, $\gamma_V$ is cocontinuous by virtue of ([**2**] III 2.5), giving the first statement of the proposition. The second statement follows by virtue of ([**2**] III 5.4). The last statement follows from the above, in particular from (VI.5.38.2).

**VI.5.39.** Let $c$ be an object of $I$, $e_c = \sigma^+(c) = f_c^+(e)$ (VI.5.32.1). The fibered site

(VI.5.39.1) $$\varpi\colon E_{/e_c} \to I_{/c}$$

is deduced from the fibered site $\pi$ by base change by the canonical functor $I_{/c} \to I$. If we endow $I_{/c}$ with the topology induced by that of $I$, the fibered site $\varpi$ satisfies the conditions of VI.5.1. By virtue of VI.5.38, the covanishing topology of $E_{/e_c}$ is induced by that of $E$ through the canonical functor $\gamma_c\colon E_{/e_c} \to E$. In particular, the topos of sheaves of $\mathbb{U}$-sets on $E_{/e_c}$ is canonically equivalent to $\widetilde{E}_{/\sigma^*(c)}$. The final object $\mathrm{id}_c$ of $I_{/c}$ then defines a morphism of topos (VI.5.32.2)

(VI.5.39.2) $$\beta_c\colon \widetilde{E}_{/\sigma^*(c)} \to \widetilde{E}_c.$$

We denote also by

(VI.5.39.3) $$\gamma_c\colon \widetilde{E}_{/\sigma^*(c)} \to \widetilde{E}$$

the localization morphism of $\widetilde{E}$ at $\sigma^*(c)$. By VI.5.38 and (VI.5.32.3), for every sheaf $F = \{i \mapsto F_i\}$ on $E$, we have a canonical isomorphism

(VI.5.39.4) $$\beta_{c*}(\gamma_c^*(\{i \mapsto F_i\})) \xrightarrow{\sim} F_c.$$

## VI.6. Morphisms with values in a generalized covanishing topos

**VI.6.1.** We keep the notation and conventions of VI.5, in particular, those introduced in VI.5.32. More precisely, $I$ denotes a $\mathbb{U}$-site and $\pi\colon E \to I$ a cleaved and normalized fibered $\mathbb{U}$-site, such that the following conditions are satisfied:

(i) Finite inverse limits are representable in $I$.
(ii) For every $i \in \mathrm{Ob}(I)$, finite inverse limits are representable in $E_i$.
(iii) For every morphism $f\colon i \to j$ of $I$, the inverse image functor $f^+\colon E_j \to E_i$ is continuous and left exact.

We endow $E$ with the covanishing topology defined by $\pi$, and denote by $\widetilde{E}$ the topos of sheaves of $\mathbb{U}$-sets on $E$. We moreover fix a $\mathbb{U}$-site $X$ and a continuous and left exact functor

(VI.6.1.1) $$\Psi^+\colon E \to X.$$

We denote by $\widetilde{X}$ the topos of sheaves of $\mathbb{U}$-sets on $X$ and by

(VI.6.1.2) $$\Psi\colon \widetilde{X} \to \widetilde{E}$$

the morphism of topos associated with $\Psi^+$ ([**2**] IV 4.9.2). We set

(VI.6.1.3) $$u^+ = \Psi^+ \circ \sigma^+\colon I \to X,$$

where $\sigma^+$ is the functor defined in (VI.5.32.4). Since the functor $u^+$ is continuous and left exact, we denote by

(VI.6.1.4) $$u = \sigma\Psi\colon \widetilde{X} \to \widetilde{I}$$

the associated morphism of topos.

Let $i$ be an object of $I$. Since $\sigma^+(i)$ is a final object of $E_i$, $\Psi^+$ induces a functor

$$\Psi_i^+ \colon E_i \to X_{/u^+(i)}. \tag{VI.6.1.5}$$

This commutes with fibered products (VI.5.6) and transforms a final object into a final object; it is therefore left exact. On the other hand, when we endow $X_{/u^+(i)}$ with the topology induced by that of $X$, $\Psi_i^+$ transforms a covering family into a covering family by ([**2**] III 1.6 and 5.2); it is therefore continuous. We denote by

$$\Psi_i \colon \widetilde{X}_{/u^*(i)} \to \widetilde{E}_i \tag{VI.6.1.6}$$

the morphism of topos defined by $\Psi_i^+$, and by

$$\jmath_i \colon \widetilde{X}_{/u^*(i)} \to \widetilde{X} \tag{VI.6.1.7}$$

the localization morphism of $\widetilde{X}$ at $u^*(i)$ ([**2**] IV 5.1). For every morphism $f \colon i' \to i$ of $I$, we denote by

$$\jmath_f \colon \widetilde{X}_{/u^*(i')} \to \widetilde{X}_{/u^*(i)} \tag{VI.6.1.8}$$

the localization morphism associated with $u^+(f) \colon u^+(i') \to u^+(i)$ ([**2**] IV 5.5).

Recall that we have fixed a final object $\iota$ of $I$ (VI.5.32). Since $u^+(\iota)$ is a final object of $X$, $\Psi_\iota$ is none other than the composition

$$\Psi_\iota = \beta\Psi \colon \widetilde{X} \to \widetilde{E}_\iota. \tag{VI.6.1.9}$$

For any $i \in \mathrm{Ob}(I)$, we denote by $f_i \colon i \to \iota$ the canonical morphism (VI.5.32.1). We identify $\jmath_i$ and $\jmath_{f_i}$.

From the relations $\Psi^*\sigma^* = u^*$ and $\Psi^*\beta^* = \Psi_\iota^*$, we obtain by adjunction morphisms

$$\sigma^* \to \Psi_* u^*, \tag{VI.6.1.10}$$

$$\beta^* \to \Psi_*\Psi_\iota^*. \tag{VI.6.1.11}$$

**Lemma VI.6.2.** (i) *For every morphism $f \colon i' \to i$ of $I$, the diagram of morphisms of topos*

$$\begin{array}{ccc} \widetilde{X}_{/u^*(i')} & \xrightarrow{\Psi_{i'}} & \widetilde{E}_{i'} \\ {\scriptstyle \jmath_f}\downarrow & & \downarrow{\scriptstyle f} \\ \widetilde{X}_{/u^*(i)} & \xrightarrow{\Psi_i} & \widetilde{E}_i \end{array} \tag{VI.6.2.1}$$

*is commutative up to a canonical isomorphism*

$$\Psi_i \jmath_f \xrightarrow{\sim} f\Psi_{i'}. \tag{VI.6.2.2}$$

(ii) *For every object $F$ of $\widetilde{X}$, we have a canonical isomorphism of $\widetilde{E}$*

$$\Psi_*(F) \xrightarrow{\sim} \{i \mapsto \Psi_{i*}(\jmath_i^* F)\}, \tag{VI.6.2.3}$$

*where for every morphism $f \colon i' \to i$ of $I$, the transition morphism*

$$\Psi_{i*}(\jmath_i^* F) \to f_*(\Psi_{i'*}(\jmath_{i'}^* F)) \tag{VI.6.2.4}$$

*is the composition*

$$\Psi_{i*}(\jmath_i^* F) \to \Psi_{i*}(\jmath_{f*}(\jmath_f^*(\jmath_i^* F))) \xrightarrow{\sim} \Psi_{i*}(\jmath_{f*}(\jmath_{i'}^* F)) \xrightarrow{\sim} f_*(\Psi_{i'*}(\jmath_{i'}^* F)), \tag{VI.6.2.5}$$

*in which the first morphism comes from the adjunction morphism* $\mathrm{id} \to \jmath_{f*}\jmath_f^*$, *the second morphism is the canonical isomorphism, and the last morphism comes from* (VI.6.2.2).

(iii) *For every sheaf $F$ of $\widetilde{E}_\iota$, the adjunction morphism*

$$\beta^*(F) \to \Psi_*(\Psi_\iota^* F)$$

(VI.6.1.11) *is induced by the morphism of presheaves on $E$*

$$\{i \mapsto f_i^* F\} \to \{i \mapsto \Psi_{i*}(\jmath_i^*(\Psi_\iota^* F))\} \tag{VI.6.2.6}$$

*defined, for every $i \in \mathrm{Ob}(I)$, by the morphism of $\widetilde{E}_i$*

$$f_i^* F \to \Psi_{i*}(\jmath_i^*(\Psi_\iota^* F)) \tag{VI.6.2.7}$$

*adjoint of the isomorphism* (VI.6.2.2).

(i) For every object $V$ of $E_i$, we have a canonical isomorphism in $E$

$$f^+(V) \xrightarrow{\sim} V \times_{\sigma^+(i)} \sigma^+(i'). \tag{VI.6.2.8}$$

Since $\Psi^+$ is left exact, we deduce from this that the diagram of morphisms of functors

$$\begin{array}{ccc} E_i & \xrightarrow{\Psi_i^+} & X_{/u^+(i)} \\ {\scriptstyle f^+}\downarrow & & \downarrow{\scriptstyle \jmath_f^+} \\ E_{i'} & \xrightarrow{\Psi_{i'}^+} & X_{/u^+(i')} \end{array} \tag{VI.6.2.9}$$

where $\jmath_f^+$ is the base change functor by $u^+(f)$, is commutative up to canonical isomorphism. The assertion follows in view of the interpretation of the functor $\jmath_f^*$ as a base change functor by the morphism $u^*(f)$ ([**2**] III 5.4).

(ii) This immediately follows from the definitions.

(iii) Set $\beta^*(F) = \{i \mapsto G_i\}$. By (ii), the morphism $\beta^*(F) \to \Psi_*(\Psi_\iota^* F)$ (VI.6.1.11) is defined, for every $i \in \mathrm{Ob}(I)$, by a morphism of $\widetilde{E}_i$

$$s_i \colon G_i \to \Psi_{i*}(\jmath_i^*(\Psi_\iota^* F)). \tag{VI.6.2.10}$$

By virtue of VI.5.34(i), we have a canonical morphism of presheaves on $E$

$$\{i \mapsto f_i^* F\} \to \{i \mapsto G_i\}, \tag{VI.6.2.11}$$

defined, for every $i \in \mathrm{Ob}(I)$, by a morphism $t_i \colon f_i^* F \to G_i$ of $\widetilde{E}_i$. The diagram

$$\begin{array}{ccccc} f_i^* F & \xrightarrow{f_i^*(t_\iota)} & f_i^*(G_\iota) & \xrightarrow{f_i^*(s_\iota)} & f_i^*(\Psi_{\iota*}(\Psi_\iota^* F)) \\ & \searrow^{t_i} & \downarrow & & \downarrow \\ & & G_i & \xrightarrow{s_i} & \Psi_{i*}(\jmath_i^*(\Psi_\iota^* F)) \end{array} \tag{VI.6.2.12}$$

where the vertical arrows are the adjoints of the transition morphisms, is commutative. On the one hand, $s_\iota$ identifies with the morphism

$$\beta_*(\beta^* F) \to \beta_*(\Psi_*(\Psi_\iota^* F)) \tag{VI.6.2.13}$$

induced by (VI.6.1.11). On the other hand, $t_\iota$ identifies with the adjunction morphism $F \to \beta_*(\beta^* F)$ by virtue of VI.5.34(i). Consequently, in view of the definition of (VI.6.1.11), $s_\iota \circ t_\iota$ is the adjunction morphism $F \to \Psi_{\iota*}(\Psi_\iota^* F)$. It then follows from (ii) and (VI.6.2.12) that the morphism $s_i \circ t_i$ is the composition

$$f_i^* F \to f_i^*(\Psi_{\iota*}(\Psi_\iota^* F)) \to \Psi_{i*}(\jmath_i^*(\Psi_\iota^* F)), \tag{VI.6.2.14}$$

where the first arrow comes from the adjunction morphism $\mathrm{id} \to \Psi_{\iota *}\Psi_{\iota}^*$ and the second arrow is the base change morphism associated with the diagram (VI.6.2.1) (for $f = f_i$). By ([**2**] XVII 2.1.3), the adjoint of $s_i \circ t_i$ is the composition

$$\Psi_i^*(f_i^*F) \to \Psi_i^*(f_i^*(\Psi_{\iota *}(\Psi_{\iota}^*F))) \xrightarrow{\sim} \jmath_i^*(\Psi_{\iota}^*(\Psi_{\iota *}(\Psi_{\iota}^*F))) \to \jmath_i^*(\Psi_{\iota}^*F), \tag{VI.6.2.15}$$

where the first arrow comes from the adjunction morphism $\mathrm{id} \to \Psi_{\iota *}\Psi_{\iota}^*$, the second arrow is the isomorphism (VI.6.2.2), and the third arrow comes from the adjunction morphism $\Psi_{\iota}^*\Psi_{\iota *} \to \mathrm{id}$. This composed morphism is equal to the composition

$$\Psi_i^*(f_i^*F) \xrightarrow{\sim} \jmath_i^*(\Psi_{\iota}^*F) \xrightarrow{u} \jmath_i^*(\Psi_{\iota}^*(\Psi_{\iota *}(\Psi_{\iota}^*F))) \xrightarrow{v} \jmath_i^*(\Psi_{\iota}^*F), \tag{VI.6.2.16}$$

where the first arrow is the isomorphism (VI.6.2.2), $u$ comes from the adjunction morphism $\mathrm{id} \to \Psi_{\iota *}\Psi_{\iota}^*$, and $v$ comes from the adjunction morphism $\Psi_{\iota}^*\Psi_{\iota *} \to \mathrm{id}$. Since $v \circ u$ is the identity, the assertion follows.

**Proposition VI.6.3.** *Suppose that, for every $i \in \mathrm{Ob}(I)$, the adjunction morphism* $\mathrm{id} \to \Psi_{i*}\Psi_i^*$ *is an isomorphism. Then:*

(i) *For every sheaf $F$ of $\widetilde{E}_{\iota}$, $\beta^*(F)$ is the sheaf on $E$ defined by $\{i \mapsto f_i^*F\}$.*
(ii) *The adjunction morphism* $\mathrm{id} \to \beta_*\beta^*$ *is an isomorphism.*
(iii) *The adjunction morphism $\beta^* \to \Psi_*\Psi_{\iota}^*$ (VI.6.1.11) is an isomorphism.*

(i) Indeed, for every $i \in \mathrm{Ob}(I)$, the morphism

$$f_i^*F \to \Psi_{i*}(\jmath_i^*(\Psi_{\iota}^*F)) \tag{VI.6.3.1}$$

defined in (VI.6.2.7), is the composition

$$f_i^*F \to \Psi_{i*}(\Psi_i^*(f_i^*F)) \xrightarrow{\sim} \Psi_{i*}(\jmath_i^*(\Psi_{\iota}^*F)),$$

where the first morphism comes from the adjunction morphism $\mathrm{id} \to \Psi_{i*}\Psi_i^*$ and the second from (VI.6.2.2). It is therefore an isomorphism. Consequently, the morphism (VI.6.2.6) is an isomorphism of $\widehat{E}$

$$\{i \mapsto f_i^*F\} \xrightarrow{\sim} \{i \mapsto \Psi_{i*}(\jmath_i^*(\Psi_{\iota}^*F))\}. \tag{VI.6.3.2}$$

Since the target of this morphism is a sheaf on $E$ (VI.6.2.3), the same holds for its source. The proposition then follows from VI.5.34(i).

(ii) This immediately follows from (i) and VI.5.34(i).

(iii) This follows from VI.6.2(iii) and from the proof of (i).

**VI.6.4.** For the remainder of this section, in addition to the general assumptions made in VI.6.1, we suppose that finite inverse limits are representable in $X$. We denote by

$$\varpi \colon D \to I \tag{VI.6.4.1}$$

the fibered $\mathbb{U}$-site associated with the functor $u^+ \colon I \to X$ (VI.6.1.3) defined in VI.5.5: the fiber of $D$ over any object $i$ of $I$ is the category $X_{/u^+(i)}$, and for every morphism $f \colon i' \to i$ of $I$, the inverse image functor $\jmath_f^+ \colon X_{/u^+(i)} \to X_{/u^+(i')}$ is the base change functor by $u^+(f)$. We endow $D$ with the covanishing topology associated with $\varpi$. We thus obtain the covanishing site associated with the functor $u^+$ defined in VI.4.1, whose topos of sheaves of $\mathbb{U}$-sets is $\widetilde{I} \overleftarrow{\times}_{\widetilde{I}} \widetilde{X}$ (VI.4.10).

By ([**37**] VI 12; cf. also [**1**] 1.1.2), the functors $\Psi_i^+$ (VI.6.1.5) for every $i \in \mathrm{Ob}(I)$ and the isomorphisms (VI.6.2.2) define a Cartesian $I$-functor

$$\rho^+ \colon E \to D. \tag{VI.6.4.2}$$

For every $V \in \mathrm{Ob}(E)$, the canonical morphism $V \mapsto \sigma^+(\pi(V))$ (VI.5.32.6) induces a morphism $\Psi^+(V) \to u^+(\pi(V))$, and we have

(VI.6.4.3) $$\rho^+(V) = (\Psi^+(V) \to \pi(V)).$$

The functor $\rho^+$ transforms a final object into a final object and commutes with fibered products; it is therefore left exact. On the other hand, it is continuous by virtue of VI.5.18. It therefore defines a morphism of topos ([**2**] IV 4.9.2)

(VI.6.4.4) $$\rho\colon \widetilde{I} \overleftarrow{\times}_{\widetilde{I}} \widetilde{X} \to \widetilde{E}.$$

For every sheaf $F = \{i \mapsto F_i\}$ on $D$, we have

(VI.6.4.5) $$\rho_*(F) = \{i \mapsto \Psi_{i*}(F_i)\}.$$

It immediately follows from the definitions that the squares of the diagram

(VI.6.4.6) $$\begin{array}{ccccc} \widetilde{I} & \xleftarrow{\mathrm{p}_1} & \widetilde{I} \overleftarrow{\times}_{\widetilde{I}} \widetilde{X} & \xrightarrow{\mathrm{p}_2} & \widetilde{X} \\ \Big\| & & \Big\downarrow{\scriptstyle \rho} & & \Big\downarrow{\scriptstyle \Psi_\iota} \\ \widetilde{I} & \xleftarrow{\sigma} & \widetilde{E} & \xrightarrow{\beta} & \widetilde{E}_\iota \end{array}$$

and the diagram

(VI.6.4.7) $$\begin{array}{ccc} \widetilde{X} & \xrightarrow{\Psi_D} & \widetilde{I} \overleftarrow{\times}_{\widetilde{I}} \widetilde{X} \\ & {\scriptstyle \Psi} \searrow & \Big\downarrow{\scriptstyle \rho} \\ & & \widetilde{E} \end{array}$$

where $\Psi_D$ is the morphism (VI.4.13.4), are commutative up to canonical isomorphisms.

**Proposition VI.6.5.** *For every sheaf $F = \{i \mapsto F_i\}$ of $\widetilde{E}$, $\rho^*(F)$ is the sheaf associated with the presheaf on $D$ defined by $\{i \mapsto \Psi_i^*(F_i)\}$, where for every morphism $f\colon i' \to i$ of $I$, the transition morphism*

(VI.6.5.1) $$\Psi_i^*(F_i) \to \jmath_{f*}(\Psi_{i'}^* F_{i'})$$

*is the adjoint of the composition*

$$\jmath_f^*(\Psi_i^* F_i) \xrightarrow{\sim} \Psi_{i'}^*(f^* F_i), \to \Psi_{i'}^* F_{i'},$$

*where the first arrow is the isomorphism* (VI.6.2.2) *and the second arrow is induced by the transition morphisms of $F$.*

Indeed, after extending $\mathbb{U}$, if necessary, we may assume that the category $E$ is $\mathbb{U}$-small ([**2**] II 3.6 and III 1.5). By ([**2**] I 5.1 and III 1.3), the sheaf $\rho^*(F)$ is the sheaf on $D$ associated with the presheaf $G$ defined, for every $U \in \mathrm{Ob}(D)$, by

(VI.6.5.2) $$G(U) = \lim_{\substack{\longrightarrow \\ (V,u) \in A_U^\circ}} F(V),$$

where $A_U$ is the category of pairs $(V, u)$ consisting of an object $V$ of $E$ and a morphism $u\colon U \to \rho^+(V)$ of $D$. We set $i = \varpi(U)$. We consider $U$ as an object of $X_{/u^+(i)}$ and we denote by $B_U$ the category of pairs $(W, v)$ consisting of an object $W$ of $E_i$ and a $u^+(i)$-morphism $v\colon U \to \Psi_i^+(W)$ of $X$. The categories $A_U$ and $B_U$ are clearly cofiltered. Every object $(W, v)$ of $B_U$ can be naturally considered as an object of $A_U$. We thus define a fully faithful functor

$$\varphi\colon B_U \to A_U.$$

For every object $(V, u)$ of $A_U$, $u$ induces a morphism $f\colon i \to \pi(V)$ of $I$ and a $u^+(i)$-morphism $v\colon U \to \Psi_i^+(f^+V)$ of $X$, so that $(f^+(V), v)$ is an object of $B_U$. Moreover, we have a canonical morphism $\varphi((f^+(V), v)) \to (V, u)$ of $A_U$. It then follows from ([**2**] I 8.1.3(c)) that the functor $\varphi^\circ$ is cofinal. Consequently, $\varphi$ induces an isomorphism

(VI.6.5.3)
$$G(U) \simeq \varinjlim_{(W,v)\in B_U^\circ} F_i(W).$$

Hence by virtue of VI.5.17, $\rho^*(F)$ is the sheaf on $D$ associated with the presheaf $\{i \mapsto \Psi_i^*(F_i)\}$ defined by the transition morphisms (VI.6.5.1).

## VI.7. Ringed total topos

**VI.7.1.** In this section, $I$ denotes a category equivalent to a $\mathbb{U}$-small category and

(VI.7.1.1)
$$\pi\colon E \to I$$

a cleaved and normalized fibered $\mathbb{U}$-site ([**2**] VI 7.2.1). For any $i \in \mathrm{Ob}(I)$, we denote by $E_i$ the fiber of $E$ over $i$, which we always consider as endowed with the topology given by $\pi$, by $\widetilde{E}_i$ the topos of sheaves of $\mathbb{U}$-sets on $E_i$, and by

(VI.7.1.2)
$$\alpha_{i!}\colon E_i \to E$$

the canonical inclusion functor. We denote by

(VI.7.1.3)
$$\mathscr{F} \to I$$

the fibered $\mathbb{U}$-topos associated with $\pi$ ([**2**] VI 7.2.6) and by

(VI.7.1.4)
$$\mathscr{F}^\vee \to I^\circ$$

the fibered category obtained by associating with each $i \in \mathrm{Ob}(I)$ the category $\widetilde{E}_i$, and with each morphism $f\colon i \to j$ of $I$ the functor $f_*\colon \widetilde{E}_i \to \widetilde{E}_j$ direct image by the morphism of topos $f\colon \widetilde{E}_i \to \widetilde{E}_j$.

We endow $E$ with the total topology associated with $\pi$ ([**2**] VI 7.4.1); it is a $\mathbb{U}$-site ([**2**] VI 7.4.3(3)). We denote by $\mathbf{Top}(E)$ the topos of sheaves of $\mathbb{U}$-sets on $E$, called the total topos associated with $\pi$. For every $i \in \mathrm{Ob}(I)$, the functor $\alpha_{i!}$ being cocontinuous ([**2**] 7.4.2), it defines a morphism of topos ([**2**] 4.7)

(VI.7.1.5)
$$\alpha_i\colon \widetilde{E}_i \to \mathbf{Top}(E).$$

Since, moreover, $\alpha_{i!}$ is continuous, the functor $\alpha_i^*$ admits a left adjoint that we denote also by

(VI.7.1.6)
$$\alpha_{i!}\colon \widetilde{E}_i \to \mathbf{Top}(E),$$

which extends the functor $\alpha_{i!}\colon E_i \to E$. Recall ([**2**] VI 7.4.5) that for every morphism $f\colon i \to j$ of $I$, the diagram

(VI.7.1.7)
$$\begin{array}{ccc} \widetilde{E}_i & \xrightarrow{\alpha_i} & \mathbf{Top}(E) \\ {\scriptstyle f}\big\downarrow & \nearrow_{\alpha_j} & \\ \widetilde{E}_j & & \end{array}$$

is not commutative in general, but there exists a 2-morphism of topos

(VI.7.1.8)
$$\alpha_i \to \alpha_j f,$$

in other words, a morphism of functors $f^* \circ \alpha_j^* \to \alpha_i^*$, or by adjunction a morphism of functors

$$\alpha_j^* \to f_* \circ \alpha_i^*. \tag{VI.7.1.9}$$

The latter satisfy a cocycle relation of type ([1] (1.1.2.2)). They therefore induce a functor

$$\begin{array}{rcl} \mathbf{Top}(E) & \to & \mathbf{Hom}_{I^\circ}(I^\circ, \mathscr{F}^\vee) \\ F & \mapsto & \{i \mapsto \alpha_i^*(F)\}, \end{array} \tag{VI.7.1.10}$$

which is in fact an equivalence of categories ([2] VI 7.4.7). From now on, we will identify $F$ with the section $\{i \mapsto F_i\}$ that is associated with it.

**Definition VI.7.2.** We say that an object $F$ of $\mathbf{Top}(E)$ is *Cartesian* if the section $\{i \mapsto F_i\}$ of $\mathscr{F}^\vee \to I^\circ$ that corresponds to it through the equivalence of categories (VI.7.1.10) is Cartesian, in other words, if for every morphism $f: i \to j$ of $I$, the transition morphism $F_j \to f_*(F_i)$ is an isomorphism.

If the category $I$ is cofiltered and $\mathbb{U}$-small, then the inverse limit of the fibered topos $\mathscr{F} \to I$ exists ([2] VI 8.2.3), and the underlying category is canonically equivalent to the subcategory of Cartesian objects of $\mathbf{Top}(E)$ ([2] VI 8.2.9).

**VI.7.3.** Let $F = \{i \mapsto F_i\}$ and $G = \{i \mapsto G_i\}$ be two objects of $\mathbf{Top}(E)$ such that $G$ is Cartesian. Then we have a canonical isomorphism

$$\mathrm{Hom}_{\mathbf{Top}(E)}(F, G) \xrightarrow{\sim} \lim_{\substack{\longleftarrow \\ i \in I}} \mathrm{Hom}_{E_i}(F_i, G_i), \tag{VI.7.3.1}$$

where for every morphism $f: i \to j$ of $I$, the transition morphism

$$d_f: \mathrm{Hom}_{E_i}(F_i, G_i) \to \mathrm{Hom}_{E_j}(F_j, G_j)$$

of the inverse system that appears in (VI.7.3.1) associates with each morphism $u: M_i \to N_i$ the morphism $d_f(u)$ composition of

$$M_j \longrightarrow f_*(M_i) \xrightarrow{f_*(u)} f_*(N_i) \xrightarrow{\sim} N_j,$$

where the first arrow is the transition morphism of $M$ and the last arrow is the inverse of the transition isomorphism $N_j \xrightarrow{\sim} f_*(N_i)$ of $N$.

**VI.7.4.** In addition to the data fixed above, we fix a sheaf of rings $R$ of $\mathbf{Top}(E)$, in other words, a ringed structure on the fibered topos $\mathscr{F}$ in the terminology of ([2] VI 8.6.1). This corresponds to fixing, for every $i \in \mathrm{Ob}(I)$, a ring $R_i$ of $\widetilde{E}_i$, and for every morphism $f: i \to j$ of $I$, a ring homomorphism $R_j \to f_*(R_i)$, these homomorphisms being subject to compatibility relations (VI.7.1.10). The morphisms of topos $f: \widetilde{E}_i \to \widetilde{E}_j$ are therefore morphisms of ringed topos (by $R_i$ and $R_j$, respectively). For modules, we use the notation $f^{-1}$ to denote the inverse image in the sense of abelian sheaves, and we keep the notation $f^*$ for the inverse image in the sense of modules.

On the other hand, for every $i \in \mathrm{Ob}(I)$, the morphism of topos $\alpha_i: \widetilde{E}_i \to \mathbf{Top}(E)$ (VI.7.1.5) is a morphism of ringed topos (by $R_i$ and $R$, respectively). Note that since $R_i = \alpha_i^*(R)$, there is no difference for modules between the inverse image in the sense of abelian sheaves and the inverse image in the sense of modules.

Giving a structure of left (resp. right) $R$-module on a sheaf $M = \{i \mapsto M_i\}$ of $\mathbf{Top}(E)$ is equivalent to giving, for every $i \in \mathrm{Ob}(I)$, a structure of left (resp. right) $R_i$-module on $M_i$ such that for every morphism $f: i \to j$ of $I$, the transition morphism $M_j \to f_*(M_i)$ is $R_j$-linear.

**Lemma VI.7.5.** *Let* $M = \{i \mapsto M_i\}$ *be a right* $R$*-module of* $\mathbf{Top}(E)$ *and* $N = \{i \mapsto N_i\}$ *a left* $R$*-module of* $\mathbf{Top}(E)$*. Then we have a bifunctorial canonical isomorphism of abelian sheaves on* $E$

(VI.7.5.1) $$M \otimes_R N \xrightarrow{\sim} \{i \mapsto M_i \otimes_{R_i} N_i\}.$$

Indeed, for every $i \in \mathrm{Ob}(I)$, we have a canonical isomorphism ([**2**] IV 13.4)

(VI.7.5.2) $$\alpha_i^*(M \otimes_R N) \xrightarrow{\sim} \alpha_i^*(M) \otimes_{\alpha_i^*(R)} \alpha_i^*(N),$$

giving the lemma.

**VI.7.6.** Let $\pi' \colon E' \to I$ be a cleaved and normalized fibered $\mathbb{U}$-site and

(VI.7.6.1) $$\Phi^+ \colon E' \to E$$

a Cartesian $I$-functor. We endow $E'$ with the total topology associated with $\pi'$, and we denote by $\mathbf{Top}(E')$ the topos of sheaves of $\mathbb{U}$-sets on $E'$. We associate with $\pi'$ objects analogous to those associated with $\pi$ (VI.7.1), which we denote by the same letters equipped with an exponent $'$. Suppose that for every $i \in \mathrm{Ob}(I)$, the functor $\Phi_i^+ \colon E'_i \to E_i$, induced by $\Phi^+$ on the fibers at $i$, is a morphism from the site $E_i$ to the site $E'_i$, and denote by

(VI.7.6.2) $$\Phi_i \colon \widetilde{E}_i \to \widetilde{E}'_i$$

the associated morphism of topos. The morphisms $\Phi_i$ define a Cartesian morphism of fibered topos ([**2**] VI 7.1.5 and 7.1.7)

(VI.7.6.3) $$\mathscr{F} \to \mathscr{F}'.$$

By virtue of ([**2**] 7.4.10), $\Phi^+$ is a morphism from the total site $E$ to the total site $E'$. It therefore defines a morphism of topos

(VI.7.6.4) $$\Psi \colon \mathbf{Top}(E) \to \mathbf{Top}(E')$$

such that for every sheaf $F = \{i \mapsto F_i\}$ on $E$, we have

(VI.7.6.5) $$\Psi_*(F) = \{i \mapsto \Phi_{i*}(F_i)\}.$$

Let $R' = \{i \mapsto R'_i\}$ be a ring of $\mathbf{Top}(E')$ and $h \colon R' \to \Psi_*(R)$ a ring homomorphism, so that $\Psi \colon \mathbf{Top}(E) \to \mathbf{Top}(E')$ is a morphism of ringed topos (by $R$ and $R'$, respectively). Giving $h$ is equivalent to giving, for every $i \in \mathrm{Ob}(I)$, a ring homomorphism $h_i \colon R'_i \to \Phi_{i*}(R_i)$ satisfying compatibility relations. In particular, for every $i \in \mathrm{Ob}(I)$, $\Phi_i \colon \widetilde{E}_i \to \widetilde{E}'_i$ is a morphism of ringed topos (by $R_i$ and $R'_i$, respectively). We denote by $\mathrm{R}^n\Psi_*$ and $\mathrm{R}^n\Phi_{i*}$ ($n \in \mathbb{N}$) the right derived functors of the functors (cf. VI.2.4)

$$\begin{aligned} \Psi_* \colon \mathbf{Mod}(R, \mathbf{Top}(E)) &\to \mathbf{Mod}(R', \mathbf{Top}(E')), \\ \Phi_{i*} \colon \mathbf{Mod}(R_i, \widetilde{E}_i) &\to \mathbf{Mod}(R'_i, \widetilde{E}'_i). \end{aligned}$$

**Proposition VI.7.7.** *Under the assumptions of* VI.7.6*, let, moreover* $M = \{i \mapsto M_i\}$ *be an* $R$*-module of* $\mathbf{Top}(E)$ *and* $n$ *an integer* $\geq 0$*. We then have a functorial and canonical* $R'$*-isomorphism*

(VI.7.7.1) $$\mathrm{R}^n\Psi_*(M) \xrightarrow{\sim} \{i \mapsto \mathrm{R}^n\Phi_{i*}(M_i)\}.$$

Indeed, for every $c \in \mathrm{Ob}(I)$, the functor

(VI.7.7.2) $$\mathbf{Mod}(R, \widetilde{E}) \to \mathbf{Mod}(R_c, \widetilde{E}_c), \quad M = \{i \mapsto M_i\} \mapsto M_c = \alpha_c^*(M)$$

is additive and exact. On the other hand, for every injective $R$-module $M = \{i \mapsto M_i\}$, $M_c$ is flabby by virtue of ([**2**] VI 8.7.2). The proposition therefore follows from (VI.7.6.5).

Let us describe explicitly the transition morphisms of the sheaf $\{i \mapsto \mathrm{R}^n\Phi_{i*}(M_i)\}$. The morphism of ringed fibered topos $(\mathscr{F}, R) \to (\mathscr{F}', R')$ induces for every morphism $f\colon i \to j$ of $I$ a diagram of morphisms of ringed topos, commutative up to canonical isomorphism ([**1**] 1.2.3)

$$(\text{VI.7.7.3})\qquad \begin{array}{ccc} (\widetilde{E}_i, R_i) & \xrightarrow{\Phi_i} & (\widetilde{E}'_i, R'_i) \\ {\scriptstyle f_E}\downarrow & & \downarrow{\scriptstyle f_{E'}} \\ (\widetilde{E}_j, R_j) & \xrightarrow{\Phi_j} & (\widetilde{E}'_j, R'_j) \end{array}$$

where the vertical arrows, usually denoted by $f$, have been equipped with an index $E$ or $E'$ to distinguish them. We leave it to the reader to verify that the transition morphism associated with $f$ is the composition

$$(\text{VI.7.7.4})\qquad \mathrm{R}^n\Phi_{j*}(M_j) \to \mathrm{R}^n\Phi_{j*}(f_{E*}M_i) \to \\ \mathrm{R}^n(\Phi_j f_E)_*(M_i) \simeq \mathrm{R}^n(f_{E'}\Phi_i)_*(M_i) \to f_{E'*}(\mathrm{R}^n\Phi_{i*}(M_i)),$$

where the first arrow comes from the transition morphism of $M$, the second arrow and the last arrow are induced by the Cartan–Leray spectral sequence ([**2**] V 5.4), and the third isomorphism comes from (VI.7.7.3).

**Proposition VI.7.8** ([**2**] VI 7.4.15). *For every $R$-module $M = \{i \mapsto M_i\}$ of $\widetilde{E}$, we have a functorial and canonical spectral sequence*

$$(\text{VI.7.8.1})\qquad \mathrm{E}_2^{p,q} = \mathrm{R}^p \varprojlim_{i\in I^\circ} \mathrm{H}^q(\widetilde{E}_i, M_i) \Rightarrow \mathrm{H}^{p+q}(\mathbf{Top}(E), M).$$

Consider the constant fibered topos $\varpi\colon \mathbf{Ens} \times I \to I$ with fiber the punctual topos **Ens** (VI.2.1). By (VI.7.1.10), the topos of sheaves of $\mathbb{U}$-sets on the total site $\mathbf{Ens} \times I$ associated with $\varpi$ is equivalent to the category $\widehat{I}$ of presheaves of $\mathbb{U}$-sets on $I$. We have a Cartesian morphism of fibered topos

$$(\text{VI.7.8.2})\qquad \Phi\colon \mathscr{F} \to \mathbf{Ens} \times I$$

whose fiber over $i \in \mathrm{Ob}(I)$ is the canonical morphism of topos $\Phi_i\colon \widetilde{E}_i \to \mathbf{Ens}$ ([**2**] IV 2.2). By ([**2**] VI 7.4.10), $\Phi$ defines a morphism of topos

$$(\text{VI.7.8.3})\qquad \Psi\colon \mathbf{Top}(E) \to \widehat{I}$$

such that for every sheaf $F = \{i \mapsto F_i\}$ on $E$, we have

$$(\text{VI.7.8.4})\qquad \Psi_*(F) = \{i \mapsto \Gamma(\widetilde{E}_i, F_i)\}.$$

We consider $\Psi$ as a morphism of ringed topos (by $R$ and $\Psi_*(R)$, respectively). By VI.7.7, for every $R$-module $M = \{i \mapsto M_i\}$ of $\mathbf{Top}(E)$ and every integer $q \geq 0$, we have a functorial and canonical isomorphism

$$(\text{VI.7.8.5})\qquad \mathrm{R}^q\Psi_*(M) \xrightarrow{\sim} \{i \mapsto \mathrm{H}^q(\widetilde{E}_i, M_i)\}.$$

On the other hand, for every abelian group $N$ of $\widehat{I}$ and every integer $p \geq 0$, we have a functorial and canonical isomorphism ([**2**] V 2.3.1)

$$(\text{VI.7.8.6})\qquad \mathrm{H}^p(\widehat{I}, N) = \mathrm{R}^p \varprojlim_{i\in I^\circ} N(i).$$

We then take for (VI.7.8.1) the Cartan–Leray spectral sequence associated with $\Psi$ ([**2**] V 5.3).

**Corollary VI.7.9.** *Suppose that $I$ admits a final object $\iota$. Then for every $R$-module $M = \{i \mapsto M_i\}$ of $\widetilde{E}$ and every integer $n \geq 0$, we have a functorial and canonical isomorphism*

$$\mathrm{H}^n(\mathbf{Top}(E), M) \xrightarrow{\sim} \mathrm{H}^n(\widetilde{E}_\iota, M_\iota). \qquad \text{(VI.7.9.1)}$$

Indeed, with the notation of the proof of VI.7.8, the functor $\widehat{I} \to \mathbf{Ens}$, $N \mapsto \Gamma(\widehat{I}, N) = N(\iota)$ is exact. Consequently, $\mathrm{R}^p \lim\limits_{\substack{\longleftarrow \\ i \in I^\circ}} = 0$ for every $p \geq 1$. The proposition then follows from VI.7.8.

**Corollary VI.7.10.** *Suppose that $I$ is the filtered category defined by the ordered set of natural numbers $\mathbb{N}$. Then for every $R$-module $M = \{i \mapsto M_i\}$ of $\widetilde{E}$ and every integer $n \geq 0$, we have a functorial and canonical exact sequence*

$$0 \to \mathrm{R}^1 \lim_{\substack{\longleftarrow \\ i \in \mathbb{N}^\circ}} \mathrm{H}^{n-1}(\widetilde{E}_i, M_i) \to \mathrm{H}^n(\mathbf{Top}(E), M) \to \lim_{\substack{\longleftarrow \\ i \in \mathbb{N}^\circ}} \mathrm{H}^n(\widetilde{E}_i, M_i) \to 0, \qquad \text{(VI.7.10.1)}$$

*where we set $\mathrm{H}^{-1}(\widetilde{E}_n, M_n) = 0$ for every $n \in \mathbb{N}$.*

This follows from VI.7.8 and from the fact that $\mathrm{R}^p \lim\limits_{\substack{\longleftarrow \\ n \in \mathbb{N}^\circ}} = 0$ for every $p \geq 2$ ([**47**] 1.4 and [**63**] 2.1).

**Definition VI.7.11.** We say that an $R$-module $M = \{i \mapsto M_i\}$ of $\mathbf{Top}(E)$ is *co-Cartesian* if for every morphism $f \colon i \to j$ of $I$, the transition morphism $f^*(M_j) \to M_i$ of $M$ is an isomorphism, where $f^*$ denotes the inverse image in the sense of modules (VI.7.4).

**Example VI.7.12.** Let $X$ be a $\mathbb{U}$-site, $\widetilde{X}$ the topos of sheaves of $\mathbb{U}$-sets on $X$, $A$ a commutative ring of $\widetilde{X}$, and $J$ an ideal of $A$. We denote also by $\mathbb{N}$ the filtered category defined by the ordered set of natural numbers $\mathbb{N}$. The total topos associated with the constant fibered $\mathbb{U}$-site $X \times \mathbb{N} \to \mathbb{N}$ with fiber $X$ is canonically equivalent to the category $\mathbf{Hom}(\mathbb{N}^\circ, \widetilde{X})$ (VI.7.1.10), which we also denote by $\widetilde{X}^{\mathbb{N}^\circ}$. We endow it with the ring $\breve{A} = \{n \mapsto A/J^{n+1}\}$. An $\breve{A}$-module $M = \{n \mapsto M_n\}$ of $\widetilde{X}^{\mathbb{N}^\circ}$ is co-Cartesian if and only if for every integer $n \geq 0$, the morphism $M_{n+1} \otimes_A (A/J^{n+1}) \to M_n$ deduced from the transition morphism $M_{n+1} \to M_n$, is an isomorphism, in other words, if the inverse system $(M_n)_{n \in \mathbb{N}}$ is $J$-adic in the sense of ([**38**] V 3.1.1).

**VI.7.13.** Let $M = \{i \mapsto M_i\}$ and $N = \{i \mapsto N_i\}$ be two $R$-modules of $\mathbf{Top}(E)$ such that $M$ is co-Cartesian. We then have a canonical isomorphism

$$\mathrm{Hom}_R(M, N) \xrightarrow{\sim} \lim_{\substack{\longleftarrow \\ i \in I^\circ}} \mathrm{Hom}_{R_i}(M_i, N_i), \qquad \text{(VI.7.13.1)}$$

where for every morphism $f \colon i \to j$ of $I$, the transition morphism

$$d_f \colon \mathrm{Hom}_{R_j}(M_j, N_j) \to \mathrm{Hom}_{R_i}(M_i, N_i)$$

of the inverse system that appears in (VI.7.13.1) associates with each $R_j$-morphism $u \colon M_j \to N_j$ the morphism $d_f(u)$ composed of

$$M_i \xrightarrow{\sim} f^*(M_j) \xrightarrow{f^*(u)} f^*(N_j) \longrightarrow N_i,$$

where the first arrow is the inverse of the transition isomorphism $f^*(M_j) \xrightarrow{\sim} M_i$ of $M$ and the last arrow is the transition morphism of $N$. In particular, if $I$ admits a final object $\iota$, we have a canonical isomorphism

$$\mathrm{Hom}_R(M, N) \xrightarrow{\sim} \mathrm{Hom}_{R_\iota}(M_\iota, N_\iota). \qquad \text{(VI.7.13.2)}$$

## VI.8. Ringed covanishing topos

**VI.8.1.** In this section, $I$ denotes a $\mathbb{U}$-site and $\pi\colon E \to I$ a cleaved and normalized fibered $\mathbb{U}$-site over the category underlying $I$, such that the following conditions are satisfied:

(i) Finite inverse limits are representable in $I$.
(ii) For every $i \in \mathrm{Ob}(I)$, finite inverse limits are representable in $E_i$.
(iii) For every morphism $f\colon i \to j$ of $I$, the inverse image functor $f^+\colon E_j \to E_i$ is continuous and left exact.

We endow $E$ with the covanishing topology defined by $\pi$, and we denote by $\widetilde{E}$ the topos of sheaves of $\mathbb{U}$-sets on $E$. We fix a final object $\iota$ of $I$ and a final object $e$ of $E_\iota$, and we take again the notation introduced in VI.5, in particular, that introduced in VI.5.32.

We also fix a ring $R$ of $\widetilde{E}$. By VI.5.10, it amounts to fixing, for every $i \in \mathrm{Ob}(I)$, a ring $R_i$ of $\widetilde{E}_i$, and for every morphism $f\colon i \to j$ of $I$, a ring homomorphism $R_j \to f_*(R_i)$, these homomorphisms being subject to compatibility (VI.5.2.1) and gluing relations (VI.5.10.1). The morphisms of topos $f\colon \widetilde{E}_i \to \widetilde{E}_j$ are therefore morphisms of ringed topos (by $R_i$ and $R_j$, respectively). For modules, we use the notation $f^{-1}$ to denote the inverse image in the sense of abelian sheaves, and we keep the notation $f^*$ for the inverse image in the sense of modules.

Giving a structure of $R$-module on a sheaf $M = \{i \mapsto M_i\}$ of $\widetilde{E}$ is equivalent to giving, for every $i \in \mathrm{Ob}(I)$, a structure of $R_i$-module on $M_i$ such that for every morphism $f\colon i \to j$ of $I$, the transition morphism $M_j \to f_*(M_i)$ is $R_j$-linear. For every $c \in \mathrm{Ob}(I)$, the functor

$$\mathbf{Mod}(R, \widetilde{E}) \to \mathbf{Mod}(R_c, \widetilde{E}_c), \quad \{i \mapsto M_i\} \mapsto M_c \tag{VI.8.1.1}$$

is clearly additive.

**Lemma VI.8.2.** *Let $u\colon \{i \mapsto M_i\} \to \{i \mapsto N_i\}$ be a morphism of $R$-modules of $\widetilde{E}$, $F$ its kernel, and $G$ its cokernel. For every $i \in \mathrm{Ob}(I)$, denote by $u_i\colon M_i \to N_i$ the $R_i$-morphism induced by $u$, and by $F_i$ (resp. $G_i$) its kernel (resp. cokernel). Then $\{i \mapsto F_i\}$ is an $R$-module of $\widetilde{E}$, $\{i \mapsto G_i\}$ is an $R$-module of $\widehat{E}$, and we have canonical $R$-isomorphisms*

$$F \xrightarrow{\sim} \{i \mapsto F_i\}, \tag{VI.8.2.1}$$

$$G \xrightarrow{\sim} \{i \mapsto G_i\}^a, \tag{VI.8.2.2}$$

*where $\{i \mapsto G_i\}^a$ is the sheaf on $E$ associated with the presheaf $\{i \mapsto G_i\}$. In particular, the functor* (VI.8.1.1) *is left exact.*

Indeed, denote by $\mathbf{Mod}(R, \widehat{E})$ the category of $R$-modules of $\widehat{E}$, by $j_R\colon \mathbf{Mod}(R, \widetilde{E}) \to \mathbf{Mod}(R, \widehat{E})$ the canonical inclusion functor, and by $a_R\colon \mathbf{Mod}(R, \widehat{E}) \to \mathbf{Mod}(R, \widetilde{E})$ the "associated sheaf" functor ([**2**] II 6.4). Then $a_R$ is left exact and commutes with direct limits, and $j_R$ commutes with inverse limits ([**2**] II 4.1). We deduce from this canonical $R$-isomorphisms

$$j_R(F) \xrightarrow{\sim} \ker(j_R(u)), \tag{VI.8.2.3}$$

$$G \xrightarrow{\sim} a_R(\mathrm{coker}(j_R(u))), \tag{VI.8.2.4}$$

and likewise for the morphisms $u_i$ for every $i \in \mathrm{Ob}(I)$. In view of (VI.5.2.1) and ([**2**] I 3.1), we have $\ker(j_R(u)) = \{i \mapsto F_i\}$. On the other hand, it follows from VI.5.17 that $\{i \mapsto G_i\}$ is an $R$-module of $\widehat{E}$ and that we have a canonical $R$-isomorphism

$$a_R(\{i \mapsto G_i\}) \xrightarrow{\sim} a_R(\mathrm{coker}(j_R(u))), \tag{VI.8.2.5}$$

giving the lemma.

**VI.8.3.** For every $c \in \mathrm{Ob}(I)$, we denote by

$$\gamma_c \colon \widetilde{E}_{/\sigma^*(c)} \to \widetilde{E} \tag{VI.8.3.1}$$

the localization morphism of $\widetilde{E}$ at $\sigma^*(c)$, and by

$$\beta_c \colon \widetilde{E}_{/\sigma^*(c)} \to \widetilde{E}_c \tag{VI.8.3.2}$$

the morphism defined in (VI.5.39.2). By (VI.5.39.4), the functor (VI.8.1.1) identifies with the composed functor $\beta_{c*} \circ \gamma_c^*$. Consequently, $\beta_c$ is a morphism of ringed topos (by $\gamma_c^*(R)$ and $R_c$, respectively). For modules, we use the notation $\beta_c^{-1}$ to denote the inverse image in the sense of abelian sheaves, and we keep the notation $\beta_c^*$ for the inverse image in the sense of modules. We denote by $\mathrm{R}^n\beta_{c*}$ ($n \in \mathbb{N}$) the right derived functors of

$$\beta_{c*} \colon \mathbf{Mod}(\gamma_c^*(R), \widetilde{E}_{/\sigma^*(c)}) \to \mathbf{Mod}(R_c, \widetilde{E}_c). \tag{VI.8.3.3}$$

The $n$th right derived functor of (VI.8.1.1) then identifies with the functor $(\mathrm{R}^n\beta_{c*}) \circ \gamma_c^*$ by virtue of ([**2**] V 4.11).

Recall that $\beta = \beta_\iota$, which we therefore consider as a morphism of ringed topos (by $R$ and $R_\iota$, respectively). For modules, we use the notation $\beta^{-1}$ to denote the inverse image in the sense of abelian sheaves, and we keep the notation $\beta^*$ for the inverse image in the sense of modules.

**VI.8.4.** Let $\mathbb{V}$ be a universe such that $\mathbb{U} \subset \mathbb{V}$ and $I \in \mathbb{V}$. The category $E$ endowed with the total topology associated with the fibered site $\pi$ is a $\mathbb{V}$-site; but it is not in general a $\mathbb{U}$-site. We denote by $\widetilde{E}_{\mathbb{V}}$ the topos of sheaves of $\mathbb{V}$-sets on the covanishing site $E$, by

$$\jmath \colon \widetilde{E} \to \widetilde{E}_{\mathbb{V}} \tag{VI.8.4.1}$$

the canonical inclusion functor, and by $\mathbf{Top}_{\mathbb{V}}(E)$ the topos of sheaves of $\mathbb{V}$-sets on the total site $E$ (VI.7.1). We consider the canonical morphism (VI.5.16.1)

$$\delta \colon \widetilde{E}_{\mathbb{V}} \to \mathbf{Top}_{\mathbb{V}}(E) \tag{VI.8.4.2}$$

as a morphism of ringed topos (by $R$ and $\delta_*(R)$, respectively). Since $\delta$ is an embedding, the adjunction morphism $\delta^*\delta_* \to \mathrm{id}$ is an isomorphism. Hence there is no difference for modules between the inverse image in the sense of abelian sheaves and the inverse image in the sense of modules. We denote by $\mathrm{R}^n\delta_*$ ($n \in \mathbb{N}$) the right derived functors of

$$\delta_* \colon \mathbf{Mod}(R, \widetilde{E}_{\mathbb{V}}) \to \mathbf{Mod}(\delta_*(R), \mathbf{Top}_{\mathbb{V}}(E)). \tag{VI.8.4.3}$$

Recall that the functor $\jmath$ is exact and fully faithful on the categories of modules and transforms injective modules into injective modules ([**2**] V 1.9). Consequently, the $n$th right derived functor of $\delta_* \circ \jmath$ is canonically isomorphic to $(\mathrm{R}^n\delta_*) \circ \jmath$.

**Proposition VI.8.5.** *Under the assumptions of* VI.8.4, *let moreover $M$ be an $R$-module of $\widetilde{E}$.*

(i) *For every integer $n \geq 0$, we have a functorial canonical isomorphism of $\delta_*(R)$-modules*

$$\mathrm{R}^n\delta_*(M) \xrightarrow{\sim} \{i \mapsto \mathrm{R}^n\beta_{i*}(\gamma_i^*(M))\}. \tag{VI.8.5.1}$$

(ii) *If $M$ is an injective $R$-module, then $\delta_*(M)$ is an injective $\delta_*(R)$-module.*
(iii) *For every integer $n > 0$, $\delta^*(\mathrm{R}^n\delta_*(M)) = 0$.*

(i) For every $i \in \mathrm{Ob}(I)$, we denote by $\widetilde{E}_{i,\mathbb{V}}$ the topos of sheaves of $\mathbb{V}$-sets on $E_i$, by $\jmath_i\colon \widetilde{E}_i \to \widetilde{E}_{i,\mathbb{V}}$ the canonical inclusion functor, and by

(VI.8.5.2) $$\alpha_i\colon \widetilde{E}_{i,\mathbb{V}} \to \mathbf{Top}_{\mathbb{V}}(E)$$

the canonical morphism (VI.7.1.5). The latter is a morphism of ringed topos (by $R_i$ and $\delta_*(R)$, respectively); recall (VI.7.4) that there is no difference for modules between the inverse image in the sense of abelian sheaves and the inverse image in the sense of modules. We therefore have a functorial canonical isomorphism

(VI.8.5.3) $$\alpha_i^*(\mathrm{R}^n\delta_*(M)) \xrightarrow{\sim} \mathrm{R}^n(\alpha_i^*\circ\delta_*)(M).$$

On the other hand, by (VI.5.16.2) and (VI.5.39.4), we have a canonical isomorphism of functors

(VI.8.5.4) $$\alpha_i^*\circ\delta_*\circ\jmath \xrightarrow{\sim} \jmath_i\circ\beta_{i*}\circ\gamma_i^*.$$

The statement follows in view of ([**2**] V 1.9 and 4.11(1)). We leave it to the reader to describe explicitly the transition morphisms of the sheaf $\{i \mapsto \mathrm{R}^n\beta_{i*}(\gamma_i^*(M))\}$ of $\mathbf{Top}_{\mathbb{V}}(E)$.

(ii) Since the module $\jmath(M)$ is injective, we may restrict to the case where $\mathbb{U}=\mathbb{V}$. The statement then follows from ([**2**] V 0.2) because the functor $\delta_*$ admits an exact left adjoint $\delta^*=\delta^{-1}$ (VI.8.4).

(iii) Since the functor $\delta^*=\delta^{-1}$ is exact, we have a functorial canonical isomorphism

(VI.8.5.5) $$\delta^*(\mathrm{R}^n\delta_*(M)) \xrightarrow{\sim} \mathrm{R}^n(\delta^*\circ\delta_*)(M).$$

On the other hand, since $\delta$ is an embedding, the adjunction morphism $\delta^*\delta_*\to\mathrm{id}$ is an isomorphism, giving the desired statement.

**Remark VI.8.6.** For every $R$-module $M$ of $\widetilde{E}$, the Cartan–Leray spectral sequence associated with $\delta$ that computes the cohomology of $M$ ([**2**] V 5.3) reduces to that associated with $\beta$

(VI.8.6.1) $$\mathrm{E}_2^{p,q}=\mathrm{H}^p(\widetilde{E}_\iota,\mathrm{R}^q\beta_*(M)) \Rightarrow \mathrm{H}^{p+q}(\widetilde{E},M),$$

by virtue of VI.7.9 and VI.8.5.

**VI.8.7.** Let $\pi'\colon E'\to I$ be a cleaved and normalized fibered $\mathbb{U}$-site satisfying conditions (ii) and (iii) of VI.8.1 and

(VI.8.7.1) $$\Phi^+\colon E'\to E$$

a Cartesian $I$-functor. We endow $E'$ with the covanishing topology defined by $\pi'$, and we denote by $\widetilde{E}'$ the topos of sheaves of $\mathbb{U}$-sets on $E'$. We associate with $\pi'$ objects analogous to those associated with $\pi$, and we denote them by the same letters equipped with an exponent $'$. Suppose that for every $i\in\mathrm{Ob}(I)$, the functor $\Phi_i^+\colon E_i'\to E_i$ induced by $\Phi^+$ on the fibers at $i$ is continuous and left exact, and denote by

(VI.8.7.2) $$\Phi_i\colon \widetilde{E}_i\to\widetilde{E}_i'$$

the associated morphism of topos. By virtue of VI.5.36, $\Phi^+$ is continuous and left exact. It therefore defines a morphism of topos

(VI.8.7.3) $$\Phi\colon \widetilde{E}\to\widetilde{E}'$$

such that for every sheaf $F=\{i\mapsto F_i\}$ on $E$, we have

(VI.8.7.4) $$\Phi_*(F)=\{i\mapsto\Phi_{i*}(F_i)\}.$$

Let $R'=\{i\mapsto R_i'\}$ be a ring of $\widetilde{E}'$ and $h\colon R'\to\Phi_*(R)$ a ring homomorphism, so that $\Phi\colon\widetilde{E}\to\widetilde{E}'$ is a morphism of ringed topos (by $R$ and $R'$, respectively). Giving $h$

amounts to giving, for every $i \in \mathrm{Ob}(I)$, a ring homomorphism $h_i \colon R'_i \to \Phi_{i*}(R_i)$ satisfying compatibility relations (VI.5.18.3). In particular, for every $i \in \mathrm{Ob}(I)$, $\Phi_i \colon \widetilde{E}_i \to \widetilde{E}'_i$ is a morphism of ringed topos (by $R_i$ and $R'_i$, respectively). We denote by $\mathrm{R}^n\Phi_*$ and $\mathrm{R}^n\Phi_{i*}$ ($n \in \mathbb{N}$) the right derived functors of

$$\begin{array}{rcl} \Phi_* \colon \mathbf{Mod}(R, \widetilde{E}) & \to & \mathbf{Mod}(R', \widetilde{E}'), \\ \Phi_{i*} \colon \mathbf{Mod}(R_i, \widetilde{E}_i) & \to & \mathbf{Mod}(R'_i, \widetilde{E}'_i). \end{array}$$

**Proposition VI.8.8.** *Under the assumptions of* VI.8.7, *for every $R$-module $M$, we have a functorial and canonical spectral sequence*

(VI.8.8.1) $$\mathrm{E}_2^{p,q} = \{i \mapsto \mathrm{R}^p\Phi_{i*}(\mathrm{R}^q\beta_{i*}(\gamma_i^* M))\}^a \Rightarrow \mathrm{R}^{p+q}\Phi_*(M),$$

*where the source denotes the sheaf associated with the presheaf* $\{i \mapsto \mathrm{R}^p\Phi_{i*}(\mathrm{R}^q\beta_{i*}(\gamma_i^* M))\}$ *on $E'$.*

After extending $\mathbb{U}$, if necessary, we may restrict to the case where $I \in \mathbb{U}$ ([**2**] II 3.6 and V 1.9). We denote by $\mathbf{Top}(E)$ and $\mathbf{Top}(E')$ the topos of sheaves of $\mathbb{U}$-sets on the total sites $E$ and $E'$, respectively (VI.7.1). The morphisms $\Phi_i$ define a Cartesian morphism of fibered topos $\mathscr{F} \to \mathscr{F}'$. On the other hand, $\Phi^+$ is a morphism from the total site $E$ to the total site $E'$ ([**2**] 7.4.10). It therefore defines a morphism of topos

(VI.8.8.2) $$\Psi \colon \mathbf{Top}(E) \to \mathbf{Top}(E')$$

such that for every sheaf $F = \{i \mapsto F_i\}$ on $E$, we have

(VI.8.8.3) $$\Psi_*(F) = \{i \mapsto \Phi_{i*}(F_i)\}.$$

In particular, the diagram of morphisms of topos

(VI.8.8.4) $$\begin{array}{ccc} \widetilde{E} & \xrightarrow{\ \delta\ } & \mathbf{Top}(E) \\ {\scriptstyle \Phi}\downarrow & & \downarrow{\scriptstyle \Psi} \\ \widetilde{E}' & \xrightarrow{\ \delta'\ } & \mathbf{Top}(E') \end{array}$$

is commutative up to canonical isomorphism. The homomorphism $h \colon R' \to \Phi_*(R)$ induces a homomorphism $h' \colon \delta'_*(R') \to \Psi_*(\delta_*(R))$ that makes $\Psi \colon \mathbf{Top}(E) \to \mathbf{Top}(E')$ into a morphism of ringed topos (by $\delta_*(R)$ and $\delta'_*(R')$, respectively).

Since $\delta'$ is an embedding, the adjunction morphism $\delta'^*\delta'_*\Phi_* \to \Phi_*$ is an isomorphism. In view of (VI.8.8.4), we deduce from this an isomorphism $\delta'^*\Psi_*\delta_* \xrightarrow{\sim} \Phi_*$. Since the functor $\delta'^* = \delta'^{-1}$ is exact (VI.8.4), the Cartan–Leray spectral sequence ([**2**] V 5.4) then induces a functorial spectral sequence in $M$

(VI.8.8.5) $$\mathrm{E}_2^{p,q} = \delta'^*(\mathrm{R}^p\Psi_*(\mathrm{R}^q\delta_*(M))) \Rightarrow \mathrm{R}^{p+q}\Phi_*(M).$$

By virtue of VI.7.7 and VI.8.5(i), we have a canonical isomorphism of $\delta'_*(R')$-modules

(VI.8.8.6) $$\mathrm{R}^p\Psi_*(\mathrm{R}^q\delta_*(M)) \xrightarrow{\sim} \{i \mapsto \mathrm{R}^p\Phi_{i*}(\mathrm{R}^q\beta_{i*}(\gamma_i^* M))\}.$$

On the other hand, for every object $G$ of $\mathbf{Top}(E')$, $\delta'^*(G)$ is canonically isomorphic to the sheaf on the covanishing site $E'$ associated with the presheaf $G$ (cf. VI.5.16), giving the proposition.

**Lemma VI.8.9.** *Let $A = \{i \mapsto A_i\}$ be a ring of $\widehat{E}$* (VI.5.2), *$M = \{i \mapsto M_i\}$ a right $A$-module of $\widehat{E}$, $N = \{i \mapsto N_i\}$ a left $A$-module of $\widehat{E}$, and $A^a$, $M^a$, and $N^a$ the sheaves on $E$ associated with $A$, $M$, and $N$, respectively. For every $i \in \mathrm{Ob}(I)$, we denote by $M_i \otimes_{A_i} N_i$ the abelian group tensor product of $M_i$ and $N_i$ in $\widehat{E}_i$. Then $\{i \mapsto M_i \otimes_{A_i} N_i\}$*

*is a presheaf on* $E$*, and denoting by* $\{i \mapsto M_i \otimes_{A_i} N_i\}^a$ *the associated sheaf on* $E$*, we have a bifunctorial canonical isomorphism of abelian sheaves on* $E$

(VI.8.9.1) $$M^a \otimes_{A^a} N^a \xrightarrow{\sim} \{i \mapsto M_i \otimes_{A_i} N_i\}^a.$$

This follows from ([**2**] IV 12.10).

**Lemma VI.8.10.** *Let* $M = \{i \mapsto M_i\}$ *be a right* $R$*-module of* $\widetilde{E}$ *and* $N = \{i \mapsto N_i\}$ *a left* $R$*-module of* $\widetilde{E}$*. Then we have a bifunctorial canonical isomorphism of abelian sheaves on* $E$

(VI.8.10.1) $$M \otimes_R N \xrightarrow{\sim} \{i \mapsto M_i \otimes_{R_i} N_i\}^a,$$

*where for every* $i \in \mathrm{Ob}(I)$*,* $M_i \otimes_{R_i} N_i$ *is the abelian group tensor product of* $M_i$ *and* $N_i$ *in* $\widetilde{E}_i$*, and* $\{i \mapsto M_i \otimes_{R_i} N_i\}^a$ *is the sheaf on* $E$ *associated with the presheaf* $\{i \mapsto M_i \otimes_{R_i} N_i\}$.

This follows from VI.5.17 and VI.8.9.

**Definition VI.8.11.** We say that an $R$-module $M = \{i \mapsto M_i\}$ of $\widetilde{E}$ is *co-Cartesian* if the $\delta_*(R)$-module $\delta_*(M)$ is co-Cartesian in the sense of (VI.7.11), in other words, if for every morphism $f \colon i \to j$ of $I$, the transition morphism $f^*(M_j) \to M_i$ of $M$ is an isomorphism, where $f^*$ denotes the inverse image in the sense of modules (VI.8.1).

**Proposition VI.8.12.** *Let* $M = \{i \mapsto M_i\}$ *and* $N = \{i \mapsto N_i\}$ *be two* $R$*-modules of* $\widetilde{E}$ *such that* $M$ *is co-Cartesian. Then we have a canonical* $R$*-isomorphism*

(VI.8.12.1) $$\mathscr{H}om_R(M, N) \xrightarrow{\sim} \{i \mapsto \mathscr{H}om_{R_i}(M_i, N_i)\}.$$

Indeed, by (VI.7.13.2), we have a canonical isomorphism

(VI.8.12.2) $$\mathrm{Hom}_R(M, N) \xrightarrow{\sim} \mathrm{Hom}_{R_\iota}(M_\iota, N_\iota).$$

Let $V$ be an object of $E$, $c = \pi(V)$, and

(VI.8.12.3) $$\varpi \colon E_{/V} \to I_{/c}$$

the functor induced by $\pi$. For every morphism $f \colon i \to c$ of $I$, the fiber of $\varpi$ over the object $(i, f)$ of $I_{/c}$ is canonically equivalent to the category $(E_i)_{/f^+(V)}$. Endowing $I_{/c}$ with the topology induced by that of $I$, and each fiber $(E_i)_{/f^+(V)}$ with the topology induced by that of $E_i$, $\varpi$ becomes a fibered site satisfying the conditions of VI.5.1. By virtue of VI.5.38, the covanishing topology of $E_{/V}$ associated with $\varpi$ is induced by that of $E$ through the canonical functor $\gamma_V \colon E_{/V} \to E$. In particular, the topos of sheaves of $\mathbb{U}$-sets on $E_{/V}$ is canonically equivalent to $\widetilde{E}_{/\varepsilon(V)}$. Moreover, we have a functorial canonical isomorphic

(VI.8.12.4) $$M|\varepsilon(V) \xrightarrow{\sim} \{(i, f) \mapsto M_i|f^*(V)\},$$

and likewise for $N$. Consequently, the $R|\varepsilon(V)$-module $M|\varepsilon(V)$ is co-Cartesian. Denoting by $\varepsilon_c \colon E_c \to \widetilde{E}_c$ the canonical functor, the isomorphisms (VI.8.12.2) and (VI.8.12.4) induce an isomorphism

(VI.8.12.5) $$\mathrm{Hom}_{R|\varepsilon(V)}(M|\varepsilon(V), N|\varepsilon(V)) \xrightarrow{\sim} \mathrm{Hom}_{R_c|\varepsilon_c(V)}(M_c|\varepsilon_c(V), N_c|\varepsilon_c(V)),$$

that is clearly functorial in $V \in \mathrm{Ob}(E_c)$, giving the proposition ([**2**] IV 12.1).

**Remark VI.8.13.** Under the assumptions of VI.8.12, it immediately follows from VI.7.13 that for every morphism $f \colon i \to j$ of $I$, the transition morphism of the $R$-module $\mathscr{H}om_R(M, N)$ is the composition

$$f^*(\mathscr{H}om_{R_j}(M_j, N_j)) \to \mathscr{H}om_{R_i}(f^*(M_j), f^*(N_j))$$
$$\xrightarrow{\sim} \mathscr{H}om_{R_i}(M_i, f^*(N_j)) \to \mathscr{H}om_{R_i}(M_i, N_i),$$

where $f^*$ denotes the inverse image in the sense of modules, the first arrow is the canonical morphism, the second arrow is induced by the transition isomorphism $f^*(M_j) \xrightarrow{\sim} M_i$ of $M$, and the last arrow is induced by the transition morphism $f^*(N_j) \to N_i$ of $N$.

**Proposition VI.8.14.** *For every locally projective $R_\iota$-module of finite type $M_\iota$ of $\widetilde{E}_\iota$ (that is, $M_\iota$ is locally a direct summand of a free $R_\iota$-module of finite type), we have a functorial canonical isomorphism*

$$\beta^*(M_\iota) \xrightarrow{\sim} \{i \mapsto f_i^*(M_\iota)\}, \tag{VI.8.14.1}$$

*where $\beta^*$ denotes the inverse image in the sense of modules* (VI.8.3) *and for every $i \in$ Ob$(I)$, $f_i \colon i \to \iota$ is the canonical morphism and $f_i^*$ denotes the inverse image in the sense of modules* (VI.8.1). *In particular, $\beta^*(M_\iota)$ is a co-Cartesian $R$-module.*

By virtue of VI.5.34(i), VI.8.9, and VI.5.17, we have a canonical isomorphism

$$\beta^*(M_\iota) \xrightarrow{\sim} \{i \mapsto f_i^*(M_\iota)\}^a, \tag{VI.8.14.2}$$

where the right-hand side is the sheaf on $E$ associated with the presheaf $\{i \mapsto f_i^*(M_\iota)\}$. It remains to show that $M = \{i \mapsto f_i^*(M_\iota)\}$ is a sheaf on $E$. Denote by $\varepsilon_\iota \colon E_\iota \to \widetilde{E}_\iota$ the canonical functor. There exists a vertical covering $(V_n \to e)_{n \in \Sigma}$ of $E$ (that is, a covering family of $E_\iota$) such that for every $n \in \Sigma$, $M_\iota | \varepsilon_\iota(V_n)$ is a direct summand of a free $(R_\iota | \varepsilon_\iota(V_n))$-module of finite type. It suffices to show that for every $n \in \Sigma$, the restriction of $M$ to the category $E_{/V_n}$ is a sheaf for the topology induced by that of $E$ (**[35]** II 3.4.4). In view of VI.5.38, we can restrict to the case where $V_n = e$, in which case there exist an integer $d \geq 0$, an $R_\iota$-module $N_\iota$ of $\widetilde{E}_\iota$, and an $R_\iota$-isomorphism $R_\iota^d \xrightarrow{\sim} M_\iota \oplus N_\iota$. We then have an isomorphism of $\widehat{E}$

$$R^d \xrightarrow{\sim} M \oplus \{i \mapsto f_i^*(N_\iota)\}, \tag{VI.8.14.3}$$

which implies that $M$ and $\{i \mapsto f_i^*(N_\iota)\}$ are sheaves on $E$.

## VI.9. Finite étale site and topos of a scheme

**VI.9.1.** For every scheme $X$, we denote by $\mathbf{\acute{E}t}_{/X}$ the *étale site* of $X$, that is, the full subcategory of $\mathbf{Sch}_{/X}$ (VI.2.1) made up of the étale schemes on $X$, endowed with the étale topology; it is a $\mathbb{U}$-site. We denote by $X_{\text{ét}}$ the *étale topos* of $X$, that is, the topos of sheaves of $\mathbb{U}$-sets on $\mathbf{\acute{E}t}_{/X}$. We denote by $\mathbf{\acute{E}t}_{\text{coh}/X}$ (resp. $\mathbf{\acute{E}t}_{\text{scoh}/X}$) the full subcategory of $\mathbf{\acute{E}t}_{/X}$ made up of the étale schemes of finite presentation over $X$ (resp. separated étale schemes of finite presentation over $X$), endowed with the topology induced by that of $\mathbf{\acute{E}t}_{/X}$; these are $\mathbb{U}$-small sites. If $X$ is quasi-separated, the restriction functor from $X_{\text{ét}}$ to the topos of sheaves of $\mathbb{U}$-sets on $\mathbf{\acute{E}t}_{\text{coh}/X}$ (resp. $\mathbf{\acute{E}t}_{\text{scoh}/X}$) is an equivalence of categories (**[2]** VII 3.1 and 3.2).

**VI.9.2.** For every scheme $X$, we call *finite étale site* of $X$ and denote by $\mathbf{\acute{E}t}_{\text{f}/X}$ the full subcategory of $\mathbf{\acute{E}t}_{/X}$ made up of the finite étale covers of $X$, endowed with the topology induced by that of $\mathbf{\acute{E}t}_{/X}$; it is a $\mathbb{U}$-small site. We call *finite étale topos* of $X$ and denote by $X_{\text{fét}}$ the topos of sheaves of $\mathbb{U}$-sets on $\mathbf{\acute{E}t}_{\text{f}/X}$. The étale topology on $\mathbf{\acute{E}t}_{\text{f}/X}$ is clearly coarser than the canonical topology. The canonical injection $\mathbf{\acute{E}t}_{\text{f}/X} \to \mathbf{\acute{E}t}_{/X}$ induces a morphism of topos

$$\rho_X \colon X_{\text{ét}} \to X_{\text{fét}}. \tag{VI.9.2.1}$$

**VI.9.3.** Let $f\colon Y \to X$ be a morphism of schemes. The inverse image functor

$$f^{\bullet}\colon \mathbf{Sch}_{/X} \to \mathbf{Sch}_{/Y}, \quad X' \mapsto X' \times_X Y \tag{VI.9.3.1}$$

induces two morphisms of topos that we denote (to distinguish them from each other) by

$$f_{\text{ét}}\colon Y_{\text{ét}} \to X_{\text{ét}}, \tag{VI.9.3.2}$$

$$f_{\text{fét}}\colon Y_{\text{fét}} \to X_{\text{fét}}. \tag{VI.9.3.3}$$

We will leave the indices out of the notation $f_{\text{ét}}$ and $f_{\text{fét}}$ when there is no risk of ambiguity. One immediately verifies that the diagram of morphisms of topos

$$\begin{array}{ccc} Y_{\text{ét}} & \xrightarrow{f_{\text{ét}}} & X_{\text{ét}} \\ \downarrow{\scriptstyle \rho_Y} & & \downarrow{\scriptstyle \rho_X} \\ Y_{\text{fét}} & \xrightarrow{f_{\text{fét}}} & X_{\text{fét}} \end{array} \tag{VI.9.3.4}$$

is commutative up to canonical isomorphism:

$$\rho_X f_{\text{ét}} \overset{\sim}{\to} f_{\text{fét}} \rho_Y. \tag{VI.9.3.5}$$

**VI.9.4.** Let $f\colon Y \to X$ be a finite étale cover and

$$\tau_f\colon \mathbf{\acute{E}t}_{\mathrm{f}/Y} \to \mathbf{\acute{E}t}_{\mathrm{f}/X} \tag{VI.9.4.1}$$

the functor defined by composition on the left with $f$. Then $\tau_f$ induces an equivalence of categories $\mathbf{\acute{E}t}_{\mathrm{f}/Y} \overset{\sim}{\to} (\mathbf{\acute{E}t}_{\mathrm{f}/X})_{/(Y,f)}$. The étale topology on $\mathbf{\acute{E}t}_{\mathrm{f}/Y}$ is induced by that on $\mathbf{\acute{E}t}_{\mathrm{f}/X}$ through the functor $\tau_f$. By virtue of ([**2**] III 5.2), $\tau_f$ is continuous and cocontinuous. It therefore induces a sequence of three adjoint functors:

$$\tau_{f!}\colon Y_{\text{fét}} \to X_{\text{fét}}, \quad \tau_f^*\colon X_{\text{fét}} \to Y_{\text{fét}}, \quad \tau_{f*}\colon Y_{\text{fét}} \to X_{\text{fét}}, \tag{VI.9.4.2}$$

in the sense that for any two consecutive functors in the sequence, the one on the right is the right adjoint of the other. By ([**2**] III 5.4), the functor $\tau_{f!}$ factors through an equivalence of categories $(Y_{\text{fét}}) \overset{\sim}{\to} (X_{\text{fét}})_{/Y}$. The pair of functors $(\tau_f^*, \tau_{f*})$ defines a morphism of topos $Y_{\text{fét}} \to X_{\text{fét}}$, called localization morphism of $X_{\text{fét}}$ at $Y$. On the other hand, $\tau_f^*$ is left adjoint to the functor $f_{\text{fét}*}$ direct image by the morphism of topos $f_{\text{fét}}$. The latter therefore identifies canonically with the localization morphism of $X_{\text{fét}}$ at $Y$. We refer to ([**2**] VII 1.6) for the analogous statements for the étale sites and topos.

**VI.9.5.** We denote by $\mathscr{R}$ the category of finite étale covers (that is, the full subcategory of the category of morphisms of **Sch** made up of the finite étale covers) and by

$$\mathscr{R} \to \mathbf{Sch} \tag{VI.9.5.1}$$

the "target functor," which makes $\mathscr{R}$ into a cleaved and normalized fibered category over **Sch**: the fiber over any scheme $X$ is canonically equivalent to the category $\mathbf{\acute{E}t}_{\mathrm{f}/X}$, and for every morphism of schemes $f\colon Y \to X$, the inverse image functor $f^+\colon \mathbf{\acute{E}t}_{\mathrm{f}/X} \to \mathbf{\acute{E}t}_{\mathrm{f}/Y}$ is none other than the base change functor by $f$. Endowing each fiber with the étale topology, $\mathscr{R}/\mathbf{Sch}$ becomes a fibered $\mathbb{U}$-site ([**2**] VI 7.2.1). We denote by

$$\mathscr{G} \to \mathbf{Sch} \tag{VI.9.5.2}$$

the fibered $\mathbb{U}$-topos associated with $\mathscr{R}/\mathbf{Sch}$ ([**2**] VI 7.2.6): the fiber of $\mathscr{G}$ over any scheme $X$ is the topos $X_{\text{fét}}$, and for every morphism of schemes $f\colon Y \to X$, the inverse image functor $f^*\colon X_{\mathrm{fét}} \to Y_{\mathrm{fét}}$ is the inverse image functor under the morphism of topos $f_{\mathrm{fét}}$ (VI.9.3.3). We denote by

$$\mathscr{G}^\vee \to \mathbf{Sch}^\circ \tag{VI.9.5.3}$$

the fibered category obtained by associating with each scheme $X$ the category $\mathscr{G}_X = X_{\mathrm{fét}}$, and with each morphism of schemes $f\colon Y \to X$, the direct image functor $f_{\mathrm{fét}*}\colon Y_{\mathrm{fét}} \to X_{\mathrm{fét}}$ by the morphism of topos $f_{\mathrm{fét}}$.

**Lemma VI.9.6.** *Let $X$ be a scheme and $\rho_X\colon X_{\mathrm{ét}} \to X_{\mathrm{fét}}$ the canonical morphism* (VI.9.2.1). *Then, the family of fiber functors of $X_{\mathrm{fét}}$ associated with the points $\rho_X(\overline{x})$, when $\overline{x}$ goes through the family of geometric points of $X$, is conservative* ([**2**] IV 6.4.0).

Indeed, for any geometric point $\overline{x}$ of $X$, to give a neighborhood of the point $\rho_X(\overline{x})$ of $X_{\mathrm{fét}}$ in the site $\mathbf{\acute{E}t}_{\mathrm{f}/X}$ is equivalent to giving an $\overline{x}$-pointed étale cover of $X$ ([**2**] IV 6.8.2 and VIII 3.9). These objects form naturally a cofiltered category, that we denote by $\mathscr{P}_{\overline{x}}$. For any object $F$ of $X_{\mathrm{fét}}$, denoting the stalk of $F$ at $\rho_X(\overline{x})$ by $F_{\overline{x}}$, we have a canonical functorial isomorphism

$$F_{\overline{x}} \xrightarrow{\sim} \varinjlim_{(U,\xi)\in\mathscr{P}_{\overline{x}}^{\circ}} F(U). \tag{VI.9.6.1}$$

Let $u\colon F \to G$ be a morphism of $X_{\mathrm{fét}}$ such that for any geometric point $\overline{x}$ of $X$, the associated morphism $u_{\overline{x}}\colon F_{\overline{x}} \to G_{\overline{x}}$ is a monomorphism. Let's prove that $u$ is a monomorphism. We need to prove that for any $X' \in \mathrm{Ob}(\mathbf{\acute{E}t}_{\mathrm{f}/X})$ and $a, b \in F(X')$ such that $u(a) = u(b)$, we have $a = b$. By VI.9.4 and (VI.9.3.4), we can assume $X' = X$. For any geometric point $\overline{x}$ of $X$, we have $a_{\overline{x}} = b_{\overline{x}}$ because $u_{\overline{x}}$ is a monomorphism. By (VI.9.6.1), there exists an object $(U_{\overline{x}}, \zeta_{\overline{x}})$ of $\mathscr{P}_{\overline{x}}$ such that $a$ and $b$ have the same image in $F(U_{\overline{x}})$. The family of morphisms $(U_{\overline{x}} \to X)_{\overline{x}}$ being clearly a covering, we deduce that $a = b$. Therefore, $u$ is a monomorphism.

Assume furthermore that for any geometric point $\overline{x}$ of $X$, the morphism $u_{\overline{x}}$ is an epimorphism. Let's prove that $u$ is an epimorphism. It is enough to show that for any $X' \in \mathrm{Ob}(\mathbf{\acute{E}t}_{\mathrm{f}/X})$ and $b \in G(X')$, there exits $a \in F(X')$ such that $b = u(a)$. We can assume again that $X' = X$. By (VI.9.6.1), for any point $\overline{x}$ of $X$, there exists an object $(U_{\overline{x}}, \xi_{\overline{x}})$ of $\mathscr{P}_{\overline{x}}$ and a section $a_{\overline{x}} \in F(U_{\overline{x}})$ whose image by $u$ in $G(U_{\overline{x}})$ is the restriction of $b$. Since $u$ is a monomorphism, the sections $a_{\overline{x}}$ coincide over $U_{\overline{x}} \times_X U_{\overline{x}'}$, for any geometric points $\overline{x}$ and $\overline{x}'$ of $X$. They are therefore induced by a section $a \in F(X)$, and we have $u(a) = b$ as the restrictions to the $U_{\overline{x}}$'s coincide.

**Definition VI.9.7.** Let $X$ be a scheme. We say that $X$ is *locally connected* if the underlying topological space is locally connected and that $X$ is *étale-locally connected* if for every étale morphism $X' \to X$, every connected component of $X'$ is an open subset of $X'$.

We can make the following remarks.

VI.9.7.1. A scheme $X$ is étale-locally connected if and only if every étale $X$-scheme is locally connected ([**12**] I § 11.6 Proposition 11).

VI.9.7.2. The set of connected components of a topological space $X$ is locally finite if and only if every connected component of $X$ is open in $X$. This is the case if $X$ is locally connected.

VI.9.7.3. Let $X$ be a scheme whose set of irreducible components is locally finite (for example a scheme whose underlying topological space is locally noetherian). Then:

(i) For every étale morphism $f\colon X' \to X$, the set of irreducible components of $X'$ is locally finite. Indeed, since the question is local on $X$ and $X'$, we can restrict to the case where $X$ and $X'$ are affine, in which case $f$ is quasi-compact and therefore quasi-finite. The assertion then follows from ([**42**] 2.3.6(iii)).

(ii) It follows from (i) and ([**39**] 0.2.1.5(i)) that $X$ is étale-locally connected.

**VI.9.8.** Let $X$ be a connected scheme and $\overline{x}$ a geometric point of $X$. We denote by

(VI.9.8.1) $$\omega_{\overline{x}}\colon \mathbf{\acute{E}t}_{\mathrm{f}/X} \to \mathbf{Ens}$$

the fiber functor at $\overline{x}$, which with each finite étale cover $Y$ of $X$ associates the set of geometric points of $Y$ over $\overline{x}$, by $\pi_1(X,\overline{x})$ the fundamental group of $X$ at $\overline{x}$, and by $\mathbf{B}_{\pi_1(X,\overline{x})}$ the classifying topos of the profinite group $\pi_1(X,\overline{x})$, that is, the category of discrete $\mathbb{U}$-sets endowed with a continuous left action of $\pi_1(X,\overline{x})$ ([**2**] IV 2.7). Then $\omega_{\overline{x}}$ induces a fully faithful functor

(VI.9.8.2) $$\mu^+_{\overline{x}}\colon \mathbf{\acute{E}t}_{\mathrm{f}/X} \to \mathbf{B}_{\pi_1(X,\overline{x})}$$

with essential image the full subcategory $\mathscr{C}(\pi_1(X,\overline{x}))$ of $\mathbf{B}_{\pi_1(X,\overline{x})}$ made up of the finite sets ([**37**] V § 4 and § 7). On the other hand, a family $(Y_\lambda \to Y)_{\lambda\in\Lambda}$ of $\mathbf{\acute{E}t}_{\mathrm{f}/X}$ is covering for the étale topology if and only if its image in $\mathbf{B}_{\pi_1(X,\overline{x})}$ is surjective, or, equivalently, covering for the canonical topology of $\mathbf{B}_{\pi_1(X,\overline{x})}$. Consequently, the étale topology on $\mathbf{\acute{E}t}_{\mathrm{f}/X}$ is induced by the canonical topology of $\mathbf{B}_{\pi_1(X,\overline{x})}$ ([**2**] III 3.3). Since the objects of $\mathscr{C}(\pi_1(X,\overline{x}))$ form a generating family of $\mathbf{B}_{\pi_1(X,\overline{x})}$, the functor

(VI.9.8.3) $$\mu_{\overline{x}}\colon \mathbf{B}_{\pi_1(X,\overline{x})} \to X_{\mathrm{f\acute{e}t}}$$

which with each object $G$ of $\mathbf{B}_{\pi_1(X,\overline{x})}$ (considered as a representable sheaf) associates its restriction to $\mathbf{\acute{E}t}_{\mathrm{f}/X}$, is an equivalence of categories by virtue of ([**2**] IV 1.2.1).

Let $(X_i)_{i\in I}$ be an inverse system on a filtered ordered set $I$ in $\mathbf{\acute{E}t}_{\mathrm{f}/X}$ that prorepresents $\omega_{\overline{x}}$, normalized by the fact that the transition morphisms $X_i \to X_j$ $(i \geq j)$ are epimorphisms and that every epimorphism $X_i \to X'$ of $\mathbf{\acute{E}t}_{\mathrm{f}/X}$ is equivalent to a suitable epimorphism $X_i \to X_j$ $(j \leq i)$. Such a pro-object is essentially unique. It is called the *normalized universal cover of* $X$ *at* $\overline{x}$ or the *normalized fundamental pro-object of* $\mathbf{\acute{E}t}_{\mathrm{f}/X}$ *at* $\overline{x}$. Note that the set $I$ is $\mathbb{U}$-small. Consider the functor

(VI.9.8.4) $$\nu_{\overline{x}}\colon X_{\mathrm{f\acute{e}t}} \to \mathbf{B}_{\pi_1(X,\overline{x})}, \quad F \mapsto \varinjlim_{i\in I} F(X_i).$$

By definition, the restriction of $\nu_{\overline{x}}$ to $\mathbf{\acute{E}t}_{\mathrm{f}/X}$ is canonically isomorphic to the functor $\mu^+_{\overline{x}}$, and we have a canonical isomorphism of functors

(VI.9.8.5) $$\nu_{\overline{x}} \circ \mu_{\overline{x}} \xrightarrow{\sim} \mathrm{id}.$$

Since $\mu_{\overline{x}}$ is an equivalence of categories, $\nu_{\overline{x}}$ is an equivalence of categories, quasi-inverse to $\mu_{\overline{x}}$. We call it *the fiber functor* of $X_{\mathrm{f\acute{e}t}}$ at $\overline{x}$.

The functor $\nu_{\overline{x}}$ induces an equivalence between the category of sheaves of groups (resp. abelian sheaves) of $X_{\mathrm{f\acute{e}t}}$ and the category of discrete $\pi_1(X,\overline{x})$-groups (resp. of discrete $\pi_1(X,\overline{x})$-$\mathbb{Z}$-modules) (cf. II.3.1). In particular, for every abelian sheaf (resp. sheaf of groups) $F$ of $X_{\mathrm{f\acute{e}t}}$ and every integer $n \geq 0$ (resp. $n \in \{0,1\}$), we have a functorial canonical isomorphism

(VI.9.8.6) $$\mathrm{H}^n(X_{\mathrm{f\acute{e}t}}, F) \simeq \mathrm{H}^n(\pi_1(X,\overline{x}), \nu_{\overline{x}}(F)).$$

**VI.9.9.** Let $X$ be a scheme whose set of connected components is locally finite, $\overline{x}$ a geometric point of $X$, $\varphi_{\overline{x}}$ the fiber functor of $X_{\text{ét}}$ associated with $\overline{x}$, and $Y$ the connected component of $X$ containing $\overline{x}$, which is then open in $X$ (VI.9.7.2). We endow $Y$ with the scheme structure induced by that of $X$ and denote by $j\colon Y \to X$ the canonical injection, by $\nu_{\overline{x}}\colon Y_{\text{fét}} \to \mathbf{B}_{\pi_1(Y,\overline{x})}$ the fiber functor of $Y_{\text{fét}}$ at $\overline{x}$ (VI.9.8.4), and by $\gamma\colon \mathbf{B}_{\pi_1(Y,\overline{x})} \to \mathbf{Ens}$ the forgetful functor of the action of $\pi_1(Y,\overline{x})$. Then we have a canonical isomorphism of functors

(VI.9.9.1) $$\varphi_{\overline{x}} \circ \rho_X^* \xrightarrow{\sim} \gamma \circ \nu_{\overline{x}} \circ j_{\text{fét}}^*.$$

Indeed, for every $V \in \text{Ob}(\mathbf{\acute{E}t}_{\text{f}/X})$, we have $\varphi_{\overline{x}}(\rho_X^* V) = \varphi_{\overline{x}}(V)$, and this set identifies with the geometric fiber $V_{\overline{x}}$ of $V$ over $\overline{x}$. Consequently, $V$ is a neighborhood of the point $\rho_X(\overline{x})$ of $X_{\text{fét}}$ (**[2]** IV 6.8.2) if and only if $V_{\overline{x}}$ is nonempty, or, equivalently, if and only if $Y$ is contained in the image of $V$ in $X$. Let $(Y_i)_{i\in I}$ be a normalized universal cover of $Y$ at $\overline{x}$. For every $i \in I$, by definition $\varphi_{\overline{x}}(Y_i)$ contains a distinguished element. We can therefore canonically consider $Y_i$ as a neighborhood of $\rho_X(\overline{x})$ in the site $\mathbf{\acute{E}t}_{\text{f}/X}$. Moreover, it follows from the equivalence of categories (VI.9.8.1) that the system $(Y_i)_{i\in I}$ is cofinal in the (opposed category of the) category of neighborhoods of $\rho_X(\overline{x})$ in the site $\mathbf{\acute{E}t}_{\text{f}/X}$. The isomorphism (VI.9.9.1) then follows from (**[2]** IV 6.8.3), VI.9.4, and from the definition (VI.9.8.4).

**VI.9.10.** Let $X$ be a connected scheme and $\Pi(X)$ the fundamental groupoid of $X$ (recall that a groupoid is a category whose morphisms are isomorphisms). The objects of $\Pi(X)$ are the geometric points of $X$. For any geometric point $\overline{x}$ of $X$, we denote by $\omega_{\overline{x}}\colon \mathbf{\acute{E}t}_{\text{f}/X} \to \mathbf{Ens}$ the corresponding fiber functor (VI.9.8.1). If $\overline{x}$ and $\overline{x}'$ are two geometric points of $X$, the set $\pi_1(X,\overline{x},\overline{x}')$ of morphisms from $\overline{x}$ to $\overline{x}'$ in $\Pi(X)$ is the set of morphisms (or, equivalently, of isomorphisms) $\omega_{\overline{x}} \to \omega_{\overline{x}'}$ of the associated fiber functors. By (**[37]** V 5.8), the functor

(VI.9.10.1) $$\mu^+\colon \mathbf{\acute{E}t}_{\text{f}/X} \to \mathbf{Hom}(\Pi(X),\mathbf{Ens}), \quad Y \mapsto (\overline{x} \mapsto \omega_{\overline{x}}(Y)),$$

induces an equivalence between the category $\mathbf{\acute{E}t}_{\text{f}/X}$ and the category of functors $\psi$ from $\Pi(X)$ to $\mathbf{Ens}$ such that for every geometric point $\overline{x}$ of $X$, $\psi(\overline{x})$ is a finite set endowed with a continuous action of $\pi_1(X,\overline{x})$.

For every geometric point $\overline{x}$ of $X$, the fiber functor $\omega_{\overline{x}}$ is left exact and transforms covering families into surjective families. It therefore extends to a fiber functor $\phi_{\overline{x}}\colon X_{\text{fét}} \to \mathbf{Ens}$ (**[2]** IV 6.3). The latter is deduced from the functor $\nu_{\overline{x}}$ defined in (VI.9.8.4) by forgetting the action of $\pi_1(X,\overline{x})$. In view of (VI.9.9.1), it therefore corresponds to the point $\rho_X(\overline{x})$ of $X_{\text{fét}}$. Interpreting $\Pi(X)$ as the opposed category of the category of normalized fundamental pro-objects of $\mathbf{\acute{E}t}_{\text{f}/X}$ (**[37]** V 5.7), for all geometric points $\overline{x}$ and $\overline{x}'$ of $X$, every morphism of $\pi_1(X,\overline{x},\overline{x}')$ induces a morphism $\phi_{\overline{x}} \to \phi_{\overline{x}'}$ of the associated fiber functors of $X_{\text{fét}}$. We deduce from this a functor

(VI.9.10.2) $$\Pi(X) \to \mathbf{Pt}(X_{\text{fét}}), \quad \overline{x} \mapsto \phi_{\overline{x}},$$

where $\mathbf{Pt}(X_{\text{fét}})$ is the category of points of $X_{\text{fét}}$. It is an equivalence of categories by virtue of (**[2]** IV 4.9.4 and 7.2.5).

**Lemma VI.9.11.** *Under the assumptions of* VI.9.10, *the functor*

(VI.9.11.1) $$\nu\colon X_{\text{fét}} \to \mathbf{Hom}(\Pi(X),\mathbf{Ens}), \quad F \mapsto (\overline{x} \mapsto \phi_{\overline{x}}(F)),$$

*deduced from* (VI.9.10.2) *induces an equivalence between the category* $X_{\text{fét}}$ *and the category* $\Phi$ *of functors* $\varphi$ *from* $\Pi(X)$ *to* **Ens** *such that for every geometric point* $\overline{x}$ *of* $X$, $\varphi(\overline{x})$ *is a discrete* $\mathbb{U}$*-set endowed with a continuous left action of* $\pi_1(X,\overline{x})$.

Indeed, let $\overline{x}$ be a geometric point of $X$ and $\mathbf{B}_{\pi_1(X,\overline{x})}$ the classifying topos of the profinite group $\pi_1(X,\overline{x})$. The functor

$$\Phi \to \mathbf{B}_{\pi_1(X,\overline{x})}, \quad \varphi \mapsto \varphi(\overline{x}), \tag{VI.9.11.2}$$

is clearly an equivalence of categories. On the other hand, the composition of the latter and the functor $\nu$ is the equivalence of categories

$$\nu_{\overline{x}} \colon X_{\text{fét}} \xrightarrow{\sim} \mathbf{B}_{\pi_1(X,\overline{x})} \tag{VI.9.11.3}$$

defined in (VI.9.8.4), giving the lemma.

**Proposition VI.9.12.** *Let $X$ be a coherent scheme (resp. a scheme with a finite number of connected components). Then every object of $\mathbf{\acute{E}t}_{\mathrm{f}/X}$ is coherent in $X_{\text{fét}}$; in particular, the topos $X_{\text{fét}}$ is coherent.*

For the first assertion, since fibered products are representable in $\mathbf{\acute{E}t}_{\mathrm{f}/X}$, it suffices to show that every object of $\mathbf{\acute{E}t}_{\mathrm{f}/X}$ is quasi-compact ([**2**] VI 2.1). If $X$ is coherent, every object of $\mathbf{\acute{E}t}_{\mathrm{f}/X}$ is a coherent scheme, and is therefore quasi-compact as an object of the site $\mathbf{\acute{E}t}_{\mathrm{f}/X}$. Next, suppose that $X$ has a finite number of connected components $X_1, \dots, X_n$. It suffices to show that for every object $Y$ of $\mathbf{\acute{E}t}_{\mathrm{f}/X}$ and every $1 \leq i \leq n$, $Y \times_X X_i$ is quasi-compact ([**2**] VI 1.3). We may therefore suppose that the image of $Y$ in $X$ is equal to $X_1$. After replacing $X$ by $X_1$ (VI.9.4), if necessary, we can restrict to the case where $X$ is connected. It then follows from the equivalence of categories (VI.9.8.1) that $Y$ is quasi-compact.

Since finite inverse limits are representable in $\mathbf{\acute{E}t}_{\mathrm{f}/X}$, the second assertion follows from the first and from ([**2**] VI 2.4.5).

**Corollary VI.9.13.** *For every scheme $X$ whose set of connected components is locally finite, the topos $X_{\text{fét}}$ is algebraic.*

Indeed, the connected components $(X_i)_{i\in I}$ of $X$ are open and closed in $X$ (VI.9.7.2). Consequently, $(X_i \to X)_{i\in I}$ is a covering of $\mathbf{\acute{E}t}_{\mathrm{f}/X}$. On the other hand, for every $i \in I$, the topos $(X_{\text{fét}})_{/X_i} = (X_i)_{\text{fét}}$ is coherent by virtue of VI.9.12, and the morphism $X_i \to X$ is clearly quasi-separated in $X_{\text{fét}}$. It then follows from the definition ([**2**] VI 2.3) that $X_{\text{fét}}$ is algebraic.

**Corollary VI.9.14.** *For every coherent scheme $X$, the morphism $\rho_X \colon X_{\text{ét}} \to X_{\text{fét}}$ is coherent.*

Indeed, for every object $Y$ of $\mathbf{\acute{E}t}_{\mathrm{f}/X}$, $\rho_X^*(Y) = Y$ is a coherent object of $X_{\text{ét}}$. Consequently, $\rho_X$ is coherent by virtue of VI.9.12 and ([**2**] VI 3.2).

**Lemma VI.9.15.** *Let $X$ be a scheme and $F$ a locally constant and constructible torsion abelian sheaf of $X_{\text{ét}}$, locally constant and constructible. Then:*

(i) *There exist a finite étale cover $Y \to X$ and, letting $\mathbb{Z}_{Y_{\text{ét}}}$ be the free $\mathbb{Z}$-module of $X_{\text{ét}}$ generated by $Y$ ([**2**] IV 11.3.3), a surjective homomorphism $u \colon \mathbb{Z}_{Y_{\text{ét}}} \to F$.*

(ii) *If $\rho_{X*}(F) = 0$, then $F = 0$.*

(i) By descent, $F$ is representable by an object $Y$ of $\mathbf{\acute{E}t}_{\mathrm{f}/X}$ ([**2**] IX 2.2). The identity of $F$ then defines a section $e$ of $F(Y)$ and consequently a homomorphism $u \colon \mathbb{Z}_{Y_{\text{ét}}} \to F$ that is clearly surjective.

(ii) Consider $Y$ and $u$ as in (i) and denote by $\mathbb{Z}_{Y_{\text{fét}}}$ the free $\mathbb{Z}$-module of $X_{\text{fét}}$ generated by $Y$. We have $\rho_X^*(\mathbb{Z}_{Y_{\text{fét}}}) \simeq \mathbb{Z}_{Y_{\text{ét}}}$ by virtue of ([**2**] IV 13.4(b)). We deduce from this a surjective homomorphism $u \colon \rho_X^*(\mathbb{Z}_{Y_{\text{fét}}}) \to F$. Let $v \colon \mathbb{Z}_{Y_{\text{fét}}} \to \rho_{X*}(F)$ be the adjoint morphism of $u$. If $\rho_{X*}(F) = 0$, then $v = 0$ and consequently $u = 0$, which implies $F = 0$.

**Lemma VI.9.16.** *Let $X$ be a scheme whose set of connected components is locally finite, $Y$ an object of $\mathbf{\acute{E}t}_{\mathrm{f}/X}$, $F$ an abelian sheaf of $X_{\mathrm{f\acute{e}t}}$, and $e \in F(Y)$. Denote by $\mathbb{Z}_{Y_{\mathrm{f\acute{e}t}}}$ the free $\mathbb{Z}$-module of $X_{\mathrm{f\acute{e}t}}$ generated by $Y$ ([**2**] IV 11.3.3) and by $u\colon \mathbb{Z}_{Y_{\mathrm{f\acute{e}t}}} \to F$ the homomorphism associated with $e$. Then:*

(i) *$Y$ is a locally constant and constructible sheaf of $\mathbb{Z}$-modules of $X_{\mathrm{\acute{e}t}}$.*
(ii) *If $u$ is an epimorphism, then $\rho_X^*(F)$ is a locally constant and constructible sheaf of $X_{\mathrm{\acute{e}t}}$.*

(i) We can restrict to the case where $X$ is connected (VI.9.7.2). By (VI.9.8.1), there exists a surjective finite étale cover $X' \to X$ such that $Y \times_X X'$ is $X'$-isomorphic to a finite disjoint sum of copies of $X'$, giving the assertion.

(ii) Proceeding as in (i), we can reduce to the case where $X$ is connected and $Y$ is $X$-isomorphic to a finite disjoint sum of copies of $X$, by virtue of (VI.9.3.4) and ([**2**] IV 13.4(b)). Hence $e$ corresponds to sections $e_1, \dots, e_n \in F(X)$ such that the induced homomorphism $\mathbb{Z}^n_{X_{\mathrm{f\acute{e}t}}} \to F$ is surjective. Consider a geometric point $\overline{x}$ of $X$, and take the notation of (VI.9.8). We deduce from this, in view of the equivalence of categories (VI.9.8.4), that $\nu_{\overline{x}}(F)$ is an abelian group of finite type, endowed with the trivial action of $\pi_1(X, \overline{x})$. Consequently, $F$ is a constant abelian sheaf of $X_{\mathrm{f\acute{e}t}}$ with value $\nu_{\overline{x}}(F)$, and the same then holds for $\rho_X^*(F)$.

**Lemma VI.9.17.** *Let $X$ be a scheme whose set of connected components is locally finite and $F$ a sheaf of $X_{\mathrm{f\acute{e}t}}$. Then the category $(\mathbf{\acute{E}t}_{\mathrm{f}/X})_{/F}$ is filtered.*

It suffices to show that the sums of two objects and the cokernels of double arrows are representable in $(\mathbf{\acute{E}t}_{\mathrm{f}/X})_{/F}$ ([**2**] I 2.7.1). If $U$ and $V$ are two objects of $(\mathbf{\acute{E}t}_{\mathrm{f}/X})_{/F}$, the disjoint sum $U \sqcup V$ is a finite étale cover of $X$ that represents the sum of $U$ and $V$ in $(\mathbf{\acute{E}t}_{\mathrm{f}/X})_{/F}$. Let $f, g\colon U \rightrightarrows V$ be a double arrow of $(\mathbf{\acute{E}t}_{\mathrm{f}/X})_{/F}$ and $G$ its cokernel in $X_{\mathrm{f\acute{e}t}}$. It suffices to show that $G$ is representable by an object of $\mathbf{\acute{E}t}_{\mathrm{f}/X}$. There exists an étale covering $(X_i \to X)_{i \in I}$ of $\mathbf{\acute{E}t}_{\mathrm{f}/X}$ such that, for every $i \in I$, $X_i$ is connected (VI.9.7.2). By descent, it suffices to show that, for every $i \in I$, $G|X_i$ is representable by an object of $\mathbf{\acute{E}t}_{\mathrm{f}/X_i}$ ([**37**] VIII 2.1 and 5.7, [**35**] II 3.4.4). We can therefore restrict to the case where $X$ is connected. We consider a geometric point $\overline{x}$ of $X$, take the notation of (VI.9.8) and identify $\mathbf{\acute{E}t}_{\mathrm{f}/X}$ with the category $\mathscr{C}(\pi_1(X, \overline{x}))$ by the equivalence of categories (VI.9.8.1). Since in every Galois category, finite direct limits are representable ([**37**] V 4.2), the cokernel $W$ of the double arrow $(f, g)$ is representable in $\mathscr{C}(\pi_1(X, \overline{x}))$. One immediately sees that $W$ is also the cokernel of $(f, g)$ in the topos $\mathbf{B}_{\pi_1(X, \overline{x})}$. Identifying $X_{\mathrm{f\acute{e}t}}$ to $\mathbf{B}_{\pi_1(X, \overline{x})}$ by the functor (VI.9.8.4), we then see that $W$ represents $G$.

**Proposition VI.9.18.** *If $X$ is a coherent scheme having a finite number of connected components, then the adjunction morphism $\mathrm{id} \to \rho_{X*}\rho_X^*$ is an isomorphism; in particular, the functor $\rho_X^*\colon X_{\mathrm{f\acute{e}t}} \to X_{\mathrm{\acute{e}t}}$ is fully faithful.*

Let $G$ be a sheaf of $X_{\mathrm{f\acute{e}t}}$. By ([**2**] II 4.1.1), the canonical morphism of $X_{\mathrm{f\acute{e}t}}$

$$\varinjlim_{(\mathbf{\acute{E}t}_{\mathrm{f}/X})_{/G}} U \to G \tag{VI.9.18.1}$$

is an isomorphism. Since the functor $\rho_X^*$ commutes with direct limits and extends the canonical injection functor $\mathbf{\acute{E}t}_{\mathrm{f}/X} \to \mathbf{\acute{E}t}_{/X}$, we deduce from this an isomorphism of $X_{\mathrm{\acute{e}t}}$

$$\varinjlim_{(\mathbf{\acute{E}t}_{\mathrm{f}/X})_{/G}} U \xrightarrow{\sim} \rho_X^*(G). \tag{VI.9.18.2}$$

On the other hand, the topos $X_{\mathrm{\acute{e}t}}$ and $X_{\mathrm{f\acute{e}t}}$ are coherent (VI.9.12) and the morphism $\rho_X$ is coherent (VI.9.14). Hence the functor $\rho_{X*}$ commutes with filtered direct limits.

Indeed, the proof of ([**2**] VI 5.1) also works for the direct image functor of sheaves of sets. Since the category $(\mathbf{\acute{E}t}_{\mathrm{f}/X})_{/G}$ is filtered by VI.9.17, we deduce from this an isomorphism

(VI.9.18.3) $$\lim_{\substack{\longrightarrow \\ (\mathbf{\acute{E}t}_{\mathrm{f}/X})/G}} U \xrightarrow{\sim} \rho_{X*}(\rho_X^*(G)).$$

By functoriality, the adjunction morphism $G \to \rho_{X*}(\rho_X^* G)$ is the direct limit of the identity morphisms $\mathrm{id}_U$, for the objects $U$ of $(\mathbf{\acute{E}t}_{\mathrm{f}/X})/G$. It is therefore an isomorphism, giving the first part of the proposition. The second part is equivalent to the first by general properties of adjoint functors.

**Remark VI.9.19.** Let $X$ be a scheme and $G$ a sheaf of $X_{\mathrm{f\acute{e}t}}$. By ([**2**] I 5.1 and III 1.3), $\rho_X^*(G)$ is the sheaf associated with the presheaf $F$ on $\mathbf{\acute{E}t}_{/X}$ defined, for every $V \in \mathrm{Ob}(\mathbf{\acute{E}t}_{/X})$, by

(VI.9.19.1) $$F(V) = \lim_{\substack{\longrightarrow \\ (U,u)\in I_V^{\circ}}} G(U),$$

where $I_V$ is the category of pairs $(U, u)$ consisting of an object $U$ of $\mathbf{\acute{E}t}_{\mathrm{f}/X}$ and an $X$-morphism $u\colon V \to U$. Moreover, the adjunction morphism $G \to \rho_{X*}(\rho_X^* G)$ is induced by the morphism of presheaves on $\mathbf{\acute{E}t}_{\mathrm{f}/X}$ defined, for every $U \in \mathrm{Ob}(\mathbf{\acute{E}t}_{\mathrm{f}/X})$, by the canonical isomorphism

(VI.9.19.2) $$G(U) \xrightarrow{\sim} F(U);$$

indeed, $I_U$ admits a final object, namely $(U, \mathrm{id}_U)$. Notice that, in general, this does not imply that the adjunction morphism $\mathrm{id} \to \rho_{X*}\rho_X^*$ is an isomorphism, because $\rho_X^*(G)$ is not in general equal to $F$.

**Corollary VI.9.20.** *Let $X$ be a coherent scheme having a finite number of connected components and $F$ a sheaf (resp. a torsion abelian sheaf) of $X_{\mathrm{\acute{e}t}}$. Then the following conditions are equivalent:*

(i) *There exist a sheaf (resp. a torsion abelian sheaf) $G$ of $X_{\mathrm{f\acute{e}t}}$ and an isomorphism $F \simeq \rho_X^*(G)$.*

(ii) *$F$ is the direct limit in $X_{\mathrm{\acute{e}t}}$ of a filtered direct system of locally constant and constructible sheaves (resp. torsion abelian sheaves).*

Let us show that (i) implies (ii). Let $G$ be a sheaf of $X_{\mathrm{f\acute{e}t}}$. By (VI.9.18.2), we have an isomorphism of $X_{\mathrm{\acute{e}t}}$

(VI.9.20.1) $$\lim_{\substack{\longrightarrow \\ (\mathbf{\acute{E}t}_{\mathrm{f}/X})/G}} U \xrightarrow{\sim} \rho_X^*(G).$$

We deduce from this by VI.9.16(i) and VI.9.17 that $\rho_X^*(G)$ satisfies the non-resp. condition of (ii).

Let $G$ be a torsion abelian sheaf of $X_{\mathrm{f\acute{e}t}}$. For every object $U$ of $(\mathbf{\acute{E}t}_{\mathrm{f}/X})_{/G}$, denote by $\mathbb{Z}_{U_{\mathrm{f\acute{e}t}}}$ the free $\mathbb{Z}$-module of $X_{\mathrm{f\acute{e}t}}$ generated by $U$ and by $H_U$ the image of the canonical morphism $\mathbb{Z}_{U_{\mathrm{f\acute{e}t}}} \to G$ ([**2**] IV 11.3.3). The isomorphism (VI.9.18.1) induces a group isomorphism of $X_{\mathrm{f\acute{e}t}}$

(VI.9.20.2) $$\lim_{\substack{\longrightarrow \\ U\in(\mathbf{\acute{E}t}_{\mathrm{f}/X})/G}} H_U \xrightarrow{\sim} G.$$

We deduce from this a group isomorphism of $X_{\mathrm{\acute{e}t}}$

(VI.9.20.3) $$\lim_{\substack{\longrightarrow \\ (\mathbf{\acute{E}t}_{\mathrm{f}/X})/G}} \rho_X^*(H_U) \to \rho_X^*(G).$$

By VI.9.16(ii) and ([**2**] IX 1.2), $\rho_X^*(H_U)$ is a locally constant and constructible torsion abelian sheaf of $X_{\text{ét}}$. Since the category $(\mathbf{\acute{E}t}_{\text{f}/X})/G$ is filtered (VI.9.17), $\rho_X^*(G)$ satisfies the resp. condition of (ii).

Let us show that (ii) implies (i). First, consider the case where $F$ is a sheaf of sets. By descent, if $F$ is locally constant and constructible, it is representable by an object $Y$ of $\mathbf{\acute{E}t}_{\text{f}/X}$ ([**2**] IX 2.2); we therefore have an isomorphism $F \xrightarrow{\sim} \rho_X^*(Y)$, which shows the required property in the case under consideration. The general case follows because $\rho_X^*$ is fully faithful (VI.9.18) and commutes with direct limits.

Next, consider the case where $F$ is a locally constant and constructible torsion abelian sheaf. By VI.9.15(i), there exist a finite étale cover $Y \to X$ and a surjective homomorphism $u\colon \mathbb{Z}_{Y_{\text{ét}}} \to F$, where $\mathbb{Z}_{Y_{\text{ét}}}$ is the free $\mathbb{Z}$-module of $X_{\text{ét}}$ generated by $Y$. Denote by $\mathbb{Z}_{Y_{\text{fét}}}$ the free $\mathbb{Z}$-module of $X_{\text{fét}}$ generated by $Y$. We have $\rho_X^*(\mathbb{Z}_{Y_{\text{fét}}}) \simeq \mathbb{Z}_{Y_{\text{ét}}}$ by virtue of ([**2**] IV 13.4(b)). Let $v\colon \mathbb{Z}_{Y_{\text{fét}}} \to \rho_{X*}(F)$ be the adjoint morphism of $u$, $G$ the image of $v$, $w\colon \rho_X^*(G) \to F$ the adjoint of the canonical morphism $G \to \rho_{X*}(F)$, and $H$ the kernel of $w$. It is clear that $w$ is surjective. On the other hand, $\rho_{X*}(F)$ is torsion by virtue of VI.9.12, VI.9.14, and ([**2**] IX 1.2(v)); the same then holds for $G$ and $\rho_X^*(G)$. By VI.9.16(ii), $\rho_X^*(G)$ is a locally constant and constructible abelian sheaf of $X_{\text{ét}}$. Consequently, $H$ is a locally constant and constructible torsion abelian sheaf, by virtue of ([**2**] IX 2.1(ii) and 2.6). On the one hand, the sequence

$$0 \to \rho_{X*}(H) \to \rho_{X*}(\rho_X^*(G)) \to \rho_{X*}(F) \tag{VI.9.20.4}$$

is exact. On the other hand, the adjunction morphism $G \to \rho_{X*}(\rho_X^*(G))$ is an isomorphism by virtue of VI.9.18. We deduce from this that $\rho_{X*}(H) = 0$ and consequently that $H = 0$ by virtue of VI.9.15(ii). Hence $w$ is an isomorphism, which shows the required property in the case under consideration.

Finally, the case where $F$ is the direct limit in $X_{\text{ét}}$ of a filtered direct system of locally constant and constructible torsion abelian sheaves follows from the previous case because $\rho_X^*$ is fully faithful (VI.9.18) and commutes with direct limits.

**Definition VI.9.21.** We say that a scheme $X$ is $K(\pi, 1)$ if for every invertible integer $n$ in $\mathscr{O}_X$ and every $(\mathbb{Z}/n\mathbb{Z})$-module $F$ of $X_{\text{fét}}$, the adjunction homomorphism $F \to \mathrm{R}\rho_{X*}(\rho_X^* F)$ is an isomorphism.

This notion seems reasonable only for schemes that satisfy the conclusion of VI.9.18, in particular for coherent schemes with a finite number of connected components.

## VI.10. Faltings site and topos

**VI.10.1.** In this section, $f\colon Y \to X$ denotes a morphism of schemes and

$$\pi\colon E \to \mathbf{\acute{E}t}_{/X} \tag{VI.10.1.1}$$

the fibered $\mathbb{U}$-site deduced from the fibered site of finite étale covers $\mathscr{R}/\mathbf{Sch}$ (VI.9.5.1) by base change by the functor

$$\mathbf{\acute{E}t}_{/X} \to \mathbf{Sch}, \quad U \mapsto U \times_X Y. \tag{VI.10.1.2}$$

We say that $\pi$ is the *Faltings fibered site* associated with $f$. We can describe the category $E$ explicitly as follows. The objects of $E$ are the morphisms of schemes $V \to U$ over $f\colon Y \to X$ such that the morphism $U \to X$ is étale and that the morphism $V \to U_Y = U \times_X Y$ is finite étale. Let $(V' \to U')$, $(V \to U)$ be two objects of $E$. A morphism from $(V' \to U')$ to $(V \to U)$ consists of an $X$-morphism $U' \to U$ and a $Y$-morphism $V' \to V$

such that the diagram

(VI.10.1.3)
$$\begin{array}{ccc} V' & \longrightarrow & U' \\ \downarrow & & \downarrow \\ V & \longrightarrow & U \end{array}$$

is commutative. The functor $\pi$ is then defined for every object $(V \to U)$ of $E$, by

(VI.10.1.4)
$$\pi(V \to U) = U.$$

Note that the fibered site $\pi$ satisfies the conditions of VI.5.1 as well as condition VI.5.32(i'). We endow $E$ with the covanishing topology associated with $\pi$ (VI.5.3), in other words, the topology generated by the coverings $\{(V_i \to U_i) \to (V \to U)\}_{i \in I}$ of the following two types:

(v) $U_i = U$ for every $i \in I$, and $(V_i \to V)_{i \in I}$ is an étale covering.
(c) $(U_i \to U)_{i \in I}$ is an étale covering and $V_i = U_i \times_U V$ for every $i \in I$.

The resulting covanishing site $E$ is also called the *Faltings site* associated with $f$; it is a $\mathbb{U}$-site. We denote by $\widehat{E}$ (resp. $\widetilde{E}$) the category of presheaves (resp. the topos of sheaves) of $\mathbb{U}$-sets on $E$. We also call $\widetilde{E}$ the *Faltings topos* associated with $f$. If $F$ is a presheaf on $E$, we denote by $F^a$ the associated sheaf.

**Remark VI.10.2.** The category $E$ was initially introduced by Faltings, but with a topology that is in general strictly finer than the covanishing topology, namely the topology generated by the families of morphisms $\{(V_i \to U_i) \to (V \to U)\}_{i \in I}$ such that $(V_i \to V)_{i \in I}$ and $(U_i \to U)_{i \in I}$ are étale coverings ([**26**] page 214). Indeed, if $Y$ is empty, $\widetilde{E}$ is the empty topos; that is, it is equivalent to the punctual category ([**2**] IV 2.2 and 4.4). This follows from VI.5.11 because for every $U \in \mathrm{Ob}(\mathbf{\acute{E}t}_{/X})$, $U_Y$ is empty, and therefore $(U_Y)_{\mathrm{f\acute{e}t}}$ is the empty topos. However, if we endow $E$ with the topology considered by Faltings, we obtain the topos $X_{\mathrm{\acute{e}t}}$. This example also shows that the topology considered by Faltings does not in general satisfy Proposition VI.5.10, which nevertheless plays an essential role in his approach, whence the need to modify it as we have done in VI.10.1. We give in III.8.18 an example that illustrates another drawback of the topology considered by Faltings.

**VI.10.3.** It follows from VI.5.6 and from the fact that $E$ admits a final object that finite inverse limits are representable in $E$, that the functor $\pi$ is left exact, and that the family of vertical (resp. Cartesian) coverings of $E$ is stable under base change. In fact, the inverse limit of a diagram

(VI.10.3.1)
$$\begin{array}{ccc} & & (V'' \to U'') \\ & & \downarrow \\ (V' \to U') & \longrightarrow & (V \to U) \end{array}$$

of $E$ is representable by the morphism $(V' \times_V V'' \to U' \times_U U'')$. Indeed, this morphism clearly represents the inverse limit of the diagram (VI.10.3.1) in the category of morphisms of schemes over $f$. It therefore suffices to show that it is an object of $E$, or, equivalently, that the morphism

(VI.10.3.2)
$$V' \times_V V'' \to U'_Y \times_{U_Y} U''_Y$$

is a finite étale cover. It is clearly a morphism of $\mathbf{\acute{E}t}_{/Y}$, and is therefore étale. On the other hand, the diagram

$$\begin{array}{ccc} V' \times_V V'' & \longrightarrow & V' \times_{U_Y} V'' \\ \downarrow & & \downarrow \\ V & \xrightarrow{\Delta_V} & V \times_{U_Y} V \end{array} \tag{VI.10.3.3}$$

where $\Delta_V$ is the diagonal embedding, is Cartesian. Since $\Delta_V$ is a closed immersion, we immediately deduce from this that the morphism (VI.10.3.2) is finite, giving our assertion.

**VI.10.4.** Let $\star$ be one of the two symbols "coh" for coherent or "scoh" for separated and coherent, introduced in VI.9.1. We denote by

$$\pi_\star\colon E_\star \to \mathbf{\acute{E}t}_{\star/X} \tag{VI.10.4.1}$$

the fibered site deduced from $\pi$ by base change by the canonical injection functor (VI.9.1)

$$\varphi\colon \mathbf{\acute{E}t}_{\star/X} \to \mathbf{\acute{E}t}_{/X}, \tag{VI.10.4.2}$$

and by

$$\Phi\colon E_\star \to E \tag{VI.10.4.3}$$

the canonical projection. We endow $E_\star$ with the covanishing topology defined by $\pi_\star$ and denote by $\widetilde{E}_\star$ the topos of sheaves of $\mathbb{U}$-sets on $E_\star$. By virtue of VI.5.21, if $X$ is quasi-separated, the functor $\Phi$ induces by restriction an equivalence of categories

$$\Phi_s\colon \widetilde{E} \xrightarrow{\sim} \widetilde{E}_\star. \tag{VI.10.4.4}$$

Moreover, under the same assumption, the covanishing topology of $E_\star$ is induced by that of $E$ through the functor $\Phi$, by VI.5.22.

**Proposition VI.10.5.** *Suppose that $X$ and $Y$ are coherent. Then:*

(i) *For every object $(V \to U)$ of $E_{\mathrm{coh}}$, $(V \to U)^a$ is a coherent object of $\widetilde{E}$.*
(ii) *The topos $\widetilde{E}$ is coherent; in particular, it has enough points.*

(i) Indeed, every object of $\mathbf{\acute{E}t}_{\mathrm{coh}/X}$ is quasi-compact. On the other hand, for every $W \in \mathrm{Ob}(\mathbf{\acute{E}t}_{\mathrm{coh}/Y})$, since $W$ is a coherent scheme, every object of $\mathbf{\acute{E}t}_{\mathrm{f}/W}$ is quasi-compact by virtue of VI.9.12. The statement then follows from (VI.10.4.4) and VI.5.27(iii) (applied to the fibered site $\pi_{\mathrm{coh}}$ (VI.10.4.1)).

(ii) This follows from VI.5.27(iv) and ([**2**] VI § 9).

**VI.10.6.** We endow $\mathbf{\acute{E}t}_{/X}$ with the final object $X$ and $E$ by the final object $(Y \to X)$. The functors $\alpha_{X!}$ (VI.5.1.2) and $\sigma^+$ (VI.5.32.4) are then explicitly defined by

$$\alpha_{X!}\colon \mathbf{\acute{E}t}_{\mathrm{f}/Y} \to E, \qquad V \mapsto (V \to X), \tag{VI.10.6.1}$$

$$\sigma^+\colon \mathbf{\acute{E}t}_{/X} \to E, \qquad U \mapsto (U_Y \to U). \tag{VI.10.6.2}$$

These are left exact and continuous (VI.5.32). Hence they define two morphisms of topos

$$\beta\colon \widetilde{E} \to Y_{\mathrm{f\acute{e}t}}, \tag{VI.10.6.3}$$

$$\sigma\colon \widetilde{E} \to X_{\mathrm{\acute{e}t}}. \tag{VI.10.6.4}$$

For every sheaf $F = \{U \mapsto F_U\}$ on $E$, we have $\beta_*(F) = F_X$.

**Lemma VI.10.7.** (i) *The functor*

(VI.10.7.1) $$\Psi^+ : E \to \mathbf{\acute{E}t}_{/Y}, \quad (V \to U) \mapsto V$$

*is continuous and left exact; it therefore defines a morphism of topos*

(VI.10.7.2) $$\Psi : Y_{\text{ét}} \to \widetilde{E}.$$

(ii) *For every sheaf $F$ of $Y_{\text{ét}}$, we have a canonical isomorphism of $\widetilde{E}$*

(VI.10.7.3) $$\Psi_*(F) \xrightarrow{\sim} \{U \mapsto \rho_{U_Y *}(F|U_Y)\},$$

*where for every object $U$ of $\mathbf{\acute{E}t}_{/X}$, $\rho_{U_Y} : (U_Y)_{\text{ét}} \to (U_Y)_{\text{fét}}$ is the canonical morphism* (VI.9.2.1), *and for every morphism $g : U' \to U$ of $\mathbf{\acute{E}t}_{/X}$, the transition morphism*

(VI.10.7.4) $$\rho_{U_Y *}(F|U_Y) \to (g_Y)_{\text{fét}*}(\rho_{U'_Y *}(F|U'_Y))$$

*is the composition*

(VI.10.7.5) $$\rho_{U_Y *}(F|U_Y) \to \rho_{U_Y *}((g_Y)_{\text{ét}*}(F|U'_Y)) \xrightarrow{\sim} (g_Y)_{\text{fét}*}(\rho_{U'_Y *}(F|U'_Y)),$$

*in which the first arrow is induced by the adjunction morphism* $\mathrm{id} \to (g_Y)_{\text{ét}*}(g_Y)^*_{\text{ét}}$ *and the second arrow by* (VI.9.3.4).

Indeed, $\Psi^+$ is clearly left exact (VI.10.3). On the other hand, for every sheaf $F$ of $Y_{\text{ét}}$, we have a canonical isomorphism of $\widehat{E}$

(VI.10.7.6) $$F \circ \Psi^+ \xrightarrow{\sim} \{U \mapsto \rho_{U_Y *}(F|U_Y)\},$$

where the right-hand side is the presheaf on $E$ defined by the transition morphisms (VI.10.7.5). Let $(U_i \to U)_{i\in I}$ be a covering of $\mathbf{\acute{E}t}_{/X}$. For every $(i,j) \in I^2$, set $V_i = U_i \times_X Y$, $U_{ij} = U_i \times_U U_j$, and $V_{ij} = U_{ij} \times_X Y$, and denote by $h_i : V_i \to U_Y$ and $h_{ij} : V_{ij} \to U_Y$ the structural morphisms. The sequence

(VI.10.7.7) $$0 \to F|U_Y \to \prod_{i\in I} (h_i)_{\text{ét}*}(F|V_i) \rightrightarrows \prod_{(i,j)\in I^2} (h_{ij})_{\text{ét}*}(F|V_{ij})$$

is exact. Since $\rho_{U_Y *}$ commutes with inverse limits, we deduce from this by (VI.9.3.4) that the sequence

(VI.10.7.8) $$0 \to \rho_{U_V *}(F|U_Y) \to \prod_{i\in I} (h_i)_{\text{fét}*}(\rho_{V_i *}(F|V_i)) \rightrightarrows \prod_{(i,j)\in I^2} (h_{ij})_{\text{fét}*}(\rho_{V_{ij} *}(F|V_{ij}))$$

is exact. Consequently, $F \circ \Psi^+$ is a sheaf on $E$ by virtue of VI.5.10, giving the lemma.

**VI.10.8.** Let us describe explicitly the constructions of (VI.6.1) for the functor $\Psi^+$ defined in (VI.10.7.1). The composed functor

(VI.10.8.1) $$\Psi^+ \circ \sigma^+ : \mathbf{\acute{E}t}_{/X} \to \mathbf{\acute{E}t}_{/Y}$$

is none other than the inverse image functor by $f : Y \to X$; we therefore have $f_{\text{ét}} = \sigma\Psi$. On the other hand, for every object $U$ of $\mathbf{\acute{E}t}_{/X}$, the functor (VI.6.1.5)

(VI.10.8.2) $$\Psi^+_U : \mathbf{\acute{E}t}_{\mathrm{f}/U_Y} \to \mathbf{\acute{E}t}_{/U_Y}$$

induced by $\Psi^+$, identifies with the canonical injection. We can therefore identify the morphism $\Psi_U$ (VI.6.1.6) with the canonical morphism $\rho_{U_Y} : (U_Y)_{\text{ét}} \to (U_Y)_{\text{fét}}$ (VI.9.2.1); in particular, we have $\beta\Psi = \rho_Y$ (VI.6.1.9). For every morphism $g : U' \to U$ of $\mathbf{\acute{E}t}_{/X}$, the diagram (VI.6.2.1) then identifies with the diagram (VI.9.3.4). From the isomorphisms $\Psi^*\sigma^* = f^*_{\text{ét}}$ and $\Psi^*\beta^* = \rho^*_Y$, we deduce, by adjunction, morphisms

(VI.10.8.3) $$\sigma^* \to \Psi_* f^*_{\text{ét}},$$

(VI.10.8.4) $$\beta^* \to \Psi_* \rho^*_Y.$$

**Proposition VI.10.9.** *Suppose that $X$ is quasi-separated, and that $Y$ is coherent and étale-locally connected* (VI.9.7)*. For every $U \in \mathrm{Ob}(\mathbf{\acute{E}t}_{/X})$, denote by $h_U \colon U_Y \to Y$ the canonical projection. Then:*

(i) *For every sheaf $F$ of $Y_{\mathrm{f\acute{e}t}}$, $\beta^*(F)$ is the sheaf on $E_{\mathrm{coh}}$ defined by $\{U \mapsto (h_U)^*_{\mathrm{f\acute{e}t}}F\}$.*
(ii) *The adjunction morphism $\mathrm{id} \to \beta_*\beta^*$ is an isomorphism.*
(iii) *The adjunction morphism $\beta^* \to \Psi_*\rho^*_Y$* (VI.10.8.4) *is an isomorphism.*

First note that the three functors

(VI.10.9.1) $$\alpha^{\mathrm{coh}}_{X!} \colon \mathbf{\acute{E}t}_{\mathrm{f}/Y} \to E_{\mathrm{coh}}, \qquad V \mapsto (V \to X),$$

(VI.10.9.2) $$\sigma^+_{\mathrm{coh}} \colon \mathbf{\acute{E}t}_{\mathrm{coh}/X} \to E_{\mathrm{coh}}, \qquad U \mapsto (U_Y \to U),$$

(VI.10.9.3) $$\Psi^+_{\mathrm{coh}} \colon E_{\mathrm{coh}} \to \mathbf{\acute{E}t}_{\mathrm{coh}/Y}, \qquad (V \to U) \mapsto V,$$

are well-defined, continuous, and left exact. The morphisms of topos they define identify with $\beta$, $\sigma$, and $\Psi$, respectively, in view of (VI.10.4.4) and VI.9.1. On the other hand, for every object $U$ of $\mathbf{\acute{E}t}_{\mathrm{coh}/X}$, the scheme $U_Y$ is coherent and locally connected. Consequently, the adjunction morphism $\mathrm{id} \to \rho_{U_Y*}\rho^*_{U_Y}$ is an isomorphism by virtue of VI.9.18. The proposition therefore follows from VI.6.3.

**Proposition VI.10.10.** *Suppose that $X$ and $Y$ are coherent. Then the morphisms $\beta$, $\sigma$, and $\Psi$ are coherent.*

Indeed, every object of $\mathbf{\acute{E}t}_{\mathrm{coh}/X}$ is coherent in $X_{\mathrm{\acute{e}t}}$; every object of $\mathbf{\acute{E}t}_{\mathrm{f}/Y}$ is coherent in $Y_{\mathrm{f\acute{e}t}}$ (VI.9.12); for every object $(V \to U)$ of $E_{\mathrm{coh}}$, $(V \to U)^a$ is a coherent object of $\widetilde{E}$ by VI.10.5(i). The proposition therefore follows from ([**2**] VI 3.3), in view of the proof of VI.10.9.

**Remark VI.10.11.** We have a 2-morphism

(VI.10.11.1) $$\tau \colon f_{\mathrm{f\acute{e}t}}\beta \to \rho_X\sigma,$$

such that for every sheaf $F$ on $E$ and every $U \in \mathrm{Ob}(\mathbf{\acute{E}t}_{\mathrm{f}/X})$,

$$f_{\mathrm{f\acute{e}t}*}(\beta_*(F))(U) \to \rho_{X*}(\sigma_*(F))(U)$$

is the canonical map $F(U_Y \to X) \to F(U_Y \to U)$.

**VI.10.12.** Consider a commutative diagram of morphisms of schemes

(VI.10.12.1) $$\begin{array}{ccc} Y' & \xrightarrow{f'} & X' \\ {\scriptstyle g'}\downarrow & & \downarrow{\scriptstyle g} \\ Y & \xrightarrow{f} & X \end{array}$$

We denote by

(VI.10.12.2) $$\pi' \colon E' \to \mathbf{\acute{E}t}_{/X'}$$

the fibered Faltings site associated with $f'$ (VI.10.1). We endow $E'$ with the covanishing topology associated with $\pi'$ and denote by $\widetilde{E}'$ the topos of sheaves of $\mathbb{U}$-sets on $E'$. For every object $(V \to U)$ of $E$, the canonical morphism $V \times_Y Y' \to U \times_X X'$ is an object of $E'$. We thus define a functor

(VI.10.12.3) $$\Phi^+ \colon E \to E', \quad (V \to U) \mapsto (V \times_Y Y' \to U \times_X X'),$$

that is clearly left exact (VI.10.3). For every sheaf $F = \{U' \mapsto F_{U'}\}$ on $E'$, $F \circ \Phi^+$ is the presheaf on $E$ defined by

(VI.10.12.4) $$\{U \mapsto (g'_U)_{\mathrm{f\acute{e}t}*}(F_{U \times_X X'})\},$$

where for every $U \in \mathrm{Ob}(\mathbf{\acute{E}t}_{/X})$, $g'_U \colon U \times_X Y' \to U \times_X Y$ is the base change by $g'$. Let $(U_i \to U)_{i \in I}$ be a covering of $\mathbf{\acute{E}t}_{/X}$. Since the functor $(g'_U)_{\mathrm{f\acute{e}t}*}$ commutes with inverse limits, the gluing relation of $F$ with regard to the covering $(U_i \times_X X' \to U \times_X X')_{i \in I}$ (VI.5.10.1) implies the analogous relation for $F \circ \Phi^+$ with regard to the covering $(U_i \to U)_{i \in I}$. Consequently, $F \circ \Phi^+$ is a sheaf on $E$ (VI.5.10), and $\Phi^+$ is continuous. The latter therefore defines a morphism of topos

(VI.10.12.5) $$\Phi \colon \widetilde{E}' \to \widetilde{E}.$$

It immediately follows from the definitions that the squares of the diagram

(VI.10.12.6) $$\begin{array}{ccccc} X'_{\mathrm{\acute{e}t}} & \xleftarrow{\sigma'} & \widetilde{E}' & \xrightarrow{\beta'} & Y'_{\mathrm{f\acute{e}t}} \\ {\scriptstyle g_{\mathrm{\acute{e}t}}}\downarrow & & \downarrow{\scriptstyle \Phi} & & \downarrow{\scriptstyle g'_{\mathrm{f\acute{e}t}}} \\ X_{\mathrm{\acute{e}t}} & \xleftarrow{\sigma} & \widetilde{E} & \xrightarrow{\beta} & Y_{\mathrm{f\acute{e}t}} \end{array}$$

where $\beta'$ and $\sigma'$ are the canonical morphisms (VI.10.6.3) and (VI.10.6.4) relative to $f'$, are commutative up to canonical isomorphisms. On the other hand, the diagram

(VI.10.12.7) $$\begin{array}{ccc} Y'_{\mathrm{\acute{e}t}} & \xrightarrow{\Psi'} & \widetilde{E}' \\ {\scriptstyle g'_{\mathrm{\acute{e}t}}}\downarrow & & \downarrow{\scriptstyle \Phi} \\ Y_{\mathrm{\acute{e}t}} & \xrightarrow{\Psi} & \widetilde{E} \end{array}$$

where $\Psi'$ is the morphism (VI.10.7.2) relative to $f'$, is commutative up to canonical isomorphism.

**Remark VI.10.13.** We keep the assumptions of VI.10.12 and let $F$ be an abelian sheaf on $\widetilde{E}$ and $i$ an integer $\geq 0$. Then the diagram

(VI.10.13.1) $$\begin{array}{ccc} \mathrm{H}^i(Y_{\mathrm{f\acute{e}t}}, \beta_* F) & \longrightarrow & \mathrm{H}^i(\widetilde{E}, F) \\ {\scriptstyle u}\downarrow & & \\ \mathrm{H}^i(Y'_{\mathrm{f\acute{e}t}}, g'^*_{\mathrm{f\acute{e}t}}(\beta_* F)) & & \downarrow{\scriptstyle w} \\ {\scriptstyle v}\downarrow & & \\ \mathrm{H}^i(Y'_{\mathrm{f\acute{e}t}}, \beta'_*(\Phi^* F)) & \longrightarrow & \mathrm{H}^i(\widetilde{E}', \Phi^* F) \end{array}$$

where the horizontal arrows come from Cartan–Leray spectral sequences ([**2**] V 5.3), $u$ and $w$ are the canonical morphisms, and $v$ is induced by the base change morphism with respect to the right square of (VI.10.12.6) ([**1**] (1.2.2.2)) is commutative.

**VI.10.14.** Let $(f' \colon Y' \to X')$ be an object of $E$. We denote by

(VI.10.14.1) $$\pi' \colon E' \to \mathbf{\acute{E}t}_{/X'}$$

the fibered Faltings site associated with $f'$. Every object of $E'$ is naturally an object of $E$. We thus define a functor

(VI.10.14.2) $$\Phi \colon E' \to E.$$

One immediately verifies that $\Phi$ factors through an equivalence of categories

(VI.10.14.3) $$E' \xrightarrow{\sim} E_{/(Y' \to X')},$$

that is even an equivalence of categories over $\mathbf{\acute{E}t}_{/X'}$ ([**37**] VI 4.3), where we consider $E_{/(Y'\to X')}$ as an $(\mathbf{\acute{E}t}_{/X'})$-category by base change by the functor $\pi$. It then follows from VI.5.38 that the covanishing topology of $E'$ is induced by that of $E$ by the functor $\Phi$. Consequently, $\Phi$ is continuous and cocontinuous ([**2**] III 5.2). It therefore defines a sequence of three adjoint functors:

(VI.10.14.4) $$\Phi_!\colon \widetilde{E}' \to \widetilde{E}, \quad \Phi^*\colon \widetilde{E} \to \widetilde{E}', \quad \Phi_*\colon \widetilde{E}' \to \widetilde{E},$$

in the sense that for any two consecutive functors in the sequence, the one on the right is right adjoint to the other. By ([**2**] III 5.4), the functor $\Phi_!$ factors through an equivalence of categories

(VI.10.14.5) $$\widetilde{E}' \xrightarrow{\sim} \widetilde{E}_{/(Y'\to X')^a}.$$

The pair of functors $(\Phi^*, \Phi_*)$ defines the localization morphism of $\widetilde{E}$ at $(Y' \to X')^a$

(VI.10.14.6) $$\Phi\colon \widetilde{E}' \to \widetilde{E}.$$

Since $\Phi\colon E' \to E$ is a left adjoint of the functor $\Phi^+\colon E \to E'$ defined in (VI.10.12.3), the morphism (VI.10.14.6) identifies with the morphism defined in (VI.10.12.5), by virtue of ([**2**] III 2.5).

**VI.10.15.** We denote by

(VI.10.15.1) $$\varpi\colon D \to \mathbf{\acute{E}t}_{/X}$$

the fibered site associated with the inverse image functor $f^+\colon \mathbf{\acute{E}t}_{/X} \to \mathbf{\acute{E}t}_{/Y}$, defined in (VI.5.5). We endow $D$ with the covanishing topology associated with $\varpi$. We thus obtain the covanishing site associated with the functor $f^+$ (VI.4.1), whose topos of sheaves of $\mathbb{U}$-sets is $X_{\text{ét}} \overleftarrow{\times}_{X_{\text{ét}}} Y_{\text{ét}}$ (VI.4.10). Every object of $E$ is naturally an object of $D$. We thus define a fully faithful and left exact functor

(VI.10.15.2) $$\rho^+\colon E \to D.$$

For every $U \in \mathrm{Ob}(\mathbf{\acute{E}t}_{/X})$, the restriction of $\rho^+$ to the fibers over $U$ is none other than the canonical injection functor $\mathbf{\acute{E}t}_{\mathrm{f}/U_Y} \to \mathbf{\acute{E}t}_{/U_Y}$, in other words, the functor $\Psi_U^+$ (VI.10.8.2). The functor $\rho^+$ therefore identifies with a functor with the same name, associated with the functor $\Psi^+$ (VI.10.7.1) and defined in (VI.6.4.2). It is continuous and left exact. It therefore defines a morphism of topos (VI.6.4.4)

(VI.10.15.3) $$\rho\colon X_{\text{ét}} \overleftarrow{\times}_{X_{\text{ét}}} Y_{\text{ét}} \to \widetilde{E}.$$

It immediately follows from the definitions that the squares of the diagram

(VI.10.15.4) $$\begin{array}{ccccc} X_{\text{ét}} & \xleftarrow{\mathrm{p}_1} & X_{\text{ét}} \overleftarrow{\times}_{X_{\text{ét}}} Y_{\text{ét}} & \xrightarrow{\mathrm{p}_2} & Y_{\text{ét}} \\ \Vert & & \downarrow{\scriptstyle \rho} & & \downarrow{\scriptstyle \rho_Y} \\ X_{\text{ét}} & \xleftarrow{\sigma} & \widetilde{E} & \xrightarrow{\beta} & Y_{\text{fét}} \end{array}$$

where $\mathrm{p}_1$ and $\mathrm{p}_2$ are the canonical projections (VI.4.11), are commutative up to canonical isomorphisms. Moreover, we have a commutative diagram

(VI.10.15.5) $$\begin{array}{ccccc} f_{\text{fét}}\beta\rho & & \xrightarrow{\tau_E * \rho} & & \rho_X\sigma\rho \\ \downarrow & & & & \downarrow \\ f_{\text{fét}}\rho_Y\mathrm{p}_2 & \longrightarrow & \rho_X f_{\text{ét}}\mathrm{p}_2 & \xrightarrow{\rho_X * \tau_D} & \rho_X\mathrm{p}_1 \end{array}$$

where $\tau_D$ is the 2-morphism (VI.4.11.5), $\tau_E$ is the 2-morphism (VI.10.11.1), the vertical arrows are the isomorphisms underlying the diagram (VI.10.15.4), and the unlabeled horizontal arrow comes from (VI.9.3.4). On the other hand, the diagram

(VI.10.15.6)
$$\begin{array}{ccc} Y_{\text{ét}} & \xrightarrow{\Psi_D} & X_{\text{ét}} \overleftarrow{\times}_{X_{\text{ét}}} Y_{\text{ét}} \\ & \underset{\Psi}{\searrow} & \downarrow{\scriptstyle\rho} \\ & & \widetilde{E} \end{array}$$

where $\Psi_D$ is the morphism (VI.4.13.4), is commutative up to canonical isomorphism.

**Proposition VI.10.16.** *Suppose that the scheme $X$ is quasi-separated and that the scheme $Y$ is coherent. Then a family $((V_i \to U_i) \to (V \to U))_{i \in I}$ of morphisms of $E_{\text{coh}}$* (VI.10.4) *is covering if and only if it is in $D$* (VI.10.15).

Fist note that since the covanishing topology of $E_{\text{coh}}$ is induced by that of $E$ (VI.10.4), a family of morphisms with the same target in $E_{\text{coh}}$ is covering in $E_{\text{coh}}$ if and only if it is in $E$ ([**2**] III 3.3). The morphism $f$ induces a functor (VI.9.1)

(VI.10.16.1)
$$f^+_{\text{coh}}\colon \mathbf{\acute{E}t}_{\text{coh}/X} \to \mathbf{\acute{E}t}_{\text{coh}/Y}.$$

We denote by $D_{\text{coh}}$ the covanishing site associated with $f^+_{\text{coh}}$ (VI.4.1). The canonical injection functor $D_{\text{coh}} \to D$ is continuous and left exact. It induces an equivalence between the associated topos, by VI.3.9 and VI.4.10. Consequently, a family of morphisms with the same target in $D_{\text{coh}}$ is covering in $D_{\text{coh}}$ if and only if it is in $D$ ([**2**] II 4.4). On the other hand, every object of $E_{\text{coh}}$ is naturally an object of $D_{\text{coh}}$. We thus define a fully faithful and left exact functor

(VI.10.16.2)
$$\rho^+_{\text{coh}}\colon E_{\text{coh}} \to D_{\text{coh}}.$$

It is a Cartesian $(\mathbf{\acute{E}t}_{\text{coh}/X})$-functor whose restriction to the fibers over any object $U$ of $\mathbf{\acute{E}t}_{\text{coh}/X}$ is none other than the canonical injection functor $\mathbf{\acute{E}t}_{\text{f}/U_Y} \to \mathbf{\acute{E}t}_{\text{coh}/U_Y}$ (cf. VI.5.5). Hence $\rho^+_{\text{coh}}$ is continuous by virtue of VI.5.18. To show the proposition, it suffices to show that a family $\mathscr{F} = ((V_i \to U_i) \to (V \to U))_{i \in I}$ of morphisms of $E_{\text{coh}}$ is covering if and only if it is in $D_{\text{coh}}$. The condition is necessary because $\rho^+_{\text{coh}}$ is continuous ([**2**] III 1.6). Conversely, suppose that the family $\mathscr{F}$ is covering in $D_{\text{coh}}$ and let us show that it is in $E_{\text{coh}}$. For every $U \in \text{Ob}(\mathbf{\acute{E}t}_{\text{coh}/X})$, the scheme $U_Y$ is coherent. Consequently, every object of $\mathbf{\acute{E}t}_{\text{coh}/U_Y}$ is quasi-compact. By VI.5.9, there consequently exists an étale covering $(U'_j \to U)_{j \in J}$ and for every $j \in J$, an étale covering $(V'_{j,k} \to U'_j \times_U V)_{k \in K_j}$ such that for all $j \in J$ and $k \in K_j$, there exist $i_{j,k} \in I$, a $U$-morphism $U'_j \to U_{i_{j,k}}$, and a $V$-morphism $V'_{j,k} \to V_{i_{j,k}}$ such that the diagram

(VI.10.16.3)
$$\begin{array}{ccc} V'_{j,k} & \longrightarrow & V_{i_{j,k}} \\ \downarrow & & \downarrow \\ U'_j & \longrightarrow & U_{i_{j,k}} \end{array}$$

is commutative. Set $W_{j,k} = U'_j \times_{U_{i_{j,k}}} V_{i_{j,k}}$. For every $j \in J$, $(W_{j,k} \to U'_j \times_U V)_{k \in K_j}$ is a covering of $\mathbf{\acute{E}t}_{\text{f}/U'_i \times_X Y}$. Consequently, $((V_i \to U_i) \to (V \to U))_{i \in I}$ is a covering of $E_{\text{coh}}$.

**Remark VI.10.17.** Under the assumptions of VI.10.12, the diagram of morphisms of topos

(VI.10.17.1)
$$\begin{array}{ccc} X'_{\text{ét}} \overleftarrow{\times}_{X'_{\text{ét}}} Y'_{\text{ét}} & \xrightarrow{\Xi} & X_{\text{ét}} \overleftarrow{\times}_{X_{\text{ét}}} Y_{\text{ét}} \\ \downarrow{\scriptstyle \rho'} & & \downarrow{\scriptstyle \rho} \\ \widetilde{E}' & \xrightarrow{\Phi} & \widetilde{E} \end{array}$$

where $\rho$ and $\rho'$ are the canonical morphisms (VI.10.15.3), $\Phi$ is the morphism (VI.10.12.5), and $\Xi$ is the morphism deduced from the functoriality of the covanishing topos (VI.3.8), is commutative up to canonical isomorphism. Indeed, by VI.4.6(i), for every $(V \to U) \in \mathrm{Ob}(D)$ (VI.10.15), we have a canonical isomorphism

(VI.10.17.2)
$$\Xi^*((V \to U)^a) \xrightarrow{\sim} (V \times_Y Y' \to U \times_X X')^a.$$

**VI.10.18.** By VI.4.20 and ([**2**] VIII 7.9), giving a point of $X_{\text{ét}} \overleftarrow{\times}_{X_{\text{ét}}} Y_{\text{ét}}$ is equivalent to giving a pair of geometric points $\overline{x}$ of $X$ and $\overline{y}$ of $Y$ and a specialization arrow $u$ from $f(\overline{y})$ to $\overline{x}$, that is, an $X$-morphism $u\colon \overline{y} \to X_{(\overline{x})}$, where $X_{(\overline{x})}$ denotes the strict localization of $X$ at $\overline{x}$. Such a point will be denoted by $(\overline{y} \rightsquigarrow \overline{x})$ or by $(u\colon \overline{y} \rightsquigarrow \overline{x})$. We denote by $\rho(\overline{y} \rightsquigarrow \overline{x})$ its image by $\rho$ (VI.10.15.3), which is therefore a point of $\widetilde{E}$. For all $F \in \mathrm{Ob}(X_{\text{ét}})$ and $G \in \mathrm{Ob}(Y_{\text{fét}})$, we have functorial canonical isomorphisms

(VI.10.18.1)
$$(\sigma^* F)_{\rho(\overline{y} \rightsquigarrow \overline{x})} \xrightarrow{\sim} F_{\overline{x}},$$

(VI.10.18.2)
$$(\beta^* G)_{\rho(\overline{y} \rightsquigarrow \overline{x})} \xrightarrow{\sim} (\rho_Y^* G)_{\overline{y}}.$$

By (VI.10.15.5), for every $H \in \mathrm{Ob}(X_{\text{fét}})$, the map

(VI.10.18.3)
$$(\sigma^*(\rho_X^* H))_{\rho(\overline{y} \rightsquigarrow \overline{x})} \to (\beta^*(f_{\text{fét}}^* H))_{\rho(\overline{y} \rightsquigarrow \overline{x})}$$

induced by $\tau$ (VI.10.11.1), identifies canonically with the specialization morphism defined by $u$

$$(\rho_X^* H)_{\overline{x}} \to (\rho_X^* H)_{f(\overline{y})}.$$

For every object $(V \to U)$ of $E$, we have a functorial canonical isomorphism

(VI.10.18.4)
$$(V \to U)^a_{\rho(\overline{y} \rightsquigarrow \overline{x})} \xrightarrow{\sim} U_{\overline{x}} \times_{U_{f(\overline{y})}} V_{\overline{y}},$$

the map $V_{\overline{y}} \to U_{f(\overline{y})}$ is induced by $V \to U$, and the map $U_{\overline{x}} \to U_{f(\overline{y})}$ is the specialization morphism defined by $u$. This follows from (VI.4.20.6) and from the fact that $\rho^*$ extends $\rho^+$ ([**2**] III 1.4).

**Remark VI.10.19.** Let $U \in \mathrm{Ob}(\mathbf{\acute{E}t}_{/X})$, $\overline{x}, \overline{z}$ be two geometric points of $X$, $u$ a specialization morphism from $\overline{z}$ to $\overline{x}$, and $u^*\colon U_{\overline{x}} \to U_{\overline{z}}$ the corresponding specialization morphism ([**2**] VIII 7.7). Then:

(i) For every $\overline{x}' \in U_{\overline{x}}$, $\overline{z}' = u^*(\overline{x}')$ is the point of $U_{\overline{z}}$ corresponding to the composition

(VI.10.19.1)
$$\overline{z} \xrightarrow{u} X_{(\overline{x})} \xrightarrow{i} U,$$

where $i$ is the $X$-morphism defined by $\overline{x}'$ ([**2**] VIII 7.3); more precisely, $i$ is induced by the inverse of the canonical isomorphism $U_{(\overline{x}')} \xrightarrow{\sim} X_{(\overline{x})}$, where $U_{(\overline{x}')}$

denotes the strict localization of $U$ at $\overline{x}'$. Hence there exists a unique specialization morphism $u'$ from $\overline{z}'$ to $\overline{x}'$ that fits into a commutative diagram

(VI.10.19.2)
$$\begin{array}{ccc} \overline{z}' & \xrightarrow{u'} & U_{(\overline{x}')} \\ \downarrow & & \downarrow \\ \overline{z} & \xrightarrow{u} & X_{(\overline{x})} \end{array}$$

where the vertical arrows are the canonical isomorphisms.

(ii) If $U$ is separated over $X$, the map $u^*$ is injective. Indeed, if $\overline{x}'$ and $\overline{x}''$ are two points of $U_{\overline{x}}$ such that $u^*(\overline{x}') = u^*(\overline{x}'') = \overline{z}'$, then we have two specializations from $\overline{z}'$ to $\overline{x}'$ and $\overline{x}''$, respectively. In particular, the image of the canonical morphism $\overline{z}' \to U \times_X X_{(\overline{x})}$ is contained in each of the open and closed subschemes $U_{(\overline{x}')}$ and $U_{(\overline{x}'')}$ of $U \times_X X_{(\overline{x})}$ (**[42]** 18.5.11), which is impossible.

(iii) If $U$ is finite over $X$, the map $u^*$ is bijective because $U \times_X X_{(\overline{x})} = \coprod_{\overline{x}' \in U_{\overline{x}}} U_{(\overline{x}')}$.

**VI.10.20.** Let $\overline{x}$ be a geometric point of $X$, $\overline{y}$ a geometric point of $Y$, $X_{(\overline{x})}$ the strict localization of $X$ at $\overline{x}$, and $u\colon \overline{y} \to X_{(\overline{x})}$ an $X$-morphism, so that $(\overline{y} \rightsquigarrow \overline{x})$ is a point of $X_{\text{ét}} \overleftarrow{\times}_{X_{\text{ét}}} Y_{\text{ét}}$ (VI.10.18). We denote by $\mathscr{P}_{\rho(\overline{y} \rightsquigarrow \overline{x})}$ the category of $\rho(\overline{y} \rightsquigarrow \overline{x})$-pointed objects of $E$, defined as follows. The objects of $\mathscr{P}_{\rho(\overline{y} \rightsquigarrow \overline{x})}$ are the triples $((V \to U), \xi, \zeta)$ consisting of an object $(V \to U)$ of $E$, an $X$-morphism $\xi\colon \overline{x} \to U$, and a $Y$-morphism $\zeta\colon \overline{y} \to V$ such that, denoting again by $\xi\colon X_{(\overline{x})} \to U$ the $X$-morphism induced by $\xi$ (**[2]** VIII 7.3), the diagram

(VI.10.20.1)
$$\begin{array}{ccc} \overline{y} & \xrightarrow{u} & X_{(\overline{x})} \\ {\scriptstyle\zeta}\downarrow & & \downarrow{\scriptstyle\xi} \\ V & \longrightarrow & U \end{array}$$

is commutative. Let $((V \to U), \xi, \zeta)$, $((V' \to U'), \xi', \zeta')$ be two objects of $\mathscr{P}_{\rho(\overline{y} \rightsquigarrow \overline{x})}$. A morphism from $((V' \to U'), \xi', \zeta')$ to $((V \to U), \xi, \zeta)$ is defined as a morphism $(g\colon U' \to U, h\colon V' \to V)$ of $E$ such that $g \circ \xi' = \xi$ and $h \circ \zeta' = \zeta$. It follows from (VI.10.18.4) and VI.10.19(i) that $\mathscr{P}_{\rho(\overline{y} \rightsquigarrow \overline{x})}$ is canonically equivalent to the category of neighborhoods of $\rho(\overline{y} \rightsquigarrow \overline{x})$ in $E$ (**[2]** IV 6.8.2). It is therefore cofiltered and for every presheaf $F = \{U \mapsto F_U\}$ on $E$, we have a functorial canonical isomorphism (**[2]** IV (6.8.4))

(VI.10.20.2)
$$(F^a)_{\rho(\overline{y} \rightsquigarrow \overline{x})} \xrightarrow{\sim} \varinjlim_{((V \to U), \xi, \zeta) \in \mathscr{P}^{\circ}_{\rho(\overline{y} \rightsquigarrow \overline{x})}} F_U(V).$$

If $X$ is quasi-separated, we can replace in the limit above $\mathscr{P}_{\rho(\overline{y} \rightsquigarrow \overline{x})}$ by the full subcategory $\mathscr{P}^{\text{coh}}_{\rho(\overline{y} \rightsquigarrow \overline{x})}$ made up of the objects $((V \to U), \xi, \zeta)$ such that $U$ is of finite presentation over $X$ (that is, is an object of $\mathbf{\acute{E}t}_{\text{coh}/X}$), which is also cofiltered (cf. VI.10.4).

**Proposition VI.10.21.** *Suppose that the schemes $X$ and $Y$ are coherent. Then when $(\overline{y} \rightsquigarrow \overline{x})$ goes through the family of points of $X_{\text{ét}} \overleftarrow{\times}_{X_{\text{ét}}} Y_{\text{ét}}$, the family of fiber functors of $\widetilde{E}$ associated with the points $\rho(\overline{y} \rightsquigarrow \overline{x})$ is conservative* (**[2]** IV 6.4.0).

Let $u\colon F \to G$ be a morphism of $\widetilde{E}$ such that for every point $(\overline{y} \rightsquigarrow \overline{x})$ of $X_{\text{ét}} \overleftarrow{\times}_{X_{\text{ét}}} Y_{\text{ét}}$, the corresponding morphism $u_{\rho(\overline{y} \rightsquigarrow \overline{x})}\colon F_{\rho(\overline{y} \rightsquigarrow \overline{x})} \to G_{\rho(\overline{y} \rightsquigarrow \overline{x})}$ is a monomorphism. Let us show that $u$ is a monomorphism. We must show that for all $(V \to U) \in \text{Ob}(E_{\text{coh}})$ (VI.10.4) and $a, b \in F_U(V)$ such that $u(a) = u(b)$, we have $a = b$. In view of VI.10.14 and VI.10.17, we may assume $(V \to U) = (Y \to X)$. For every point $(\overline{y} \rightsquigarrow \overline{x})$ of

$X_{\text{ét}} \overleftarrow{\times}_{X_{\text{ét}}} Y_{\text{ét}}$, we have $a_{\rho(\overline{y} \rightsquigarrow \overline{x})} = b_{\rho(\overline{y} \rightsquigarrow \overline{x})}$ because $u_{\rho(\overline{y} \rightsquigarrow \overline{x})}$ is a monomorphism. By (VI.10.20.2), there exists an object $((U_{(\overline{y} \rightsquigarrow \overline{x})} \to V_{(\overline{y} \rightsquigarrow \overline{x})}), \xi_{(\overline{y} \rightsquigarrow \overline{x})}, \zeta_{(\overline{y} \rightsquigarrow \overline{x})})$ of $\mathscr{P}^{\text{coh}}_{\rho(\overline{y} \rightsquigarrow \overline{x})}$ such that $a$ and $b$ have the same images in $F_{U_{(\overline{y} \rightsquigarrow \overline{x})}}(V_{(\overline{y} \rightsquigarrow \overline{x})})$. On the other hand, since the topos $X_{\text{ét}}$ and $Y_{\text{ét}}$ are coherent and the morphism $f\colon Y_{\text{ét}} \to X_{\text{ét}}$ is coherent ([**2**] VI 3.3), the family of fiber functors of $X_{\text{ét}} \overleftarrow{\times}_{X_{\text{ét}}} Y_{\text{ét}}$ associated with the points $(\overline{y} \rightsquigarrow \overline{x})$ is conservative, by VI.5.30 and VI.10.18. We deduce from this that the family of morphisms $((U_{(\overline{y} \rightsquigarrow \overline{x})} \to V_{(\overline{y} \rightsquigarrow \overline{x})}) \to (Y \to X))_{(\overline{y} \rightsquigarrow \overline{x})}$ is covering in $D$ (VI.10.15) ([**2**] IV 6.5). It is therefore covering in $E_{\text{coh}}$ by virtue of VI.10.16. Consequently, $a = b$ and $u$ is a monomorphism.

Suppose, moreover, that for every point $(\overline{y} \rightsquigarrow \overline{x})$ of $X_{\text{ét}} \overleftarrow{\times}_{X_{\text{ét}}} Y_{\text{ét}}$, the morphism $u_{\rho(\overline{y} \rightsquigarrow \overline{x})}$ is an epimorphism and let us show that the same holds for $u$. It suffices to show that for all $(V \to U) \in \text{Ob}(E_{\text{coh}})$ and $b \in G_U(V)$, there exists $a \in F_U(V)$ such that $b = u(a)$. We may still assume that $(V \to U) = (Y \to X)$. By (VI.10.20.2), for every point $(\overline{y} \rightsquigarrow \overline{x})$ of $X_{\text{ét}} \overleftarrow{\times}_{X_{\text{ét}}} Y_{\text{ét}}$, there exist an object $((U_{(\overline{y} \rightsquigarrow \overline{x})} \to V_{(\overline{y} \rightsquigarrow \overline{x})}), \xi_{(\overline{y} \rightsquigarrow \overline{x})}, \zeta_{(\overline{y} \rightsquigarrow \overline{x})})$ of $\mathscr{P}^{\text{coh}}_{\rho(\overline{y} \rightsquigarrow \overline{x})}$ and a section $a_{(\overline{y} \rightsquigarrow \overline{x})} \in F_{U_{(\overline{y} \rightsquigarrow \overline{x})}}(V_{(\overline{y} \rightsquigarrow \overline{x})})$ whose image by $u$ in $G_{U_{(\overline{y} \rightsquigarrow \overline{x})}}(V_{(\overline{y} \rightsquigarrow \overline{x})})$ is the restriction of $b$. Since $u$ is a monomorphism, the sections $a_{(\overline{y} \rightsquigarrow \overline{x})}$ coincide on $(V_{(\overline{y}' \rightsquigarrow \overline{x}')} \times_Y V_{(\overline{y} \rightsquigarrow \overline{x})} \to U_{(\overline{y}' \rightsquigarrow \overline{x}')} \times_X U_{(\overline{y} \rightsquigarrow \overline{x})})$, for all points $(\overline{y} \rightsquigarrow \overline{x})$ and $(\overline{y}' \rightsquigarrow \overline{x}')$ of $X_{\text{ét}} \overleftarrow{\times}_{X_{\text{ét}}} Y_{\text{ét}}$. They therefore come from a section $a \in F_X(Y)$, and we have $u(a) = b$ because the restrictions on the $(U_{(\overline{y} \rightsquigarrow \overline{x})} \to V_{(\overline{y} \rightsquigarrow \overline{x})})$ coincide.

**VI.10.22.** Suppose that $X$ is strictly local with closed point $x$. For every $U \in \text{Ob}(\mathbf{\acute{E}t}_{\text{scoh}/X})$ (VI.9.1), we denote by $U^{\text{f}}$ the disjoint sum of the strict localizations of $U$ at the points of $U_x$; it is an open and closed subscheme of $U$, and is finite over $X$ ([**42**] 18.5.11). The correspondence $U \mapsto U^{\text{f}}$ defines a functor

(VI.10.22.1) $$\iota_x^+ \colon \mathbf{\acute{E}t}_{\text{scoh}/X} \to \mathbf{\acute{E}t}_{\text{f}/X},$$

that is clearly left exact and continuous. The associated morphism of topos

(VI.10.22.2) $$\iota_x \colon X_{\text{fét}} \to X_{\text{ét}}$$

identifies with the morphism $\mathbf{Ens} \to X_{\text{ét}}$ defined by $x$. Indeed, the fiber functor at $x$ induces an equivalence between the topos $X_{\text{fét}}$ and $\mathbf{Ens}$ because the group $\pi_1(X, x)$ is trivial (VI.9.8.1), and for every $U \in \text{Ob}(\mathbf{\acute{E}t}_{\text{scoh}/X})$, the canonical map $U^{\text{f}}_x \to U_x$ is bijective.

Let us consider the fibered site

(VI.10.22.3) $$\pi_{\text{scoh}} \colon E_{\text{scoh}} \to \mathbf{\acute{E}t}_{\text{scoh}/X}$$

defined in (VI.10.4), and endow $E_{\text{scoh}}$ with the covanishing topology associated with $\pi_{\text{scoh}}$. Recall that the topos of sheaves of $\mathbb{U}$-sets on $E_{\text{scoh}}$ identifies canonically with $\widetilde{E}$ (VI.10.4.4). For every object $(V \to U)$ of $E_{\text{scoh}}$, $V \times_U U^{\text{f}} = V \times_{U_Y} U^{\text{f}}_Y$ is a finite étale cover of $Y$. We thus obtain a functor

(VI.10.22.4) $$\theta^+ \colon E_{\text{scoh}} \to \mathbf{\acute{E}t}_{\text{f}/Y}, \quad (V \to U) \mapsto V \times_U U^{\text{f}}.$$

**Lemma VI.10.23.** *We keep the assumptions of* VI.10.22.

(i) *The functor $\theta^+$ is continuous and left exact; it therefore defines a morphism of topos*

(VI.10.23.1) $$\theta \colon Y_{\text{fét}} \to \widetilde{E}.$$

(ii) *For every sheaf $F$ of $Y_{\text{fét}}$, we have a canonical isomorphism of sheaves on $E_{\text{scoh}}$*

(VI.10.23.2) $$\theta_*(F) \xrightarrow{\sim} \{U \mapsto (j_U)_{\text{fét}*}(F|U^{\text{f}}_Y)\},$$

*where for every* $U \in \mathrm{Ob}(\mathbf{\acute{E}t}_{\mathrm{scoh}/X})$, $j_U \colon U_Y^{\mathrm{f}} \to U_Y$ *denotes the canonical injection, and for every morphism* $g \colon U' \to U$ *of* $\mathbf{\acute{E}t}_{\mathrm{scoh}/X}$, *the transition morphism*

$$(j_U)_{\mathrm{f\acute{e}t}*}(F|U_Y^{\mathrm{f}}) \to (g_Y)_{\mathrm{f\acute{e}t}*}((j_{U'})_{\mathrm{f\acute{e}t}*}(F|U_Y'^{\mathrm{f}})) \tag{VI.10.23.3}$$

*is the composition*

$$(j_U)_{\mathrm{f\acute{e}t}*}(F|U_Y^{\mathrm{f}}) \to (j_U)_{\mathrm{f\acute{e}t}*}((g_Y^{\mathrm{f}})_{\mathrm{f\acute{e}t}*}(F|U_Y'^{\mathrm{f}})) \xrightarrow{\sim} (g_Y)_{\mathrm{f\acute{e}t}*}((j_{U'})_{\mathrm{f\acute{e}t}*}(F|U_Y'^{\mathrm{f}})), \tag{VI.10.23.4}$$

*where* $g^{\mathrm{f}} \colon U'^{\mathrm{f}} \to U^{\mathrm{f}}$ *is the image of* $g$ *by the functor* (VI.10.22.1), *the first arrow is induced by the adjunction morphism* $\mathrm{id} \to (g_Y^{\mathrm{f}})_{\mathrm{f\acute{e}t}*}(g_Y^{\mathrm{f}})^*_{\mathrm{f\acute{e}t}}$, *and the second arrow is the canonical isomorphism.*

One easily verifies that $\theta^+$ commutes with fibered products and transforms final objects into final objects. It is therefore left exact. On the other hand, for every sheaf $F$ of $Y_{\mathrm{f\acute{e}t}}$, we have a canonical isomorphism of $\widehat{E}_{\mathrm{scoh}}$

$$F \circ \theta^+ \xrightarrow{\sim} \{U \mapsto (j_U)_{\mathrm{f\acute{e}t}*}(F|U_Y^{\mathrm{f}})\}, \tag{VI.10.23.5}$$

where the right-hand side is the presheaf on $E_{\mathrm{scoh}}$ defined by the transition morphisms (VI.10.23.4). Let $(U_i \to U)_{i\in I}$ be a covering of $\mathbf{\acute{E}t}_{\mathrm{scoh}/X}$. For every $(i,j) \in I^2$, set $V_i = U_i \times_X Y$, $W_i = U_i^{\mathrm{f}} \times_X Y$, $U_{ij} = U_i \times_U U_j$, $V_{ij} = U_{ij} \times_X Y$, and $W_{ij} = U_{ij}^{\mathrm{f}} \times_X Y$, and denote by $h_i \colon V_i \to U_Y$, $g_i \colon W_i \to U_Y^{\mathrm{f}}$, $h_{ij} \colon V_{ij} \to U_Y$, and $g_{ij} \colon W_{ij} \to U_Y^{\mathrm{f}}$ the structural morphisms. Since the functor (VI.10.22.1) is left exact and continuous, the sequence

$$0 \to F|U_Y^{\mathrm{f}} \to \prod_{i\in I} (g_i)_{\mathrm{f\acute{e}t}*}(F|W_i) \rightrightarrows \prod_{(i,j)\in I^2} (g_{ij})_{\mathrm{f\acute{e}t}*}(F|W_{ij}) \tag{VI.10.23.6}$$

is exact. Since $(j_U)_{\mathrm{f\acute{e}t}*}$ commutes with inverse limits, we deduce from this that the sequence

(VI.10.23.7)

$$0 \to (j_U)_{\mathrm{f\acute{e}t}*}(F|U_Y^{\mathrm{f}}) \to \prod_{i\in I} (h_i)_{\mathrm{f\acute{e}t}*}((j_{U_i})_{\mathrm{f\acute{e}t}*}(F|W_i)) \rightrightarrows \prod_{(i,j)\in I^2} (h_{ij})_{\mathrm{f\acute{e}t}*}((j_{U_{ij}})_{\mathrm{f\acute{e}t}*}(F|W_{ij}))$$

is exact. Consequently, $F \circ \theta^+$ is a sheaf by virtue of VI.5.10. Hence $\theta^+$ is continuous, giving the lemma.

**VI.10.24.** We keep the assumptions of VI.10.22 and describe explicitly the constructions of (VI.6.1) for the functor $\theta^+$ defined in (VI.10.22.4). The composed functor

$$\theta^+ \circ \sigma^+ \colon \mathbf{\acute{E}t}_{\mathrm{scoh}/X} \to \mathbf{\acute{E}t}_{\mathrm{f}/Y} \tag{VI.10.24.1}$$

is none other than the functor $U \mapsto U_Y^{\mathrm{f}}$; we therefore have $\sigma\theta = \iota_x f_{\mathrm{f\acute{e}t}}$, where $\iota_x$ is the morphism (VI.10.22.2). On the other hand, for every object $U$ of $\mathbf{\acute{E}t}_{\mathrm{scoh}/X}$, the functor (VI.6.1.5)

$$\theta_U^+ \colon \mathbf{\acute{E}t}_{\mathrm{f}/U_Y} \to \mathbf{\acute{E}t}_{\mathrm{f}/U_Y^{\mathrm{f}}} \tag{VI.10.24.2}$$

induced by $\theta^+$ is none other than the inverse image under the canonical morphism $j_U \colon U_Y^{\mathrm{f}} \to U_Y$. In particular, $\theta$ is a section of $\beta$; that is, we have a canonical isomorphism (VI.6.1.9)

$$\beta\theta \xrightarrow{\sim} \mathrm{id}_{Y_{\mathrm{f\acute{e}t}}}. \tag{VI.10.24.3}$$

We obtain a base change morphism

$$\beta_* \to \theta^*, \tag{VI.10.24.4}$$

composition of $\beta_* \to \beta_*\theta_*\theta^* \xrightarrow{\sim} \theta^*$, where the first arrow is induced by the adjunction morphism $\mathrm{id} \to \theta_*\theta^*$, and the second arrow by (VI.10.24.3).

**VI.10.25.** We keep the assumptions of VI.10.22. The canonical morphism $\rho_X\colon X_{\text{ét}} \to X_{\text{fét}}$ (VI.9.2.1) identifies with the unique morphism of topos $X_{\text{ét}} \to \mathbf{Ens}$ ([**2**] IV 4.3). On the other hand, the composed morphism $\iota_x\rho_X\colon X_{\text{ét}} \to X_{\text{ét}}$ is defined by the morphism of sites

(VI.10.25.1) $$\mathbf{\acute{E}t}_{\text{scoh}/X} \to \mathbf{\acute{E}t}_{\text{scoh}/X}, \quad U \mapsto U^{\text{f}}.$$

The canonical injection $U^{\text{f}} \to U$, for $U \in \text{Ob}(\mathbf{\acute{E}t}_{\text{scoh}/X})$, then defines a 2-morphism

(VI.10.25.2) $$\text{id}_{X_{\text{ét}}} \to \iota_x\rho_X.$$

By VI.3.7, the morphisms of topos $\iota_x\rho_X f_{\text{ét}}\colon Y_{\text{ét}} \to X_{\text{ét}}$ and $\text{id}_{Y_{\text{ét}}}$, and the 2-morphism $f_{\text{ét}} \to \iota_x\rho_X f_{\text{ét}}$ induced by (VI.10.25.2), define a morphism of topos

(VI.10.25.3) $$\gamma\colon Y_{\text{ét}} \to X_{\text{ét}} \overleftarrow{\times}_{X_{\text{ét}}} Y_{\text{ét}}$$

such that $\text{p}_1\gamma = \iota_x\rho_X f_{\text{ét}}$, $\text{p}_2\gamma = \text{id}_{Y_{\text{ét}}}$, and $\tau * \gamma$ is induced by (VI.10.25.2):

(VI.10.25.4) $$\begin{array}{ccccc} X_{\text{ét}} & \xleftarrow{f_{\text{ét}}} & Y_{\text{ét}} & & \\ \downarrow{\scriptstyle \iota_x\rho_X} & & \downarrow{\scriptstyle \gamma} & \searrow^{\text{id}} & \\ X_{\text{ét}} & \xleftarrow{\text{p}_1} & X_{\text{ét}} \overleftarrow{\times}_{X_{\text{ét}}} Y_{\text{ét}} & \xrightarrow{\text{p}_2} & Y_{\text{ét}} \\ & \searrow & & \swarrow_{f_{\text{ét}}} & \\ & & X_{\text{ét}} & & \end{array}$$

We take again the notation of (VI.10.15) and denote by $D_{\text{scoh}}$ the fibered site over $\mathbf{\acute{E}t}_{\text{scoh}/X}$ deduced from $D$ (VI.10.15.1) by base change by the canonical injection functor $\mathbf{\acute{E}t}_{\text{scoh}/X} \to \mathbf{\acute{E}t}_{/X}$ (VI.10.4). It follows from VI.4.6(i) that for every $(V \to U) \in \text{Ob}(D_{\text{scoh}})$, we have a canonical isomorphism

(VI.10.25.5) $$\gamma^*((V \to U)^a) \xrightarrow{\sim} V \times_U U^{\text{f}}.$$

We deduce from this that the diagram

(VI.10.25.6) $$\begin{array}{ccc} Y_{\text{ét}} & \xrightarrow{\gamma} & X_{\text{ét}} \overleftarrow{\times}_{X_{\text{ét}}} Y_{\text{ét}} \\ \downarrow{\scriptstyle \rho_Y} & & \downarrow{\scriptstyle \rho} \\ Y_{\text{fét}} & \xrightarrow{\theta} & \widetilde{E} \end{array}$$

where $\rho$ is the canonical morphism (VI.10.15.3), is commutative up to canonical isomorphism.

**Proposition VI.10.26.** *Under the assumptions of* VI.10.22, *for every sheaf $F$ of $\widetilde{E}$ and every geometric point $\overline{y}$ of $Y$, the map*

(VI.10.26.1) $$(\beta_*F)_{\rho_Y(\overline{y})} \to (\theta^*F)_{\rho_Y(\overline{y})}$$

*induced by the base change morphism* (VI.10.24.4) *is bijective.*

First, note that there exists a unique specialization from $f(\overline{y})$ to $x$; we can therefore consider the point $(\overline{y} \rightsquigarrow x)$ of $X_{\text{ét}} \overleftarrow{\times}_{X_{\text{ét}}} Y_{\text{ét}}$ (VI.10.18). It is clear that the points $\gamma(\overline{y})$ and $(\overline{y} \rightsquigarrow x)$ are canonically isomorphic. It then follows from (VI.10.25.6) that the points $\theta(\rho_Y(\overline{y}))$ and $\rho(\overline{y} \rightsquigarrow x)$ of $\widetilde{E}$ are canonically isomorphic. We denote by $\mathscr{C}_{\overline{y}}$ the category of $\overline{y}$-pointed finite étale $Y$-schemes, which we identify with the category of neighborhoods of $\rho_Y(\overline{y})$ in the site $\mathbf{\acute{E}t}_{\text{f}/Y}$ ([**2**] IV 6.8.2), and by $\mathscr{P}^{\text{scoh}}_{\rho(\overline{y} \rightsquigarrow x)}$ the full subcategory of the category $\mathscr{P}_{\rho(\overline{y}\rightsquigarrow x)}$ (VI.10.20) made up of the $\rho(\overline{y} \rightsquigarrow x)$-pointed objects $((V \to U), \xi, \zeta)$ of $E$ such that $U$ is separated and of finite presentation over $X$ (that is, is an object of $\mathbf{\acute{E}t}_{\mathrm{scoh}/X}$). It follows from (VI.10.18.4) and VI.10.19(i) that $\mathscr{P}^{\mathrm{scoh}}_{\rho(\overline{y}\rightsquigarrow \overline{x})}$ is canonically equivalent to the category of neighborhoods of the point $\rho(y \rightsquigarrow x)$ in the site $E_{\mathrm{scoh}}$. It is therefore cofiltered. Denote by $\xi_0 \colon x \to X$ the canonical injection. We have a fully faithful functor

$$\jmath_{\overline{y}} \colon \mathscr{C}_{\overline{y}} \to \mathscr{P}^{\mathrm{scoh}}_{\rho(\overline{y}\rightsquigarrow x)}, \quad (V, \zeta \colon \overline{y} \to V) \mapsto ((V \to X), \xi_0, \zeta), \tag{VI.10.26.2}$$

compatible with the canonical functor (VI.10.6.1)

$$\alpha^{\mathrm{scoh}}_{X!} \colon \mathbf{\acute{E}t}_{\mathrm{f}/Y} \to E_{\mathrm{scoh}}, \quad V \mapsto (V \to X). \tag{VI.10.26.3}$$

The adjunction morphism $F \to \theta_*(\theta^* F)$ is defined, for every $(V \to U) \in \mathrm{Ob}(E_{\mathrm{scoh}})$, by the canonical map

$$F(V \to U) \to (\theta^* F)(V \times_U U^{\mathrm{f}}).$$

Consequently, (VI.10.26.1) identifies with the map

$$\varinjlim_{(W,\zeta)\in \mathscr{C}^{\circ}_{\overline{y}}} F(W \to X) \to F_{\rho(\overline{y}\rightsquigarrow x)} \tag{VI.10.26.4}$$

induced by the functor $\jmath^{\circ}_{\overline{y}}$. Hence it suffices to show that $\jmath^{\circ}_{\overline{y}}$ is cofinal. Let $((V \to U), \xi, \zeta)$ be an object of $\mathscr{P}^{\mathrm{scoh}}_{\rho(\overline{y}\rightsquigarrow x)}$. Denote also by $\xi \colon X \to U$ the $X$-morphism induced by $\xi$, so that the diagram

$$\begin{array}{ccc} \overline{y} & \longrightarrow & X \\ {\scriptstyle\zeta}\downarrow & & \downarrow{\scriptstyle\xi} \\ V & \longrightarrow & U \end{array} \tag{VI.10.26.5}$$

is commutative. We deduce from this a $Y$-morphism $\zeta' \colon \overline{y} \to V \times_{U,\xi} X$ that fits into a commutative diagram

$$\begin{array}{ccccc} \overline{y} & \xrightarrow{\zeta'} & V \times_{U,\xi} X & \longrightarrow & X \\ & {\scriptstyle\zeta}\searrow & \downarrow & & \downarrow{\scriptstyle\xi} \\ & & V & \longrightarrow & U \end{array} \tag{VI.10.26.6}$$

Consequently, $(V \times_{U,\xi} X, \zeta')$ is an object of $\mathscr{C}_{\overline{y}}$, and the diagram (VI.10.26.6) induces a morphism

$$\jmath_{\overline{y}}(V \times_{U,\xi} X, \zeta') = ((V \times_{U,\xi} X \to X), \xi_0, \zeta') \to ((V \to U), \xi, \zeta)$$

of $\mathscr{P}^{\mathrm{scoh}}_{\rho(\overline{y}\rightsquigarrow x)}$. We deduce from this that $\jmath^{\circ}_{\overline{y}}$ is cofinal by ([**2**] I 8.1.3(c)), giving the proposition.

**Corollary VI.10.27.** *Suppose that $X$ is strictly local, and denote by $\theta \colon Y_{\mathrm{f\acute{e}t}} \to \widetilde{E}$ the morphism of topos defined in* (VI.10.23.1). *Then, the base change morphism $\beta_* \to \theta^*$* (VI.10.24.4) *is an isomorphism; in particular, the functor $\beta_*$ is exact.*

It follows from VI.10.26 and VI.9.6.

**Corollary VI.10.28.** *Suppose that $X$ is strictly local, and denote by $\theta \colon Y_{\mathrm{f\acute{e}t}} \to \widetilde{E}$ the morphism of topos defined in* (VI.10.23.1). *Then:*

(i) *For every sheaf $F$ of $\widetilde{E}$, the canonical map*

$$\Gamma(\widetilde{E}, F) \to \Gamma(Y_{\mathrm{f\acute{e}t}}, \theta^* F) \tag{VI.10.28.1}$$

*is bijective.*

(ii) *For every abelian sheaf $F$ of $\widetilde{E}$, the canonical map*

(VI.10.28.2)
$$\mathrm{H}^i(\widetilde{E},F)\to \mathrm{H}^i(Y_{\text{fét}},\theta^*F)$$

*is bijective for every $i\geq 0$.*

(i) Indeed, the diagram

(VI.10.28.3)
$$\begin{array}{ccc}
\Gamma(\widetilde{E},F) & \xleftarrow[\sim]{\ w\ } & \Gamma(Y_{\text{fét}},\beta_*F)\\
\Big\downarrow{\scriptstyle v} & & \Big\downarrow{\scriptstyle u}\\
\Gamma(\widetilde{E},\theta_*(\theta^*F)) & \xleftarrow[\sim]{\ w'\ } & \Gamma(Y_{\text{fét}},\beta_*(\theta_*(\theta^*F)))\\
 & \searrow{\scriptstyle v'} & \Big\downarrow{\scriptstyle u'}\\
 & & \Gamma(Y_{\text{fét}},\theta^*F)
\end{array}$$

where $w$, $w'$, and $v'$ are the canonical bijections, $u$ and $v$ are induced by the adjunction morphism $\mathrm{id}\to\theta_*\theta^*$, and $u'$ is induced by the isomorphism (VI.10.24.3), is commutative. Since $u'\circ u$ is bijective by virtue of VI.10.27, the same holds for $v'\circ v$, giving the assertion.

(ii) Indeed, the diagram

(VI.10.28.4)
$$\begin{array}{ccc}
\mathrm{H}^i(\widetilde{E},F) & \xleftarrow{\ w\ } & \mathrm{H}^i(Y_{\text{fét}},\beta_*F)\\
\Big\downarrow{\scriptstyle v} & & \Big\downarrow{\scriptstyle u}\\
\mathrm{H}^i(\widetilde{E},\theta_*(\theta^*F)) & \xleftarrow{\ w'\ } & \mathrm{H}^i(Y_{\text{fét}},\beta_*(\theta_*(\theta^*F)))\\
 & \searrow{\scriptstyle v'} & \Big\downarrow{\scriptstyle u'}\\
 & & \mathrm{H}^i(Y_{\text{fét}},\theta^*F)
\end{array}$$

where $w$, $w'$, and $v'$ are induced by the Cartan–Leray spectral sequence ([**2**] V 5.3), $u$ and $v$ are induced by the adjunction morphism $\mathrm{id}\to\theta_*\theta^*$, and $u'$ is induced by the isomorphism (VI.10.24.3), is commutative. On the other hand, $u'\circ u$ is bijective, and the functor $\beta_*$ is exact by virtue of VI.10.27; hence $w$ is bijective, giving the assertion.

**VI.10.29.** Let $\overline{x}$ be a geometric point of $X$, $\underline{X}$ the strict localization of $X$ at $\overline{x}$, $\underline{Y}=Y\times_X\underline{X}$, and $\underline{f}\colon\underline{Y}\to\underline{X}$ the canonical projection. We denote by $\underline{\widetilde{E}}$ the Faltings topos associated with $\underline{f}$ (VI.10.1) and by

(VI.10.29.1)
$$\theta\colon\underline{Y}_{\text{fét}}\to\underline{\widetilde{E}}$$

the morphism defined in (VI.10.23.1). The canonical morphism $\underline{X}\to X$ induces by functoriality a morphism (VI.10.12.5)

(VI.10.29.2)
$$\Phi\colon\underline{\widetilde{E}}\to\widetilde{E}.$$

We denote by

(VI.10.29.3)
$$\varphi_{\overline{x}}\colon\widetilde{E}\to\underline{Y}_{\text{fét}}$$

the composed functor $\theta^*\circ\Phi^*$.

**Proposition VI.10.30.** *We keep the assumptions of* VI.10.29 *and moreover suppose that $f$ is coherent. Then:*

(i) *For every sheaf $F$ of $\widetilde{E}$, we have a functorial canonical isomorphism*

(VI.10.30.1)
$$\sigma_*(F)_{\overline{x}}\xrightarrow{\sim}\Gamma(\underline{Y}_{\text{fét}},\varphi_{\overline{x}}(F)).$$

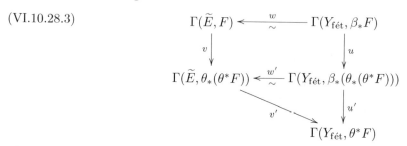

(ii) *For every abelian sheaf $F$ of $\widetilde{E}$ and every integer $i \geq 0$, we have a functorial canonical isomorphism*

$$\mathrm{R}^i\sigma_*(F)_{\overline{x}} \xrightarrow{\sim} \mathrm{H}^i(\underline{Y}_{\text{fét}}, \varphi_{\overline{x}}(F)). \tag{VI.10.30.2}$$

(iii) *For every exact sequence of abelian sheaves $0 \to F' \to F \to F'' \to 0$ of $\widetilde{E}$ and every integer $i \geq 0$, the diagram*

$$\begin{array}{ccc} \mathrm{R}^i\sigma_*(F'')_{\overline{x}} & \longrightarrow & \mathrm{R}^{i+1}\sigma_*(F')_{\overline{x}} \\ \downarrow & & \downarrow \\ \mathrm{H}^i(\underline{Y}_{\text{fét}}, \varphi_{\overline{x}}(F'')) & \longrightarrow & \mathrm{H}^{i+1}(\underline{Y}_{\text{fét}}, \varphi_{\overline{x}}(F')) \end{array} \tag{VI.10.30.3}$$

*where the vertical arrows are the canonical isomorphisms* (VI.10.30.2) *and the horizontal arrows are the boundary maps of the long exact sequences of cohomology, is commutative.*

This proposition will be proved in VI.11.14.

**Remark VI.10.31.** We keep the assumptions and notation of VI.10.29 and moreover let $\overline{y}$ be a geometric point of $Y$ and $u\colon \overline{y} \to \underline{X}$ an $X$-morphism, so that $(\overline{y} \rightsquigarrow \overline{x})$ is a point of $X_{\text{ét}} \overleftarrow{\times}_{X_{\text{ét}}} Y_{\text{ét}}$ (VI.10.18). Denote by $\widetilde{x}$ the closed point of $\underline{X}$, by $v\colon \overline{y} \to \underline{Y}$ the morphism induced by $u$, by $\widetilde{y}$ the geometric point of $\underline{Y}$ defined by $v$, by $\rho_{\underline{Y}}\colon \underline{Y}_{\text{ét}} \to \underline{Y}_{\text{fét}}$ the canonical morphism (VI.9.2.1), and by $\psi_{\widetilde{y}}\colon \underline{Y}_{\text{fét}} \to \mathbf{Ens}$ the fiber functor associated with the point $\rho_{\underline{Y}}(\widetilde{y})$ of $\underline{Y}_{\text{fét}}$. The composed functor

$$\psi_{\widetilde{y}} \circ \varphi_{\overline{x}}\colon \widetilde{E} \to \mathbf{Ens} \tag{VI.10.31.1}$$

is then canonically isomorphic to the fiber functor associated with the point $\rho(\overline{y} \rightsquigarrow \overline{x})$ of $\widetilde{E}$ (VI.10.15.3). Indeed, the squares of the diagram of morphisms of topos

$$\begin{array}{ccccc} \underline{Y}_{\text{ét}} & \xrightarrow{\gamma} & \underline{X}_{\text{ét}} \overleftarrow{\times}_{\underline{X}_{\text{ét}}} \underline{Y}_{\text{ét}} & \xrightarrow{\Xi} & X_{\text{ét}} \overleftarrow{\times}_{X_{\text{ét}}} Y_{\text{ét}} \\ {\scriptstyle \rho_{\underline{Y}}}\downarrow & & {\scriptstyle \underline{\rho}}\downarrow & & \downarrow{\scriptstyle \rho} \\ \underline{Y}_{\text{fét}} & \xrightarrow{\theta} & \underline{\widetilde{E}} & \xrightarrow{\Phi} & \widetilde{E} \end{array} \tag{VI.10.31.2}$$

where $\rho$ and $\underline{\rho}$ are the canonical morphisms (VI.10.15.3), $\gamma$ is the morphism (VI.10.25.3), and $\Xi$ is the morphism deduced from the functoriality of the covanishing topos (VI.3.8), are commutative up to canonical isomorphisms: the left square corresponds to the diagram (VI.10.25.6) and the right square to the diagram (VI.10.17.1). On the other hand, there exists a unique specialization arrow from $\underline{f}(\widetilde{y})$ to $\widetilde{x}$; we may therefore consider the point $(\widetilde{y} \rightsquigarrow \widetilde{x})$ of $\underline{X}_{\text{ét}} \overleftarrow{\times}_{\underline{X}_{\text{ét}}} \underline{Y}_{\text{ét}}$. It is clear that $\gamma(\widetilde{y})$ is canonically isomorphic to $(\widetilde{y} \rightsquigarrow \widetilde{x})$ and that $\Xi(\widetilde{y} \rightsquigarrow \widetilde{x}) = (\overline{y} \rightsquigarrow \overline{x})$. Consequently, $\rho(\overline{y} \rightsquigarrow \overline{x})$ is canonically isomorphic to $\Phi(\theta(\rho_{\underline{Y}}(\widetilde{y})))$, giving the assertion.

**Proposition VI.10.32.** *Suppose that the schemes $X$ and $Y$ are coherent. For every geometric point $\overline{x}$ of $X$, let $X_{(\overline{x})}$ be the strict localization of $X$ at $\overline{x}$, $Y_{(\overline{x})} = Y \times_X X_{(\overline{x})}$, and*

$$\varphi_{\overline{x}}\colon \widetilde{E} \to (Y_{(\overline{x})})_{\text{fét}} \tag{VI.10.32.1}$$

*the functor defined in* (VI.10.29.3). *Then, the family of functors $(\varphi_{\overline{x}})$, when $\overline{x}$ goes through the set of geometric points of $X$, is conservative.*

Indeed, the family of fiber functors of $\widetilde{E}$ associated with the points of the form $\rho(\overline{y} \rightsquigarrow \overline{x})$, when $(\overline{y} \rightsquigarrow \overline{x})$ goes through the points of $X_{\text{ét}} \overleftarrow{\times}_{X_{\text{ét}}} Y_{\text{ét}}$, is conservative by virtue of VI.10.21. The proposition follows in view of VI.10.31.

**Corollary VI.10.33.** *Under the assumptions of* VI.10.32, *a morphism $u$ of $\widetilde{E}$ is a monomorphism (resp. epimorphism) if and only if for every geometric point $\overline{x}$ of $X$, $\varphi_{\overline{x}}(u)$ is one* (VI.10.32.1).

This follows from VI.10.32 and ([**2**] I 6.2(ii)).

**VI.10.34.** We take the assumptions and notation of VI.10.29 and moreover denote by $\mathscr{C}_{\overline{x}}$ the category of $\overline{x}$-pointed étale $X$-schemes ([**2**] VIII 3.9), which we identify with the category of neighborhoods of $\overline{x}$ in the site $\mathbf{\acute{E}t}_{/X}$ ([**2**] IV 6.8.2). It is a cofiltered category. For every object $(U, \xi\colon x \to U)$ of $\mathscr{C}_{\overline{x}}$, we denote again by $\xi\colon \underline{X} \to U$ the morphism deduced from $\xi$ ([**2**] VIII 7.3) and by

(VI.10.34.1) $$\xi_Y : \underline{Y} \to U_Y$$

its base change by $f$. The topos $\widetilde{E}_{/(U_Y \to U)^a}$ is canonically equivalent to the Faltings topos associated with the morphism $f_U : U_Y \to U$ by (VI.10.14.5). Denote by

(VI.10.34.2) $$\jmath_U : \widetilde{E}_{/(U_Y \to U)^a} \to \widetilde{E}$$

the localization morphism of $\widetilde{E}$ at $(U_Y \to U)^a$. The morphism $\xi\colon \underline{X} \to U$ then induces by functoriality a morphism of topos (VI.10.12.5)

(VI.10.34.3) $$\Phi_\xi : \underline{\widetilde{E}} \to \widetilde{E}_{/(U_Y \to U)^a}.$$

By (VI.10.12.6), the diagram

(VI.10.34.4) $$\begin{array}{ccc} \underline{\widetilde{E}} & \xrightarrow{\Phi_\xi} & \widetilde{E}_{/(U_Y \to U)^a} \\ {\scriptstyle \underline{\beta}}\downarrow & & \downarrow{\scriptstyle \beta_U} \\ \underline{Y}_{\text{fét}} & \xrightarrow{(\xi_Y)_{\text{fét}}} & (U_Y)_{\text{fét}} \end{array}$$

where $\underline{\beta}$ and $\beta_U$ are the canonical morphisms (VI.10.6.3), is commutative up to canonical isomorphism. We deduce from this a base change morphism

(VI.10.34.5) $$(\xi_Y)^*_{\text{fét}} \beta_{U*} \to \underline{\beta}_* \Phi^*_\xi.$$

By ([**1**] 1.2.4(i)), the composition

(VI.10.34.6) $$(\xi_Y)^*_{\text{fét}} \beta_{U*} \to \underline{\beta}_* \Phi^*_\xi \to \theta^* \Phi^*_\xi,$$

where the second arrow is induced by (VI.10.24.4), is the base change morphism deduced from the canonical isomorphism (VI.10.24.3)

(VI.10.34.7) $$\beta_U \circ \Phi_\xi \circ \theta \xrightarrow{\sim} (\xi_Y)_{\text{fét}}.$$

Let $F = \{U \mapsto F_U\}$ be an object of $\widehat{E}$, $F^a = \{U \mapsto G_U\}$ the sheaf of $\widetilde{E}$ associated with $F$, and for every $U \in \mathrm{Ob}(\mathbf{\acute{E}t}_{/X})$, $F^a_U$ the sheaf of $(U_Y)_{\text{fét}}$ associated with $F_U$. By VI.5.17, $\{U \mapsto F^a_U\}$ is a presheaf on $E$ and we have a canonical morphism $\{U \mapsto F_U\} \to \{U \mapsto F^a_U\}$ of $\widehat{E}$, inducing an isomorphism between the associated sheaves. We associate with the presheaf $\{U \mapsto F^a_U\}$ the functor

(VI.10.34.8) $$\mathscr{C}^\circ_{\overline{x}} \to \underline{Y}_{\text{fét}}, \quad (U, \xi) \mapsto (\xi_Y)^*_{\text{fét}}(F^a_U),$$

that with each morphism $t\colon (U', \xi') \to (U, \xi)$ to $\mathscr{C}_{\overline{x}}$ associates the composed morphism

$$(\xi_Y)^*_{\text{fét}}(F^a_U) \xrightarrow{\sim} (\xi'_Y)^*_{\text{fét}}((t_Y)^*_{\text{fét}} F^a_U) \to (\xi'_Y)^*_{\text{fét}}(F^a_{U'}),$$

where the first arrow is induced by the relation $\xi = t \circ \xi'$ and the second arrow comes from the transition morphism $F_U^a \to (t_Y)_{\text{fét}*}(F_{U'}^a)$ of $\{U \mapsto F_U^a\}$. Likewise, we associate with the sheaf $F^a = \{U \mapsto G_U\}$ the functor

$$\mathscr{C}_{\overline{x}}^{\circ} \to \underline{Y}_{\text{fét}}, \quad (U,\xi) \mapsto (\xi_Y)_{\text{fét}}^*(G_U). \tag{VI.10.34.9}$$

The canonical morphism $\{U \mapsto F_U^a\} \to \{U \mapsto G_U\}$ then induces a morphism of functors from $\mathscr{C}_{\overline{x}}^{\circ}$ to $\underline{Y}_{\text{fét}}$ :

$$(\xi_Y)_{\text{fét}}^*(F_U^a) \to (\xi_Y)_{\text{fét}}^*(G_U), \quad (U,\xi) \in \mathrm{Ob}(\mathscr{C}_{\overline{x}}). \tag{VI.10.34.10}$$

By ([2] III 5.3), we have a canonical isomorphism

$$\beta_{U*}(j_U^*(F^a)) \xrightarrow{\sim} G_U. \tag{VI.10.34.11}$$

The morphisms (VI.10.34.5) and (VI.10.34.6) therefore induce two functorial morphisms

$$(\xi_Y)_{\text{fét}}^*(G_U) \to \underline{\beta}_*(\Phi^* F^a), \tag{VI.10.34.12}$$

$$(\xi_Y)_{\text{fét}}^*(G_U) \to \varphi_{\overline{x}}(F^a). \tag{VI.10.34.13}$$

By ([1] 1.2.4(i)), these are morphisms of inverse systems on the category $\mathscr{C}_{\overline{x}}^{\circ}$ (VI.10.34.9). In view of (VI.10.34.10), we deduce from this two functorial morphisms in $F$

$$\varinjlim_{(U,\xi)\in\mathscr{C}_{\overline{x}}^{\circ}} (\xi_Y)_{\text{fét}}^*(F_U^a) \to \underline{\beta}_*(\Phi^* F^a), \tag{VI.10.34.14}$$

$$\varinjlim_{(U,\xi)\in\mathscr{C}_{\overline{x}}^{\circ}} (\xi_Y)_{\text{fét}}^*(F_U^a) \to \varphi_{\overline{x}}(F^a). \tag{VI.10.34.15}$$

**Remarks VI.10.35.** We keep the assumptions of VI.10.34.

(i) By ([2] XVII 2.1.3), the morphism (VI.10.34.6) is equal to the composition

$$(\xi_Y)_{\text{fét}}^*\beta_{U*} \xrightarrow{\sim} \theta^*\Phi_\xi^*\beta_U^*\beta_{U*} \to \theta^*\Phi_\xi^*, \tag{VI.10.35.1}$$

where the first arrow is induced by (VI.10.34.7) and the second arrow is induced by the adjunction morphism $\beta_U^*\beta_{U*} \to \mathrm{id}$.

(ii) We choose an affine object $(X_0, \xi_0)$ of $\mathscr{C}_{\overline{x}}$ and denote by $I$ the category of $\xi_0$-pointed étale $X_0$-schemes that are affine over $X_0$. The canonical functor $I \to \mathscr{C}_{\overline{x}}$ is then cofinal ([2] VIII 4.5). We may therefore replace in the direct limits in (VI.10.34.14) and (VI.10.34.15) the category $\mathscr{C}_{\overline{x}}$ by $I$.

**Proposition VI.10.36.** *Under the assumptions of* VI.10.34, *let moreover $\overline{y}$ be a geometric point of $Y$ and $u\colon \overline{y} \to \underline{X}$ an $X$-morphism, so that $(\overline{y} \rightsquigarrow \overline{x})$ is a point of $X_{\text{ét}} \overleftarrow{\times}_{X_{\text{ét}}} Y_{\text{ét}}$* (VI.10.18). *We denote by $v\colon \overline{y} \to \underline{Y}$ the $Y$-morphism induced by $u$, by $\widetilde{y}$ the geometric point of $\underline{Y}$ defined by $v$, by $\rho_{\underline{Y}}\colon \underline{Y}_{\text{ét}} \to \underline{Y}_{\text{fét}}$ the canonical morphism* (VI.9.2.1), *and by $\psi_{\widetilde{y}}\colon \underline{Y}_{\text{fét}} \to$ **Ens** the fiber functor associated with the point $\rho_{\underline{Y}}(\widetilde{y})$. Let $F = \{U \mapsto F_U\}$ be an object of $\widehat{E}$, $F^a$ the sheaf of $\widetilde{E}$ associated with $F$, and for every $U \in \mathrm{Ob}(\mathbf{\acute{E}t}_{/X})$, $F_U^a$ the sheaf of $(U_Y)_{\text{fét}}$ associated with $F_U$. Then, we have a functorial and canonical isomorphism*

$$(F^a)_{\rho(\overline{y}\rightsquigarrow\overline{x})} \xrightarrow{\sim} \varinjlim_{(U,\xi)\in\mathscr{C}_{\overline{x}}^{\circ}} \psi_{\widetilde{y}}((\xi_Y)_{\text{fét}}^*(F_U^a)), \tag{VI.10.36.1}$$

*whose inverse identifies with the image of the canonical morphism* (VI.10.34.15) *by the functor $\psi_{\widetilde{y}}$.*

First, note that $\{U \mapsto F_U^a\}$ is naturally an object of $\widehat{E}$ and that the canonical morphism $\{U \mapsto F_U\} \to \{U \mapsto F_U^a\}$ induces an isomorphism between the associated sheaves, by VI.5.17. Let $\mathscr{P}_{\rho(\overline{y}\rightsquigarrow\overline{x})}$ be the category of $\rho(\overline{y}\rightsquigarrow\overline{x})$-pointed objects of $E$ (VI.10.20). We have a functor

$$\phi\colon \mathscr{P}_{\rho(\overline{y}\rightsquigarrow\overline{x})}\to\mathscr{C}_{\overline{x}},\quad ((V\to U),\xi,\zeta)\mapsto(U,\xi). \tag{VI.10.36.2}$$

For every $(U,\xi)\in\mathrm{Ob}(\mathscr{C}_{\overline{x}})$, the fiber of $\phi$ over $(U,\xi)$ is canonically equivalent to the category $\mathscr{D}^{\widetilde{y}}_{(U,\xi)}$ of $\xi_Y(\widetilde{y})$-pointed finite étale $U_Y$-schemes (VI.10.34.1). The isomorphism (VI.10.20.2) therefore induces a functorial and canonical isomorphism

$$(F^a)_{\rho(\overline{y}\rightsquigarrow\overline{x})}\xrightarrow{\sim}\varinjlim_{(U,\xi)\in\mathscr{C}^{\circ}_{\overline{x}}}\ \varinjlim_{(V,\zeta)\in(\mathscr{D}^{\widetilde{y}}_{(U,\xi)})^{\circ}} F_U(V). \tag{VI.10.36.3}$$

In view of (VI.9.3.4) and ([**2**] IV (6.8.4)), for every $(U,\xi)\in\mathrm{Ob}(\mathscr{C}_{\overline{x}})$, we have a functorial canonical isomorphism

$$\psi_{\widetilde{y}}((\xi_Y)^*_{\text{fét}}(F^a_U))\xrightarrow{\sim}\varinjlim_{(V,\zeta)\in(\mathscr{D}^{\widetilde{y}}_{(U,\xi)})^{\circ}} F_U(V), \tag{VI.10.36.4}$$

giving the isomorphism (VI.10.36.1). On the other hand, $\psi_{\widetilde{y}}\circ\varphi_{\overline{x}}$ is the fiber functor of $\widetilde{E}$ associated with the point $\rho(\overline{y}\rightsquigarrow\overline{x})$, by virtue of VI.10.31. To establish the second assertion, it therefore suffices to show that for every object $((V\to U),\xi,\zeta)$ of $\mathscr{P}_{\rho(\overline{y}\rightsquigarrow\overline{x})}$, the canonical map (VI.10.36.3)

$$F_U(V)\to(F^a)_{\rho(\overline{y}\rightsquigarrow\overline{x})} \tag{VI.10.36.5}$$

is the composition

$$F_U(V)\to\psi_{\widetilde{y}}((\xi_Y)^*_{\text{fét}}(F^a_U))\to\psi_{\widetilde{y}}(\varphi_{\overline{x}}(F^a)), \tag{VI.10.36.6}$$

where the first arrow is defined by the object $(V,\zeta)$ of $\mathscr{D}^{\widetilde{y}}_{(U,\xi)}$ and the second arrow is the image by $\psi_{\widetilde{y}}$ of the composition of (VI.10.34.10) and (VI.10.34.13). We may restrict to the case where $F$ is a sheaf of $\widetilde{E}$, so that $F^a_U=F_U$. By localization (VI.10.14), we may assume $U=X$. Denote by $\mathrm{pr}_Y\colon \underline{Y}\to Y$ the canonical projection. By VI.10.35(i), the image by $\psi_{\widetilde{y}}$ of the morphism $(\mathrm{pr}_Y)^*_{\text{fét}}(F_X)\to\varphi_{\overline{x}}(F)$ (VI.10.34.13) coincides with the map

$$(F_X)_{\rho_Y(\overline{y})}\to F_{\rho(\overline{y}\rightsquigarrow\overline{x})} \tag{VI.10.36.7}$$

induced by the adjunction morphism $\beta^*(\beta_*(F))\to F$ and the isomorphism (VI.10.18.2), giving the desired assertion.

**Corollary VI.10.37.** *We keep the assumptions of* VI.10.34, *and moreover let* $F=\{U\mapsto F_U\}$ *be an object of* $\widehat{E}$, $F^a$ *the sheaf of* $\widetilde{E}$ *associated with* $F$, *and for every* $U\in\mathrm{Ob}(\mathbf{\acute{E}t}_{/X})$, $F^a_U$ *the sheaf of* $(U_Y)_{\text{fét}}$ *associated with* $F_U$. *Then, the canonical morphism* (VI.10.34.15)

$$\varinjlim_{(U,\xi)\in\mathscr{C}^{\circ}_{\overline{x}}}(\xi_Y)^*_{\text{fét}}(F^a_U)\to\varphi_{\overline{x}}(F^a) \tag{VI.10.37.1}$$

*is an isomorphism.*

It follows from VI.10.36 and VI.9.6.

**Proposition VI.10.38.** *Under the assumptions of* VI.10.34, *for every sheaf* $F=\{U\mapsto F_U\}$ *of* $\widetilde{E}$, *the canonical morphism* (VI.10.34.14)

$$\varinjlim_{(U,\xi)\in\mathscr{C}^{\circ}_{\overline{x}}}(\xi_Y)^*_{\text{fét}}(F_U)\to\underline{\beta}_*(\Phi^*F) \tag{VI.10.38.1}$$

*is an isomorphism.*

It follows from VI.10.27 and VI.10.37.

**VI.10.39.** Let $F = \{U \mapsto F_U\}$ be a presheaf of abelian groups on $E$ such that for every $U \in \mathrm{Ob}(\mathbf{\acute{E}t}_{/X})$, $F_U$ is a sheaf (of abelian groups) of $(U_Y)_{\text{fét}}$ (for example, $F$ is an abelian sheaf of $\widetilde{E}$), and let $i$ be an integer $\geq 0$. We denote by $\mathscr{H}^i(F)$ the sheaf of $X_{\text{ét}}$ associated with the presheaf on $\mathbf{\acute{E}t}_{/X}$ defined for every $U \in \mathrm{Ob}(\mathbf{\acute{E}t}_{/X})$ by the group

$$\mathrm{H}^i((U_Y)_{\text{fét}}, F_U), \tag{VI.10.39.1}$$

and for every morphism $g\colon U' \to U$ of $\mathbf{\acute{E}t}_{/X}$, by the composed map

$$\mathrm{H}^i((U_Y)_{\text{fét}}, F_U) \to \mathrm{H}^i((U_Y)_{\text{fét}}, (g_Y)_{\text{fét}*}(F_{U'})) \to \mathrm{H}^i((U'_Y)_{\text{fét}}, F_{U'}), \tag{VI.10.39.2}$$

where the first arrow is induced by the transition morphism of $F$ and the second arrow is induced by the Cartan–Leray spectral sequence ([**2**] V 5.3).

We denote by $F^a = \{U \mapsto G_U\}$ the sheaf (of abelian groups) of $\widetilde{E}$ associated with $F$ ([**2**] III 6.4). For every $U \in \mathrm{Ob}(\mathbf{\acute{E}t}_{/X})$, the topos $\widetilde{E}_{/(U_Y \to U)^a}$ is canonically equivalent to the Faltings topos associated with the morphism $U_Y \to U$ by (VI.10.14.5). We therefore have a canonical morphism of topos (VI.10.6.3)

$$\beta_U : \widetilde{E}_{/(U_Y \to U)^a} \to (U_Y)_{\text{fét}}. \tag{VI.10.39.3}$$

By definition of the restriction functor ([**2**] III 5.3), we have a canonical isomorphism

$$\beta_{U*}(F^a|(U_Y \to U)^a) \xrightarrow{\sim} G_U. \tag{VI.10.39.4}$$

Consequently, the Cartan–Leray spectral sequence induces a functorial map in $F$

$$\mathrm{H}^i((U_Y)_{\text{fét}}, G_U) \to \mathrm{H}^i((U_Y \to U)^a, F^a). \tag{VI.10.39.5}$$

Composing with the map $\mathrm{H}^i((U_Y)_{\text{fét}}, F_U) \to \mathrm{H}^i((U_Y)_{\text{fét}}, G_U)$ induced by the canonical morphism $F \to F^a$, we obtain a functorial map in $F$

$$\mathrm{H}^i((U_Y)_{\text{fét}}, F_U) \to \mathrm{H}^i((U_Y \to U)^a, F^a). \tag{VI.10.39.6}$$

Let $g\colon U' \to U$ be a morphism of $\mathbf{\acute{E}t}_{/X}$. The diagram of morphisms of topos

$$\begin{array}{ccc} \widetilde{E}_{/(U'_Y \to U')^a} & \xrightarrow{\beta_{U'}} & (U'_Y)_{\text{fét}} \\ {\scriptstyle j}\downarrow & & \downarrow{\scriptstyle (g_Y)_{\text{fét}}} \\ \widetilde{E}_{/(U_Y \to U)^a} & \xrightarrow{\beta_U} & (U_Y)_{\text{fét}} \end{array} \tag{VI.10.39.7}$$

where $j$ is the localization morphism of $\widetilde{E}_{/(U_Y \to U)^a}$ at $(U'_Y \to U')^a$, is commutative, up to canonical isomorphism. One immediately verifies this on the corresponding diagram of morphisms of sites. On the other hand, the diagram

(VI.10.39.8)

$$\begin{array}{ccccc} \beta_{U*}(F^a|(U_Y \to U)^a) & \xrightarrow{u} & \beta_{U*}(j_*(F^a|(U'_Y \to U')^a)) & \xrightarrow[\sim]{v} & (g_Y)_{\text{fét}*}(\beta_{U'*}(F^a|(U'_Y \to U')^a)) \\ \| & & & & \| \\ G_U & & \xrightarrow{\quad w \quad} & & (g_Y)_{\text{fét}*}(G_{U'}) \end{array}$$

where $u$ is induced by the adjunction morphism $\mathrm{id} \to j_*j^*$, $v$ is induced by the diagram (VI.10.39.7), $w$ is the transition morphism of $F^a$, and the vertical identifications come from the isomorphism (VI.10.39.4), is commutative. We deduce from this that the diagram

(VI.10.39.9)
$$\begin{array}{ccc}
\mathrm{H}^i((U_Y)_{\text{fét}}, G_U) & \longrightarrow & \mathrm{H}^i((U_Y \to U)^a, F^a) \\
\downarrow & & \\
\mathrm{H}^i((U_Y)_{\text{fét}}, (g_Y)_{\text{fét}*}(G_{U'})) & & \downarrow \\
\downarrow & & \\
\mathrm{H}^i((U'_Y)_{\text{fét}}, G_{U'}) & \longrightarrow & \mathrm{H}^i((U'_Y \to U')^a, F^a)
\end{array}$$

where the horizontal arrows are the maps (VI.10.39.5), the left vertical arrows are defined in (VI.10.39.2) (for $F^a$), and the right vertical arrow is the canonical map, is commutative. Consequently, the map (VI.10.39.6) defines a morphism of presheaves on $\mathbf{\acute{E}t}_{/X}$. Taking the associated sheaves, we obtain a morphism of abelian groups of $X_{\text{ét}}$

(VI.10.39.10)
$$\mathscr{H}^i(F) \to \mathrm{R}^i\sigma_*(F^a).$$

**Theorem VI.10.40.** *Let $F = \{U \mapsto F_U\}$ be a presheaf of abelian groups on $E$, $\overline{x}$ a geometric point of $X$, and $X_{(\overline{x})}$ the strict localization of $X$ at $\overline{x}$. Suppose that $f$ is coherent, and that for every $U \in \mathrm{Ob}(\mathbf{\acute{E}t}_{/X})$, $F_U$ is a sheaf of $(U_Y)_{\text{fét}}$. Then, for every integer $i \geq 0$, the stalk $\mathscr{H}^i(F)_{\overline{x}} \to \mathrm{R}^i\sigma_*(F^a)_{\overline{x}}$ of the morphism* (VI.10.39.10) *at $\overline{x}$ is an isomorphism.*

This theorem will be proved in VI.11.15.

## VI.11. Inverse limit of Faltings topos

**VI.11.1.** We denote by $\mathfrak{M}$ the category of morphisms of schemes and by $\mathfrak{D}$ the category of morphisms of $\mathfrak{M}$. The objects of $\mathfrak{D}$ are therefore commutative diagrams of morphisms of schemes

(VI.11.1.1)
$$\begin{array}{ccc}
V & \longrightarrow & U \\
\downarrow & & \downarrow \\
Y & \longrightarrow & X
\end{array}$$

where we consider the horizontal arrows as objects of $\mathfrak{M}$ and the vertical arrows as morphisms of $\mathfrak{M}$; such an object will be denoted by $(V, U, Y, X)$. We denote by $\mathfrak{E}$ the full subcategory of $\mathfrak{D}$ made up of the objects $(V, U, Y, X)$ such that the morphism $U \to X$ is étale of finite presentation and that the morphism $V \to U_Y = U \times_X Y$ is finite étale. The "target functor"

(VI.11.1.2)
$$\mathfrak{E} \to \mathfrak{M}, \quad (V, U, Y, X) \mapsto (Y \to X),$$

makes $\mathfrak{E}$ into a cleaved and normalized fibered category. The fiber over an object $f\colon Y \to X$ of $\mathfrak{M}$ is the category denoted by $E_{\text{coh}}$ in VI.10.4. For every commutative diagram of morphisms of schemes

(VI.11.1.3)
$$\begin{array}{ccc}
Y' & \xrightarrow{f'} & X' \\
{\scriptstyle g'}\downarrow & & \downarrow{\scriptstyle g} \\
Y & \xrightarrow{f} & X
\end{array}$$

the inverse image functor of (VI.11.1.2) associated with the morphism $(g', g)$ of $\mathfrak{M}$ is the functor (VI.10.12.3)

$$\Phi^+ \colon \mathfrak{E}_f \to \mathfrak{E}_{f'}, \quad (V \to U) \mapsto (V \times_Y Y' \to U \times_X X'). \tag{VI.11.1.4}$$

Endowing each fiber of $\mathfrak{E}/\mathfrak{M}$ with the covanishing topology (VI.10.4), $\mathfrak{E}$ becomes a fibered $\mathbb{U}$-site (cf. VI.10.12 and [**2**] VI 7.2.4). We denote by

$$\mathfrak{F} \to \mathfrak{M} \tag{VI.11.1.5}$$

the fibered $\mathbb{U}$-topos associated with $\mathfrak{E}/\mathfrak{M}$ ([**2**] VI 7.2.6): the fiber of $\mathfrak{F}$ over an object $f \colon Y \to X$ of $\mathfrak{M}$ is the topos $\widetilde{\mathfrak{E}}_f$ of sheaves of $\mathbb{U}$-sets on the covanishing site $\mathfrak{E}_f$, and the inverse image functor relative to the morphism defined by the diagram (VI.11.1.3) is the functor $\Phi^* \colon \widetilde{\mathfrak{E}}_f \to \widetilde{\mathfrak{E}}_{f'}$, inverse image under the morphism of topos $\Phi \colon \widetilde{\mathfrak{E}}_{f'} \to \widetilde{\mathfrak{E}}_f$ associated with the morphism of sites $\Phi^+$ (VI.11.1.4). We denote by

$$\mathfrak{F}^\vee \to \mathfrak{M}^\circ \tag{VI.11.1.6}$$

the fibered category obtained by associating with each object $f \colon Y \to X$ of $\mathfrak{M}$ the category $\mathfrak{F}_f = \widetilde{\mathfrak{E}}_f$, and with each morphism defined by a diagram (VI.11.1.3) the functor $\Phi_* \colon \widetilde{\mathfrak{E}}_{f'} \to \widetilde{\mathfrak{E}}_f$, direct image by the morphism of topos $\Phi \colon \widetilde{\mathfrak{E}}_{f'} \to \widetilde{\mathfrak{E}}_f$.

**VI.11.2.** Let $I$ be an essentially small cofiltered category ([**2**] I 2.7 and 8.1.8) and

$$\varphi \colon I \to \mathfrak{M}, \quad i \mapsto (f_i \colon Y_i \to X_i) \tag{VI.11.2.1}$$

a functor such that for every morphism $j \to i$ of $I$, the morphisms $Y_j \to Y_i$ and $X_j \to X_i$ are affine. We suppose that there exists $i_0 \in \mathrm{Ob}(I)$ such that $X_{i_0}$ and $Y_{i_0}$ are coherent. We denote by

$$\mathfrak{E}_\varphi \to I \tag{VI.11.2.2}$$

$$\mathfrak{F}_\varphi \to I \tag{VI.11.2.3}$$

$$\mathfrak{F}^\vee_\varphi \to I^\circ \tag{VI.11.2.4}$$

the site, topos, and fibered category deduced from $\mathfrak{E}$ (VI.11.1.2), $\mathfrak{F}$ (VI.11.1.5), and $\mathfrak{F}^\vee$ (VI.11.1.6), respectively, by base change by the functor $\varphi$. Note that $\mathfrak{F}_\varphi$ is the fibered topos associated with $\mathfrak{E}_\varphi$ ([**2**] VI 7.2.6.8). By ([**42**] 8.2.3), the inverse limits

$$X = \lim_{\substack{\longleftarrow \\ i \in \mathrm{Ob}(I)}} X_i \quad \text{and} \quad Y = \lim_{\substack{\longleftarrow \\ i \in \mathrm{Ob}(I)}} Y_i \tag{VI.11.2.5}$$

are representable in the category of schemes. The morphisms $(f_i)_{i \in I}$ induce a morphism $f \colon Y \to X$, that represents the inverse limit of the functor (VI.11.2.1).

For every $i \in \mathrm{Ob}(I)$, we have a canonical commutative diagram

$$\begin{array}{ccc} Y & \xrightarrow{f} & X \\ \downarrow & & \downarrow \\ Y_i & \xrightarrow{f_i} & X_i \end{array} \tag{VI.11.2.6}$$

There corresponds to it an inverse image functor (VI.11.1.4)

$$\Phi_i^+ \colon \mathfrak{E}_{f_i} \to \mathfrak{E}_f, \tag{VI.11.2.7}$$

that is continuous and left exact, and consequently a morphism of topos

$$\Phi_i \colon \widetilde{\mathfrak{E}}_f \to \widetilde{\mathfrak{E}}_{f_i}. \tag{VI.11.2.8}$$

We have a natural functor

$$\mathfrak{E}_\varphi \to \mathfrak{E}_f, \tag{VI.11.2.9}$$

whose restriction to the fiber over any $i \in \mathrm{Ob}(I)$ is the functor $\Phi_i^+$ (VI.11.2.7). This functor transforms Cartesian morphisms into isomorphisms. It therefore factors uniquely through a functor ([**2**] VI 6.3)

$$\lim_{\overrightarrow{I^\circ}} \mathfrak{E}_\varphi \to \mathfrak{E}_f. \tag{VI.11.2.10}$$

The $I$-functor $\mathfrak{E}_\varphi \to \mathfrak{E}_f \times I$ deduced from (VI.11.2.9) is a Cartesian morphism of fibered sites ([**2**] VI 7.2.2). It therefore induces a Cartesian morphism of fibered topos ([**2**] VI 7.2.7)

$$\widetilde{\mathfrak{E}}_f \times I \to \mathfrak{F}_\varphi. \tag{VI.11.2.11}$$

**Proposition VI.11.3.** *The topos $\widetilde{\mathfrak{E}}_f$ equipped with the morphism* (VI.11.2.11) *is an inverse limit of the fibered topos $\mathfrak{F}_\varphi/I$* ([**2**] VI 8.1.1).

First, note that the functor (VI.11.2.10) is an equivalence of categories by virtue of ([**42**] 8.8.2, 8.10.5 and 17.7.8). Let $T$ be a $\mathbb{U}$-topos,

$$h\colon T \times I \to \mathfrak{F}_\varphi \tag{VI.11.3.1}$$

a Cartesian morphism of fibered topos over $I$. Denote by $\varepsilon_I\colon \mathfrak{E}_\varphi \to \mathfrak{F}_\varphi$ the canonical Cartesian functor ([**2**] VI (7.2.6.7)), and set

$$h^+ = h^* \circ \varepsilon_I\colon \mathfrak{E}_\varphi \to T \times I. \tag{VI.11.3.2}$$

For every $i \in \mathrm{Ob}(I)$, we denote by

$$h_i^+\colon \mathfrak{E}_{f_i} \to T \tag{VI.11.3.3}$$

the restriction of $h^+$ to the fiber over $i$. By the equivalence of categories (VI.11.2.10) and ([**2**] VI 6.2), there exists essentially a unique functor

$$g^+\colon \mathfrak{E}_f \to T \tag{VI.11.3.4}$$

such that $h^+$ is isomorphic to the composition

$$\mathfrak{E}_\varphi \longrightarrow \mathfrak{E}_f \times I \xrightarrow{g^+ \times \mathrm{id}_I} T \times I, \tag{VI.11.3.5}$$

where the first arrow is the functor deduced from (VI.11.2.9). Let us show that $g^+$ is a morphism of sites. For every object $(V \to U)$ of $\mathfrak{E}_f$, there exist $i \in \mathrm{Ob}(I)$, an object $(V_i \to U_i)$ of $\mathfrak{E}_{f_i}$, and an isomorphism of $\mathfrak{E}_f$

$$(V \to U) \xrightarrow{\sim} \Phi_i^+(V_i \to U_i). \tag{VI.11.3.6}$$

Since the functors $h_i^+$ and $\Phi_i^+$ are left exact, we deduce from this that $g^+$ is left exact. On the other hand, every *finite* Cartesian (resp. vertical) covering of $\mathfrak{E}_f$ (VI.5.3) is the inverse image of a Cartesian (resp. vertical) covering of $\mathfrak{E}_{f_i}$ for an object $i \in I$, by virtue of ([**42**] 8.10.5(vi)). Since the schemes $X$ and $Y$ are coherent, we deduce from this that $g^+$ transforms Cartesian (resp. vertical) coverings of $\mathfrak{E}_f$ into epimorphic families of $T$. Consequently, $g^+$ is continuous by virtue of VI.5.10. It therefore defines a morphism of topos

$$g\colon T \to \widetilde{\mathfrak{E}}_f \tag{VI.11.3.7}$$

such that $h$ is isomorphic to the composition

$$T \times I \xrightarrow{g \times \mathrm{id}_I} \widetilde{\mathfrak{E}}_f \times I \longrightarrow \mathfrak{F}_\varphi, \tag{VI.11.3.8}$$

where the second arrow is the morphism (VI.11.2.11). Such a morphism $g$ is essentially unique because the "restriction" $g^+\colon \mathfrak{E}_f \to T$ of the functor $g^*$ is essentially unique by the above, giving the proposition.

**VI.11.4.** We endow $\mathfrak{E}_\varphi$ with the total topology ([**2**] VI 7.4.1) and denote by $\mathbf{Top}(\mathfrak{E}_\varphi)$ the topos of sheaves of $\mathbb{U}$-sets on $\mathfrak{E}_\varphi$. By ([**2**] VI 7.4.7), we have a canonical equivalence of categories (VI.2.2)

$$\mathbf{Top}(\mathfrak{E}_\varphi) \xrightarrow{\sim} \mathbf{Hom}_{I^\circ}(I^\circ, \mathfrak{F}_\varphi^\vee). \tag{VI.11.4.1}$$

On the other hand, the natural functor $\mathfrak{E}_\varphi \to \mathfrak{E}_f$ (VI.11.2.9) is a morphism of sites ([**2**] VI 7.4.4) and therefore defines a morphism of topos

$$\varpi \colon \widetilde{\mathfrak{E}}_f \to \mathbf{Top}(\mathfrak{E}_\varphi). \tag{VI.11.4.2}$$

By virtue of VI.11.3 and ([**2**] VI 8.2.9), there exists an equivalence of categories $\Theta$ that fits into a commutative diagram

$$\begin{array}{ccc} \widetilde{\mathfrak{E}}_f & \xrightarrow[\sim]{\Theta} & \mathbf{Hom}_{\mathrm{cart}/I^\circ}(I^\circ, \mathfrak{F}_\varphi^\vee) \\ {\scriptstyle\varpi_*}\downarrow & & \downarrow \\ \mathbf{Top}(\mathfrak{E}_\varphi) & \xrightarrow{\sim} & \mathbf{Hom}_{I^\circ}(I^\circ, \mathfrak{F}_\varphi^\vee) \end{array} \tag{VI.11.4.3}$$

where the bottom horizontal arrow is the equivalence of categories (VI.11.4.1) and the right vertical arrow is the canonical injection.

For every object $F \in \mathrm{Ob}(\mathbf{Top}(\mathfrak{E}_\varphi))$, if $\{i \mapsto F_i\}$ is the corresponding section of $\mathbf{Hom}_{I^\circ}(I^\circ, \mathfrak{F}_\varphi^\vee)$, we have a functorial canonical isomorphism ([**2**] VI 8.5.2)

$$\varpi^*(F) \xrightarrow{\sim} \varinjlim_{i \in I^\circ} \Phi_i^*(F_i). \tag{VI.11.4.4}$$

**Corollary VI.11.5.** *Let $F$ be a sheaf of $\mathbf{Top}(\mathfrak{E}_\varphi)$ and $\{i \mapsto F_i\}$ the associated section of $\mathbf{Hom}_{I^\circ}(I^\circ, \mathfrak{F}_\varphi^\vee)$ under the equivalence of categories* (VI.11.4.1). *Then we have a functorial canonical isomorphism*

$$\varinjlim_{i \in I^\circ} \Gamma(\widetilde{\mathfrak{E}}_{f_i}, F_i) \xrightarrow{\sim} \Gamma(\widetilde{\mathfrak{E}}_f, \varinjlim_{i \in I^\circ} \Phi_i^*(F_i)). \tag{VI.11.5.1}$$

**Corollary VI.11.6.** *Let $F$ be an abelian sheaf of $\mathbf{Top}(\mathfrak{E}_\varphi)$ and $\{i \mapsto F_i\}$ the section of $\mathbf{Hom}_{I^\circ}(I^\circ, \mathfrak{F}_\varphi^\vee)$ associated with it by the equivalence of categories* (VI.11.4.1). *Then for every integer $q \geq 0$, we have a functorial canonical isomorphism*

$$\varinjlim_{i \in I^\circ} \mathrm{H}^q(\widetilde{\mathfrak{E}}_{f_i}, F_i) \xrightarrow{\sim} \mathrm{H}^q(\widetilde{\mathfrak{E}}_f, \varinjlim_{i \in I^\circ} \Phi_i^*(F_i)). \tag{VI.11.6.1}$$

Corollaries VI.11.5 and VI.11.6 follow from VI.11.3 and ([**2**] VI 8.7.7). Note that the conditions required in ([**2**] VI 8.7.1 and 8.7.7) are satisfied by virtue of VI.10.5 and ([**2**] VI 3.3, 5.1, and 5.2).

**VI.11.7.** We denote by

$$\mathscr{R}_\varphi \to I, \tag{VI.11.7.1}$$

$$\mathscr{G}_\varphi \to I, \tag{VI.11.7.2}$$

$$\mathscr{G}_\varphi^\vee \to I^\circ \tag{VI.11.7.3}$$

the fibered site, topos, and category deduced, respectively, from the fibered site of finite étale covers $\mathscr{R}/\mathbf{Sch}$ (VI.9.5.1), the fibered topos $\mathscr{G}/\mathbf{Sch}$ (VI.9.5.2), and the fibered category $\mathscr{G}^\vee/\mathbf{Sch}^\circ$ (VI.9.5.3), by base change by the functor

$$I \to \mathbf{Sch}, \quad i \mapsto Y_i \tag{VI.11.7.4}$$

induced by $\varphi$ (VI.11.2.1). For every $i \in \mathrm{Ob}(I)$, we denote by

$$t_i \colon Y \to Y_i \tag{VI.11.7.5}$$

the canonical morphism (VI.11.2.5). We have a natural functor

$$\mathscr{R}_\varphi \to \mathscr{R}_Y \tag{VI.11.7.6}$$

whose restriction to the fiber over $i \in \mathrm{Ob}(I)$ is given by the base change functor by the morphism $t_i$

$$\mathscr{R}_{Y_i} \to \mathscr{R}_Y, \quad Y'_i \mapsto Y'_i \times_{Y_i} Y.$$

This functor transforms Cartesian morphisms into isomorphisms, and therefore factors uniquely through a functor

$$\varinjlim_{I^\circ} \mathscr{R}_\varphi \to \mathscr{R}_Y. \tag{VI.11.7.7}$$

The $I$-functor $\mathscr{R}_\varphi \to \mathscr{R}_Y \times I$ deduced from (VI.11.7.6) is a Cartesian morphism of fibered sites ([**2**] VI 7.2.2). It therefore induces a morphism of fibered topos

$$Y_{\text{fét}} \times I \to \mathscr{G}_\varphi. \tag{VI.11.7.8}$$

**Lemma VI.11.8.** *The functor* (VI.11.7.7) *is an equivalence of sites when we endow the source with the topology of the direct limit of the fibered site* $\mathscr{R}_\varphi$ ([**2**] VI 8.2.5) *and the target with the étale topology.*

Indeed, the functor (VI.11.7.7) is an equivalence of categories by virtue of ([**42**] 8.8.2, 8.10.5, and 17.7.8). Let $i \in \mathrm{Ob}(I)$, $g_i : Y'_i \to Y_i$ be a finite étale cover and $g : Y' \to Y$ the finite étale cover deduced from $g_i$ by base change by the morphism $Y \to Y_i$. By ([**42**] 8.10.5), $g$ is surjective if and only if there exists a morphism $j \to i$ of $I$ such that the finite étale cover $g_j : Y'_j \to Y_j$ deduced from $g_i$ by base change is surjective. The assertion concerning the topologies follows in view of ([**2**] VI 8.2.2 and III 1.6).

**Proposition VI.11.9.** *The topos* $Y_{\text{fét}}$ *equipped with the morphism* (VI.11.7.8) *is an inverse limit of the fibered topos* $\mathscr{G}_\varphi/I$.

This follows from VI.11.8 and ([**2**] VI 8.2.3).

**Corollary VI.11.10.** *Let* $F$ *be an abelian sheaf of the total topos of* $\mathscr{R}_\varphi$ *and* $\{i \mapsto F_i\}$ *the section of* $\mathbf{Hom}_{I^\circ}(I^\circ, \mathscr{G}_\varphi^\vee)$ *associated with it. Then for every integer* $q \geq 0$, *we have a functorial canonical isomorphism*

$$\varinjlim_{i \in I^\circ} \mathrm{H}^q((Y_i)_{\text{fét}}, F_i) \xrightarrow{\sim} \mathrm{H}^q(Y_{\text{fét}}, \varinjlim_{i \in I^\circ} (t_i)^*_{\text{fét}}(F_i)). \tag{VI.11.10.1}$$

This follows from VI.11.9 and ([**2**] VI 8.7.7). Note that the conditions required in ([**2**] VI 8.7.1 and 8.7.7) are satisfied by virtue of VI.9.12 and ([**2**] VI 3.3, 5.1, and 5.2).

**VI.11.11.** We denote by

$$\beta : \widetilde{\mathfrak{E}}_f \to Y_{\text{fét}} \tag{VI.11.11.1}$$

and, for every $i \in \mathrm{Ob}(I)$, by

$$\beta_i : \widetilde{\mathfrak{E}}_{f_i} \to (Y_i)_{\text{fét}} \tag{VI.11.11.2}$$

the canonical morphisms (VI.10.6.3). It follows from (VI.10.12.6) and ([**37**] VI 12; cf. also [**1**] 1.1.2) that there exists essentially one Cartesian morphism of fibered topos

$$\beta_\varphi : \mathfrak{F}_\varphi \to \mathscr{G}_\varphi \tag{VI.11.11.3}$$

whose fiber over each $i \in \mathrm{Ob}(I)$ is the morphism $\beta_i$. Moreover, the diagram of morphisms of fibered topos

(VI.11.11.4)
$$\begin{array}{ccc} \widetilde{\mathfrak{E}}_f \times I & \xrightarrow{\beta\times\mathrm{id}} & Y_{\text{fét}} \times I \\ \downarrow & & \downarrow \\ \mathfrak{F}_\varphi & \xrightarrow{\beta_\varphi} & \mathscr{G}_\varphi \end{array}$$

where the vertical arrows are the morphisms (VI.11.2.11) and (VI.11.7.8), is commutative up to canonical isomorphism. We can therefore identify $\beta$ with the morphism deduced from $\beta_\varphi$ by taking the inverse limit in the sense of ([**2**] VI 8.1.4).

**Proposition VI.11.12.** *Let $F$ be a sheaf of* $\mathbf{Top}(\mathfrak{E}_\varphi)$ *and* $\{i \mapsto F_i\}$ *the section of* $\mathbf{Hom}_{I^\circ}(I^\circ, \mathfrak{F}_\varphi^\vee)$ *associated with it by the equivalence of categories* (VI.11.4.1). *Then we have a functorial canonical isomorphism*

(VI.11.12.1)
$$\lim_{\substack{\longrightarrow \\ i\in I^\circ}} (t_i)^*_{\text{fét}}(\beta_{i*}(F_i)) \xrightarrow{\sim} \beta_*(\varpi^* F),$$

*where $\varpi$ is the morphism* (VI.11.4.2) *and $t_i$ is the morphism* (VI.11.7.5).

This follows from VI.11.11 and ([**2**] VI 8.5.9). Note that the conditions required in loc. cit. are satisfied by virtue of VI.10.5, VI.10.10, and ([**2**] VI 3.3 and 5.1).

**Remark VI.11.13.** Let $F$ be an abelian sheaf of $\mathbf{Top}(\mathfrak{E}_\varphi)$, $\{i \mapsto F_i\}$ the object of $\mathbf{Hom}_{I^\circ}(I^\circ, \mathfrak{F}_\varphi^\vee)$ associated with it by the equivalence of categories (VI.11.4.1), and $q$ an integer $\geq 0$. It then follows from VI.10.13 that the diagram

(VI.11.13.1)
$$\begin{array}{ccc} \displaystyle\lim_{\substack{\longrightarrow \\ i\in I^\circ}} \mathrm{H}^q((Y_i)_{\text{fét}}, \beta_{i*}(F_i)) & \longrightarrow & \displaystyle\lim_{\substack{\longrightarrow \\ i\in I^\circ}} \mathrm{H}^q(\widetilde{\mathfrak{E}}_{f_i}, F_i) \\ \downarrow{\scriptstyle u} & & \\ \displaystyle\lim_{\substack{\longrightarrow \\ i\in I^\circ}} \mathrm{H}^q(Y_{\text{fét}}, (t_i)^*_{\text{fét}}(\beta_{i*}(F_i))) & & \downarrow{\scriptstyle w} \\ \downarrow{\scriptstyle v} & & \\ \mathrm{H}^q(Y_{\text{fét}}, \beta_*(\varpi^* F)) & \longrightarrow & \mathrm{H}^q(\widetilde{\mathfrak{E}}_f, \varpi^* F) \end{array}$$

where the horizontal arrows come from the Cartan–Leray spectral sequences, $u$ is the canonical morphism, $v$ is induced by (VI.11.12.1), and $w$ is induced by (VI.11.4.4), is commutative.

**VI.11.14.** We can now prove Proposition VI.10.30. We choose an affine étale neighborhood $X_0$ of $\overline{x}$ in $X$. We denote by $I$ the category of $\overline{x}$-pointed étale $X_0$-schemes that are affine over $X_0$ (cf. [**2**] VIII 3.9 and 4.5), and by $\varphi\colon I \to \mathfrak{M}$ the functor that with each object $U$ of $I$, associates the canonical projection $f_U\colon U_Y \to U$. Then, $\underline{f}$ identifies canonically with the inverse limit of the functor $\varphi$. For every $U \in \mathrm{Ob}(I)$, the topos $\widetilde{E}_{/(U_Y\to U)^a}$ is canonically equivalent to the Faltings topos associated with the morphism $f_U$ by (VI.10.14.5). Consequently, with the notation of this section, for every sheaf $F$ of $\widetilde{E}$, $\{U \mapsto F|(U_Y \to U)^a\}$ is naturally a section of $\mathbf{Hom}_{I^\circ}(I^\circ, \mathfrak{F}_\varphi^\vee)$. It therefore defines a sheaf of $\mathbf{Top}(\mathfrak{E}_\varphi)$ (VI.11.4.1). We have a functorial canonical isomorphism (VI.11.4.4)

(VI.11.14.1)
$$\Phi^*(F) \xrightarrow{\sim} \varpi^*(\{U \mapsto F|(U_Y \to U)^a\}).$$

(i) By ([**2**] VIII 3.9 and 4.5), we have a canonical isomorphism

$$\sigma_*(F)_{\overline{x}} \xrightarrow{\sim} \varinjlim_{U \in I^\circ} \Gamma((U_Y \to U)^a, F). \tag{VI.11.14.2}$$

This induces a functorial isomorphism

$$\sigma_*(F)_{\overline{x}} \xrightarrow{\sim} \Gamma(\widetilde{\underline{E}}, \Phi^*(F)), \tag{VI.11.14.3}$$

by virtue of VI.11.5. The statement follows in view of VI.10.28(i).

(ii) This follows, like (i), from VI.11.6, VI.10.28(ii), and the canonical isomorphism ([**2**] V 5.1(1))

$$\mathrm{R}^i\sigma_*(F)_{\overline{x}} \xrightarrow{\sim} \varinjlim_{U \in I^\circ} \mathrm{H}^i((U_Y \to U)^a, F). \tag{VI.11.14.4}$$

(iii) This immediately follows from the proof of (ii).

**VI.11.15.** We can finally prove Theorem VI.10.40. We choose an affine étale neighborhood $X_0$ of $\overline{x}$ in $X$. We denote by $I$ the category of $\overline{x}$-pointed étale $X_0$-schemes that are affine over $X_0$, and by $\varphi\colon I \to \mathfrak{M}$ the functor that with each object $U$ of $I$, associates the canonical projection $f_U\colon U_Y \to U$. By ([**2**] IV (6.8.4)) and VI.11.10, we have canonical isomorphisms

$$\mathscr{H}^i(F)_{\overline{x}} \xrightarrow{\sim} \varinjlim_{U \in I^\circ} \mathrm{H}^i((U_Y)_{\text{fét}}, F_U) \xrightarrow{\sim} \mathrm{H}^i(\underline{Y}_{\text{fét}}, \varinjlim_{U \in I^\circ} (\xi_Y)^*_{\text{fét}} F_U). \tag{VI.11.15.1}$$

By virtue of VI.10.30(ii), we have an isomorphism

$$\mathrm{R}^i\sigma_*(F^a)_{\overline{x}} \xrightarrow{\sim} \mathrm{H}^i(\underline{Y}_{\text{fét}}, \varphi_{\overline{x}}(F^a)). \tag{VI.11.15.2}$$

On the other hand, it follows from VI.11.13 and from the definitions that the diagram

$$\begin{array}{ccc} \mathscr{H}^i(F)_{\overline{x}} & \xrightarrow{\sim} & \mathrm{H}^i(\underline{Y}_{\text{fét}}, \varinjlim_{U \in I^\circ} (\xi_Y)^*_{\text{fét}} F_U) \\ {\scriptstyle u}\downarrow & & \downarrow{\scriptstyle v} \\ \mathrm{R}^i\sigma_*(F^a)_{\overline{x}} & \xrightarrow{\sim} & \mathrm{H}^i(\underline{Y}_{\text{fét}}, \varphi_{\overline{x}}(F^a)) \end{array} \tag{VI.11.15.3}$$

where the horizontal arrows are the isomorphisms (VI.11.15.1) and (VI.11.15.2), $u$ is the stalk of the morphism (VI.10.39.10) at $\overline{x}$, and $v$ is induced by the isomorphism (VI.10.37.1), is commutative. Consequently, $u$ is an isomorphism, giving the theorem.

# Facsimile : A $p$-adic Simpson correspondence

by Gerd Faltings
Advances in Mathematics **198** (2005), 847–862

Available online at www.sciencedirect.com
SCIENCE DIRECT
ELSEVIER
Advances in Mathematics 198 (2005) 847–862
ADVANCES IN Mathematics
www.elsevier.com/locate/aim

# A $p$-adic Simpson correspondence

Gerd Faltings*

*Max-Planck-Institut für Mathematik, Vivatsgasse 7, 53111 Bonn, Germany*

Received 28 May 2004; accepted 25 May 2005

Communicated by Johan De Jong

Dedicated to M. Artin on the occasion of his 70th birthday

**Abstract**

For curves over a $p$-adic field we construct an equivalence between the category of Higgs-bundles and that of "generalised representations" of the étale fundamental group. The definition of "generalised representations" uses $p$-adic Hodge theory and almost étale coverings, and it includes usual representations which form a full subcategory. The equivalence depends on the choice of an exponential function for the multiplicative group.


*Keywords:* Almost étale extensions; Higgs-bundles; $p$-adic Hodge theory

## 1. Introduction

The purpose of this note is to construct Higgs-bundles associated to representations of the geometric fundamental group of a curve over a $p$-adic field $K$. It thus can be considered a $p$-adic analogue of the results of Simpson and Corlette (see [12]). The functor is fully faithful but it is difficult to characterise its image: namely the resulting Higgs-bundles are semistable of slope zero, but we do not know whether any such Higgs-bundle lies in the image (this is true for line-bundles on curves over $p$-adic local fields). Conversely we can construct for all Higgs-bundles so-called "generalised representations", which form a category containing the usual representations as full subcategory. However, we do not know which of those come from genuine representations.

* Fax: +49 228 402 277.
*E-mail address:* gerd@mpim-bonn.mpg.de.

0001-8708/$ - see front matter © 2005 Elsevier Inc. All rights reserved.
doi:10.1016/j.aim.2005.05.026

Over a local field $K$ and for line-bundles one can check that one gets all line-bundles of degree zero, so one can hope that over local fields all semistable Higgs-bundles of degree zero lie in the image.

Recall that a Higgs-bundle on an algebraic manifold $X$ is a pair $(\mathcal{E}, \theta)$, where $\mathcal{E}$ is a vectorbundle on $X$ and $\theta$ a global section of $\mathcal{E}nd(E) \otimes \Omega_X$ satisfying $\theta \wedge \theta = 0$ (that is in local coordinates the components of $\theta$ commute). We also use variants where $X$ is only logsmooth, or where $\theta$ has coefficients in some Tate-twist. The latter corresponds (I assume) to a factor $2\pi i$ in the classical complex setup.

We should note that our functors depend on certain choices, the most important being that of an exponential function for the multiplicative group. This induces exponential functions on all commutative group schemes over $K$. Another choice is that we have to choose lifts to certain types of dual numbers, and different lifts amount to twists by "Higgs-line-bundles". That is we do not obtain just one functor but a whole family of them, all related by such twists.

The method used in the proofs is the theory of almost étale extensions (see [6], for a more systematic treatment [11]) which was developed by the author for applications in $p$-adic Hodge-theory. We develop a nonabelian Hodge–Tate-theory. What is still missing is an appropriate role for connections and Frobenius, which might result in a more powerful theory generalising Fontaine's ideas.

This work was inspired by the workshop on nonabelian Hodge-theory at MSRI Berkeley, at Easter 2002. We also mention the preprint [9] which uses similar techniques in a different setting.

## 2. Generalised representations

We denote by $V$ a complete discrete valuation-ring with perfect residue-field $k$ of characteristic $p > 0$ and fraction-field $K$ of characteristic 0. $\bar{K}$ is the algebraic closure of $K$ and $\bar{V}$ the integral closure of $V$ in $\bar{K}$. $X$ is a proper $V$-scheme which has toroidal singularities (as explained in [6, Chapter 2, Appendix 1]), for example $X$ could be smooth or have semistable singularities. Further more $D \subset X$ is a divisor which satisfies the conditions in [6]. Especially the generic fibre $X_K$ is smooth and $D_K$ a divisor with normal crossings. As in [6] $X^\circ = X - D$.

We have a topos $\mathcal{X}^\circ$ of sheaves on the situs whose objects consists of finite étale coverings of the generic fibres $U_K^\circ$ of schemes $U \to X$ which are étale over $X$. The normalisation of $\mathcal{O}_U$ in such covers defines a sheaf $\bar{\mathcal{O}}$ on $\mathcal{X}^\circ$. Furthermore if $\mathbb{L}$ is the locally constant sheaf on $\mathcal{X}_K^\circ$ associated to a representation of the fundamental group $\pi_1(X_K^\circ)$ on a finite $V$-module the associated maps

$$H^i(X_{\bar{K}}^\circ, \mathbb{L}) \otimes \bar{V} \to H^i(\mathcal{X}_{\bar{K}}^\circ, \mathbb{L} \otimes \bar{\mathcal{O}})$$

are almost isomorphisms, that is their kernels and cokernels are annihilated by the maximal ideal of $\bar{V}$ (see [6, Chapter 4, Theorem 9] for curves also [4]). This remains true if $\mathbb{L}$ is only a locally constant étale sheaf on the geometric fibre $X_{\bar{K}}^\circ$ because such a sheaf is already defined over a finite extension of $K$. It follows that the functor $\mathbb{L} \otimes \bar{\mathcal{O}}$ is fully faithful as a functor from continuous representations of $\pi_1(X^\circ_{\bar{K}})$ on finitely presented torsion $\bar{V}$-modules, to $\bar{\mathcal{O}}$-modules up to almost isomorphisms. (In the first category almost maps coincide with usual maps). Also the essential image is closed under extensions and deformations. Deformations arise if we have a family of representations over a complete local ring, and consider various base-changes to $\bar{V}/(p^s)$. If one of them lies in the essential image so do all. In the following we restrict to representations on free modules over $\bar{V}/(p^s)$ which correspond to vectorbundles over $\bar{O}/(p^s)$. We call the latter generalised representations.

## 3. The local structure of generalised representations

Now assume given an affine $U = Spec(R) \subset X$ which is small, that is $R$ is étale over a toroidal model. By adjoining roots of characters of the torus we obtain a subextension $R_\infty$ of $\bar{R}$ (the integral closure of $R$ in the maximal étale cover of $U^\circ_K$) which is Galois over $R_1 = R \otimes_V \bar{V}$ with group $\Delta_\infty = \hat{\mathbb{Z}}(1)^d$. We write it as the union of algebras $R_n$ which have themselves toroidal singularities. Furthermore $\bar{R}$ is almost étale over $R_\infty$ ([6, Section 2, Theorem 4]). This implies that each $\bar{R}$-module with a continuous semilinear action of $\Delta = Gal(\bar{R}/R)$ is almost induced from an $R_\infty$-module with $\Delta_\infty$-action. Also we can compute the Galois-cohomology $H^i(\Delta, \bar{R}/(p^s))$ (with $s$ any positive rational number): namely it is almost isomorphic to $H^i(\Delta_\infty, R_\infty/(p^s))$ which in turn is the direct sum of $H^i(\Delta_\infty, R \otimes_V \bar{V}/(p^s))$ and of a direct summand annihilated by $p^{1/(p-1)}$ (see [6, p. 206]). This results from the decomposition of $R_\infty$ into $\Delta_\infty$-eigenspaces where the contributions from nontrivial eigenspaces are annihilated by $\zeta - 1$, $\zeta$ a nontrivial root of unity. Finally the (interesting) first summand is canonically identified with the logarithmic differentials $\tilde{\Omega}^i_{R/V} \otimes_V \bar{V}/(p^s)$. Examples of such (locally defined) generalised representations are given by homomorphisms

$$\Delta \twoheadrightarrow \Delta_\infty \to GL(r, R \otimes_V \bar{V}/(p^s)),$$

we show that many others are close to these.

**Lemma 1.** *Suppose $\alpha > 1/(p-1)$ is a rational number, and $\bar{M} \cong \bar{R}^r/(p^s)$ a generalised representation (it admits a semilinear $\Delta$-operation).*

(i) *Suppose that $\bar{M}$ is trivial modulo $p^{2\alpha}$. Then its reduction modulo $p^{s-\alpha}$ is given by a representation $\Delta_\infty \to GL(r.R \otimes_V \bar{V}/(p^{s-\alpha}))$, and this representation is trivial modulo $p^\alpha$.*

(ii) *Suppose given two representations $\Delta_\infty \to GL(r_i, R \otimes_V \bar{V}/(p^s))$, trivial modulo $p^\alpha$, and an $\bar{R} - \Delta$-linear map between the associated generalised representations. Then its reduction modulo $p^{s-\alpha}$ is given by an $R_1 - \Delta_\infty$-linear map of representations.*

**Proof.** For (ii) we consider the representation on the Hom-space, which we call $M$, and have to show that $\Delta$-invariants in $\bar{M} = M \otimes \bar{R}$ come from $\Delta_\infty$-invariants in $M$. This follows because $H^0(\Delta, \bar{M})$ is almost isomorphic to $H^0(\Delta_\infty, M \otimes R_\infty)$, and the latter decomposes into a direct sum corresponding to the eigenspace decomposition of $R_\infty$. On nontrivial eigenspaces some element of $\Delta_\infty$ operates as the sum of a nontrivial root of unity $\zeta$ and of an endomorphism divisible by $p^\alpha$ and the corresponding contribution is annihilated by $p^\alpha$. For (i) we choose a positive rational $\varepsilon$ with $\alpha > 3\varepsilon + 1/(p-1)$, and show by induction over $n$ that the assertion holds for the representation modulo $p^{2\alpha+n\varepsilon}$. We may assume that $s \geqslant 3\alpha + n\varepsilon$.

For $n = 0$ $\bar{M}$ modulo $p^{2\alpha}$ is by assumption induced from a (trivial) $M$. Assume we have found such an $M$ modulo $p^{2\alpha+n\varepsilon}$, inducing $\bar{M}$. If we try to lift $M$ to a representation modulo $p^{3\alpha+n\varepsilon}$ we encounter an obstruction in $H^2(\Delta_\infty, End(M)/(p^\alpha))$ whose image in $H^2(\Delta, End(\bar{M})/(p^\alpha))$ vanishes because $\bar{M}$ lifts. As the induced map is almost injective (over $R_\infty$ it is a direct summand) the obstruction vanishes after multiplication by $p^\varepsilon$, that is $M$ modulo $p^{2\alpha+(n-1)\varepsilon}$ lifts to a representation modulo $p^{3\alpha+(n-1)\varepsilon}$. Over $\bar{R}$ the induced generalised representation differs from $\bar{M}$ by a class in $H^1(\Delta, End(\bar{M})/(p^\alpha))$. Again this cohomology is almost isomorphic to the direct sum of $H^1(\Delta_\infty, End(M)/(p^\alpha))$ and terms annihilated by $p^{1/(p-1)}$. Hence our class becomes "constant" after multiplication by $p^{\alpha-2\varepsilon}$ and vanishes after modifying the lift of $M$, which is now a lift from coefficients modulo $p^{\alpha+(n+1)\varepsilon}$ to coefficients modulo $p^{2\alpha+(n+1)\varepsilon}$. This finishes the proof. □

**Remarks.** (i) The result extends to $p$-adic representations: this follows from the inductive method of liftings.

(ii) For $\alpha > 1/(p-1)$ the exponential and logarithmic series converge for arguments divisible by $p^\alpha$. Applying the logarithm to the images of generators of $\Delta_\infty$ then defines endomorphisms of $M$ or an element of $End(M) \otimes \tilde{\Omega}_{R/V} \otimes \bar{V}(-1)$, divisible by $p^\alpha$, and with commuting components. We shall see that this element is independent of the choices involved in the construction of $R_\infty$, by defining an inverse functor which associates to such "Higgs-bundles" a generalised representation.

Namely consider Fontaine's rings $A_{inf}(R)$ and $A_{inf}(V)$ associated to $R$ and $V$ (see [8]). $A_{inf}(V)$ surjects onto the $p$-adic completion $\hat{\bar{V}}$ and the kernel is a principal ideal with generator $\xi$. Here we only need the quotient $A_2(V) = A_{inf}(V)/(\xi^2)$, which is an extension of $\hat{\bar{V}}$ by $\hat{\bar{V}}\xi$. The latter contains canonically $\hat{\bar{V}}(1) = p^{1/(p-1)}\hat{\bar{V}}\xi$. Similar assertions (with $\bar{V}$ replaced by $\bar{R}$) hold for $A_2(R)$. Also $A_2(V)$ and $A_2(R)$ have natural toroidal (or logarithmic) structures.

Now we first lift $R \otimes_V \hat{\bar{V}}$ to a log-smooth algebra $\tilde{R}$ over $A_2(V)$. Two such lifts are isomorphic but not canonically isomorphic. Namely the automorphism group of a lift is the group of logarithmic derivations $Hom_R(\tilde{\Omega}_{R/V}, R \otimes_V \hat{\bar{V}}\xi)$. Also an étale map from $Spec(R)$ to a toroidal model defines such an $\tilde{R}$ induced from the toroidal model.

Next we lift $R \subset \hat{\bar{R}}$ to an $A_2(V)$-linear $\tilde{R} \to A_2(R)$. Again two lifts differ by a logarithmic derivation into $\hat{\bar{R}}\xi$. If $M \cong R \otimes_V \bar{V}^r/(p^s)$ denotes a free module together with an endomorphism $\theta \in End(M) \otimes_R \tilde{\Omega}_{R/V}$ such that $\theta \wedge \theta = 0$ (that is $\theta$ defines a Higgs-bundle, or the components of $\theta$ commute), and $\theta$ is divisible by $p^\alpha$ for some $\alpha > 1/(p-1)$, then the two pushforwards of $M$ via different lifts $\tilde{R} \to A_2(R)$ are canonically isomorphic:

Use the fact that $\theta$ behaves like a connection. For example if $R$ is smooth over $V$, $t_i \in R$ form local coordinates with associated derivations $\partial_i = \partial/\partial t_i$, and if the two lifts differ on $t_i$ by $u_i\xi \in \hat{\bar{R}}\xi$, then an isomorphism is given by the Taylor-series

$$\sum_I \theta(\partial)^I(m)/I! \otimes u^I.$$

Here $m \in M$, the sum is over all multi-indices $I = (i_1, \ldots, i_d)$, and there is an obvious divided power structure to explain how to divide by the factorials.

For more general logarithmic coordinates one has to use the logarithmic Taylor-series: for one variable $t$ with derivative $\tilde{\partial} = t\partial/\partial t$ and two lifts differing by $tu\xi$ one obtains

$$\sum \theta(\tilde{\partial})^n(m)/n! \otimes u^n,$$

and the same for several variables. Note that the pushforward is always the same module $M \otimes \hat{\bar{R}}$ but that different lifts of $\tilde{R}$ induce nontrivial automorphisms of this module. Also $\theta$ should be considered as an element of $End(M) \otimes \hat{\bar{R}}\xi^{-1}$. Finally everything extends to $p$-adic Higgs-bundles, by passing to a $p$-adic limit.

Now the Galois-group $\Delta$ acts on lifts $\tilde{R} \to A_2(R)$ and thus semilinearly on $M \otimes \hat{\bar{R}}$. If we choose an étale map from $Spec(R)$ to a toroidal model we obtain a lift $\tilde{R}$ mapping to $A_2(R)$ by extracting $p$-power roots out of characters of the torus (which map to elements of $R$). Also we have an $R_\infty$ and $\Delta$ acts on the preferred lifting via its quotient $\Delta_\infty = \hat{\mathbb{Z}}(1)^d$. That group acts in turn on the $i$th coordinate via its $i$th projection to $\hat{\mathbb{Z}}(1)$ and the inclusion $\hat{\bar{R}}(1) \subset \hat{\bar{R}}\xi$. It follows that $\theta$ induces the previous element of $End(M) \otimes_R \tilde{\Omega}_{R/V}(-1)$ via $\hat{\bar{R}}\xi^{-1} \subset \hat{\bar{R}}(-1)$. As conclusion we get that for a Higgs-bundle modulo $p^s$ with $\theta$ divisible by $p^\alpha$ $(\alpha > 1/(p-1))$ we get an associated generalised representation modulo $p^s$ which will be trivial modulo $p^\beta$ for any $\beta < \alpha + 1/(p-1)$. Conversely, if we start with a generalised representation modulo $(p^s)$ which is trivial modulo $p^\alpha$ we obtain a bundle with endomorphism modulo $(p^{s-\alpha})$, and dividing this endomorphism by $p^{1/(p-1)}$ gives a Higgs-bundle modulo $p^{s-\alpha-1/(p-1)}$. These procedures are inverse up to a loss of exponents $s$ which can be bounded by any $\beta > 2/(p-1)$. Passing to the limit $s \to \infty$ we obtain an equivalence of categories between generalised $p$-adic representations which are trivial modulo $p^\beta$ for some $\beta > 2/(p-1)$, and $p$-adic Higgs-bundles with $\theta$ divisible by $p^\alpha$ for some $\alpha > 1/(p-1)$. We formalise this as follows:

**Definition 2.** A Higgs-module $(M, \theta)$ ($M$ a finitely generated free $R$-module, $\theta \in End_R(M) \otimes_R \Omega_{R/V} \otimes_R \hat{\bar{R}}\xi^{-1}$) is called small if $\theta$ is divisible by $p^\alpha$ for some $\alpha > 1/(p-1)$. A generalised representation $\bar{M}$ is called small if it is trivial modulo $p^{2\alpha}$.

Thus our local theory gives a bijection between small Higgs-modules and small generalised representations.

There also exists a $\mathbb{Q}_p$-theory where we consider continuous $\Delta$-representations on finitely generated projective $\hat{\bar{R}}[1/p]$-modules and projective $\hat{R}_1[1/p]$-modules with an endomorphism $\theta$ with coefficients $\tilde{\Omega}_{X/V}(-1)$, $\theta\wedge\theta=0$. For simplicity we do this only for the case of curves, that is for relative dimension $d=1$.

Suppose first that we are given a finitely generated projective $R\hat{\otimes}_V\bar{V}[1/p]$-module $M$ with an $\tilde{\Omega}_{X/V}(-1)$-valued endomorphism $\theta$. Then the coefficients $\lambda_i$ of the characteristic polynomial of $\theta$ (that is the traces of $\bigwedge^i\theta$ on $\bigwedge^i M$) are elements of $\tilde{\Omega}_{X/V}^{\otimes i}\otimes R\hat{\otimes}_V\bar{V}[1/p]$. If we assume that they are integral and divisible by $p^{2i\alpha}$ for some $\alpha>1/(p-1)$ we can find a finitely generated $R\hat{\otimes}_V\bar{V}$-submodule $M^\circ\subset M$ which is stable under $\theta/p^{2\alpha}$ and which generates $M$. The previous constructions (using local lifts of $R$ to $A_2(V)$ and to $A_2(R)$) then define a representation of $\Delta$ on the $p$-adic completion of $\bar{R}\cdot M$. Less canonically the quotient $\Delta_\infty=\hat{\mathbb{Z}}(1)$ acts by exponentiating $\theta$. Thus we get a functor from small $\mathbb{Q}_p$-Higgs-bundles ("small" is now defined in terms of divisibility of the coefficients $\lambda_i$) to generalised $\mathbb{Q}_p$-representations.

For the converse assume $\Delta$ operates semilinearly on a projective $\hat{\bar{R}}[1/p]$-module $\bar{M}$ and assume it is generated by a finitely generated $\hat{\bar{R}}$-submodule $\bar{M}^\circ$ such that $\bar{M}^\circ$ is generated by elements which are $\Delta$-invariant modulo $p^{2\alpha}\bar{M}^\circ$, for some $\alpha>1/(p-1)$. Replacing $\alpha$ by a slightly smaller $\alpha'$ we may replace $\bar{M}^\circ$ by its invariants under $Gal(\bar{R}/R_\infty)$ and consider the problem with coefficients $\hat{R}_\infty$. Then lifting the action of a generator of $\Delta_\infty=\hat{\mathbb{Z}}(1)$ on generators of $\bar{M}^\circ$ we find a small action of $\Delta_\infty$ on some $\hat{R}_\infty^n$ and a $\Delta_\infty$-linear surjection of this module onto $\bar{M}^\circ$. That is $\bar{M}$ becomes the quotient of the $\mathbb{Q}_p$-object defined by an integral small generalised representation. Applying the same reasoning to the kernel we get a resolution by integral small generalised representations. The cokernel of the induced map on associated Higgs-bundles then defines the inverse functor.

Both functors are fully faithful: using resolutions and duals we reduce to the previous results for integral objects. Thus:

**Theorem 3.** *The construction above defines, for small toroidal affines, an equivalence of categories between small generalised representations and small Higgs-bundles, or generalised $\mathbb{Q}_p$-representations and Higgs-bundles (both assumed small) on the generic fibre.*

We can use these methods to compare cohomologies, first in a locals setting but in such a way that it will globalise later. Suppose $M$ is a finitely generated projective $\hat{R}_1$-module with a Higgs-field $\theta$ such that all invariants $\lambda_i$ are divisible by $p^{2i\alpha}$, $\alpha>1/(p-1)$. Then we can find a sublattice $M^0\subset M$ with $\theta(M^0)\subseteq p^{2\alpha}M^0\otimes\tilde{\Omega}_{R/V}\xi^{-1}$, and a representation of $\Delta$ on $M^0$ which induces a generalised representation $\bar{M}$. The cohomology-groups $H^i(\Delta,\bar{M})$ can be, up to almost isomorphism, computed by reduction to $R_\infty$ and the action of $\Delta_\infty=\hat{\mathbb{Z}}^d(1)$. Decomposing $R_\infty$ into eigenspaces we see that the contribution of nontrivial eigenspaces is annihilated by $p^{1/(p-1)}$, while the trivial eigenspace has the same cohomology as the Koszul-complex $\bigwedge^*(\tilde{\Omega}_{R/V} \otimes \hat{R}_0\xi^{-1}) \otimes M^0$, with exterior multiplication by $\theta$ as differential. A map of complexes inducing this isomorphism can be constructed as follows:

Consider the symmetric algebra $\mathcal{S} = \bigoplus_{n \geqslant 0} S^n(\tilde{\Omega}_{R/V} \otimes \hat{R}_0\xi^{-1})$. It has a Higgs-field $\theta = -\sum \partial_i \otimes \omega_i$, where the $\omega_i$ are a local basis of $\tilde{\Omega}_{R/V}$ and the $\partial_i$ the dual derivations (of degree $-1$) on the symmetric algebra. The Koszul-complex of $-\theta_{\mathcal{S}}$ is a resolution of $R_0$. Also the Higgs-field $\theta_{\mathcal{S}}$ has divided powers $\theta_{\mathcal{S}}^n/n!$, and the associated exponential series is finite when applied to elements of $\mathcal{S}$.

The completed tensorproduct $M^0 \hat{\otimes} \hat{\mathcal{S}}$ also has a Higgs-field and by the usual procedure we obtain $\Delta$-representations on the $p$-adic completion of the tensorproduct with $\bar{R}$. This gives a resolution of $\bar{M}$ in the category of continuous $\Delta$-representations, and as for any such resolution its $\Delta$-invariants map in the derived category to the complex representing cohomology (in fact the analogue also holds for all coefficients $M^0/p^s M^0$, thus one can avoid continuous cohomology with $\mathbb{Q}_p$-coefficients). Finally the Koszul-complex for $M^0$ maps into the complex of invariants, by sending $m$ to the sum $\sum_{n \geqslant 0} \theta^n(m)/n!$ in the $p$-adic completion of $M^0 \otimes \mathcal{S}$.

For general $\mathbb{Q}_p$-representations we denote $R_0 = R \otimes_V \bar{V}$ and by $R_n$ the result of adjoining all roots of order $n!$ of the toroidal coordinates, so that $R_0 \subset R_n \subset R_\infty$, and the $R_n$ are still toroidal over $\bar{V}$. We now pass to an $R_n$ where our representation becomes small, thus obtaining invariants $\lambda_i$ in the $p$-adic completion of $\tilde{\Omega}_{R_n/V}^{\otimes i}\xi^{-i}$. These are invariant under $Gal(R_n/R_0)$ and thus define elements in $\hat{\tilde{\Omega}}_{R/V}^{\otimes i} \otimes_R R_0[1/p]\xi^{-i}$. We claim that they are independent of the choice of the $R_n$:

**Lemma 4.** *The invariants $\lambda_i$ do not depend on the choice of $R_\infty$ and commute with basechange.*

**Proof.** We use that we have found a lift of $R$ to $A_2(V)$ and to $A_2(R)$ such that for each $\delta \in \Delta$ and $a$ in the lift we have $\delta(a) - a \in \hat{R}_0\xi$, and for $\delta$ in a subgroup $\Delta_n$ of finite index this is divisible by $p^{n+\alpha}$. Assume more generally that we have a finite subextension $R_0 \subset S \subset \bar{R}$ such that $\delta(a) - a \in p^{n+\alpha}\hat{S}\xi$ if $\delta$ lies in the subgroup $\Delta_S$ fixing $S$, and an element $\theta \in p^{-n}\tilde{\Omega}_{R/V} \otimes End(S^r)\xi^{-1}$. Then we can define an action of $\Delta_S$ on $\hat{S}^r$ and on $\hat{\bar{R}}^r$ by the Taylor-series above, which converges because of the assumptions on divisibility. We also assume that there are subgroups $\Delta_{S,m} \subset \Delta$, of finite index, such that $\delta(a) - a$ is divisible by $p^{m+n+\alpha}$ if $\delta \in \Delta_{S,m}$.

Now assume that two such data define representations which become isomorphic after tensoring with $\mathbb{Q}_p$. We then claim that the two $\theta$'s have the same coefficients $\lambda_i$ in their characteristic polynomials. To show this we are allowed to enlarge the $S$'s and may assume that they coincide. The two lifts of $R$ to $A_2(R)$ then differ by a logarithmic derivation $R \to \hat{\bar{R}}\xi$. Its reduction modulo $p^{m+n+\alpha}$ is invariant under $\Delta_{S,m}$. As the discriminant of the compositum of $S$ and $R_s$ over $R_s$ (our original sequence) divides $p$ for big $s$ we can pass to this compositum and then have that after multiplication by $p$ our $\Delta_{S,m}$-invariants are $S$-linear combinations of invariants under

$Gal(\bar{R}/R_s)$. However, from the explicit decomposition of $R_\infty$ into $\Delta_\infty$-eigenspaces it follows that such invariants modulo $p^{m+n+\alpha}$ are almost sums of elements of $R_s\xi$ and elements annihilated by $p^{1/(p-1)}$. Thus multiplying again by (say) $p^2$ we see that our invariant lifts. That means after replacing $\Delta_S$ by a smaller subgroup we may assume that the two lifts differ by the sum of a derivation with values in $\hat{S}\xi$ and of a derivation with values in $p^{n+\alpha}\hat{\bar{R}}\xi$. Changing one of the lifts we get rid of the first term.

But then previous arguments imply that the two lifts give isomorphic representations, so we may assume that the two lifts coincide and only the $\theta$'s might differ. But then one concludes that an isomorphism between the two representations has coefficients in $\hat{S}\otimes\mathbb{Q}_p$ and conjugates the two $\theta$'s. Namely to show that the coefficients are invariant under $\Delta_S$ localise and reduce to $R$ a discrete valuation-ring, then replace $R$ by $S$. This finally shows the claim. □

For a generalised $\mathbb{Q}_p$-representation over $R$ we can pass to some $R_n$ where it is given by $\theta$ acting on a projective $\hat{R}_n[1/p]$-module, then add a direct summand (with trivial $\theta$) to make it free of rank $r$. Thus our reasoning applies also to such modules and shows that the $\lambda_i$ are canonical.

Finally the same type of reasoning shows that the definition of $\lambda_i$'s commutes with basechange: assume $R \to R'$ is a logarithmic map, and we are given a generalised $\mathbb{Q}_p$-representation over $R$ which induces such a representation over $R'$. Then we can choose $S$ and $S'$ as above, and assume that the map extends to $S \to S'$. Next we compare the two maps from an $A_2(V)$-lift of $R$ to $A_2(R')$ obtained by either mapping to $A_2(R) \to A_2(R')$, or first to a lift of $R'$ and then to $A_2(R')$. For any element $a$ of the lift we obtain two 1-cocycles on $\Delta'_{S'}$ with values in $\hat{S}'\xi$ which are divisible by $p^n$, and whose difference is a boundary. As before we then can assume that it is the boundary of an element divisible by $p^n$.

The restriction to $\Delta'_{S'}$ of the pushforward is defined by the first cocycle, and a Higgs-field $\theta \in End(S^r)\otimes\tilde{\Omega}_{R/V}\xi^{-1}$, and (after basechange $S \to S'$) isomorphic to the representation given by the second cocycle. However, this second cocyle is defined for all lifts of elements of $R'$, not just of $R$, and we get the representation defined by the pushforward of $\theta$. But this is our claim.

## 4. Globalisation

Again $X$ is proper over $Spec(V)$, with toroidal singularities. A small generalised representation of its fundamental group is defined as a compatible system of small generalised representations on a covering of $X$ by small affines. For example, it can be defined by a representation of $\pi_1(X^\circ \otimes_V \bar{K})$ on $\hat{\bar{V}}^r$ which is trivial module $p^{2\alpha}$, $\alpha > 1/(p-1)$. However, a representation may induce locally small generalised representations without being trivial modulo some $p$-power. A small Higgs-bundle on $X$ is a vector-bundle $\mathcal{E}$ on $X$ together with an endomorphism $\theta \in End(\mathcal{E})\otimes\tilde{\Omega}_{X/V}\otimes\hat{\bar{V}}\xi^{-1}$ which is divisible by $p^\alpha$ for some $\alpha > 1/(p-1)$, such that $\theta\wedge\theta = 0$. Also $\hat{X}$ and

$\hat{\mathcal{E}}$ denote $p$-adic completions over $\hat{\bar{V}}$, that is $\hat{X}$ is a formal scheme and $\hat{\mathcal{E}}$ a formal vector-bundle on it. For any open $\hat{U} \subset \hat{X}$ and any formal vectorfield $\vartheta \in \Gamma(\hat{U}, \tilde{T}_{X/V})\xi$ we obtain an automorphism $\exp(\theta(\vartheta))$ of $\hat{\mathcal{E}}$ on $\hat{U}$. Hence for any open covering $\hat{U}_i$ of $\hat{X}$ and any 1-Cech-cocycle $\vartheta_{ij} \in \Gamma(\hat{U}_i \cap \hat{U}_j, \tilde{T}_{X/V})\xi$ we get a functor $\exp(\theta(\vartheta_{ij}))$ by formal Higgs-bundles by twisting $\hat{\mathcal{E}}$ by the cocycle $\exp(\theta(\vartheta_{ij}))$. These functors are multiplicative in $\{\vartheta_{ij}\}$, up to canonical isomorphism. Furthermore if $\vartheta_{ij} = \vartheta_i - \vartheta_j$ is a boundary the $\exp(\theta(\vartheta_i))$ define an isomorphism between the identity and the functor given by $\{\vartheta_{ij}\}$.

If we choose a covering $U_i$ of $X$ by small affines we have over each $U_i$ equivalences of categories between small Higgs-bundles and small generalised representations. However, these depend on a choice of lifting $U_i$ to $A_2(V)$ since automorphisms of such a lift act on small Higgs-bundles via our $\exp(\theta(\vartheta))$-construction. Only if we choose a lift of $X$ to $A_2(V)$ we get a global equivalence of categories which depends on the lift: for two lifts we can choose local isomorphisms over each $U_i$. On the overlaps they differ by a vectorfield $\vartheta_{ij}$, and $\exp(\theta(\vartheta_{ij}))$ describes the difference between the associated equivalences of categories. Furthermore change of local isomorphisms results in modifying $\{\vartheta_{ij}\}$ by a coboundary. We also remark that after inverting $p$ $K$ lifts to $A_2(V)$, and thus if $X$ is defined over $K$ we can lift it if we invert $p$. We then could extend the theory to any such $X$ if we only consider Higgs-bundles with $\theta$ divisible by a sufficiently high $p$-power.

Similarly for functoriality: if $f : X \to Y$ denotes a (logarithmic) map we have natural pullbacks $f^*$ for Higgs-bundles as well as for generalised representations. If we lift $X$ and $Y$ to $A_2(V)$ and choose local lifts $f_i$ of $f$ these will differ on the overlaps by vectorfields $\vartheta_{ij} \in f^*(\tilde{T}_{Y/S})$ which act on the pullbacks $f^*(\mathcal{F}, \theta)$ of small Higgs-bundles on $Y$. Then the two functors $f^*$ differ by twisting by the corresponding cocycle.

All in all we get:

**Theorem 5.** *This procedure defines (for liftable schemes) an equivalence of categories between small generalised representations and small Higgs-bundles. This also extends to the $\mathbb{Q}_p$-theory, and there the functor induces an isomorphism on cohomology (we have shown this locally, but the construction gives a global map).*

We cannot drop the adjective "small" as is demonstrated by explicit calculations for rank one bundles on curves:

Namely suppose $X$ is a smooth proper (geometrically connected) curve over $V$. Then the abelianised fundamental group of $X \otimes_V \bar{K}$ is free and isomorphic to $\hat{\mathbb{Z}}^{2g}$ where $g$ denotes the genus of $X$. Its continuous $\hat{\bar{K}}$-representations are parametrised by the images of the generators which are elements of $\hat{\bar{V}}^*$ whose reduction modulo the maximal ideal is torsion, that is lie in the algebraic closure of $\mathbb{F}_p$. This is always the case if $k$ is finite, that is if $K$ is a local field. The logarithm maps such elements surjectively onto $\hat{\bar{K}}$, with kernel the roots of unity $\mu(\bar{K})$, and we obtain an exact sequence

$$0 \to Hom(\pi_1(X_{\bar{K}}), \mu(\bar{K})) \to Hom(\pi_1(X_{\bar{K}}), \hat{\bar{K}}^*) \to Hom(\pi_1(X_{\bar{K}}), \hat{\bar{K}}) \to 0.$$

The first term coincides with the torsion-points in the Jacobian $J$ of $X$, and the third with the homomorphisms from the Tate-module $T_p(J)$ into $\hat{\bar{K}}$. By Hodge–Tate theory [3] the latter coincides with the direct sum $Lie(J) \otimes_V \hat{\bar{K}} \oplus \Gamma(X, \Omega_{X/V}) \otimes_V \hat{\bar{K}}(-1)$. On the other hand from the logarithm-sequence for the Jacobian we obtain an exact sequence

$$0 \to J(\bar{K})_{tors} \to J(\hat{\bar{K}})' \to Lie(J) \otimes_V \hat{\bar{K}} \to 0,$$

where the middle term is the preimage of the torsion in $J(\bar{k})$. This exact sequence turns out to be the restriction of the first sequence to the direct summand $Lie(J) \otimes_V \hat{\bar{K}}$ (proofs will follow from the considerations below. They amount to functoriality of the logarithm map for the homomorphism $J[p^\infty] \to \mu_{p^\infty}$ defined by an element of the Tate-module of $J$, or better of its dual which, however, coincides with $J$).

Now assume that $K$ is a local field, that is its residue-field is finite. If we choose a splitting of this exact sequence over the second summand $\Gamma(X, \Omega_{X/V}) \otimes_V \hat{\bar{K}}(-1)$ we obtain an isomorphism between one-dimensional representations of $\pi_1(X_{\bar{K}})$ and the product of $J(\hat{\bar{K}})$ and $\Gamma(X, \Omega_{X/V}) \otimes_V \hat{\bar{K}}(-1)$, that is a bijection (in rank one) between representations and Higgs-bundles. Unfortunately there is no canonical splitting because the first exact sequence realises the universal extension of $T_p(J)[1/p]$ by $\hat{\bar{K}} - Gal(\bar{K}/K)$-modules, so it cannot be trivial on a direct summand (see [10]). That is there exists no continuous $Gal(\bar{K}/K)$-invariant bijection between representations and Higgs-bundles. Nevertheless we can construct such bijections if we choose an exponential map for $\hat{\bar{K}}$:

Namely the logarithm defines an exact sequence

$$0 \to \mu(\bar{K}) \to \mathbb{G}_m(\hat{\bar{V}})' \to \hat{\bar{K}} \to 0,$$

where $\mathbb{G}_m(\hat{\bar{V}})'$ denotes the elements of $\hat{\bar{V}}^*$ which are torsion modulo the maximal ideal. An exponential map is a continuous right inverse of the logarithm, or a continuous splitting of this extension. It induces such a splitting for all commutative algebraic groups over $K$:

Suppose $G$ is such an algebraic group. Then the exponential and logarithm induce isomorphisms between sufficiently small open subgroups of $G(\hat{\bar{K}})$ and its Lie-algebra $\mathfrak{g} \otimes_K \hat{\bar{K}}$. If $G(\hat{\bar{K}})' \subseteq G(\hat{\bar{K}})$ denotes the subgroup of elements $g$ for which some multiple $g^n$ ($n \in \mathbb{N}$ nonzero) lies in this neighbourhood, we can extend the logarithm to $G(\hat{\bar{K}})'$ and obtain an exact sequence

$$0 \to G(\hat{\bar{K}})_{tors} \to G(\hat{\bar{K}})' \to \mathfrak{g} \otimes_K \hat{\bar{K}}.$$

We show that the last map is surjective and construct a right inverse, as follows:

First of all the logarithm is an isomorphism for additive or unipotent groups. We thus may divide the connected component of the identity of $G$ by its maximal unipotent subgroup and may assume that $G$ is semiabelian, using Rosenlicht's theorem that $G$ is an extension of an abelian variety by a linear group. Namely this operation does not change the torsion-points and thus allows us to lift splittings. Then passing to a finite extension of $K$ we may assume that $G$ has split semistable reduction and is over $V$ the quotient of a semiabelian $\tilde{G}$ by a group of periods $\iota : Y \to \tilde{G}(K)$. We may replace $G$ by $\tilde{G}$ and may assume that $G$ is semiabelian over $V$, that is $G$ is an extension

$$0 \to T \to G \to A \to 0,$$

with $A$ an abelian variety and $T$ a split torus. If $T_p(G)$ denotes its Tate-module, then any element $\rho \in Hom(T_p(G), \mathbb{Z}_p(1))$ in the Tate-module of its dual group defines over $\hat{\bar{V}}$ a homomorphism of $p$-divisible groups $G[p^\infty] \to \mu_{p^\infty}$ and thus also a map from the kernel of the reduction map $G(\hat{\bar{K}}) \to G(\bar{k})$ to the 1-units in $\hat{\bar{V}}^*$. We thus obtain a transformation from the kernel of the reduction-map to the kernel of the reduction map on $T_p(G)(-1) \otimes \hat{\bar{V}}^*$, and this induces an isomorphism on $p$-torsion. On the level of Lie-algebras we get a map from $\mathfrak{g} \otimes_K \hat{\bar{K}}$ to $T_p(G) \otimes \hat{\bar{K}}(-1)$. These maps are compatible via the logarithm, by naturality.

Now by Hodge–Tate theory $T_p(G) \otimes \hat{\bar{K}}(-1)$ is isomorphic to $\mathfrak{g} \otimes_K \hat{\bar{K}} \oplus Lie(A^t) \otimes_K \hat{\bar{K}}(-1)$. Strictly speaking this has been shown as it stands only if $G$ is defined over a finite extension of $V$, not if $G$ is defined over $\hat{\bar{V}}$. In general ($G = A$ only defined over $\hat{\bar{K}}$) we obtain a subspace $\mathfrak{g} \otimes_K \hat{\bar{K}}$ with quotient $Lie(A^t) \otimes_K \hat{\bar{K}}(-1)$, without a canonical lift. This follows from the proof in [3] by applying it to a universal family of abelian schemes $A$, and then specialising. By Galois-invariance (or for many other reasons) our map lands in the first factor and it is known to be an isomorphism onto the first direct summand. As the logarithm-map is surjective for $G$ (it is $p$-divisible) and as we have an isomorphism on torsion-groups we get all in all an isomorphism from the kernel of reduction on $G(\hat{\bar{K}})$ onto the preimage of $\mathfrak{g} \otimes_K \hat{\bar{K}}$ in the kernel of reduction of $T_p(G)(-1) \otimes \hat{\bar{V}}^*$. Thus finally the exponential map on the multiplicative group induces such a map on $G$. These maps are functorial.

We also remark that we obtain a certain uniformity in $G$, as follows: call an element of the Lie-algebra $\mathfrak{g}$ small if there exists a rigid analytic homomorphism from the closed rigid unit disk (with addition) into $G$, with derivative at the origin the given element. Such a map is unique, has values in $G(\hat{\bar{K}})'$, and one checks that for the multiplicative group $\mathbb{G}_m$ an element is small if and only if it has valuation $> 1/(p-1)$ (the exponential series has to converge). It then follows that there exists an integer $n$ such that the exponential map is rigid analytic on any product of $p^n$ with a small element: reduce to $G$ semiabelian over $\hat{\bar{V}}$. Then an analytic homomorphism from the closed unit disk into $G$ coincides with a map from the $p$-adic completion of the additive group into $G$, is trivial on the special fibre, the map from $G(\hat{\bar{V}})'$ to $T_p(G)(-1) \otimes \hat{\bar{V}}^*$ is analytic, and everything reduces to the case of the multiplicative group.

As an application we can exponentiate cohomology-classes to line-bundles: consider a semistable proper curve $X$ over $V$ and a sequence of elements $\alpha_i \in \Gamma(X, \tilde{\Omega}^{\otimes i}_{X/V}) \otimes \hat{\bar{K}}(-i)$, for $1 \leqslant i \leqslant r$. For example the $\alpha_i$ could be the coefficients of $det(Tid - \theta)$ for a Higgs-field $\theta$, on a vectorbundle of rank $r$ on the generic fibre of $X$. Associated to these invariants is an algebra $\mathcal{A}$ over $X_{\hat{\bar{K}}}$, commutative and locally free of rank $r$. Namely it is the quotient of the symmetric algebra in the logarithmic tangent-bundle $\tilde{\mathcal{T}}_{X/V} \otimes \hat{\bar{K}}(1)$, under the monic polynomial with coefficients $\lambda_i$. As a module $\mathcal{A}$ is the direct sum of powers $\tilde{\mathcal{T}}^{\otimes i}_{X/V} \otimes \hat{\bar{K}}(i)$, for $0 \leqslant i \leqslant r-1$. For example if the $\lambda_i$ come from a Higgs-field on $\mathcal{E}$ then $\mathcal{E}_{\hat{\bar{K}}}$ becomes naturally an $\mathcal{A}$-module.

Now the Picard-group $H^1(X_{\hat{\bar{K}}}, \mathcal{A}^*)$ consists of the $\hat{\bar{K}}$-points of a smooth algebraic group over $\hat{\bar{K}}$, with tangent-space $H^1(X_{\hat{\bar{K}}}, \mathcal{A})$, so we can exponentiate classes in the latter to get line-bundles. To avoid the complications due to automorphisms we first consider line-bundles trivialised at finitely many $\hat{\bar{V}}$-points of $X$, up to isomorphism. If the number of points is positive and $> (r-1)(2-2g)$ then $\mathcal{A}$ has no global sections vanishing at all these points, so these objects have no automorphisms, and the moduli-problem is representable by a smooth locally algebraic group over $\hat{\bar{K}}$ with tangent space the first cohomology of $X_{\hat{\bar{K}}}$ with coefficients in the local sections of $\mathcal{A}$ which vanish in the prescribed points. Let $G$ denote the connected component of the identity of this moduli-space.

The product of the fibres of $\mathcal{A}^*$ maps to $G$ (via change of trivialisations) and has as image a closed subgroup $H$ (the kernel is given by the global sections of $\mathcal{A}^*$). If $\mathfrak{g}/\mathfrak{h} = H^1(X_{\hat{\bar{K}}}, \mathcal{A})$ denotes the Lie-algebra of the quotient $G/H$, choose a section $\mathfrak{g}/\mathfrak{h} \to \mathfrak{g}$ of the projection, and the exponential map for $G$, to define a family of $\mathcal{A}$-line-bundles on $X_{\hat{\bar{K}}}$ parametrised by $\mathfrak{g}/\mathfrak{h}$. This family is $p$-adically continuous and additive in the sense that the line-bundle corresponding to a sum is canonically isomorphic to the tensorproduct of the line-bundles corresponding to the factors, and these isomorphisms are compatible with associativity.

Finally the difference between two sections of $\mathfrak{g} \to \mathfrak{g}/\mathfrak{h}$ lifts to a linear map from $\mathfrak{g}$ into the product of the fibres of $\mathcal{A}$ in the prescribed points, and then exponentiates to a coherent family of isomorphisms between the associated line-bundles. Finally this coherent family of isomorphisms is canonical up to a multiplicative family of invertible global sections of $\mathcal{A}$. Also for $p$-adically small elements of $\mathfrak{g}/\mathfrak{h}$ the exponential coincides with the map defined by the exponential for some model for $\mathcal{A}$ on the $p$-adic formal scheme defined by $X$, by continuity.

This construction allows us to construct a twisted pullback for Higgs-bundles on $X_{\hat{\bar{K}}}$: namely for a map $f : X \to Y$ and a Higgs-bundle $(\mathcal{E}, \theta_Y)$ on $Y_{\hat{\bar{K}}}$ the usual pullback $f^*(\mathcal{E}), f^*(\theta_Y)$ admits an action of the algebra $f^*(\mathcal{A}_Y)$ ($\mathcal{A}_Y$ is defined like the previous $\mathcal{A}$, with $X$ replaced by $Y$). Also the obstruction to lift $f$ to $A_2(V)$ is an element of $H^1(X, f^*(\tilde{\mathcal{T}}_{Y/V}))$ which exponentiates to an $f^*(\mathcal{A}_Y)$-line-bundle. The twisted pullback $f^\circ(\mathcal{E}, \theta_Y)$ is defined as the tensorproduct of $f^*(\mathcal{E}, \theta_Y)$ with this line-bundle. This functor depends on certain choices (trivialisations in suitable points) but different families give isomorphic functors. Also for compositions $(fg)^\circ$ is naturally isomorphic to $g^\circ f^\circ$, and we leave it to the reader to check the various compatibilities involved (later we consider actions of a finite group on $X$ where compatibilities become easy to check because one can make all arbitrary choices invariant under this group). Finally there exists an integer $n \geqslant 2$, independent of $X$, such that if the $\lambda_i$ extend to a semistable model of $X$ over $\hat{\bar{V}}$ (so $\mathcal{A}$ has an extension to this model) and if a class in $H^1(X, \tilde{T}_{X/V}) \otimes_V \hat{\bar{V}}(1)$ is divisible by $p^n$, then our exponential coincides with the usual exponential of the class on the formal scheme defined by $X$.

Now we come to our main result.

**Theorem 6.** *There exists an equivalence of categories between Higgs-bundles and generalised representations, if we allow* $\hat{\bar{K}}$*-coefficients.*

**Proof.** For this we use the equivalence of categories for small representations and a descent argument. The latter uses the following construction of coverings of $X$.

After passing to an extension of $V$ choose a $V$-rational point in $X$ and use it to embed the generic fibre $X_K$ into its Jacobian $J$. Then multiplication by $p^n$ on $J$ induces a covering $X_{n,K}$ of $X_K$ which has (after finite extension of $K$) a semistable model $X_n$ mapping to $X$. If $J_{n,K}$ denotes the Jacobian of $X_{n,K}$ the induced map $J_{n,K} \to J_K$ is divisible by $p^n$, and so is by duality the pullback $J_K \to J_{n,K}$. Next consider the induced map on differentials on Néron-models which is identified with the pullback $\Gamma(X, \tilde{\Omega}_{X/V}) \to \Gamma(X_n, \tilde{\Omega}_{X_n/V})$. Thus the pullback on differentials is divisible by $p^n$ on global sections. As global sections generate $\tilde{\Omega}_{X/V}$ on the generic fibre there exists an integer $n_0$ such that the subsheaf of $\tilde{\Omega}_{X/V}$ generated by global sections contains $p^{n_0}\tilde{\Omega}_{X/V}$, and so the pullback-map on differentials is divisible by $p^{n-n_0}$. Thus we can make generalised representations or Higgs-bundles small by pullback along such a covering.

Now if we have a generalised representation on $X$ we choose a finite Galois covering $Y \to X$ such that it becomes small on $Y$, that is it is given by a small Higgs-bundle there. The covering group acts via the $f^\circ$-action on this Higgs-bundle, and via the usual action if we twist by the obstruction to lift $Y \to X$ to $A_2(V)$. By descent we get a Higgs-bundle on $X$. Conversely given such a Higgs-bundle on $X$ we choose $Y$ such that it becomes small on $Y$, then twist the pullback by the inverse of the obstruction-class, and get a generalised representation on $Y$ which descends to $X$. □

Things are slightly more complicated if we allow open curves, that is the divisor at infinity $D$ is nonempty: namely then the covering may ramify over $D$, and an equivariant action of the covering group does not always allow descent. We need that the (finite) stabiliser of a point acts trivially on the fibre in this point. In the $\mathcal{A}$-linebundles used for the twisting, the action of inertia is given by the exponential of its derivative (the inertia-group is $\hat{\mathbb{Z}}(1)$), and to get descent we need the same for the action of inertia on the fibre of our generalised representation. The same reasoning gives the converse, so we get an equivalence between Higgs-bundles and representations satisfying the following condition on inertia.

The action of the inertia group $\hat{\mathbb{Z}}(1)$ on the fibre at a point in the boundary has a derivative $Res(\theta)$ (it is the residue of the associated $\theta$). It then must be equal to the exponential (in the multiplicative group of the algebra generated by $Res(\theta)$) of its derivative. Especially the action factors over $\mathbb{Z}_p(1)$.

Our functors induce isomorphisms between the $\mathcal{X}^0$-cohomology of generalised representations and Higgs-cohomology of Higgs-bundles. This has been shown for small objects. The general result follows by descent once we show that Higgs-cohomology is invariant under our twists by $\mathcal{A}[1/p]$-line-bundles. We only use that these line-bundles come from twists by vectorfields, and that they are in the image of the exponential map. The latter is used as follows.

$\mathcal{A}$ is a quotient of the symmetric algebra $S(\tilde{\mathcal{T}})$ under a monic polynomial (naturally an element $P \in S(\tilde{\mathcal{T}}) \otimes \tilde{\Omega}^{\otimes r}$) with coefficients $\lambda_i$. Denote by $\mathcal{B}$ the algebra (of rank $r+1$) defined by multiplying this polynomial by $T$. Then our line-bundles lift to compatible systems of $\mathcal{B}[1/p]$-line-bundles, trivialised on $\mathcal{A}/(\theta) = \mathcal{O}_X$. We claim that the Koszul-complex $\mathcal{E} \to \mathcal{E} \otimes \tilde{\Omega}_{X/V}$ is up to quasi-isomorphism invariant under tensoring $\mathcal{E}$ with $\mathcal{B}$-line-bundles $\mathcal{M}$ which are trivialised modulo $\theta$:

**Proposition 1.** *For two $\mathcal{B}$-line-bundles $\mathcal{M}_1$ and $\mathcal{M}_2$ any isomorphism $\mathcal{M}_1/(\theta) \cong \mathcal{M}_2/(\theta)$ induces a quasi-isomorphism on Koszul-complexes for $\mathcal{E} \otimes_{\mathcal{B}} \mathcal{M}_i$.*

**Proof.** Denote by $\mathcal{N} \subset \mathcal{M}_1 \oplus \mathcal{M}_2$ the elements which lie modulo $\theta$ in the graph of the isomorphism. Then multiplication by $\theta$ induces a map from $\mathcal{M}_1 \oplus \mathcal{M}_2$ to $\mathcal{N} \otimes \tilde{\Omega}_{X/V}$, and after tensoring (over $\mathcal{B}$) with $\mathcal{A}$ the resulting complex becomes locally free over $\mathcal{A}$, and the two projections induce quasi-isomorphisms with the Koszul-complexes of the $\mathcal{M}_i$. Tensoring with $\mathcal{E}$ gives the desired zigzag of quasi-isomorphisms. Also the construction is transitive if we have three bundles $\mathcal{M}_i$ and compatible isomorphisms between their reductions modulo $\theta$ (do the construction above with three summands). □

## 5. Examples and open questions

A natural question to ask is which Higgs-bundles come from actual representations of the fundamental group. If this representation is trivial modulo a sufficiently high $p$-power the associated Higgs-bundle has a model which is trivial modulo some positive $p$-power, so its restriction to the generic fibre is semistable (even without the Higgs-field) of slope zero. By descent it follows that in general the associated Higgs-bundle is semistable of slope zero (now we need the Higgs-field to construct the twisted pullback $f^\circ$). Furthermore if a Higgs-bundle (on the generic fibre) has an integral model which is the trivial bundle modulo some $p^\alpha$ for $\alpha > 0$ we can (by pullback) assume that a Higgs-field is divisible by a high $p$-power. It then follows by deformation-theory that it is associated to an $\pi_1$-representation: for some small $\varepsilon$ (to take care of the "almost") we first get this modulo $p^{2\alpha-\varepsilon}$, by considering extensions of trivial representations or generalised representations. Then continue. This generalises a result in [1,2].

Also if for a dominant (i.e. nonconstant) map $Y_K \to X_K$ the pullback to $Y$ of a generalised representation comes from a real representations, then the same holds already on $X$: we may assume that the cover is Galois, with group $G$. Then $G$ operates on the locally constant system of $Y$, and the inertia acts trivially on the fibres at fixed-points (by checking for the generalised representations). Thus the system descends.

This suffices to show that all rank-one Higgs-bundles $(L, \theta)$ with $\mathcal{L} \in J(\hat{\bar{K}})'$ come from $\pi_1$-representations: if $\mathcal{L}$ is torsion it can be trivialised by pullback. Thus we reduce to $\mathcal{L}$ which is trivial on the special fibre.

The resulting homomorphism

$$J(\hat{\bar{K}})' \times \Gamma(X, \Omega_X) \otimes_K \hat{\bar{K}}(-1) \to Hom(\pi_1(X_{\bar{K}}), \hat{\bar{V}}^*)$$

coincides on the first factor with the previously defined map: namely this is true on the torsion, and this determines the map by $Gal(\bar{K}/K)$-invariance. If $X$ is only defined over $\hat{\bar{K}}$ the same follows by an approximation argument. For the second factor one can show that there the map is the exponential of a $\hat{\bar{K}}$-linear map into the Lie-algebra of the group of representations, that is into $H^1(X, \mathcal{O}_X) \otimes_K \hat{\bar{K}} \oplus \Gamma(X, \Omega_X) \otimes_K \hat{\bar{K}}(-1)$. Furthermore the second component of our map is the identity, while the first one depends on the choice of lifting $X$ to $A_2(V)$.

**Example.** Sometimes $\mathbb{Q}_p$-representations $\mathbb{L}$ of the fundamental group are associated to filtered Frobenius-crystals $\mathcal{E}$, that is there are functorial isomorphisms $B_{crys}(R) \otimes \mathcal{L} \cong B_{crys}(R) \otimes \mathcal{E}$ respecting Galois-action, filtration, and Frobenius (for small $R$'s). The induced isomorphism on $gr_F^0$ is $\hat{\bar{R}} \otimes \mathbb{L} \cong \oplus_i \hat{\bar{R}}[1/p](-i) \otimes gr_F^i(\mathcal{E})$ shows that $\mathbb{L}$ corresponds to the graded Higgs-bundle $gr_F(\mathcal{E})$ "with Tate-twists".

Unipotent representations of the fundamental group correspond to unipotent Higgs-bundles. Here "unipotent" means in both cases that the object is a successive extension of the trivial representation, respectively, Higgs-bundle. That follows easily from the comparison-theorem for cohomology as extensions are classified by $H^1$.

In fact using the method in [5] (proof of Theorem 5) one can even show that unipotent representations are (logarithmically) crystalline (see also [13]): for simplicity assume that our curve $X$ has two $V$-rational points $O$ and $\infty$. We consider representations of the fundamental-group of $X - \infty$ which has the advantage of being free. If $T_p(J)$ denotes the Tate-module of the Jacobian $J$ of $X$ we construct by induction unipotent smooth $p$-adic systems $\mathbb{L}_n$ on the generic fibre $(X - \{\infty\})_K$ such that we have exact sequences

$$0 \to T_p(J)^{\otimes n} \to \mathbb{L}_n \to \mathbb{L}_{n-1} \to 0.$$

Furthermore in the fibre over 0 the quotient $\mathbb{L}_0 = \mathbb{Z}_p$ lifts to a subsheaf of $\mathbb{L}_n$, the projection $\mathbb{L}_n \to \mathbb{L}_0 = \mathbb{Z}_p$ induces an isomorphism on homomorphisms into $\mathbb{Z}_p$ (over $(X - \{\infty\})_{\bar{K}}$), and the inclusion $T_p(J)^{\otimes n} \to \mathbb{L}_n$ an isomorphism on extensions by $\mathbb{Z}_p$.

All these hold for $\mathbb{L}_0 = \mathbb{Z}_p$. If we have constructed $\mathbb{L}_{n-1}$ we know that its extensions (over $X_{\bar{K}}$) by the constant sheaf $T_p(J)^{\otimes n}$ are classified by the corresponding $H^1$ which is isomorphic to

$$\begin{aligned} &H^1((X - \{\infty\})_{\bar{K}}, \mathcal{H}om(T_p(J)^{\otimes n-1}, T_p(J)^{\otimes n})) \\ &= Hom(T_p(J)^{\otimes n}, T_p(J)^{\otimes n}), \end{aligned}$$

and $\mathbb{L}_n$ is the class corresponding to the identity. From the long exact sequence of Ext-groups one derives the assertion about $Ext^i((X - \{\infty\})_{\bar{K}}; \mathbb{L}_n, \mathbb{Z}_p)$. Finally, the automorphisms of the extension are $T_p(J)^{\otimes n}$ and they act simply transitively on the splittings in 0. Thus we get that $\mathbb{L}_n$ is unique up to unique isomorphism, and thus already defined over $K$.

The same type of argument works on the crystalline side and constructs filtered (with degrees $\leqslant 0$) crystals $\mathcal{E}_n$, on the logarithmic crystalline site of $X$ relative to $\mathbb{Z}_p$, with splittings at 0 as before. These are unique, either filtered or unfiltered, and thus Frobenius-crystals. Finally, the two objects correspond via the comparison theory. As any unipotent representation of the fundamental-group is a quotient of a direct sum of $\mathbb{L}_n$'s the assertion follows.

## References

[1] C. Deninger, A. Werner, Vector bundles and $p$-adic representations I, Mŭnster, 2003, preprint.
[2] C. Deninger, A. Werner, Bundles on $p$-adic curves and parallel transport, Mŭnster, 2004, preprint.
[3] G. Faltings, Hodge–Tate structures and modular forms, Math. Ann. 278 (1987) 133–149.
[4] G. Faltings, Crystalline cohomology of semistable curves, and $p$-adic Galois-representations, J. Algebraic Geom. 1 (1992) 61–82.
[5] G. Faltings, Curves and their fundamental groups (following Grothendieck, Tamagawa and Mochizuki), Sem. Bourbaki 840, Astérisque 252 (1998) 131–150.
[6] G. Faltings, Almost étale extensions, Astérisque 279 (2002) 185–270.
[8] J.M. Fontaine, Le corps de périodes $p$-adiques, in: Périodes $p$-adiques, Astérisque, vol. 223, 1994.
[9] J.M. Fontaine, Arithmétique des représentations galoisiennes $p$-adiques, Orsay 2000-24, preprint.
[10] J.M. Fontaine, Presque $\mathbb{C}_p$-représentations, Orsay 2002-12, preprint.
[11] O. Gabber, L. Ramero, Almost ring theory, Springer Lecture Notes, vol. 1800, Springer, Berlin, 2003.
[12] C.T. Simpson, Higgs bundles and local systems, Publ. Math. IHES 75 (1992) 5–95.
[13] V. Vologodsky, Hodge structure on the fundamental group and its application to $p$-adic integration, Moscow Math. J. 3 (2003) 205–247.

# Bibliography


[1] A. Abbes, Éléments de géométrie rigide. Volume I. Construction et étude géométrique des espaces rigides, Progress in Mathematics Vol. **286**, Birkhäuser (2010).

[2] M. Artin, A. Grothendieck, J. L. Verdier, Théorie des topos et cohomologie étale des schémas, SGA 4, Lecture Notes in Math. Tome 1, **269** (1972); Tome 2, **270** (1972); Tome 3, **305** (1973), Springer-Verlag.

[3] P. Berthelot, Cohomologie Cristalline des Schémas de Caractéristique $p > 0$, Lecture Notes in Math. **407** (1974), Springer-Verlag.

[4] P. Berthelot, A. Grothendieck, L. Illusie, Théorie des intersections et théorème de Riemann-Roch, SGA 6, Lecture Notes in Math. **225** (1971), Springer-Verlag.

[5] P. Berthelot, A. Ogus, Notes on Crystalline Cohomology, Mathematical Notes **21** (1978), Princeton University Press.

[6] R. Berger, R. Kiehl, E. Kunz, H.-J. Nastold, Differentialrechnung in der analytischen geometry, Lecture Notes in Math. **38** (1967), Springer-Verlag.

[7] S. Bosch, U. Güntzer, R. Remmert, Non-archimedean analysis, Springer-Verlag (1984).

[8] S. Bosch, W. Lütkebohmert, M. Raynaud, *Formal and rigid geometry, IV. The reduced fiber theorem*, Invent. Math. **119** (1995), 361–398.

[9] N. Bourbaki, Algèbre, Chapitres 1–3, Masson (1970).

[10] N. Bourbaki, Algèbre, Chapitre 10, Masson (1980).

[11] N. Bourbaki, Algèbre commutative, Chapitres 1–9, Hermann (1985).

[12] N. Bourbaki, Topologie générale, Chapitres 1–4, Hermann (1971).

[13] O. Brinon, *Fibré de Higgs et structure locale d'une petite représentation généralisée (d'après Faltings)*, Lecture notes for a workshop on the *p-adic Simpson Correspondence following Faltings*, Rennes (2009).

[14] O. Brinon, *Représentations p-adiques cristallines et de de Rham dans le cas relatif*, Mémoire de la Soc. Math. France **112** (2008).

[15] P. Deligne, *Théorie de Hodge : II*, Publ. Math. IHÉS **40** (1971), 5–57.

[16] P. Deligne, Letter to L. Illusie of Mars 31, 1995.

[17] M. Demazure, A. Grothendieck, Schémas en Groupes (SGA 3) I : Propriétés Générales des Schémas en Groupes, Lecture Notes in Math. **151** (1970), Springer-Verlag.

[18] C. Deninger, *Representations attached to vector bundles on curves over finite and p-adic fields, a comparison*, Münster J. Math. **3** (2010), 29–41.

[19] C. Deninger, A. Werner, *Vector bundles on p-adic curves and parallel transport*, Annales Scientifiques ENS **38** (2005), 553–597.

[20] C. Deninger, A. Werner, *Vector Bundles on p-adic curves II*, in *Algebraic and arithmetic structures of moduli spaces*, Adv. Stud. Pure Math. **58** (2010), Math. Soc. Japan, 1–26.

[21] S. Donaldson, *Infinite determinants, stable bundles and curvature*, Duke Math. J. **54** (1987), 231–247.

[22] T. Ekedahl, *On the multiplicative properties of the de Rham-Witt complex. II*, Ark. Mat. **23** (1985), 53–102.

[23] G. Faltings, *Hodge-Tate structures and modular forms*, Math. Ann. **278** (1987), 133-149.

[24] G. Faltings, *p-adic Hodge theory*, J. Amer. Math. Soc. **1** (1988), 255–299.

[25] G. Faltings, *Crystalline cohomology and p-adic Galois representations*, in Algebraic analysis, geometry, and number theory (Baltimore, 1988), Johns Hopkins Univ. Press, Baltimore (1989), 25–80.

[26] G. Faltings, *Almost étale extensions*, in Cohomologies $p$-adiques et applications arithmétiques. II, Astérisque **279** (2002), 185–270.

[27] G. Faltings, *A p-adic Simpson correspondence*, Adv. Math. **198** (2005), 847–862; reprinted as a facsimile on page 577 of this volume.

[28] G. Faltings, *A p-adic Simpson correspondence II: small representations*, Pure Appl. Math. Q. **7** (2011), Special Issue: In memory of Eckart Viehweg, 1241–1264.

[29] J.-M. Fontaine, *Sur certains types de représentations p-adiques du groupe de Galois d'un corps local; construction d'un anneau de Barsotti-Tate*, Annals of Math. **115** (1982), 529–577.

[30] J.-M. Fontaine, *Formes différentielles et Modules de Tate des variétés abéliennes sur les corps locaux*, Inv. Math. **65** (1982), 379–409.

[31] J.-M. Fontaine, *Le corps des périodes p-adiques*, in Périodes $p$-adiques, Séminaire de Bures, 1988, Astérisque **223** (1994), 59–111.

[32] O. Gabber, L. Ramero, Almost Ring Theory, Lecture Notes in Mathematics **1800** (2003), Springer-Verlag.

[33] O. Gabber, L. Ramero, Foundations of $p$-adic Hodge theory (fourth release), preprint (2009).

[34] P. Gabriel, *Des catégories abéliennes*, Bull. Soc. Math. France **90** (1962), 323–448.

[35] J. Giraud, Cohomologie non abélienne, Springer-Verlag (1971).

[36] A. Grothendieck, Groupes de Barsotti-Tate et cristaux de Dieudonné, SMS (1974) Montréal.

[37] A. Grothendieck, Revêtements étales et groupe fondamental, SGA 1, Lecture Notes in Mathematics **224** (1971), Springer-Verlag.

[38] A. Grothendieck, Cohomologie $\ell$-adique et fonctions $L$, SGA 5, *Lecture Notes in Mathematics* **589** (1977), Springer-Verlag.

[39] A. Grothendieck, J.A. Dieudonné, Éléments de Géométrie Algébrique I, Seconde édition, Springer-Verlag (1971).

[40] A. Grothendieck, J.A. Dieudonné, Éléments de Géométrie Algébrique, II Étude globale élémentaire de quelques classes de morphismes, Pub. Math. IHÉS **8** (1961).

[41] A. Grothendieck, J.A. Dieudonné, Éléments de Géométrie Algébrique, III Étude cohomologique des faisceaux cohérents, Pub. Math. IHÉS **11** (1961), **17** (1963).

[42] A. Grothendieck, J.A. Dieudonné, Éléments de Géométrie Algébrique, IV Étude locale des schémas et des morphismes de schémas, Pub. Math. IHÉS **20** (1964), **24** (1965), **28** (1966), **32** (1967).

[43] O. Hyodo, *On variation of Hodge-Tate structures*, Math. Ann. **284** (1989), 7–22.

[44] O. Hyodo, *On the Hodge-Tate decomposition in the imperfect residue field case*, J. reine angew. Math. **365** (1986), 97–113.

[45] L. Illusie, Complexe cotangent et déformations. I, Lecture Notes in Math. **239** (1971), Springer-Verlag.

[46] L. Illusie, *Produits orientés*, in Travaux de Gabber sur l'uniformisation locale et la cohomologie étale des schémas quasi-excellents, Séminaire à l'École Polytechnique dirigé par L. Illusie, Y. Laszlo et F. Orgogozo (2006-2008), Astérisque **363-364** (2014), 213–234.

[47] U. Jannsen, *Continuous étale Cohomology*, Math. Ann. **280** (1988), 207–245.

[48] U. Jannsen, *The splitting of the Hochschild-Serre spectral sequence for a product of groups*, Canad. Math. Bull. **33** (1990), 181–183.

[49] F. Kato, *Log smooth deformation theory*, Tôhoku Math. J. **48** (1996), 317–354.

[50] K. Kato, *Logarithmic structures of Fontaine-Illusie*, Algebraic analysis, geometry, and number theory, Johns Hopkins UP, Baltimore (1989), 191–224.

[51] K. Kato, *Toric singularities*, American Journal of Math. **116** (1994), 1073–1099.

[52] G. Lan, M. Sheng, K. Zuo, *Semistable Higgs bundles, periodic Higgs bundles and representations of algebraic fundamental groups*, preprint (2013), arXiv:1311.6424.

[53] G. Laumon, *Vanishing cycles over a base of dimension* $\geq 1$, in *Algebraic geometry* (Tokyo/Kyoto, 1982), Lecture Notes in Math. **1016** (1983), Springer-Verlag, 143–150.

[54] J. Le Potier, *Fibrés de Higgs et systèmes locaux*, Séminaire Bourbaki, Exp. **737** (1991), Astérisque **201-203** (1992), 221–268.

[55] P. Monsky, G. Washnitzer, *Formal cohomology: I*, Ann. of Math. **88** (1968), 181–217.

[56] M.S. Narasimhan, C.S. Seshadri, *Stable and unitary vector bundles on a compact Riemann surface*, Ann. of Math. **82** (1965), 540–567.

[57] W. Niziol, *Toric singularities : log blow-ups and global resolutions*, J. of Alg. Geom. **15** (2006), 1–29.

[58] A. Ogus, Lectures on logarithmic algebraic geometry, book in preparation (version of December 11, 2013).

[59] A. Ogus, V. Vologodsky, *Non abelian Hodge theory in characteristic p*, Pub. Math. IHÉS **106** (2007), 1–138.

[60] F. Orgogozo, *Modifications et cycles proches sur une base générale*, Int. Math. Res. Not. (2006), 1–38.

[61] H. Oyama, *PD-Higgs crystals and Higgs cohomology in characteristic p*, preprint (2014).

[62] M. Raynaud, Anneaux locaux henséliens, Lecture Notes in Mathematics **169** (1970), Springer-Verlag.

[63] J.-E. Roos, *Derived functors of inverse limits revisited*, J. London Math. Soc. (2) **73** (2006), 65–83.

[64] P. Scholze, *Perfectoid spaces*, Publ. Math. IHÉS **116** (2012), 245–313.

[65] J.-P. Serre, Cohomologie galoisienne, Cinquième édition révisée et complétée, Lecture Notes in Math. **5** (1997), Springer-Verlag.

[66] A. Shiho, *Notes on generalizations of local Ogus-Vologodsky correspondence*, preprint (2012), arXiv:1206.5907.

[67] C. Simpson, *Constructing variations of Hodge structure using Yang-Mills theory and applications to uniformization*, J. Amer. Math. Soc. **1** (1988), no. 4, 867–918.

[68] C. Simpson, *Higgs bundles and local systems*, Pub. Math. IHÉS **75** (1992), 5–95.

[69] C. Simpson, *Moduli of representations of the fundamental group of a smooth variety I*, Pub. Math. IHÉS **79** (1994), 47–129.

[70] C. Simpson, *Moduli of representations of the fundamental group of a smooth projective variety II*, Publ. Math. IHÉS **80** (1994), 5–79.

[71] J. Tate, *p-divisible groups*, in Proceedings of a conference on local fields, Driebergen 1966, Springer-Verlag (1967), 158–183.

[72] J. Tate, *Relations between $K_2$ and Galois cohomology*, Invent. Math. **36** (1976), 257–274.

[73] T. Tsuji, *p-adic étale cohomology and crystalline cohomology in the semi-stable reduction case*, Invent. Math. **137** (1999), 233–411.

[74] T. Tsuji, *Saturated morphisms of logarithmic schemes*, preprint (1997).

[75] T. Tsuji, *Notes on p-adic Simpson correspondence and Galois cohomology*, preprint (2009).

# Indexes

## Index of Symbols

## Alphabetical Index